To the Student

As you begin, you may feel anxious about the number of theorems, definitions, procedures, and equations. You may wonder if you can learn it all in time. Don't worry—your concerns are normal. This textbook was written with you in mind. If you attend class, work hard, and read and study this text, you will build the knowledge and skills you need to be successful. Here's how you can use the text to your benefit.

Read Carefully

When you get busy, it's easy to skip reading and go right to the problems. Don't . . . the text has a large number of examples and clear explanations to help you break down the mathematics into easy-to-understand steps. Reading will provide you with a clearer understanding, beyond simple memorization. Read before class (not after) so you can ask questions about anything you didn't understand. You'll be amazed at how much more you'll get out of class if you do this.

Use the Features

I use many different methods in the classroom to communicate. Those methods, when incorporated into the text, are called "features." The features serve many purposes, from providing timely review of material you learned before (just when you need it) to providing organized review sessions to help you prepare for quizzes and tests. Take advantage of the features and you will master the material.

To make this easier, we've provided a brief guide to getting the most from this text. Refer to "Prepare for Class," "Practice," and "Review" on the following three pages. Spend fifteen minutes reviewing the guide and familiarizing yourself with the features by flipping to the page numbers provided. Then, as you read, use them. This is the best way to make the most of your text.

Please do not hesitate to contact us, through Pearson Education, with any questions, comments, or suggestions for improving this text. I look forward to hearing from you, and good luck with all of your studies.

Best Wishes!

Michael Sullivan

Prepare for Class "Read the Book"

Feature	Description	Benefit	Page
Every Chapter Opener begins with . . .			
Chapter-Opening Topic & Project	Each chapter begins with a discussion of a topic of current interest and ends with a related project.	The Project lets you apply what you learned to solve a problem related to the topic.	99
Internet-Based Projects	The projects allow for the integration of spreadsheet technology that you will need to be a productive member of the workforce.	The projects give you an opportunity to collaborate and use mathematics to deal with issues of current interest.	186
Every Section begins with . . .			
Learning Objectives 2	Each section begins with a list of objectives. Objectives also appear in the text where the objective is covered.	These focus your studying by emphasizing what's most important and where to find it.	114
Sections contain . . .			
PREPARING FOR THIS SECTION	Most sections begin with a list of key concepts to review with page numbers.	Ever forget what you've learned? This feature highlights previously learned material to be used in this section. Review it, and you'll always be prepared to move forward.	114
Now Work the **'Are You Prepared?' Problems**	Problems that assess whether you have the prerequisite knowledge for the upcoming section.	Not sure you need the Preparing for This Section review? Work the 'Are You Prepared?' problems. If you get one wrong, you'll know exactly what you need to review and where to review it!	114, 126
Now Work PROBLEMS	These follow most examples and direct you to a related exercise.	We learn best by doing. You'll solidify your understanding of examples if you try a similar problem right away, to be sure you understand what you've just read.	116, 119, 123
WARNING	Warnings are provided in the text.	These point out common mistakes and help you to avoid them.	116, 124
Exploration and **Seeing the Concept**	These graphing utility activities foreshadow a concept or solidify a concept just presented.	You will obtain a deeper and more intuitive understanding of theorems and definitions.	69, 149
In Words	These provide alternative descriptions of select definitions and theorems.	Does math ever look foreign to you? This feature translates math into plain English.	115, 134
Calculus	These appear next to information essential for the study of calculus.	Pay attention–if you spend extra time now, you'll do better later!	103, 128
SHOWCASE EXAMPLES	These examples provide "how-to" instruction by offering a guided, step-by-step approach to solving a problem.	With each step presented on the left and the mathematics displayed on the right, you can immediately see how each step is employed.	152, 170
Model It! Examples and Problems	These examples and problems require you to build a mathematical model from either a verbal description or data. The homework Model It! problems are marked by purple headings.	It is rare for a problem to come in the form *"Solve the following equation."* Rather, the equation must be developed based on an explanation of the problem. These problems require you to develop models that will allow you to describe the problem mathematically and suggest a solution to the problem.	111, 159, 171

Practice "Work the Problems"

Feature	Description	Benefit	Page
'Are You Prepared?' Problems	These assess your retention of the prerequisite material you'll need. Answers are given at the end of the section exercises. This feature is related to the Preparing for This Section feature.	Do you always remember what you've learned? Working these problems is the best way to find out. If you get one wrong, you'll know exactly what you need to review and where to review it!	145, 155
Concepts and Vocabulary	These short-answer questions, mainly Fill-in-the-Blank, Multiple-Choice and True/False items, assess your understanding of key definitions and concepts in the current section.	It is difficult to learn math without knowing the language of mathematics. These problems test your understanding of the formulas and vocabulary.	155
Skill Building	Correlated with section examples, these problems provide straightforward practice.	It's important to dig in and develop your skills. These problems provide you with ample opportunity to do so.	156–158
Mixed Practice	These problems offer comprehensive assessment of the skills learned in the section by asking problems that relate to more than one concept or objective. These problems may also require you to utilize skills learned in previous sections.	Learning mathematics is a building process. Many concepts are interrelated. These problems help you see how mathematics builds on itself and also see how the concepts tie together.	158
Applications and Extensions	These problems allow you to apply your skills to real-world problems. They also allow you to extend concepts learned in the section.	You will see that the material learned within the section has many uses in everyday life.	158–159
Explaining Concepts: Discussion and Writing	"Discussion and Writing" problems are colored red. They support class discussion, verbalization of mathematical ideas, and writing and research projects.	To verbalize an idea, or to describe it clearly in writing, shows real understanding. These problems nurture that understanding. Many are challenging, but you'll get out what you put in.	159
NEW! **Retain Your Knowledge**	These problems allow you to practice content learned earlier in the course.	Remembering how to solve all the different kinds of problems that you encounter throughout the course is difficult. This practice helps you remember.	159
Now Work PROBLEMS	Many examples refer you to a related homework problem. These related problems are marked by a pencil and orange numbers.	If you get stuck while working problems, look for the closest Now Work problem, and refer to the related example to see if it helps.	146, 147, 151
Review Exercises	Every chapter concludes with a comprehensive list of exercises to pratice. Use the list of objectives to determine the objective and examples that correspond to the problems.	Work these problems to ensure that you understand all the skills and concepts of the chapter. Think of it as a comprehensive review of the chapter.	182–184

Review "Study for Quizzes and Tests"

Feature	Description	Benefit	Page
The Chapter Review at the end of each chapter contains . . .			
Things to Know	A detailed list of important theorems, formulas, and definitions from the chapter.	Review these and you'll know the most important material in the chapter!	178–181
You Should Be Able to . . .	Contains a complete list of objectives by section, examples that illustrate the objective, and practice exercises that test your understanding of the objective.	Do the recommended exercises and you'll have mastered the key material. If you get something wrong, go back and work through the example listed and try again.	181–182
Review Exercises	These provide comprehensive review and practice of key skills, matched to the Learning Objectives for each section.	Practice makes perfect. These problems combine exercises from all sections, giving you a comprehensive review in one place.	182–184
Chapter Test	About 15–20 problems that can be taken as a Chapter Test. Be sure to take the Chapter Test under test conditions—no notes!	Be prepared. Take the sample practice test under test conditions. This will get you ready for your instructor's test. If you get a problem wrong, you can watch the Chapter Test Prep Video.	184–185
Cumulative Review	These problem sets appear at the end of each chapter, beginning with Chapter 2. They combine problems from previous chapters, providing an ongoing cumulative review. When you use them in conjunction with the Retain Your Knowledge problems, you will be ready for the final exam.	These problem sets are really important. Completing them will ensure that you are not forgetting anything as you go. This will go a long way toward keeping you primed for the final exam.	185
Chapter Projects	The Chapter Projects apply to what you've learned in the chapter. Additional projects are available on the Instructor's Resource Center (IRC).	The Chapter Projects give you an opportunity to apply what you've learned in the chapter to the opening topic. If your instructor allows, these make excellent opportunities to work in a group, which is often the best way of learning math.	186–187
Internet-Based Projects	In selected chapters, a Web-based project is given.	These projects give you an opportunity to collaborate and use mathematics to deal with issues of current interest by using the Internet to research and collect data.	186

Achieve Your Potential

The author, Michael Sullivan, has developed specific content in MyMathLab® to ensure you have many resources to help you achieve success in mathematics - and beyond! The MyMathLab features described here will help you:

- **Review** math skills and concepts you may have forgotten
- **Retain** new concepts as you move through your math course
- **Develop** skills that will help with your transition to college

Adaptive Study Plan

The Study Plan will help you study more efficiently and effectively.

Your performance and activity are assessed continually in real time, providing a personalized experience based on your individual needs.

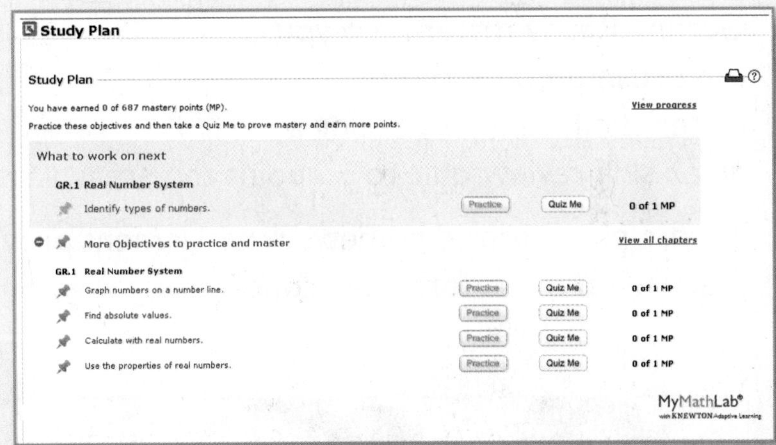

Skills for Success

The Skills for Success Modules support your continued success in college. These modules provide tutorials and guidance on a variety of topics, including transitioning to college, online learning, time management, and more.

Additional content is provided to help with the development of professional skills such as resume writing and interview preparation.

Getting Ready

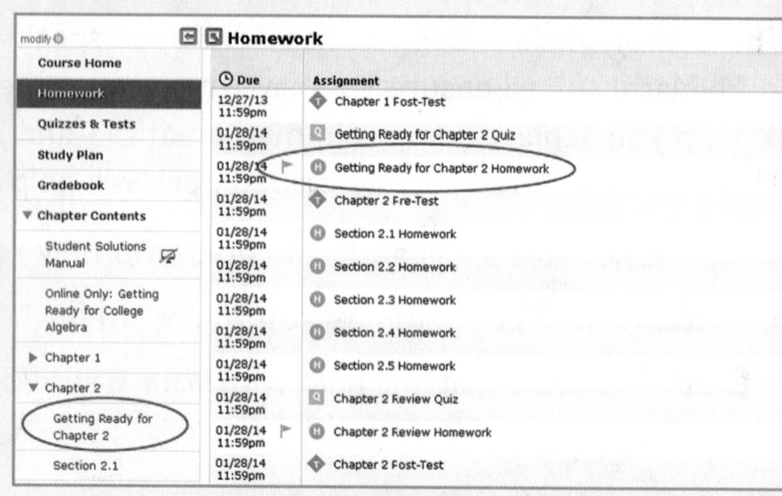

Are you frustrated when you know you learned a math concept in the past, but you can't quite remember the skill when it's time to use it? Don't worry!

The author has included Getting Ready material so you can brush up on forgotten material efficiently by taking a quick skill review quiz to pinpoint the areas where you need help.

Then, a personalized homework assignment provides additional practice on those forgotten concepts, right when you need it.

Retain Your Knowledge

As you work through your math course, these MyMathLab® exercises support ongoing review to help you maintain essential skills.

The ability to recall important math concepts as you continually acquire new mathematical skills will help you be successful in this math course and in your future math courses.

Trigonometry

A Unit Circle Approach

Tenth Edition

Michael Sullivan
Chicago State University

PEARSON

Boston Columbus Indianapolis New York San Francisco Hoboken
Amsterdam Cape Town Dubai London Madrid Milan Munich Paris Montreal Toronto
Delhi Mexico City São Paulo Sydney Hong Kong Seoul Singapore Taipei Tokyo

Editor in Chief: *Anne Kelly*
Acquisitions Editor: *Dawn Murrin*
Assistant Editor: *Joseph Colella*
Program Team Lead: *Karen Wernholm*
Program Manager: *Chere Bemelmans*
Project Team Lead: *Peter Silvia*
Project Manager: *Peggy McMahon*
Associate Media Producer: *Marielle Guiney*
Senior Project Manager, MyMathLab: *Kristina Evans*
QA Manager, Assessment Content: *Marty Wright*
Senior Marketing Manager: *Michelle Cook*
Marketing Manager: *Peggy Sue Lucas*
Marketing Assistant: *Justine Goulart*
Senior Author Support/Technology Specialist: *Joe Vetere*
Procurement Manager: *Vincent Scelta*
Procurement Specialist: *Carol Melville*
Text Design: *Tamara Newnam*
Production Coordination,
Composition, Illustrations: *Cenveo® Publisher Services*
Associate Director of Design,
USHE EMSS/HSC/EDU: *Andrea Nix*
Project Manager, Rights and Permissions: *Diahanne Lucas Dowridge*
Art Director: *Heather Scott*
Cover Design and Cover Illustration: *Tamara Newnam*

Acknowledgments of third-party content appear on page C1, which constitutes an extension of this copyright page.

The student edition of this text has been cataloged as follows:
Library of Congress Cataloging-in-Publication Data
Sullivan, Michael, 1942-
Trigonometry : a unit circle approach / Michael Sullivan, Chicago State University. -- Tenth edition.
pages cm
Includes index.
ISBN 978-0-321-97860-8
1. Trigonometry--Textbooks. I. Title.
QA531.S87 2016
516.24--dc23

2014045813

2 3 4 5 6 7 8 9 10—CRK—18 17 16 15

www.pearsonhighered.com

ISBN-10: **0-321-97860-9**
ISBN-13: **978-0-321-97860-8**

Contents

Three Distinct Series xv

The Contemporary Series xvi

Preface to the Instructor xvii

Resources for Success xxii

Applications Index xxiv

1 Graphs and Functions 1

1.1 **The Distance and Midpoint Formulas** 2
Use the Distance Formula • Use the Midpoint Formula

1.2 **Graphs of Equations in Two Variables; Circles** 9
Graph Equations by Plotting Points • Find Intercepts from a Graph • Find Intercepts from an Equation • Test an Equation for Symmetry with Respect to the x-Axis, the y-Axis, and the Origin • Know How to Graph Key Equations Write the Standard Form of the Equation of a Circle• Graph a Circle • Work with the General Form of the Equation of a Circle

1.3 **Functions and Their Graphs** 23
Determine Whether a Relation Represents a Function • Find the Value of a Function • Find the Difference Quotient of a Function • Find the Domain of a Function Defined by an Equation • Identify the Graph of a Function • Obtain Information from or about the Graph of a Function

1.4 **Properties of Functions** 42
Determine Even and Odd Functions from a Graph • Identify Even and Odd Functions from an Equation • Use a Graph to Determine Where a Function Is Increasing, Decreasing, or Constant • Use a Graph to Locate Local Maxima and Local Minima • Use a Graph to Locate the Absolute Maximum and the Absolute Minimum • Use a Graphing Utility to Approximate Local Maxima and Local Minima and to Determine Where a Function Is Increasing or Decreasing • Find the Average Rate of Change of a Function

1.5 **Library of Functions; Piecewise-defined Functions** 55
Graph the Functions Listed in the Library of Functions • Graph Piecewise-defined Functions

1.6 **Graphing Techniques: Transformations** 64
Graph Functions Using Vertical and Horizontal Shifts • Graph Functions Using Compressions and Stretches • Graph Functions Using Reflections about the x-Axis and the y-Axis

1.7 **One-to-One Functions; Inverse Functions** 78
Determine Whether a Function Is One-to-One • Determine the Inverse of a Function Defined by a Map or a Set of Ordered Pairs • Obtain the Graph of the Inverse Function from the Graph of the Function • Find the Inverse of a Function Defined by an Equation

Chapter Review 91

Chapter Test 96

Chapter Projects 97

2 Trigonometric Functions 99

2.1 Angles and Their Measure 100
Convert between Decimals and Degrees, Minutes, Seconds Measures for Angles • Find the Length of an Arc of a Circle • Convert from Degrees to Radians and from Radians to Degrees • Find the Area of a Sector of a Circle • Find the Linear Speed of an Object Traveling in Circular Motion

2.2 Trigonometric Functions: Unit Circle Approach 114
Find the Exact Values of the Trigonometric Functions Using a Point on the Unit Circle • Find the Exact Values of the Trigonometric Functions of Quadrantal Angles • Find the Exact Values of the Trigonometric Functions of $\frac{\pi}{4} = 45°$ • Find the Exact Values of the Trigonometric Functions of $\frac{\pi}{6} = 30°$ and $\frac{\pi}{3} = 60°$ • Find the Exact Values of the Trigonometric Functions for Integer Multiples of $\frac{\pi}{6} = 30°$, $\frac{\pi}{4} = 45°$, and $\frac{\pi}{3} = 60°$ • Use a Calculator to Approximate the Value of a Trigonometric Function • Use a Circle of Radius r to Evaluate the Trigonometric Functions

2.3 Properties of the Trigonometric Functions 131
Determine the Domain and the Range of the Trigonometric Functions • Determine the Period of the Trigonometric Functions • Determine the Signs of the Trigonometric Functions in a Given Quadrant • Find the Values of the Trigonometric Functions Using Fundamental Identities • Find the Exact Values of the Trigonometric Functions of an Angle Given One of the Functions and the Quadrant of the Angle • Use Even–Odd Properties to Find the Exact Values of the Trigonometric Functions

2.4 Graphs of the Sine and Cosine Functions 145
Graph Functions of the Form $y = A \sin(\omega x)$ Using Transformations • Graph Functions of the Form $y = A \cos(\omega x)$ Using Transformations • Determine the Amplitude and Period of Sinusoidal Functions • Graph Sinusoidal Functions Using Key Points • Find an Equation for a Sinusoidal Graph

2.5 Graphs of the Tangent, Cotangent, Cosecant, and Secant Functions 160
Graph Functions of the Form $y = A \tan(\omega x) + B$ and $y = A \cot(\omega x) + B$ • Graph Functions of the Form $y = A \csc(\omega x) + B$ and $y = A \sec(\omega x) + B$

2.6 Phase Shift; Sinusoidal Curve Fitting 168
Graph Sinusoidal Functions of the Form $y = A \sin(\omega x - \phi) + B$ • Build Sinusoidal Models from Data

Chapter Review 178

Chapter Test 184

Cumulative Review 185

Chapter Projects 186

3 Analytic Trigonometry 188

3.1 The Inverse Sine, Cosine, and Tangent Functions 189
Find the Exact Value of an Inverse Sine Function • Find an Approximate Value of an Inverse Sine Function • Use Properties of Inverse Functions to Find Exact Values of Certain Composite Functions • Find the Inverse Function of a Trigonometric Function • Solve Equations Involving Inverse Trigonometric Functions

3.2 The Inverse Trigonometric Functions (Continued) 202
Find the Exact Value of Expressions Involving the Inverse Sine, Cosine, and Tangent Functions • Define the Inverse Secant, Cosecant, and Cotangent Functions • Use a Calculator to Evaluate $\sec^{-1} x$, $\csc^{-1} x$, and $\cot^{-1} x$ • Write a Trigonometric Expression as an Algebraic Expression

3.3 Trigonometric Equations 208
Solve Equations Involving a Single Trigonometric Function • Solve Trigonometric Equations Using a Calculator • Solve Trigonometric Equations Quadratic in Form • Solve Trigonometric Equations Using Fundamental Identities • Solve Trigonometric Equations Using a Graphing Utility

3.4 Trigonometric Identities 218
Use Algebra to Simplify Trigonometric Expressions • Establish Identities

3.5 Sum and Difference Formulas 226
Use Sum and Difference Formulas to Find Exact Values • Use Sum and Difference Formulas to Establish Identities • Use Sum and Difference Formulas Involving Inverse Trigonometric Functions • Solve Trigonometric Equations Linear in Sine and Cosine

3.6 Double-angle and Half-angle Formulas 238
Use Double-angle Formulas to Find Exact Values • Use Double-angle Formulas to Establish Identities • Use Half-angle Formulas to Find Exact Values

3.7 Product-to-Sum and Sum-to-Product Formulas 248
Express Products as Sums • Express Sums as Products

Chapter Review 252

Chapter Test 255

Cumulative Review 256

Chapter Projects 257

4 Applications of Trigonometric Functions 258

4.1 Right Triangle Trigonometry; Applications 259
Find the Value of Trigonometric Functions of Acute Angles Using Right Triangles • Use the Complementary Angle Theorem • Solve Right Triangles • Solve Applied Problems

4.2 The Law of Sines 272
Solve SAA or ASA Triangles • Solve SSA Triangles • Solve Applied Problems

4.3 The Law of Cosines 282
Solve SAS Triangles • Solve SSS Triangles • Solve Applied Problems

4.4 Area of a Triangle 289
Find the Area of SAS Triangles • Find the Area of SSS Triangles

4.5 Simple Harmonic Motion; Damped Motion; Combining Waves 295
Build a Model for an Object in Simple Harmonic Motion • Analyze Simple Harmonic Motion • Analyze an Object in Damped Motion • Graph the Sum of Two Functions

Chapter Review 304

Chapter Test 307

Cumulative Review 308

Chapter Projects 309

5 Polar Coordinates; Vectors 311

5.1 Polar Coordinates 312
Plot Points Using Polar Coordinates • Convert from Polar Coordinates to Rectangular Coordinates • Convert from Rectangular Coordinates to Polar Coordinates • Transform Equations between Polar and Rectangular Forms

5.2 Polar Equations and Graphs 321
Identify and Graph Polar Equations by Converting to Rectangular Equations • Test Polar Equations for Symmetry • Graph Polar Equations by Plotting Points

5.3 The Complex Plane; De Moivre's Theorem 336
Plot Points in the Complex Plane • Convert a Complex Number between Rectangular Form and Polar Form • Find Products and Quotients of Complex Numbers in Polar Form • Use De Moivre's Theorem • Find Complex Roots

5.4 Vectors 344
Graph Vectors • Find a Position Vector • Add and Subtract Vectors Algebraically • Find a Scalar Multiple and the Magnitude of a Vector • Find a Unit Vector • Find a Vector from Its Direction and Magnitude • Model with Vectors

5.5 The Dot Product 358
Find the Dot Product of Two Vectors • Find the Angle between Two Vectors • Determine Whether Two Vectors Are Parallel • Determine Whether Two Vectors Are Orthogonal • Decompose a Vector into Two Orthogonal Vectors • Compute Work

5.6 Vectors in Space 365
Find the Distance between Two Points in Space • Find Position Vectors in Space • Perform Operations on Vectors • Find the Dot Product • Find the Angle between Two Vectors • Find the Direction Angles of a Vector

5.7 The Cross Product 375
Find the Cross Product of Two Vectors • Know Algebraic Properties of the Cross Product • Know Geometric Properties of the Cross Product • Find a Vector Orthogonal to Two Given Vectors • Find the Area of a Parallelogram

Chapter Review 381

Chapter Test 384

Cumulative Review 385

Chapter Projects 385

6 Analytic Geometry 386

6.1 Conics 387
Know the Names of the Conics

6.2 The Parabola 388
Analyze Parabolas with Vertex at the Origin • Analyze Parabolas with Vertex at (h, k) • Solve Applied Problems Involving Parabolas

6.3 The Ellipse 397
Analyze Ellipses with Center at the Origin • Analyze Ellipses with Center at (h, k) • Solve Applied Problems Involving Ellipses

6.4 The Hyperbola 407
Analyze Hyperbolas with Center at the Origin • Find the Asymptotes of a Hyperbola • Analyze Hyperbolas with Center at (h, k) • Solve Applied Problems Involving Hyperbolas

6.5 Rotation of Axes; General Form of a Conic 420
Identify a Conic • Use a Rotation of Axes to Transform Equations • Analyze an Equation Using a Rotation of Axes • Identify Conics without a Rotation of Axes

6.6 Polar Equations of Conics 428
Analyze and Graph Polar Equations of Conics • Convert the Polar Equation of a Conic to a Rectangular Equation

6.7 Plane Curves and Parametric Equations 434
Graph Parametric Equations • Find a Rectangular Equation for a Curve Defined Parametrically • Use Time as a Parameter in Parametric Equations • Find Parametric Equations for Curves Defined by Rectangular Equations

Chapter Review 446

Chapter Test 449

Cumulative Review 449

Chapter Projects 450

7 Exponential and Logarithmic Functions 452

7.1 Exponential Functions 453
Evaluate Exponential Functions • Graph Exponential Functions • Define the Number *e* • Solve Exponential Equations

7.2 Logarithmic Functions 470
Change Exponential Statements to Logarithmic Statements and Logarithmic Statements to Exponential Statements • Evaluate Logarithmic Expressions • Determine the Domain of a Logarithmic Function • Graph Logarithmic Functions • Solve Logarithmic Equations

7.3 Properties of Logarithms 482
Work with the Properties of Logarithms • Write a Logarithmic Expression as a Sum or Difference of Logarithms • Write a Logarithmic Expression as a Single Logarithm • Evaluate Logarithms Whose Base Is Neither 10 Nor *e*

7.4 Logarithmic and Exponential Equations 491
Solve Logarithmic Equations • Solve Exponential Equations • Solve Logarithmic and Exponential Equations Using a Graphing Utility

7.5 Financial Models 498
Determine the Future Value of a Lump Sum of Money • Calculate Effective Rates of Return • Determine the Present Value of a Lump Sum of Money • Determine the Rate of Interest or the Time Required to Double a Lump Sum of Money

7.6 Exponential Growth and Decay Models; Newton's Law; Logistic Growth and Decay Models 508
Find Equations of Populations That Obey the Law of Uninhibited Growth • Find Equations of Populations That Obey the Law of Decay • Use Newton's Law of Cooling • Use Logistic Models

7.7 Building Exponential, Logarithmic, and Logistic Models from Data 519
Build an Exponential Model from Data • Build a Logarithmic Model from Data • Build a Logistic Model from Data

Chapter Review 526

Chapter Test 530

Cumulative Review 531

Chapter Projects 532

Appendix A Review A1

A.1 Algebra Essentials A1
Work with Sets • Graph Inequalities • Find Distance on the Real Number Line • Evaluate Algebraic Expressions • Determine the Domain of a Variable • Use the Laws of Exponents • Evaluate Square Roots • Use a Calculator to Evaluate Exponents

A.2 Geometry Essentials A14
Use the Pythagorean Theorem and Its Converse • Know Geometry Formulas • Understand Congruent Triangles and Similar Triangles

A.3 Factoring Polynomials; Completing the Square A22
Know Formulas for Special Products • Factor Polynomials • Complete the Square

A.4 Solving Equations A27
Solve Equations by Factoring • Solve Equations Involving Absolute Value • Solve a Quadratic Equation by Factoring • Solve a Quadratic Equation by Completing the Square • Solve a Quadratic Equation Using the Quadratic Formula

A.5 Complex Numbers; Quadratic Equations in the Complex Number System A37
Add, Subtract, Multiply, and Divide Complex Numbers • Solve Quadratic Equations in the Complex Number System

A.6 Interval Notation; Solving Inequalities A46
Use Interval Notation • Use Properties of Inequalities • Solve Inequalities • Solve Combined Inequalities • Solve Inequalities Involving Absolute Value

A.7 *n*th Roots; Rational Exponents A56
Work with *n*th Roots • Simplify Radicals • Rationalize Denominators • Solve Radical Equations • Simplify Expressions with Rational Exponents

A.8 Lines A64
Calculate and Interpret the Slope of a Line • Graph Lines Given a Point and the Slope • Find the Equation of a Vertical Line • Use the Point–Slope Form of a Line; Identify Horizontal Lines • Find the Equation of a Line Given Two Points • Write the Equation of a Line in Slope–Intercept Form • Identify the Slope and *y*-Intercept of a Line from Its Equation • Graph Lines Written in General Form Using Intercepts • Find Equations of Parallel Lines • Find Equations of Perpendicular Lines

A.9 Building Linear Models from Data A79
Draw and Interpret Scatter Diagrams • Distinguish between Linear and Nonlinear Relations • Use a Graphing Utility to Find the Line of Best Fit

Appendix B Graphing Utilities B1

B.1 The Viewing Rectangle B1

B.2 Using a Graphing Utility to Graph Equations B3

B.3 Using a Graphing Utility to Locate Intercepts and Check for Symmetry B5

B.4 Using a Graphing Utility to Solve Equations B6

B.5 Square Screens B8

B.6 Using a Graphing Utility to Graph a Polar Equation B9

B.7 Using a Graphing Utility to Graph Parametric Equations B9

Answers AN1

Credits C1

Index I1

To the Memory of Joe and Rita
and My Sister Maryrose

Three Distinct Series

Students have different goals, learning styles, and levels of preparation. Instructors have different teaching philosophies, styles, and techniques. Rather than write one series to fit all, the Sullivans have written three distinct series. All share the same goal—to develop a high level of mathematical understanding and an appreciation for the way mathematics can describe the world around us. The manner of reaching that goal, however, differs from series to series.

Contemporary Series, Tenth Edition

The Contemporary Series is the most traditional in approach yet modern in its treatment of precalculus mathematics. Graphing utility coverage is optional and can be included or excluded at the discretion of the instructor: *College Algebra, Algebra & Trigonometry, Trigonometry: A Unit Circle Approach, Precalculus.*

Enhanced with Graphing Utilities Series, Sixth Edition

This series provides a thorough integration of graphing utilities into topics, allowing students to explore mathematical concepts and encounter ideas usually studied in later courses. Using technology, the approach to solving certain problems differs from the Contemporary Series, while the emphasis on understanding concepts and building strong skills does not: *College Algebra, Algebra & Trigonometry, Precalculus.*

Concepts through Functions Series, Third Edition

This series differs from the others, utilizing a functions approach that serves as the organizing principle tying concepts together. Functions are introduced early in various formats. This approach supports the Rule of Four, which states that functions are represented symbolically, numerically, graphically, and verbally. Each chapter introduces a new type of function and then develops all concepts pertaining to that particular function. The solutions of equations and inequalities, instead of being developed as stand-alone topics, are developed in the context of the underlying functions. Graphing utility coverage is optional and can be included or excluded at the discretion of the instructor: *College Algebra; Precalculus, with a Unit Circle Approach to Trigonometry; Precalculus, with a Right Triangle Approach to Trigonometry*.

The Contemporary Series

College Algebra, Tenth Edition

This text provides a contemporary approach to college algebra, with three chapters of review material preceding the chapters on functions. Graphing calculator usage is provided, but is optional. After completing this book, a student will be adequately prepared for trigonometry, finite mathematics, and business calculus.

Algebra & Trigonometry, Tenth Edition

This text contains all the material in *College Algebra*, but also develops the trigonometric functions using a right triangle approach and showing how it relates to the unit circle approach. Graphing techniques are emphasized, including a thorough discussion of polar coordinates, parametric equations, and conics using polar coordinates. Graphing calculator usage is provided, but is optional. After completing this book, a student will be adequately prepared for finite mathematics, business calculus, and engineering calculus.

Precalculus, Tenth Edition

This text contains one review chapter before covering the traditional precalculus topic of functions and their graphs, polynomial and rational functions, and exponential and logarithmic functions. The trigonometric functions are introduced using a unit circle approach and showing how it relates to the right triangle approach. Graphing techniques are emphasized, including a thorough discussion of polar coordinates, parametric equations, and conics using polar coordinates. Graphing calculator usage is provided, but is optional. The final chapter provides an introduction to calculus, with a discussion of the limit, the derivative, and the integral of a function. After completing this book, a student will be adequately prepared for finite mathematics, business calculus, and engineering calculus.

Trigonometry: a Unit Circle Approach, Tenth Edition

This text, designed for stand-alone courses in trigonometry, develops the trigonometric functions using a unit circle approach and showing how it relates to the right triangle approach. Graphing techniques are emphasized, including a thorough discussion of polar coordinates, parametric equations, and conics using polar coordinates. Graphing calculator usage is provided, but is optional. After completing this book, a student will be adequately prepared for finite mathematics, business calculus, and engineering calculus.

Preface to the Instructor

As a professor of mathematics at an urban public university for 35 years, I understand the varied needs of students taking trigonometry. Students range from being underprepared, with little mathematical background and a fear of mathematics, to being highly prepared and motivated. For some, this is their final course in mathematics. For others, it is preparation for future mathematics courses. I have written this text with both groups in mind.

A tremendous benefit of authoring a successful series is the broad-based feedback I receive from teachers and students who have used previous editions. I am sincerely grateful for their support. Virtually every change to this edition is the result of their thoughtful comments and suggestions. I hope that I have been able to take their ideas and, building upon a successful foundation of the ninth edition, make this series an even better learning and teaching tool for students and teachers.

Features in the Tenth Edition

A descriptive list of the many special features of *Trigonometry* can be found on the endpapers in the front of this text.

This list places the features in their proper context, as building blocks of an overall learning system that has been carefully crafted over the years to help students get the most out of the time they put into studying. Please take the time to review this and to discuss it with your students at the beginning of your course. My experience has been that when students utilize these features, they are more successful in the course.

New to the Tenth Edition

- **Retain Your Knowledge** This new category of problems in the exercise set are based on the article "To Retain New Learning, Do the Math" published in the *Edurati Review*. In this article, Kevin Washburn suggests that "the more students are required to recall new content or skills, the better their memory will be." It is frustrating when students cannot recall skills learned earlier in the course. To alleviate this recall problem, we have created "Retain Your Knowledge" problems. These are problems considered to be "final exam material" that students can use to maintain their skills. Answers to all these problems appear in the back of the Student Edition, and all are programmed in MyMathLab.
- **Guided Lecture Notes** Ideal for online, emporium/redesign courses, inverted classrooms, or traditional lecture classrooms. These lecture notes help students take thorough, organized, and understandable notes as they watch the Author in Action videos. They ask students to complete definitions, procedures, and examples based on the content of the videos and text. In addition, experience suggests that students learn by doing and understanding the why/how of the concept or property. Therefore, many sections will have an exploration activity to motivate student learning. These explorations introduce the topic and/or connect it to either a real-world application or a previous section. For example, when the vertical-line test is discussed in Section 1.3, after the theorem statement, the notes ask the students to explain why the vertical-line test works by using the definition of a function. This challenge helps students process the information at a higher level of understanding.
- **Illustrations** Many of the figures now have captions to help connect the illustrations to the explanations in the body of the text.
- **TI Screen Shots** In this edition we have replaced all the screen shots from the ninth edition with screen shots using TI-84Plus C. These updated screen shots help students visualize concepts clearly and help make stronger connections between equations, data, and graphs in full color.
- **Chapter Projects,** which apply the concepts of each chapter to a real-world situation, have been enhanced to give students an up-to-the-minute experience. Many projects are new and Internet-based, requiring the student to research information online in order to solve problems.
- **Exercise Sets** All the exercises in the text have been reviewed and analyzed for this edition, some have been removed, and new ones have been added. All time-sensitive problems have been updated to the most recent information available. The problem sets remain classified according to purpose.

  The ***'Are You Prepared?'*** problems have been improved to better serve their purpose as a just-in-time review of concepts that the student will need to apply in the upcoming section.

  The ***Concepts and Vocabulary*** problems have been expanded and now include multiple-choice exercises. Together with the fill-in-the-blank and True/False problems, these exercises have been written to serve as reading quizzes.

  Skill Building problems develop the student's computational skills with a large selection of exercises that are directly related to the objectives of the section. ***Mixed Practice*** problems offer a comprehensive assessment of skills that relate to more than one objective. Often these require skills learned earlier in the course.

  Applications and Extensions problems have been updated. Further, many new application-type exercises have been added, especially ones involving information and data drawn from sources the student will recognize, to improve relevance and timeliness.

  The ***Explaining Concepts: Discussion and Writing*** exercises have been improved and expanded to provide more opportunity for classroom discussion and group projects.

New to this edition, ***Retain Your Knowledge*** exercises consist of a collection of four problems in each exercise set that are based on material learned earlier in the course. They serve to keep information that has already been learned "fresh" in the mind of the student. Answers to all these problems appear in the Student Edition.

The ***Review Exercises*** in the Chapter Review have been streamlined, but they remain tied to the clearly expressed objectives of the chapter. Answers to all these problems appear in the Student Edition.

- **Annotated Instructor's Edition** As a guide, the author's suggestions for homework assignments are indicated by a blue underscore below the problem number. These problems are assignable in the MyMathLab as part of a "Ready-to-Go" course.
- **Section 1.3** The objective Find the Difference Quotient of a Function has been added.

Using the Tenth Edition Effectively with Your Syllabus

To meet the varied needs of diverse syllabi, this text contains more content than is likely to be covered in a *Trigonometry* course. As the chart illustrates, this text has been organized with flexibility of use in mind. Within a given chapter, certain sections are optional (see the details that follow the figure below) and can be omitted without loss of continuity.

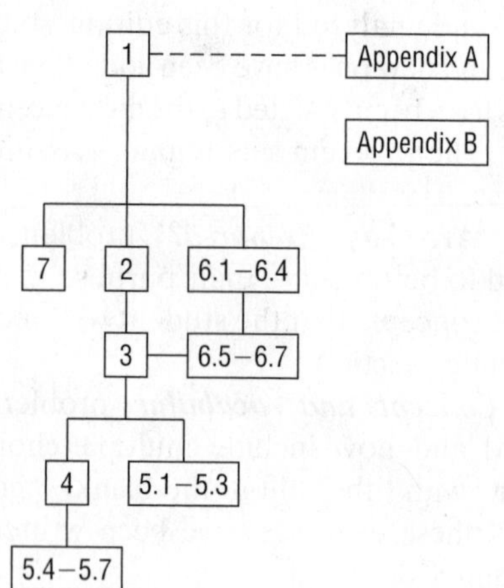

Appendix A Review
This chapter consists of review material. It may be used as the first part of the course or later as a just-in-time review when the content is required. Specific references to this chapter occur throughout the book to assist in the review process.

Chapter 1 Graphs and Functions
This chapter lays the foundation for Chapters 2, 6, and 7.

Chapter 2 Trigonometric Functions
Section 2.6 may be omitted in a brief course.

Chapter 3 Analytic Trigonometry
Section 3.7 may be omitted in a brief course.

Chapter 4 Applications of Trigonometric Functions
Sections 4.4 and 4.5 may be omitted in a brief course.

Chapter 5 Polar Coordinates; Vectors
Sections 5.1–5.3 and Sections 5.4–5.7 are independent and may be covered separately.

Chapter 6 Analytic Geometry
Sections 6.1–6.4 follow in sequence. Sections 6.5, 6.6, and 6.7 are independent of each other, but each requires Sections 6.1–6.4.

Chapter 7 Exponential and Logarithmic Functions
Sections 7.1–7.6 follow in sequence. Sections 7.7, 7.8, and 7.9 are optional.

Acknowledgments

Textbooks are written by authors, but evolve from an idea to final form through the efforts of many people. It was Don Dellen who first suggested this text and series to me. Don is remembered for his extensive contributions to publishing and mathematics.

Thanks are due to the following people for their assistance and encouragement to the preparation of this edition:

- From Pearson Education: Anne Kelly for her substantial contributions, ideas, and enthusiasm; Dawn Murrin, for her unmatched talent at getting the details right; Joseph Colella for always getting the reviews and pages to me on time; Peggy McMahon for directing the always difficult production process; Rose Kernan for handling liaison between the compositor and author; Peggy Lucas for her genuine interest in marketing this text; Chris Hoag for her continued support and genuine interest; Paul Corey for his leadership and commitment to excellence; and the Pearson Math and Science Sales team, for their continued confidence and personal support of our texts.
- Accuracy checkers: C. Brad Davis, who read the entire manuscript and accuracy checked answers. His attention to detail is amazing; Timothy Britt, for creating the Solutions Manuals and accuracy checking answers.

Finally, I offer my grateful thanks to the dedicated users and reviewers of my texts, whose collective insights form the backbone of each textbook revision.

James Africh, College of DuPage
Steve Agronsky, Cal Poly State University
Gererdo Aladro, Florida International University
Grant Alexander, Joliet Junior College
Dave Anderson, South Suburban College
Richard Andrews, Florida A&M University
Joby Milo Anthony, University of Central Florida
James E. Arnold, University of Wisconsin-Milwaukee
Adel Arshaghi, Center for Educational Merit
Carolyn Autray, University of West Georgia
Agnes Azzolino, Middlesex County College
Wilson P. Banks, Illinois State University
Sudeshna Basu, Howard University
Dale R. Bedgood, East Texas State University
Beth Beno, South Suburban College
Carolyn Bernath, Tallahassee Community College
Rebecca Berthiaume, Edison State College
William H. Beyer, University of Akron
Annette Blackwelder, Florida State University
Richelle Blair, Lakeland Community College
Kevin Bodden, Lewis and Clark College
Jeffrey Boerner, University of Wisconsin-Stout
Barry Booten, Florida Atlantic University
Larry Bouldin, Roane State Community College
Bob Bradshaw, Ohlone College
Trudy Bratten, Grossmont College
Tim Bremer, Broome Community College
Tim Britt, Jackson State Community College
Michael Brook, University of Delaware
Joanne Brunner, Joliet Junior College
Warren Burch, Brevard Community College
Mary Butler, Lincoln Public Schools
Melanie Butler, West Virginia University
Jim Butterbach, Joliet Junior College
William J. Cable, University of Wisconsin-Stevens Point
Lois Calamia, Brookdale Community College
Jim Campbell, Lincoln Public Schools
Roger Carlsen, Moraine Valley Community College
Elena Catoiu, Joliet Junior College
Mathews Chakkanakuzhi, Palomar College
Tim Chappell, Penn Valley Community College
John Collado, South Suburban College
Alicia Collins, Mesa Community College
Nelson Collins, Joliet Junior College
Rebecca Connell, Troy University
Jim Cooper, Joliet Junior College
Denise Corbett, East Carolina University
Carlos C. Corona, San Antonio College
Theodore C. Coskey, South Seattle Community College
Rebecca Connell, Troy University
Donna Costello, Plano Senior High School
Paul Crittenden, University of Nebraska at Lincoln
John Davenport, East Texas State University
Faye Dang, Joliet Junior College
Antonio David, Del Mar College
Stephanie Deacon, Liberty University
Duane E. Deal, Ball State University
Jerry DeGroot, Purdue North Central
Timothy Deis, University of Wisconsin-Platteville
Joanna DelMonaco, Middlesex Community College
Vivian Dennis, Eastfield College
Deborah Dillon, R. L. Turner High School
Guesna Dohrman, Tallahassee Community College
Cheryl Doolittle, Iowa State University
Karen R. Dougan, University of Florida
Jerrett Dumouchel, Florida Community College at Jacksonville
Louise Dyson, Clark College
Paul D. East, Lexington Community College
Don Edmondson, University of Texas-Austin
Erica Egizio, Joliet Junior College
Jason Eltrevoog, Joliet Junior College
Christopher Ennis, University of Minnesota
Kathy Eppler, Salt Lake Community College
Ralph Esparza, Jr., Richland College
Garret J. Etgen, University of Houston
Scott Fallstrom, Shoreline Community College
Pete Falzone, Pensacola Junior College
Arash Farahmand, Skyline College
W.A. Ferguson, University of Illinois-Urbana/Champaign
Iris B. Fetta, Clemson University
Mason Flake, student at Edison Community College
Timothy W. Flood, Pittsburg State University
Robert Frank, Westmoreland County Community College
Merle Friel, Humboldt State University
Richard A. Fritz, Moraine Valley Community College
Dewey Furness, Ricks College
Mary Jule Gabiou, North Idaho College
Randy Gallaher, Lewis and Clark College
Tina Garn, University of Arizona
Dawit Getachew, Chicago State University
Wayne Gibson, Rancho Santiago College
Loran W. Gierhart, University of Texas at San Antonio and Palo Alto College
Robert Gill, University of Minnesota Duluth
Nina Girard, University of Pittsburgh at Johnstown
Sudhir Kumar Goel, Valdosta State University
Adrienne Goldstein, Miami Dade College, Kendall Campus
Joan Goliday, Sante Fe Community College
Lourdes Gonzalez, Miami Dade College, Kendall Campus
Frederic Gooding, Goucher College
Donald Goral, Northern Virginia Community College
Sue Graupner, Lincoln Public Schools
Mary Beth Grayson, Liberty University
Jennifer L. Grimsley, University of Charleston
Ken Gurganus, University of North Carolina
James E. Hall, University of Wisconsin-Madison
Judy Hall, West Virginia University
Edward R. Hancock, DeVry Institute of Technology
Julia Hassett, DeVry Institute, Dupage
Christopher Hay-Jahans, University of South Dakota
Michah Heibel, Lincoln Public Schools
LaRae Helliwell, San Jose City College
Celeste Hernandez, Richland College
Gloria P. Hernandez, Louisiana State University at Eunice
Brother Herron, Brother Rice High School
Robert Hoburg, Western Connecticut State University
Lynda Hollingsworth, Northwest Missouri State University
Deltrye Holt, Augusta State University
Charla Holzbog, Denison High School
Lee Hruby, Naperville North High School
Miles Hubbard, St. Cloud State University
Kim Hughes, California State College-San Bernardino
Stanislav, Jabuka, University of Nevada, Reno
Ron Jamison, Brigham Young University
Richard A. Jensen, Manatee Community College
Glenn Johnson, Middlesex Community College
Sandra G. Johnson, St. Cloud State University
Tuesday Johnson, New Mexico State University
Susitha Karunaratne, Purdue University North Central
Moana H. Karsteter, Tallahassee Community College
Donna Katula, Joliet Junior College

Arthur Kaufman, College of Staten Island
Thomas Kearns, North Kentucky University
Jack Keating, Massasoit Community College
Shelia Kellenbarger, Lincoln Public Schools
Rachael Kenney, North Carolina State University
John B. Klassen, North Idaho College
Debra Kopcso, Louisiana State University
Lynne Kowski, Raritan Valley Community College
Yelena Kravchuk, University of Alabama at Birmingham
Ray S. Kuan, Skyline College
Keith Kuchar, Manatee Community College
Tor Kwembe, Chicago State University
Linda J. Kyle, Tarrant Country Jr. College
H.E. Lacey, Texas A & M University
Harriet Lamm, Coastal Bend College
James Lapp, Fort Lewis College
Matt Larson, Lincoln Public Schools
Christopher Lattin, Oakton Community College
Julia Ledet, Lousiana State University
Adele LeGere, Oakton Community College
Kevin Leith, University of Houston
JoAnn Lewin, Edison College
Jeff Lewis, Johnson County Community College
Janice C. Lyon, Tallahassee Community College
Jean McArthur, Joliet Junior College
Virginia McCarthy, Iowa State University
Karla McCavit, Albion College
Michael McClendon, University of Central Oklahoma
Tom McCollow, DeVry Institute of Technology
Marilyn McCollum, North Carolina State University
Jill McGowan, Howard University
Will McGowant, Howard University
Angela McNulty, Joliet Junior College
Laurence Maher, North Texas State University
Jay A. Malmstrom, Oklahoma City Community College
Rebecca Mann, Apollo High School
Lynn Marecek, Santa Ana College
Sherry Martina, Naperville North High School
Alec Matheson, Lamar University
Nancy Matthews, University of Oklahoma
James Maxwell, Oklahoma State University-Stillwater
Marsha May, Midwestern State University
James McLaughlin, West Chester University
Judy Meckley, Joliet Junior College
David Meel, Bowling Green State University
Carolyn Meitler, Concordia University
Samia Metwali, Erie Community College
Rich Meyers, Joliet Junior College
Eldon Miller, University of Mississippi
James Miller, West Virginia University
Michael Miller, Iowa State University
Kathleen Miranda, SUNY at Old Westbury
Chris Mirbaha, The Community College of Baltimore County
Val Mohanakumar, Hillsborough Community College
Thomas Monaghan, Naperville North High School
Miguel Montanez, Miami Dade College, Wolfson Campus
Maria Montoya, Our Lady of the Lake University
Susan Moosai, Florida Atlantic University
Craig Morse, Naperville North High School
Samad Mortabit, Metropolitan State University
Pat Mower, Washburn University
Tammy Muhs, University of Central Florida
A. Muhundan, Manatee Community College
Jane Murphy, Middlesex Community College
Richard Nadel, Florida International University
Gabriel Nagy, Kansas State University
Bill Naegele, South Suburban College
Karla Neal, Lousiana State University
Lawrence E. Newman, Holyoke Community College
Dwight Newsome, Pasco-Hernando Community College
Denise Nunley, Maricopa Community Colleges
James Nymann, University of Texas-El Paso
Mark Omodt, Anoka-Ramsey Community College
Seth F. Oppenheimer, Mississippi State University
Leticia Oropesa, University of Miami
Linda Padilla, Joliet Junior College
Sanja Pantic, University of Illinois at Chicago
E. James Peake, Iowa State University
Kelly Pearson, Murray State University
Dashamir Petrela, Florida Atlantic University
Philip Pina, Florida Atlantic University
Charlotte Pisors, Baylor University
Michael Prophet, University of Northern Iowa
Laura Pyzdrowski, West Virginia University
Carrie Quesnell, Weber State University
Neal C. Raber, University of Akron
Thomas Radin, San Joaquin Delta College
Aibeng Serene Radulovic, Florida Atlantic University
Ken A. Rager, Metropolitan State College
Kenneth D. Reeves, San Antonio College
Elsi Reinhardt, Truckee Meadows Community College
Jose Remesar, Miami Dade College, Wolfson Campus
Jane Ringwald, Iowa State University
Douglas F. Robertson, University of Minnesota, MPLS
Stephen Rodi, Austin Community College
William Rogge, Lincoln Northeast High School
Howard L. Rolf, Baylor University
Mike Rosenthal, Florida International University
Phoebe Rouse, Lousiana State University
Edward Rozema, University of Tennessee at Chattanooga
Dennis C. Runde, Manatee Community College
Alan Saleski, Loyola University of Chicago
Susan Sandmeyer, Jamestown Community College
Brenda Santistevan, Salt Lake Community College
Linda Schmidt, Greenville Technical College
Ingrid Scott, Montgomery College
A.K. Shamma, University of West Florida
Zachery Sharon, University of Texas at San Antonio
Martin Sherry, Lower Columbia College
Carmen Shershin, Florida International University
Tatrana Shubin, San Jose State University
Anita Sikes, Delgado Community College
Timothy Sipka, Alma College
Charlotte Smedberg, University of Tampa
Lori Smellegar, Manatee Community College
Gayle Smith, Loyola Blakefield
Cindy Soderstrom, Salt Lake Community College
Leslie Soltis, Mercyhurst College
John Spellman, Southwest Texas State University
Karen Spike, University of North Carolina
Rajalakshmi Sriram, Okaloosa-Walton Community College
Katrina Staley, North Carolina Agricultural and Technical State University
Becky Stamper, Western Kentucky University
Judy Staver, Florida Community College-South
Robin Steinberg, Pima Community College
Neil Stephens, Hinsdale South High School
Sonya Stephens, Florida A&M Univeristy
Patrick Stevens, Joliet Junior College
John Sumner, University of Tampa
Matthew TenHuisen, University of North Carolina, Wilmington
Christopher Terry, Augusta State University
Diane Tesar, South Suburban College
Tommy Thompson, Brookhaven College
Martha K. Tietze, Shawnee Mission Northwest High School
Richard J. Tondra, Iowa State University
Florentina Tone, University of West Florida
Suzanne Topp, Salt Lake Community College
Marilyn Toscano, University of Wisconsin, Superior
Marvel Townsend, University of Florida
Jim Trudnowski, Carroll College
Robert Tuskey, Joliet Junior College
Mihaela Vajiac, Chapman University-Orange
Julia Varbalow, Thomas Nelson Community College-Leesville
Richard G. Vinson, University of South Alabama
Jorge Viola-Prioli, Florida Atlantic University
Mary Voxman, University of Idaho
Jennifer Walsh, Daytona Beach Community College
Donna Wandke, Naperville North High School
Timothy L. Warkentin, Cloud County Community College
Melissa J. Watts, Virginia State University
Hayat Weiss, Middlesex Community College
Kathryn Wetzel, Amarillo College
Darlene Whitkenack, Northern Illinois University
Suzanne Williams, Central Piedmont Community College
Larissa Williamson, University of Florida
Christine Wilson, West Virginia University
Brad Wind, Florida International University
Anna Wiodarczyk, Florida International University
Mary Wolyniak, Broome Community College
Canton Woods, Auburn University
Tamara S. Worner, Wayne State College
Terri Wright, New Hampshire Community Technical College, Manchester
Aletheia Zambesi, University of West Florida
George Zazi, Chicago State University
Steve Zuro, Joliet Junior College

Michael Sullivan
Chicago State University

Resources for Success

MyMathLab® Online Course (access code required)

MyMathLab delivers **proven results** in helping individual students succeed. It provides **engaging experiences** that personalize, stimulate, and measure learning for each student. And it comes from an **experienced partner** with educational expertise and an eye on the future. MyMathLab helps prepare students and gets them thinking more conceptually and visually through the following features:

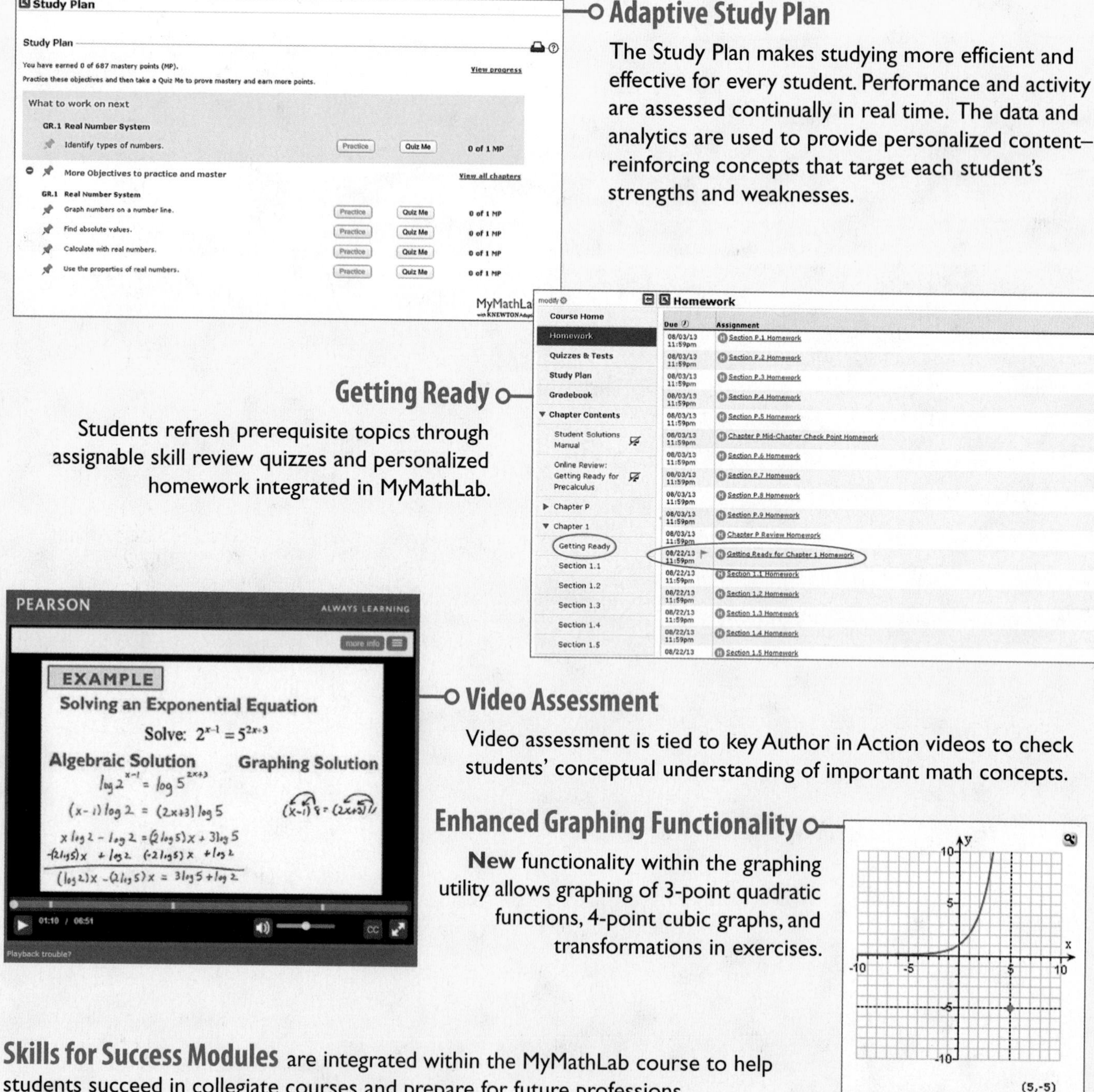

Adaptive Study Plan

The Study Plan makes studying more efficient and effective for every student. Performance and activity are assessed continually in real time. The data and analytics are used to provide personalized content–reinforcing concepts that target each student's strengths and weaknesses.

Getting Ready

Students refresh prerequisite topics through assignable skill review quizzes and personalized homework integrated in MyMathLab.

Video Assessment

Video assessment is tied to key Author in Action videos to check students' conceptual understanding of important math concepts.

Enhanced Graphing Functionality

New functionality within the graphing utility allows graphing of 3-point quadratic functions, 4-point cubic graphs, and transformations in exercises.

Skills for Success Modules are integrated within the MyMathLab course to help students succeed in collegiate courses and prepare for future professions.

Retain Your Knowledge These new exercises support ongoing review at the course level and help students maintain essential skills.

Instructor Resources

Additional resources can be downloaded from **www.pearsonhighered.com** or hardcopy resources can be ordered from your sales representative.

Ready to Go MyMathLab® Course

Now it is even easier to get started with MyMathLab. The Ready to Go MyMathLab course option includes author-chosen preassigned homework, integrated review, and more.

TestGen®

TestGen® (www.pearsoned.com/testgen) enables instructors to build, edit, print, and administer tests using a computerized bank of questions developed to cover all the objectives of the text.

PowerPoint® Lecture Slides

Fully editable slides correlated with the text.

Annotated Instructor's Edition

Shorter answers are on the page beside the exercises. Longer answers are in the back of the text.

Instructor Solutions Manual

Includes fully worked solutions to all exercises in the text.

Mini Lecture Notes

Includes additional examples and helpful teaching tips, by section.

Online Chapter Projects

Additional projects that give students an opportunity to apply what they learned in the chapter.

Student Resources

Additional resources to enhance student success:

Lecture Video

Author in Action videos are actual classroom lectures with fully worked out examples presented by Michael Sullivan, III. All video is assignable within MyMathlab.

Chapter Test Prep Videos

Students can watch instructors work through step-by-step solutions to all chapter test exercises from the text. These are available in MyMathlab and on YouTube.

Student Solutions Manual

Provides detailed worked-out solutions to odd-numbered exercises.

Guided Lecture Notes

These lecture notes assist students in taking thorough, organized, and understandable notes while watching Author in Action videos. Students actively participate in learning the how/why of important concepts through explorations and activities. The Guided Lecture Notes are available as PDF's and customizable Word files in MyMathLab. They can also be packaged with the text and the MyMathLab access code.

Algebra Review

Four chapters of Intermediate Algebra review. Perfect for a slower-paced course or for individual review.

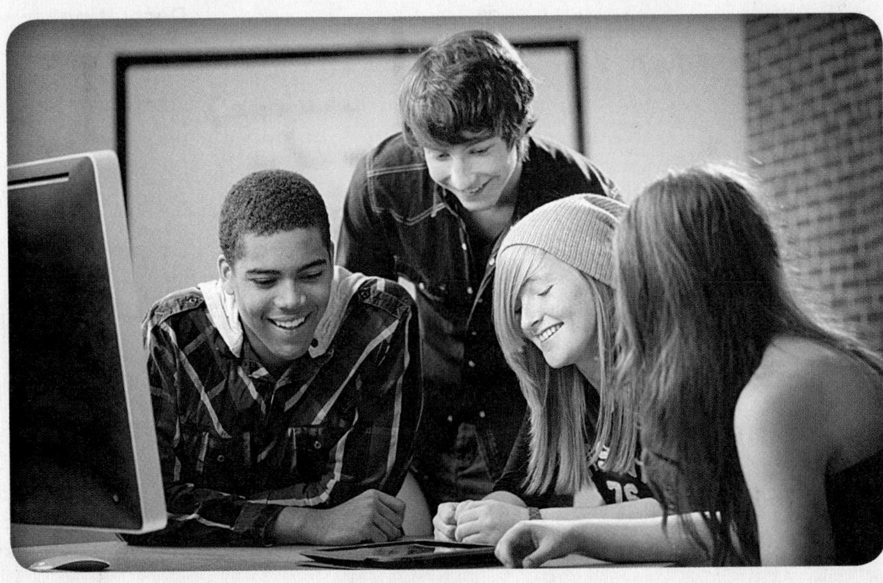

Applications Index

Acoustics
amplifying sound, 529
loudness of sound, 481
loudspeaker, 302
tuning fork, 302
whispering galleries, 403–404

Aerodynamics
modeling aircraft motion, 385

Aeronautics
Challenger disaster, 518

Agriculture
farm worker population trends, 517
milk production, 524
removing stump, 357

Air travel
bearing of aircraft, 269, 279
frequent flyer miles, 279
holding pattern, 216
revising a flight plan, 287
speed and direction of aircraft, 351–352, 355
ticket price and flight time, A85

Archaeology
age of ancient tools, 510–511
age of fossil, 516
age of tree, 516
date of prehistoric man's death, 529

Architecture
Burj Khalifa building, A15
Norman window, A20
racetrack design, 406

Area
of Bermuda Triangle, 293
under a curve, 201
of isosceles triangle, 247
of region, 112
of sector of circle, 107, 110
of segment of circle, 305

Art
fine decorative pieces, 129

Astronomy
angle of elevation of Sun, 268
planetary orbits, 403
  Earth, 406
  elliptical, 406
  Jupiter, 406
  Mars, 406
  Mercury, 433
  Neptune, 451
  Pluto, 406, 451
radius of the moon, 130

Aviation
modeling aircraft motion, 385

Biology
alcohol and driving, 477, 482
bacterial growth, 509–510, 523
  E-coli, 54
healing of wounds, 467, 481
maternal age versus Down syndrome, A85
muscle force, 356
yeast biomass as function of time, 522

Business
cable rates, 524
cost
  of manufacturing, 40, A13
  of production, 53
  of transporting goods, 63
cost equation, A77
drive-thru rate
  at Burger King, 463
  at Citibank, 467, 481
  at McDonald's, 468
Jiffy Lube's car arrival rate, 467, 481
new-car markup, A55
precision ball bearings, A13
price of plane tickets, A85
product promotion, A78
rate of return on, 505
revenue from digital music, 77
salary, gross, 38
sales
  commission on, A54
  net, 8
salvage value, 529
truck rentals, A77
wages of car salesperson, A77

Calculus
area under a curve, 78, 201
area under graph, 53
carrying a ladder around a corner, 216–217
index of refraction, 217
maximizing rain gutter construction, 247
projectile motion, 247

Carpentry
pitch, A79

Chemistry
alpha particles, 419
decomposition reactions, 517
ethanol production, 523
pH, 480
radioactive decay, 516, 523–524, 529
radioactivity from Chernobyl, 517
volume of gas, A54

Communications
cell phone towers, 525
data plan cost, 40
satellite dish, 393–394, 395
spreading of rumors, 467, 481
tablet service, 62
Touch-Tone phones, 251, 303
wireless data plan, 1, 97

Computers and computing
cost of manufacturing, 40
graphics, 357
home computer ownership, 517
mind-mapping software, 257
Word users, 517

Construction
of flashlight, 395
of garden walk, 185
of headlight, 395
of highway, 269, 280, 306
pitch of roof, 270
of rain gutter, 122, 246–247, 260–261
of ramp, 279
  access ramp, A78
of swimming pool, A20, A21
of swing set, 288
of tent, 292
TV dish, 395
vent pipe installation, 406

Crime
income *vs.* rate of, 525

Decorating
Christmas tree, A16

Demographics
diversity index, 480
life expectancy, A54
mosquito colony growth, 516

Design
of awning, 281
of fine decorative pieces, 129
of Little League Field, 113
of water sprinkler, 111

Direction
of aircraft, 351–352, 355
compass heading, 356
for crossing a river, 355, 356
of fireworks display, 418

of lightning strikes, 418
of motorboat, 355
of swimmer, 384

Distance

Bermuda Triangle, A21
bicycle riding, 41
from Chicago to Honolulu, 201
circumference of Earth, 113
between cities, 106–107, 111
between Earth and Mercury, 281
between Earth and Venus, 281
from Earth to a star, 268–269
of explosion, 419
height
  of aircraft, 279, 281
  of bridge, 279
  of building, 268, 269
  of cloud, 264
  of Eiffel Tower, 268
  of embankment, 269
  of Ferris Wheel rider, 216
  of Great Pyramid of Cheops, 281, A21
  of helicopter, 306
  of hot-air balloon, 269
  of Lincoln's caricature on Mt. Rushmore, 269
  of mountain, 276, 279
  of statue on a building, 264–265
  of tower, 270
  of tree, 279
  of Washington Monument, 269
  of Willis Tower, 269
from home, 41
from Honolulu to Melbourne, Australia, 201
of hot-air balloon
  to airport, 307
  from intersection, 8
length
  of guy wire, 287
  of mountain trail, 269
  of ski lift, 279
limiting magnitude of telescope, 529
to the Moon, 280
nautical miles, 112
to plateau, 268
across a pond, 268
reach of ladder, 268
of rotating beacon, 167
between runners, 279
at sea, 280
to shore, 268, 280, 305
between skyscrapers, 270, 271
to tower, 281
traveled by wheel, A20
tree height, 130
between two moving vehicles, 8
between two objects, 268, 269
viewing, 130
visibility of Gibb's Hill Lighthouse beam, 265–266, 271, A21
visual, A21
walking, 41
width
  of gorge, 267
  of Mississippi River, 270
  of river, 263, 305

Economics

Consumer Price Index (CPI), 507
federal stimulus package of 2009, 506
inflation, 506
per capita federal debt, 506
poverty threshold, 8

Education

college costs, 506
funding a college education, 529
grade computation, A55
IQ tests, A55
learning curve, 468, 481
tuition for colleges and universities, 529–530
video games and grade-point average, A84

Electricity

alternating current (ac), 183, 237
alternating current (ac) circuits, 158, 176
alternating current (ac) generators, 158–159
charging a capacitor, 303
current in *RC* circuit, 468
current in *RL* circuit, 468, 481
impedance, A45
parallel circuits, A45
rates for, A54, A77
voltage
  foreign, A13
  U.S., A13

Electronics

loudspeakers, 302
microphones, 21
sawtooth curve, 247, 303

Energy

nuclear power plant, 418–419
solar, 21, 364
solar heat, 396
thermostat control, 77

Engineering

bridges
  clearance, 159
  parabolic arch, 396
  semielliptical arch, 405, 406, 448
  suspension, 395–396
drive wheel, 306
Gateway Arch (St. Louis), 396
grade of road, A79
lean of Leaning Tower of Pisa, 280
moment of inertia, 251
piston engines, 129
product of inertia, 247
road system, 320
robotic arm, 374
rods and pistons, 288
searchlight, 225, 396, 448
whispering galleries, 405

Entertainment

cable rates, 524
Demon Roller Coaster customer rate, 468
movie theater, 200–201

Environment

endangered species, 467

Exercise and fitness

for weight loss, A54

Finance

balancing a checkbook, A13
college costs, 506
computer system purchase, 505
cost
  car rental, 63
  data plan, 53
  of driving car, A77
  of natural gas, 62
  of trans-Atlantic travel, 39
  of triangular lot, 292
credit cards
  interest on, 505
  payment, 63
depreciation, 467
  of car, 497, 532
electricity rates, A77
federal income tax, A54
federal stimulus package of 2009, 506
funding a college education, 529
gross salary, 38
income *vs.* crime rate, 525
loans
  mortgage fees, 63
  repayment of, 505
mortgages
  interest rates on, 506
  second, 506
price appreciation of homes, 505
saving for a car, 505
savings accounts interest, 505
taxes
  e-filing returns, 54
  federal income, 63, 90
used-car purchase, 505
water bills, A55

Food and nutrition

candy, A84
cooling time of pizza, 516
"light" foods, A55
raisins, A84
warming time of Beer stein, 517

Forestry
wood product classification, 515

Geography
area of Bermuda Triangle, 293
area of lake, 293, 306
inclination of mountain trail, 263, 305

Geology
earthquakes, 481–482

Geometry
angle between two lines, 237
circle
  area of, 293
  circumference of, A13
  length of chord of, 288
cube
  surface area of, A13
  volume of, A13
ladder angle, 307
quadrilateral area, 308
rectangle
  area of, 38, A13
  inscribed in ellipse, 406
  inscribed in semicircle, 247
  perimeter of, A13
semicircle area, 293, 308
sphere
  surface area of, A13
  volume of, A13
square, area of, A20
surface area
  of cube, A13
  of sphere, A13
triangle
  area of, 292, 293, 308, A13
  circumscribing, 282
  equilateral, A13
  isosceles, 38, 308
  perimeter of, A13
  right, 267
  sides of, 308
volume of parallelepiped, 380

Government
federal income tax, 63, 90, A54
  e-filing returns, 54
federal stimulus package of 2009, 506
first-class mail, 64
national debt, 53
per capita federal debt, 506

Health. *See also* **Medicine**
blood pressure, 216
cigarette use among teens, A78
ideal body weight, 90

Housing
number of rooms in, 39
price appreciation of homes, 505

Investment(s)
in zero-coupon bonds, 503, 506
compound interest on, 498–499, 500, 501–502, 505
doubling of, 503–504, 506
finance charges, 505
in fixed-income securities, 506
growth rate for, 505
IRA, 506
money market account, 502
mutual fund growth rate, 520
return on, 505
savings account, 501–502
stock appreciation, 505
time to reach goal, 505, 507
tripling of, 504, 506

Landscaping
building a walk, 185
height of tree, 279
removing stump, 357
watering lawn, 111

Leisure and recreation
Ferris wheel, 22, 216, 281, 302
video games and grade-point average, A84

Measurement
optical methods of, 225
of rainfall, 364

Mechanics. *See* **Physics**

Medicine. *See also* **Health**
blood pressure, 216
breast cancer, 523
drug concentration, 53
drug medication, 467, 481
healing of wounds, 467, 481
pancreatic cancer, 467
spreading of disease, 530

Meteorology
weather balloon height and atmospheric pressure, 521

Miscellaneous
biorhythms, 159
birthdays shared in roomful of people, 517
carrying a ladder around a corner, 167, 216–217
coffee container, 532
cross-sectional area of beam, 39
drafting error, 8
grazing area for cow, 294
land dimensions, 279
Mandelbrot sets, 343
surveillance satellites, 270–271

Motion, 303. *See also* **Physics**
catching a train, 448
on a circle, 111
of Ferris Wheel rider, 216
of golf ball, 39–40
minute hand of clock, 110, 183
objects approaching intersection, 445
of pendulum, 303, 307
revolutions of circular disk, A20
simulating, 439
uniform, 445

Motor vehicles
alcohol and driving, 477, 482
approaching intersection, 445
braking load, 364, 384
crankshafts, 280
depreciation of, 497, 532
with Global Positioning System (GPS), 529
new-car markup, A55
parking fee, 62
rental costs, 63
revolutions per minute of wheel, 497
spin balancing tires, 112
stopping distance, 90
tailgating, 130
used-car purchase, 505
windshield wiper, area cleaned by, 111

Music
revenues from, 77

Navigation
avoiding a tropical storm, 287
bearing, 265–266, 286
  of aircraft, 269, 279
  of ship, 269, 306
charting a course, 356
commercial, 279
compass heading, 356
crossing a river, 355, 356
error in
  correcting, 284–285, 306
  time lost due to, 279
rescue at sea, 276–277, 279–280
revising a flight plan, 287

Oceanography
tides, 177

Optics
angle of refraction, 217
bending light, 218
Brewster's Law, 218
index of refraction, 217
laser beam, 268
laser projection, 247
light obliterated through glass, 467
mirrors, 419
optical measurement, 225
reflecting telescope, 396

Pediatrics
height vs. head circumference, 90, A84

Photography
camera distance, 269

Physics
angle of elevation of Sun, 268
damped motion, 298, 306–307
force, 355
  to hold a wagon on a hill, 361–362
  muscle, 356
  resultant, 355
gravity
  on Earth, 39, 90
  on Jupiter, 39
harmonic motion, 297, 306–307
heat transfer, 216
inclination of mountain trail, 263
incline angle, 364
inclined ramp, 356
moment of inertia, 251
motion of object, 297, 306–307
pendulum motion, 110, 303, 307, A64
  period, 78, 90
product of inertia, 247
projectile motion, 128–129, 130, 216, 217, 242, 247, 251, 437–438, 444–445, 448
  artillery, 207
  thrown object, 351, 444
simulating motion, 439
static equilibrium, 352–353, 356, 357, 384
static friction, 356
tension, 352–353, 356, 384
truck pulls, 357
uniform motion, 445, 448
velocity down inclined planes, A63
weight
  of a boat, 355
  of a car, 355
  of a piano, 352
work, 374

Play
swinging, 308
wagon pulling, 355, 362

Population. *See also* **Demographics**
bacterial, 516, 523
decline in, 516
E-coli growth, 54
of endangered species, 517–518
of fruit fly, 514
as function of age, 38
growth in, 516, 517, 518
insect, 516
of United States, 497, 524
of world, 497, 525, 529

Probability
exponential, 463, 467, 481
Poisson, 468

Pyrotechnics
fireworks display, 418

Rate. *See also* **Speed**
of car, 111
catching a bus, 444
catching a train, 444
to keep up with the Sun, 112
revolutions per minute
  of bicycle wheels, 111
  of pulleys, 113

Real estate
commission schedule, A54
cost of triangular lot, 292

Recreation
Demon Roller Coaster customer rate, 468

Security
security cameras, 268

Seismology
calibrating instruments, 448

Speed
of aircraft, 355
angular, 111, 183
  of Ferris wheel, 112
of current, 112
as function of time, 41
of glider, 305
linear, 108–109
  on Earth, 111, 112
  of Ferris wheel, 112
  of International Space Station (ISS), 238
of Moon, 111
revolutions per minute, 111, 112
of rotation of lighthouse beacons, 183
of swimmer, 384
of truck, 268
of wheel pulling cable cars, 112

Sports
baseball, 444, 445
  diamond, 7
  dimensions of home plate, 292
  field, 287, 288
  Little League, 7, 113
  on-base percentage, A79–A80
  stadium, 287
basketball
  free throws, 39, 270
  granny shots, 39
  pool shots, 271
distance between runners, 279
football, 406
golf, 39–40, 437–438, 444, 526
  distance to the green, 286
  sand bunkers, 207
hammer throw, 185
swimming, 308, 384

Statistics. *See* **Probability**

Temperature
body, A13
conversion of, 90
cooling time of pizza, 516
measuring, A77
monthly, 176–177, 183
relationship between scales, 77
sinusoidal function from, 172–173
of skillet, 529
thermostat control, 77
warming time of Beer stein, 517
wind chill factor, 530

Tests and testing
IQ, A55

Time
for Beer stein to warm, 517
for block to slide down inclined plane, 129
Ferris Wheel rider height as function of, 216
hours of daylight, 99, 174–175, 178, 186–187, 200
for pizza to cool, 516
of sunrise, 112, 200
of trip, 129, 144

Transportation
deicing salt, 207
Niagara Falls Incline Railway, 269

Travel. *See also* **Air travel; Navigation**
bearing, 279, 306
drivers stopped by the police, 531

Volume
of gasoline in tank, A63

Weapons
artillery, 207

Weather
atmospheric pressure, 467, 481
avoiding a tropical storm, 287
hurricanes, 176
lightning strikes, 415–416, 418
rainfall measurement, 364
relative humidity, 468
wind chill, 63–64, 530

Work, 362
computing, 362, 363, 384
pulling a wagon, 362
ramp angle, 364
wheelbarrow push, 355

Trigonometry

A Unit Circle Approach

Tenth Edition

Graphs and Functions

1

Choosing a Wireless Data Plan

Most consumers choose a cellular provider first and then select an appropriate data plan from that provider. The choice as to the type of plan selected depends on your use of the device. For example, is online gaming important? Do you want to stream audio or video? The mathematics learned in this chapter can help you decide what plan is best suited to your particular needs.

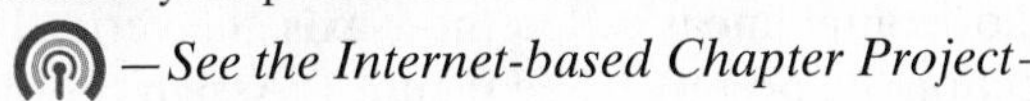
—See the Internet-based Chapter Project—

••• A Look Back

Appendix A reviews skills from intermediate algebra.

A Look Ahead •••

Here we connect algebra and geometry using the rectangular coordinate system to graph equations in two variables. Then we look at a special type of equation involving two variables called a *function*. This chapter deals with what a function is, how to graph functions, properties of functions, and how functions are used in applications. The word *function* apparently was introduced by René Descartes in 1637. For him, a function was simply any positive integral power of a variable *x*. Gottfried Wilhelm Leibniz (1646–1716), who always emphasized the geometric side of mathematics, used the word function to denote any quantity associated with a curve, such as the coordinates of a point on the curve. Leonhard Euler (1707–1783) employed the word to mean any equation or formula involving variables and constants. His idea of a function is similar to the one most often seen in courses that precede calculus. Later, the use of functions in investigating heat flow equations led to a very broad definition that originated with Lejeune Dirichlet (1805–1859), which describes a function as a correspondence between two sets. That is the definition used in this text.

Outline

1.1 The Distance and Midpoint Formulas
1.2 Graphs of Equations in Two Variables; Circles
1.3 Functions and Their Graphs
1.4 Properties of Functions
1.5 Library of Functions; Piecewise-defined Functions
1.6 Graphing Techniques: Transformations
1.7 One-to-One Functions; Inverse Functions
Chapter Review
Chapter Test
Chapter Projects

1.1 The Distance and Midpoint Formulas

PREPARING FOR THIS SECTION *Before getting started, review the following:*

- Algebra Essentials (Appendix A, Section A.1, pp. A1–A10)
- Geometry Essentials (Appendix A, Section A.2, pp. A14–A18)

Now Work the **'Are You Prepared?'** problems on page 5.

OBJECTIVES 1 Use the Distance Formula (p. 3)
2 Use the Midpoint Formula (p. 5)

Rectangular Coordinates

We locate a point on the real number line by assigning it a single real number, called the *coordinate of the point*. For work in a two-dimensional plane, we locate points by using two numbers.

Begin with two real number lines located in the same plane: one horizontal and the other vertical. The horizontal line is called the ***x*-axis**, the vertical line the ***y*-axis**, and the point of intersection the **origin *O***. See Figure 1. Assign coordinates to every point on these number lines using a convenient scale. In mathematics, we usually use the same scale on each axis, but in applications, different scales appropriate to the application may be used.

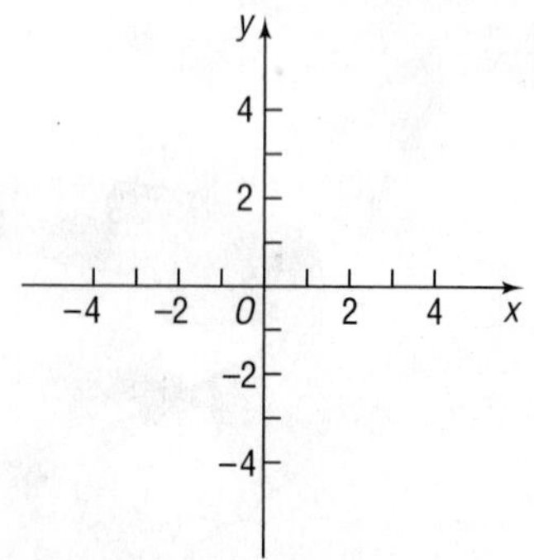

Figure 1 *xy*-plane

The origin O has a value of 0 on both the x-axis and the y-axis. Points on the x-axis to the right of O are associated with positive real numbers, and those to the left of O are associated with negative real numbers. Points on the y-axis above O are associated with positive real numbers, and those below O are associated with negative real numbers. In Figure 1, the x-axis and y-axis are labeled as x and y, respectively, and an arrow at the end of each axis is used to denote the positive direction.

The coordinate system described here is called a **rectangular** or **Cartesian*** **coordinate system**. The plane formed by the x-axis and y-axis is sometimes called the ***xy*-plane**, and the x-axis and y-axis are referred to as the **coordinate axes**.

Any point P in the xy-plane can be located by using an **ordered pair** (x, y) of real numbers. Let x denote the signed distance of P from the y-axis (*signed* means that if P is to the right of the y-axis, then $x > 0$, and if P is to the left of the y-axis, then $x < 0$); and let y denote the signed distance of P from the x-axis. The ordered pair (x, y), also called the **coordinates** of P, gives us enough information to locate the point P in the plane.

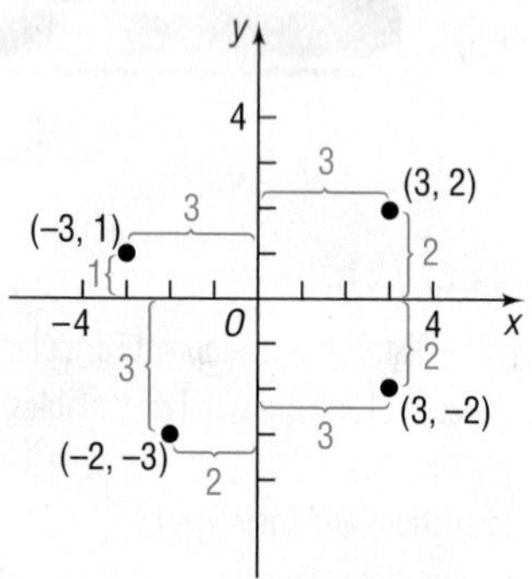

Figure 2

For example, to locate the point whose coordinates are $(-3, 1)$, go 3 units along the x-axis to the left of O and then go straight up 1 unit. We **plot** this point by placing a dot at this location. See Figure 2, in which the points with coordinates $(-3, 1)$, $(-2, -3)$, $(3, -2)$, and $(3, 2)$ are plotted.

The origin has coordinates $(0, 0)$. Any point on the x-axis has coordinates of the form $(x, 0)$, and any point on the y-axis has coordinates of the form $(0, y)$.

If (x, y) are the coordinates of a point P, then x is called the ***x*-coordinate**, or **abscissa**, of P, and y is the ***y*-coordinate**, or **ordinate**, of P. We identify the point P by its coordinates (x, y) by writing $P = (x, y)$. Usually, we will simply say "the point (x, y)" rather than "the point whose coordinates are (x, y)."

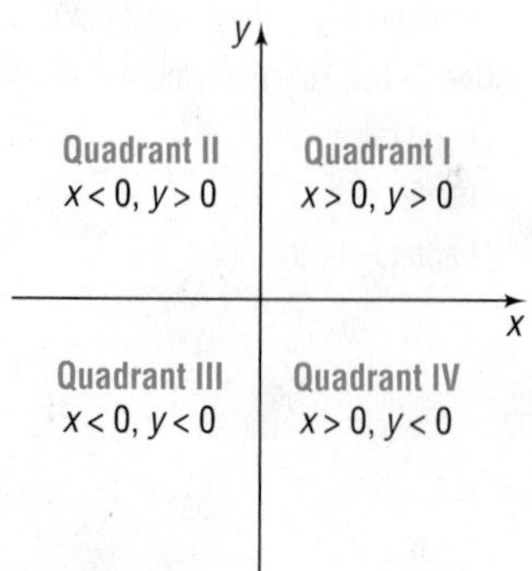

Figure 3

The coordinate axes divide the xy-plane into four sections called **quadrants**, as shown in Figure 3. In quadrant I, both the x-coordinate and the y-coordinate of all points are positive; in quadrant II, x is negative and y is positive; in quadrant III, both x and y are negative; and in quadrant IV, x is positive and y is negative. Points on the coordinate axes belong to no quadrant.

Now Work PROBLEM 15

* Named after René Descartes (1596–1650), a French mathematician, philosopher, and theologian.

COMMENT On a graphing calculator, you can set the scale on each axis. Once this has been done, you obtain the **viewing rectangle.** See Figure 4 for a typical viewing rectangle. You should now read Section B.1, *The Viewing Rectangle,* in Appendix B.

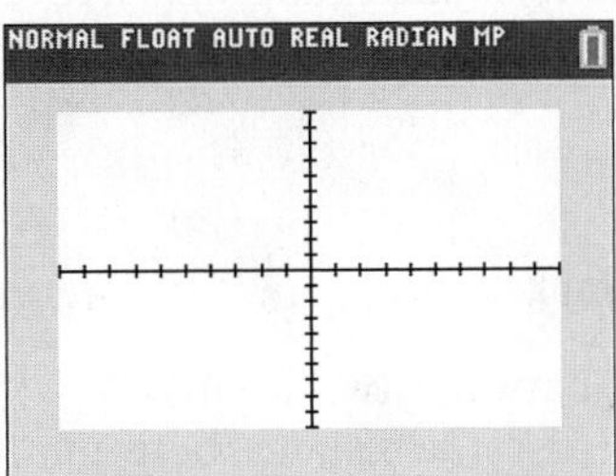

Figure 4 TI-84 Plus C Standard Viewing Rectangle

1 Use the Distance Formula

If the same units of measurement (such as inches, centimeters, and so on) are used for both the x-axis and y-axis, then all distances in the xy-plane can be measured using this unit of measurement.

EXAMPLE 1

Finding the Distance between Two Points

Find the distance d between the points $(1, 3)$ and $(5, 6)$.

Solution First plot the points $(1, 3)$ and $(5, 6)$ and connect them with a straight line. See Figure 5(a). To find the length d, begin by drawing a horizontal line from $(1, 3)$ to $(5, 3)$ and a vertical line from $(5, 3)$ to $(5, 6)$, forming a right triangle, as shown in Figure 5(b). One leg of the triangle is of length 4 (since $|5 - 1| = 4$), and the other is of length 3 (since $|6 - 3| = 3$). By the Pythagorean Theorem, the square of the distance d that we seek is

$$d^2 = 4^2 + 3^2 = 16 + 9 = 25$$
$$d = \sqrt{25} = 5$$

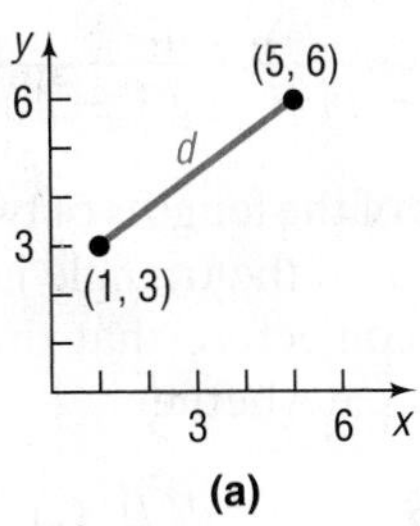

(a)

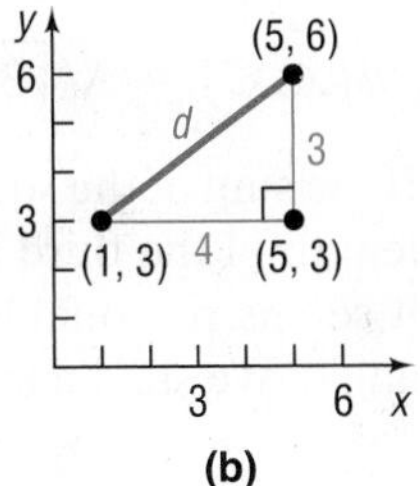

(b)

Figure 5

In Words
To compute the distance between two points, find the difference of the x-coordinates, square it, and add this to the square of the difference of the y-coordinates. The square root of this sum is the distance.

The **distance formula** provides a straightforward method for computing the distance between two points.

THEOREM

Distance Formula

The distance between two points $P_1 = (x_1, y_1)$ and $P_2 = (x_2, y_2)$, denoted by $d(P_1, P_2)$, is

$$d(P_1, P_2) = \sqrt{(x_2 - x_1)^2 + (y_2 - y_1)^2} \qquad (1)$$

EXAMPLE 2 Using the Distance Formula

Find the distance d between the points $(-4, 5)$ and $(3, 2)$.

Solution Using the distance formula, equation (1), reveals that the distance d is

$$d = \sqrt{[3 - (-4)]^2 + (2 - 5)^2} = \sqrt{7^2 + (-3)^2}$$
$$= \sqrt{49 + 9} = \sqrt{58} \approx 7.62$$

Now Work PROBLEMS 19 AND 23

The distance between two points $P_1 = (x_1, y_1)$ and $P_2 = (x_2, y_2)$ is never a negative number. Also, the distance between two points is 0 only when the points are identical—that is, when $x_1 = x_2$ and $y_1 = y_2$. And, because $(x_2 - x_1)^2 = (x_1 - x_2)^2$ and $(y_2 - y_1)^2 = (y_1 - y_2)^2$, it makes no difference whether the distance is computed from P_1 to P_2 or from P_2 to P_1; that is, $d(P_1, P_2) = d(P_2, P_1)$.

The introduction to this chapter mentioned that rectangular coordinates enable us to translate geometry problems into algebra problems, and vice versa. The next example shows how algebra (the distance formula) can be used to solve geometry problems.

EXAMPLE 3 Using Algebra to Solve Geometry Problems

Consider the three points $A = (-2, 1)$, $B = (2, 3)$, and $C = (3, 1)$.

(a) Plot each point and form the triangle ABC.
(b) Find the length of each side of the triangle.
(c) Show that the triangle is a right triangle.
(d) Find the area of the triangle.

Solution (a) Figure 6 shows the points A, B, C and the triangle ABC.

(b) To find the length of each side of the triangle, use the distance formula, equation (1).

$$d(A, B) = \sqrt{[2 - (-2)]^2 + (3 - 1)^2} = \sqrt{16 + 4} = \sqrt{20} = 2\sqrt{5}$$
$$d(B, C) = \sqrt{(3 - 2)^2 + (1 - 3)^2} = \sqrt{1 + 4} = \sqrt{5}$$
$$d(A, C) = \sqrt{[3 - (-2)]^2 + (1 - 1)^2} = \sqrt{25 + 0} = 5$$

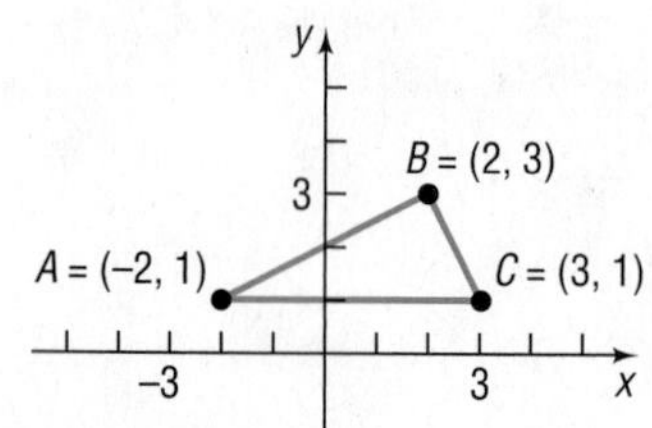

Figure 6

(c) If the sum of the squares of the lengths of two of the sides equals the square of the length of the third side, then the triangle is a right triangle. Looking at Figure 6, it seems reasonable to conjecture that the angle at vertex B might be a right angle. We shall check to see whether

$$[d(A, B)]^2 + [d(B, C)]^2 = [d(A, C)]^2$$

Using the results in part (b) yields

$$[d(A, B)]^2 + [d(B, C)]^2 = (2\sqrt{5})^2 + (\sqrt{5})^2$$
$$= 20 + 5 = 25 = [d(A, C)]^2$$

It follows from the converse of the Pythagorean Theorem that triangle ABC is a right triangle.

(d) Because the right angle is at vertex B, the sides AB and BC form the base and height of the triangle. Its area is

$$\text{Area} = \frac{1}{2}(\text{Base})(\text{Height}) = \frac{1}{2}(2\sqrt{5})(\sqrt{5}) = 5 \text{ square units}$$

Now Work PROBLEM 31

2 Use the Midpoint Formula

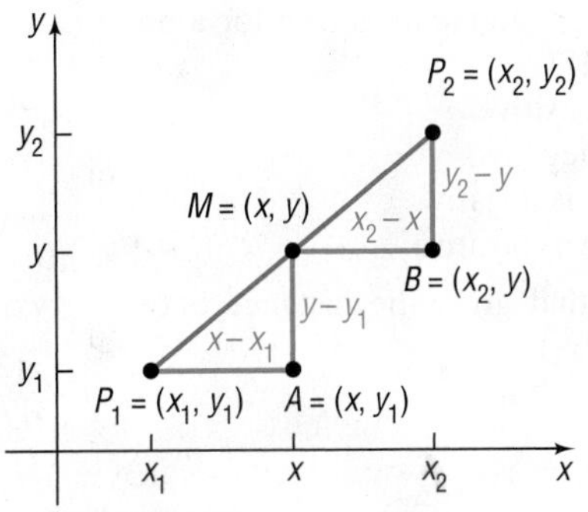

We now derive a formula for the coordinates of the **midpoint of a line segment.** Let $P_1 = (x_1, y_1)$ and $P_2 = (x_2, y_2)$ be the endpoints of a line segment, and let $M = (x, y)$ be the point on the line segment that is the same distance from P_1 as it is from P_2. See Figure 7. The triangles P_1AM and MBP_2 are congruent. [Do you see why? $d(P_1, M) = d(M, P_2)$ is given; also, $\angle AP_1M = \angle BMP_2$* and $\angle P_1MA = \angle MP_2B$. Thus, we have angle–side–angle.] Because triangles P_1AM and MBP_2 are congruent, corresponding sides are equal in length. That is,

Figure 7

$$x - x_1 = x_2 - x \qquad \text{and} \qquad y - y_1 = y_2 - y$$
$$2x = x_1 + x_2 \qquad\qquad 2y = y_1 + y_2$$
$$x = \frac{x_1 + x_2}{2} \qquad\qquad y = \frac{y_1 + y_2}{2}$$

THEOREM

Midpoint Formula

The midpoint $M = (x, y)$ of the line segment from $P_1 = (x_1, y_1)$ to $P_2 = (x_2, y_2)$ is

$$M = (x, y) = \left(\frac{x_1 + x_2}{2}, \frac{y_1 + y_2}{2}\right) \qquad \textbf{(2)}$$

In Words

To find the midpoint of a line segment, average the x-coordinates of the endpoints, and average the y-coordinates of the endpoints.

EXAMPLE 4 **Finding the Midpoint of a Line Segment**

Find the midpoint of the line segment from $P_1 = (-5, 5)$ to $P_2 = (3, 1)$. Plot the points P_1 and P_2 and their midpoint.

Solution Apply the midpoint formula (2) using $x_1 = -5, y_1 = 5, x_2 = 3$, and $y_2 = 1$. Then the coordinates (x, y) of the midpoint M are

$$x = \frac{x_1 + x_2}{2} = \frac{-5 + 3}{2} = -1 \quad \text{and} \quad y = \frac{y_1 + y_2}{2} = \frac{5 + 1}{2} = 3$$

That is, $M = (-1, 3)$. See Figure 8. ●

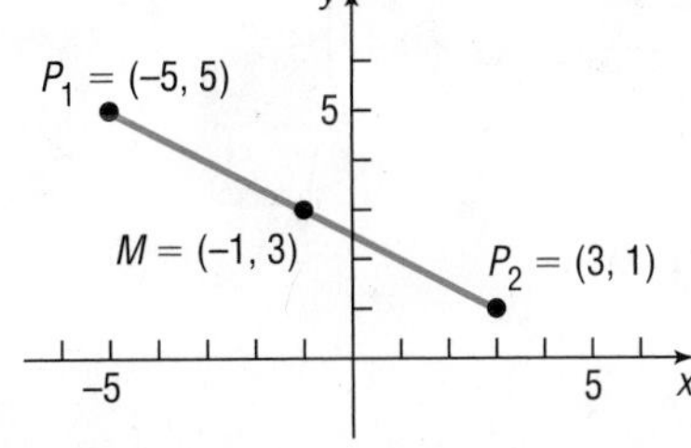

Figure 8

Now Work PROBLEM 37

1.1 Assess Your Understanding

'Are You Prepared?' *Answers are given at the end of these exercises. If you get a wrong answer, read the pages listed in red.*

1. On the real number line, the origin is assigned the number ________. (p. A4)

2. If −3 and 5 are the coordinates of two points on the real number line, the distance between these points is ________. (pp. A5–A6)

3. If 3 and 4 are the legs of a right triangle, the hypotenuse is ________. (p. A14)

4. Use the converse of the Pythagorean Theorem to show that a triangle whose sides are of lengths 11, 60, and 61 is a right triangle. (pp. A14–A15)

5. The area A of a triangle whose base is b and whose altitude is h is $A =$ ______. (p. A15)

6. ***True or False*** Two triangles are congruent if two angles and the included side of one equals two angles and the included side of the other. (pp. A16–A17).

*A postulate from geometry states that the transversal $\overline{P_1P_2}$ forms congruent corresponding angles with the parallel line segments $\overline{P_1A}$ and $\overline{MB}$.

Concepts and Vocabulary

7. If (x, y) are the coordinates of a point P in the xy-plane, then x is called the ______________ of P, and y is the ______________ of P.

8. The coordinate axes divide the xy-plane into four sections called ______________.

9. If three distinct points P, Q, and R all lie on a line, and if $d(P, Q) = d(Q, R)$, then Q is called the ______________ of the line segment from P to R.

10. *True or False* The distance between two points is sometimes a negative number.

11. *True or False* The point $(-1, 4)$ lies in quadrant IV of the Cartesian plane.

12. *True or False* The midpoint of a line segment is found by averaging the x-coordinates and averaging the y-coordinates of the endpoints.

13. Which of the following statements is true for a point (x, y) that lies in quadrant III?
(a) Both x and y are positive.
(b) Both x and y are negative.
(c) x is positive, and y is negative.
(d) x is negative, and y is positive.

14. Choose the formula that gives the distance between two points (x_1, y_1) and (x_2, y_2).
(a) $\sqrt{(x_2 - x_1)^2 + (y_2 - y_1)^2}$
(b) $\sqrt{(x_2 + x_1)^2 - (y_2 + y_1)^2}$
(c) $\sqrt{(x_2 - x_1)^2 - (y_2 - y_1)^2}$
(d) $\sqrt{(x_2 + x_1)^2 + (y_2 + y_1)^2}$

Skill Building

In Problems 15 and 16, plot each point in the xy-plane. Tell in which quadrant or on what coordinate axis each point lies.

15. (a) $A = (-3, 2)$ (b) $B = (6, 0)$ (c) $C = (-2, -2)$ (d) $D = (6, 5)$ (e) $E = (0, -3)$ (f) $F = (6, -3)$

16. (a) $A = (1, 4)$ (b) $B = (-3, -4)$ (c) $C = (-3, 4)$ (d) $D = (4, 1)$ (e) $E = (0, 1)$ (f) $F = (-3, 0)$

17. Plot the points $(2, 0)$, $(2, -3)$, $(2, 4)$, $(2, 1)$, and $(2, -1)$. Describe the set of all points of the form $(2, y)$, where y is a real number.

18. Plot the points $(0, 3)$, $(1, 3)$, $(-2, 3)$, $(5, 3)$, and $(-4, 3)$. Describe the set of all points of the form $(x, 3)$, where x is a real number.

In Problems 19–30, find the distance $d(P_1, P_2)$ between the points P_1 and P_2.

19.
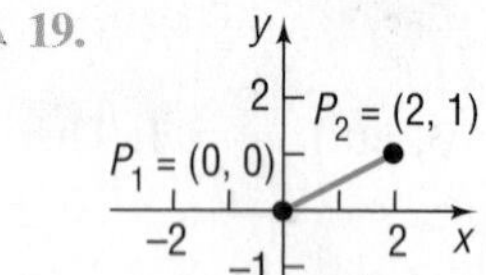

20.
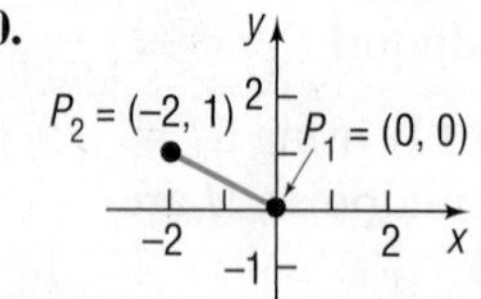

21.
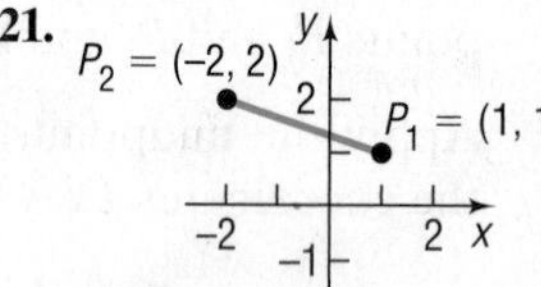

22.
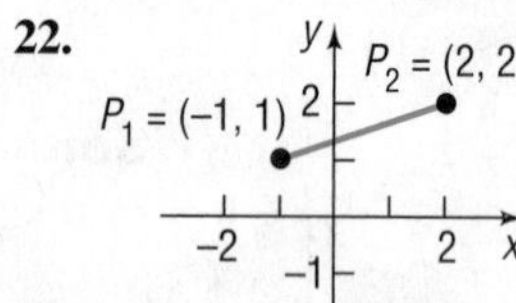

23. $P_1 = (3, -4)$; $P_2 = (5, 4)$

24. $P_1 = (-1, 0)$; $P_2 = (2, 4)$

25. $P_1 = (-3, 2)$; $P_2 = (6, 0)$

26. $P_1 = (2, -3)$; $P_2 = (4, 2)$

27. $P_1 = (4, -3)$; $P_2 = (6, 4)$

28. $P_1 = (-4, -3)$; $P_2 = (6, 2)$

29. $P_1 = (a, b)$; $P_2 = (0, 0)$

30. $P_1 = (a, a)$; $P_2 = (0, 0)$

In Problems 31–36, plot each point and form the triangle ABC. Show that the triangle is a right triangle. Find its area.

31. $A = (-2, 5)$; $B = (1, 3)$; $C = (-1, 0)$

32. $A = (-2, 5)$; $B = (12, 3)$; $C = (10, -11)$

33. $A = (-5, 3)$; $B = (6, 0)$; $C = (5, 5)$

34. $A = (-6, 3)$; $B = (3, -5)$; $C = (-1, 5)$

35. $A = (4, -3)$; $B = (0, -3)$; $C = (4, 2)$

36. $A = (4, -3)$; $B = (4, 1)$; $C = (2, 1)$

In Problems 37–44, find the midpoint of the line segment joining the points P_1 and P_2.

37. $P_1 = (3, -4)$; $P_2 = (5, 4)$

38. $P_1 = (-2, 0)$; $P_2 = (2, 4)$

39. $P_1 = (-3, 2)$; $P_2 = (6, 0)$

40. $P_1 = (2, -3)$; $P_2 = (4, 2)$

41. $P_1 = (4, -3)$; $P_2 = (6, 1)$

42. $P_1 = (-4, -3)$; $P_2 = (2, 2)$

43. $P_1 = (a, b)$; $P_2 = (0, 0)$

44. $P_1 = (a, a)$; $P_2 = (0, 0)$

Applications and Extensions

45. If the point $(2, 5)$ is shifted 3 units to the right and 2 units down, what are its new coordinates?

46. If the point $(-1, 6)$ is shifted 2 units to the left and 4 units up, what are its new coordinates?

47. Find all points having an x-coordinate of 3 whose distance from the point $(-2, -1)$ is 13.
(a) By using the Pythagorean Theorem.
(b) By using the distance formula.

48. Find all points having a y-coordinate of -6 whose distance from the point $(1, 2)$ is 17.
(a) By using the Pythagorean Theorem.
(b) By using the distance formula.

49. Find all points on the x-axis that are 6 units from the point $(4, -3)$.

50. Find all points on the y-axis that are 6 units from the point $(4, -3)$.

51. Suppose that $A = (2, 5)$ are the coordinates of a point in the xy-plane.
(a) Find the coordinates of the point if A is shifted 3 units to the left and 4 units down.
(b) Find the coordinates of the point if A is shifted 2 units to the left and 8 units up.

52. Plot the points $A = (-1, 8)$ and $M = (2, 3)$ in the xy-plane. If M is the midpoint of a line segment AB, find the coordinates of B.

53. The midpoint of the line segment from P_1 to P_2 is $(-1, 4)$. If $P_1 = (-3, 6)$, what is P_2?

54. The midpoint of the line segment from P_1 to P_2 is $(5, -4)$. If $P_2 = (7, -2)$, what is P_1?

55. Geometry The **medians** of a triangle are the line segments from each vertex to the midpoint of the opposite side (see the figure). Find the lengths of the medians of the triangle with vertices at $A = (0, 0)$, $B = (6, 0)$, and $C = (4, 4)$.

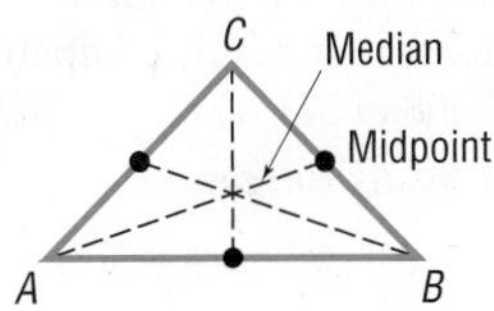

56. Geometry An **equilateral triangle** is one in which all three sides are of equal length. If two vertices of an equilateral triangle are $(0, 4)$ and $(0, 0)$, find the third vertex. How many of these triangles are possible?

57. Geometry Find the midpoint of each diagonal of a square with side of length s. Draw the conclusion that the diagonals of a square intersect at their midpoints.
[**Hint:** Use $(0, 0)$, $(0, s)$, $(s, 0)$, and (s, s) as the vertices of the square.]

58. Geometry Verify that the points $(0,0)$, $(a,0)$, and $\left(\frac{a}{2}, \frac{\sqrt{3}a}{2}\right)$ are the vertices of an equilateral triangle. Then show that the midpoints of the three sides are the vertices of a second equilateral triangle (refer to Problem 56).

*In Problems 59–62, find the length of each side of the triangle determined by the three points P_1, P_2, and P_3. State whether the triangle is an isosceles triangle, a right triangle, neither of these, or both. (An **isosceles triangle** is one in which at least two of the sides are of equal length.)*

59. $P_1 = (2, 1)$; $P_2 = (-4, 1)$; $P_3 = (-4, -3)$

60. $P_1 = (-1, 4)$; $P_2 = (6, 2)$; $P_3 = (4, -5)$

61. $P_1 = (-2, -1)$; $P_2 = (0, 7)$; $P_3 = (3, 2)$

62. $P_1 = (7, 2)$; $P_2 = (-4, 0)$; $P_3 = (4, 6)$

63. Baseball A major league baseball "diamond" is actually a square 90 feet on a side (see the figure). What is the distance directly from home plate to second base (the diagonal of the square)?

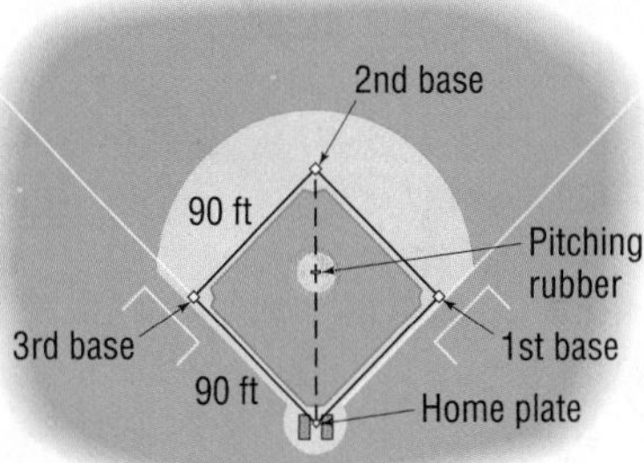

64. Little League Baseball The layout of a Little League playing field is a square 60 feet on a side. How far is it directly from home plate to second base (the diagonal of the square)?
Source: Little League Baseball, Official Regulations and Playing Rules, 2014.

65. Baseball Refer to Problem 63. Overlay a rectangular coordinate system on a major league baseball diamond so that the origin is at home plate, the positive x-axis lies in the direction from home plate to first base, and the positive y-axis lies in the direction from home plate to third base.
(a) What are the coordinates of first base, second base, and third base? Use feet as the unit of measurement.
(b) If the right fielder is located at $(310, 15)$, how far is it from the right fielder to second base?
(c) If the center fielder is located at $(300, 300)$, how far is it from the center fielder to third base?

66. Little League Baseball Refer to Problem 64. Overlay a rectangular coordinate system on a Little League baseball diamond so that the origin is at home plate, the positive x-axis lies in the direction from home plate to first base, and the positive y-axis lies in the direction from home plate to third base.
(a) What are the coordinates of first base, second base, and third base? Use feet as the unit of measurement.
(b) If the right fielder is located at $(180, 20)$, how far is it from the right fielder to second base?
(c) If the center fielder is located at $(220, 220)$, how far is it from the center fielder to third base?

67. Distance between Moving Objects A Ford Focus and a Freightliner truck leave an intersection at the same time. The Focus heads east at an average speed of 30 miles per hour, while the truck heads south at an average speed of 40 miles per hour. Find an expression for their distance d (in miles) apart at the end of t hours.

68. Distance of a Moving Object from a Fixed Point A hot-air balloon, headed due east at an average speed of 15 miles per hour and at a constant altitude of 100 feet, passes over an intersection (see the figure). Find an expression for the distance d (measured in feet) from the balloon to the intersection t seconds later.

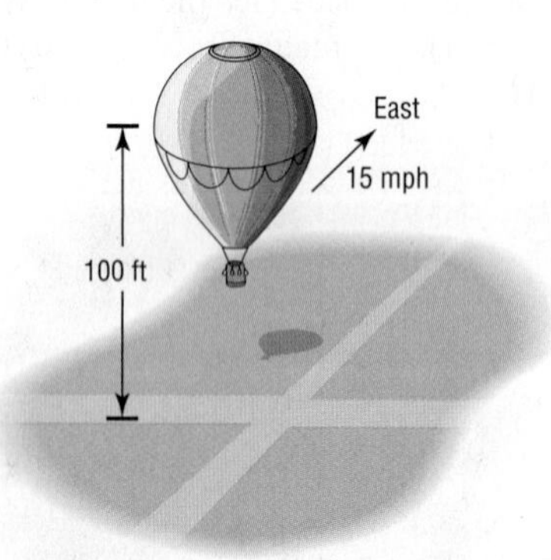

69. Drafting Error When a draftsman draws three lines that are to intersect at one point, the lines may not intersect as intended and subsequently will form an **error triangle**. If this error triangle is long and thin, one estimate for the location of the desired point is the midpoint of the shortest side. The figure shows one such error triangle.

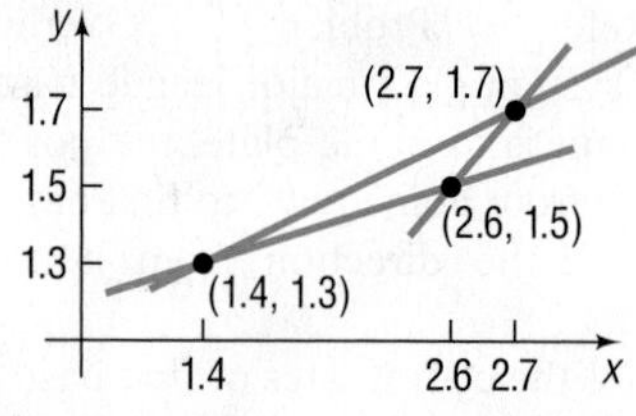

(a) Find an estimate for the desired intersection point.

(b) Find the length of the median for the midpoint found in part (a). See Problem 55.

70. Net Sales The figure below illustrates how net sales of Wal-Mart Stores, Inc., grew from 2007 through 2013. Use the midpoint formula to estimate the net sales of Wal-Mart Stores, Inc., in 2010. How does your result compare to the reported value of \$405 billion?

Source: Wal-Mart Stores, Inc., 2013 Annual Report

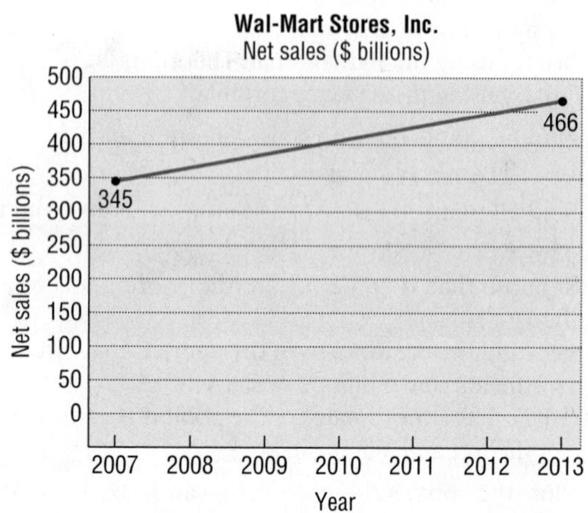

71. Poverty Threshold Poverty thresholds are determined by the U.S. Census Bureau. A poverty threshold represents the minimum annual household income for a family not to be considered poor. In 2003, the poverty threshold for a family of four with two children under the age of 18 years was \$18,660. In 2013, the poverty threshold for a family of four with two children under the age of 18 years was \$23,624. Assuming that poverty thresholds increase in a straight-line fashion, use the midpoint formula to estimate the poverty threshold for a family of four with two children under the age of 18 in 2008. How does your result compare to the actual poverty threshold in 2008 of \$21,834?

Source: U.S. Census Bureau

Explaining Concepts: Discussion and Writing

72. Write a paragraph that describes a Cartesian plane. Then write a second paragraph that describes how to plot points in the Cartesian plane. Your paragraphs should include the terms "coordinate axes," "ordered pair," "coordinates," "plot," "x-coordinate," and "y-coordinate."

'Are You Prepared?' Answers

1. 0 **2.** 8 **3.** 5 **4.** $11^2 + 60^2 = 121 + 3600 = 3721 = 61^2$ **5.** $A = \frac{1}{2}bh$ **6.** True

1.2 Graphs of Equations in Two Variables; Circles

PREPARING FOR THIS SECTION *Before getting started, review the following:*

- Solving Equations (Appendix A, Section A.4, pp. A27–A34)
- Complete the Square (Appendix A, Section A.3, pp. A24–A25)

Now Work the 'Are You Prepared?' problems on page 18.

OBJECTIVES
1. Graph Equations by Plotting Points (p. 9)
2. Find Intercepts from a Graph (p. 10)
3. Find Intercepts from an Equation (p. 11)
4. Test an Equation for Symmetry with Respect to the x-Axis, the y-Axis, and the Origin (p. 12)
5. Know How to Graph Key Equations (p. 14)
6. Write the Standard Form of the Equation of a Circle (p. 15)
7. Graph a Circle (p. 16)
8. Work with the General Form of the Equation of a Circle (p. 17)

1 Graph Equations by Plotting Points

An **equation in two variables**, say x and y, is a statement in which two expressions involving x and y are equal. The expressions are called the **sides** of the equation. Since an equation is a statement, it may be true or false, depending on the value of the variables. Any pair of values for x and y that result in a true statement are said to **satisfy** the equation.

For example, the following are all equations in two variables x and y:

$$x^2 + y^2 = 5 \qquad 2x - y = 6 \qquad y = 2x + 5 \qquad x^2 = y$$

The first of these, $x^2 + y^2 = 5$, is satisfied for $x = 1, y = 2$, since $1^2 + 2^2 = 5$. Other choices of x and y, such as $x = -1$, $y = -2$, also satisfy this equation. It is not satisfied for $x = 2$ and $y = 3$, since $2^2 + 3^2 = 4 + 9 = 13 \neq 5$.

The **graph of an equation in two variables** x and y consists of the set of points in the xy-plane whose coordinates (x, y) satisfy the equation.

EXAMPLE 1 **Determining Whether a Point Is on the Graph of an Equation**

Determine if the following points are on the graph of the equation $2x - y = 6$.

(a) $(2, 3)$ (b) $(2, -2)$

Solution (a) For the point $(2, 3)$, check to see whether $x = 2, y = 3$ satisfies the equation $2x - y = 6$.

$$2x - y = 2(2) - 3 = 4 - 3 = 1 \neq 6$$

The equation is not satisfied, so the point $(2, 3)$ is not on the graph of $2x - y = 6$.

(b) For the point $(2, -2)$,

$$2x - y = 2(2) - (-2) = 4 + 2 = 6$$

The equation is satisfied, so the point $(2, -2)$ is on the graph of $2x - y = 6$. ●

Now Work PROBLEM 13

EXAMPLE 2 **Graphing an Equation by Plotting Points**

Graph the equation: $y = 2x + 5$

Solution The graph consists of all points (x, y) that satisfy the equation. To locate some of these points (and get an idea of the pattern of the graph), assign some numbers to x, and find corresponding values for y.

If	Then	Point on Graph
$x = 0$	$y = 2(0) + 5 = 5$	$(0, 5)$
$x = 1$	$y = 2(1) + 5 = 7$	$(1, 7)$
$x = -5$	$y = 2(-5) + 5 = -5$	$(-5, -5)$
$x = 10$	$y = 2(10) + 5 = 25$	$(10, 25)$

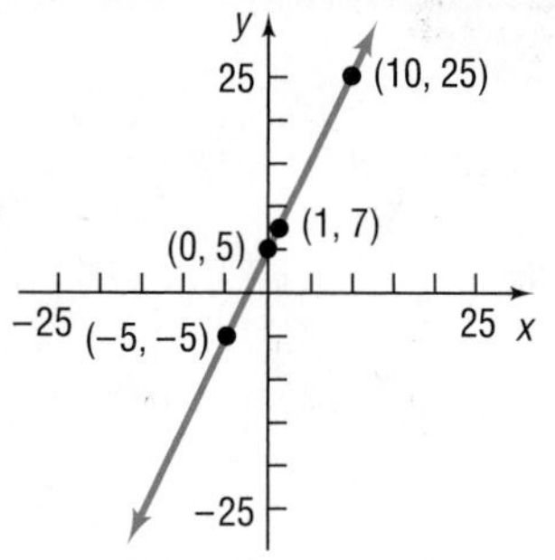

Figure 9 $y = 2x + 5$

By plotting these points and then connecting them, we obtain the graph of the equation (a *line*)*, as shown in Figure 9. ●

EXAMPLE 3

Graphing an Equation by Plotting Points

Graph the equation: $y = x^2$

Solution Table 1 provides several points on the graph. Plotting these points and connecting them with a smooth curve gives the graph (a *parabola*) shown in Figure 10.

Table 1

x	$y = x^2$	(x, y)
−4	16	(−4, 16)
−3	9	(−3, 9)
−2	4	(−2, 4)
−1	1	(−1, 1)
0	0	(0, 0)
1	1	(1, 1)
2	4	(2, 4)
3	9	(3, 9)
4	16	(4, 16)

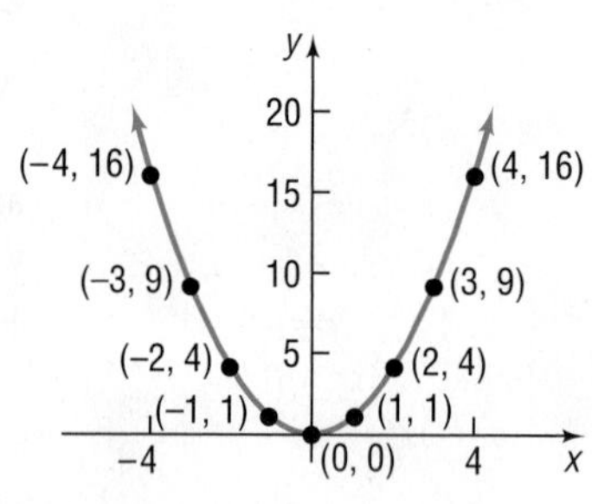

Figure 10 $y = x^2$

●

The graphs of the equations shown in Figures 9 and 10 do not show all points. For example, in Figure 9, the point $(20, 45)$ is a part of the graph of $y = 2x + 5$, but it is not shown. Since the graph of $y = 2x + 5$ can be extended out indefinitely, we use arrows to indicate that the pattern shown continues. It is important when illustrating a graph to present enough of the graph so that any viewer of the illustration will "see" the rest of it as an obvious continuation of what is actually there. This is referred to as a **complete graph**.

One way to obtain the complete graph of an equation is to plot enough points on the graph for a pattern to become evident. Then these points are connected with a smooth curve following the suggested pattern. But how many points are sufficient? Sometimes knowledge about the equation tells us. For example, if an equation is of the form $y = mx + b$, then its graph is a line.* In this case, only two points are needed to obtain the graph.

COMMENT Another way to obtain the graph of an equation is to use a graphing utility. Read Section B.2, *Using a Graphing Utility to Graph Equations*, in Appendix B. ■

One purpose of this text is to investigate the properties of equations in order to decide whether a graph is complete. Sometimes we shall graph equations by plotting points. Shortly, we shall investigate various techniques that will enable us to graph an equation without plotting so many points.

Two techniques that sometimes reduce the number of points required to graph an equation involve finding *intercepts* and checking for *symmetry*.

2 Find Intercepts from a Graph

The points, if any, at which a graph crosses or touches the coordinate axes are called the **intercepts**. See Figure 11. The x-coordinate of a point at which the graph crosses or touches the x-axis is an ***x*-intercept**, and the y-coordinate of a point at which the graph crosses or touches the y-axis is a ***y*-intercept**.

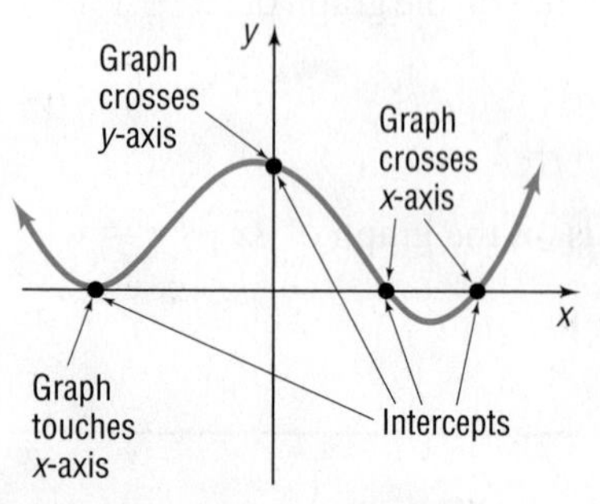

Figure 11

*Lines are discussed in detail in Appendix A, Section A.8.

EXAMPLE 4 **Finding Intercepts from a Graph**

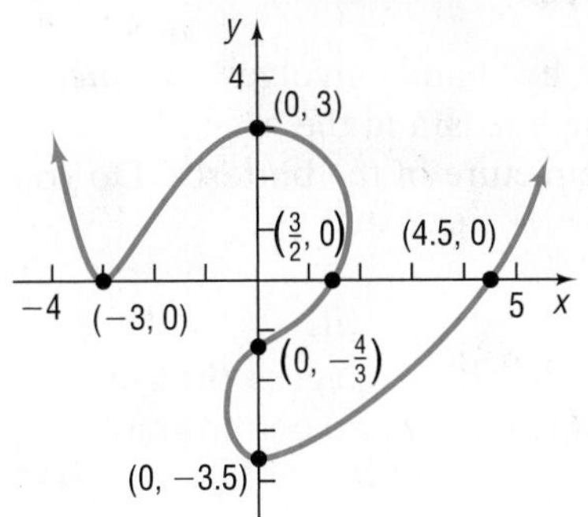

Figure 12

Find the intercepts of the graph in Figure 12. What are its x-intercepts? What are its y-intercepts?

Solution The intercepts of the graph are the points

$$(-3, 0) \quad (0, 3) \quad \left(\frac{3}{2}, 0\right) \quad \left(0, -\frac{4}{3}\right) \quad (0, -3.5) \quad (4.5, 0)$$

The x-intercepts are $-3, \frac{3}{2}$, and 4.5; the y-intercepts are $-3.5, -\frac{4}{3}$, and 3. ●

In Example 4, note the following usage: If the type of intercept (x- versus y-) is not specified, then report the intercept as an ordered pair. However, if the type of intercept is specified, then report the coordinate of the specified intercept. For x-intercepts, report the x-coordinate of the intercept; for y-intercepts, report the y-coordinate of the intercept.

Now Work PROBLEM 41(a)

3 Find Intercepts from an Equation

The intercepts of a graph can be found from its equation by using the fact that points on the x-axis have y-coordinates equal to 0, and points on the y-axis have x-coordinates equal to 0.

COMMENT For many equations, finding intercepts may not be so easy. In such cases, a graphing utility can be used. Read the first part of Section B.3, *Using a Graphing Utility to Locate Intercepts and Check for Symmetry*, in Appendix B, to find out how to locate intercepts using a graphing utility. ■

Procedure for Finding Intercepts

1. To find the x-intercept(s), if any, of the graph of an equation, let $y = 0$ in the equation and solve for x, where x is a real number.
2. To find the y-intercept(s), if any, of the graph of an equation, let $x = 0$ in the equation and solve for y, where y is a real number.

EXAMPLE 5 **Finding Intercepts from an Equation**

Find the x-intercept(s) and the y-intercept(s) of the graph of $y = x^2 - 4$. Then graph $y = x^2 - 4$ by plotting points.

Solution To find the x-intercept(s), let $y = 0$ and obtain the equation

$$\begin{aligned} x^2 - 4 &= 0 && y = x^2 - 4 \text{ with } y = 0 \\ (x + 2)(x - 2) &= 0 && \text{Factor.} \\ x + 2 = 0 \quad \text{or} \quad x - 2 &= 0 && \text{Zero-Product Property} \\ x = -2 \quad \text{or} \quad x &= 2 && \text{Solve.} \end{aligned}$$

The equation has two solutions, -2 and 2. The x-intercepts are -2 and 2.

To find the y-intercept(s), let $x = 0$ in the equation.

$$\begin{aligned} y &= x^2 - 4 \\ &= 0^2 - 4 = -4 \end{aligned}$$

The y-intercept is -4.

Since $x^2 \geq 0$ for all x, we deduce from the equation $y = x^2 - 4$ that $y \geq -4$ for all x. This information, the intercepts, and the points from Table 2 enable us to graph $y = x^2 - 4$. See Figure 13.

Figure 13 $y = x^2 - 4$

Table 2

x	$y = x^2 - 4$	(x, y)
−3	5	(−3, 5)
−1	−3	(−1, −3)
1	−3	(1, −3)
3	5	(3, 5)

●

Now Work PROBLEM 23

4 Test an Equation for Symmetry with Respect to the x-Axis, the y-Axis, and the Origin

Another helpful tool for graphing equations by hand involves *symmetry*, particularly symmetry with respect to the *x*-axis, the *y*-axis, and the origin.

Symmetry often occurs in nature. Consider the picture of the butterfly. Do you see the symmetry?

DEFINITION A graph is said to be **symmetric with respect to the *x*-axis** if, for every point (x, y) on the graph, the point $(x, -y)$ is also on the graph.

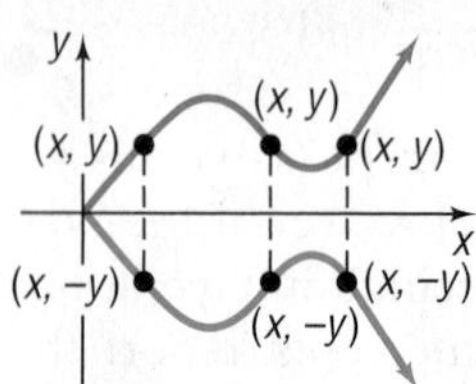

Figure 14
Symmetry with respect to the *x*-axis

Figure 14 illustrates the definition. Note that when a graph is symmetric with respect to the *x*-axis, the part of the graph above the *x*-axis is a reflection (or mirror image) of the part below it, and vice versa.

EXAMPLE 6 **Points Symmetric with Respect to the x-Axis**

If a graph is symmetric with respect to the *x*-axis, and the point $(3, 2)$ is on the graph, then the point $(3, -2)$ is also on the graph. •

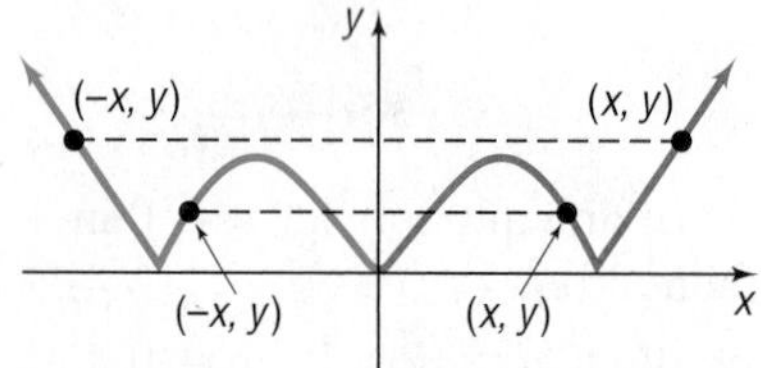

Figure 15
Symmetry with respect to the *y*-axis

DEFINITION A graph is said to be **symmetric with respect to the *y*-axis** if, for every point (x, y) on the graph, the point $(-x, y)$ is also on the graph.

Figure 15 illustrates the definition. When a graph is symmetric with respect to the *y*-axis, the part of the graph to the right of the *y*-axis is a reflection of the part to the left of it, and vice versa.

EXAMPLE 7 **Points Symmetric with Respect to the y-Axis**

If a graph is symmetric with respect to the *y*-axis and the point $(5, 8)$ is on the graph, then the point $(-5, 8)$ is also on the graph. •

DEFINITION A graph is said to be **symmetric with respect to the origin** if, for every point (x, y) on the graph, the point $(-x, -y)$ is also on the graph.

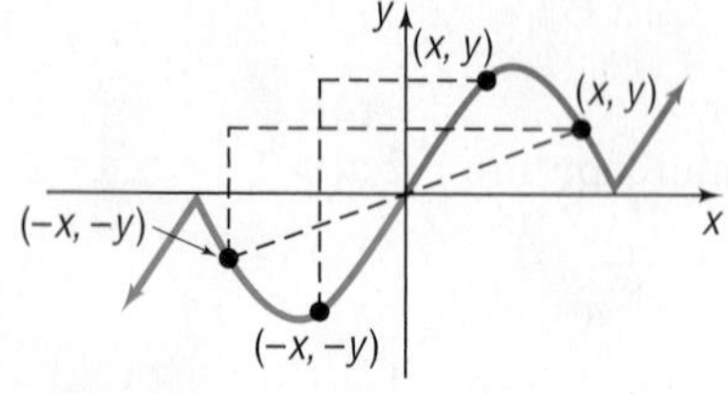

Figure 16
Symmetry with respect to the origin

Figure 16 illustrates the definition. Symmetry with respect to the origin may be viewed in three ways:

1. As a reflection about the *y*-axis, followed by a reflection about the *x*-axis
2. As a projection along a line through the origin so that the distances from the origin are equal
3. As half of a complete revolution about the origin

EXAMPLE 8 **Points Symmetric with Respect to the Origin**

If a graph is symmetric with respect to the origin, and the point $(4, 2)$ is on the graph, then the point $(-4, -2)$ is also on the graph. •

Now Work PROBLEMS 31 AND 41(b)

When the graph of an equation is symmetric with respect to a coordinate axis or the origin, the number of points that you need to plot in order to see the pattern is reduced. For example, if the graph of an equation is symmetric with respect to the

y-axis, then once points to the right of the y-axis are plotted, an equal number of points on the graph can be obtained by reflecting them about the y-axis. Because of this, before graphing an equation, first determine whether it has any symmetry. The following tests are used for this purpose.

Tests for Symmetry

To test the graph of an equation for symmetry with respect to the

x-Axis Replace y by $-y$ in the equation and simplify. If an equivalent equation results, the graph of the equation is symmetric with respect to the x-axis.

y-Axis Replace x by $-x$ in the equation and simplify. If an equivalent equation results, the graph of the equation is symmetric with respect to the y-axis.

Origin Replace x by $-x$ and y by $-y$ in the equation and simplify. If an equivalent equation results, the graph of the equation is symmetric with respect to the origin.

EXAMPLE 9

Testing an Equation for Symmetry

Test $y = \dfrac{4x^2}{x^2 + 1}$ for symmetry.

Solution *x-Axis:* To test for symmetry with respect to the x-axis, replace y by $-y$. Since $-y = \dfrac{4x^2}{x^2 + 1}$ is not equivalent to $y = \dfrac{4x^2}{x^2 + 1}$, the graph of the equation is not symmetric with respect to the x-axis.

y-Axis: To test for symmetry with respect to the y-axis, replace x by $-x$. Since $y = \dfrac{4(-x)^2}{(-x)^2 + 1} = \dfrac{4x^2}{x^2 + 1}$ is equivalent to $y = \dfrac{4x^2}{x^2 + 1}$, the graph of the equation is symmetric with respect to the y-axis.

Origin: To test for symmetry with respect to the origin, replace x by $-x$ and y by $-y$.

$$-y = \frac{4(-x)^2}{(-x)^2 + 1} \quad \text{Replace } x \text{ by } -x \text{ and } y \text{ by } -y.$$

$$-y = \frac{4x^2}{x^2 + 1} \quad \text{Simplify.}$$

$$y = -\frac{4x^2}{x^2 + 1} \quad \text{Multiply both sides by } -1.$$

Since the result is not equivalent to the original equation, the graph of the equation $y = \dfrac{4x^2}{x^2 + 1}$ is not symmetric with respect to the origin. ●

Seeing the Concept

Figure 17 shows the graph of $y = \dfrac{4x^2}{x^2 + 1}$ using a graphing utility. Do you see the symmetry with respect to the y-axis?

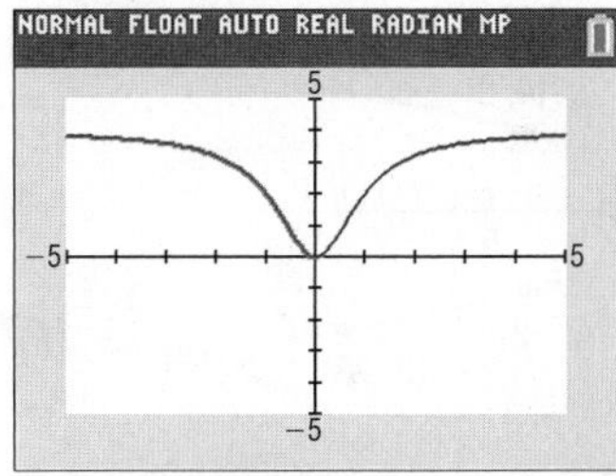

Figure 17 $y = \dfrac{4x^2}{x^2 + 1}$

Now Work PROBLEM 61

5 Know How to Graph Key Equations

The next three examples use intercepts, symmetry, and point plotting to obtain the graphs of key equations. It is important to know the graphs of these key equations because we use them later. The first of these is $y = x^3$.

EXAMPLE 10 **Graphing the Equation $y = x^3$ by Finding Intercepts, Checking for Symmetry, and Plotting Points**

Graph the equation $y = x^3$ by plotting points. Find any intercepts and check for symmetry first.

Solution First, find the intercepts. When $x = 0$, then $y = 0$; and when $y = 0$, then $x = 0$. The origin $(0, 0)$ is the only intercept. Now test for symmetry.

x-Axis: Replace y by $-y$. Since $-y = x^3$ is not equivalent to $y = x^3$, the graph is not symmetric with respect to the x-axis.

y-Axis: Replace x by $-x$. Since $y = (-x)^3 = -x^3$ is not equivalent to $y = x^3$, the graph is not symmetric with respect to the y-axis.

Origin: Replace x by $-x$ and y by $-y$. Since $-y = (-x)^3 = -x^3$ is equivalent to $y = x^3$ (multiply both sides by -1), the graph is symmetric with respect to the origin.

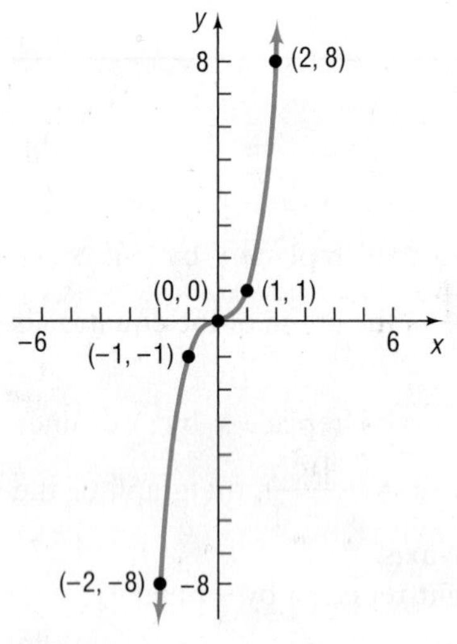

Figure 18 $y = x^3$

To graph $y = x^3$, use the equation to obtain several points on the graph. Because of the symmetry, we need to locate only points on the graph for which $x \geq 0$. See Table 3. Since $(1, 1)$ is on the graph, and the graph is symmetric with respect to the origin, the point $(-1, -1)$ is also on the graph. Plot the points from Table 3 and use the symmetry. Figure 18 shows the graph.

Table 3

x	$y = x^3$	(x, y)
0	0	(0, 0)
1	1	(1, 1)
2	8	(2, 8)
3	27	(3, 27)

●

EXAMPLE 11 **Graphing the Equation $x = y^2$**

(a) Graph the equation $x = y^2$. Find any intercepts and check for symmetry first.
(b) Graph $x = y^2$, $y \geq 0$.

Solution (a) The lone intercept is $(0, 0)$. The graph is symmetric with respect to the x-axis. (Do you see why? Replace y by $-y$.) Figure 19 shows the graph.

(b) If we restrict y so that $y \geq 0$, the equation $x = y^2$, $y \geq 0$, may be written equivalently as $y = \sqrt{x}$. The portion of the graph of $x = y^2$ in quadrant I is therefore the graph of $y = \sqrt{x}$. See Figure 20.

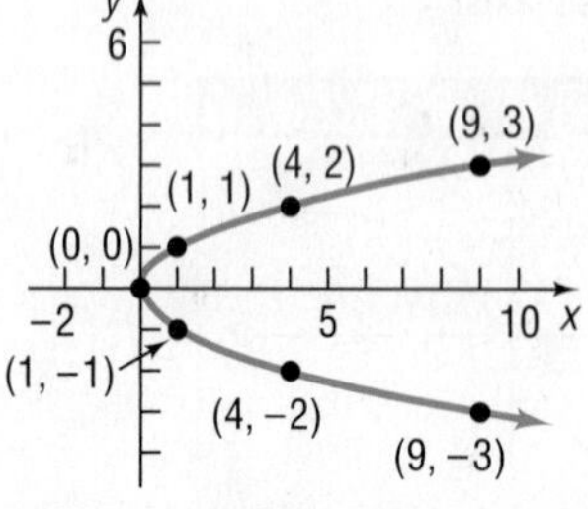

Figure 19 $x = y^2$

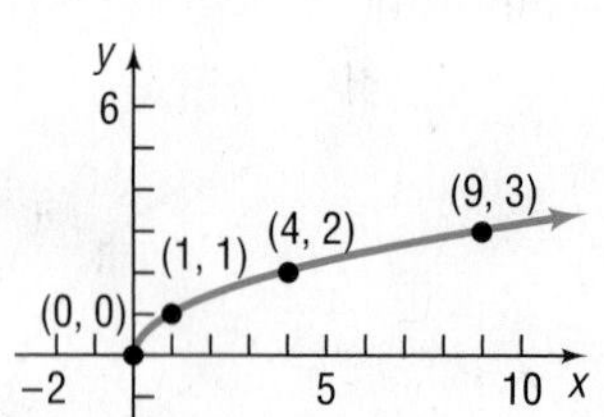

Figure 20 $y = \sqrt{x}$

●

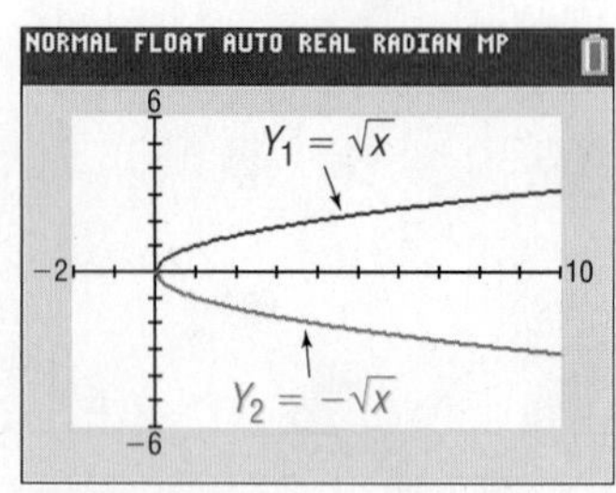

Figure 21

COMMENT To see the graph of the equation $x = y^2$ on a graphing calculator, you will need to graph two equations: $Y_1 = \sqrt{x}$ and $Y_2 = -\sqrt{x}$. See Figure 21. ■

EXAMPLE 12

Graphing the Equation $y = \dfrac{1}{x}$

Graph the equation $y = \dfrac{1}{x}$. First, find any intercepts and check for symmetry.

Solution Check for intercepts first. Letting $x = 0$, we obtain 0 in the denominator, which makes y undefined. We conclude that there is no y-intercept. Letting $y = 0$, we get the equation $\dfrac{1}{x} = 0$, which has no solution. We conclude that there is no x-intercept. The graph of $y = \dfrac{1}{x}$ does not cross or touch the coordinate axes.

Next check for symmetry:

x-Axis: Replacing y by $-y$ yields $-y = \dfrac{1}{x}$, which is not equivalent to $y = \dfrac{1}{x}$.

y-Axis: Replacing x by $-x$ yields $y = \dfrac{1}{-x} = -\dfrac{1}{x}$, which is not equivalent to $y = \dfrac{1}{x}$.

Origin: Replacing x by $-x$ and y by $-y$ yields $-y = -\dfrac{1}{x}$, which is equivalent to $y = \dfrac{1}{x}$. The graph is symmetric with respect to the origin.

Table 4

x	$y = \dfrac{1}{x}$	(x, y)
$\dfrac{1}{10}$	10	$\left(\dfrac{1}{10}, 10\right)$
$\dfrac{1}{3}$	3	$\left(\dfrac{1}{3}, 3\right)$
$\dfrac{1}{2}$	2	$\left(\dfrac{1}{2}, 2\right)$
1	1	$(1, 1)$
2	$\dfrac{1}{2}$	$\left(2, \dfrac{1}{2}\right)$
3	$\dfrac{1}{3}$	$\left(3, \dfrac{1}{3}\right)$
10	$\dfrac{1}{10}$	$\left(10, \dfrac{1}{10}\right)$

Now set up Table 4, listing several points on the graph. Because of the symmetry with respect to the origin, we use only positive values of x. From Table 4 we infer that if x is a large and positive number, then $y = \dfrac{1}{x}$ is a positive number close to 0. We also infer that if x is a positive number close to 0, then $y = \dfrac{1}{x}$ is a large and positive number. Using this information, we can graph the equation.

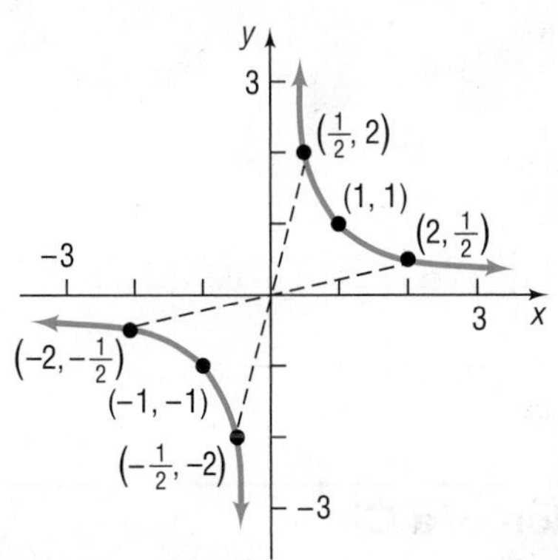

Figure 22 $y = \dfrac{1}{x}$

Figure 22 illustrates some of these points and the graph of $y = \dfrac{1}{x}$. Observe how the absence of intercepts and the existence of symmetry with respect to the origin were utilized. ●

COMMENT Refer to Example 2 in Appendix B, Section B.3, for the graph of $y = \dfrac{1}{x}$ found using a graphing utility. ■

6 Write the Standard Form of the Equation of a Circle

One advantage of a coordinate system is that it enables us to translate a geometric statement into an algebraic statement, and vice versa. Consider, for example, the following geometric statement that defines a circle.

DEFINITION

A **circle** is a set of points in the xy-plane that are a fixed distance r from a fixed point (h, k). The fixed distance r is called the **radius**, and the fixed point (h, k) is called the **center** of the circle.

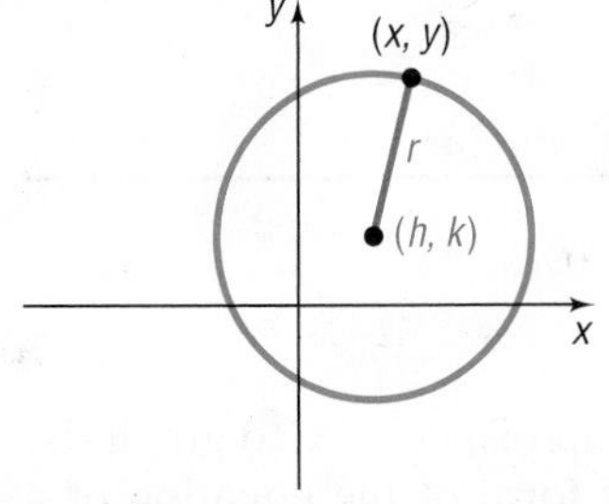

Figure 23 $(x - h)^2 + (y - k)^2 = r^2$

Figure 23 shows the graph of a circle. To find the equation, let (x, y) represent the coordinates of any point on a circle with radius r and center (h, k). Then the distance between the points (x, y) and (h, k) must always equal r. That is, by the distance formula,

$$\sqrt{(x - h)^2 + (y - k)^2} = r$$

or, equivalently,

$$(x - h)^2 + (y - k)^2 = r^2$$

DEFINITION

The **standard form of an equation of a circle** with radius r and center (h, k) is

$$(x - h)^2 + (y - k)^2 = r^2 \quad (1)$$

THEOREM

The standard form of an equation of a circle of radius r with center at the origin $(0, 0)$ is

$$x^2 + y^2 = r^2$$

DEFINITION

If the radius $r = 1$, the circle whose center is at the origin is called the **unit circle** and has the equation

$$x^2 + y^2 = 1$$

See Figure 24. Notice that the graph of the unit circle is symmetric with respect to the x-axis, the y-axis, and the origin.

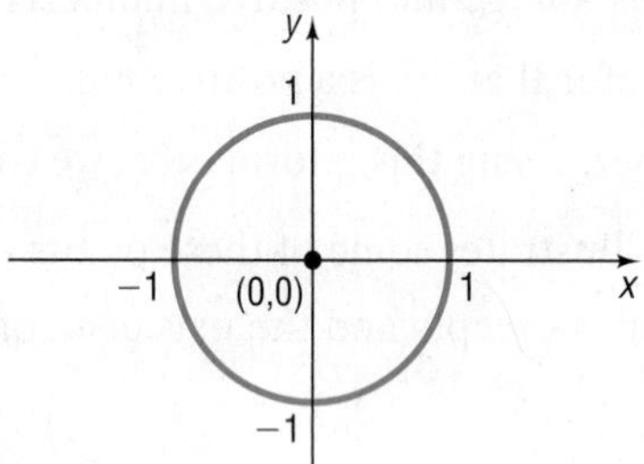

Figure 24
Unit circle $x^2 + y^2 = 1$

EXAMPLE 13 **Writing the Standard Form of the Equation of a Circle**

Write the standard form of the equation of the circle with radius 5 and center $(-3, 6)$.

Solution Substitute the values $r = 5$, $h = -3$, and $k = 6$ into equation (1).

$$(x - h)^2 + (y - k)^2 = r^2$$

$$(x + 3)^2 + (y - 6)^2 = 25$$

●

Now Work PROBLEM 79

7 Graph a Circle

EXAMPLE 14 **Graphing a Circle**

Graph the equation: $(x + 3)^2 + (y - 2)^2 = 16$

Solution Since the equation is in the form of equation (1), its graph is a circle. To graph the equation, compare the given equation to the standard form of the equation of a circle. The comparison yields information about the circle.

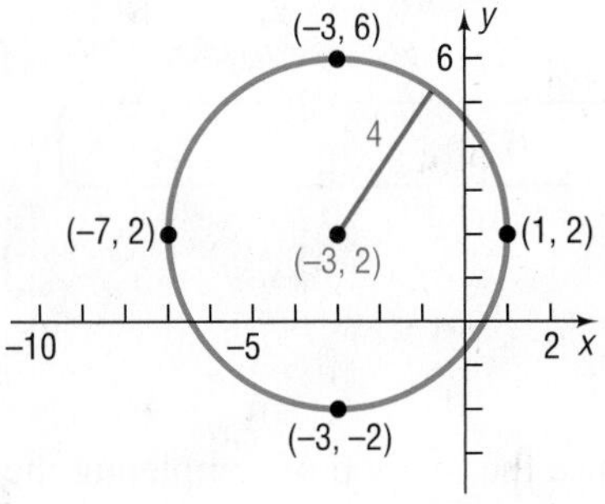

$$(x + 3)^2 + (y - 2)^2 = 16$$
$$(x - (-3))^2 + (y - 2)^2 = 4^2$$
$$(x - h)^2 + (y - k)^2 = r^2$$

We see that $h = -3$, $k = 2$, and $r = 4$. The circle has center $(-3, 2)$ and a radius of 4 units. To graph this circle, first plot the center $(-3, 2)$. Since the radius is 4, we can locate four points on the circle by plotting points 4 units to the left, to the right, up, and down from the center. These four points can then be used as guides to obtain the graph. See Figure 25. ●

(-3, 6) y 6 4 (-7, 2) (1, 2) (-3, 2) -10 -5 2 x (-3, -2)

Figure 25 $(x + 3)^2 + (y - 2)^2 = 16$

Now Work PROBLEMS 95(a) AND (b)

EXAMPLE 15 **Finding the Intercepts of a Circle**

For the circle $(x + 3)^2 + (y - 2)^2 = 16$, find the intercepts, if any, of its graph.

Solution This is the equation discussed and graphed in Example 14. To find the x-intercepts, if any, let $y = 0$. Then

$$\begin{aligned}(x + 3)^2 + (y - 2)^2 &= 16 \\ (x + 3)^2 + (0 - 2)^2 &= 16 && y = 0 \\ (x + 3)^2 + 4 &= 16 && \text{Simplify.} \\ (x + 3)^2 &= 12 && \text{Simplify.} \\ x + 3 &= \pm\sqrt{12} && \text{Apply the Square Root Method.} \\ x &= -3 \pm 2\sqrt{3} && \text{Solve for } x.\end{aligned}$$

> **In Words**
> The symbol $\pm$ is read "plus or minus." It means to add and subtract the quantity following the $\pm$ symbol. For example, 5 ± 2 means "$5 - 2 = 3$ or $5 + 2 = 7$."

The x-intercepts are $-3 - 2\sqrt{3} \approx -6.46$ and $-3 + 2\sqrt{3} \approx 0.46$.

To find the y-intercepts, if any, let $x = 0$. Then

$$\begin{aligned}(x + 3)^2 + (y - 2)^2 &= 16 \\ (0 + 3)^2 + (y - 2)^2 &= 16 \\ 9 + (y - 2)^2 &= 16 \\ (y - 2)^2 &= 7 \\ y - 2 &= \pm\sqrt{7} \\ y &= 2 \pm \sqrt{7}\end{aligned}$$

The y-intercepts are $2 - \sqrt{7} \approx -0.65$ and $2 + \sqrt{7} \approx 4.65$.

Look back at Figure 25 to verify the approximate locations of the intercepts. ●

Now Work PROBLEM 95(c)

8 Work with the General Form of the Equation of a Circle

If we eliminate the parentheses from the standard form of the equation of the circle given in Example 15, we get

$$(x + 3)^2 + (y - 2)^2 = 16$$
$$x^2 + 6x + 9 + y^2 - 4y + 4 = 16$$

which simplifies to

$$x^2 + y^2 + 6x - 4y - 3 = 0$$

It can be shown that any equation of the form

$$x^2 + y^2 + ax + by + c = 0$$

has a graph that is a circle, is a point, or has no graph at all. For example, the graph of the equation $x^2 + y^2 = 0$ is the single point $(0, 0)$. The equation $x^2 + y^2 + 5 = 0$, or $x^2 + y^2 = -5$, has no graph, because sums of squares of real numbers are never negative.

DEFINITION

When its graph is a circle, the equation

$$x^2 + y^2 + ax + by + c = 0$$

is the **general form of the equation of a circle.**

Now Work PROBLEM 85

If an equation of a circle is in general form, we use the method of completing the square to put the equation in standard form so that we can identify its center and radius.

EXAMPLE 16 Graphing a Circle Whose Equation Is in General Form

Graph the equation: $x^2 + y^2 + 4x - 6y + 12 = 0$

Solution Group the terms involving x, group the terms involving y, and put the constant on the right side of the equation. The result is

$$(x^2 + 4x) + (y^2 - 6y) = -12$$

Next, complete the square of each expression in parentheses. Remember that any number added on the left side of the equation must also be added on the right.

$$(x^2 + 4x + 4) + (y^2 - 6y + 9) = -12 + 4 + 9$$

$$\left(\frac{4}{2}\right)^2 = 4 \qquad \left(\frac{-6}{2}\right)^2 = 9$$

$$(x + 2)^2 + (y - 3)^2 = 1 \quad \text{Factor.}$$

(−2, 4) (−1, 3) (−3, 3) (−2, 3) (−2, 2) 1 4 −3 1 x y

Figure 26 $(x + 2)^2 + (y - 3)^2 = 1$

This equation is the standard form of the equation of a circle with radius 1 and center $(-2, 3)$. To graph the equation, use the center $(-2, 3)$ and the radius 1. See Figure 26.

Now Work PROBLEM 99

EXAMPLE 17 Using a Graphing Utility to Graph a Circle

Graph the equation: $x^2 + y^2 = 4$

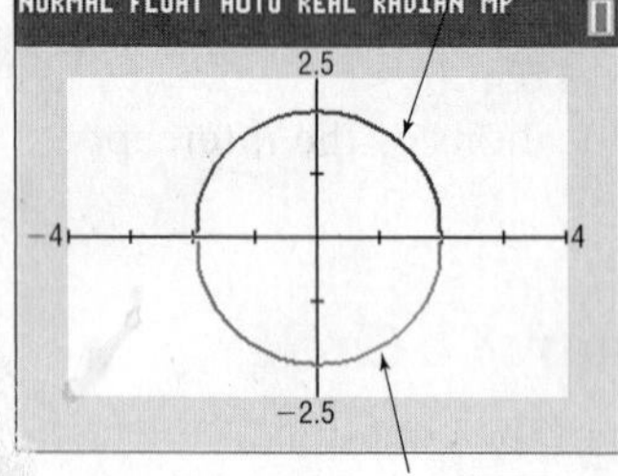

Figure 27 $x^2 + y^2 = 4$

Solution This is the equation of a circle with center at the origin and radius 2. To graph this equation, solve for y.

$$x^2 + y^2 = 4$$

$$y^2 = 4 - x^2 \quad \text{Subtract } x^2 \text{ from each side.}$$

$$y = \pm\sqrt{4 - x^2} \quad \text{Apply the Square Root Method to solve for } y.$$

There are two equations to graph: first graph $Y_1 = \sqrt{4 - x^2}$ and then graph $Y_2 = -\sqrt{4 - x^2}$ on the same square screen. (Your circle will appear oval if you do not use a square screen.*) See Figure 27.

1.2 Assess Your Understanding

Are You Prepared? *Answers are given at the end of these exercises. If you get a wrong answer, read the pages listed in red.*

1. To complete the square of $x^2 + 10x$, you would _____ (*add/subtract*) the number _____. (pp. A24–A25)

2. Use the Square Root Method to solve the equation $(x - 2)^2 = 9$. (p. A31)

*The square screen ratio for the TI-84 Plus C calculator is 8:5.

Concepts and Vocabulary

3. The points, if any, at which a graph crosses or touches the coordinate axes are called __________.

4. If for every point (x, y) on the graph of an equation the point $(-x, y)$ is also on the graph, then the graph is symmetric with respect to the ________.

5. If the graph of an equation is symmetric with respect to the origin and $(3, -4)$ is a point on the graph, then ________ is also a point on the graph.

6. ***True or False*** To find the y-intercepts of the graph of an equation, let $x = 0$ and solve for y.

7. ***True or False*** If a graph is symmetric with respect to the x-axis, then it cannot be symmetric with respect to the y-axis.

8. For a circle, the ________ is the distance from the center to any point on the circle.

9. ***True or False*** The radius of the circle $x^2 + y^2 = 9$ is 3.

10. Given that the intercepts of a graph are $(-4, 0)$ and $(0, 5)$, choose the statement that is true.
(a) The y-intercept is -4, and the x-intercept is 5.
(b) The y-intercepts are -4 and 5.
(c) The x-intercepts are -4 and 5.
(d) The x-intercept is -4, and the y-intercept is 5.

11. ***True or False*** The center of the circle

$$(x + 3)^2 + (y - 2)^2 = 13$$

is $(3, -2)$.

12. The equation of a circle can be changed from general form to standard from by doing which of the following?
(a) completing the squares
(b) solving for x
(c) solving for y
(d) squaring both sides

Skill Building

In Problems 13–18, determine which of the given points are on the graph of the equation.

13. Equation: $y = x^4 - \sqrt{x}$
Points: $(0, 0)$; $(1, 1)$; $(2, 4)$

14. Equation: $y = x^3 - 2\sqrt{x}$
Points: $(0, 0)$; $(1, 1)$; $(1, -1)$

15. Equation: $y^2 = x^2 + 9$
Points: $(0, 3)$; $(3, 0)$; $(-3, 0)$

16. Equation: $y^3 = x + 1$
Points: $(1, 2)$; $(0, 1)$; $(-1, 0)$

17. Equation: $x^2 + y^2 = 4$
Points: $(0, 2)$; $(-2, 2)$; $\left(\sqrt{2}, \sqrt{2}\right)$

18. Equation: $x^2 + 4y^2 = 4$
Points: $(0, 1)$; $(2, 0)$; $\left(2, \frac{1}{2}\right)$

In Problems 19–30, find the intercepts and graph each equation by plotting points. Be sure to label the intercepts.

19. $y = x + 2$
20. $y = x - 6$
21. $y = 2x + 8$
22. $y = 3x - 9$
23. $y = x^2 - 1$
24. $y = x^2 - 9$
25. $y = -x^2 + 4$
26. $y = -x^2 + 1$
27. $2x + 3y = 6$
28. $5x + 2y = 10$
29. $9x^2 + 4y = 36$
30. $4x^2 + y = 4$

In Problems 31–40, plot each point. Then plot the point that is symmetric to it with respect to (a) the x-axis; (b) the y-axis; (c) the origin.

31. $(3, 4)$
32. $(5, 3)$
33. $(-2, 1)$
34. $(4, -2)$
35. $(5, -2)$
36. $(-1, -1)$
37. $(-3, -4)$
38. $(4, 0)$
39. $(0, -3)$
40. $(-3, 0)$

In Problems 41–52, the graph of an equation is given. (a) Find the intercepts. (b) Indicate whether the graph is symmetric with respect to the x-axis, the y-axis, or the origin.

41.

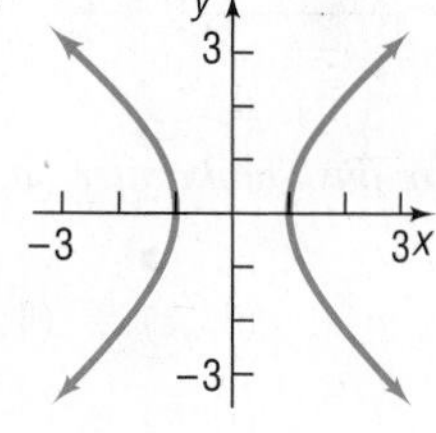

42.

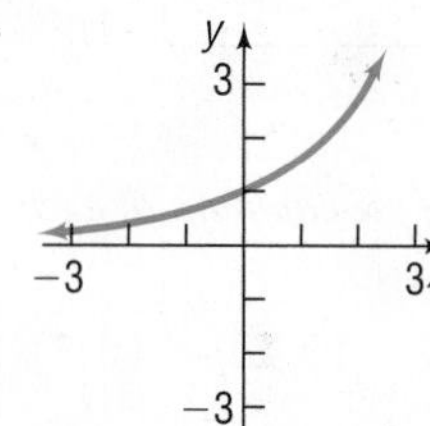

43.

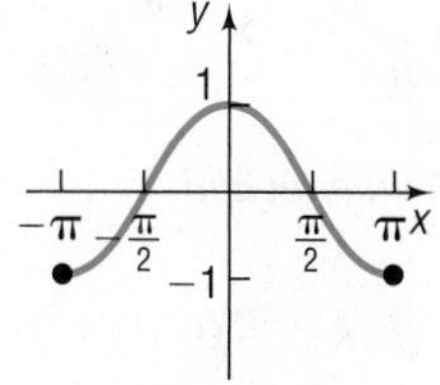

44.

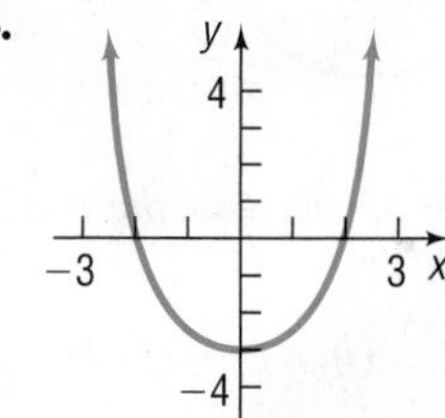

45.

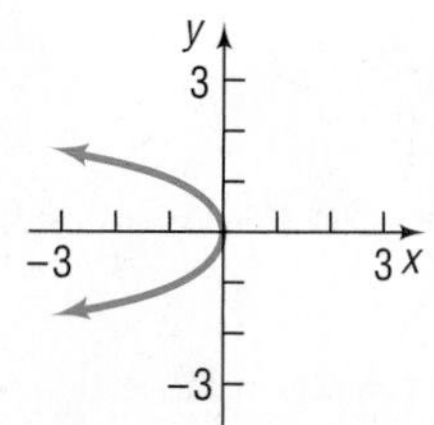

46.

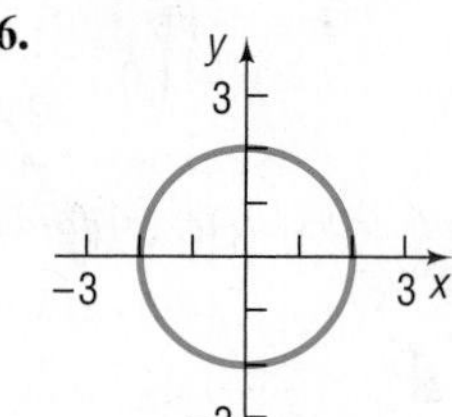

47.

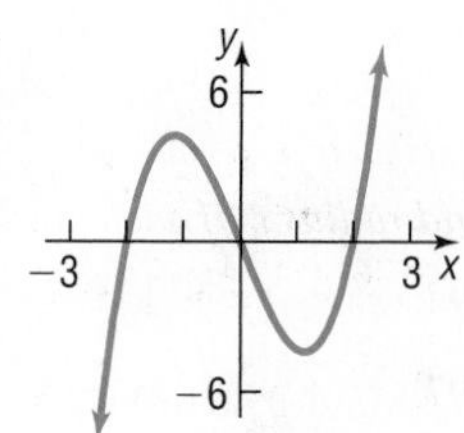

48.

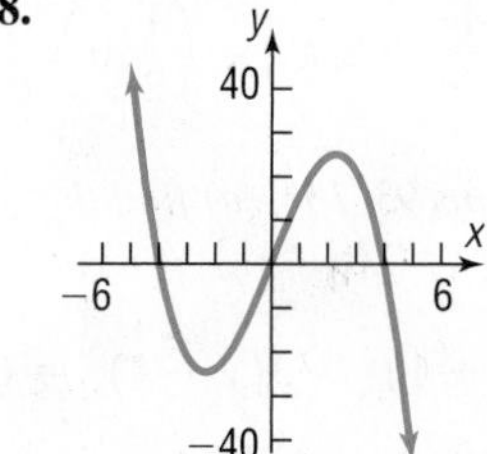

49.

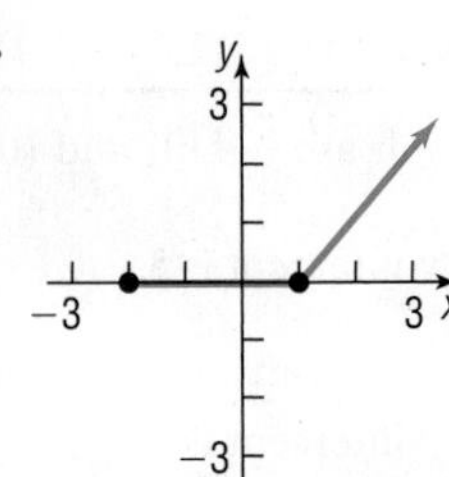

50.

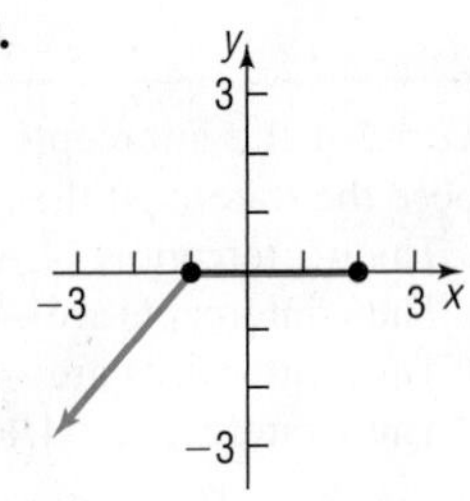

51.

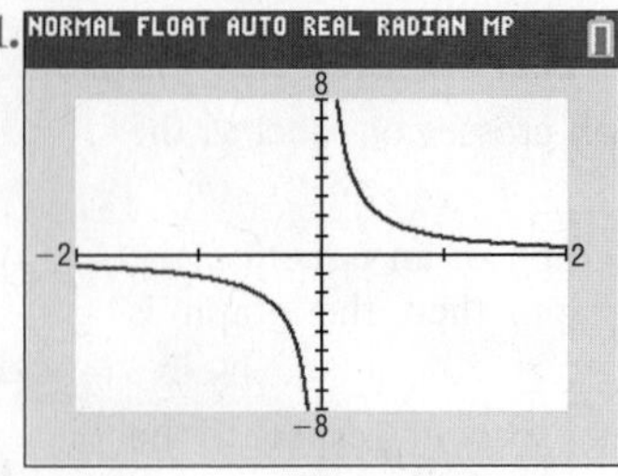

52.

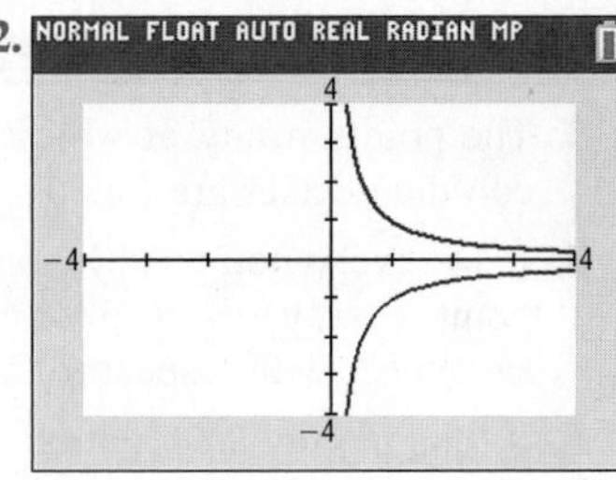

In Problems 53–56, draw a complete graph so that it has the type of symmetry indicated.

53. y-axis

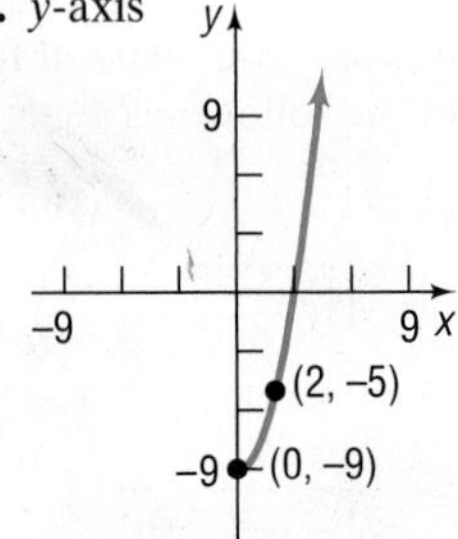

54. x-axis

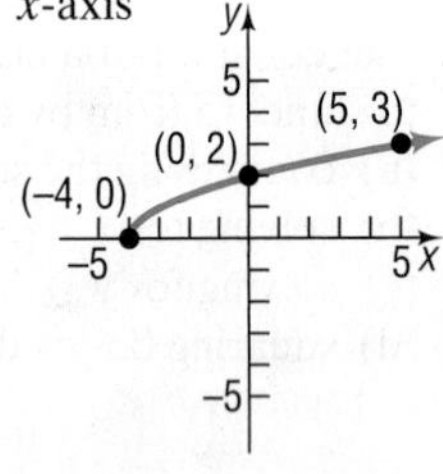

55. Origin

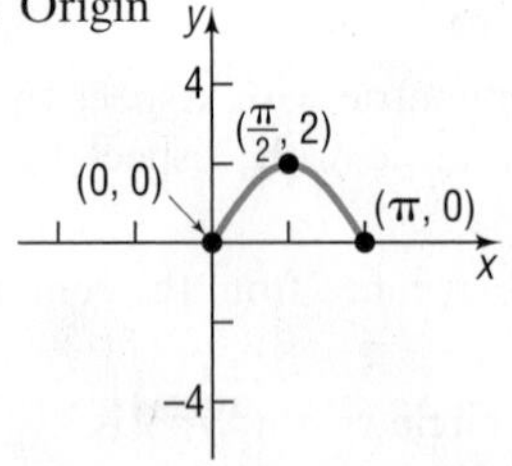

56. y-axis

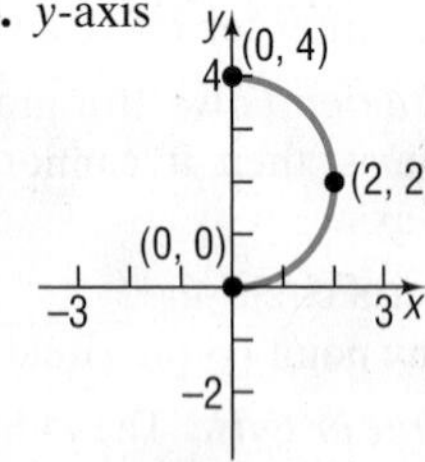

In Problems 57–72, list the intercepts and test for symmetry.

57. $y^2 = x + 4$ **58.** $y^2 = x + 9$ **59.** $y = \sqrt[3]{x}$ **60.** $y = \sqrt[5]{x}$

61. $x^2 + y - 9 = 0$ **62.** $x^2 - y - 4 = 0$ **63.** $9x^2 + 4y^2 = 36$ **64.** $4x^2 + y^2 = 4$

65. $y = x^3 - 27$ **66.** $y = x^4 - 1$ **67.** $y = x^2 - 3x - 4$ **68.** $y = x^2 + 4$

69. $y = \dfrac{3x}{x^2 + 9}$ **70.** $y = \dfrac{x^2 - 4}{2x}$ **71.** $y = \dfrac{-x^3}{x^2 - 9}$ **72.** $y = \dfrac{x^4 + 1}{2x^5}$

In Problems 73–76, draw a quick sketch of the graph of each equation.

73. $y = x^3$ **74.** $x = y^2$ **75.** $y = \sqrt{x}$ **76.** $y = \dfrac{1}{x}$

77. If $(a, 4)$ is a point on the graph of $y = x^2 + 3x$, what is a?

78. If $(a, -5)$ is a point on the graph of $y = x^2 + 6x$, what is a?

In Problems 79–82, find the center and radius of each circle. Write the standard form of the equation.

79. y; (0, 1); (2, 1); x

80. y; (1, 2); (1, 0); x

81.

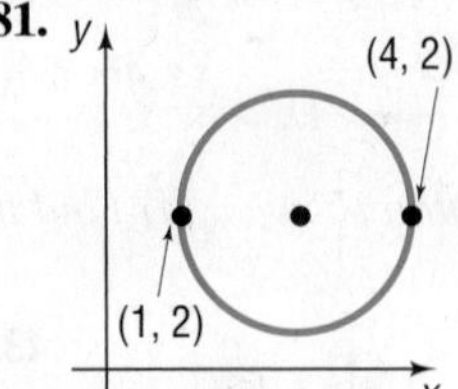

82. y; (2, 3); (0, 1); x

In Problems 83–92, write the standard form of the equation and the general form of the equation of each circle of radius r and center (h, k). Graph each circle.

83. $r = 2;\ (h, k) = (0, 0)$ **84.** $r = 3;\ (h, k) = (0, 0)$ **85.** $r = 2;\ (h, k) = (0, 2)$ **86.** $r = 3;\ (h, k) = (1, 0)$

87. $r = 5;\ (h, k) = (4, -3)$ **88.** $r = 4;\ (h, k) = (2, -3)$ **89.** $r = 4;\ (h, k) = (-2, 1)$ **90.** $r = 7;\ (h, k) = (-5, -2)$

91. $r = \dfrac{1}{2};\ (h, k) = \left(\dfrac{1}{2}, 0\right)$ **92.** $r = \dfrac{1}{2};\ (h, k) = \left(0, -\dfrac{1}{2}\right)$

In Problems 93–106, (a) find the center (h, k) and radius r of each circle; (b) graph each circle; (c) find the intercepts, if any.

93. $x^2 + y^2 = 4$ **94.** $x^2 + (y - 1)^2 = 1$ **95.** $2(x - 3)^2 + 2y^2 = 8$

96. $3(x + 1)^2 + 3(y - 1)^2 = 6$ **97.** $x^2 + y^2 - 2x - 4y - 4 = 0$ **98.** $x^2 + y^2 + 4x + 2y - 20 = 0$

99. $x^2 + y^2 + 4x - 4y - 1 = 0$

100. $x^2 + y^2 - 6x + 2y + 9 = 0$

101. $x^2 + y^2 - x + 2y + 1 = 0$

102. $x^2 + y^2 + x + y - \frac{1}{2} = 0$

103. $2x^2 + 2y^2 - 12x + 8y - 24 = 0$

104. $2x^2 + 2y^2 + 8x + 7 = 0$

105. $2x^2 + 8x + 2y^2 = 0$

106. $3x^2 + 3y^2 - 12y = 0$

In Problems 107–114, find the standard form of the equation of each circle.

107. Center at the origin and containing the point $(-2, 3)$

108. Center $(1, 0)$ and containing the point $(-3, 2)$

109. Center $(2, 3)$ and tangent to the x-axis

110. Center $(-3, 1)$ and tangent to the y-axis

111. With endpoints of a diameter at $(1, 4)$ and $(-3, 2)$

112. With endpoints of a diameter at $(4, 3)$ and $(0, 1)$

113. Center $(-1, 3)$ and tangent to the line $y = 2$

114. Center $(4, -2)$ and tangent to the line $x = 1$

Applications and Extensions

115. Given that the point (1, 2) is on the graph of an equation that is symmetric with respect to the origin, what other point is on the graph?

116. If the graph of an equation is symmetric with respect to the y-axis and 6 is an x-intercept of this graph, name another x-intercept.

117. If the graph of an equation is symmetric with respect to the origin and -4 is an x-intercept of this graph, name another x-intercept.

118. If the graph of an equation is symmetric with respect to the x-axis and 2 is a y-intercept, name another y-intercept.

119. **Microphones** In studios and on stages, cardioid microphones are often preferred for the richness they add to voices and for their ability to reduce the level of sound from the sides and rear of the microphone. Suppose one such cardioid pattern is given by the equation $(x^2 + y^2 - x)^2 = x^2 + y^2$.

(a) Find the intercepts of the graph of the equation.
(b) Test for symmetry with respect to the x-axis, the y-axis, and the origin.

Source: www.notaviva.com

120. **Solar Energy** The solar electric generating systems at Kramer Junction, California, use parabolic troughs to heat a heat-transfer fluid to a high temperature. This fluid is used to generate steam that drives a power conversion system to produce electricity. For troughs 7.5 feet wide, an equation for the cross section is $16y^2 = 120x - 225$.
(a) Find the intercepts of the graph of the equation.
(b) Test for symmetry with respect to the x-axis, the y-axis, and the origin.

Source: U.S. Department of Energy

121. Find the area of the square in the figure.

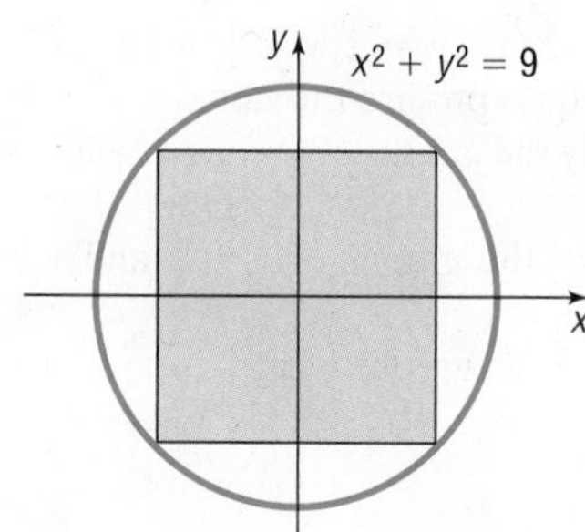

122. Find the area of the blue shaded region in the figure, assuming the quadrilateral inside the circle is a square.

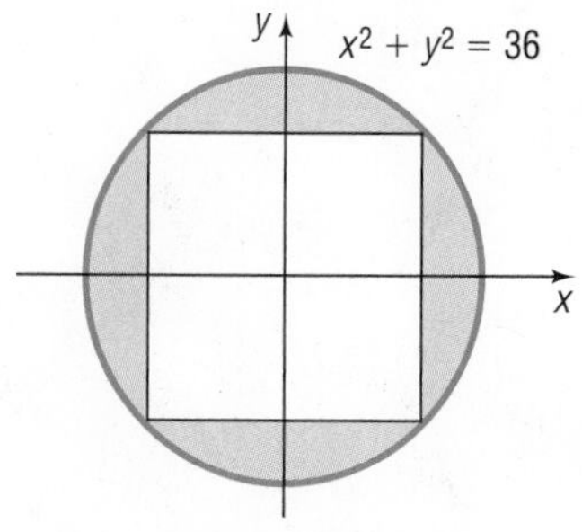

123. Ferris Wheel The original Ferris wheel was built in 1893 by Pittsburgh, Pennsylvania bridge builder George W. Ferris. The Ferris wheel was originally built for the 1893 World's Fair in Chicago, but it was also later reconstructed for the 1904 World's Fair in St. Louis. It had a maximum height of 264 feet and a wheel diameter of 250 feet. Find an equation for the wheel if the center of the wheel is on the y-axis.

Source: guinnessworldrecords.com

124. Ferris Wheel Opening in 2014 in Las Vegas, The High Roller observation wheel has a maximum height of 550 feet and a diameter of 520 feet, with one full rotation taking approximately 30 minutes. Find an equation for the wheel if the center of the wheel is on the y-axis.

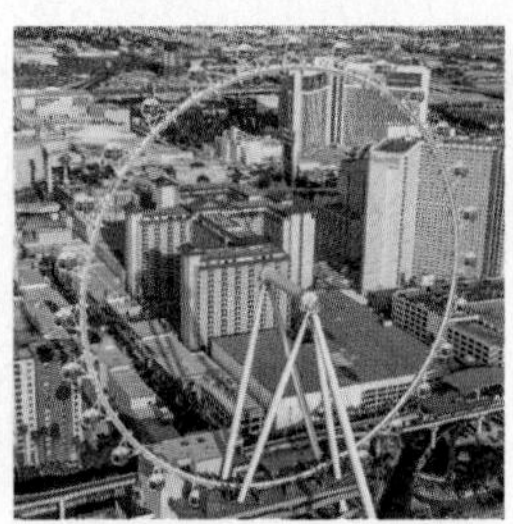

Source: Las Vegas Review Journal

Explaining Concepts: Discussion and Writing

125. Which of the following equations might have the graph shown? (More than one answer is possible.)

(a) $(x-2)^2 + (y+3)^2 = 13$
(b) $(x-2)^2 + (y-2)^2 = 8$
(c) $(x-2)^2 + (y-3)^2 = 13$
(d) $(x+2)^2 + (y-2)^2 = 8$
(e) $x^2 + y^2 - 4x - 9y = 0$
(f) $x^2 + y^2 + 4x - 2y = 0$
(g) $x^2 + y^2 - 9x - 4y = 0$
(h) $x^2 + y^2 - 4x - 4y = 4$

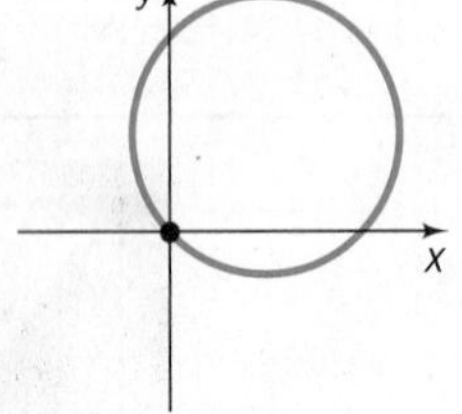

126. Which of the following equations might have the graph shown? (More than one answer is possible.)

(a) $(x-2)^2 + y^2 = 3$
(b) $(x+2)^2 + y^2 = 3$
(c) $x^2 + (y-2)^2 = 3$
(d) $(x+2)^2 + y^2 = 4$
(e) $x^2 + y^2 + 10x + 16 = 0$
(f) $x^2 + y^2 + 10x - 2y = 1$
(g) $x^2 + y^2 + 9x + 10 = 0$
(h) $x^2 + y^2 - 9x - 10 = 0$

y
x

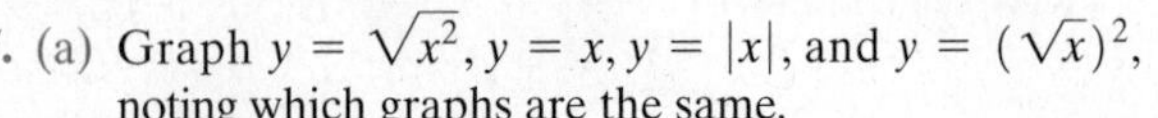

127. (a) Graph $y = \sqrt{x^2}$, $y = x$, $y = |x|$, and $y = (\sqrt{x})^2$, noting which graphs are the same.
(b) Explain why the graphs of $y = \sqrt{x^2}$ and $y = |x|$ are the same.
(c) Explain why the graphs of $y = x$ and $y = (\sqrt{x})^2$ are not the same.
(d) Explain why the graphs of $y = \sqrt{x^2}$ and $y = x$ are not the same.

128. Draw a graph of an equation that contains two x-intercepts; at one the graph crosses the x-axis, and at the other the graph touches the x-axis.

129. Make up an equation with the intercepts $(2, 0)$, $(4, 0)$, and $(0, 1)$. Compare your equation with a friend's equation. Comment on any similarities.

130. Draw a graph that contains the points $(-2, -1)$, $(0, 1)$, $(1, 3)$, and $(3, 5)$. Compare your graph with those of other students. Are most of the graphs almost straight lines? How many are "curved"? Discuss the various ways in which these points might be connected.

131. An equation is being tested for symmetry with respect to the x-axis, the y-axis, and the origin. Explain why, if two of these symmetries are present, the remaining one must also be present.

132. Draw a graph that contains the points $(-2, 5)$, $(-1, 3)$, and $(0, 2)$ and is symmetric with respect to the y-axis. Compare your graph with those of other students; comment on any similarities. Can a graph contain these points and be symmetric with respect to the x-axis? the origin? Why or why not?

133. Explain how the center and radius of a circle can be used to graph the circle.

134. What Went Wrong? A student stated that the center and radius of the graph whose equation is $(x+3)^2 + (y-2)^2 = 16$ are $(3, -2)$ and 4, respectively. Why is this incorrect?

'Are You Prepared?' Answers

1. add; 25 **2.** $\{-1, 5\}$

1.3 Functions and Their Graphs

PREPARING FOR THIS SECTION *Before getting started, review the following:*

- Interval Notation (Appendix A, Section A.6, pp. A46–A47)
- Solving Inequalities (Appendix A, Section A.6, pp. A49–A51)
- Evaluating Algebraic Expressions, Domain of a Variable (Appendix A, Section A.1, pp. A6–A7)

Now Work the **'Are You Prepared?'** problems on page 35.

OBJECTIVES
1. Determine Whether a Relation Represents a Function (p. 23)
2. Find the Value of a Function (p. 26)
3. Find the Difference Quotient of a Function (p. 29)
4. Find the Domain of a Function Defined by an Equation (p. 30)
5. Identify the Graph of a Function (p. 32)
6. Obtain Information from or about the Graph of a Function (p. 33)

1 Determine Whether a Relation Represents a Function

Often there are situations where the value of one variable is somehow linked to the value of another variable. For example, an individual's level of education is linked to annual income. Engine size is linked to gas mileage. When the value of one variable is related to the value of a second variable, we have a *relation*. A **relation** is a correspondence between two sets. If x and y are two elements, one from each of these sets, and if a relation exists between x and y, then we say that x **corresponds** to y or that y **depends on** x, and we write $x \rightarrow y$.

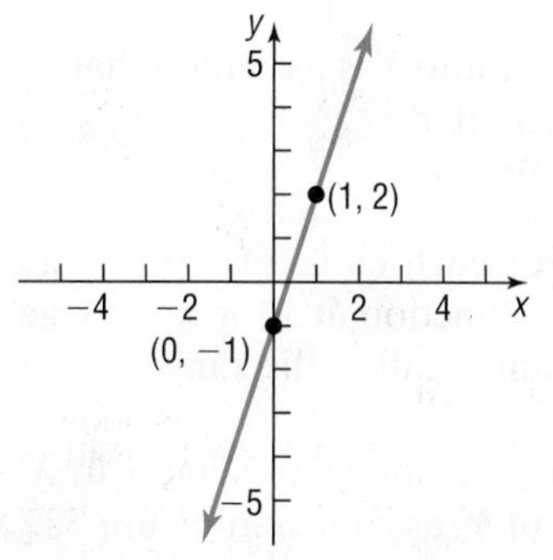

Figure 28 $y = 3x - 1$

There are a number of ways to express relations between two sets. For example, the equation $y = 3x - 1$ shows a relation between x and y. It says that if we take some number x, multiply it by 3, and then subtract 1, we obtain the corresponding value of y. In this sense, x serves as the **input** to the relation, and y is the **output** of the relation. This relation, expressed as a graph, is shown in Figure 28.

The set of all inputs for a relation is called the **domain** of the relation, and the set of all outputs is called the **range**.

In addition to being expressed as equations and graphs, relations can be expressed through a technique called *mapping*. A **map** illustrates a relation as a set of inputs with an arrow drawn from each element in the set of inputs to the corresponding element in the set of outputs. **Ordered pairs** can be used to represent $x \rightarrow y$ as (x, y).

EXAMPLE 1 Maps and Ordered Pairs as Relations

Figure 29 shows a relation between states and the number of representatives each state has in the House of Representatives. (***Source:*** www.house.gov). The relation might be named "number of representatives."

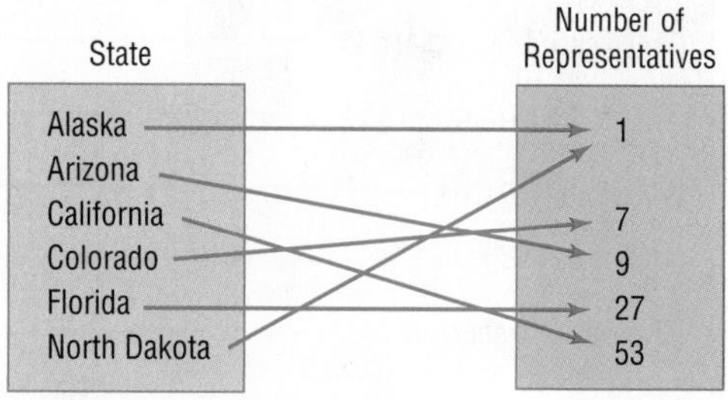

Figure 29 Number of representatives

In this relation, Alaska corresponds to 1, Arizona corresponds to 9, and so on. Using ordered pairs, this relation would be expressed as

$$\{(\text{Alaska}, 1), (\text{Arizona}, 9), (\text{California}, 53), (\text{Colorado}, 7), (\text{Florida}, 27), (\text{North Dakota}, 1)\}$$

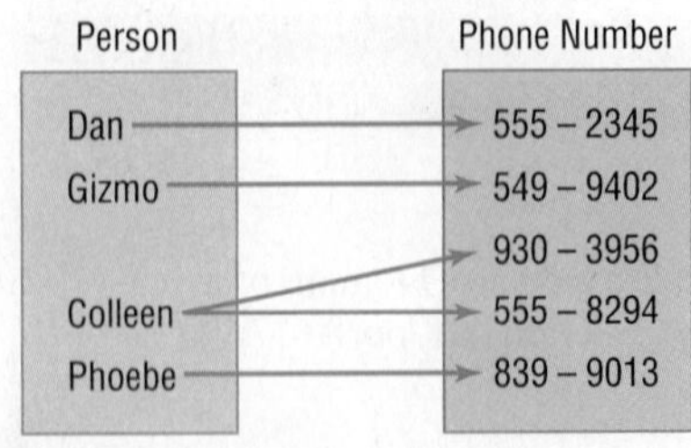

Figure 30 Phone numbers

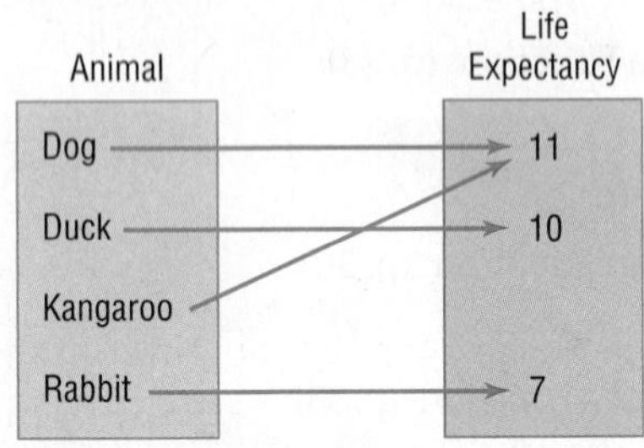

Figure 31 Animal life expectancy

The domain of the relation is {Alaska, Arizona, California, Colorado, Florida, North Dakota}, and the range is {1, 7, 9, 27, 53}. Note that the output "1" is listed only once in the range. ●

One of the most important concepts in algebra is the *function*. A function is a special type of relation. To understand the idea behind a function, let's revisit the relation presented in Example 1. If we were to ask, "How many representatives does Alaska have?" you would respond "1." In fact, each input *state* corresponds to a single output *number of representatives*.

Let's consider a second relation, one that involves a correspondence between four people and their phone numbers. See Figure 30. Notice that Colleen has two telephone numbers. There is no single answer to the question "What is Colleen's phone number?"

Let's look at one more relation. Figure 31 is a relation that shows a correspondence between type of *animal* and *life expectancy*. If asked to determine the life expectancy of a dog, we would all respond, "11 years." If asked to determine the life expectancy of a rabbit, we would all respond, "7 years."

Notice that the relations presented in Figures 29 and 31 have something in common. What is it? In both of these relations, each input corresponds to exactly one output. This leads to the definition of a *function*.

DEFINITION

Let X and Y be two nonempty sets.* A **function** from X into Y is a relation that associates with each element of X exactly one element of Y.

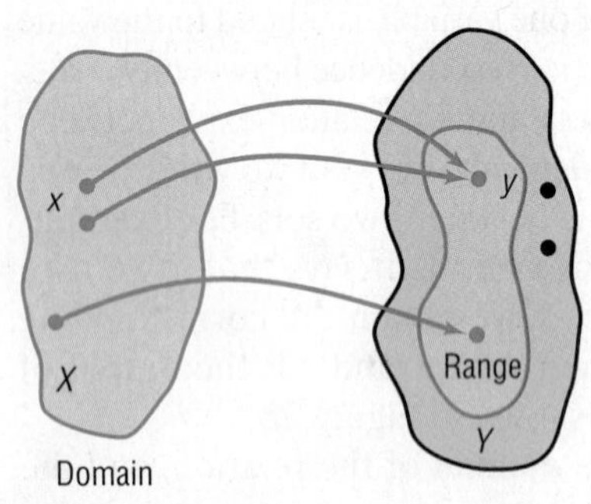

Figure 32

The set X is called the **domain** of the function. For each element x in X, the corresponding element y in Y is called the **value** of the function at x, or the **image** of x. The set of all images of the elements in the domain is called the **range** of the function. See Figure 32.

Since there may be some elements in Y that are not the image of some x in X, it follows that the range of a function may be a subset of Y, as shown in Figure 32.

Not all relations between two sets are functions. The next example shows how to determine whether a relation is a function.

EXAMPLE 2

Determining Whether a Relation Is a Function

For each relation in Figures 33, 34, and 35, state the domain and range. Then determine whether the relation is a function.

(a) See Figure 33. For this relation, the input is the number of calories in a fast-food sandwich, and the output is the fat content (in grams).

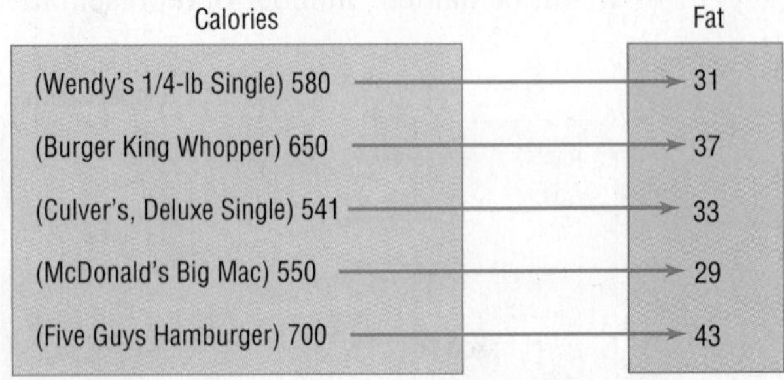

Figure 33 Fat content
Source: Each company's Web site

*The sets X and Y will usually be sets of real numbers, in which case a (real) function results. The two sets can also be sets of complex numbers, and then we have defined a complex function. In the broad definition (proposed by Lejeune Dirichlet), X and Y can be any two sets.

(b) See Figure 34. For this relation, the inputs are gasoline stations in Harris County, Texas, and the outputs are the price per gallon of unleaded regular in January 2015.

(c) See Figure 35. For this relation, the inputs are the weight (in carats) of pear-cut diamonds and the outputs are the price (in dollars).

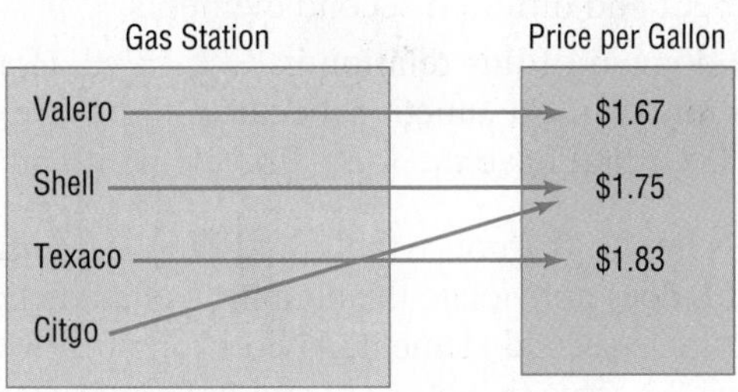

Figure 34 Unleaded price per gallon

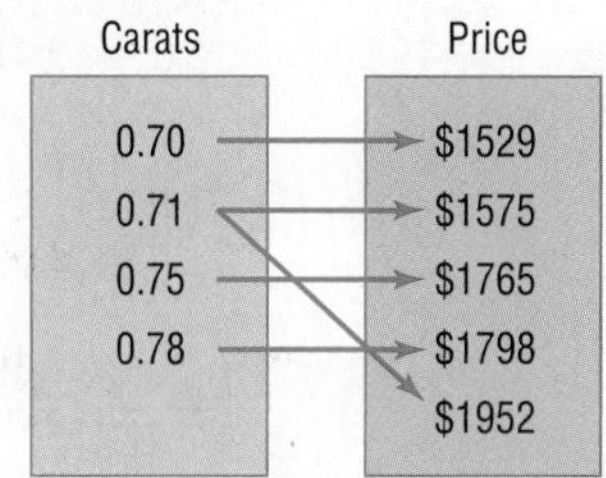

Figure 35 Diamond price

Source: Used with permission of Diamonds.com

Solution

(a) The domain of the relation is {541, 550, 580, 650, 700}, and the range of the relation is {29, 31, 33, 37, 43}. The relation in Figure 33 is a function because each element in the domain corresponds to exactly one element in the range.

(b) The domain of the relation is {Citgo, Shell, Texaco, Valero}. The range of the relation is {1.67, 1.75, 1.83}. The relation in Figure 34 is a function because each element in the domain corresponds to exactly one element in the range. Notice that it is okay for more than one element in the domain to correspond to the same element in the range (Shell and Citgo both sold gas for $1.75 a gallon).

(c) The domain of the relation is {0.70, 0.71, 0.75, 0.78} and the range is {$1529, $1575, $1765, $1798, $1952}. The relation in Figure 35 is not a function because not every element in the domain corresponds to exactly one element in the range. If a 0.71-carat diamond is chosen from the domain, a single price cannot be assigned to it. ●

Now Work PROBLEM 19

The idea behind a function is its predictability. If the input is known, we can use the function to determine the output. With "nonfunctions," we don't have this predictability. Look back at Figure 33. If asked, "How many grams of fat are in a 580-calorie sandwich?" we could use the correspondence to answer, "31." Now consider Figure 35. If asked, "What is the price of a 0.71-carat diamond?" we could not give a single response because two outputs result from the single input "0.71." For this reason, the relation in Figure 35 is not a function.

> **In Words**
> For a function, no input has more than one output. The domain of a function is the set of all inputs; the range is the set of all outputs.

We may also think of a function as a set of ordered pairs (x, y) in which no ordered pairs have the same first element and different second elements. The set of all first elements x is the domain of the function, and the set of all second elements y is its range. Each element x in the domain corresponds to exactly one element y in the range.

EXAMPLE 3 **Determining Whether a Relation Is a Function**

For each relation, state the domain and range. Then determine whether the relation is a function.

(a) $\{(1, 4), (2, 5), (3, 6), (4, 7)\}$

(b) $\{(1, 4), (2, 4), (3, 5), (6, 10)\}$

(c) $\{(-3, 9), (-2, 4), (0, 0), (1, 1), (-3, 8)\}$

Solution (a) The domain of this relation is $\{1, 2, 3, 4\}$, and its range is $\{4, 5, 6, 7\}$. This relation is a function because there are no ordered pairs with the same first element and different second elements.

(b) The domain of this relation is $\{1, 2, 3, 6\}$, and its range is $\{4, 5, 10\}$. This relation is a function because there are no ordered pairs with the same first element and different second elements.

(c) The domain of this relation is $\{-3, -2, 0, 1\}$, and its range is $\{0, 1, 4, 8, 9\}$. This relation is not a function because there are two ordered pairs, $(-3, 9)$ and $(-3, 8)$, that have the same first element and different second elements. ●

In Example 3(b), notice that 1 and 2 in the domain both have the same image in the range. This does not violate the definition of a function; two different first elements can have the same second element. A violation of the definition occurs when two ordered pairs have the same first element and different second elements, as in Example 3(c).

Now Work PROBLEM 23

Up to now we have shown how to identify when a relation is a function for relations defined by mappings (Example 2) and ordered pairs (Example 3). But relations can also be expressed as equations. The circumstances under which equations are functions are discussed next.

To determine whether an equation, where y depends on x, is a function, it is often easiest to solve the equation for y. If any value of x in the domain corresponds to more than one y, the equation does not define a function; otherwise, it does define a function.

EXAMPLE 4 Determining Whether an Equation Is a Function

Determine whether the equation $y = 2x - 5$ defines y as a function of x.

Solution The equation tells us to take an input x, multiply it by 2, and then subtract 5. For any input x, these operations yield only one output y, so the equation is a function. For example, if $x = 1$, then $y = 2(1) - 5 = -3$. If $x = 3$, then $y = 2(3) - 5 = 1$. The graph of the equation $y = 2x - 5$ is a line with slope 2 and y-intercept -5. The function is called a *linear function*. ●

EXAMPLE 5 Determining Whether an Equation Is a Function

Determine whether the equation $x^2 + y^2 = 1$ defines y as a function of x.

Solution To determine whether the equation $x^2 + y^2 = 1$, which defines the unit circle, is a function, solve the equation for y.

$$x^2 + y^2 = 1$$
$$y^2 = 1 - x^2$$
$$y = \pm\sqrt{1 - x^2}$$

For values of x for which $-1 < x < 1$, two values of y result. For example, if $x = 0$, then $y = \pm 1$, so two different outputs result from the same input. This means that the equation $x^2 + y^2 = 1$ does not define a function. ●

Now Work PROBLEM 37

2 Find the Value of a Function

Functions are often denoted by letters such as f, F, g, G, and others. If f is a function, then for each number x in its domain, the corresponding image in the range is designated by the symbol $f(x)$, read as "f of x" or as "f at x." We refer to $f(x)$ as the **value of f at the number x**; $f(x)$ is the number that results when x is given and the function f is applied; $f(x)$ is the output corresponding to x or $f(x)$ is the image of x.

CAUTION
$f(x)$ does not mean "f times x".

For example, the function given in Example 4 may be written as $y = f(x) = 2x - 5$. Then $f(1) = -3$ and $f(3) = 1$. ■

Figure 36 illustrates some other functions. Notice that in every function, for each x in the domain, there is exactly one value in the range.

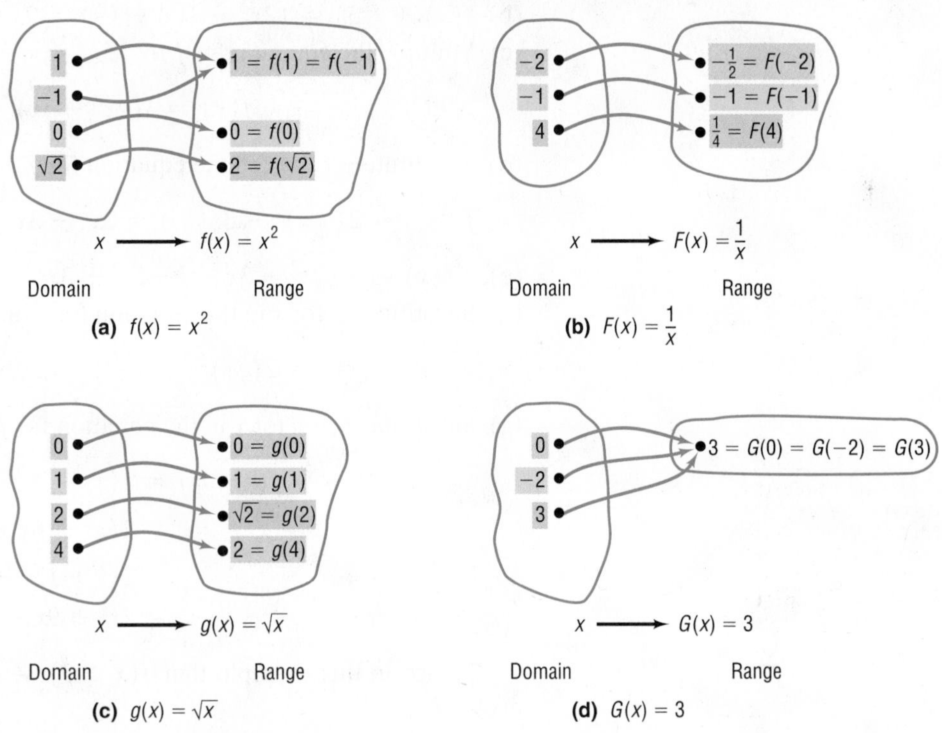

Figure 36

Sometimes it is helpful to think of a function f as a machine that receives as input a number from the domain, manipulates it, and outputs a value. See Figure 37.

The restrictions on this input/output machine are as follows:

1. It accepts only numbers from the domain of the function.
2. For each input, there is exactly one output (which may be repeated for different inputs).

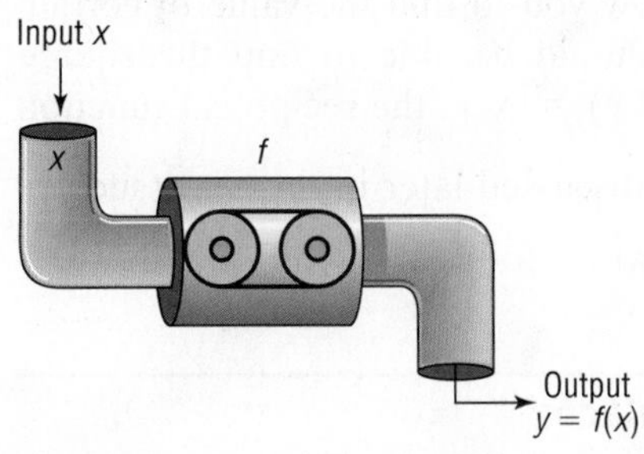

Figure 37 Input/output machine

For a function $y = f(x)$, the variable x is called the **independent variable**, because it can be assigned any of the permissible numbers from the domain. The variable y is called the **dependent variable**, because its value depends on x.

Any symbols can be used to represent the independent and dependent variables. For example, if f is the *cube function,* then f can be given by $f(x) = x^3$ or $f(t) = t^3$ or $f(z) = z^3$. All three functions are the same. Each says to cube the independent variable to get the output. In practice, the symbols used for the independent and dependent variables are based on common usage, such as using C for cost in business.

The independent variable is also called the **argument** of the function. Thinking of the independent variable as an argument can sometimes make it easier to find the value of a function. For example, if f is the function defined by $f(x) = x^3$, then f tells us to cube the argument. Thus $f(2)$ means to cube 2, $f(a)$ means to cube the number a, and $f(x + h)$ means to cube the quantity $x + h$.

EXAMPLE 6

Finding Values of a Function

For the function f defined by $f(x) = 2x^2 - 3x$, evaluate

(a) $f(3)$ (b) $f(x) + f(3)$ (c) $3f(x)$ (d) $f(-x)$
(e) $-f(x)$ (f) $f(3x)$ (g) $f(x + 3)$

Solution (a) Substitute 3 for x in the equation for f, $f(x) = 2x^2 - 3x$, to get

$$f(3) = 2(3)^2 - 3(3) = 18 - 9 = 9$$

The image of 3 is 9.

(b) $f(x) + f(3) = (2x^2 - 3x) + (9) = 2x^2 - 3x + 9$

(c) Multiply the equation for f by 3.

$$3f(x) = 3(2x^2 - 3x) = 6x^2 - 9x$$

(d) Substitute $-x$ for x in the equation for f and simplify.

$$f(-x) = 2(-x)^2 - 3(-x) = 2x^2 + 3x$$ Notice the use of parentheses here.

(e) $-f(x) = -(2x^2 - 3x) = -2x^2 + 3x$

(f) Substitute $3x$ for x in the equation for f and simplify.

$$f(3x) = 2(3x)^2 - 3(3x) = 2(9x^2) - 9x = 18x^2 - 9x$$

(g) Substitute $x + 3$ for x in the equation for f and simplify.

$$\begin{aligned} f(x+3) &= 2(x+3)^2 - 3(x+3) \\ &= 2(x^2 + 6x + 9) - 3x - 9 \\ &= 2x^2 + 12x + 18 - 3x - 9 \\ &= 2x^2 + 9x + 9 \end{aligned}$$

Notice in this example that $f(x+3) \neq f(x) + f(3)$, $f(-x) \neq -f(x)$, and $3f(x) \neq f(3x)$.

Now Work PROBLEM 43

Most calculators have special keys that allow you to find the value of certain commonly used functions. For example, you should be able to find the square function $f(x) = x^2$, the square root function $f(x) = \sqrt{x}$, the reciprocal function $f(x) = \frac{1}{x} = x^{-1}$, and many others that will be discussed later in this text (such as $\ln x$ and $\log x$). Verify the results of Example 7, which follows, on your calculator.

EXAMPLE 7 Finding Values of a Function on a Calculator

(a) $f(x) = x^2$ $\quad f(1.234) = 1.234^2 = 1.522756$

(b) $F(x) = \frac{1}{x}$ $\quad F(1.234) = \frac{1}{1.234} \approx 0.8103727715$

(c) $g(x) = \sqrt{x}$ $\quad g(1.234) = \sqrt{1.234} \approx 1.110855526$

COMMENT Graphing calculators can be used to evaluate any function. Figure 38 shows the result obtained in Example 6(a) on a TI-84 Plus C graphing calculator with the function to be evaluated, $f(x) = 2x^2 - 3x$, in Y_1.

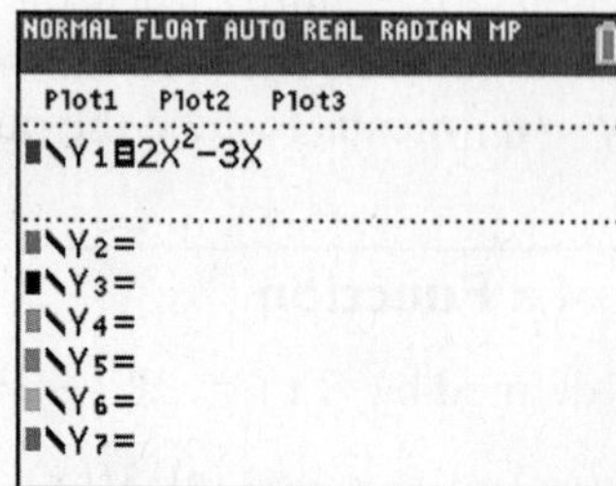

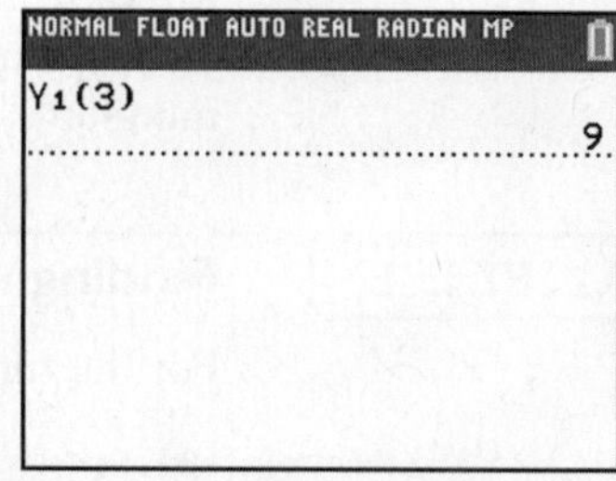

Figure 38 Evaluating $f(x) = 2x^2 - 3x$ for $x = 3$

Implicit Form of a Function

COMMENT The explicit form of a function is the form required by a graphing calculator. ■

In general, when a function f is defined by an equation in x and y, we say that the function f is given **implicitly**. If it is possible to solve the equation for y in terms of x, then we write $y = f(x)$ and say that the function is given **explicitly**. For example,

Implicit Form	**Explicit Form**
$3x + y = 5$	$y = f(x) = -3x + 5$
$x^2 - y = 6$	$y = f(x) = x^2 - 6$
$xy = 4$	$y = f(x) = \frac{4}{x}$

SUMMARY

Important Facts about Functions

(a) For each x in the domain of a function f, there is exactly one image $f(x)$ in the range; however, an element in the range can result from more than one x in the domain.

(b) f is the symbol that we use to denote the function. It is symbolic of the equation (rule) that we use to get from an x in the domain to $f(x)$ in the range.

(c) If $y = f(x)$, then x is called the independent variable or argument of f, and y is called the dependent variable or the value of f at x.

3 Find the Difference Quotient of a Function

An important concept in calculus involves looking at a certain quotient. For a given function $y = f(x)$, the inputs x and $x + h$, $h \neq 0$, result in the images $f(x)$ and $f(x + h)$. The quotient of their differences

$$\frac{f(x + h) - f(x)}{(x + h) - x} = \frac{f(x + h) - f(x)}{h}$$

with $h \neq 0$, is called the *difference quotient of f at x.*

DEFINITION

The **difference quotient** of a function f at x is given by

$$\frac{f(x + h) - f(x)}{h} \qquad h \neq 0 \qquad \textbf{(1)}$$

The difference quotient is used in calculus to define the derivative, which leads to applications such as the velocity of an object and optimization of resources.

When finding a difference quotient, it is necessary to simplify the expression in order to cancel the h in the denominator, as illustrated in the following example.

EXAMPLE 8 **Finding the Difference Quotient of a Function**

Find the difference quotient of each function.

(a) $f(x) = 2x^2 - 3x$

(b) $f(x) = \frac{4}{x}$

(c) $f(x) = \sqrt{x}$

Solution (a)

$$\frac{f(x+h)-f(x)}{h} = \frac{[2(x+h)^2 - 3(x+h)] - [2x^2 - 3x]}{h}$$

$\uparrow$ $f(x+h) = 2(x+h)^2 - 3(x+h)$

$$= \frac{2(x^2 + 2xh + h^2) - 3x - 3h - 2x^2 + 3x}{h} \quad \text{Simplify.}$$

$$= \frac{2x^2 + 4xh + 2h^2 - 3h - 2x^2}{h} \quad \text{Distribute and combine like terms.}$$

$$= \frac{4xh + 2h^2 - 3h}{h} \quad \text{Combine like terms.}$$

$$= \frac{h(4x + 2h - 3)}{h} \quad \text{Factor out } h.$$

$$= 4x + 2h - 3 \quad \text{Divide out the factor } h.$$

(b)

$$\frac{f(x+h)-f(x)}{h} = \frac{\frac{4}{x+h} - \frac{4}{x}}{h} \quad f(x+h) = \frac{4}{x+h}$$

$$= \frac{\frac{4x - 4(x+h)}{x(x+h)}}{h} \quad \text{Subtract.}$$

$$= \frac{4x - 4x - 4h}{x(x+h)h} \quad \text{Divide and distribute.}$$

$$= \frac{-4h}{x(x+h)h} \quad \text{Simplify.}$$

$$= -\frac{4}{x(x+h)} \quad \text{Divide out the factor } h.$$

(c)

$$\frac{f(x+h)-f(x)}{h} = \frac{\sqrt{x+h} - \sqrt{x}}{h} \quad f(x+h) = \sqrt{x+h}$$

$$= \frac{\sqrt{x+h} - \sqrt{x}}{h} \cdot \frac{\sqrt{x+h} + \sqrt{x}}{\sqrt{x+h} + \sqrt{x}} \quad \text{Rationalize the numerator.}$$

$$= \frac{(\sqrt{x+h})^2 - (\sqrt{x})^2}{h(\sqrt{x+h} + \sqrt{x})} \quad (A-B)(A+B) = A^2 - B^2$$

$$= \frac{h}{h(\sqrt{x+h} + \sqrt{x})} \quad (\sqrt{x+h})^2 - (\sqrt{x})^2 = x + h - x = h$$

$$= \frac{1}{\sqrt{x+h} + \sqrt{x}} \quad \text{Divide out the factor } h.$$

Now Work PROBLEM 67

4 Find the Domain of a Function Defined by an Equation

Often the domain of a function f is not specified; instead, only the equation defining the function is given. In such cases, we agree that the **domain of f** is the largest set of real numbers for which the value $f(x)$ is a real number. The domain of a function f is the same as the domain of the variable x in the expression $f(x)$.

EXAMPLE 9 **Finding the Domain of a Function**

Find the domain of each of the following functions.

(a) $f(x) = x^2 + 5x$

(b) $g(x) = \dfrac{3x}{x^2 - 4}$

(c) $h(t) = \sqrt{4 - 3t}$

(d) $F(x) = \dfrac{\sqrt{3x + 12}}{x - 5}$

Solution (a) The function says to square a number and then add five times the number. Since these operations can be performed on any real number, the domain of f is the set of all real numbers.

(b) The function g says to divide $3x$ by $x^2 - 4$. Since division by 0 is not defined, the denominator $x^2 - 4$ can never be 0, so x can never equal -2 or 2. The domain of the function g is $\{x | x \neq -2, x \neq 2\}$.

In Words
The domain of g found in Example 9(b) is $\{x | x \neq -2, x \neq 2\}$. This notation is read, "The domain of the function g is the set of all real numbers x such that x does not equal -2 and x does not equal 2."

(c) The function h says to take the square root of $4 - 3t$. But only nonnegative numbers have real square roots, so the expression under the square root (the radicand) must be nonnegative (greater than or equal to zero). This requires that

$$4 - 3t \geq 0$$
$$-3t \geq -4$$
$$t \leq \frac{4}{3}$$

The domain of h is $\left\{t \middle| t \leq \frac{4}{3}\right\}$, or the interval $\left(-\infty, \frac{4}{3}\right]$.

(d) The function F says to take the square root of $3x + 12$ and divide this result by $x - 5$. This requires that $3x + 12 \geq 0$, so $x \geq -4$, and also that $x - 5 \neq 0$, so $x \neq 5$. Combining these two restrictions, the domain of F is

$$\{x | x \geq -4, \quad x \neq 5\}.$$

The following steps may prove helpful for finding the domain of a function that is defined by an equation and whose domain is a subset of the real numbers.

Finding the Domain of a Function Defined by an Equation

1. Start with the domain as the set of real numbers.
2. If the equation has a denominator, exclude any numbers that give a zero denominator.
3. If the equation has a radical of even index, exclude any numbers that cause the expression inside the radical (the radicand) to be negative.

Now Work PROBLEM 55

If x is in the domain of a function f, we shall say that **f is defined at x**, or **$f(x)$ exists**. If x is not in the domain of f, we say that **f is not defined at x**, or **$f(x)$ does not exist**. For example, if $f(x) = \frac{x}{x^2 - 1}$, then $f(0)$ exists, but $f(1)$ and $f(-1)$ do not exist. (Do you see why?)

We have not said much about finding the range of a function. We will say more about finding the range when we look at the graph of a function. When a function is defined by an equation, it can be difficult to find the range. Therefore, we shall usually be content to find just the domain of a function when the function is defined by an equation. We shall express the domain of a function using inequalities, interval notation, set notation, or words, whichever is most convenient.

When we use functions in applications, the domain may be restricted by physical or geometric considerations. For example, the domain of the function f defined by $f(x) = x^2$ is the set of all real numbers. However, if f is used to obtain the area of a square when the length x of a side is known, then we must restrict the domain of f to the positive real numbers, since the length of a side can never be 0 or negative.

EXAMPLE 10 **Finding the Domain in an Application**

Express the area of a circle as a function of its radius. Find the domain.

Solution See Figure 39. The formula for the area A of a circle of radius r is $A = \pi r^2$. Using r to represent the independent variable and A to represent the dependent variable, the function expressing this relationship is

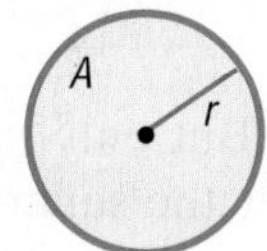

Figure 39 Circle of radius r

$$A(r) = \pi r^2$$

In this setting, the domain is $\{r \mid r > 0\}$. (Do you see why?) ●

Observe, in the solution to Example 10, that the symbol A is used in two ways: It is used to name the function, and it is used to symbolize the dependent variable. This double use is common in applications and should not cause any difficulty.

Now Work PROBLEM 103

The Graph of a Function

In applications, a graph often demonstrates more clearly the relationship between two variables than, say, an equation or table. For example, Table 5 shows the average price of gasoline in the United States for the years 1985–2014 (adjusted for inflation, based on 2014 dollars). If we plot these data and then connect the points, we obtain Figure 40.

Table 5

Year	Price	Year	Price	Year	Price
1985	2.55	1995	1.72	2005	2.74
1986	1.90	1996	1.80	2006	3.01
1987	1.89	1997	1.76	2007	3.19
1988	1.81	1998	1.49	2008	3.56
1989	1.87	1999	1.61	2009	2.58
1990	2.03	2000	2.03	2010	3.00
1991	1.91	2001	1.90	2011	3.69
1992	1.82	2002	1.76	2012	3.72
1993	1.74	2003	1.99	2013	3.54
1994	1.71	2004	2.31	2014	3.43

Source: U.S. Energy Information Administration

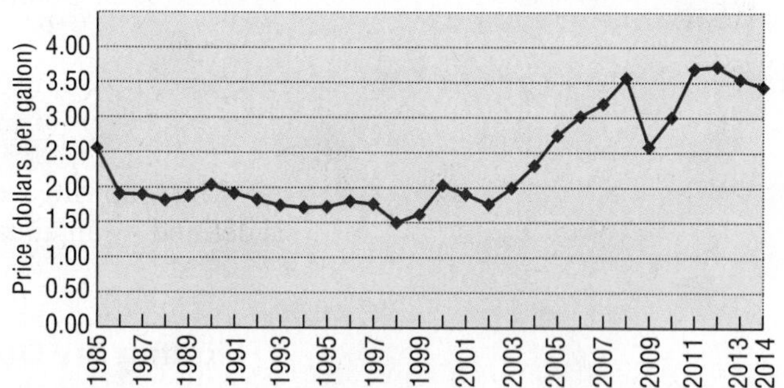

Figure 40 Average retail price of gasoline (2014 dollars)
Source: U.S. Energy Information Administration

We can see from the graph that the price of gasoline (adjusted for inflation) stayed roughly the same from 1986 to 1991 and rose rapidly from 2002 to 2008. The graph also shows that the lowest price occurred in 1998. To learn information such as this from an equation requires that some calculations be made.

Look again at Figure 40. The graph shows that for each date on the horizontal axis, there is only one price on the vertical axis. The graph represents a function, although the exact rule for getting from date to price is not given.

When a function is defined by an equation in x and y, the **graph of the function** is the graph of the equation; that is, it is the set of points (x, y) in the xy-plane that satisfy the equation.

5 Identify the Graph of a Function

> In Words
> If any vertical line intersects a graph at more than one point, the graph is not the graph of a function.

Not every collection of points in the xy-plane represents the graph of a function. Remember, for a function, each number x in the domain has exactly one image y in the range. This means that the graph of a function cannot contain two points with the same x-coordinate and different y-coordinates. Therefore, the graph of a function must satisfy the following **vertical-line test**.

THEOREM **Vertical-Line Test**

A set of points in the xy-plane is the graph of a function if and only if every vertical line intersects the graph in at most one point.

EXAMPLE 11

Identifying the Graph of a Function

Which of the graphs in Figure 41 are graphs of functions?

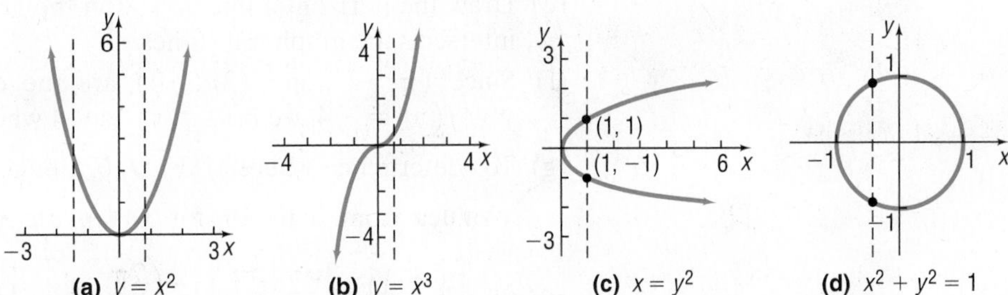

Figure 41

Solution The graphs in Figures 41(a) and 41(b) are graphs of functions, because every vertical line intersects each graph in at most one point. The graphs in Figures 41(c) and 41(d) are not graphs of functions, because there is a vertical line that intersects each graph in more than one point. Notice in Figure 41(c) that the input 1 corresponds to two outputs, −1 and 1. This is why the graph does not represent a function. ●

Now Work PROBLEMS 83 AND 85

6 Obtain Information from or about the Graph of a Function

If (x, y) is a point on the graph of a function *f*, then *y* is the value of *f* at *x*; that is, $y = f(x)$. Also if $y = f(x)$, then (x, y) is a point on the graph of *f*. For example, if $(-2, 7)$ is on the graph of *f*, then $f(-2) = 7$, and if $f(5) = 8$, then the point $(5, 8)$ is on the graph of $y = f(x)$. The next example illustrates how to obtain information about a function if its graph is given.

EXAMPLE 12

Obtaining Information from the Graph of a Function

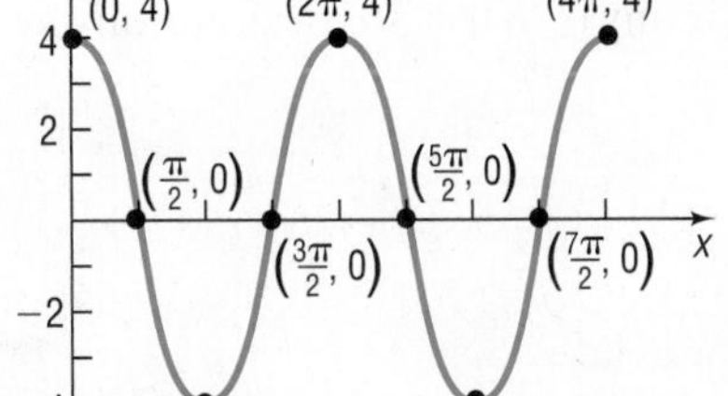

Figure 42

Let f be the function whose graph is given in Figure 42. (The graph of f might represent the distance *y* that the bob of a pendulum is from its *at-rest* position at time *x*. Negative values of *y* mean that the pendulum is to the left of the at-rest position, and positive values of *y* mean that the pendulum is to the right of the at-rest position.)

(a) What are $f(0), f\left(\frac{3\pi}{2}\right)$, and $f(3\pi)$?
(b) What is the domain of f?
(c) What is the range of f?
(d) List the intercepts. (Recall that these are the points, if any, where the graph crosses or touches the coordinate axes.)
(e) How many times does the line $y = 2$ intersect the graph?
(f) For what values of *x* does $f(x) = -4$?
(g) For what values of *x* is $f(x) > 0$?

Solution

(a) Since $(0, 4)$ is on the graph of *f*, the *y*-coordinate 4 is the value of f at the *x*-coordinate 0; that is, $f(0) = 4$. In a similar way, when $x = \frac{3\pi}{2}$, then $y = 0$, so $f\left(\frac{3\pi}{2}\right) = 0$. When $x = 3\pi$, then $y = -4$, so $f(3\pi) = -4$.
(b) To determine the domain of *f*, notice that the points on the graph of f have *x*-coordinates between 0 and 4π, inclusive; and for each number *x* between 0 and 4π, there is a point $(x, f(x))$ on the graph. The domain of *f* is $\{x \mid 0 \le x \le 4\pi\}$ or the interval $[0, 4\pi]$.
(c) The points on the graph all have *y*-coordinates between −4 and 4, inclusive; and for each such number *y*, there is at least one number *x* in the domain. The range of *f* is $\{y \mid -4 \le y \le 4\}$ or the interval $[-4, 4]$.

(d) The intercepts are the points

$$(0,4), \left(\frac{\pi}{2}, 0\right), \left(\frac{3\pi}{2}, 0\right), \left(\frac{5\pi}{2}, 0\right), \text{ and } \left(\frac{7\pi}{2}, 0\right)$$

(e) Draw the horizontal line $y = 2$ on the graph in Figure 42. Notice that the line intersects the graph four times.

(f) Since $(\pi, -4)$ and $(3\pi, -4)$ are the only points on the graph for which $y = f(x) = -4$, we have $f(x) = -4$ when $x = \pi$ and $x = 3\pi$.

(g) To determine where $f(x) > 0$, look at Figure 42 and determine the x-values from 0 to 4π for which the y-coordinate is positive. This occurs on $\left[0, \frac{\pi}{2}\right) \cup \left(\frac{3\pi}{2}, \frac{5\pi}{2}\right) \cup \left(\frac{7\pi}{2}, 4\pi\right]$. Using inequality notation, $f(x) > 0$ for $0 \le x < \frac{\pi}{2}$ or $\frac{3\pi}{2} < x < \frac{5\pi}{2}$ or $\frac{7\pi}{2} < x \le 4\pi$. ●

When the graph of a function is given, its domain may be viewed as the shadow created by the graph on the x-axis by vertical beams of light. Its range can be viewed as the shadow created by the graph on the y-axis by horizontal beams of light. Try this technique with the graph given in Figure 42.

Now Work PROBLEM 79

EXAMPLE 13 **Obtaining Information about the Graph of a Function**

Consider the function: $f(x) = \dfrac{x+1}{x+2}$

(a) Find the domain of f.

(b) Is the point $\left(1, \frac{1}{2}\right)$ on the graph of f?

(c) If $x = 2$, what is $f(x)$? What point is on the graph of f?

(d) If $f(x) = 2$, what is x? What point is on the graph of f?

(e) What are the x-intercepts of the graph of f (if any)? What point(s) are on the graph of f?

Solution (a) The domain of f is $\{x \mid x \neq -2\}$.

(b) When $x = 1$, then

$$f(1) = \frac{1+1}{1+2} = \frac{2}{3} \qquad f(x) = \frac{x+1}{x+2}$$

The point $\left(1, \frac{2}{3}\right)$ is on the graph of f; the point $\left(1, \frac{1}{2}\right)$ is not.

(c) If $x = 2$, then

$$f(2) = \frac{2+1}{2+2} = \frac{3}{4}$$

The point $\left(2, \frac{3}{4}\right)$ is on the graph of f.

(d) If $f(x) = 2$, then

$$\frac{x+1}{x+2} = 2 \qquad f(x) = 2$$

$$x + 1 = 2(x+2) \qquad \text{Multiply both sides by } x + 2.$$

$$x + 1 = 2x + 4 \qquad \text{Distribute.}$$

$$x = -3 \qquad \text{Solve for } x.$$

If $f(x) = 2$, then $x = -3$. The point $(-3, 2)$ is on the graph of f.

(e) The x-intercepts of the graph of f are the real solutions of the equation $f(x) = 0$ that are in the domain of f.

$$\frac{x+1}{x+2} = 0$$

$$x + 1 = 0 \quad \text{Multiply both sides by } x + 2.$$

$$x = -1 \quad \text{Subtract 1 from both sides.}$$

The only real solution of the equation $f(x) = \frac{x+1}{x+2} = 0$ is $x = -1$, so -1 is the only x-intercept. Since $f(-1) = 0$, the point $(-1, 0)$ is on the graph of f. ●

Now Work PROBLEM 95

SUMMARY

Function	A relation between two sets of real numbers so that each number x in the first set, the domain, has corresponding to it exactly one number y in the second set, the range. A set of ordered pairs (x, y) or $(x, f(x))$ in which no first element is paired with two different second elements. The range is the set of y values of the function that are the images of the x values in the domain. A function f may be defined implicitly by an equation involving x and y or explicitly by writing $y = f(x)$.
Unspecified Domain	If a function f is defined by an equation and no domain is specified, then the domain will be taken to be the largest set of real numbers for which the equation defines a real number.
Function Notation	$y = f(x)$ f is a symbol for the function. x is the independent variable, or argument. y is the dependent variable. $f(x)$ is the value of the function at x, or the image of x.
Graph of a Function	The collection of points (x, y) that satisfies the equation $y = f(x)$.
Vertical-Line Test	A collection of points is the graph of a function provided that every vertical line intersects the graph in at most one point.

1.3 Assess Your Understanding

'Are You Prepared?' *Answers are given at the end of these exercises. If you get a wrong answer, read the pages listed in red.*

1. The inequality $-1 < x < 3$ can be written in interval notation as ________. (pp. A46–A47)

2. If $x = -2$, the value of the expression $3x^2 - 5x + \frac{1}{x}$ is ______. (pp. A6–A7)

3. The domain of the variable in the expression $\frac{x-3}{x+4}$ is ____________. (p. A7)

4. Solve the inequality: $3 - 2x > 5$. (pp. A49–A52)

Concepts and Vocabulary

5. If f is a function defined by the equation $y = f(x)$, then x is called the ____________ variable, and y is the __________ variable.

6. ***True or False*** Every relation is a function.

7. ***True or False*** If no domain is specified for a function f, then the domain of f is taken to be the set of real numbers.

8. ***True or False*** The domain of the function $f(x) = \frac{x^2 - 4}{x}$ is $\{x \mid x \neq \pm 2\}$.

9. The set of all images of the elements in the domain of a function is called the ______.
(a) range (b) domain (c) solution set (d) function

10. The independent variable is sometimes referred to as the ______ of the function.
(a) range (b) value (c) argument (d) definition

11. The expression $\frac{f(x+h)-f(x)}{h}$ is called the ______ of f.
(a) radicand (b) image
(c) correspondence (d) difference quotient

12. When written as $y = f(x)$, a function is said to be defined ______.
(a) explicitly (b) consistently
(c) implicitly (d) rationally

13. A set of points in the xy-plane is the graph of a function if and only if every _______ line intersects the graph in at most one point.

14. If the point $(5, -3)$ is a point on the graph of f, then $f(\_\_) = \_\_\_\_$.

15. Find a so that the point $(-1, 2)$ is on the graph of $f(x) = ax^2 + 4$.

16. ***True or False*** The graph of a function $y = f(x)$ always crosses the y-axis.

17. ***True or False*** The y-intercept of the graph of the function $y = f(x)$, whose domain is all real numbers, is $f(0)$.

18. The graph of a function $y = f(x)$ can have more than one of which type of intercept?
(a) x-intercept (b) y-intercept
(c) both (d) neither

Skill Building

In Problems 19–30, state the domain and range for each relation. Then determine whether each relation represents a function.

19.
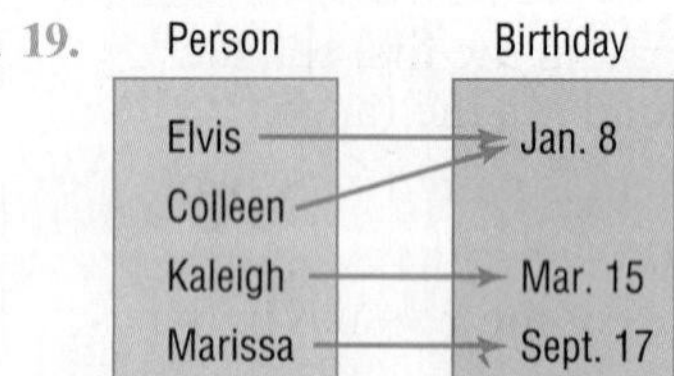

20.
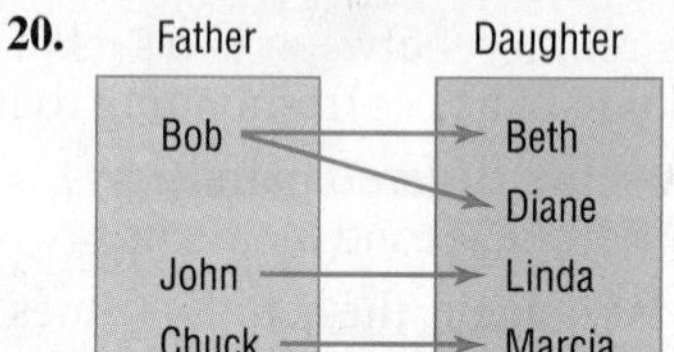

21.
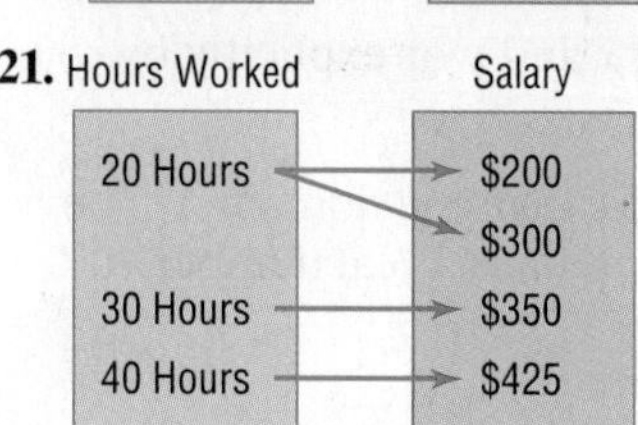

22.
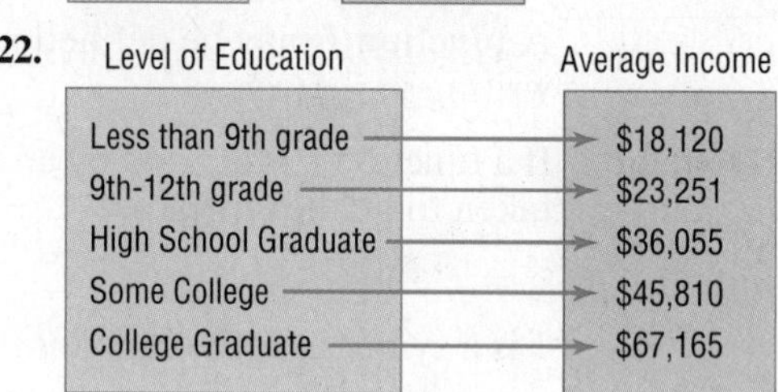

23. $\{(2,6), (-3,6), (4,9), (2,10)\}$ **24.** $\{(-2,5), (-1,3), (3,7), (4,12)\}$ **25.** $\{(1,3), (2,3), (3,3), (4,3)\}$

26. $\{(0,-2), (1,3), (2,3), (3,7)\}$ **27.** $\{(-2,4), (-2,6), (0,3), (3,7)\}$ **28.** $\{(-4,4), (-3,3), (-2,2), (-1,1), (-4,0)\}$

29. $\{(-2,4), (-1,1), (0,0), (1,1)\}$ **30.** $\{(-2,16), (-1,4), (0,3), (1,4)\}$

In Problems 31–42, determine whether the equation defines y as a function of x.

31. $y = 2x^2 - 3x + 4$ **32.** $y = x^3$ **33.** $y = \frac{1}{x}$ **34.** $y = |x|$

35. $y^2 = 4 - x^2$ **36.** $y = \pm\sqrt{1-2x}$ 37. $x = y^2$ **38.** $x + y^2 = 1$

39. $y = \sqrt[3]{x}$ **40.** $y = \frac{3x-1}{x+2}$ **41.** $2x^2 + 3y^2 = 1$ **42.** $x^2 - 4y^2 = 1$

In Problems 43–50, find the following for each function:
(a) $f(0)$ *(b)* $f(1)$ *(c)* $f(-1)$ *(d)* $f(-x)$ *(e)* $-f(x)$ *(f)* $f(x+1)$ *(g)* $f(2x)$ *(h)* $f(x+h)$

43. $f(x) = 3x^2 + 2x - 4$ **44.** $f(x) = -2x^2 + x - 1$ **45.** $f(x) = \frac{x}{x^2+1}$ **46.** $f(x) = \frac{x^2-1}{x+4}$

47. $f(x) = |x| + 4$ **48.** $f(x) = \sqrt{x^2+x}$ **49.** $f(x) = \frac{2x+1}{3x-5}$ **50.** $f(x) = 1 - \frac{1}{(x+2)^2}$

In Problems 51–66, find the domain of each function.

51. $f(x) = -5x + 4$ **52.** $f(x) = x^2 + 2$ **53.** $f(x) = \frac{x}{x^2+1}$ **54.** $f(x) = \frac{x^2}{x^2+1}$

55. $g(x) = \frac{x}{x^2-16}$ **56.** $h(x) = \frac{2x}{x^2-4}$ **57.** $F(x) = \frac{x-2}{x^3+x}$ **58.** $G(x) = \frac{x+4}{x^3-4x}$

59. $h(x) = \sqrt{3x - 12}$ **60.** $G(x) = \sqrt{1 - x}$ **61.** $p(x) = \sqrt{\dfrac{2}{x - 1}}$ **62.** $f(x) = \dfrac{4}{\sqrt{x - 9}}$

63. $f(x) = \dfrac{x}{\sqrt{x - 4}}$ **64.** $f(x) = \dfrac{-x}{\sqrt{-x - 2}}$ **65.** $P(t) = \dfrac{\sqrt{t - 4}}{3t - 21}$ **66.** $h(z) = \dfrac{\sqrt{z + 3}}{z - 2}$

In Problems 67–78, find the difference quotient of f; that is, find $\dfrac{f(x + h) - f(x)}{h}$, $h \neq 0$, *for each function. Be sure to simplify.*

67. $f(x) = 4x + 3$ **68.** $f(x) = -3x + 1$ **69.** $f(x) = x^2 - 4$ **70.** $f(x) = 3x^2 + 2$

71. $f(x) = x^2 - x + 4$ **72.** $f(x) = 3x^2 - 2x + 6$ **73.** $f(x) = \dfrac{1}{x^2}$ **74.** $f(x) = \dfrac{1}{x + 3}$

75. $f(x) = \dfrac{2x}{x + 3}$ **76.** $f(x) = \dfrac{5x}{x - 4}$ **77.** $f(x) = \sqrt{x - 2}$ **78.** $f(x) = \sqrt{x + 1}$

[**Hint:** Rationalize the numerator.]

79. Use the given graph of the function f to answer parts (a)–(n).

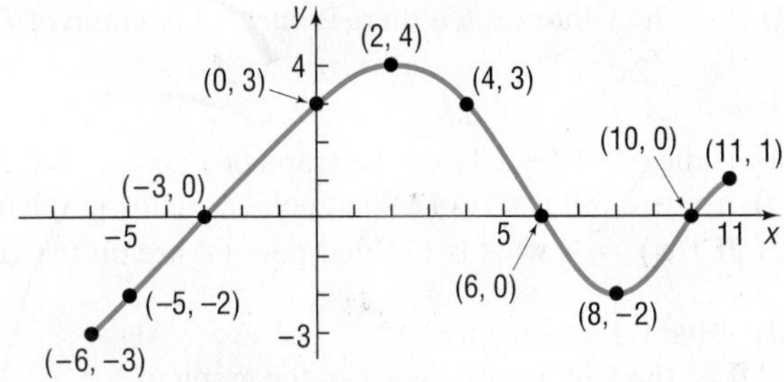

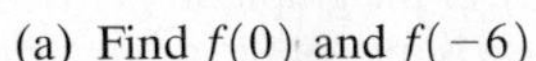

(a) Find $f(0)$ and $f(-6)$.
(b) Find $f(6)$ and $f(11)$.
(c) Is $f(3)$ positive or negative?
(d) Is $f(-4)$ positive or negative?
(e) For what values of x is $f(x) = 0$?
(f) For what values of x is $f(x) > 0$?
(g) What is the domain of f?
(h) What is the range of f?
(i) What are the x-intercepts?
(j) What is the y-intercept?
(k) How often does the line $y = \dfrac{1}{2}$ intersect the graph?
(l) How often does the line $x = 5$ intersect the graph?
(m) For what values of x does $f(x) = 3$?
(n) For what values of x does $f(x) = -2$?

80. Use the given graph of the function f to answer parts (a)–(n).

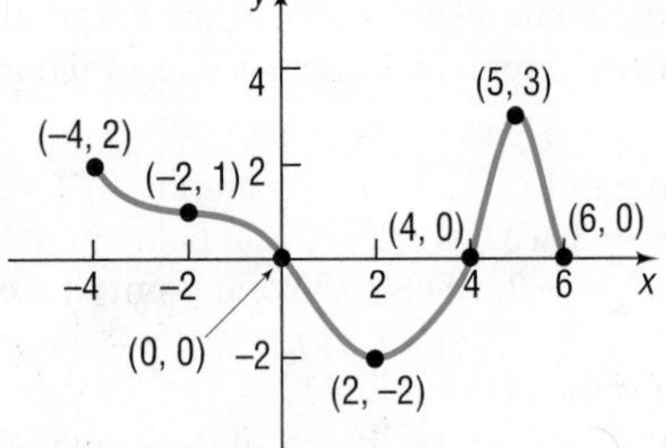

(a) Find $f(0)$ and $f(6)$.
(b) Find $f(2)$ and $f(-2)$.
(c) Is $f(3)$ positive or negative?
(d) Is $f(-1)$ positive or negative?
(e) For what values of x is $f(x) = 0$?
(f) For what values of x is $f(x) < 0$?
(g) What is the domain of f?
(h) What is the range of f?
(i) What are the x-intercepts?
(j) What is the y-intercept?
(k) How often does the line $y = -1$ intersect the graph?
(l) How often does the line $x = 1$ intersect the graph?
(m) For what value of x does $f(x) = 3$?
(n) For what value of x does $f(x) = -2$?

In Problems 81–92, determine whether the graph is that of a function by using the vertical-line test. If it is, use the graph to find:
(a) The domain and range *(b) The intercepts, if any* *(c) Any symmetry with respect to the x-axis, the y-axis, or the origin*

81.

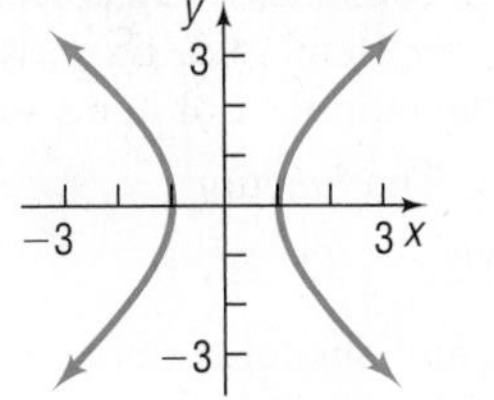

82.

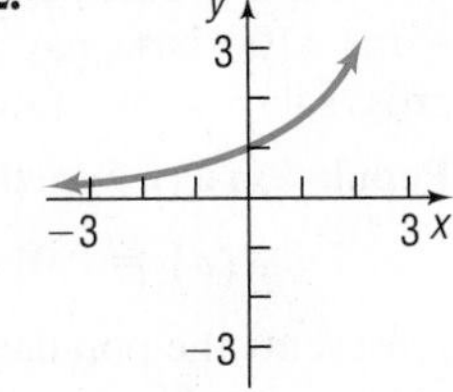

83.

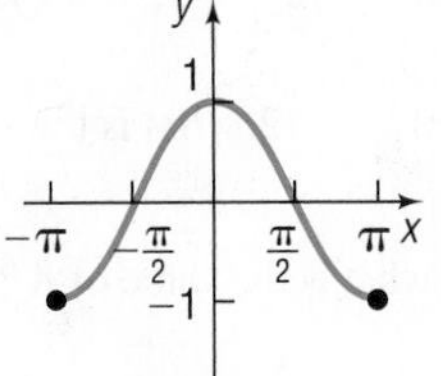

84.

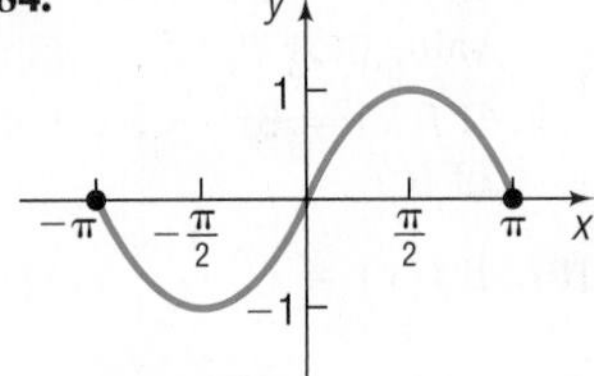

85.

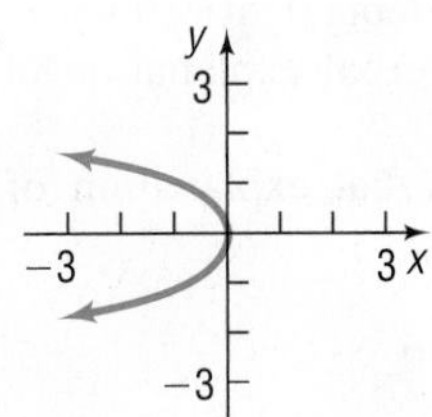

86.

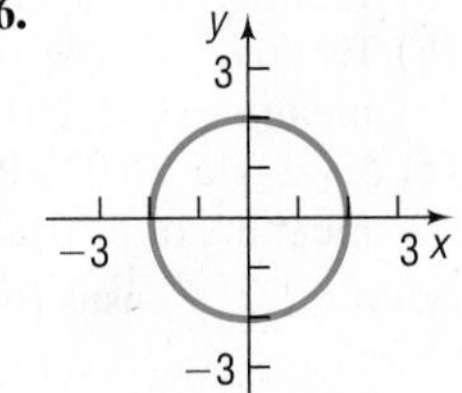

87.

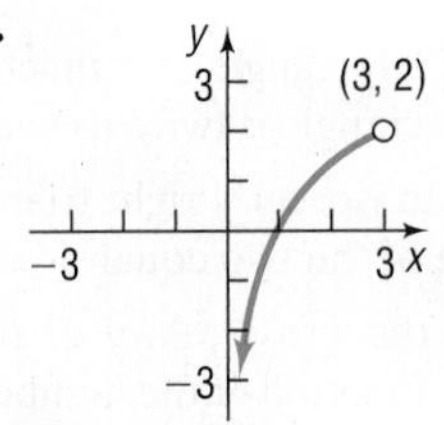

88.

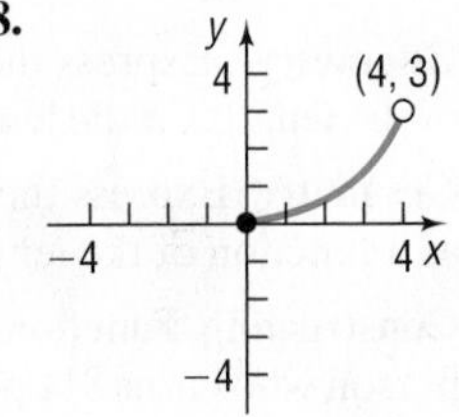

89.

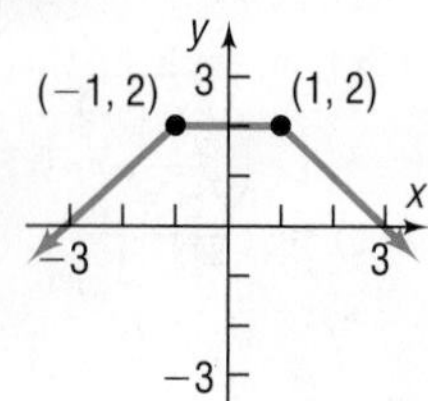

90.

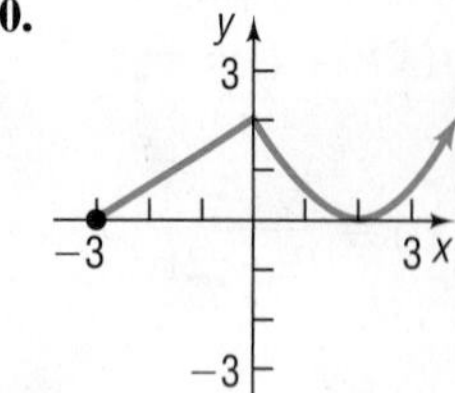

91.

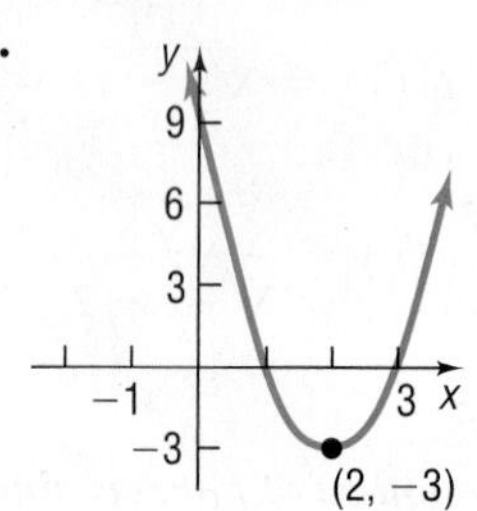

92.

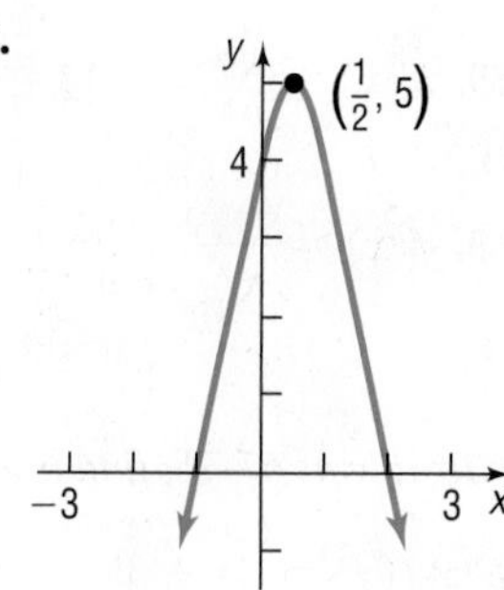

In Problems 93–98, answer the questions about the given function.

93. $f(x) = 2x^2 - x - 1$
(a) Is the point $(-1, 2)$ on the graph of f?
(b) If $x = -2$, what is $f(x)$? What point is on the graph of f?
(c) If $f(x) = -1$, what is x? What point(s) are on the graph of f?
(d) What is the domain of f?
(e) List the x-intercepts, if any, of the graph of f.
(f) List the y-intercept, if there is one, of the graph of f.

94. $f(x) = -3x^2 + 5x$
(a) Is the point $(-1, 2)$ on the graph of f?
(b) If $x = -2$, what is $f(x)$? What point is on the graph of f?
(c) If $f(x) = -2$, what is x? What point(s) are on the graph of f?
(d) What is the domain of f?
(e) List the x-intercepts, if any, of the graph of f.
(f) List the y-intercept, if there is one, of the graph of f.

95. $f(x) = \dfrac{x + 2}{x - 6}$
(a) Is the point $(3, 14)$ on the graph of f?
(b) If $x = 4$, what is $f(x)$? What point is on the graph of f?
(c) If $f(x) = 2$, what is x? What point(s) are on the graph of f?
(d) What is the domain of f?
(e) List the x-intercepts, if any, of the graph of f.
(f) List the y-intercept, if there is one, of the graph of f.

96. $f(x) = \dfrac{x^2 + 2}{x + 4}$
(a) Is the point $\left(1, \dfrac{3}{5}\right)$ on the graph of f?
(b) If $x = 0$, what is $f(x)$? What point is on the graph of f?
(c) If $f(x) = \dfrac{1}{2}$, what is x? What point(s) are on the graph of f?
(d) What is the domain of f?
(e) List the x-intercepts, if any, of the graph of f.
(f) List the y-intercept, if there is one, of the graph of f.

97. $f(x) = \dfrac{2x^2}{x^4 + 1}$
(a) Is the point $(-1, 1)$ on the graph of f?
(b) If $x = 2$, what is $f(x)$? What point is on the graph of f?
(c) If $f(x) = 1$, what is x? What point(s) are on the graph of f?
(d) What is the domain of f?
(e) List the x-intercepts, if any, of the graph of f.
(f) List the y-intercept, if there is one, of the graph of f.

98. $f(x) = \dfrac{2x}{x - 2}$
(a) Is the point $\left(\dfrac{1}{2}, -\dfrac{2}{3}\right)$ on the graph of f?
(b) If $x = 4$, what is $f(x)$? What point is on the graph of f?
(c) If $f(x) = 1$, what is x? What point(s) are on the graph of f?
(d) What is the domain of f?
(e) List the x-intercepts, if any, of the graph of f.
(f) List the y-intercept, if there is one, of the graph of f.

Applications and Extensions

99. If $f(x) = 2x^3 + Ax^2 + 4x - 5$ and $f(2) = 5$, what is the value of A?

100. If $f(x) = 3x^2 - Bx + 4$ and $f(-1) = 12$, what is the value of B?

101. If $f(x) = \dfrac{3x + 8}{2x - A}$ and $f(0) = 2$, what is the value of A?

102. If $f(x) = \dfrac{2x - B}{3x + 4}$ and $f(2) = \dfrac{1}{2}$, what is the value of B?

103. Geometry Express the area A of a rectangle as a function of the length x if the length of the rectangle is twice its width.

104. Geometry Express the area A of an isosceles right triangle as a function of the length x of one of the two equal sides.

105. Constructing Functions Express the gross salary G of a person who earns \$14 per hour as a function of the number x of hours worked.

106. Constructing Functions Tiffany, a commissioned salesperson, earns \$100 base pay plus \$10 per item sold. Express her gross salary G as a function of the number x of items sold.

107. Population as a Function of Age The function

$$P(a) = 0.014a^2 - 5.073a + 327.287$$

represents the population P (in millions) of Americans that are a years of age or older in 2012.
(a) Identify the dependent and independent variables.
(b) Evaluate $P(20)$. Provide a verbal explanation of the meaning of $P(20)$.
(c) Evaluate $P(0)$. Provide a verbal explanation of the meaning of $P(0)$.

Source: U.S. Census Bureau

108. Number of Rooms The function

$$N(r) = -1.35r^2 + 15.45r - 20.71$$

represents the number N of housing units (in millions) in 2012 that had r rooms, where r is an integer and $2 \le r \le 9$.
(a) Identify the dependent and independent variables.
(b) Evaluate $N(3)$. Provide a verbal explanation of the meaning of $N(3)$.

109. Effect of Gravity on Earth If a rock falls from a height of 20 meters on Earth, the height H (in meters) after x seconds is approximately

$$H(x) = 20 - 4.9x^2$$

(a) What is the height of the rock when $x = 1$ second? $x = 1.1$ seconds? $x = 1.2$ seconds? $x = 1.3$ seconds?
(b) When is the height of the rock 15 meters? When is it 10 meters? When is it 5 meters?
(c) When does the rock strike the ground?

110. Effect of Gravity on Jupiter If a rock falls from a height of 20 meters on the planet Jupiter, its height H (in meters) after x seconds is approximately

$$H(x) = 20 - 13x^2$$

(a) What is the height of the rock when $x = 1$ second? $x = 1.1$ seconds? $x = 1.2$ seconds?
(b) When is the height of the rock 15 meters? When is it 10 meters? When is it 5 meters?
(c) When does the rock strike the ground?

111. Cost of Trans-Atlantic Travel A Boeing 747 crosses the Atlantic Ocean (3000 miles) with an airspeed of 500 miles per hour. The cost C (in dollars) per passenger is given by

$$C(x) = 100 + \frac{x}{10} + \frac{36{,}000}{x}$$

where x is the ground speed (airspeed $\pm$ wind).
(a) What is the cost per passenger for quiescent (no wind) conditions?
(b) What is the cost per passenger with a head wind of 50 miles per hour?
(c) What is the cost per passenger with a tail wind of 100 miles per hour?
(d) What is the cost per passenger with a head wind of 100 miles per hour?

112. Cross-sectional Area The cross-sectional area of a beam cut from a log with radius 1 foot is given by the function $A(x) = 4x\sqrt{1 - x^2}$, where x represents the length, in feet, of half the base of the beam. See the figure. Determine the cross-sectional area of the beam if the length of half the base of the beam is as follows:
(a) One-third of a foot
(b) One-half of a foot
(c) Two-thirds of a foot

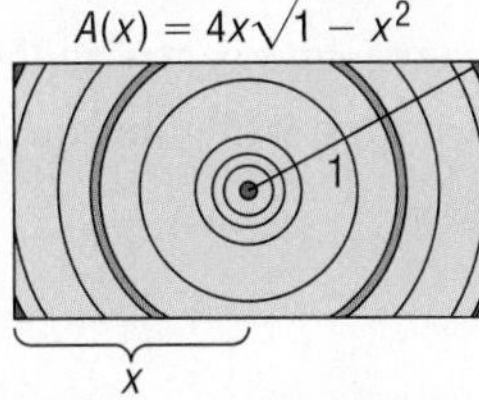

113. Free-throw Shots According to physicist Peter Brancazio, the key to a successful foul shot in basketball lies in the arc of the shot. Brancazio determined the optimal angle of the arc from the free-throw line to be 45 degrees. The arc also depends on the velocity with which the ball is shot. If a player shoots a foul shot, releasing the ball at a 45-degree angle from a position 6 feet above the floor, then the path of the ball can be modeled by the function

$$h(x) = -\frac{44x^2}{v^2} + x + 6$$

where h is the height of the ball above the floor, x is the forward distance of the ball in front of the foul line, and v is the initial velocity with which the ball is shot in feet per second. Suppose a player shoots a ball with an initial velocity of 28 feet per second.
(a) Determine the height of the ball after it has traveled 8 feet in front of the foul line.
(b) Determine the height of the ball after it has traveled 12 feet in front of the foul line.
(c) Find additional points and graph the path of the basketball.
(d) The center of the hoop is 10 feet above the floor and 15 feet in front of the foul line. Will the ball go through the hoop? Why or why not? If not, with what initial velocity must the ball be shot in order for the ball to go through the hoop?

Source: The Physics of Foul Shots, Discover, *Vol. 21, No. 10, October 2000*

114. Granny Shots The last player in the NBA to use an underhand foul shot (a "granny" shot) was Hall of Fame forward Rick Barry, who retired in 1980. Barry believes that current NBA players could increase their free-throw percentage if they were to use an underhand shot. Since underhand shots are released from a lower position, the angle of the shot must be increased. If a player shoots an underhand foul shot, releasing the ball at a 70-degree angle from a position 3.5 feet above the floor, then the path of the ball can be modeled by the function $h(x) = -\frac{136x^2}{v^2} + 2.7x + 3.5$, where h is the height of the ball above the floor, x is the forward distance of the ball in front of the foul line, and v is the initial velocity with which the ball is shot in feet per second.
(a) The center of the hoop is 10 feet above the floor and 15 feet in front of the foul line. Determine the initial velocity with which the ball must be shot in order for the ball to go through the hoop.
(b) Write the function for the path of the ball using the velocity found in part (a).
(c) Determine the height of the ball after it has traveled 9 feet in front of the foul line.
(d) Find additional points and graph the path of the basketball.

Source: The Physics of Foul Shots, Discover, *Vol. 21, No. 10, October 2000*

115. Motion of a Golf Ball A golf ball is hit with an initial velocity of 130 feet per second at an inclination of 45° to the horizontal. In physics, it is established that the height h of the golf ball is given by the function

$$h(x) = \frac{-32x^2}{130^2} + x$$

where x is the horizontal distance that the golf ball has traveled.

(a) Determine the height of the golf ball after it has traveled 100 feet.
(b) What is the height after it has traveled 300 feet?
(c) What is the height after it has traveled 500 feet?
(d) How far was the golf ball hit?
(e) Use a graphing utility to graph the function $h = h(x)$.
(f) Use a graphing utility to determine the distance that the ball has traveled when the height of the ball is 90 feet.
(g) Create a TABLE with TblStart = 0 and ΔTbl = 25. To the nearest 25 feet, how far does the ball travel before it reaches a maximum height? What is the maximum height?
(h) Adjust the value of ΔTbl until you determine the distance, to within 1 foot, that the ball travels before it reaches its maximum height.

116. Reading and Interpreting Graphs Let C be the function whose graph is given at the top of the next column. This graph represents the cost C of manufacturing q computers in a day.
(a) Determine $C(0)$. Interpret this value.
(b) Determine $C(10)$. Interpret this value.
(c) Determine $C(50)$. Interpret this value.
(d) What is the domain of C? What does this domain imply in terms of daily production?
(e) Describe the shape of the graph.
(f) The point (30, 32 000) is called an *inflection point*. Describe the behavior of the graph around the inflection point.

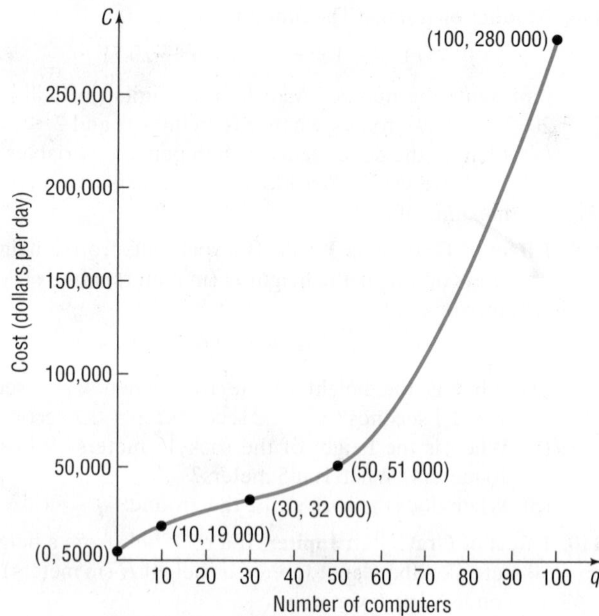

117. Reading and Interpreting Graphs Let C be the function whose graph is given below. This graph represents the cost C of using g gigabytes of data in a month for a shared-data family plan.
(a) Determine $C(0)$. Interpret this value.
(b) Determine $C(5)$. Interpret this value.
(c) Determine $C(15)$. Interpret this value.
(d) What is the domain of C? What does this domain imply in terms of the number of gigabytes?
(e) Describe the shape of the graph.

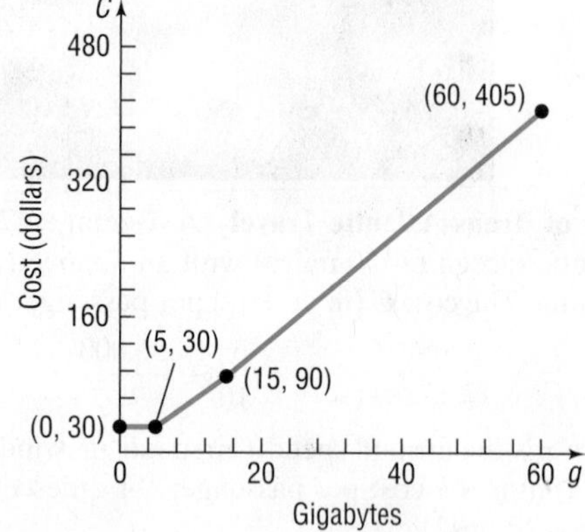

118. Some functions f have the property that $f(a + b) = f(a) + f(b)$ for all real numbers a and b. Which of the following functions have this property?
(a) $h(x) = 2x$ (b) $g(x) = x^2$
(c) $F(x) = 5x - 2$ (d) $G(x) = \dfrac{1}{x}$

Explaining Concepts: Discussion and Writing

119. Describe how you would find the domain and range of a function if you were given its graph. How would your strategy change if you were given the equation defining the function instead of its graph?

120. How many x-intercepts can the graph of a function have? How many y-intercepts can the graph of a function have?

121. Is a graph that consists of a single point the graph of a function? Can you write the equation of such a function?

122. Match each of the following functions with the graph that best describes the situation.

(a) The cost of building a house as a function of its square footage
(b) The height of an egg dropped from a 300-foot building as a function of time
(c) The height of a human as a function of time
(d) The demand for Big Macs as a function of price
(e) The height of a child on a swing as a function of time

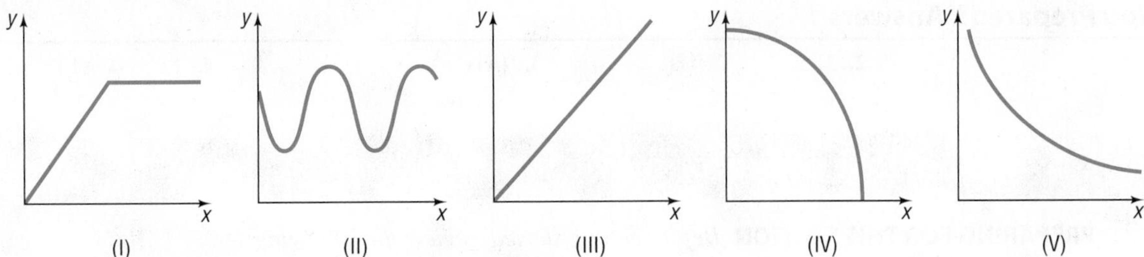

123. Match each of the following functions with the graph that best describes the situation.

(a) The temperature of a bowl of soup as a function of time
(b) The number of hours of daylight per day over a 2-year period
(c) The population of Florida as a function of time
(d) The distance traveled by a car going at a constant velocity as a function of time
(e) The height of a golf ball hit with a 7-iron as a function of time

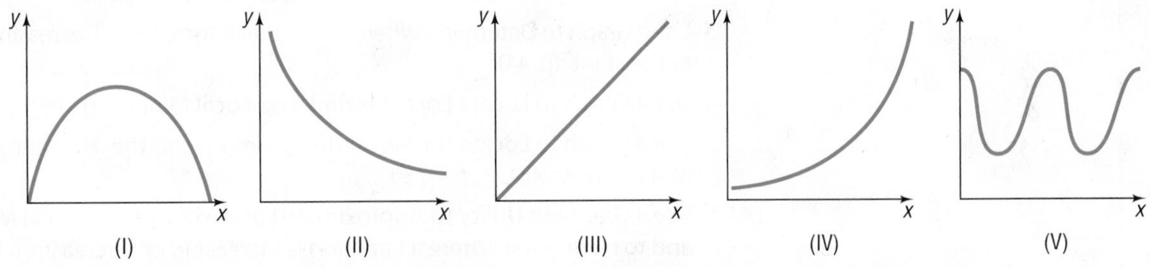

124. Consider the following scenario: Barbara decides to take a walk. She leaves home, walks 2 blocks in 5 minutes at a constant speed, and realizes that she forgot to lock the door. So Barbara runs home in 1 minute. While at her doorstep, it takes her 1 minute to find her keys and lock the door. Barbara walks 5 blocks in 15 minutes and then decides to jog home. It takes her 7 minutes to get home. Draw a graph of Barbara's distance from home (in blocks) as a function of time.

125. Consider the following scenario: Jayne enjoys riding her bicycle through the woods. At the forest preserve, she gets on her bicycle and rides up a 2000-foot incline in 10 minutes. She then travels down the incline in 3 minutes. The next 5000 feet is level terrain, and she covers the distance in 20 minutes. She rests for 15 minutes. Jayne then travels 10,000 feet in 30 minutes. Draw a graph of Jayne's distance traveled (in feet) as a function of time.

126. The following sketch represents the distance d (in miles) that Kevin was from home as a function of time t (in hours). Answer the questions by referring to the graph. In parts (a)–(g), how many hours elapsed and how far was Kevin from home during this time?

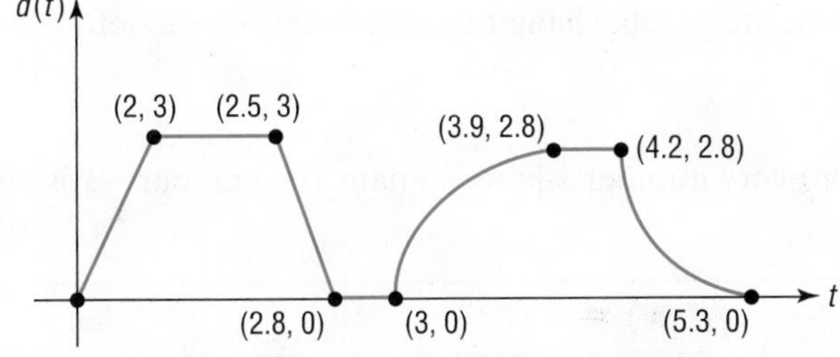

(a) From $t = 0$ to $t = 2$
(b) From $t = 2$ to $t = 2.5$
(c) From $t = 2.5$ to $t = 2.8$
(d) From $t = 2.8$ to $t = 3$
(e) From $t = 3$ to $t = 3.9$
(f) From $t = 3.9$ to $t = 4.2$
(g) From $t = 4.2$ to $t = 5.3$
(h) What is the farthest distance that Kevin was from home?
(i) How many times did Kevin return home?

127. The following sketch represents the speed v (in miles per hour) of Michael's car as a function of time t (in minutes).

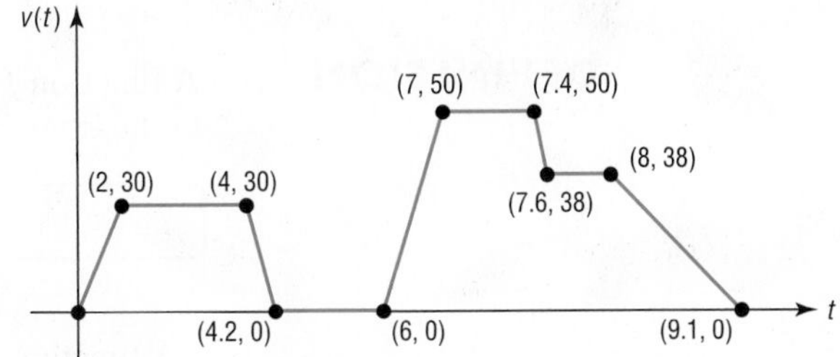

(a) Over what interval of time was Michael traveling fastest?
(b) Over what interval(s) of time was Michael's speed zero?
(c) What was Michael's speed between 0 and 2 minutes?
(d) What was Michael's speed between 4.2 and 6 minutes?
(e) What was Michael's speed between 7 and 7.4 minutes?
(f) When was Michael's speed constant?

128. Draw the graph of a function whose domain is $\{x \mid -3 \le x \le 8, \ x \ne 5\}$ and whose range is $\{y \mid -1 \le y \le 2, \ y \ne 0\}$. What point(s) in the rectangle $-3 \le x \le 8, -1 \le y \le 2$ cannot be on the graph? Compare your graph with those of other students. What differences do you see?

129. Is there a function whose graph is symmetric with respect to the x-axis? Explain.

130. Explain why the vertical-line test works.

'Are You Prepared?' Answers

1. $(-1, 3)$ **2.** 21.5 **3.** $\{x \mid x \ne -4\}$ **4.** $\{x \mid x < -1\}$

1.4 Properties of Functions

PREPARING FOR THIS SECTION *Before getting started, review the following:*

- Interval Notation (Appendix A, Section A.6, pp. A46–A47)
- Intercepts (Section 1.2, pp. 10–11)
- Slope of a Line (Appendix A, Section A.8, pp. A64–A66)
- Point–Slope Form of a Line (Appendix A, Section A.8, p. A68)
- Symmetry (Section 1.2, pp. 12–13)

Now Work the **'Are You Prepared?'** problems on page 50.

OBJECTIVES
1 Determine Even and Odd Functions from a Graph (p. 42)
2 Identify Even and Odd Functions from an Equation (p. 43)
3 Use a Graph to Determine Where a Function Is Increasing, Decreasing, or Constant (p. 44)
4 Use a Graph to Locate Local Maxima and Local Minima (p. 45)
5 Use a Graph to Locate the Absolute Maximum and the Absolute Minimum (p. 46)

6 Use a Graphing Utility to Approximate Local Maxima and Local Minima and to Determine Where a Function Is Increasing or Decreasing (p. 48)
7 Find the Average Rate of Change of a Function (p. 48)

To obtain the graph of a function $y = f(x)$, it is often helpful to know certain properties that the function has and the impact of these properties on the way the graph will look.

1 Determine Even and Odd Functions from a Graph

The words *even* and *odd*, when applied to a function f, describe the symmetry that exists for the graph of the function.

A function f is even if and only if, whenever the point (x, y) is on the graph of f, the point $(-x, y)$ is also on the graph. Using function notation, we define an even function as follows:

DEFINITION

A function f is **even** if, for every number x in its domain, the number $-x$ is also in the domain and

$$f(-x) = f(x)$$

A function f is odd if and only if, whenever the point (x, y) is on the graph of f, the point $(-x, -y)$ is also on the graph. Using function notation, we define an odd function as follows:

DEFINITION

A function f is **odd** if, for every number x in its domain, the number $-x$ is also in the domain and

$$f(-x) = -f(x)$$

Refer to page 13, where the tests for symmetry are listed. The following results are then evident.

THEOREM A function is even if and only if its graph is symmetric with respect to the y-axis. A function is odd if and only if its graph is symmetric with respect to the origin.

EXAMPLE 1 **Determining Even and Odd Functions from the Graph**

Determine whether each graph given in Figure 43 is the graph of an even function, an odd function, or a function that is neither even nor odd.

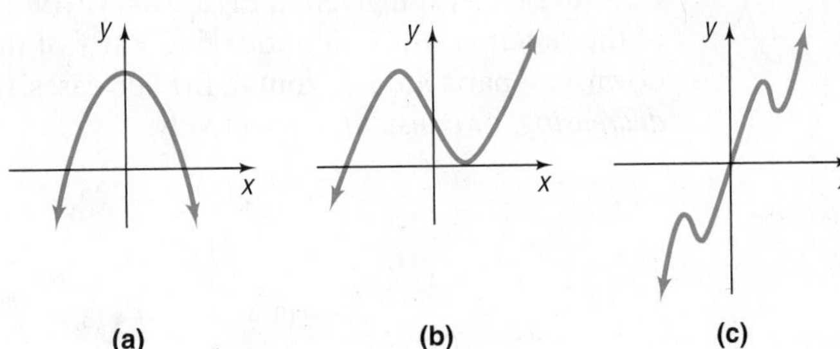

Figure 43 (a) (b) (c)

Solution (a) The graph in Figure 43(a) is that of an even function, because the graph is symmetric with respect to the y-axis.

(b) The function whose graph is given in Figure 43(b) is neither even nor odd, because the graph is neither symmetric with respect to the y-axis nor symmetric with respect to the origin.

(c) The function whose graph is given in Figure 43(c) is odd, because its graph is symmetric with respect to the origin. ●

Now Work PROBLEMS 25(a), (b), AND (d)

2 Identify Even and Odd Functions from an Equation

EXAMPLE 2 **Identifying Even and Odd Functions Algebraically**

Determine whether each of the following functions is even, odd, or neither. Then determine whether the graph is symmetric with respect to the y-axis, with respect to the origin, or neither.

(a) $f(x) = x^2 - 5$ (b) $g(x) = x^3 - 1$

(c) $h(x) = 5x^3 - x$ (d) $F(x) = |x|$

Solution (a) To determine whether f is even, odd, or neither, replace x by $-x$ in $f(x) = x^2 - 5$.

$$f(-x) = (-x)^2 - 5 = x^2 - 5 = f(x)$$

Since $f(-x) = f(x)$, the function is even, and the graph of f is symmetric with respect to the y-axis.

(b) Replace x by $-x$ in $g(x) = x^3 - 1$.

$$g(-x) = (-x)^3 - 1 = -x^3 - 1$$

Since $g(-x) \neq g(x)$ and $g(-x) \neq -g(x) = -(x^3 - 1) = -x^3 + 1$, the function is neither even nor odd. The graph of g is not symmetric with respect to the y-axis, nor is it symmetric with respect to the origin.

(c) Replace x by $-x$ in $h(x) = 5x^3 - x$.

$$h(-x) = 5(-x)^3 - (-x) = -5x^3 + x = -(5x^3 - x) = -h(x)$$

Since $h(-x) = -h(x)$, h is an odd function, and the graph of h is symmetric with respect to the origin.

(d) Replace x by $-x$ in $F(x) = |x|$.

$$F(-x) = |-x| = |-1| \cdot |x| = |x| = F(x)$$

Since $F(-x) = F(x)$, F is an even function, and the graph of F is symmetric with respect to the y-axis. ●

Now Work PROBLEM 37

3 Use a Graph to Determine Where a Function Is Increasing, Decreasing, or Constant

Consider the graph given in Figure 44. If you look from left to right along the graph of the function, you will notice that parts of the graph are going up, parts are going down, and parts are horizontal. In such cases, the function is described as *increasing, decreasing,* or *constant,* respectively.

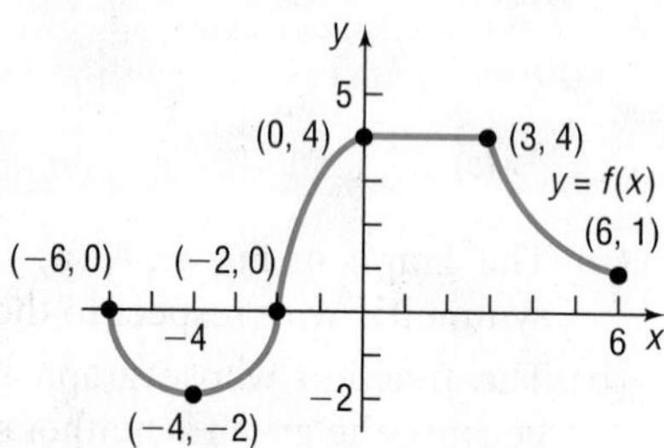

Figure 44

EXAMPLE 3 Determining Where a Function Is Increasing, Decreasing, or Constant from Its Graph

Determine the values of x for which the function in Figure 44 is increasing. Where is it decreasing? Where is it constant?

Solution When determining where a function is increasing, where it is decreasing, and where it is constant, we use strict inequalities involving the independent variable x, or we use open intervals* of x-coordinates. The function whose graph is given in Figure 44 is increasing on the open interval $(-4, 0)$, or for $-4 < x < 0$. The function is decreasing on the open intervals $(-6, -4)$ and $(3, 6)$, or for $-6 < x < -4$ and $3 < x < 6$. The function is constant on the open interval $(0, 3)$, or for $0 < x < 3$. ●

WARNING Describe the behavior of a graph in terms of its x-values. Do not say the graph in Figure 44 is increasing from the point $(-4, -2)$ to the point $(0, 4)$. Rather, say it is increasing on the interval $(-4, 0)$. ■

More precise definitions follow:

DEFINITIONS A function f is **increasing** on an open interval I if, for any choice of x_1 and x_2 in I, with $x_1 < x_2$, we have $f(x_1) < f(x_2)$.

A function f is **decreasing** on an open interval I if, for any choice of x_1 and x_2 in I, with $x_1 < x_2$, we have $f(x_1) > f(x_2)$.

A function f is **constant** on an open interval I if, for all choices of x in I, the values $f(x)$ are equal.

In Words
If a function is decreasing, then as the values of x get bigger, the values of the function get smaller. If a function is increasing, then as the values of x get bigger, the values of the function also get bigger. If a function is constant, then as the values of x get bigger, the values of the function remain unchanged.

Figure 45 illustrates the definitions. The graph of an increasing function goes up from left to right, the graph of a decreasing function goes down from left to right, and the graph of a constant function remains at a fixed height.

*The open interval (a, b) consists of all real numbers x for which $a < x < b$.

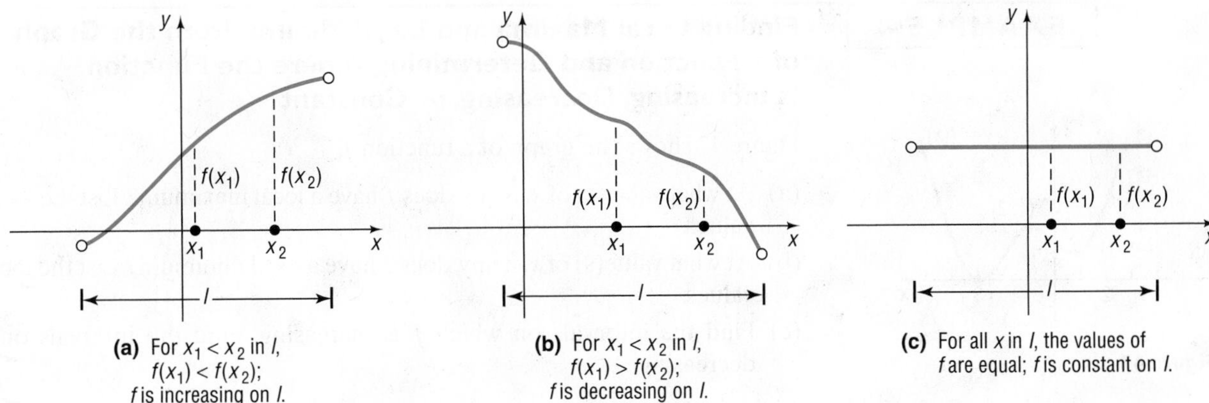

Figure 45 (a) For $x_1 < x_2$ in I, $f(x_1) < f(x_2)$; f is increasing on I. (b) For $x_1 < x_2$ in I, $f(x_1) > f(x_2)$; f is decreasing on I. (c) For all x in I, the values of f are equal; f is constant on I.

Now Work PROBLEMS 13, 15, 17, AND 25(c)

4 Use a Graph to Locate Local Maxima and Local Minima

Suppose f is a function defined on an open interval I containing c. If the value of f at c is greater than or equal to the values of f on I, then f has a *local maximum* at c.* See Figure 46(a).

If the value of f at c is less than or equal to the values of f on I, then f has a *local minimum* at c. See Figure 46(b).

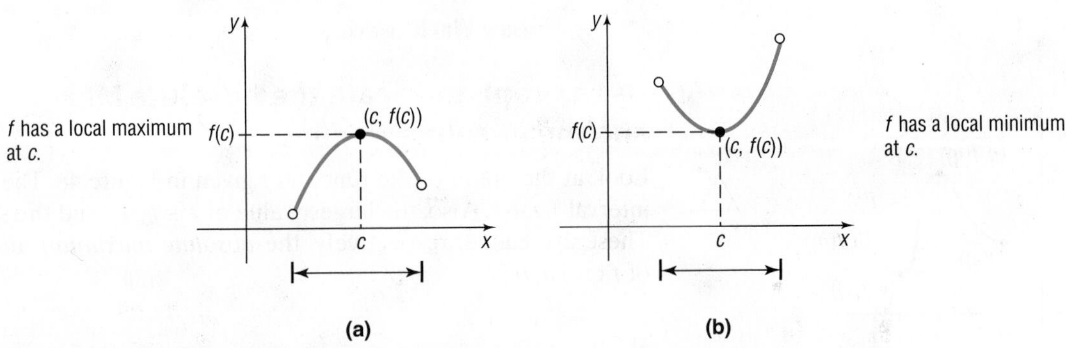

Figure 46 Local maximum and local minimum

DEFINITIONS

Let f be a function defined on some interval I.

A function f has a **local maximum** at c if there is an open interval in I containing c so that, for all x in this open interval, we have $f(x) \le f(c)$. We call $f(c)$ a **local maximum value of f.**

A function f has a **local minimum** at c if there is an open interval in I containing c so that, for all x in this open interval, we have $f(x) \ge f(c)$. We call $f(c)$ a **local minimum value of f.**

If f has a local maximum at c, then the value of f at c is greater than or equal to the values of f near c. If f has a local minimum at c, then the value of f at c is less than or equal to the values of f near c. The word *local* is used to suggest that it is only near c, not necessarily over the entire domain, that the value $f(c)$ has these properties.

*Some texts use the term *relative* instead of *local*.

EXAMPLE 4 Finding Local Maxima and Local Minima from the Graph of a Function and Determining Where the Function Is Increasing, Decreasing, or Constant

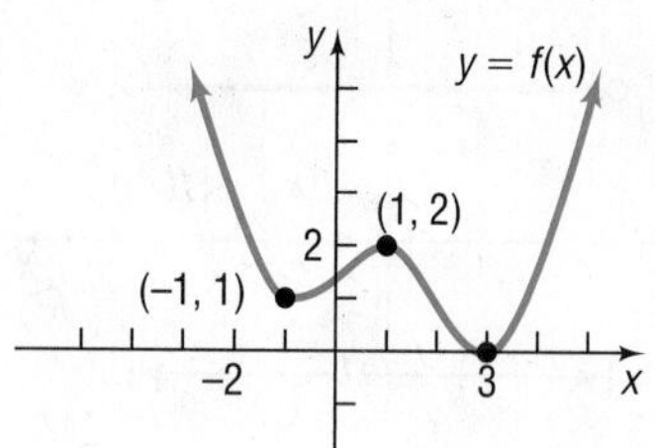

Figure 47

Figure 47 shows the graph of a function f.

(a) At what value(s) of x, if any, does f have a local maximum? List the local maximum values.

(b) At what value(s) of x, if any, does f have a local minimum? List the local minimum values.

(c) Find the intervals on which f is increasing. Find the intervals on which f is decreasing.

Solution The domain of f is the set of real numbers.

(a) f has a local maximum at 1, since for all x close to 1, we have $f(x) \le f(1)$. The local maximum value is $f(1) = 2$.

(b) f has local minima at -1 and at 3. The local minimum values are $f(-1) = 1$ and $f(3) = 0$.

(c) The function whose graph is given in Figure 47 is increasing for all values of x between -1 and 1 and for all values of x greater than 3. That is, the function is increasing on the intervals $(-1, 1)$ and $(3, \infty)$, or for $-1 < x < 1$ and $x > 3$. The function is decreasing for all values of x less than -1 and for all values of x between 1 and 3. That is, the function is decreasing on the intervals $(-\infty, -1)$ and $(1, 3)$, or for $x < -1$ and $1 < x < 3$. ●

WARNING The y-value is the local maximum value or local minimum value, and it occurs at some x-value. For example, in Figure 47, we say f has a local maximum at 1 and the local maximum value is 2. ■

Now Work PROBLEMS 19 AND 21

5 Use a Graph to Locate the Absolute Maximum and the Absolute Minimum

Look at the graph of the function f given in Figure 48. The domain of f is the closed interval $[a, b]$. Also, the largest value of f is $f(u)$ and the smallest value of f is $f(v)$. These are called, respectively, the *absolute maximum* and the *absolute minimum* of f on $[a, b]$.

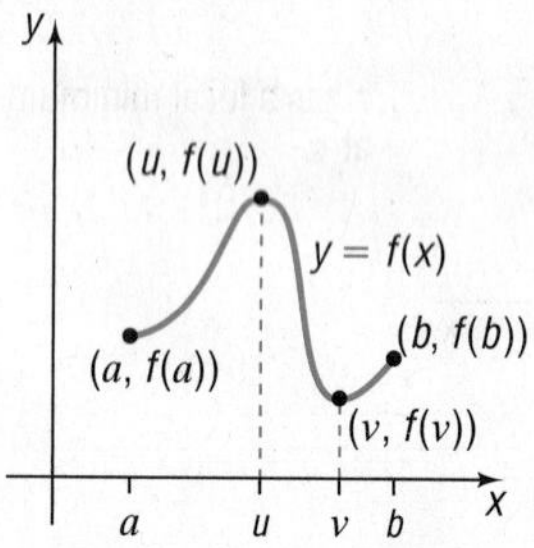

domain: $[a, b]$
for all x in $[a, b]$, $f(x) \le f(u)$
for all x in $[a, b]$, $f(x) \ge f(v)$
absolute maximum: $f(u)$
absolute minimum: $f(v)$

Figure 48

DEFINITION Let f be a function defined on some interval I. If there is a number u in I for which $f(x) \le f(u)$ for all x in I, then f has an **absolute maximum at u**, and the number $f(u)$ is the **absolute maximum of f on I**.

If there is a number v in I for which $f(x) \ge f(v)$ for all x in I, then f has an **absolute minimum at v**, and the number $f(v)$ is the **absolute minimum of f on I**.

The absolute maximum and absolute minimum of a function f are sometimes called the **extreme values** of f on I.

The absolute maximum or absolute minimum of a function f may not exist. Let's look at some examples.

EXAMPLE 5 Finding the Absolute Maximum and the Absolute Minimum from the Graph of a Function

For each graph of a function $y = f(x)$ in Figure 49, find the absolute maximum and the absolute minimum, if they exist. Also, find any local maxima or local minima.

Solution (a) The function f whose graph is given in Figure 49(a) has the closed interval $[0, 5]$ as its domain. The largest value of f is $f(3) = 6$, the absolute maximum. The smallest value of f is $f(0) = 1$, the absolute minimum. The function has a local maximum of 6 at $x = 3$ and a local minimum of 4 at $x = 4$.

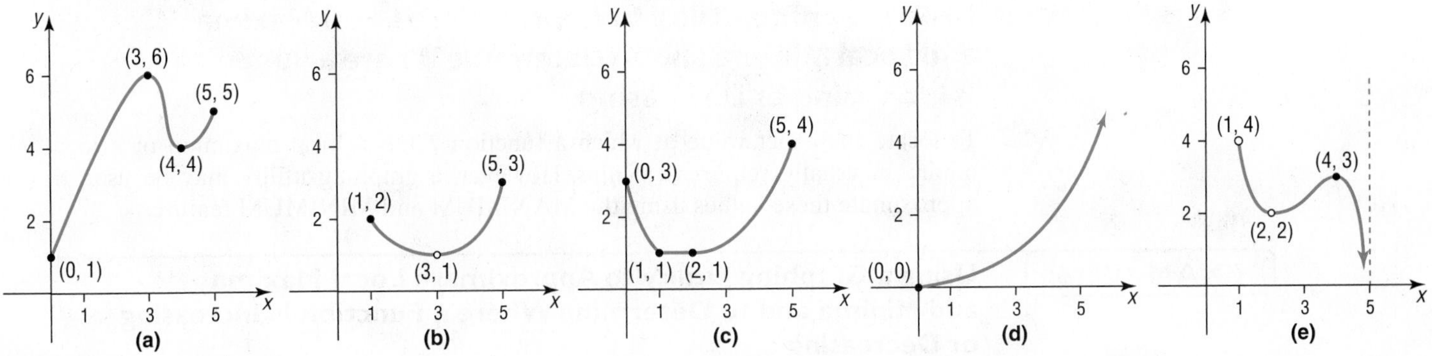

Figure 49

WARNING A function may have an absolute maximum or an absolute minimum at an endpoint but not a local maximum or a local minimum. Why? Local maxima and local minima are found over some open interval *I*, and this interval cannot be created around an endpoint. ■

(b) The function f whose graph is given in Figure 49(b) has the domain $\{x|1 \leq x \leq 5, x \neq 3\}$. Note that we exclude 3 from the domain because of the "hole" at (3, 1). The largest value of f on its domain is $f(5) = 3$, the absolute maximum. There is no absolute minimum. Do you see why? As you trace the graph, getting closer to the point (3, 1), there is no single smallest value. [As soon as you claim a smallest value, we can trace closer to (3, 1) and get a smaller value!] The function has no local maxima or minima.

(c) The function f whose graph is given in Figure 49(c) has the interval [0, 5] as its domain. The absolute maximum of f is $f(5) = 4$. The absolute minimum is 1. Notice that the absolute minimum 1 occurs at any number in the interval [1, 2]. The function has a local minimum value of 1 at every x in the interval [1, 2], and has a local maximum value of 1 at every x in the interval (1, 2).

(d) The function f given in Figure 49(d) has the interval $[0, \infty)$ as its domain. The function has no absolute maximum; the absolute minimum is $f(0) = 0$. The function has no local maximum or local minimum.

(e) The function f in Figure 49(e) has the domain $\{x|1 < x < 5, x \neq 2\}$. The function has no absolute maximum and no absolute minimum. Do you see why? The function has a local maximum value of 3 at $x = 4$, but no local minimum value. ●

In calculus, there is a theorem with conditions that guarantee a function will have an absolute maximum and an absolute minimum.

THEOREM

Extreme Value Theorem

If f is a continuous function* whose domain is a closed interval $[a, b]$, then f has an absolute maximum and an absolute minimum on $[a, b]$.

The absolute maximum (minimum) can be found by selecting the largest (smallest) value of f from the following list:

1. The values of f at any local maxima or local minima of f in $[a, b]$.
2. The value of f at each endpoint of $[a, b]$—that is, $f(a)$ and $f(b)$.

For example, the graph of the function f given in Figure 49(a) is continuous on the closed interval [0, 5]. The Extreme Value Theorem guarantees that f has extreme values on [0, 5]. To find them, we list

1. The value of f at the local extrema: $f(3) = 6, f(4) = 4$
2. The value of f at the endpoints: $f(0) = 1, f(5) = 5$

The largest of these, 6, is the absolute maximum; the smallest of these, 1, is the absolute minimum.

Now Work PROBLEM 49

*Although a precise definition requires calculus, we'll agree for now that a continuous function is one whose graph has no gaps or holes and can be traced without lifting the pencil from the paper.

6 Use a Graphing Utility to Approximate Local Maxima and Local Minima and to Determine Where a Function Is Increasing or Decreasing

To locate the exact value at which a function f has a local maximum or a local minimum usually requires calculus. However, a graphing utility may be used to approximate these values using the MAXIMUM and MINIMUM features.

EXAMPLE 6 **Using a Graphing Utility to Approximate Local Maxima and Minima and to Determine Where a Function Is Increasing or Decreasing**

(a) Use a graphing utility to graph $f(x) = 6x^3 - 12x + 5$ for $-2 < x < 2$. Approximate where f has a local maximum and where f has a local minimum.

(b) Determine where f is increasing and where it is decreasing.

Solution (a) Graphing utilities have a feature that finds the maximum or minimum point of a graph within a given interval. Graph the function f for $-2 < x < 2$. The MAXIMUM and MINIMUM commands require us to first determine the open interval I. The graphing utility will then approximate the maximum or minimum value in the interval. Using MAXIMUM, we find that the local maximum value is 11.53 and that it occurs at $x = -0.82$, rounded to two decimal places. See Figure 50(a). Using MINIMUM, we find that the local minimum value is -1.53 and that it occurs at $x = 0.82$, rounded to two decimal places. See Figure 50(b).

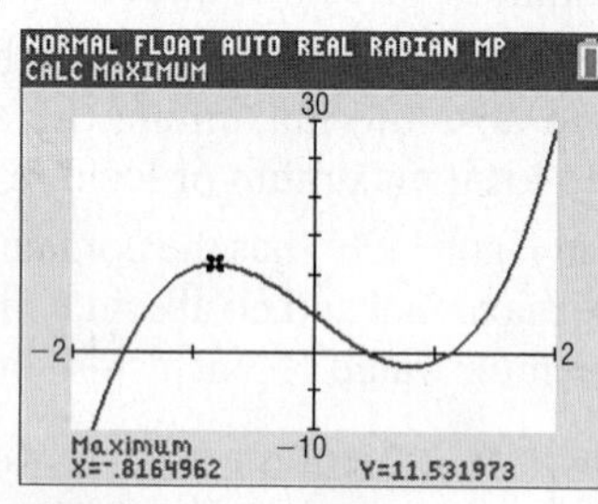

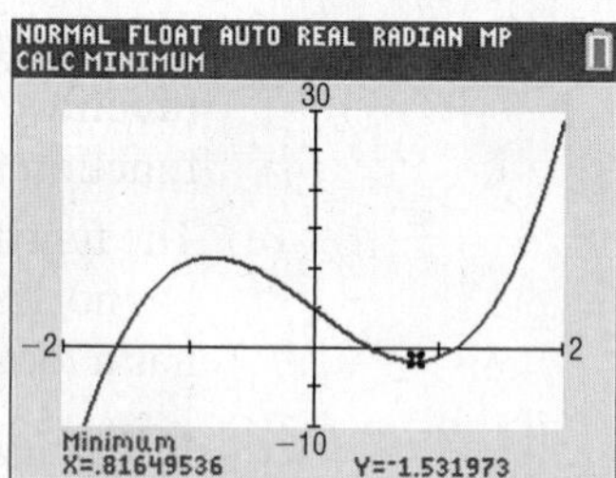

Figure 50 **(a)** Local maximum **(b)** Local minimum

(b) Looking at Figures 50(a) and (b), we see that the graph of f is increasing from $x = -2$ to $x = -0.82$ and from $x = 0.82$ to $x = 2$, so f is increasing on the intervals $(-2, -0.82)$ and $(0.82, 2)$, or for $-2 < x < -0.82$ and $0.82 < x < 2$. The graph is decreasing from $x = -0.82$ to $x = 0.82$, so f is decreasing on the interval $(-0.82, 0.82)$, or for $-0.82 < x < 0.82$. ●

Now Work PROBLEM 57

7 Find the Average Rate of Change of a Function

In Appendix A, we said that the slope of a line can be interpreted as the average rate of change. To find the average rate of change of a function between any two points on its graph, calculate the slope of the line containing the two points.

DEFINITION

If a and b, $a \neq b$, are in the domain of a function $y = f(x)$, the **average rate of change of f from a to b** is defined as

$$\text{Average rate of change} = \frac{\Delta y}{\Delta x} = \frac{f(b) - f(a)}{b - a} \qquad a \neq b \qquad (1)$$

In Words
The symbol Δ is the Greek capital letter delta and is read "change in."

The symbol Δy in equation (1) is the "change in y," and Δx is the "change in x." The average rate of change of f is the change in y divided by the change in x.

EXAMPLE 7 **Finding the Average Rate of Change**

Find the average rate of change of $f(x) = 3x^2$:

(a) From 1 to 3 (b) From 1 to 5 (c) From 1 to 7

Solution (a) The average rate of change of $f(x) = 3x^2$ from 1 to 3 is

$$\frac{\Delta y}{\Delta x} = \frac{f(3) - f(1)}{3 - 1} = \frac{27 - 3}{3 - 1} = \frac{24}{2} = 12$$

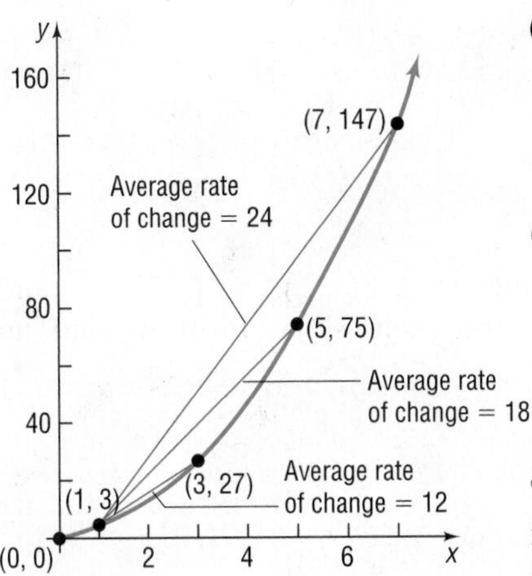

Figure 51 $f(x) = 3x^2$

(b) The average rate of change of $f(x) = 3x^2$ from 1 to 5 is

$$\frac{\Delta y}{\Delta x} = \frac{f(5) - f(1)}{5 - 1} = \frac{75 - 3}{5 - 1} = \frac{72}{4} = 18$$

(c) The average rate of change of $f(x) = 3x^2$ from 1 to 7 is

$$\frac{\Delta y}{\Delta x} = \frac{f(7) - f(1)}{7 - 1} = \frac{147 - 3}{7 - 1} = \frac{144}{6} = 24$$

See Figure 51 for a graph of $f(x) = 3x^2$. The function f is increasing for $x > 0$. The fact that the average rate of change is positive for any $x_1, x_2, x_1 \neq x_2$, in the interval $(1, 7)$ indicates that the graph is increasing on $1 < x < 7$. Further, the average rate of change is consistently getting larger for $1 < x < 7$, which indicates that the graph is increasing at an increasing rate.

Now Work PROBLEM 65

The Secant Line

The average rate of change of a function has an important geometric interpretation. Look at the graph of $y = f(x)$ in Figure 52. Two points are labeled on the graph: $(a, f(a))$ and $(b, f(b))$. The line containing these two points is called the **secant line**; its slope is

$$m_{\text{sec}} = \frac{f(b) - f(a)}{b - a}$$

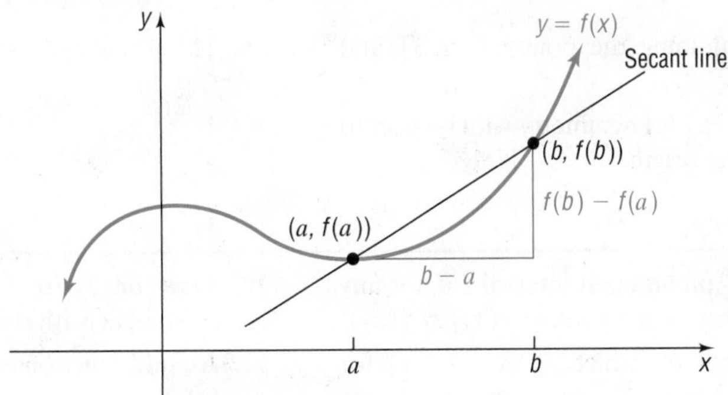

Figure 52 Secant line

THEOREM **Slope of the Secant Line**

The average rate of change of a function from a to b equals the slope of the secant line containing the two points $(a, f(a))$ and $(b, f(b))$ on its graph.

EXAMPLE 8 **Finding an Equation of a Secant Line**

Suppose that $g(x) = 3x^2 - 2x + 3$.

(a) Find the average rate of change of g from -2 to 1.

(b) Find an equation of the secant line containing $(-2, g(-2))$ and $(1, g(1))$.

(c) Using a graphing utility, draw the graph of g and the secant line obtained in part (b) on the same screen.

Solution (a) The average rate of change of $g(x) = 3x^2 - 2x + 3$ from -2 to 1 is

$$\text{Average rate of change} = \frac{g(1) - g(-2)}{1 - (-2)}$$

$$= \frac{4 - 19}{3} \qquad g(1) = 3(1)^2 - 2(1) + 3 = 4,\quad g(-2) = 3(-2)^2 - 2(-2) + 3 = 19$$

$$= -\frac{15}{3} = -5$$

(b) The slope of the secant line containing $(-2, g(-2)) = (-2, 19)$ and $(1, g(1)) = (1, 4)$ is $m_{sec} = -5$. Use the point–slope form to find an equation of the secant line.

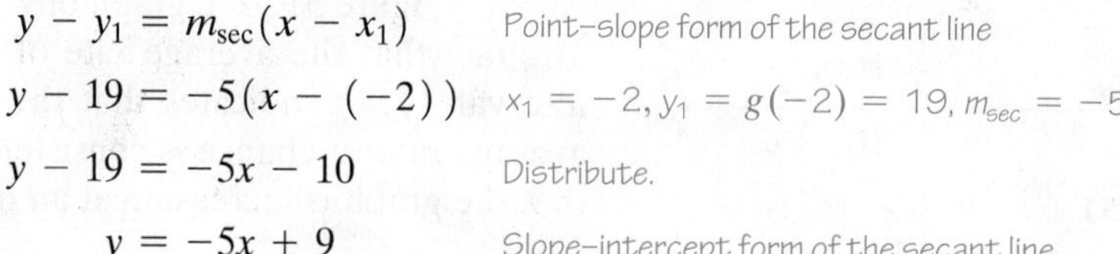

$$y - y_1 = m_{sec}(x - x_1) \qquad \text{Point–slope form of the secant line}$$

$$y - 19 = -5(x - (-2)) \qquad x_1 = -2, y_1 = g(-2) = 19, m_{sec} = -5$$

$$y - 19 = -5x - 10 \qquad \text{Distribute.}$$

$$y = -5x + 9 \qquad \text{Slope-intercept form of the secant line}$$

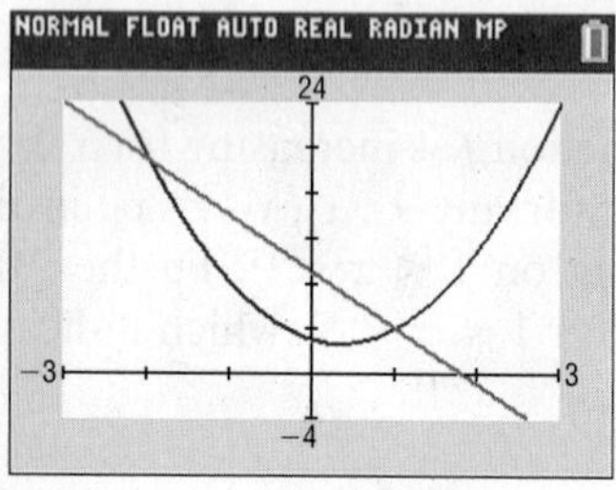

Figure 53 Graph of g and the secant line

(c) Figure 53 shows the graph of g along with the secant line $y = -5x + 9$. ●

Now Work PROBLEM 71

1.4 Assess Your Understanding

'Are You Prepared?' *Answers are given at the end of these exercises. If you get a wrong answer, read the pages listed in red.*

1. The interval $(2, 5)$ can be written as the inequality _____. (pp. A46–A47)
2. The slope of the line containing the points $(-2, 3)$ and $(3, 8)$ is ___. (pp. 64–66)
3. Test the equation $y = 5x^2 - 1$ for symmetry with respect to the x-axis, the y-axis, and the origin. (pp. 12–13)
4. Write the point–slope form of the line with slope 5 containing the point $(3, -2)$. (p. 68)
5. The intercepts of the equation $y = x^2 - 9$ are _____. (pp. 10–11)

Concepts and Vocabulary

6. A function f is _________ on an open interval I if, for any choice of x_1 and x_2 in I, with $x_1 < x_2$, we have $f(x_1) < f(x_2)$.
7. A(n) _____ function f is one for which $f(-x) = f(x)$ for every x in the domain of f; a(n) ____ function f is one for which $f(-x) = -f(x)$ for every x in the domain of f.
8. ***True or False*** A function f is decreasing on an open interval I if, for any choice of x_1 and x_2 in I, with $x_1 < x_2$, we have $f(x_1) > f(x_2)$.
9. ***True or False*** A function f has a local maximum at c if there is an open interval I containing c such that for all x in I, $f(x) \le f(c)$.
10. ***True or False*** Even functions have graphs that are symmetric with respect to the origin.
11. An odd function is symmetric with respect to _____.
    (a) the x-axis (b) the y-axis
    (c) the origin (d) the line $y = x$
12. Which of the following intervals is required to guarantee a continuous function will have both an absolute maximum and an absolute minimum?
    (a) (a, b) (b) $(a, b]$
    (c) $[a, b)$ (d) $[a, b]$

Skill Building

In Problems 13–24, use the graph of the function f given.

13. Is f increasing on the interval $(-8, -2)$?
14. Is f decreasing on the interval $(-8, -4)$?
15. Is f increasing on the interval $(-2, 6)$?
16. Is f decreasing on the interval $(2, 5)$?
17. List the interval(s) on which f is increasing.
18. List the interval(s) on which f is decreasing.
19. Is there a local maximum at 2? If yes, what is it?
20. Is there a local maximum at 5? If yes, what is it?
21. List the number(s) at which f has a local maximum. What are the local maximum values?
22. List the number(s) at which f has a local minimum. What are the local minimum values?
23. Find the absolute minimum of f on $[-10, 7]$.
24. Find the absolute maximum of f on $[-10, 7]$.

In Problems 25–32, the graph of a function is given. Use the graph to find:
(a) The intercepts, if any
(b) The domain and range
(c) The intervals on which the function is increasing, decreasing, or constant
(d) Whether the function is even, odd, or neither

25.
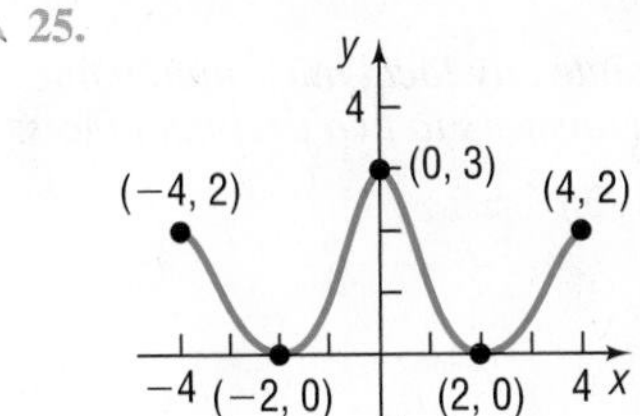

26.
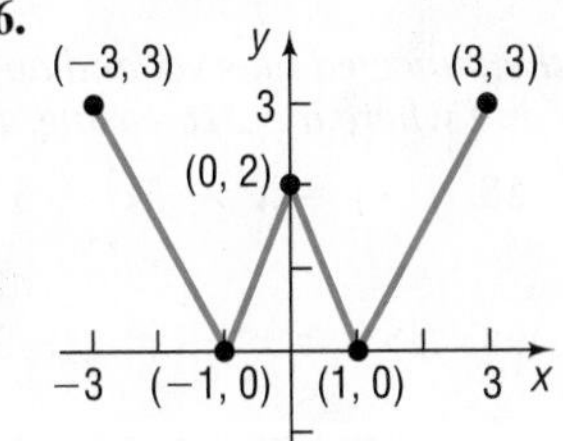

27.
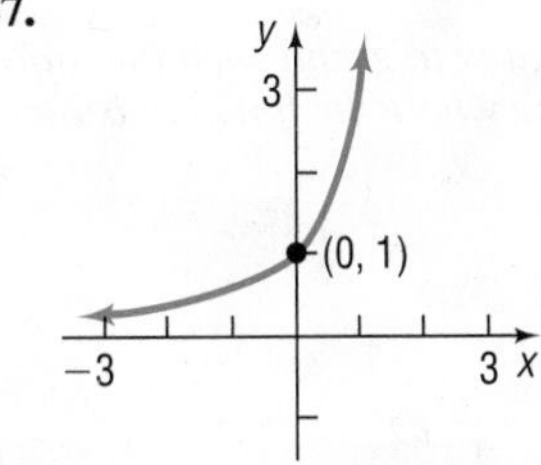

28.
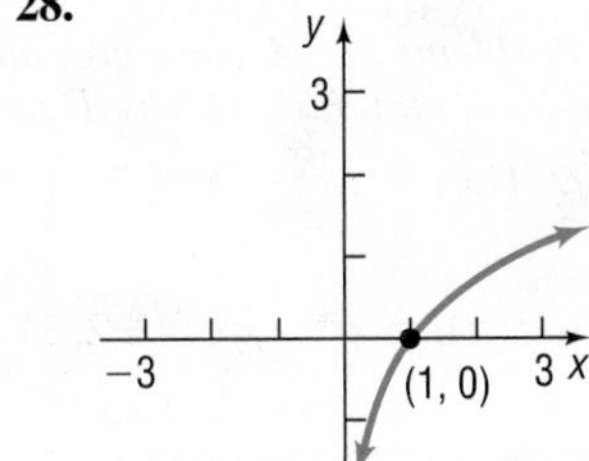

29.
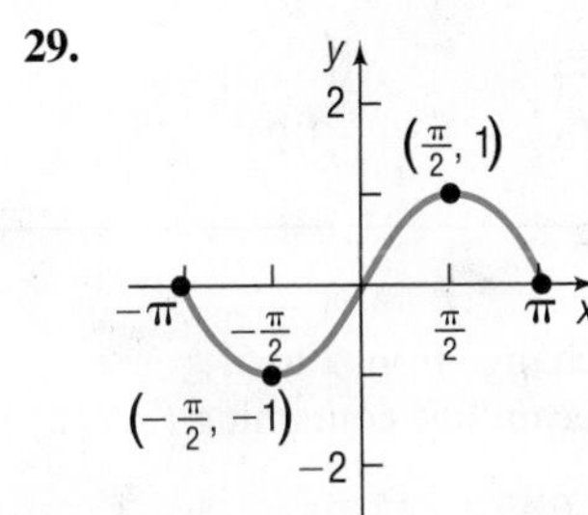

30.
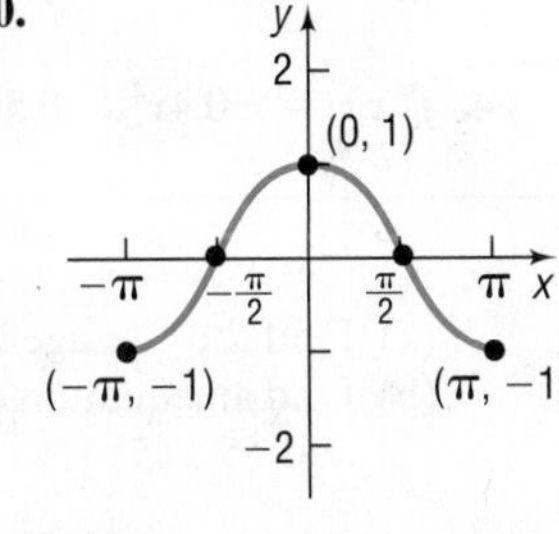

31.
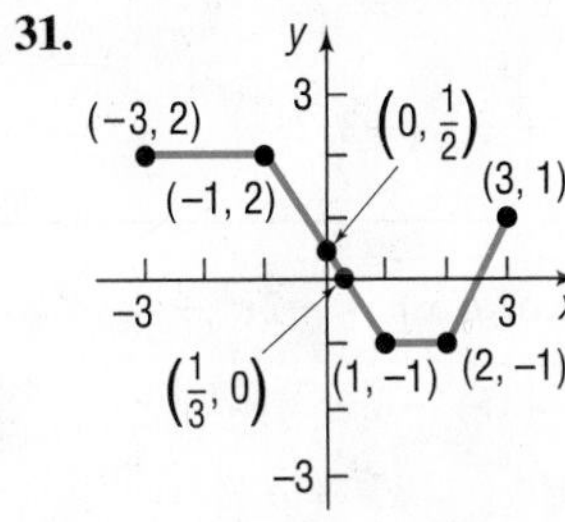

32.
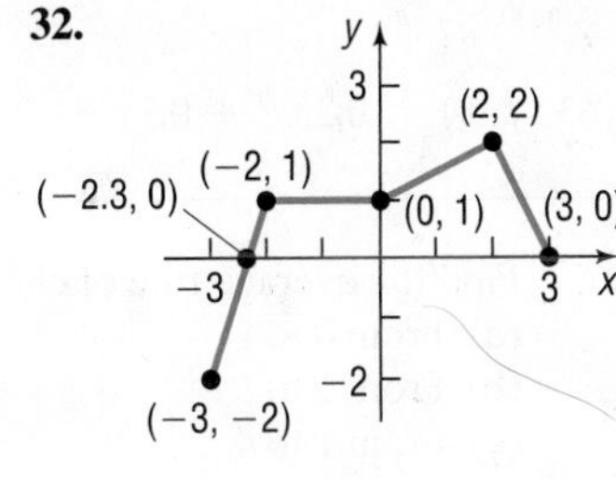

In Problems 33–36, the graph of a function f is given. Use the graph to find:
(a) The numbers, if any, at which f has a local maximum. What are the local maximum values?
(b) The numbers, if any, at which f has a local minimum. What are the local minimum values?

33.
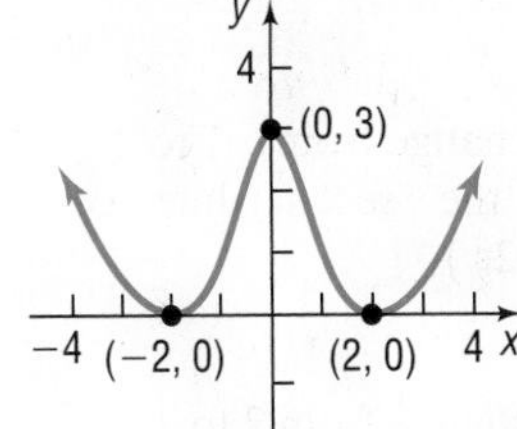

34.
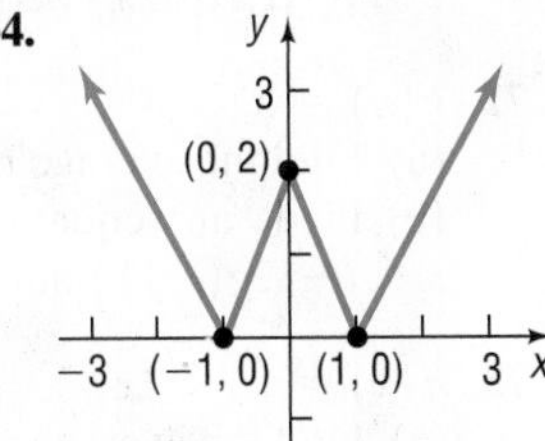

35.
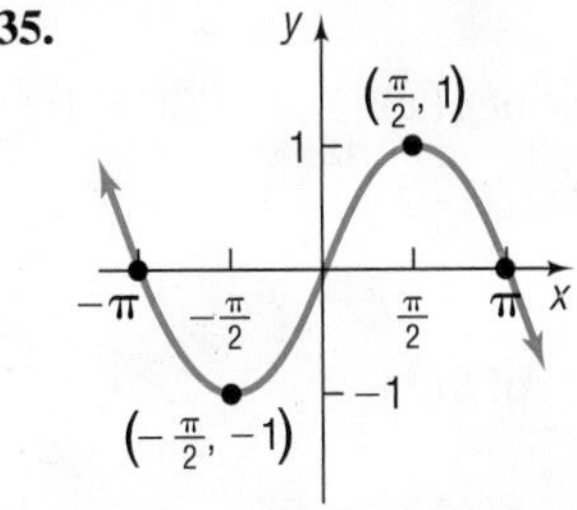

36.
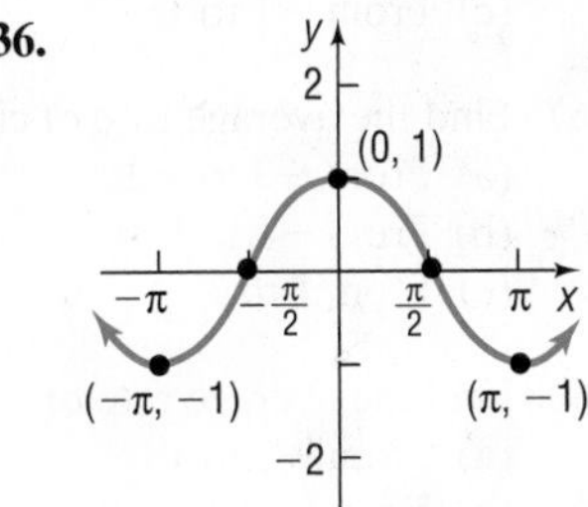

In Problems 37–48, determine algebraically whether each function is even, odd, or neither.

37. $f(x) = 4x^3$
38. $f(x) = 2x^4 - x^2$
39. $g(x) = -3x^2 - 5$
40. $h(x) = 3x^3 + 5$
41. $F(x) = \sqrt[3]{x}$
42. $G(x) = \sqrt{x}$
43. $f(x) = x + |x|$
44. $f(x) = \sqrt[3]{2x^2 + 1}$
45. $g(x) = \dfrac{1}{x^2}$
46. $h(x) = \dfrac{x}{x^2 - 1}$
47. $h(x) = \dfrac{-x^3}{3x^2 - 9}$
48. $F(x) = \dfrac{2x}{|x|}$

In Problems 49–56, for each graph of a function $y = f(x)$, find the absolute maximum and the absolute minimum, if they exist. Identify any local maximum values or local minimum values.

49.

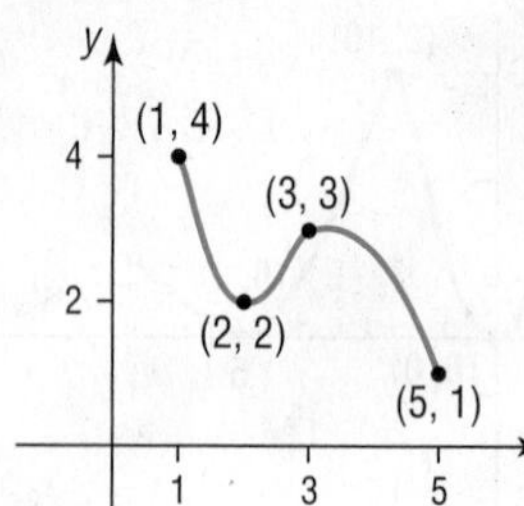

50.

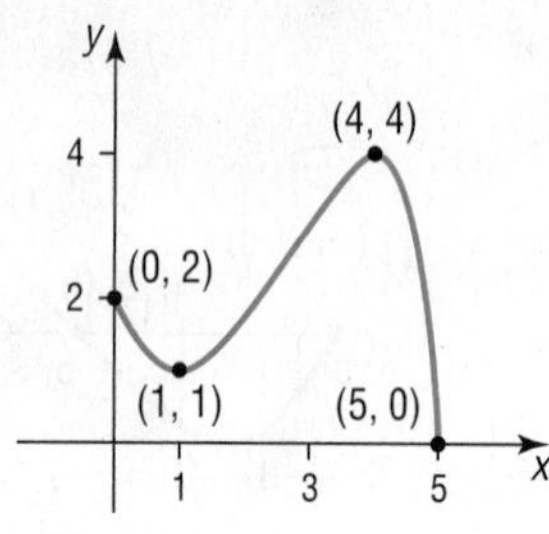

51.

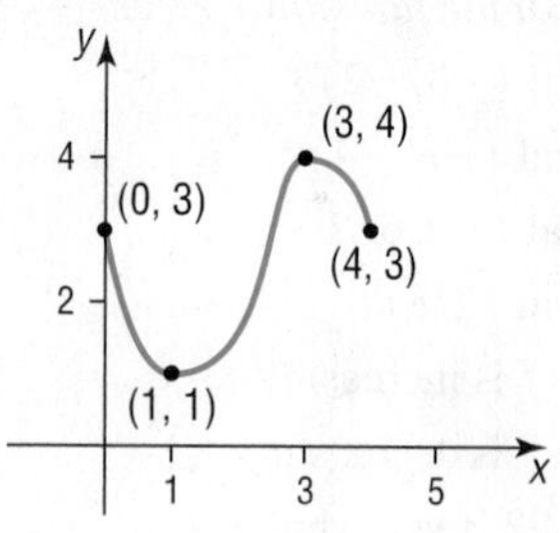

52.

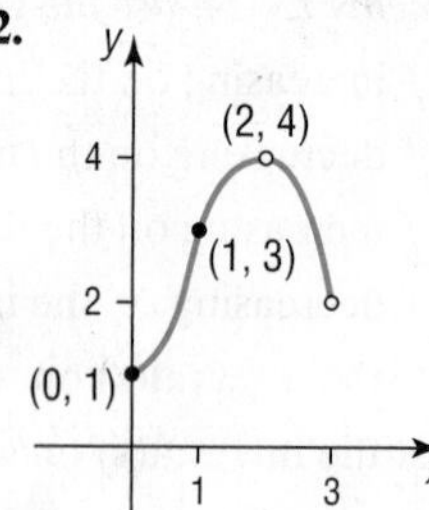

53.

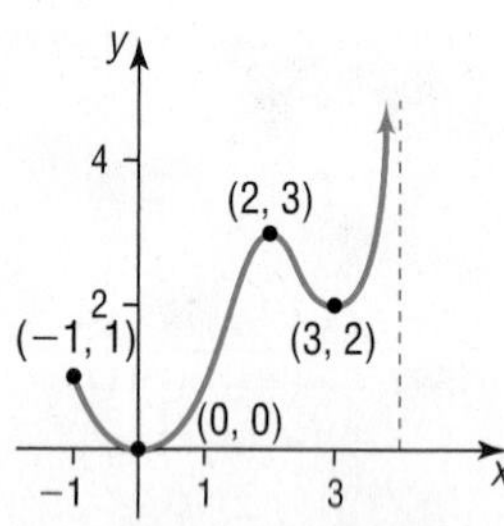

54.

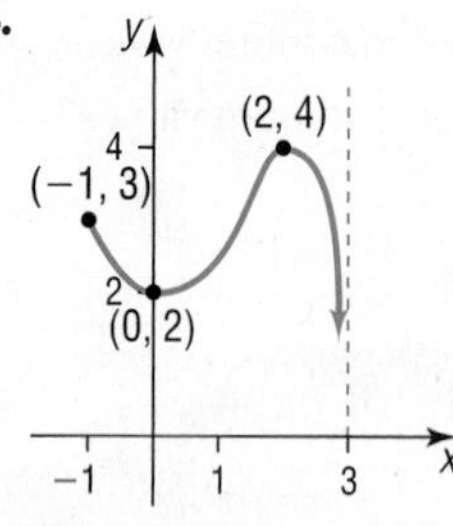

55. **56.**

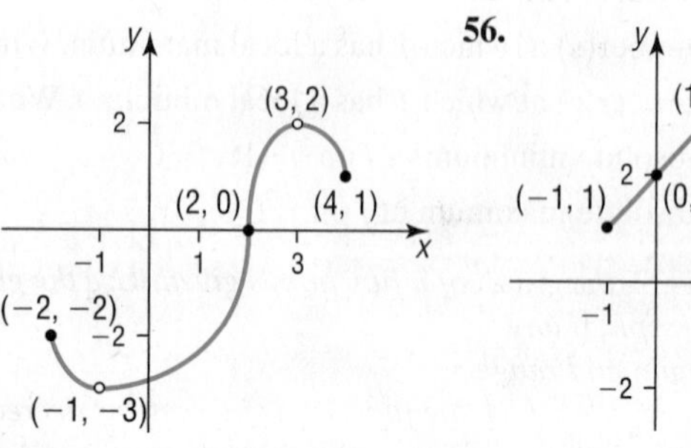

In Problems 57–64, use a graphing utility to graph each function over the indicated interval and approximate any local maximum values and local minimum values. Determine where the function is increasing and where it is decreasing. Round answers to two decimal places.

57. $f(x) = x^3 - 3x + 2 \quad (-2, 2)$

58. $f(x) = x^3 - 3x^2 + 5 \quad (-1, 3)$

59. $f(x) = x^5 - x^3 \quad (-2, 2)$

60. $f(x) = x^4 - x^2 \quad (-2, 2)$

61. $f(x) = -0.2x^3 - 0.6x^2 + 4x - 6 \quad (-6, 4)$

62. $f(x) = -0.4x^3 + 0.6x^2 + 3x - 2 \quad (-4, 5)$

63. $f(x) = 0.25x^4 + 0.3x^3 - 0.9x^2 + 3 \quad (-3, 2)$

64. $f(x) = -0.4x^4 - 0.5x^3 + 0.8x^2 - 2 \quad (-3, 2)$

---

65. Find the average rate of change of $f(x) = -2x^2 + 4$:
(a) From 0 to 2
(b) From 1 to 3
(c) From 1 to 4

66. Find the average rate of change of $f(x) = -x^3 + 1$:
(a) From 0 to 2
(b) From 1 to 3
(c) From -1 to 1

67. Find the average rate of change of $g(x) = x^3 - 2x + 1$:
(a) From -3 to -2
(b) From -1 to 1
(c) From 1 to 3

68. Find the average rate of change of $h(x) = x^2 - 2x + 3$:
(a) From -1 to 1
(b) From 0 to 2
(c) From 2 to 5

69. $f(x) = 5x - 2$
(a) Find the average rate of change from 1 to 3.
(b) Find an equation of the secant line containing $(1, f(1))$ and $(3, f(3))$.

70. $f(x) = -4x + 1$
(a) Find the average rate of change from 2 to 5.
(b) Find an equation of the secant line containing $(2, f(2))$ and $(5, f(5))$.

71. $g(x) = x^2 - 2$
(a) Find the average rate of change from -2 to 1.
(b) Find an equation of the secant line containing $(-2, g(-2))$ and $(1, g(1))$.

72. $g(x) = x^2 + 1$
(a) Find the average rate of change from -1 to 2.
(b) Find an equation of the secant line containing $(-1, g(-1))$ and $(2, g(2))$.

73. $h(x) = x^2 - 2x$
(a) Find the average rate of change from 2 to 4.
(b) Find an equation of the secant line containing $(2, h(2))$ and $(4, h(4))$.

74. $h(x) = -2x^2 + x$
(a) Find the average rate of change from 0 to 3.
(b) Find an equation of the secant line containing $(0, h(0))$ and $(3, h(3))$.

Mixed Practice

75. $g(x) = x^3 - 27x$
(a) Determine whether g is even, odd, or neither.
(b) There is a local minimum value of -54 at 3. Determine the local maximum value.

76. $f(x) = -x^3 + 12x$
(a) Determine whether f is even, odd, or neither.
(b) There is a local maximum value of 16 at 2. Determine the local minimum value.

77. $F(x) = -x^4 + 8x^2 + 8$
(a) Determine whether F is even, odd, or neither.
(b) There is a local maximum value of 24 at $x = 2$. Determine a second local maximum value.
(c) Suppose the area under the graph of F between $x = 0$ and $x = 3$ that is bounded from below by the x-axis is 47.4 square units. Using the result from part (a), determine the area under the graph of F between $x = -3$ and $x = 0$ that is bounded from below by the x-axis.

78. $G(x) = -x^4 + 32x^2 + 144$
(a) Determine whether G is even, odd, or neither.
(b) There is a local maximum value of 400 at $x = 4$. Determine a second local maximum value.
(c) Suppose the area under the graph of G between $x = 0$ and $x = 6$ that is bounded from below by the x-axis is 1612.8 square units. Using the result from part (a), determine the area under the graph of G between $x = -6$ and $x = 0$ that is bounded from below by the x-axis.

Applications and Extensions

79. Minimum Average Cost The average cost per hour in dollars, $\overline{C}$, of producing x riding lawn mowers can be modeled by the function

$$\overline{C}(x) = 0.3x^2 + 21x - 251 + \frac{2500}{x}$$

(a) Use a graphing utility to graph $\overline{C} = \overline{C}(x)$.
(b) Determine the number of riding lawn mowers to produce in order to minimize average cost.
(c) What is the minimum average cost?

80. Medicine Concentration The concentration C of a medication in the bloodstream t hours after being administered is modeled by the function

$$C(t) = -0.002t^4 + 0.039t^3 - 0.285t^2 + 0.766t + 0.085$$

(a) After how many hours will the concentration be highest?
(b) A woman nursing a child must wait until the concentration is below 0.5 before she can feed him. After taking the medication, how long must she wait before feeding her child?

81. Data Plan Cost The monthly cost C, in dollars, for wireless data plans with x gigabytes of data included is shown in the table below. Since each input value for x corresponds to exactly one output value for C, the plan cost is a function of the number of data gigabytes. Thus $C(x)$ represents the monthly cost for a wireless data plan with x gigabytes included.

GB	Cost ($)	GB	Cost ($)
4	70	20	150
6	80	30	225
10	100	40	300
15	130	50	375

(a) Plot the points (4, 70), (6, 80), (10, 100), and so on in a Cartesian plane.
(b) Draw a line segment from the point (10, 100) to (30, 225). What does the slope of this line segment represent?
(c) Find the average rate of change of the monthly cost from 4 to 10 gigabytes.
(d) Find the average rate of change of the monthly cost from 10 to 30 gigabytes.
(e) Find the average rate of change of the monthly cost from 30 to 50 gigabytes.
(f) What is happening to the average rate of change as the gigabytes of data increase?

82. National Debt The size of the total debt owed by the United States federal government continues to grow. In fact, according to the Department of the Treasury, the debt per person living in the United States is approximately $53,000 (or over $140,000 per U.S. household). The following data represent the U.S. debt for the years 2001–2013. Since the debt D depends on the year y, and each input corresponds to exactly one output, the debt is a function of the year. So $D(y)$ represents the debt for each year y.

Year	Debt (billions of dollars)	Year	Debt (billions of dollars)
2001	5807	2008	10,025
2002	6228	2009	11,910
2003	6783	2010	13,562
2004	7379	2011	14,790
2005	7933	2012	16,066
2006	8507	2013	16,738
2007	9008		

Source: *www.treasurydirect.gov*

(a) Plot the points (2001, 5807), (2002, 6228), and so on in a Cartesian plane.
(b) Draw a line segment from the point (2001, 5807) to (2006, 8507). What does the slope of this line segment represent?
(c) Find the average rate of change of the debt from 2002 to 2004.
(d) Find the average rate of change of the debt from 2006 to 2008.
(e) Find the average rate of change of the debt from 2010 to 2012.
(f) What appears to be happening to the average rate of change as time passes?

83. *E. coli* Growth A strain of *E. coli* Beu 397-recA441 is placed into a nutrient broth at 30° Celsius and allowed to grow. The data shown in the table are collected. The population is measured in grams and the time in hours. Since population P depends on time t, and each input corresponds to exactly one output, we can say that population is a function of time. Thus $P(t)$ represents the population at time t.

Time (hours), t	Population (grams), P
0	0.09
2.5	0.18
3.5	0.26
4.5	0.35
6	0.50

(a) Find the average rate of change of the population from 0 to 2.5 hours.
(b) Find the average rate of change of the population from 4.5 to 6 hours.
(c) What is happening to the average rate of change as time passes?

84. e-Filing Tax Returns The Internal Revenue Service Restructuring and Reform Act (RRA) was signed into law by President Bill Clinton in 1998. A major objective of the RRA was to promote electronic filing of tax returns. The data in the table (top, right) show the percentage of individual income tax returns filed electronically for filing years 2004–2012. Since the percentage P of returns filed electronically depends on the filing year y, and each input corresponds to exactly one output, the percentage of returns filed electronically is a function of the filing year; so $P(y)$ represents the percentage of returns filed electronically for filing year y.
(a) Find the average rate of change of the percentage of e-filed returns from 2004 to 2006.
(b) Find the average rate of change of the percentage of e-filed returns from 2007 to 2009.
(c) Find the average rate of change of the percentage of e-filed returns from 2010 to 2012.
(d) What is happening to the average rate of change as time passes?

Year	Percentage of returns e-filed
2004	46.5
2005	51.1
2006	53.8
2007	57.1
2008	58.5
2009	67.2
2010	69.8
2011	77.2
2012	82.7

Source: Internal Revenue Service

85. For the function $f(x) = x^2$, compute the average rate of change:
(a) From 0 to 1
(b) From 0 to 0.5
(c) From 0 to 0.1
(d) From 0 to 0.01
(e) From 0 to 0.001
(f) Use a graphing utility to graph each of the secant lines along with f.
(g) What do you think is happening to the secant lines?
(h) What is happening to the slopes of the secant lines? Is there some number that they are getting closer to? What is that number?

86. For the function $f(x) = x^2$, compute the average rate of change:
(a) From 1 to 2
(b) From 1 to 1.5
(c) From 1 to 1.1
(d) From 1 to 1.01
(e) From 1 to 1.001
(f) Use a graphing utility to graph each of the secant lines along with f.
(g) What do you think is happening to the secant lines?
(h) What is happening to the slopes of the secant lines? Is there some number that they are getting closer to? What is that number?

Problems 87–94 require the following discussion of a secant line. The slope of the secant line containing the two points $(x, f(x))$ and $(x + h, f(x + h))$ on the graph of a function $y = f(x)$ may be given as

$$m_{sec} = \frac{f(x+h) - f(x)}{(x+h) - x} = \frac{f(x+h) - f(x)}{h}, \qquad h \neq 0$$

In calculus, this expression is called the **difference quotient of *f*.**
(a) Express the slope of the secant line of each function in terms of x and h. Be sure to simplify your answer.
(b) Find m_{sec} for $h = 0.5, 0.1,$ and 0.01 at $x = 1$. What value does m_{sec} approach as h approaches 0?
(c) Find an equation for the secant line at $x = 1$ with $h = 0.01$.
(d) Use a graphing utility to graph f and the secant line found in part (c) in the same viewing window.

87. $f(x) = 2x + 5$
88. $f(x) = -3x + 2$
89. $f(x) = x^2 + 2x$
90. $f(x) = 2x^2 + x$
91. $f(x) = 2x^2 - 3x + 1$
92. $f(x) = -x^2 + 3x - 2$
93. $f(x) = \dfrac{1}{x}$
94. $f(x) = \dfrac{1}{x^2}$

Explaining Concepts: Discussion and Writing

95. Draw the graph of a function that has the following properties: domain: all real numbers; range: all real numbers; intercepts: $(0, -3)$ and $(3, 0)$; a local maximum value of -2 is at -1; a local minimum value of -6 is at 2. Compare your graph with those of others. Comment on any differences.

96. Redo Problem 95 with the following additional information: increasing on $(-\infty, -1), (2, \infty)$; decreasing on $(-1, 2)$. Again compare your graph with others and comment on any differences.

97. How many x-intercepts can a function defined on an interval have if it is increasing on that interval? Explain.

98. Suppose that a friend of yours does not understand the idea of increasing and decreasing functions. Provide an explanation, complete with graphs, that clarifies the idea.

99. Can a function be both even and odd? Explain.

100. Using a graphing utility, graph $y = 5$ on the interval $(-3, 3)$. Use MAXIMUM to find the local maximum values on $(-3, 3)$. Comment on the result provided by the calculator.

101. A function f has a positive average rate of change on the interval $[2, 5]$. Is f increasing on $[2, 5]$? Explain.

102. Show that a constant function $f(x) = b$ has an average rate of change of 0. Compute the average rate of change of $y = \sqrt{4 - x^2}$ on the interval $[-2, 2]$. Explain how this can happen.

'Are You Prepared?' Answers

1. $2 < x < 5$ **2.** 1 **3.** symmetric with respect to the y-axis **4.** $y + 2 = 5(x - 3)$ **5.** $(-3, 0), (3, 0), (0, -9)$

1.5 Library of Functions; Piecewise-defined Functions

PREPARING FOR THIS SECTION *Before getting started, review the following:*

- Intercepts (Section 1.2, pp. 10–11)
- Lines (Appendix A, Section A.8, pp. A64–A74)
- Graphs of Key Equations (Section 1.2: Example 3, p. 10; Example 10, p. 14; Example 11, p. 14; Example 12, p. 15)

Now Work the 'Are You Prepared?' problems on page 61.

OBJECTIVES 1 Graph the Functions Listed in the Library of Functions (p. 55)
2 Graph Piecewise-defined Functions (p. 59)

1 Graph the Functions Listed in the Library of Functions

First we introduce a few more functions, beginning with the *square root function*.

On page 14, we graphed the equation $y = \sqrt{x}$. Figure 54 shows a graph of the function $f(x) = \sqrt{x}$. Based on the graph, we have the following properties:

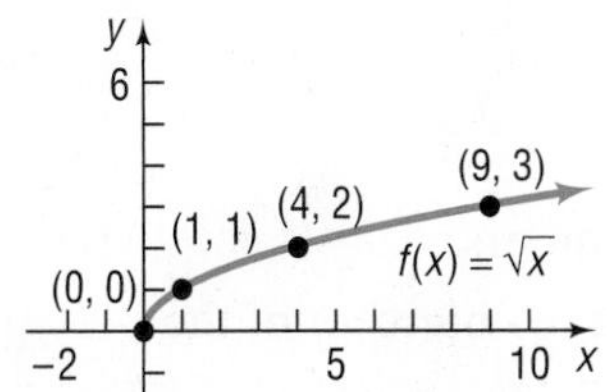

Figure 54 Square root function

Properties of $f(x) = \sqrt{x}$

1. The domain and the range are the set of nonnegative real numbers.
2. The x-intercept of the graph of $f(x) = \sqrt{x}$ is 0. The y-intercept of the graph of $f(x) = \sqrt{x}$ is also 0.
3. The function is neither even nor odd.
4. The function is increasing on the interval $(0, \infty)$.
5. The function has an absolute minimum of 0 at $x = 0$.

EXAMPLE 1 **Graphing the Cube Root Function**

(a) Determine whether $f(x) = \sqrt[3]{x}$ is even, odd, or neither. State whether the graph of f is symmetric with respect to the y-axis or symmetric with respect to the origin.

(b) Determine the intercepts, if any, of the graph of $f(x) = \sqrt[3]{x}$.

(c) Graph $f(x) = \sqrt[3]{x}$.

Solution (a) Because

$$f(-x) = \sqrt[3]{-x} = -\sqrt[3]{x} = -f(x)$$

the function is odd. The graph of f is symmetric with respect to the origin.

(b) The y-intercept is $f(0) = \sqrt[3]{0} = 0$. The x-intercept is found by solving the equation $f(x) = 0$.

$$f(x) = 0$$

$$\sqrt[3]{x} = 0 \quad f(x) = \sqrt[3]{x}$$

$$x = 0 \quad \text{Cube both sides of the equation.}$$

The x-intercept is also 0.

(c) Use the function to form Table 6 and obtain some points on the graph. Because of the symmetry with respect to the origin, we find only points (x, y) for which $x \geq 0$. Figure 55 shows the graph of $f(x) = \sqrt[3]{x}$.

Table 6

x	$y = f(x) = \sqrt[3]{x}$	(x, y)
0	0	$(0, 0)$
$\frac{1}{8}$	$\frac{1}{2}$	$\left(\frac{1}{8}, \frac{1}{2}\right)$
1	1	$(1, 1)$
2	$\sqrt[3]{2} \approx 1.26$	$(2, \sqrt[3]{2})$
8	2	$(8, 2)$

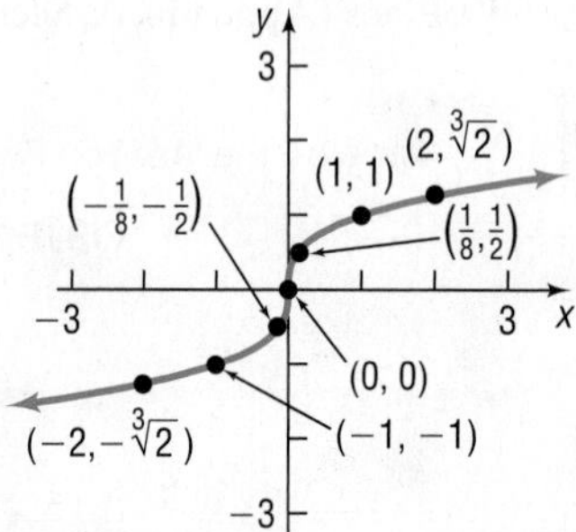

Figure 55 Cube Root Function

From the results of Example 1 and Figure 55, we have the following properties of the cube root function.

Properties of $f(x) = \sqrt[3]{x}$

1. The domain and the range are the set of all real numbers.
2. The x-intercept of the graph of $f(x) = \sqrt[3]{x}$ is 0. The y-intercept of the graph of $f(x) = \sqrt[3]{x}$ is also 0.
3. The function is odd. The graph is symmetric with respect to the origin.
4. The function is increasing on the interval $(-\infty, \infty)$.
5. The function does not have any local minima or any local maxima.

EXAMPLE 2

Graphing the Absolute Value Function

(a) Determine whether $f(x) = |x|$ is even, odd, or neither. State whether the graph of f is symmetric with respect to the y-axis, symmetric with respect to the origin, or neither.

(b) Determine the intercepts, if any, of the graph of $f(x) = |x|$.

(c) Graph $f(x) = |x|$.

Solution (a) Because

$$f(-x) = |-x| \\ = |x| = f(x)$$

the function is even. The graph of f is symmetric with respect to the y-axis.

(b) The y-intercept is $f(0) = |0| = 0$. The x-intercept is found by solving the equation $f(x) = 0$, or $|x| = 0$. The x-intercept is 0.

(c) Use the function to form Table 7 and obtain some points on the graph. Because of the symmetry with respect to the y-axis, we only need to find points (x, y) for which $x \geq 0$. Figure 56 shows the graph of $f(x) = |x|$.

Table 7

x	y = f(x) = |x|	(x, y)
0	0	(0, 0)
1	1	(1, 1)
2	2	(2, 2)
3	3	(3, 3)

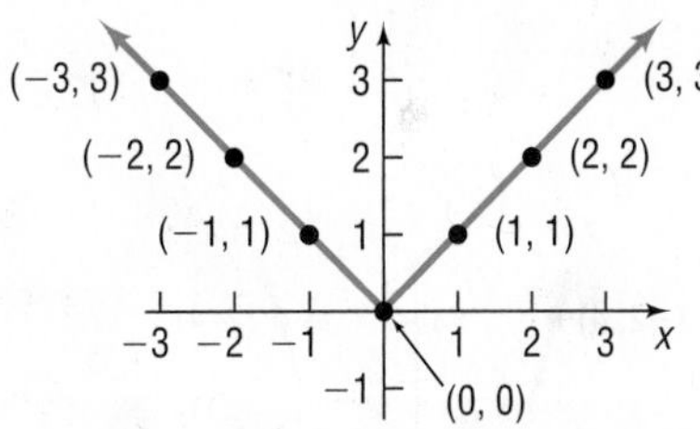

Figure 56 Absolute Value Function

From the results of Example 2 and Figure 56, we have the following properties of the absolute value function.

Properties of $f(x) = |x|$

1. The domain is the set of all real numbers. The range of f is $\{y \mid y \geq 0\}$.
2. The x-intercept of the graph of $f(x) = |x|$ is 0. The y-intercept of the graph of $f(x) = |x|$ is also 0.
3. The function is even. The graph is symmetric with respect to the y-axis.
4. The function is decreasing on the interval $(-\infty, 0)$. It is increasing on the interval $(0, \infty)$.
5. The function has an absolute minimum of 0 at $x = 0$.

Seeing the Concept

Graph $y = |x|$ on a square screen and compare what you see with Figure 56. Note that some graphing calculators use abs(x) for absolute value.

Next is a list of the key functions that we have discussed. In going through this list, pay special attention to the properties of each function, particularly to the shape of each graph. Knowing these graphs, along with key points on each graph, will lay the foundation for further graphing techniques.

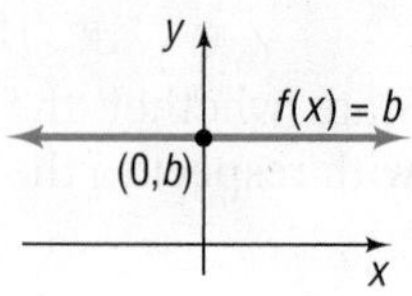

Figure 57 Constant Function

Constant Function

$$f(x) = b \qquad b \text{ is a real number}$$

See Figure 57.

The domain of a **constant function** is the set of all real numbers; its range is the set consisting of a single number b. Its graph is a horizontal line whose y-intercept is b. The constant function is an even function.

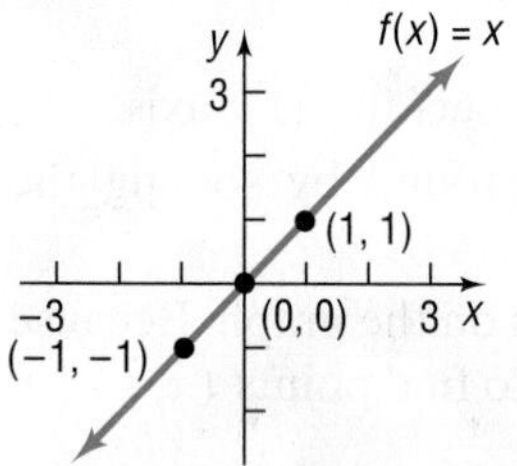

Figure 58 Identity Function

Identity Function

$$f(x) = x$$

See Figure 58.

The domain and the range of the **identity function** are the set of all real numbers. Its graph is a line whose slope is 1 and whose y-intercept is 0. The line consists of all points for which the x-coordinate equals the y-coordinate. The identity function is an odd function that is increasing over its domain. Note that the graph bisects quadrants I and III.

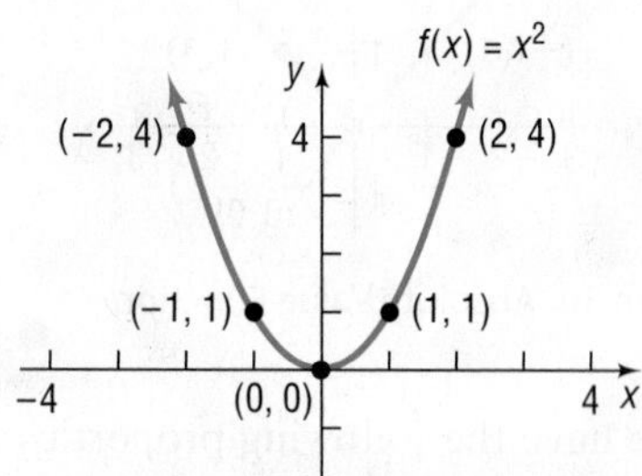

Figure 59 Square Function

Square Function

$$f(x) = x^2$$

See Figure 59.

The domain of the **square function** is the set of all real numbers; its range is the set of nonnegative real numbers. The graph of this function is a parabola whose intercept is at $(0, 0)$. The square function is an even function that is decreasing on the interval $(-\infty, 0)$ and increasing on the interval $(0, \infty)$.

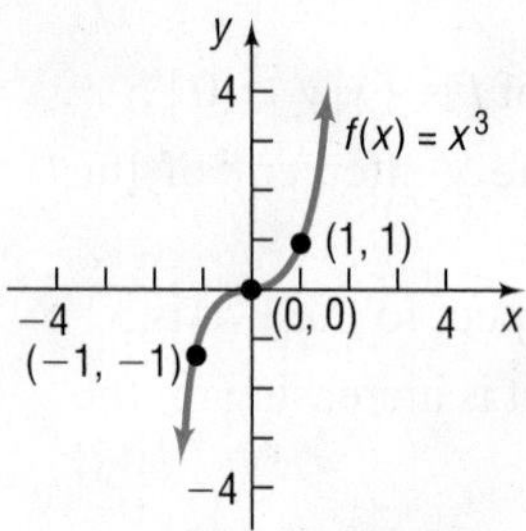

Figure 60 Cube Function

Cube Function

$$f(x) = x^3$$

See Figure 60.

The domain and the range of the **cube function** are the set of all real numbers. The intercept of the graph is at $(0, 0)$. The cube function is odd and is increasing on the interval $(-\infty, \infty)$.

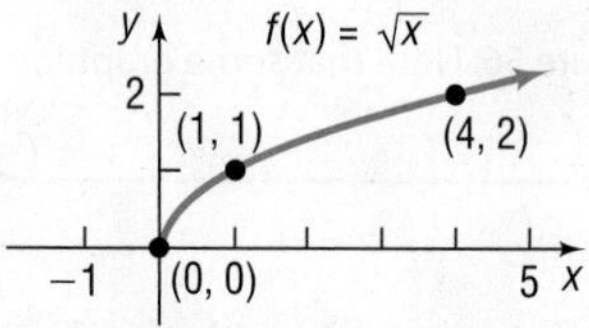

Figure 61 Square Root Function

Square Root Function

$$f(x) = \sqrt{x}$$

See Figure 61.

The domain and the range of the **square root function** are the set of nonnegative real numbers. The intercept of the graph is at $(0, 0)$. The square root function is neither even nor odd and is increasing on the interval $(0, \infty)$.

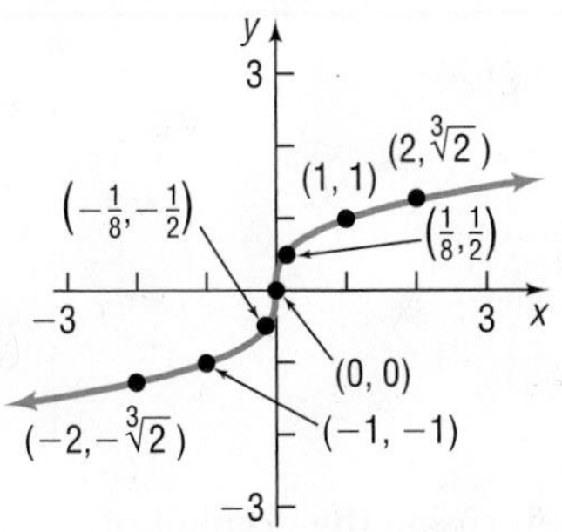

Figure 62 Cube Root Function

Cube Root Function

$$f(x) = \sqrt[3]{x}$$

See Figure 62.

The domain and the range of the **cube root function** are the set of all real numbers. The intercept of the graph is at $(0, 0)$. The cube root function is an odd function that is increasing on the interval $(-\infty, \infty)$.

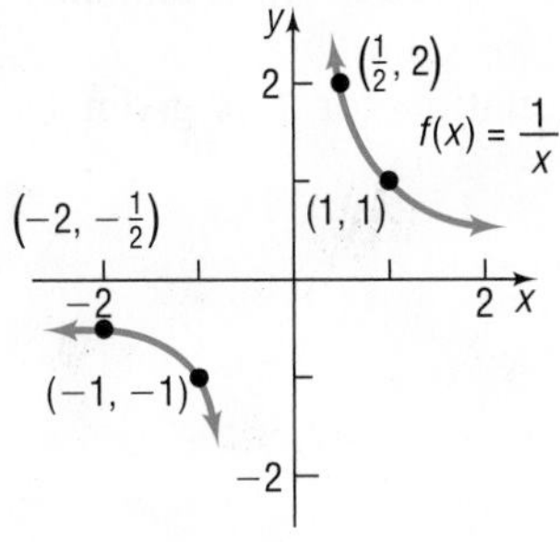

Figure 63 Reciprocal Function

Reciprocal Function

$$f(x) = \frac{1}{x}$$

Refer to Example 12, page 15, for a discussion of the equation $y = \frac{1}{x}$. See Figure 63.

The domain and the range of the **reciprocal function** are the set of all nonzero real numbers. The graph has no intercepts. The reciprocal function is decreasing on the intervals $(-\infty, 0)$ and $(0, \infty)$ and is an odd function.

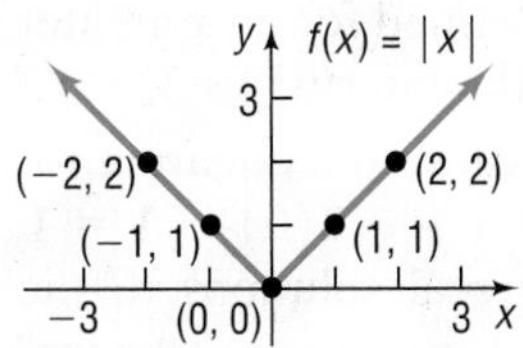

Figure 64 Absolute Value Function

Absolute Value Function

$$f(x) = |x|$$

See Figure 64.

The domain of the **absolute value function** is the set of all real numbers; its range is the set of nonnegative real numbers. The intercept of the graph is at $(0, 0)$. If $x \geq 0$, then $f(x) = x$, and the graph of f is part of the line $y = x$; if $x < 0$, then $f(x) = -x$, and the graph of f is part of the line $y = -x$. The absolute value function is an even function; it is decreasing on the interval $(-\infty, 0)$ and increasing on the interval $(0, \infty)$.

The functions discussed so far are basic. Whenever you encounter one of them, you should see a mental picture of its graph. For example, if you encounter the function $f(x) = x^2$, you should see in your mind's eye a picture like Figure 59.

Now Work PROBLEMS 11 THROUGH 18

2 Graph Piecewise-defined Functions

Sometimes a function is defined using different equations on different parts of its domain. For example, the absolute value function $f(x) = |x|$ is actually defined by two equations: $f(x) = x$ if $x \geq 0$ and $f(x) = -x$ if $x < 0$. For convenience, these equations are generally combined into one expression as

$$f(x) = |x| = \begin{cases} x & \text{if } x \geq 0 \\ -x & \text{if } x < 0 \end{cases}$$

When a function is defined by different equations on different parts of its domain, it is called a **piecewise-defined** function.

EXAMPLE 3

Analyzing a Piecewise-defined Function

The function f is defined as

$$f(x) = \begin{cases} -2x + 1 & \text{if } -3 \le x < 1 \\ 2 & \text{if } x = 1 \\ x^2 & \text{if } x > 1 \end{cases}$$

(a) Find $f(-2), f(1)$, and $f(2)$. (b) Determine the domain of f.
(c) Locate any intercepts. (d) Graph f.
(e) Use the graph to find the range of f. (f) Is f continuous on its domain?

Solution (a) To find $f(-2)$, observe that when $x = -2$, the equation for f is given by $f(x) = -2x + 1$. So

$$f(-2) = -2(-2) + 1 = 5$$

When $x = 1$, the equation for f is $f(x) = 2$. So,

$$f(1) = 2$$

When $x = 2$, the equation for f is $f(x) = x^2$. So

$$f(2) = 2^2 = 4$$

(b) To find the domain of f, look at its definition. Since f is defined for all x greater than or equal to -3, the domain of f is $\{x \mid x \ge -3\}$, or the interval $[-3, \infty)$.

(c) The y-intercept of the graph of the function is $f(0)$. Because the equation for f when $x = 0$ is $f(x) = -2x + 1$, the y-intercept is $f(0) = -2(0) + 1 = 1$. The x-intercepts of the graph of a function f are the real solutions to the equation $f(x) = 0$. To find the x-intercepts of f, solve $f(x) = 0$ for each "piece" of the function, and then determine what values of x, if any, satisfy the condition that defines the piece.

$$\begin{aligned} f(x) &= 0 \\ -2x + 1 &= 0 \quad -3 \le x < 1 \\ -2x &= -1 \\ x &= \frac{1}{2} \end{aligned} \qquad \begin{aligned} f(x) &= 0 \\ 2 &= 0 \quad x = 1 \\ &\text{No solution} \end{aligned} \qquad \begin{aligned} f(x) &= 0 \\ x^2 &= 0 \quad x > 1 \\ x &= 0 \end{aligned}$$

The first potential x-intercept, $x = \frac{1}{2}$, satisfies the condition $-3 \le x < 1$, so $x = \frac{1}{2}$ is an x-intercept. The second potential x-intercept, $x = 0$, does not satisfy the condition $x > 1$, so $x = 0$ is not an x-intercept. The only x-intercept is $\frac{1}{2}$. The intercepts are $(0, 1)$ and $\left(\frac{1}{2}, 0\right)$.

(d) To graph f, graph each "piece." First graph the line $y = -2x + 1$ and keep only the part for which $-3 \le x < 1$. Then plot the point $(1, 2)$ because, when $x = 1, f(x) = 2$. Finally, graph the parabola $y = x^2$ and keep only the part for which $x > 1$. See Figure 65.

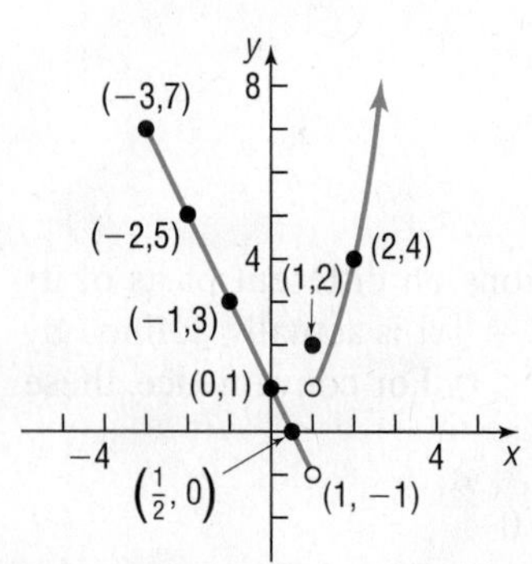

Figure 65

(e) From the graph, we conclude that the range of f is $\{y \mid y > -1\}$, or the interval $(-1, \infty)$.

(f) The function f is not continuous because there is a "jump" in the graph at $x = 1$. ●

Now Work PROBLEM 31

1.5 Assess Your Understanding

'Are You Prepared?' *Answers are given at the end of these exercises. If you get a wrong answer, read the pages listed in red.*

1. Sketch the graph of $y = \sqrt{x}$. (p. 14)

2. Sketch the graph of $y = \dfrac{1}{x}$. (p. 15)

3. List the intercepts of the equation $y = x^3 - 8$. (pp. 10–11)

Concepts and Vocabulary

4. The function $f(x) = x^2$ is decreasing on the interval ________.

5. When functions are defined by more than one equation, they are called ______________ functions.

6. ***True or False*** The cube function is odd and is increasing on the interval $(-\infty, \infty)$.

7. ***True or False*** The cube root function is odd and is decreasing on the interval $(-\infty, \infty)$.

8. ***True or False*** The domain and the range of the reciprocal function are the set of all real numbers.

9. Which of the following functions has a graph that is symmetric about the y-axis?
(a) $y = \sqrt{x}$ (b) $y = |x|$ (c) $y = x^3$ (d) $y = \dfrac{1}{x}$

10. Consider the following function.

$$f(x) = \begin{cases} 3x - 2 & \text{if } \quad x < 2 \\ x^2 + 5 & \text{if } \quad 2 \le x < 10 \\ 3 & \text{if } \quad x \ge 10 \end{cases}$$

Which "piece(s)" should be used to find the y-intercept?
(a) $3x - 2$ (b) $x^2 + 5$ (c) 3 (d) all three

Skill Building

In Problems 11–18, match each graph to its function.

A. Constant function *B. Identity function* *C. Square function* *D. Cube function*
E. Square root function *F. Reciprocal function* *G. Absolute value function* *H. Cube root function*

11.
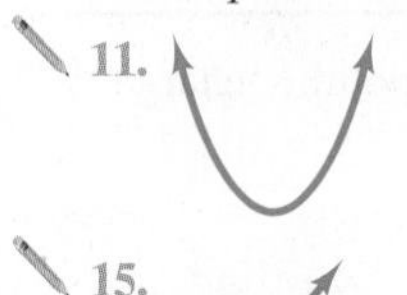

12.
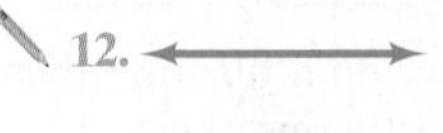

13.
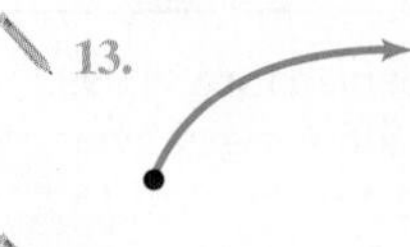

14.
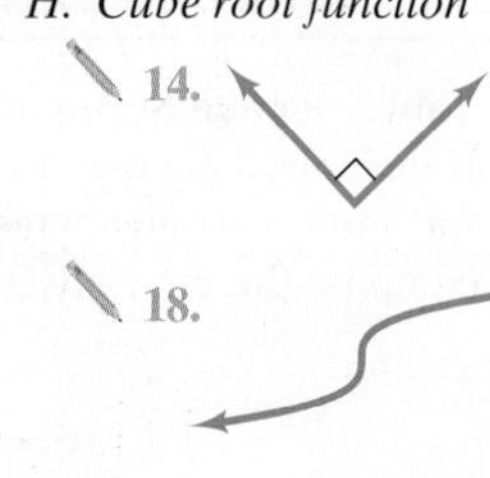

15.
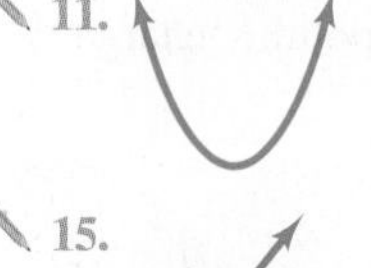

16.
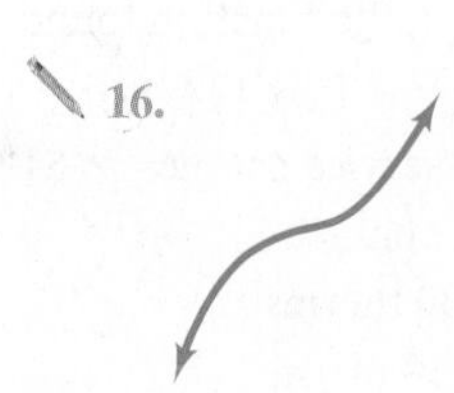

17.

18.

In Problems 19–26, sketch the graph of each function. Be sure to label three points on the graph.

19. $f(x) = x$ **20.** $f(x) = x^2$ **21.** $f(x) = x^3$ **22.** $f(x) = \sqrt{x}$

23. $f(x) = \dfrac{1}{x}$ **24.** $f(x) = |x|$ **25.** $f(x) = \sqrt[3]{x}$ **26.** $f(x) = 3$

27. If $f(x) = \begin{cases} x^2 & \text{if } x < 0 \\ 2 & \text{if } x = 0 \\ 2x + 1 & \text{if } x > 0 \end{cases}$

find: (a) $f(-2)$ (b) $f(0)$ (c) $f(2)$

28. If $f(x) = \begin{cases} -3x & \text{if } x < -1 \\ 0 & \text{if } x = -1 \\ 2x^2 + 1 & \text{if } x > -1 \end{cases}$

find: (a) $f(-2)$ (b) $f(-1)$ (c) $f(0)$

29. If $f(x) = \begin{cases} 2x - 4 & \text{if } -1 \le x \le 2 \\ x^3 - 2 & \text{if } 2 < x \le 3 \end{cases}$

find: (a) $f(0)$ (b) $f(1)$ (c) $f(2)$ (d) $f(3)$

30. If $f(x) = \begin{cases} x^3 & \text{if } -2 \le x < 1 \\ 3x + 2 & \text{if } 1 \le x \le 4 \end{cases}$

find: (a) $f(-1)$ (b) $f(0)$ (c) $f(1)$ (d) $f(3)$

In Problems 31–40:

(a) Find the domain of each function. *(b) Locate any intercepts.* *(c) Graph each function.*
(d) Based on the graph, find the range. *(e) Is f continuous on its domain?*

31. $f(x) = \begin{cases} 2x & \text{if } x \ne 0 \\ 1 & \text{if } x = 0 \end{cases}$

32. $f(x) = \begin{cases} 3x & \text{if } x \ne 0 \\ 4 & \text{if } x = 0 \end{cases}$

33. $f(x) = \begin{cases} -2x + 3 & \text{if } x < 1 \\ 3x - 2 & \text{if } x \ge 1 \end{cases}$

34. $f(x) = \begin{cases} x + 3 & \text{if } x < -2 \\ -2x - 3 & \text{if } x \geq -2 \end{cases}$

35. $f(x) = \begin{cases} x + 3 & \text{if } -2 \leq x < 1 \\ 5 & \text{if } x = 1 \\ -x + 2 & \text{if } x > 1 \end{cases}$

36. $f(x) = \begin{cases} 2x + 5 & \text{if } -3 \leq x < 0 \\ -3 & \text{if } x = 0 \\ -5x & \text{if } x > 0 \end{cases}$

37. $f(x) = \begin{cases} 1 + x & \text{if } x < 0 \\ x^2 & \text{if } x \geq 0 \end{cases}$

38. $f(x) = \begin{cases} \dfrac{1}{x} & \text{if } x < 0 \\ \sqrt[3]{x} & \text{if } x \geq 0 \end{cases}$

39. $f(x) = \begin{cases} |x| & \text{if } -2 \leq x < 0 \\ x^3 & \text{if } x > 0 \end{cases}$

40. $f(x) = \begin{cases} 2 - x & \text{if } -3 \leq x < 1 \\ \sqrt{x} & \text{if } x > 1 \end{cases}$

In Problems 41–44, the graph of a piecewise-defined function is given. Write a definition for each function.

41.

y
2
(−1, 1)
(2, 1)
−2
(0, 0)
2 x

42.

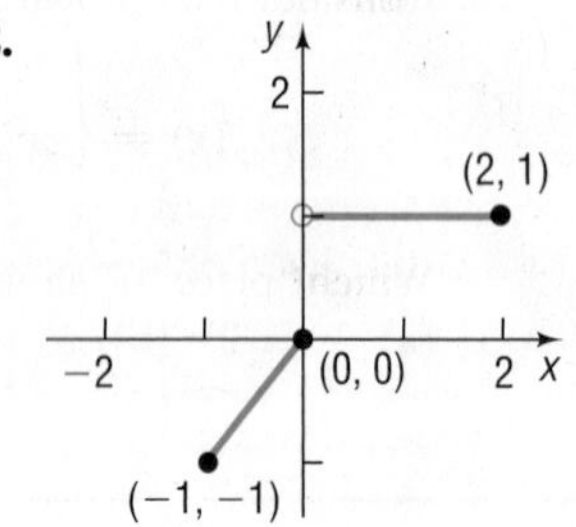

43.

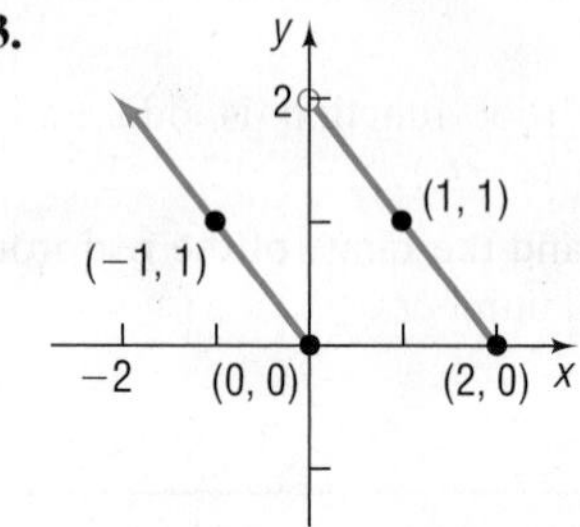

44.

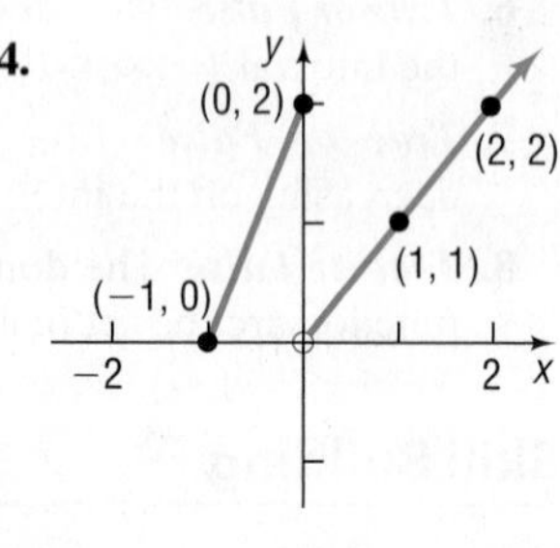

Applications and Extensions

45. Tablet Service Sprint offers a monthly tablet plan for \$34.99. It includes 3 gigabytes of data and charges \$15 per gigabyte for additional gigabytes. The following function is used to compute the monthly cost for a subscriber.

$$C(x) = \begin{cases} 34.99 & \text{if } \quad 0 \leq x \leq 3 \\ 15x - 10.01 & \text{if } \quad x > 3 \end{cases}$$

Compute the monthly cost for each of the following gigabytes of use.

(a) 2 (b) 5 (c) 13

Source: Sprint, April 2014

46. Parking at O'Hare International Airport The short-term (no more than 24 hours) parking fee F (in dollars) for parking x hours on a weekday at O'Hare International Airport's main parking garage can be modeled by the function

$$F(x) = \begin{cases} 3 & \text{if } 0 < x \leq 3 \\ 5x + 5 & \text{if } 3 < x < 9 \\ 50 & \text{if } 9 \leq x \leq 24 \end{cases}$$

where fractions of a dollar are rounded upto the next dollar.

Determine the fee for parking in the short-term parking garage for

(a) 2 hours (b) 7 hours
(c) 15 hours (d) 8 hours and 10 minutes

Source: O'Hare International Airport

47. Cost of Natural Gas In March 2014, Laclede Gas had the rate schedule (top, right) for natural gas usage in single-family residences.

(a) What is the charge for using 20 therms in a month?
(b) What is the charge for using 150 therms in a month?
(c) Develop a function that models the monthly charge C for x therms of gas.
(d) Graph the function found in part (c).

Monthly service charge	\$19.50
Delivery charge	
First 30 therms	\$0.91686/therm
Over 30 therms	\$0
Natural gas cost	
First 30 therms	\$0.3313/therm
Over 30 therms	\$0.5757/therm

Source: Laclede Gas

48. Cost of Natural Gas In April 2014, Nicor Gas had the following rate schedule for natural gas usage in small businesses.

Monthly customer charge	\$72.60
Distribution charge	
1st 150 therms	\$0.1201/therm
Next 4850 therms	\$0.0549/therm
Over 5000 therms	\$0.0482/therm
Gas supply charge	\$0.68/therm

(a) What is the charge for using 1000 therms in a month?
(b) What is the charge for using 6000 therms in a month?
(c) Develop a function that models the monthly charge C for x therms of gas.
(d) Graph the function found in part (c).

Source: Nicor Gas, 2014

49. **Federal Income Tax** Two 2014 Tax Rate Schedules are given in the accompanying table. If x equals taxable income and y equals the tax due, construct a function $y = f(x)$ for Schedule X.

2014 Tax Rate Schedules

Schedule X—Single						Schedule Y-1—Married Filing Jointly or Qualified Widow(er)					
If Taxable Income is Over	But Not Over	The Tax is This Amount		Plus This %	Of the Excess Over	If Taxable Income is Over	But Not Over	The Tax is This Amount		Plus This %	Of the Excess Over
$0	$9,075	$0	+	10%	$0	$0	$18,150	$0	+	10%	$0
9,075	36,900	907.50	+	15%	9,075	18,150	73,800	1,815	+	15%	18,150
36,900	89,350	5,081.25	+	25%	36,900	73,800	148,850	10,162.50	+	25%	73,800
89,350	186,350	18,193.75	+	28%	89,350	148,850	226,850	28,925.00	+	28%	148,850
186,350	405,100	45,353.75	+	33%	186,350	226,850	405,100	50,765.00	+	33%	226,850
405,100	406,750	117,541.25	+	35%	405,100	405,100	457,600	109,587.50	+	35%	405,100
406,750	—	118,188.75	+	39.6%	406,750	457,600	—	127,962.50	+	39.6%	457,600

50. **Federal Income Tax** Refer to the 2014 tax rate schedules. If x equals taxable income and y equals the tax due, construct a function $y = f(x)$ for Schedule Y-1.

51. **Cost of Transporting Goods** A trucking company transports goods between Chicago and New York, a distance of 960 miles. The company's policy is to charge, for each pound, \$0.50 per mile for the first 100 miles, \$0.40 per mile for the next 300 miles, \$0.25 per mile for the next 400 miles, and no charge for the remaining 160 miles.
(a) Graph the relationship between the cost of transportation in dollars and mileage over the entire 960-mile route.
(b) Find the cost as a function of mileage for hauls between 100 and 400 miles from Chicago.
(c) Find the cost as a function of mileage for hauls between 400 and 800 miles from Chicago.

52. **Car Rental Costs** An economy car rented in Florida from Enterprise® on a weekly basis costs \$185 per week. Extra days cost \$37 per day until the day rate exceeds the weekly rate, in which case the weekly rate applies. Also, any part of a day used counts as a full day. Find the cost C of renting an economy car as a function of the number x of days used, where $7 \le x \le 14$. Graph this function.

53. **Mortgage Fees** Fannie Mae charges a loan-level price adjustment (LLPA) on all mortgages, which represents a fee homebuyers seeking a loan must pay. The rate paid depends on the credit score of the borrower, the amount borrowed, and the loan-to-value (LTV) ratio. The LTV ratio is the ratio of amount borrowed to appraised value of the home. For example, a homebuyer who wishes to borrow \$250,000 with a credit score of 730 and an LTV ratio of 80% will pay 0.5% (0.005) of \$250,000, or \$1250. The table shows the LLPA for various credit scores and an LTV ratio of 80%.

Credit Score	Loan-Level Price Adjustment Rate
≤659	3.00%
660–679	2.50%
680–699	1.75%
700–719	1%
720–739	0.5%
≥740	0.25%

Source: Fannie Mae

(a) Construct a function $C = C(s)$, where C is the loan-level price adjustment (LLPA) and s is the credit score of an individual who wishes to borrow \$300,000 with an 80% LTV ratio.
(b) What is the LLPA on a \$300,000 loan with an 80% LTV ratio for a borrower whose credit score is 725?
(c) What is the LLPA on a \$300,000 loan with an 80% LTV ratio for a borrower whose credit score is 670?

54. **Minimum Payments for Credit Cards** Holders of credit cards issued by banks, department stores, oil companies, and so on receive bills each month that state minimum amounts that must be paid by a certain due date. The minimum due depends on the total amount owed. One such credit card company uses the following rules: For a bill of less than \$10, the entire amount is due. For a bill of at least \$10 but less than \$500, the minimum due is \$10. A minimum of \$30 is due on a bill of at least \$500 but less than \$1000, a minimum of \$50 is due on a bill of at least \$1000 but less than \$1500, and a minimum of \$70 is due on bills of \$1500 or more. Find the function f that describes the minimum payment due on a bill of x dollars. Graph f.

55. **Wind Chill** The wind chill factor represents the air temperature at a standard wind speed that would produce the same heat loss as the given temperature and wind speed. One formula for computing the equivalent temperature is

$$W = \begin{cases} t & 0 \le v < 1.79 \\ 33 - \dfrac{(10.45 + 10\sqrt{v} - v)(33 - t)}{22.04} & 1.79 \le v \le 20 \\ 33 - 1.5958(33 - t) & v > 20 \end{cases}$$

where v represents the wind speed (in meters per second) and t represents the air temperature (°C). Compute the wind chill for the following:
(a) An air temperature of 10°C and a wind speed of 1 meter per second (m/sec)
(b) An air temperature of 10°C and a wind speed of 5 m/sec
(c) An air temperature of 10°C and a wind speed of 15 m/sec
(d) An air temperature of 10°C and a wind speed of 25 m/sec
(e) Explain the physical meaning of the equation corresponding to $0 \le v < 1.79$.
(f) Explain the physical meaning of the equation corresponding to $v > 20$.

56. **Wind Chill** Redo Problem 55(a)–(d) for an air temperature of $-10°C$.

57. **First-class Mail** In 2014 the U.S. Postal Service charged $0.98 postage for first-class mail retail flats (such as an 8.5" by 11" envelope) weighing up to 1 ounce, plus $0.21 for each additional ounce up to 13 ounces. First-class rates do not apply to flats weighing more than 13 ounces. Develop a model that relates C, the first-class postage charged, for a flat weighing x ounces. Graph the function.
Source: United States Postal Service

Explaining Concepts: Discussion and Writing

In Problems 58–65, use a graphing utility.

58. **Exploration** Graph $y = x^2$. Then on the same screen graph $y = x^2 + 2$, followed by $y = x^2 + 4$, followed by $y = x^2 - 2$. What pattern do you observe? Can you predict the graph of $y = x^2 - 4$? Of $y = x^2 + 5$?

59. **Exploration** Graph $y = x^2$. Then on the same screen graph $y = (x - 2)^2$, followed by $y = (x - 4)^2$, followed by $y = (x + 2)^2$. What pattern do you observe? Can you predict the graph of $y = (x + 4)^2$? Of $y = (x - 5)^2$?

60. **Exploration** Graph $y = |x|$. Then on the same screen graph $y = 2|x|$, followed by $y = 4|x|$, followed by $y = \frac{1}{2}|x|$. What pattern do you observe? Can you predict the graph of $y = \frac{1}{4}|x|$? Of $y = 5|x|$?

61. **Exploration** Graph $y = x^2$. Then on the same screen graph $y = -x^2$. Now try $y = |x|$ and $y = -|x|$. What do you conclude?

62. **Exploration** Graph $y = \sqrt{x}$. Then on the same screen graph $y = \sqrt{-x}$. Now try $y = 2x + 1$ and $y = 2(-x) + 1$. What do you conclude?

63. **Exploration** Graph $y = x^3$. Then on the same screen graph $y = (x - 1)^3 + 2$. Could you have predicted the result?

64. **Exploration** Graph $y = x^2$, $y = x^4$, and $y = x^6$ on the same screen. What do you notice is the same about each graph? What do you notice is different?

65. **Exploration** Graph $y = x^3$, $y = x^5$, and $y = x^7$ on the same screen. What do you notice is the same about each graph? What do you notice is different?

66. Consider the equation

$$y = \begin{cases} 1 & \text{if } x \text{ is rational} \\ 0 & \text{if } x \text{ is irrational} \end{cases}$$

Is this a function? What is its domain? What is its range? What is its y-intercept, if any? What are its x-intercepts, if any? Is it even, odd, or neither? How would you describe its graph?

67. Define some functions that pass through $(0, 0)$ and $(1, 1)$ and are increasing for $x \geq 0$. Begin your list with $y = \sqrt{x}$, $y = x$, and $y = x^2$. Can you propose a general result about such functions?

'Are You Prepared?' Answers

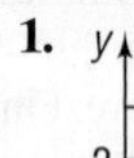

1.

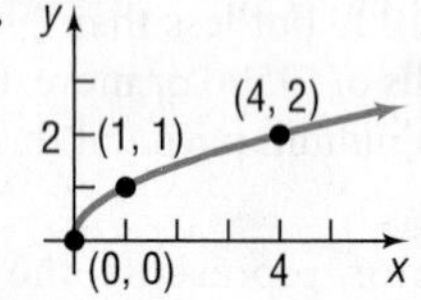

2.

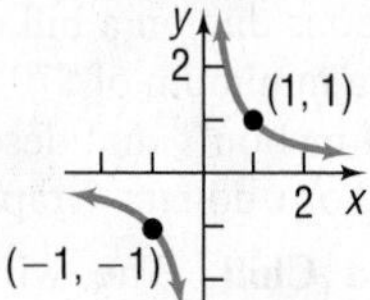

3. $(0, -8)$, $(2, 0)$

1.6 Graphing Techniques: Transformations

OBJECTIVES
1 Graph Functions Using Vertical and Horizontal Shifts (p. 65)
2 Graph Functions Using Compressions and Stretches (p. 68)
3 Graph Functions Using Reflections about the x-Axis and the y-Axis (p. 70)

At this stage, if you were asked to graph any of the functions defined by $y = x$, $y = x^2$, $y = x^3$, $y = \sqrt{x}$, $y = \sqrt[3]{x}$, $y = \frac{1}{x}$, or $y = |x|$, your response should be, "Yes, I recognize these functions and know the general shapes of their graphs." (If this is not your answer, review the previous section, Figures 58 through 64.)

Sometimes we are asked to graph a function that is "almost" like one that we already know how to graph. In this section, we develop techniques for graphing such functions. Collectively, these techniques are referred to as **transformations**.

1 Graph Functions Using Vertical and Horizontal Shifts

EXAMPLE 1

Vertical Shift Up

Use the graph of $f(x) = x^2$ to obtain the graph of $g(x) = x^2 + 3$. Find the domain and range of g.

Solution Begin by obtaining some points on the graphs of f and g. For example, when $x = 0$, then $y = f(0) = 0$ and $y = g(0) = 3$. When $x = 1$, then $y = f(1) = 1$ and $y = g(1) = 4$. Table 8 lists these and a few other points on each graph. Notice that each y-coordinate of a point on the graph of g is 3 units larger than the y-coordinate of the corresponding point on the graph of f. We conclude that the graph of g is identical to that of f, except that it is shifted vertically up 3 units. See Figure 66.

Table 8

x	$y = f(x) = x^2$	$y = g(x) = x^2 + 3$
−2	4	7
−1	1	4
0	0	3
1	1	4
2	4	7

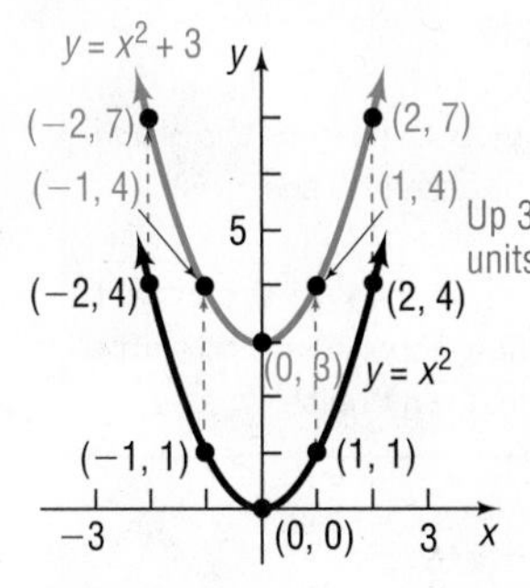

Figure 66

The domain of g is all real numbers, or $(-\infty, \infty)$. The range of g is $[3, \infty)$. ●

EXAMPLE 2

Vertical Shift Down

Use the graph of $f(x) = x^2$ to obtain the graph of $g(x) = x^2 - 4$. Find the domain and range of g.

Solution Table 9 lists some points on the graphs of f and g. Notice that each y-coordinate of g is 4 units less than the corresponding y-coordinate of f.

To obtain the graph of g from the graph of f, subtract 4 from each y-coordinate on the graph of f. The graph of g is identical to that of f, except that it is shifted down 4 units. See Figure 67.

Table 9

x	$y = f(x) = x^2$	$y = g(x) = x^2 - 4$
−2	4	0
−1	1	−3
0	0	−4
1	1	−3
2	4	0

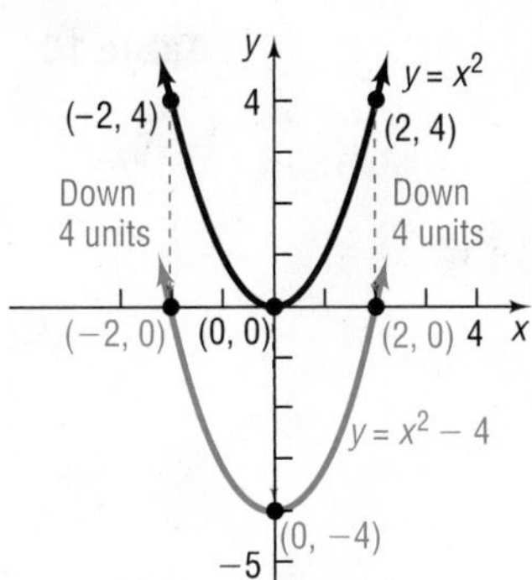

Figure 67

The domain of g is all real numbers, or $(-\infty, \infty)$. The range of g is $[-4, \infty)$. ●

Note that a vertical shift affects only the range of a function, not the domain. For example, the range of $f(x) = x^2$ is $[0, \infty)$. In Example 1 the range of g is $[3, \infty)$, whereas in Example 2 the range of g is $[-4, \infty)$. The domain for all three functions is all real numbers.

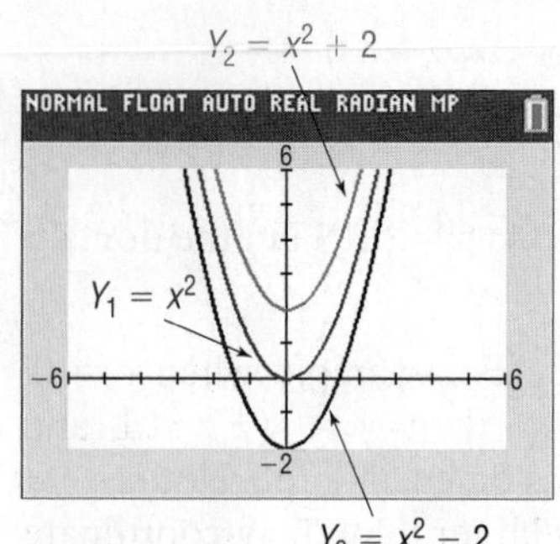

Figure 68

Exploration On the same screen, graph each of the following functions:

$$Y_1 = x^2$$
$$Y_2 = x^2 + 2$$
$$Y_3 = x^2 - 2$$

Figure 68 illustrates the graphs. You should have observed a general pattern. With $Y_1 = x^2$ on the screen, the graph of $Y_2 = x^2 + 2$ is identical to that of $Y_1 = x^2$, except that it is shifted vertically up 2 units. The graph of $Y_3 = x^2 - 2$ is identical to that of $Y_1 = x^2$, except that it is shifted vertically down 2 units.

In Words

For $y = f(x) + k, k > 0$, add k to each y-coordinate on the graph of $y = f(x)$ to shift the graph up k units.

For $y = f(x) - k, k > 0$, subtract k from each y-coordinate to shift the graph down k units.

We are led to the following conclusions:

If a positive real number k is added to the output of a function $y = f(x)$, the graph of the new function $y = f(x) + k$ is the graph of f **shifted vertically up** k units.

If a positive real number k is subtracted from the output of a function $y = f(x)$, the graph of the new function $y = f(x) - k$ is the graph of f **shifted vertically down** k units.

Now Work PROBLEM 41

EXAMPLE 3 Horizontal Shift to the Right

Use the graph of $f(x) = \sqrt{x}$ to obtain the graph of $g(x) = \sqrt{x - 2}$. Find the domain and range of g.

Solution The function $g(x) = \sqrt{x - 2}$ is basically a square root function. Table 10 lists some points on the graphs of f and g. Note that when $f(x) = 0$, then $x = 0$, and when $g(x) = 0$, then $x = 2$. Also, when $f(x) = 2$, then $x = 4$, and when $g(x) = 2$, then $x = 6$. Notice that the x-coordinates on the graph of g are 2 units larger than the corresponding x-coordinates on the graph of f for any given y-coordinate. We conclude that the graph of g is identical to that of f, except that it is shifted horizontally 2 units to the right. See Figure 69.

Table 10

x	$y = f(x) = \sqrt{x}$	x	$y = g(x) = \sqrt{x - 2}$
0	0	2	0
1	1	3	1
4	2	6	2
9	3	11	3

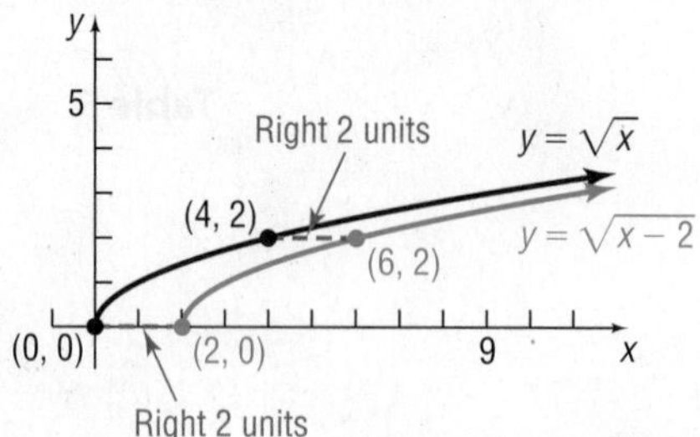

Figure 69

The domain of g is $[2, \infty)$ and the range is $[0, \infty)$. ●

EXAMPLE 4 Horizontal Shift to the Left

Use the graph of $f(x) = \sqrt{x}$ to obtain the graph of $g(x) = \sqrt{x + 4}$. Find the domain and range of g.

Solution The function $g(x) = \sqrt{x + 4}$ is basically a square root function. Its graph is the same as that of f, except that it is shifted horizontally 4 units to the left. See Figure 70.

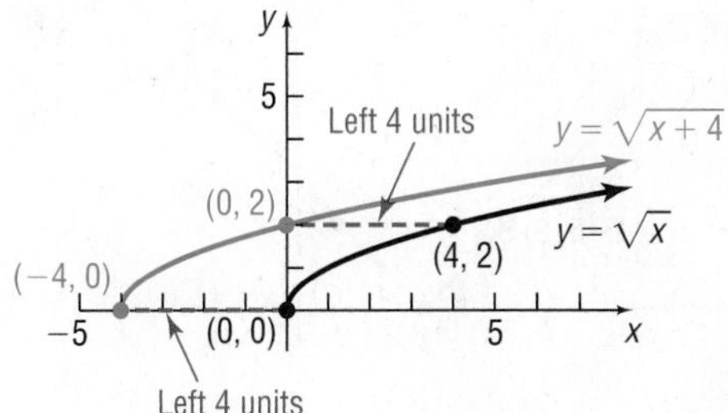

Figure 70

The domain of g is $[-4, \infty)$ and the range is $[0, \infty)$. ●

Now Work PROBLEM 45

Note that a horizontal shift affects only the domain of a function, not the range. For example, the domain of $f(x) = \sqrt{x}$ is $[0, \infty)$. In Example 3 the domain of g is $[2, \infty)$, whereas in Example 4 the domain of g is $[-4, \infty)$. The range for all three functions is $[0, \infty)$.

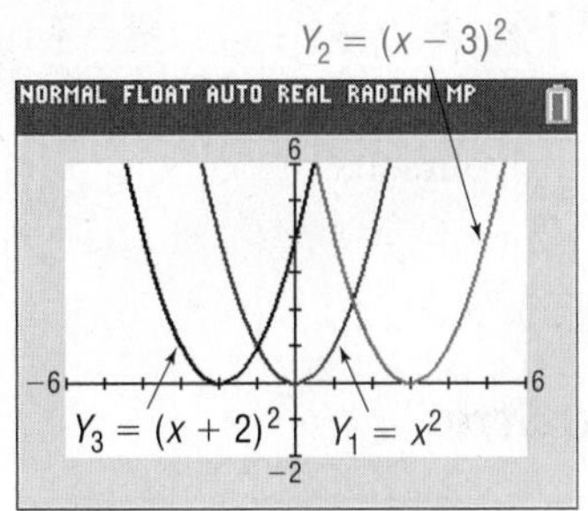

Figure 71

Exploration On the same screen, graph each of the following functions:

$$Y_1 = x^2$$
$$Y_2 = (x-3)^2$$
$$Y_3 = (x+2)^2$$

Figure 71 illustrates the graphs.

You should have observed the following pattern. With the graph of $Y_1 = x^2$ on the screen, the graph of $Y_2 = (x-3)^2$ is identical to that of $Y_1 = x^2$, except that it is shifted horizontally to the right 3 units. The graph of $Y_3 = (x+2)^2$ is identical to that of $Y_1 = x^2$, except that it is shifted horizontally to the left 2 units.

We are led to the following conclusions:

> **In Words**
> For $y = f(x-h)$, $h > 0$, add h to each x-coordinate on the graph of $y = f(x)$ to shift the graph right h units. For $y = f(x+h)$, $h > 0$, subtract h from each x-coordinate on the graph of $y = f(x)$ to shift the graph left h units.

If the argument x of a function f is replaced by $x - h$, $h > 0$, the graph of the new function $y = f(x-h)$ is the graph of f **shifted horizontally right** h units.

If the argument x of a function f is replaced by $x + h$, $h > 0$, the graph of the new function $y = f(x+h)$ is the graph of f **shifted horizontally left** h units.

Observe the distinction between vertical and horizontal shifts. The graph of $f(x) = x^3 + 2$ is obtained by shifting the graph of $y = x^3$ *up* 2 units, because we evaluate the cube function first and then add 2. The graph of $g(x) = (x+2)^3$ is obtained by shifting the graph of $y = x^3$ *left* 2 units, because we add 2 to x before we evaluate the cube function.

Vertical and horizontal shifts are sometimes combined.

EXAMPLE 5 Combining Vertical and Horizontal Shifts

Graph the function $f(x) = |x+3| - 5$. Find the domain and range of f.

Solution We graph f in steps. First, note that the rule for f is basically an absolute value function, so begin with the graph of $y = |x|$ as shown in Figure 72(a) on page 68. Next, to get the graph of $y = |x+3|$, shift the graph of $y = |x|$ horizontally 3 units to the left. See Figure 72(b). Finally, to get the graph of $y = |x+3| - 5$, shift the graph of $y = |x+3|$ vertically down 5 units. See Figure 72(c). Note the points plotted on each graph. Using key points can be helpful in keeping track of the transformation that has taken place.

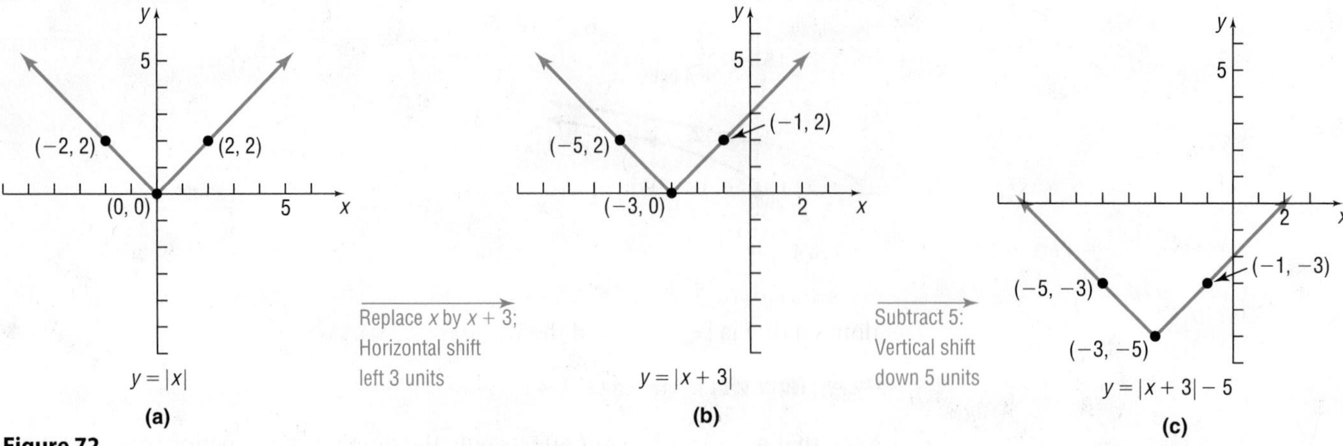

Figure 72

The domain of f is all real numbers, or $(-\infty, \infty)$. The range of f is $[-5, \infty)$. ●

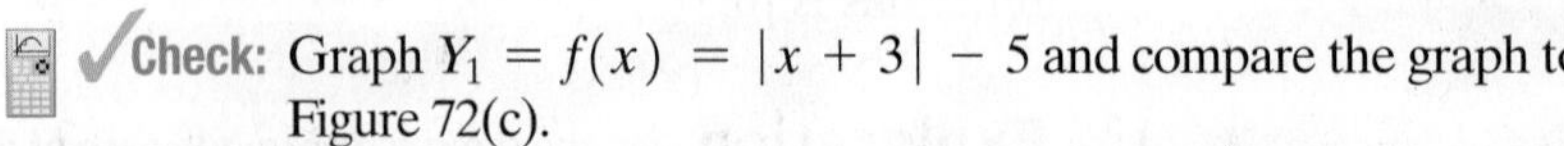

✓Check: Graph $Y_1 = f(x) = |x + 3| - 5$ and compare the graph to Figure 72(c).

In Example 5, if the vertical shift had been done first, followed by the horizontal shift, the final graph would have been the same. Try it for yourself.

Now Work PROBLEMS 47 AND 71

2 Graph Functions Using Compressions and Stretches

EXAMPLE 6

Vertical Stretch

Use the graph of $f(x) = \sqrt{x}$ to obtain the graph of $g(x) = 2\sqrt{x}$.

Solution To see the relationship between the graphs of f and g, we form Table 11, listing points on each graph. For each x, the y-coordinate of a point on the graph of g is 2 times as large as the corresponding y-coordinate on the graph of f. The graph of $f(x) = \sqrt{x}$ is vertically stretched by a factor of 2 to obtain the graph of $g(x) = 2\sqrt{x}$. For example, $(1, 1)$ is on the graph of f, but $(1, 2)$ is on the graph of g. See Figure 73.

Table 11

x	$y = f(x) = \sqrt{x}$	$y = g(x) = 2\sqrt{x}$
0	0	0
1	1	2
4	2	4
9	3	6

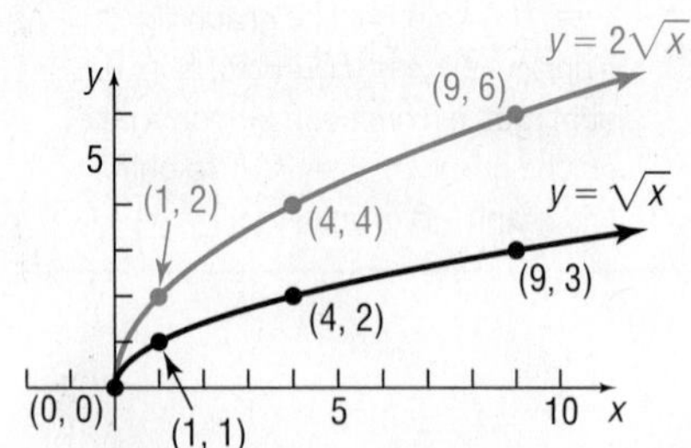

Figure 73 ●

EXAMPLE 7

Vertical Compression

Use the graph of $f(x) = |x|$ to obtain the graph of $g(x) = \frac{1}{2}|x|$.

Solution For each x, the y-coordinate of a point on the graph of g is $\frac{1}{2}$ as large as the corresponding y-coordinate on the graph of f. The graph of $f(x) = |x|$ is vertically compressed by a factor of $\frac{1}{2}$ to obtain the graph of $g(x) = \frac{1}{2}|x|$. For example, $(2, 2)$ is on the graph of f, but $(2, 1)$ is on the graph of g. See Table 12 and Figure 74.

Table 12

x	$y = f(x) = \lvert x \rvert$	$y = g(x) = \frac{1}{2}\lvert x \rvert$
−2	2	1
−1	1	$\frac{1}{2}$
0	0	0
1	1	$\frac{1}{2}$
2	2	1

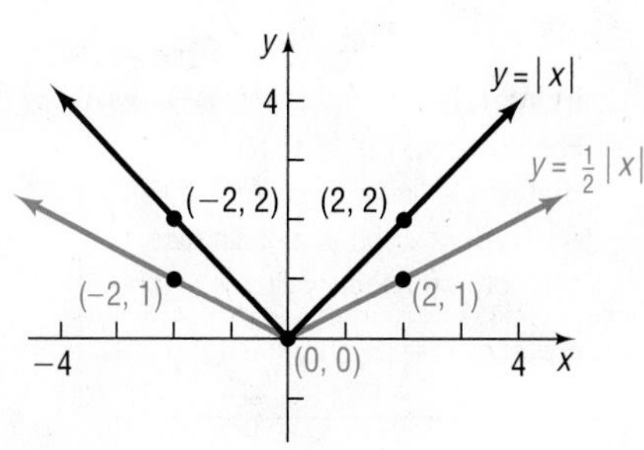

Figure 74

> **In Words**
> For $y = af(x)$, $a > 0$, the factor a is "outside" the function, so it affects the y-coordinates. Multiply each y-coordinate on the graph of $y = f(x)$ by a.

When the right side of a function $y = f(x)$ is multiplied by a positive number a, the graph of the new function $y = af(x)$ is obtained by multiplying each y-coordinate on the graph of $y = f(x)$ by a. The new graph is a **vertically compressed** (if $0 < a < 1$) or a **vertically stretched** (if $a > 1$) version of the graph of $y = f(x)$.

 **Now Work** PROBLEM 49

What happens if the argument x of a function $y = f(x)$ is multiplied by a positive number a, creating a new function $y = f(ax)$? To find the answer, look at the following Exploration.

Exploration On the same screen, graph each of the following functions:

$$Y_1 = f(x) = \sqrt{x} \qquad Y_2 = f(2x) = \sqrt{2x} \qquad Y_3 = f\left(\frac{1}{2}x\right) = \sqrt{\frac{1}{2}x} = \sqrt{\frac{x}{2}}$$

Create a table of values to explore the relation between the x- and y-coordinates of each function.

Result You should have obtained the graphs in Figure 75. Look at Table 13(a). Note that (1, 1), (4, 2), and (9, 3) are points on the graph of $Y_1 = \sqrt{x}$. Also, (0.5, 1), (2, 2), and (4.5, 3) are points on the graph of $Y_2 = \sqrt{2x}$. For a given y-coordinate, the x-coordinate on the graph of Y_2 is $\frac{1}{2}$ of the x-coordinate on Y_1.

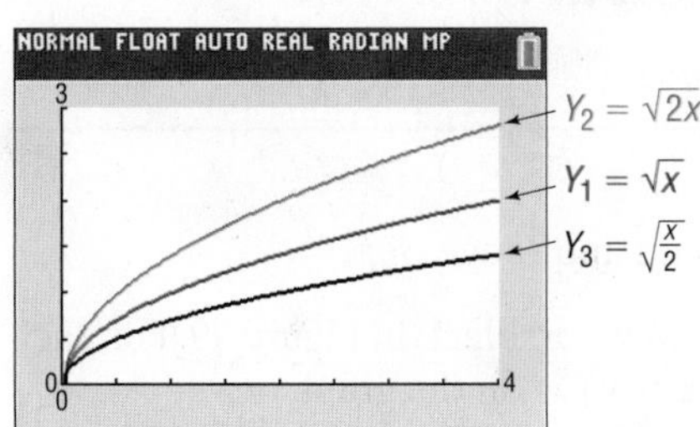

Figure 75

Table 13

NORMAL FLOAT AUTO REAL RADIAN MP
PRESS ENTER TO EDIT

X	Y1	Y2
0	0	0
.5	.70711	1
1	1	1.4142
2	1.4142	2
4	2	2.8284
4.5	2.1213	3
8	2.8284	4
9	3	4.2426
16	4	5.6569
12.5	3.5355	5
25	5	7.0711

Y2=√(2X)

(a)

NORMAL FLOAT AUTO REAL RADIAN MP
PRESS ENTER TO EDIT

X	Y1	Y3
0	0	0
1	1	.70711
2	1.4142	1
4	2	1.4142
8	2.8284	2
9	3	2.1213
16	4	2.8284
18	4.2426	3
25	5	3.5355
32	5.6569	4
50	7.0711	5

Y3=√(X/2)

(b)

We conclude that the graph of $Y_2 = \sqrt{2x}$ is obtained by multiplying the x-coordinate of each point on the graph of $Y_1 = \sqrt{x}$ by $\frac{1}{2}$. The graph of $Y_2 = \sqrt{2x}$ is the graph of $Y_1 = \sqrt{x}$ *compressed* horizontally.

Look at Table 13(b). Notice that (1, 1), (4, 2), and (9, 3) are points on the graph of $Y_1 = \sqrt{x}$. Also notice that (2, 1), (8, 2), and (18, 3) are points on the graph of $Y_3 = \sqrt{\frac{x}{2}}$. For a given y-coordinate, the x-coordinate on the graph of Y_3 is 2 times the x-coordinate on Y_1. We conclude that the graph of $Y_3 = \sqrt{\frac{x}{2}}$ is obtained by multiplying the x-coordinate of each point on the graph of $Y_1 = \sqrt{x}$ by 2. The graph of $Y_3 = \sqrt{\frac{x}{2}}$ is the graph of $Y_1 = \sqrt{x}$ *stretched* horizontally.

> **In Words**
> For $y = f(ax)$, $a > 0$, the factor a is "inside" the function, so it affects the x-coordinates. Multiply each x-coordinate on the graph of $y = f(x)$ by $\frac{1}{a}$.

Based on the Exploration, we have the following result:

> If the argument x of a function $y = f(x)$ is multiplied by a positive number a, then the graph of the new function $y = f(ax)$ is obtained by multiplying each x-coordinate on the graph of $y = f(x)$ by $\frac{1}{a}$. A **horizontal compression** results if $a > 1$, and a **horizontal stretch** results if $0 < a < 1$.

Let's look at an example.

EXAMPLE 8

Graphing Using Stretches and Compressions

The graph of $y = f(x)$ is given in Figure 76. Use this graph to find the graphs of

(a) $y = 2f(x)$ (b) $y = f(3x)$

Solution (a) The graph of $y = 2f(x)$ is obtained by multiplying each y-coordinate of $y = f(x)$ by 2. See Figure 77.

(b) The graph of $y = f(3x)$ is obtained from the graph of $y = f(x)$ by multiplying each x-coordinate of $y = f(x)$ by $\frac{1}{3}$. See Figure 78.

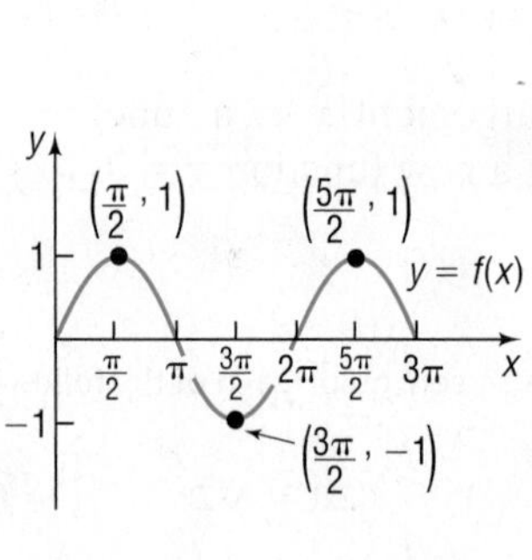

Figure 76 $y = f(x)$

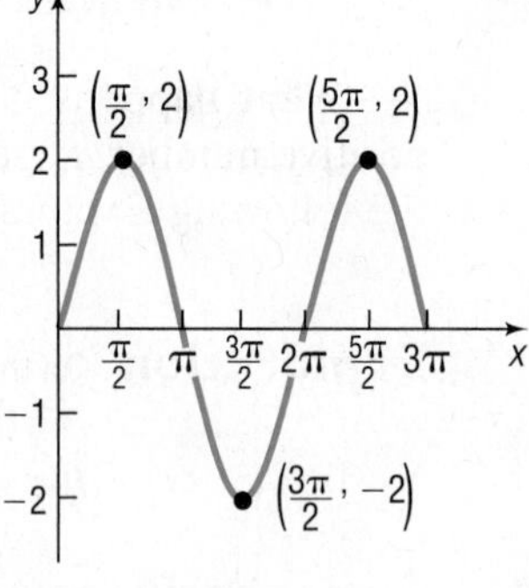

Figure 77 $y = 2f(x)$

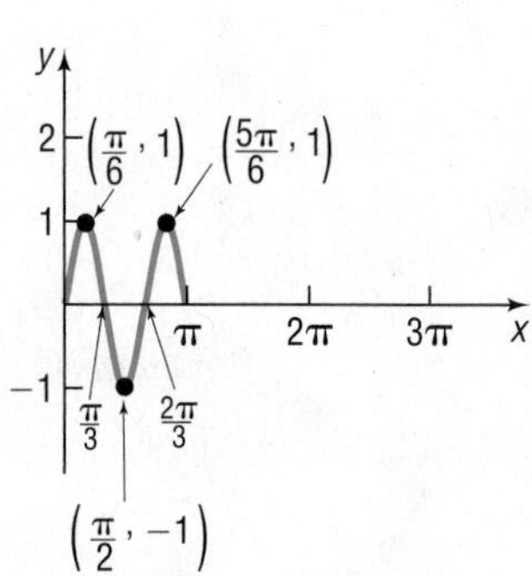

Figure 78 $y = f(3x)$

Now Work PROBLEMS 65(e) AND (g)

3 Graph Functions Using Reflections about the x-Axis and the y-Axis

EXAMPLE 9

Reflection about the x-Axis

Graph the function $f(x) = -x^2$. Find the domain and range of f.

Solution Begin with the graph of $y = x^2$, as shown in black in Figure 79. For each point (x, y) on the graph of $y = x^2$, the point $(x, -y)$ is on the graph of $y = -x^2$, as indicated in Table 14. Draw the graph of $y = -x^2$ by reflecting the graph of $y = x^2$ about the x-axis. See Figure 79.

Table 14

x	$y = x^2$	$y = -x^2$
−2	4	−4
−1	1	−1
0	0	0
1	1	−1
2	4	−4

Figure 79

The domain of f is all real numbers, or $(-\infty, \infty)$. The range of f is $(-\infty, 0]$.

When the right side of the function $y = f(x)$ is multiplied by -1, the graph of the new function $y = -f(x)$ is the **reflection about the x-axis** of the graph of the function $y = f(x)$.

Now Work PROBLEM 51

EXAMPLE 10 Reflection about the y-Axis

Graph the function $f(x) = \sqrt{-x}$. Find the domain and range of f.

Solution To get the graph of $f(x) = \sqrt{-x}$, begin with the graph of $y = \sqrt{x}$, as shown in Figure 80. For each point (x, y) on the graph of $y = \sqrt{x}$, the point $(-x, y)$ is on the graph of $y = \sqrt{-x}$. Obtain the graph of $y = \sqrt{-x}$ by reflecting the graph of $y = \sqrt{x}$ about the y-axis. See Figure 80.

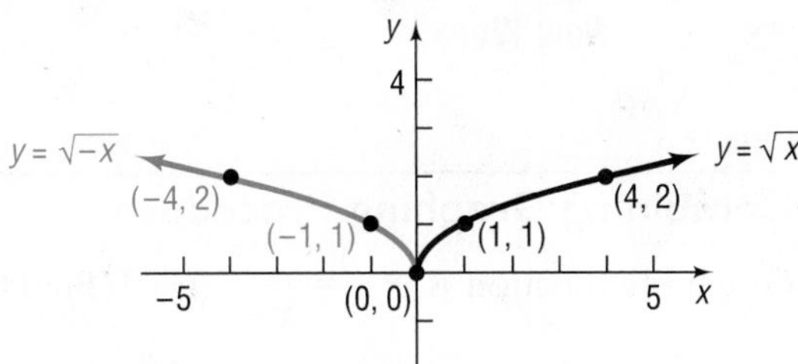

Figure 80

The domain of f is $(-\infty, 0]$. The range of f is the set of all nonnegative real numbers, or $[0, \infty)$. ●

In Words
For $y = -f(x)$, multiply each y-coordinate on the graph of $y = f(x)$ by -1.
For $y = f(-x)$, multiply each x-coordinate by -1.

When the graph of the function $y = f(x)$ is known, the graph of the new function $y = f(-x)$ is the **reflection about the y-axis** of the graph of the function $y = f(x)$.

SUMMARY OF GRAPHING TECHNIQUES

To Graph:	Draw the Graph of f and:	Functional Change to $f(x)$
Vertical shifts		
$y = f(x) + k, \quad k > 0$	Raise the graph of f by k units.	Add k to $f(x)$.
$y = f(x) - k, \quad k > 0$	Lower the graph of f by k units.	Subtract k from $f(x)$.
Horizontal shifts		
$y = f(x + h), \quad h > 0$	Shift the graph of f to the left h units.	Replace x by $x + h$.
$y = f(x - h), \quad h > 0$	Shift the graph of f to the right h units.	Replace x by $x - h$.
Compressing or stretching		
$y = af(x), \quad a > 0$	Multiply each y-coordinate of $y = f(x)$ by a. Stretch the graph of f vertically if $a > 1$. Compress the graph of f vertically if $0 < a < 1$.	Multiply $f(x)$ by a.
$y = f(ax), \quad a > 0$	Multiply each x-coordinate of $y = f(x)$ by $\frac{1}{a}$. Stretch the graph of f horizontally if $0 < a < 1$. Compress the graph of f horizontally if $a > 1$.	Replace x by ax.
Reflection about the x-axis		
$y = -f(x)$	Reflect the graph of f about the x-axis.	Multiply $f(x)$ by -1.
Reflection about the y-axis		
$y = f(-x)$	Reflect the graph of f about the y-axis.	Replace x by $-x$.

EXAMPLE 11 **Determining the Function Obtained from a Series of Transformations**

Find the function that is finally graphed after the following three transformations are applied to the graph of $y = |x|$.

1. Shift left 2 units
2. Shift up 3 units
3. Reflect about the y-axis

Solution

1. Shift left 2 units: Replace x by $x + 2$. $\quad y = |x + 2|$
2. Shift up 3 units: Add 3. $\quad y = |x + 2| + 3$
3. Reflect about the y-axis: Replace x by $-x$. $\quad y = |-x + 2| + 3$

Now Work PROBLEM 29

EXAMPLE 12 **Combining Graphing Procedures**

Graph the function $R(x) = \dfrac{3}{x-2} + 1$. Find the domain and range of R.

Solution It is helpful to write R as $R(x) = 3\left(\dfrac{1}{x-2}\right) + 1$. Now use the following steps to obtain the graph of R.

STEP 1: $y = \dfrac{1}{x}$ — Reciprocal function

STEP 2: $y = 3 \cdot \left(\dfrac{1}{x}\right) = \dfrac{3}{x}$ — Multiply by 3; vertical stretch by a factor of 3.

STEP 3: $y = \dfrac{3}{x-2}$ — Replace x by $x - 2$; horizontal shift to the right 2 units.

STEP 4: $y = \dfrac{3}{x-2} + 1$ — Add 1; vertical shift up 1 unit.

See Figure 81.

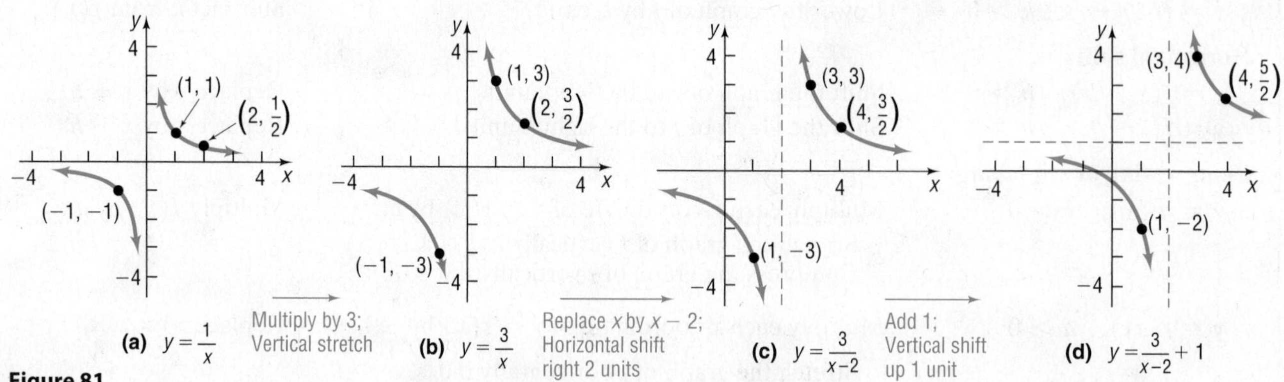

Figure 81

The domain of $y = \dfrac{1}{x}$ is $\{x \mid x \neq 0\}$ and its range is $\{y \mid y \neq 0\}$. Because we shifted right 2 units and up 1 unit to obtain the graph of R, the domain of R is $\{x \mid x \neq 2\}$ and its range is $\{y \mid y \neq 1\}$.

HINT: Although the order in which transformations are performed can be altered, consider using the following order for consistency:

1. Reflections
2. Compressions and stretches
3. Shifts

Other orderings of the steps shown in Example 12 would also result in the graph of R. For example, try this one:

STEP 1: $y = \dfrac{1}{x}$ — Reciprocal function

STEP 2: $y = \dfrac{1}{x-2}$ — Replace x by $x - 2$; horizontal shift to the right 2 units.

STEP 3: $y = \dfrac{3}{x-2}$ — Multiply by 3; vertical stretch by a factor of 3.

STEP 4: $y = \dfrac{3}{x-2} + 1$ — Add 1; vertical shift up 1 unit.

EXAMPLE 13

Combining Graphing Procedures

Graph the function $f(x) = \sqrt{1-x} + 2$. Find the domain and range of f.

Solution Because horizontal shifts require the form $x - h$, begin by rewriting $f(x)$ as $f(x) = \sqrt{1-x} + 2 = \sqrt{-(x-1)} + 2$. Now use the following steps.

STEP 1: $y = \sqrt{x}$ — Square root function

STEP 2: $y = \sqrt{-x}$ — Replace x by $-x$; reflect about the y-axis.

STEP 3: $y = \sqrt{-(x-1)} = \sqrt{1-x}$ — Replace x by $x - 1$; horizontal shift to the right 1 unit.

STEP 4: $y = \sqrt{1-x} + 2$ — Add 2; vertical shift up 2 units.

See Figure 82.

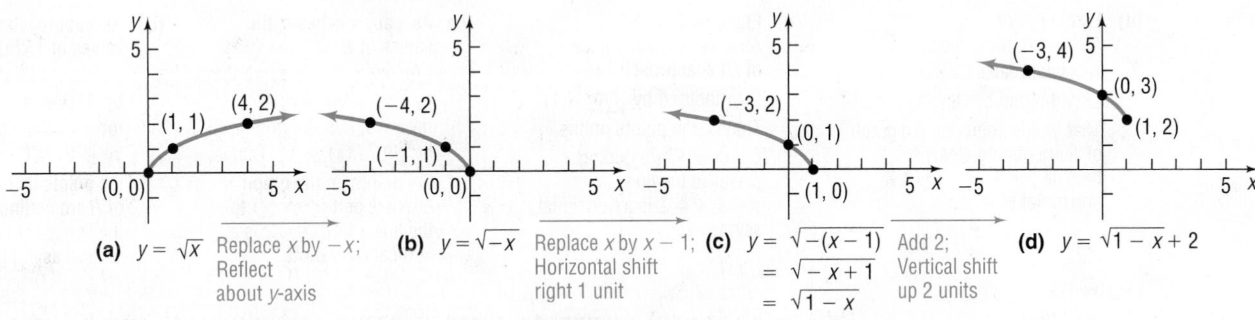

Figure 82

The domain of f is $(-\infty, 1]$ and the range is $[2, \infty)$. ●

Now Work PROBLEM 57

Asymptotes

Look back at Figure 81(d). Notice that as the values of x become more negative, that is, as x becomes **unbounded in the negative direction** ($x \to -\infty$, read as "**x approaches negative infinity**"), the values $R(x)$ approach 1. Also, as the values of x become **unbounded in the positive direction** ($x \to \infty$, read as "**x approaches infinity**"), the values of $R(x)$ also approach 1.That is,

As $x \to -\infty$, the values $R(x)$ approach 1. $\lim_{x \to -\infty} R(x) = 1$

As $x \to \infty$, the values $R(x)$ approach 1. $\lim_{x \to \infty} R(x) = 1$

This behavior of the graph is depicted by the horizontal line $y = 1$, called *a horizontal asymptote* of the graph.

DEFINITION

Let R denote a function. If, as $x \to -\infty$ or as $x \to \infty$, the values of $R(x)$ approach some fixed number, L, then the line $y = L$ is a **horizontal asymptote** of the graph of R.

Now look again at Figure 81(d). Notice that as x gets closer to 2, but remains less than 2, values of R are becoming unbounded in the negative direction; that is, $R(x) \to -\infty$. And as x gets closer to 2, but remains larger than 2, the values of R are becoming unbounded in the positive direction; that is $R(x) \to \infty$. We conclude that

$$\text{As } x \to 2, \text{ the values of } |R(x)| \to \infty \qquad \lim_{x \to 2} |R(x)| = \infty$$

This behavior of the graph is depicted by the vertical line $x = 2$, which is called a *vertical asymptote* of the graph.

DEFINITION

If, as x approaches some number c, the values $|R(x)| \to \infty$, then the line $x = c$ is a **vertical asymptote** of the graph of R. The graph of a function never intersects a vertical asymptote.

Even though the asymptotes of a function are not part of the graph of the function, they provide information about how the graph looks. Figure 83 illustrates some of the possibilities.

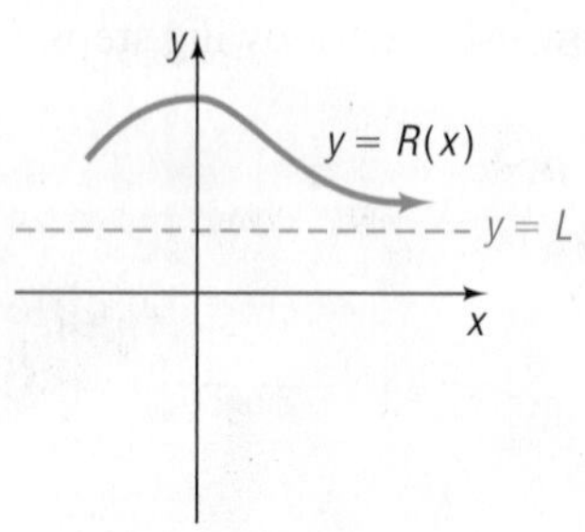

(a) End behavior: As $x \to \infty$, the values of $R(x)$ approach L [symbolized by $\lim_{x \to \infty} R(x) = L$]. That is, the points on the graph of R are getting closer to the line $y = L$; $y = L$ is a horizontal asymptote.

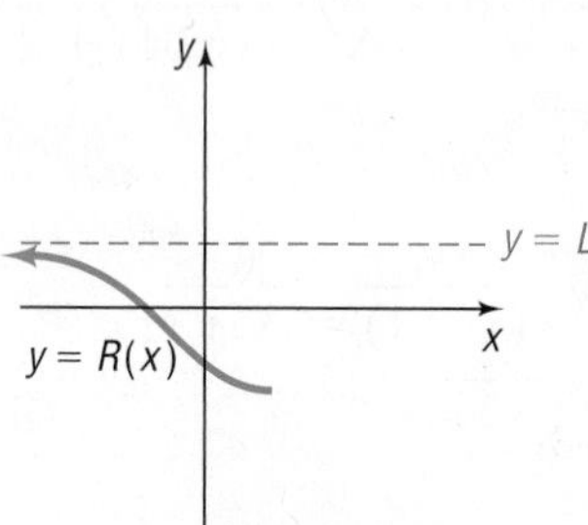

(b) End behavior: As $x \to -\infty$, the values of $R(x)$ approach L [symbolized by $\lim_{x \to -\infty} R(x) = L$]. That is, the points on the graph of R are getting closer to the line $y = L$; $y = L$ is a horizontal asymptote.

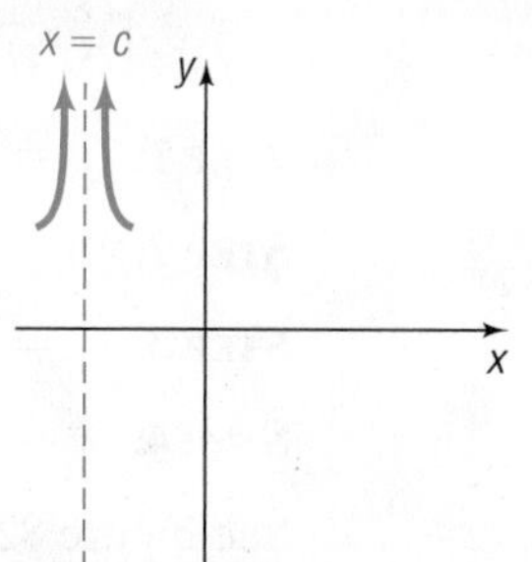

(c) As x approaches c, the values of $R(x) \to \infty$ [for $x < c$, this is symbolized by $\lim_{x \to c^-} R(x) = \infty$; for $x > c$, this is symbolized by $\lim_{x \to c^+} R(x) = \infty$]. That is, the points on the graph of R are getting closer to the line $x = c$; $x = c$ is a vertical asymptote.

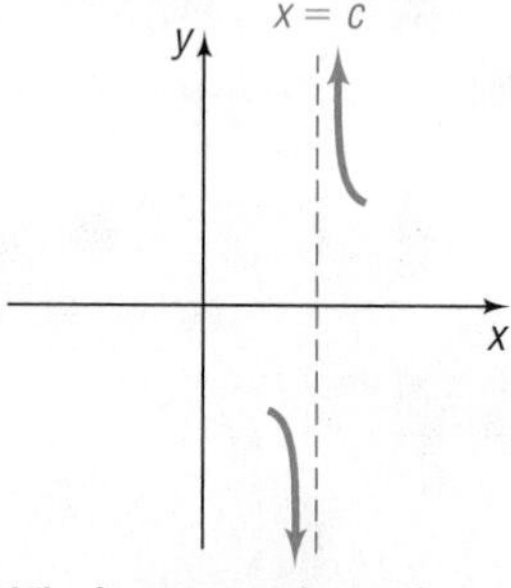

(d) As x approaches c, the values of $|R(x)| \to \infty$ [for $x < c$, this is symbolized by $\lim_{x \to c^-} R(x) = -\infty$; for $x > c$, this is symbolized by $\lim_{x \to c^+} R(x) = \infty$]. That is, the points on the graph of R are getting closer to the line $x = c$; $x = c$ is a vertical asymptote.

Figure 83

A horizontal asymptote, when it occurs, describes the end behavior of the graph as $x \to \infty$ or as $x \to -\infty$. **The graph of a function may intersect a horizontal asymptote.**

A vertical asymptote, when it occurs, describes the behavior of the graph when x is close to some number c. **The graph of a function will never intersect a vertical asymptote.**

1.6 Assess Your Understanding

Concepts and Vocabulary

1. Suppose that the graph of a function f is known. Then the graph of $y = f(x - 2)$ may be obtained by a(n) ________ shift of the graph of f to the _____ a distance of 2 units.
2. Suppose that the graph of a function f is known. Then the graph of $y = f(-x)$ may be obtained by a reflection about the _____-axis of the graph of the function $y = f(x)$.
3. ***True or False*** The graph of $y = \frac{1}{3}g(x)$ is the graph of $y = g(x)$ vertically stretched by a factor of 3.
4. ***True or False*** The graph of $y = -f(x)$ is the reflection about the x-axis of the graph of $y = f(x)$.
5. Which of the following functions has a graph that is the graph of $y = \sqrt{x}$ shifted down 3 units?
   (a) $y = \sqrt{x + 3}$ (b) $y = \sqrt{x} - 3$
   (c) $y = \sqrt{x} + 3$ (d) $y = \sqrt{x - 3}$
6. Which of the following functions has a graph that is the graph of $y = f(x)$ compressed horizontally by a factor of 4?
   (a) $y = f(4x)$ (b) $y = f\left(\frac{1}{4}x\right)$
   (c) $y = 4f(x)$ (d) $y = \frac{1}{4}f(x)$

7. ***True or False*** The graph of a function may cross a vertical asymptote.

8. If the line $y = L$ is a horizontal asymptote of the graph of a function R, then as $x \to \infty$, __________.

(a) the values of $|R(x)| \to -\infty$
(b) the values of $R(x) \to L$
(c) the values of $|R(x)| \to \infty$
(d) the values of $R(x) \to 0$

Skill Building

In Problems 9–20, match each graph to one of the following functions:

A. $y = x^2 + 2$ B. $y = -x^2 + 2$ C. $y = |x| + 2$ D. $y = -|x| + 2$
E. $y = (x - 2)^2$ F. $y = -(x + 2)^2$ G. $y = |x - 2|$ H. $y = -|x + 2|$
I. $y = 2x^2$ J. $y = -2x^2$ K. $y = 2|x|$ L. $y = -2|x|$

9. 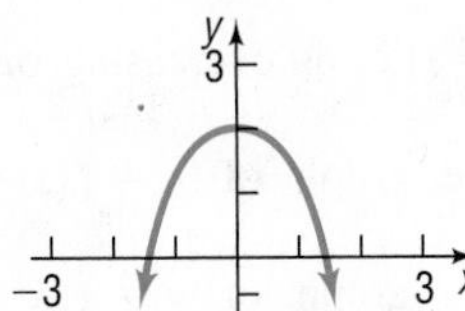

10. 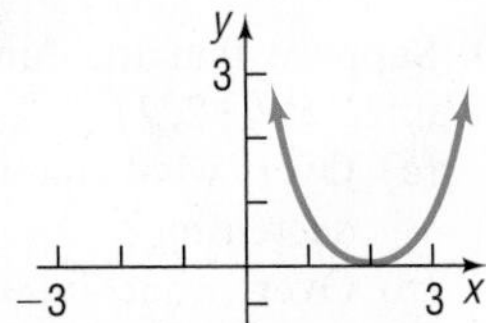

11. 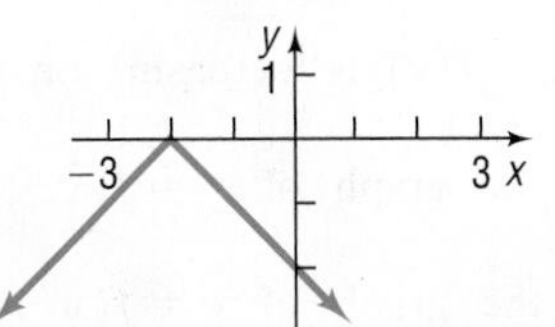

12. 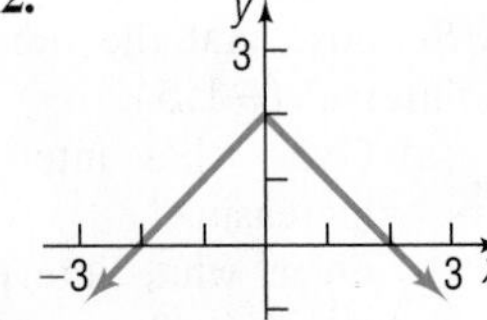

13. 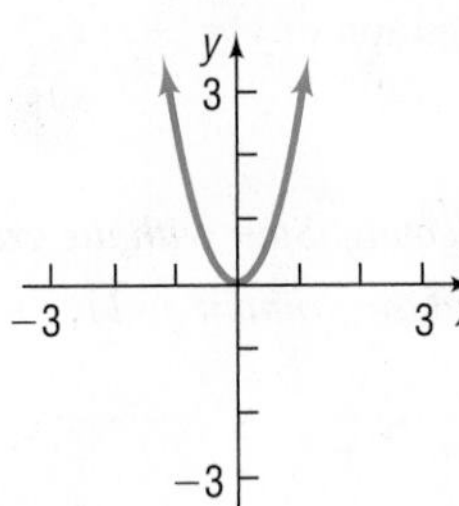

14. 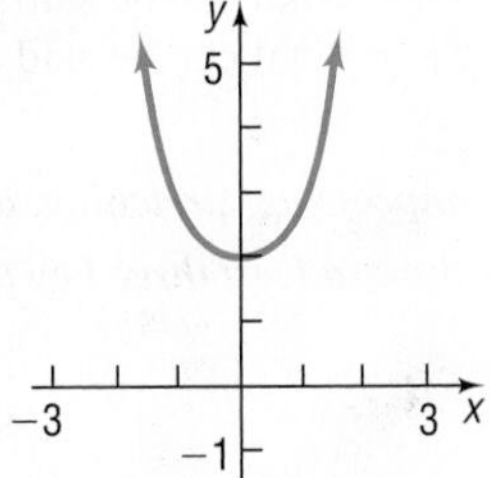

15. 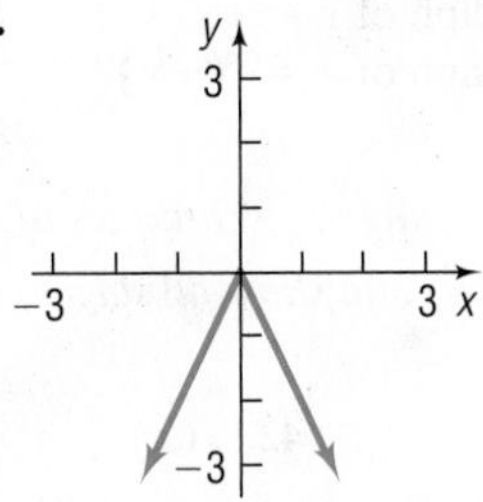

16. 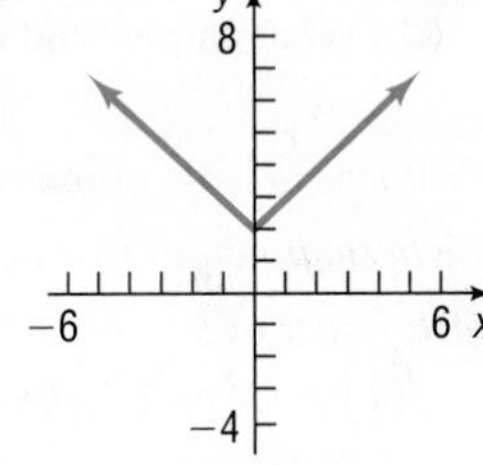

17. 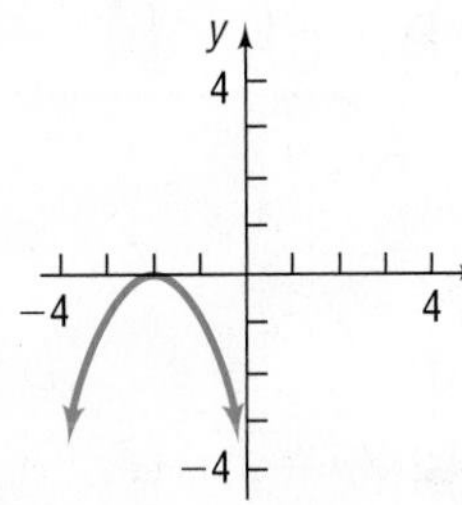

18. 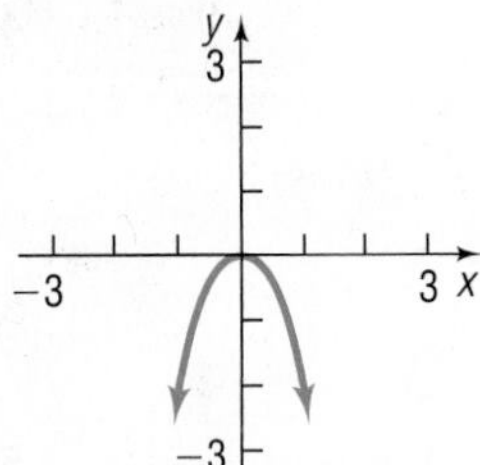

19. 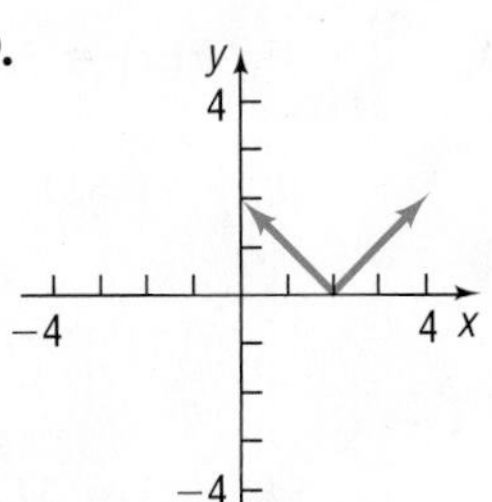

20. 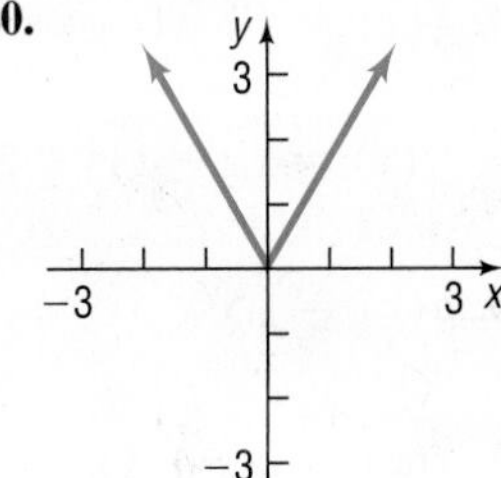

In Problems 21–28, write the function whose graph is the graph of $y = x^3$, but is:

21. Shifted to the right 4 units
22. Shifted to the left 4 units
23. Shifted up 4 units
24. Shifted down 4 units
25. Reflected about the y-axis
26. Reflected about the x-axis
27. Vertically stretched by a factor of 4
28. Horizontally stretched by a factor of 4

In Problems 29–32, find the function that is finally graphed after each of the following transformations is applied to the graph of $y = \sqrt{x}$ in the order stated.

29. (1) Shift up 2 units
(2) Reflect about the x-axis
(3) Reflect about the y-axis

30. (1) Reflect about the x-axis
(2) Shift right 3 units
(3) Shift down 2 units

31. (1) Reflect about the x-axis
(2) Shift up 2 units
(3) Shift left 3 units

32. (1) Shift up 2 units
(2) Reflect about the y-axis
(3) Shift left 3 units

33. If $(3, 6)$ is a point on the graph of $y = f(x)$, which of the following points must be on the graph of $y = -f(x)$?
(a) $(6, 3)$ (b) $(6, -3)$
(c) $(3, -6)$ (d) $(-3, 6)$

34. If $(3, 6)$ is a point on the graph of $y = f(x)$, which of the following points must be on the graph of $y = f(-x)$?
(a) $(6, 3)$ (b) $(6, -3)$
(c) $(3, -6)$ (d) $(-3, 6)$

35. If $(1, 3)$ is a point on the graph of $y = f(x)$, which of the following points must be on the graph of $y = 2f(x)$?

(a) $\left(1, \frac{3}{2}\right)$ (b) $(2, 3)$
(c) $(1, 6)$ (d) $\left(\frac{1}{2}, 3\right)$

36. If $(4, 2)$ is a point on the graph of $y = f(x)$, which of the following points must be on the graph of $y = f(2x)$?

(a) $(4, 1)$ (b) $(8, 2)$
(c) $(2, 2)$ (d) $(4, 4)$

37. Suppose that the x-intercepts of the graph of $y = f(x)$ are -5 and 3.
(a) What are the x-intercepts of the graph of $y = f(x+2)$?
(b) What are the x-intercepts of the graph of $y = f(x-2)$?
(c) What are the x-intercepts of the graph of $y = 4f(x)$?
(d) What are the x-intercepts of the graph of $y = f(-x)$?

38. Suppose that the x-intercepts of the graph of $y = f(x)$ are -8 and 1.
(a) What are the x-intercepts of the graph of $y = f(x+4)$?
(b) What are the x-intercepts of the graph of $y = f(x-3)$?
(c) What are the x-intercepts of the graph of $y = 2f(x)$?
(d) What are the x-intercepts of the graph of $y = f(-x)$?

39. Suppose that the function $y = f(x)$ is increasing on the interval $(-1, 5)$.
(a) Over what interval is the graph of $y = f(x + 2)$ increasing?
(b) Over what interval is the graph of $y = f(x - 5)$ increasing?
(c) What can be said about the graph of $y = -f(x)$?
(d) What can be said about the graph of $y = f(-x)$?

40. Suppose that the function $y = f(x)$ is decreasing on the interval $(-2, 7)$.
(a) Over what interval is the graph of $y = f(x + 2)$ decreasing?
(b) Over what interval is the graph of $y = f(x - 5)$ decreasing?
(c) What can be said about the graph of $y = -f(x)$?
(d) What can be said about the graph of $y = f(-x)$?

In Problems 41–64, graph each function using the techniques of shifting, compressing, stretching, and/or reflecting. Start with the graph of the basic function (for example, $y = x^2$) and show all stages. Be sure to show at least three key points. Find the domain and the range of each function.

41. $f(x) = x^2 - 1$
42. $f(x) = x^2 + 4$
43. $g(x) = x^3 + 1$
44. $g(x) = x^3 - 1$
45. $h(x) = \sqrt{x + 2}$
46. $h(x) = \sqrt{x + 1}$
47. $f(x) = (x - 1)^3 + 2$
48. $f(x) = (x + 2)^3 - 3$
49. $g(x) = 4\sqrt{x}$
50. $g(x) = \frac{1}{2}\sqrt{x}$
51. $f(x) = -\sqrt[3]{x}$
52. $f(x) = -\sqrt{x}$
53. $f(x) = 2(x + 1)^2 - 3$
54. $f(x) = 3(x - 2)^2 + 1$
55. $g(x) = 2\sqrt{x - 2} + 1$
56. $g(x) = 3|x + 1| - 3$
57. $h(x) = \sqrt{-x} - 2$
58. $h(x) = \frac{4}{x} + 2$
59. $f(x) = -(x + 1)^3 - 1$
60. $f(x) = -4\sqrt{x - 1}$
61. $g(x) = 2|1 - x|$
62. $g(x) = 4\sqrt{2 - x}$
63. $h(x) = \frac{1}{2x}$
64. $h(x) = \sqrt[3]{x - 1} + 3$

In Problems 65–68, the graph of a function f is illustrated. Use the graph of f as the first step toward graphing each of the following functions:

(a) $F(x) = f(x) + 3$ *(b)* $G(x) = f(x + 2)$ *(c)* $P(x) = -f(x)$ *(d)* $H(x) = f(x + 1) - 2$
(e) $Q(x) = \frac{1}{2}f(x)$ *(f)* $g(x) = f(-x)$ *(g)* $h(x) = f(2x)$

65.

y
4
(0, 2) (2, 2)
(4, 0)
−4 2 x
−2
(−4, −2)

66.

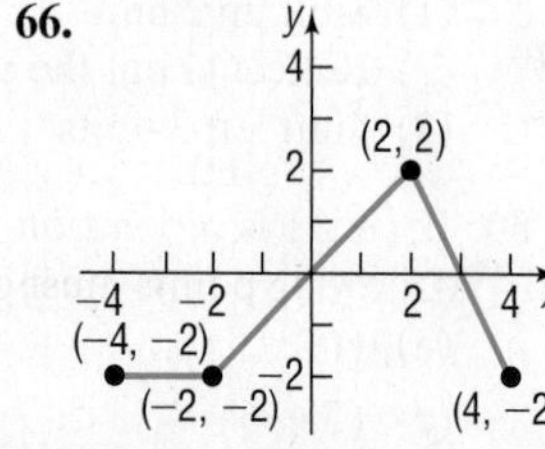

67.

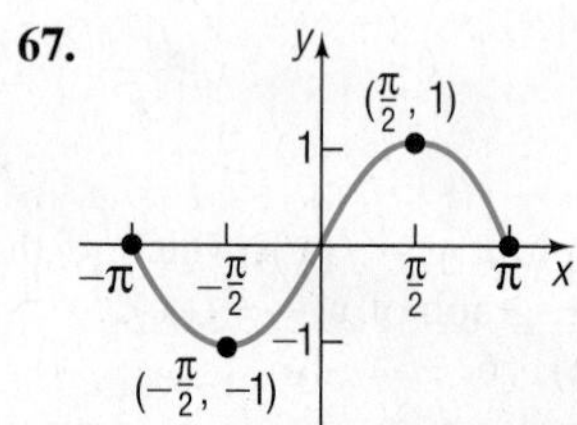

68.

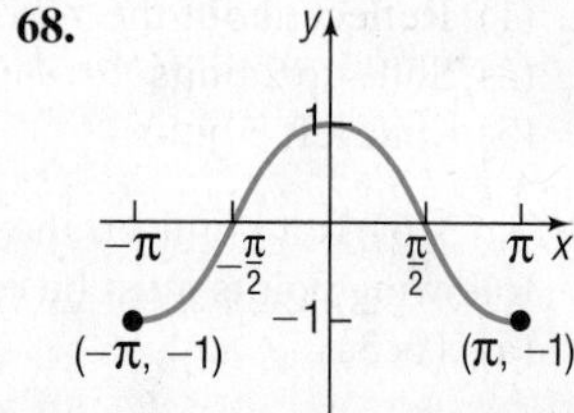

Mixed Practice

In Problems 69–76, complete the square of each quadratic expression. Then graph each function using the technique of shifting. (If necessary, refer to Appendix A, Section A.3 to review completing the square.)

69. $f(x) = x^2 + 2x$ **70.** $f(x) = x^2 - 6x$ **71.** $f(x) = x^2 - 8x + 1$ **72.** $f(x) = x^2 + 4x + 2$

73. $f(x) = 2x^2 - 12x + 19$ **74.** $f(x) = 3x^2 + 6x + 1$ **75.** $f(x) = -3x^2 - 12x - 17$ **76.** $f(x) = -2x^2 - 12x - 13$

Applications and Extensions

77. The graph of a function f is illustrated in the figure.
(a) Draw the graph of $y = |f(x)|$.
(b) Draw the graph of $y = f(|x|)$.

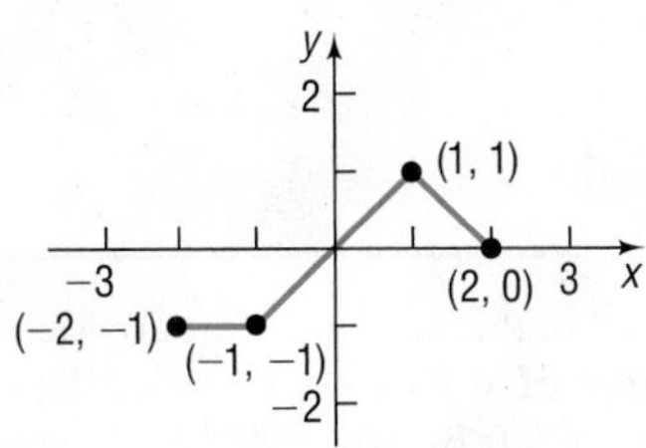

78. The graph of a function f is illustrated in the figure.
(a) Draw the graph of $y = |f(x)|$.
(b) Draw the graph of $y = f(|x|)$.

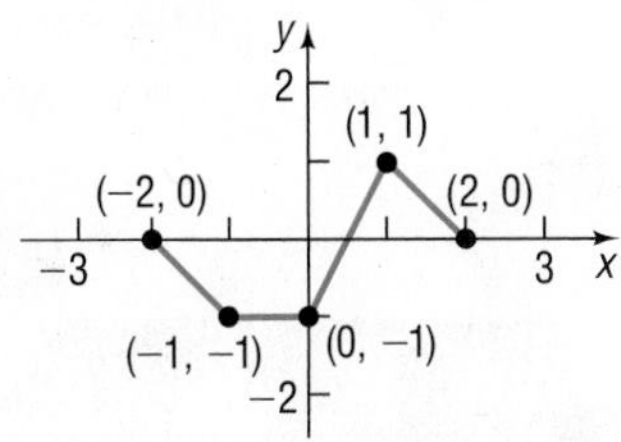

79. Suppose $(1, 3)$ is a point on the graph of $y = f(x)$.
(a) What point is on the graph of $y = f(x + 3) - 5$?
(b) What point is on the graph of $y = -2f(x - 2) + 1$?
(c) What point is on the graph of $y = f(2x + 3)$?

80. Suppose $(-3, 5)$ is a point on the graph of $y = g(x)$.
(a) What point is on the graph of $y = g(x + 1) - 3$?
(b) What point is on the graph of $y = -3g(x - 4) + 3$?
(c) What point is on the graph of $y = g(3x + 9)$?

81. (a) Graph $f(x) = |x - 3| - 3$ using transformations.
(b) Find the area of the region that is bounded by f and the x-axis and lies below the x-axis.

82. (a) Graph $f(x) = -2|x - 4| + 4$ using transformations.
(b) Find the area of the region that is bounded by f and the x-axis and lies above the x-axis.

83. Thermostat Control Energy conservation experts estimate that homeowners can save 5% to 10% on winter heating bills by programming their thermostats 5 to 10 degrees lower while sleeping. In the graph (top, right), the temperature T (in degrees Fahrenheit) of a home is given as a function of time t (in hours after midnight) over a 24-hour period.
(a) At what temperature is the thermostat set during daytime hours? At what temperature is the thermostat set overnight?

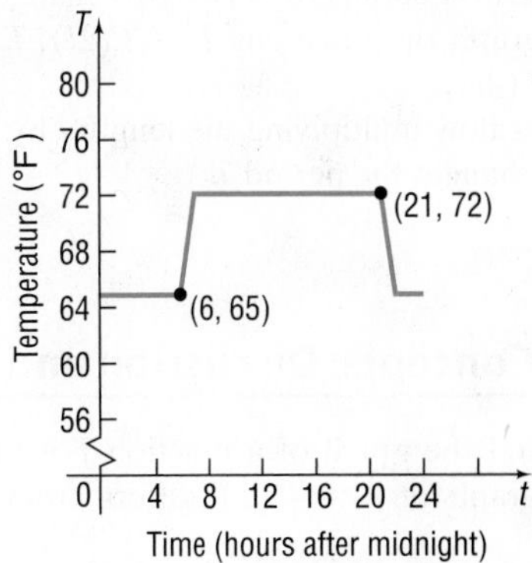

(b) The homeowner reprograms the thermostat to $y = T(t) - 2$. Explain how this affects the temperature in the house. Graph this new function.
(c) The homeowner reprograms the thermostat to $y = T(t + 1)$. Explain how this affects the temperature in the house. Graph this new function.

Source: Roger Albright, *547 Ways to Be Fuel Smart,* 2000

84. Digital Music Revenues The total projected worldwide digital music revenues R, in millions of dollars, for the years 2012 through 2017 can be estimated by the function

$$R(x) = 28.6x^2 + 300x + 4843$$

where x is the number of years after 2012.
(a) Find $R(0)$, $R(3)$, and $R(5)$ and explain what each value represents.
(b) Find $r(x) = R(x - 2)$.
(c) Find $r(2)$, $r(5)$, and $r(7)$ and explain what each value represents.
(d) In the model $r = r(x)$, what does x represent?
(e) Would there be an advantage in using the model r when estimating the projected revenues for a given year instead of the model R?

Source: IFPI Digital Music Report 2013

85. Temperature Measurements The relationship between the Celsius (°C) and Fahrenheit (°F) scales for measuring temperature is given by the equation

$$F = \frac{9}{5}C + 32$$

The relationship between the Celsius (°C) and Kelvin (K) scales is $K = C + 273$. Graph the equation $F = \frac{9}{5}C + 32$ using degrees Fahrenheit on the y-axis and degrees Celsius on the x-axis. Use the techniques introduced in this section to obtain the graph showing the relationship between Kelvin and Fahrenheit temperatures.

86. Period of a Pendulum The period T (in seconds) of a simple pendulum is a function of its length l (in feet) defined by the equation

$$T = 2\pi\sqrt{\frac{l}{g}}$$

where $g \approx 32.2$ feet per second per second is the acceleration due to gravity.

(a) Use a graphing utility to graph the function $T = T(l)$.
(b) Now graph the functions $T = T(l + 1)$, $T = T(l + 2)$, and $T = T(l + 3)$.
(c) Discuss how adding to the length l changes the period T.
(d) Now graph the functions $T = T(2l)$, $T = T(3l)$, and $T = T(4l)$.
(e) Discuss how multiplying the length l by factors of 2, 3, and 4 changes the period T.

87. The equation $y = (x - c)^2$ defines a *family of parabolas*, one parabola for each value of c. On one set of coordinate axes, graph the members of the family for $c = 0$, $c = 3$, and $c = -2$.

88. Repeat Problem 87 for the family of parabolas $y = x^2 + c$.

Explaining Concepts: Discussion and Writing

89. Suppose that the graph of a function f is known. Explain how the graph of $y = 4f(x)$ differs from the graph of $y = f(4x)$.

90. Suppose that the graph of a function f is known. Explain how the graph of $y = f(x) - 2$ differs from the graph of $y = f(x - 2)$.

91. The area under the curve $y = \sqrt{x}$ bounded from below by the x-axis and on the right by $x = 4$ is $\frac{16}{3}$ square units. Using the ideas presented in this section, what do you think is the area under the curve of $y = \sqrt{-x}$ bounded from below by the x-axis and on the left by $x = -4$? Justify your answer.

92. Explain how the range of the function $f(x) = x^2$ compares to the range of $g(x) = f(x) + k$.

93. Explain how the domain of $g(x) = \sqrt{x}$ compares to the domain of $g(x - k)$, where $k \geq 0$.

1.7 One-to-One Functions; Inverse Functions

PREPARING FOR THIS SECTION *Before getting started, review the following:*

- Functions and Their Graphs (Section 1.3, pp. 23–35)

Now Work the 'Are You Prepared?' problems on page 87.

OBJECTIVES
1 Determine Whether a Function Is One-to-One (p. 78)
2 Determine the Inverse of a Function Defined by a Map or a Set of Ordered Pairs (p. 81)
3 Obtain the Graph of the Inverse Function from the Graph of the Function (p. 83)
4 Find the Inverse of a Function Defined by an Equation (p. 84)

1 Determine Whether a Function Is One-to-One

Section 1.3 presented four different ways to represent a function: (1) a map, (2) a set of ordered pairs, (3) a graph, and (4) an equation. For example, Figures 84 and 85 illustrate two different functions represented as mappings. The function in Figure 84 shows the correspondence between states and their populations (in millions). The function in Figure 85 shows a correspondence between animals and life expectancies (in years).

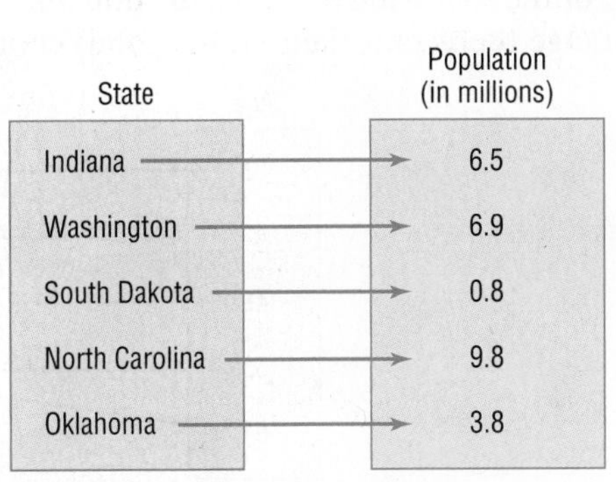

Figure 84

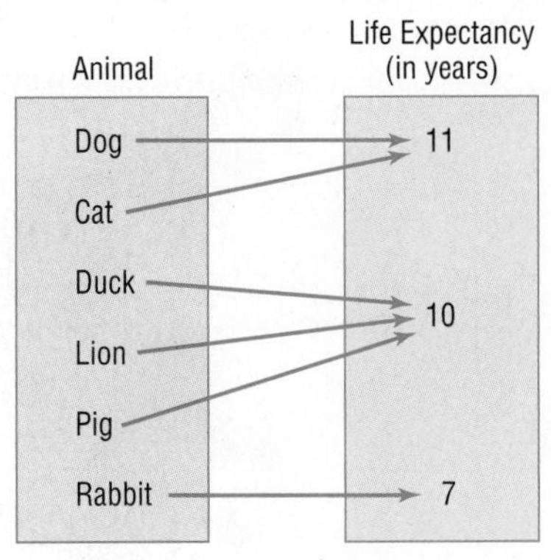

Figure 85

Suppose several people are asked to name a state that has a population of 0.8 million based on the function in Figure 84. Everyone will respond "South Dakota." Now, if the same people are asked to name an animal whose life expectancy is 11 years based on the function in Figure 85, some may respond "dog," while others may respond "cat." What is the difference between the functions in Figures 84 and 85? In Figure 84, no two elements in the domain correspond to the same element in the range. In Figure 85, this is not the case: Different elements in the domain correspond to the same element in the range. Functions such as the one in Figure 84 are given a special name.

DEFINITION

A function is **one-to-one** if any two different inputs in the domain correspond to two different outputs in the range. That is, if x_1 and x_2 are two different inputs of a function f, then f is one-to-one if $f(x_1) \neq f(x_2)$.

> **In Words**
> A function is not one-to-one if two different inputs correspond to the same output.

Put another way, a function f is one-to-one if no y in the range is the image of more than one x in the domain. A function is not one-to-one if any two (or more) different elements in the domain correspond to the same element in the range. So the function in Figure 85 is not one-to-one because two different elements in the domain, *dog* and *cat*, both correspond to 11 (and also because three different elements in the domain correspond to 10). Figure 86 illustrates the distinction among one-to-one functions, functions that are not one-to-one, and relations that are not functions.

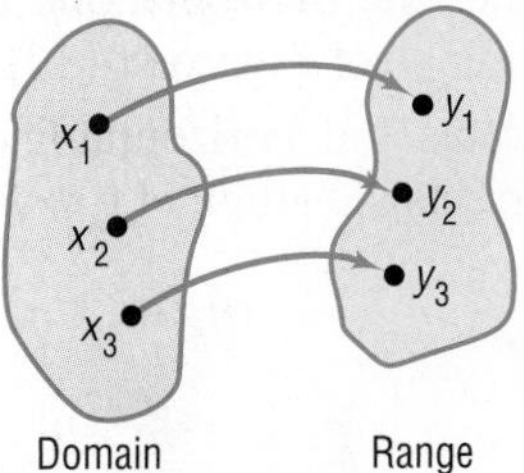

Domain Range

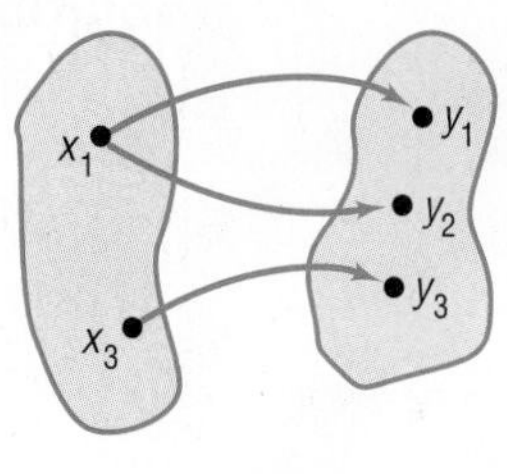

Figure 86

(a) One-to-one function: Each x in the domain has one and only one image in the range.

Domain Range

(b) Not a one-to-one function: y_1 is the image of both x_1 and x_2.

(c) Not a function: x_1 has two images, y_1 and y_2.

EXAMPLE 1 Determining Whether a Function Is One-to-One

Determine whether the functions given on page 80 are one-to-one.

(a) For the following function, the domain represents the ages of five males, and the range represents their HDL (good) cholesterol scores (mg/dL).

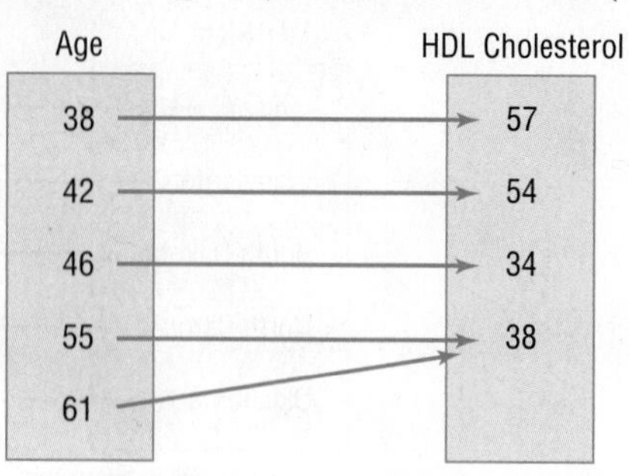

(b) $\{(-2, 6), (-1, 3), (0, 2), (1, 5), (2, 8)\}$

Solution (a) The function is not one-to-one because there are two different inputs, 55 and 61, that correspond to the same output, 38.

(b) The function is one-to-one because no two distinct inputs correspond to the same output. ●

Now Work PROBLEMS 13 AND 17

For functions defined by an equation $y = f(x)$ and for which the graph of f is known, there is a simple test, called the **horizontal-line test**, to determine whether f is one-to-one.

THEOREM

Horizontal-line Test

If every horizontal line intersects the graph of a function f in at most one point, then f is one-to-one.

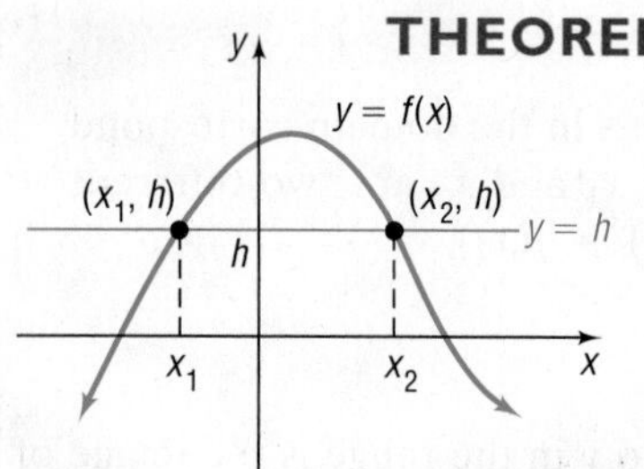

Figure 87
$f(x_1) = f(x_2) = h$ and $x_1 \neq x_2$; f is not a one-to-one function.

The reason why this test works can be seen in Figure 87, where the horizontal line $y = h$ intersects the graph at two distinct points, (x_1, h) and (x_2, h). Since h is the image of both x_1 and x_2 and $x_1 \neq x_2$, f is not one-to-one. Based on Figure 87, we can state the horizontal-line test in another way: If the graph of any horizontal line intersects the graph of a function f at more than one point, then f is not one-to-one.

EXAMPLE 2 Using the Horizontal-line Test

For each function, use its graph to determine whether the function is one-to-one.

(a) $f(x) = x^2$ (b) $g(x) = x^3$

Solution (a) Figure 88(a) illustrates the horizontal-line test for $f(x) = x^2$. The horizontal line $y = 1$ intersects the graph of f twice, at $(1, 1)$ and at $(-1, 1)$, so f is not one-to-one.

(b) Figure 88(b) illustrates the horizontal-line test for $g(x) = x^3$. Because every horizontal line intersects the graph of g exactly once, it follows that g is one-to-one.

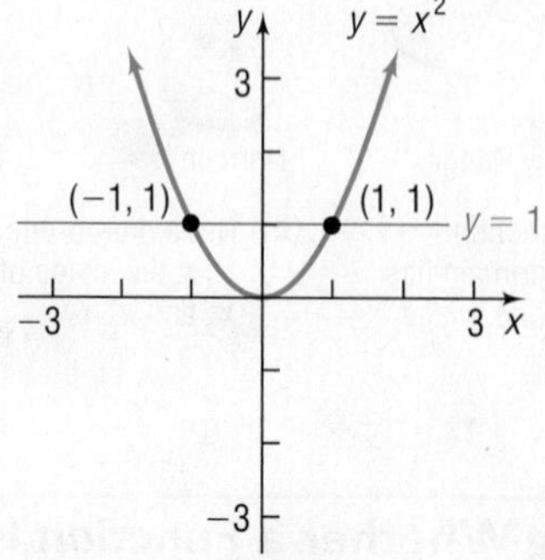

Figure 88 **(a)** A horizontal line intersects the graph twice; f is not one-to-one. **(b)** Every horizontal line intersects the graph exactly once; g is one-to-one. ●

Now Work PROBLEM 21

Look more closely at the one-to-one function $g(x) = x^3$. This function is an increasing function. Because an increasing (or decreasing) function will always have different y-values for unequal x-values, it follows that a function that is increasing (or decreasing) over its domain is also a one-to-one function.

THEOREM

A function that is increasing on an interval I is a one-to-one function on I.
A function that is decreasing on an interval I is a one-to-one function on I.

2 Determine the Inverse of a Function Defined by a Map or a Set of Ordered Pairs

DEFINITION

Suppose that f is a one-to-one function. Then, corresponding to each x in the domain of f, there is exactly one y in the range (because f is a function); and corresponding to each y in the range of f, there is exactly one x in the domain (because f is one-to-one). The correspondence from the range of f back to the domain of f is called the **inverse function of f**. The symbol f^{-1} is used to denote the inverse function of f.

In Words

Suppose that f is a one-to-one function so that the input 5 corresponds to the output 10. In the inverse function f^{-1}, the input 10 will correspond to the output 5.

We will discuss how to find inverses for all four representations of functions: (1) maps, (2) sets of ordered pairs, (3) graphs, and (4) equations. We begin with finding inverses of functions represented by maps or sets of ordered pairs.

EXAMPLE 3 **Finding the Inverse of a Function Defined by a Map**

Find the inverse of the function defined by the map below. Let the domain of the function represent certain states, and let the range represent the states' populations (in millions). Find the domain and the range of the inverse function.

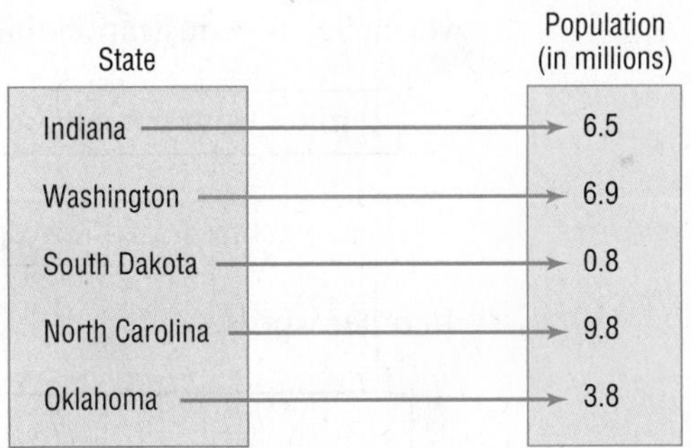

Solution The function is one-to-one. To find the inverse function, interchange the elements in the domain with the elements in the range. For example, the function receives as input Indiana and outputs 6.5 million. So the inverse receives as input 6.5 million and outputs Indiana. The inverse function is shown next.

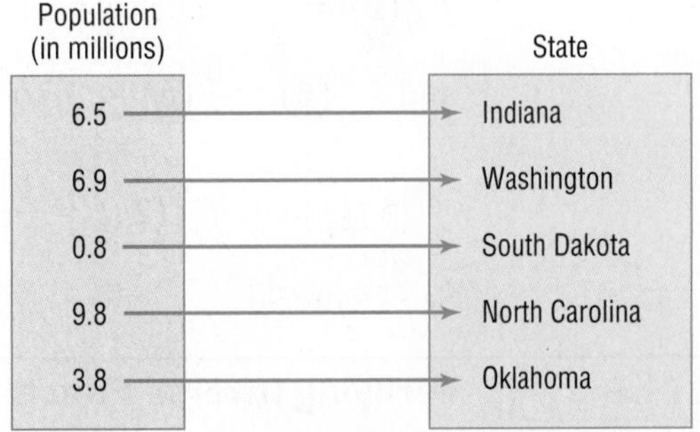

The domain of the inverse function is $\{6.5, 6.9, 0.8, 9.8, 3.8\}$. The range of the inverse function is $\{$Indiana, Washington, South Dakota, North Carolina, Oklahoma$\}$. ●

If the function f is a set of ordered pairs (x, y), then the inverse function of f, denoted f^{-1}, is the set of ordered pairs (y, x).

EXAMPLE 4 **Finding the Inverse of a Function Defined by a Set of Ordered Pairs**

Find the inverse of the following one-to-one function:

$$\{(-3,-27),(-2,-8),(-1,-1),(0,0),(1,1),(2,8),(3,27)\}$$

State the domain and the range of the function and its inverse.

Solution The inverse of the given function is found by interchanging the entries in each ordered pair and so is given by

$$\{(-27,-3),(-8,-2),(-1,-1),(0,0),(1,1),(8,2),(27,3)\}$$

The domain of the function is $\{-3,-2,-1,0,1,2,3\}$. The range of the function is $\{-27,-8,-1,0,1,8,27\}$. The domain of the inverse function is $\{-27,-8,-1,0,1,8,27\}$. The range of the inverse function is $\{-3,-2,-1,0,1,2,3\}$. ●

Now Work PROBLEMS 27 AND 31

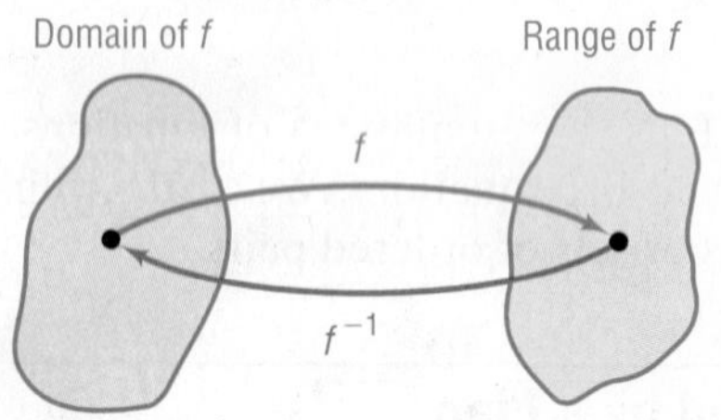

Figure 89

Remember, if f is a one-to-one function, it has an inverse function, f^{-1}. See Figure 89.

Based on the results of Example 4 and Figure 89, two facts are now apparent about a one-to-one function f and its inverse f^{-1}.

Domain of f = Range of f^{-1} Range of f = Domain of f^{-1}

Look again at Figure 89 to visualize the relationship. Starting with x, applying f, and then applying f^{-1} gets x back again. Starting with x, applying f^{-1}, and then applying f gets the number x back again. To put it simply, what f does, f^{-1} undoes, and vice versa. See the illustration that follows.

WARNING Be careful! f^{-1} is a symbol for the inverse function of f. The -1 used in f^{-1} is not an exponent. That is, f^{-1} *does not* mean the reciprocal of f; $f^{-1}(x)$ is not equal to $\dfrac{1}{f(x)}$. ■

Input x from domain of f $\xrightarrow{\text{Apply } f}$ $f(x)$ $\xrightarrow{\text{Apply } f^{-1}}$ $f^{-1}(f(x)) = x$

Input x from domain of f^{-1} $\xrightarrow{\text{Apply } f^{-1}}$ $f^{-1}(x)$ $\xrightarrow{\text{Apply } f}$ $f(f^{-1}(x)) = x$

In other words,

$$f^{-1}(f(x)) = x \quad \text{where } x \text{ is in the domain of } f$$
$$f(f^{-1}(x)) = x \quad \text{where } x \text{ is in the domain of } f^{-1}$$

Consider the function $f(x) = 2x$, which multiplies the argument x by 2. The inverse function f^{-1} undoes whatever f does. So the inverse function of f is $f^{-1}(x) = \frac{1}{2}x$, which divides the argument by 2. For example, $f(3) = 2(3) = 6$ and $f^{-1}(6) = \frac{1}{2}(6) = 3$, so f^{-1} undoes what f did. This is verified by showing that

$$f^{-1}(f(x)) = f^{-1}(2x) = \frac{1}{2}(2x) = x \quad \text{and} \quad f(f^{-1}(x)) = f\left(\frac{1}{2}x\right) = 2\left(\frac{1}{2}x\right) = x$$

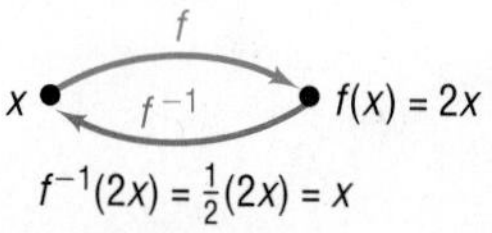

See Figure 90.

x $\xrightarrow{f}$ $f(x) = 2x$; $f^{-1}(2x) = \frac{1}{2}(2x) = x$

Figure 90

EXAMPLE 5 **Verifying Inverse Functions**

(a) Verify that the inverse of $g(x) = x^3$ is $g^{-1}(x) = \sqrt[3]{x}$.

(b) Verify that the inverse of $f(x) = 2x + 3$ is $f^{-1}(x) = \frac{1}{2}(x - 3)$.

Solution (a) $g^{-1}(g(x)) = g^{-1}(x^3) = \sqrt[3]{x^3} = x$ for all x in the domain of g

$g(g^{-1}(x)) = g(\sqrt[3]{x}) = (\sqrt[3]{x})^3 = x$ for all x in the domain of g^{-1}

(b) $f^{-1}(f(x)) = f^{-1}(2x + 3) = \frac{1}{2}[(2x + 3) - 3] = \frac{1}{2}(2x) = x$ for all x in the domain of f

$f(f^{-1}(x)) = f\left(\frac{1}{2}(x - 3)\right) = 2\left[\frac{1}{2}(x - 3)\right] + 3 = (x - 3) + 3 = x$ for all x in the domain of f^{-1}

EXAMPLE 6 **Verifying Inverse Functions**

Verify that the inverse of $f(x) = \frac{1}{x - 1}$ is $f^{-1}(x) = \frac{1}{x} + 1$. For what values of x is $f^{-1}(f(x)) = x$? For what values of x is $f(f^{-1}(x)) = x$?

Solution The domain of f is $\{x|x \neq 1\}$ and the domain of f^{-1} is $\{x|x \neq 0\}$. Now

$$f^{-1}(f(x)) = f^{-1}\left(\frac{1}{x - 1}\right) = \frac{1}{\frac{1}{x - 1}} + 1 = x - 1 + 1 = x \quad \text{provided } x \neq 1$$

$$f(f^{-1}(x)) = f\left(\frac{1}{x} + 1\right) = \frac{1}{\left(\frac{1}{x} + 1\right) - 1} = \frac{1}{\frac{1}{x}} = x \quad \text{provided } x \neq 0$$

Now Work PROBLEMS 35 AND 39

3 Obtain the Graph of the Inverse Function from the Graph of the Function

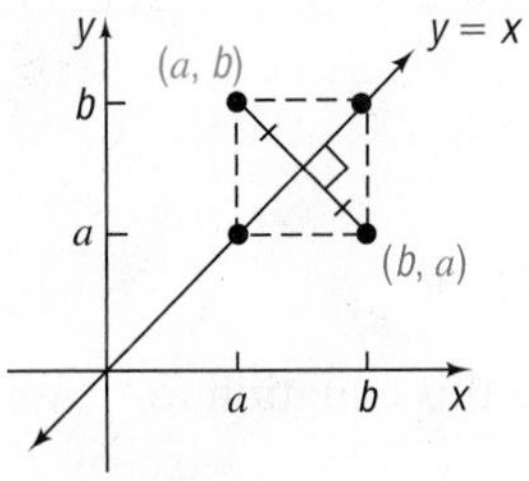

Suppose that (a, b) is a point on the graph of a one-to-one function f defined by $y = f(x)$. Then $b = f(a)$. This means that $a = f^{-1}(b)$, so (b, a) is a point on the graph of the inverse function f^{-1}. The relationship between the point (a, b) on f and the point (b, a) on f^{-1} is shown in Figure 91. The line segment with endpoints (a, b) and (b, a) is perpendicular to the line $y = x$ and is bisected by the line $y = x$. (Do you see why?) It follows that the point (b, a) on f^{-1} is the reflection about the line $y = x$ of the point (a, b) on f.

Figure 91

THEOREM The graph of a one-to-one function f and the graph of its inverse function f^{-1} are symmetric with respect to the line $y = x$.

Figure 92 illustrates this result. Once the graph of f is known, the graph of f^{-1} may be obtained by reflecting the graph of f about the line $y = x$.

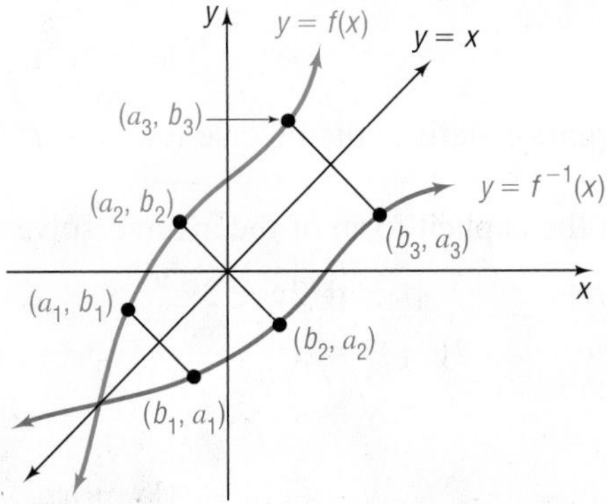

Figure 92

EXAMPLE 7 **Graphing the Inverse Function**

The graph in Figure 93(a) is that of a one-to-one function $y = f(x)$. Draw the graph of its inverse.

Solution Begin by adding the graph of $y = x$ to Figure 93(a). Since the points $(-2, -1)$, $(-1, 0)$, and $(2, 1)$ are on the graph of f, the points $(-1, -2)$, $(0, -1)$, and $(1, 2)$ must be on the graph of f^{-1}. Keeping in mind that the graph of f^{-1} is the reflection about the line $y = x$ of the graph of f, draw the graph of f^{-1}. See Figure 93(b).

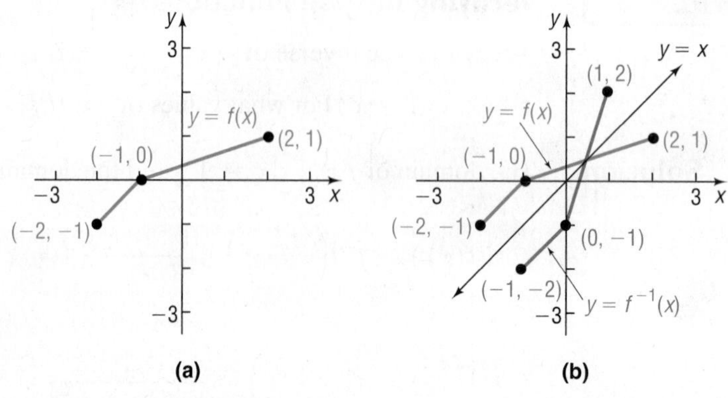

Figure 93 (a) (b)

Now Work PROBLEM 45

4 Find the Inverse of a Function Defined by an Equation

The fact that the graphs of a one-to-one function f and its inverse function f^{-1} are symmetric with respect to the line $y = x$ tells us more. It says that we can obtain f^{-1} by interchanging the roles of x and y in f. Look again at Figure 92. If f is defined by the equation

$$y = f(x)$$

then f^{-1} is defined by the equation

$$x = f(y)$$

The equation $x = f(y)$ defines f^{-1} *implicitly*. If we can solve this equation for y, we will have the *explicit* form of f^{-1}, that is,

$$y = f^{-1}(x)$$

Let's use this procedure to find the inverse of $f(x) = 2x + 3$. (Because f is a linear function and is increasing, f is one-to-one and so has an inverse function.)

EXAMPLE 8 **How to Find the Inverse Function**

Step-by-Step Solution

Find the inverse of $f(x) = 2x + 3$. Graph f and f^{-1} on the same coordinate axes.

Step 1: Replace $f(x)$ with y. In $y = f(x)$, interchange the variables x and y to obtain $x = f(y)$. This equation defines the inverse function f^{-1} implicitly.

Replace $f(x)$ with y in $f(x) = 2x + 3$ and obtain $y = 2x + 3$. Now interchange the variables x and y to obtain

$$x = 2y + 3$$

This equation defines the inverse function f^{-1} implicitly.

Step 2: If possible, solve the implicit equation for y in terms of x to obtain the explicit form of f^{-1}, $y = f^{-1}(x)$.

To find the explicit form of the inverse, solve $x = 2y + 3$ for y.

$$x = 2y + 3$$

$$2y + 3 = x \quad \text{Reflexive Property; If } a = b, \text{ then } b = a.$$

$$2y = x - 3 \quad \text{Subtract 3 from both sides.}$$

$$y = \frac{1}{2}(x - 3) \quad \text{Multiply both sides by } \frac{1}{2}.$$

The explicit form of the inverse function f^{-1} is

$$f^{-1}(x) = \frac{1}{2}(x - 3)$$

Step 3: Check the result by showing that $f^{-1}(f(x)) = x$ and $f(f^{-1}(x)) = x$.

We verified that f and f^{-1} are inverses in Example 5(b).

The graphs of $f(x) = 2x + 3$ and its inverse $f^{-1}(x) = \frac{1}{2}(x - 3)$ are shown in Figure 94. Note the symmetry of the graphs with respect to the line $y = x$. ●

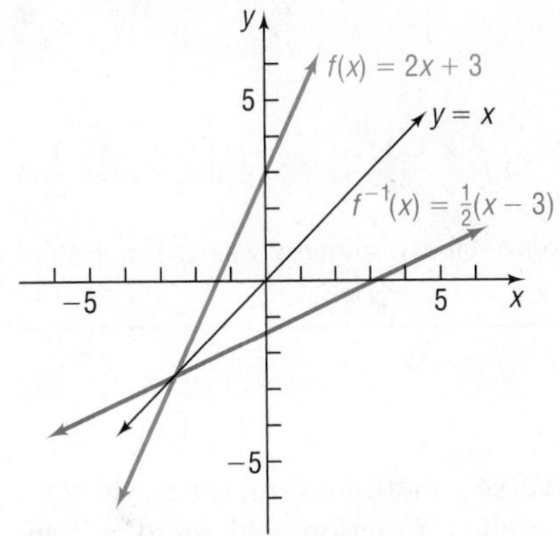

Figure 94

Procedure for Finding the Inverse of a One-to-One Function

STEP 1: In $y = f(x)$, interchange the variables x and y to obtain

$$x = f(y)$$

This equation defines the inverse function f^{-1} implicitly.

STEP 2: If possible, solve the implicit equation for y in terms of x to obtain the explicit form of f^{-1}:

$$y = f^{-1}(x)$$

STEP 3: Check the result by showing that

$$f^{-1}(f(x)) = x \quad \text{and} \quad f(f^{-1}(x)) = x$$

EXAMPLE 9 **Finding the Inverse Function**

The function

$$f(x) = \frac{2x + 1}{x - 1} \qquad x \neq 1$$

is one-to-one. Find its inverse function and check the result.

Solution **STEP 1:** Replace $f(x)$ with y and interchange the variables x and y in

$$y = \frac{2x + 1}{x - 1}$$

to obtain

$$x = \frac{2y + 1}{y - 1}$$

STEP 2: Solve for y.

$$\begin{aligned} x &= \frac{2y + 1}{y - 1} \\ x(y - 1) &= 2y + 1 && \text{Multiply both sides by } y - 1. \\ xy - x &= 2y + 1 && \text{Use the Distributive Property.} \\ xy - 2y &= x + 1 && \text{Subtract } 2y \text{ from both sides; add } x \text{ to both sides.} \\ (x - 2)y &= x + 1 && \text{Factor.} \\ y &= \frac{x + 1}{x - 2} && \text{Divide by } x - 2. \end{aligned}$$

The inverse function is

$$f^{-1}(x) = \frac{x + 1}{x - 2} \qquad x \neq 2 \qquad \text{Replace } y \text{ by } f^{-1}(x).$$

STEP 3: ✓Check:

$$f^{-1}(f(x)) = f^{-1}\left(\frac{2x+1}{x-1}\right) = \frac{\frac{2x+1}{x-1}+1}{\frac{2x+1}{x-1}-2} = \frac{2x+1+x-1}{2x+1-2(x-1)} = \frac{3x}{3} = x, \quad x \neq 1$$

$$f(f^{-1}(x)) = f\left(\frac{x+1}{x-2}\right) = \frac{2\left(\frac{x+1}{x-2}\right)+1}{\frac{x+1}{x-2}-1} = \frac{2(x+1)+x-2}{x+1-(x-2)} = \frac{3x}{3} = x, \quad x \neq 2$$

●

Exploration

In Example 9, we found that if $f(x) = \frac{2x+1}{x-1}$, then $f^{-1}(x) = \frac{x+1}{x-2}$. Compare the vertical and horizontal asymptotes of f and f^{-1}.

Result The vertical asymptote of f is $x = 1$, and the horizontal asymptote is $y = 2$. The vertical asymptote of f^{-1} is $x = 2$, and the horizontal asymptote is $y = 1$.

Now Work PROBLEMS 53 AND 67

If a function is not one-to-one, it has no inverse function. Sometimes, though, an appropriate restriction on the domain of such a function will yield a new function that *is* one-to-one. Then the function defined on the restricted domain has an inverse function. Let's look at an example of this common practice.

EXAMPLE 10

Finding the Inverse of a Domain-restricted Function

Find the inverse of $y = f(x) = x^2$ if $x \geq 0$. Graph f and f^{-1}.

Solution The function $y = x^2$ is not one-to-one. [Refer to Example 2(a).] However, restricting the domain of this function to $x \geq 0$, as indicated, results in a new function that is increasing and therefore is one-to-one. Consequently, the function defined by $y = f(x) = x^2, x \geq 0$, has an inverse function, f^{-1}.

Follow the steps given previously to find f^{-1}.

STEP 1: In the equation $y = x^2, x \geq 0$, interchange the variables x and y. The result is

$$x = y^2 \qquad y \geq 0$$

This equation defines the inverse function implicitly.

STEP 2: Solve for y to get the explicit form of the inverse. Because $y \geq 0$, only one solution for y is obtained: $y = \sqrt{x}$. So $f^{-1}(x) = \sqrt{x}$.

STEP 3: ✓Check: $f^{-1}(f(x)) = f^{-1}(x^2) = \sqrt{x^2} = |x| = x$ because $x \geq 0$

$$f(f^{-1}(x)) = f(\sqrt{x}) = (\sqrt{x})^2 = x$$

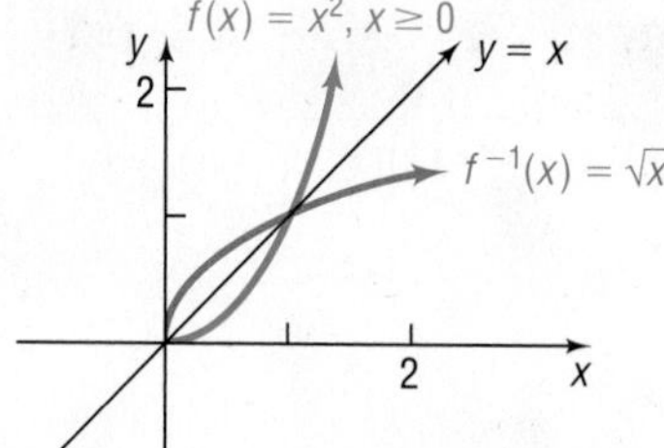

Figure 95

Figure 95 illustrates the graphs of $f(x) = x^2, x \geq 0$, and $f^{-1}(x) = \sqrt{x}$. ●

SUMMARY

1. If a function f is one-to-one, then it has an inverse function f^{-1}.
2. Domain of f = Range of f^{-1}; Range of f = Domain of f^{-1}.
3. To verify that f^{-1} is the inverse of f, show that $f^{-1}(f(x)) = x$ for every x in the domain of f and that $f(f^{-1}(x)) = x$ for every x in the domain of f^{-1}.
4. The graphs of f and f^{-1} are symmetric with respect to the line $y = x$.

1.7 Assess Your Understanding

'Are You Prepared?' *Answers are given at the end of these exercises. If you get a wrong answer, read the pages listed in red.*

1. Is the set of ordered pairs $\{(1,3), (2,3), (-1,2)\}$ a function? Why or why not? (pp. 24–26)
2. What is the range of the relation $\{(2,-1), (4,5), (3,8), (-4,0)\}$? (pp. 24–26)
3. What is the domain of $f(x) = \dfrac{x+5}{x^2+3x-18}$? (pp. 30–31)
4. Find the function $y = f(x)$ that is defined implicitly by the equation $x - 3y = -8$. (p. 29)

Concepts and Vocabulary

5. If x_1 and x_2 are two different inputs of a function f, then f is one-to-one if ________.
6. If every horizontal line intersects the graph of a function f at no more than one point, then f is a(n) ________ function.
7. If f is a one-to-one function and $f(3) = 8$, then $f^{-1}(8) =$ ________.
8. If f^{-1} denotes the inverse of a function f, then the graphs of f and f^{-1} are symmetric with respect to the line ________.
9. If the domain of a one-to-one function f is $[4, \infty)$, then the range of its inverse function f^{-1} is ________.
10. ***True or False*** If f and g are inverse functions, then the domain of f is the same as the range of g.
11. If $(-2, 3)$ is a point on the graph of a one-to-one function f, which of the following points is on the graph of f^{-1}?
(a) $(3,-2)$ (b) $(2,-3)$ (c) $(-3,2)$ (d) $(-2,-3)$
12. Suppose f is a one-to-one function with a domain of $\{x \mid x \neq 3\}$ and a range of $\left\{x \,\middle|\, x \neq \frac{2}{3}\right\}$. Which of the following is the domain of f^{-1}?
(a) $\{x \mid x \neq 3\}$ (b) All real numbers
(c) $\left\{x \,\middle|\, x \neq \frac{2}{3}, x \neq 3\right\}$ (d) $\left\{x \,\middle|\, x \neq \frac{2}{3}\right\}$

Skill Building

In Problems 13–20, determine whether the function is one-to-one.

13.
Domain → Range
20 Hours → $200
25 Hours → $300
30 Hours → $350
40 Hours → $425

14.
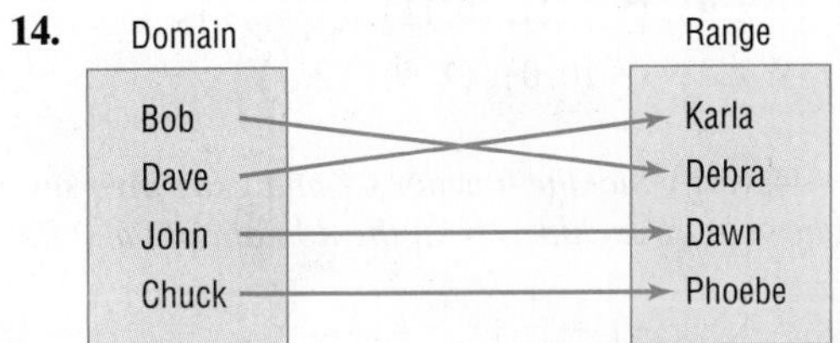

15.
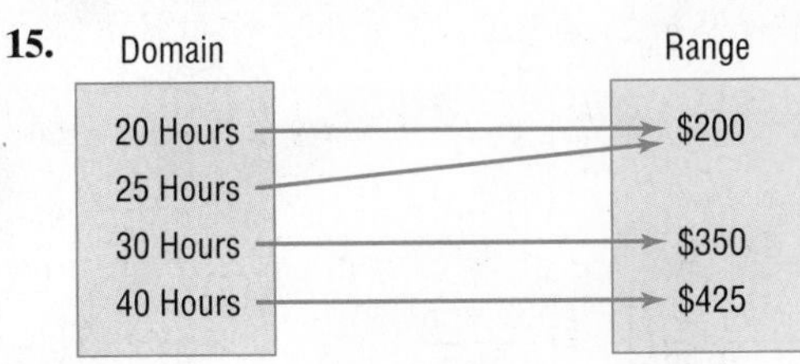

16.
Domain → Range
Bob → Karla
Dave → Debra
John → Phoebe
Chuck → Phoebe

17. $\{(2,6), (-3,6), (4,9), (1,10)\}$
18. $\{(-2,5), (-1,3), (3,7), (4,12)\}$
19. $\{(0,0), (1,1), (2,16), (3,81)\}$
20. $\{(1,2), (2,8), (3,18), (4,32)\}$

In Problems 21–26, the graph of a function f is given. Use the horizontal-line test to determine whether f is one-to-one.

21.
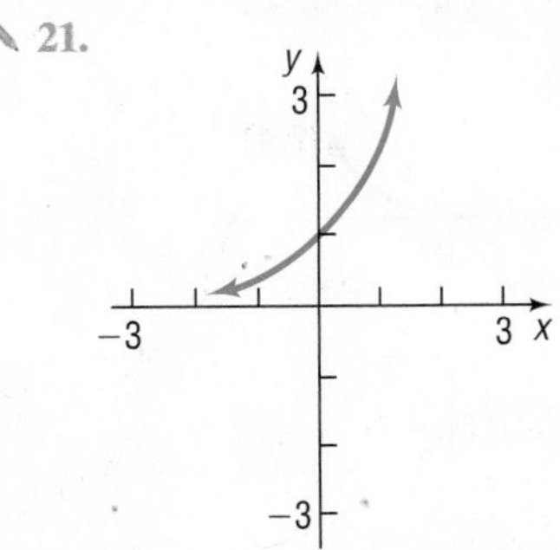

22.
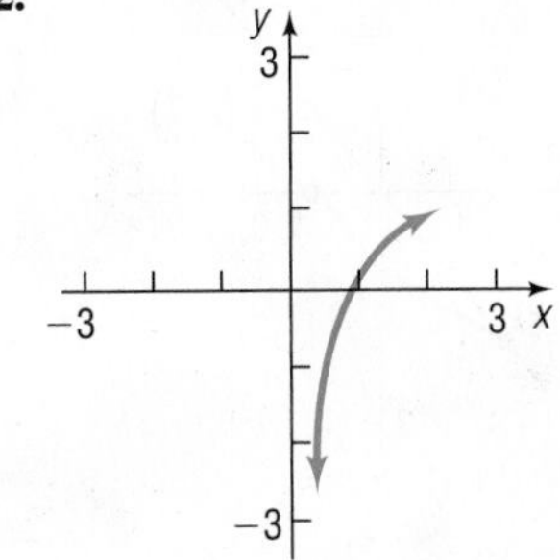

23.
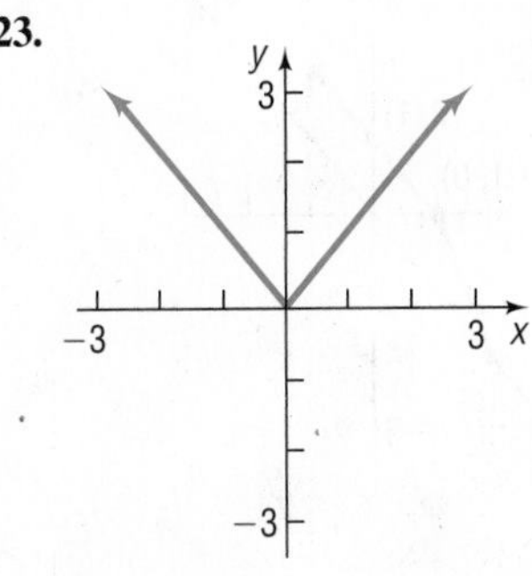

24.

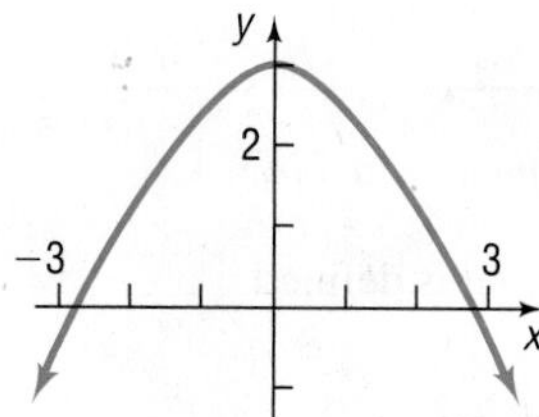

25.

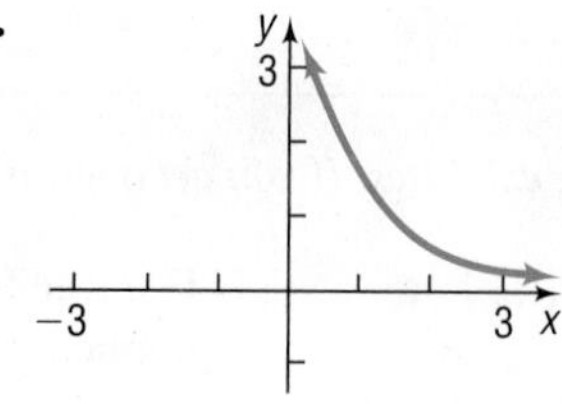

26.

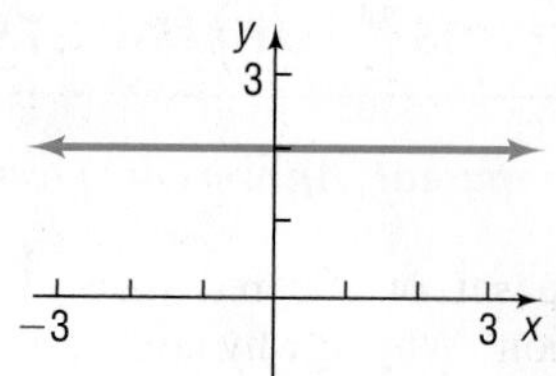

In Problems 27–34, find the inverse of each one-to-one function. State the domain and the range of each inverse function.

27.

Location	Annual Precipitation (inches)
Atlanta, GA	49.7
Boston, MA	43.8
Las Vegas, NV	4.2
Miami, FL	61.9
Los Angeles, CA	12.8

Source: currentresults.com

28.

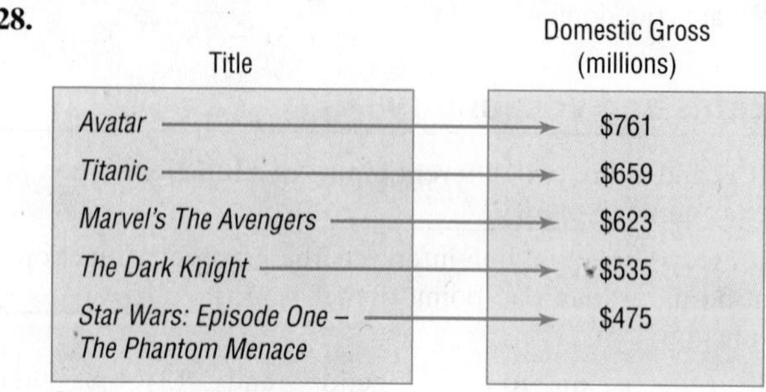

Source: boxofficemojo.com

29.

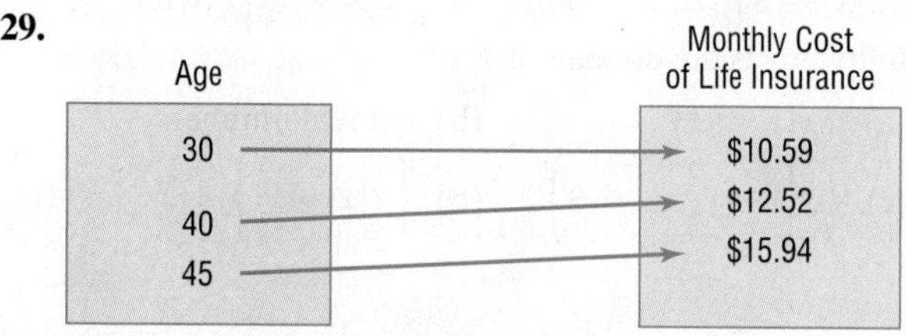

Source: tiaa-cref.org

30.

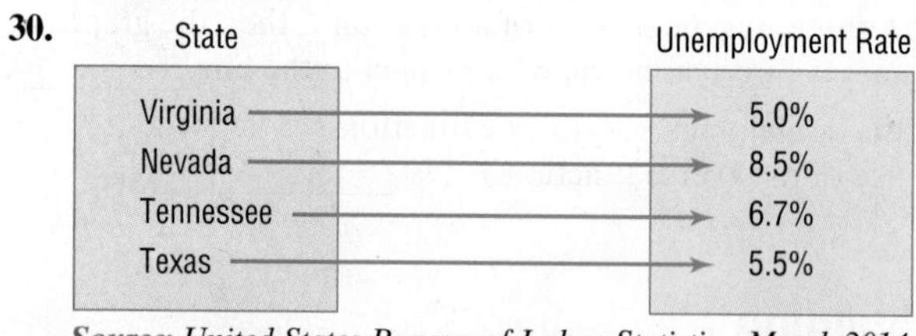

Source: United States Bureau of Labor Statistics, March 2014

31. $\{(-3,5),(-2,9),(-1,2),(0,11),(1,-5)\}$

32. $\{(-2,2),(-1,6),(0,8),(1,-3),(2,9)\}$

33. $\{(-2,1),(-3,2),(-10,0),(1,9),(2,4)\}$

34. $\{(-2,-8),(-1,-1),(0,0),(1,1),(2,8)\}$

In Problems 35–44, verify that the functions f and g are inverses of each other by showing that $f(g(x)) = x$ and $g(f(x)) = x$. Give any values of x that need to be excluded from the domain of f and the domain of g.

35. $f(x) = 3x + 4;\ \ g(x) = \frac{1}{3}(x - 4)$

36. $f(x) = 3 - 2x;\ \ g(x) = -\frac{1}{2}(x - 3)$

37. $f(x) = 4x - 8;\ \ g(x) = \frac{x}{4} + 2$

38. $f(x) = 2x + 6;\ \ g(x) = \frac{1}{2}x - 3$

39. $f(x) = x^3 - 8;\ \ g(x) = \sqrt[3]{x + 8}$

40. $f(x) = (x - 2)^2, x \geq 2;\ \ g(x) = \sqrt{x} + 2$

41. $f(x) = \frac{1}{x};\ \ g(x) = \frac{1}{x}$

42. $f(x) = x;\ \ g(x) = x$

43. $f(x) = \frac{2x + 3}{x + 4};\ \ g(x) = \frac{4x - 3}{2 - x}$

44. $f(x) = \frac{x - 5}{2x + 3};\ \ g(x) = \frac{3x + 5}{1 - 2x}$

In Problems 45–50, the graph of a one-to-one function f is given. Draw the graph of the inverse function f^{-1}.

45.

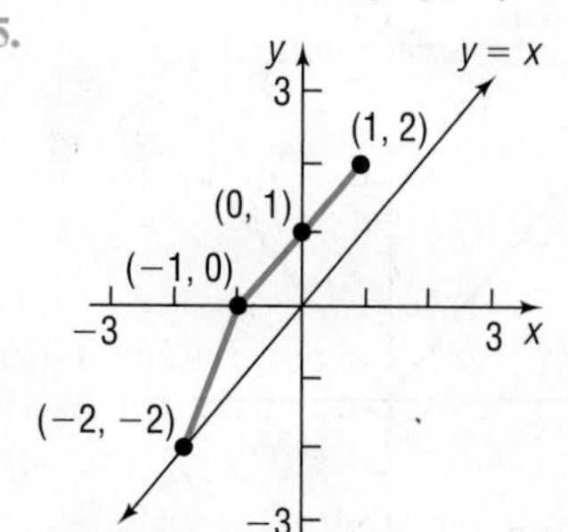

46.

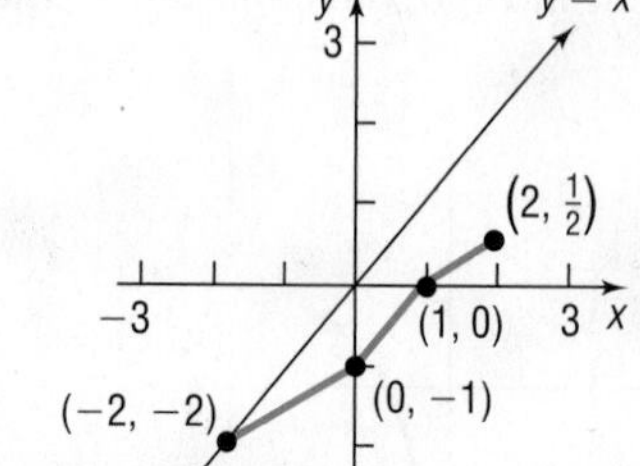

47.

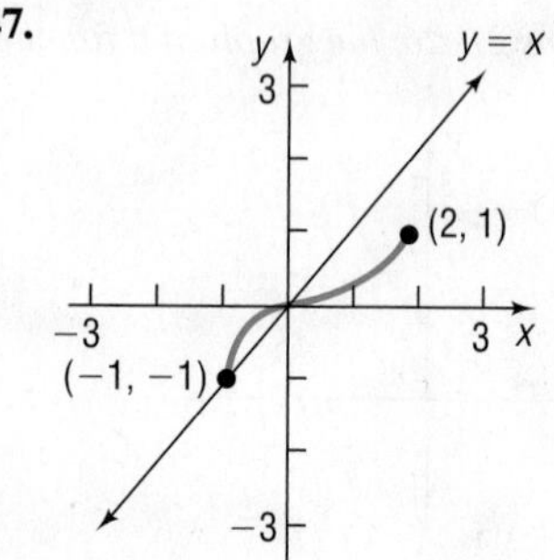

48.

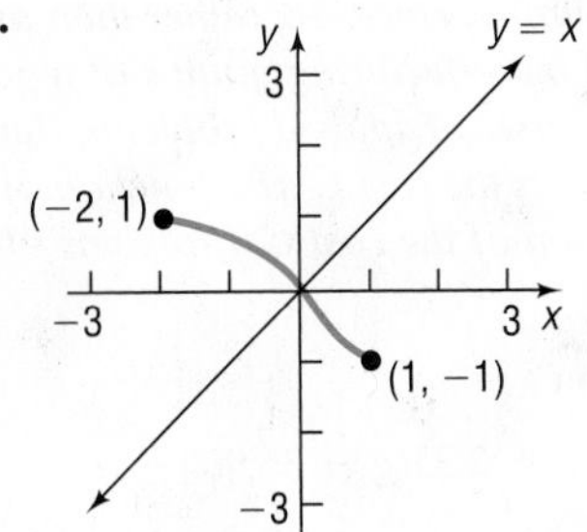

49.

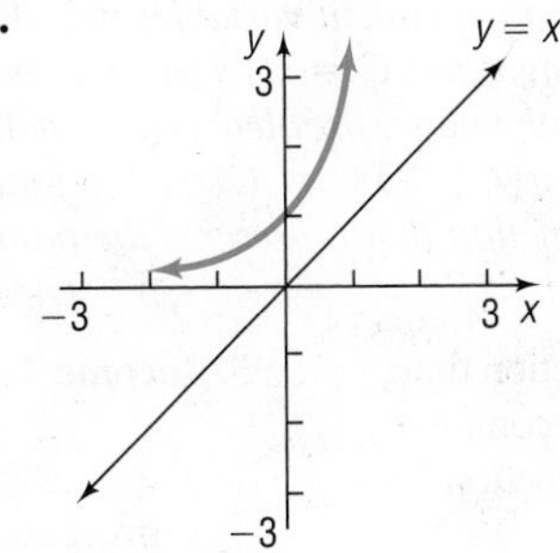

50.

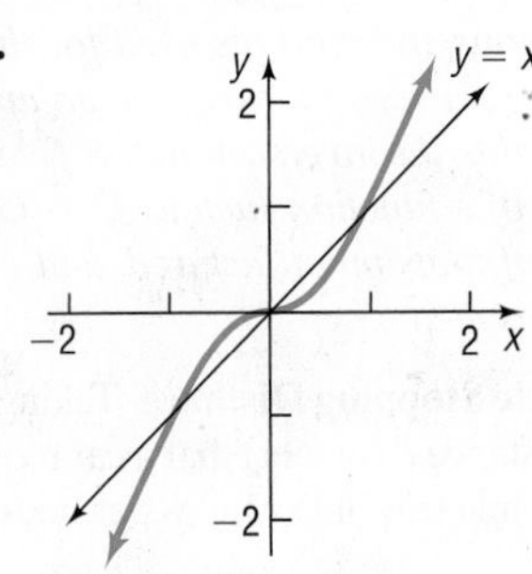

In Problems 51–62, the function f is one-to-one. (a) Find its inverse function f^{-1} and check your answer. (b) Find the domain and the range of f and f^{-1}. (c) Graph f, f^{-1}, and $y = x$ on the same coordinate axes.

51. $f(x) = 3x$ **52.** $f(x) = -4x$ **53.** $f(x) = 4x + 2$

54. $f(x) = 1 - 3x$ **55.** $f(x) = x^3 - 1$ **56.** $f(x) = x^3 + 1$

57. $f(x) = x^2 + 4, \quad x \ge 0$ **58.** $f(x) = x^2 + 9, \quad x \ge 0$ **59.** $f(x) = \dfrac{4}{x}$

60. $f(x) = -\dfrac{3}{x}$ **61.** $f(x) = \dfrac{1}{x - 2}$ **62.** $f(x) = \dfrac{4}{x + 2}$

In Problems 63–74, the function f is one-to-one. (a) Find its inverse function f^{-1} and check your answer. (b) Find the domain and the range of f and f^{-1}.

63. $f(x) = \dfrac{2}{3 + x}$ **64.** $f(x) = \dfrac{4}{2 - x}$ **65.** $f(x) = \dfrac{3x}{x + 2}$

66. $f(x) = -\dfrac{2x}{x - 1}$ **67.** $f(x) = \dfrac{2x}{3x - 1}$ **68.** $f(x) = -\dfrac{3x + 1}{x}$

69. $f(x) = \dfrac{3x + 4}{2x - 3}$ **70.** $f(x) = \dfrac{2x - 3}{x + 4}$ **71.** $f(x) = \dfrac{2x + 3}{x + 2}$

72. $f(x) = \dfrac{-3x - 4}{x - 2}$ **73.** $f(x) = \dfrac{x^2 - 4}{2x^2}, \quad x > 0$ **74.** $f(x) = \dfrac{x^2 + 3}{3x^2} \quad x > 0$

Applications and Extensions

75. Use the graph of $y = f(x)$ given in Problem 45 to evaluate the following:
(a) $f(-1)$ (b) $f(1)$ (c) $f^{-1}(1)$ (d) $f^{-1}(2)$

76. Use the graph of $y = f(x)$ given in Problem 46 to evaluate the following:
(a) $f(2)$ (b) $f(1)$ (c) $f^{-1}(0)$ (d) $f^{-1}(-1)$

77. If $f(7) = 13$ and f is one-to-one, what is $f^{-1}(13)$?

78. If $g(-5) = 3$ and g is one-to-one, what is $g^{-1}(3)$?

79. The domain of a one-to-one function f is $[5, \infty)$, and its range is $[-2, \infty)$. State the domain and the range of f^{-1}.

80. The domain of a one-to-one function f is $[0, \infty)$, and its range is $[5, \infty)$. State the domain and the range of f^{-1}.

81. The domain of a one-to-one function g is $(-\infty, 0]$, and its range is $[0, \infty)$. State the domain and the range of g^{-1}.

82. The domain of a one-to-one function g is $[0, 15]$, and its range is $(0, 8)$. State the domain and the range of g^{-1}.

83. A function $y = f(x)$ is increasing on the interval $(0, 5)$. What conclusions can you draw about the graph of $y = f^{-1}(x)$?

84. A function $y = f(x)$ is decreasing on the interval $(0, 5)$. What conclusions can you draw about the graph of $y = f^{-1}(x)$?

85. Find the inverse of the linear function

$$f(x) = mx + b, \quad m \neq 0$$

86. Find the inverse of the function

$$f(x) = \sqrt{r^2 - x^2}, \quad 0 \le x \le r$$

87. A function f has an inverse function f^{-1}. If the graph of f lies in quadrant I, in which quadrant does the graph of f^{-1} lie?

88. A function f has an inverse function f^{-1}. If the graph of f lies in quadrant II, in which quadrant does the graph of f^{-1} lie?

89. The function $f(x) = |x|$ is not one-to-one. Find a suitable restriction on the domain of f so that the new function that results is one-to-one. Then find the inverse of the new function.

90. The function $f(x) = x^4$ is not one-to-one. Find a suitable restriction on the domain of f so that the new function that results is one-to-one. Then find the inverse of the new function.

In applications, the symbols used for the independent and dependent variables are often based on common usage. So, rather than using $y = f(x)$ to represent a function, an applied problem might use $C = C(q)$ to represent the cost C of manufacturing q units of a good. Because of this, the inverse notation f^{-1} used in a pure mathematics problem is not used when finding inverses of applied problems. Rather, the inverse of a function such as $C = C(q)$ will be $q = q(C)$. So $C = C(q)$ is a function that represents the cost C as a function of the number q of units manufactured, and $q = q(C)$ is a function that represents the number q as a function of the cost C. Problems 91–94 illustrate this idea.

91. Vehicle Stopping Distance Taking into account reaction time, the distance d (in feet) that a car requires to come to a complete stop while traveling r miles per hour is given by the function

$$d(r) = 6.97r - 90.39$$

(a) Express the speed r at which the car is traveling as a function of the distance d required to come to a complete stop.
(b) Verify that $r = r(d)$ is the inverse of $d = d(r)$ by showing that $r(d(r)) = r$ and $d(r(d)) = d$.
(c) Predict the speed that a car was traveling if the distance required to stop was 300 feet.

92. Height and Head Circumference The head circumference C of a child is related to the height H of the child (both in inches) through the function

$$H(C) = 2.15C - 10.53$$

(a) Express the head circumference C as a function of height H.
(b) Verify that $C = C(H)$ is the inverse of $H = H(C)$ by showing that $H(C(H)) = H$ and $C(H(C)) = C$.
(c) Predict the head circumference of a child who is 26 inches tall.

93. Ideal Body Weight One model for the ideal body weight W for men (in kilograms) as a function of height h (in inches) is given by the function

$$W(h) = 50 + 2.3(h - 60)$$

(a) What is the ideal weight of a 6-foot male?
(b) Express the height h as a function of weight W.
(c) Verify that $h = h(W)$ is the inverse of $W = W(h)$ by showing that $h(W(h)) = h$ and $W(h(W)) = W$.
(d) What is the height of a male who is at his ideal weight of 80 kilograms?

[**Note:** The ideal body weight W for women (in kilograms) as a function of height h (in inches) is given by $W(h) = 45.5 + 2.3(h - 60)$.]

94. Temperature Conversion The function $F(C) = \frac{9}{5}C + 32$ converts a temperature from C degrees Celsius to F degrees Fahrenheit.

(a) Express the temperature in degrees Celsius C as a function of the temperature in degrees Fahrenheit F.
(b) Verify that $C = C(F)$ is the inverse of $F = F(C)$ by showing that $C(F(C)) = C$ and $F(C(F)) = F$.
(c) What is the temperature in degrees Celsius if it is 70 degrees Fahrenheit?

95. Income Taxes The function

$$T(g) = 5081.25 + 0.25(g - 36{,}900)$$

represents the 2014 federal income tax T (in dollars) due for a "single" filer whose modified adjusted gross income is g dollars, where $36{,}900 \le g \le 89{,}350$.

(a) What is the domain of the function T?
(b) Given that the tax due T is an increasing linear function of modified adjusted gross income g, find the range of the function T.
(c) Find adjusted gross income g as a function of federal income tax T. What are the domain and the range of this function?

96. Income Taxes The function

$$T(g) = 1815 + 0.15(g - 18{,}150)$$

represents the 2014 federal income tax T (in dollars) due for a "married filing jointly" filer whose modified adjusted gross income is g dollars, where $18{,}150 \le g \le 73{,}800$.

(a) What is the domain of the function T?
(b) Given that the tax due T is an increasing linear function of modified adjusted gross income g, find the range of the function T.
(c) Find adjusted gross income g as a function of federal income tax T. What are the domain and the range of this function?

97. Gravity on Earth If a rock falls from a height of 100 meters on Earth, the height H (in meters) after t seconds is approximately

$$H(t) = 100 - 4.9t^2$$

(a) In general, quadratic functions are not one-to-one. However, the function H is one-to-one. Why?
(b) Find the inverse of H and verify your result.
(c) How long will it take a rock to fall 80 meters?

98. Period of a Pendulum The period T (in seconds) of a simple pendulum as a function of its length l (in feet) is given by

$$T(l) = 2\pi\sqrt{\frac{l}{32.2}}$$

(a) Express the length l as a function of the period T.
(b) How long is a pendulum whose period is 3 seconds?

99. Given

$$f(x) = \frac{ax + b}{cx + d}$$

find $f^{-1}(x)$. If $c \neq 0$, under what conditions on $a, b, c,$ and d is $f = f^{-1}$?

Explaining Concepts: Discussion and Writing

100. Can a one-to-one function and its inverse be equal? What must be true about the graph of f for this to happen? Give some examples to support your conclusion.

101. Draw the graph of a one-to-one function that contains the points $(-2, -3)$, $(0, 0)$, and $(1, 5)$. Now draw the graph of its inverse. Compare your graph to those of other students. Discuss any similarities. What differences do you see?

102. Give an example of a function whose domain is the set of real numbers and that is neither increasing nor decreasing on its domain, but is one-to-one.

[**Hint:** Use a piecewise-defined function.]

103. Is every odd function one-to-one? Explain.

104. Suppose that $C(g)$ represents the cost C, in dollars, of manufacturing g cars. Explain what $C^{-1}(800{,}000)$ represents.

105. Explain why the horizontal-line test can be used to identify one-to-one functions from a graph.

106. Explain why a function must be one-to-one in order to have an inverse that is a function. Use the function $y = x^2$ to support your explanation.

'Are You Prepared?' Answers

1. Yes; for each input x there is one output y.

2. $\{-1, 0, 5, 8\}$

3. $\{x \mid x \neq -6, x \neq 3\}$

4. $y = \dfrac{x + 8}{3}$ or $y = \dfrac{1}{3}x + \dfrac{8}{3}$

Chapter Review

Library of Functions

Constant function (p. 58)

$f(x) = b$

The graph is a horizontal line with y-intercept b.

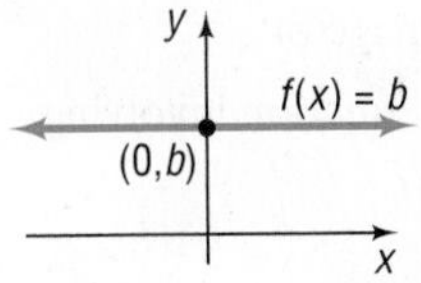

Identity function (p. 58)

$f(x) = x$

The graph is a line with slope 1 and y-intercept 0.

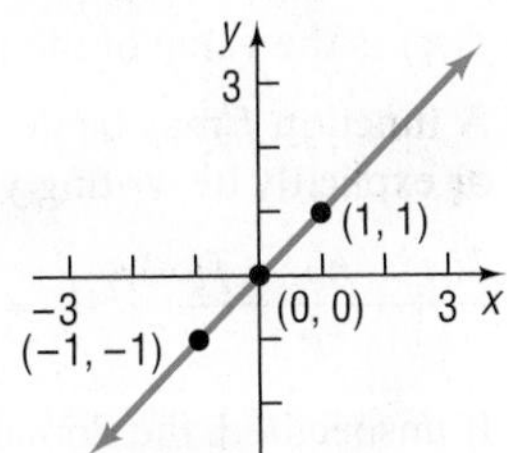

Linear function

$f(x) = mx + b$

The graph is a line with slope m and y-intercept b.

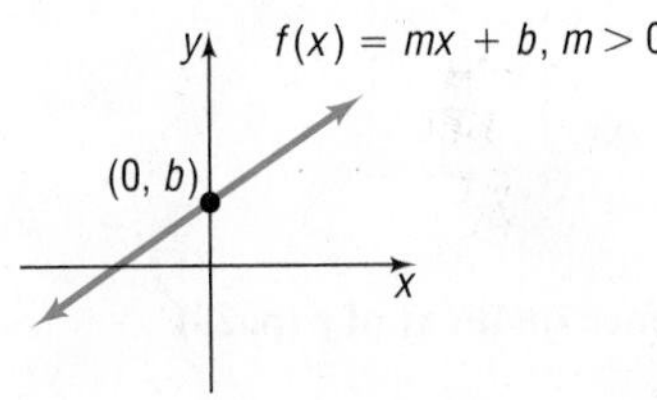

Square function (p. 58)

$f(x) = x^2$

The graph is a parabola with intercept at $(0, 0)$.

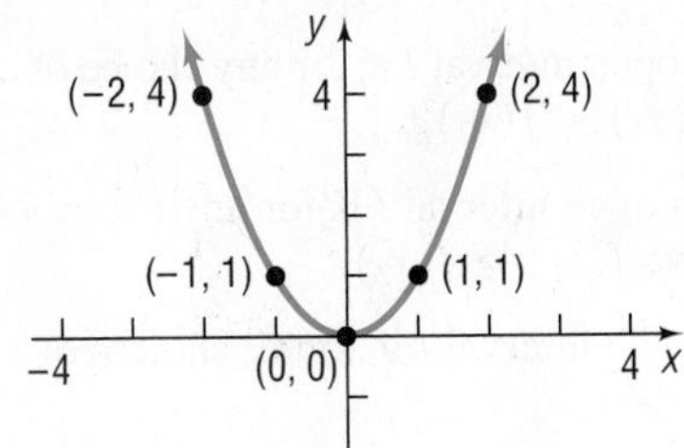

Cube function (p. 58)

$f(x) = x^3$

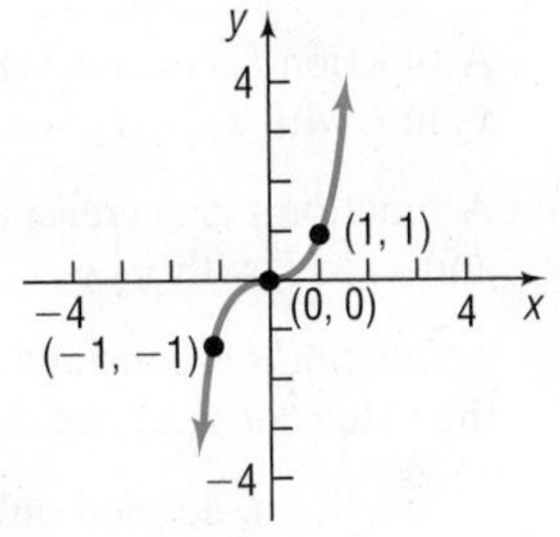

Square root function (p. 58)

$f(x) = \sqrt{x}$

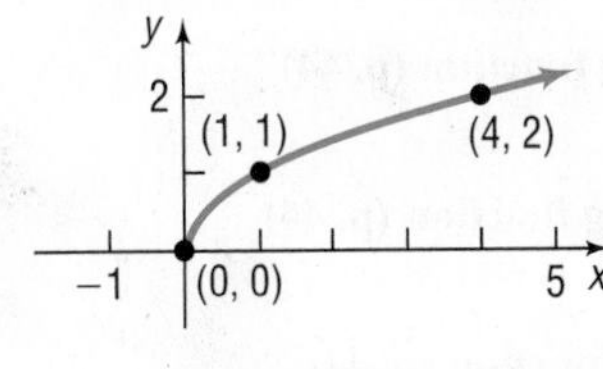

Cube root function (p. 59)

$f(x) = \sqrt[3]{x}$

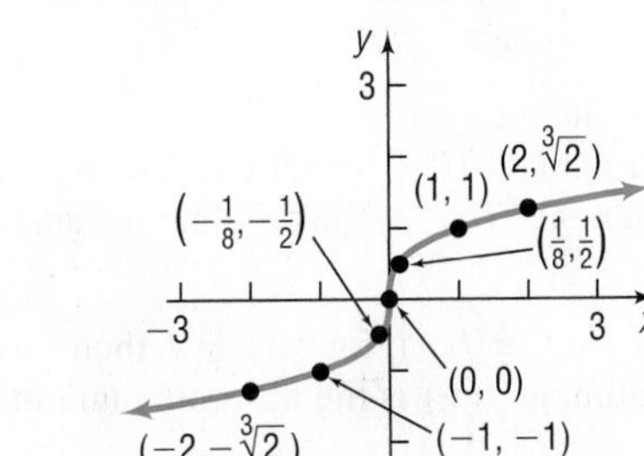

Reciprocal function (p. 59)

$f(x) = \dfrac{1}{x}$

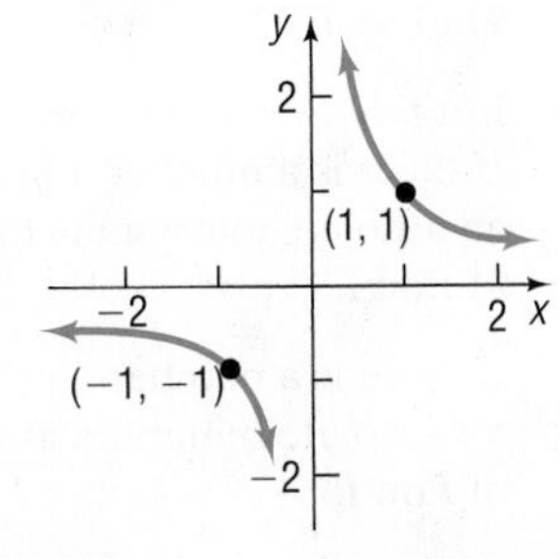

Absolute value function (p. 59)

$f(x) = |x|$

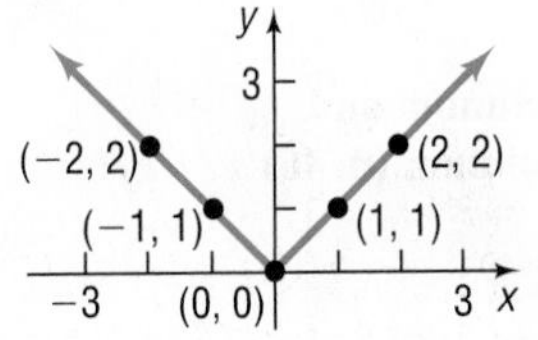

Things to Know

Formulas

Distance formula (p. 3)	$d = \sqrt{(x_2 - x_1)^2 + (y_2 - y_1)^2}$
Midpoint formula (p. 5)	$(x, y) = \left(\frac{x_1 + x_2}{2}, \frac{y_1 + y_2}{2}\right)$
Equations of Circles	
Standard form of the equation of a circle (p. 16)	$(x - h)^2 + (y - k)^2 = r^2$; r is the radius of the circle, (h, k) is the center of the circle.
Equation of the unit circle (p. 16)	$x^2 + y^2 = 1$
General form of the equation of a circle (p. 18)	$x^2 + y^2 + ax + by + c = 0$ with restrictions on a, b, and c
Function (pp. 23–26)	A relation between two sets so that each element x in the first set, the domain, has corresponding to it exactly one element y in the second set, the range. The range is the set of y values of the function for the x values in the domain. A function can also be characterized as a set of ordered pairs (x, y) in which no first element is paired with two different second elements.
Function notation (pp. 26–29)	$y = f(x)$ f is a symbol for the function. x is the argument, or independent variable. y is the dependent variable. $f(x)$ is the value of the function at x, or the image of x. A function f may be defined implicitly by an equation involving x and y or explicitly by writing $y = f(x)$.
Difference quotient of f (p. 29)	$\frac{f(x + h) - f(x)}{h} \quad h \neq 0$
Domain (pp. 30–32)	If unspecified, the domain of a function f defined by an equation is the largest set of real numbers for which $f(x)$ is a real number.
Vertical-line test (p. 32)	A set of points in the plane is the graph of a function if and only if every vertical line intersects the graph in at most one point.
Even function (p. 42)	$f(-x) = f(x)$ for every x in the domain ($-x$ must also be in the domain).
Odd function (p. 42)	$f(-x) = -f(x)$ for every x in the domain ($-x$ must also be in the domain).
Increasing function (p. 44)	A function f is increasing on an open interval I if, for any choice of x_1 and x_2 in I, with $x_1 < x_2$, we have $f(x_1) < f(x_2)$.
Decreasing function (p. 44)	A function f is decreasing on an open interval I if, for any choice of x_1 and x_2 in I, with $x_1 < x_2$, we have $f(x_1) > f(x_2)$.
Constant function (p. 44)	A function f is constant on an open interval I if, for all choices of x in I, the values of $f(x)$ are equal.
Local maximum (p. 45)	A function f, defined on some interval I, has a local maximum at c if there is an open interval in I containing c such that, for all x in this open interval, $f(x) \leq f(c)$.
Local minimum (p. 45)	A function f, defined on some interval I, has a local minimum at c if there is an open interval in I containing c such that, for all x in this open interval, $f(x) \geq f(c)$.
Absolute maximum and Absolute minimum (p. 46)	Let f denote a function defined on some interval I. If there is a number u in I for which $f(x) \leq f(u)$ for all x in I, then f has an absolute maximum at u, and the number $f(u)$ is the absolute maximum of f on I. If there is a number v in I for which $f(x) \geq f(v)$, for all x in I, then f has an absolute minimum at v and the number $f(v)$ is the absolute minimum of f on I.

Average rate of change of a function (p. 48)	The average rate of change of f from a to b is $\frac{\Delta y}{\Delta x} = \frac{f(b) - f(a)}{b - a} \quad a \neq b$
One-to-one function f (p. 79)	A function for which any two different inputs in the domain correspond to two different outputs in the range For any choice of elements x_1, x_2 in the domain of f, if $x_1 \neq x_2$, then $f(x_1) \neq f(x_2)$
Horizontal-line test (p. 80)	If every horizontal line intersects the graph of a function f in at most one point, f is one-to-one.
Inverse function f^{-1} of f (pp. 81–83)	Domain of f = range of f^{-1}; range of f = domain of f^{-1} $f^{-1}(f(x)) = x$ for all x in the domain of f $f(f^{-1}(x)) = x$ for all x in the domain of f^{-1} The graphs of f and f^{-1} are symmetric with respect to the line $y = x$.

Objectives

Section		You should be able to . . .	Examples	Review Exercises
1.1	1	Use the distance formula (p. 3)	1–3	1(a)–3(a), 16, 17
	2	Use the midpoint formula (p. 5)	4	1(b)–3(b), 17
1.2	1	Graph equations by plotting points (p. 9)	1–3	4
	2	Find intercepts from a graph (p. 10)	4	5
	3	Find intercepts from an equation (p. 11)	5	6–10
	4	Test an equation for symmetry with respect to the x-axis, the y-axis, and the origin (p. 12)	6–9	6–10
	5	Know how to graph key equations (p. 14)	10–12	43, 44
	6	Write the standard form of the equation of a circle (p. 15)	13	11, 12, 17
	7	Graph a circle (p. 16)	14, 15	13–15
	8	Work with the general form of the equation of a circle (p. 17)	16	14, 15
1.3	1	Determine whether a relation represents a function (p. 23)	1–5	18, 19
	2	Find the value of a function (p. 26)	6, 7	20–22, 53
	3	Find the difference quotient of a function (p. 29)	8	29
	4	Find the domain of a function defined by an equation (p. 30)	9, 10	23–28
	5	Identify the graph of a function (p. 32)	11	41, 42
	6	Obtain information from or about the graph of a function (p. 33)	12, 13	30(a)–(e), 31(a), 31(e), 31(g)
1.4	1	Determine even and odd functions from a graph (p. 42)	1	31(f)
	2	Identify even and odd functions from the equation (p. 43)	2	32–35
	3	Use a graph to determine where a function is increasing, decreasing, or constant (p. 44)	3	31(b)
	4	Use a graph to locate local maxima and local minima (p. 45)	4	31(c)
	5	Use a graph to locate the absolute maximum and the absolute minimum (p. 46)	5	31(d)
	6	Use a graphing utility to approximate local maxima and local minima and to determine where a function is increasing or decreasing (p. 48)	6	36, 37
	7	Find the average rate of change of a function (p. 48)	7, 8	38–40
1.5	1	Graph the functions listed in the library of functions (p. 55)	1, 2	43, 44
	2	Graph piecewise-defined functions (p. 59)	3	51, 52
1.6	1	Graph functions using vertical and horizontal shifts (p. 65)	1–5	30(f), 47–50
	2	Graph functions using compressions and stretches (p. 68)	6–8, 12	30(g), 46, 50
	3	Graph functions using reflections about the x-axis or y-axis (p. 70)	9–11, 13	30(h), 46, 48, 50

Section	You should be able to . . .	Examples	Review Exercises
1.7	1 Determine whether a function is one-to-one (p. 78)	1, 2	54(a), 55
	2 Determine the inverse of a function defined by a map or a set of ordered pairs (p. 81)	3, 4	54(b)
	3 Obtain the graph of the inverse function from the graph of the function (p. 83)	7	55
	4 Find the inverse of a function defined by an equation (p. 84)	8–10	56–59

Review Exercises

In Problems 1–3, find the following for each pair of points:
(a) The distance between the points
(b) The midpoint of the line segment connecting the points

1. $(0, 0)$; $(4, 2)$

2. $(1, -1)$; $(-2, 3)$

3. $(4, -4)$; $(4, 8)$

4. Graph $y = x^2 + 4$ by plotting points.

5. List the intercepts of the graph below.

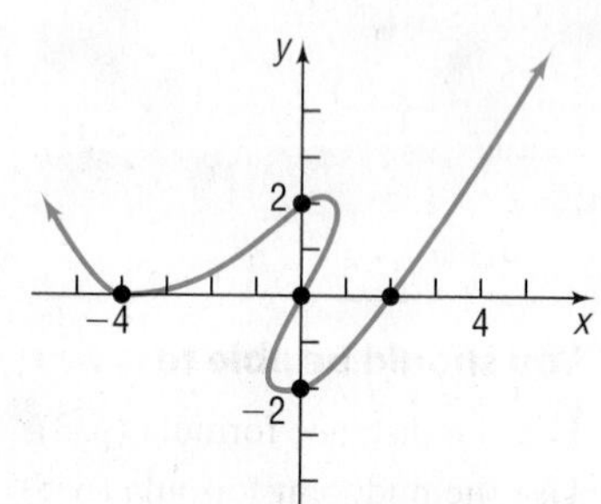

In Problems 6–10, list the intercepts and test for symmetry with respect to the x-axis, the y-axis, and the origin.

6. $2x = 3y^2$ **7.** $x^2 + 4y^2 = 16$ **8.** $y = x^4 + 2x^2 + 1$

9. $y = x^3 - x$ **10.** $x^2 + x + y^2 + 2y = 0$

In Problems 11 and 12, find the standard form of the equation of the circle whose center and radius are given.

11. $(h, k) = (-2, 3); r = 4$ **12.** $(h, k) = (-1, -2); r = 1$

In Problems 13–15, find the center and radius of each circle. Graph each circle. Find the intercepts, if any, of each circle.

13. $x^2 + (y - 1)^2 = 4$ **14.** $x^2 + y^2 - 2x + 4y - 4 = 0$ **15.** $3x^2 + 3y^2 - 6x + 12y = 0$

16. Show that the points $A = (3, 4)$, $B = (1, 1)$, and $C = (-2, 3)$ are the vertices of an isosceles triangle.

17. The endpoints of the diameter of a circle are $(-3, 2)$ and $(5, -6)$. Find the center and radius of the circle. Write the standard equation of this circle.

In Problems 18 and 19, determine whether each relation represents a function. For each function, state the domain and range.

18. $\{(-1, 0), (2, 3), (4, 0)\}$ **19.** $\{(4, -1), (2, 1), (4, 2)\}$

In Problems 20–22, find the following for each function:

(a) $f(2)$ (b) $f(-2)$ (c) $f(-x)$ (d) $-f(x)$ (e) $f(x - 2)$ (f) $f(2x)$

20. $f(x) = \dfrac{3x}{x^2 - 1}$ **21.** $f(x) = \sqrt{x^2 - 4}$ **22.** $f(x) = \dfrac{x^2 - 4}{x^2}$

In Problems 23–28, find the domain of each function.

23. $f(x) = \dfrac{x}{x^2 - 9}$ **24.** $f(x) = \sqrt{2 - x}$ **25.** $g(x) = \dfrac{|x|}{x}$

26. $f(x) = \dfrac{x}{x^2 + 2x - 3}$ **27.** $f(x) = \dfrac{\sqrt{x + 1}}{x^2 - 4}$ **28.** $g(x) = \dfrac{x}{\sqrt{x + 8}}$

29. Find the difference quotient of $f(x) = -2x^2 + x + 1$; that is, find $\dfrac{f(x+h) - f(x)}{h}$, $h \neq 0$.

30. Consider the graph of the function f on the right.
(a) Find the domain and the range of f.
(b) List the intercepts.
(c) Find $f(-2)$.
(d) For what value of x does $f(x) = -3$?
(e) Solve $f(x) > 0$.
(f) Graph $y = f(x-3)$.
(g) Graph $y = f\left(\frac{1}{2}x\right)$.
(h) Graph $y = -f(x)$.

31. Use the graph of the function f shown to find:
(a) The domain and the range of f.
(b) The intervals on which f is increasing, decreasing, or constant.
(c) The local minimum values and local maximum values.
(d) The absolute maximum and absolute minimum.
(e) Whether the graph is symmetric with respect to the x-axis, the y-axis, or the origin.
(f) Whether the function is even, odd, or neither.
(g) The intercepts, if any.

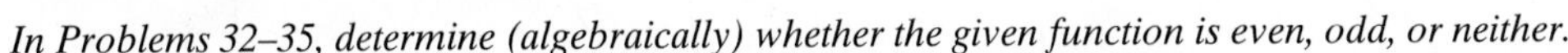

In Problems 32–35, determine (algebraically) whether the given function is even, odd, or neither.

32. $f(x) = x^3 - 4x$ **33.** $g(x) = \dfrac{4 + x^2}{1 + x^4}$ **34.** $G(x) = 1 - x + x^3$ **35.** $f(x) = \dfrac{x}{1 + x^2}$

In Problems 36 and 37, use a graphing utility to graph each function over the indicated interval. Approximate any local maximum values and local minimum values. Determine where the function is increasing and where it is decreasing.

36. $f(x) = 2x^3 - 5x + 1 \quad (-3, 3)$ **37.** $f(x) = 2x^4 - 5x^3 + 2x + 1 \quad (-2, 3)$

38. Find the average rate of change of $f(x) = 8x^2 - x$:
(a) From 1 to 2 (b) From 0 to 1 (c) From 2 to 4

In Problems 39 and 40, find the average rate of change from 2 to 3 for each function f. Be sure to simplify.

39. $f(x) = 2 - 5x$ **40.** $f(x) = 3x - 4x^2$

In Problems 41 and 42, is the graph shown the graph of a function?

41.

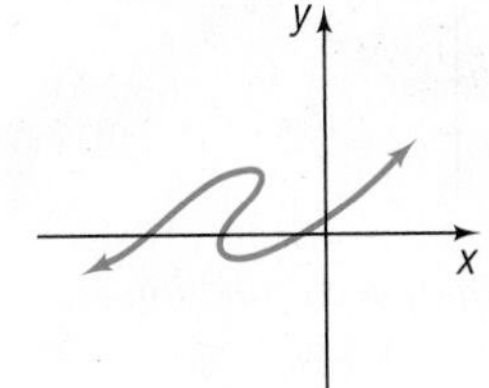

42.

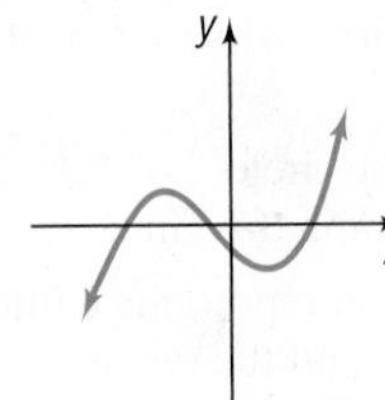

In Problems 43 and 44, graph each function. Be sure to label at least three points.

43. $f(x) = |x|$ **44.** $f(x) = \sqrt{x}$

In Problems 45–50, graph each function using the techniques of shifting, compressing or stretching, and reflections. Identify any intercepts of the graph. State the domain and, based on the graph, find the range.

45. $F(x) = |x| - 4$ **46.** $g(x) = -2|x|$ **47.** $h(x) = \sqrt{x - 1}$

48. $f(x) = \sqrt{1 - x}$ **49.** $h(x) = (x - 1)^2 + 2$ **50.** $g(x) = -2(x + 2)^3 - 8$

In Problems 51 and 52:
(a) Find the domain of each function. *(b) Locate any intercepts.* *(c) Graph each function.*
(d) Based on the graph, find the range. *(e) Is f continuous on its domain?*

51. $f(x) = \begin{cases} 3x & \text{if } -2 < x \le 1 \\ x + 1 & \text{if } x > 1 \end{cases}$ **52.** $f(x) = \begin{cases} x & \text{if } -4 \le x < 0 \\ 1 & \text{if } x = 0 \\ 3x & \text{if } x > 0 \end{cases}$

53. A function f is defined by

$$f(x) = \frac{Ax + 5}{6x - 2}$$

If $f(1) = 4$, find A.

54. (a) Verify that the function below is one-to-one; (b) find its inverse.

$\{(1, 2), (3, 5), (5, 8), (6, 10)\}$

55. State why the graph of the function is one-to-one. Then draw the graph of the inverse function f^{-1}.

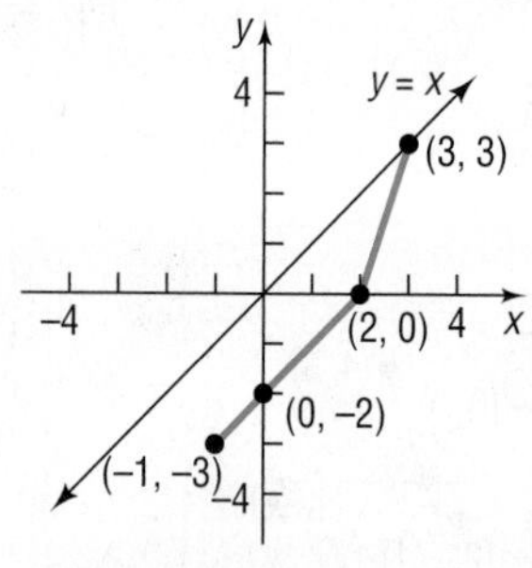

In Problems 56–59, each function is one-to-one. Find the inverse of each function and check your answer. State the domain and range of f and f^{-1}.

56. $f(x) = \dfrac{2x + 3}{5x - 2}$

57. $f(x) = \dfrac{1}{x - 1}$

58. $f(x) = \sqrt{x - 2}$

59. $f(x) = x^{1/3} + 1$

Chapter Test

CHAPTER **Test Prep** VIDEOS

The Chapter Test Prep Videos are step-by-step solutions available in MyMathLab®, or on this text's YouTube™ Channel. Flip back to the Resources for Success page for a link to this text's YouTube channel.

In Problems 1–2, use $P_1 = (-1, 3)$ and $P_2 = (5, -1)$.

1. Find the distance from P_1 to P_2.

2. Find the midpoint of the line segment joining P_1 and P_2.

3. Graph $y = x^2 - 9$ by plotting points.

4. Sketch the graph of $y^2 = x$.

5. List the intercepts and test for symmetry: $x^2 + y = 9$.

6. Write the general form of the circle with center $(4, -3)$ and radius 5.

7. Find the center and radius of the circle $x^2 + y^2 + 4x - 2y - 4 = 0$. Graph this circle.

8. Determine whether each relation represents a function. For each function, state the domain and the range.

(a) $\{(2, 5), (4, 6), (6, 7), (8, 8)\}$

(b) $\{(1, 3), (4, -2), (-3, 5), (1, 7)\}$

(c)

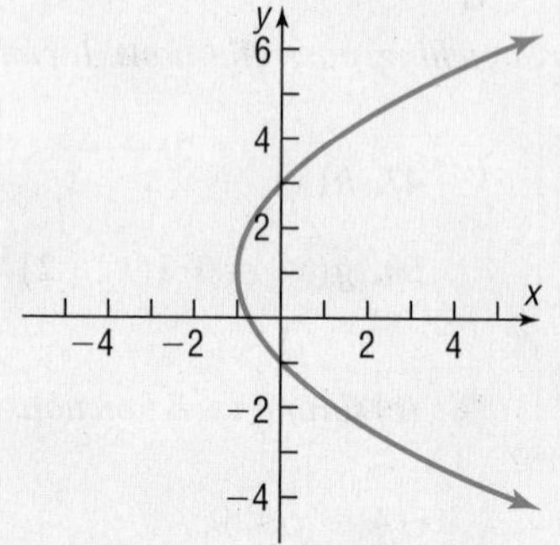

(d)

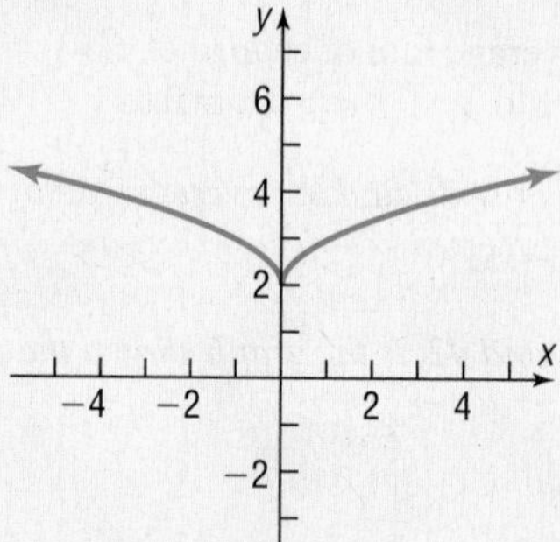

In Problems 9–11, find the domain of each function and evaluate each function at $x = -1$.

9. $f(x) = \sqrt{4 - 5x}$

10. $g(x) = \dfrac{x + 2}{|x + 2|}$

11. $h(x) = \dfrac{x - 4}{x^2 + 5x - 36}$

12. Consider the graph of the function f:

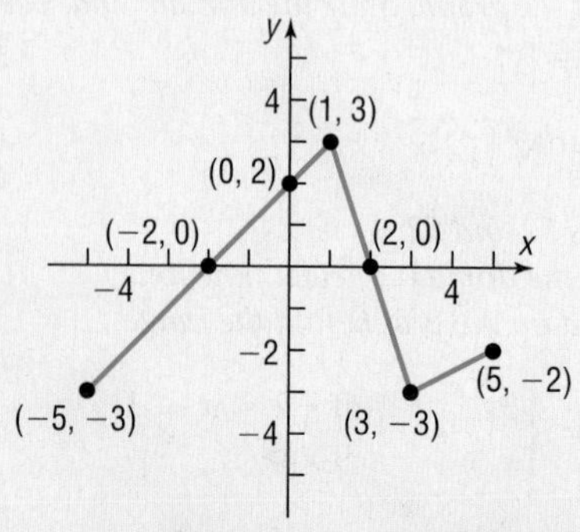

(a) Find the domain and the range of f.
(b) List the intercepts.
(c) Find $f(1)$.
(d) For what value(s) of x does $f(x) = -3$?
(e) Solve $f(x) < 0$.

13. Use a graphing utility to graph the function $f(x) = -x^4 + 2x^3 + 4x^2 - 2$ on the interval $(-5, 5)$. Approximate any local maximum values and local minimum values rounded to two decimal places. Determine where the function is increasing and where it is decreasing.

14. Consider the function $g(x) = \begin{cases} 2x + 1 & \text{if } x < -1 \\ x - 4 & \text{if } x \geq -1 \end{cases}$
(a) Graph the function.
(b) List the intercepts.
(c) Find $g(-5)$.
(d) Find $g(2)$.

15. For the function $f(x) = 3x^2 - 2x + 4$, find the average rate of change of f from 3 to 4.

16. Graph each function using the techniques of shifting, compressing or stretching, and reflections. Start with the graph of the basic function and show all stages.
(a) $h(x) = -2(x + 1)^3 + 3$
(b) $g(x) = |x + 4| + 2$

17. Find the inverse of $f(x) = \dfrac{2}{3x - 5}$ and check your answer. State the domain and the range of f and f^{-1}.

18. If the point $(3, -5)$ is on the graph of a one-to-one function f, what point must be on the graph of f^{-1}?

Chapter Projects

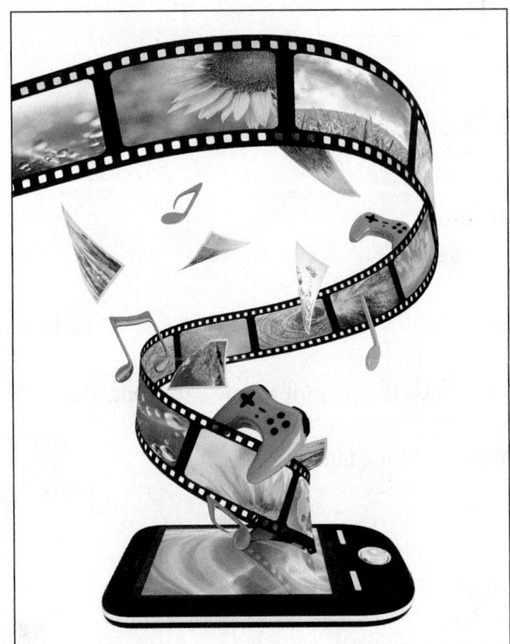

Internet-based Project

I. Choosing a Wireless Data Plan Collect information from your family, friends, or consumer agencies such as Consumer Reports. Then decide on a cellular provider, choosing the company that you feel offers the best service. Once you have selected a service provider, research the various types of individual plans offered by the company by visiting the provider's website. Many providers offer family plans that include unlimited talk and text. The monthly cost is primarily determined by the amount of data used and the number of devices.

1. Suppose you expect to use 10 gigabytes of data for a single smartphone. What would be the monthly cost of each plan you are considering?
2. Suppose you expect to use 30 gigabytes of data and want a personal hotspot, but you still have only a single smartphone. What would be the monthly cost of each plan you are considering?
3. Suppose you expect to use 20 gigabytes of data with three smartphones sharing the data. What would be the monthly cost of each plan you are considering?
4. Suppose you expect to use 20 gigabytes of data with a single smartphone and a personal hotspot. What would be the monthly cost of each plan you are considering?
5. Build a model that describes the monthly cost C, in dollars, as a function of the number of data gigabytes used, g, assuming a single smartphone and a personal hotspot for each plan you are considering.
6. Graph each function from Problem 5.
7. Based on your particular usage, which plan is best for you?
8. Now, develop an Excel spreadsheet to analyze the various plans you are considering. Suppose you want a family plan with unlimited talk and text that offers 10 gigabytes of shared data and costs \$100 per month. Additional gigabytes of data cost \$15 per gigabyte, extra phones can be added to the plan for \$15 each per month, and each hotspot costs \$20 per month. Because wireless

data plans have a cost structure based on piecewise-defined functions, we need an "if/then" statement within Excel to analyze the cost of the plan. Use the accompanying Excel spreadsheet as a guide in developing your spreadsheet. Enter into your spreadsheet a variety of possible amounts of data and various numbers of additional phones and hotspots.

	A	B	C	D	
1					
2	Monthly fee	$100			
3	Allotted data per month (GB)	10			
4	Data used (GB)	12			
5	Cost per additional GB of data	$15			
6					
7	Monthly cost of hotspot	$20			
8	Number of hotspots	1			
9	Monthly cost of additional phone	$15			
10	Number of additional phones	2			
11					
12	Cost of data	=IF(B4<B3,B2,B2+B5*(B4-B3))			
13	Cost of additional devices/hotspots	=B8*B7+B10*B9			
14					
15	Total Cost	=B12+B13			
16					

9. Write a paragraph supporting the choice in plans that best meets your needs.

10. How are "if/then" loops similar to a piecewise-defined function?

Citation: Excel © 2013 Microsoft Corporation. Used with permission from Microsoft.

The following projects are available on the Instructor's Resource Center (IRC).

II. Project at Motorola: *Wireless Internet Service* Use functions and their graphs to analyze the total cost of various wireless Internet service plans.

III. Cost of Cable When government regulations and customer preference influence the path of a new cable line, the Pythagorean Theorem can be used to assess the cost of installation.

IV. Oil Spill Functions are used to analyze the size and spread of an oil spill from a leaking tanker.

Trigonometric Functions

2

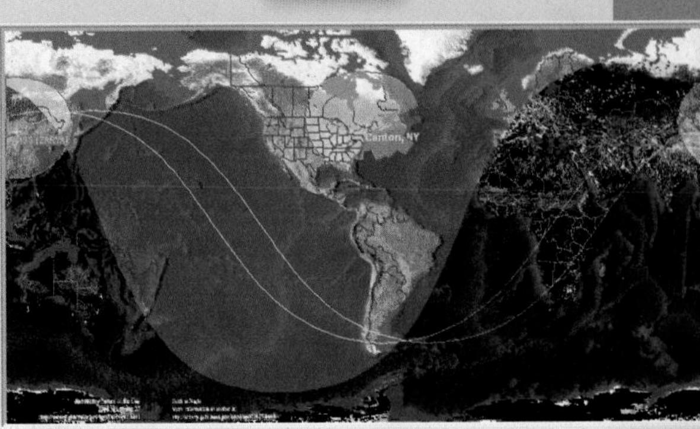

Length of Day

The length of a day depends upon the day of the year as well as the latitude of the location. Latitude gives the location of a point on Earth north or south of the equator. In the Internet Project at the end of this chapter, we will find a model that describes the relation between the length of day and day of the year for a specific latitude.

 *—See the Internet-based Chapter Project I—*

••• A Look Back

In Chapter 1, we began our discussion of functions. We defined domain and range and independent and dependent variables; we found the value of a function and graphed functions. We continued our study of functions by listing properties that a function might have, like being even or odd, and we created a library of functions, naming key functions and listing their properties, including the graph.

A Look Ahead •••

In this chapter we define the trigonometric functions, six functions that have wide application. We shall talk about their domain and range, see how to find values, graph them, and develop a list of their properties.

There are two widely accepted approaches to the development of the trigonometric functions: one uses right triangles; the other uses circles, especially the unit circle. In this book, we develop the trigonometric functions using the unit circle. In Chapter 4, we present right triangle trigonometry.

Outline

2.1 Angles and Their Measure
2.2 Trigonometric Functions: Unit Circle Approach
2.3 Properties of the Trigonometric Functions
2.4 Graphs of the Sine and Cosine Functions
2.5 Graphs of the Tangent, Cotangent, Cosecant, and Secant Functions
2.6 Phase Shift; Sinusoidal Curve Fitting
Chapter Review
Chapter Test
Cumulative Review
Chapter Projects

2.1 Angles and Their Measure

PREPARING FOR THIS SECTION *Before getting started, review the following:*

- Area and Circumference of a Circle (Appendix A, Section A.2, p. A16)

Now Work the 'Are You Prepared?' problems on page 109.

OBJECTIVES
1. Convert between Decimals and Degrees, Minutes, Seconds Measures for Angles (p. 102)
2. Find the Length of an Arc of a Circle (p. 103)
3. Convert from Degrees to Radians and from Radians to Degrees (p. 104)
4. Find the Area of a Sector of a Circle (p. 107)
5. Find the Linear Speed of an Object Traveling in Circular Motion (p. 108)

A **ray**, or **half-line**, is that portion of a line that starts at a point V on the line and extends indefinitely in one direction. The starting point V of a ray is called its **vertex**. See Figure 1.

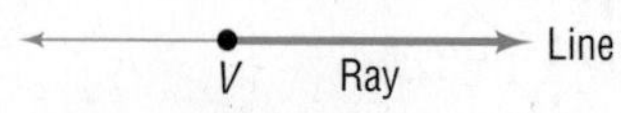

Figure 1 A ray or half-line

If two rays are drawn with a common vertex, they form an **angle**. We call one ray of an angle the **initial side** and the other the **terminal side**. The angle formed is identified by showing the direction and amount of rotation from the initial side to the terminal side. If the rotation is in the counterclockwise direction, the angle is **positive;** if the rotation is clockwise, the angle is **negative**. See Figure 2.

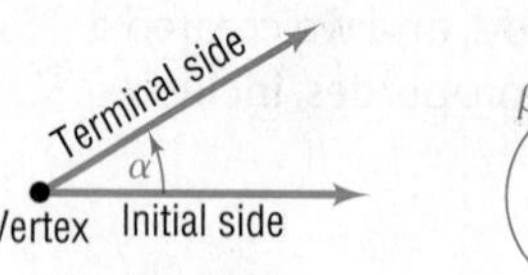

(a) Counterclockwise rotation Positive angle

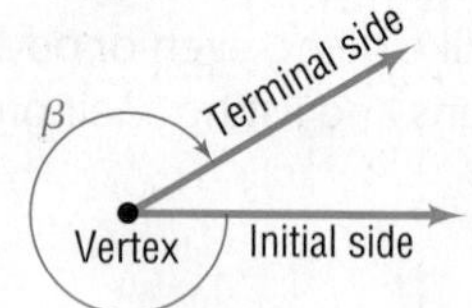

(b) Clockwise rotation Negative angle

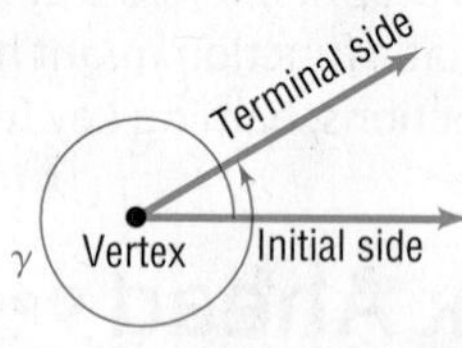

(c) Counterclockwise rotation Positive angle

Figure 2

Lowercase Greek letters, such as α (alpha), β (beta), γ (gamma), and θ (theta), will often be used to denote angles. Notice in Figure 2(a) that the angle α is positive because the direction of the rotation from the initial side to the terminal side is counterclockwise. The angle β in Figure 2(b) is negative because the rotation is clockwise. The angle γ in Figure 2(c) is positive. Notice that the angle α in Figure 2(a) and the angle γ in Figure 2(c) have the same initial side and the same terminal side. However, α and γ are unequal, because the amount of rotation required to go from the initial side to the terminal side is greater for angle γ than for angle α.

An angle θ is said to be in **standard position** if its vertex is at the origin of a rectangular coordinate system and its initial side coincides with the positive x-axis. See Figure 3.

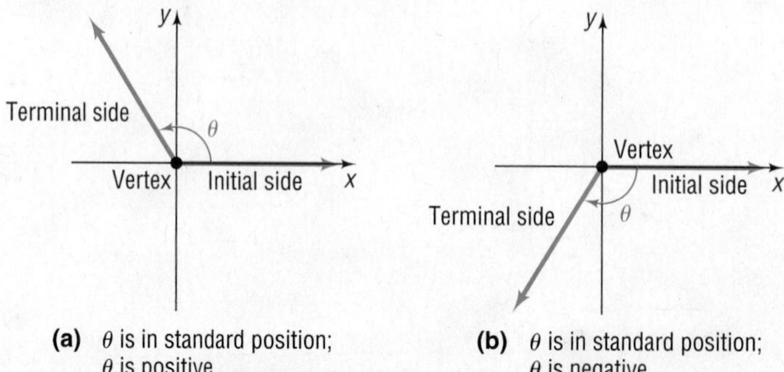

(a) θ is in standard position; θ is positive

(b) θ is in standard position; θ is negative

Figure 3 Standard position of an angle

When an angle θ is in standard position, the terminal side will lie either in a quadrant, in which case we say that θ **lies in that quadrant**, or the terminal side will lie on the x-axis or the y-axis, in which case we say that θ is a **quadrantal angle**. For example, the angle θ in Figure 4(a) lies in quadrant II, the angle θ in Figure 4(b) lies in quadrant IV, and the angle θ in Figure 4(c) is a quadrantal angle.

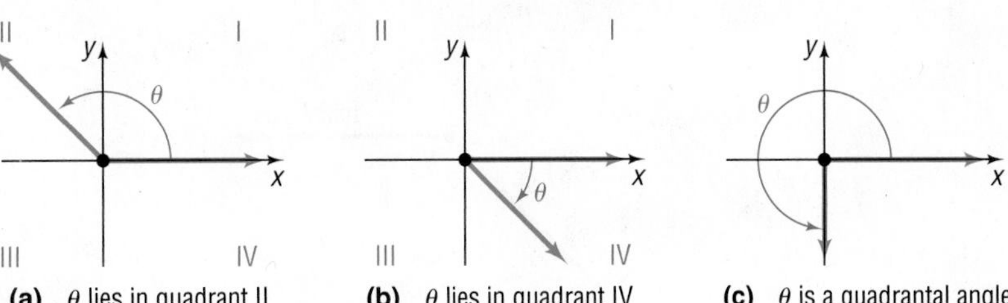

Figure 4 **(a)** θ lies in quadrant II **(b)** θ lies in quadrant IV **(c)** θ is a quadrantal angle

Angles are measured by determining the amount of rotation needed for the initial side to become coincident with the terminal side. The two commonly used measures for angles are *degrees* and *radians*.

Degrees

HISTORICAL NOTE One counterclockwise rotation was said to measure 360° because the Babylonian year had 360 days. ■

The angle formed by rotating the initial side exactly once in the counterclockwise direction until it coincides with itself (1 revolution) is said to measure 360 degrees, abbreviated 360°. **One degree, 1°,** is $\frac{1}{360}$ revolution. A **right angle** is an angle that measures 90°, or $\frac{1}{4}$ revolution; a **straight angle** is an angle that measures 180°, or $\frac{1}{2}$ revolution. See Figure 5. As Figure 5(b) shows, it is customary to indicate a right angle by using the symbol ∟.

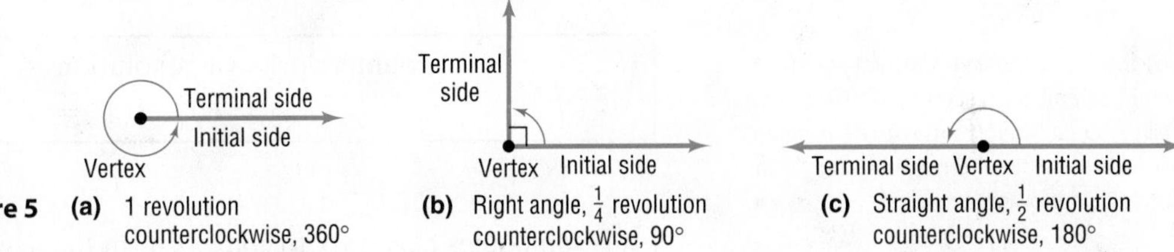

Figure 5 **(a)** 1 revolution counterclockwise, 360° **(b)** Right angle, $\frac{1}{4}$ revolution counterclockwise, 90° **(c)** Straight angle, $\frac{1}{2}$ revolution counterclockwise, 180°

It is also customary to refer to an angle that measures θ degrees as an angle *of* θ degrees.

EXAMPLE 1

Drawing an Angle

Draw each angle.

(a) 45° (b) −90° (c) 225° (d) 405°

Solution (a) An angle of 45° is $\frac{1}{2}$ of a right angle. See Figure 6.

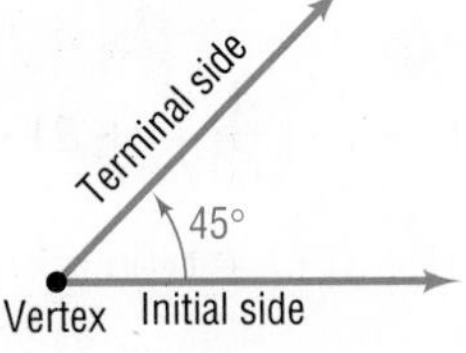

Figure 6 45° angle

(b) An angle of −90° is $\frac{1}{4}$ revolution in the clockwise direction. See Figure 7.

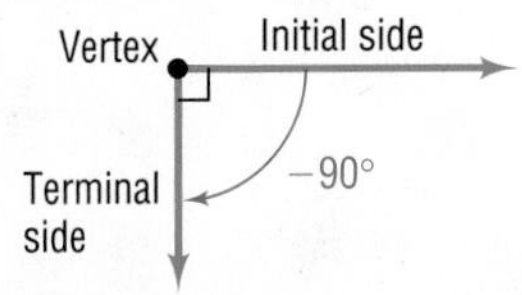

Figure 7 −90° angle

(c) An angle of 225° consists of a rotation through 180° followed by a rotation through 45°. See Figure 8.

(d) An angle of 405° consists of 1 revolution (360°) followed by a rotation through 45°. See Figure 9.

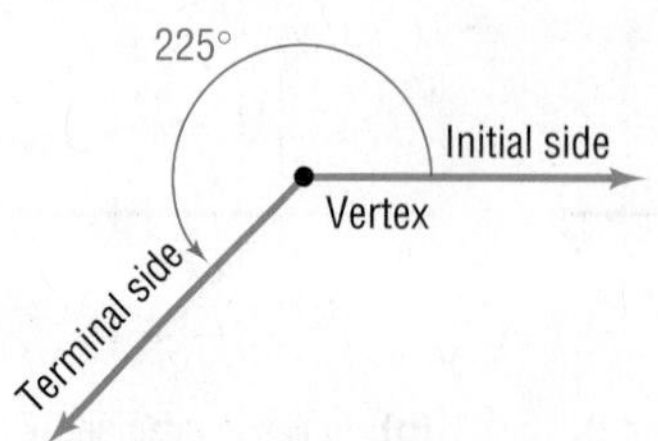

Figure 8 225° angle

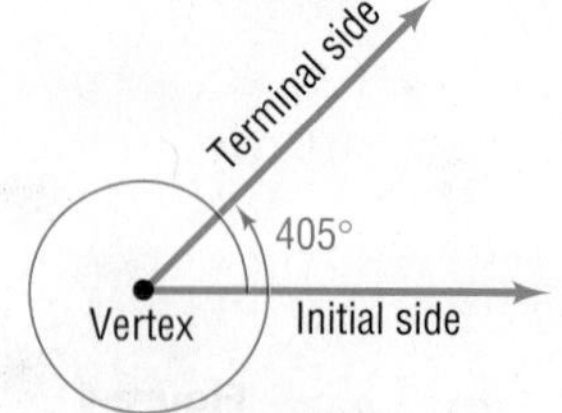

Figure 9 405° angle

1 Convert between Decimals and Degrees, Minutes, Seconds Measures for Angles

Although subdivisions of a degree may be obtained by using decimals, the terms *minute* and *second* are also used. **One minute**, denoted by **1′**, is defined as $\frac{1}{60}$ degree. **One second**, denoted by **1″**, is defined as $\frac{1}{60}$ minute or, equivalently, as $\frac{1}{3600}$ degree. An angle of, say, 30 degrees, 40 minutes, 10 seconds is written compactly as 30°40′10″. To summarize:

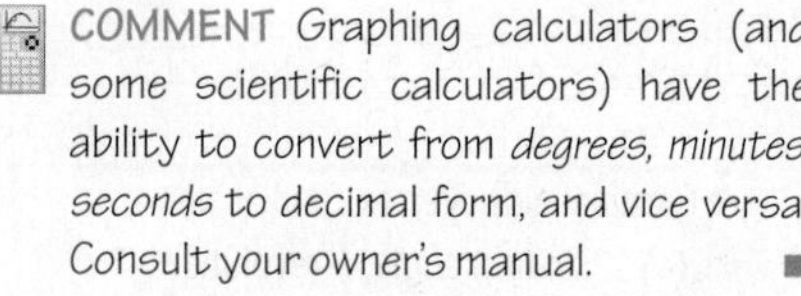
COMMENT Graphing calculators (and some scientific calculators) have the ability to convert from *degrees, minutes, seconds* to decimal form, and vice versa. Consult your owner's manual. ■

$$1 \text{ counterclockwise revolution} = 360^\circ$$
$$1^\circ = 60' \qquad 1' = 60'' \tag{1}$$

It is sometimes necessary to convert from the degrees, minutes, seconds notation (D°M′S″) to a decimal form, and vice versa.

EXAMPLE 2 **Converting between Degrees, Minutes, Seconds, and Decimal Forms**

(a) Convert 50°6′21″ to a decimal in degrees. Round the answer to four decimal places.

(b) Convert 21.256° to the D°M′S″ form. Round the answer to the nearest second.

Solution (a) Because $1' = \left(\frac{1}{60}\right)^\circ$ and $1'' = \left(\frac{1}{60}\right)' = \left(\frac{1}{60}\cdot\frac{1}{60}\right)^\circ$, convert as follows:

$$
\begin{aligned}
50^\circ 6' 21'' &= 50^\circ + 6' + 21'' = 50^\circ + 6\cdot 1' + 21\cdot 1'' \\
&= 50^\circ + 6\cdot\left(\frac{1}{60}\right)^\circ + 21\cdot\left(\frac{1}{60}\cdot\frac{1}{60}\right)^\circ \\
&\approx 50^\circ + 0.1^\circ + 0.0058^\circ \\
&= 50.1058^\circ
\end{aligned}
$$

Convert minutes and seconds to degrees.

(b) Because $1° = 60'$ and $1' = 60''$, proceed as follows:

$$\begin{aligned} 21.256° &= 21° + 0.256° \\ &= 21° + (0.256) \cdot 1° \\ &= 21° + (0.256)(60') && \text{Convert fraction of degree to minutes; } 1° = 60'. \\ &= 21° + 15.36' \\ &= 21° + 15' + 0.36' \\ &= 21° + 15' + (0.36) \cdot 1' \\ &= 21° + 15' + (0.36)(60'') && \text{Convert fraction of minute to seconds; } 1' = 60''. \\ &= 21° + 15' + 21.6'' \\ &\approx 21°15'22'' && \text{Round to the nearest second.} \end{aligned}$$

Now Work PROBLEMS 23 AND 29

In many applications, such as describing the exact location of a star or the precise position of a ship at sea, angles measured in degrees, minutes, and even seconds are used. For calculation purposes, these are transformed to decimal form. In other applications, especially those in calculus, angles are measured using *radians*.

Radians

A **central angle** is a positive angle whose vertex is at the center of a circle. The rays of a central angle subtend (intersect) an arc on the circle. If the radius of the circle is r and the length of the arc subtended by the central angle is also r, then the measure of the angle is **1 radian**. See Figure 10(a).

For a circle of radius 1, the rays of a central angle with measure 1 radian subtend an arc of length 1. For a circle of radius 3, the rays of a central angle with measure 1 radian subtend an arc of length 3. See Figure 10(b).

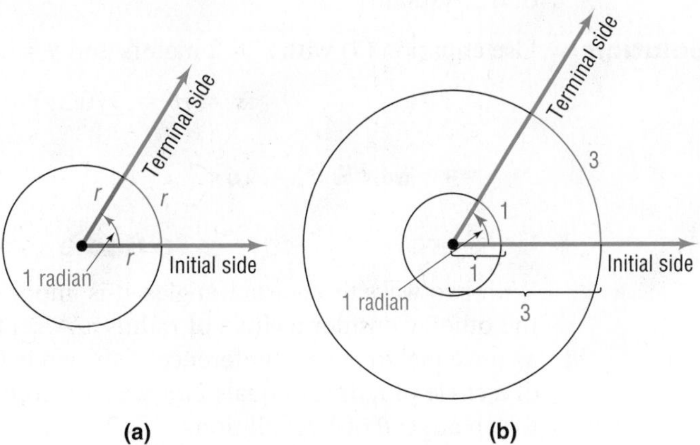

Figure 10 (a) (b)

2 Find the Length of an Arc of a Circle

Now consider a circle of radius r and two central angles, θ and θ_1, measured in radians. Suppose that these central angles subtend arcs of lengths s and s_1, respectively, as shown in Figure 11. From geometry, the ratio of the measures of the angles equals the ratio of the corresponding lengths of the arcs subtended by these angles; that is,

$$\frac{\theta}{\theta_1} = \frac{s}{s_1} \qquad (2)$$

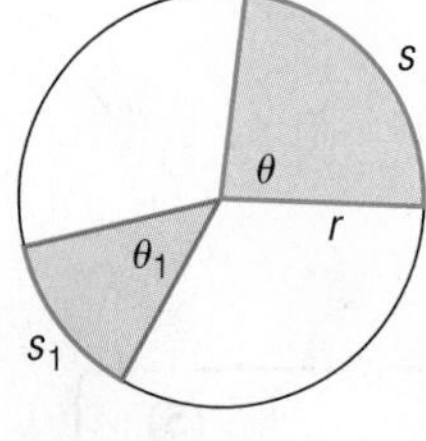

Figure 11 $\frac{\theta}{\theta_1} = \frac{s}{s_1}$

Suppose that $\theta_1 = 1$ radian. Refer again to Figure 10(a). The length s_1 of the arc subtended by the central angle $\theta_1 = 1$ radian equals the radius r of the circle. Then $s_1 = r$, so equation (2) reduces to

$$\frac{\theta}{1} = \frac{s}{r} \quad \text{or} \quad s = r\theta \tag{3}$$

THEOREM

Arc Length

For a circle of radius r, a central angle of θ radians subtends an arc whose length s is

$$s = r\theta \tag{4}$$

NOTE Formulas must be consistent with regard to the units used. In equation (4), we write

$$s = r\theta$$

To see the units, however, we must go back to equation (3) and write

$$\frac{\theta \text{ radians}}{1 \text{ radian}} = \frac{s \text{ length units}}{r \text{ length units}}$$

$$s \text{ length units} = r \text{ length units} \frac{\theta \text{ radians}}{1 \text{ radian}}$$

The radians divide out, leaving

$$s \text{ length units} = (r \text{ length units})\theta \qquad s = r\theta$$

where θ appears to be "dimensionless" but, in fact, is measured in radians. So, in using the formula $s = r\theta$, the dimension for θ is radians, and any convenient unit of length (such as inches or meters) may be used for s and r. ■

EXAMPLE 3 **Finding the Length of an Arc of a Circle**

Find the length of the arc of a circle of radius 2 meters subtended by a central angle of 0.25 radian.

Solution Use equation (4) with $r = 2$ meters and $\theta = 0.25$. The length s of the arc is

$$s = r\theta = 2(0.25) = 0.5 \text{ meter}$$ ●

Now Work PROBLEM 71

3 Convert from Degrees to Radians and from Radians to Degrees

With two ways to measure angles, it is important to be able to convert from one to the other. Consider a circle of radius r. A central angle of 1 revolution will subtend an arc equal to the circumference of the circle (Figure 12). Because the circumference of a circle of radius r equals $2\pi r$, we substitute $2\pi r$ for s in equation (4) to find that, for an angle θ of 1 revolution,

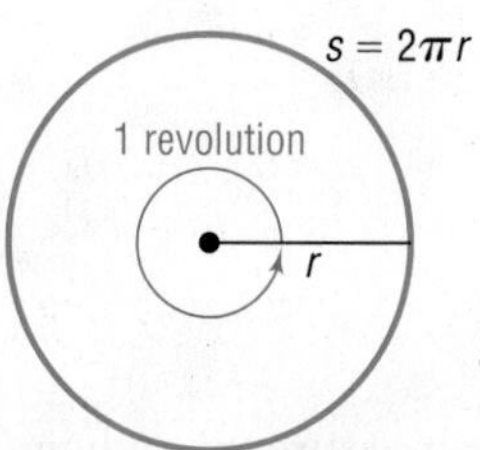

Figure 12
1 revolution = 2π radians

$$s = r\theta$$

$$2\pi r = r\theta \qquad \theta = 1 \text{ revolution}; s = 2\pi r$$

$$\theta = 2\pi \text{ radians} \qquad \text{Solve for } \theta.$$

From this, we have

$$1 \text{ revolution} = 2\pi \text{ radians} \tag{5}$$

Since 1 revolution $= 360°$, we have

$$360° = 2\pi \text{ radians}$$

Dividing both sides by 2 yields

$$180° = \pi \text{ radians} \qquad (6)$$

Divide both sides of equation (6) by 180. Then

$$1 \text{ degree} = \frac{\pi}{180} \text{ radian}$$

Divide both sides of (6) by π. Then

$$\frac{180}{\pi} \text{ degrees} = 1 \text{ radian}$$

We have the following two conversion formulas:*

$$1 \text{ degree} = \frac{\pi}{180} \text{ radian} \qquad 1 \text{ radian} = \frac{180}{\pi} \text{ degrees} \qquad (7)$$

EXAMPLE 4

Converting from Degrees to Radians

Convert each angle in degrees to radians.

(a) $60°$ (b) $150°$ (c) $-45°$ (d) $90°$ (e) $107°$

Solution

(a) $60° = 60 \cdot 1 \text{ degree} = 60 \cdot \frac{\pi}{180} \text{ radian} = \frac{\pi}{3} \text{ radians}$

(b) $150° = 150 \cdot 1° = 150 \cdot \frac{\pi}{180} \text{ radian} = \frac{5\pi}{6} \text{ radians}$

(c) $-45° = -45 \cdot \frac{\pi}{180} \text{ radian} = -\frac{\pi}{4} \text{ radian}$

(d) $90° = 90 \cdot \frac{\pi}{180} \text{ radian} = \frac{\pi}{2} \text{ radians}$

(e) $107° = 107 \cdot \frac{\pi}{180} \text{ radian} \approx 1.868 \text{ radians}$ •

Example 4, parts (a)–(d), illustrates that angles that are "nice" fractions of a revolution are expressed in radian measure as fractional multiples of π, rather than as decimals. For example, a right angle, as in Example 4(d), is left in the form $\frac{\pi}{2}$ radians, which is exact, rather than using the approximation $\frac{\pi}{2} \approx \frac{3.1416}{2} = 1.5708$ radians. When the fractions are not "nice," we use the decimal approximation of the angle, as in Example 4(e).

Now Work PROBLEMS 35 AND 61

*Some students prefer instead to use the proportion $\frac{\text{Degree}}{180°} = \frac{\text{Radian}}{\pi}$, then substitute for what is given, and solve for the measurement sought.

EXAMPLE 5 Converting Radians to Degrees

Convert each angle in radians to degrees.

(a) $\frac{\pi}{6}$ radian (b) $\frac{3\pi}{2}$ radians (c) $-\frac{3\pi}{4}$ radians

(d) $\frac{7\pi}{3}$ radians (e) 5 radians

Solution

(a) $\frac{\pi}{6}\text{ radian} = \frac{\pi}{6}\cdot 1\text{ radian} = \frac{\pi}{6}\cdot\frac{180}{\pi}\text{ degrees} = 30°$

(b) $\frac{3\pi}{2}\text{ radians} = \frac{3\pi}{2}\cdot\frac{180}{\pi}\text{ degrees} = 270°$

(c) $-\frac{3\pi}{4}\text{ radians} = -\frac{3\pi}{4}\cdot\frac{180}{\pi}\text{ degrees} = -135°$

(d) $\frac{7\pi}{3}\text{ radians} = \frac{7\pi}{3}\cdot\frac{180}{\pi}\text{ degrees} = 420°$

(e) $5\text{ radians} = 5\cdot\frac{180}{\pi}\text{ degrees} \approx 286.48°$

Now Work PROBLEM 47

Table 1 lists the degree and radian measures of some commonly encountered angles. You should learn to feel equally comfortable using either measure for these angles.

Table 1

Degrees	0°	30°	45°	60°	90°	120°	135°	150°	180°
Radians	0	$\frac{\pi}{6}$	$\frac{\pi}{4}$	$\frac{\pi}{3}$	$\frac{\pi}{2}$	$\frac{2\pi}{3}$	$\frac{3\pi}{4}$	$\frac{5\pi}{6}$	π
Degrees		210°	225°	240°	270°	300°	315°	330°	360°
Radians		$\frac{7\pi}{6}$	$\frac{5\pi}{4}$	$\frac{4\pi}{3}$	$\frac{3\pi}{2}$	$\frac{5\pi}{3}$	$\frac{7\pi}{4}$	$\frac{11\pi}{6}$	2π

EXAMPLE 6 Finding the Distance between Two Cities

The latitude of a location L is the measure of the angle formed by a ray drawn from the center of Earth to the equator and a ray drawn from the center of Earth to L. See Figure 13(a). Sioux Falls, South Dakota, is due north of Dallas, Texas. Find the distance between Sioux Falls (43° 33′ north latitude) and Dallas (32° 46′ north latitude). See Figure 13(b). Assume that the radius of Earth is 3960 miles.

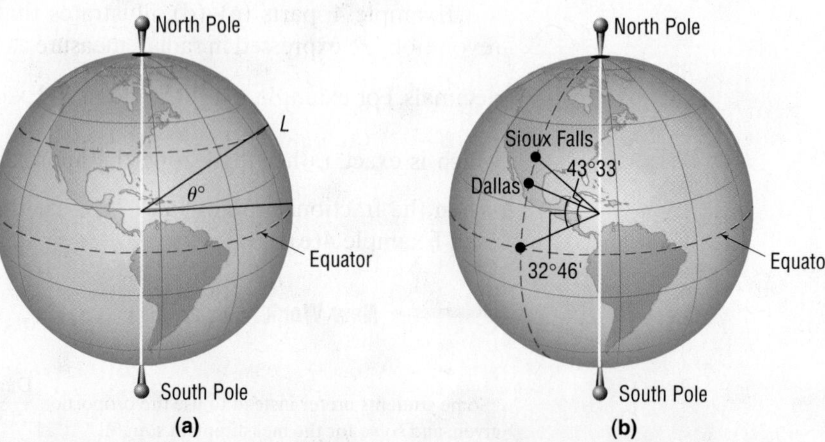

Figure 13 (a) (b)

Solution The measure of the central angle between the two cities is $43° 33' - 32° 46' = 10° 47'$. Use equation (4), $s = r\theta$. But remember to first convert the angle of $10° 47'$ to radians.

$$\theta = 10°47' \approx 10.7833° = 10.7833 \cdot \frac{\pi}{180} \text{ radian} \approx 0.188 \text{ radian}$$

$$\uparrow$$

$$47' = 47\left(\frac{1}{60}\right)^{\circ}$$

Use $\theta = 0.188$ radian and $r = 3960$ miles in equation (4). The distance between the two cities is

$$s = r\theta = 3960 \cdot 0.188 \approx 744 \text{ miles}$$

NOTE If the measure of an angle is given as 5, it is understood to mean 5 radians; if the measure of an angle is given as 5°, it means 5 degrees. ■

When an angle is measured in degrees, the degree symbol will always be shown. However, when an angle is measured in radians, the usual practice is to omit the word *radians*. So if the measure of an angle is given as $\frac{\pi}{6}$, it is understood to mean $\frac{\pi}{6}$ radian.

Now Work PROBLEM 107

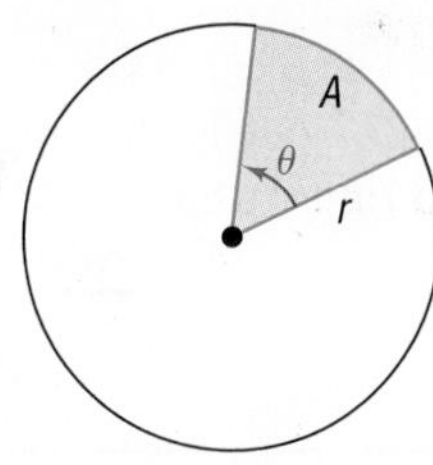

Figure 14 Sector of a Circle

4 Find the Area of a Sector of a Circle

Consider a circle of radius r. Suppose that θ, measured in radians, is a central angle of this circle. See Figure 14. We seek a formula for the area A of the sector (shown in blue) formed by the angle θ.

Now consider a circle of radius r and two central angles θ and θ_1, both measured in radians. See Figure 15. From geometry, we know that the ratio of the measures of the angles equals the ratio of the corresponding areas of the sectors formed by these angles. That is,

$$\frac{\theta}{\theta_1} = \frac{A}{A_1}$$

Figure 15 $\frac{\theta}{\theta_1} = \frac{A}{A_1}$

Suppose that $\theta_1 = 2\pi$ radians. Then $A_1 =$ area of the circle $= \pi r^2$. Solving for A, we find

$$A = A_1 \frac{\theta}{\theta_1} = \pi r^2 \frac{\theta}{2\pi} = \frac{1}{2} r^2 \theta$$

$$\uparrow$$

$$A_1 = \pi r^2$$
$$\theta_1 = 2\pi$$

THEOREM **Area of a Sector**

The area A of the sector of a circle of radius r formed by a central angle of θ radians is

$$A = \frac{1}{2} r^2 \theta \qquad (8)$$

EXAMPLE 7 **Finding the Area of a Sector of a Circle**

Find the area of the sector of a circle of radius 2 feet formed by an angle of 30°. Round the answer to two decimal places.

Solution Use equation (8) with $r = 2$ feet and $\theta = 30° = \frac{\pi}{6}$ radian. [Remember, in equation (8), θ must be in radians.]

$$A = \frac{1}{2} r^2 \theta = \frac{1}{2} (2)^2 \frac{\pi}{6} = \frac{\pi}{3} \approx 1.05$$

The area A of the sector is 1.05 square feet, rounded to two decimal places.

Now Work PROBLEM 79

5 Find the Linear Speed of an Object Traveling in Circular Motion

The average speed of an object equals the distance traveled divided by the elapsed time. For motion along a circle, we distinguish between *linear speed* and *angular speed.*

DEFINITION

Suppose that an object moves around a circle of radius r at a constant speed. If s is the distance traveled in time t around this circle, then the **linear speed** v of the object is defined as

$$v = \frac{s}{t} \quad \textbf{(9)}$$

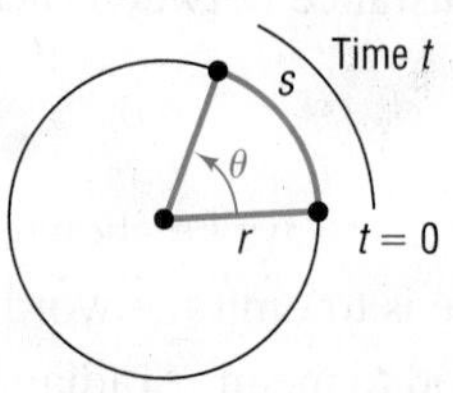

Figure 16 $v = \frac{s}{t}$

As this object travels around the circle, suppose that θ (measured in radians) is the central angle swept out in time t. See Figure 16.

DEFINITION

The **angular speed** ω (the Greek letter omega) of this object is the angle θ (measured in radians) swept out, divided by the elapsed time t; that is,

$$\omega = \frac{\theta}{t} \quad \textbf{(10)}$$

Angular speed is the way the turning rate of an engine is described. For example, an engine idling at 900 rpm (revolutions per minute) is one that rotates at an angular speed of

$$900\,\frac{\text{revolutions}}{\text{minute}} = 900\,\frac{\cancel{\text{revolutions}}}{\text{minute}} \cdot 2\pi\,\frac{\text{radians}}{\cancel{\text{revolution}}} = 1800\pi\,\frac{\text{radians}}{\text{minute}}$$

There is an important relationship between linear speed and angular speed:

$$\text{linear speed} = v = \underset{(9)}{\frac{s}{t}} = \underset{s = r\theta}{\frac{r\theta}{t}} = r\underset{(10)}{\left(\frac{\theta}{t}\right)} = r \cdot \omega$$

$$v = r\omega \quad \textbf{(11)}$$

where ω is measured in radians per unit time.

When using equation (11), remember that $v = \frac{s}{t}$ (the linear speed) has the dimensions of length per unit of time (such as feet per second or miles per hour), r (the radius of the circular motion) has the same length dimension as s, and ω (the angular speed) has the dimensions of radians per unit of time. If the angular speed is given in terms of *revolutions* per unit of time (as is often the case), be sure to convert it to *radians* per unit of time using the fact that 1 revolution $= 2\pi$ radians before attempting to use equation (11).

EXAMPLE 8 **Finding Linear Speed**

A child is spinning a rock at the end of a 2-foot rope at a rate of 180 revolutions per minute (rpm). Find the linear speed of the rock when it is released.

Solution Look at Figure 17. The rock is moving around a circle of radius $r = 2$ feet. The angular speed ω of the rock is

$$\omega = 180\,\frac{\text{revolutions}}{\text{minute}} = 180\,\frac{\cancel{\text{revolutions}}}{\text{minute}} \cdot 2\pi\,\frac{\text{radians}}{\cancel{\text{revolution}}} = 360\pi\,\frac{\text{radians}}{\text{minute}}$$

Figure 17

From equation (11), the linear speed v of the rock is

$$v = r\omega = 2 \text{ feet} \cdot 360\pi \frac{\text{radians}}{\text{minute}} = 720\pi \frac{\text{feet}}{\text{minute}} \approx 2262 \frac{\text{feet}}{\text{minute}}$$

The linear speed of the rock when it is released is 2262 ft/min $\approx$ 25.7 mi/h. ●

Now Work PROBLEM 99

Historical Feature

Trigonometry was developed by Greek astronomers, who regarded the sky as the inside of a sphere, so it was natural that triangles on a sphere were investigated early (by Menelaus of Alexandria about AD 100) and that triangles in the plane were studied much later. The first book containing a systematic treatment of plane and spherical trigonometry was written by the Persian astronomer Nasir Eddin (about AD 1250).

Regiomontanus (1436–1476) is the person most responsible for moving trigonometry from astronomy into mathematics. His work was improved by Copernicus (1473–1543) and Copernicus's student Rhaeticus (1514–1576). Rhaeticus's book was the first to define the six trigonometric functions as ratios of sides of triangles, although he did not give the functions their present names. Credit for this is due to Thomas Finck (1583), but Finck's notation was by no means universally accepted at the time. The notation was finally stabilized by the textbooks of Leonhard Euler (1707–1783).

Trigonometry has since evolved from its use by surveyors, navigators, and engineers to present applications involving ocean tides, the rise and fall of food supplies in certain ecologies, brain wave patterns, and many other phenomena.

2.1 Assess Your Understanding

'Are You Prepared?' *Answers are given at the end of these exercises. If you get a wrong answer, read the pages listed in red.*

1. What is the formula for the circumference C of a circle of radius r? (p. A16)

2. The area of a semicircle of radius 2 is ________. (p. A16)

Concepts and Vocabulary

3. An angle θ is in ________ ________ if its vertex is at the origin of a rectangular coordinate system and its initial side coincides with the positive x-axis.

4. A ________ ________ is a positive angle whose vertex is at the center of a circle.

5. If the radius of a circle is r and the length of the arc subtended by a central angle is also r, then the measure of the angle is 1 ______.
(a) degree (b) minute (c) second (d) radian

6. On a circle of radius r, a central angle of θ radians subtends an arc of length $s =$ ___; the area of the sector formed by this angle θ is $A =$ ______.

7. $180° =$ ______ radians
(a) $\frac{\pi}{2}$ (b) π (c) $\frac{3\pi}{2}$ (d) 2π

8. An object travels around a circle of radius r with constant speed. If s is the distance traveled in time t around the circle and θ is the central angle (in radians) swept out in time t, then the linear speed of the object is $v =$ __ and the angular speed of the object is $\omega =$ __.

9. ***True or False*** The angular speed ω of an object traveling around a circle of radius r is the angle θ (measured in radians) swept out, divided by the elapsed time t.

10. ***True or False*** For circular motion on a circle of radius r, linear speed equals angular speed divided by r.

Skill Building

In Problems 11–22, draw each angle.

11. 30° **12.** 60° **13.** 135° **14.** −120° **15.** 450° **16.** 540°

17. $\frac{3\pi}{4}$ **18.** $\frac{4\pi}{3}$ **19.** $-\frac{\pi}{6}$ **20.** $-\frac{2\pi}{3}$ **21.** $\frac{16\pi}{3}$ **22.** $\frac{21\pi}{4}$

In Problems 23–28, convert each angle to a decimal in degrees. Round your answer to two decimal places.

23. 40°10′25″ **24.** 61°42′21″ **25.** 50°14′20″ **26.** 73°40′40″ **27.** 9°9′9″ **28.** 98°22′45″

In Problems 29–34, convert each angle to D°M′S″ form. Round your answer to the nearest second.

29. 40.32° **30.** 61.24° **31.** 18.255° **32.** 29.411° **33.** 19.99° **34.** 44.01°

In Problems 35–46, convert each angle in degrees to radians. Express your answer as a multiple of π.

35. 30° **36.** 120° **37.** 240° **38.** 330° **39.** $-60°$ **40.** $-30°$

41. 180° **42.** 270° **43.** $-135°$ **44.** $-225°$ **45.** $-90°$ **46.** $-180°$

In Problems 47–58, convert each angle in radians to degrees.

47. $\frac{\pi}{3}$ **48.** $\frac{5\pi}{6}$ **49.** $-\frac{5\pi}{4}$ **50.** $-\frac{2\pi}{3}$ **51.** $\frac{\pi}{2}$ **52.** 4π

53. $\frac{\pi}{12}$ **54.** $\frac{5\pi}{12}$ **55.** $-\frac{\pi}{2}$ **56.** $-\pi$ **57.** $-\frac{\pi}{6}$ **58.** $-\frac{3\pi}{4}$

In Problems 59–64, convert each angle in degrees to radians. Express your answer in decimal form, rounded to two decimal places.

59. 17° **60.** 73° **61.** $-40°$ **62.** $-51°$ **63.** 125° **64.** 350°

In Problems 65–70, convert each angle in radians to degrees. Express your answer in decimal form, rounded to two decimal places.

65. 3.14 **66.** 0.75 **67.** 2 **68.** 3 **69.** 6.32 **70.** $\sqrt{2}$

In Problems 71–78, s denotes the length of the arc of a circle of radius r subtended by the central angle θ. Find the missing quantity. Round answers to three decimal places.

71. $r = 10$ meters, $\theta = \frac{1}{2}$ radian, $s = ?$

72. $r = 6$ feet, $\theta = 2$ radians, $s = ?$

73. $\theta = \frac{1}{3}$ radian, $s = 2$ feet, $r = ?$

74. $\theta = \frac{1}{4}$ radian, $s = 6$ centimeters, $r = ?$

75. $r = 5$ miles, $s = 3$ miles, $\theta = ?$

76. $r = 6$ meters, $s = 8$ meters, $\theta = ?$

77. $r = 2$ inches, $\theta = 30°$, $s = ?$

78. $r = 3$ meters, $\theta = 120°$, $s = ?$

In Problems 79–86, A denotes the area of the sector of a circle of radius r formed by the central angle θ. Find the missing quantity. Round answers to three decimal places.

79. $r = 10$ meters, $\theta = \frac{1}{2}$ radian, $A = ?$

80. $r = 6$ feet, $\theta = 2$ radians, $A = ?$

81. $\theta = \frac{1}{3}$ radian, $A = 2$ square feet, $r = ?$

82. $\theta = \frac{1}{4}$ radian, $A = 6$ square centimeters, $r = ?$

83. $r = 5$ miles, $A = 3$ square miles, $\theta = ?$

84. $r = 6$ meters, $A = 8$ square meters, $\theta = ?$

85. $r = 2$ inches, $\theta = 30°$, $A = ?$

86. $r = 3$ meters, $\theta = 120°$, $A = ?$

In Problems 87–90, find the length s and area A. Round answers to three decimal places.

87.

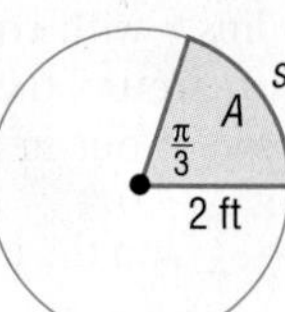

88.

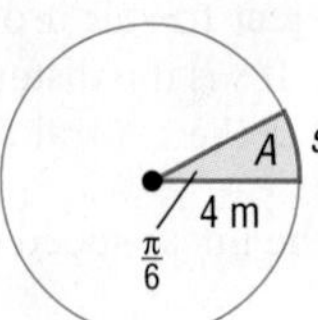

89.

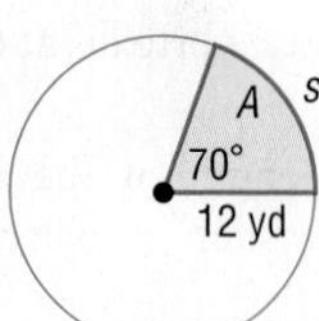

90.

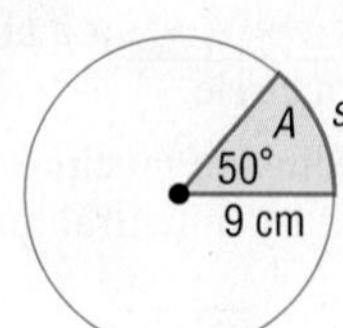

Applications and Extensions

91. Movement of a Minute Hand The minute hand of a clock is 6 inches long. How far does the tip of the minute hand move in 15 minutes? How far does it move in 25 minutes? Round answers to two decimal places.

92. Movement of a Pendulum A pendulum swings through an angle of 20° each second. If the pendulum is 40 inches long, how far does its tip move each second? Round answers to two decimal places.

93. Area of a Sector Find the area of the sector of a circle of radius 4 meters formed by an angle of 45°. Round the answer to two decimal places.

94. Area of a Sector Find the area of the sector of a circle of radius 3 centimeters formed by an angle of 60°. Round the answer to two decimal places.

95. Watering a Lawn A water sprinkler sprays water over a distance of 30 feet while rotating through an angle of 135°. What area of lawn receives water?

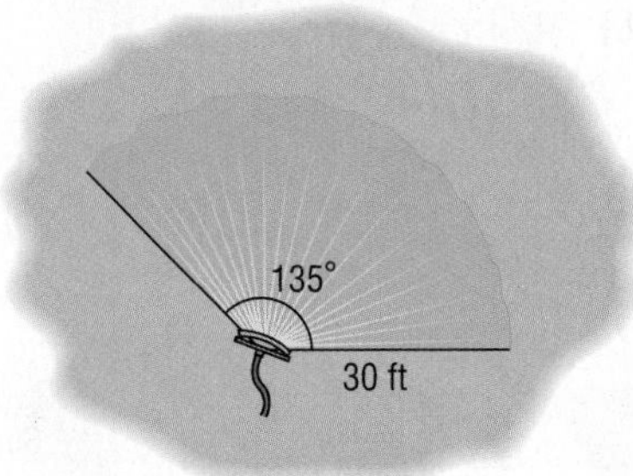

96. Designing a Water Sprinkler An engineer is asked to design a water sprinkler that will cover a field of 100 square yards that is in the shape of a sector of a circle of radius 15 yards. Through what angle should the sprinkler rotate?

97. Windshield Wiper The arm and blade of a windshield wiper have a total length of 34 inches. If the blade is 25 inches long and the wiper sweeps out an angle of 120°, how much window area can the blade clean?

98. Windshield Wiper The arm and blade of a windshield wiper have a total length of 30 inches. If the blade is 24 inches long and the wiper sweeps out an angle of 125°, how much window area can the blade clean?

99. Motion on a Circle An object is traveling around a circle with a radius of 5 centimeters. If in 20 seconds a central angle of $\frac{1}{3}$ radian is swept out, what is the angular speed of the object? What is its linear speed?

100. Motion on a Circle An object is traveling around a circle with a radius of 2 meters. If in 20 seconds the object travels 5 meters, what is its angular speed? What is its linear speed?

101. Amusement Park Ride A gondola on an amusement park ride, similar to the Spin Cycle at Silverwood Theme Park, spins at a speed of 13 revolutions per minute. If the gondola is 25 feet from the ride's center, what is the linear speed of the gondola in miles per hour?

102. Amusement Park Ride A centrifugal force ride, similar to the Gravitron, spins at a speed of 22 revolutions per minute. If the diameter of the ride is 13 meters, what is the linear speed of the passengers in kilometers per hour?

103. Blu-ray Drive A Blu-ray drive has a maximum speed of 10,000 revolutions per minute. If a Blu-ray disc has a diameter of 12 cm, what is the linear speed, in km/h, of a point 4 cm from the center if the disc is spinning at a rate of 8000 revolutions per minute?

104. DVD Drive A DVD drive has a maximum speed of 7200 revolutions per minute. If a DVD has a diameter of 12 cm, what is the linear speed, in km/h, of a point 5 cm from the disc's center if it is spinning at a rate of 5400 revolutions per minute?

105. Bicycle Wheels The diameter of each wheel of a bicycle is 26 inches. If you are traveling at a speed of 35 miles per hour on this bicycle, through how many revolutions per minute are the wheels turning?

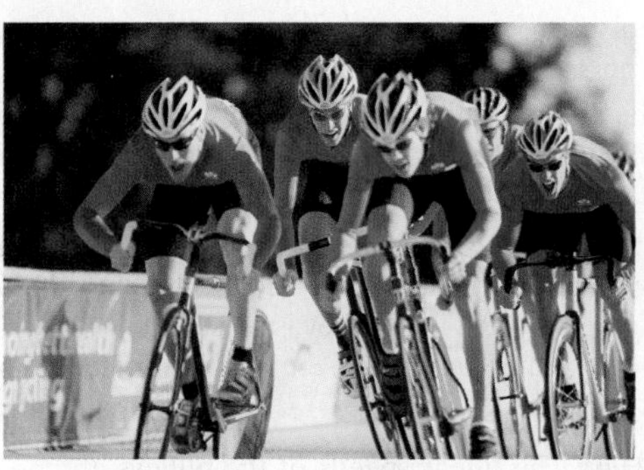

106. Car Wheels The radius of each wheel of a car is 15 inches. If the wheels are turning at a rate of 3 revolutions per second, how fast is the car moving? Express your answer in inches per second and in miles per hour.

In Problems 107–110, the latitude of a location L is the angle formed by a ray drawn from the center of Earth to the Equator and a ray drawn from the center of Earth to L. See the figure.

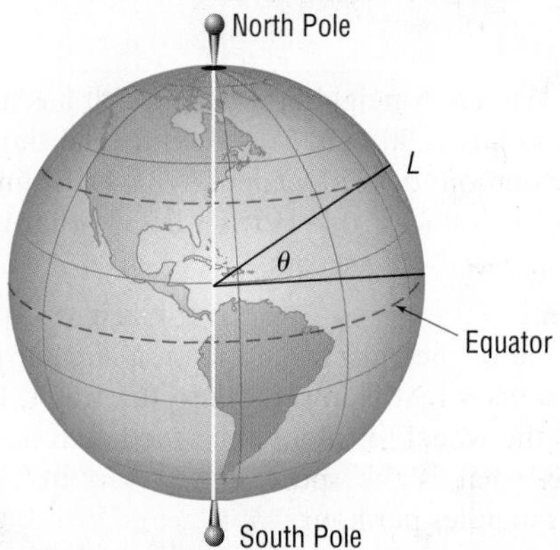

107. Distance between Cities Memphis, Tennessee, is due north of New Orleans, Louisiana. Find the distance between Memphis (35°9′ north latitude) and New Orleans (29°57′ north latitude). Assume that the radius of Earth is 3960 miles.

108. Distance between Cities Charleston, West Virginia, is due north of Jacksonville, Florida. Find the distance between Charleston (38°21′ north latitude) and Jacksonville (30°20′ north latitude). Assume that the radius of Earth is 3960 miles.

109. Linear Speed on Earth Earth rotates on an axis through its poles. The distance from the axis to a location on Earth at 30° north latitude is about 3429.5 miles. Therefore, a location on Earth at 30° north latitude is spinning on a circle of radius 3429.5 miles. Compute the linear speed on the surface of Earth at 30° north latitude.

110. Linear Speed on Earth Earth rotates on an axis through its poles. The distance from the axis to a location on Earth at 40° north latitude is about 3033.5 miles. Therefore, a location on Earth at 40° north latitude is spinning on a circle of radius 3033.5 miles. Compute the linear speed on the surface of Earth at 40° north latitude.

111. Speed of the Moon The mean distance of the moon from Earth is 2.39×10^5 miles. Assuming that the orbit of the moon around Earth is circular and that 1 revolution takes 27.3 days, find the linear speed of the moon. Express your answer in miles per hour.

112. Speed of Earth The mean distance of Earth from the Sun is 9.29×10^7 miles. Assuming that the orbit of Earth around the Sun is circular and that 1 revolution takes 365 days, find the linear speed of Earth. Express your answer in miles per hour.

113. Pulleys Two pulleys, one with radius 2 inches and the other with radius 8 inches, are connected by a belt. (See the figure.) If the 2-inch pulley is caused to rotate at 3 revolutions per minute, determine the revolutions per minute of the 8-inch pulley.

[**Hint:** The linear speeds of the pulleys are the same; both equal the speed of the belt.]

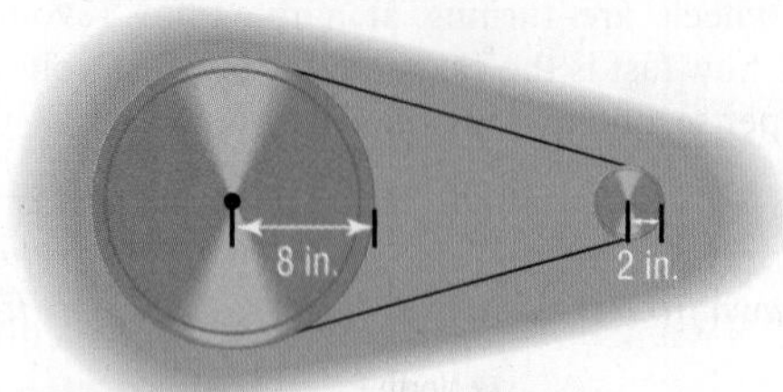

114. Ferris Wheels A neighborhood carnival has a Ferris wheel whose radius is 30 feet. You measure the time it takes for one revolution to be 70 seconds. What is the linear speed (in feet per second) of this Ferris wheel? What is the angular speed in radians per second?

115. Computing the Speed of a River Current To approximate the speed of the current of a river, a circular paddle wheel with radius 4 feet is lowered into the water. If the current causes the wheel to rotate at a speed of 10 revolutions per minute, what is the speed of the current? Express your answer in miles per hour.

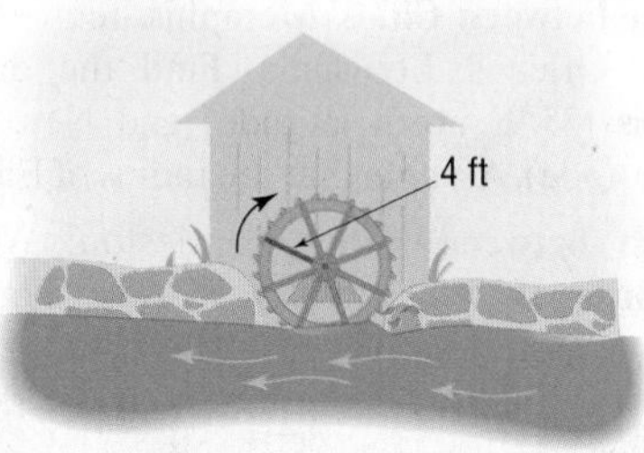

116. Spin Balancing Tires A spin balancer rotates the wheel of a car at 480 revolutions per minute. If the diameter of the wheel is 26 inches, what road speed is being tested? Express your answer in miles per hour. At how many revolutions per minute should the balancer be set to test a road speed of 80 miles per hour?

117. The Cable Cars of San Francisco At the Cable Car Museum you can see the four cable lines that are used to pull cable cars up and down the hills of San Francisco. Each cable travels at a speed of 9.55 miles per hour, caused by a rotating wheel whose diameter is 8.5 feet. How fast is the wheel rotating? Express your answer in revolutions per minute.

118. Difference in Time of Sunrise Naples, Florida, is about 90 miles due west of Ft. Lauderdale. How much sooner would a person in Ft. Lauderdale first see the rising Sun than a person in Naples? See the hint.

[**Hint:** Consult the figure. When a person at Q sees the first rays of the Sun, a person at P is still in the dark. The person at P sees the first rays after Earth has rotated so that P is at the location Q. Now use the fact that at the latitude of Ft. Lauderdale in 24 hours an arc of length $2\pi(3559)$ miles is subtended.]

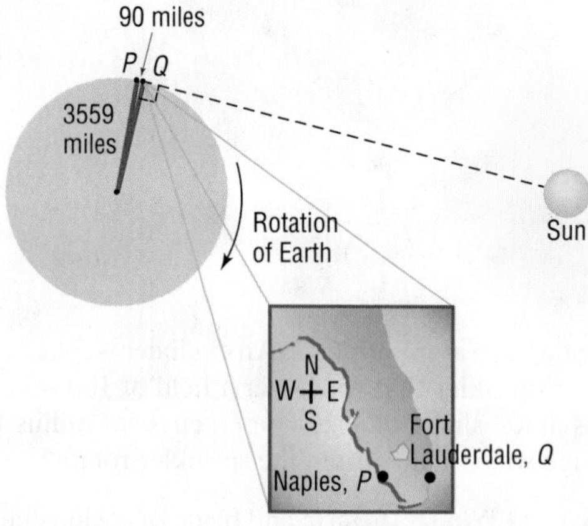

119. Let the Dog Roam A dog is attached to a 9-foot rope fastened to the outside corner of a fenced-in garden that measures 6 feet by 10 feet. Assuming that the dog cannot enter the garden, compute the exact area that the dog can wander. Write the exact area in square feet.*

120. Area of a Region The measure of arc $\widehat{BE}$ is 2π. Find the exact area of the portion of the rectangle $ABCD$ that falls outside of the circle whose center is at A.*

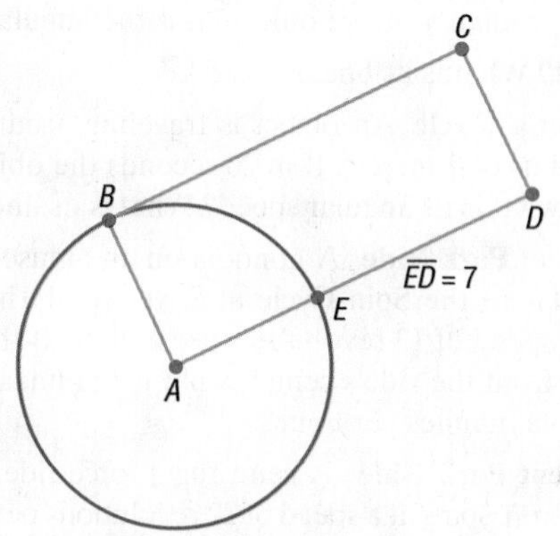

*Courtesy of the Joliet Junior College Mathematics Department

121. Keeping Up with the Sun How fast would you have to travel on the surface of Earth at the equator to keep up with the Sun (that is, so that the Sun would appear to remain in the same position in the sky)?

122. Nautical Miles A **nautical mile** equals the length of arc subtended by a central angle of 1 minute on a great circle† on the surface of Earth. See the figure on the next page. If the radius of Earth is taken as 3960 miles, express 1 nautical mile in terms of ordinary, or **statute**, miles.

†Any circle drawn on the surface of Earth that divides Earth into two equal hemispheres.

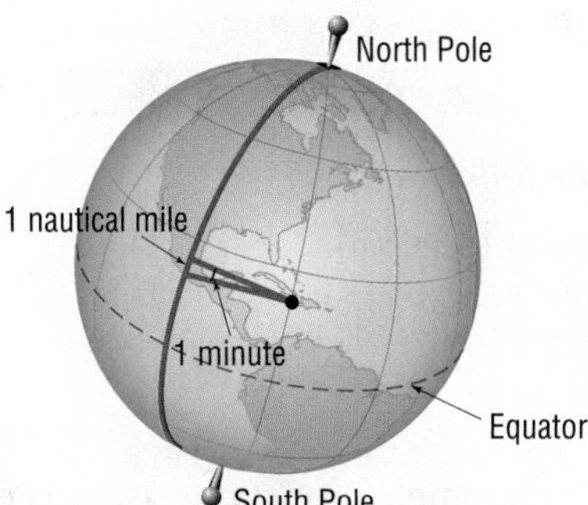

123. Approximating the Circumference of Earth Eratosthenes of Cyrene (276–195 BC) was a Greek scholar who lived and worked in Cyrene and Alexandria. One day while visiting in Syene he noticed that the Sun's rays shone directly down a well. On this date 1 year later, in Alexandria, which is 500 miles due north of Syene he measured the angle of the Sun to be about 7.2 degrees. See the figure. Use this information to approximate the radius and circumference of Earth.

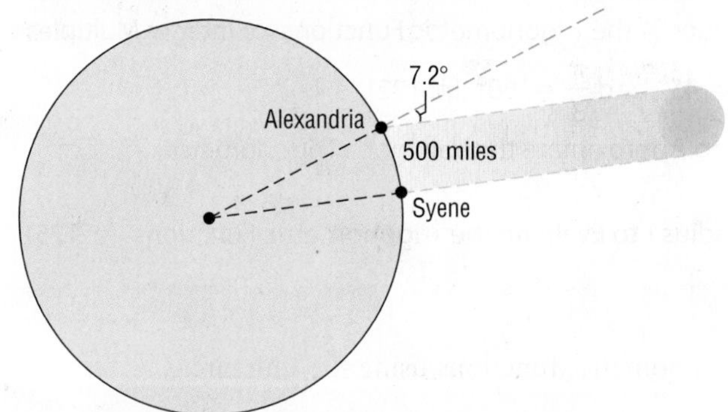

124. Designing a Little League Field For a 60-foot Little League Baseball field, the distance from home base to the nearest fence (or other obstruction) in fair territory should be a minimum of 200 feet. The commissioner of parks and recreation is making plans for a new 60-foot field. Because of limited ground availability, he will use the minimum required distance to the outfield fence. To increase safety, however, he plans to include a 10-foot-wide warning track on the inside of the fence. To further increase safety, the fence and warning track will extend both directions into foul territory. In total, the arc formed by the outfield fence (including the extensions into the foul territories) will be subtended by a central angle at home plate measuring 96°, as illustrated.

(a) Determine the length of the outfield fence.

(b) Determine the area of the warning track.

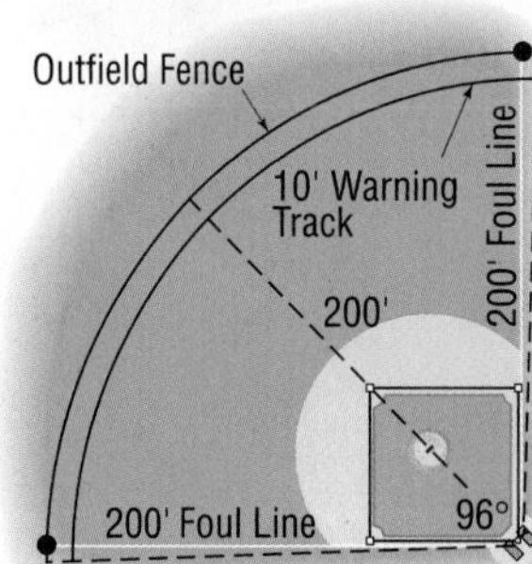

Source: www.littleleague.org

[**Note:** There is a 90° angle between the two foul lines. Then there are two 3° angles between the foul lines and the dotted lines shown. The angle between the two dotted lines outside the 200-foot foul lines is 96°.]

125. Pulleys Two pulleys, one with radius r_1 and the other with radius r_2, are connected by a belt. The pulley with radius r_1 rotates at ω_1 revolutions per minute, whereas the pulley with radius r_2 rotates at ω_2 revolutions per minute. Show that $\frac{r_1}{r_2} = \frac{\omega_2}{\omega_1}$.

Explaining Concepts: Discussion and Writing

126. Do you prefer to measure angles using degrees or radians? Provide justification and a rationale for your choice.

127. What is 1 radian? What is 1 degree?

128. Which angle has the larger measure: 1 degree or 1 radian? Or are they equal?

129. Explain the difference between linear speed and angular speed.

130. For a circle of radius r, a central angle of θ degrees subtends an arc whose length s is $s = \frac{\pi}{180} r\theta$. Discuss whether this statement is true or false. Defend your position.

131. Discuss why ships and airplanes use nautical miles to measure distance. Explain the difference between a nautical mile and a statute mile.

132. Investigate the way that speed bicycles work. In particular, explain the differences and similarities between 5-speed and 9-speed derailleurs. Be sure to include a discussion of linear speed and angular speed.

133. In Example 6, we found that the distance between Dallas, Texas and Sioux Falls, South Dakota is approximately 744 miles. According to *mapquest.com*, the distance is approximately 850 miles. What might account for the difference?

Retain Your Knowledge

Problems 134–137 are based on material learned earlier in the course. The purpose of these problems is to keep the material fresh in your mind so that you are better prepared for the final exam.

134. Find the x-intercepts of $f(x) = 3x + 7$, if any.

135. Find the domain of $h(x) = \frac{3x}{x^2 - 9}$.

136. Write the function that is finally graphed if all of the following transformations are applied to the graph of $y = |x|$.
(a) Shift left 3 units. (b) Reflect about the x-axis. (c) Shift down 4 units.

137. Find the midpoint of the line segment joining the points $P_1 = (3, 6)$ and $P_2 = (-4, 2)$.

'Are You Prepared?' Answers

1. $C = 2\pi r$ **2.** 2π

2.2 Trigonometric Functions: Unit Circle Approach

PREPARING FOR THIS SECTION *Before getting started, review the following:*

- Geometry Essentials (Appendix A, Section A.2, pp. A14–A18)
- Unit Circle (Section 1.2, p. 16)
- Symmetry (Section 1.2, pp. 12–13)
- Functions (Section 1.3, pp. 23–35)

Now Work the 'Are You Prepared?' problems on page 126.

OBJECTIVES
1 Find the Exact Values of the Trigonometric Functions Using a Point on the Unit Circle (p. 116)
2 Find the Exact Values of the Trigonometric Functions of Quadrantal Angles (p. 117)
3 Find the Exact Values of the Trigonometric Functions of $\frac{\pi}{4} = 45°$ (p. 119)
4 Find the Exact Values of the Trigonometric Functions of $\frac{\pi}{6} = 30°$ and $\frac{\pi}{3} = 60°$ (p. 120)
5 Find the Exact Values of the Trigonometric Functions for Integer Multiples of $\frac{\pi}{6} = 30°$, $\frac{\pi}{4} = 45°$, and $\frac{\pi}{3} = 60°$ (p. 122)
6 Use a Calculator to Approximate the Value of a Trigonometric Function (p. 124)
7 Use a Circle of Radius r to Evaluate the Trigonometric Functions (p. 125)

We now introduce the trigonometric functions using the unit circle.

The Unit Circle

Recall that the unit circle is a circle whose radius is 1 and whose center is at the origin of a rectangular coordinate system. Also recall that any circle of radius r has circumference of length $2\pi r$. Therefore, the unit circle (radius $= 1$) has a circumference of length 2π. In other words, for 1 revolution around the unit circle the length of the arc is 2π units.

The following discussion sets the stage for defining the trigonometric functions using the unit circle.

Let t be any real number. Position the t-axis so that it is vertical with the positive direction up. Place this t-axis in the xy-plane so that $t = 0$ is located at the point $(1, 0)$ in the xy-plane.

If $t \geq 0$, let s be the distance from the origin to t on the t-axis. See the red portion of Figure 18(a). Beginning at the point $(1, 0)$ on the unit circle, travel

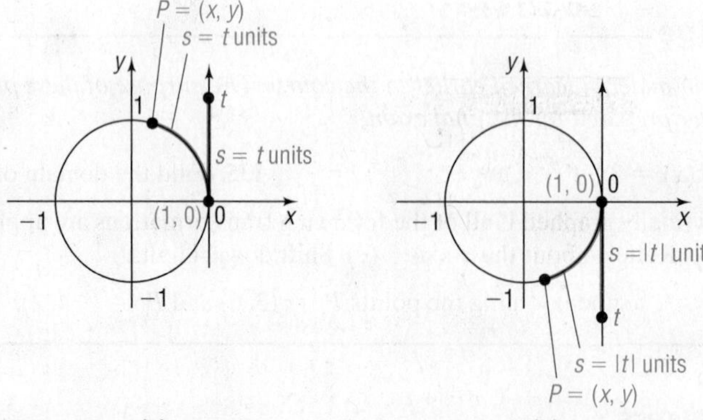

Figure 18 (a) (b)

$s = t$ units in the counterclockwise direction along the circle, to arrive at the point $P = (x, y)$. In this sense, the length $s = t$ units is being **wrapped** around the unit circle.

If $t < 0$, we begin at the point $(1, 0)$ on the unit circle and travel $s = |t|$ units in the clockwise direction to arrive at the point $P = (x, y)$. See Figure 18(b).

If $t > 2\pi$ or if $t < -2\pi$, it will be necessary to travel around the unit circle more than once before arriving at the point P. Do you see why?

Let's describe this process another way. Picture a string of length $s = |t|$ units being wrapped around a circle of radius 1 unit. Start wrapping the string around the circle at the point $(1, 0)$. If $t \geq 0$, wrap the string in the counterclockwise direction; if $t < 0$, wrap the string in the clockwise direction. The point $P = (x, y)$ is the point where the string ends.

This discussion tells us that, for any real number t, we can locate a unique point $P = (x, y)$ on the unit circle. We call P **the point on the unit circle that corresponds to *t*.** This is the important idea here. No matter what real number t is chosen, there is a unique point P on the unit circle corresponding to it. The coordinates of the point $P = (x, y)$ on the unit circle corresponding to the real number t are used to define the **six trigonometric functions of *t*.**

DEFINITION

Let t be a real number and let $P = (x, y)$ be the point on the unit circle that corresponds to t.

The **sine function** associates with t the y-coordinate of P and is denoted by

$$\sin t = y$$

The **cosine function** associates with t the x-coordinate of P and is denoted by

$$\cos t = x$$

If $x \neq 0$, the **tangent function** associates with t the ratio of the y-coordinate to the x-coordinate of P and is denoted by

$$\tan t = \frac{y}{x}$$

If $y \neq 0$, the **cosecant function** is defined as

$$\csc t = \frac{1}{y}$$

If $x \neq 0$, the **secant function** is defined as

$$\sec t = \frac{1}{x}$$

If $y \neq 0$, the **cotangent function** is defined as

$$\cot t = \frac{x}{y}$$

In Words

The point $P = (x, y)$ on the unit circle corresponding to a real number t is given by $(\cos t, \sin t)$.

Notice in these definitions that if $x = 0$, that is, if the point P is on the y-axis, then the tangent function and the secant function are undefined. Also, if $y = 0$, that is, if the point P is on the x-axis, then the cosecant function and the cotangent function are undefined.

Because we use the unit circle in these definitions of the trigonometric functions, they are sometimes referred to as **circular functions**.

1 Find the Exact Values of the Trigonometric Functions Using a Point on the Unit Circle

EXAMPLE 1

Finding the Values of the Six Trigonometric Functions Using a Point on the Unit Circle

Let t be a real number and let $P = \left(-\frac{1}{2}, \frac{\sqrt{3}}{2}\right)$ be the point on the unit circle that corresponds to t. Find the values of $\sin t$, $\cos t$, $\tan t$, $\csc t$, $\sec t$, and $\cot t$.

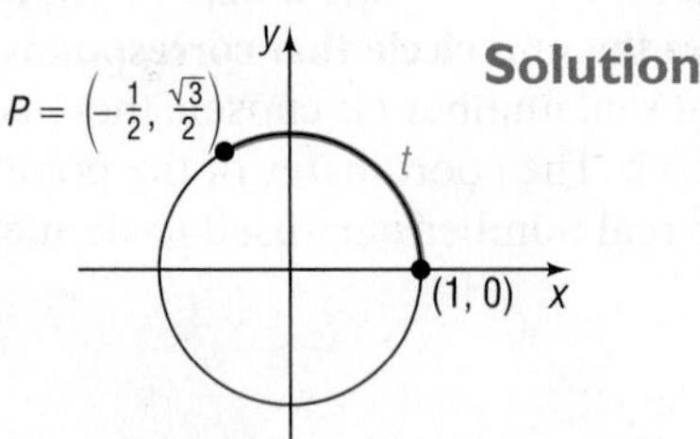

Figure 19

Solution See Figure 19. We follow the definition of the six trigonometric functions, using $P = \left(-\frac{1}{2}, \frac{\sqrt{3}}{2}\right) = (x, y)$. Then, with $x = -\frac{1}{2}$ and $y = \frac{\sqrt{3}}{2}$, we have

$$\sin t = y = \frac{\sqrt{3}}{2} \qquad \cos t = x = -\frac{1}{2} \qquad \tan t = \frac{y}{x} = \frac{\frac{\sqrt{3}}{2}}{-\frac{1}{2}} = -\sqrt{3}$$

$$\csc t = \frac{1}{y} = \frac{1}{\frac{\sqrt{3}}{2}} = \frac{2\sqrt{3}}{3} \qquad \sec t = \frac{1}{x} = \frac{1}{-\frac{1}{2}} = -2 \qquad \cot t = \frac{x}{y} = \frac{-\frac{1}{2}}{\frac{\sqrt{3}}{2}} = -\frac{\sqrt{3}}{3}$$

WARNING When writing the values of the trigonometric functions, do not forget the argument of the function.

$\sin t = \frac{\sqrt{3}}{2}$ ✓

$\sin = \frac{\sqrt{3}}{2}$ ✗

Now Work PROBLEM 13

Trigonometric Functions of Angles

Let $P = (x, y)$ be the point on the unit circle corresponding to the real number t. See Figure 20(a). Let θ be the angle in standard position, measured in radians, whose terminal side is the ray from the origin through P. Suppose θ subtends an arc of length s on the unit circle. See Figure 20(b). Since the unit circle has radius 1 unit, if $s = |t|$ units, then from the arc length formula $s = r|\theta|$, we have $\theta = t$ radians. See Figures 20(c) and (d).

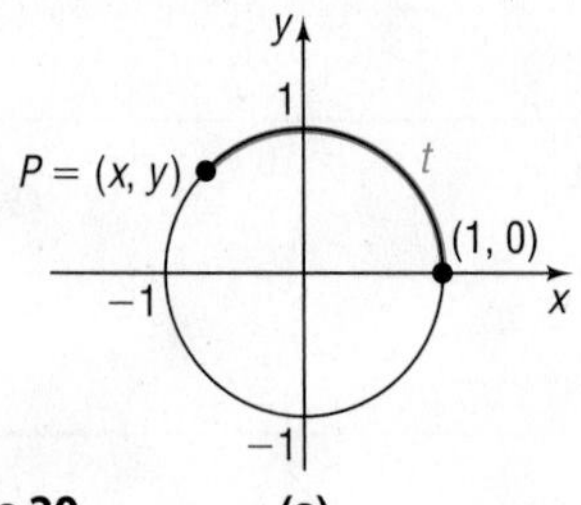

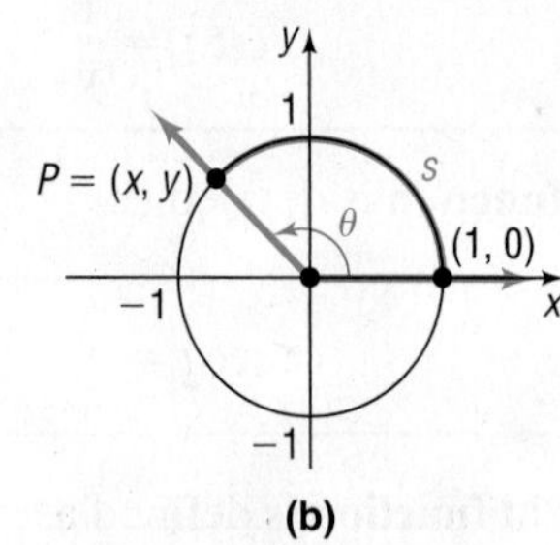

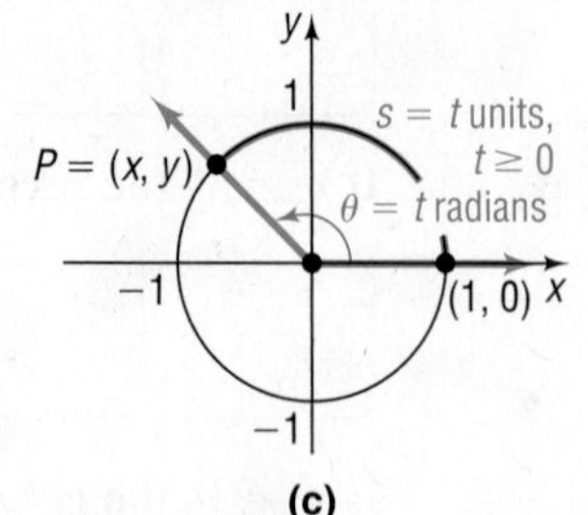

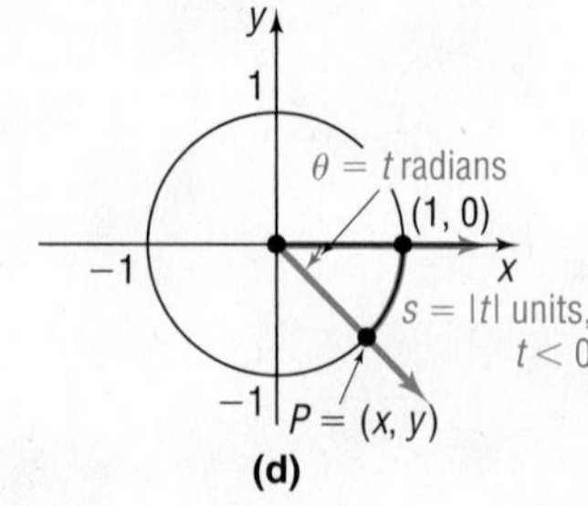

Figure 20 (a) (b) (c) (d)

The point $P = (x, y)$ on the unit circle that corresponds to the real number t is also the point P on the terminal side of the angle $\theta = t$ radians. As a result, we can say that

$$\underset{\text{Real number}}{\sin t} = \underset{\theta = t \text{ radians}}{\sin \theta}$$

and so on. We can now define the trigonometric functions of the angle θ.

DEFINITION

If $\theta = t$ radians, the **six trigonometric functions of the angle θ** are defined as

$\sin \theta = \sin t$	$\cos \theta = \cos t$	$\tan \theta = \tan t$
$\csc \theta = \csc t$	$\sec \theta = \sec t$	$\cot \theta = \cot t$

Even though the trigonometric functions can be viewed both as functions of real numbers and as functions of angles, it is customary to refer to trigonometric functions of real numbers and trigonometric functions of angles collectively as the *trigonometric functions*. We shall follow this practice from now on.

If an angle θ is measured in degrees, we shall use the degree symbol when writing a trigonometric function of θ, as, for example, in $\sin 30°$ and $\tan 45°$. If an angle θ is measured in radians, then no symbol is used when writing a trigonometric function of θ, as, for example, in $\cos \pi$ and $\sec \frac{\pi}{3}$.

Finally, since the values of the trigonometric functions of an angle θ are determined by the coordinates of the point $P = (x, y)$ on the unit circle corresponding to θ, the units used to measure the angle θ are irrelevant. For example, it does not matter whether we write $\theta = \frac{\pi}{2}$ radians or $\theta = 90°$. The point on the unit circle corresponding to this angle is $P = (0, 1)$. As a result,

$$\sin \frac{\pi}{2} = \sin 90° = 1 \quad \text{and} \quad \cos \frac{\pi}{2} = \cos 90° = 0$$

2 Find the Exact Values of the Trigonometric Functions of Quadrantal Angles

To find the exact value of a trigonometric function of an angle θ or a real number t requires that we locate the point $P = (x, y)$ on the unit circle that corresponds to t. This is not always easy to do. In the examples that follow, we will evaluate the trigonometric functions of certain angles or real numbers for which this process is relatively easy. A calculator will be used to evaluate the trigonometric functions of most other angles.

EXAMPLE 2

Finding the Exact Values of the Six Trigonometric Functions of Quadrantal Angles

Find the exact values of the six trigonometric functions of:

(a) $\theta = 0 = 0°$ (b) $\theta = \frac{\pi}{2} = 90°$

(c) $\theta = \pi = 180°$ (d) $\theta = \frac{3\pi}{2} = 270°$

Solution (a) The point on the unit circle that corresponds to $\theta = 0 = 0°$ is $P = (1, 0)$. See Figure 21(a). Then using $x = 1$ and $y = 0$

$$\sin 0 = \sin 0° = y = 0 \qquad \cos 0 = \cos 0° = x = 1$$

$$\tan 0 = \tan 0° = \frac{y}{x} = 0 \qquad \sec 0 = \sec 0° = \frac{1}{x} = 1$$

Since the y-coordinate of P is 0, $\csc 0$ and $\cot 0$ are not defined.

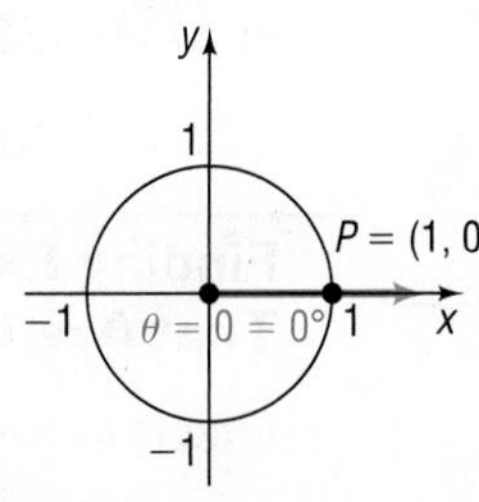

Figure 21 **(a)** $\theta = 0 = 0°$

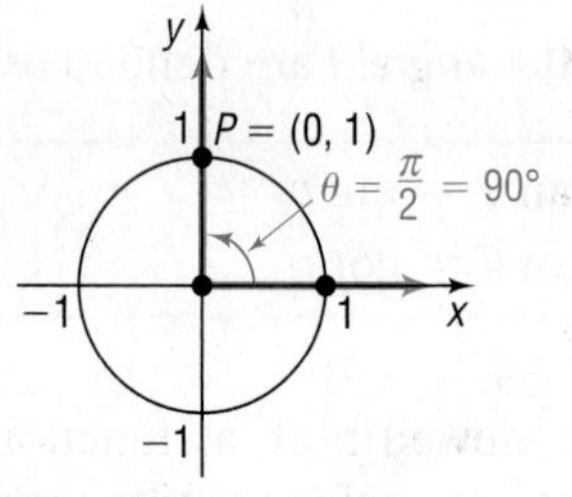

Figure 21 (b) $\theta = \frac{\pi}{2} = 90°$

(b) The point on the unit circle that corresponds to $\theta = \frac{\pi}{2} = 90°$ is $P = (0, 1)$. See Figure 21(b). Then

$$\sin\frac{\pi}{2} = \sin 90° = y = 1 \qquad \cos\frac{\pi}{2} = \cos 90° = x = 0$$

$$\csc\frac{\pi}{2} = \csc 90° = \frac{1}{y} = 1 \qquad \cot\frac{\pi}{2} = \cot 90° = \frac{x}{y} = 0$$

Since the x-coordinate of P is 0, $\tan\frac{\pi}{2}$ and $\sec\frac{\pi}{2}$ are not defined.

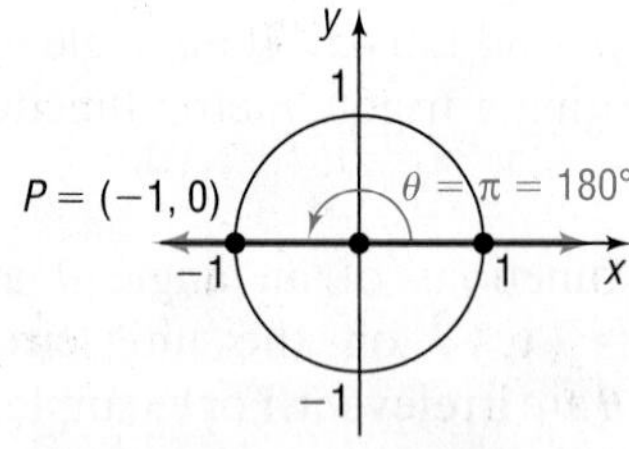

Figure 21 (c) $\theta = \pi = 180°$

(c) The point on the unit circle that corresponds to $\theta = \pi = 180°$ is $P = (-1, 0)$. See Figure 21(c). Then

$$\sin\pi = \sin 180° = y = 0 \qquad \cos\pi = \cos 180° = x = -1$$

$$\tan\pi = \tan 180° = \frac{y}{x} = 0 \qquad \sec\pi = \sec 180° = \frac{1}{x} = -1$$

Since the y-coordinate of P is 0, $\csc\pi$ and $\cot\pi$ are not defined.

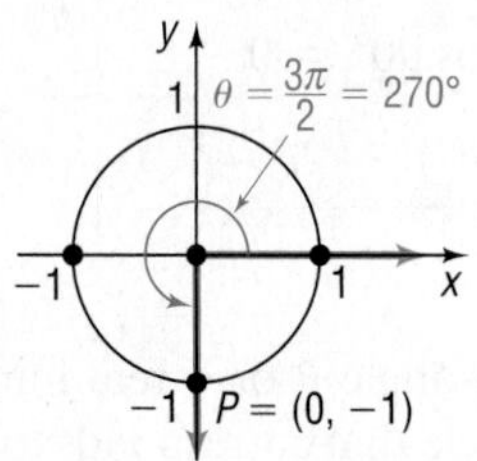

Figure 21 (d) $\theta = \frac{3\pi}{2} = 270°$

(d) The point on the unit circle that corresponds to $\theta = \frac{3\pi}{2} = 270°$ is $P = (0, -1)$. See Figure 21(d). Then

$$\sin\frac{3\pi}{2} = \sin 270° = y = -1 \qquad \cos\frac{3\pi}{2} = \cos 270° = x = 0$$

$$\csc\frac{3\pi}{2} = \csc 270° = \frac{1}{y} = -1 \qquad \cot\frac{3\pi}{2} = \cot 270° = \frac{x}{y} = 0$$

Since the x-coordinate of P is 0, $\tan\frac{3\pi}{2}$ and $\sec\frac{3\pi}{2}$ are not defined. ●

Table 2 summarizes the values of the trigonometric functions found in Example 2.

Table 2

Quadrantal Angles							
θ (Radians)	**θ (Degrees)**	**sin θ**	**cos θ**	**tan θ**	**csc θ**	**sec θ**	**cot θ**
0	0°	0	1	0	Not defined	1	Not defined
$\frac{\pi}{2}$	90°	1	0	Not defined	1	Not defined	0
π	180°	0	−1	0	Not defined	−1	Not defined
$\frac{3\pi}{2}$	270°	−1	0	Not defined	−1	Not defined	0

There is no need to memorize Table 2. To find the value of a trigonometric function of a quadrantal angle, draw the angle and apply the definition, as we did in Example 2.

EXAMPLE 3

Finding Exact Values of the Trigonometric Functions of Angles That Are Integer Multiples of Quadrantal Angles

Find the exact value of:

(a) $\sin(3\pi)$ (b) $\cos(-270°)$

Solution (a) See Figure 22(a). The point P on the unit circle that corresponds to $\theta = 3\pi$ is $P = (-1, 0)$, so $\sin(3\pi) = y = 0$.

(b) See Figure 22(b). The point P on the unit circle that corresponds to $\theta = -270°$ is $P = (0, 1)$, so $\cos(-270°) = x = 0$.

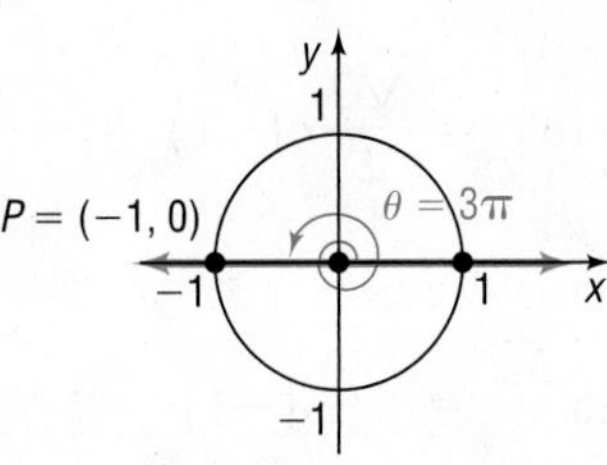

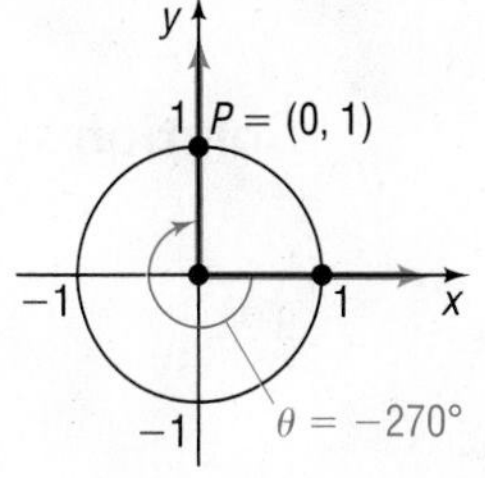

Figure 22 **(a)** $\theta = 3\pi$ **(b)** $\theta = -270°$

Now Work PROBLEMS 21 AND 61

3 Find the Exact Values of the Trigonometric Functions of $\frac{\pi}{4} = 45°$

EXAMPLE 4 **Finding the Exact Values of the Trigonometric Functions of $\frac{\pi}{4} = 45°$**

Find the exact values of the six trigonometric functions of $\frac{\pi}{4} = 45°$.

Solution We seek the coordinates of the point $P = (x, y)$ on the unit circle that corresponds to $\theta = \frac{\pi}{4} = 45°$. See Figure 23. First, observe that P lies on the line $y = x$. (Do you see why? Since $\theta = 45° = \frac{1}{2} \cdot 90°$, P must lie on the line that bisects quadrant I.) Since $P = (x, y)$ also lies on the unit circle, $x^2 + y^2 = 1$, it follows that

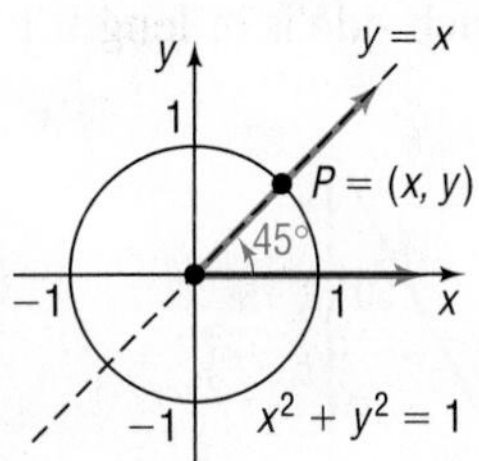

Figure 23 $\theta = 45°$

$$x^2 + y^2 = 1$$

$$x^2 + x^2 = 1 \quad y = x, x > 0, y > 0$$

$$2x^2 = 1$$

$$x = \frac{1}{\sqrt{2}} = \frac{\sqrt{2}}{2} \qquad y = \frac{\sqrt{2}}{2}$$

Then

$$\sin\frac{\pi}{4} = \sin 45° = \frac{\sqrt{2}}{2} \qquad \cos\frac{\pi}{4} = \cos 45° = \frac{\sqrt{2}}{2} \qquad \tan\frac{\pi}{4} = \tan 45° = \frac{\frac{\sqrt{2}}{2}}{\frac{\sqrt{2}}{2}} = 1$$

$$\csc\frac{\pi}{4} = \csc 45° = \frac{1}{\frac{\sqrt{2}}{2}} = \sqrt{2} \qquad \sec\frac{\pi}{4} = \sec 45° = \frac{1}{\frac{\sqrt{2}}{2}} = \sqrt{2} \qquad \cot\frac{\pi}{4} = \cot 45° = \frac{\frac{\sqrt{2}}{2}}{\frac{\sqrt{2}}{2}} = 1$$

EXAMPLE 5 **Finding the Exact Value of a Trigonometric Expression**

Find the exact value of each expression.

(a) $\sin 45° \cos 180°$ (b) $\tan \frac{\pi}{4} - \sin \frac{3\pi}{2}$ (c) $\left(\sec \frac{\pi}{4}\right)^2 + \csc \frac{\pi}{2}$

Solution (a) $\sin 45° \cos 180° = \frac{\sqrt{2}}{2} \cdot (-1) = -\frac{\sqrt{2}}{2}$

From Example 4 ($\frac{\sqrt{2}}{2}$); From Table 2 (-1)

(b) $\tan \frac{\pi}{4} - \sin \frac{3\pi}{2} = 1 - (-1) = 2$

From Example 4 (1); From Table 2 (-1)

(c) $\left(\sec \frac{\pi}{4}\right)^2 + \csc \frac{\pi}{2} = (\sqrt{2})^2 + 1 = 2 + 1 = 3$ ●

Now Work PROBLEM 35

4 Find the Exact Values of the Trigonometric Functions of $\frac{\pi}{6} = 30°$ and $\frac{\pi}{3} = 60°$

Consider a right triangle in which one of the angles is $\frac{\pi}{6} = 30°$. It then follows that the third angle is $\frac{\pi}{3} = 60°$. Figure 24(a) illustrates such a triangle with hypotenuse of length 1. Our problem is to determine a and b.

Begin by placing next to the triangle in Figure 24(a) another triangle congruent to the first, as shown in Figure 24(b). Notice that we now have a triangle whose three angles each equal 60°. This triangle is therefore equilateral, so each side is of length 1.

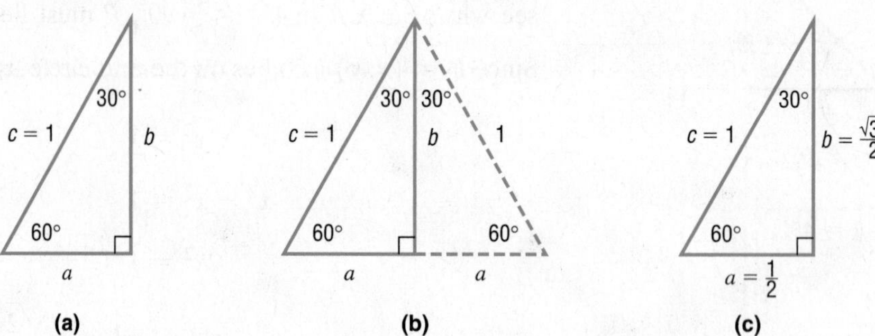

Figure 24 30°–60°–90° triangle

This means the base is $2a = 1$, and so $a = \frac{1}{2}$. By the Pythagorean Theorem, b satisfies the equation $a^2 + b^2 = c^2$, so we have

$$a^2 + b^2 = c^2$$

$$\frac{1}{4} + b^2 = 1 \qquad a = \frac{1}{2}, c = 1$$

$$b^2 = 1 - \frac{1}{4} = \frac{3}{4}$$

$$b = \frac{\sqrt{3}}{2} \qquad b > 0 \text{ because } b \text{ is the length of the side of a triangle.}$$

This results in Figure 24(c).

EXAMPLE 6

Finding the Exact Values of the Trigonometric Functions of $\frac{\pi}{3} = 60°$

Find the exact values of the six trigonometric functions of $\frac{\pi}{3} = 60°$.

Solution Position the triangle in Figure 24(c) so that the $60°$ angle is in standard position. See Figure 25. The point on the unit circle that corresponds to $\theta = \frac{\pi}{3} = 60°$ is $P = \left(\frac{1}{2}, \frac{\sqrt{3}}{2}\right)$. Then

Figure 25 $\theta = \frac{\pi}{3} = 60°$

$$\sin\frac{\pi}{3} = \sin 60° = \frac{\sqrt{3}}{2} \qquad \cos\frac{\pi}{3} = \cos 60° = \frac{1}{2}$$

$$\csc\frac{\pi}{3} = \csc 60° = \frac{1}{\frac{\sqrt{3}}{2}} = \frac{2}{\sqrt{3}} = \frac{2\sqrt{3}}{3} \qquad \sec\frac{\pi}{3} = \sec 60° = \frac{1}{\frac{1}{2}} = 2$$

$$\tan\frac{\pi}{3} = \tan 60° = \frac{\frac{\sqrt{3}}{2}}{\frac{1}{2}} = \sqrt{3} \qquad \cot\frac{\pi}{3} = \cot 60° = \frac{\frac{1}{2}}{\frac{\sqrt{3}}{2}} = \frac{1}{\sqrt{3}} = \frac{\sqrt{3}}{3}$$

●

EXAMPLE 7

Finding the Exact Values of the Trigonometric Functions of $\frac{\pi}{6} = 30°$

Find the exact values of the trigonometric functions of $\frac{\pi}{6} = 30°$.

Solution Position the triangle in Figure 24(c) so that the $30°$ angle is in standard position. See Figure 26. The point on the unit circle that corresponds to $\theta = \frac{\pi}{6} = 30°$ is $P = \left(\frac{\sqrt{3}}{2}, \frac{1}{2}\right)$. Then

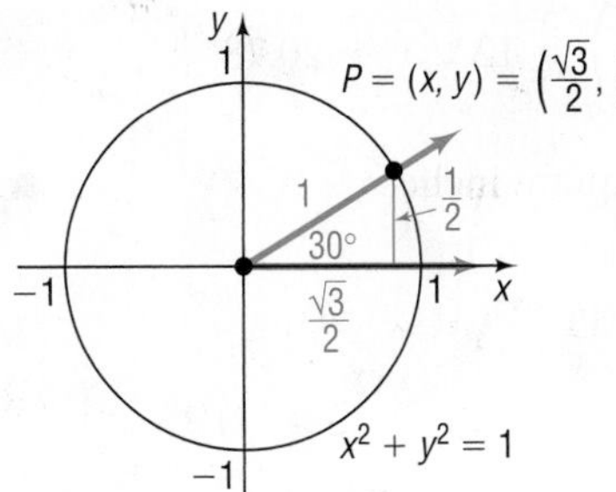

Figure 26 $\theta = \frac{\pi}{6} = 30°$

$$\sin\frac{\pi}{6} = \sin 30° = \frac{1}{2} \qquad \cos\frac{\pi}{6} = \cos 30° = \frac{\sqrt{3}}{2}$$

$$\csc\frac{\pi}{6} = \csc 30° = \frac{1}{\frac{1}{2}} = 2 \qquad \sec\frac{\pi}{6} = \sec 30° = \frac{1}{\frac{\sqrt{3}}{2}} = \frac{2}{\sqrt{3}} = \frac{2\sqrt{3}}{3}$$

$$\tan\frac{\pi}{6} = \tan 30° = \frac{\frac{1}{2}}{\frac{\sqrt{3}}{2}} = \frac{1}{\sqrt{3}} = \frac{\sqrt{3}}{3} \qquad \cot\frac{\pi}{6} = \cot 30° = \frac{\frac{\sqrt{3}}{2}}{\frac{1}{2}} = \sqrt{3}$$

●

Table 3 on the next page summarizes the information just derived for $\frac{\pi}{6} = 30°$, $\frac{\pi}{4} = 45°$, and $\frac{\pi}{3} = 60°$. Until you memorize the entries in Table 3, you should draw an appropriate diagram to determine the values given in the table.

Table 3

θ (Radians)	θ (Degrees)	$\sin\theta$	$\cos\theta$	$\tan\theta$	$\csc\theta$	$\sec\theta$	$\cot\theta$
$\frac{\pi}{6}$	30°	$\frac{1}{2}$	$\frac{\sqrt{3}}{2}$	$\frac{\sqrt{3}}{3}$	2	$\frac{2\sqrt{3}}{3}$	$\sqrt{3}$
$\frac{\pi}{4}$	45°	$\frac{\sqrt{2}}{2}$	$\frac{\sqrt{2}}{2}$	1	$\sqrt{2}$	$\sqrt{2}$	1
$\frac{\pi}{3}$	60°	$\frac{\sqrt{3}}{2}$	$\frac{1}{2}$	$\sqrt{3}$	$\frac{2\sqrt{3}}{3}$	2	$\frac{\sqrt{3}}{3}$

Now Work PROBLEM 41

EXAMPLE 8 **Constructing a Rain Gutter**

A rain gutter is to be constructed of aluminum sheets 12 inches wide. After marking off a length of 4 inches from each edge, this length is bent up at an angle θ. See Figure 27. The area A of the opening may be expressed as a function of θ as

$$A(\theta) = 16\sin\theta(\cos\theta + 1)$$

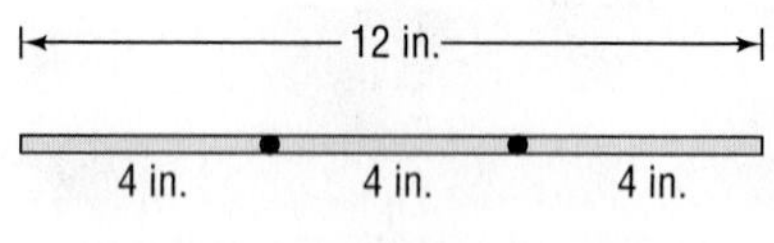

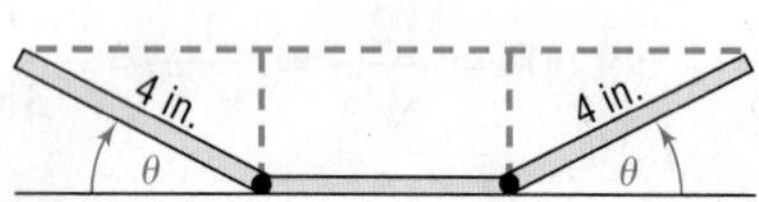

Figure 27

Find the area A of the opening for $\theta = 30°$, $\theta = 45°$, and $\theta = 60°$.

Solution For $\theta = 30°$: $A(30°) = 16\sin 30°(\cos 30° + 1)$

$$= 16\left(\frac{1}{2}\right)\left(\frac{\sqrt{3}}{2} + 1\right) = 4\sqrt{3} + 8 \approx 14.93$$

The area of the opening for $\theta = 30°$ is about 14.93 square inches.

For $\theta = 45°$: $A(45°) = 16\sin 45°(\cos 45° + 1)$

$$= 16\left(\frac{\sqrt{2}}{2}\right)\left(\frac{\sqrt{2}}{2} + 1\right) = 8 + 8\sqrt{2} \approx 19.31$$

The area of the opening for $\theta = 45°$ is about 19.31 square inches.

For $\theta = 60°$: $A(60°) = 16\sin 60°(\cos 60° + 1)$

$$= 16\left(\frac{\sqrt{3}}{2}\right)\left(\frac{1}{2} + 1\right) = 12\sqrt{3} \approx 20.78$$

The area of the opening for $\theta = 60°$ is about 20.78 square inches. ●

5 Find the Exact Values of the Trigonometric Functions for Integer Multiples of $\frac{\pi}{6} = 30°, \frac{\pi}{4} = 45°$, and $\frac{\pi}{3} = 60°$

We know the exact values of the trigonometric functions of $\frac{\pi}{4} = 45°$. Using symmetry, we can find the exact values of the trigonometric functions of $\frac{3\pi}{4} = 135°$, $\frac{5\pi}{4} = 225°$, and $\frac{7\pi}{4} = 315°$. The point on the unit circle corresponding to $\frac{\pi}{4} = 45°$ is $\left(\frac{\sqrt{2}}{2}, \frac{\sqrt{2}}{2}\right)$. See Figure 28. Using symmetry with respect to the y-axis, the point $\left(-\frac{\sqrt{2}}{2}, \frac{\sqrt{2}}{2}\right)$ is the point on the unit circle that corresponds to the angle $\frac{3\pi}{4} = 135°$. Similarly, using symmetry with respect to the origin, the

Figure 28

point $\left(-\frac{\sqrt{2}}{2}, -\frac{\sqrt{2}}{2}\right)$ is the point on the unit circle that corresponds to the angle $\frac{5\pi}{4} = 225°$. Finally, using symmetry with respect to the x-axis, the point $\left(\frac{\sqrt{2}}{2}, -\frac{\sqrt{2}}{2}\right)$ is the point on the unit circle that corresponds to the angle $\frac{7\pi}{4} = 315°$.

EXAMPLE 9 **Finding Exact Values for Multiples of $\frac{\pi}{4} = 45°$**

Find the exact value of each expression.

(a) $\cos \frac{5\pi}{4}$ (b) $\sin 135°$ (c) $\tan 315°$ (d) $\sin\left(-\frac{\pi}{4}\right)$ (e) $\cos \frac{11\pi}{4}$

Solution (a) From Figure 28, we see the point $\left(-\frac{\sqrt{2}}{2}, -\frac{\sqrt{2}}{2}\right)$ corresponds to $\frac{5\pi}{4}$, so $\cos \frac{5\pi}{4} = x = -\frac{\sqrt{2}}{2}$.

(b) Since $135° = \frac{3\pi}{4}$, the point $\left(-\frac{\sqrt{2}}{2}, \frac{\sqrt{2}}{2}\right)$ corresponds to 135°, so $\sin 135° = \frac{\sqrt{2}}{2}$.

(c) Since $315° = \frac{7\pi}{4}$, the point $\left(\frac{\sqrt{2}}{2}, -\frac{\sqrt{2}}{2}\right)$ corresponds to 315°, so

$$\tan 315° = \frac{-\frac{\sqrt{2}}{2}}{\frac{\sqrt{2}}{2}} = -1.$$

(d) The point $\left(\frac{\sqrt{2}}{2}, -\frac{\sqrt{2}}{2}\right)$ corresponds to $-\frac{\pi}{4}$, so $\sin\left(-\frac{\pi}{4}\right) = -\frac{\sqrt{2}}{2}$.

(e) The point $\left(-\frac{\sqrt{2}}{2}, \frac{\sqrt{2}}{2}\right)$ corresponds to $\frac{11\pi}{4}$, so $\cos \frac{11\pi}{4} = -\frac{\sqrt{2}}{2}$. ●

Now Work PROBLEMS 51 AND 55

The use of symmetry also provides information about certain integer multiples of the angles $\frac{\pi}{6} = 30°$ and $\frac{\pi}{3} = 60°$. See Figures 29 and 30.

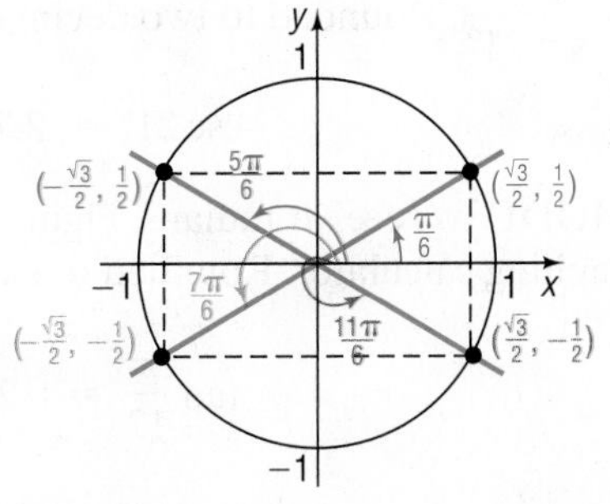

Figure 29

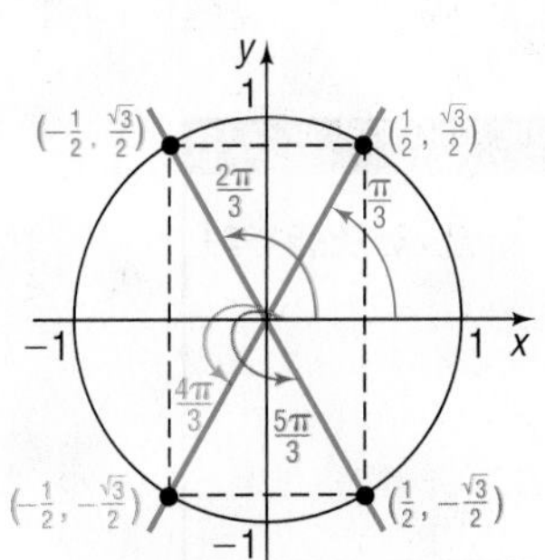

Figure 30

EXAMPLE 10 **Finding Exact Values for Multiples of $\frac{\pi}{6} = 30°$ or $\frac{\pi}{3} = 60°$**

Based on Figures 29 and 30, we see that

(a) $\cos 210° = \cos \frac{7\pi}{6} = -\frac{\sqrt{3}}{2}$ (b) $\sin(-60°) = \sin\left(-\frac{\pi}{3}\right) = -\frac{\sqrt{3}}{2}$

(c) $\tan \frac{5\pi}{3} = \frac{-\frac{\sqrt{3}}{2}}{\frac{1}{2}} = -\sqrt{3}$ (d) $\cos \frac{8\pi}{3} = \cos \frac{2\pi}{3} = -\frac{1}{2}$ ●

Now Work PROBLEM 47

6 Use a Calculator to Approximate the Value of a Trigonometric Function

Before getting started, you must first decide whether to enter the angle in the calculator using radians or degrees and then set the calculator to the correct MODE. Check your instruction manual to find out how your calculator handles degrees and radians. Your calculator has keys marked [sin], [cos], and [tan]. To find the values of the remaining three trigonometric functions, secant, cosecant, and cotangent, use the fact that, if $P = (x, y)$ is a point on the unit circle on the terminal side of θ, then

$$\sec \theta = \frac{1}{x} = \frac{1}{\cos \theta} \qquad \csc \theta = \frac{1}{y} = \frac{1}{\sin \theta} \qquad \cot \theta = \frac{x}{y} = \frac{1}{\frac{y}{x}} = \frac{1}{\tan \theta}$$

WARNING On your calculator the second functions $\sin^{-1}$, $\cos^{-1}$, and $\tan^{-1}$ do not represent the reciprocal of sin, cos, and tan. ■

EXAMPLE 11 **Using a Calculator to Approximate the Value of a Trigonometric Function**

Use a calculator to find the approximate value of:

(a) $\cos 48°$ (b) $\csc 21°$ (c) $\tan \frac{\pi}{12}$

Express your answer rounded to two decimal places.

Solution (a) First, set the MODE to receive degrees. Rounded to two decimal places,

$$\cos 48° = 0.6691306 \approx 0.67$$

(b) Most calculators do not have a csc key. The manufacturers assume that the user knows some trigonometry. To find the value of csc 21°, use the fact that $\csc 21° = \frac{1}{\sin 21°}$. Rounded to two decimal places,

$$\csc 21° \approx 2.79$$

(c) Set the MODE to receive radians. Figure 31 shows the solution using a TI-84 Plus C graphing calculator. Rounded to two decimal places,

$$\tan \frac{\pi}{12} \approx 0.27$$ ●

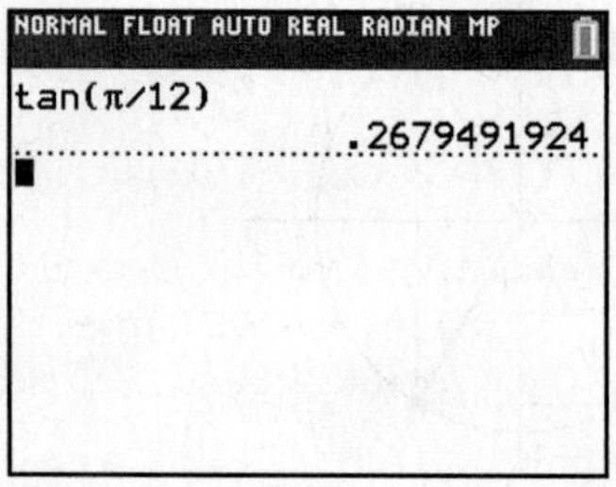

Figure 31

Now Work PROBLEM 65

7 Use a Circle of Radius r to Evaluate the Trigonometric Functions

Until now, finding the exact value of a trigonometric function of an angle θ required that we locate the corresponding point $P = (x, y)$ on the unit circle. In fact, though, any circle whose center is at the origin can be used.

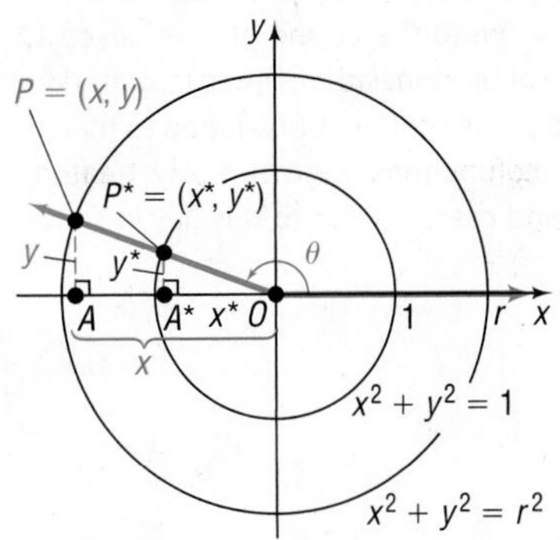

Figure 32

Let θ be any nonquadrantal angle placed in standard position. Let $P = (x, y)$ be the point on the circle $x^2 + y^2 = r^2$ that corresponds to θ, and let $P^* = (x^*, y^*)$ be the point on the unit circle that corresponds to θ. See Figure 32, where θ is shown in quadrant II.

Notice that the triangles OA^*P^* and OAP are similar; as a result, the ratios of corresponding sides are equal.

$$\frac{y^*}{1} = \frac{y}{r} \qquad \frac{x^*}{1} = \frac{x}{r} \qquad \frac{y^*}{x^*} = \frac{y}{x}$$

$$\frac{1}{y^*} = \frac{r}{y} \qquad \frac{1}{x^*} = \frac{r}{x} \qquad \frac{x^*}{y^*} = \frac{x}{y}$$

These results lead us to formulate the following theorem:

THEOREM

For an angle θ in standard position, let $P = (x, y)$ be the point on the terminal side of θ that is also on the circle $x^2 + y^2 = r^2$. Then

$$\sin\theta = \frac{y}{r} \qquad \cos\theta = \frac{x}{r} \qquad \tan\theta = \frac{y}{x} \quad x \neq 0$$

$$\csc\theta = \frac{r}{y} \quad y \neq 0 \qquad \sec\theta = \frac{r}{x} \quad x \neq 0 \qquad \cot\theta = \frac{x}{y} \quad y \neq 0$$

EXAMPLE 12 **Finding the Exact Values of the Six Trigonometric Functions**

Find the exact values of each of the six trigonometric functions of an angle θ if $(4, -3)$ is a point on its terminal side in standard position.

Solution See Figure 33. The point $(4, -3)$ is on a circle that has a radius of $r = \sqrt{4^2 + (-3)^2} = \sqrt{16 + 9} = \sqrt{25} = 5$ with center at the origin.

For the point $(x, y) = (4, -3)$, we have $x = 4$ and $y = -3$. Since $r = 5$, we find

$$\sin\theta = \frac{y}{r} = -\frac{3}{5} \qquad \cos\theta = \frac{x}{r} = \frac{4}{5} \qquad \tan\theta = \frac{y}{x} = -\frac{3}{4}$$

$$\csc\theta = \frac{r}{y} = -\frac{5}{3} \qquad \sec\theta = \frac{r}{x} = \frac{5}{4} \qquad \cot\theta = \frac{x}{y} = -\frac{4}{3}$$

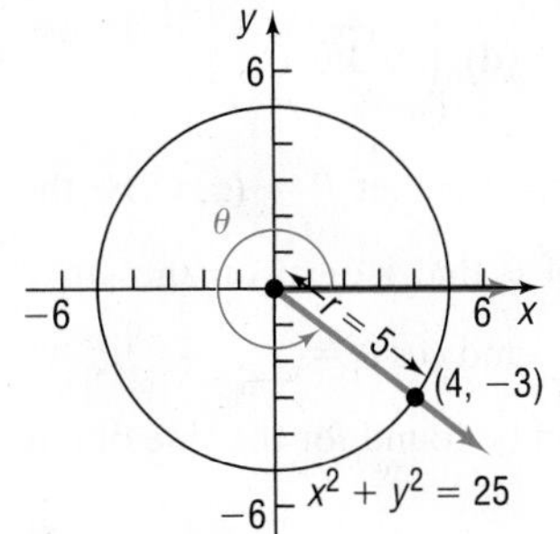

Figure 33

Now Work PROBLEM 77

Historical Feature

The name *sine* for the sine function arose from a medieval confusion. The name comes from the Sanskrit word *jīva* (meaning chord), first used in India by Araybhata the Elder (AD 510). He really meant half-chord, but abbreviated it. This was brought into Arabic as *jība,* which was meaningless. Because the proper Arabic word *jaib* would be written the same way (short vowels are not written out in Arabic), *jība* was pronounced as jaib, which meant bosom or hollow, and *jība* remains as the Arabic word for sine to this day. Scholars translating the Arabic works into Latin found that the word *sinus* also meant bosom or hollow, and from *sinus* we get *sine.*

The name *tangent,* due to Thomas Finck (1583), can be understood by looking at Figure 34 on page 126. The line segment $\overline{DC}$ is tangent to the circle at C. If $d(O, B) = d(O, C) = 1$, then the length of the line segment $\overline{DC}$ is

$$d(D, C) = \frac{d(D, C)}{1} = \frac{d(D, C)}{d(O, C)} = \tan\alpha$$

continued

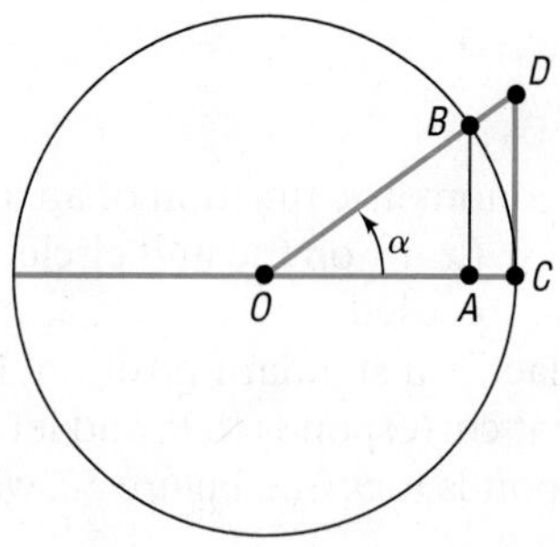

Figure 34

The old name for the tangent is *umbra versa* (meaning turned shadow), referring to the use of the tangent in solving height problems with shadows.

The names of the remaining functions came about as follows. If α and β are complementary angles, then $\cos\alpha = \sin\beta$. Because β is the complement of α, it was natural to write the cosine of α as *sin co* α. Probably for reasons involving ease of pronunciation, the *co* migrated to the front, and then cosine received a three-letter abbreviation to match sin, sec, and tan. The two other cofunctions were similarly treated, except that the long forms *cotan* and *cosec* survive to this day in some countries.

2.2 Assess Your Understanding

'Are You Prepared?' *Answers are given at the end of these exercises. If you get a wrong answer, read the pages listed in red.*

1. In a right triangle, with legs a and b and hypotenuse c, the Pythagorean Theorem states that __________. (p. A14)
2. The value of the function $f(x) = 3x - 7$ at 5 is ____. (pp. 23–35)
3. ***True or False*** For a function $y = f(x)$, for each x in the domain, there is exactly one element y in the range. (pp. 23–35)
4. If two triangles are similar, then corresponding angles are ______ and the lengths of corresponding sides are __________. (pp. A14–A18)
5. What point is symmetric with respect to the y-axis to the point $\left(\frac{1}{2}, \frac{\sqrt{3}}{2}\right)$? (pp. 12–13)
6. If (x, y) is a point on the unit circle in quadrant IV and if $x = \frac{\sqrt{3}}{2}$, what is y? (p. 16)

Concepts and Vocabulary

7. Which function takes as input a real number t that corresponds to a point $P = (x, y)$ on the unit circle and outputs the x-coordinate?
(a) sine (b) cosine (c) tangent (d) secant
8. The point on the unit circle that corresponds to $\theta = \frac{\pi}{2}$ is $P =$ ______.
9. The point on the unit circle that corresponds to $\theta = \frac{\pi}{4}$ is $P =$ __________.
10. The point on the unit circle that corresponds to $\theta = \frac{\pi}{3}$ is
(a) $\left(\frac{1}{2}, \frac{\sqrt{3}}{2}\right)$ (b) $\left(\frac{\sqrt{2}}{2}, \frac{\sqrt{2}}{2}\right)$
(c) $\left(\frac{\sqrt{3}}{2}, \frac{1}{2}\right)$ (d) $\left(\sqrt{3}, \frac{2\sqrt{3}}{3}\right)$
11. For any angle θ in standard position, let $P = (x, y)$ be the point on the terminal side of θ that is also on the circle $x^2 + y^2 = r^2$. Then, $\sin\theta =$ __ and $\cos\theta =$ __.
12. ***True or False*** Exact values can be found for the sine of any angle.

Skill Building

In Problems 13–20, $P = (x, y)$ is the point on the unit circle that corresponds to a real number t. Find the exact values of the six trigonometric functions of t.

13. $\left(\frac{\sqrt{3}}{2}, \frac{1}{2}\right)$
14. $\left(\frac{1}{2}, -\frac{\sqrt{3}}{2}\right)$
15. $\left(-\frac{2}{5}, \frac{\sqrt{21}}{5}\right)$
16. $\left(-\frac{1}{5}, \frac{2\sqrt{6}}{5}\right)$
17. $\left(-\frac{\sqrt{2}}{2}, \frac{\sqrt{2}}{2}\right)$
18. $\left(\frac{\sqrt{2}}{2}, \frac{\sqrt{2}}{2}\right)$
19. $\left(\frac{2\sqrt{2}}{3}, -\frac{1}{3}\right)$
20. $\left(-\frac{\sqrt{5}}{3}, -\frac{2}{3}\right)$

In Problems 21–30, find the exact value. Do not use a calculator.

21. $\sin \frac{11\pi}{2}$ **22.** $\cos(7\pi)$ **23.** $\tan(6\pi)$ **24.** $\cot \frac{7\pi}{2}$ **25.** $\csc \frac{11\pi}{2}$

26. $\sec(8\pi)$ **27.** $\cos\left(-\frac{3\pi}{2}\right)$ **28.** $\sin(-3\pi)$ **29.** $\sec(-\pi)$ **30.** $\tan(-3\pi)$

In Problems 31–46, find the exact value of each expression. Do not use a calculator.

31. $\sin 45° + \cos 60°$ **32.** $\sin 30° - \cos 45°$ **33.** $\sin 90° + \tan 45°$ **34.** $\cos 180° - \sin 180°$

35. $\sin 45° \cos 45°$ **36.** $\tan 45° \cos 30°$ **37.** $\csc 45° \tan 60°$ **38.** $\sec 30° \cot 45°$

39. $4 \sin 90° - 3 \tan 180°$ **40.** $5 \cos 90° - 8 \sin 270°$ **41.** $2 \sin \frac{\pi}{3} - 3 \tan \frac{\pi}{6}$ **42.** $2 \sin \frac{\pi}{4} + 3 \tan \frac{\pi}{4}$

43. $2 \sec \frac{\pi}{4} + 4 \cot \frac{\pi}{3}$ **44.** $3 \csc \frac{\pi}{3} + \cot \frac{\pi}{4}$ **45.** $\csc \frac{\pi}{2} + \cot \frac{\pi}{2}$ **46.** $\sec \pi - \csc \frac{\pi}{2}$

In Problems 47–64, find the exact values of the six trigonometric functions of the given angle. If any are not defined, say "not defined." Do not use a calculator.

47. $\frac{2\pi}{3}$ **48.** $\frac{5\pi}{6}$ **49.** $210°$ **50.** $240°$ **51.** $\frac{3\pi}{4}$ **52.** $\frac{11\pi}{4}$

53. $\frac{8\pi}{3}$ **54.** $\frac{13\pi}{6}$ **55.** $405°$ **56.** $390°$ **57.** $-\frac{\pi}{6}$ **58.** $-\frac{\pi}{3}$

59. $-135°$ **60.** $-240°$ **61.** $\frac{5\pi}{2}$ **62.** 5π **63.** $-\frac{14\pi}{3}$ **64.** $-\frac{13\pi}{6}$

In Problems 65–76, use a calculator to find the approximate value of each expression rounded to two decimal places.

65. $\sin 28°$ **66.** $\cos 14°$ **67.** $\sec 21°$ **68.** $\cot 70°$

69. $\tan \frac{\pi}{10}$ **70.** $\sin \frac{\pi}{8}$ **71.** $\cot \frac{\pi}{12}$ **72.** $\csc \frac{5\pi}{13}$

73. $\sin 1$ **74.** $\tan 1$ **75.** $\sin 1°$ **76.** $\tan 1°$

In Problems 77–84, a point on the terminal side of an angle θ in standard position is given. Find the exact value of each of the six trigonometric functions of θ.

77. $(-3, 4)$ **78.** $(5, -12)$ **79.** $(2, -3)$ **80.** $(-1, -2)$

81. $(-2, -2)$ **82.** $(-1, 1)$ **83.** $\left(\frac{1}{3}, \frac{1}{4}\right)$ **84.** $(0.3, 0.4)$

---

85. Find the exact value of:

$$\sin 45° + \sin 135° + \sin 225° + \sin 315°$$

86. Find the exact value of:

$$\tan 60° + \tan 150°$$

87. Find the exact value of:

$$\sin 40° + \sin 130° + \sin 220° + \sin 310°$$

88. Find the exact value of:

$$\tan 40° + \tan 140°$$

89. If $f(\theta) = \sin \theta = 0.1$, find $f(\theta + \pi)$.

90. If $f(\theta) = \cos \theta = 0.3$, find $f(\theta + \pi)$.

91. If $f(\theta) = \tan \theta = 3$, find $f(\theta + \pi)$.

92. If $f(\theta) = \cot \theta = -2$, find $f(\theta + \pi)$.

93. If $\sin \theta = \frac{1}{5}$, find $\csc \theta$.

94. If $\cos \theta = \frac{2}{3}$, find $\sec \theta$.

In Problems 95–106, $f(\theta) = \sin \theta$ and $g(\theta) = \cos \theta$. Find the exact value of each function below if $\theta = 60°$. Do not use a calculator.

95. $f(\theta)$ **96.** $g(\theta)$ **97.** $f\left(\frac{\theta}{2}\right)$ **98.** $g\left(\frac{\theta}{2}\right)$

99. $[f(\theta)]^2$ **100.** $[g(\theta)]^2$ **101.** $f(2\theta)$ **102.** $g(2\theta)$

103. $2f(\theta)$ **104.** $2g(\theta)$ **105.** $f(-\theta)$ **106.** $g(-\theta)$

Mixed Practice

In Problems 107–112, $f(x) = \sin x$, $g(x) = \cos x$, $h(x) = 2x$, and $p(x) = \frac{x}{2}$. Find the value of each of the following:

107. $f\left(h\left(\frac{\pi}{6}\right)\right)$

108. $g(p(60°))$

109. $p(g(315°))$

110. $h\left(f\left(\frac{5\pi}{6}\right)\right)$

111. (a) Find $f\left(\frac{\pi}{4}\right)$. What point is on the graph of f?
(b) Assuming f is one-to-one*, use the result of part (a) to find a point on the graph of f^{-1}.
(c) What point is on the graph of $y = f\left(x + \frac{\pi}{4}\right) - 3$ if $x = \frac{\pi}{4}$?

112. (a) Find $g\left(\frac{\pi}{6}\right)$. What point is on the graph of g?
(b) Assuming g is one-to-one*, use the result of part (a) to find a point on the graph of g^{-1}.
(c) What point is on the graph of $y = 2g\left(x - \frac{\pi}{6}\right)$ if $x = \frac{\pi}{6}$?

Applications and Extensions

113. Find two negative and three positive angles, expressed in radians, for which the point on the unit circle that corresponds to each angle is $\left(\frac{1}{2}, \frac{\sqrt{3}}{2}\right)$.

114. Find two negative and three positive angles, expressed in radians, for which the point on the unit circle that corresponds to each angle is $\left(-\frac{\sqrt{2}}{2}, \frac{\sqrt{2}}{2}\right)$.

115. Use a calculator in radian mode to complete the following table.

What can you conclude about the value of $f(\theta) = \frac{\sin \theta}{\theta}$ as θ approaches 0?

θ	0.5	0.4	0.2	0.1	0.01	0.001	0.0001	0.00001
$\sin \theta$								
$f(\theta) = \frac{\sin \theta}{\theta}$								

116. Use a calculator in radian mode to complete the following table.

What can you conclude about the value of $g(\theta) = \frac{\cos \theta - 1}{\theta}$ as θ approaches 0?

θ	0.5	0.4	0.2	0.1	0.01	0.001	0.0001	0.00001
$\cos \theta - 1$								
$g(\theta) = \frac{\cos \theta - 1}{\theta}$								

For Problems 117–120, use the following discussion.

Projectile Motion *The path of a projectile fired at an inclination θ to the horizontal with initial speed v_0 is a parabola (see the figure).*

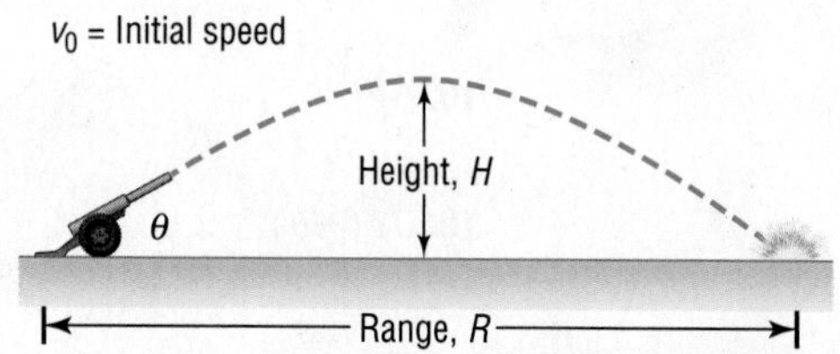

The range R of the projectile, that is, the horizontal distance that the projectile travels, is found by using the function

$$R(\theta) = \frac{v_0^2 \sin(2\theta)}{g}$$

where $g \approx 32.2$ feet per second per second ≈ 9.8 meters per second per second is the acceleration due to gravity. The maximum height H of the projectile is given by the function

$$H(\theta) = \frac{v_0^2 (\sin \theta)^2}{2g}$$

*In Section 3.1, we discuss the necessary domain restriction so that the function is one-to-one.

In Problems 117–120, find the range R and maximum height H.

117. The projectile is fired at an angle of 45° to the horizontal with an initial speed of 100 feet per second.

118. The projectile is fired at an angle of 30° to the horizontal with an initial speed of 150 meters per second.

119. The projectile is fired at an angle of 25° to the horizontal with an initial speed of 500 meters per second.

120. The projectile is fired at an angle of 50° to the horizontal with an initial speed of 200 feet per second.

121. Inclined Plane See the figure.

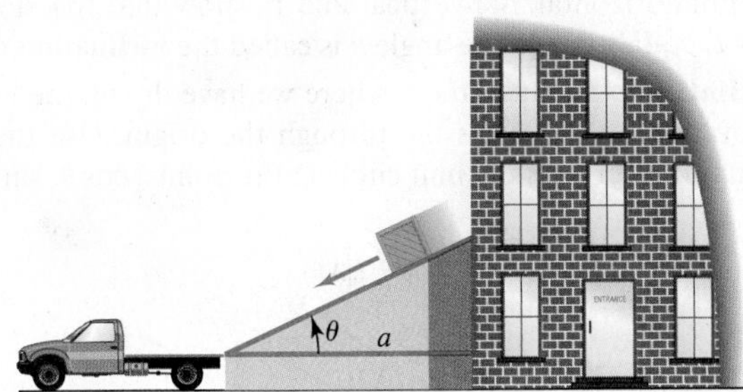

If friction is ignored, the time t (in seconds) required for a block to slide down an inclined plane is given by the function

$$t(\theta) = \sqrt{\frac{2a}{g \sin\theta \cos\theta}}$$

where a is the length (in feet) of the base and $g \approx 32$ feet per second per second is the acceleration due to gravity. How long does it take a block to slide down an inclined plane with base $a = 10$ feet when:

(a) $\theta = 30°$?
(b) $\theta = 45°$?
(c) $\theta = 60°$?

122. Piston Engines In a certain piston engine, the distance x (in centimeters) from the center of the drive shaft to the head of the piston is given by the function

$$x(\theta) = \cos\theta + \sqrt{16 + 0.5\cos(2\theta)}$$

where θ is the angle between the crank and the path of the piston head. See the figure. Find x when $\theta = 30°$ and when $\theta = 45°$.

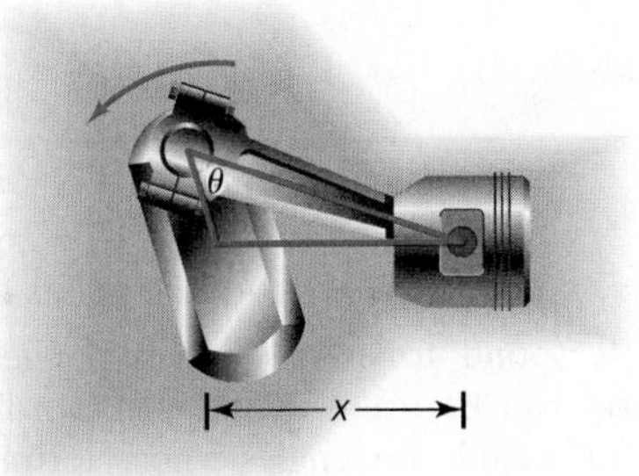

123. Calculating the Time of a Trip Two oceanfront homes are located 8 miles apart on a straight stretch of beach, each a distance of 1 mile from a paved road that parallels the ocean. See the figure.

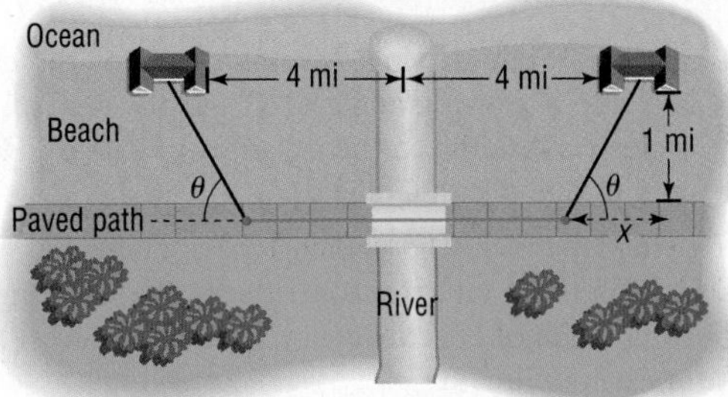

Sally can jog 8 miles per hour along the paved road, but only 3 miles per hour in the sand on the beach. Because of a river directly between the two houses, it is necessary to jog in the sand to the road, continue on the road, and then jog directly back in the sand to get from one house to the other. The time T to get from one house to the other as a function of the angle θ shown in the illustration is

$$T(\theta) = 1 + \frac{2}{3\sin\theta} - \frac{1}{4\tan\theta}, \quad 0° < \theta < 90°$$

(a) Calculate the time T for $\theta = 30°$. How long is Sally on the paved road?
(b) Calculate the time T for $\theta = 45°$. How long is Sally on the paved road?
(c) Calculate the time T for $\theta = 60°$. How long is Sally on the paved road?
(d) Calculate the time T for $\theta = 90°$. Describe the path taken. Why can't the formula for T be used?

124. Designing Fine Decorative Pieces A designer of decorative art plans to market solid gold spheres encased in clear crystal cones. Each sphere is of fixed radius R and will be enclosed in a cone of height h and radius r. See the illustration. Many cones can be used to enclose the sphere, each having a different slant angle θ. The volume V of the cone can be expressed as a function of the slant angle θ of the cone as

$$V(\theta) = \frac{1}{3}\pi R^3 \frac{(1 + \sec\theta)^3}{(\tan\theta)^2} \quad 0° < \theta < 90°$$

What volume V is required to enclose a sphere of radius 2 centimeters in a cone whose slant angle θ is 30°? 45°? 60°?

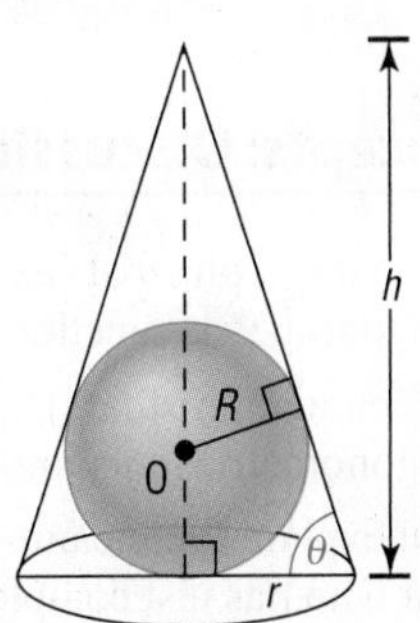

Use the following to answer Problems 125–128. The viewing angle, θ, of an object is the angle the object forms at the lens of the viewer's eye. This is also known as the perceived or angular size of the object. The viewing angle is related to the object's height, H, and distance from the viewer, D, through the formula $\tan\frac{\theta}{2} = \frac{H}{2D}$.

125. Tailgating While driving, Arletha observes the car in front of her with a viewing angle of 22°. If the car is 6 feet wide, how close is Arletha to the car in front of her? Round your answer to one decimal place.

126. Viewing Distance The Washington Monument in Washington, D.C. is 555 feet tall. If a tourist sees the monument with a viewing angle of 8°, how far away, to the nearest foot, is she from the monument?

127. Tree Height A forest ranger views a tree that is 200 feet away with a viewing angle of 20°. How tall is the tree to the nearest foot?

128. Radius of the Moon An astronomer observes the moon with a viewing angle of 0.52°. If the moon's average distance from Earth is 384,400 km, what is its radius to the nearest kilometer?

129. Let θ be the measure of an angle, in radians, in standard position with $\pi < \theta < \frac{3\pi}{2}$. Find the exact y-coordinate of the intersection of the terminal side of θ with the unit circle, given $\cos\theta + \sin^2\theta = \frac{41}{49}$. State the answer as a single fraction, completely simplified, with rationalized denominator.

130. Let θ be the measure of an angle, in radians, in standard position with $\frac{\pi}{2} < \theta < \pi$. Find the exact x-coordinate of the intersection of the terminal side of θ with the unit circle, given $\cos^2\theta - \sin\theta = -\frac{1}{9}$. State the answer as a single fraction, completely simplified, with rationalized denominator.

131. Projectile Motion An object is propelled upward at an angle θ, $45° < \theta < 90°$, to the horizontal with an initial velocity of v_0 feet per second from the base of an inclined plane that makes an angle of 45° with the horizontal. See the illustration.

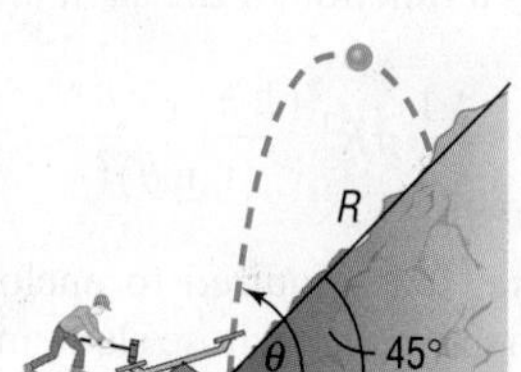

If air resistance is ignored, the distance R that it travels up the inclined plane as a function of θ is given by

$$R(\theta) = \frac{v_0^2\sqrt{2}}{32}\left[\sin(2\theta) - \cos(2\theta) - 1\right]$$

(a) Find the distance R that the object travels along the inclined plane if the initial velocity is 32 feet per second and $\theta = 60°$.

(b) Graph $R = R(\theta)$ if the initial velocity is 32 feet per second.

(c) What value of θ makes R largest?

132. If θ, $0 < \theta < \pi$, is the angle between the positive x-axis and a nonhorizontal, nonvertical line L, show that the slope m of L equals $\tan\theta$. The angle θ is called the **inclination** of L. [**Hint:** See the illustration, where we have drawn the line M parallel to L and passing through the origin. Use the fact that M intersects the unit circle at the point $(\cos\theta, \sin\theta)$.]

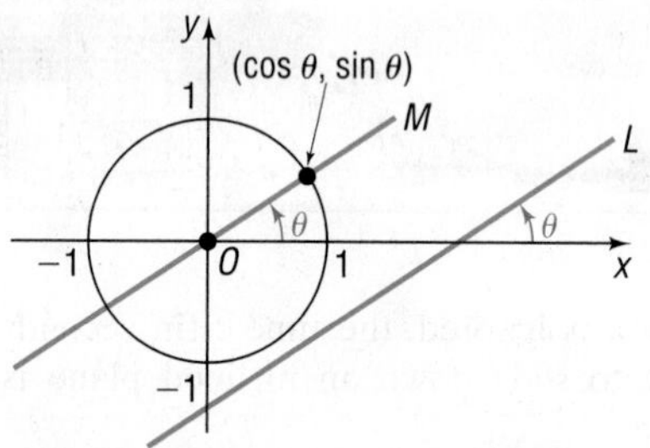

In Problems 133 and 134, use the figure to approximate the value of the six trigonometric functions at t to the nearest tenth. Then use a calculator to approximate each of the six trigonometric functions at t.

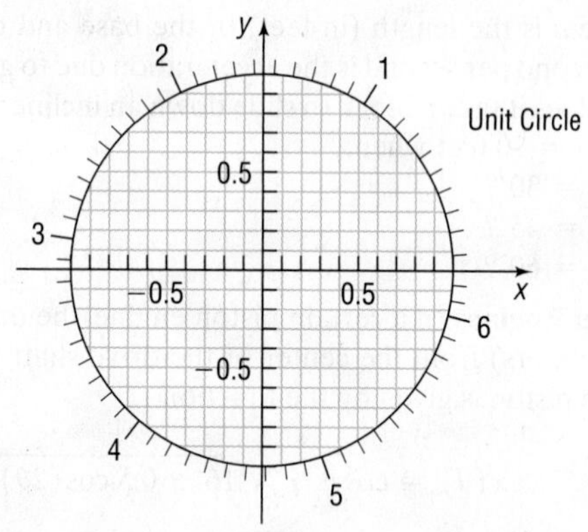

133. (a) $t = 1$ (b) $t = 5.1$

134. (a) $t = 2$ (b) $t = 4$

Explaining Concepts: Discussion and Writing

135. Write a brief paragraph that explains how to quickly compute the trigonometric functions of 30°, 45°, and 60°.

136. Write a brief paragraph that explains how to quickly compute the trigonometric functions of 0°, 90°, 180°, and 270°.

137. How would you explain the meaning of the sine function to a fellow student who has just completed college algebra?

138. Draw a unit circle. Label the angles $0, \frac{\pi}{6}, \frac{\pi}{4}, \frac{\pi}{3}, \ldots, \frac{7\pi}{4}, \frac{11\pi}{6}, 2\pi$ and the coordinates of the points on the unit circle that correspond to each of these angles. Explain how symmetry can be used to find the coordinates of points on the unit circle for angles whose terminal sides are in quadrants II, III, and IV.

Retain Your Knowledge

Problems 139–142 are based on material learned earlier in the course. The purpose of these problems is to keep the material fresh in your mind so that you are better prepared for the final exam.

139. Find the inverse of the one-to-one function $f(x) = \dfrac{2}{x-7}$.

140. Convert $-\dfrac{13\pi}{3}$ radians to degrees.

141. Find $f(x+3)$ given $f(x) = \dfrac{x-1}{x^2+2}$.

142. Find the distance between the points $(3, -6)$ and $(1, 0)$.

'Are You Prepared?' Answers

1. $c^2 = a^2 + b^2$ **2.** 8 **3.** True **4.** equal; proportional **5.** $\left(-\dfrac{1}{2}, \dfrac{\sqrt{3}}{2}\right)$ **6.** $-\dfrac{1}{2}$

2.3 Properties of the Trigonometric Functions

PREPARING FOR THIS SECTION *Before getting started, review the following:*

- Functions (Section 1.3, pp. 23–35)
- Identity (Appendix A, Section A.4, p. A27)
- Even and Odd Functions (Section 1.4, pp. 42–44)

Now Work the **'Are You Prepared?'** problems on page 142.

OBJECTIVES
1. Determine the Domain and the Range of the Trigonometric Functions (p. 131)
2. Determine the Period of the Trigonometric Functions (p. 133)
3. Determine the Signs of the Trigonometric Functions in a Given Quadrant (p. 135)
4. Find the Values of the Trigonometric Functions Using Fundamental Identities (p. 136)
5. Find the Exact Values of the Trigonometric Functions of an Angle Given One of the Functions and the Quadrant of the Angle (p. 138)
6. Use Even–Odd Properties to Find the Exact Values of the Trigonometric Functions (p. 141)

1 Determine the Domain and the Range of the Trigonometric Functions

Let θ be an angle in standard position, and let $P = (x, y)$ be the point on the unit circle that corresponds to θ. See Figure 35. Then, by the definition given earlier,

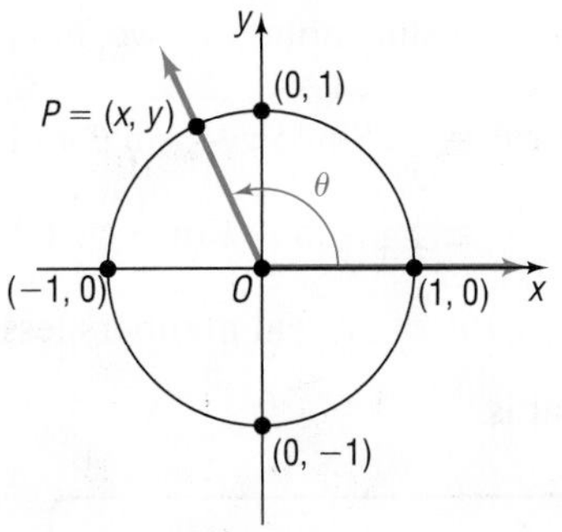

$$\sin\theta = y \qquad \cos\theta = x \qquad \tan\theta = \frac{y}{x} \quad x \neq 0$$

$$\csc\theta = \frac{1}{y} \quad y \neq 0 \qquad \sec\theta = \frac{1}{x} \quad x \neq 0 \qquad \cot\theta = \frac{x}{y} \quad y \neq 0$$

Figure 35

For $\sin\theta$ and $\cos\theta$, there is no concern about dividing by 0, so θ can be any angle. It follows that the domain of the sine function and cosine function is the set of all real numbers.

> The domain of the sine function is the set of all real numbers.
> The domain of the cosine function is the set of all real numbers.

For the tangent function and secant function, the x-coordinate of $P = (x, y)$ cannot be 0 since this results in division by 0. See Figure 35. On the unit circle, there are two such points, $(0, 1)$ and $(0, -1)$. These two points correspond to the angles $\frac{\pi}{2}$ (90°) and $\frac{3\pi}{2}$ (270°) or, more generally, to any angle that is an odd integer multiple of $\frac{\pi}{2}$ (90°), such as $\pm\frac{\pi}{2}$ ($\pm 90°$), $\pm\frac{3\pi}{2}$ ($\pm 270°$), $\pm\frac{5\pi}{2}$ ($\pm 450°$), and so on. Such angles must be excluded from the domain of the tangent function and secant function.

> The domain of the tangent function is the set of all real numbers, except odd integer multiples of $\frac{\pi}{2}$ (90°).
> The domain of the secant function is the set of all real numbers, except odd integer multiples of $\frac{\pi}{2}$ (90°).

For the cotangent function and cosecant function, the y-coordinate of $P = (x, y)$ cannot be 0 since this results in division by 0. See Figure 35. On the unit circle, there are two such points, $(1, 0)$ and $(-1, 0)$. These two points correspond to the angles 0 (0°) and π (180°) or, more generally, to any angle that is an integer multiple of π (180°), such as 0 (0°), $\pm\pi$ ($\pm 180°$), $\pm 2\pi$ ($\pm 360°$), $\pm 3\pi$ ($\pm 540°$), and so on. Such angles must therefore be excluded from the domain of the cotangent function and cosecant function.

> The domain of the cotangent function is the set of all real numbers, except integer multiples of π (180°).
> The domain of the cosecant function is the set of all real numbers, except integer multiples of π (180°).

Next we determine the range of each of the six trigonometric functions. Refer again to Figure 35. Let $P = (x, y)$ be the point on the unit circle that corresponds to the angle θ. It follows that $-1 \le x \le 1$ and $-1 \le y \le 1$. Since $\sin\theta = y$ and $\cos\theta = x$, we have

$$-1 \le \sin\theta \le 1 \qquad -1 \le \cos\theta \le 1$$

The range of both the sine function and the cosine function consists of all real numbers between -1 and 1, inclusive. Using absolute value notation, we have $|\sin\theta| \le 1$ and $|\cos\theta| \le 1$.

If θ is not an integer multiple of π (180°), then $\csc\theta = \frac{1}{y}$. Since $y = \sin\theta$ and $|y| = |\sin\theta| \le 1$, it follows that $|\csc\theta| = \frac{1}{|\sin\theta|} = \frac{1}{|y|} \ge 1 \left(\frac{1}{y} \le -1 \text{ or } \frac{1}{y} \ge 1\right)$. Since $\csc\theta = \frac{1}{y}$, the range of the cosecant function consists of all real numbers less than or equal to -1 or greater than or equal to 1. That is,

$$\csc\theta \le -1 \quad \text{or} \quad \csc\theta \ge 1$$

If θ is not an odd integer multiple of $\frac{\pi}{2}$ $(90°)$, then $\sec\theta = \frac{1}{x}$. Since $x = \cos\theta$ and $|x| = |\cos\theta| \le 1$, it follows that $|\sec\theta| = \frac{1}{|\cos\theta|} = \frac{1}{|x|} \ge 1 \left(\frac{1}{x} \le -1 \text{ or } \frac{1}{x} \ge 1\right)$. Since $\sec\theta = \frac{1}{x}$, the range of the secant function consists of all real numbers less than or equal to -1 or greater than or equal to 1. That is,

$$\sec\theta \le -1 \quad \text{or} \quad \sec\theta \ge 1$$

The range of both the tangent function and the cotangent function is the set of all real numbers.

$$-\infty < \tan\theta < \infty \qquad -\infty < \cot\theta < \infty$$

You are asked to prove this in Problems 121 and 122.

Table 4 summarizes these results.

Table 4

Function	Symbol	Domain	Range
sine	$f(\theta) = \sin\theta$	All real numbers	All real numbers from -1 to 1, inclusive
cosine	$f(\theta) = \cos\theta$	All real numbers	All real numbers from -1 to 1, inclusive
tangent	$f(\theta) = \tan\theta$	All real numbers, except odd integer multiples of $\frac{\pi}{2}$ $(90°)$	All real numbers
cosecant	$f(\theta) = \csc\theta$	All real numbers, except integer multiples of π $(180°)$	All real numbers greater than or equal to 1 or less than or equal to -1
secant	$f(\theta) = \sec\theta$	All real numbers, except odd integer multiples of $\frac{\pi}{2}$ $(90°)$	All real numbers greater than or equal to 1 or less than or equal to -1
cotangent	$f(\theta) = \cot\theta$	All real numbers, except integer multiples of π $(180°)$	All real numbers

Now Work PROBLEM 97

2 Determine the Period of the Trigonometric Functions

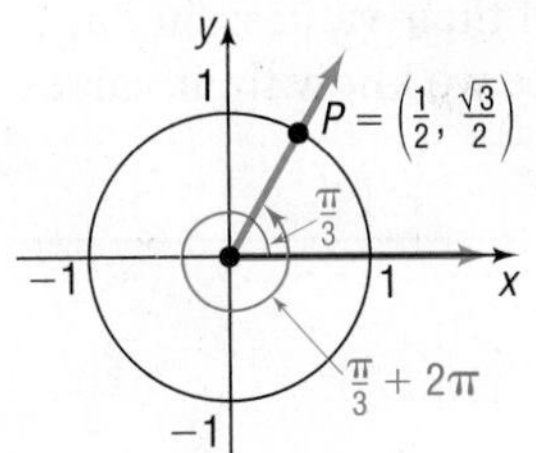

Figure 36 $\sin\left(\frac{\pi}{3} + 2\pi\right) = \sin\frac{\pi}{3}$; $\cos\left(\frac{\pi}{3} + 2\pi\right) = \cos\frac{\pi}{3}$

Look at Figure 36. This figure shows that for an angle of $\frac{\pi}{3}$ radians the corresponding point P on the unit circle is $\left(\frac{1}{2}, \frac{\sqrt{3}}{2}\right)$. Notice that, for an angle of $\frac{\pi}{3} + 2\pi$ radians, the corresponding point P on the unit circle is also $\left(\frac{1}{2}, \frac{\sqrt{3}}{2}\right)$. Then

$$\sin\frac{\pi}{3} = \frac{\sqrt{3}}{2} \quad \text{and} \quad \sin\left(\frac{\pi}{3} + 2\pi\right) = \frac{\sqrt{3}}{2}$$

$$\cos\frac{\pi}{3} = \frac{1}{2} \quad \text{and} \quad \cos\left(\frac{\pi}{3} + 2\pi\right) = \frac{1}{2}$$

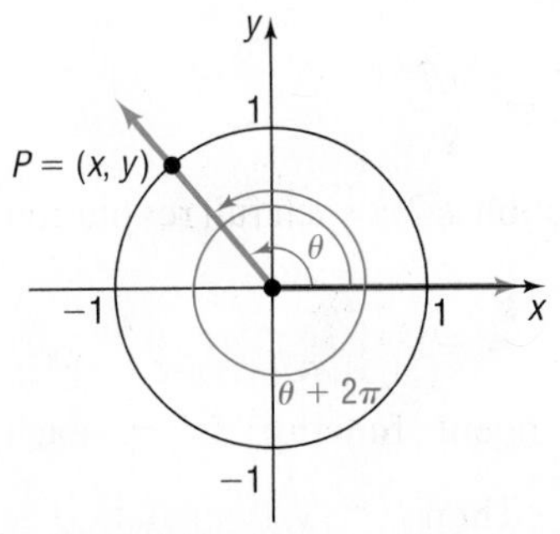

Figure 37 $\sin(\theta + 2\pi k) = \sin\theta$; $\cos(\theta + 2\pi k) = \cos\theta$

This example illustrates a more general situation. For a given angle θ, measured in radians, suppose that we know the corresponding point $P = (x, y)$ on the unit circle. Now add 2π to θ. The point on the unit circle corresponding to $\theta + 2\pi$ is identical to the point P on the unit circle corresponding to θ. See Figure 37. The values of the trigonometric functions of $\theta + 2\pi$ are equal to the values of the corresponding trigonometric functions of θ.

If we add (or subtract) integer multiples of 2π to θ, the values of the sine and cosine function remain unchanged.

That is, for all θ

$$\sin(\theta + 2\pi k) = \sin\theta \qquad \cos(\theta + 2\pi k) = \cos\theta \tag{1}$$
where k is any integer.

Functions that exhibit this kind of behavior are called *periodic functions.*

DEFINITION

A function f is called **periodic** if there is a positive number p such that, whenever θ is in the domain of f, so is $\theta + p$, and

$$f(\theta + p) = f(\theta)$$

If there is a smallest such number p, this smallest value is called the **(fundamental) period** of f.

Based on equation (1), the sine and cosine functions are periodic. In fact, the sine and cosine functions have period 2π. You are asked to prove this fact in Problems 123 and 124. The secant and cosecant functions are also periodic with period 2π, and the tangent and cotangent functions are periodic with period π. You are asked to prove these statements in Problems 125 through 128.

These facts are summarized as follows:

In Words
Tangent and cotangent have period π; the others have period 2π.

Periodic Properties

$$\sin(\theta + 2\pi) = \sin\theta \qquad \cos(\theta + 2\pi) = \cos\theta \qquad \tan(\theta + \pi) = \tan\theta$$
$$\csc(\theta + 2\pi) = \csc\theta \qquad \sec(\theta + 2\pi) = \sec\theta \qquad \cot(\theta + \pi) = \cot\theta$$

Because the sine, cosine, secant, and cosecant functions have period 2π, once we know their values over an interval of length 2π, we know all their values; similarly, since the tangent and cotangent functions have period π, once we know their values over an interval of length π, we know all their values.

EXAMPLE 1

Finding Exact Values Using Periodic Properties

Find the exact value of:

(a) $\sin\dfrac{17\pi}{4}$ (b) $\cos(5\pi)$ (c) $\tan\dfrac{5\pi}{4}$

Solution

(a) It is best to sketch the angle first, as shown in Figure 38(a). Since the period of the sine function is 2π, each full revolution can be ignored leaving the angle $\dfrac{\pi}{4}$. Then

$$\sin\frac{17\pi}{4} = \sin\left(\frac{\pi}{4} + 4\pi\right) = \sin\frac{\pi}{4} = \frac{\sqrt{2}}{2}$$

(b) See Figure 38(b). Since the period of the cosine function is 2π, each full revolution can be ignored leaving the angle π. Then

$$\cos(5\pi) = \cos(\pi + 4\pi) = \cos\pi = -1$$

(c) See Figure 38(c). Since the period of the tangent function is π, each half-revolution can be ignored leaving the angle $\dfrac{\pi}{4}$. Then

$$\tan\frac{5\pi}{4} = \tan\left(\frac{\pi}{4} + \pi\right) = \tan\frac{\pi}{4} = 1$$

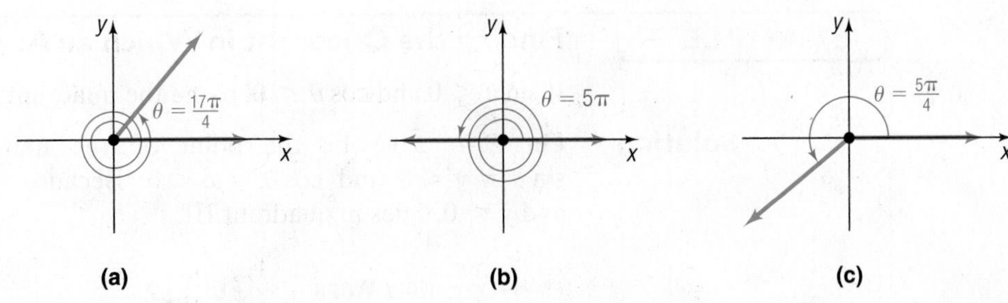

Figure 38

The periodic properties of the trigonometric functions will be very helpful to us when we study their graphs later in the chapter.

Now Work PROBLEM 11

3 Determine the Signs of the Trigonometric Functions in a Given Quadrant

Let $P = (x, y)$ be the point on the unit circle that corresponds to the angle θ. If we know in which quadrant the point P lies, then we can determine the signs of the trigonometric functions of θ. For example, if $P = (x, y)$ lies in quadrant IV, as shown in Figure 39, then we know that $x > 0$ and $y < 0$. Consequently,

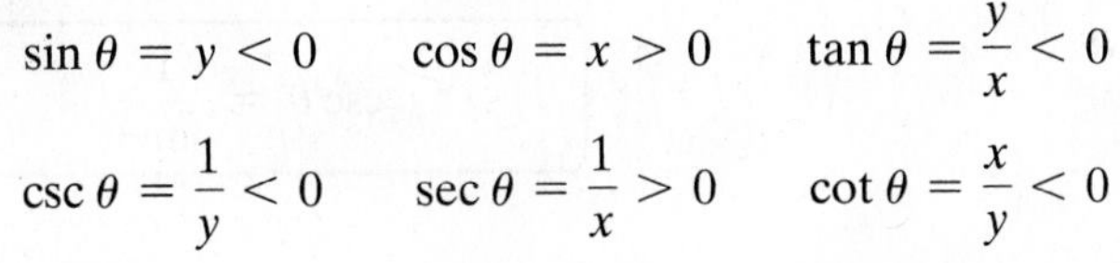

$$\sin\theta = y < 0 \qquad \cos\theta = x > 0 \qquad \tan\theta = \frac{y}{x} < 0$$

$$\csc\theta = \frac{1}{y} < 0 \qquad \sec\theta = \frac{1}{x} > 0 \qquad \cot\theta = \frac{x}{y} < 0$$

Figure 39
θ in quadrant IV, $x > 0, y < 0$

Table 5 lists the signs of the six trigonometric functions for each quadrant. See also Figure 40.

Table 5

Quadrant of *P*	$\sin\theta$, $\csc\theta$	$\cos\theta$, $\sec\theta$	$\tan\theta$, $\cot\theta$
I	Positive	Positive	Positive
II	Positive	Negative	Negative
III	Negative	Negative	Positive
IV	Negative	Positive	Negative

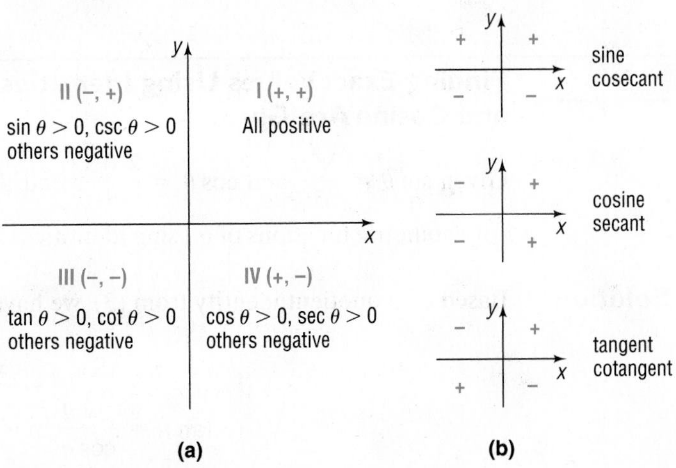

Figure 40 Signs of the trigonometric functions

EXAMPLE 2 **Finding the Quadrant in Which an Angle θ Lies**

If $\sin\theta < 0$ and $\cos\theta < 0$, name the quadrant in which the angle θ lies.

Solution Let $P = (x, y)$ be the point on the unit circle corresponding to θ. Then $\sin\theta = y < 0$ and $\cos\theta = x < 0$. Because points in quadrant III have $x < 0$ and $y < 0$, θ lies in quadrant III. ●

Now Work PROBLEM 27

4 Find the Values of the Trigonometric Functions Using Fundamental Identities

If $P = (x, y)$ is the point on the unit circle corresponding to θ, then

$$\sin\theta = y \qquad \cos\theta = x \qquad \tan\theta = \frac{y}{x} \text{ if } x \neq 0$$

$$\csc\theta = \frac{1}{y} \text{ if } y \neq 0 \qquad \sec\theta = \frac{1}{x} \text{ if } x \neq 0 \qquad \cot\theta = \frac{x}{y} \text{ if } y \neq 0$$

Based on these definitions, we have the **reciprocal identities**:

Reciprocal Identities

$$\csc\theta = \frac{1}{\sin\theta} \qquad \sec\theta = \frac{1}{\cos\theta} \qquad \cot\theta = \frac{1}{\tan\theta} \tag{2}$$

Two other fundamental identities are the **quotient identities.**

Quotient Identities

$$\tan\theta = \frac{\sin\theta}{\cos\theta} \qquad \cot\theta = \frac{\cos\theta}{\sin\theta} \tag{3}$$

The proofs of identities (2) and (3) follow from the definitions of the trigonometric functions. (See Problems 129 and 130.)

If $\sin\theta$ and $\cos\theta$ are known, identities (2) and (3) make it easy to find the values of the remaining trigonometric functions.

EXAMPLE 3 **Finding Exact Values Using Identities When Sine and Cosine Are Given**

Given $\sin\theta = \dfrac{\sqrt{5}}{5}$ and $\cos\theta = \dfrac{2\sqrt{5}}{5}$, find the exact values of the four remaining trigonometric functions of θ using identities.

Solution Based on a quotient identity from (3), we have

$$\tan\theta = \frac{\sin\theta}{\cos\theta} = \frac{\dfrac{\sqrt{5}}{5}}{\dfrac{2\sqrt{5}}{5}} = \frac{1}{2}$$

Then we use the reciprocal identities from (2) to get

$$\csc\theta = \frac{1}{\sin\theta} = \frac{1}{\frac{\sqrt{5}}{5}} = \frac{5}{\sqrt{5}} = \sqrt{5} \qquad \sec\theta = \frac{1}{\cos\theta} = \frac{1}{\frac{2\sqrt{5}}{5}} = \frac{5}{2\sqrt{5}} = \frac{\sqrt{5}}{2} \qquad \cot\theta = \frac{1}{\tan\theta} = \frac{1}{\frac{1}{2}} = 2$$

Now Work PROBLEM 35

The equation of the unit circle is $x^2 + y^2 = 1$ or, equivalently,

$$y^2 + x^2 = 1$$

If $P = (x, y)$ is the point on the unit circle that corresponds to the angle θ, then $y = \sin\theta$ and $x = \cos\theta$, so we have

$$(\sin\theta)^2 + (\cos\theta)^2 = 1 \qquad \textbf{(4)}$$

It is customary to write $\sin^2\theta$ instead of $(\sin\theta)^2$, $\cos^2\theta$ instead of $(\cos\theta)^2$, and so on. With this notation, we can rewrite equation (4) as

$$\sin^2\theta + \cos^2\theta = 1 \qquad \textbf{(5)}$$

If $\cos\theta \neq 0$, we can divide each side of equation (5) by $\cos^2\theta$.

$$\frac{\sin^2\theta}{\cos^2\theta} + \frac{\cos^2\theta}{\cos^2\theta} = \frac{1}{\cos^2\theta}$$

$$\left(\frac{\sin\theta}{\cos\theta}\right)^2 + 1 = \left(\frac{1}{\cos\theta}\right)^2$$

Now use identities (2) and (3) to get

$$\tan^2\theta + 1 = \sec^2\theta \qquad \textbf{(6)}$$

Similarly, if $\sin\theta \neq 0$, we can divide equation (5) by $\sin^2\theta$ and use identities (2) and (3) to get $1 + \cot^2\theta = \csc^2\theta$, which we write as

$$\cot^2\theta + 1 = \csc^2\theta \qquad \textbf{(7)}$$

Collectively, the identities in (5), (6), and (7) are referred to as the **Pythagorean identities.**

Fundamental Identities

$$\tan\theta = \frac{\sin\theta}{\cos\theta} \qquad \cot\theta = \frac{\cos\theta}{\sin\theta}$$

$$\csc\theta = \frac{1}{\sin\theta} \qquad \sec\theta = \frac{1}{\cos\theta} \qquad \cot\theta = \frac{1}{\tan\theta}$$

$$\sin^2\theta + \cos^2\theta = 1 \qquad \tan^2\theta + 1 = \sec^2\theta \qquad \cot^2\theta + 1 = \csc^2\theta$$

EXAMPLE 4

Finding the Exact Value of a Trigonometric Expression Using Identities

Find the exact value of each expression. Do not use a calculator.

(a) $\tan 20^\circ - \dfrac{\sin 20^\circ}{\cos 20^\circ}$ (b) $\sin^2\dfrac{\pi}{12} + \dfrac{1}{\sec^2\dfrac{\pi}{12}}$

Solution (a) $\tan 20° - \dfrac{\sin 20°}{\cos 20°} = \tan 20° - \tan 20° = 0$

$\uparrow$ $\dfrac{\sin\theta}{\cos\theta} = \tan\theta$

(b) $\sin^2 \dfrac{\pi}{12} + \dfrac{1}{\sec^2 \dfrac{\pi}{12}} = \sin^2 \dfrac{\pi}{12} + \cos^2 \dfrac{\pi}{12} = 1$

$\uparrow$ $\cos\theta = \dfrac{1}{\sec\theta}$ $\qquad \uparrow$ $\sin^2\theta + \cos^2\theta = 1$

Now Work PROBLEM 79

5 Find the Exact Values of the Trigonometric Functions of an Angle Given One of the Functions and the Quadrant of the Angle

Many problems require finding the exact values of the remaining trigonometric functions when the value of one of them is known and the quadrant in which θ lies can be found. There are two approaches to solving such problems. One approach uses a circle of radius r; the other uses identities.

When using identities, sometimes a rearrangement is required. For example, the Pythagorean identity

$$\sin^2\theta + \cos^2\theta = 1$$

can be solved for $\sin\theta$ in terms of $\cos\theta$ (or vice versa) as follows:

$$\sin^2\theta = 1 - \cos^2\theta$$
$$\sin\theta = \pm\sqrt{1 - \cos^2\theta}$$

where the + sign is used if $\sin\theta > 0$ and the − sign is used if $\sin\theta < 0$. Similarly, in $\tan^2\theta + 1 = \sec^2\theta$, we can solve for $\tan\theta$ (or $\sec\theta$), and in $\cot^2\theta + 1 = \csc^2\theta$, we can solve for $\cot\theta$ (or $\csc\theta$).

EXAMPLE 5 **Finding Exact Values Given One Value and the Sign of Another**

Given that $\sin\theta = \dfrac{1}{3}$ and $\cos\theta < 0$, find the exact value of each of the remaining five trigonometric functions.

Option I Using a Circle

Suppose that $P = (x, y)$ is the point on a circle that corresponds to θ. Since $\sin\theta = \dfrac{1}{3} > 0$ and $\cos\theta < 0$, the point $P = (x, y)$ is in quadrant II. Because $\sin\theta = \dfrac{1}{3} = \dfrac{y}{r}$, we let $y = 1$ and $r = 3$. The point $P = (x, y) = (x, 1)$ that corresponds to θ lies on the circle of radius 3 centered at the origin: $x^2 + y^2 = 9$. See Figure 41.

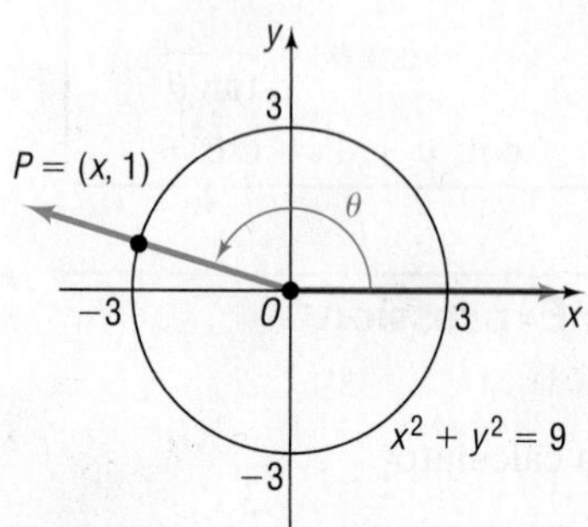

Figure 41

To find x, we use the fact that $x^2 + y^2 = 9$, $y = 1$, and P is in quadrant II (so $x < 0$).

$$x^2 + y^2 = 9$$
$$x^2 + 1^2 = 9 \qquad y = 1$$
$$x^2 = 8$$
$$x = -2\sqrt{2} \qquad x < 0$$

Since $x = -2\sqrt{2}$, $y = 1$, and $r = 3$, we find that

$$\cos\theta = \frac{x}{r} = -\frac{2\sqrt{2}}{3} \qquad \tan\theta = \frac{y}{x} = \frac{1}{-2\sqrt{2}} = -\frac{\sqrt{2}}{4}$$

$$\csc\theta = \frac{r}{y} = \frac{3}{1} = 3 \qquad \sec\theta = \frac{r}{x} = \frac{3}{-2\sqrt{2}} = -\frac{3\sqrt{2}}{4} \qquad \cot\theta = \frac{x}{y} = \frac{-2\sqrt{2}}{1} = -2\sqrt{2}$$

Option 2 Using Identities First, solve the identity $\sin^2\theta + \cos^2\theta = 1$ for $\cos\theta$.

$$\sin^2\theta + \cos^2\theta = 1$$
$$\cos^2\theta = 1 - \sin^2\theta$$
$$\cos\theta = \pm\sqrt{1 - \sin^2\theta}$$

Because $\cos\theta < 0$, choose the minus sign and use the fact that $\sin\theta = \frac{1}{3}$.

$$\cos\theta = -\sqrt{1 - \sin^2\theta} = -\sqrt{1 - \frac{1}{9}} = -\sqrt{\frac{8}{9}} = -\frac{2\sqrt{2}}{3}$$

$\uparrow$ $\sin\theta = \frac{1}{3}$

Now we know the values of $\sin\theta$ and $\cos\theta$, so we can use quotient and reciprocal identities to get

$$\tan\theta = \frac{\sin\theta}{\cos\theta} = \frac{\frac{1}{3}}{\frac{-2\sqrt{2}}{3}} = \frac{1}{-2\sqrt{2}} = -\frac{\sqrt{2}}{4} \qquad \cot\theta = \frac{1}{\tan\theta} = -2\sqrt{2}$$

$$\sec\theta = \frac{1}{\cos\theta} = \frac{1}{\frac{-2\sqrt{2}}{3}} = \frac{-3}{2\sqrt{2}} = -\frac{3\sqrt{2}}{4} \qquad \csc\theta = \frac{1}{\sin\theta} = \frac{1}{\frac{1}{3}} = 3$$

●

Finding the Values of the Trigonometric Functions of θ When the Value of One Function Is Known and the Quadrant of θ Is Known

Given the value of one trigonometric function and the quadrant in which θ lies, the exact value of each of the remaining five trigonometric functions can be found in either of two ways.

Option 1 Using a Circle of Radius r

STEP 1: Draw a circle centered at the origin showing the location of the angle θ and the point $P = (x, y)$ that corresponds to θ. The radius of the circle that contains $P = (x, y)$ is $r = \sqrt{x^2 + y^2}$.

STEP 2: Assign a value to two of the three variables x, y, r based on the value of the given trigonometric function and the location of P.

STEP 3: Use the fact that P lies on the circle $x^2 + y^2 = r^2$ to find the value of the missing variable.

STEP 4: Apply the theorem on page 125 to find the values of the remaining trigonometric functions.

Option 2 Using Identities

Use appropriately selected identities to find the value of each remaining trigonometric function.

EXAMPLE 6

Given the Value of One Trigonometric Function and the Sign of Another, Find the Values of the Remaining Ones

Given that $\tan\theta = \frac{1}{2}$ and $\sin\theta < 0$, find the exact value of each of the remaining five trigonometric functions of θ.

Option 1 Using a Circle

STEP 1: Since $\tan\theta = \frac{1}{2} > 0$ and $\sin\theta < 0$, the point $P = (x, y)$ that corresponds to θ lies in quadrant III. See Figure 42.

STEP 2: Since $\tan\theta = \frac{1}{2} = \frac{y}{x}$ and θ lies in quadrant III, let $x = -2$ and $y = -1$.

STEP 3: With $x = -2$ and $y = -1$, then $r = \sqrt{x^2 + y^2} = \sqrt{(-2)^2 + (-1)^2} = \sqrt{5}$, so P lies on the circle $x^2 + y^2 = 5$.

STEP 4: Apply the theorem on page 125 using $x = -2$, $y = -1$, and $r = \sqrt{5}$.

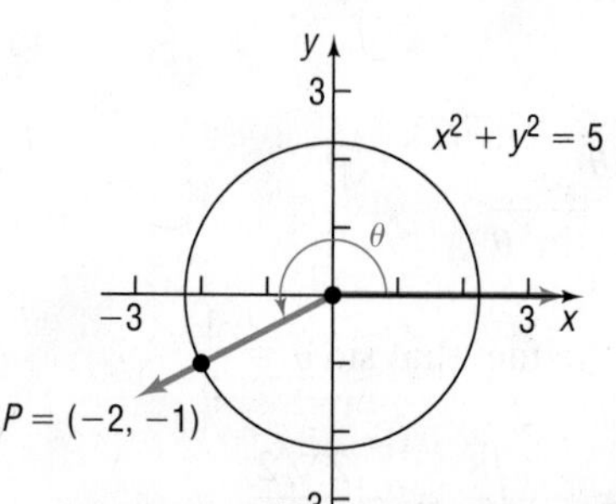

Figure 42 $\tan\theta = \frac{1}{2}$; θ in quadrant III

$$\sin\theta = \frac{y}{r} = \frac{-1}{\sqrt{5}} = -\frac{\sqrt{5}}{5} \qquad \cos\theta = \frac{x}{r} = \frac{-2}{\sqrt{5}} = -\frac{2\sqrt{5}}{5}$$

$$\csc\theta = \frac{r}{y} = \frac{\sqrt{5}}{-1} = -\sqrt{5} \qquad \sec\theta = \frac{r}{x} = \frac{\sqrt{5}}{-2} = -\frac{\sqrt{5}}{2} \qquad \cot\theta = \frac{x}{y} = \frac{-2}{-1} = 2$$

●

Option 2 Using Identities

Because the value of $\tan\theta$ is known, use the Pythagorean identity that involves $\tan\theta$, that is, $\tan^2\theta + 1 = \sec^2\theta$. Since $\tan\theta = \frac{1}{2} > 0$ and $\sin\theta < 0$, then θ lies in quadrant III, where $\sec\theta < 0$.

$$\tan^2\theta + 1 = \sec^2\theta \qquad \text{Pythagorean identity}$$

$$\left(\frac{1}{2}\right)^2 + 1 = \sec^2\theta \qquad \tan\theta = \frac{1}{2}$$

$$\sec^2\theta = \frac{1}{4} + 1 = \frac{5}{4} \qquad \text{Proceed to solve for } \sec\theta.$$

$$\sec\theta = -\frac{\sqrt{5}}{2} \qquad \sec\theta < 0$$

Now we know $\tan\theta = \frac{1}{2}$ and $\sec\theta = -\frac{\sqrt{5}}{2}$. Using reciprocal identities, we find

$$\cos\theta = \frac{1}{\sec\theta} = \frac{1}{-\frac{\sqrt{5}}{2}} = -\frac{2}{\sqrt{5}} = -\frac{2\sqrt{5}}{5}$$

$$\cot\theta = \frac{1}{\tan\theta} = \frac{1}{\frac{1}{2}} = 2$$

To find $\sin\theta$, use the following reasoning:

$$\tan\theta = \frac{\sin\theta}{\cos\theta} \quad \text{so} \quad \sin\theta = (\tan\theta)(\cos\theta) = \left(\frac{1}{2}\right)\cdot\left(-\frac{2\sqrt{5}}{5}\right) = -\frac{\sqrt{5}}{5}$$

$$\csc\theta = \frac{1}{\sin\theta} = \frac{1}{-\frac{\sqrt{5}}{5}} = -\frac{5}{\sqrt{5}} = -\sqrt{5}$$

●

Now Work PROBLEM 43

6 Use Even–Odd Properties to Find the Exact Values of the Trigonometric Functions

Recall that a function f is even if $f(-\theta) = f(\theta)$ for all θ in the domain of f; a function f is odd if $f(-\theta) = -f(\theta)$ for all θ in the domain of f. We will now show that the trigonometric functions sine, tangent, cotangent, and cosecant are odd functions and the functions cosine and secant are even functions.

In Words

Cosine and secant are even functions; the others are odd functions.

Even–Odd Properties

$$\sin(-\theta) = -\sin\theta \qquad \cos(-\theta) = \cos\theta \qquad \tan(-\theta) = -\tan\theta$$
$$\csc(-\theta) = -\csc\theta \qquad \sec(-\theta) = \sec\theta \qquad \cot(-\theta) = -\cot\theta$$

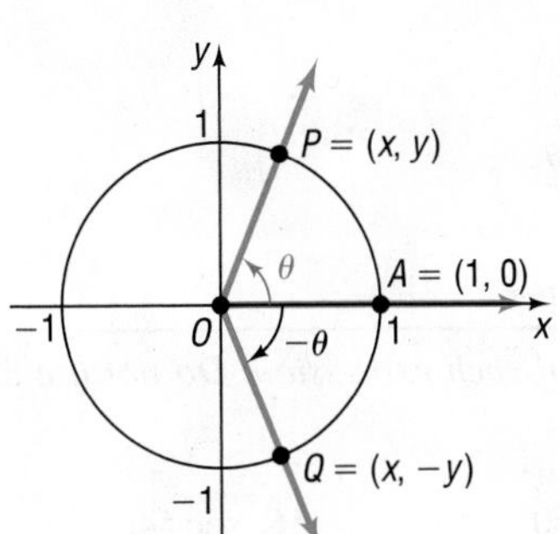

Figure 43

Proof Let $P = (x, y)$ be the point on the unit circle that corresponds to the angle θ. See Figure 43. Using symmetry, the point Q on the unit circle that corresponds to the angle $-\theta$ will have coordinates $(x, -y)$. Using the definition of the trigonometric functions, we have

$$\sin\theta = y \qquad \sin(-\theta) = -y \qquad \cos\theta = x \qquad \cos(-\theta) = x$$

so

$$\sin(-\theta) = -y = -\sin\theta \qquad \cos(-\theta) = x = \cos\theta$$

Now, using these results and some of the fundamental identities, we have

$$\tan(-\theta) = \frac{\sin(-\theta)}{\cos(-\theta)} = \frac{-\sin\theta}{\cos\theta} = -\tan\theta \qquad \cot(-\theta) = \frac{1}{\tan(-\theta)} = \frac{1}{-\tan\theta} = -\cot\theta$$

$$\sec(-\theta) = \frac{1}{\cos(-\theta)} = \frac{1}{\cos\theta} = \sec\theta \qquad \csc(-\theta) = \frac{1}{\sin(-\theta)} = \frac{1}{-\sin\theta} = -\csc\theta$$

■

EXAMPLE 7 **Finding Exact Values Using Even–Odd Properties**

Find the exact value of:

(a) $\sin(-45°)$ (b) $\cos(-\pi)$ (c) $\cot\left(-\frac{3\pi}{2}\right)$ (d) $\tan\left(-\frac{37\pi}{4}\right)$

Solution

(a) $\sin(-45°) = -\sin 45° = -\frac{\sqrt{2}}{2}$ (Odd function)

(b) $\cos(-\pi) = \cos\pi = -1$ (Even function)

(c) $\cot\left(-\frac{3\pi}{2}\right) = -\cot\frac{3\pi}{2} = 0$ (Odd function)

(d) $\tan\left(-\frac{37\pi}{4}\right) = -\tan\frac{37\pi}{4} = -\tan\left(\frac{\pi}{4} + 9\pi\right) = -\tan\frac{\pi}{4} = -1$ (Odd function; Period is π.)

●

Now Work PROBLEM 59

2.3 Assess Your Understanding

'Are You Prepared?' *Answers are given at the end of these exercises. If you get a wrong answer, read the pages listed in red.*

1. Find the domain of the function $f(x) = \dfrac{x+1}{2x+1}$. (pp. 23–35)

2. A function for which $f(x) = f(-x)$ for all x in the domain of f is called a(n) _____ function. (pp. 42–44)

3. ***True or False*** The function $f(x) = \sqrt{x}$ is even. (pp. 42–44)

4. ***True or False*** The equation $x^2 + 2x = (x+1)^2 - 1$ is an identity. (p. A27)

Concepts and Vocabulary

5. The sine, cosine, cosecant, and secant functions have period ___; the tangent and cotangent functions have period __.

6. The domain of the tangent function is _________.

7. Which of the following is not in the range of the sine function?
(a) $\dfrac{\pi}{4}$ (b) $\dfrac{3}{2}$ (c) -0.37 (d) -1

8. Which of the following functions is even?
(a) cosine (b) sine (c) tangent (d) cosecant

9. $\sin^2\theta + \cos^2\theta =$ __.

10. ***True or False*** $\sec\theta = \dfrac{1}{\sin\theta}$

Skill Building

In Problems 11–26, use the fact that the trigonometric functions are periodic to find the exact value of each expression. Do not use a calculator.

11. $\sin 405°$ **12.** $\cos 420°$ **13.** $\tan 405°$ **14.** $\sin 390°$ **15.** $\csc 450°$ **16.** $\sec 540°$

17. $\cot 390°$ **18.** $\sec 420°$ **19.** $\cos\dfrac{33\pi}{4}$ **20.** $\sin\dfrac{9\pi}{4}$ **21.** $\tan(21\pi)$ **22.** $\csc\dfrac{9\pi}{2}$

23. $\sec\dfrac{17\pi}{4}$ **24.** $\cot\dfrac{17\pi}{4}$ **25.** $\tan\dfrac{19\pi}{6}$ **26.** $\sec\dfrac{25\pi}{6}$

In Problems 27–34, name the quadrant in which the angle θ lies.

27. $\sin\theta > 0,\ \cos\theta < 0$ **28.** $\sin\theta < 0,\ \cos\theta > 0$ **29.** $\sin\theta < 0,\ \tan\theta < 0$ **30.** $\cos\theta > 0,\ \tan\theta > 0$

31. $\cos\theta > 0,\ \tan\theta < 0$ **32.** $\cos\theta < 0,\ \tan\theta > 0$ **33.** $\sec\theta < 0,\ \sin\theta > 0$ **34.** $\csc\theta > 0,\ \cos\theta < 0$

In Problems 35–42, sin θ and cos θ are given. Find the exact value of each of the four remaining trigonometric functions.

35. $\sin\theta = -\dfrac{3}{5},\ \cos\theta = \dfrac{4}{5}$ **36.** $\sin\theta = \dfrac{4}{5},\ \cos\theta = -\dfrac{3}{5}$ **37.** $\sin\theta = \dfrac{2\sqrt{5}}{5},\ \cos\theta = \dfrac{\sqrt{5}}{5}$

38. $\sin\theta = -\dfrac{\sqrt{5}}{5},\ \cos\theta = -\dfrac{2\sqrt{5}}{5}$ **39.** $\sin\theta = \dfrac{1}{2},\ \cos\theta = \dfrac{\sqrt{3}}{2}$ **40.** $\sin\theta = \dfrac{\sqrt{3}}{2},\ \cos\theta = \dfrac{1}{2}$

41. $\sin\theta = -\dfrac{1}{3},\ \cos\theta = \dfrac{2\sqrt{2}}{3}$ **42.** $\sin\theta = \dfrac{2\sqrt{2}}{3},\ \cos\theta = -\dfrac{1}{3}$

In Problems 43–58, find the exact value of each of the remaining trigonometric functions of θ.

43. $\sin\theta = \dfrac{12}{13},\ \theta$ in quadrant II **44.** $\cos\theta = \dfrac{3}{5},\ \theta$ in quadrant IV **45.** $\cos\theta = -\dfrac{4}{5},\ \theta$ in quadrant III

46. $\sin\theta = -\dfrac{5}{13},\ \theta$ in quadrant III **47.** $\sin\theta = \dfrac{5}{13},\ 90° < \theta < 180°$ **48.** $\cos\theta = \dfrac{4}{5},\ 270° < \theta < 360°$

49. $\cos\theta = -\dfrac{1}{3},\ \dfrac{\pi}{2} < \theta < \pi$ **50.** $\sin\theta = -\dfrac{2}{3},\ \pi < \theta < \dfrac{3\pi}{2}$ **51.** $\sin\theta = \dfrac{2}{3},\ \tan\theta < 0$

52. $\cos\theta = -\dfrac{1}{4},\ \tan\theta > 0$ **53.** $\sec\theta = 2,\ \sin\theta < 0$ **54.** $\csc\theta = 3,\ \cot\theta < 0$

55. $\tan\theta = \dfrac{3}{4},\ \sin\theta < 0$ **56.** $\cot\theta = \dfrac{4}{3},\ \cos\theta < 0$ **57.** $\tan\theta = -\dfrac{1}{3},\ \sin\theta > 0$

58. $\sec\theta = -2,\ \tan\theta > 0$

In Problems 59–76, use the even–odd properties to find the exact value of each expression. Do not use a calculator.

59. $\sin(-60°)$
60. $\cos(-30°)$
61. $\tan(-30°)$
62. $\sin(-135°)$
63. $\sec(-60°)$
64. $\csc(-30°)$
65. $\sin(-90°)$
66. $\cos(-270°)$
67. $\tan\left(-\frac{\pi}{4}\right)$
68. $\sin(-\pi)$
69. $\cos\left(-\frac{\pi}{4}\right)$
70. $\sin\left(-\frac{\pi}{3}\right)$
71. $\tan(-\pi)$
72. $\sin\left(-\frac{3\pi}{2}\right)$
73. $\csc\left(-\frac{\pi}{4}\right)$
74. $\sec(-\pi)$
75. $\sec\left(-\frac{\pi}{6}\right)$
76. $\csc\left(-\frac{\pi}{3}\right)$

In Problems 77–88, use properties of the trigonometric functions to find the exact value of each expression. Do not use a calculator.

77. $\sin^2 40° + \cos^2 40°$
78. $\sec^2 18° - \tan^2 18°$
79. $\sin 80° \csc 80°$
80. $\tan 10° \cot 10°$
81. $\tan 40° - \frac{\sin 40°}{\cos 40°}$
82. $\cot 20° - \frac{\cos 20°}{\sin 20°}$
83. $\cos 400° \cdot \sec 40°$
84. $\tan 200° \cdot \cot 20°$
85. $\sin\left(-\frac{\pi}{12}\right)\csc\frac{25\pi}{12}$
86. $\sec\left(-\frac{\pi}{18}\right)\cdot\cos\frac{37\pi}{18}$
87. $\frac{\sin(-20°)}{\cos 380°} + \tan 200°$
88. $\frac{\sin 70°}{\cos(-430°)} + \tan(-70°)$

89. If $\sin\theta = 0.3$, find the value of:
$$\sin\theta + \sin(\theta + 2\pi) + \sin(\theta + 4\pi)$$

90. If $\cos\theta = 0.2$, find the value of:
$$\cos\theta + \cos(\theta + 2\pi) + \cos(\theta + 4\pi)$$

91. If $\tan\theta = 3$, find the value of:
$$\tan\theta + \tan(\theta + \pi) + \tan(\theta + 2\pi)$$

92. If $\cot\theta = -2$, find the value of:
$$\cot\theta + \cot(\theta - \pi) + \cot(\theta - 2\pi)$$

93. Find the exact value of:
$$\sin 1° + \sin 2° + \sin 3° + \cdots + \sin 358° + \sin 359°$$

94. Find the exact value of:
$$\cos 1° + \cos 2° + \cos 3° + \cdots + \cos 358° + \cos 359°$$

95. What is the domain of the sine function?

96. What is the domain of the cosine function?

97. For what numbers θ is $f(\theta) = \tan\theta$ not defined?

98. For what numbers θ is $f(\theta) = \cot\theta$ not defined?

99. For what numbers θ is $f(\theta) = \sec\theta$ not defined?

100. For what numbers θ is $f(\theta) = \csc\theta$ not defined?

101. What is the range of the sine function?

102. What is the range of the cosine function?

103. What is the range of the tangent function?

104. What is the range of the cotangent function?

105. What is the range of the secant function?

106. What is the range of the cosecant function?

107. Is the sine function even, odd, or neither? Is its graph symmetric? With respect to what?

108. Is the cosine function even, odd, or neither? Is its graph symmetric? With respect to what?

109. Is the tangent function even, odd, or neither? Is its graph symmetric? With respect to what?

110. Is the cotangent function even, odd, or neither? Is its graph symmetric? With respect to what?

111. Is the secant function even, odd, or neither? Is its graph symmetric? With respect to what?

112. Is the cosecant function even, odd, or neither? Is its graph symmetric? With respect to what?

Applications and Extensions

In Problems 113–118, use the periodic and even–odd properties.

113. If $f(\theta) = \sin\theta$ and $f(a) = \frac{1}{3}$, find the exact value of:
(a) $f(-a)$ (b) $f(a) + f(a + 2\pi) + f(a + 4\pi)$

114. If $f(\theta) = \cos\theta$ and $f(a) = \frac{1}{4}$, find the exact value of:
(a) $f(-a)$ (b) $f(a) + f(a + 2\pi) + f(a - 2\pi)$

115. If $f(\theta) = \tan\theta$ and $f(a) = 2$, find the exact value of:
(a) $f(-a)$ (b) $f(a) + f(a + \pi) + f(a + 2\pi)$

116. If $f(\theta) = \cot\theta$ and $f(a) = -3$, find the exact value of:
(a) $f(-a)$ (b) $f(a) + f(a + \pi) + f(a + 4\pi)$

117. If $f(\theta) = \sec\theta$ and $f(a) = -4$, find the exact value of:
(a) $f(-a)$ (b) $f(a) + f(a + 2\pi) + f(a + 4\pi)$

118. If $f(\theta) = \csc\theta$ and $f(a) = 2$, find the exact value of:
(a) $f(-a)$ (b) $f(a) + f(a + 2\pi) + f(a + 4\pi)$

119. Calculating the Time of a Trip From a parking lot, you want to walk to a house on the beach. The house is located 1500 feet down a paved path that parallels the ocean, which is 500 feet away. See the illustration. Along the path you can walk 300 feet per minute, but in the sand on the beach you can only walk 100 feet per minute.

The time T to get from the parking lot to the beach house can be expressed as a function of the angle θ shown in the illustration is

$$T(\theta) = 5 - \frac{5}{3\tan\theta} + \frac{5}{\sin\theta}, \quad 0 < \theta < \frac{\pi}{2}$$

Calculate the time T if you walk directly from the parking lot to the house.

[**Hint:** $\tan\theta = \frac{500}{1500}$.]

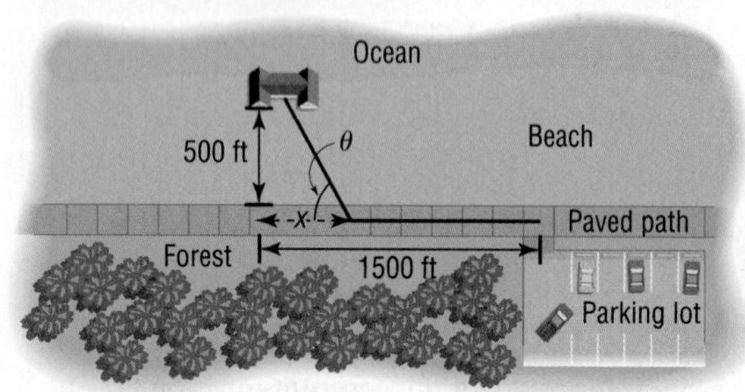

120. Calculating the Time of a Trip Two oceanfront homes are located 8 miles apart on a straight stretch of beach, each a distance of 1 mile from a paved path that parallels the ocean. Sally can jog 8 miles per hour on the paved path, but only 3 miles per hour in the sand on the beach. Because a river flows directly between the two houses, it is necessary to jog in the sand to the road, continue on the path, and then jog directly back in the sand to get from one house to the other. See the illustration. The time T to get from one house to the other as a function of the angle θ shown in the illustration is

$$T(\theta) = 1 + \frac{2}{3\sin\theta} - \frac{1}{4\tan\theta} \quad 0 < \theta < \frac{\pi}{2}$$

(a) Calculate the time T for $\tan\theta = \frac{1}{4}$.
(b) Describe the path taken.
(c) Explain why θ must be larger than 14°.

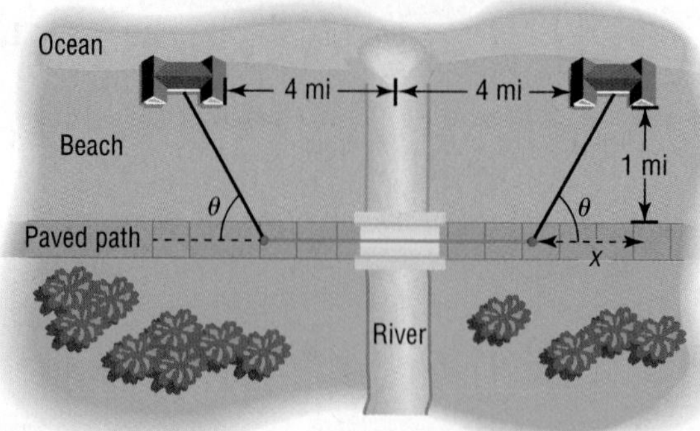

121. Show that the range of the tangent function is the set of all real numbers.

122. Show that the range of the cotangent function is the set of all real numbers.

123. Show that the period of $f(\theta) = \sin\theta$ is 2π.
[**Hint:** Assume that $0 < p < 2\pi$ exists so that $\sin(\theta + p) = \sin\theta$ for all θ. Let $\theta = 0$ to find p. Then let $\theta = \frac{\pi}{2}$ to obtain a contradiction.]

124. Show that the period of $f(\theta) = \cos\theta$ is 2π.

125. Show that the period of $f(\theta) = \sec\theta$ is 2π.

126. Show that the period of $f(\theta) = \csc\theta$ is 2π.

127. Show that the period of $f(\theta) = \tan\theta$ is π.

128. Show that the period of $f(\theta) = \cot\theta$ is π.

129. Prove the reciprocal identities given in formula (2).

130. Prove the quotient identities given in formula (3).

131. Establish the identity:

$$(\sin\theta\cos\phi)^2 + (\sin\theta\sin\phi)^2 + \cos^2\theta = 1$$

Explaining Concepts: Discussion and Writing

132. Write down five properties of the tangent function. Explain the meaning of each.

133. Describe your understanding of the meaning of a periodic function.

134. Explain how to find the value of sin 390° using periodic properties.

135. Explain how to find the value of $\cos(-45°)$ using even–odd properties.

136. Explain how to find the value of sin 390° and $\cos(-45°)$ using the unit circle.

Retain Your Knowledge

Problems 137–140 are based on material learned earlier in the course. The purpose of these problems is to keep the material fresh in your mind so that you are better prepared for the final exam.

137. Determine if $f(x) = \dfrac{x^4 + 3}{x^2 - 5}$ is even, odd, or neither.

138. Graph $f(x) = -2x^2 + 12x - 13$ using transformations.

139. Is (25, 18) on the graph of $f(x) = 3\sqrt{x - 9} + 6$?

140. Find the y-intercept of $f(x) = x^3 - 9x^2 + 3x - 27$.

'Are You Prepared?' Answers

1. $\left\{x \mid x \neq -\dfrac{1}{2}\right\}$ **2.** even **3.** False **4.** True

2.4 Graphs of the Sine and Cosine Functions*

PREPARING FOR THIS SECTION *Before getting started, review the following:*

- Graphing Techniques: Transformations (Section 1.6, pp. 64–73)

Now Work the **'Are You Prepared?'** problems on page 155.

OBJECTIVES
1 Graph Functions of the Form $y = A \sin(\omega x)$ Using Transformations (p. 146)
2 Graph Functions of the Form $y = A \cos(\omega x)$ Using Transformations (p. 148)
3 Determine the Amplitude and Period of Sinusoidal Functions (p. 149)
4 Graph Sinusoidal Functions Using Key Points (p. 151)
5 Find an Equation for a Sinusoidal Graph (p. 154)

Since we want to graph the trigonometric functions in the xy-plane, we shall use the traditional symbols x for the independent variable (or argument) and y for the dependent variable (or value at x) for each function. So the six trigonometric functions can be written as

$$y = f(x) = \sin x \qquad y = f(x) = \cos x \qquad y = f(x) = \tan x$$
$$y = f(x) = \csc x \qquad y = f(x) = \sec x \qquad y = f(x) = \cot x$$

Here the independent variable x represents an angle, measured in radians. In calculus, x will usually be treated as a real number. As noted earlier, these are equivalent ways of viewing x.

The Graph of the Sine Function $y = \sin x$

Because the sine function has period 2π, it is only necessary to graph $y = \sin x$ on the interval $[0, 2\pi]$. The remainder of the graph will consist of repetitions of this portion of the graph.

To begin, consider Table 6 on page 146, which lists some points on the graph of $y = \sin x, 0 \le x \le 2\pi$. As the table shows, the graph of $y = \sin x, 0 \le x \le 2\pi$, begins at the origin. As x increases from 0 to $\frac{\pi}{2}$, the value of $y = \sin x$ increases from 0 to 1; as x increases from $\frac{\pi}{2}$ to π to $\frac{3\pi}{2}$, the value of y decreases from 1 to 0 to -1;

*For those who wish to include phase shifts here, Section 2.6 can be covered immediately after Section 2.4 without loss of continuity.

Table 6

x	$y = \sin x$	(x, y)
0	0	$(0, 0)$
$\frac{\pi}{6}$	$\frac{1}{2}$	$\left(\frac{\pi}{6}, \frac{1}{2}\right)$
$\frac{\pi}{2}$	1	$\left(\frac{\pi}{2}, 1\right)$
$\frac{5\pi}{6}$	$\frac{1}{2}$	$\left(\frac{5\pi}{6}, \frac{1}{2}\right)$
π	0	$(\pi, 0)$
$\frac{7\pi}{6}$	$-\frac{1}{2}$	$\left(\frac{7\pi}{6}, -\frac{1}{2}\right)$
$\frac{3\pi}{2}$	-1	$\left(\frac{3\pi}{2}, -1\right)$
$\frac{11\pi}{6}$	$-\frac{1}{2}$	$\left(\frac{11\pi}{6}, -\frac{1}{2}\right)$
2π	0	$(2\pi, 0)$

as x increases from $\frac{3\pi}{2}$ to 2π, the value of y increases from -1 to 0. Plotting the points listed in Table 6 and connecting them with a smooth curve yields the graph shown in Figure 44.

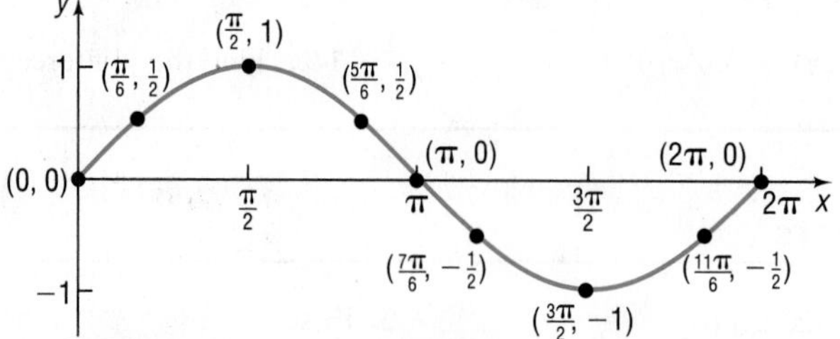

Figure 44 $y = \sin x, 0 \leq x \leq 2\pi$

The graph in Figure 44 is one period, or **cycle**, of the graph of $y = \sin x$. To obtain a more complete graph of $y = \sin x$, continue the graph in each direction, as shown in Figure 45.

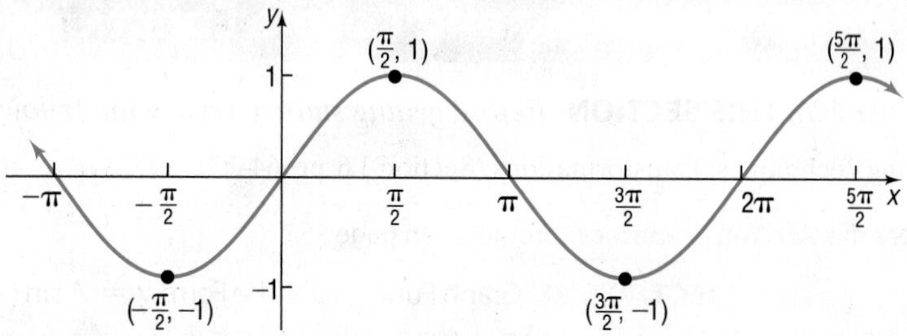

Figure 45 $y = \sin x, -\infty < x < \infty$

The graph of $y = \sin x$ illustrates some of the facts already discussed about the sine function.

Properties of the Sine Function $y = \sin x$

1. The domain is the set of all real numbers.
2. The range consists of all real numbers from -1 to 1, inclusive.
3. The sine function is an odd function, as the symmetry of the graph with respect to the origin indicates.
4. The sine function is periodic, with period 2π.
5. The x-intercepts are $\ldots, -2\pi, -\pi, 0, \pi, 2\pi, 3\pi, \ldots$; the y-intercept is 0.
6. The maximum value is 1 and occurs at $x = \ldots, -\frac{3\pi}{2}, \frac{\pi}{2}, \frac{5\pi}{2}, \frac{9\pi}{2}, \ldots$; the minimum value is -1 and occurs at $x = \ldots, -\frac{\pi}{2}, \frac{3\pi}{2}, \frac{7\pi}{2}, \frac{11\pi}{2}, \ldots$.

Now Work PROBLEM 11

1 Graph Functions of the Form $y = A \sin(\omega x)$ Using Transformations

EXAMPLE 1

Graphing Functions of the Form $y = A \sin(\omega x)$ Using Transformations

Graph $y = 3 \sin x$ using transformations. Use the graph to determine the domain and the range of the function.

Solution Figure 46 illustrates the steps.

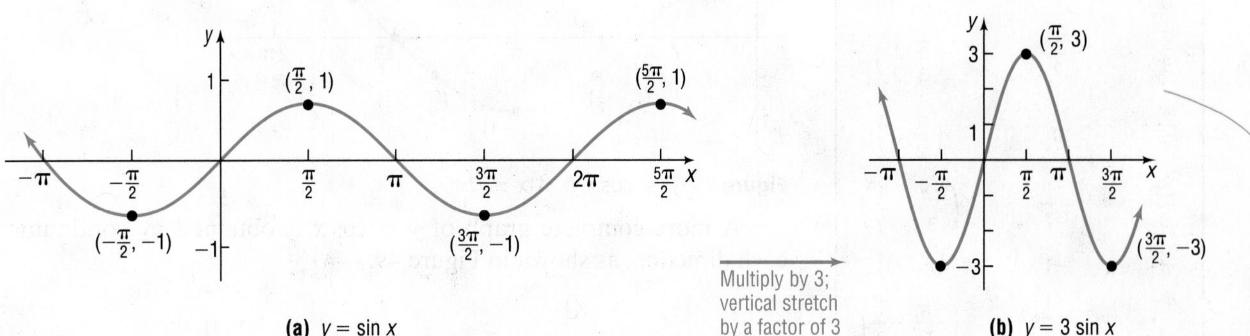

Figure 46 (a) $y = \sin x$ (b) $y = 3 \sin x$

The domain of $y = 3 \sin x$ is the set of all real numbers, or $(-\infty, \infty)$. The range is $\{y | -3 \leq y \leq 3\}$, or $[-3, 3]$.

EXAMPLE 2

Graphing Functions of the Form $y = A \sin(\omega x)$ Using Transformations

Graph $y = -\sin(2x)$ using transformations. Use the graph to determine the domain and the range of the function.

Solution Figure 47 illustrates the steps.

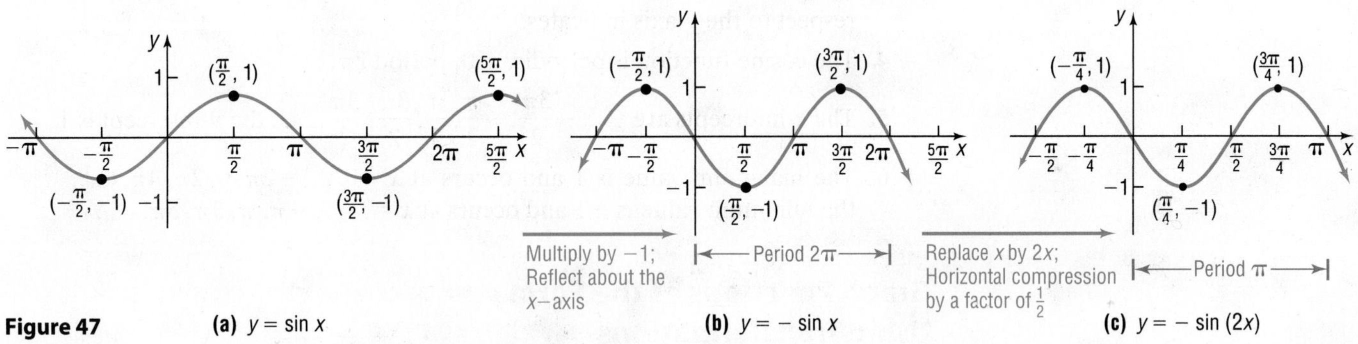

Figure 47 (a) $y = \sin x$ (b) $y = -\sin x$ (c) $y = -\sin(2x)$

The domain of $y = -\sin(2x)$ is the set of all real numbers, or $(-\infty, \infty)$. The range is $\{y | -1 \leq y \leq 1\}$, or $[-1, 1]$.

Note in Figure 47(c) that the period of the function $y = -\sin(2x)$ is π because of the horizontal compression of the original period 2π by a factor of $\frac{1}{2}$.

Now Work PROBLEM 35 USING TRANSFORMATIONS

The Graph of the Cosine Function $y = \cos x$

The cosine function also has period 2π. Proceed as with the sine function by constructing Table 7 on page 148, which lists some points on the graph of $y = \cos x$, $0 \leq x \leq 2\pi$. As the table shows, the graph of $y = \cos x$, $0 \leq x \leq 2\pi$, begins at the point $(0, 1)$. As x increases from 0 to $\frac{\pi}{2}$ to π, the value of y decreases from 1 to 0 to -1; as x increases from π to $\frac{3\pi}{2}$ to 2π, the value of y increases from -1 to 0 to 1. As before, plot the points in Table 7 to get one period or cycle of the graph. See Figure 48.

Table 7

x	$y = \cos x$	(x, y)
0	1	(0, 1)
$\frac{\pi}{3}$	$\frac{1}{2}$	$\left(\frac{\pi}{3}, \frac{1}{2}\right)$
$\frac{\pi}{2}$	0	$\left(\frac{\pi}{2}, 0\right)$
$\frac{2\pi}{3}$	$-\frac{1}{2}$	$\left(\frac{2\pi}{3}, -\frac{1}{2}\right)$
π	-1	$(\pi, -1)$
$\frac{4\pi}{3}$	$-\frac{1}{2}$	$\left(\frac{4\pi}{3}, -\frac{1}{2}\right)$
$\frac{3\pi}{2}$	0	$\left(\frac{3\pi}{2}, 0\right)$
$\frac{5\pi}{3}$	$\frac{1}{2}$	$\left(\frac{5\pi}{3}, \frac{1}{2}\right)$
2π	1	$(2\pi, 1)$

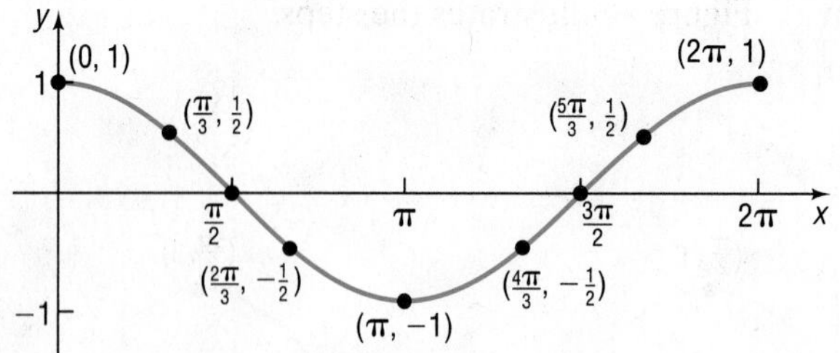

Figure 48 $y = \cos x, 0 \le x \le 2\pi$

A more complete graph of $y = \cos x$ is obtained by continuing the graph in each direction, as shown in Figure 49.

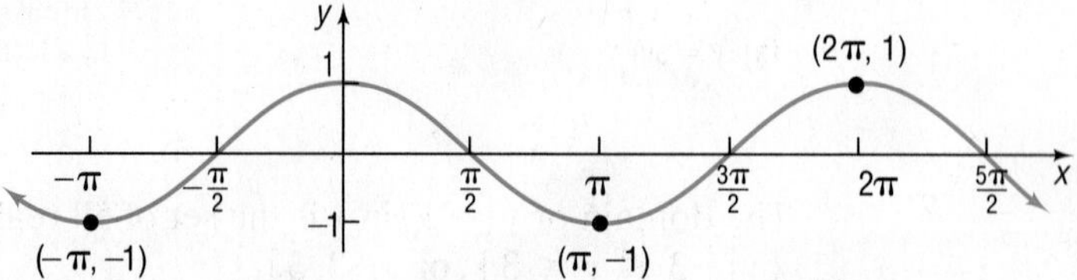

Figure 49 $y = \cos x, -\infty < x < \infty$

The graph of $y = \cos x$ illustrates some of the facts already discussed about the cosine function.

Properties of the Cosine Function

1. The domain is the set of all real numbers.
2. The range consists of all real numbers from -1 to 1, inclusive.
3. The cosine function is an even function, as the symmetry of the graph with respect to the y-axis indicates.
4. The cosine function is periodic, with period 2π.
5. The x-intercepts are $\ldots, -\frac{3\pi}{2}, -\frac{\pi}{2}, \frac{\pi}{2}, \frac{3\pi}{2}, \frac{5\pi}{2}, \ldots$; the y-intercept is 1.
6. The maximum value is 1 and occurs at $x = \ldots, -2\pi, 0, 2\pi, 4\pi, 6\pi, \ldots$; the minimum value is -1 and occurs at $x = \ldots, -\pi, \pi, 3\pi, 5\pi, \ldots$.

2 Graph Functions of the Form $y = A\cos(\omega x)$ Using Transformations

EXAMPLE 3

Graphing Functions of the Form $y = A\cos(\omega x)$ Using Transformations

Graph $y = 2\cos(3x)$ using transformations. Use the graph to determine the domain and the range of the function.

Solution Figure 50 shows the steps.

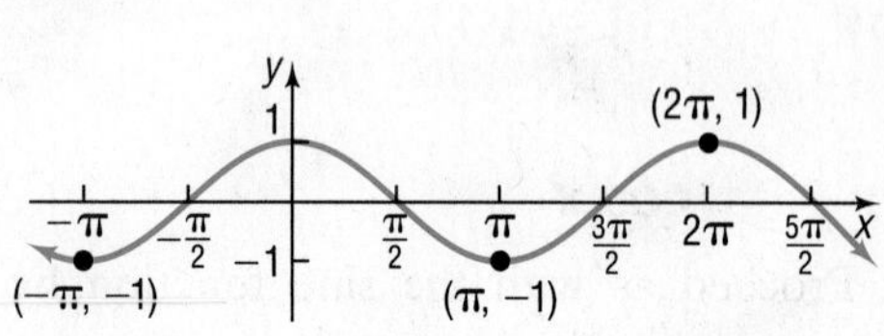

(a) $y = \cos x$

Multiply by 2; Vertical stretch by a factor of 2

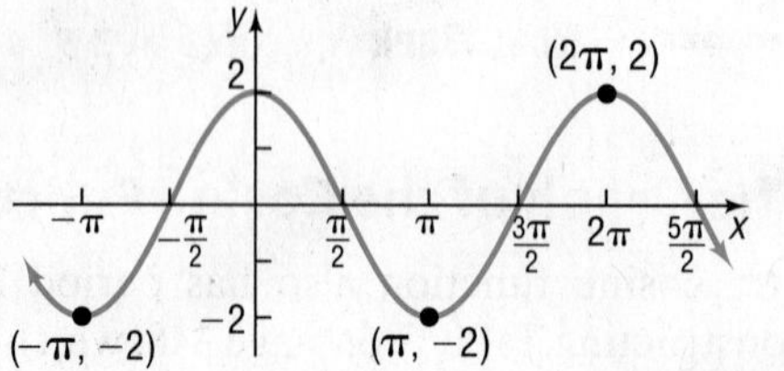

(b) $y = 2\cos x$

Replace x by 3x; Horizontal compression by a factor of $\frac{1}{3}$

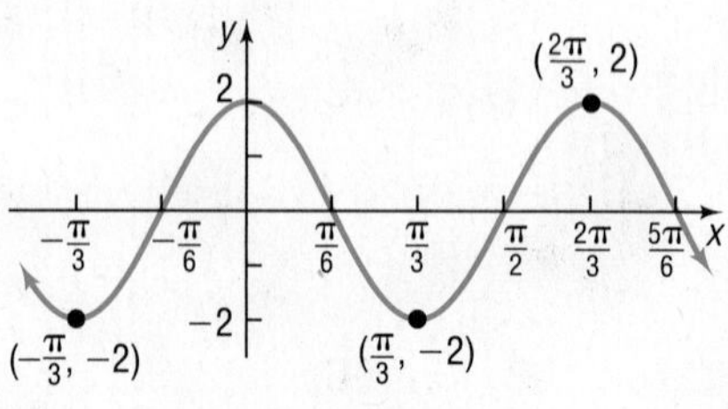

(c) $y = 2\cos(3x)$

Figure 50

The domain of $y = 2\cos(3x)$ is the set of all real numbers, or $(-\infty, \infty)$. The range is $\{y \mid -2 \leq y \leq 2\}$, or $[-2, 2]$. ●

Notice in Figure 50(c) that the period of the function $y = 2\cos(3x)$ is $\dfrac{2\pi}{3}$ because of the compression of the original period 2π by a factor of $\dfrac{1}{3}$.

Now Work PROBLEM 43 USING TRANSFORMATIONS

Sinusoidal Graphs

Shift the graph of $y = \cos x$ to the right $\dfrac{\pi}{2}$ units to obtain the graph of $y = \cos\left(x - \dfrac{\pi}{2}\right)$. See Figure 51(a). Now look at the graph of $y = \sin x$ in Figure 51(b). Notice that the graph of $y = \sin x$ is the same as the graph of $y = \cos\left(x - \dfrac{\pi}{2}\right)$.

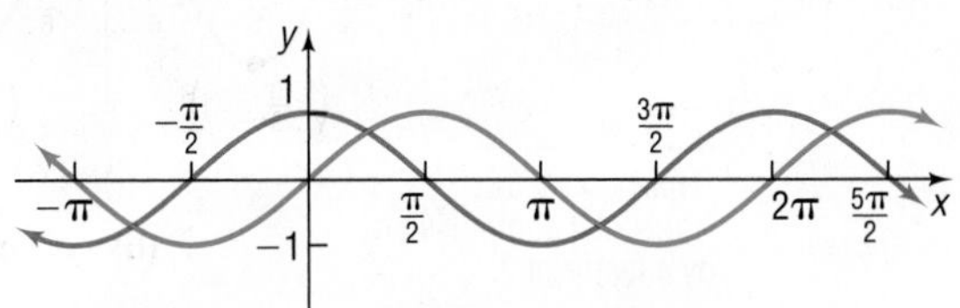

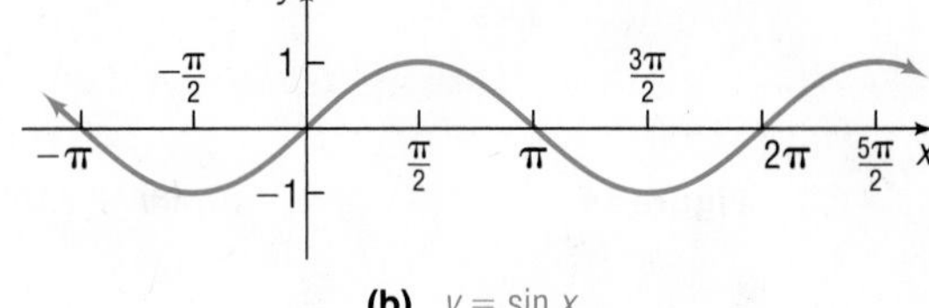

Figure 51 **(a)** $y = \cos x$, $y = \cos\left(x - \frac{\pi}{2}\right)$ **(b)** $y = \sin x$

Seeing the Concept

Graph $Y_1 = \sin x$ and $Y_2 = \cos\left(x - \dfrac{\pi}{2}\right)$. How many graphs do you see?

Based on Figure 51, we conjecture that

$$\sin x = \cos\left(x - \frac{\pi}{2}\right)$$

(We shall prove this fact in Chapter 3.) Because of this relationship, the graphs of functions of the form $y = A\sin(\omega x)$ or $y = A\cos(\omega x)$ are referred to as **sinusoidal graphs**.

3 Determine the Amplitude and Period of Sinusoidal Functions

Figure 52 uses transformations to obtain the graph of $y = 2\cos x$. Note that the values of $y = 2\cos x$ lie between -2 and 2, inclusive.

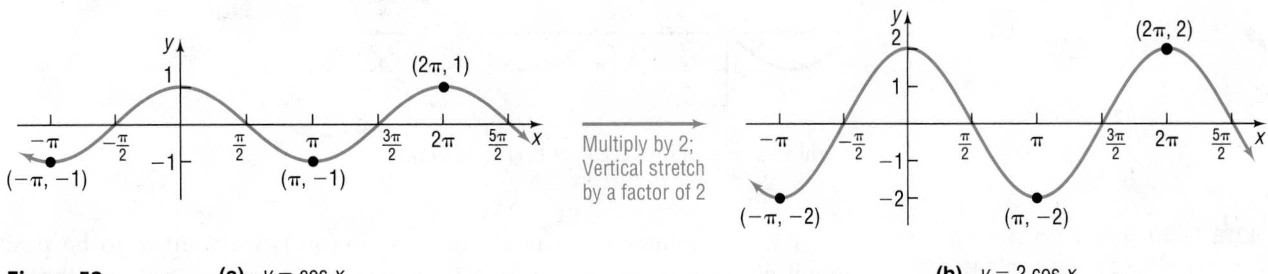

Figure 52 **(a)** $y = \cos x$ **(b)** $y = 2\cos x$

In general, the values of the functions $y = A\sin x$ and $y = A\cos x$, where $A \neq 0$, will always satisfy the inequalities

$$-|A| \leq A\sin x \leq |A| \quad \text{and} \quad -|A| \leq A\cos x \leq |A|$$

respectively. The number $|A|$ is called the **amplitude** of $y = A\sin x$ or $y = A\cos x$. See Figure 53 on page 150.

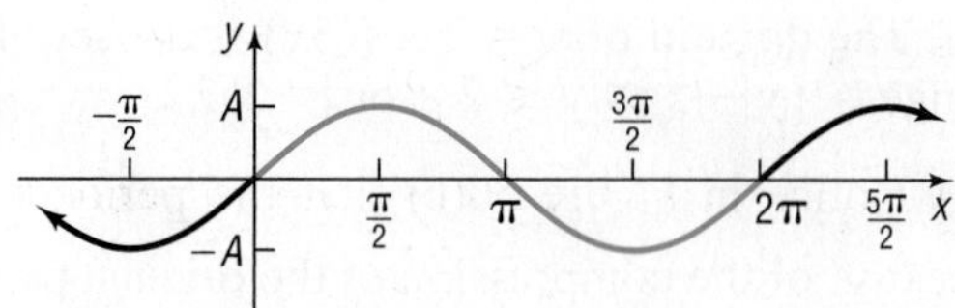

Figure 53 $y = A \sin x, A > 0$; period $= 2\pi$

Figure 54 uses transformations to obtain the graph of $y = \cos(3x)$. Note that the period of this function is $\frac{2\pi}{3}$, because of the horizontal compression of the original period 2π by a factor of $\frac{1}{3}$.

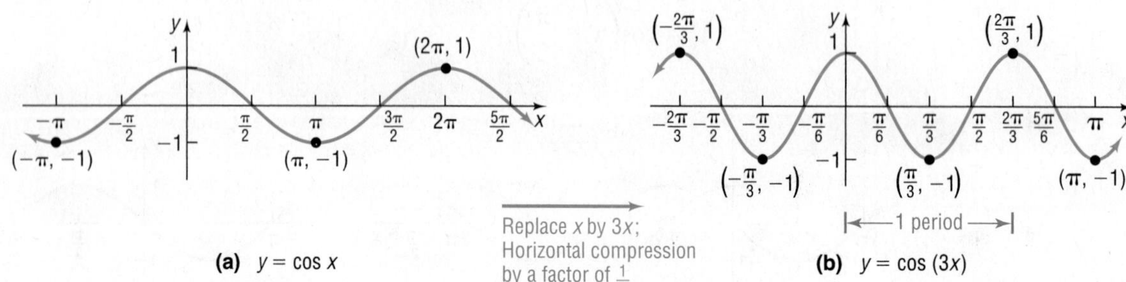

Figure 54 **(a)** $y = \cos x$ **(b)** $y = \cos(3x)$

If $\omega > 0$, the functions $y = \sin(\omega x)$ and $y = \cos(\omega x)$ will have period $T = \frac{2\pi}{\omega}$. To see why, recall that the graph of $y = \sin(\omega x)$ is obtained from the graph of $y = \sin x$ by performing a horizontal compression or stretch by a factor $\frac{1}{\omega}$. This horizontal compression replaces the interval $[0, 2\pi]$, which contains one period of the graph of $y = \sin x$, by the interval $\left[0, \frac{2\pi}{\omega}\right]$, which contains one period of the graph of $y = \sin(\omega x)$. So the function $y = \cos(3x)$, graphed in Figure 54(b), with $\omega = 3$, has period $\frac{2\pi}{\omega} = \frac{2\pi}{3}$.

One period of the graph of $y = \sin(\omega x)$ or $y = \cos(\omega x)$ is called a **cycle**. Figure 55 illustrates the general situation. The blue portion of the graph is one cycle.

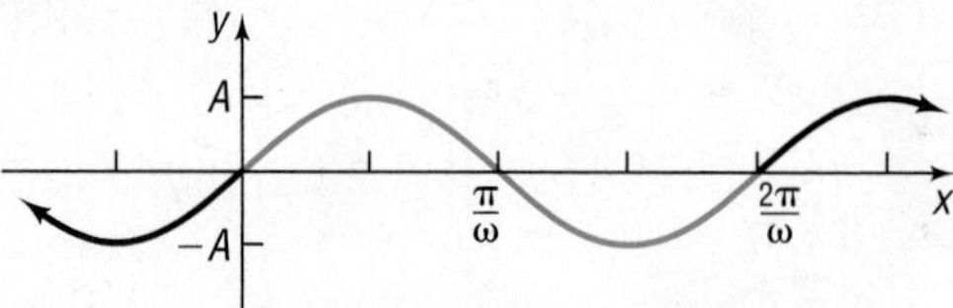

Figure 55 $y = A \sin(\omega x), A > 0, \omega > 0$; period $= \frac{2\pi}{\omega}$

NOTE Recall that a function f is even if $f(-x) = f(x)$; a function f is odd if $f(-x) = -f(x)$. Since the sine function is odd, $\sin(-x) = -\sin x$; since the cosine function is even, $\cos(-x) = \cos x$. ■

When graphing $y = \sin(\omega x)$ or $y = \cos(\omega x)$, we want ω to be positive. To graph either $y = \sin(-\omega x), \omega > 0$, or $y = \cos(-\omega x), \omega > 0$, use the even–odd properties of the sine and cosine functions as follows:

$$\sin(-\omega x) = -\sin(\omega x) \quad \text{and} \quad \cos(-\omega x) = \cos(\omega x)$$

This provides an equivalent form in which the coefficient of x in the argument is positive. For example,

$$\sin(-2x) = -\sin(2x) \quad \text{and} \quad \cos(-\pi x) = \cos(\pi x)$$

Because of this, we can assume that $\omega > 0$.

THEOREM If $\omega > 0$, the amplitude and period of $y = A\sin(\omega x)$ and $y = A\cos(\omega x)$ are given by

$$\text{Amplitude} = |A| \qquad \text{Period} = T = \frac{2\pi}{\omega} \qquad \textbf{(1)}$$

EXAMPLE 4 **Finding the Amplitude and Period of a Sinusoidal Function**

Determine the amplitude and period of $y = 3\sin(4x)$.

Solution Comparing $y = 3\sin(4x)$ to $y = A\sin(\omega x)$, note that $A = 3$ and $\omega = 4$. From equation (1),

$$\text{Amplitude} = |A| = 3 \qquad \text{Period} = T = \frac{2\pi}{\omega} = \frac{2\pi}{4} = \frac{\pi}{2}$$

Now Work PROBLEM 17

4 Graph Sinusoidal Functions Using Key Points

So far, we have graphed functions of the form $y = A\sin(\omega x)$ or $y = A\cos(\omega x)$ using transformations. We now introduce another method that can be used to graph these functions.

Figure 56 shows one cycle of the graphs of $y = \sin x$ and $y = \cos x$ on the interval $[0, 2\pi]$. Notice that each graph consists of four parts corresponding to the four subintervals:

$$\left[0, \frac{\pi}{2}\right], \quad \left[\frac{\pi}{2}, \pi\right], \quad \left[\pi, \frac{3\pi}{2}\right], \quad \left[\frac{3\pi}{2}, 2\pi\right]$$

Each subinterval is of length $\frac{\pi}{2}$ (the period 2π divided by 4, the number of parts), and the endpoints of these intervals $x = 0, x = \frac{\pi}{2}, x = \pi, x = \frac{3\pi}{2}, x = 2\pi$ give rise to five key points on each graph:

$$\text{For } y = \sin x: \quad (0, 0), \left(\frac{\pi}{2}, 1\right), (\pi, 0), \left(\frac{3\pi}{2}, -1\right), (2\pi, 0)$$

$$\text{For } y = \cos x: \quad (0, 1), \left(\frac{\pi}{2}, 0\right), (\pi, -1), \left(\frac{3\pi}{2}, 0\right), (2\pi, 1)$$

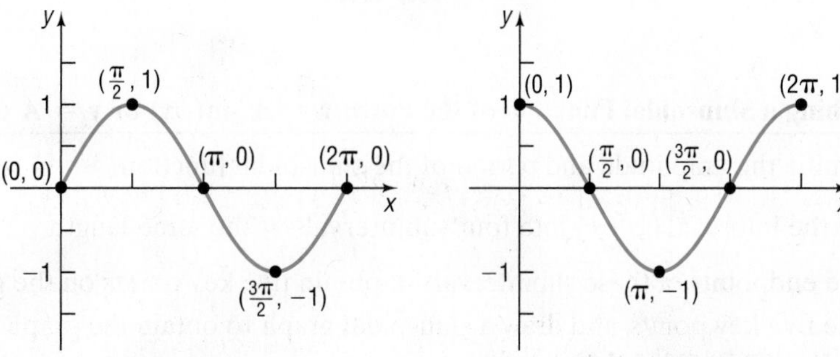

Figure 56 **(a)** $y = \sin x$ **(b)** $y = \cos x$

EXAMPLE 5 Graphing a Sinusoidal Function Using Key Points

Graph $y = 3\sin(4x)$ using key points.

Step-by-Step Solution

Step 1: Determine the amplitude and period of the sinusoidal function.

Comparing $y = 3\sin(4x)$ to $y = A\sin(\omega x)$, note that $A = 3$ and $\omega = 4$, so the amplitude is $|A| = 3$ and the period is $\frac{2\pi}{\omega} = \frac{2\pi}{4} = \frac{\pi}{2}$. Because the amplitude is 3, the graph of $y = 3\sin(4x)$ will lie between -3 and 3 on the y-axis. Because the period is $\frac{\pi}{2}$, one cycle will begin at $x = 0$ and end at $x = \frac{\pi}{2}$.

Step 2: Divide the interval $\left[0, \frac{2\pi}{\omega}\right]$ into four subintervals of the same length.

Divide the interval $\left[0, \frac{\pi}{2}\right]$ into four subintervals, each of length $\frac{\pi}{2} \div 4 = \frac{\pi}{8}$, as follows:

$$\left[0, \frac{\pi}{8}\right] \quad \left[\frac{\pi}{8}, \frac{\pi}{8} + \frac{\pi}{8}\right] = \left[\frac{\pi}{8}, \frac{\pi}{4}\right] \quad \left[\frac{\pi}{4}, \frac{\pi}{4} + \frac{\pi}{8}\right] = \left[\frac{\pi}{4}, \frac{3\pi}{8}\right] \quad \left[\frac{3\pi}{8}, \frac{3\pi}{8} + \frac{\pi}{8}\right] = \left[\frac{3\pi}{8}, \frac{\pi}{2}\right]$$

The endpoints of the subintervals are $0, \frac{\pi}{8}, \frac{\pi}{4}, \frac{3\pi}{8}, \frac{\pi}{2}$. These values represent the x-coordinates of the five key points on the graph.

Step 3: Use the endpoints of the subintervals from Step 2 to obtain five key points on the graph.

NOTE The five key points could also be obtained by evaluating $y = 3\sin(4x)$ at each endpoint. ■

To obtain the y-coordinates of the five key points of $y = 3\sin(4x)$, multiply the y-coordinates of the five key points for $y = \sin x$ in Figure 56(a) by $A = 3$. The five key points are

$$(0, 0) \quad \left(\frac{\pi}{8}, 3\right) \quad \left(\frac{\pi}{4}, 0\right) \quad \left(\frac{3\pi}{8}, -3\right) \quad \left(\frac{\pi}{2}, 0\right)$$

Step 4: Plot the five key points and draw a sinusoidal graph to obtain the graph of one cycle. Extend the graph in each direction to make it complete.

Plot the five key points obtained in Step 3, and fill in the graph of the sine curve as shown in Figure 57(a). Extend the graph in each direction to obtain the complete graph shown in Figure 57(b). Notice that additional key points appear every $\frac{\pi}{8}$ radian.

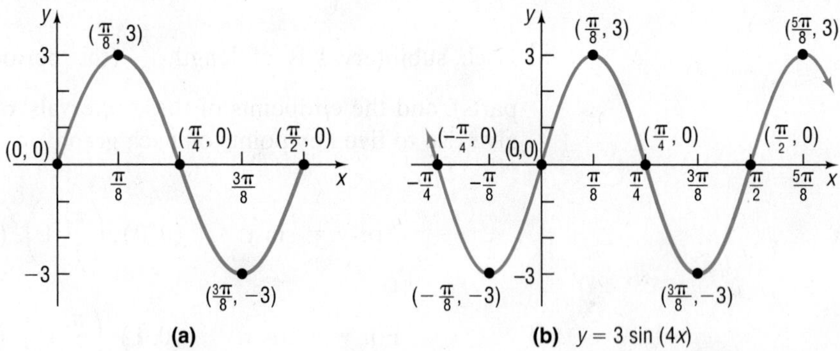

Figure 57 (a) (b) $y = 3\sin(4x)$ ●

✓**Check:** Graph $y = 3\sin(4x)$ using transformations. Which graphing method do you prefer?

Now Work PROBLEM 35 USING KEY POINTS

SUMMARY

Steps for Graphing a Sinusoidal Function of the Form $y = A\sin(\omega x)$ or $y = A\cos(\omega x)$ Using Key Points

STEP 1: Determine the amplitude and period of the sinusoidal function.

STEP 2: Divide the interval $\left[0, \frac{2\pi}{\omega}\right]$ into four subintervals of the same length.

STEP 3: Use the endpoints of these subintervals to obtain five key points on the graph.

STEP 4: Plot the five key points, and draw a sinusoidal graph to obtain the graph of one cycle. Extend the graph in each direction to make it complete.

EXAMPLE 6

Graphing a Sinusoidal Function Using Key Points

Graph $y = 2\sin\left(-\frac{\pi}{2}x\right)$ using key points.

Solution Since the sine function is odd, use the equivalent form:

$$y = -2\sin\left(\frac{\pi}{2}x\right)$$

Step 1: Comparing $y = -2\sin\left(\frac{\pi}{2}x\right)$ to $y = A\sin(\omega x)$, note that $A = -2$ and $\omega = \frac{\pi}{2}$. The amplitude is $|A| = |-2| = 2$, and the period is $T = \frac{2\pi}{\omega} = \frac{2\pi}{\frac{\pi}{2}} = 4$. The graph of $y = -2\sin\left(\frac{\pi}{2}x\right)$ lies between -2 and 2 on the y-axis. One cycle will begin at $x = 0$ and end at $x = 4$.

Step 2: Divide the interval $[0, 4]$ into four subintervals, each of length $4 \div 4 = 1$. The x-coordinates of the five key points are

0	$0 + 1 = 1$	$1 + 1 = 2$	$2 + 1 = 3$	$3 + 1 = 4$
1st x-coordinate	2nd x-coordinate	3rd x-coordinate	4th x-coordinate	5th x-coordinate

Step 3: Since $y = -2\sin\left(\frac{\pi}{2}x\right)$, multiply the y-coordinates of the five key points in Figure 56(a) by $A = -2$. The five key points on the graph are

$$(0, 0) \quad (1, -2) \quad (2, 0) \quad (3, 2) \quad (4, 0)$$

Step 4: Plot these five points, and fill in the graph of the sine function as shown in Figure 58(a). Extend the graph in each direction to obtain Figure 58(b).

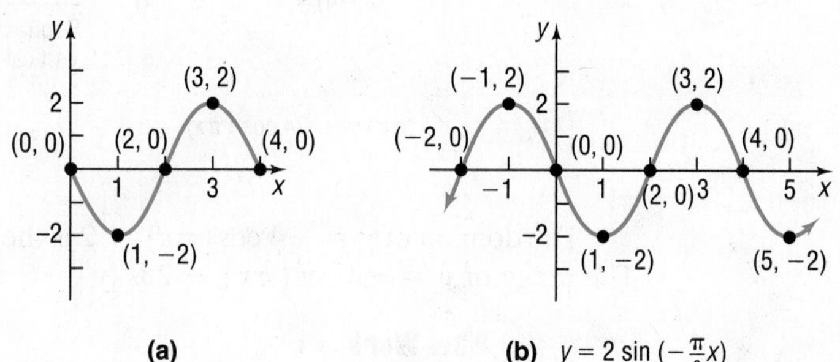

Figure 58 (a) (b) $y = 2\sin\left(-\frac{\pi}{2}x\right)$

COMMENT To graph a sinusoidal function of the form $y = A\sin(\omega x)$ or $y = A\cos(\omega x)$ using a graphing utility, use the amplitude to set Y_{min} and Y_{max}, and use the period to set X_{min} and X_{max}.

Check: Graph $y = 2\sin\left(-\frac{\pi}{2}x\right)$ using transformations. Which graphing method do you prefer?

Now Work PROBLEM 39 USING KEY POINTS

If the function to be graphed is of the form $y = A\sin(\omega x) + B$ [or $y = A\cos(\omega x) + B$], first graph $y = A\sin(\omega x)$ [or $y = A\cos(\omega x)$] and then use a vertical shift.

EXAMPLE 7

Graphing a Sinusoidal Function Using Key Points

Graph $y = -4\cos(\pi x) - 2$ using key points. Use the graph to determine the domain and the range of $y = -4\cos(\pi x) - 2$.

Solution Begin by graphing the function $y = -4\cos(\pi x)$. Comparing $y = -4\cos(\pi x)$ with $y = A\cos(\omega x)$, note that $A = -4$ and $\omega = \pi$. The amplitude is $|A| = |-4| = 4$, and the period is $T = \dfrac{2\pi}{\omega} = \dfrac{2\pi}{\pi} = 2$.

The graph of $y = -4\cos(\pi x)$ will lie between -4 and 4 on the y-axis. One cycle will begin at $x = 0$ and end at $x = 2$.

Divide the interval $[0, 2]$ into four subintervals, each of length $2 \div 4 = \dfrac{1}{2}$. The x-coordinates of the five key points are

0	$0 + \frac{1}{2} = \frac{1}{2}$	$\frac{1}{2} + \frac{1}{2} = 1$	$1 + \frac{1}{2} = \frac{3}{2}$	$\frac{3}{2} + \frac{1}{2} = 2$
1st x-coordinate	2nd x-coordinate	3rd x-coordinate	4th x-coordinate	5th x-coordinate

Since $y = -4\cos(\pi x)$, multiply the y-coordinates of the five key points of $y = \cos x$ shown in Figure 56(b) by $A = -4$ to obtain the five key points on the graph of $y = -4\cos(\pi x)$:

$$(0, -4) \quad \left(\frac{1}{2}, 0\right) \quad (1, 4) \quad \left(\frac{3}{2}, 0\right) \quad (2, -4)$$

Plot these five points, and fill in the graph of the cosine function as shown in Figure 59(a). Extend the graph in each direction to obtain Figure 59(b), the graph of $y = -4\cos(\pi x)$.

A vertical shift down 2 units gives the graph of $y = -4\cos(\pi x) - 2$, as shown in Figure 59(c).

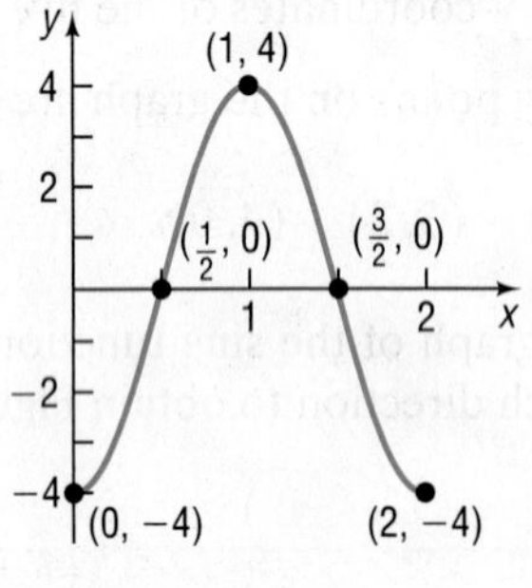

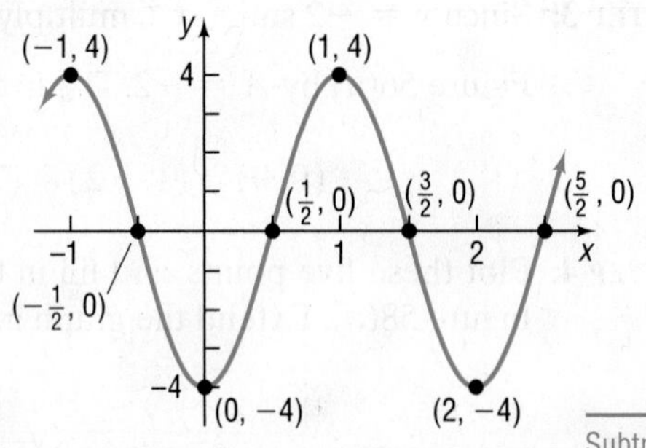

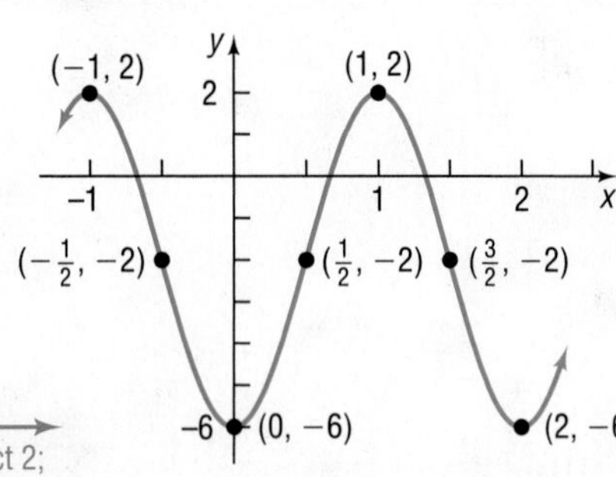

Figure 59 (a) (b) $y = -4\cos(\pi x)$ (c) $y = -4\cos(\pi x) - 2$

The domain of $y = -4\cos(\pi x) - 2$ is the set of all real numbers, or $(-\infty, \infty)$. The range of $y = -4\cos(\pi x) - 2$ is $\{y \mid -6 \le y \le 2\}$, or $[-6, 2]$. ●

Now Work PROBLEM 49

5 Find an Equation for a Sinusoidal Graph

EXAMPLE 8 **Finding an Equation for a Sinusoidal Graph**

Find an equation for the graph shown in Figure 60.

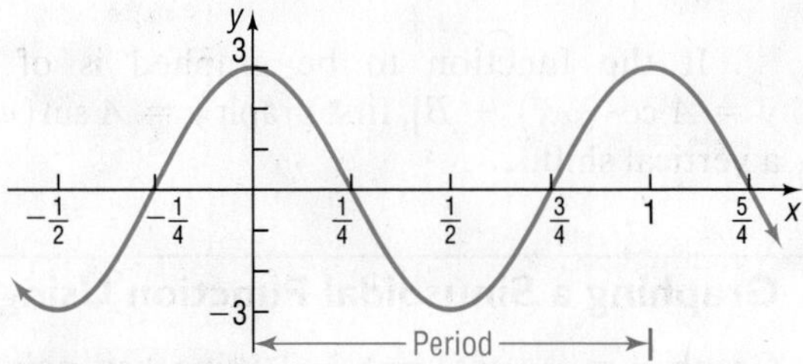

Figure 60

Solution The graph has the characteristics of a cosine function. Do you see why? The maximum value, 3, occurs at $x = 0$. So the equation can be viewed as a cosine function $y = A\cos(\omega x)$ with $A = 3$ and period $T = 1$. Then $\frac{2\pi}{\omega} = 1$, so $\omega = 2\pi$. The cosine function whose graph is given in Figure 60 is

$$y = A\cos(\omega x) = 3\cos(2\pi x)$$

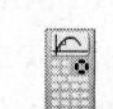 **Check:** Graph $Y_1 = 3\cos(2\pi x)$ and compare the result with Figure 60.

EXAMPLE 9 **Finding an Equation for a Sinusoidal Graph**

Find an equation for the graph shown in Figure 61.

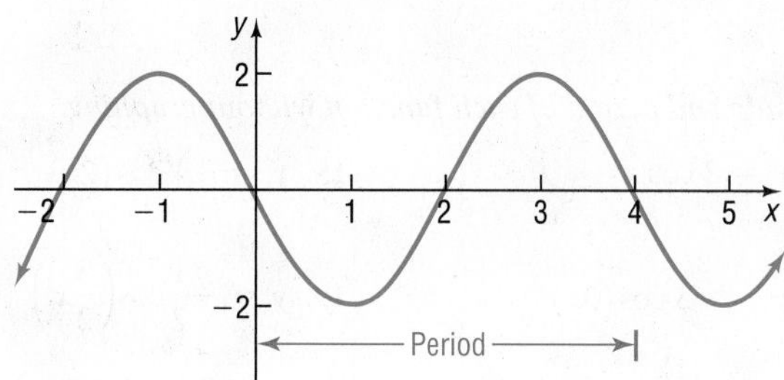

Figure 61

Solution The graph is sinusoidal, with amplitude $|A| = 2$. The period is 4, so $\frac{2\pi}{\omega} = 4$, or $\omega = \frac{\pi}{2}$. Since the graph passes through the origin, it is easier to view the equation as a sine function,* but note that the graph is actually the reflection of a sine function about the x-axis (since the graph is decreasing near the origin). This requires that $A = -2$. The sine function whose graph is given in Figure 61 is

$$y = A\sin(\omega x) = -2\sin\left(\frac{\pi}{2}x\right)$$

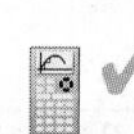 **Check:** Graph $Y_1 = -2\sin\left(\frac{\pi}{2}x\right)$ and compare the result with Figure 61.

 **Now Work** PROBLEMS 57 AND 61

2.4 Assess Your Understanding

'Are You Prepared?' *Answers are given at the end of these exercises. If you get a wrong answer, read the pages listed in red.*

1. Use transformations to graph $y = 3x^2$. (pp. 64–73)
2. Use transformations to graph $y = \sqrt{2x}$. (pp. 64–73)

Concepts and Vocabulary

3. The maximum value of $y = \sin x$, $0 \le x \le 2\pi$, is _____ and occurs at $x =$ _____.
4. The function $y = A\sin(\omega x)$, $A > 0$, has amplitude 3 and period 2; then $A =$ _____ and $\omega =$ _____.
5. The function $y = 3\cos(6x)$ has amplitude _____ and period ___ .
6. ***True or False*** The graphs of $y = \sin x$ and $y = \cos x$ are identical except for a horizontal shift.
7. ***True or False*** For $y = 2\sin(\pi x)$, the amplitude is 2 and the period is $\frac{\pi}{2}$.
8. ***True or False*** The graph of the sine function has infinitely many x-intercepts.
9. One period of the graph of $y = \sin(\omega x)$ or $y = \cos(\omega x)$ is called a(n) _____.
   (a) amplitude (b) phase shift
   (c) transformation (d) cycle
10. To graph $y = 3\sin(-2x)$ using key points, the equivalent form _____ could be graphed instead.
   (a) $y = -3\sin(-2x)$ (b) $y = -2\sin(3x)$
   (c) $y = 3\sin(2x)$ (d) $y = -3\sin(2x)$

*The equation could also be viewed as a cosine function with a horizontal shift, but viewing it as a sine function is easier.

Skill Building

11. $f(x) = \sin x$
    (a) What is the y-intercept of the graph of f?
    (b) For what numbers x, $-\pi \le x \le \pi$, is the graph of f increasing?
    (c) What is the absolute maximum of f?
    (d) For what numbers x, $0 \le x \le 2\pi$, does $f(x) = 0$?
    (e) For what numbers x, $-2\pi \le x \le 2\pi$, does $f(x) = 1$? Where does $f(x) = -1$?
    (f) For what numbers x, $-2\pi \le x \le 2\pi$, does $f(x) = -\frac{1}{2}$?
    (g) What are the x-intercepts of f?

12. $g(x) = \cos x$
    (a) What is the y-intercept of the graph of g?
    (b) For what numbers x, $-\pi \le x \le \pi$, is the graph of g decreasing?
    (c) What is the absolute minimum of g?
    (d) For what numbers x, $0 \le x \le 2\pi$, does $g(x) = 0$?
    (e) For what numbers x, $-2\pi \le x \le 2\pi$, does $g(x) = 1$? Where does $g(x) = -1$?
    (f) For what numbers x, $-2\pi \le x \le 2\pi$, does $g(x) = \frac{\sqrt{3}}{2}$?
    (g) What are the x-intercepts of g?

In Problems 13–22, determine the amplitude and period of each function without graphing.

13. $y = 2 \sin x$ **14.** $y = 3 \cos x$ **15.** $y = -4 \cos(2x)$ **16.** $y = -\sin\left(\frac{1}{2}x\right)$

17. $y = 6 \sin(\pi x)$ **18.** $y = -3 \cos(3x)$ **19.** $y = -\frac{1}{2}\cos\left(\frac{3}{2}x\right)$ **20.** $y = \frac{4}{3}\sin\left(\frac{2}{3}x\right)$

21. $y = \frac{5}{3}\sin\left(-\frac{2\pi}{3}x\right)$ **22.** $y = \frac{9}{5}\cos\left(-\frac{3\pi}{2}x\right)$

In Problems 23–32, match the given function to one of the graphs (A)–(J).

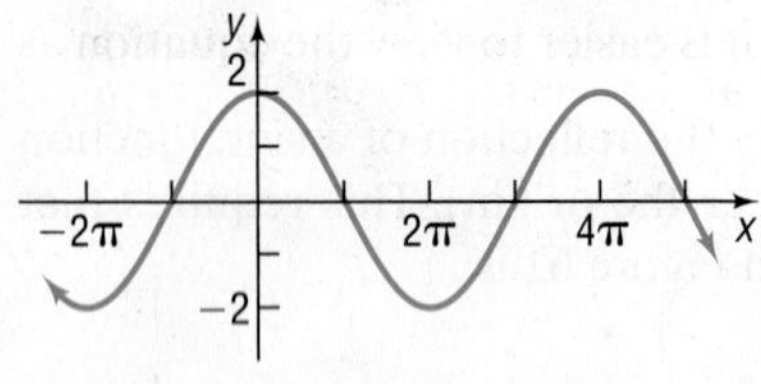

(A)

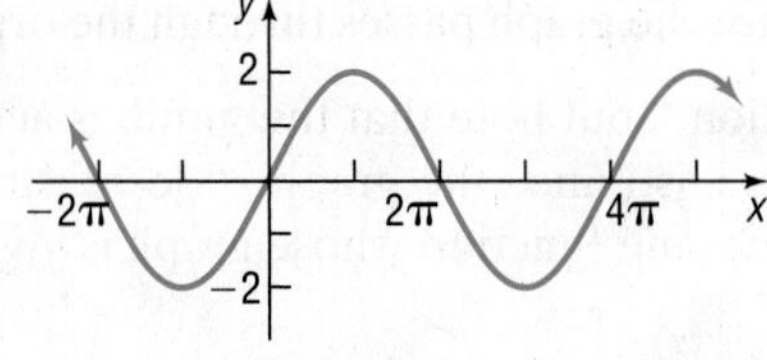

(B)

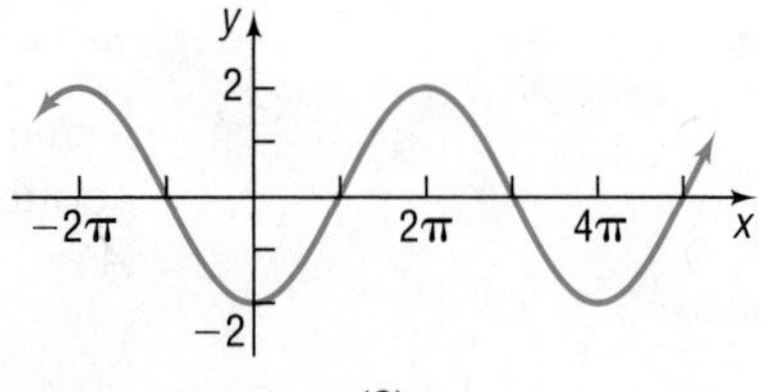

(C)

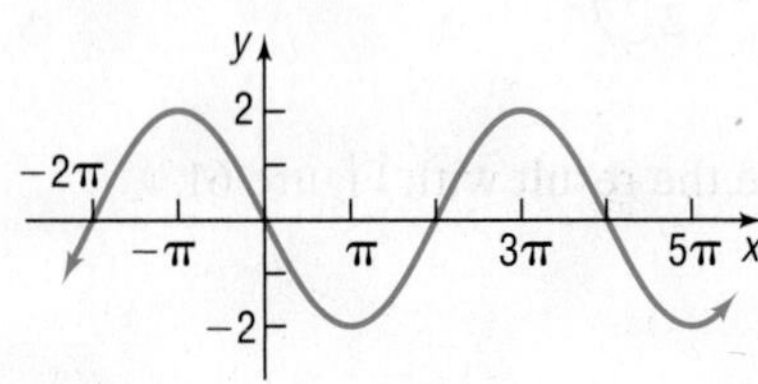

(D)

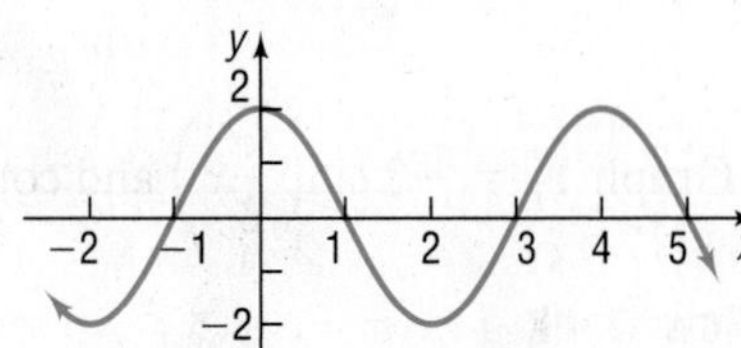

(E)

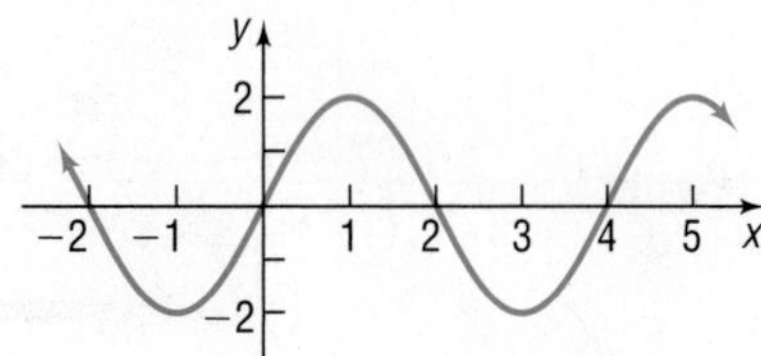

(F)

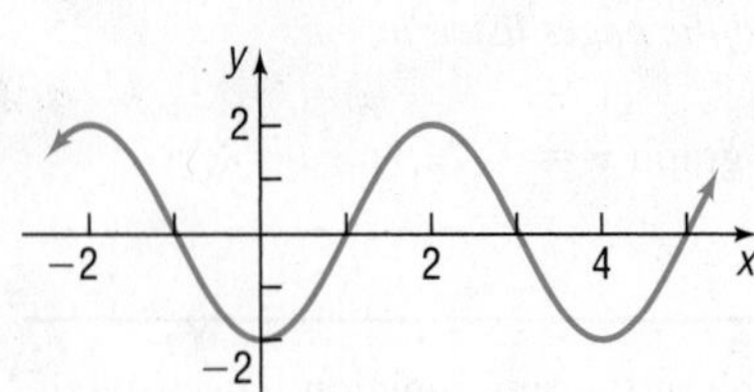

(G)

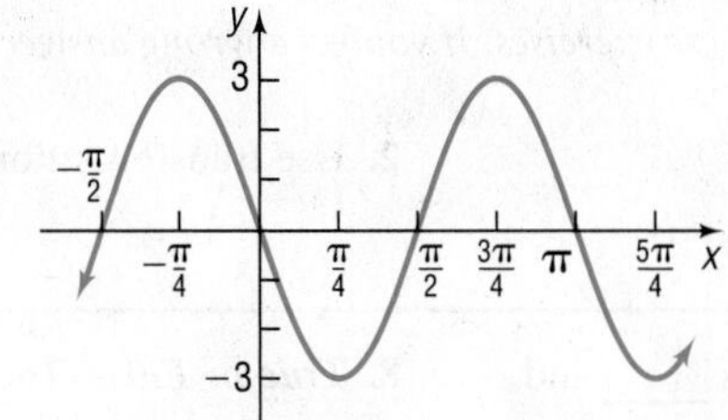

(H)

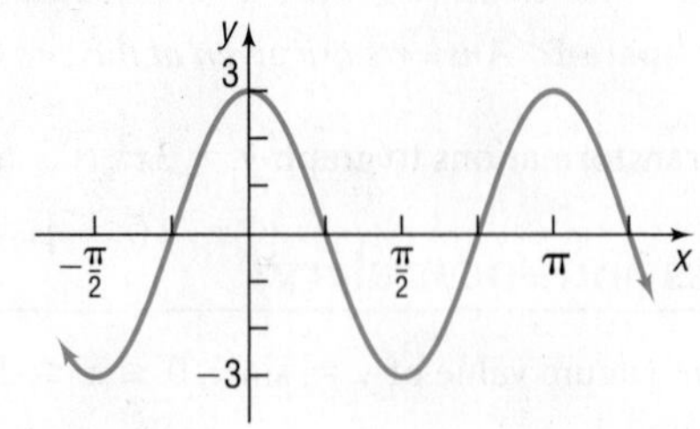

(I)

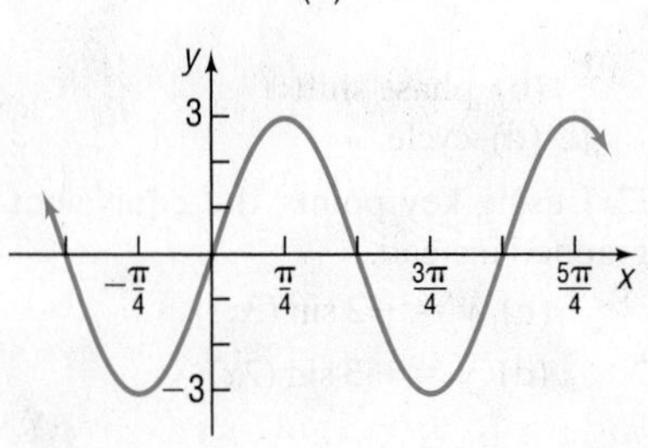

(J)

23. $y = 2 \sin\left(\frac{\pi}{2}x\right)$ **24.** $y = 2 \cos\left(\frac{\pi}{2}x\right)$ **25.** $y = 2 \cos\left(\frac{1}{2}x\right)$

26. $y = 3 \cos(2x)$ **27.** $y = -3 \sin(2x)$ **28.** $y = 2 \sin\left(\frac{1}{2}x\right)$

29. $y = -2 \cos\left(\frac{1}{2}x\right)$ **30.** $y = -2 \cos\left(\frac{\pi}{2}x\right)$ **31.** $y = 3 \sin(2x)$

32. $y = -2 \sin\left(\frac{1}{2}x\right)$

In Problems 33–56, graph each function using transformations or the method of key points. Be sure to label key points and show at least two cycles. Use the graph to determine the domain and the range of each function.

33. $y = 4\cos x$
34. $y = 3\sin x$
35. $y = -4\sin x$
36. $y = -3\cos x$

37. $y = \cos(4x)$
38. $y = \sin(3x)$
39. $y = \sin(-2x)$
40. $y = \cos(-2x)$

41. $y = 2\sin\left(\frac{1}{2}x\right)$
42. $y = 2\cos\left(\frac{1}{4}x\right)$
43. $y = -\frac{1}{2}\cos(2x)$
44. $y = -4\sin\left(\frac{1}{8}x\right)$

45. $y = 2\sin x + 3$
46. $y = 3\cos x + 2$
47. $y = 5\cos(\pi x) - 3$
48. $y = 4\sin\left(\frac{\pi}{2}x\right) - 2$

49. $y = -6\sin\left(\frac{\pi}{3}x\right) + 4$
50. $y = -3\cos\left(\frac{\pi}{4}x\right) + 2$
51. $y = 5 - 3\sin(2x)$
52. $y = 2 - 4\cos(3x)$

53. $y = \frac{5}{3}\sin\left(-\frac{2\pi}{3}x\right)$
54. $y = \frac{9}{5}\cos\left(-\frac{3\pi}{2}x\right)$
55. $y = -\frac{3}{2}\cos\left(\frac{\pi}{4}x\right) + \frac{1}{2}$
56. $y = -\frac{1}{2}\sin\left(\frac{\pi}{8}x\right) + \frac{3}{2}$

In Problems 57–60, write the equation of a sine function that has the given characteristics.

57. Amplitude: 3
Period: π

58. Amplitude: 2
Period: 4π

59. Amplitude: 3
Period: 2

60. Amplitude: 4
Period: 1

In Problems 61–74, find an equation for each graph.

61.

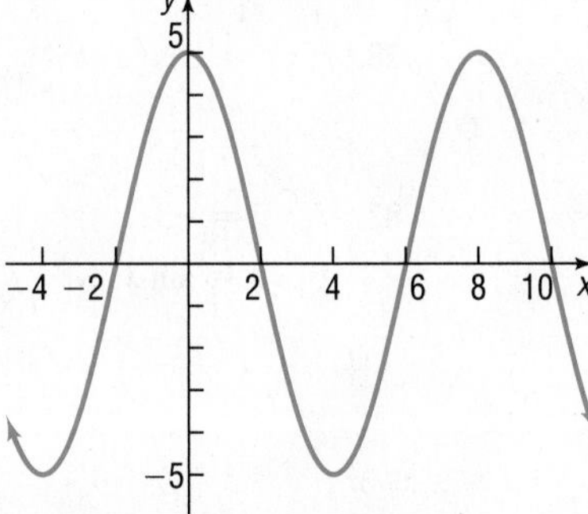

62.

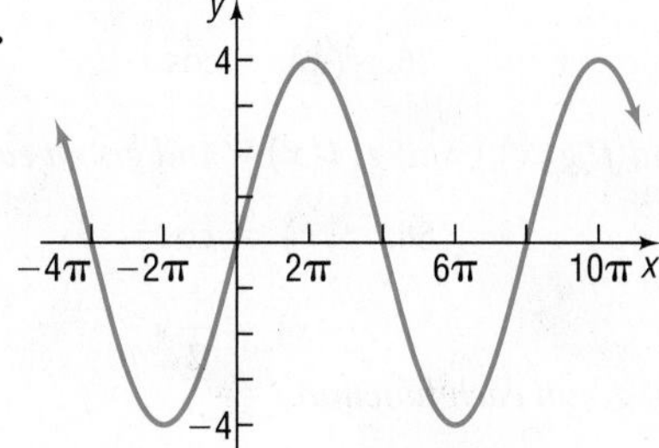

63.

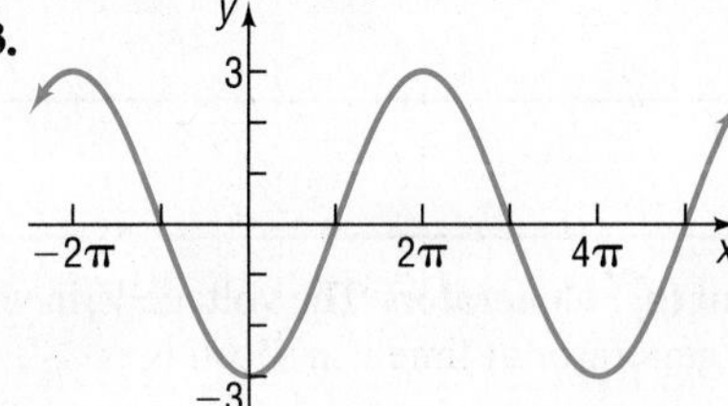

64.

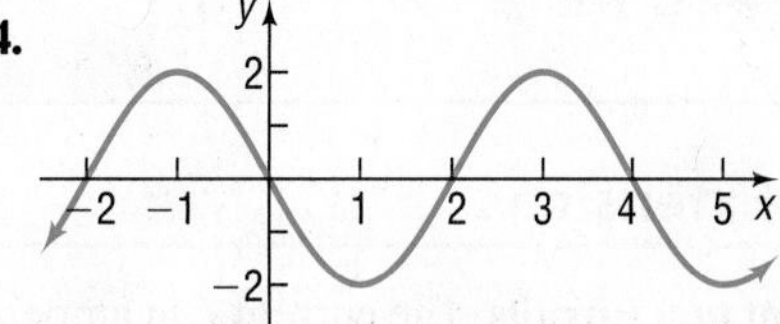

65.

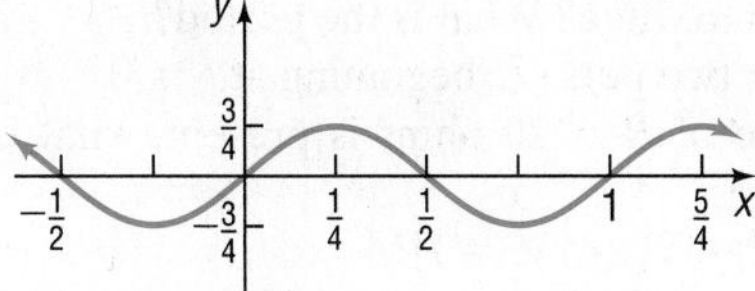

66.

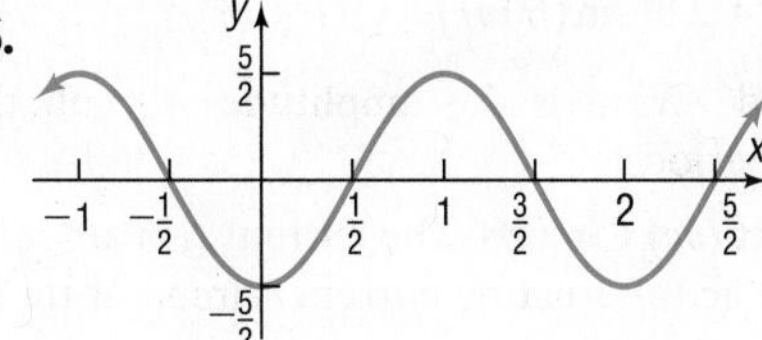

67.

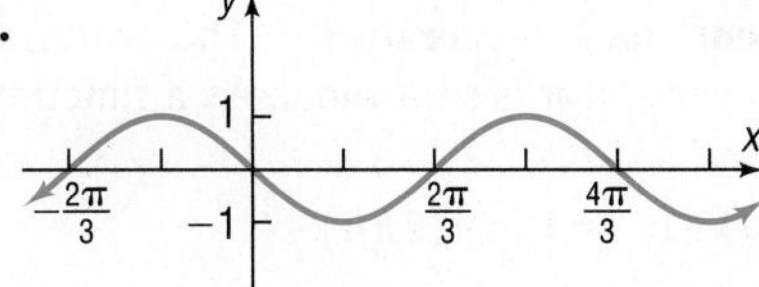

68.

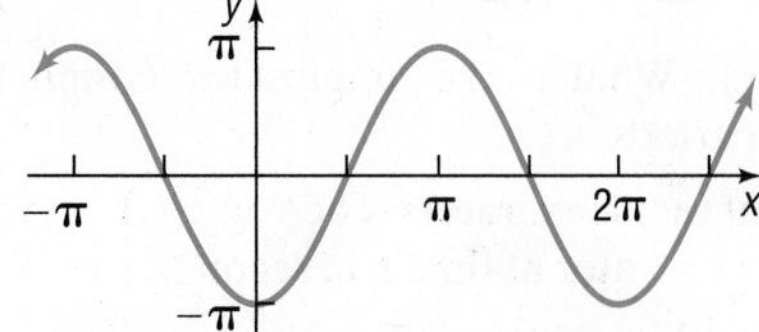

69.

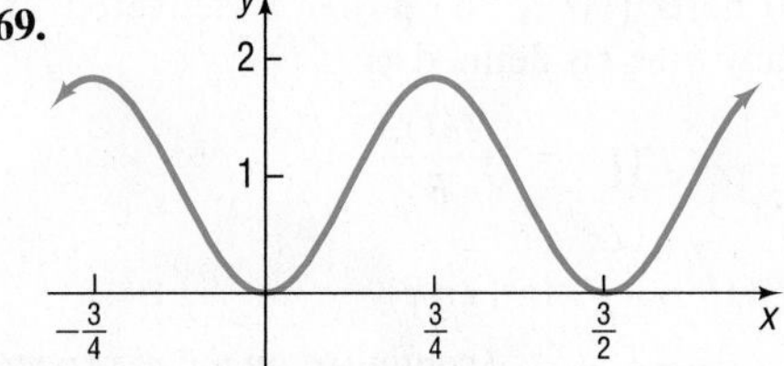

70.

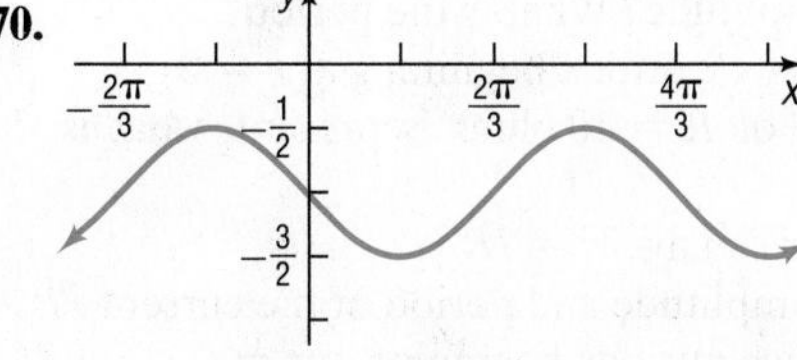

71.
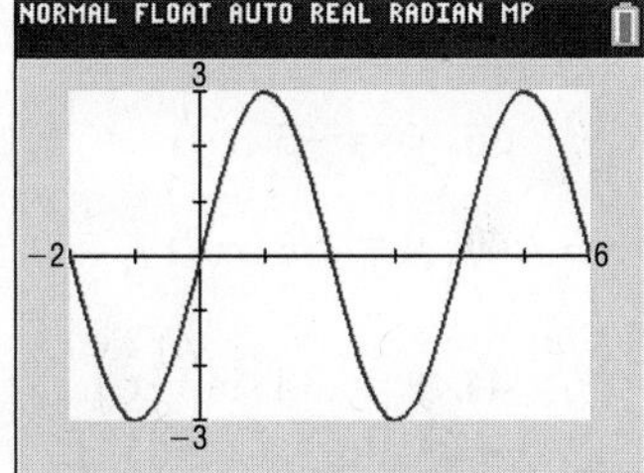

72.
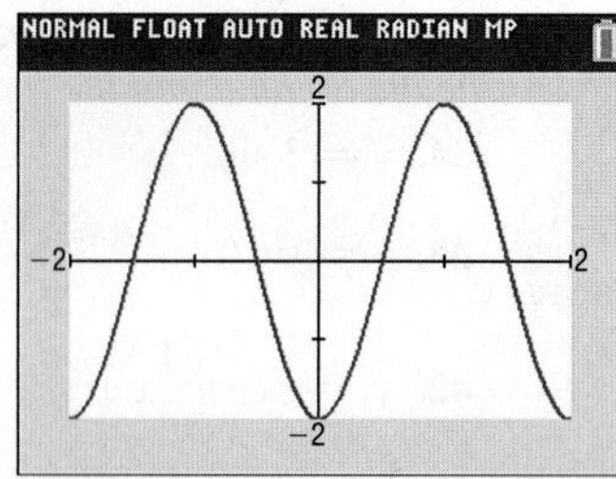

73.
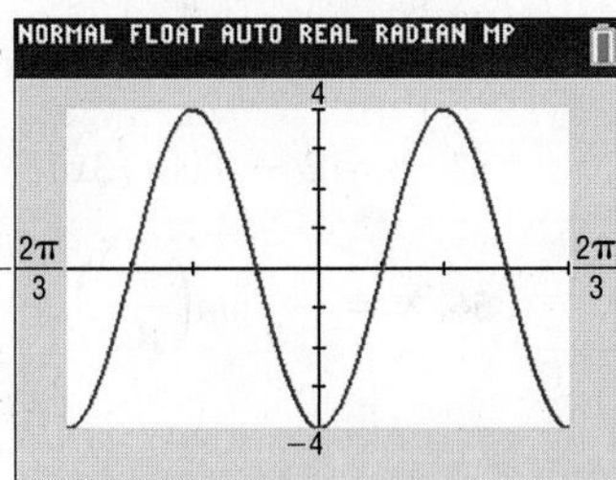

74.
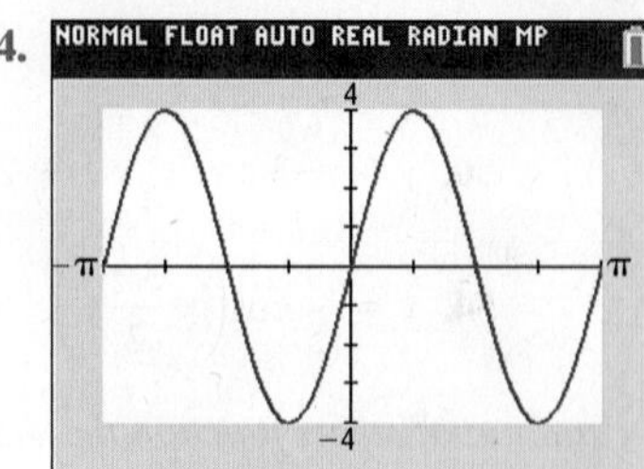

Mixed Practice

In Problems 75–78, find the average rate of change of f from 0 to $\frac{\pi}{2}$.

75. $f(x) = \sin x$

76. $f(x) = \cos x$

77. $f(x) = \sin \frac{x}{2}$

78. $f(x) = \cos(2x)$

In Problems 79–82, find $f(g(x))$ *and* $g(f(x))$, *and graph each of these functions.*

79. $f(x) = \sin x$
$g(x) = 4x$

80. $f(x) = \cos x$
$g(x) = \frac{1}{2}x$

81. $f(x) = -2x$
$g(x) = \cos x$

82. $f(x) = -3x$
$g(x) = \sin x$

In Problems 83 and 84, graph each function.

83. $f(x) = \begin{cases} \sin x & 0 \le x < \frac{5\pi}{4} \\ \cos x & \frac{5\pi}{4} \le x \le 2\pi \end{cases}$

84. $g(x) = \begin{cases} 2\sin x & 0 \le x \le \pi \\ \cos x + 1 & \pi < x \le 2\pi \end{cases}$

Applications and Extensions

85. Alternating Current (ac) Circuits The current I, in amperes, flowing through an ac (alternating current) circuit at time t, in seconds, is

$$I(t) = 220\sin(60\pi t) \qquad t \ge 0$$

What is the period? What is the amplitude? Graph this function over two periods.

86. Alternating Current (ac) Circuits The current I, in amperes, flowing through an ac (alternating current) circuit at time t, in seconds, is

$$I(t) = 120\sin(30\pi t) \qquad t \ge 0$$

What is the period? What is the amplitude? Graph this function over two periods.

87. Alternating Current (ac) Generators The voltage V, in volts, produced by an ac generator at time t, in seconds, is

$$V(t) = 220\sin(120\pi t)$$

(a) What is the amplitude? What is the period?
(b) Graph V over two periods, beginning at $t = 0$.
(c) If a resistance of $R = 10$ ohms is present, what is the current I?
[Hint: Use Ohm's Law, $V = IR$.]
(d) What are the amplitude and period of the current I?
(e) Graph I over two periods, beginning at $t = 0$.

88. Alternating Current (ac) Generators The voltage V, in volts, produced by an ac generator at time t, in seconds, is

$$V(t) = 120\sin(120\pi t)$$

(a) What is the amplitude? What is the period?
(b) Graph V over two periods, beginning at $t = 0$.
(c) If a resistance of $R = 20$ ohms is present, what is the current I?
[Hint: Use Ohm's Law, $V = IR$.]
(d) What are the amplitude and period of the current I?
(e) Graph I over two periods, beginning at $t = 0$.

89. Alternating Current (ac) Generators The voltage V produced by an ac generator is sinusoidal. As a function of time, the voltage V is

$$V(t) = V_0\sin(2\pi ft)$$

where f is the **frequency**, the number of complete oscillations (cycles) per second. [In the United States and Canada, f is 60 hertz (Hz).] The **power** P delivered to a resistance R at any time t is defined as

$$P(t) = \frac{[V(t)]^2}{R}$$

(a) Show that $P(t) = \frac{V_0^2}{R}\sin^2(2\pi ft)$.

(continued on the next page)

(b) The graph of P is shown in the figure. Express P as a sinusoidal function.

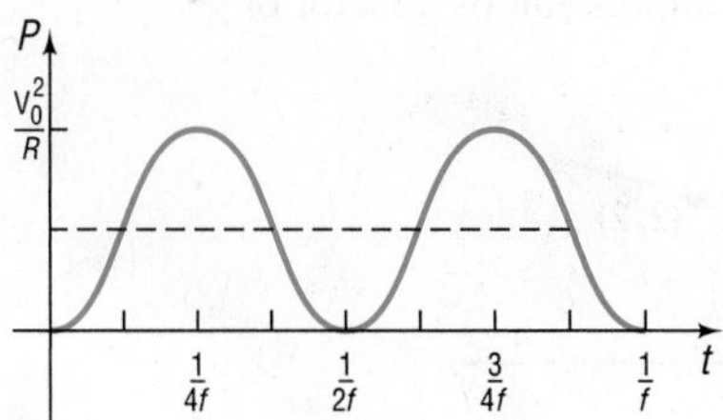

Power in an ac generator

(c) Deduce that

$$\sin^2(2\pi ft) = \frac{1}{2}[1 - \cos(4\pi ft)]$$

90. Bridge Clearance A one-lane highway runs through a tunnel in the shape of one-half a sine curve cycle. The opening is 28 feet wide at road level and is 15 feet tall at its highest point.

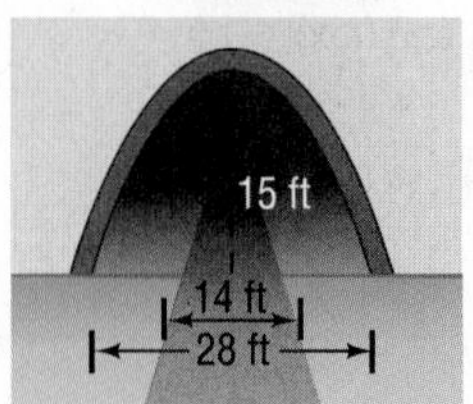

(a) Find an equation for the sine curve that fits the opening. Place the origin at the left end of the opening.
(b) If the road is 14 feet wide with 7-foot shoulders on each side, what is the height of the tunnel at the edge of the road?

Source: *en.wikipedia.org/wiki/Interstate_Highway_standards* and *Ohio Revised Code*

91. Biorhythms In the theory of biorhythms, a sine function of the form

$$P(t) = 50\sin(\omega t) + 50$$

is used to measure the percent P of a person's potential at time t, where t is measured in days and $t = 0$ is the day the person is born. Three characteristics are commonly measured:

Physical potential: period of 23 days
Emotional potential: period of 28 days
Intellectual potential: period of 33 days

(a) Find ω for each characteristic.
(b) Using a graphing utility, graph all three functions on the same screen.
(c) Is there a time t when all three characteristics have 100% potential? When is it?
(d) Suppose that you are 20 years old today ($t = 7305$ days). Describe your physical, emotional, and intellectual potential for the next 30 days.

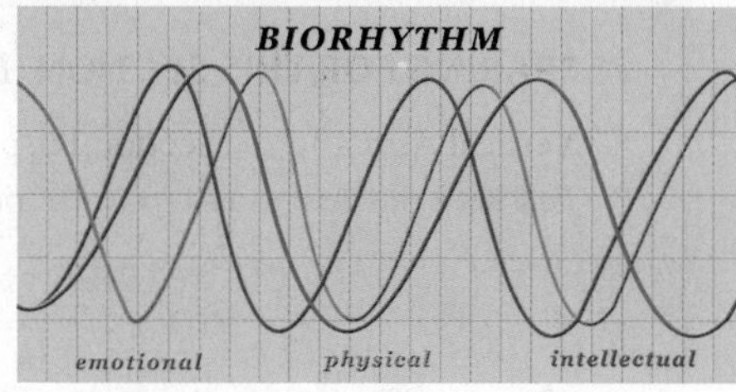

92. Graph $y = |\cos x|, -2\pi \le x \le 2\pi$.

93. Graph $y = |\sin x|, -2\pi \le x \le 2\pi$.

In Problems 94–97, the graphs of the given pairs of functions intersect infinitely many times. In each problem, find four of these points of intersection.

94. $y = \sin x$, $y = \frac{1}{2}$

95. $y = \cos x$, $y = \frac{1}{2}$

96. $y = 2\sin x$, $y = -2$

97. $y = \tan x$, $y = 1$

Explaining Concepts: Discussion and Writing

98. Explain how you would scale the x-axis and y-axis before graphing $y = 3\cos(\pi x)$.

99. Explain the term *amplitude* as it relates to the graph of a sinusoidal function.

100. Explain the term *period* as it relates to the graph of a sinusoidal function.

101. Explain how the amplitude and period of a sinusoidal graph are used to establish the scale on each coordinate axis.

102. Find an application in your major field that leads to a sinusoidal graph. Write a summary of your findings.

Retain Your Knowledge

Problems 103–106 are based on material learned earlier in the course. The purpose of these problems is to keep the material fresh in your mind so that you are better prepared for the final exam.

103. If $f(x) = x^2 - 5x + 1$, find $\dfrac{f(x+h) - f(x)}{h}$.

104. Find the exact value of $\sin\left(-\dfrac{11\pi}{4}\right)$ without using a calculator.

105. Find the intercepts of the graph of $h(x) = 3|x + 2| - 1$.

106. Find the area of the sector of a circle of radius 5 subtended by a central angle of 40°.

'Are You Prepared?' Answers

1. Vertical stretch by a factor of 3

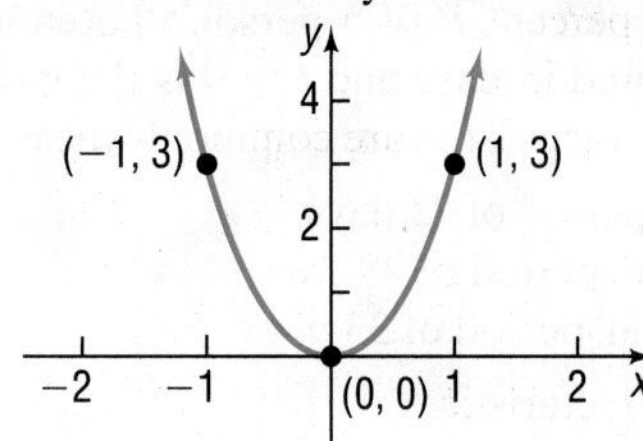

2. Horizontal compression by a factor of $\frac{1}{2}$

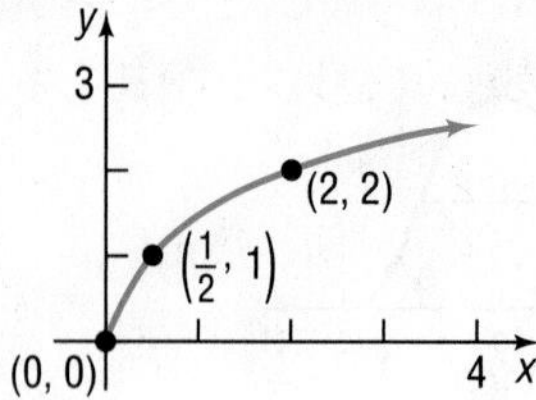

2.5 Graphs of the Tangent, Cotangent, Cosecant, and Secant Functions

PREPARING FOR THIS SECTION *Before getting started, review the following:*

- Vertical Asymptotes (Section 1.6, pp. 73–74)

Now Work the 'Are You Prepared?' problems on page 165.

OBJECTIVES 1 Graph Functions of the Form $y = A\tan(\omega x) + B$ and $y = A\cot(\omega x) + B$ (p. 162)

2 Graph Functions of the Form $y = A\csc(\omega x) + B$ and $y = A\sec(\omega x) + B$ (p. 164)

The Graph of the Tangent Function $y = \tan x$

Because the tangent function has period π, we only need to determine the graph over some interval of length π. The rest of the graph will consist of repetitions of that graph. Because the tangent function is not defined at $\ldots, -\frac{3\pi}{2}, -\frac{\pi}{2}, \frac{\pi}{2}, \frac{3\pi}{2}, \ldots$, we will concentrate on the interval $\left(-\frac{\pi}{2}, \frac{\pi}{2}\right)$, of length π, and construct Table 8, which lists some points on the graph of $y = \tan x$, $-\frac{\pi}{2} < x < \frac{\pi}{2}$. We plot the points in the table and connect them with a smooth curve. See Figure 62 for a partial graph of $y = \tan x$, where $-\frac{\pi}{3} \le x \le \frac{\pi}{3}$.

Table 8

x	$y = \tan x$	(x, y)
$-\frac{\pi}{3}$	$-\sqrt{3} \approx -1.73$	$\left(-\frac{\pi}{3}, -\sqrt{3}\right)$
$-\frac{\pi}{4}$	-1	$\left(-\frac{\pi}{4}, -1\right)$
$-\frac{\pi}{6}$	$-\frac{\sqrt{3}}{3} \approx -0.58$	$\left(-\frac{\pi}{6}, -\frac{\sqrt{3}}{3}\right)$
0	0	$(0, 0)$
$\frac{\pi}{6}$	$\frac{\sqrt{3}}{3} \approx 0.58$	$\left(\frac{\pi}{6}, \frac{\sqrt{3}}{3}\right)$
$\frac{\pi}{4}$	1	$\left(\frac{\pi}{4}, 1\right)$
$\frac{\pi}{3}$	$\sqrt{3} \approx 1.73$	$\left(\frac{\pi}{3}, \sqrt{3}\right)$

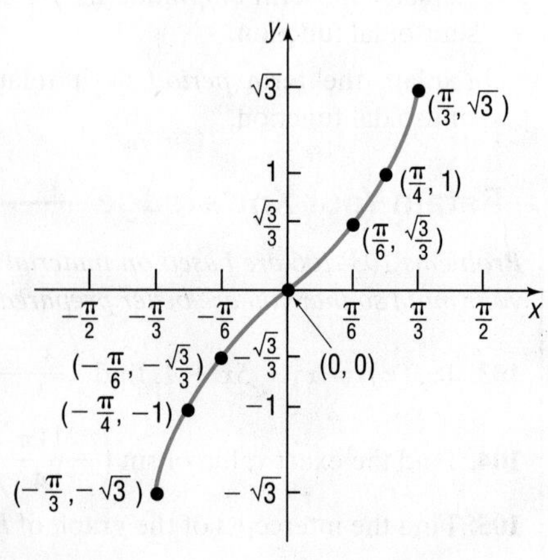

Figure 62 $y = \tan x$, $-\frac{\pi}{3} \le x \le \frac{\pi}{3}$

To complete one period of the graph of $y = \tan x$, we need to investigate the behavior of the function as x approaches $-\frac{\pi}{2}$ and $\frac{\pi}{2}$. We must be careful, though, because $y = \tan x$ is not defined at these numbers. To determine this behavior, we use the identity

$$\tan x = \frac{\sin x}{\cos x}$$

See Table 9. If x is close to $\frac{\pi}{2} \approx 1.5708$ but remains less than $\frac{\pi}{2}$, then $\sin x$ will be close to 1, and $\cos x$ will be positive and close to 0. (To see this, refer to the graphs of the sine function and the cosine function.) So the ratio $\frac{\sin x}{\cos x}$ will be positive and large. In fact, the closer x gets to $\frac{\pi}{2}$, the closer $\sin x$ gets to 1 and $\cos x$ gets to 0, so $\tan x$ approaches ∞ $\left(\lim_{x \to \frac{\pi}{2}^-} \tan x = \infty\right)$. In other words, the vertical line $x = \frac{\pi}{2}$ is a vertical asymptote to the graph of $y = \tan x$.

Table 9

x	$\sin x$	$\cos x$	$y = \tan x$
$\frac{\pi}{3} \approx 1.05$	$\frac{\sqrt{3}}{2}$	$\frac{1}{2}$	$\sqrt{3} \approx 1.73$
1.5	0.9975	0.0707	14.1
1.57	0.9999	7.96×10^{-4}	1255.8
1.5707	0.9999	9.6×10^{-5}	10,381
$\frac{\pi}{2} \approx 1.5708$	1	0	Undefined

If x is close to $-\frac{\pi}{2}$ but remains greater than $-\frac{\pi}{2}$, then $\sin x$ is close to -1, and $\cos x$ is positive and close to 0. The ratio $\frac{\sin x}{\cos x}$ approaches $-\infty$ $\left(\lim_{x \to -\frac{\pi}{2}^+} \tan x = -\infty\right)$. In other words, the vertical line $x = -\frac{\pi}{2}$ is also a vertical asymptote to the graph.

With these observations, one period of the graph can be completed. Obtain the complete graph of $y = \tan x$ by repeating this period, as shown in Figure 63.

Check: Graph $Y_1 = \tan x$ and compare the result with Figure 63. Use TRACE to see what happens as x gets close to $\frac{\pi}{2}$ but remains less than $\frac{\pi}{2}$.

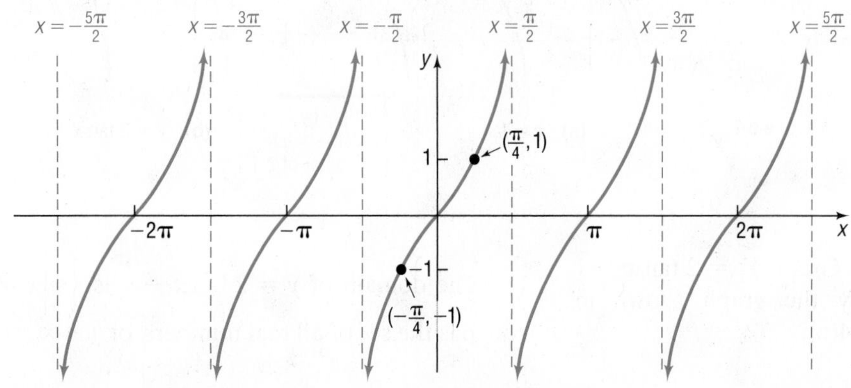

Figure 63 $y = \tan x$, $-\infty < x < \infty$, x not equal to odd multiples of $\frac{\pi}{2}$, $-\infty < y < \infty$

The graph of $y = \tan x$ in Figure 63 illustrates the following properties.

Properties of the Tangent Function

1. The domain is the set of all real numbers, except odd multiples of $\frac{\pi}{2}$.
2. The range is the set of all real numbers.
3. The tangent function is an odd function, as the symmetry of the graph with respect to the origin indicates.
4. The tangent function is periodic, with period π.
5. The x-intercepts are $\dots, -2\pi, -\pi, 0, \pi, 2\pi, 3\pi, \dots$; the y-intercept is 0.
6. Vertical asymptotes occur at $x = \dots, -\frac{3\pi}{2}, -\frac{\pi}{2}, \frac{\pi}{2}, \frac{3\pi}{2}, \dots$.

Now Work PROBLEMS 7 AND 15

1 Graph Functions of the Form $y = A\tan(\omega x) + B$ and $y = A\cot(\omega x) + B$

For tangent functions, there is no concept of amplitude since the range of the tangent function is $(-\infty, \infty)$. The role of A in $y = A\tan(\omega x) + B$ is to provide the magnitude of the vertical stretch. The period of $y = \tan x$ is π, so the period of $y = A\tan(\omega x) + B$ is $\frac{\pi}{\omega}$, caused by the horizontal compression of the graph by a factor of $\frac{1}{\omega}$. Finally, the presence of B indicates that a vertical shift is required.

EXAMPLE 1 **Graphing Functions of the Form $y = A\tan(\omega x) + B$**

Graph $y = 2\tan x - 1$. Use the graph to determine the domain and the range of the function $y = 2\tan x - 1$.

Solution Figure 64 shows the steps using transformations.

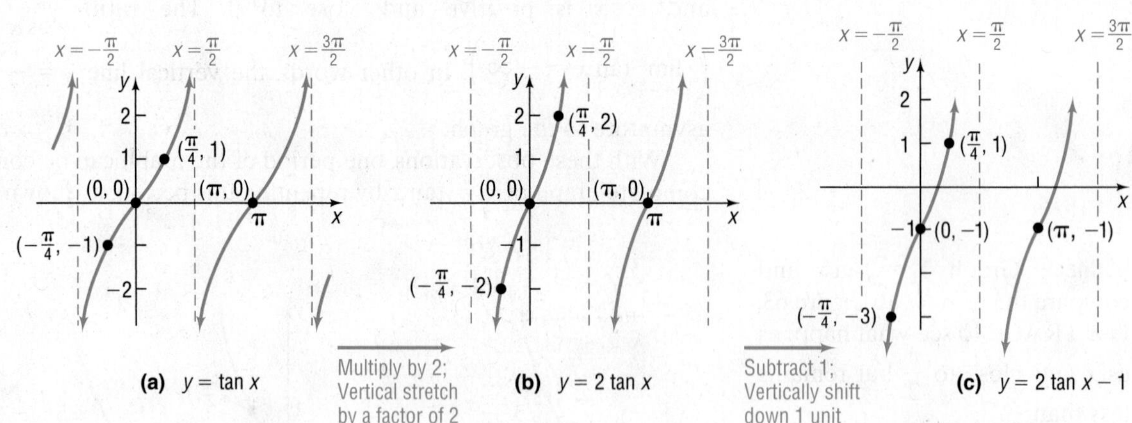

Figure 64 (a) $y = \tan x$ (b) $y = 2\tan x$ (c) $y = 2\tan x - 1$

Check: Graph $Y_1 = 2\tan x - 1$ to verify the graph shown in Figure 64(c).

The domain of $y = 2\tan x - 1$ is $\left\{x \,\middle|\, x \neq \frac{k\pi}{2}, k \text{ is an odd integer}\right\}$, and the range is the set of all real numbers, or $(-\infty, \infty)$.

EXAMPLE 2 **Graphing Functions of the Form $y = A\tan(\omega x) + B$**

Graph $y = 3\tan(2x)$. Use the graph to determine the domain and the range of $y = 3\tan(2x)$.

Solution Figure 65 shows the steps using transformations.

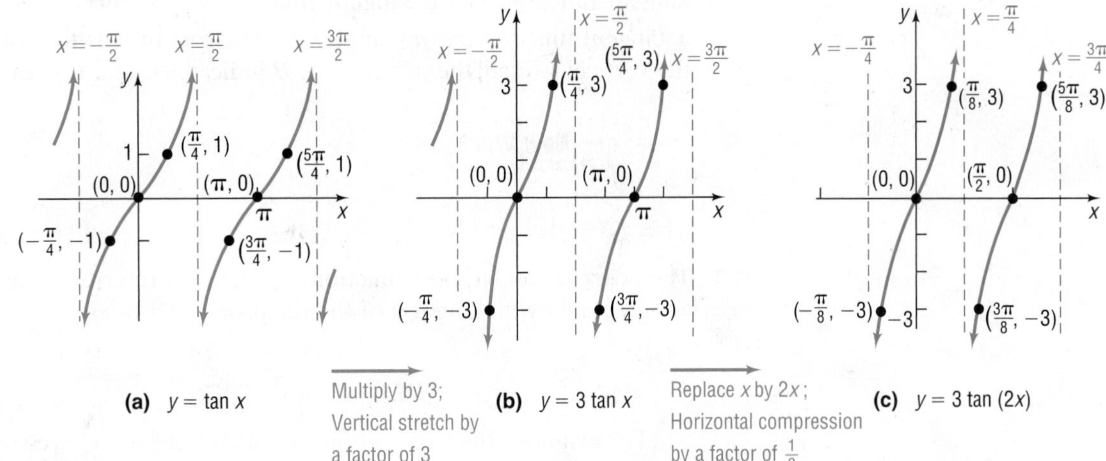

Figure 65 (a) $y = \tan x$ → Multiply by 3; Vertical stretch by a factor of 3 → (b) $y = 3\tan x$ → Replace x by $2x$; Horizontal compression by a factor of $\frac{1}{2}$ → (c) $y = 3\tan(2x)$

The domain of $y = 3\tan(2x)$ is $\left\{x \,\middle|\, x \neq \frac{k\pi}{4}, k \text{ is an odd integer}\right\}$, and the range is the set of all real numbers, or $(-\infty, \infty)$.

Check: Graph $Y_1 = 3\tan(2x)$ to verify the graph in Figure 65(c).

Note in Figure 65(c) that the period of $y = 3\tan(2x)$ is $\frac{\pi}{2}$ because of the compression of the original period π by a factor of $\frac{1}{2}$. Notice that the asymptotes are $x = -\frac{\pi}{4}, x = \frac{\pi}{4}, x = \frac{3\pi}{4}$, and so on, also because of the compression.

Now Work PROBLEM 21

The Graph of the Cotangent Function $y = \cot x$

The graph of $y = \cot x$ can be obtained in the same manner as the graph of $y = \tan x$. The period of $y = \cot x$ is π. Because the cotangent function is not defined for integer multiples of π, concentrate on the interval $(0, \pi)$. Table 10 lists some points on the graph of $y = \cot x$, $0 < x < \pi$. As x approaches 0 but remains greater than 0, the value of $\cos x$ will be close to 1, and the value of $\sin x$ will be positive and close to 0. The ratio $\frac{\cos x}{\sin x} = \cot x$ will be positive and large; so as x approaches 0, with $x > 0$, $\cot x$ approaches ∞ $\left(\lim_{x\to 0^+} \cot x = \infty\right)$. Similarly, as x approaches π but remains less than π, the value of $\cos x$ will be close to -1, and the value of $\sin x$ will be positive and close to 0. So the ratio $\frac{\cos x}{\sin x} = \cot x$ will be negative and will approach $-\infty$ as x approaches π $\left(\lim_{x\to \pi^-} \cot x = -\infty\right)$. Figure 66 shows the graph.

Table 10

x	$y = \cot x$	(x, y)
$\frac{\pi}{6}$	$\sqrt{3}$	$\left(\frac{\pi}{6}, \sqrt{3}\right)$
$\frac{\pi}{4}$	1	$\left(\frac{\pi}{4}, 1\right)$
$\frac{\pi}{3}$	$\frac{\sqrt{3}}{3}$	$\left(\frac{\pi}{3}, \frac{\sqrt{3}}{3}\right)$
$\frac{\pi}{2}$	0	$\left(\frac{\pi}{2}, 0\right)$
$\frac{2\pi}{3}$	$-\frac{\sqrt{3}}{3}$	$\left(\frac{2\pi}{3}, -\frac{\sqrt{3}}{3}\right)$
$\frac{3\pi}{4}$	-1	$\left(\frac{3\pi}{4}, -1\right)$
$\frac{5\pi}{6}$	$-\sqrt{3}$	$\left(\frac{5\pi}{6}, -\sqrt{3}\right)$

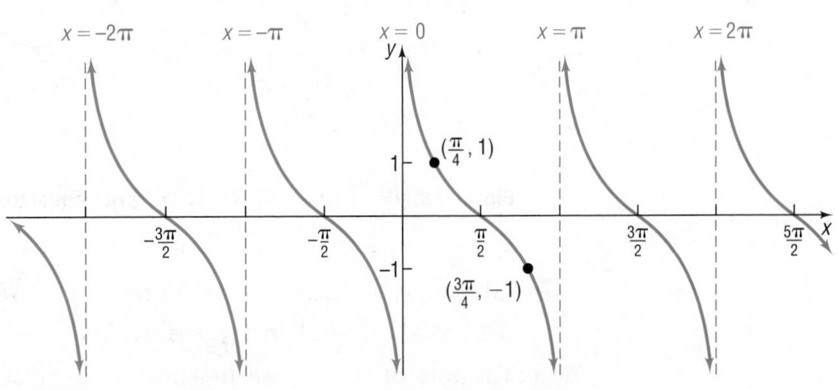

Figure 66 $y = \cot x, -\infty < x < \infty$, x not equal to integer multiples of π, $-\infty < y < \infty$

The graph of $y = A\cot(\omega x) + B$ has characteristics similar to those of the tangent function. The cotangent function $y = A\cot(\omega x) + B$ has period $\frac{\pi}{\omega}$. The cotangent function has no amplitude. The role of A is to provide the magnitude of the vertical stretch; the presence of B indicates that a vertical shift is required.

Now Work PROBLEM 23

The Graphs of the Cosecant Function and the Secant Function

The cosecant and secant functions, sometimes referred to as **reciprocal functions**, are graphed by making use of the reciprocal identities

$$\csc x = \frac{1}{\sin x} \quad \text{and} \quad \sec x = \frac{1}{\cos x}$$

For example, the value of the cosecant function $y = \csc x$ at a given number x equals the reciprocal of the corresponding value of the sine function, provided that the value of the sine function is not 0. If the value of $\sin x$ is 0, then x is an integer multiple of π. At such numbers, the cosecant function is not defined. In fact, the graph of the cosecant function has vertical asymptotes at integer multiples of π. Figure 67 shows the graph.

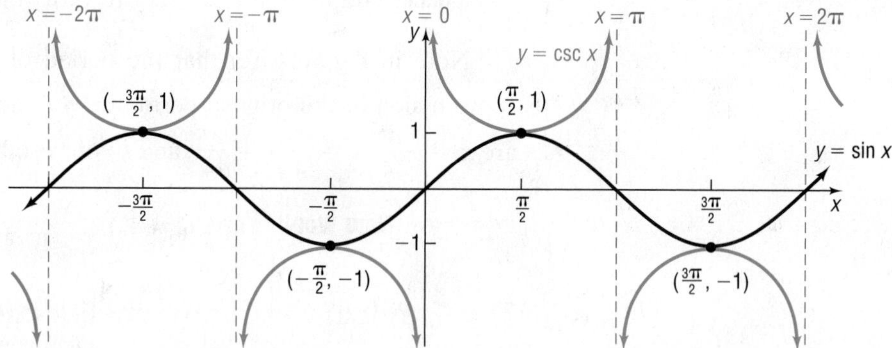

Figure 67 $y = \csc x, -\infty < x < \infty$, x not equal to integer multiples of π, $|y| \geq 1$

Using the idea of reciprocals, the graph of $y = \sec x$ can be obtained in a similar manner. See Figure 68.

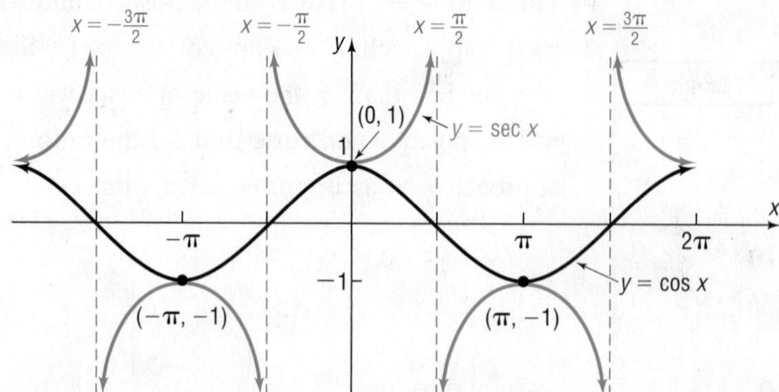

Figure 68 $y = \sec x, -\infty < x < \infty$, x not equal to odd multiples of $\frac{\pi}{2}$, $|y| \geq 1$

2 Graph Functions of the Form $y = A\csc(\omega x) + B$ and $y = A\sec(\omega x) + B$

The role of A in these functions is to set the range. The range of $y = \csc x$ is $\{y \mid y \leq -1 \text{ or } y \geq 1\}$ or $\{y \mid |y| \geq 1\}$; the range of $y = A\csc x$ is $\{y \mid |y| \geq |A|\}$

because of the vertical stretch of the graph by a factor of $|A|$. Just as with the sine and cosine functions, the period of $y = \csc(\omega x)$ and $y = \sec(\omega x)$ becomes $\frac{2\pi}{\omega}$ because of the horizontal compression of the graph by a factor of $\frac{1}{\omega}$. The presence of B indicates that a vertical shift is required.

EXAMPLE 3 **Graphing Functions of the Form $y = A\csc(\omega x) + B$**

Graph $y = 2\csc x - 1$. Use the graph to determine the domain and the range of $y = 2\csc x - 1$.

Solution We use transformations. Figure 69 shows the required steps.

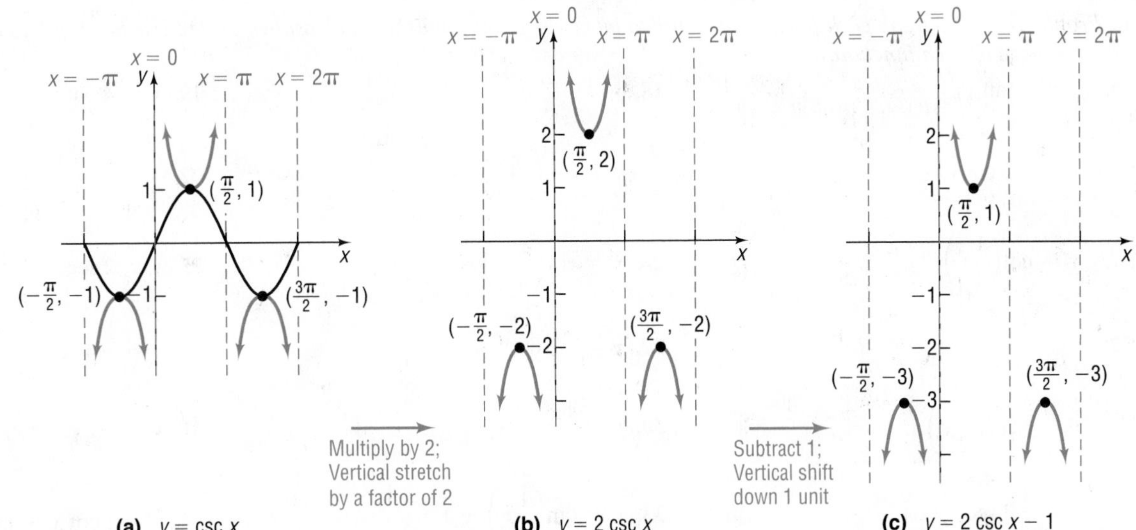

Figure 69 (a) $y = \csc x$ (b) $y = 2\csc x$ (c) $y = 2\csc x - 1$

The domain of $y = 2\csc x - 1$ is $\{x | x \neq k\pi, k \text{ is an integer}\}$ and the range is $\{y | y \leq -3 \text{ or } y \geq 1\}$, or, using interval notation, $(-\infty, -3] \cup [1, \infty)$. ●

✓Check: Graph $Y_1 = 2\csc x - 1$ to verify the graph shown in Figure 69.

Now Work PROBLEM 29

2.5 Assess Your Understanding

'Are You Prepared?' *Answers are given at the end of these exercises. If you get a wrong answer, read the pages listed in red.*

1. The graph of $y = \frac{3x - 6}{x - 4}$ has a vertical asymptote. What is it? (pp. 73–74)

2. ***True or False*** If $x = 3$ is a vertical asymptote of a rational function R, then $\lim_{x \to 3} |R(x)| = \infty$. (pp. 73–74)

Concepts and Vocabulary

3. The graph of $y = \tan x$ is symmetric with respect to the ______ and has vertical asymptotes at ____________.

4. The graph of $y = \sec x$ is symmetric with respect to the ______ and has vertical asymptotes at ____________.

5. It is easiest to graph $y = \sec x$ by first sketching the graph of ______.
(a) $y = \sin x$ (b) $y = \cos x$ (c) $y = \tan x$ (d) $y = \csc x$

6. ***True or False*** The graphs of $y = \tan x$, $y = \cot x$, $y = \sec x$, and $y = \csc x$ each have infinitely many vertical asymptotes.

Skill Building

In Problems 7–16, if necessary, refer to the graphs of the functions to answer each question.

7. What is the y-intercept of $y = \tan x$?

8. What is the y-intercept of $y = \cot x$?

9. What is the y-intercept of $y = \sec x$?

10. What is the y-intercept of $y = \csc x$?

11. For what numbers x, $-2\pi \leq x \leq 2\pi$, does $\sec x = 1$? For what numbers x does $\sec x = -1$?

12. For what numbers x, $-2\pi \leq x \leq 2\pi$, does $\csc x = 1$? For what numbers x does $\csc x = -1$?

13. For what numbers x, $-2\pi \leq x \leq 2\pi$, does the graph of $y = \sec x$ have vertical asymptotes?

14. For what numbers x, $-2\pi \leq x \leq 2\pi$, does the graph of $y = \csc x$ have vertical asymptotes?

15. For what numbers x, $-2\pi \leq x \leq 2\pi$, does the graph of $y = \tan x$ have vertical asymptotes?

16. For what numbers x, $-2\pi \leq x \leq 2\pi$, does the graph of $y = \cot x$ have vertical asymptotes?

In Problems 17–40, graph each function. Be sure to label key points and show at least two cycles. Use the graph to determine the domain and the range of each function.

17. $y = 3\tan x$
18. $y = -2\tan x$
19. $y = 4\cot x$
20. $y = -3\cot x$
21. $y = \tan\left(\frac{\pi}{2}x\right)$
22. $y = \tan\left(\frac{1}{2}x\right)$
23. $y = \cot\left(\frac{1}{4}x\right)$
24. $y = \cot\left(\frac{\pi}{4}x\right)$
25. $y = 2\sec x$
26. $y = \frac{1}{2}\csc x$
27. $y = -3\csc x$
28. $y = -4\sec x$
29. $y = 4\sec\left(\frac{1}{2}x\right)$
30. $y = \frac{1}{2}\csc(2x)$
31. $y = -2\csc(\pi x)$
32. $y = -3\sec\left(\frac{\pi}{2}x\right)$
33. $y = \tan\left(\frac{1}{4}x\right) + 1$
34. $y = 2\cot x - 1$
35. $y = \sec\left(\frac{2\pi}{3}x\right) + 2$
36. $y = \csc\left(\frac{3\pi}{2}x\right)$
37. $y = \frac{1}{2}\tan\left(\frac{1}{4}x\right) - 2$
38. $y = 3\cot\left(\frac{1}{2}x\right) - 2$
39. $y = 2\csc\left(\frac{1}{3}x\right) - 1$
40. $y = 3\sec\left(\frac{1}{4}x\right) + 1$

Mixed Practice

In Problems 41–44, find the average rate of change of f from 0 to $\frac{\pi}{6}$.

41. $f(x) = \tan x$
42. $f(x) = \sec x$
43. $f(x) = \tan(2x)$
44. $f(x) = \sec(2x)$

In Problems 45–48, find $f(g(x))$ *and* $g(f(x))$*, and graph each of these functions.*

45. $f(x) = \tan x$, $g(x) = 4x$
46. $f(x) = 2\sec x$, $g(x) = \frac{1}{2}x$
47. $f(x) = -2x$, $g(x) = \cot x$
48. $f(x) = \frac{1}{2}x$, $g(x) = 2\csc x$

In Problems 49 and 50, graph each function.

49. $f(x) = \begin{cases} \tan x & 0 \leq x < \frac{\pi}{2} \\ 0 & x = \frac{\pi}{2} \\ \sec x & \frac{\pi}{2} < x \leq \pi \end{cases}$

50. $g(x) = \begin{cases} \csc x & 0 < x < \pi \\ 0 & x = \pi \\ \cot x & \pi < x < 2\pi \end{cases}$

Applications and Extensions

51. Carrying a Ladder around a Corner Two hallways, one of width 3 feet, the other of width 4 feet, meet at a right angle. See the illustration.

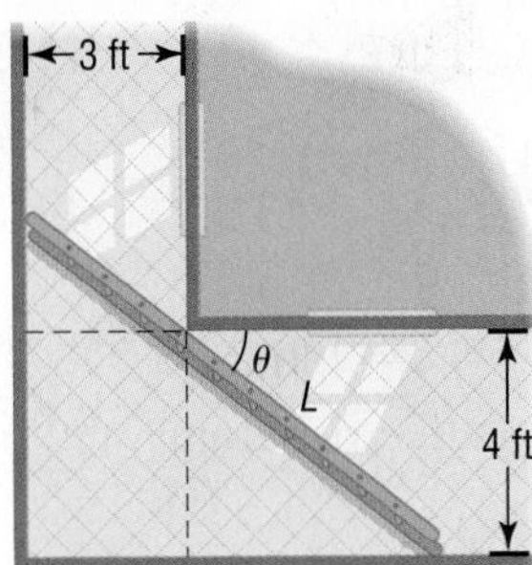

(a) Show that the length L of the ladder shown as a function of the angle θ is

$$L(\theta) = 3\sec\theta + 4\csc\theta$$

(b) Graph $L = L(\theta), 0 < \theta < \frac{\pi}{2}$.

(c) For what value of θ is L the least?

(d) What is the length of the longest ladder that can be carried around the corner? Why is this also the least value of L?

52. A Rotating Beacon Suppose that a fire truck is parked in front of a building as shown in the figure.

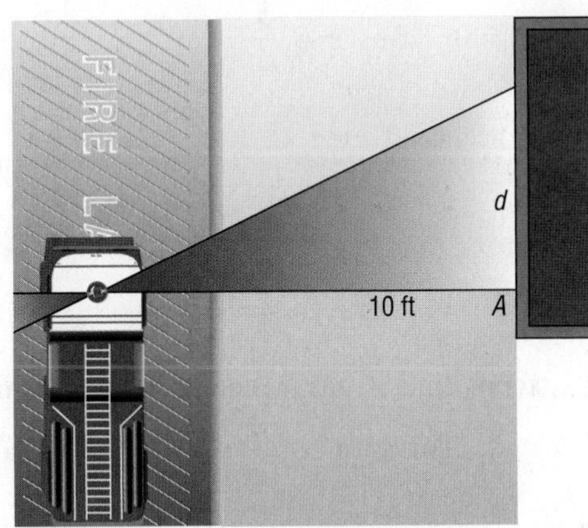

The beacon light on top of the fire truck is located 10 feet from the wall and has a light on each side. If the beacon light rotates 1 revolution every 2 seconds, then a model for determining the distance d, in feet, that the beacon of light is from point A on the wall after t seconds is given by

$$d(t) = |10\tan(\pi t)|$$

(a) Graph $d(t) = |10\tan(\pi t)|$ for $0 \le t \le 2$.

(b) For what values of t is the function undefined? Explain what this means in terms of the beam of light on the wall.

(c) Fill in the following table.

t	0	0.1	0.2	0.3	0.4
$d(t) = 10\tan(\pi t)$					

(d) Compute $\frac{d(0.1) - d(0)}{0.1 - 0}, \frac{d(0.2) - d(0.1)}{0.2 - 0.1}$, and so on, for each consecutive value of t. These are called **first differences**.

(e) Interpret the first differences found in part (d). What is happening to the speed of the beam of light as d increases?

53. Exploration Graph

$$y = \tan x \quad \text{and} \quad y = -\cot\left(x + \frac{\pi}{2}\right)$$

Do you think that $\tan x = -\cot\left(x + \frac{\pi}{2}\right)$?

Retain Your Knowledge

Problems 54–57 are based on material learned earlier in the course. The purpose of these problems is to keep the material fresh in your mind so that you are better prepared for the final exam.

54. Test $y^2 = x - 4$ for symmetry with respect to the x-axis, y-axis, and origin.

55. Write the standard form of the equation of a circle with radius 3 and center $(-3, 1)$.

56. Given $h(x) = \frac{\sqrt{x} - 6}{x^2 + 10}$, find $h(4)$.

57. Determine if the relation $x^2 + (y - 4)^2 = 16$ is a function.

'Are You Prepared?' Answers

1. $x = 4$ **2.** True

2.6 Phase Shift; Sinusoidal Curve Fitting

OBJECTIVES 1 Graph Sinusoidal Functions of the Form $y = A\sin(\omega x - \phi) + B$ (p. 168)
2 Build Sinusoidal Models from Data (p. 171)

1 Graph Sinusoidal Functions of the Form $y = A\sin(\omega x - \phi) + B$

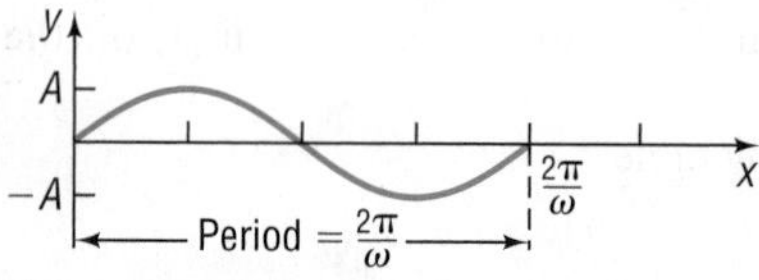

Figure 70 One cycle of $y = A\sin(\omega x)$, $A > 0, \omega > 0$

We have seen that the graph of $y = A\sin(\omega x)$, $\omega > 0$, has amplitude $|A|$ and period $T = \frac{2\pi}{\omega}$. One cycle can be drawn as x varies from 0 to $\frac{2\pi}{\omega}$, or, equivalently, as ωx varies from 0 to 2π. See Figure 70.

Now consider the graph of

$$y = A\sin(\omega x - \phi)$$

which may also be written as

$$y = A\sin\left[\omega\left(x - \frac{\phi}{\omega}\right)\right]$$

NOTE The beginning and end of the period can also be found by solving the inequality:

$$0 \le \omega x - \phi \le 2\pi$$
$$\phi \le \omega x \le 2\pi + \phi$$
$$\frac{\phi}{\omega} \le x \le \frac{2\pi}{\omega} + \frac{\phi}{\omega}$$ ■

where $\omega > 0$ and ϕ (the Greek letter phi) are real numbers. The graph is a sine curve with amplitude $|A|$. As $\omega x - \phi$ varies from 0 to 2π, one period will be traced out. This period begins when

$$\omega x - \phi = 0 \quad \text{or} \quad x = \frac{\phi}{\omega}$$

and ends when

$$\omega x - \phi = 2\pi \quad \text{or} \quad x = \frac{\phi}{\omega} + \frac{2\pi}{\omega}$$

Figure 71 One cycle of $y = A\sin(\omega x - \phi)$, $A > 0$, $\omega > 0, \phi > 0$

See Figure 71.

Notice that the graph of $y = A\sin(\omega x - \phi) = A\sin\left[\omega\left(x - \frac{\phi}{\omega}\right)\right]$ is the same as the graph of $y = A\sin(\omega x)$, except that it has been shifted $\left|\frac{\phi}{\omega}\right|$ units (to the right if $\phi > 0$ and to the left if $\phi < 0$). This number $\frac{\phi}{\omega}$ is called the **phase shift** of the graph of $y = A\sin(\omega x - \phi)$.

For the graphs of $y = A\sin(\omega x - \phi)$ or $y = A\cos(\omega x - \phi)$, $\omega > 0$,

Amplitude $= |A|$	Period $= T = \frac{2\pi}{\omega}$	Phase shift $= \frac{\phi}{\omega}$

The phase shift is to the left if $\phi < 0$ and to the right if $\phi > 0$.

EXAMPLE 1 **Finding the Amplitude, Period, and Phase Shift of a Sinusoidal Function and Graphing It**

Find the amplitude, period, and phase shift of $y = 3\sin(2x - \pi)$, and graph the function.

Solution Use the same four steps used to graph sinusoidal functions of the form $y = A\sin(\omega x)$ or $y = A\cos(\omega x)$ given on page 152.

STEP 1: Compare

$$y = 3\sin(2x - \pi) = 3\sin\left[2\left(x - \frac{\pi}{2}\right)\right]$$

to

$$y = A\sin(\omega x - \phi) = A\sin\left[\omega\left(x - \frac{\phi}{\omega}\right)\right]$$

Note that $A = 3$, $\omega = 2$, and $\phi = \pi$. The graph is a sine curve with amplitude $|A| = 3$, period $T = \frac{2\pi}{\omega} = \frac{2\pi}{2} = \pi$, and phase shift $= \frac{\phi}{\omega} = \frac{\pi}{2}$.

NOTE The interval defining one cycle can also be found by solving the inequality

$$0 \le 2x - \pi \le 2\pi$$

Then

$$\pi \le 2x \le 3\pi$$
$$\frac{\pi}{2} \le x \le \frac{3\pi}{2}$$

■

STEP 2: The graph of $y = 3\sin(2x - \pi)$ will lie between -3 and 3 on the y-axis. One cycle will begin at $x = \frac{\phi}{\omega} = \frac{\pi}{2}$ and end at $x = \frac{\phi}{\omega} + \frac{2\pi}{\omega} = \frac{\pi}{2} + \pi = \frac{3\pi}{2}$. To find the five key points, divide the interval $\left[\frac{\pi}{2}, \frac{3\pi}{2}\right]$ into four subintervals, each of length $\pi \div 4 = \frac{\pi}{4}$, by finding the following values of x:

$\frac{\pi}{2}$	$\frac{\pi}{2} + \frac{\pi}{4} = \frac{3\pi}{4}$	$\frac{3\pi}{4} + \frac{\pi}{4} = \pi$	$\pi + \frac{\pi}{4} = \frac{5\pi}{4}$	$\frac{5\pi}{4} + \frac{\pi}{4} = \frac{3\pi}{2}$
1st x-coordinate	2nd x-coordinate	3rd x-coordinate	4th x-coordinate	5th x-coordinate

STEP 3: Use these values of x to determine the five key points on the graph:

$$\left(\frac{\pi}{2}, 0\right) \quad \left(\frac{3\pi}{4}, 3\right) \quad (\pi, 0) \quad \left(\frac{5\pi}{4}, -3\right) \quad \left(\frac{3\pi}{2}, 0\right)$$

STEP 4: Plot these five points and fill in the graph of the sine function as shown in Figure 72(a). Extend the graph in each direction to obtain Figure 72(b).

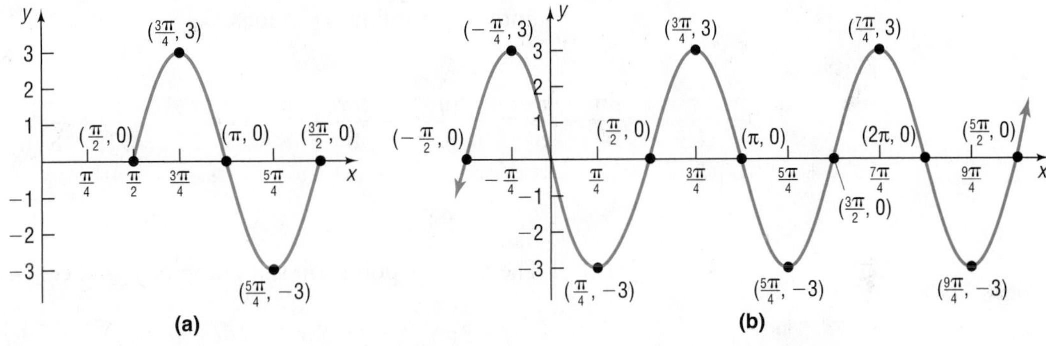

Figure 72 (a) (b)

The graph of $y = 3\sin(2x - \pi) = 3\sin\left[2\left(x - \frac{\pi}{2}\right)\right]$ may also be obtained using transformations. See Figure 73.

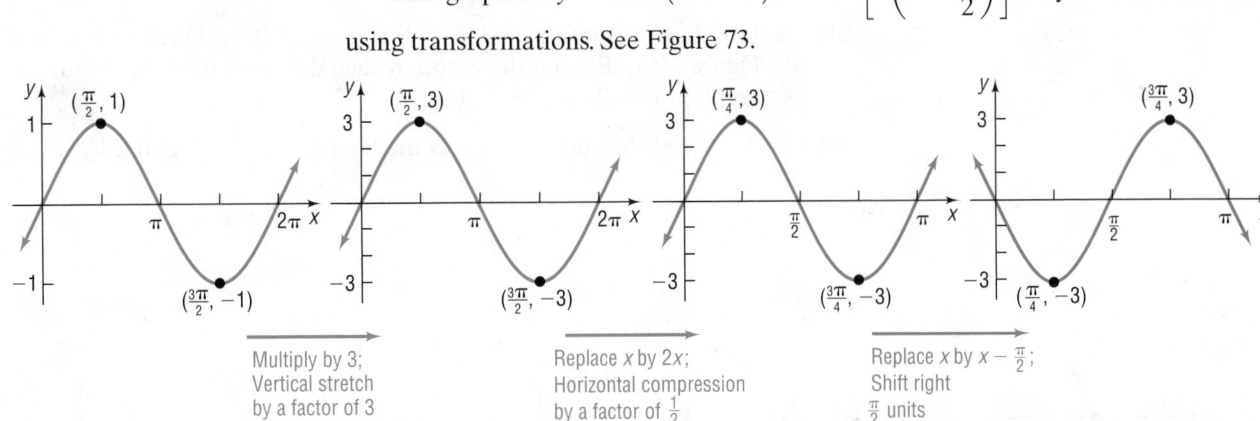

Figure 73 (a) $y = \sin x$ (b) $y = 3\sin x$ (c) $y = 3\sin(2x)$ (d) $y = 3\sin\left[2\left(x - \frac{\pi}{2}\right)\right] = 3\sin(2x - \pi)$ ●

Check: Graph $Y_1 = 3\sin(2x - \pi)$ using a graphing utility to verify Figure 73(d).

To graph a sinusoidal function of the form $y = A\sin(\omega x - \phi) + B$, first graph the function $y = A\sin(\omega x - \phi)$ and then apply a vertical shift.

EXAMPLE 2 **Finding the Amplitude, Period, and Phase Shift of a Sinusoidal Function and Graphing It**

Find the amplitude, period, and phase shift of $y = 2\cos(4x + 3\pi) + 1$, and graph the function.

Solution **STEP 1:** Compare

$$y = 2\cos(4x + 3\pi) = 2\cos\left[4\left(x + \frac{3\pi}{4}\right)\right]$$

to

$$y = A\cos(\omega x - \phi) = A\cos\left[\omega\left(x - \frac{\phi}{\omega}\right)\right]$$

Note that $A = 2$, $\omega = 4$, and $\phi = -3\pi$. The graph is a cosine curve with amplitude $|A| = 2$, period $T = \frac{2\pi}{\omega} = \frac{2\pi}{4} = \frac{\pi}{2}$, and phase shift $= \frac{\phi}{\omega} = -\frac{3\pi}{4}$.

STEP 2: The graph of $y = 2\cos(4x + 3\pi)$ will lie between -2 and 2 on the y-axis. One cycle begins at $x = \frac{\phi}{\omega} = -\frac{3\pi}{4}$ and ends at $x = \frac{\phi}{\omega} + \frac{2\pi}{\omega} = -\frac{3\pi}{4} + \frac{\pi}{2} = -\frac{\pi}{4}$. To find the five key points, divide the interval $\left[-\frac{3\pi}{4}, -\frac{\pi}{4}\right]$ into four subintervals, each of the length $\frac{\pi}{2} \div 4 = \frac{\pi}{8}$, by finding the following values.

NOTE The interval defining one cycle can also be found by solving the inequality

$$0 \le 4x + 3\pi \le 2\pi$$

Then

$$-3\pi \le 4x \le -\pi$$

$$-\frac{3\pi}{4} \le x \le -\frac{\pi}{4}$$

■

$-\frac{3\pi}{4}$	$-\frac{3\pi}{4} + \frac{\pi}{8} = -\frac{5\pi}{8}$	$-\frac{5\pi}{8} + \frac{\pi}{8} = -\frac{\pi}{2}$	$-\frac{\pi}{2} + \frac{\pi}{8} = -\frac{3\pi}{8}$	$-\frac{3\pi}{8} + \frac{\pi}{8} = -\frac{\pi}{4}$
1st x-coordinate	2nd x-coordinate	3rd x-coordinate	4th x-coordinate	5th x-coordinate

STEP 3: The five key points on the graph of $y = 2\cos(4x + 3\pi)$ are

$$\left(-\frac{3\pi}{4}, 2\right) \left(-\frac{5\pi}{8}, 0\right) \left(-\frac{\pi}{2}, -2\right) \left(-\frac{3\pi}{8}, 0\right) \left(-\frac{\pi}{4}, 2\right)$$

STEP 4: Plot these five points and fill in the graph of the cosine function as shown in Figure 74(a). Extend the graph in each direction to obtain Figure 74(b), the graph of $y = 2\cos(4x + 3\pi)$.

STEP 5: A vertical shift up 1 unit gives the final graph. See Figure 74(c).

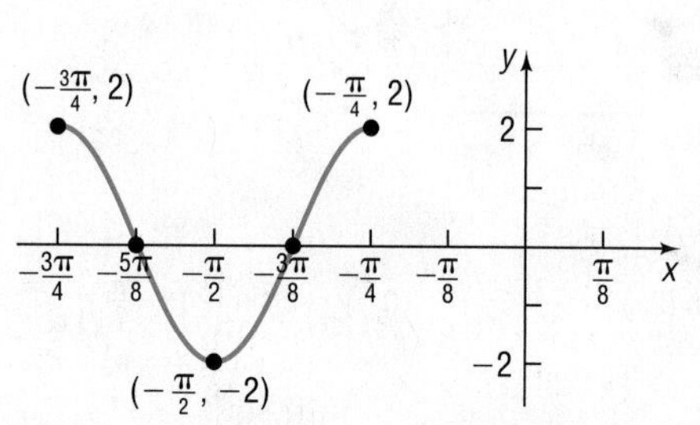

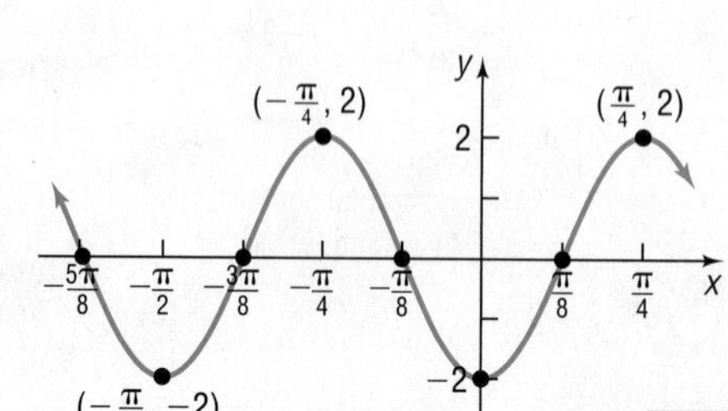

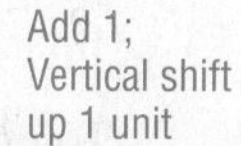

Figure 74 (a) (b) $y = 2\cos(4x + 3\pi)$ (c) $y = 2\cos(4x + 3\pi) + 1$

●

The graph of $y = 2\cos(4x + 3\pi) + 1 = 2\cos\left[4\left(x + \frac{3\pi}{4}\right)\right] + 1$ may also be obtained using transformations. See Figure 75.

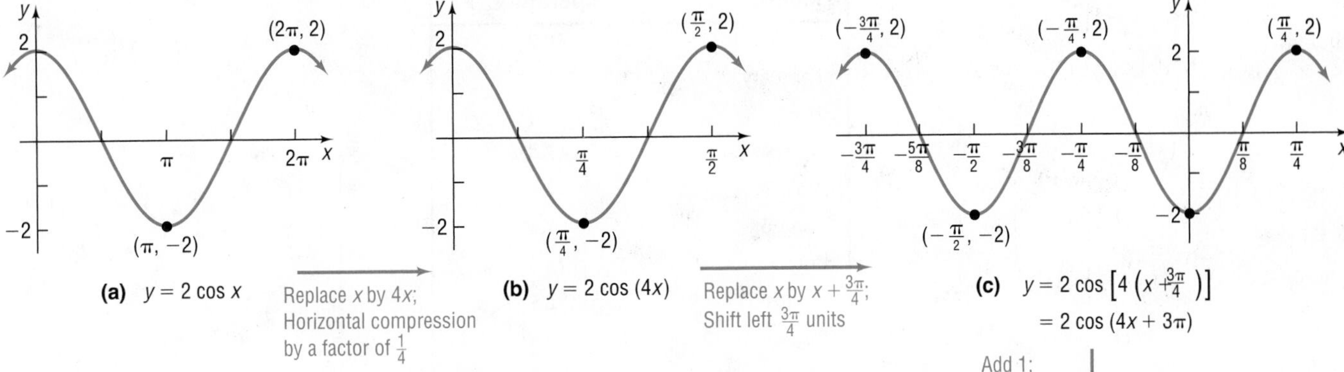

(a) $y = 2\cos x$

(b) $y = 2\cos(4x)$

(c) $y = 2\cos\left[4\left(x + \frac{3\pi}{4}\right)\right] = 2\cos(4x + 3\pi)$

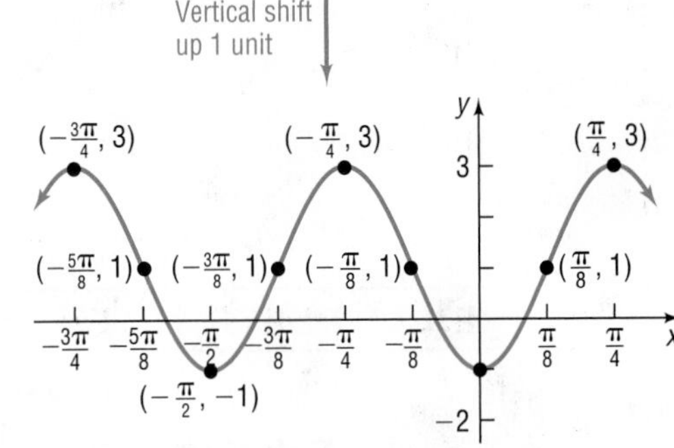

(d) $y = 2\cos(4x + 3\pi) + 1$

Figure 75

Now Work PROBLEM 3

SUMMARY

Steps for Graphing Sinusoidal Functions $y = A\sin(\omega x - \phi) + B$ or $y = A\cos(\omega x - \phi) + B$

STEP 1: Determine the amplitude $|A|$, period $T = \frac{2\pi}{\omega}$, and phase shift $\frac{\phi}{\omega}$.

STEP 2: Determine the starting point of one cycle of the graph, $\frac{\phi}{\omega}$. Determine the ending point of one cycle of the graph, $\frac{\phi}{\omega} + \frac{2\pi}{\omega}$. Divide the interval $\left[\frac{\phi}{\omega}, \frac{\phi}{\omega} + \frac{2\pi}{\omega}\right]$ into four subintervals, each of length $\frac{2\pi}{\omega} \div 4$.

STEP 3: Use the endpoints of the subintervals to find the five key points on the graph.

STEP 4: Plot the five key points, and connect them with a sinusoidal graph to obtain one cycle of the graph. Extend the graph in each direction to make it complete.

STEP 5: If $B \neq 0$, apply a vertical shift.

2 Build Sinusoidal Models from Data

Scatter diagrams of data sometimes resemble the graph of a sinusoidal function. Let's look at an example.

The data given in Table 11 on the next page represent the average monthly temperatures in Denver, Colorado. Since the data represent *average* monthly temperatures collected over many years, the data will not vary much from year to year and so will essentially repeat each year. In other words, the data are periodic. Figure 76 shows the scatter diagram of these data, where $x = 1$ represents January, $x = 2$ represents February, and so on.

Notice that the scatter diagram looks like the graph of a sinusoidal function. We choose to fit the data to a sine function of the form

$$y = A\sin(\omega x - \phi) + B$$

where A, B, ω, and ϕ are constants.

Table 11

Month, x	Average Monthly Temperature, °F
January, 1	30.7
February, 2	32.5
March, 3	40.4
April, 4	47.4
May, 5	57.1
June, 6	67.4
July, 7	74.2
August, 8	72.5
September, 9	62.4
October, 10	50.9
November, 11	38.3
December, 12	30.0

Source: U.S. National Oceanic and Atmospheric Administration

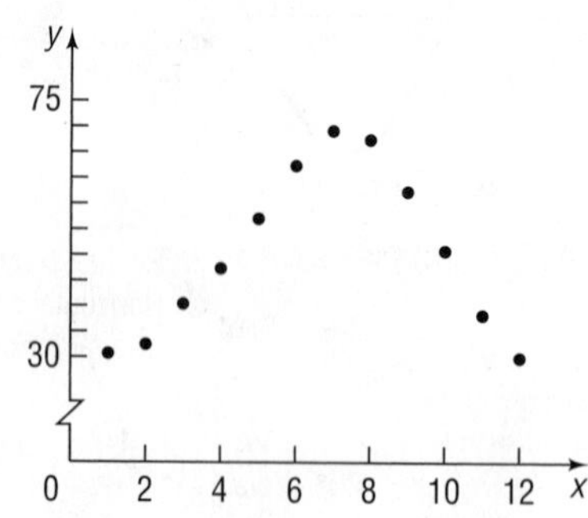

Figure 76 Denver average monthly temperature

EXAMPLE 3

Finding a Sinusoidal Function from Temperature Data

Fit a sine function to the data in Table 11.

Solution Begin with a scatter diagram of the data for one year. See Figure 77. The data will be fitted to a sine function of the form

$$y = A \sin(\omega x - \phi) + B$$

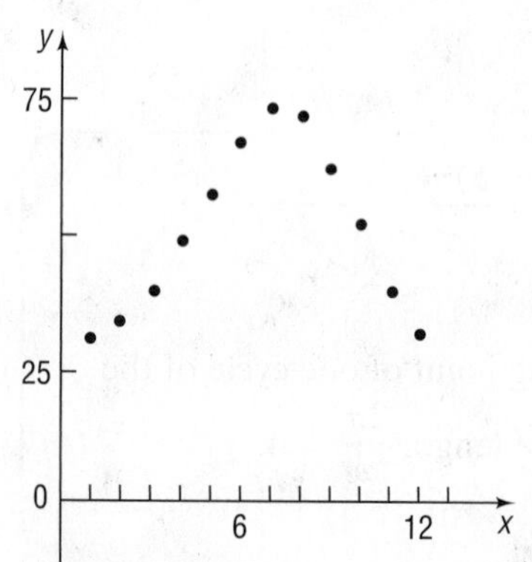

Figure 77

Step 1: To find the amplitude A, compute

$$\text{Amplitude} = \frac{\text{largest data value} - \text{smallest data value}}{2}$$

$$= \frac{74.2 - 30.0}{2} = 22.1$$

To see the remaining steps in this process, superimpose the graph of the function $y = 22.1 \sin x$, where x represents months, on the scatter diagram.

Figure 78 shows the two graphs. To fit the data, the graph needs to be shifted vertically, shifted horizontally, and stretched horizontally.

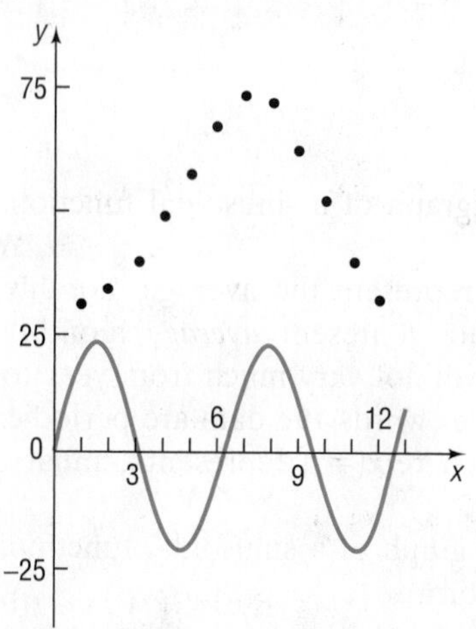

Figure 78

Step 2: Determine the vertical shift by finding the average of the highest and lowest data values.

$$\text{Vertical shift} = \frac{74.2 + 30.0}{2} = 52.1$$

Now superimpose the graph of $y = 22.1 \sin x + 52.1$ on the scatter diagram. See Figure 79 on the next page.

We see that the graph needs to be shifted horizontally and stretched horizontally.

Step 3: It is easier to find the horizontal stretch factor first. Since the temperatures repeat every 12 months, the period of the function is $T = 12$. Because $T = \dfrac{2\pi}{\omega} = 12$,

$$\omega = \frac{2\pi}{12} = \frac{\pi}{6}$$

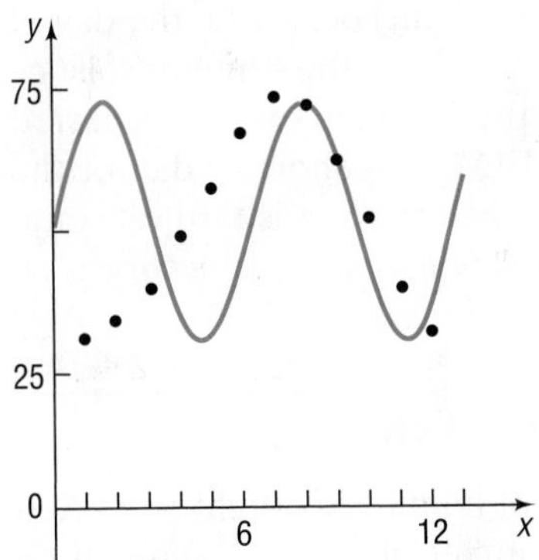

Figure 79

Now superimpose the graph of $y = 22.1 \sin\left(\frac{\pi}{6}x\right) + 52.1$ on the scatter diagram. See Figure 80, where it is clear that the graph still needs to be shifted horizontally.

STEP 4: To determine the horizontal shift, use the period $T = 12$ and divide the interval $[0, 12]$ into four subintervals of length $12 \div 4 = 3$:

$$[0, 3], \quad [3, 6], \quad [6, 9], \quad [9, 12]$$

The sine curve is increasing on the interval $(0, 3)$ and is decreasing on the interval $(3, 9)$, so a local maximum occurs at $x = 3$. The data indicate that a maximum occurs at $x = 7$ (corresponding to July's temperature), so the graph of the function must be shifted 4 units to the right by replacing x by $x - 4$. Doing this yields

$$y = 22.1 \sin\left(\frac{\pi}{6}(x - 4)\right) + 52.1$$

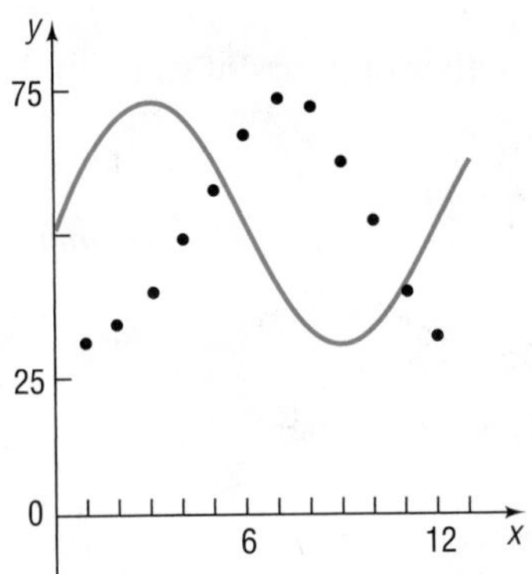

Figure 80

Multiplying out reveals that a sine function of the form $y = A \sin(\omega x - \phi) + B$ that fits the data is

$$y = 22.1 \sin\left(\frac{\pi}{6}x - \frac{2\pi}{3}\right) + 52.1$$

The graph of $y = 22.1 \sin\left(\frac{\pi}{6}x - \frac{2\pi}{3}\right) + 52.1$ and the scatter diagram of the data are shown in Figure 81.

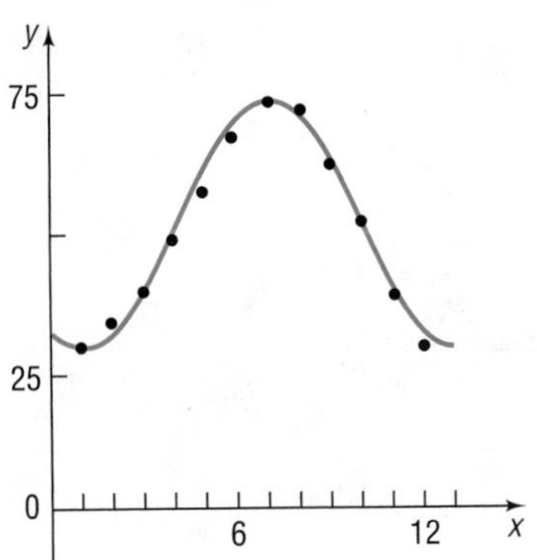

Figure 81

The steps to fit a sine function

$$y = A \sin(\omega x - \phi) + B$$

to sinusoidal data follow.

Steps for Fitting a Sine Function $y = A \sin(\omega x - \phi) + B$ to Data

STEP 1: Determine A, the amplitude of the function.

$$\text{Amplitude} = \frac{\text{largest data value} - \text{smallest data value}}{2}$$

STEP 2: Determine B, the vertical shift of the function.

$$\text{Vertical shift} = \frac{\text{largest data value} + \text{smallest data value}}{2}$$

STEP 3: Determine ω. Since the period T, the time it takes for the data to repeat, is $T = \frac{2\pi}{\omega}$, we have

$$\omega = \frac{2\pi}{T}$$

STEP 4: Determine the horizontal shift of the function by using the period of the data. Divide the period into four subintervals of equal length. Determine the x-coordinate for the maximum of the sine function and the x-coordinate for the maximum value of the data. Use this information to determine the value of the phase shift, $\frac{\phi}{\omega}$.

Now Work PROBLEMS 29(a)–(c)

Let's look at another example. Since the number of hours of sunlight in a day cycles annually, the number of hours of sunlight in a day for a given location can be modeled by a sinusoidal function.

The longest day of the year (in terms of hours of sunlight) occurs on the day of the summer solstice. For locations in the Northern Hemisphere, the summer solstice is the time when the Sun is farthest north. In 2014, the summer solstice occurred on June 21 (the 172nd day of the year) at 6:51 AM EDT. The shortest day of the year occurs on the day of the winter solstice, the time when the Sun is farthest south (for locations in the Northern Hemisphere). In 2014, the winter solstice occurred on December 21 (the 355th day of the year) at 6:03 PM (EST).

EXAMPLE 4

Finding a Sinusoidal Function for Hours of Daylight

According to the *Old Farmer's Almanac,* the number of hours of sunlight in Boston on the day of the summer solstice is 15.30, and the number of hours of sunlight on the day of the winter solstice is 9.08.

(a) Find a sinusoidal function of the form $y = A\sin(\omega x - \phi) + B$ that fits the data.

(b) Use the function found in part (a) to predict the number of hours of sunlight in Boston on April 1, the 91st day of the year.

(c) Graph the function found in part (a).

(d) Look up the number of hours of sunlight for April 1 in the *Old Farmer's Almanac* and compare it to the results found in part (b).

Source: The Old Farmer's Almanac, www.almanac.com/rise

Solution (a) **STEP 1:** Amplitude $= \dfrac{\text{largest data value} - \text{smallest data value}}{2}$

$$= \frac{15.30 - 9.08}{2} = 3.11$$

STEP 2: Vertical shift $= \dfrac{\text{largest data value} + \text{smallest data value}}{2}$

$$= \frac{15.30 + 9.08}{2} = 12.19$$

STEP 3: The data repeat every 365 days. Since $T = \dfrac{2\pi}{\omega} = 365$, we find

$$\omega = \frac{2\pi}{365}$$

So far, we have $y = 3.11\sin\left(\dfrac{2\pi}{365}x - \phi\right) + 12.19$.

STEP 4: To determine the horizontal shift, use the period $T = 365$ and divide the interval $[0, 365]$ into four subintervals of length $365 \div 4 = 91.25$:

$$[0, 91.25],\quad [91.25, 182.5],\quad [182.5, 273.75],\quad [273.75, 365]$$

The sine curve is increasing on the interval $(0, 91.25)$ and is decreasing on the interval $(91.25, 273.75)$, so a local maximum occurs at $x = 91.25$. Since the maximum occurs on the summer solstice at $x = 172$, we must shift the graph of the function $172 - 91.25 = 80.75$ units to the right by replacing x by $x - 80.75$. Doing this yields

$$y = 3.11\sin\left(\frac{2\pi}{365}(x - 80.75)\right) + 12.19$$

Next, multiply out to obtain a sine function of the form $y = A\sin(\omega x - \phi) + B$ that fits the data.

$$y = 3.11\sin\left(\frac{2\pi}{365}x - \frac{323\pi}{730}\right) + 12.19$$

(b) To predict the number of hours of daylight on April 1, let $x = 91$ in the function found in part (a) and obtain

$$y = 3.11 \sin\left(\frac{2\pi}{365} \cdot 91 - \frac{323}{730}\pi\right) + 12.19$$

$$\approx 12.74$$

The prediction is that there will be about 12.74 hours = 12 hours, 44 minutes of sunlight on April 1 in Boston.

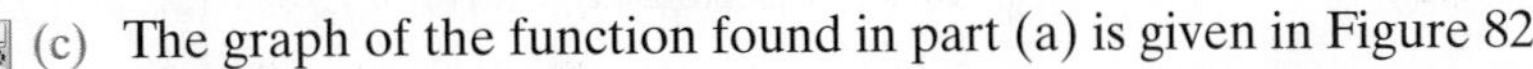

(c) The graph of the function found in part (a) is given in Figure 82.

(d) According to the *Old Farmer's Almanac*, there will be 12 hours 45 minutes of sunlight on April 1 in Boston. ●

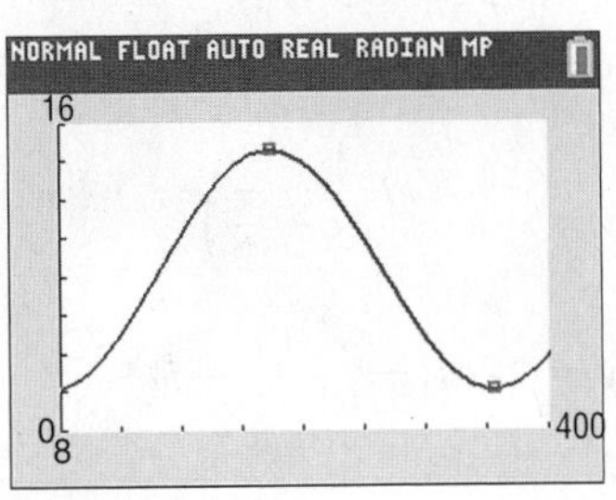

Figure 82

Now Work PROBLEM 35

Certain graphing utilities (such as the TI-83, TI-84 Plus C, and TI-89) have the capability of finding the sine function of best fit for sinusoidal data. At least four data points are required for this process.

EXAMPLE 5 Finding the Sine Function of Best Fit

Use a graphing utility to find the sine function of best fit for the data in Table 11. Graph this function with the scatter diagram of the data.

Solution Enter the data from Table 11 and execute the SINe REGression program. The result is shown in Figure 83.

The output that the graphing utility provides shows the equation

$$y = a \sin(bx + c) + d$$

The sinusoidal function of best fit is

$$y = 21.43 \sin(0.56x - 2.44) + 51.71$$

where x represents the month and y represents the average temperature.

Figure 84 shows the graph of the sinusoidal function of best fit on the scatter diagram.

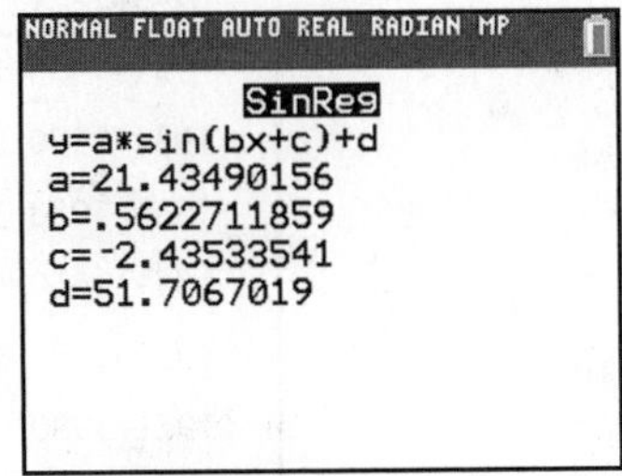

Figure 83

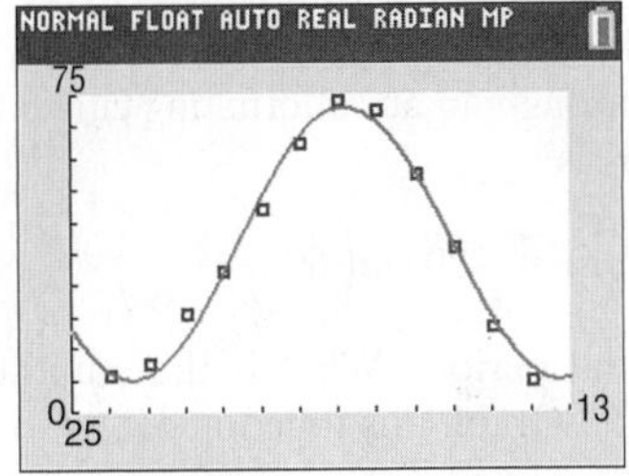

Figure 84 ●

Now Work PROBLEMS 29(d) AND (e)

2.6 Assess Your Understanding

Concepts and Vocabulary

1. For the graph of $y = A \sin(\omega x - \phi)$, the number $\frac{\phi}{\omega}$ is called the ______ ______.

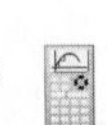

2. ***True or False*** A graphing utility requires only two data points to find the sine function of best fit.

Skill Building

In Problems 3–14, find the amplitude, period, and phase shift of each function. Graph each function. Be sure to label key points. Show at least two periods.

3. $y = 4\sin(2x - \pi)$
4. $y = 3\sin(3x - \pi)$
5. $y = 2\cos\left(3x + \frac{\pi}{2}\right)$
6. $y = 3\cos(2x + \pi)$
7. $y = -3\sin\left(2x + \frac{\pi}{2}\right)$
8. $y = -2\cos\left(2x - \frac{\pi}{2}\right)$
9. $y = 4\sin(\pi x + 2) - 5$
10. $y = 2\cos(2\pi x + 4) + 4$
11. $y = 3\cos(\pi x - 2) + 5$
12. $y = 2\cos(2\pi x - 4) - 1$
13. $y = -3\sin\left(-2x + \frac{\pi}{2}\right)$
14. $y = -3\cos\left(-2x + \frac{\pi}{2}\right)$

In Problems 15–18, write the equation of a sine function that has the given characteristics.

15. Amplitude: 2
Period: π
Phase shift: $\frac{1}{2}$

16. Amplitude: 3
Period: $\frac{\pi}{2}$
Phase shift: 2

17. Amplitude: 3
Period: 3π
Phase shift: $-\frac{1}{3}$

18. Amplitude: 2
Period: π
Phase shift: -2

Mixed Practice

In Problems 19–26, apply the methods of this and the previous section to graph each function. Be sure to label key points and show at least two periods.

19. $y = 2\tan(4x - \pi)$
20. $y = \frac{1}{2}\cot(2x - \pi)$
21. $y = 3\csc\left(2x - \frac{\pi}{4}\right)$
22. $y = \frac{1}{2}\sec(3x - \pi)$
23. $y = -\cot\left(2x + \frac{\pi}{2}\right)$
24. $y = -\tan\left(3x + \frac{\pi}{2}\right)$
25. $y = -\sec(2\pi x + \pi)$
26. $y = -\csc\left(-\frac{1}{2}\pi x + \frac{\pi}{4}\right)$

Applications and Extensions

27. **Alternating Current (ac) Circuits** The current I, in amperes, flowing through an ac (alternating current) circuit at time t, in seconds, is

$$I(t) = 120\sin\left(30\pi t - \frac{\pi}{3}\right) \qquad t \geq 0$$

What is the period? What is the amplitude? What is the phase shift? Graph this function over two periods.

28. **Alternating Current (ac) Circuits** The current I, in amperes, flowing through an ac (alternating current) circuit at time t, in seconds, is

$$I(t) = 220\sin\left(60\pi t - \frac{\pi}{6}\right) \qquad t \geq 0$$

What is the period? What is the amplitude? What is the phase shift? Graph this function over two periods.

29. **Hurricanes** Hurricanes are categorized using the Saffir-Simpson Hurricane Scale, with winds 111–130 miles per hour (mph) corresponding to a category 3 hurricane, winds 131–155 mph corresponding to a category 4 hurricane, and winds in excess of 155 mph corresponding to a category 5 hurricane. The following data represent the number of major hurricanes in the Atlantic Basin (category 3, 4, or 5) each decade from 1921 to 2010.
(a) Draw a scatter diagram of the data.
(b) Find a sinusoidal function of the form $y = A\sin(\omega x - \phi) + B$ that models the data.
(c) Draw the sinusoidal function found in part (b) on the scatter diagram.

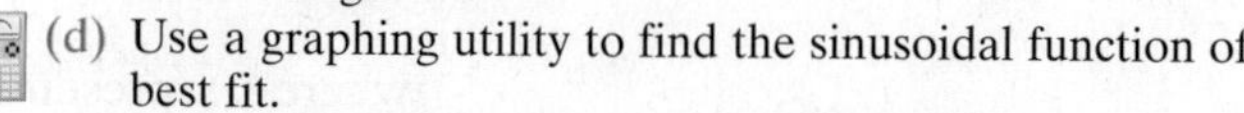

(d) Use a graphing utility to find the sinusoidal function of best fit.
(e) Graph the sinusoidal function of best fit on a scatter diagram of the data.

Decade, x	Major Hurricanes, H
1921–1930, 1	17
1931–1940, 2	16
1941–1950, 3	29
1951–1960, 4	33
1961–1970, 5	27
1971–1980, 6	16
1981–1990, 7	16
1991–2000, 8	27
2001–2010, 9	33

Source: U.S. National Oceanic and Atmospheric Administration

30. **Monthly Temperature** The data on the next page represent the average monthly temperatures for Washington, D.C.
(a) Draw a scatter diagram of the data for one period.
(b) Find a sinusoidal function of the form $y = A\sin(\omega x - \phi) + B$ that models the data.

(c) Draw the sinusoidal function found in part (b) on the scatter diagram.

(d) Use a graphing utility to find the sinusoidal function of best fit.

(e) Graph the sinusoidal function of best fit on a scatter diagram of the data.

Month, x	Average Monthly Temperature, °F
January, 1	36.0
February, 2	39.0
March, 3	46.8
April, 4	56.8
May, 5	66.0
June, 6	75.2
July, 7	79.8
August, 8	78.1
September, 9	71.0
October, 10	59.5
November, 11	49.6
December, 12	39.7

Source: U.S. National Oceanic and Atmospheric Administration

31. Monthly Temperature The following data represent the average monthly temperatures for Indianapolis, Indiana.

Month, x	Average Monthly Temperature, °F
January, 1	28.1
February, 2	32.1
March, 3	42.2
April, 4	53.0
May, 5	62.7
June, 6	72.0
July, 7	75.4
August, 8	74.2
September, 9	66.9
October, 10	55.0
November, 11	43.6
December, 12	31.6

Source: U.S. National Oceanic and Atmospheric Administration

(a) Draw a scatter diagram of the data for one period.

(b) Find a sinusoidal function of the form

$$y = A\sin(\omega x - \phi) + B$$

that models the data.

(c) Draw the sinusoidal function found in part (b) on the scatter diagram.

(d) Use a graphing utility to find the sinusoidal function of best fit.

(e) Graph the sinusoidal function of best fit on a scatter diagram of the data.

32. Monthly Temperature The following data represent the average monthly temperatures for Baltimore, Maryland.

(a) Draw a scatter diagram of the data for one period.

(b) Find a sinusoidal function of the form $y = A\sin(\omega x - \phi) + B$ that models the data.

(c) Draw the sinusoidal function found in part (b) on the scatter diagram.

(d) Use a graphing utility to find the sinusoidal function of best fit.

(e) Graph the sinusoidal function of best fit on a scatter diagram of the data.

Month, x	Average Monthly Temperature, °F
January, 1	32.9
February, 2	35.8
March, 3	43.6
April, 4	53.7
May, 5	62.9
June, 6	72.4
July, 7	77.0
August, 8	75.1
September, 9	67.8
October, 10	56.1
November, 11	46.5
December, 12	36.7

Source: U.S. National Oceanic and Atmospheric Administration

33. Tides The length of time between consecutive high tides is 12 hours and 25 minutes. According to the National Oceanic and Atmospheric Administration, on Saturday, April 26, 2014, in Charleston, South Carolina, high tide occurred at 6:30 AM (6.5 hours) and low tide occurred at 12:24 PM (12.4 hours). Water heights are measured as the amounts above or below the mean lower low water. The height of the water at high tide was 5.86 feet, and the height of the water at low tide was −0.38 foot.

(a) Approximately when will the next high tide occur?

(b) Find a sinusoidal function of the form

$$y = A\sin(\omega x - \phi) + B$$

that models the data.

(c) Use the function found in part (b) to predict the height of the water at 3 PM on April 26, 2014.

34. Tides The length of time between consecutive high tides is 12 hours and 25 minutes. According to the National Oceanic and Atmospheric Administration, on Saturday, April 26, 2014, in Sitka, Alaska, high tide occurred at 12:06 AM (0.10 hours) and low tide occurred at 6:24 AM (6.4 hours). Water heights are measured as the amounts above or below the mean lower low water. The height of the water at high tide was 9.97 feet, and the height of the water at low tide was 0.59 feet.

(a) Approximately when will the next high tide occur?

(b) Find a sinusoidal function of the form

$$y = A\sin(\omega x - \phi) + B$$

that models the data.

(c) Use the function found in part (b) to predict the height of the water at 6 PM.

35. **Hours of Daylight** According to the *Old Farmer's Almanac*, in Miami, Florida, the number of hours of sunlight on the summer solstice of 2014 was 13.75, and the number of hours of sunlight on the winter solstice was 10.52.
(a) Find a sinusoidal function of the form
$$y = A\sin(\omega x - \phi) + B$$
that models the data.
(b) Use the function found in part (a) to predict the number of hours of sunlight on April 1, the 91st day of the year.
(c) Draw a graph of the function found in part (a).
(d) Look up the number of hours of sunlight for April 1 in the *Old Farmer's Almanac*, and compare the actual hours of daylight to the results found in part (b).

36. **Hours of Daylight** According to the *Old Farmer's Almanac*, in Detroit, Michigan, the number of hours of sunlight on the summer solstice of 2014 was 15.27, and the number of hours of sunlight on the winter solstice was 9.07.
(a) Find a sinusoidal function of the form
$$y = A\sin(\omega x - \phi) + B$$
that models the data.
(b) Use the function found in part (a) to predict the number of hours of sunlight on April 1, the 91st day of the year.
(c) Draw a graph of the function found in part (a).
(d) Look up the number of hours of sunlight for April 1 in the *Old Farmer's Almanac*, and compare the actual hours of daylight to the results found in part (b).

37. **Hours of Daylight** According to the *Old Farmer's Almanac*, in Anchorage, Alaska, the number of hours of sunlight on the summer solstice of 2014 was 19.37, and the number of hours of sunlight on the winter solstice was 5.45.
(a) Find a sinusoidal function of the form
$$y = A\sin(\omega x - \phi) + B$$
that models the data.
(b) Use the function found in part (a) to predict the number of hours of sunlight on April 1, the 91st day of the year.
(c) Draw a graph of the function found in part (a).
(d) Look up the number of hours of sunlight for April 1 in the *Old Farmer's Almanac*, and compare the actual hours of daylight to the results found in part (b).

38. **Hours of Daylight** According to the *Old Farmer's Almanac*, in Honolulu, Hawaii, the number of hours of sunlight on the summer solstice of 2014 was 13.42, and the number of hours of sunlight on the winter solstice was 10.83.
(a) Find a sinusoidal function of the form
$$y = A\sin(\omega x - \phi) + B$$
that models the data.
(b) Use the function found in part (a) to predict the number of hours of sunlight on April 1, the 91st day of the year.
(c) Draw a graph of the function found in part (a).
(d) Look up the number of hours of sunlight for April 1 in the *Old Farmer's Almanac*, and compare the actual hours of daylight to the results found in part (b).

Discussion and Writing

39. Explain how the amplitude and period of a sinusoidal graph are used to establish the scale on each coordinate axis.

40. Find an application in your major field that leads to a sinusoidal graph. Write an account of your findings.

Retain Your Knowledge

Problems 41–44 are based on material learned earlier in the course. The purpose of these problems is to keep the material fresh in your mind so that you are better prepared for the final exam.

41. Use a graphing utility to approximate the local maxima and local minima of $f(x) = 2x^3 - 3x^2 - 8x + 14$.

42. Find the average rate of change of $f(x) = -16x^2 + 5x$ on the interval $[-1, 2]$.

43. Convert the angle $\theta = \dfrac{7\pi}{12}$ to degrees.

44. Given $g(x) = \begin{cases} -3x + 7 & x \geq 0 \\ x^2 - 9 & x < 0 \end{cases}$, find all values for x where $g(x) = -5$.

Chapter Review

Things to Know

Definitions

Angle in standard position (p. 100)
Vertex is at the origin; initial side is along the positive x-axis.

1 Degree (1°) (p. 101)
$1° = \dfrac{1}{360}$ revolution

1 Radian (p. 103)
The measure of a central angle of a circle whose rays subtend an arc that is the same length as the radius of the circle.

Trigonometric functions (p. 115)
$P = (x, y)$ is the point on the unit circle corresponding to $\theta = t$ radians.

$\sin t = \sin\theta = y$ $\quad \cos t = \cos\theta = x$ $\quad \tan t = \tan\theta = \dfrac{y}{x} \quad x \neq 0$

$\csc t = \csc\theta = \dfrac{1}{y} \quad y \neq 0$ $\quad \sec t = \sec\theta = \dfrac{1}{x} \quad x \neq 0$ $\quad \cot t = \cot\theta = \dfrac{x}{y} \quad y \neq 0$

Trigonometric functions using a circle of radius r (p. 125)	For an angle θ in standard position, $P = (x, y)$ is the point on the terminal side of θ that is also on the circle $x^2 + y^2 = r^2$.
	$\sin\theta = \frac{y}{r}$ $\quad$ $\cos\theta = \frac{x}{r}$ $\quad$ $\tan\theta = \frac{y}{x}\quad x \neq 0$
	$\csc\theta = \frac{r}{y}\quad y \neq 0$ $\quad$ $\sec\theta = \frac{r}{x}\quad x \neq 0$ $\quad$ $\cot\theta = \frac{x}{y}\quad y \neq 0$
Periodic function (p. 134)	$f(\theta + p) = f(\theta)$, for all θ, $p > 0$, where the smallest such p is the fundamental period.

Formulas

1 counterclockwise revolution $= 360°$ (p. 101–102) $= 2\pi$ radians (p. 104)	$1° = \frac{\pi}{180}$ radian (p. 105); 1 radian $= \frac{180}{\pi}$ degrees (p. 105)
Arc length: $s = r\theta$ (p. 104)	θ is measured in radians; s is the length of the arc subtended by the central angle θ of the circle of radius r.
Area of a sector: $A = \frac{1}{2}r^2\theta$ (p. 107)	A is the area of the sector of a circle of radius r formed by a central angle of θ radians.
Linear speed: $v = \frac{s}{t}$ (p. 108)	v is the linear speed along the circle of radius r; ω is the angular speed (measured in radians per unit time).
Angular speed: $\omega = \frac{\theta}{t}$ (p. 108) $\quad v = r\omega$ (p. 108)	

Table of Values (pp. 118 and 122)

θ (Radians)	θ (Degrees)	$\sin\theta$	$\cos\theta$	$\tan\theta$	$\csc\theta$	$\sec\theta$	$\cot\theta$
0	0°	0	1	0	Not defined	1	Not defined
$\frac{\pi}{6}$	30°	$\frac{1}{2}$	$\frac{\sqrt{3}}{2}$	$\frac{\sqrt{3}}{3}$	2	$\frac{2\sqrt{3}}{3}$	$\sqrt{3}$
$\frac{\pi}{4}$	45°	$\frac{\sqrt{2}}{2}$	$\frac{\sqrt{2}}{2}$	1	$\sqrt{2}$	$\sqrt{2}$	1
$\frac{\pi}{3}$	60°	$\frac{\sqrt{3}}{2}$	$\frac{1}{2}$	$\sqrt{3}$	$\frac{2\sqrt{3}}{3}$	2	$\frac{\sqrt{3}}{3}$
$\frac{\pi}{2}$	90°	1	0	Not defined	1	Not defined	0
π	180°	0	−1	0	Not defined	−1	Not defined
$\frac{3\pi}{2}$	270°	−1	0	Not defined	−1	Not defined	0

The Unit Circle (pp. 122–123)

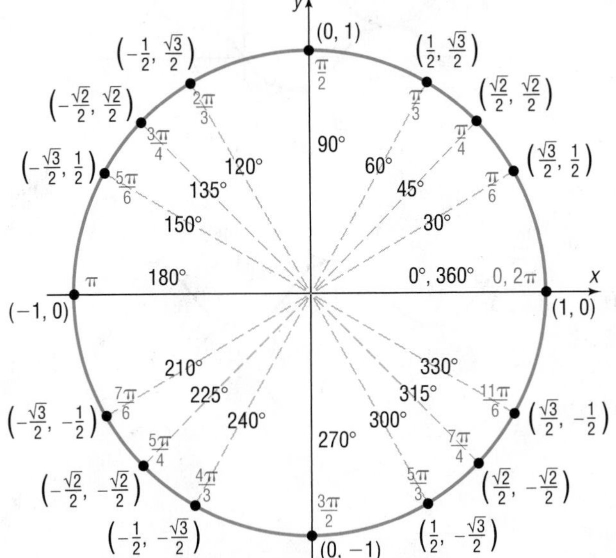

Fundamental Identities (pp. 136–137)

$$\tan\theta = \frac{\sin\theta}{\cos\theta} \qquad \cot\theta = \frac{\cos\theta}{\sin\theta}$$

$$\csc\theta = \frac{1}{\sin\theta} \qquad \sec\theta = \frac{1}{\cos\theta} \qquad \cot\theta = \frac{1}{\tan\theta}$$

$$\sin^2\theta + \cos^2\theta = 1 \qquad \tan^2\theta + 1 = \sec^2\theta \qquad \cot^2\theta + 1 = \csc^2\theta$$

Properties of the Trigonometric Functions (pp. 131–134, 141)

$y = \sin x$ (p. 146)

Domain: $-\infty < x < \infty$

Range: $-1 \le y \le 1$

Periodic: period $= 2\pi\,(360°)$

Odd function

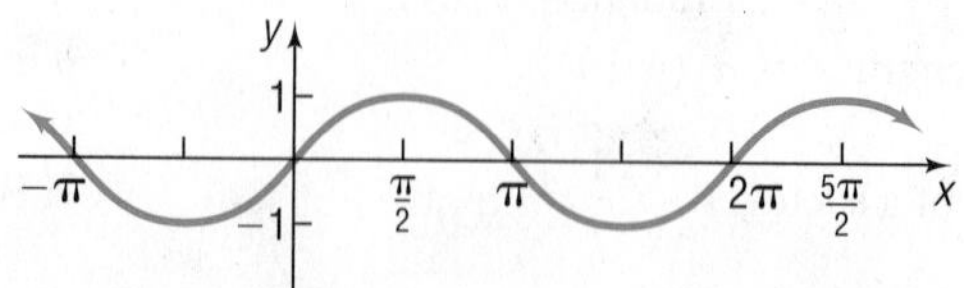

$y = \cos x$ (p. 148)

Domain: $-\infty < x < \infty$

Range: $-1 \le y \le 1$

Periodic: period $= 2\pi\,(360°)$

Even function

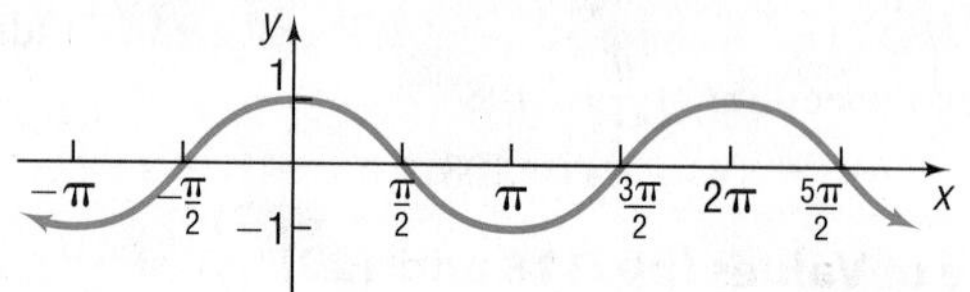

$y = \tan x$ (pp. 161–162)

Domain: $-\infty < x < \infty$, except odd integer multiples of $\frac{\pi}{2}$ (90°)

Range: $-\infty < y < \infty$

Periodic: period $= \pi\,(180°)$

Odd function

Vertical asymptotes at odd integer multiples of $\frac{\pi}{2}$

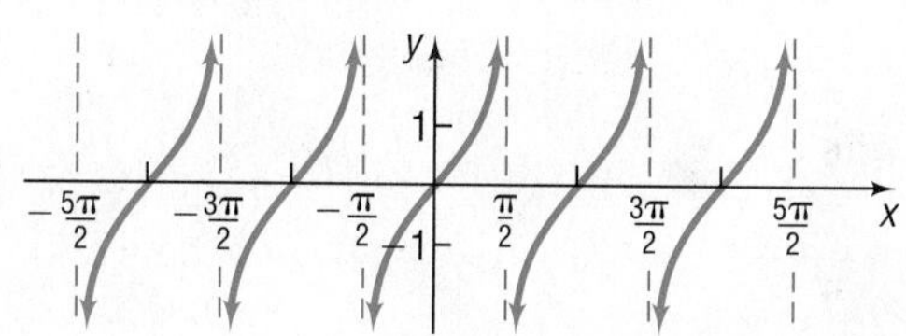

$y = \cot x$ (p. 163)

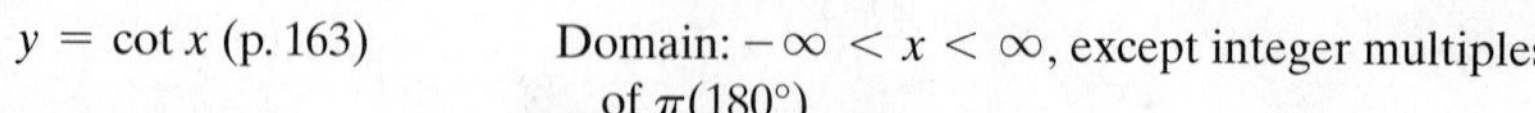

Domain: $-\infty < x < \infty$, except integer multiples of π(180°)

Range: $-\infty < y < \infty$

Periodic: period $= \pi\,(180°)$

Odd function

Vertical asymptotes at integer multiples of π

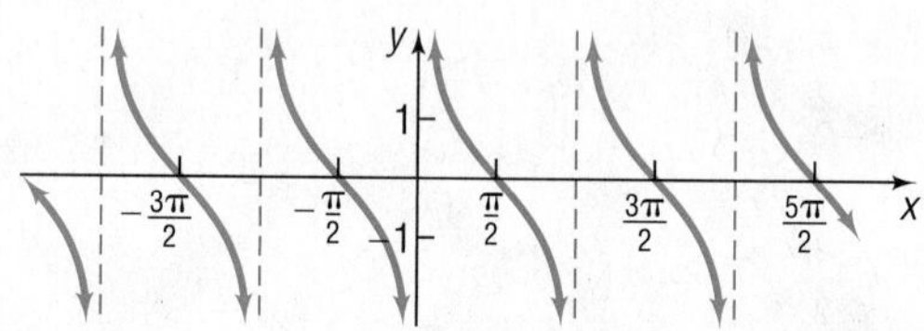

$y = \csc x$ (p. 164)

Domain: $-\infty < x < \infty$, except integer multiples of $\pi\,(180°)$

Range: $|y| \ge 1$ ($y \le -1$ or $y \ge 1$)

Periodic: period $= 2\pi\,(360°)$

Odd function

Vertical asymptotes at integer multiples of π

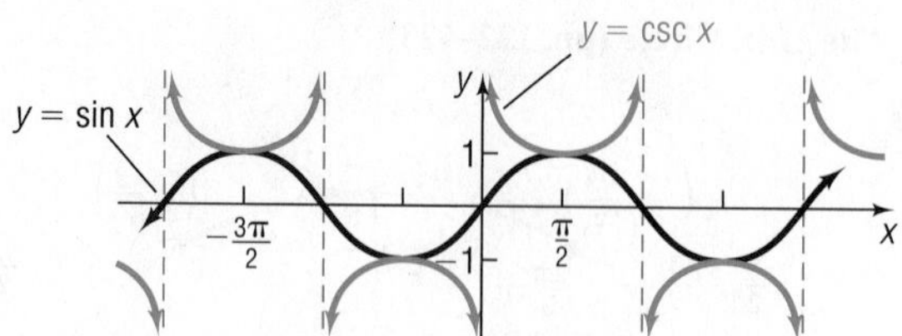

$y = \sec x$ (p. 164)

Domain: $-\infty < x < \infty$, except odd integer multiples of $\frac{\pi}{2}$ (90°)

Range: $|y| \ge 1$ ($y \le -1$ or $y \ge 1$)

Periodic: period $= 2\pi\,(360°)$

Even function

Vertical asymptotes at odd integer multiples of $\frac{\pi}{2}$

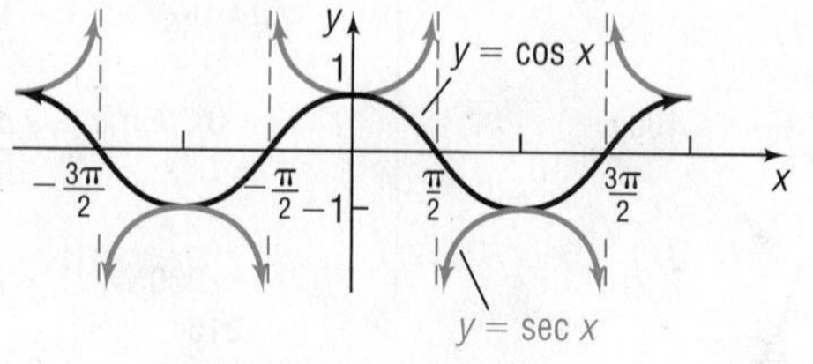

Sinusoidal Graphs

$y = A\sin(\omega x) + B,\quad \omega > 0$	Period $= \dfrac{2\pi}{\omega}$ (pp. 151, 168)
$y = A\cos(\omega x) + B,\quad \omega > 0$	Amplitude $= \lvert A \rvert$ (pp. 151, 168)
$y = A\sin(\omega x - \phi) + B = A\sin\left[\omega\left(x - \dfrac{\phi}{\omega}\right)\right] + B$	Phase shift $= \dfrac{\phi}{\omega}$ (p. 168)
$y = A\cos(\omega x - \phi) + B = A\cos\left[\omega\left(x - \dfrac{\phi}{\omega}\right)\right] + B$	

Objectives

Section		You should be able to...	Example(s)	Review Exercises
2.1	1	Convert between decimals and degrees, minutes, seconds measures for angles (p. 102)	2	48
	2	Find the length of an arc of a circle (p. 103)	3	49, 50
	3	Convert from degrees to radians and from radians to degrees (p. 104)	4–6	1–4
	4	Find the area of a sector of a circle (p. 107)	7	49
	5	Find the linear speed of an object traveling in circular motion (p. 108)	8	51, 52
2.2	1	Find the exact values of the trigonometric functions using a point on the unit circle (p. 116)	1	45, 55
	2	Find the exact values of the trigonometric functions of quadrantal angles (p. 117)	2, 3	9, 55
	3	Find the exact values of the trigonometric functions of $\frac{\pi}{4} = 45°$ (p. 119)	4, 5	5, 6
	4	Find the exact values of the trigonometric functions of $\frac{\pi}{6} = 30°$ and $\frac{\pi}{3} = 60°$ (p. 120)	6–8	5, 6
	5	Find the exact values of the trigonometric functions for integer multiples of $\frac{\pi}{6} = 30°$, $\frac{\pi}{4} = 45°$, and $\frac{\pi}{3} = 60°$ (p. 122)	9, 10	7, 8, 10, 55
	6	Use a calculator to approximate the value of a trigonometric function (p. 124)	11	41, 42
	7	Use a circle of radius r to evaluate the trigonometric functions (p. 125)	12	46
2.3	1	Determine the domain and the range of the trigonometric functions (p. 131)	pp. 131–133	47
	2	Determine the period of the trigonometric functions (p. 133)	1	47
	3	Determine the signs of the trigonometric functions in a given quadrant (p. 135)	2	43, 44
	4	Find the values of the trigonometric functions using fundamental identities (p. 136)	3, 4	11–15
	5	Find the exact values of the trigonometric functions of an angle given one of the functions and the quadrant of the angle (p. 138)	5, 6	16–23
	6	Use even–odd properties to find the exact values of the trigonometric functions (p. 141)	7	13–15
2.4	1	Graph functions of the form $y = A\sin(\omega x)$ using transformations (p. 146)	1, 2	24
	2	Graph functions of the form $y = A\cos(\omega x)$ using transformations (p. 148)	3	25

(continued)

Section		Objectives	Examples	Review Exercises
	3	Determine the amplitude and period of sinusoidal functions (p. 149)	4	33–38
	4	Graph sinusoidal functions using key points (p. 151)	5–7	24, 25, 35
	5	Find an equation for a sinusoidal graph (p. 154)	8, 9	39, 40
2.5	1	Graph functions of the form $y = A\tan(\omega x) + B$ and $y = A\cot(\omega x) + B$ (p. 162)	1, 2	27
	2	Graph functions of the form $y = A\csc(\omega x) + B$ and $y = A\sec(\omega x) + B$ (p. 164)	3	29
2.6	1	Graph sinusoidal functions of the form $y = A\sin(\omega x - \phi) + B$ (p. 168)	1, 2	31, 35–38, 53(d)
	2	Build sinusoidal models from data (p. 171)	3–5	54

Review Exercises

In Problems 1 and 2, convert each angle in degrees to radians. Express your answer as a multiple of π.

1. $135°$ **2.** $18°$

In Problems 3 and 4, convert each angle in radians to degrees.

3. $\frac{3\pi}{4}$ **4.** $-\frac{5\pi}{2}$

In Problems 5–15, find the exact value of each expression. Do not use a calculator.

5. $\tan\frac{\pi}{4} - \sin\frac{\pi}{6}$ **6.** $3\sin 45° - 4\tan\frac{\pi}{6}$ **7.** $6\cos\frac{3\pi}{4} + 2\tan\left(-\frac{\pi}{3}\right)$

8. $\sec\left(-\frac{\pi}{3}\right) - \cot\left(-\frac{5\pi}{4}\right)$ **9.** $\tan\pi + \sin\pi$ **10.** $\cos 540° - \tan(-405°)$

11. $\sin^2 20° + \frac{1}{\sec^2 20°}$ **12.** $\sec 50° \cos 50°$ **13.** $\frac{\cos(-40°)}{\cos 40°}$

14. $\frac{\sin(-40°)}{\sin 40°}$ **15.** $\sin 310° \csc(-50°)$

In Problems 16–23, find the exact value of each of the remaining trigonometric functions.

16. $\sin\theta = \frac{4}{5}$, θ is acute **17.** $\tan\theta = \frac{12}{5}$, $\sin\theta < 0$ **18.** $\sec\theta = -\frac{5}{4}$, $\tan\theta < 0$

19. $\sin\theta = \frac{12}{13}$, θ in quadrant II **20.** $\sin\theta = -\frac{5}{13}$, $\frac{3\pi}{2} < \theta < 2\pi$ **21.** $\tan\theta = \frac{1}{3}$, $180° < \theta < 270°$

22. $\sec\theta = 3$, $\frac{3\pi}{2} < \theta < 2\pi$ **23.** $\cot\theta = -2$, $\frac{\pi}{2} < \theta < \pi$

In Problems 24–32, graph each function. Each graph should contain at least two periods. Use the graph to determine the domain and the range of each function.

24. $y = 2\sin(4x)$ **25.** $y = -3\cos(2x)$ **26.** $y = \tan(x + \pi)$

27. $y = -2\tan(3x)$ **28.** $y = \cot\left(x + \frac{\pi}{4}\right)$ **29.** $y = 4\sec(2x)$

30. $y = \csc\left(x + \frac{\pi}{4}\right)$ **31.** $y = 4\sin(2x + 4) - 2$ **32.** $y = 5\cot\left(\frac{x}{3} - \frac{\pi}{4}\right)$

In Problems 33 and 34, determine the amplitude and period of each function without graphing.

33. $y = \sin(2x)$ **34.** $y = -2\cos(3\pi x)$

In Problems 35–38, find the amplitude, period, and phase shift of each function. Graph each function. Show at least two periods.

35. $y = 4\sin(3x)$ **36.** $y = -\cos\left(\frac{1}{2}x + \frac{\pi}{2}\right)$ **37.** $y = \frac{1}{2}\sin\left(\frac{3}{2}x - \pi\right)$ **38.** $y = -\frac{2}{3}\cos(\pi x - 6)$

In Problems 39 and 40, find a function whose graph is given.

39.

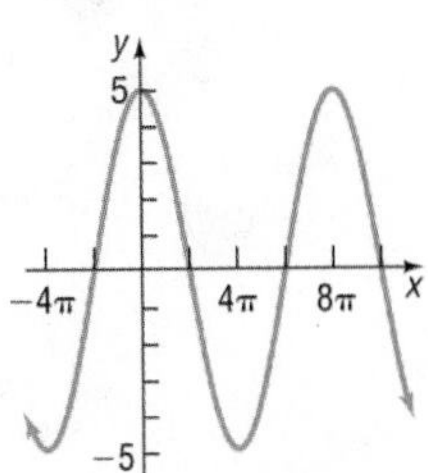

40.

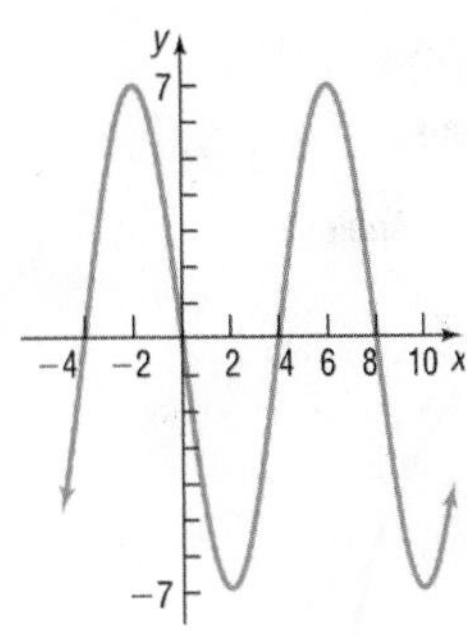

41. Use a calculator to approximate $\sin \frac{\pi}{8}$. Round the answer to two decimal places.

42. Use a calculator to approximate sec 10°. Round the answer to two decimal places.

43. Determine the signs of the six trigonometric functions of an angle θ whose terminal side is in quadrant III.

44. Name the quadrant θ lies in if $\cos \theta > 0$ and $\tan \theta < 0$.

45. Find the exact values of the six trigonometric functions of t if $P = \left(-\frac{1}{3}, \frac{2\sqrt{2}}{3}\right)$ is the point on the unit circle that corresponds to t.

46. Find the exact value of $\sin t$, $\cos t$, and $\tan t$ if $P = (-2, 5)$ is the point on the circle that corresponds to t.

47. What are the domain and the range of the secant function? What is the period?

48. (a) Convert the angle 32°20′35″ to a decimal in degrees. Round the answer to two decimal places.
(b) Convert the angle 63.18° to D°M′S″ form. Express the answer to the nearest second.

49. Find the length of the arc subtended by a central angle of 30° on a circle of radius 2 feet. What is the area of the sector?

50. The minute hand of a clock is 8 inches long. How far does the tip of the minute hand move in 30 minutes? How far does it move in 20 minutes?

51. Angular Speed of a Race Car A race car is driven around a circular track at a constant speed of 180 miles per hour. If the diameter of the track is $\frac{1}{2}$ mile, what is the angular speed of the car? Express your answer in revolutions per hour (which is equivalent to laps per hour).

52. Lighthouse Beacons The Montauk Point Lighthouse on Long Island has dual beams (two light sources opposite each other). Ships at sea observe a blinking light every 5 seconds. What rotation speed is required to achieve this?

53. Alternating Current The current I, in amperes, flowing through an ac (alternating current) circuit at time t is

$$I(t) = 220 \sin\left(30\pi t + \frac{\pi}{6}\right), \quad t \geq 0$$

(a) What is the period?
(b) What is the amplitude?
(c) What is the phase shift?
(d) Graph this function over two periods.

54. Monthly Temperature The data below represent the average monthly temperatures for Phoenix, Arizona.
(a) Draw a scatter diagram of the data for one period.
(b) Find a sinusoidal function of the form $y = A \sin(\omega x - \phi) + B$ that fits the data.
(c) Draw the sinusoidal function found in part (b) on the scatter diagram.

(d) Use a graphing utility to find the sinusoidal function of best fit.
(e) Graph the sinusoidal function of best fit on the scatter diagram.

Month, m	Average Monthly Temperature, T
January, 1	56
February, 2	60
March, 3	65
April, 4	73
May, 5	82
June, 6	91
July, 7	95
August, 8	94
September, 9	88
October, 10	77
November, 11	64
December, 12	55

Source: U.S. National Oceanic and Atmospheric Administration

55. Unit Circle On the given unit circle, fill in the missing angles ($0 \le \theta \le 2\pi$) and the corresponding points P.

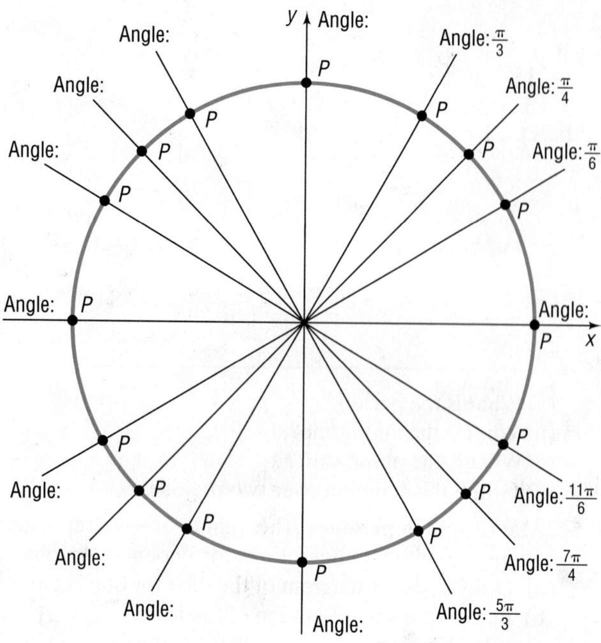

Chapter Test

CHAPTER

The Chapter Test Prep Videos are step-by-step solutions available in MyMathLab®, or on this text's YouTube Channel. Flip back to the Resources for Success page for a link to this text's YouTube channel.

In Problems 1–3, convert each angle in degrees to radians. Express your answer as a multiple of π.

1. $260°$ **2.** $-400°$ **3.** $13°$

In Problems 4–6, convert each angle in radians to degrees.

4. $-\frac{\pi}{8}$ **5.** $\frac{9\pi}{2}$ **6.** $\frac{3\pi}{4}$

In Problems 7–12, find the exact value of each expression.

7. $\sin\frac{\pi}{6}$ **8.** $\cos\left(-\frac{5\pi}{4}\right) - \cos\frac{3\pi}{4}$

9. $\cos(-120°)$ **10.** $\tan 330°$

11. $\sin\frac{\pi}{2} - \tan\frac{19\pi}{4}$ **12.** $2\sin^2 60° - 3\cos 45°$

In Problems 13–16, use a calculator to evaluate each expression. Round your answer to three decimal places.

13. $\sin 17°$

14. $\cos\frac{2\pi}{5}$

15. $\sec 229°$

16. $\cot\frac{28\pi}{9}$

17. Fill in each table entry with the sign of each function.

	sin θ	cos θ	tan θ	sec θ	csc θ	cot θ
θ in QI						
θ in QII						
θ in QIII						
θ in QIV						

18. If $f(x) = \sin x$ and $f(a) = \frac{3}{5}$, find $f(-a)$.

In Problems 19–21, find the value of the remaining five trigonometric functions of θ.

19. $\sin\theta = \frac{5}{7}$, θ in quadrant II **20.** $\cos\theta = \frac{2}{3}$, $\frac{3\pi}{2} < \theta < 2\pi$

21. $\tan\theta = -\frac{12}{5}$, $\frac{\pi}{2} < \theta < \pi$

In Problems 22–24, the point (x, y) *is on the terminal side of angle* θ *in standard position. Find the exact value of the given trigonometric function.*

22. $(2, 7)$, $\sin\theta$ **23.** $(-5, 11)$, $\cos\theta$

24. $(6, -3)$, $\tan\theta$

In Problems 25 and 26, graph the function.

25. $y = 2\sin\left(\frac{x}{3} - \frac{\pi}{6}\right)$ **26.** $y = \tan\left(-x + \frac{\pi}{4}\right) + 2$

27. Write an equation for a sinusoidal graph with the following properties:

$$A = -3 \quad \text{period} = \frac{2\pi}{3} \quad \text{phase shift} = -\frac{\pi}{4}$$

28. Logan has a garden in the shape of a sector of a circle; the outer rim of the garden is 25 feet long and the central angle of the sector is 50°. She wants to add a 3-foot-wide walk to the outer rim. How many square feet of paving blocks will she need to build the walk?

29. Hungarian Adrian Annus won the gold medal for the hammer throw at the 2004 Olympics in Athens with a winning distance of 83.19 meters.* The event consists of swinging a 16-pound weight attached to a wire 190 centimeters long in a circle and then releasing it. Assuming his release is at a 45° angle to the ground, the hammer will travel a distance of $\frac{v_0^2}{g}$ meters, where $g = 9.8$ meters/second^2 and v_0 is the linear speed of the hammer when released. At what rate (rpm) was he swinging the hammer upon release?

*Annus was stripped of his medal after refusing to cooperate with postmedal drug testing.

Cumulative Review

1. Find the real solutions, if any, of the equation $2x^2 + x - 1 = 0$.

2. Find an equation for the line with slope -3 containing the point $(-2, 5)$.

3. Find an equation for a circle of radius 4 and center at the point $(0, -2)$.

4. Discuss the equation $2x - 3y = 12$. Graph it.

5. Discuss the equation $x^2 + y^2 - 2x + 4y - 4 = 0$. Graph it.

6. Use transformations to graph the function $y = (x - 3)^2 + 2$.

7. Sketch a graph of each of the following functions. Label at least three points on each graph.

(a) $y = x^2$ (b) $y = x^3$

(c) $y = \sin x$ (d) $y = \tan x$

8. Find the inverse function of $f(x) = 3x - 2$.

9. Find the exact value of $(\sin 14°)^2 + (\cos 14°)^2 - 3$.

10. Graph $y = 3\sin(2x)$.

11. Find the exact value of $\tan\frac{\pi}{4} - 3\cos\frac{\pi}{6} + \csc\frac{\pi}{6}$.

12. Find a sinusoidal function for the following graph.

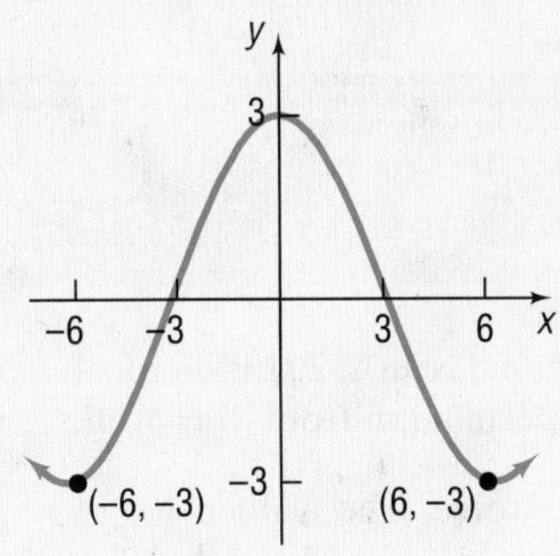

Chapter Projects

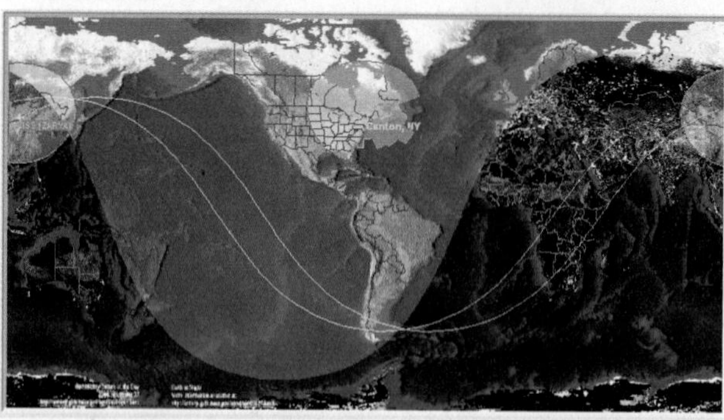

Internet-based Project

I. Length of Day Go to *http://en.wikipedia.org/wiki/latitude* and read about latitude. Then go to *http://www.orchidculture.com/COD/daylength.html.*

1. For a particular latitude, record in a table the length of day for the various days of the year. For January 1, use 1 as the day, for January 16, use 16 as the day, for February 1, use 32 as the day, and so on. Enter the data into an Excel spreadsheet using column B for the day of the year and column C for the length of day.
2. Draw a scatter diagram of the data with day of the year as the independent variable and length of day as the dependent variable using Excel.
3. Determine the sinusoidal function of best fit, $y = A\sin(Bx + C) + D$, as follows:
   (a) Enter initial guesses for the values of A, B, C, and D into column A with the value of A in cell A1, B in cell A2, C in cell A3, and D in cell A4.
   (b) Enter "=A$1*sin(A$2*B1+A$3)+A$4" into cell D1. Copy this cell entry into the cells below D1 for as many rows as there are data. For example, if column C goes to row 23, then column D should also go to row 23.
   (c) Enter "=(D1−C1)^2" into cell E1. Copy this entry below as described in part 3(b).
   (d) The idea behind curve fitting is to make the sum of the squared differences between what is predicted and actual observations as small as possible. Enter "=sum(E1..E#)" into cell A6, where # represents the row number of the last data point. For example, if you have 23 rows of data, enter "=sum(E1..E23)" in cell A6.
   (e) Now, install the Solver feature of Excel. To do this, click the File tab, and then select Options. Select Add-Ins. In the drop-down menu entitled "Manage," choose Excel Add-ins, and then click Go. . . . Check the box entitled "Solver Add-in" and click OK. The Solver add-in is now available in the Data tab. Choose Solver. Fill in the screen below:

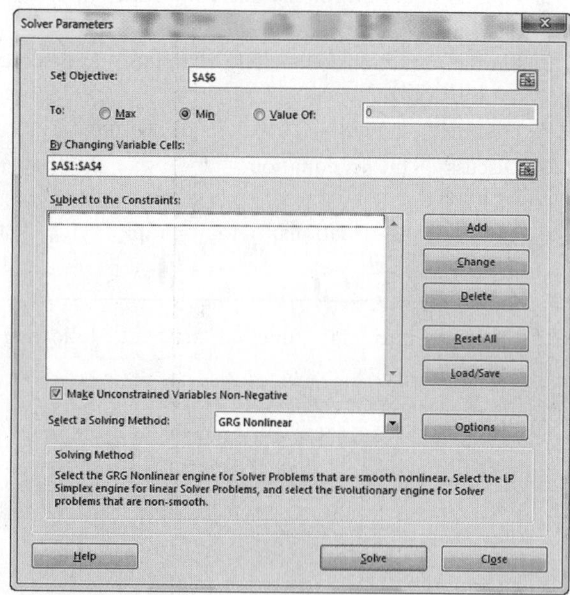

Click Solve. The values for A, B, C, and D are located in cells A1–A4. What is the sinusoidal function of best fit?

(continued)

4. Determine the longest day of the year according to your model. What is the day length on the longest day of the year? Determine the shortest day of the year according to your model. What is the day length on the shortest day of the year?
5. On which days is the day length exactly 12 hours according to your model?
6. Look up the day on which the vernal equinox and autumnal equinox occur. How do they match up with the results obtained in part 5?
7. Do you think your model accurately describes the relation between day of the year and length of day?
8. Use your model to predict the hours of daylight for the latitude you selected for various days of the year. Go to the *Old Farmer's Almanac* or other website (such as *http://astro.unl.edu/classaction/animations/coordsmotion/daylighthoursexplorer.html*) to determine the hours of daylight for the latitude you selected. How do the two compare?

The following projects are available on the Instructor's Resource Center (IRC):

II. Tides Data from a tide table are used to build a sine function that models tides.

III. Project at Motorola *Digital Transmission over the Air* Learn how Motorola Corporation transmits digital sequences by modulating the phase of the carrier waves.

IV. Identifying Mountain Peaks in Hawaii The visibility of a mountain is affected by its altitude, its distance from the viewer, and the curvature of Earth's surface. Trigonometry can be used to determine whether a distant object can be seen.

V. CBL Experiment Technology is used to model and study the effects of damping on sound waves.

3 Analytic Trigonometry

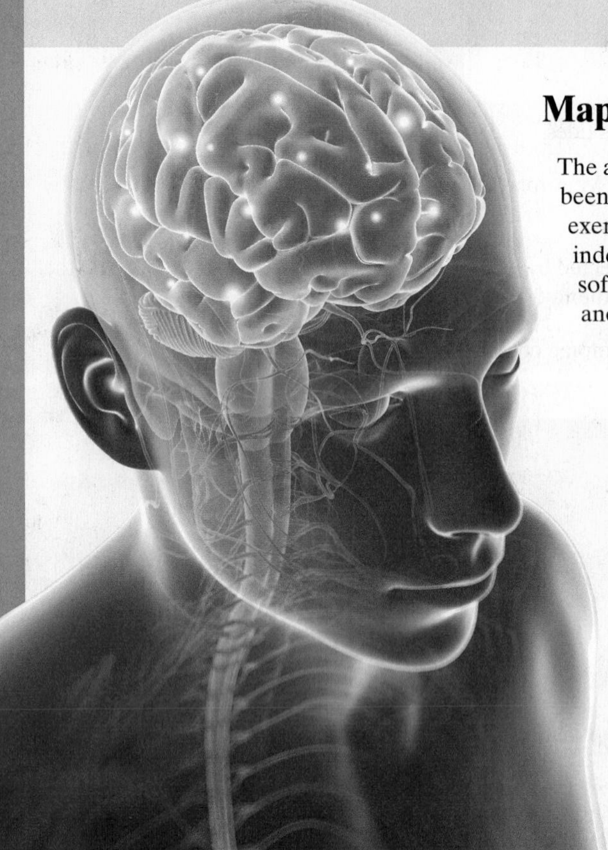

Mapping Your Mind

The ability to organize material in your mind is key to understanding. You have been exposed to a lot of concepts at this point in the course, and it is a worthwhile exercise to organize the material. In the past, we might organize material using index cards or an outline. But in today's digital world, we can use interesting software that enables us to digitally organize the material that is in our mind and share it with anyone on the Web.

—*See the Internet-based Chapter Project I*—

Outline

3.1 The Inverse Sine, Cosine, and Tangent Functions
3.2 The Inverse Trigonometric Functions (Continued)
3.3 Trigonometric Equations
3.4 Trigonometric Identities
3.5 Sum and Difference Formulas
3.6 Double-angle and Half-angle Formulas
3.7 Product-to-Sum and Sum-to-Product Formulas
Chapter Review
Chapter Test
Cumulative Review
Chapter Projects

••• A Look Back

Chapter 1 introduced inverse functions and developed their properties, particularly the relationship between the domain and range of a function and those of its inverse. We learned that the graphs of a function and its inverse are symmetric with respect to the line $y = x$.

Chapter 2 defined the six trigonometric functions and discussed their properties.

A Look Ahead •••

The first two sections of this chapter define the six inverse trigonometric functions and investigate their properties. Section 3.3 discusses equations that contain trigonometric functions. Sections 3.4 through 3.7 continue the derivation of identities. These identities play an important role in calculus, the physical and life sciences, and economics, where they are used to simplify complicated expressions.

3.1 The Inverse Sine, Cosine, and Tangent Functions

PREPARING FOR THIS SECTION *Before getting started, review the following:*

- Inverse Functions (Section 1.7, pp. 78–86)
- Values of the Trigonometric Functions (Section 2.2, pp. 116–125)
- Properties of the Sine, Cosine, and Tangent Functions (Section 2.3, pp. 131–141)
- Graphs of the Sine, Cosine, and Tangent Functions (Sections 2.4, pp. 145–149 and 2.5, pp. 160–163)

Now Work the **'Are You Prepared?'** problems on page 198.

OBJECTIVES
1. Find the Exact Value of an Inverse Sine Function (p. 190)
2. Find an Approximate Value of an Inverse Sine Function (p. 191)
3. Use Properties of Inverse Functions to Find Exact Values of Certain Composite Functions (p. 192)
4. Find the Inverse Function of a Trigonometric Function (p. 197)
5. Solve Equations Involving Inverse Trigonometric Functions (p. 198)

In Section 1.7 we discussed inverse functions, and we concluded that if a function is one-to-one, it will have an inverse function. We also observed that if a function is not one-to-one, it may be possible to restrict its domain in some suitable manner so that the restricted function is one-to-one. For example, the function $y = x^2$ is not one-to-one; however, if the domain is restricted to $x \geq 0$, the function is one-to-one.

Other properties of a one-to-one function f and its inverse function f^{-1} that were discussed in Section 1.7 are summarized next.

1. $f^{-1}(f(x)) = x$ for every x in the domain of f, and $f(f^{-1}(x)) = x$ for every x in the domain of f^{-1}.
2. The domain of $f =$ the range of f^{-1}, and the range of $f =$ the domain of f^{-1}.
3. The graph of f and the graph of f^{-1} are reflections of one another about the line $y = x$.
4. If a function $y = f(x)$ has an inverse function, the implicit equation of the inverse function is $x = f(y)$. If we solve this equation for y, we obtain the explicit equation $y = f^{-1}(x)$.

The Inverse Sine Function

Figure 1 shows the graph of $y = \sin x$. Because every horizontal line $y = b$, where b is between -1 and 1, inclusive, intersects the graph of $y = \sin x$ infinitely many times, it follows from the horizontal-line test that the function $y = \sin x$ is not one-to-one.

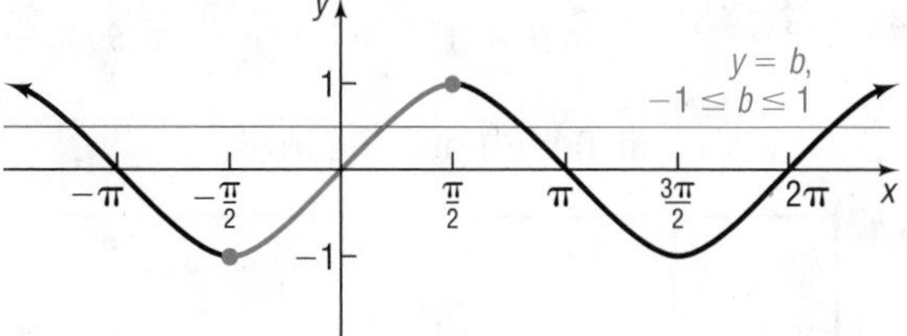

Figure 1 $y = \sin x, -\infty < x < \infty, -1 \leq y \leq 1$

However, if the domain of $y = \sin x$ is restricted to the interval $\left[-\frac{\pi}{2}, \frac{\pi}{2}\right]$, the restricted function

$$y = \sin x \qquad -\frac{\pi}{2} \leq x \leq \frac{\pi}{2}$$

is one-to-one and has an inverse function.* See Figure 2.

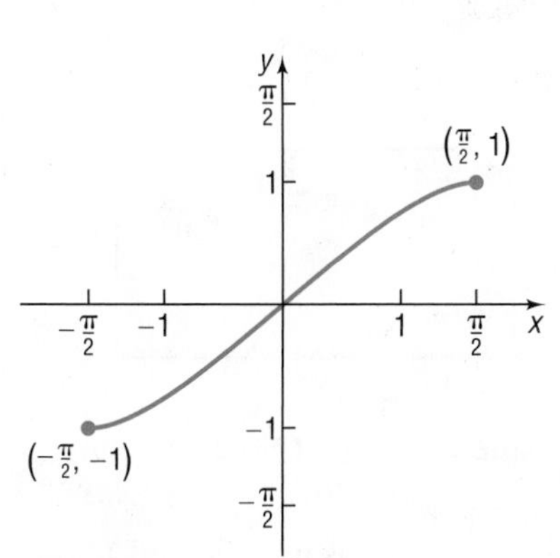

Figure 2 $y = \sin x, -\frac{\pi}{2} \leq x \leq \frac{\pi}{2}, -1 \leq y \leq 1$

*Although there are many other ways to restrict the domain and obtain a one-to-one function, mathematicians have agreed to use the interval $\left[-\frac{\pi}{2}, \frac{\pi}{2}\right]$ to define the inverse of $y = \sin x$.

NOTE Remember, the domain of a function f equals the range of its inverse, function f^{-1}, and the range of a function f equals the domain of its inverse function f^{-1}. Because the restricted domain of the sine function is $\left[-\frac{\pi}{2}, \frac{\pi}{2}\right]$, the range of inverse sine function is $\left[-\frac{\pi}{2}, \frac{\pi}{2}\right]$, and because the range of the sine function is $[-1, 1]$, the domain of the inverse sine function is $[-1, 1]$. ■

An equation for the inverse of $y = f(x) = \sin x$ is obtained by interchanging x and y. The implicit form of the inverse function is $x = \sin y$, $-\frac{\pi}{2} \le y \le \frac{\pi}{2}$. The explicit form is called the **inverse sine** of x and is symbolized by $y = f^{-1}(x) = \sin^{-1} x$.

$$y = \sin^{-1} x \quad \text{means} \quad x = \sin y$$
$$\text{where} \quad -1 \le x \le 1 \quad \text{and} \quad -\frac{\pi}{2} \le y \le \frac{\pi}{2} \tag{1}$$

Because $y = \sin^{-1} x$ means $x = \sin y$, $y = \sin^{-1} x$ is read as "y is the angle or real number whose sine equals x," or, alternatively, is read "y is the inverse sine of x." Be careful about the notation used. The superscript -1 that appears in $y = \sin^{-1} x$ is not an exponent but the symbol used to denote the inverse function f^{-1} of f. (To avoid confusion, some texts use the notation $y = \text{Arcsin}\, x$ instead of $y = \sin^{-1} x$.)

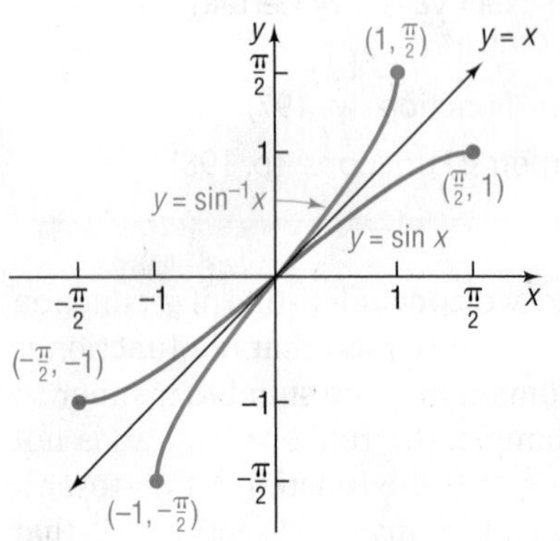

Figure 3
$y = \sin^{-1} x$, $-1 \le x \le 1$, $-\frac{\pi}{2} \le y \le \frac{\pi}{2}$

The inverse of a function f receives as input an element from the range of f and returns as output an element in the domain of f. The restricted sine function, $y = f(x) = \sin x$, receives as input an angle or real number x in the interval $\left[-\frac{\pi}{2}, \frac{\pi}{2}\right]$ and outputs a real number in the interval $[-1, 1]$. Therefore, the inverse sine function, $y = \sin^{-1} x$, receives as input a real number in the interval $[-1, 1]$ or $-1 \le x \le 1$, its domain, and outputs an angle or real number in the interval $\left[-\frac{\pi}{2}, \frac{\pi}{2}\right]$ or $-\frac{\pi}{2} \le y \le \frac{\pi}{2}$, its range.

The graph of the inverse sine function can be obtained by reflecting the restricted portion of the graph of $y = f(x) = \sin x$ about the line $y = x$, as shown in Figure 3.

Check: Graph $Y_1 = \sin x$ and $Y_2 = \sin^{-1} x$. Compare the result with Figure 3.

1 Find the Exact Value of an Inverse Sine Function

For some numbers x, it is possible to find the exact value of $y = \sin^{-1} x$.

EXAMPLE 1 **Finding the Exact Value of an Inverse Sine Function**

Find the exact value of: $\sin^{-1} 1$

Solution Let $\theta = \sin^{-1} 1$. Then θ is the angle, $-\frac{\pi}{2} \le \theta \le \frac{\pi}{2}$, whose sine equals 1.

$$\theta = \sin^{-1} 1 \qquad -\frac{\pi}{2} \le \theta \le \frac{\pi}{2}$$

$$\sin \theta = 1 \qquad -\frac{\pi}{2} \le \theta \le \frac{\pi}{2} \qquad \text{By definition of } y = \sin^{-1} x$$

Now look at Table 1 and Figure 4.

Table 1

θ	$-\frac{\pi}{2}$	$-\frac{\pi}{3}$	$-\frac{\pi}{4}$	$-\frac{\pi}{6}$	0	$\frac{\pi}{6}$	$\frac{\pi}{4}$	$\frac{\pi}{3}$	$\frac{\pi}{2}$
$\sin \theta$	−1	$-\frac{\sqrt{3}}{2}$	$-\frac{\sqrt{2}}{2}$	$-\frac{1}{2}$	0	$\frac{1}{2}$	$\frac{\sqrt{2}}{2}$	$\frac{\sqrt{3}}{2}$	1

The only angle θ in the interval $\left[-\frac{\pi}{2}, \frac{\pi}{2}\right]$ whose sine is 1 is $\frac{\pi}{2}$. (Note that $\sin \frac{5\pi}{2}$ also equals 1, but $\frac{5\pi}{2}$ lies outside the interval $\left[-\frac{\pi}{2}, \frac{\pi}{2}\right]$, which is not allowed.) So

$$\sin^{-1} 1 = \frac{\pi}{2}$$

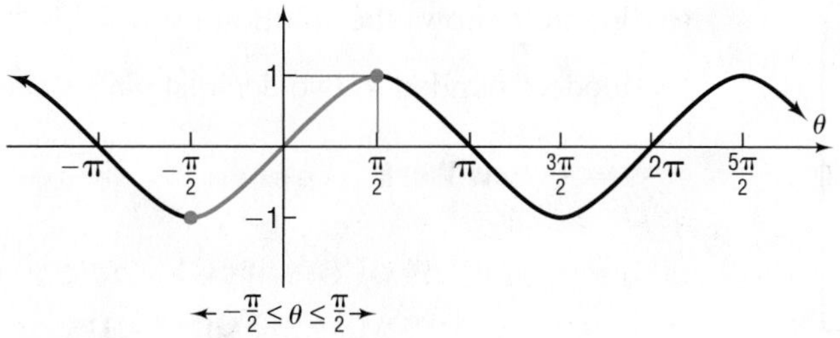

Figure 4

Now Work PROBLEM 15

EXAMPLE 2

Finding the Exact Value of an Inverse Sine Function

Find the exact value of: $\sin^{-1}\left(-\frac{1}{2}\right)$

Solution Let $\theta = \sin^{-1}\left(-\frac{1}{2}\right)$. Then θ is the angle, $-\frac{\pi}{2} \le \theta \le \frac{\pi}{2}$, whose sine equals $-\frac{1}{2}$.

$$\theta = \sin^{-1}\left(-\frac{1}{2}\right) \qquad -\frac{\pi}{2} \le \theta \le \frac{\pi}{2}$$

$$\sin\theta = -\frac{1}{2} \qquad -\frac{\pi}{2} \le \theta \le \frac{\pi}{2}$$

(Refer to Table 1 and Figure 4, if necessary.) The only angle in the interval $\left[-\frac{\pi}{2}, \frac{\pi}{2}\right]$ whose sine is $-\frac{1}{2}$ is $-\frac{\pi}{6}$, so

$$\sin^{-1}\left(-\frac{1}{2}\right) = -\frac{\pi}{6}$$

Now Work PROBLEM 21

2 Find an Approximate Value of an Inverse Sine Function

For most numbers x, the value $y = \sin^{-1} x$ must be approximated.

EXAMPLE 3

Finding an Approximate Value of an Inverse Sine Function

Find an approximate value of each of the following functions:

(a) $\sin^{-1}\frac{1}{3}$

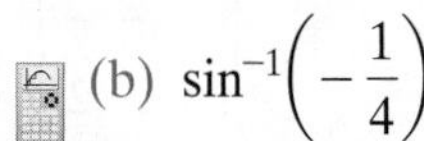

(b) $\sin^{-1}\left(-\frac{1}{4}\right)$

Express the answer in radians rounded to two decimal places.

Solution (a) Because the angle is to be measured in radians, first set the mode of the calculator to radians.* Rounded to two decimal places,

$$\sin^{-1}\frac{1}{3} = 0.34$$

*On most calculators, the inverse sine is obtained by pressing SHIFT or 2nd, followed by sin. On some calculators, $\sin^{-1}$ is pressed first, then 1/3 is entered; on others, this sequence is reversed. Consult your owner's manual for the correct sequence.

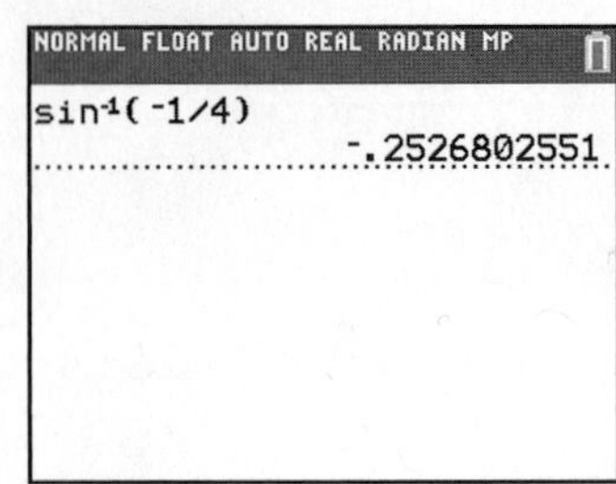

Figure 5

(b) Figure 5 shows the solution using a TI-84 Plus C graphing calculator in radian mode. Rounded to two decimal places, $\sin^{-1}\left(-\frac{1}{4}\right) = -0.25$. ●

Now Work PROBLEM 27

3 Use Properties of Inverse Functions to Find Exact Values of Certain Composite Functions

Recall from the discussion of functions and their inverses in Section 1.7, that $f^{-1}(f(x)) = x$ for all x in the domain of f and that $f(f^{-1}(x)) = x$ for all x in the domain of f^{-1}. In terms of the sine function and its inverse, these properties are of the form

$$f^{-1}(f(x)) = \sin^{-1}(\sin x) = x \quad \text{where } -\frac{\pi}{2} \le x \le \frac{\pi}{2} \qquad \textbf{(2a)}$$

$$f(f^{-1}(x)) = \sin(\sin^{-1} x) = x \quad \text{where } -1 \le x \le 1 \qquad \textbf{(2b)}$$

EXAMPLE 4

Finding the Exact Value of Certain Composite Functions

Find the exact value of each of the following composite functions.

(a) $\sin^{-1}\left(\sin\frac{\pi}{8}\right)$ (b) $\sin^{-1}\left(\sin\frac{5\pi}{8}\right)$

Solution (a) The composite function $\sin^{-1}\left(\sin\frac{\pi}{8}\right)$ follows the form of property (2a). Because $\frac{\pi}{8}$ is in the interval $\left[-\frac{\pi}{2}, \frac{\pi}{2}\right]$, property (2a) can be used. Then

$$\sin^{-1}\left(\sin\frac{\pi}{8}\right) = \frac{\pi}{8}$$

(b) The composite function $\sin^{-1}\left(\sin\frac{5\pi}{8}\right)$ follows the form of property (2a), but $\frac{5\pi}{8}$ is not in the interval $\left[-\frac{\pi}{2}, \frac{\pi}{2}\right]$. To use (2a), first find an angle θ in the interval $\left[-\frac{\pi}{2}, \frac{\pi}{2}\right]$ for which $\sin\theta = \sin\frac{5\pi}{8}$. Then, using (2a),

$$\sin^{-1}\left(\sin\frac{5\pi}{8}\right) = \sin^{-1}(\sin\theta) = \theta$$

Figure 6 illustrates that $\sin\frac{5\pi}{8} = y = \sin\frac{3\pi}{8}$. Since $\frac{3\pi}{8}$ is in the interval $\left[-\frac{\pi}{2}, \frac{\pi}{2}\right]$, this means

$$\sin^{-1}\left(\sin\frac{5\pi}{8}\right) = \sin^{-1}\left(\sin\frac{3\pi}{8}\right) = \frac{3\pi}{8}$$

Use property (2a). ●

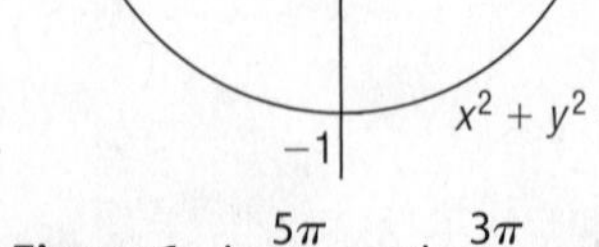

Figure 6 $\sin\frac{5\pi}{8} = \sin\frac{3\pi}{8}$

Now Work PROBLEM 43

EXAMPLE 5

Finding the Exact Value of Certain Composite Functions

Find the exact value, if any, of each composite function.

(a) $\sin(\sin^{-1} 0.5)$ (b) $\sin(\sin^{-1} 1.8)$

Solution (a) The composite function $\sin(\sin^{-1} 0.5)$ follows the form of property (2b), and 0.5 is in the interval $[-1, 1]$. Using (2b) reveals that

$$\sin(\sin^{-1} 0.5) = 0.5$$

(b) The composite function $\sin(\sin^{-1} 1.8)$ follows the form of property (2b). Since the domain of the inverse sine function is $[-1, 1]$, $\sin^{-1} 1.8$ is not defined. Therefore, $\sin(\sin^{-1} 1.8)$ is also not defined. ●

Now Work PROBLEM 51

The Inverse Cosine Function

Figure 7 shows the graph of $y = \cos x$. Because every horizontal line $y = b$, where b is between -1 and 1, inclusive, intersects the graph of $y = \cos x$ infinitely many times, it follows that the cosine function is not one-to-one.

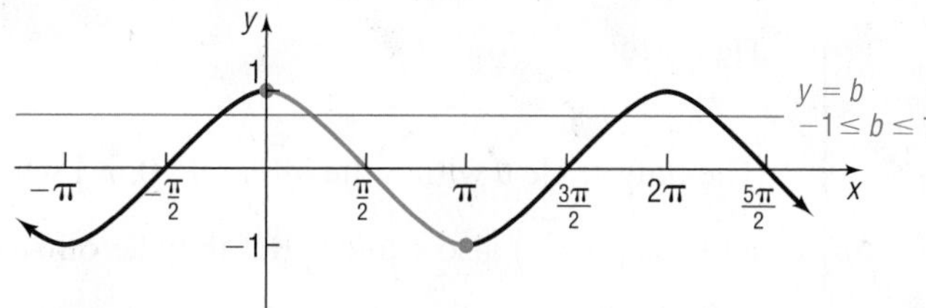

Figure 7 $y = \cos x, -\infty < x < \infty, -1 \le y \le 1$

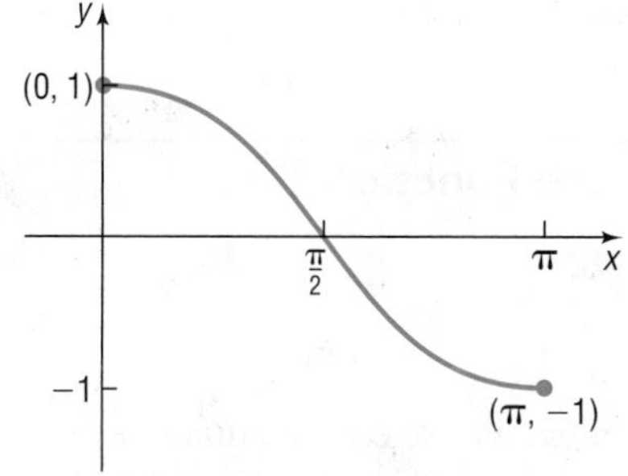

Figure 8
$y = \cos x, 0 \le x \le \pi, -1 \le y \le 1$

However, if the domain of $y = \cos x$ is restricted to the interval $[0, \pi]$, the restricted function

$$y = \cos x \qquad 0 \le x \le \pi$$

is one-to-one and has an inverse function.* See Figure 8.

An equation for the inverse of $y = f(x) = \cos x$ is obtained by interchanging x and y. The implicit form of the inverse function is $x = \cos y$, $0 \le y \le \pi$. The explicit form is called the **inverse cosine** of x and is symbolized by $y = f^{-1}(x) = \cos^{-1} x$ (or by $y = \text{Arccos}\, x$).

DEFINITION

$$y = \cos^{-1} x \quad \text{means} \quad x = \cos y$$
$$\text{where} \quad -1 \le x \le 1 \quad \text{and} \quad 0 \le y \le \pi \qquad \textbf{(3)}$$

Here y is the angle whose cosine is x. Because the range of the cosine function, $y = \cos x$, is $-1 \le y \le 1$, the domain of the inverse function $y = \cos^{-1} x$ is $-1 \le x \le 1$. Because the restricted domain of the cosine function, $y = \cos x$, is $0 \le x \le \pi$, the range of the inverse function $y = \cos^{-1} x$ is $0 \le y \le \pi$.

The graph of $y = \cos^{-1} x$ can be obtained by reflecting the restricted portion of the graph of $y = \cos x$ about the line $y = x$, as shown in Figure 9.

Figure 9
$y = \cos^{-1} x, -1 \le x \le 1, 0 \le y \le \pi$

✓**Check:** Graph $Y_1 = \cos x$ and $Y_2 = \cos^{-1} x$. Compare the result with Figure 9.

*This is the generally accepted restriction to define the inverse cosine function.

EXAMPLE 6 **Finding the Exact Value of an Inverse Cosine Function**

Find the exact value of: $\cos^{-1} 0$

Solution Let $\theta = \cos^{-1} 0$. Then θ is the angle, $0 \le \theta \le \pi$, whose cosine equals 0.

$$\theta = \cos^{-1} 0 \qquad 0 \le \theta \le \pi$$
$$\cos\theta = 0 \qquad 0 \le \theta \le \pi$$

Table 2

θ	$\cos\theta$
0	1
$\frac{\pi}{6}$	$\frac{\sqrt{3}}{2}$
$\frac{\pi}{4}$	$\frac{\sqrt{2}}{2}$
$\frac{\pi}{3}$	$\frac{1}{2}$
$\frac{\pi}{2}$	0
$\frac{2\pi}{3}$	$-\frac{1}{2}$
$\frac{3\pi}{4}$	$-\frac{\sqrt{2}}{2}$
$\frac{5\pi}{6}$	$-\frac{\sqrt{3}}{2}$
π	-1

Look at Table 2 and Figure 10.

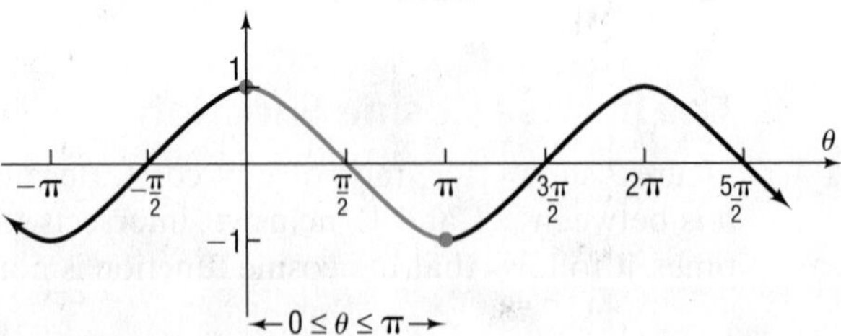

Figure 10

The only angle θ within the interval $[0, \pi]$ whose cosine is 0 is $\frac{\pi}{2}$. [Note that $\cos\frac{3\pi}{2}$ and $\cos\left(-\frac{\pi}{2}\right)$ also equal 0, but they lie outside the interval $[0, \pi]$, so these values are not allowed.] Therefore,

$$\cos^{-1} 0 = \frac{\pi}{2}$$

EXAMPLE 7 **Finding the Exact Value of an Inverse Cosine Function**

Find the exact value of: $\cos^{-1}\left(-\frac{\sqrt{2}}{2}\right)$

Solution Let $\theta = \cos^{-1}\left(-\frac{\sqrt{2}}{2}\right)$. Then θ is the angle, $0 \le \theta \le \pi$, whose cosine equals $-\frac{\sqrt{2}}{2}$.

$$\theta = \cos^{-1}\left(-\frac{\sqrt{2}}{2}\right) \qquad 0 \le \theta \le \pi$$
$$\cos\theta = -\frac{\sqrt{2}}{2} \qquad 0 \le \theta \le \pi$$

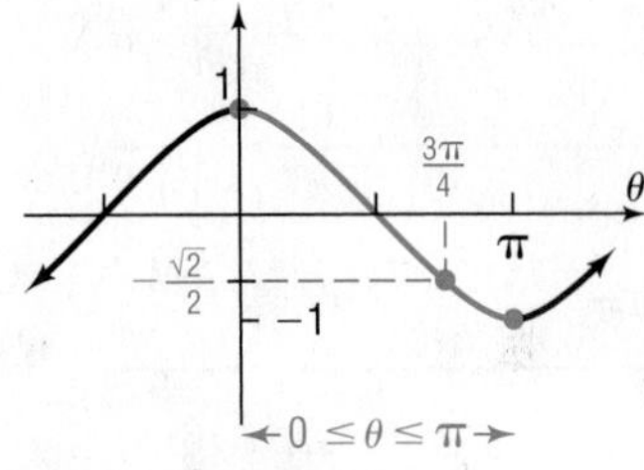

Figure 11

Look at Table 2 and Figure 11.

The only angle θ within the interval $[0, \pi]$ whose cosine is $-\frac{\sqrt{2}}{2}$ is $\frac{3\pi}{4}$, so

$$\cos^{-1}\left(-\frac{\sqrt{2}}{2}\right) = \frac{3\pi}{4}$$

Now Work PROBLEM 25

For the cosine function and its inverse, the following properties hold.

$$f^{-1}(f(x)) = \cos^{-1}(\cos x) = x \quad \text{where } 0 \le x \le \pi \qquad \textbf{(4a)}$$
$$f(f^{-1}(x)) = \cos(\cos^{-1} x) = x \quad \text{where } -1 \le x \le 1 \qquad \textbf{(4b)}$$

EXAMPLE 8 **Using Properties of Inverse Functions to Find the Exact Value of Certain Composite Functions**

Find the exact value of:

(a) $\cos^{-1}\left(\cos\dfrac{\pi}{12}\right)$ (b) $\cos\left[\cos^{-1}(-0.4)\right]$

(c) $\cos^{-1}\left[\cos\left(-\dfrac{2\pi}{3}\right)\right]$ (d) $\cos(\cos^{-1}\pi)$

Solution (a) $\cos^{-1}\left(\cos\dfrac{\pi}{12}\right) = \dfrac{\pi}{12}$ $\dfrac{\pi}{12}$ is in the interval $[0, \pi]$; use property (4a).

(b) $\cos\left[\cos^{-1}(-0.4)\right] = -0.4$ -0.4 is in the interval $[-1, 1]$; use property (4b).

(c) The angle $-\dfrac{2\pi}{3}$ is not in the interval $[0, \pi]$ so property (4a) cannot be used. However, because the cosine function is even, $\cos\left(-\dfrac{2\pi}{3}\right) = \cos\dfrac{2\pi}{3}$. Because $\dfrac{2\pi}{3}$ is in the interval $[0, \pi]$, property (4a) can be used, and

$$\cos^{-1}\left[\cos\left(-\frac{2\pi}{3}\right)\right] = \cos^{-1}\left(\cos\frac{2\pi}{3}\right) = \frac{2\pi}{3}$$ $\dfrac{2\pi}{3}$ is in the interval $[0, \pi]$; use (4a).

(d) Because π is not in the interval $[-1, 1]$, the domain of the inverse cosine function, $\cos^{-1}\pi$ is not defined. This means the composite function $\cos(\cos^{-1}\pi)$ is also not defined. ●

Now Work PROBLEMS 39 AND 55

The Inverse Tangent Function

Figure 12 shows the graph of $y = \tan x$. Because every horizontal line intersects the graph infinitely many times, it follows that the tangent function is not one-to-one.

However, if the domain of $y = \tan x$ is restricted to the interval $\left(-\dfrac{\pi}{2}, \dfrac{\pi}{2}\right)$, the restricted function

$$y = \tan x \qquad -\frac{\pi}{2} < x < \frac{\pi}{2}$$

is one-to-one and so has an inverse function.* See Figure 13.

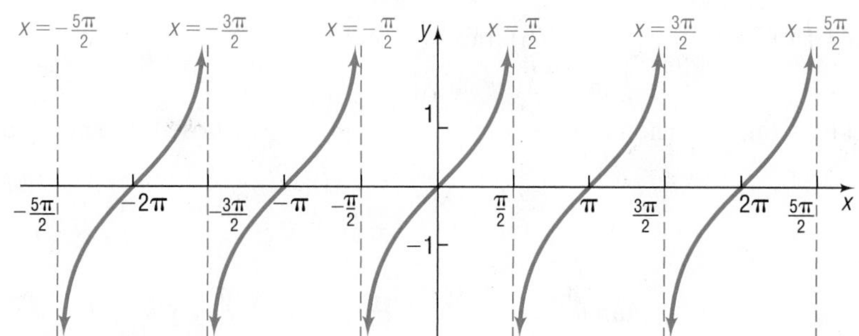

Figure 12
$y = \tan x, -\infty < x < \infty$, x not equal to odd multiples of $\dfrac{\pi}{2}$, $-\infty < y < \infty$

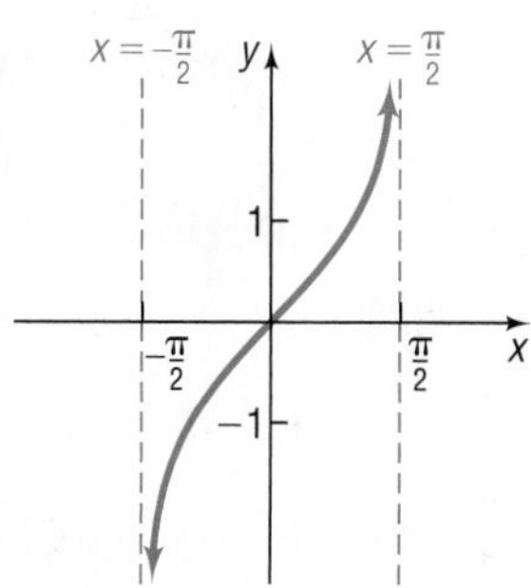

Figure 13
$y = \tan x, -\dfrac{\pi}{2} < x < \dfrac{\pi}{2}, -\infty < y < \infty$

*This is the generally accepted restriction.

An equation for the inverse of $y = f(x) = \tan x$ is obtained by interchanging x and y. The implicit form of the inverse function is $x = \tan y$, $-\frac{\pi}{2} < y < \frac{\pi}{2}$. The explicit form is called the **inverse tangent** of x and is symbolized by $y = f^{-1}(x) = \tan^{-1} x$ (or by $y = \text{Arctan}\, x$).

DEFINITION

$$y = \tan^{-1} x \quad \text{means} \quad x = \tan y$$
$$\text{where} \quad -\infty < x < \infty \quad \text{and} \quad -\frac{\pi}{2} < y < \frac{\pi}{2} \tag{5}$$

Here y is the angle whose tangent is x. The domain of the function $y = \tan^{-1} x$ is $-\infty < x < \infty$, and its range is $-\frac{\pi}{2} < y < \frac{\pi}{2}$. The graph of $y = \tan^{-1} x$ can be obtained by reflecting the restricted portion of the graph of $y = \tan x$ about the line $y = x$, as shown in Figure 14.

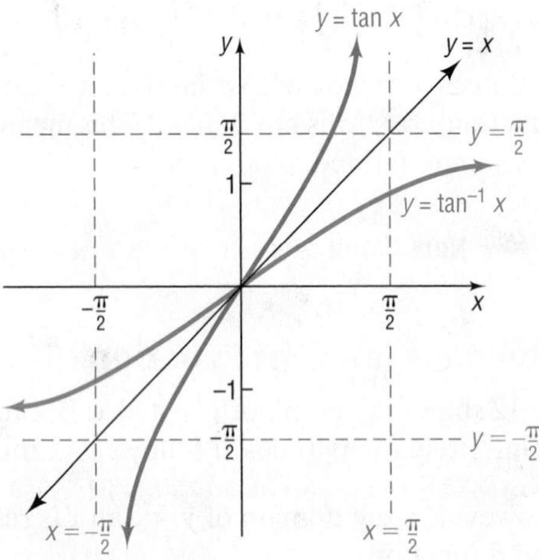

✓**Check:** Graph $Y_1 = \tan x$ and $Y_2 = \tan^{-1} x$. Compare the result with Figure 14.

Figure 14 $y = \tan^{-1} x, -\infty < x < \infty, -\frac{\pi}{2} < y < \frac{\pi}{2}$

EXAMPLE 9 Finding the Exact Value of an Inverse Tangent Function

Find the exact value of:

(a) $\tan^{-1} 1$ (b) $\tan^{-1}(-\sqrt{3})$

Solution (a) Let $\theta = \tan^{-1} 1$. Then θ is the angle, $-\frac{\pi}{2} < \theta < \frac{\pi}{2}$, whose tangent equals 1.

$$\theta = \tan^{-1} 1 \qquad -\frac{\pi}{2} < \theta < \frac{\pi}{2}$$

$$\tan \theta = 1 \qquad -\frac{\pi}{2} < \theta < \frac{\pi}{2}$$

Look at Table 3. The only angle θ within the interval $\left(-\frac{\pi}{2}, \frac{\pi}{2}\right)$ whose tangent is 1 is $\frac{\pi}{4}$, so

$$\tan^{-1} 1 = \frac{\pi}{4}$$

Table 3

θ	$\tan\theta$
$-\frac{\pi}{2}$	Undefined
$-\frac{\pi}{3}$	$-\sqrt{3}$
$-\frac{\pi}{4}$	-1
$-\frac{\pi}{6}$	$-\frac{\sqrt{3}}{3}$
0	0
$\frac{\pi}{6}$	$\frac{\sqrt{3}}{3}$
$\frac{\pi}{4}$	1
$\frac{\pi}{3}$	$\sqrt{3}$
$\frac{\pi}{2}$	Undefined

(b) Let $\theta = \tan^{-1}(-\sqrt{3})$. Then θ is the angle, $-\frac{\pi}{2} < \theta < \frac{\pi}{2}$, whose tangent equals $-\sqrt{3}$.

$$\theta = \tan^{-1}(-\sqrt{3}) \qquad -\frac{\pi}{2} < \theta < \frac{\pi}{2}$$

$$\tan\theta = -\sqrt{3} \qquad -\frac{\pi}{2} < \theta < \frac{\pi}{2}$$

Look at Table 3. The only angle θ within the interval $\left(-\frac{\pi}{2}, \frac{\pi}{2}\right)$ whose tangent is $-\sqrt{3}$ is $-\frac{\pi}{3}$, so

$$\tan^{-1}(-\sqrt{3}) = -\frac{\pi}{3}$$

●

Now Work PROBLEM 19

For the tangent function and its inverse, the following properties hold.

$$f^{-1}(f(x)) = \tan^{-1}(\tan x) = x \qquad \text{where } -\frac{\pi}{2} < x < \frac{\pi}{2}$$

$$f(f^{-1}(x)) = \tan(\tan^{-1} x) = x \qquad \text{where } -\infty < x < \infty$$

Now Work PROBLEM 45

4 Find the Inverse Function of a Trigonometric Function

EXAMPLE 10 **Finding the Inverse Function of a Trigonometric Function**

(a) Find the inverse function f^{-1} of $f(x) = 2\sin x - 1$, $-\frac{\pi}{2} \le x \le \frac{\pi}{2}$.

(b) Find the range of f and the domain and range of f^{-1}.

Solution (a) The function f is one-to-one and so has an inverse function. Follow the steps on page 85 for finding the inverse function.

$$y = 2\sin x - 1$$

$$x = 2\sin y - 1 \qquad \text{Interchange } x \text{ and } y.$$

$$x + 1 = 2\sin y \qquad \text{Solve for } y.$$

$$\sin y = \frac{x+1}{2}$$

$$y = \sin^{-1}\frac{x+1}{2} \qquad \text{Apply the definition (1).}$$

The inverse function is $f^{-1}(x) = \sin^{-1}\frac{x+1}{2}$.

(b) To find the range of f, solve $y = 2\sin x - 1$ for $\sin x$, and use the fact that $-1 \le \sin x \le 1$.

$$y = 2\sin x - 1$$

$$\sin x = \frac{y+1}{2}$$

$$-1 \le \frac{y+1}{2} \le 1$$
$$-2 \le y + 1 \le 2$$
$$-3 \le y \le 1$$

NOTE The range of f also can be found using transformations. The range of $y = \sin x$ is $[-1, 1]$. The range of $y = 2 \sin x$ is $[-2, 2]$ due to the vertical stretch by a factor of 2. The range of $f(x) = 2 \sin x - 1$ is $[-3, 1]$ due to the shift down of 1 unit. ■

The range of f is $\{y \mid -3 \le y \le 1\}$, or $[-3, 1]$ using interval notation.

The domain of f^{-1} equals the range of f, $[-3, 1]$.

The range of f^{-1} equals the domain of f, $\left[-\frac{\pi}{2}, \frac{\pi}{2}\right]$. ●

Now Work PROBLEM 61

5 Solve Equations Involving Inverse Trigonometric Functions

Equations that contain inverse trigonometric functions are called **inverse trigonometric equations.**

EXAMPLE 11 Solving an Equation Involving an Inverse Trigonometric Function

Solve the equation: $3 \sin^{-1} x = \pi$

Solution To solve an equation involving a single inverse trigonometric function, first isolate the inverse trigonometric function.

$$3 \sin^{-1} x = \pi$$
$$\sin^{-1} x = \frac{\pi}{3} \quad \text{Divide both sides by 3.}$$
$$x = \sin \frac{\pi}{3} \quad y = \sin^{-1} x \text{ means } x = \sin y.$$
$$x = \frac{\sqrt{3}}{2}$$

The solution set is $\left\{\frac{\sqrt{3}}{2}\right\}$. ●

Now Work PROBLEM 67

3.1 Assess Your Understanding

'Are You Prepared?' *Answers are given at the end of these exercises. If you get a wrong answer, read the pages listed in red.*

1. What are the domain and the range of $y = \sin x$? (pp. 131–141)
2. A suitable restriction on the domain of the function $f(x) = (x - 1)^2$ to make it one-to-one would be ______. (pp. 78–86)
3. If the domain of a one-to-one function is $[3, \infty)$, the range of its inverse is ______. (pp. 78–86)
4. ***True or False*** The graph of $y = \cos x$ is decreasing on the interval $(0, \pi)$. (pp. 145–149)
5. $\tan \frac{\pi}{4} =$ ______; $\sin \frac{\pi}{3} =$ ______ (pp. 116–125)
6. $\sin\left(-\frac{\pi}{6}\right) =$ ______; $\cos \pi =$ ______. (pp. 116–125)

Concepts and Vocabulary

7. $y = \sin^{-1} x$ means ______, where $-1 \le x \le 1$ and $-\frac{\pi}{2} \le y \le \frac{\pi}{2}$.
8. $\cos^{-1}(\cos x) = x$, where ______.
9. $\tan(\tan^{-1} x) = x$, where ______.
10. ***True or False*** The domain of $y = \sin^{-1} x$ is $-\frac{\pi}{2} \le x \le \frac{\pi}{2}$.
11. ***True or False*** $\sin(\sin^{-1} 0) = 0$ and $\cos(\cos^{-1} 0) = 0$.
12. ***True or False*** $y = \tan^{-1} x$ means $x = \tan y$, where $-\infty < x < \infty$ and $-\frac{\pi}{2} < y < \frac{\pi}{2}$.

13. Which of the following inequalities describes where $\sin^{-1}(\sin x) = x$?
(a) $-\infty < x < \infty$ (b) $0 \le x \le \pi$
(c) $-1 \le x \le 1$ (d) $-\frac{\pi}{2} \le x \le \frac{\pi}{2}$

14. Choose the inverse function f^{-1} of
$$f(x) = \frac{1}{2}\tan x, \ -\frac{\pi}{2} < x < \frac{\pi}{2}$$
(a) $f^{-1}(x) = \tan^{-1}(2x); -\infty < x < \infty$
(b) $f^{-1}(x) = 2\tan^{-1} x; -\infty < x < \infty$
(c) $f^{-1}(x) = \frac{1}{2}\tan^{-1} x; -\infty < x < \infty$
(d) $f^{-1}(x) = \tan^{-1}\left(\frac{1}{2}x\right); -\infty < x < \infty$

Skill Building

In Problems 15–26, find the exact value of each expression.

15. $\sin^{-1} 0$
16. $\cos^{-1} 1$
17. $\sin^{-1}(-1)$
18. $\cos^{-1}(-1)$
19. $\tan^{-1} 0$
20. $\tan^{-1}(-1)$
21. $\sin^{-1}\frac{\sqrt{2}}{2}$
22. $\tan^{-1}\frac{\sqrt{3}}{3}$
23. $\tan^{-1}\sqrt{3}$
24. $\sin^{-1}\left(-\frac{\sqrt{3}}{2}\right)$
25. $\cos^{-1}\left(-\frac{\sqrt{3}}{2}\right)$
26. $\sin^{-1}\left(-\frac{\sqrt{2}}{2}\right)$

In Problems 27–38, use a calculator to find the value of each expression rounded to two decimal places.

27. $\sin^{-1} 0.1$
28. $\cos^{-1} 0.6$
29. $\tan^{-1} 5$
30. $\tan^{-1} 0.2$
31. $\cos^{-1}\frac{7}{8}$
32. $\sin^{-1}\frac{1}{8}$
33. $\tan^{-1}(-0.4)$
34. $\tan^{-1}(-3)$
35. $\sin^{-1}(-0.12)$
36. $\cos^{-1}(-0.44)$
37. $\cos^{-1}\frac{\sqrt{2}}{3}$
38. $\sin^{-1}\frac{\sqrt{3}}{5}$

In Problems 39–58, find the exact value, if any, of each composite function. If there is no value, say it is "not defined." Do not use a calculator.

39. $\cos^{-1}\left(\cos\frac{4\pi}{5}\right)$
40. $\sin^{-1}\left[\sin\left(-\frac{\pi}{10}\right)\right]$
41. $\tan^{-1}\left[\tan\left(-\frac{3\pi}{8}\right)\right]$
42. $\sin^{-1}\left[\sin\left(-\frac{3\pi}{7}\right)\right]$
43. $\sin^{-1}\left(\sin\frac{9\pi}{8}\right)$
44. $\cos^{-1}\left[\cos\left(-\frac{5\pi}{3}\right)\right]$
45. $\tan^{-1}\left(\tan\frac{4\pi}{5}\right)$
46. $\tan^{-1}\left[\tan\left(-\frac{2\pi}{3}\right)\right]$
47. $\cos^{-1}\left[\cos\left(-\frac{\pi}{4}\right)\right]$
48. $\sin^{-1}\left[\sin\left(-\frac{3\pi}{4}\right)\right]$
49. $\tan^{-1}\left[\tan\left(\frac{\pi}{2}\right)\right]$
50. $\tan^{-1}\left[\tan\left(-\frac{3\pi}{2}\right)\right]$
51. $\sin\left(\sin^{-1}\frac{1}{4}\right)$
52. $\cos\left[\cos^{-1}\left(-\frac{2}{3}\right)\right]$
53. $\tan(\tan^{-1} 4)$
54. $\tan[\tan^{-1}(-2)]$
55. $\cos(\cos^{-1} 1.2)$
56. $\sin[\sin^{-1}(-2)]$
57. $\tan(\tan^{-1}\pi)$
58. $\sin[\sin^{-1}(-1.5)]$

In Problems 59–66, find the inverse function f^{-1} of each function f. Find the range of f and the domain and range of f^{-1}.

59. $f(x) = 5\sin x + 2; -\frac{\pi}{2} \le x \le \frac{\pi}{2}$
60. $f(x) = 2\tan x - 3; -\frac{\pi}{2} < x < \frac{\pi}{2}$
61. $f(x) = -2\cos(3x); 0 \le x \le \frac{\pi}{3}$
62. $f(x) = 3\sin(2x); -\frac{\pi}{4} \le x \le \frac{\pi}{4}$
63. $f(x) = -\tan(x+1) - 3; -1 - \frac{\pi}{2} < x < \frac{\pi}{2} - 1$
64. $f(x) = \cos(x+2) + 1; -2 \le x \le \pi - 2$
65. $f(x) = 3\sin(2x+1); -\frac{1}{2} - \frac{\pi}{4} \le x \le -\frac{1}{2} + \frac{\pi}{4}$
66. $f(x) = 2\cos(3x+2); -\frac{2}{3} \le x \le -\frac{2}{3} + \frac{\pi}{3}$

In Problems 67–74, find the exact solution of each equation.

67. $4\sin^{-1} x = \pi$
68. $2\cos^{-1} x = \pi$
69. $3\cos^{-1}(2x) = 2\pi$
70. $-6\sin^{-1}(3x) = \pi$
71. $3\tan^{-1} x = \pi$
72. $-4\tan^{-1} x = \pi$
73. $4\cos^{-1} x - 2\pi = 2\cos^{-1} x$
74. $5\sin^{-1} x - 2\pi = 2\sin^{-1} x - 3\pi$

Applications and Extensions

In Problems 75–80, use the following discussion. The formula

$$D = 24\left[1 - \frac{\cos^{-1}(\tan i \tan \theta)}{\pi}\right]$$

can be used to approximate the number of hours of daylight D when the declination of the Sun is i° at a location θ° north latitude for any date between the vernal equinox and autumnal equinox. The declination of the Sun is defined as the angle i between the equatorial plane and any ray of light from the Sun. The latitude of a location is the angle θ between the Equator and the location on the surface of Earth, with the vertex of the angle located at the center of Earth. See the figure. To use the formula, $\cos^{-1}(\tan i \tan \theta)$ *must be expressed in radians.*

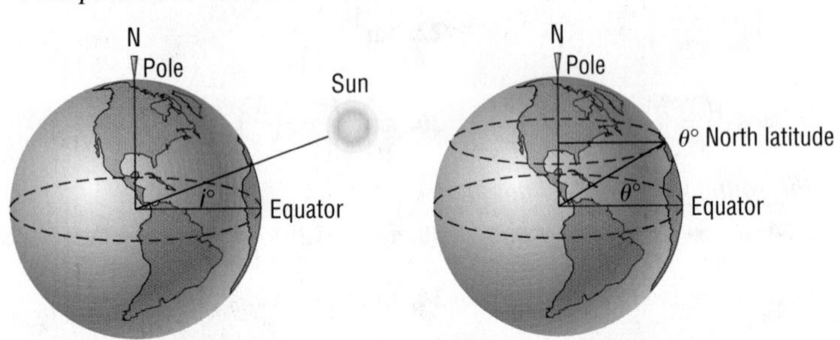

75. Approximate the number of hours of daylight in Houston, Texas (29°45′ north latitude), for the following dates:
(a) Summer solstice $(i = 23.5°)$
(b) Vernal equinox $(i = 0°)$
(c) July 4 $(i = 22°48')$

76. Approximate the number of hours of daylight in New York, New York (40°45′ north latitude), for the following dates:
(a) Summer solstice $(i = 23.5°)$
(b) Vernal equinox $(i = 0°)$
(c) July 4 $(i = 22°48')$

77. Approximate the number of hours of daylight in Honolulu, Hawaii (21°18′ north latitude), for the following dates:
(a) Summer solstice $(i = 23.5°)$
(b) Vernal equinox $(i = 0°)$
(c) July 4 $(i = 22°48')$

78. Approximate the number of hours of daylight in Anchorage, Alaska (61°10′ north latitude), for the following dates:
(a) Summer solstice $(i = 23.5°)$
(b) Vernal equinox $(i = 0°)$
(c) July 4 $(i = 22°48')$

79. Approximate the number of hours of daylight at the Equator (0° north latitude) for the following dates:
(a) Summer solstice $(i = 23.5°)$
(b) Vernal equinox $(i = 0°)$
(c) July 4 $(i = 22°48')$
(d) What do you conclude about the number of hours of daylight throughout the year for a location at the Equator?

80. Approximate the number of hours of daylight for any location that is 66°30′ north latitude for the following dates:
(a) Summer solstice $(i = 23.5°)$
(b) Vernal equinox $(i = 0°)$
(c) July 4 $(i = 22°48')$
(d) Thanks to the symmetry of the orbital path of Earth around the Sun, the number of hours of daylight on the winter solstice may be found by computing the number of hours of daylight on the summer solstice and subtracting this result from 24 hours. Compute the number of hours of daylight for this location on the winter solstice. What do you conclude about daylight for a location at 66°30′ north latitude?

81. Being the First to See the Rising Sun Cadillac Mountain, elevation 1530 feet, is located in Acadia National Park, Maine, and is the highest peak on the east coast of the United States. It is said that a person standing on the summit will be the first person in the United States to see the rays of the rising Sun. How much sooner would a person atop Cadillac Mountain see the first rays than a person standing below, at sea level?

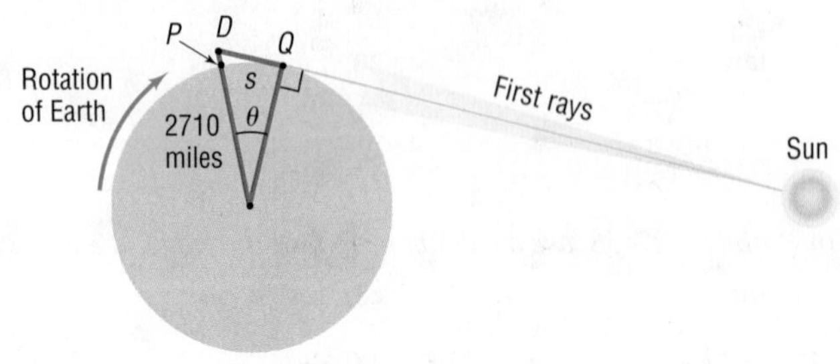

[**Hint:** Consult the figure. When the person at D sees the first rays of the Sun, the person at P does not. The person at P sees the first rays of the Sun only after Earth has rotated so that P is at location Q. Compute the length of the arc subtended by the central angle θ. Then use the fact that at the latitude of Cadillac Mountain, in 24 hours a length of $2\pi(2710) \approx 17027.4$ miles is subtended.]

82. Movie Theater Screens Suppose that a movie theater has a screen that is 28 feet tall. When you sit down, the bottom of the screen is 6 feet above your eye level. The angle formed by drawing a line from your eye to the bottom of the screen and another line from your eye to the top of the screen is called the **viewing angle**. In the figure, θ is the viewing angle. Suppose that you sit x feet from the screen. The viewing angle θ is given by the function

$$\theta(x) = \tan^{-1}\left(\frac{34}{x}\right) - \tan^{-1}\left(\frac{6}{x}\right)$$

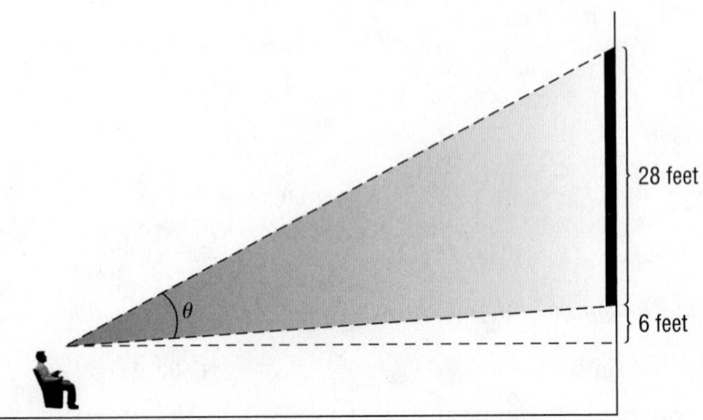

(a) What is your viewing angle if you sit 10 feet from the screen? 15 feet? 20 feet?
(b) If there are 5 feet between the screen and the first row of seats and there are 3 feet between each row and the row behind it, which row results in the largest viewing angle?

(c) Using a graphing utility, graph

$$\theta(x) = \tan^{-1}\left(\frac{34}{x}\right) - \tan^{-1}\left(\frac{6}{x}\right)$$

What value of x results in the largest viewing angle?

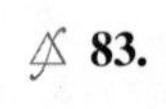
83. Area under a Curve The area under the graph of $y = \dfrac{1}{1 + x^2}$ and above the x-axis between $x = a$ and $x = b$ is given by

$$\tan^{-1} b - \tan^{-1} a$$

See the figure.

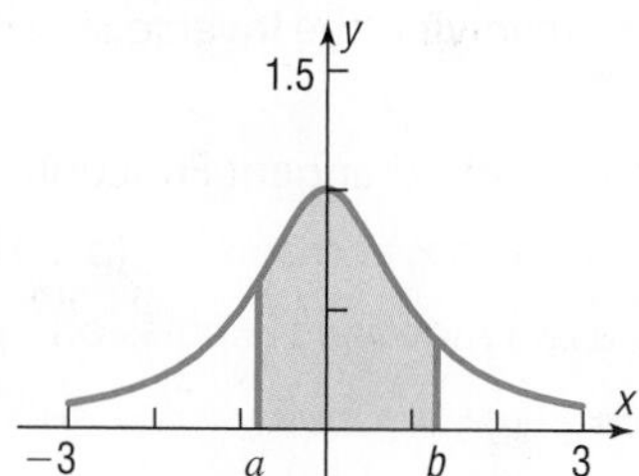

(a) Find the exact area under the graph of $y = \dfrac{1}{1 + x^2}$ and above the x-axis between $x = 0$ and $x = \sqrt{3}$.

(b) Find the exact area under the graph of $y = \dfrac{1}{1 + x^2}$ and above the x-axis between $x = -\dfrac{\sqrt{3}}{3}$ and $x = 1$.

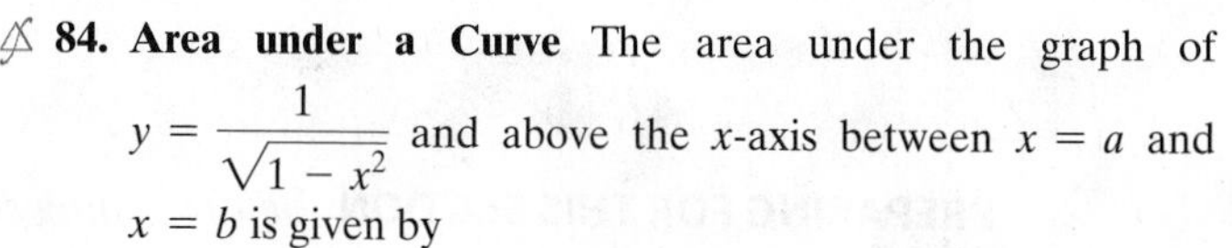
84. Area under a Curve The area under the graph of $y = \dfrac{1}{\sqrt{1 - x^2}}$ and above the x-axis between $x = a$ and $x = b$ is given by

$$\sin^{-1} b - \sin^{-1} a$$

See the figure.

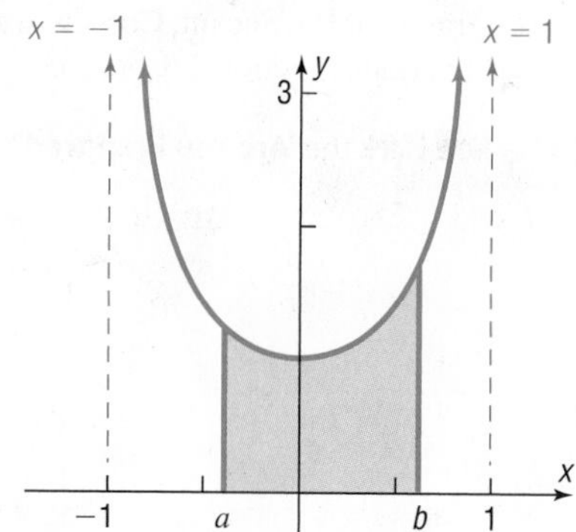

(a) Find the exact area under the graph of $y = \dfrac{1}{\sqrt{1 - x^2}}$ and above the x-axis between $x = 0$ and $x = \dfrac{\sqrt{3}}{2}$.

(b) Find the exact area under the graph of $y = \dfrac{1}{\sqrt{1 - x^2}}$ and above the x-axis between $x = -\dfrac{1}{2}$ and $x = \dfrac{1}{2}$.

Problems 85 and 86 require the following discussion:
The shortest distance between two points on Earth's surface can be determined from the latitude and longitude of the two locations. For example, if location 1 has (lat, lon) $= (\alpha_1, \beta_1)$ *and location 2 has* (lat, lon) $= (\alpha_2, \beta_2)$, *the shortest distance between the two locations is approximately*

$$d = r\cos^{-1}[(\cos\alpha_1\cos\beta_1\cos\alpha_2\cos\beta_2) + (\cos\alpha_1\sin\beta_1\cos\alpha_2\sin\beta_2) + (\sin\alpha_1\sin\alpha_2)],$$

where r = *radius of Earth* $\approx$ 3960 *miles and the inverse cosine function is expressed in radians. Also, N latitude and E longitude are positive angles, and S latitude and W longitude are negative angles.*

City	Latitude	Longitude
Chicago, IL	41°50′N	87°37′W
Honolulu, HI	21°18′N	157°50′W
Melbourne, Australia	37°47′S	144°58′E

Source: www.infoplease.com

85. Shortest Distance from Chicago to Honolulu Find the shortest distance from Chicago, latitude 41°50′N, longitude 87°37′W, to Honolulu, latitude 21°18′N, longitude 157°50′W. Round your answer to the nearest mile.

86. Shortest Distance from Honolulu to Melbourne, Australia Find the shortest distance from Honolulu to Melbourne, Australia, latitude 37°47′S, longitude 144°58′E. Round your answer to the nearest mile.

Retain Your Knowledge

Problems 87–90 are based on material learned earlier in the course. The purpose of these problems is to keep the material fresh in your mind so that you are better prepared for the final exam.

87. If $f(x) = \dfrac{2x + B}{x - 3}$ and $f(5) = 8$, what is the value of B?

88. State why the graph of the function f shown to the right is one-to-one. Then draw the graph of the inverse function f^{-1}.
Hint: The graph of $y = x$ is given.

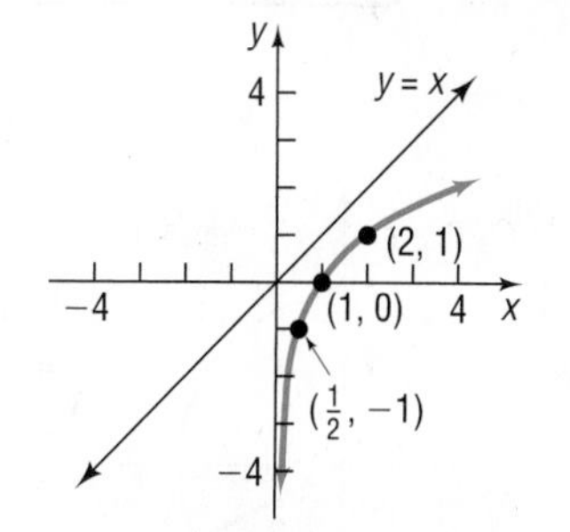

89. Find the average rate of change of $f(x) = 3x^2 - 5x$ from 1 to 3.

90. Find the exact value: $\sin\dfrac{\pi}{3}\cos\dfrac{\pi}{3}$

'Are You Prepared?' Answers

1. Domain: the set of all real numbers; range: $-1 \le y \le 1$ **2.** Two answers are possible: $x \le 1$ or $x \ge 1$

3. $[3, \infty)$ **4.** True **5.** $1; \dfrac{\sqrt{3}}{2}$ **6.** $-\dfrac{1}{2}; -1$

3.2 The Inverse Trigonometric Functions (Continued)

PREPARING FOR THIS SECTION *Before getting started, review the following concepts:*

- Finding Exact Values Given the Value of a Trigonometric Function and the Quadrant of the Angle (Section 2.3, pp. 138–140)
- Graphs of the Secant, Cosecant, and Cotangent Functions (Section 2.5, pp. 163–165)
- Domain and Range of the Secant, Cosecant, and Cotangent Functions (Section 2.3, pp. 131–133)

Now Work the **'Are You Prepared?'** problems on page 205.

OBJECTIVES
1 Find the Exact Value of Expressions Involving the Inverse Sine, Cosine, and Tangent Functions (p. 202)
2 Define the Inverse Secant, Cosecant, and Cotangent Functions (p. 203)
3 Use a Calculator to Evaluate $\sec^{-1} x$, $\csc^{-1} x$, and $\cot^{-1} x$ (p. 204)
4 Write a Trigonometric Expression as an Algebraic Expression (p. 205)

1 Find the Exact Value of Expressions Involving the Inverse Sine, Cosine, and Tangent Functions

EXAMPLE 1 **Finding the Exact Value of Expressions Involving Inverse Trigonometric Functions**

Find the exact value of: $\sin\left(\tan^{-1}\frac{1}{2}\right)$

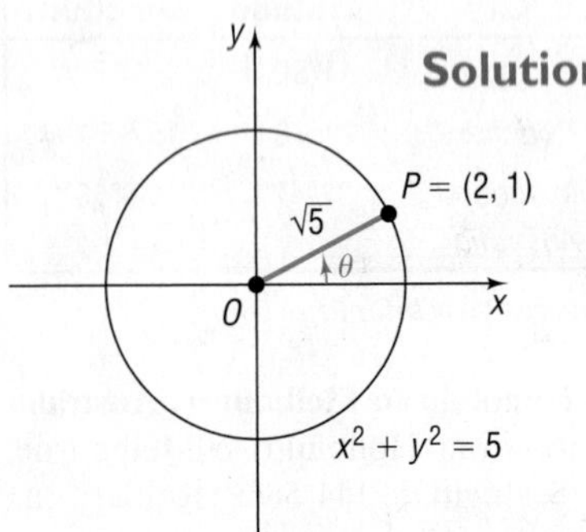

Figure 15 $\tan\theta = \frac{1}{2}$

Solution Let $\theta = \tan^{-1}\frac{1}{2}$. Then $\tan\theta = \frac{1}{2}$, where $-\frac{\pi}{2} < \theta < \frac{\pi}{2}$. We seek $\sin\theta$. Because $\tan\theta > 0$, it follows that $0 < \theta < \frac{\pi}{2}$, so θ lies in quadrant I. Because $\tan\theta = \frac{1}{2} = \frac{y}{x}$, let $x = 2$ and $y = 1$. Since $r = d(O, P) = \sqrt{2^2 + 1^2} = \sqrt{5}$, the point $P = (x, y) = (2, 1)$ is on the circle $x^2 + y^2 = 5$. See Figure 15. Then, with $x = 2$, $y = 1$, and $r = \sqrt{5}$, it follows that

$$\sin\left(\tan^{-1}\frac{1}{2}\right) = \sin\theta = \underset{\uparrow\ \sin\theta = \frac{y}{r}}{\frac{1}{\sqrt{5}}} = \frac{\sqrt{5}}{5}$$

EXAMPLE 2 **Finding the Exact Value of Expressions Involving Inverse Trigonometric Functions**

Find the exact value of: $\cos\left[\sin^{-1}\left(-\frac{1}{3}\right)\right]$

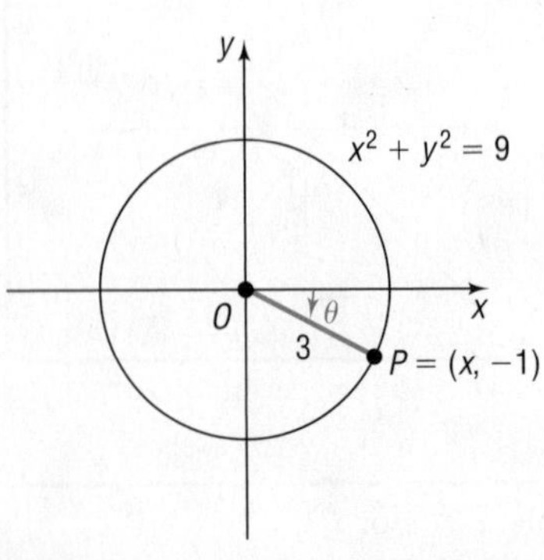

Figure 16 $\sin\theta = -\frac{1}{3}$

Solution Let $\theta = \sin^{-1}\left(-\frac{1}{3}\right)$. Then $\sin\theta = -\frac{1}{3}$ and $-\frac{\pi}{2} \le \theta \le \frac{\pi}{2}$. We seek $\cos\theta$. Because $\sin\theta < 0$, it follows that $-\frac{\pi}{2} \le \theta < 0$, so θ lies in quadrant IV. Since $\sin\theta = \frac{-1}{3} = \frac{y}{r}$, let $y = -1$ and $r = 3$. The point $P = (x, y) = (x, -1), x > 0$, is on a circle of radius 3, $x^2 + y^2 = 9$. See Figure 16. Then

$$x^2 + y^2 = 9$$
$$x^2 + (-1)^2 = 9 \qquad y = -1$$
$$x^2 = 8$$
$$x = 2\sqrt{2} \qquad x > 0$$

Using $x = 2\sqrt{2}$, $y = -1$, and $r = 3$ gives the result

$$\cos\left[\sin^{-1}\left(-\frac{1}{3}\right)\right] = \cos\theta = \frac{2\sqrt{2}}{3}$$

$\uparrow \cos\theta = \frac{x}{r}$

EXAMPLE 3 **Finding the Exact Value of Expressions Involving Inverse Trigonometric Functions**

Find the exact value of: $\tan\left[\cos^{-1}\left(-\frac{1}{3}\right)\right]$

Solution Let $\theta = \cos^{-1}\left(-\frac{1}{3}\right)$. Then $\cos\theta = -\frac{1}{3}$ and $0 \le \theta \le \pi$. We seek $\tan\theta$. Because $\cos\theta < 0$, it follows that $\frac{\pi}{2} < \theta \le \pi$, so θ lies in quadrant II. Since $\cos\theta = \frac{-1}{3} = \frac{x}{r}$, let $x = -1$ and $r = 3$. The point $P = (x, y) = (-1, y)$, $y > 0$, is on a circle of radius $r = 3$, $x^2 + y^2 = 9$. See Figure 17. Then

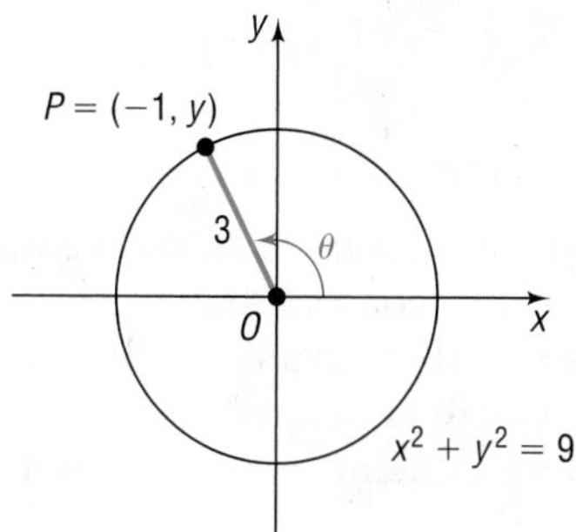

Figure 17 $\cos\theta = -\frac{1}{3}$

$$x^2 + y^2 = 9$$

$$(-1)^2 + y^2 = 9 \qquad x = -1$$

$$y^2 = 8$$

$$y = 2\sqrt{2} \qquad y > 0$$

Then, $x = -1$, $y = 2\sqrt{2}$, and $r = 3$, which means

$$\tan\left[\cos^{-1}\left(-\frac{1}{3}\right)\right] = \tan\theta = \frac{2\sqrt{2}}{-1} = -2\sqrt{2}$$

$\uparrow \tan\theta = \frac{y}{x}$

Now Work PROBLEMS 9 AND 27

2 Define the Inverse Secant, Cosecant, and Cotangent Functions

The inverse secant, inverse cosecant, and inverse cotangent functions are defined as follows:

DEFINITION

$$y = \sec^{-1} x \quad \text{means} \quad x = \sec y \tag{1}$$
$$\text{where} \quad |x| \ge 1 \quad \text{and} \quad 0 \le y \le \pi, \quad y \ne \frac{\pi}{2}^{*}$$

$$y = \csc^{-1} x \quad \text{means} \quad x = \csc y \tag{2}$$
$$\text{where} \quad |x| \ge 1 \quad \text{and} \quad -\frac{\pi}{2} \le y \le \frac{\pi}{2}, \quad y \ne 0^{\dagger}$$

$$y = \cot^{-1} x \quad \text{means} \quad x = \cot y \tag{3}$$
$$\text{where} \quad -\infty < x < \infty \quad \text{and} \quad 0 < y < \pi$$

You are encouraged to review the graphs of the cotangent, cosecant, and secant functions in Figures 66, 67, and 68 in Section 2.5 to help you to see the basis for these definitions.

*Most books use this definition. A few use the restriction $0 \le y < \frac{\pi}{2}, \pi \le y < \frac{3\pi}{2}$.

†Most books use this definition. A few use the restriction $-\pi < y \le -\frac{\pi}{2}, 0 < y \le \frac{\pi}{2}$.

EXAMPLE 4 **Finding the Exact Value of an Inverse Cosecant Function**

Find the exact value of: $\csc^{-1} 2$

Solution Let $\theta = \csc^{-1} 2$. We seek the angle θ, $-\frac{\pi}{2} \le \theta \le \frac{\pi}{2}$, $\theta \ne 0$, whose cosecant equals 2 $\left(\text{or, equivalently, whose sine equals } \frac{1}{2}\right)$.

$$\theta = \csc^{-1} 2 \qquad -\frac{\pi}{2} \le \theta \le \frac{\pi}{2}, \quad \theta \ne 0$$

$$\csc \theta = 2 \qquad -\frac{\pi}{2} \le \theta \le \frac{\pi}{2}, \quad \theta \ne 0 \quad \sin\theta = \frac{1}{2}$$

The only angle θ in the interval $-\frac{\pi}{2} \le \theta \le \frac{\pi}{2}$, $\theta \ne 0$, whose cosecant is 2 $\left[\sin\theta = \frac{1}{2}\right]$ is $\frac{\pi}{6}$, so $\csc^{-1} 2 = \frac{\pi}{6}$. ●

Now Work PROBLEM 39

3 Use a Calculator to Evaluate $\sec^{-1} x$, $\csc^{-1} x$, and $\cot^{-1} x$

Most calculators do not have keys for evaluating the inverse cotangent, cosecant, and secant functions. The easiest way to evaluate them is to convert to an inverse trigonometric function whose range is the same as the one to be evaluated. In this regard, notice that $y = \cot^{-1} x$ and $y = \sec^{-1} x$, except where undefined, each have the same range as $y = \cos^{-1} x$; $y = \csc^{-1} x$, except where undefined, has the same range as $y = \sin^{-1} x$.

NOTE Remember that the range of $y = \sin^{-1} x$ is $\left[-\frac{\pi}{2}, \frac{\pi}{2}\right]$; the range of $y = \cos^{-1} x$ is $[0, \pi]$. ■

EXAMPLE 5 **Approximating the Value of Inverse Trigonometric Functions**

Use a calculator to approximate each expression in radians rounded to two decimal places.

(a) $\sec^{-1} 3$ (b) $\csc^{-1}(-4)$ (c) $\cot^{-1}\frac{1}{2}$ (d) $\cot^{-1}(-2)$

Solution First, set your calculator to radian mode.

(a) Let $\theta = \sec^{-1} 3$. Then $\sec\theta = 3$ and $0 \le \theta \le \pi$, $\theta \ne \frac{\pi}{2}$. Now find $\cos\theta$ because $y = \cos^{-1} x$ has the same range as $y = \sec^{-1} x$, except where undefined. Because $\sec\theta = \frac{1}{\cos\theta} = 3$, this means $\cos\theta = \frac{1}{3}$. Then $\theta = \cos^{-1}\frac{1}{3}$ and

$$\sec^{-1} 3 = \theta = \cos^{-1}\frac{1}{3} \approx 1.23$$

Use a calculator.

(b) Let $\theta = \csc^{-1}(-4)$. Then $\csc\theta = -4$, $-\frac{\pi}{2} \le \theta \le \frac{\pi}{2}$, $\theta \ne 0$. Now find $\sin\theta$ because $y = \sin^{-1} x$ has the same range as $y = \csc^{-1} x$, except where undefined. Because $\csc\theta = \frac{1}{\sin\theta} = -4$, this means $\sin\theta = -\frac{1}{4}$. Then $\theta = \sin^{-1}\left(-\frac{1}{4}\right)$, and

$$\csc^{-1}(-4) = \theta = \sin^{-1}\left(-\frac{1}{4}\right) \approx -0.25$$

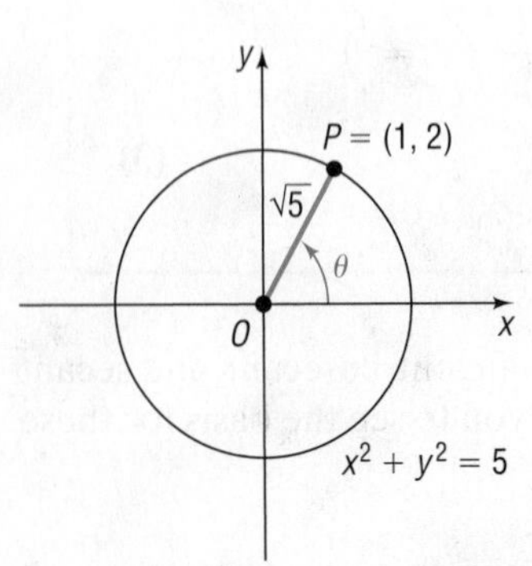

Figure 18 $\cot\theta = \frac{1}{2}, 0 < \theta < \pi$

(c) Let $\theta = \cot^{-1}\frac{1}{2}$. Then $\cot\theta = \frac{1}{2}$, $0 < \theta < \pi$. So θ lies in quadrant I. Now find $\cos\theta$ because $y = \cos^{-1} x$ has the same range as $y = \cot^{-1} x$, except where undefined. Use Figure 18 to find that $\cos\theta = \frac{1}{\sqrt{5}}$, $0 < \theta < \frac{\pi}{2}$.

So $\theta = \cos^{-1}\left(\frac{1}{\sqrt{5}}\right)$, and

$$\cot^{-1}\frac{1}{2} = \theta = \cos^{-1}\left(\frac{1}{\sqrt{5}}\right) \approx 1.11$$

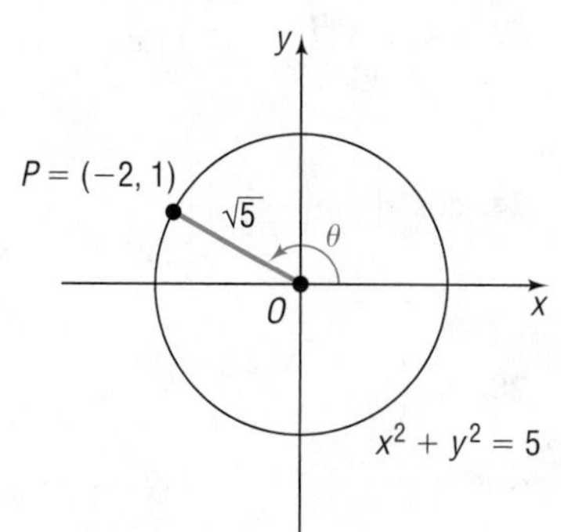

Figure 19 $\cot\theta = -2, 0 < \theta < \pi$

(d) Let $\theta = \cot^{-1}(-2)$. Then $\cot\theta = -2, 0 < \theta < \pi$. These facts indicate that θ lies in quadrant II. Now find $\cos\theta$. Use Figure 19 to find that $\cos\theta = -\frac{2}{\sqrt{5}}, \frac{\pi}{2} < \theta < \pi$. This means $\theta = \cos^{-1}\left(-\frac{2}{\sqrt{5}}\right)$, and

$$\cot^{-1}(-2) = \theta = \cos^{-1}\left(-\frac{2}{\sqrt{5}}\right) \approx 2.68$$

Now Work PROBLEM 45

4 Write a Trigonometric Expression as an Algebraic Expression

EXAMPLE 6 **Writing a Trigonometric Expression as an Algebraic Expression**

Write $\sin(\tan^{-1} u)$ as an algebraic expression containing u.

Solution Let $\theta = \tan^{-1} u$ so that $\tan\theta = u, -\frac{\pi}{2} < \theta < \frac{\pi}{2}, -\infty < u < \infty$. This means $\sec\theta > 0$. Then

$$\sin(\tan^{-1} u) = \sin\theta = \sin\theta \cdot \frac{\cos\theta}{\cos\theta} = \tan\theta\cos\theta = \frac{\tan\theta}{\sec\theta} = \frac{\tan\theta}{\sqrt{1+\tan^2\theta}} = \frac{u}{\sqrt{1+u^2}}$$

Multiply by 1: $\frac{\cos\theta}{\cos\theta}$; $\frac{\sin\theta}{\cos\theta} = \tan\theta$; $\sec^2\theta = 1 + \tan^2\theta$, $\sec\theta > 0$

Now Work PROBLEM 57

3.2 Assess Your Understanding

'Are You Prepared?' *Answers are given at the end of these exercises. If you get a wrong answer, read the pages listed in red.*

1. What is the domain and the range of $y = \sec x$? (pp. 131–133)
2. ***True or False*** The graph of $y = \sec x$ is one-to-one on the interval $\left[0, \frac{\pi}{2}\right)$ and on the interval $\left(\frac{\pi}{2}, \pi\right]$. (pp. 163–165)
3. If $\tan\theta = \frac{1}{2}, -\frac{\pi}{2} < \theta < \frac{\pi}{2}$, then $\sin\theta =$ ______. (pp. 138–140)

Concepts and Vocabulary

4. $y = \sec^{-1} x$ means ________, where $|x|$ _____ and _____ $\leq y \leq$ _____, $y \neq \frac{\pi}{2}$.
5. To find the inverse secant of a real number x such that $|x| \geq 1$, convert the inverse secant to an inverse ______.
6. ***True or False*** It is impossible to obtain exact values for the inverse secant function.
7. ***True or False*** $\csc^{-1} 0.5$ is not defined.
8. ***True or False*** The domain of the inverse cotangent function is the set of real numbers.

Skill Building

In Problems 9–36, find the exact value of each expression.

9. $\cos\left(\sin^{-1}\frac{\sqrt{2}}{2}\right)$
10. $\sin\left(\cos^{-1}\frac{1}{2}\right)$
11. $\tan\left[\cos^{-1}\left(-\frac{\sqrt{3}}{2}\right)\right]$
12. $\tan\left[\sin^{-1}\left(-\frac{1}{2}\right)\right]$

13. $\sec\left(\cos^{-1}\frac{1}{2}\right)$ **14.** $\cot\left[\sin^{-1}\left(-\frac{1}{2}\right)\right]$ **15.** $\csc(\tan^{-1} 1)$ **16.** $\sec(\tan^{-1}\sqrt{3})$

17. $\sin[\tan^{-1}(-1)]$ **18.** $\cos\left[\sin^{-1}\left(-\frac{\sqrt{3}}{2}\right)\right]$ **19.** $\sec\left[\sin^{-1}\left(-\frac{1}{2}\right)\right]$ **20.** $\csc\left[\cos^{-1}\left(-\frac{\sqrt{3}}{2}\right)\right]$

21. $\cos^{-1}\left(\sin\frac{5\pi}{4}\right)$ **22.** $\tan^{-1}\left(\cot\frac{2\pi}{3}\right)$ **23.** $\sin^{-1}\left[\cos\left(-\frac{7\pi}{6}\right)\right]$ **24.** $\cos^{-1}\left[\tan\left(-\frac{\pi}{4}\right)\right]$

25. $\tan\left(\sin^{-1}\frac{1}{3}\right)$ **26.** $\tan\left(\cos^{-1}\frac{1}{3}\right)$ 27. $\sec\left(\tan^{-1}\frac{1}{2}\right)$ **28.** $\cos\left(\sin^{-1}\frac{\sqrt{2}}{3}\right)$

29. $\cot\left[\sin^{-1}\left(-\frac{\sqrt{2}}{3}\right)\right]$ **30.** $\csc[\tan^{-1}(-2)]$ **31.** $\sin[\tan^{-1}(-3)]$ **32.** $\cot\left[\cos^{-1}\left(-\frac{\sqrt{3}}{3}\right)\right]$

33. $\sec\left(\sin^{-1}\frac{2\sqrt{5}}{5}\right)$ **34.** $\csc\left(\tan^{-1}\frac{1}{2}\right)$ **35.** $\sin^{-1}\left(\cos\frac{3\pi}{4}\right)$ **36.** $\cos^{-1}\left(\sin\frac{7\pi}{6}\right)$

In Problems 37–44, find the exact value of each expression.

37. $\cot^{-1}\sqrt{3}$ **38.** $\cot^{-1} 1$ 39. $\csc^{-1}(-1)$ **40.** $\csc^{-1}\sqrt{2}$

41. $\sec^{-1}\frac{2\sqrt{3}}{3}$ **42.** $\sec^{-1}(-2)$ **43.** $\cot^{-1}\left(-\frac{\sqrt{3}}{3}\right)$ **44.** $\csc^{-1}\left(-\frac{2\sqrt{3}}{3}\right)$

In Problems 45–56, use a calculator to find the value of each expression rounded to two decimal places.

45. $\sec^{-1} 4$ **46.** $\csc^{-1} 5$ **47.** $\cot^{-1} 2$ **48.** $\sec^{-1}(-3)$

49. $\csc^{-1}(-3)$ **50.** $\cot^{-1}\left(-\frac{1}{2}\right)$ **51.** $\cot^{-1}(-\sqrt{5})$ **52.** $\cot^{-1}(-8.1)$

53. $\csc^{-1}\left(-\frac{3}{2}\right)$ **54.** $\sec^{-1}\left(-\frac{4}{3}\right)$ **55.** $\cot^{-1}\left(-\frac{3}{2}\right)$ **56.** $\cot^{-1}(-\sqrt{10})$

In Problems 57–66, write each trigonometric expression as an algebraic expression in u.

57. $\cos(\tan^{-1} u)$ **58.** $\sin(\cos^{-1} u)$ **59.** $\tan(\sin^{-1} u)$ **60.** $\tan(\cos^{-1} u)$ **61.** $\sin(\sec^{-1} u)$

62. $\sin(\cot^{-1} u)$ **63.** $\cos(\csc^{-1} u)$ **64.** $\cos(\sec^{-1} u)$ **65.** $\tan(\cot^{-1} u)$ **66.** $\tan(\sec^{-1} u)$

Mixed Practice

In Problems 67–78, $f(x) = \sin x,\ -\frac{\pi}{2} \le x \le \frac{\pi}{2}$, $g(x) = \cos x,\ 0 \le x \le \pi$, *and* $h(x) = \tan x,\ -\frac{\pi}{2} < x < \frac{\pi}{2}$. *Find the exact value of each composite function.*

67. $g\left(f^{-1}\left(\frac{12}{13}\right)\right)$ **68.** $f\left(g^{-1}\left(\frac{5}{13}\right)\right)$ **69.** $g^{-1}\left(f\left(\frac{7\pi}{4}\right)\right)$ **70.** $f^{-1}\left(g\left(\frac{5\pi}{6}\right)\right)$

71. $h\left(f^{-1}\left(-\frac{3}{5}\right)\right)$ **72.** $h\left(g^{-1}\left(-\frac{4}{5}\right)\right)$ **73.** $g\left(h^{-1}\left(\frac{12}{5}\right)\right)$ **74.** $f\left(h^{-1}\left(\frac{5}{12}\right)\right)$

75. $g^{-1}\left(f\left(-\frac{4\pi}{3}\right)\right)$ **76.** $g^{-1}\left(f\left(-\frac{5\pi}{6}\right)\right)$ **77.** $h\left(g^{-1}\left(-\frac{1}{4}\right)\right)$ **78.** $h\left(f^{-1}\left(-\frac{2}{5}\right)\right)$

Applications and Extensions

Problems 79 and 80 require the following discussion: When granular materials are allowed to fall freely, they form conical (cone-shaped) piles. The naturally occurring angle of slope, measured from the horizontal, at which the loose material comes to rest is called the **angle of repose** *and varies for different materials. The angle of repose θ is related to the height h and base radius r of the conical pile by the equation $\theta = \cot^{-1}\frac{r}{h}$. See the illustration.*

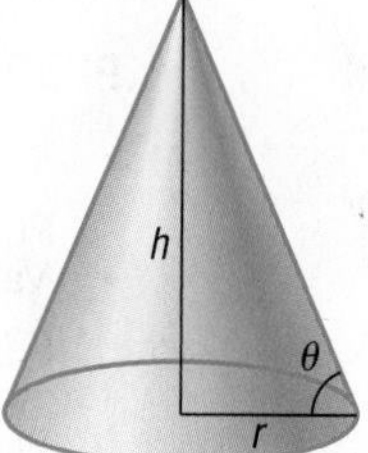

79. Angle of Repose: Deicing Salt Due to potential transportation issues (for example, frozen waterways) deicing salt used by highway departments in the Midwest must be ordered early and stored for future use. When deicing salt is stored in a pile 14 feet high, the diameter of the base of the pile is 45 feet.
(a) Find the angle of repose for deicing salt.
(b) What is the base diameter of a pile that is 17 feet high?
(c) What is the height of a pile that has a base diameter of approximately 122 feet?
Source: Salt Institute, *The Salt Storage Handbook,* 2013

80. Angle of Repose: Bunker Sand The steepness of sand bunkers on a golf course is affected by the angle of repose of the sand (a larger angle of repose allows for steeper bunkers). A freestanding pile of loose sand from a United States Golf Association (USGA) bunker had a height of 4 feet and a base diameter of approximately 6.68 feet.
(a) Find the angle of repose for USGA bunker sand.
(b) What is the height of such a pile if the diameter of the base is 8 feet?
(c) A 6-foot-high pile of loose Tour Grade 50/50 sand has a base diameter of approximately 8.44 feet. Which type of sand (USGA or Tour Grade 50/50) would be better suited for steep bunkers?
Source: 2004 Annual Report, Purdue University Turfgrass Science Program

81. Artillery A projectile fired into the first quadrant from the origin of a coordinate system will pass through the point (x, y) at time t according to the relationship $\cot\theta = \frac{2x}{2y + gt^2}$, where $\theta =$ the angle of elevation of the launcher and $g =$ the acceleration due to gravity $= 32.2$ feet/second^2. An artilleryman is firing at an enemy bunker located 2450 feet up the side of a hill that is 6175 feet away. He fires a round, and exactly 2.27 seconds later he scores a direct hit.
(a) What angle of elevation did he use?
(b) If the angle of elevation is also given by $\sec\theta = \frac{v_0 t}{x}$, where v_0 is the muzzle velocity of the weapon, find the muzzle velocity of the artillery piece he used.
Source: www.egwald.com/geometry/projectile3d.php

82. Using a graphing utility, graph $y = \cot^{-1}x$.
83. Using a graphing utility, graph $y = \sec^{-1}x$.
84. Using a graphing utility, graph $y = \csc^{-1}x$.

Explaining Concepts: Discussion and Writing

85. Explain in your own words how you would use your calculator to find the value of $\cot^{-1} 10$.

86. Consult three books on calculus and write down the definition in each of $y = \sec^{-1} x$ and $y = \csc^{-1} x$. Compare these with the definitions given in this book.

Retain Your Knowledge

Problems 87–90 are based on material learned earlier in the course. The purpose of these problems is to keep the material fresh in your mind so that you are better prepared for the final exam.

87. Find the domain of $f(x) = \frac{x - 1}{x^2 - 25}$.
88. Determine algebraically whether $f(x) = x^3 + x^2 - x$ is even, odd, or neither.
89. Convert 315° to radians.
90. Find the length of the arc subtended by a central angle of 75° on a circle of radius 6 inches. Give both the exact length and an approximation rounded to two decimal places.

'Are You Prepared?' Answers

1. Domain: $\left\{x \middle| x \neq \text{odd integer multiples of } \frac{\pi}{2}\right\}$; range: $\{y \le -1 \text{ or } y \ge 1\}$ **2.** True **3.** $\frac{\sqrt{5}}{5}$

3.3 Trigonometric Equations

PREPARING FOR THIS SECTION *Before getting started, review the following:*

- Solving Equations (Appendix A, Section A.4, pp. A27–A34)
- Values of the Trigonometric Functions (Section 2.2, pp. 116–125)
- Using a Graphing Utility to Solve Equations (Appendix B, Section B.4, pp. B6–B7)

Now Work the 'Are You Prepared?' problems on page 213.

OBJECTIVES
1. Solve Equations Involving a Single Trigonometric Function (p. 208)
2. Solve Trigonometric Equations Using a Calculator (p. 211)
3. Solve Trigonometric Equations Quadratic in Form (p. 211)
4. Solve Trigonometric Equations Using Fundamental Identities (p. 212)
5. Solve Trigonometric Equations Using a Graphing Utility (p. 213)

1 Solve Equations Involving a Single Trigonometric Function

In this section, we discuss **trigonometric equations**—that is, equations involving trigonometric functions that are satisfied only by some values of the variable (or, possibly, are not satisfied by any values of the variable). The values that satisfy the equation are called **solutions** of the equation.

EXAMPLE 1 Checking Whether a Given Number Is a Solution of a Trigonometric Equation

Determine whether $\theta = \frac{\pi}{4}$ is a solution of the equation $2 \sin \theta - 1 = 0$. Is $\theta = \frac{\pi}{6}$ a solution?

Solution Replace θ by $\frac{\pi}{4}$ in the given equation. The result is

$$2 \sin \frac{\pi}{4} - 1 = 2 \cdot \frac{\sqrt{2}}{2} - 1 = \sqrt{2} - 1 \neq 0$$

Therefore, $\frac{\pi}{4}$ is not a solution.

Next replace θ by $\frac{\pi}{6}$ in the equation. The result is

$$2 \sin \frac{\pi}{6} - 1 = 2 \cdot \frac{1}{2} - 1 = 0$$

Therefore, $\frac{\pi}{6}$ is a solution of the given equation. ●

The equation given in Example 1 has other solutions besides $\theta = \frac{\pi}{6}$. For example, $\theta = \frac{5\pi}{6}$ is also a solution, as is $\theta = \frac{13\pi}{6}$. (You should check this for yourself.) In fact, the equation has an infinite number of solutions due to the periodicity of the sine function, as can be seen in Figure 20 where we graph $y = 2 \sin x - 1$. Each x-intercept of the graph represents a solution to the equation $2 \sin x - 1 = 0$.

Unless the domain of the variable is restricted, we need to find *all* the solutions of a trigonometric equation. As the next example illustrates, finding all the solutions can be accomplished by first finding solutions over an interval whose length equals the period of the function and then adding multiples of that period to the solutions found.

Figure 20

EXAMPLE 2 Finding All the Solutions of a Trigonometric Equation

Solve the equation: $\cos\theta = \dfrac{1}{2}$

Give a general formula for all the solutions. List eight of the solutions.

In Words

Solving the equation $\cos\theta = \dfrac{1}{2}$ means finding all the angles θ whose cosine is $\dfrac{1}{2}$.

Solution The period of the cosine function is 2π. In the interval $[0, 2\pi)$, there are two angles θ for which $\cos\theta = \dfrac{1}{2}$: $\theta = \dfrac{\pi}{3}$ and $\theta = \dfrac{5\pi}{3}$. See Figure 21. Because the cosine function has period 2π, all the solutions of $\cos\theta = \dfrac{1}{2}$ may be given by the general formula

$$\theta = \frac{\pi}{3} + 2k\pi \quad \text{or} \quad \theta = \frac{5\pi}{3} + 2k\pi \qquad k \text{ any integer}$$

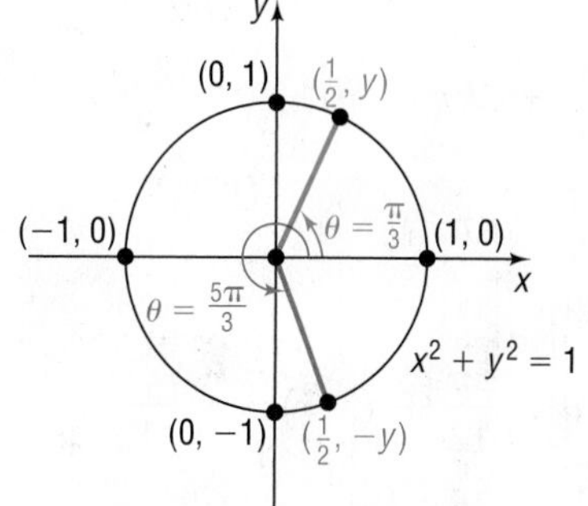

Figure 21

Eight of the solutions are

$$\underbrace{-\frac{5\pi}{3}, -\frac{\pi}{3}}_{k=-1}, \underbrace{\frac{\pi}{3}, \frac{5\pi}{3}}_{k=0}, \underbrace{\frac{7\pi}{3}, \frac{11\pi}{3}}_{k=1}, \underbrace{\frac{13\pi}{3}, \frac{17\pi}{3}}_{k=2}$$

Check: To verify the solutions, graph $Y_1 = \cos x$ and $Y_2 = \dfrac{1}{2}$ and determine where the graphs intersect. (Be sure to graph in radian mode.) See Figure 22. The graph of Y_1 intersects the graph of Y_2 at $x = 1.05\left(\approx\dfrac{\pi}{3}\right)$, $5.24\left(\approx\dfrac{5\pi}{3}\right)$, $7.33\left(\approx\dfrac{7\pi}{3}\right)$, and $11.52\left(\approx\dfrac{11\pi}{3}\right)$, rounded to two decimal places.

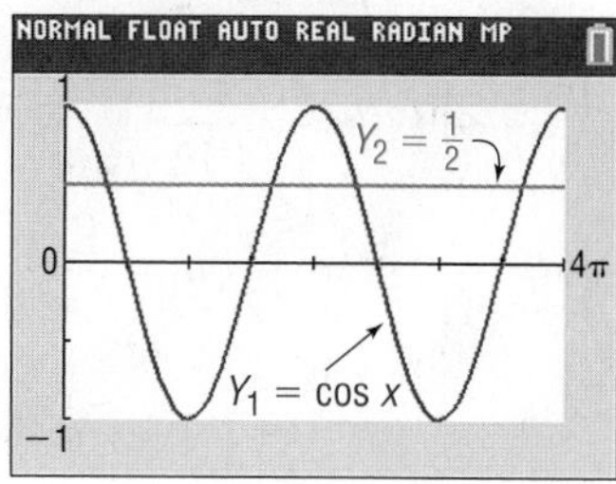

Figure 22

Now Work PROBLEM 37

In most of the work we do, we shall be interested only in finding solutions of trigonometric equations for $0 \le \theta < 2\pi$.

EXAMPLE 3 Solving a Linear Trigonometric Equation

Solve the equation: $2\sin\theta + \sqrt{3} = 0, \quad 0 \le \theta < 2\pi$

Solution First solve the equation for $\sin\theta$.

$$\begin{aligned} 2\sin\theta + \sqrt{3} &= 0 \\ 2\sin\theta &= -\sqrt{3} && \text{Subtract } \sqrt{3} \text{ from both sides.} \\ \sin\theta &= -\frac{\sqrt{3}}{2} && \text{Divide both sides by 2.} \end{aligned}$$

In the interval $[0, 2\pi)$, there are two angles θ for which $\sin\theta = -\frac{\sqrt{3}}{2}$: $\theta = \frac{4\pi}{3}$ and $\theta = \frac{5\pi}{3}$. The solution set is $\left\{\frac{4\pi}{3}, \frac{5\pi}{3}\right\}$. ●

Now Work PROBLEM 13

When the argument of the trigonometric function in an equation is a multiple of θ, the general formula must be used to solve the equation.

EXAMPLE 4 Solving a Trigonometric Equation

Solve the equation: $\sin(2\theta) = \frac{1}{2}, \quad 0 \le \theta < 2\pi$

Solution In the interval $[0, 2\pi)$, the sine function equals $\frac{1}{2}$ at $\frac{\pi}{6}$ and $\frac{5\pi}{6}$. See Figure 23(a). Therefore, 2θ must equal $\frac{\pi}{6}$ or $\frac{5\pi}{6}$. Here's the problem, however. The period of $y = \sin(2\theta)$ is $\frac{2\pi}{2} = \pi$. So, in the interval $[0, 2\pi)$, the graph of $y = \sin(2\theta)$ completes two cycles, and the graph of $y = \sin(2\theta)$ intersects the graph of $y = \frac{1}{2}$ four times. See Figure 23(b). For this reason, there are four solutions to the equation $\sin(2\theta) = \frac{1}{2}$ in $[0, 2\pi)$. To find these solutions, write the general formula that gives all the solutions.

$$2\theta = \frac{\pi}{6} + 2k\pi \quad \text{or} \quad 2\theta = \frac{5\pi}{6} + 2k\pi \qquad k \text{ any integer}$$

$$\theta = \frac{\pi}{12} + k\pi \quad \text{or} \quad \theta = \frac{5\pi}{12} + k\pi \qquad \text{Divide by 2.}$$

Then

$$\theta = \frac{\pi}{12} + (-1)\pi = \frac{-11\pi}{12} \qquad k = -1 \qquad \theta = \frac{5\pi}{12} + (-1)\pi = \frac{-7\pi}{12}$$

$$\theta = \frac{\pi}{12} + (0)\pi = \frac{\pi}{12} \qquad k = 0 \qquad \theta = \frac{5\pi}{12} + (0)\pi = \frac{5\pi}{12}$$

$$\theta = \frac{\pi}{12} + (1)\pi = \frac{13\pi}{12} \qquad k = 1 \qquad \theta = \frac{5\pi}{12} + (1)\pi = \frac{17\pi}{12}$$

$$\theta = \frac{\pi}{12} + (2)\pi = \frac{25\pi}{12} \qquad k = 2 \qquad \theta = \frac{5\pi}{12} + (2)\pi = \frac{29\pi}{12}$$

In the interval $[0, 2\pi)$, the solutions of $\sin(2\theta) = \frac{1}{2}$ are $\theta = \frac{\pi}{12}, \theta = \frac{5\pi}{12}, \theta = \frac{13\pi}{12}$, and $\theta = \frac{17\pi}{12}$. The solution set is $\left\{\frac{\pi}{12}, \frac{5\pi}{12}, \frac{13\pi}{12}, \frac{17\pi}{12}\right\}$.

This means the graph of $y = \sin(2\theta)$ intersects the line $y = \frac{1}{2}$ at the points $\left(\frac{\pi}{12}, \frac{1}{2}\right), \left(\frac{5\pi}{12}, \frac{1}{2}\right), \left(\frac{13\pi}{12}, \frac{1}{2}\right)$, and $\left(\frac{17\pi}{12}, \frac{1}{2}\right)$ in the interval $[0, 2\pi)$. ●

✓Check: Verify these solutions by graphing $Y_1 = \sin(2x)$ and $Y_2 = \frac{1}{2}$ for $0 \le x \le 2\pi$.

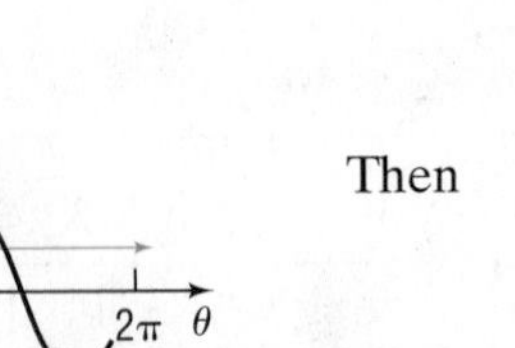

Figure 23

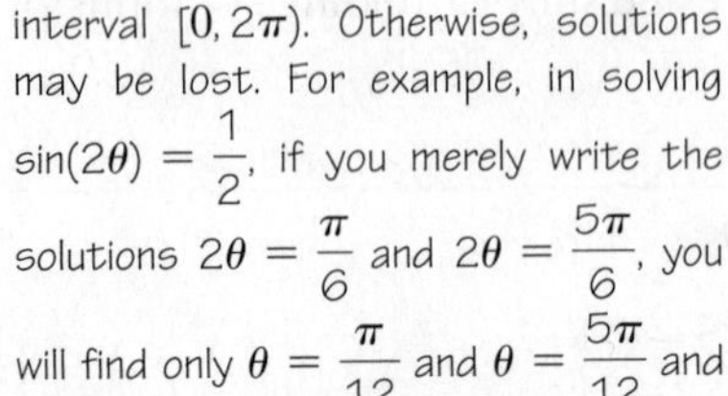
WARNING In solving a trigonometric equation for $\theta, 0 \le \theta < 2\pi$, in which the argument is not θ (as in Example 4), you must write down all the solutions first and then list those that are in the interval $[0, 2\pi)$. Otherwise, solutions may be lost. For example, in solving $\sin(2\theta) = \frac{1}{2}$, if you merely write the solutions $2\theta = \frac{\pi}{6}$ and $2\theta = \frac{5\pi}{6}$, you will find only $\theta = \frac{\pi}{12}$ and $\theta = \frac{5\pi}{12}$ and miss the other solutions. ■

EXAMPLE 5 Solving a Trigonometric Equation

Solve the equation: $\tan\left(\theta - \frac{\pi}{2}\right) = 1, \quad 0 \le \theta < 2\pi$

Solution The period of the tangent function is π. In the interval $[0, \pi)$, the tangent function has the value 1 when the argument is $\frac{\pi}{4}$. Because the argument is $\theta - \frac{\pi}{2}$ in the given equation, write the general formula that gives all the solutions.

$$\theta - \frac{\pi}{2} = \frac{\pi}{4} + k\pi \quad \text{k any integer}$$

$$\theta = \frac{3\pi}{4} + k\pi$$

In the interval $[0, 2\pi)$, $\theta = \frac{3\pi}{4}$ and $\theta = \frac{3\pi}{4} + \pi = \frac{7\pi}{4}$ are the only solutions. The solution set is $\left\{\frac{3\pi}{4}, \frac{7\pi}{4}\right\}$.

Now Work PROBLEM 27

2 Solve Trigonometric Equations Using a Calculator

The next example illustrates how to solve trigonometric equations using a calculator. Remember that the function keys on a calculator will only give values consistent with the definition of the function.

EXAMPLE 6 **Solving a Trigonometric Equation with a Calculator**

Use a calculator to solve the equation $\tan\theta = -2, 0 \le \theta < 2\pi$. Express any solutions in radians, rounded to two decimal places.

Solution To solve $\tan\theta = -2$ on a calculator, first set the mode to radians. Then use the $\boxed{\tan^{-1}}$ key to obtain

$$\theta = \tan^{-1}(-2) \approx -1.1071487$$

Rounded to two decimal places, $\theta = \tan^{-1}(-2) = -1.11$ radian. Because of the definition of $y = \tan^{-1} x$, the angle θ that is obtained is the angle $-\frac{\pi}{2} < \theta < \frac{\pi}{2}$ for which $\tan\theta = -2$. Because we seek solutions for which $0 \le \theta < 2\pi$, we express the angle as $2\pi - 1.11$. See the blue portion of Figure 24.

Another angle for which $\tan\theta = -2$ is $\pi - 1.11$. See the red portion of Figure 24. The angle $\pi - 1.11$ is the angle in quadrant II, where $\tan\theta = -2$. The solutions for $\tan\theta = -2, 0 \le \theta < 2\pi$, are

$$\theta = 2\pi - 1.11 \approx 5.17 \text{ radians} \quad \text{and} \quad \theta = \pi - 1.11 \approx 2.03 \text{ radians}$$

The solution set is {5.17, 2.03}.

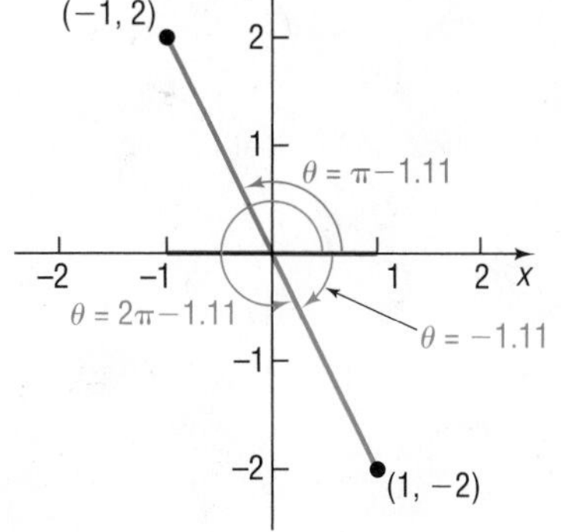

Figure 24 $\tan\theta = -2$

WARNING Example 6 illustrates that caution must be exercised when solving trigonometric equations on a calculator. Remember that the calculator supplies an angle only within the restrictions of the definition of the inverse trigonometric function. To find the remaining solutions, you must identify other quadrants, if any, in which a solution may be located.

Now Work PROBLEM 47

3 Solve Trigonometric Equations Quadratic in Form

Many trigonometric equations can be solved by applying techniques that we already know, such as applying the quadratic formula (if the equation is a second-degree polynomial) or factoring.

EXAMPLE 7 **Solving a Trigonometric Equation Quadratic in Form**

Solve the equation: $2\sin^2\theta - 3\sin\theta + 1 = 0, \quad 0 \le \theta < 2\pi$

Solution This equation is a quadratic equation (in $\sin\theta$) that can be factored.

$$2\sin^2\theta - 3\sin\theta + 1 = 0 \qquad 2x^2 - 3x + 1 = 0, \quad x = \sin\theta$$
$$(2\sin\theta - 1)(\sin\theta - 1) = 0 \qquad (2x-1)(x-1) = 0$$
$$2\sin\theta - 1 = 0 \quad \text{or} \quad \sin\theta - 1 = 0 \qquad \text{Use the Zero-Product Property.}$$
$$\sin\theta = \frac{1}{2} \quad \text{or} \quad \sin\theta = 1$$

Solving each equation in the interval $[0, 2\pi)$ yields

$$\theta = \frac{\pi}{6} \qquad \theta = \frac{5\pi}{6} \qquad \theta = \frac{\pi}{2}$$

The solution set is $\left\{\frac{\pi}{6}, \frac{5\pi}{6}, \frac{\pi}{2}\right\}$.

Now Work PROBLEM 61

4 Solve Trigonometric Equations Using Fundamental Identities

When a trigonometric equation contains more than one trigonometric function, identities sometimes can be used to obtain an equivalent equation that contains only one trigonometric function.

EXAMPLE 8 **Solving a Trigonometric Equation Using Identities**

Solve the equation: $3\cos\theta + 3 = 2\sin^2\theta, \quad 0 \le \theta < 2\pi$

Solution The equation in its present form contains a sine and a cosine. However, a form of the Pythagorean Identity, $\sin^2\theta + \cos^2\theta = 1$, can be used to transform the equation into an equivalent one containing only cosines.

$$3\cos\theta + 3 = 2\sin^2\theta$$
$$3\cos\theta + 3 = 2(1 - \cos^2\theta) \qquad \sin^2\theta = 1 - \cos^2\theta$$
$$3\cos\theta + 3 = 2 - 2\cos^2\theta$$
$$2\cos^2\theta + 3\cos\theta + 1 = 0 \qquad \text{Quadratic in } \cos\theta$$
$$(2\cos\theta + 1)(\cos\theta + 1) = 0 \qquad \text{Factor.}$$
$$2\cos\theta + 1 = 0 \quad \text{or} \quad \cos\theta + 1 = 0 \qquad \text{Use the Zero-Product Property.}$$
$$\cos\theta = -\frac{1}{2} \quad \text{or} \quad \cos\theta = -1$$

Solving each equation in the interval $[0, 2\pi)$ yields

$$\theta = \frac{2\pi}{3} \qquad \theta = \frac{4\pi}{3} \qquad \theta = \pi$$

The solution set is $\left\{\frac{2\pi}{3}, \pi, \frac{4\pi}{3}\right\}$.

Check: Graph $Y_1 = 3\cos x + 3$ and $Y_2 = 2\sin^2 x$, $0 \le x \le 2\pi$, and find the points of intersection. How close are your approximate solutions to the exact ones found in Example 8?

EXAMPLE 9 **Solving a Trigonometric Equation Using Identities**

Solve the equation: $\cos^2\theta + \sin\theta = 2, \quad 0 \le \theta < 2\pi$

Solution This equation involves two trigonometric functions: sine and cosine. By using a Pythagorean Identity, we can express the equation in terms of just sine functions.

$$\cos^2 \theta + \sin \theta = 2$$

$$(1 - \sin^2 \theta) + \sin \theta = 2 \quad \cos^2 \theta = 1 - \sin^2 \theta$$

$$\sin^2 \theta - \sin \theta + 1 = 0$$

This is a quadratic equation in $\sin \theta$. The discriminant is

$$b^2 - 4ac = 1 - 4 = -3 < 0.$$

Therefore, the equation has no real solution. The solution set is the empty set, $\varnothing$. ●

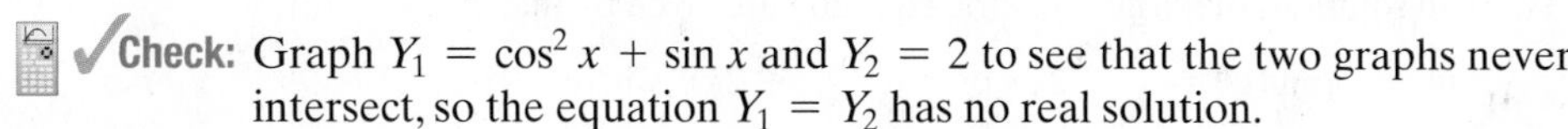

Check: Graph $Y_1 = \cos^2 x + \sin x$ and $Y_2 = 2$ to see that the two graphs never intersect, so the equation $Y_1 = Y_2$ has no real solution.

5 Solve Trigonometric Equations Using a Graphing Utility

The techniques introduced in this section apply only to certain types of trigonometric equations. Solutions for other types are usually studied in calculus, using numerical methods.

EXAMPLE 10 **Solving a Trigonometric Equation Using a Graphing Utility**

Solve: $5 \sin x + x = 3$

Express the solution(s) rounded to two decimal places.

Solution This type of trigonometric equation cannot be solved by previous methods. A graphing utility, though, can be used here. Each solution of this equation is the x-coordinate of a point of intersection of the graphs of $Y_1 = 5 \sin x + x$ and $Y_2 = 3$. See Figure 25.

There are three points of intersection; the x-coordinates provide the solutions. Use INTERSECT, to find

$$x = 0.52 \qquad x = 3.18 \qquad x = 5.71$$

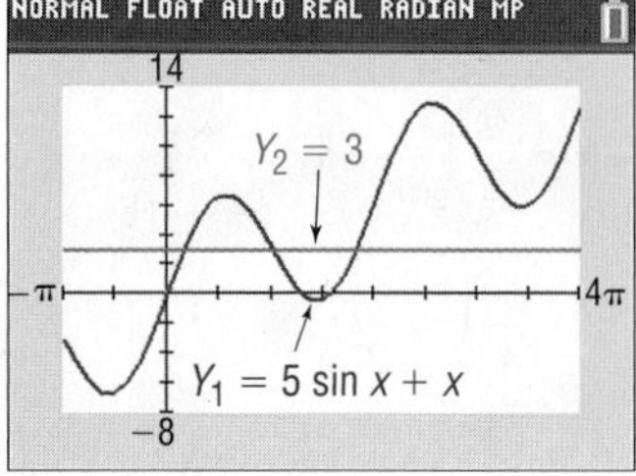

Figure 25

The solution set is $\{0.52, 3.18, 5.71\}$. ●

Now Work PROBLEM 83

3.3 Assess Your Understanding

'Are You Prepared?' *Answers are given at the end of these exercises. If you get a wrong answer, read the pages listed in red.*

1. Solve: $3x - 5 = -x + 1$ (pp. A27–A34)
2. $\sin\left(\frac{\pi}{4}\right) =$ ____; $\cos\left(\frac{8\pi}{3}\right) =$ ____ (pp. 116–125)
3. Find the real solutions of $4x^2 - x - 5 = 0$. (pp. A27–A34)
4. Find the real solutions of $x^2 - x - 1 = 0$. (pp. A27–A34)
5. Find the real solutions of $(2x - 1)^2 - 3(2x - 1) - 4 = 0$. (pp. A27–A34)
6. Use a graphing utility to solve $5x^3 - 2 = x - x^2$. Round answers to two decimal places. (pp. B6–B7)

Concepts and Vocabulary

7. ***True or False*** Most trigonometric equations have unique solutions.

8. ***True or False*** Two solutions of the equation $\sin\theta = \frac{1}{2}$ are $\frac{\pi}{6}$ and $\frac{5\pi}{6}$.

9. ***True or False*** The set of all solutions of the equation $\tan\theta = 1$ is given by $\left\{\theta \middle| \theta = \frac{\pi}{4} + k\pi, k \text{ is any integer}\right\}$.

10. ***True or False*** The equation $\sin\theta = 2$ has a real solution that can be found using a calculator.

11. If all solutions of a trigonometric equation are given by the general formula $\theta = \frac{\pi}{6} + 2k\pi$ or $\theta = \frac{11\pi}{6} + 2k\pi$, where k is any integer, then which of the following is *not* a solution of the equation?

(a) $\frac{35\pi}{6}$ (b) $\frac{23\pi}{6}$ (c) $\frac{13\pi}{6}$ (d) $\frac{7\pi}{6}$

12. Suppose $\theta = \frac{\pi}{2}$ is the only solution of a trigonometric equation in the interval $0 \le \theta < 2\pi$. Assuming a period of 2π, which of the following formulas gives all solutions of the equation, where k is any integer?

(a) $\theta = \frac{\pi}{2} + 2k\pi$ (b) $\theta = \frac{\pi}{2} + k\pi$

(c) $\theta = \frac{k\pi}{2}$ (d) $\theta = \frac{\pi + k\pi}{2}$

Skill Building

In Problems 13–36, solve each equation on the interval $0 \le \theta < 2\pi$.

13. $2\sin\theta + 3 = 2$ **14.** $1 - \cos\theta = \frac{1}{2}$ **15.** $2\sin\theta + 1 = 0$

16. $\cos\theta + 1 = 0$ **17.** $\tan\theta + 1 = 0$ **18.** $\sqrt{3}\cot\theta + 1 = 0$

19. $4\sec\theta + 6 = -2$ **20.** $5\csc\theta - 3 = 2$ **21.** $3\sqrt{2}\cos\theta + 2 = -1$

22. $4\sin\theta + 3\sqrt{3} = \sqrt{3}$ **23.** $4\cos^2\theta = 1$ **24.** $\tan^2\theta = \frac{1}{3}$

25. $2\sin^2\theta - 1 = 0$ **26.** $4\cos^2\theta - 3 = 0$ **27.** $\sin(3\theta) = -1$

28. $\tan\frac{\theta}{2} = \sqrt{3}$ **29.** $\cos(2\theta) = -\frac{1}{2}$ **30.** $\tan(2\theta) = -1$

31. $\sec\frac{3\theta}{2} = -2$ **32.** $\cot\frac{2\theta}{3} = -\sqrt{3}$ **33.** $\cos\left(2\theta - \frac{\pi}{2}\right) = -1$

34. $\sin\left(3\theta + \frac{\pi}{18}\right) = 1$ **35.** $\tan\left(\frac{\theta}{2} + \frac{\pi}{3}\right) = 1$ **36.** $\cos\left(\frac{\theta}{3} - \frac{\pi}{4}\right) = \frac{1}{2}$

In Problems 37–46, solve each equation. Give a general formula for all the solutions. List six solutions.

37. $\sin\theta = \frac{1}{2}$ **38.** $\tan\theta = 1$ **39.** $\tan\theta = -\frac{\sqrt{3}}{3}$ **40.** $\cos\theta = -\frac{\sqrt{3}}{2}$

41. $\cos\theta = 0$ **42.** $\sin\theta = \frac{\sqrt{2}}{2}$ **43.** $\cos(2\theta) = -\frac{1}{2}$ **44.** $\sin(2\theta) = -1$

45. $\sin\frac{\theta}{2} = -\frac{\sqrt{3}}{2}$ **46.** $\tan\frac{\theta}{2} = -1$

In Problems 47–58, use a calculator to solve each equation on the interval $0 \le \theta < 2\pi$. Round answers to two decimal places.

47. $\sin\theta = 0.4$ **48.** $\cos\theta = 0.6$ **49.** $\tan\theta = 5$ **50.** $\cot\theta = 2$

51. $\cos\theta = -0.9$ **52.** $\sin\theta = -0.2$ **53.** $\sec\theta = -4$ **54.** $\csc\theta = -3$

55. $5\tan\theta + 9 = 0$ **56.** $4\cot\theta = -5$ **57.** $3\sin\theta - 2 = 0$ **58.** $4\cos\theta + 3 = 0$

In Problems 59–82, solve each equation on the interval $0 \le \theta < 2\pi$.

59. $2\cos^2\theta + \cos\theta = 0$

60. $\sin^2\theta - 1 = 0$

61. $2\sin^2\theta - \sin\theta - 1 = 0$

62. $2\cos^2\theta + \cos\theta - 1 = 0$

63. $(\tan\theta - 1)(\sec\theta - 1) = 0$

64. $(\cot\theta + 1)\left(\csc\theta - \frac{1}{2}\right) = 0$

65. $\sin^2\theta - \cos^2\theta = 1 + \cos\theta$

66. $\cos^2\theta - \sin^2\theta + \sin\theta = 0$

67. $\sin^2\theta = 6(\cos(-\theta) + 1)$

68. $2\sin^2\theta = 3(1 - \cos(-\theta))$

69. $\cos\theta = -\sin(-\theta)$

70. $\cos\theta - \sin(-\theta) = 0$

71. $\tan\theta = 2\sin\theta$

72. $\tan\theta = \cot\theta$

73. $1 + \sin\theta = 2\cos^2\theta$

74. $\sin^2\theta = 2\cos\theta + 2$

75. $2\sin^2\theta - 5\sin\theta + 3 = 0$

76. $2\cos^2\theta - 7\cos\theta - 4 = 0$

77. $3(1 - \cos\theta) = \sin^2\theta$

78. $4(1 + \sin\theta) = \cos^2\theta$

79. $\tan^2\theta = \frac{3}{2}\sec\theta$

80. $\csc^2\theta = \cot\theta + 1$

81. $\sec^2\theta + \tan\theta = 0$

82. $\sec\theta = \tan\theta + \cot\theta$

In Problems 83–94, use a graphing utility to solve each equation. Express the solution(s) rounded to two decimal places.

83. $x + 5\cos x = 0$

84. $x - 4\sin x = 0$

85. $22x - 17\sin x = 3$

86. $19x + 8\cos x = 2$

87. $\sin x + \cos x = x$

88. $\sin x - \cos x = x$

89. $x^2 - 2\cos x = 0$

90. $x^2 + 3\sin x = 0$

91. $x^2 - 2\sin(2x) = 3x$

92. $x^2 = x + 3\cos(2x)$

93. $6\sin x - e^x = 2, \quad x > 0$

94. $4\cos(3x) - e^x = 1, \quad x > 0$

Mixed Practice

95. What are the zeros of $f(x) = 4\sin^2 x - 3$ on the interval $[0, 2\pi]$?

96. What are the zeros of $f(x) = 2\cos(3x) + 1$ on the interval $[0, \pi]$?

97. $f(x) = 3\sin x$
   (a) Find the zeros of f on the interval $[-2\pi, 4\pi]$.
   (b) Graph $f(x) = 3\sin x$ on the interval $[-2\pi, 4\pi]$.
   (c) Solve $f(x) = \frac{3}{2}$ on the interval $[-2\pi, 4\pi]$. What points are on the graph of f? Label these points on the graph drawn in part (b).
   (d) Use the graph drawn in part (b) along with the results of part (c) to determine the values of x such that $f(x) > \frac{3}{2}$ on the interval $[-2\pi, 4\pi]$.

98. $f(x) = 2\cos x$
   (a) Find the zeros of f on the interval $[-2\pi, 4\pi]$.
   (b) Graph $f(x) = 2\cos x$ on the interval $[-2\pi, 4\pi]$.
   (c) Solve $f(x) = -\sqrt{3}$ on the interval $[-2\pi, 4\pi]$. What points are on the graph of f? Label these points on the graph drawn in part (b).
   (d) Use the graph drawn in part (b) along with the results of part (c) to determine the values of x such that $f(x) < -\sqrt{3}$ on the interval $[-2\pi, 4\pi]$.

99. $f(x) = 4\tan x$
   (a) Solve $f(x) = -4$.
   (b) For what values of x is $f(x) < -4$ on the interval $\left(-\frac{\pi}{2}, \frac{\pi}{2}\right)$?

100. $f(x) = \cot x$
   (a) Solve $f(x) = -\sqrt{3}$.
   (b) For what values of x is $f(x) > -\sqrt{3}$ on the interval $(0, \pi)$?

101. (a) Graph $f(x) = 3\sin(2x) + 2$ and $g(x) = \frac{7}{2}$ on the same Cartesian plane for the interval $[0, \pi]$.
(b) Solve $f(x) = g(x)$ on the interval $[0, \pi]$, and label the points of intersection on the graph drawn in part (a).
(c) Solve $f(x) > g(x)$ on the interval $[0, \pi]$.
(d) Shade the region bounded by $f(x) = 3\sin(2x) + 2$ and $g(x) = \frac{7}{2}$ between the two points found in part (b) on the graph drawn in part (a).

102. (a) Graph $f(x) = 2\cos\frac{x}{2} + 3$ and $g(x) = 4$ on the same Cartesian plane for the interval $[0, 4\pi]$.
(b) Solve $f(x) = g(x)$ on the interval $[0, 4\pi]$, and label the points of intersection on the graph drawn in part (a).
(c) Solve $f(x) < g(x)$ on the interval $[0, 4\pi]$.
(d) Shade the region bounded by $f(x) = 2\cos\frac{x}{2} + 3$ and $g(x) = 4$ between the two points found in part (b) on the graph drawn in part (a).

103. (a) Graph $f(x) = -4\cos x$ and $g(x) = 2\cos x + 3$ on the same Cartesian plane for the interval $[0, 2\pi]$.
(b) Solve $f(x) = g(x)$ on the interval $[0, 2\pi]$, and label the points of intersection on the graph drawn in part (a).
(c) Solve $f(x) > g(x)$ on the interval $[0, 2\pi]$.
(d) Shade the region bounded by $f(x) = -4\cos x$ and $g(x) = 2\cos x + 3$ between the two points found in part (b) on the graph drawn in part (a).

104. (a) Graph $f(x) = 2\sin x$ and $g(x) = -2\sin x + 2$ on the same Cartesian plane for the interval $[0, 2\pi]$.
(b) Solve $f(x) = g(x)$ on the interval $[0, 2\pi]$, and label the points of intersection on the graph drawn in part (a).
(c) Solve $f(x) > g(x)$ on the interval $[0, 2\pi]$.
(d) Shade the region bounded by $f(x) = 2\sin x$ and $g(x) = -2\sin x + 2$ between the two points found in part (b) on the graph drawn in part (a).

Applications and Extensions

105. Blood Pressure Blood pressure is a way of measuring the amount of force exerted on the walls of blood vessels. It is measured using two numbers: systolic (as the heart beats) blood pressure and diastolic (as the heart rests) blood pressure. Blood pressures vary substantially from person to person, but a typical blood pressure is 120/80, which means the systolic blood pressure is 120 mmHg and the diastolic blood pressure is 80 mmHg. Assuming that a person's heart beats 70 times per minute, the blood pressure P of an individual after t seconds can be modeled by the function

$$P(t) = 100 + 20\sin\left(\frac{7\pi}{3}t\right)$$

(a) In the interval [0, 1], determine the times at which the blood pressure is 100 mmHg.
(b) In the interval [0, 1], determine the times at which the blood pressure is 120 mmHg.
(c) In the interval [0, 1], determine the times at which the blood pressure is between 100 and 105 mmHg.

106. The Ferris Wheel In 1893, George Ferris engineered the Ferris wheel. It was 250 feet in diameter. If a Ferris wheel makes 1 revolution every 40 seconds, then the function

$$h(t) = 125\sin\left(0.157t - \frac{\pi}{2}\right) + 125$$

represents the height h, in feet, of a seat on the wheel as a function of time t, where t is measured in seconds. The ride begins when $t = 0$.
(a) During the first 40 seconds of the ride, at what time t is an individual on the Ferris wheel exactly 125 feet above the ground?
(b) During the first 80 seconds of the ride, at what time t is an individual on the Ferris wheel exactly 250 feet above the ground?
(c) During the first 40 seconds of the ride, over what interval of time t is an individual on the Ferris wheel more than 125 feet above the ground?

107. Holding Pattern An airplane is asked to stay within a holding pattern near Chicago's O'Hare International Airport. The function $d(x) = 70\sin(0.65x) + 150$ represents the distance d, in miles, of the airplane from the airport at time x, in minutes.
(a) When the plane enters the holding pattern, $x = 0$, how far is it from O'Hare?
(b) During the first 20 minutes after the plane enters the holding pattern, at what time x is the plane exactly 100 miles from the airport?
(c) During the first 20 minutes after the plane enters the holding pattern, at what time x is the plane more than 100 miles from the airport?
(d) While the plane is in the holding pattern, will it ever be within 70 miles of the airport? Why?

108. Projectile Motion A golfer hits a golf ball with an initial velocity of 100 miles per hour. The range R of the ball as a function of the angle θ to the horizontal is given by $R(\theta) = 672\sin(2\theta)$, where R is measured in feet.
(a) At what angle θ should the ball be hit if the golfer wants the ball to travel 450 feet (150 yards)?
(b) At what angle θ should the ball be hit if the golfer wants the ball to travel 540 feet (180 yards)?
(c) At what angle θ should the ball be hit if the golfer wants the ball to travel at least 480 feet (160 yards)?
(d) Can the golfer hit the ball 720 feet (240 yards)?

109. Heat Transfer In the study of heat transfer, the equation $x + \tan x = 0$ occurs. Graph $Y_1 = -x$ and $Y_2 = \tan x$ for $x \geq 0$. Conclude that there are an infinite number of points of intersection of these two graphs. Now find the first two positive solutions of $x + \tan x = 0$ rounded to two decimal places.

110. Carrying a Ladder around a Corner Two hallways, one of width 3 feet, the other of width 4 feet, meet at a right angle. See the illustration on page 217. It can be shown that the length L of the ladder as a function of θ is $L(\theta) = 4\csc\theta + 3\sec\theta$.
(a) In calculus, you will be asked to find the length of the longest ladder that can turn the corner by solving the equation

$$3\sec\theta\tan\theta - 4\csc\theta\cot\theta = 0 \quad 0° < \theta < 90°$$

Solve this equation for θ.

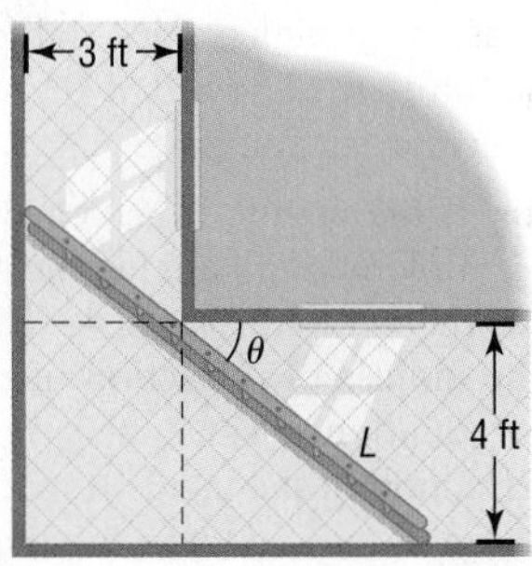

(b) What is the length of the longest ladder that can be carried around the corner?

(c) Graph $L = L(\theta), 0° \leq \theta \leq 90°$, and find the angle θ that minimizes the length L.

(d) Compare the result with the one found in part (a). Explain why the two answers are the same.

111. Projectile Motion The horizontal distance that a projectile will travel in the air (ignoring air resistance) is given by the equation

$$R(\theta) = \frac{v_0^2 \sin(2\theta)}{g}$$

where v_0 is the initial velocity of the projectile, θ is the angle of elevation, and g is acceleration due to gravity (9.8 meters per second squared).

(a) If you can throw a baseball with an initial speed of 34.8 meters per second, at what angle of elevation θ should you direct the throw so that the ball travels a distance of 107 meters before striking the ground?

(b) Determine the maximum distance that you can throw the ball.

(c) Graph $R = R(\theta)$, with $v_0 = 34.8$ meters per second.

(d) Verify the results obtained in parts (a) and (b) using a graphing utility.

112. Projectile Motion Refer to Problem 111.

(a) If you can throw a baseball with an initial speed of 40 meters per second, at what angle of elevation θ should you direct the throw so that the ball travels a distance of 110 meters before striking the ground?

(b) Determine the maximum distance that you can throw the ball.

(c) Graph $R = R(\theta)$, with $v_0 = 40$ meters per second.

(d) Verify the results obtained in parts (a) and (b) using a graphing utility.

The following discussion of Snell's Law of Refraction (named after Willebrord Snell, 1580–1626) is needed for Problems 113–120. Light, sound, and other waves travel at different speeds, depending on the medium (air, water, wood, and so on) through which they pass. Suppose that light travels from a point A in one medium, where its speed is v_1, to a point B in another medium, where its speed is v_2. Refer to the figure, where the angle θ_1 is called the angle* of incidence *and the angle θ_2 is the* angle of refraction. *Snell's Law, which can be proved using calculus, states that*

$$\frac{\sin \theta_1}{\sin \theta_2} = \frac{v_1}{v_2}$$

The ratio $\frac{v_1}{v_2}$ is called the index of refraction. *Some values are given in the table shown to the right.*

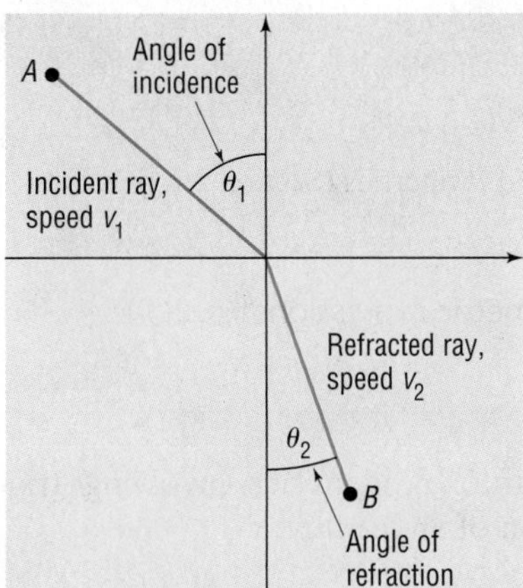

Some Indexes of Refraction	
Medium	**Index of Refraction**[†]
Water	1.33
Ethyl alcohol (20°C)	1.36
Carbon disulfide	1.63
Air (1 atm and 0°C)	1.00029
Diamond	2.42
Fused quartz	1.46
Glass, crown	1.52
Glass, dense flint	1.66
Sodium chloride	1.54

113. The index of refraction of light in passing from a vacuum into water is 1.33. If the angle of incidence is 40°, determine the angle of refraction.

114. The index of refraction of light in passing from a vacuum into dense flint glass is 1.66. If the angle of incidence is 50°, determine the angle of refraction.

115. Ptolemy, who lived in the city of Alexandria in Egypt during the second century AD, gave the measured values in the following table for the angle of incidence θ_1 and the angle of refraction θ_2 for a light beam passing from air into water. Do these values agree with Snell's Law? If so, what index of refraction results? (These data are of interest as the oldest recorded physical measurements.)

θ_1	θ_2	θ_1	θ_2
10°	8°	50°	35°0′
20°	15°30′	60°	40°30′
30°	22°30′	70°	45°30′
40°	29°0′	80°	50°0′

*Because this law was also deduced by René Descartes in France, it is also known as Descartes' Law.

[†] For light of wavelength 589 nanometers, measured with respect to a vacuum. The index with respect to air is negligibly different in most cases.

116. **Bending Light** The speed of yellow sodium light (wavelength, 589 nanometers) in a certain liquid is measured to be 1.92×10^8 meters per second. What is the index of refraction of this liquid, with respect to air, for sodium light?*
[**Hint:** The speed of light in air is approximately 2.998×10^8 meters per second.]

117. **Bending Light** A beam of light with a wavelength of 589 nanometers traveling in air makes an angle of incidence of 40° on a slab of transparent material, and the refracted beam makes an angle of refraction of 26°. Find the index of refraction of the material.*

118. **Bending Light** A light ray with a wavelength of 589 nanometers (produced by a sodium lamp) traveling through air makes an angle of incidence of 30° on a smooth, flat slab of crown glass. Find the angle of refraction.*

119. A light beam passes through a thick slab of material whose index of refraction is n_2. Show that the emerging beam is parallel to the incident beam.*

120. **Brewster's Law** If the angle of incidence and the angle of refraction are complementary angles, the angle of incidence is referred to as the Brewster angle θ_B. The Brewster angle is related to the indices of refraction of the two media, n_1 and n_2, by the equation $n_1 \sin \theta_B = n_2 \cos \theta_B$, where n_1 is the index of refraction of the incident medium and n_2 is the index of refraction of the refractive medium. Determine the Brewster angle for a light beam traveling through water (at 20°C) that makes an angle of incidence with a smooth, flat slab of crown glass.

*Adapted from Halliday, Resnick, and Walker, *Fundamentals of Physics*, 10th ed., 2014, John Wiley & Sons.

Explaining Concepts: Discussion and Writing

121. Explain in your own words how you would use your calculator to solve the equation $\cos x = -0.6, 0 \le x < 2\pi$. How would you modify your approach to solve the equation $\cot x = 5, 0 < x < 2\pi$?

122. Explain why no further points of intersection (and therefore no further solutions) exist in Figure 25 for $x < -\pi$ or $x > 4\pi$.

Retain Your Knowledge

Problems 123–126 are based on material learned earlier in the course. The purpose of these problems is to keep the material fresh in your mind so that you are better prepared for the final exam.

123. Find the center and radius of the circle $x^2 + y^2 - 10x + 4y + 20 = 0$.

124. Write the function whose graph is the graph of $y = \sqrt{x}$, but reflected about the y-axis.

125. Given $\sin \theta = -\dfrac{\sqrt{10}}{10}$ and $\cos \theta = \dfrac{3\sqrt{10}}{10}$, find the exact value of each of the four remaining trigonometric functions.

126. Determine the amplitude, period, and phase shift of the function $y = 2 \sin (2x - \pi)$. Graph the function. Show at least two periods.

'Are You Prepared?' Answers

1. $\left\{\dfrac{3}{2}\right\}$ **2.** $\dfrac{\sqrt{2}}{2}; -\dfrac{1}{2}$ **3.** $\left\{-1, \dfrac{5}{4}\right\}$ **4.** $\left\{\dfrac{1-\sqrt{5}}{2}, \dfrac{1+\sqrt{5}}{2}\right\}$ **5.** $\left\{0, \dfrac{5}{2}\right\}$ **6.** $\{0.76\}$

3.4 Trigonometric Identities

PREPARING FOR THIS SECTION *Before getting started, review the following:*

- Fundamental Identities (Section 2.3, p. 137)
- Even–Odd Properties (Section 2.3, p. 141)

Now Work the **'Are You Prepared?'** problems on page 223.

OBJECTIVES 1 Use Algebra to Simplify Trigonometric Expressions (p. 219)
2 Establish Identities (p. 220)

This section establishes some additional identities involving trigonometric functions. First, let's review the definition of an *identity*.

DEFINITION

Two functions f and g are **identically equal** if

$$f(x) = g(x)$$

for every value of x for which both functions are defined. Such an equation is referred to as an **identity**. An equation that is not an identity is called a **conditional equation**.

For example, the following are identities:

$$(x + 1)^2 = x^2 + 2x + 1 \qquad \sin^2 x + \cos^2 x = 1 \qquad \csc x = \frac{1}{\sin x}$$

The following are conditional equations:

$2x + 5 = 0$ True only if $x = -\frac{5}{2}$

$\sin x = 0$ True only if $x = k\pi$, k an integer

$\sin x = \cos x$ True only if $x = \frac{\pi}{4} + 2k\pi$ or $x = \frac{5\pi}{4} + 2k\pi$, k an integer

The following summarizes the trigonometric identities that have been established thus far.

Quotient Identities

$$\tan\theta = \frac{\sin\theta}{\cos\theta} \qquad \cot\theta = \frac{\cos\theta}{\sin\theta}$$

Reciprocal Identities

$$\csc\theta = \frac{1}{\sin\theta} \qquad \sec\theta = \frac{1}{\cos\theta} \qquad \cot\theta = \frac{1}{\tan\theta}$$

Pythagorean Identities

$$\sin^2\theta + \cos^2\theta = 1 \qquad \tan^2\theta + 1 = \sec^2\theta$$
$$\cot^2\theta + 1 = \csc^2\theta$$

Even–Odd Identities

$$\sin(-\theta) = -\sin\theta \qquad \cos(-\theta) = \cos\theta \qquad \tan(-\theta) = -\tan\theta$$
$$\csc(-\theta) = -\csc\theta \qquad \sec(-\theta) = \sec\theta \qquad \cot(-\theta) = -\cot\theta$$

This list comprises what shall be referred to as the **basic trigonometric identities**. These identities should not merely be memorized, but should be *known* (just as you know your name rather than have it memorized). In fact, minor variations of a basic identity are often used. For example,

$$\sin^2\theta = 1 - \cos^2\theta \quad \text{or} \quad \cos^2\theta = 1 - \sin^2\theta$$

might be used instead of $\sin^2\theta + \cos^2\theta = 1$. For this reason, among others, it is very important to know these relationships and be comfortable with variations of them.

1 Use Algebra to Simplify Trigonometric Expressions

The ability to use algebra to manipulate trigonometric expressions is a key skill that one must have to establish identities. Four basic algebraic techniques are used to establish identities:

1. Rewriting a trigonometric expression in terms of sine and cosine only
2. Multiplying the numerator and denominator of a ratio by a "well-chosen 1"
3. Writing sums of trigonometric ratios as a single ratio
4. Factoring

EXAMPLE 1

Using Algebraic Techniques to Simplify Trigonometric Expressions

(a) Simplify $\dfrac{\cot\theta}{\csc\theta}$ by rewriting each trigonometric function in terms of sine and cosine functions.

(b) Show that $\dfrac{\cos\theta}{1+\sin\theta} = \dfrac{1-\sin\theta}{\cos\theta}$ by multiplying the numerator and denominator by $1 - \sin\theta$.

(c) Simplify $\dfrac{1+\sin u}{\sin u} + \dfrac{\cot u - \cos u}{\cos u}$ by rewriting the expression as a single ratio.

(d) Simplify $\dfrac{\sin^2 v - 1}{\tan v \sin v - \tan v}$ by factoring.

Solution

(a) $$\frac{\cot\theta}{\csc\theta} = \frac{\dfrac{\cos\theta}{\sin\theta}}{\dfrac{1}{\sin\theta}} = \frac{\cos\theta}{\sin\theta}\cdot\frac{\sin\theta}{1} = \cos\theta$$

(b) $$\frac{\cos\theta}{1+\sin\theta} = \frac{\cos\theta}{1+\sin\theta}\cdot\frac{1-\sin\theta}{1-\sin\theta} = \frac{\cos\theta(1-\sin\theta)}{1-\sin^2\theta}$$

Multiply by a well-chosen 1: $\dfrac{1-\sin\theta}{1-\sin\theta}$.

$$= \frac{\cos\theta(1-\sin\theta)}{\cos^2\theta} = \frac{1-\sin\theta}{\cos\theta}$$

(c) $$\frac{1+\sin u}{\sin u} + \frac{\cot u - \cos u}{\cos u} = \frac{1+\sin u}{\sin u}\cdot\frac{\cos u}{\cos u} + \frac{\cot u - \cos u}{\cos u}\cdot\frac{\sin u}{\sin u}$$

$$= \frac{\cos u + \sin u\cos u + \cot u\sin u - \cos u\sin u}{\sin u\cos u} = \frac{\cos u + \cot u\sin u}{\sin u\cos u}$$

$$= \frac{\cos u + \dfrac{\cos u}{\sin u}\cdot\sin u}{\sin u\cos u} = \frac{\cos u + \cos u}{\sin u\cos u} = \frac{2\cos u}{\sin u\cos u} = \frac{2}{\sin u}$$

$\cot u = \dfrac{\cos u}{\sin u}$

(d) $$\frac{\sin^2 v - 1}{\tan v\sin v - \tan v} = \frac{(\sin v + 1)(\sin v - 1)}{\tan v(\sin v - 1)} = \frac{\sin v + 1}{\tan v}$$

●

Now Work PROBLEMS 11, 13, AND 15

2 Establish Identities

In the examples that follow, the directions read "Establish the identity...." This is accomplished by starting with one side of the given equation (usually the side containing the more complicated expression) and, using appropriate basic identities and algebraic manipulations, arriving at the other side. The selection of appropriate basic identities to obtain the desired result is learned only through experience and lots of practice.

EXAMPLE 2

Establishing an Identity

Establish the identity: $\csc\theta\cdot\tan\theta = \sec\theta$

Solution Start with the left side, because it contains the more complicated expression, and apply a reciprocal identity and a quotient identity.

$$\csc\theta \cdot \tan\theta = \frac{1}{\cancel{\sin\theta}} \cdot \frac{\cancel{\sin\theta}}{\cos\theta} = \frac{1}{\cos\theta} = \sec\theta$$

The right side has been reached, so the identity is established. ●

NOTE A graphing utility can be used to provide evidence of an identity. For example, if we graph $Y_1 = \csc\theta \cdot \tan\theta$ and $Y_2 = \sec\theta$, the graphs appear to be the same. This provides evidence that $Y_1 = Y_2$. However, it does not prove their equality. A graphing utility *cannot be used to establish an identity*—identities must be established algebraically. ■

Now Work PROBLEM 21

EXAMPLE 3 Establishing an Identity

Establish the identity: $\sin^2(-\theta) + \cos^2(-\theta) = 1$

Solution Begin with the left side and, because the arguments are $-\theta$, apply Even–Odd Identities.

$$\begin{aligned} \sin^2(-\theta) + \cos^2(-\theta) &= [\sin(-\theta)]^2 + [\cos(-\theta)]^2 \\ &= (-\sin\theta)^2 + (\cos\theta)^2 && \text{Even–Odd Identities} \\ &= (\sin\theta)^2 + (\cos\theta)^2 \\ &= 1 && \text{Pythagorean Identity} \end{aligned}$$

●

EXAMPLE 4 Establishing an Identity

Establish the identity: $\dfrac{\sin^2(-\theta) - \cos^2(-\theta)}{\sin(-\theta) - \cos(-\theta)} = \cos\theta - \sin\theta$

Solution Observe that the left side contains the more complicated expression. Also, the left side contains expressions with the argument $-\theta$, whereas the right side contains expressions with the argument θ. So start with the left side and apply Even–Odd Identities.

$$\begin{aligned} \frac{\sin^2(-\theta) - \cos^2(-\theta)}{\sin(-\theta) - \cos(-\theta)} &= \frac{[\sin(-\theta)]^2 - [\cos(-\theta)]^2}{\sin(-\theta) - \cos(-\theta)} \\ &= \frac{(-\sin\theta)^2 - (\cos\theta)^2}{-\sin\theta - \cos\theta} && \text{Even–Odd Identities} \\ &= \frac{(\sin\theta)^2 - (\cos\theta)^2}{-\sin\theta - \cos\theta} && \text{Simplify.} \\ &= \frac{(\sin\theta - \cos\theta)\cancel{(\sin\theta + \cos\theta)}}{-\cancel{(\sin\theta + \cos\theta)}} && \text{Factor.} \\ &= \cos\theta - \sin\theta && \text{Divide out and simplify.} \end{aligned}$$

●

EXAMPLE 5 Establishing an Identity

Establish the identity: $\dfrac{1 + \tan u}{1 + \cot u} = \tan u$

Solution

$$\begin{aligned} \frac{1 + \tan u}{1 + \cot u} &= \frac{1 + \tan u}{1 + \dfrac{1}{\tan u}} = \frac{1 + \tan u}{\dfrac{\tan u + 1}{\tan u}} \\ &= \frac{\tan u\cancel{(1 + \tan u)}}{\cancel{\tan u + 1}} = \tan u \end{aligned}$$

●

Now Work PROBLEMS 25 AND 29

When sums or differences of quotients appear, it is usually best to rewrite them as a single quotient, especially if the other side of the identity consists of only one term.

EXAMPLE 6 **Establishing an Identity**

Establish the identity: $\dfrac{\sin\theta}{1+\cos\theta} + \dfrac{1+\cos\theta}{\sin\theta} = 2\csc\theta$

Solution The left side is more complicated. Start with it and add.

$$\frac{\sin\theta}{1+\cos\theta} + \frac{1+\cos\theta}{\sin\theta} = \frac{\sin^2\theta + (1+\cos\theta)^2}{(1+\cos\theta)(\sin\theta)} \quad \text{Add the quotients.}$$

$$= \frac{\sin^2\theta + 1 + 2\cos\theta + \cos^2\theta}{(1+\cos\theta)(\sin\theta)} \quad \text{Multiply out in the numerator.}$$

$$= \frac{(\sin^2\theta + \cos^2\theta) + 1 + 2\cos\theta}{(1+\cos\theta)(\sin\theta)} \quad \text{Regroup.}$$

$$= \frac{2 + 2\cos\theta}{(1+\cos\theta)(\sin\theta)} \quad \text{Pythagorean Identity}$$

$$= \frac{2\cancel{(1+\cos\theta)}}{\cancel{(1+\cos\theta)}(\sin\theta)} \quad \text{Factor and cancel.}$$

$$= \frac{2}{\sin\theta}$$

$$= 2\csc\theta \quad \text{Reciprocal Identity}$$

Now Work PROBLEM 51

Sometimes it helps to write one side in terms of sine and cosine functions only.

EXAMPLE 7 **Establishing an Identity**

Establish the identity: $\dfrac{\tan v + \cot v}{\sec v \csc v} = 1$

Solution

$$\frac{\tan v + \cot v}{\sec v \csc v} \underset{\uparrow}{=} \frac{\dfrac{\sin v}{\cos v} + \dfrac{\cos v}{\sin v}}{\dfrac{1}{\cos v}\cdot\dfrac{1}{\sin v}} \underset{\uparrow}{=} \frac{\dfrac{\sin^2 v + \cos^2 v}{\cos v \sin v}}{\dfrac{1}{\cos v \sin v}}$$

Change to sines and cosines. — Add the quotients in the numerator.

$$\underset{\uparrow}{=} \frac{1}{\cos v \sin v}\cdot\frac{\cos v \sin v}{1} = 1$$

Divide the quotients; $\sin^2 v + \cos^2 v = 1$.

Now Work PROBLEM 71

Sometimes, multiplying the numerator and the denominator by an appropriate factor simplifies an expression.

EXAMPLE 8 Establishing an Identity

Establish the identity: $\dfrac{1 - \sin\theta}{\cos\theta} = \dfrac{\cos\theta}{1 + \sin\theta}$

Solution Start with the left side and multiply the numerator and the denominator by $1 + \sin\theta$. (Alternatively, we could multiply the numerator and the denominator of the right side by $1 - \sin\theta$.)

$$\frac{1 - \sin\theta}{\cos\theta} = \frac{1 - \sin\theta}{\cos\theta} \cdot \frac{1 + \sin\theta}{1 + \sin\theta} \quad \text{Multiply the numerator and the denominator by } 1 + \sin\theta.$$

$$= \frac{1 - \sin^2\theta}{\cos\theta(1 + \sin\theta)}$$

$$= \frac{\cos^2\theta}{\cos\theta(1 + \sin\theta)} \quad 1 - \sin^2\theta = \cos^2\theta$$

$$= \frac{\cos\theta}{1 + \sin\theta} \quad \text{Divide out } \cos\theta.$$

●

Now Work PROBLEM 55

Although a lot of practice is the only real way to learn how to establish identities, the following guidelines should prove helpful.

WARNING Be careful not to handle identities to be established as if they were conditional equations. You *cannot* establish an identity by such methods as adding the same expression to each side and obtaining a true statement. This practice is not allowed, because the original statement is precisely the one that you are trying to establish. You do not know until it has been established that it is, in fact, true. ■

Guidelines for Establishing Identities

1. It is almost always preferable to start with the side containing the more complicated expression.
2. Rewrite sums or differences of quotients as a single quotient.
3. Sometimes it will help to rewrite one side in terms of sine and cosine functions only.
4. Always keep the goal in mind. As you manipulate one side of the expression, keep in mind the form of the expression on the other side.

3.4 Assess Your Understanding

'Are You Prepared?' *Answers are given at the end of these exercises. If you get a wrong answer, read the pages listed in red.*

1. ***True or False*** $\sin^2\theta = 1 - \cos^2\theta$. (p. 137)

2. ***True or False*** $\sin(-\theta) + \cos(-\theta) = \cos\theta - \sin\theta$. (p. 141)

Concepts and Vocabulary

3. Suppose that f and g are two functions with the same domain. If $f(x) = g(x)$ for every x in the domain, the equation is called a(n) ________. Otherwise, it is called a(n) ________ equation.

4. $\tan^2\theta - \sec^2\theta =$ ________.

5. $\cos(-\theta) - \cos\theta =$ ________.

6. ***True or False*** $\sin(-\theta) + \sin\theta = 0$ for any value of θ.

7. ***True or False*** In establishing an identity, it is often easiest to just multiply both sides by a well-chosen nonzero expression involving the variable.

8. ***True or False*** $\tan\theta \cdot \cos\theta = \sin\theta$ for any $\theta \neq (2k + 1)\dfrac{\pi}{2}$.

9. Which of the following equations is *not* an identity?
(a) $\cot^2\theta + 1 = \csc^2\theta$ (b) $\tan(-\theta) = -\tan\theta$
(c) $\tan\theta = \dfrac{\cos\theta}{\sin\theta}$ (d) $\csc\theta = \dfrac{1}{\sin\theta}$

10. The expression $\dfrac{1}{1 - \sin\theta} + \dfrac{1}{1 + \sin\theta}$ simplifies to which of the following?
(a) $2\cos^2\theta$ (b) $2\sec^2\theta$ (c) $2\sin^2\theta$ (d) $2\csc^2\theta$

Skill Building

In Problems 11–20, simplify each trigonometric expression by following the indicated direction.

11. Rewrite in terms of sine and cosine functions: $\tan\theta \cdot \csc\theta$.

12. Rewrite in terms of sine and cosine functions: $\cot\theta \cdot \sec\theta$.

13. Multiply $\dfrac{\cos\theta}{1-\sin\theta}$ by $\dfrac{1+\sin\theta}{1+\sin\theta}$.

14. Multiply $\dfrac{\sin\theta}{1+\cos\theta}$ by $\dfrac{1-\cos\theta}{1-\cos\theta}$.

15. Rewrite over a common denominator: $\dfrac{\sin\theta+\cos\theta}{\cos\theta}+\dfrac{\cos\theta-\sin\theta}{\sin\theta}$

16. Rewrite over a common denominator: $\dfrac{1}{1-\cos v}+\dfrac{1}{1+\cos v}$

17. Multiply and simplify: $\dfrac{(\sin\theta+\cos\theta)(\sin\theta+\cos\theta)-1}{\sin\theta\cos\theta}$

18. Multiply and simplify: $\dfrac{(\tan\theta+1)(\tan\theta+1)-\sec^2\theta}{\tan\theta}$

19. Factor and simplify: $\dfrac{3\sin^2\theta+4\sin\theta+1}{\sin^2\theta+2\sin\theta+1}$

20. Factor and simplify: $\dfrac{\cos^2\theta-1}{\cos^2\theta-\cos\theta}$

In Problems 21–100, establish each identity.

21. $\csc\theta\cdot\cos\theta=\cot\theta$

22. $\sec\theta\cdot\sin\theta=\tan\theta$

23. $1+\tan^2(-\theta)=\sec^2\theta$

24. $1+\cot^2(-\theta)=\csc^2\theta$

25. $\cos\theta(\tan\theta+\cot\theta)=\csc\theta$

26. $\sin\theta(\cot\theta+\tan\theta)=\sec\theta$

27. $\tan u\cot u-\cos^2 u=\sin^2 u$

28. $\sin u\csc u-\cos^2 u=\sin^2 u$

29. $(\sec\theta-1)(\sec\theta+1)=\tan^2\theta$

30. $(\csc\theta-1)(\csc\theta+1)=\cot^2\theta$

31. $(\sec\theta+\tan\theta)(\sec\theta-\tan\theta)=1$

32. $(\csc\theta+\cot\theta)(\csc\theta-\cot\theta)=1$

33. $\cos^2\theta(1+\tan^2\theta)=1$

34. $(1-\cos^2\theta)(1+\cot^2\theta)=1$

35. $(\sin\theta+\cos\theta)^2+(\sin\theta-\cos\theta)^2=2$

36. $\tan^2\theta\cos^2\theta+\cot^2\theta\sin^2\theta=1$

37. $\sec^4\theta-\sec^2\theta=\tan^4\theta+\tan^2\theta$

38. $\csc^4\theta-\csc^2\theta=\cot^4\theta+\cot^2\theta$

39. $\sec u-\tan u=\dfrac{\cos u}{1+\sin u}$

40. $\csc u-\cot u=\dfrac{\sin u}{1+\cos u}$

41. $3\sin^2\theta+4\cos^2\theta=3+\cos^2\theta$

42. $9\sec^2\theta-5\tan^2\theta=5+4\sec^2\theta$

43. $1-\dfrac{\cos^2\theta}{1+\sin\theta}=\sin\theta$

44. $1-\dfrac{\sin^2\theta}{1-\cos\theta}=-\cos\theta$

45. $\dfrac{1+\tan v}{1-\tan v}=\dfrac{\cot v+1}{\cot v-1}$

46. $\dfrac{\csc v-1}{\csc v+1}=\dfrac{1-\sin v}{1+\sin v}$

47. $\dfrac{\sec\theta}{\csc\theta}+\dfrac{\sin\theta}{\cos\theta}=2\tan\theta$

48. $\dfrac{\csc\theta-1}{\cot\theta}=\dfrac{\cot\theta}{\csc\theta+1}$

49. $\dfrac{1+\sin\theta}{1-\sin\theta}=\dfrac{\csc\theta+1}{\csc\theta-1}$

50. $\dfrac{\cos\theta+1}{\cos\theta-1}=\dfrac{1+\sec\theta}{1-\sec\theta}$

51. $\dfrac{1-\sin v}{\cos v}+\dfrac{\cos v}{1-\sin v}=2\sec v$

52. $\dfrac{\cos v}{1+\sin v}+\dfrac{1+\sin v}{\cos v}=2\sec v$

53. $\dfrac{\sin\theta}{\sin\theta-\cos\theta}=\dfrac{1}{1-\cot\theta}$

54. $1-\dfrac{\sin^2\theta}{1+\cos\theta}=\cos\theta$

55. $\dfrac{1-\sin\theta}{1+\sin\theta}=(\sec\theta-\tan\theta)^2$

56. $\dfrac{1-\cos\theta}{1+\cos\theta}=(\csc\theta-\cot\theta)^2$

57. $\dfrac{\cos\theta}{1-\tan\theta}+\dfrac{\sin\theta}{1-\cot\theta}=\sin\theta+\cos\theta$

58. $\dfrac{\cot\theta}{1-\tan\theta}+\dfrac{\tan\theta}{1-\cot\theta}=1+\tan\theta+\cot\theta$

59. $\tan\theta+\dfrac{\cos\theta}{1+\sin\theta}=\sec\theta$

60. $\dfrac{\sin\theta\cos\theta}{\cos^2\theta-\sin^2\theta}=\dfrac{\tan\theta}{1-\tan^2\theta}$

61. $\dfrac{\tan\theta+\sec\theta-1}{\tan\theta-\sec\theta+1}=\tan\theta+\sec\theta$

62. $\dfrac{\sin\theta-\cos\theta+1}{\sin\theta+\cos\theta-1}=\dfrac{\sin\theta+1}{\cos\theta}$

63. $\dfrac{\tan\theta-\cot\theta}{\tan\theta+\cot\theta}=\sin^2\theta-\cos^2\theta$

64. $\dfrac{\sec\theta-\cos\theta}{\sec\theta+\cos\theta}=\dfrac{\sin^2\theta}{1+\cos^2\theta}$

65. $\dfrac{\tan u-\cot u}{\tan u+\cot u}+1=2\sin^2 u$

66. $\dfrac{\tan u-\cot u}{\tan u+\cot u}+2\cos^2 u=1$

67. $\dfrac{\sec\theta+\tan\theta}{\cot\theta+\cos\theta}=\tan\theta\sec\theta$

68. $\dfrac{\sec\theta}{1+\sec\theta}=\dfrac{1-\cos\theta}{\sin^2\theta}$

69. $\dfrac{1-\tan^2\theta}{1+\tan^2\theta}+1=2\cos^2\theta$

70. $\dfrac{1-\cot^2\theta}{1+\cot^2\theta}+2\cos^2\theta=1$

71. $\dfrac{\sec\theta-\csc\theta}{\sec\theta\csc\theta}=\sin\theta-\cos\theta$

72. $\dfrac{\sin^2\theta-\tan\theta}{\cos^2\theta-\cot\theta}=\tan^2\theta$

73. $\sec\theta-\cos\theta=\sin\theta\tan\theta$

74. $\tan\theta+\cot\theta=\sec\theta\csc\theta$

75. $\dfrac{1}{1-\sin\theta}+\dfrac{1}{1+\sin\theta}=2\sec^2\theta$

76. $\dfrac{1+\sin\theta}{1-\sin\theta}-\dfrac{1-\sin\theta}{1+\sin\theta}=4\tan\theta\sec\theta$

77. $\dfrac{\sec\theta}{1-\sin\theta} = \dfrac{1+\sin\theta}{\cos^3\theta}$

78. $\dfrac{1+\sin\theta}{1-\sin\theta} = (\sec\theta + \tan\theta)^2$

79. $\dfrac{(\sec v - \tan v)^2 + 1}{\csc v(\sec v - \tan v)} = 2\tan v$

80. $\dfrac{\sec^2 v - \tan^2 v + \tan v}{\sec v} = \sin v + \cos v$

81. $\dfrac{\sin\theta + \cos\theta}{\cos\theta} - \dfrac{\sin\theta - \cos\theta}{\sin\theta} = \sec\theta\csc\theta$

82. $\dfrac{\sin\theta + \cos\theta}{\sin\theta} - \dfrac{\cos\theta - \sin\theta}{\cos\theta} = \sec\theta\csc\theta$

83. $\dfrac{\sin^3\theta + \cos^3\theta}{\sin\theta + \cos\theta} = 1 - \sin\theta\cos\theta$

84. $\dfrac{\sin^3\theta + \cos^3\theta}{1 - 2\cos^2\theta} = \dfrac{\sec\theta - \sin\theta}{\tan\theta - 1}$

85. $\dfrac{\cos^2\theta - \sin^2\theta}{1 - \tan^2\theta} = \cos^2\theta$

86. $\dfrac{\cos\theta + \sin\theta - \sin^3\theta}{\sin\theta} = \cot\theta + \cos^2\theta$

87. $\dfrac{(2\cos^2\theta - 1)^2}{\cos^4\theta - \sin^4\theta} = 1 - 2\sin^2\theta$

88. $\dfrac{1 - 2\cos^2\theta}{\sin\theta\cos\theta} = \tan\theta - \cot\theta$

89. $\dfrac{1 + \sin\theta + \cos\theta}{1 + \sin\theta - \cos\theta} = \dfrac{1 + \cos\theta}{\sin\theta}$

90. $\dfrac{1 + \cos\theta + \sin\theta}{1 + \cos\theta - \sin\theta} = \sec\theta + \tan\theta$

91. $(a\sin\theta + b\cos\theta)^2 + (a\cos\theta - b\sin\theta)^2 = a^2 + b^2$

92. $(2a\sin\theta\cos\theta)^2 + a^2(\cos^2\theta - \sin^2\theta)^2 = a^2$

93. $\dfrac{\tan\alpha + \tan\beta}{\cot\alpha + \cot\beta} = \tan\alpha\tan\beta$

94. $(\tan\alpha + \tan\beta)(1 - \cot\alpha\cot\beta) + (\cot\alpha + \cot\beta)(1 - \tan\alpha\tan\beta) = 0$

95. $(\sin\alpha + \cos\beta)^2 + (\cos\beta + \sin\alpha)(\cos\beta - \sin\alpha) = 2\cos\beta(\sin\alpha + \cos\beta)$

96. $(\sin\alpha - \cos\beta)^2 + (\cos\beta + \sin\alpha)(\cos\beta - \sin\alpha) = -2\cos\beta(\sin\alpha - \cos\beta)$

97. $\ln|\sec\theta| = -\ln|\cos\theta|$

98. $\ln|\tan\theta| = \ln|\sin\theta| - \ln|\cos\theta|$

99. $\ln|1 + \cos\theta| + \ln|1 - \cos\theta| = 2\ln|\sin\theta|$

100. $\ln|\sec\theta + \tan\theta| + \ln|\sec\theta - \tan\theta| = 0$

In Problems 101–104, show that the functions f and g are identically equal.

101. $f(x) = \sin x \cdot \tan x \quad g(x) = \sec x - \cos x$

102. $f(x) = \cos x \cdot \cot x \quad g(x) = \csc x - \sin x$

103. $f(\theta) = \dfrac{1 - \sin\theta}{\cos\theta} - \dfrac{\cos\theta}{1 + \sin\theta} \quad g(\theta) = 0$

104. $f(\theta) = \tan\theta + \sec\theta \quad g(\theta) = \dfrac{\cos\theta}{1 - \sin\theta}$

Applications and Extensions

105. **Searchlights** A searchlight at the grand opening of a new car dealership casts a spot of light on a wall located 75 meters from the searchlight. The acceleration $\ddot{r}$ of the spot of light is found to be $\ddot{r} = 1200\sec\theta(2\sec^2\theta - 1)$. Show that this is equivalent to $\ddot{r} = 1200\left(\dfrac{1 + \sin^2\theta}{\cos^3\theta}\right)$.

Source: Adapted from Hibbeler, *Engineering Mechanics: Dynamics,* 13th ed., Pearson © 2013.

106. **Optical Measurement** Optical methods of measurement often rely on the interference of two light waves. If two light waves, identical except for a phase lag, are mixed together, the resulting intensity, or irradiance, is given by

$$I_t = 4A^2\frac{(\csc\theta - 1)(\sec\theta + \tan\theta)}{\csc\theta\sec\theta}.$$

Show that this is equivalent to $I_t = (2A\cos\theta)^2$.

Source: Experimental Techniques, July/August 2002

Explaining Concepts: Discussion and Writing

107. Write a few paragraphs outlining your strategy for establishing identities.

108. Write down the three Pythagorean Identities.

109. Why do you think it is usually preferable to start with the side containing the more complicated expression when establishing an identity?

110. Make up an identity that is not a basic identity.

Retain Your Knowledge

Problems 111–114 are based on material learned earlier in the course. The purpose of these problems is to keep the material fresh in your mind so that you are better prepared for the final exam.

111. Find the distance between the points $P_1 = (4, -7)$ and $P_2 = (-1, 5)$.

112. Find the equation of the circle with center $(-6, 0)$ and radius $r = \sqrt{7}$.

113. Find the exact values of the six trigonometric functions of an angle θ in standard position if $(-12, 5)$ is a point on its terminal side.

114. Find the average rate of change of $f(x) = \cos x$ from 0 to $\dfrac{\pi}{2}$.

'Are You Prepared?' Answers

1. True 2. True

3.5 Sum and Difference Formulas

PREPARING FOR THIS SECTION *Before getting started, review the following:*

- Distance Formula (Section 1.1, p. 3)
- Values of the Trigonometric Functions (Section 2.2, pp. 116–125)
- Finding Exact Values Given the Value of a Trigonometric Function and the Quadrant of the Angle (Section 2.3, p. 138–140)

Now Work the 'Are You Prepared?' problems on page 235.

OBJECTIVES
1. Use Sum and Difference Formulas to Find Exact Values (p. 227)
2. Use Sum and Difference Formulas to Establish Identities (p. 230)
3. Use Sum and Difference Formulas Involving Inverse Trigonometric Functions (p. 232)
4. Solve Trigonometric Equations Linear in Sine and Cosine (p. 233)

This section continues the derivation of trigonometric identities by obtaining formulas that involve the sum or the difference of two angles, such as $\cos(\alpha + \beta)$, $\cos(\alpha - \beta)$, and $\sin(\alpha + \beta)$. These formulas are referred to as the **sum and difference formulas**. We begin with the formulas for $\cos(\alpha + \beta)$ and $\cos(\alpha - \beta)$.

THEOREM **Sum and Difference Formulas for the Cosine Function**

$$\cos(\alpha + \beta) = \cos\alpha\cos\beta - \sin\alpha\sin\beta \qquad (1)$$
$$\cos(\alpha - \beta) = \cos\alpha\cos\beta + \sin\alpha\sin\beta \qquad (2)$$

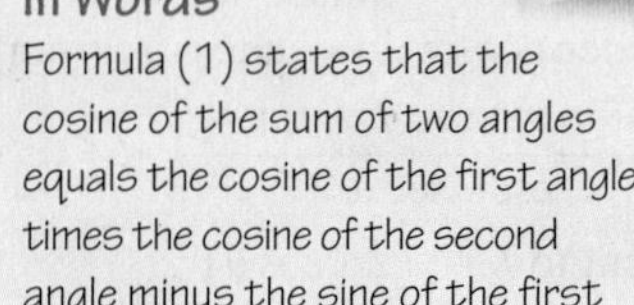

In Words
Formula (1) states that the cosine of the sum of two angles equals the cosine of the first angle times the cosine of the second angle minus the sine of the first angle times the sine of the second angle.

Proof We will prove formula (2) first. Although this formula is true for all numbers α and β, we shall assume in our proof that $0 < \beta < \alpha < 2\pi$. Begin with the unit circle and place the angles α and β in standard position, as shown in Figure 26(a). The point P_1 lies on the terminal side of β, so its coordinates are $(\cos\beta, \sin\beta)$; and the point P_2 lies on the terminal side of α, so its coordinates are $(\cos\alpha, \sin\alpha)$.

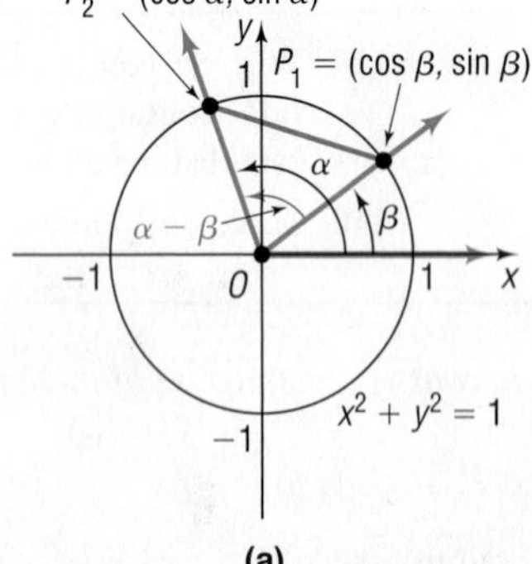

(a)

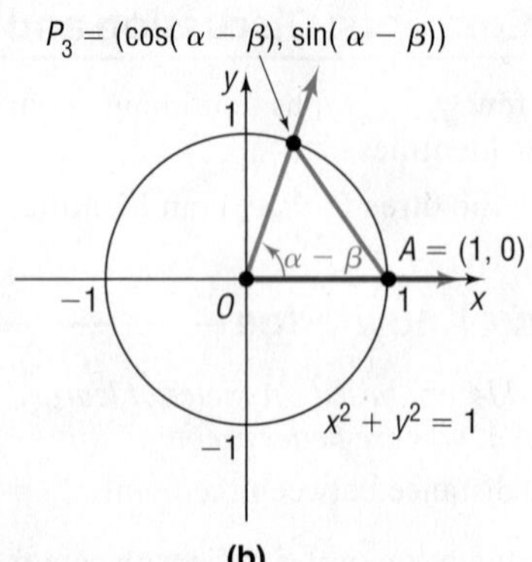

(b)

Figure 26

Now place the angle $\alpha - \beta$ in standard position, as shown in Figure 26(b). The point A has coordinates $(1, 0)$, and the point P_3 is on the terminal side of the angle $\alpha - \beta$, so its coordinates are $(\cos(\alpha - \beta), \sin(\alpha - \beta))$.

Looking at triangle OP_1P_2 in Figure 26(a) and triangle OAP_3 in Figure 26(b), note that these triangles are congruent. (Do you see why? SAS: two sides and the included angle, $\alpha - \beta$, are equal.) As a result, the unknown side of triangle $OP_1 P_2$ and the unknown side of triangle OAP_3 must be equal; that is,

$$d(A, P_3) = d(P_1, P_2)$$

Now use the distance formula to obtain

$$\sqrt{[\cos(\alpha - \beta) - 1]^2 + [\sin(\alpha - \beta) - 0]^2} = \sqrt{(\cos\alpha - \cos\beta)^2 + (\sin\alpha - \sin\beta)^2} \quad d(A, P_3) = d(P_1, P_2)$$

$$[\cos(\alpha - \beta) - 1]^2 + \sin^2(\alpha - \beta) = (\cos\alpha - \cos\beta)^2 + (\sin\alpha - \sin\beta)^2 \quad \text{Square both sides.}$$

$$\cos^2(\alpha - \beta) - 2\cos(\alpha - \beta) + 1 + \sin^2(\alpha - \beta) = \cos^2\alpha - 2\cos\alpha\cos\beta + \cos^2\beta + \sin^2\alpha - 2\sin\alpha\sin\beta + \sin^2\beta \quad \text{Multiply out the squared terms.}$$

$$2 - 2\cos(\alpha - \beta) = 2 - 2\cos\alpha\cos\beta - 2\sin\alpha\sin\beta \quad \text{Apply a Pythagorean Identity (3 times).}$$

$$-2\cos(\alpha - \beta) = -2\cos\alpha\cos\beta - 2\sin\alpha\sin\beta \quad \text{Subtract 2 from each side.}$$

$$\cos(\alpha - \beta) = \cos\alpha\cos\beta + \sin\alpha\sin\beta \quad \text{Divide each side by } -2.$$

This is formula (2). ■

The proof of formula (1) follows from formula (2) and the Even–Odd Identities. Because $\alpha + \beta = \alpha - (-\beta)$, it follows that

$$\begin{aligned}\cos(\alpha + \beta) &= \cos[\alpha - (-\beta)] \\ &= \cos\alpha\cos(-\beta) + \sin\alpha\sin(-\beta) && \text{Use formula (2).} \\ &= \cos\alpha\cos\beta - \sin\alpha\sin\beta && \text{Even–Odd Identities}\end{aligned}$$

1 Use Sum and Difference Formulas to Find Exact Values

One use of formulas (1) and (2) is to obtain the exact value of the cosine of an angle that can be expressed as the sum or difference of angles whose sine and cosine are known exactly.

EXAMPLE 1 Using the Sum Formula to Find an Exact Value

Find the exact value of $\cos 75°$.

Solution Because $75° = 45° + 30°$, use formula (1) to obtain

$$\cos 75° = \cos(45° + 30°) \underset{\text{Formula (1)}}{=} \cos 45° \cos 30° - \sin 45° \sin 30°$$

$$= \frac{\sqrt{2}}{2} \cdot \frac{\sqrt{3}}{2} - \frac{\sqrt{2}}{2} \cdot \frac{1}{2} = \frac{1}{4}(\sqrt{6} - \sqrt{2})$$

●

EXAMPLE 2 Using the Difference Formula to Find an Exact Value

Find the exact value of $\cos \dfrac{\pi}{12}$.

Solution

$$\cos\frac{\pi}{12} = \cos\left(\frac{3\pi}{12} - \frac{2\pi}{12}\right) = \cos\left(\frac{\pi}{4} - \frac{\pi}{6}\right)$$

$$= \cos\frac{\pi}{4}\cos\frac{\pi}{6} + \sin\frac{\pi}{4}\sin\frac{\pi}{6} \quad \text{Use formula (2).}$$

$$= \frac{\sqrt{2}}{2} \cdot \frac{\sqrt{3}}{2} + \frac{\sqrt{2}}{2} \cdot \frac{1}{2} = \frac{1}{4}(\sqrt{6} + \sqrt{2})$$

●

Now Work PROBLEMS 13 AND 19

Another use of formulas (1) and (2) is to establish other identities. Two important identities conjectured earlier, in Section 2.4, are given next.

Seeing the Concept

Graph $Y_1 = \cos\left(\frac{\pi}{2} - x\right)$ and $Y_2 = \sin x$ on the same screen. Does doing this demonstrate the result 3(a)? How would you demonstrate the result 3(b)?

$$\cos\left(\frac{\pi}{2} - \theta\right) = \sin\theta \qquad \textbf{(3a)}$$

$$\sin\left(\frac{\pi}{2} - \theta\right) = \cos\theta \qquad \textbf{(3b)}$$

Proof To prove formula (3a), use the formula for $\cos(\alpha - \beta)$ with $\alpha = \frac{\pi}{2}$ and $\beta = \theta$.

$$\begin{aligned}\cos\left(\frac{\pi}{2} - \theta\right) &= \cos\frac{\pi}{2}\cos\theta + \sin\frac{\pi}{2}\sin\theta \\ &= 0\cdot\cos\theta + 1\cdot\sin\theta \\ &= \sin\theta\end{aligned}$$

To prove formula (3b), make use of the identity (3a) just established.

$$\sin\left(\frac{\pi}{2} - \theta\right) \underset{\text{Use (3a).}}{=} \cos\left[\frac{\pi}{2} - \left(\frac{\pi}{2} - \theta\right)\right] = \cos\theta$$

■

Also, because

$$\cos\left(\frac{\pi}{2} - \theta\right) = \cos\left[-\left(\theta - \frac{\pi}{2}\right)\right] \underset{\text{Even Property of Cosine}}{=} \cos\left(\theta - \frac{\pi}{2}\right)$$

and because

$$\cos\left(\frac{\pi}{2} - \theta\right) \underset{3(a)}{=} \sin\theta$$

it follows that $\cos\left(\theta - \frac{\pi}{2}\right) = \sin\theta$. The graphs of $y = \cos\left(\theta - \frac{\pi}{2}\right)$ and $y = \sin\theta$ are identical.

Having established the identities in formulas (3a) and (3b), we now can derive the sum and difference formulas for $\sin(\alpha + \beta)$ and $\sin(\alpha - \beta)$.

Proof

$$\begin{aligned}\sin(\alpha + \beta) &= \cos\left[\frac{\pi}{2} - (\alpha + \beta)\right] && \text{Formula (3a)} \\ &= \cos\left[\left(\frac{\pi}{2} - \alpha\right) - \beta\right] \\ &= \cos\left(\frac{\pi}{2} - \alpha\right)\cos\beta + \sin\left(\frac{\pi}{2} - \alpha\right)\sin\beta && \text{Formula (2)} \\ &= \sin\alpha\cos\beta + \cos\alpha\sin\beta && \text{Formulas (3a) and (3b)}\end{aligned}$$

$$\begin{aligned}\sin(\alpha - \beta) &= \sin[\alpha + (-\beta)] \\ &= \sin\alpha\cos(-\beta) + \cos\alpha\sin(-\beta) && \text{Use the sum formula for sine just obtained.} \\ &= \sin\alpha\cos\beta + \cos\alpha(-\sin\beta) && \text{Even–Odd Identities} \\ &= \sin\alpha\cos\beta - \cos\alpha\sin\beta\end{aligned}$$

■

In Words

Formula (4) states that the sine of the sum of two angles equals the sine of the first angle times the cosine of the second angle plus the cosine of the first angle times the sine of the second angle.

THEOREM

Sum and Difference Formulas for the Sine Function

$$\sin(\alpha + \beta) = \sin\alpha\cos\beta + \cos\alpha\sin\beta \qquad (4)$$

$$\sin(\alpha - \beta) = \sin\alpha\cos\beta - \cos\alpha\sin\beta \qquad (5)$$

EXAMPLE 3 Using the Sum Formula to Find an Exact Value

Find the exact value of $\sin\dfrac{7\pi}{12}$.

Solution

$$\sin\frac{7\pi}{12} = \sin\left(\frac{3\pi}{12} + \frac{4\pi}{12}\right) = \sin\left(\frac{\pi}{4} + \frac{\pi}{3}\right)$$

$$= \sin\frac{\pi}{4}\cos\frac{\pi}{3} + \cos\frac{\pi}{4}\sin\frac{\pi}{3} \qquad \text{Formula (4)}$$

$$= \frac{\sqrt{2}}{2}\cdot\frac{1}{2} + \frac{\sqrt{2}}{2}\cdot\frac{\sqrt{3}}{2} = \frac{1}{4}(\sqrt{2} + \sqrt{6})$$

Now Work PROBLEM 21

EXAMPLE 4 Using the Difference Formula to Find an Exact Value

Find the exact value of $\sin 80^\circ \cos 20^\circ - \cos 80^\circ \sin 20^\circ$.

Solution The form of the expression $\sin 80^\circ \cos 20^\circ - \cos 80^\circ \sin 20^\circ$ is that of the right side of formula (5) for $\sin(\alpha - \beta)$ with $\alpha = 80^\circ$ and $\beta = 20^\circ$. That is,

$$\sin 80^\circ \cos 20^\circ - \cos 80^\circ \sin 20^\circ = \sin(80^\circ - 20^\circ) = \sin 60^\circ = \frac{\sqrt{3}}{2}$$

Now Work PROBLEMS 27 AND 31

EXAMPLE 5 Finding Exact Values

Say it is known that $\sin\alpha = \dfrac{4}{5}$, $\dfrac{\pi}{2} < \alpha < \pi$, and that $\sin\beta = -\dfrac{2}{\sqrt{5}} = -\dfrac{2\sqrt{5}}{5}$, $\pi < \beta < \dfrac{3\pi}{2}$. Find the exact value of each of the following.

(a) $\cos\alpha$ (b) $\cos\beta$ (c) $\cos(\alpha + \beta)$ (d) $\sin(\alpha + \beta)$

Solution (a) Because $\sin\alpha = \dfrac{4}{5} = \dfrac{y}{r}$ and $\dfrac{\pi}{2} < \alpha < \pi$, let $y = 4$ and $r = 5$, and place α in quadrant II. The point $P = (x, y) = (x, 4)$, $x < 0$, is on a circle of radius 5, $x^2 + y^2 = 25$. See Figure 27. Then

$$x^2 + y^2 = 25$$

$$x^2 + 16 = 25 \qquad y = 4$$

$$x^2 = 25 - 16 = 9$$

$$x = -3 \qquad x < 0$$

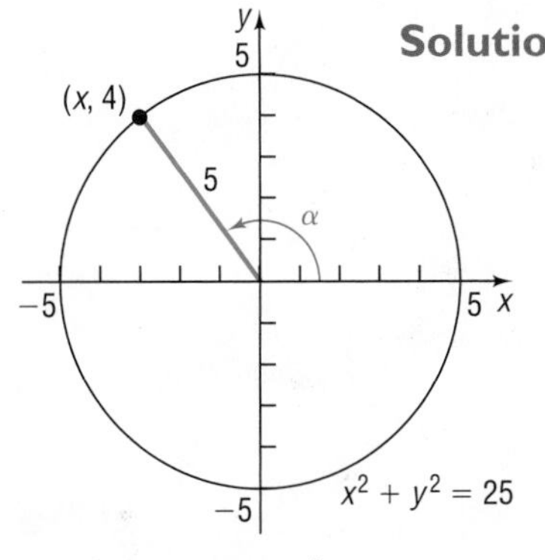

Figure 27 $\sin\alpha = \dfrac{4}{5}, \dfrac{\pi}{2} < \alpha < \pi$

Then

$$\cos\alpha = \frac{x}{r} = -\frac{3}{5}$$

Alternatively, $\cos \alpha$ can be found using identities, as follows:

$$\cos \alpha = -\sqrt{1 - \sin^2 \alpha} = -\sqrt{1 - \frac{16}{25}} = -\sqrt{\frac{9}{25}} = -\frac{3}{5}$$

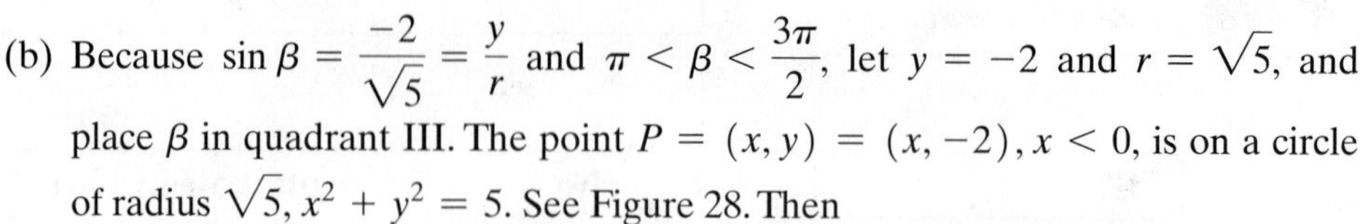

(b) Because $\sin \beta = \frac{-2}{\sqrt{5}} = \frac{y}{r}$ and $\pi < \beta < \frac{3\pi}{2}$, let $y = -2$ and $r = \sqrt{5}$, and place β in quadrant III. The point $P = (x, y) = (x, -2), x < 0$, is on a circle of radius $\sqrt{5}$, $x^2 + y^2 = 5$. See Figure 28. Then

$$x^2 + y^2 = 5$$
$$x^2 + 4 = 5 \quad y = -2$$
$$x^2 = 1$$
$$x = -1 \quad x < 0$$

Figure 28 $\sin \beta = \frac{-2}{\sqrt{5}}, \pi < \beta < \frac{3\pi}{2}$

Then

$$\cos \beta = \frac{a}{r} = \frac{-1}{\sqrt{5}} = -\frac{\sqrt{5}}{5}$$

Alternatively, $\cos \beta$ can be found using identities, as follows:

$$\cos \beta = -\sqrt{1 - \sin^2 \beta} = -\sqrt{1 - \frac{4}{5}} = -\sqrt{\frac{1}{5}} = -\frac{\sqrt{5}}{5}$$

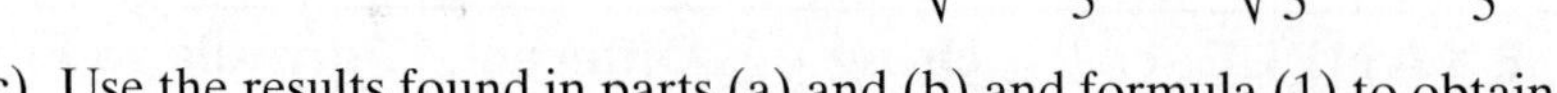

(c) Use the results found in parts (a) and (b) and formula (1) to obtain

$$\cos(\alpha + \beta) = \cos \alpha \cos \beta - \sin \alpha \sin \beta$$
$$= -\frac{3}{5}\left(-\frac{\sqrt{5}}{5}\right) - \frac{4}{5}\left(-\frac{2\sqrt{5}}{5}\right) = \frac{11\sqrt{5}}{25}$$

(d) $\sin(\alpha + \beta) = \sin \alpha \cos \beta + \cos \alpha \sin \beta$

$$= \frac{4}{5}\left(-\frac{\sqrt{5}}{5}\right) + \left(-\frac{3}{5}\right)\left(-\frac{2\sqrt{5}}{5}\right) = \frac{2\sqrt{5}}{25}$$

Now Work PROBLEMS 35(a), (b), AND (c)

2 Use Sum and Difference Formulas to Establish Identities

EXAMPLE 6 **Establishing an Identity**

Establish the identity: $\frac{\cos(\alpha - \beta)}{\sin \alpha \sin \beta} = \cot \alpha \cot \beta + 1$

Solution

$$\frac{\cos(\alpha - \beta)}{\sin \alpha \sin \beta} = \frac{\cos \alpha \cos \beta + \sin \alpha \sin \beta}{\sin \alpha \sin \beta}$$
$$= \frac{\cos \alpha \cos \beta}{\sin \alpha \sin \beta} + \frac{\sin \alpha \sin \beta}{\sin \alpha \sin \beta}$$
$$= \frac{\cos \alpha}{\sin \alpha} \cdot \frac{\cos \beta}{\sin \beta} + 1$$
$$= \cot \alpha \cot \beta + 1$$

Now Work PROBLEMS 49 AND 61

Use the identity $\tan\theta = \dfrac{\sin\theta}{\cos\theta}$ and the sum formulas for $\sin(\alpha+\beta)$ and $\cos(\alpha+\beta)$ to derive a formula for $\tan(\alpha+\beta)$.

Proof $\tan(\alpha+\beta) = \dfrac{\sin(\alpha+\beta)}{\cos(\alpha+\beta)} = \dfrac{\sin\alpha\cos\beta + \cos\alpha\sin\beta}{\cos\alpha\cos\beta - \sin\alpha\sin\beta}$

Now divide the numerator and the denominator by $\cos\alpha\cos\beta$.

$$\tan(\alpha+\beta) = \frac{\dfrac{\sin\alpha\cos\beta + \cos\alpha\sin\beta}{\cos\alpha\cos\beta}}{\dfrac{\cos\alpha\cos\beta - \sin\alpha\sin\beta}{\cos\alpha\cos\beta}} = \frac{\dfrac{\sin\alpha\cancel{\cos\beta}}{\cos\alpha\cancel{\cos\beta}} + \dfrac{\cancel{\cos\alpha}\sin\beta}{\cancel{\cos\alpha}\cos\beta}}{\dfrac{\cancel{\cos\alpha}\,\cancel{\cos\beta}}{\cancel{\cos\alpha}\,\cancel{\cos\beta}} - \dfrac{\sin\alpha\sin\beta}{\cos\alpha\cos\beta}}$$

$$= \frac{\dfrac{\sin\alpha}{\cos\alpha} + \dfrac{\sin\beta}{\cos\beta}}{1 - \dfrac{\sin\alpha}{\cos\alpha}\cdot\dfrac{\sin\beta}{\cos\beta}} = \frac{\tan\alpha + \tan\beta}{1 - \tan\alpha\tan\beta}$$

■

Proof Use the sum formula for $\tan(\alpha+\beta)$ and Even–Odd Properties to get the difference formula.

$$\tan(\alpha-\beta) = \tan[\alpha + (-\beta)] = \frac{\tan\alpha + \tan(-\beta)}{1 - \tan\alpha\tan(-\beta)} = \frac{\tan\alpha - \tan\beta}{1 + \tan\alpha\tan\beta}$$ ■

We have proved the following results:

THEOREM

Sum and Difference Formulas for the Tangent Function

$$\tan(\alpha+\beta) = \frac{\tan\alpha + \tan\beta}{1 - \tan\alpha\tan\beta} \qquad (6)$$

$$\tan(\alpha-\beta) = \frac{\tan\alpha - \tan\beta}{1 + \tan\alpha\tan\beta} \qquad (7)$$

In Words

Formula (6) states that the tangent of the sum of two angles equals the tangent of the first angle plus the tangent of the second angle, all divided by 1 minus their product.

Now Work PROBLEMS 15 AND 35(d)

EXAMPLE 7

Establishing an Identity

Prove the identity: $\tan(\theta+\pi) = \tan\theta$

Solution

$$\tan(\theta+\pi) = \frac{\tan\theta + \tan\pi}{1 - \tan\theta\tan\pi} = \frac{\tan\theta + 0}{1 - \tan\theta\cdot 0} = \tan\theta$$ ●

Example 7 verifies that the tangent function is periodic with period π, a fact that was discussed earlier.

EXAMPLE 8

Establishing an Identity

Prove the identity: $\tan\left(\theta + \dfrac{\pi}{2}\right) = -\cot\theta$

Solution

Formula (6) cannot be used because $\tan\dfrac{\pi}{2}$ is not defined. Instead, proceed as follows:

$$\tan\left(\theta+\frac{\pi}{2}\right) = \frac{\sin\left(\theta+\dfrac{\pi}{2}\right)}{\cos\left(\theta+\dfrac{\pi}{2}\right)} = \frac{\sin\theta\cos\dfrac{\pi}{2} + \cos\theta\sin\dfrac{\pi}{2}}{\cos\theta\cos\dfrac{\pi}{2} - \sin\theta\sin\dfrac{\pi}{2}}$$

$$= \frac{(\sin\theta)(0) + (\cos\theta)(1)}{(\cos\theta)(0) - (\sin\theta)(1)} = \frac{\cos\theta}{-\sin\theta} = -\cot\theta$$ ●

WARNING Be careful when using formulas (6) and (7). These formulas can be used only for angles α and β for which $\tan\alpha$ and $\tan\beta$ are defined. That is, they can be used for all angles except odd integer multiples of $\dfrac{\pi}{2}$. ■

3 Use Sum and Difference Formulas Involving Inverse Trigonometric Functions

EXAMPLE 9 **Finding the Exact Value of an Expression Involving Inverse Trigonometric Functions**

Find the exact value of: $\sin\left(\cos^{-1}\frac{1}{2} + \sin^{-1}\frac{3}{5}\right)$

Solution We seek the sine of the sum of two angles, $\alpha = \cos^{-1}\frac{1}{2}$ and $\beta = \sin^{-1}\frac{3}{5}$. Then

$$\cos\alpha = \frac{1}{2} \quad 0 \le \alpha \le \pi \quad \text{and} \quad \sin\beta = \frac{3}{5} \quad -\frac{\pi}{2} \le \beta \le \frac{\pi}{2}$$

Use Pythagorean Identities to obtain $\sin\alpha$ and $\cos\beta$. Because $\sin\alpha \ge 0$ and $\cos\beta \ge 0$ (do you know why?), this means that

$$\sin\alpha = \sqrt{1 - \cos^2\alpha} = \sqrt{1 - \frac{1}{4}} = \sqrt{\frac{3}{4}} = \frac{\sqrt{3}}{2}$$

$$\cos\beta = \sqrt{1 - \sin^2\beta} = \sqrt{1 - \frac{9}{25}} = \sqrt{\frac{16}{25}} = \frac{4}{5}$$

As a result,

$$\sin\left(\cos^{-1}\frac{1}{2} + \sin^{-1}\frac{3}{5}\right) = \sin(\alpha + \beta) = \sin\alpha\cos\beta + \cos\alpha\sin\beta$$

$$= \frac{\sqrt{3}}{2}\cdot\frac{4}{5} + \frac{1}{2}\cdot\frac{3}{5} = \frac{4\sqrt{3} + 3}{10}$$

NOTE In Example 9 $\sin\alpha$ also can be found by using $\cos\alpha = \frac{1}{2} = \frac{x}{r}$, so $x = 1$ and $r = 2$. Then $y = \sqrt{3}$ and $\sin\alpha = \frac{y}{r} = \frac{\sqrt{3}}{2}$. Also, $\cos\beta$ can be found in a similar fashion. ■

Now Work PROBLEM 77

EXAMPLE 10 **Writing a Trigonometric Expression as an Algebraic Expression**

Write $\sin(\sin^{-1}u + \cos^{-1}v)$ as an algebraic expression containing u and v (that is, without any trigonometric functions). Give the restrictions on u and v.

Solution First, for $\sin^{-1}u$, the restriction on u is $-1 \le u \le 1$, and for $\cos^{-1}v$, the restriction on v is $-1 \le v \le 1$. Now let $\alpha = \sin^{-1}u$ and $\beta = \cos^{-1}v$. Then

$$\sin\alpha = u \quad -\frac{\pi}{2} \le \alpha \le \frac{\pi}{2} \quad -1 \le u \le 1$$

$$\cos\beta = v \quad 0 \le \beta \le \pi \quad -1 \le v \le 1$$

Because $-\frac{\pi}{2} \le \alpha \le \frac{\pi}{2}$, this means that $\cos\alpha \ge 0$. As a result,

$$\cos\alpha = \sqrt{1 - \sin^2\alpha} = \sqrt{1 - u^2}$$

Similarly, because $0 \le \beta \le \pi$, this means that $\sin\beta \ge 0$. Then

$$\sin\beta = \sqrt{1 - \cos^2\beta} = \sqrt{1 - v^2}$$

As a result,

$$\sin(\sin^{-1}u + \cos^{-1}v) = \sin(\alpha + \beta) = \sin\alpha\cos\beta + \cos\alpha\sin\beta$$

$$= uv + \sqrt{1 - u^2}\sqrt{1 - v^2}$$

Now Work PROBLEM 87

4 Solve Trigonometric Equations Linear in Sine and Cosine

Sometimes it is necessary to square both sides of an equation to obtain expressions that allow the use of identities. Remember, squaring both sides of an equation may introduce extraneous solutions. As a result, apparent solutions must be checked.

EXAMPLE 11 Solving a Trigonometric Equation Linear in Sine and Cosine

Solve the equation: $\sin\theta + \cos\theta = 1, \quad 0 \le \theta < 2\pi$

Option 1 Attempts to use available identities do not lead to equations that are easy to solve. (Try it yourself.) So, given the form of this equation, square each side.

$$\begin{aligned} \sin\theta + \cos\theta &= 1 \\ (\sin\theta + \cos\theta)^2 &= 1 && \text{Square each side.} \\ \sin^2\theta + 2\sin\theta\cos\theta + \cos^2\theta &= 1 && \text{Remove parentheses.} \\ 2\sin\theta\cos\theta &= 0 && \sin^2\theta + \cos^2\theta = 1 \\ \sin\theta\cos\theta &= 0 \end{aligned}$$

Setting each factor equal to zero leads to

$$\sin\theta = 0 \quad \text{or} \quad \cos\theta = 0$$

The apparent solutions are

$$\theta = 0 \qquad \theta = \pi \qquad \theta = \frac{\pi}{2} \qquad \theta = \frac{3\pi}{2}$$

Because both sides of the original equation were squared, these apparent solutions must be checked to see whether any are extraneous.

$$\begin{aligned} &\theta = 0: && \sin 0 + \cos 0 = 0 + 1 = 1 && \text{A solution} \\ &\theta = \pi: && \sin\pi + \cos\pi = 0 + (-1) = -1 && \text{Not a solution} \\ &\theta = \frac{\pi}{2}: && \sin\frac{\pi}{2} + \cos\frac{\pi}{2} = 1 + 0 = 1 && \text{A solution} \\ &\theta = \frac{3\pi}{2}: && \sin\frac{3\pi}{2} + \cos\frac{3\pi}{2} = -1 + 0 = -1 && \text{Not a solution} \end{aligned}$$

The values $\theta = \pi$ and $\theta = \frac{3\pi}{2}$ are extraneous. The solution set is $\left\{0, \frac{\pi}{2}\right\}$. ●

Option 2 Start with the equation

$$\sin\theta + \cos\theta = 1$$

and divide each side by $\sqrt{2}$. (The reason for this choice will become apparent shortly.) Then

$$\frac{1}{\sqrt{2}}\sin\theta + \frac{1}{\sqrt{2}}\cos\theta = \frac{1}{\sqrt{2}}$$

The left side now resembles the formula for the sine of the sum of two angles, one of which is θ. The other angle is unknown (call it ϕ.) Then

$$\sin(\theta + \phi) = \sin\theta\cos\phi + \cos\theta\sin\phi = \frac{1}{\sqrt{2}} = \frac{\sqrt{2}}{2} \tag{8}$$

where

$$\cos\phi = \frac{1}{\sqrt{2}} = \frac{\sqrt{2}}{2} \qquad \sin\phi = \frac{1}{\sqrt{2}} = \frac{\sqrt{2}}{2} \qquad 0 \le \phi < 2\pi$$

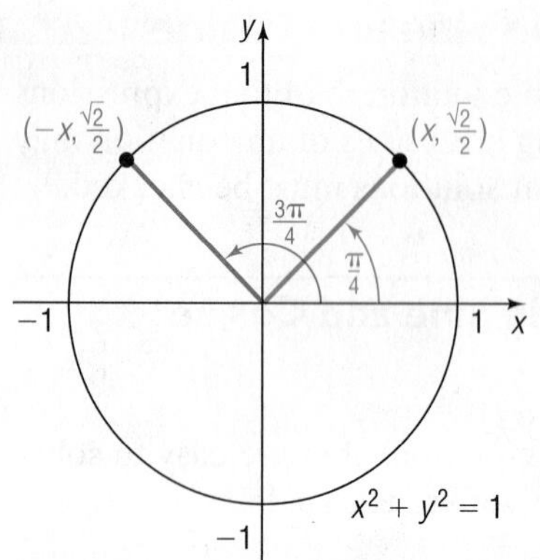

Figure 29

The angle ϕ is therefore $\frac{\pi}{4}$. As a result, equation (8) becomes

$$\sin\left(\theta + \frac{\pi}{4}\right) = \frac{\sqrt{2}}{2}$$

In the interval $[0, 2\pi)$, there are two angles whose sine is $\frac{\sqrt{2}}{2}$: $\frac{\pi}{4}$ and $\frac{3\pi}{4}$. See Figure 29. As a result,

$$\theta + \frac{\pi}{4} = \frac{\pi}{4} \quad \text{or} \quad \theta + \frac{\pi}{4} = \frac{3\pi}{4}$$

$$\theta = 0 \quad \text{or} \quad \theta = \frac{\pi}{2}$$

The solution set is $\left\{0, \frac{\pi}{2}\right\}$.

This second option can be used to solve any linear equation in the variables $\sin\theta$ and $\cos\theta$.

EXAMPLE 12 Solving a Trigonometric Equation Linear in $\sin\theta$ and $\cos\theta$

Solve:

$$a\sin\theta + b\cos\theta = c \tag{9}$$

where a, b, and c are constants and either $a \neq 0$ or $b \neq 0$.

Solution Divide each side of equation (9) by $\sqrt{a^2 + b^2}$. Then

$$\frac{a}{\sqrt{a^2+b^2}}\sin\theta + \frac{b}{\sqrt{a^2+b^2}}\cos\theta = \frac{c}{\sqrt{a^2+b^2}} \tag{10}$$

There is a unique angle ϕ, $0 \leq \phi < 2\pi$, for which

$$\cos\phi = \frac{a}{\sqrt{a^2+b^2}} \quad \text{and} \quad \sin\phi = \frac{b}{\sqrt{a^2+b^2}} \tag{11}$$

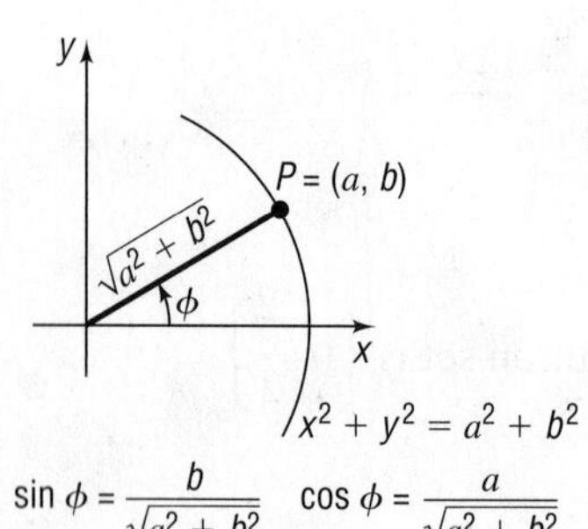

Figure 30

Figure 30 shows the situation for $a > 0$ and $b > 0$. Equation (10) may be written as

$$\sin\theta\cos\phi + \cos\theta\sin\phi = \frac{c}{\sqrt{a^2+b^2}}$$

or, equivalently,

$$\sin(\theta + \phi) = \frac{c}{\sqrt{a^2+b^2}} \tag{12}$$

where ϕ satisfies equation (11).

If $|c| > \sqrt{a^2+b^2}$, then $\sin(\theta+\phi) > 1$ or $\sin(\theta+\phi) < -1$, and equation (12) has no solution.

If $|c| \leq \sqrt{a^2+b^2}$, then the solutions of equation (12) are

$$\theta + \phi = \sin^{-1}\frac{c}{\sqrt{a^2+b^2}} \quad \text{or} \quad \theta + \phi = \pi - \sin^{-1}\frac{c}{\sqrt{a^2+b^2}}$$

Because the angle ϕ is determined by equations (11), these give the solutions to equation (9).

Now Work PROBLEM 95

SUMMARY

Sum and Difference Formulas

$\cos(\alpha + \beta) = \cos\alpha\cos\beta - \sin\alpha\sin\beta$

$\cos(\alpha - \beta) = \cos\alpha\cos\beta + \sin\alpha\sin\beta$

$\sin(\alpha + \beta) = \sin\alpha\cos\beta + \cos\alpha\sin\beta$

$\sin(\alpha - \beta) = \sin\alpha\cos\beta - \cos\alpha\sin\beta$

$\tan(\alpha + \beta) = \dfrac{\tan\alpha + \tan\beta}{1 - \tan\alpha\tan\beta}$

$\tan(\alpha - \beta) = \dfrac{\tan\alpha - \tan\beta}{1 + \tan\alpha\tan\beta}$

3.5 Assess Your Understanding

'Are You Prepared?' *Answers are given at the end of these exercises. If you get a wrong answer, read the pages listed in red.*

1. The distance d from the point $(2, -3)$ to the point $(5, 1)$ is ____. (p. 3)

2. If $\sin\theta = \dfrac{4}{5}$ and θ is in quadrant II, then $\cos\theta =$ ____. (pp. 138–140)

3. (a) $\sin\dfrac{\pi}{4}\cdot\cos\dfrac{\pi}{3} =$ ____. (pp. 116–125)
   (b) $\tan\dfrac{\pi}{4} - \sin\dfrac{\pi}{6} =$ ____. (pp. 116–125)

4. If $\tan\alpha = \dfrac{4}{3}$, $\pi < \alpha < \dfrac{3\pi}{2}$, then $\sin\alpha =$ ____. (pp. 138–140)

Concepts and Vocabulary

5. $\cos(\alpha + \beta) = \cos\alpha\cos\beta$ ____ $\sin\alpha\sin\beta$
6. $\sin(\alpha - \beta) = \sin\alpha\cos\beta$ ____ $\cos\alpha\sin\beta$
7. ***True or False*** $\sin(\alpha + \beta) = \sin\alpha + \sin\beta + 2\sin\alpha\sin\beta$
8. ***True or False*** $\tan 75° = \tan 30° + \tan 45°$
9. ***True or False*** $\cos\left(\dfrac{\pi}{2} - \theta\right) = \cos\theta$
10. ***True or False*** If $f(x) = \sin x$ and $g(x) = \cos x$, then $g(\alpha + \beta) = g(\alpha)g(\beta) - f(\alpha)f(\beta)$
11. Choose the expression that completes the sum formula for tangent functions: $\tan(\alpha + \beta) =$ ____
    (a) $\dfrac{\tan\alpha + \tan\beta}{1 - \tan\alpha\tan\beta}$ (b) $\dfrac{\tan\alpha - \tan\beta}{1 + \tan\alpha\tan\beta}$
    (c) $\dfrac{\tan\alpha + \tan\beta}{1 + \tan\alpha\tan\beta}$ (d) $\dfrac{\tan\alpha - \tan\beta}{1 - \tan\alpha\tan\beta}$
12. Choose the expression that is equivalent to $\sin 60°\cos 20° + \cos 60°\sin 20°$.
    (a) $\cos 40°$ (b) $\sin 40°$ (c) $\cos 80°$ (d) $\sin 80°$

Skill Building

In Problems 13–24, find the exact value of each expression.

13. $\cos 165°$
14. $\sin 105°$
15. $\tan 15°$
16. $\tan 195°$
17. $\sin\dfrac{5\pi}{12}$
18. $\sin\dfrac{\pi}{12}$
19. $\cos\dfrac{7\pi}{12}$
20. $\tan\dfrac{7\pi}{12}$
21. $\sin\dfrac{17\pi}{12}$
22. $\tan\dfrac{19\pi}{12}$
23. $\sec\left(-\dfrac{\pi}{12}\right)$
24. $\cot\left(-\dfrac{5\pi}{12}\right)$

In Problems 25–34, find the exact value of each expression.

25. $\sin 20°\cos 10° + \cos 20°\sin 10°$
26. $\sin 20°\cos 80° - \cos 20°\sin 80°$
27. $\cos 70°\cos 20° - \sin 70°\sin 20°$
28. $\cos 40°\cos 10° + \sin 40°\sin 10°$
29. $\dfrac{\tan 20° + \tan 25°}{1 - \tan 20°\tan 25°}$
30. $\dfrac{\tan 40° - \tan 10°}{1 + \tan 40°\tan 10°}$
31. $\sin\dfrac{\pi}{12}\cos\dfrac{7\pi}{12} - \cos\dfrac{\pi}{12}\sin\dfrac{7\pi}{12}$
32. $\cos\dfrac{5\pi}{12}\cos\dfrac{7\pi}{12} - \sin\dfrac{5\pi}{12}\sin\dfrac{7\pi}{12}$
33. $\cos\dfrac{\pi}{12}\cos\dfrac{5\pi}{12} + \sin\dfrac{5\pi}{12}\sin\dfrac{\pi}{12}$
34. $\sin\dfrac{\pi}{18}\cos\dfrac{5\pi}{18} + \cos\dfrac{\pi}{18}\sin\dfrac{5\pi}{18}$

In Problems 35–40, find the exact value of each of the following under the given conditions:

(a) $\sin(\alpha + \beta)$ *(b)* $\cos(\alpha + \beta)$ *(c)* $\sin(\alpha - \beta)$ *(d)* $\tan(\alpha - \beta)$

35. $\sin\alpha = \dfrac{3}{5}, 0 < \alpha < \dfrac{\pi}{2}$; $\cos\beta = \dfrac{2\sqrt{5}}{5}, -\dfrac{\pi}{2} < \beta < 0$
36. $\cos\alpha = \dfrac{\sqrt{5}}{5}, 0 < \alpha < \dfrac{\pi}{2}$; $\sin\beta = -\dfrac{4}{5}, -\dfrac{\pi}{2} < \beta < 0$

37. $\tan \alpha = -\frac{4}{3}, \frac{\pi}{2} < \alpha < \pi;\ \cos \beta = \frac{1}{2}, 0 < \beta < \frac{\pi}{2}$

38. $\tan \alpha = \frac{5}{12}, \pi < \alpha < \frac{3\pi}{2};\ \sin \beta = -\frac{1}{2}, \pi < \beta < \frac{3\pi}{2}$

39. $\sin \alpha = \frac{5}{13}, -\frac{3\pi}{2} < \alpha < -\pi;\ \tan \beta = -\sqrt{3}, \frac{\pi}{2} < \beta < \pi$

40. $\cos \alpha = \frac{1}{2}, -\frac{\pi}{2} < \alpha < 0;\ \sin \beta = \frac{1}{3}, 0 < \beta < \frac{\pi}{2}$

41. If $\sin \theta = \frac{1}{3}$, θ in quadrant II, find the exact value of:

(a) $\cos \theta$ (b) $\sin\left(\theta + \frac{\pi}{6}\right)$

(c) $\cos\left(\theta - \frac{\pi}{3}\right)$ (d) $\tan\left(\theta + \frac{\pi}{4}\right)$

42. If $\cos \theta = \frac{1}{4}$, θ in quadrant IV, find the exact value of:

(a) $\sin \theta$ (b) $\sin\left(\theta - \frac{\pi}{6}\right)$

(c) $\cos\left(\theta + \frac{\pi}{3}\right)$ (d) $\tan\left(\theta - \frac{\pi}{4}\right)$

In Problems 43–48, use the figures to evaluate each function if $f(x) = \sin x$, $g(x) = \cos x$, and $h(x) = \tan x$.

43. $f(\alpha + \beta)$ **44.** $g(\alpha + \beta)$

45. $g(\alpha - \beta)$ **46.** $f(\alpha - \beta)$

47. $h(\alpha + \beta)$ **48.** $h(\alpha - \beta)$

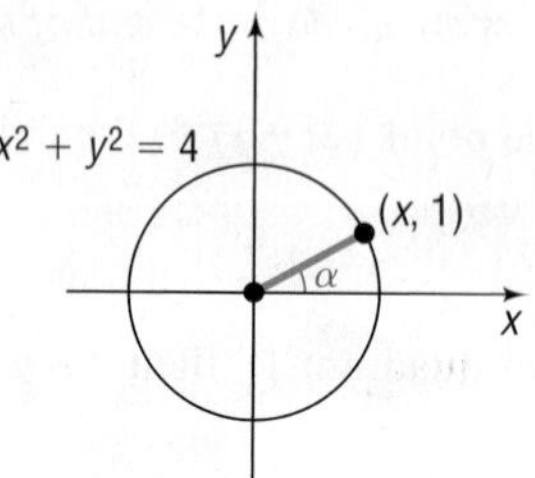

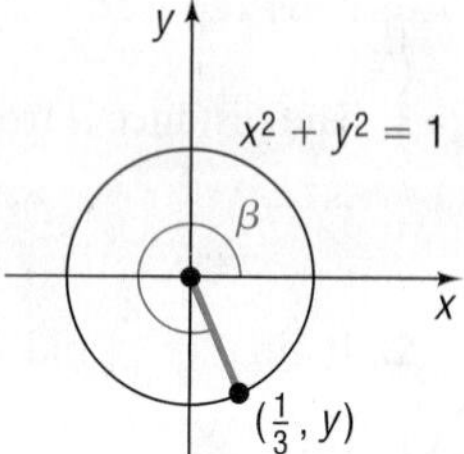

In Problems 49–74, establish each identity.

49. $\sin\left(\frac{\pi}{2} + \theta\right) = \cos \theta$

50. $\cos\left(\frac{\pi}{2} + \theta\right) = -\sin \theta$

51. $\sin(\pi - \theta) = \sin \theta$

52. $\cos(\pi - \theta) = -\cos \theta$

53. $\sin(\pi + \theta) = -\sin \theta$

54. $\cos(\pi + \theta) = -\cos \theta$

55. $\tan(\pi - \theta) = -\tan \theta$

56. $\tan(2\pi - \theta) = -\tan \theta$

57. $\sin\left(\frac{3\pi}{2} + \theta\right) = -\cos \theta$

58. $\cos\left(\frac{3\pi}{2} + \theta\right) = \sin \theta$

59. $\sin(\alpha + \beta) + \sin(\alpha - \beta) = 2 \sin \alpha \cos \beta$

60. $\cos(\alpha + \beta) + \cos(\alpha - \beta) = 2 \cos \alpha \cos \beta$

61. $\frac{\sin(\alpha + \beta)}{\sin \alpha \cos \beta} = 1 + \cot \alpha \tan \beta$

62. $\frac{\sin(\alpha + \beta)}{\cos \alpha \cos \beta} = \tan \alpha + \tan \beta$

63. $\frac{\cos(\alpha + \beta)}{\cos \alpha \cos \beta} = 1 - \tan \alpha \tan \beta$

64. $\frac{\cos(\alpha - \beta)}{\sin \alpha \cos \beta} = \cot \alpha + \tan \beta$

65. $\frac{\sin(\alpha + \beta)}{\sin(\alpha - \beta)} = \frac{\tan \alpha + \tan \beta}{\tan \alpha - \tan \beta}$

66. $\frac{\cos(\alpha + \beta)}{\cos(\alpha - \beta)} = \frac{1 - \tan \alpha \tan \beta}{1 + \tan \alpha \tan \beta}$

67. $\cot(\alpha + \beta) = \frac{\cot \alpha \cot \beta - 1}{\cot \beta + \cot \alpha}$

68. $\cot(\alpha - \beta) = \frac{\cot \alpha \cot \beta + 1}{\cot \beta - \cot \alpha}$

69. $\sec(\alpha + \beta) = \frac{\csc \alpha \csc \beta}{\cot \alpha \cot \beta - 1}$

70. $\sec(\alpha - \beta) = \frac{\sec \alpha \sec \beta}{1 + \tan \alpha \tan \beta}$

71. $\sin(\alpha - \beta) \sin(\alpha + \beta) = \sin^2 \alpha - \sin^2 \beta$

72. $\cos(\alpha - \beta) \cos(\alpha + \beta) = \cos^2 \alpha - \sin^2 \beta$

73. $\sin(\theta + k\pi) = (-1)^k \sin \theta$, k any integer

74. $\cos(\theta + k\pi) = (-1)^k \cos \theta$, k any integer

In Problems 75–86, find the exact value of each expression.

75. $\sin\left(\sin^{-1} \frac{1}{2} + \cos^{-1} 0\right)$

76. $\sin\left(\sin^{-1} \frac{\sqrt{3}}{2} + \cos^{-1} 1\right)$

77. $\sin\left[\sin^{-1} \frac{3}{5} - \cos^{-1}\left(-\frac{4}{5}\right)\right]$

78. $\sin\left[\sin^{-1}\left(-\frac{4}{5}\right) - \tan^{-1} \frac{3}{4}\right]$

79. $\cos\left(\tan^{-1} \frac{4}{3} + \cos^{-1} \frac{5}{13}\right)$

80. $\cos\left[\tan^{-1} \frac{5}{12} - \sin^{-1}\left(-\frac{3}{5}\right)\right]$

81. $\cos\left(\sin^{-1}\frac{5}{13} - \tan^{-1}\frac{3}{4}\right)$

82. $\cos\left(\tan^{-1}\frac{4}{3} + \cos^{-1}\frac{12}{13}\right)$

83. $\tan\left(\sin^{-1}\frac{3}{5} + \frac{\pi}{6}\right)$

84. $\tan\left(\frac{\pi}{4} - \cos^{-1}\frac{3}{5}\right)$

85. $\tan\left(\sin^{-1}\frac{4}{5} + \cos^{-1}1\right)$

86. $\tan\left(\cos^{-1}\frac{4}{5} + \sin^{-1}1\right)$

In Problems 87–92, write each trigonometric expression as an algebraic expression containing u and v. Give the restrictions required on u and v.

87. $\cos(\cos^{-1}u + \sin^{-1}v)$

88. $\sin(\sin^{-1}u - \cos^{-1}v)$

89. $\sin(\tan^{-1}u - \sin^{-1}v)$

90. $\cos(\tan^{-1}u + \tan^{-1}v)$

91. $\tan(\sin^{-1}u - \cos^{-1}v)$

92. $\sec(\tan^{-1}u + \cos^{-1}v)$

In Problems 93–98, solve each equation on the interval $0 \le \theta < 2\pi$.

93. $\sin\theta - \sqrt{3}\cos\theta = 1$

94. $\sqrt{3}\sin\theta + \cos\theta = 1$

95. $\sin\theta + \cos\theta = \sqrt{2}$

96. $\sin\theta - \cos\theta = -\sqrt{2}$

97. $\tan\theta + \sqrt{3} = \sec\theta$

98. $\cot\theta + \csc\theta = -\sqrt{3}$

Applications and Extensions

99. Show that $\sin^{-1}v + \cos^{-1}v = \frac{\pi}{2}$.

100. Show that $\tan^{-1}v + \cot^{-1}v = \frac{\pi}{2}$.

101. Show that $\tan^{-1}\left(\frac{1}{v}\right) = \frac{\pi}{2} - \tan^{-1}v$, if $v > 0$.

102. Show that $\cot^{-1}e^{v} = \tan^{-1}e^{-v}$.

103. Show that $\sin(\sin^{-1}v + \cos^{-1}v) = 1$.

104. Show that $\cos(\sin^{-1}v + \cos^{-1}v) = 0$.

105. Calculus Show that the difference quotient for $f(x) = \sin x$ is given by

$$\frac{f(x+h) - f(x)}{h} = \frac{\sin(x+h) - \sin x}{h} = \cos x \cdot \frac{\sin h}{h} - \sin x \cdot \frac{1 - \cos h}{h}$$

106. Calculus Show that the difference quotient for $f(x) = \cos x$ is given by

$$\frac{f(x+h) - f(x)}{h} = \frac{\cos(x+h) - \cos x}{h} = -\sin x \cdot \frac{\sin h}{h} - \cos x \cdot \frac{1 - \cos h}{h}$$

107. One, Two, Three
(a) Show that $\tan(\tan^{-1}1 + \tan^{-1}2 + \tan^{-1}3) = 0$.
(b) Conclude from part (a) that

$$\tan^{-1}1 + \tan^{-1}2 + \tan^{-1}3 = \pi$$

Source: College Mathematics Journal, Vol. 37, No. 3, May 2006

108. Electric Power In an alternating current (ac) circuit, the instantaneous power p at time t is given by

$$p(t) = V_m I_m \cos\phi \sin^2(\omega t) - V_m I_m \sin\phi \sin(\omega t)\cos(\omega t)$$

Show that this is equivalent to

$$p(t) = V_m I_m \sin(\omega t)\sin(\omega t - \phi)$$

Source: HyperPhysics, hosted by Georgia State University

109. Geometry: Angle between Two Lines Let L_1 and L_2 denote two nonvertical intersecting lines, and let θ denote the acute angle between L_1 and L_2 (see the figure). Show that

$$\tan\theta = \frac{m_2 - m_1}{1 + m_1 m_2}$$

where m_1 and m_2 are the slopes of L_1 and L_2, respectively. [**Hint:** Use the facts that $\tan\theta_1 = m_1$ and $\tan\theta_2 = m_2$.]

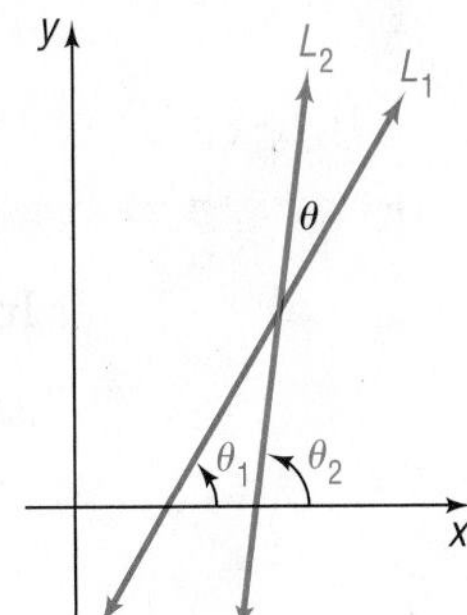

110. If $\alpha + \beta + \gamma = 180°$ and

$$\cot\theta = \cot\alpha + \cot\beta + \cot\gamma \quad 0 < \theta < 90°$$

show that

$$\sin^3\theta = \sin(\alpha - \theta)\sin(\beta - \theta)\sin(\gamma - \theta)$$

111. If $\tan\alpha = x + 1$ and $\tan\beta = x - 1$, show that

$$2\cot(\alpha - \beta) = x^2$$

Explaining Concepts: Discussion and Writing

112. Discuss the following derivation:

$$\tan\left(\theta + \frac{\pi}{2}\right) = \frac{\tan\theta + \tan\frac{\pi}{2}}{1 - \tan\theta\tan\frac{\pi}{2}} = \frac{\frac{\tan\theta}{\tan\frac{\pi}{2}} + 1}{\frac{1}{\tan\frac{\pi}{2}} - \tan\theta} = \frac{0+1}{0-\tan\theta} = \frac{1}{-\tan\theta} = -\cot\theta$$

Can you justify each step?

113. Explain why formula (7) cannot be used to show that

$$\tan\left(\frac{\pi}{2} - \theta\right) = \cot\theta$$

Establish this identity by using formulas (3a) and (3b).

Retain Your Knowledge

Problems 114–117 are based on material learned earlier in the course. The purpose of these problems is to keep the material fresh in your mind so that you are better prepared for the final exam.

114. The International Space Station (ISS) completes about 15.5 orbits of Earth per day. Assuming the orbit is circular with an average radius of 4200 miles, compute the linear speed of the ISS in miles per hour. Round the answer to the nearest whole number.

115. Convert $\frac{17\pi}{6}$ to degrees.

116. Find the area of the sector of a circle of radius 6 centimeters formed by an angle of 45°. Give both the exact area and an approximation rounded to two decimal places.

117. Given $\tan\theta = -2$, $270° < \theta < 360°$, find the exact value of the remaining five trigonometric functions.

'Are You Prepared?' Answers

1. 5 **2.** $-\frac{3}{5}$ **3.** (a) $\frac{\sqrt{2}}{4}$ (b) $\frac{1}{2}$ **4.** $-\frac{4}{5}$

3.6 Double-angle and Half-angle Formulas

OBJECTIVES
1 Use Double-angle Formulas to Find Exact Values (p. 239)
2 Use Double-angle Formulas to Establish Identities (p. 239)
3 Use Half-angle Formulas to Find Exact Values (p. 242)

In this section, formulas for $\sin(2\theta)$, $\cos(2\theta)$, $\sin\left(\frac{1}{2}\theta\right)$, and $\cos\left(\frac{1}{2}\theta\right)$ are established in terms of $\sin\theta$ and $\cos\theta$. They are derived using the sum formulas.

In the sum formulas for $\sin(\alpha+\beta)$ and $\cos(\alpha+\beta)$, let $\alpha = \beta = \theta$. Then

$$\begin{aligned}\sin(\alpha+\beta) &= \sin\alpha\cos\beta + \cos\alpha\sin\beta\\ \sin(\theta+\theta) &= \sin\theta\cos\theta + \cos\theta\sin\theta\\ \sin(2\theta) &= 2\sin\theta\cos\theta\end{aligned}$$

and

$$\begin{aligned}\cos(\alpha+\beta) &= \cos\alpha\cos\beta - \sin\alpha\sin\beta\\ \cos(\theta+\theta) &= \cos\theta\cos\theta - \sin\theta\sin\theta\\ \cos(2\theta) &= \cos^2\theta - \sin^2\theta\end{aligned}$$

An application of the Pythagorean Identity $\sin^2\theta + \cos^2\theta = 1$ results in two other ways to express $\cos(2\theta)$.

$$\cos(2\theta) = \cos^2\theta - \sin^2\theta = (1 - \sin^2\theta) - \sin^2\theta = 1 - 2\sin^2\theta$$

and

$$\cos(2\theta) = \cos^2\theta - \sin^2\theta = \cos^2\theta - (1 - \cos^2\theta) = 2\cos^2\theta - 1$$

The following theorem summarizes the **Double-angle Formulas.**

THEOREM

Double-angle Formulas

$$\sin(2\theta) = 2\sin\theta\cos\theta \quad (1)$$

$$\cos(2\theta) = \cos^2\theta - \sin^2\theta \quad (2)$$

$$\cos(2\theta) = 1 - 2\sin^2\theta \quad (3)$$

$$\cos(2\theta) = 2\cos^2\theta - 1 \quad (4)$$

1 Use Double-angle Formulas to Find Exact Values

EXAMPLE 1

Finding Exact Values Using Double-angle Formulas

If $\sin\theta = \frac{3}{5}$, $\frac{\pi}{2} < \theta < \pi$, find the exact value of:

(a) $\sin(2\theta)$ (b) $\cos(2\theta)$

Solution (a) Because $\sin(2\theta) = 2\sin\theta\cos\theta$ and because $\sin\theta = \frac{3}{5}$ is known, begin by finding $\cos\theta$. Since $\sin\theta = \frac{3}{5} = \frac{y}{r}$, $\frac{\pi}{2} < \theta < \pi$, let $y = 3$ and $r = 5$, and place θ in quadrant II. The point $P = (x, y) = (x, 3)$ is on a circle of radius 5, $x^2 + y^2 = 25$. See Figure 31. Then

$$\begin{aligned} x^2 + y^2 &= 25 \\ x^2 &= 25 - 9 = 16 \quad y = 3 \\ x &= -4 \quad x < 0 \end{aligned}$$

This means that $\cos\theta = \frac{x}{r} = \frac{-4}{5}$. Now use formula (1) to obtain

$$\sin(2\theta) = 2\sin\theta\cos\theta = 2\left(\frac{3}{5}\right)\left(-\frac{4}{5}\right) = -\frac{24}{25}$$

(b) Because $\sin\theta = \frac{3}{5}$ is given, it is easiest to use formula (3) to get $\cos(2\theta)$.

$$\cos(2\theta) = 1 - 2\sin^2\theta = 1 - 2\left(\frac{9}{25}\right) = 1 - \frac{18}{25} = \frac{7}{25}$$

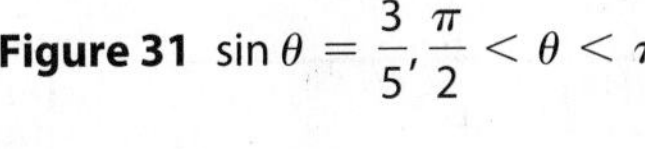

Figure 31 $\sin\theta = \frac{3}{5}, \frac{\pi}{2} < \theta < \pi$

WARNING In finding $\cos(2\theta)$ in Example 1(b), a version of the Double-angle Formula, formula (3), was used. Note that it is not possible to use the Pythagorean identity $\cos(2\theta) = \pm\sqrt{1 - \sin^2(2\theta)}$, with $\sin(2\theta) = -\frac{24}{25}$, because there is no way of knowing which sign to choose.

Now Work PROBLEMS 9(a) AND (b)

2 Use Double-angle Formulas to Establish Identities

EXAMPLE 2

Establishing Identities

(a) Develop a formula for $\tan(2\theta)$ in terms of $\tan\theta$.

(b) Develop a formula for $\sin(3\theta)$ in terms of $\sin\theta$ and $\cos\theta$.

Solution (a) In the sum formula for $\tan(\alpha + \beta)$, let $\alpha = \beta = \theta$. Then

$$\tan(\alpha + \beta) = \frac{\tan\alpha + \tan\beta}{1 - \tan\alpha\tan\beta}$$

$$\tan(\theta + \theta) = \frac{\tan\theta + \tan\theta}{1 - \tan\theta\tan\theta}$$

$$\tan(2\theta) = \frac{2\tan\theta}{1 - \tan^2\theta} \quad (5)$$

(b) To get a formula for $\sin(3\theta)$, write 3θ as $2\theta + \theta$, and use the sum formula.

$$\sin(3\theta) = \sin(2\theta + \theta) = \sin(2\theta)\cos\theta + \cos(2\theta)\sin\theta$$

Now use the Double-angle Formulas to get

$$\begin{aligned}\sin(3\theta) &= (2\sin\theta\cos\theta)(\cos\theta) + (\cos^2\theta - \sin^2\theta)(\sin\theta)\\ &= 2\sin\theta\cos^2\theta + \sin\theta\cos^2\theta - \sin^3\theta\\ &= 3\sin\theta\cos^2\theta - \sin^3\theta\end{aligned}$$

The formula obtained in Example 2(b) also can be written as

$$\begin{aligned}\sin(3\theta) = 3\sin\theta\cos^2\theta - \sin^3\theta &= 3\sin\theta(1 - \sin^2\theta) - \sin^3\theta\\ &= 3\sin\theta - 4\sin^3\theta\end{aligned}$$

That is, $\sin(3\theta)$ is a third-degree polynomial in the variable $\sin\theta$. In fact, $\sin(n\theta)$, n a positive odd integer, can always be written as a polynomial of degree n in the variable $\sin\theta$.*

Now Work PROBLEM 67

Rearranging the Double-angle Formulas (3) and (4) leads to other formulas that will be used later in this section.

Begin with formula (3) and solve for $\sin^2\theta$.

$$\begin{aligned}\cos(2\theta) &= 1 - 2\sin^2\theta\\ 2\sin^2\theta &= 1 - \cos(2\theta)\end{aligned}$$

$$\sin^2\theta = \frac{1 - \cos(2\theta)}{2} \quad (6)$$

Similarly, using formula (4), solve for $\cos^2\theta$.

$$\begin{aligned}\cos(2\theta) &= 2\cos^2\theta - 1\\ 2\cos^2\theta &= 1 + \cos(2\theta)\end{aligned}$$

$$\cos^2\theta = \frac{1 + \cos(2\theta)}{2} \quad (7)$$

Formulas (6) and (7) can be used to develop a formula for $\tan^2\theta$.

$$\tan^2\theta = \frac{\sin^2\theta}{\cos^2\theta} = \frac{\dfrac{1 - \cos(2\theta)}{2}}{\dfrac{1 + \cos(2\theta)}{2}}$$

$$\tan^2\theta = \frac{1 - \cos(2\theta)}{1 + \cos(2\theta)} \quad (8)$$

Formulas (6) through (8) do not have to be memorized since their derivations are so straightforward.

Formulas (6) and (7) are important in calculus. The next example illustrates a problem that arises in calculus requiring the use of formula (7).

*Because of the work done by P. L. Chebyshëv, these polynomials are sometimes called *Chebyshëv polynomials*.

EXAMPLE 3 **Establishing an Identity**

Write an equivalent expression for $\cos^4\theta$ that does not involve any powers of sine or cosine greater than 1.

Solution The idea here is to apply formula (7) twice.

$$\begin{aligned}
\cos^4\theta = (\cos^2\theta)^2 &= \left(\frac{1+\cos(2\theta)}{2}\right)^2 && \text{Formula (7)}\\
&= \frac{1}{4}[1 + 2\cos(2\theta) + \cos^2(2\theta)]\\
&= \frac{1}{4} + \frac{1}{2}\cos(2\theta) + \frac{1}{4}\cos^2(2\theta)\\
&= \frac{1}{4} + \frac{1}{2}\cos(2\theta) + \frac{1}{4}\left\{\frac{1+\cos[2(2\theta)]}{2}\right\} && \text{Formula (7)}\\
&= \frac{1}{4} + \frac{1}{2}\cos(2\theta) + \frac{1}{8}[1+\cos(4\theta)]\\
&= \frac{3}{8} + \frac{1}{2}\cos(2\theta) + \frac{1}{8}\cos(4\theta)
\end{aligned}$$

Now Work PROBLEM 43

EXAMPLE 4 **Solving a Trigonometric Equation Using Identities**

Solve the equation: $\sin\theta\cos\theta = -\frac{1}{2}$, $0 \le \theta < 2\pi$

Solution The left side of the given equation is in the form of the Double-angle Formula $2\sin\theta\cos\theta = \sin(2\theta)$, except for a factor of 2. Multiply each side by 2.

$$\begin{aligned}
\sin\theta\cos\theta &= -\frac{1}{2}\\
2\sin\theta\cos\theta &= -1 && \text{Multiply each side by 2.}\\
\sin(2\theta) &= -1 && \text{Double-angle Formula}
\end{aligned}$$

The argument here is 2θ. Write the general formula that gives all the solutions of this equation, and then list those that are in the interval $[0, 2\pi)$. Because $\sin\left(\frac{3\pi}{2} + 2\pi k\right) = -1$, for any integer k, this means that

$$\begin{aligned}
2\theta &= \frac{3\pi}{2} + 2k\pi && k \text{ any integer}\\
\theta &= \frac{3\pi}{4} + k\pi
\end{aligned}$$

$$\underset{k=-1}{\theta = \frac{3\pi}{4} + (-1)\pi = -\frac{\pi}{4}}, \quad \underset{k=0}{\theta = \frac{3\pi}{4} + (0)\pi = \frac{3\pi}{4}}, \quad \underset{k=1}{\theta = \frac{3\pi}{4} + (1)\pi = \frac{7\pi}{4}}, \quad \underset{k=2}{\theta = \frac{3\pi}{4} + (2)\pi = \frac{11\pi}{4}}$$

The solutions in the interval $[0, 2\pi)$ are

$$\theta = \frac{3\pi}{4}, \quad \theta = \frac{7\pi}{4}$$

The solution set is $\left\{\frac{3\pi}{4}, \frac{7\pi}{4}\right\}$.

Now Work PROBLEM 71

EXAMPLE 5 **Projectile Motion**

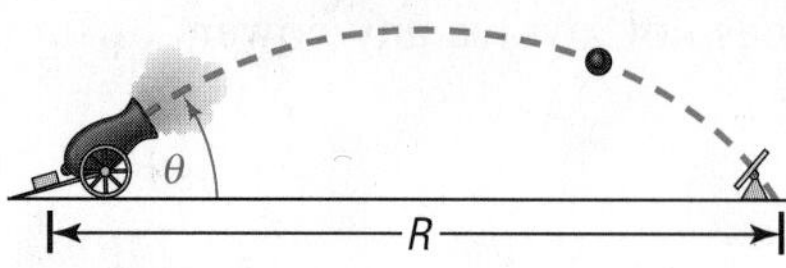

Figure 32

An object is propelled upward at an angle θ to the horizontal with an initial velocity of v_0 feet per second. See Figure 32. If air resistance is ignored, the **range** R—the horizontal distance that the object travels—is given by the function

$$R(\theta) = \frac{1}{16}v_0^2 \sin\theta\cos\theta$$

(a) Show that $R(\theta) = \frac{1}{32}v_0^2\sin(2\theta)$.

(b) Find the angle θ for which R is a maximum.

Solution (a) Rewrite the given expression for the range using the Double-angle Formula $\sin(2\theta) = 2\sin\theta\cos\theta$. Then

$$R(\theta) = \frac{1}{16}v_0^2\sin\theta\cos\theta = \frac{1}{16}v_0^2\frac{2\sin\theta\cos\theta}{2} = \frac{1}{32}v_0^2\sin(2\theta)$$

(b) In this form, the largest value for the range R can be found. For a fixed initial speed v_0, the angle θ of inclination to the horizontal determines the value of R. The largest value of a sine function is 1, which occurs when the argument 2θ is 90°. For maximum R, it follows that

$$2\theta = 90^\circ$$
$$\theta = 45^\circ$$

An inclination to the horizontal of 45° results in the maximum range. ●

3 Use Half-angle Formulas to Find Exact Values

Another important use of formulas (6) through (8) is to prove the *Half-angle Formulas*. In formulas (6) through (8), let $\theta = \frac{\alpha}{2}$. Then

$$\sin^2\frac{\alpha}{2} = \frac{1-\cos\alpha}{2} \qquad \cos^2\frac{\alpha}{2} = \frac{1+\cos\alpha}{2} \qquad \tan^2\frac{\alpha}{2} = \frac{1-\cos\alpha}{1+\cos\alpha} \quad \textbf{(9)}$$

The identities in box (9) will prove useful in integral calculus.

Solving for the trigonometric functions on the left sides of equations (9) gives the Half-angle Formulas.

THEOREM **Half-angle Formulas**

$$\sin\frac{\alpha}{2} = \pm\sqrt{\frac{1-\cos\alpha}{2}} \quad \textbf{(10)}$$

$$\cos\frac{\alpha}{2} = \pm\sqrt{\frac{1+\cos\alpha}{2}} \quad \textbf{(11)}$$

$$\tan\frac{\alpha}{2} = \pm\sqrt{\frac{1-\cos\alpha}{1+\cos\alpha}} \quad \textbf{(12)}$$

where the + or − sign is determined by the quadrant of the angle $\frac{\alpha}{2}$.

EXAMPLE 6 **Finding Exact Values Using Half-angle Formulas**

Use a Half-angle Formula to find the exact value of:

(a) $\cos 15^\circ$ (b) $\sin(-15^\circ)$

Solution (a) Because $15^\circ = \frac{30^\circ}{2}$, use the Half-angle Formula for $\cos\frac{\alpha}{2}$ with $\alpha = 30^\circ$. Also, because 15° is in quadrant I, $\cos 15^\circ > 0$, so choose the + sign in using formula (11):

$$\cos 15° = \cos\frac{30°}{2} = \sqrt{\frac{1+\cos 30°}{2}}$$
$$= \sqrt{\frac{1+\sqrt{3}/2}{2}} = \sqrt{\frac{2+\sqrt{3}}{4}} = \frac{\sqrt{2+\sqrt{3}}}{2}$$

(b) Use the fact that $\sin(-15°) = -\sin 15°$, and then apply formula (10a).

$$\sin(-15°) = -\sin\frac{30°}{2} = -\sqrt{\frac{1-\cos 30°}{2}}$$
$$= -\sqrt{\frac{1-\sqrt{3}/2}{2}} = -\sqrt{\frac{2-\sqrt{3}}{4}} = -\frac{\sqrt{2-\sqrt{3}}}{2}$$ ●

It is interesting to compare the answer found in Example 6(a) with the answer to Example 2 of Section 3.5. There it was calculated that

$$\cos\frac{\pi}{12} = \cos 15° = \frac{1}{4}(\sqrt{6}+\sqrt{2})$$

Based on this and the result of Example 6(a),

$$\frac{1}{4}(\sqrt{6}+\sqrt{2}) \quad \text{and} \quad \frac{\sqrt{2+\sqrt{3}}}{2}$$

are equal. (Since each expression is positive, you can verify this equality by squaring each expression.) Two very different-looking, yet correct, answers can be obtained, depending on the approach taken to solve a problem.

Now Work PROBLEM 21

EXAMPLE 7 **Finding Exact Values Using Half-angle Formulas**

If $\cos\alpha = -\frac{3}{5}$, $\pi < \alpha < \frac{3\pi}{2}$, find the exact value of:

(a) $\sin\frac{\alpha}{2}$ (b) $\cos\frac{\alpha}{2}$ (c) $\tan\frac{\alpha}{2}$

Solution First, observe that if $\pi < \alpha < \frac{3\pi}{2}$, then $\frac{\pi}{2} < \frac{\alpha}{2} < \frac{3\pi}{4}$. As a result, $\frac{\alpha}{2}$ lies in quadrant II.

(a) Because $\frac{\alpha}{2}$ lies in quadrant II, $\sin\frac{\alpha}{2} > 0$, so use the + sign in formula (10) to get

$$\sin\frac{\alpha}{2} = \sqrt{\frac{1-\cos\alpha}{2}} = \sqrt{\frac{1-\left(-\frac{3}{5}\right)}{2}}$$
$$= \sqrt{\frac{\frac{8}{5}}{2}} = \sqrt{\frac{4}{5}} = \frac{2}{\sqrt{5}} = \frac{2\sqrt{5}}{5}$$

(b) Because $\frac{\alpha}{2}$ lies in quadrant II, $\cos\frac{\alpha}{2} < 0$, so use the − sign in formula (11) to get

$$\cos\frac{\alpha}{2} = -\sqrt{\frac{1+\cos\alpha}{2}} = -\sqrt{\frac{1+\left(-\frac{3}{5}\right)}{2}}$$
$$= -\sqrt{\frac{\frac{2}{5}}{2}} = -\frac{1}{\sqrt{5}} = -\frac{\sqrt{5}}{5}$$

(c) Because $\frac{\alpha}{2}$ lies in quadrant II, $\tan\frac{\alpha}{2} < 0$, so use the $-$ sign in formula (12) to get

$$\tan\frac{\alpha}{2} = -\sqrt{\frac{1-\cos\alpha}{1+\cos\alpha}} = -\sqrt{\frac{1-\left(-\frac{3}{5}\right)}{1+\left(-\frac{3}{5}\right)}} = -\sqrt{\frac{\frac{8}{5}}{\frac{2}{5}}} = -2$$

Another way to solve Example 7(c) is to use the results of parts (a) and (b).

$$\tan\frac{\alpha}{2} = \frac{\sin\frac{\alpha}{2}}{\cos\frac{\alpha}{2}} = \frac{\frac{2\sqrt{5}}{5}}{-\frac{\sqrt{5}}{5}} = -2$$

Now Work PROBLEMS 9(c) AND (d)

There is a formula for $\tan\frac{\alpha}{2}$ that does not contain $+$ and $-$ signs, making it more useful than formula (12). To derive it, use the formulas

$$1 - \cos\alpha = 2\sin^2\frac{\alpha}{2} \qquad \text{Formula (9)}$$

and

$$\sin\alpha = \sin\left[2\left(\frac{\alpha}{2}\right)\right] = 2\sin\frac{\alpha}{2}\cos\frac{\alpha}{2} \qquad \text{Double-angle Formula (1)}$$

Then

$$\frac{1-\cos\alpha}{\sin\alpha} = \frac{2\sin^2\frac{\alpha}{2}}{2\sin\frac{\alpha}{2}\cos\frac{\alpha}{2}} = \frac{\sin\frac{\alpha}{2}}{\cos\frac{\alpha}{2}} = \tan\frac{\alpha}{2}$$

Because it also can be shown that

$$\frac{1-\cos\alpha}{\sin\alpha} = \frac{\sin\alpha}{1+\cos\alpha}$$

this results in the following two Half-angle Formulas:

Half-angle Formulas for $\tan\frac{\alpha}{2}$

$$\tan\frac{\alpha}{2} = \frac{1-\cos\alpha}{\sin\alpha} = \frac{\sin\alpha}{1+\cos\alpha} \qquad \textbf{(13)}$$

With this formula, the solution to Example 7(c) can be obtained as follows:

$$\cos\alpha = -\frac{3}{5} \qquad \pi < \alpha < \frac{3\pi}{2}$$

$$\sin\alpha = -\sqrt{1-\cos^2\alpha} = -\sqrt{1-\frac{9}{25}} = -\sqrt{\frac{16}{25}} = -\frac{4}{5}$$

Then, by equation (13),

$$\tan\frac{\alpha}{2} = \frac{1-\cos\alpha}{\sin\alpha} = \frac{1-\left(-\frac{3}{5}\right)}{-\frac{4}{5}} = \frac{\frac{8}{5}}{-\frac{4}{5}} = -2$$

3.6 Assess Your Understanding

Concepts and Vocabulary

1. $\cos(2\theta) = \cos^2\theta -$ _____ $=$ _____ $- 1 = 1 -$ _____.

2. $\sin^2\dfrac{\theta}{2} = \dfrac{\text{\_\_\_\_\_}}{2}$.

3. $\tan\dfrac{\theta}{2} = \dfrac{1-\cos\theta}{\text{\_\_\_\_\_}}$.

4. ***True or False*** $\tan(2\theta) = \dfrac{2\tan\theta}{1-\tan^2\theta}$

5. ***True or False*** $\sin(2\theta)$ has two equivalent forms:
$2\sin\theta\cos\theta$ and $\sin^2\theta - \cos^2\theta$

6. ***True or False*** $\tan(2\theta) + \tan(2\theta) = \tan(4\theta)$

7. Choose the expression that completes the Half-angle Formula for cosine functions: $\cos\dfrac{\alpha}{2} =$ _____

(a) $\pm\sqrt{\dfrac{1-\cos\alpha}{2}}$ (b) $\pm\sqrt{\dfrac{1+\cos\alpha}{2}}$

(c) $\pm\sqrt{\dfrac{\cos\alpha-\sin\alpha}{2}}$ (d) $\pm\sqrt{\dfrac{1-\cos\alpha}{1+\cos\alpha}}$

8. If $\sin\alpha = \pm\sqrt{\dfrac{1-\cos\theta}{2}}$, then which of the following describes how the value of θ is related to the value of α?

(a) $\theta = \alpha$ (b) $\theta = \dfrac{\alpha}{2}$ (c) $\theta = 2\alpha$ (d) $\theta = \alpha^2$

Skill Building

In Problems 9–20, use the information given about the angle θ, $0 \le \theta < 2\pi$, to find the exact value of:

(a) $\sin(2\theta)$ *(b)* $\cos(2\theta)$ *(c)* $\sin\dfrac{\theta}{2}$ *(d)* $\cos\dfrac{\theta}{2}$

9. $\sin\theta = \dfrac{3}{5}$, $0 < \theta < \dfrac{\pi}{2}$

10. $\cos\theta = \dfrac{3}{5}$, $0 < \theta < \dfrac{\pi}{2}$

11. $\tan\theta = \dfrac{4}{3}$, $\pi < \theta < \dfrac{3\pi}{2}$

12. $\tan\theta = \dfrac{1}{2}$, $\pi < \theta < \dfrac{3\pi}{2}$

13. $\cos\theta = -\dfrac{\sqrt{6}}{3}$, $\dfrac{\pi}{2} < \theta < \pi$

14. $\sin\theta = -\dfrac{\sqrt{3}}{3}$, $\dfrac{3\pi}{2} < \theta < 2\pi$

15. $\sec\theta = 3$, $\sin\theta > 0$

16. $\csc\theta = -\sqrt{5}$, $\cos\theta < 0$

17. $\cot\theta = -2$, $\sec\theta < 0$

18. $\sec\theta = 2$, $\csc\theta < 0$

19. $\tan\theta = -3$, $\sin\theta < 0$

20. $\cot\theta = 3$, $\cos\theta < 0$

In Problems 21–30, use the Half-angle Formulas to find the exact value of each expression.

21. $\sin 22.5°$

22. $\cos 22.5°$

23. $\tan\dfrac{7\pi}{8}$

24. $\tan\dfrac{9\pi}{8}$

25. $\cos 165°$

26. $\sin 195°$

27. $\sec\dfrac{15\pi}{8}$

28. $\csc\dfrac{7\pi}{8}$

29. $\sin\left(-\dfrac{\pi}{8}\right)$

30. $\cos\left(-\dfrac{3\pi}{8}\right)$

In Problems 31–42, use the figures to evaluate each function, given that $f(x) = \sin x$, $g(x) = \cos x$, and $h(x) = \tan x$.

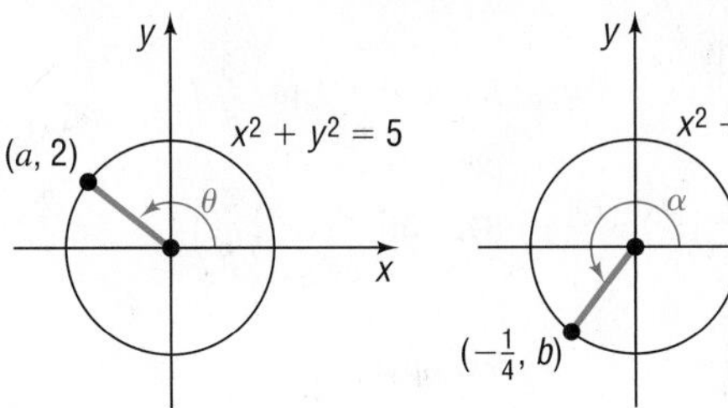

31. $f(2\theta)$

32. $g(2\theta)$

33. $g\left(\dfrac{\theta}{2}\right)$

34. $f\left(\dfrac{\theta}{2}\right)$

35. $h(2\theta)$

36. $h\left(\dfrac{\theta}{2}\right)$

37. $g(2\alpha)$

38. $f(2\alpha)$

39. $f\left(\dfrac{\alpha}{2}\right)$

40. $g\left(\dfrac{\alpha}{2}\right)$

41. $h\left(\dfrac{\alpha}{2}\right)$

42. $h(2\alpha)$

43. Show that $\sin^4\theta = \dfrac{3}{8} - \dfrac{1}{2}\cos(2\theta) + \dfrac{1}{8}\cos(4\theta)$.

44. Show that $\sin(4\theta) = (\cos\theta)(4\sin\theta - 8\sin^3\theta)$.

45. Develop a formula for $\cos(3\theta)$ as a third-degree polynomial in the variable $\cos\theta$.

46. Develop a formula for $\cos(4\theta)$ as a fourth-degree polynomial in the variable $\cos\theta$.

47. Find an expression for $\sin(5\theta)$ as a fifth-degree polynomial in the variable $\sin\theta$.

48. Find an expression for $\cos(5\theta)$ as a fifth-degree polynomial in the variable $\cos\theta$.

In Problems 49–70, establish each identity.

49. $\cos^4\theta - \sin^4\theta = \cos(2\theta)$

50. $\dfrac{\cot\theta - \tan\theta}{\cot\theta + \tan\theta} = \cos(2\theta)$

51. $\cot(2\theta) = \dfrac{\cot^2\theta - 1}{2\cot\theta}$

52. $\cot(2\theta) = \dfrac{1}{2}(\cot\theta - \tan\theta)$

53. $\sec(2\theta) = \dfrac{\sec^2\theta}{2 - \sec^2\theta}$

54. $\csc(2\theta) = \dfrac{1}{2}\sec\theta\csc\theta$

55. $\cos^2(2u) - \sin^2(2u) = \cos(4u)$

56. $(4\sin u\cos u)(1 - 2\sin^2 u) = \sin(4u)$

57. $\dfrac{\cos(2\theta)}{1 + \sin(2\theta)} = \dfrac{\cot\theta - 1}{\cot\theta + 1}$

58. $\sin^2\theta\cos^2\theta = \dfrac{1}{8}[1 - \cos(4\theta)]$

59. $\sec^2\dfrac{\theta}{2} = \dfrac{2}{1 + \cos\theta}$

60. $\csc^2\dfrac{\theta}{2} = \dfrac{2}{1 - \cos\theta}$

61. $\cot^2\dfrac{v}{2} = \dfrac{\sec v + 1}{\sec v - 1}$

62. $\tan\dfrac{v}{2} = \csc v - \cot v$

63. $\cos\theta = \dfrac{1 - \tan^2\dfrac{\theta}{2}}{1 + \tan^2\dfrac{\theta}{2}}$

64. $1 - \dfrac{1}{2}\sin(2\theta) = \dfrac{\sin^3\theta + \cos^3\theta}{\sin\theta + \cos\theta}$

65. $\dfrac{\sin(3\theta)}{\sin\theta} - \dfrac{\cos(3\theta)}{\cos\theta} = 2$

66. $\dfrac{\cos\theta + \sin\theta}{\cos\theta - \sin\theta} - \dfrac{\cos\theta - \sin\theta}{\cos\theta + \sin\theta} = 2\tan(2\theta)$

67. $\tan(3\theta) = \dfrac{3\tan\theta - \tan^3\theta}{1 - 3\tan^2\theta}$

68. $\tan\theta + \tan(\theta + 120°) + \tan(\theta + 240°) = 3\tan(3\theta)$

69. $\ln|\sin\theta| = \dfrac{1}{2}(\ln|1 - \cos(2\theta)| - \ln 2)$

70. $\ln|\cos\theta| = \dfrac{1}{2}(\ln|1 + \cos(2\theta)| - \ln 2)$

In Problems 71–80, solve each equation on the interval $0 \le \theta < 2\pi$.

71. $\cos(2\theta) + 6\sin^2\theta = 4$

72. $\cos(2\theta) = 2 - 2\sin^2\theta$

73. $\cos(2\theta) = \cos\theta$

74. $\sin(2\theta) = \cos\theta$

75. $\sin(2\theta) + \sin(4\theta) = 0$

76. $\cos(2\theta) + \cos(4\theta) = 0$

77. $3 - \sin\theta = \cos(2\theta)$

78. $\cos(2\theta) + 5\cos\theta + 3 = 0$

79. $\tan(2\theta) + 2\sin\theta = 0$

80. $\tan(2\theta) + 2\cos\theta = 0$

Mixed Practice

In Problems 81–92, find the exact value of each expression.

81. $\sin\left(2\sin^{-1}\dfrac{1}{2}\right)$

82. $\sin\left[2\sin^{-1}\dfrac{\sqrt{3}}{2}\right]$

83. $\cos\left(2\sin^{-1}\dfrac{3}{5}\right)$

84. $\cos\left(2\cos^{-1}\dfrac{4}{5}\right)$

85. $\tan\left[2\cos^{-1}\left(-\dfrac{3}{5}\right)\right]$

86. $\tan\left(2\tan^{-1}\dfrac{3}{4}\right)$

87. $\sin\left(2\cos^{-1}\dfrac{4}{5}\right)$

88. $\cos\left[2\tan^{-1}\left(-\dfrac{4}{3}\right)\right]$

89. $\sin^2\left(\dfrac{1}{2}\cos^{-1}\dfrac{3}{5}\right)$

90. $\cos^2\left(\dfrac{1}{2}\sin^{-1}\dfrac{3}{5}\right)$

91. $\sec\left(2\tan^{-1}\dfrac{3}{4}\right)$

92. $\csc\left[2\sin^{-1}\left(-\dfrac{3}{5}\right)\right]$

In Problems 93–95, find the real zeros of each trigonometric function on the interval $0 \le \theta < 2\pi$.

93. $f(x) = \sin(2x) - \sin x$

94. $f(x) = \cos(2x) + \cos x$

95. $f(x) = \cos(2x) + \sin^2 x$

Applications and Extensions

96. Constructing a Rain Gutter A rain gutter is to be constructed of aluminum sheets 12 inches wide. After marking off a length of 4 inches from each edge, the builder bends this length up at an angle θ. See the illustration. The area A of the opening as a function of θ is given by

$$A(\theta) = 16\sin\theta(\cos\theta + 1) \quad 0° < \theta < 90°$$

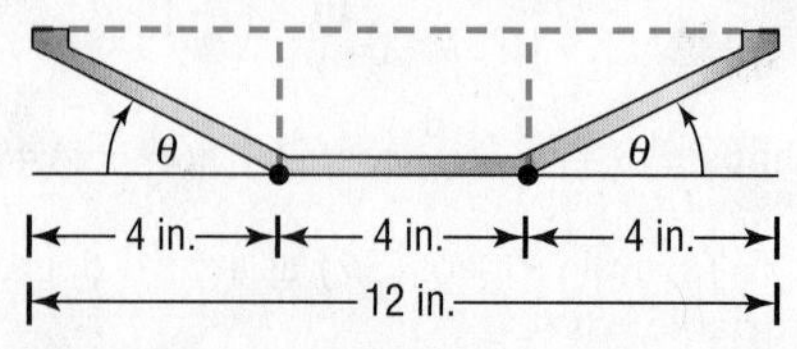

(a) In calculus, you will be asked to find the angle θ that maximizes A by solving the equation

$$\cos(2\theta) + \cos\theta = 0, \quad 0° < \theta < 90°$$

Solve this equation for θ.

(b) What is the maximum area A of the opening?

(c) Graph $A = A(\theta)$, $0° \leq \theta \leq 90°$, and find the angle θ that maximizes the area A. Also find the maximum area.

97. Laser Projection In a laser projection system, the **optical angle** or **scanning angle** θ is related to the throw distance D from the scanner to the screen and the projected image width W by the equation

$$D = \frac{\frac{1}{2}W}{\csc\theta - \cot\theta}$$

(a) Show that the projected image width is given by

$$W = 2D\tan\frac{\theta}{2}$$

(b) Find the optical angle if the throw distance is 15 feet and the projected image width is 6.5 feet.

Source: Pangolin Laser Systems, Inc.

98. Product of Inertia The **product of inertia** for an area about inclined axes is given by the formula

$$I_{uv} = I_x \sin\theta\cos\theta - I_y \sin\theta\cos\theta + I_{xy}(\cos^2\theta - \sin^2\theta)$$

Show that this is equivalent to

$$I_{uv} = \frac{I_x - I_y}{2}\sin(2\theta) + I_{xy}\cos(2\theta)$$

Source: Adapted from Hibbeler, *Engineering Mechanics: Statics,* 13th ed., Pearson © 2013.

99. Projectile Motion An object is propelled upward at an angle θ, $45° < \theta < 90°$, to the horizontal with an initial velocity of v_0 feet per second from the base of a plane that makes an angle of 45° with the horizontal. See the illustration. If air resistance is ignored, the distance R that it travels up the inclined plane is given by the function

$$R(\theta) = \frac{v_0^2\sqrt{2}}{16}\cos\theta(\sin\theta - \cos\theta)$$

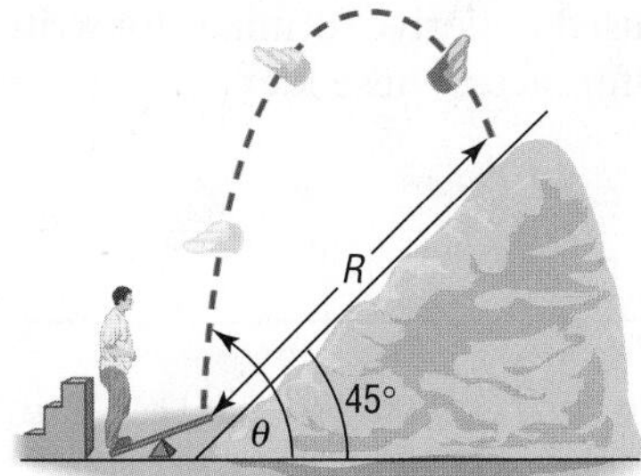

(a) Show that

$$R(\theta) = \frac{v_0^2\sqrt{2}}{32}[\sin(2\theta) - \cos(2\theta) - 1]$$

(b) In calculus, you will be asked to find the angle θ that maximizes R by solving the equation

$$\sin(2\theta) + \cos(2\theta) = 0$$

Solve this equation for θ.

(c) What is the maximum distance R if $v_0 = 32$ feet per second?

(d) Graph $R = R(\theta)$, $45° \leq \theta \leq 90°$, and find the angle θ that maximizes the distance R. Also find the maximum distance. Use $v_0 = 32$ feet per second. Compare the results with the answers found in parts (b) and (c).

100. Sawtooth Curve An oscilloscope often displays a sawtooth curve. This curve can be approximated by sinusoidal curves of varying periods and amplitudes. A first approximation to the sawtooth curve is given by

$$y = \frac{1}{2}\sin(2\pi x) + \frac{1}{4}\sin(4\pi x)$$

Show that $y = \sin(2\pi x)\cos^2(\pi x)$.

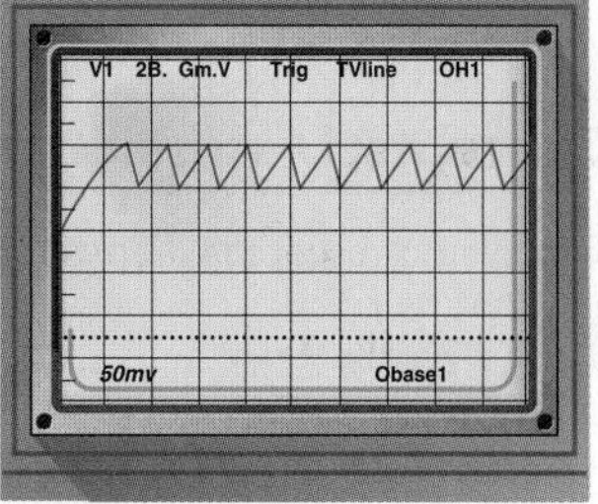

101. Area of an Isosceles Triangle Show that the area A of an isosceles triangle whose equal sides are of length s, and where θ is the angle between them, is

$$A = \frac{1}{2}s^2\sin\theta$$

[**Hint:** See the illustration. The height h bisects the angle θ and is the perpendicular bisector of the base.]

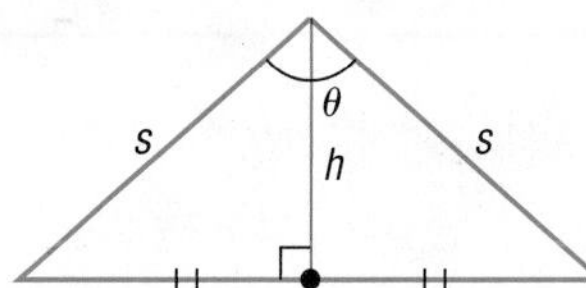

102. Geometry A rectangle is inscribed in a semicircle of radius 1. See the illustration.

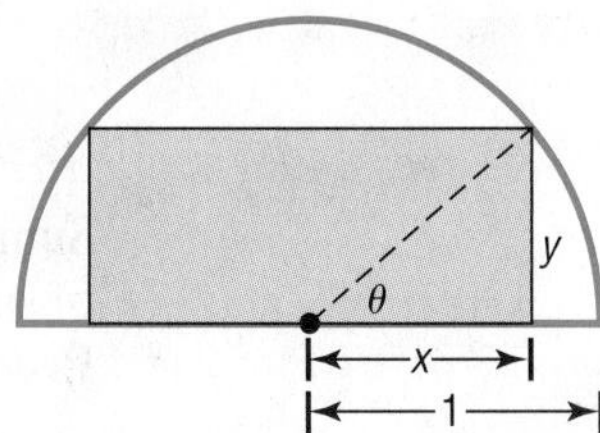

(a) Express the area A of the rectangle as a function of the angle θ shown in the illustration.

(b) Show that $A(\theta) = \sin(2\theta)$.

(c) Find the angle θ that results in the largest area A.

(d) Find the dimensions of this largest rectangle.

103. If $x = 2\tan\theta$, express $\sin(2\theta)$ as a function of x.

104. If $x = 2\tan\theta$, express $\cos(2\theta)$ as a function of x.

105. Find the value of the number C:

$$\frac{1}{2}\sin^2 x + C = -\frac{1}{4}\cos(2x)$$

106. Find the value of the number C:

$$\frac{1}{2}\cos^2 x + C = \frac{1}{4}\cos(2x)$$

107. If $z = \tan\frac{\alpha}{2}$, show that $\sin\alpha = \frac{2z}{1+z^2}$.

108. If $z = \tan\frac{\alpha}{2}$, show that $\cos\alpha = \frac{1-z^2}{1+z^2}$.

109. Graph $f(x) = \sin^2 x = \frac{1-\cos(2x)}{2}$ for $0 \le x \le 2\pi$ by using transformations.

110. Repeat Problem 109 for $g(x) = \cos^2 x$.

111. Use the fact that

$$\cos\frac{\pi}{12} = \frac{1}{4}(\sqrt{6}+\sqrt{2})$$

to find $\sin\frac{\pi}{24}$ and $\cos\frac{\pi}{24}$.

112. Show that

$$\cos\frac{\pi}{8} = \frac{\sqrt{2+\sqrt{2}}}{2}$$

and use it to find $\sin\frac{\pi}{16}$ and $\cos\frac{\pi}{16}$.

113. Show that

$$\sin^3\theta + \sin^3(\theta+120°) + \sin^3(\theta+240°) = -\frac{3}{4}\sin(3\theta)$$

114. If $\tan\theta = a\tan\frac{\theta}{3}$, express $\tan\frac{\theta}{3}$ in terms of a.

115. For $\cos(2x) + (2m-1)\sin x + m - 1 = 0$, find m such that there is exactly one real solution for x, $-\frac{\pi}{2} \le x \le \frac{\pi}{2}$.[†]

[†]Courtesy of Joliet Junior College Mathematics Department

Explaining Concepts: Discussion and Writing

116. Go to the library and research Chebyshëv polynomials. Write a report on your findings.

Retain Your Knowledge

Problems 117–120 are based on material learned earlier in the course. The purpose of these problems is to keep the material fresh in your mind so that you are better prepared for the final exam.

117. Determine the domain of $f(x) = \sqrt{10-2x}$.

118. Determine algebraically whether $f(x) = \frac{-x}{x^2+9}$ is even, odd, or neither.

119. Find the exact value of $\sin\left(\frac{2\pi}{3}\right) - \cos\left(\frac{4\pi}{3}\right)$.

120. Graph $y = -2\cos\left(\frac{\pi}{2}x\right)$. Show at least two periods.

3.7 Product-to-Sum and Sum-to-Product Formulas

OBJECTIVES
1 Express Products as Sums (p. 248)
2 Express Sums as Products (p. 249)

1 Express Products as Sums

Sum and difference formulas can be used to derive formulas for writing the products of sines and/or cosines as sums or differences. These identities are usually called the **Product-to-Sum Formulas.**

THEOREM

Product-to-Sum Formulas

$$\sin\alpha\sin\beta = \frac{1}{2}[\cos(\alpha-\beta) - \cos(\alpha+\beta)] \quad \textbf{(1)}$$

$$\cos\alpha\cos\beta = \frac{1}{2}[\cos(\alpha-\beta) + \cos(\alpha+\beta)] \quad \textbf{(2)}$$

$$\sin\alpha\cos\beta = \frac{1}{2}[\sin(\alpha+\beta) + \sin(\alpha-\beta)] \quad \textbf{(3)}$$

These formulas do not have to be memorized. Instead, remember how they are derived. Then, when you want to use them, either look them up or derive them, as needed.

To derive formulas (1) and (2), write down the sum and difference formulas for the cosine:

$$\cos(\alpha - \beta) = \cos\alpha\cos\beta + \sin\alpha\sin\beta \quad \textbf{(4)}$$
$$\cos(\alpha + \beta) = \cos\alpha\cos\beta - \sin\alpha\sin\beta \quad \textbf{(5)}$$

To derive formula (1), subtract equation (5) from equation (4) to get

$$\cos(\alpha - \beta) - \cos(\alpha + \beta) = 2\sin\alpha\sin\beta$$

from which

$$\sin\alpha\sin\beta = \frac{1}{2}[\cos(\alpha - \beta) - \cos(\alpha + \beta)]$$

To derive formula (2), add equations (4) and (5) to get

$$\cos(\alpha - \beta) + \cos(\alpha + \beta) = 2\cos\alpha\cos\beta$$

from which

$$\cos\alpha\cos\beta = \frac{1}{2}[\cos(\alpha - \beta) + \cos(\alpha + \beta)]$$

To derive Product-to-Sum Formula (3), use the sum and difference formulas for sine in a similar way. (You are asked to do this in Problem 53.)

EXAMPLE 1 **Expressing Products as Sums**

Express each of the following products as a sum containing only sines or only cosines.

(a) $\sin(6\theta)\sin(4\theta)$ (b) $\cos(3\theta)\cos\theta$ (c) $\sin(3\theta)\cos(5\theta)$

Solution (a) Use formula (1) to get

$$\sin(6\theta)\sin(4\theta) = \frac{1}{2}[\cos(6\theta - 4\theta) - \cos(6\theta + 4\theta)]$$
$$= \frac{1}{2}[\cos(2\theta) - \cos(10\theta)]$$

(b) Use formula (2) to get

$$\cos(3\theta)\cos\theta = \frac{1}{2}[\cos(3\theta - \theta) + \cos(3\theta + \theta)]$$
$$= \frac{1}{2}[\cos(2\theta) + \cos(4\theta)]$$

(c) Use formula (3) to get

$$\sin(3\theta)\cos(5\theta) = \frac{1}{2}[\sin(3\theta + 5\theta) + \sin(3\theta - 5\theta)]$$
$$= \frac{1}{2}[\sin(8\theta) + \sin(-2\theta)] = \frac{1}{2}[\sin(8\theta) - \sin(2\theta)]$$

●

Now Work PROBLEM 7

2 Express Sums as Products

The **Sum-to-Product Formulas** are given next.

THEOREM **Sum-to-Product Formulas**

$$\sin\alpha + \sin\beta = 2\sin\frac{\alpha + \beta}{2}\cos\frac{\alpha - \beta}{2} \quad \textbf{(6)}$$

$$\sin\alpha - \sin\beta = 2\sin\frac{\alpha - \beta}{2}\cos\frac{\alpha + \beta}{2} \quad \textbf{(7)}$$

$$\cos\alpha + \cos\beta = 2\cos\frac{\alpha + \beta}{2}\cos\frac{\alpha - \beta}{2} \quad \textbf{(8)}$$

$$\cos\alpha - \cos\beta = -2\sin\frac{\alpha + \beta}{2}\sin\frac{\alpha - \beta}{2} \quad \textbf{(9)}$$

Formula (6) is derived here. The derivations of formulas (7) through (9) are left as exercises (see Problems 54 through 56).

Proof

$$2\sin\frac{\alpha+\beta}{2}\cos\frac{\alpha-\beta}{2} = 2\cdot\frac{1}{2}\left[\sin\left(\frac{\alpha+\beta}{2}+\frac{\alpha-\beta}{2}\right)+\sin\left(\frac{\alpha+\beta}{2}-\frac{\alpha-\beta}{2}\right)\right]$$

Product-to-Sum Formula (3)

$$= \sin\frac{2\alpha}{2} + \sin\frac{2\beta}{2} = \sin\alpha + \sin\beta$$

■

EXAMPLE 2 **Expressing Sums (or Differences) as Products**

Express each sum or difference as a product of sines and/or cosines.

(a) $\sin(5\theta) - \sin(3\theta)$ (b) $\cos(3\theta) + \cos(2\theta)$

Solution (a) Use formula (7) to get

$$\sin(5\theta) - \sin(3\theta) = 2\sin\frac{5\theta-3\theta}{2}\cos\frac{5\theta+3\theta}{2} = 2\sin\theta\cos(4\theta)$$

(b)
$$\cos(3\theta) + \cos(2\theta) = 2\cos\frac{3\theta+2\theta}{2}\cos\frac{3\theta-2\theta}{2} \quad \text{Formula (8)}$$
$$= 2\cos\frac{5\theta}{2}\cos\frac{\theta}{2}$$

●

Now Work PROBLEM 17

3.7 Assess Your Understanding

Skill Building

In Problems 1–6, find the exact value of each expression.

1. $\sin 195°\cdot\cos 75°$
2. $\cos 285°\cdot\cos 195°$
3. $\sin 285°\cdot\sin 75°$
4. $\sin 75° + \sin 15°$
5. $\cos 255° - \cos 195°$
6. $\sin 255° - \sin 15°$

In Problems 7–16, express each product as a sum containing only sines or only cosines.

7. $\sin(4\theta)\sin(2\theta)$
8. $\cos(4\theta)\cos(2\theta)$
9. $\sin(4\theta)\cos(2\theta)$
10. $\sin(3\theta)\sin(5\theta)$
11. $\cos(3\theta)\cos(5\theta)$
12. $\sin(4\theta)\cos(6\theta)$
13. $\sin\theta\sin(2\theta)$
14. $\cos(3\theta)\cos(4\theta)$
15. $\sin\frac{3\theta}{2}\cos\frac{\theta}{2}$
16. $\sin\frac{\theta}{2}\cos\frac{5\theta}{2}$

In Problems 17–24, express each sum or difference as a product of sines and/or cosines.

17. $\sin(4\theta) - \sin(2\theta)$
18. $\sin(4\theta) + \sin(2\theta)$
19. $\cos(2\theta) + \cos(4\theta)$
20. $\cos(5\theta) - \cos(3\theta)$
21. $\sin\theta + \sin(3\theta)$
22. $\cos\theta + \cos(3\theta)$
23. $\cos\frac{\theta}{2} - \cos\frac{3\theta}{2}$
24. $\sin\frac{\theta}{2} - \sin\frac{3\theta}{2}$

In Problems 25–42, establish each identity.

25. $\dfrac{\sin\theta + \sin(3\theta)}{2\sin(2\theta)} = \cos\theta$
26. $\dfrac{\cos\theta + \cos(3\theta)}{2\cos(2\theta)} = \cos\theta$
27. $\dfrac{\sin(4\theta) + \sin(2\theta)}{\cos(4\theta) + \cos(2\theta)} = \tan(3\theta)$
28. $\dfrac{\cos\theta - \cos(3\theta)}{\sin(3\theta) - \sin\theta} = \tan(2\theta)$
29. $\dfrac{\cos\theta - \cos(3\theta)}{\sin\theta + \sin(3\theta)} = \tan\theta$
30. $\dfrac{\cos\theta - \cos(5\theta)}{\sin\theta + \sin(5\theta)} = \tan(2\theta)$
31. $\sin\theta[\sin\theta + \sin(3\theta)] = \cos\theta[\cos\theta - \cos(3\theta)]$
32. $\sin\theta[\sin(3\theta) + \sin(5\theta)] = \cos\theta[\cos(3\theta) - \cos(5\theta)]$
33. $\dfrac{\sin(4\theta) + \sin(8\theta)}{\cos(4\theta) + \cos(8\theta)} = \tan(6\theta)$
34. $\dfrac{\sin(4\theta) - \sin(8\theta)}{\cos(4\theta) - \cos(8\theta)} = -\cot(6\theta)$
35. $\dfrac{\sin(4\theta) + \sin(8\theta)}{\sin(4\theta) - \sin(8\theta)} = -\dfrac{\tan(6\theta)}{\tan(2\theta)}$
36. $\dfrac{\cos(4\theta) - \cos(8\theta)}{\cos(4\theta) + \cos(8\theta)} = \tan(2\theta)\tan(6\theta)$

37. $\dfrac{\sin \alpha + \sin \beta}{\sin \alpha - \sin \beta} = \tan \dfrac{\alpha + \beta}{2} \cot \dfrac{\alpha - \beta}{2}$

38. $\dfrac{\cos \alpha + \cos \beta}{\cos \alpha - \cos \beta} = -\cot \dfrac{\alpha + \beta}{2} \cot \dfrac{\alpha - \beta}{2}$

39. $\dfrac{\sin \alpha + \sin \beta}{\cos \alpha + \cos \beta} = \tan \dfrac{\alpha + \beta}{2}$

40. $\dfrac{\sin \alpha - \sin \beta}{\cos \alpha - \cos \beta} = -\cot \dfrac{\alpha + \beta}{2}$

41. $1 + \cos(2\theta) + \cos(4\theta) + \cos(6\theta) = 4\cos\theta\cos(2\theta)\cos(3\theta)$

42. $1 - \cos(2\theta) + \cos(4\theta) - \cos(6\theta) = 4\sin\theta\cos(2\theta)\sin(3\theta)$

In Problems 43–46, solve each equation on the interval $0 \le \theta < 2\pi$.

43. $\sin(2\theta) + \sin(4\theta) = 0$

44. $\cos(2\theta) + \cos(4\theta) = 0$

45. $\cos(4\theta) - \cos(6\theta) = 0$

46. $\sin(4\theta) - \sin(6\theta) = 0$

Applications and Extensions

47. **Touch-Tone Phones** On a Touch-Tone phone, each button produces a unique sound. The sound produced is the sum of two tones, given by

$$y = \sin(2\pi l t) \quad \text{and} \quad y = \sin(2\pi h t)$$

where l and h are the low and high frequencies (cycles per second) shown on the illustration. For example, if you touch 7, the low frequency is $l = 852$ cycles per second and the high frequency is $h = 1209$ cycles per second. The sound emitted when you touch 7 is

$$y = \sin[2\pi(852)t] + \sin[2\pi(1209)t]$$

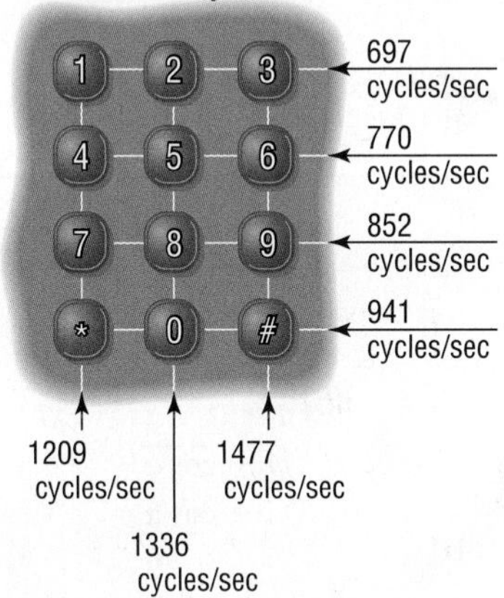

(a) Write this sound as a product of sines and/or cosines.
(b) Determine the maximum value of y.
(c) Graph the sound emitted when 7 is touched.

48. **Touch-Tone Phones**
(a) Write, as a product of sines and/or cosines, the sound emitted when the # key is touched.
(b) Determine the maximum value of y.
(c) Graph the sound emitted when the # key is touched.

49. **Moment of Inertia** The moment of inertia I of an object is a measure of how easy it is to rotate the object about some fixed point. In engineering mechanics, it is sometimes necessary to compute moments of inertia with respect to a set of rotated axes. These moments are given by the equations

$$I_u = I_x \cos^2\theta + I_y \sin^2\theta - 2I_{xy}\sin\theta\cos\theta$$
$$I_v = I_x \sin^2\theta + I_y \cos^2\theta + 2I_{xy}\sin\theta\cos\theta$$

Use Product-to-Sum Formulas to show that

$$I_u = \frac{I_x + I_y}{2} + \frac{I_x - I_y}{2}\cos(2\theta) - I_{xy}\sin(2\theta)$$

and

$$I_v = \frac{I_x + I_y}{2} - \frac{I_x - I_y}{2}\cos(2\theta) + I_{xy}\sin(2\theta)$$

Source: Adapted from Hibbeler, *Engineering Mechanics: Statics,* 13th ed., Pearson © 2013.

50. **Projectile Motion** The range R of a projectile propelled downward from the top of an inclined plane at an angle θ to the inclined plane is given by

$$R(\theta) = \frac{2v_0^2 \sin\theta\cos(\theta - \phi)}{g\cos^2\phi}$$

where v_0 is the initial velocity of the projectile, ϕ is the angle the plane makes with respect to the horizontal, and g is acceleration due to gravity.

(a) Show that for fixed v_0 and ϕ, the maximum range down the incline is given by $R_{\max} = \dfrac{v_0^2}{g(1 - \sin\phi)}$.
(b) Determine the maximum range if the projectile has an initial velocity of 50 meters/second, the angle of the plane is $\phi = 35°$, and $g = 9.8$ meters/second^2.

51. If $\alpha + \beta + \gamma = \pi$, show that

$$\sin(2\alpha) + \sin(2\beta) + \sin(2\gamma) = 4\sin\alpha\sin\beta\sin\gamma$$

52. If $\alpha + \beta + \gamma = \pi$, show that

$$\tan\alpha + \tan\beta + \tan\gamma = \tan\alpha\tan\beta\tan\gamma$$

53. Derive formula (3).

54. Derive formula (7).

55. Derive formula (8).

56. Derive formula (9).

Retain Your Knowledge

Problems 57–60 are based on material learned earlier in the course. The purpose of these problems is to keep the material fresh in your mind so that you are better prepared for the final exam.

57. Find the intercepts of the graph of $3x^2 - 4y = 48$.

58. Find the exact value of $\cos\left(\csc^{-1}\dfrac{7}{5}\right)$.

59. For $y = 5\cos(4x - \pi)$, find the amplitude, the period, and the phase shift.

60. Find the inverse function f^{-1} of $f(x) = 3\sin x - 5$, $-\dfrac{\pi}{2} \le x \le \dfrac{\pi}{2}$. Find the range of f and the domain and range of f^{-1}.

Chapter Review

Things to Know

Definitions of the six inverse trigonometric functions

$y = \sin^{-1} x$ means $x = \sin y$ where $-1 \le x \le 1,\ -\frac{\pi}{2} \le y \le \frac{\pi}{2}$ (p. 190)

$y = \cos^{-1} x$ means $x = \cos y$ where $-1 \le x \le 1,\ 0 \le y \le \pi$ (p. 193)

$y = \tan^{-1} x$ means $x = \tan y$ where $-\infty < x < \infty,\ -\frac{\pi}{2} < y < \frac{\pi}{2}$ (p. 196)

$y = \sec^{-1} x$ means $x = \sec y$ where $|x| \ge 1,\ 0 \le y \le \pi,\ y \ne \frac{\pi}{2}$ (p. 203)

$y = \csc^{-1} x$ means $x = \csc y$ where $|x| \ge 1,\ -\frac{\pi}{2} \le y \le \frac{\pi}{2},\ y \ne 0$ (p. 203)

$y = \cot^{-1} x$ means $x = \cot y$ where $-\infty < x < \infty,\ 0 < y < \pi$ (p. 203)

Sum and Difference Formulas (pp. 226, 229, and 231)

$$\cos(\alpha + \beta) = \cos\alpha\cos\beta - \sin\alpha\sin\beta \qquad \cos(\alpha - \beta) = \cos\alpha\cos\beta + \sin\alpha\sin\beta$$

$$\sin(\alpha + \beta) = \sin\alpha\cos\beta + \cos\alpha\sin\beta \qquad \sin(\alpha - \beta) = \sin\alpha\cos\beta - \cos\alpha\sin\beta$$

$$\tan(\alpha + \beta) = \frac{\tan\alpha + \tan\beta}{1 - \tan\alpha\tan\beta} \qquad \tan(\alpha - \beta) = \frac{\tan\alpha - \tan\beta}{1 + \tan\alpha\tan\beta}$$

Double-angle Formulas (pp. 239 and 240)

$$\sin(2\theta) = 2\sin\theta\cos\theta \qquad \cos(2\theta) = \cos^2\theta - \sin^2\theta \qquad \tan(2\theta) = \frac{2\tan\theta}{1 - \tan^2\theta}$$

$$\cos(2\theta) = 2\cos^2\theta - 1 \qquad \cos(2\theta) = 1 - 2\sin^2\theta$$

Half-angle Formulas (pp. 240, 242, and 244)

$$\sin^2\frac{\alpha}{2} = \frac{1 - \cos\alpha}{2} \qquad \cos^2\frac{\alpha}{2} = \frac{1 + \cos\alpha}{2} \qquad \tan^2\frac{\alpha}{2} = \frac{1 - \cos\alpha}{1 + \cos\alpha}$$

$$\sin\frac{\alpha}{2} = \pm\sqrt{\frac{1 - \cos\alpha}{2}} \qquad \cos\frac{\alpha}{2} = \pm\sqrt{\frac{1 + \cos\alpha}{2}} \qquad \tan\frac{\alpha}{2} = \pm\sqrt{\frac{1 - \cos\alpha}{1 + \cos\alpha}} = \frac{1 - \cos\alpha}{\sin\alpha} = \frac{\sin\alpha}{1 + \cos\alpha}$$

where the + or − sign is determined by the quadrant of $\frac{\alpha}{2}$.

Product-to-Sum Formulas (p. 248)

$$\sin\alpha\sin\beta = \frac{1}{2}[\cos(\alpha - \beta) - \cos(\alpha + \beta)]$$

$$\cos\alpha\cos\beta = \frac{1}{2}[\cos(\alpha - \beta) + \cos(\alpha + \beta)]$$

$$\sin\alpha\cos\beta = \frac{1}{2}[\sin(\alpha + \beta) + \sin(\alpha - \beta)]$$

Sum-to-Product Formulas (p. 249)

$$\sin\alpha + \sin\beta = 2\sin\frac{\alpha + \beta}{2}\cos\frac{\alpha - \beta}{2} \qquad \sin\alpha - \sin\beta = 2\sin\frac{\alpha - \beta}{2}\cos\frac{\alpha + \beta}{2}$$

$$\cos\alpha + \cos\beta = 2\cos\frac{\alpha + \beta}{2}\cos\frac{\alpha - \beta}{2} \qquad \cos\alpha - \cos\beta = -2\sin\frac{\alpha + \beta}{2}\sin\frac{\alpha - \beta}{2}$$

Objectives

Section		You should be able to ...	Example(s)	Review Exercises
3.1	1	Find the exact value of an inverse sine function (p. 190)	1, 2, 6, 7, 9	1–6
	2	Find an approximate value of an inverse sine function (p. 191)	3	76–78
	3	Use properties of inverse functions to find exact values of certain composite functions (p. 192)	4, 5, 8	9–17
	4	Find the inverse function of a trigonometric function (p. 197)	10	24, 25
	5	Solve equations involving inverse trigonometric functions (p. 198)	11	84, 85
3.2	1	Find the exact value of expressions involving the inverse sine, cosine, and tangent functions (p. 202)	1–3	18–21, 23
	2	Define the inverse secant, cosecant, and cotangent functions (p. 203)	4	7, 8, 22
	3	Use a calculator to evaluate $\sec^{-1} x$, $\csc^{-1} x$, and $\cot^{-1} x$ (p. 204)	5	79, 80
	4	Write a trigonometric expression as an algebraic expression (p. 205)	6	26, 27
3.3	1	Solve equations involving a single trigonometric function (p. 208)	1–5	64–68
	2	Solve trigonometric equations using a calculator (p. 211)	6	69
	3	Solve trigonometric equations quadratic in form (p. 211)	7	72
	4	Solve trigonometric equations using fundamental identities (p. 212)	8, 9	70, 71, 73
	5	Solve trigonometric equations using a graphing utility (p. 213)	10	81–83
3.4	1	Use algebra to simplify trigonometric expressions (p. 219)	1	28–44
	2	Establish identities (p. 220)	2–8	28–36
3.5	1	Use sum and difference formulas to find exact values (p. 227)	1–5	45–50, 53–57(a)–(d), 86
	2	Use sum and difference formulas to establish identities (p. 230)	6–8	37, 38
	3	Use sum and difference formulas involving inverse trigonometric functions (p. 232)	9, 10	58–61
	4	Solve trigonometric equations linear in sine and cosine (p. 233)	11, 12	75
3.6	1	Use double-angle formulas to find exact values (p. 239)	1	53–57(e), (f), 62, 63, 87
	2	Use double-angle formulas to establish identities (p. 239)	2–5	40, 41, 74
	3	Use half-angle formulas to find exact values (p. 242)	6, 7	51, 52, 53–57(g), (h), 86
3.7	1	Express products as sums (p. 248)	1	42
	2	Express sums as products (p. 249)	2	43, 44

Review Exercises

In Problems 1–8, find the exact value of each expression. Do not use a calculator.

1. $\sin^{-1} 1$ **2.** $\cos^{-1} 0$ **3.** $\tan^{-1} 1$ **4.** $\sin^{-1}\left(-\frac{1}{2}\right)$

5. $\cos^{-1}\left(-\frac{\sqrt{3}}{2}\right)$ **6.** $\tan^{-1}(-\sqrt{3})$ **7.** $\sec^{-1}\sqrt{2}$ **8.** $\cot^{-1}(-1)$

In Problems 9–23, find the exact value, if any, of each composite function. If there is no value, say it is "not defined." Do not use a calculator.

9. $\sin^{-1}\left(\sin\frac{3\pi}{8}\right)$ **10.** $\cos^{-1}\left(\cos\frac{3\pi}{4}\right)$ **11.** $\tan^{-1}\left(\tan\frac{2\pi}{3}\right)$ **12.** $\cos^{-1}\left(\cos\frac{15\pi}{7}\right)$

13. $\sin^{-1}\left[\sin\left(-\frac{8\pi}{9}\right)\right]$ **14.** $\sin(\sin^{-1} 0.9)$ **15.** $\cos(\cos^{-1} 0.6)$ **16.** $\tan[\tan^{-1} 5]$

17. $\cos[\cos^{-1}(-1.6)]$ **18.** $\sin^{-1}\left(\cos\frac{2\pi}{3}\right)$ **19.** $\cos^{-1}\left(\tan\frac{3\pi}{4}\right)$ **20.** $\tan\left[\sin^{-1}\left(-\frac{\sqrt{3}}{2}\right)\right]$

21. $\sec\left(\tan^{-1}\frac{\sqrt{3}}{3}\right)$

22. $\sin\left(\cot^{-1}\frac{3}{4}\right)$

23. $\tan\left[\sin^{-1}\left(-\frac{4}{5}\right)\right]$

In Problems 24 and 25, find the inverse function f^{-1} of each function f. Find the range of f and the domain and range of f^{-1}.

24. $f(x) = 2\sin(3x) \quad -\frac{\pi}{6} \le x \le \frac{\pi}{6}$

25. $f(x) = -\cos x + 3 \quad 0 \le x \le \pi$

In Problems 26 and 27, write each trigonometric expression as an algebraic expression in u.

26. $\cos(\sin^{-1} u)$

27. $\tan(\csc^{-1} u)$

In Problems 28–44, establish each identity.

28. $\tan\theta\cot\theta - \sin^2\theta = \cos^2\theta$

29. $\sin^2\theta(1 + \cot^2\theta) = 1$

30. $5\cos^2\theta + 3\sin^2\theta = 3 + 2\cos^2\theta$

31. $\frac{1 - \cos\theta}{\sin\theta} + \frac{\sin\theta}{1 - \cos\theta} = 2\csc\theta$

32. $\frac{\cos\theta}{\cos\theta - \sin\theta} = \frac{1}{1 - \tan\theta}$

33. $\frac{\csc\theta}{1 + \csc\theta} = \frac{1 - \sin\theta}{\cos^2\theta}$

34. $\csc\theta - \sin\theta = \cos\theta\cot\theta$

35. $\frac{1 - \sin\theta}{\sec\theta} = \frac{\cos^3\theta}{1 + \sin\theta}$

36. $\frac{1 - 2\sin^2\theta}{\sin\theta\cos\theta} = \cot\theta - \tan\theta$

37. $\frac{\cos(\alpha + \beta)}{\cos\alpha\sin\beta} = \cot\beta - \tan\alpha$

38. $\frac{\cos(\alpha - \beta)}{\cos\alpha\cos\beta} = 1 + \tan\alpha\tan\beta$

39. $(1 + \cos\theta)\tan\frac{\theta}{2} = \sin\theta$

40. $2\cot\theta\cot(2\theta) = \cot^2\theta - 1$

41. $1 - 8\sin^2\theta\cos^2\theta = \cos(4\theta)$

42. $\frac{\sin(3\theta)\cos\theta - \sin\theta\cos(3\theta)}{\sin(2\theta)} = 1$

43. $\frac{\sin(2\theta) + \sin(4\theta)}{\cos(2\theta) + \cos(4\theta)} = \tan(3\theta)$

44. $\frac{\cos(2\theta) - \cos(4\theta)}{\cos(2\theta) + \cos(4\theta)} - \tan\theta\tan(3\theta) = 0$

In Problems 45–52, find the exact value of each expression.

45. $\sin 165°$

46. $\tan 105°$

47. $\cos\frac{5\pi}{12}$

48. $\sin\left(-\frac{\pi}{12}\right)$

49. $\cos 80°\cos 20° + \sin 80°\sin 20°$

50. $\sin 70°\cos 40° - \cos 70°\sin 40°$

51. $\tan\frac{\pi}{8}$

52. $\sin\frac{5\pi}{8}$

In Problems 53–57, use the information given about the angles α and β to find the exact value of:

(a) $\sin(\alpha + \beta)$ (b) $\cos(\alpha + \beta)$ (c) $\sin(\alpha - \beta)$ (d) $\tan(\alpha + \beta)$

(e) $\sin(2\alpha)$ (f) $\cos(2\beta)$ (g) $\sin\frac{\beta}{2}$ (h) $\cos\frac{\alpha}{2}$

53. $\sin\alpha = \frac{4}{5}, 0 < \alpha < \frac{\pi}{2}; \sin\beta = \frac{5}{13}, \frac{\pi}{2} < \beta < \pi$

54. $\sin\alpha = -\frac{3}{5}, \pi < \alpha < \frac{3\pi}{2}; \cos\beta = \frac{12}{13}, \frac{3\pi}{2} < \beta < 2\pi$

55. $\tan\alpha = \frac{3}{4}, \pi < \alpha < \frac{3\pi}{2}; \tan\beta = \frac{12}{5}, 0 < \beta < \frac{\pi}{2}$

56. $\sec\alpha = 2, -\frac{\pi}{2} < \alpha < 0; \sec\beta = 3, \frac{3\pi}{2} < \beta < 2\pi$

57. $\sin\alpha = -\frac{2}{3}, \pi < \alpha < \frac{3\pi}{2}; \cos\beta = -\frac{2}{3}, \pi < \beta < \frac{3\pi}{2}$

In Problems 58–63, find the exact value of each expression.

58. $\cos\left(\sin^{-1}\frac{3}{5} - \cos^{-1}\frac{1}{2}\right)$

59. $\sin\left(\cos^{-1}\frac{5}{13} - \cos^{-1}\frac{4}{5}\right)$

60. $\tan\left[\sin^{-1}\left(-\frac{1}{2}\right) - \tan^{-1}\frac{3}{4}\right]$

61. $\cos\left[\tan^{-1}(-1) + \cos^{-1}\left(-\frac{4}{5}\right)\right]$

62. $\sin\left[2\cos^{-1}\left(-\frac{3}{5}\right)\right]$

63. $\cos\left(2\tan^{-1}\frac{4}{3}\right)$

In Problems 64–75, solve each equation on the interval $0 \le \theta < 2\pi$.

64. $\cos\theta = \frac{1}{2}$

65. $\tan\theta + \sqrt{3} = 0$

66. $\sin(2\theta) + 1 = 0$

67. $\tan(2\theta) = 0$

68. $\sec^2\theta = 4$

69. $0.2\sin\theta = 0.05$

70. $\sin\theta + \sin(2\theta) = 0$

71. $\sin(2\theta) - \cos\theta - 2\sin\theta + 1 = 0$

72. $2\sin^2\theta - 3\sin\theta + 1 = 0$

73. $4\sin^2\theta = 1 + 4\cos\theta$

74. $\sin(2\theta) = \sqrt{2}\cos\theta$

75. $\sin\theta - \cos\theta = 1$

In Problems 76–80, use a calculator to find an approximate value for each expression, rounded to two decimal places.

76. $\sin^{-1} 0.7$

77. $\tan^{-1}(-2)$

78. $\cos^{-1}(-0.2)$

79. $\sec^{-1} 3$

80. $\cot^{-1}(-4)$

In Problems 81–83, use a graphing utility to solve each equation on the interval $0 \le x \le 2\pi$. Approximate any solutions rounded to two decimal places.

81. $2x = 5\cos x$

82. $2\sin x + 3\cos x = 4x$

83. $\sin x = \ln x$

In Problems 84 and 85, find the exact solution of each equation.

84. $-3\sin^{-1} x = \pi$

85. $2\cos^{-1} x + \pi = 4\cos^{-1} x$

86. Use a half-angle formula to find the exact value of sin 15°. Then use a difference formula to find the exact value of sin 15°. Show that the answers you found are the same.

87. If you are given the value of $\cos\theta$ and want the exact value of $\cos(2\theta)$, what form of the Double-angle Formula for $\cos(2\theta)$ is most efficient to use?

Chapter Test

CHAPTER **Test Prep** VIDEOS

The Chapter Test Prep Videos are step-by-step solutions available in MyMathLab®, or on this text's YouTube Channel. Flip back to the Resources for Success page for a link to this text's YouTube channel.

In Problems 1–6, find the exact value of each expression. Express angles in radians.

1. $\sec^{-1}\left(\frac{2}{\sqrt{3}}\right)$

2. $\sin^{-1}\left(-\frac{\sqrt{2}}{2}\right)$

3. $\sin^{-1}\left(\sin\frac{11\pi}{5}\right)$

4. $\tan\left(\tan^{-1}\frac{7}{3}\right)$

5. $\cot\left(\csc^{-1}\sqrt{10}\right)$

6. $\sec\left(\cos^{-1}\left(-\frac{3}{4}\right)\right)$

In Problems 7–10, use a calculator to evaluate each expression. Express angles in radians rounded to two decimal places.

7. $\sin^{-1} 0.382$

8. $\sec^{-1} 1.4$

9. $\tan^{-1} 3$

10. $\cot^{-1} 5$

In Problems 11–16 establish each identity.

11. $\dfrac{\csc\theta + \cot\theta}{\sec\theta + \tan\theta} = \dfrac{\sec\theta - \tan\theta}{\csc\theta - \cot\theta}$

12. $\sin\theta\tan\theta + \cos\theta = \sec\theta$

13. $\tan\theta + \cot\theta = 2\csc(2\theta)$

14. $\dfrac{\sin(\alpha + \beta)}{\tan\alpha + \tan\beta} = \cos\alpha\cos\beta$

15. $\sin(3\theta) = 3\sin\theta - 4\sin^3\theta$

16. $\dfrac{\tan\theta - \cot\theta}{\tan\theta + \cot\theta} = 1 - 2\cos^2\theta$

In Problems 17–24 use sum, difference, product, or half-angle formulas to find the exact value of each expression.

17. $\cos 15°$

18. $\tan 75°$

19. $\sin\left(\frac{1}{2}\cos^{-1}\frac{3}{5}\right)$

20. $\tan\left(2\sin^{-1}\frac{6}{11}\right)$

21. $\cos\left(\sin^{-1}\frac{2}{3} + \tan^{-1}\frac{3}{2}\right)$

22. $\sin 75° \cos 15°$

23. $\sin 75° + \sin 15°$

24. $\cos 65° \cos 20° + \sin 65° \sin 20°$

In Problems 25–29, solve each equation on $0 \le \theta < 2\pi$.

25. $4\sin^2\theta - 3 = 0$

26. $-3\cos\left(\frac{\pi}{2} - \theta\right) = \tan\theta$

27. $\cos^2\theta + 2\sin\theta\cos\theta - \sin^2\theta = 0$

28. $\sin(\theta + 1) = \cos\theta$

29. $4\sin^2\theta + 7\sin\theta = 2$

Cumulative Review

1. Find the real solutions, if any, of the equation $3x^2 + x - 1 = 0$.

2. Find an equation for the line containing the points $(-2, 5)$ and $(4, -1)$. What is the distance between these points? What is their midpoint?

3. Test the equation $3x + y^2 = 9$ for symmetry with respect to the x-axis, y-axis, and origin. List the intercepts.

4. Use transformations to graph the equation $y = |x - 3| + 2$.

5. Use transformations to graph the equation

$$y = \cos\left(x - \frac{\pi}{2}\right) - 1$$

6. Graph each of the following functions. Label at least three points on each graph. Name the inverse function of each and show its graph.

(a) $y = x^3$

(b) $y = \sin x, \quad -\frac{\pi}{2} \le x \le \frac{\pi}{2}$

(c) $y = \cos x, \quad 0 \le x \le \pi$

7. If $\sin\theta = -\frac{1}{3}$ and $\pi < \theta < \frac{3\pi}{2}$, find the exact value of:

(a) $\cos\theta$ (b) $\tan\theta$ (c) $\sin(2\theta)$

(d) $\cos(2\theta)$ (e) $\sin\left(\frac{1}{2}\theta\right)$ (f) $\cos\left(\frac{1}{2}\theta\right)$

8. Find the exact value of $\cos(\tan^{-1} 2)$.

9. If $\sin\alpha = \frac{1}{3}, \frac{\pi}{2} < \alpha < \pi$, and $\cos\beta = -\frac{1}{3}, \pi < \beta < \frac{3\pi}{2}$, find the exact value of:

(a) $\cos\alpha$ (b) $\sin\beta$ (c) $\cos(2\alpha)$

(d) $\cos(\alpha + \beta)$ (e) $\sin\frac{\beta}{2}$

Chapter Projects

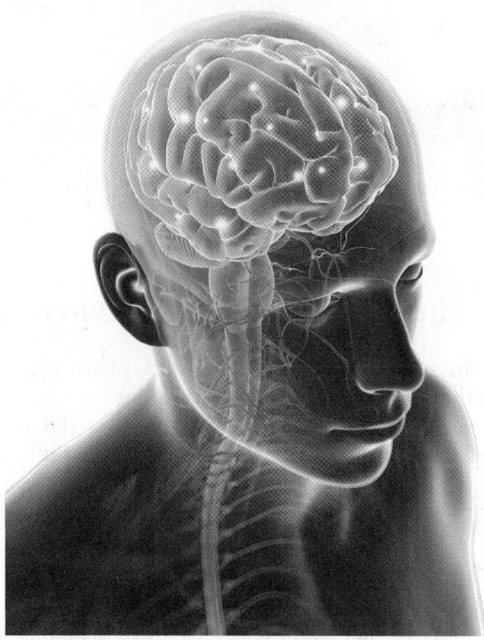

Internet-based Project

I. Mapping Your Mind The goal of this project is to organize the material learned in Chapters 2 and 3 in our minds. To do this, we will use mind-mapping software called Mindomo. Mindomo is free software that enables you to organize your thoughts digitally and share these thoughts with anyone on the Web. By organizing your thoughts, you can see the big picture and then communicate this big picture to others. You are also able to see how various concepts are related to each other.

1. Go to *http://www.mindomo.com* and register. Learn how to use Mindomo. A video on using Mindomo can be found at *http://www.screencast.com/t/ZPwJQDs4*
2. Use an Internet search engine to research Mind Mapping. Write a few paragraphs that explain the history and benefit of mind mapping.
3. Create a MindMap that explains the following:
   (a) The six trigonometric functions and their properties (including the inverses of these functions)
   (b) The fundamental trigonometric identities
   When creating your map, be creative! Perhaps you can share ideas about when a particular identity might be used, or when a particular identity cannot be used.
4. Share the MindMap so that students in your class can view it.

The following projects are available on the Instructor's Resource Center (IRC):

II. Waves Wave motion is described by a sinusoidal equation. The Principle of Superposition of two waves is discussed.

III. Project at Motorola *Sending Pictures Wirelessly* The electronic transmission of pictures is made practical by image compression, mathematical methods that greatly reduce the number of bits of data used to compose the picture.

IV. Calculus of Differences Finding consecutive difference quotients is called finding finite differences and is used to analyze the graph of an unknown function.

Applications of Trigonometric Functions

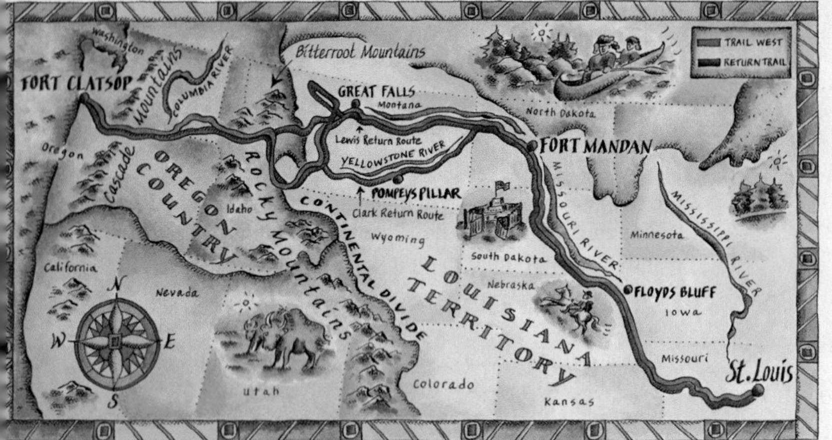

The Lewis and Clark Expedition

In today's world of GPS and smart phone apps that can precisely track one's whereabouts, it is difficult to fathom the magnitude of the challenge that confronted Meriwether Lewis and William Clark in 1804.

But Lewis and Clark managed. Commissioned by President Thomas Jefferson to explore the newly purchased Louisiana Territory, the co-captains led their expedition—the Corps of Discovery—on a journey that took nearly two and a half years and carried them more than 7000 miles. Starting at St. Louis, Missouri, they traveled up the Missouri River, across the Great Plains, over the Rocky Mountains, down the Columbia River to the Pacific Ocean, and then back. Along the way, using limited tools such as a compass and octant, they created more than 130 maps of the area with remarkable detail and accuracy.

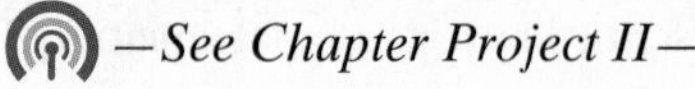
—See Chapter Project II—

Outline

4.1 Right Triangle Trigonometry; Applications
4.2 The Law of Sines
4.3 The Law of Cosines
4.4 Area of a Triangle
4.5 Simple Harmonic Motion; Damped Motion; Combining Waves
Chapter Review
Chapter Test
Cumulative Review
Chapter Projects

••• A Look Back

In Chapter 2, we defined the six trigonometric functions using the unit circle. In particular, we learned to evaluate the trigonometric functions. We also learned how to graph sinusoidal functions. In Chapter 3, we defined the inverse trigonometric functions and solved equations involving the trigonometric functions.

A Look Ahead •••

In this chapter, we define the trigonometric functions using right triangles and then use the trigonometric functions to solve applied problems. The first four sections deal with applications involving right triangles and *oblique triangles,* triangles that do not have a right angle. To solve problems involving oblique triangles, we will develop the Law of Sines and the Law of Cosines. We will also develop formulas for finding the area of a triangle.

The final section deals with applications of sinusoidal functions involving simple harmonic motion and damped motion.

4.1 Right Triangle Trigonometry; Applications

PREPARING FOR THIS SECTION *Before getting started, review the following:*

- Pythagorean Theorem (Appendix A, Section A.2, pp. A14–A15)
- Trigonometric Equations (Section 3.3, pp. 208–213)

Now Work the **'Are You Prepared?'** problems on page 266.

OBJECTIVES
1 Find the Value of Trigonometric Functions of Acute Angles Using Right Triangles (p. 259)
2 Use the Complementary Angle Theorem (p. 261)
3 Solve Right Triangles (p. 261)
4 Solve Applied Problems (p. 262)

1 Find the Value of Trigonometric Functions of Acute Angles Using Right Triangles

A triangle in which one angle is a right angle $(90°)$ is called a **right triangle**. Recall that the side opposite the right angle is called the **hypotenuse**, and the remaining two sides are called the **legs** of the triangle. In Figure 1(a), the hypotenuse is labeled as c to indicate that its length is c units, and, in a like manner, the legs are labeled as a and b. Because the triangle is a right triangle, the Pythagorean Theorem tells us that

$$a^2 + b^2 = c^2$$

Figure 1(a) also shows the angle θ. The angle θ is an **acute angle**: that is, $0° < \theta < 90°$ for θ measured in degrees and $0 < \theta < \frac{\pi}{2}$ for θ measured in radians.

Place θ in standard position, as shown in Figure 1(b). Then the coordinates of the point P are (a, b). Also, P is a point on the terminal side of θ that is on the circle $x^2 + y^2 = c^2$. (Do you see why?)

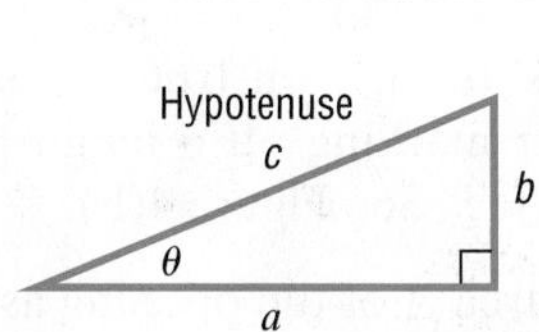

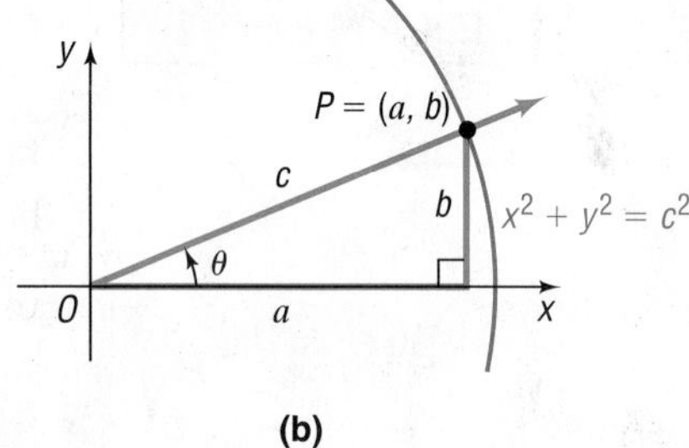

Figure 1 Right triangle with acute angle θ

(a) (b)

Now apply the theorem on page 125 for evaluating trigonometric functions using a circle of radius c, $x^2 + y^2 = c^2$. By referring to the lengths of the sides of the triangle by the names hypotenuse (c), opposite (b), and adjacent (a), as indicated in Figure 2, the trigonometric functions of θ can be expressed as ratios of the sides of a right triangle.

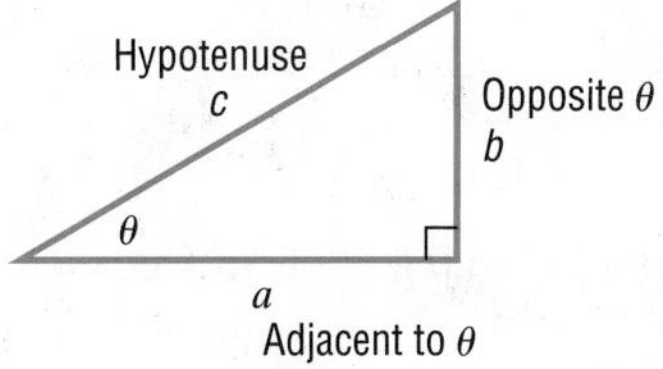

Figure 2 Right triangle

$$\sin\theta = \frac{\text{Opposite}}{\text{Hypotenuse}} = \frac{b}{c} \qquad \csc\theta = \frac{\text{Hypotenuse}}{\text{Opposite}} = \frac{c}{b}$$
$$\cos\theta = \frac{\text{Adjacent}}{\text{Hypotenuse}} = \frac{a}{c} \qquad \sec\theta = \frac{\text{Hypotenuse}}{\text{Adjacent}} = \frac{c}{a} \qquad \textbf{(1)}$$
$$\tan\theta = \frac{\text{Opposite}}{\text{Adjacent}} = \frac{b}{a} \qquad \cot\theta = \frac{\text{Adjacent}}{\text{Opposite}} = \frac{a}{b}$$

Notice that each trigonometric function of the acute angle θ is positive.

EXAMPLE 1 Finding the Value of Trigonometric Functions from a Right Triangle

Find the exact value of the six trigonometric functions of the angle θ in Figure 3.

Solution In Figure 3 the two given sides of the triangle are

$$c = \text{Hypotenuse} = 5, \quad a = \text{Adjacent} = 3$$

5

Opposite

θ

3

Figure 3

To find the length of the opposite side, use the Pythagorean Theorem.

$$(\text{Adjacent})^2 + (\text{Opposite})^2 = (\text{Hypotenuse})^2$$
$$3^2 + (\text{Opposite})^2 = 5^2$$
$$(\text{Opposite})^2 = 25 - 9 = 16$$
$$\text{Opposite} = 4$$

Now that the lengths of the three sides are known, use the ratios in equations (1) to find the value of each of the six trigonometric functions.

$$\sin\theta = \frac{\text{Opposite}}{\text{Hypotenuse}} = \frac{4}{5} \quad \cos\theta = \frac{\text{Adjacent}}{\text{Hypotenuse}} = \frac{3}{5} \quad \tan\theta = \frac{\text{Opposite}}{\text{Adjacent}} = \frac{4}{3}$$

$$\csc\theta = \frac{\text{Hypotenuse}}{\text{Opposite}} = \frac{5}{4} \quad \sec\theta = \frac{\text{Hypotenuse}}{\text{Adjacent}} = \frac{5}{3} \quad \cot\theta = \frac{\text{Adjacent}}{\text{Opposite}} = \frac{3}{4}$$

●

Now Work PROBLEM 9

The values of the trigonometric functions of an acute angle are ratios of the lengths of the sides of a right triangle. This way of viewing the trigonometric functions leads to many applications and, in fact, was the point of view used by early mathematicians (before calculus) in studying the subject of trigonometry.

EXAMPLE 2 Constructing a Rain Gutter

A rain gutter is to be constructed of aluminum sheets 12 inches wide. See Figure 4(a). After marking off a length of 4 inches from each edge, the sides are bent up at an angle θ. See Figure 4(b).

(a) Express the area A of the opening as a function of θ.

[Hint: Let b denote the vertical height of the bend.]

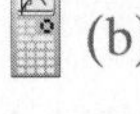
(b) Graph $A = A(\theta)$. Find the angle θ that makes A largest. (This bend will allow the most water to flow through the gutter.)

Solution (a) Look again at Figure 4(b). The area A of the opening is the sum of the areas of two congruent right triangles and one rectangle. Look at Figure 4(c), which shows the triangle on the right in Figure 4(b) redrawn. We see that

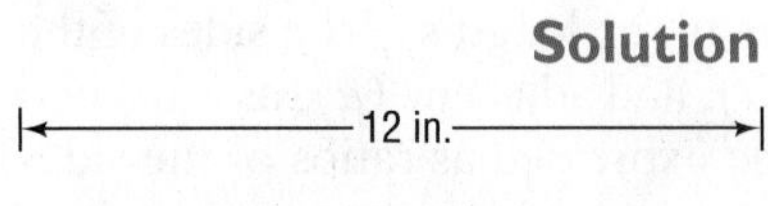

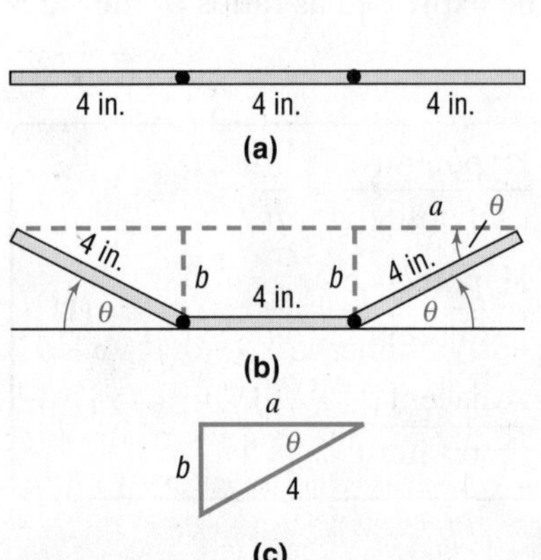

Figure 4

$$\cos\theta = \frac{a}{4} \quad \text{so} \quad a = 4\cos\theta \qquad \sin\theta = \frac{b}{4} \quad \text{so} \quad b = 4\sin\theta$$

The area of the triangle is

$$\text{area} = \frac{1}{2}(\text{base})(\text{height}) = \frac{1}{2}ab = \frac{1}{2}(4\cos\theta)(4\sin\theta) = 8\sin\theta\cos\theta$$

So the area of the two congruent triangles together is $16\sin\theta\cos\theta$.

The rectangle has length 4 and height b, so its area is

$$4b = 4(4\sin\theta) = 16\sin\theta$$

The area A of the opening is

$$A = \text{area of the two triangles} + \text{area of the rectangle}$$

$$A(\theta) = 16 \sin\theta \cos\theta + 16 \sin\theta = 16 \sin\theta(\cos\theta + 1)$$

(b) Figure 5 shows the graph of $A = A(\theta)$. Using MAXIMUM, the angle θ that makes A largest is 60°. ●

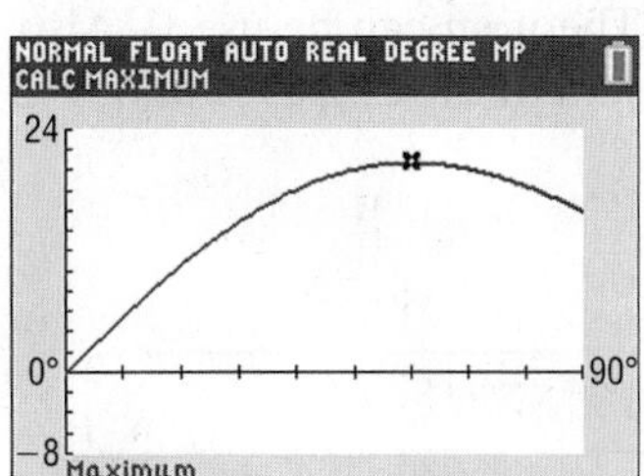

Figure 5

2 Use the Complementary Angle Theorem

Two acute angles are called **complementary** if their sum is a right angle, or 90°. Because the sum of the angles of any triangle is 180°, it follows that, for a right triangle, the sum of the acute angles is 90°, so the two acute angles are complementary.

Refer now to Figure 6, which labels the angle opposite side b as B and the angle opposite side a as A. Notice that side b is adjacent to angle A and side a is adjacent to angle B. As a result,

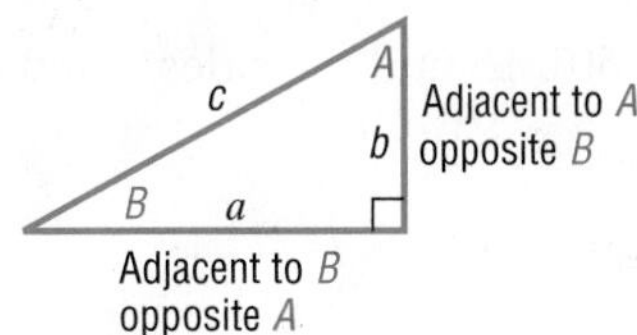

Figure 6

$$\sin B = \frac{b}{c} = \cos A \qquad \cos B = \frac{a}{c} = \sin A \qquad \tan B = \frac{b}{a} = \cot A$$

$$\csc B = \frac{c}{b} = \sec A \qquad \sec B = \frac{c}{a} = \csc A \qquad \cot B = \frac{a}{b} = \tan A \tag{2}$$

Because of these relationships, the functions sine and cosine, tangent and cotangent, and secant and cosecant are called **cofunctions** of each other. The identities (2) may be expressed in words as follows:

THEOREM

Complementary Angle Theorem

Cofunctions of complementary angles are equal.

Examples of this theorem are given next:

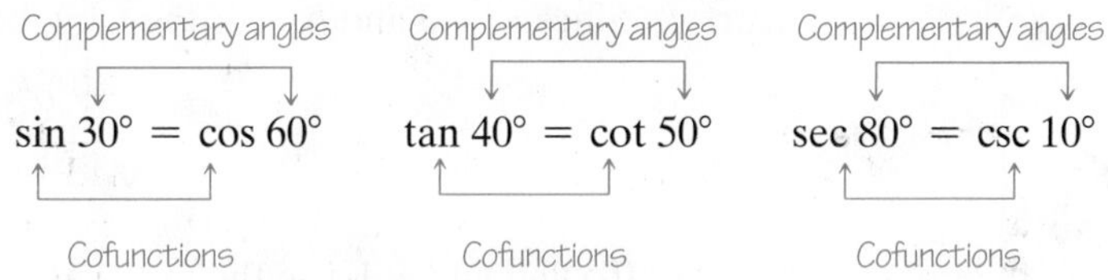

EXAMPLE 3

Using the Complementary Angle Theorem

(a) $\sin 62° = \cos(90° - 62°) = \cos 28°$

(b) $\tan\dfrac{\pi}{12} = \cot\left(\dfrac{\pi}{2} - \dfrac{\pi}{12}\right) = \cot\dfrac{5\pi}{12}$

(c) $\sin^2 40° + \sin^2 50° = \sin^2 40° + \cos^2 40° = 1$

↑ $\sin 50° = \cos 40°$ ●

Now Work PROBLEM 19

3 Solve Right Triangles

Figure 7 Right triangle

In the discussion that follows, we will always label a right triangle so that side a is opposite angle A, side b is opposite angle B, and side c is the hypotenuse, as shown in Figure 7. **To solve a right triangle** means to find the missing lengths of its sides and the measurements of its angles. We shall follow the practice of expressing the lengths of the sides rounded to two decimal places and expressing angles in degrees rounded to one decimal place. (Be sure that your calculator is in degree mode.)

To solve a right triangle, we need to know one of the acute angles A or B and a side, or else two sides (in which case the Pythagorean Theorem can be used). Also, because the sum of the angles of a triangle is 180°, the sum of the angles A and B in a right triangle must be 90°.

THEOREM For the right triangle shown in Figure 7, we have

$$c^2 = a^2 + b^2 \qquad A + B = 90°$$

EXAMPLE 4 **Solving a Right Triangle**

Use Figure 8. If $b = 2$ and $A = 40°$, find a, c, and B.

Solution Since $A = 40°$ and $A + B = 90°$, it follows that $B = 50°$. To find the sides a and c, use the facts that

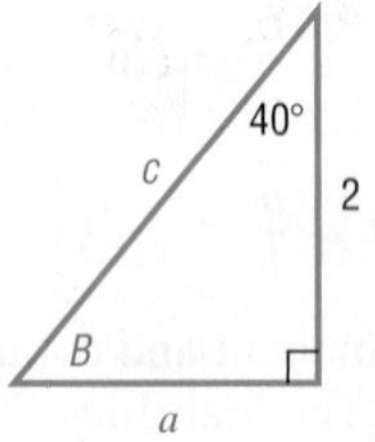

Figure 8

$$\tan 40° = \frac{a}{2} \quad \text{and} \quad \cos 40° = \frac{2}{c}$$

Now solve for a and c.

$$a = 2 \tan 40° \approx 1.68 \quad \text{and} \quad c = \frac{2}{\cos 40°} \approx 2.61$$

Now Work PROBLEM 29

EXAMPLE 5 **Solving a Right Triangle**

Use Figure 9. If $a = 3$ and $b = 2$, find c, A, and B.

Solution Since $a = 3$ and $b = 2$, then, by the Pythagorean Theorem, we have

$$c^2 = a^2 + b^2 = 3^2 + 2^2 = 9 + 4 = 13$$
$$c = \sqrt{13} \approx 3.61$$

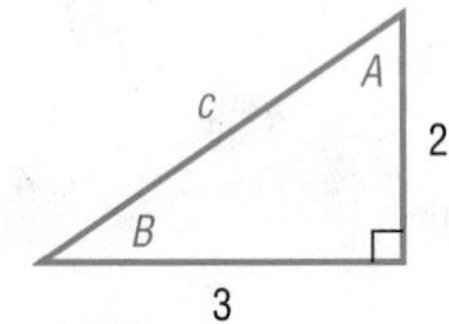

Figure 9

To find angle A, use the fact that

$$\tan A = \frac{3}{2} \quad \text{so} \quad A = \tan^{-1}\frac{3}{2}$$

Use a calculator with the mode set to degrees to find that $A = 56.3°$ rounded to one decimal place. Since $A + B = 90°$, this means that $B = 33.7°$.

NOTE To avoid round-off errors when using a calculator, we will store unrounded values in memory for use in subsequent calculations. ■

Now Work PROBLEM 39

4 Solve Applied Problems*

In addition to developing models using right triangles, we can use right triangle trigonometry to measure heights and distances that are either awkward or impossible to measure by ordinary means. When using right triangles to solve these problems, pay attention to the known measures. This will indicate what trigonometric function to use. For example, if you know the measure of an angle and the length of the side adjacent to the angle, and wish to find the length of the opposite side, you would use the tangent function. Do you know why?

*In applied problems, it is important that answers be reported with both justifiable accuracy and appropriate significant figures. In this chapter we shall assume that the problem data are accurate to the number of significant digits resulting in sides being rounded to two decimal places and angles being rounded to one decimal place.

EXAMPLE 6

Finding the Width of a River

A surveyor can measure the width of a river by setting up a transit* at a point C on one side of the river and taking a sighting of a point A on the other side. Refer to Figure 10. After turning through an angle of 90° at C, the surveyor walks a distance of 200 meters to point B. Using the transit at B, the angle θ is measured and found to be 20°. What is the width of the river rounded to the nearest meter?

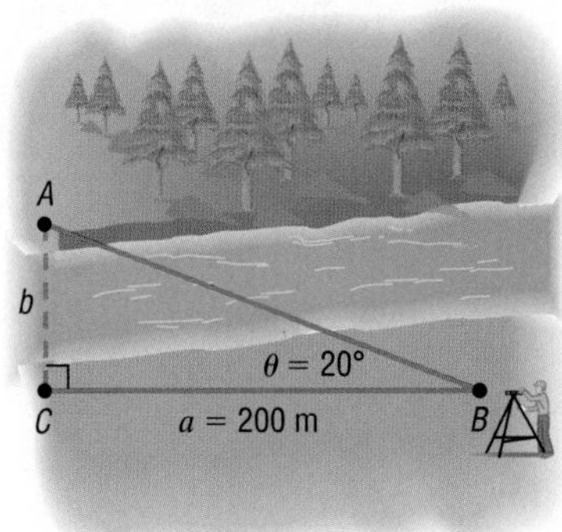

Figure 10

Solution As seen in Figure 10, the width of the river is the length of side b, and a and θ are known. Use the facts that b is opposite θ and a is adjacent to θ and write

$$\tan \theta = \frac{b}{a}$$

which leads to

$$\tan 20° = \frac{b}{200}$$

$$b = 200 \tan 20° \approx 72.79 \text{ meters}$$

The width of the river is 73 meters, rounded to the nearest meter. ●

Now Work PROBLEM 49

EXAMPLE 7

Finding the Inclination of a Mountain Trail

A straight trail leads from the Alpine Hotel, elevation 8000 feet, to a scenic overlook, elevation 11,100 feet. The length of the trail is 14,100 feet. What is the inclination (grade) of the trail? That is, what is the angle B in Figure 11?

Solution Figure 11 shows that the length of the side opposite angle B is 11,100 − 8000 = 3100 feet, and the length of the hypotenuse is 14,100 feet. The angle B obeys the equation

$$\sin B = \frac{3100}{14{,}100}$$

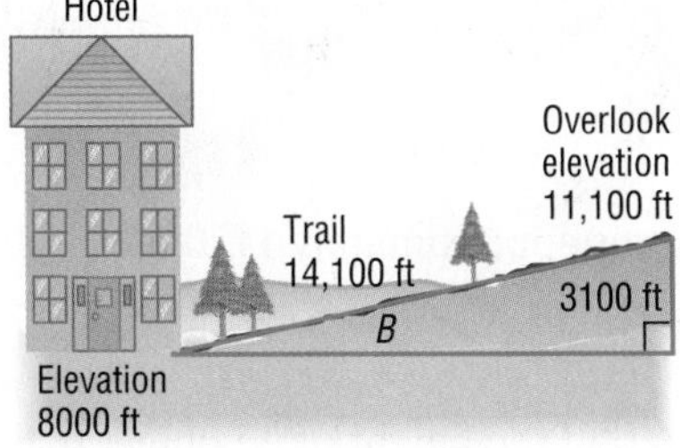

Figure 11

Using a calculator,

$$B = \sin^{-1} \frac{3100}{14{,}100} \approx 12.7°$$

The inclination (grade) of the trail is approximately 12.7°. ●

Now Work PROBLEM 55

Vertical heights can sometimes be measured using either the *angle of elevation* or the *angle of depression*. If a person is looking up at an object, the acute angle measured from the horizontal to a line of sight to the object is called the **angle of elevation**. See Figure 12(a) on the next page.

*An instrument used in surveying to measure angles.

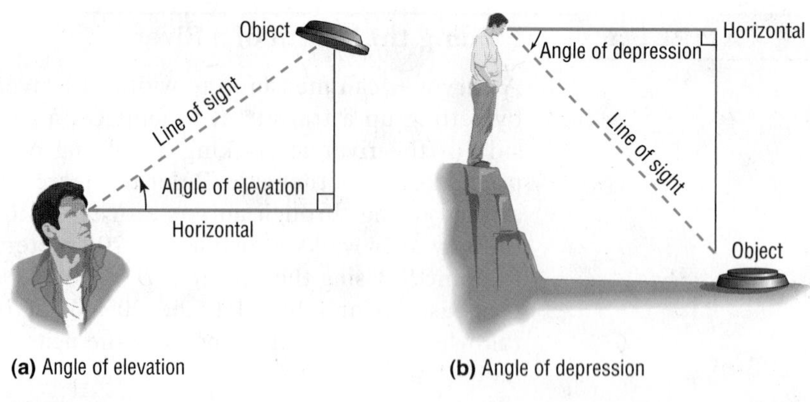

(a) Angle of elevation

(b) Angle of depression

Figure 12

If a person is looking down at an object, the acute angle made by the line of sight to the object and the horizontal is called the **angle of depression**. See Figure 12(b).

EXAMPLE 8

Finding the Height of a Cloud

Meteorologists find the height of a cloud using an instrument called a **ceilometer**. A ceilometer consists of a **light projector** that directs a vertical light beam up to the cloud base and a **light detector** that scans the cloud to detect the light beam. See Figure 13(a). At Midway Airport in Chicago, a ceilometer was employed to find the height of the cloud cover. It was set up with its light detector 300 feet from its light projector. If the angle of elevation from the light detector to the base of the cloud was 75°, what was the height of the cloud cover?

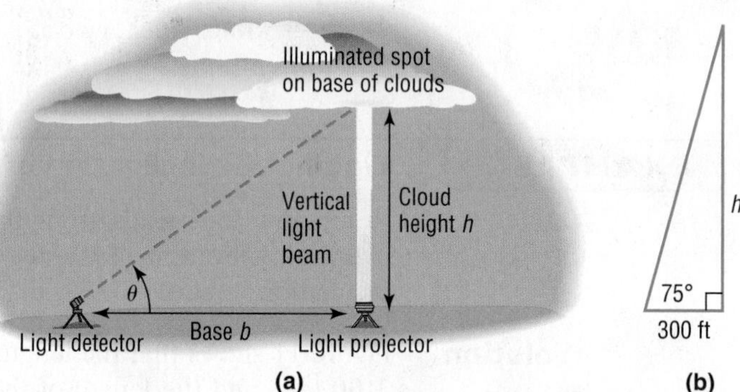

Figure 13 (a) (b)

Solution Figure 13(b) illustrates the situation. To find the height h, use the fact that $\tan 75° = \dfrac{h}{300}$, so

$$h = 300 \tan 75° \approx 1120 \text{ feet}$$

The ceiling (height to the base of the cloud cover) was approximately 1120 feet. •

 **Now Work** PROBLEM 51

The idea behind Example 8 can also be used to find the height of an object that is positioned above ground level.

EXAMPLE 9

Finding the Height of a Statue on a Building

Adorning the top of the Board of Trade building in Chicago is a statue of Ceres, the Roman goddess of wheat. From street level, two observations are taken 400 feet from the center of the building. The angle of elevation to the base of the statue is found to be 55.1° and the angle of elevation to the top of the statue is 56.5°. See Figure 14(a). What is the height of the statue?

Solution Figure 14(b) shows two triangles that replicate Figure 14(a). The height of the statue of Ceres will be $b' - b$. To find b and b', refer to Figure 14(b).

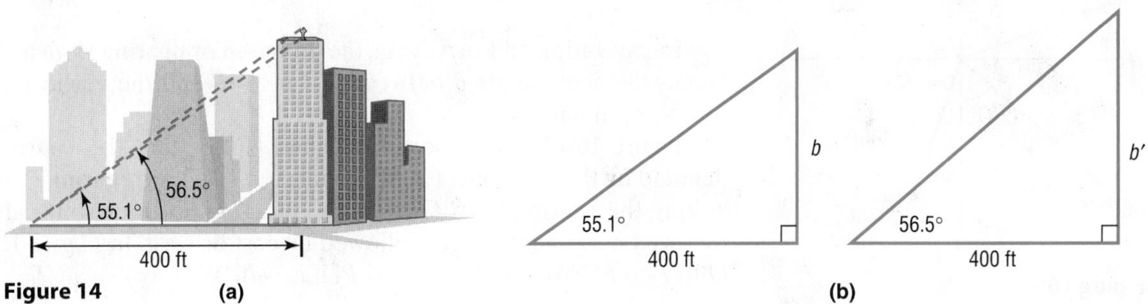

Figure 14 (a) (b)

$$\tan 55.1° = \frac{b}{400} \qquad \tan 56.5° = \frac{b'}{400}$$

$$b = 400 \tan 55.1° \approx 573.39 \qquad b' = 400 \tan 56.5° \approx 604.33$$

The height of the statue is approximately $604.33 - 573.39 = 30.94$ feet ≈ 31 feet.

Now Work PROBLEM 71

EXAMPLE 10 The Gibb's Hill Lighthouse, Southampton, Bermuda

In operation since 1846, the Gibb's Hill Lighthouse stands 117 feet high on a hill 245 feet high, so its beam of light is 362 feet above sea level. A brochure states that the light can be seen on the horizon about 26 miles distant. Verify the accuracy of this statement.

Solution Figure 15 illustrates the situation. The central angle θ, positioned at the center of Earth, radius 3960 miles, obeys the equation

$$\cos\theta = \frac{3960}{3960 + \frac{362}{5280}} \approx 0.999982687 \qquad \text{1 mile} = 5280 \text{ feet}$$

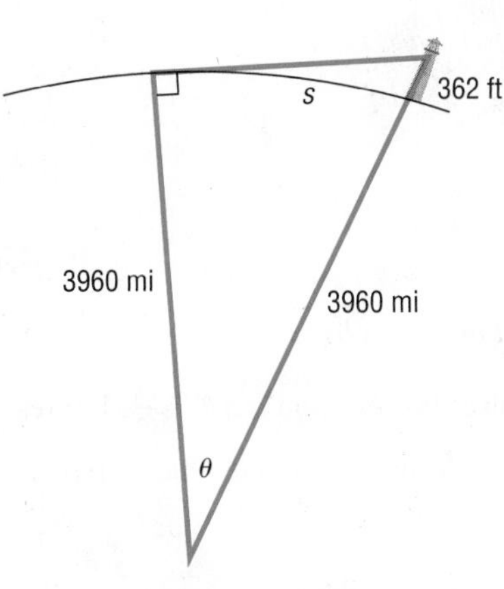

Figure 15

Solving for θ yields

$$\theta \approx \cos^{-1}(0.999982687) \approx 0.33715° \approx 20.23'$$

The brochure does not indicate whether the distance is measured in nautical miles or statute miles. Let's calculate both distances.

The distance s in nautical miles (refer to Problem 122, p. 112) is the measure of the angle θ in minutes, so $s \approx 20.23$ nautical miles.

The distance s in statute miles is given by the formula $s = r\theta$, where θ is measured in radians. Then, since

$$\theta \approx 0.33715° \approx 0.00588 \text{ radian}$$

$$\uparrow$$

$$1° = \frac{\pi}{180} \text{ radian}$$

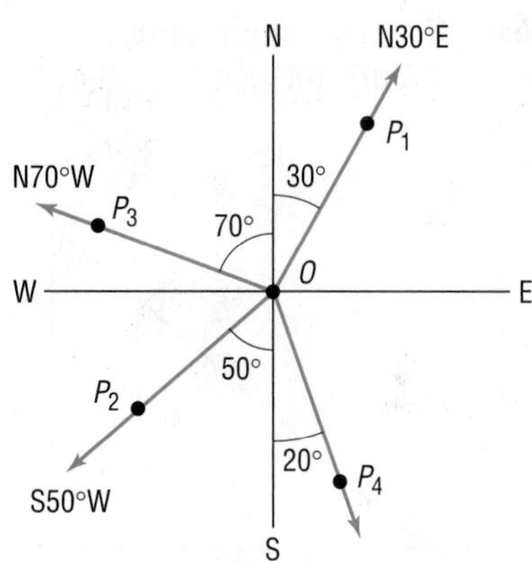

Figure 16

this means that

$$s = r\theta \approx (3960)(0.00588) \approx 23.3 \text{ miles}$$

In either case, it would seem that the brochure overstated the distance somewhat. ●

In navigation and surveying, the **direction** or **bearing** from a point O to a point P equals the acute angle θ between the ray OP and the vertical line through O, the north–south line.

Figure 16 illustrates some bearings. Notice that the bearing from O to P_1 is denoted by the symbolism N30°E, indicating that the bearing is 30° east of north. In writing the bearing from O to P, the direction north or south always appears first, followed by an acute angle, followed by east or west. In Figure 16, the bearing from O to P_2 is S50°W, and from O to P_3 it is N70°W.

EXAMPLE 11

Finding the Bearing of an Object

In Figure 16, what is the bearing from O to an object at P_4?

Solution The acute angle between the ray OP_4 and the north–south line through O is 20°. The bearing from O to P_4 is S20°E. ●

EXAMPLE 12

Finding the Bearing of an Airplane

A Boeing 777 aircraft takes off from O'Hare Airport on runway 2 LEFT, which has a bearing of N20°E.* After flying for 1 mile, the pilot of the aircraft requests permission to turn 90° and head toward the northwest. The request is granted. After the plane goes 2 miles in this direction, what bearing should the control tower use to locate the aircraft?

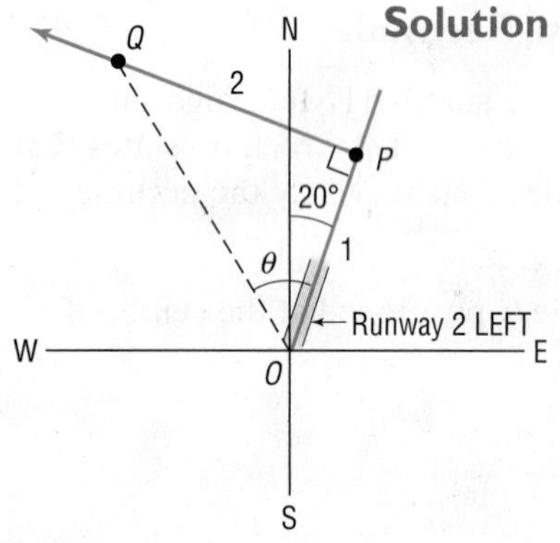

Solution Figure 17 illustrates the situation. After flying 1 mile from the airport O (the control tower), the aircraft is at P. After turning 90° toward the northwest and flying 2 miles, the aircraft is at the point Q. In triangle OPQ, the angle θ obeys the equation

$$\tan\theta = \frac{2}{1} = 2 \quad \text{so} \quad \theta = \tan^{-1} 2 \approx 63.4°$$

The acute angle between north and the ray OQ is $63.4° - 20° = 43.4°$. The bearing of the aircraft from O to Q is N43.4°W. ●

Figure 17

Now Work PROBLEM 63

4.1 Assess Your Understanding

'Are You Prepared?' *Answers are given at the end of these exercises. If you get a wrong answer, read the pages listed in red.*

1. In a right triangle, if the length of the hypotenuse is 5 and the length of one of the other sides is 3, what is the length of the third side? (pp. A14–A15)

2. If θ is an acute angle, solve the equation $\tan\theta = \frac{1}{2}$. Express your answer in degrees, rounded to one decimal place. (pp. 208–213)

3. If θ is an acute angle, solve the equation $\sin\theta = \frac{1}{2}$. (pp. 208–213)

*In air navigation, the term **azimuth** denotes the positive angle measured clockwise from the north (N) to a ray OP. In Figure 16, the azimuth from O to P_1 is 30°; the azimuth from O to P_2 is 230°; the azimuth from O to P_3 is 290°. In naming runways, the units digit is left off the azimuth. Runway 2 LEFT means the left runway with a direction of azimuth 20° (bearing N20°E). Runway 23 is the runway with azimuth 230° and bearing S50°W.

Concepts and Vocabulary

4. ***True or False*** $\sin 52° = \cos 48°$.

5. The sum of the measures of the two acute angles in a right triangle is ______.
(a) 45° (b) 90° (c) 180° (d) 360°

6. When you look up at an object, the acute angle measured from the horizontal to a line-of-sight observation of the object is called the ______ ___ ________.

7. ***True or False*** In a right triangle, if two sides are known, we can solve the triangle.

8. ***True or False*** In a right triangle, if we know the two acute angles, we can solve the triangle.

Skill Building

In Problems 9–18, find the exact value of the six trigonometric functions of the angle θ in each figure.

9.
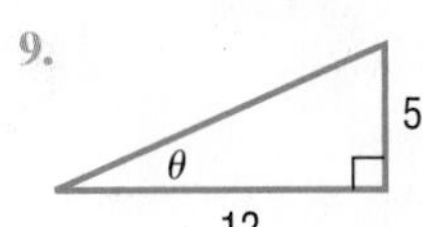

10.
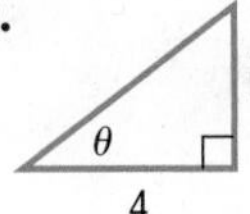

11.
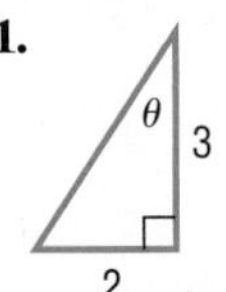

12.
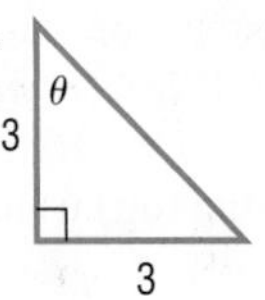

13.
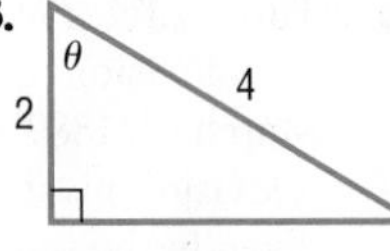

14.
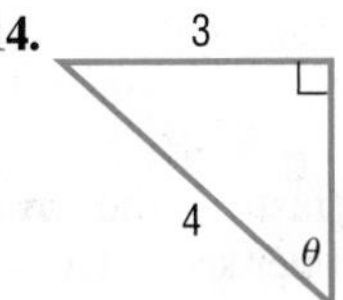

15.
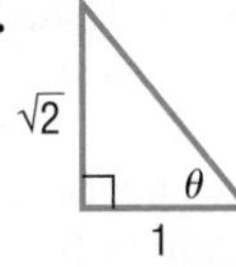

16.
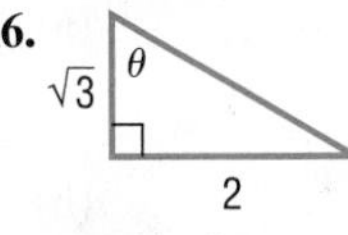

17.
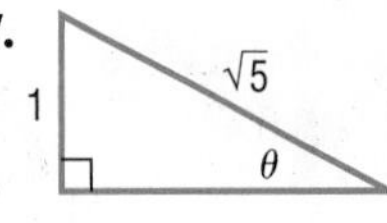

18.
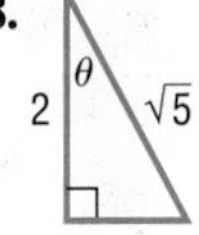

In Problems 19–28, find the exact value of each expression. Do not use a calculator.

19. $\sin 38° - \cos 52°$

20. $\tan 12° - \cot 78°$

21. $\dfrac{\cos 10°}{\sin 80°}$

22. $\dfrac{\cos 40°}{\sin 50°}$

23. $1 - \cos^2 20° - \cos^2 70°$

24. $1 + \tan^2 5° - \csc^2 85°$

25. $\tan 20° - \dfrac{\cos 70°}{\cos 20°}$

26. $\cot 40° - \dfrac{\sin 50°}{\sin 40°}$

27. $\cos 35° \sin 55° + \sin 35° \cos 55°$

28. $\sec 35° \csc 55° - \tan 35° \cot 55°$

In Problems 29–42, use the right triangle shown below. Then, using the given information, solve the triangle.

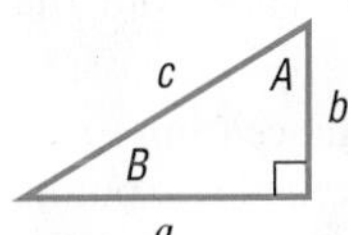

29. $b = 5$, $B = 20°$; find a, c, and A

30. $b = 4$, $B = 10°$; find a, c, and A

31. $a = 6$, $B = 40°$; find b, c, and A

32. $a = 7$, $B = 50°$; find b, c, and A

33. $b = 4$, $A = 10°$; find a, c, and B

34. $b = 6$, $A = 20°$; find a, c, and B

35. $a = 5$, $A = 25°$; find b, c, and B

36. $a = 6$, $A = 40°$; find b, c, and B

37. $c = 9$, $B = 20°$; find b, a, and A

38. $c = 10$, $A = 40°$; find b, a, and B

39. $a = 5$, $b = 3$; find c, A, and B

40. $a = 2$, $b = 8$; find c, A, and B

*41. $a = 2$, $c = 5$; find b, A, and B

42. $b = 4$, $c = 6$; find a, A, and B

Applications and Extensions

43. **Geometry** The hypotenuse of a right triangle is 5 inches. If one leg is 2 inches, find the degree measure of each angle.

44. **Geometry** The hypotenuse of a right triangle is 3 feet. If one leg is 1 foot, find the degree measure of each angle.

45. **Geometry** A right triangle has a hypotenuse of length 8 inches. If one angle is 35°, find the length of each leg.

46. **Geometry** A right triangle has a hypotenuse of length 10 centimeters. If one angle is 40°, find the length of each leg.

47. **Geometry** A right triangle contains a 25° angle.
(a) If one leg is of length 5 inches, what is the length of the hypotenuse?
(b) There are two answers. How is this possible?

48. **Geometry** A right triangle contains an angle of $\dfrac{\pi}{8}$ radian.
(a) If one leg is of length 3 meters, what is the length of the hypotenuse?
(b) There are two answers. How is this possible?

49. **Finding the Width of a Gorge** Find the distance from A to C across the gorge illustrated in the figure.

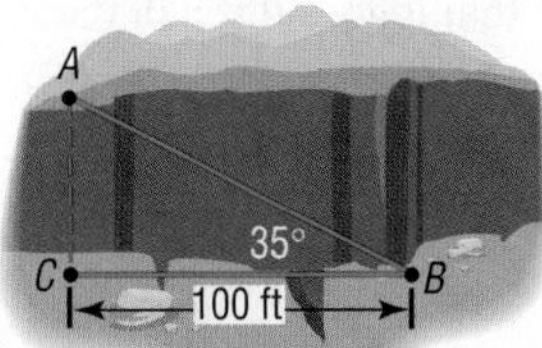

50. **Finding the Distance across a Pond** Find the distance from A to C across the pond illustrated in the figure.

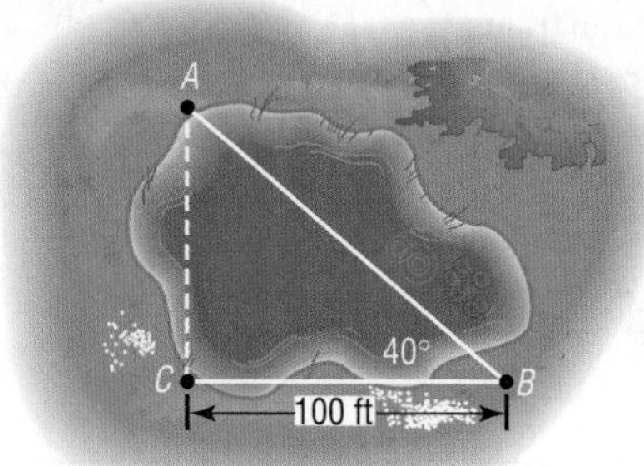

51. **The Eiffel Tower** The tallest tower built before the era of television masts, the Eiffel Tower was completed on March 31, 1889. Find the height of the Eiffel Tower (before a television mast was added to the top) using the information given in the illustration.

52. **Finding the Distance of a Ship from Shore** A person in a small boat, offshore from a vertical cliff known to be 100 feet in height, takes a sighting of the top of the cliff. If the angle of elevation is found to be 25°, how far offshore is the boat?

53. **Finding the Distance to a Plateau** Suppose that you are headed toward a plateau 50 meters high. If the angle of elevation to the top of the plateau is 20°, how far are you from the base of the plateau?

54. **Finding the Reach of a Ladder** A 22-foot extension ladder leaning against a building makes a 70° angle with the ground. How far up the building does the ladder touch?

55. **Finding the Angle of Elevation of the Sun** At 10 AM on April 26, 2009, a building 300 feet high cast a shadow 50 feet long. What was the angle of elevation of the Sun?

56. **Directing a Laser Beam** A laser beam is to be directed through a small hole in the center of a circle of radius 10 feet. The origin of the beam is 35 feet from the circle (see the figure). At what angle of elevation should the beam be aimed to ensure that it goes through the hole?

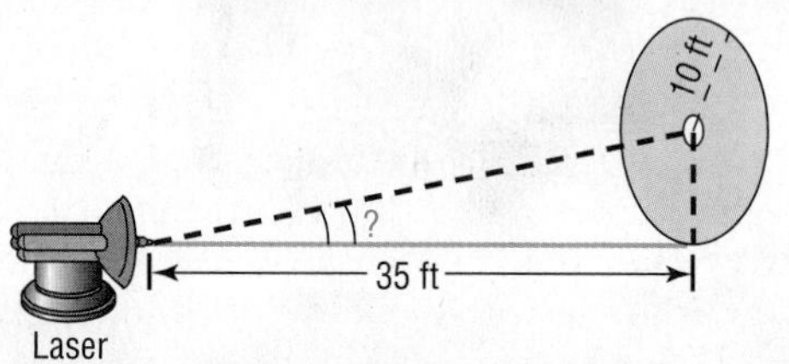

57. **Finding the Speed of a Truck** A state trooper is hidden 30 feet from a highway. One second after a truck passes, the angle θ between the highway and the line of observation from the patrol car to the truck is measured. See the illustration.

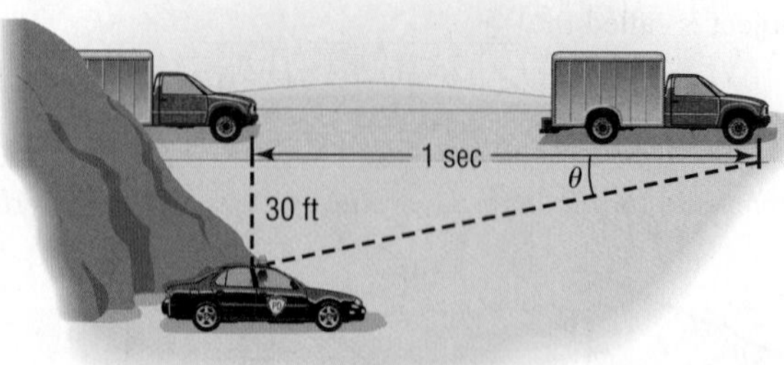

(a) If the angle measures 15°, how fast is the truck traveling? Express the answer in feet per second and in miles per hour.
(b) If the angle measures 20°, how fast is the truck traveling? Express the answer in feet per second and in miles per hour.
(c) If the speed limit is 55 miles per hour and a speeding ticket is issued for speeds of 5 miles per hour or more over the limit, for what angles should the trooper issue a ticket?

58. **Security** A security camera in a neighborhood bank is mounted on a wall 9 feet above the floor. What angle of depression should be used if the camera is to be directed to a spot 6 feet above the floor and 12 feet from the wall?

59. **Parallax** One method of measuring the distance from Earth to a star is the parallax method. The idea behind computing this distance is to measure the angle formed between the Earth and the star at two different points in time. Typically, the measurements are taken so that the side opposite the angle is as large as possible. Therefore, the optimal approach is to measure the angle when Earth is on opposite sides of the Sun, as shown in the figure.

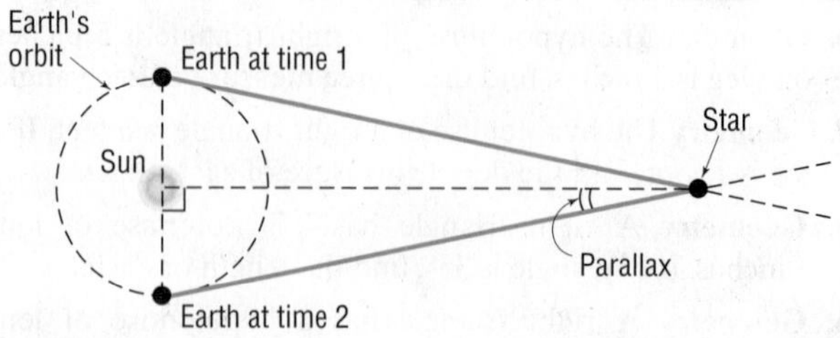

(a) Proxima Centauri is 4.22 light-years from Earth. If 1 light-year is about 5.9 trillion miles, how many miles is Proxima Centauri from Earth?
(b) The mean distance from Earth to the Sun is 93,000,000 miles. What is the parallax of Proxima Centauri?

60. **Parallax** See Problem 59. 61 Cygni, sometimes called Bessel's Star (after Friedrich Bessel, who measured the distance from Earth to the star in 1838), is a star in the constellation Cygnus.
   (a) 61 Cygni is 11.14 light-years from Earth. If 1 light-year is about 5.9 trillion miles, how many miles is 61 Cygni from Earth?
   (b) The mean distance from Earth to the Sun is 93,000,000 miles. What is the parallax of 61 Cygni?

61. **Washington Monument** The angle of elevation of the Sun is 35.1° at the instant the shadow cast by the Washington Monument is 789 feet long. Use this information to calculate the height of the monument.

62. **Finding the Length of a Mountain Trail** A straight trail with an inclination of 17° leads from a hotel at an elevation of 9000 feet to a mountain lake at an elevation of 11,200 feet. What is the length of the trail?

63. **Finding the Bearing of an Aircraft** A DC-9 aircraft leaves Midway Airport from runway 4 RIGHT, whose bearing is N40°E. After flying for $\frac{1}{2}$ mile, the pilot requests permission to turn 90° and head toward the southeast. The permission is granted. After the airplane goes 1 mile in this direction, what bearing should the control tower use to locate the aircraft?

64. **Finding the Bearing of a Ship** A ship leaves the port of Miami with a bearing of S80°E and a speed of 15 knots. After 1 hour, the ship turns 90° toward the south. After 2 hours, maintaining the same speed, what is the bearing to the ship from port?

65. **Niagara Falls Incline Railway** Situated between Portage Road and the Niagara Parkway directly across from the Canadian Horseshoe Falls, the Falls Incline Railway is a funicular that carries passengers up an embankment to Table Rock Observation Point. If the length of the track is 51.8 meters and the angle of inclination is 36°2′, determine the height of the embankment.

   Source: www.niagaraparks.com

66. **Willis Tower** Willis Tower in Chicago is the second tallest building in the United States and is topped by a high antenna. A surveyor on the ground makes the following measurement:
   1. The angle of elevation from his position to the top of the building is 34°.
   2. The distance from his position to the top of the building is 2593 feet.
   3. The distance from his position to the top of the antenna is 2743 feet.

   (a) How far away from the (base of the) building is the surveyor located?
   (b) How tall is the building?
   (c) What is the angle of elevation from the surveyor to the top of the antenna?
   (d) How tall is the antenna?

   Source: Council on Tall Buildings and Urban Habitat

67. **Constructing a Highway** A highway whose primary directions are north–south is being constructed along the west coast of Florida. Near Naples, a bay obstructs the straight path of the road. Since the cost of a bridge is prohibitive, engineers decide to go around the bay. The illustration shows the path that they decide on and the measurements taken. What is the length of highway needed to go around the bay?

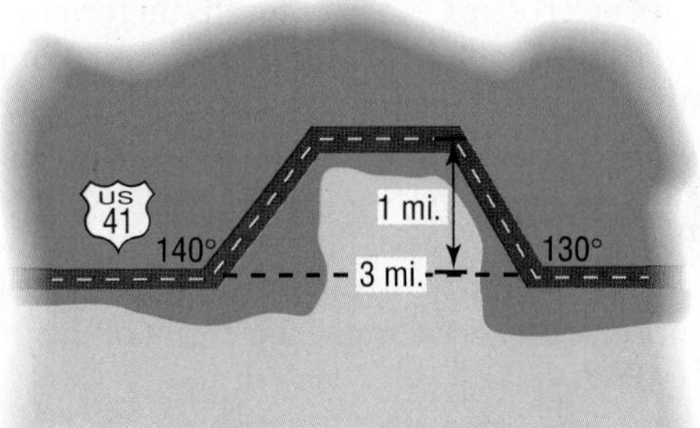

68. **Photography** A camera is mounted on a tripod 4 feet high at a distance of 10 feet from George, who is 6 feet tall. See the illustration. If the camera lens has angles of depression and elevation of 20°, will George's feet and head be seen by the lens? If not, how far back will the camera need to be moved to include George's feet and head?

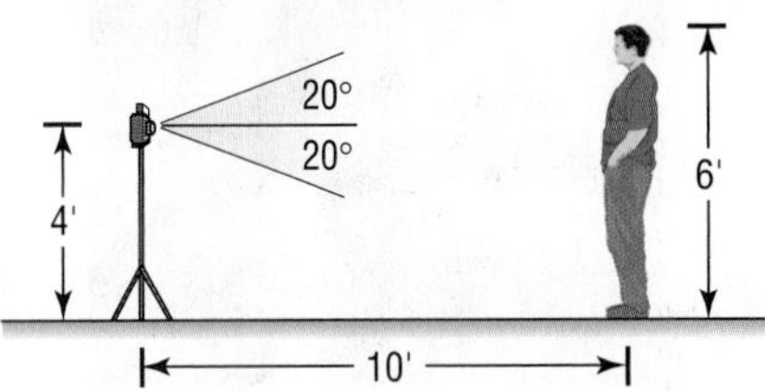

69. **Finding the Distance between Two Objects** A blimp, suspended in the air at a height of 500 feet, lies directly over a line from Soldier Field to the Adler Planetarium on Lake Michigan (see the figure). If the angle of depression from the blimp to the stadium is 32° and from the blimp to the planetarium is 23°, find the distance between Soldier Field and the Adler Planetarium.

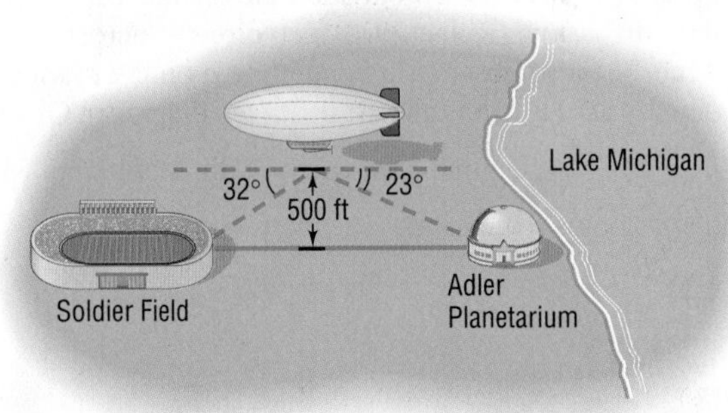

70. **Hot-Air Balloon** While taking a ride in a hot-air balloon in Napa Valley, Francisco wonders how high he is. To find out, he chooses a landmark that is to the east of the balloon and measures the angle of depression to be 54°. A few minutes later, after traveling 100 feet east, the angle of depression to the same landmark is determined to be 61°. Use this information to determine the height of the balloon.

71. **Mt. Rushmore** To measure the height of Lincoln's caricature on Mt. Rushmore, two sightings 800 feet from the base of the mountain are taken. If the angle of elevation to the bottom of Lincoln's face is 32° and the angle of elevation to the top is 35°, what is the height of Lincoln's face?

72. **The CN Tower** The CN Tower, located in Toronto, Canada, is the tallest structure in the Americas. While visiting Toronto, a tourist wondered what the height of the tower above the top of the Sky Pod is. While standing 4000 feet from the tower, she measured the angle to the top of the Sky Pod to be 20.1°. At this same distance, the angle of elevation to the top of the tower was found to be 24.4°. Use this information to determine the height of the tower above the Sky Pod.

73. **Chicago Skyscrapers** The angle of inclination from the base of the John Hancock Center to the top of the main structure of the Willis Tower is approximately 10.3°. If the main structure of the Willis Tower is 1451 feet tall, how far apart are the two skyscrapers? Assume the bases of the two buildings are at the same elevation.

Source: www.emporis.com

74. **Estimating the Width of the Mississippi River** A tourist at the top of the Gateway Arch (height, 630 feet) in St. Louis, Missouri, observes a boat moored on the Illinois side of the Mississippi River 2070 feet directly across from the Arch. She also observes a boat moored on the Missouri side directly across from the first boat (see diagram). Given that $B = \cot^{-1}\frac{67}{55}$, estimate the width of the Mississippi River at the St. Louis riverfront.

Source: U.S. Army Corps of Engineers

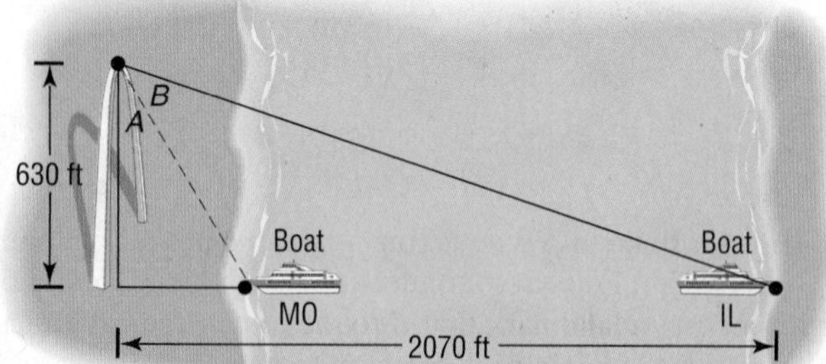

75. **Finding the Pitch of a Roof** A carpenter is preparing to put a roof on a garage that is 20 feet by 40 feet by 20 feet. A steel support beam 46 feet in length is positioned in the center of the garage. To support the roof, another beam will be attached to the top of the center beam (see the figure). At what angle of elevation is the new beam? In other words, what is the pitch of the roof?

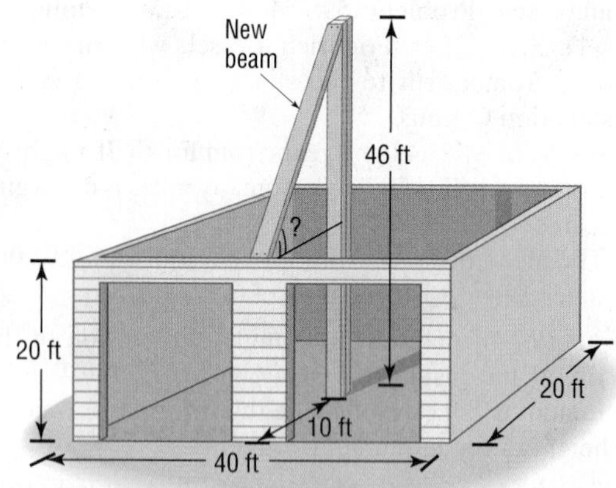

76. **Shooting Free Throws in Basketball** The eyes of a basketball player are 6 feet above the floor. The player is at the free-throw line, which is 15 feet from the center of the basket rim (see the figure). What is the angle of elevation from the player's eyes to the center of the rim?

[**Hint:** The rim is 10 feet above the floor.]

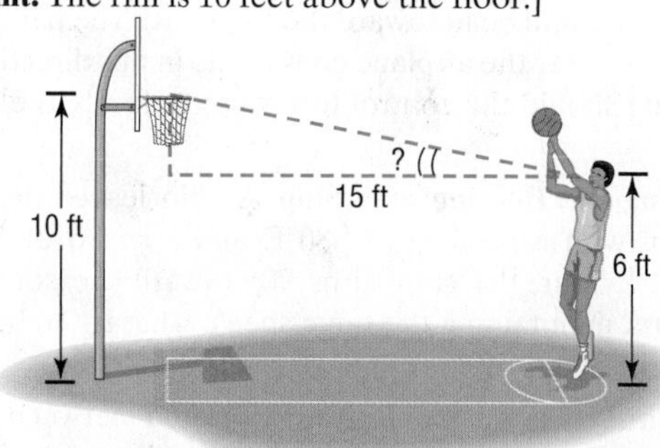

77. **Geometry** Find the value of the angle θ in degrees rounded to the nearest tenth of a degree.

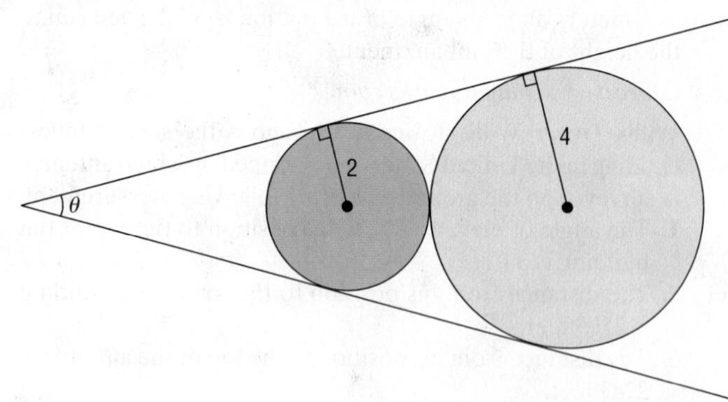

78. **Surveillance Satellites** A surveillance satellite circles Earth at a height of h miles above the surface. Suppose that d is the distance, in miles, on the surface of Earth that can be observed from the satellite. See the illustration on the following page.

(a) Find an equation that relates the central angle θ to the height h.

(b) Find an equation that relates the observable distance d and θ.

(c) Find an equation that relates d and h.

(d) If d is to be 2500 miles, how high must the satellite orbit above Earth?

(e) If the satellite orbits at a height of 300 miles, what distance d on the surface can be observed?

79. Calculating Pool Shots A pool player located at **X** wants to shoot the white ball off the top cushion and hit the red ball dead center. He knows from physics that the white ball will come off a cushion at the same angle as it hits the cushion. Where on the top cushion should he hit the white ball?

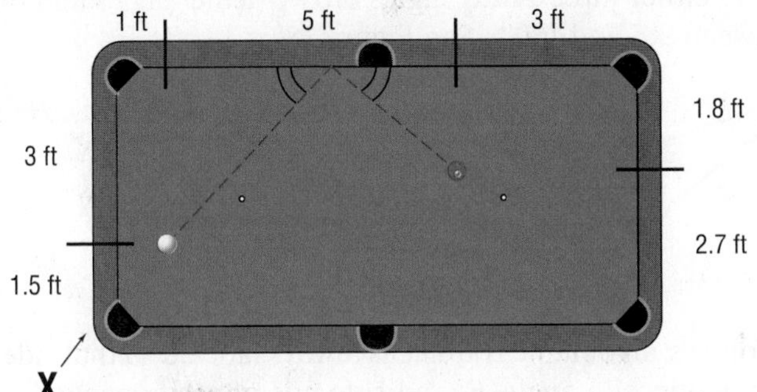

80. One World Trade Center One World Trade Center (1WTC) is the centerpiece of the rebuilding of the World Trade Center in New York City. The tower is 1776 feet tall (including its spire). The angle of elevation from the base of an office building to the tip of the spire is 34°. The angle of elevation from the helipad on the roof of the office building to the tip of the spire is 20°.

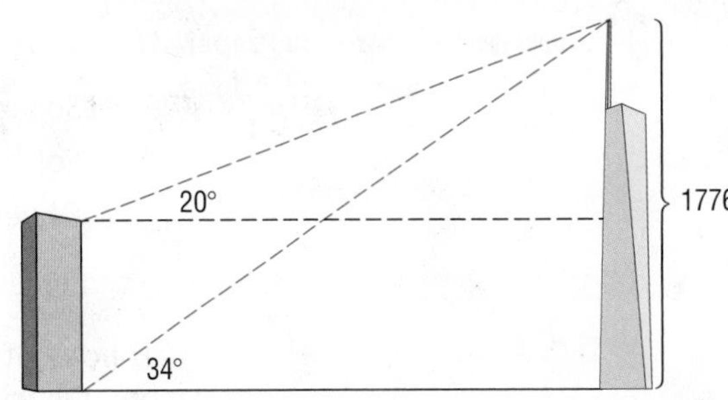

(a) How far away is the office building from One World Trade Center? Assume the side of the tower is vertical. Round to the nearest foot.

(b) How tall is the office building? Round to the nearest foot.

Explaining Concepts: Discussion and Writing

81. Explain how you would measure the width of the Grand Canyon from a point on its ridge.

82. Explain how you would measure the height of a TV tower that is on the roof of a tall building.

83. The Gibb's Hill Lighthouse, Southampton, Bermuda In operation since 1846, the Gibb's Hill Lighthouse stands 117 feet high on a hill 245 feet high, so its beam of light is 362 feet above sea level. A brochure states that ships 40 miles away can see the light and planes flying at 10,000 feet can see it 120 miles away. Verify the accuracy of these statements. What assumption did the brochure make about the height of the ship?

Retain Your Knowledge

Problems 84–87 are based on material learned earlier in the course. The purpose of these problems is to keep the material fresh in your mind so that you are better prepared for the final exam.

84. Determine the amplitude and period of $y = -8\sin(6x)$.

85. Find the exact value of $\sin 15°$.
Hint: $15° = 45° - 30°$

86. Evaluate $\dfrac{f(x) - f(4)}{x - 4}$, where $f(x) = \sqrt{x}$ for $x = 5, 4.5,$ and 4.1. Round results to three decimal places.

87. Solve $2\sin^2\theta - \sin\theta - 1 = 0$ for $0 \le \theta < 2\pi$.

'Are You Prepared?' Answers

1. 4 **2.** 26.6° **3.** 30°

4.2 The Law of Sines

PREPARING FOR THIS SECTION *Before getting started, review the following:*

- Trigonometric Equations (Section 3.3, pp. 208–213)
- Difference Formula for the Sine Function (Section 3.5, p. 229)
- Geometry Essentials (Appendix A, Section A.2, pp. A14–A18)

Now Work the **'Are You Prepared?'** problems on page 278.

OBJECTIVES
1. Solve SAA or ASA Triangles (p. 273)
2. Solve SSA Triangles (p. 274)
3. Solve Applied Problems (p. 276)

If none of the angles of a triangle is a right angle, the triangle is called **oblique**. An oblique triangle will have either three acute angles or two acute angles and one obtuse angle (an angle between 90° and 180°). See Figure 18.

Obtuse angle

Figure 18 **(a)** All angles are acute **(b)** Two acute angles and one obtuse angle

In the discussion that follows, an oblique triangle is always labeled so that side a is opposite angle A, side b is opposite angle B, and side c is opposite angle C, as shown in Figure 19.

Figure 19 Oblique triangle

To **solve an oblique triangle** means to find the lengths of its sides and the measurements of its angles. To do this, we need to know the length of one side,* along with (i) two angles, (ii) one angle and one other side, or (iii) the other two sides. There are four possibilities to consider.

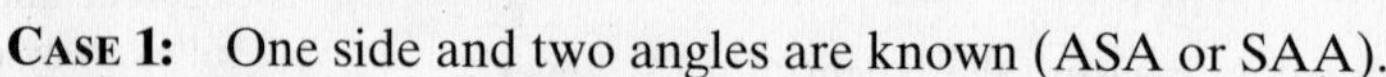

Case 1: One side and two angles are known (ASA or SAA).
Case 2: Two sides and the angle opposite one of them are known (SSA).
Case 3: Two sides and the included angle are known (SAS).
Case 4: Three sides are known (SSS).

Figure 20 illustrates the four cases, where the known measurements are shown in blue.

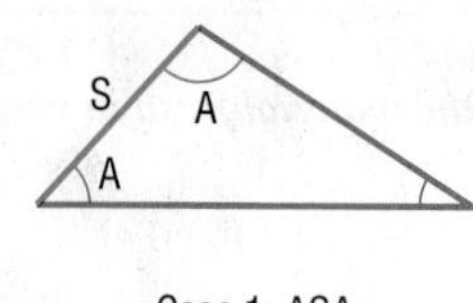

Case 1: ASA

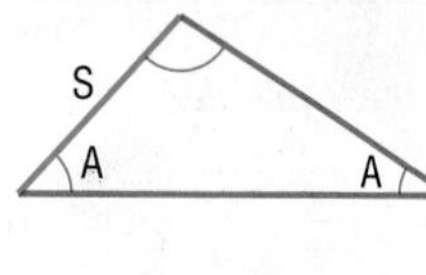

Case 1: SAA

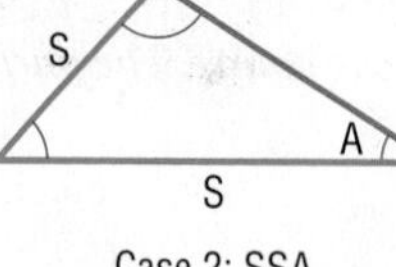

Case 2: SSA

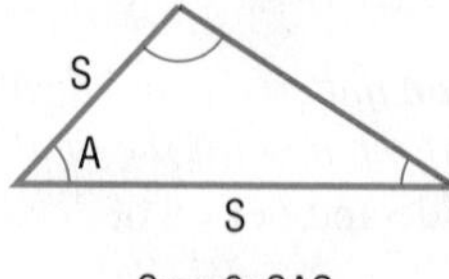

Case 3: SAS

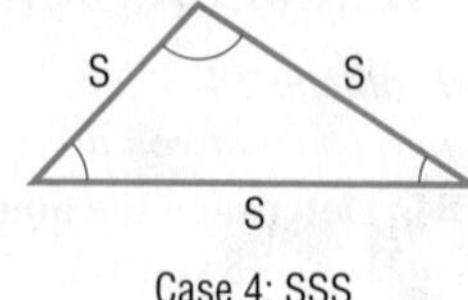

Case 4: SSS

Figure 20

The **Law of Sines** is used to solve triangles for which Case 1 or 2 holds. Cases 3 and 4 are considered when we study the Law of Cosines in the next section.

THEOREM

Law of Sines

For a triangle with sides a, b, c and opposite angles A, B, C, respectively,

$$\frac{\sin A}{a} = \frac{\sin B}{b} = \frac{\sin C}{c} \quad (1)$$

WARNING *Oblique triangles cannot be solved using the methods of Section 4.1. Do you know why?* ■

*The length of one side must be known because knowing only the angles will reveal only a family of *similar triangles*.

A proof of the Law of Sines is given at the end of this section.

The Law of Sines actually consists of three equalities:

$$\frac{\sin A}{a} = \frac{\sin B}{b} \qquad \frac{\sin A}{a} = \frac{\sin C}{c} \qquad \frac{\sin B}{b} = \frac{\sin C}{c}$$

Formula (1) is a compact way to write these three equations.

Typically, applying the Law of Sines to solve triangles uses the fact that the sum of the angles of any triangle equals 180°; that is,

$$A + B + C = 180° \qquad \textbf{(2)}$$

1 Solve SAA or ASA Triangles

The first two examples show how to solve a triangle when one side and two angles are known (Case 1: SAA or ASA).

EXAMPLE 1 Using the Law of Sines to Solve an SAA Triangle

Solve the triangle: $A = 40°, B = 60°, a = 4$

Solution Figure 21 shows the triangle to be solved. The third angle C is found using equation (2).

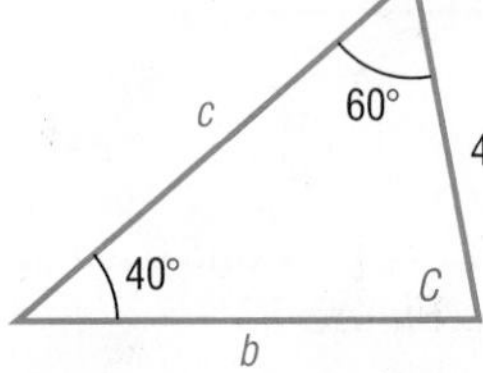

Figure 21

NOTE Although it is not a check, the reasonableness of answers can be verified by determining whether the longest side is opposite the largest angle and the shortest side is opposite the smallest angle. ■

$$A + B + C = 180°$$
$$40° + 60° + C = 180°$$
$$C = 80°$$

Now use the Law of Sines (twice) to find the unknown sides b and c.

$$\frac{\sin A}{a} = \frac{\sin B}{b} \qquad \frac{\sin A}{a} = \frac{\sin C}{c}$$

Because $a = 4$, $A = 40°$, $B = 60°$, and $C = 80°$, we have

$$\frac{\sin 40°}{4} = \frac{\sin 60°}{b} \qquad \frac{\sin 40°}{4} = \frac{\sin 80°}{c}$$

Solving for b and c yields

$$b = \frac{4 \sin 60°}{\sin 40°} \approx 5.39 \qquad c = \frac{4 \sin 80°}{\sin 40°} \approx 6.13$$ ●

Notice in Example 1 that b and c are found by working with the given side a. This is better than finding b first and working with a rounded value of b to find c.

Now Work PROBLEM 9

EXAMPLE 2 Using the Law of Sines to Solve an ASA Triangle

Solve the triangle: $A = 35°, B = 15°, c = 5$

Solution Figure 22 illustrates the triangle to be solved. Two angles are known ($A = 35°$ and $B = 15°$). Find the third angle using equation (2):

$$A + B + C = 180°$$
$$35° + 15° + C = 180°$$
$$C = 130°$$

Figure 22

Now the three angles and one side ($c = 5$) of the triangle are known. To find the remaining two sides a and b, use the Law of Sines (twice).

$$\frac{\sin A}{a} = \frac{\sin C}{c} \qquad \frac{\sin B}{b} = \frac{\sin C}{c}$$

$$\frac{\sin 35°}{a} = \frac{\sin 130°}{5} \qquad \frac{\sin 15°}{b} = \frac{\sin 130°}{5}$$

$$a = \frac{5 \sin 35°}{\sin 130°} \approx 3.74 \qquad b = \frac{5 \sin 15°}{\sin 130°} \approx 1.69$$

Now Work PROBLEM 23

2 Solve SSA Triangles

Figure 23 $\sin A = \frac{h}{b}$

Case 2 (SSA), which applies to triangles for which two sides and the angle opposite one of them are known, is referred to as the **ambiguous case**, because the known information may result in one triangle, two triangles, or no triangle at all. Suppose that sides a and b and angle A are given, as illustrated in Figure 23. The key to determining how many triangles, if any, may be formed from the given information lies primarily with the relative size of side a, the height h, and the fact that $h = b \sin A$.

No Triangle If $a < h = b \sin A$, then side a is not sufficiently long to form a triangle. See Figure 24.

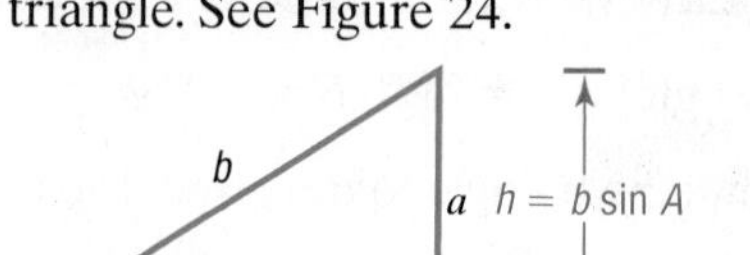

Figure 24 $a < h = b \sin A$

One Right Triangle If $a = h = b \sin A$, then side a is just long enough to form a right triangle. See Figure 25.

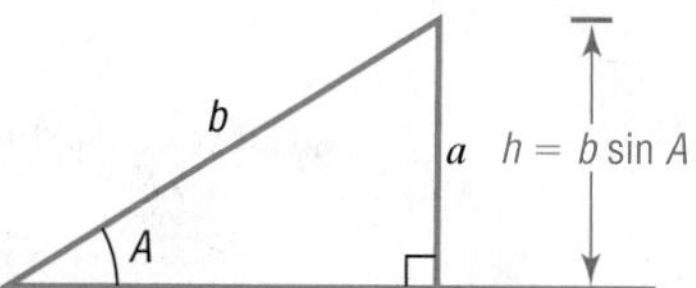

Figure 25 $a = h = b \sin A$

Two Triangles If $h = b \sin A < a$ and $a < b$, then two distinct triangles can be formed from the given information. See Figure 26.

One Triangle If $a \geq b$, only one triangle can be formed. See Figure 27.

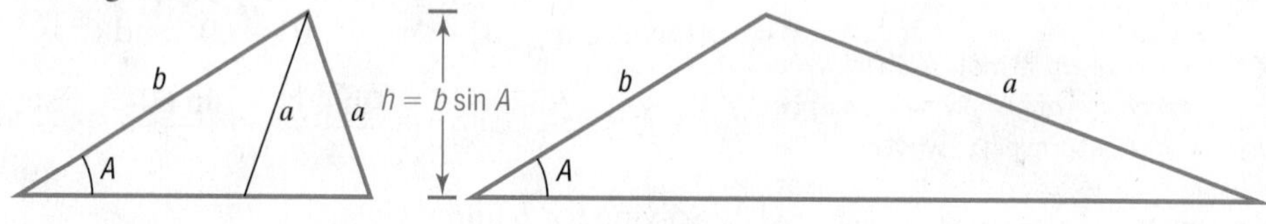

Figure 26 $b \sin A < a$ and $a < b$

Figure 27 $a \geq b$

Fortunately, it is not necessary to rely on an illustration or on complicated relationships to draw the correct conclusion in the ambiguous case. The Law of Sines will lead us to the correct determination. Let's see how.

EXAMPLE 3 Using the Law of Sines to Solve an SSA Triangle (No Solution)

Solve the triangle: $a = 2, c = 1, C = 50°$

Solution Because $a = 2$, $c = 1$, and $C = 50°$ are known, use the Law of Sines to find the angle A.

$$\frac{\sin A}{a} = \frac{\sin C}{c}$$

$$\frac{\sin A}{2} = \frac{\sin 50°}{1}$$

$$\sin A = 2 \sin 50° \approx 1.53$$

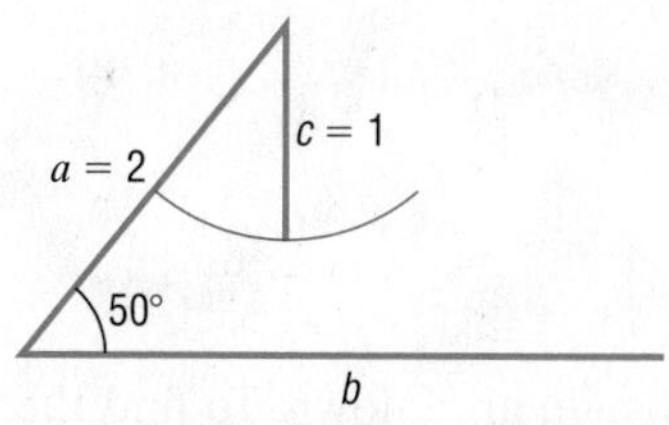

Figure 28

Since there is no angle A for which $\sin A > 1$, there can be no triangle with the given measurements. Figure 28 illustrates the measurements given. Note that no matter how side c is positioned, it will never touch side b to form a triangle.

EXAMPLE 4 Using the Law of Sines to Solve an SSA Triangle (One Solution)

Solve the triangle: $a = 3, b = 2, A = 40°$

Solution See Figure 29(a). Because $a = 3, b = 2$, and $A = 40°$ are known, use the Law of Sines to find the angle B.

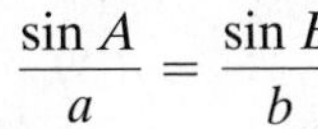

$$\frac{\sin A}{a} = \frac{\sin B}{b}$$

Then

$$\frac{\sin 40°}{3} = \frac{\sin B}{2}$$

$$\sin B = \frac{2 \sin 40°}{3} \approx 0.43$$

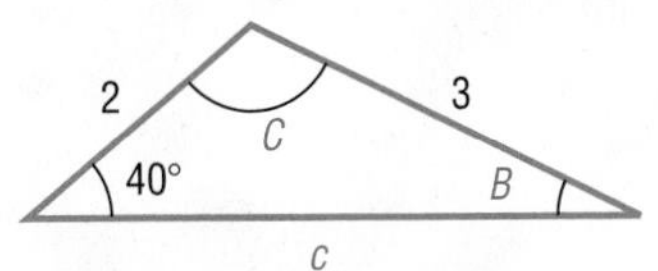

Figure 29(a)

NOTE The angle B_1 was determined by finding the value of $\sin^{-1}\left(\frac{2 \sin 40°}{3}\right)$. Using the rounded value and evaluating $\sin^{-1}(0.43)$ will yield a slightly different result. ■

There are two angles B, $0° < B < 180°$, for which $\sin B \approx 0.43$.

$$B_1 \approx 25.4° \quad \text{and} \quad B_2 \approx 180° - 25.4° = 154.6°$$

The second possibility, $B_2 \approx 154.6°$, is ruled out, because $A = 40°$ makes $A + B_2 \approx 194.6° > 180°$. Now use $B_1 \approx 25.4°$ to find that

$$C = 180° - A - B_1 \approx 180° - 40° - 25.4° = 114.6°$$

The third side c may now be determined using the Law of Sines.

$$\frac{\sin A}{a} = \frac{\sin C}{c}$$

$$\frac{\sin 40°}{3} = \frac{\sin 114.6°}{c}$$

$$c = \frac{3 \sin 114.6°}{\sin 40°} \approx 4.24$$

Figure 29(b)

Figure 29(b) illustrates the solved triangle. ●

EXAMPLE 5 Using the Law of Sines to Solve an SSA Triangle (Two Solutions)

Solve the triangle: $a = 6, b = 8, A = 35°$

Solution Because $a = 6, b = 8$, and $A = 35°$ are known, use the Law of Sines to find the angle B.

$$\frac{\sin A}{a} = \frac{\sin B}{b}$$

Then

$$\frac{\sin 35°}{6} = \frac{\sin B}{8}$$

$$\sin B = \frac{8 \sin 35°}{6} \approx 0.76$$

$$B_1 \approx 49.9° \quad \text{or} \quad B_2 \approx 180° - 49.9° = 130.1°$$

Figure 30(a)

Both choices of B result in $A + B < 180°$. There are two triangles, one containing the angle $B_1 \approx 49.9°$ and the other containing the angle $B_2 \approx 130.1°$. See Figure 30(a).

The third angle C is either

$$C_1 = 180° - A - B_1 \approx 95.1° \quad \text{or} \quad C_2 = 180° - A - B_2 \approx 14.9°$$

$A = 35°$, $B_1 = 49.9°$ (for C_1); $A = 35°$, $B_2 = 130.1°$ (for C_2)

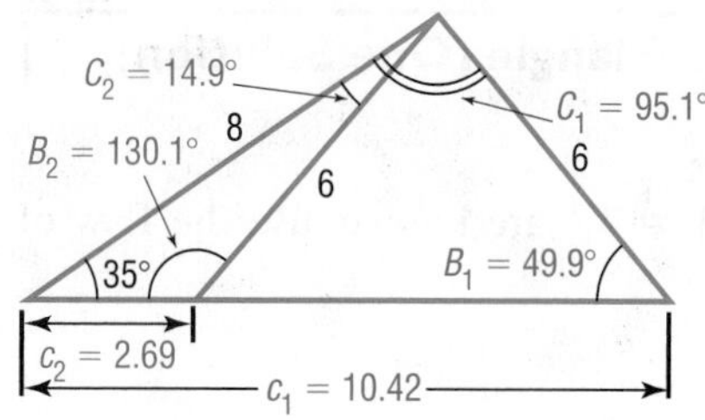

Figure 30(b)

The third side c obeys the Law of Sines, so

$$\frac{\sin A}{a} = \frac{\sin C_1}{c_1} \qquad \frac{\sin A}{a} = \frac{\sin C_2}{c_2}$$

$$\frac{\sin 35°}{6} = \frac{\sin 95.1°}{c_1} \qquad \frac{\sin 35°}{6} = \frac{\sin 14.9°}{c_2}$$

$$c_1 = \frac{6 \sin 95.1°}{\sin 35°} \approx 10.42 \qquad c_2 = \frac{6 \sin 14.9°}{\sin 35°} \approx 2.69$$

The two solved triangles are illustrated in Figure 30(b).

Now Work PROBLEMS 25 AND 31

3 Solve Applied Problems

EXAMPLE 6 **Finding the Height of a Mountain**

To measure the height of a mountain, a surveyor takes two sightings of the peak at a distance 900 meters apart on a direct line to the mountain.* See Figure 31(a). The first observation results in an angle of elevation of 47°, and the second results in an angle of elevation of 35°. If the transit is 2 meters high, what is the height h of the mountain?

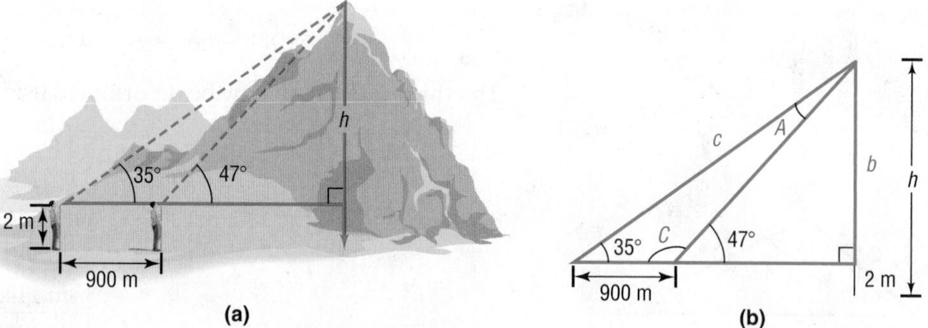

Figure 31

Solution Figure 31(b) shows the triangles that replicate the illustration in Figure 31(a). Since $C + 47° = 180°$, this means that $C = 133°$. Also, since $A + C + 35° = 180°$, this means that $A = 180° - 35° - C = 145° - 133° = 12°$. Use the Law of Sines to find c.

$$\frac{\sin A}{a} = \frac{\sin C}{c} \qquad A = 12°, C = 133°, a = 900$$

$$c = \frac{900 \sin 133°}{\sin 12°} \approx 3165.86$$

Using the larger right triangle gives

$$\sin 35° = \frac{b}{c}$$

$$b = 3165.86 \sin 35° \approx 1815.86 \approx 1816 \text{ meters}$$

The height of the peak from ground level is approximately $1816 + 2 = 1818$ meters.

Now Work PROBLEM 37

EXAMPLE 7 **Rescue at Sea**

Coast Guard Station Zulu is located 120 miles due west of Station X-ray. A ship at sea sends an SOS call that is received by each station. The call to Station Zulu indicates that the bearing of the ship from Zulu is N40°E (40° east of north). The call to Station X-ray indicates that the bearing of the ship from X-ray is N30°W (30° west of north).

(a) How far is each station from the ship?

(b) If a helicopter capable of flying 200 miles per hour is dispatched from the nearest station to the ship, how long will it take to reach the ship?

* For simplicity, assume that these sightings are at the same level.

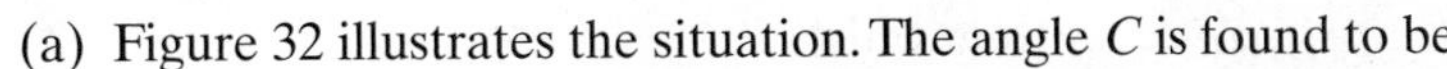

Solution (a) Figure 32 illustrates the situation. The angle C is found to be

$$C = 180° - 50° - 60° = 70°$$

Figure 32

The Law of Sines can now be used to find the two distances a and b that are needed.

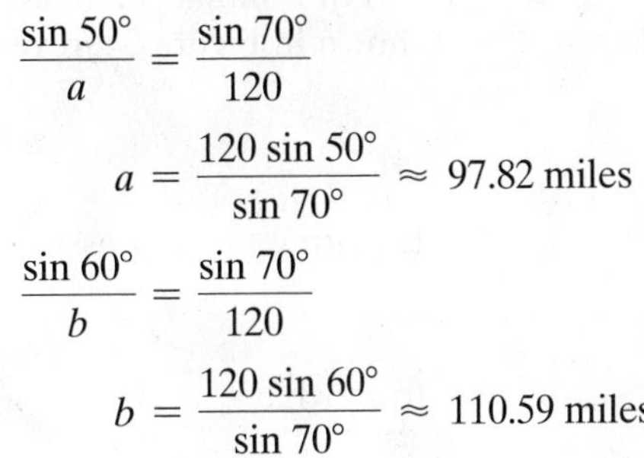

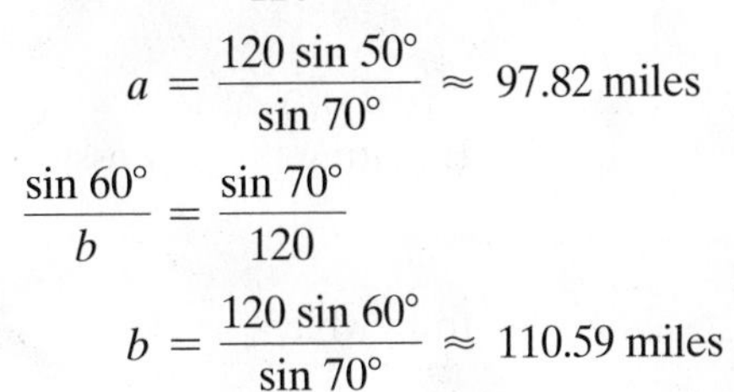

$$\frac{\sin 50°}{a} = \frac{\sin 70°}{120}$$

$$a = \frac{120 \sin 50°}{\sin 70°} \approx 97.82 \text{ miles}$$

$$\frac{\sin 60°}{b} = \frac{\sin 70°}{120}$$

$$b = \frac{120 \sin 60°}{\sin 70°} \approx 110.59 \text{ miles}$$

Station Zulu is about 111 miles from the ship, and Station X-ray is about 98 miles from the ship.

(b) The time t needed for the helicopter to reach the ship from Station X-ray is found by using the formula

$$(\text{Rate}, r)(\text{Time}, t) = \text{Distance}, a$$

Then

$$t = \frac{a}{r} = \frac{97.82}{200} \approx 0.49 \text{ hour} \approx 29 \text{ minutes}$$

It will take about 29 minutes for the helicopter to reach the ship. ●

Now Work PROBLEM 47

Proof of the Law of Sines To prove the Law of Sines, construct an altitude of length h from one of the vertices of a triangle. Figure 33(a) shows h for a triangle with three acute angles, and Figure 33(b) shows h for a triangle with an obtuse angle. In each case, the altitude is drawn from the vertex at B. Using either illustration

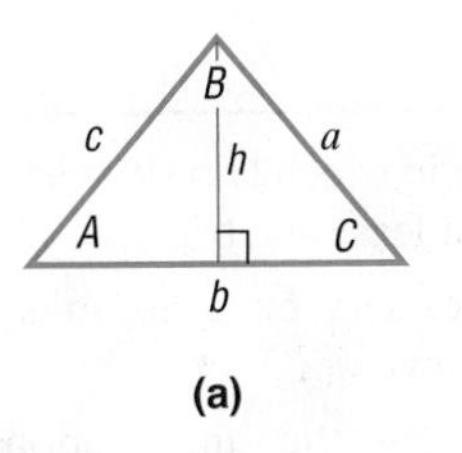

(a)

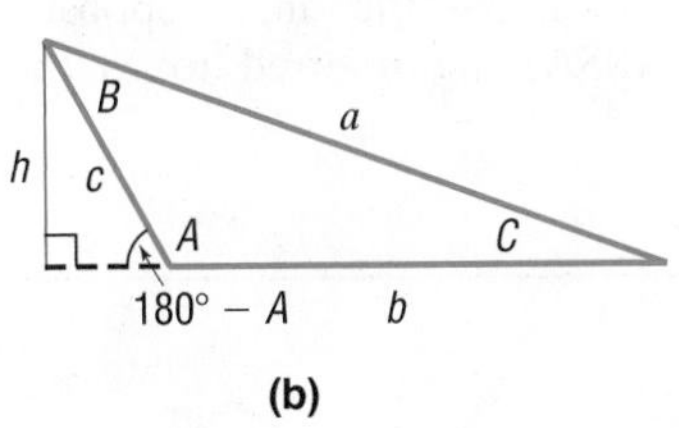

(b)

Figure 33

$$\sin C = \frac{h}{a}$$

from which

$$h = a \sin C \tag{3}$$

From Figure 33(a), it also follows that

$$\sin A = \frac{h}{c}$$

from which

$$h = c \sin A \tag{4}$$

From Figure 33(b), it follows that

$$\sin(180° - A) = \sin A = \frac{h}{c}$$

$\sin(180° - A) = \sin 180° \cos A - \cos 180° \sin A = \sin A$

which again gives

$$h = c \sin A$$

Thus, whether the triangle has three acute angles or has two acute angles and one obtuse angle, equations (3) and (4) hold. As a result, the expressions for h in equations (3) and (4) are equal. That is,

$$a \sin C = c \sin A$$

from which

$$\frac{\sin A}{a} = \frac{\sin C}{c} \tag{5}$$

In a similar manner, constructing the altitude h' from the vertex of angle A, as shown in Figure 34, reveals that

$$\sin B = \frac{h'}{c} \quad \text{and} \quad \sin C = \frac{h'}{b}$$

Equating the expressions for h' gives

$$h' = c \sin B = b \sin C$$

from which

$$\frac{\sin B}{b} = \frac{\sin C}{c} \tag{6}$$

When equations (5) and (6) are combined, the result is equation (1), the Law of Sines. ■

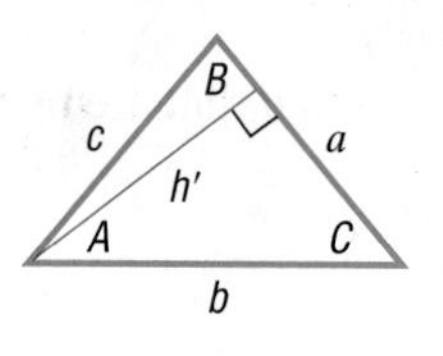

(a)

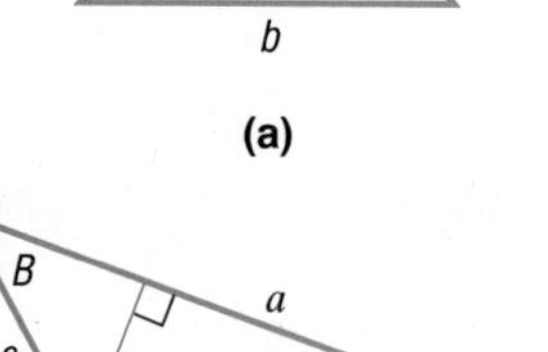

(b)

Figure 34

4.2 Assess Your Understanding

'Are You Prepared?' *Answers are given at the end of these exercises. If you get a wrong answer, read the pages listed in red.*

1. The difference formula for the sine function is $\sin(A - B) =$ ______. (p. 229)
2. If θ is an acute angle, solve the equation $\cos\theta = \frac{\sqrt{3}}{2}$. (pp. 208–213)
3. The two triangles shown are similar. Find the missing length. (pp. A14–A18)

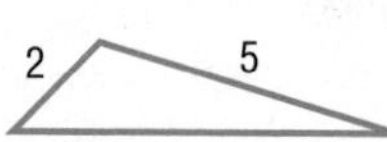

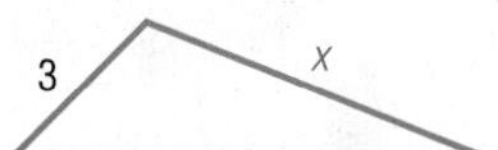

Concepts and Vocabulary

4. If none of the angles of a triangle is a right angle, the triangle is called ________.
   (a) oblique (b) obtuse (c) acute (d) scalene
5. For a triangle with sides a, b, c and opposite angles A, B, C, the Law of Sines states that ________________.
6. ***True or False*** An oblique triangle in which two sides and an angle are given always results in at least one triangle.
7. ***True or False*** The Law of Sines can be used to solve triangles where three sides are known.
8. Triangles for which two sides and the angle opposite one of them are known (SSA) are referred to as the __________ ______.

Skill Building

In Problems 9–16, solve each triangle.

9.
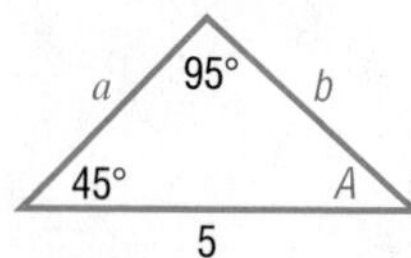

10.
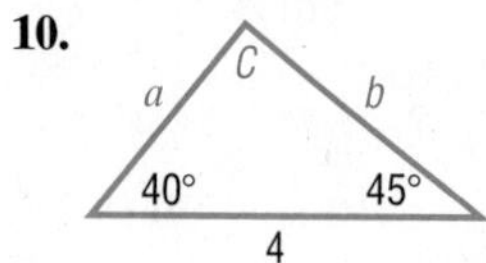

11.
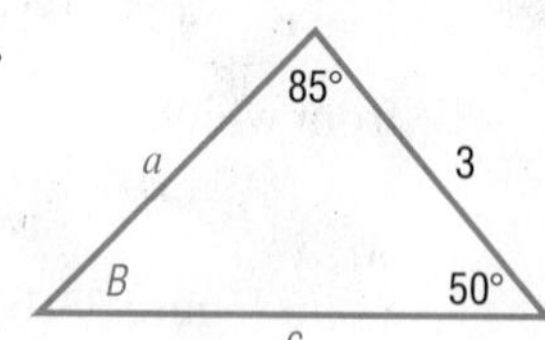

12.
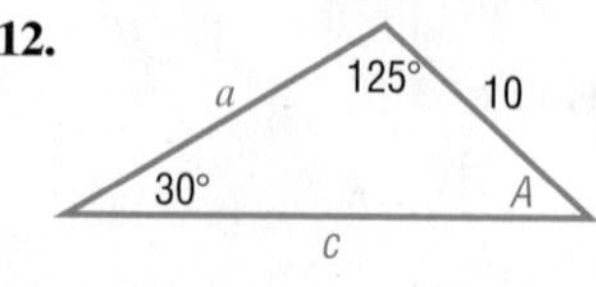

13.
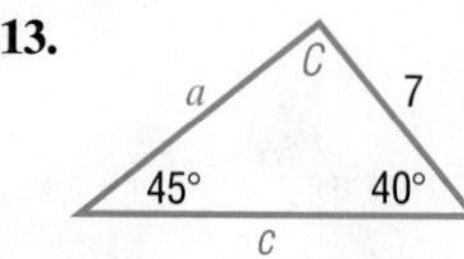

14.
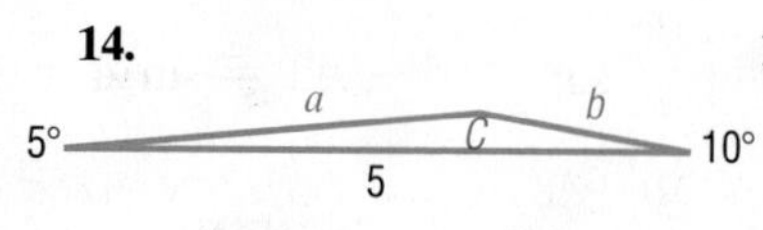

15.
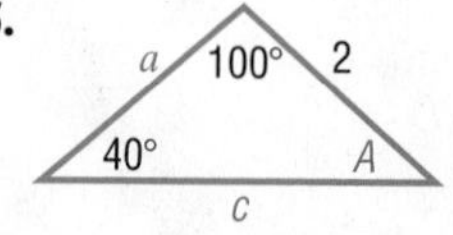

16.
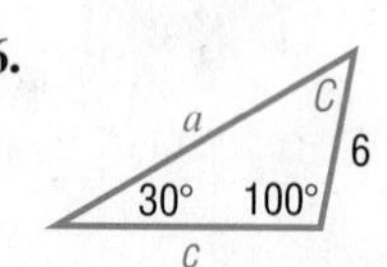

In Problems 17–24, solve each triangle.

17. $A = 40°$, $B = 20°$, $a = 2$
18. $A = 50°$, $C = 20°$, $a = 3$
19. $B = 70°$, $C = 10°$, $b = 5$
20. $A = 70°$, $B = 60°$, $c = 4$
21. $A = 110°$, $C = 30°$, $c = 3$
22. $B = 10°$, $C = 100°$, $b = 2$
23. $A = 40°$, $B = 40°$, $c = 2$
24. $B = 20°$, $C = 70°$, $a = 1$

In Problems 25–36, two sides and an angle are given. Determine whether the given information results in one triangle, two triangles, or no triangle at all. Solve any resulting triangle(s).

25. $a = 3, \quad b = 2, \quad A = 50°$
26. $b = 4, \quad c = 3, \quad B = 40°$
27. $b = 5, \quad c = 3, \quad B = 100°$
28. $a = 2, \quad c = 1, \quad A = 120°$
29. $a = 4, \quad b = 5, \quad A = 60°$
30. $b = 2, \quad c = 3, \quad B = 40°$
31. $b = 4, \quad c = 6, \quad B = 20°$
32. $a = 3, \quad b = 7, \quad A = 70°$
33. $a = 2, \quad c = 1, \quad C = 100°$
34. $b = 4, \quad c = 5, \quad B = 95°$
35. $a = 2, \quad c = 1, \quad C = 25°$
36. $b = 4, \quad c = 5, \quad B = 40°$

Applications and Extensions

37. Finding the Length of a Ski Lift Consult the figure. To find the length of the span of a proposed ski lift from P to Q, a surveyor measures $\angle DPQ$ to be 25° and then walks back a distance of 1000 feet to R and measures $\angle PRQ$ to be 15°. What is the distance from P to Q?

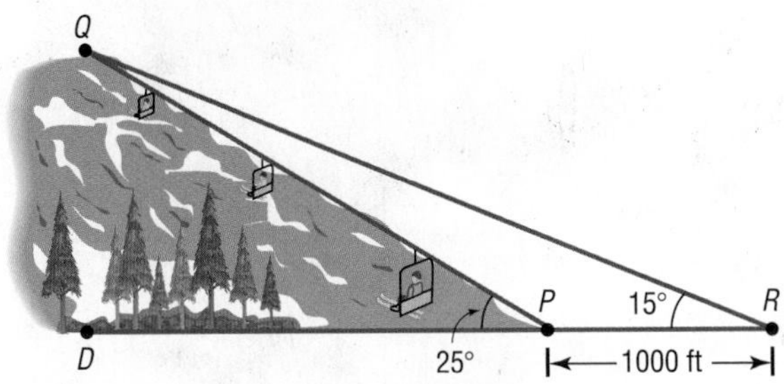

38. Finding the Height of a Mountain Use the illustration in Problem 37 to find the height QD of the mountain.

39. Finding the Height of an Airplane An aircraft is spotted by two observers who are 1000 feet apart. As the airplane passes over the line joining them, each observer takes a sighting of the angle of elevation to the plane, as indicated in the figure. How high is the airplane?

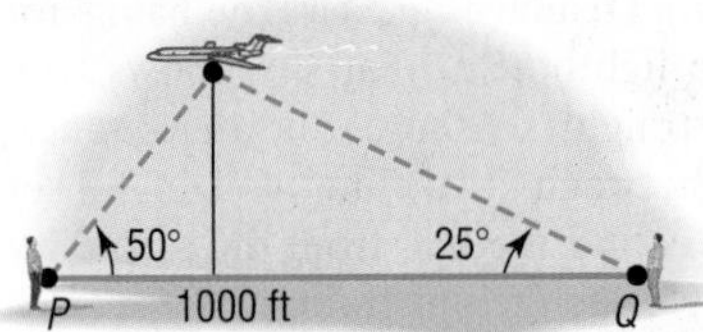

40. Finding the Height of the Bridge over the Royal Gorge The highest bridge in the world is the bridge over the Royal Gorge of the Arkansas River in Colorado. Sightings to the same point at water level directly under the bridge are taken from each side of the 880-foot-long bridge, as indicated in the figure. How high is the bridge?

Source: Guinness Book of World Records

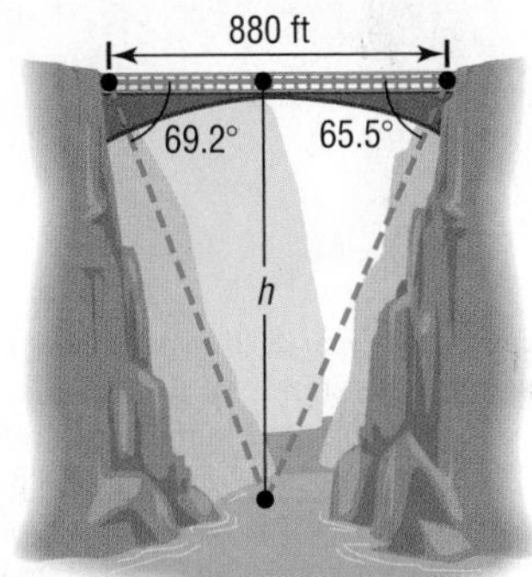

41. Land Dimensions A triangular plot of land has one side along a straight road measuring 200 feet. A second side makes a 50° angle with the road, and the third side makes a 43° angle with the road. How long are the other two sides?

42. Distance between Runners Two runners in a marathon determine that the angles of elevation of a news helicopter covering the race are 38° and 45°. If the helicopter is 1700 feet directly above the finish line, how far apart are the runners?

43. Landscaping Pat needs to determine the height of a tree before cutting it down to be sure that it will not fall on a nearby fence. The angle of elevation of the tree from one position on a flat path from the tree is 30°, and from a second position 40 feet farther along this path it is 20°. What is the height of the tree?

44. Construction A loading ramp 10 feet long that makes an angle of 18° with the horizontal is to be replaced by one that makes an angle of 12° with the horizontal. How long is the new ramp?

45. Commercial Navigation Adam must fly home to St. Louis from a business meeting in Oklahoma City. One flight option flies directly to St. Louis, a distance of about 461.1 miles. A second flight option flies first to Kansas City and then connects to St. Louis. The bearing from Oklahoma City to Kansas City is N29.6°E, and the bearing from Oklahoma City to St. Louis is N57.7°E. The bearing from St. Louis to Oklahoma City is S57.7°W, and the bearing from St. Louis to Kansas City is N79.4°W. How many more frequent flyer miles will Adam receive if he takes the connecting flight rather than the direct flight?

Source: www.landings.com

46. Time Lost to a Navigation Error In attempting to fly from city P to city Q, an aircraft followed a course that was 10° in error, as indicated in the figure. After flying a distance of 50 miles, the pilot corrected the course by turning at point R and flying 300 miles farther. If the constant speed of the aircraft was 250 miles per hour, how much time was lost due to the error?

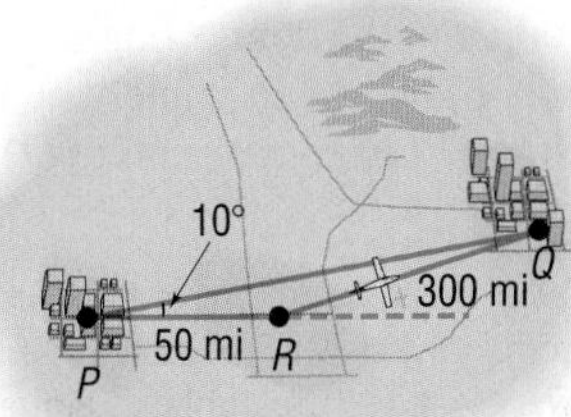

47. Rescue at Sea Coast Guard Station Able is located 150 miles due south of Station Baker. A ship at sea sends an SOS call that is received by each station. The call to Station Able indicates that the ship is located N55°E; the call to Station Baker indicates that the ship is located S60°E.

(continued on next page)

(a) How far is each station from the ship?
(b) If a helicopter capable of flying 200 miles per hour is dispatched from the station nearest the ship, how long will it take to reach the ship?

48. Distance to the Moon At exactly the same time, Tom and Alice measured the angle of elevation to the moon while standing exactly 300 km apart. The angle of elevation to the moon for Tom was 49.8974°, and the angle of elevation to the moon for Alice was 49.9312°. See the figure. To the nearest 1000 km, how far was the moon from Earth when the measurement was obtained?

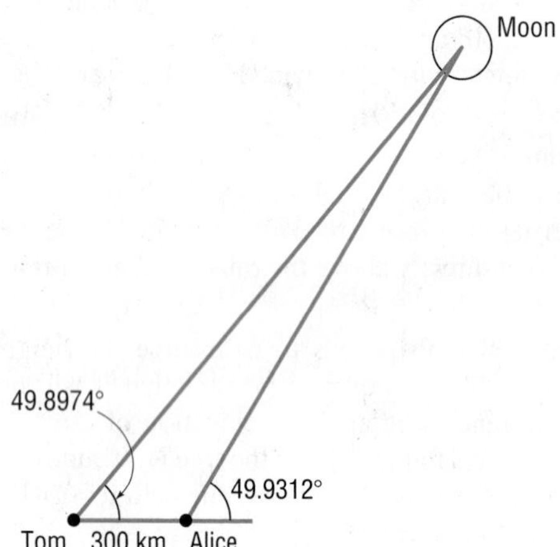

49. Finding the Lean of the Leaning Tower of Pisa The famous Leaning Tower of Pisa was originally 184.5 feet high.* At a distance of 123 feet from the base of the tower, the angle of elevation to the top of the tower is found to be 60°. Find $\angle RPQ$ indicated in the figure. Also, find the perpendicular distance from R to PQ.

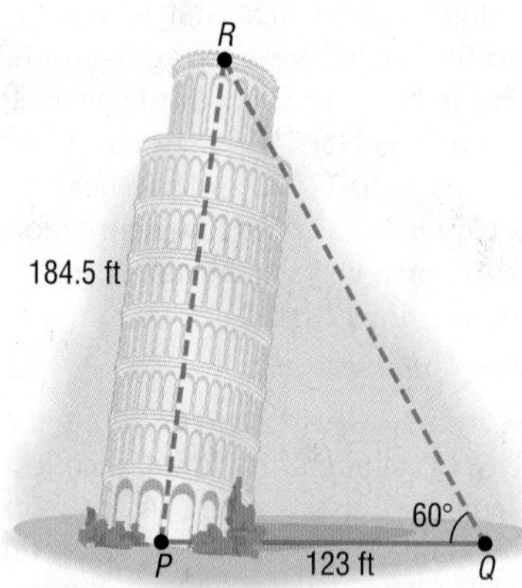

50. Crankshafts on Cars On a certain automobile, the crankshaft is 3 inches long and the connecting rod is 9 inches long (see the figure, top, right). At the time when $\angle OPQ$ is 15°, how far is the piston (P) from the center (O) of the crankshaft?

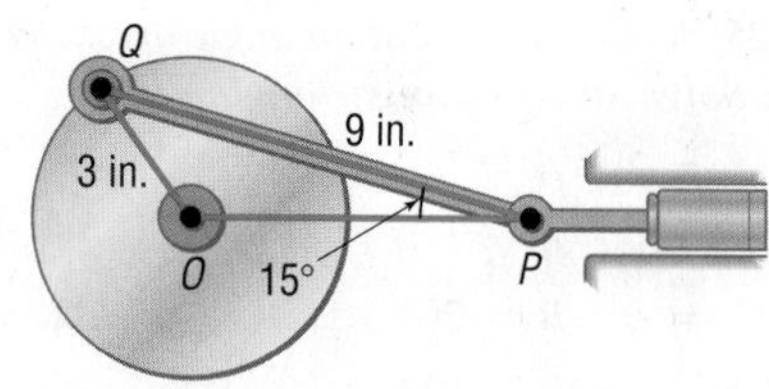

51. Constructing a Highway U.S. 41, a highway whose primary directions are north–south, is being constructed along the west coast of Florida. Near Naples, a bay obstructs the straight path of the road. Since the cost of a bridge is prohibitive, engineers decide to go around the bay. The illustration shows the path that they decide on and the measurements taken. What is the length of highway needed to go around the bay?

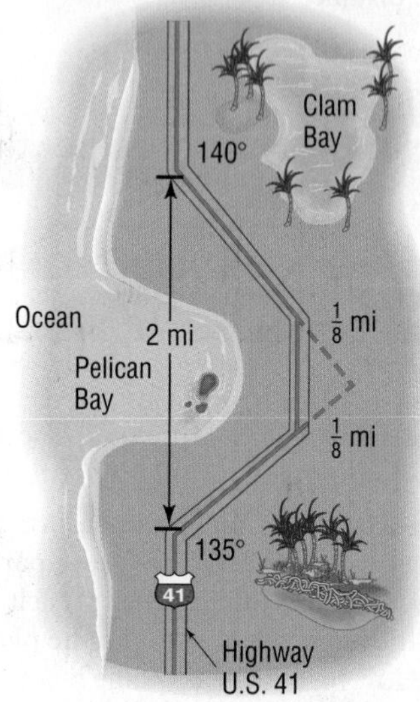

52. Calculating Distances at Sea The navigator of a ship at sea spots two lighthouses that she knows to be 3 miles apart along a straight seashore. She determines that the angles formed between two line-of-sight observations of the lighthouses and the line from the ship directly to shore are 15° and 35°. See the illustration.
(a) How far is the ship from lighthouse P?
(b) How far is the ship from lighthouse Q?
(c) How far is the ship from shore?

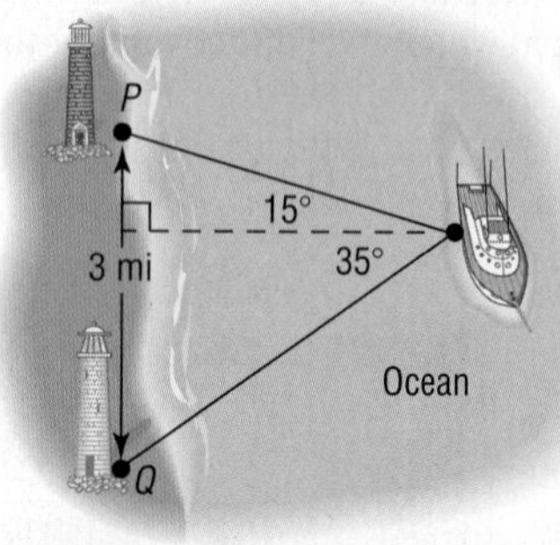

* On February 27, 1964, the government of Italy requested aid in preventing the tower from toppling. A multinational task force of engineers, mathematicians, and historians was assigned and met on the Azores islands to discuss stabilization methods. After over two decades of work on the subject, the tower was closed to the public in January 1990. During the time that the tower was closed, the bells were removed to relieve it of some weight, and cables were cinched around the third level and anchored several hundred meters away. Apartments and houses in the path of the tower were vacated for safety concerns. After a decade of corrective reconstruction and stabilization efforts, the tower was reopened to the public on December 15, 2001. Many methods were proposed to stabilize the tower, including the addition of 800 metric tons of lead counterweights to the raised end of the base. The final solution was to remove 38 cubic meters of soil from underneath the raised end. The tower has been declared stable for at least another 300 years.

***Source:** http://en.wikipedia.org/wiki/Leaning_Tower_of_Pisa*

53. Designing an Awning An awning that covers a sliding glass door that is 88 inches tall forms an angle of 50° with the wall. The purpose of the awning is to prevent sunlight from entering the house when the angle of elevation of the Sun is more than 65°. See the figure. Find the length L of the awning.

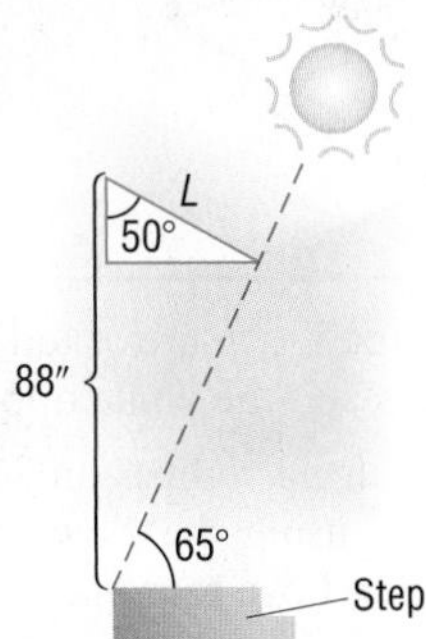

54. Finding Distances A forest ranger is walking on a path inclined at 5° to the horizontal directly toward a 100-foot-tall fire observation tower. The angle of elevation from the path to the top of the tower is 40°. How far is the ranger from the tower at this time?

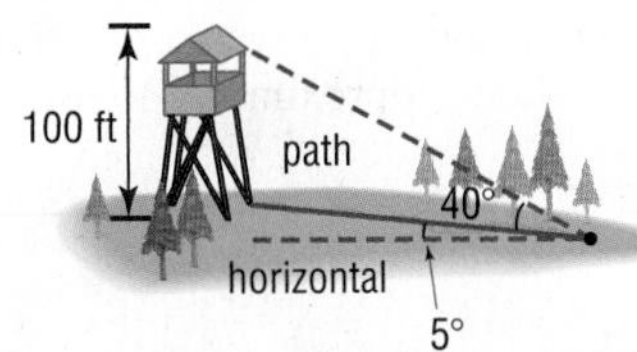

55. Great Pyramid of Cheops One of the original Seven Wonders of the World, the Great Pyramid of Cheops was built about 2580 BC. Its original height was 480 feet 11 inches, but owing to the loss of its topmost stones, it is now shorter. Find the current height of the Great Pyramid using the information given in the illustration.

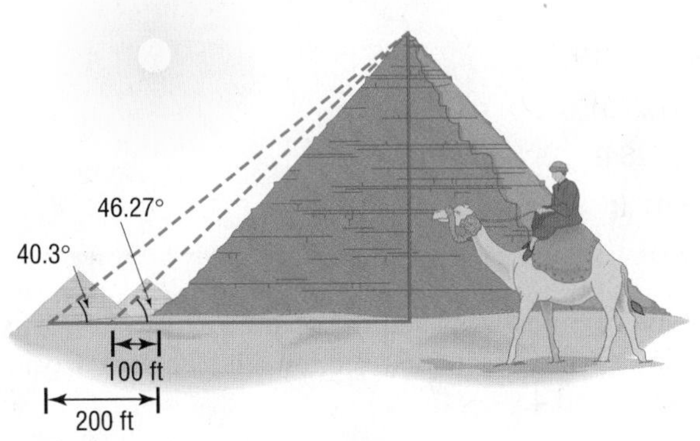

Source: Guinness Book of World Records

56. Determining the Height of an Aircraft Two sensors are spaced 700 feet apart along the approach to a small airport. When an aircraft is nearing the airport, the angle of elevation from the first sensor to the aircraft is 20°, and from the second sensor to the aircraft it is 15°. Determine how high the aircraft is at this time.

57. Mercury The distance from the Sun to Earth is approximately 149,600,000 kilometers (km). The distance from the Sun to Mercury is approximately 57,910,000 km. The **elongation angle** α is the angle formed between the line of sight from Earth to the Sun and the line of sight from Earth to Mercury. See the figure. Suppose that the elongation angle for Mercury is 15°. Use this information to find the possible distances between Earth and Mercury.

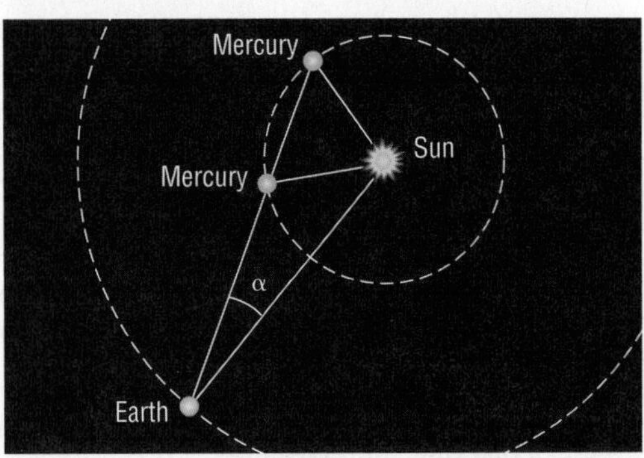

58. Venus The distance from the Sun to Earth is approximately 149,600,000 km. The distance from the Sun to Venus is approximately 108,200,000 km. The elongation angle α is the angle formed between the line of sight from Earth to the Sun and the line of sight from Earth to Venus. Suppose that the elongation angle for Venus is 10°. Use this information to find the possible distances between Earth and Venus.

59. The Original Ferris Wheel George Washington Gale Ferris, Jr., designed the original Ferris wheel for the 1893 World's Columbian Exposition in Chicago, Illinois. The wheel had 36 equally spaced cars each the size of a school bus. The distance between adjacent cars was approximately 22 feet. Determine the diameter of the wheel to the nearest foot.

Source: Carnegie Library of Pittsburgh, www.clpgh.org

60. Mollweide's Formula For any triangle, **Mollweide's Formula** (named after Karl Mollweide, 1774–1825) states that

$$\frac{a+b}{c} = \frac{\cos\left[\frac{1}{2}(A-B)\right]}{\sin\left(\frac{1}{2}C\right)}$$

Derive it.

[**Hint:** Use the Law of Sines and then a Sum-to-Product Formula. Notice that this formula involves all six parts of a triangle. As a result, it is sometimes used to check the solution of a triangle.]

61. Mollweide's Formula Another form of Mollweide's Formula is

$$\frac{a-b}{c} = \frac{\sin\left[\frac{1}{2}(A-B)\right]}{\cos\left(\frac{1}{2}C\right)}$$

Derive it.

62. For any triangle, derive the formula

$$a = b\cos C + c\cos B$$

[**Hint:** Use the fact that $\sin A = \sin(180° - B - C)$.]

63. Law of Tangents For any triangle, derive the **Law of Tangents:**

$$\frac{a-b}{a+b} = \frac{\tan\left[\frac{1}{2}(A-B)\right]}{\tan\left[\frac{1}{2}(A+B)\right]}$$

[**Hint:** Use Mollweide's Formula.]

64. Circumscribing a Triangle Show that

$$\frac{\sin A}{a} = \frac{\sin B}{b} = \frac{\sin C}{c} = \frac{1}{2r}$$

where r is the radius of the circle circumscribing the triangle PQR whose sides are a, b, and c, as shown in the figure. [**Hint:** Draw the diameter PP'. Then $B = \angle PQR = \angle PP'R$, and angle $\angle PRP' = 90°$.]

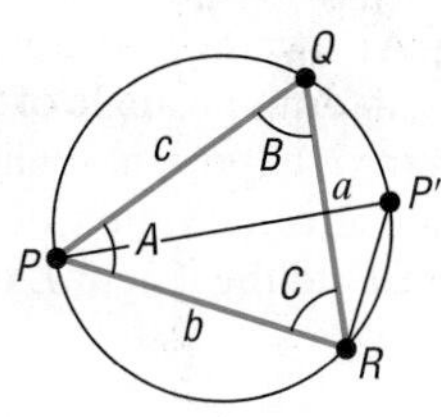

Explaining Concepts: Discussion and Writing

65. Make up three problems involving oblique triangles. One should result in one triangle, the second in two triangles, and the third in no triangle.

66. What do you do first if you are asked to solve a triangle and are given one side and two angles?

67. What do you do first if you are asked to solve a triangle and are given two sides and the angle opposite one of them?

68. Solve Example 6 using right-triangle geometry. Comment on which solution, using the Law of Sines or using right triangles, you prefer. Give reasons.

Retain Your Knowledge

Problems 69–72 are based on material learned earlier in the course. The purpose of these problems is to keep the material fresh in your mind so that you are better prepared for the final exam.

69. Find the center (h, k) and radius r of the circle $x^2 + 2x + y^2 - 6y = 6$.

70. Graph $y = 4\sin\left(\frac{1}{2}x\right)$. Show at least two periods.

71. Find the exact value of $\tan\left[\cos^{-1}\left(-\frac{7}{8}\right)\right]$.

72. Find the exact distance between $P_1 = (-1, -7)$ and $P_2 = (2, -1)$. Then approximate the distance to two decimal places.

'Are You Prepared?' Answers

1. $\sin A\cos B - \cos A\sin B$ **2.** $30°$ or $\frac{\pi}{6}$ **3.** $\frac{15}{2}$

4.3 The Law of Cosines

PREPARING FOR THIS SECTION *Before getting started, review the following:*

- Trigonometric Equations (Section 3.3, pp. 208–213)
- Distance Formula (Section 1.1, p. 3)

Now Work the 'Are You Prepared?' problems on page 285.

OBJECTIVES 1 Solve SAS Triangles (p. 283)
2 Solve SSS Triangles (p. 284)
3 Solve Applied Problems (p. 284)

In the previous section, the Law of Sines was used to solve Case 1 (SAA or ASA) and Case 2 (SSA) of an oblique triangle. In this section, the Law of Cosines is derived and used to solve Cases 3 and 4.

CASE 3: Two sides and the included angle are known (SAS).
CASE 4: Three sides are known (SSS).

THEOREM

Law of Cosines

For a triangle with sides a, b, c and opposite angles A, B, C, respectively,

$$c^2 = a^2 + b^2 - 2ab\cos C \quad \textbf{(1)}$$
$$b^2 = a^2 + c^2 - 2ac\cos B \quad \textbf{(2)}$$
$$a^2 = b^2 + c^2 - 2bc\cos A \quad \textbf{(3)}$$

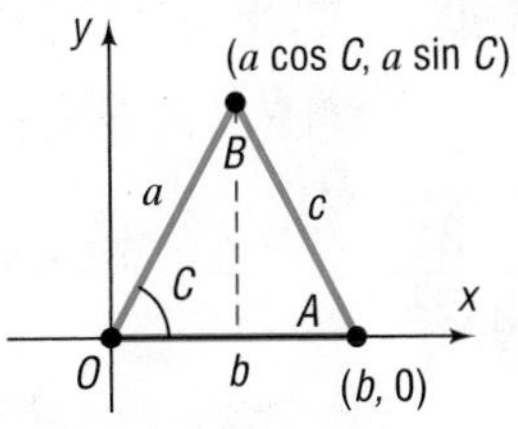

(a) Angle C is acute

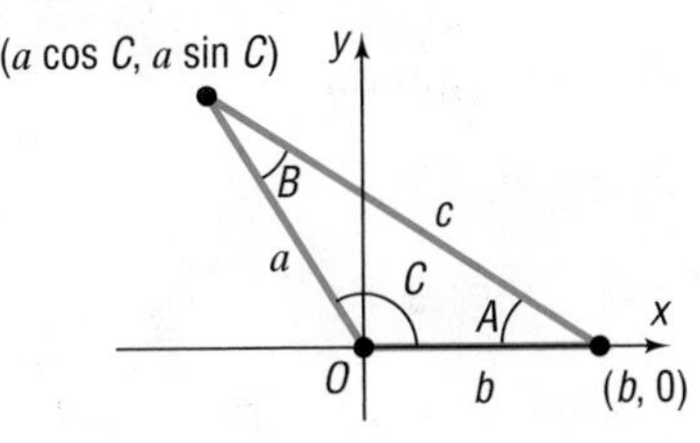

(b) Angle C is obtuse

Figure 35

Proof Only formula (1) is proved here. Formulas (2) and (3) may be proved using the same argument.

Begin by strategically placing a triangle on a rectangular coordinate system so that the vertex of angle C is at the origin and side b lies along the positive x-axis. Regardless of whether C is acute, as in Figure 35(a), or obtuse, as in Figure 35(b), the vertex of angle B has coordinates $(a \cos C, a \sin C)$. The vertex of angle A has coordinates $(b, 0)$.

Use the distance formula to compute c^2.

$$
\begin{aligned}
c^2 &= (b - a\cos C)^2 + (0 - a \sin C)^2 \\
&= b^2 - 2ab\cos C + a^2\cos^2 C + a^2 \sin^2 C \\
&= b^2 - 2ab\cos C + a^2(\cos^2 C + \sin^2 C) \\
&= a^2 + b^2 - 2ab\cos C
\end{aligned}
$$
■

Each of formulas (1), (2), and (3) may be stated in words as follows:

THEOREM

Law of Cosines

The square of one side of a triangle equals the sum of the squares of the other two sides, minus twice their product times the cosine of their included angle.

Observe that if the triangle is a right triangle (so that, say, $C = 90°$), formula (1) becomes the familiar Pythagorean Theorem: $c^2 = a^2 + b^2$. The Pythagorean Theorem is a special case of the Law of Cosines!

1 Solve SAS Triangles

The Law of Cosines is used to solve Case 3 (SAS), which applies to triangles for which two sides and the included angle are known.

EXAMPLE 1

Using the Law of Cosines to Solve an SAS Triangle

Solve the triangle: $a = 2, \quad b = 3, \quad C = 60°$

Solution See Figure 36. Because two sides, a and b, and the included angle, $C = 60°$, are known, the Law of Cosines makes it easy to find the third side, c.

$$
\begin{aligned}
c^2 &= a^2 + b^2 - 2ab\cos C \\
&= 2^2 + 3^2 - 2\cdot 2\cdot 3\cdot \cos 60° \quad a = 2, b = 3, C = 60° \\
&= 13 - \left(12\cdot\frac{1}{2}\right) = 7 \\
c &= \sqrt{7}
\end{aligned}
$$

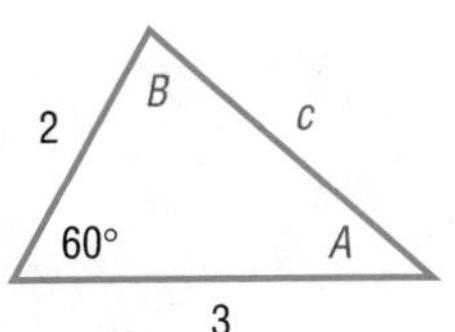

Figure 36

Side c is of length $\sqrt{7}$. To find the angles A and B, either the Law of Sines or the Law of Cosines may be used. It is preferable to use the Law of Cosines because it will lead to an equation with one solution. Using the Law of Sines would lead to an equation with two solutions that would need to be checked to determine which solution fits the given data.* We choose to use formulas (2) and (3) of the Law of Cosines to find A and B.

For A:

$$
\begin{aligned}
a^2 &= b^2 + c^2 - 2bc\cos A \\
2bc\cos A &= b^2 + c^2 - a^2 \\
\cos A &= \frac{b^2 + c^2 - a^2}{2bc} = \frac{9 + 7 - 4}{2\cdot 3\sqrt{7}} = \frac{12}{6\sqrt{7}} = \frac{2\sqrt{7}}{7} \\
A &= \cos^{-1}\frac{2\sqrt{7}}{7} \approx 40.9°
\end{aligned}
$$

*The Law of Sines can be used if the angle sought is opposite the smaller side, thus ensuring that it is acute. (In Figure 36, use the Law of Sines to find A, the angle opposite the smaller side.)

For B:

$$b^2 = a^2 + c^2 - 2ac \cos B$$

$$\cos B = \frac{a^2 + c^2 - b^2}{2ac} = \frac{4 + 7 - 9}{4\sqrt{7}} = \frac{2}{4\sqrt{7}} = \frac{\sqrt{7}}{14}$$

$$B = \cos^{-1} \frac{\sqrt{7}}{14} \approx 79.1°$$

NOTE The angle B could also have been found by using the fact that the sum $A + B + C = 180°$, so $B = 180° - 40.9° - 60° = 79.1°$. However, using the Law of Cosines twice allows for a check. ■

Notice that $A + B + C = 40.9° + 79.1° + 60° = 180°$, as required. ●

Now Work PROBLEM 9

2 Solve SSS Triangles

The next example illustrates how the Law of Cosines is used when three sides of a triangle are known, Case 4 (SSS).

EXAMPLE 2 Using the Law of Cosines to Solve an SSS Triangle

Solve the triangle: $a = 4, b = 3, c = 6$

Solution See Figure 37. To find the angles A, B, and C, proceed as in the solution to Example 1.

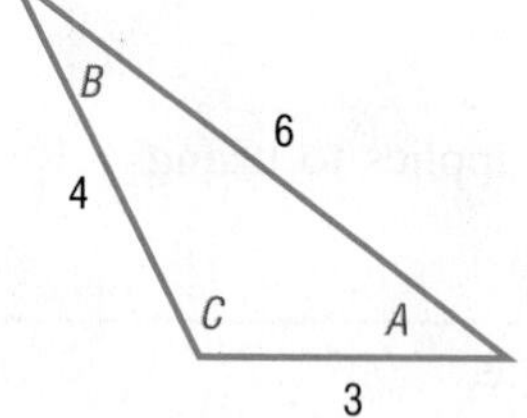

Figure 37

For A:

$$\cos A = \frac{b^2 + c^2 - a^2}{2bc} = \frac{9 + 36 - 16}{2 \cdot 3 \cdot 6} = \frac{29}{36}$$

$$A = \cos^{-1} \frac{29}{36} \approx 36.3°$$

For B:

$$\cos B = \frac{a^2 + c^2 - b^2}{2ac} = \frac{16 + 36 - 9}{2 \cdot 4 \cdot 6} = \frac{43}{48}$$

$$B = \cos^{-1} \frac{43}{48} \approx 26.4°$$

Now use A and B to find C:

$$C = 180° - A - B \approx 180° - 36.3° - 26.4° = 117.3°$$ ●

Now Work PROBLEM 15

3 Solve Applied Problems

EXAMPLE 3 Correcting a Navigational Error

A motorized sailboat leaves Naples, Florida, bound for Key West, 150 miles away. Maintaining a constant speed of 15 miles per hour, but encountering heavy crosswinds and strong currents, the crew finds, after 4 hours, that the sailboat is off course by 20°.

(a) How far is the sailboat from Key West at this time?
(b) Through what angle should the sailboat turn to correct its course?
(c) How much time has been added to the trip because of this? (Assume that the speed remains at 15 miles per hour.)

Solution See Figure 38. With a speed of 15 miles per hour, the sailboat has gone 60 miles after 4 hours. The distance x of the sailboat from Key West is to be found, along with the angle θ that the sailboat should turn through to correct its course.

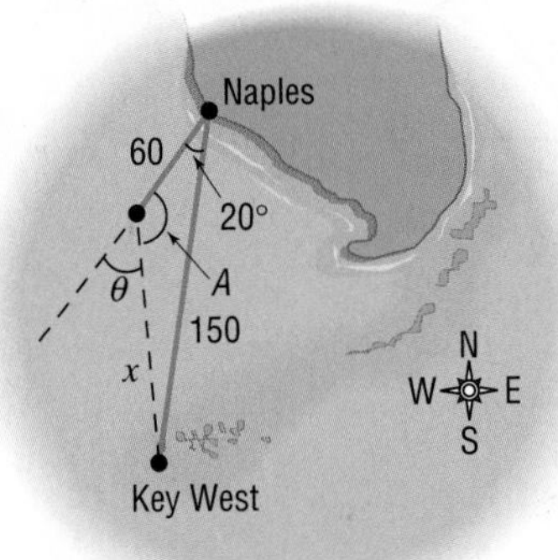

Figure 38

(a) To find x, use the Law of Cosines, because two sides and the included angle are known.

$$x^2 = 150^2 + 60^2 - 2(150)(60)\cos 20° \approx 9185.53$$
$$x \approx 95.8$$

The sailboat is about 96 miles from Key West.

(b) With all three sides of the triangle now known, use the Law of Cosines again to find the angle A opposite the side of length 150 miles.

$$150^2 = 96^2 + 60^2 - 2(96)(60)\cos A$$
$$9684 = -11{,}520 \cos A$$
$$\cos A \approx -0.8406$$
$$A \approx 147.2°$$

So,

$$\theta = 180° - A \approx 180° - 147.2° = 32.8°$$

The sailboat should turn through an angle of about 33° to correct its course.

(c) The total length of the trip is now $60 + 96 = 156$ miles. The extra 6 miles will only require about 0.4 hour, or 24 minutes, more if the speed of 15 miles per hour is maintained.

Now Work PROBLEM 45

Historical Feature

The Law of Sines was known vaguely long before it was explicitly stated by Nasir Eddin (about AD 1250). Ptolemy (about AD 150) was aware of it in a form using a chord function instead of the sine function. But it was first clearly stated in Europe by Regiomontanus, writing in 1464.

The Law of Cosines appears first in Euclid's *Elements* (Book II), but in a well-disguised form in which squares built on the sides of triangles are added and a rectangle representing the cosine term is subtracted. It was thus known to all mathematicians because of their familiarity with Euclid's work. An early modern form of the Law of Cosines, that for finding the angle when the sides are known, was stated by François Viète (in 1593).

The Law of Tangents (see Problem 63 in Section 4.2) has become obsolete. In the past it was used in place of the Law of Cosines, because the Law of Cosines was very inconvenient for calculation with logarithms or slide rules. Mixing of addition and multiplication is now very easy on a calculator, however, and the Law of Tangents has been shelved along with the slide rule.

4.3 Assess Your Understanding

'Are You Prepared?' *Answers are given at the end of these exercises. If you get a wrong answer, read the pages listed in red.*

1. Write the formula for the distance d from $P_1 = (x_1, y_1)$ to $P_2 = (x_2, y_2)$. (p. 3)

2. If θ is an acute angle, solve the equation $\cos\theta = \frac{\sqrt{2}}{2}$. (pp. 208–213)

Concepts and Vocabulary

3. If three sides of a triangle are given, the Law of ________ is used to solve the triangle.

4. If one side and two angles of a triangle are given, which law can be used to solve the triangle?
(a) Law of Sines (b) Law of Cosines
(c) Either a or b (d) The triangle cannot be solved.

5. If two sides and the included angle of a triangle are given, which law can be used to solve the triangle?
(a) Law of Sines (b) Law of Cosines
(c) Either a or b (d) The triangle cannot be solved.

6. ***True or False*** Given only the three sides of a triangle, there is insufficient information to solve the triangle.

7. ***True or False*** The Law of Cosines states that the square of one side of a triangle equals the sum of the squares of the other two sides, minus twice their product.

8. ***True or False*** A special case of the Law of Cosines is the Pythagorean Theorem.

Skill Building

In Problems 9–16, solve each triangle.

9.

10.

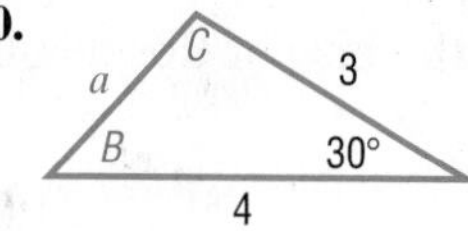

11.

14.

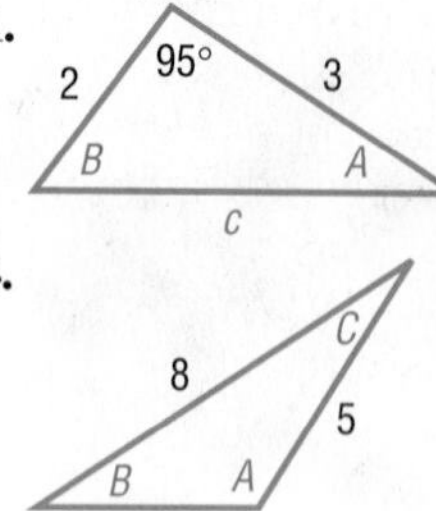

12.

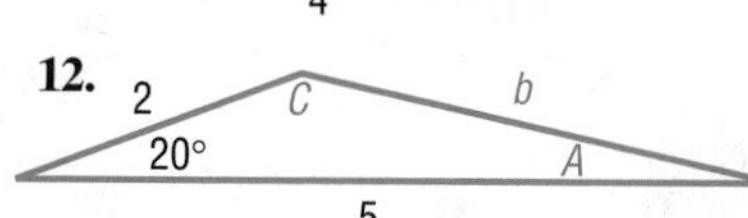

13.

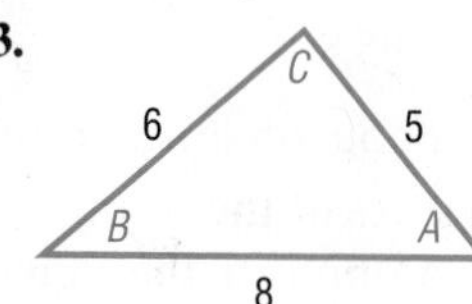

15.

16.

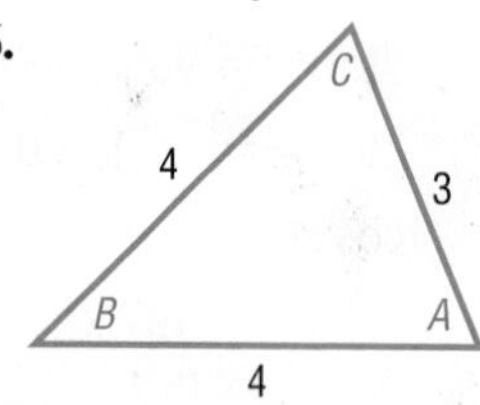

In Problems 17–32, solve each triangle.

17. $a = 3, \quad b = 4, \quad C = 40°$

18. $a = 2, \quad c = 1, \quad B = 10°$

19. $b = 1, \quad c = 3, \quad A = 80°$

20. $a = 6, \quad b = 4, \quad C = 60°$

21. $a = 3, \quad c = 2, \quad B = 110°$

22. $b = 4, \quad c = 1, \quad A = 120°$

23. $a = 2, \quad b = 2, \quad C = 50°$

24. $a = 3, \quad c = 2, \quad B = 90°$

25. $a = 12, \quad b = 13, \quad c = 5$

26. $a = 4, \quad b = 5, \quad c = 3$

27. $a = 2, \quad b = 2, \quad c = 2$

28. $a = 3, \quad b = 3, \quad c = 2$

29. $a = 5, \quad b = 8, \quad c = 9$

30. $a = 4, \quad b = 3, \quad c = 6$

31. $a = 10, \quad b = 8, \quad c = 5$

32. $a = 9, \quad b = 7, \quad c = 10$

Mixed Practice

In Problems 33–42, solve each triangle using either the Law of Sines or the Law of Cosines.

33. $B = 20°, C = 75°, b = 5$

34. $A = 50°, B = 55°, c = 9$

35. $a = 6, b = 8, c = 9$

36. $a = 14, b = 7, A = 85°$

37. $B = 35°, C = 65°, a = 15$

38. $a = 4, c = 5, B = 55°$

39. $A = 10°, a = 3, b = 10$

40. $A = 65°, B = 72°, b = 7$

41. $b = 5, c = 12, A = 60°$

42. $a = 10, b = 10, c = 15$

Applications and Extensions

43. Distance to the Green A golfer hits an errant tee shot that lands in the rough. A marker in the center of the fairway is 150 yards from the center of the green. While standing on the marker and facing the green, the golfer turns 110° toward his ball. He then paces off 35 yards to his ball. See the figure. How far is the ball from the center of the green?

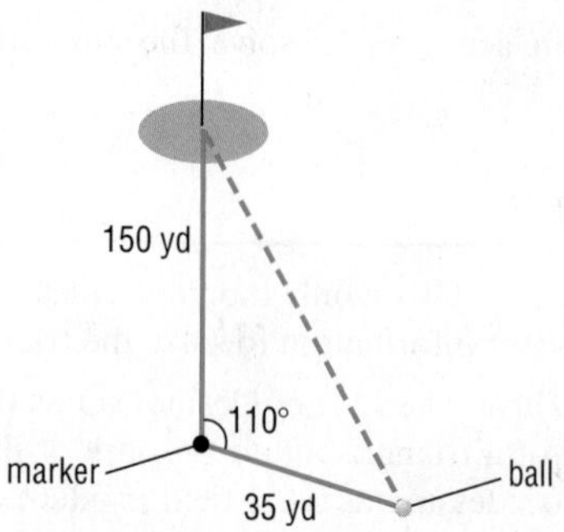

44. Navigation An airplane flies due north from Ft. Myers to Sarasota, a distance of 150 miles, and then turns through an angle of 50° and flies to Orlando, a distance of 100 miles. See the figure.

(a) How far is it directly from Ft. Myers to Orlando?

(b) What bearing should the pilot use to fly directly from Ft. Myers to Orlando?

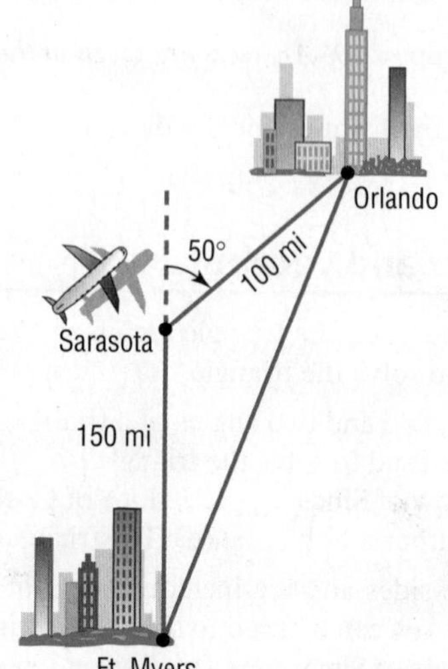

45. **Avoiding a Tropical Storm** A cruise ship maintains an average speed of 15 knots in going from San Juan, Puerto Rico, to Barbados, West Indies, a distance of 600 nautical miles. To avoid a tropical storm, the captain heads out of San Juan in a direction of 20° off a direct heading to Barbados. The captain maintains the 15-knot speed for 10 hours, after which time the path to Barbados becomes clear of storms.
   (a) Through what angle should the captain turn to head directly to Barbados?
   (b) Once the turn is made, how long will it be before the ship reaches Barbados if the same 15-knot speed is maintained?

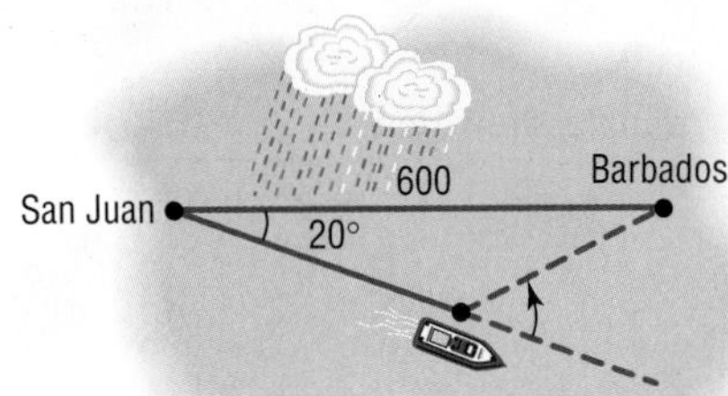

46. **Revising a Flight Plan** In attempting to fly from Chicago to Louisville, a distance of 330 miles, a pilot inadvertently took a course that was 10° in error, as indicated in the figure.
   (a) If the aircraft maintains an average speed of 220 miles per hour, and if the error in direction is discovered after 15 minutes, through what angle should the pilot turn to head toward Louisville?
   (b) What new average speed should the pilot maintain so that the total time of the trip is 90 minutes?

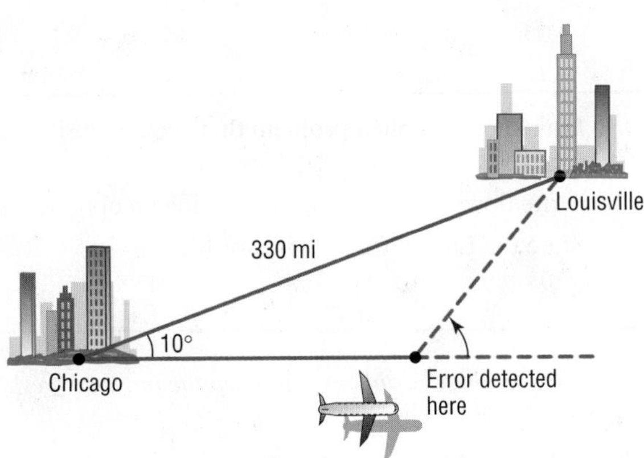

47. **Major League Baseball Field** A major league baseball diamond is actually a square 90 feet on a side. The pitching rubber is located 60.5 feet from home plate on a line joining home plate and second base.
   (a) How far is it from the pitching rubber to first base?
   (b) How far is it from the pitching rubber to second base?
   (c) If a pitcher faces home plate, through what angle does he need to turn to face first base?

48. **Little League Baseball Field** According to Little League baseball official regulations, the diamond is a square 60 feet on a side. The pitching rubber is located 46 feet from home plate on a line joining home plate and second base.
   (a) How far is it from the pitching rubber to first base?
   (b) How far is it from the pitching rubber to second base?
   (c) If a pitcher faces home plate, through what angle does he need to turn to face first base?

49. **Finding the Length of a Guy Wire** The height of a radio tower is 500 feet, and the ground on one side of the tower slopes upward at an angle of 10° (see the figure).
   (a) How long should a guy wire be if it is to connect to the top of the tower and be secured at a point on the sloped side 100 feet from the base of the tower?
   (b) How long should a second guy wire be if it is to connect to the middle of the tower and be secured at a position 100 feet from the base on the flat side?

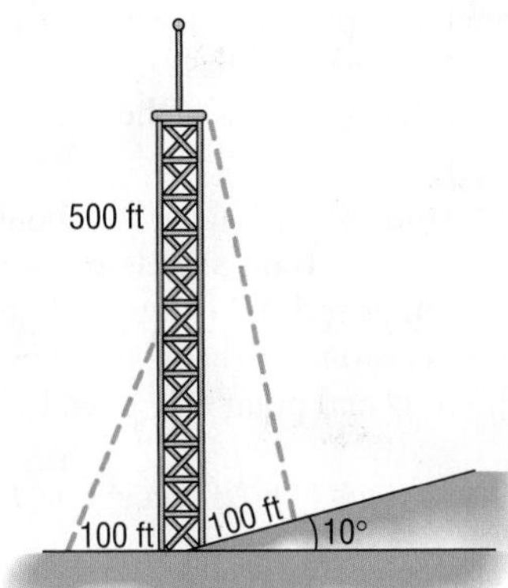

50. **Finding the Length of a Guy Wire** A radio tower 500 feet high is located on the side of a hill with an inclination to the horizontal of 5°. See the figure. How long should two guy wires be if they are to connect to the top of the tower and be secured at two points 100 feet directly above and directly below the base of the tower?

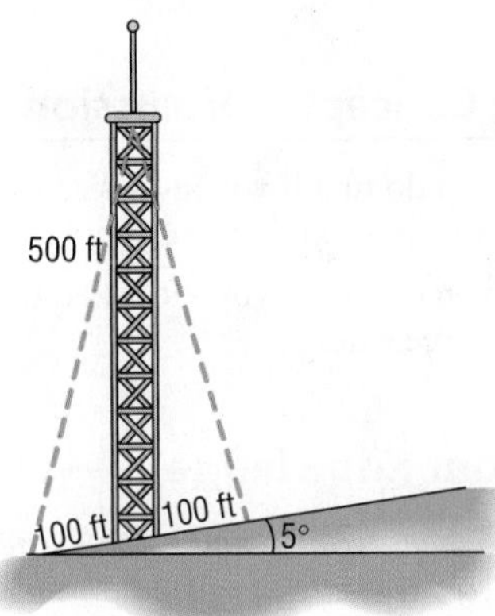

51. **Wrigley Field, Home of the Chicago Cubs** The distance from home plate to the fence in dead center in Wrigley Field is 400 feet (see the figure). How far is it from the fence in dead center to third base?

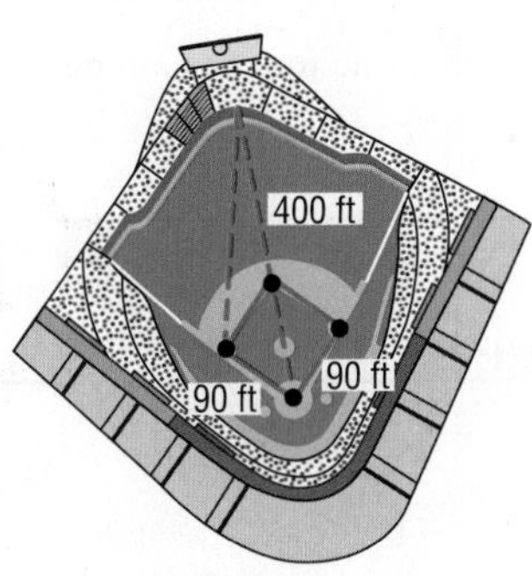

52. Little League Baseball The distance from home plate to the fence in dead center at the Oak Lawn Little League field is 280 feet. How far is it from the fence in dead center to third base?

[**Hint:** The distance between the bases in Little League is 60 feet.]

53. Building a Swing Set Clint is building a wooden swing set for his children. Each supporting end of the swing set is to be an A-frame constructed with two 10-foot-long 4 by 4's joined at a 45° angle. To prevent the swing set from tipping over, Clint wants to secure the base of each A-frame to concrete footings. How far apart should the footings for each A-frame be?

54. Rods and Pistons Rod OA rotates about the fixed point O so that point A travels on a circle of radius r. Connected to point A is another rod AB of length $L > 2r$, and point B is connected to a piston. See the figure. Show that the distance x between point O and point B is given by

$$x = r\cos\theta + \sqrt{r^2\cos^2\theta + L^2 - r^2}$$

where θ is the angle of rotation of rod OA.

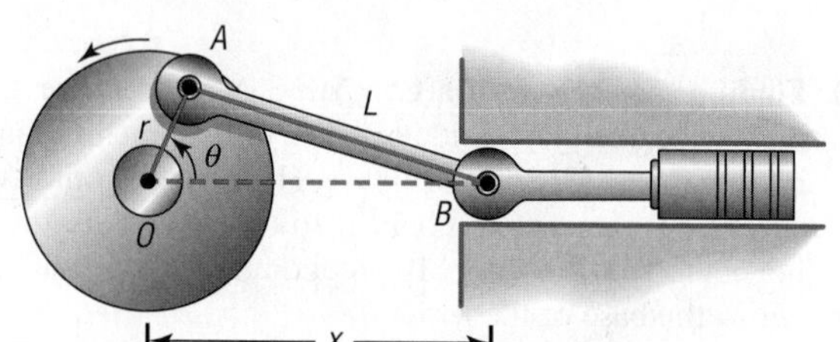

55. Geometry Show that the length d of a chord of a circle of radius r is given by the formula

$$d = 2r\sin\frac{\theta}{2}$$

where θ is the central angle formed by the radii to the ends of the chord. See the figure. Use this result to derive the fact that $\sin\theta < \theta$, where $\theta > 0$ is measured in radians.

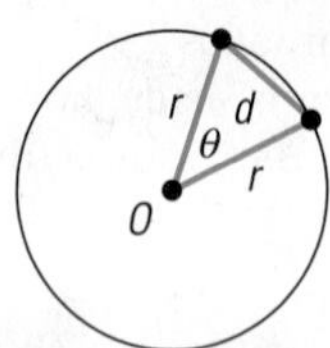

56. For any triangle, show that

$$\cos\frac{C}{2} = \sqrt{\frac{s(s-c)}{ab}}$$

where $s = \frac{1}{2}(a+b+c)$.

[**Hint:** Use a Half-angle Formula and the Law of Cosines.]

57. For any triangle, show that

$$\sin\frac{C}{2} = \sqrt{\frac{(s-a)(s-b)}{ab}}$$

where $s = \frac{1}{2}(a+b+c)$.

58. Use the Law of Cosines to prove the identity

$$\frac{\cos A}{a} + \frac{\cos B}{b} + \frac{\cos C}{c} = \frac{a^2+b^2+c^2}{2abc}$$

Explaining Concepts: Discussion and Writing

59. What do you do first if you are asked to solve a triangle and are given two sides and the included angle?

60. What do you do first if you are asked to solve a triangle and are given three sides?

61. Make up an applied problem that requires using the Law of Cosines.

62. Write down your strategy for solving an oblique triangle.

63. State the Law of Cosines in words.

Retain Your Knowledge

Problems 64–67 are based on material learned earlier in the course. The purpose of these problems is to keep the material fresh in your mind so that you are better prepared for the final exam.

64. If $(4, -5)$ is a point on the graph of $y = f(x)$, what point must be on the graph of $y = -f(2x)$?

65. Find the average rate of change of $f(x) = \cot x$ from $\frac{\pi}{6}$ to $\frac{\pi}{2}$.

66. Given $\tan\theta = -\frac{2\sqrt{6}}{5}$ and $\cos\theta = -\frac{5}{7}$, find the exact value of each of the four remaining trigonometric functions.

67. Find an equation for the graph.

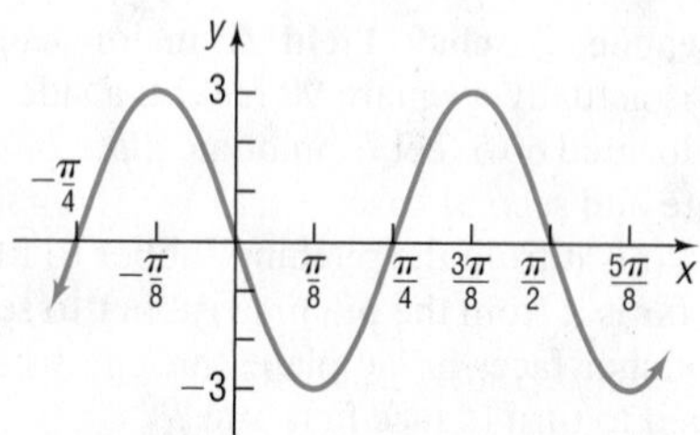

'Are You Prepared?' Answers

1. $d = \sqrt{(x_2 - x_1)^2 + (y_2 - y_1)^2}$ **2.** $\theta = 45°$ or $\frac{\pi}{4}$

4.4 Area of a Triangle

PREPARING FOR THIS SECTION *Before getting started, review the following:*

- Geometry Essentials (Appendix A, Section A.2, pp. A14–A18)

Now Work the 'Are You Prepared?' problem on page 291.

OBJECTIVES 1 Find the Area of SAS Triangles (p. 289)
2 Find the Area of SSS Triangles (p. 290)

In this section, several formulas for calculating the area of a triangle are derived. The most familiar of these follows.

THEOREM

The area K of a triangle is

$$K = \frac{1}{2}bh \qquad (1)$$

where b is the base and h is an altitude drawn to that base.

NOTE Typically, A is used for area. However, because A is also used as the measure of an angle, K is used here for area to avoid confusion. ■

Proof Look at the triangle in Figure 39. Around the triangle construct a rectangle of height h and base b, as shown in Figure 40.

Triangles 1 and 2 in Figure 40 are equal in area, as are triangles 3 and 4. Consequently, the area of the triangle with base b and altitude h is exactly half the area of the rectangle, which is bh.

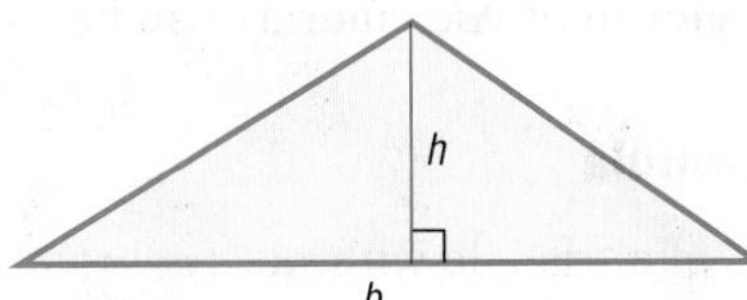

Figure 39

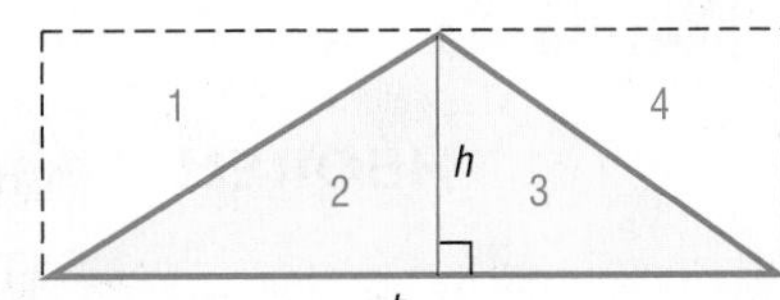

Figure 40 ■

1 Find the Area of SAS Triangles

If the base b and the altitude h to that base are known, then the area of such a triangle can be found using formula (1). Usually, though, the information required to use formula (1) is not given. Suppose, for example, that two sides a and b and the included angle C are known. See Figure 41. Then the altitude h can be found by noting that

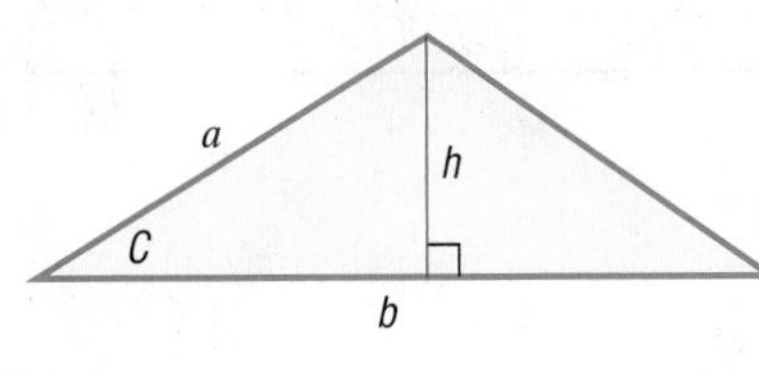

Figure 41

$$\frac{h}{a} = \sin C$$

so

$$h = a \sin C$$

Using this fact in formula (1) produces

$$K = \frac{1}{2}bh = \frac{1}{2}b(a \sin C) = \frac{1}{2}ab \sin C$$

The area K of the triangle is given by the formula

$$K = \frac{1}{2}ab \sin C \qquad (2)$$

Dropping altitudes from the other two vertices of the triangle leads to the following corresponding formulas:

$$K = \frac{1}{2}bc \sin A \qquad (3)$$

$$K = \frac{1}{2}ac \sin B \qquad (4)$$

It is easiest to remember these formulas by using the following wording:

THEOREM

The area K of a triangle equals one-half the product of two of its sides times the sine of their included angle.

EXAMPLE 1 **Finding the Area of an SAS Triangle**

Find the area K of the triangle for which $a = 8$, $b = 6$, and $C = 30°$.

Solution See Figure 42. Use formula (2) to get

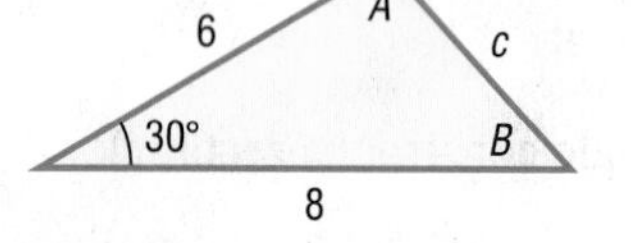

Figure 42

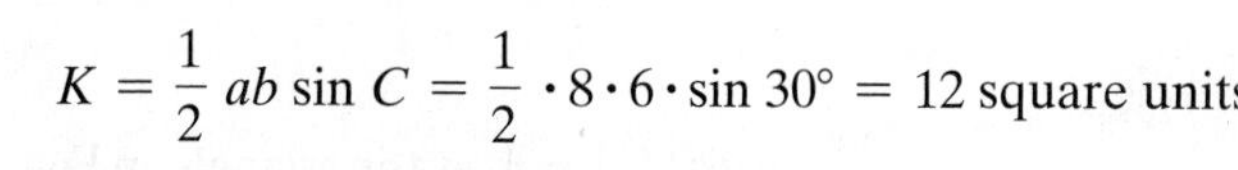

$$K = \frac{1}{2}ab \sin C = \frac{1}{2} \cdot 8 \cdot 6 \cdot \sin 30° = 12 \text{ square units}$$

Now Work PROBLEM 7

2 Find the Area of SSS Triangles

If the three sides of a triangle are known, another formula, called **Heron's Formula** (named after Heron of Alexandria), can be used to find the area of a triangle.

THEOREM

Heron's Formula

The area K of a triangle with sides a, b, and c is

$$K = \sqrt{s(s-a)(s-b)(s-c)} \qquad (5)$$

where $s = \frac{1}{2}(a + b + c)$.

EXAMPLE 2 **Finding the Area of an SSS Triangle**

Find the area of a triangle whose sides are 4, 5, and 7.

Solution Let $a = 4$, $b = 5$, and $c = 7$. Then

$$s = \frac{1}{2}(a + b + c) = \frac{1}{2}(4 + 5 + 7) = 8$$

Heron's Formula gives the area K as

$$K = \sqrt{s(s-a)(s-b)(s-c)} = \sqrt{8 \cdot 4 \cdot 3 \cdot 1} = \sqrt{96} = 4\sqrt{6} \text{ square units}$$

Now Work PROBLEM 13

Proof of Heron's Formula The proof given here uses the Law of Cosines and is quite different from the proof given by Heron.

From the Law of Cosines,

$$c^2 = a^2 + b^2 - 2ab \cos C$$

and the Half-angle Formula,

$$\cos^2\frac{C}{2} = \frac{1 + \cos C}{2}$$

it follows that

$$\cos^2\frac{C}{2} = \frac{1 + \cos C}{2} = \frac{1 + \dfrac{a^2 + b^2 - c^2}{2ab}}{2}$$

$$= \frac{a^2 + 2ab + b^2 - c^2}{4ab} = \frac{(a + b)^2 - c^2}{4ab}$$

$$= \frac{(a + b - c)(a + b + c)}{4ab} = \frac{2(s - c)\cdot 2s}{4ab} = \frac{s(s - c)}{ab} \tag{6}$$

↑ Factor. ↑ $a + b - c = a + b + c - 2c = 2s - 2c = 2(s - c)$

Similarly, using $\sin^2\frac{C}{2} = \frac{1 - \cos C}{2}$, it follows that

$$\sin^2\frac{C}{2} = \frac{(s - a)(s - b)}{ab} \tag{7}$$

Now use formula (2) for the area.

$$K = \frac{1}{2}ab\sin C$$

$$= \frac{1}{2}ab\cdot 2\sin\frac{C}{2}\cos\frac{C}{2} \qquad \sin C = \sin\left[2\left(\frac{C}{2}\right)\right] = 2\sin\frac{C}{2}\cos\frac{C}{2}$$

$$= ab\sqrt{\frac{(s - a)(s - b)}{ab}}\sqrt{\frac{s(s - c)}{ab}} \qquad \text{Use equations (6) and (7).}$$

$$= \sqrt{s(s - a)(s - b)(s - c)}$$

■

Historical Feature

Heron's Formula (also known as *Hero's Formula*) was first expressed by Heron of Alexandria (first century AD), who had, besides his mathematical talents, a good deal of engineering skills. In various temples his mechanical devices produced effects that seemed supernatural, and visitors presumably were thus moved to generosity. Heron's book *Metrica*, on making such devices, has survived and was discovered in 1896 in the city of Constantinople.

Heron's Formulas for the area of a triangle caused some mild discomfort in Greek mathematics, because a product with two factors was an area and one with three factors was a volume, but four factors seemed contradictory in Heron's time.

4.4 Assess Your Understanding

'Are You Prepared?' *The answer is given at the end of these exercises. If you get the wrong answer, read the pages listed in red.*

1. The area K of a triangle whose base is b and whose height is h is ________. (pp. A14–A18)

Concepts and Vocabulary

2. If two sides a and b and the included angle C are known in a triangle, then the area K is found using the formula $K =$ __________.

3. The area K of a triangle with sides a, b, and c is $K =$ ____________________, where $s =$ ___________.

4. ***True or False*** The area of a triangle equals one-half the product of the lengths of two of its sides times the sine of their included angle.

5. Given two sides of a triangle, b and c, and the included angle A, the altitude h from angle B to side b is given by _______.
(a) $\frac{1}{2}ab\sin A$ (b) $b\sin A$ (c) $c\sin A$ (d) $\frac{1}{2}bc\sin A$

6. Heron's Formula is used to find the area of ________ triangles.
(a) ASA (b) SAS (c) SSS (d) AAS

Skill Building

In Problems 7–14, find the area of each triangle. Round answers to two decimal places.

7.

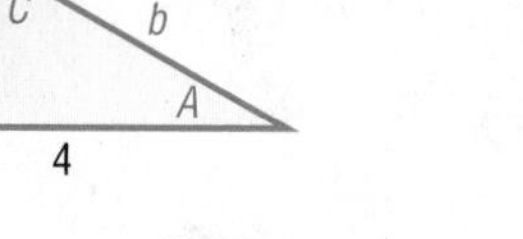

8.

9.

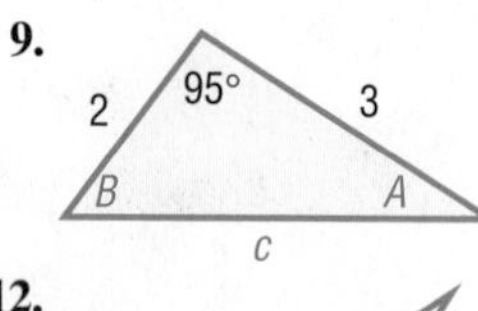

10.

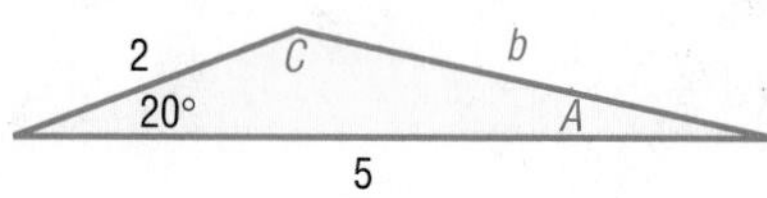

11.

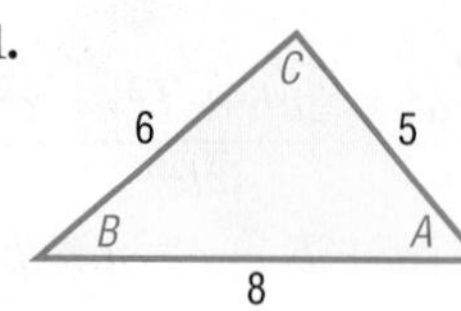

12.

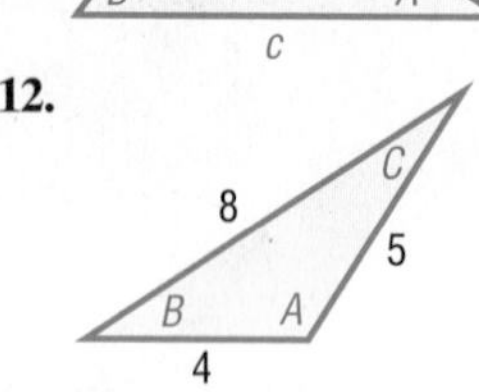

13.

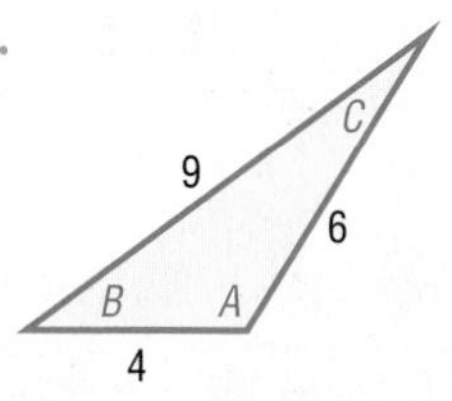

14.

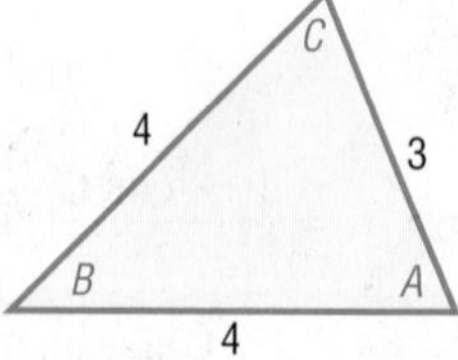

In Problems 15–26, find the area of each triangle. Round answers to two decimal places.

15. $a = 3, \quad b = 4, \quad C = 40°$

16. $a = 2, \quad c = 1, \quad B = 10°$

17. $b = 1, \quad c = 3, \quad A = 80°$

18. $a = 6, \quad b = 4, \quad C = 60°$

19. $a = 3, \quad c = 2, \quad B = 110°$

20. $b = 4, \quad c = 1, \quad A = 120°$

21. $a = 12, \quad b = 13, \quad c = 5$

22. $a = 4, \quad b = 5, \quad c = 3$

23. $a = 2, \quad b = 2, \quad c = 2$

24. $a = 3, \quad b = 3, \quad c = 2$

25. $a = 5, \quad b = 8, \quad c = 9$

26. $a = 4, \quad b = 3, \quad c = 6$

Applications and Extensions

27. Area of an ASA Triangle If two angles and the included side are given, the third angle is easy to find. Use the Law of Sines to show that the area K of a triangle with side a and angles A, B, and C is

$$K = \frac{a^2 \sin B \sin C}{2 \sin A}$$

28. Area of a Triangle Prove the two other forms of the formula given in Problem 27.

$$K = \frac{b^2 \sin A \sin C}{2 \sin B} \quad \text{and} \quad K = \frac{c^2 \sin A \sin B}{2 \sin C}$$

In Problems 29–34, use the results of Problem 27 or 28 to find the area of each triangle. Round answers to two decimal places.

29. $A = 40°, \quad B = 20°, \quad a = 2$

30. $A = 50°, \quad C = 20°, \quad a = 3$

31. $B = 70°, \quad C = 10°, \quad b = 5$

32. $A = 70°, \quad B = 60°, \quad c = 4$

33. $A = 110°, \quad C = 30°, \quad c = 3$

34. $B = 10°, \quad C = 100°, \quad b = 2$

35. Area of a Segment Find the area of the segment (shaded in blue in the figure) of a circle whose radius is 8 feet, formed by a central angle of 70°.

[**Hint:** Subtract the area of the triangle from the area of the sector to obtain the area of the segment.]

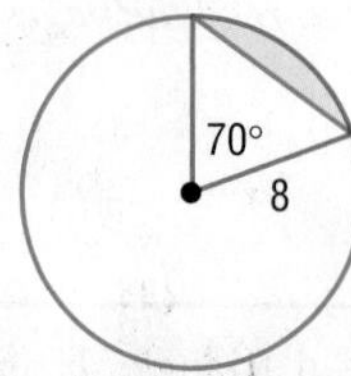

36. Area of a Segment Find the area of the segment of a circle whose radius is 5 inches, formed by a central angle of 40°.

37. Cost of a Triangular Lot The dimensions of a triangular lot are 100 feet by 50 feet by 75 feet. If the price of such land is $3 per square foot, how much does the lot cost?

38. Amount of Material to Make a Tent A cone-shaped tent is made from a circular piece of canvas 24 feet in diameter by removing a sector with central angle 100° and connecting the ends. What is the surface area of the tent?

39. Dimensions of Home Plate The dimensions of home plate at any major league baseball stadium are shown. Find the area of home plate.

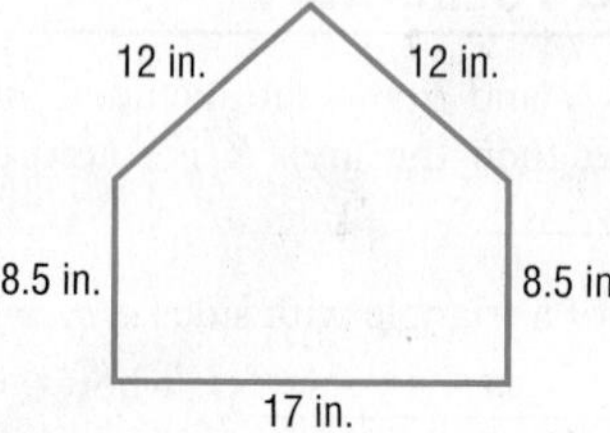

40. Computing Areas See the figure. Find the area of the shaded region enclosed in a semicircle of diameter 10 inches. The length of the chord PQ is 8 inches.
[**Hint:** Triangle PQR is a right triangle.]

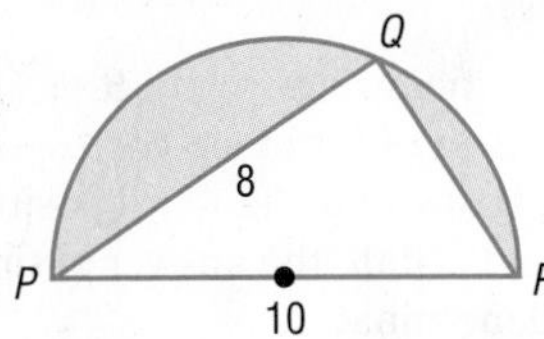

41. Geometry See the figure, which shows a circle of radius r with center at O. Find the area K of the shaded region as a function of the central angle θ.

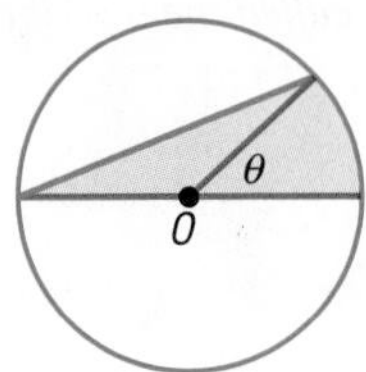

42. Approximating the Area of a Lake To approximate the area of a lake, a surveyor walks around the perimeter of the lake, taking the measurements shown in the illustration. Using this technique, what is the approximate area of the lake?
[**Hint:** Use the Law of Cosines on the three triangles shown, and then find the sum of their areas.]

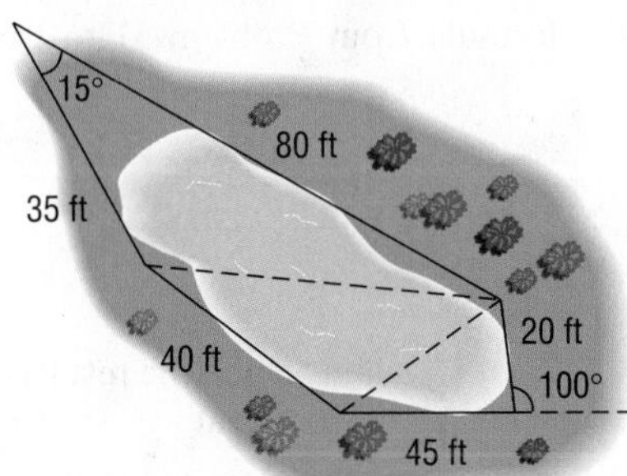

43. The Flatiron Building Completed in 1902 in New York City, the Flatiron Building is triangular shaped and bounded by 22nd Street, Broadway, and 5th Avenue. The building measures approximately 87 feet on the 22nd Street side, 190 feet on the Broadway side, and 173 feet on the 5th Avenue side. Approximate the ground area covered by the building.

Source: Sarah Bradford Landau and Carl W. Condit, *Rise of the New York Skyscraper: 1865–1913*. New Haven, CT: Yale University Press, 1996

44. Bermuda Triangle The Bermuda Triangle is roughly defined by Hamilton, Bermuda; San Juan, Puerto Rico; and Fort Lauderdale, Florida. The distances from Hamilton to Fort Lauderdale, Fort Lauderdale to San Juan, and San Juan to Hamilton are approximately 1028, 1046, and 965 miles, respectively. Ignoring the curvature of Earth, approximate the area of the Bermuda Triangle.
Source: www.worldatlas.com

45. Geometry Refer to the figure. If $|OA| = 1$, show that:

(a) Area $\Delta OAC = \frac{1}{2}\sin\alpha\cos\alpha$

(b) Area $\Delta OCB = \frac{1}{2}|OB|^2\sin\beta\cos\beta$

(c) Area $\Delta OAB = \frac{1}{2}|OB|\sin(\alpha+\beta)$

(d) $|OB| = \dfrac{\cos\alpha}{\cos\beta}$

(e) $\sin(\alpha+\beta) = \sin\alpha\cos\beta + \cos\alpha\sin\beta$

[**Hint:** Area ΔOAB = Area ΔOAC + Area ΔOCB]

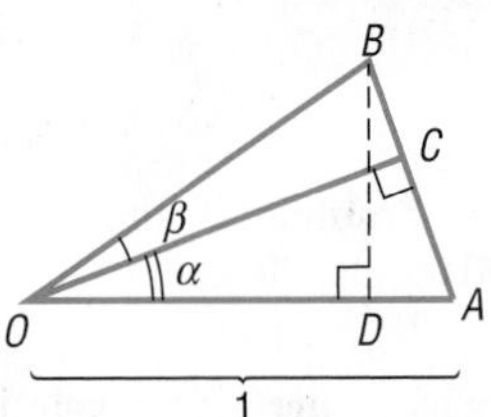

46. Geometry Refer to the figure, in which a unit circle is drawn. The line segment DB is tangent to the circle and θ is acute.
(a) Express the area of ΔOBC in terms of $\sin\theta$ and $\cos\theta$.
(b) Express the area of ΔOBD in terms of $\sin\theta$ and $\cos\theta$.
(c) The area of the sector $\widehat{OBC}$ of the circle is $\frac{1}{2}\theta$, where θ is measured in radians. Use the results of parts (a) and (b) and the fact that

$$\text{Area } \Delta OBC < \text{Area } \widehat{OBC} < \text{Area } \Delta OBD$$

to show that

$$1 < \frac{\theta}{\sin\theta} < \frac{1}{\cos\theta}$$

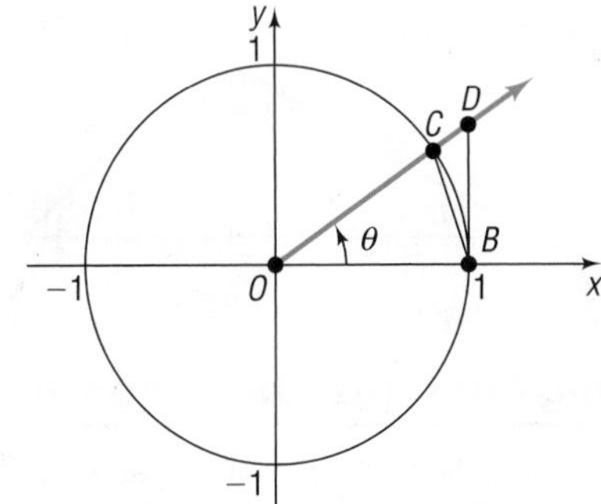

47. **The Cow Problem*** A cow is tethered to one corner of a square barn, 10 feet by 10 feet, with a rope 100 feet long. What is the maximum grazing area for the cow? [See the illustration that follows.]

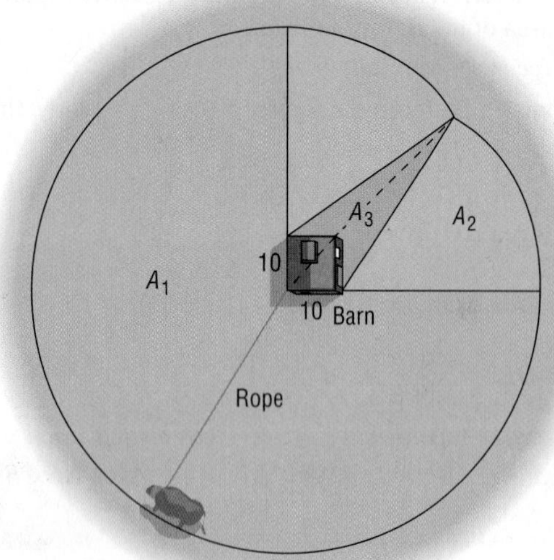

48. **Another Cow Problem** If the barn in Problem 47 is rectangular, 10 feet by 20 feet, what is the maximum grazing area for the cow?

49. **Perfect Triangles** A **perfect triangle** is one having integers for sides for which the area is numerically equal to the perimeter. Show that the triangles with the given side lengths are perfect.
(a) 9, 10, 17 (b) 6, 25, 29
Source: M.V. Bonsangue, G. E. Gannon, E. Buchman, and N. Gross, "In Search of Perfect Triangles," *Mathematics Teacher,* Vol. 92, No. 1, 1999: 56–61

50. If h_1, h_2, and h_3 are the altitudes dropped from P, Q, and R, respectively, in a triangle (see the figure), show that

$$\frac{1}{h_1}+\frac{1}{h_2}+\frac{1}{h_3}=\frac{s}{K}$$

where K is the area of the triangle and $s=\frac{1}{2}(a+b+c)$. [**Hint:** $h_1=\frac{2K}{a}$.]

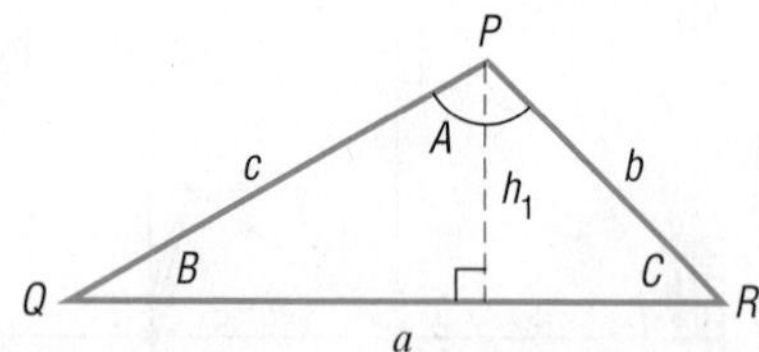

51. Show that a formula for the altitude h from a vertex to the opposite side a of a triangle is

$$h=\frac{a\sin B\sin C}{\sin A}$$

52. A triangle has vertices $A=(0,0)$, $B=(1,0)$, and C, where C is the point on the unit circle corresponding to an angle of 105° when it is drawn in standard position. Find the exact area of the triangle. State the answer in simplified form with a rationalized denominator.

Inscribed Circle *For Problems 53–56, the lines that bisect each angle of a triangle meet in a single point O, and the perpendicular distance r from O to each side of the triangle is the same. The circle with center at O and radius r is called the inscribed circle of the triangle (see the figure).*

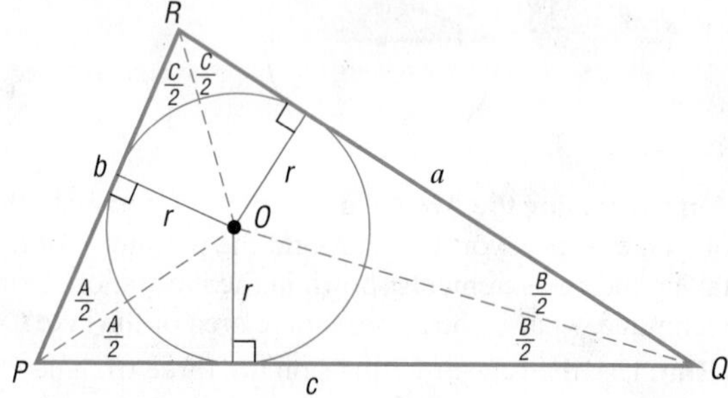

53. Apply the formula from Problem 51 to triangle OPQ to show that

$$r=\frac{c\sin\frac{A}{2}\sin\frac{B}{2}}{\cos\frac{C}{2}}$$

54. Use the result of Problem 53 and the results of Problems 56 and 57 in Section 4.3 to show that

$$\cot\frac{C}{2}=\frac{s-c}{r}$$

where $s=\frac{1}{2}(a+b+c)$.

55. Show that

$$\cot\frac{A}{2}+\cot\frac{B}{2}+\cot\frac{C}{2}=\frac{s}{r}$$

56. Show that the area K of triangle PQR is $K=rs$, where $s=\frac{1}{2}(a+b+c)$. Then show that

$$r=\sqrt{\frac{(s-a)(s-b)(s-c)}{s}}$$

Explaining Concepts: Discussion and Writing

57. What do you do first if you are asked to find the area of a triangle and are given two sides and the included angle?

58. What do you do first if you are asked to find the area of a triangle and are given three sides?

59. State the area of an SAS triangle in words.

* Suggested by Professor Teddy Koukounas of Suffolk Community College, who learned of it from an old farmer in Virginia. Solution provided by Professor Kathleen Miranda of SUNY at Old Westbury.

Retain Your Knowledge

Problems 60–63 are based on material learned earlier in the course. The purpose of these problems is to keep the material fresh in your mind so that you are better prepared for the final exam.

60. Determine the phase shift of $y = -\frac{1}{2}\sin\left(\frac{5}{2}x - \frac{\pi}{4}\right) + 10$.

61. Find the domain of $f(x) = \frac{x+1}{x^2 - 9}$.

62. $P = \left(-\frac{\sqrt{7}}{3}, \frac{\sqrt{2}}{3}\right)$ is the point on the unit circle that corresponds to a real number t. Find the exact values of the six trigonometric functions of t.

63. Establish the identity: $\csc\theta - \sin\theta = \cos\theta\cot\theta$

'Are You Prepared?' Answer

1. $K = \frac{1}{2}bh$

4.5 Simple Harmonic Motion; Damped Motion; Combining Waves

PREPARING FOR THIS SECTION *Before getting started, review the following:*

- Sinusoidal Graphs (Section 2.4, pp. 149–155)

Now Work the 'Are You Prepared?' problem on page 301.

OBJECTIVES
1 Build a Model for an Object in Simple Harmonic Motion (p. 295)
2 Analyze Simple Harmonic Motion (p. 297)
3 Analyze an Object in Damped Motion (p. 298)
4 Graph the Sum of Two Functions (p. 299)

1 Build a Model for an Object in Simple Harmonic Motion

Many physical phenomena can be described as simple harmonic motion. Radio and television waves, light waves, sound waves, and water waves exhibit motion that is simple harmonic.

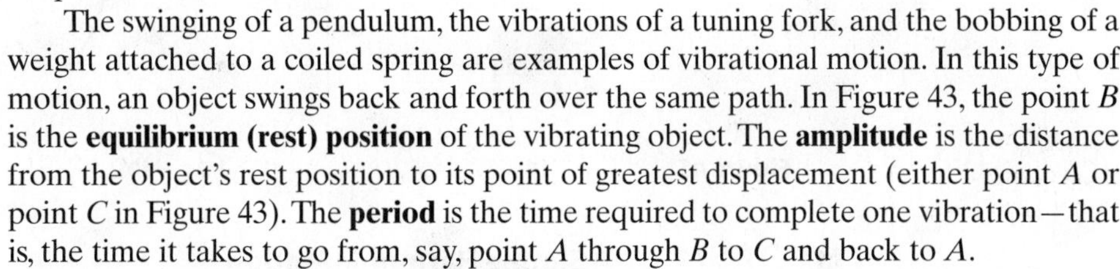

The swinging of a pendulum, the vibrations of a tuning fork, and the bobbing of a weight attached to a coiled spring are examples of vibrational motion. In this type of motion, an object swings back and forth over the same path. In Figure 43, the point B is the **equilibrium (rest) position** of the vibrating object. The **amplitude** is the distance from the object's rest position to its point of greatest displacement (either point A or point C in Figure 43). The **period** is the time required to complete one vibration—that is, the time it takes to go from, say, point A through B to C and back to A.

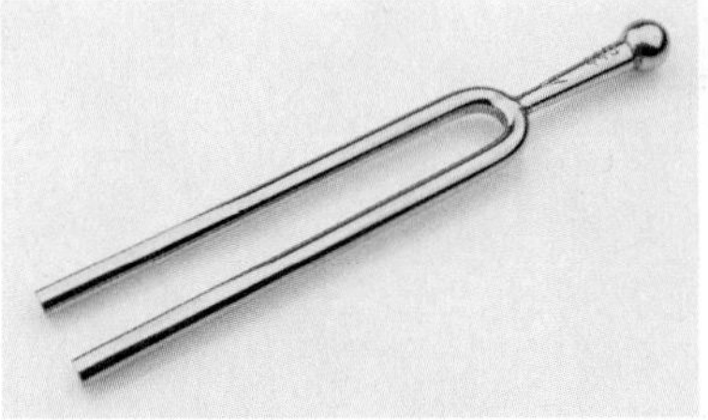

Tuning fork

Simple harmonic motion is a special kind of vibrational motion in which the acceleration a of the object is directly proportional to the negative of its displacement d from its rest position. That is, $a = -kd$, $k > 0$.

A Compressed
B Rest
C Stretched
Amplitude
Amplitude

Figure 43 Coiled spring

For example, when the mass hanging from the spring in Figure 43 is pulled down from its rest position B to the point C, the force of the spring tries to restore the mass to its rest position. Assuming that there is no frictional force* to retard the motion, the amplitude will remain constant. The force increases in direct proportion to the distance that the mass is pulled from its rest position. Since the force increases directly, the acceleration of the mass of the object must do likewise, because (by Newton's Second Law of Motion) force is directly proportional to acceleration. As a result, the acceleration of the object varies directly with its displacement, and the motion is an example of simple harmonic motion.

*If friction is present, the amplitude will decrease with time to 0. This type of motion is an example of **damped motion**, which is discussed later in this section.

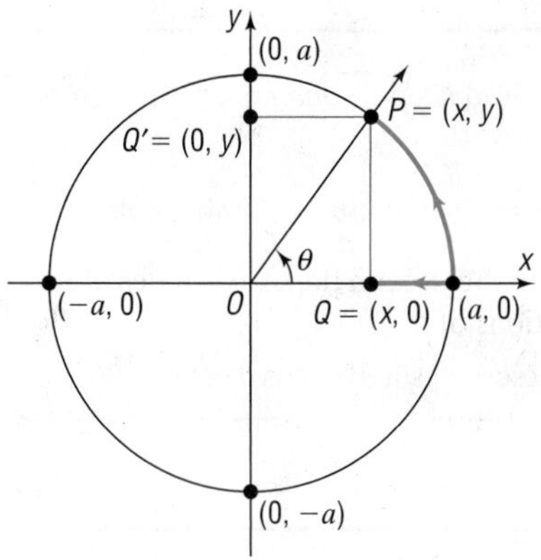

Figure 44

Simple harmonic motion is related to circular motion. To see this relationship, consider a circle of radius a, with center at $(0, 0)$. See Figure 44. Suppose that an object initially placed at $(a, 0)$ moves counterclockwise around the circle at a constant angular speed ω. Suppose further that after time t has elapsed the object is at the point $P = (x, y)$ on the circle. The angle θ, in radians, swept out by the ray $\overrightarrow{OP}$ in this time t is

$$\theta = \omega t \qquad \omega = \frac{\theta}{t}$$

The coordinates of the point P at time t are

$$x = a\cos\theta = a\cos(\omega t)$$
$$y = a\sin\theta = a\sin(\omega t)$$

Corresponding to each position $P = (x, y)$ of the object moving about the circle, there is the point $Q = (x, 0)$, called the **projection of P on the x-axis**. As P moves around the circle at a constant rate, the point Q moves back and forth between the points $(a, 0)$ and $(-a, 0)$ along the x-axis with a motion that is simple harmonic. Similarly, for each point P there is a point $Q' = (0, y)$, called the **projection of P on the y-axis**. As P moves around the circle, the point Q' moves back and forth between the points $(0, a)$ and $(0, -a)$ on the y-axis with a motion that is simple harmonic. Simple harmonic motion can be described as the projection of constant circular motion on a coordinate axis.

To put it another way, again consider a mass hanging from a spring where the mass is pulled down from its rest position to the point C and then released. See Figure 45(a). The graph shown in Figure 45(b) describes the displacement d of the object from its rest position as a function of time t, assuming that no frictional force is present.

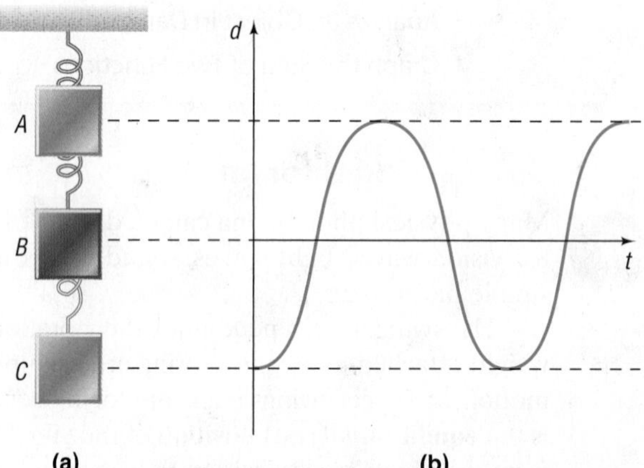

Figure 45 (a) (b)

THEOREM

Simple Harmonic Motion

An object that moves on a coordinate axis so that the displacement d from its rest position at time t is given by either

$$d = a\cos(\omega t) \quad \text{or} \quad d = a\sin(\omega t)$$

where a and $\omega > 0$ are constants, moves with simple harmonic motion. The motion has amplitude $|a|$ and period $\dfrac{2\pi}{\omega}$.

The **frequency** f of an object in simple harmonic motion is the number of oscillations per unit time. Since the period is the time required for one oscillation, it follows that the frequency is the reciprocal of the period; that is,

$$f = \frac{\omega}{2\pi} \qquad \omega > 0$$

EXAMPLE 1

Build a Model for an Object in Harmonic Motion

Suppose that an object attached to a coiled spring is pulled down a distance of 5 inches from its rest position and then released. If the time for one oscillation is 3 seconds, develop a model that relates the displacement d of the object from its rest position after time t (in seconds). Assume no friction.

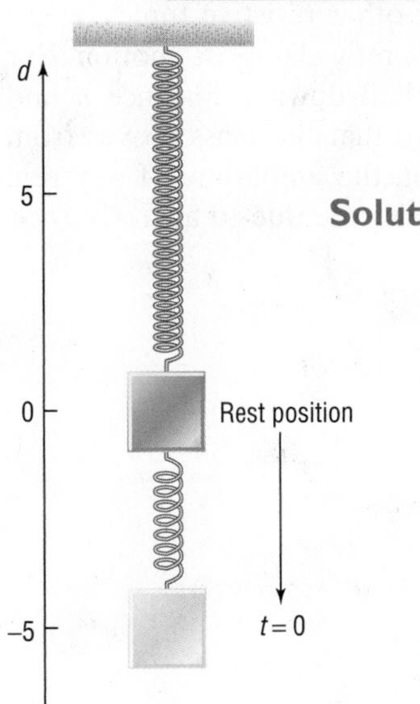

Figure 46

NOTE In the solution to Example 1, $a = -5$ because the object is initially pulled down. (If the initial direction is up, then use $a = 5$.)

Solution The motion of the object is simple harmonic. See Figure 46. When the object is released $(t = 0)$, the displacement of the object from the rest position is -5 units (since the object was pulled down). Because $d = -5$ when $t = 0$, it is easier to use the cosine function*

$$d = a\cos(\omega t)$$

to describe the motion. Now the amplitude is $|-5| = 5$ and the period is 3, so

$$a = -5 \quad \text{and} \quad \frac{2\pi}{\omega} = \text{period} = 3, \quad \text{so } \omega = \frac{2\pi}{3}$$

An equation that models the motion of the object is

$$d = -5\cos\left[\frac{2\pi}{3}t\right]$$

Now Work PROBLEM 5

2 Analyze Simple Harmonic Motion

EXAMPLE 2

Analyzing the Motion of an Object

Suppose that the displacement d (in meters) of an object at time t (in seconds) satisfies the equation

$$d = 10\sin(5t)$$

(a) Describe the motion of the object.
(b) What is the maximum displacement from its resting position?
(c) What is the time required for one oscillation?
(d) What is the frequency?

Solution Observe that the given equation is of the form

$$d = a\sin(\omega t) \qquad d = 10\sin(5t)$$

where $a = 10$ and $\omega = 5$.

(a) The motion is simple harmonic.
(b) The maximum displacement of the object from its resting position is the amplitude: $|a| = 10$ meters.
(c) The time required for one oscillation is the period:

$$\text{Period} = \frac{2\pi}{\omega} = \frac{2\pi}{5} \text{ seconds}$$

(d) The frequency is the reciprocal of the period.

$$\text{Frequency} = f = \frac{5}{2\pi} \text{ oscillation per second}$$

Now Work PROBLEM 13

* No phase shift is required when a cosine function is used.

3 Analyze an Object in Damped Motion*

Most physical phenomena are affected by friction or other resistive forces. These forces remove energy from a moving system and thereby damp its motion. For example, when a mass hanging from a spring is pulled down a distance a and released, the friction in the spring causes the distance that the mass moves from its at-rest position to decrease over time. As a result, the amplitude of any real oscillating spring or swinging pendulum decreases with time due to air resistance, friction, and so forth. See Figure 47.

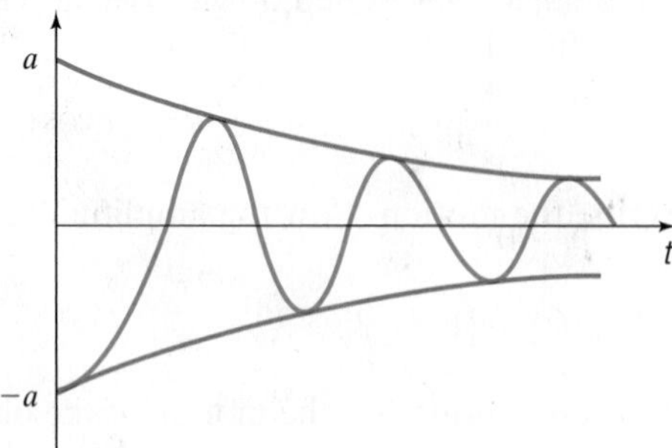

Figure 47

A model that describes this phenomenon maintains a sinusoidal component, but the amplitude of this component will decrease with time to account for the damping effect. In addition, the period of the oscillating component will be affected by the damping. The next result, from physics, describes damped motion.

THEOREM

Damped Motion

The displacement d of an oscillating object from its at-rest position at time t is given by

$$d(t) = ae^{-bt/(2m)} \cos\left(\sqrt{\omega^2 - \frac{b^2}{4m^2}}\, t\right)$$

where b is the **damping factor** or **damping coefficient** and m is the mass of the oscillating object. Here $|a|$ is the displacement at $t = 0$, and $\frac{2\pi}{\omega}$ is the period under simple harmonic motion (no damping).

Notice that for $b = 0$ (zero damping), we have the formula for simple harmonic motion with amplitude $|a|$ and period $\frac{2\pi}{\omega}$.

EXAMPLE 3 **Analyzing a Damped Vibration Curve**

Analyze the damped vibration curve

$$d(t) = e^{-t/\pi} \cos t \quad t \geq 0$$

Solution The displacement d is the product of $y = e^{-t/\pi}$ and $y = \cos t$. Using properties of absolute value and the fact that $|\cos t| \leq 1$, it follows that

$$|d(t)| = |e^{-t/\pi} \cos t| = |e^{-t/\pi}||\cos t| \leq |e^{-t/\pi}| \underset{\uparrow \atop e^{-t/\pi} > 0}{=} e^{-t/\pi}$$

As a result,

$$-e^{-t/\pi} \leq d(t) \leq e^{-t/\pi}$$

This means that the graph of d will lie between the graphs of $y = e^{-t/\pi}$ and $y = -e^{-t/\pi}$, called the **bounding curves** of d.

*Requires Section 7.1, Exponential Functions

Also, the graph of d will touch these graphs when $|\cos t| = 1$; that is, when $t = 0, \pi, 2\pi$, and so on. The x-intercepts of the graph of d occur when $\cos t = 0$, that is, at $\frac{\pi}{2}, \frac{3\pi}{2}, \frac{5\pi}{2}$, and so on. See Table 1.

Table 1

t	0	$\frac{\pi}{2}$	π	$\frac{3\pi}{2}$	2π
$e^{-t/\pi}$	1	$e^{-1/2}$	e^{-1}	$e^{-3/2}$	e^{-2}
$\cos t$	1	0	-1	0	1
$d(t) = e^{-t/\pi} \cos t$	1	0	$-e^{-1}$	0	e^{-2}
Point on graph of d	$(0, 1)$	$\left(\frac{\pi}{2}, 0\right)$	$(\pi, -e^{-1})$	$\left(\frac{3\pi}{2}, 0\right)$	$(2\pi, e^{-2})$

The graphs of $y = \cos t$, $y = e^{-t/\pi}$, $y = -e^{-t/\pi}$, and $d(t) = e^{-t/\pi} \cos t$ are shown in in Figure 48.

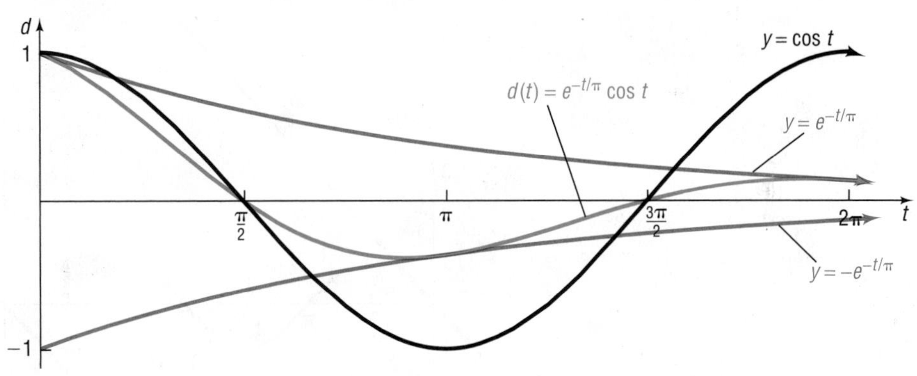

Figure 48

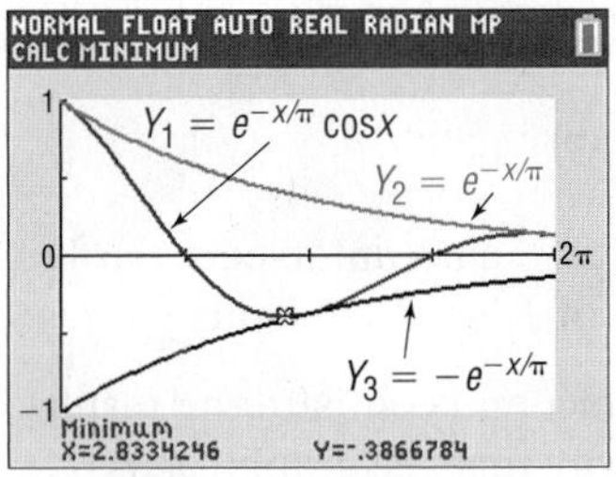

NORMAL FLOAT AUTO REAL RADIAN MP
CALC MINIMUM
$Y_1 = e^{-x/\pi} \cos x$
$Y_2 = e^{-x/\pi}$
$Y_3 = -e^{-x/\pi}$
Minimum
X=2.8334246 Y=-.3866784

Figure 49

Exploration

Graph $Y_1 = e^{-x/\pi} \cos x$, along with $Y_2 = e^{-x/\pi}$ and $Y_3 = -e^{-x/\pi}$, for $0 \le x \le 2\pi$. Determine where Y_1 has its first turning point (local minimum). Compare this to where Y_1 intersects Y_3.

Result Figure 49 shows the graphs of $Y_1 = e^{-x/\pi} \cos x$, $Y_2 = e^{-x/\pi}$, and $Y_3 = -e^{-x/\pi}$. Using MINIMUM, the first turning point occurs at $x \approx 2.83$; Y_1 INTERSECTS Y_3 at $x = \pi \approx 3.14$.

Now Work PROBLEM 21

4 Graph the Sum of Two Functions

Many physical and biological applications require the graph of the sum of two functions, such as

$$f(x) = x + \sin x \quad \text{or} \quad g(x) = \sin x + \cos(2x)$$

For example, if two tones are emitted, the sound produced is the sum of the waves produced by the two tones. See Problem 57 for an explanation of Touch-Tone phones.

To graph the sum of two (or more) functions, use the method of adding y-coordinates described next.

EXAMPLE 4 **Graphing the Sum of Two Functions**

Use the method of adding y-coordinates to graph $f(x) = x + \sin x$.

Solution First, graph the component functions,

$$y = f_1(x) = x \qquad y = f_2(x) = \sin x$$

on the same coordinate system. See Figure 50(a). Now, select several values of x say $x = 0$, $x = \frac{\pi}{2}$, $x = \pi$, $x = \frac{3\pi}{2}$, and $x = 2\pi$, and use them to compute $f(x) = f_1(x) + f_2(x)$. Table 2 shows the computations. Plot these points and connect them to get the graph, as shown in Figure 50(b).

Table 2

x	0	$\frac{\pi}{2}$	π	$\frac{3\pi}{2}$	2π
$y = f_1(x) = x$	0	$\frac{\pi}{2}$	π	$\frac{3\pi}{2}$	2π
$y = f_2(x) = \sin x$	0	1	0	-1	0
$f(x) = x + \sin x$	0	$\frac{\pi}{2} + 1 \approx 2.57$	π	$\frac{3\pi}{2} - 1 \approx 3.71$	2π
Point on graph of f	$(0, 0)$	$\left(\frac{\pi}{2}, 2.57\right)$	(π, π)	$\left(\frac{3\pi}{2}, 3.71\right)$	$(2\pi, 2\pi)$

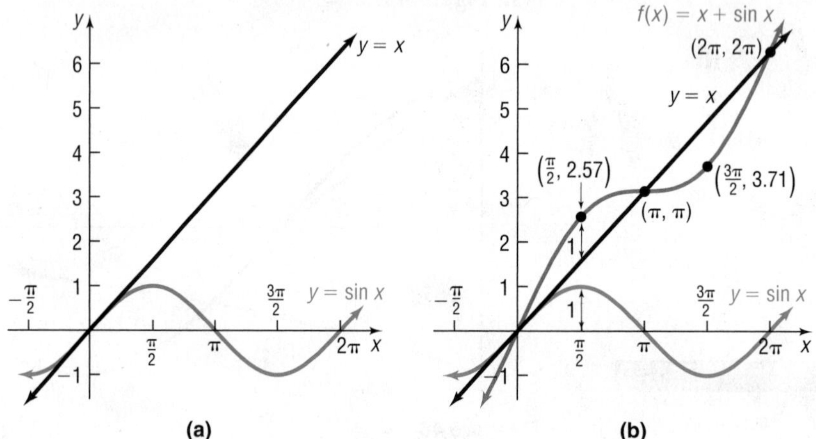

Figure 50

In Figure 50(b), notice that the graph of $f(x) = x + \sin x$ intersects the line $y = x$ whenever $\sin x = 0$. Also, notice that the graph of f is not periodic. ●

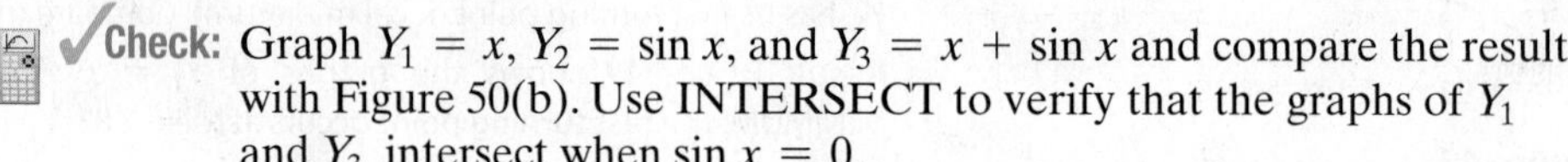

✓**Check:** Graph $Y_1 = x$, $Y_2 = \sin x$, and $Y_3 = x + \sin x$ and compare the result with Figure 50(b). Use INTERSECT to verify that the graphs of Y_1 and Y_3 intersect when $\sin x = 0$.

The next example shows a periodic graph.

EXAMPLE 5 Graphing the Sum of Two Sinusoidal Functions

Use the method of adding y-coordinates to graph

$$f(x) = \sin x + \cos(2x)$$

Solution Table 3 shows the steps for computing several points on the graph of f. Figure 51 on page 301 illustrates the graphs of the component functions, $y = f_1(x) = \sin x$ and $y = f_2(x) = \cos(2x)$, and the graph of $f(x) = \sin x + \cos(2x)$, which is shown in red.

Table 3

x	$-\frac{\pi}{2}$	0	$\frac{\pi}{2}$	π	$\frac{3\pi}{2}$	2π
$y = f_1(x) = \sin x$	-1	0	1	0	-1	0
$y = f_2(x) = \cos(2x)$	-1	1	-1	1	-1	1
$f(x) = \sin x + \cos(2x)$	-2	1	0	1	-2	1
Point on graph of f	$\left(-\frac{\pi}{2}, -2\right)$	$(0, 1)$	$\left(\frac{\pi}{2}, 0\right)$	$(\pi, 1)$	$\left(\frac{3\pi}{2}, -2\right)$	$(2\pi, 1)$

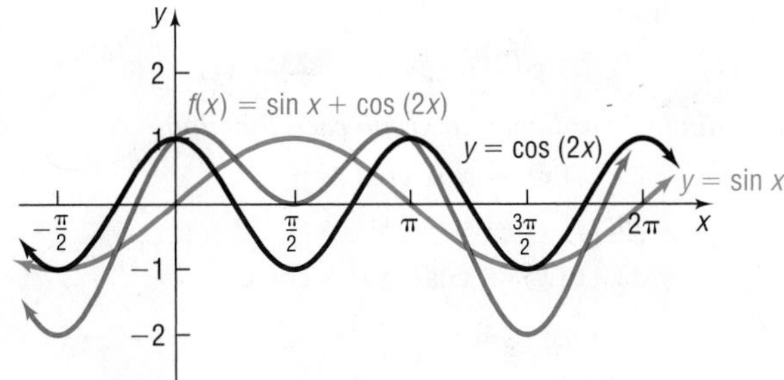

Figure 51

Notice that f is periodic, with period 2π. ●

✓**Check:** Graph $Y_1 = \sin x$, $Y_2 = \cos(2x)$, and $Y_3 = \sin x + \cos(2x)$ and compare the result with Figure 51.

Now Work PROBLEM 25

4.5 Assess Your Understanding

'**Are You Prepared?**' *The answer is given at the end of these exercises. If you get a wrong answer, read the pages listed in red.*

1. The amplitude A and period T of $f(x) = 5\sin(4x)$ are ___ and _____. (pp. 149–155)

Concepts and Vocabulary

2. The motion of an object obeys the equation $d = 4\cos(6t)$. Such motion is described as ______ _________. The number 4 is called the __________.

3. When a mass hanging from a spring is pulled down and then released, the motion is called ______ _________ if there is no frictional force to retard the motion, and the motion is called __________ if there is such friction.

4. ***True or False*** If the distance d of an object from its rest position at time t is given by a sinusoidal graph, the motion of the object is simple harmonic motion.

Skill Building

In Problems 5–8, an object attached to a coiled spring is pulled down a distance a from its rest position and then released. Assuming that the motion is simple harmonic with period T, write an equation that relates the displacement d of the object from its rest position after t seconds. Also assume that the positive direction of the motion is up.

5. $a = 5$; $T = 2$ seconds

6. $a = 10$; $T = 3$ seconds

7. $a = 6$; $T = \pi$ seconds

8. $a = 4$; $T = \frac{\pi}{2}$ seconds

9. Rework Problem 5 under the same conditions, except that at time $t = 0$, the object is at its resting position and moving down.

10. Rework Problem 6 under the same conditions, except that at time $t = 0$, the object is at its resting position and moving down.

11. Rework Problem 7 under the same conditions, except that at time $t = 0$, the object is at its resting position and moving down.

12. Rework Problem 8 under the same conditions, except that at time $t = 0$, the object is at its resting position and moving down.

In Problems 13–20, the displacement d (in meters) of an object at time t (in seconds) is given.
(a) Describe the motion of the object.
(b) What is the maximum displacement from its resting position?
(c) What is the time required for one oscillation?
(d) What is the frequency?

13. $d = 5\sin(3t)$

14. $d = 4\sin(2t)$

15. $d = 6\cos(\pi t)$

16. $d = 5\cos\left(\frac{\pi}{2}t\right)$

17. $d = -3\sin\left(\frac{1}{2}t\right)$

18. $d = -2\cos(2t)$

19. $d = 6 + 2\cos(2\pi t)$

20. $d = 4 + 3\sin(\pi t)$

In Problems 21–24, graph each damped vibration curve for $0 \le t \le 2\pi$.

21. $d(t) = e^{-t/\pi}\cos(2t)$
22. $d(t) = e^{-t/2\pi}\cos(2t)$
23. $d(t) = e^{-t/2\pi}\cos t$
24. $d(t) = e^{-t/4\pi}\cos t$

In Problems 25–32, use the method of adding y-coordinates to graph each function.

25. $f(x) = x + \cos x$
26. $f(x) = x + \cos(2x)$
27. $f(x) = x - \sin x$
28. $f(x) = x - \cos x$
29. $f(x) = \sin x + \cos x$
30. $f(x) = \sin(2x) + \cos x$
31. $g(x) = \sin x + \sin(2x)$
32. $g(x) = \cos(2x) + \cos x$

Mixed Practice

In Problems 33–38, (a) use the Product-to-Sum Formulas to express each product as a sum, and (b) use the method of adding y-coordinates to graph each function on the interval $[0, 2\pi]$.

33. $f(x) = \sin(2x)\sin x$
34. $F(x) = \sin(3x)\sin x$
35. $G(x) = \cos(4x)\cos(2x)$
36. $h(x) = \cos(2x)\cos x$
37. $H(x) = 2\sin(3x)\cos x$
38. $g(x) = 2\sin x\cos(3x)$

Applications and Extensions

In Problems 39–44, an object of mass m (in grams) attached to a coiled spring with damping factor b (in grams per second) is pulled down a distance a (in centimeters) from its rest position and then released. Assume that the positive direction of the motion is up and the period is T (in seconds) under simple harmonic motion.

(a) Write an equation that relates the displacement d of the object from its rest position after t seconds.

(b) Graph the equation found in part (a) for 5 oscillations using a graphing utility.

39. $m = 25, \quad a = 10, \quad b = 0.7, \quad T = 5$
40. $m = 20, \quad a = 15, \quad b = 0.75, \quad T = 6$
41. $m = 30, \quad a = 18, \quad b = 0.6, \quad T = 4$
42. $m = 15, \quad a = 16, \quad b = 0.65, \quad T = 5$
43. $m = 10, \quad a = 5, \quad b = 0.8, \quad T = 3$
44. $m = 10, \quad a = 5, \quad b = 0.7, \quad T = 3$

In Problems 45–50, the distance d (in meters) of the bob of a pendulum of mass m (in kilograms) from its rest position at time t (in seconds) is given. The bob is released from the left of its rest position and represents a negative direction.

(a) Describe the motion of the object. Be sure to give the mass and damping factor.

(b) What is the initial displacement of the bob? That is, what is the displacement at $t = 0$?

(c) Graph the motion using a graphing utility.

(d) What is the displacement of the bob at the start of the second oscillation?

(e) What happens to the displacement of the bob as time increases without bound?

45. $d = -20e^{-0.7t/40}\cos\left(\sqrt{\left(\frac{2\pi}{5}\right)^2 - \frac{0.49}{1600}}\,t\right)$
46. $d = -20e^{-0.8t/40}\cos\left(\sqrt{\left(\frac{2\pi}{5}\right)^2 - \frac{0.64}{1600}}\,t\right)$
47. $d = -30e^{-0.6t/80}\cos\left(\sqrt{\left(\frac{2\pi}{7}\right)^2 - \frac{0.36}{6400}}\,t\right)$
48. $d = -30e^{-0.5t/70}\cos\left(\sqrt{\left(\frac{\pi}{2}\right)^2 - \frac{0.25}{4900}}\,t\right)$
49. $d = -15e^{-0.9t/30}\cos\left(\sqrt{\left(\frac{\pi}{3}\right)^2 - \frac{0.81}{900}}\,t\right)$
50. $d = -10e^{-0.8t/50}\cos\left(\sqrt{\left(\frac{2\pi}{3}\right)^2 - \frac{0.64}{2500}}\,t\right)$

51. **Loudspeaker** A loudspeaker diaphragm is oscillating in simple harmonic motion described by the equation $d = a\cos(\omega t)$ with a frequency of 520 hertz (cycles per second) and a maximum displacement of 0.80 millimeter. Find ω and then determine the equation that describes the movement of the diaphragm.

52. **Colossus** Added to Six Flags St. Louis in 1986, the Colossus is a giant Ferris wheel. Its diameter is 165 feet, it rotates at a rate of about 1.6 revolutions per minute, and the bottom of the wheel is 15 feet above the ground. Determine an equation that relates a rider's height above the ground at time *t*. Assume the passenger begins the ride at the bottom of the wheel.

    Source: Six Flags Theme Parks, Inc.

53. **Tuning Fork** The end of a tuning fork moves in simple harmonic motion described by the equation $d = a\sin(\omega t)$. If a tuning fork for the note A above middle C on an even-tempered scale (A_4, the tone by which an orchestra tunes itself) has a frequency of 440 hertz (cycles per second), find ω. If the maximum displacement of the end of the tuning fork is 0.01 millimeter, determine the equation that describes the movement of the tuning fork.

    Source: David Lapp. *Physics of Music and Musical Instruments.* Medford, MA: Tufts University, 2003

54. **Tuning Fork** The end of a tuning fork moves in simple harmonic motion described by the equation $d = a\sin(\omega t)$. If a tuning fork for the note E above middle C on an even-tempered scale (E_4) has a frequency of approximately 329.63 hertz (cycles per second), find ω. If the maximum displacement of the end of the tuning fork is 0.025 millimeter, determine the equation that describes the movement of the tuning fork.

    Source: David Lapp. *Physics of Music and Musical Instruments.* Medford, MA: Tufts University, 2003

55. Charging a Capacitor See the illustration. If a charged capacitor is connected to a coil by closing a switch, energy is transferred to the coil and then back to the capacitor in an oscillatory motion. The voltage V (in volts) across the capacitor will gradually diminish to 0 with time t (in seconds).

(a) Graph the function relating V and t:

$$V(t) = e^{-t/3}\cos(\pi t) \qquad 0 \le t \le 3$$

(b) At what times t will the graph of V touch the graph of $y = e^{-t/3}$? When does the graph of V touch the graph of $y = -e^{-t/3}$?

(c) When will the voltage V be between -0.4 and 0.4 volt?

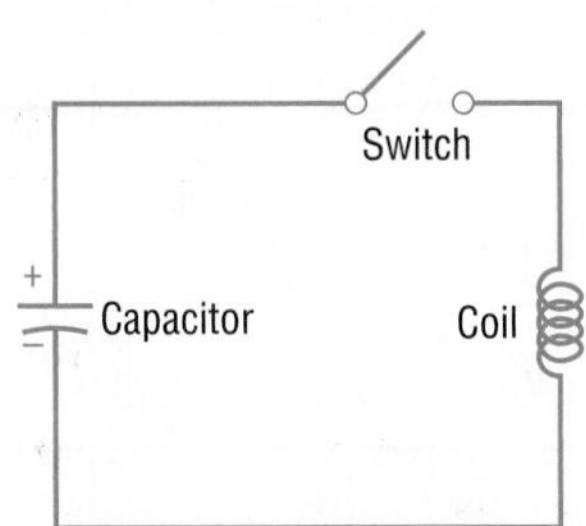

56. The Sawtooth Curve An oscilloscope often displays a *sawtooth curve*. This curve can be approximated by sinusoidal curves of varying periods and amplitudes.

(a) Use a graphing utility to graph the following function, which can be used to approximate the sawtooth curve.

$$f(x) = \frac{1}{2}\sin(2\pi x) + \frac{1}{4}\sin(4\pi x) \qquad 0 \le x \le 4$$

(b) A better approximation to the sawtooth curve is given by

$$f(x) = \frac{1}{2}\sin(2\pi x) + \frac{1}{4}\sin(4\pi x) + \frac{1}{8}\sin(8\pi x)$$

Use a graphing utility to graph this function for $0 \le x \le 4$ and compare the result to the graph obtained in part (a).

(c) A third and even better approximation to the sawtooth curve is given by

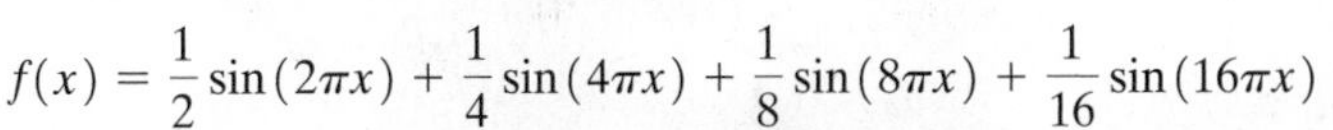

$$f(x) = \frac{1}{2}\sin(2\pi x) + \frac{1}{4}\sin(4\pi x) + \frac{1}{8}\sin(8\pi x) + \frac{1}{16}\sin(16\pi x)$$

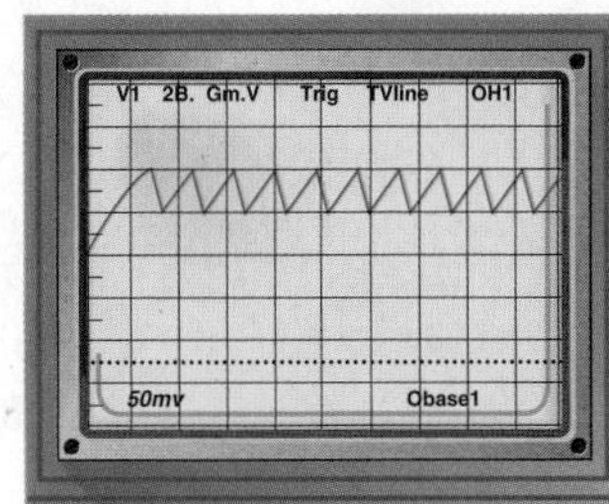

Use a graphing utility to graph this function for $0 \le x \le 4$ and compare the result to the graphs obtained in parts (a) and (b).

(d) What do you think the next approximation to the sawtooth curve is?

57. Touch-Tone Phones On a Touch-Tone phone, each button produces a unique sound. The sound produced is the sum of two tones, given by

$$y = \sin(2\pi l t) \quad \text{and} \quad y = \sin(2\pi h t)$$

where l and h are the low and high frequencies (cycles per second) shown in the illustration. For example, if you touch 7, the low frequency is $l = 852$ cycles per second and the high frequency is $h = 1209$ cycles per second. The sound emitted by touching 7 is

$$y = \sin[2\pi(852)t] + \sin[2\pi(1209)t]$$

Use a graphing utility to graph the sound emitted by touching 7.

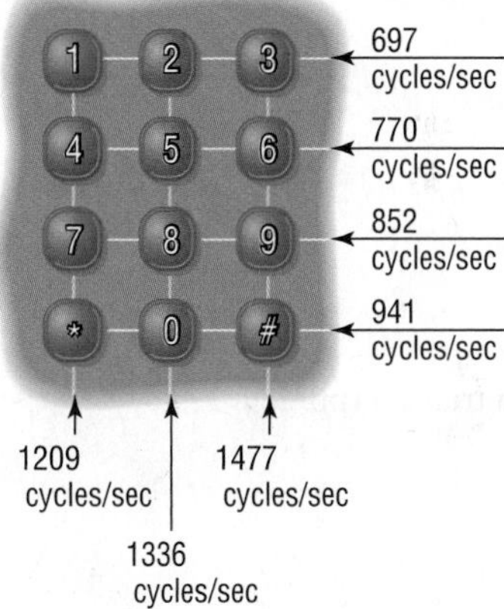

58. Use a graphing utility to graph the sound emitted by the * key on a Touch-Tone phone. See Problem 57.

59. CBL Experiment Pendulum motion is analyzed to estimate simple harmonic motion. A plot is generated with the position of the pendulum over time. The graph is used to find a sinusoidal curve of the form $y = A\cos[B(x - C)] + D$. Determine the amplitude, period, and frequency. (Activity 16, Real-World Math with the CBL System.)

60. CBL Experiment The sound from a tuning fork is collected over time. A model of the form $y = A\cos[B(x - C)]$ is fitted to the data. Determine the amplitude, frequency, and period of the graph. (Activity 23, Real-World Math with the CBL System.)

Explaining Concepts: Discussion and Writing

61. Use a graphing utility to graph the function $f(x) = \dfrac{\sin x}{x}$, $x > 0$. Based on the graph, what do you conjecture about the value of $\dfrac{\sin x}{x}$ for x close to 0?

62. Use a graphing utility to graph $y = x\sin x$, $y = x^2\sin x$, and $y = x^3\sin x$ for $x > 0$. What patterns do you observe?

63. Use a graphing utility to graph $y = \dfrac{1}{x}\sin x$, $y = \dfrac{1}{x^2}\sin x$, and $y = \dfrac{1}{x^3}\sin x$ for $x > 0$. What patterns do you observe?

64. How would you explain to a friend what simple harmonic motion is? How would you explain damped motion?

Retain Your Knowledge

Problems 65–68 are based on material learned earlier in the course. The purpose of these problems is to keep the material fresh in your mind so that you are better prepared for the final exam.

65. The function $f(x) = \frac{x-3}{x-4}$, $x \neq 4$, is one-to-one. Find its inverse function.

66. Find the exact value of $\csc\left(\cos^{-1}\frac{2}{5}\right)$.

67. Solve the triangle where $a = 20$, $B = 110°$, and $b = 55$.

68. Given $\cos\alpha = \frac{4}{5}$, $0 < \alpha < \frac{\pi}{2}$, find the exact value of:

(a) $\cos\frac{\alpha}{2}$ (b) $\sin\frac{\alpha}{2}$ (c) $\tan\frac{\alpha}{2}$

'Are You Prepared?' Answer

1. $A = 5$; $T = \frac{\pi}{2}$

Chapter Review

Things to Know

Formulas

Law of Sines (p. 272) $\frac{\sin A}{a} = \frac{\sin B}{b} = \frac{\sin C}{c}$

Law of Cosines (p. 282)

$$c^2 = a^2 + b^2 - 2ab\cos C$$
$$b^2 = a^2 + c^2 - 2ac\cos B$$
$$a^2 = b^2 + c^2 - 2bc\cos A$$

Area of a triangle (pp. 289–290)

$$K = \frac{1}{2}bh \quad K = \frac{1}{2}ab\sin C \quad K = \frac{1}{2}bc\sin A \quad K = \frac{1}{2}ac\sin B$$

$$K = \sqrt{s(s-a)(s-b)(s-c)} \quad \text{where} \quad s = \frac{1}{2}(a+b+c)$$

Objectives

Section		You should be able to . . .	Example(s)	Review Exercises
4.1	1	Find the value of trigonometric functions of acute angles using right triangles (p. 259)	1, 2	1, 2, 27
	2	Use the complementary angle theorem (p. 261)	3	3–5
	3	Solve right triangles (p. 261)	4, 5	6, 7, 27
	4	Solve applied problems (p. 262)	6–12	28–31, 36–38
4.2	1	Solve SAA or ASA triangles (p. 273)	1, 2	8, 19
	2	Solve SSA triangles (p. 274)	3–5	9, 10, 12, 16, 18
	3	Solve applied problems (p. 276)	6, 7	32, 33
4.3	1	Solve SAS triangles (p. 283)	1	11, 15, 20
	2	Solve SSS triangles (p. 284)	2	13, 14, 17
	3	Solve applied problems (p. 284)	3	34
4.4	1	Find the area of SAS triangles (p. 289)	1	21, 22, 26, 35
	2	Find the area of SSS triangles (p. 290)	2	23, 24
4.5	1	Build a model for an object in simple harmonic motion (p. 295)	1	39
	2	Analyze simple harmonic motion (p. 297)	2	40, 41
	3	Analyze an object in damped motion (p. 298)	3	42, 43
	4	Graph the sum of two functions (p. 299)	4, 5	44

Review Exercises

In Problems 1 and 2, find the exact value of the six trigonometric functions of the angle θ in each figure.

1.

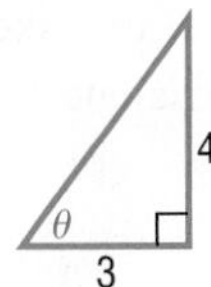

2.

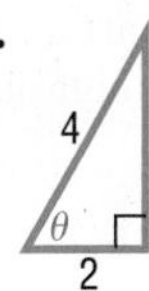

In Problems 3–5, find the exact value of each expression. Do not use a calculator.

3. $\cos 62° - \sin 28°$ **4.** $\dfrac{\sec 55°}{\csc 35°}$ **5.** $\cos^2 40° + \cos^2 50°$

In Problems 6 and 7, solve each triangle.

6.

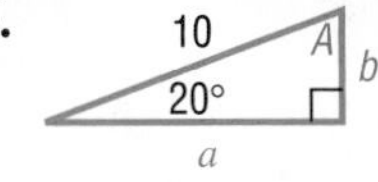

7.

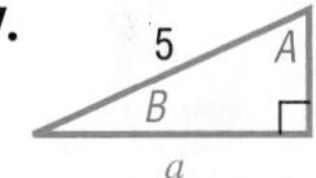

In Problems 8–20, find the remaining angle(s) and side(s) of each triangle, if it (they) exists. If no triangle exists, say "No triangle."

8. $A = 50°,\ B = 30°,\ a = 1$ **9.** $A = 100°,\ a = 5,\ c = 2$ **10.** $a = 3,\ c = 1,\ C = 110°$

11. $a = 3,\ c = 1,\ B = 100°$ **12.** $a = 3,\ b = 5,\ B = 80°$ **13.** $a = 2,\ b = 3,\ c = 1$

14. $a = 10,\ b = 7,\ c = 8$ **15.** $a = 1,\ b = 3,\ C = 40°$ **16.** $a = 5,\ b = 3,\ A = 80°$

17. $a = 1,\ b = \frac{1}{2},\ c = \frac{4}{3}$ **18.** $a = 3,\ A = 10°,\ b = 4$ **19.** $a = 4,\ A = 20°,\ B = 100°$

20. $c = 5,\ b = 4,\ A = 70°$

In Problems 21–25, find the area of each triangle.

21. $a = 2,\ b = 3,\ C = 40°$ **22.** $b = 4,\ c = 10,\ A = 70°$ **23.** $a = 4,\ b = 3,\ c = 5$

24. $a = 4,\ b = 2,\ c = 5$ **25.** $A = 50°,\ B = 30°,\ a = 1$

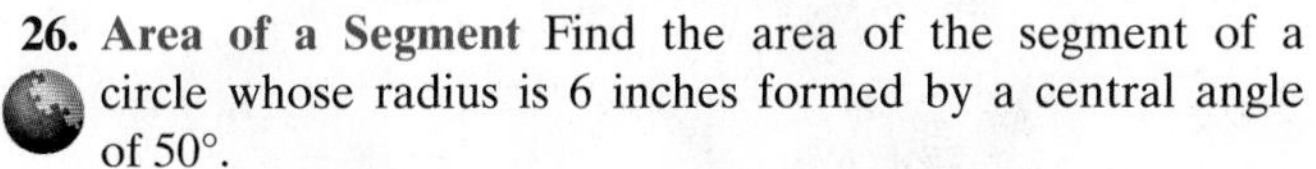

26. Area of a Segment Find the area of the segment of a circle whose radius is 6 inches formed by a central angle of 50°.

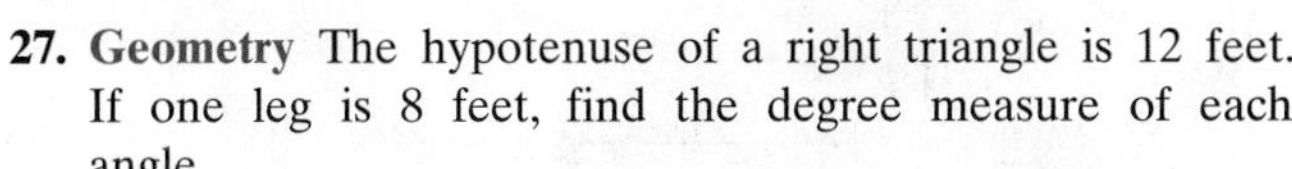

27. Geometry The hypotenuse of a right triangle is 12 feet. If one leg is 8 feet, find the degree measure of each angle.

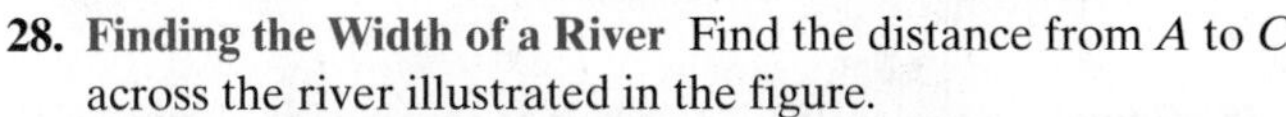

28. Finding the Width of a River Find the distance from A to C across the river illustrated in the figure.

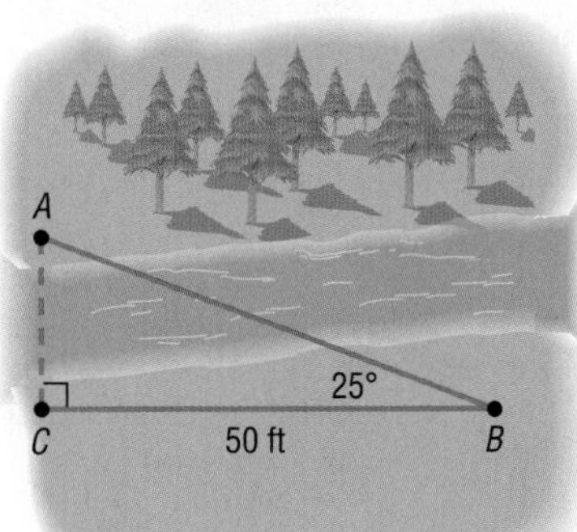

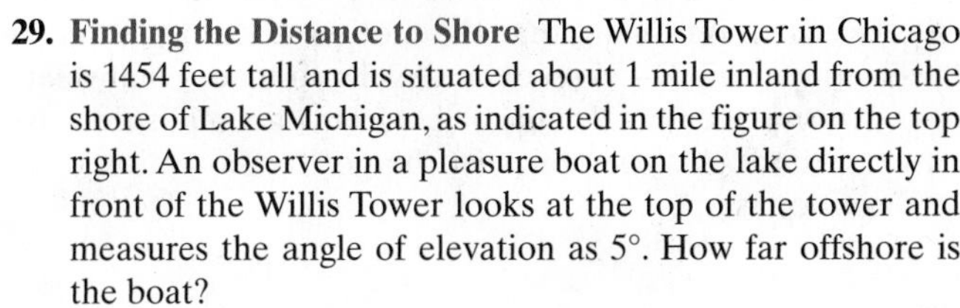

29. Finding the Distance to Shore The Willis Tower in Chicago is 1454 feet tall and is situated about 1 mile inland from the shore of Lake Michigan, as indicated in the figure on the top right. An observer in a pleasure boat on the lake directly in front of the Willis Tower looks at the top of the tower and measures the angle of elevation as 5°. How far offshore is the boat?

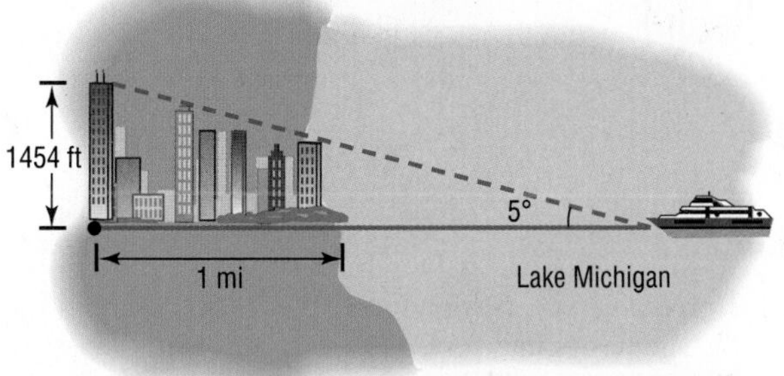

30. Finding the Speed of a Glider From a glider 200 feet above the ground, two sightings of a stationary object directly in front are taken 1 minute apart (see the figure). What is the speed of the glider?

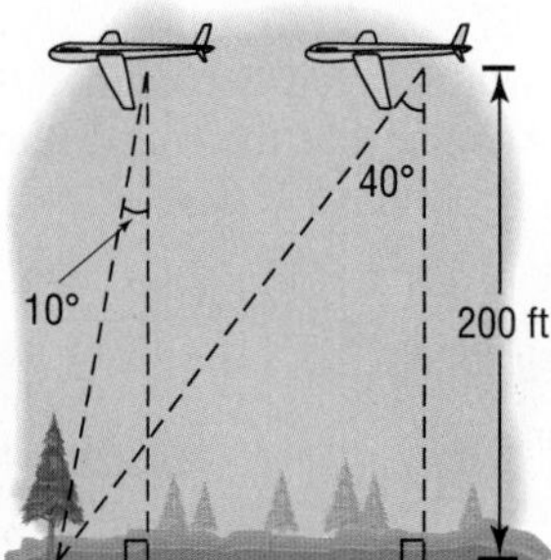

31. Finding the Grade of a Mountain Trail A straight trail with a uniform inclination leads from a hotel, elevation 5000 feet, to a lake in a valley, elevation 4100 feet. The length of the trail is 4100 feet. What is the inclination (grade) of the trail?

32. Finding the Height of a Helicopter Two observers simultaneously measure the angle of elevation of a helicopter. One angle is measured as 25°, the other as 40° (see the figure). If the observers are 100 feet apart and the helicopter lies over the line joining them, how high is the helicopter?

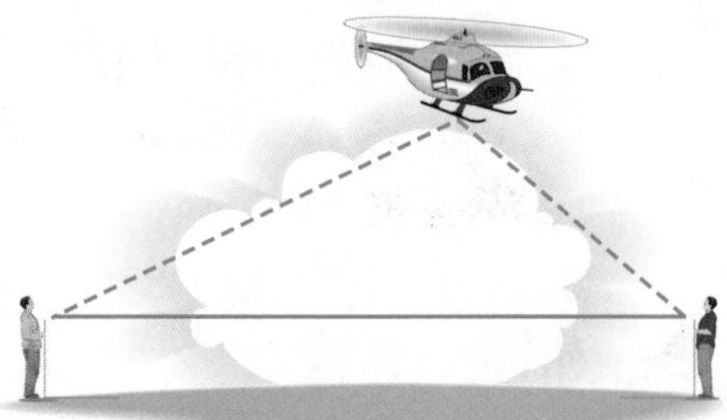

33. Constructing a Highway A highway whose primary directions are north–south is being constructed along the west coast of Florida. Near Naples, a bay obstructs the straight path of the road. Since the cost of a bridge is prohibitive, engineers decide to go around the bay. The illustration shows the path that they decide on and the measurements taken. What is the length of highway needed to go around the bay?

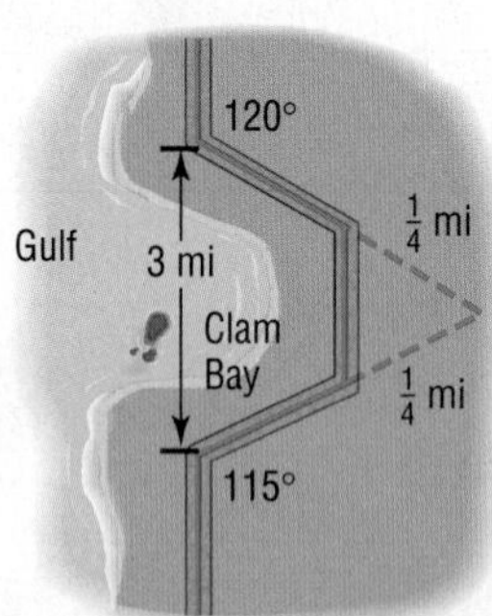

34. Correcting a Navigational Error A sailboat leaves St. Thomas bound for an island in the British West Indies, 200 miles away. Maintaining a constant speed of 18 miles per hour, but encountering heavy crosswinds and strong currents, the crew finds, after 4 hours, that the sailboat is off course by 15°.

(a) How far is the sailboat from the island at this time?

(b) Through what angle should the sailboat turn to correct its course?

(c) How much time has been added to the trip because of this? (Assume that the speed remains at 18 miles per hour.)

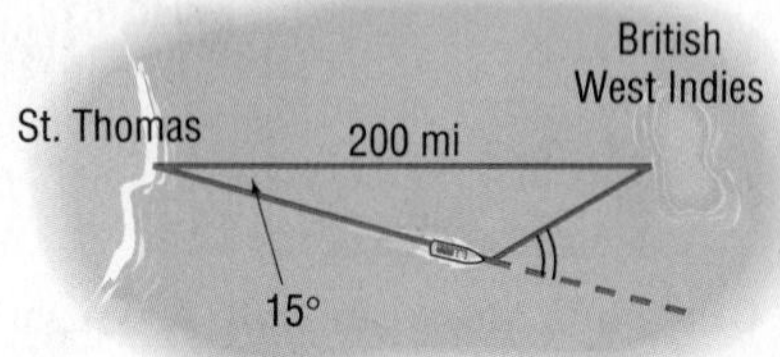

35. Approximating the Area of a Lake To approximate the area of a lake, Cindy walks around the perimeter of the lake, taking the measurements shown in the illustration. Using this technique, what is the approximate area of the lake?

[**Hint:** Use the Law of Cosines on the three triangles shown and then find the sum of their areas.]

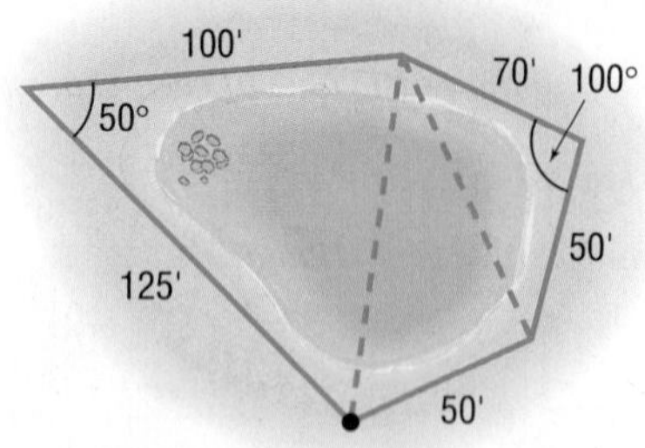

36. Finding the Bearing of a Ship The *Majesty* leaves the Port at Boston for Bermuda with a bearing of S80°E at an average speed of 10 knots. After 1 hour, the ship turns 90° toward the southwest. After 2 hours at an average speed of 20 knots, what is the bearing of the ship from Boston?

37. Drive Wheels of an Engine The drive wheel of an engine is 13 inches in diameter, and the pulley on the rotary pump is 5 inches in diameter. If the shafts of the drive wheel and the pulley are 2 feet apart, what length of belt is required to join them as shown in the figure?

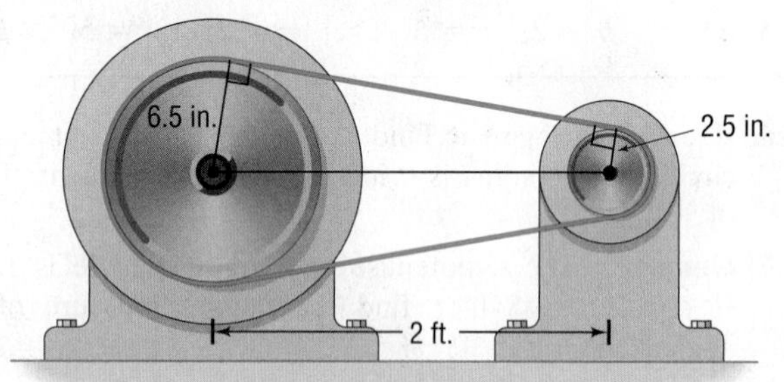

38. Rework Problem 37 if the belt is crossed, as shown in the figure.

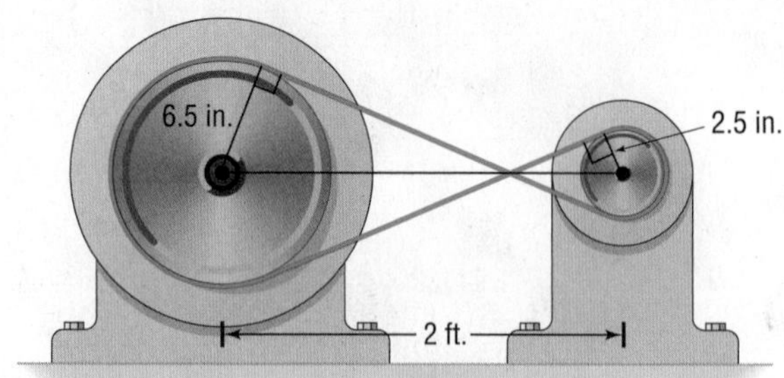

39. An object attached to a coiled spring is pulled down a distance $a = 3$ units from its rest position and then released. Assuming that the motion is simple harmonic with period $T = 4$ seconds, develop a model that relates the displacement d of the object from its rest position after t seconds. Also assume that the positive direction of the motion is up.

In Problems 40 and 41, the displacement d (in feet) of an object from its rest position at time t (in seconds) is given.
(a) Describe the motion of the object.
(b) What is the maximum displacement from its rest position?
(c) What is the time required for one oscillation?
(d) What is the frequency?

40. $d = 6\sin(2t)$

41. $d = -2\cos(\pi t)$

42. An object of mass $m = 40$ grams attached to a coiled spring with damping factor $b = 0.75$ gram/second is pulled down a distance $a = 15$ centimeters from its rest position and then released. Assume that the positive direction of the motion is up and the period is $T = 5$ seconds under simple harmonic motion.
(a) Develop a model that relates the displacement d of the object from its rest position after t seconds.
(b) Graph the equation found in part (a) for 5 oscillations.

43. The displacement d (in meters) of the bob of a pendulum of mass m (in kilograms) from its rest position at time t (in seconds) is given as $d = -15e^{-0.6t/40}\cos\left(\sqrt{\left(\frac{2\pi}{5}\right)^2 - \frac{0.36}{1600}}\,t\right)$
(a) Describe the motion of the object.
(b) What is the initial displacement of the bob? That is, what is the displacement at $t = 0$?
(c) Graph the motion using a graphing utility.
(d) What is the displacement of the bob at the start of the second oscillation?
(e) What happens to the displacement of the bob as time increases without bound?

44. Use the method of adding y-coordinates to graph $y = 2\sin x + \cos(2x)$.

Chapter Test

CHAPTER **Test Prep** VIDEOS

The Chapter Test Prep Videos are step-by-step solutions available in MyMathLab®, or on this text's YouTube Channel. Flip back to the Resources for Success page for a link to this text's YouTube channel.

1. Find the exact value of the six trigonometric functions of the angle θ in the figure.

2. Find the exact value of $\sin 40° - \cos 50°$.

In Problems 3–5, use the given information to determine the three remaining parts of each triangle.

3.

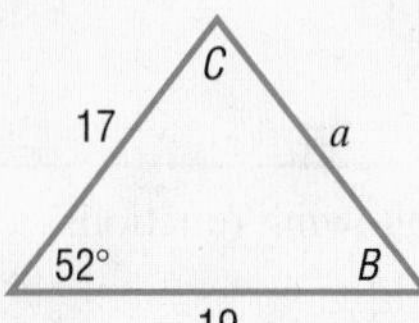

4.

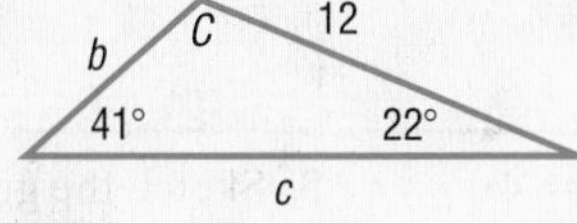

5.

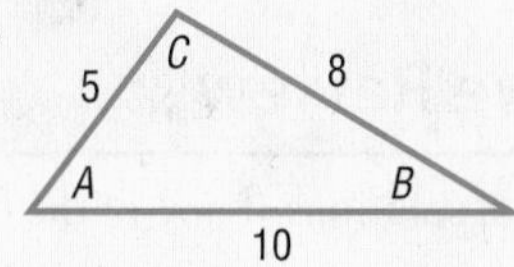

In Problems 6–8, solve each triangle.

6. $A = 55°,\ C = 20°,\ a = 4$

7. $a = 3,\ b = 7,\ A = 40°$

8. $a = 8,\ b = 4,\ C = 70°$

9. Find the area of the triangle described in Problem 8.

10. Find the area of the triangle described in Problem 5.

11. A 12-foot ladder leans against a building. The top of the ladder leans against the wall 10.5 feet from the ground. What is the angle formed by the ground and the ladder?

12. A hot-air balloon is flying at a height of 600 feet and is directly above the Marshall Space Flight Center in Huntsville, Alabama. The pilot of the balloon looks down at the airport that is known to be 5 miles from the Marshall Space Flight Center. What is the angle of depression from the balloon to the airport?

13. Find the area of the shaded region enclosed in a semicircle of diameter 8 centimeters. The length of the chord AB is 6 centimeters.

[**Hint:** Triangle ABC is a right triangle.]

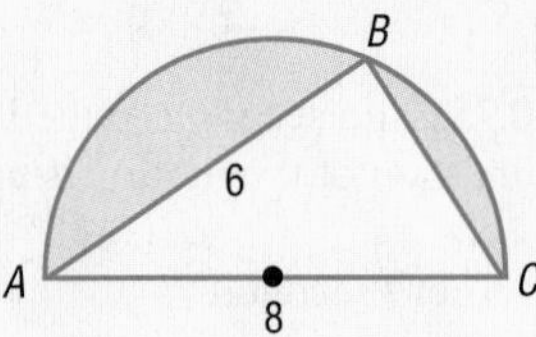

14. Find the area of the quadrilateral shown.

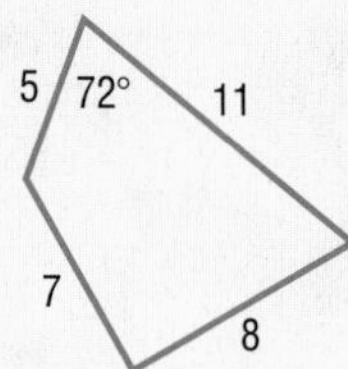

15. Madison wants to swim across Lake William from the fishing lodge (A) to the boat ramp (B), but she wants to know the distance first. Highway 20 goes right past the boat ramp and County Road 3 goes to the lodge. The two roads intersect at point (C), 4.2 miles from the ramp and 3.5 miles from the lodge. Madison uses a transit to measure the angle of intersection of the two roads to be 32°. How far will she need to swim?

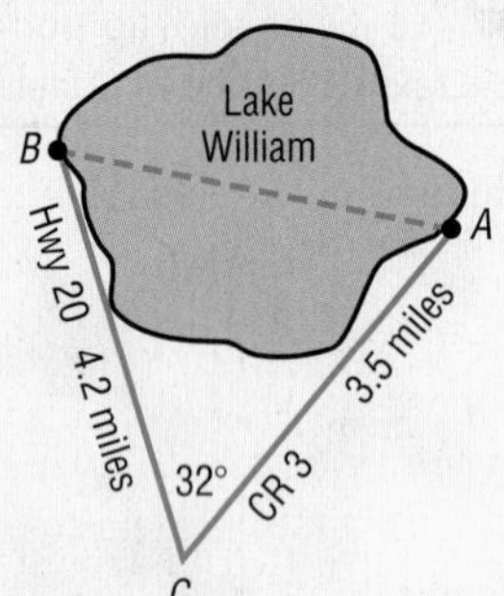

16. Given that $\triangle OAB$ is an isosceles triangle and the shaded sector is a semicircle, find the area of the entire region. Express your answer as a decimal rounded to two places.

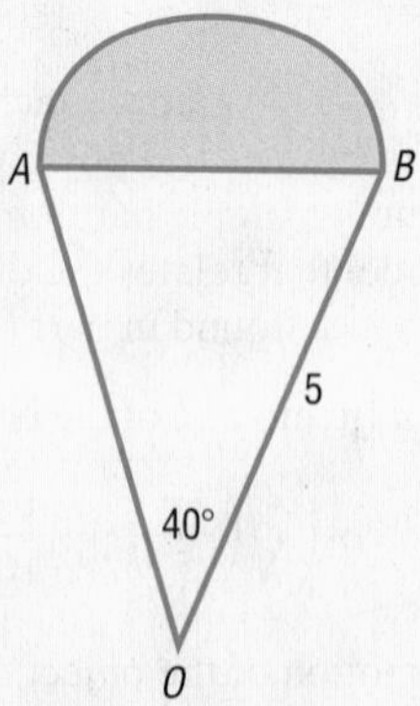

17. The area of the triangle shown below is $54\sqrt{6}$ square units; find the lengths of the sides.

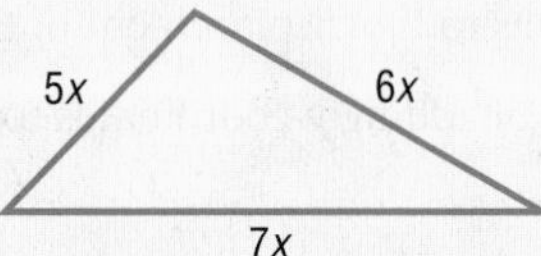

18. Logan is playing on her swing. One full swing (front to back to front) takes 6 seconds and at the peak of her swing she is at an angle of 42° with the vertical. If her swing is 5 feet long, and we ignore all resistive forces, build a model that relates her horizontal displacement (from the rest position) after time t.

Cumulative Review

1. Find the real solutions, if any, of the equation $3x^2 + 1 = 4x$.

2. Find an equation for the circle with center at the point $(-5, 1)$ and radius 3. Graph this circle.

3. Determine the domain of the function

$$f(x) = \sqrt{x^2 - 3x - 4}$$

4. Graph the function $y = 3\sin(\pi x)$.

5. Graph the function $y = -2\cos(2x - \pi)$.

6. If $\tan\theta = -2$ and $\frac{3\pi}{2} < \theta < 2\pi$, find the exact value of:

(a) $\sin\theta$ (b) $\cos\theta$ (c) $\sin(2\theta)$

(d) $\cos(2\theta)$ (e) $\sin\left(\frac{1}{2}\theta\right)$ (f) $\cos\left(\frac{1}{2}\theta\right)$

7. Graph each of the following functions on the interval $[0, 4]$:

(a) $y = \sin x$ (b) $y = 2x + \sin x$

8. Sketch the graph of each of the following functions:

(a) $y = x$ (b) $y = x^2$

(c) $y = \sqrt{x}$ (d) $y = x^3$

(e) $y = \sin x$ (f) $y = \cos x$

(g) $y = \tan x$

9. Solve the triangle in which side a is 20, side c is 15, and angle C is 40°.

10. On the interval $0 \le \theta < 2\pi$, solve the equation $\cos(3\theta) = 1$.

11. What is the length of the arc subtended by a central angle of 60° on a circle of radius 3 feet?

Chapter Projects

I. Spherical Trigonometry When the distance between two locations on the surface of Earth is small, the distance can be computed in statutory miles. Using this assumption, the Law of Sines and the Law of Cosines can be used to approximate distances and angles. However, Earth is a sphere, so as the distance between two points on its surface increases, the linear distance gets less accurate because of curvature. Under this circumstance, the curvature of Earth must be considered when using the Law of Sines and the Law of Cosines.

1. Draw a spherical triangle and label the vertices A, B, and C. Then connect each vertex by a radius to the center O of the sphere. Now draw tangent lines to the sides a and b of the triangle that go through C. Extend the lines OA and OB to intersect the tangent lines at P and Q, respectively. See the figure. List the plane right triangles. Determine the measures of the central angles.

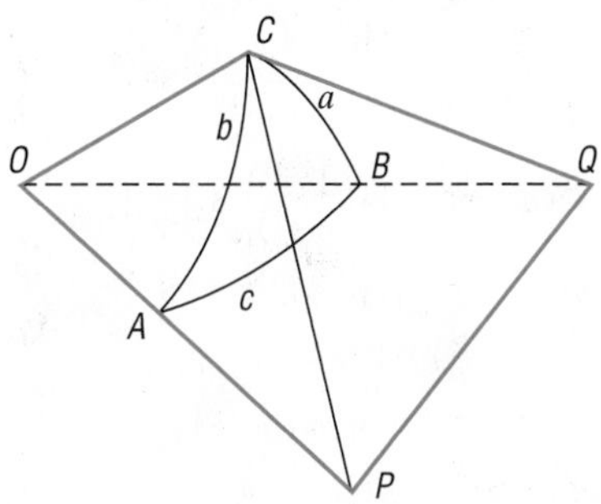

2. Apply the Law of Cosines to triangles OPQ and CPQ to find two expressions for the length of PQ.
3. Subtract the expressions in part (2) from each other. Solve for the term containing $\cos c$.
4. Use the Pythagorean Theorem to find another value for $OQ^2 - CQ^2$ and $OP^2 - CP^2$. Now solve for $\cos c$.
5. Replacing the ratios in part (4) by the cosines of the sides of the spherical triangle, you should now have the Law of Cosines for spherical triangles:

$$\cos c = \cos a \cos b + \sin a \sin b \cos C$$

Source: For the spherical Law of Cosines; see *Mathematics from the Birth of Numbers* by Jan Gullberg. W. W. Norton &Co., Publishers, 1996, pp. 491–494.

II. The Lewis and Clark Expedition Lewis and Clark followed several rivers in their trek from what is now Great Falls, Montana, to the Pacific coast. First, they went down the Missouri and Jefferson rivers from Great Falls to Lemhi, Idaho. Because the two cities are at different longitudes and different latitudes, the curvature of Earth must be accounted for when computing the distance that they traveled. Assume that the radius of Earth is 3960 miles.

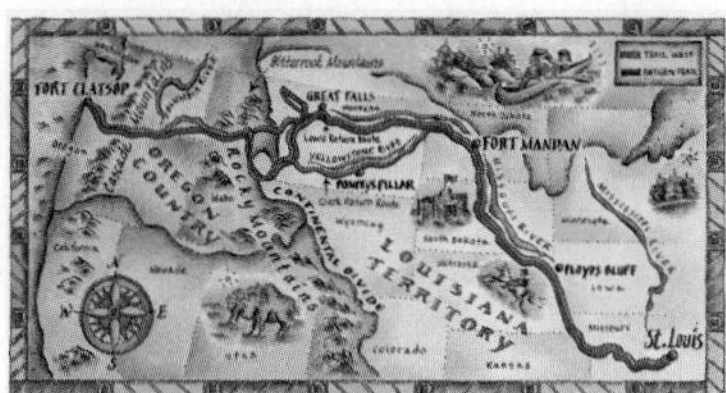

1. Great Falls is at approximately 47.5°N and 111.3°W. Lemhi is at approximately 45.5°N and 113.5°W. (Assume that the rivers flow straight from Great Falls to Lemhi on the surface of Earth.) This line is called a geodesic line. Apply the Law of Cosines for a spherical triangle to find the angle between Great Falls and Lemhi. (The central angles are found by using the differences in the latitudes and longitudes of the towns. See the diagram.) Then find the length of the arc joining the two towns. (Recall $s = r\theta$.)

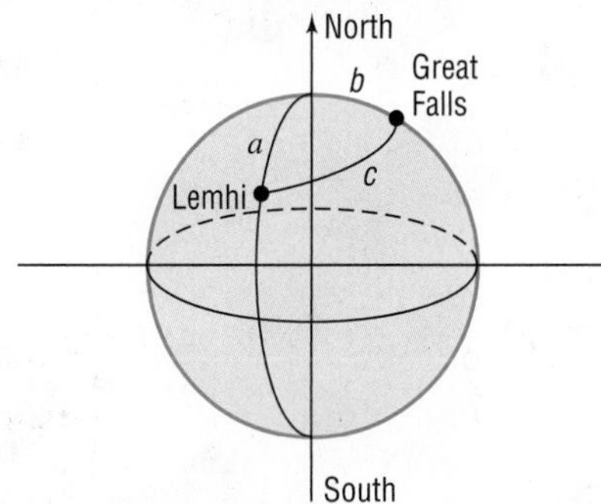

2. From Lemhi, they went up the Bitteroot River and the Snake River to what is now Lewiston and Clarkston on the border of Idaho and Washington. Although this is not really a side to a triangle, make a side that goes from Lemhi to Lewiston and Clarkston. If Lewiston and Clarkston are at about 46.5°N 117.0°W, find the distance from Lemhi using the Law of Cosines for a spherical triangle and the arc length.
3. How far did the explorers travel just to get that far?
4. Draw a plane triangle connecting the three towns. If the distance from Lewiston to Great Falls is 282 miles, and if the angle at Great Falls is 42° and the angle at Lewiston is 48.5°, find the distance from Great Falls to Lemhi and from Lemhi to Lewiston. How do these distances compare with the ones computed in parts (1) and (2)?

Source: For Lewis and Clark Expedition: *American Journey: The Quest for Liberty to 1877, Texas Edition.* Prentice Hall, 1992, p. 345.

Source: For map coordinates: *National Geographic Atlas of the World,* published by National Geographic Society, 1981, pp. 74–75.

The following projects are available at the Instructor's Resource Center (IRC):

III. Project at Motorola: ***How Can You Build or Analyze a Vibration Profile?*** Fourier functions not only are important to analyze vibrations, but they are also what a mathematician would call interesting. Complete the project to see why.

IV. Leaning Tower of Pisa Trigonometry is used to analyze the apparent height and tilt of the Leaning Tower of Pisa.

V. Locating Lost Treasure Clever treasure seekers who know the Law of Sines are able to find a buried treasure efficiently.

VI. Jacob's Field Angles of elevation and the Law of Sines are used to determine the height of the stadium wall and the distance from home plate to the top of the wall.

Polar Coordinates; Vectors

5

How Airplanes Fly

Four aerodynamic forces act on an airplane in flight: **lift**, **drag**, **thrust**, and **weight** (gravity).

Drag is the resistance of air molecules hitting the airplane (the *backward* force), thrust is the power of the airplane's engine (the *forward* force), lift is the *upward* force, and weight is the *downward* force. So for airplanes to fly and stay airborne, the thrust must be greater than the drag, and the lift must be greater than the weight (*so, drag opposes thrust, and lift opposes weight*).

This is certainly the case when an airplane takes off or climbs. However, when it is in straight and level flight, the opposing forces of lift and weight are balanced. During a descent, weight exceeds lift, and to slow the airplane, drag has to overcome thrust.

Thrust is generated by the airplane's engine (propeller or jet), weight is created by the natural force of gravity acting on the airplane, and drag comes from friction as the plane moves through air molecules. Drag is also a *reaction* to lift, and this lift must be generated by the airplane in flight. This is done by the **wings** of the airplane.

A cross section of a typical airplane wing shows the top surface to be more curved than the bottom surface. This shaped profile is called an **airfoil** (or aerofoil), and the shape is used because an airfoil generates significantly more lift than opposing drag. In other words, it is very **efficient** at generating lift.

During flight, air naturally flows over and beneath the wing and is deflected upward over the top surface and downward beneath the lower surface. Any difference in deflection causes a difference in air pressure (pressure gradient), and because of the airfoil shape, the pressure of the deflected air is lower above the airfoil than below it. As a result the wing is "pushed" upward by the higher pressure beneath, or, you can argue, it is "sucked" upward by the lower pressure above.

Source: Adapted from Pete Carpenter. How Airplanes Fly—The Basic Principles of Flight http://www.rc-airplane-world.com/how-airplanes-fly.html, accessed June 2014. © rc-airplane-world.com

—*See Chapter Project I*—

••• A Look Back, A Look Ahead •••

This chapter is in two parts: Polar Coordinates (Sections 5.1–5.3) and Vectors (Sections 5.4–5.7). They are independent of each other and may be covered in either order.

Sections 5.1–5.3: In Chapter 1 we introduced rectangular coordinates (the *xy*-plane) and discussed the graph of an equation in two variables involving *x* and *y*. In Sections 5.1 and 5.2, we introduce polar coordinates, an alternative to rectangular coordinates, and discuss graphing equations that involve polar coordinates. In Section 5.3, we discuss raising a complex number to a real power. As it turns out, polar coordinates are useful for the discussion.

Sections 5.4–5.7: We have seen in many chapters that we are often required to solve an equation to obtain a solution to applied problems. In the last four sections of this chapter, we develop the notion of a vector and show how it can be used to model applied problems in physics and engineering.

Outline

5.1 Polar Coordinates
5.2 Polar Equations and Graphs
5.3 The Complex Plane; De Moivre's Theorem
5.4 Vectors
5.5 The Dot Product
5.6 Vectors in Space
5.7 The Cross Product
Chapter Review
Chapter Test
Cumulative Review
Chapter Projects

5.1 Polar Coordinates

PREPARING FOR THIS SECTION *Before getting started, review the following:*

- Rectangular Coordinates (Section 1.1, pp. 2–4)
- Definition of the Trigonometric Functions (Section 2.2, pp. 115 and 125)
- Inverse Tangent Function (Section 3.1, pp. 195–197)
- Complete the Square (Appendix A, Section A.3, pp. A24–A25)

Now Work the 'Are You Prepared?' problems on page 319.

OBJECTIVES
1 Plot Points Using Polar Coordinates (p. 312)
2 Convert from Polar Coordinates to Rectangular Coordinates (p. 314)
3 Convert from Rectangular Coordinates to Polar Coordinates (p. 316)
4 Transform Equations between Polar and Rectangular Forms (p. 318)

So far, we have always used a system of rectangular coordinates to plot points in the plane. Now we are ready to describe another system, called *polar coordinates*. In many instances, polar coordinates offer certain advantages over rectangular coordinates.

In a rectangular coordinate system, you will recall, a point in the plane is represented by an ordered pair of numbers (x, y), where x and y equal the signed distances of the point from the y-axis and the x-axis, respectively. In a polar coordinate system, we select a point, called the **pole**, and then a ray with vertex at the pole, called the **polar axis**. See Figure 1. Comparing the rectangular and polar coordinate systems, note that the origin in rectangular coordinates coincides with the pole in polar coordinates, and the positive x-axis in rectangular coordinates coincides with the polar axis in polar coordinates.

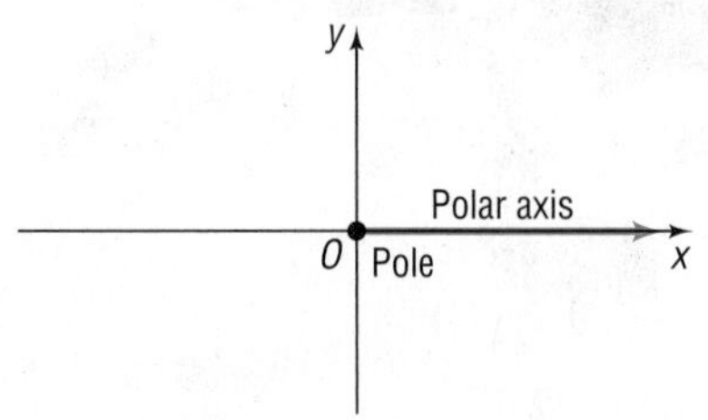

Figure 1

1 Plot Points Using Polar Coordinates

A point P in a polar coordinate system is represented by an ordered pair of numbers (r, θ). If $r > 0$, then r is the distance of the point from the pole; θ is an angle (in degrees or radians) formed by the polar axis and a ray from the pole through the point. We call the ordered pair (r, θ) the **polar coordinates** of the point. See Figure 2.

As an example, suppose that a point P has polar coordinates $\left(2, \frac{\pi}{4}\right)$. Locate P by first drawing an angle of $\frac{\pi}{4}$ radian, placing its vertex at the pole and its initial side along the polar axis. Then go out a distance of 2 units along the terminal side of the angle to reach the point P. See Figure 3.

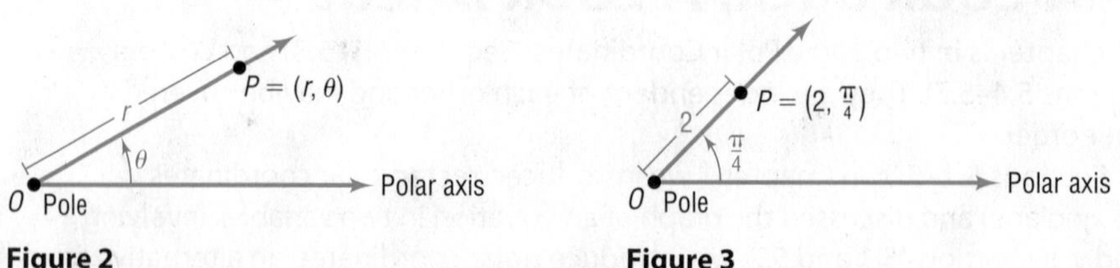

Figure 2 Figure 3

In using polar coordinates (r, θ), it is possible for r to be negative. When this happens, instead of the point being on the terminal side of θ, it is on the ray from the pole extending in the direction *opposite* the terminal side of θ at a distance $|r|$ units from the pole. See Figure 4 for an illustration.

For example, to plot the point $\left(-3, \frac{2\pi}{3}\right)$, use the ray in the opposite direction of $\frac{2\pi}{3}$ and go out $|-3| = 3$ units along that ray. See Figure 5.

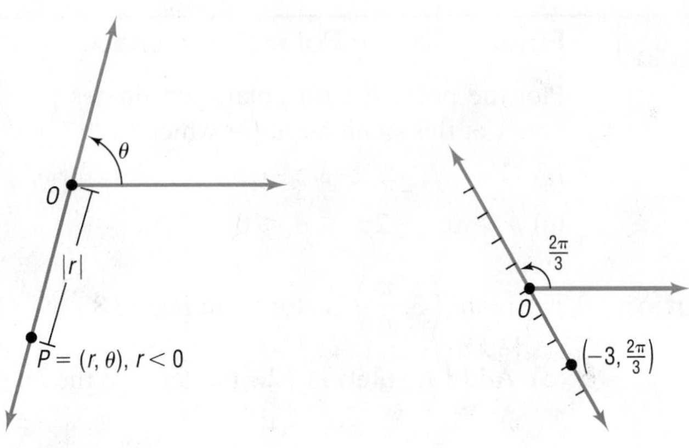

Figure 4

Figure 5

EXAMPLE 1

Plotting Points Using Polar Coordinates

Plot the points with the following polar coordinates:

(a) $\left(3, \dfrac{5\pi}{3}\right)$ (b) $\left(2, -\dfrac{\pi}{4}\right)$ (c) $(3, 0)$ (d) $\left(-2, \dfrac{\pi}{4}\right)$

Solution Figure 6 shows the points.

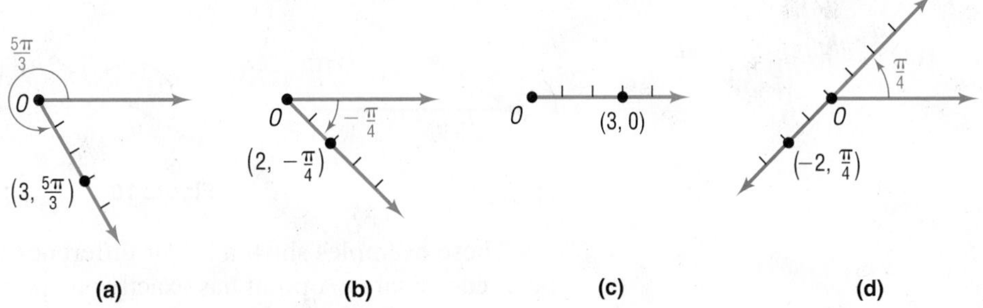

Figure 6 (a) (b) (c) (d)

Now Work PROBLEMS 11, 19, AND 29

Recall that an angle measured counterclockwise is positive and an angle measured clockwise is negative. This convention has some interesting consequences related to polar coordinates.

EXAMPLE 2

Finding Several Polar Coordinates of a Single Point

Consider again the point P with polar coordinates $\left(2, \dfrac{\pi}{4}\right)$, as shown in Figure 7(a). Because $\dfrac{\pi}{4}$, $\dfrac{9\pi}{4}$, and $-\dfrac{7\pi}{4}$ all have the same terminal side, this point P also can be located by using the polar coordinates $\left(2, \dfrac{9\pi}{4}\right)$ or the polar coordinates $\left(2, -\dfrac{7\pi}{4}\right)$, as shown in Figures 7(b) and (c). The point $\left(2, \dfrac{\pi}{4}\right)$ can also be represented by the polar coordinates $\left(-2, \dfrac{5\pi}{4}\right)$. See Figure 7(d).

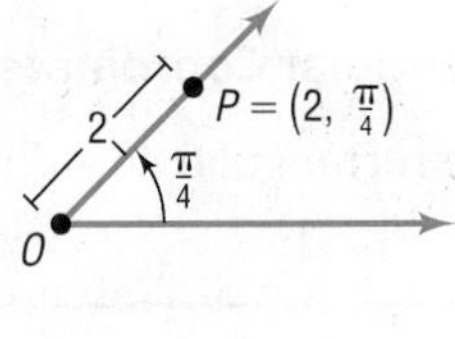

P = (2, 9π/4)

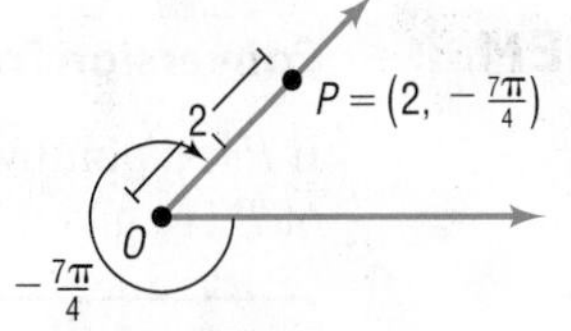

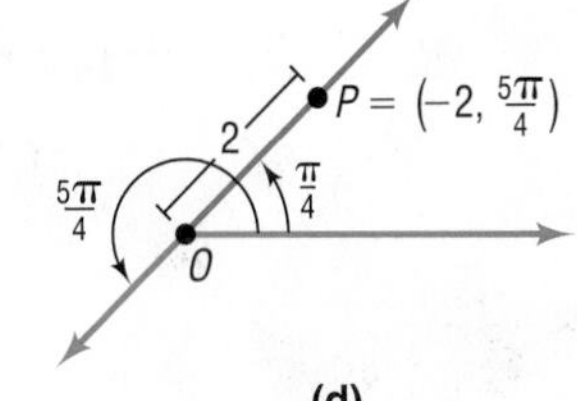

Figure 7 (a) (b) (c) (d)

EXAMPLE 3 **Finding Other Polar Coordinates of a Given Point**

Plot the point P with polar coordinates $\left(3, \frac{\pi}{6}\right)$, and find other polar coordinates (r, θ) of this same point for which:

(a) $r > 0, \quad 2\pi \le \theta < 4\pi$ (b) $r < 0, \quad 0 \le \theta < 2\pi$

(c) $r > 0, \quad -2\pi \le \theta < 0$

Solution The point $\left(3, \frac{\pi}{6}\right)$ is plotted in Figure 8.

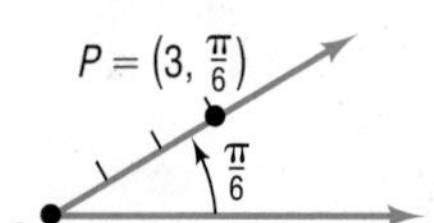

Figure 8

(a) Add 1 revolution (2π radians) to the angle $\frac{\pi}{6}$ to get

$$P = \left(3, \frac{\pi}{6} + 2\pi\right) = \left(3, \frac{13\pi}{6}\right).$$

See Figure 9.

(b) Add $\frac{1}{2}$ revolution (π radians) to the angle $\frac{\pi}{6}$, and replace 3 by -3 to get $P = \left(-3, \frac{\pi}{6} + \pi\right) = \left(-3, \frac{7\pi}{6}\right)$. See Figure 10.

(c) Subtract 2π from the angle $\frac{\pi}{6}$ to get $P = \left(3, \frac{\pi}{6} - 2\pi\right) = \left(3, -\frac{11\pi}{6}\right)$. See Figure 11.

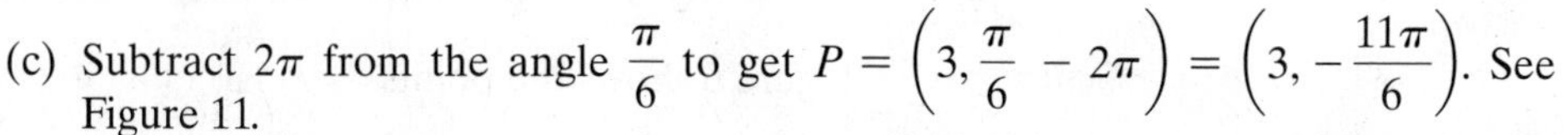

Figure 9: $P = \left(3, \frac{13\pi}{6}\right)$, angle $\frac{13\pi}{6}$

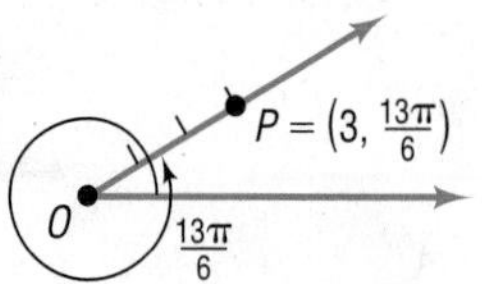

Figure 10

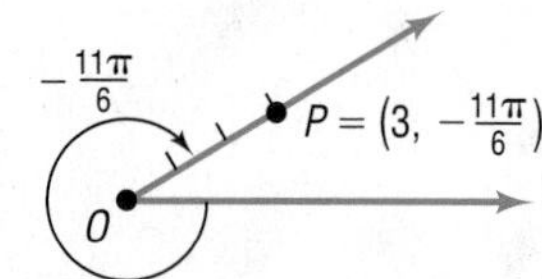

Figure 11

These examples show a major difference between rectangular coordinates and polar coordinates. A point has exactly one pair of rectangular coordinates; however, a point has infinitely many pairs of polar coordinates.

SUMMARY

A point with polar coordinates (r, θ), θ in radians, can also be represented by either of the following:

$$(r, \theta + 2\pi k) \quad \text{or} \quad (-r, \theta + \pi + 2\pi k) \qquad k \text{ any integer}$$

The polar coordinates of the pole are $(0, \theta)$, where θ can be any angle.

Now Work PROBLEM 33

2 Convert from Polar Coordinates to Rectangular Coordinates

Sometimes it is necessary to convert coordinates or equations in rectangular form to polar form, and vice versa. To do this, recall that the origin in rectangular coordinates is the pole in polar coordinates and that the positive x-axis in rectangular coordinates is the polar axis in polar coordinates.

THEOREM **Conversion from Polar Coordinates to Rectangular Coordinates**

If P is a point with polar coordinates (r, θ), the rectangular coordinates (x, y) of P are given by

$$x = r\cos\theta \qquad y = r\sin\theta \tag{1}$$

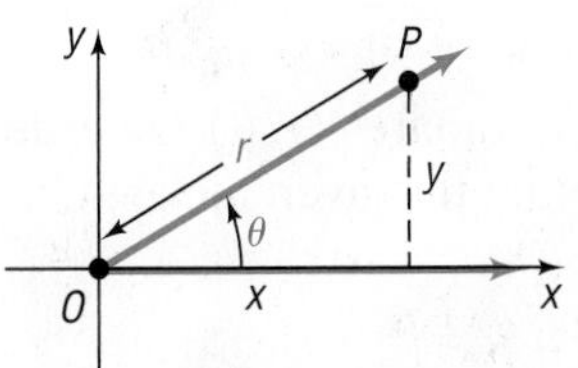

Figure 12

Proof Suppose that P has the polar coordinates (r, θ). We seek the rectangular coordinates (x, y) of P. Refer to Figure 12.

If $r = 0$, then, regardless of θ, the point P is the pole, for which the rectangular coordinates are $(0, 0)$. Formula (1) is valid for $r = 0$.

If $r > 0$, the point P is on the terminal side of θ, and $r = d(O, P) = \sqrt{x^2 + y^2}$. Because

$$\cos\theta = \frac{x}{r} \qquad \sin\theta = \frac{y}{r}$$

this means

$$x = r\cos\theta \qquad y = r\sin\theta$$

If $r < 0$ and θ is in radians, the point $P = (r, \theta)$ can be represented as $(-r, \pi + \theta)$, where $-r > 0$. Because

$$\cos(\pi + \theta) = -\cos\theta = \frac{x}{-r} \qquad \sin(\pi + \theta) = -\sin\theta = \frac{y}{-r}$$

this means

$$x = r\cos\theta \qquad y = r\sin\theta$$ ■

EXAMPLE 4

Converting from Polar Coordinates to Rectangular Coordinates

Find the rectangular coordinates of the points with the following polar coordinates:

(a) $\left(6, \frac{\pi}{6}\right)$ (b) $\left(-4, -\frac{\pi}{4}\right)$

Solution Use formula (1): $x = r\cos\theta$ and $y = r\sin\theta$.

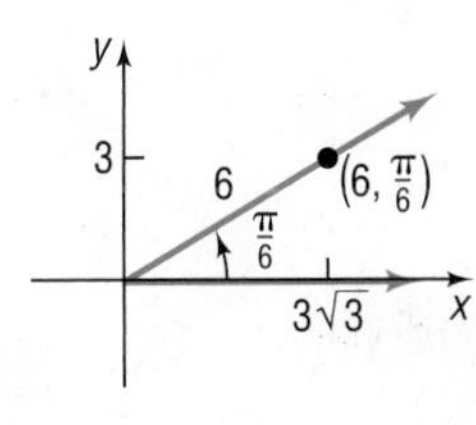

(a)

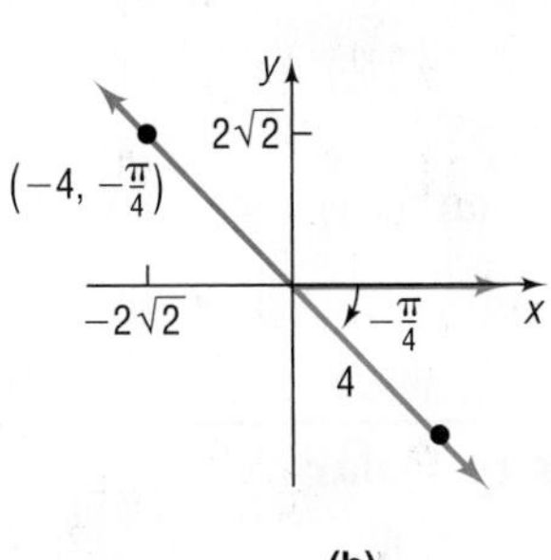

(b)

Figure 13

(a) Figure 13(a) shows $\left(6, \frac{\pi}{6}\right)$ plotted. Notice that $\left(6, \frac{\pi}{6}\right)$ lies in quadrant I of the rectangular coordinate system. So both the x-coordinate and the y-coordinate will be positive. Substituting $r = 6$ and $\theta = \frac{\pi}{6}$ gives

$$x = r\cos\theta = 6\cos\frac{\pi}{6} = 6\cdot\frac{\sqrt{3}}{2} = 3\sqrt{3}$$

$$y = r\sin\theta = 6\sin\frac{\pi}{6} = 6\cdot\frac{1}{2} = 3$$

The rectangular coordinates of the point $\left(6, \frac{\pi}{6}\right)$ are $(3\sqrt{3}, 3)$, which lies in quadrant I, as expected.

(b) Figure 13(b) shows $\left(-4, -\frac{\pi}{4}\right)$ plotted. Notice that $\left(-4, -\frac{\pi}{4}\right)$ lies in quadrant II of the rectangular coordinate system. Substituting $r = -4$ and $\theta = -\frac{\pi}{4}$ gives

$$x = r\cos\theta = -4\cos\left(-\frac{\pi}{4}\right) = -4\cdot\frac{\sqrt{2}}{2} = -2\sqrt{2}$$

$$y = r\sin\theta = -4\sin\left(-\frac{\pi}{4}\right) = -4\left(-\frac{\sqrt{2}}{2}\right) = 2\sqrt{2}$$

The rectangular coordinates of the point $\left(-4, -\frac{\pi}{4}\right)$ are $(-2\sqrt{2}, 2\sqrt{2})$, which lies in quadrant II, as expected. ●

COMMENT Many calculators have the capability of converting from polar coordinates to rectangular coordinates. Consult your owner's manual for the proper keystrokes. In most cases this procedure is tedious, so you will probably find that using formula (1) is faster. ■

Now Work PROBLEMS 41 AND 53

3 Convert from Rectangular Coordinates to Polar Coordinates

Converting from rectangular coordinates (x, y) to polar coordinates (r, θ) is a little more complicated. Notice that each solution begins by plotting the given rectangular coordinates.

EXAMPLE 5 How to Convert from Rectangular Coordinates to Polar Coordinates with the Point on a Coordinate Axis

Find polar coordinates of a point whose rectangular coordinates are $(0, 3)$.

Step-by-Step Solution

Step 1: Plot the point (x, y) and note the quadrant the point lies in or the coordinate axis the point lies on.

Plot the point $(0, 3)$ in a rectangular coordinate system. See Figure 14. The point lies on the positive y-axis.

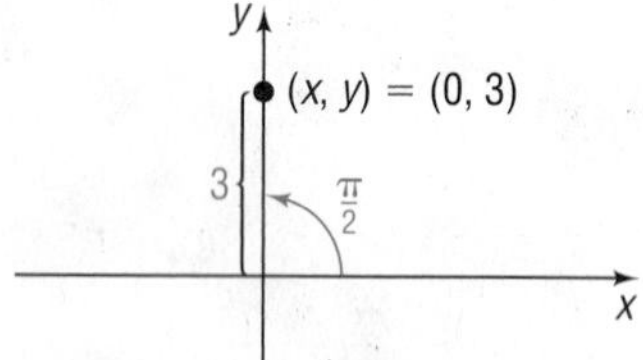

Figure 14

Step 2: Determine the distance r from the origin to the point.

The point $(0, 3)$ lies on the y-axis a distance of 3 units from the origin (pole), so $r = 3$.

Step 3: Determine θ.

A ray with vertex at the pole through $(0, 3)$ forms an angle $\theta = \dfrac{\pi}{2}$ with the polar axis.

Polar coordinates for this point can be given by $\left(3, \dfrac{\pi}{2}\right)$. Other possible representations include $\left(-3, -\dfrac{\pi}{2}\right)$ and $\left(3, \dfrac{5\pi}{2}\right)$. ●

Figure 15 shows polar coordinates of points that lie on either the x-axis or the y-axis. In each illustration, $a > 0$.

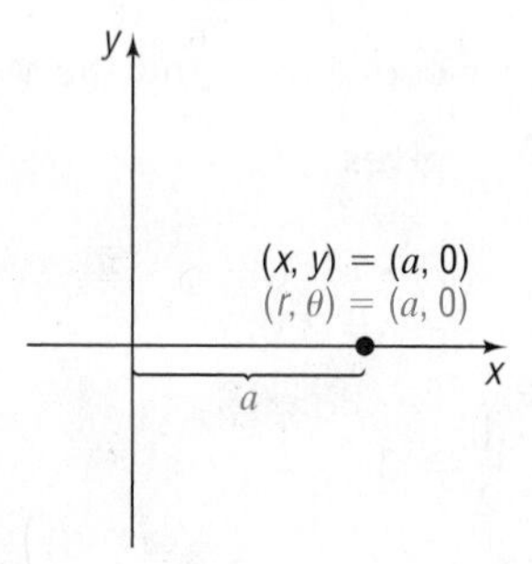

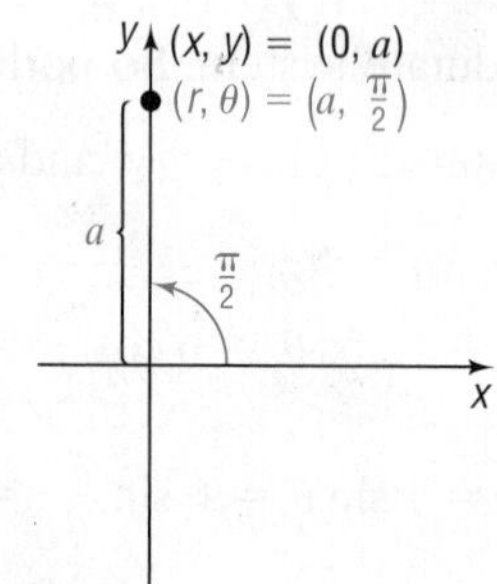

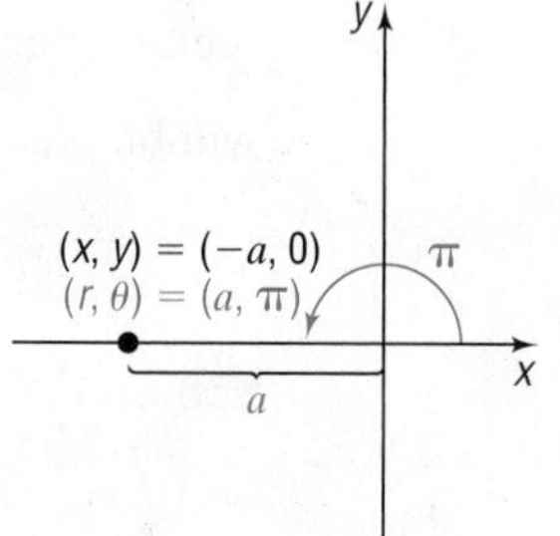

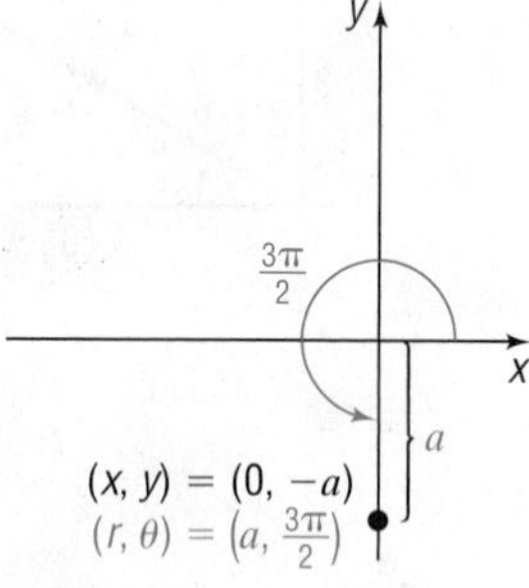

Figure 15 (a) $(x, y) = (a, 0), a > 0$ (b) $(x, y) = (0, a), a > 0$ (c) $(x, y) = (-a, 0), a > 0$ (d) $(x, y) = (0, -a), a > 0$

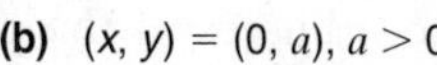

Now Work PROBLEM 57

EXAMPLE 6 How to Convert from Rectangular Coordinates to Polar Coordinates with the Point in a Quadrant

Find the polar coordinates of a point whose rectangular coordinates are $(2, -2)$.

Step-by-Step Solution

Step 1: Plot the point (x, y) and note the quadrant the point lies in or the coordinate axis the point lies on.

Plot the point $(2, -2)$ in a rectangular coordinate system. See Figure 16. The point lies in quadrant IV.

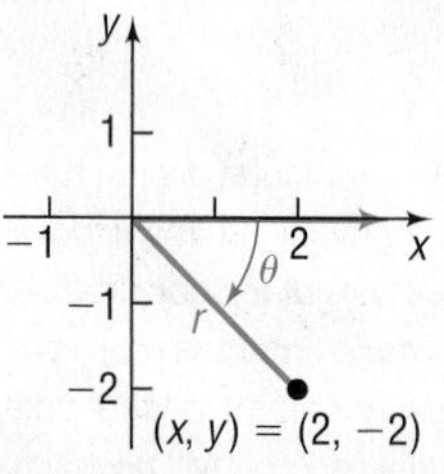

Figure 16

Step 2: Determine the distance r from the origin to the point using $r = \sqrt{x^2 + y^2}$.

$$r = \sqrt{x^2 + y^2} = \sqrt{(2)^2 + (-2)^2} = \sqrt{8} = 2\sqrt{2}$$

Step 3: Determine θ.

Find θ by recalling that $\tan\theta = \frac{y}{x}$, so $\theta = \tan^{-1}\frac{y}{x}$, $-\frac{\pi}{2} < \theta < \frac{\pi}{2}$. Because $(2, -2)$ lies in quadrant IV, this means that $-\frac{\pi}{2} < \theta < 0$. As a result,

$$\theta = \tan^{-1}\frac{y}{x} = \tan^{-1}\left(\frac{-2}{2}\right) = \tan^{-1}(-1) = -\frac{\pi}{4}$$

COMMENT Many calculators have the capability of converting from rectangular coordinates to polar coordinates. Consult your owner's manual for the proper keystrokes. ■

A set of polar coordinates for the point $(2, -2)$ is $\left(2\sqrt{2}, -\frac{\pi}{4}\right)$. Other possible representations include $\left(2\sqrt{2}, \frac{7\pi}{4}\right)$ and $\left(-2\sqrt{2}, \frac{3\pi}{4}\right)$. ●

EXAMPLE 7

Converting from Rectangular Coordinates to Polar Coordinates

Find polar coordinates of a point whose rectangular coordinates are $(-1, -\sqrt{3})$.

Solution

STEP 1: See Figure 17. The point lies in quadrant III.

STEP 2: The distance r from the origin to the point $(-1, -\sqrt{3})$ is

$$r = \sqrt{(-1)^2 + (-\sqrt{3})^2} = \sqrt{4} = 2$$

$(x, y) = (-1, -\sqrt{3})$

Figure 17

STEP 3: To find θ, use $\alpha = \tan^{-1}\frac{y}{x} = \tan^{-1}\frac{-\sqrt{3}}{-1} = \tan^{-1}\sqrt{3}$, $-\frac{\pi}{2} < \alpha < \frac{\pi}{2}$.

Since the point $(-1, -\sqrt{3})$ lies in quadrant III and the inverse tangent function gives an angle in quadrant I, add π to the result to obtain an angle in quadrant III.

$$\theta = \pi + \tan^{-1}\left(\frac{-\sqrt{3}}{-1}\right) = \pi + \tan^{-1}\sqrt{3} = \pi + \frac{\pi}{3} = \frac{4\pi}{3}$$

A set of polar coordinates for this point is $\left(2, \frac{4\pi}{3}\right)$. Other possible representations include $\left(-2, \frac{\pi}{3}\right)$ and $\left(2, -\frac{2\pi}{3}\right)$. ●

Figure 18 shows how to find polar coordinates of a point that lies in a quadrant when its rectangular coordinates (x, y) are given.

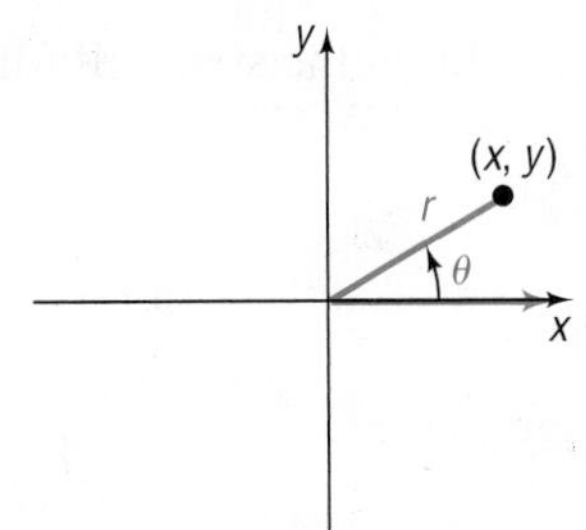

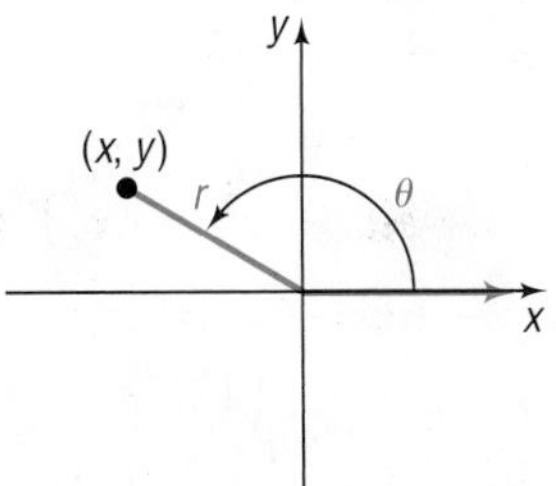

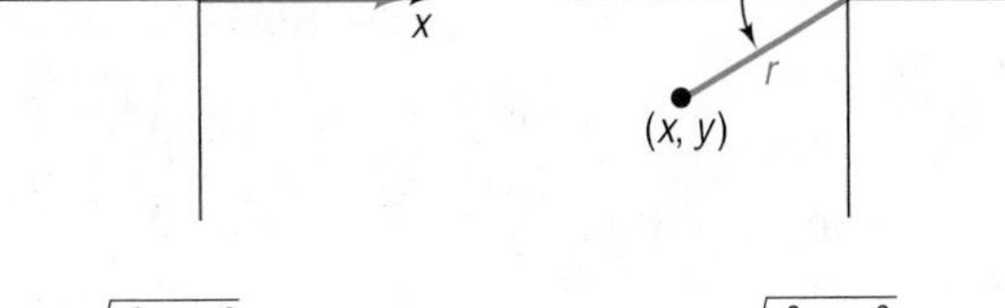

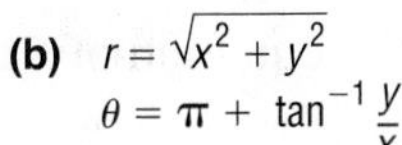

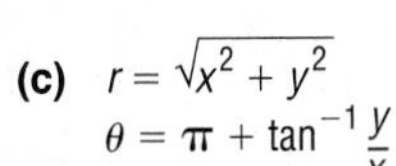

Figure 18 (a) $r = \sqrt{x^2 + y^2}$, $\theta = \tan^{-1}\frac{y}{x}$ (b) $r = \sqrt{x^2 + y^2}$, $\theta = \pi + \tan^{-1}\frac{y}{x}$ (c) $r = \sqrt{x^2 + y^2}$, $\theta = \pi + \tan^{-1}\frac{y}{x}$ (d) $r = \sqrt{x^2 + y^2}$, $\theta = \tan^{-1}\frac{y}{x}$

The preceding discussion provides the formulas

$$r^2 = x^2 + y^2 \qquad \tan\theta = \frac{y}{x} \quad \text{if } x \neq 0 \qquad \textbf{(2)}$$

To use formula (2) effectively, follow these steps:

Steps for Converting from Rectangular to Polar Coordinates

STEP 1: Always plot the point (x, y) first, as shown in Examples 5, 6, and 7. Note the quadrant the point lies in or the coordinate axis the point lies on.

STEP 2: If $x = 0$ or $y = 0$, use your illustration to find r.
If $x \neq 0$ and $y \neq 0$, then $r = \sqrt{x^2 + y^2}$.

STEP 3: Find θ. If $x = 0$ or $y = 0$, use your illustration to find θ.
If $x \neq 0$ and $y \neq 0$, note the quadrant in which the point lies.

$$\text{Quadrant I or IV:} \quad \theta = \tan^{-1}\frac{y}{x}$$

$$\text{Quadrant II or III:} \quad \theta = \pi + \tan^{-1}\frac{y}{x}$$

Now Work PROBLEM 61

4 Transform Equations between Polar and Rectangular Forms

Formulas (1) and (2) can also be used to transform equations from polar form to rectangular form, and vice versa. Two common techniques for transforming an equation from polar form to rectangular form are

1. Multiplying both sides of the equation by r
2. Squaring both sides of the equation

EXAMPLE 8 Transforming an Equation from Polar to Rectangular Form

Transform the equation $r = 6\cos\theta$ from polar coordinates to rectangular coordinates, and identify the graph.

Solution Multiplying each side by r makes it easier to apply formulas (1) and (2).

$$r = 6\cos\theta$$

$$r^2 = 6r\cos\theta \quad \text{Multiply each side by } r.$$

$$x^2 + y^2 = 6x \quad r^2 = x^2 + y^2;\ x = r\cos\theta$$

This is the equation of a circle. Complete the square to obtain the standard form of the equation.

$$x^2 + y^2 = 6x$$

$$(x^2 - 6x) + y^2 = 0 \quad \text{General form}$$

$$(x^2 - 6x + 9) + y^2 = 9 \quad \text{Complete the square in } x.$$

$$(x - 3)^2 + y^2 = 9 \quad \text{Factor.}$$

This is the standard form of the equation of a circle with center $(3, 0)$ and radius 3.

●

Now Work PROBLEM 77

EXAMPLE 9 Transforming an Equation from Rectangular to Polar Form

Transform the equation $4xy = 9$ from rectangular coordinates to polar coordinates.

Solution Use formula (1): $x = r\cos\theta$ and $y = r\sin\theta$.

$$\begin{aligned} 4xy &= 9 \\ 4(r\cos\theta)(r\sin\theta) &= 9 \quad x = r\cos\theta, y = r\sin\theta \\ 4r^2\cos\theta\sin\theta &= 9 \end{aligned}$$

This is the polar form of the equation. It can be simplified as follows:

$$\begin{aligned} 2r^2(2\sin\theta\cos\theta) &= 9 \quad \text{Factor out } 2r^2. \\ 2r^2\sin(2\theta) &= 9 \quad \text{Double-angle Formula} \end{aligned}$$

Now Work PROBLEM 71

5.1 Assess Your Understanding

'Are You Prepared?' *Answers are given at the end of these exercises. If you get a wrong answer, read the pages listed in red.*

1. Plot the point whose rectangular coordinates are $(3, -1)$. What quadrant does the point lie in? (pp. 2–4)
2. To complete the square of $x^2 + 6x$, add ______. (pp. A24–A25)
3. If $P = (x, y)$ is a point on the terminal side of the angle θ at a distance r from the origin, then $\tan\theta =$ ______. (p. 125)
4. $\tan^{-1}(-1) =$ ______. (pp. 195–197)

Concepts and Vocabulary

5. The origin in rectangular coordinates coincides with the ______ in polar coordinates; the positive x-axis in rectangular coordinates coincides with the ______ ______ in polar coordinates.
6. If P is a point with polar coordinates (r, θ), the rectangular coordinates (x, y) of P are given by $x =$ ______ and $y =$ ______.
7. For the point with polar coordinates $\left(1, -\frac{\pi}{2}\right)$, which of the following best describes the location of the point in a rectangular coordinate system?
   (a) in quadrant IV (b) on the y-axis
   (c) in quadrant II (d) on the x-axis
8. The point $\left(5, \frac{\pi}{6}\right)$ can also be represented by which of the following polar coordinates?
   (a) $\left(5, -\frac{\pi}{6}\right)$ (b) $\left(-5, \frac{13\pi}{6}\right)$
   (c) $\left(5, -\frac{5\pi}{6}\right)$ (d) $\left(-5, \frac{7\pi}{6}\right)$
9. ***True or False*** In the polar coordinates (r, θ), r can be negative.
10. ***True or False*** The polar coordinates of a point are unique.

Skill Building

In Problems 11–18, match each point in polar coordinates with either A, B, C, or D on the graph.

11. $\left(2, -\frac{11\pi}{6}\right)$ 12. $\left(-2, -\frac{\pi}{6}\right)$ 13. $\left(-2, \frac{\pi}{6}\right)$ 14. $\left(2, \frac{7\pi}{6}\right)$

15. $\left(2, \frac{5\pi}{6}\right)$ 16. $\left(-2, \frac{5\pi}{6}\right)$ 17. $\left(-2, \frac{7\pi}{6}\right)$ 18. $\left(2, \frac{11\pi}{6}\right)$

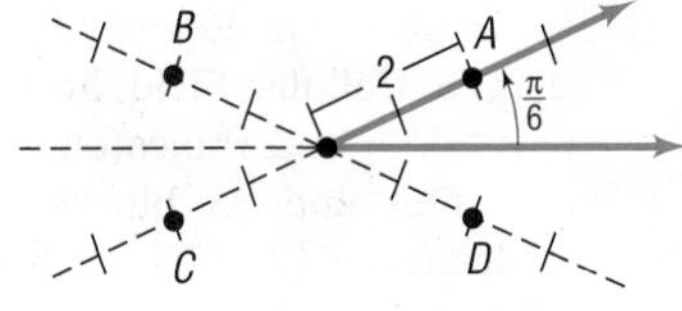

In Problems 19–32, plot each point given in polar coordinates.

19. $(3, 90°)$ 20. $(4, 270°)$ 21. $(-2, 0)$ 22. $(-3, \pi)$ 23. $\left(6, \frac{\pi}{6}\right)$

24. $\left(5, \frac{5\pi}{3}\right)$ 25. $(-2, 135°)$ 26. $(-3, 120°)$ 27. $\left(4, -\frac{2\pi}{3}\right)$ 28. $\left(2, -\frac{5\pi}{4}\right)$

29. $\left(-1, -\frac{\pi}{3}\right)$ 30. $\left(-3, -\frac{3\pi}{4}\right)$ 31. $(-2, -\pi)$ 32. $\left(-3, -\frac{\pi}{2}\right)$

In Problems 33–40, plot each point given in polar coordinates, and find other polar coordinates (r, θ) of the point for which:
(a) $r > 0, \; -2\pi \le \theta < 0$ (b) $r < 0, \; 0 \le \theta < 2\pi$ (c) $r > 0, \; 2\pi \le \theta < 4\pi$

33. $\left(5, \frac{2\pi}{3}\right)$ 34. $\left(4, \frac{3\pi}{4}\right)$ 35. $(-2, 3\pi)$ 36. $(-3, 4\pi)$

37. $\left(1, \frac{\pi}{2}\right)$ 38. $(2, \pi)$ 39. $\left(-3, -\frac{\pi}{4}\right)$ 40. $\left(-2, -\frac{2\pi}{3}\right)$

In Problems 41–56, polar coordinates of a point are given. Find the rectangular coordinates of each point.

41. $\left(3, \frac{\pi}{2}\right)$ 42. $\left(4, \frac{3\pi}{2}\right)$ 43. $(-2, 0)$ 44. $(-3, \pi)$

45. $(6, 150°)$ 46. $(5, 300°)$ 47. $\left(-2, \frac{3\pi}{4}\right)$ 48. $\left(-2, \frac{2\pi}{3}\right)$

49. $\left(-1, -\frac{\pi}{3}\right)$ 50. $\left(-3, -\frac{3\pi}{4}\right)$ 51. $(-2, -180°)$ 52. $(-3, -90°)$

53. $(7.5, 110°)$ 54. $(-3.1, 182°)$ 55. $(6.3, 3.8)$ 56. $(8.1, 5.2)$

In Problems 57–68, the rectangular coordinates of a point are given. Find polar coordinates for each point.

57. $(3, 0)$ 58. $(0, 2)$ 59. $(-1, 0)$ 60. $(0, -2)$

61. $(1, -1)$ 62. $(-3, 3)$ 63. $(\sqrt{3}, 1)$ 64. $(-2, -2\sqrt{3})$

65. $(1.3, -2.1)$ 66. $(-0.8, -2.1)$ 67. $(8.3, 4.2)$ 68. $(-2.3, 0.2)$

In Problems 69–76, the letters x and y represent rectangular coordinates. Write each equation using polar coordinates (r, θ).

69. $2x^2 + 2y^2 = 3$ 70. $x^2 + y^2 = x$ 71. $x^2 = 4y$ 72. $y^2 = 2x$

73. $2xy = 1$ 74. $4x^2 y = 1$ 75. $x = 4$ 76. $y = -3$

In Problems 77–84, the letters r and θ represent polar coordinates. Write each equation using rectangular coordinates (x, y).

77. $r = \cos\theta$ 78. $r = \sin\theta + 1$ 79. $r^2 = \cos\theta$ 80. $r = \sin\theta - \cos\theta$

81. $r = 2$ 82. $r = 4$ 83. $r = \frac{4}{1 - \cos\theta}$ 84. $r = \frac{3}{3 - \cos\theta}$

Applications and Extensions

85. **Chicago** In Chicago, the road system is set up like a Cartesian plane, where streets are indicated by the number of blocks they are from Madison Street and State Street. For example, Wrigley Field in Chicago is located at 1060 West Addison, which is 10 blocks west of State Street and 36 blocks north of Madison Street. Treat the intersection of Madison Street and State Street as the origin of a coordinate system, with east being the positive x-axis.
(a) Write the location of Wrigley Field using rectangular coordinates.
(b) Write the location of Wrigley Field using polar coordinates. Use the east direction for the polar axis. Express θ in degrees.
(c) U.S. Cellular Field, home of the White Sox, is located at 35th and Princeton, which is 3 blocks west of State Street and 35 blocks south of Madison. Write the location of U.S. Cellular Field using rectangular coordinates.
(d) Write the location of U.S. Cellular Field using polar coordinates. Use the east direction for the polar axis. Express θ in degrees.

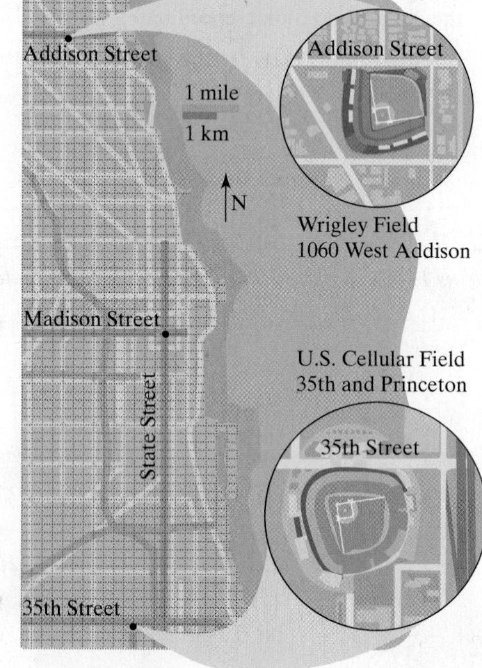

86. Show that the formula for the distance d between two points $P_1 = (r_1, \theta_1)$ and $P_2 = (r_2, \theta_2)$ is

$$d = \sqrt{r_1^2 + r_2^2 - 2r_1 r_2 \cos(\theta_2 - \theta_1)}$$

Explaining Concepts: Discussion and Writing

87. In converting from polar coordinates to rectangular coordinates, what formulas will you use?

88. Explain how to convert from rectangular coordinates to polar coordinates.

89. Is the street system in your town based on a rectangular coordinate system, a polar coordinate system, or some other system? Explain.

Retain Your Knowledge

Problems 90–93 are based on material learned earlier in the course. The purpose of these problems is to keep the material fresh in your mind so that you are better prepared for the final exam.

90. Find the midpoint of the line segment connecting the points $(-3, 7)$ and $\left(\frac{1}{2}, 2\right)$.

91. Given that the point (3, 8) is on the graph of $y = f(x)$, what is the corresponding point on the graph of $y = -2f(x + 3) + 5$?

92. The point $(-8, 15)$ is on the terminal side of an angle θ in standard position. Find the exact value of each of the six trigonometric functions of θ.

93. Determine the amplitude and period of $y = 4\cos\left(\frac{2}{3}x\right)$ without graphing.

'Are You Prepared?' Answers

1. (3, −1); quadrant IV **2.** 9 **3.** $\frac{y}{x}$ **4.** $-\frac{\pi}{4}$

5.2 Polar Equations and Graphs

PREPARING FOR THIS SECTION *Before getting started, review the following:*

- Symmetry (Section 1.2, pp. 12–13)
- Circles (Section 1.2, pp. 15–18)
- Even–Odd Properties of Trigonometric Functions (Section 2.3, p. 141)
- Difference Formulas for Sine and Cosine (Section 3.5, pp. 226 and 229)
- Values of the Sine and Cosine Functions at Certain Angles (Section 2.2, pp. 117–124)

Now Work the **'Are You Prepared?'** problems on page 333.

OBJECTIVES
1. Identify and Graph Polar Equations by Converting to Rectangular Equations (p. 322)
2. Test Polar Equations for Symmetry (p. 325)
3. Graph Polar Equations by Plotting Points (p. 326)

Just as a rectangular grid may be used to plot points given by rectangular coordinates, such as the points (–3, 1) and (1, 2) shown in Figure 19(a), a grid consisting of concentric circles (with centers at the pole) and rays (with vertices at the pole) can be used to plot points given by polar coordinates, such as the points $\left(4, \frac{5\pi}{4}\right)$ and $\left(2, \frac{\pi}{4}\right)$ shown in Figure 19(b). Such **polar grids** are used to graph *polar equations*.

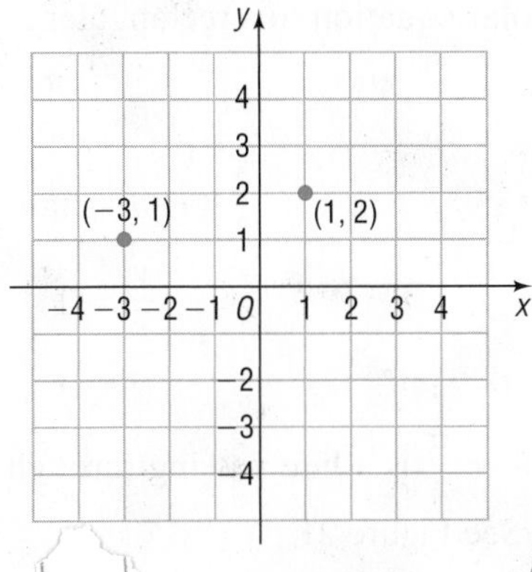

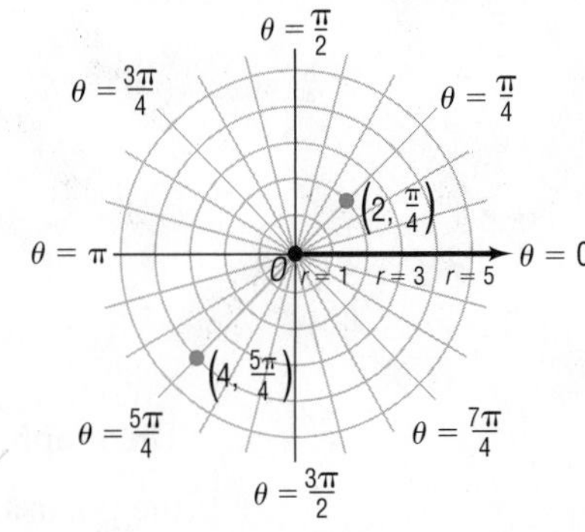

Figure 19 **(a)** Rectangular grid **(b)** Polar grid

DEFINITION An equation whose variables are polar coordinates is called a **polar equation**. The **graph of a polar equation** consists of all points whose polar coordinates satisfy the equation.

1 Identify and Graph Polar Equations by Converting to Rectangular Equations

One method that can be used to graph a polar equation is to convert the equation to rectangular coordinates. In the following discussion, (x, y) represents the rectangular coordinates of a point P, and (r, θ) represents polar coordinates of the point P.

EXAMPLE 1 Identifying and Graphing a Polar Equation (Circle)

Identify and graph the equation: $r = 3$

Solution Convert the polar equation to a rectangular equation.

$$r = 3$$

$$r^2 = 9 \quad \text{Square both sides.}$$

$$x^2 + y^2 = 9 \quad r^2 = x^2 + y^2$$

The graph of $r = 3$ is a circle, with center at the pole and radius 3. See Figure 20.

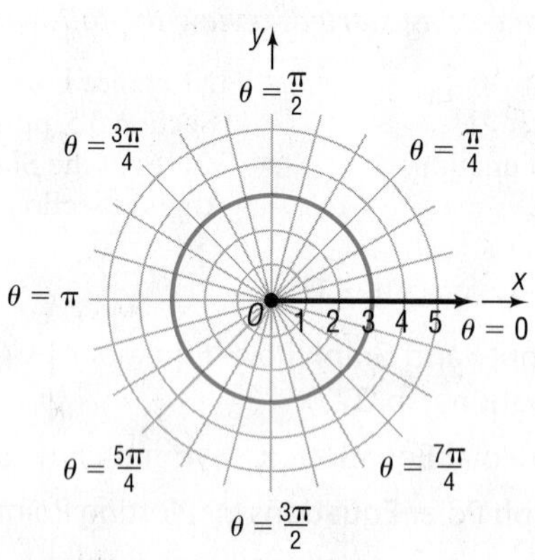

Figure 20 $r = 3$, or $x^2 + y^2 = 9$

Now Work PROBLEM 15

EXAMPLE 2 Identifying and Graphing a Polar Equation (Line)

Identify and graph the equation: $\theta = \dfrac{\pi}{4}$

Solution Convert the polar equation to a rectangular equation.

$$\theta = \frac{\pi}{4}$$

$$\tan \theta = \tan \frac{\pi}{4} \quad \text{Take the tangent of both sides.}$$

$$\frac{y}{x} = 1 \quad \tan \theta = \frac{y}{x};\ \tan \frac{\pi}{4} = 1$$

$$y = x$$

The graph of $\theta = \dfrac{\pi}{4}$ is a line passing through the pole making an angle of $\dfrac{\pi}{4}$ with the polar axis. See Figure 21.

Figure 21 $\theta = \dfrac{\pi}{4}$, or $y = x$

Now Work PROBLEM 17

EXAMPLE 3 **Identifying and Graphing a Polar Equation (Horizontal Line)**

Identify and graph the equation: $r \sin \theta = 2$

Solution Because $y = r \sin \theta$, we can write the equation as

$$y = 2$$

Therefore, the graph of $r \sin \theta = 2$ is a horizontal line 2 units above the pole. See Figure 22.

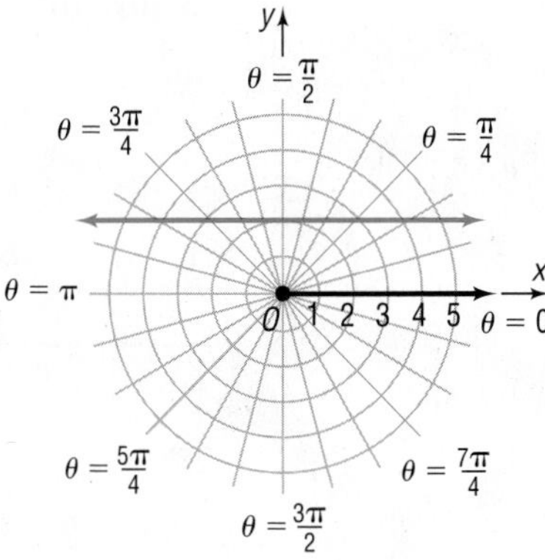

Figure 22 $r \sin \theta = 2$, or $y = 2$

COMMENT A graphing utility can be used to graph polar equations. Read Using a Graphing Utility to Graph a Polar Equation, Appendix B, Section B.6. ■

EXAMPLE 4 **Identifying and Graphing a Polar Equation (Vertical Line)**

Identify and graph the equation: $r \cos \theta = -3$

Solution Since $x = r \cos \theta$, we can write the equation as

$$x = -3$$

Therefore, the graph of $r \cos \theta = -3$ is a vertical line 3 units to the left of the pole. See Figure 23.

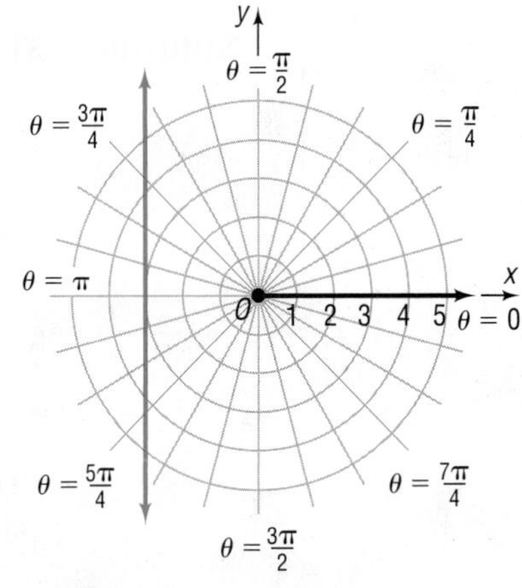

Figure 23 $r \cos \theta = -3$, or $x = -3$

Examples 3 and 4 lead to the following results. (The proofs are left as exercises. See Problems 83 and 84.)

THEOREM

Let a be a real number. Then the graph of the equation

$$r \sin \theta = a$$

is a horizontal line. It lies a units above the pole if $a \geq 0$ and lies $|a|$ units below the pole if $a < 0$.

The graph of the equation

$$r \cos \theta = a$$

is a vertical line. It lies a units to the right of the pole if $a \geq 0$ and lies $|a|$ units to the left of the pole if $a < 0$.

Now Work PROBLEM 21

EXAMPLE 5 Identifying and Graphing a Polar Equation (Circle)

Identify and graph the equation: $r = 4 \sin \theta$

Solution To transform the equation to rectangular coordinates, multiply each side by r.

$$r^2 = 4r \sin \theta$$

Now use the facts that $r^2 = x^2 + y^2$ and $y = r \sin \theta$. Then

$$\begin{aligned} x^2 + y^2 &= 4y \\ x^2 + (y^2 - 4y) &= 0 \\ x^2 + (y^2 - 4y + 4) &= 4 \quad \text{Complete the square in } y. \\ x^2 + (y - 2)^2 &= 4 \quad \text{Factor.} \end{aligned}$$

This is the standard equation of a circle with center at $(0, 2)$ in rectangular coordinates and radius 2. See Figure 24.

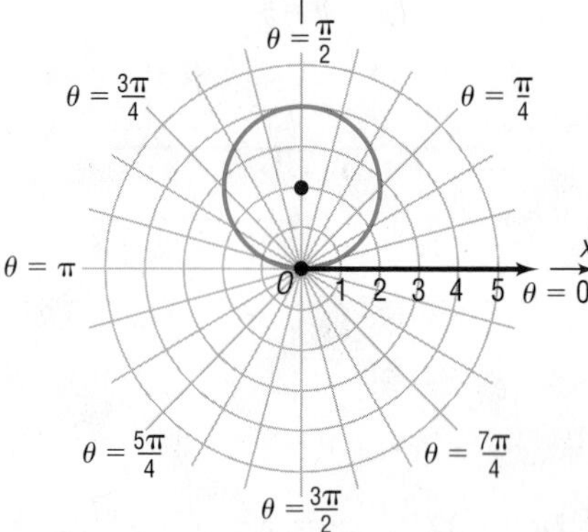

Figure 24
$r = 4 \sin \theta$, or $x^2 + (y - 2)^2 = 4$

EXAMPLE 6 Identifying and Graphing a Polar Equation (Circle)

Identify and graph the equation: $r = -2 \cos \theta$

Solution To transform the equation to rectangular coordinates, multiply each side by r.

$$\begin{aligned} r^2 &= -2r \cos \theta \quad \text{Multiply both sides by } r. \\ x^2 + y^2 &= -2x \quad r^2 = x^2 + y^2;\ x = r\cos\theta \\ x^2 + 2x + y^2 &= 0 \\ (x^2 + 2x + 1) + y^2 &= 1 \quad \text{Complete the square in } x. \\ (x + 1)^2 + y^2 &= 1 \quad \text{Factor.} \end{aligned}$$

This is the standard equation of a circle with center at $(-1, 0)$ in rectangular coordinates and radius 1. See Figure 25.

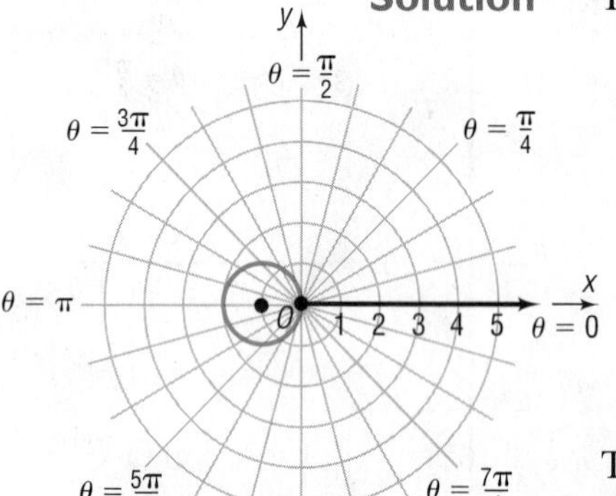

Figure 25
$r = -2 \cos \theta$, or $(x + 1)^2 + y^2 = 1$

Exploration

Using a square screen, graph $r_1 = \sin \theta$, $r_2 = 2 \sin \theta$, and $r_3 = 3 \sin \theta$. Do you see the pattern? Clear the screen and graph $r_1 = -\sin \theta$, $r_2 = -2 \sin \theta$, and $r_3 = -3 \sin \theta$. Do you see the pattern? Clear the screen and graph $r_1 = \cos \theta$, $r_2 = 2 \cos \theta$, and $r_3 = 3 \cos \theta$. Do you see the pattern? Clear the screen and graph $r_1 = -\cos \theta$, $r_2 = -2 \cos \theta$, and $r_3 = -3 \cos \theta$. Do you see the pattern?

Based on Examples 5 and 6 and the Exploration above, we are led to the following results. (The proofs are left as exercises. See Problems 85–88.)

THEOREM

Let a be a positive real number. Then

	Equation	**Description**
(a)	$r = 2a \sin \theta$	Circle: radius a; center at $(0, a)$ in rectangular coordinates
(b)	$r = -2a \sin \theta$	Circle: radius a; center at $(0, -a)$ in rectangular coordinates
(c)	$r = 2a \cos \theta$	Circle: radius a; center at $(a, 0)$ in rectangular coordinates
(d)	$r = -2a \cos \theta$	Circle: radius a; center at $(-a, 0)$ in rectangular coordinates

Each circle passes through the pole.

Now Work PROBLEM 23

The method of converting a polar equation to an identifiable rectangular equation to obtain the graph is not always helpful, nor is it always necessary. Usually, a table is created that lists several points on the graph. By checking for symmetry, it may be possible to reduce the number of points needed to draw the graph.

2 Test Polar Equations for Symmetry

In polar coordinates, the points (r, θ) and $(r, -\theta)$ are symmetric with respect to the polar axis (and to the x-axis). See Figure 26(a). The points (r, θ) and $(r, \pi - \theta)$ are symmetric with respect to the line $\theta = \frac{\pi}{2}$ (the y-axis). See Figure 26(b). The points (r, θ) and $(-r, \theta)$ are symmetric with respect to the pole (the origin). See Figure 26(c).

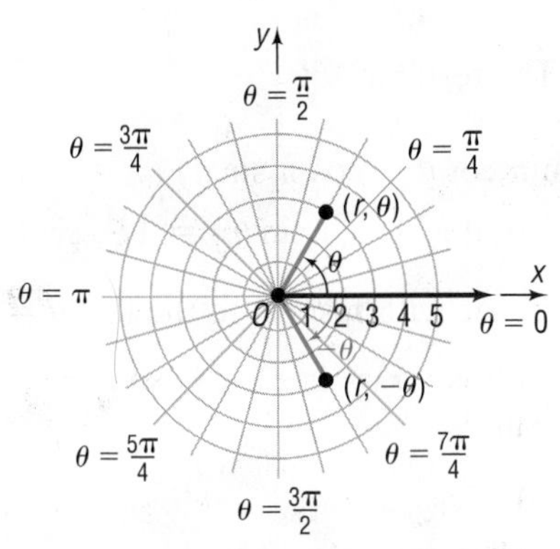

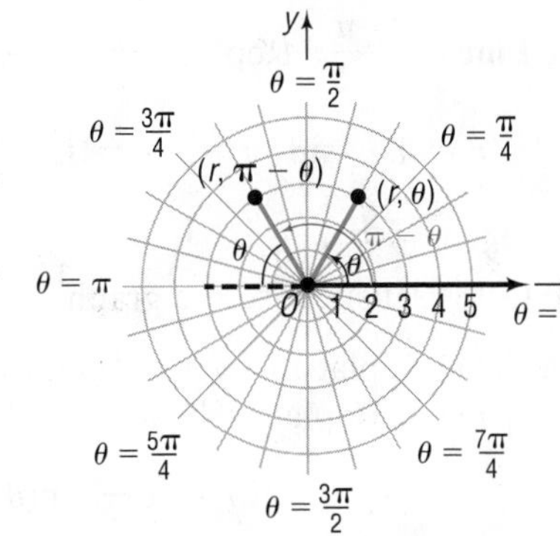

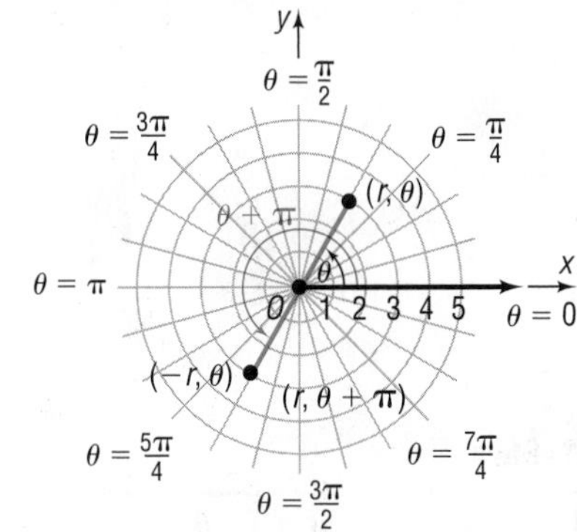

Figure 26 **(a)** Points symmetric with respect to the polar axis **(b)** Points symmetric with respect to the line $\theta = \frac{\pi}{2}$ **(c)** Points symmetric with respect to the pole

The following tests are a consequence of these observations.

THEOREM

Tests for Symmetry

Symmetry with Respect to the Polar Axis (x-Axis)

In a polar equation, replace θ by $-\theta$. If an equivalent equation results, the graph is symmetric with respect to the polar axis.

Symmetry with Respect to the Line $\theta = \frac{\pi}{2}$ (y-Axis)

In a polar equation, replace θ by $\pi - \theta$. If an equivalent equation results, the graph is symmetric with respect to the line $\theta = \frac{\pi}{2}$.

Symmetry with Respect to the Pole (Origin)

In a polar equation, replace r by $-r$ or θ by $\theta + \pi$. If an equivalent equation results, the graph is symmetric with respect to the pole.

The three tests for symmetry given here are *sufficient* conditions for symmetry, but they are not *necessary* conditions. That is, an equation may fail these tests and still have a graph that is symmetric with respect to the polar axis, the line $\theta = \frac{\pi}{2}$, or the pole. For example, the graph of $r = \sin(2\theta)$ turns out to be symmetric with respect to the polar axis, the line $\theta = \frac{\pi}{2}$, and the pole, but only the test for symmetry with respect to the pole (replace θ by $\theta + \pi$) works. See also Problems 89–91.

3 Graph Polar Equations by Plotting Points

EXAMPLE 7 Graphing a Polar Equation (Cardioid)

Graph the equation: $r = 1 - \sin\theta$

Solution Check for symmetry first.

Polar Axis: Replace θ by $-\theta$. The result is

$$r = 1 - \sin(-\theta) = 1 + \sin\theta \quad \sin(-\theta) = -\sin\theta$$

The test fails, so the graph may or may not be symmetric with respect to the polar axis.

The Line $\theta = \frac{\pi}{2}$: Replace θ by $\pi - \theta$. The result is

$$r = 1 - \sin(\pi - \theta) = 1 - (\sin\pi\cos\theta - \cos\pi\sin\theta)$$
$$= 1 - [0\cdot\cos\theta - (-1)\sin\theta] = 1 - \sin\theta$$

The test is satisfied, so the graph is symmetric with respect to the line $\theta = \frac{\pi}{2}$.

The Pole: Replace r by $-r$. Then the result is $-r = 1 - \sin\theta$, so $r = -1 + \sin\theta$. The test fails. Replace θ by $\theta + \pi$. The result is

$$r = 1 - \sin(\theta + \pi)$$
$$= 1 - [\sin\theta\cos\pi + \cos\theta\sin\pi]$$
$$= 1 - [\sin\theta\cdot(-1) + \cos\theta\cdot 0]$$
$$= 1 + \sin\theta$$

This test also fails, so the graph may or may not be symmetric with respect to the pole.

Next, identify points on the graph by assigning values to the angle θ and calculating the corresponding values of r. Due to the periodicity of the sine function and the symmetry with respect to the line $\theta = \frac{\pi}{2}$, just assign values to θ from $-\frac{\pi}{2}$ to $\frac{\pi}{2}$, as given in Table 1.

Now plot the points (r, θ) from Table 1 and trace out the graph, beginning at the point $\left(2, -\frac{\pi}{2}\right)$ and ending at the point $\left(0, \frac{\pi}{2}\right)$. Then reflect this portion of the graph about the line $\theta = \frac{\pi}{2}$ (the y-axis) to obtain the complete graph. See Figure 27.

Table 1

θ	$r = 1 - \sin\theta$
$-\frac{\pi}{2}$	$1 - (-1) = 2$
$-\frac{\pi}{3}$	$1 - \left(-\frac{\sqrt{3}}{2}\right) \approx 1.87$
$-\frac{\pi}{6}$	$1 - \left(-\frac{1}{2}\right) = \frac{3}{2}$
0	$1 - 0 = 1$
$\frac{\pi}{6}$	$1 - \frac{1}{2} = \frac{1}{2}$
$\frac{\pi}{3}$	$1 - \frac{\sqrt{3}}{2} \approx 0.13$
$\frac{\pi}{2}$	$1 - 1 = 0$

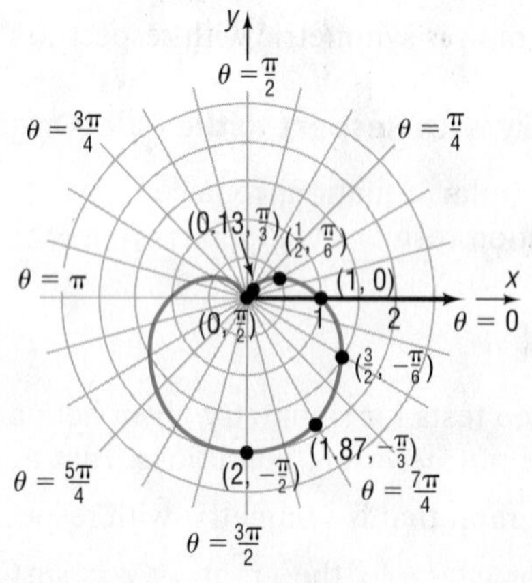

Figure 27 $r = 1 - \sin\theta$

The curve in Figure 27 is an example of a *cardioid* (a heart-shaped curve).

Exploration

Graph $r_1 = 1 + \sin\theta$. Clear the screen and graph $r_1 = 1 - \cos\theta$. Clear the screen and graph $r_1 = 1 + \cos\theta$. Do you see a pattern?

DEFINITION

Cardioids are characterized by equations of the form

$$r = a(1 + \cos\theta) \qquad r = a(1 + \sin\theta)$$
$$r = a(1 - \cos\theta) \qquad r = a(1 - \sin\theta)$$

where $a > 0$. The graph of a cardioid passes through the pole.

Now Work PROBLEM 39

EXAMPLE 8 Graphing a Polar Equation (Limaçon without an Inner Loop)

Graph the equation: $r = 3 + 2\cos\theta$

Solution Check for symmetry first.

Polar Axis: Replace θ by $-\theta$. The result is

$$r = 3 + 2\cos(-\theta) = 3 + 2\cos\theta \quad \cos(-\theta) = \cos\theta$$

The test is satisfied, so the graph is symmetric with respect to the polar axis.

The Line $\theta = \frac{\pi}{2}$: Replace θ by $\pi - \theta$. The result is

$$r = 3 + 2\cos(\pi - \theta) = 3 + 2(\cos\pi\cos\theta + \sin\pi\sin\theta)$$
$$= 3 - 2\cos\theta$$

The test fails, so the graph may or may not be symmetric with respect to the line $\theta = \frac{\pi}{2}$.

The Pole: Replace r by $-r$. The test fails, so the graph may or may not be symmetric with respect to the pole. Replace θ by $\theta + \pi$. The test fails, so the graph may or may not be symmetric with respect to the pole.

Next, identify points on the graph by assigning values to the angle θ and calculating the corresponding values of r. Due to the periodicity of the cosine function and the symmetry with respect to the polar axis, just assign values to θ from 0 to π, as given in Table 2.

Now plot the points (r, θ) from Table 2 and trace out the graph, beginning at the point $(5, 0)$ and ending at the point $(1, \pi)$. Then reflect this portion of the graph about the polar axis (the x-axis) to obtain the complete graph. See Figure 28.

Table 2

θ	$r = 3 + 2\cos\theta$
0	$3 + 2(1) = 5$
$\frac{\pi}{6}$	$3 + 2\left(\frac{\sqrt{3}}{2}\right) \approx 4.73$
$\frac{\pi}{3}$	$3 + 2\left(\frac{1}{2}\right) = 4$
$\frac{\pi}{2}$	$3 + 2(0) = 3$
$\frac{2\pi}{3}$	$3 + 2\left(-\frac{1}{2}\right) = 2$
$\frac{5\pi}{6}$	$3 + 2\left(-\frac{\sqrt{3}}{2}\right) \approx 1.27$
π	$3 + 2(-1) = 1$

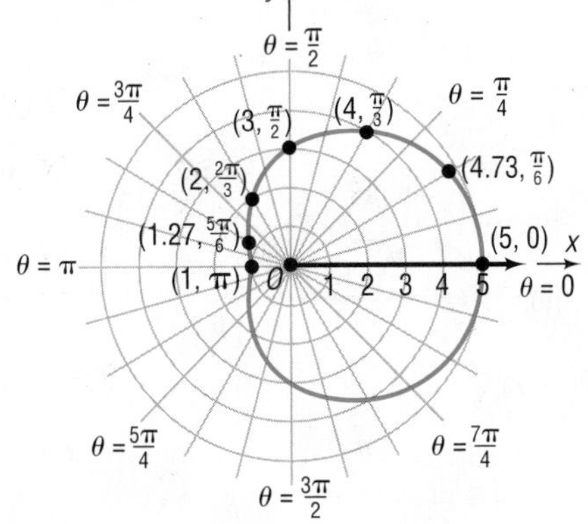

Figure 28 $r = 3 + 2\cos\theta$

The curve in Figure 28 is an example of a *limaçon* (a French word for *snail*) *without an inner loop.*

Exploration

Graph $r_1 = 3 - 2\cos\theta$. Clear the screen and graph $r_1 = 3 + 2\sin\theta$. Clear the screen and graph $r_1 = 3 - 2\sin\theta$. Do you see a pattern?

DEFINITION

Limaçons without an inner loop are characterized by equations of the form

$$r = a + b\cos\theta \qquad r = a + b\sin\theta$$
$$r = a - b\cos\theta \qquad r = a - b\sin\theta$$

where $a > 0$, $b > 0$, and $a > b$. The graph of a limaçon without an inner loop does not pass through the pole.

Now Work PROBLEM 45

EXAMPLE 9 **Graphing a Polar Equation (Limaçon with an Inner Loop)**

Graph the equation: $r = 1 + 2\cos\theta$

Solution First, check for symmetry.

Polar Axis: Replace θ by $-\theta$. The result is

$$r = 1 + 2\cos(-\theta) = 1 + 2\cos\theta$$

The test is satisfied, so the graph is symmetric with respect to the polar axis.

The Line $\theta = \frac{\pi}{2}$: Replace θ by $\pi - \theta$. The result is

$$r = 1 + 2\cos(\pi - \theta) = 1 + 2(\cos\pi\cos\theta + \sin\pi\sin\theta)$$
$$= 1 - 2\cos\theta$$

The test fails, so the graph may or may not be symmetric with respect to the line $\theta = \frac{\pi}{2}$.

The Pole: Replace r by $-r$. The test fails, so the graph may or may not be symmetric with respect to the pole. Replace θ by $\theta + \pi$. The test fails, so the graph may or may not be symmetric with respect to the pole.

Next, identify points on the graph of $r = 1 + 2\cos\theta$ by assigning values to the angle θ and calculating the corresponding values of r. Due to the periodicity of the cosine function and the symmetry with respect to the polar axis, just assign values to θ from 0 to π, as given in Table 3.

Table 3

θ	$r = 1 + 2\cos\theta$
0	$1 + 2(1) = 3$
$\frac{\pi}{6}$	$1 + 2\left(\frac{\sqrt{3}}{2}\right) \approx 2.73$
$\frac{\pi}{3}$	$1 + 2\left(\frac{1}{2}\right) = 2$
$\frac{\pi}{2}$	$1 + 2(0) = 1$
$\frac{2\pi}{3}$	$1 + 2\left(-\frac{1}{2}\right) = 0$
$\frac{5\pi}{6}$	$1 + 2\left(-\frac{\sqrt{3}}{2}\right) \approx -0.73$
π	$1 + 2(-1) = -1$

Now plot the points (r, θ) from Table 3, beginning at $(3, 0)$ and ending at $(-1, \pi)$. See Figure 29(a). Finally, reflect this portion of the graph about the polar axis (the x-axis) to obtain the complete graph. See Figure 29(b).

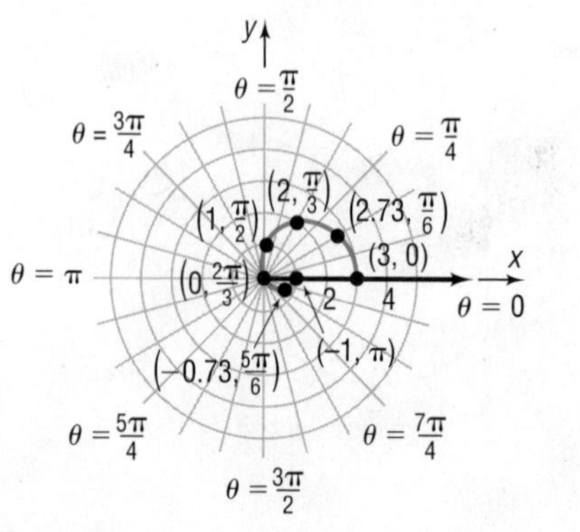

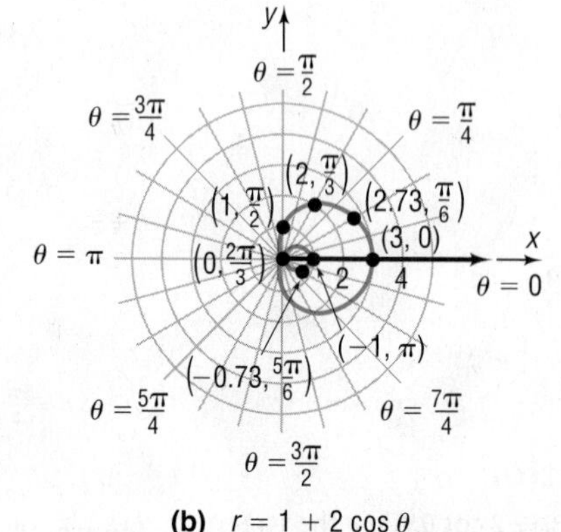

Figure 29 (a) (b) $r = 1 + 2\cos\theta$

Exploration

Graph $r_1 = 1 - 2\cos\theta$. Clear the screen and graph $r_1 = 1 + 2\sin\theta$. Clear the screen and graph $r_1 = 1 - 2\sin\theta$. Do you see a pattern?

The curve in Figure 29(b) is an example of a *limaçon with an inner loop*.

DEFINITION

Limaçons with an inner loop are characterized by equations of the form

$$r = a + b\cos\theta \qquad r = a + b\sin\theta$$
$$r = a - b\cos\theta \qquad r = a - b\sin\theta$$

where $a > 0$, $b > 0$, and $a < b$. The graph of a limaçon with an inner loop passes through the pole twice.

Now Work PROBLEM 47

EXAMPLE 10 Graphing a Polar Equation (Rose)

Graph the equation: $r = 2\cos(2\theta)$

Solution Check for symmetry.

Polar Axis: Replace θ by $-\theta$. The result is

$$r = 2\cos[2(-\theta)] = 2\cos(2\theta)$$

The test is satisfied, so the graph is symmetric with respect to the polar axis.

The Line $\theta = \frac{\pi}{2}$: Replace θ by $\pi - \theta$. The result is

$$r = 2\cos[2(\pi - \theta)] = 2\cos(2\pi - 2\theta) = 2\cos(2\theta)$$

The test is satisfied, so the graph is symmetric with respect to the line $\theta = \frac{\pi}{2}$.

The Pole: Since the graph is symmetric with respect to both the polar axis and the line $\theta = \frac{\pi}{2}$, it must be symmetric with respect to the pole.

Next, construct Table 4. Because of the periodicity of the cosine function and the symmetry with respect to the polar axis, the line $\theta = \frac{\pi}{2}$, and the pole, consider only values of θ from 0 to $\frac{\pi}{2}$.

Table 4

θ	$r = 2\cos(2\theta)$
0	$2(1) = 2$
$\frac{\pi}{6}$	$2\left(\frac{1}{2}\right) = 1$
$\frac{\pi}{4}$	$2(0) = 0$
$\frac{\pi}{3}$	$2\left(-\frac{1}{2}\right) = -1$
$\frac{\pi}{2}$	$2(-1) = -2$

Plot and connect these points as shown in Figure 30(a). Finally, because of symmetry, reflect this portion of the graph first about the polar axis (the x-axis) and then about the line $\theta = \frac{\pi}{2}$ (the y-axis) to obtain the complete graph. See Figure 30(b).

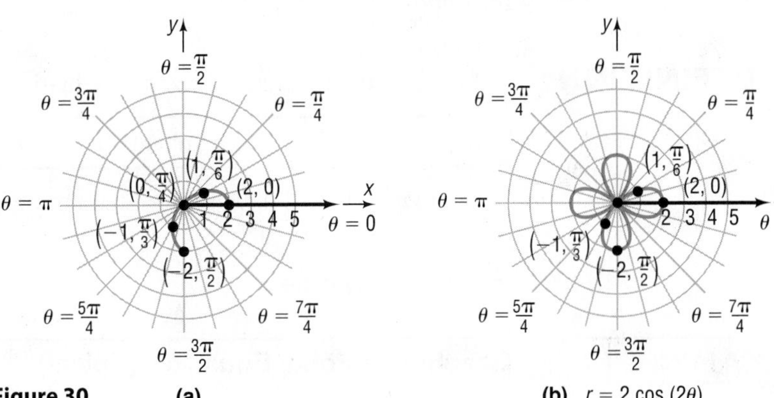

Figure 30 (a) (b) $r = 2\cos(2\theta)$

Exploration

Graph $r_1 = 2\cos(4\theta)$; clear the screen and graph $r_1 = 2\cos(6\theta)$. How many petals did each of these graphs have?

Clear the screen and graph, in order, each on a clear screen, $r_1 = 2\cos(3\theta)$, $r_1 = 2\cos(5\theta)$, and $r_1 = 2\cos(7\theta)$. What do you notice about the number of petals?

The curve in Figure 30(b) is called a *rose* with four petals.

DEFINITION

Rose curves are characterized by equations of the form

$$r = a\cos(n\theta) \qquad r = a\sin(n\theta) \qquad a \neq 0$$

and have graphs that are rose shaped. If $n \neq 0$ is even, the rose has $2n$ petals; if $n \neq \pm 1$ is odd, the rose has n petals.

Now Work PROBLEM 51

EXAMPLE 11 Graphing a Polar Equation (Lemniscate)

Graph the equation: $r^2 = 4\sin(2\theta)$

Solution We leave it to you to verify that the graph is symmetric with respect to the pole. Because of the symmetry with respect to the pole, consider only those values of θ between $\theta = 0$ and $\theta = \pi$. Note that there are no points on the graph for $\frac{\pi}{2} < \theta < \pi$ (quadrant II), since $r^2 < 0$ for such values. Table 5 lists points on the graph for values of $\theta = 0$ through $\theta = \frac{\pi}{2}$. The points from Table 5 where $r \geq 0$ are plotted in Figure 31(a). The remaining points on the graph may be obtained by using symmetry. Figure 31(b) shows the final graph drawn.

Table 5

θ	$r^2 = 4\sin(2\theta)$	r
0	$4(0) = 0$	0
$\frac{\pi}{6}$	$4\left(\frac{\sqrt{3}}{2}\right) = 2\sqrt{3}$	± 1.9
$\frac{\pi}{4}$	$4(1) = 4$	± 2
$\frac{\pi}{3}$	$4\left(\frac{\sqrt{3}}{2}\right) = 2\sqrt{3}$	± 1.9
$\frac{\pi}{2}$	$4(0) = 0$	0

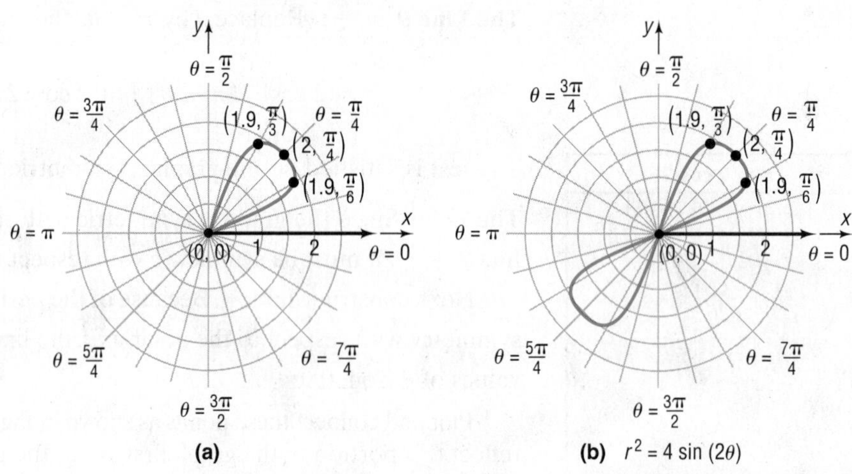

Figure 31 (a) (b) $r^2 = 4\sin(2\theta)$

The curve in Figure 31(b) is an example of a *lemniscate* (from the Greek word for *ribbon*).

DEFINITION

Lemniscates are characterized by equations of the form

$$r^2 = a^2\sin(2\theta) \qquad r^2 = a^2\cos(2\theta)$$

where $a \neq 0$, and have graphs that are propeller shaped.

Now Work PROBLEM 55

EXAMPLE 12 Graphing a Polar Equation (Spiral)

Graph the equation: $r = e^{\theta/5}$

Solution The tests for symmetry with respect to the pole, the polar axis, and the line $\theta = \frac{\pi}{2}$ fail. Furthermore, there is no number θ for which $r = 0$, so the graph does not pass through the pole. Observe that r is positive for all θ, r increases as θ increases, $r \rightarrow 0$

Table 6

θ	$r = e^{\theta/5}$
$-\dfrac{3\pi}{2}$	0.39
$-\pi$	0.53
$-\dfrac{\pi}{2}$	0.73
$-\dfrac{\pi}{4}$	0.85
0	1
$\dfrac{\pi}{4}$	1.17
$\dfrac{\pi}{2}$	1.37
π	1.87
$\dfrac{3\pi}{2}$	2.57
2π	3.51

as $\theta \to -\infty$, and $r \to \infty$ as $\theta \to \infty$. With the help of a calculator, the values in Table 6 can be obtained. See Figure 32.

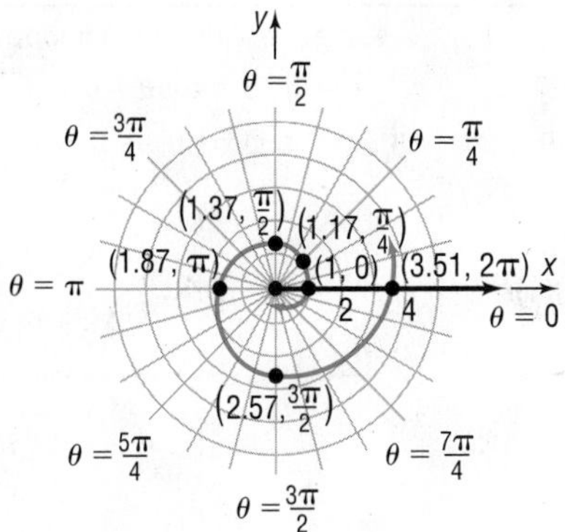

Figure 32 $r = e^{\theta/5}$

The curve in Figure 32 is called a **logarithmic spiral**, since its equation may be written as $\theta = 5 \ln r$ and it spirals infinitely both toward the pole and away from it.

Classification of Polar Equations

The equations of some lines and circles in polar coordinates and their corresponding equations in rectangular coordinates are given in Table 7. Also included are the names and graphs of a few of the more frequently encountered polar equations.

Table 7

Lines

Description	Line passing through the pole making an angle α with the polar axis	Vertical line	Horizontal line
Rectangular equation	$y = (\tan \alpha)x$	$x = a$	$y = b$
Polar equation	$\theta = \alpha$	$r \cos \theta = a$	$r \sin \theta = b$
Typical graph			

Circles

Description	Center at the pole, radius a	Passing through the pole, tangent to the line $\theta = \dfrac{\pi}{2}$, center on the polar axis, radius a	Passing through the pole, tangent to the polar axis, center on the line $\theta = \dfrac{\pi}{2}$, radius a
Rectangular equation	$x^2 + y^2 = a^2, \quad a > 0$	$x^2 + y^2 = \pm 2ax, \quad a > 0$	$x^2 + y^2 = \pm 2ay, \quad a > 0$
Polar equation	$r = a, \quad a > 0$	$r = \pm 2a \cos \theta, \quad a > 0$	$r = \pm 2a \sin \theta, \quad a > 0$
Typical graph			

(continued)

Table 7 (Continued)

Other Equations			
Name	Cardioid	Limaçon without inner loop	Limaçon with inner loop
Polar equations	$r = a \pm a\cos\theta,\quad a > 0$	$r = a \pm b\cos\theta,\quad 0 < b < a$	$r = a \pm b\cos\theta,\quad 0 < a < b$
	$r = a \pm a\sin\theta,\quad a > 0$	$r = a \pm b\sin\theta,\quad 0 < b < a$	$r = a \pm b\sin\theta,\quad 0 < a < b$
Typical graph			
Name	Lemniscate	Rose with three petals	Rose with four petals
Polar equations	$r^2 = a^2\cos(2\theta),\quad a \neq 0$	$r = a\sin(3\theta),\quad a > 0$	$r = a\sin(2\theta),\quad a > 0$
	$r^2 = a^2\sin(2\theta),\quad a \neq 0$	$r = a\cos(3\theta),\quad a > 0$	$r = a\cos(2\theta),\quad a > 0$
Typical graph			

Sketching Quickly

If a polar equation involves only a sine (or cosine) function, you can quickly obtain its graph by making use of Table 7, periodicity, and a short table.

EXAMPLE 13 Sketching the Graph of a Polar Equation Quickly

Graph the equation: $r = 2 + 2\sin\theta$

Solution You should recognize the polar equation: Its graph is a cardioid. The period of $\sin\theta$ is 2π, so form a table using $0 \le \theta \le 2\pi$, compute r, plot the points (r, θ), and sketch the graph of a cardioid as θ varies from 0 to 2π. See Table 8 and Figure 33.

Table 8

θ	$r = 2 + 2\sin\theta$
0	$2 + 2(0) = 2$
$\frac{\pi}{2}$	$2 + 2(1) = 4$
π	$2 + 2(0) = 2$
$\frac{3\pi}{2}$	$2 + 2(-1) = 0$
2π	$2 + 2(0) = 2$

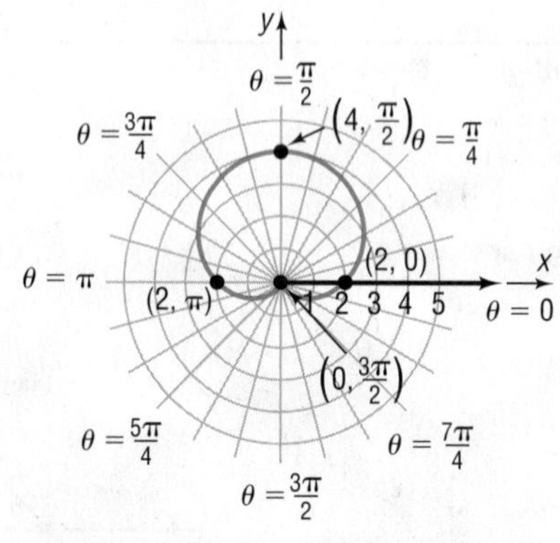

Figure 33 $r = 2 + 2\sin\theta$

Calculus Comment For those of you who are planning to study calculus, a comment about one important role of polar equations is in order.

In rectangular coordinates, the equation $x^2 + y^2 = 1$, whose graph is the unit circle, is not the graph of a function. In fact, it requires two functions to obtain the graph of the unit circle:

$$y_1 = \sqrt{1 - x^2} \quad \text{Upper semicircle} \qquad y_2 = -\sqrt{1 - x^2} \quad \text{Lower semicircle}$$

In polar coordinates, the equation $r = 1$, whose graph is also the unit circle, does define a function. For each choice of θ, there is only one corresponding value of r, that is, $r = 1$. Since many problems in calculus require the use of functions, the opportunity to express nonfunctions in rectangular coordinates as functions in polar coordinates becomes extremely useful.

Note also that the vertical-line test for functions is valid only for equations in rectangular coordinates.

Historical Feature

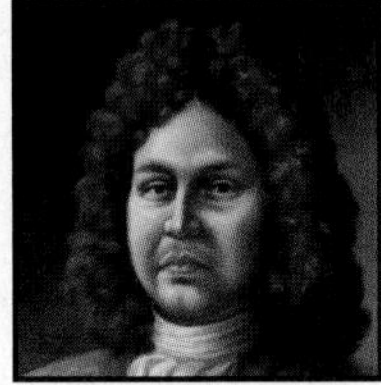

Jakob Bernoulli (1654–1705)

Polar coordinates seem to have been invented by Jakob Bernoulli (1654–1705) in about 1691, although, as with most such ideas, earlier traces of the notion exist. Early users of calculus remained committed to rectangular coordinates, and polar coordinates did not become widely used until the early 1800s. Even then, it was mostly geometers who used them for describing odd curves. Finally, about the mid-1800s, applied mathematicians realized the tremendous simplification that polar coordinates make possible in the description of objects with circular or cylindrical symmetry. From then on, their use became widespread.

5.2 Assess Your Understanding

'Are You Prepared?' *Answers are given at the end of these exercises. If you get a wrong answer, read the pages listed in red.*

1. If the rectangular coordinates of a point are $(4, -6)$, the point symmetric to it with respect to the origin is ________. (pp. 12–13)

2. The difference formula for cosine is $\cos(A - B) =$ _____. (p. 226)

3. The standard equation of a circle with center at $(-2, 5)$ and radius 3 is _____. (pp. 15–18)

4. Is the sine function even, odd, or neither? (p. 141)

5. $\sin \frac{5\pi}{4} =$ _____. (pp. 117–124)

6. $\cos \frac{2\pi}{3} =$ ____. (pp. 117–124)

Concepts and Vocabulary

7. An equation whose variables are polar coordinates is called a(n) _____ ________.

8. ***True or False*** The tests for symmetry in polar coordinates are always conclusive.

9. To test whether the graph of a polar equation may be symmetric with respect to the polar axis, replace θ by ____.

10. To test whether the graph of a polar equation may be symmetric with respect to the line $\theta = \frac{\pi}{2}$, replace θ by _______.

11. ***True or False*** A cardioid passes through the pole.

12. Rose curves are characterized by equations of the form $r = a\cos(n\theta)$ or $r = a\sin(n\theta)$, $a \neq 0$. If $n \neq 0$ is even, the rose has ____ petals; if $n \neq \pm 1$ is odd, the rose has _____ petals.

13. For a positive real number a, the graph of which of the following polar equations is a circle with radius a and center at $(a, 0)$ in rectangular coordinates?
(a) $r = 2a\sin\theta$ (b) $r = -2a\sin\theta$
(c) $r = 2a\cos\theta$ (d) $r = -2a\cos\theta$

14. In polar coordinates, the points (r, θ) and $(-r, \theta)$ are symmetric with respect to which of the following?
(a) the polar axis (or x-axis) (b) the pole (or origin)
(c) the line $\theta = \frac{\pi}{2}$ (or y-axis) (d) the line $\theta = \frac{\pi}{4}$ (or $y = x$)

Skill Building

In Problems 15–30, transform each polar equation to an equation in rectangular coordinates. Then identify and graph the equation.

15. $r = 4$
16. $r = 2$
17. $\theta = \frac{\pi}{3}$
18. $\theta = -\frac{\pi}{4}$
19. $r\sin\theta = 4$
20. $r\cos\theta = 4$
21. $r\cos\theta = -2$
22. $r\sin\theta = -2$

23. $r = 2\cos\theta$ **24.** $r = 2\sin\theta$ **25.** $r = -4\sin\theta$ **26.** $r = -4\cos\theta$

27. $r\sec\theta = 4$ **28.** $r\csc\theta = 8$ **29.** $r\csc\theta = -2$ **30.** $r\sec\theta = -4$

In Problems 31–38, match each of the graphs (A) through (H) to one of the following polar equations.

31. $r = 2$ **32.** $\theta = \dfrac{\pi}{4}$ **33.** $r = 2\cos\theta$ **34.** $r\cos\theta = 2$

35. $r = 1 + \cos\theta$ **36.** $r = 2\sin\theta$ **37.** $\theta = \dfrac{3\pi}{4}$ **38.** $r\sin\theta = 2$

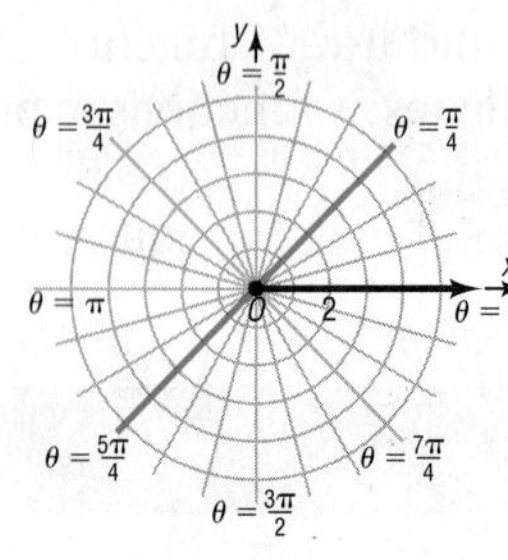
(A)

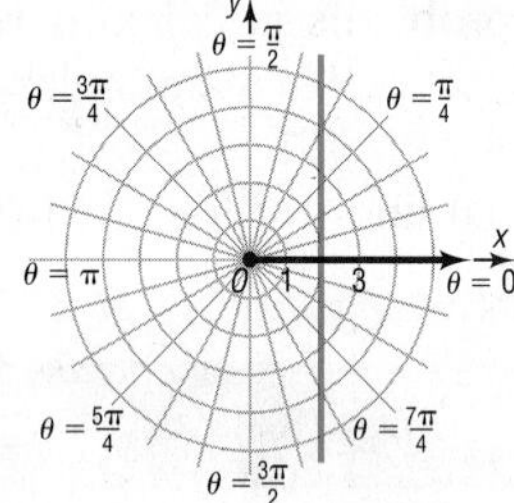
(B)

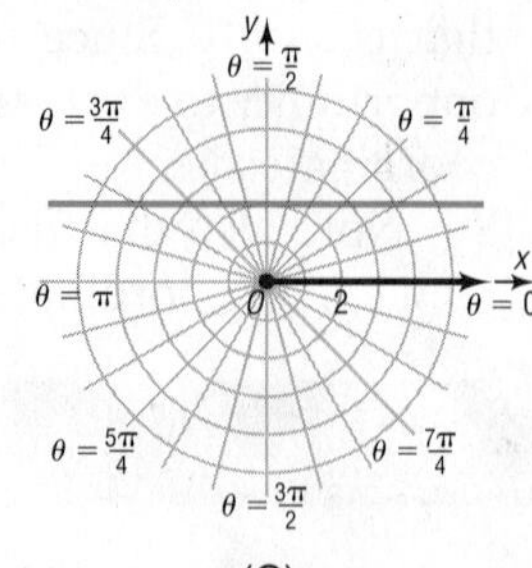
(C)

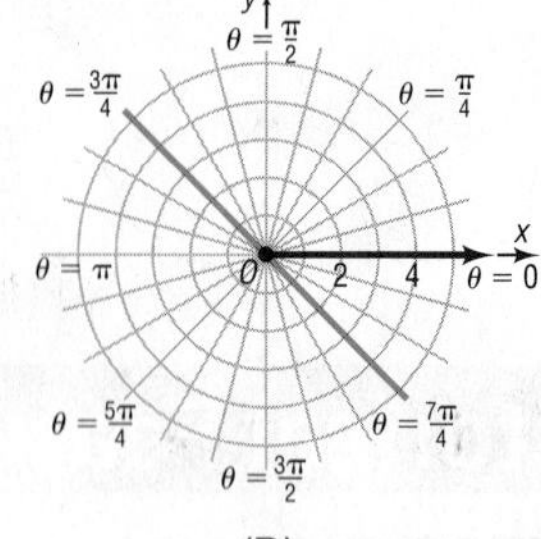
(D)

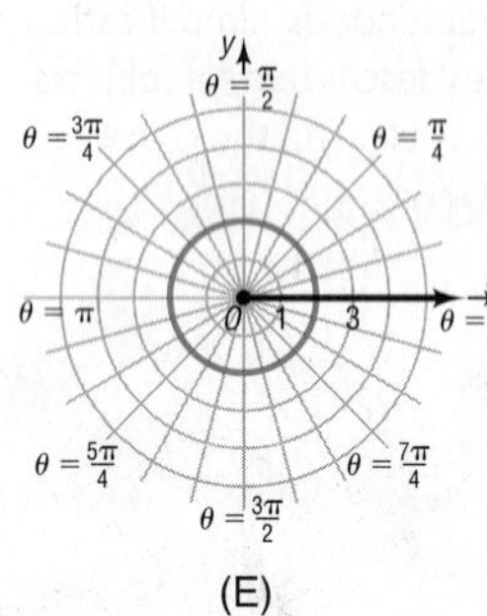
(E)

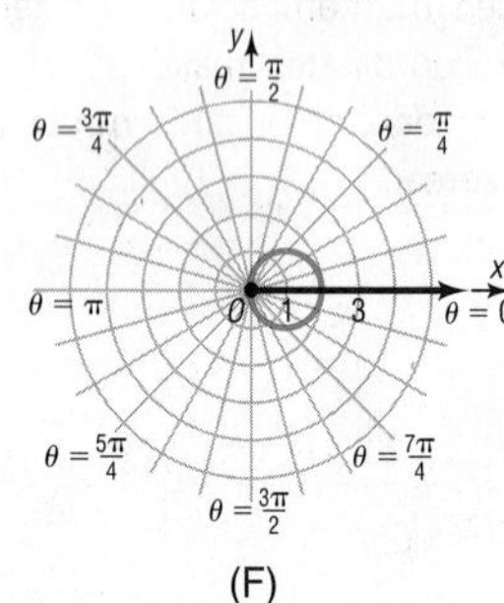
(F)

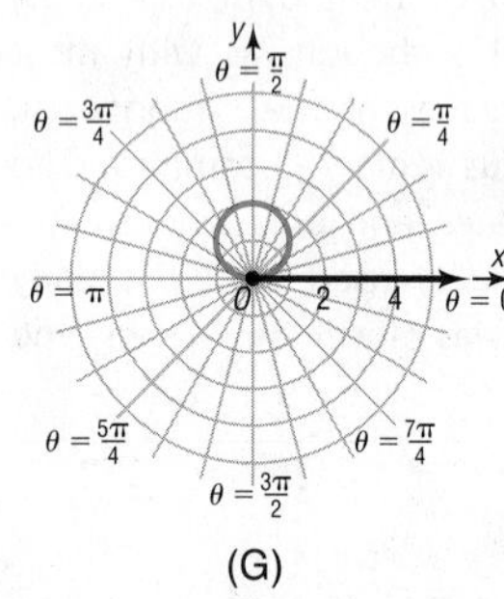
(G)

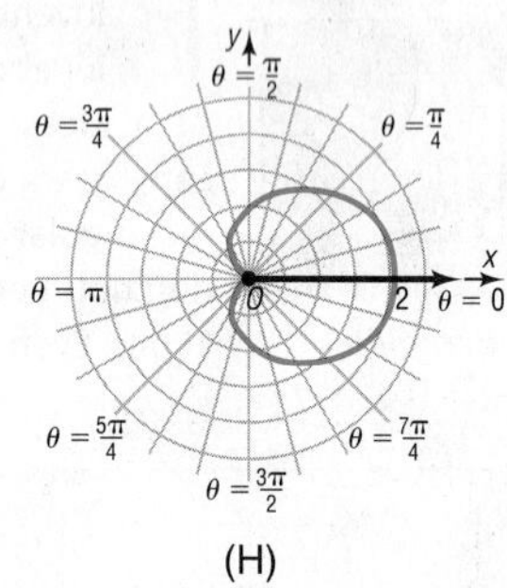
(H)

In Problems 39–62, identify and graph each polar equation.

39. $r = 2 + 2\cos\theta$ **40.** $r = 1 + \sin\theta$ **41.** $r = 3 - 3\sin\theta$ **42.** $r = 2 - 2\cos\theta$

43. $r = 2 + \sin\theta$ **44.** $r = 2 - \cos\theta$ **45.** $r = 4 - 2\cos\theta$ **46.** $r = 4 + 2\sin\theta$

47. $r = 1 + 2\sin\theta$ **48.** $r = 1 - 2\sin\theta$ **49.** $r = 2 - 3\cos\theta$ **50.** $r = 2 + 4\cos\theta$

51. $r = 3\cos(2\theta)$ **52.** $r = 2\sin(3\theta)$ **53.** $r = 4\sin(5\theta)$ **54.** $r = 3\cos(4\theta)$

55. $r^2 = 9\cos(2\theta)$ **56.** $r^2 = \sin(2\theta)$ **57.** $r = 2^\theta$ **58.** $r = 3^\theta$

59. $r = 1 - \cos\theta$ **60.** $r = 3 + \cos\theta$ **61.** $r = 1 - 3\cos\theta$ **62.** $r = 4\cos(3\theta)$

Mixed Practice

In Problems 63–68, graph each pair of polar equations on the same polar grid. Find the polar coordinates of the point(s) of intersection and label the point(s) on the graph.

63. $r = 8\cos\theta;\ r = 2\sec\theta$ **64.** $r = 8\sin\theta;\ r = 4\csc\theta$ **65.** $r = \sin\theta;\ r = 1 + \cos\theta$

66. $r = 3;\ r = 2 + 2\cos\theta$ **67.** $r = 1 + \sin\theta;\ r = 1 + \cos\theta$ **68.** $r = 1 + \cos\theta;\ r = 3\cos\theta$

Applications and Extensions

In Problems 69–72, the polar equation for each graph is either $r = a + b\cos\theta$ or $r = a + b\sin\theta$, $a > 0$. Select the correct equation and find the values of a and b.

69.

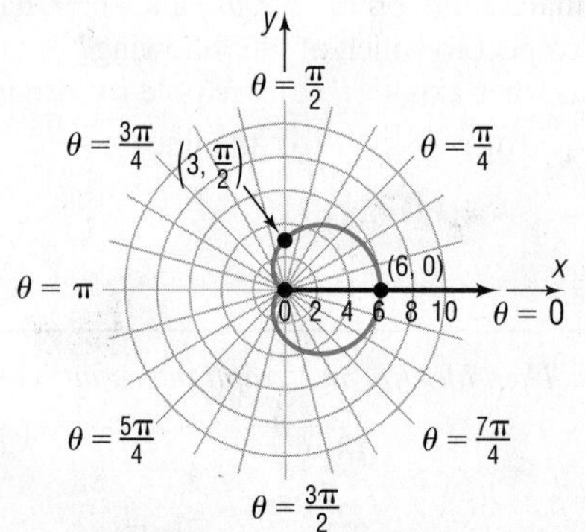

70.

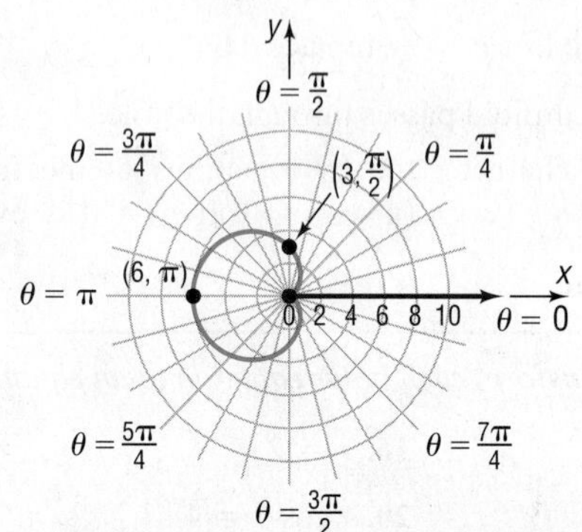

71.

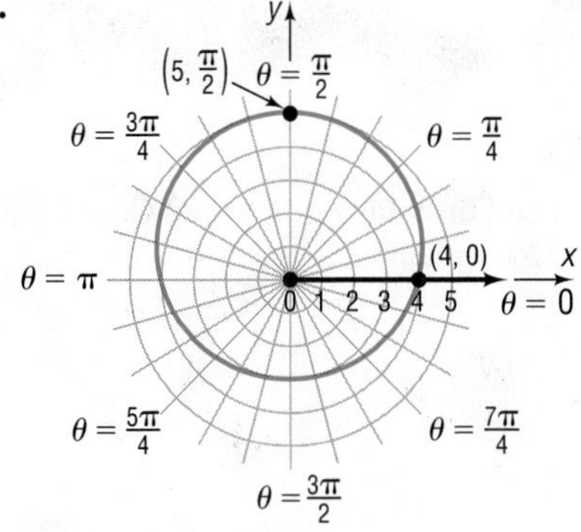

72.

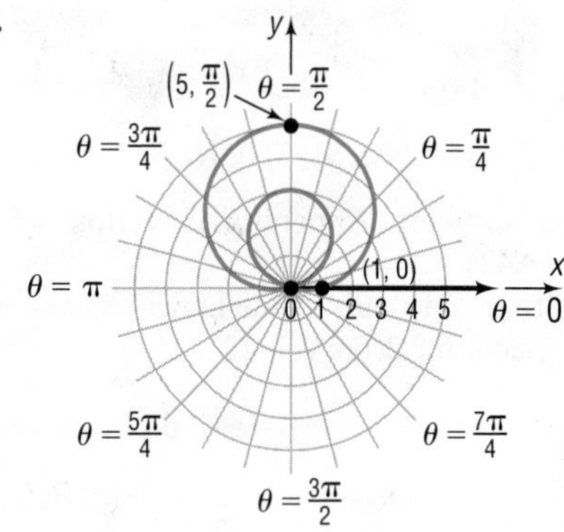

In Problems 73–82, graph each polar equation.

73. $r = \dfrac{2}{1 - \cos\theta}$ *(parabola)*

74. $r = \dfrac{2}{1 - 2\cos\theta}$ *(hyperbola)*

75. $r = \dfrac{1}{3 - 2\cos\theta}$ *(ellipse)*

76. $r = \dfrac{1}{1 - \cos\theta}$ *(parabola)*

77. $r = \theta, \quad \theta \geq 0$ *(spiral of Archimedes)*

78. $r = \dfrac{3}{\theta}$ *(reciprocal spiral)*

79. $r = \csc\theta - 2, \quad 0 < \theta < \pi$ *(conchoid)*

80. $r = \sin\theta\tan\theta$ *(cissoid)*

81. $r = \tan\theta, \quad -\dfrac{\pi}{2} < \theta < \dfrac{\pi}{2}$ *(kappa curve)*

82. $r = \cos\dfrac{\theta}{2}$

83. Show that the graph of the equation $r\sin\theta = a$ is a horizontal line a units above the pole if $a \geq 0$ and $|a|$ units below the pole if $a < 0$.

84. Show that the graph of the equation $r\cos\theta = a$ is a vertical line a units to the right of the pole if $a \geq 0$ and $|a|$ units to the left of the pole if $a < 0$.

85. Show that the graph of the equation $r = 2a\sin\theta, a > 0$, is a circle of radius a with center at $(0, a)$ in rectangular coordinates.

86. Show that the graph of the equation $r = -2a\sin\theta, a > 0$, is a circle of radius a with center at $(0, -a)$ in rectangular coordinates.

87. Show that the graph of the equation $r = 2a\cos\theta, a > 0$, is a circle of radius a with center at $(a, 0)$ in rectangular coordinates.

88. Show that the graph of the equation $r = -2a\cos\theta, a > 0$, is a circle of radius a with center at $(-a, 0)$ in rectangular coordinates.

Explaining Concepts: Discussion and Writing

89. Explain why the following test for symmetry is valid: Replace r by $-r$ and θ by $-\theta$ in a polar equation. If an equivalent equation results, the graph is symmetric with respect to the line $\theta = \dfrac{\pi}{2}$ (y-axis).

(a) Show that the test on page 325 fails for $r^2 = \cos\theta$, yet this new test works.

(b) Show that the test on page 325 works for $r^2 = \sin\theta$, yet this new test fails.

90. Write down two different tests for symmetry with respect to the polar axis. Find examples in which one test works and the other fails. Which test do you prefer to use? Justify your answer.

91. The tests for symmetry given on page 325 are sufficient, but not necessary. Explain what this means.

92. Explain why the vertical-line test used to identify functions in rectangular coordinates does not work for equations expressed in polar coordinates.

Retain Your Knowledge

Problems 93–96 are based on material learned earlier in the course. The purpose of these problems is to keep the material fresh in your mind so that you are better prepared for the final exam.

93. Given $f(x) = x^2 - 3x$, find $f(x + 2)$.

94. Convert $\dfrac{7\pi}{3}$ radians to degrees.

95. Determine the amplitude and period of $y = -2\sin(5x)$ without graphing.

96. Solve triangle ABC: $a = 5, A = 38°$, and $B = 57°$

'Are You Prepared?' Answers

1. $(-4, 6)$ **2.** $\cos A\cos B + \sin A\sin B$ **3.** $(x + 2)^2 + (y - 5)^2 = 9$ **4.** Odd **5.** $-\dfrac{\sqrt{2}}{2}$ **6.** $-\dfrac{1}{2}$

5.3 The Complex Plane; De Moivre's Theorem

PREPARING FOR THIS SECTION *Before getting started, review the following:*

- Complex Numbers (Appendix A, Section A.5, pp. A37–A42)
- Values of the Sine and Cosine Functions at Certain Angles (Section 2.2, pp. 117–124)
- Sum and Difference Formulas for Sine and Cosine (Section 3.5, pp. 226 and 229)

Now Work the 'Are You Prepared?' problems on page 342.

OBJECTIVES
1. Plot Points in the Complex Plane (p. 336)
2. Convert a Complex Number between Rectangular Form and Polar Form (p. 337)
3. Find Products and Quotients of Complex Numbers in Polar Form (p. 338)
4. Use De Moivre's Theorem (p. 339)
5. Find Complex Roots (p. 340)

1 Plot Points in the Complex Plane

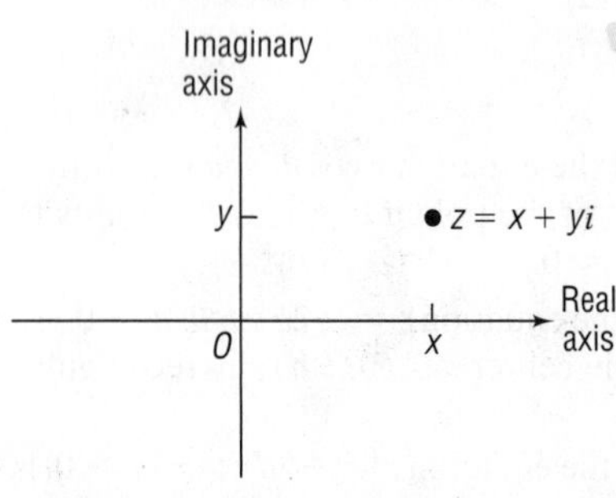

Figure 34 Complex plane

Complex numbers are discussed in Appendix A, Section A.5. In that discussion, we were not prepared to give a geometric interpretation of a complex number. Now we are ready.

A complex number $z = x + yi$ can be interpreted geometrically as the point (x, y) in the xy-plane. Each point in the plane corresponds to a complex number, and conversely, each complex number corresponds to a point in the plane. The collection of such points is referred to as the **complex plane**. The x-axis is referred to as the **real axis**, because any point that lies on the real axis is of the form $z = x + 0i = x$, a real number. The y-axis is called the **imaginary axis**, because any point that lies on it is of the form $z = 0 + yi = yi$, a pure imaginary number. See Figure 34.

EXAMPLE 1 **Plotting a Point in the Complex Plane**

Plot the point corresponding to $z = \sqrt{3} - i$ in the complex plane.

Solution The point corresponding to $z = \sqrt{3} - i$ has the rectangular coordinates $(\sqrt{3}, -1)$. This point, located in quadrant IV, is plotted in Figure 35.

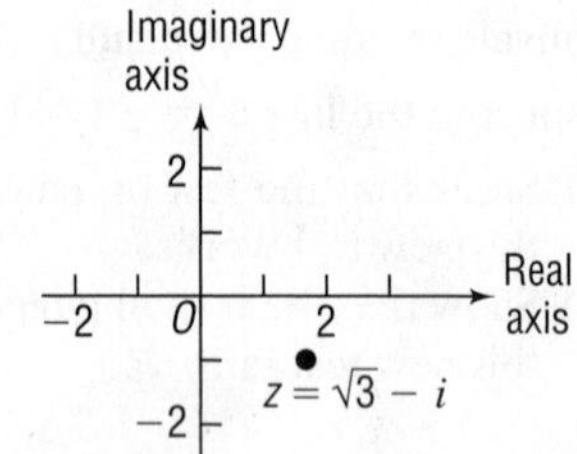

Figure 35

DEFINITION

Let $z = x + yi$ be a complex number. The **magnitude** or **modulus** of z, denoted by $|z|$, is defined as the distance from the origin to the point (x, y). That is,

$$|z| = \sqrt{x^2 + y^2} \tag{1}$$

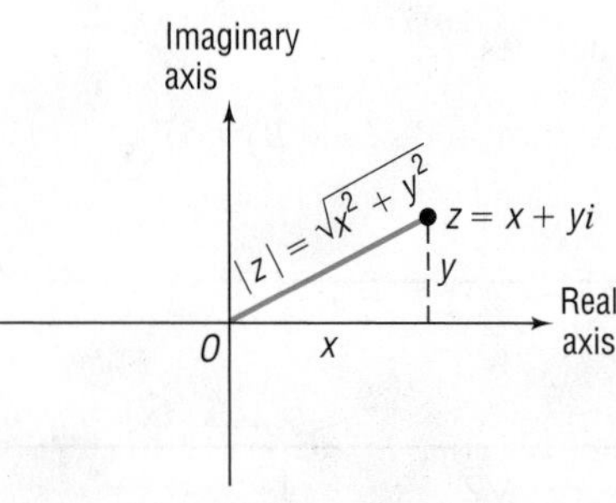

Figure 36

See Figure 36 for an illustration.

This definition for $|z|$ is consistent with the definition for the absolute value of a real number: If $z = x + yi$ is real, then $z = x + 0i$ and

$$|z| = \sqrt{x^2 + 0^2} = \sqrt{x^2} = |x|$$

For this reason, the magnitude of z is sometimes called the **absolute value of z**.

Recall that if $z = x + yi$, then its **conjugate**, denoted by $\bar{z}$, is $\bar{z} = x - yi$. Because $z\bar{z} = x^2 + y^2$, which is a nonnegative real number, it follows from equation (1) that the magnitude of z can be written as

$$|z| = \sqrt{z\bar{z}} \qquad \textbf{(2)}$$

2 Convert a Complex Number between Rectangular Form and Polar Form

When a complex number is written in the standard form $z = x + yi$, it is in **rectangular**, or **Cartesian**, **form**, because (x, y) are the rectangular coordinates of the corresponding point in the complex plane. Suppose that (r, θ) are polar coordinates of this point. Then

$$x = r\cos\theta \qquad y = r\sin\theta \qquad \textbf{(3)}$$

DEFINITION

If $r \geq 0$ and $0 \leq \theta < 2\pi$, the complex number $z = x + yi$ may be written in **polar form** as

$$z = x + yi = (r\cos\theta) + (r\sin\theta)i = r(\cos\theta + i\sin\theta) \qquad \textbf{(4)}$$

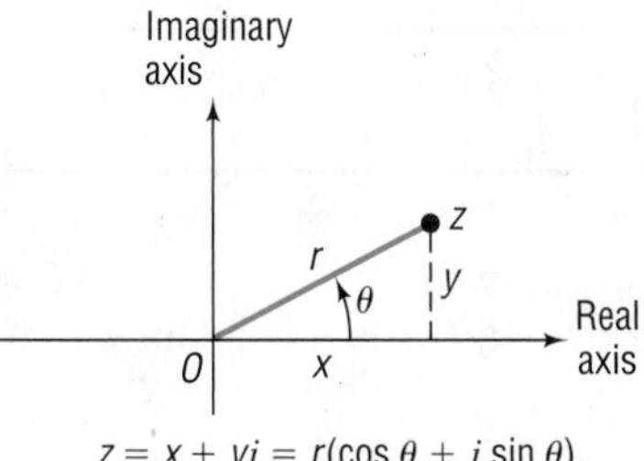

$z = x + yi = r(\cos\theta + i\sin\theta)$, $r \geq 0$, $0 \leq \theta < 2\pi$

Figure 37

See Figure 37.

If $z = r(\cos\theta + i\sin\theta)$ is the polar form of a complex number,* the angle θ, $0 \leq \theta < 2\pi$, is called the **argument of** z.

Also, because $r \geq 0$, we have $r = \sqrt{x^2 + y^2}$. From equation (1), it follows that the magnitude of $z = r(\cos\theta + i\sin\theta)$ is

$$|z| = r$$

EXAMPLE 2 Writing a Complex Number in Polar Form

Write an expression for $z = \sqrt{3} - i$ in polar form.

Solution The point, located in quadrant IV, is plotted in Figure 35. Because $x = \sqrt{3}$ and $y = -1$, it follows that

$$r = \sqrt{x^2 + y^2} = \sqrt{(\sqrt{3})^2 + (-1)^2} = \sqrt{4} = 2$$

so

$$\sin\theta = \frac{y}{r} = \frac{-1}{2} \qquad \cos\theta = \frac{x}{r} = \frac{\sqrt{3}}{2} \qquad 0 \leq \theta < 2\pi$$

The angle θ, $0 \leq \theta < 2\pi$, that satisfies both equations is $\theta = \frac{11\pi}{6}$. With $\theta = \frac{11\pi}{6}$ and $r = 2$, the polar form of $z = \sqrt{3} - i$ is

$$z = r(\cos\theta + i\sin\theta) = 2\left(\cos\frac{11\pi}{6} + i\sin\frac{11\pi}{6}\right)$$

Now Work PROBLEM 13

EXAMPLE 3 Plotting a Point in the Complex Plane and Converting from Polar to Rectangular Form

Plot the point corresponding to $z = 2(\cos 30° + i\sin 30°)$ in the complex plane, and write an expression for z in rectangular form.

*Some texts abbreviate the polar form using $z = r(\cos\theta + i\sin\theta) = r\,\text{cis}\,\theta$.

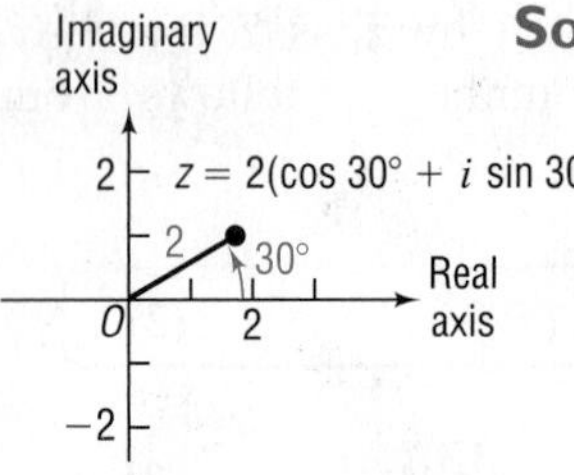

Figure 38 $z = \sqrt{3} + i$

Solution To plot the complex number $z = 2(\cos 30° + i \sin 30°)$, plot the point whose polar coordinates are $(r, \theta) = (2, 30°)$, as shown in Figure 38. In rectangular form,

$$z = 2(\cos 30° + i \sin 30°) = 2\left(\frac{\sqrt{3}}{2} + \frac{1}{2}i\right) = \sqrt{3} + i$$

Now Work PROBLEM 25

3 Find Products and Quotients of Complex Numbers in Polar Form

The polar form of a complex number provides an alternative method for finding products and quotients of complex numbers.

THEOREM

Let $z_1 = r_1(\cos \theta_1 + i \sin \theta_1)$ and $z_2 = r_2(\cos \theta_2 + i \sin \theta_2)$ be two complex numbers. Then

$$z_1 z_2 = r_1 r_2 [\cos(\theta_1 + \theta_2) + i \sin(\theta_1 + \theta_2)] \quad \textbf{(5)}$$

If $z_2 \neq 0$, then

$$\frac{z_1}{z_2} = \frac{r_1}{r_2}[\cos(\theta_1 - \theta_2) + i \sin(\theta_1 - \theta_2)] \quad \textbf{(6)}$$

In Words

The magnitude of a complex number z is r, and its argument is θ, so when

$$z = r(\cos \theta + i \sin \theta)$$

the magnitude of the product (quotient) of two complex numbers equals the product (quotient) of their magnitudes; the argument of the product (quotient) of two complex numbers is determined by the sum (difference) of their arguments.

Proof We will prove formula (5). The proof of formula (6) is left as an exercise (see Problem 68).

$$\begin{aligned} z_1 z_2 &= [r_1(\cos \theta_1 + i \sin \theta_1)][r_2(\cos \theta_2 + i \sin \theta_2)] \\ &= r_1 r_2[(\cos \theta_1 + i \sin \theta_1)(\cos \theta_2 + i \sin \theta_2)] \\ &= r_1 r_2[(\cos \theta_1 \cos \theta_2 - \sin \theta_1 \sin \theta_2) + i(\sin \theta_1 \cos \theta_2 + \cos \theta_1 \sin \theta_2)] \\ &= r_1 r_2[\cos(\theta_1 + \theta_2) + i \sin(\theta_1 + \theta_2)] \end{aligned}$$

■

Let's look at an example of how this theorem can be used.

EXAMPLE 4 **Finding Products and Quotients of Complex Numbers in Polar Form**

If $z = 3(\cos 20° + i \sin 20°)$ and $w = 5(\cos 100° + i \sin 100°)$, find the following (leave your answers in polar form).

(a) zw (b) $\dfrac{z}{w}$

Solution

(a)
$$\begin{aligned} zw &= [3(\cos 20° + i \sin 20°)][5(\cos 100° + i \sin 100°)] \\ &= (3 \cdot 5)[\cos(20° + 100°) + i \sin(20° + 100°)] && \text{Apply equation (5).} \\ &= 15(\cos 120° + i \sin 120°) \end{aligned}$$

(b)
$$\begin{aligned} \frac{z}{w} &= \frac{3(\cos 20° + i \sin 20°)}{5(\cos 100° + i \sin 100°)} \\ &= \frac{3}{5}[\cos(20° - 100°) + i \sin(20° - 100°)] && \text{Apply equation (6).} \\ &= \frac{3}{5}[\cos(-80°) + i \sin(-80°)] \\ &= \frac{3}{5}(\cos 280° + i \sin 280°) && \text{The argument must lie between } 0° \text{ and } 360°. \end{aligned}$$

Now Work PROBLEM 35

4 Use De Moivre's Theorem

De Moivre's Theorem, stated by Abraham De Moivre (1667–1754) in 1730, but already known to many people by 1710, is important for the following reason: The fundamental processes of algebra are the four operations of addition, subtraction, multiplication, and division, together with powers and the extraction of roots. De Moivre's Theorem allows the last two fundamental algebraic operations to be applied to complex numbers.

De Moivre's Theorem, in its most basic form, is a formula for raising a complex number z to the power n, where $n \geq 1$ is a positive integer. Let's try to conjecture the form of the result.

Let $z = r(\cos\theta + i\sin\theta)$ be a complex number. Then equation (5) yields

$$n = 2:\quad z^2 = r^2[\cos(2\theta) + i\sin(2\theta)] \qquad \text{Equation (5)}$$

$$n = 3:\quad z^3 = z^2 \cdot z$$

$$= \{r^2[\cos(2\theta) + i\sin(2\theta)]\}[r(\cos\theta + i\sin\theta)]$$

$$= r^3[\cos(3\theta) + i\sin(3\theta)] \qquad \text{Equation (5)}$$

$$n = 4:\quad z^4 = z^3 \cdot z$$

$$= \{r^3[\cos(3\theta) + i\sin(3\theta)]\}[r(\cos\theta + i\sin\theta)]$$

$$= r^4[\cos(4\theta) + i\sin(4\theta)] \qquad \text{Equation (5)}$$

Do you see the pattern?

THEOREM

De Moivre's Theorem

If $z = r(\cos\theta + i\sin\theta)$ is a complex number, then

$$z^n = r^n[\cos(n\theta) + i\sin(n\theta)] \qquad \textbf{(7)}$$

where $n \geq 1$ is a positive integer.

The proof of De Moivre's Theorem requires mathematical induction (which is not discussed in this text), so it is omitted here. The theorem is actually true for all integers, n. You are asked to prove this in Problem 69.

EXAMPLE 5 **Using De Moivre's Theorem**

Write $[2(\cos 20° + i\sin 20°)]^3$ in the standard form $a + bi$.

Solution

$$[2(\cos 20° + i\sin 20°)]^3 = 2^3[\cos(3\cdot 20°) + i\sin(3\cdot 20°)] \qquad \text{Apply De Moivre's Theorem.}$$

$$= 8(\cos 60° + i\sin 60°)$$

$$= 8\left(\frac{1}{2} + \frac{\sqrt{3}}{2}i\right) = 4 + 4\sqrt{3}i$$

●

Now Work PROBLEM 43

EXAMPLE 6 **Using De Moivre's Theorem**

Write $(1 + i)^5$ in the standard form $a + bi$.

Solution

To apply De Moivre's Theorem, first write the complex number in polar form. Since the magnitude of $1 + i$ is $\sqrt{1^2 + 1^2} = \sqrt{2}$, begin by writing

$$1 + i = \sqrt{2}\left(\frac{1}{\sqrt{2}} + \frac{1}{\sqrt{2}}i\right) = \sqrt{2}\left(\cos\frac{\pi}{4} + i\sin\frac{\pi}{4}\right)$$

NOTE In the solution of Example 6, the approach used in Example 2 could also be used to write $1 + i$ in polar form. ■

Now

$$\begin{aligned}(1+i)^5 &= \left[\sqrt{2}\left(\cos\frac{\pi}{4} + i\sin\frac{\pi}{4}\right)\right]^5\\ &= (\sqrt{2})^5\left[\cos\left(5\cdot\frac{\pi}{4}\right) + i\sin\left(5\cdot\frac{\pi}{4}\right)\right]\\ &= 4\sqrt{2}\left(\cos\frac{5\pi}{4} + i\sin\frac{5\pi}{4}\right)\\ &= 4\sqrt{2}\left[-\frac{1}{\sqrt{2}} + \left(-\frac{1}{\sqrt{2}}\right)i\right] = -4-4i\end{aligned}$$

5 Find Complex Roots

Let w be a given complex number, and let $n \geq 2$ denote a positive integer. Any complex number z that satisfies the equation

$$z^n = w$$

is a **complex *n*th root** of w. In keeping with previous usage, if $n = 2$, the solutions of the equation $z^2 = w$ are called **complex square roots** of w, and if $n = 3$, the solutions of the equation $z^3 = w$ are called **complex cube roots** of w.

THEOREM

Finding Complex Roots

Let $w = r(\cos\theta_0 + i\sin\theta_0)$ be a complex number, and let $n \geq 2$ be an integer. If $w \neq 0$, there are n distinct complex nth roots of w, given by the formula

$$z_k = \sqrt[n]{r}\left[\cos\left(\frac{\theta_0}{n} + \frac{2k\pi}{n}\right) + i\sin\left(\frac{\theta_0}{n} + \frac{2k\pi}{n}\right)\right] \quad \textbf{(8)}$$

where $k = 0, 1, 2, \ldots, n-1$.

Proof (Outline) We will not prove this result in its entirety. Instead, we shall show only that each z_k in equation (8) satisfies the equation $z_k^n = w$, proving that each z_k is a complex nth root of w.

$$\begin{aligned}z_k^n &= \left\{\sqrt[n]{r}\left[\cos\left(\frac{\theta_0}{n} + \frac{2k\pi}{n}\right) + i\sin\left(\frac{\theta_0}{n} + \frac{2k\pi}{n}\right)\right]\right\}^n\\ &= (\sqrt[n]{r})^n\left\{\cos\left[n\left(\frac{\theta_0}{n} + \frac{2k\pi}{n}\right)\right] + i\sin\left[n\left(\frac{\theta_0}{n} + \frac{2k\pi}{n}\right)\right]\right\} && \text{Apply De Moivre's Theorem.}\\ &= r[\cos(\theta_0 + 2k\pi) + i\sin(\theta_0 + 2k\pi)] && \text{Simplify.}\\ &= r(\cos\theta_0 + i\sin\theta_0) = w && \text{Periodic Property}\end{aligned}$$

So each z_k, $k = 0, 1, \ldots, n-1$, is a complex nth root of w. To complete the proof, we would need to show that each z_k, $k = 0, 1, \ldots, n-1$, is, in fact, distinct and that there are no complex nth roots of w other than those given by equation (8). ■

EXAMPLE 7 **Finding Complex Cube Roots**

Find the complex cube roots of $-1 + \sqrt{3}i$. Leave your answers in polar form, with the argument in degrees.

Solution First, express $-1 + \sqrt{3}i$ in polar form using degrees.

$$-1 + \sqrt{3}i = 2\left(-\frac{1}{2} + \frac{\sqrt{3}}{2}i\right) = 2(\cos 120° + i\sin 120°)$$

The three complex cube roots of $-1 + \sqrt{3}i = 2(\cos 120° + i \sin 120°)$ are

$$\begin{aligned} z_k &= \sqrt[3]{2}\left[\cos\left(\frac{120°}{3} + \frac{360°k}{3}\right) + i \sin\left(\frac{120°}{3} + \frac{360°k}{3}\right)\right] \\ &= \sqrt[3]{2}\,[\cos(40° + 120°k) + i \sin(40° + 120°k)] \qquad k = 0, 1, 2 \end{aligned}$$

so

WARNING Most graphing utilities will provide only the answer z_0 to the calculation $(-1 + \sqrt{3}i) \wedge (1/3)$. The paragraph following Example 7 explains how to obtain z_1 and z_2 from z_0. ■

$$z_0 = \sqrt[3]{2}\,[\cos(40° + 120° \cdot 0) + i \sin(40° + 120° \cdot 0)] = \sqrt[3]{2}\,(\cos 40° + i \sin 40°)$$

$$z_1 = \sqrt[3]{2}\,[\cos(40° + 120° \cdot 1) + i \sin(40° + 120° \cdot 1)] = \sqrt[3]{2}\,(\cos 160° + i \sin 160°)$$

$$z_2 = \sqrt[3]{2}\,[\cos(40° + 120° \cdot 2) + i \sin(40° + 120° \cdot 2)] = \sqrt[3]{2}\,(\cos 280° + i \sin 280°)$$

●

Notice that all of the three complex cube roots of $-1 + \sqrt{3}i$ have the same magnitude, $\sqrt[3]{2}$. This means that the points corresponding to each cube root lie the same distance from the origin; that is, the three points lie on a circle with center at the origin and radius $\sqrt[3]{2}$. Furthermore, the arguments of these cube roots are 40°, 160°, and 280°, the difference of consecutive pairs being $120° = \frac{360°}{3}$. This means that the three points are equally spaced on the circle, as shown in Figure 39. These results are not coincidental. In fact, you are asked to show that these results hold for complex nth roots in Problems 65 through 67.

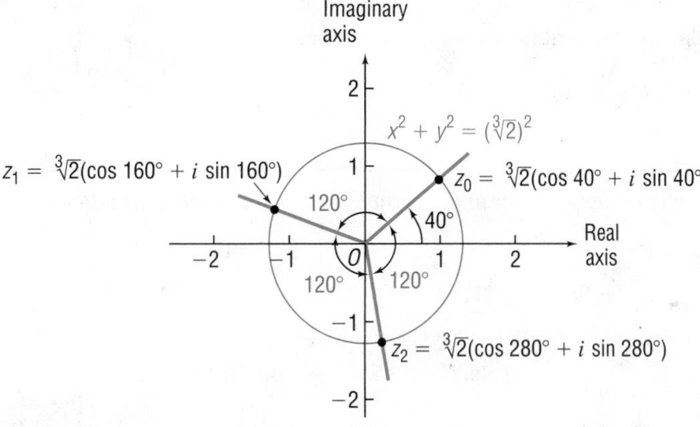

Figure 39

Now Work PROBLEM 55

Historical Feature

John Wallis

The Babylonians, Greeks, and Arabs considered square roots of negative quantities to be impossible and equations with complex solutions to be unsolvable. The first hint that there was some connection between real solutions of equations and complex numbers came when Girolamo Cardano (1501–1576) and Tartaglia (1499–1557) found *real* roots of cubic equations by taking cube roots of *complex* quantities. For centuries thereafter, mathematicians worked with complex numbers without much belief in their actual existence. In 1673, John Wallis appears to have been the first to suggest the graphical representation of complex numbers, a truly significant idea that was not pursued further until about 1800. Several people, including Karl Friedrich Gauss (1777–1855), then rediscovered the idea, and graphical representation helped to establish complex numbers as equal members of the number family. In practical applications, complex numbers have found their greatest uses in the study of alternating current, where they are a commonplace tool, and in the field of subatomic physics.

Historical Problems

1. The quadratic formula works perfectly well if the coefficients are complex numbers. Solve the following.

(a) $z^2 - (2 + 5i)z - 3 + 5i = 0$ **(b)** $z^2 - (1 + i)z - 2 - i = 0$

5.3 Assess Your Understanding

'Are You Prepared?' Answers are given at the end of these exercises. If you get a wrong answer, read the pages listed in red.

1. The conjugate of $-4 - 3i$ is ________. (pp. A37–A42)

2. The sum formula for the sine function is $\sin(A + B) =$ ______. (p. 229)

3. The sum formula for the cosine function is $\cos(A + B) =$ ______. (p. 226)

4. $\sin 120° =$ ________; $\cos 240° =$ ________. (pp. 117–124)

Concepts and Vocabulary

5. In the complex plane, the x-axis is referred to as the ______ axis, and the y-axis is called the __________ axis.

6. When a complex number z is written in the polar form $z = r(\cos\theta + i\sin\theta)$, the nonnegative number r is the __________ or __________ of z, and the angle θ, $0 \le \theta < 2\pi$, is the __________ of z.

7. Let $z_1 = r_1(\cos\theta_1 + i\sin\theta_1)$ and $z_2 = r_2(\cos\theta_2 + i\sin\theta_2)$ be two complex numbers. Then

$z_1 z_2 =$ ____ [cos (________) + i sin (________)].

8. If $z = r(\cos\theta + i\sin\theta)$ is a complex number, then $z^n =$ ____ [cos(____) + i sin(____)].

9. Every nonzero complex number will have exactly ______ distinct complex cube roots.

10. ***True or False*** The polar form of a nonzero complex number is unique.

11. If $z = x + yi$ is a complex number, then $|z|$ equals which of the following?

(a) $x^2 + y^2$ (b) $|x| + |y|$

(c) $\sqrt{x^2 + y^2}$ (d) $\sqrt{|x| + |y|}$

12. If $z_1 = r_1(\cos\theta_1 + i\sin\theta_1)$ and $z_2 = r_2(\cos\theta_2 + i\sin\theta_2)$ are complex numbers, then $\frac{z_1}{z_2}$, $z_2 \ne 0$, equals which of the following?

(a) $\frac{r_1}{r_2}[\cos(\theta_1 - \theta_2) + i\sin(\theta_1 - \theta_2)]$

(b) $\frac{r_1}{r_2}\left[\cos\left(\frac{\theta_1}{\theta_2}\right) + i\sin\left(\frac{\theta_1}{\theta_2}\right)\right]$

(c) $\frac{r_1}{r_2}[\cos(\theta_1 + \theta_2) - i\sin(\theta_1 + \theta_2)]$

(d) $\frac{r_1}{r_2}\left[\cos\left(\frac{\theta_1}{\theta_2}\right) - i\sin\left(\frac{\theta_1}{\theta_2}\right)\right]$

Skill Building

In Problems 13–24, plot each complex number in the complex plane and write it in polar form. Express the argument in degrees.

13. $1 + i$ **14.** $-1 + i$ **15.** $\sqrt{3} - i$ **16.** $1 - \sqrt{3}i$ **17.** $-3i$ **18.** -2

19. $4 - 4i$ **20.** $9\sqrt{3} + 9i$ **21.** $3 - 4i$ **22.** $2 + \sqrt{3}i$ **23.** $-2 + 3i$ **24.** $\sqrt{5} - i$

In Problems 25–34, write each complex number in rectangular form.

25. $2(\cos 120° + i\sin 120°)$ **26.** $3(\cos 210° + i\sin 210°)$ **27.** $4\left(\cos\frac{7\pi}{4} + i\sin\frac{7\pi}{4}\right)$

28. $2\left(\cos\frac{5\pi}{6} + i\sin\frac{5\pi}{6}\right)$ **29.** $3\left(\cos\frac{3\pi}{2} + i\sin\frac{3\pi}{2}\right)$ **30.** $4\left(\cos\frac{\pi}{2} + i\sin\frac{\pi}{2}\right)$

31. $0.2(\cos 100° + i\sin 100°)$ **32.** $0.4(\cos 200° + i\sin 200°)$

33. $2\left(\cos\frac{\pi}{18} + i\sin\frac{\pi}{18}\right)$ **34.** $3\left(\cos\frac{\pi}{10} + i\sin\frac{\pi}{10}\right)$

In Problems 35–42, find zw and $\frac{z}{w}$. Leave your answers in polar form.

35. $z = 2(\cos 40° + i\sin 40°)$, $w = 4(\cos 20° + i\sin 20°)$

36. $z = \cos 120° + i\sin 120°$, $w = \cos 100° + i\sin 100°$

37. $z = 3(\cos 130° + i\sin 130°)$, $w = 4(\cos 270° + i\sin 270°)$

38. $z = 2(\cos 80° + i\sin 80°)$, $w = 6(\cos 200° + i\sin 200°)$

39. $z = 2\left(\cos\frac{\pi}{8} + i\sin\frac{\pi}{8}\right)$, $w = 2\left(\cos\frac{\pi}{10} + i\sin\frac{\pi}{10}\right)$

40. $z = 4\left(\cos\frac{3\pi}{8} + i\sin\frac{3\pi}{8}\right)$, $w = 2\left(\cos\frac{9\pi}{16} + i\sin\frac{9\pi}{16}\right)$

41. $z = 2 + 2i$, $w = \sqrt{3} - i$

42. $z = 1 - i$, $w = 1 - \sqrt{3}i$

In Problems 43–54, write each expression in the standard form $a + bi$.

43. $[4(\cos 40° + i\sin 40°)]^3$ **44.** $[3(\cos 80° + i\sin 80°)]^3$ **45.** $\left[2\left(\cos\frac{\pi}{10} + i\sin\frac{\pi}{10}\right)\right]^5$

46. $\left[\sqrt{2}\left(\cos\frac{5\pi}{16} + i\sin\frac{5\pi}{16}\right)\right]^4$ **47.** $[\sqrt{3}(\cos 10° + i\sin 10°)]^6$ **48.** $\left[\frac{1}{2}(\cos 72° + i\sin 72°)\right]^5$

49. $\left[\sqrt{5}\left(\cos\frac{3\pi}{16} + i\sin\frac{3\pi}{16}\right)\right]^4$

50. $\left[\sqrt{3}\left(\cos\frac{5\pi}{18} + i\sin\frac{5\pi}{18}\right)\right]^6$

51. $(1 - i)^5$

52. $(\sqrt{3} - i)^6$

53. $(\sqrt{2} - i)^6$

54. $(1 - \sqrt{5}i)^8$

In Problems 55–62, find all the complex roots. Leave your answers in polar form with the argument in degrees.

55. The complex cube roots of $1 + i$

56. The complex fourth roots of $\sqrt{3} - i$

57. The complex fourth roots of $4 - 4\sqrt{3}i$

58. The complex cube roots of $-8 - 8i$

59. The complex fourth roots of $-16i$

60. The complex cube roots of -8

61. The complex fifth roots of i

62. The complex fifth roots of $-i$

Applications and Extensions

63. Find the four complex fourth roots of unity (1) and plot them.

64. Find the six complex sixth roots of unity (1) and plot them.

65. Show that each complex nth root of a nonzero complex number w has the same magnitude.

66. Use the result of Problem 65 to draw the conclusion that each complex nth root lies on a circle with center at the origin. What is the radius of this circle?

67. Refer to Problem 66. Show that the complex nth roots of a nonzero complex number w are equally spaced on the circle.

68. Prove formula (6).

69. Prove that De Moivre's Theorem is true for *all* integers n by assuming it is true for integers $n \geq 1$ and then showing it is true for 0 and for negative integers.

Hint: Multiply the numerator and the denominator by the conjugate of the denominator, and use even-odd properties.

70. **Mandelbrot Sets**

(a) Consider the expression $a_n = (a_{n-1})^2 + z$, where z is some complex number (called the **seed**) and $a_0 = z$. Compute $a_1 (= a_0^2 + z)$, $a_2 (= a_1^2 + z)$, $a_3 (= a_2^2 + z)$, a_4, a_5, and a_6 for the following seeds: $z_1 = 0.1 - 0.4i$, $z_2 = 0.5 + 0.8i$, $z_3 = -0.9 + 0.7i$, $z_4 = -1.1 + 0.1i$, $z_5 = 0 - 1.3i$, and $z_6 = 1 + 1i$.

(b) The dark portion of the graph represents the set of all values $z = x + yi$ that are in the Mandelbrot set. Determine which complex numbers in part (a) are in this set by plotting them on the graph. Do the complex numbers that are not in the Mandelbrot set have any common characteristics regarding the values of a_6 found in part (a)?

(c) Compute $|z| = \sqrt{x^2 + y^2}$ for each of the complex numbers in part (a). Now compute $|a_6|$ for each of the complex numbers in part (a). For which complex numbers is $|a_6| \leq |z|$ and $|z| \leq 2$? Conclude that the criterion for a complex number to be in the Mandelbrot set is that $|a_n| \leq |z|$ and $|z| \leq 2$.

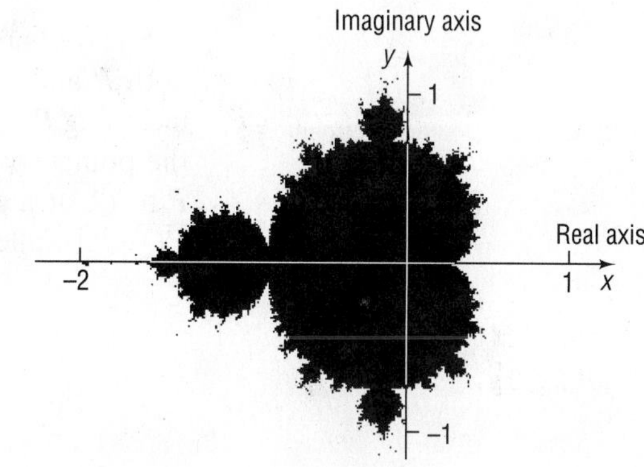

Retain Your Knowledge

Problems 71–74 are based on material learned earlier in the course. The purpose of these problems is to keep the material fresh in your mind so that you are better prepared for the final exam.

71. Find the area of the triangle with $a = 8$, $b = 11$, and $C = 113°$.

72. Convert 240° to radians. Express your answer as a multiple of π.

73. Find the exact distance between the points $(-3, 4)$ and $(2, -1)$.

74. Given $\cos\theta = \frac{1}{4}$ and $\frac{3\pi}{2} < \theta < 2\pi$, find the exact values of the five remaining trigonometric functions of θ.

'Are You Prepared?' Answers

1. $-4 + 3i$ 2. $\sin A\cos B + \cos A\sin B$ 3. $\cos A\cos B - \sin A\sin B$ 4. $\frac{\sqrt{3}}{2}; -\frac{1}{2}$

5.4 Vectors

OBJECTIVES 1 Graph Vectors (p. 346)
2 Find a Position Vector (p. 346)
3 Add and Subtract Vectors Algebraically (p. 348)
4 Find a Scalar Multiple and the Magnitude of a Vector (p. 349)
5 Find a Unit Vector (p. 349)
6 Find a Vector from Its Direction and Magnitude (p. 350)
7 Model with Vectors (p. 351)

In simple terms, a **vector** (derived from the Latin *vehere,* meaning "to carry") is a quantity that has both magnitude and direction. It is customary to represent a vector by using an arrow. The length of the arrow represents the **magnitude** of the vector, and the arrowhead indicates the **direction** of the vector.

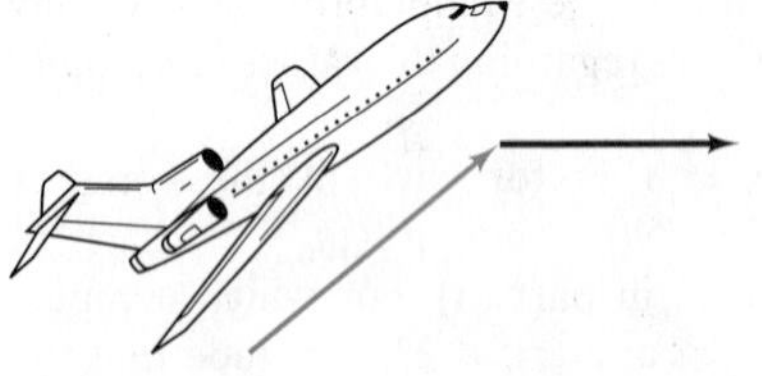

Figure 40

Many quantities in physics can be represented by vectors. For example, the velocity of an aircraft can be represented by an arrow that points in the direction of movement; the length of the arrow represents the speed. If the aircraft speeds up, we lengthen the arrow; if the aircraft changes direction, we introduce an arrow in the new direction. See Figure 40. Based on this representation, it is not surprising that vectors and *directed line segments* are somehow related.

Geometric Vectors

If P and Q are two distinct points in the xy-plane, there is exactly one line containing both P and Q [Figure 41(a)]. The points on that part of the line that joins P to Q, including P and Q, form what is called the **line segment** $\overline{PQ}$ [Figure 41(b)]. Ordering the points so that they proceed from P to Q results in a **directed line segment** from P to Q, or a **geometric vector**, which is denoted by $\overrightarrow{PQ}$. In a directed line segment $\overrightarrow{PQ}$, P is called the **initial point** and Q the **terminal point**, as indicated in Figure 41(c).

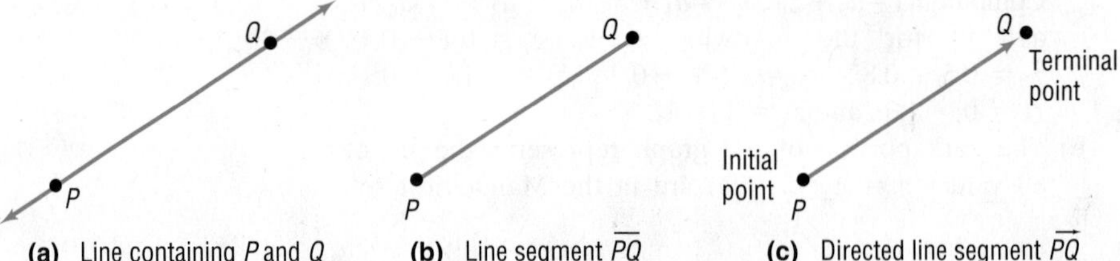

Figure 41 (a) Line containing P and Q (b) Line segment $\overline{PQ}$ (c) Directed line segment $\overrightarrow{PQ}$

The magnitude of the directed line segment $\overrightarrow{PQ}$ is the distance from the point P to the point Q; that is, it is the length of the line segment. The direction of $\overrightarrow{PQ}$ is from P to Q. If a vector **v*** has the same magnitude and the same direction as the directed line segment $\overrightarrow{PQ}$, write

$$\mathbf{v} = \overrightarrow{PQ}$$

The vector **v** whose magnitude is 0 is called the **zero vector, 0.** The zero vector is assigned no direction.

Two vectors **v** and **w** are **equal**, written

$$\mathbf{v} = \mathbf{w}$$

if they have the same magnitude and the same direction.

For example, the three vectors shown in Figure 42 have the same magnitude and the same direction, so they are equal, even though they have different initial points and different terminal points. As a result, it is useful to think of a vector simply as an arrow, keeping in mind that two arrows (vectors) are equal if they have the same direction and the same magnitude (length).

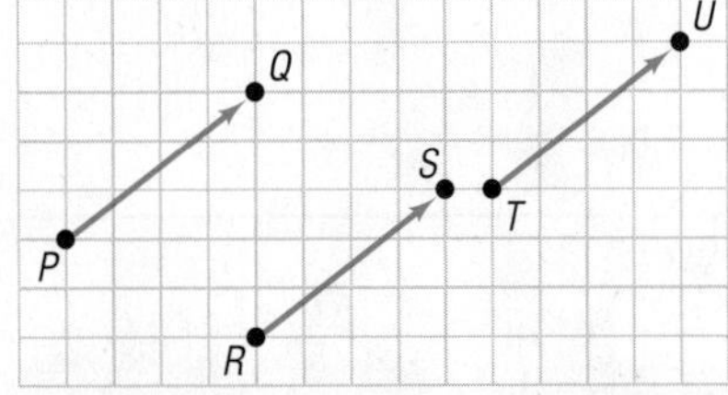

Figure 42 Equal vectors

*Boldface letters will be used to denote vectors, to distinguish them from numbers. For handwritten work, an arrow is placed over the letter to signify a vector. For example, write a vector by hand as $\vec{v}$.

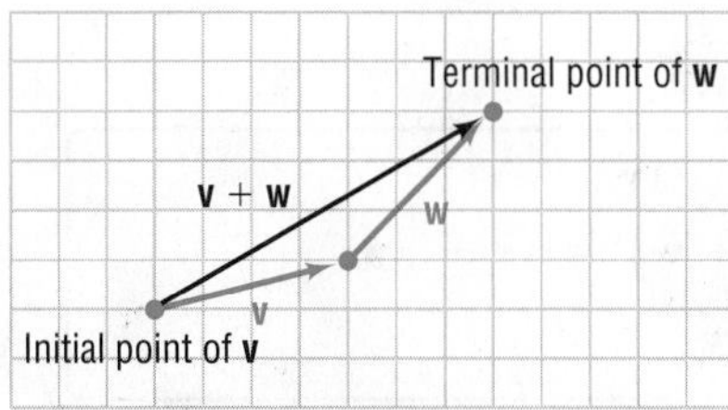

Figure 43 Adding vectors

Adding Vectors Geometrically

The **sum** $\mathbf{v} + \mathbf{w}$ of two vectors is defined as follows: Position the vectors **v** and **w** so that the terminal point of **v** coincides with the initial point of **w**, as shown in Figure 43. The vector $\mathbf{v} + \mathbf{w}$ is then the unique vector whose initial point coincides with the initial point of **v** and whose terminal point coincides with the terminal point of **w**.

Vector addition is **commutative**. That is, if **v** and **w** are any two vectors, then

$$\mathbf{v} + \mathbf{w} = \mathbf{w} + \mathbf{v}$$

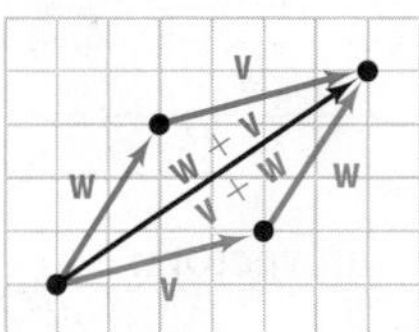

Figure 44 $\mathbf{v} + \mathbf{w} = \mathbf{w} + \mathbf{v}$

Figure 44 illustrates this fact. (Observe that the commutative property is another way of saying that opposite sides of a parallelogram are equal and parallel.)

Vector addition is also **associative**. That is, if **u**, **v**, and **w** are vectors, then

$$\mathbf{u} + (\mathbf{v} + \mathbf{w}) = (\mathbf{u} + \mathbf{v}) + \mathbf{w}$$

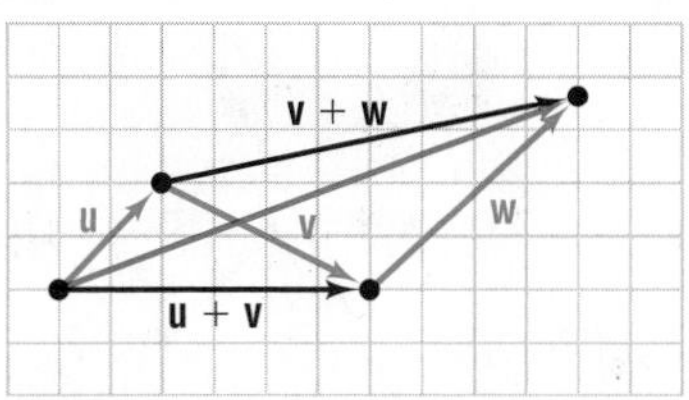

Figure 45
$(\mathbf{u} + \mathbf{v}) + \mathbf{w} = \mathbf{u} + (\mathbf{v} + \mathbf{w})$

Figure 45 illustrates the associative property for vectors.

The zero vector **0** has the property that

$$\mathbf{v} + \mathbf{0} = \mathbf{0} + \mathbf{v} = \mathbf{v}$$

for any vector **v**.

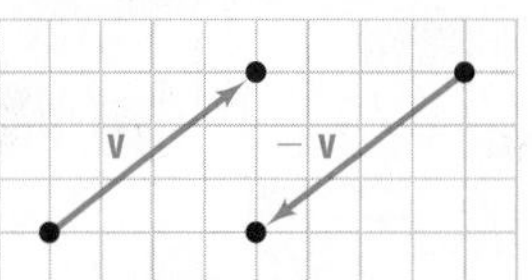

Figure 46 Opposite vectors

If **v** is a vector, then $-\mathbf{v}$ is the vector that has the same magnitude as **v**, but whose direction is opposite to **v**, as shown in Figure 46.

Furthermore,

$$\mathbf{v} + (-\mathbf{v}) = \mathbf{0}$$

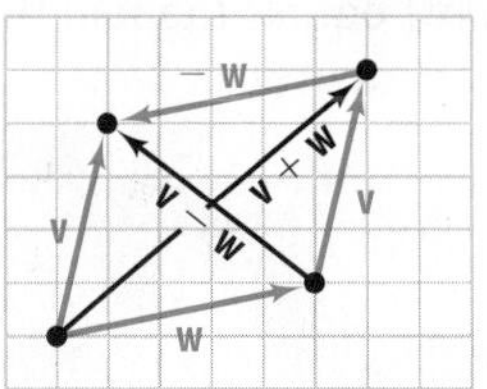

Figure 47

If **v** and **w** are two vectors, then the **difference** $\mathbf{v} - \mathbf{w}$ is defined as

$$\mathbf{v} - \mathbf{w} = \mathbf{v} + (-\mathbf{w})$$

Figure 47 illustrates the relationships among **v**, **w**, $\mathbf{v} + \mathbf{w}$, and $\mathbf{v} - \mathbf{w}$.

Multiplying Vectors by Numbers Geometrically

When dealing with vectors, real numbers are referred to as **scalars**. Scalars are quantities that have only magnitude. Examples of scalar quantities from physics are temperature, speed, and time. We now define how to multiply a vector by a scalar.

DEFINITION

If α is a scalar and **v** is a vector, the **scalar multiple** $\alpha\mathbf{v}$ is defined as follows:

1. If $\alpha > 0$, $\alpha\mathbf{v}$ is the vector whose magnitude is α times the magnitude of **v** and whose direction is the same as that of **v**.
2. If $\alpha < 0$, $\alpha\mathbf{v}$ is the vector whose magnitude is $|\alpha|$ times the magnitude of **v** and whose direction is opposite that of **v**.
3. If $\alpha = 0$ or if $\mathbf{v} = \mathbf{0}$, then $\alpha\mathbf{v} = \mathbf{0}$.

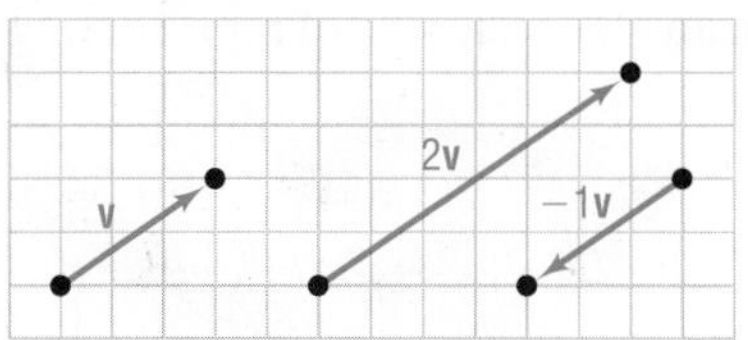

Figure 48 Scalar multiples

See Figure 48 for some illustrations.

For example, if **a** is the acceleration of an object of mass m due to a force **F** being exerted on it, then, by Newton's second law of motion, $\mathbf{F} = m\mathbf{a}$. Here, $m\mathbf{a}$ is the product of the scalar m and the vector **a**.

Scalar multiples have the following properties:

$$0\mathbf{v} = \mathbf{0} \qquad 1\mathbf{v} = \mathbf{v} \qquad -1\mathbf{v} = -\mathbf{v}$$
$$(\alpha + \beta)\mathbf{v} = \alpha\mathbf{v} + \beta\mathbf{v} \qquad \alpha(\mathbf{v} + \mathbf{w}) = \alpha\mathbf{v} + \alpha\mathbf{w}$$
$$\alpha(\beta\mathbf{v}) = (\alpha\beta)\mathbf{v}$$

1 Graph Vectors

EXAMPLE 1 **Graphing Vectors**

Use the vectors illustrated in Figure 49 to graph each of the following vectors:

(a) $\mathbf{v} - \mathbf{w}$ (b) $2\mathbf{v} + 3\mathbf{w}$ (c) $2\mathbf{v} - \mathbf{w} + \mathbf{u}$

Figure 49

Solution Figure 50 shows each graph.

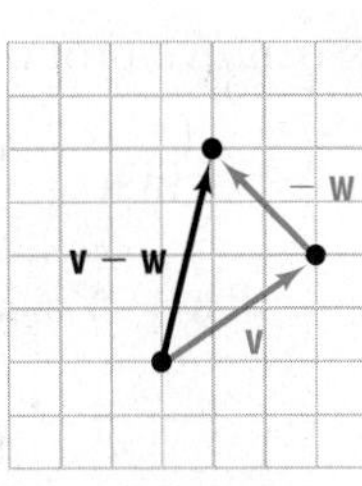

(a) v − w

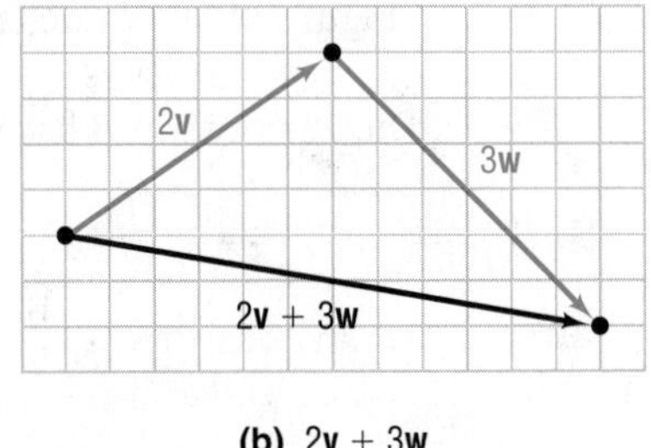

(b) 2v + 3w

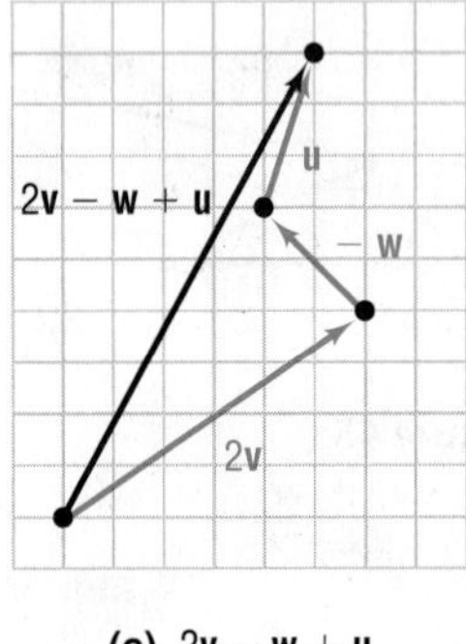

(c) 2v − w + u

Figure 50

Now Work PROBLEMS 11 AND 13

Magnitude of Vectors

The symbol $\|\mathbf{v}\|$ represents the **magnitude** of a vector $\mathbf{v}$. Since $\|\mathbf{v}\|$ equals the length of a directed line segment, it follows that $\|\mathbf{v}\|$ has the following properties:

THEOREM **Properties of $\|\mathbf{v}\|$**

If $\mathbf{v}$ is a vector and if α is a scalar, then

(a) $\|\mathbf{v}\| \geq 0$ (b) $\|\mathbf{v}\| = 0$ if and only if $\mathbf{v} = \mathbf{0}$

(c) $\|-\mathbf{v}\| = \|\mathbf{v}\|$ (d) $\|\alpha\mathbf{v}\| = |\alpha|\,\|\mathbf{v}\|$

Property (a) is a consequence of the fact that distance is a nonnegative number. Property (b) follows because the length of the directed line segment $\overrightarrow{PQ}$ is positive unless P and Q are the same point, in which case the length is 0. Property (c) follows because the length of the line segment $\overline{PQ}$ equals the length of the line segment $\overline{QP}$. Property (d) is a direct consequence of the definition of a scalar multiple.

DEFINITION A vector $\mathbf{u}$ for which $\|\mathbf{u}\| = 1$ is called a **unit vector**.

2 Find a Position Vector

To compute the magnitude and direction of a vector, an algebraic way of representing vectors is needed.

DEFINITION An **algebraic vector v** is represented as

$$\mathbf{v} = \langle a, b \rangle$$

where a and b are real numbers (scalars) called the **components** of the vector $\mathbf{v}$.

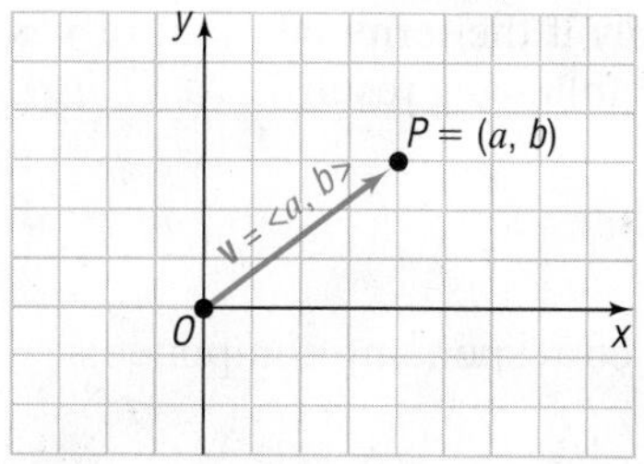

Figure 51 Position vector **v**

A rectangular coordinate system is used to represent algebraic vectors in the plane. If $\mathbf{v} = \langle a, b \rangle$ is an algebraic vector whose initial point is at the origin, then **v** is called a **position vector**. See Figure 51. Notice that the terminal point of the position vector $\mathbf{v} = \langle a, b \rangle$ is $P = (a, b)$.

The next result states that any vector whose initial point is not at the origin is equal to a unique position vector.

THEOREM

Suppose that **v** is a vector with initial point $P_1 = (x_1, y_1)$, not necessarily the origin, and terminal point $P_2 = (x_2, y_2)$. If $\mathbf{v} = \overrightarrow{P_1P_2}$, then **v** is equal to the position vector

$$\mathbf{v} = \langle x_2 - x_1, y_2 - y_1 \rangle \qquad (1)$$

In Words

An algebraic vector represents "driving directions" to get from the initial point to the terminal point of a vector. So if $\mathbf{v} = \langle 5, 4 \rangle$, travel 5 units right and 4 units up from the initial point to arrive at the terminal point.

To see why this is true, look at Figure 52.

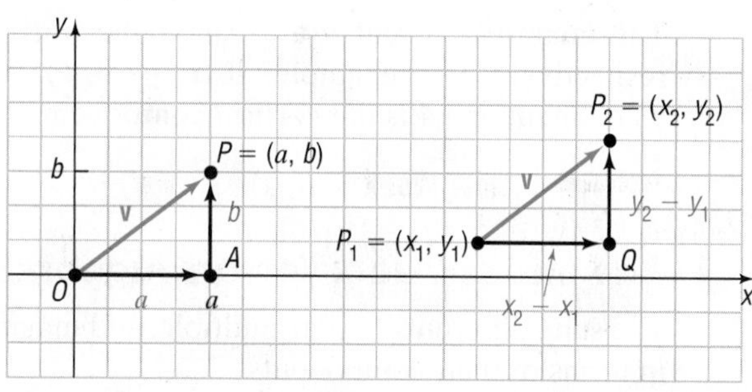

Figure 52 $\mathbf{v} = \langle a, b \rangle = \langle x_2 - x_1, y_2 - y_1 \rangle$

Triangle OPA and triangle P_1P_2Q are congruent. [Do you see why? The line segments have the same magnitude, so $d(O, P) = d(P_1, P_2)$; and they have the same direction, so $\angle POA = \angle P_2P_1Q$. Since the triangles are right triangles, we have angle–side–angle.] It follows that corresponding sides are equal. As a result, $x_2 - x_1 = a$ and $y_2 - y_1 = b$, so **v** may be written as

$$\mathbf{v} = \langle a, b \rangle = \langle x_2 - x_1, y_2 - y_1 \rangle$$

Because of this result, any algebraic vector can be replaced by a unique position vector, and vice versa. This flexibility is one of the main reasons for the wide use of vectors.

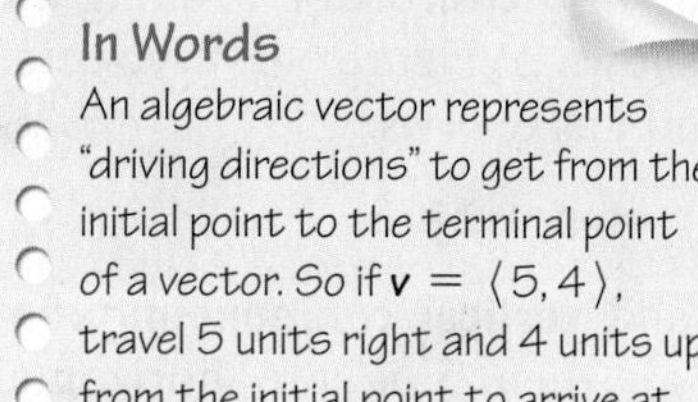

EXAMPLE 2

Finding a Position Vector

Find the position vector of the vector $\mathbf{v} = \overrightarrow{P_1P_2}$ if $P_1 = (-1, 2)$ and $P_2 = (4, 6)$.

Solution By equation (1), the position vector equal to **v** is

$$\mathbf{v} = \langle 4 - (-1), 6 - 2 \rangle = \langle 5, 4 \rangle$$

See Figure 53.

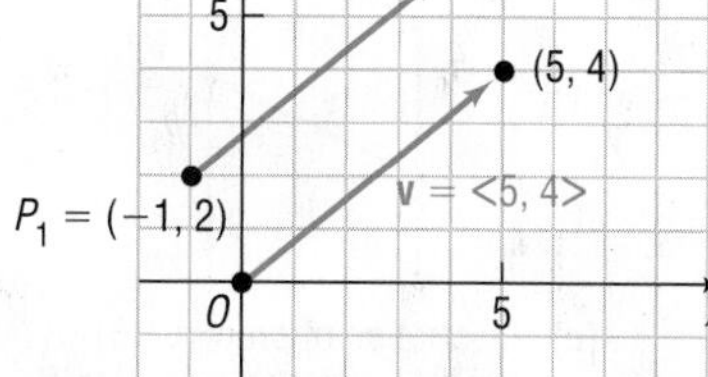

Figure 53

Two position vectors **v** and **w** are equal if and only if the terminal point of **v** is the same as the terminal point of **w**. This leads to the following result:

THEOREM

Equality of Vectors

Two vectors **v** and **w** are equal if and only if their corresponding components are equal. That is,

$$\text{If } \mathbf{v} = \langle a_1, b_1 \rangle \quad \text{and} \quad \mathbf{w} = \langle a_2, b_2 \rangle$$
$$\text{then} \quad \mathbf{v} = \mathbf{w} \quad \text{if and only if} \quad a_1 = a_2 \quad \text{and} \quad b_1 = b_2.$$

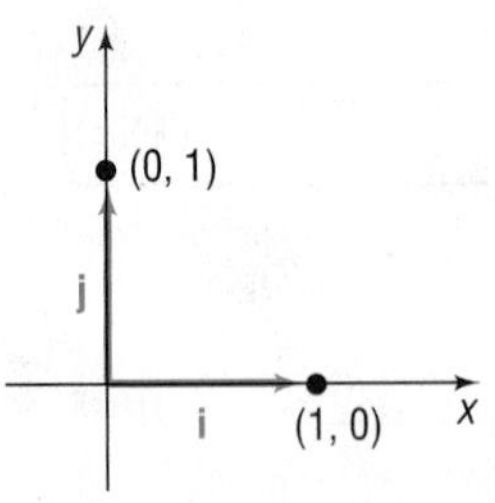

Figure 54 Unit vectors **i** and **j**

We now present an alternative representation of a vector in the plane that is common in the physical sciences. Let **i** denote the unit vector whose direction is along the positive x-axis; let **j** denote the unit vector whose direction is along the positive y-axis. Then $\mathbf{i} = \langle 1, 0 \rangle$ and $\mathbf{j} = \langle 0, 1 \rangle$, as shown in Figure 54. Any vector $\mathbf{v} = \langle a, b \rangle$ can be written using the unit vectors **i** and **j** as follows:

$$\mathbf{v} = \langle a, b \rangle = a\langle 1, 0 \rangle + b\langle 0, 1 \rangle = a\mathbf{i} + b\mathbf{j}$$

The quantities a and b are called the **horizontal** and **vertical components** of **v**, respectively. For example, if $\mathbf{v} = \langle 5, 4 \rangle = 5\mathbf{i} + 4\mathbf{j}$, then 5 is the horizontal component and 4 is the vertical component.

Now Work PROBLEM 31

3 Add and Subtract Vectors Algebraically

The sum, difference, scalar multiple, and magnitude of algebraic vectors are defined in terms of their components.

DEFINITION

Let $\mathbf{v} = a_1\mathbf{i} + b_1\mathbf{j} = \langle a_1, b_1 \rangle$ and $\mathbf{w} = a_2\mathbf{i} + b_2\mathbf{j} = \langle a_2, b_2 \rangle$ be two vectors, and let α be a scalar. Then

$$\mathbf{v} + \mathbf{w} = (a_1 + a_2)\mathbf{i} + (b_1 + b_2)\mathbf{j} = \langle a_1 + a_2, b_1 + b_2 \rangle \quad (2)$$
$$\mathbf{v} - \mathbf{w} = (a_1 - a_2)\mathbf{i} + (b_1 - b_2)\mathbf{j} = \langle a_1 - a_2, b_1 - b_2 \rangle \quad (3)$$
$$\alpha\mathbf{v} = (\alpha a_1)\mathbf{i} + (\alpha b_1)\mathbf{j} = \langle \alpha a_1, \alpha b_1 \rangle \quad (4)$$
$$\|\mathbf{v}\| = \sqrt{a_1^2 + b_1^2} \quad (5)$$

In Words
To add two vectors, add corresponding components. To subtract two vectors, subtract corresponding components.

These definitions are compatible with the geometric definitions given earlier in this section. See Figure 55.

Figure 55

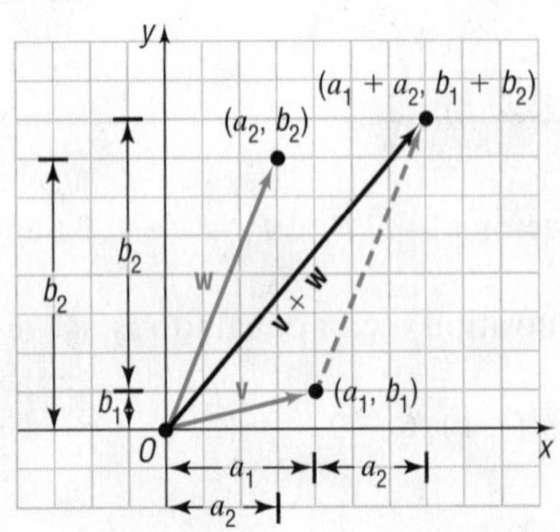

(a) Illustration of property (2)

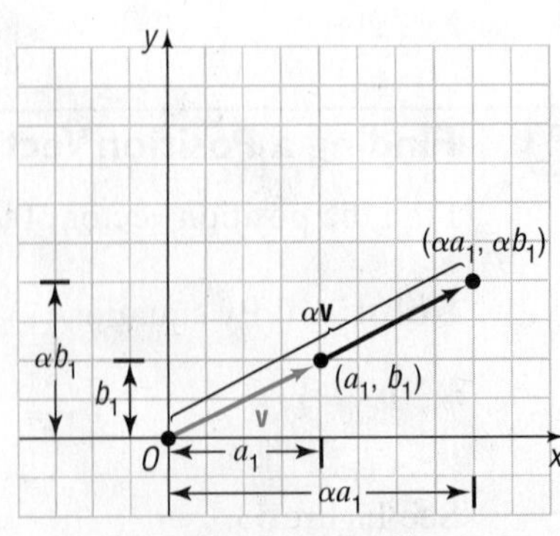

(b) Illustration of property (4), $\alpha > 0$

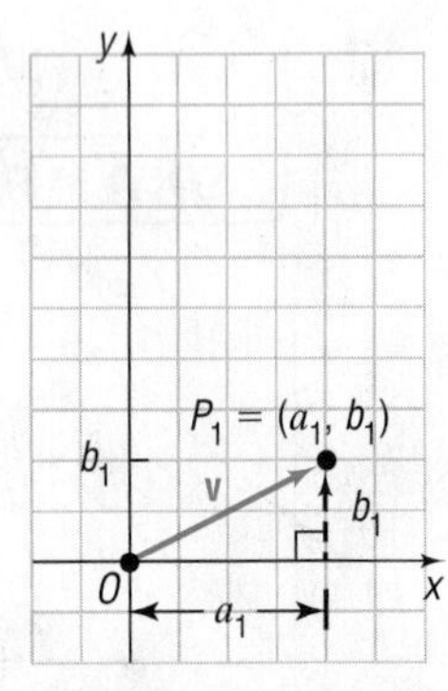

(c) Illustration of property (5): $\|\mathbf{v}\|$ = Distance from O to P_1
$\|\mathbf{v}\| = \sqrt{a_1^2 + b_1^2}$

EXAMPLE 3 **Adding and Subtracting Vectors**

If $\mathbf{v} = 2\mathbf{i} + 3\mathbf{j} = \langle 2, 3 \rangle$ and $\mathbf{w} = 3\mathbf{i} - 4\mathbf{j} = \langle 3, -4 \rangle$, find:

(a) $\mathbf{v} + \mathbf{w}$ (b) $\mathbf{v} - \mathbf{w}$

Solution (a) $\mathbf{v} + \mathbf{w} = (2\mathbf{i} + 3\mathbf{j}) + (3\mathbf{i} - 4\mathbf{j}) = (2 + 3)\mathbf{i} + (3 - 4)\mathbf{j} = 5\mathbf{i} - \mathbf{j}$

or

$\mathbf{v} + \mathbf{w} = \langle 2, 3 \rangle + \langle 3, -4 \rangle = \langle 2 + 3, 3 + (-4) \rangle = \langle 5, -1 \rangle$

(b) $\mathbf{v} - \mathbf{w} = (2\mathbf{i} + 3\mathbf{j}) - (3\mathbf{i} - 4\mathbf{j}) = (2 - 3)\mathbf{i} + [3 - (-4)]\mathbf{j} = -\mathbf{i} + 7\mathbf{j}$

or

$\mathbf{v} - \mathbf{w} = \langle 2, 3 \rangle - \langle 3, -4 \rangle = \langle 2 - 3, 3 - (-4) \rangle = \langle -1, 7 \rangle$ •

4 Find a Scalar Multiple and the Magnitude of a Vector

EXAMPLE 4 **Finding Scalar Multiples and Magnitudes of Vectors**

If $\mathbf{v} = 2\mathbf{i} + 3\mathbf{j} = \langle 2, 3 \rangle$ and $\mathbf{w} = 3\mathbf{i} - 4\mathbf{j} = \langle 3, -4 \rangle$, find:

(a) $3\mathbf{v}$ (b) $2\mathbf{v} - 3\mathbf{w}$ (c) $\|\mathbf{v}\|$

Solution (a) $3\mathbf{v} = 3(2\mathbf{i} + 3\mathbf{j}) = 6\mathbf{i} + 9\mathbf{j}$

or

$3\mathbf{v} = 3\langle 2, 3 \rangle = \langle 6, 9 \rangle$

(b) $$2\mathbf{v} - 3\mathbf{w} = 2(2\mathbf{i} + 3\mathbf{j}) - 3(3\mathbf{i} - 4\mathbf{j}) = 4\mathbf{i} + 6\mathbf{j} - 9\mathbf{i} + 12\mathbf{j}$$
$$= -5\mathbf{i} + 18\mathbf{j}$$

or

$$2\mathbf{v} - 3\mathbf{w} = 2\langle 2, 3 \rangle - 3\langle 3, -4 \rangle = \langle 4, 6 \rangle - \langle 9, -12 \rangle$$
$$= \langle 4 - 9, 6 - (-12) \rangle = \langle -5, 18 \rangle$$

(c) $\|\mathbf{v}\| = \|2\mathbf{i} + 3\mathbf{j}\| = \sqrt{2^2 + 3^2} = \sqrt{13}$ •

Now Work PROBLEMS 37 AND 43

For the remainder of the section, we will express a vector $\mathbf{v}$ in the form $a\mathbf{i} + b\mathbf{j}$.

5 Find a Unit Vector

Recall that a unit vector $\mathbf{u}$ is a vector for which $\|\mathbf{u}\| = 1$. In many applications, it is useful to be able to find a unit vector $\mathbf{u}$ that has the same direction as a given vector $\mathbf{v}$.

THEOREM **Unit Vector in the Direction of v**

For any nonzero vector $\mathbf{v}$, the vector

$$\mathbf{u} = \frac{\mathbf{v}}{\|\mathbf{v}\|} \tag{6}$$

is a unit vector that has the same direction as $\mathbf{v}$.

Proof Let $\mathbf{v} = a\mathbf{i} + b\mathbf{j}$. Then $\|\mathbf{v}\| = \sqrt{a^2 + b^2}$ and

$$\mathbf{u} = \frac{\mathbf{v}}{\|\mathbf{v}\|} = \frac{a\mathbf{i} + b\mathbf{j}}{\sqrt{a^2 + b^2}} = \frac{a}{\sqrt{a^2 + b^2}}\mathbf{i} + \frac{b}{\sqrt{a^2 + b^2}}\mathbf{j}$$

The vector **u** is in the same direction as **v**, since $\|\mathbf{v}\| > 0$. Furthermore,

$$\|\mathbf{u}\| = \sqrt{\frac{a^2}{a^2 + b^2} + \frac{b^2}{a^2 + b^2}} = \sqrt{\frac{a^2 + b^2}{a^2 + b^2}} = 1$$

That is, **u** is a unit vector in the direction of **v**. ■

As a consequence of this theorem, if **u** is a unit vector in the same direction as a vector **v**, then **v** may be expressed as

$$\mathbf{v} = \|\mathbf{v}\|\mathbf{u} \qquad \textbf{(7)}$$

This way of expressing a vector is useful in many applications.

EXAMPLE 5 Finding a Unit Vector

Find a unit vector in the same direction as $\mathbf{v} = 4\mathbf{i} - 3\mathbf{j}$.

Solution Find $\|\mathbf{v}\|$ first.

$$\|\mathbf{v}\| = \|4\mathbf{i} - 3\mathbf{j}\| = \sqrt{16 + 9} = 5$$

Now multiply **v** by the scalar $\dfrac{1}{\|\mathbf{v}\|} = \dfrac{1}{5}$. A unit vector in the same direction as **v** is

$$\frac{\mathbf{v}}{\|\mathbf{v}\|} = \frac{4\mathbf{i} - 3\mathbf{j}}{5} = \frac{4}{5}\mathbf{i} - \frac{3}{5}\mathbf{j}$$

✓**Check:** This vector is indeed a unit vector because

$$\left\|\frac{\mathbf{v}}{\|\mathbf{v}\|}\right\| = \sqrt{\left(\frac{4}{5}\right)^2 + \left(-\frac{3}{5}\right)^2} = \sqrt{\frac{16}{25} + \frac{9}{25}} = \sqrt{\frac{25}{25}} = 1$$ ●

Now Work PROBLEM 53

6 Find a Vector from Its Direction and Magnitude

If a vector represents the speed and direction of an object, it is called a **velocity vector**. If a vector represents the direction and amount of a force acting on an object, it is called a **force vector**. In many applications, a vector is described in terms of its magnitude and direction, rather than in terms of its components. For example, a ball thrown with an initial speed of 25 miles per hour at an angle of 30° to the horizontal is a velocity vector.

Suppose that we are given the magnitude $\|\mathbf{v}\|$ of a nonzero vector **v** and the **direction angle** α, $0° \le \alpha < 360°$, between **v** and **i**. To express **v** in terms of $\|\mathbf{v}\|$ and α, first find the unit vector **u** having the same direction as **v**.

Look at Figure 56. The coordinates of the terminal point of **u** are $(\cos\alpha, \sin\alpha)$. Then $\mathbf{u} = \cos\alpha\,\mathbf{i} + \sin\alpha\,\mathbf{j}$ and, from equation (7),

$$\mathbf{v} = \|\mathbf{v}\|(\cos\alpha\,\mathbf{i} + \sin\alpha\,\mathbf{j}) \qquad \textbf{(8)}$$

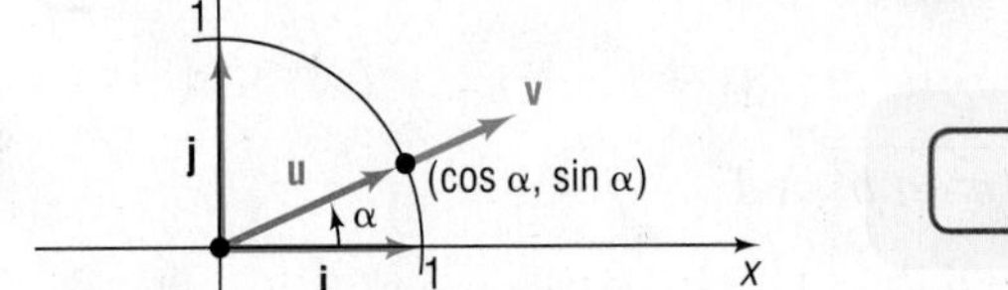

Figure 56 $\mathbf{v} = \|\mathbf{v}\|(\cos\alpha\,\mathbf{i} + \sin\alpha\,\mathbf{j})$

where α is the direction angle between **v** and **i**.

EXAMPLE 6 Finding a Vector When Its Magnitude and Direction Are Given

A ball is thrown with an initial speed of 25 miles per hour in a direction that makes an angle of 30° with the positive x-axis. Express the velocity vector $\mathbf{v}$ in terms of $\mathbf{i}$ and $\mathbf{j}$. What is the initial speed in the horizontal direction? What is the initial speed in the vertical direction?

Solution The magnitude of $\mathbf{v}$ is $\|\mathbf{v}\| = 25$ miles per hour, and the angle between the direction of $\mathbf{v}$ and $\mathbf{i}$, the positive x-axis, is $\alpha = 30°$. By equation (8),

$$\mathbf{v} = \|\mathbf{v}\|(\cos\alpha\mathbf{i} + \sin\alpha\mathbf{j}) = 25(\cos 30°\mathbf{i} + \sin 30°\mathbf{j})$$

$$= 25\left(\frac{\sqrt{3}}{2}\mathbf{i} + \frac{1}{2}\mathbf{j}\right) = \frac{25\sqrt{3}}{2}\mathbf{i} + \frac{25}{2}\mathbf{j}$$

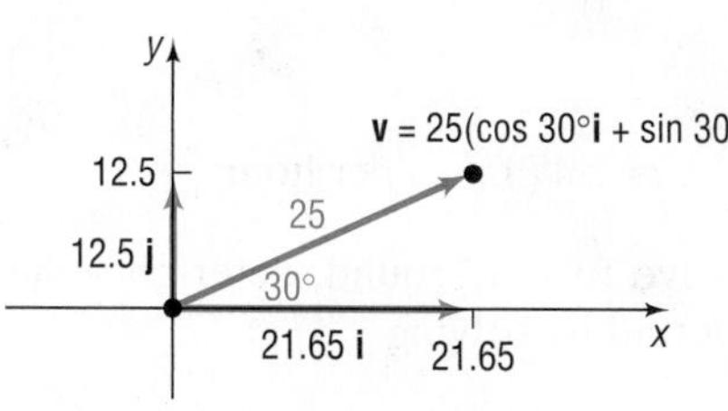

Figure 57

The initial speed of the ball in the horizontal direction is the horizontal component of $\mathbf{v}$, $\frac{25\sqrt{3}}{2} \approx 21.65$ miles per hour. The initial speed in the vertical direction is the vertical component of $\mathbf{v}$, $\frac{25}{2} = 12.5$ miles per hour. See Figure 57. ●

Now Work PROBLEM 59

EXAMPLE 7 Finding the Direction Angle of a Vector

Find the direction angle α of $\mathbf{v} = 4\mathbf{i} - 4\mathbf{j}$.

Solution See Figure 58. The direction angle α of $\mathbf{v} = 4\mathbf{i} - 4\mathbf{j}$ can be found by solving

$$\tan\alpha = \frac{-4}{4} = -1$$

Because $0° \leq \alpha < 360°$, the direction angle is $\alpha = 315°$. ●

Figure 58

Now Work PROBLEM 65

7 Model with Vectors

Because forces can be represented by vectors, two forces "combine" the way that vectors "add." If $\mathbf{F}_1$ and $\mathbf{F}_2$ are two forces simultaneously acting on an object, the vector sum $\mathbf{F}_1 + \mathbf{F}_2$ is the **resultant force**. The resultant force produces the same effect on the object as that obtained when the two forces $\mathbf{F}_1$ and $\mathbf{F}_2$ act on the object. See Figure 59.

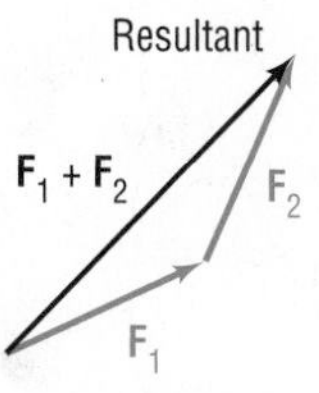

Figure 59 Resultant force

EXAMPLE 8 Finding the Actual Speed and Direction of an Aircraft

A Boeing 737 aircraft maintains a constant airspeed of 500 miles per hour headed due south. The jet stream is 80 miles per hour in the northeasterly direction.

(a) Express the velocity $\mathbf{v}_a$ of the 737 relative to the air and the velocity $\mathbf{v}_w$ of the jet stream in terms of $\mathbf{i}$ and $\mathbf{j}$.

(b) Find the velocity of the 737 relative to the ground.

(c) Find the actual speed and direction of the 737 relative to the ground.

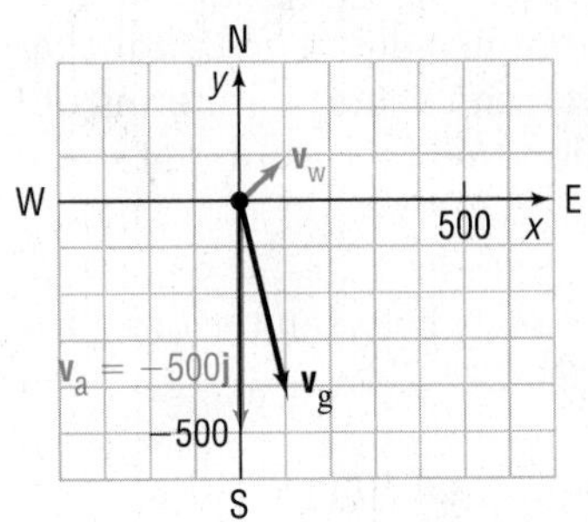

Figure 60

Solution (a) Set up a coordinate system in which north (N) is along the positive y-axis. See Figure 60. The velocity of the 737 relative to the air is $\mathbf{v}_a = -500\mathbf{j}$. The velocity of the jet stream $\mathbf{v}_w$ has magnitude 80 and direction NE (northeast), so the angle between $\mathbf{v}_w$ and $\mathbf{i}$ is 45°. Express $\mathbf{v}_w$ in terms of $\mathbf{i}$ and $\mathbf{j}$ as

$$\mathbf{v}_w = 80(\cos 45°\mathbf{i} + \sin 45°\mathbf{j}) = 80\left(\frac{\sqrt{2}}{2}\mathbf{i} + \frac{\sqrt{2}}{2}\mathbf{j}\right) = 40\sqrt{2}\,(\mathbf{i} + \mathbf{j})$$

(b) The velocity of the 737 relative to the ground $\mathbf{v}_g$ is

$$\mathbf{v}_g = \mathbf{v}_a + \mathbf{v}_w = -500\mathbf{j} + 40\sqrt{2}(\mathbf{i} + \mathbf{j}) = 40\sqrt{2}\mathbf{i} + \left(40\sqrt{2} - 500\right)\mathbf{j}$$

(c) The actual speed of the 737 is

$$\|\mathbf{v}_g\| = \sqrt{(40\sqrt{2})^2 + (40\sqrt{2} - 500)^2} \approx 447 \text{ miles per hour}$$

To find the actual direction of the 737 relative to the ground, determine the direction angle of $\mathbf{v}_g$. The direction angle is found by solving

$$\tan\alpha = \frac{40\sqrt{2} - 500}{40\sqrt{2}}$$

Then $\alpha \approx -82.7°$. The 737 is traveling S 7.3° E.

Now Work PROBLEM 77

EXAMPLE 9 Finding the Weight of a Piano

Two movers require a magnitude of force of 300 pounds to push a piano up a ramp inclined at an angle 20° from the horizontal. How much does the piano weigh?

Solution Let $\mathbf{F}_1$ represent the force of gravity, $\mathbf{F}_2$ represent the force required to move the piano up the ramp, and $\mathbf{F}_3$ represent the force of the piano against the ramp. See Figure 61. The angle between the ground and the ramp is the same as the angle between $\mathbf{F}_1$ and $\mathbf{F}_3$ because triangles ABC and BDE are similar, so $\angle BAC = \angle DBE = 20°$. To find the magnitude of $\mathbf{F}_1$ (the weight of the piano), calculate

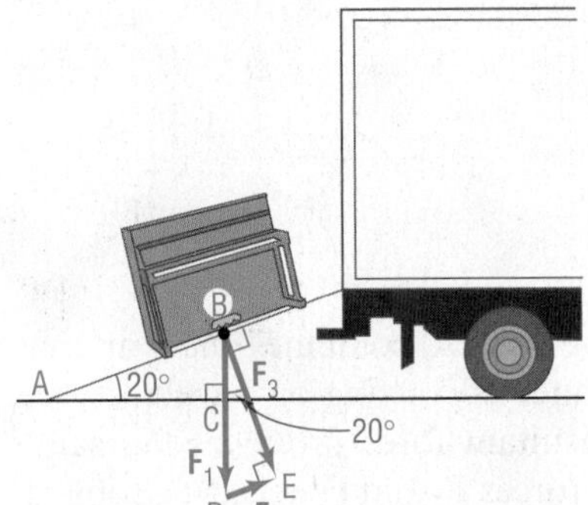

Figure 61

$$\sin 20° = \frac{\|\mathbf{F}_2\|}{\|\mathbf{F}_1\|} = \frac{300}{\|\mathbf{F}_1\|}$$

$$\|\mathbf{F}_1\| = \frac{300 \text{ lb}}{\sin 20°} \approx 877 \text{ lb}$$

The piano weighs approximately 877 pounds.

An object is said to be in **static equilibrium** if the object is at rest and the sum of all forces acting on the object is zero—that is, if the resultant force is **0**.

EXAMPLE 10 Analyzing an Object in Static Equilibrium

A box of supplies that weighs 1200 pounds is suspended by two cables attached to the ceiling, as shown in Figure 62. What are the tensions in the two cables?

Solution Draw a force diagram using the vectors as shown in Figure 63. The tensions in the cables are the magnitudes $\|\mathbf{F}_1\|$ and $\|\mathbf{F}_2\|$ of the force vectors $\mathbf{F}_1$ and $\mathbf{F}_2$. The magnitude of the force vector $\mathbf{F}_3$ equals 1200 pounds, the weight of the box. Now write each force vector in terms of the unit vectors $\mathbf{i}$ and $\mathbf{j}$. For $\mathbf{F}_1$ and $\mathbf{F}_2$, use equation (8). Remember that α is the angle between the vector and the positive x-axis.

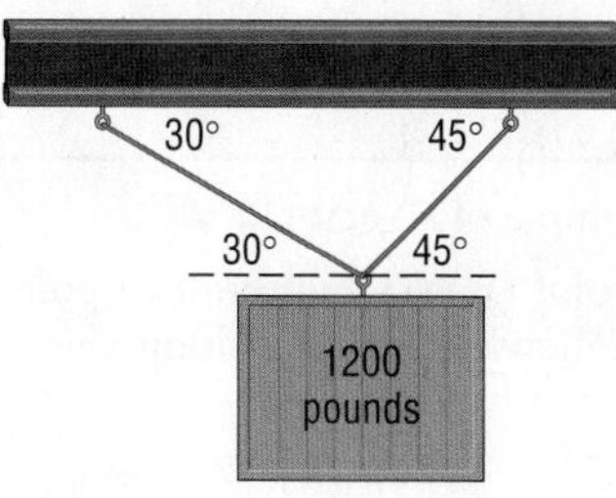

Figure 62

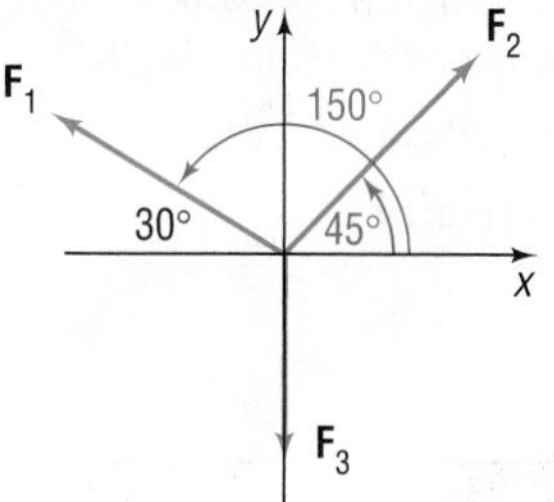

Figure 63 Force diagram

$$\mathbf{F}_1 = \|\mathbf{F}_1\|(\cos 150°\mathbf{i} + \sin 150°\mathbf{j}) = \|\mathbf{F}_1\|\left(-\frac{\sqrt{3}}{2}\mathbf{i} + \frac{1}{2}\mathbf{j}\right) = -\frac{\sqrt{3}}{2}\|\mathbf{F}_1\|\mathbf{i} + \frac{1}{2}\|\mathbf{F}_1\|\mathbf{j}$$

$$\mathbf{F}_2 = \|\mathbf{F}_2\|(\cos 45°\mathbf{i} + \sin 45°\mathbf{j}) = \|\mathbf{F}_2\|\left(\frac{\sqrt{2}}{2}\mathbf{i} + \frac{\sqrt{2}}{2}\mathbf{j}\right) = \frac{\sqrt{2}}{2}\|\mathbf{F}_2\|\mathbf{i} + \frac{\sqrt{2}}{2}\|\mathbf{F}_2\|\mathbf{j}$$

$$\mathbf{F}_3 = -1200\mathbf{j}$$

For static equilibrium, the sum of the force vectors must equal zero.

$$\mathbf{F}_1 + \mathbf{F}_2 + \mathbf{F}_3 = -\frac{\sqrt{3}}{2}\|\mathbf{F}_1\|\mathbf{i} + \frac{1}{2}\|\mathbf{F}_1\|\mathbf{j} + \frac{\sqrt{2}}{2}\|\mathbf{F}_2\|\mathbf{i} + \frac{\sqrt{2}}{2}\|\mathbf{F}_2\|\mathbf{j} - 1200\mathbf{j} = \mathbf{0}$$

The **i** component and **j** component will each equal zero. This results in the two equations

$$-\frac{\sqrt{3}}{2}\|\mathbf{F}_1\| + \frac{\sqrt{2}}{2}\|\mathbf{F}_2\| = 0 \tag{9}$$

$$\frac{1}{2}\|\mathbf{F}_1\| + \frac{\sqrt{2}}{2}\|\mathbf{F}_2\| - 1200 = 0 \tag{10}$$

Solve equation (9) for $\|\mathbf{F}_2\|$ to obtain

$$\|\mathbf{F}_2\| = \frac{\sqrt{3}}{\sqrt{2}}\|\mathbf{F}_1\| \tag{11}$$

Substituting into equation (10) and solving for $\|\mathbf{F}_1\|$ yields

$$\frac{1}{2}\|\mathbf{F}_1\| + \frac{\sqrt{2}}{2}\left(\frac{\sqrt{3}}{\sqrt{2}}\|\mathbf{F}_1\|\right) - 1200 = 0$$

$$\frac{1}{2}\|\mathbf{F}_1\| + \frac{\sqrt{3}}{2}\|\mathbf{F}_1\| - 1200 = 0$$

$$\frac{1 + \sqrt{3}}{2}\|\mathbf{F}_1\| = 1200$$

$$\|\mathbf{F}_1\| = \frac{2400}{1 + \sqrt{3}} \approx 878.5 \text{ pounds}$$

Substituting this value into equation (11) gives $\|\mathbf{F}_2\|$.

$$\|\mathbf{F}_2\| = \frac{\sqrt{3}}{\sqrt{2}}\|\mathbf{F}_1\| = \frac{\sqrt{3}}{\sqrt{2}} \cdot \frac{2400}{1 + \sqrt{3}} \approx 1075.9 \text{ pounds}$$

The left cable has tension of approximately 878.5 pounds, and the right cable has tension of approximately 1075.9 pounds. ●

Now Work PROBLEM 87

Historical Feature

Josiah Gibbs (1839–1903)

The history of vectors is surprisingly complicated for such a natural concept. In the *xy*-plane, complex numbers do a good job of imitating vectors. About 1840, mathematicians became interested in finding a system that would do for three dimensions what the complex numbers do for two dimensions. Hermann Grassmann (1809–1877), in Germany, and William Rowan Hamilton (1805–1865), in Ireland, both attempted to find solutions.

Hamilton's system was the *quaternions,* which are best thought of as a real number plus a vector; they do for four dimensions what complex numbers do for two dimensions. In this system the order of multiplication matters; that is, ab ≠ ba. Also, two products of vectors emerged, the scalar product (or dot product) and the vector product (or cross product).

Grassmann's abstract style, although easily read today, was almost impenetrable during the nineteenth century, and only a few of his ideas were appreciated. Among those few were the same scalar and vector products that Hamilton had found.

About 1880, the American physicist Josiah Willard Gibbs (1839–1903) worked out an algebra involving only the simplest concepts: the vectors and the two products. He then added some calculus, and the resulting system was simple, flexible, and well adapted to expressing a large number of physical laws. This system remains in use essentially unchanged. Hamilton's and Grassmann's more extensive systems each gave birth to much interesting mathematics, but little of it is seen at elementary levels.

5.4 Assess Your Understanding

Concepts and Vocabulary

1. A ______ is a quantity that has both magnitude and direction.
2. If **v** is a vector, then $\mathbf{v} + (-\mathbf{v}) =$ ____.
3. A vector **u** for which $\|\mathbf{u}\| = 1$ is called a(n) ______ vector.
4. If $\mathbf{v} = \langle a, b \rangle$ is an algebraic vector whose initial point is the origin, then **v** is called a(n) ________ vector.
5. If $\mathbf{v} = a\mathbf{i} + b\mathbf{j}$, then a is called the __________ component of **v** and b is called the ________ component of **v**.
6. If $\mathbf{F}_1$ and $\mathbf{F}_2$ are two forces simultaneously acting on an object, the vector sum $\mathbf{F}_1 + \mathbf{F}_2$ is called the ________ force.
7. ***True or False*** Force is an example of a vector.
8. ***True or False*** Mass is an example of a vector.
9. If **v** is a vector with initial point (x_1, y_1) and terminal point (x_2, y_2), then which of the following is the position vector that equals **v**?
   (a) $\langle x_2 - x_1, y_2 - y_1 \rangle$ (b) $\langle x_1 - x_2, y_1 - y_2 \rangle$
   (c) $\left\langle \frac{x_2 - x_1}{2}, \frac{y_2 - y_1}{2} \right\rangle$ (d) $\left\langle \frac{x_1 + x_2}{2}, \frac{y_1 + y_2}{2} \right\rangle$
10. If **v** is a nonzero vector with direction angle α, $0° \le \alpha < 360°$, between **v** and **i**, then **v** equals which of the following?
    (a) $\|\mathbf{v}\|(\cos\alpha\mathbf{i} - \sin\alpha\mathbf{j})$ (b) $\|\mathbf{v}\|(\cos\alpha\mathbf{i} + \sin\alpha\mathbf{j})$
    (c) $\|\mathbf{v}\|(\sin\alpha\mathbf{i} - \cos\alpha\mathbf{j})$ (d) $\|\mathbf{v}\|(\sin\alpha\mathbf{i} + \cos\alpha\mathbf{j})$

Skill Building

In Problems 11–18, use the vectors in the figure at the right to graph each of the following vectors.

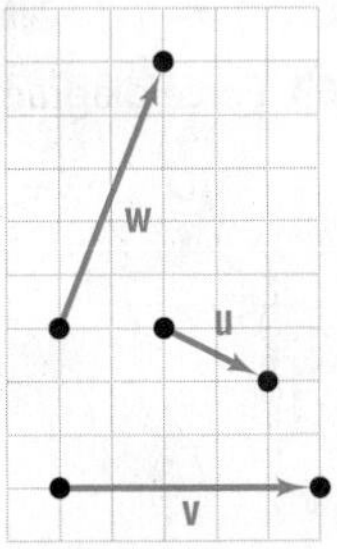

11. **v** + **w**
12. **u** + **v**
13. 3**v**
14. 2**w**
15. **v** − **w**
16. **u** − **v**
17. 3**v** + **u** − 2**w**
18. 2**u** − 3**v** + **w**

In Problems 19–26, use the figure at the right. Determine whether each statement given is true or false.

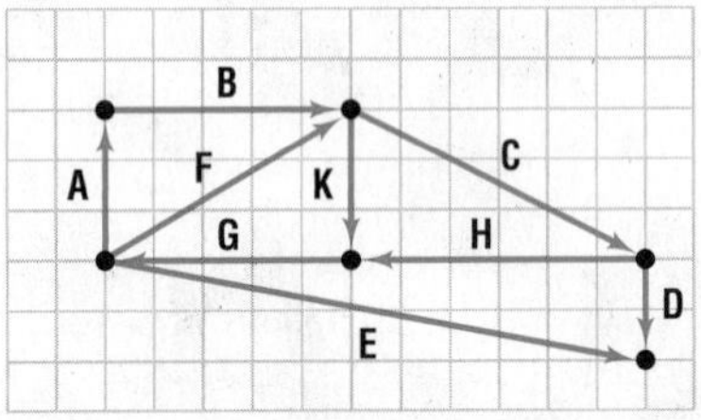

19. **A** + **B** = **F**
20. **K** + **G** = **F**
21. **C** = **D** − **E** + **F**
22. **G** + **H** + **E** = **D**
23. **E** + **D** = **G** + **H**
24. **H** − **C** = **G** − **F**
25. **A** + **B** + **K** + **G** = **0**
26. **A** + **B** + **C** + **H** + **G** = **0**

27. If $\|\mathbf{v}\| = 4$, what is $\|3\mathbf{v}\|$?
28. If $\|\mathbf{v}\| = 2$, what is $\|-4\mathbf{v}\|$?

*In Problems 29–36, the vector **v** has initial point P and terminal point Q. Write **v** in the form $a\mathbf{i} + b\mathbf{j}$; that is, find its position vector.*

29. $P = (0, 0);\ Q = (3, 4)$
30. $P = (0, 0);\ Q = (-3, -5)$
31. $P = (3, 2);\ Q = (5, 6)$
32. $P = (-3, 2);\ Q = (6, 5)$
33. $P = (-2, -1);\ Q = (6, -2)$
34. $P = (-1, 4);\ Q = (6, 2)$
35. $P = (1, 0);\ Q = (0, 1)$
36. $P = (1, 1);\ Q = (2, 2)$

In Problems 37–42, find $\|\mathbf{v}\|$.

37. **v** = 3**i** − 4**j**
38. **v** = −5**i** + 12**j**
39. **v** = **i** − **j**
40. **v** = −**i** − **j**
41. **v** = −2**i** + 3**j**
42. **v** = 6**i** + 2**j**

*In Problems 43–48, find each quantity if **v** = 3**i** − 5**j** and **w** = −2**i** + 3**j**.*

43. 2**v** + 3**w**
44. 3**v** − 2**w**
45. $\|\mathbf{v} - \mathbf{w}\|$
46. $\|\mathbf{v} + \mathbf{w}\|$
47. $\|\mathbf{v}\| - \|\mathbf{w}\|$
48. $\|\mathbf{v}\| + \|\mathbf{w}\|$

*In Problems 49–54, find the unit vector in the same direction as **v**.*

49. **v** = 5**i**
50. **v** = −3**j**
51. **v** = 3**i** − 4**j**
52. **v** = −5**i** + 12**j**
53. **v** = **i** − **j**
54. **v** = 2**i** − **j**

55. Find a vector $\mathbf{v}$ whose magnitude is 4 and whose component in the $\mathbf{i}$ direction is twice the component in the $\mathbf{j}$ direction.

56. Find a vector $\mathbf{v}$ whose magnitude is 3 and whose component in the $\mathbf{i}$ direction is equal to the component in the $\mathbf{j}$ direction.

57. If $\mathbf{v} = 2\mathbf{i} - \mathbf{j}$ and $\mathbf{w} = x\mathbf{i} + 3\mathbf{j}$, find all numbers x for which $\|\mathbf{v} + \mathbf{w}\| = 5$.

58. If $P = (-3, 1)$ and $Q = (x, 4)$, find all numbers x such that the vector represented by $\overrightarrow{PQ}$ has length 5.

In Problems 59–64, write the vector $\mathbf{v}$ *in the form* $a\mathbf{i} + b\mathbf{j}$, *given its magnitude* $\|\mathbf{v}\|$ *and the angle* α *it makes with the positive x-axis.*

59. $\|\mathbf{v}\| = 5, \quad \alpha = 60°$

60. $\|\mathbf{v}\| = 8, \quad \alpha = 45°$

61. $\|\mathbf{v}\| = 14, \quad \alpha = 120°$

62. $\|\mathbf{v}\| = 3, \quad \alpha = 240°$

63. $\|\mathbf{v}\| = 25, \quad \alpha = 330°$

64. $\|\mathbf{v}\| = 15, \quad \alpha = 315°$

In Problems 65–72, find the direction angle of $\mathbf{v}$.

65. $\mathbf{v} = 3\mathbf{i} + 3\mathbf{j}$

66. $\mathbf{v} = \mathbf{i} + \sqrt{3}\mathbf{j}$

67. $\mathbf{v} = -3\sqrt{3}\mathbf{i} + 3\mathbf{j}$

68. $\mathbf{v} = -5\mathbf{i} - 5\mathbf{j}$

69. $\mathbf{v} = 4\mathbf{i} - 2\mathbf{j}$

70. $\mathbf{v} = 6\mathbf{i} - 4\mathbf{j}$

71. $\mathbf{v} = -\mathbf{i} - 5\mathbf{j}$

72. $\mathbf{v} = -\mathbf{i} + 3\mathbf{j}$

Applications and Extensions

73. **Force Vectors** A child pulls a wagon with a force of 40 pounds. The handle of the wagon makes an angle of 30° with the ground. Express the force vector $\mathbf{F}$ in terms of $\mathbf{i}$ and $\mathbf{j}$.

74. **Force Vectors** A man pushes a wheelbarrow up an incline of 20° with a force of 100 pounds. Express the force vector $\mathbf{F}$ in terms of $\mathbf{i}$ and $\mathbf{j}$.

75. **Resultant Force** Two forces of magnitude 40 newtons (N) and 60 N act on an object at angles of 30° and −45° with the positive x-axis, as shown in the figure. Find the direction and magnitude of the resultant force; that is, find $\mathbf{F}_1 + \mathbf{F}_2$.

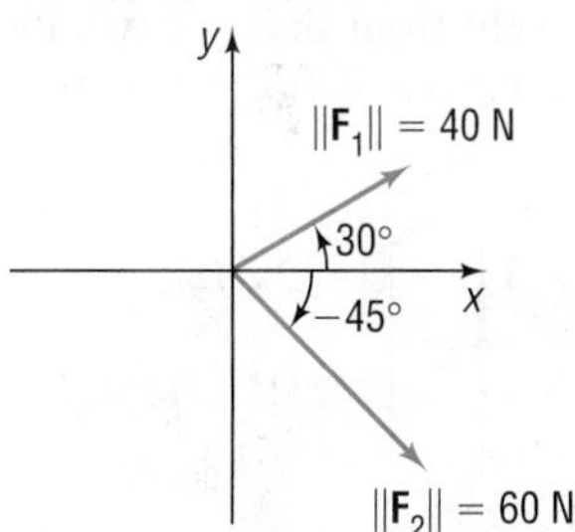

76. **Resultant Force** Two forces of magnitude 30 newtons (N) and 70 N act on an object at angles of 45° and 120° with the positive x-axis, as shown in the figure. Find the direction and magnitude of the resultant force; that is, find $\mathbf{F}_1 + \mathbf{F}_2$.

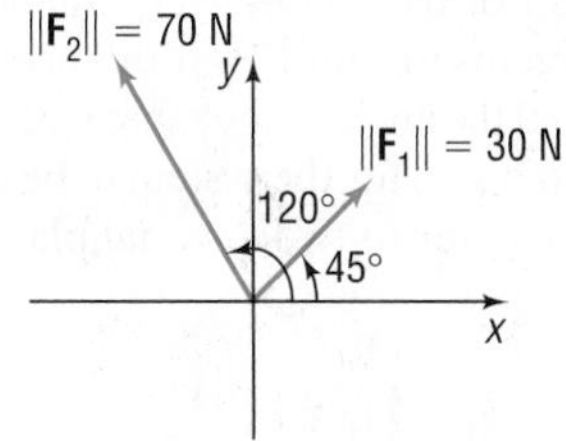

77. **Finding the Actual Speed and Direction of an Aircraft** A Boeing 747 jumbo jet maintains a constant airspeed of 550 miles per hour (mph) headed due north. The jet stream is 100 mph in the northeasterly direction.
   (a) Express the velocity $\mathbf{v}_a$ of the 747 relative to the air and the velocity $\mathbf{v}_w$ of the jet stream in terms of $\mathbf{i}$ and $\mathbf{j}$.
   (b) Find the velocity of the 747 relative to the ground.
   (c) Find the actual speed and direction of the 747 relative to the ground.

78. **Finding the Actual Speed and Direction of an Aircraft** An Airbus A320 jet maintains a constant airspeed of 500 mph headed due west. The jet stream is 100 mph in the southeasterly direction.
   (a) Express the velocity $\mathbf{v}_a$ of the A320 relative to the air and the velocity $\mathbf{v}_w$ of the jet stream in terms of $\mathbf{i}$ and $\mathbf{j}$.
   (b) Find the velocity of the A320 relative to the ground.
   (c) Find the actual speed and direction of the A320 relative to the ground.

79. **Ground Speed and Direction of an Airplane** An airplane has an airspeed of 500 kilometers per hour (km/h) bearing N45°E. The wind velocity is 60 km/h in the direction N30°W. Find the resultant vector representing the path of the plane relative to the ground. What is the groundspeed of the plane? What is its direction?

80. **Ground Speed and Direction of an Airplane** An airplane has an airspeed of 600 km/h bearing S30°E. The wind velocity is 40 km/h in the direction S45°E. Find the resultant vector representing the path of the plane relative to the ground. What is the groundspeed of the plane? What is its direction?

81. **Weight of a Boat** A magnitude of 700 pounds of force is required to hold a boat and its trailer in place on a ramp whose incline is 10° to the horizontal. What is the combined weight of the boat and its trailer?

82. **Weight of a Car** A magnitude of 1200 pounds of force is required to prevent a car from rolling down a hill whose incline is 15° to the horizontal. What is the weight of the car?

83. **Correct Direction for Crossing a River** A river has a constant current of 3 km/h. At what angle to a boat dock should a motorboat capable of maintaining a constant speed of 20 km/h be headed in order to reach a point directly opposite the dock? If the river is $\frac{1}{2}$ kilometer wide, how long will it take to cross?

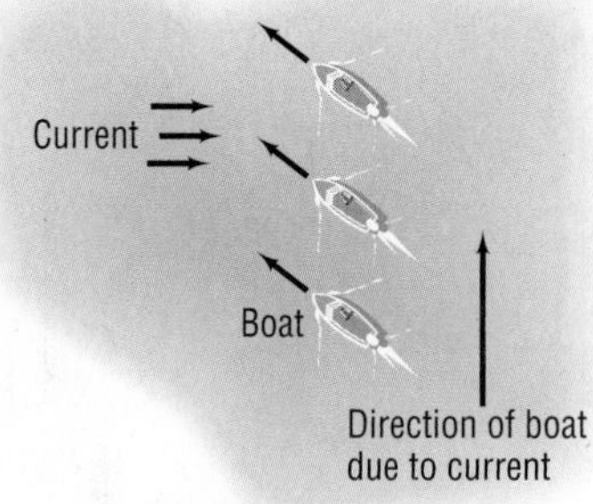

84. Finding the Correct Compass Heading The pilot of an aircraft wishes to head directly east but is faced with a wind speed of 40 mph from the northwest. If the pilot maintains an airspeed of 250 mph, what compass heading should be maintained to head directly east? What is the actual speed of the aircraft?

85. Charting a Course A helicopter pilot needs to travel to a regional airport 25 miles away. She flies at an actual heading of N16.26°E with an airspeed of 120 mph, and there is a wind blowing directly east at 20 mph.

(a) Determine the compass heading that the pilot needs to reach her destination.

(b) How long will it take her to reach her destination? Round to the nearest minute.

86. Crossing a River A captain needs to pilot a boat across a river that is 2 km wide. The current in the river is 2 km/h and the speed of the boat in still water is 10 km/h. The desired landing point on the other side is 1 km upstream.

(a) Determine the direction in which the captain should aim the boat.

(b) How long will the trip take?

87. Static Equilibrium A weight of 1000 pounds is suspended from two cables, as shown in the figure. What are the tensions in the two cables?

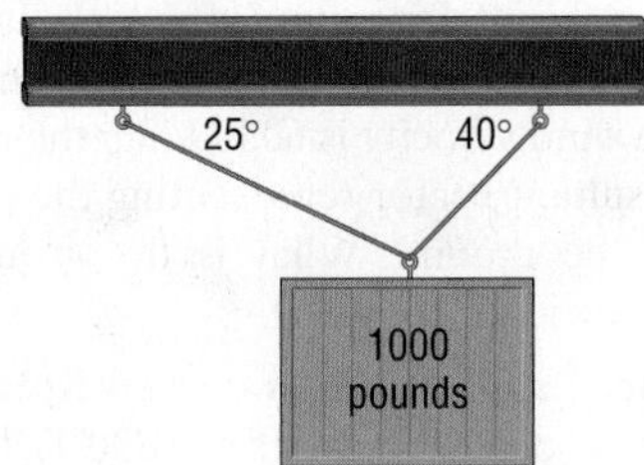

88. Static Equilibrium A weight of 800 pounds is suspended from two cables, as shown in the figure. What are the tensions in the two cables?

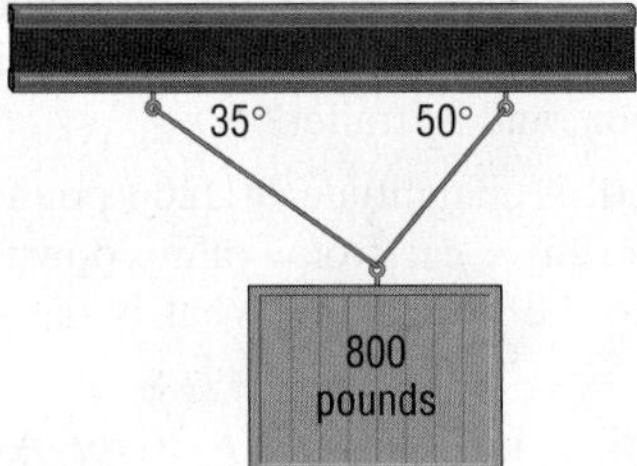

89. Static Equilibrium A tightrope walker located at a certain point deflects the rope as indicated in the figure. If the weight of the tightrope walker is 150 pounds, how much tension is in each part of the rope?

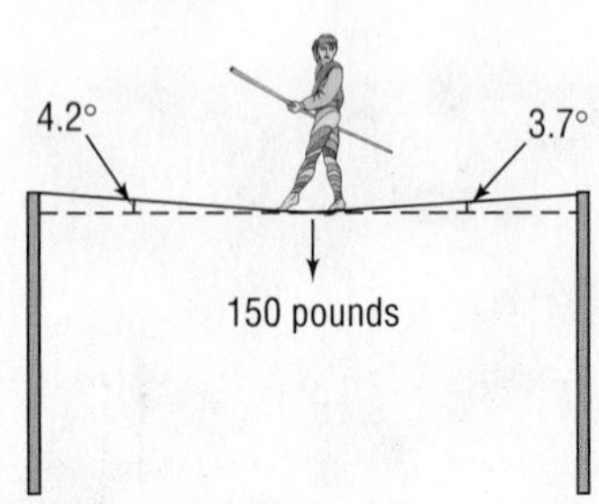

90. Static Equilibrium Repeat Problem 89 if the angle on the left is 3.8°, the angle on the right is 2.6°, and the weight of the tightrope walker is 135 pounds.

91. Static Friction A 20-pound box sits at rest on a horizontal surface, and there is friction between the box and the surface. One side of the surface is raised slowly to create a ramp. The friction force **f** opposes the direction of motion and is proportional to the normal force $\mathbf{F}_N$ exerted by the surface on the box. The proportionality constant is called the **coefficient of friction**, μ. When the angle of the ramp, θ, reaches 20°, the box begins to slide. Find the value of μ to two decimal places.

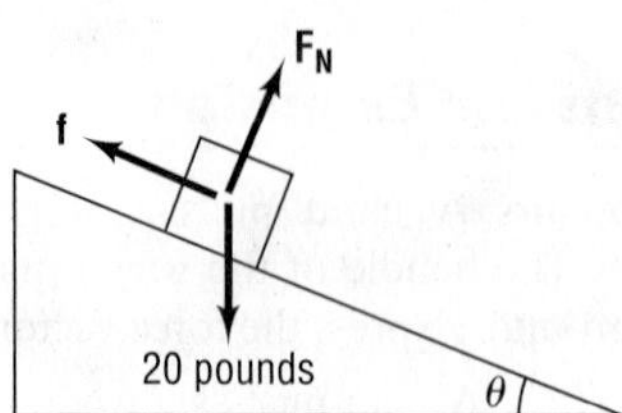

92. Inclined Ramp A 2-pound weight is attached to a 3-pound weight by a rope that passes over an ideal pulley. The smaller weight hangs vertically, while the larger weight sits on a frictionless inclined ramp with angle θ. The rope exerts a tension force **T** on both weights along the direction of the rope. Find the angle measure for θ that is needed to keep the larger weight from sliding down the ramp. Round your answer to the nearest tenth of a degree.

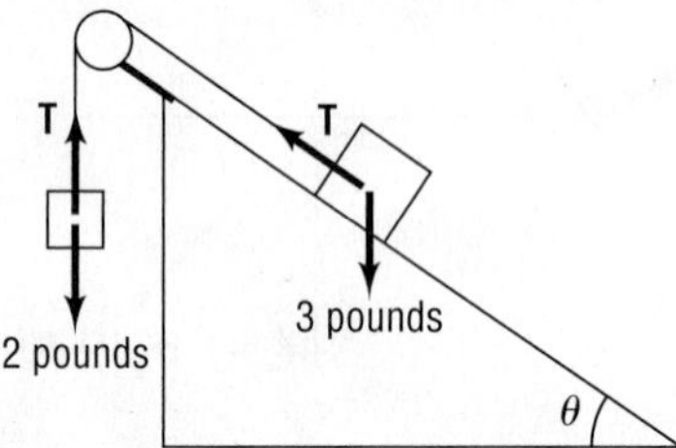

93. Inclined Ramp A box sitting on a horizontal surface is attached to a second box sitting on an inclined ramp by a rope that passes over an ideal pulley. The rope exerts a tension force **T** on both weights along the direction of the rope, and the coefficient of friction between the surface and boxes is 0.6 (see Problems 91 and 92). If the box on the right weighs 100 pounds and the angle of the ramp is 35°, how much must the box on the left weigh for the system to be in static equilibrium? Round your answer to two decimal places.

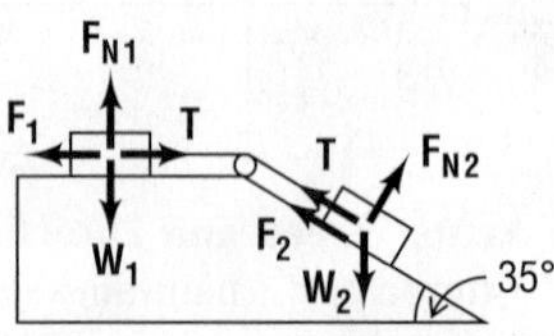

94. Muscle Force Two muscles exert force on a bone at the same point. The first muscle exerts a force of 800 N at a 10° angle with the bone. The second muscle exerts a force of 710 N at a 35° angle with the bone. What are the direction and magnitude of the resulting force on the bone?

95. Truck Pull At a county fair truck pull, two pickup trucks are attached to the back end of a monster truck as illustrated in the figure. One of the pickups pulls with a force of 2000 pounds, and the other pulls with a force of 3000 pounds. There is an angle of 45° between them. With how much force must the monster truck pull in order to remain unmoved? [**Hint:** Find the resultant force of the two trucks.]

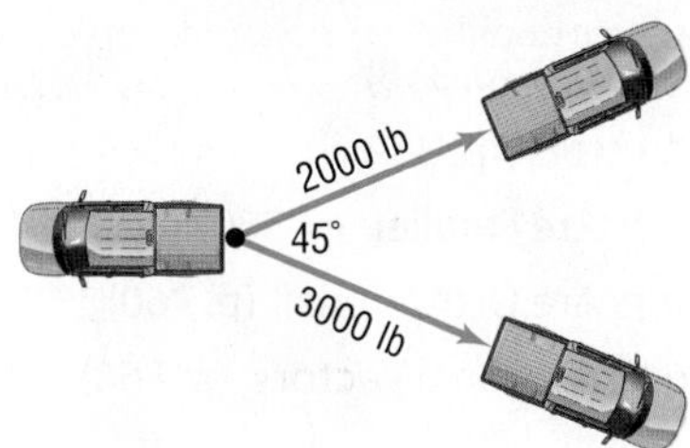

96. Removing a Stump A farmer wishes to remove a stump from a field by pulling it out with his tractor. Having removed many stumps before, he estimates that he will need 6 tons (12,000 pounds) of force to remove the stump. However, his tractor is only capable of pulling with a force of 7000 pounds, so he asks his neighbor to help. His neighbor's tractor can pull with a force of 5500 pounds. They attach the two tractors to the stump with a 40° angle between the forces, as shown in the figure.

(a) Assuming the farmer's estimate of a needed 6-ton force is correct, will the farmer be successful in removing the stump?

(b) Had the farmer arranged the tractors with a 25° angle between the forces, would he have been successful in removing the stump?

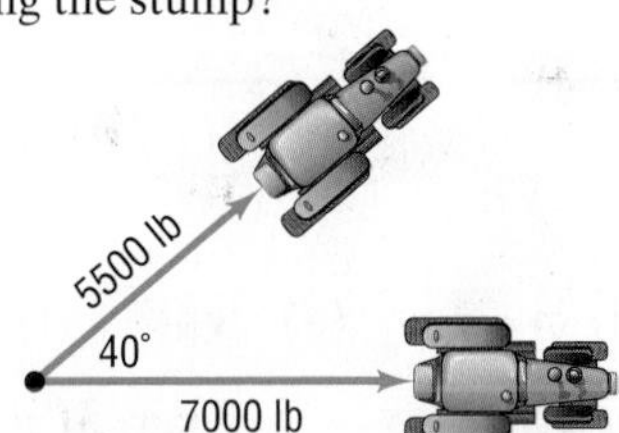

97. Computer Graphics The field of computer graphics utilizes vectors to compute translations of points. For example, if the point $(-3, 2)$ is to be translated by $\mathbf{v} = \langle 5, 2 \rangle$, then the new location will be $\mathbf{u}' = \mathbf{u} + \mathbf{v} = \langle -3, 2 \rangle + \langle 5, 2 \rangle = \langle 2, 4 \rangle$.

As illustrated in the figure, the point $(-3, 2)$ is translated to $(2, 4)$ by $\mathbf{v}$.

(a) Determine the new coordinates of $(3, -1)$ if it is translated by $\mathbf{v} = \langle -4, 5 \rangle$.

(b) Illustrate this translation graphically.

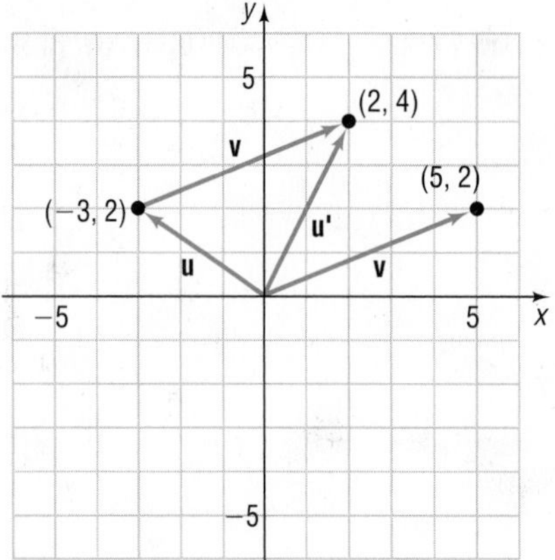

Source: Phil Dadd. Vectors and Matrices: A Primer. *www.gamedev.net/reference/articles/article1832.asp*

98. Computer Graphics Refer to Problem 97. The points $(-3, 0)$, $(-1, -2)$, $(3, 1)$, and $(1, 3)$ are the vertices of a parallelogram $ABCD$.

(a) Find the new vertices of a parallelogram $A'B'C'D'$ if it is translated by $\mathbf{v} = \langle 3, -2 \rangle$.

(b) Find the new vertices of a parallelogram $A'B'C'D'$ if it is translated by $-\frac{1}{2}\mathbf{v}$.

99. Static Equilibrium Show on the following graph the force needed for the object at P to be in static equilibrium.

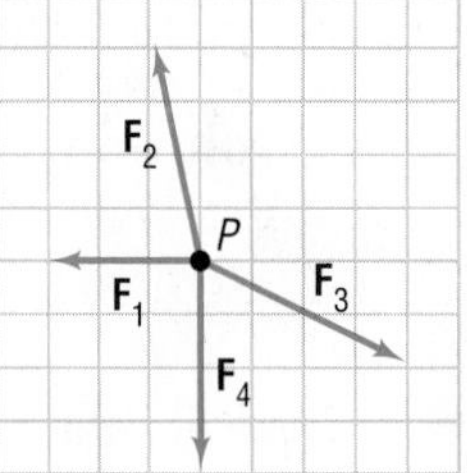

Explaining Concepts: Discussion and Writing

100. Explain in your own words what a vector is. Give an example of a vector.

101. Write a brief paragraph comparing the algebra of complex numbers and the algebra of vectors.

102. Explain the difference between an algebraic vector and a position vector.

Retain Your Knowledge

Problems 103–106 are based on material learned earlier in the course. The purpose of these problems is to keep the material fresh in your mind so that you are better prepared for the final exam.

103. Solve the equation $4\sin^2\theta - 3 = 0$ on the interval $0 \le \theta < 2\pi$.

104. Solve triangle ABC: $a = 4, b = 1$, and $C = 100°$

105. Find the exact value of $\tan\left[\cos^{-1}\left(\frac{1}{2}\right)\right]$.

106. Find the amplitude, period, and phase shift of $y = \frac{3}{2}\cos(6x + 3\pi)$. Graph the function, showing at least two periods.

5.5 The Dot Product

PREPARING FOR THIS SECTION *Before getting started, review the following:*

- Law of Cosines (Section 4.3, p. 282)

Now Work the **'Are You Prepared?'** problem on page 363.

OBJECTIVES
1. Find the Dot Product of Two Vectors (p. 358)
2. Find the Angle between Two Vectors (p. 359)
3. Determine Whether Two Vectors Are Parallel (p. 360)
4. Determine Whether Two Vectors Are Orthogonal (p. 360)
5. Decompose a Vector into Two Orthogonal Vectors (p. 360)
6. Compute Work (p. 362)

1 Find the Dot Product of Two Vectors

The definition for a product of two vectors is somewhat unexpected. However, such a product has meaning in many geometric and physical applications.

DEFINITION

If $\mathbf{v} = a_1\mathbf{i} + b_1\mathbf{j}$ and $\mathbf{w} = a_2\mathbf{i} + b_2\mathbf{j}$ are two vectors, the **dot product** $\mathbf{v}\cdot\mathbf{w}$ is defined as

$$\mathbf{v}\cdot\mathbf{w} = a_1a_2 + b_1b_2 \tag{1}$$

EXAMPLE 1 **Finding Dot Products**

If $\mathbf{v} = 2\mathbf{i} - 3\mathbf{j}$ and $\mathbf{w} = 5\mathbf{i} + 3\mathbf{j}$, find:

(a) $\mathbf{v}\cdot\mathbf{w}$ (b) $\mathbf{w}\cdot\mathbf{v}$ (c) $\mathbf{v}\cdot\mathbf{v}$ (d) $\mathbf{w}\cdot\mathbf{w}$ (e) $\|\mathbf{v}\|$ (f) $\|\mathbf{w}\|$

Solution

(a) $\mathbf{v}\cdot\mathbf{w} = 2(5) + (-3)3 = 1$ (b) $\mathbf{w}\cdot\mathbf{v} = 5(2) + 3(-3) = 1$

(c) $\mathbf{v}\cdot\mathbf{v} = 2(2) + (-3)(-3) = 13$ (d) $\mathbf{w}\cdot\mathbf{w} = 5(5) + 3(3) = 34$

(e) $\|\mathbf{v}\| = \sqrt{2^2 + (-3)^2} = \sqrt{13}$ (f) $\|\mathbf{w}\| = \sqrt{5^2 + 3^2} = \sqrt{34}$ ●

COMMENT A scalar multiple $\alpha\mathbf{v}$ is a vector. A dot product $\mathbf{u}\cdot\mathbf{v}$ is a scalar (real number). ■

Since the dot product $\mathbf{v}\cdot\mathbf{w}$ of two vectors $\mathbf{v}$ and $\mathbf{w}$ is a real number (a scalar), it is sometimes referred to as the **scalar product**.

The results obtained in Example 1 suggest some general properties of the dot product.

THEOREM

Properties of the Dot Product

If **u**, **v**, and **w** are vectors, then

Commutative Property

$$\mathbf{u}\cdot\mathbf{v} = \mathbf{v}\cdot\mathbf{u} \tag{2}$$

Distributive Property

$$\mathbf{u}\cdot(\mathbf{v} + \mathbf{w}) = \mathbf{u}\cdot\mathbf{v} + \mathbf{u}\cdot\mathbf{w} \tag{3}$$

$$\mathbf{v}\cdot\mathbf{v} = \|\mathbf{v}\|^2 \tag{4}$$

$$\mathbf{0}\cdot\mathbf{v} = 0 \tag{5}$$

Proof We prove properties (2) and (4) here and leave properties (3) and (5) as exercises (see Problems 38 and 39).

To prove property (2), let $\mathbf{u} = a_1\mathbf{i} + b_1\mathbf{j}$ and $\mathbf{v} = a_2\mathbf{i} + b_2\mathbf{j}$. Then

$$\mathbf{u}\cdot\mathbf{v} = a_1a_2 + b_1b_2 = a_2a_1 + b_2b_1 = \mathbf{v}\cdot\mathbf{u}$$

To prove property (4), let $\mathbf{v} = a\mathbf{i} + b\mathbf{j}$. Then

$$\mathbf{v}\cdot\mathbf{v} = a^2 + b^2 = \|\mathbf{v}\|^2$$

■

2 Find the Angle between Two Vectors

One use of the dot product is to calculate the angle between two vectors.

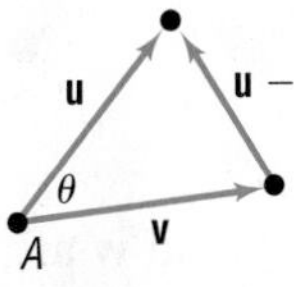

Let **u** and **v** be two vectors with the same initial point A. Then the vectors **u**, **v**, and $\mathbf{u} - \mathbf{v}$ form a triangle. The angle θ at vertex A of the triangle is the angle between the vectors **u** and **v**. See Figure 64. We wish to find a formula for calculating the angle θ.

Figure 64

The sides of the triangle have lengths $\|\mathbf{v}\|, \|\mathbf{u}\|$, and $\|\mathbf{u} - \mathbf{v}\|$, and θ is the included angle between the sides of length $\|\mathbf{v}\|$ and $\|\mathbf{u}\|$. The Law of Cosines (Section 4.3) can be used to find the cosine of the included angle.

$$\|\mathbf{u} - \mathbf{v}\|^2 = \|\mathbf{u}\|^2 + \|\mathbf{v}\|^2 - 2\|\mathbf{u}\|\|\mathbf{v}\|\cos\theta$$

Now use property (4) to rewrite this equation in terms of dot products.

$$(\mathbf{u} - \mathbf{v})\cdot(\mathbf{u} - \mathbf{v}) = \mathbf{u}\cdot\mathbf{u} + \mathbf{v}\cdot\mathbf{v} - 2\|\mathbf{u}\|\|\mathbf{v}\|\cos\theta \qquad \textbf{(6)}$$

Then apply the distributive property (3) twice on the left side of (6) to obtain

$$\begin{aligned}(\mathbf{u} - \mathbf{v})\cdot(\mathbf{u} - \mathbf{v}) &= \mathbf{u}\cdot(\mathbf{u} - \mathbf{v}) - \mathbf{v}\cdot(\mathbf{u} - \mathbf{v})\\ &= \mathbf{u}\cdot\mathbf{u} - \mathbf{u}\cdot\mathbf{v} - \mathbf{v}\cdot\mathbf{u} + \mathbf{v}\cdot\mathbf{v}\\ &= \mathbf{u}\cdot\mathbf{u} + \mathbf{v}\cdot\mathbf{v} - 2\,\mathbf{u}\cdot\mathbf{v} \qquad \textbf{(7)}\end{aligned}$$

↑ Property (2)

Combining equations (6) and (7) gives

$$\mathbf{u}\cdot\mathbf{u} + \mathbf{v}\cdot\mathbf{v} - 2\,\mathbf{u}\cdot\mathbf{v} = \mathbf{u}\cdot\mathbf{u} + \mathbf{v}\cdot\mathbf{v} - 2\|\mathbf{u}\|\|\mathbf{v}\|\cos\theta$$

$$\mathbf{u}\cdot\mathbf{v} = \|\mathbf{u}\|\|\mathbf{v}\|\cos\theta$$

THEOREM

Angle between Vectors

If **u** and **v** are two nonzero vectors, the angle θ, $0 \le \theta \le \pi$, between **u** and **v** is determined by the formula

$$\cos\theta = \frac{\mathbf{u}\cdot\mathbf{v}}{\|\mathbf{u}\|\|\mathbf{v}\|} \qquad \textbf{(8)}$$

EXAMPLE 2 **Finding the Angle θ between Two Vectors**

Find the angle θ between $\mathbf{u} = 4\mathbf{i} - 3\mathbf{j}$ and $\mathbf{v} = 2\mathbf{i} + 5\mathbf{j}$.

Solution Find $\mathbf{u}\cdot\mathbf{v}$, $\|\mathbf{u}\|$, and $\|\mathbf{v}\|$.

$$\mathbf{u}\cdot\mathbf{v} = 4(2) + (-3)(5) = -7$$

$$\|\mathbf{u}\| = \sqrt{4^2 + (-3)^2} = 5$$

$$\|\mathbf{v}\| = \sqrt{2^2 + 5^2} = \sqrt{29}$$

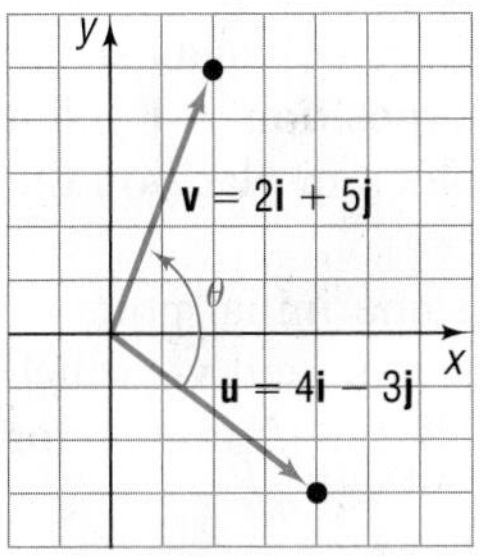

By formula (8), if θ is the angle between **u** and **v**, then

$$\cos\theta = \frac{\mathbf{u}\cdot\mathbf{v}}{\|\mathbf{u}\|\|\mathbf{v}\|} = \frac{-7}{5\sqrt{29}} \approx -0.26$$

Therefore, $\theta \approx \cos^{-1}(-0.26) \approx 105°$. See Figure 65. ●

Figure 65

Now Work PROBLEMS 9(a) AND (b)

3 Determine Whether Two Vectors Are Parallel

Two vectors **v** and **w** are said to be **parallel** if there is a nonzero scalar α so that $\mathbf{v} = \alpha\mathbf{w}$. In this case, the angle θ between **v** and **w** is 0 or π.

EXAMPLE 3 Determining Whether Two Vectors Are Parallel

The vectors $\mathbf{v} = 3\mathbf{i} - \mathbf{j}$ and $\mathbf{w} = 6\mathbf{i} - 2\mathbf{j}$ are parallel, since $\mathbf{v} = \frac{1}{2}\mathbf{w}$. Furthermore, since

$$\cos\theta = \frac{\mathbf{v}\cdot\mathbf{w}}{\|\mathbf{v}\|\,\|\mathbf{w}\|} = \frac{18+2}{\sqrt{10}\,\sqrt{40}} = \frac{20}{\sqrt{400}} = 1$$

the angle θ between **v** and **w** is 0. ●

4 Determine Whether Two Vectors Are Orthogonal

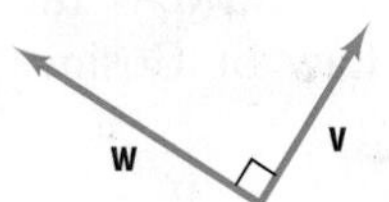

Figure 66 **v** is orthogonal to **w**.

If the angle θ between two nonzero vectors **v** and **w** is $\frac{\pi}{2}$, the vectors **v** and **w** are called **orthogonal**.* See Figure 66.

Since $\cos\frac{\pi}{2} = 0$, it follows from formula (8) that if the vectors **v** and **w** are orthogonal, then $\mathbf{v}\cdot\mathbf{w} = 0$.

On the other hand, if $\mathbf{v}\cdot\mathbf{w} = 0$, then $\mathbf{v} = \mathbf{0}$ or $\mathbf{w} = \mathbf{0}$ or $\cos\theta = 0$. If $\cos\theta = 0$, then $\theta = \frac{\pi}{2}$, and **v** and **w** are orthogonal. If **v** or **w** is the zero vector, then, since the zero vector has no specific direction, we adopt the convention that the zero vector is orthogonal to every vector.

THEOREM

Two vectors **v** and **w** are orthogonal if and only if

$$\mathbf{v}\cdot\mathbf{w} = 0$$

EXAMPLE 4 Determining Whether Two Vectors Are Orthogonal

The vectors

$$\mathbf{v} = 2\mathbf{i} - \mathbf{j} \quad \text{and} \quad \mathbf{w} = 3\mathbf{i} + 6\mathbf{j}$$

are orthogonal, since

$$\mathbf{v}\cdot\mathbf{w} = 6 - 6 = 0$$

See Figure 67. ●

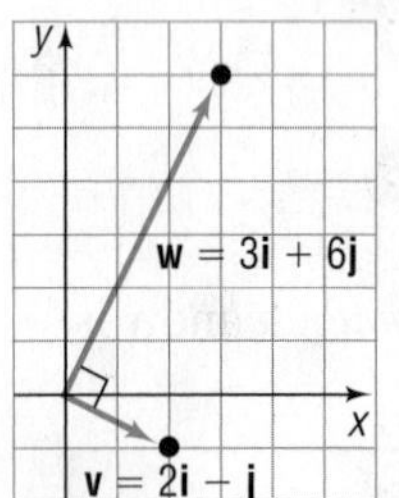

Figure 67

Now Work PROBLEM 9(c)

5 Decompose a Vector into Two Orthogonal Vectors

Figure 68

The last section discussed how to add two vectors to find the resultant vector. Now, the reverse problem is discussed, namely that of decomposing a vector into the sum of two components.

In many physical applications, it is necessary to find "how much" of a vector is applied in a given direction. Look at Figure 68. The force **F** due to gravity is pulling straight down (toward the center of Earth) on the block. To study the effect of gravity on the block, it is necessary to determine how much of **F** is actually pushing the block down the incline ($\mathbf{F}_1$) and how much is pressing the block against the incline ($\mathbf{F}_2$), at a right angle to the incline. Knowing the **decomposition** of **F** often enables us to determine when friction (the force holding the block in place on the incline) is overcome and the block will slide down the incline.

Suppose that **v** and **w** are two nonzero vectors with the same initial point *P*. We seek to decompose **v** into two vectors: $\mathbf{v}_1$, which is parallel to **w**, and $\mathbf{v}_2$, which is orthogonal to **w**. See Figure 69(a) and (b). The vector $\mathbf{v}_1$ is called the **vector projection of v onto w**.

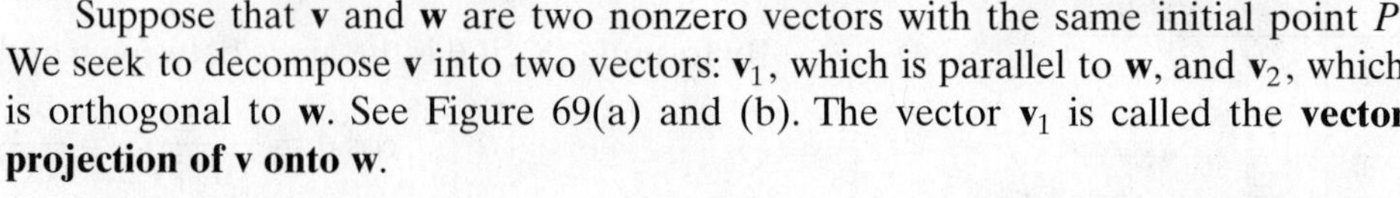

Orthogonal, perpendicular, and *normal* are all terms that mean "meet at a right angle." It is customary to refer to two vectors as being *orthogonal*, to two lines as being *perpendicular*, and to a line and a plane or a vector and a plane as being *normal*.

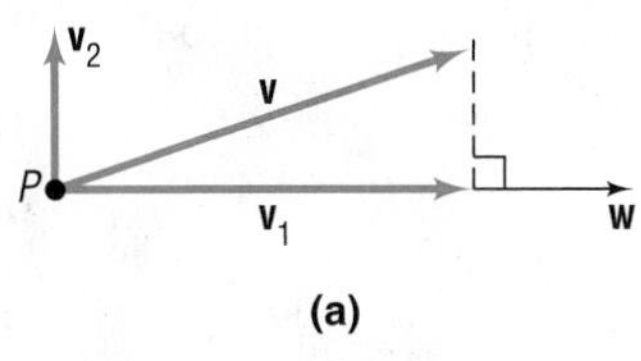

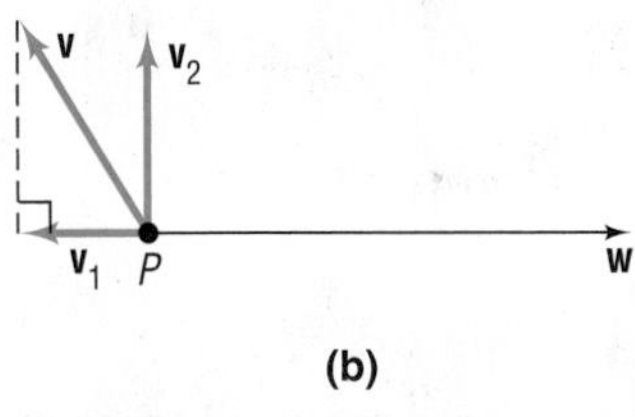

Figure 69

The vector $\mathbf{v}_1$ is obtained as follows: From the terminal point of $\mathbf{v}$, drop a perpendicular to the line containing $\mathbf{w}$. The vector $\mathbf{v}_1$ is the vector from P to the foot of this perpendicular. The vector $\mathbf{v}_2$ is given by $\mathbf{v}_2 = \mathbf{v} - \mathbf{v}_1$. Note that $\mathbf{v} = \mathbf{v}_1 + \mathbf{v}_2$, the vector $\mathbf{v}_1$ is parallel to $\mathbf{w}$, and the vector $\mathbf{v}_2$ is orthogonal to $\mathbf{w}$. This is the decomposition of $\mathbf{v}$ that was sought.

Now we seek a formula for $\mathbf{v}_1$ that is based on a knowledge of the vectors $\mathbf{v}$ and $\mathbf{w}$. Since $\mathbf{v} = \mathbf{v}_1 + \mathbf{v}_2$, we have

$$\mathbf{v}\cdot\mathbf{w} = (\mathbf{v}_1 + \mathbf{v}_2)\cdot\mathbf{w} = \mathbf{v}_1\cdot\mathbf{w} + \mathbf{v}_2\cdot\mathbf{w} \qquad (9)$$

Since $\mathbf{v}_2$ is orthogonal to $\mathbf{w}$, we have $\mathbf{v}_2\cdot\mathbf{w} = 0$. Since $\mathbf{v}_1$ is parallel to $\mathbf{w}$, we have $\mathbf{v}_1 = \alpha\mathbf{w}$ for some scalar α. Equation (9) can be written as

$$\mathbf{v}\cdot\mathbf{w} = \alpha\mathbf{w}\cdot\mathbf{w} = \alpha\|\mathbf{w}\|^2 \qquad \mathbf{v}_1 = \alpha\mathbf{w};\ \mathbf{v}_2\cdot\mathbf{w} = 0$$

$$\alpha = \frac{\mathbf{v}\cdot\mathbf{w}}{\|\mathbf{w}\|^2}$$

Then

$$\mathbf{v}_1 = \alpha\mathbf{w} = \frac{\mathbf{v}\cdot\mathbf{w}}{\|\mathbf{w}\|^2}\,\mathbf{w}$$

THEOREM If $\mathbf{v}$ and $\mathbf{w}$ are two nonzero vectors, the vector projection of $\mathbf{v}$ onto $\mathbf{w}$ is

$$\mathbf{v}_1 = \frac{\mathbf{v}\cdot\mathbf{w}}{\|\mathbf{w}\|^2}\,\mathbf{w} \qquad (10)$$

The decomposition of $\mathbf{v}$ into $\mathbf{v}_1$ and $\mathbf{v}_2$, where $\mathbf{v}_1$ is parallel to $\mathbf{w}$, and $\mathbf{v}_2$ is orthogonal to $\mathbf{w}$, is

$$\mathbf{v}_1 = \frac{\mathbf{v}\cdot\mathbf{w}}{\|\mathbf{w}\|^2}\,\mathbf{w} \qquad \mathbf{v}_2 = \mathbf{v} - \mathbf{v}_1 \qquad (11)$$

EXAMPLE 5 Decomposing a Vector into Two Orthogonal Vectors

Find the vector projection of $\mathbf{v} = \mathbf{i} + 3\mathbf{j}$ onto $\mathbf{w} = \mathbf{i} + \mathbf{j}$. Decompose $\mathbf{v}$ into two vectors, $\mathbf{v}_1$ and $\mathbf{v}_2$, where $\mathbf{v}_1$ is parallel to $\mathbf{w}$, and $\mathbf{v}_2$ is orthogonal to $\mathbf{w}$.

Solution Use formulas (10) and (11).

$$\mathbf{v}_1 = \frac{\mathbf{v}\cdot\mathbf{w}}{\|\mathbf{w}\|^2}\mathbf{w} = \frac{1+3}{(\sqrt{2})^2}\mathbf{w} = 2\mathbf{w} = 2(\mathbf{i} + \mathbf{j})$$

$$\mathbf{v}_2 = \mathbf{v} - \mathbf{v}_1 = (\mathbf{i} + 3\mathbf{j}) - 2(\mathbf{i} + \mathbf{j}) = -\mathbf{i} + \mathbf{j}$$

See Figure 70.

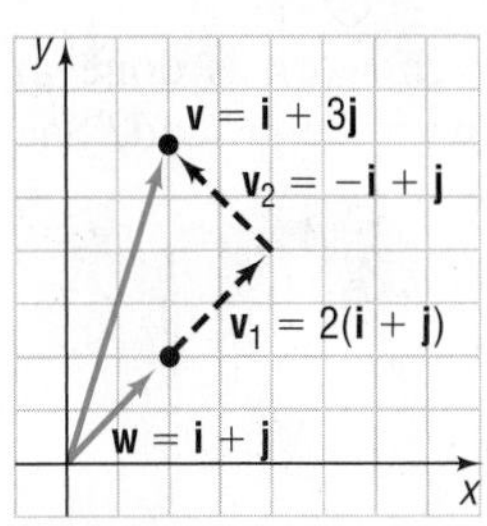

Figure 70

Now Work PROBLEM 21

EXAMPLE 6 Finding the Force Required to Hold a Wagon on a Hill

A wagon with two small children as occupants weighs 100 pounds and is on a hill with a grade of 20°. What is the magnitude of the force that is required to keep the wagon from rolling down the hill?

Solution See Figure 71. We wish to find the magnitude of the force $\mathbf{v}$ that is acting to cause the wagon to roll down the hill. A force with the same magnitude in the opposite direction of $\mathbf{v}$ will keep the wagon from rolling down the hill. The force of gravity is orthogonal to the level ground, so the force of the wagon due to gravity can be represented by the vector

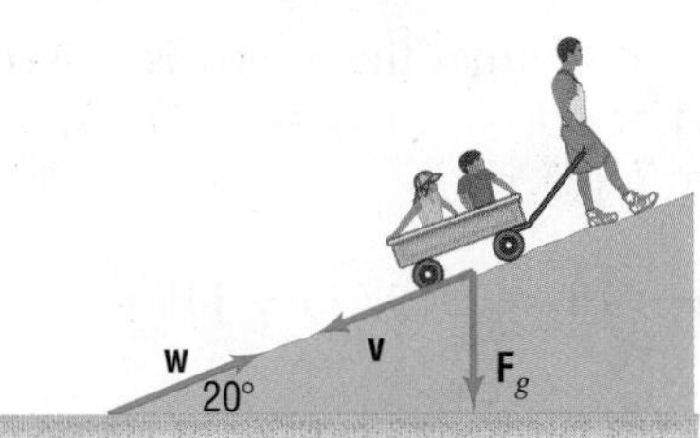

Figure 71

$$\mathbf{F}_g = -100\mathbf{j}$$

Determine the vector projection of $\mathbf{F}_g$ onto $\mathbf{w}$, which is the force parallel to the hill. The vector $\mathbf{w}$ is given by

$$\mathbf{w} = \cos 20°\mathbf{i} + \sin 20°\mathbf{j}$$

The vector projection of $\mathbf{F}_g$ onto $\mathbf{w}$ is

$$\begin{aligned}\mathbf{v} &= \frac{\mathbf{F}_g \cdot \mathbf{w}}{\|\mathbf{w}\|^2}\mathbf{w} \\ &= \frac{-100(\sin 20°)}{\left(\sqrt{\cos^2 20° + \sin^2 20°}\right)^2}(\cos 20°\mathbf{i} + \sin 20°\mathbf{j}) \\ &= -34.2(\cos 20°\mathbf{i} + \sin 20°\mathbf{j})\end{aligned}$$

The magnitude of $\mathbf{v}$ is 34.2 pounds, so the magnitude of the force required to keep the wagon from rolling down the hill is 34.2 pounds. ●

6 Compute Work

In elementary physics, the **work** W done by a constant force $\mathbf{F}$ in moving an object from a point A to a point B is defined as

$$W = (\text{magnitude of force})(\text{distance}) = \|\mathbf{F}\|\|\overrightarrow{AB}\|$$

Work is commonly measured in foot-pounds or in newton-meters (joules).

In this definition, it is assumed that the force $\mathbf{F}$ is applied along the line of motion. If the constant force $\mathbf{F}$ is not along the line of motion, but instead is at an angle θ to the direction of the motion, as illustrated in Figure 72, then the **work** W **done by F** in moving an object from A to B is defined as

$$W = \mathbf{F} \cdot \overrightarrow{AB} \qquad \textbf{(12)}$$

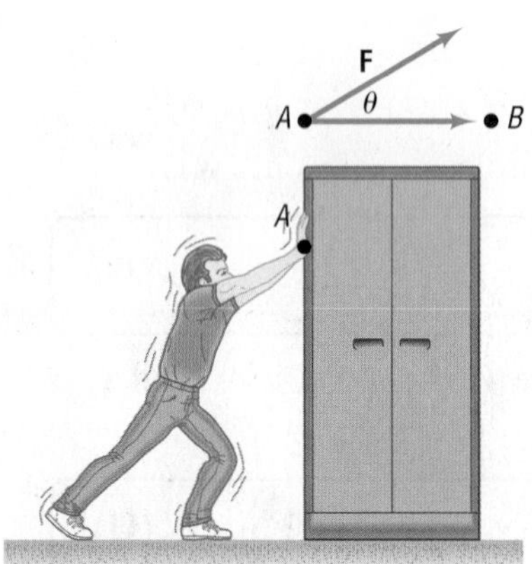

Figure 72

This definition is compatible with the force-times-distance definition, since

$$\begin{aligned}W &= (\text{amount of force in the direction of } \overrightarrow{AB})(\text{distance}) \\ &= \|\text{projection of } \mathbf{F} \text{ on } AB\|\|\overrightarrow{AB}\| \underset{\uparrow}{=} \frac{\mathbf{F}\cdot\overrightarrow{AB}}{\|\overrightarrow{AB}\|^2}\|\overrightarrow{AB}\|\|\overrightarrow{AB}\| = \mathbf{F}\cdot\overrightarrow{AB}\end{aligned}$$

Use formula (10).

EXAMPLE 7 **Computing Work**

A girl is pulling a wagon with a force of 50 pounds. How much work is done in moving the wagon 100 feet if the handle makes an angle of 30° with the ground? See Figure 73(a).

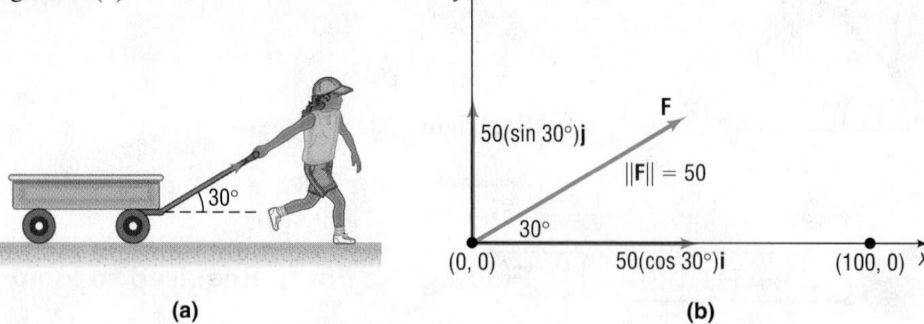

Figure 73 (a) (b)

Solution Position the vectors in a coordinate system in such a way that the wagon is moved from $(0, 0)$ to $(100, 0)$. The motion is from $A = (0, 0)$ to $B = (100, 0)$, so $\overrightarrow{AB} = 100\mathbf{i}$. The force vector $\mathbf{F}$, as shown in Figure 73(b), is

$$\mathbf{F} = 50(\cos 30°\mathbf{i} + \sin 30°\mathbf{j}) = 50\left(\frac{\sqrt{3}}{2}\mathbf{i} + \frac{1}{2}\mathbf{j}\right) = 25(\sqrt{3}\mathbf{i} + \mathbf{j})$$

By formula (12), the work done is

$$W = \mathbf{F}\cdot\overrightarrow{AB} = 25(\sqrt{3}\mathbf{i} + \mathbf{j}) \cdot 100\mathbf{i} = 2500\sqrt{3} \text{ foot-pounds}$$ ●

Now Work PROBLEM 29

Historical Feature

We stated in the Historical Feature in Section 5.4 that complex numbers were used as vectors in the plane before the general notion of a vector was clarified. Suppose that we make the correspondence

$$\text{Vector} \leftrightarrow \text{Complex number}$$
$$a\mathbf{i} + b\mathbf{j} \leftrightarrow a + bi$$
$$c\mathbf{i} + d\mathbf{j} \leftrightarrow c + di$$

Show that

$$(a\mathbf{i} + b\mathbf{j}) \cdot (c\mathbf{i} + d\mathbf{j}) = \text{real part } [(\overline{a + bi})(c + di)]$$

This is how the dot product was found originally. The imaginary part is also interesting. It is a determinant and represents the area of the parallelogram whose edges are the vectors. See Section 5.7. This is close to some of Hermann Grassmann's ideas and is also connected with the scalar triple product of three-dimensional vectors.

5.5 Assess Your Understanding

'Are You Prepared?' *The answer is given at the end of these exercises. If you get the wrong answer, read the page listed in red.*

1. In a triangle with sides a, b, c and angles A, B, C, the Law of Cosines states that ______. (p. 282)

Concepts and Vocabulary

2. If $\mathbf{v} = a_1\mathbf{i} + b_1\mathbf{j}$ and $\mathbf{w} = a_2\mathbf{i} + b_2\mathbf{j}$ are two vectors, then the _____ _________ is defined as $\mathbf{v} \cdot \mathbf{w} = a_1a_2 + b_1b_2$.
3. If $\mathbf{v} \cdot \mathbf{w} = 0$, then the two vectors $\mathbf{v}$ and $\mathbf{w}$ are ___________.
4. If $\mathbf{v} = 3\mathbf{w}$, then the two vectors $\mathbf{v}$ and $\mathbf{w}$ are _________.
5. ***True or False*** Given two nonzero vectors $\mathbf{v}$ and $\mathbf{w}$, it is always possible to decompose $\mathbf{v}$ into two vectors, one parallel to $\mathbf{w}$ and the other orthogonal to $\mathbf{w}$.
6. ***True or False*** Work is a physical example of a vector.
7. The angle θ, $0 \le \theta \le \pi$, between two nonzero vectors $\mathbf{u}$ and $\mathbf{v}$ can be found using which of the following formulas?
   (a) $\sin\theta = \dfrac{\|\mathbf{u}\|}{\|\mathbf{v}\|}$ (b) $\cos\theta = \dfrac{\|\mathbf{u}\|}{\|\mathbf{v}\|}$
   (c) $\sin\theta = \dfrac{\mathbf{u} \cdot \mathbf{v}}{\|\mathbf{u}\|\,\|\mathbf{v}\|}$ (d) $\cos\theta = \dfrac{\mathbf{u} \cdot \mathbf{v}}{\|\mathbf{u}\|\,\|\mathbf{v}\|}$
8. If two nonzero vectors $\mathbf{v}$ and $\mathbf{w}$ are orthogonal, then the angle between them has which of the following measures?
   (a) π (b) $\dfrac{\pi}{2}$ (c) $\dfrac{3\pi}{2}$ (d) 2π

Skill Building

In Problems 9–18, (a) find the dot product $\mathbf{v} \cdot \mathbf{w}$; *(b) find the angle between* $\mathbf{v}$ *and* $\mathbf{w}$; *(c) state whether the vectors are parallel, orthogonal, or neither.*

9. $\mathbf{v} = \mathbf{i} - \mathbf{j}$, $\mathbf{w} = \mathbf{i} + \mathbf{j}$
10. $\mathbf{v} = \mathbf{i} + \mathbf{j}$, $\mathbf{w} = -\mathbf{i} + \mathbf{j}$
11. $\mathbf{v} = 2\mathbf{i} + \mathbf{j}$, $\mathbf{w} = \mathbf{i} - 2\mathbf{j}$
12. $\mathbf{v} = 2\mathbf{i} + 2\mathbf{j}$, $\mathbf{w} = \mathbf{i} + 2\mathbf{j}$
13. $\mathbf{v} = \sqrt{3}\mathbf{i} - \mathbf{j}$, $\mathbf{w} = \mathbf{i} + \mathbf{j}$
14. $\mathbf{v} = \mathbf{i} + \sqrt{3}\mathbf{j}$, $\mathbf{w} = \mathbf{i} - \mathbf{j}$
15. $\mathbf{v} = 3\mathbf{i} + 4\mathbf{j}$, $\mathbf{w} = -6\mathbf{i} - 8\mathbf{j}$
16. $\mathbf{v} = 3\mathbf{i} - 4\mathbf{j}$, $\mathbf{w} = 9\mathbf{i} - 12\mathbf{j}$
17. $\mathbf{v} = 4\mathbf{i}$, $\mathbf{w} = \mathbf{j}$
18. $\mathbf{v} = \mathbf{i}$, $\mathbf{w} = -3\mathbf{j}$
19. Find a so that the vectors $\mathbf{v} = \mathbf{i} - a\mathbf{j}$ and $\mathbf{w} = 2\mathbf{i} + 3\mathbf{j}$ are orthogonal.
20. Find b so that the vectors $\mathbf{v} = \mathbf{i} + \mathbf{j}$ and $\mathbf{w} = \mathbf{i} + b\mathbf{j}$ are orthogonal.

In Problems 21–26, decompose $\mathbf{v}$ *into two vectors* $\mathbf{v}_1$ *and* $\mathbf{v}_2$, *where* $\mathbf{v}_1$ *is parallel to* $\mathbf{w}$, *and* $\mathbf{v}_2$ *is orthogonal to* $\mathbf{w}$.

21. $\mathbf{v} = 2\mathbf{i} - 3\mathbf{j}$, $\mathbf{w} = \mathbf{i} - \mathbf{j}$
22. $\mathbf{v} = -3\mathbf{i} + 2\mathbf{j}$, $\mathbf{w} = 2\mathbf{i} + \mathbf{j}$
23. $\mathbf{v} = \mathbf{i} - \mathbf{j}$, $\mathbf{w} = -\mathbf{i} - 2\mathbf{j}$
24. $\mathbf{v} = 2\mathbf{i} - \mathbf{j}$, $\mathbf{w} = \mathbf{i} - 2\mathbf{j}$
25. $\mathbf{v} = 3\mathbf{i} + \mathbf{j}$, $\mathbf{w} = -2\mathbf{i} - \mathbf{j}$
26. $\mathbf{v} = \mathbf{i} - 3\mathbf{j}$, $\mathbf{w} = 4\mathbf{i} - \mathbf{j}$

Applications and Extensions

27. Given vectors $\mathbf{u} = \mathbf{i} + 5\mathbf{j}$ and $\mathbf{v} = 4\mathbf{i} + y\mathbf{j}$, find y so that the angle between the vectors is 60°.
28. Given vectors $\mathbf{u} = x\mathbf{i} + 2\mathbf{j}$ and $\mathbf{v} = 7\mathbf{i} - 3\mathbf{j}$, find x so that the angle between the vectors is 30°.
29. **Computing Work** Find the work done by a force of 3 pounds acting in the direction 60° to the horizontal in moving an object 6 feet from $(0, 0)$ to $(6, 0)$.
30. **Computing Work** A wagon is pulled horizontally by exerting a force of 20 pounds on the handle at an angle of 30° with the horizontal. How much work is done in moving the wagon 100 feet?

31. **Solar Energy** The amount of energy collected by a solar panel depends on the intensity of the sun's rays and the area of the panel. Let the vector **I** represent the intensity, in watts per square centimeter, having the direction of the sun's rays. Let the vector **A** represent the area, in square centimeters, whose direction is the orientation of a solar panel. See the figure. The total number of watts collected by the panel is given by $W = |\mathbf{I} \cdot \mathbf{A}|$.

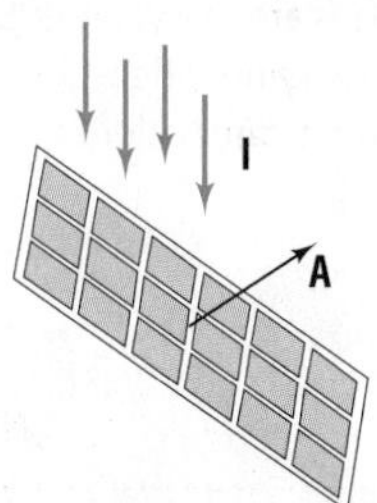

Suppose that $\mathbf{I} = \langle -0.02, -0.01 \rangle$ and $\mathbf{A} = \langle 300, 400 \rangle$.
(a) Find $\|\mathbf{I}\|$ and $\|\mathbf{A}\|$, and interpret the meaning of each.
(b) Compute W and interpret its meaning.
(c) If the solar panel is to collect the maximum number of watts, what must be true about **I** and **A**?

32. **Rainfall Measurement** Let the vector **R** represent the amount of rainfall, in inches, whose direction is the inclination of the rain to a rain gauge. Let the vector **A** represent the area, in square inches, whose direction is the orientation of the opening of the rain gauge. See the figure. The volume of rain collected in the gauge, in cubic inches, is given by $V = |\mathbf{R} \cdot \mathbf{A}|$, even when the rain falls in a slanted direction or the gauge is not perfectly vertical.

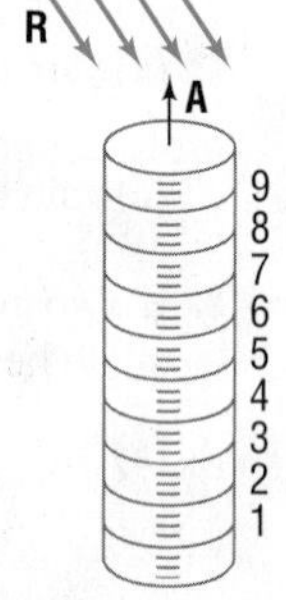

Suppose that $\mathbf{R} = \langle 0.75, -1.75 \rangle$ and $\mathbf{A} = \langle 0.3, 1 \rangle$.
(a) Find $\|\mathbf{R}\|$ and $\|\mathbf{A}\|$, and interpret the meaning of each.
(b) Compute V and interpret its meaning.
(c) If the gauge is to collect the maximum volume of rain, what must be true about **R** and **A**?

33. **Braking Load** A Toyota Sienna with a gross weight of 5300 pounds is parked on a street with an 8° grade. See the figure. Find the magnitude of the force required to keep the Sienna from rolling down the hill. What is the magnitude of the force perpendicular to the hill?

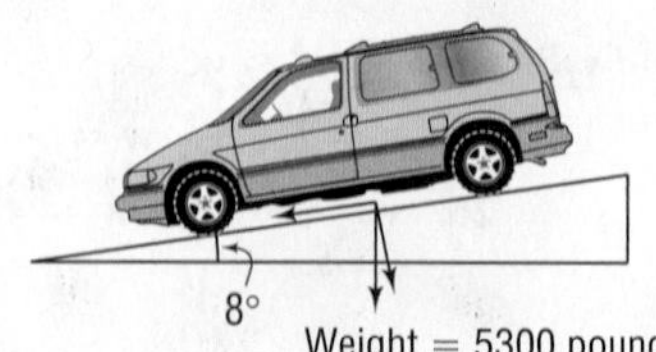

34. **Braking Load** A Chevrolet Silverado with a gross weight of 4500 pounds is parked on a street with a 10° grade. Find the magnitude of the force required to keep the Silverado from rolling down the hill. What is the magnitude of the force perpendicular to the hill?

35. **Ramp Angle** Billy and Timmy are using a ramp to load furniture into a truck. While rolling a 250-pound piano up the ramp, they discover that the truck is too full of other furniture for the piano to fit. Timmy holds the piano in place on the ramp while Billy repositions other items to make room for it in the truck. If the angle of inclination of the ramp is 20°, how many pounds of force must Timmy exert to hold the piano in position?

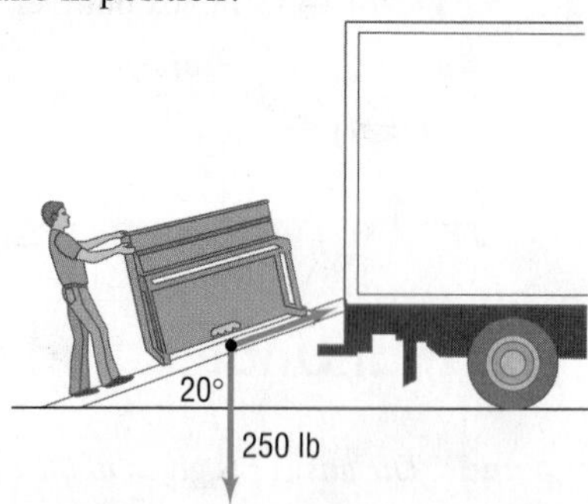

36. **Incline Angle** A bulldozer exerts 1000 pounds of force to prevent a 5000-pound boulder from rolling down a hill. Determine the angle of inclination of the hill.

37. Find the acute angle that a constant unit force vector makes with the positive x-axis if the work done by the force in moving a particle from $(0, 0)$ to $(4, 0)$ equals 2.

38. Prove the distributive property:
$$\mathbf{u} \cdot (\mathbf{v} + \mathbf{w}) = \mathbf{u} \cdot \mathbf{v} + \mathbf{u} \cdot \mathbf{w}$$

39. Prove property (5): $\mathbf{0} \cdot \mathbf{v} = 0$.

40. If **v** is a unit vector and the angle between **v** and **i** is α, show that $\mathbf{v} = \cos \alpha \mathbf{i} + \sin \alpha \mathbf{j}$.

41. Suppose that **v** and **w** are unit vectors. If the angle between **v** and **i** is α and the angle between **w** and **i** is β, use the idea of the dot product $\mathbf{v} \cdot \mathbf{w}$ to prove that
$$\cos(\alpha - \beta) = \cos \alpha \cos \beta + \sin \alpha \sin \beta$$

42. Show that the projection of **v** onto **i** is $(\mathbf{v} \cdot \mathbf{i})\mathbf{i}$. Then show that we can always write a vector **v** as
$$\mathbf{v} = (\mathbf{v} \cdot \mathbf{i})\mathbf{i} + (\mathbf{v} \cdot \mathbf{j})\mathbf{j}$$

43. (a) If **u** and **v** have the same magnitude, show that $\mathbf{u} + \mathbf{v}$ and $\mathbf{u} - \mathbf{v}$ are orthogonal.
(b) Use this to prove that an angle inscribed in a semicircle is a right angle (see the figure).

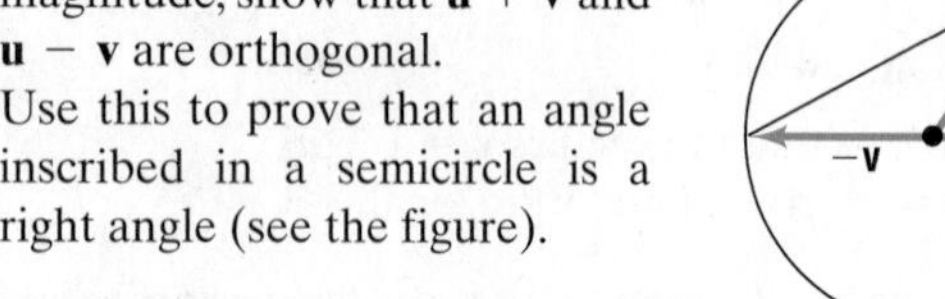

44. Let **v** and **w** denote two nonzero vectors. Show that the vector $\mathbf{v} - \alpha\mathbf{w}$ is orthogonal to **w** if $\alpha = \dfrac{\mathbf{v} \cdot \mathbf{w}}{\|\mathbf{w}\|^2}$.

45. Let **v** and **w** denote two nonzero vectors. Show that the vectors $\|\mathbf{w}\|\mathbf{v} + \|\mathbf{v}\|\mathbf{w}$ and $\|\mathbf{w}\|\mathbf{v} - \|\mathbf{v}\|\mathbf{w}$ are orthogonal.

46. In the definition of work given in this section, what is the work done if **F** is orthogonal to $\overrightarrow{AB}$?

47. Prove the **polarization identity**,
$$\|\mathbf{u} + \mathbf{v}\|^2 - \|\mathbf{u} - \mathbf{v}\|^2 = 4(\mathbf{u} \cdot \mathbf{v})$$

Explaining Concepts: Discussion and Writing

48. Create an application (different from any found in the text) that requires a dot product.

Retain Your Knowledge

Problems 49–52 are based on material learned earlier in the course. The purpose of these problems is to keep the material fresh in your mind so that you are better prepared for the final exam.

49. Find the average rate of change of $f(x) = x^3 - 5x^2 + 27$ from -3 to 2.

50. Find the exact value of $5 \cos 60° + 2 \tan \frac{\pi}{4}$. Do not use a calculator.

51. Establish the identity: $(1 - \sin^2 \theta)(1 + \tan^2 \theta) = 1$

52. Solve triangle ABC: $a = 8, b = 11$, and $c = 16$

'Are You Prepared?' Answer

1. $c^2 = a^2 + b^2 - 2ab \cos C$

5.6 Vectors in Space

PREPARING FOR THIS SECTION *Before getting started, review the following:*

- Distance Formula (Section 1.1, p. 3)

Now Work the **'Are You Prepared?'** problem on page 372.

OBJECTIVES
1. Find the Distance between Two Points in Space (p. 366)
2. Find Position Vectors in Space (p. 367)
3. Perform Operations on Vectors (p. 367)
4. Find the Dot Product (p. 369)
5. Find the Angle between Two Vectors (p. 369)
6. Find the Direction Angles of a Vector (p. 370)

Rectangular Coordinates in Space

In the plane, each point is associated with an ordered pair of real numbers. In space, each point is associated with an ordered triple of real numbers. Through a fixed point, called the **origin** O, draw three mutually perpendicular lines: the x-axis, the y-axis, and the z-axis. On each of these axes, select an appropriate scale and the positive direction. See Figure 74.

The direction chosen for the positive z-axis in Figure 74 makes the system *right-handed*. This conforms to the *right-hand rule,* which states that if the index finger of the right hand points in the direction of the positive x-axis and the middle finger points in the direction of the positive y-axis, then the thumb will point in the direction of the positive z-axis. See Figure 75.

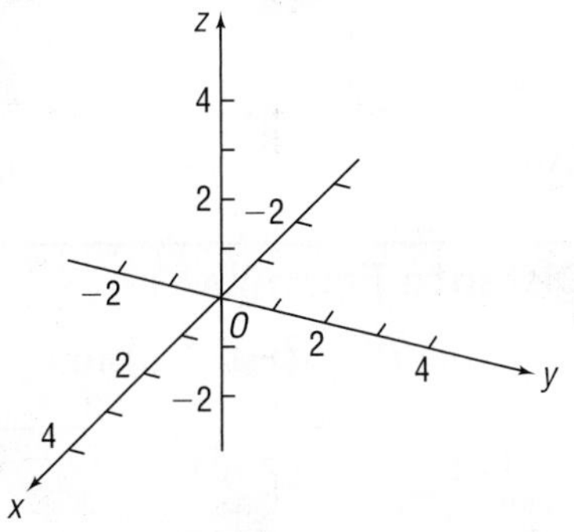

Figure 74

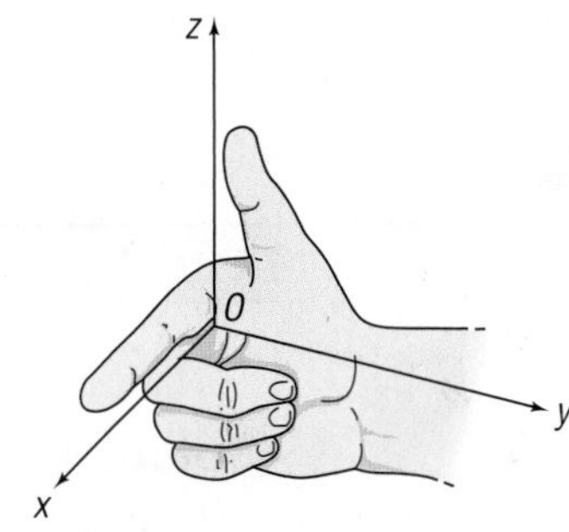

Figure 75

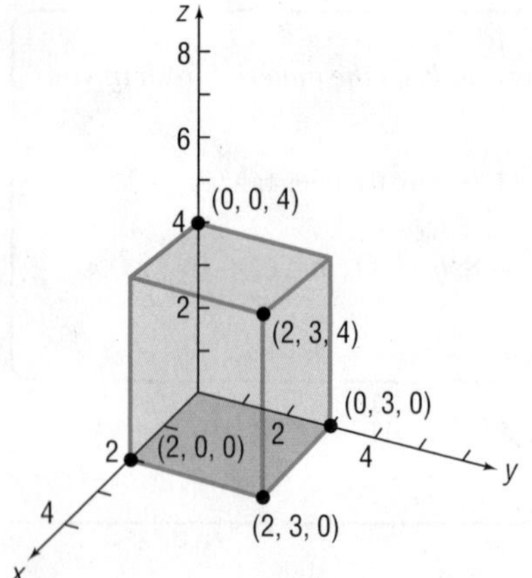

Figure 76

Associate with each point P an ordered triple (x, y, z) of real numbers, the **coordinates of *P***. For example, the point $(2, 3, 4)$ is located by starting at the origin and moving 2 units along the positive x-axis, 3 units in the direction of the positive y-axis, and 4 units in the direction of the positive z-axis. See Figure 76.

Figure 76 also shows the location of the points $(2, 0, 0)$, $(0, 3, 0)$, $(0, 0, 4)$, and $(2, 3, 0)$. Points of the form $(x, 0, 0)$ lie on the x-axis, and points of the forms $(0, y, 0)$ and $(0, 0, z)$ lie on the y-axis and z-axis, respectively. Points of the form $(x, y, 0)$ lie in a plane called the ***xy*-plane**. Its equation is $z = 0$. Similarly, points of the form $(x, 0, z)$ lie in the ***xz*-plane** (equation $y = 0$), and points of the form $(0, y, z)$ lie in the ***yz*-plane** (equation $x = 0$). See Figure 77(a). By extension of these ideas, all points obeying the equation $z = 3$ will lie in a plane parallel to and 3 units above the xy-plane. The equation $y = 4$ represents a plane parallel to the xz-plane and 4 units to the right of the plane $y = 0$. See Figure 77(b).

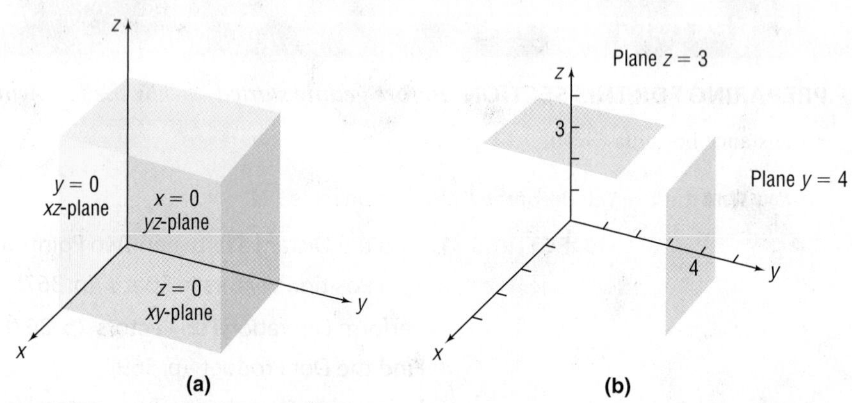

Figure 77 (a) (b)

Now Work PROBLEM 9

1 Find the Distance between Two Points in Space

The formula for the distance between two points in space is an extension of the Distance Formula for points in the plane given in Section 1.1.

THEOREM **Distance Formula in Space**

If $P_1 = (x_1, y_1, z_1)$ and $P_2 = (x_2, y_2, z_2)$ are two points in space, the distance d from P_1 to P_2 is

$$d = \sqrt{(x_2 - x_1)^2 + (y_2 - y_1)^2 + (z_2 - z_1)^2} \quad (1)$$

The proof, which we omit, utilizes a double application of the Pythagorean Theorem.

EXAMPLE 1 **Using the Distance Formula**

Find the distance from $P_1 = (-1, 3, 2)$ to $P_2 = (4, -2, 5)$.

Solution $d = \sqrt{[4 - (-1)]^2 + [-2 - 3]^2 + [5 - 2]^2} = \sqrt{25 + 25 + 9} = \sqrt{59}$ •

Now Work PROBLEM 15

2 Find Position Vectors in Space

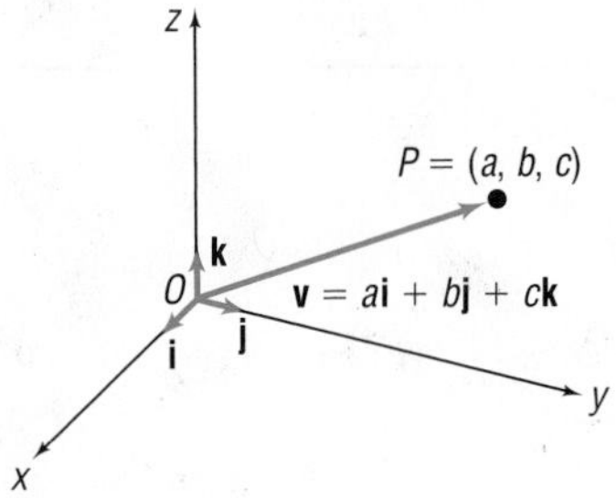

Figure 78

To represent vectors in space, we introduce the unit vectors **i**, **j**, and **k** whose directions are along the positive x-axis, the positive y-axis, and the positive z-axis, respectively. If **v** is a vector with initial point at the origin O and terminal point at $P = (a, b, c)$, then we can represent **v** in terms of the vectors **i**, **j**, and **k** as

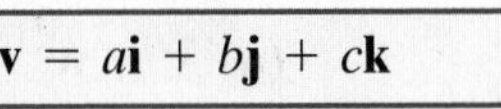

$$\mathbf{v} = a\mathbf{i} + b\mathbf{j} + c\mathbf{k}$$

See Figure 78.

The scalars $a, b,$ and c are called the **components** of the vector $\mathbf{v} = a\mathbf{i} + b\mathbf{j} + c\mathbf{k}$, with a being the component in the direction **i**, b the component in the direction **j**, and c the component in the direction **k**.

A vector whose initial point is at the origin is called a **position vector.** The next result states that any vector whose initial point is not at the origin is equal to a unique position vector.

THEOREM

Suppose that **v** is a vector with initial point $P_1 = (x_1, y_1, z_1)$, not necessarily the origin, and terminal point $P_2 = (x_2, y_2, z_2)$. If $\mathbf{v} = \overrightarrow{P_1P_2}$, then **v** is equal to the position vector

$$\mathbf{v} = (x_2 - x_1)\mathbf{i} + (y_2 - y_1)\mathbf{j} + (z_2 - z_1)\mathbf{k} \quad (2)$$

Figure 79 illustrates this result.

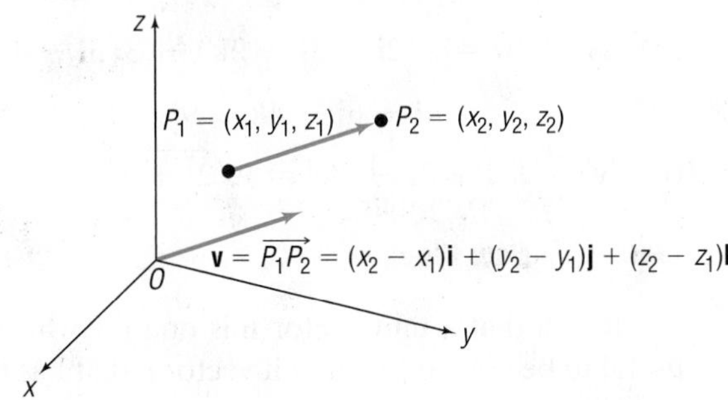

Figure 79

EXAMPLE 2 **Finding a Position Vector**

Find the position vector of the vector $\mathbf{v} = \overrightarrow{P_1P_2}$ if $P_1 = (-1, 2, 3)$ and $P_2 = (4, 6, 2)$.

Solution By equation (2), the position vector equal to **v** is

$$\mathbf{v} = [4 - (-1)]\mathbf{i} + (6 - 2)\mathbf{j} + (2 - 3)\mathbf{k} = 5\mathbf{i} + 4\mathbf{j} - \mathbf{k}$$

Now Work PROBLEM 29

3 Perform Operations on Vectors

Equality, addition, subtraction, scalar product, and magnitude can be defined in terms of the components of a vector.

DEFINITION

Let $\mathbf{v} = a_1\mathbf{i} + b_1\mathbf{j} + c_1\mathbf{k}$ and $\mathbf{w} = a_2\mathbf{i} + b_2\mathbf{j} + c_2\mathbf{k}$ be two vectors, and let α be a scalar. Then

$$\mathbf{v} = \mathbf{w} \text{ if and only if } a_1 = a_2, b_1 = b_2, \text{ and } c_1 = c_2$$
$$\mathbf{v} + \mathbf{w} = (a_1 + a_2)\mathbf{i} + (b_1 + b_2)\mathbf{j} + (c_1 + c_2)\mathbf{k}$$
$$\mathbf{v} - \mathbf{w} = (a_1 - a_2)\mathbf{i} + (b_1 - b_2)\mathbf{j} + (c_1 - c_2)\mathbf{k}$$
$$\alpha\mathbf{v} = (\alpha a_1)\mathbf{i} + (\alpha b_1)\mathbf{j} + (\alpha c_1)\mathbf{k}$$
$$\|\mathbf{v}\| = \sqrt{a_1^2 + b_1^2 + c_1^2}$$

These definitions are compatible with the geometric definitions given in Section 5.4 for vectors in a plane.

EXAMPLE 3

Adding and Subtracting Vectors

If $\mathbf{v} = 2\mathbf{i} + 3\mathbf{j} - 2\mathbf{k}$ and $\mathbf{w} = 3\mathbf{i} - 4\mathbf{j} + 5\mathbf{k}$, find:

(a) $\mathbf{v} + \mathbf{w}$ (b) $\mathbf{v} - \mathbf{w}$

Solution

(a)
$$\begin{aligned}\mathbf{v} + \mathbf{w} &= (2\mathbf{i} + 3\mathbf{j} - 2\mathbf{k}) + (3\mathbf{i} - 4\mathbf{j} + 5\mathbf{k})\\ &= (2 + 3)\mathbf{i} + (3 - 4)\mathbf{j} + (-2 + 5)\mathbf{k}\\ &= 5\mathbf{i} - \mathbf{j} + 3\mathbf{k}\end{aligned}$$

(b)
$$\begin{aligned}\mathbf{v} - \mathbf{w} &= (2\mathbf{i} + 3\mathbf{j} - 2\mathbf{k}) - (3\mathbf{i} - 4\mathbf{j} + 5\mathbf{k})\\ &= (2 - 3)\mathbf{i} + [3 - (-4)]\mathbf{j} + [-2 - 5]\mathbf{k}\\ &= -\mathbf{i} + 7\mathbf{j} - 7\mathbf{k}\end{aligned}$$

EXAMPLE 4

Finding Scalar Products and Magnitudes

If $\mathbf{v} = 2\mathbf{i} + 3\mathbf{j} - 2\mathbf{k}$ and $\mathbf{w} = 3\mathbf{i} - 4\mathbf{j} + 5\mathbf{k}$, find:

(a) $3\mathbf{v}$ (b) $2\mathbf{v} - 3\mathbf{w}$ (c) $\|\mathbf{v}\|$

Solution

(a) $3\mathbf{v} = 3(2\mathbf{i} + 3\mathbf{j} - 2\mathbf{k}) = 6\mathbf{i} + 9\mathbf{j} - 6\mathbf{k}$

(b)
$$\begin{aligned}2\mathbf{v} - 3\mathbf{w} &= 2(2\mathbf{i} + 3\mathbf{j} - 2\mathbf{k}) - 3(3\mathbf{i} - 4\mathbf{j} + 5\mathbf{k})\\ &= 4\mathbf{i} + 6\mathbf{j} - 4\mathbf{k} - 9\mathbf{i} + 12\mathbf{j} - 15\mathbf{k} = -5\mathbf{i} + 18\mathbf{j} - 19\mathbf{k}\end{aligned}$$

(c) $\|\mathbf{v}\| = \|2\mathbf{i} + 3\mathbf{j} - 2\mathbf{k}\| = \sqrt{2^2 + 3^2 + (-2)^2} = \sqrt{17}$

Now Work PROBLEMS 33 AND 39

Recall that a unit vector $\mathbf{u}$ is one for which $\|\mathbf{u}\| = 1$. In many applications, it is useful to be able to find a unit vector $\mathbf{u}$ that has the same direction as a given vector $\mathbf{v}$.

THEOREM

Unit Vector in the Direction of v

For any nonzero vector $\mathbf{v}$, the vector

$$\mathbf{u} = \frac{\mathbf{v}}{\|\mathbf{v}\|}$$

is a unit vector that has the same direction as $\mathbf{v}$.

As a consequence of this theorem, if $\mathbf{u}$ is a unit vector in the same direction as a vector $\mathbf{v}$, then $\mathbf{v}$ may be expressed as

$$\mathbf{v} = \|\mathbf{v}\|\mathbf{u}$$

EXAMPLE 5

Finding a Unit Vector

Find the unit vector in the same direction as $\mathbf{v} = 2\mathbf{i} - 3\mathbf{j} - 6\mathbf{k}$.

Solution Find $\|\mathbf{v}\|$ first.

$$\|\mathbf{v}\| = \|2\mathbf{i} - 3\mathbf{j} - 6\mathbf{k}\| = \sqrt{4 + 9 + 36} = \sqrt{49} = 7$$

Now multiply **v** by the scalar $\frac{1}{\|\mathbf{v}\|} = \frac{1}{7}$. The result is the unit vector

$$\mathbf{u} = \frac{\mathbf{v}}{\|\mathbf{v}\|} = \frac{2\mathbf{i} - 3\mathbf{j} - 6\mathbf{k}}{7} = \frac{2}{7}\mathbf{i} - \frac{3}{7}\mathbf{j} - \frac{6}{7}\mathbf{k}$$

Now Work PROBLEM 47

4 Find the Dot Product

The definition of *dot product* in space is an extension of the definition given for vectors in a plane.

DEFINITION

If $\mathbf{v} = a_1\mathbf{i} + b_1\mathbf{j} + c_1\mathbf{k}$ and $\mathbf{w} = a_2\mathbf{i} + b_2\mathbf{j} + c_2\mathbf{k}$ are two vectors, the **dot product $\mathbf{v} \cdot \mathbf{w}$** is defined as

$$\mathbf{v} \cdot \mathbf{w} = a_1a_2 + b_1b_2 + c_1c_2 \qquad (3)$$

EXAMPLE 6 **Finding Dot Products**

If $\mathbf{v} = 2\mathbf{i} - 3\mathbf{j} + 6\mathbf{k}$ and $\mathbf{w} = 5\mathbf{i} + 3\mathbf{j} - \mathbf{k}$, find:

(a) $\mathbf{v} \cdot \mathbf{w}$ (b) $\mathbf{w} \cdot \mathbf{v}$ (c) $\mathbf{v} \cdot \mathbf{v}$
(d) $\mathbf{w} \cdot \mathbf{w}$ (e) $\|\mathbf{v}\|$ (f) $\|\mathbf{w}\|$

Solution

(a) $\mathbf{v} \cdot \mathbf{w} = 2(5) + (-3)3 + 6(-1) = -5$
(b) $\mathbf{w} \cdot \mathbf{v} = 5(2) + 3(-3) + (-1)(6) = -5$
(c) $\mathbf{v} \cdot \mathbf{v} = 2(2) + (-3)(-3) + 6(6) = 49$
(d) $\mathbf{w} \cdot \mathbf{w} = 5(5) + 3(3) + (-1)(-1) = 35$
(e) $\|\mathbf{v}\| = \sqrt{2^2 + (-3)^2 + 6^2} = \sqrt{49} = 7$
(f) $\|\mathbf{w}\| = \sqrt{5^2 + 3^2 + (-1)^2} = \sqrt{35}$

The dot product in space has the same properties as the dot product in the plane.

THEOREM

Properties of the Dot Product

If **u**, **v**, and **w** are vectors, then

Commutative Property

$$\mathbf{u} \cdot \mathbf{v} = \mathbf{v} \cdot \mathbf{u}$$

Distributive Property

$$\mathbf{u} \cdot (\mathbf{v} + \mathbf{w}) = \mathbf{u} \cdot \mathbf{v} + \mathbf{u} \cdot \mathbf{w}$$

$$\mathbf{v} \cdot \mathbf{v} = \|\mathbf{v}\|^2$$

$$\mathbf{0} \cdot \mathbf{v} = 0$$

5 Find the Angle between Two Vectors

The angle θ between two vectors in space follows the same formula as for two vectors in the plane.

THEOREM

Angle between Vectors

If **u** and **v** are two nonzero vectors, the angle θ, $0 \le \theta \le \pi$, between **u** and **v** is determined by the formula

$$\cos\theta = \frac{\mathbf{u}\cdot\mathbf{v}}{\|\mathbf{u}\|\,\|\mathbf{v}\|} \qquad (4)$$

EXAMPLE 7

Finding the Angle between Two Vectors

Find the angle θ between $\mathbf{u} = 2\mathbf{i} - 3\mathbf{j} + 6\mathbf{k}$ and $\mathbf{v} = 2\mathbf{i} + 5\mathbf{j} - \mathbf{k}$.

Solution Compute the quantities $\mathbf{u}\cdot\mathbf{v}$, $\|\mathbf{u}\|$, and $\|\mathbf{v}\|$.

$$\mathbf{u}\cdot\mathbf{v} = 2(2) + (-3)(5) + 6(-1) = -17$$

$$\|\mathbf{u}\| = \sqrt{2^2 + (-3)^2 + 6^2} = \sqrt{49} = 7$$

$$\|\mathbf{v}\| = \sqrt{2^2 + 5^2 + (-1)^2} = \sqrt{30}$$

By formula (4), if θ is the angle between **u** and **v**, then

$$\cos\theta = \frac{\mathbf{u}\cdot\mathbf{v}}{\|\mathbf{u}\|\,\|\mathbf{v}\|} = \frac{-17}{7\sqrt{30}} \approx -0.443$$

Thus, $\theta \approx \cos^{-1}(-0.443) \approx 116.3°$.

Now Work PROBLEM 51

6 Find the Direction Angles of a Vector

A nonzero vector **v** in space can be described by specifying its magnitude and its three **direction angles** α, β, and γ. These direction angles are defined as

α = the angle between **v** and **i**, the positive x-axis, $0 \le \alpha \le \pi$

β = the angle between **v** and **j**, the positive y-axis, $0 \le \beta \le \pi$

γ = the angle between **v** and **k**, the positive z-axis, $0 \le \gamma \le \pi$

See Figure 80.

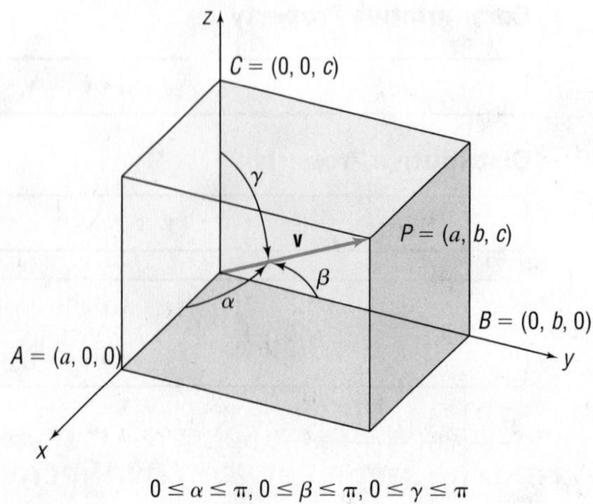

Figure 80 Direction angles

Our first goal is to find expressions for α, β, and γ in terms of the components of a vector. Let $\mathbf{v} = a\mathbf{i} + b\mathbf{j} + c\mathbf{k}$ denote a nonzero vector. The angle α between $\mathbf{v}$ and $\mathbf{i}$, the positive x-axis, obeys

$$\cos \alpha = \frac{\mathbf{v} \cdot \mathbf{i}}{\|\mathbf{v}\| \, \|\mathbf{i}\|} = \frac{a}{\|\mathbf{v}\|}$$

Similarly,

$$\cos \beta = \frac{b}{\|\mathbf{v}\|} \quad \text{and} \quad \cos \gamma = \frac{c}{\|\mathbf{v}\|}$$

Since $\|\mathbf{v}\| = \sqrt{a^2 + b^2 + c^2}$, the following result is obtained.

THEOREM

Direction Angles

If $\mathbf{v} = a\mathbf{i} + b\mathbf{j} + c\mathbf{k}$ is a nonzero vector in space, the direction angles α, β, and γ obey

$$\cos \alpha = \frac{a}{\sqrt{a^2 + b^2 + c^2}} = \frac{a}{\|\mathbf{v}\|} \qquad \cos \beta = \frac{b}{\sqrt{a^2 + b^2 + c^2}} = \frac{b}{\|\mathbf{v}\|}$$

$$\cos \gamma = \frac{c}{\sqrt{a^2 + b^2 + c^2}} = \frac{c}{\|\mathbf{v}\|} \qquad (5)$$

The numbers $\cos \alpha$, $\cos \beta$, and $\cos \gamma$ are called the **direction cosines** of the vector $\mathbf{v}$.

EXAMPLE 8

Finding the Direction Angles of a Vector

Find the direction angles of $\mathbf{v} = -3\mathbf{i} + 2\mathbf{j} - 6\mathbf{k}$.

Solution $\|\mathbf{v}\| = \sqrt{(-3)^2 + 2^2 + (-6)^2} = \sqrt{49} = 7$

Using the formulas in equation (5), we have

$$\cos \alpha = \frac{-3}{7} \qquad \cos \beta = \frac{2}{7} \qquad \cos \gamma = \frac{-6}{7}$$

$$\alpha \approx 115.4° \qquad \beta \approx 73.4° \qquad \gamma \approx 149.0°$$

THEOREM

Property of the Direction Cosines

If α, β, and γ are the direction angles of a nonzero vector $\mathbf{v}$ in space, then

$$\cos^2 \alpha + \cos^2 \beta + \cos^2 \gamma = 1 \qquad (6)$$

The proof is a direct consequence of the equations in (5).

Based on equation (6), when two direction cosines are known, the third is determined up to its sign. Knowing two direction cosines is not sufficient to uniquely determine the direction of a vector in space.

EXAMPLE 9

Finding a Direction Angle of a Vector

The vector $\mathbf{v}$ makes an angle of $\alpha = \frac{\pi}{3}$ with the positive x-axis, an angle of $\beta = \frac{\pi}{3}$ with the positive y-axis, and an acute angle γ with the positive z-axis. Find γ.

Solution By equation (6), we have

$$\cos^2\left(\frac{\pi}{3}\right) + \cos^2\left(\frac{\pi}{3}\right) + \cos^2\gamma = 1 \qquad 0 < \gamma < \frac{\pi}{2}$$

$$\left(\frac{1}{2}\right)^2 + \left(\frac{1}{2}\right)^2 + \cos^2\gamma = 1$$

$$\cos^2\gamma = \frac{1}{2}$$

$$\cos\gamma = \frac{\sqrt{2}}{2} \quad \text{or} \quad \cos\gamma = -\frac{\sqrt{2}}{2}$$

$$\gamma = \frac{\pi}{4} \quad \text{or} \quad \gamma = \frac{3\pi}{4}$$

Since γ must be acute, $\gamma = \frac{\pi}{4}$. ●

The direction cosines of a vector give information about only the direction of the vector; they provide no information about its magnitude. For example, *any* vector that is parallel to the xy-plane and makes an angle of $\frac{\pi}{4}$ radian with the positive x-axis and y-axis has direction cosines

$$\cos\alpha = \frac{\sqrt{2}}{2} \qquad \cos\beta = \frac{\sqrt{2}}{2} \qquad \cos\gamma = 0$$

However, if the direction angles *and* the magnitude of a vector are known, the vector is uniquely determined.

EXAMPLE 10 **Writing a Vector in Terms of Its Magnitude and Direction Cosines**

Show that any nonzero vector **v** in space can be written in terms of its magnitude and direction cosines as

$$\mathbf{v} = \|\mathbf{v}\|[(\cos\alpha)\mathbf{i} + (\cos\beta)\mathbf{j} + (\cos\gamma)\mathbf{k}] \qquad \textbf{(7)}$$

Solution Let $\mathbf{v} = a\mathbf{i} + b\mathbf{j} + c\mathbf{k}$. From the equations in (5), note that

$$a = \|\mathbf{v}\|\cos\alpha \qquad b = \|\mathbf{v}\|\cos\beta \qquad c = \|\mathbf{v}\|\cos\gamma$$

Substituting gives

$$\begin{aligned}\mathbf{v} = a\mathbf{i} + b\mathbf{j} + c\mathbf{k} &= \|\mathbf{v}\|(\cos\alpha)\mathbf{i} + \|\mathbf{v}\|(\cos\beta)\mathbf{j} + \|\mathbf{v}\|(\cos\gamma)\mathbf{k} \\ &= \|\mathbf{v}\|[(\cos\alpha)\mathbf{i} + (\cos\beta)\mathbf{j} + (\cos\gamma)\mathbf{k}]\end{aligned}$$ ●

Now Work PROBLEM 59

Example 10 shows that the direction cosines of a vector **v** are also the components of the unit vector in the direction of **v**.

5.6 Assess Your Understanding

'Are You Prepared?' *The answer is given at the end of these exercises. If you get the wrong answer, read the page listed in red.*

1. The distance d from $P_1 = (x_1, y_1)$ to $P_2 = (x_2, y_2)$ is $d =$ ______. (p. 3)

Concepts and Vocabulary

2. In space, points of the form $(x, y, 0)$ lie in a plane called the _______.

3. If $\mathbf{v} = a\mathbf{i} + b\mathbf{j} + c\mathbf{k}$ is a vector in space, the scalars a, b, c are called the __________ of $\mathbf{v}$.

4. The squares of the direction cosines of a vector in space add up to _____.

5. ***True or False*** In space, the dot product of two vectors is a positive number.

6. ***True or False*** A vector in space may be described by specifying its magnitude and its direction angles.

Skill Building

In Problems 7–14, describe the set of points (x, y, z) defined by the equation(s).

7. $y = 0$ **8.** $x = 0$ 9. $z = 2$ **10.** $y = 3$

11. $x = -4$ **12.** $z = -3$ **13.** $x = 1$ and $y = 2$ **14.** $x = 3$ and $z = 1$

In Problems 15–20, find the distance from P_1 to P_2.

15. $P_1 = (0, 0, 0)$ and $P_2 = (4, 1, 2)$

16. $P_1 = (0, 0, 0)$ and $P_2 = (1, -2, 3)$

17. $P_1 = (-1, 2, -3)$ and $P_2 = (0, -2, 1)$

18. $P_1 = (-2, 2, 3)$ and $P_2 = (4, 0, -3)$

19. $P_1 = (4, -2, -2)$ and $P_2 = (3, 2, 1)$

20. $P_1 = (2, -3, -3)$ and $P_2 = (4, 1, -1)$

In Problems 21–26, opposite vertices of a rectangular box whose edges are parallel to the coordinate axes are given. List the coordinates of the other six vertices of the box.

21. $(0, 0, 0)$; $(2, 1, 3)$ **22.** $(0, 0, 0)$; $(4, 2, 2)$ **23.** $(1, 2, 3)$; $(3, 4, 5)$

24. $(5, 6, 1)$; $(3, 8, 2)$ **25.** $(-1, 0, 2)$; $(4, 2, 5)$ **26.** $(-2, -3, 0)$; $(-6, 7, 1)$

In Problems 27–32, the vector $\mathbf{v}$ has initial point P and terminal point Q. Write $\mathbf{v}$ in the form $a\mathbf{i} + b\mathbf{j} + c\mathbf{k}$; that is, find its position vector.

27. $P = (0, 0, 0)$; $Q = (3, 4, -1)$

28. $P = (0, 0, 0)$; $Q = (-3, -5, 4)$

29. $P = (3, 2, -1)$; $Q = (5, 6, 0)$

30. $P = (-3, 2, 0)$; $Q = (6, 5, -1)$

31. $P = (-2, -1, 4)$; $Q = (6, -2, 4)$

32. $P = (-1, 4, -2)$; $Q = (6, 2, 2)$

In Problems 33–38, find $\|\mathbf{v}\|$.

33. $\mathbf{v} = 3\mathbf{i} - 6\mathbf{j} - 2\mathbf{k}$ **34.** $\mathbf{v} = -6\mathbf{i} + 12\mathbf{j} + 4\mathbf{k}$ **35.** $\mathbf{v} = \mathbf{i} - \mathbf{j} + \mathbf{k}$

36. $\mathbf{v} = -\mathbf{i} - \mathbf{j} + \mathbf{k}$ **37.** $\mathbf{v} = -2\mathbf{i} + 3\mathbf{j} - 3\mathbf{k}$ **38.** $\mathbf{v} = 6\mathbf{i} + 2\mathbf{j} - 2\mathbf{k}$

In Problems 39–44, find each quantity if $\mathbf{v} = 3\mathbf{i} - 5\mathbf{j} + 2\mathbf{k}$ and $\mathbf{w} = -2\mathbf{i} + 3\mathbf{j} - 2\mathbf{k}$.

39. $2\mathbf{v} + 3\mathbf{w}$ **40.** $3\mathbf{v} - 2\mathbf{w}$ **41.** $\|\mathbf{v} - \mathbf{w}\|$

42. $\|\mathbf{v} + \mathbf{w}\|$ **43.** $\|\mathbf{v}\| - \|\mathbf{w}\|$ **44.** $\|\mathbf{v}\| + \|\mathbf{w}\|$

In Problems 45–50, find the unit vector in the same direction as $\mathbf{v}$.

45. $\mathbf{v} = 5\mathbf{i}$ **46.** $\mathbf{v} = -3\mathbf{j}$ 47. $\mathbf{v} = 3\mathbf{i} - 6\mathbf{j} - 2\mathbf{k}$

48. $\mathbf{v} = -6\mathbf{i} + 12\mathbf{j} + 4\mathbf{k}$ **49.** $\mathbf{v} = \mathbf{i} + \mathbf{j} + \mathbf{k}$ **50.** $\mathbf{v} = 2\mathbf{i} - \mathbf{j} + \mathbf{k}$

In Problems 51–58, find the dot product $\mathbf{v} \cdot \mathbf{w}$ and the angle between $\mathbf{v}$ and $\mathbf{w}$.

51. $\mathbf{v} = \mathbf{i} - \mathbf{j}$, $\mathbf{w} = \mathbf{i} + \mathbf{j} + \mathbf{k}$

52. $\mathbf{v} = \mathbf{i} + \mathbf{j}$, $\mathbf{w} = -\mathbf{i} + \mathbf{j} - \mathbf{k}$

53. $\mathbf{v} = 2\mathbf{i} + \mathbf{j} - 3\mathbf{k}$, $\mathbf{w} = \mathbf{i} + 2\mathbf{j} + 2\mathbf{k}$

54. $\mathbf{v} = 2\mathbf{i} + 2\mathbf{j} - \mathbf{k}$, $\mathbf{w} = \mathbf{i} + 2\mathbf{j} + 3\mathbf{k}$

55. $\mathbf{v} = 3\mathbf{i} - \mathbf{j} + 2\mathbf{k}$, $\mathbf{w} = \mathbf{i} + \mathbf{j} - \mathbf{k}$

56. $\mathbf{v} = \mathbf{i} + 3\mathbf{j} + 2\mathbf{k}$, $\mathbf{w} = \mathbf{i} - \mathbf{j} + \mathbf{k}$

57. $\mathbf{v} = 3\mathbf{i} + 4\mathbf{j} + \mathbf{k}$, $\mathbf{w} = 6\mathbf{i} + 8\mathbf{j} + 2\mathbf{k}$

58. $\mathbf{v} = 3\mathbf{i} - 4\mathbf{j} + \mathbf{k}$, $\mathbf{w} = 6\mathbf{i} - 8\mathbf{j} + 2\mathbf{k}$

In Problems 59 – 66, find the direction angles of each vector. Write each vector in the form of equation (7).

59. $\mathbf{v} = 3\mathbf{i} - 6\mathbf{j} - 2\mathbf{k}$ **60.** $\mathbf{v} = -6\mathbf{i} + 12\mathbf{j} + 4\mathbf{k}$ **61.** $\mathbf{v} = \mathbf{i} + \mathbf{j} + \mathbf{k}$ **62.** $\mathbf{v} = \mathbf{i} - \mathbf{j} - \mathbf{k}$

63. $\mathbf{v} = \mathbf{i} + \mathbf{j}$ **64.** $\mathbf{v} = \mathbf{j} + \mathbf{k}$ **65.** $\mathbf{v} = 3\mathbf{i} - 5\mathbf{j} + 2\mathbf{k}$ **66.** $\mathbf{v} = 2\mathbf{i} + 3\mathbf{j} - 4\mathbf{k}$

Applications and Extensions

67. Robotic Arm Consider the double-jointed robotic arm shown in the figure. Let the lower arm be modeled by $\mathbf{a} = \langle 2, 3, 4 \rangle$, the middle arm be modeled by $\mathbf{b} = \langle 1, -1, 3 \rangle$, and the upper arm be modeled by $\mathbf{c} = \langle 4, -1, -2 \rangle$, where units are in feet.

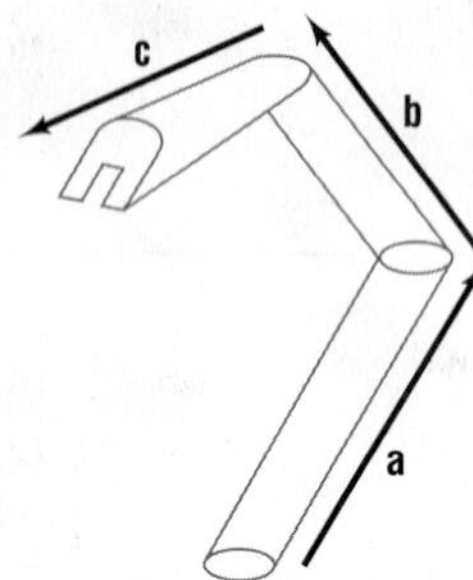

(a) Find a vector **d** that represents the position of the hand.
(b) Determine the distance of the hand from the origin.

68. The Sphere In space, the collection of all points that are the same distance from some fixed point is called a **sphere**. See the illustration. The constant distance is called the **radius**, and the fixed point is the **center** of the sphere. Show that an equation of a sphere with center at (x_0, y_0, z_0) and radius r is

$$(x - x_0)^2 + (y - y_0)^2 + (z - z_0)^2 = r^2$$

[**Hint:** Use the Distance Formula (1).]

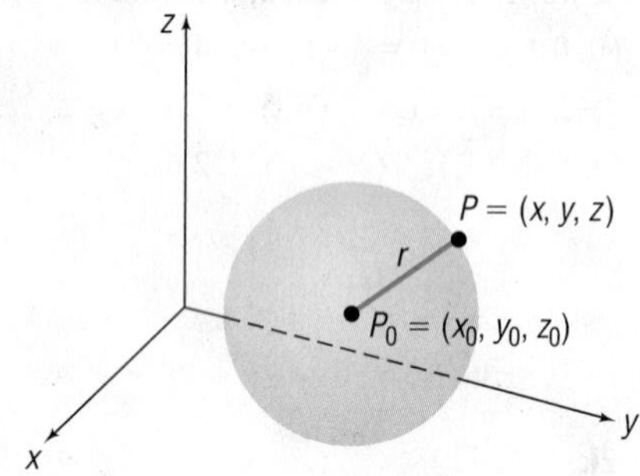

In Problems 69 and 70, find an equation of a sphere with radius r and center P_0.

69. $r = 1;\ P_0 = (3, 1, 1)$

70. $r = 2;\ P_0 = (1, 2, 2)$

In Problems 71–76, find the radius and center of each sphere. [**Hint:** *Complete the square in each variable.*]

71. $x^2 + y^2 + z^2 + 2x - 2y = 2$

72. $x^2 + y^2 + z^2 + 2x - 2z = -1$

73. $x^2 + y^2 + z^2 - 4x + 4y + 2z = 0$

74. $x^2 + y^2 + z^2 - 4x = 0$

75. $2x^2 + 2y^2 + 2z^2 - 8x + 4z = -1$

76. $3x^2 + 3y^2 + 3z^2 + 6x - 6y = 3$

The **work** *W done by a constant force* **F** *in moving an object from a point A in space to a point B in space is defined as* $W = \mathbf{F} \cdot \overrightarrow{AB}$. *Use this definition in Problems 77–79.*

77. Work Find the work done by a force of 3 newtons acting in the direction $2\mathbf{i} + \mathbf{j} + 2\mathbf{k}$ in moving an object 2 meters from $(0, 0, 0)$ to $(0, 2, 0)$.

78. Work Find the work done by a force of 1 newton acting in the direction $2\mathbf{i} + 2\mathbf{j} + \mathbf{k}$ in moving an object 3 meters from $(0, 0, 0)$ to $(1, 2, 2)$.

79. Work Find the work done in moving an object along a vector $\mathbf{u} = 3\mathbf{i} + 2\mathbf{j} - 5\mathbf{k}$ if the applied force is $\mathbf{F} = 2\mathbf{i} - \mathbf{j} - \mathbf{k}$. Use meters for distance and newtons for force.

Retain Your Knowledge

Problems 80–83 are based on material learned earlier in the course. The purpose of these problems is to keep the material fresh in your mind so that you are better prepared for the final exam.

80. Find the inverse f^{-1} of $f(x) = 3\cos x + 5,\ 0 \le x \le \pi$. Find the range of f and the domain and range of f^{-1}.

81. If $\mathbf{v} = -3\mathbf{i} + 6\mathbf{j}$, find $\|\mathbf{v}\|$.

82. Find the exact value of $\sin 80° \cos 50° - \cos 80° \sin 50°$.

83. Solve the triangle.

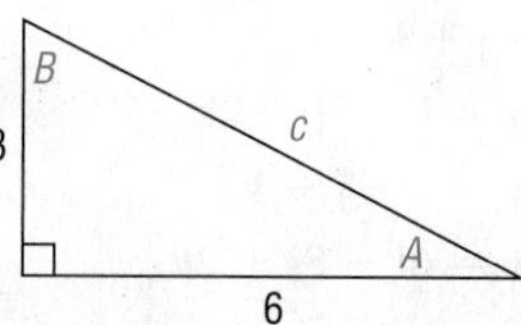

'Are You Prepared?' Answer

1. $\sqrt{(x_2 - x_1)^2 + (y_2 - y_1)^2}$

5.7 The Cross Product

OBJECTIVES 1 Find the Cross Product of Two Vectors (p. 375)
2 Know Algebraic Properties of the Cross Product (p. 376)
3 Know Geometric Properties of the Cross Product (p. 377)
4 Find a Vector Orthogonal to Two Given Vectors (p. 378)
5 Find the Area of a Parallelogram (p. 378)

1 Find the Cross Product of Two Vectors

For vectors in space, and only for vectors in space, a second product of two vectors is defined, called the *cross product*. The cross product of two vectors in space is also a vector that has applications in both geometry and physics.

DEFINITION

If $\mathbf{v} = a_1\mathbf{i} + b_1\mathbf{j} + c_1\mathbf{k}$ and $\mathbf{w} = a_2\mathbf{i} + b_2\mathbf{j} + c_2\mathbf{k}$ are two vectors in space, the **cross product** $\mathbf{v} \times \mathbf{w}$ is defined as the vector

$$\mathbf{v} \times \mathbf{w} = (b_1c_2 - b_2c_1)\mathbf{i} - (a_1c_2 - a_2c_1)\mathbf{j} + (a_1b_2 - a_2b_1)\mathbf{k} \qquad (1)$$

Notice that the cross product $\mathbf{v} \times \mathbf{w}$ of two vectors is a vector. Because of this, it is sometimes referred to as the **vector product**.

EXAMPLE 1 **Finding a Cross Product Using Equation (1)**

If $\mathbf{v} = 2\mathbf{i} + 3\mathbf{j} + 5\mathbf{k}$ and $\mathbf{w} = \mathbf{i} + 2\mathbf{j} + 3\mathbf{k}$, find $\mathbf{v} \times \mathbf{w}$.

Solution

$$\begin{aligned}\mathbf{v} \times \mathbf{w} &= (3\cdot 3 - 2\cdot 5)\mathbf{i} - (2\cdot 3 - 1\cdot 5)\mathbf{j} + (2\cdot 2 - 1\cdot 3)\mathbf{k} \quad \text{Equation (1)}\\ &= (9 - 10)\mathbf{i} - (6 - 5)\mathbf{j} + (4 - 3)\mathbf{k}\\ &= -\mathbf{i} - \mathbf{j} + \mathbf{k}\end{aligned}$$

Determinants may be used as an aid in computing cross products. A **2 by 2 determinant**, symbolized by

$$\begin{vmatrix} a_1 & b_1 \\ a_2 & b_2 \end{vmatrix}$$

has the value $a_1b_2 - a_2b_1$; that is,

$$\begin{vmatrix} a_1 & b_1 \\ a_2 & b_2 \end{vmatrix} = a_1b_2 - a_2b_1$$

A **3 by 3 determinant** has the value

$$\begin{vmatrix} A & B & C \\ a_1 & b_1 & c_1 \\ a_2 & b_2 & c_2 \end{vmatrix} = \begin{vmatrix} b_1 & c_1 \\ b_2 & c_2 \end{vmatrix} A - \begin{vmatrix} a_1 & c_1 \\ a_2 & c_2 \end{vmatrix} B + \begin{vmatrix} a_1 & b_1 \\ a_2 & b_2 \end{vmatrix} C$$

EXAMPLE 2

Evaluating Determinants

(a) $\begin{vmatrix} 2 & 3 \\ 1 & 2 \end{vmatrix} = 2 \cdot 2 - 1 \cdot 3 = 4 - 3 = 1$

(b) $$\begin{vmatrix} A & B & C \\ 2 & 3 & 5 \\ 1 & 2 & 3 \end{vmatrix} = \begin{vmatrix} 3 & 5 \\ 2 & 3 \end{vmatrix} A - \begin{vmatrix} 2 & 5 \\ 1 & 3 \end{vmatrix} B + \begin{vmatrix} 2 & 3 \\ 1 & 2 \end{vmatrix} C$$
$$= (9 - 10)A - (6 - 5)B + (4 - 3)C$$
$$= -A - B + C$$

Now Work PROBLEM 7

The cross product of the vectors $\mathbf{v} = a_1\mathbf{i} + b_1\mathbf{j} + c_1\mathbf{k}$ and $\mathbf{w} = a_2\mathbf{i} + b_2\mathbf{j} + c_2\mathbf{k}$, that is,

$$\mathbf{v} \times \mathbf{w} = (b_1c_2 - b_2c_1)\mathbf{i} - (a_1c_2 - a_2c_1)\mathbf{j} + (a_1b_2 - a_2b_1)\mathbf{k}$$

may be written symbolically using determinants as

$$\mathbf{v} \times \mathbf{w} = \begin{vmatrix} \mathbf{i} & \mathbf{j} & \mathbf{k} \\ a_1 & b_1 & c_1 \\ a_2 & b_2 & c_2 \end{vmatrix} = \begin{vmatrix} b_1 & c_1 \\ b_2 & c_2 \end{vmatrix}\mathbf{i} - \begin{vmatrix} a_1 & c_1 \\ a_2 & c_2 \end{vmatrix}\mathbf{j} + \begin{vmatrix} a_1 & b_1 \\ a_2 & b_2 \end{vmatrix}\mathbf{k}$$

EXAMPLE 3

Using Determinants to Find Cross Products

If $\mathbf{v} = 2\mathbf{i} + 3\mathbf{j} + 5\mathbf{k}$ and $\mathbf{w} = \mathbf{i} + 2\mathbf{j} + 3\mathbf{k}$, find:

(a) $\mathbf{v} \times \mathbf{w}$ (b) $\mathbf{w} \times \mathbf{v}$ (c) $\mathbf{v} \times \mathbf{v}$ (d) $\mathbf{w} \times \mathbf{w}$

Solution

(a) $$\mathbf{v} \times \mathbf{w} = \begin{vmatrix} \mathbf{i} & \mathbf{j} & \mathbf{k} \\ 2 & 3 & 5 \\ 1 & 2 & 3 \end{vmatrix} = \begin{vmatrix} 3 & 5 \\ 2 & 3 \end{vmatrix}\mathbf{i} - \begin{vmatrix} 2 & 5 \\ 1 & 3 \end{vmatrix}\mathbf{j} + \begin{vmatrix} 2 & 3 \\ 1 & 2 \end{vmatrix}\mathbf{k} = -\mathbf{i} - \mathbf{j} + \mathbf{k}$$

(b) $$\mathbf{w} \times \mathbf{v} = \begin{vmatrix} \mathbf{i} & \mathbf{j} & \mathbf{k} \\ 1 & 2 & 3 \\ 2 & 3 & 5 \end{vmatrix} = \begin{vmatrix} 2 & 3 \\ 3 & 5 \end{vmatrix}\mathbf{i} - \begin{vmatrix} 1 & 3 \\ 2 & 5 \end{vmatrix}\mathbf{j} + \begin{vmatrix} 1 & 2 \\ 2 & 3 \end{vmatrix}\mathbf{k} = \mathbf{i} + \mathbf{j} - \mathbf{k}$$

(c) $$\mathbf{v} \times \mathbf{v} = \begin{vmatrix} \mathbf{i} & \mathbf{j} & \mathbf{k} \\ 2 & 3 & 5 \\ 2 & 3 & 5 \end{vmatrix} = \begin{vmatrix} 3 & 5 \\ 3 & 5 \end{vmatrix}\mathbf{i} - \begin{vmatrix} 2 & 5 \\ 2 & 5 \end{vmatrix}\mathbf{j} + \begin{vmatrix} 2 & 3 \\ 2 & 3 \end{vmatrix}\mathbf{k} = 0\mathbf{i} - 0\mathbf{j} + 0\mathbf{k} = \mathbf{0}$$

(d) $$\mathbf{w} \times \mathbf{w} = \begin{vmatrix} \mathbf{i} & \mathbf{j} & \mathbf{k} \\ 1 & 2 & 3 \\ 1 & 2 & 3 \end{vmatrix}$$
$$= \begin{vmatrix} 2 & 3 \\ 2 & 3 \end{vmatrix}\mathbf{i} - \begin{vmatrix} 1 & 3 \\ 1 & 3 \end{vmatrix}\mathbf{j} + \begin{vmatrix} 1 & 2 \\ 1 & 2 \end{vmatrix}\mathbf{k} = 0\mathbf{i} - 0\mathbf{j} + 0\mathbf{k} = \mathbf{0}$$

Now Work PROBLEM 15

2 Know Algebraic Properties of the Cross Product

Notice in Examples 3(a) and (b) that $\mathbf{v} \times \mathbf{w}$ and $\mathbf{w} \times \mathbf{v}$ are negatives of one another. From Examples 3(c) and (d), one might conjecture that the cross product of a vector with itself is the zero vector. These and other algebraic properties of the cross product are given next.

THEOREM

Algebraic Properties of the Cross Product

If **u**, **v**, and **w** are vectors in space and if α is a scalar, then

$$\mathbf{u} \times \mathbf{u} = \mathbf{0} \tag{2}$$

$$\mathbf{u} \times \mathbf{v} = -(\mathbf{v} \times \mathbf{u}) \tag{3}$$

$$\alpha(\mathbf{u} \times \mathbf{v}) = (\alpha\mathbf{u}) \times \mathbf{v} = \mathbf{u} \times (\alpha\mathbf{v}) \tag{4}$$

$$\mathbf{u} \times (\mathbf{v} + \mathbf{w}) = (\mathbf{u} \times \mathbf{v}) + (\mathbf{u} \times \mathbf{w}) \tag{5}$$

Proof We will prove properties (2) and (4) here and leave properties (3) and (5) as exercises (see Problems 60 and 61).

To prove property (2), let $\mathbf{u} = a_1\mathbf{i} + b_1\mathbf{j} + c_1\mathbf{k}$. Then

$$\mathbf{u} \times \mathbf{u} = \begin{vmatrix} \mathbf{i} & \mathbf{j} & \mathbf{k} \\ a_1 & b_1 & c_1 \\ a_1 & b_1 & c_1 \end{vmatrix} = \begin{vmatrix} b_1 & c_1 \\ b_1 & c_1 \end{vmatrix}\mathbf{i} - \begin{vmatrix} a_1 & c_1 \\ a_1 & c_1 \end{vmatrix}\mathbf{j} + \begin{vmatrix} a_1 & b_1 \\ a_1 & b_1 \end{vmatrix}\mathbf{k}$$
$$= 0\mathbf{i} - 0\mathbf{j} + 0\mathbf{k} = \mathbf{0}$$

To prove property (4), let $\mathbf{u} = a_1\mathbf{i} + b_1\mathbf{j} + c_1\mathbf{k}$ and $\mathbf{v} = a_2\mathbf{i} + b_2\mathbf{j} + c_2\mathbf{k}$. Then

$$\alpha(\mathbf{u} \times \mathbf{v}) = \alpha[(b_1c_2 - b_2c_1)\mathbf{i} - (a_1c_2 - a_2c_1)\mathbf{j} + (a_1b_2 - a_2b_1)\mathbf{k}]$$

Apply (1).

$$= \alpha(b_1c_2 - b_2c_1)\mathbf{i} - \alpha(a_1c_2 - a_2c_1)\mathbf{j} + \alpha(a_1b_2 - a_2b_1)\mathbf{k} \tag{6}$$

Since $\alpha\mathbf{u} = \alpha a_1\mathbf{i} + \alpha b_1\mathbf{j} + \alpha c_1\mathbf{k}$, we have

$$(\alpha\mathbf{u}) \times \mathbf{v} = (\alpha b_1c_2 - b_2\alpha c_1)\mathbf{i} - (\alpha a_1c_2 - a_2\alpha c_1)\mathbf{j} + (\alpha a_1b_2 - a_2\alpha b_1)\mathbf{k}$$
$$= \alpha(b_1c_2 - b_2c_1)\mathbf{i} - \alpha(a_1c_2 - a_2c_1)\mathbf{j} + \alpha(a_1b_2 - a_2b_1)\mathbf{k} \tag{7}$$

Based on equations (6) and (7), the first part of property (4) follows. The second part can be proved in like fashion. ■

Now Work PROBLEM 17

3 Know Geometric Properties of the Cross Product

THEOREM

Geometric Properties of the Cross Product

Let **u** and **v** be vectors in space.

$\mathbf{u} \times \mathbf{v}$ is orthogonal to both **u** and **v**. **(8)**

$\|\mathbf{u} \times \mathbf{v}\| = \|\mathbf{u}\|\,\|\mathbf{v}\| \sin\theta$, **(9)**

where θ is the angle between **u** and **v**.

$\|\mathbf{u} \times \mathbf{v}\|$ is the area of the parallelogram

having $\mathbf{u} \neq \mathbf{0}$ and $\mathbf{v} \neq \mathbf{0}$ as adjacent sides. **(10)**

$\mathbf{u} \times \mathbf{v} = \mathbf{0}$ if and only if **u** and **v** are parallel. **(11)**

Proof of Property (8) Let $\mathbf{u} = a_1\mathbf{i} + b_1\mathbf{j} + c_1\mathbf{k}$ and $\mathbf{v} = a_2\mathbf{i} + b_2\mathbf{j} + c_2\mathbf{k}$. Then

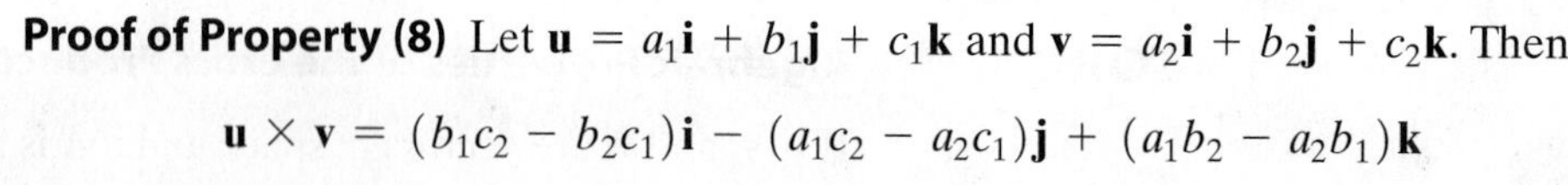

$$\mathbf{u} \times \mathbf{v} = (b_1c_2 - b_2c_1)\mathbf{i} - (a_1c_2 - a_2c_1)\mathbf{j} + (a_1b_2 - a_2b_1)\mathbf{k}$$

Now compute the dot product $\mathbf{u} \cdot (\mathbf{u} \times \mathbf{v})$.

$$\begin{aligned}\mathbf{u} \cdot (\mathbf{u} \times \mathbf{v}) &= (a_1\mathbf{i} + b_1\mathbf{j} + c_1\mathbf{k}) \cdot [(b_1c_2 - b_2c_1)\mathbf{i} - (a_1c_2 - a_2c_1)\mathbf{j} + (a_1b_2 - a_2b_1)\mathbf{k}] \\ &= a_1(b_1c_2 - b_2c_1) - b_1(a_1c_2 - a_2c_1) + c_1(a_1b_2 - a_2b_1) = 0\end{aligned}$$

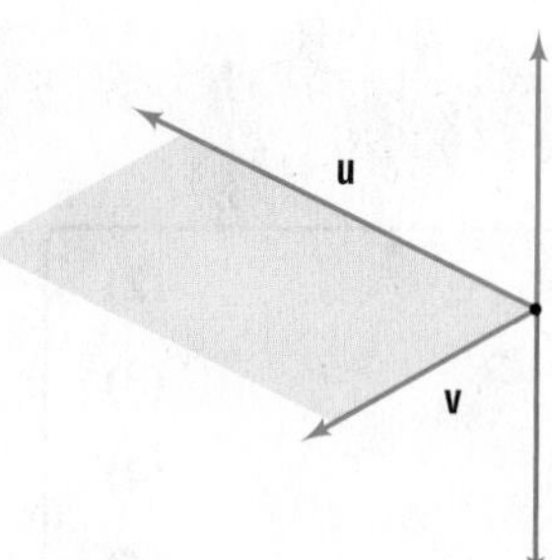

Figure 81

Since two vectors are orthogonal if their dot product is zero, it follows that $\mathbf{u}$ and $\mathbf{u} \times \mathbf{v}$ are orthogonal. Similarly, $\mathbf{v} \cdot (\mathbf{u} \times \mathbf{v}) = 0$, so $\mathbf{v}$ and $\mathbf{u} \times \mathbf{v}$ are orthogonal. ■

4 Find a Vector Orthogonal to Two Given Vectors

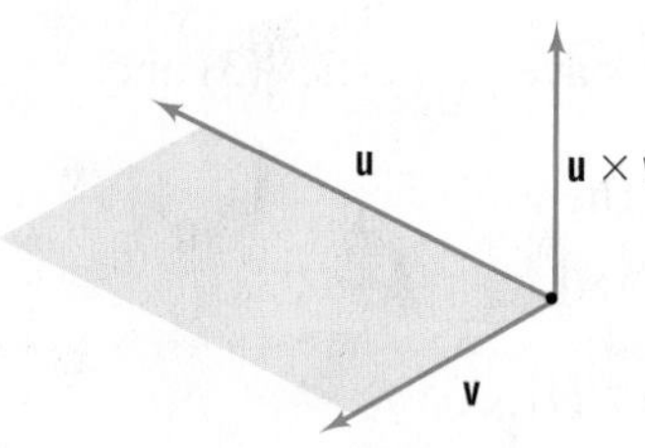

Figure 82

As long as the vectors $\mathbf{u}$ and $\mathbf{v}$ are not parallel, they will form a plane in space. See Figure 81. Based on property (8), the vector $\mathbf{u} \times \mathbf{v}$ is normal to this plane. As Figure 81 illustrates, there are essentially (without regard to magnitude) two vectors normal to the plane containing $\mathbf{u}$ and $\mathbf{v}$. It can be shown that the vector $\mathbf{u} \times \mathbf{v}$ is the one determined by the thumb of the right hand when the other fingers of the right hand are cupped so that they point in a direction from $\mathbf{u}$ to $\mathbf{v}$. See Figure 82.*

EXAMPLE 4 Finding a Vector Orthogonal to Two Given Vectors

Find a vector that is orthogonal to $\mathbf{u} = 3\mathbf{i} - 2\mathbf{j} + \mathbf{k}$ and $\mathbf{v} = -\mathbf{i} + 3\mathbf{j} - \mathbf{k}$.

Solution Based on property (8), such a vector is $\mathbf{u} \times \mathbf{v}$.

$$\mathbf{u} \times \mathbf{v} = \begin{vmatrix} \mathbf{i} & \mathbf{j} & \mathbf{k} \\ 3 & -2 & 1 \\ -1 & 3 & -1 \end{vmatrix} = (2 - 3)\mathbf{i} - [-3 - (-1)]\mathbf{j} + (9 - 2)\mathbf{k} = -\mathbf{i} + 2\mathbf{j} + 7\mathbf{k}$$

The vector $-\mathbf{i} + 2\mathbf{j} + 7\mathbf{k}$ is orthogonal to both $\mathbf{u}$ and $\mathbf{v}$.

Check: Two vectors are orthogonal if their dot product is zero.

$$\mathbf{u} \cdot (-\mathbf{i} + 2\mathbf{j} + 7\mathbf{k}) = (3\mathbf{i} - 2\mathbf{j} + \mathbf{k}) \cdot (-\mathbf{i} + 2\mathbf{j} + 7\mathbf{k}) = -3 - 4 + 7 = 0$$

$$\mathbf{v} \cdot (-\mathbf{i} + 2\mathbf{j} + 7\mathbf{k}) = (-\mathbf{i} + 3\mathbf{j} - \mathbf{k}) \cdot (-\mathbf{i} + 2\mathbf{j} + 7\mathbf{k}) = 1 + 6 - 7 = 0$$

Now Work PROBLEM 41 ●

The proof of property (9) is left as an exercise. See Problem 62.

5 Find the Area of a Parallelogram

Proof of Property (10) Suppose that $\mathbf{u}$ and $\mathbf{v}$ are adjacent sides of a parallelogram. See Figure 83. Then the lengths of these sides are $\|\mathbf{u}\|$ and $\|\mathbf{v}\|$. If θ is the angle between $\mathbf{u}$ and $\mathbf{v}$, then the height of the parallelogram is $\|\mathbf{v}\| \sin\theta$ and its area is

$$\text{Area of parallelogram} = \text{Base} \times \text{Height} = \|\mathbf{u}\|[\|\mathbf{v}\| \sin\theta] \underset{\substack{\uparrow \\ \text{Property (9)}}}{=} \|\mathbf{u} \times \mathbf{v}\|$$

■

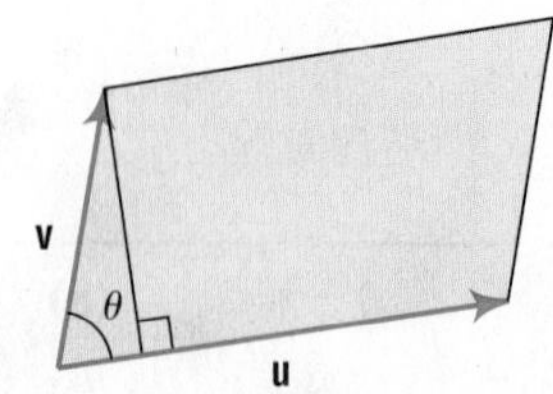

Figure 83

EXAMPLE 5 Finding the Area of a Parallelogram

Find the area of the parallelogram whose vertices are $P_1 = (0, 0, 0)$, $P_2 = (3, -2, 1)$, $P_3 = (-1, 3, -1)$, and $P_4 = (2, 1, 0)$.

*This is a consequence of using a "right-handed" coordinate system.

Solution Two adjacent sides of this parallelogram are

$$\mathbf{u} = \overrightarrow{P_1P_2} = 3\mathbf{i} - 2\mathbf{j} + \mathbf{k} \quad \text{and} \quad \mathbf{v} = \overrightarrow{P_1P_3} = -\mathbf{i} + 3\mathbf{j} - \mathbf{k}$$

WARNING Not all pairs of vertices give rise to a side. For example, $\overrightarrow{P_1P_4}$ is a diagonal of the parallelogram since $\overrightarrow{P_1P_3} + \overrightarrow{P_3P_4} = \overrightarrow{P_1P_4}$. Also, $\overrightarrow{P_1P_3}$ and $\overrightarrow{P_2P_4}$ are not adjacent sides; they are parallel sides. ■

Since $\mathbf{u} \times \mathbf{v} = -\mathbf{i} + 2\mathbf{j} + 7\mathbf{k}$ (Example 4), the area of the parallelogram is

$$\text{Area of parallelogram} = \|\mathbf{u} \times \mathbf{v}\| = \sqrt{1 + 4 + 49} = \sqrt{54} = 3\sqrt{6} \text{ square units}$$

●

Now Work PROBLEM 49

Proof of Property (11) The proof requires two parts. If **u** and **v** are parallel, then there is a scalar α such that $\mathbf{u} = \alpha\mathbf{v}$. Then

$$\mathbf{u} \times \mathbf{v} = (\alpha\mathbf{v}) \times \mathbf{v} \underset{\substack{\uparrow \\ \text{Property (4)}}}{=} \alpha(\mathbf{v} \times \mathbf{v}) \underset{\substack{\uparrow \\ \text{Property (2)}}}{=} \mathbf{0}$$

If $\mathbf{u} \times \mathbf{v} = \mathbf{0}$, then, by property (9), we have

$$\|\mathbf{u} \times \mathbf{v}\| = \|\mathbf{u}\|\,\|\mathbf{v}\| \sin\theta = 0$$

Since $\mathbf{u} \neq \mathbf{0}$ and $\mathbf{v} \neq \mathbf{0}$, we must have $\sin\theta = 0$, so $\theta = 0$ or $\theta = \pi$. In either case, since θ is the angle between **u** and **v**, then **u** and **v** are parallel. ■

5.7 Assess Your Understanding

Concepts and Vocabulary

1. ***True or False*** If **u** and **v** are parallel vectors, then $\mathbf{u} \times \mathbf{v} = \mathbf{0}$.
2. ***True or False*** For any vector **v**, $\mathbf{v} \times \mathbf{v} = \mathbf{0}$.
3. ***True or False*** If **u** and **v** are vectors, then $\mathbf{u} \times \mathbf{v} + \mathbf{v} \times \mathbf{u} = \mathbf{0}$.
4. ***True or False*** $\mathbf{u} \times \mathbf{v}$ is a vector that is parallel to both **u** and **v**.
5. ***True or False*** $\|\mathbf{u} \times \mathbf{v}\| = \|\mathbf{u}\|\,\|\mathbf{v}\| \cos\theta$, where θ is the angle between **u** and **v**.
6. ***True or False*** The area of the parallelogram having **u** and **v** as adjacent sides is the magnitude of the cross product of **u** and **v**.

Skill Building

In Problems 7–14, find the value of each determinant.

7. $\begin{vmatrix} 3 & 4 \\ 1 & 2 \end{vmatrix}$

8. $\begin{vmatrix} -2 & 5 \\ 2 & -3 \end{vmatrix}$

9. $\begin{vmatrix} 6 & 5 \\ -2 & -1 \end{vmatrix}$

10. $\begin{vmatrix} -4 & 0 \\ 5 & 3 \end{vmatrix}$

11. $\begin{vmatrix} A & B & C \\ 2 & 1 & 4 \\ 1 & 3 & 1 \end{vmatrix}$

12. $\begin{vmatrix} A & B & C \\ 0 & 2 & 4 \\ 3 & 1 & 3 \end{vmatrix}$

13. $\begin{vmatrix} A & B & C \\ -1 & 3 & 5 \\ 5 & 0 & -2 \end{vmatrix}$

14. $\begin{vmatrix} A & B & C \\ 1 & -2 & -3 \\ 0 & 2 & -2 \end{vmatrix}$

In Problems 15–22, find (a) $\mathbf{v} \times \mathbf{w}$, *(b)* $\mathbf{w} \times \mathbf{v}$, *(c)* $\mathbf{w} \times \mathbf{w}$, *and (d)* $\mathbf{v} \times \mathbf{v}$.

15. $\mathbf{v} = 2\mathbf{i} - 3\mathbf{j} + \mathbf{k}$, $\mathbf{w} = 3\mathbf{i} - 2\mathbf{j} - \mathbf{k}$
16. $\mathbf{v} = -\mathbf{i} + 3\mathbf{j} + 2\mathbf{k}$, $\mathbf{w} = 3\mathbf{i} - 2\mathbf{j} - \mathbf{k}$
17. $\mathbf{v} = \mathbf{i} + \mathbf{j}$, $\mathbf{w} = 2\mathbf{i} + \mathbf{j} + \mathbf{k}$
18. $\mathbf{v} = \mathbf{i} - 4\mathbf{j} + 2\mathbf{k}$, $\mathbf{w} = 3\mathbf{i} + 2\mathbf{j} + \mathbf{k}$
19. $\mathbf{v} = 2\mathbf{i} - \mathbf{j} + 2\mathbf{k}$, $\mathbf{w} = \mathbf{j} - \mathbf{k}$
20. $\mathbf{v} = 3\mathbf{i} + \mathbf{j} + 3\mathbf{k}$, $\mathbf{w} = \mathbf{i} - \mathbf{k}$
21. $\mathbf{v} = \mathbf{i} - \mathbf{j} - \mathbf{k}$, $\mathbf{w} = 4\mathbf{i} - 3\mathbf{k}$
22. $\mathbf{v} = 2\mathbf{i} - 3\mathbf{j}$, $\mathbf{w} = 3\mathbf{j} - 2\mathbf{k}$

In Problems 23–44, use the given vectors **u**, **v**, *and* **w** *to find each expression.*

$$\mathbf{u} = 2\mathbf{i} - 3\mathbf{j} + \mathbf{k} \qquad \mathbf{v} = -3\mathbf{i} + 3\mathbf{j} + 2\mathbf{k} \qquad \mathbf{w} = \mathbf{i} + \mathbf{j} + 3\mathbf{k}$$

23. $\mathbf{u} \times \mathbf{v}$

24. $\mathbf{v} \times \mathbf{w}$

25. $\mathbf{v} \times \mathbf{u}$

26. $\mathbf{w} \times \mathbf{v}$

27. $\mathbf{v} \times \mathbf{v}$

28. $\mathbf{w} \times \mathbf{w}$

29. $(3\mathbf{u}) \times \mathbf{v}$

30. $\mathbf{v} \times (4\mathbf{w})$

31. $\mathbf{u} \times (2\mathbf{v})$

32. $(-3\mathbf{v}) \times \mathbf{w}$

33. $\mathbf{u} \cdot (\mathbf{u} \times \mathbf{v})$

34. $\mathbf{v} \cdot (\mathbf{v} \times \mathbf{w})$

35. $\mathbf{u} \cdot (\mathbf{v} \times \mathbf{w})$

36. $(\mathbf{u} \times \mathbf{v}) \cdot \mathbf{w}$

37. $\mathbf{v} \cdot (\mathbf{u} \times \mathbf{w})$

38. $(\mathbf{v} \times \mathbf{u}) \cdot \mathbf{w}$

39. $\mathbf{u} \times (\mathbf{v} \times \mathbf{v})$

40. $(\mathbf{w} \times \mathbf{w}) \times \mathbf{v}$

41. Find a vector orthogonal to both **u** and **v**.

42. Find a vector orthogonal to both **u** and **w**.

43. Find a vector orthogonal to both **u** and $\mathbf{i} + \mathbf{j}$.

44. Find a vector orthogonal to both **u** and $\mathbf{j} + \mathbf{k}$.

In Problems 45–48, find the area of the parallelogram with one corner at P_1 and adjacent sides $\overrightarrow{P_1P_2}$ and $\overrightarrow{P_1P_3}$.

45. $P_1 = (0, 0, 0),\ P_2 = (1, 2, 3),\ P_3 = (-2, 3, 0)$

46. $P_1 = (0, 0, 0),\ P_2 = (2, 3, 1),\ P_3 = (-2, 4, 1)$

47. $P_1 = (1, 2, 0),\ P_2 = (-2, 3, 4),\ P_3 = (0, -2, 3)$

48. $P_1 = (-2, 0, 2),\ P_2 = (2, 1, -1),\ P_3 = (2, -1, 2)$

In Problems 49–52, find the area of the parallelogram with vertices P_1, P_2, P_3, and P_4.

49. $P_1 = (1, 1, 2),\ P_2 = (1, 2, 3),\ P_3 = (-2, 3, 0),\ P_4 = (-2, 4, 1)$

50. $P_1 = (2, 1, 1),\ P_2 = (2, 3, 1),\ P_3 = (-2, 4, 1),\ P_4 = (-2, 6, 1)$

51. $P_1 = (1, 2, -1),\ P_2 = (4, 2, -3),\ P_3 = (6, -5, 2),\ P_4 = (9, -5, 0)$

52. $P_1 = (-1, 1, 1),\ P_2 = (-1, 2, 2),\ P_3 = (-3, 4, -5),\ P_4 = (-3, 5, -4)$

Applications and Extensions

53. Find a unit vector normal to the plane containing $\mathbf{v} = \mathbf{i} + 3\mathbf{j} - 2\mathbf{k}$ and $\mathbf{w} = -2\mathbf{i} + \mathbf{j} + 3\mathbf{k}$.

54. Find a unit vector normal to the plane containing $\mathbf{v} = 2\mathbf{i} + 3\mathbf{j} - \mathbf{k}$ and $\mathbf{w} = -2\mathbf{i} - 4\mathbf{j} - 3\mathbf{k}$.

55. Volume of a Parallelepiped A **parallelepiped** is a prism whose faces are all parallelograms. Let **A**, **B**, and **C** be the vectors that define the parallelepiped shown in the figure. The volume V of the parallelepiped is given by the formula $V = |(\mathbf{A} \times \mathbf{B}) \cdot \mathbf{C}|$.

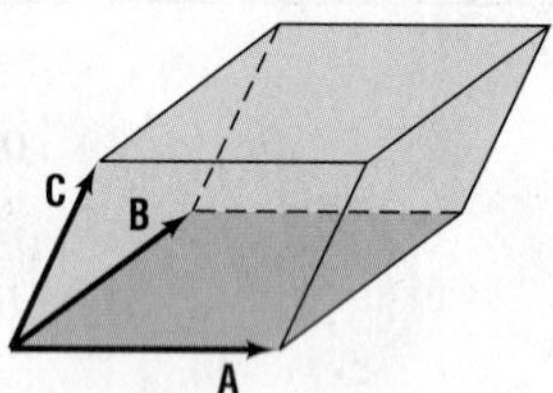

Find the volume of a parallelepiped if the defining vectors are $\mathbf{A} = 3\mathbf{i} - 2\mathbf{j} + 4\mathbf{k}$, $\mathbf{B} = 2\mathbf{i} + \mathbf{j} - 2\mathbf{k}$, and $\mathbf{C} = 3\mathbf{i} - 6\mathbf{j} - 2\mathbf{k}$.

56. Volume of a Parallelepiped Refer to Problem 55. Find the volume of a parallelepiped whose defining vectors are $\mathbf{A} = \mathbf{i} + 6\mathbf{k}$, $\mathbf{B} = 2\mathbf{i} + 3\mathbf{j} - 8\mathbf{k}$, and $\mathbf{C} = 8\mathbf{i} - 5\mathbf{j} + 6\mathbf{k}$.

57. Prove for vectors **u** and **v** that

$$\|\mathbf{u} \times \mathbf{v}\|^2 = \|\mathbf{u}\|^2\|\mathbf{v}\|^2 - (\mathbf{u} \cdot \mathbf{v})^2$$

[**Hint:** Proceed as in the proof of property (4), computing first the left side and then the right side.]

58. Show that if **u** and **v** are orthogonal, then

$$\|\mathbf{u} \times \mathbf{v}\| = \|\mathbf{u}\|\,\|\mathbf{v}\|$$

59. Show that if **u** and **v** are orthogonal unit vectors, then $\mathbf{u} \times \mathbf{v}$ is also a unit vector.

60. Prove property (3).

61. Prove property (5).

62. Prove property (9).

[**Hint:** Use the result of Problem 57 and the fact that if θ is the angle between **u** and **v**, then $\mathbf{u} \cdot \mathbf{v} = \|\mathbf{u}\|\,\|\mathbf{v}\| \cos \theta$.]

Discussion and Writing

63. If $\mathbf{u} \cdot \mathbf{v} = 0$ and $\mathbf{u} \times \mathbf{v} = \mathbf{0}$, what, if anything, can you conclude about **u** and **v**?

Retain Your Knowledge

Problems 64–67 are based on material learned earlier in the course. The purpose of these problems is to keep the material fresh in your mind so that you are better prepared for the final exam.

64. Find the exact value of $\cos^{-1}\left(\frac{1}{\sqrt{2}}\right)$.

65. Find two pairs of polar coordinates (r, θ), one with $r > 0$ and the other with $r < 0$, for the point with rectangular coordinates $(-8, -15)$. Express θ in radians.

66. Solve the equation $\tan(3\theta) = 1$ on the interval $0 \le \theta < 2\pi$.

67. Given $\tan\theta = -\frac{20}{21}$ and $\sin\theta = \frac{20}{29}$, find the exact values of the four remaining trigonometric functions of θ.

Chapter Review

Things to Know

Polar Coordinates (pp. 312–319)

Relationship between polar coordinates (r, θ) and rectangular coordinates (x, y) (pp. 314 and 317)	$x = r\cos\theta,\ y = r\sin\theta$ $r^2 = x^2 + y^2,\ \tan\theta = \frac{y}{x},\quad x \neq 0$
Polar form of a complex number (p. 337)	If $z = x + yi$, then $z = r(\cos\theta + i\sin\theta)$, where $r = \lvert z\rvert = \sqrt{x^2 + y^2}$, $\sin\theta = \frac{y}{r}$, $\cos\theta = \frac{x}{r}$, $0 \le \theta < 2\pi$.
De Moivre's Theorem (p. 339)	If $z = r(\cos\theta + i\sin\theta)$, then $z^n = r^n[\cos(n\theta) + i\sin(n\theta)]$, where $n \ge 1$ is a positive integer.
nth root of a complex number $w = r(\cos\theta_0 + i\sin\theta_0)$ (p. 340)	$z_k = \sqrt[n]{r}\left[\cos\left(\frac{\theta_0}{n} + \frac{2k\pi}{n}\right) + i\sin\left(\frac{\theta_0}{n} + \frac{2k\pi}{n}\right)\right],\quad k = 0, \dots, n-1,$ where $n \ge 2$ is an integer
Vectors (pp. 344–353)	Quantity having magnitude and direction; equivalent to a directed line segment $\overrightarrow{PQ}$
Position vector (pp. 347 and 367)	Vector whose initial point is at the origin
Unit vector (pp. 346 and 368)	Vector whose magnitude is 1
Dot product (pp. 358 and 369)	If $\mathbf{v} = a_1\mathbf{i} + b_1\mathbf{j}$ and $\mathbf{w} = a_2\mathbf{i} + b_2\mathbf{j}$, then $\mathbf{v}\cdot\mathbf{w} = a_1a_2 + b_1b_2$. If $\mathbf{v} = a_1\mathbf{i} + b_1\mathbf{j} + c_1\mathbf{k}$ and $\mathbf{w} = a_2\mathbf{i} + b_2\mathbf{j} + c_2\mathbf{k}$, then $\mathbf{v}\cdot\mathbf{w} = a_1a_2 + b_1b_2 + c_1c_2$.
Angle θ between two nonzero vectors $\mathbf{u}$ and $\mathbf{v}$ (pp. 359 and 370)	$\cos\theta = \frac{\mathbf{u}\cdot\mathbf{v}}{\lVert\mathbf{u}\rVert\,\lVert\mathbf{v}\rVert}$
Direction angles of a vector in space (p. 371)	If $\mathbf{v} = a\mathbf{i} + b\mathbf{j} + c\mathbf{k}$, then $\mathbf{v} = \lVert\mathbf{v}\rVert[(\cos\alpha)\mathbf{i} + (\cos\beta)\mathbf{j} + (\cos\gamma)\mathbf{k}]$, where $\cos\alpha = \frac{a}{\lVert\mathbf{v}\rVert}$, $\cos\beta = \frac{b}{\lVert\mathbf{v}\rVert}$, and $\cos\gamma = \frac{c}{\lVert\mathbf{v}\rVert}$.
Cross product (p. 375)	If $\mathbf{v} = a_1\mathbf{i} + b_1\mathbf{j} + c_1\mathbf{k}$ and $\mathbf{w} = a_2\mathbf{i} + b_2\mathbf{j} + c_2\mathbf{k}$, then $\mathbf{v}\times\mathbf{w} = [b_1c_2 - b_2c_1]\mathbf{i} - [a_1c_2 - a_2c_1]\mathbf{j} + [a_1b_2 - a_2b_1]\mathbf{k}$.
Area of a parallelogram (p. 377)	$\lVert\mathbf{u}\times\mathbf{v}\rVert = \lVert\mathbf{u}\rVert\,\lVert\mathbf{v}\rVert\sin\theta$, where θ is the angle between the two adjacent sides $\mathbf{u}$ and $\mathbf{v}$.

Objectives

Section		You should be able to...	Example(s)	Review Exercises
5.1	1	Plot points using polar coordinates (p. 312)	1–3	1–3
	2	Convert from polar coordinates to rectangular coordinates (p. 314)	4	1–3
	3	Convert from rectangular coordinates to polar coordinates (p. 316)	5–7	4–6
	4	Transform equations between polar and rectangular forms (p. 318)	8, 9	7(a)–10(a)
5.2	1	Identify and graph polar equations by converting to rectangular equations (p. 322)	1–6	7(b)–10(b)
	2	Test polar equations for symmetry (p. 325)	7–10	11–13
	3	Graph polar equations by plotting points (p. 326)	7–13	11–13
5.3	1	Plot points in the complex plane (p. 336)	1	16–18
	2	Convert a complex number between rectangular form and polar form (p. 337)	2, 3	14–18
	3	Find products and quotients of complex numbers in polar form (p. 338)	4	19–21
	4	Use De Moivre's Theorem (p. 339)	5, 6	22–25
	5	Find complex roots (p. 340)	7	26
5.4	1	Graph vectors (p. 346)	1	27, 28
	2	Find a position vector (p. 346)	2	29, 30
	3	Add and subtract vectors algebraically (p. 348)	3	31
	4	Find a scalar multiple and the magnitude of a vector (p. 349)	4	29, 30, 32–34
	5	Find a unit vector (p. 349)	5	35
	6	Find a vector from its direction and magnitude (p. 350)	6	36, 37
	7	Model with vectors (p. 351)	8–10	59, 60
5.5	1	Find the dot product of two vectors (p. 358)	1	46, 47
	2	Find the angle between two vectors (p. 359)	2	46, 47
	3	Determine whether two vectors are parallel (p. 360)	3	50–52
	4	Determine whether two vectors are orthogonal (p. 360)	4	50–52
	5	Decompose a vector into two orthogonal vectors (p. 360)	5, 6	53, 54, 62
	6	Compute work (p. 362)	7	61
5.6	1	Find the distance between two points in space (p. 366)	1	38
	2	Find position vectors in space (p. 367)	2	39
	3	Perform operations on vectors (p. 367)	3–5	40–42
	4	Find the dot product (p. 369)	6	48, 49
	5	Find the angle between two vectors (p. 369)	7	48, 49
	6	Find the direction angles of a vector (p. 370)	8–10	55
5.7	1	Find the cross product of two vectors (p. 375)	1–3	43, 44
	2	Know algebraic properties of the cross product (p. 376)	p. 377	57, 58
	3	Know geometric properties of the cross product (p. 377)	p. 377	56
	4	Find a vector orthogonal to two given vectors (p. 378)	4	45
	5	Find the area of a parallelogram (p. 378)	5	56

Review Exercises

In Problems 1–3, plot each point given in polar coordinates, and find its rectangular coordinates.

1. $\left(3, \frac{\pi}{6}\right)$ **2.** $\left(-2, \frac{4\pi}{3}\right)$ **3.** $\left(-3, -\frac{\pi}{2}\right)$

In Problems 4–6, the rectangular coordinates of a point are given. Find two pairs of polar coordinates (r, θ) for each point, one with $r > 0$ and the other with $r < 0$. Express θ in radians.

4. $(-3, 3)$ **5.** $(0, -2)$ **6.** $(3, 4)$

In Problems 7–10, the variables r and θ represent polar coordinates. (a) Write each polar equation as an equation in rectangular coordinates (x, y). (b) Identify the equation and graph it.

7. $r = 2 \sin \theta$ **8.** $r = 5$ **9.** $\theta = \dfrac{\pi}{4}$ **10.** $r^2 + 4r \sin \theta - 8r \cos \theta = 5$

In Problems 11–13, sketch the graph of each polar equation. Be sure to test for symmetry.

11. $r = 4 \cos \theta$ **12.** $r = 3 - 3 \sin \theta$ **13.** $r = 4 - \cos \theta$

In Problems 14 and 15, write each complex number in polar form. Express each argument in degrees.

14. $-1 - i$ **15.** $4 - 3i$

In Problems 16–18, write each complex number in the standard form $a + bi$, and plot each in the complex plane.

16. $2(\cos 150° + i \sin 150°)$ **17.** $3\left(\cos \dfrac{2\pi}{3} + i \sin \dfrac{2\pi}{3}\right)$ **18.** $0.1(\cos 350° + i \sin 350°)$

In Problems 19–21, find zw and $\dfrac{z}{w}$. Leave your answers in polar form.

19. $z = \cos 80° + i \sin 80°$
$w = \cos 50° + i \sin 50°$

20. $z = 3\left(\cos \dfrac{9\pi}{5} + i \sin \dfrac{9\pi}{5}\right)$
$w = 2\left(\cos \dfrac{\pi}{5} + i \sin \dfrac{\pi}{5}\right)$

21. $z = 5(\cos 10° + i \sin 10°)$
$w = \cos 355° + i \sin 355°$

In Problems 22–25, write each expression in the standard form $a + bi$.

22. $[3(\cos 20° + i \sin 20°)]^3$ **23.** $\left[\sqrt{2}\left(\cos \dfrac{5\pi}{8} + i \sin \dfrac{5\pi}{8}\right)\right]^4$ **24.** $(1 - \sqrt{3}i)^6$

25. $(3 + 4i)^4$ **26.** Find all the complex cube roots of 27.

In Problems 27 and 28, use the figure to the right to graph each of the following:

27. $\mathbf{u} + \mathbf{v}$ **28.** $2\mathbf{u} + 3\mathbf{v}$

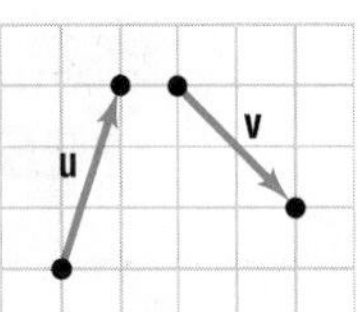

In Problems 29 and 30, the vector **v** *is represented by the directed line segment $\overrightarrow{PQ}$. Write* **v** *in the form $a\mathbf{i} + b\mathbf{j}$ and find $\|\mathbf{v}\|$.*

29. $P = (1, -2);\ Q = (3, -6)$ **30.** $P = (0, -2);\ Q = (-1, 1)$

In Problems 31–35, use the vectors $\mathbf{v} = -2\mathbf{i} + \mathbf{j}$ and $\mathbf{w} = 4\mathbf{i} - 3\mathbf{j}$ to find:

31. $\mathbf{v} + \mathbf{w}$ **32.** $4\mathbf{v} - 3\mathbf{w}$ **33.** $\|\mathbf{v}\|$

34. $\|\mathbf{v}\| + \|\mathbf{w}\|$ **35.** Find a unit vector in the same direction as **v**.

36. Find the vector **v** in the xy-plane with magnitude 3 if the direction angle of **v** is 60°.

37. Find the direction angle between **i** and $\mathbf{v} = -\mathbf{i} + \sqrt{3}\,\mathbf{j}$.

38. Find the distance from $P_1 = (1, 3, -2)$ to $P_2 = (4, -2, 1)$.

39. A vector **v** has initial point $P = (1, 3, -2)$ and terminal point $Q = (4, -2, 1)$. Write **v** in the form $\mathbf{v} = a\mathbf{i} + b\mathbf{j} + c\mathbf{k}$.

In Problems 40–45, use the vectors $\mathbf{v} = 3\mathbf{i} + \mathbf{j} - 2\mathbf{k}$ and $\mathbf{w} = -3\mathbf{i} + 2\mathbf{j} - \mathbf{k}$ to find each expression.

40. $4\mathbf{v} - 3\mathbf{w}$ **41.** $\|\mathbf{v} - \mathbf{w}\|$ **42.** $\|\mathbf{v}\| - \|\mathbf{w}\|$

43. $\mathbf{v} \times \mathbf{w}$ **44.** $\mathbf{v} \cdot (\mathbf{v} \times \mathbf{w})$ **45.** Find a unit vector orthogonal to both **v** and **w**.

In Problems 46–49, find the dot product $\mathbf{v} \cdot \mathbf{w}$ and the angle between **v** *and* **w**.

46. $\mathbf{v} = -2\mathbf{i} + \mathbf{j},\ \mathbf{w} = 4\mathbf{i} - 3\mathbf{j}$ **47.** $\mathbf{v} = \mathbf{i} - 3\mathbf{j},\ \mathbf{w} = -\mathbf{i} + \mathbf{j}$

48. $\mathbf{v} = \mathbf{i} + \mathbf{j} + \mathbf{k},\ \mathbf{w} = \mathbf{i} - \mathbf{j} + \mathbf{k}$ **49.** $\mathbf{v} = 4\mathbf{i} - \mathbf{j} + 2\mathbf{k},\ \mathbf{w} = \mathbf{i} - 2\mathbf{j} - 3\mathbf{k}$

In Problems 50–52, determine whether **v** *and* **w** *are parallel, orthogonal, or neither.*

50. $\mathbf{v} = 2\mathbf{i} + 3\mathbf{j};\ \mathbf{w} = -4\mathbf{i} - 6\mathbf{j}$ **51.** $\mathbf{v} = -2\mathbf{i} + 2\mathbf{j};\ \mathbf{w} = -3\mathbf{i} + 2\mathbf{j}$ **52.** $\mathbf{v} = 3\mathbf{i} - 2\mathbf{j};\ \mathbf{w} = 4\mathbf{i} + 6\mathbf{j}$

In Problems 53 and 54, decompose **v** *into two vectors, one parallel to* **w** *and the other orthogonal to* **w**.

53. $\mathbf{v} = 2\mathbf{i} + \mathbf{j};\ \mathbf{w} = -4\mathbf{i} + 3\mathbf{j}$ **54.** $\mathbf{v} = 2\mathbf{i} + 3\mathbf{j};\ \mathbf{w} = 3\mathbf{i} + \mathbf{j}$

55. Find the direction angles of the vector $\mathbf{v} = 3\mathbf{i} - 4\mathbf{j} + 2\mathbf{k}$.

56. Find the area of the parallelogram with vertices $P_1 = (1, 1, 1)$, $P_2 = (2, 3, 4)$, $P_3 = (6, 5, 2)$, and $P_4 = (7, 7, 5)$.

57. If $\mathbf{u} \times \mathbf{v} = 2\mathbf{i} - 3\mathbf{j} + \mathbf{k}$, what is $\mathbf{v} \times \mathbf{u}$?

58. Suppose that $\mathbf{u} = 3\mathbf{v}$. What is $\mathbf{u} \times \mathbf{v}$?

59. Actual Speed and Direction of a Swimmer A swimmer can maintain a constant speed of 5 miles per hour. If the swimmer heads directly across a river that has a current moving at the rate of 2 miles per hour, what is the actual speed of the swimmer? (See the figure.) If the river is 1 mile wide, how far downstream will the swimmer end up from the point directly across the river from the starting point?

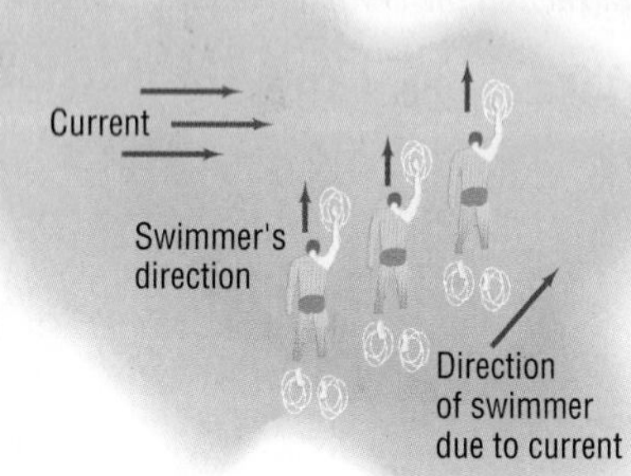

60. Static Equilibrium A weight of 2000 pounds is suspended from two cables, as shown in the figure. What are the tensions in the two cables?

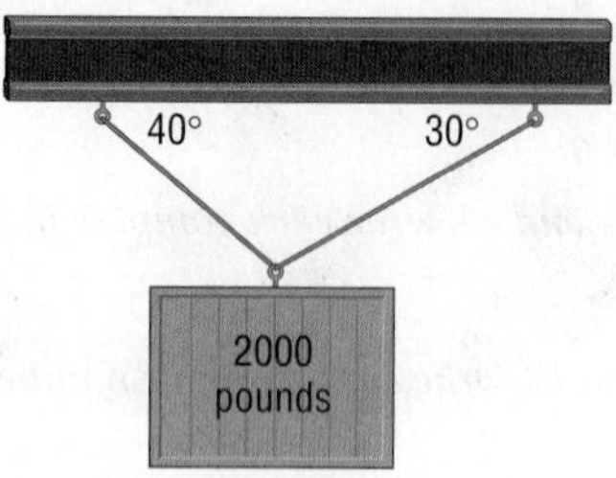

61. Computing Work Find the work done by a force of 5 pounds acting in the direction 60° to the horizontal in moving an object 20 feet from $(0, 0)$ to $(20, 0)$.

62. Braking Load A moving van with a gross weight of 8000 pounds is parked on a street with a 5° grade. Find the magnitude of the force required to keep the van from rolling down the hill. What is the magnitude of the force perpendicular to the hill?

Chapter Test

CHAPTER **Test Prep** VIDEOS — The Chapter Test Prep Videos are step-by-step solutions available in MyMathLab®, or on this text's YouTube Channel. Flip back to the Resources for Success page for a link to this text's YouTube channel.

In Problems 1–3, plot each point given in polar coordinates.

1. $\left(2, \frac{3\pi}{4}\right)$ **2.** $\left(3, -\frac{\pi}{6}\right)$ **3.** $\left(-4, \frac{\pi}{3}\right)$

4. Convert $(2, 2\sqrt{3})$ from rectangular coordinates to polar coordinates (r, θ), where $r > 0$ and $0 \le \theta < 2\pi$.

In Problems 5–7, convert the polar equation to a rectangular equation. Graph the equation.

5. $r = 7$ **6.** $\tan\theta = 3$ **7.** $r\sin^2\theta + 8\sin\theta = r$

In Problems 8 and 9, test the polar equation for symmetry with respect to the pole, the polar axis, and the line $\theta = \frac{\pi}{2}$.

8. $r^2\cos\theta = 5$ **9.** $r = 5\sin\theta\cos^2\theta$

In Problems 10–12, perform the given operation, where $z = 2(\cos 85° + i\sin 85°)$ *and* $w = 3(\cos 22° + i\sin 22°)$. *Write your answer in polar form.*

10. $z \cdot w$ **11.** $\frac{w}{z}$ **12.** w^5

13. Find all the complex cube roots of $-8 + 8\sqrt{3}i$. Then plot them in the complex plane.

In Problems 14–18, $P_1 = (3\sqrt{2}, 7\sqrt{2})$ *and* $P_2 = (8\sqrt{2}, 2\sqrt{2})$.

14. Find the position vector **v** equal to $\overrightarrow{P_1P_2}$.

15. Find $\|\mathbf{v}\|$.

16. Find the unit vector in the direction of **v**.

17. Find the direction angle of **v**.

18. Decompose **v** into its vertical and horizontal components.

In Problems 19–22, $\mathbf{v}_1 = \langle 4, 6\rangle$, $\mathbf{v}_2 = \langle -3, -6\rangle$, $\mathbf{v}_3 = \langle -8, 4\rangle$, *and* $\mathbf{v}_4 = \langle 10, 15\rangle$.

19. Find the vector $\mathbf{v}_1 + 2\mathbf{v}_2 - \mathbf{v}_3$.

20. Which two vectors are parallel?

21. Which two vectors are orthogonal?

22. Find the angle between the vectors $\mathbf{v}_1$ and $\mathbf{v}_2$.

In Problems 23–25, use the vectors $\mathbf{u} = 2\mathbf{i} - 3\mathbf{j} + \mathbf{k}$ *and* $\mathbf{v} = -\mathbf{i} + 3\mathbf{j} + 2\mathbf{k}$.

23. Find $\mathbf{u} \times \mathbf{v}$.

24. Find the direction angles for **u**.

25. Find the area of the parallelogram that has **u** and **v** as adjacent sides.

26. A 1200-pound chandelier is to be suspended over a large ballroom; the chandelier will be hung on two cables of equal length whose ends will be attached to the ceiling, 16 feet apart. The chandelier will be free-hanging so that the ends of the cable will make equal angles with the ceiling. If the top of the chandelier is to be 16 feet from the ceiling, what is the minimum tension each cable must be able to endure?

Cumulative Review

1. Find the real solutions, if any, of the equation $x^2 - 9 = 0$.
2. Find an equation for the line containing the origin that makes an angle of 30° with the positive x-axis.
3. Find an equation for the circle with center at the point $(0, 1)$ and radius 3. Graph this circle.
4. What is the domain of the function $f(x) = \dfrac{1}{1 - 2x}$?
5. Test the equation $x^2 + y^3 = 2x^4$ for symmetry with respect to the x-axis, the y-axis, and the origin.
6. Graph the function $y = |\sin x|$.
7. Graph the function $y = \sin|x|$.
8. Find the exact value of $\sin^{-1}\left(-\dfrac{1}{2}\right)$.
9. Graph the equations $x = 3$ and $y = 4$ on the same set of rectangular axes.
10. Graph the equations $r = 2$ and $\theta = \dfrac{\pi}{3}$ on the same set of polar axes.
11. What are the amplitude and period of $y = -4\cos(\pi x)$?

Chapter Projects

I. Modeling Aircraft Motion Four aerodynamic forces act on an airplane in flight: lift, weight, thrust, and drag. While an aircraft is in flight, these four forces continuously battle each other. Weight opposes lift, and drag opposes thrust. See the diagram below. In balanced flight at constant speed, the lift and weight are equal, and the thrust and drag are equal.

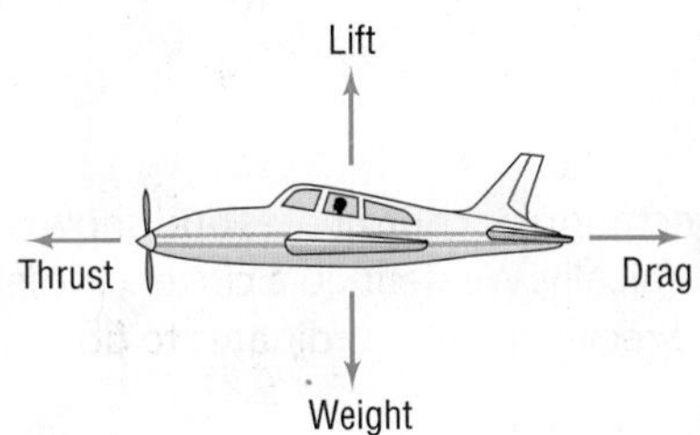

Source: www.aeromuseum.org

1. What will happen to the aircraft if the lift is held constant while the weight is decreased (say, from burning off fuel)?
2. What will happen to the aircraft if the lift is decreased while the weight is held constant?
3. What will happen to the aircraft if the thrust is increased while the drag is held constant?
4. What will happen to the aircraft if the drag is increased while the thrust is held constant?

In 1903 the Wright brothers made the first controlled powered flight. The weight of their plane was approximately 700 pounds (lb). Newton's Second Law of Motion states that force = mass × acceleration ($F = ma$). If the mass is measured in kilograms (kg) and acceleration in meters per second squared (m/sec^2), then the force will be measured in newtons (N). [**Note:** $1\text{ N} = 1\text{ kg}\cdot\text{m/sec}^2$.]

5. If 1 kg = 2.205 lb, convert the weight of the Wright brothers' plane to kilograms.
6. If acceleration due to gravity is $a = 9.80\text{ m/sec}^2$, determine the force due to weight on the Wright brothers' plane.
7. What must be true about the lift force of the Wright brothers' plane for it to get off the ground?
8. The weight of a fully loaded Cessna 170B is 2200 lb. What lift force is required to get this plane off the ground?
9. The maximum gross weight of a Boeing 747 is 255,000 lb. What lift force is required to get this jet off the ground?

The following projects are available at the Instructors' Resource Center (IRC):

II. Project at Motorola ***Signal Fades Due to Interference*** Complex trigonometric functions are used to ensure that a cellphone has optimal reception as the user travels up and down an elevator.

III. Compound Interest The effect of continuously compounded interest is analyzed using polar coordinates.

IV. Complex Equations Analysis of complex equations illustrates the connections between complex and real equations. At times, using complex equations is more efficient for proving mathematical theorems.

6 Analytic Geometry

The Orbit of Comet Hale-Bopp

The orbits of Comet Hale-Bopp and Earth can be modeled using *ellipses,* the subject of Section 6.3. The Internet-based Project at the end of this chapter explores the possibility of Comet Hale-Bopp colliding with Earth.

—See the Internet-based Chapter Project I—

Outline

6.1 Conics
6.2 The Parabola
6.3 The Ellipse
6.4 The Hyperbola
6.5 Rotation of Axes; General Form of a Conic
6.6 Polar Equations of Conics
6.7 Plane Curves and Parametric Equations
Chapter Review
Chapter Test
Cumulative Review
Chapter Projects

••• A Look Back

In Chapter 1, we introduced rectangular coordinates and showed how geometry problems can be solved algebraically. We defined a circle geometrically and then used the distance formula and rectangular coordinates to obtain an equation for a circle.

A Look Ahead •••

In this chapter, geometric definitions are given for the *conics,* and the distance formula and rectangular coordinates are used to obtain their equations.

Historically, Apollonius (200 BC) was among the first to study conics and discover some of their interesting properties. Today, conics are still studied because of their many uses. *Paraboloids of revolution* (parabolas rotated about their axes of symmetry) are used as signal collectors (the satellite dishes used with radar and dish TV, for example), as solar energy collectors, and as reflectors (telescopes, light projection, and so on). The planets circle the Sun in approximately *elliptical* orbits. Elliptical surfaces can be used to reflect signals such as light and sound from one place to another. A third conic, the *hyperbola,* can be used to determine the location of lightning strikes.

The Greeks used Euclidean geometry to study conics. However, we shall use the more powerful methods of analytic geometry, which uses both algebra and geometry, for our study of conics.

6.1 Conics

OBJECTIVE 1 Know the Names of the Conics (p. 387)

1 Know the Names of the Conics

The word *conic* derives from the word *cone,* which is a geometric figure that can be constructed in the following way: Let a and g be two distinct lines that intersect at a point V. Keep the line a fixed. Now rotate the line g about a, while maintaining the same angle between a and g. The collection of points swept out (generated) by the line g is called a **right circular cone**. See Figure 1. The fixed line a is called the **axis** of the cone; the point V is its **vertex**; the lines that pass through V and make the same angle with a as g are **generators** of the cone. Each generator is a line that lies entirely on the cone. The cone consists of two parts, called **nappes**, that intersect at the vertex.

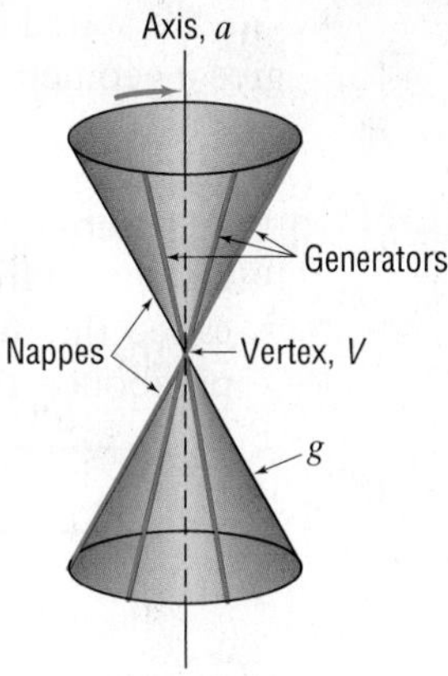

Figure 1 Right circular cone

Conics, an abbreviation for **conic sections**, are curves that result from the intersection of a right circular cone and a plane. The conics we shall study arise when the plane does not contain the vertex, as shown in Figure 2. These conics are **circles** when the plane is perpendicular to the axis of the cone and intersects each generator; **ellipses** when the plane is tilted slightly so that it intersects each generator, but intersects only one nappe of the cone; **parabolas** when the plane is tilted farther so that it is parallel to one (and only one) generator and intersects only one nappe of the cone; and **hyperbolas** when the plane intersects both nappes.

If the plane contains the vertex, the intersection of the plane and the cone is a point, a line, or a pair of intersecting lines. These are usually called **degenerate conics.**

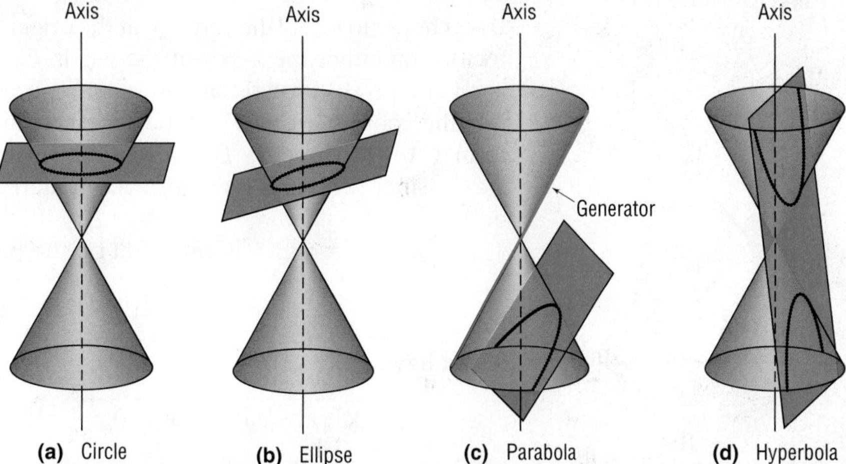

Figure 2 **(a)** Circle **(b)** Ellipse **(c)** Parabola **(d)** Hyperbola

Conic sections are used in modeling many different applications. For example, parabolas are used in describing searchlights and telescopes (see Figures 14 and 15 on page 393). Ellipses are used to model the orbits of planets and whispering galleries (see pages 403–404). And hyperbolas are used to locate lightning strikes and model nuclear cooling towers (see Problems 76 and 77 in Section 6.4).

6.2 The Parabola

PREPARING FOR THIS SECTION *Before getting started, review the following:*

- Distance Formula (Section 1.1, p. 3)
- Symmetry (Section 1.2, pp. 12–13)
- Square Root Method (Appendix A, Section A.4, p. A31)
- Complete the Square (Appendix A, Section A.3, pp. A24–A25)
- Graphing Techniques: Transformations (Section 1.6, pp. 64–73)

Now Work the 'Are You Prepared?' problems on page 394.

OBJECTIVES
1. Analyze Parabolas with Vertex at the Origin (p. 388)
2. Analyze Parabolas with Vertex at (h, k) (p. 391)
3. Solve Applied Problems Involving Parabolas (p. 393)

We stated earlier that the graph of $f(x) = x^2$ is a parabola. In this section, we give a geometric definition of a parabola and use it to obtain an equation.

DEFINITION

A **parabola** is the collection of all points P in the plane that are the same distance d from a fixed point F as they are from a fixed line D. The point F is called the **focus** of the parabola, and the line D is its **directrix**. As a result, a parabola is the set of points P for which

$$d(F, P) = d(P, D) \qquad \textbf{(1)}$$

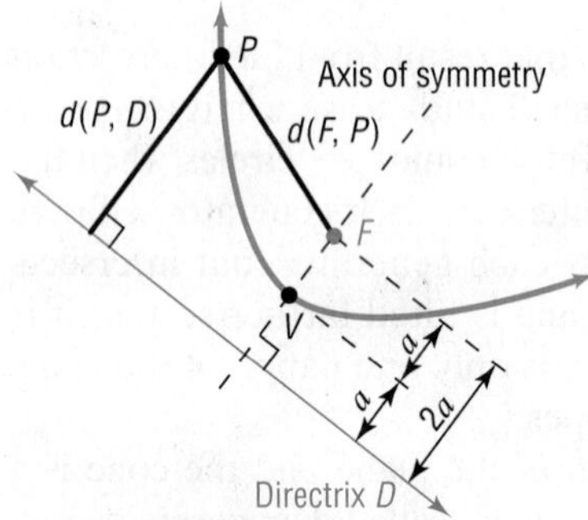

Figure 3 Parabola

Figure 3 shows a parabola (in blue). The line through the focus F and perpendicular to the directrix D is called the **axis of symmetry** of the parabola. The point of intersection of the parabola with its axis of symmetry is called the **vertex** V.

Because the vertex V lies on the parabola, it must satisfy equation (1): $d(F, V) = d(V, D)$. The vertex is midway between the focus and the directrix. We shall let a equal the distance $d(F, V)$ from F to V. Now we are ready to derive an equation for a parabola. To do this, we use a rectangular system of coordinates positioned so that the vertex V, focus F, and directrix D of the parabola are conveniently located.

1 Analyze Parabolas with Vertex at the Origin

If we choose to locate the vertex V at the origin $(0, 0)$, we can conveniently position the focus F on either the x-axis or the y-axis. First, consider the case where the focus F is on the positive x-axis, as shown in Figure 4. Because the distance from F to V is a, the coordinates of F will be $(a, 0)$ with $a > 0$. Similarly, because the distance from V to the directrix D is also a, and because D must be perpendicular to the x-axis (since the x-axis is the axis of symmetry), the equation of the directrix D must be $x = -a$.

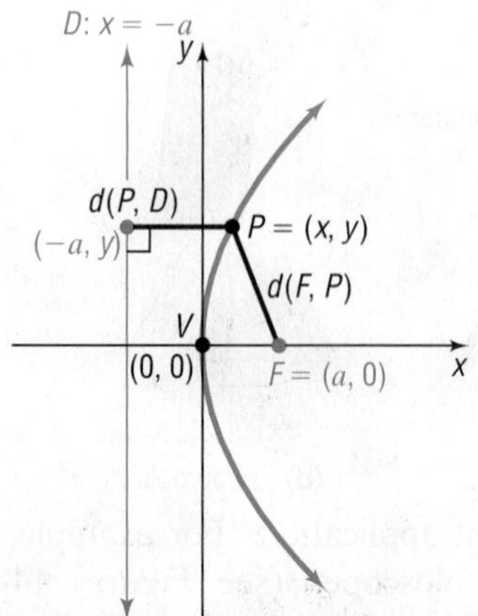

Figure 4

Now, if $P = (x, y)$ is any point on the parabola, P must satisfy equation (1):

$$d(F, P) = d(P, D)$$

So we have

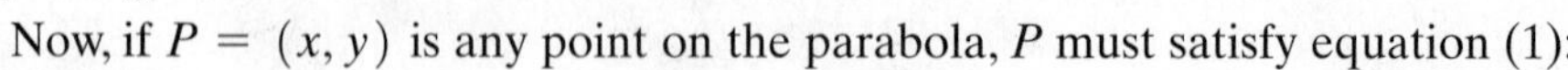

$$\sqrt{(x-a)^2 + (y-0)^2} = |x + a| \qquad \text{Use the Distance Formula.}$$

$$(x-a)^2 + y^2 = (x+a)^2 \qquad \text{Square both sides.}$$

$$x^2 - 2ax + a^2 + y^2 = x^2 + 2ax + a^2 \qquad \text{Multiply out.}$$

$$y^2 = 4ax \qquad \text{Simplify.}$$

THEOREM

Equation of a Parabola: Vertex at (0, 0), Focus at (a, 0), a > 0

The equation of a parabola with vertex at $(0, 0)$, focus at $(a, 0)$, and directrix $x = -a$, $a > 0$, is

$$y^2 = 4ax \quad (2)$$

Recall that a is the distance from the vertex to the focus of a parabola. When graphing the parabola $y^2 = 4ax$ it is helpful to determine the "opening" by finding the points that lie directly above or below the focus $(a, 0)$. This is done by letting $x = a$ in $y^2 = 4ax$, so $y^2 = 4a(a) = 4a^2$, or $y = \pm 2a$. The line segment joining the two points, $(a, 2a)$ and $(a, -2a)$, is called the **latus rectum**; its length is $4a$.

EXAMPLE 1 Finding the Equation of a Parabola and Graphing It

Find an equation of the parabola with vertex at $(0, 0)$ and focus at $(3, 0)$. Graph the equation.

Solution The distance from the vertex $(0, 0)$ to the focus $(3, 0)$ is $a = 3$. Based on equation (2), the equation of this parabola is

$$y^2 = 4ax$$
$$y^2 = 12x \qquad a = 3$$

To graph this parabola, find the two points that determine the latus rectum by letting $x = 3$. Then

$$y^2 = 12x = 12(3) = 36$$
$$y = \pm 6 \qquad \text{Solve for } y.$$

The points $(3, 6)$ and $(3, -6)$ determine the latus rectum. These points help graph the parabola because they determine the "opening." See Figure 5. ●

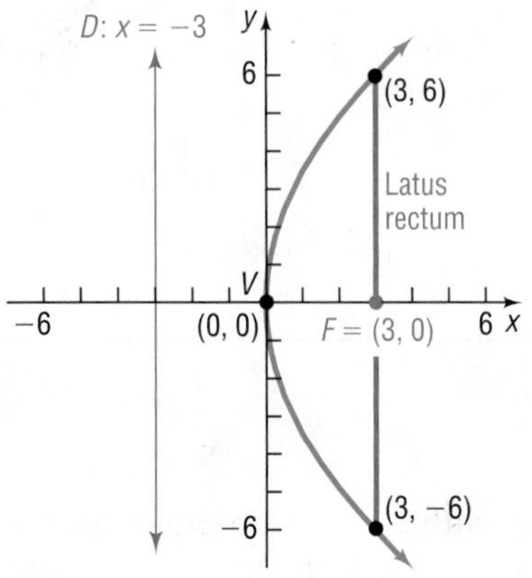

D: x = −3, y, 6, (3, 6), Latus rectum, V, −6, (0, 0), F = (3, 0), 6 x, (3, −6), −6

Figure 5 $y^2 = 12x$

Now Work PROBLEM 21

COMMENT To graph the parabola $y^2 = 12x$ discussed in Example 1, graph the two functions $Y_1 = \sqrt{12x}$ and $Y_2 = -\sqrt{12x}$. Do this and compare what you see with Figure 5. ■

By reversing the steps used to obtain equation (2), it follows that the graph of an equation of the form of equation (2), $y^2 = 4ax$, is a parabola; its vertex is at $(0, 0)$, its focus is at $(a, 0)$, its directrix is the line $x = -a$, and its axis of symmetry is the x-axis.

For the remainder of this section, the direction **"Analyze the equation"** will mean to find the vertex, focus, and directrix of the parabola and graph it.

EXAMPLE 2 Analyzing the Equation of a Parabola

Analyze the equation: $y^2 = 8x$

Solution The equation $y^2 = 8x$ is of the form $y^2 = 4ax$, where $4a = 8$, so $a = 2$. Consequently, the graph of the equation is a parabola with vertex at $(0, 0)$ and focus on the positive x-axis at $(a, 0) = (2, 0)$. The directrix is the vertical line $x = -2$. The two points that determine the latus rectum are obtained by letting $x = 2$. Then $y^2 = 16$, so $y = \pm 4$. The points $(2, -4)$ and $(2, 4)$ determine the latus rectum. See Figure 6 for the graph. ●

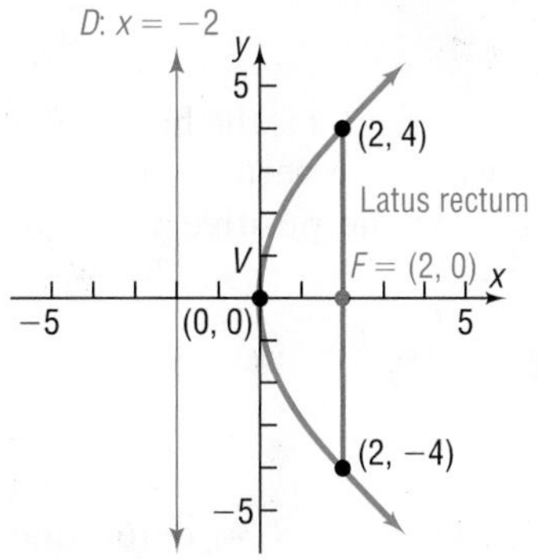

D: x = −2, y, 5, (2, 4), Latus rectum, V, F = (2, 0), x, −5, (0, 0), 5, (2, −4), −5

Figure 6 $y^2 = 8x$

Recall that we obtained equation (2) after placing the focus on the positive x-axis. If the focus is placed on the negative x-axis, positive y-axis, or negative y-axis, a different form of the equation for the parabola results. The four forms of the equation of a parabola with vertex at $(0, 0)$ and focus on a coordinate axis a distance a from $(0, 0)$ are given in Table 1, and their graphs are given in Figure 7. Notice that each graph is symmetric with respect to its axis of symmetry.

Table 1

Equations of a Parabola: Vertex at (0, 0); Focus on an Axis; $a > 0$

Vertex	Focus	Directrix	Equation	Description
(0, 0)	$(a, 0)$	$x = -a$	$y^2 = 4ax$	Axis of symmetry is the x-axis, opens right
(0, 0)	$(-a, 0)$	$x = a$	$y^2 = -4ax$	Axis of symmetry is the x-axis, opens left
(0, 0)	$(0, a)$	$y = -a$	$x^2 = 4ay$	Axis of symmetry is the y-axis, opens up
(0, 0)	$(0, -a)$	$y = a$	$x^2 = -4ay$	Axis of symmetry is the y-axis, opens down

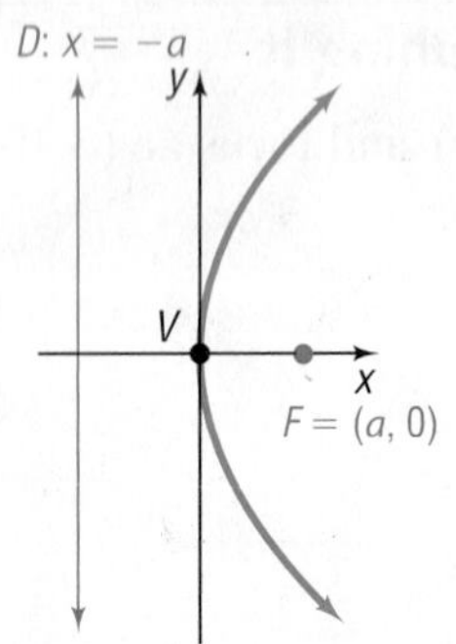

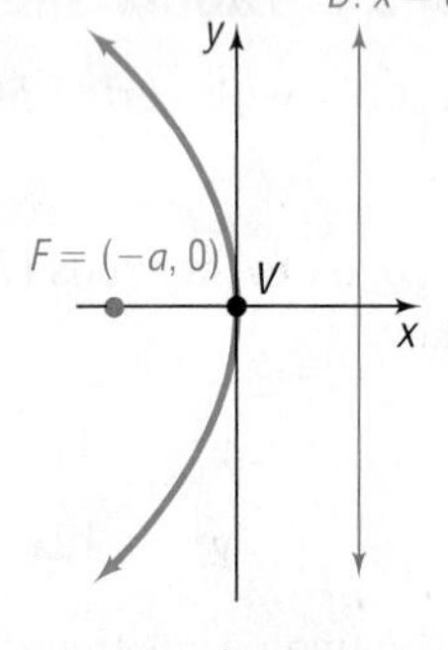

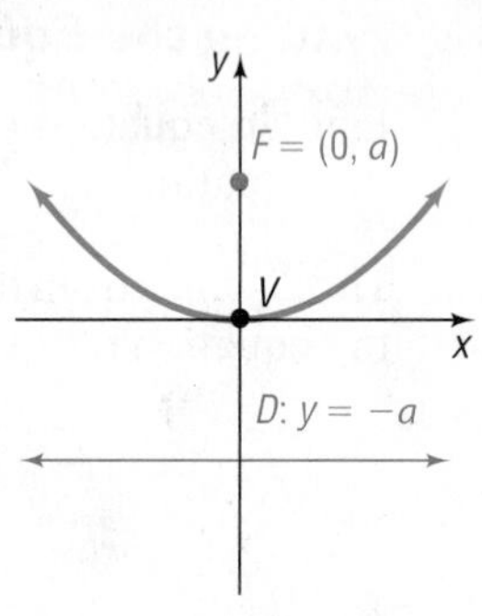

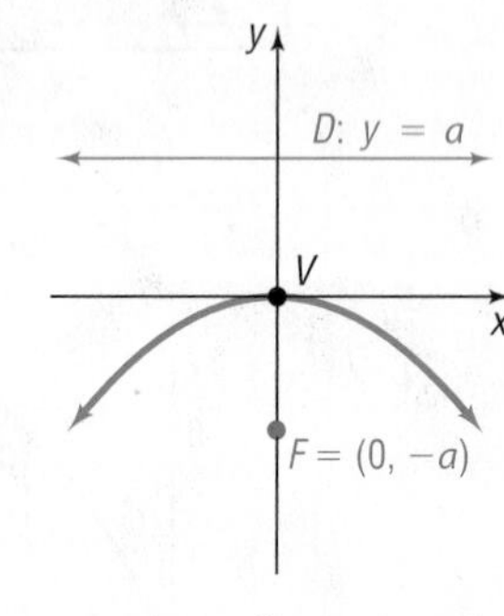

Figure 7 (a) $y^2 = 4ax$ (b) $y^2 = -4ax$ (c) $x^2 = 4ay$ (d) $x^2 = -4ay$

EXAMPLE 3

Analyzing the Equation of a Parabola

Analyze the equation: $x^2 = -12y$

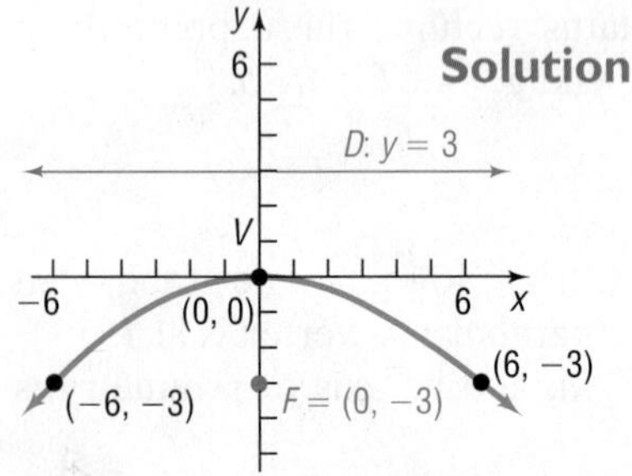

Figure 8 $x^2 = -12y$

Solution The equation $x^2 = -12y$ is of the form $x^2 = -4ay$, with $a = 3$. Consequently, the graph of the equation is a parabola with vertex at $(0, 0)$, focus at $(0, -3)$, and directrix the line $y = 3$. The parabola opens down, and its axis of symmetry is the y-axis. To obtain the points defining the latus rectum, let $y = -3$. Then $x^2 = 36$, so $x = \pm 6$. The points $(-6, -3)$ and $(6, -3)$ determine the latus rectum. See Figure 8 for the graph. ●

Now Work PROBLEM 41

EXAMPLE 4

Finding the Equation of a Parabola

Find the equation of the parabola with focus at $(0, 4)$ and directrix the line $y = -4$. Graph the equation.

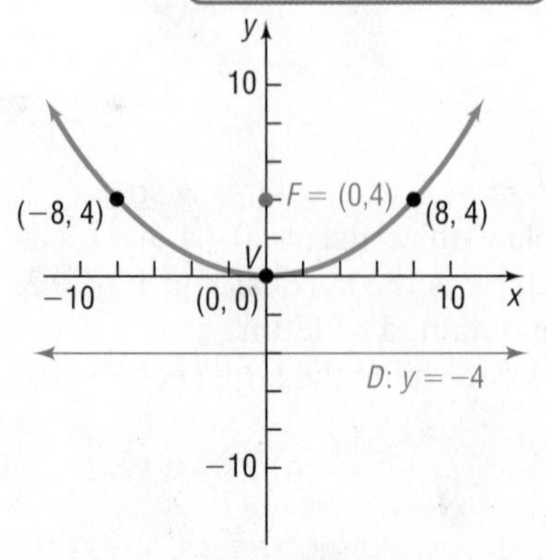

Figure 9 $x^2 = 16y$

Solution A parabola whose focus is at $(0, 4)$ and whose directrix is the horizontal line $y = -4$ will have its vertex at $(0, 0)$. (Do you see why? The vertex is midway between the focus and the directrix.) Since the focus is on the positive y-axis at $(0, 4)$, the equation of this parabola is of the form $x^2 = 4ay$, with $a = 4$. That is,

$$x^2 = 4ay = 4(4)y = 16y$$
$$\uparrow$$
$$a = 4$$

Letting $y = 4$ yields $x^2 = 64$, so $x = \pm 8$. The points $(8, 4)$ and $(-8, 4)$ determine the latus rectum. Figure 9 shows the graph of $x^2 = 16y$. ●

EXAMPLE 5

Finding the Equation of a Parabola

Find the equation of a parabola with vertex at $(0, 0)$ if its axis of symmetry is the x-axis and its graph contains the point $\left(-\frac{1}{2}, 2\right)$. Find its focus and directrix, and graph the equation.

Solution The vertex is at the origin, the axis of symmetry is the x-axis, and the graph contains a point in the second quadrant, so the parabola opens to the left. From Table 1, note that the form of the equation is

$$y^2 = -4ax$$

Because the point $\left(-\frac{1}{2}, 2\right)$ is on the parabola, the coordinates $x = -\frac{1}{2}, y = 2$ must satisfy $y^2 = -4ax$. Substituting $x = -\frac{1}{2}$ and $y = 2$ into this equation leads to

$$4 = -4a\left(-\frac{1}{2}\right) \quad y^2 = -4ax; x = -\frac{1}{2}, y = 2$$

$$a = 2$$

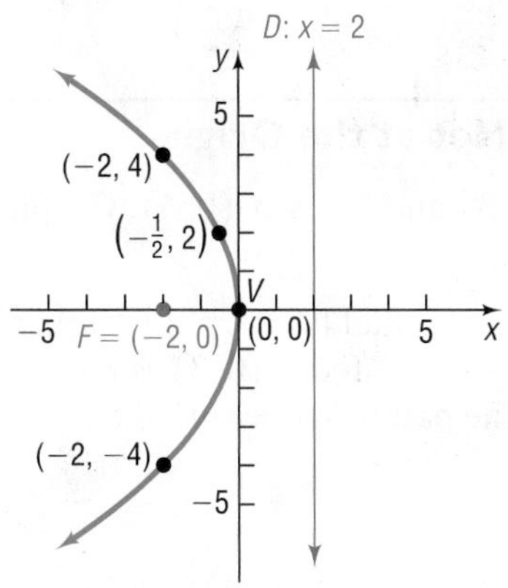

Figure 10 $y^2 = -8x$

The equation of the parabola is

$$y^2 = -4(2)x = -8x$$

The focus is at $(-2, 0)$ and the directrix is the line $x = 2$. Letting $x = -2$ gives $y^2 = 16$, so $y = \pm 4$. The points $(-2, 4)$ and $(-2, -4)$ determine the latus rectum. See Figure 10. ●

Now Work PROBLEM 29

2 Analyze Parabolas with Vertex at (h, k)

If a parabola with vertex at the origin and axis of symmetry along a coordinate axis is shifted horizontally h units and then vertically k units, the result is a parabola with vertex at (h, k) and axis of symmetry parallel to a coordinate axis. The equations of such parabolas have the same forms as those in Table 1, but with x replaced by $x - h$ (the horizontal shift) and y replaced by $y - k$ (the vertical shift). Table 2 gives the forms of the equations of such parabolas. Figures 11(a)–(d) on page 392 illustrate the graphs for $h > 0, k > 0$.

NOTE It is not recommended that Table 2 be memorized. Rather, use transformations (shift horizontally h units, vertically k units), along with the fact that a represents the distance from the vertex to the focus, to determine the various components of a parabola. It is also helpful to remember that parabolas of the form "$x^2 =$ " open up or down, while parabolas of the form "$y^2 =$ " open left or right. ■

Table 2 Equations of a Parabola: Vertex at (h, k); Axis of Symmetry Parallel to a Coordinate Axis; $a > 0$

Vertex	Focus	Directrix	Equation	Description
(h, k)	$(h + a, k)$	$x = h - a$	$(y - k)^2 = 4a(x - h)$	Axis of symmetry is parallel to the x-axis, opens right
(h, k)	$(h - a, k)$	$x = h + a$	$(y - k)^2 = -4a(x - h)$	Axis of symmetry is parallel to the x-axis, opens left
(h, k)	$(h, k + a)$	$y = k - a$	$(x - h)^2 = 4a(y - k)$	Axis of symmetry is parallel to the y-axis, opens up
(h, k)	$(h, k - a)$	$y = k + a$	$(x - h)^2 = -4a(y - k)$	Axis of symmetry is parallel to the y-axis, opens down

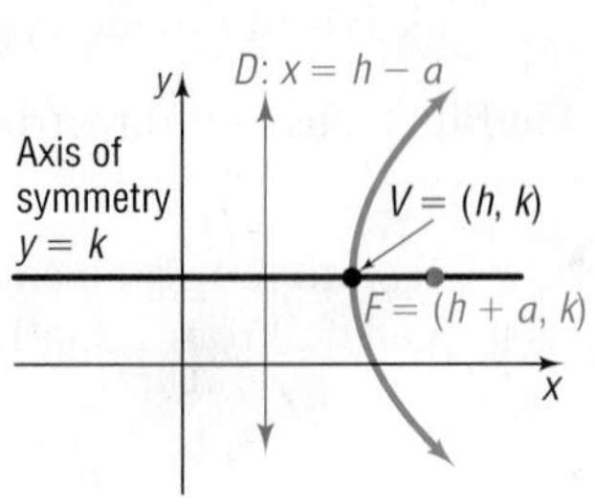

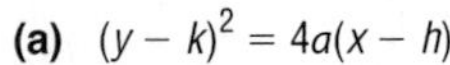
(a) $(y - k)^2 = 4a(x - h)$

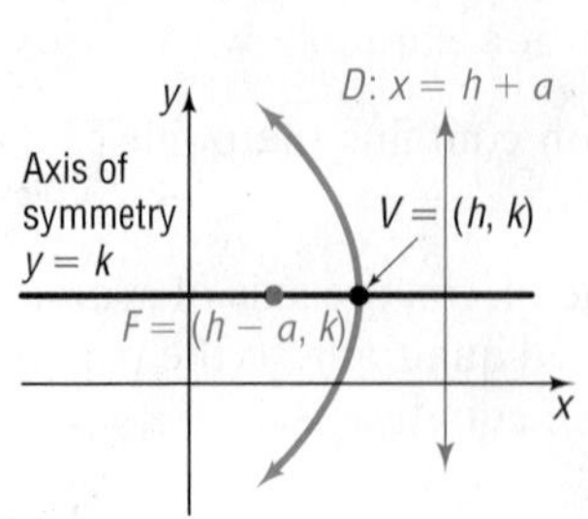

(b) $(y - k)^2 = -4a(x - h)$

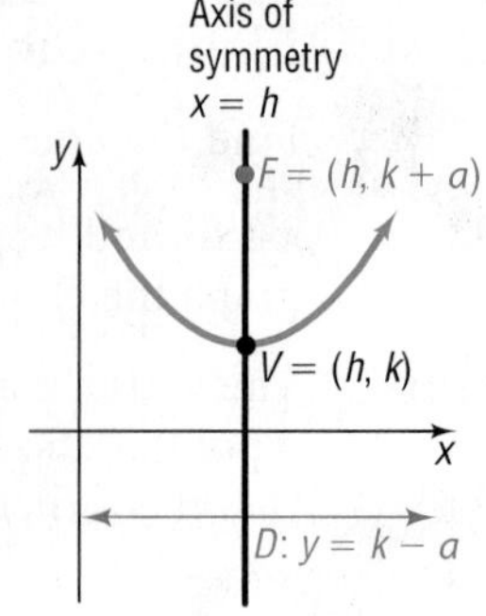

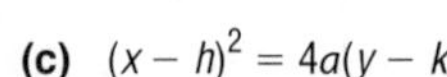
(c) $(x - h)^2 = 4a(y - k)$

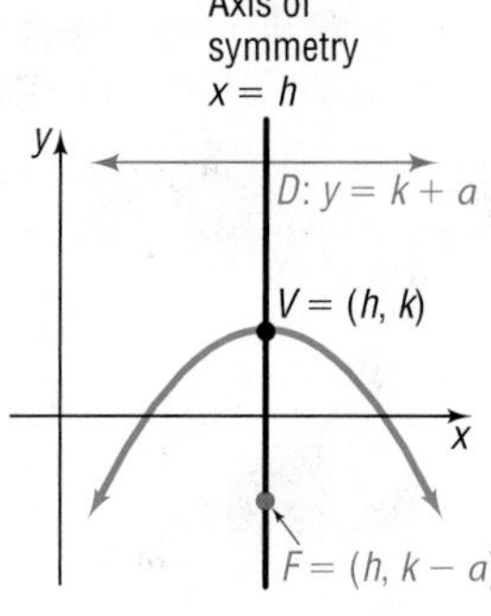

(d) $(x - h)^2 = -4a(y - k)$

Figure 11

EXAMPLE 6 **Finding the Equation of a Parabola, Vertex Not at the Origin**

Find an equation of the parabola with vertex at $(-2, 3)$ and focus at $(0, 3)$. Graph the equation.

Solution The vertex $(-2, 3)$ and focus $(0, 3)$ both lie on the horizontal line $y = 3$ (the axis of symmetry). The distance a from the vertex $(-2, 3)$ to the focus $(0, 3)$ is $a = 2$. Also, because the focus lies to the right of the vertex, the parabola opens to the right. Consequently, the form of the equation is

$$(y - k)^2 = 4a(x - h)$$

where $(h, k) = (-2, 3)$ and $a = 2$. Therefore, the equation is

$$(y - 3)^2 = 4 \cdot 2[x - (-2)]$$
$$(y - 3)^2 = 8(x + 2)$$

To find the points that define the latus rectum, let $x = 0$, so that $(y - 3)^2 = 16$. Then $y - 3 = \pm 4$, so $y = -1$ or $y = 7$. The points $(0, -1)$ and $(0, 7)$ determine the latus rectum; the line $x = -4$ is the directrix. See Figure 12. ●

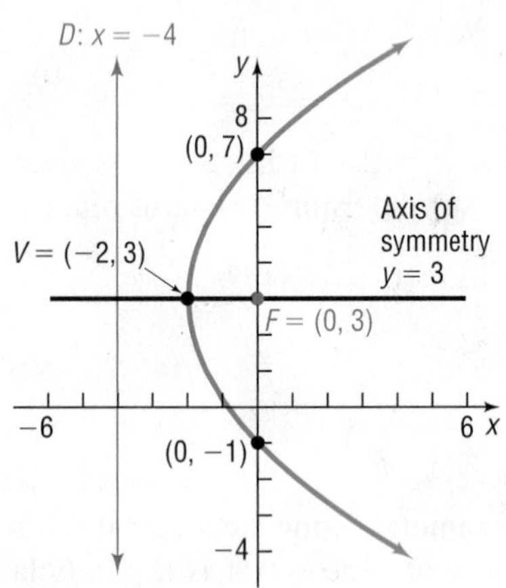

Figure 12 $(y - 3)^2 = 8(x + 2)$

Now Work PROBLEM 31

Polynomial equations define parabolas whenever they involve two variables that are quadratic in one variable and linear in the other.

EXAMPLE 7 **Analyzing the Equation of a Parabola**

Analyze the equation: $x^2 + 4x - 4y = 0$

Solution To analyze the equation $x^2 + 4x - 4y = 0$, complete the square involving the variable x.

$$x^2 + 4x - 4y = 0$$
$$x^2 + 4x = 4y \quad \text{Isolate the terms involving } x \text{ on the left side.}$$
$$x^2 + 4x + 4 = 4y + 4 \quad \text{Complete the square on the left side.}$$
$$(x + 2)^2 = 4(y + 1) \quad \text{Factor.}$$

This equation is of the form $(x - h)^2 = 4a(y - k)$, with $h = -2$, $k = -1$, and $a = 1$. The graph is a parabola with vertex at $(h, k) = (-2, -1)$ that opens up. The focus is at $(-2, 0)$, and the directrix is the line $y = -2$. See Figure 13. ●

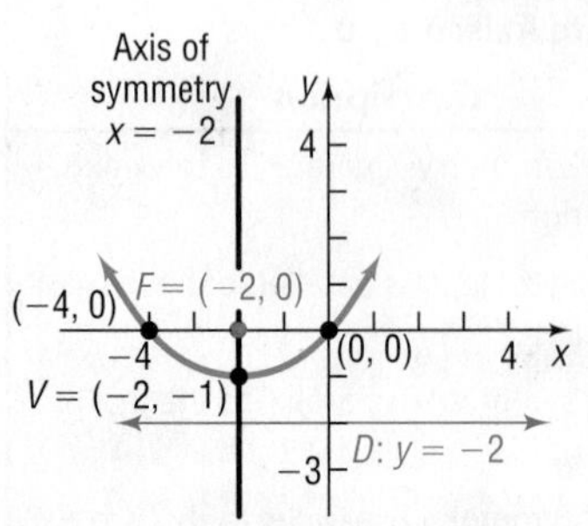

Figure 13 $x^2 + 4x - 4y = 0$

Now Work PROBLEM 49

3 Solve Applied Problems Involving Parabolas

Parabolas find their way into many applications. For example, suspension bridges have cables in the shape of a parabola. Another property of parabolas that is used in applications is their reflecting property.

Suppose that a mirror is shaped like a **paraboloid of revolution**, a surface formed by rotating a parabola about its axis of symmetry. If a light (or any other emitting source) is placed at the focus of the parabola, all the rays emanating from the light will reflect off the mirror in lines parallel to the axis of symmetry. This principle is used in the design of searchlights, flashlights, certain automobile headlights, and other such devices. See Figure 14.

Conversely, suppose that rays of light (or other signals) emanate from a distant source so that they are essentially parallel. When these rays strike the surface of a parabolic mirror whose axis of symmetry is parallel to these rays, they are reflected to a single point at the focus. This principle is used in the design of some solar energy devices, satellite dishes, and the mirrors used in some types of telescopes. See Figure 15.

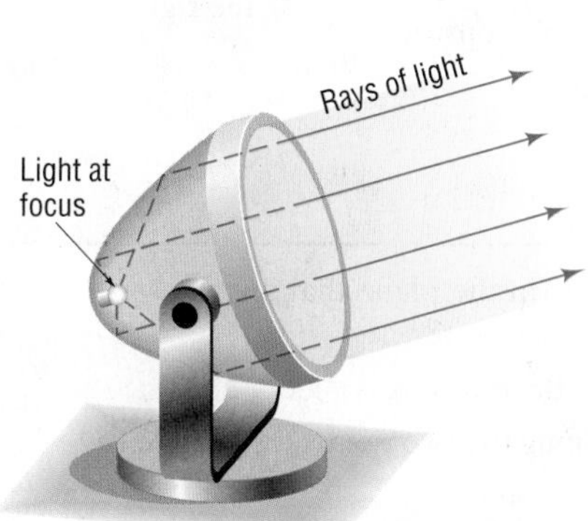

Figure 14 Searchlight

Figure 15 Telescope

EXAMPLE 8 **Satellite Dish**

A satellite dish is shaped like a paraboloid of revolution. The signals that emanate from a satellite strike the surface of the dish and are reflected to a single point, where the receiver is located. If the dish is 8 feet across at its opening and 3 feet deep at its center, at what position should the receiver be placed? That is, where is the focus?

Solution Figure 16(a) shows the satellite dish. On a rectangular coordinate system, draw the parabola used to form the dish so that the vertex of the parabola is at the origin and its focus is on the positive y-axis. See Figure 16(b).

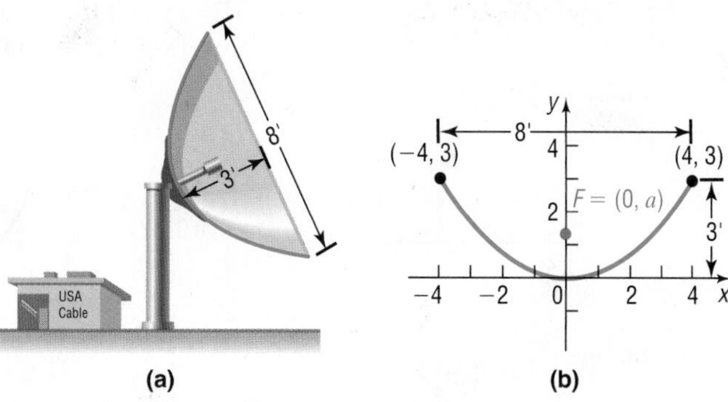

Figure 16

The form of the equation of the parabola is

$$x^2 = 4ay$$

and its focus is at $(0, a)$. Since $(4, 3)$ is a point on the graph, this gives

$$4^2 = 4a(3) \quad x^2 = 4ay; x = 4, y = 3$$

$$a = \frac{4}{3} \quad \text{Solve for } a.$$

The receiver should be located $1\frac{1}{3}$ feet (1 foot, 4 inches) from the base of the dish, along its axis of symmetry. ●

Now Work PROBLEM 65

6.2 Assess Your Understanding

'Are You Prepared?' *Answers are given at the end of these exercises. If you get a wrong answer, read the pages listed in red.*

1. The formula for the distance d from $P_1 = (x_1, y_1)$ to $P_2 = (x_2, y_2)$ is $d =$ ________________. (p. 3)
2. To complete the square of $x^2 - 4x$, add ____. (pp. A24–A25)
3. Use the Square Root Method to find the real solutions of $(x + 4)^2 = 9$. (p. A31)
4. The point that is symmetric with respect to the x-axis to the point $(-2, 5)$ is ______. (pp. 12–13)
5. To graph $y = (x - 3)^2 + 1$, shift the graph of $y = x^2$ to the right _____ units and then _____ 1 unit. (pp. 64–73)

Concepts and Vocabulary

6. A(n) __________ is the collection of all points in the plane that are the same distance from a fixed point as they are from a fixed line.
7. The line through the focus and perpendicular to the directrix is called the ____________ of the parabola.
8. For the parabola $y^2 = 4ax$, the line segment joining the two points $(a, 2a)$ and $(a, -2a)$ is called the ___________.

Answer Problems 9–12 using the figure.

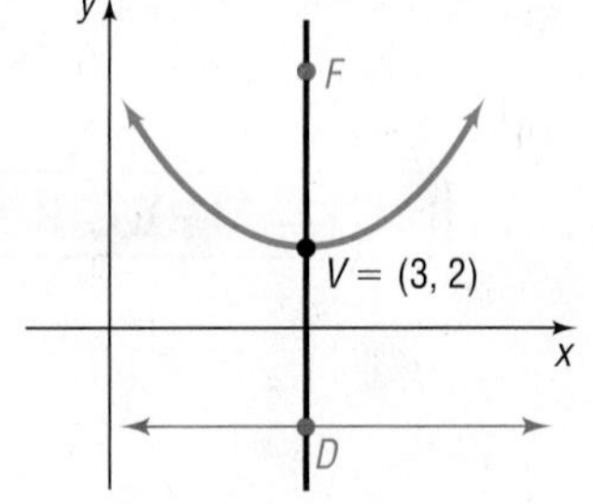

9. If $a > 0$, the equation of the parabola is of the form
   (a) $(y - k)^2 = 4a(x - h)$ (b) $(y - k)^2 = -4a(x - h)$
   (c) $(x - h)^2 = 4a(y - k)$ (d) $(x - h)^2 = -4a(y - k)$
10. The coordinates of the vertex are ________.
11. If $a = 4$, then the coordinates of the focus are ________.
    (a) $(-1, 2)$ (b) $(3, -2)$ (c) $(7, 2)$ (d) $(3, 6)$
12. If $a = 4$, then the equation of the directrix is ________.
    (a) $x = -3$ (b) $x = 3$ (c) $y = -2$ (d) $y = 2$

Skill Building

In Problems 13–20, the graph of a parabola is given. Match each graph to its equation.

(A) $y^2 = 4x$
(B) $x^2 = 4y$
(C) $y^2 = -4x$
(D) $x^2 = -4y$
(E) $(y - 1)^2 = 4(x - 1)$
(F) $(x + 1)^2 = 4(y + 1)$
(G) $(y - 1)^2 = -4(x - 1)$
(H) $(x + 1)^2 = -4(y + 1)$

13.
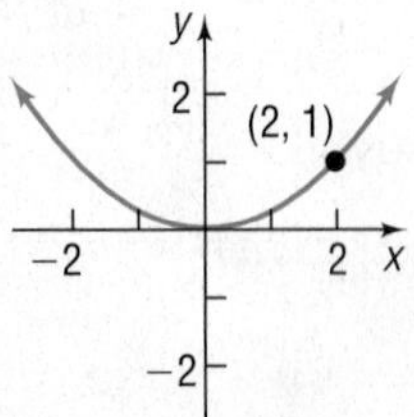

14.
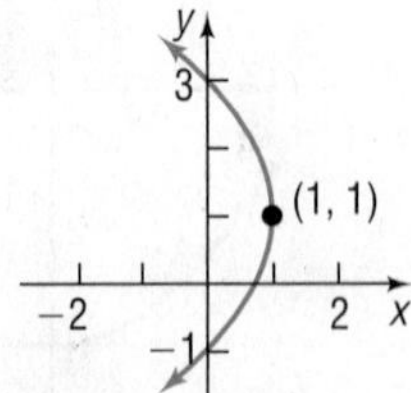

15.
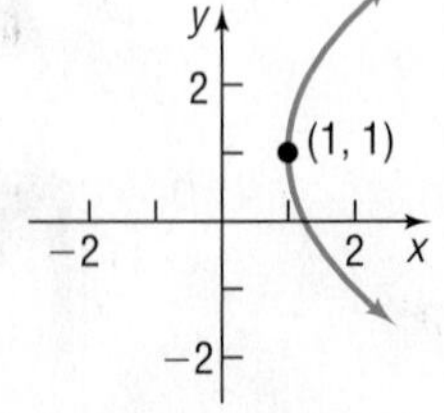

16.
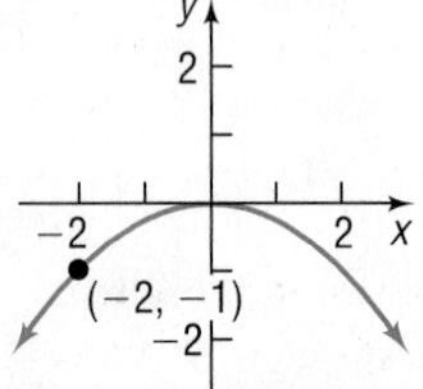

17.
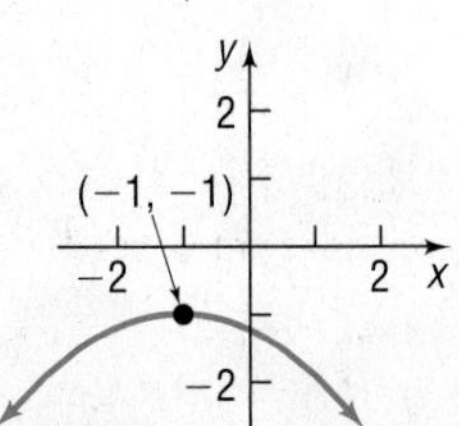

18.
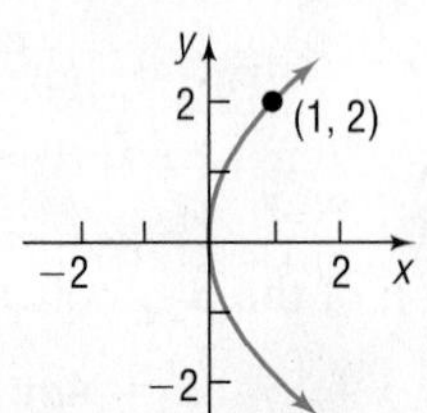

19.
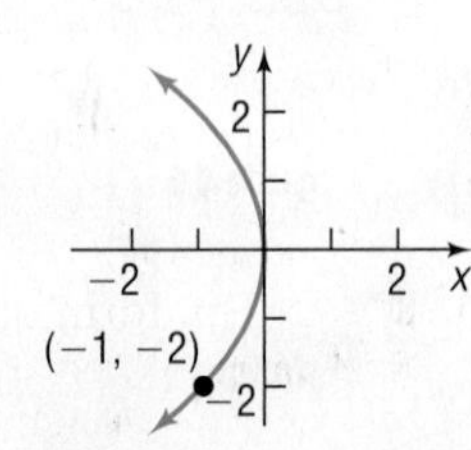

20.
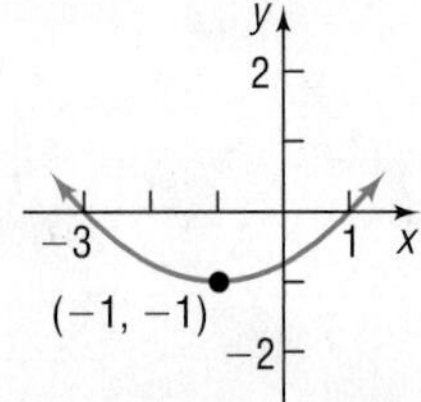

In Problems 21–38, find the equation of the parabola described. Find the two points that define the latus rectum, and graph the equation.

21. Focus at $(4, 0)$; vertex at $(0, 0)$
22. Focus at $(0, 2)$; vertex at $(0, 0)$
23. Focus at $(0, -3)$; vertex at $(0, 0)$
24. Focus at $(-4, 0)$; vertex at $(0, 0)$
25. Focus at $(-2, 0)$; directrix the line $x = 2$
26. Focus at $(0, -1)$; directrix the line $y = 1$
27. Directrix the line $y = -\frac{1}{2}$; vertex at $(0, 0)$
28. Directrix the line $x = -\frac{1}{2}$; vertex at $(0, 0)$
29. Vertex at $(0, 0)$; axis of symmetry the y-axis; containing the point $(2, 3)$
30. Vertex at $(0, 0)$; axis of symmetry the x-axis; containing the point $(2, 3)$
31. Vertex at $(2, -3)$; focus at $(2, -5)$
32. Vertex at $(4, -2)$; focus at $(6, -2)$
33. Vertex at $(-1, -2)$; focus at $(0, -2)$
34. Vertex at $(3, 0)$; focus at $(3, -2)$
35. Focus at $(-3, 4)$; directrix the line $y = 2$
36. Focus at $(2, 4)$; directrix the line $x = -4$
37. Focus at $(-3, -2)$; directrix the line $x = 1$
38. Focus at $(-4, 4)$; directrix the line $y = -2$

In Problems 39–56, find the vertex, focus, and directrix of each parabola. Graph the equation.

39. $x^2 = 4y$
40. $y^2 = 8x$
41. $y^2 = -16x$
42. $x^2 = -4y$
43. $(y - 2)^2 = 8(x + 1)$
44. $(x + 4)^2 = 16(y + 2)$
45. $(x - 3)^2 = -(y + 1)$
46. $(y + 1)^2 = -4(x - 2)$
47. $(y + 3)^2 = 8(x - 2)$
48. $(x - 2)^2 = 4(y - 3)$
49. $y^2 - 4y + 4x + 4 = 0$
50. $x^2 + 6x - 4y + 1 = 0$
51. $x^2 + 8x = 4y - 8$
52. $y^2 - 2y = 8x - 1$
53. $y^2 + 2y - x = 0$
54. $x^2 - 4x = 2y$
55. $x^2 - 4x = y + 4$
56. $y^2 + 12y = -x + 1$

In Problems 57–64, write an equation for each parabola.

57.
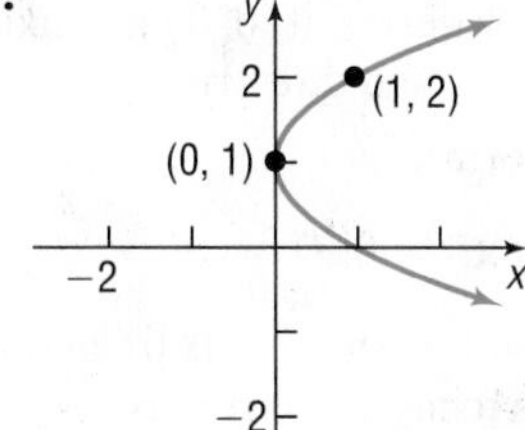

58.
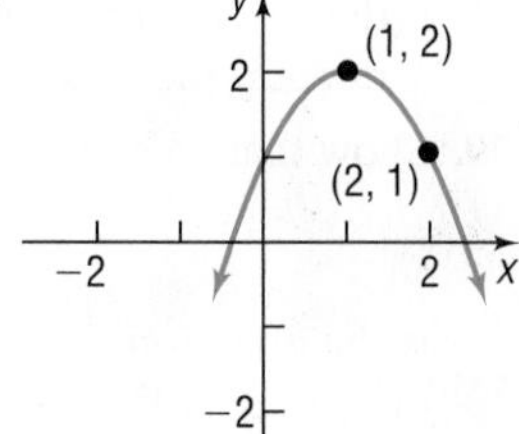

59.
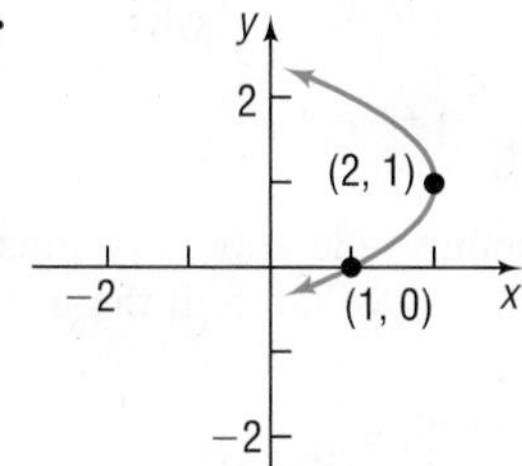

60.
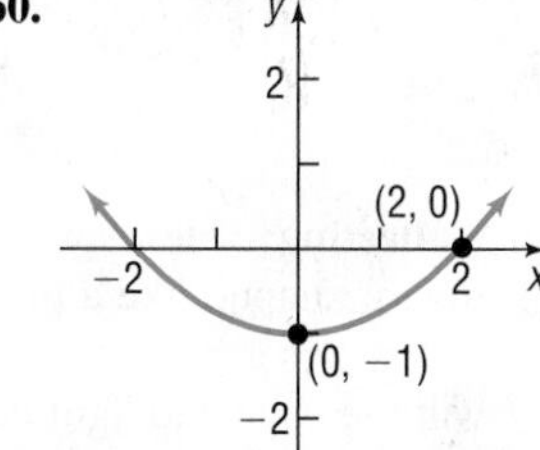

61.
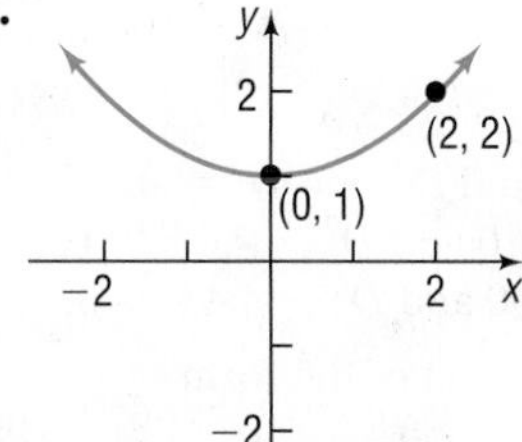

62.
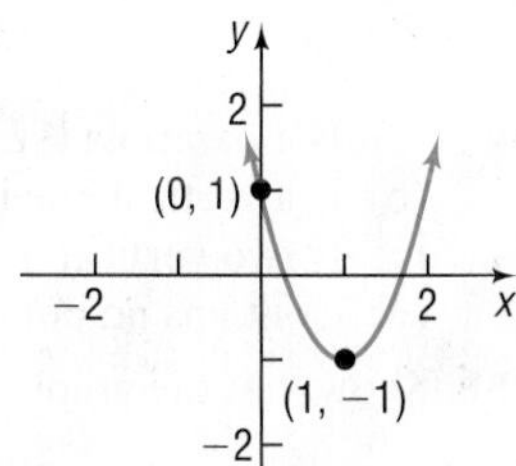

63.
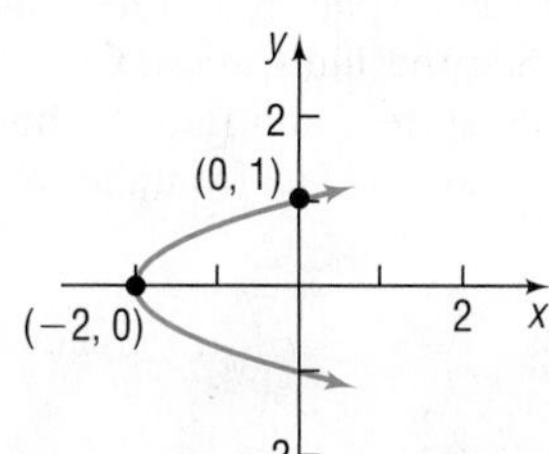

64.
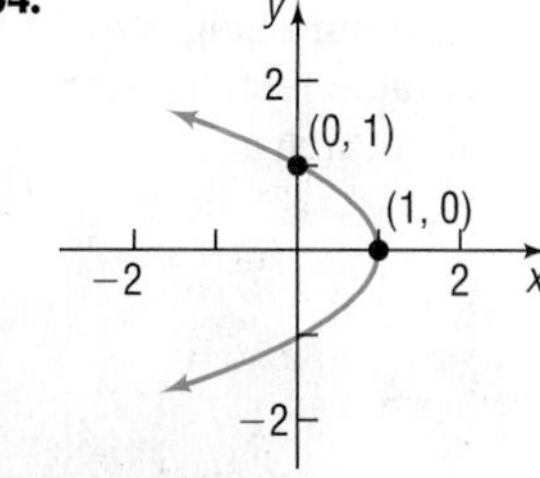

Applications and Extensions

65. **Satellite Dish** A satellite dish is shaped like a paraboloid of revolution. The signals that emanate from a satellite strike the surface of the dish and are reflected to a single point, where the receiver is located. If the dish is 10 feet across at its opening and 4 feet deep at its center, at what position should the receiver be placed?

66. **Constructing a TV Dish** A cable TV receiving dish is in the shape of a paraboloid of revolution. Find the location of the receiver, which is placed at the focus, if the dish is 6 feet across at its opening and 2 feet deep.

67. **Constructing a Flashlight** The reflector of a flashlight is in the shape of a paraboloid of revolution. Its diameter is 4 inches and its depth is 1 inch. How far from the vertex should the light bulb be placed so that the rays will be reflected parallel to the axis?

68. **Constructing a Headlight** A sealed-beam headlight is in the shape of a paraboloid of revolution. The bulb, which is placed at the focus, is 1 inch from the vertex. If the depth is to be 2 inches, what is the diameter of the headlight at its opening?

69. **Suspension Bridge** The cables of a suspension bridge are in the shape of a parabola, as shown in the figure. The towers supporting the cable are 600 feet apart and 80 feet high. If the cables touch the road surface midway between the towers, what is the height of the cable from the road at a point 150 feet from the center of the bridge?

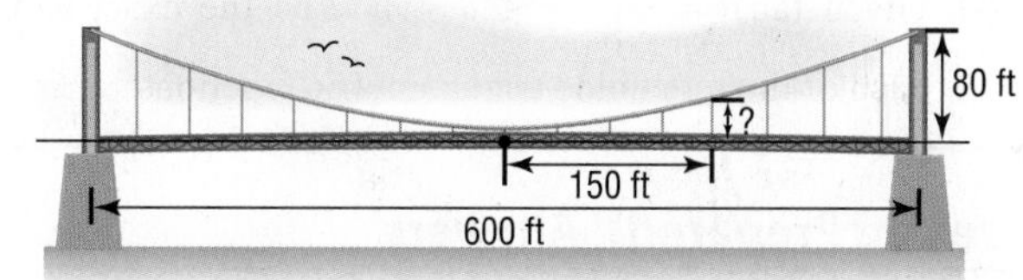

70. Suspension Bridge The cables of a suspension bridge are in the shape of a parabola. The towers supporting the cable are 400 feet apart and 100 feet high. If the cables are at a height of 10 feet midway between the towers, what is the height of the cable at a point 50 feet from the center of the bridge?

71. Searchlight A searchlight is shaped like a paraboloid of revolution. If the light source is located 2 feet from the base along the axis of symmetry and the opening is 5 feet across, how deep should the searchlight be?

72. Searchlight A searchlight is shaped like a paraboloid of revolution. If the light source is located 2 feet from the base along the axis of symmetry and the depth of the searchlight is 4 feet, what should the width of the opening be?

73. Solar Heat A mirror is shaped like a paraboloid of revolution and will be used to concentrate the rays of the sun at its focus, creating a heat source. See the figure. If the mirror is 20 feet across at its opening and is 6 feet deep, where will the heat source be concentrated?

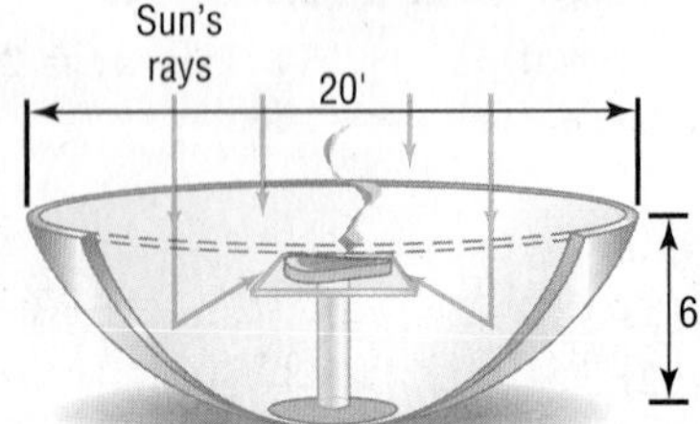

74. Reflecting Telescope A reflecting telescope contains a mirror shaped like a paraboloid of revolution. If the mirror is 4 inches across at its opening and is 3 inches deep, where will the collected light be concentrated?

75. Parabolic Arch Bridge A bridge is built in the shape of a parabolic arch. The bridge has a span of 120 feet and a maximum height of 25 feet. See the illustration. Choose a suitable rectangular coordinate system and find the height of the arch at distances of 10, 30, and 50 feet from the center.

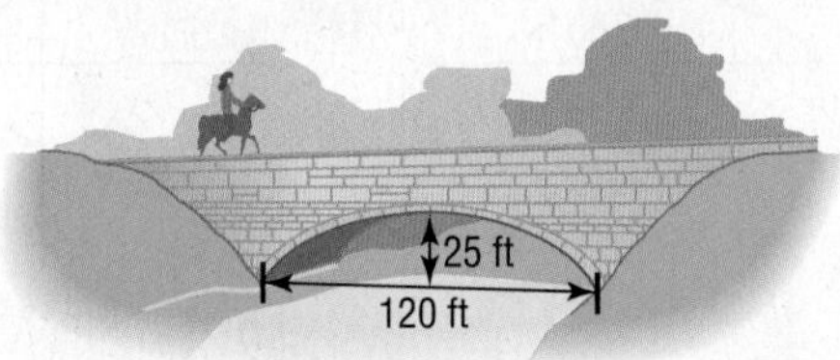

76. Parabolic Arch Bridge A bridge is to be built in the shape of a parabolic arch and is to have a span of 100 feet. The height of the arch a distance of 40 feet from the center is to be 10 feet. Find the height of the arch at its center.

77. Gateway Arch The Gateway Arch in St. Louis is often mistaken to be parabolic in shape. In fact, it is a *catenary*, which has a more complicated formula than a parabola. The Arch is 630 feet high and 630 feet wide at its base.

(a) Find the equation of a parabola with the same dimensions. Let x equal the horizontal distance from the center of the arch.

(b) The table below gives the height of the Arch at various widths; find the corresponding heights for the parabola found in (a).

Width (ft)	Height (ft)
567	100
478	312.5
308	525

(c) Do the data support the notion that the Arch is in the shape of a parabola?

Source: gatewayarch.com

78. Show that an equation of the form

$$Ax^2 + Ey = 0 \qquad A \neq 0, E \neq 0$$

is the equation of a parabola with vertex at $(0, 0)$ and axis of symmetry the y-axis. Find its focus and directrix.

79. Show that an equation of the form

$$Cy^2 + Dx = 0 \qquad C \neq 0, D \neq 0$$

is the equation of a parabola with vertex at $(0, 0)$ and axis of symmetry the x-axis. Find its focus and directrix.

80. Show that the graph of an equation of the form

$$Ax^2 + Dx + Ey + F = 0 \qquad A \neq 0$$

(a) Is a parabola if $E \neq 0$.
(b) Is a vertical line if $E = 0$ and $D^2 - 4AF = 0$.
(c) Is two vertical lines if $E = 0$ and $D^2 - 4AF > 0$.
(d) Contains no points if $E = 0$ and $D^2 - 4AF < 0$.

81. Show that the graph of an equation of the form

$$Cy^2 + Dx + Ey + F = 0 \qquad C \neq 0$$

(a) Is a parabola if $D \neq 0$.
(b) Is a horizontal line if $D = 0$ and $E^2 - 4CF = 0$.
(c) Is two horizontal lines if $D = 0$ and $E^2 - 4CF > 0$.
(d) Contains no points if $D = 0$ and $E^2 - 4CF < 0$.

Retain Your Knowledge

Problems 82–85 are based on material learned earlier in the course. The purpose of these problems is to keep the material fresh in your mind so that you are better prepared for the final exam.

82. For $x = 9y^2 - 36$, list the intercepts and test for symmetry.

83. Convert $-\frac{7\pi}{18}$ from radians to degrees.

84. Given $\tan\theta = -\frac{5}{8}, \frac{\pi}{2} < \theta < \pi$, find the exact value of each of the remaining trigonometric functions.

85. Find the exact value: $\tan\left[\cos^{-1}\left(-\frac{3}{7}\right)\right]$

'Are You Prepared?' Answers

1. $\sqrt{(x_2 - x_1)^2 + (y_2 - y_1)^2}$ **2.** 4 **3.** $\{-7, -1\}$ **4.** $(-2, -5)$ **5.** 3; up

6.3 The Ellipse

PREPARING FOR THIS SECTION *Before getting started, review the following:*

- Distance Formula (Section 1.1, p. 3)
- Complete the Square (Appendix A, Section A.3, pp. A24–A25)
- Intercepts (Section 1.2, pp. 10–11)
- Symmetry (Section 1.2, pp. 12–13)
- Circles (Section 1.2, pp. 15–18)
- Graphing Techniques: Transformations (Section 1.6, pp. 64–73)

Now Work the 'Are You Prepared?' problems on page 404.

OBJECTIVES
1 Analyze Ellipses with Center at the Origin (p. 397)
2 Analyze Ellipses with Center at (h, k) (p. 401)
3 Solve Applied Problems Involving Ellipses (p. 403)

DEFINITION An **ellipse** is the collection of all points in the plane the sum of whose distances from two fixed points, called the **foci**, is a constant.

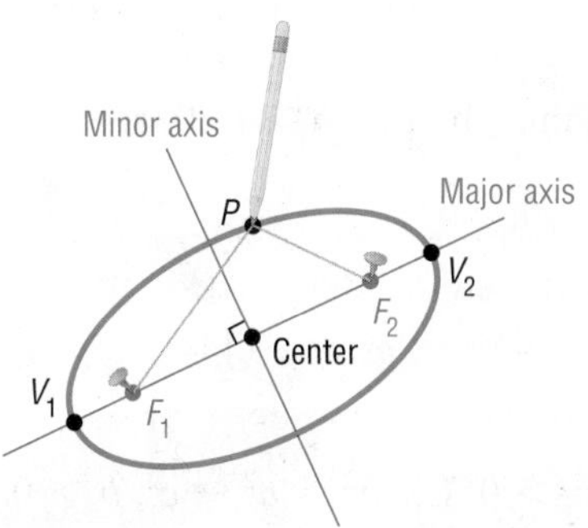

Figure 17 Ellipse

The definition contains within it a physical means for drawing an ellipse. Find a piece of string (the length of this string is the constant referred to in the definition). Then take two thumbtacks (the foci) and stick them into a piece of cardboard so that the distance between them is less than the length of the string. Now attach the ends of the string to the thumbtacks and, using the point of a pencil, pull the string taut. See Figure 17. Keeping the string taut, rotate the pencil around the two thumbtacks. The pencil traces out an ellipse, as shown in Figure 17.

In Figure 17, the foci are labeled F_1 and F_2. The line containing the foci is called the **major axis**. The midpoint of the line segment joining the foci is the **center** of the ellipse. The line through the center and perpendicular to the major axis is the **minor axis**.

The two points of intersection of the ellipse and the major axis are the **vertices**, V_1 and V_2, of the ellipse. The distance from one vertex to the other is the **length of the major axis**. The ellipse is symmetric with respect to its major axis, with respect to its minor axis, and with respect to its center.

1 Analyze Ellipses with Center at the Origin

With these ideas in mind, we are ready to find the equation of an ellipse in a rectangular coordinate system. First, place the center of the ellipse at the origin. Second, position the ellipse so that its major axis coincides with a coordinate axis, say the x-axis, as shown in Figure 18. If c is the distance from the center to a focus, one focus will be at $F_1 = (-c, 0)$ and the other at $F_2 = (c, 0)$. As we shall see, it is

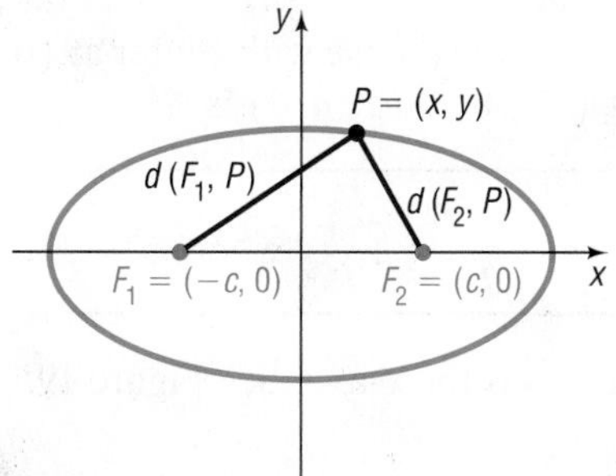

Figure 18

convenient to let $2a$ denote the constant distance referred to in the definition. Then, if $P = (x, y)$ is any point on the ellipse,

$$d(F_1, P) + d(F_2, P) = 2a \quad \text{The sum of the distances from } P \text{ to the foci equals a constant, } 2a.$$

$$\sqrt{(x+c)^2 + y^2} + \sqrt{(x-c)^2 + y^2} = 2a \quad \text{Use the Distance Formula.}$$

$$\sqrt{(x+c)^2 + y^2} = 2a - \sqrt{(x-c)^2 + y^2} \quad \text{Isolate one radical.}$$

$$(x+c)^2 + y^2 = 4a^2 - 4a\sqrt{(x-c)^2 + y^2} + (x-c)^2 + y^2 \quad \text{Square both sides.}$$

$$x^2 + 2cx + c^2 + y^2 = 4a^2 - 4a\sqrt{(x-c)^2 + y^2} + x^2 - 2cx + c^2 + y^2 \quad \text{Multiply out.}$$

$$4cx - 4a^2 = -4a\sqrt{(x-c)^2 + y^2} \quad \text{Simplify; isolate the radical.}$$

$$cx - a^2 = -a\sqrt{(x-c)^2 + y^2} \quad \text{Divide each side by 4.}$$

$$(cx - a^2)^2 = a^2[(x-c)^2 + y^2] \quad \text{Square both sides again.}$$

$$c^2x^2 - 2a^2cx + a^4 = a^2(x^2 - 2cx + c^2 + y^2) \quad \text{Multiply out.}$$

$$(c^2 - a^2)x^2 - a^2y^2 = a^2c^2 - a^4 \quad \text{Rearrange the terms.}$$

$$(a^2 - c^2)x^2 + a^2y^2 = a^2(a^2 - c^2) \quad \text{Multiply each side by } -1; \text{ factor } a^2 \text{ on the right side.} \quad \textbf{(1)}$$

To obtain points on the ellipse off the x-axis, it must be that $a > c$. To see why, look again at Figure 18. Then

$$d(F_1, P) + d(F_2, P) > d(F_1, F_2) \quad \text{The sum of the lengths of two sides of a triangle is greater than the length of the third side.}$$

$$2a > 2c \quad d(F_1, P) + d(F_2, P) = 2a,\ d(F_1, F_2) = 2c$$

$$a > c$$

Because $a > c > 0$, this means $a^2 > c^2$, so $a^2 - c^2 > 0$. Let $b^2 = a^2 - c^2$, $b > 0$. Then $a > b$ and equation (1) can be written as

$$b^2x^2 + a^2y^2 = a^2b^2$$

$$\frac{x^2}{a^2} + \frac{y^2}{b^2} = 1 \quad \text{Divide each side by } a^2b^2.$$

As you can verify, the graph of this equation has symmetry with respect to the x-axis, the y-axis, and the origin.

Because the major axis is the x-axis, find the vertices of this ellipse by letting $y = 0$. The vertices satisfy the equation $\frac{x^2}{a^2} = 1$, the solutions of which are $x = \pm a$. Consequently, the vertices of this ellipse are $V_1 = (-a, 0)$ and $V_2 = (a, 0)$. The y-intercepts of the ellipse, found by letting $x = 0$, have coordinates $(0, -b)$ and $(0, b)$. The four intercepts, $(a, 0)$, $(-a, 0)$, $(0, b)$, and $(0, -b)$, are used to graph the ellipse.

THEOREM

Equation of an Ellipse: Center at (0, 0); Major Axis along the *x*-Axis

An equation of the ellipse with center at $(0, 0)$, foci at $(-c, 0)$ and $(c, 0)$, and vertices at $(-a, 0)$ and $(a, 0)$ is

$$\frac{x^2}{a^2} + \frac{y^2}{b^2} = 1 \quad \text{where } a > b > 0 \text{ and } b^2 = a^2 - c^2 \qquad \textbf{(2)}$$

The major axis is the x-axis. See Figure 19.

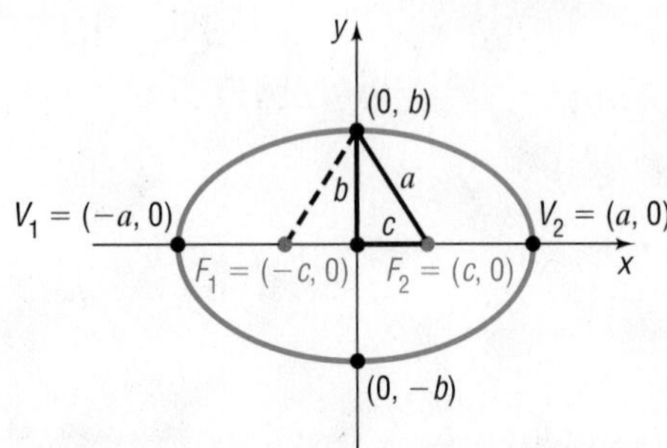

Figure 19

Notice in Figure 19 the points $(0, 0)$, $(c, 0)$, and $(0, b)$ form a right triangle. Because $b^2 = a^2 - c^2$ (or $b^2 + c^2 = a^2$), the distance from the focus at $(c, 0)$ to the point $(0, b)$ is a.

This can be seen another way. Look at the two right triangles in Figure 19. They are congruent. Do you see why? Because the sum of the distances from the foci to a point on the ellipse is $2a$, it follows that the distance from $(c, 0)$ to $(0, b)$ is a.

EXAMPLE 1 Finding an Equation of an Ellipse

Find an equation of the ellipse with center at the origin, one focus at $(3, 0)$, and a vertex at $(-4, 0)$. Graph the equation.

Solution The ellipse has its center at the origin, and since the given focus and vertex lie on the x-axis, the major axis is the x-axis. The distance from the center, $(0, 0)$, to one of the foci, $(3, 0)$, is $c = 3$. The distance from the center, $(0, 0)$, to one of the vertices, $(-4, 0)$, is $a = 4$. From equation (2), it follows that

$$b^2 = a^2 - c^2 = 16 - 9 = 7$$

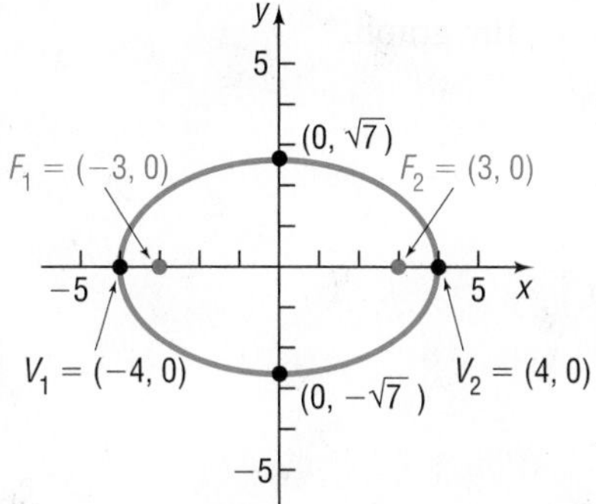

Figure 20 $\frac{x^2}{16} + \frac{y^2}{7} = 1$

so an equation of the ellipse is

$$\frac{x^2}{16} + \frac{y^2}{7} = 1$$

Figure 20 shows the graph. ●

In Figure 20, the intercepts of the equation are used to graph the ellipse. Following this practice will make it easier for you to obtain an accurate graph of an ellipse.

Now Work PROBLEM 27

COMMENT The intercepts of the ellipse also provide information about how to set the viewing rectangle for graphing an ellipse. To graph the ellipse

$$\frac{x^2}{16} + \frac{y^2}{7} = 1$$

discussed in Example 1, set the viewing rectangle using a square screen that includes the intercepts, perhaps $-4.8 \le x \le 4.8$, $-3 \le y \le 3$. Then solve the equation for y:

$$\frac{x^2}{16} + \frac{y^2}{7} = 1$$

$$\frac{y^2}{7} = 1 - \frac{x^2}{16} \quad \text{Subtract } \frac{x^2}{16} \text{ from each side.}$$

$$y^2 = 7\left(1 - \frac{x^2}{16}\right) \quad \text{Multiply both sides by 7.}$$

$$y = \pm\sqrt{7\left(1 - \frac{x^2}{16}\right)} \quad \text{Take the square root of each side.}$$

Now graph the two functions

$$Y_1 = \sqrt{7\left(1 - \frac{x^2}{16}\right)} \text{ and } Y_2 = -\sqrt{7\left(1 - \frac{x^2}{16}\right)}$$

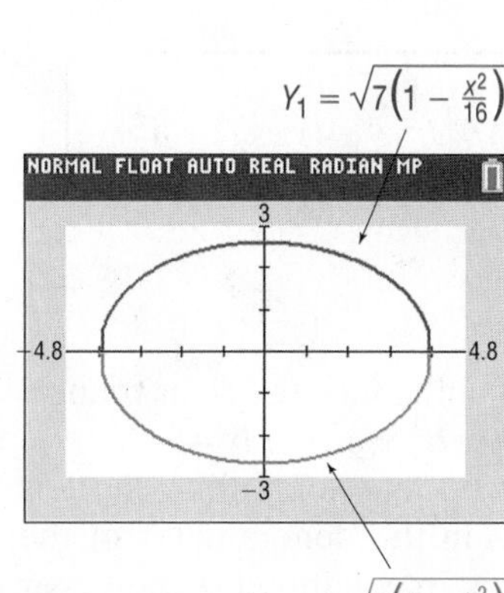

Figure 21

Figure 21 shows the result. ■

An equation of the form of equation (2), with $a^2 > b^2$, is the equation of an ellipse with center at the origin, foci on the x-axis at $(-c, 0)$ and $(c, 0)$, where $c^2 = a^2 - b^2$, and major axis along the x-axis.

For the remainder of this section, the direction **"Analyze the equation"** will mean to find the center, major axis, foci, and vertices of the ellipse and graph it.

EXAMPLE 2 Analyzing the Equation of an Ellipse

Analyze the equation: $\dfrac{x^2}{25} + \dfrac{y^2}{9} = 1$

Solution The equation is of the form of equation (2), with $a^2 = 25$ and $b^2 = 9$. The equation is that of an ellipse with center $(0, 0)$ and major axis along the x-axis. The vertices are at $(\pm a, 0) = (\pm 5, 0)$. Because $b^2 = a^2 - c^2$, this means

$$c^2 = a^2 - b^2 = 25 - 9 = 16$$

The foci are at $(\pm c, 0) = (\pm 4, 0)$. Figure 22 shows the graph.

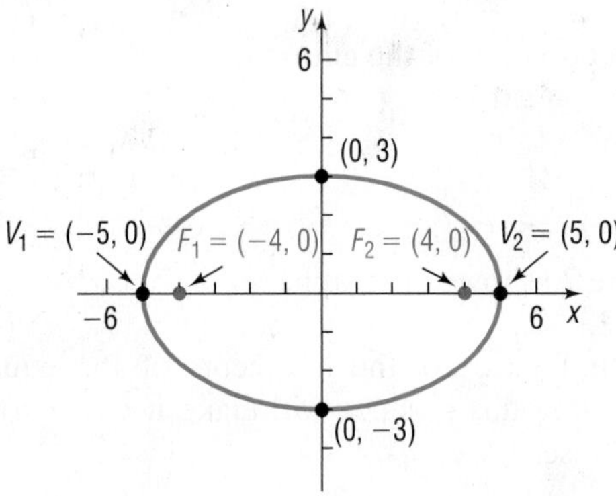

Figure 22 $\dfrac{x^2}{25} + \dfrac{y^2}{9} = 1$

Now Work PROBLEM 17

If the major axis of an ellipse with center at $(0, 0)$ lies on the y-axis, the foci are at $(0, -c)$ and $(0, c)$. Using the same steps as before, the definition of an ellipse leads to the following result.

THEOREM

Equation of an Ellipse: Center at (0, 0); Major Axis along the *y*-Axis

An equation of the ellipse with center at $(0, 0)$, foci at $(0, -c)$ and $(0, c)$, and vertices at $(0, -a)$ and $(0, a)$ is

$$\frac{x^2}{b^2} + \frac{y^2}{a^2} = 1 \quad \text{where } a > b > 0 \text{ and } b^2 = a^2 - c^2 \tag{3}$$

The major axis is the y-axis.

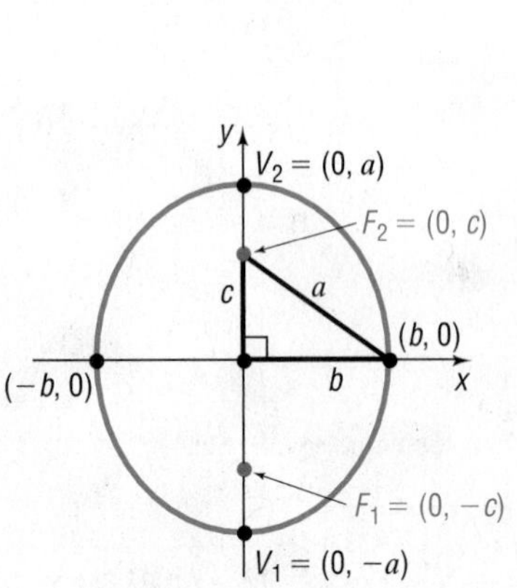

Figure 23 $\dfrac{x^2}{b^2} + \dfrac{y^2}{a^2} = 1, a > b > 0$

Figure 23 illustrates the graph of such an ellipse. Again, notice the right triangle formed by the points at $(0, 0)$, $(b, 0)$, and $(0, c)$, so that $a^2 = b^2 + c^2$ (or $b^2 = a^2 - c^2$).

Look closely at equations (2) and (3). Although they may look alike, there is a difference! In equation (2), the larger number, a^2, is in the denominator of the x^2-term, so the major axis of the ellipse is along the x-axis. In equation (3), the larger number, a^2, is in the denominator of the y^2-term, so the major axis is along the y-axis.

EXAMPLE 3

Analyzing the Equation of an Ellipse

Analyze the equation: $9x^2 + y^2 = 9$

Solution To put the equation in proper form, divide each side by 9.

$$x^2 + \frac{y^2}{9} = 1$$

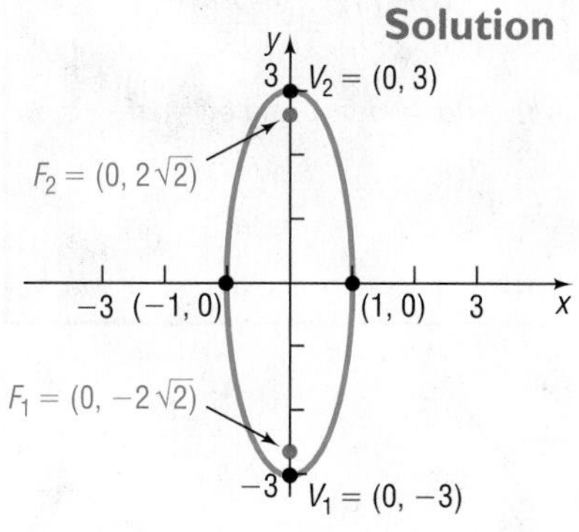

Figure 24 $9x^2 + y^2 = 9$

The larger denominator, 9, is in the y^2-term so, based on equation (3), this is the equation of an ellipse with center at the origin and major axis along the y-axis. Also, $a^2 = 9$, $b^2 = 1$, and $c^2 = a^2 - b^2 = 9 - 1 = 8$. The vertices are at $(0, \pm a) = (0, \pm 3)$, and the foci are at $(0, \pm c) = (0, \pm 2\sqrt{2})$. The x-intercepts are at $(\pm b, 0) = (\pm 1, 0)$. Figure 24 shows the graph. ●

Now Work PROBLEM 21

EXAMPLE 4

Finding an Equation of an Ellipse

Find an equation of the ellipse having one focus at $(0, 2)$ and vertices at $(0, -3)$ and $(0, 3)$. Graph the equation.

Solution Plot the given focus and vertices, and note that the major axis is the y-axis. Because the vertices are at $(0, -3)$ and $(0, 3)$, the center of this ellipse is at their midpoint, the origin. The distance from the center, $(0, 0)$, to the given focus, $(0, 2)$, is $c = 2$. The distance from the center, $(0, 0)$, to one of the vertices, $(0, 3)$, is $a = 3$. So $b^2 = a^2 - c^2 = 9 - 4 = 5$. The form of the equation of this ellipse is given by equation (3).

$$\frac{x^2}{b^2} + \frac{y^2}{a^2} = 1$$

$$\frac{x^2}{5} + \frac{y^2}{9} = 1$$

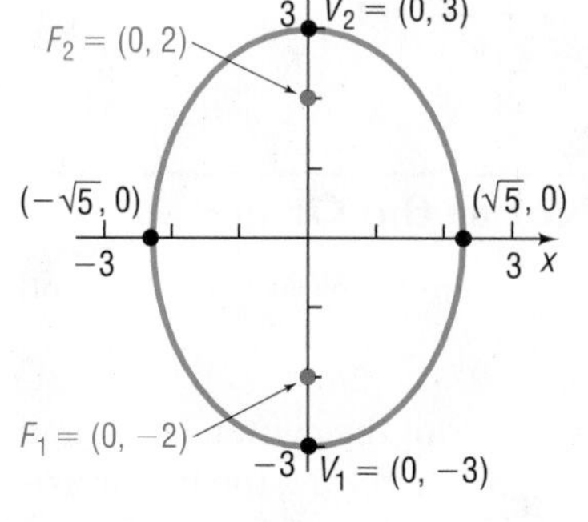

Figure 25 $\frac{x^2}{5} + \frac{y^2}{9} = 1$

Figure 25 shows the graph. ●

Now Work PROBLEM 29

A circle may be considered a special kind of ellipse. To see why, let $a = b$ in equation (2) or (3). Then

$$\frac{x^2}{a^2} + \frac{y^2}{a^2} = 1$$

$$x^2 + y^2 = a^2$$

This is the equation of a circle with center at the origin and radius a. The value of c is

$$c^2 = a^2 - \underset{\substack{\uparrow \\ a = b}}{b^2} = 0$$

This indicates that the closer the two foci of an ellipse are to the center, the more the ellipse will look like a circle.

2 Analyze Ellipses with Center at (h, k)

If an ellipse with center at the origin and major axis coinciding with a coordinate axis is shifted horizontally h units and then vertically k units, the result is an ellipse with center at (h, k) and major axis parallel to a coordinate axis. The equations of such ellipses have the same forms as those given in equations (2) and (3), except that x is replaced by $x - h$ (the horizontal shift) and y is replaced by $y - k$ (the vertical shift). Table 3 (on the next page) gives the forms of the equations of such ellipses, and Figure 26 shows their graphs.

Table 3

Equations of an Ellipse: Center at (h, k); Major Axis Parallel to a Coordinate Axis				
Center	**Major Axis**	**Foci**	**Vertices**	**Equation**
(h, k)	Parallel to the x-axis	$(h + c, k)$	$(h + a, k)$	$\frac{(x-h)^2}{a^2} + \frac{(y-k)^2}{b^2} = 1$
		$(h - c, k)$	$(h - a, k)$	$a > b > 0$ and $b^2 = a^2 - c^2$
(h, k)	Parallel to the y-axis	$(h, k + c)$	$(h, k + a)$	$\frac{(x-h)^2}{b^2} + \frac{(y-k)^2}{a^2} = 1$
		$(h, k - c)$	$(h, k - a)$	$a > b > 0$ and $b^2 = a^2 - c^2$

NOTE It is not recommended that Table 3 be memorized. Rather, use transformations (shift horizontally h units, vertically k units), along with the fact that a represents the distance from the center to the vertices, c represents the distance from the center to the foci, and $b^2 = a^2 - c^2$ (or $c^2 = a^2 - b^2$). ■

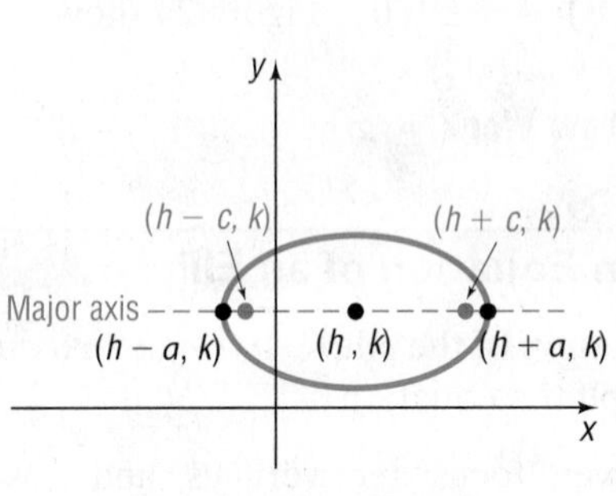

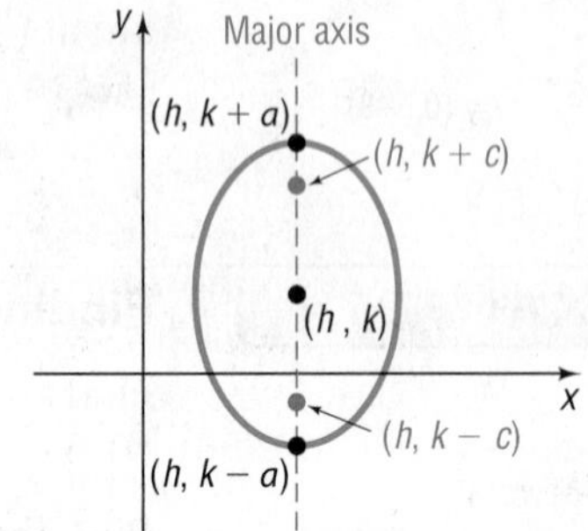

Figure 26 (a) $\frac{(x-h)^2}{a^2} + \frac{(y-k)^2}{b^2} = 1$ (b) $\frac{(x-h)^2}{b^2} + \frac{(y-k)^2}{a^2} = 1$

EXAMPLE 5 **Finding an Equation of an Ellipse, Center Not at the Origin**

Find an equation for the ellipse with center at $(2, -3)$, one focus at $(3, -3)$, and one vertex at $(5, -3)$. Graph the equation.

Solution The center is at $(h, k) = (2, -3)$, so $h = 2$ and $k = -3$. Plot the center, focus, and vertex, and note that the points all lie on the line $y = -3$. Therefore, the major axis is parallel to the x-axis. The distance from the center $(2, -3)$ to a focus $(3, -3)$ is $c = 1$; the distance from the center $(2, -3)$ to a vertex $(5, -3)$ is $a = 3$. Then $b^2 = a^2 - c^2 = 9 - 1 = 8$. The form of the equation is

$$\frac{(x-h)^2}{a^2} + \frac{(y-k)^2}{b^2} = 1 \qquad h = 2, k = -3, a = 3, b = 2\sqrt{2}$$

$$\frac{(x-2)^2}{9} + \frac{(y+3)^2}{8} = 1$$

To graph the equation, use the center $(h, k) = (2, -3)$ to locate the vertices. The major axis is parallel to the x-axis, so the vertices are $a = 3$ units left and right of the center $(2, -3)$. Therefore, the vertices are

$$V_1 = (2 - 3, -3) = (-1, -3) \quad \text{and} \quad V_2 = (2 + 3, -3) = (5, -3)$$

Since $c = 1$ and the major axis is parallel to the x-axis, the foci are 1 unit left and right of the center. Therefore, the foci are

$$F_1 = (2 - 1, -3) = (1, -3) \quad \text{and} \quad F_2 = (2 + 1, -3) = (3, -3)$$

Finally, use the value of $b = 2\sqrt{2}$ to find the two points above and below the center.

$$(2, -3 - 2\sqrt{2}) \quad \text{and} \quad (2, -3 + 2\sqrt{2})$$

Figure 27 shows the graph. ●

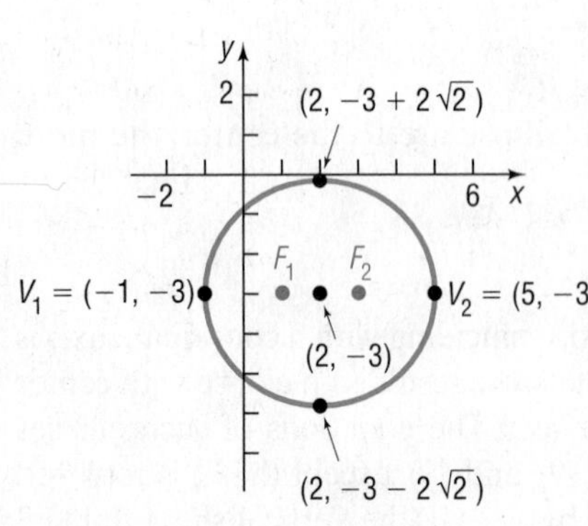

Figure 27 $\frac{(x-2)^2}{9} + \frac{(y+3)^2}{8} = 1$

Now Work PROBLEM 55

EXAMPLE 6

Analyzing the Equation of an Ellipse

Analyze the equation: $4x^2 + y^2 - 8x + 4y + 4 = 0$

Solution Complete the squares in x and in y.

$$4x^2 + y^2 - 8x + 4y + 4 = 0$$

$$4x^2 - 8x + y^2 + 4y = -4 \quad \text{Group like variables; place the constant on the right side.}$$

$$4(x^2 - 2x) + (y^2 + 4y) = -4 \quad \text{Factor out 4 from the first two terms.}$$

$$4(x^2 - 2x + 1) + (y^2 + 4y + 4) = -4 + 4 + 4 \quad \text{Complete each square.}$$

$$4(x - 1)^2 + (y + 2)^2 = 4 \quad \text{Factor.}$$

$$(x - 1)^2 + \frac{(y + 2)^2}{4} = 1 \quad \text{Divide each side by 4.}$$

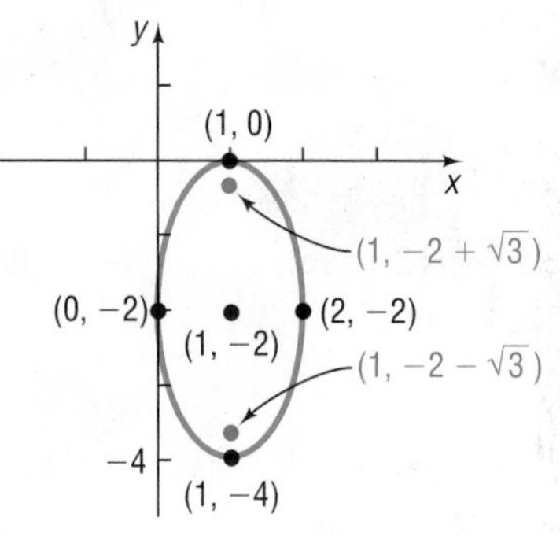

Figure 28
$4x^2 + y^2 - 8x + 4y + 4 = 0$

This is the equation of an ellipse with center at $(1, -2)$ and major axis parallel to the y-axis. Since $a^2 = 4$ and $b^2 = 1$, we have $c^2 = a^2 - b^2 = 4 - 1 = 3$. The vertices are at $(h, k \pm a) = (1, -2 \pm 2)$, or $(1, -4)$ and $(1, 0)$. The foci are at $(h, k \pm c) = (1, -2 \pm \sqrt{3})$, or $(1, -2 - \sqrt{3})$ and $(1, -2 + \sqrt{3})$. Figure 28 shows the graph. ●

Now Work PROBLEM 47

3 Solve Applied Problems Involving Ellipses

Ellipses are found in many applications in science and engineering. For example, the orbits of the planets around the Sun are elliptical, with the Sun's position at a focus. See Figure 29.

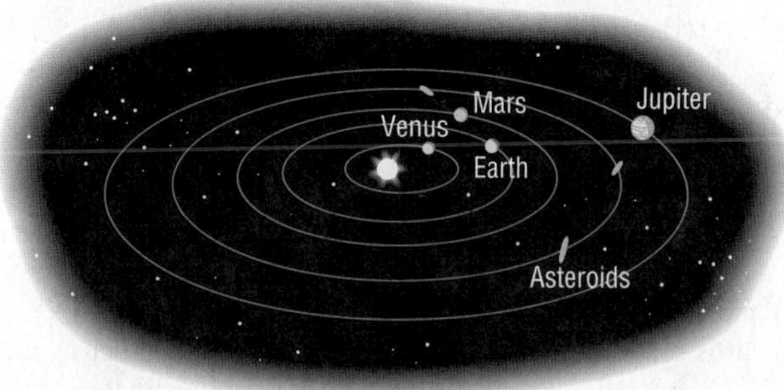

Figure 29

Stone and concrete bridges are often shaped as semielliptical arches. Elliptical gears are used in machinery when a variable rate of motion is required.

Ellipses also have an interesting reflection property. If a source of light (or sound) is placed at one focus, the waves transmitted by the source will reflect off the ellipse and concentrate at the other focus. This is the principle behind *whispering galleries,* which are rooms designed with elliptical ceilings. A person standing at one focus of the ellipse can whisper and be heard by a person standing at the other focus, because all the sound waves that reach the ceiling are reflected to the other person.

EXAMPLE 7

A Whispering Gallery

The whispering gallery in the Museum of Science and Industry in Chicago is 47.3 feet long. The distance from the center of the room to the foci is 20.3 feet. Find an equation that describes the shape of the room. How high is the room at its center?

Source: *Chicago Museum of Science and Industry Web site; www.msichicago.org*

Solution Set up a rectangular coordinate system so that the center of the ellipse is at the origin and the major axis is along the x-axis. The equation of the ellipse is

$$\frac{x^2}{a^2} + \frac{y^2}{b^2} = 1$$

Since the length of the room is 47.3 feet, the distance from the center of the room to each vertex (the end of the room) will be $\frac{47.3}{2} = 23.65$ feet; so $a = 23.65$ feet. The distance from the center of the room to each focus is $c = 20.3$ feet. See Figure 30.

Because $b^2 = a^2 - c^2$, this means that $b^2 = 23.65^2 - 20.3^2 = 147.2325$. An equation that describes the shape of the room is given by

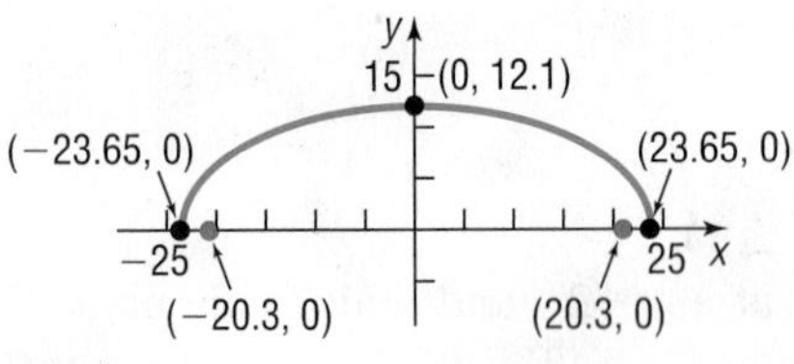

Figure 30

$$\frac{x^2}{23.65^2} + \frac{y^2}{147.2325} = 1$$

The height of the room at its center is $b = \sqrt{147.2325} \approx 12.1$ feet. ●

Now Work PROBLEM 71

6.3 Assess Your Understanding

'Are You Prepared?' *Answers are given at the end of these exercises. If you get a wrong answer, read the pages listed in red.*

1. The distance d from $P_1 = (2, -5)$ to $P_2 = (4, -2)$ is $d =$ ______. (p. 3)
2. To complete the square of $x^2 - 3x$, add ______. (pp. A24–A25)
3. Find the intercepts of the equation $y^2 = 16 - 4x^2$. (pp. 10–11)
4. The point that is symmetric with respect to the y-axis to the point $(-2, 5)$ is ______. (pp. 12–13)
5. To graph $y = (x + 1)^2 - 4$, shift the graph of $y = x^2$ to the (left/right) ______ unit(s) and then (up/down) ______ unit(s). (pp. 64–73)
6. The standard equation of a circle with center at $(2, -3)$ and radius 1 is ______. (pp. 15–18)

Concepts and Vocabulary

7. A(n) ________ is the collection of all points in the plane the sum of whose distances from two fixed points is a constant.
8. For an ellipse, the foci lie on a line called the ______.
   (a) minor axis (b) major axis
   (c) directrix (d) latus rectum
9. For the ellipse $\frac{x^2}{4} + \frac{y^2}{25} = 1$, the vertices are the points ________ and ________.
10. For the ellipse $\frac{x^2}{25} + \frac{y^2}{9} = 1$, the value of a is ____, the value of b is ____, and the major axis is the ____ -axis.
11. If the center of an ellipse is $(2, -3)$, the major axis is parallel to the x-axis, and the distance from the center of the ellipse to its vertices is $a = 4$ units, then the coordinates of the vertices are ________ and ________.
12. If the foci of an ellipse are $(-4, 4)$ and $(6, 4)$, then the coordinates of the center of the ellipse are ______.
    (a) $(1, 4)$ (b) $(4, 1)$
    (c) $(1, 0)$ (d) $(5, 4)$

Skill Building

In Problems 13–16, the graph of an ellipse is given. Match each graph to its equation.

(A) $\frac{x^2}{4} + y^2 = 1$ (B) $x^2 + \frac{y^2}{4} = 1$ (C) $\frac{x^2}{16} + \frac{y^2}{4} = 1$ (D) $\frac{x^2}{4} + \frac{y^2}{16} = 1$

13.
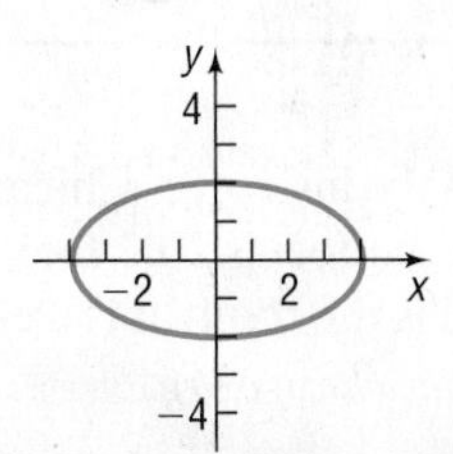

14.
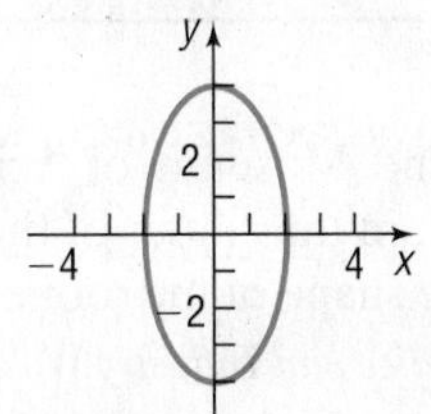

15.
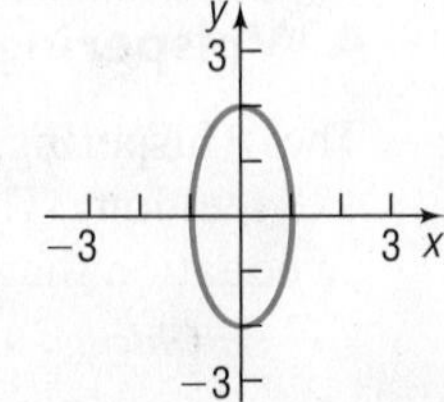

16.
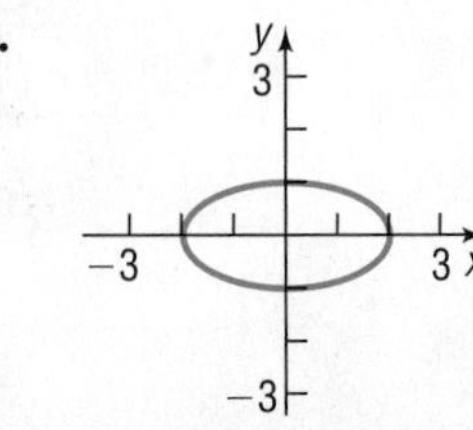

In Problems 17–26, analyze each equation. That is, find the center, vertices, and foci of each ellipse and graph it.

17. $\dfrac{x^2}{25} + \dfrac{y^2}{4} = 1$ **18.** $\dfrac{x^2}{9} + \dfrac{y^2}{4} = 1$ **19.** $\dfrac{x^2}{9} + \dfrac{y^2}{25} = 1$ **20.** $x^2 + \dfrac{y^2}{16} = 1$ **21.** $4x^2 + y^2 = 16$

22. $x^2 + 9y^2 = 18$ **23.** $4y^2 + x^2 = 8$ **24.** $4y^2 + 9x^2 = 36$ **25.** $x^2 + y^2 = 16$ **26.** $x^2 + y^2 = 4$

In Problems 27–38, find an equation for each ellipse. Graph the equation.

27. Center at $(0, 0)$; focus at $(3, 0)$; vertex at $(5, 0)$

28. Center at $(0, 0)$; focus at $(-1, 0)$; vertex at $(3, 0)$

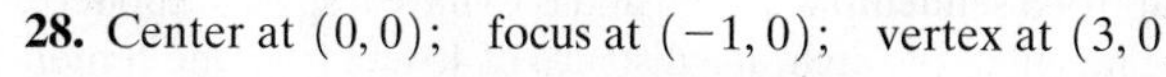

29. Center at $(0, 0)$; focus at $(0, -4)$; vertex at $(0, 5)$

30. Center at $(0, 0)$; focus at $(0, 1)$; vertex at $(0, -2)$

31. Foci at $(\pm 2, 0)$; length of the major axis is 6

32. Foci at $(0, \pm 2)$; length of the major axis is 8

33. Focus at $(-4, 0)$; vertices at $(\pm 5, 0)$

34. Focus at $(0, -4)$; vertices at $(0, \pm 8)$

35. Foci at $(0, \pm 3)$; x-intercepts are ± 2

36. Vertices at $(\pm 4, 0)$; y-intercepts are ± 1

37. Center at $(0, 0)$; vertex at $(0, 4)$; $b = 1$

38. Vertices at $(\pm 5, 0)$; $c = 2$

In Problems 39–42, write an equation for each ellipse.

39.

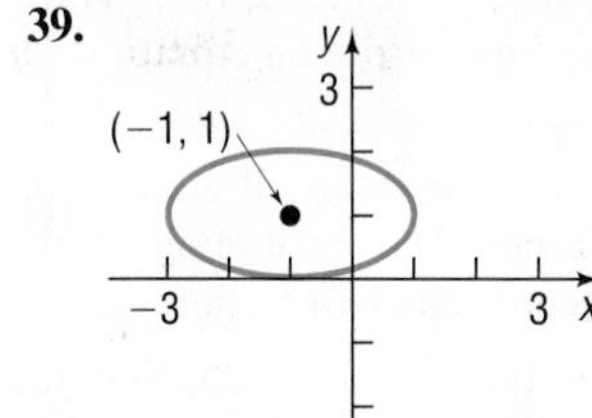

40.

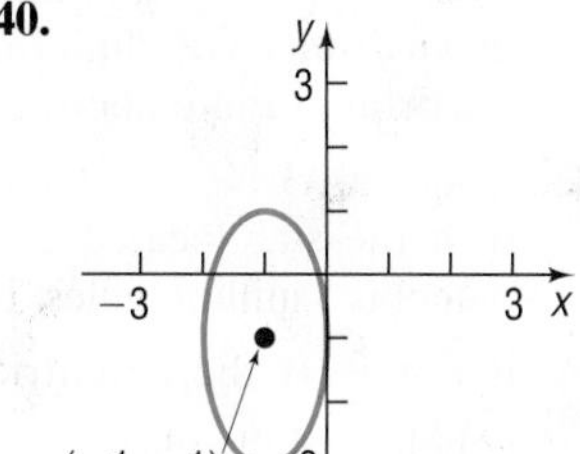

41.

42.

In Problems 43–54, analyze each equation; that is, find the center, foci, and vertices of each ellipse. Graph each equation.

43. $\dfrac{(x-3)^2}{4} + \dfrac{(y+1)^2}{9} = 1$

44. $\dfrac{(x+4)^2}{9} + \dfrac{(y+2)^2}{4} = 1$

45. $(x+5)^2 + 4(y-4)^2 = 16$

46. $9(x-3)^2 + (y+2)^2 = 18$

47. $x^2 + 4x + 4y^2 - 8y + 4 = 0$

48. $x^2 + 3y^2 - 12y + 9 = 0$

49. $2x^2 + 3y^2 - 8x + 6y + 5 = 0$

50. $4x^2 + 3y^2 + 8x - 6y = 5$

51. $9x^2 + 4y^2 - 18x + 16y - 11 = 0$

52. $x^2 + 9y^2 + 6x - 18y + 9 = 0$

53. $4x^2 + y^2 + 4y = 0$

54. $9x^2 + y^2 - 18x = 0$

In Problems 55–64, find an equation for each ellipse. Graph the equation.

55. Center at $(2, -2)$; vertex at $(7, -2)$; focus at $(4, -2)$

56. Center at $(-3, 1)$; vertex at $(-3, 3)$; focus at $(-3, 0)$

57. Vertices at $(4, 3)$ and $(4, 9)$; focus at $(4, 8)$

58. Foci at $(1, 2)$ and $(-3, 2)$; vertex at $(-4, 2)$

59. Foci at $(5, 1)$ and $(-1, 1)$; length of the major axis is 8

60. Vertices at $(2, 5)$ and $(2, -1)$; $c = 2$

61. Center at $(1, 2)$; focus at $(4, 2)$; contains the point $(1, 3)$

62. Center at $(1, 2)$; focus at $(1, 4)$; contains the point $(2, 2)$

63. Center at $(1, 2)$; vertex at $(4, 2)$; contains the point $(1, 5)$

64. Center at $(1, 2)$; vertex at $(1, 4)$; contains the point $(1 + \sqrt{3}, 3)$

In Problems 65–68, graph each function. Be sure to label all the intercepts. **[Hint:** *Notice that each function is half an ellipse.*]

65. $f(x) = \sqrt{16 - 4x^2}$ **66.** $f(x) = \sqrt{9 - 9x^2}$ **67.** $f(x) = -\sqrt{64 - 16x^2}$ **68.** $f(x) = -\sqrt{4 - 4x^2}$

Applications and Extensions

69. Semielliptical Arch Bridge An arch in the shape of the upper half of an ellipse is used to support a bridge that is to span a river 20 meters wide. The center of the arch is 6 meters above the center of the river. See the figure. Write an equation for the ellipse in which the x-axis coincides with the water level and the y-axis passes through the center of the arch.

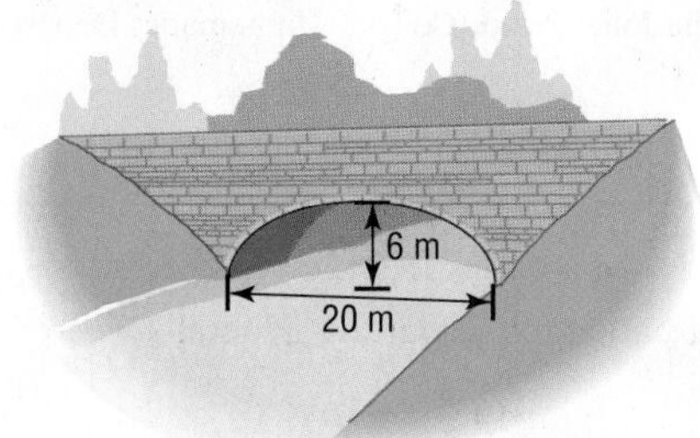

70. Semielliptical Arch Bridge The arch of a bridge is a semiellipse with a horizontal major axis. The span is 30 feet, and the top of the arch is 10 feet above the major axis. The roadway is horizontal and is 2 feet above the top of the arch. Find the vertical distance from the roadway to the arch at 5-foot intervals along the roadway.

71. Whispering Gallery A hall 100 feet in length is to be designed as a whispering gallery. If the foci are located 25 feet from the center, how high will the ceiling be at the center?

72. Whispering Gallery Jim, standing at one focus of a whispering gallery, is 6 feet from the nearest wall. His friend is standing at the other focus, 100 feet away. What is the length of this whispering gallery? How high is its elliptical ceiling at the center?

73. **Semielliptical Arch Bridge** A bridge is built in the shape of a semielliptical arch. The bridge has a span of 120 feet and a maximum height of 25 feet. Choose a suitable rectangular coordinate system and find the height of the arch at distances of 10, 30, and 50 feet from the center.

74. **Semielliptical Arch Bridge** A bridge is to be built in the shape of a semielliptical arch and is to have a span of 100 feet. The height of the arch, at a distance of 40 feet from the center, is to be 10 feet. Find the height of the arch at its center.

75. **Racetrack Design** Consult the figure. A racetrack is in the shape of an ellipse 100 feet long and 50 feet wide. What is the width 10 feet from a vertex?

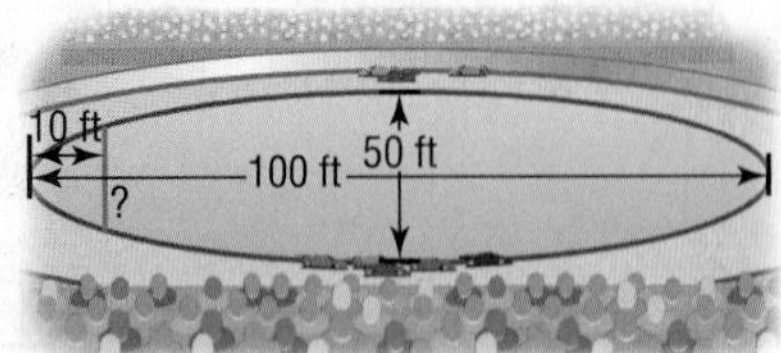

76. **Semielliptical Arch Bridge** An arch for a bridge over a highway is in the form of half an ellipse. The top of the arch is 20 feet above the ground level (the major axis). The highway has four lanes, each 12 feet wide; a center safety strip 8 feet wide; and two side strips, each 4 feet wide. What should the span of the bridge be (the length of its major axis) if the height 28 feet from the center is to be 13 feet?

77. **Installing a Vent Pipe** A homeowner is putting in a fireplace that has a 4-inch-radius vent pipe. He needs to cut an elliptical hole in his roof to accommodate the pipe. If the pitch of his roof is $\frac{5}{4}$ (a rise of 5, run of 4), what are the dimensions of the hole?

Source: www.doe.virginia.gov

78. **Volume of a Football** A football is in the shape of a **prolate spheroid**, which is simply a solid obtained by rotating an ellipse $\left(\frac{x^2}{a^2}+\frac{y^2}{b^2}=1\right)$ about its major axis. An inflated NFL football averages 11.125 inches in length and 28.25 inches in center circumference. If the volume of a prolate spheroid is $\frac{4}{3}\pi ab^2$, how much air does the football contain? (Neglect material thickness).

Source: www.answerbag.com

*In Problems 79–83, use the fact that the orbit of a planet about the Sun is an ellipse, with the Sun at one focus. The **aphelion** of a planet is its greatest distance from the Sun, and the **perihelion** is its shortest distance. The **mean distance** of a planet from the Sun is the length of the semimajor axis of the elliptical orbit. See the illustration.*

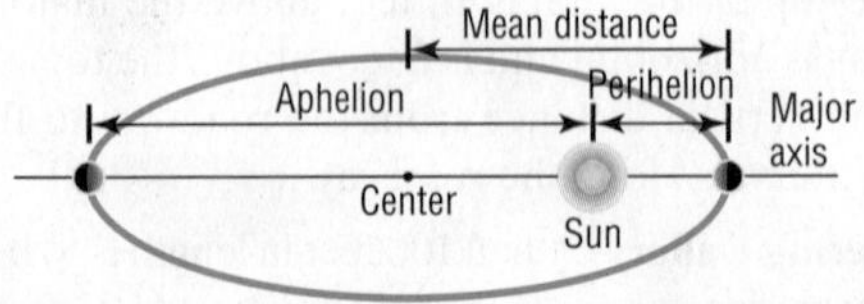

79. **Earth** The mean distance of Earth from the Sun is 93 million miles. If the aphelion of Earth is 94.5 million miles, what is the perihelion? Write an equation for the orbit of Earth around the Sun.

80. **Mars** The mean distance of Mars from the Sun is 142 million miles. If the perihelion of Mars is 128.5 million miles, what is the aphelion? Write an equation for the orbit of Mars about the Sun.

81. **Jupiter** The aphelion of Jupiter is 507 million miles. If the distance from the center of its elliptical orbit to the Sun is 23.2 million miles, what is the perihelion? What is the mean distance? Write an equation for the orbit of Jupiter around the Sun.

82. **Pluto** The perihelion of Pluto is 4551 million miles, and the distance from the center of its elliptical orbit to the Sun is 897.5 million miles. Find the aphelion of Pluto. What is the mean distance of Pluto from the Sun? Write an equation for the orbit of Pluto about the Sun.

83. **Elliptical Orbit** A planet orbits a star in an elliptical orbit with the star located at one focus. The perihelion of the planet is 5 million miles. The **eccentricity** e of a conic section is $e=\frac{c}{a}$. If the eccentricity of the orbit is 0.75, find the aphelion of the planet.*

84. A rectangle is inscribed in an ellipse with major axis of length 14 meters and minor axis of length 4 meters. Find the maximum area of a rectangle inscribed in the ellipse. Round your answer to two decimal places.*

85. Let D be the line given by the equation $x+5=0$. Let E be the conic section given by the equation $x^2+5y^2=20$. Let the point C be the vertex of E with the smaller x-coordinate, and let B be the endpoint of the minor axis of E with the larger y-coordinate. Determine the exact y-coordinate of the point M on D that is equidistant from points B and C.*

86. Show that an equation of the form

$$Ax^2+Cy^2+F=0 \qquad A\neq 0, C\neq 0, F\neq 0$$

where A and C are of the same sign and F is of opposite sign,
(a) is the equation of an ellipse with center at $(0,0)$ if $A\neq C$.
(b) is the equation of a circle with center $(0,0)$ if $A=C$.

87. Show that the graph of an equation of the form

$$Ax^2+Cy^2+Dx+Ey+F=0 \qquad A\neq 0, C\neq 0$$

where A and C are of the same sign,
(a) is an ellipse if $\frac{D^2}{4A}+\frac{E^2}{4C}-F$ is the same sign as A.
(b) is a point if $\frac{D^2}{4A}+\frac{E^2}{4C}-F=0$.
(c) contains no points if $\frac{D^2}{4A}+\frac{E^2}{4C}-F$ is of opposite sign to A.

*Courtesy of the Joliet Junior College Mathematics Department

Discussion and Writing

88. The **eccentricity** e of an ellipse is defined as the number $\frac{c}{a}$, where a is the distance of a vertex from the center and c is the distance of a focus from the center. Because $a>c$, it follows that $e<1$. Write a brief paragraph about the general shape of each of the following ellipses. Be sure to justify your conclusions.

(a) Eccentricity close to 0 (b) Eccentricity $=0.5$ (c) Eccentricity close to 1

Retain Your Knowledge

Problems 89–92 are based on material learned earlier in the course. The purpose of these problems is to keep the material fresh in your mind so that you are better prepared for the final exam.

89. What are the x-intercepts, if any, of the graph of the function $f(x) = (x-5)^2 - 12$?

90. Find the domain of the rational function $f(x) = \dfrac{2x-3}{x-5}$. Find any horizontal or vertical asymptotes.

91. Find the value of $\csc^{-1}(-3.6)$ rounded to two decimal places.

92. Solve the right triangle shown.

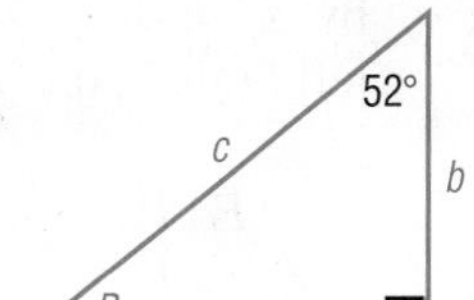

'Are You Prepared?' Answers

1. $\sqrt{13}$ **2.** $\dfrac{9}{4}$ **3.** $(-2,0),(2,0),(0,-4),(0,4)$ **4.** $(2,5)$ **5.** left; 1; down; 4 **6.** $(x-2)^2 + (y+3)^2 = 1$

6.4 The Hyperbola

PREPARING FOR THIS SECTION *Before getting started, review the following:*

- Distance Formula (Section 1.1, p. 3)
- Complete the Square (Appendix A, Section A.3, pp. A24–A25)
- Intercepts (Section 1.2, pp. 10–11)
- Symmetry (Section 1.2, pp. 12–13)
- Asymptotes (Section 1.6, pp. 73–74)
- Graphing Techniques: Transformations (Section 1.6, pp. 64–73)
- Square Root Method (Appendix A, Section A.4, p. A31)

Now Work the **'Are You Prepared?'** problems on page 417.

OBJECTIVES
1 Analyze Hyperbolas with Center at the Origin (p. 407)
2 Find the Asymptotes of a Hyperbola (p. 412)
3 Analyze Hyperbolas with Center at (h, k) (p. 414)
4 Solve Applied Problems Involving Hyperbolas (p. 415)

DEFINITION A **hyperbola** is the collection of all points in the plane the difference of whose distances from two fixed points, called the **foci**, is a constant.

Figure 31 illustrates a hyperbola with foci F_1 and F_2. The line containing the foci is called the **transverse axis**. The midpoint of the line segment joining the foci is the **center** of the hyperbola. The line through the center and perpendicular to the transverse axis is the **conjugate axis**. The hyperbola consists of two separate curves, called **branches**, that are symmetric with respect to the transverse axis, conjugate axis, and center. The two points of intersection of the hyperbola and the transverse axis are the **vertices**, V_1 and V_2, of the hyperbola.

Figure 31 Hyperbola

1 Analyze Hyperbolas with Center at the Origin

With these ideas in mind, we are now ready to find the equation of a hyperbola in the rectangular coordinate system. First, place the center at the origin. Next, position

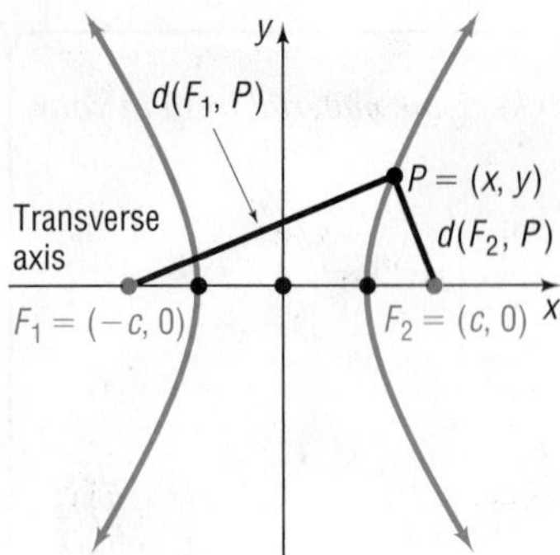

Figure 32
$d(F_1, P) - d(F_2, P) = \pm 2a$

the hyperbola so that its transverse axis coincides with a coordinate axis. Suppose that the transverse axis coincides with the x-axis, as shown in Figure 32.

If c is the distance from the center to a focus, one focus will be at $F_1 = (-c, 0)$ and the other at $F_2 = (c, 0)$. Now we let the constant difference of the distances from any point $P = (x, y)$ on the hyperbola to the foci F_1 and F_2 be denoted by $\pm 2a$, where $a > 0$. (If P is on the right branch, the + sign is used; if P is on the left branch, the − sign is used.) The coordinates of P must satisfy the equation

$$d(F_1, P) - d(F_2, P) = \pm 2a \qquad \text{Difference of the distances from } P \text{ to the foci equals } \pm 2a.$$

$$\sqrt{(x+c)^2 + y^2} - \sqrt{(x-c)^2 + y^2} = \pm 2a \qquad \text{Use the Distance Formula.}$$

$$\sqrt{(x+c)^2 + y^2} = \pm 2a + \sqrt{(x-c)^2 + y^2} \qquad \text{Isolate one radical.}$$

$$(x+c)^2 + y^2 = 4a^2 \pm 4a\sqrt{(x-c)^2 + y^2} + (x-c)^2 + y^2 \qquad \text{Square both sides.}$$

Next multiply out.

$$x^2 + 2cx + c^2 + y^2 = 4a^2 \pm 4a\sqrt{(x-c)^2 + y^2} + x^2 - 2cx + c^2 + y^2$$

$$4cx - 4a^2 = \pm 4a\sqrt{(x-c)^2 + y^2} \qquad \text{Simplify; isolate the radical.}$$

$$cx - a^2 = \pm a\sqrt{(x-c)^2 + y^2} \qquad \text{Divide each side by 4.}$$

$$(cx - a^2)^2 = a^2[(x-c)^2 + y^2] \qquad \text{Square both sides.}$$

$$c^2x^2 - 2ca^2x + a^4 = a^2(x^2 - 2cx + c^2 + y^2) \qquad \text{Multiply out.}$$

$$c^2x^2 + a^4 = a^2x^2 + a^2c^2 + a^2y^2 \qquad \text{Distribute and simplify.}$$

$$(c^2 - a^2)x^2 - a^2y^2 = a^2c^2 - a^4 \qquad \text{Rearrange terms.}$$

$$(c^2 - a^2)x^2 - a^2y^2 = a^2(c^2 - a^2) \qquad \text{Factor } a^2 \text{ on the right side.} \qquad \textbf{(1)}$$

To obtain points on the hyperbola off the x-axis, it must be that $a < c$. To see why, look again at Figure 32.

$$d(F_1, P) < d(F_2, P) + d(F_1, F_2) \qquad \text{Use triangle } F_1PF_2.$$

$$d(F_1, P) - d(F_2, P) < d(F_1, F_2)$$

$$2a < 2c \qquad P \text{ is on the right branch, so } d(F_1, P) - d(F_2, P) = 2a;\ d(F_1, F_2) = 2c.$$

$$a < c$$

Since $a < c$, we also have $a^2 < c^2$, so $c^2 - a^2 > 0$. Let $b^2 = c^2 - a^2$, $b > 0$. Then equation (1) can be written as

$$b^2x^2 - a^2y^2 = a^2b^2$$

$$\frac{x^2}{a^2} - \frac{y^2}{b^2} = 1 \qquad \text{Divide each side by } a^2b^2.$$

To find the vertices of the hyperbola defined by this equation, let $y = 0$. The vertices satisfy the equation $\frac{x^2}{a^2} = 1$, the solutions of which are $x = \pm a$. Consequently, the vertices of the hyperbola are $V_1 = (-a, 0)$ and $V_2 = (a, 0)$. Notice that the distance from the center $(0, 0)$ to either vertex is a.

THEOREM

Equation of a Hyperbola: Center at (0, 0); Transverse Axis along the *x*-Axis

An equation of the hyperbola with center at $(0, 0)$, foci at $(-c, 0)$ and $(c, 0)$, and vertices at $(-a, 0)$ and $(a, 0)$ is

$$\frac{x^2}{a^2} - \frac{y^2}{b^2} = 1 \qquad \text{where } b^2 = c^2 - a^2 \qquad \textbf{(2)}$$

The transverse axis is the x-axis.

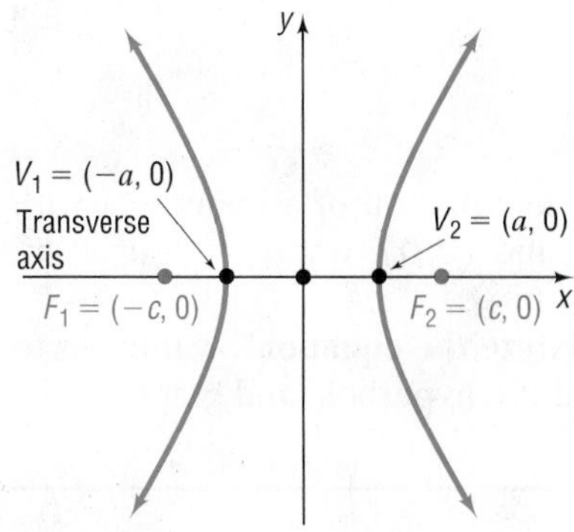

Figure 33 $\frac{x^2}{a^2} - \frac{y^2}{b^2} = 1, \; b^2 = c^2 - a^2$

See Figure 33. As you can verify, the hyperbola defined by equation (2) is symmetric with respect to the x-axis, y-axis, and origin. To find the y-intercepts, if any, let $x = 0$ in equation (2). This results in the equation $\frac{y^2}{b^2} = -1$, which has no real solution, so the hyperbola defined by equation (2) has no y-intercepts. In fact, since $\frac{x^2}{a^2} - 1 = \frac{y^2}{b^2} \geq 0$, it follows that $\frac{x^2}{a^2} \geq 1$. There are no points on the graph for $-a < x < a$.

EXAMPLE 1

Finding and Graphing an Equation of a Hyperbola

Find an equation of the hyperbola with center at the origin, one focus at $(3, 0)$, and one vertex at $(-2, 0)$. Graph the equation.

Solution The hyperbola has its center at the origin. Plot the center, focus, and vertex. Since they all lie on the x-axis, the transverse axis coincides with the x-axis. One focus is at $(c, 0) = (3, 0)$, so $c = 3$. One vertex is at $(-a, 0) = (-2, 0)$, so $a = 2$. From equation (2), it follows that $b^2 = c^2 - a^2 = 9 - 4 = 5$, so an equation of the hyperbola is

$$\frac{x^2}{4} - \frac{y^2}{5} = 1$$

To graph a hyperbola, it is helpful to locate and plot other points on the graph. For example, to find the points above and below the foci, let $x = \pm 3$. Then

$$\frac{x^2}{4} - \frac{y^2}{5} = 1$$

$$\frac{(\pm 3)^2}{4} - \frac{y^2}{5} = 1 \qquad x = \pm 3$$

$$\frac{9}{4} - \frac{y^2}{5} = 1$$

$$\frac{y^2}{5} = \frac{5}{4}$$

$$y^2 = \frac{25}{4}$$

$$y = \pm\frac{5}{2}$$

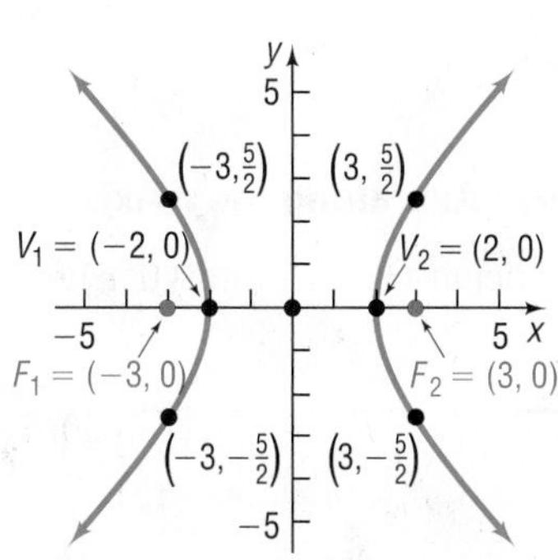

Figure 34 $\frac{x^2}{4} - \frac{y^2}{5} = 1$

The points above and below the foci are $\left(\pm 3, \frac{5}{2}\right)$ and $\left(\pm 3, -\frac{5}{2}\right)$. These points determine the "opening" of the hyperbola. See Figure 34.

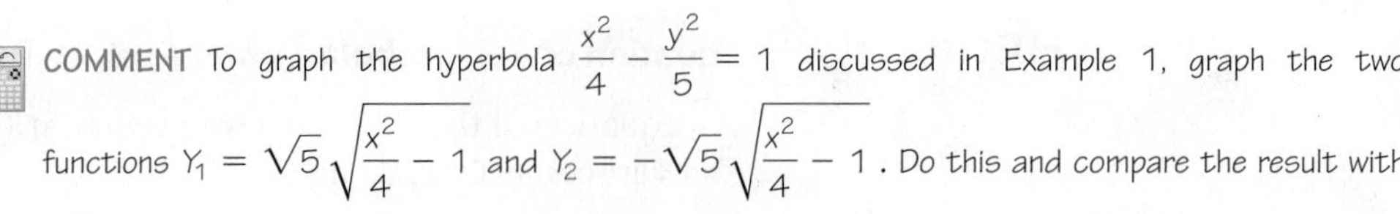

COMMENT To graph the hyperbola $\frac{x^2}{4} - \frac{y^2}{5} = 1$ discussed in Example 1, graph the two functions $Y_1 = \sqrt{5}\sqrt{\frac{x^2}{4} - 1}$ and $Y_2 = -\sqrt{5}\sqrt{\frac{x^2}{4} - 1}$. Do this and compare the result with Figure 34. ■

Now Work PROBLEM 19

An equation of the form of equation (2) is the equation of a hyperbola with center at the origin, foci on the x-axis at $(-c, 0)$ and $(c, 0)$, where $c^2 = a^2 + b^2$, and transverse axis along the x-axis.

For the next two examples, the direction **"Analyze the equation"** will mean to find the center, transverse axis, vertices, and foci of the hyperbola and graph it.

EXAMPLE 2 Analyzing the Equation of a Hyperbola

Analyze the equation: $\frac{x^2}{16} - \frac{y^2}{4} = 1$

Solution The given equation is of the form of equation (2), with $a^2 = 16$ and $b^2 = 4$. The graph of the equation is a hyperbola with center at $(0, 0)$ and transverse axis along the x-axis. Also, $c^2 = a^2 + b^2 = 16 + 4 = 20$. The vertices are at $(\pm a, 0) = (\pm 4, 0)$, and the foci are at $(\pm c, 0) = (\pm 2\sqrt{5}, 0)$.

To locate the points on the graph above and below the foci, let $x = \pm 2\sqrt{5}$. Then

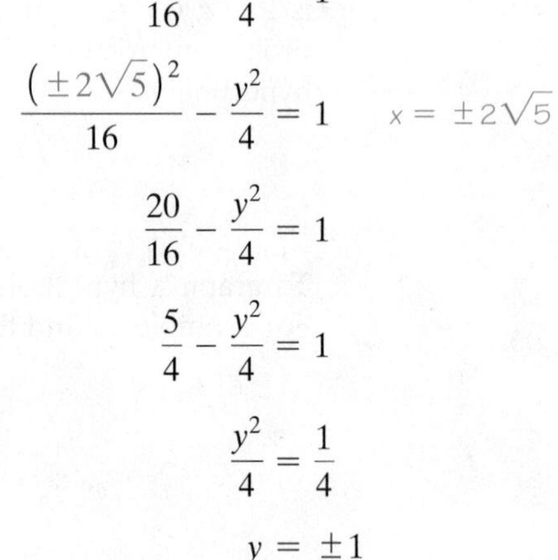

$$\frac{x^2}{16} - \frac{y^2}{4} = 1$$

$$\frac{(\pm 2\sqrt{5})^2}{16} - \frac{y^2}{4} = 1 \qquad x = \pm 2\sqrt{5}$$

$$\frac{20}{16} - \frac{y^2}{4} = 1$$

$$\frac{5}{4} - \frac{y^2}{4} = 1$$

$$\frac{y^2}{4} = \frac{1}{4}$$

$$y = \pm 1$$

The points above and below the foci are $(\pm 2\sqrt{5}, 1)$ and $(\pm 2\sqrt{5}, -1)$. See Figure 35. ●

Figure 35 $\frac{x^2}{16} - \frac{y^2}{4} = 1$

THEOREM

Equation of a Hyperbola: Center at (0, 0); Transverse Axis along the y-Axis

An equation of the hyperbola with center at $(0, 0)$, foci at $(0, -c)$ and $(0, c)$, and vertices at $(0, -a)$ and $(0, a)$ is

$$\frac{y^2}{a^2} - \frac{x^2}{b^2} = 1 \quad \text{where } b^2 = c^2 - a^2 \tag{3}$$

The transverse axis is the y-axis.

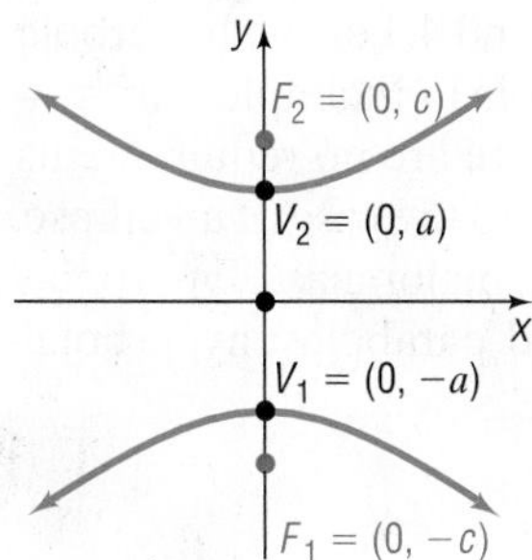

Figure 36 $\dfrac{y^2}{a^2} - \dfrac{x^2}{b^2} = 1,\ b^2 = c^2 - a^2$

Figure 36 shows the graph of a typical hyperbola defined by equation (3).

An equation of the form of equation (2), $\dfrac{x^2}{a^2} - \dfrac{y^2}{b^2} = 1$, is the equation of a hyperbola with center at the origin, foci on the x-axis at $(-c, 0)$ and $(c, 0)$, where $c^2 = a^2 + b^2$, and transverse axis along the x-axis.

An equation of the form of equation (3), $\dfrac{y^2}{a^2} - \dfrac{x^2}{b^2} = 1$, is the equation of a hyperbola with center at the origin, foci on the y-axis at $(0, -c)$ and $(0, c)$, where $c^2 = a^2 + b^2$, and transverse axis along the y-axis.

Notice the difference in the forms of equations (2) and (3). When the y^2-term is subtracted from the x^2-term, the transverse axis is along the x-axis. When the x^2-term is subtracted from the y^2-term, the transverse axis is along the y-axis.

EXAMPLE 3

Analyzing the Equation of a Hyperbola

Analyze the equation: $4y^2 - x^2 = 4$

Solution To put the equation in proper form, divide each side by 4:

$$y^2 - \frac{x^2}{4} = 1$$

Since the x^2-term is subtracted from the y^2-term, the equation is that of a hyperbola with center at the origin and transverse axis along the y-axis. Also, comparing the above equation to equation (3), note that $a^2 = 1, b^2 = 4$, and $c^2 = a^2 + b^2 = 5$. The vertices are at $(0, \pm a) = (0, \pm 1)$, and the foci are at $(0, \pm c) = (0, \pm\sqrt{5})$.

To locate points on the graph to the left and right of the foci, let $y = \pm\sqrt{5}$. Then

$$\begin{aligned} 4y^2 - x^2 &= 4 \\ 4(\pm\sqrt{5})^2 - x^2 &= 4 \qquad y = \pm\sqrt{5} \\ 20 - x^2 &= 4 \\ x^2 &= 16 \\ x &= \pm 4 \end{aligned}$$

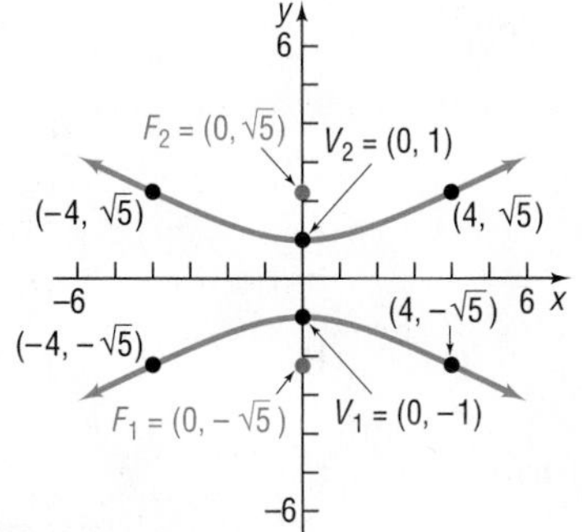

Figure 37 $4y^2 - x^2 = 4$

Four other points on the graph are $(\pm 4, \sqrt{5})$ and $(\pm 4, -\sqrt{5})$. See Figure 37.

EXAMPLE 4

Finding an Equation of a Hyperbola

Find an equation of the hyperbola that has one vertex at $(0, 2)$ and foci at $(0, -3)$ and $(0, 3)$. Graph the equation.

Solution Since the foci are at $(0, -3)$ and $(0, 3)$, the center of the hyperbola, which is at their midpoint, is the origin. Also, the transverse axis is along the y-axis. The given information also reveals that $c = 3, a = 2$, and $b^2 = c^2 - a^2 = 9 - 4 = 5$. The form of the equation of the hyperbola is given by equation (3):

$$\frac{y^2}{a^2} - \frac{x^2}{b^2} = 1$$

$$\frac{y^2}{4} - \frac{x^2}{5} = 1$$

Let $y = \pm 3$ to obtain points on the graph on either side of each focus. See Figure 38.

Figure 38 $\dfrac{y^2}{4} - \dfrac{x^2}{5} = 1$

Now Work PROBLEM 21

Look at the equations of the hyperbolas in Examples 2 and 4. For the hyperbola in Example 2, $a^2 = 16$ and $b^2 = 4$, so $a > b$; for the hyperbola in Example 4, $a^2 = 4$ and $b^2 = 5$, so $a < b$. This indicates that for hyperbolas, there are no requirements involving the relative sizes of a and b. Contrast this situation to the case of an ellipse, in which the relative sizes of a and b dictate which axis is the major axis. Hyperbolas have another feature to distinguish them from ellipses and parabolas: hyperbolas have asymptotes.

2 Find the Asymptotes of a Hyperbola

An **oblique asymptote** of the graph of a funtion $y = R(x)$ is the line $y = mx + b$, $m \neq 0$, if $R(x)$ approaches $mx + b$ as $x \to -\infty$ or as $x \to \infty$. Such asymptotes provide information about the end behavior of the graph of a hyperbola.

THEOREM **Asymptotes of a Hyperbola**

The hyperbola $\dfrac{x^2}{a^2} - \dfrac{y^2}{b^2} = 1$ has the two oblique asymptotes

$$y = \frac{b}{a}x \quad \text{and} \quad y = -\frac{b}{a}x \tag{4}$$

Proof Begin by solving for y in the equation of the hyperbola.

$$\frac{x^2}{a^2} - \frac{y^2}{b^2} = 1$$

$$\frac{y^2}{b^2} = \frac{x^2}{a^2} - 1$$

$$y^2 = b^2\left(\frac{x^2}{a^2} - 1\right)$$

Since $x \neq 0$, the right side can be rearranged in the form

$$y^2 = \frac{b^2 x^2}{a^2}\left(1 - \frac{a^2}{x^2}\right)$$

$$y = \pm\frac{bx}{a}\sqrt{1 - \frac{a^2}{x^2}}$$

Now, as $x \to -\infty$ or as $x \to \infty$, the term $\dfrac{a^2}{x^2}$ approaches 0, so the expression under the radical approaches 1. So as $x \to -\infty$ or as $x \to \infty$, the value of y approaches $\pm\dfrac{bx}{a}$; that is, the graph of the hyperbola approaches the lines

$$y = -\frac{b}{a}x \quad \text{and} \quad y = \frac{b}{a}x$$

These lines are oblique asymptotes of the hyperbola. ■

The asymptotes of a hyperbola are not part of the hyperbola, but they do serve as a guide for graphing a hyperbola. For example, suppose that we want to graph the equation

$$\frac{x^2}{a^2} - \frac{y^2}{b^2} = 1$$

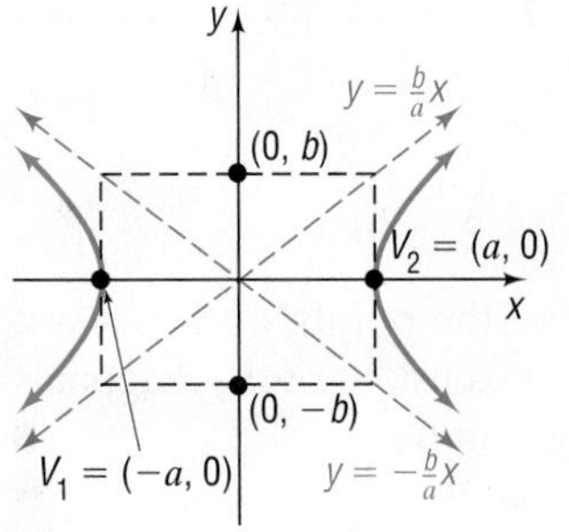

Figure 39 $\frac{x^2}{a^2} - \frac{y^2}{b^2} = 1$

Begin by plotting the vertices $(-a, 0)$ and $(a, 0)$. Then plot the points $(0, -b)$ and $(0, b)$ and use these four points to construct a rectangle, as shown in Figure 39. The diagonals of this rectangle have slopes $\frac{b}{a}$ and $-\frac{b}{a}$, and their extensions are the asymptotes of the hyperbola, $y = \frac{b}{a}x$ and $y = -\frac{b}{a}x$. If we graph the asymptotes, we can use them to establish the "opening" of the hyperbola and avoid plotting other points.

THEOREM

Asymptotes of a Hyperbola

The hyperbola $\frac{y^2}{a^2} - \frac{x^2}{b^2} = 1$ has the two oblique asymptotes

$$y = \frac{a}{b}x \quad \text{and} \quad y = -\frac{a}{b}x \tag{5}$$

You are asked to prove this result in Problem 84.

For the remainder of this section, the direction **"Analyze the equation"** will mean to find the center, transverse axis, vertices, foci, and asymptotes of the hyperbola and graph it.

EXAMPLE 5 Analyzing the Equation of a Hyperbola

Analyze the equation: $\frac{y^2}{4} - x^2 = 1$

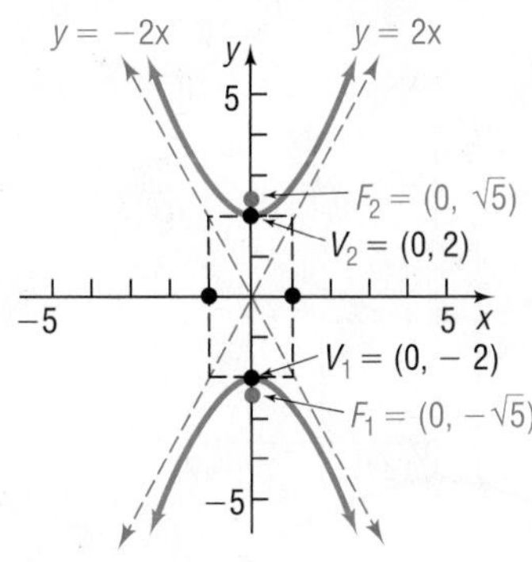

Figure 40 $\frac{y^2}{4} - x^2 = 1$

Solution Since the x^2-term is subtracted from the y^2-term, the equation is of the form of equation (3) and is a hyperbola with center at the origin and transverse axis along the y-axis. Comparing this equation to equation (3), note that $a^2 = 4$, $b^2 = 1$, and $c^2 = a^2 + b^2 = 5$. The vertices are at $(0, \pm a) = (0, \pm 2)$, and the foci are at $(0, \pm c) = (0, \pm\sqrt{5})$. Using equation (5) with $a = 2$ and $b = 1$, the asymptotes are the lines $y = \frac{a}{b}x = 2x$ and $y = -\frac{a}{b}x = -2x$. Form the rectangle containing the points $(0, \pm a) = (0, \pm 2)$ and $(\pm b, 0) = (\pm 1, 0)$. The extensions of the diagonals of this rectangle are the asymptotes. Now graph the asymptotes and the hyperbola. See Figure 40. ●

EXAMPLE 6 Analyzing the Equation of a Hyperbola

Analyze the equation: $9x^2 - 4y^2 = 36$

Solution Divide each side of the equation by 36 to put the equation in proper form.

$$\frac{x^2}{4} - \frac{y^2}{9} = 1$$

The center of the hyperbola is the origin. Since the y^2-term is subtracted from the x^2-term, the transverse axis is along the x-axis, and the vertices and foci will lie on the x-axis. Using equation (2), note that $a^2 = 4$, $b^2 = 9$, and $c^2 = a^2 + b^2 = 13$. The vertices are $a = 2$ units left and right of the center at $(\pm a, 0) = (\pm 2, 0)$, the foci

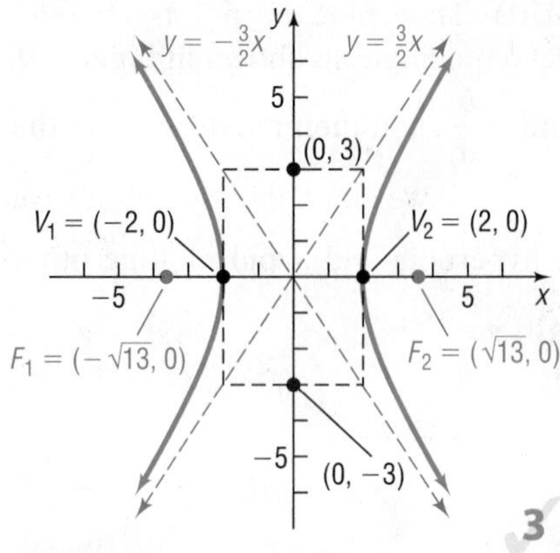

Figure 41 $9x^2 - 4y^2 = 36$

are $c = \sqrt{13}$ units left and right of the center at $(\pm c, 0) = (\pm\sqrt{13}, 0)$, and the asymptotes have the equations

$$y = \frac{b}{a}x = \frac{3}{2}x \quad \text{and} \quad y = -\frac{b}{a}x = -\frac{3}{2}x$$

To graph the hyperbola, form the rectangle containing the points $(\pm a, 0)$ and $(0, \pm b)$, that is, $(-2, 0)$, $(2, 0)$, $(0, -3)$, and $(0, 3)$. The extensions of the diagonals of this rectangle are the asymptotes. See Figure 41 for the graph. ●

Now Work PROBLEM 31

3 Analyze Hyperbolas with Center at (h, k)

If a hyperbola with center at the origin and transverse axis coinciding with a coordinate axis is shifted horizontally h units and then vertically k units, the result is a hyperbola with center at (h, k) and transverse axis parallel to a coordinate axis. The equations of such hyperbolas have the same forms as those given in equations (2) and (3), except that x is replaced by $x - h$ (the horizontal shift) and y is replaced by $y - k$ (the vertical shift). Table 4 gives the forms of the equations of such hyperbolas. See Figure 42 for typical graphs.

Table 4 Equations of a Hyperbola: Center at (h, k); Transverse Axis Parallel to a Coordinate Axis

Center	Transverse Axis	Foci	Vertices	Equation	Asymptotes
(h, k)	Parallel to the x-axis	$(h \pm c, k)$	$(h \pm a, k)$	$\frac{(x-h)^2}{a^2} - \frac{(y-k)^2}{b^2} = 1, \; b^2 = c^2 - a^2$	$y - k = \pm\frac{b}{a}(x - h)$
(h, k)	Parallel to the y-axis	$(h, k \pm c)$	$(h, k \pm a)$	$\frac{(y-k)^2}{a^2} - \frac{(x-h)^2}{b^2} = 1, \; b^2 = c^2 - a^2$	$y - k = \pm\frac{a}{b}(x - h)$

NOTE It is not recommended that Table 4 be memorized. Rather, use transformations (shift horizontally h units, vertically k units), along with the fact that a represents the distance from the center to the vertices, c represents the distance from the center to the foci, and $b^2 = c^2 - a^2$ (or $c^2 = a^2 + b^2$). ■

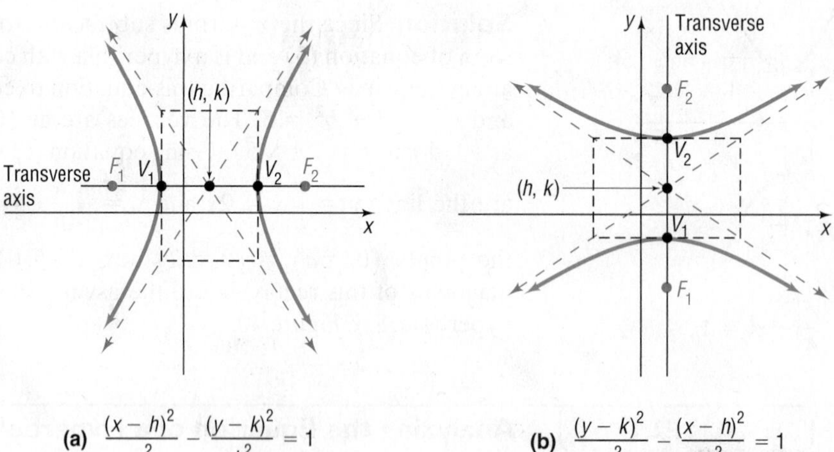

Figure 42 (a) $\frac{(x-h)^2}{a^2} - \frac{(y-k)^2}{b^2} = 1$ (b) $\frac{(y-k)^2}{a^2} - \frac{(x-h)^2}{b^2} = 1$

EXAMPLE 7 **Finding an Equation of a Hyperbola, Center Not at the Origin**

Find an equation for the hyperbola with center at $(1, -2)$, one focus at $(4, -2)$, and one vertex at $(3, -2)$. Graph the equation.

Solution The center is at $(h, k) = (1, -2)$, so $h = 1$ and $k = -2$. Since the center, focus, and vertex all lie on the line $y = -2$, the transverse axis is parallel to the x-axis. The distance from the center $(1, -2)$ to the focus $(4, -2)$ is $c = 3$; the distance from

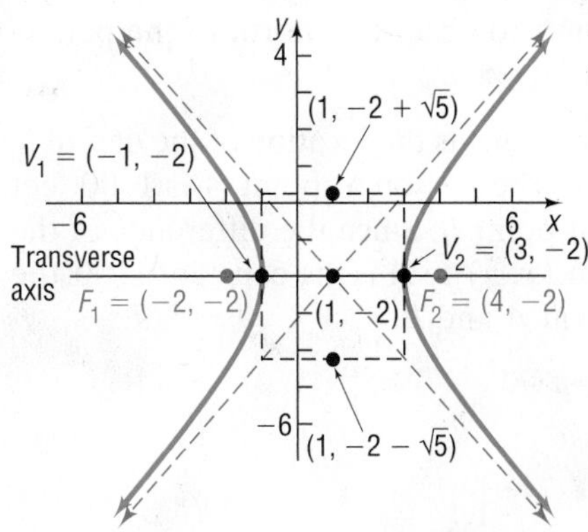

Figure 43 $\frac{(x-1)^2}{4} - \frac{(y+2)^2}{5} = 1$

the center $(1, -2)$ to the vertex $(3, -2)$ is $a = 2$. Then $b^2 = c^2 - a^2 = 9 - 4 = 5$. The equation is

$$\frac{(x-h)^2}{a^2} - \frac{(y-k)^2}{b^2} = 1$$

$$\frac{(x-1)^2}{4} - \frac{(y+2)^2}{5} = 1$$

See Figure 43.

Now Work PROBLEM 41

EXAMPLE 8 Analyzing the Equation of a Hyperbola

Analyze the equation: $-x^2 + 4y^2 - 2x - 16y + 11 = 0$

Solution Complete the squares in x and in y.

$$-x^2 + 4y^2 - 2x - 16y + 11 = 0$$

$$-(x^2 + 2x) + 4(y^2 - 4y) = -11 \quad \text{Group terms.}$$

$$-(x^2 + 2x + 1) + 4(y^2 - 4y + 4) = -11 - 1 + 16 \quad \text{Complete each square.}$$

$$-(x+1)^2 + 4(y-2)^2 = 4 \quad \text{Factor.}$$

$$(y-2)^2 - \frac{(x+1)^2}{4} = 1 \quad \text{Divide each side by 4.}$$

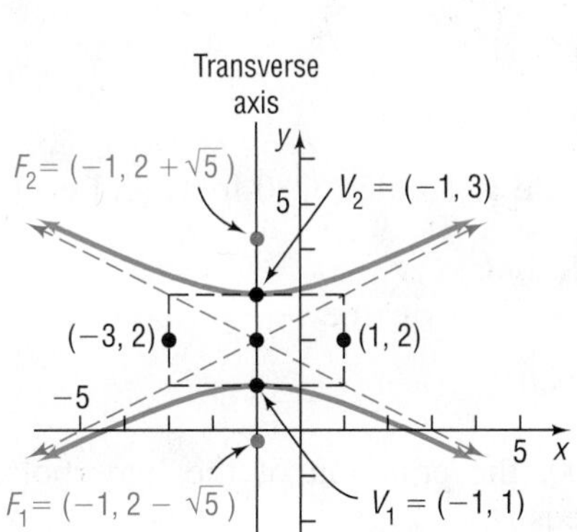

Figure 44
$-x^2 + 4y^2 - 2x - 16y + 11 = 0$

This is the equation of a hyperbola with center at $(-1, 2)$ and transverse axis parallel to the y-axis. Also, $a^2 = 1$ and $b^2 = 4$, so $c^2 = a^2 + b^2 = 5$. Since the transverse axis is parallel to the y-axis, the vertices and foci are located a and c units above and below the center, respectively. The vertices are at $(h, k \pm a) = (-1, 2 \pm 1)$, or $(-1, 1)$ and $(-1, 3)$. The foci are at $(h, k \pm c) = (-1, 2 \pm \sqrt{5})$. The asymptotes are $y - 2 = \frac{1}{2}(x+1)$ and $y - 2 = -\frac{1}{2}(x+1)$. Figure 44 shows the graph.

Now Work PROBLEM 55

4 Solve Applied Problems Involving Hyperbolas

Look at Figure 45. Suppose that three microphones are located at points O_1, O_2, and O_3 (the foci of the two hyperbolas). In addition, suppose that a gun is fired at S and the microphone at O_1 records the gunshot 1 second after the microphone at O_2. Because sound travels at about 1100 feet per second, we conclude that the microphone at O_1 is 1100 feet farther from the gunshot than O_2. We can model this situation by saying that S lies on a branch of a hyperbola with foci at O_1 and O_2. (Do you see why? The difference of the distances from S to O_1 and from S to O_2 is the constant 1100.) If the third microphone at O_3 records the gunshot 2 seconds after O_1, then S will lie on a branch of a second hyperbola with foci at O_1 and O_3. In this case, the constant difference will be 2200. The intersection of the two hyperbolas will identify the location of S.

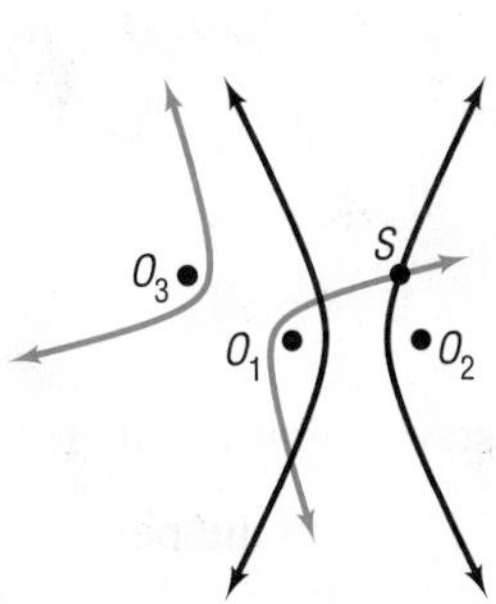

Figure 45

EXAMPLE 9 Lightning Strikes

Suppose that two people standing 1 mile apart both see a flash of lightning. After a period of time, the person standing at point A hears the thunder. One second later, the person standing at point B hears the thunder. If the person at B is due west of the person at A and the lightning strike is known to occur due north of the person standing at point A, where did the lightning strike occur?

Solution See Figure 46, in which the ordered pair (x, y) represents the location of the lightning strike. Sound travels at 1100 feet per second, so the person at point A is 1100 feet closer to the lightning strike than the person at point B. Since the difference of the distance from (x, y) to B and the distance from (x, y) to A is the constant 1100, the point (x, y) lies on a hyperbola whose foci are at A and B.

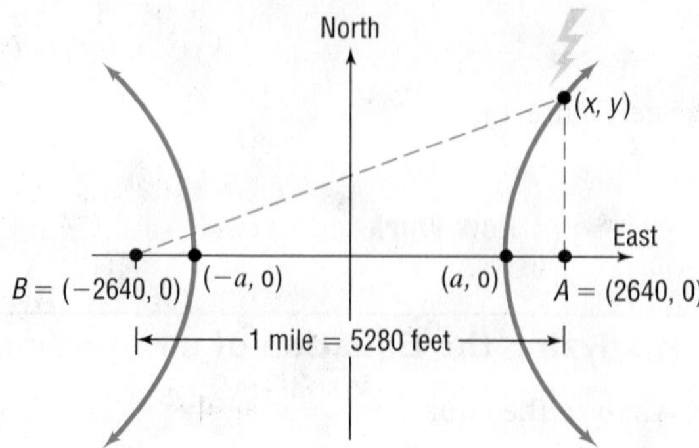

Figure 46

An equation of the hyperbola is

$$\frac{x^2}{a^2} - \frac{y^2}{b^2} = 1$$

where $2a = 1100$, so $a = 550$.

Because the distance between the two people is 1 mile (5280 feet) and each person is at a focus of the hyperbola, this means

$$2c = 5280$$

$$c = \frac{5280}{2} = 2640$$

Since $b^2 = c^2 - a^2 = 2640^2 - 550^2 = 6{,}667{,}100$, the equation of the hyperbola that describes the location of the lightning strike is

$$\frac{x^2}{550^2} - \frac{y^2}{6{,}667{,}100} = 1$$

Refer to Figure 46. Since the lightning strike occurred due north of the individual at the point $A = (2640, 0)$, let $x = 2640$ and solve the resulting equation.

$$\frac{2640^2}{550^2} - \frac{y^2}{6{,}667{,}100} = 1 \qquad x = 2640$$

$$-\frac{y^2}{6{,}667{,}100} = -22.04 \qquad \text{Subtract } \frac{2640^2}{550^2} \text{ from both sides.}$$

$$y^2 = 146{,}942{,}884 \qquad \text{Multiply both sides by } -6{,}667{,}100.$$

$$y = 12{,}122 \qquad y > 0 \text{ since the lightning strike occurred in quadrant I.}$$

The lightning strike occurred 12,122 feet north of the person standing at point A.

Check: The difference between the distance from $(2640, 12122)$ to the person at the point $B = (-2640, 0)$ and the distance from $(2640, 12122)$ to the person at the point $A = (2640, 0)$ should be 1100. Using the distance formula, the difference of the distances is

$$\sqrt{[2640 - (-2640)]^2 + (12{,}122 - 0)^2} - \sqrt{(2640 - 2640)^2 + (12{,}122 - 0)^2} = 1100$$

as required.

Now Work PROBLEM 75

6.4 Assess Your Understanding

'Are You Prepared?' Answers are given at the end of these exercises. If you get a wrong answer, read the pages listed in red.

1. The distance d from $P_1 = (3, -4)$ to $P_2 = (-2, 1)$ is $d =$ ____. (p. 3)
2. To complete the square of $x^2 + 5x$, add ____. (pp. A24–A25)
3. Find the intercepts of the equation $y^2 = 9 + 4x^2$. (pp. 10–11)
4. ***True or False*** The equation $y^2 = 9 + x^2$ is symmetric with respect to the x-axis, the y-axis, and the origin. (pp. 12–13)
5. To graph $y = (x - 5)^3 - 4$, shift the graph of $y = x^3$ to the (left/right) ___ unit(s) and then (up/down) ___ unit(s). (pp. 64–73)
6. Find the vertical asymptotes, if any, and the horizontal asymptote, if any, of $y = \dfrac{x^2 - 9}{x^2 - 4}$. (pp. 73–74)

Concepts and Vocabulary

7. A(n) ________ is the collection of points in the plane the difference of whose distances from two fixed points is a constant.
8. For a hyperbola, the foci lie on a line called the ________ _____.

Answer Problems 9–11 using the figure to the right.

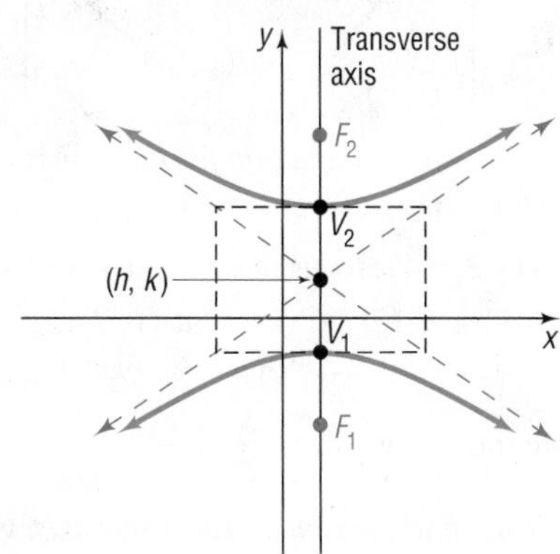

9. The equation of the hyperbola is of the form
   (a) $\dfrac{(x-h)^2}{a^2} - \dfrac{(y-k)^2}{b^2} = 1$
   (b) $\dfrac{(y-k)^2}{a^2} - \dfrac{(x-h)^2}{b^2} = 1$
   (c) $\dfrac{(x-h)^2}{a^2} + \dfrac{(y-k)^2}{b^2} = 1$
   (d) $\dfrac{(x-h)^2}{b^2} + \dfrac{(y-k)^2}{a^2} = 1$
10. If the center of the hyperbola is (2, 1) and $a = 3$, then the coordinates of the vertices are _____ and ______.
11. If the center of the hyperbola is (2, 1) and $c = 5$, then the coordinates of the foci are _____ and ______.
12. In a hyperbola, if $a = 3$ and $c = 5$, then $b =$ _____.
    (a) 1 (b) 2 (c) 4 (d) 8
13. For the hyperbola $\dfrac{x^2}{4} - \dfrac{y^2}{9} = 1$, the value of a is ____, the value of b is ____, and the transverse axis is the ____ -axis.
14. For the hyperbola $\dfrac{y^2}{16} - \dfrac{x^2}{81} = 1$, the asymptotes are

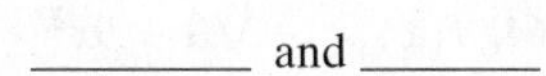
________ and _______.

Skill Building

In Problems 15–18, the graph of a hyperbola is given. Match each graph to its equation.

(A) $\dfrac{x^2}{4} - y^2 = 1$ (B) $x^2 - \dfrac{y^2}{4} = 1$ (C) $\dfrac{y^2}{4} - x^2 = 1$ (D) $y^2 - \dfrac{x^2}{4} = 1$

15. 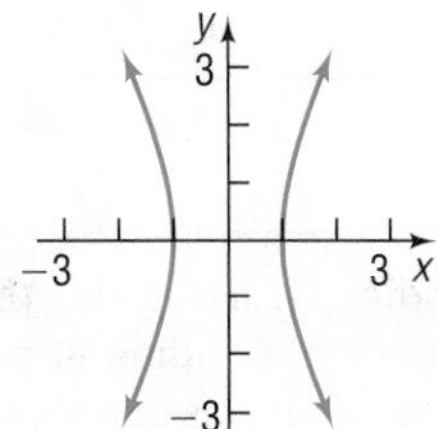

16. 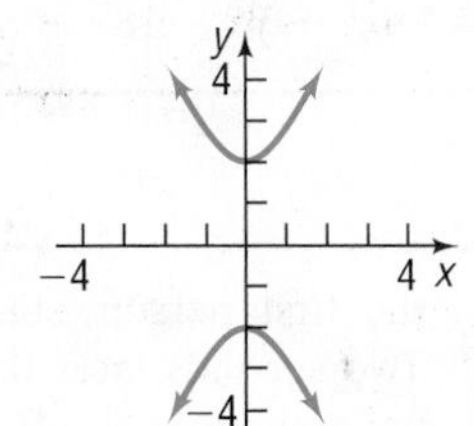

17. 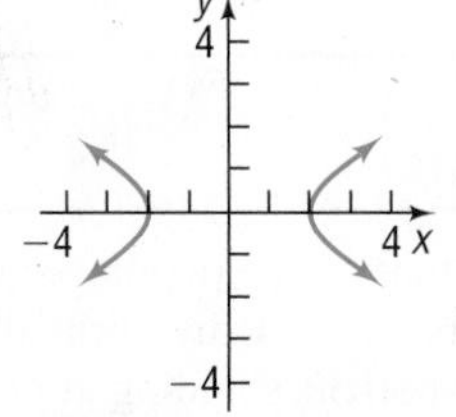

18. 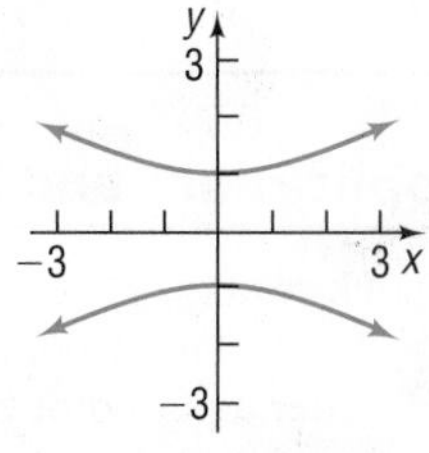

In Problems 19–28, find an equation for the hyperbola described. Graph the equation.

19. Center at $(0, 0)$; focus at $(3, 0)$; vertex at $(1, 0)$
20. Center at $(0, 0)$; focus at $(0, 5)$; vertex at $(0, 3)$
21. Center at $(0, 0)$; focus at $(0, -6)$; vertex at $(0, 4)$
22. Center at $(0, 0)$; focus at $(-3, 0)$; vertex at $(2, 0)$
23. Foci at $(-5, 0)$ and $(5, 0)$; vertex at $(3, 0)$
24. Focus at $(0, 6)$; vertices at $(0, -2)$ and $(0, 2)$

25. Vertices at $(0, -6)$ and $(0, 6)$; asymptote the line $y = 2x$

26. Vertices at $(-4, 0)$ and $(4, 0)$; asymptote the line $y = 2x$

27. Foci at $(-4, 0)$ and $(4, 0)$; asymptote the line $y = -x$

28. Foci at $(0, -2)$ and $(0, 2)$; asymptote the line $y = -x$

In Problems 29–36, find the center, transverse axis, vertices, foci, and asymptotes. Graph each equation.

29. $\dfrac{x^2}{25} - \dfrac{y^2}{9} = 1$ **30.** $\dfrac{y^2}{16} - \dfrac{x^2}{4} = 1$ **31.** $4x^2 - y^2 = 16$ **32.** $4y^2 - x^2 = 16$

33. $y^2 - 9x^2 = 9$ **34.** $x^2 - y^2 = 4$ **35.** $y^2 - x^2 = 25$ **36.** $2x^2 - y^2 = 4$

In Problems 37–40, write an equation for each hyperbola.

37.

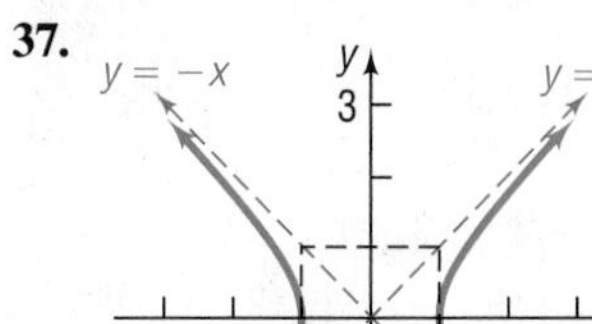

38.

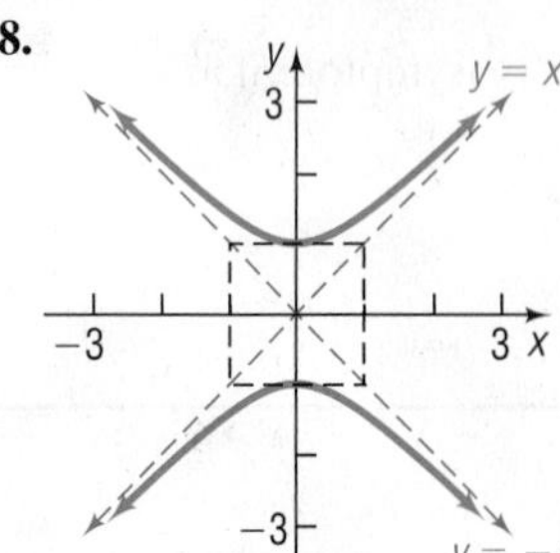

39.

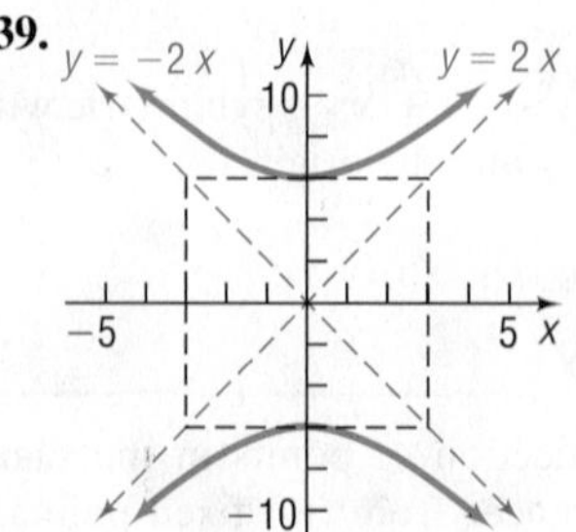

40.

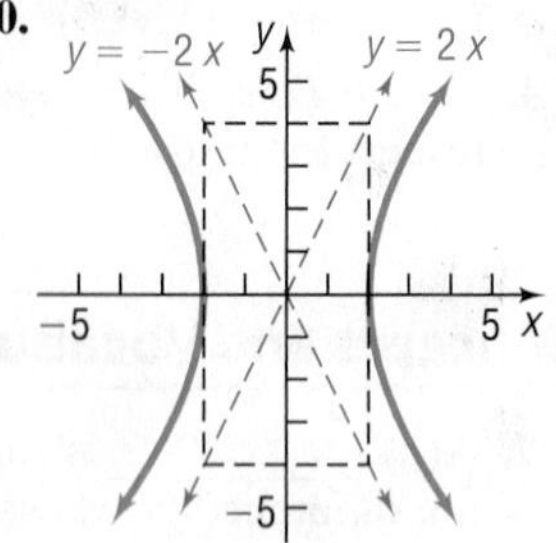

In Problems 41–48, find an equation for the hyperbola described. Graph the equation.

41. Center at $(4, -1)$; focus at $(7, -1)$; vertex at $(6, -1)$

42. Center at $(-3, 1)$; focus at $(-3, 6)$; vertex at $(-3, 4)$

43. Center at $(-3, -4)$; focus at $(-3, -8)$; vertex at $(-3, -2)$

44. Center at $(1, 4)$; focus at $(-2, 4)$; vertex at $(0, 4)$

45. Foci at $(3, 7)$ and $(7, 7)$; vertex at $(6, 7)$

46. Focus at $(-4, 0)$ vertices at $(-4, 4)$ and $(-4, 2)$

47. Vertices at $(-1, -1)$ and $(3, -1)$; asymptote the line $y + 1 = \dfrac{3}{2}(x - 1)$

48. Vertices at $(1, -3)$ and $(1, 1)$; asymptote the line $y + 1 = \dfrac{3}{2}(x - 1)$

In Problems 49–62, find the center, transverse axis, vertices, foci, and asymptotes. Graph each equation.

49. $\dfrac{(x-2)^2}{4} - \dfrac{(y+3)^2}{9} = 1$ **50.** $\dfrac{(y+3)^2}{4} - \dfrac{(x-2)^2}{9} = 1$ **51.** $(y-2)^2 - 4(x+2)^2 = 4$

52. $(x+4)^2 - 9(y-3)^2 = 9$ **53.** $(x+1)^2 - (y+2)^2 = 4$ **54.** $(y-3)^2 - (x+2)^2 = 4$

55. $x^2 - y^2 - 2x - 2y - 1 = 0$ **56.** $y^2 - x^2 - 4y + 4x - 1 = 0$ **57.** $y^2 - 4x^2 - 4y - 8x - 4 = 0$

58. $2x^2 - y^2 + 4x + 4y - 4 = 0$ **59.** $4x^2 - y^2 - 24x - 4y + 16 = 0$ **60.** $2y^2 - x^2 + 2x + 8y + 3 = 0$

61. $y^2 - 4x^2 - 16x - 2y - 19 = 0$ **62.** $x^2 - 3y^2 + 8x - 6y + 4 = 0$

In Problems 63–66, graph each function. Be sure to label any intercepts. [**Hint:** *Notice that each function is half a hyperbola.*]

63. $f(x) = \sqrt{16 + 4x^2}$ **64.** $f(x) = -\sqrt{9 + 9x^2}$ **65.** $f(x) = -\sqrt{-25 + x^2}$ **66.** $f(x) = \sqrt{-1 + x^2}$

Mixed Practice

In Problems 67–74, analyze each equation.

67. $\dfrac{(x-3)^2}{4} - \dfrac{y^2}{25} = 1$ **68.** $\dfrac{(y+2)^2}{16} - \dfrac{(x-2)^2}{4} = 1$ **69.** $x^2 = 16(y - 3)$

70. $y^2 = -12(x + 1)$ **71.** $25x^2 + 9y^2 - 250x + 400 = 0$ **72.** $x^2 + 36y^2 - 2x + 288y + 541 = 0$

73. $x^2 - 6x - 8y - 31 = 0$ **74.** $9x^2 - y^2 - 18x - 8y - 88 = 0$

Applications and Extensions

75. Fireworks Display Suppose that two people standing 2 miles apart both see the burst from a fireworks display. After a period of time the first person, standing at point A, hears the burst. One second later the second person, standing at point B, hears the burst. If the person at point B is due west of the person at point A, and if the display is known to occur due north of the person at point A, where did the fireworks display occur?

76. Lightning Strikes Suppose that two people standing 1 mile apart both see a flash of lightning. After a period of time the first person, standing at point A, hears the thunder. Two seconds later the second person, standing at point B, hears the thunder. If the person at point B is due west of the person at point A, and if the lightning strike is known to occur due north of the person standing at point A, where did the lightning strike occur?

77. Nuclear Power Plant Some nuclear power plants utilize "natural draft" cooling towers in the shape of a **hyperboloid**, a solid obtained by rotating a hyperbola about its conjugate axis. Suppose that such a cooling tower has a base diameter of 400 feet and the diameter at its narrowest point, 360 feet above the ground, is 200 feet. If the diameter at the top of the tower is 300 feet, how tall is the tower?

Source: Bay Area Air Quality Management District

78. An Explosion Two recording devices are set 2400 feet apart, with the device at point A to the west of the device at point B. At a point between the devices 300 feet from point B, a small amount of explosive is detonated. The recording devices record the time until the sound reaches each. How far directly north of point B should a second explosion be done so that the measured time difference recorded by the devices is the same as that for the first detonation?

79. Rutherford's Experiment In May 1911, Ernest Rutherford published a paper in *Philosophical Magazine*. In this article, he described the motion of alpha particles as they are shot at a piece of gold foil 0.00004 cm thick. Before conducting this experiment, Rutherford expected that the alpha particles would shoot through the foil just as a bullet would shoot through snow. Instead, a small fraction of the alpha particles bounced off the foil. This led to the conclusion that the nucleus of an atom is dense, while the remainder of the atom is sparse. Only the density of the nucleus could cause the alpha particles to deviate from their path. The figure shows a diagram from Rutherford's paper that indicates that the deflected alpha particles follow the path of one branch of a hyperbola.

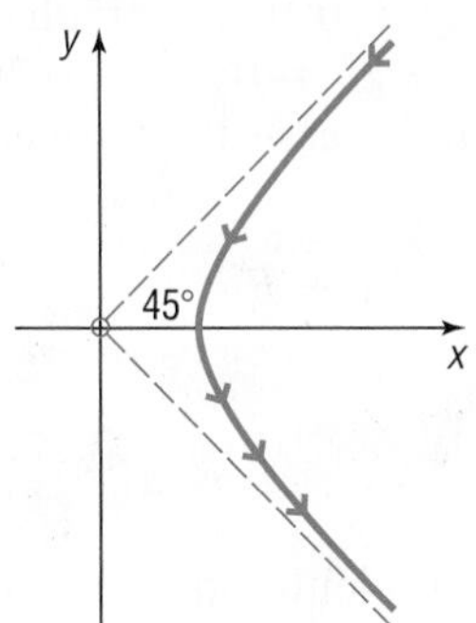

(a) Find an equation of the asymptotes under this scenario.
(b) If the vertex of the path of the alpha particles is 10 cm from the center of the hyperbola, find a model that describes the path of the particle.

80. Hyperbolic Mirrors Hyperbolas have interesting reflective properties that make them useful for lenses and mirrors. For example, if a ray of light strikes a convex hyperbolic mirror on a line that would (theoretically) pass through its rear focus, it is reflected through the front focus. This property, and that of the parabola, were used to develop the *Cassegrain* telescope in 1672. The focus of the parabolic mirror and the rear focus of the hyperbolic mirror are the same point. The rays are collected by the parabolic mirror, then are reflected toward the (common) focus, and thus are reflected by the hyperbolic mirror through the opening to its front focus, where the eyepiece is located. If the equation of the hyperbola is $\frac{y^2}{9} - \frac{x^2}{16} = 1$ and the focal length (distance from the vertex to the focus) of the parabola is 6, find the equation of the parabola.

Source: www.enchantedlearning.com

81. The **eccentricity** e of a hyperbola is defined as the number $\frac{c}{a}$, where a is the distance of a vertex from the center and c is the distance of a focus from the center. Because $c > a$, it follows that $e > 1$. Describe the general shape of a hyperbola whose eccentricity is close to 1. What is the shape if e is very large?

82. A hyperbola for which $a = b$ is called an **equilateral hyperbola**. Find the eccentricity e of an equilateral hyperbola.

[**Note:** The eccentricity of a hyperbola is defined in Problem 81.]

83. Two hyperbolas that have the same set of asymptotes are called **conjugate**. Show that the hyperbolas

$$\frac{x^2}{4} - y^2 = 1 \quad \text{and} \quad y^2 - \frac{x^2}{4} = 1$$

are conjugate. Graph each hyperbola on the same set of coordinate axes.

84. Prove that the hyperbola

$$\frac{y^2}{a^2} - \frac{x^2}{b^2} = 1$$

has the two oblique asymptotes

$$y = \frac{a}{b}x \quad \text{and} \quad y = -\frac{a}{b}x$$

85. Show that the graph of an equation of the form

$$Ax^2 + Cy^2 + F = 0 \qquad A \neq 0, C \neq 0, F \neq 0$$

where A and C are opposite in sign, is a hyperbola with center at $(0, 0)$.

86. Show that the graph of an equation of the form

$$Ax^2 + Cy^2 + Dx + Ey + F = 0 \qquad A \neq 0, C \neq 0$$

where A and C are opposite in sign,

(a) is a hyperbola if $\frac{D^2}{4A} + \frac{E^2}{4C} - F \neq 0$.
(b) is two intersecting lines if $\frac{D^2}{4A} + \frac{E^2}{4C} - F = 0$.

Retain Your Knowledge

Problems 87–90 are based on material learned earlier in the course. The purpose of these problems is to keep the material fresh in your mind so that you are better prepared for the final exam.

87. For $y = \frac{3}{2}\cos(6x + 3\pi)$, find the amplitude, period, and phase shift. Then graph the function, showing at least two periods.

88. Solve the triangle described: $a = 7, b = 10,$ and $C = 100°$.

89. Find the rectangular coordinates of the point with the polar coordinates $\left(12, -\frac{\pi}{3}\right)$.

90. Transform the polar equation $r = 6\sin\theta$ to an equation in rectangular coordinates. Then identify and graph the equation.

'Are You Prepared?' Answers

1. $5\sqrt{2}$ **2.** $\frac{25}{4}$ **3.** $(0, -3), (0, 3)$ **4.** True **5.** right; 5; down; 4 **6.** Vertical: $x = -2, x = 2$; horizontal: $y = 1$

6.5 Rotation of Axes; General Form of a Conic

PREPARING FOR THIS SECTION *Before getting started, review the following:*

- Sum Formulas for Sine and Cosine (Section 3.5, pp. 229 and 226)
- Half-angle Formulas for Sine and Cosine (Section 3.6, p. 242)
- Double-angle Formulas for Sine and Cosine (Section 3.6, p. 239)

Now Work the 'Are You Prepared?' problems on page 426.

OBJECTIVES
1 Identify a Conic (p. 420)
2 Use a Rotation of Axes to Transform Equations (p. 421)
3 Analyze an Equation Using a Rotation of Axes (p. 423)
4 Identify Conics without a Rotation of Axes (p. 425)

In this section, we show that the graph of a general second-degree polynomial equation containing two variables x and y—that is, an equation of the form

$$Ax^2 + Bxy + Cy^2 + Dx + Ey + F = 0 \quad \textbf{(1)}$$

where A, B, and C are not simultaneously 0—is a conic. We shall not concern ourselves here with the degenerate cases of equation (1), such as $x^2 + y^2 = 0$, whose graph is a single point $(0, 0)$; or $x^2 + 3y^2 + 3 = 0$, whose graph contains no points; or $x^2 - 4y^2 = 0$, whose graph is two lines, $x - 2y = 0$ and $x + 2y = 0$.

We begin with the case where $B = 0$. In this case, the term containing xy is not present, so equation (1) has the form

$$Ax^2 + Cy^2 + Dx + Ey + F = 0$$

where either $A \neq 0$ or $C \neq 0$.

1 Identify a Conic

We have already discussed the procedure for identifying the graph of this kind of equation; we complete the squares of the quadratic expressions in x or y, or both. Once this has been done, the conic can be identified by comparing it to one of the forms studied in Sections 6.2 through 6.4.

In fact, though, the conic can be identified directly from the equation without completing the squares.

THEOREM **Identifying Conics without Completing the Squares**

Excluding degenerate cases, the equation

$$Ax^2 + Cy^2 + Dx + Ey + F = 0 \quad \textbf{(2)}$$

where A and C are not both equal to zero:
(a) Defines a parabola if $AC = 0$.
(b) Defines an ellipse (or a circle) if $AC > 0$.
(c) Defines a hyperbola if $AC < 0$.

Proof

(a) If $AC = 0$, then either $A = 0$ or $C = 0$, but not both, so the form of equation (2) is either

$$Ax^2 + Dx + Ey + F = 0 \qquad A \neq 0$$

or

$$Cy^2 + Dx + Ey + F = 0 \qquad C \neq 0$$

Using the results of Problems 80 and 81 at the end of Section 6.2, it follows that, except for the degenerate cases, the equation is a parabola.

(b) If $AC > 0$, then A and C have the same sign. Using the results of Problem 87 at the end of Section 6.3, except for the degenerate cases, the equation is an ellipse.

(c) If $AC < 0$, then A and C have opposite signs. Using the results of Problem 86 at the end of Section 6.4, except for the degenerate cases, the equation is a hyperbola. ■

We will not be concerned with the degenerate cases of equation (2). However, in practice, you should be alert to the possibility of degeneracy.

EXAMPLE 1 **Identifying a Conic without Completing the Squares**

Identify the graph of each equation without completing the squares.

(a) $3x^2 + 6y^2 + 6x - 12y = 0$ (b) $2x^2 - 3y^2 + 6y + 4 = 0$

(c) $y^2 - 2x + 4 = 0$

Solution (a) Compare the given equation to equation (2), and conclude that $A = 3$ and $C = 6$. Since $AC = 18 > 0$, the equation defines an ellipse.

(b) Here $A = 2$ and $C = -3$, so $AC = -6 < 0$. The equation defines a hyperbola.

(c) Here $A = 0$ and $C = 1$, so $AC = 0$. The equation defines a parabola. ●

Now Work PROBLEM 11

Although we can now identify the type of conic represented by any equation of the form of equation (2) without completing the squares, we still need to complete the squares if we desire additional information about the conic, such as its graph.

2 Use a Rotation of Axes to Transform Equations

Now let's turn our attention to equations of the form of equation (1), where $B \neq 0$. To discuss this case, we introduce a new procedure: *rotation of axes.*

In a **rotation of axes**, the origin remains fixed while the x-axis and y-axis are rotated through an angle θ to a new position; the new positions of the x-axis and the y-axis are denoted by x' and y', respectively, as shown in Figure 47(a).

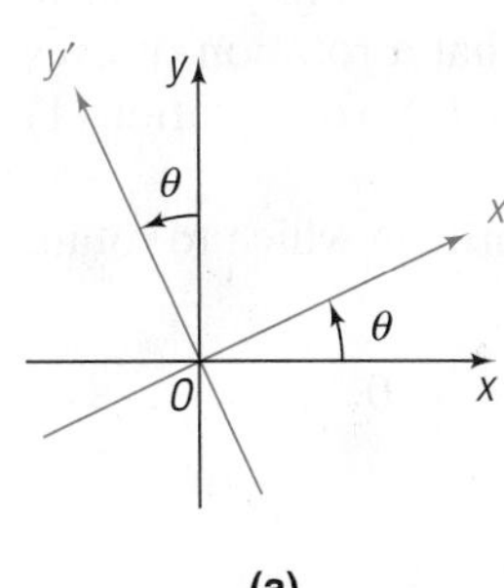

(a)

(b)

Figure 47

Now look at Figure 47(b). There the point P has the coordinates (x, y) relative to the xy-plane, while the same point P has coordinates (x', y') relative to the $x'y'$-plane. We seek relationships that will enable us to express x and y in terms of x', y', and θ.

As Figure 47(b) shows, r denotes the distance from the origin O to the point P, and α denotes the angle between the positive x'-axis and the ray from O through P. Then, using the definitions of sine and cosine, we have

$$x' = r\cos\alpha \qquad y' = r\sin\alpha \tag{3}$$

$$x = r\cos(\theta + \alpha) \qquad y = r\sin(\theta + \alpha) \tag{4}$$

Now

$$\begin{aligned} x &= r\cos(\theta + \alpha) \\ &= r(\cos\theta\cos\alpha - \sin\theta\sin\alpha) && \text{Apply the Sum Formula for cosine.} \\ &= (r\cos\alpha)(\cos\theta) - (r\sin\alpha)(\sin\theta) \\ &= x'\cos\theta - y'\sin\theta && \text{By equation (3)} \end{aligned}$$

Similarly,

$$\begin{aligned} y &= r\sin(\theta + \alpha) \\ &= r(\sin\theta\cos\alpha + \cos\theta\sin\alpha) && \text{Apply the Sum Formula for sine.} \\ &= x'\sin\theta + y'\cos\theta && \text{By equation (3)} \end{aligned}$$

THEOREM **Rotation Formulas**

If the x- and y-axes are rotated through an angle θ, the coordinates (x, y) of a point P relative to the xy-plane and the coordinates (x', y') of the same point relative to the new x'- and y'-axes are related by the formulas

$$x = x' \cos \theta - y' \sin \theta \qquad y = x' \sin \theta + y' \cos \theta \tag{5}$$

EXAMPLE 2 Rotating Axes

Express the equation $xy = 1$ in terms of new $x'y'$-coordinates by rotating the axes through a 45° angle. Discuss the new equation.

Solution Let $\theta = 45°$ in equation (5). Then

$$x = x' \cos 45° - y' \sin 45° = x' \frac{\sqrt{2}}{2} - y' \frac{\sqrt{2}}{2} = \frac{\sqrt{2}}{2}(x' - y')$$

$$y = x' \sin 45° + y' \cos 45° = x' \frac{\sqrt{2}}{2} + y' \frac{\sqrt{2}}{2} = \frac{\sqrt{2}}{2}(x' + y')$$

Substituting these expressions for x and y in $xy = 1$ gives

$$\left[\frac{\sqrt{2}}{2}(x' - y')\right]\left[\frac{\sqrt{2}}{2}(x' + y')\right] = 1$$

$$\frac{1}{2}(x'^2 - y'^2) = 1$$

$$\frac{x'^2}{2} - \frac{y'^2}{2} = 1$$

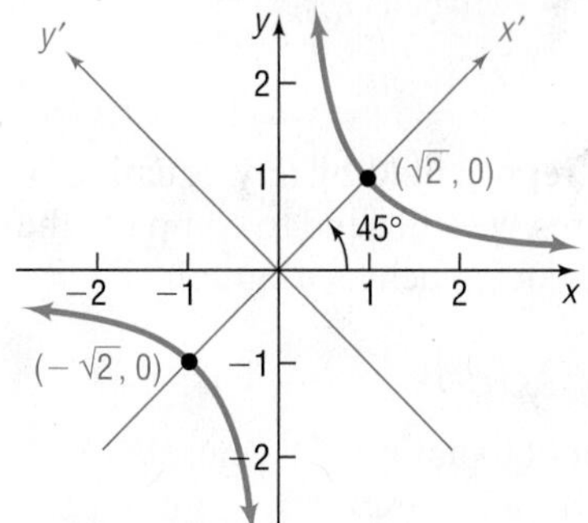

Figure 48 $xy = 1$

This is the equation of a hyperbola with center at $(0, 0)$ and transverse axis along the x'-axis. The vertices are at $(\pm\sqrt{2}, 0)$ on the x'-axis; the asymptotes are $y' = x'$ and $y' = -x'$ (which correspond to the original x- and y-axes). See Figure 48 for the graph. ●

As Example 2 illustrates, a rotation of axes through an appropriate angle can transform a second-degree equation in x and y containing an xy-term into one in x' and y' in which no $x'y'$-term appears. In fact, we will show that a rotation of axes through an appropriate angle will transform any equation of the form of equation (1) into an equation in x' and y' without an $x'y'$-term.

To find the formula for choosing an appropriate angle θ through which to rotate the axes, begin with equation (1),

$$Ax^2 + Bxy + Cy^2 + Dx + Ey + F = 0 \qquad B \neq 0$$

Next rotate through an angle θ using the rotation formulas (5).

$$\begin{aligned} A(x' \cos \theta - y' \sin \theta)^2 &+ B(x' \cos \theta - y' \sin \theta)(x' \sin \theta + y' \cos \theta) \\ &+ C(x' \sin \theta + y' \cos \theta)^2 + D(x' \cos \theta - y' \sin \theta) \\ &+ E(x' \sin \theta + y' \cos \theta) + F = 0 \end{aligned}$$

Expanding and collecting like terms gives

$$\begin{aligned} (A \cos^2 \theta + B \sin \theta \cos \theta + C \sin^2 \theta)x'^2 &+ [B(\cos^2 \theta - \sin^2 \theta) + 2(C - A)(\sin \theta \cos \theta)]x'y' \\ &+ (A \sin^2 \theta - B \sin \theta \cos \theta + C \cos^2 \theta)y'^2 \\ &+ (D \cos \theta + E \sin \theta)x' \\ &+ (-D \sin \theta + E \cos \theta)y' + F = 0 \end{aligned} \tag{6}$$

In equation (6), the coefficient of $x'y'$ is

$$B(\cos^2 \theta - \sin^2 \theta) + 2(C - A)(\sin \theta \cos \theta)$$

Since we want to eliminate the $x'y'$-term, we select an angle θ so that this coefficient is 0.

$$B(\cos^2\theta - \sin^2\theta) + 2(C - A)(\sin\theta\cos\theta) = 0$$
$$B\cos(2\theta) + (C - A)\sin(2\theta) = 0 \qquad \text{Double-angle Formulas}$$
$$B\cos(2\theta) = (A - C)\sin(2\theta)$$
$$\cot(2\theta) = \frac{A - C}{B} \qquad B \neq 0$$

THEOREM

To transform the equation

$$Ax^2 + Bxy + Cy^2 + Dx + Ey + F = 0 \qquad B \neq 0$$

into an equation in x' and y' without an $x'y'$-term, rotate the axes through an angle θ that satisfies the equation

$$\cot(2\theta) = \frac{A - C}{B} \qquad \textbf{(7)}$$

WARNING Be careful if you use a calculator to solve equation (7).

1. If $\cot(2\theta) = 0$, then $2\theta = 90°$ and $\theta = 45°$.
2. If $\cot(2\theta) \neq 0$, first find $\cos(2\theta)$. Then use the inverse cosine function key(s) to obtain 2θ, $0° < 2\theta < 180°$. Finally, divide by 2 to obtain the correct acute angle θ. ■

Equation (7) has an infinite number of solutions for θ. We shall adopt the convention of choosing the acute angle θ that satisfies (7). There are two possibilities:

If $\cot(2\theta) \geq 0$, then $0° < 2\theta \leq 90°$, so $0° < \theta \leq 45°$.
If $\cot(2\theta) < 0$, then $90° < 2\theta < 180°$, so $45° < \theta < 90°$.

Each of these results in a counterclockwise rotation of the axes through an acute angle θ.*

3 Analyze an Equation Using a Rotation of Axes

For the remainder of this section, the direction **"Analyze the equation"** will mean to transform the given equation so that it contains no xy-term and to graph the equation.

EXAMPLE 3

Analyzing an Equation Using a Rotation of Axes

Analyze the equation: $x^2 + \sqrt{3}\,xy + 2y^2 - 10 = 0$

Solution Since an xy-term is present, the axes must rotate. Using $A = 1$, $B = \sqrt{3}$, and $C = 2$ in equation (7), the appropriate acute angle θ through which to rotate the axes satisfies the equation

$$\cot(2\theta) = \frac{A - C}{B} = \frac{-1}{\sqrt{3}} = -\frac{\sqrt{3}}{3} \qquad 0° < 2\theta < 180°$$

Since $\cot(2\theta) = -\dfrac{\sqrt{3}}{3}$, this means $2\theta = 120°$, so $\theta = 60°$. Using $\theta = 60°$ in the rotation formulas (5) yields

$$x = x'\cos 60° - y'\sin 60° = \frac{1}{2}x' - \frac{\sqrt{3}}{2}y' = \frac{1}{2}(x' - \sqrt{3}\,y')$$
$$y = x'\sin 60° + y'\cos 60° = \frac{\sqrt{3}}{2}x' + \frac{1}{2}y' = \frac{1}{2}(\sqrt{3}\,x' + y')$$

*Any rotation through an angle θ that satisfies $\cot(2\theta) = \dfrac{A - C}{B}$ will eliminate the $x'y'$-term. However, the final form of the transformed equation may be different (but equivalent), depending on the angle chosen.

Substituting these values into the original equation and simplifying gives

$$x^2 + \sqrt{3}\,xy + 2y^2 - 10 = 0$$

$$\frac{1}{4}(x' - \sqrt{3}\,y')^2 + \sqrt{3}\left[\frac{1}{2}(x' - \sqrt{3}\,y')\right]\left[\frac{1}{2}(\sqrt{3}\,x' + y')\right] + 2\left[\frac{1}{4}(\sqrt{3}\,x' + y')^2\right] = 10$$

Multiply both sides by 4 and expand to obtain

$$x'^2 - 2\sqrt{3}\,x'y' + 3y'^2 + \sqrt{3}\,(\sqrt{3}\,x'^2 - 2x'y' - \sqrt{3}\,y'^2) + 2(3x'^2 + 2\sqrt{3}\,x'y' + y'^2) = 40$$

$$10x'^2 + 2y'^2 = 40$$

$$\frac{x'^2}{4} + \frac{y'^2}{20} = 1$$

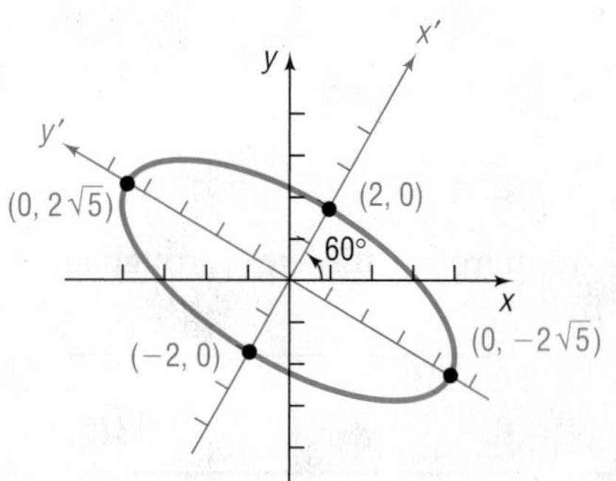

Figure 49 $\frac{x'^2}{4} + \frac{y'^2}{20} = 1$

This is the equation of an ellipse with center at $(0, 0)$ and major axis along the y'-axis. The vertices are at $(0, \pm 2\sqrt{5})$ on the y'-axis. See Figure 49 for the graph. ●

Now Work PROBLEM 31

In Example 3, the acute angle θ through which to rotate the axes was easy to find because of the numbers used in the given equation. In general, the equation $\cot(2\theta) = \frac{A - C}{B}$ will not have such a "nice" solution. As the next example shows, we can still find the appropriate rotation formulas without using a calculator approximation by applying Half-angle Formulas.

EXAMPLE 4 Analyzing an Equation Using a Rotation of Axes

Analyze the equation: $4x^2 - 4xy + y^2 + 5\sqrt{5}\,x + 5 = 0$

Solution Letting $A = 4$, $B = -4$, and $C = 1$ in equation (7), the appropriate angle θ through which to rotate the axes satisfies

$$\cot(2\theta) = \frac{A - C}{B} = \frac{3}{-4} = -\frac{3}{4}$$

To use the rotation formulas (5), we need to know the values of $\sin\theta$ and $\cos\theta$. Because we seek an acute angle θ, we know that $\sin\theta > 0$ and $\cos\theta > 0$. Use the Half-angle Formulas in the form

$$\sin\theta = \sqrt{\frac{1 - \cos(2\theta)}{2}} \qquad \cos\theta = \sqrt{\frac{1 + \cos(2\theta)}{2}}$$

Now we need to find the value of $\cos(2\theta)$. Because $\cot(2\theta) = -\frac{3}{4}$, then $90° < 2\theta < 180°$ (Do you know why?), so $\cos(2\theta) = -\frac{3}{5}$. Then

$$\sin\theta = \sqrt{\frac{1 - \cos(2\theta)}{2}} = \sqrt{\frac{1 - \left(-\frac{3}{5}\right)}{2}} = \sqrt{\frac{4}{5}} = \frac{2}{\sqrt{5}} = \frac{2\sqrt{5}}{5}$$

$$\cos\theta = \sqrt{\frac{1 + \cos(2\theta)}{2}} = \sqrt{\frac{1 + \left(-\frac{3}{5}\right)}{2}} = \sqrt{\frac{1}{5}} = \frac{1}{\sqrt{5}} = \frac{\sqrt{5}}{5}$$

With these values, the rotation formulas (5) are

$$x = \frac{\sqrt{5}}{5}x' - \frac{2\sqrt{5}}{5}y' = \frac{\sqrt{5}}{5}(x' - 2y')$$

$$y = \frac{2\sqrt{5}}{5}x' + \frac{\sqrt{5}}{5}y' = \frac{\sqrt{5}}{5}(2x' + y')$$

Substituting these values in the original equation and simplifying gives

$$4x^2 - 4xy + y^2 + 5\sqrt{5}x + 5 = 0$$

$$4\left[\frac{\sqrt{5}}{5}(x' - 2y')\right]^2 - 4\left[\frac{\sqrt{5}}{5}(x' - 2y')\right]\left[\frac{\sqrt{5}}{5}(2x' + y')\right] + \left[\frac{\sqrt{5}}{5}(2x' + y')\right]^2 + 5\sqrt{5}\left[\frac{\sqrt{5}}{5}(x' - 2y')\right] = -5$$

Multiply both sides by 5 and expand to obtain

$$4(x'^2 - 4x'y' + 4y'^2) - 4(2x'^2 - 3x'y' - 2y'^2) + 4x'^2 + 4x'y' + y'^2 + 25(x' - 2y') = -25$$

$$25y'^2 - 50y' + 25x' = -25 \quad \text{Combine like terms.}$$

$$y'^2 - 2y' + x' = -1 \quad \text{Divide by 25.}$$

$$y'^2 - 2y' + 1 = -x' \quad \text{Complete the square in } y'.$$

$$(y' - 1)^2 = -x'$$

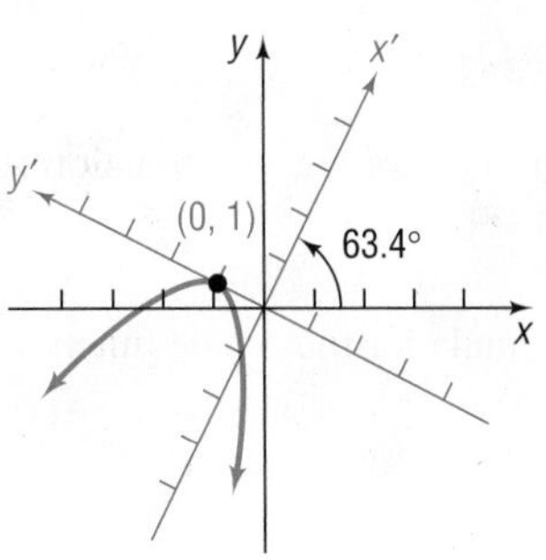

Figure 50 $(y' - 1)^2 = -x'$

This is the equation of a parabola with vertex at $(0, 1)$ in the $x'y'$-plane. The axis of symmetry is parallel to the x'-axis. Use a calculator to solve $\sin\theta = \frac{2\sqrt{5}}{5}$, and find that $\theta \approx 63.4°$. See Figure 50 for the graph. ●

Now Work PROBLEM 37

4 Identify Conics without a Rotation of Axes

Suppose that we are required only to identify (rather than analyze) the graph of an equation of the form

$$Ax^2 + Bxy + Cy^2 + Dx + Ey + F = 0 \qquad B \neq 0 \tag{8}$$

Applying the rotation formulas (5) to this equation gives an equation of the form

$$A'x'^2 + B'x'y' + C'y'^2 + D'x' + E'y' + F' = 0 \tag{9}$$

where A', B', C', D', E', and F' can be expressed in terms of A, B, C, D, E, F and the angle θ of rotation (see Problem 53). It can be shown that the value of $B^2 - 4AC$ in equation (8) and the value of $B'^2 - 4A'C'$ in equation (9) are equal no matter what angle θ of rotation is chosen (see Problem 55). In particular, if the angle θ of rotation satisfies equation (7), then $B' = 0$ in equation (9), and $B^2 - 4AC = -4A'C'$. Since equation (9) then has the form of equation (2),

$$A'x'^2 + C'y'^2 + D'x' + E'y' + F' = 0$$

we can identify its graph without completing the squares, as we did in the beginning of this section. In fact, now we can identify the conic described by any equation of the form of equation (8) without a rotation of axes.

THEOREM

Identifying Conics without a Rotation of Axes

Except for degenerate cases, the equation

$$Ax^2 + Bxy + Cy^2 + Dx + Ey + F = 0$$

(a) Defines a parabola if $B^2 - 4AC = 0$.
(b) Defines an ellipse (or a circle) if $B^2 - 4AC < 0$.
(c) Defines a hyperbola if $B^2 - 4AC > 0$.

You are asked to prove this theorem in Problem 56.

EXAMPLE 5 **Identifying a Conic without a Rotation of Axes**

Identify the graph of the equation: $8x^2 - 12xy + 17y^2 - 4\sqrt{5}x - 2\sqrt{5}y - 15 = 0$

Solution Here $A = 8$, $B = -12$, and $C = 17$, so $B^2 - 4AC = -400$. Since $B^2 - 4AC < 0$, the equation defines an ellipse. ●

Now Work PROBLEM 43

6.5 Assess Your Understanding

'Are You Prepared?' *Answers are given at the end of these exercises. If you get a wrong answer, read the pages listed in red.*

1. The sum formula for the sine function is $\sin(A + B) =$ ______. (p. 229)
2. The Double-angle Formula for the sine function is $\sin(2\theta) =$ ______. (p. 239)
3. If θ is acute, the Half-angle Formula for the sine function is $\sin\frac{\theta}{2} =$ ______. (p. 242)
4. If θ is acute, the Half-angle Formula for the cosine function is $\cos\frac{\theta}{2} =$ ______. (p. 242)

Concepts and Vocabulary

5. To transform the equation
$$Ax^2 + Bxy + Cy^2 + Dx + Ey + F = 0 \qquad B \neq 0$$
into one in x' and y' without an $x'y'$-term, rotate the axes through an acute angle θ that satisfies the equation ______.
6. Except for degenerate cases, the equation
$$Ax^2 + Bxy + Cy^2 + Dx + Ey + F = 0$$
defines a(n) ________ if $B^2 - 4AC = 0$.
7. ***True or False*** The equation $ax^2 + 6y^2 - 12y = 0$ defines an ellipse if $a > 0$.
8. Except for degenerate cases, the equation
$$Ax^2 + Bxy + Cy^2 + Dx + Ey + F = 0$$
defines an ellipse if ____________.
9. ***True or False*** The equation $3x^2 + Bxy + 12y^2 = 10$ defines a parabola if $B = -12$.
10. ***True or False*** To eliminate the xy-term from the equation $x^2 - 2xy + y^2 - 2x + 3y + 5 = 0$, rotate the axes through an angle θ, where $\cot\theta = B^2 - 4AC$.

Skill Building

In Problems 11–20, identify the graph of each equation without completing the squares.

11. $x^2 + 4x + y + 3 = 0$
12. $2y^2 - 3y + 3x = 0$
13. $6x^2 + 3y^2 - 12x + 6y = 0$
14. $2x^2 + y^2 - 8x + 4y + 2 = 0$
15. $3x^2 - 2y^2 + 6x + 4 = 0$
16. $4x^2 - 3y^2 - 8x + 6y + 1 = 0$
17. $2y^2 - x^2 - y + x = 0$
18. $y^2 - 8x^2 - 2x - y = 0$
19. $x^2 + y^2 - 8x + 4y = 0$
20. $2x^2 + 2y^2 - 8x + 8y = 0$

In Problems 21–30, determine the appropriate rotation formulas to use so that the new equation contains no xy-term.

21. $x^2 + 4xy + y^2 - 3 = 0$
22. $x^2 - 4xy + y^2 - 3 = 0$
23. $5x^2 + 6xy + 5y^2 - 8 = 0$
24. $3x^2 - 10xy + 3y^2 - 32 = 0$
25. $13x^2 - 6\sqrt{3}xy + 7y^2 - 16 = 0$
26. $11x^2 + 10\sqrt{3}xy + y^2 - 4 = 0$
27. $4x^2 - 4xy + y^2 - 8\sqrt{5}x - 16\sqrt{5}y = 0$
28. $x^2 + 4xy + 4y^2 + 5\sqrt{5}y + 5 = 0$
29. $25x^2 - 36xy + 40y^2 - 12\sqrt{13}x - 8\sqrt{13}y = 0$
30. $34x^2 - 24xy + 41y^2 - 25 = 0$

In Problems 31–42, rotate the axes so that the new equation contains no xy-term. Analyze and graph the new equation. Refer to Problems 21–30 for Problems 31–40.

31. $x^2 + 4xy + y^2 - 3 = 0$
32. $x^2 - 4xy + y^2 - 3 = 0$
33. $5x^2 + 6xy + 5y^2 - 8 = 0$
34. $3x^2 - 10xy + 3y^2 - 32 = 0$

35. $13x^2 - 6\sqrt{3}xy + 7y^2 - 16 = 0$

36. $11x^2 + 10\sqrt{3}xy + y^2 - 4 = 0$

37. $4x^2 - 4xy + y^2 - 8\sqrt{5}x - 16\sqrt{5}y = 0$

38. $x^2 + 4xy + 4y^2 + 5\sqrt{5}y + 5 = 0$

39. $25x^2 - 36xy + 40y^2 - 12\sqrt{13}x - 8\sqrt{13}y = 0$

40. $34x^2 - 24xy + 41y^2 - 25 = 0$

41. $16x^2 + 24xy + 9y^2 - 130x + 90y = 0$

42. $16x^2 + 24xy + 9y^2 - 60x + 80y = 0$

In Problems 43–52, identify the graph of each equation without applying a rotation of axes.

43. $x^2 + 3xy - 2y^2 + 3x + 2y + 5 = 0$

44. $2x^2 - 3xy + 4y^2 + 2x + 3y - 5 = 0$

45. $x^2 - 7xy + 3y^2 - y - 10 = 0$

46. $2x^2 - 3xy + 2y^2 - 4x - 2 = 0$

47. $9x^2 + 12xy + 4y^2 - x - y = 0$

48. $10x^2 + 12xy + 4y^2 - x - y + 10 = 0$

49. $10x^2 - 12xy + 4y^2 - x - y - 10 = 0$

50. $4x^2 + 12xy + 9y^2 - x - y = 0$

51. $3x^2 - 2xy + y^2 + 4x + 2y - 1 = 0$

52. $3x^2 + 2xy + y^2 + 4x - 2y + 10 = 0$

Applications and Extensions

In Problems 53–56, apply the rotation formulas (5) to

$$Ax^2 + Bxy + Cy^2 + Dx + Ey + F = 0$$

to obtain the equation

$$A'x'^2 + B'x'y' + C'y'^2 + D'x' + E'y' + F' = 0$$

53. Express A', B', C', D', E', and F' in terms of A, B, C, D, E, F, and the angle θ of rotation.
[Hint: Refer to equation (6).]

54. Show that $A + C = A' + C'$, which proves that $A + C$ is **invariant**; that is, its value does not change under a rotation of axes.

55. Refer to Problem 54. Show that $B^2 - 4AC$ is invariant.

56. Prove that, except for degenerate cases, the equation

$$Ax^2 + Bxy + Cy^2 + Dx + Ey + F = 0$$

(a) Defines a parabola if $B^2 - 4AC = 0$.
(b) Defines an ellipse (or a circle) if $B^2 - 4AC < 0$.
(c) Defines a hyperbola if $B^2 - 4AC > 0$.

57. Use the rotation formulas (5) to show that distance is invariant under a rotation of axes. That is, show that the distance from $P_1 = (x_1, y_1)$ to $P_2 = (x_2, y_2)$ in the xy-plane equals the distance from $P_1 = (x'_1, y'_1)$ to $P_2 = (x'_2, y'_2)$ in the $x'y'$-plane.

58. Show that the graph of the equation $x^{1/2} + y^{1/2} = a^{1/2}$ is part of the graph of a parabola.

Explaining Concepts: Discussion and Writing

59. Formulate a strategy for analyzing and graphing an equation of the form

$$Ax^2 + Cy^2 + Dx + Ey + F = 0$$

60. Explain how your strategy presented in Problem 59 changes if the equation is of the form

$$Ax^2 + Bxy + Cy^2 + Dx + Ey + F = 0$$

Retain Your Knowledge

Problems 61–64 are based on material learned earlier in the course. The purpose of these problems is to keep the material fresh in your mind so that you are better prepared for the final exam.

61. Solve the triangle whose sides are:

$$a = 7, b = 9, \text{and } c = 11.$$

62. Find the area of the triangle described: $a = 14$, $b = 11$, and $C = 30°$.

63. Transform the equation $xy = 1$ from rectangular coordinates to polar coordinates.

64. Write the complex number $2 - 5i$ in polar form.

'Are You Prepared?' Answers

1. $\sin A \cos B + \cos A \sin B$
2. $2 \sin \theta \cos \theta$
3. $\sqrt{\dfrac{1 - \cos \theta}{2}}$
4. $\sqrt{\dfrac{1 + \cos \theta}{2}}$

6.6 Polar Equations of Conics

PREPARING FOR THIS SECTION *Before getting started, review the following:*

- Polar Coordinates (Section 5.1, pp. 312–319)

Now Work the **'Are You Prepared?'** problems on page 432.

OBJECTIVES 1 Analyze and Graph Polar Equations of Conics (p. 428)
2 Convert the Polar Equation of a Conic to a Rectangular Equation (p. 432)

1 Analyze and Graph Polar Equations of Conics

In Sections 6.2 through 6.4, we gave separate definitions for the parabola, ellipse, and hyperbola based on geometric properties and the distance formula. This section presents an alternative definition that simultaneously defines *all* these conics. As we shall see, this approach is well suited to polar coordinate representation. (Refer to Section 5.1.)

DEFINITION

Let D denote a fixed line called the **directrix**; let F denote a fixed point called the **focus**, which is not on D; and let e be a fixed positive number called the **eccentricity**. A **conic** is the set of points P in the plane such that the ratio of the distance from F to P to the distance from D to P equals e. That is, a conic is the collection of points P for which

$$\frac{d(F,P)}{d(D,P)} = e \qquad \textbf{(1)}$$

If $e = 1$, the conic is a **parabola**.
If $e < 1$, the conic is an **ellipse**.
If $e > 1$, the conic is a **hyperbola**.

Observe that if $e = 1$, the definition of a parabola in equation (1) is exactly the same as the definition used earlier in Section 6.2.

In the case of an ellipse, the **major axis** is a line through the focus perpendicular to the directrix. In the case of a hyperbola, the **transverse axis** is a line through the focus perpendicular to the directrix. For both an ellipse and a hyperbola, the eccentricity e satisfies

$$e = \frac{c}{a} \qquad \textbf{(2)}$$

where c is the distance from the center to the focus, and a is the distance from the center to a vertex.

Just as we did earlier using rectangular coordinates, we derive equations for the conics in polar coordinates by choosing a convenient position for the focus F and the directrix D. The focus F is positioned at the pole, and the directrix D is either parallel or perpendicular to the polar axis.

Suppose that we start with the directrix D perpendicular to the polar axis at a distance p units to the left of the pole (the focus F). See Figure 51.

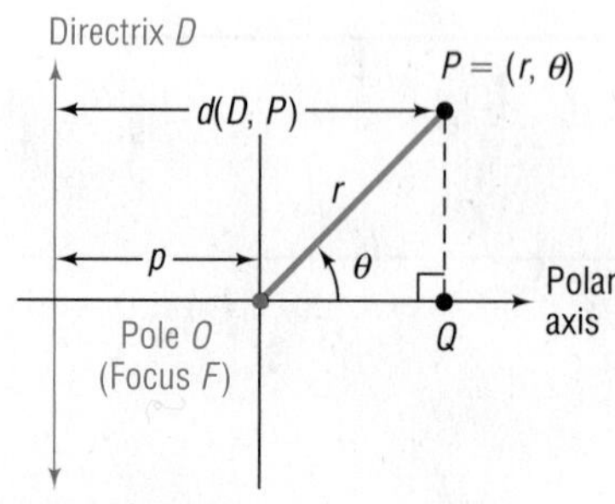

Figure 51

If $P = (r, \theta)$ is any point on the conic, then, by equation (1),

$$\frac{d(F,P)}{d(D,P)} = e \quad \text{or} \quad d(F,P) = e \cdot d(D,P) \qquad \textbf{(3)}$$

Now use the point Q obtained by dropping the perpendicular from P to the polar axis to calculate $d(D, P)$.

$$d(D, P) = p + d(O, Q) = p + r\cos\theta$$

Using this expression and the fact that $d(F, P) = d(O, P) = r$ in equation (3) gives

$$\begin{aligned} d(F, P) &= e \cdot d(D, P) \\ r &= e(p + r\cos\theta) \\ r &= ep + er\cos\theta \\ r - er\cos\theta &= ep \\ r(1 - e\cos\theta) &= ep \\ r &= \frac{ep}{1 - e\cos\theta} \end{aligned}$$

THEOREM

Polar Equation of a Conic; Focus at the Pole; Directrix Perpendicular to the Polar Axis a Distance *p* to the Left of the Pole

The polar equation of a conic with focus at the pole and directrix perpendicular to the polar axis at a distance p to the left of the pole is

$$r = \frac{ep}{1 - e\cos\theta} \quad \textbf{(4)}$$

where e is the eccentricity of the conic.

EXAMPLE 1 Analyzing and Graphing the Polar Equation of a Conic

Analyze and graph the equation: $r = \dfrac{4}{2 - \cos\theta}$

Solution The given equation is not quite in the form of equation (4), since the first term in the denominator is 2 instead of 1. Divide the numerator and denominator by 2 to obtain

$$r = \frac{2}{1 - \frac{1}{2}\cos\theta} \qquad r = \frac{ep}{1 - e\cos\theta}$$

This equation is in the form of equation (4), with

$$e = \frac{1}{2} \quad \text{and} \quad ep = 2$$

Then

$$\frac{1}{2}p = 2, \quad \text{so} \quad p = 4$$

Since $e = \frac{1}{2} < 1$, the conic is an ellipse. One focus is at the pole, and the directrix is perpendicular to the polar axis, a distance of $p = 4$ units to the left of the pole. It follows that the major axis is along the polar axis. To find the vertices, let $\theta = 0$ and $\theta = \pi$. The vertices of the ellipse are $(4, 0)$ and $\left(\frac{4}{3}, \pi\right)$. The midpoint of the vertices, $\left(\frac{4}{3}, 0\right)$ in polar coordinates, is the center of the ellipse.

[Do you see why? The vertices $(4, 0)$ and $\left(\frac{4}{3}, \pi\right)$ in polar coordinates are $(4, 0)$ and $\left(-\frac{4}{3}, 0\right)$ in rectangular coordinates. The midpoint in rectangular coordinates is $\left(\frac{4}{3}, 0\right)$, which is also $\left(\frac{4}{3}, 0\right)$ in polar coordinates.] Then a = distance from the center to a vertex $= \frac{8}{3}$. Using $a = \frac{8}{3}$ and $e = \frac{1}{2}$ in equation (2), $e = \frac{c}{a}$ yields $c = ae = \frac{4}{3}$. Finally, using $a = \frac{8}{3}$ and $c = \frac{4}{3}$ in $b^2 = a^2 - c^2$ yields

$$b^2 = a^2 - c^2 = \frac{64}{9} - \frac{16}{9} = \frac{48}{9}$$

$$b = \frac{4\sqrt{3}}{3}$$

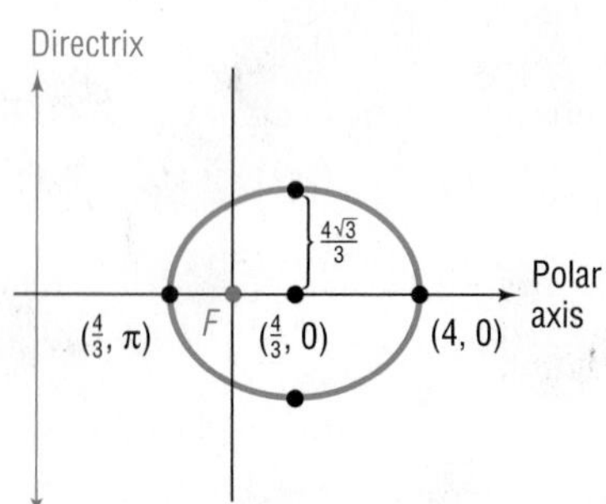

Figure 52 $r = \dfrac{4}{2 - \cos\theta}$

Figure 52 shows the graph.

Exploration

Graph $r_1 = \dfrac{4}{2 + \cos\theta}$ and compare the result with Figure 52. What do you conclude? Clear the screen and graph $r_1 = \dfrac{4}{2 - \sin\theta}$ and then $r_1 = \dfrac{4}{2 + \sin\theta}$. Compare each of these graphs with Figure 52. What do you conclude?

Now Work PROBLEM 11

Equation (4) was obtained under the assumption that the directrix was perpendicular to the polar axis at a distance p units to the left of the pole. A similar derivation (see Problem 43), in which the directrix is perpendicular to the polar axis at a distance p units to the right of the pole, results in the equation

$$r = \frac{ep}{1 + e\cos\theta}$$

In Problems 44 and 45, you are asked to derive the polar equations of conics with focus at the pole and directrix parallel to the polar axis. Table 5 summarizes the polar equations of conics.

Table 5 Polar Equations of Conics (Focus at the Pole, Eccentricity e)

Equation	Description
(a) $r = \dfrac{ep}{1 - e\cos\theta}$	Directrix is perpendicular to the polar axis at a distance p units to the left of the pole.
(b) $r = \dfrac{ep}{1 + e\cos\theta}$	Directrix is perpendicular to the polar axis at a distance p units to the right of the pole.
(c) $r = \dfrac{ep}{1 + e\sin\theta}$	Directrix is parallel to the polar axis at a distance p units above the pole.
(d) $r = \dfrac{ep}{1 - e\sin\theta}$	Directrix is parallel to the polar axis at a distance p units below the pole.

Eccentricity

If $e = 1$, the conic is a parabola; the axis of symmetry is perpendicular to the directrix.

If $e < 1$, the conic is an ellipse; the major axis is perpendicular to the directrix.

If $e > 1$, the conic is a hyperbola; the transverse axis is perpendicular to the directrix.

EXAMPLE 2 **Analyzing and Graphing the Polar Equation of a Conic**

Analyze and graph the equation: $r = \dfrac{6}{3 + 3\sin\theta}$

Solution To place the equation in proper form, divide the numerator and denominator by 3 to get

$$r = \frac{2}{1 + \sin\theta}$$

Referring to Table 5, conclude that this equation is in the form of equation (c) with

$$e = 1 \quad \text{and} \quad ep = 2$$

$$p = 2 \quad e = 1$$

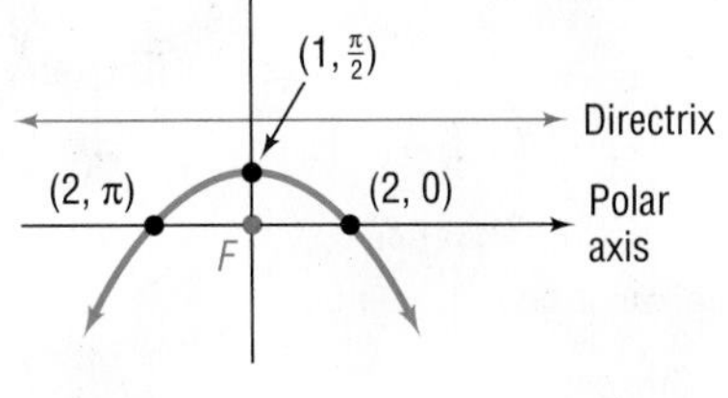

Figure 53 $r = \dfrac{6}{3 + 3\sin\theta}$

The conic is a parabola with focus at the pole. The directrix is parallel to the polar axis at a distance 2 units above the pole; the axis of symmetry is perpendicular to the polar axis. The vertex of the parabola is at $\left(1, \dfrac{\pi}{2}\right)$. (Do you see why?) See Figure 53 for the graph. Notice that we plotted two additional points, $(2, 0)$ and $(2, \pi)$, to assist in graphing. ●

Now Work PROBLEM 13

EXAMPLE 3 **Analyzing and Graphing the Polar Equation of a Conic**

Analyze and graph the equation: $r = \dfrac{3}{1 + 3\cos\theta}$

Solution This equation is in the form of equation (b) in Table 5. This means that

$$e = 3 \quad \text{and} \quad ep = 3$$

$$p = 1 \quad e = 3$$

This is the equation of a hyperbola with a focus at the pole. The directrix is perpendicular to the polar axis, 1 unit to the right of the pole. The transverse axis is along the polar axis. To find the vertices, let $\theta = 0$ and $\theta = \pi$. The vertices are $\left(\dfrac{3}{4}, 0\right)$ and $\left(-\dfrac{3}{2}, \pi\right)$. The center, which is at the midpoint of $\left(\dfrac{3}{4}, 0\right)$ and $\left(-\dfrac{3}{2}, \pi\right)$, is $\left(\dfrac{9}{8}, 0\right)$. Then c = distance from the center to a focus $= \dfrac{9}{8}$. Since $e = 3$, it follows from equation (2), $e = \dfrac{c}{a}$, that $a = \dfrac{3}{8}$. Finally, using $a = \dfrac{3}{8}$ and $c = \dfrac{9}{8}$ in $b^2 = c^2 - a^2$ gives

$$b^2 = c^2 - a^2 = \frac{81}{64} - \frac{9}{64} = \frac{72}{64} = \frac{9}{8}$$

$$b = \frac{3}{2\sqrt{2}} = \frac{3\sqrt{2}}{4}$$

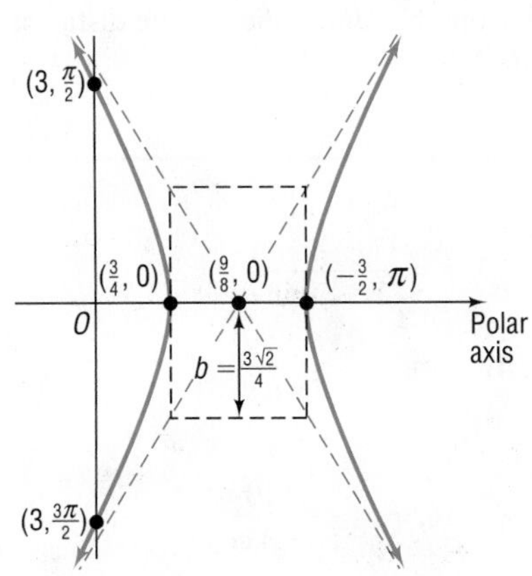

Figure 54 $r = \dfrac{3}{1 + 3\cos\theta}$

Figure 54 shows the graph. Notice that we plotted two additional points, $\left(3, \dfrac{\pi}{2}\right)$ and $\left(3, \dfrac{3\pi}{2}\right)$, on the left branch and used symmetry to obtain the right branch. The asymptotes of this hyperbola were found in the usual way by constructing the rectangle shown. ●

Now Work PROBLEM 17

2 Convert the Polar Equation of a Conic to a Rectangular Equation

EXAMPLE 4 **Converting a Polar Equation to a Rectangular Equation**

Convert the polar equation

$$r = \frac{1}{3 - 3\cos\theta}$$

to a rectangular equation.

Solution The strategy here is to rearrange the equation and square each side before using the transformation equations.

$$\begin{aligned} r &= \frac{1}{3 - 3\cos\theta} \\ 3r - 3r\cos\theta &= 1 \\ 3r &= 1 + 3r\cos\theta && \text{Rearrange the equation.} \\ 9r^2 &= (1 + 3r\cos\theta)^2 && \text{Square each side.} \\ 9(x^2 + y^2) &= (1 + 3x)^2 && x^2 + y^2 = r^2;\ x = r\cos\theta \\ 9x^2 + 9y^2 &= 9x^2 + 6x + 1 \\ 9y^2 &= 6x + 1 \end{aligned}$$

This is the equation of a parabola in rectangular coordinates. ●

Now Work PROBLEM 25

6.6 Assess Your Understanding

'Are You Prepared?' *Answers are given at the end of these exercises. If you get a wrong answer, read the pages listed in red.*

1. If (x, y) are the rectangular coordinates of a point P and (r, θ) are its polar coordinates, then $x =$ ___ and $y =$ ___. (pp. 312–319)
2. Transform the equation $r = 6\cos\theta$ from polar coordinates to rectangular coordinates. (pp. 312–319)

Concepts and Vocabulary

3. A ________ is the set of points P in the plane such that the ratio of the distance from a fixed point called the ________ to P to the distance from a fixed line called the ________ to P equals a constant e.
4. The eccentricity e of a parabola is ________, of an ellipse it is ________, and of a hyperbola it is ________.
5. ***True or False*** If (r, θ) are polar coordinates, the equation $r = \frac{2}{2 + 3\sin\theta}$ defines a hyperbola.
6. ***True or False*** The eccentricity e of an ellipse is $\frac{c}{a}$, where a is the distance of a vertex from the center and c is the distance of a focus from the center.

Skill Building

In Problems 7–12, identify the conic that each polar equation represents. Also give the position of the directrix.

7. $r = \frac{1}{1 + \cos\theta}$
8. $r = \frac{3}{1 - \sin\theta}$
9. $r = \frac{4}{2 - 3\sin\theta}$
10. $r = \frac{2}{1 + 2\cos\theta}$
11. $r = \frac{3}{4 - 2\cos\theta}$
12. $r = \frac{6}{8 + 2\sin\theta}$

In Problems 13–24, analyze each equation and graph it.

13. $r = \frac{1}{1 + \cos\theta}$
14. $r = \frac{3}{1 - \sin\theta}$
15. $r = \frac{8}{4 + 3\sin\theta}$
16. $r = \frac{10}{5 + 4\cos\theta}$
17. $r = \frac{9}{3 - 6\cos\theta}$
18. $r = \frac{12}{4 + 8\sin\theta}$
19. $r = \frac{8}{2 - \sin\theta}$
20. $r = \frac{8}{2 + 4\cos\theta}$
21. $r(3 - 2\sin\theta) = 6$
22. $r(2 - \cos\theta) = 2$
23. $r = \frac{6\sec\theta}{2\sec\theta - 1}$
24. $r = \frac{3\csc\theta}{\csc\theta - 1}$

In Problems 25–36, convert each polar equation to a rectangular equation.

25. $r = \dfrac{1}{1 + \cos\theta}$

26. $r = \dfrac{3}{1 - \sin\theta}$

27. $r = \dfrac{8}{4 + 3\sin\theta}$

28. $r = \dfrac{10}{5 + 4\cos\theta}$

29. $r = \dfrac{9}{3 - 6\cos\theta}$

30. $r = \dfrac{12}{4 + 8\sin\theta}$

31. $r = \dfrac{8}{2 - \sin\theta}$

32. $r = \dfrac{8}{2 + 4\cos\theta}$

33. $r(3 - 2\sin\theta) = 6$

34. $r(2 - \cos\theta) = 2$

35. $r = \dfrac{6\sec\theta}{2\sec\theta - 1}$

36. $r = \dfrac{3\csc\theta}{\csc\theta - 1}$

In Problems 37–42, find a polar equation for each conic. For each, a focus is at the pole.

37. $e = 1$; directrix is parallel to the polar axis, 1 unit above the pole.

38. $e = 1$; directrix is parallel to the polar axis, 2 units below the pole.

39. $e = \frac{4}{5}$; directrix is perpendicular to the polar axis, 3 units to the left of the pole.

40. $e = \frac{2}{3}$; directrix is parallel to the polar axis, 3 units above the pole.

41. $e = 6$; directrix is parallel to the polar axis, 2 units below the pole.

42. $e = 5$; directrix is perpendicular to the polar axis, 5 units to the right of the pole.

Applications and Extensions

43. Derive equation (b) in Table 5:
$$r = \frac{ep}{1 + e\cos\theta}$$

44. Derive equation (c) in Table 5:
$$r = \frac{ep}{1 + e\sin\theta}$$

45. Derive equation (d) in Table 5:
$$r = \frac{ep}{1 - e\sin\theta}$$

46. **Orbit of Mercury** The planet Mercury travels around the Sun in an elliptical orbit given approximately by
$$r = \frac{(3.442)10^7}{1 - 0.206\cos\theta}$$
where r is measured in miles and the Sun is at the pole. Find the distance from Mercury to the Sun at *aphelion* (greatest distance from the Sun) and at *perihelion* (shortest distance from the Sun). See the figure. Use the aphelion and perihelion to graph the orbit of Mercury using a graphing utility.

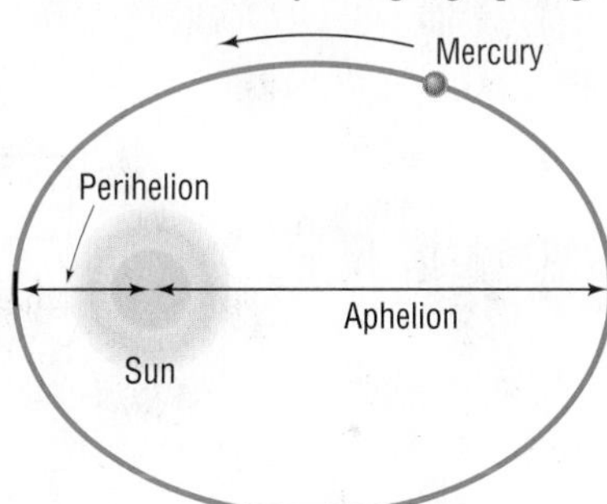

Retain Your Knowledge

Problems 47–50 are based on material learned earlier in the course. The purpose of these problems is to keep the material fresh in your mind so that you are better prepared for the final exam.

47. Find the area of the triangle whose sides are:
$$a = 7, b = 8, \text{and } c = 10.$$
Round the answer to two decimal places.

48. Without graphing, determine the amplitude and period of $y = 4\cos\left(\frac{1}{5}x\right)$.

49. Solve $2\cos^2 x + \cos x - 1 = 0,\ 0 \le x < 2\pi$

50. For $\mathbf{v} = 10\mathbf{i} - 24\mathbf{j}$, find $\|\mathbf{v}\|$.

'Are You Prepared?' Answers

1. $r\cos\theta$; $r\sin\theta$ 2. $x^2 + y^2 = 6x$ or $(x - 3)^2 + y^2 = 9$

6.7 Plane Curves and Parametric Equations

PREPARING FOR THIS SECTION *Before getting started, review the following:*

- Amplitude and Period of Sinusoidal Graphs (Section 2.4, pp. 149–151)

Now Work the 'Are You Prepared?' problem on page 443.

OBJECTIVES
1 Graph Parametric Equations (p. 434)
2 Find a Rectangular Equation for a Curve Defined Parametrically (p. 435)
3 Use Time as a Parameter in Parametric Equations (p. 437)
4 Find Parametric Equations for Curves Defined by Rectangular Equations (p. 440)

Equations of the form $y = f(x)$, where f is a function, have graphs that are intersected no more than once by any vertical line. The graphs of many of the conics and certain other, more complicated graphs do not have this characteristic. Yet each graph, like the graph of a function, is a collection of points (x, y) in the xy-plane; that is, each is a *plane curve*. This section discusses another way of representing such graphs.

Let $x = f(t)$ and $y = g(t)$, where f and g are two functions whose common domain is some interval I. The collection of points defined by

$$(x, y) = (f(t), g(t))$$

is called a **plane curve**. The equations

$$x = f(t) \qquad y = g(t)$$

where t is in I, are called **parametric equations** of the curve. The variable t is called a **parameter**.

1 Graph Parametric Equations

Parametric equations are particularly useful in describing movement along a curve. Suppose that a curve is defined by the parametric equations

$$x = f(t) \qquad y = g(t) \qquad a \le t \le b$$

where f and g are each defined over the interval $a \le t \le b$. For a given value of t, we can find the value of $x = f(t)$ and $y = g(t)$, obtaining a point (x, y) on the curve. In fact, as t varies over the interval from $t = a$ to $t = b$, successive values of t give rise to a directed movement along the curve; that is, the curve is traced out in a certain direction by the corresponding succession of points (x, y). See Figure 55. The arrows show the direction, or **orientation**, along the curve as t varies from a to b.

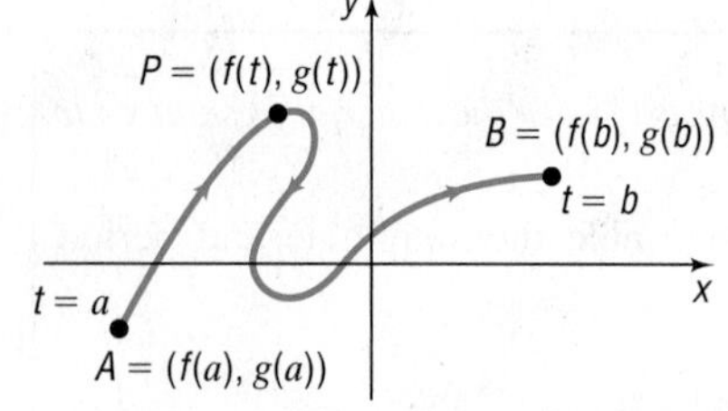

Figure 55

EXAMPLE 1

Graphing a Curve Defined by Parametric Equations

Graph the curve defined by the parametric equations

$$x = 3t^2 \qquad y = 2t \qquad -2 \le t \le 2 \tag{1}$$

Solution For each number t, $-2 \le t \le 2$, there corresponds a number x and a number y. For example, when $t = -2$, then $x = 3(-2)^2 = 12$ and $y = 2(-2) = -4$. When $t = 0$, then $x = 0$ and $y = 0$. Set up a table listing various choices of the parameter t and the corresponding values for x and y, as shown in Table 6. Plotting these points and connecting them with a smooth curve leads to Figure 56. The arrows in Figure 56 are used to indicate the orientation.

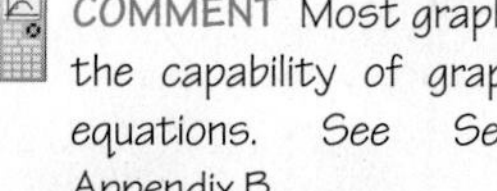
COMMENT Most graphing utilities have the capability of graphing parametric equations. See Section B.7 in Appendix B. ■

Table 6

t	x	y	(x, y)
−2	12	−4	(12, −4)
−1	3	−2	(3, −2)
0	0	0	(0, 0)
1	3	2	(3, 2)
2	12	4	(12, 4)

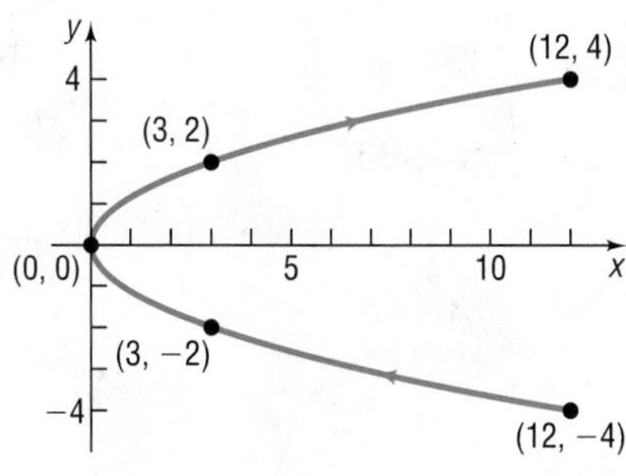

Figure 56 $x = 3t^2, y = 2t, -2 \leq t \leq 2$

Exploration

Graph the following parametric equations using a graphing utility with $Xmin = 0, Xmax = 15$, $Ymin = -5$, $Ymax = 5$, and $Tstep = 0.1$.

1. $x = \dfrac{3t^2}{4}, y = t, -4 \leq t \leq 4$
2. $x = 3t^2 + 12t + 12, y = 2t + 4, -4 \leq t \leq 0$
3. $x = 3t^{\frac{2}{3}}, y = 2\sqrt[3]{t}, -8 \leq t \leq 8$

Compare these graphs to Figure 56. Conclude that parametric equations defining a curve are not unique; that is, different parametric equations can represent the same graph.

2 Find a Rectangular Equation for a Curve Defined Parametrically

The curve given in Example 1 should be familiar. To identify it accurately, find the corresponding rectangular equation by eliminating the parameter t from the parametric equations given in Example 1:

$$x = 3t^2 \qquad y = 2t \qquad -2 \leq t \leq 2$$

Solve for t in $y = 2t$, obtaining $t = \dfrac{y}{2}$, and substitute this expression in the other equation to get

$$x = 3t^2 = 3\left(\frac{y}{2}\right)^2 = \frac{3y^2}{4}$$

$$\uparrow\; t = \frac{y}{2}$$

This equation, $x = \dfrac{3y^2}{4}$, is the equation of a parabola with vertex at $(0, 0)$ and axis of symmetry along the x-axis.

Note that the parameterized curve defined by equation (1) and shown in Figure 56 is only a part of the parabola $x = \dfrac{3y^2}{4}$. The graph of the rectangular equation obtained by eliminating the parameter will, in general, contain more points than the original parameterized curve. Care must therefore be taken when a parameterized curve is graphed after eliminating the parameter. Even so, eliminating the parameter t of a parameterized curve to identify it accurately is sometimes a better approach than plotting points. However, the elimination process sometimes requires a little ingenuity.

Exploration

Graph $x = \dfrac{3y^2}{4}\left(Y_1 = \sqrt{\dfrac{4x}{3}} \text{ and } Y_2 = -\sqrt{\dfrac{4x}{3}}\right)$ in FUNction mode with $Xmin = 0$, $Xmax = 15$, $Ymin = -5$, $Ymax = 5$. Compare this graph with Figure 56. Why do the graphs differ?

EXAMPLE 2 **Finding the Rectangular Equation of a Curve Defined Parametrically**

Find the rectangular equation of the curve whose parametric equations are

$$x = a\cos t \qquad y = a\sin t \qquad -\infty < t < \infty$$

where $a > 0$ is a constant. Graph this curve, indicating its orientation.

Solution The presence of sines and cosines in the parametric equations suggests using a Pythagorean Identity. In fact, since

$$\cos t = \frac{x}{a} \qquad \sin t = \frac{y}{a}$$

this means that

$$\cos^2 t + \sin^2 t = 1$$
$$\left(\frac{x}{a}\right)^2 + \left(\frac{y}{a}\right)^2 = 1$$
$$x^2 + y^2 = a^2$$

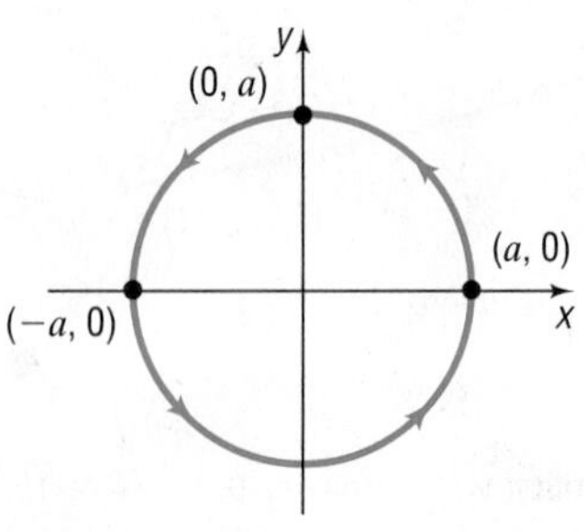

Figure 57 $x = a\cos t, y = a\sin t$

The curve is a circle with center at $(0, 0)$ and radius a. As the parameter t increases, say from $t = 0$ [the point $(a, 0)$] to $t = \frac{\pi}{2}$ [the point $(0, a)$] to $t = \pi$ [the point $(-a, 0)$], note that the corresponding points are traced in a counterclockwise direction around the circle. The orientation is as indicated in Figure 57. ●

Now Work PROBLEMS 7 AND 19

Let's analyze the curve in Example 2 further. The domain of each parametric equation is $-\infty < t < \infty$. That means the graph in Figure 57 is actually being repeated each time that t increases by 2π.

If we wanted the curve to consist of exactly 1 revolution in the counterclockwise direction, we could write

$$x = a\cos t \qquad y = a\sin t \qquad 0 \le t \le 2\pi$$

This curve starts at $t = 0$ [the point $(a, 0)$] and, proceeding counterclockwise around the circle, ends at $t = 2\pi$ [also the point $(a, 0)$].

If we wanted the curve to consist of exactly three revolutions in the counterclockwise direction, we could write

$$x = a\cos t \qquad y = a\sin t \qquad -2\pi \le t \le 4\pi$$

or

$$x = a\cos t \qquad y = a\sin t \qquad 0 \le t \le 6\pi$$

or

$$x = a\cos t \qquad y = a\sin t \qquad 2\pi \le t \le 8\pi$$

EXAMPLE 3

Describing Parametric Equations

Find rectangular equations for the following curves defined by parametric equations. Graph each curve.

(a) $x = a\cos t \quad y = a\sin t \quad 0 \le t \le \pi, \quad a > 0$
(b) $x = -a\sin t \quad y = -a\cos t \quad 0 \le t \le \pi, \quad a > 0$

Solution (a) Eliminate the parameter t using a Pythagorean Identity.

$$\cos^2 t + \sin^2 t = 1$$
$$\left(\frac{x}{a}\right)^2 + \left(\frac{y}{a}\right)^2 = 1$$
$$x^2 + y^2 = a^2$$

The curve defined by these parametric equations lies on a circle with radius a and center at $(0, 0)$. The curve begins at the point $(a, 0)$, when $t = 0$; passes through the point $(0, a)$, when $t = \frac{\pi}{2}$; and ends at the point $(-a, 0)$, when $t = \pi$.

The parametric equations define the upper semicircle of a circle of radius a with a counterclockwise orientation. See Figure 58. The rectangular equation is

$$y = \sqrt{a^2 - x^2} \qquad -a \le x \le a$$

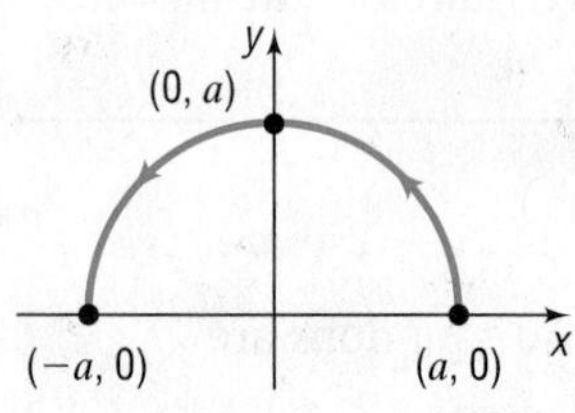

Figure 58
$x = a\cos t, y = a\sin t, 0 \le t \le \pi, a > 0$

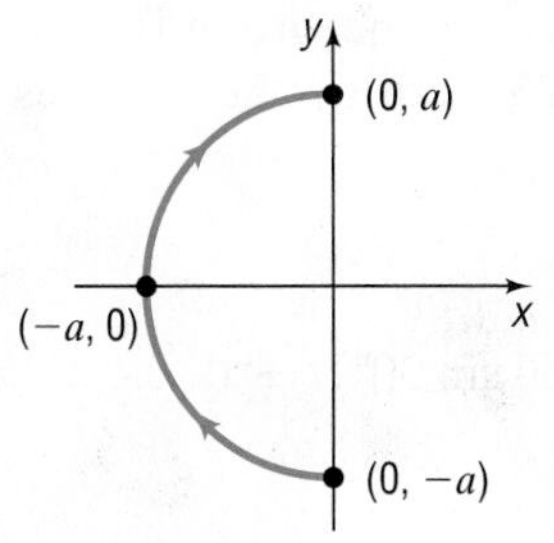

Figure 59 $x = -a \sin t, y = -a \cos t,$
$0 \le t \le \pi, a > 0$

(b) Eliminate the parameter t using a Pythagorean Identity.

$$\sin^2 t + \cos^2 t = 1$$
$$\left(\frac{x}{-a}\right)^2 + \left(\frac{y}{-a}\right)^2 = 1$$
$$x^2 + y^2 = a^2$$

The curve defined by these parametric equations lies on a circle with radius a and center at $(0, 0)$. The curve begins at the point $(0, -a)$, when $t = 0$; passes through the point $(-a, 0)$, when $t = \frac{\pi}{2}$; and ends at the point $(0, a)$, when $t = \pi$. The parametric equations define the left semicircle of a circle of radius a with a clockwise orientation. See Figure 59. The rectangular equation is

$$x = -\sqrt{a^2 - y^2} \qquad -a \le y \le a$$

Seeing the Concept

Graph $x = \cos t, y = \sin t$ for $0 \le t \le 2\pi$. Compare to Figure 57. Graph $x = \cos t$, $y = \sin t$ for $0 \le t \le \pi$. Compare to Figure 58. Graph $x = -\sin t, y = -\cos t$ for $0 \le t \le \pi$. Compare to Figure 59.

Example 3 illustrates the versatility of parametric equations for replacing complicated rectangular equations, while providing additional information about orientation. These characteristics make parametric equations very useful in applications, such as projectile motion.

3 Use Time as a Parameter in Parametric Equations

If we think of the parameter t as time, then the parametric equations $x = f(t)$ and $y = g(t)$ of a curve C specify how the x- and y-coordinates of a moving point vary with time.

For example, we can use parametric equations to model the motion of an object, sometimes referred to as **curvilinear motion**. Using parametric equations, we can specify not only where the object travels—that is, its location (x, y)—but also when it gets there—that is, the time t.

When an object is propelled upward at an inclination θ to the horizontal with initial speed v_0, the resulting motion is called **projectile motion**. See Figure 60(a).

In calculus it is shown that the parametric equations of the path of a projectile fired at an inclination θ to the horizontal, with an initial speed v_0, from a height h above the horizontal, are

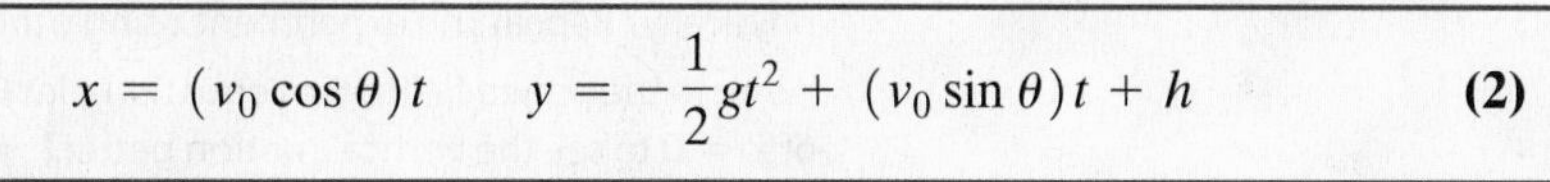

$$x = (v_0 \cos \theta)t \qquad y = -\frac{1}{2}gt^2 + (v_0 \sin \theta)t + h \qquad \textbf{(2)}$$

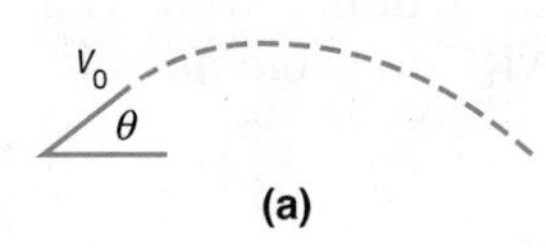

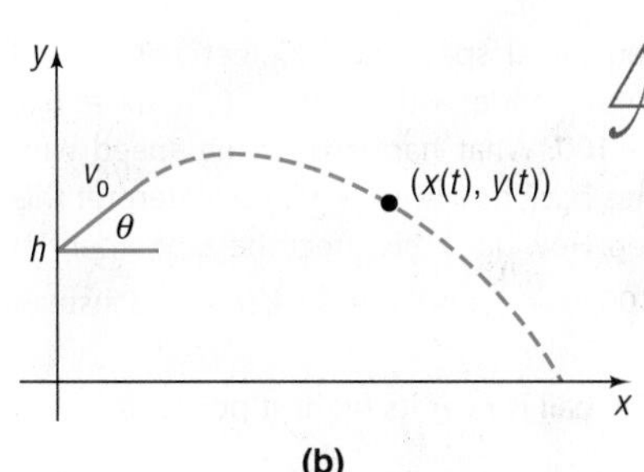

Figure 60

where t is the time and g is the constant acceleration due to gravity (approximately 32 ft/sec/sec, or 9.8 m/sec/sec). See Figure 60(b).

EXAMPLE 4 Projectile Motion

Suppose that Jim hit a golf ball with an initial velocity of 150 feet per second at an angle of 30° to the horizontal. See Figure 61.

(a) Find parametric equations that describe the position of the ball as a function of time.
(b) How long was the golf ball in the air?
(c) When was the ball at its maximum height? Determine the maximum height of the ball.
(d) Determine the distance that the ball traveled.
(e) Using a graphing utility, simulate the motion of the golf ball by simultaneously graphing the equations found in part (a).

30°

Figure 61

Solution (a) We have $v_0 = 150$ ft/sec, $\theta = 30°$, $h = 0$ ft (the ball is on the ground), and $g = 32$ ft/sec^2 (since the units are in feet and seconds). Substitute these values into equations (2) to get

$$x = (v_0 \cos\theta)t = (150 \cos 30°)t = 75\sqrt{3}t$$
$$y = -\frac{1}{2}gt^2 + (v_0 \sin\theta)t + h = -\frac{1}{2}(32)t^2 + (150 \sin 30°)t + 0$$
$$= -16t^2 + 75t$$

(b) To determine the length of time that the ball was in the air, solve the equation $y = 0$.

$$-16t^2 + 75t = 0$$
$$t(-16t + 75) = 0$$
$$t = 0 \text{ sec} \quad \text{or} \quad t = \frac{75}{16} = 4.6875 \text{ sec}$$

The ball struck the ground after 4.6875 seconds.

(c) Notice that the height y of the ball is a quadratic function of t, so the maximum height of the ball can be found by determining the vertex of $y = -16t^2 + 75t$. The value of t at the vertex is

$$t = \frac{-b}{2a} = \frac{-75}{-32} = 2.34375 \text{ sec}$$

The ball was at its maximum height after 2.34375 seconds. The maximum height of the ball is found by evaluating the function y at $t = 2.34375$ seconds.

$$\text{Maximum height} = -16(2.34375)^2 + 75(2.34375) \approx 87.89 \text{ feet}$$

(d) Since the ball was in the air for 4.6875 seconds, the horizontal distance that the ball traveled is

$$x = (75\sqrt{3})4.6875 \approx 608.92 \text{ feet}$$

(e) Enter the equations from part (a) into a graphing utility with Tmin = 0, Tmax = 4.7, and Tstep = 0.1. Use ZOOM-SQUARE to avoid any distortion to the angle of elevation. See Figure 62. ●

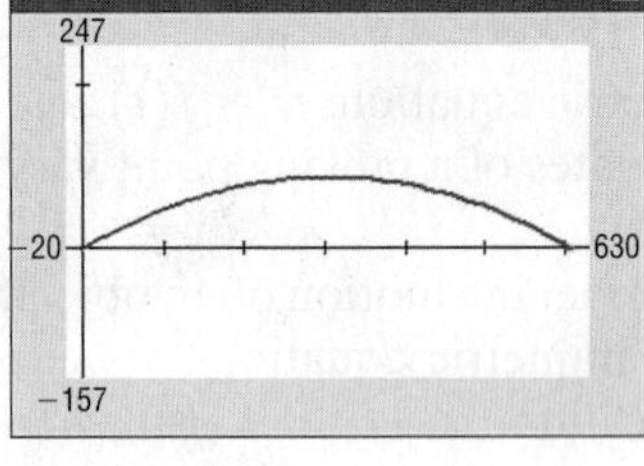

Figure 62

Exploration

Simulate the motion of a ball thrown straight up with an initial speed of 100 feet per second from a height of 5 feet above the ground. Use PARametric mode with Tmin = 0, Tmax = 6.5, Tstep = 0.1, Xmin = 0, Xmax = 5, Ymin = 0, and Ymax = 180. What happens to the speed with which the graph is drawn as the ball goes up and then comes back down? How do you interpret this physically? Repeat the experiment using other values for Tstep. How does this affect the experiment?

[**Hint:** In the projectile motion equations, let $\theta = 90°$, $v_0 = 100$, $h = 5$, and $g = 32$. Use $x = 3$ instead of $x = 0$ to see the vertical motion better.]

Result In Figure 63(a) the ball is going up. In Figure 63(b) the ball is near its highest point. Finally, in Figure 63(c), the ball is coming back down.

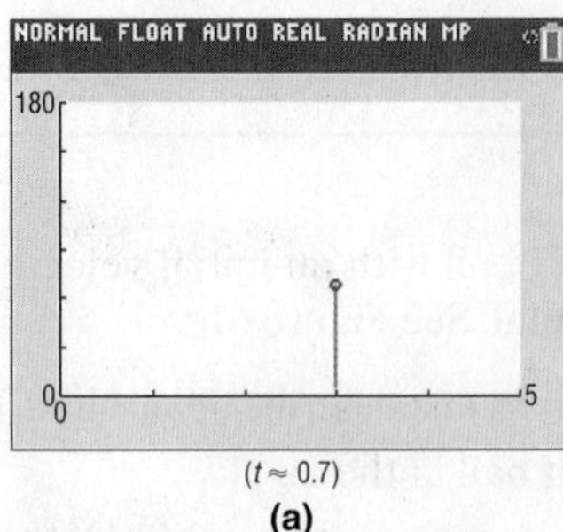

($t \approx 0.7$)
(a)

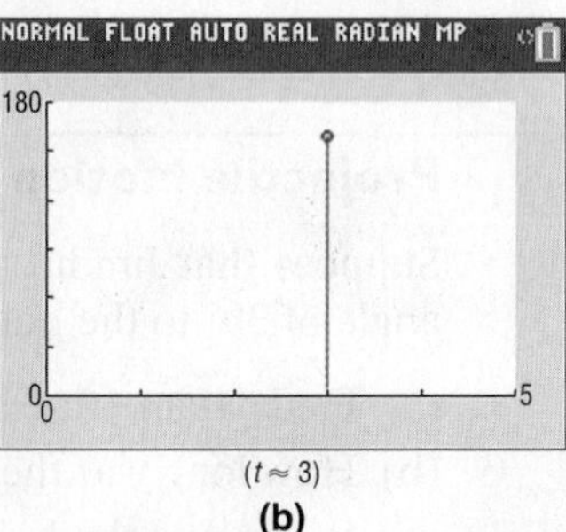

($t \approx 3$)
(b)

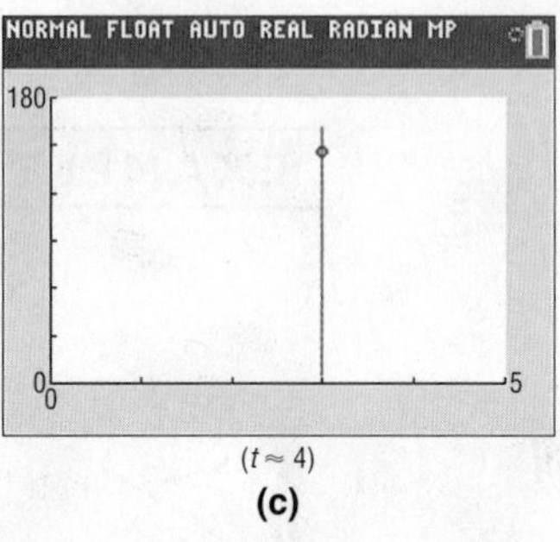

($t \approx 4$)
(c)

Figure 63

Notice that as the ball goes up, its speed decreases, until at the highest point it is zero. Then the speed increases as the ball comes back down.

Now Work PROBLEM 49

A graphing utility can be used to simulate other kinds of motion as well.

EXAMPLE 5

Simulating Motion

Tanya, who is a long-distance runner, runs at an average speed of 8 miles per hour. Two hours after Tanya leaves your house, you leave in your Honda and follow the same route. If your average speed is 40 miles per hour, how long will it be before you catch up to Tanya? See Figure 64. Use a simulation of the two motions to verify the answer.

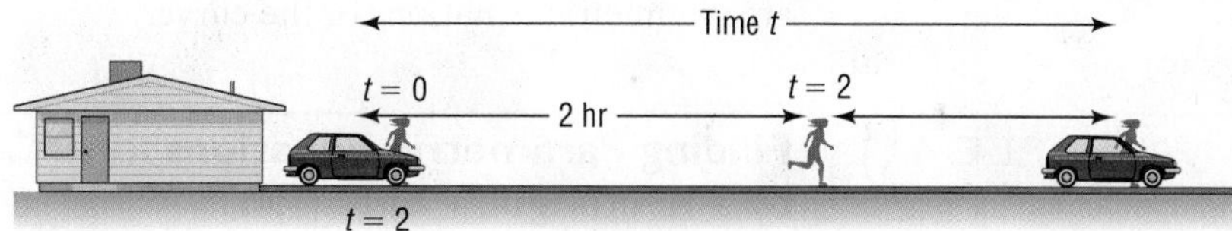

Figure 64

Solution Begin with two sets of parametric equations: one to describe Tanya's motion, the other to describe the motion of the Honda. We choose time $t = 0$ to be when Tanya leaves the house. If we choose $y_1 = 2$ as Tanya's path, then we can use $y_2 = 4$ as the parallel path of the Honda. The horizontal distances traversed in time t (Distance = Rate × Time) are

$$\text{Tanya: } x_1 = 8t \qquad \text{Honda: } x_2 = 40(t - 2)$$

The Honda catches up to Tanya when $x_1 = x_2$.

$$\begin{aligned} 8t &= 40(t - 2) \\ 8t &= 40t - 80 \\ -32t &= -80 \\ t &= \frac{-80}{-32} = 2.5 \end{aligned}$$

The Honda catches up to Tanya 2.5 hours after Tanya leaves the house.

In PARametric mode with Tstep = 0.01, simultaneously graph

$$\begin{aligned} \text{Tanya: } x_1 &= 8t & \text{Honda: } x_2 &= 40(t - 2) \\ y_1 &= 2 & y_2 &= 4 \end{aligned}$$

for $0 \leq t \leq 3$.

Figure 65 shows the relative positions of Tanya and the Honda for $t = 0$, $t = 2$, $t = 2.25$, $t = 2.5$, and $t = 2.75$.

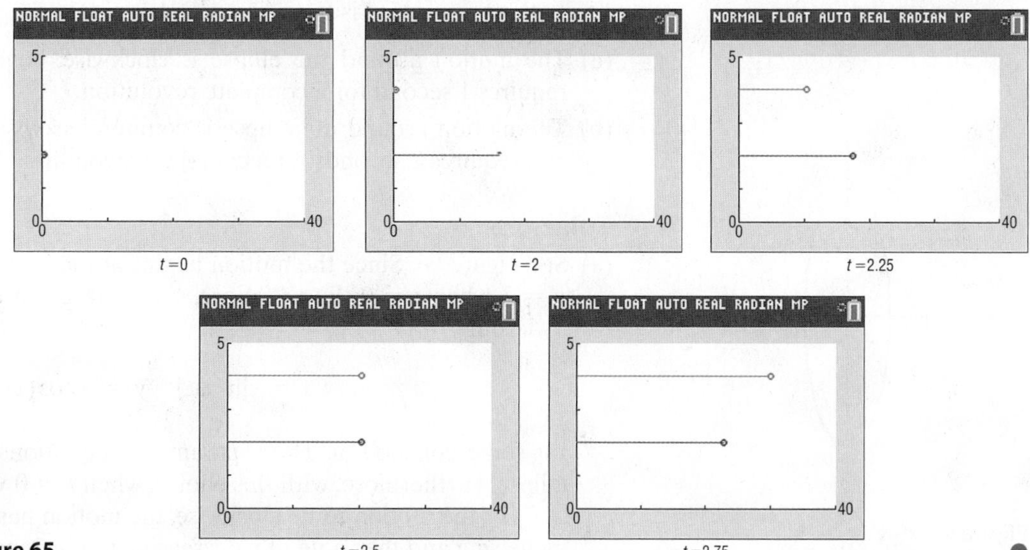

Figure 65

4 Find Parametric Equations for Curves Defined by Rectangular Equations

We now take up the question of how to find parametric equations of a given curve.

If a curve is defined by the equation $y = f(x)$, where f is a function, one way of finding parametric equations is to let $x = t$. Then $y = f(t)$ and

$$x = t \quad y = f(t) \qquad t \text{ in the domain of } f$$

are parametric equations of the curve.

EXAMPLE 6 **Finding Parametric Equations for a Curve Defined by a Rectangular Equation**

Find two different pairs of parametric equations for the equation $y = x^2 - 4$.

Solution For the first pair of parametric equations, let $x = t$. Then the parametric equations are

$$x = t \quad y = t^2 - 4 \qquad -\infty < t < \infty$$

A second pair of parametric equations is found by letting $x = t^3$. Then the parametric equations become

$$x = t^3 \quad y = t^6 - 4 \qquad -\infty < t < \infty$$

●

Care must be taken when using the second approach in Example 6, since the substitution for x must be a function that allows x to take on all the values stipulated by the domain of f. For example, letting $x = t^2$ so that $y = t^4 - 4$ does not result in equivalent parametric equations for $y = x^2 - 4$, since only points for which $x \geq 0$ are obtained; yet the domain of $y = x^2 - 4$ is $\{x \mid x \text{ is any real number}\}$.

Now Work PROBLEM 33

EXAMPLE 7 **Finding Parametric Equations for an Object in Motion**

Find parametric equations for the ellipse

$$x^2 + \frac{y^2}{9} = 1$$

where the parameter t is time (in seconds) and

(a) The motion around the ellipse is clockwise, begins at the point $(0, 3)$, and requires 1 second for a complete revolution.

(b) The motion around the ellipse is counterclockwise, begins at the point $(1, 0)$, and requires 2 seconds for a complete revolution.

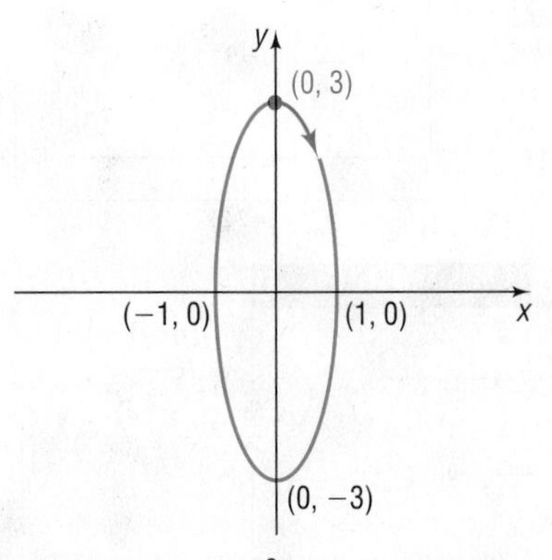

Figure 66 $x^2 + \frac{y^2}{9} = 1$

Solution

(a) See Figure 66. Since the motion begins at the point $(0, 3)$, we want $x = 0$ and $y = 3$ when $t = 0$. Furthermore, since the given equation is an ellipse, begin by letting

$$x = \sin(\omega t) \quad y = 3\cos(\omega t)$$

for some constant ω. These parametric equations satisfy the equation of the ellipse. Furthermore, with this choice, when $t = 0$ we have $x = 0$ and $y = 3$.

For the motion to be clockwise, the motion has to begin with the value of x increasing and the value of y decreasing as t increases. This requires that $\omega > 0$.

[Do you know why? If $\omega > 0$, then $x = \sin(\omega t)$ is increasing when $t > 0$ is near zero, and $y = 3\cos(\omega t)$ is decreasing when $t > 0$ is near zero.] See the red part of the graph in Figure 66.

Finally, since 1 revolution requires 1 second, the period $\frac{2\pi}{\omega} = 1$, so $\omega = 2\pi$. Parametric equations that satisfy the conditions stipulated are

$$x = \sin(2\pi t) \qquad y = 3\cos(2\pi t) \qquad 0 \le t \le 1 \tag{3}$$

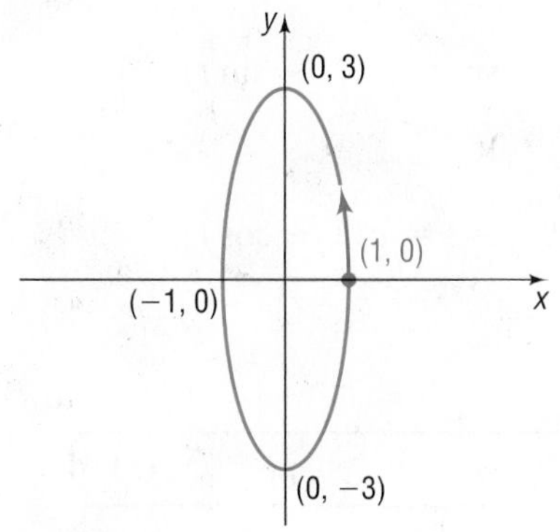

Figure 67 $x^2 + \frac{y^2}{9} = 1$

(b) See Figure 67. Since the motion begins at the point $(1, 0)$, we want $x = 1$ and $y = 0$ when $t = 0$. The given equation is an ellipse, so begin by letting

$$x = \cos(\omega t) \qquad y = 3\sin(\omega t)$$

for some constant ω. These parametric equations satisfy the equation of the ellipse. Furthermore, with this choice, when $t = 0$ we have $x = 1$ and $y = 0$.

For the motion to be counterclockwise, the motion has to begin with the value of x decreasing and the value of y increasing as t increases. This requires that $\omega > 0$. (Do you know why?) Finally, since 1 revolution requires 2 seconds, the period is $\frac{2\pi}{\omega} = 2$, so $\omega = \pi$. The parametric equations that satisfy the conditions stipulated are

$$x = \cos(\pi t) \qquad y = 3\sin(\pi t) \qquad 0 \le t \le 2 \tag{4}$$

Either equations (3) or equations (4) can serve as parametric equations for the ellipse $x^2 + \frac{y^2}{9} = 1$ given in Example 7. The direction of the motion, the beginning point, and the time for 1 revolution give a particular parametric representation.

Now Work PROBLEM 39

The Cycloid

Suppose that a circle of radius a rolls along a horizontal line without slipping. As the circle rolls along the line, a point P on the circle will trace out a curve called a **cycloid** (see Figure 68). We now seek parametric equations* for a cycloid.

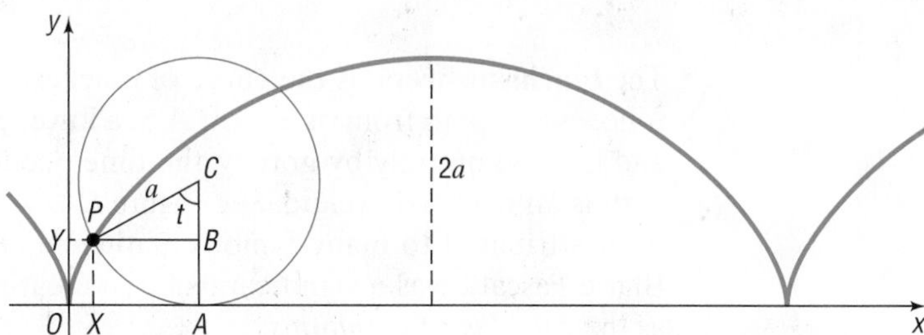

Figure 68 Cycloid

We begin with a circle of radius a and take the fixed line on which the circle rolls as the x-axis. Let the origin be one of the points at which the point P comes in contact with the x-axis. Figure 68 illustrates the position of this point P after the circle has rolled somewhat. The angle t (in radians) measures the angle through which the circle has rolled.

*Any attempt to derive the rectangular equation of a cycloid would soon demonstrate how complicated the task is.

Since we require no slippage, it follows that

$$\text{Arc } AP = d(O, A)$$

The length of the arc AP is given by $s = r\theta$, where $r = a$ and $\theta = t$ radians. Then

$$at = d(O, A) \qquad s = r\theta, \text{ where } r = a \text{ and } \theta = t$$

The x-coordinate of the point P is

$$d(O, X) = d(O, A) - d(X, A) = at - a\sin t = a(t - \sin t)$$

The y-coordinate of the point P is

$$d(O, Y) = d(A, C) - d(B, C) = a - a\cos t = a(1 - \cos t)$$

Exploration

Graph $x = t - \sin t$, $y = 1 - \cos t$, $0 \le t \le 3\pi$, using your graphing utility with Tstep $= \frac{\pi}{36}$ and a square screen. Compare your results with Figure 68.

The parametric equations of the cycloid are

$$x = a(t - \sin t) \qquad y = a(1 - \cos t) \qquad \textbf{(5)}$$

Applications to Mechanics

If a is negative in equation (5), we obtain an inverted cycloid, as shown in Figure 69(a). The inverted cycloid occurs as a result of some remarkable applications in the field of mechanics. We shall mention two of them: the *brachistochrone* and the *tautochrone*.*

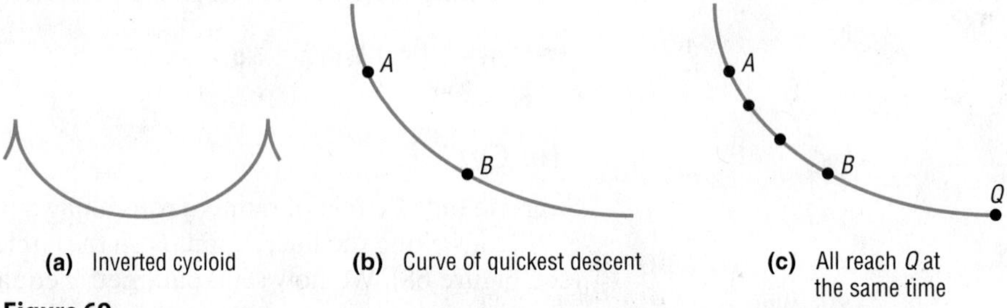

(a) Inverted cycloid (b) Curve of quickest descent (c) All reach Q at the same time

Figure 69

The **brachistochrone** is the curve of quickest descent. If a particle is constrained to follow some path from one point A to a lower point B (not on the same vertical line) and is acted on only by gravity, the time needed to make the descent is least if the path is an inverted cycloid. See Figure 69(b). This remarkable discovery, which has been attributed to many famous mathematicians (including Johann Bernoulli and Blaise Pascal), was a significant step in creating the branch of mathematics known as the *calculus of variations*.

To define the **tautochrone**, let Q be the lowest point on an inverted cycloid. If several particles placed at various positions on an inverted cycloid simultaneously begin to slide down the cycloid, they will reach the point Q at the same time, as indicated in Figure 69(c). The tautochrone property of the cycloid was used by Christiaan Huygens (1629–1695), the Dutch mathematician, physicist, and astronomer, to construct a pendulum clock with a bob that swings along a cycloid (see Figure 70). In Huygens's clock, the bob was made to swing along a cycloid by suspending the bob on a thin wire constrained by two plates shaped like cycloids. In a clock of this design, the period of the pendulum is independent of its amplitude.

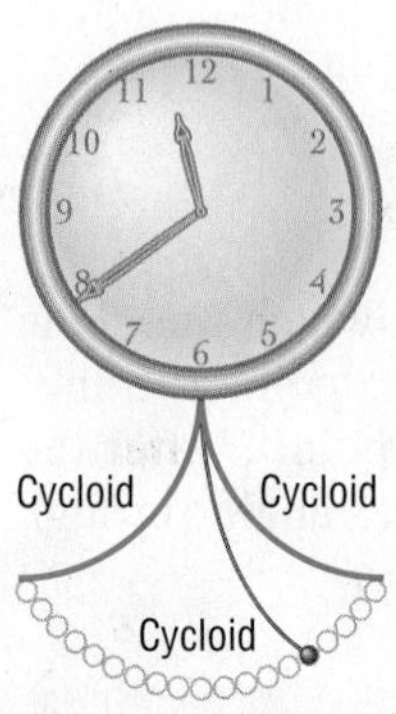

Figure 70

* In Greek, *brachistochrone* means "the shortest time" and *tautochrone* means "equal time."

6.7 Assess Your Understanding

'Are You Prepared?' *The answer is given at the end of these exercises. If you get a wrong answer, read the pages listed in red.*

1. The function $f(x) = 3\sin(4x)$ has amplitude ____ and period ____. (pp. 149–151)

Concepts and Vocabulary

2. Let $x = f(t)$ and $y = g(t)$, where f and g are two functions whose common domain is some interval I. The collection of points defined by $(x, y) = (f(t), g(t))$ is called a(n) ______. The variable t is called a(n) ______.

3. The parametric equations $x = 2\sin t$, $y = 3\cos t$ define a(n) ______.

4. If a circle rolls along a horizontal line without slippage, a fixed point P on the circle will trace out a curve called a(n) ______.

5. ***True or False*** Parametric equations defining a curve are unique.

6. ***True or False*** Curves defined using parametric equations have an orientation.

Skill Building

In Problems 7–26, graph the curve whose parametric equations are given, and show its orientation. Find the rectangular equation of each curve.

7. $x = 3t + 2, \quad y = t + 1; \quad 0 \le t \le 4$

8. $x = t - 3, \quad y = 2t + 4; \quad 0 \le t \le 2$

9. $x = t + 2, \quad y = \sqrt{t}; \quad t \ge 0$

10. $x = \sqrt{2t}, \quad y = 4t; \quad t \ge 0$

11. $x = t^2 + 4, \quad y = t^2 - 4; \quad -\infty < t < \infty$

12. $x = \sqrt{t} + 4, \quad y = \sqrt{t} - 4; \quad t \ge 0$

13. $x = 3t^2, \quad y = t + 1; \quad -\infty < t < \infty$

14. $x = 2t - 4, \quad y = 4t^2; \quad -\infty < t < \infty$

15.⁺ $x = 2e^t, \quad y = 1 + e^t; \quad t \ge 0$

16.⁺ $x = e^t, \quad y = e^{-t}; \quad t \ge 0$

17. $x = \sqrt{t}, \quad y = t^{3/2}; \quad t \ge 0$

18. $x = t^{3/2} + 1, \quad y = \sqrt{t}; \quad t \ge 0$

19. $x = 2\cos t, \quad y = 3\sin t; \quad 0 \le t \le 2\pi$

20. $x = 2\cos t, \quad y = 3\sin t; \quad 0 \le t \le \pi$

21. $x = 2\cos t, \quad y = 3\sin t; \quad -\pi \le t \le 0$

22. $x = 2\cos t, \quad y = \sin t; \quad 0 \le t \le \frac{\pi}{2}$

23. $x = \sec t, \quad y = \tan t; \quad 0 \le t \le \frac{\pi}{4}$

24. $x = \csc t, \quad y = \cot t; \quad \frac{\pi}{4} \le t \le \frac{\pi}{2}$

25. $x = \sin^2 t, \quad y = \cos^2 t; \quad 0 \le t \le 2\pi$

26.⁺⁺ $x = t^2, \quad y = \ln t; \quad t > 0$

In Problems 27–34, find two different pairs of parametric equations for each rectangular equation.

27. $y = 4x - 1$

28. $y = -8x + 3$

29. $y = x^2 + 1$

30. $y = -2x^2 + 1$

31. $y = x^3$

32. $y = x^4 + 1$

33. $x = y^{3/2}$

34. $x = \sqrt{y}$

In Problems 35–38, find parametric equations that define the curve shown.

35.

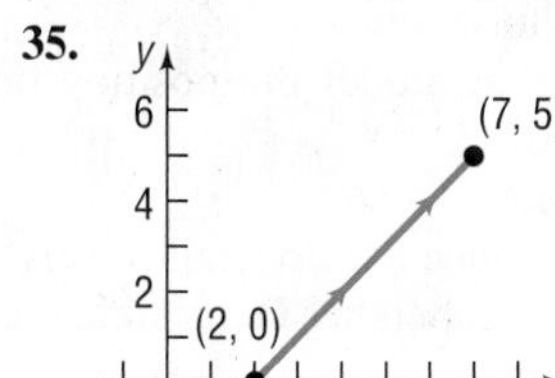

36.

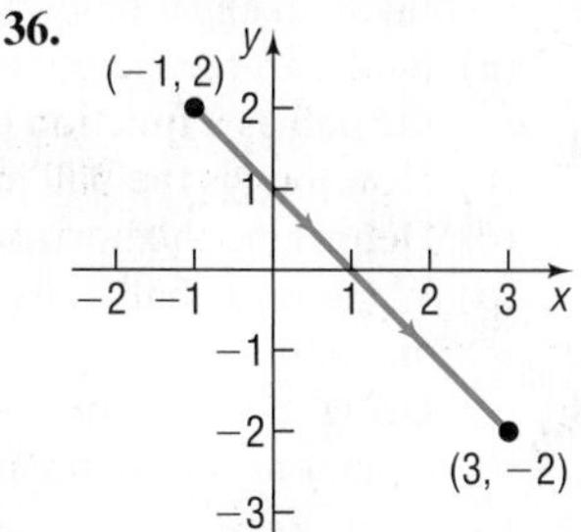

37.

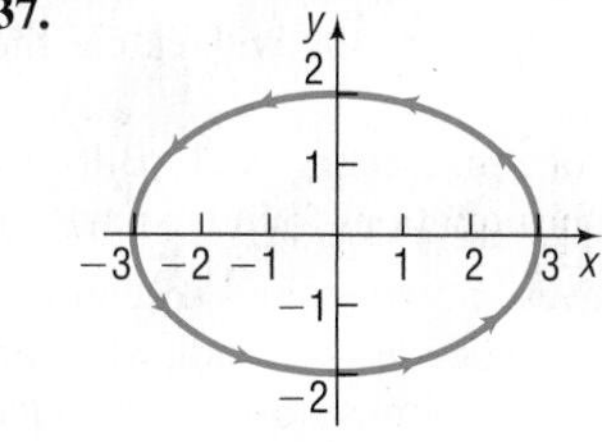

38.

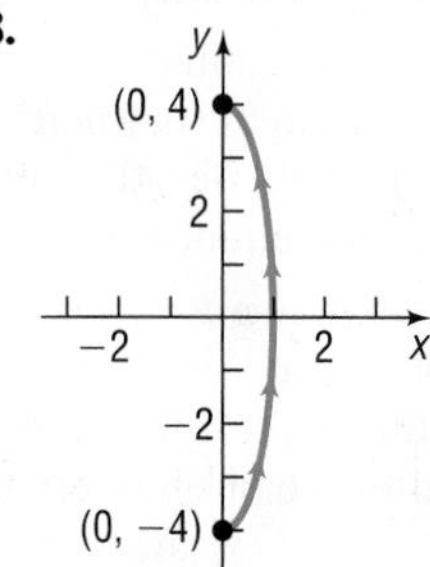

In Problems 39–42, find parametric equations for an object that moves along the ellipse $\frac{x^2}{4} + \frac{y^2}{9} = 1$ *with the motion described.*

39. The motion begins at $(2, 0)$, is clockwise, and requires 2 seconds for a complete revolution.

40. The motion begins at $(0, 3)$, is counterclockwise, and requires 1 second for a complete revolution.

41. The motion begins at $(0, 3)$, is clockwise, and requires 1 second for a complete revolution.

42. The motion begins at $(2, 0)$, is counterclockwise, and requires 3 seconds for a complete revolution.

⁺Requires Section 7.1, Exponential Functions
⁺⁺Requires Section 7.2, Logarithmic Functions

In Problems 43 and 44, the parametric equations of four curves are given. Graph each of them, indicating the orientation.

43. $C_1: \; x = t, \; y = t^2; \; -4 \le t \le 4$
$C_2: \; x = \cos t, \; y = 1 - \sin^2 t; \; 0 \le t \le \pi$
*$C_3: \; x = e^t, \; y = e^{2t}; \; 0 \le t \le \ln 4$
$C_4: \; x = \sqrt{t}, \; y = t; \; 0 \le t \le 16$

44. $C_1: \; x = t, \; y = \sqrt{1 - t^2}; \; -1 \le t \le 1$
$C_2: \; x = \sin t, \; y = \cos t; \; 0 \le t \le 2\pi$
$C_3: \; x = \cos t, \; y = \sin t; \; 0 \le t \le 2\pi$
$C_4: \; x = \sqrt{1 - t^2}, \; y = t; \; -1 \le t \le 1$

 *In Problems 45–48, use a graphing utility to graph the curve defined by the given parametric equations.*

45. $x = t \sin t, \; y = t \cos t, \; t > 0$

46. $x = \sin t + \cos t, \; y = \sin t - \cos t$

47. $x = 4 \sin t - 2 \sin(2t)$
$y = 4 \cos t - 2 \cos(2t)$

48. $x = 4 \sin t + 2 \sin(2t)$
$y = 4 \cos t + 2 \cos(2t)$

Applications and Extensions

49. Projectile Motion Bob throws a ball straight up with an initial speed of 50 feet per second from a height of 6 feet.

(a) Find parametric equations that model the motion of the ball as a function of time.
(b) How long is the ball in the air?
(c) When is the ball at its maximum height? Determine the maximum height of the ball.
 (d) Simulate the motion of the ball by graphing the equations found in part (a).

50. Projectile Motion Alice throws a ball straight up with an initial speed of 40 feet per second from a height of 5 feet.
(a) Find parametric equations that model the motion of the ball as a function of time.
(b) How long is the ball in the air?
(c) When is the ball at its maximum height? Determine the maximum height of the ball.
 (d) Simulate the motion of the ball by graphing the equations found in part (a).

51. Catching a Train Bill's train leaves at 8:06 AM and accelerates at the rate of 2 meters per second per second. Bill, who can run 5 meters per second, arrives at the train station 5 seconds after the train has left and runs for the train.
(a) Find parametric equations that model the motions of the train and Bill as a function of time.
[**Hint:** The position s at time t of an object having acceleration a is $s = \frac{1}{2}at^2$.]
(b) Determine algebraically whether Bill will catch the train. If so, when?
 (c) Simulate the motion of the train and Bill by simultaneously graphing the equations found in part (a).

52. Catching a Bus Jodi's bus leaves at 5:30 PM and accelerates at the rate of 3 meters per second per second. Jodi, who can run 5 meters per second, arrives at the bus station 2 seconds after the bus has left and runs for the bus.
(a) Find parametric equations that model the motions of the bus and Jodi as a function of time.
[**Hint:** The position s at time t of an object having acceleration a is $s = \frac{1}{2}at^2$.]
(b) Determine algebraically whether Jodi will catch the bus. If so, when?
 (c) Simulate the motion of the bus and Jodi by graphing simultaneously the equations found in part (a).

*Requires Section 7.1, Exponential Functions and Section 7.2, Logarithmic Functions

53. Projectile Motion Ichiro throws a baseball with an initial speed of 145 feet per second at an angle of 20° to the horizontal. The ball leaves Ichiro's hand at a height of 5 feet.
(a) Find parametric equations that model the position of the ball as a function of time.
(b) How long is the ball in the air?
(c) Determine the horizontal distance that the ball travels.
(d) When is the ball at its maximum height? Determine the maximum height of the ball.
 (e) Using a graphing utility, simultaneously graph the equations found in part (a).

54. Projectile Motion Mark Texeira hit a baseball with an initial speed of 125 feet per second at an angle of 40° to the horizontal. The ball was hit at a height of 3 feet above the ground.
(a) Find parametric equations that model the position of the ball as a function of time.
(b) How long was the ball in the air?
(c) Determine the horizontal distance that the ball traveled.
(d) When was the ball at its maximum height? Determine the maximum height of the ball.
 (e) Using a graphing utility, simultaneously graph the equations found in part (a).

55. Projectile Motion Suppose that Adam hits a golf ball off a cliff 300 meters high with an initial speed of 40 meters per second at an angle of 45° to the horizontal.
(a) Find parametric equations that model the position of the ball as a function of time.
(b) How long is the ball in the air?
(c) Determine the horizontal distance that the ball travels.
(d) When is the ball at its maximum height? Determine the maximum height of the ball.
 (e) Using a graphing utility, simultaneously graph the equations found in part (a).

56. Projectile Motion Suppose that Karla hits a golf ball off a cliff 300 meters high with an initial speed of 40 meters per second at an angle of 45° to the horizontal on the Moon (gravity on the Moon is one-sixth of that on Earth).
(a) Find parametric equations that model the position of the ball as a function of time.
(b) How long is the ball in the air?
(c) Determine the horizontal distance that the ball travels.
(d) When is the ball at its maximum height? Determine the maximum height of the ball.
 (e) Using a graphing utility, simultaneously graph the equations found in part (a).

57. Uniform Motion A Toyota Camry (traveling east at 40 mph) and a Chevy Impala (traveling north at 30 mph) are heading toward the same intersection. The Camry is 5 miles from the intersection when the Impala is 4 miles from the intersection. See the figure.

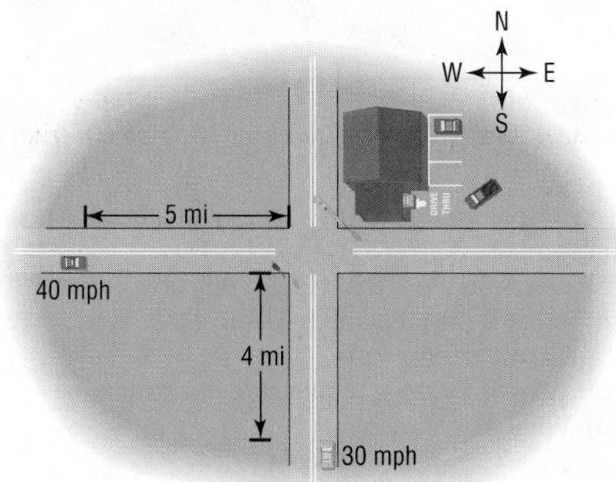

(a) Find parametric equations that model the motion of the Camry and the Impala.
(b) Find a formula for the distance between the cars as a function of time.
(c) Graph the function in part (b) using a graphing utility.
(d) What is the minimum distance between the cars? When are the cars closest?
(e) Simulate the motion of the cars by simultaneously graphing the equations found in part (a).

58. Uniform Motion A Cessna (heading south at 120 mph) and a Boeing 747 (heading west at 600 mph) are flying toward the same point at the same altitude. The Cessna is 100 miles from the point where the flight patterns intersect, and the 747 is 550 miles from this intersection point. See the figure.

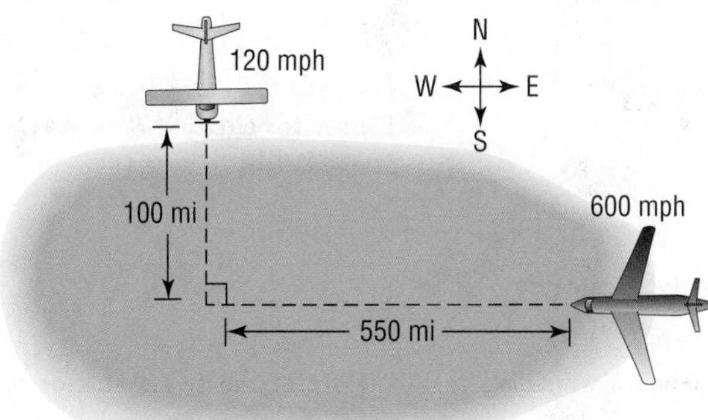

(a) Find parametric equations that model the motion of the Cessna and the 747.
(b) Find a formula for the distance between the planes as a function of time.
(c) Graph the function in part (b) using a graphing utility.
(d) What is the minimum distance between the planes? When are the planes closest?
(e) Simulate the motion of the planes by simultaneously graphing the equations found in part (a).

59. The Green Monster The left field wall at Fenway Park is 310 feet from home plate; the wall itself (affectionately named the Green Monster) is 37 feet high. A batted ball must clear the wall to be a home run. Suppose a ball leaves the bat 3 feet above the ground, at an angle of 45°. Use $g = 32$ feet per second^2 as the acceleration due to gravity, and ignore any air resistance.
(a) Find parametric equations that model the position of the ball as a function of time.
(b) What is the maximum height of the ball if it leaves the bat with a speed of 90 miles per hour? Give your answer in feet.
(c) How far is the ball from home plate at its maximum height? Give your answer in feet.
(d) If the ball is hit straight down the left field line, will it clear the Green Monster? If it does, by how much does it clear the wall?

Source: The Boston Red Sox

60. Projectile Motion The position of a projectile fired with an initial velocity v_0 feet per second and at an angle θ to the horizontal at the end of t seconds is given by the parametric equations

$$x = (v_0 \cos \theta)t \qquad y = (v_0 \sin \theta)t - 16t^2$$

See the illustration.

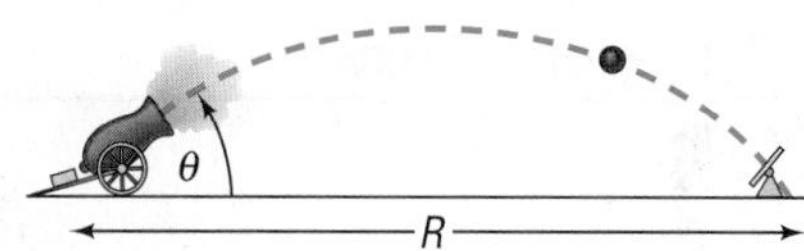

(a) Obtain the rectangular equation of the trajectory, and identify the curve.
(b) Show that the projectile hits the ground $(y = 0)$ when $t = \frac{1}{16} v_0 \sin \theta$.
(c) How far has the projectile traveled (horizontally) when it strikes the ground? In other words, find the range R.
(d) Find the time t when $x = y$. Next find the horizontal distance x and the vertical distance y traveled by the projectile in this time. Then compute $\sqrt{x^2 + y^2}$. This is the distance R, the range, that the projectile travels up a plane inclined at 45° to the horizontal $(x = y)$. See the following illustration. (See also Problem 99 in Section 3.6.)

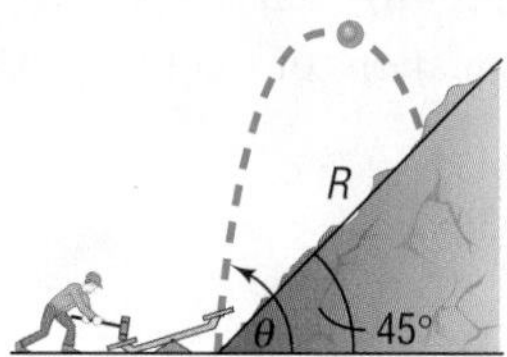

61. Show that the parametric equations for a line passing through the points (x_1, y_1) and (x_2, y_2) are

$$x = (x_2 - x_1)t + x_1$$
$$y = (y_2 - y_1)t + y_1 \qquad -\infty < t < \infty$$

What is the orientation of this line?

62. Hypocycloid The hypocycloid is a curve defined by the parametric equations

$$x(t) = \cos^3 t \quad y(t) = \sin^3 t \quad 0 \le t \le 2\pi$$

(a) Graph the hypocycloid using a graphing utility.
(b) Find a rectangular equation of the hypocycloid.

Explaining Concepts: Discussion and Writing

63. In Problem 62, we graphed the hypocycloid. Now graph the rectangular equations of the hypocycloid. Did you obtain a complete graph? If not, experiment until you do.

64. Look up the curves called *hypocycloid* and *epicycloid*. Write a report on what you find. Be sure to draw comparisons with the cycloid.

Retain Your Knowledge

Problems 65–68 are based on material learned earlier in the course. The purpose of these problems is to keep the material fresh in your mind so that you are better prepared for the final exam.

65. Graph the equation $3x - 4y = 8$.

66. Graph $y = 2\cos(2x) + \sin\left(\frac{x}{2}\right)$.

67. The International Space Station (ISS) orbits Earth at a height of approximately 248 miles above the surface. What is the distance, in miles, on the surface of Earth that can be observed from the ISS? Assume that Earth's radius is 3960 miles.
Source: nasa.gov

68. The displacement d (in meters) of an object at time t (in seconds) is given by $d = 2\cos(4t)$.
(a) Describe the motion of the object.
(b) What is the maximum displacement of the object from its rest position?
(c) What is the time required for 1 oscillation?
(d) What is the frequency?

'Are You Prepared?' Answer

1. $3; \frac{\pi}{2}$

Chapter Review

Things to Know

Equations

Parabola (pp. 388–394)	See Tables 1 and 2 (pp. 390 and 391).	
Ellipse (pp. 397–404)	See Table 3 (p. 402).	
Hyperbola (pp. 407–416)	See Table 4 (p. 414).	
General equation of a conic (p. 425)	$Ax^2 + Bxy + Cy^2 + Dx + Ey + F = 0$	Parabola if $B^2 - 4AC = 0$ Ellipse (or circle) if $B^2 - 4AC < 0$ Hyperbola if $B^2 - 4AC > 0$
Polar equations of a conic with focus at the pole (pp. 428–432)	See Table 5 (p. 430).	
Parametric equations of a curve (p. 434)	$x = f(t), y = g(t)$, t is the parameter	

Definitions

Parabola (p. 388)	Set of points P in the plane for which $d(F, P) = d(P, D)$, where F is the focus and D is the directrix	
Ellipse (p. 397)	Set of points P in the plane the sum of whose distances from two fixed points (the foci) is a constant	
Hyperbola (p. 407)	Set of points P in the plane the difference of whose distances from two fixed points (the foci) is a constant	
Conic in polar coordinates (p. 428)	$\frac{d(F, P)}{d(D, P)} = e$	Parabola if $e = 1$ Ellipse if $e < 1$ Hyperbola if $e > 1$

Formulas

Rotation formulas (p. 422)	$x = x'\cos\theta - y'\sin\theta$ $y = x'\sin\theta + y'\cos\theta$	
Angle θ of rotation that eliminates the $x'y'$-term (p. 423)	$\cot(2\theta) = \frac{A - C}{B}$	$0° < \theta < 90°$

Objectives

Section		You should be able to . . .	Example(s)	Review Exercises
6.1	1	Know the names of the conics (p. 387)		1–16
6.2	1	Analyze parabolas with vertex at the origin (p. 388)	1–5	1, 11
	2	Analyze parabolas with vertex at (h, k) (p. 391)	6, 7	4, 6, 9, 14
	3	Solve applied problems involving parabolas (p. 393)	8	39
6.3	1	Analyze ellipses with center at the origin (p. 397)	1–4	3, 13
	2	Analyze ellipses with center at (h, k) (p. 401)	5, 6	8, 10, 16, 38
	3	Solve applied problems involving ellipses (p. 403)	7	40
6.4	1	Analyze hyperbolas with center at the origin (p. 407)	1–4	2, 5, 12, 37
	2	Find the asymptotes of a hyperbola (p. 412)	5, 6	2, 5, 7
	3	Analyze hyperbolas with center at (h, k) (p. 414)	7, 8	7, 15, 17, 18
	4	Solve applied problems involving hyperbolas (p. 415)	9	41
6.5	1	Identify a conic (p. 420)	1	19, 20
	2	Use a rotation of axes to transform equations (p. 421)	2	24–26
	3	Analyze an equation using a rotation of axes (p. 423)	3, 4	24–26, 44
	4	Identify conics without a rotation of axes (p. 425)	5	21–23
6.6	1	Analyze and graph polar equations of conics (p. 428)	1–3	27–29
	2	Convert the polar equation of a conic to a rectangular equation (p. 432)	4	30, 31
6.7	1	Graph parametric equations (p. 434)	1	32–34
	2	Find a rectangular equation for a curve defined parametrically (p. 435)	2, 3	32–34
	3	Use time as a parameter in parametric equations (p. 437)	4, 5	42, 43
	4	Find parametric equations for curves defined by rectangular equations (p. 440)	6, 7	35, 36

Review Exercises

In Problems 1–10, identify each equation. If it is a parabola, give its vertex, focus, and directrix; if it is an ellipse, give its center, vertices, and foci; if it is a hyperbola, give its center, vertices, foci, and asymptotes.

1. $y^2 = -16x$

2. $\dfrac{x^2}{25} - y^2 = 1$

3. $\dfrac{y^2}{25} + \dfrac{x^2}{16} = 1$

4. $x^2 + 4y = 4$

5. $4x^2 - y^2 = 8$

6. $x^2 - 4x = 2y$

7. $y^2 - 4y - 4x^2 + 8x = 4$

8. $4x^2 + 9y^2 - 16x - 18y = 11$

9. $4x^2 - 16x + 16y + 32 = 0$

10. $9x^2 + 4y^2 - 18x + 8y = 23$

In Problems 11–18, find an equation of the conic described. Graph the equation.

11. Parabola; focus at $(-2, 0)$; directrix the line $x = 2$

12. Hyperbola; center at $(0, 0)$; focus at $(0, 4)$; vertex at $(0, -2)$

13. Ellipse; foci at $(-3, 0)$ and $(3, 0)$; vertex at $(4, 0)$

14. Parabola; vertex at $(2, -3)$; focus at $(2, -4)$

15. Hyperbola; center at $(-2, -3)$; focus at $(-4, -3)$; vertex at $(-3, -3)$

16. Ellipse; foci at $(-4, 2)$ and $(-4, 8)$; vertex at $(-4, 10)$

17. Center at $(-1, 2)$; $a = 3$; $c = 4$; transverse axis parallel to the x-axis

18. Vertices at $(0, 1)$ and $(6, 1)$; asymptote the line $3y + 2x = 9$

In Problems 19–23, identify each conic without completing the squares and without applying a rotation of axes.

19. $y^2 + 4x + 3y - 8 = 0$

20. $x^2 + 2y^2 + 4x - 8y + 2 = 0$

21. $9x^2 - 12xy + 4y^2 + 8x + 12y = 0$

22. $4x^2 + 10xy + 4y^2 - 9 = 0$

23. $x^2 - 2xy + 3y^2 + 2x + 4y - 1 = 0$

In Problems 24–26, rotate the axes so that the new equation contains no xy-term. Analyze and graph the new equation.

24. $2x^2 + 5xy + 2y^2 - \frac{9}{2} = 0$

25. $6x^2 + 4xy + 9y^2 - 20 = 0$

26. $4x^2 - 12xy + 9y^2 + 12x + 8y = 0$

In Problems 27–29, identify the conic that each polar equation represents, and graph it.

27. $r = \dfrac{4}{1 - \cos\theta}$

28. $r = \dfrac{6}{2 - \sin\theta}$

29. $r = \dfrac{8}{4 + 8\cos\theta}$

In Problems 30 and 31, convert each polar equation to a rectangular equation.

30. $r = \dfrac{4}{1 - \cos\theta}$

31. $r = \dfrac{8}{4 + 8\cos\theta}$

In Problems 32–34, graph the curve whose parametric equations are given, and show its orientation. Find the rectangular equation of each curve.

32. $x = 4t - 2, \quad y = 1 - t; \quad -\infty < t < \infty$

33. $x = 3\sin t, \quad y = 4\cos t + 2; \quad 0 \le t \le 2\pi$

34. $x = \sec^2 t, \quad y = \tan^2 t; \quad 0 \le t \le \dfrac{\pi}{4}$

35. Find two different pairs of parametric equations for $y = -2x + 4$.

36. Find parametric equations for an object that moves along the ellipse $\dfrac{x^2}{16} + \dfrac{y^2}{9} = 1$, where the motion begins at $(4, 0)$, is counterclockwise, and requires 4 seconds for a complete revolution.

---

37. Find an equation of the hyperbola whose foci are the vertices of the ellipse $4x^2 + 9y^2 = 36$ and whose vertices are the foci of this ellipse.

38. Describe the collection of points in a plane so that the distance from each point to the point $(3, 0)$ is three-fourths of its distance from the line $x = \dfrac{16}{3}$.

39. **Searchlight** A searchlight is shaped like a paraboloid of revolution. If a light source is located 1 foot from the vertex along the axis of symmetry and the opening is 2 feet across, how deep should the mirror be in order to reflect the light rays parallel to the axis of symmetry?

40. **Semielliptical Arch Bridge** A bridge is built in the shape of a semielliptical arch. The bridge has a span of 60 feet and a maximum height of 20 feet. Find the height of the arch at distances of 5, 10, and 20 feet from the center.

41. **Calibrating Instruments** In a test of their recording devices, a team of seismologists positioned two of the devices 2000 feet apart, with the device at point A to the west of the device at point B. At a point between the devices and 200 feet from point B, a small amount of explosive was detonated and a note made of the time at which the sound reached each device. A second explosion is to be carried out at a point directly north of point B. How far north should the site of the second explosion be chosen so that the measured time difference recorded by the devices for the second detonation is the same as that recorded for the first detonation?

42. **Uniform Motion** Mary's train leaves at 7:15 AM and accelerates at the rate of 3 meters per second per second. Mary, who can run 6 meters per second, arrives at the train station 2 seconds after the train has left.
    (a) Find parametric equations that model the motion of the train and Mary as a function of time.

    [**Hint:** The position s at time t of an object having acceleration a is $s = \frac{1}{2}at^2$.]

    (b) Determine algebraically whether Mary will catch the train. If so, when?
    (c) Simulate the motions of the train and Mary by simultaneously graphing the equations found in part (a).

43. **Projectile Motion** Drew Brees throws a football with an initial speed of 80 feet per second at an angle of 35° to the horizontal. The ball leaves Brees's hand at a height of 6 feet.
    (a) Find parametric equations that model the position of the ball as a function of time.
    (b) How long is the ball in the air?
    (c) When is the ball at its maximum height? Determine the maximum height of the ball.
    (d) Determine the horizontal distance that the ball travels.
    (e) Using a graphing utility, simultaneously graph the equations found in part (a).

44. Formulate a strategy for discussing and graphing an equation of the form

$$Ax^2 + Bxy + Cy^2 + Dx + Ey + F = 0$$

Chapter Test

CHAPTER **Test Prep** VIDEOS

The Chapter Test Prep Videos are step-by-step solutions available in MyMathLab®, or on this text's YouTube™ Channel. Flip back to the Resources for Success page for a link to this text's YouTube channel.

In Problems 1–3, identify each equation. If it is a parabola, give its vertex, focus, and directrix; if an ellipse, give its center, vertices, and foci; if a hyperbola, give its center, vertices, foci, and asymptotes.

1. $\dfrac{(x+1)^2}{4} - \dfrac{y^2}{9} = 1$

2. $8y = (x-1)^2 - 4$

3. $2x^2 + 3y^2 + 4x - 6y = 13$

In Problems 4–6, find an equation of the conic described; graph the equation.

4. Parabola: focus $(-1, 4.5)$, vertex $(-1, 3)$

5. Ellipse: center $(0, 0)$, vertex $(0, -4)$, focus $(0, 3)$

6. Hyperbola: center $(2, 2)$, vertex $(2, 4)$, contains the point $\left(2 + \sqrt{10}, 5\right)$

In Problems 7–9, identify each conic without completing the square or rotating axes.

7. $2x^2 + 5xy + 3y^2 + 3x - 7 = 0$

8. $3x^2 - xy + 2y^2 + 3y + 1 = 0$

9. $x^2 - 6xy + 9y^2 + 2x - 3y - 2 = 0$

10. Given the equation $41x^2 - 24xy + 34y^2 - 25 = 0$, rotate the axes so that there is no xy-term. Analyze and graph the new equation.

11. Identify the conic represented by the polar equation $r = \dfrac{3}{1 - 2\cos\theta}$. Find the rectangular equation.

12. Graph the curve whose parametric equations are given, and show its orientation. Find the rectangular equation for the curve.

$$x = 3t - 2 \quad y = 1 - \sqrt{t} \quad 0 \le t \le 9$$

13. A parabolic reflector (paraboloid of revolution) is used by TV crews at football games to pick up the referee's announcements, quarterback signals, and so on. A microphone is placed at the focus of the parabola. If a certain reflector is 4 feet wide and 1.5 feet deep, where should the microphone be placed?

Cumulative Review

1. For $f(x) = -3x^2 + 5x - 2$, find

$$\frac{f(x+h) - f(x)}{h} \qquad h \neq 0$$

2. Find an equation for each of the following graphs.

(a) Line:

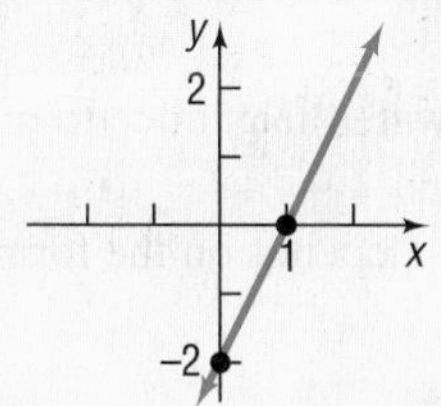

(b) Circle:

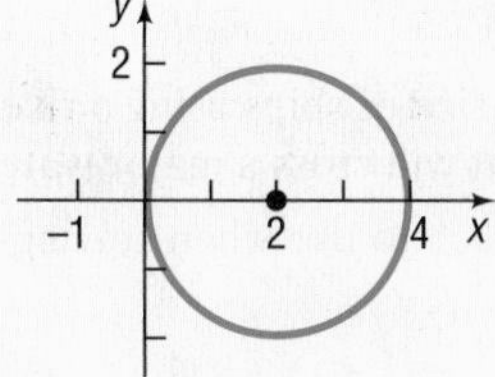

(c) Ellipse:

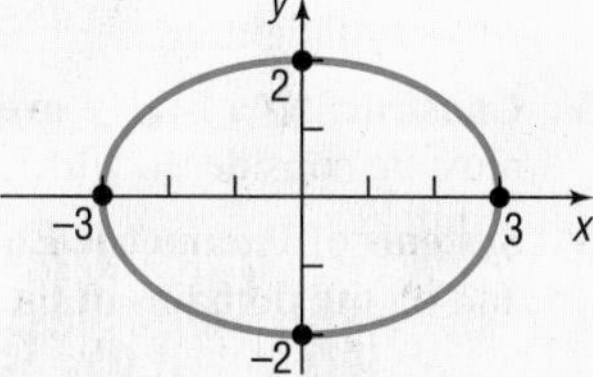

(d) Parabola:

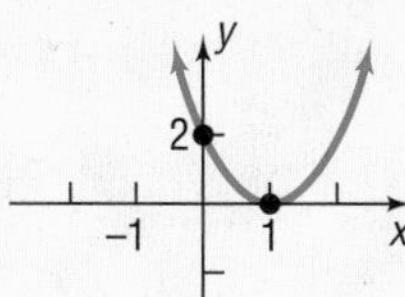

(e) Hyperbola:

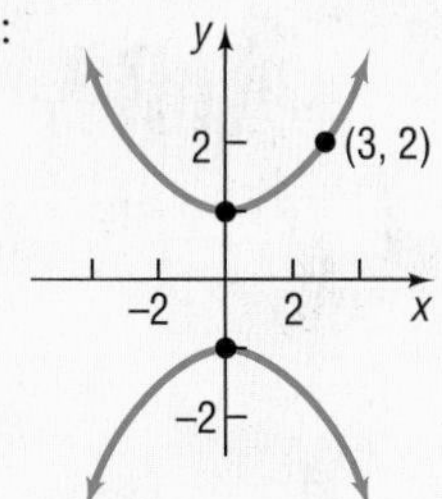

3. Find all the solutions of the equation $\sin(2\theta) = 0.5$.

4. Find a polar equation for the line containing the origin that makes an angle of 30° with the positive x-axis.

5. Find a polar equation for the circle with center at the point $(0, 4)$ and radius 4. Graph this circle.

6. What is the domain of the function $f(x) = \dfrac{3}{\sin x + \cos x}$?

7. Solve the equation $\cot(2\theta) = 1$, where $0° < \theta < 90°$.

8. Find the rectangular equation of the curve

$$x = 5\tan t \quad y = 5\sec^2 t \quad -\frac{\pi}{2} < t < \frac{\pi}{2}$$

Chapter Projects

Internet-based Project

I. Comet Hale-Bopp The orbits of planets and some comets about the Sun are ellipses, with the Sun at one focus. The **aphelion** of a planet is its greatest distance from the Sun, and the **perihelion** is its shortest distance. The **mean distance** of a planet from the Sun is the length of the semimajor axis of the elliptical orbit. See the illustration.

1. Research the history of Comet Hale-Bopp on the Internet. In particular, determine the aphelion and perihelion. Often these values are given in terms of astronomical units. What is an astronomical unit? What is it equivalent to in miles? In kilometers? What is the orbital period of Comet Hale-Bopp? When will it next be visible from Earth? How close does it come to Earth?
2. Find a model for the orbit of Comet Hale-Bopp around the Sun. Use the x-axis as the major axis.
3. Comet Hale-Bopp has an orbit that is roughly perpendicular to that of Earth. Find a model for the orbit of Earth using the y-axis as the major axis.
4. Use a graphing utility or some other graphing technology to graph the paths of the orbits. Based on the graphs, do the paths of the orbits intersect? Does this mean that Comet Hale-Bopp will collide with Earth?

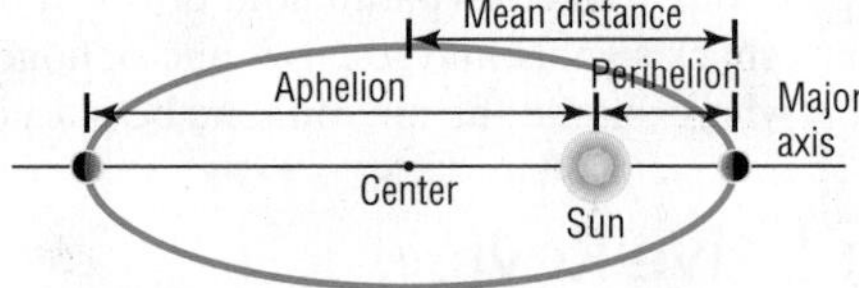

The following projects can be found at the Instructor's Resource Center (IRC):

II. The Orbits of Neptune and Pluto The astronomical body known as Pluto and the planet Neptune travel around the Sun in elliptical orbits. Pluto, at times, comes closer to the Sun than Neptune, the outermost planet. This project examines and analyzes the two orbits.

III. Project at Motorola ***Distorted Deployable Space Reflector Antennas*** An engineer designs an antenna that will deploy in space to collect sunlight.

IV. Constructing a Bridge over the East River The size of ships using a river and fluctuations in water height due to tides or flooding must be considered when designing a bridge that will cross a major waterway.

V. Systems of Parametric Equations Which approach to use when solving a system of equations depends on the form of the system and on the domains of the equations.

7 Exponential and Logarithmic Functions

Depreciation of Cars

You are ready to buy that first new car. You know that cars lose value over time due to depreciation and that different cars have different rates of depreciation. So you will research the depreciation rates for the cars you are thinking of buying. After all, for cars that sell for about the same price, the lower the depreciation rate, the more the car will be worth each year.

—See the Internet-based Chapter Project I—

Outline

7.1 Exponential Functions
7.2 Logarithmic Functions
7.3 Properties of Logarithms
7.4 Logarithmic and Exponential Equations
7.5 Financial Models
7.6 Exponential Growth and Decay Models; Newton's Law; Logistic Growth and Decay Models
7.7 Building Exponential, Logarithmic, and Logistic Models from Data
Chapter Review
Chapter Test
Cumulative Review
Chapter Projects

A Look Back

In Chapter 1, we created a library of functions, naming key functions and listing their properties, including their graphs. These functions belong to the class of **algebraic functions**—that is, functions that can be expressed in terms of sums, differences, products, quotients, powers, or roots of polynomials. Functions that are not algebraic are termed **transcendental** (they transcend, or go beyond, algebraic functions).

A Look Ahead

In this chapter, we study two transcendental functions: the exponential function and the logarithmic function. These functions occur frequently in a wide variety of applications, such as biology, chemistry, economics, and psychology.

7.1 Exponential Functions

PREPARING FOR THIS SECTION *Before getting started, review the following:*

- Exponents (Appendix A, Section A.1, pp. A8–A9, and Section A.7, pp. A59–A61)
- Graphing Techniques: Transformations (Section 1.6, pp. 64–73)
- Solving Equations (Appendix A, Section A.4, pp. A27–A34)
- Average Rate of Change (Section 1.4, pp. 48–49)
- Asymptotes (Section 1.6, pp. 73–74)
- Average Rate of Change of a Line (Appendix A, Section A.8, p. A65)

Now Work the 'Are You Prepared?' problems on page 464.

OBJECTIVES
1 Evaluate Exponential Functions (p. 453)
2 Graph Exponential Functions (p. 457)
3 Define the Number *e* (p. 460)
4 Solve Exponential Equations (p. 462)

1 Evaluate Exponential Functions

Appendix A, Section A.7 gives a definition for raising a real number a to a rational power. That discussion provides meaning to expressions of the form

$$a^r$$

where the base a is a positive real number and the exponent r is a rational number.

But what is the meaning of a^x, where the base a is a positive real number and the exponent x is an irrational number? Although a rigorous definition requires methods discussed in calculus, the basis for the definition is easy to follow: Select a rational number r that is formed by truncating (removing) all but a finite number of digits from the irrational number x. Then it is reasonable to expect that

$$a^x \approx a^r$$

For example, take the irrational number $\pi = 3.14159\ldots$. Then an approximation to a^π is

$$a^\pi \approx a^{3.14}$$

where the digits after the hundredths position have been removed from the value for π. A better approximation would be

$$a^\pi \approx a^{3.14159}$$

where the digits after the hundred-thousandths position have been removed. Continuing in this way, we can obtain approximations to a^π to any desired degree of accuracy.

Most calculators have an $\boxed{x^y}$ key or a caret key $\boxed{\wedge}$ for working with exponents. To evaluate expressions of the form a^x, enter the base a, then press the $\boxed{x^y}$ key (or the $\boxed{\wedge}$ key), enter the exponent x, and press $\boxed{=}$ (or $\boxed{\text{ENTER}}$).

EXAMPLE 1 **Using a Calculator to Evaluate Powers of 2**

Using a calculator, evaluate:

(a) $2^{1.4}$ (b) $2^{1.41}$ (c) $2^{1.414}$ (d) $2^{1.4142}$ (e) $2^{\sqrt{2}}$

Solution (a) $2^{1.4} \approx 2.639015822$ (b) $2^{1.41} \approx 2.657371628$
(c) $2^{1.414} \approx 2.66474965$ (d) $2^{1.4142} \approx 2.665119089$
(e) $2^{\sqrt{2}} \approx 2.665144143$

 **Now Work** PROBLEM 15

It can be shown that the familiar laws for rational exponents hold for real exponents.

THEOREM

Laws of Exponents

If s, t, a, and b are real numbers with $a > 0$ and $b > 0$, then

$$a^s \cdot a^t = a^{s+t} \qquad (a^s)^t = a^{st} \qquad (ab)^s = a^s \cdot b^s$$
$$1^s = 1 \qquad a^{-s} = \frac{1}{a^s} = \left(\frac{1}{a}\right)^s \qquad a^0 = 1 \tag{1}$$

Introduction to Exponential Growth

Suppose a function f has the following two properties:

1. The value of f doubles with every 1-unit increase in the independent variable x.
2. The value of f at $x = 0$ is 5, so $f(0) = 5$.

Table 1 shows values of the function f for $x = 0, 1, 2, 3$, and 4.

Table 1

x	f(x)
0	5
1	10
2	20
3	40
4	80

Let's find an equation $y = f(x)$ that describes this function f. The key fact is that the value of f doubles for every 1-unit increase in x.

$f(0) = 5$
$f(1) = 2f(0) = 2 \cdot 5 = 5 \cdot 2^1$ — Double the value of f at 0 to get the value at 1.
$f(2) = 2f(1) = 2(5 \cdot 2) = 5 \cdot 2^2$ — Double the value of f at 1 to get the value at 2.
$f(3) = 2f(2) = 2(5 \cdot 2^2) = 5 \cdot 2^3$
$f(4) = 2f(3) = 2(5 \cdot 2^3) = 5 \cdot 2^4$

The pattern leads to

$$f(x) = 2f(x-1) = 2(5 \cdot 2^{x-1}) = 5 \cdot 2^x$$

DEFINITION

An **exponential function** is a function of the form

$$f(x) = Ca^x$$

where a is a positive real number $(a > 0)$, $a \neq 1$, and $C \neq 0$ is a real number. The domain of f is the set of all real numbers. The base a is the **growth factor**, and, because $f(0) = Ca^0 = C$, C is called the **initial value.**

WARNING It is important to distinguish a power function, $g(x) = ax^n$, $n \geq 2$ an integer, from an exponential function, $f(x) = Ca^x$, $a \neq 1$, $a > 0$. In a power function, the base is a variable and the exponent is a constant. In an exponential function, the base is a constant and the exponent is a variable. ■

In the definition of an exponential function, the base $a = 1$ is excluded because this function is simply the constant function $f(x) = C \cdot 1^x = C$. Bases that are negative are also excluded; otherwise, many values of x would have to be excluded from the domain, such as $x = \frac{1}{2}$ and $x = \frac{3}{4}$. [Recall that $(-2)^{1/2} = \sqrt{-2}$, $(-3)^{3/4} = \sqrt[4]{(-3)^3} = \sqrt[4]{-27}$, and so on, are not defined in the set of real numbers.]

Transformations (vertical shifts, horizontal shifts, reflections, and so on) of a function of the form $f(x) = Ca^x$ also represent exponential functions. Some examples of exponential functions are

$$f(x) = 2^x \qquad F(x) = \left(\frac{1}{3}\right)^x + 5 \qquad G(x) = 2 \cdot 3^{x-3}$$

For each function, note that the base of the exponential expression is a constant and the exponent contains a variable.

In the function $f(x) = 5 \cdot 2^x$, notice that the ratio of consecutive outputs is constant for 1-unit increases in the input. This ratio equals the constant 2, the base of the exponential function. In other words,

$$\frac{f(1)}{f(0)} = \frac{5 \cdot 2^1}{5} = 2 \qquad \frac{f(2)}{f(1)} = \frac{5 \cdot 2^2}{5 \cdot 2^1} = 2 \qquad \frac{f(3)}{f(2)} = \frac{5 \cdot 2^3}{5 \cdot 2^2} = 2 \quad \text{and so on}$$

This leads to the following result.

THEOREM

For an exponential function $f(x) = Ca^x$, $a > 0$, $a \neq 1$, and $C \neq 0$, if x is any real number, then

$$\frac{f(x+1)}{f(x)} = a \quad \text{or} \quad f(x+1) = af(x)$$

In Words

For 1-unit changes in the input x of an exponential function $f(x) = Ca^x$, the ratio of consecutive outputs is the constant a.

Proof

$$\frac{f(x+1)}{f(x)} = \frac{Ca^{x+1}}{Ca^x} = a^{x+1-x} = a^1 = a$$ ■

EXAMPLE 2 Identifying Linear or Exponential Functions

Determine whether the given function is linear, exponential, or neither. For those that are linear, find a linear function that models the data. For those that are exponential, find an exponential function that models the data.

(a)

x	y
−1	5
0	2
1	−1
2	−4
3	−7

(b)

x	y
−1	32
0	16
1	8
2	4
3	2

(c)

x	y
−1	2
0	4
1	7
2	11
3	16

Solution For each function, compute the average rate of change of y with respect to x and the ratio of consecutive outputs. If the average rate of change is constant, then the function is linear. If the ratio of consecutive outputs is constant, then the function is exponential.

Table 2

x	y	Average Rate of Change	Ratio of Consecutive Outputs
-1	5		
		$\frac{\Delta y}{\Delta x} = \frac{2-5}{0-(-1)} = -3$	$\frac{2}{5}$
0	2		
		$\frac{-1-2}{1-0} = -3$	$\frac{-1}{2} = -\frac{1}{2}$
1	-1		
		$\frac{-4-(-1)}{2-1} = -3$	$\frac{-4}{-1} = 4$
2	-4		
		$\frac{-7-(-4)}{3-2} = -3$	$\frac{-7}{-4} = \frac{7}{4}$
3	-7		

(a)

x	y	Average Rate of Change	Ratio of Consecutive Outputs
-1	32		
		$\frac{\Delta y}{\Delta x} = \frac{16-32}{0-(-1)} = -16$	$\frac{16}{32} = \frac{1}{2}$
0	16		
		-8	$\frac{8}{16} = \frac{1}{2}$
1	8		
		-4	$\frac{4}{8} = \frac{1}{2}$
2	4		
		-2	$\frac{2}{4} = \frac{1}{2}$
3	2		

(b)

x	y	Average Rate of Change	Ratio of Consecutive Outputs
-1	2		
		$\frac{\Delta y}{\Delta x} = \frac{4-2}{0-(-1)} = 2$	2
0	4		
		3	$\frac{7}{4}$
1	7		
		4	$\frac{11}{7}$
2	11		
		5	$\frac{16}{11}$
3	16		

(c)

(a) See Table 2(a). The average rate of change for every 1-unit increase in x is -3. Therefore, the function is a linear function. In a linear function the average rate of change is the slope m, so $m = -3$. The y-intercept b is the value of the function at $x = 0$, so $b = 2$. The linear function that models the data is $f(x) = mx + b = -3x + 2$.

(b) See Table 2(b). For this function, the average rate of change from -1 to 0 is -16, and the average rate of change from 0 to 1 is -8. Because the average rate of change is not constant, the function is not a linear function. The ratio of consecutive outputs for a 1-unit increase in the inputs is a constant, $\frac{1}{2}$. Because the ratio of consecutive outputs is constant, the function is an exponential function with growth factor $a = \frac{1}{2}$. The initial value C of the exponential function is $C = 16$, the value of the function at 0. Therefore, the exponential function that models the data is $g(x) = Ca^x = 16 \cdot \left(\frac{1}{2}\right)^x$.

(c) See Table 2(c). For this function, the average rate of change from -1 to 0 is 2, and the average rate of change from 0 to 1 is 3. Because the average rate of change is not constant, the function is not a linear function. The ratio of consecutive outputs from -1 to 0 is 2, and the ratio of consecutive outputs from 0 to 1 is $\frac{7}{4}$. Because the ratio of consecutive outputs is not a constant, the function is not an exponential function. ●

Now Work PROBLEM 27

2 Graph Exponential Functions

If we know how to graph an exponential function of the form $f(x) = a^x$, then we can use transformations (shifting, stretching, and so on) to obtain the graph of any exponential function.

First, let's graph the exponential function $f(x) = 2^x$.

EXAMPLE 3

Graphing an Exponential Function

Graph the exponential function: $f(x) = 2^x$

Solution The domain of $f(x) = 2^x$ is the set of all real numbers. Begin by locating some points on the graph of $f(x) = 2^x$, as listed in Table 3.

Because $2^x > 0$ for all x, the range of f is $(0, \infty)$. Therefore, the graph has no x-intercepts, and in fact the graph will lie above the x-axis for all x. As Table 3 indicates, the y-intercept is 1. Table 3 also indicates that as $x \to -\infty$, the value of $f(x) = 2^x$ gets closer and closer to 0. Therefore, the x-axis $(y = 0)$ is a horizontal asymptote to the graph as $x \to -\infty$. This provides the end behavior for x large and negative.

To determine the end behavior for x large and positive, look again at Table 3. As $x \to \infty$, $f(x) = 2^x$ grows very quickly, causing the graph of $f(x) = 2^x$ to rise very rapidly. It is apparent that f is an increasing function and so is one-to-one.

Using all this information, plot some of the points from Table 3 and connect them with a smooth, continuous curve, as shown in Figure 1.

Table 3

x	$f(x) = 2^x$
-10	$2^{-10} \approx 0.00098$
-3	$2^{-3} = \frac{1}{8}$
-2	$2^{-2} = \frac{1}{4}$
-1	$2^{-1} = \frac{1}{2}$
0	$2^0 = 1$
1	$2^1 = 2$
2	$2^2 = 4$
3	$2^3 = 8$
10	$2^{10} = 1024$

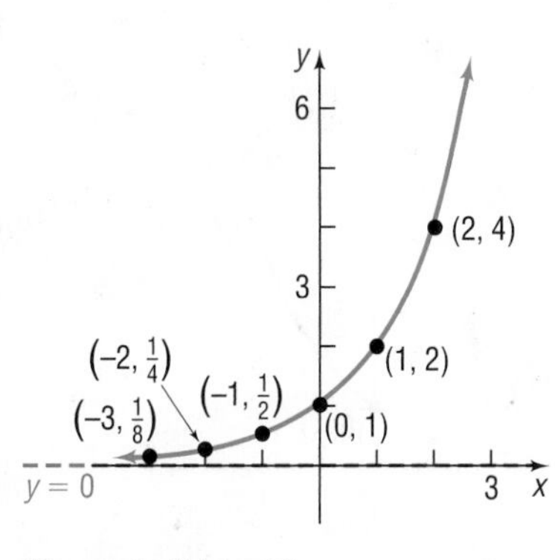

Figure 1 $f(x) = 2^x$ ●

Graphs that look like the one in Figure 1 occur very frequently in a variety of situations. For example, the graph in Figure 2 on the next page illustrates the number of Facebook subscribers by year from 2004 to 2013. One might conclude from this graph that the number of Facebook subscribers is growing *exponentially*.

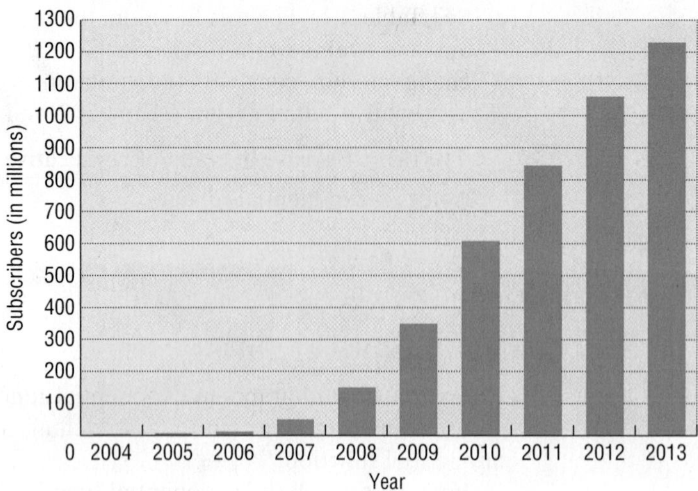

Source: Facebook Newsroom

Figure 2

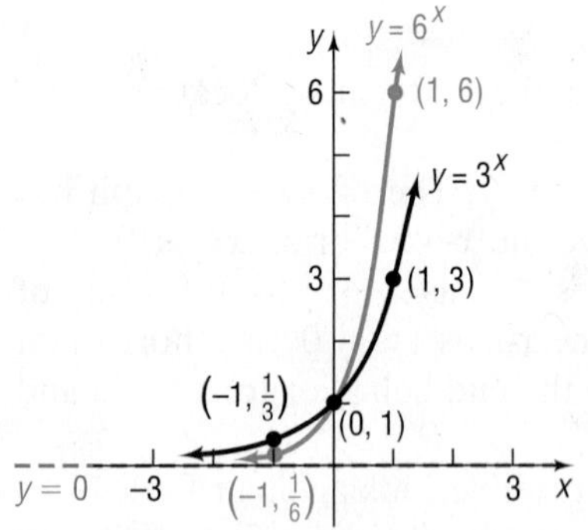

Figure 3

Later in this chapter, more will be said about situations that lead to exponential growth. For now, let's continue to explore properties of exponential functions.

The graph of $f(x) = 2^x$ in Figure 1 is typical of all exponential functions of the form $f(x) = a^x$ with $a > 1$. Such functions are increasing functions and hence are one-to-one. Their graphs lie above the x-axis, pass through the point $(0, 1)$, and thereafter rise rapidly as $x \to \infty$. As $x \to -\infty$, the x-axis $(y = 0)$ is a horizontal asymptote. There are no vertical asymptotes. Finally, the graphs are smooth and continuous with no corners or gaps.

Figure 3 illustrates the graphs of two more exponential functions whose bases are larger than 1. Notice that the larger the base, the steeper the graph is when $x > 0$, and when $x < 0$, the larger the base, the closer the graph of the equation is to the x-axis.

Seeing the Concept

Graph $Y_1 = 2^x$ and compare what you see to Figure 1. Clear the screen, graph $Y_1 = 3^x$ and $Y_2 = 6^x$, and compare what you see to Figure 3. Clear the screen and graph $Y_1 = 10^x$ and $Y_2 = 100^x$.

Properties of the Exponential Function $f(x) = a^x, a > 1$

1. The domain is the set of all real numbers, or $(-\infty, \infty)$ using interval notation; the range is the set of positive real numbers, or $(0, \infty)$ using interval notation.
2. There are no x-intercepts; the y-intercept is 1.
3. The x-axis $(y = 0)$ is a horizontal asymptote as $x \to -\infty$. $\left[\lim_{x \to -\infty} a^x = 0\right]$.
4. $f(x) = a^x, a > 1$, is an increasing function and is one-to-one.
5. The graph of f contains the points $\left(-1, \frac{1}{a}\right)$, $(0, 1)$, and $(1, a)$.
6. The graph of f is smooth and continuous, with no corners or gaps. See Figure 4.

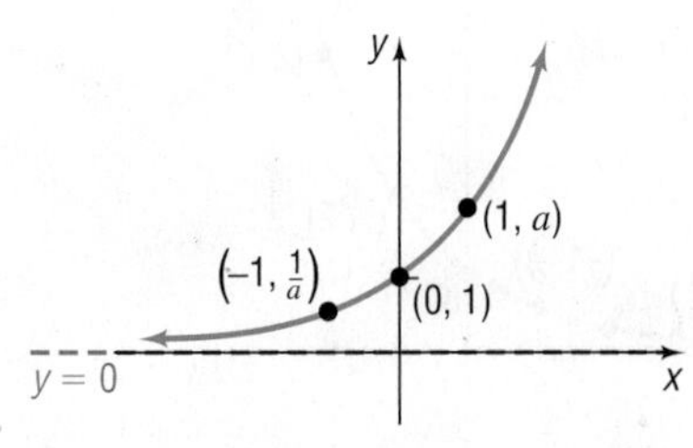

Figure 4 $f(x) = a^x, a > 1$

Now consider $f(x) = a^x$ when $0 < a < 1$.

EXAMPLE 4 Graphing an Exponential Function

Graph the exponential function: $f(x) = \left(\frac{1}{2}\right)^x$

Solution The domain of $f(x) = \left(\frac{1}{2}\right)^x$ consists of all real numbers. As before, locate some points on the graph as shown in Table 4. Because $\left(\frac{1}{2}\right)^x > 0$ for all x, the range of f is the interval $(0, \infty)$. The graph lies above the x-axis and has no x-intercepts. The y-intercept is 1. As $x \to -\infty$, $f(x) = \left(\frac{1}{2}\right)^x$ grows very quickly. As $x \to \infty$, the values of $f(x)$ approach 0. The x-axis ($y = 0$) is a horizontal asymptote as $x \to \infty$. It is apparent that f is a decreasing function and so is one-to-one. Figure 5 illustrates the graph.

Table 4

x	$f(x) = \left(\frac{1}{2}\right)^x$
-10	$\left(\frac{1}{2}\right)^{-10} = 1024$
-3	$\left(\frac{1}{2}\right)^{-3} = 8$
-2	$\left(\frac{1}{2}\right)^{-2} = 4$
-1	$\left(\frac{1}{2}\right)^{-1} = 2$
0	$\left(\frac{1}{2}\right)^{0} = 1$
1	$\left(\frac{1}{2}\right)^{1} = \frac{1}{2}$
2	$\left(\frac{1}{2}\right)^{2} = \frac{1}{4}$
3	$\left(\frac{1}{2}\right)^{3} = \frac{1}{8}$
10	$\left(\frac{1}{2}\right)^{10} \approx 0.00098$

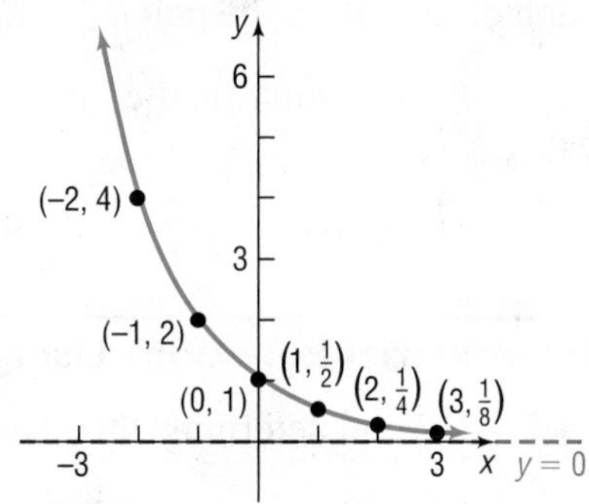

Figure 5 $f(x) = \left(\frac{1}{2}\right)^x$

The graph of $y = \left(\frac{1}{2}\right)^x$ also can be obtained from the graph of $y = 2^x$ using transformations. The graph of $y = \left(\frac{1}{2}\right)^x = 2^{-x}$ is a reflection about the y-axis of the graph of $y = 2^x$ (replace x by $-x$). See Figures 6(a) and (b).

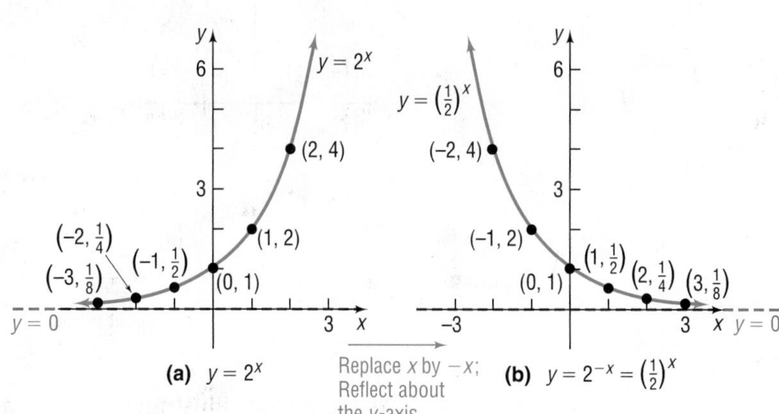

Figure 6 **(a)** $y = 2^x$ **(b)** $y = 2^{-x} = \left(\frac{1}{2}\right)^x$

The graph of $f(x) = \left(\frac{1}{2}\right)^x$ in Figure 5 is typical of all exponential functions of the form $f(x) = a^x$ with $0 < a < 1$. Such functions are decreasing and one-to-one. Their graphs lie above the x-axis and pass through the point $(0, 1)$. The graphs rise rapidly as $x \to -\infty$. As $x \to \infty$, the x-axis ($y = 0$) is a horizontal asymptote. There are no vertical asymptotes. Finally, the graphs are smooth and continuous, with no corners or gaps.

Seeing the Concept

Using a graphing utility, simultaneously graph:

(a) $Y_1 = 3^x, Y_2 = \left(\frac{1}{3}\right)^x$

(b) $Y_1 = 6^x, Y_2 = \left(\frac{1}{6}\right)^x$

Conclude that the graph of $Y_2 = \left(\frac{1}{a}\right)^x$, for $a > 0$, is the reflection about the y-axis of the graph of $Y_1 = a^x$.

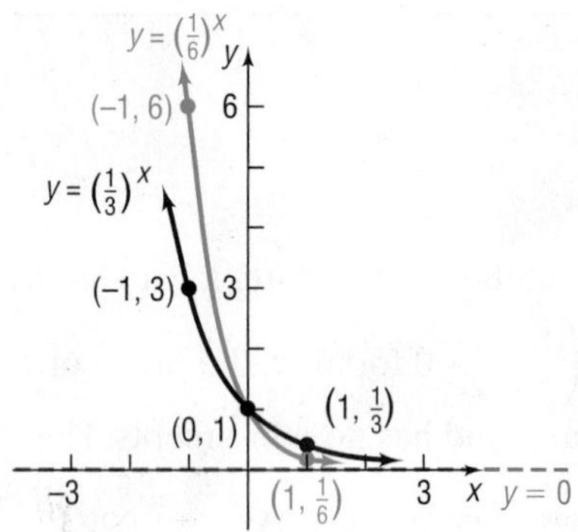

Figure 7

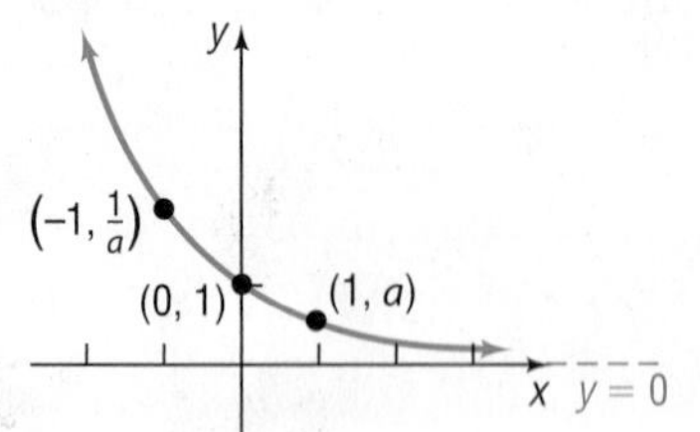

Figure 8 $f(x) = a^x, 0 < a < 1$

Figure 7 illustrates the graphs of two more exponential functions whose bases are between 0 and 1. Notice that the smaller base results in a graph that is steeper when $x < 0$. When $x > 0$, the graph of the equation with the smaller base is closer to the x-axis.

Properties of the Exponential Function $f(x) = a^x, 0 < a < 1$

1. The domain is the set of all real numbers, or $(-\infty, \infty)$ using interval notation; the range is the set of positive real numbers, or $(0, \infty)$ using interval notation.
2. There are no x-intercepts; the y-intercept is 1.
3. The x-axis $(y = 0)$ is a horizontal asymptote as $x \to \infty$ $\left[\lim_{x\to\infty} a^x = 0\right]$.
4. $f(x) = a^x, 0 < a < 1$, is a decreasing function and is one-to-one.
5. The graph of f contains the points $\left(-1, \frac{1}{a}\right)$, $(0, 1)$, and $(1, a)$.
6. The graph of f is smooth and continuous, with no corners or gaps. See Figure 8.

EXAMPLE 5 **Graphing Exponential Functions Using Transformations**

Graph $f(x) = 2^{-x} - 3$ and determine the domain, range, and horizontal asymptote of f.

Solution Begin with the graph of $y = 2^x$. Figure 9 shows the stages.

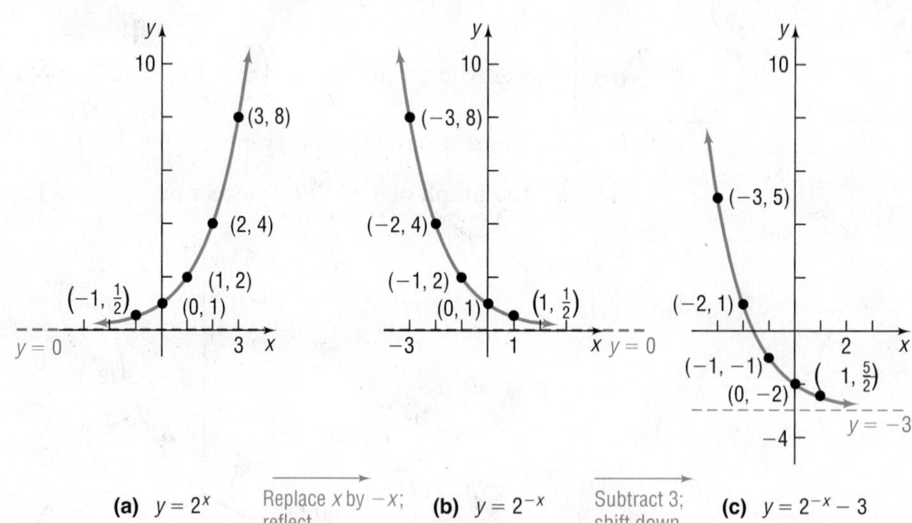

Figure 9 (a) $y = 2^x$ (b) $y = 2^{-x}$ (c) $y = 2^{-x} - 3$

As Figure 9(c) illustrates, the domain of $f(x) = 2^{-x} - 3$ is the interval $(-\infty, \infty)$ and the range is the interval $(-3, \infty)$. The horizontal asymptote of f is the line $y = -3$. ●

Now Work PROBLEM 43

3 Define the Number e

Many problems that occur in nature require the use of an exponential function whose base is a certain irrational number, symbolized by the letter e.

One way of arriving at this important number e is given next.

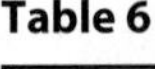

DEFINITION

The **number e** is defined as the number that the expression

$$\left(1 + \frac{1}{n}\right)^n \quad \text{(2)}$$

approaches as $n \to \infty$. In calculus, this is expressed, using limit notation, as

$$e = \lim_{n\to\infty}\left(1 + \frac{1}{n}\right)^n$$

Table 5 illustrates what happens to the defining expression (2) as n takes on increasingly large values. The last number in the right column in the table approximates e correct to nine decimal places. That is, $e = 2.718281827\ldots$. Remember, the three dots indicate that the decimal places continue. Because these decimal places continue but do not repeat, e is an irrational number. The number e is often expressed as a decimal rounded to a specific number of places. For example, $e \approx 2.71828$ is rounded to five decimal places.

Table 5

n	$\frac{1}{n}$	$1 + \frac{1}{n}$	$\left(1 + \frac{1}{n}\right)^n$
1	1	2	2
2	0.5	1.5	2.25
5	0.2	1.2	2.48832
10	0.1	1.1	2.59374246
100	0.01	1.01	2.704813829
1,000	0.001	1.001	2.716923932
10,000	0.0001	1.0001	2.718145927
100,000	0.00001	1.00001	2.718268237
1,000,000	0.000001	1.000001	2.718280469
1,000,000,000	10^{-9}	$1 + 10^{-9}$	2.718281827

Table 6

x	e^x
−2	$e^{-2} \approx 0.14$
−1	$e^{-1} \approx 0.37$
0	$e^0 = 1$
1	$e^1 \approx 2.72$
2	$e^2 \approx 7.39$

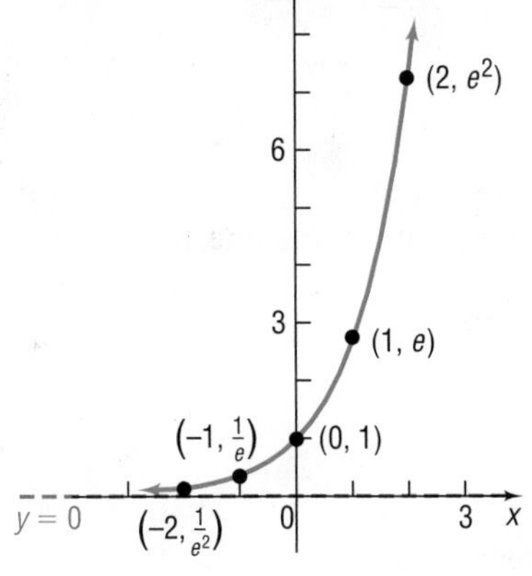

Figure 10 $y = e^x$

The exponential function $f(x) = e^x$, whose base is the number e, occurs with such frequency in applications that it is usually referred to as *the* exponential function. Indeed, most calculators have the key $\boxed{e^x}$ or $\boxed{\exp(x)}$, which may be used to evaluate the exponential function for a given value of x.*

Now use your calculator to approximate e^x for $x = -2, x = -1, x = 0, x = 1$, and $x = 2$. See Table 6. The graph of the exponential function $f(x) = e^x$ is given in Figure 10. Since $2 < e < 3$, the graph of $y = e^x$ lies between the graphs of $y = 2^x$ and $y = 3^x$. Do you see why? (Refer to Figures 1 and 3.)

Seeing the Concept

Graph $Y_1 = e^x$ and compare what you see to Figure 10. Use eVALUEate or TABLE to verify the points on the graph shown in Figure 10. Now graph $Y_2 = 2^x$ and $Y_3 = 3^x$ on the same screen as $Y_1 = e^x$. Notice that the graph of $Y_1 = e^x$ lies between these two graphs.

EXAMPLE 6

Graphing Exponential Functions Using Transformations

Graph $f(x) = -e^{x-3}$ and determine the domain, range, and horizontal asymptote of f.

*If your calculator does not have one of these keys, refer to your owner's manual.

Solution Begin with the graph of $y = e^x$. Figure 11 shows the stages.

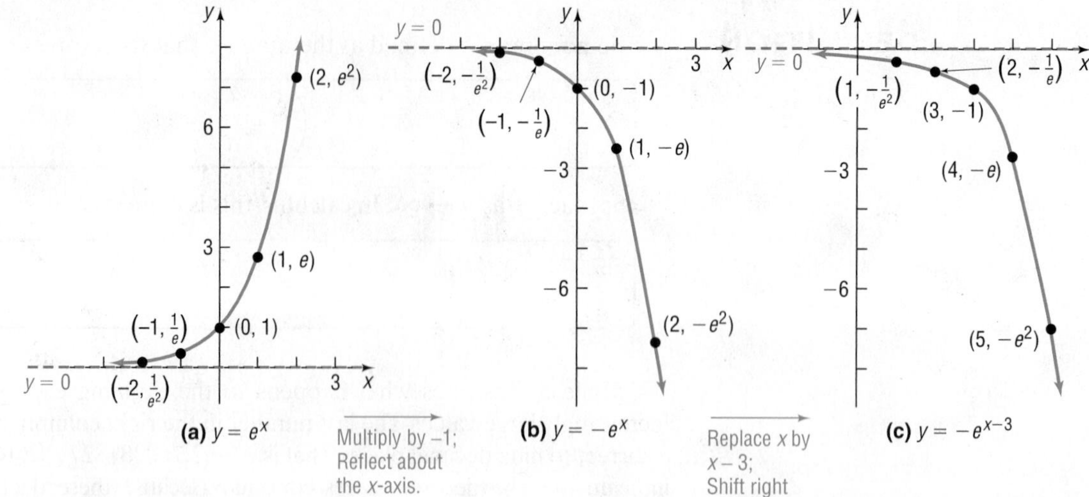

Figure 11

As Figure 11(c) illustrates, the domain of $f(x) = -e^{x-3}$ is the interval $(-\infty, \infty)$, and the range is the interval $(-\infty, 0)$. The horizontal asymptote is the line $y = 0$.

Now Work PROBLEM 55

4 Solve Exponential Equations

Equations that involve terms of the form a^x, where $a > 0$ and $a \neq 1$, are referred to as **exponential equations**. Such equations can sometimes be solved by appropriately applying the Laws of Exponents and property (3):

> **In Words**
> When two exponential expressions with the same base are equal, then their exponents are equal.

$$\text{If} \quad a^u = a^v, \quad \text{then} \quad u = v. \tag{3}$$

Property (3) is a consequence of the fact that exponential functions are one-to-one. To use property (3), each side of the equality must be written with the same base.

EXAMPLE 7 **Solving Exponential Equations**

Solve each exponential equation.

(a) $3^{x+1} = 81$ (b) $4^{2x-1} = 8^{x+3}$

Solution (a) Since $81 = 3^4$, write the equation as

$$3^{x+1} = 81 = 3^4$$

Now the expressions on both sides of the equation have the same base, 3. Set the exponents equal to each other to obtain

$$x + 1 = 4$$
$$x = 3$$

The solution set is {3}.

(b)
$$4^{2x-1} = 8^{x+3}$$
$$(2^2)^{(2x-1)} = (2^3)^{(x+3)} \quad 4 = 2^2; 8 = 2^3$$
$$2^{2(2x-1)} = 2^{3(x+3)} \quad (a^r)^s = a^{rs}$$
$$2(2x - 1) = 3(x + 3) \quad \text{If } a^u = a^v, \text{ then } u = v.$$
$$4x - 2 = 3x + 9$$
$$x = 11$$

The solution set is {11}.

Now Work PROBLEMS 65 AND 75

EXAMPLE 8

Solving an Exponential Equation

Solve: $e^{-x^2} = (e^x)^2 \cdot \dfrac{1}{e^3}$

Solution Use the Laws of Exponents first to get a single expression with the base e on the right side.

$$(e^x)^2 \cdot \frac{1}{e^3} = e^{2x} \cdot e^{-3} = e^{2x-3}$$

As a result,

$$\begin{aligned} e^{-x^2} &= e^{2x-3} \\ -x^2 &= 2x - 3 && \text{Apply property (3).} \\ x^2 + 2x - 3 &= 0 && \text{Place the quadratic equation in standard form.} \\ (x+3)(x-1) &= 0 && \text{Factor.} \\ x = -3 \quad &\text{or} \quad x = 1 && \text{Use the Zero-Product Property.} \end{aligned}$$

The solution set is $\{-3, 1\}$.

Now Work PROBLEM 81

EXAMPLE 9

Exponential Probability

Between 9:00 PM and 10:00 PM, cars arrive at Burger King's drive-thru at the rate of 12 cars per hour (0.2 car per minute). The following formula from statistics can be used to determine the probability that a car will arrive within t minutes of 9:00 PM.

$$F(t) = 1 - e^{-0.2t}$$

(a) Determine the probability that a car will arrive within 5 minutes of 9 PM (that is, before 9:05 PM).

(b) Determine the probability that a car will arrive within 30 minutes of 9 PM (before 9:30 PM).

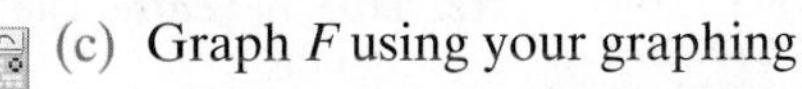

(c) Graph F using your graphing utility.

(d) What value does F approach as t increases without bound in the positive direction?

Solution (a) The probability that a car will arrive within 5 minutes is found by evaluating $F(t)$ at $t = 5$.

$$F(5) = 1 - e^{-0.2(5)} \approx 0.63212$$

Use a calculator.

There is a 63% probability that a car will arrive within 5 minutes.

(b) The probability that a car will arrive within 30 minutes is found by evaluating $F(t)$ at $t = 30$.

$$F(30) = 1 - e^{-0.2(30)} \approx 0.9975$$

Use a calculator.

There is a 99.75% probability that a car will arrive within 30 minutes.

Figure 12 $F(t) = 1 - e^{-0.2t}$

(c) See Figure 12 for the graph of F.

(d) As time passes, the probability that a car will arrive increases. The value that F approaches can be found by letting $t \to \infty$. Since $e^{-0.2t} = \dfrac{1}{e^{0.2t}}$, it follows that $e^{-0.2t} \to 0$ as $t \to \infty$. Therefore, F approaches 1 as t gets large. The algebraic analysis is confirmed by Figure 12.

Now Work PROBLEM 113

SUMMARY

Properties of the Exponential Function

$f(x) = a^x,\ a > 1$ — Domain: the interval $(-\infty, \infty)$; range: the interval $(0, \infty)$
x-intercepts: none; y-intercept: 1
Horizontal asymptote: x-axis $(y = 0)$ as $x \to -\infty$
Increasing; one-to-one; smooth; continuous
See Figure 4 for a typical graph.

$f(x) = a^x,\ 0 < a < 1$ — Domain: the interval $(-\infty, \infty)$; range: the interval $(0, \infty)$
x-intercepts: none; y-intercept: 1
Horizontal asymptote: x-axis $(y = 0)$ as $x \to \infty$
Decreasing; one-to-one; smooth; continuous
See Figure 8 for a typical graph.

If $a^u = a^v$, then $u = v$.

7.1 Assess Your Understanding

'Are You Prepared?' *Answers are given at the end of these exercises. If you get a wrong answer, read the pages listed in red.*

1. $4^3 =$ _____ ; $8^{2/3} =$ _____ ; $3^{-2} =$ _____ . (pp. A8–A9 and pp. A59–A61)
2. Solve: $x^2 + 3x = 4$ (pp. A27–A34)
3. ***True or False*** To graph $y = (x - 2)^3$, shift the graph of $y = x^3$ to the left 2 units. (pp. 64–73)
4. Find the average rate of change of $f(x) = 3x - 5$ from $x = 0$ to $x = 4$. (pp. 48–49)
5. ***True or False*** The function $f(x) = \dfrac{2x}{x - 3}$ has $y = 2$ as a horizontal asymptote. (pp. 73–74)

Concepts and Vocabulary

6. A(n) __________ is a function of the form $f(x) = Ca^x$, where $a > 0$, $a \neq 1$, and $C \neq 0$ are real numbers. The base a is the _____ _____ and C is the _____ _____.
7. For an exponential function $f(x) = Ca^x$, $\dfrac{f(x+1)}{f(x)} =$ ___.
8. ***True or False*** The domain of the exponential function $f(x) = a^x$, where $a > 0$ and $a \neq 1$, is the set of all real numbers.
9. ***True or False*** The graph of the exponential function $f(x) = a^x$, where $a > 0$ and $a \neq 1$, has no x-intercept.
10. The graph of every exponential function $f(x) = a^x$, where $a > 0$ and $a \neq 1$, passes through three points: ____, ____, and ____.
11. If $3^x = 3^4$, then $x =$ ____.
12. ***True or False*** The graphs of $y = 3^x$ and $y = \left(\dfrac{1}{3}\right)^x$ are identical.
13. Which of the following exponential functions is an increasing function?
    (a) $f(x) = 0.5^x$ (b) $f(x) = \left(\dfrac{5}{2}\right)^x$
    (c) $f(x) = \left(\dfrac{2}{3}\right)^x$ (d) $f(x) = 0.9^x$
14. Which of the following is the range of the exponential function $f(x) = a^x$, $a > 0$ and $a \neq 1$?
    (a) $(-\infty, \infty)$ (b) $(-\infty, 0)$
    (c) $(0, \infty)$ (d) $(-\infty, 0) \cup (0, \infty)$

Skill Building

In Problems 15–26, approximate each number using a calculator. Express your answer rounded to three decimal places.

15. (a) $2^{3.14}$ (b) $2^{3.141}$ (c) $2^{3.1415}$ (d) 2^{π}
16. (a) $2^{2.7}$ (b) $2^{2.71}$ (c) $2^{2.718}$ (d) 2^{e}
17. (a) $3.1^{2.7}$ (b) $3.14^{2.71}$ (c) $3.141^{2.718}$ (d) π^{e}
18. (a) $2.7^{3.1}$ (b) $2.71^{3.14}$ (c) $2.718^{3.141}$ (d) e^{π}
19. $(1 + 0.04)^6$
20. $\left(1 + \dfrac{0.09}{12}\right)^{24}$
21. $8.4\left(\dfrac{1}{3}\right)^{2.9}$
22. $158\left(\dfrac{5}{6}\right)^{8.63}$
23. $e^{1.2}$
24. $e^{-1.3}$
25. $125e^{0.026(7)}$
26. $83.6e^{-0.157(9.5)}$

In Problems 27–34, determine whether the given function is linear, exponential, or neither. For those that are linear functions, find a linear function that models the data; for those that are exponential, find an exponential function that models the data.

27.

x	$f(x)$
−1	3
0	6
1	12
2	18
3	30

28.

x	$g(x)$
−1	2
0	5
1	8
2	11
3	14

29.

x	$H(x)$
−1	$\frac{1}{4}$
0	1
1	4
2	16
3	64

30.

x	$F(x)$
−1	$\frac{2}{3}$
0	1
1	$\frac{3}{2}$
2	$\frac{9}{4}$
3	$\frac{27}{8}$

31.

x	$f(x)$
−1	$\frac{3}{2}$
0	3
1	6
2	12
3	24

32.

x	$g(x)$
−1	6
0	1
1	0
2	3
3	10

33.

x	$H(x)$
−1	2
0	4
1	6
2	8
3	10

34.

x	$F(x)$
−1	$\frac{1}{2}$
0	$\frac{1}{4}$
1	$\frac{1}{8}$
2	$\frac{1}{16}$
3	$\frac{1}{32}$

In Problems 35–42, the graph of an exponential function is given. Match each graph to one of the following functions.

(A) $y = 3^x$ (B) $y = 3^{-x}$ (C) $y = -3^x$ (D) $y = -3^{-x}$
(E) $y = 3^x - 1$ (F) $y = 3^{x-1}$ (G) $y = 3^{1-x}$ (H) $y = 1 - 3^x$

35.
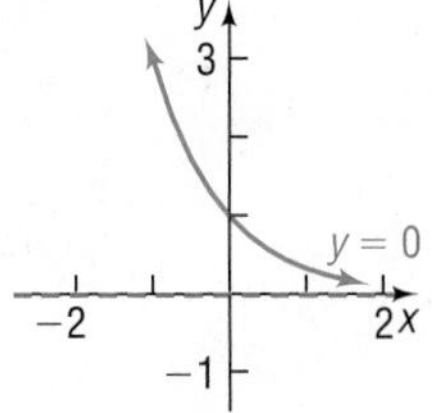

36.
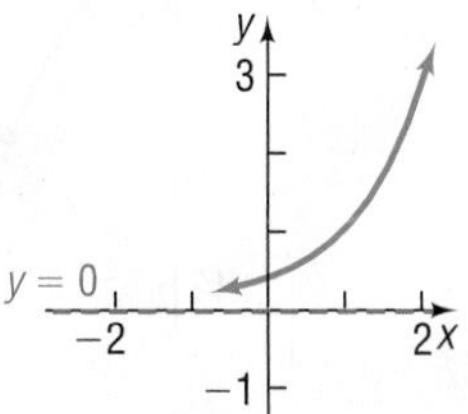

37.
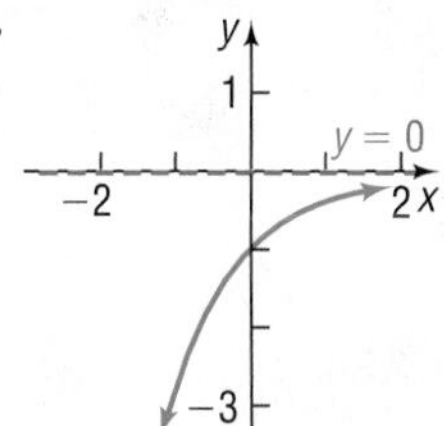

38.
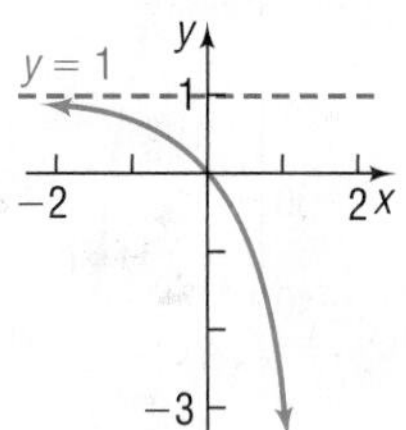

39.
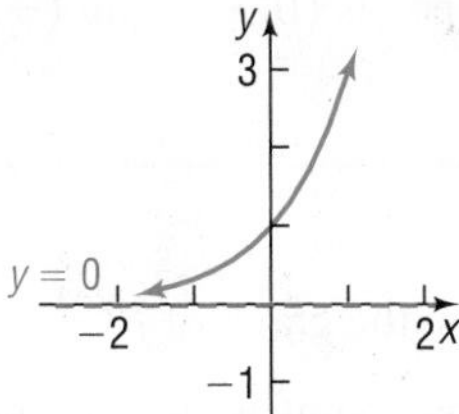

40.
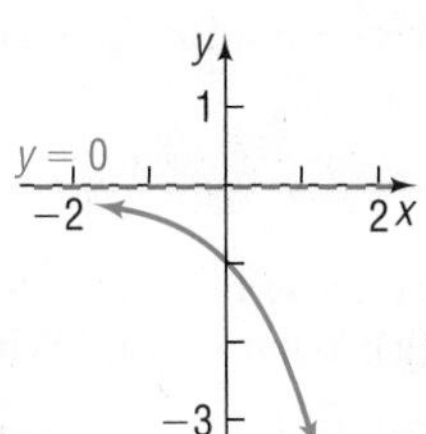

41.
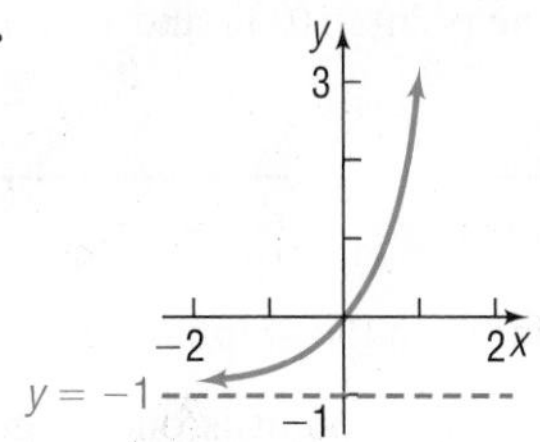

42.
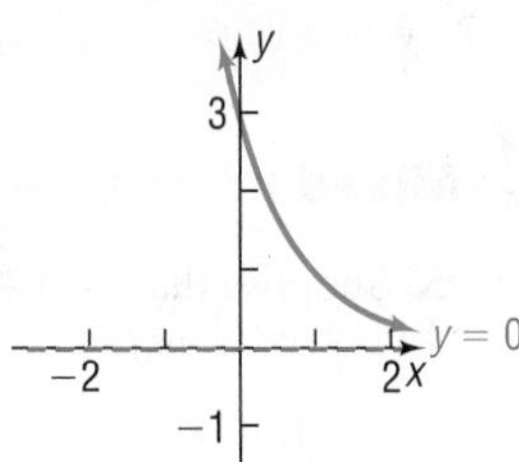

In Problems 43–54, use transformations to graph each function. Determine the domain, range, and horizontal asymptote of each function.

43. $f(x) = 2^x + 1$
44. $f(x) = 3^x - 2$
45. $f(x) = 3^{x-1}$
46. $f(x) = 2^{x+2}$
47. $f(x) = 3 \cdot \left(\frac{1}{2}\right)^x$
48. $f(x) = 4 \cdot \left(\frac{1}{3}\right)^x$
49. $f(x) = 3^{-x} - 2$
50. $f(x) = -3^x + 1$
51. $f(x) = 2 + 4^{x-1}$
52. $f(x) = 1 - 2^{x+3}$
53. $f(x) = 2 + 3^{x/2}$
54. $f(x) = 1 - 2^{-x/3}$

In Problems 55–62, begin with the graph of $y = e^x$ (Figure 10) and use transformations to graph each function. Determine the domain, range, and horizontal asymptote of each function.

55. $f(x) = e^{-x}$
56. $f(x) = -e^x$
57. $f(x) = e^{x+2}$
58. $f(x) = e^x - 1$
59. $f(x) = 5 - e^{-x}$
60. $f(x) = 9 - 3e^{-x}$
61. $f(x) = 2 - e^{-x/2}$
62. $f(x) = 7 - 3e^{2x}$

In Problems 63–82, solve each equation.

63. $7^x = 7^3$

64. $5^x = 5^{-6}$

65. $2^{-x} = 16$

66. $3^{-x} = 81$

67. $\left(\frac{1}{5}\right)^x = \frac{1}{25}$

68. $\left(\frac{1}{4}\right)^x = \frac{1}{64}$

69. $2^{2x-1} = 4$

70. $5^{x+3} = \frac{1}{5}$

71. $3^{x^3} = 9^x$

72. $4^{x^2} = 2^x$

73. $8^{-x+14} = 16^x$

74. $9^{-x+15} = 27^x$

75. $3^{x^2-7} = 27^{2x}$

76. $5^{x^2+8} = 125^{2x}$

77. $4^x \cdot 2^{x^2} = 16^2$

78. $9^{2x} \cdot 27^{x^2} = 3^{-1}$

79. $e^x = e^{3x+8}$

80. $e^{3x} = e^{2-x}$

81. $e^{x^2} = e^{3x} \cdot \frac{1}{e^2}$

82. $(e^4)^x \cdot e^{x^2} = e^{12}$

83. If $4^x = 7$, what does 4^{-2x} equal?

84. If $2^x = 3$, what does 4^{-x} equal?

85. If $3^{-x} = 2$, what does 3^{2x} equal?

86. If $5^{-x} = 3$, what does 5^{3x} equal?

87. If $9^x = 25$, what does 3^x equal?

88. If $2^{-3x} = \frac{1}{1000}$, what does 2^x equal?

In Problems 89–92, determine the exponential function whose graph is given.

89.

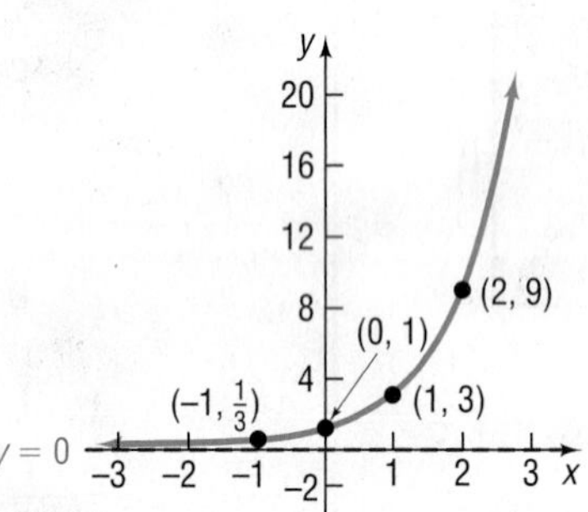

90.

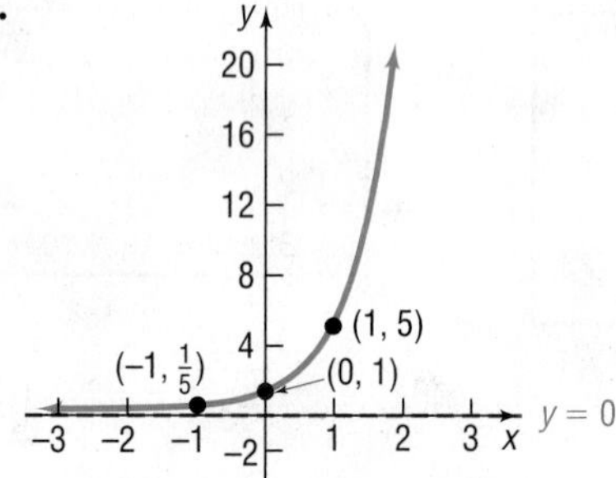

91.

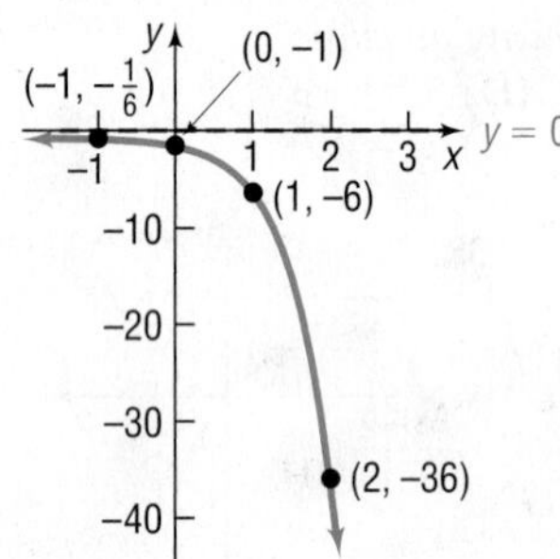

92.

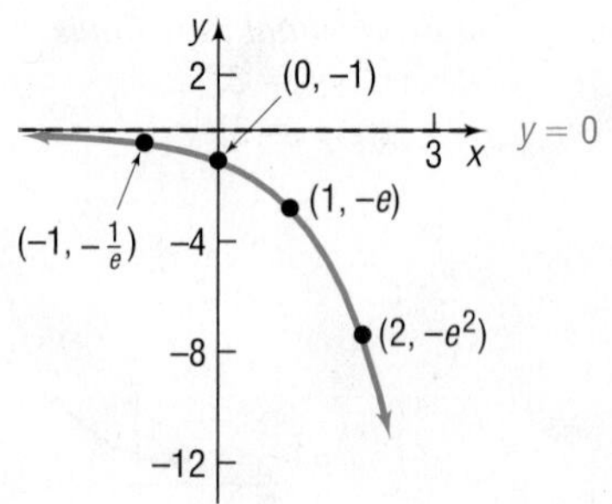

93. Find an exponential function with horizontal asymptote $y = 2$ whose graph contains the points $(0, 3)$ and $(1, 5)$.

94. Find an exponential function with horizontal asymptote $y = -3$ whose graph contains the points $(0, -2)$ and $(-2, 1)$.

Mixed Practice

95. Suppose that $f(x) = 2^x$.
(a) What is $f(4)$? What point is on the graph of f?
(b) If $f(x) = \frac{1}{16}$, what is x? What point is on the graph of f?

96. Suppose that $f(x) = 3^x$.
(a) What is $f(4)$? What point is on the graph of f?
(b) If $f(x) = \frac{1}{9}$, what is x? What point is on the graph of f?

97. Suppose that $g(x) = 4^x + 2$.
(a) What is $g(-1)$? What point is on the graph of g?
(b) If $g(x) = 66$, what is x? What point is on the graph of g?

98. Suppose that $g(x) = 5^x - 3$.
(a) What is $g(-1)$? What point is on the graph of g?
(b) If $g(x) = 122$, what is x? What point is on the graph of g?

99. Suppose that $H(x) = \left(\frac{1}{2}\right)^x - 4$.
(a) What is $H(-6)$? What point is on the graph of H?
(b) If $H(x) = 12$, what is x? What point is on the graph of H?
(c) Find the zero of H.

100. Suppose that $F(x) = \left(\frac{1}{3}\right)^x - 3$.
(a) What is $F(-5)$? What point is on the graph of F?
(b) If $F(x) = 24$, what is x? What point is on the graph of F?
(c) Find the zero of F.

In Problems 101–104, graph each function. Based on the graph, state the domain and the range, and find any intercepts.

101. $f(x) = \begin{cases} e^{-x} & \text{if } x < 0 \\ e^{x} & \text{if } x \geq 0 \end{cases}$

102. $f(x) = \begin{cases} e^{x} & \text{if } x < 0 \\ e^{-x} & \text{if } x \geq 0 \end{cases}$

103. $f(x) = \begin{cases} -e^{x} & \text{if } x < 0 \\ -e^{-x} & \text{if } x \geq 0 \end{cases}$

104. $f(x) = \begin{cases} -e^{-x} & \text{if } x < 0 \\ -e^{x} & \text{if } x \geq 0 \end{cases}$

Applications and Extensions

105. Optics If a single pane of glass obliterates 3% of the light passing through it, the percent p of light that passes through n successive panes is given approximately by the function

$$p(n) = 100(0.97)^n$$

(a) What percent of light will pass through 10 panes?
(b) What percent of light will pass through 25 panes?
(c) Explain the meaning of the base 0.97 in this problem.

106. Atmospheric Pressure The atmospheric pressure p on a balloon or airplane decreases with increasing height. This pressure, measured in millimeters of mercury, is related to the height h (in kilometers) above sea level by the function

$$p(h) = 760e^{-0.145h}$$

(a) Find the atmospheric pressure at a height of 2 km (over a mile).
(b) What is it at a height of 10 kilometers (over 30,000 feet)?

107. Depreciation The price p, in dollars, of a Honda Civic EX-L sedan that is x years old is modeled by

$$p(x) = 22{,}265(0.90)^x$$

(a) How much should a 3-year-old Civic EX-L sedan cost?
(b) How much should a 9-year-old Civic EX-L sedan cost?
(c) Explain the meaning of the base 0.90 in this problem.

108. Healing of Wounds The normal healing of wounds can be modeled by an exponential function. If A_0 represents the original area of the wound and if A equals the area of the wound, then the function

$$A(n) = A_0e^{-0.35n}$$

describes the area of a wound after n days following an injury when no infection is present to retard the healing. Suppose that a wound initially had an area of 100 square millimeters.
(a) If healing is taking place, how large will the area of the wound be after 3 days?
(b) How large will it be after 10 days?

109. Advanced-Stage Pancreatic Cancer The percentage of patients P who have survived t years after initial diagnosis of advanced-stage pancreatic cancer is modeled by the function

$$P(t) = 100(0.3)^t$$

(a) According to the model, what percent of patients survive 1 year after initial diagnosis?
(b) What percent of patients survive 2 years after initial diagnosis?
(c) Explain the meaning of the base 0.3 in the context of this problem.
Source: Cancer Treatment Centers of America

110. Endangered Species In a protected environment, the population P of a certain endangered species recovers over time t (in years) according to the model

$$P(t) = 30(1.149)^t$$

(a) What is the size of the initial population of the species?
(b) According to the model, what will be the population of the species in 5 years?
(c) According to the model, what will be the population of the species in 10 years?
(d) According to the model, what will be the population of the species in 15 years?
(e) What is happening to the population every 5 years?

111. Drug Medication The function

$$D(h) = 5e^{-0.4h}$$

can be used to find the number of milligrams D of a certain drug that is in a patient's bloodstream h hours after the drug has been administered. How many milligrams will be present after 1 hour? After 6 hours?

112. Spreading of Rumors A model for the number N of people in a college community who have heard a certain rumor is

$$N = P(1 - e^{-0.15d})$$

where P is the total population of the community and d is the number of days that have elapsed since the rumor began. In a community of 1000 students, how many students will have heard the rumor after 3 days?

113. Exponential Probability Between 12:00 PM and 1:00 PM, cars arrive at Citibank's drive-thru at the rate of 6 cars per hour (0.1 car per minute). The following formula from probability can be used to determine the probability that a car will arrive within t minutes of 12:00 PM.

$$F(t) = 1 - e^{-0.1t}$$

(a) Determine the probability that a car will arrive within 10 minutes of 12:00 PM (that is, before 12:10 PM).
(b) Determine the probability that a car will arrive within 40 minutes of 12:00 PM (before 12:40 PM).
(c) What value does F approach as t becomes unbounded in the positive direction?
(d) Graph F using a graphing utility.
(e) Using INTERSECT, determine how many minutes are needed for the probability to reach 50%.

114. Exponential Probability Between 5:00 PM and 6:00 PM, cars arrive at Jiffy Lube at the rate of 9 cars per hour (0.15 car per minute). This formula from probability can be used to determine the probability that a car will arrive within t minutes of 5:00 PM:

$$F(t) = 1 - e^{-0.15t}$$

(a) Determine the probability that a car will arrive within 15 minutes of 5:00 PM (that is, before 5:15 PM).
(b) Determine the probability that a car will arrive within 30 minutes of 5:00 PM (before 5:30 PM).
(c) What value does F approach as t becomes unbounded in the positive direction?
(d) Graph F using a graphing utility.
(e) Using INTERSECT, determine how many minutes are needed for the probability to reach 60%.

115. Poisson Probability Between 5:00 PM and 6:00 PM, cars arrive at a McDonald's drive-thru at the rate of 20 cars per hour. The following formula from probability can be used to determine the probability that x cars will arrive between 5:00 PM and 6:00 PM.

$$P(x) = \frac{20^x e^{-20}}{x!}$$

where

$$x! = x \cdot (x - 1) \cdot (x - 2) \cdot \cdots \cdot 3 \cdot 2 \cdot 1$$

(a) Determine the probability that $x = 15$ cars will arrive between 5:00 PM and 6:00 PM.
(b) Determine the probability that $x = 20$ cars will arrive between 5:00 PM and 6:00 PM.

116. Poisson Probability People enter a line for the *Demon Roller Coaster* at the rate of 4 per minute. The following formula from probability can be used to determine the probability that x people will arrive within the next minute.

$$P(x) = \frac{4^x e^{-4}}{x!}$$

where

$$x! = x \cdot (x - 1) \cdot (x - 2) \cdot \cdots \cdot 3 \cdot 2 \cdot 1$$

(a) Determine the probability that $x = 5$ people will arrive within the next minute.
(b) Determine the probability that $x = 8$ people will arrive within the next minute.

117. Relative Humidity The relative humidity is the ratio (expressed as a percent) of the amount of water vapor in the air to the maximum amount that the air can hold at a specific temperature. The relative humidity, R, is found using the following formula:

$$R = 10^{\left(\frac{4221}{T+459.4} - \frac{4221}{D+459.4} + 2\right)}$$

where T is the air temperature (in °F) and D is the dew point temperature (in °F).
(a) Determine the relative humidity if the air temperature is 50° Fahrenheit and the dew point temperature is 41° Fahrenheit.
(b) Determine the relative humidity if the air temperature is 68° Fahrenheit and the dew point temperature is 59° Fahrenheit.
(c) What is the relative humidity if the air temperature and the dew point temperature are the same?

118. Learning Curve Suppose that a student has 500 vocabulary words to learn. If the student learns 15 words after 5 minutes, the function

$$L(t) = 500(1 - e^{-0.0061t})$$

approximates the number of words L that the student will have learned after t minutes.
(a) How many words will the student have learned after 30 minutes?
(b) How many words will the student have learned after 60 minutes?

119. Current in an RL Circuit The equation governing the amount of current I (in amperes) after time t (in seconds) in a single RL circuit consisting of a resistance R (in ohms), an inductance L (in henrys), and an electromotive force E (in volts) is

$$I = \frac{E}{R}\left[1 - e^{-(R/L)t}\right]$$

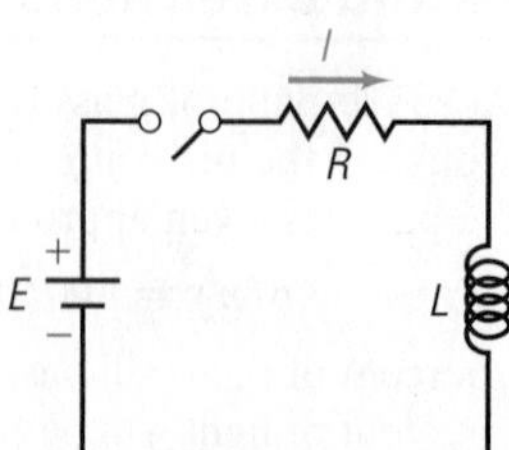

(a) If $E = 120$ volts, $R = 10$ ohms, and $L = 5$ henrys, how much current I_1 is flowing after 0.3 second? After 0.5 second? After 1 second?
(b) What is the maximum current?
(c) Graph this function $I = I_1(t)$, measuring I along the y-axis and t along the x-axis.
(d) If $E = 120$ volts, $R = 5$ ohms, and $L = 10$ henrys, how much current I_2 is flowing after 0.3 second? After 0.5 second? After 1 second?
(e) What is the maximum current?
(f) Graph the function $I = I_2(t)$ on the same coordinate axes as $I_1(t)$.

120. Current in an RC Circuit The equation governing the amount of current I (in amperes) after time t (in microseconds) in a single RC circuit consisting of a resistance R (in ohms), a capacitance C (in microfarads), and an electromotive force E (in volts) is

$$I = \frac{E}{R}e^{-t/(RC)}$$

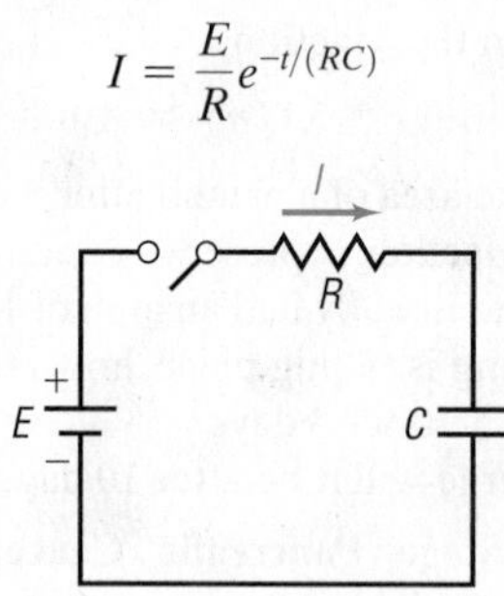

(a) If $E = 120$ volts, $R = 2000$ ohms, and $C = 1.0$ microfarad, how much current I_1 is flowing initially ($t = 0$)? After 1000 microseconds? After 3000 microseconds?
(b) What is the maximum current?
(c) Graph the function $I = I_1(t)$, measuring I along the y-axis and t along the x-axis.
(d) If $E = 120$ volts, $R = 1000$ ohms, and $C = 2.0$ microfarads, how much current I_2 is flowing initially? After 1000 microseconds? After 3000 microseconds?
(e) What is the maximum current?
(f) Graph the function $I = I_2(t)$ on the same coordinate axes as $I_1(t)$.

121. If f is an exponential function of the form $f(x) = Ca^x$ with growth factor 3, and if $f(6) = 12$, what is $f(7)$?

122. Another Formula for *e* Use a calculator to compute the values of

$$2 + \frac{1}{2!} + \frac{1}{3!} + \cdots + \frac{1}{n!}$$

for $n = 4, 6, 8$, and 10. Compare each result with e.
[**Hint:** $1! = 1, 2! = 2 \cdot 1, 3! = 3 \cdot 2 \cdot 1,$
$n! = n(n-1) \cdot \cdots \cdot (3)(2)(1)$.]

123. Another Formula for *e* Use a calculator to compute the various values of the expression. Compare the values to e.

$$2 + \cfrac{1}{1 + \cfrac{1}{2 + \cfrac{2}{3 + \cfrac{3}{4 + \cfrac{4}{\text{etc.}}}}}}$$

124. Difference Quotient If $f(x) = a^x$, show that

$$\frac{f(x+h) - f(x)}{h} = a^x \cdot \frac{a^h - 1}{h} \quad h \neq 0$$

125. If $f(x) = a^x$, show that $f(A + B) = f(A) \cdot f(B)$.

126. If $f(x) = a^x$, show that $f(-x) = \dfrac{1}{f(x)}$.

127. If $f(x) = a^x$, show that $f(\alpha x) = [f(x)]^\alpha$.

Problems 128 and 129 provide definitions for two other transcendental functions.

128. The **hyperbolic sine function,** designated by sinh x, is defined as

$$\sinh x = \frac{1}{2}(e^x - e^{-x})$$

(a) Show that $f(x) = \sinh x$ is an odd function.
(b) Graph $f(x) = \sinh x$ using a graphing utility.

129. The **hyperbolic cosine function,** designated by cosh x, is defined as

$$\cosh x = \frac{1}{2}(e^x + e^{-x})$$

(a) Show that $f(x) = \cosh x$ is an even function.
(b) Graph $f(x) = \cosh x$ using a graphing utility.
(c) Refer to Problem 128. Show that, for every x,

$$(\cosh x)^2 - (\sinh x)^2 = 1$$

130. Historical Problem Pierre de Fermat (1601–1665) conjectured that the function

$$f(x) = 2^{(2^x)} + 1$$

for $x = 1, 2, 3, \ldots$, would always have a value equal to a prime number. But Leonhard Euler (1707–1783) showed that this formula fails for $x = 5$. Use a calculator to determine the prime numbers produced by f for $x = 1, 2, 3, 4$. Then show that $f(5) = 641 \times 6{,}700{,}417$, which is not prime.

Explaining Concepts: Discussion and Writing

131. The bacteria in a 4-liter container double every minute. After 60 minutes the container is full. How long did it take to fill half the container?

132. Explain in your own words what the number e is. Provide at least two applications that use this number.

133. Do you think that there is a power function that increases more rapidly than an exponential function whose base is greater than 1? Explain.

134. As the base a of an exponential function $f(x) = a^x$, where $a > 1$, increases, what happens to the behavior of its graph for $x > 0$? What happens to the behavior of its graph for $x < 0$?

135. The graphs of $y = a^{-x}$ and $y = \left(\dfrac{1}{a}\right)^x$ are identical. Why?

Retain Your Knowledge

Problems 136–139 are based on material learned earlier in the course. The purpose of these problems is to keep the material fresh in your mind so that you are better prepared for the final exam.

136. Convert 100° to radians.

137. Determine the amplitude and period of $y = -4\cos(6x)$.

138. Find the exact value of $\sin^{-1}\left[\sin\left(\dfrac{7\pi}{10}\right)\right]$.

139. Find the rectangular coordinates of the point whose polar coordinates are $\left(8, \dfrac{5\pi}{6}\right)$.

'Are You Prepared?' Answers

1. $64; 4; \frac{1}{9}$ **2.** $\{-4, 1\}$ **3.** False **4.** 3 **5.** True

7.2 Logarithmic Functions

PREPARING FOR THIS SECTION *Before getting started, review the following:*

- Solving Inequalities (Appendix A, Section A.6, pp. A49–A52)
- One-to-One Functions; Inverse Functions (Section 1.7, pp. 78–86)
- Solving Equations (Appendix A, Section A.4, pp. A27–A34)

Now Work the 'Are You Prepared?' problems on page 478.

OBJECTIVES
1. Change Exponential Statements to Logarithmic Statements and Logarithmic Statements to Exponential Statements (p. 470)
2. Evaluate Logarithmic Expressions (p. 471)
3. Determine the Domain of a Logarithmic Function (p. 471)
4. Graph Logarithmic Functions (p. 472)
5. Solve Logarithmic Equations (p. 476)

Recall that a one-to-one function $y = f(x)$ has an inverse function that is defined implicitly by the equation $x = f(y)$. In particular, the exponential function $y = f(x) = a^x$, where $a > 0$ and $a \neq 1$, is one-to-one and hence has an inverse function that is defined implicitly by the equation

$$x = a^y \qquad a > 0 \qquad a \neq 1$$

This inverse function is so important that it is given a name, the *logarithmic function*.

DEFINITION

The **logarithmic function with base *a***, where $a > 0$ and $a \neq 1$, is denoted by $y = \log_a x$ (read as "y is the logarithm with base a of x") and is defined by

$$y = \log_a x \quad \text{if and only if} \quad x = a^y$$

The domain of the logarithmic function $y = \log_a x$ is $x > 0$.

In Words
When you need to evaluate $\log_a x$, think to yourself "a raised to what power gives me x?"

As this definition illustrates, **a logarithm is a name for a certain exponent**. So $\log_a x$ represents the exponent to which a must be raised to obtain x.

EXAMPLE 1 **Relating Logarithms to Exponents**

(a) If $y = \log_3 x$, then $x = 3^y$. For example, the logarithmic statement $4 = \log_3 81$ is equivalent to the exponential statement $81 = 3^4$.

(b) If $y = \log_5 x$, then $x = 5^y$. For example, $-1 = \log_5\left(\frac{1}{5}\right)$ is equivalent to $\frac{1}{5} = 5^{-1}$.

1 Change Exponential Statements to Logarithmic Statements and Logarithmic Statements to Exponential Statements

The definition of a logarithm can be used to convert from exponential form to logarithmic form, and vice versa, as the following two examples illustrate.

EXAMPLE 2 **Changing Exponential Statements to Logarithmic Statements**

Change each exponential statement to an equivalent statement involving a logarithm.

(a) $1.2^3 = m$ (b) $e^b = 9$ (c) $a^4 = 24$

Solution Use the fact that $y = \log_a x$ and $x = a^y$, where $a > 0$ and $a \neq 1$, are equivalent.

(a) If $1.2^3 = m$, then $3 = \log_{1.2} m$. (b) If $e^b = 9$, then $b = \log_e 9$.
(c) If $a^4 = 24$, then $4 = \log_a 24$.

Now Work PROBLEM 11

EXAMPLE 3 **Changing Logarithmic Statements to Exponential Statements**

Change each logarithmic statement to an equivalent statement involving an exponent.

(a) $\log_a 4 = 5$ (b) $\log_e b = -3$ (c) $\log_3 5 = c$

Solution

(a) If $\log_a 4 = 5$, then $a^5 = 4$.
(b) If $\log_e b = -3$, then $e^{-3} = b$.
(c) If $\log_3 5 = c$, then $3^c = 5$. ●

Now Work PROBLEM 19

2 Evaluate Logarithmic Expressions

To find the exact value of a logarithm, write the logarithm in exponential notation using the fact that $y = \log_a x$ is equivalent to $a^y = x$, and use the fact that if $a^u = a^v$, then $u = v$.

EXAMPLE 4 **Finding the Exact Value of a Logarithmic Expression**

Find the exact value of:

(a) $\log_2 16$ (b) $\log_3 \frac{1}{27}$

Solution

(a) To evaluate $\log_2 16$, think "2 raised to what power yields 16?" Then,

$y = \log_2 16$

$2^y = 16$ Change to exponential form.

$2^y = 2^4$ $16 = 2^4$

$y = 4$ Equate exponents.

Therefore, $\log_2 16 = 4$.

(b) To evaluate $\log_3 \frac{1}{27}$, think "3 raised to what power yields $\frac{1}{27}$?" Then,

$y = \log_3 \frac{1}{27}$

$3^y = \frac{1}{27}$ Change to exponential form.

$3^y = 3^{-3}$ $\frac{1}{27} = \frac{1}{3^3} = 3^{-3}$

$y = -3$ Equate exponents.

Therefore, $\log_3 \frac{1}{27} = -3$. ●

Now Work PROBLEM 27

3 Determine the Domain of a Logarithmic Function

The logarithmic function $y = \log_a x$ has been defined as the inverse of the exponential function $y = a^x$. That is, if $f(x) = a^x$, then $f^{-1}(x) = \log_a x$. Based on the discussion in Section 1.7 on inverse functions, for a function f and its inverse f^{-1},

$$\text{Domain of } f^{-1} = \text{Range of } f \quad \text{and} \quad \text{Range of } f^{-1} = \text{Domain of } f$$

Consequently, it follows that

> Domain of the logarithmic function = Range of the exponential function = $(0, \infty)$
> Range of the logarithmic function = Domain of the exponential function = $(-\infty, \infty)$

The next box summarizes some properties of the logarithmic function.

> $y = \log_a x$ (defining equation: $x = a^y$)
> Domain: $0 < x < \infty$ Range: $-\infty < y < \infty$

The domain of a logarithmic function consists of the *positive* real numbers, so the argument of a logarithmic function must be greater than zero.

EXAMPLE 5 **Finding the Domain of a Logarithmic Function**

Find the domain of each logarithmic function.

(a) $F(x) = \log_2(x + 3)$ (b) $g(x) = \log_5\left(\dfrac{1 + x}{1 - x}\right)$ (c) $h(x) = \log_{1/2}|x|$

Solution (a) The domain of F consists of all x for which $x + 3 > 0$, that is, $x > -3$. Using interval notation, the domain of F is $(-3, \infty)$.

(b) The domain of g is restricted to

$$\frac{1 + x}{1 - x} > 0$$

Solve this inequality to find that the domain of g consists of all x between -1 and 1, that is, $-1 < x < 1$, or, using interval notation, $(-1, 1)$.

(c) Since $|x| > 0$, provided that $x \neq 0$, the domain of h consists of all real numbers except zero, or, using interval notation, $(-\infty, 0) \cup (0, \infty)$. ●

Now Work PROBLEMS 41 AND 47

4 Graph Logarithmic Functions

Because exponential functions and logarithmic functions are inverses of each other, the graph of the logarithmic function $y = \log_a x$ is the reflection about the line $y = x$ of the graph of the exponential function $y = a^x$, as shown in Figure 13.

For example, to graph $y = \log_2 x$, graph $y = 2^x$ and reflect it about the line $y = x$. See Figure 14. To graph $y = \log_{1/3} x$, graph $y = \left(\dfrac{1}{3}\right)^x$ and reflect it about the line $y = x$. See Figure 15.

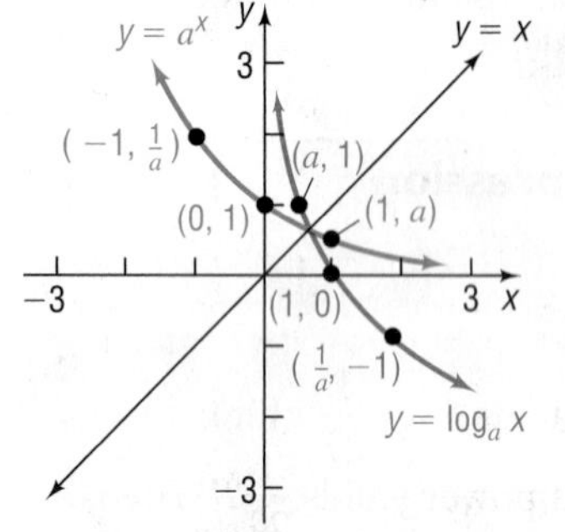

(a) $0 < a < 1$

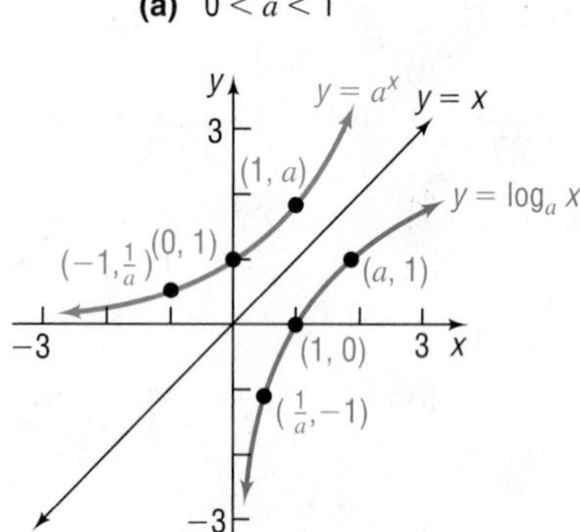

(b) $a > 1$

Figure 13

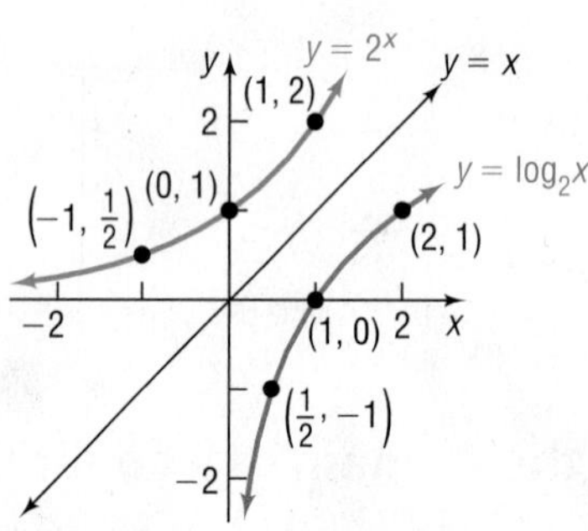

Figure 14

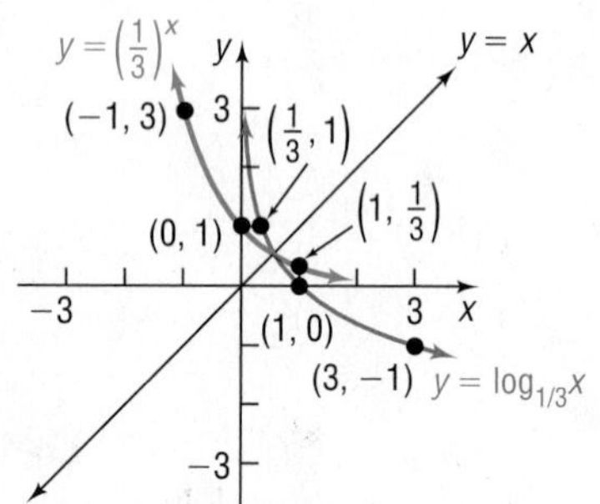

Figure 15

Now Work PROBLEM 61

The graphs of $y = \log_a x$ in Figures 13(a) and (b) lead to the following properties.

Properties of the Logarithmic Function $f(x) = \log_a x; a > 0, a \neq 1$

1. The domain is the set of positive real numbers, or $(0, \infty)$ using interval notation; the range is the set of all real numbers, or $(-\infty, \infty)$ using interval notation.
2. The x-intercept of the graph is 1. There is no y-intercept.
3. The y-axis $(x = 0)$ is a vertical asymptote of the graph.
4. A logarithmic function is decreasing if $0 < a < 1$ and is increasing if $a > 1$.
5. The graph of f contains the points $(1, 0)$, $(a, 1)$, and $\left(\dfrac{1}{a}, -1\right)$.
6. The graph is smooth and continuous, with no corners or gaps.

If the base of a logarithmic function is the number e, the result is the **natural logarithm function.** This function occurs so frequently in applications that it is given a special symbol, **ln** (from the Latin, *logarithmus naturalis*). That is,

$$y = \ln x \quad \text{if and only if} \quad x = e^y \qquad \textbf{(1)}$$

Because $y = \ln x$ and the exponential function $y = e^x$ are inverse functions, the graph of $y = \ln x$ can be obtained by reflecting the graph of $y = e^x$ about the line $y = x$. See Figure 16.

Using a calculator with an ln key, we can obtain other points on the graph of $f(x) = \ln x$. See Table 7.

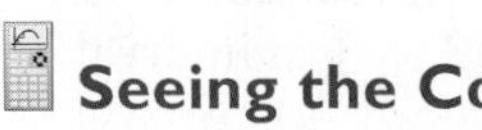

Seeing the Concept

Graph $Y_1 = e^x$ and $Y_2 = \ln x$ on the same square screen. Use eVALUEate to verify the points on the graph given in Figure 16. Do you see the symmetry of the two graphs with respect to the line $y = x$?

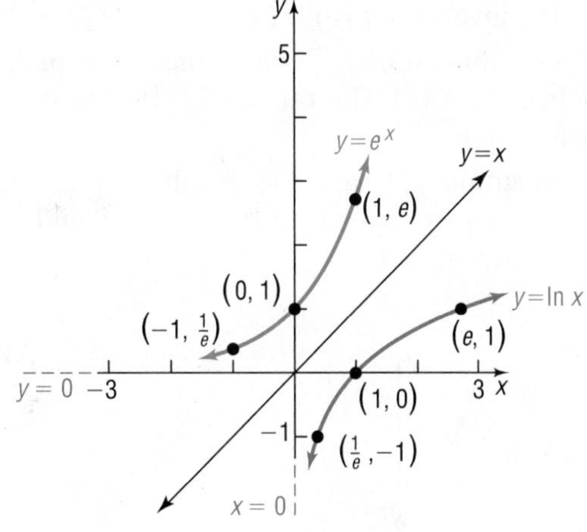

Figure 16

Table 7

x	$\ln x$
$\frac{1}{2}$	−0.69
2	0.69
3	1.10

EXAMPLE 6 Graphing a Logarithmic Function and Its Inverse

(a) Find the domain of the logarithmic function $f(x) = -\ln(x - 2)$.
(b) Graph f.
(c) From the graph, determine the range and vertical asymptote of f.
(d) Find f^{-1}, the inverse of f.
(e) Find the domain and the range of f^{-1}.
(f) Graph f^{-1}.

Solution

(a) The domain of f consists of all x for which $x - 2 > 0$, or equivalently, $x > 2$. The domain of f is $\{x | x > 2\}$, or $(2, \infty)$ in interval notation.

(b) To obtain the graph of $y = -\ln(x - 2)$, begin with the graph of $y = \ln x$ and use transformations. See Figure 17.

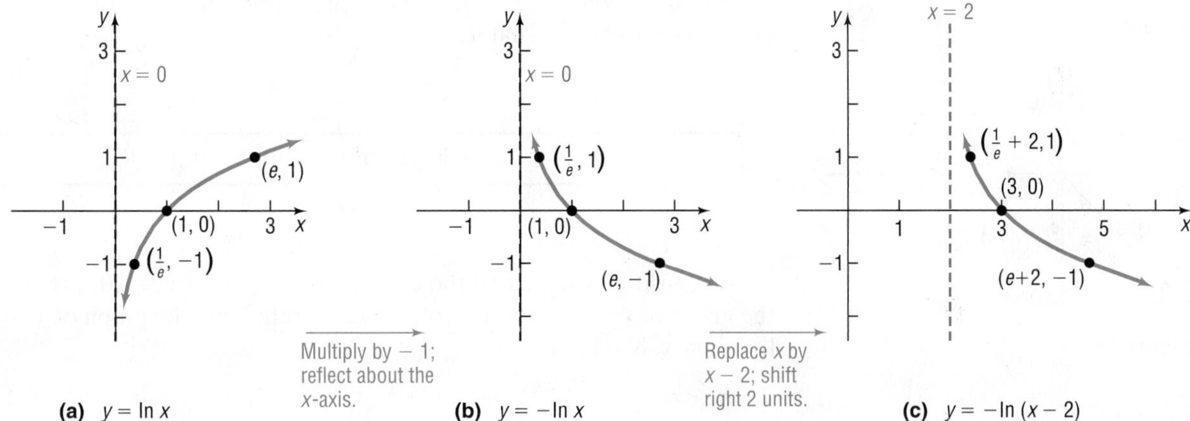

Figure 17 (a) $y = \ln x$ (b) $y = -\ln x$ (c) $y = -\ln(x - 2)$

(c) The range of $f(x) = -\ln(x - 2)$ is the set of all real numbers. The vertical asymptote is $x = 2$. [Do you see why? The original asymptote $(x = 0)$ is shifted to the right 2 units.]

(d) To find f^{-1}, begin with $y = -\ln(x - 2)$. The inverse function is defined implicitly by the equation

$$x = -\ln(y - 2)$$

Now solve for y.

$$-x = \ln(y - 2) \quad \text{Isolate the logarithm.}$$
$$e^{-x} = y - 2 \quad \text{Change to exponential form.}$$
$$y = e^{-x} + 2 \quad \text{Solve for } y.$$

The inverse of f is $f^{-1}(x) = e^{-x} + 2$.

(e) The domain of f^{-1} equals the range of f, which is the set of all real numbers, from part (c). The range of f^{-1} is the domain of f, which is $(2, \infty)$ in interval notation.

(f) To graph f^{-1}, use the graph of f in Figure 17(c) and reflect it about the line $y = x$. See Figure 18. We could also graph $f^{-1}(x) = e^{-x} + 2$ using transformations.

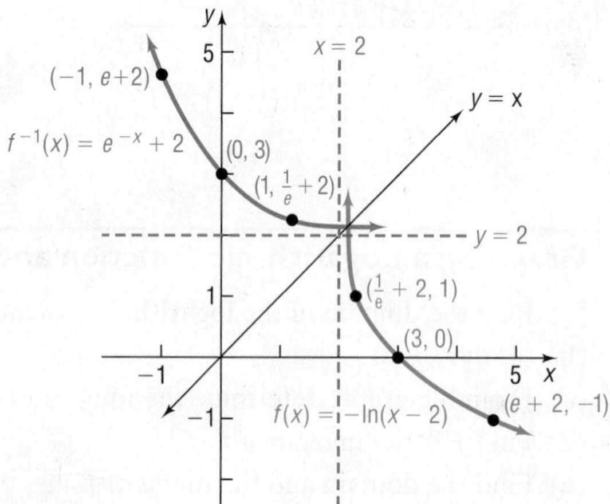

Figure 18

If the base of a logarithmic function is the number 10, the result is the **common logarithm function**. If the base a of the logarithmic function is not indicated, it is understood to be 10. That is,

$$y = \log x \quad \text{if and only if} \quad x = 10^y$$

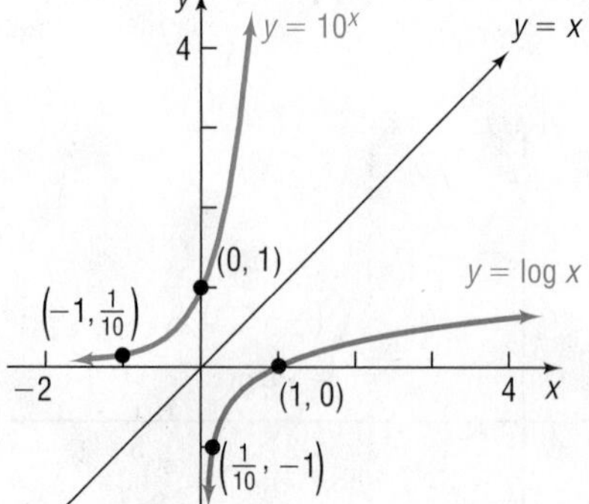

Figure 19

Because $y = \log x$ and the exponential function $y = 10^x$ are inverse functions, the graph of $y = \log x$ can be obtained by reflecting the graph of $y = 10^x$ about the line $y = x$. See Figure 19.

EXAMPLE 7 **Graphing a Logarithmic Function and Its Inverse**

(a) Find the domain of the logarithmic function $f(x) = 3\log(x-1)$.
(b) Graph f.
(c) From the graph, determine the range and vertical asymptote of f.
(d) Find f^{-1}, the inverse of f.
(e) Find the domain and the range of f^{-1}.
(f) Graph f^{-1}.

Solution (a) The domain of f consists of all x for which $x - 1 > 0$, or equivalently, $x > 1$. The domain of f is $\{x|x > 1\}$, or $(1, \infty)$ in interval notation.

(b) To obtain the graph of $y = 3\log(x-1)$, begin with the graph of $y = \log x$ and use transformations. See Figure 20.

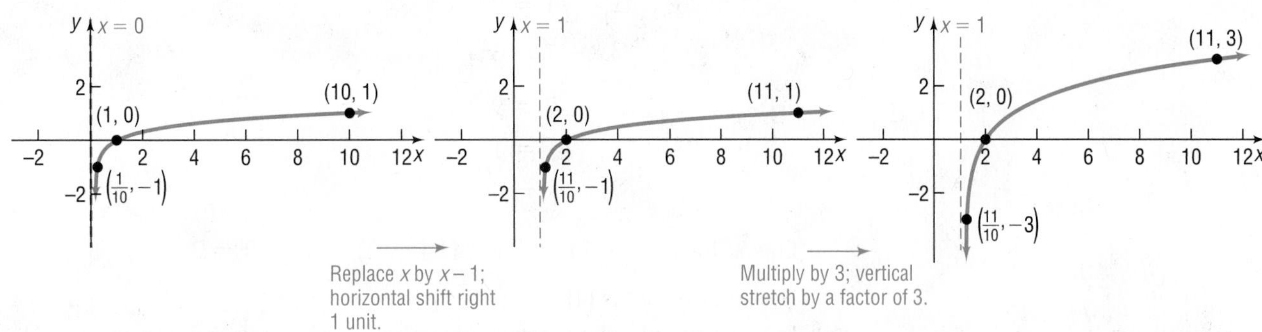

Figure 20 **(a)** $y = \log x$ **(b)** $y = \log(x-1)$ **(c)** $y = 3\log(x-1)$

(c) The range of $f(x) = 3\log(x-1)$ is the set of all real numbers. The vertical asymptote is $x = 1$.

(d) Begin with $y = 3\log(x-1)$. The inverse function is defined implicitly by the equation

$$x = 3\log(y-1)$$

Proceed to solve for y.

$$\frac{x}{3} = \log(y-1) \quad \text{Isolate the logarithm.}$$

$$10^{x/3} = y - 1 \quad \text{Change to exponential form.}$$

$$y = 10^{x/3} + 1 \quad \text{Solve for } y.$$

The inverse of f is $f^{-1}(x) = 10^{x/3} + 1$.

(e) The domain of f^{-1} is the range of f, which is the set of all real numbers, from part (c). The range of f^{-1} is the domain of f, which is $(1, \infty)$ in interval notation.

(f) To graph f^{-1}, use the graph of f in Figure 20(c) and reflect it about the line $y = x$. See Figure 21. We could also graph $f^{-1}(x) = 10^{x/3} + 1$ using transformations. ●

Figure 21

Now Work PROBLEM 81

5 Solve Logarithmic Equations

Equations that contain logarithms are called **logarithmic equations**. Care must be taken when solving logarithmic equations algebraically. In the expression $\log_a M$, remember that a and M are positive and $a \neq 1$. Be sure to check each apparent solution in the original equation and discard any that are extraneous.

Some logarithmic equations can be solved by changing the logarithmic equation to exponential form using the fact that $y = \log_a x$ means $a^y = x$.

EXAMPLE 8 Solving Logarithmic Equations

Solve:

(a) $\log_3(4x - 7) = 2$ (b) $\log_x 64 = 2$

Solution (a) To solve, change the logarithmic equation to exponential form.

$$\log_3(4x - 7) = 2$$

$$4x - 7 = 3^2 \quad \text{Change to exponential form.}$$

$$4x - 7 = 9$$

$$4x = 16$$

$$x = 4$$

Check: $\log_3(4x - 7) = \log_3(4 \cdot 4 - 7) = \log_3 9 = 2 \quad 3^2 = 9$

The solution set is {4}.

(b) To solve, change the logarithmic equation to exponential form.

$$\log_x 64 = 2$$

$$x^2 = 64 \quad \text{Change to exponential form.}$$

$$x = \pm\sqrt{64} = \pm 8 \quad \text{Square Root Method}$$

Because the base of a logarithm must be positive, discard -8. Check the potential solution 8.

Check: $\log_8 64 = 2 \quad 8^2 = 64$

The solution set is {8}. ●

EXAMPLE 9 Using Logarithms to Solve an Exponential Equation

Solve: $e^{2x} = 5$

Solution To solve, change the exponential equation to logarithmic form.

$$e^{2x} = 5$$

$$\ln 5 = 2x \quad \text{Change to logarithmic form.}$$

$$x = \frac{\ln 5}{2} \quad \text{Exact solution}$$

$$\approx 0.805 \quad \text{Approximate solution}$$

The solution set is $\left\{\frac{\ln 5}{2}\right\}$. ●

Now Work PROBLEMS 89 AND 101

EXAMPLE 10 **Alcohol and Driving**

Blood alcohol concentration (BAC) is a measure of the amount of alcohol in a person's bloodstream. A BAC of 0.04% means that a person has 4 parts alcohol per 10,000 parts blood in the body. Relative risk is defined as the likelihood of one event occurring divided by the likelihood of a second event occurring. For example, if an individual with a BAC of 0.02% is 1.4 times as likely to have a car accident as an individual who has not been drinking, the relative risk of an accident with a BAC of 0.02% is 1.4. Recent medical research suggests that the relative risk R of having an accident while driving a car can be modeled by an equation of the form

$$R = e^{kx}$$

where x is the percent concentration of alcohol in the bloodstream and k is a constant.

(a) Research indicates that the relative risk of a person having an accident with a BAC of 0.02% is 1.4. Find the constant k in the equation.

(b) Using this value of k, what is the relative risk if the concentration is 0.17%?

(c) Using this same value of k, what BAC corresponds to a relative risk of 100?

(d) If the law asserts that anyone with a relative risk of 4 or more should not have driving privileges, at what concentration of alcohol in the bloodstream should a driver be arrested and charged with DUI (driving under the influence)?

Solution (a) For a concentration of alcohol in the blood of 0.02% and a relative risk of 1.4, let $x = 0.02$ and $R = 1.4$ in the equation and solve for k.

$$R = e^{kx}$$

$$1.4 = e^{k(0.02)} \quad R = 1.4; x = 0.02$$

$$0.02k = \ln 1.4 \quad \text{Change to a logarithmic statement.}$$

$$k = \frac{\ln 1.4}{0.02} \approx 16.82 \quad \text{Solve for } k.$$

(b) A concentration of 0.17% means $x = 0.17$. Use $k = 16.82$ in the equation to find the relative risk R:

$$R = e^{kx} = e^{(16.82)(0.17)} \approx 17.5$$

For a concentration of alcohol in the blood of 0.17%, the relative risk of an accident is about 17.5. That is, a person with a BAC of 0.17% is 17.5 times as likely to have a car accident as a person with no alcohol in the bloodstream.

(c) A relative risk of 100 means $R = 100$. Use $k = 16.82$ in the equation $R = e^{kx}$. The concentration x of alcohol in the blood obeys

$$100 = e^{16.82x} \quad R = e^{kx}, R = 100, k = 16.82$$

$$16.82x = \ln 100 \quad \text{Change to a logarithmic statement.}$$

$$x = \frac{\ln 100}{16.82} \approx 0.27 \quad \text{Solve for } x.$$

For a concentration of alcohol in the blood of 0.27%, the relative risk of an accident is 100.

NOTE A BAC of 0.30% results in a loss of consciousness in most people. ■

(d) A relative risk of 4 means $R = 4$. Use $k = 16.82$ in the equation $R = e^{kx}$. The concentration x of alcohol in the bloodstream obeys

$$4 = e^{16.82x}$$

$$16.82x = \ln 4$$

$$x = \frac{\ln 4}{16.82} \approx 0.082$$

A driver with a BAC of 0.082% or more should be arrested and charged with DUI. ●

NOTE In most states, the blood alcohol content at which a DUI citation is given is 0.08%. ■

SUMMARY

Properties of the Logarithmic Function

$f(x) = \log_a x, \quad a > 1$
($y = \log_a x$ means $x = a^y$)

Domain: the interval $(0, \infty)$; Range: the interval $(-\infty, \infty)$

x-intercept: 1; y-intercept: none; vertical asymptote: $x = 0$ (y-axis); increasing; one-to-one

See Figure 22(a) for a typical graph.

$f(x) = \log_a x, \quad 0 < a < 1$
($y = \log_a x$ means $x = a^y$)

Domain: the interval $(0, \infty)$; Range: the interval $(-\infty, \infty)$

x-intercept: 1; y-intercept: none; vertical asymptote: $x = 0$ (y-axis); decreasing; one-to-one

See Figure 22(b) for a typical graph.

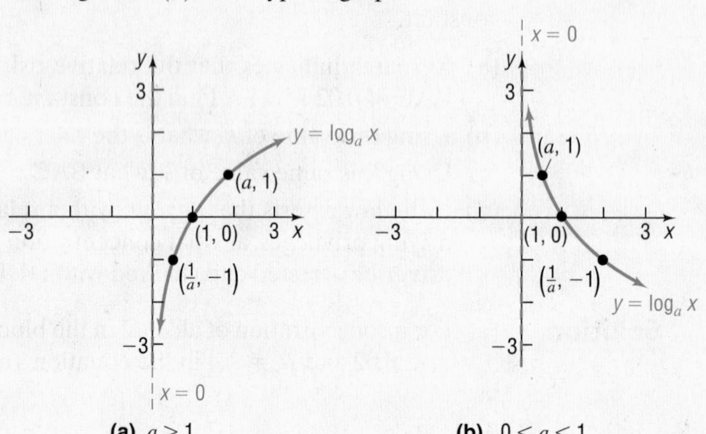

Figure 22 (a) $a > 1$ (b) $0 < a < 1$

7.2 Assess Your Understanding

'Are You Prepared?' *Answers are given at the end of these exercises. If you get a wrong answer, read the pages listed in red.*

1. Solve the inequality: $3x - 7 \le 8 - 2x$ (pp. A49–A52)

2. Solve: $2x + 3 = 9$ (pp. A27–A34)

3. ***True or False*** A one-to-one function $y = f(x)$ has an inverse function that is defined by the equation $x = f(y)$. (pp. 78–86)

Concepts and Vocabulary

4. The domain of the logarithmic function $f(x) = \log_a x$ is ______.

5. The graph of every logarithmic function $f(x) = \log_a x$, where $a > 0$ and $a \neq 1$, passes through three points: ______, ______, and ______.

6. If the graph of a logarithmic function $f(x) = \log_a x$, where $a > 0$ and $a \neq 1$, is increasing, then its base must be larger than ______.

7. ***True or False*** If $y = \log_a x$, then $y = a^x$.

8. ***True or False*** The graph of $f(x) = \log_a x$, where $a > 0$ and $a \neq 1$, has an x-intercept equal to 1 and no y-intercept.

9. Select the answer that completes the statement: $y = \ln x$ if and only if ______.
(a) $x = e^y$ (b) $y = e^x$ (c) $x = 10^y$ (d) $y = 10^x$

10. Choose the domain of $f(x) = \log_3(x + 2)$.
(a) $(-\infty, \infty)$ (b) $(2, \infty)$ (c) $(-2, \infty)$ (d) $(0, \infty)$

Skill Building

In Problems 11–18, change each exponential statement to an equivalent statement involving a logarithm.

11. $9 = 3^2$ **12.** $16 = 4^2$ **13.** $a^2 = 1.6$ **14.** $a^3 = 2.1$

15. $2^x = 7.2$ **16.** $3^x = 4.6$ **17.** $e^x = 8$ **18.** $e^{2.2} = M$

In Problems 19–26, change each logarithmic statement to an equivalent statement involving an exponent.

19. $\log_2 8 = 3$ **20.** $\log_3\left(\frac{1}{9}\right) = -2$ **21.** $\log_a 3 = 6$ **22.** $\log_b 4 = 2$

23. $\log_3 2 = x$ **24.** $\log_2 6 = x$ **25.** $\ln 4 = x$ **26.** $\ln x = 4$

In Problems 27–38, find the exact value of each logarithm without using a calculator.

27. $\log_2 1$
28. $\log_8 8$
29. $\log_5 25$
30. $\log_3\left(\frac{1}{9}\right)$
31. $\log_{1/2} 16$
32. $\log_{1/3} 9$
33. $\log_{10} \sqrt{10}$
34. $\log_5 \sqrt[3]{25}$
35. $\log_{\sqrt{2}} 4$
36. $\log_{\sqrt{3}} 9$
37. $\ln \sqrt{e}$
38. $\ln e^3$

In Problems 39–50, find the domain of each function.

39. $f(x) = \ln(x - 3)$
40. $g(x) = \ln(x - 1)$
41. $F(x) = \log_2 x^2$
42. $H(x) = \log_5 x^3$
43. $f(x) = 3 - 2\log_4\left(\frac{x}{2} - 5\right)$
44. $g(x) = 8 + 5\ln(2x + 3)$
45. $f(x) = \ln\left(\frac{1}{x + 1}\right)$
46. $g(x) = \ln\left(\frac{1}{x - 5}\right)$
47. $g(x) = \log_5\left(\frac{x + 1}{x}\right)$
48. $h(x) = \log_3\left(\frac{x}{x - 1}\right)$
49. $f(x) = \sqrt{\ln x}$
50. $g(x) = \frac{1}{\ln x}$

In Problems 51–58, use a calculator to evaluate each expression. Round your answer to three decimal places.

51. $\ln \frac{5}{3}$
52. $\frac{\ln 5}{3}$
53. $\frac{\ln \frac{10}{3}}{0.04}$
54. $\frac{\ln \frac{2}{3}}{-0.1}$
55. $\frac{\ln 4 + \ln 2}{\log 4 + \log 2}$
56. $\frac{\log 15 + \log 20}{\ln 15 + \ln 20}$
57. $\frac{2\ln 5 + \log 50}{\log 4 - \ln 2}$
58. $\frac{3\log 80 - \ln 5}{\log 5 + \ln 20}$

59. Find a so that the graph of $f(x) = \log_a x$ contains the point $(2, 2)$.

60. Find a so that the graph of $f(x) = \log_a x$ contains the point $\left(\frac{1}{2}, -4\right)$.

In Problems 61–64, graph each function and its inverse on the same set of axes.

61. $f(x) = 3^x; f^{-1}(x) = \log_3 x$
62. $f(x) = 4^x; f^{-1}(x) = \log_4 x$
63. $f(x) = \left(\frac{1}{2}\right)^x; f^{-1}(x) = \log_{1/2} x$
64. $f(x) = \left(\frac{1}{3}\right)^x; f^{-1}(x) = \log_{1/3} x$

In Problems 65–72, the graph of a logarithmic function is given. Match each graph to one of the following functions:

(A) $y = \log_3 x$ (B) $y = \log_3(-x)$ (C) $y = -\log_3 x$ (D) $y = -\log_3(-x)$
(E) $y = \log_3 x - 1$ (F) $y = \log_3(x - 1)$ (G) $y = \log_3(1 - x)$ (H) $y = 1 - \log_3 x$

65.

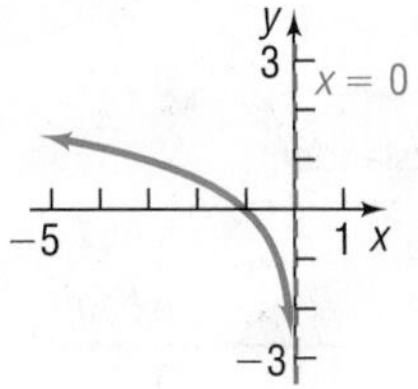

66.

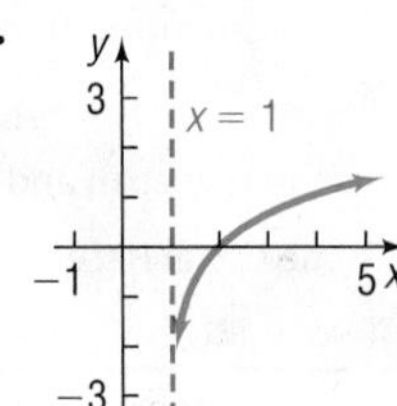

67.

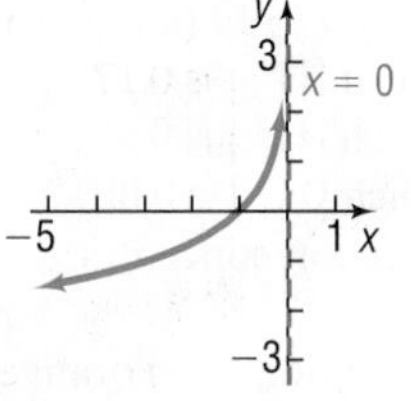

68.

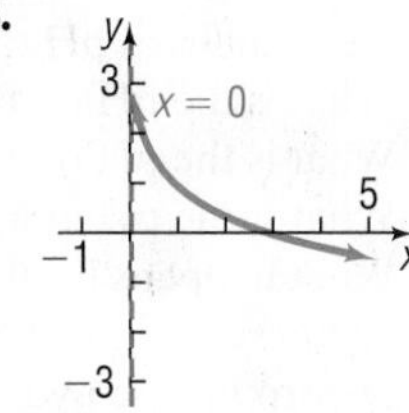

69.

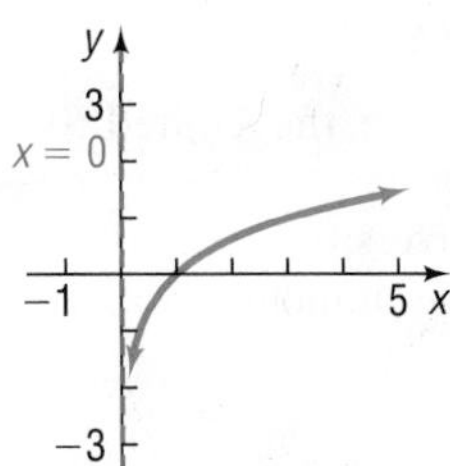

70.

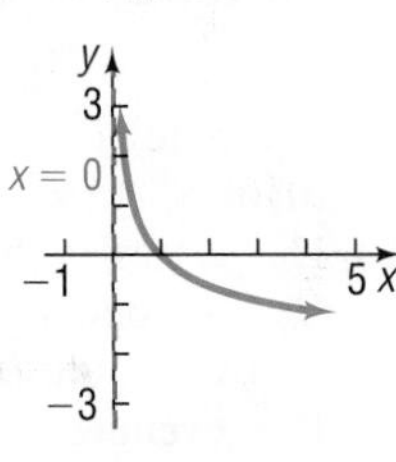

71.

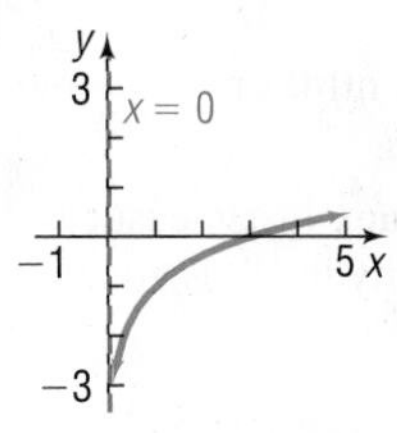

72.

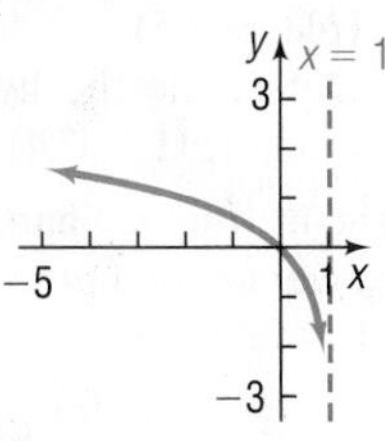

In Problems 73–88, use the given function f.

(a) Find the domain of f. (b) Graph f. (c) From the graph, determine the range and any asymptotes of f.
(d) Find f^{-1}, the inverse of f. (e) Find the domain and the range of f^{-1}. (f) Graph f^{-1}.

73. $f(x) = \ln(x + 4)$
74. $f(x) = \ln(x - 3)$
75. $f(x) = 2 + \ln x$
76. $f(x) = -\ln(-x)$
77. $f(x) = \ln(2x) - 3$
78. $f(x) = -2\ln(x + 1)$
79. $f(x) = \log(x - 4) + 2$
80. $f(x) = \frac{1}{2}\log x - 5$
81. $f(x) = \frac{1}{2}\log(2x)$
82. $f(x) = \log(-2x)$
83. $f(x) = 3 + \log_3(x + 2)$
84. $f(x) = 2 - \log_3(x + 1)$
85. $f(x) = e^{x+2} - 3$
86. $f(x) = 3e^x + 2$
87. $f(x) = 2^{x/3} + 4$
88. $f(x) = -3^{x+1}$

In Problems 89–112, solve each equation.

89. $\log_3 x = 2$ **90.** $\log_5 x = 3$ **91.** $\log_2(2x+1) = 3$ **92.** $\log_3(3x-2) = 2$

93. $\log_x 4 = 2$ **94.** $\log_x\left(\frac{1}{8}\right) = 3$ **95.** $\ln e^x = 5$ **96.** $\ln e^{-2x} = 8$

97. $\log_4 64 = x$ **98.** $\log_5 625 = x$ **99.** $\log_3 243 = 2x + 1$ **100.** $\log_6 36 = 5x + 3$

101. $e^{3x} = 10$ **102.** $e^{-2x} = \frac{1}{3}$ **103.** $e^{2x+5} = 8$ **104.** $e^{-2x+1} = 13$

105. $\log_3(x^2+1) = 2$ **106.** $\log_5(x^2+x+4) = 2$ **107.** $\log_2 8^x = -3$ **108.** $\log_3 3^x = -1$

109. $5e^{0.2x} = 7$ **110.** $8 \cdot 10^{2x-7} = 3$ **111.** $2 \cdot 10^{2-x} = 5$ **112.** $4e^{x+1} = 5$

Mixed Practice

113. Suppose that $G(x) = \log_3(2x+1) - 2$.
(a) What is the domain of G?
(b) What is $G(40)$? What point is on the graph of G?
(c) If $G(x) = 3$, what is x? What point is on the graph of G?
(d) What is the zero of G?

114. Suppose that $F(x) = \log_2(x+1) - 3$.
(a) What is the domain of F?
(b) What is $F(7)$? What point is on the graph of F?
(c) If $F(x) = -1$, what is x? What point is on the graph of F?
(d) What is the zero of F?

In Problems 115–118, graph each function. Based on the graph, state the domain and the range, and find any intercepts.

115. $f(x) = \begin{cases} \ln(-x) & \text{if } x < 0 \\ \ln x & \text{if } x > 0 \end{cases}$

116. $f(x) = \begin{cases} \ln(-x) & \text{if } x \le -1 \\ -\ln(-x) & \text{if } -1 < x < 0 \end{cases}$

117. $f(x) = \begin{cases} -\ln x & \text{if } 0 < x < 1 \\ \ln x & \text{if } x \ge 1 \end{cases}$

118. $f(x) = \begin{cases} \ln x & \text{if } 0 < x < 1 \\ -\ln x & \text{if } x \ge 1 \end{cases}$

Applications and Extensions

119. Chemistry The pH of a chemical solution is given by the formula

$$\text{pH} = -\log_{10}[\text{H}^+]$$

where $[\text{H}^+]$ is the concentration of hydrogen ions in moles per liter. Values of pH range from 0 (acidic) to 14 (alkaline).
(a) What is the pH of a solution for which $[\text{H}^+]$ is 0.1?
(b) What is the pH of a solution for which $[\text{H}^+]$ is 0.01?
(c) What is the pH of a solution for which $[\text{H}^+]$ is 0.001?
(d) What happens to pH as the hydrogen ion concentration decreases?
(e) Determine the hydrogen ion concentration of an orange (pH = 3.5).
(f) Determine the hydrogen ion concentration of human blood (pH = 7.4).

120. Diversity Index Shannon's diversity index is a measure of the diversity of a population. The diversity index is given by the formula

$$H = -(p_1 \log p_1 + p_2 \log p_2 + \cdots + p_n \log p_n)$$

where p_1 is the proportion of the population that is species 1, p_2 is the proportion of the population that is species 2, and so on. In this problem, the population is people in the United States and the species is race.
(a) According to the U.S. Census Bureau, the distribution of race in the United States in 2010 was as follows:

Race	Proportion
White	0.724
Black or African American	0.126
American Indian and Alaska Native	0.009
Asian	0.048
Native Hawaiian and Other Pacific Islander	0.002
Some Other Race	0.062
Two or More Races	0.029

Source: *U.S. Census Bureau*

Compute the diversity index of the United States in 2010.
(b) The largest value of the diversity index is given by $H_{max} = \log(S)$, where S is the number of categories of race. Compute H_{max}.
(c) The **evenness ratio** is given by $E_H = \frac{H}{H_{max}}$, where $0 \le E_H \le 1$. If $E_H = 1$, there is complete evenness. Compute the evenness ratio for the United States.
(d) Obtain the distribution of race for the United States in 2000 from the Census Bureau. Compute Shannon's diversity index. Is the United States becoming more diverse? Why?

121. Atmospheric Pressure The atmospheric pressure p on an object decreases with increasing height. This pressure, measured in millimeters of mercury, is related to the height h (in kilometers) above sea level by the function

$$p(h) = 760e^{-0.145h}$$

(a) Find the height of an aircraft if the atmospheric pressure is 320 millimeters of mercury.
(b) Find the height of a mountain if the atmospheric pressure is 667 millimeters of mercury.

122. Healing of Wounds The normal healing of wounds can be modeled by an exponential function. If A_0 represents the original area of the wound, and if A equals the area of the wound, then the function

$$A(n) = A_0e^{-0.35n}$$

describes the area of a wound after n days following an injury when no infection is present to retard the healing. Suppose that a wound initially had an area of 100 square millimeters.
(a) If healing is taking place, after how many days will the wound be one-half its original size?
(b) How long before the wound is 10% of its original size?

123. Exponential Probability Between 12:00 PM and 1:00 PM, cars arrive at Citibank's drive-thru at the rate of 6 cars per hour (0.1 car per minute). The following formula from statistics can be used to determine the probability that a car will arrive within t minutes of 12:00 PM.

$$F(t) = 1 - e^{-0.1t}$$

(a) Determine how many minutes are needed for the probability to reach 50%.
(b) Determine how many minutes are needed for the probability to reach 80%.
(c) Is it possible for the probability to equal 100%? Explain.

124. Exponential Probability Between 5:00 PM and 6:00 PM, cars arrive at Jiffy Lube at the rate of 9 cars per hour (0.15 car per minute). The following formula from statistics can be used to determine the probability that a car will arrive within t minutes of 5:00 PM.

$$F(t) = 1 - e^{-0.15t}$$

(a) Determine how many minutes are needed for the probability to reach 50%.
(b) Determine how many minutes are needed for the probability to reach 80%.

125. Drug Medication The formula

$$D = 5e^{-0.4h}$$

can be used to find the number of milligrams D of a certain drug that is in a patient's bloodstream h hours after the drug was administered. When the number of milligrams reaches 2, the drug is to be administered again. What is the time between injections?

126. Spreading of Rumors A model for the number N of people in a college community who have heard a certain rumor is

$$N = P(1 - e^{-0.15d})$$

where P is the total population of the community and d is the number of days that have elapsed since the rumor began. In a community of 1000 students, how many days will elapse before 450 students have heard the rumor?

127. Current in an RL Circuit The equation governing the amount of current I (in amperes) after time t (in seconds) in a simple RL circuit consisting of a resistance R (in ohms), an inductance L (in henrys), and an electromotive force E (in volts) is

$$I = \frac{E}{R}[1 - e^{-(R/L)t}]$$

If $E = 12$ volts, $R = 10$ ohms, and $L = 5$ henrys, how long does it take to obtain a current of 0.5 ampere? Of 1.0 ampere? Graph the equation.

128. Learning Curve Psychologists sometimes use the function

$$L(t) = A(1 - e^{-kt})$$

to measure the amount L learned at time t. Here A represents the amount to be learned, and the number k measures the rate of learning. Suppose that a student has an amount A of 200 vocabulary words to learn. A psychologist determines that the student has learned 20 vocabulary words after 5 minutes.

(a) Determine the rate of learning k.
(b) Approximately how many words will the student have learned after 10 minutes?
(c) After 15 minutes?
(d) How long does it take for the student to learn 180 words?

Loudness of Sound *Problems 129–132 use the following discussion: The* ***loudness*** $L(x)$*, measured in decibels (dB), of a sound of intensity x, measured in watts per square meter, is defined as* $L(x) = 10 \log \frac{x}{I_0}$*, where* $I_0 = 10^{-12}$ *watt per square meter is the least intense sound that a human ear can detect. Determine the loudness, in decibels, of each of the following sounds.*

129. Normal conversation: intensity of $x = 10^{-7}$ watt per square meter.

130. Amplified rock music: intensity of 10^{-1} watt per square meter.

131. Heavy city traffic: intensity of $x = 10^{-3}$ watt per square meter.

132. Diesel truck traveling 40 miles per hour 50 feet away: intensity 10 times that of a passenger car traveling 50 miles per hour 50 feet away, whose loudness is 70 decibels.

The Richter Scale *Problems 133 and 134 on the next page use the following discussion: The* ***Richter scale*** *is one way of converting seismographic readings into numbers that provide an easy reference for measuring the magnitude M of an earthquake. All earthquakes are compared to a* ***zero-level earthquake*** *whose seismographic reading measures 0.001 millimeter at a distance of 100 kilometers from the epicenter. An earthquake whose seismographic reading measures x millimeters has* ***magnitude*** $M(x)$*, given by*

$$M(x) = \log\left(\frac{x}{x_0}\right)$$

where $x_0 = 10^{-3}$ *is the reading of a zero-level earthquake the same distance from its epicenter. In Problems 133 and 134, determine the magnitude of each earthquake.*

133. Magnitude of an Earthquake Mexico City in 1985: seismographic reading of 125,892 millimeters 100 kilometers from the center

134. Magnitude of an Earthquake San Francisco in 1906: seismographic reading of 50,119 millimeters 100 kilometers from the center

135. Alcohol and Driving The concentration of alcohol in a person's bloodstream is measurable. Suppose that the relative risk R of having an accident while driving a car can be modeled by an equation of the form

$$R = e^{kx}$$

where x is the percent concentration of alcohol in the bloodstream and k is a constant.

(a) Suppose that a concentration of alcohol in the bloodstream of 0.03 percent results in a relative risk of an accident of 1.4. Find the constant k in the equation.

(b) Using this value of k, what is the relative risk if the concentration is 0.17 percent?

(c) Using the same value of k, what concentration of alcohol corresponds to a relative risk of 100?

(d) If the law asserts that anyone with a relative risk of having an accident of 5 or more should not have driving privileges, at what concentration of alcohol in the bloodstream should a driver be arrested and charged with a DUI?

(e) Compare this situation with that of Example 10. If you were a lawmaker, which situation would you support? Give your reasons.

Explaining Concepts: Discussion and Writing

136. Is there any function of the form $y = x^{\alpha}, 0 < \alpha < 1$, that increases more slowly than a logarithmic function whose base is greater than 1? Explain.

137. In the definition of the logarithmic function, the base a is not allowed to equal 1. Why?

138. **Critical Thinking** In buying a new car, one consideration might be how well the price of the car holds up over time. Different makes of cars have different depreciation rates. One way to compute a depreciation rate for a car is given here. Suppose that the current prices of a certain automobile are as shown in the table.

	Age in Years				
New	1	2	3	4	5
\$38,000	\$36,600	\$32,400	\$28,750	\$25,400	\$21,200

Use the formula New $=$ Old(e^{Rt}) to find R, the annual depreciation rate, for a specific time t. When might be the best time to trade in the car? Consult the NADA ("blue") book and compare two like models that you are interested in. Which has the better depreciation rate?

Retain Your Knowledge

Problems 139–142 are based on material learned earlier in the course. The purpose of these problems is to keep the material fresh in your mind so that you are better prepared for the final exam.

139. Find the average rate of change of $f(x) = 9^x$ from $\frac{1}{2}$ to 1.

140. Find the exact value of $\sin\left(-\frac{5\pi}{2}\right)$. Do not use a calculator.

141. Solve the equation $\cos(3\theta) = -1$ on the interval $0 \le \theta < 2\pi$.

142. For $\mathbf{v} = -\mathbf{i} + 2\mathbf{j}$ and $\mathbf{w} = \mathbf{i} + 3\mathbf{j}$, find the dot product $\mathbf{v} \cdot \mathbf{w}$ and the angle θ between $\mathbf{v}$ and $\mathbf{w}$.

'Are You Prepared?' Answers

1. $x \le 3$ **2.** $x < -4$ or $x > 1$ **3.** $\{3\}$

7.3 Properties of Logarithms

OBJECTIVES
1 Work with the Properties of Logarithms (p. 482)
2 Write a Logarithmic Expression as a Sum or Difference of Logarithms (p. 484)
3 Write a Logarithmic Expression as a Single Logarithm (p. 485)
4 Evaluate Logarithms Whose Base Is Neither 10 Nor e (p. 487)

1 Work with the Properties of Logarithms

Logarithms have some very useful properties that can be derived directly from the definition and the laws of exponents.

EXAMPLE 1 **Establishing Properties of Logarithms**

(a) Show that $\log_a 1 = 0$. (b) Show that $\log_a a = 1$.

Solution (a) This fact was established when we graphed $y = \log_a x$ (see Figure 13 on page 472). To show the result algebraically, let $y = \log_a 1$. Then

$$\begin{aligned} y &= \log_a 1 \\ a^y &= 1 && \text{Change to exponential form.} \\ a^y &= a^0 && a^0 = 1 \text{ since } a > 0, a \neq 1 \\ y &= 0 && \text{Equate exponents.} \\ \log_a 1 &= 0 && y = \log_a 1 \end{aligned}$$

(b) Let $y = \log_a a$. Then

$$\begin{aligned} y &= \log_a a \\ a^y &= a && \text{Change to exponential form.} \\ a^y &= a^1 && a = a^1 \\ y &= 1 && \text{Equate exponents.} \\ \log_a a &= 1 && y = \log_a a \end{aligned}$$

To summarize:

$$\log_a 1 = 0 \qquad \log_a a = 1$$

THEOREM

Properties of Logarithms

In the properties given next, M and a are positive real numbers, $a \neq 1$, and r is any real number.

The number $\log_a M$ is the exponent to which a must be raised to obtain M. That is,

$$a^{\log_a M} = M \qquad \textbf{(1)}$$

The logarithm with base a of a raised to a power equals that power. That is,

$$\log_a a^r = r \qquad \textbf{(2)}$$

The proof uses the fact that $y = a^x$ and $y = \log_a x$ are inverse functions.

Proof of Property (1) For inverse functions,

$$f(f^{-1}(x)) = x \quad \text{for all } x \text{ in the domain of } f^{-1}$$

Use $f(x) = a^x$ and $f^{-1}(x) = \log_a x$ to find

$$f(f^{-1}(x)) = a^{\log_a x} = x \quad \text{for } x > 0$$

Now let $x = M$ to obtain $a^{\log_a M} = M$, where $M > 0$. ■

Proof of Property (2) For inverse functions,

$$f^{-1}(f(x)) = x \quad \text{for all } x \text{ in the domain of } f$$

Use $f(x) = a^x$ and $f^{-1}(x) = \log_a x$ to find

$$f^{-1}(f(x)) = \log_a a^x = x \quad \text{for all real numbers } x$$

Now let $x = r$ to obtain $\log_a a^r = r$, where r is any real number. ■

EXAMPLE 2 **Using Properties (1) and (2)**

(a) $2^{\log_2 \pi} = \pi$ (b) $\log_{0.2} 0.2^{-\sqrt{2}} = -\sqrt{2}$ (c) $\ln e^{kt} = kt$

Now Work PROBLEM 15

Other useful properties of logarithms are given next.

THEOREM **Properties of Logarithms**

In the following properties, M, N, and a are positive real numbers, $a \neq 1$, and r is any real number.

The Log of a Product Equals the Sum of the Logs

$$\log_a(MN) = \log_a M + \log_a N \quad (3)$$

The Log of a Quotient Equals the Difference of the Logs

$$\log_a\left(\frac{M}{N}\right) = \log_a M - \log_a N \quad (4)$$

The Log of a Power Equals the Product of the Power and the Log

$$\log_a M^r = r \log_a M \quad (5)$$

$$a^r = e^{r \ln a} \quad (6)$$

We shall derive properties (3), (5), and (6) and leave the derivation of property (4) as an exercise (see Problem 109).

Proof of Property (3) Let $A = \log_a M$ and let $B = \log_a N$. These expressions are equivalent to the exponential expressions

$$a^A = M \quad \text{and} \quad a^B = N$$

Now

$$\begin{aligned} \log_a(MN) = \log_a(a^A a^B) &= \log_a a^{A+B} && \text{Law of Exponents} \\ &= A + B && \text{Property (2) of logarithms} \\ &= \log_a M + \log_a N \end{aligned}$$

Proof of Property (5) Let $A = \log_a M$. This expression is equivalent to

$$a^A = M$$

Now

$$\begin{aligned} \log_a M^r = \log_a(a^A)^r &= \log_a a^{rA} && \text{Law of Exponents} \\ &= rA && \text{Property (2) of logarithms} \\ &= r \log_a M \end{aligned}$$

Proof of Property (6) Property (1), with $a = e$, gives

$$e^{\ln M} = M$$

Now let $M = a^r$ and apply property (5).

$$e^{\ln a^r} = e^{r \ln a} = a^r$$

Now Work PROBLEM 19

2 Write a Logarithmic Expression as a Sum or Difference of Logarithms

Logarithms can be used to transform products into sums, quotients into differences, and powers into factors. Such transformations prove useful in certain types of calculus problems.

EXAMPLE 3 **Writing a Logarithmic Expression as a Sum of Logarithms**

Write $\log_a(x\sqrt{x^2+1})$, $x > 0$, as a sum of logarithms. Express all powers as factors.

Solution

$$\begin{aligned}\log_a(x\sqrt{x^2+1}) &= \log_a x + \log_a \sqrt{x^2+1} && \log_a(M \cdot N) = \log_a M + \log_a N\\ &= \log_a x + \log_a(x^2+1)^{1/2}\\ &= \log_a x + \frac{1}{2}\log_a(x^2+1) && \log_a M^r = r\log_a M\end{aligned}$$

EXAMPLE 4 **Writing a Logarithmic Expression as a Difference of Logarithms**

Write

$$\ln\frac{x^2}{(x-1)^3} \qquad x > 1$$

as a difference of logarithms. Express all powers as factors.

Solution

$$\ln\frac{x^2}{(x-1)^3} = \ln x^2 - \ln(x-1)^3 = 2\ln x - 3\ln(x-1)$$

$$\uparrow \log_a\left(\frac{M}{N}\right) = \log_a M - \log_a N \qquad \uparrow \log_a M^r = r\log_a M$$

EXAMPLE 5 **Writing a Logarithmic Expression as a Sum and Difference of Logarithms**

Write

$$\log_a\frac{\sqrt{x^2+1}}{x^3(x+1)^4} \qquad x > 0$$

as a sum and difference of logarithms. Express all powers as factors.

Solution

$$\begin{aligned}\log_a\frac{\sqrt{x^2+1}}{x^3(x+1)^4} &= \log_a\sqrt{x^2+1} - \log_a[x^3(x+1)^4] && \text{Property (4)}\\ &= \log_a\sqrt{x^2+1} - [\log_a x^3 + \log_a(x+1)^4] && \text{Property (3)}\\ &= \log_a(x^2+1)^{1/2} - \log_a x^3 - \log_a(x+1)^4\\ &= \frac{1}{2}\log_a(x^2+1) - 3\log_a x - 4\log_a(x+1) && \text{Property (5)}\end{aligned}$$

WARNING In using properties (3) through (5), be careful about the values that the variable may assume. For example, the domain of the variable for $\log_a x$ is $x > 0$ and for $\log_a(x-1)$ is $x > 1$. If these functions are added, the domain is $x > 1$. That is, the equality

$$\log_a x + \log_a(x-1) = \log_a[x(x-1)]$$

is true only for $x > 1$.

Now Work PROBLEM 51

3 Write a Logarithmic Expression as a Single Logarithm

Another use of properties (3) through (5) is to write sums and/or differences of logarithms with the same base as a single logarithm. This skill will be needed to solve certain logarithmic equations discussed in the next section.

EXAMPLE 6 **Writing Expressions as a Single Logarithm**

Write each of the following as a single logarithm.

(a) $\log_a 7 + 4\log_a 3$ (b) $\frac{2}{3}\ln 8 - \ln(5^2 - 1)$

(c) $\log_a x + \log_a 9 + \log_a(x^2+1) - \log_a 5$

Solution (a)
$$\begin{aligned}\log_a 7 + 4\log_a 3 &= \log_a 7 + \log_a 3^4 && r\log_a M = \log_a M^r\\ &= \log_a 7 + \log_a 81\\ &= \log_a(7\cdot 81) && \log_a M + \log_a N = \log_a(M\cdot N)\\ &= \log_a 567\end{aligned}$$

(b)
$$\begin{aligned}\frac{2}{3}\ln 8 - \ln(5^2 - 1) &= \ln 8^{2/3} - \ln(25 - 1) && r\log_a M = \log_a M^r\\ &= \ln 4 - \ln 24 && 8^{2/3} = (\sqrt[3]{8})^2 = 2^2 = 4\\ &= \ln\left(\frac{4}{24}\right) && \log_a M - \log_a N = \log_a\left(\frac{M}{N}\right)\\ &= \ln\left(\frac{1}{6}\right)\\ &= \ln 1 - \ln 6\\ &= -\ln 6 && \ln 1 = 0\end{aligned}$$

(c)
$$\begin{aligned}\log_a x + \log_a 9 + \log_a(x^2+1) - \log_a 5 &= \log_a(9x) + \log_a(x^2+1) - \log_a 5\\ &= \log_a[9x(x^2+1)] - \log_a 5\\ &= \log_a\left[\frac{9x(x^2+1)}{5}\right]\end{aligned}$$

●

WARNING A common error that some students make is to express the logarithm of a sum as the sum of logarithms.

$$\log_a(M+N) \quad \text{is not equal to} \quad \log_a M + \log_a N$$
$$\log_a(MN) = \log_a M + \log_a N$$

Another common error is to express the difference of logarithms as the quotient of logarithms.

$$\log_a M - \log_a N \quad \text{is not equal to} \quad \frac{\log_a M}{\log_a N}$$
$$\log_a M - \log_a N = \log_a\left(\frac{M}{N}\right)$$

A third common error is to express a logarithm raised to a power as the product of the power times the logarithm.

$$(\log_a M)^r \quad \text{is not equal to} \quad r\log_a M$$
$$\log_a M^r = r\log_a M$$

■

Now Work PROBLEMS 57 AND 63

Two other important properties of logarithms are consequences of the fact that the logarithmic function $y = \log_a x$ is a one-to-one function.

THEOREM

Properties of Logarithms

In the following properties, M, N, and a are positive real numbers, $a \neq 1$.

If $M = N$, then $\log_a M = \log_a N$. **(7)**

If $\log_a M = \log_a N$, then $M = N$. **(8)**

Property (7) is used as follows: Starting with the equation $M = N$, "take the logarithm of both sides" to obtain $\log_a M = \log_a N$.

Properties (7) and (8) are useful for solving *exponential and logarithmic equations*, a topic discussed in the next section.

4 Evaluate Logarithms Whose Base Is Neither 10 Nor *e*

Logarithms with base 10—common logarithms—were used to facilitate arithmetic computations before the widespread use of calculators. (See the Historical Feature at the end of this section.) Natural logarithms—that is, logarithms whose base is the number e—remain very important because they arise frequently in the study of natural phenomena.

Common logarithms are usually abbreviated by writing **log**, with the base understood to be 10, just as natural logarithms are abbreviated by **ln**, with the base understood to be e.

Most calculators have both [log] and [ln] keys to calculate the common logarithm and the natural logarithm of a number, respectively. Let's look at an example to see how to approximate logarithms having a base other than 10 or e.

EXAMPLE 7 **Approximating a Logarithm Whose Base Is Neither 10 Nor e**

Approximate $\log_2 7$. Round the answer to four decimal places.

Solution Remember, evaluating $\log_2 7$ means answering the question "2 raised to what exponent equals 7?" Let $y = \log_2 7$. Then $2^y = 7$. Because $2^2 = 4$ and $2^3 = 8$, the value of $\log_2 7$ is between 2 and 3.

$$\begin{aligned} 2^y &= 7 \\ \ln 2^y &= \ln 7 && \text{Property (7)} \\ y \ln 2 &= \ln 7 && \text{Property (5)} \\ y &= \frac{\ln 7}{\ln 2} && \text{Exact value} \\ y &\approx 2.8074 && \text{Approximate value rounded to four decimal places} \end{aligned}$$

Example 7 shows how to approximate a logarithm whose base is 2 by changing to logarithms involving the base e. In general, the **Change-of-Base Formula** is used.

THEOREM **Change-of-Base Formula**

If $a \neq 1, b \neq 1$, and M are positive real numbers, then

$$\log_a M = \frac{\log_b M}{\log_b a} \qquad \textbf{(9)}$$

Proof Let $y = \log_a M$. Then

$$\begin{aligned} a^y &= M \\ \log_b a^y &= \log_b M && \text{Property (7)} \\ y \log_b a &= \log_b M && \text{Property (5)} \\ y &= \frac{\log_b M}{\log_b a} && \text{Solve for } y. \\ \log_a M &= \frac{\log_b M}{\log_b a} && y = \log_a M \end{aligned}$$

Because calculators have keys only for [log] and [ln], in practice, the Change-of-Base Formula uses either $b = 10$ or $b = e$. That is,

$$\log_a M = \frac{\log M}{\log a} \quad \text{and} \quad \log_a M = \frac{\ln M}{\ln a} \qquad \textbf{(10)}$$

EXAMPLE 8 **Using the Change-of-Base Formula**

Approximate:

(a) $\log_5 89$ (b) $\log_{\sqrt{2}} \sqrt{5}$

Round answers to four decimal places.

Solution

(a) $\log_5 89 = \dfrac{\log 89}{\log 5} \approx \dfrac{1.949390007}{0.6989700043} \approx 2.7889$

or

$\log_5 89 = \dfrac{\ln 89}{\ln 5} \approx \dfrac{4.48863637}{1.609437912} \approx 2.7889$

(b) $\log_{\sqrt{2}} \sqrt{5} = \dfrac{\log \sqrt{5}}{\log \sqrt{2}} = \dfrac{\frac{1}{2}\log 5}{\frac{1}{2}\log 2} = \dfrac{\log 5}{\log 2} \approx 2.3219$

or

$\log_{\sqrt{2}} \sqrt{5} = \dfrac{\ln \sqrt{5}}{\ln \sqrt{2}} = \dfrac{\frac{1}{2}\ln 5}{\frac{1}{2}\ln 2} = \dfrac{\ln 5}{\ln 2} \approx 2.3219$

●

Now Work PROBLEMS 23 AND 71

COMMENT To graph logarithmic functions when the base is different from e or 10 requires the Change-of-Base Formula. For example, to graph $y = \log_2 x$, graph either $y = \dfrac{\ln x}{\ln 2}$ or $y = \dfrac{\log x}{\log 2}$. ■

Now Work PROBLEM 79

SUMMARY

Properties of Logarithms

In the list that follows, a, b, M, N, and r are real numbers. Also, $a > 0$, $a \neq 1$, $b > 0$, $b \neq 1$, $M > 0$, and $N > 0$.

Definition $y = \log_a x$ means $x = a^y$

Properties of Logarithms

$\log_a 1 = 0$ $\quad \log_a a = 1$ $\quad \log_a M^r = r \log_a M$

$a^{\log_a M} = M$ $\quad \log_a a^r = r$ $\quad a^r = e^{r \ln a}$

$\log_a (MN) = \log_a M + \log_a N$ $\quad \log_a\left(\dfrac{M}{N}\right) = \log_a M - \log_a N$

If $M = N$, then $\log_a M = \log_a N$. If $\log_a M = \log_a N$, then $M = N$.

Change-of-Base Formula $\log_a M = \dfrac{\log_b M}{\log_b a}$

Historical Feature

John Napier (1550–1617)

Logarithms were invented about 1590 by John Napier (1550–1617) and Joost Bürgi (1552–1632), working independently. Napier, whose work had the greater influence, was a Scottish lord, a secretive man whose neighbors were inclined to believe him to be in league with the devil. His approach to logarithms was very different from ours; it was based on the relationship between arithmetic and geometric sequences, and not on the inverse function relationship of logarithms to exponential functions (described in Section 7.2). Napier's tables, published in 1614, listed what would now be called *natural logarithms* of sines and were rather difficult to use. A London professor, Henry Briggs, became interested in the tables and visited Napier. In their conversations, they developed the idea of common logarithms, which were published in 1617. The importance of this tool for calculation was immediately recognized, and by 1650 common logarithms were being printed as far away as China. They remained an important calculation tool until the advent of the inexpensive handheld calculator about 1972, which has decreased their calculational—but not their theoretical—importance.

A side effect of the invention of logarithms was the popularization of the decimal system of notation for real numbers.

7.3 Assess Your Understanding

Concepts and Vocabulary

1. $\log_a 1 =$ ____
2. $a^{\log_a M} =$ ____
3. $\log_a a^r =$ ____
4. $\log_a (MN) =$ ________ + ________
5. $\log_a\left(\frac{M}{N}\right) =$ ________ − ________
6. $\log_a M^r =$ ________
7. If $\log_8 M = \frac{\log_5 7}{\log_5 8}$, then $M =$ ____.
8. ***True or False*** $\ln(x+3) - \ln(2x) = \frac{\ln(x+3)}{\ln(2x)}$
9. ***True or False*** $\log_2(3x^4) = 4\log_2(3x)$
10. ***True or False*** $\log\left(\frac{2}{3}\right) = \frac{\log 2}{\log 3}$
11. Choose the expression equivalent to 2^x.
    (a) e^{2x} (b) $e^{x\ln 2}$ (c) $e^{\log_2 x}$ (d) $e^{2\ln x}$
12. Writing $\log_a x - \log_a y + 2\log_a z$ as a single logarithm results in which of the following?
    (a) $\log_a(x - y + 2z)$ (b) $\log_a\left(\frac{xz^2}{y}\right)$
    (c) $\log_a\left(\frac{2xz}{y}\right)$ (d) $\log_a\left(\frac{x}{yz^2}\right)$

Skill Building

In Problems 13–28, use properties of logarithms to find the exact value of each expression. Do not use a calculator.

13. $\log_3 3^{71}$
14. $\log_2 2^{-13}$
15. $\ln e^{-4}$
16. $\ln e^{\sqrt{2}}$
17. $2^{\log_2 7}$
18. $e^{\ln 8}$
19. $\log_8 2 + \log_8 4$
20. $\log_6 9 + \log_6 4$
21. $\log_6 18 - \log_6 3$
22. $\log_8 16 - \log_8 2$
23. $\log_2 6 \cdot \log_6 8$
24. $\log_3 8 \cdot \log_8 9$
25. $3^{\log_3 5 - \log_3 4}$
26. $5^{\log_5 6 + \log_5 7}$
27. $e^{\log_{e^2} 16}$
28. $e^{\log_{e^2} 9}$

In Problems 29–36, suppose that $\ln 2 = a$ *and* $\ln 3 = b$. *Use properties of logarithms to write each logarithm in terms of a and b.*

29. $\ln 6$
30. $\ln \frac{2}{3}$
31. $\ln 1.5$
32. $\ln 0.5$
33. $\ln 8$
34. $\ln 27$
35. $\ln \sqrt[5]{6}$
36. $\ln \sqrt[4]{\frac{2}{3}}$

In Problems 37–56, write each expression as a sum and/or difference of logarithms. Express powers as factors.

37. $\log_5(25x)$
38. $\log_3 \frac{x}{9}$
39. $\log_2 z^3$
40. $\log_7 x^5$
41. $\ln(ex)$
42. $\ln \frac{e}{x}$
43. $\ln \frac{x}{e^x}$
44. $\ln(xe^x)$
45. $\log_a(u^2v^3) \quad u > 0, v > 0$
46. $\log_2\left(\frac{a}{b^2}\right) \quad a > 0, b > 0$
47. $\ln(x^2\sqrt{1-x}) \quad 0 < x < 1$
48. $\ln(x\sqrt{1+x^2}) \quad x > 0$
49. $\log_2\left(\frac{x^3}{x-3}\right) \quad x > 3$
50. $\log_5\left(\frac{\sqrt[3]{x^2+1}}{x^2-1}\right) \quad x > 1$
51. $\log\left[\frac{x(x+2)}{(x+3)^2}\right] \quad x > 0$
52. $\log\left[\frac{x^3\sqrt{x+1}}{(x-2)^2}\right] \quad x > 2$
53. $\ln\left[\frac{x^2-x-2}{(x+4)^2}\right]^{1/3} \quad x > 2$
54. $\ln\left[\frac{(x-4)^2}{x^2-1}\right]^{2/3} \quad x > 4$
55. $\ln\frac{5x\sqrt{1+3x}}{(x-4)^3} \quad x > 4$
56. $\ln\left[\frac{5x^2\sqrt[3]{1-x}}{4(x+1)^2}\right] \quad 0 < x < 1$

In Problems 57–70, write each expression as a single logarithm.

57. $3\log_5 u + 4\log_5 v$
58. $2\log_3 u - \log_3 v$
59. $\log_3\sqrt{x} - \log_3 x^3$
60. $\log_2\left(\frac{1}{x}\right) + \log_2\left(\frac{1}{x^2}\right)$
61. $\log_4(x^2-1) - 5\log_4(x+1)$
62. $\log(x^2+3x+2) - 2\log(x+1)$
63. $\ln\left(\frac{x}{x-1}\right) + \ln\left(\frac{x+1}{x}\right) - \ln(x^2-1)$
64. $\log\left(\frac{x^2+2x-3}{x^2-4}\right) - \log\left(\frac{x^2+7x+6}{x+2}\right)$
65. $8\log_2\sqrt{3x-2} - \log_2\left(\frac{4}{x}\right) + \log_2 4$
66. $21\log_3\sqrt[3]{x} + \log_3(9x^2) - \log_3 9$
67. $2\log_a(5x^3) - \frac{1}{2}\log_a(2x+3)$
68. $\frac{1}{3}\log(x^3+1) + \frac{1}{2}\log(x^2+1)$
69. $2\log_2(x+1) - \log_2(x+3) - \log_2(x-1)$
70. $3\log_5(3x+1) - 2\log_5(2x-1) - \log_5 x$

In Problems 71–78, use the Change-of-Base Formula and a calculator to evaluate each logarithm. Round your answer to three decimal places.

71. $\log_3 21$ **72.** $\log_5 18$ **73.** $\log_{1/3} 71$ **74.** $\log_{1/2} 15$

75. $\log_{\sqrt{2}} 7$ **76.** $\log_{\sqrt{5}} 8$ **77.** $\log_\pi e$ **78.** $\log_\pi \sqrt{2}$

In Problems 79–84, graph each function using a graphing utility and the Change-of-Base Formula.

79. $y = \log_4 x$ **80.** $y = \log_5 x$ **81.** $y = \log_2(x + 2)$

82. $y = \log_4(x - 3)$ **83.** $y = \log_{x-1}(x + 1)$ **84.** $y = \log_{x+2}(x - 2)$

Mixed Practice

85. If $f(x) = \ln x, g(x) = e^x$, and $h(x) = x^2$, find:
(a) $(f \circ g)(x)$. What is the domain of $f \circ g$?
(b) $(g \circ f)(x)$. What is the domain of $g \circ f$?
(c) $(f \circ g)(5)$
(d) $(f \circ h)(x)$. What is the domain of $f \circ h$?
(e) $(f \circ h)(e)$

86. If $f(x) = \log_2 x, g(x) = 2^x$, and $h(x) = 4x$, find:
(a) $(f \circ g)(x)$. What is the domain of $f \circ g$?
(b) $(g \circ f)(x)$. What is the domain of $g \circ f$?
(c) $(f \circ g)(3)$
(d) $(f \circ h)(x)$. What is the domain of $f \circ h$?
(e) $(f \circ h)(8)$

Applications and Extensions

In Problems 87–96, express y as a function of x. The constant C is a positive number.

87. $\ln y = \ln x + \ln C$

88. $\ln y = \ln(x + C)$

89. $\ln y = \ln x + \ln(x + 1) + \ln C$

90. $\ln y = 2 \ln x - \ln(x + 1) + \ln C$

91. $\ln y = 3x + \ln C$

92. $\ln y = -2x + \ln C$

93. $\ln(y - 3) = -4x + \ln C$

94. $\ln(y + 4) = 5x + \ln C$

95. $3 \ln y = \frac{1}{2}\ln(2x + 1) - \frac{1}{3}\ln(x + 4) + \ln C$

96. $2 \ln y = -\frac{1}{2}\ln x + \frac{1}{3}\ln(x^2 + 1) + \ln C$

97. Find the value of $\log_2 3 \cdot \log_3 4 \cdot \log_4 5 \cdot \log_5 6 \cdot \log_6 7 \cdot \log_7 8$.

98. Find the value of $\log_2 4 \cdot \log_4 6 \cdot \log_6 8$.

99. Find the value of $\log_2 3 \cdot \log_3 4 \cdot \cdots \cdot \log_n(n + 1) \cdot \log_{n+1} 2$.

100. Find the value of $\log_2 2 \cdot \log_2 4 \cdot \cdots \cdot \log_2 2^n$.

101. Show that $\log_a(x + \sqrt{x^2 - 1}) + \log_a(x - \sqrt{x^2 - 1}) = 0$.

102. Show that $\log_a(\sqrt{x} + \sqrt{x - 1}) + \log_a(\sqrt{x} - \sqrt{x - 1}) = 0$.

103. Show that $\ln(1 + e^{2x}) = 2x + \ln(1 + e^{-2x})$.

104. Difference Quotient If $f(x) = \log_a x$, show that $\dfrac{f(x + h) - f(x)}{h} = \log_a\left(1 + \dfrac{h}{x}\right)^{1/h}, \quad h \neq 0$.

105. If $f(x) = \log_a x$, show that $-f(x) = \log_{1/a} x$.

106. If $f(x) = \log_a x$, show that $f(AB) = f(A) + f(B)$.

107. If $f(x) = \log_a x$, show that $f\left(\dfrac{1}{x}\right) = -f(x)$.

108. If $f(x) = \log_a x$, show that $f(x^\alpha) = \alpha f(x)$.

109. Show that $\log_a\left(\dfrac{M}{N}\right) = \log_a M - \log_a N$, where a, M, and N are positive real numbers and $a \neq 1$.

110. Show that $\log_a\left(\dfrac{1}{N}\right) = -\log_a N$, where a and N are positive real numbers and $a \neq 1$.

Explaining Concepts: Discussion and Writing

111. Graph $Y_1 = \log(x^2)$ and $Y_2 = 2\log(x)$ using a graphing utility. Are they equivalent? What might account for any differences in the two functions?

112. Write an example that illustrates why $(\log_a x)^r \neq r\log_a x$.

113. Write an example that illustrates why $\log_2(x + y) \neq \log_2 x + \log_2 y$.

114. Does $3^{\log_3(-5)} = -5$? Why or why not?

Retain Your Knowledge

Problems 115–118 are based on material learned earlier in the course. The purpose of these problems is to keep the material fresh in your mind so that you are better prepared for the final exam.

115. Graph $f(x) = \sqrt{2 - x}$ using the techniques of shifting, compressing or stretching, and reflecting. State the domain and the range of f.

116. The point $(12, -5)$ is on the terminal side of an angle θ in standard position. Find the exact value of each of the six trigonometric functions of θ.

117. Find the exact value of $\cos^{-1}\left(\cos \dfrac{7\pi}{6}\right)$.

118. For the equation $x^2 + y^2 = 2y$, the variables x and y represent rectangular coordinates. Write the equation using polar coordinates (r, θ).

7.4 Logarithmic and Exponential Equations

PREPARING FOR THIS SECTION *Before getting started, review the following:*

- Solving Equations Using a Graphing Utility (Appendix B, Section B.4, pp. B6–B7)
- Solving Quadratic Equations (Appendix A, Section A.4, pp. A30–A34)

Now Work the **'Are You Prepared?'** problems on page 495.

OBJECTIVES
1. Solve Logarithmic Equations (p. 491)
2. Solve Exponential Equations (p. 493)
3. Solve Logarithmic and Exponential Equations Using a Graphing Utility (p. 494)

1 Solve Logarithmic Equations

In Section 7.2 we solved logarithmic equations by changing a logarithmic expression to an exponential expression. That is, we used the definition of a logarithm:

$$y = \log_a x \quad \text{is equivalent to} \quad x = a^y \qquad a > 0 \quad a \neq 1$$

For example, to solve the equation $\log_2(1 - 2x) = 3$, write the logarithmic equation as an equivalent exponential equation $1 - 2x = 2^3$ and solve for x.

$$\log_2(1 - 2x) = 3$$
$$1 - 2x = 2^3 \qquad \text{Change to exponential form.}$$
$$-2x = 7 \qquad \text{Simplify.}$$
$$x = -\frac{7}{2} \qquad \text{Solve.}$$

You should check this solution for yourself.

For most logarithmic equations, some manipulation of the equation (usually using properties of logarithms) is required to obtain a solution. Also, to avoid extraneous solutions with logarithmic equations, determine the domain of the variable first.

Let's begin with an example of a logarithmic equation that requires using the fact that a logarithmic function is a one-to-one function:

$$\text{If } \log_a M = \log_a N, \text{ then } M = N \qquad M, N, \text{ and } a \text{ are positive and } a \neq 1$$

EXAMPLE 1 **Solving a Logarithmic Equation**

Solve: $2 \log_5 x = \log_5 9$

Solution The domain of the variable in this equation is $x > 0$. Note that each logarithm has the same base, 5. Then find the exact solution as follows:

$$2 \log_5 x = \log_5 9$$
$$\log_5 x^2 = \log_5 9 \qquad r\log_a M = \log_a M^r$$
$$x^2 = 9 \qquad \text{If } \log_a M = \log_a N, \text{ then } M = N.$$
$$x = 3 \quad \text{or} \quad x = -3$$

Recall that the domain of the variable is $x > 0$. Therefore, -3 is extraneous and must be discarded.

✓**Check:** $2\log_5 3 \stackrel{?}{=} \log_5 9$

$$\log_5 3^2 \stackrel{?}{=} \log_5 9 \qquad r\log_a M = \log_a M^r$$

$$\log_5 9 = \log_5 9$$

The solution set is $\{3\}$.

Now Work PROBLEM 13

Often one or more properties of logarithms are needed to rewrite the equation as a single logarithm. In the next example, the log of a product property is used.

EXAMPLE 2

Solving a Logarithmic Equation

Solve: $\log_5(x+6) + \log_5(x+2) = 1$

Solution The domain of the variable requires that $x + 6 > 0$ and $x + 2 > 0$, so $x > -6$ and $x > -2$. This means any solution must satisfy $x > -2$. To obtain an exact solution, first express the left side as a single logarithm. Then change the equation to an equivalent exponential equation.

$$\log_5(x+6) + \log_5(x+2) = 1$$

$$\log_5[(x+6)(x+2)] = 1 \qquad \log_a M + \log_a N = \log_a(MN)$$

$$(x+6)(x+2) = 5^1 = 5 \qquad \text{Change to exponential form.}$$

$$x^2 + 8x + 12 = 5 \qquad \text{Multiply out.}$$

$$x^2 + 8x + 7 = 0 \qquad \text{Place the quadratic equation in standard form.}$$

$$(x+7)(x+1) = 0 \qquad \text{Factor.}$$

$$x = -7 \quad \text{or} \quad x = -1 \qquad \text{Zero-Product Property}$$

WARNING A negative solution is not automatically extraneous. You must determine whether the potential solution causes the argument of any logarithmic expression in the equation to be negative or 0. ■

Only $x = -1$ satisfies the restriction that $x > -2$, so $x = -7$ is extraneous. The solution set is $\{-1\}$, which you should check.

Now Work PROBLEM 21

EXAMPLE 3

Solving a Logarithmic Equation

Solve: $\ln x = \ln(x+6) - \ln(x-4)$

Solution The domain of the variable requires that $x > 0$, $x + 6 > 0$, and $x - 4 > 0$. As a result, the domain of the variable here is $x > 4$. Begin the solution using the log of a difference property.

$$\ln x = \ln(x+6) - \ln(x-4)$$

$$\ln x = \ln\left(\frac{x+6}{x-4}\right) \qquad \ln M - \ln N = \ln\left(\frac{M}{N}\right)$$

$$x = \frac{x+6}{x-4} \qquad \text{If } \ln M = \ln N, \text{ then } M = N.$$

$$x(x-4) = x + 6 \qquad \text{Multiply both sides by } x - 4.$$

$$x^2 - 4x = x + 6 \qquad \text{Multiply out.}$$

$$x^2 - 5x - 6 = 0 \qquad \text{Place the quadratic equation in standard form.}$$

$$(x-6)(x+1) = 0 \qquad \text{Factor.}$$

$$x = 6 \quad \text{or} \quad x = -1 \qquad \text{Zero-Product Property}$$

Because the domain of the variable is $x > 4$, discard -1 as extraneous. The solution set is $\{6\}$, which you should check.

WARNING In using properties of logarithms to solve logarithmic equations, avoid using the property $\log_a x^r = r\log_a x$, when r is even. The reason can be seen in this example:

Solve: $\log_3 x^2 = 4$

Solution: The domain of the variable x is all real numbers except 0.

(a)
$$\begin{aligned}\log_3 x^2 &= 4\\ x^2 &= 3^4 = 81\\ x &= -9 \text{ or } x = 9\end{aligned}$$

(b)
$$\begin{aligned}\log_3 x^2 &= 4\\ 2\log_3 x &= 4 \qquad x > 0\\ \log_3 x &= 2\\ x &= 9\end{aligned}$$

Both -9 and 9 are solutions of $\log_3 x^2 = 4$ (as you can verify). The solution in part (b) does not find the solution -9 because the domain of the variable was further restricted due to the application of the property $\log_a x^r = r\log_a x$. ■

Now Work PROBLEM 31

2 Solve Exponential Equations

In Sections 7.1 and 7.2, we solved exponential equations algebraically by expressing each side of the equation using the same base. That is, we used the one-to-one property of the exponential function:

$$\text{If } a^u = a^v, \text{ then } u = v \qquad a > 0 \quad a \neq 1$$

For example, to solve the exponential equation $4^{2x+1} = 16$, notice that $16 = 4^2$ and apply the property above to obtain the equation $2x + 1 = 2$, from which we find $x = \frac{1}{2}$.

Not all exponential equations can be readily expressed so that each side of the equation has the same base. For such equations, algebraic techniques often can be used to obtain exact solutions.

EXAMPLE 4 **Solving Exponential Equations**

Solve: (a) $2^x = 5$ (b) $8 \cdot 3^x = 5$

Solution (a) Because 5 cannot be written as an integer power of 2 ($2^2 = 4$ and $2^3 = 8$), write the exponential equation as the equivalent logarithmic equation.

$$\begin{aligned}2^x &= 5\\ x &= \log_2 5 = \frac{\ln 5}{\ln 2}\end{aligned}$$

↑ Change-of-Base Formula (10), Section 7.3

Alternatively, the equation $2^x = 5$ can be solved by taking the natural logarithm (or common logarithm) of each side.

$$\begin{aligned}2^x &= 5\\ \ln 2^x &= \ln 5 && \text{If } M = N, \text{ then } \ln M = \ln N.\\ x\ln 2 &= \ln 5 && \ln M^r = r\ln M\\ x &= \frac{\ln 5}{\ln 2} && \text{Exact solution}\\ &\approx 2.322 && \text{Approximate solution}\end{aligned}$$

The solution set is $\left\{\frac{\ln 5}{\ln 2}\right\}$.

(b)
$$\begin{aligned}8 \cdot 3^x &= 5\\ 3^x &= \frac{5}{8} && \text{Solve for } 3^x.\end{aligned}$$

$$x = \log_3\left(\frac{5}{8}\right) = \frac{\ln\left(\frac{5}{8}\right)}{\ln 3} \quad \text{Exact solution}$$

$$\approx -0.428 \quad \text{Approximate solution}$$

The solution set is $\left\{\dfrac{\ln\left(\frac{5}{8}\right)}{\ln 3}\right\}$.

Now Work PROBLEM 43

EXAMPLE 5 **Solving an Exponential Equation**

Solve: $5^{x-2} = 3^{3x+2}$

Solution Because the bases are different, first apply property (7), Section 7.3 (take the natural logarithm of each side), and then use a property of logarithms. The result is an equation in x that can be solved.

$$5^{x-2} = 3^{3x+2}$$

$$\ln 5^{x-2} = \ln 3^{3x+2} \quad \text{If } M = N, \ln M = \ln N.$$

$$(x-2)\ln 5 = (3x+2)\ln 3 \quad \ln M^r = r\ln M$$

$$(\ln 5)x - 2\ln 5 = (3\ln 3)x + 2\ln 3 \quad \text{Distribute.}$$

$$(\ln 5)x - (3\ln 3)x = 2\ln 3 + 2\ln 5 \quad \text{Place terms involving } x \text{ on the left.}$$

$$(\ln 5 - 3\ln 3)x = 2(\ln 3 + \ln 5) \quad \text{Factor.}$$

$$x = \frac{2(\ln 3 + \ln 5)}{\ln 5 - 3\ln 3} \quad \text{Exact solution}$$

$$\approx -3.212 \quad \text{Approximate solution}$$

The solution set is $\left\{\dfrac{2(\ln 3 + \ln 5)}{\ln 5 - 3\ln 3}\right\}$.

NOTE: Because of the properties of logarithms, exact solutions involving logarithms often can be expressed in multiple ways. For example, the solution to $5^{x-2} = 3^{3x+2}$ from Example 5 can be expressed equivalently as $\dfrac{2\ln 15}{\ln 5 - \ln 27}$ or as $\dfrac{\ln 225}{\ln(5/27)}$, among others. Do you see why?

Now Work PROBLEM 53

EXAMPLE 6 **Solving an Exponential Equation That Is Quadratic in Form**

Solve: $4^x - 2^x - 12 = 0$

Solution Note that $4^x = (2^2)^x = 2^{(2x)} = (2^x)^2$, so the equation is quadratic in form and can be written as

$$(2^x)^2 - 2^x - 12 = 0 \quad \text{Let } u = 2^x\text{; then } u^2 - u - 12 = 0.$$

Now factor as usual.

$$(2^x - 4)(2^x + 3) = 0 \quad (u-4)(u+3) = 0$$

$$2^x - 4 = 0 \quad \text{or} \quad 2^x + 3 = 0 \quad u - 4 = 0 \text{ or } u + 3 = 0$$

$$2^x = 4 \qquad 2^x = -3 \quad u = 2^x = 4 \qquad u = 2^x = -3$$

The equation on the left has the solution $x = 2$, since $2^x = 4 = 2^2$; the equation on the right has no solution, since $2^x > 0$ for all x. The only solution is 2. The solution set is $\{2\}$.

Now Work PROBLEM 61

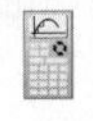

3 Solve Logarithmic and Exponential Equations Using a Graphing Utility

The algebraic techniques introduced in this section to obtain exact solutions apply only to certain types of logarithmic and exponential equations. Solutions for other types are generally studied in calculus, using numerical methods. For such types, we can use a graphing utility to approximate the solution.

EXAMPLE 7

Solving Equations Using a Graphing Utility

Solve: $x + e^x = 2$

Express the solution(s) rounded to two decimal places.

Solution The solution is found by graphing $Y_1 = x + e^x$ and $Y_2 = 2$. Since Y_1 is an increasing function (do you know why?), there is only one point of intersection for Y_1 and Y_2. Figure 23 shows the graphs of Y_1 and Y_2. Using the INTERSECT command reveals that the solution is 0.44, rounded to two decimal places. ●

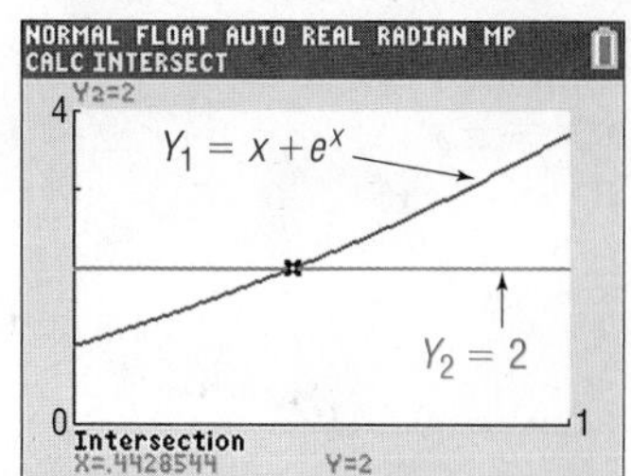

Figure 23

Now Work PROBLEM 71

7.4 Assess Your Understanding

'Are You Prepared?' *Answers are given at the end of these exercises. If you get a wrong answer, read the pages listed in red.*

1. Solve $x^2 - 7x - 30 = 0$. (pp. A30–A34)

2. Solve $(x + 3)^2 - 4(x + 3) + 3 = 0$. (pp. A30–A34)

 **3.** Approximate the solution(s) to $x^3 = x^2 - 5$ using a graphing utility. (pp. B6–B7)

 **4.** Approximate the solution(s) to $x^3 - 2x + 2 = 0$ using a graphing utility. (pp. B6–B7)

Skill Building

In Problems 5–40, solve each logarithmic equation. Express irrational solutions in exact form and as a decimal rounded to three decimal places.

5. $\log_4 x = 2$

6. $\log (x + 6) = 1$

7. $\log_2(5x) = 4$

8. $\log_3(3x - 1) = 2$

9. $\log_4(x + 2) = \log_4 8$

10. $\log_5(2x + 3) = \log_5 3$

11. $\frac{1}{2}\log_3 x = 2\log_3 2$

12. $-2\log_4 x = \log_4 9$

13. $3\log_2 x = -\log_2 27$

14. $2\log_5 x = 3\log_5 4$

15. $3\log_2(x - 1) + \log_2 4 = 5$

16. $2\log_3(x + 4) - \log_3 9 = 2$

17. $\log x + \log(x + 15) = 2$

18. $\log x + \log (x - 21) = 2$

19. $\log(2x + 1) = 1 + \log(x - 2)$

20. $\log(2x) - \log(x - 3) = 1$

21. $\log_2(x + 7) + \log_2(x + 8) = 1$

22. $\log_6(x + 4) + \log_6(x + 3) = 1$

23. $\log_8(x + 6) = 1 - \log_8(x + 4)$

24. $\log_5(x + 3) = 1 - \log_5(x - 1)$

25. $\ln x + \ln(x + 2) = 4$

26. $\ln(x + 1) - \ln x = 2$

27. $\log_3(x + 1) + \log_3(x + 4) = 2$

28. $\log_2(x + 1) + \log_2(x + 7) = 3$

29. $\log_{1/3}(x^2 + x) - \log_{1/3}(x^2 - x) = -1$

30. $\log_4(x^2 - 9) - \log_4(x + 3) = 3$

31. $\log_a(x - 1) - \log_a(x + 6) = \log_a(x - 2) - \log_a(x + 3)$

32. $\log_a x + \log_a(x - 2) = \log_a(x + 4)$

33. $2\log_5 (x - 3) - \log_5 8 = \log_5 2$

34. $\log_3 x - 2\log_3 5 = \log_3 (x + 1) - 2\log_3 10$

35. $2\log_6 (x + 2) = 3\log_6 2 + \log_6 4$

36. $3(\log_7 x - \log_7 2) = 2\log_7 4$

37. $2\log_{13} (x + 2) = \log_{13} (4x + 7)$

38. $\log (x - 1) = \frac{1}{3}\log 2$

39. $(\log_3 x)^2 - 5(\log_3 x) = 6$

40. $\ln x - 3\sqrt{\ln x} + 2 = 0$

In Problems 41–68, solve each exponential equation. Express irrational solutions in exact form and as a decimal rounded to three decimal places.

41. $2^{x-5} = 8$

42. $5^{-x} = 25$

43. $2^x = 10$

44. $3^x = 14$

45. $8^{-x} = 1.2$

46. $2^{-x} = 1.5$

47. $5(2^{3x}) = 8$

48. $0.3(4^{0.2x}) = 0.2$

49. $3^{1-2x} = 4^x$

50. $2^{x+1} = 5^{1-2x}$

51. $\left(\frac{3}{5}\right)^x = 7^{1-x}$

52. $\left(\frac{4}{3}\right)^{1-x} = 5^x$

53. $1.2^x = (0.5)^{-x}$

54. $0.3^{1+x} = 1.7^{2x-1}$

55. $\pi^{1-x} = e^x$

56. $e^{x+3} = \pi^x$

57. $2^{2x} + 2^x - 12 = 0$ **58.** $3^{2x} + 3^x - 2 = 0$ **59.** $3^{2x} + 3^{x+1} - 4 = 0$ **60.** $2^{2x} + 2^{x+2} - 12 = 0$

61. $16^x + 4^{x+1} - 3 = 0$ **62.** $9^x - 3^{x+1} + 1 = 0$ **63.** $25^x - 8 \cdot 5^x = -16$ **64.** $36^x - 6 \cdot 6^x = -9$

65. $3 \cdot 4^x + 4 \cdot 2^x + 8 = 0$ **66.** $2 \cdot 49^x + 11 \cdot 7^x + 5 = 0$ **67.** $4^x - 10 \cdot 4^{-x} = 3$ **68.** $3^x - 14 \cdot 3^{-x} = 5$

In Problems 69–82, use a graphing utility to solve each equation. Express your answer rounded to two decimal places.

69. $\log_5(x+1) - \log_4(x-2) = 1$ **70.** $\log_2(x-1) - \log_6(x+2) = 2$

71. $e^x = -x$ **72.** $e^{2x} = x + 2$ **73.** $e^x = x^2$ **74.** $e^x = x^3$

75. $\ln x = -x$ **76.** $\ln(2x) = -x + 2$ **77.** $\ln x = x^3 - 1$ **78.** $\ln x = -x^2$

79. $e^x + \ln x = 4$ **80.** $e^x - \ln x = 4$ **81.** $e^{-x} = \ln x$ **82.** $e^{-x} = -\ln x$

Mixed Practice

In Problems 83–94, solve each equation. Express irrational solutions in exact form and as a decimal rounded to three decimal places.

83. $\log_2(x+1) - \log_4 x = 1$
[**Hint:** Change $\log_4 x$ to base 2.]

84. $\log_2(3x+2) - \log_4 x = 3$

85. $\log_{16} x + \log_4 x + \log_2 x = 7$

86. $\log_9 x + 3\log_3 x = 14$

87. $(\sqrt[3]{2})^{2-x} = 2^{x^2}$

88. $\log_2 x^{\log_2 x} = 4$

89. $\dfrac{e^x + e^{-x}}{2} = 1$
[**Hint:** Multiply each side by e^x.]

90. $\dfrac{e^x + e^{-x}}{2} = 3$

91. $\dfrac{e^x - e^{-x}}{2} = 2$

92. $\dfrac{e^x - e^{-x}}{2} = -2$

93. $\log_5 x + \log_3 x = 1$
[**Hint:** Use the Change-of-Base Formula.]

94. $\log_2 x + \log_6 x = 3$

95. $f(x) = \log_2(x+3)$ and $g(x) = \log_2(3x+1)$.
(a) Solve $f(x) = 3$. What point is on the graph of f?
(b) Solve $g(x) = 4$. What point is on the graph of g?
(c) Solve $f(x) = g(x)$. Do the graphs of f and g intersect? If so, where?
(d) Solve $(f+g)(x) = 7$.
(e) Solve $(f-g)(x) = 2$.

96. $f(x) = \log_3(x+5)$ and $g(x) = \log_3(x-1)$.
(a) Solve $f(x) = 2$. What point is on the graph of f?
(b) Solve $g(x) = 3$. What point is on the graph of g?
(c) Solve $f(x) = g(x)$. Do the graphs of f and g intersect? If so, where?
(d) Solve $(f+g)(x) = 3$.
(e) Solve $(f-g)(x) = 2$.

97. (a) If $f(x) = 3^{x+1}$ and $g(x) = 2^{x+2}$, graph f and g on the same Cartesian plane.
(b) Find the point(s) of intersection of the graphs of f and g by solving $f(x) = g(x)$. Round answers to three decimal places. Label any intersection points on the graph drawn in part (a).
(c) Based on the graph, solve $f(x) > g(x)$.

98. (a) If $f(x) = 5^{x-1}$ and $g(x) = 2^{x+1}$, graph f and g on the same Cartesian plane.
(b) Find the point(s) of intersection of the graphs of f and g by solving $f(x) = g(x)$. Label any intersection points on the graph drawn in part (a).
(c) Based on the graph, solve $f(x) > g(x)$.

99. (a) Graph $f(x) = 3^x$ and $g(x) = 10$ on the same Cartesian plane.
(b) Shade the region bounded by the y-axis, $f(x) = 3^x$, and $g(x) = 10$ on the graph drawn in part (a).
(c) Solve $f(x) = g(x)$ and label the point of intersection on the graph drawn in part (a).

100. (a) Graph $f(x) = 2^x$ and $g(x) = 12$ on the same Cartesian plane.
(b) Shade the region bounded by the y-axis, $f(x) = 2^x$, and $g(x) = 12$ on the graph drawn in part (a).
(c) Solve $f(x) = g(x)$ and label the point of intersection on the graph drawn in part (a).

101. (a) Graph $f(x) = 2^{x+1}$ and $g(x) = 2^{-x+2}$ on the same Cartesian plane.
(b) Shade the region bounded by the y-axis, $f(x) = 2^{x+1}$, and $g(x) = 2^{-x+2}$ on the graph drawn in part (a).
(c) Solve $f(x) = g(x)$ and label the point of intersection on the graph drawn in part (a).

102. (a) Graph $f(x) = 3^{-x+1}$ and $g(x) = 3^{x-2}$ on the same Cartesian plane.
(b) Shade the region bounded by the y-axis, $f(x) = 3^{-x+1}$, and $g(x) = 3^{x-2}$ on the graph drawn in part (a).
(c) Solve $f(x) = g(x)$ and label the point of intersection on the graph drawn in part (a).

103. (a) Graph $f(x) = 2^x - 4$.
(b) Find the zero of f.
(c) Based on the graph, solve $f(x) < 0$.

104. (a) Graph $g(x) = 3^x - 9$.
(b) Find the zero of g.
(c) Based on the graph, solve $g(x) > 0$.

Applications and Extensions

105. A Population Model The resident population of the United States in 2014 was 317 million people and was growing at a rate of 0.7% per year. Assuming that this growth rate continues, the model $P(t) = 317(1.007)^{t-2014}$ represents the population P (in millions of people) in year t.

(a) According to this model, when will the population of the United States be 400 million people?

(b) According to this model, when will the population of the United States be 435 million people?

Source: U.S. Census Bureau

106. A Population Model The population of the world in 2014 was 7.14 billion people and was growing at a rate of 1.1% per year. Assuming that this growth rate continues, the model $P(t) = 7.14(1.011)^{t-2014}$ represents the population P (in billions of people) in year t.

(a) According to this model, when will the population of the world be 9 billion people?

(b) According to this model, when will the population of the world be 12.5 billion people?

Source: U.S. Census Bureau

107. Depreciation The value V of a Chevy Cruze that is t years old can be modeled by $V(t) = 18{,}700(0.84)^t$.

(a) According to the model, when will the car be worth \$9000?

(b) According to the model, when will the car be worth \$6000?

(c) According to the model, when will the car be worth \$2000?

Source: Kelley Blue Book

108. Depreciation The value V of a Honda Civic LX that is t years old can be modeled by $V(t) = 18{,}955(0.905)^t$.

(a) According to the model, when will the car be worth \$16,000?

(b) According to the model, when will the car be worth \$10,000?

(c) According to the model, when will the car be worth \$7500?

Source: Kelley Blue Book

Explaining Concepts: Discussion and Writing

109. Fill in the reason for each step in the following two solutions.

Solve: $\log_3(x-1)^2 = 2$

Solution A	**Solution B**
$\log_3(x-1)^2 = 2$	$\log_3(x-1)^2 = 2$
$(x-1)^2 = 3^2 = 9$ ______	$2\log_3(x-1) = 2$ ______
$(x-1) = \pm 3$ ______	$\log_3(x-1) = 1$ ______
$x-1 = -3$ or $x-1 = 3$ ______	$x-1 = 3^1 = 3$ ______
$x = -2$ or $x = 4$ ______	$x = 4$ ______

Both solutions given in Solution A check. Explain what caused the solution $x = -2$ to be lost in Solution B.

Retain Your Knowledge

Problems 110–113 are based on material learned earlier in the course. The purpose of these problems is to keep the material fresh in your mind so that you are better prepared for the final exam.

110. Determine whether the function $\{(0,-4), (2,-2), (4,0), (6,2)\}$ is one-to-one.

111. Each wheel of a truck has a radius of 17 inches. If the truck is traveling at 65 miles per hour, through how many revolutions per minute are the wheels traveling?

112. Write $-4 + 4i$ in polar form. Express the argument in degrees.

113. For the equation $2r\cos\theta - 5r\sin\theta = 10$, the variables r and θ represent polar coordinates. Write the equation using rectangular coordinates (x, y).

'Are You Prepared?' Answers

1. $\{-3, 10\}$ **2.** $\{-2, 0\}$ **3.** $\{-1.43\}$ **4.** $\{-1.77\}$

7.5 Financial Models

OBJECTIVES
1 Determine the Future Value of a Lump Sum of Money (p. 498)
2 Calculate Effective Rates of Return (p. 501)
3 Determine the Present Value of a Lump Sum of Money (p. 502)
4 Determine the Rate of Interest or the Time Required to Double a Lump Sum of Money (p. 503)

1 Determine the Future Value of a Lump Sum of Money

Interest is money paid for the use of money. The total amount borrowed (whether by an individual from a bank in the form of a loan or by a bank from an individual in the form of a savings account) is called the **principal**. The **rate of interest**, expressed as a percent, is the amount charged for the use of the principal for a given period of time, usually on a yearly (that is, per annum) basis.

THEOREM

Simple Interest Formula

If a principal of P dollars is borrowed for a period of t years at a per annum interest rate r, expressed as a decimal, the interest I charged is

$$I = Prt \qquad \textbf{(1)}$$

Interest charged according to formula (1) is called **simple interest**.

In problems involving interest, the term **payment period** is defined as follows.

Annually:	Once per year	**Monthly:**	12 times per year
Semiannually:	Twice per year	**Daily:**	365 times per year*
Quarterly:	Four times per year		

When the interest due at the end of a payment period is added to the principal so that the interest computed at the end of the next payment period is based on this new principal amount (old principal + interest), the interest is said to have been **compounded**. **Compound interest** is interest paid on the principal and on previously earned interest.

EXAMPLE 1

Computing Compound Interest

A credit union pays interest of 2% per annum compounded quarterly on a certain savings plan. If \$1000 is deposited in such a plan and the interest is left to accumulate, how much is in the account after 1 year?

Solution Use the simple interest formula, $I = Prt$. The principal P is \$1000 and the rate of interest is $2\% = 0.02$. After the first quarter of a year, the time t is $\frac{1}{4}$ year, so the interest earned is

$$I = Prt = (\$1000)(0.02)\left(\frac{1}{4}\right) = \$5$$

* Most banks use a 360-day "year." Why do you think they do?

The new principal is $P + I = \$1000 + \$5 = \$1005$. At the end of the second quarter, the interest on this principal is

$$I = (\$1005)(0.02)\left(\frac{1}{4}\right) = \$5.03$$

At the end of the third quarter, the interest on the new principal of $\$1005 + \$5.03 = \$1010.03$ is

$$I = (\$1010.03)(0.02)\left(\frac{1}{4}\right) = \$5.05$$

Finally, after the fourth quarter, the interest is

$$I = (\$1015.08)(0.02)\left(\frac{1}{4}\right) = \$5.08$$

After 1 year the account contains $\$1015.08 + \$5.08 = \$1020.16$. ●

The pattern of the calculations performed in Example 1 leads to a general formula for compound interest. For this purpose, let P represent the principal to be invested at a per annum interest rate r that is compounded n times per year, so the time of each compounding period is $\frac{1}{n}$ years. (For computing purposes, r is expressed as a decimal.) The interest earned after each compounding period is given by formula (1).

$$\text{Interest} = \text{principal} \times \text{rate} \times \text{time} = P \cdot r \cdot \frac{1}{n} = P \cdot \left(\frac{r}{n}\right)$$

The amount A after one compounding period is

$$A = P + P \cdot \left(\frac{r}{n}\right) = P \cdot \left(1 + \frac{r}{n}\right)$$

After two compounding periods, the amount A, based on the new principal $P \cdot \left(1 + \frac{r}{n}\right)$, is

$$A = \underbrace{P \cdot \left(1 + \frac{r}{n}\right)}_{\text{New principal}} + \underbrace{P \cdot \left(1 + \frac{r}{n}\right)\left(\frac{r}{n}\right)}_{\text{Interest on new principal}} \underset{\substack{\uparrow \\ \text{Factor out } P \cdot \left(1 + \frac{r}{n}\right)}}{=} P \cdot \left(1 + \frac{r}{n}\right) \cdot \left(1 + \frac{r}{n}\right) = P \cdot \left(1 + \frac{r}{n}\right)^2$$

After three compounding periods, the amount A is

$$A = P \cdot \left(1 + \frac{r}{n}\right)^2 + P \cdot \left(1 + \frac{r}{n}\right)^2 \left(\frac{r}{n}\right) = P \cdot \left(1 + \frac{r}{n}\right)^2 \cdot \left(1 + \frac{r}{n}\right) = P \cdot \left(1 + \frac{r}{n}\right)^3$$

Continuing this way, after n compounding periods (1 year), the amount A is

$$A = P \cdot \left(1 + \frac{r}{n}\right)^n$$

Because t years will contain $n \cdot t$ compounding periods, the amount after t years is

$$A = P \cdot \left(1 + \frac{r}{n}\right)^{nt}$$

THEOREM

Compound Interest Formula

The amount A after t years due to a principal P invested at an annual interest rate r, expressed as a decimal, compounded n times per year is

$$A = P \cdot \left(1 + \frac{r}{n}\right)^{nt} \tag{2}$$

Exploration

To observe the effects of compounding interest monthly on an initial deposit of $1, graph $Y_1 = \left(1 + \frac{r}{12}\right)^{12x}$ with $r = 0.06$ and $r = 0.12$ for $0 \le x \le 30$. What is the future value of $1 in 30 years when the interest rate per annum is $r = 0.06$ (6%)? What is the future value of $1 in 30 years when the interest rate per annum is $r = 0.12$ (12%)? Does doubling the interest rate double the future value?

For example, to rework Example 1, use $P = \$1000$, $r = 0.02$, $n = 4$ (quarterly compounding), and $t = 1$ year to obtain

$$A = P\cdot\left(1 + \frac{r}{n}\right)^{nt} = 1000\left(1 + \frac{0.02}{4}\right)^{4\cdot 1} = \$1020.16$$

In equation (2), the amount A is typically referred to as the **future value** of the account, and P is called the **present value.**

Now Work PROBLEM 7

EXAMPLE 2

Comparing Investments Using Different Compounding Periods

Investing $1000 at an annual rate of 10% compounded annually, semiannually, quarterly, monthly, and daily will yield the following amounts after 1 year:

Annual compounding ($n = 1$): $A = P\cdot(1 + r)$

$$= (\$1000)(1 + 0.10) = \$1100.00$$

Semiannual compounding ($n = 2$): $A = P\cdot\left(1 + \frac{r}{2}\right)^2$

$$= (\$1000)(1 + 0.05)^2 = \$1102.50$$

Quarterly compounding ($n = 4$): $A = P\cdot\left(1 + \frac{r}{4}\right)^4$

$$= (\$1000)(1 + 0.025)^4 = \$1103.81$$

Monthly compounding ($n = 12$): $A = P\cdot\left(1 + \frac{r}{12}\right)^{12}$

$$= (\$1000)\left(1 + \frac{0.10}{12}\right)^{12} = \$1104.71$$

Daily compounding ($n = 365$): $A = P\cdot\left(1 + \frac{r}{365}\right)^{365}$

$$= (\$1000)\left(1 + \frac{0.10}{365}\right)^{365} = \$1105.16$$

From Example 2, note that the effect of compounding more frequently is that the amount after 1 year is higher: $1000 compounded 4 times a year at 10% results in $1103.81, $1000 compounded 12 times a year at 10% results in $1104.71, and $1000 compounded 365 times a year at 10% results in $1105.16. This leads to the following question: What would happen to the amount after 1 year if the number of times that the interest is compounded were increased without bound?

Let's find the answer. Suppose that P is the principal, r is the per annum interest rate, and n is the number of times that the interest is compounded each year. The amount A after 1 year is

$$A = P\cdot\left(1 + \frac{r}{n}\right)^n$$

Rewrite this expression as follows:

$$A = P\cdot\left(1 + \frac{r}{n}\right)^n = P\cdot\left(1 + \frac{1}{\frac{n}{r}}\right)^n = P\cdot\left[\left(1 + \frac{1}{\frac{n}{r}}\right)^{n/r}\right]^r = P\cdot\left[\left(1 + \frac{1}{h}\right)^h\right]^r \quad (3)$$

$$h = \frac{n}{r}$$

Now suppose that the number n of times that the interest is compounded per year gets larger and larger; that is, suppose that $n \to \infty$. Then $h = \frac{n}{r} \to \infty$, and the expression in brackets in equation (3) equals e. That is, $A \to Pe^r$.

Table 8 compares $\left(1 + \frac{r}{n}\right)^n$, for large values of n, to e^r for $r = 0.05$, $r = 0.10$, $r = 0.15$, and $r = 1$. The larger that n gets, the closer $\left(1 + \frac{r}{n}\right)^n$ gets to e^r. No matter how frequent the compounding, the amount after 1 year has the definite ceiling Pe^r.

Table 8

	$\left(1 + \frac{r}{n}\right)^n$			
	$n = 100$	$n = 1000$	$n = 10{,}000$	e^r
$r = 0.05$	1.0512580	1.0512698	1.051271	1.0512711
$r = 0.10$	1.1051157	1.1051654	1.1051704	1.1051709
$r = 0.15$	1.1617037	1.1618212	1.1618329	1.1618342
$r = 1$	2.7048138	2.7169239	2.7181459	2.7182818

When interest is compounded so that the amount after 1 year is Pe^r, the interest is said to be **compounded continuously**.

THEOREM

Continuous Compounding

The amount A after t years due to a principal P invested at an annual interest rate r compounded continuously is

$$A = Pe^{rt} \quad (4)$$

EXAMPLE 3

Using Continuous Compounding

The amount A that results from investing a principal P of \$1000 at an annual rate r of 10% compounded continuously for a time t of 1 year is

$$A = \$1000e^{0.10} = (\$1000)(1.10517) = \$1105.17$$

●

Now Work PROBLEM 13

2 Calculate Effective Rates of Return

Suppose that you have \$1000 and a bank offers to pay you 3% annual interest on a savings account with interest compounded monthly. What annual interest rate must be earned for you to have the same amount at the end of the year as if the interest had been compounded annually (once per year)? To answer this question, first determine the value of the \$1000 in the account that earns 3% compounded monthly.

$$A = \$1000\left(1 + \frac{0.03}{12}\right)^{12} \quad \text{Use } A = P\left(1 + \frac{r}{n}\right)^n \text{ with } P = \$1000, r = 0.03, n = 12.$$

$$= \$1030.42$$

So the interest earned is \$30.42. Using $I = Prt$ with $t = 1, I = \$30.42$, and $P = \$1000$, the annual simple interest rate is $0.03042 = 3.042\%$. This interest rate is known as the *effective rate of interest.*

The **effective rate of interest** is the annual simple interest rate that would yield the same amount as compounding n times per year, or continuously, after 1 year.

THEOREM **Effective Rate of Interest**

The effective rate of interest r_e of an investment earning an annual interest rate r is given by

$$\text{Compounding } n \text{ times per year: } r_e = \left(1 + \frac{r}{n}\right)^n - 1$$

$$\text{Continuous compounding: } r_e = e^r - 1$$

EXAMPLE 4 **Computing the Effective Rate of Interest—Which Is the Best Deal?**

Suppose you want to buy a 5-year certificate of deposit (CD). You visit three banks to determine their CD rates. American Express offers you 2.15% annual interest compounded monthly, and First Internet Bank offers you 2.20% compounded quarterly. Discover offers 2.12% compounded daily. Determine which bank is offering the best deal.

Solution The bank that offers the best deal is the one with the highest effective interest rate.

American Express	**First Internet Bank**	**Discover**
$r_e = \left(1 + \frac{0.0215}{12}\right)^{12} - 1$	$r_e = \left(1 + \frac{0.022}{4}\right)^{4} - 1$	$r_e = \left(1 + \frac{0.0212}{365}\right)^{365} - 1$
$\approx 1.02171 - 1$	$\approx 1.02218 - 1$	$\approx 1.02143 - 1$
$= 0.02171$	$= 0.02218$	$= 0.02143$
$= 2.171\%$	$= 2.218\%$	$= 2.143\%$

The effective rate of interest is highest for First Internet Bank, so First Internet Bank is offering the best deal. ●

Now Work PROBLEM 23

3 Determine the Present Value of a Lump Sum of Money

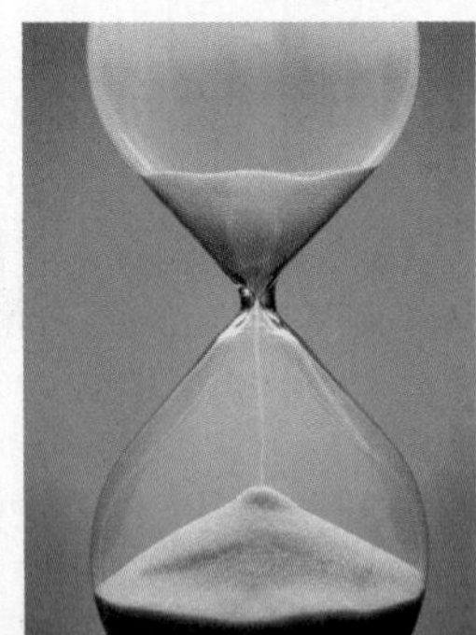

When people in finance speak of the "time value of money," they are usually referring to the *present value* of money. The **present value** of A dollars to be received at a future date is the principal that you would need to invest now so that it will grow to A dollars in the specified time period. The present value of money to be received at a future date is always less than the amount to be received, since the amount to be received will equal the present value (money invested now) *plus* the interest accrued over the time period.

The compound interest formula (2) is used to develop a formula for present value. If P is the present value of A dollars to be received after t years at a per annum interest rate r compounded n times per year, then, by formula (2),

$$A = P \cdot \left(1 + \frac{r}{n}\right)^{nt}$$

To solve for P, divide both sides by $\left(1 + \frac{r}{n}\right)^{nt}$. The result is

$$\frac{A}{\left(1 + \frac{r}{n}\right)^{nt}} = P \quad \text{or} \quad P = A \cdot \left(1 + \frac{r}{n}\right)^{-nt}$$

THEOREM

Present Value Formulas

The present value P of A dollars to be received after t years, assuming a per annum interest rate r compounded n times per year, is

$$P = A \cdot \left(1 + \frac{r}{n}\right)^{-nt} \qquad (5)$$

If the interest is compounded continuously, then

$$P = Ae^{-rt} \qquad (6)$$

To derive (6), solve formula (4) for P.

EXAMPLE 5

Computing the Value of a Zero-Coupon Bond

A zero-coupon (noninterest-bearing) bond can be redeemed in 10 years for \$1000. How much should you be willing to pay for it now if you want a return of

(a) 8% compounded monthly? (b) 7% compounded continuously?

Solution (a) To find the present value of \$1000, use formula (5) with $A = \$1000$, $n = 12$, $r = 0.08$, and $t = 10$.

$$P = A \cdot \left(1 + \frac{r}{n}\right)^{-nt} = \$1000\left(1 + \frac{0.08}{12}\right)^{-12(10)} = \$450.52$$

For a return of 8% compounded monthly, pay \$450.52 for the bond.

(b) Here use formula (6) with $A = \$1000$, $r = 0.07$, and $t = 10$.

$$P = Ae^{-rt} = \$1000e^{-(0.07)(10)} = \$496.59$$

For a return of 7% compounded continuously, pay \$496.59 for the bond. ●

Now Work PROBLEM 15

4 Determine the Rate of Interest or the Time Required to Double a Lump Sum of Money

EXAMPLE 6

Rate of Interest Required to Double an Investment

What annual rate of interest compounded annually is needed in order to double an investment in 5 years?

Solution If P is the principal and P is to double, then the amount A will be $2P$. Use the compound interest formula with $n = 1$ and $t = 5$ to find r.

$$A = P \cdot \left(1 + \frac{r}{n}\right)^{nt}$$

$$2P = P \cdot (1 + r)^5 \qquad A = 2P, n = 1, t = 5$$

$$2 = (1 + r)^5 \qquad \text{Divide both sides by } P.$$

$$1 + r = \sqrt[5]{2} \qquad \text{Take the fifth root of each side.}$$

$$r = \sqrt[5]{2} - 1 \approx 1.148698 - 1 = 0.148698$$

The annual rate of interest needed to double the principal in 5 years is 14.87%. ●

Now Work PROBLEM 31

EXAMPLE 7 **Time Required to Double or Triple an Investment**

(a) How long will it take for an investment to double in value if it earns 5% compounded continuously?

(b) How long will it take to triple at this rate?

Solution (a) If P is the initial investment and P is to double, then the amount A will be $2P$. Use formula (4) for continuously compounded interest with $r = 0.05$.

$$
\begin{aligned}
A &= Pe^{rt} \\
2P &= Pe^{0.05t} && A = 2P, r = 0.05 \\
2 &= e^{0.05t} && \text{Divide out the } P\text{'s.} \\
0.05t &= \ln 2 && \text{Rewrite as a logarithm.} \\
t &= \frac{\ln 2}{0.05} \approx 13.86 && \text{Solve for } t.
\end{aligned}
$$

It will take about 14 years to double the investment.

(b) To triple the investment, let $A = 3P$ in formula (4).

$$
\begin{aligned}
A &= Pe^{rt} \\
3P &= Pe^{0.05t} && A = 3P, r = 0.05 \\
3 &= e^{0.05t} && \text{Divide out the } P\text{'s.} \\
0.05t &= \ln 3 && \text{Rewrite as a logarithm.} \\
t &= \frac{\ln 3}{0.05} \approx 21.97 && \text{Solve for } t.
\end{aligned}
$$

It will take about 22 years to triple the investment. ●

Now Work PROBLEM 35

7.5 Assess Your Understanding

Concepts and Vocabulary

1. What is the interest due when \$500 is borrowed for 6 months at a simple interest rate of 6% per annum?

2. If you borrow \$5000 and, after 9 months, pay off the loan in the amount of \$5500, what per annum rate of interest was charged?

3. The total amount borrowed (whether by an individual from a bank in the form of a loan or by a bank from an individual in the form of a savings account) is called the ________.

4. If a principal of P dollars is borrowed for a period of t years at a per annum interest rate r, expressed as a decimal, the interest I charged is ____ = ______. Interest charged according to this formula is called __________.

5. In problems involving interest, if the payment period of the interest is quarterly, then interest is paid ____ times per year.

6. The ________ ____ __ _______ is the annual simple interest rate that would yield the same amount as compounding n times per year, or continuously, after 1 year.

Skill Building

In Problems 7–14, find the amount that results from each investment.

7. \$100 invested at 4% compounded quarterly after a period of 2 years

8. \$50 invested at 6% compounded monthly after a period of 3 years

9. \$500 invested at 8% compounded quarterly after a period of $2\frac{1}{2}$ years

10. \$300 invested at 12% compounded monthly after a period of $1\frac{1}{2}$ years

11. \$600 invested at 5% compounded daily after a period of 3 years

12. \$700 invested at 6% compounded daily after a period of 2 years

13. \$1000 invested at 11% compounded continuously after a period of 2 years

14. \$400 invested at 7% compounded continuously after a period of 3 years

In Problems 15–22, find the principal needed now to get each amount; that is, find the present value.

15. To get \$100 after 2 years at 6% compounded monthly

16. To get \$75 after 3 years at 8% compounded quarterly

17. To get \$1000 after $2\frac{1}{2}$ years at 6% compounded daily

18. To get \$800 after $3\frac{1}{2}$ years at 7% compounded monthly

19. To get \$600 after 2 years at 4% compounded quarterly

20. To get \$300 after 4 years at 3% compounded daily

21. To get \$80 after $3\frac{1}{4}$ years at 9% compounded continuously

22. To get \$800 after $2\frac{1}{2}$ years at 8% compounded continuously

In Problems 23–26, find the effective rate of interest.

23. For 5% compounded quarterly

24. For 6% compounded monthly

25. For 5% compounded continuously

26. For 6% compounded continuously

In Problems 27–30, determine the rate that represents the better deal.

27. 6% compounded quarterly or $6\frac{1}{4}$% compounded annually

28. 9% compounded quarterly or $9\frac{1}{4}$% compounded annually

29. 9% compounded monthly or 8.8% compounded daily

30. 8% compounded semiannually or 7.9% compounded daily

31. What rate of interest compounded annually is required to double an investment in 3 years?

32. What rate of interest compounded annually is required to double an investment in 6 years?

33. What rate of interest compounded annually is required to triple an investment in 5 years?

34. What rate of interest compounded annually is required to triple an investment in 10 years?

35. (a) How long does it take for an investment to double in value if it is invested at 8% compounded monthly?
(b) How long does it take if the interest is compounded continuously?

36. (a) How long does it take for an investment to triple in value if it is invested at 6% compounded monthly?
(b) How long does it take if the interest is compounded continuously?

37. What rate of interest compounded quarterly will yield an effective interest rate of 7%?

38. What rate of interest compounded continuously will yield an effective interest rate of 6%?

Applications and Extensions

39. **Time Required to Reach a Goal** If Tanisha has \$100 to invest at 4% per annum compounded monthly, how long will it be before she has \$150? If the compounding is continuous, how long will it be?

40. **Time Required to Reach a Goal** If Angela has \$100 to invest at 2.5% per annum compounded monthly, how long will it be before she has \$175? If the compounding is continuous, how long will it be?

41. **Time Required to Reach a Goal** How many years will it take for an initial investment of \$10,000 to grow to \$25,000? Assume a rate of interest of 6% compounded continuously.

42. **Time Required to Reach a Goal** How many years will it take for an initial investment of \$25,000 to grow to \$80,000? Assume a rate of interest of 7% compounded continuously.

43. **Price Appreciation of Homes** What will a \$90,000 condominium cost 5 years from now if the price appreciation for condos over that period averages 3% compounded annually?

44. **Credit Card Interest** A department store charges 1.25% per month on the unpaid balance for customers with charge accounts (interest is compounded monthly). A customer charges \$200 and does not pay her bill for 6 months. What is the bill at that time?

45. **Saving for a Car** Jerome will be buying a used car for \$15,000 in 3 years. How much money should he ask his parents for now so that, if he invests it at 5% compounded continuously, he will have enough to buy the car?

46. **Paying off a Loan** John requires \$3000 in 6 months to pay off a loan that has no prepayment privileges. If he has the \$3000 now, how much of it should he save in an account paying 3% compounded monthly so that in 6 months he will have exactly \$3000?

47. **Return on a Stock** George contemplates the purchase of 100 shares of a stock selling for \$15 per share. The stock pays no dividends. The history of the stock indicates that it should grow at an annual rate of 15% per year. How much should the 100 shares of stock be worth in 5 years?

48. **Return on an Investment** A business purchased for \$650,000 in 2010 is sold in 2013 for \$850,000. What is the annual rate of return for this investment?

49. **Comparing Savings Plans** Jim places \$1000 in a bank account that pays 5.6% compounded continuously. After 1 year, will he have enough money to buy a computer system that costs \$1060? If another bank will pay Jim 5.9% compounded monthly, is this a better deal?

50. **Savings Plans** On January 1, Kim places \$1000 in a certificate of deposit that pays 6.8% compounded continuously and matures in 3 months. Then Kim places the \$1000 and the interest in a passbook account that pays 5.25% compounded monthly. How much does Kim have in the passbook account on May 1?

51. **Comparing IRA Investments** Will invests \$2000 of the money in his IRA in a bond trust that pays 9% interest compounded semiannually. His friend Henry invests \$2000 in his IRA in a certificate of deposit that pays $8\frac{1}{2}\%$ compounded continuously. Who has more money after 20 years, Will or Henry?

52. **Comparing Two Alternatives** Suppose that April has access to an investment that will pay 10% interest compounded continuously. Which is better: to be given \$1000 now so that she can take advantage of this investment opportunity or to be given \$1325 after 3 years?

53. **College Costs** The average annual cost of college at 4-year private colleges was \$30,094 in the 2013–2014 academic year. This was a 3.8% increase from the previous year.
    (a) If the cost of college increases by 3.8% each year, what will be the average cost of college at a 4-year private college for the 2033–2034 academic year?
    (b) College savings plans, such as a 529 plan, allow individuals to put money aside now to help pay for college later. If one such plan offers a rate of 2% compounded continuously, how much should be put in a college savings plan in 2015 to pay for 1 year of the cost of college at a 4-year private college for an incoming freshman in 2033?

    Source: The College Board

54. **Analyzing Interest Rates on a Mortgage** Colleen and Bill have just purchased a house for \$650,000, with the seller holding a second mortgage of \$100,000. They promise to pay the seller \$100,000 plus all accrued interest 5 years from now. The seller offers them three interest options on the second mortgage:
    (a) Simple interest at 6% per annum
    (b) 5.5% interest compounded monthly
    (c) 5.25% interest compounded continuously

    Which option is best? That is, which results in paying the least interest on the loan?

55. **2009 Federal Stimulus Package** In February 2009, President Obama signed into law a \$787 billion federal stimulus package. At that time, 20-year Series EE bonds had a fixed rate of 1.3% compounded semiannually. If the federal government financed the stimulus through EE bonds, how much would it have to pay back in 2029? How much interest was paid to finance the stimulus?

    Source: U.S. Treasury Department

56. **Per Capita Federal Debt** In 2014, the federal debt was about \$17.5 trillion. In 2014, the U.S. population was about 317 million. Assuming that the federal debt is increasing about 6.4% per year and the U.S. population is increasing about 0.7% per year, determine the per capita debt (total debt divided by population) in 2030.

Problems 57–62 require the following discussion. ***Inflation*** *is a term used to describe the erosion of the purchasing power of money. For example, if the annual inflation rate is 3%, then \$1000 worth of purchasing power now will have only \$970 worth of purchasing power in 1 year because 3% of the original \$1000 ($0.03 \times 1000 = 30$) has been eroded due to inflation. In general, if the rate of inflation averages r% per annum over n years, the amount A that \$P will purchase after n years is*

$$A = P \cdot (1 - r)^n$$

where r is expressed as a decimal.

57. **Inflation** If the inflation rate averages 3%, what will be the purchasing power of \$1000 in 2 years?

58. **Inflation** If the inflation rate averages 2%, what will be the purchasing power of \$1000 in 3 years?

59. **Inflation** If the purchasing power of \$1000 is only \$950 after 2 years, what was the average inflation rate?

60. **Inflation** If the purchasing power of \$1000 is only \$930 after 2 years, what was the average inflation rate?

61. **Inflation** If the average inflation rate is 2%, how long is it until purchasing power is cut in half?

62. **Inflation** If the average inflation rate is 4%, how long is it until purchasing power is cut in half?

Problems 63–66 involve zero-coupon bonds. A ***zero-coupon bond*** *is a bond that is sold now at a discount and will pay its face value at the time when it matures; no interest payments are made.*

63. **Zero-Coupon Bonds** A zero-coupon bond can be redeemed in 20 years for \$10,000. How much should you be willing to pay for it now if you want a return of:
    (a) 5% compounded monthly?
    (b) 5% compounded continuously?

64. **Zero-Coupon Bonds** A child's grandparents are considering buying an \$80,000 face-value, zero-coupon bond at her birth so that she will have enough money for her college education 17 years later. If they want a rate of return of 6% compounded annually, what should they pay for the bond?

65. **Zero-Coupon Bonds** How much should a \$10,000 face-value, zero-coupon bond, maturing in 10 years, be sold for now if its rate of return is to be 4.5% compounded annually?

66. **Zero-Coupon Bonds** If Pat pays \$15,334.65 for a \$25,000 face-value, zero-coupon bond that matures in 8 years, what is his annual rate of return?

67. **Time to Double or Triple an Investment** The formula

$$t = \frac{\ln m}{n \ln\left(1 + \frac{r}{n}\right)}$$

can be used to find the number of years t required to multiply an investment m times when r is the per annum interest rate compounded n times a year.
    (a) How many years will it take to double the value of an IRA that compounds annually at the rate of 6%?
    (b) How many years will it take to triple the value of a savings account that compounds quarterly at an annual rate of 5%?
    (c) Give a derivation of this formula.

68. Time to Reach an Investment Goal The formula

$$t = \frac{\ln A - \ln P}{r}$$

can be used to find the number of years t required for an investment P to grow to a value A when compounded continuously at an annual rate r.

(a) How long will it take to increase an initial investment of \$1000 to \$4500 at an annual rate of 5.75%?
(b) What annual rate is required to increase the value of a \$2000 IRA to \$30,000 in 35 years?
(c) Give a derivation of this formula.

Problems 69–72 require the following discussion. The ***consumer price index (CPI)*** *indicates the relative change in price over time for a fixed basket of goods and services. It is a cost-of-living index that helps measure the effect of inflation on the cost of goods and services. The CPI uses the base period 1982–1984 for comparison (the CPI for this period is 100). The CPI for March 2014 was 236.29. This means that \$100 in the period 1982–1984 had the same purchasing power as \$236.29 in March 2014. In general, if the rate of inflation averages r% per annum over n years, then the CPI index after n years is*

$$\text{CPI} = \text{CPI}_0\left(1 + \frac{r}{100}\right)^n$$

where CPI_0 *is the CPI index at the beginning of the n-year period.*
Source: *U.S. Bureau of Labor Statistics*

69. Consumer Price Index
(a) The CPI was 215.3 for 2008 and 233.0 for 2013. Assuming that annual inflation remained constant for this time period, determine the average annual inflation rate.
(b) Using the inflation rate from part (a), in what year will the CPI reach 300?

70. Consumer Price Index If the current CPI is 234.2 and the average annual inflation rate is 2.8%, what will be the CPI in 5 years?

71. Consumer Price Index If the average annual inflation rate is 3.1%, how long will it take for the CPI index to double? (A doubling of the CPI index means purchasing power is cut in half.)

72. Consumer Price Index The base period for the CPI changed in 1998. Under the previous weight and item structure, the CPI for 1995 was 456.5. If the average annual inflation rate was 5.57%, what year was used as the base period for the CPI?

Explaining Concepts: Discussion and Writing

73. Explain in your own words what the term *compound interest* means. What does *continuous compounding* mean?

74. Explain in your own words the meaning of *present value*.

75. **Critical Thinking** You have just contracted to buy a house and will seek financing in the amount of \$100,000. You go to several banks. Bank 1 will lend you \$100,000 at the rate of 4.125% amortized over 30 years with a loan origination fee of 0.45%. Bank 2 will lend you \$100,000 at the rate of 3.375% amortized over 15 years with a loan origination fee of 0.95%. Bank 3 will lend you \$100,000 at the rate of 4.25% amortized over 30 years with no loan origination fee. Bank 4 will lend you \$100,000 at the rate of 3.625% amortized over 15 years with no loan origination fee. Which loan would you take? Why? Be sure to have sound reasons for your choice. Use the information in the table to assist you. If the amount of the monthly payment does not matter to you, which loan would you take? Again, have sound reasons for your choice. Compare your final decision with others in the class. Discuss.

	Monthly Payment	Loan Origination Fee
Bank 1	\$485	\$450
Bank 2	\$709	\$950
Bank 3	\$492	\$0
Bank 4	\$721	\$0

Retain Your Knowledge

Problems 76–79 are based on material learned earlier in the course. The purpose of these problems is to keep the material fresh in your mind so that you are better prepared for the final exam.

76. The function $f(x) = \dfrac{x}{x-2}$ is one-to-one. Find f^{-1}.

77. Given $\sin\theta = -\dfrac{3\sqrt{7}}{8}$ and $\cos\theta = -\dfrac{1}{8}$, find the exact values of the four remaining trigonometric functions of θ.

78. Solve triangle ABC: $A = 40°$, $B = 60°$, and $c = 12$

79. Solve: $\log_2(x+3) = 2\log_2(x-3)$

7.6 Exponential Growth and Decay Models; Newton's Law; Logistic Growth and Decay Models

OBJECTIVES 1 Find Equations of Populations That Obey the Law of Uninhibited Growth (p. 508)
2 Find Equations of Populations That Obey the Law of Decay (p. 510)
3 Use Newton's Law of Cooling (p. 511)
4 Use Logistic Models (p. 513)

1 Find Equations of Populations That Obey the Law of Uninhibited Growth

Many natural phenomena have been found to follow the law that an amount A varies with time t according to the function

$$A(t) = A_0 e^{kt} \qquad \textbf{(1)}$$

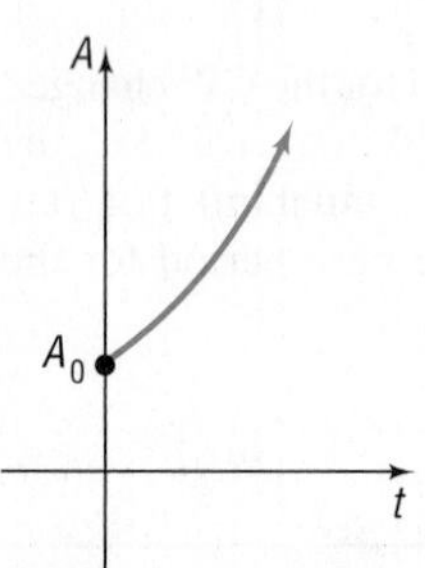

(a) $A(t) = A_0e^{kt}, k > 0$
Exponential growth

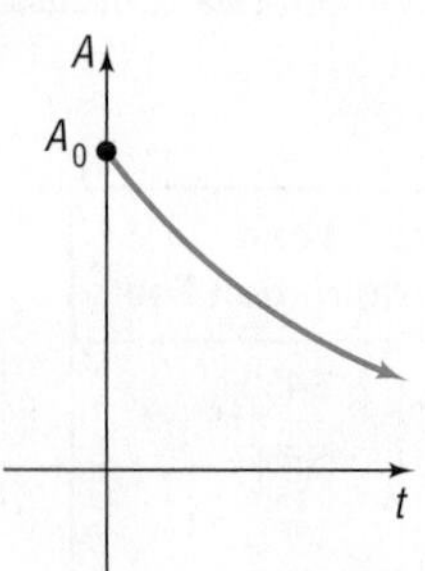

(b) $A(t) = A_0e^{kt}, k < 0$
Exponential decay

Figure 24

Here A_0 is the original amount $(t = 0)$ and $k \neq 0$ is a constant.

If $k > 0$, then equation (1) states that the amount A is increasing over time; if $k < 0$, the amount A is decreasing over time. In either case, when an amount A varies over time according to equation (1), it is said to follow the **exponential law**, or the **law of uninhibited growth** $(k > 0)$ **or decay** $(k < 0)$. See Figure 24.

For example, in Section 7.5, continuously compounded interest was shown to follow the law of uninhibited growth. In this section we shall look at some additional phenomena that follow the exponential law.

Cell division is the growth process of many living organisms, such as amoebas, plants, and human skin cells. Based on an ideal situation in which no cells die and no by-products are produced, the number of cells present at a given time follows the law of uninhibited growth. Actually, however, after enough time has passed, growth at an exponential rate will cease as a consequence of factors such as lack of living space and dwindling food supply. The law of uninhibited growth accurately models only the early stages of the cell division process.

The cell division process begins with a culture containing N_0 cells. Each cell in the culture grows for a certain period of time and then divides into two identical cells. Assume that the time needed for each cell to divide in two is constant and does not change as the number of cells increases. These new cells then grow, and eventually each divides in two, and so on.

Uninhibited Growth of Cells

A model that gives the number N of cells in a culture after a time t has passed (in the early stages of growth) is

$$N(t) = N_0 e^{kt} \qquad k > 0 \qquad \textbf{(2)}$$

where N_0 is the initial number of cells and k is a positive constant that represents the growth rate of the cells.

Using formula (2) to model the growth of cells employs a function that yields positive real numbers, even though the number of cells being counted must be an integer. This is a common practice in many applications.

EXAMPLE 1

Bacterial Growth

A colony of bacteria that grows according to the law of uninhibited growth is modeled by the function $N(t) = 100e^{0.045t}$, where N is measured in grams and t is measured in days.

(a) Determine the initial amount of bacteria.
(b) What is the growth rate of the bacteria?
(c) What is the population after 5 days?
(d) How long will it take for the population to reach 140 grams?
(e) What is the doubling time for the population?

Solution (a) The initial amount of bacteria, N_0, is obtained when $t = 0$, so

$$N_0 = N(0) = 100e^{0.045(0)} = 100 \text{ grams}$$

(b) Compare $N(t) = 100e^{0.045t}$ to $N(t) = N_0 e^{kt}$. The value of k, 0.045, indicates a growth rate of 4.5%.

(c) The population after 5 days is $N(5) = 100e^{0.045(5)} \approx 125.2$ grams.

(d) To find how long it takes for the population to reach 140 grams, solve the equation $N(t) = 140$.

$$100e^{0.045t} = 140$$

$$e^{0.045t} = 1.4 \qquad \text{Divide both sides of the equation by 100.}$$

$$0.045t = \ln 1.4 \qquad \text{Rewrite as a logarithm.}$$

$$t = \frac{\ln 1.4}{0.045} \qquad \text{Divide both sides of the equation by 0.045.}$$

$$\approx 7.5 \text{ days}$$

The population reaches 140 grams in about 7.5 days.

(e) The population doubles when $N(t) = 200$ grams, so the doubling time is found by solving the equation $200 = 100e^{0.045t}$ for t.

$$200 = 100e^{0.045t}$$

$$2 = e^{0.045t} \qquad \text{Divide both sides of the equation by 100.}$$

$$\ln 2 = 0.045t \qquad \text{Rewrite as a logarithm.}$$

$$t = \frac{\ln 2}{0.045} \qquad \text{Divide both sides of the equation by 0.045.}$$

$$\approx 15.4 \text{ days}$$

The population doubles approximately every 15.4 days. ●

Now Work PROBLEM 1

EXAMPLE 2

Bacterial Growth

A colony of bacteria increases according to the law of uninhibited growth.

(a) If N is the number of cells and t is the time in hours, express N as a function of t.
(b) If the number of bacteria doubles in 3 hours, find the function that gives the number of cells in the culture.
(c) How long will it take for the size of the colony to triple?
(d) How long will it take for the population to double a second time (that is, to increase four times)?

Solution (a) Using formula (2), the number N of cells at time t is

$$N(t) = N_0 e^{kt}$$

where N_0 is the initial number of bacteria present and k is a positive number.

(b) To find the growth rate k, note that the number of cells doubles in 3 hours, so

$$N(3) = 2N_0$$

But $N(3) = N_0 e^{k(3)}$, so

$$N_0 e^{k(3)} = 2N_0$$

$$e^{3k} = 2 \quad \text{Divide both sides by } N_0.$$

$$3k = \ln 2 \quad \text{Write the exponential equation as a logarithm.}$$

$$k = \frac{1}{3}\ln 2 \approx 0.23105$$

The function that models this growth process is therefore

$$N(t) = N_0\, e^{0.23105t}$$

(c) The time t needed for the size of the colony to triple requires that $N = 3N_0$. Substitute $3N_0$ for N to get

$$3N_0 = N_0\, e^{0.23105t}$$

$$3 = e^{0.23105t}$$

$$0.23105t = \ln 3$$

$$t = \frac{\ln 3}{0.23105} \approx 4.755 \text{ hours}$$

It will take about 4.755 hours, or 4 hours and 45 minutes, for the size of the colony to triple.

(d) If a population doubles in 3 hours, it will double a second time in 3 more hours, for a total time of 6 hours. ●

2 Find Equations of Populations That Obey the Law of Decay

Radioactive materials follow the law of uninhibited decay.

Uninhibited Radioactive Decay

The amount A of a radioactive material present at time t is given by

$$A(t) = A_0 e^{kt} \qquad k < 0 \qquad \textbf{(3)}$$

where A_0 is the original amount of radioactive material and k is a negative number that represents the rate of decay.

All radioactive substances have a specific **half-life,** which is the time required for half of the radioactive substance to decay. **Carbon dating** uses the fact that all living organisms contain two kinds of carbon, carbon-12 (a stable carbon) and carbon-14 (a radioactive carbon with a half-life of 5730 years). While an organism is living, the ratio of carbon-12 to carbon-14 is constant. But when an organism dies, the original amount of carbon-12 present remains unchanged, whereas the amount of carbon-14 begins to decrease. This change in the amount of carbon-14 present relative to the amount of carbon-12 present makes it possible to calculate when the organism died.

EXAMPLE 3

Estimating the Age of Ancient Tools

Traces of burned wood along with ancient stone tools in an archeological dig in Chile were found to contain approximately 1.67% of the original amount of carbon-14. If the half-life of carbon-14 is 5730 years, approximately when was the tree cut and burned?

Solution Using formula (3), the amount A of carbon-14 present at time t is

$$A(t) = A_0 e^{kt}$$

where A_0 is the original amount of carbon-14 present and k is a negative number. We first seek the number k. To find it, we use the fact that after 5730 years, half of the original amount of carbon-14 remains, so $A(5730) = \frac{1}{2}A_0$. Then

$$\frac{1}{2}A_0 = A_0 e^{k(5730)}$$

$$\frac{1}{2} = e^{5730k} \qquad \text{Divide both sides of the equation by } A_0.$$

$$5730k = \ln\frac{1}{2} \qquad \text{Rewrite as a logarithm.}$$

$$k = \frac{1}{5730}\ln\frac{1}{2} \approx -0.000120968$$

Formula (3) therefore becomes

$$A(t) = A_0\, e^{-0.000120968t}$$

If the amount A of carbon-14 now present is 1.67% of the original amount, it follows that

$$0.0167A_0 = A_0\, e^{-0.000120968t}$$

$$0.0167 = e^{-0.000120968t} \qquad \text{Divide both sides of the equation by } A_0.$$

$$-0.000120968t = \ln 0.0167 \qquad \text{Rewrite as a logarithm.}$$

$$t = \frac{\ln 0.0167}{-0.000120968} \approx 33{,}830 \text{ years}$$

The tree was cut and burned about 33,830 years ago. Some archeologists use this conclusion to argue that humans lived in the Americas nearly 34,000 years ago, much earlier than is generally accepted. ●

Now Work PROBLEM 3

3 Use Newton's Law of Cooling

Newton's Law of Cooling* states that the temperature of a heated object decreases exponentially over time toward the temperature of the surrounding medium.

Newton's Law of Cooling

The temperature u of a heated object at a given time t can be modeled by the following function:

$$u(t) = T + (u_0 - T)e^{kt} \qquad k < 0 \qquad \textbf{(4)}$$

where T is the constant temperature of the surrounding medium, u_0 is the initial temperature of the heated object, and k is a negative constant.

EXAMPLE 4 **Using Newton's Law of Cooling**

An object is heated to 100°C (degrees Celsius) and is then allowed to cool in a room whose air temperature is 30°C.

(a) If the temperature of the object is 80°C after 5 minutes, when will its temperature be 50°C?

(b) Determine the elapsed time before the temperature of the object is 35°C.

(c) What do you notice about the temperature as time passes?

*Named after Sir Isaac Newton (1643–1727), one of the cofounders of calculus.

Solution (a) Using formula (4) with $T = 30$ and $u_0 = 100$, the temperature $u(t)$ (in degrees Celsius) of the object at time t (in minutes) is

$$u(t) = 30 + (100 - 30)e^{kt} = 30 + 70e^{kt} \qquad (5)$$

where k is a negative constant. To find k, use the fact that $u = 80$ when $t = 5$. Then

$$\begin{aligned} u(t) &= 30 + 70e^{kt} \\ 80 &= 30 + 70e^{k(5)} && u(5) = 80 \\ 50 &= 70e^{5k} && \text{Simplify.} \\ e^{5k} &= \frac{50}{70} && \text{Solve for } e^{5k}. \\ 5k &= \ln\frac{5}{7} && \text{Rewrite as a logarithm.} \\ k &= \frac{1}{5}\ln\frac{5}{7} \approx -0.0673 && \text{Solve for } k. \end{aligned}$$

Formula (5) therefore becomes

$$u(t) = 30 + 70e^{-0.0673t} \qquad (6)$$

To find t when $u = 50°\text{C}$, solve the equation

$$\begin{aligned} 50 &= 30 + 70e^{-0.0673t} \\ 20 &= 70e^{-0.0673t} && \text{Simplify.} \\ e^{-0.0673t} &= \frac{20}{70} \\ -0.0673t &= \ln\frac{2}{7} && \text{Rewrite as a logarithm.} \\ t &= \frac{\ln\frac{2}{7}}{-0.0673} \approx 18.6 \text{ minutes} && \text{Solve for } t. \end{aligned}$$

The temperature of the object will be 50°C after about 18.6 minutes, or 18 minutes, 36 seconds.

(b) Use equation (6) to find t when $u = 35°\text{C}$.

$$\begin{aligned} 35 &= 30 + 70e^{-0.0673t} \\ 5 &= 70e^{-0.0673t} && \text{Simplify.} \\ e^{-0.0673t} &= \frac{5}{70} \\ -0.0673t &= \ln\frac{5}{70} && \text{Rewrite as a logarithm.} \\ t &= \frac{\ln\frac{5}{70}}{-0.0673} \approx 39.2 \text{ minutes} && \text{Solve for } t. \end{aligned}$$

The object will reach a temperature of 35°C after about 39.2 minutes.

(c) Look at equation (6). As t increases, the exponent $-0.0673t$ becomes unbounded in the negative direction. As a result, the value of $e^{-0.0673t}$ approaches zero, so the value of u, the temperature of the object, approaches 30°C, the air temperature of the room. ●

Now Work PROBLEM 13

4 Use Logistic Models

The exponential growth model $A(t) = A_0e^{kt}$, $k > 0$, assumes uninhibited growth, meaning that the value of the function grows without limit. Recall that cell division could be modeled using this function, assuming that no cells die and no by-products are produced. However, cell division eventually is limited by factors such as living space and food supply. The **logistic model**, given next, can describe situations where the growth or decay of the dependent variable is limited.

Logistic Model

In a logistic model, the population P after time t is given by the function

$$P(t) = \frac{c}{1 + ae^{-bt}} \qquad (7)$$

where a, b, and c are constants with $a > 0$ and $c > 0$. The model is a growth model if $b > 0$; the model is a decay model if $b < 0$.

The number c is called the **carrying capacity** (for growth models) because the value $P(t)$ approaches c as t approaches infinity; that is, $\lim_{t\to\infty} P(t) = c$. The number $|b|$ is the growth rate for $b > 0$ and the decay rate for $b < 0$. Figure 25(a) shows the graph of a typical logistic growth function, and Figure 25(b) shows the graph of a typical logistic decay function.

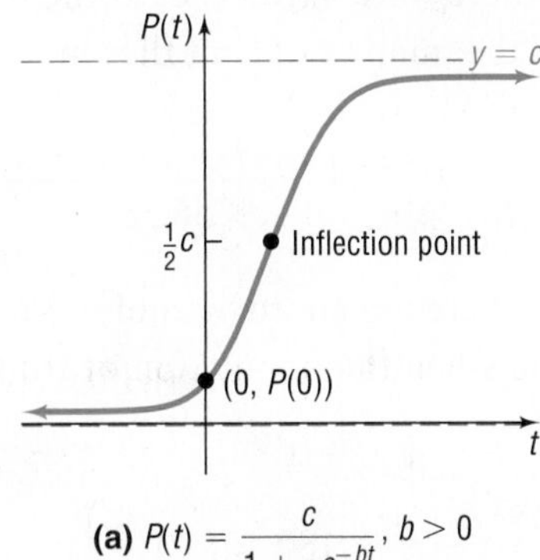

Figure 25

(a) $P(t) = \dfrac{c}{1 + ae^{-bt}}$, $b > 0$
Logistic growth

(b) $P(t) = \dfrac{c}{1 + ae^{-bt}}$, $b < 0$
Logistic decay

Based on the figures, the following properties of logistic functions emerge.

Properties of the Logistic Model, Equation (7)

1. The domain is the set of all real numbers. The range is the interval $(0, c)$, where c is the carrying capacity.
2. There are no x-intercepts; the y-intercept is $P(0)$.
3. There are two horizontal asymptotes: $y = 0$ and $y = c$.
4. $P(t)$ is an increasing function if $b > 0$ and a decreasing function if $b < 0$.
5. There is an **inflection point** where $P(t)$ equals $\frac{1}{2}$ of the carrying capacity. The inflection point is the point on the graph where the graph changes from being curved upward to being curved downward for growth functions, and the point where the graph changes from being curved downward to being curved upward for decay functions.
6. The graph is smooth and continuous, with no corners or gaps.

EXAMPLE 5 Fruit Fly Population

Fruit flies are placed in a half-pint milk bottle with a banana (for food) and yeast plants (for food and to provide a stimulus to lay eggs). Suppose that the fruit fly population after t days is given by

$$P(t) = \frac{230}{1 + 56.5e^{-0.37t}}$$

(a) State the carrying capacity and the growth rate.
(b) Determine the initial population.
(c) What is the population after 5 days?
(d) How long does it take for the population to reach 180?
(e) Use a graphing utility to determine how long it takes for the population to reach one-half of the carrying capacity.

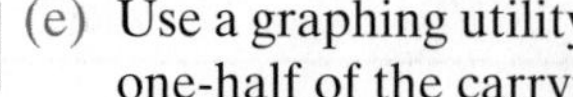

Solution

(a) As $t \to \infty$, $e^{-0.37t} \to 0$ and $P(t) \to \frac{230}{1}$. The carrying capacity of the half-pint bottle is 230 fruit flies. The growth rate is $|b| = |0.37| = 37\%$ per day.

(b) To find the initial number of fruit flies in the half-pint bottle, evaluate $P(0)$.

$$P(0) = \frac{230}{1 + 56.5e^{-0.37(0)}} = \frac{230}{1 + 56.5} = 4$$

So, initially, there were 4 fruit flies in the half-pint bottle.

(c) After 5 days the number of fruit flies in the half-pint bottle is

$$P(5) = \frac{230}{1 + 56.5e^{-0.37(5)}} \approx 23 \text{ fruit flies}$$

After 5 days, there are approximately 23 fruit flies in the bottle.

(d) To determine when the population of fruit flies will be 180, solve the equation $P(t) = 180$.

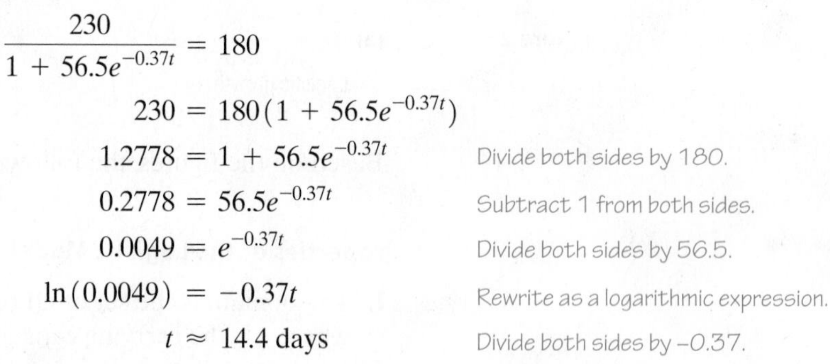

$$\frac{230}{1 + 56.5e^{-0.37t}} = 180$$

$$230 = 180(1 + 56.5e^{-0.37t})$$

$$1.2778 = 1 + 56.5e^{-0.37t} \quad \text{Divide both sides by 180.}$$

$$0.2778 = 56.5e^{-0.37t} \quad \text{Subtract 1 from both sides.}$$

$$0.0049 = e^{-0.37t} \quad \text{Divide both sides by 56.5.}$$

$$\ln(0.0049) = -0.37t \quad \text{Rewrite as a logarithmic expression.}$$

$$t \approx 14.4 \text{ days} \quad \text{Divide both sides by } -0.37.$$

It will take approximately 14.4 days (14 days, 10 hours) for the population to reach 180 fruit flies.

(e) One-half of the carrying capacity is 115 fruit flies. Solve $P(t) = 115$ by graphing $Y_1 = \frac{230}{1 + 56.5e^{-0.37t}}$ and $Y_2 = 115$ and using INTERSECT. See Figure 26. The population will reach one-half of the carrying capacity in about 10.9 days (10 days, 22 hours).

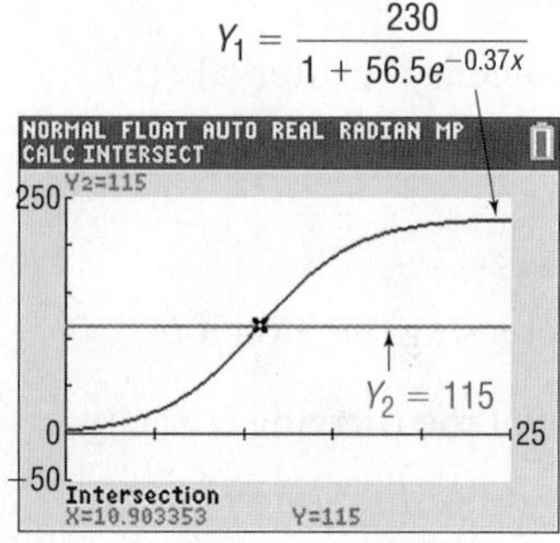

Figure 26

Look at Figure 26. Notice the point where the graph reaches 115 fruit flies (one-half of the carrying capacity): The graph changes from being curved upward to being curved downward. Using the language of calculus, we say the graph changes from increasing at an increasing rate to increasing at a decreasing rate. For any logistic growth function, when the population reaches one-half the carrying capacity, the population growth starts to slow down.

Now Work PROBLEM 23

Exploration

On the same viewing rectangle, graph

$$Y_1 = \frac{500}{1 + 24e^{-0.03t}} \quad \text{and} \quad Y_2 = \frac{500}{1 + 24e^{-0.08t}}$$

What effect does the growth rate $|b|$ have on the logistic growth function?

EXAMPLE 6 Wood Products

The EFISCEN wood product model classifies wood products according to their life-span. There are four classifications: short (1 year), medium short (4 years), medium long (16 years), and long (50 years). Based on data obtained from the European Forest Institute, the percentage of remaining wood products after t years for wood products with long life-spans (such as those used in the building industry) is given by

$$P(t) = \frac{100.3952}{1 + 0.0316e^{0.0581t}}$$

(a) What is the decay rate?
(b) What is the percentage of remaining wood products after 10 years?
(c) How long does it take for the percentage of remaining wood products to reach 50%?
(d) Explain why the numerator given in the model is reasonable.

Solution

(a) The decay rate is $|b| = |-0.0581| = 5.81\%$ per year.
(b) Evaluate $P(10)$.

$$P(10) = \frac{100.3952}{1 + 0.0316e^{0.0581(10)}} \approx 95.0$$

So 95% of long-life-span wood products remain after 10 years.

(c) Solve the equation $P(t) = 50$.

$$\frac{100.3952}{1 + 0.0316e^{0.0581t}} = 50$$

$$100.3952 = 50(1 + 0.0316e^{0.0581t})$$

$$2.0079 = 1 + 0.0316e^{0.0581t} \qquad \text{Divide both sides by 50.}$$

$$1.0079 = 0.0316e^{0.0581t} \qquad \text{Subtract 1 from both sides.}$$

$$31.8956 = e^{0.0581t} \qquad \text{Divide both sides by 0.0316.}$$

$$\ln(31.8956) = 0.0581t \qquad \text{Rewrite as a logarithmic expression.}$$

$$t \approx 59.6 \text{ years} \qquad \text{Divide both sides by 0.0581.}$$

It will take approximately 59.6 years for the percentage of long-life-span wood products remaining to reach 50%.

(d) The numerator of 100.3952 is reasonable because the maximum percentage of wood products remaining that is possible is 100%. ●

7.6 Assess Your Understanding

Applications and Extensions

1. **Growth of an Insect Population** The size P of a certain insect population at time t (in days) obeys the law of uninhibited growth $P(t) = 500e^{0.02t}$.
   (a) Determine the number of insects at $t = 0$ days.
   (b) What is the growth rate of the insect population?
   (c) What is the population after 10 days?
   (d) When will the insect population reach 800?
   (e) When will the insect population double?

2. **Growth of Bacteria** The number N of bacteria present in a culture at time t (in hours) obeys the law of uninhibited growth $N(t) = 1000e^{0.01t}$.
   (a) Determine the number of bacteria at $t = 0$ hours.
   (b) What is the growth rate of the bacteria?
   (c) What is the population after 4 hours?
   (d) When will the number of bacteria reach 1700?
   (e) When will the number of bacteria double?

3. **Radioactive Decay** Strontium-90 is a radioactive material that decays according to the function $A(t) = A_0e^{-0.0244t}$, where A_0 is the initial amount present and A is the amount present at time t (in years). Assume that a scientist has a sample of 500 grams of strontium-90.
   (a) What is the decay rate of strontium-90?
   (b) How much strontium-90 is left after 10 years?
   (c) When will 400 grams of strontium-90 be left?
   (d) What is the half-life of strontium-90?

4. **Radioactive Decay** Iodine-131 is a radioactive material that decays according to the function $A(t) = A_0e^{-0.087t}$, where A_0 is the initial amount present and A is the amount present at time t (in days). Assume that a scientist has a sample of 100 grams of iodine-131.
   (a) What is the decay rate of iodine-131?
   (b) How much iodine-131 is left after 9 days?
   (c) When will 70 grams of iodine-131 be left?
   (d) What is the half-life of iodine-131?

5. **Growth of a Colony of Mosquitoes** The population of a colony of mosquitoes obeys the law of uninhibited growth.
   (a) If N is the population of the colony and t is the time in days, express N as a function of t.
   (b) If there are 1000 mosquitoes initially and there are 1800 after 1 day, what is the size of the colony after 3 days?
   (c) How long is it until there are 10,000 mosquitoes?

6. **Bacterial Growth** A culture of bacteria obeys the law of uninhibited growth.
   (a) If N is the number of bacteria in the culture and t is the time in hours, express N as a function of t.
   (b) If 500 bacteria are present initially and there are 800 after 1 hour, how many will be present in the culture after 5 hours?
   (c) How long is it until there are 20,000 bacteria?

7. **Population Growth** The population of a southern city follows the exponential law.
   (a) If N is the population of the city and t is the time in years, express N as a function of t.
   (b) If the population doubled in size over an 18-month period and the current population is 10,000, what will the population be 2 years from now?

8. **Population Decline** The population of a midwestern city follows the exponential law.
   (a) If N is the population of the city and t is the time in years, express N as a function of t.
   (b) If the population decreased from 900,000 to 800,000 from 2008 to 2010, what will the population be in 2012?

9. **Radioactive Decay** The half-life of radium is 1690 years. If 10 grams is present now, how much will be present in 50 years?

10. **Radioactive Decay** The half-life of radioactive potassium is 1.3 billion years. If 10 grams is present now, how much will be present in 100 years? In 1000 years?

11. **Estimating the Age of a Tree** A piece of charcoal is found to contain 30% of the carbon-14 that it originally had. When did the tree die from which the charcoal came? Use 5730 years as the half-life of carbon-14.

12. **Estimating the Age of a Fossil** A fossilized leaf contains 70% of its normal amount of carbon-14. How old is the fossil?

13. **Cooling Time of a Pizza Pan** A pizza pan is removed at 5:00 PM from an oven whose temperature is fixed at 450°F into a room that is a constant 70°F. After 5 minutes, the temperature of the pan is 300°F.
   (a) At what time is the temperature of the pan 135°F?
   (b) Determine the time that needs to elapse before the temperature of the pan is 160°F.
   (c) What do you notice about the temperature as time passes?

14. **Newton's Law of Cooling** A thermometer reading 72°F is placed in a refrigerator where the temperature is a constant 38°F.
   (a) If the thermometer reads 60°F after 2 minutes, what will it read after 7 minutes?
   (b) How long will it take before the thermometer reads 39°F?
   (c) Determine the time that must elapse before the thermometer reads 45°F.
   (d) What do you notice about the temperature as time passes?

15. **Newton's Law of Heating** A thermometer reading 8°C is brought into a room with a constant temperature of 35°C. If the thermometer reads 15°C after 3 minutes, what will it read after being in the room for 5 minutes? For 10 minutes?
   [**Hint:** You need to construct a formula similar to equation (4).]

16. **Warming Time of a Beer Stein** A beer stein has a temperature of 28°F. It is placed in a room with a constant temperature of 70°F. After 10 minutes, the temperature of the stein has risen to 35°F. What will the temperature of the stein be after 30 minutes? How long will it take the stein to reach a temperature of 45°F? (See the hint given for Problem 15.)

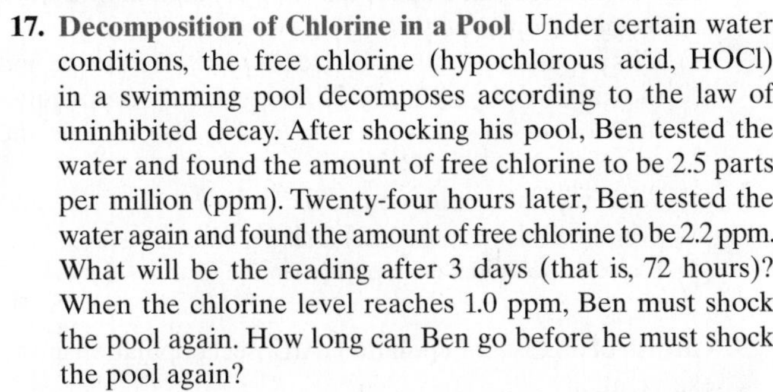

17. **Decomposition of Chlorine in a Pool** Under certain water conditions, the free chlorine (hypochlorous acid, HOCl) in a swimming pool decomposes according to the law of uninhibited decay. After shocking his pool, Ben tested the water and found the amount of free chlorine to be 2.5 parts per million (ppm). Twenty-four hours later, Ben tested the water again and found the amount of free chlorine to be 2.2 ppm. What will be the reading after 3 days (that is, 72 hours)? When the chlorine level reaches 1.0 ppm, Ben must shock the pool again. How long can Ben go before he must shock the pool again?

18. **Decomposition of Dinitrogen Pentoxide** At 45°C, dinitrogen pentoxide (N_2O_5) decomposes into nitrous dioxide (NO_2) and oxygen (O_2) according to the law of uninhibited decay. An initial amount of 0.25 M N_2O_5 (M is a measure of concentration known as molarity) decomposes to 0.15 M N_2O_5 in 17 minutes. What concentration of N_2O_5 will remain after 30 minutes? How long will it take until only 0.01 M N_2O_5 remains?

19. **Decomposition of Sucrose** Reacting with water in an acidic solution at 35°C, sucrose ($C_{12}H_{22}O_{11}$) decomposes into glucose ($C_6H_{12}O_6$) and fructose ($C_6H_{12}O_6$)* according to the law of uninhibited decay. An initial concentration of 0.40 M of sucrose decomposes to 0.36 M sucrose in 30 minutes. What concentration of sucrose will remain after 2 hours? How long will it take until only 0.10 M sucrose remains?

20. **Decomposition of Salt in Water** Salt (NaCl) decomposes in water into sodium (Na^+) and chloride (Cl^-) ions according to the law of uninhibited decay. If the initial amount of salt is 25 kilograms and, after 10 hours, 15 kilograms of salt is left, how much salt is left after 1 day? How long does it take until $\frac{1}{2}$ kilogram of salt is left?

21. **Radioactivity from Chernobyl** After the release of radioactive material into the atmosphere from a nuclear power plant at Chernobyl (Ukraine) in 1986, the hay in Austria was contaminated by iodine 131 (half-life 8 days). If it is safe to feed the hay to cows when 10% of the iodine 131 remains, how long did the farmers need to wait to use this hay?

22. **Word Users** According to a survey by Olsten Staffing Services, the percentage of companies reporting usage of Microsoft Word t years since 1984 is given by

$$P(t) = \frac{99.744}{1 + 3.014e^{-0.799t}}$$

(a) What is the growth rate in the percentage of Microsoft Word users?
(b) Use a graphing utility to graph $P = P(t)$.
(c) What was the percentage of Microsoft Word users in 1990?
(d) During what year did the percentage of Microsoft Word users reach 90%?
(e) Explain why the numerator given in the model is reasonable. What does it imply?

23. **Home Computers** The logistic model

$$P(t) = \frac{95.4993}{1 + 0.0405e^{0.1968t}}$$

represents the percentage of households that do not own a personal computer t years since 1984.
(a) Evaluate and interpret $P(0)$.
(b) Use a graphing utility to graph $P = P(t)$.
(c) What percentage of households did not own a personal computer in 1995?
(d) In what year did the percentage of households that do not own a personal computer reach 10%?

Source: U.S. Department of Commerce

24. **Farmers** The logistic model

$$W(t) = \frac{14{,}656{,}248}{1 + 0.059e^{0.057t}}$$

represents the number of farm workers in the United States t years after 1910.
(a) Evaluate and interpret $W(0)$.
(b) Use a graphing utility to graph $W = W(t)$.
(c) How many farm workers were there in the United States in 2010?
(d) When did the number of farm workers in the United States reach 10,000,000?
(e) According to this model, what happens to the number of farm workers in the United States as t approaches ∞? Based on this result, do you think that it is reasonable to use this model to predict the number of farm workers in the United States in 2060? Why?

Source: U.S. Department of Agriculture

25. **Birthdays** The logistic model

$$P(n) = \frac{113.3198}{1 + 0.115e^{0.0912n}}$$

models the probability that, in a room of n people, no two people share the same birthday.
(a) Use a graphing utility to graph $P = P(n)$.
(b) In a room of $n = 15$ people, what is the probability that no two share the same birthday?
(c) How many people must be in a room before the probability that no two people share the same birthday falls below 10%?
(d) What happens to the probability as n increases? Explain what this result means.

26. **Population of an Endangered Species** Environmentalists often capture an endangered species and transport the species to a controlled environment where the species can produce offspring and regenerate its population. Suppose that six American bald eagles are captured, transported to Montana, and set free. Based on experience, the environmentalists expect the population to grow according to the model

$$P(t) = \frac{500}{1 + 83.33e^{-0.162t}}$$

where t is measured in years. (*continued on the next page*)

*Author's Note: Surprisingly, the chemical formulas for glucose and fructose are the same: This is not a typo.

(a) Determine the carrying capacity of the environment.
(b) What is the growth rate of the bald eagle?
(c) What is the population after 3 years?
(d) When will the population be 300 eagles?
(e) How long does it take for the population to reach one-half of the carrying capacity?

27. The *Challenger* Disaster After the *Challenger* disaster in 1986, a study was made of the 23 launches that preceded the fatal flight. A mathematical model was developed involving the relationship between the Fahrenheit temperature x around the O-rings and the number y of eroded or leaky primary O-rings. The model stated that

$$y = \frac{6}{1 + e^{-(5.085 - 0.1156x)}}$$

where the number 6 indicates the 6 primary O-rings on the spacecraft.

(a) What is the predicted number of eroded or leaky primary O-rings at a temperature of 100°F?
(b) What is the predicted number of eroded or leaky primary O-rings at a temperature of 60°F?
(c) What is the predicted number of eroded or leaky primary O-rings at a temperature of 30°F?
(d) Graph the equation using a graphing utility. At what temperature is the predicted number of eroded or leaky O-rings 1? 3? 5?

Source: Linda Tappin, "Analyzing Data Relating to the Challenger *Disaster,"* Mathematics Teacher, *Vol. 87, No. 6, September 1994, pp. 423–426.*

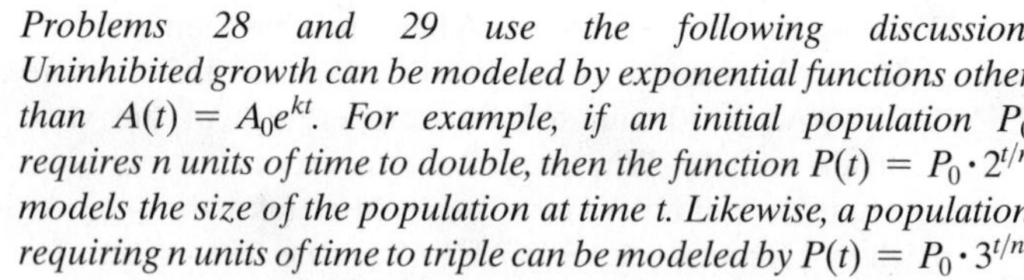

Problems 28 and 29 use the following discussion: Uninhibited growth can be modeled by exponential functions other than $A(t) = A_0e^{kt}$. For example, if an initial population P_0 requires n units of time to double, then the function $P(t) = P_0 \cdot 2^{t/n}$ models the size of the population at time t. Likewise, a population requiring n units of time to triple can be modeled by $P(t) = P_0 \cdot 3^{t/n}$.

28. Growth of a Human Population The population of a town is growing exponentially.

(a) If its population doubled in size over an 8-year period and the current population is 25,000, write an exponential function of the form $P(t) = P_0 \cdot 2^{t/n}$ that models the population.
(b) What will the population be in 3 years?
(c) When will the population reach 80,000?
(d) Express the model from part (a) in the form $A(t) = A_0e^{kt}$.

29. Growth of an Insect Population An insect population grows exponentially.

(a) If the population triples in 20 days, and 50 insects are present initially, write an exponential function of the form $P(t) = P_0 \cdot 3^{t/n}$ that models the population.
(b) What will the population be in 47 days?
(c) When will the population reach 700?
(d) Express the model from part (a) in the form $A(t) = A_0e^{kt}$.

Retain Your Knowledge

Problems 30–33 are based on material learned earlier in the course. The purpose of these problems is to keep the material fresh in your mind so that you are better prepared for the final exam.

30. Given $\sin\theta = \frac{3}{4}$ and $\cos\theta < 0$, find the exact value of each of the remaining five trigonometric functions of θ.

31. Solve triangle ABC: $a = 5, b = 4$, and $C = 110°$

32. Find the area of a triangle whose sides are 8, 11, and 13. Round the answer to two decimal places.

33. Write the logarithmic expression $\ln\left(\frac{x^2\sqrt{y}}{z}\right)$ as the sum and/or difference of logarithms. Express powers as factors.

7.7 Building Exponential, Logarithmic, and Logistic Models from Data

PREPARING FOR THIS SECTION *Before getting started, review the following:*

- Building Linear Models from Data (Appendix A, Section A.9, pp. A79–A83)

OBJECTIVES
1 Build an Exponential Model from Data (p. 519)
2 Build a Logarithmic Model from Data (p. 521)
3 Build a Logistic Model from Data (p. 521)

In Appendix A, Section A.9 we discussed how to find the linear function of best fit $(y = ax + b)$.

In this section we discuss how to use a graphing utility to find equations of best fit that describe the relation between two variables when the relation is thought to be exponential $(y = ab^x)$, logarithmic $(y = a + b \ln x)$, or logistic $\left(y = \dfrac{c}{1 + ae^{-bx}}\right)$. As before, we draw a scatter diagram of the data to help to determine the appropriate model to use.

Figure 27 shows scatter diagrams that will typically be observed for the three models. Below each scatter diagram are any restrictions on the values of the parameters.

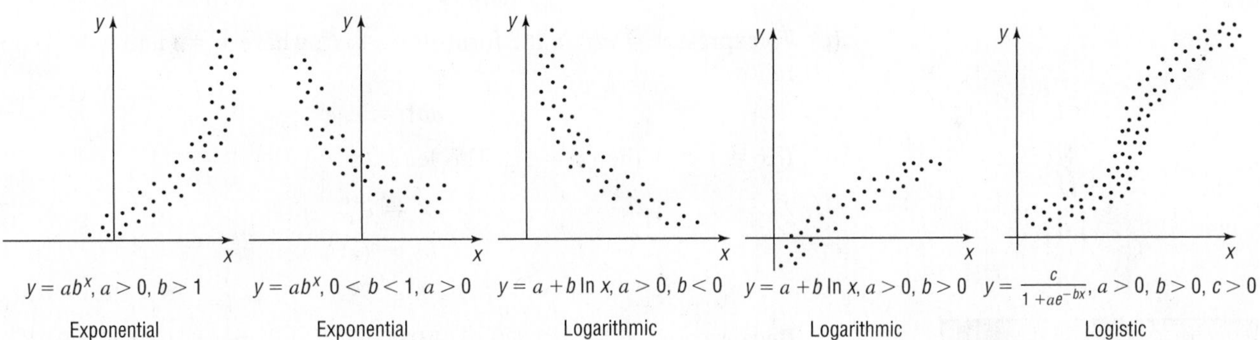

Figure 27

Most graphing utilities have REGression options that fit data to a specific type of curve. Once the data have been entered and a scatter diagram obtained, the type of curve that you want to fit to the data is selected. Then that REGression option is used to obtain the curve of *best fit* of the type selected.

The correlation coefficient r will appear only if the model can be written as a linear expression. As it turns out, r will appear for the linear, power, exponential, and logarithmic models, since these models can be written as a linear expression. Remember, the closer $|r|$ is to 1, the better the fit.

1 Build an Exponential Model from Data

We saw in Section 7.5 that the future value of money behaves exponentially, and we saw in Section 7.6 that growth and decay models also behave exponentially. The next example shows how data can lead to an exponential model.

EXAMPLE 1 Fitting an Exponential Function to Data

Table 9

Year, x	Account Value, y
0	20,000
1	21,516
2	23,355
3	24,885
4	27,484
5	30,053
6	32,622

Mariah deposited \$20,000 in a well-diversified mutual fund 6 years ago. The data in Table 9 represent the value of the account at the beginning of each year for the last 7 years.

(a) Using a graphing utility, draw a scatter diagram with year as the independent variable.

(b) Using a graphing utility, build an exponential model from the data.

(c) Express the function found in part (b) in the form $A = A_0e^{kt}$.

(d) Graph the exponential function found in part (b) or (c) on the scatter diagram.

(e) Using the solution to part (b) or (c), predict the value of the account after 10 years.

(f) Interpret the value of k found in part (c).

Solution

(a) Enter the data into the graphing utility and draw the scatter diagram as shown in Figure 28.

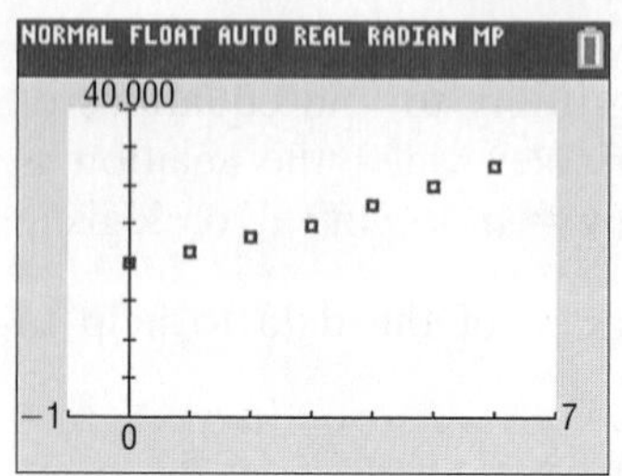

Figure 28

(b) A graphing utility fits the data in Table 9 to an exponential model of the form $y = ab^x$ using the EXPonential REGression option. Figure 29 shows that $y = ab^x = 19{,}820.43(1.085568)^x$. Notice that $|r| = 0.999$, which is close to 1, indicating a good fit.

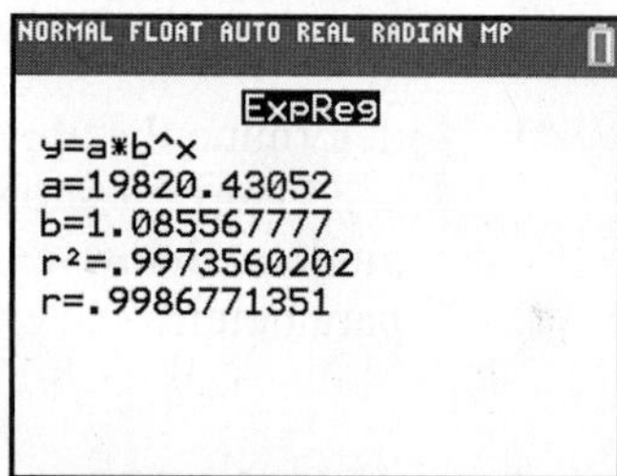

Figure 29

(c) To express $y = ab^x$ in the form $A = A_0e^{kt}$, where $x = t$ and $y = A$, proceed as follows:

$$ab^x = A_0e^{kt}$$

If $x = t = 0$, then $a = A_0$. This leads to

$$b^x = e^{kt}$$
$$b^x = (e^k)^t$$
$$b = e^k \qquad x = t$$

Because $y = ab^x = 19{,}820.43(1.085568)^x$, this means that $a = 19{,}820.43$ and $b = 1.085568$.

$$a = A_0 = 19{,}820.43 \quad \text{and} \quad b = e^k = 1.085568$$

To find k, rewrite $e^k = 1.085568$ as a logarithm to obtain

$$k = \ln(1.085568) \approx 0.08210$$

As a result, $A = A_0e^{kt} = 19{,}820.43e^{0.08210t}$.

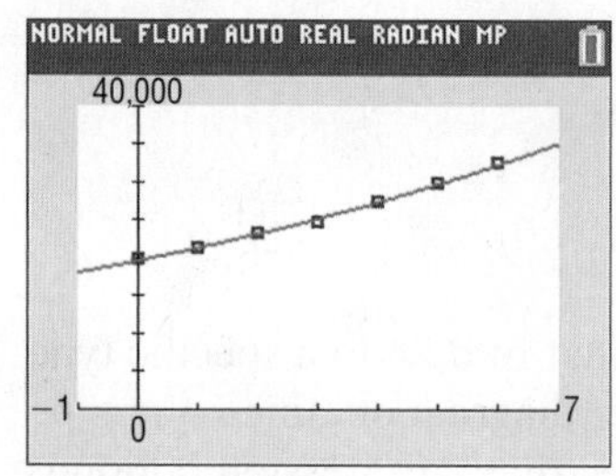

Figure 30

(d) See Figure 30 for the graph of the exponential function of best fit.

(e) Let $t = 10$ in the function found in part (c). The predicted value of the account after 10 years is

$$A = A_0e^{kt} = 19{,}820.43e^{0.08210(10)} \approx \$45{,}047$$

(f) The value of $k = 0.08210 = 8.210\%$ represents the annual growth rate of the account. It represents the rate of interest earned, assuming the account is growing continuously. ●

Now Work PROBLEM 1

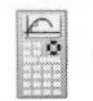

2 Build a Logarithmic Model from Data

Some relations between variables follow a logarithmic model.

EXAMPLE 2 Fitting a Logarithmic Function to Data

Jodi, a meteorologist, is interested in finding a function that explains the relation between the height of a weather balloon (in kilometers) and the atmospheric pressure (measured in millimeters of mercury) on the balloon. She collects the data shown in Table 10.

Table 10

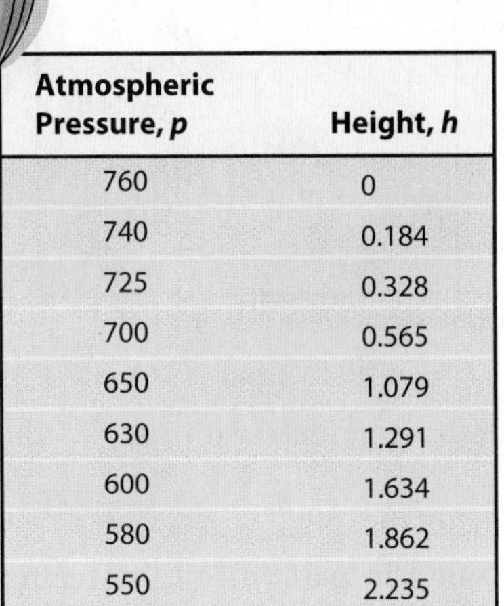

Atmospheric Pressure, p	Height, h
760	0
740	0.184
725	0.328
700	0.565
650	1.079
630	1.291
600	1.634
580	1.862
550	2.235

(a) Using a graphing utility, draw a scatter diagram of the data with atmospheric pressure as the independent variable.

(b) It is known that the relation between atmospheric pressure and height follows a logarithmic model. Using a graphing utility, build a logarithmic model from the data.

(c) Draw the logarithmic function found in part (b) on the scatter diagram.

(d) Use the function found in part (b) to predict the height of the weather balloon if the atmospheric pressure is 560 millimeters of mercury.

Solution

(a) Enter the data into the graphing utility, and draw the scatter diagram. See Figure 31.

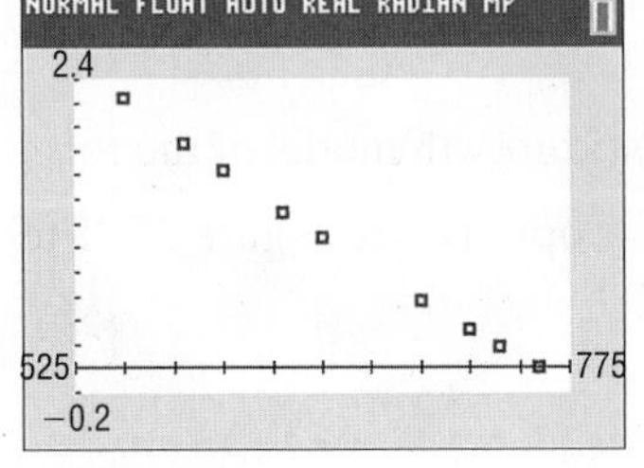

Figure 31

(b) A graphing utility fits the data in Table 10 to a logarithmic function of the form $y = a + b \ln x$ by using the LOGarithm REGression option. See Figure 32. The logarithmic model from the data is

$$h(p) = 45.7863 - 6.9025 \ln p$$

where h is the height of the weather balloon and p is the atmospheric pressure. Notice that $|r|$ is close to 1, indicating a good fit.

(c) Figure 33 shows the graph of $h(p) = 45.7863 - 6.9025 \ln p$ on the scatter diagram.

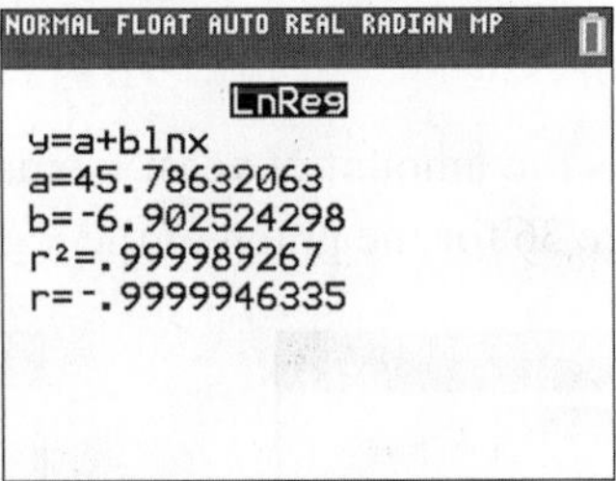

Figure 32

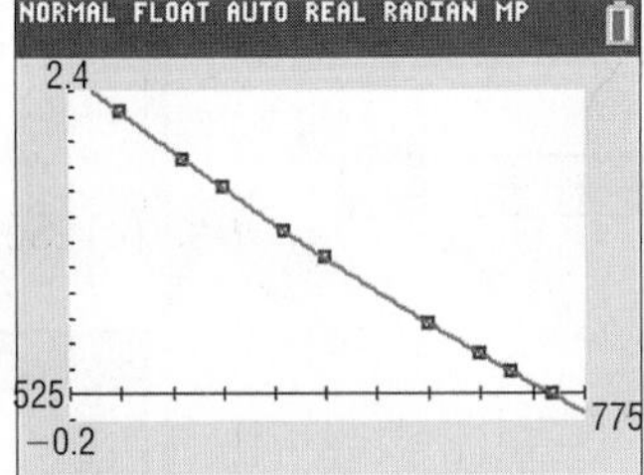

Figure 33

(d) Using the function found in part (b), Jodi predicts the height of the weather balloon when the atmospheric pressure is 560 to be

$$\begin{aligned} h(560) &= 45.7863 - 6.9025 \ln 560 \\ &\approx 2.108 \text{ kilometers} \end{aligned}$$

 **Now Work** PROBLEM 5

3 Build a Logistic Model from Data

Logistic growth models can be used to model situations for which the value of the dependent variable is limited. Many real-world situations conform to this scenario. For example, the population of the human race is limited by the availability of natural resources such as food and shelter. When the value of the dependent variable is limited, a logistic growth model is often appropriate.

EXAMPLE 3 Fitting a Logistic Function to Data

The data in Table 11 represent the amount of yeast biomass in a culture after t hours.

Table 11

Time (hours)	Yeast Biomass	Time (hours)	Yeast Biomass	Time (hours)	Yeast Biomass
0	9.6	7	257.3	14	640.8
1	18.3	8	350.7	15	651.1
2	29.0	9	441.0	16	655.9
3	47.2	10	513.3	17	659.6
4	71.1	11	559.7	18	661.8
5	119.1	12	594.8		
6	174.6	13	629.4		

Source: Tor Carlson (Über Geschwindigkeit and Grösse der Hefevermehrung in Würze, Biochemische Zeitschrift, Bd. 57, pp. 313–334, 1913)

(a) Using a graphing utility, draw a scatter diagram of the data with time as the independent variable.

(b) Using a graphing utility, build a logistic model from the data.

(c) Using a graphing utility, graph the function found in part (b) on the scatter diagram.

(d) What is the predicted carrying capacity of the culture?

(e) Use the function found in part (b) to predict the population of the culture at $t = 19$ hours.

Solution

(a) See Figure 34 for a scatter diagram of the data.

Figure 34

(b) A graphing utility fits the data in Table 11 to a logistic growth model of the form $y = \dfrac{c}{1 + ae^{-bx}}$ by using the LOGISTIC regression option. See Figure 35. The logistic model from the data is

$$y = \frac{663.0}{1 + 71.6e^{-0.5470x}}$$

where y is the amount of yeast biomass in the culture and x is the time.

(c) See Figure 36 for the graph of the logistic model.

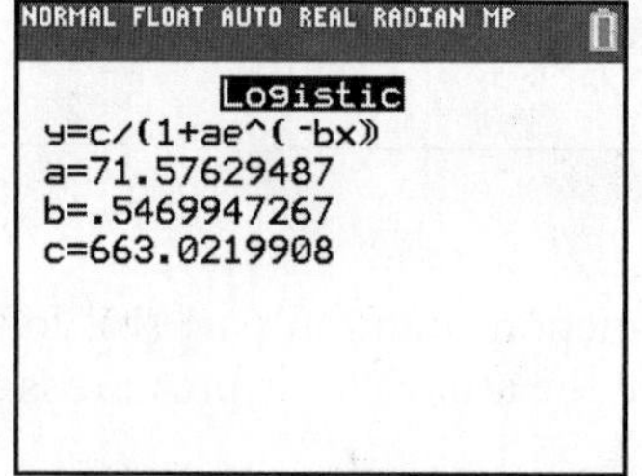

Figure 35

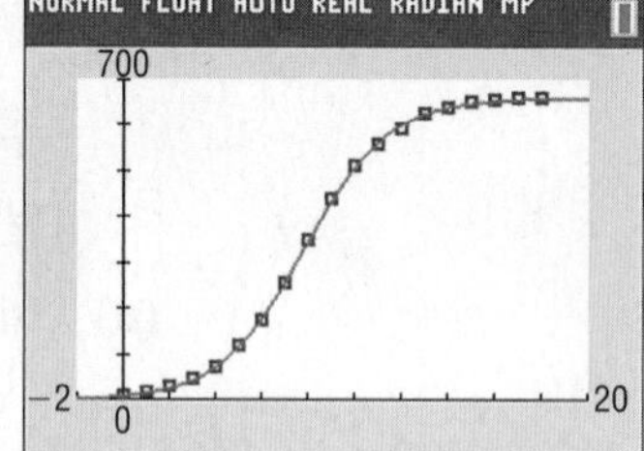

Figure 36

(d) Based on the logistic growth model found in part (b), the carrying capacity of the culture is 663.

(e) Using the logistic growth model found in part (b), the predicted amount of yeast biomass at $t = 19$ hours is

$$y = \frac{663.0}{1 + 71.6e^{-0.5470(19)}} \approx 661.5$$

Now Work PROBLEM 7

7.7 Assess Your Understanding

Applications and Extensions

1. **Biology** A strain of E-coli Beu 397-recA441 is placed into a nutrient broth at 30° Celsius and allowed to grow. The following data are collected. Theory states that the number of bacteria in the petri dish will initially grow according to the law of uninhibited growth. The population is measured using an optical device in which the amount of light that passes through the petri dish is measured.

Time (hours), x	Population, y
0	0.09
2.5	0.18
3.5	0.26
4.5	0.35
6	0.50

Source: Dr. Polly Lavery, Joliet Junior College

(a) Draw a scatter diagram treating time as the independent variable.
(b) Using a graphing utility, build an exponential model from the data.
(c) Express the function found in part (b) in the form $N(t) = N_0e^{kt}$.
(d) Graph the exponential function found in part (b) or (c) on the scatter diagram.
(e) Use the exponential function from part (b) or (c) to predict the population at $x = 7$ hours.
(f) Use the exponential function from part (b) or (c) to predict when the population will reach 0.75.

2. **Ethanol Production** The data in the table below represent ethanol production (in billions of gallons) in the United States from 2000 to 2013.

Year	Ethanol Produced (billion gallons)	Year	Ethanol Produced (billion gallons)
2000 ($x = 0$)	1.6	2007 ($x = 7$)	6.5
2001 ($x = 1$)	1.8	2008 ($x = 8$)	9.3
2002 ($x = 2$)	2.1	2009 ($x = 9$)	10.9
2003 ($x = 3$)	2.8	2010 ($x = 10$)	13.3
2004 ($x = 4$)	3.4	2011 ($x = 11$)	13.9
2005 ($x = 5$)	3.9	2012 ($x = 12$)	13.2
2006 ($x = 6$)	4.9	2013 ($x = 13$)	13.3

Source: Renewable Fuels Association, 2014

(a) Using a graphing utility, draw a scatter diagram of the data using 0 for 2000, 1 for 2001, and so on, as the independent variable.
(b) Using a graphing utility, build an exponential model from the data.
(c) Express the function found in part (b) in the form $A(t) = A_0e^{kt}$.
(d) Graph the exponential function found in part (b) or (c) on the scatter diagram
(e) Use the model to predict the amount of ethanol that will be produced in 2015.
(f) Interpret the meaning of k in the function found in part (c).

3. **Advanced-Stage Breast Cancer** The data in the table below represents the percentage of patients who have survived after diagnosis of advanced-stage breast cancer at 6-month intervals of time.

Time after Diagnosis (years)	Percentage Surviving
0.5	95.7
1	83.6
1.5	74.0
2	58.6
2.5	47.4
3	41.9
3.5	33.6

Source: Cancer Treatment Centers of America

(a) Using a graphing utility, draw a scatter diagram of the data with time after diagnosis as the independent variable.
(b) Using a graphing utility, build an exponential model from the data.
(c) Express the function found in part (b) in the form $A(t) = A_0e^{kt}$.
(d) Graph the exponential function found in part (b) or (c) on the scatter diagram.
(e) What percentage of patients diagnosed with advanced-stage cancer are expected to survive for 4 years after initial diagnosis?
(f) Interpret the meaning of k in the function found in part (c).

4. **Chemistry** A chemist has a 100-gram sample of a radioactive material. He records the amount of radioactive material every week for 7 weeks and obtains the following data: (*continued on the next page*)

Week	Weight (in grams)
0	100.0
1	88.3
2	75.9
3	69.4
4	59.1
5	51.8
6	45.5

(a) Using a graphing utility, draw a scatter diagram with week as the independent variable.
(b) Using a graphing utility, build an exponential model from the data.
(c) Express the function found in part (b) in the form $A(t) = A_0 e^{kt}$.
(d) Graph the exponential function found in part (b) or (c) on the scatter diagram.
(e) From the result found in part (b), determine the half-life of the radioactive material.
(f) How much radioactive material will be left after 50 weeks?
(g) When will there be 20 grams of radioactive material?

5. **Milk Production** The data in the table below represent the number of dairy farms (in thousands) and the amount of milk produced (in billions of pounds) in the United States for various years.

Year	Dairy Farms (thousands)	Milk Produced (billion pounds)
1980	334	128
1985	269	143
1990	193	148
1995	140	155
2000	105	167
2005	78	177
2010	63	193

Source: Statistical Abstract of the United States, 2012

(a) Using a graphing utility, draw a scatter diagram of the data with the number of dairy farms as the independent variable.
(b) Using a graphing utility, build a logarithmic model from the data.
(c) Graph the logarithmic function found in part (b) on the scatter diagram.
(d) In 2008, there were 67 thousand dairy farms in the United States. Use the function in part (b) to predict the amount of milk produced in 2008.
(e) The actual amount of milk produced in 2008 was 190 billion pounds. How does your prediction in part (d) compare to this?

6. **Cable Rates** The data (top, right) represent the average monthly rate charged for expanded basic cable television in the United States from 1995 to 2012. A market researcher believes that external factors, such as the growth of satellite television and internet programming, have affected the cost of basic cable. She is interested in building a model that will describe the average monthly cost of basic cable.
(a) Using a graphing utility, draw a scatter diagram of the data using 0 for 1995, 1 for 1996, and so on, as the independent variable and average monthly rate as the dependent variable.
(b) Using a graphing utility, build a logistic model from the data.
(c) Graph the logistic function found in part (b) on the scatter diagram.
(d) Based on the model found in part (b), what is the maximum possible average monthly rate for basic cable?
(e) Use the model to predict the average rate for basic cable in 2017.

Year	Average Monthly Rate (dollars)
1995 ($x = 0$)	22.35
1996 ($x = 1$)	24.28
1997 ($x = 2$)	26.31
1998 ($x = 3$)	27.88
1999 ($x = 4$)	28.94
2000 ($x = 5$)	31.22
2001 ($x = 6$)	33.75
2002 ($x = 7$)	36.47
2003 ($x = 8$)	38.95
2004 ($x = 9$)	41.04
2005 ($x = 10$)	43.04
2006 ($x = 11$)	45.26
2007 ($x = 12$)	47.27
2008 ($x = 13$)	49.65
2009 ($x = 14$)	52.37
2010 ($x = 15$)	54.44
2011 ($x = 16$)	57.46
2012 ($x = 17$)	61.63

Source: Federal Communications Commission, 2013

7. **Population Model** The following data represent the population of the United States. An ecologist is interested in building a model that describes the population of the United States.

Year	Population
1900	76,212,168
1910	92,228,496
1920	106,021,537
1930	123,202,624
1940	132,164,569
1950	151,325,798
1960	179,323,175
1970	203,302,031
1980	226,542,203
1990	248,709,873
2000	281,421,906
2010	308,745,538

Source: U.S. Census Bureau

(a) Using a graphing utility, draw a scatter diagram of the data using years since 1900 as the independent variable and population as the dependent variable.
(b) Using a graphing utility, build a logistic model from the data.
(c) Using a graphing utility, draw the function found in part (b) on the scatter diagram.
(d) Based on the function found in part (b), what is the carrying capacity of the United States?
(e) Use the function found in part (b) to predict the population of the United States in 2012.
(f) When will the United States population be 350,000,000?
(g) Compare actual U.S. Census figures to the predictions found in parts (e) and (f). Discuss any differences.

8. Population Model The data on the right represent the world population. An ecologist is interested in building a model that describes the world population.

(a) Using a graphing utility, draw a scatter diagram of the data using years since 2000 as the independent variable and population as the dependent variable.
(b) Using a graphing utility, build a logistic model from the data.
(c) Using a graphing utility, draw the function found in part (b) on the scatter diagram.
(d) Based on the function found in part (b), what is the carrying capacity of the world?
(e) Use the function found in part (b) to predict the population of the world in 2020.
(f) When will world population be 10 billion?

Year	Population (billions)	Year	Population (billions)
2001	6.17	2008	6.71
2002	6.24	2009	6.79
2003	6.32	2010	6.86
2004	6.40	2011	6.94
2005	6.47	2012	7.02
2006	6.55	2013	7.10
2007	6.63		

Source: U.S. Census Bureau

9. Cell Phone Towers The following data represent the number of cell sites in service in the United States from 1985 to 2012 at the end of June each year.

Year	Cell Sites (thousands)	Year	Cell Sites (thousands)	Year	Cell Sites (thousands)
1985 ($x = 1$)	0.6	1995 ($x = 11$)	19.8	2004 ($x = 20$)	174.4
1986 ($x = 2$)	1.2	1996 ($x = 12$)	24.8	2005 ($x = 21$)	178.0
1987 ($x = 3$)	1.7	1997 ($x = 13$)	38.7	2006 ($x = 22$)	197.6
1988 ($x = 4$)	2.8	1998 ($x = 14$)	57.7	2007 ($x = 23$)	210.4
1989 ($x = 5$)	3.6	1999 ($x = 15$)	74.2	2008 ($x = 24$)	220.5
1990 ($x = 6$)	4.8	2000 ($x = 16$)	95.7	2009 ($x = 25$)	245.9
1991 ($x = 7$)	6.7	2001 ($x = 17$)	114.1	2010 ($x = 26$)	251.6
1992 ($x = 8$)	8.9	2002 ($x = 18$)	131.4	2011 ($x = 27$)	256.9
1993 ($x = 9$)	11.6	2003 ($x = 19$)	147.7	2012 ($x = 28$)	285.6
1994 ($x = 10$)	14.7				

Source: ©2013 CTIA-The Wireless Association®. All Rights Reserved.

(a) Using a graphing utility, draw a scatter diagram of the data using 1 for 1985, 2 for 1986, and so on, as the independent variable and number of cell sites as the dependent variable.
(b) Using a graphing utility, build a logistic model from the data.
(c) Graph the logistic function found in part (b) on the scatter diagram.
(d) What is the predicted carrying capacity for cell sites in the United States?
(e) Use the model to predict the number of cell sites in the United States at the end of June 2017.

Mixed Practice

10. Income versus Crime Rate The data on the right represent property crime rate against individuals (crimes per 1000 households) and their household income (in dollars) in the United States in 2009.

(a) Using a graphing utility, draw a scatter diagram of the data using income, x, as the independent variable and crime rate, y, as the dependent variable.
(b) Based on the scatter diagram drawn in part (a), decide on a model (exponential, logarithmic, or logistic) that you think best describes the relation between income and crime rate. Be sure to justify your choice of model.
(c) Using a graphing utility, find the model of best fit.
(d) Using a graphing utility, draw the model of best fit on the scatter diagram you drew in part (a).

Income Level	Property Crime Rate
5000	201.1
11,250	157.0
20,000	141.6
30,000	134.1
42,500	139.7
62,500	120.0

Source: Statistical Abstract of the United States, 2012

(e) Use your model to predict the crime rate of a household whose income is $55,000.

11. Golfing The data on the right represent the expected percentage of putts that will be made by professional golfers on the PGA Tour depending on distance. For example, it is expected that 99.3% of 2-foot putts will be made.

(a) Using a graphing utility, draw a scatter diagram of the data with distance as the independent variable.

(b) Based on the scatter diagram drawn in part (a), decide on a model (exponential, logarithmic, or logistic) that you think best describes the relation between distance and expected percentage. Be sure to justify your choice of model.

(c) Using a graphing utility, find the model of best fit.

(d) Graph the function found in part (c) on the scatter diagram.

(e) Use the function found in part (c) to predict what percentage of 30-foot putts will be made.

Distance (feet)	Expected Percentage	Distance (feet)	Expected Percentage
2	99.3	14	25.0
3	94.8	15	22.0
4	85.8	16	20.0
5	74.7	17	19.0
6	64.7	18	17.0
7	55.6	19	16.0
8	48.5	20	14.0
9	43.4	21	13.0
10	38.3	22	12.0
11	34.2	23	11.0
12	30.1	24	11.0
13	27.0	25	10.0

Source: TheSandTrap.com

Retain Your Knowledge

Problems 12–15 are based on material learned earlier in the course. The purpose of these problems is to keep the material fresh in your mind so that you are better prepared for the final exam.

12. Graph the equation $(x - 3)^2 + y^2 = 25$.

13. Find the exact value of $\sin^2 \frac{\pi}{8} + \cos^2 \frac{\pi}{8}$. Do not use a calculator.

14. Solve the equation $2 \sin(2\theta) = 1$ in the interval $0 \leq \theta < 2\pi$.

15. Solve: $\log_5(x - 8) = 3$

Chapter Review

Things to Know

Properties of the exponential function (pp. 458, 460)

$f(x) = Ca^x \quad a > 1, C > 0$

Domain: the interval $(-\infty, \infty)$
Range: the interval $(0, \infty)$
x-intercepts: none; y-intercept: C
Horizontal asymptote: x-axis $(y = 0)$ as $x \rightarrow -\infty$
Increasing; one-to-one; smooth; continuous
See Figure 4 for a typical graph.

$f(x) = Ca^x \quad 0 < a < 1, C > 0$

Domain: the interval $(-\infty, \infty)$
Range: the interval $(0, \infty)$
x-intercepts: none; y-intercept: C
Horizontal asymptote: x-axis $(y = 0)$ as $x \rightarrow \infty$
Decreasing; one-to-one; smooth; continuous
See Figure 8 for a typical graph.

Number *e* (p. 461)

Number approached by the expression $\left(1 + \frac{1}{n}\right)^n$ as $n \rightarrow \infty$; that is, $\lim_{n \to \infty} \left(1 + \frac{1}{n}\right)^n = e$.

Property of exponents (p. 462)

If $a^u = a^v$, then $u = v$.

Properties of the logarithmic function (pp. 471, 472, 478)

$f(x) = \log_a x \quad a > 1$
$(y = \log_a x$ means $x = a^y)$

Domain: the interval $(0, \infty)$
Range: the interval $(-\infty, \infty)$
x-intercept: 1; y-intercept: none
Vertical asymptote: $x = 0$ (y-axis)

$f(x) = \log_a x \quad 0 < a < 1$ $(y = \log_a x \text{ means } x = a^y)$	Increasing; one-to-one; smooth; continuous See Figure 22(a) for a typical graph. Domain: the interval $(0, \infty)$ Range: the interval $(-\infty, \infty)$ x-intercept: 1; y-intercept: none Vertical asymptote: $x = 0$ (y-axis) Decreasing; one-to-one; smooth; continuous See Figure 22(b) for a typical graph.

Natural logarithm (p. 473) $y = \ln x$ means $x = e^y$.

Properties of logarithms (pp. 483–484, 486) $\log_a 1 = 0 \quad \log_a a = 1 \quad a^{\log_a M} = M \quad \log_a a^r = r \quad a^r = e^{r \ln a}$

$\log_a(MN) = \log_a M + \log_a N \quad \log_a\left(\frac{M}{N}\right) = \log_a M - \log_a N$

$\log_a M^r = r \log_a M$

If $M = N$, then $\log_a M = \log_a N$.

If $\log_a M = \log_a N$, then $M = N$.

Formulas

Change-of-Base Formula (p. 487) $\log_a M = \dfrac{\log_b M}{\log_b a}$

Compound Interest Formula (p. 499) $A = P\cdot\left(1 + \dfrac{r}{n}\right)^{nt}$

Continuous compounding (p. 501) $A = Pe^{rt}$

Effective rate of interest (p. 502) Compounding n times per year: $r_e = \left(1 + \dfrac{r}{n}\right)^n - 1$

Continuous compounding: $r_e = e^r - 1$

Present Value Formulas (p. 503) $P = A\cdot\left(1 + \dfrac{r}{n}\right)^{-nt}$ or $P = Ae^{-rt}$

Uninhibited Growth and decay (pp. 508, 510) $A(t) = A_0 e^{kt}$

Newton's Law of Cooling (p. 511) $u(t) = T + (u_0 - T)e^{kt} \quad k < 0$

Logistic model (p. 513) $P(t) = \dfrac{c}{1 + ae^{-bt}}$

Objectives

Section		You should be able to . . .	Example(s)	Review Exercises
7.1	1	Evaluate exponential functions (p. 453)	1	1(a), (c), 34(a)
	2	Graph exponential functions (p. 457)	3–6	18–20
	3	Define the number e (p. 460)	p. 461	
	4	Solve exponential equations (p. 462)	7, 8	22, 23, 26, 28, 34(b)
7.2	1	Change exponential statements to logarithmic statements and logarithmic statements to exponential statements (p. 470)	2, 3	2, 3
	2	Evaluate logarithmic expressions (p. 471)	4	1(b), (d), 6, 33(b), 35(a), 36(a)
	3	Determine the domain of a logarithmic function (p. 471)	5	4, 5, 21(a)
	4	Graph logarithmic functions (p. 472)	6, 7	17, 21(b), 33(a)
	5	Solve logarithmic equations (p. 476)	8, 9	24, 27, 33(c), 35(b)
7.3	1	Work with the properties of logarithms (p. 482)	1, 2	7, 8
	2	Write a logarithmic expression as a sum or difference of logarithms (p. 484)	3–5	9–12
	3	Write a logarithmic expression as a single logarithm (p. 485)	6	13–15
	4	Evaluate logarithms whose base is neither 10 nor e (p. 487)	7, 8	16

Section		You should be able to . . .	Example(s)	Review Exercises
7.4	1	Solve logarithmic equations (p. 491)	1–3	24, 30
	2	Solve exponential equations (p. 493)	4–6	25, 29, 31, 32
	3	Solve logarithmic and exponential equations using a graphing utility (p. 494)	7	22–32
7.5	1	Determine the future value of a lump sum of money (p. 498)	1–3	37
	2	Calculate effective rates of return (p. 501)	4	37
	3	Determine the present value of a lump sum of money (p. 502)	5	38
	4	Determine the rate of interest or the time required to double a lump sum of money (p. 503)	6, 7	37
7.6	1	Find equations of populations that obey the law of uninhibited growth (p. 508)	1, 2	41
	2	Find equations of populations that obey the law of decay (p. 510)	3	39, 42
	3	Use Newton's Law of Cooling (p. 511)	4	40
	4	Use logistic models (p. 513)	5, 6	43
7.7	1	Build an exponential model from data (p. 519)	1	44
	2	Build a logarithmic model from data (p. 521)	2	45
	3	Build a logistic model from data (p. 521)	3	46

Review Exercises

In Problem 1, $f(x) = 3^x$ and $g(x) = \log_3 x$.

1. Evaluate: (a) $f(4)$ (b) $g(9)$ (c) $f(-2)$ (d) $g\left(\frac{1}{27}\right)$

2. Change $5^2 = z$ to an equivalent statement involving a logarithm.

3. Change $\log_5 u = 13$ to an equivalent statement involving an exponent.

In Problems 4 and 5, find the domain of each logarithmic function.

4. $f(x) = \log(3x - 2)$ **5.** $H(x) = \log_2(x^2 - 3x + 2)$

In Problems 6–8, find the exact value of each expression. Do not use a calculator.

6. $\log_2\left(\frac{1}{8}\right)$ **7.** $\ln e^{\sqrt{2}}$ **8.** $2^{\log_2 0.4}$

In Problems 9–12, write each expression as the sum and/or difference of logarithms. Express powers as factors.

9. $\log_3\left(\frac{uv^2}{w}\right)$ $u > 0, v > 0, w > 0$ **10.** $\log_2(a^2\sqrt{b})^4$ $a > 0, b > 0$

11. $\log(x^2\sqrt{x^3 + 1})$ $x > 0$ **12.** $\ln\left(\frac{2x + 3}{x^2 - 3x + 2}\right)^2$ $x > 2$

In Problems 13–15, write each expression as a single logarithm.

13. $3\log_4 x^2 + \frac{1}{2}\log_4\sqrt{x}$ **14.** $\ln\left(\frac{x - 1}{x}\right) + \ln\left(\frac{x}{x + 1}\right) - \ln(x^2 - 1)$

15. $\frac{1}{2}\ln(x^2 + 1) - 4\ln\frac{1}{2} - \frac{1}{2}[\ln(x - 4) + \ln x]$

16. Use the Change-of-Base Formula and a calculator to evaluate $\log_4 19$. Round your answer to three decimal places.

17. Graph $y = \log_3 x$ using a graphing utility and the Change-of-Base Formula.

In Problems 18–21, use the given function f to:
(a) Find the domain of f. (b) Graph f. (c) From the graph, determine the range and any asymptotes of f.
(d) Find f^{-1}, the inverse function of f. (e) Find the domain and the range of f^{-1}. (f) Graph f^{-1}.

18. $f(x) = 2^{x-3}$ **19.** $f(x) = 1 + 3^{-x}$ **20.** $f(x) = 3e^{x-2}$ **21.** $f(x) = \frac{1}{2}\ln(x + 3)$

In Problems 22–32, solve each equation. Express any irrational solution in exact form and as a decimal rounded to three decimal places.

22. $8^{6+3x} = 4$ **23.** $3^{x^2+x} = \sqrt{3}$ **24.** $\log_x 64 = -3$ **25.** $5^x = 3^{x+2}$

26. $25^{2x} = 5^{x^2-12}$ **27.** $\log_3 \sqrt{x-2} = 2$ **28.** $8 = 4^{x^2} \cdot 2^{5x}$ **29.** $2^x \cdot 5 = 10^x$

30. $\log_6(x+3) + \log_6(x+4) = 1$ **31.** $e^{1-x} = 5$ **32.** $9^x + 4 \cdot 3^x - 3 = 0$

33. Suppose that $f(x) = \log_2(x-2) + 1$.
(a) Graph f.
(b) What is $f(6)$? What point is on the graph of f?
(c) Solve $f(x) = 4$. What point is on the graph of f?
(d) Based on the graph drawn in part (a), solve $f(x) > 0$.
(e) Find $f^{-1}(x)$. Graph f^{-1} on the same Cartesian plane as f.

34. Amplifying Sound An amplifier's power output P (in watts) is related to its decibel voltage gain d by the formula

$$P = 25e^{0.1d}$$

(a) Find the power output for a decibel voltage gain of 4 decibels.
(b) For a power output of 50 watts, what is the decibel voltage gain?

35. Limiting Magnitude of a Telescope A telescope is limited in its usefulness by the brightness of the star that it is aimed at and by the diameter of its lens. One measure of a star's brightness is its *magnitude;* the dimmer the star, the larger its magnitude. A formula for the limiting magnitude L of a telescope—that is, the magnitude of the dimmest star that it can be used to view—is given by

$$L = 9 + 5.1 \log d$$

where d is the diameter (in inches) of the lens.
(a) What is the limiting magnitude of a 3.5-inch telescope?
(b) What diameter is required to view a star of magnitude 14?

36. Salvage Value The number of years n for a piece of machinery to depreciate to a known salvage value can be found using the formula

$$n = \frac{\log s - \log i}{\log(1-d)}$$

where s is the salvage value of the machinery, i is its initial value, and d is the annual rate of depreciation.
(a) How many years will it take for a piece of machinery to decline in value from \$90,000 to \$10,000 if the annual rate of depreciation is 0.20 (20%)?
(b) How many years will it take for a piece of machinery to lose half of its value if the annual rate of depreciation is 15%?

37. Funding a College Education A child's grandparents purchase a \$10,000 bond fund that matures in 18 years to be used for her college education. The bond fund pays 4% interest compounded semiannually. How much will the bond fund be worth at maturity? What is the effective rate of interest? How long will it take the bond to double in value under these terms?

38. Funding a College Education A child's grandparents wish to purchase a bond that matures in 18 years to be used for her college education. The bond pays 4% interest compounded semiannually. How much should they pay so that the bond will be worth \$85,000 at maturity?

39. Estimating the Date When a Prehistoric Man Died The bones of a prehistoric man found in the desert of New Mexico contain approximately 5% of the original amount of carbon-14. If the half-life of carbon-14 is 5730 years, approximately how long ago did the man die?

40. Temperature of a Skillet A skillet is removed from an oven where the temperature is 450°F and placed in a room where the temperature is 70°F. After 5 minutes, the temperature of the skillet is 400°F. How long will it be until its temperature is 150°F?

41. World Population The annual growth rate of the world's population in 2014 was $k = 1.1\% = 0.011$. The population of the world in 2014 was 7,137,577,750. Letting $t = 0$ represent 2014, use the uninhibited growth model to predict the world's population in the year 2024.

Source: U.S. Census Bureau

42. Radioactive Decay The half-life of radioactive cobalt is 5.27 years. If 100 grams of radioactive cobalt is present now, how much will be present in 20 years? In 40 years?

43. Logistic Growth The logistic growth model

$$P(t) = \frac{0.8}{1 + 1.67e^{-0.16t}}$$

represents the proportion of new cars with a global positioning system (GPS). Let $t = 0$ represent 2006, $t = 1$ represent 2007, and so on.
(a) What proportion of new cars in 2006 had a GPS?
(b) Determine the maximum proportion of new cars that have a GPS.
(c) Using a graphing utility, graph $P = P(t)$.
(d) When will 75% of new cars have a GPS?

44. Rising Tuition The following data represent the average in-state tuition and fees (in 2013 dollars) at public four-year colleges and universities in the United States from the academic year 1983–84 to the academic year 2013–14.

Academic Year	Tuition and Fees (2013 dollars)
1983–84 ($x = 0$)	2684
1988–89 ($x = 5$)	3111
1993–94 ($x = 10$)	4101
1998–99 ($x = 15$)	4648
2003–04 ($x = 20$)	5900
2008–09 ($x = 25$)	7008
2013–14 ($x = 30$)	8893

Source: The College Board

(continued)

(a) Using a graphing utility, draw a scatter diagram with academic year as the independent variable.
(b) Using a graphing utility, build an exponential model from the data.
(c) Express the function found in part (b) in the form $A(t) = A_0e^{kt}$.
(d) Graph the exponential function found in part (b) or (c) on the scatter diagram.
(e) Predict the academic year when the average tuition will reach $12,000.

45. Wind Chill Factor The data below represent the wind speed (mph) and the wind chill factor at an air temperature of 15°F.

Wind Speed (mph)	Wind Chill Factor (°F)
5	7
10	3
15	0
20	−2
25	−4
30	−5
35	−7

Source: U.S. National Weather Service

(a) Using a graphing utility, draw a scatter diagram with wind speed as the independent variable.
(b) Using a graphing utility, build a logarithmic model from the data.
(c) Using a graphing utility, draw the logarithmic function found in part (b) on the scatter diagram.
(d) Use the function found in part (b) to predict the wind chill factor if the air temperature is 15°F and the wind speed is 23 mph.

46. Spreading of a Disease Jack and Diane live in a small town of 50 people. Unfortunately, both Jack and Diane have a cold. Those who come in contact with someone who has this cold will themselves catch the cold. The following data represent the number of people in the small town who have caught the cold after t days.

Days, t	Number of People with Cold, C
0	2
1	4
2	8
3	14
4	22
5	30
6	37
7	42
8	44

(a) Using a graphing utility, draw a scatter diagram of the data. Comment on the type of relation that appears to exist between the number of days that have passed and the number of people with a cold.
(b) Using a graphing utility, build a logistic model from the data.
(c) Graph the function found in part (b) on the scatter diagram.
(d) According to the function found in part (b), what is the maximum number of people who will catch the cold? In reality, what is the maximum number of people who could catch the cold?
(e) Sometime between the second day and the third day, 10 people in the town had a cold. According to the model found in part (b), when did 10 people have a cold?
(f) How long will it take for 46 people to catch the cold?

Chapter Test

The Chapter Test Prep Videos are step-by-step solutions available in MyMathLab®, or on this text's YouTube Channel. Flip back to the Resources for Success page for a link to this text's YouTube channel.

In Problems 1–3, solve each equation.

1. $3^x = 243$ **2.** $\log_b 16 = 2$
3. $\log_5 x = 4$

In Problems 4–7, use a calculator to evaluate each expression. Round your answer to three decimal places.

4. $e^3 + 2$ **5.** $\log 20$
6. $\log_3 21$ **7.** $\ln 133$

In Problems 8 and 9, use the given function f.

(a) Find the domain of f.
(b) Graph f.
(c) From the graph, determine the range and any asymptotes of f.
(d) Find f^{-1}, the inverse of f.
(e) Find the domain and the range of f^{-1}.
(f) Graph f^{-1}.

8. $f(x) = 4^{x+1} - 2$ **9.** $f(x) = 1 - \log_5(x - 2)$

In Problems 10–15, solve each equation.

10. $5^{x+2} = 125$

11. $\log(x + 9) = 2$

12. $8 - 2e^{-x} = 4$

13. $\log(x^2 + 3) = \log(x + 6)$

14. $7^{x+3} = e^x$

15. $\log_2(x - 4) + \log_2(x + 4) = 3$

16. Write $\log_2\left(\dfrac{4x^3}{x^2 - 3x - 18}\right)$ as the sum and/or difference of logarithms. Express powers as factors.

17. A 50-mg sample of a radioactive substance decays to 34 mg after 30 days. How long will it take for there to be 2 mg remaining?

18. (a) If $1000 is invested at 5% compounded monthly, how much is there after 8 months?

(b) If you want to have $1000 in 9 months, how much do you need to place in a savings account now that pays 5% compounded quarterly?

(c) How long does it take to double your money if you can invest it at 6% compounded annually?

19. The decibel level, D, of sound is given by the equation $D = 10\log\left(\dfrac{I}{I_0}\right)$, where I is the intensity of the sound and $I_0 = 10^{-12}$ watt per square meter.

(a) If the shout of a single person measures 80 decibels, how loud would the sound be if two people shouted at the same time? That is, how loud would the sound be if the intensity doubled?

(b) The pain threshold for sound is 125 decibels. If the Athens Olympic Stadium 2004 (Olympiako Stadio Athinas 'Spyros Louis') can seat 74,400 people, how many people in the crowd need to shout at the same time for the resulting sound level to meet or exceed the pain threshold? (Ignore any possible sound dampening.)

Cumulative Review

1. Is the following graph the graph of a function? If it is, is the function one-to-one?

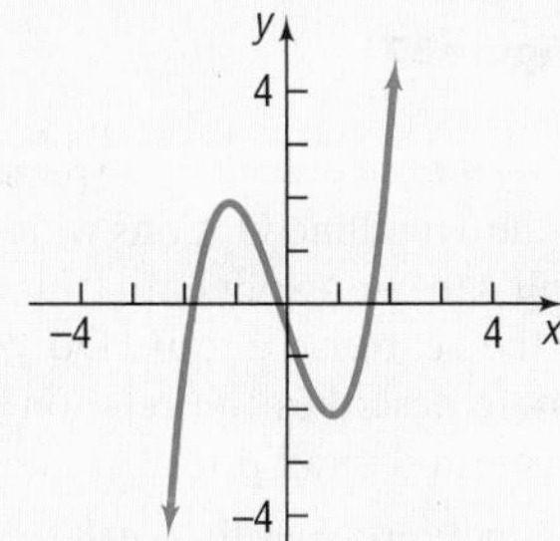

2. For the function $f(x) = 2x^2 - 3x + 1$, find the following:

(a) $f(3)$ (b) $f(-x)$ (c) $f(x + h)$

3. Determine which of the following points are on the graph of $x^2 + y^2 = 1$.

(a) $\left(\dfrac{1}{2}, \dfrac{1}{2}\right)$ (b) $\left(\dfrac{1}{2}, \dfrac{\sqrt{3}}{2}\right)$

4. Solve the equation $3(x - 2) = 4(x + 5)$.

5. Graph the line $2x - 4y = 16$.

6. Graph $f(x) = 3(x + 1)^3 - 2$ using transformations.

7. For the function $g(x) = 3^x + 2$:

(a) Graph g using transformations. State the domain, range, and horizontal asymptote of g.

(b) Determine the inverse of g. State the domain, range, and vertical asymptote of g^{-1}.

(c) On the same graph as g, graph g^{-1}.

8. Solve the equation: $4^{x-3} = 8^{2x}$

9. Solve the equation: $\log_3(x + 1) + \log_3(2x - 3) = \log_9 9$

10. Suppose that $f(x) = \log_3(x + 2)$. Solve:

(a) $f(x) = 0$ (b) $f(x) > 0$

(c) $f(x) = 3$

11. Data Analysis The following data represent the percent of all drivers by age who have been stopped by the police for any reason within the past year. The median age represents the midpoint of the upper and lower limit for the age range.

Age Range	Median Age, x	Percent Stopped, y
16–19	17.5	18.2
20–29	24.5	16.8
30–39	34.5	11.3
40–49	44.5	9.4
50–59	54.5	7.7
≥60	69.5	3.8

(a) Using your graphing utility, draw a scatter diagram of the data treating median age, x, as the independent variable.

(b) What type of model do you feel best describes the relation between median age and percent stopped? You may choose from among linear, exponential, logarithmic, and logistic models.

(c) Provide a justification for the model that you selected in part (b).

Chapter Projects

Internet-based Project

I. Depreciation of Cars Kelley Blue Book is a guide that provides the current retail price of cars. You can access the Kelley Blue Book at your library or online at *www.kbb.com.*

1. Identify three cars that you are considering purchasing, and find the Kelley Blue Book value of the cars for 0 (brand new), 1, 2, 3, 4, and 5 years of age. Online, the value of the car can be found by selecting What should I pay for a used car? Enter the year, make, and model of the car you are selecting. To be consistent, we will assume the cars will be driven 12,000 miles per year, so a 1-year-old car will have 12,000 miles, a 2-year-old car will have 24,000 miles, and so on. Choose the same options for each year, and finally determine the suggested retail price for cars that are in Excellent, Good, and Fair shape. You should have a total of 16 observations (1 for a brand new car, 3 for a 1-year-old car, 3 for a 2-year-old car, and so on).
2. Draw a scatter diagram of the data with age as the independent variable and value as the dependent variable using Excel, a TI-graphing calculator, or some other spreadsheet.
3. Determine the exponential function of best fit. Graph the exponential function of best fit on the scatter diagram. To do this in Excel, click on any data point in the scatter diagram. Now select the Chart Element icon (+). Check the box for Trendline, select the arrow to the right, and choose More Options. Select the Exponential radio button and select Display Equation on Chart. See Figure 37. Move the Trendline Options window off to the side, and you will see the exponential function of best fit displayed on the scatter diagram. Do you think the function accurately describes the relation between age of the car and suggested retail price?

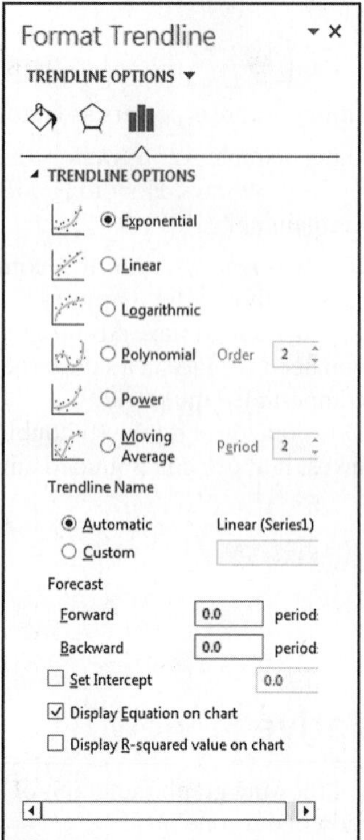

Figure 37

4. The exponential function of best fit is of the form $y = Ce^{rx}$, where y is the suggested retail value of the car and x is the age of the car (in years). What does the value of C represent? What does the value of r represent? What is the depreciation rate for each car that you are considering?
5. Write a report detailing which car you would purchase based on the depreciation rate you found for each car.

The following projects are available on the Instructor's Resource Center (IRC):

II. Hot Coffee A fast-food restaurant wants a special container to hold coffee. The restaurant wishes the container to quickly cool the coffee from 200° to 130°F and keep the liquid between 110° and 130°F as long as possible. The restaurant has three containers to select from. Which one should be purchased?

III. Project at Motorola ***Thermal Fatigue of Solder Connections*** Product reliability is a major concern of a manufacturer. Here a logarithmic transformation is used to simplify the analysis of a cell phone's ability to withstand temperature change.

Appendix A
Review

Outline

A.1 Algebra Essentials
A.2 Geometry Essentials
A.3 Factoring Polynomials; Completing the Square
A.4 Solving Equations
A.5 Complex Numbers; Quadratic Equations in the Complex Number System
A.6 Interval Notation; Solving Inequalities
A.7 *n*th Roots; Rational Exponents
A.8 Lines
A.9 Building Linear Models from Data

A.1 Algebra Essentials

PREPARING FOR THIS SECTION *Before getting started, read "To the Student" at the beginning of this book.*

OBJECTIVES
1 Work with Sets (p. A1)
2 Graph Inequalities (p. A5)
3 Find Distance on the Real Number Line (p. A5)
4 Evaluate Algebraic Expressions (p. A6)
5 Determine the Domain of a Variable (p. A7)
6 Use the Laws of Exponents (p. A8)
7 Evaluate Square Roots (p. A9)
8 Use a Calculator to Evaluate Exponents (p. A10)

1 Work with Sets

A **set** is a well-defined collection of distinct objects. The objects of a set are called its **elements**. By **well-defined**, we mean that there is a rule that enables us to determine whether a given object is an element of the set. If a set has no elements, it is called the **empty set**, or **null set**, and is denoted by the symbol $\varnothing$.

For example, the set of *digits* consists of the collection of numbers 0, 1, 2, 3, 4, 5, 6, 7, 8, and 9. If we use the symbol D to denote the set of digits, then we can write

$$D = \{0, 1, 2, 3, 4, 5, 6, 7, 8, 9\}$$

In this notation, the braces { } are used to enclose the objects, or **elements**, in the set. This method of denoting a set is called the **roster method**. A second way to denote a set is to use **set-builder notation**, where the set D of digits is written as

$$D = \{\quad x \quad | \; x \text{ is a digit}\}$$

Read as "D is the set of all x such that x is a digit."

EXAMPLE 1

Using Set-builder Notation and the Roster Method

(a) $E = \{x|x \text{ is an even digit}\} = \{0, 2, 4, 6, 8\}$
(b) $O = \{x|x \text{ is an odd digit}\} = \{1, 3, 5, 7, 9\}$ ●

Because the elements of a set are distinct, we never repeat elements. For example, we would never write $\{1, 2, 3, 2\}$; the correct listing is $\{1, 2, 3\}$. Because a set is a collection, the order in which the elements are listed is immaterial. $\{1, 2, 3\}$, $\{1, 3, 2\}$, $\{2, 1, 3\}$, and so on, all represent the same set.

If every element of a set A is also an element of a set B, then A is a **subset** of B, which is denoted $A \subseteq B$. If two sets A and B have the same elements, then A **equals** B, which is denoted $A = B$.

For example, $\{1, 2, 3\} \subseteq \{1, 2, 3, 4, 5\}$ and $\{1, 2, 3\} = \{2, 3, 1\}$.

DEFINITION

If A and B are sets, the **intersection** of A with B, denoted $A \cap B$, is the set consisting of elements that belong to both A and B. The **union** of A with B, denoted $A \cup B$, is the set consisting of elements that belong to either A or B, or both.

EXAMPLE 2

Finding the Intersection and Union of Sets

Let $A = \{1, 3, 5, 8\}$, $B = \{3, 5, 7\}$, and $C = \{2, 4, 6, 8\}$. Find:

(a) $A \cap B$ (b) $A \cup B$ (c) $B \cap (A \cup C)$

Solution

(a) $A \cap B = \{1, 3, 5, 8\} \cap \{3, 5, 7\} = \{3, 5\}$
(b) $A \cup B = \{1, 3, 5, 8\} \cup \{3, 5, 7\} = \{1, 3, 5, 7, 8\}$
(c) $B \cap (A \cup C) = \{3, 5, 7\} \cap (\{1, 3, 5, 8\} \cup \{2, 4, 6, 8\})$
$= \{3, 5, 7\} \cap \{1, 2, 3, 4, 5, 6, 8\} = \{3, 5\}$ ●

Now Work PROBLEM 15

When working with sets, it is common practice to designate a **universal set** U, the set consisting of all the elements to be considered. With a universal set designated, elements of the universal set not found in a given set can be considered.

DEFINITION

If A is a set, the **complement** of A, denoted $\overline{A}$, is the set consisting of all the elements in the universal set that are not in A.*

EXAMPLE 3

Finding the Complement of a Set

If the universal set is $U = \{1, 2, 3, 4, 5, 6, 7, 8, 9\}$ and if $A = \{1, 3, 5, 7, 9\}$, then $\overline{A} = \{2, 4, 6, 8\}$. ●

It follows from the definition of complement that $A \cup \overline{A} = U$ and $A \cap \overline{A} = \varnothing$. Do you see why?

Now Work PROBLEM 19

It is often helpful to draw pictures of sets. Such pictures, called **Venn diagrams**, represent sets as circles enclosed in a rectangle. The rectangle represents the universal set. Such diagrams often make it easier to visualize various relationships among sets. See Figure 1.

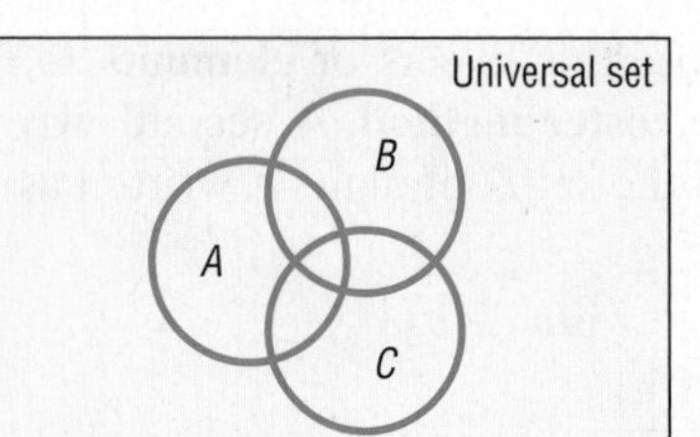

Figure 1

*Some books use the notation A' or A^c for the complement of A.

The Venn diagram in Figure 2(a) illustrates that $A \subseteq B$. The Venn diagram in Figure 2(b) illustrates that A and B have no elements in common, that is, that $A \cap B = \varnothing$. The sets A and B in Figure 2(b) are **disjoint**.

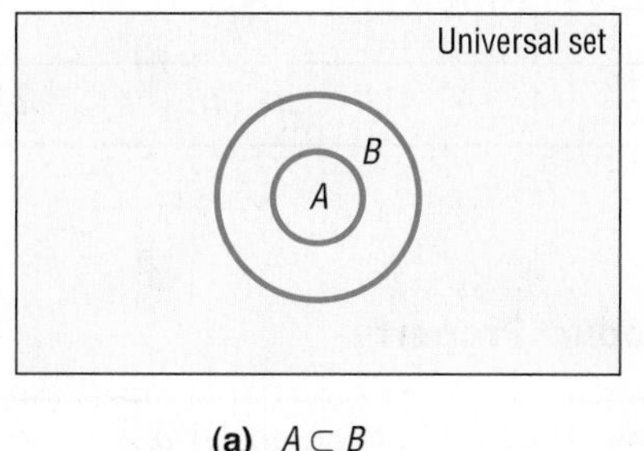

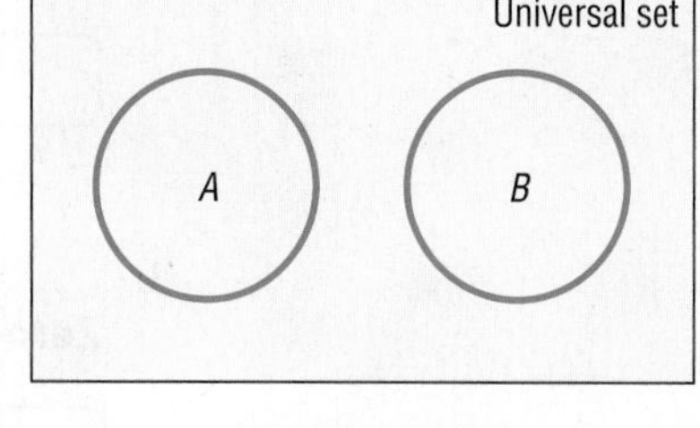

Figure 2 (a) $A \subseteq B$ subset (b) $A \cap B = \varnothing$ disjoint sets

Figures 3(a), 3(b), and 3(c) use Venn diagrams to illustrate the definitions of *intersection*, *union*, and *complement*, respectively.

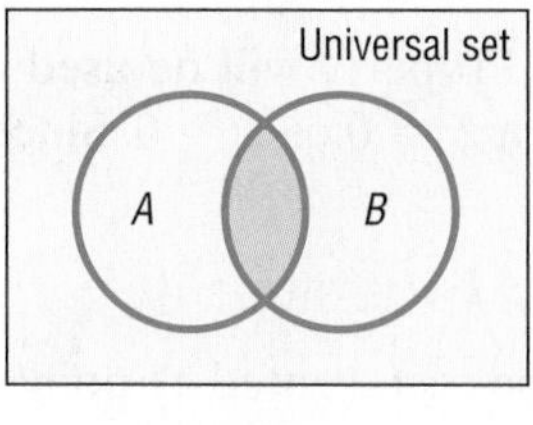

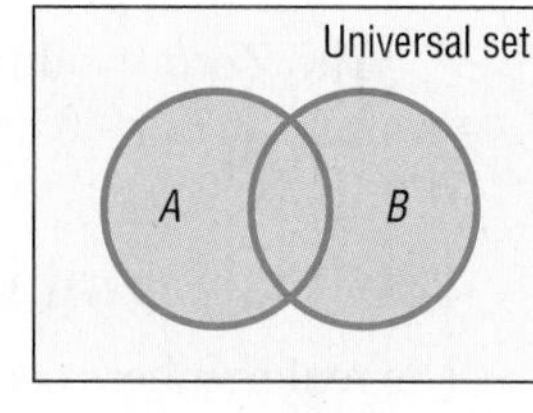

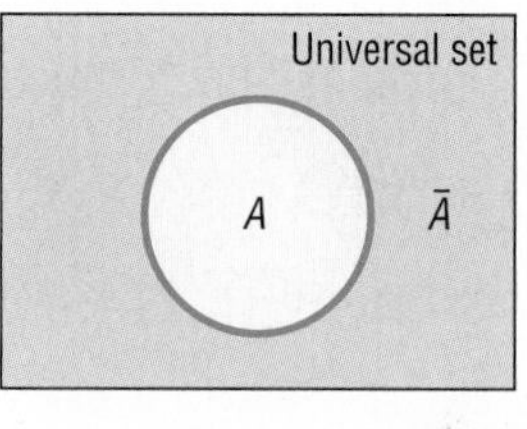

Figure 3 (a) $A \cap B$ intersection (b) $A \cup B$ union (c) $\bar{A}$ complement

Real Numbers

Real numbers are represented by symbols such as

$$25, \quad 0, \quad -3, \quad \frac{1}{2}, \quad -\frac{5}{4}, \quad 0.125, \quad \sqrt{2}, \quad \pi, \quad \sqrt[3]{-2}, \quad 0.666\ldots$$

The set of **counting numbers**, or **natural numbers**, contains the numbers in the set $\{1, 2, 3, 4, \ldots\}$. (The three dots, called an **ellipsis**, indicate that the pattern continues indefinitely.) The set of **integers** contains the numbers in the set $\{\ldots, -3, -2, -1, 0, 1, 2, 3, \ldots\}$. A **rational number** is a number that can be expressed as a *quotient* $\frac{a}{b}$ of two integers, where the integer b cannot be 0. Examples of rational numbers are $\frac{3}{4}, \frac{5}{2}, \frac{0}{4}$, and $-\frac{2}{3}$. Since $\frac{a}{1} = a$ for any integer a, every integer is also a rational number. Real numbers that are not rational are called **irrational**. Examples of irrational numbers are $\sqrt{2}$ and π (the Greek letter pi), which equals the constant ratio of the circumference to the diameter of a circle. See Figure 4.

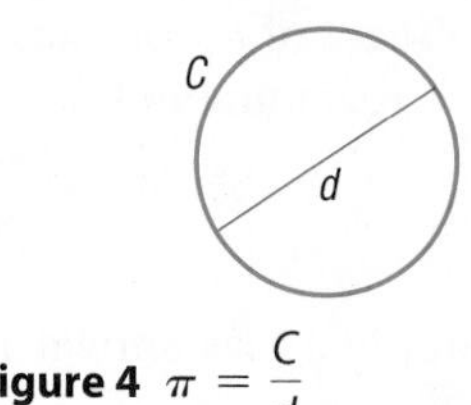

Figure 4 $\pi = \frac{C}{d}$

Real numbers can be represented as **decimals**. Rational real numbers have decimal representations that either **terminate** or are nonterminating with **repeating** blocks of digits. For example, $\frac{3}{4} = 0.75$, which terminates; and $\frac{2}{3} = 0.666\ldots$, in which the digit 6 repeats indefinitely. Irrational real numbers have decimal representations that neither repeat nor terminate. For example, $\sqrt{2} = 1.414213\ldots$ and $\pi = 3.14159\ldots$. In practice, the decimal representation of an irrational number is given as an approximation. We use the symbol $\approx$ (read as "approximately equal to") to write $\sqrt{2} \approx 1.4142$ and $\pi \approx 3.1416$.

Two frequently used properties of real numbers are given next. Suppose that a, b, and c are real numbers.

Distributive Property

$$a \cdot (b + c) = ab + ac$$

Zero-Product Property

If $ab = 0$, then either $a = 0$ or $b = 0$ or both equal 0.

In Words

If a product equals 0, then one or both of the factors is 0.

The Distributive Property can be used to remove parentheses:

$$2(x + 3) = 2x + 2 \cdot 3 = 2x + 6$$

The Zero-Product Property will be used to solve equations (Section A.4). For example, if $2x = 0$, then $2 = 0$ or $x = 0$. Since $2 \neq 0$, it follows that $x = 0$.

The Real Number Line

The real numbers can be represented by points on a line called the **real number line**. There is a one-to-one correspondence between real numbers and points on a line. That is, every real number corresponds to a point on the line, and each point on the line has a unique real number associated with it.

Pick a point on the line somewhere in the center, and label it O. This point, called the **origin**, corresponds to the real number 0. See Figure 5. The point 1 unit to the right of O corresponds to the number 1. The distance between 0 and 1 determines the **scale** of the number line. For example, the point associated with the number 2 is twice as far from O as 1. Notice that an arrowhead on the right end of the line indicates the direction in which the numbers increase. Points to the left of the origin correspond to the real numbers -1, -2, and so on. Figure 5 also shows the points associated with the rational numbers $-\frac{1}{2}$ and $\frac{1}{2}$ and with the irrational numbers $\sqrt{2}$ and π.

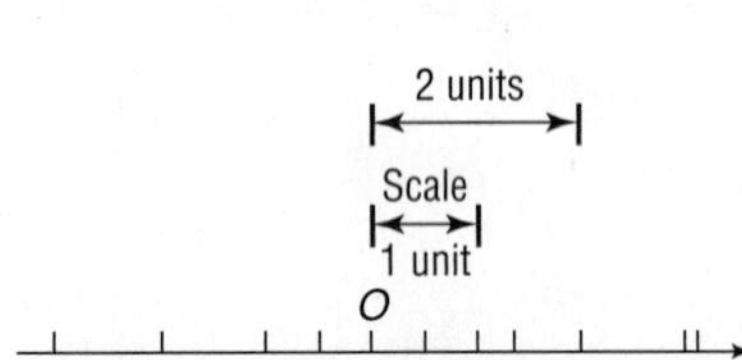

Figure 5 Real number line

DEFINITION

The real number associated with a point P is called the **coordinate** of P, and the line whose points have been assigned coordinates is called the **real number line**.

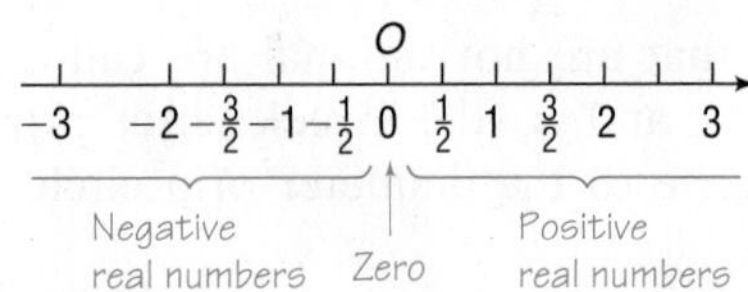

Figure 6

The real number line consists of three classes of real numbers, as shown in Figure 6.

1. The **negative real numbers** are the coordinates of points to the left of the origin O.
2. The real number **zero** is the coordinate of the origin O.
3. The **positive real numbers** are the coordinates of points to the right of the origin O.

Now Work PROBLEM 23

2 Graph Inequalities

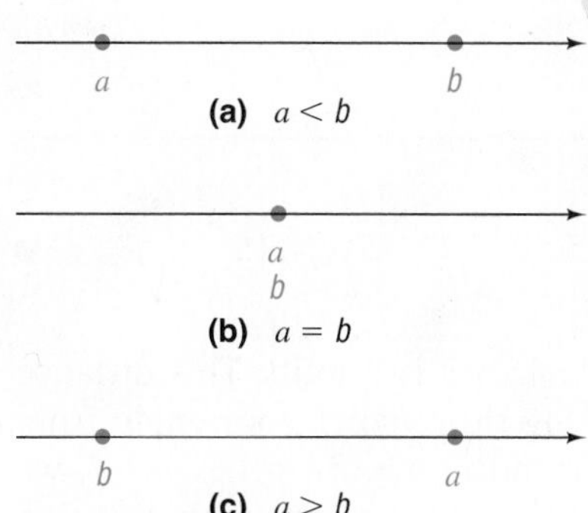

Figure 7

An important property of the real number line follows from the fact that, given two numbers (points) a and b, either a is to the left of b, or a is at the same location as b, or a is to the right of b. See Figure 7.

If a is to the left of b, then "a is less than b," which is written $a < b$. If a is to the right of b, then "a is greater than b," which is written $a > b$. If a is at the same location as b, then $a = b$. If a is either less than or equal to b, then $a \leq b$. Similarly, $a \geq b$ means that a is either greater than or equal to b. Collectively, the symbols $<, >, \leq$, and $\geq$ are called **inequality symbols**.

Note that $a < b$ and $b > a$ mean the same thing. It does not matter whether we write $2 < 3$ or $3 > 2$.

Furthermore, if $a < b$ or if $b > a$, then the difference $b - a$ is positive. Do you see why?

An **inequality** is a statement in which two expressions are related by an inequality symbol. The expressions are referred to as the **sides** of the inequality. Inequalities of the form $a < b$ or $b > a$ are called **strict inequalities**, whereas inequalities of the form $a \leq b$ or $b \geq a$ are called **nonstrict inequalities**.

The following conclusions are based on the discussion so far.

$a > 0$ is equivalent to a is positive

$a < 0$ is equivalent to a is negative

The inequality $a > 0$ is sometimes read as "a is positive." If $a \geq 0$, then either $a > 0$ or $a = 0$, and this is read as "a is nonnegative."

Now Work PROBLEMS 27 AND 37

EXAMPLE 4 **Graphing Inequalities**

(a) On the real number line, graph all numbers x for which $x > 4$.
(b) On the real number line, graph all numbers x for which $x \leq 5$.

Solution

(a) See Figure 8. Notice that we use a left parenthesis to indicate that the number 4 is *not* part of the graph.

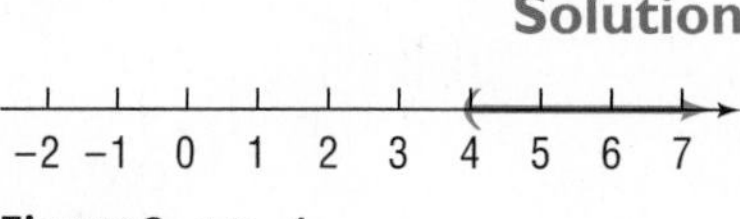

Figure 8 $x > 4$

(b) See Figure 9. Notice that we use a right bracket to indicate that the number 5 *is* part of the graph. ●

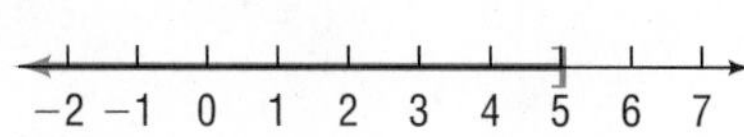

Figure 9 $x \leq 5$

Now Work PROBLEM 43

3 Find Distance on the Real Number Line

The *absolute value* of a number a is the distance from 0 to a on the number line. For example, -4 is 4 units from 0, and 3 is 3 units from 0. See Figure 10. That is, the absolute value of -4 is 4, and the absolute value of 3 is 3.

4 units 3 units

−5 −4 −3 −2 −1 0 1 2 3 4

Figure 10

A more formal definition of absolute value is given next.

DEFINITION

The **absolute value** of a real number a, denoted by the symbol $|a|$, is defined by the rules

$$|a| = a \quad \text{if } a \geq 0 \qquad \text{and} \qquad |a| = -a \quad \text{if } a < 0$$

For example, because $-4 < 0$, the second rule must be used to get $|-4| = -(-4) = 4$.

EXAMPLE 5 **Computing Absolute Value**

(a) $|8| = 8$ (b) $|0| = 0$ (c) $|-15| = -(-15) = 15$ ●

Look again at Figure 10. The distance from -4 to 3 is 7 units. This distance is the difference $3 - (-4)$, obtained by subtracting the smaller coordinate from the larger. However, since $|3 - (-4)| = |7| = 7$ and $|-4 - 3| = |-7| = 7$, we can use absolute value to calculate the distance between two points without being concerned about which is smaller.

DEFINITION If P and Q are two points on a real number line with coordinates a and b, respectively, the **distance between P and Q**, denoted by $d(P, Q)$, is

$$d(P, Q) = |b - a|$$

Since $|b - a| = |a - b|$, it follows that $d(P, Q) = d(Q, P)$.

EXAMPLE 6 **Finding Distance on a Number Line**

Let P, Q, and R be points on a real number line with coordinates -5, 7, and -3, respectively. Find the distance

(a) between P and Q (b) between Q and R

Solution See Figure 11.

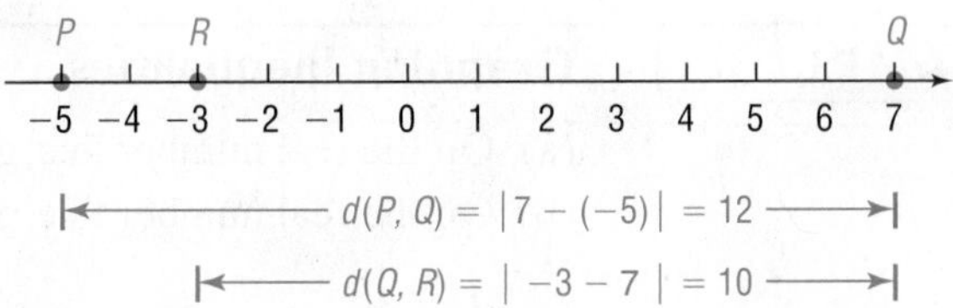

Figure 11

(a) $d(P, Q) = |7 - (-5)| = |12| = 12$
(b) $d(Q, R) = |-3 - 7| = |-10| = 10$ ●

Now Work PROBLEM 49

4 Evaluate Algebraic Expressions

Remember, in algebra we use letters such as x, y, a, b, and c to represent numbers. If the letter used is to represent *any* number from a given set of numbers, it is called a **variable**. A **constant** is either a fixed number, such as 5 or $\sqrt{3}$, or a letter that represents a fixed (possibly unspecified) number.

Constants and variables are combined using the operations of addition, subtraction, multiplication, and division to form *algebraic expressions*. Examples of algebraic expressions include

$$x + 3 \qquad \frac{3}{1 - t} \qquad 7x - 2y$$

To evaluate an algebraic expression, substitute a numerical value for each variable.

EXAMPLE 7

Evaluating an Algebraic Expression

Evaluate each expression if $x = 3$ and $y = -1$.

(a) $x + 3y$ (b) $5xy$ (c) $\dfrac{3y}{2 - 2x}$ (d) $|-4x + y|$

Solution (a) Substitute 3 for x and -1 for y in the expression $x + 3y$.

$$x + 3y = 3 + 3(-1) = 3 + (-3) = 0$$

$\uparrow$ $x = 3, y = -1$

(b) If $x = 3$ and $y = -1$, then

$$5xy = 5(3)(-1) = -15$$

(c) If $x = 3$ and $y = -1$, then

$$\frac{3y}{2 - 2x} = \frac{3(-1)}{2 - 2(3)} = \frac{-3}{2 - 6} = \frac{-3}{-4} = \frac{3}{4}$$

(d) If $x = 3$ and $y = -1$, then

$$|-4x + y| = |-4(3) + (-1)| = |-12 + (-1)| = |-13| = 13$$

Now Work PROBLEMS 51 AND 59

5 Determine the Domain of a Variable

In working with expressions or formulas involving variables, the variables may be allowed to take on values from only a certain set of numbers. For example, in the formula for the area A of a circle of radius r, $A = \pi r^2$, the variable r is necessarily restricted to the positive real numbers. In the expression $\dfrac{1}{x}$, the variable x cannot take on the value 0, since division by 0 is not defined.

DEFINITION The set of values that a variable may assume is called the **domain of the variable.**

EXAMPLE 8

Finding the Domain of a Variable

The domain of the variable x in the expression

$$\frac{5}{x - 2}$$

is $\{x|x \neq 2\}$ since, if $x = 2$, the denominator becomes 0, which is not defined.

EXAMPLE 9

Circumference of a Circle

In the formula for the circumference C of a circle of radius r,

$$C = 2\pi r$$

the domain of the variable r, representing the radius of the circle, is the set of positive real numbers, $\{r|r > 0\}$. The domain of the variable C, representing the circumference of the circle, is also the set of positive real numbers, $\{C|C > 0\}$.

In describing the domain of a variable, we may use either set notation or words, whichever is more convenient.

Now Work PROBLEM 69

6 Use the Laws of Exponents

Integer exponents provide a shorthand notation for representing repeated multiplications of a real number. For example,

$$2^3 = 2 \cdot 2 \cdot 2 = 8 \qquad 3^4 = 3 \cdot 3 \cdot 3 \cdot 3 = 81$$

DEFINITION

If a is a real number and n is a positive integer, then the symbol $\boldsymbol{a^n}$ represents the product of n factors of a. That is,

$$a^n = \underbrace{a \cdot a \cdot \ldots \cdot a}_{n \text{ factors}} \tag{1}$$

WARNING Be careful with minus signs and exponents.

$$-2^4 = -1 \cdot 2^4 = -16$$

whereas

$$(-2)^4 = (-2)(-2)(-2)(-2) = 16$$

■

In the definition it is understood that $a^1 = a$. Furthermore, $a^2 = a \cdot a$, $a^3 = a \cdot a \cdot a$, and so on. In the expression a^n, a is called the **base** and n is called the **exponent**, or **power**. We read a^n as "a raised to the power n" or as "a to the nth power." We usually read a^2 as "a squared" and a^3 as "a cubed."

In working with exponents, the operation of *raising to a power* is performed before any other operation. As examples,

$$4 \cdot 3^2 = 4 \cdot 9 = 36 \qquad 2^2 + 3^2 = 4 + 9 = 13$$

$$-2^4 = -16 \qquad 5 \cdot 3^2 + 2 \cdot 4 = 5 \cdot 9 + 2 \cdot 4 = 45 + 8 = 53$$

Parentheses are used to indicate operations to be performed first. For example,

$$(-2)^4 = (-2)(-2)(-2)(-2) = 16 \qquad (2 + 3)^2 = 5^2 = 25$$

DEFINITION

If $a \neq 0$, then

$$a^0 = 1$$

DEFINITION

If $a \neq 0$ and if n is a positive integer, then

$$a^{-n} = \frac{1}{a^n}$$

Whenever you encounter a negative exponent, think "reciprocal."

EXAMPLE 10 **Evaluating Expressions Containing Negative Exponents**

(a) $2^{-3} = \frac{1}{2^3} = \frac{1}{8}$ (b) $x^{-4} = \frac{1}{x^4}$ (c) $\left(\frac{1}{5}\right)^{-2} = \frac{1}{\left(\frac{1}{5}\right)^2} = \frac{1}{\frac{1}{25}} = 25$ ●

Now Work PROBLEMS 87 AND 107

The following properties, called the **Laws of Exponents**, can be proved using the preceding definitions. In the list, a and b are real numbers, and m and n are integers.

THEOREM

Laws of Exponents

$$a^m a^n = a^{m+n} \qquad (a^m)^n = a^{mn} \qquad (ab)^n = a^n b^n$$

$$\frac{a^m}{a^n} = a^{m-n} = \frac{1}{a^{n-m}} \quad \text{if } a \neq 0 \qquad \left(\frac{a}{b}\right)^n = \frac{a^n}{b^n} \quad \text{if } b \neq 0$$

EXAMPLE 11

Using the Laws of Exponents

Write each expression so that all exponents are positive.

(a) $\dfrac{x^5 y^{-2}}{x^3 y} \quad x \neq 0, \quad y \neq 0$ (b) $\left(\dfrac{x^{-3}}{3y^{-1}}\right)^{-2} \quad x \neq 0, \quad y \neq 0$

Solution

(a) $$\frac{x^5 y^{-2}}{x^3 y} = \frac{x^5}{x^3} \cdot \frac{y^{-2}}{y} = x^{5-3} \cdot y^{-2-1} = x^2 y^{-3} = x^2 \cdot \frac{1}{y^3} = \frac{x^2}{y^3}$$

(b) $$\left(\frac{x^{-3}}{3y^{-1}}\right)^{-2} = \frac{(x^{-3})^{-2}}{(3y^{-1})^{-2}} = \frac{x^6}{3^{-2}(y^{-1})^{-2}} = \frac{x^6}{\frac{1}{9}y^2} = \frac{9x^6}{y^2}$$

Now Work PROBLEMS 89 AND 99

7 Evaluate Square Roots

> **In Words**
> $\sqrt{36}$ means "give me the nonnegative number whose square is 36."

A real number is squared when it is raised to the power 2. The inverse of squaring is finding a **square root**. For example, since $6^2 = 36$ and $(-6)^2 = 36$, the numbers 6 and -6 are square roots of 36.

The symbol $\sqrt{}$, called a **radical sign**, is used to denote the **principal**, or nonnegative, square root. For example, $\sqrt{36} = 6$.

DEFINITION

If a is a nonnegative real number, the nonnegative number b such that $b^2 = a$ is the **principal square root** of a, and is denoted by $b = \sqrt{a}$.

The following comments are noteworthy:

1. Negative numbers do not have square roots (in the real number system), because the square of any real number is *nonnegative*. For example, $\sqrt{-4}$ is not a real number, because there is no real number whose square is -4.
2. The principal square root of 0 is 0, since $0^2 = 0$. That is, $\sqrt{0} = 0$.
3. The principal square root of a positive number is positive.
4. If $c \geq 0$, then $(\sqrt{c})^2 = c$. For example, $(\sqrt{2})^2 = 2$ and $(\sqrt{3})^2 = 3$.

EXAMPLE 12

Evaluating Square Roots

(a) $\sqrt{64} = 8$ (b) $\sqrt{\dfrac{1}{16}} = \dfrac{1}{4}$ (c) $(\sqrt{1.4})^2 = 1.4$

Examples 12(a) and (b) are examples of square roots of perfect squares, since $64 = 8^2$ and $\dfrac{1}{16} = \left(\dfrac{1}{4}\right)^2$.

Consider the expression $\sqrt{a^2}$. Since $a^2 \geq 0$, the principal square root of a^2 is defined whether $a > 0$ or $a < 0$. However, since the principal square root is nonnegative, we need an absolute value to ensure the nonnegative result. That is,

$$\sqrt{a^2} = |a| \qquad a \text{ any real number} \qquad (2)$$

EXAMPLE 13

Using Equation (2)

(a) $\sqrt{(2.3)^2} = |2.3| = 2.3$ (b) $\sqrt{(-2.3)^2} = |-2.3| = 2.3$

(c) $\sqrt{x^2} = |x|$

Now Work PROBLEM 95

Calculators

Calculators are incapable of displaying decimals that contain a large number of digits. For example, some calculators are capable of displaying only eight digits. When a number requires more than eight digits, the calculator either truncates or rounds. To see how your calculator handles decimals, divide 2 by 3. How many digits do you see? Is the last digit a 6 or a 7? If it is a 6, your calculator truncates; if it is a 7, your calculator rounds.

There are different kinds of calculators. An **arithmetic** calculator can only add, subtract, multiply, and divide numbers; therefore, this type is not adequate for this course. **Scientific** calculators have all the capabilities of arithmetic calculators and also contain **function keys** labeled ln, log, sin, cos, tan, x^y, inv, and so on. **Graphing** calculators have all the capabilities of scientific calculators and contain a screen on which graphs can be displayed. We use the symbol whenever a graphing calculator needs to be used. In this text the use of a graphing calculator is optional.

8 Use a Calculator to Evaluate Exponents

Your calculator has either a caret key, ^, or an x^y key, that is used for computations involving exponents.

EXAMPLE 14

Exponents on a Graphing Calculator

Evaluate: $(2.3)^5$

Solution Figure 12 shows the result using a TI-84 Plus C graphing calculator.

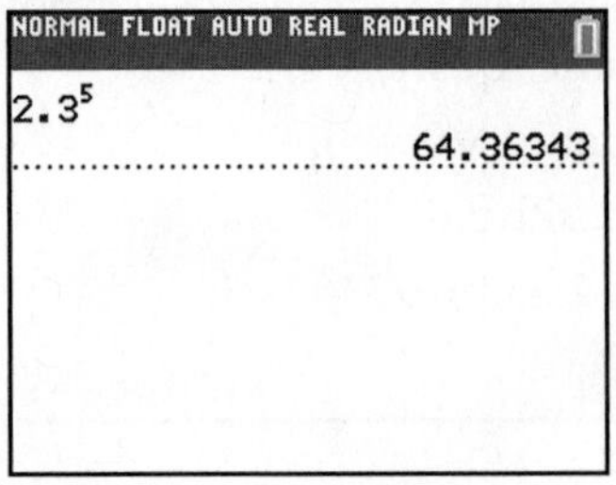

NORMAL FLOAT AUTO REAL RADIAN MP

2.3^5

64.36343

Figure 12

Now Work PROBLEM 125

A.1 Assess Your Understanding

Concepts and Vocabulary

1. A(n) ________ is a letter used in algebra to represent any number from a given set of numbers.

2. On the real number line, the real number zero is the coordinate of the ______.

3. An inequality of the form $a > b$ is called a(n) ______ inequality.

4. In the expression 2^4, the number 2 is called the _____ and 4 is called the _______________.

5. If a is a nonnegative real number, then which inequality statement best describes a?
(a) $a < 0$ (b) $a > 0$ (c) $a \leq 0$ (d) $a \geq 0$

6. Let a and b be nonzero real numbers and m and n be integers. Which of the following is not a law of exponents?
(a) $\left(\frac{a}{b}\right)^n = \frac{a^n}{b^n}$ (b) $(a^m)^n = a^{m+n}$
(c) $\frac{a^m}{a^n} = a^{m-n}$ (d) $(ab)^n = a^n b^n$

7. ***True or False*** The product of two negative real numbers is always greater than zero.

8. ***True or False*** The distance between two distinct points on the real number line is always greater than zero.

9. ***True or False*** The absolute value of a real number is always greater than zero.

10. ***True or False*** To multiply two expressions having the same base, retain the base and multiply the exponents.

Skill Building

In Problems 11–22, use U = *universal set* = $\{0, 1, 2, 3, 4, 5, 6, 7, 8, 9\}$, $A = \{1, 3, 4, 5, 9\}$, $B = \{2, 4, 6, 7, 8\}$, *and* $C = \{1, 3, 4, 6\}$ *to find each set.*

11. $A \cup B$ **12.** $A \cup C$ **13.** $A \cap B$ **14.** $A \cap C$

15. $(A \cup B) \cap C$ **16.** $(A \cap B) \cup C$ **17.** $\overline{A}$ **18.** $\overline{C}$

19. $\overline{A \cap B}$ **20.** $\overline{B \cup C}$ **21.** $\overline{A} \cup \overline{B}$ **22.** $\overline{B} \cap \overline{C}$

23. On the real number line, label the points with coordinates $0, 1, -1, \frac{5}{2}, -2.5, \frac{3}{4}$, and 0.25.

24. Repeat Problem 23 for the coordinates $0, -2, 2, -1.5, \frac{3}{2}, \frac{1}{3}$, and $\frac{2}{3}$.

In Problems 25–34, replace the question mark by $<$, $>$, *or* $=$, *whichever is correct.*

25. $\frac{1}{2}\ ?\ 0$ **26.** $5\ ?\ 6$ 27. $-1\ ?\ -2$ **28.** $-3\ ?\ -\frac{5}{2}$ **29.** $\pi\ ?\ 3.14$

30. $\sqrt{2}\ ?\ 1.41$ **31.** $\frac{1}{2}\ ?\ 0.5$ **32.** $\frac{1}{3}\ ?\ 0.33$ **33.** $\frac{2}{3}\ ?\ 0.67$ **34.** $\frac{1}{4}\ ?\ 0.25$

In Problems 35–40, write each statement as an inequality.

35. x is positive **36.** z is negative 37. x is less than 2

38. y is greater than -5 **39.** x is less than or equal to 1 **40.** x is greater than or equal to 2

In Problems 41–44, graph the numbers x on the real number line.

41. $x \geq -2$ **42.** $x < 4$ 43. $x > -1$ **44.** $x \leq 7$

In Problems 45–50, use the given real number line to compute each distance.

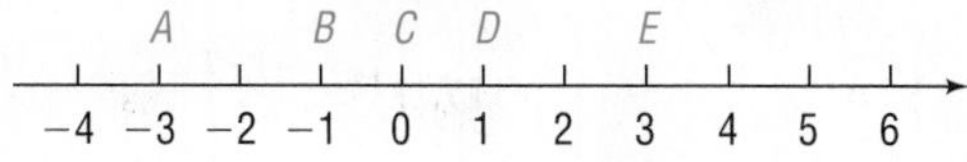

45. $d(C, D)$ **46.** $d(C, A)$ **47.** $d(D, E)$ **48.** $d(C, E)$ 49. $d(A, E)$ **50.** $d(D, B)$

In Problems 51–58, evaluate each expression if $x = -2$ *and* $y = 3$.

51. $x + 2y$ **52.** $3x + y$ **53.** $5xy + 2$ **54.** $-2x + xy$

55. $\frac{2x}{x - y}$ **56.** $\frac{x + y}{x - y}$ **57.** $\frac{3x + 2y}{2 + y}$ **58.** $\frac{2x - 3}{y}$

In Problems 59–68, find the value of each expression if $x = 3$ and $y = -2$.

59. $|x + y|$
60. $|x - y|$
61. $|x| + |y|$
62. $|x| - |y|$
63. $\dfrac{|x|}{x}$
64. $\dfrac{|y|}{y}$
65. $|4x - 5y|$
66. $|3x + 2y|$
67. $||4x| - |5y||$
68. $3|x| + 2|y|$

In Problems 69–76, determine which of the values (a) through (d), if any, must be excluded from the domain of the variable in each expression.

(a) $x = 3$ (b) $x = 1$ (c) $x = 0$ (d) $x = -1$

69. $\dfrac{x^2 - 1}{x}$
70. $\dfrac{x^2 + 1}{x}$
71. $\dfrac{x}{x^2 - 9}$
72. $\dfrac{x}{x^2 + 9}$
73. $\dfrac{x^2}{x^2 + 1}$
74. $\dfrac{x^3}{x^2 - 1}$
75. $\dfrac{x^2 + 5x - 10}{x^3 - x}$
76. $\dfrac{-9x^2 - x + 1}{x^3 + x}$

In Problems 77–80, determine the domain of the variable x in each expression.

77. $\dfrac{4}{x - 5}$
78. $\dfrac{-6}{x + 4}$
79. $\dfrac{x}{x + 4}$
80. $\dfrac{x - 2}{x - 6}$

In Problems 81–84, use the formula $C = \dfrac{5}{9}(F - 32)$ for converting degrees Fahrenheit into degrees Celsius to find the Celsius measure of each Fahrenheit temperature.

81. $F = 32°$
82. $F = 212°$
83. $F = 77°$
84. $F = -4°$

In Problems 85–96, simplify each expression.

85. $(-4)^2$
86. -4^2
87. 4^{-2}
88. -4^{-2}
89. $3^{-6} \cdot 3^4$
90. $4^{-2} \cdot 4^3$
91. $(3^{-2})^{-1}$
92. $(2^{-1})^{-3}$
93. $\sqrt{25}$
94. $\sqrt{36}$
95. $\sqrt{(-4)^2}$
96. $\sqrt{(-3)^2}$

In Problems 97–106, simplify each expression. Express the answer so that all exponents are positive. Whenever an exponent is 0 or negative, we assume that the base is not 0.

97. $(8x^3)^2$
98. $(-4x^2)^{-1}$
99. $(x^2y^{-1})^2$
100. $(x^{-1}y)^3$
101. $\dfrac{x^2y^3}{xy^4}$
102. $\dfrac{x^{-2}y}{xy^2}$
103. $\dfrac{(-2)^3x^4(yz)^2}{3^2xy^3z}$
104. $\dfrac{4x^{-2}(yz)^{-1}}{2^3x^4y}$
105. $\left(\dfrac{3x^{-1}}{4y^{-1}}\right)^{-2}$
106. $\left(\dfrac{5x^{-2}}{6y^{-2}}\right)^{-3}$

In Problems 107–118, find the value of each expression if $x = 2$ and $y = -1$.

107. $2xy^{-1}$
108. $-3x^{-1}y$
109. $x^2 + y^2$
110. x^2y^2
111. $(xy)^2$
112. $(x + y)^2$
113. $\sqrt{x^2}$
114. $(\sqrt{x})^2$
115. $\sqrt{x^2 + y^2}$
116. $\sqrt{x^2} + \sqrt{y^2}$
117. x^y
118. y^x

119. Find the value of the expression $2x^3 - 3x^2 + 5x - 4$ if $x = 2$. What is the value if $x = 1$?

120. Find the value of the expression $4x^3 + 3x^2 - x + 2$ if $x = 1$. What is the value if $x = 2$?

121. What is the value of $\dfrac{(666)^4}{(222)^4}$?

122. What is the value of $(0.1)^3(20)^3$?

In Problems 123–130, use a calculator to evaluate each expression. Round your answer to three decimal places.

123. $(8.2)^6$
124. $(3.7)^5$
125. $(6.1)^{-3}$
126. $(2.2)^{-5}$
127. $(-2.8)^6$
128. $-(2.8)^6$
129. $(-8.11)^{-4}$
130. $-(8.11)^{-4}$

Applications and Extensions

In Problems 131–140, express each statement as an equation involving the indicated variables.

131. Area of a Rectangle The area A of a rectangle is the product of its length l and its width w.

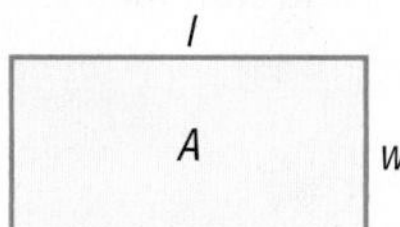

132. Perimeter of a Rectangle The perimeter P of a rectangle is twice the sum of its length l and its width w.

133. Circumference of a Circle The circumference C of a circle is the product of π and its diameter d.

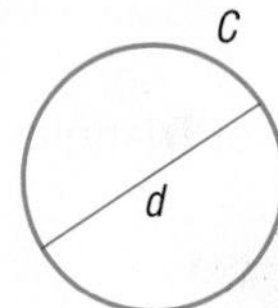

134. Area of a Triangle The area A of a triangle is one-half the product of its base b and its height h.

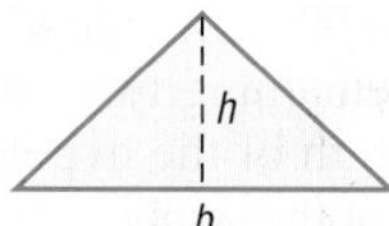

135. Area of an Equilateral Triangle The area A of an equilateral triangle is $\frac{\sqrt{3}}{4}$ times the square of the length x of one side.

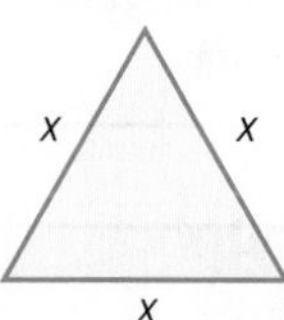

136. Perimeter of an Equilateral Triangle The perimeter P of an equilateral triangle is 3 times the length x of one side.

137. Volume of a Sphere The volume V of a sphere is $\frac{4}{3}$ times π times the cube of the radius r.

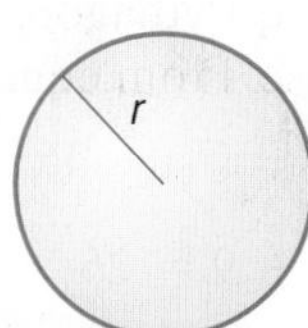

138. Surface Area of a Sphere The surface area S of a sphere is 4 times π times the square of the radius r.

139. Volume of a Cube The volume V of a cube is the cube of the length x of a side.

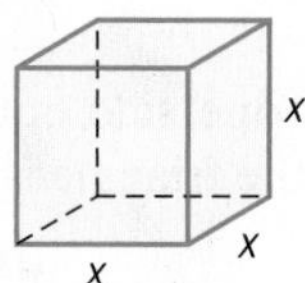

140. Surface Area of a Cube The surface area S of a cube is 6 times the square of the length x of a side.

141. Manufacturing Cost The weekly production cost C of manufacturing x watches is given by the formula $C = 4000 + 2x$, where the variable C is in dollars.
(a) What is the cost of producing 1000 watches?
(b) What is the cost of producing 2000 watches?

142. Balancing a Checkbook At the beginning of the month, Mike had a balance of \$210 in his checking account. During the next month, he deposited \$80, wrote a check for \$120, made another deposit of \$25, and wrote two checks: one for \$60 and the other for \$32. He was also assessed a monthly service charge of \$5. What was his balance at the end of the month?

In Problems 143 and 144, write an inequality using an absolute value to describe each statement.

143. x is at least 6 units from 4.

144. x is more than 5 units from 2.

145. U.S. Voltage In the United States, normal household voltage is 110 volts. It is acceptable for the actual voltage x to differ from normal by at most 5 volts. A formula that describes this is

$$|x - 110| \le 5$$

(a) Show that a voltage of 108 volts is acceptable.
(b) Show that a voltage of 104 volts is not acceptable.

146. Foreign Voltage In other countries, normal household voltage is 220 volts. It is acceptable for the actual voltage x to differ from normal by at most 8 volts. A formula that describes this is

$$|x - 220| \le 8$$

(a) Show that a voltage of 214 volts is acceptable.
(b) Show that a voltage of 209 volts is not acceptable.

147. Making Precision Ball Bearings The FireBall Company manufactures ball bearings for precision equipment. One of its products is a ball bearing with a stated radius of 3 centimeters (cm). Only ball bearings with a radius within 0.01 cm of this stated radius are acceptable. If x is the radius of a ball bearing, a formula describing this situation is

$$|x - 3| \le 0.01$$

(a) Is a ball bearing of radius $x = 2.999$ acceptable?
(b) Is a ball bearing of radius $x = 2.89$ acceptable?

148. Body Temperature Normal human body temperature is 98.6°F. A temperature x that differs from normal by at least 1.5°F is considered unhealthy. A formula that describes this is

$$|x - 98.6| \ge 1.5$$

(a) Show that a temperature of 97°F is unhealthy.
(b) Show that a temperature of 100°F is not unhealthy.

149. Does $\frac{1}{3}$ equal 0.333? If not, which is larger? By how much?

150. Does $\frac{2}{3}$ equal 0.666? If not, which is larger? By how much?

Explaining Concepts: Discussion and Writing

151. Is there a positive real number "closest" to 0?

152. **Number game** I'm thinking of a number! It lies between 1 and 10; its square is rational and lies between 1 and 10. The number is larger than π. Correct to two decimal places (that is, truncated to two decimal places), name the number. Now think of your own number, describe it, and challenge a fellow student to name it.

153. Write a brief paragraph that illustrates the similarities and differences between "less than" ($<$) and "less than or equal to" ($\leq$).

154. Give a reason why the statement $5 < 8$ is true.

A.2 Geometry Essentials

OBJECTIVES 1 Use the Pythagorean Theorem and Its Converse (p. A14)
2 Know Geometry Formulas (p. A15)
3 Understand Congruent Triangles and Similar Triangles (p. A16)

1 Use the Pythagorean Theorem and Its Converse

The *Pythagorean Theorem* is a statement about *right triangles*. A **right triangle** is one that contains a **right angle**—that is, an angle of 90°. The side of the triangle opposite the 90° angle is called the **hypotenuse**; the remaining two sides are called **legs**. In Figure 13 we have used c to represent the length of the hypotenuse and a and b to represent the lengths of the legs. Notice the use of the symbol ⊿ to show the 90° angle. We now state the Pythagorean Theorem.

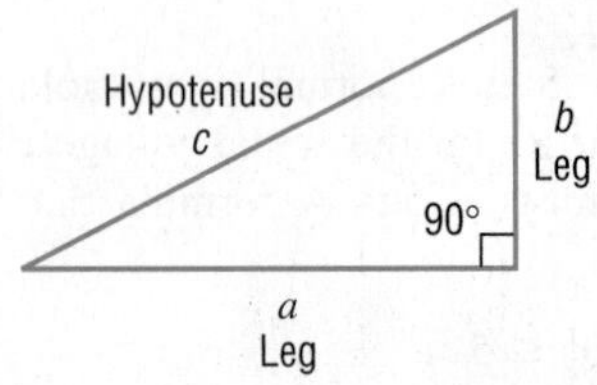

Figure 13 A right triangle

PYTHAGOREAN THEOREM

In a right triangle, the square of the length of the hypotenuse is equal to the sum of the squares of the lengths of the legs. That is, in the right triangle shown in Figure 13,

$$c^2 = a^2 + b^2 \quad (1)$$

EXAMPLE 1 **Finding the Hypotenuse of a Right Triangle**

In a right triangle, one leg has length 4 and the other has length 3. What is the length of the hypotenuse?

Solution Since the triangle is a right triangle, we use the Pythagorean Theorem with $a = 4$ and $b = 3$ to find the length c of the hypotenuse. From equation (1),

$$c^2 = a^2 + b^2$$
$$c^2 = 4^2 + 3^2 = 16 + 9 = 25$$
$$c = \sqrt{25} = 5$$

Now Work PROBLEM 15

The converse of the Pythagorean Theorem is also true.

CONVERSE OF THE PYTHAGOREAN THEOREM

In a triangle, if the square of the length of one side equals the sum of the squares of the lengths of the other two sides, the triangle is a right triangle. The 90° angle is opposite the longest side.

EXAMPLE 2 Verifying That a Triangle Is a Right Triangle

Show that a triangle whose sides are of lengths 5, 12, and 13 is a right triangle. Identify the hypotenuse.

Figure 14

Solution Square the lengths of the sides.

$$5^2 = 25 \qquad 12^2 = 144 \qquad 13^2 = 169$$

Notice that the sum of the first two squares (25 and 144) equals the third square (169). That is, because $5^2 + 12^2 = 13^2$, the triangle is a right triangle. The longest side, 13, is the hypotenuse. See Figure 14.

Now Work PROBLEM 23

EXAMPLE 3 Applying the Pythagorean Theorem

The tallest building in the world is Burj Khalifa in Dubai, United Arab Emirates, at 2717 feet and 163 floors. The observation deck is 1483 feet above ground level. How far can a person standing on the observation deck see (with the aid of a telescope)? Use 3960 miles for the radius of Earth.
Source: Council on Tall Buildings and Urban Habitat

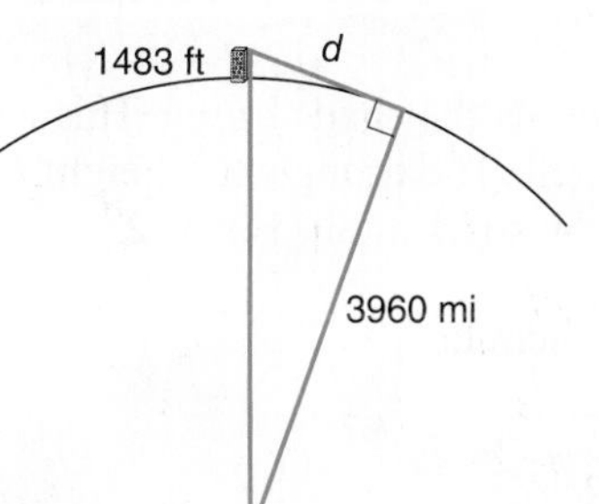

Figure 15

Solution From the center of Earth, draw two radii: one through Burj Khalifa and the other to the farthest point a person can see from the observation deck. See Figure 15. Apply the Pythagorean Theorem to the right triangle.

Since 1 mile = 5280 feet, 1483 feet $= \dfrac{1483}{5280}$ mile. Then

$$d^2 + (3960)^2 = \left(3960 + \frac{1483}{5280}\right)^2$$

$$d^2 = \left(3960 + \frac{1483}{5280}\right)^2 - (3960)^2 \approx 2224.58$$

$$d \approx 47.17$$

A person can see more than 47 miles from the observation deck.

Now Work PROBLEM 55

2 Know Geometry Formulas

Certain formulas from geometry are useful in solving algebra problems.

For a rectangle of length l and width w,

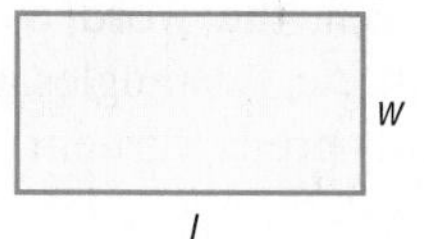

$$\text{Area} = lw \qquad \text{Perimeter} = 2l + 2w$$

For a triangle with base b and altitude h,

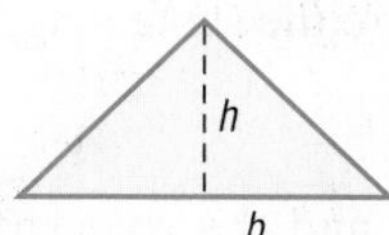

$$\text{Area} = \frac{1}{2}bh$$

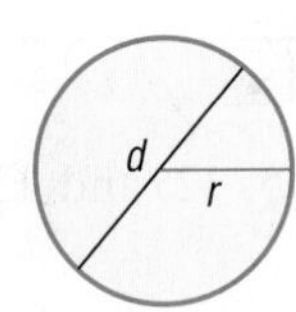

For a circle of radius r (diameter $d = 2r$),

$$\text{Area} = \pi r^2 \qquad \text{Circumference} = 2\pi r = \pi d$$

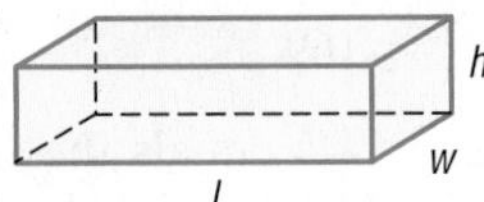

For a closed rectangular box of length l, width w, and height h,

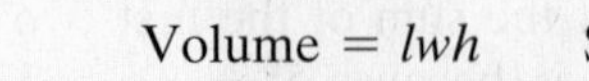

$$\text{Volume} = lwh \qquad \text{Surface area} = 2lh + 2wh + 2lw$$

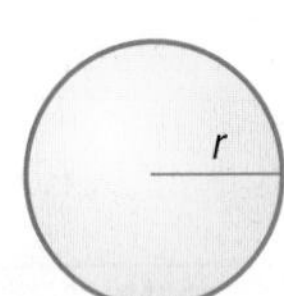

For a sphere of radius r,

$$\text{Volume} = \frac{4}{3}\pi r^3 \qquad \text{Surface area} = 4\pi r^2$$

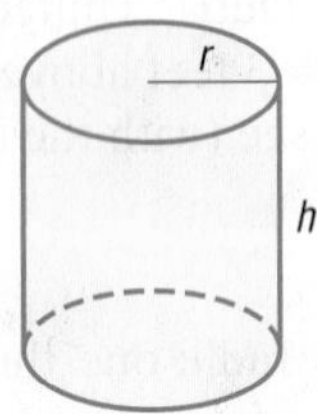

For a closed right circular cylinder of height h and radius r,

$$\text{Volume} = \pi r^2 h \qquad \text{Surface area} = 2\pi r^2 + 2\pi rh$$

Now Work PROBLEM 31

EXAMPLE 4 **Using Geometry Formulas**

A Christmas tree ornament is in the shape of a semicircle on top of a triangle. How many square centimeters (cm²) of copper is required to make the ornament if the height of the triangle is 6 cm and the base is 4 cm?

Solution See Figure 16. The amount of copper required equals the shaded area. This area is the sum of the areas of the triangle and the semicircle. The triangle has height $h = 6$ and base $b = 4$. The semicircle has diameter $d = 4$, so its radius is $r = 2$.

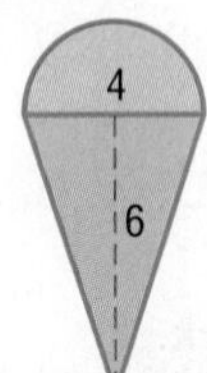

Figure 16

$$\begin{aligned}\text{Area} &= \text{Area of triangle} + \text{Area of semicircle}\\ &= \frac{1}{2}bh + \frac{1}{2}\pi r^2 = \frac{1}{2}(4)(6) + \frac{1}{2}\pi \cdot 2^2 \qquad b = 4; h = 6; r = 2\\ &= 12 + 2\pi \approx 18.28 \text{ cm}^2\end{aligned}$$

About 18.28 cm² of copper is required.

Now Work PROBLEM 49

3 Understand Congruent Triangles and Similar Triangles

> **In Words**
> Two triangles are congruent if they have the same size and shape.

Throughout the text we will make reference to triangles. We begin with a discussion of *congruent* triangles. According to dictionary.com, the word **congruent** means "coinciding exactly when superimposed." For example, two angles are congruent if they have the same measure, and two line segments are congruent if they have the same length.

DEFINITION Two triangles are **congruent** if each pair of corresponding angles have the same measure and each pair of corresponding sides are the same length.

In Figure 17, corresponding angles are equal and the corresponding sides are equal in length: $a = d$, $b = e$, and $c = f$. As a result, these triangles are congruent.

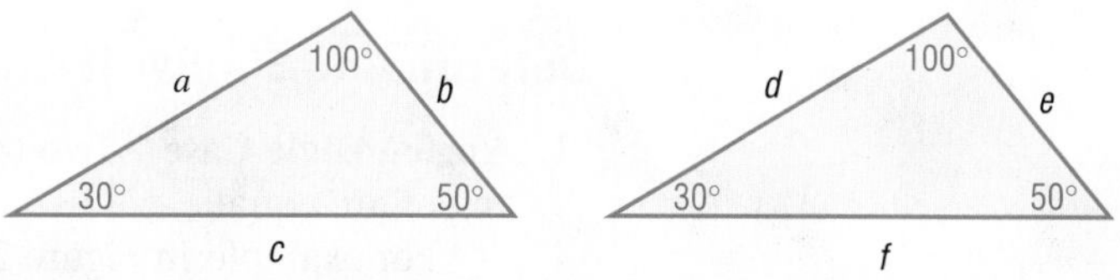

Figure 17 Congruent triangles

Actually, it is not necessary to verify that all three angles and all three sides are the same measure to determine whether two triangles are congruent.

Determining Congruent Triangles

1. **Angle–Side–Angle Case** Two triangles are congruent if two of the angles are equal and the lengths of the corresponding sides between the two angles are equal.

   For example, in Figure 18(a), the two triangles are congruent because two angles and the included side are equal.

2. **Side–Side–Side Case** Two triangles are congruent if the lengths of the corresponding sides of the triangles are equal.

   For example, in Figure 18(b), the two triangles are congruent because the three corresponding sides are all equal.

3. **Side–Angle–Side Case** Two triangles are congruent if the lengths of two corresponding sides are equal and the angles between the two sides are the same.

   For example, in Figure 18(c), the two triangles are congruent because two sides and the included angle are equal.

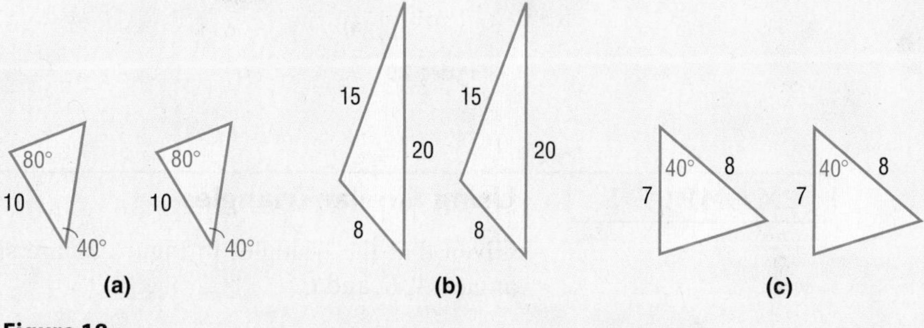

Figure 18

We contrast congruent triangles with *similar* triangles.

DEFINITION

Two triangles are **similar** if the corresponding angles are equal and the lengths of the corresponding sides are proportional.

In Words

Two triangles are similar if they have the same shape, but (possibly) different sizes.

For example, the triangles in Figure 19 are similar because the corresponding angles are equal. In addition, the lengths of the corresponding sides are proportional because each side in the triangle on the right is twice as long as each corresponding side in the triangle on the left. That is, the ratio of the corresponding sides is a constant: $\dfrac{d}{a} = \dfrac{e}{b} = \dfrac{f}{c} = 2$.

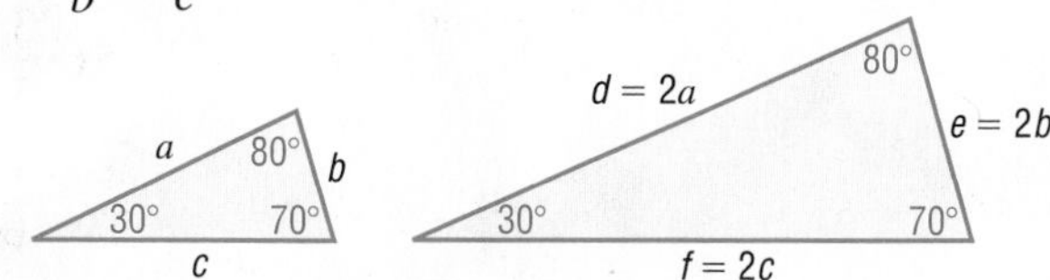

Figure 19 Similar triangles

It is not necessary to verify that all three angles are equal and all three sides are proportional to determine whether two triangles are similar.

Determining Similar Triangles

1. **Angle–Angle Case** Two triangles are similar if two of the corresponding angles are equal.
   For example, in Figure 20(a), the two triangles are similar because two angles are equal.
2. **Side–Side–Side Case** Two triangles are similar if the lengths of all three sides of each triangle are proportional.
   For example, in Figure 20(b), the two triangles are similar because
   $$\frac{10}{30} = \frac{5}{15} = \frac{6}{18} = \frac{1}{3}$$
3. **Side–Angle–Side Case** Two triangles are similar if two corresponding sides are proportional and the angles between the two sides are equal.
   For example, in Figure 20(c), the two triangles are similar because $\frac{4}{6} = \frac{12}{18} = \frac{2}{3}$ and the angles between the sides are equal.

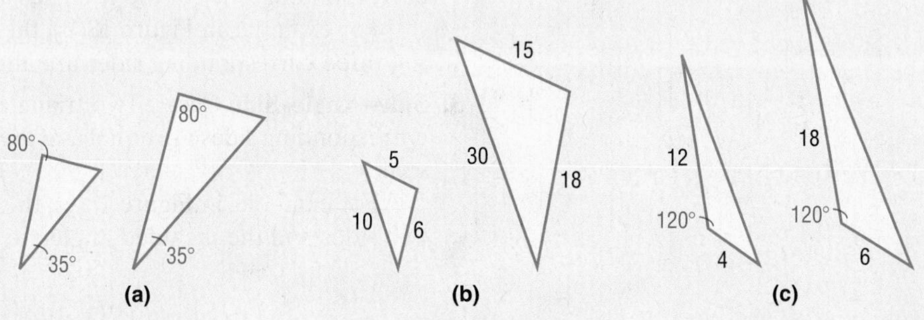

Figure 20

EXAMPLE 5 Using Similar Triangles

Given that the triangles in Figure 21 are similar, find the missing length x and the angles A, B, and C.

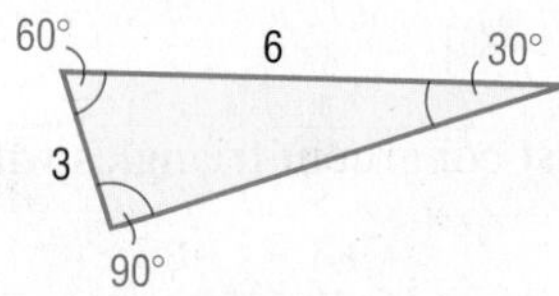

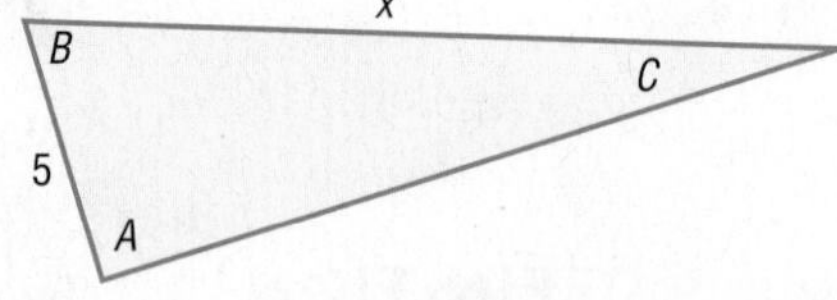

Figure 21

Solution Because the triangles are similar, corresponding angles are equal. So $A = 90°$, $B = 60°$, and $C = 30°$. Also, the corresponding sides are proportional. That is, $\frac{3}{5} = \frac{6}{x}$. We solve this equation for x.

$$\frac{3}{5} = \frac{6}{x}$$

$$5x \cdot \frac{3}{5} = 5x \cdot \frac{6}{x} \quad \text{Multiply both sides by } 5x.$$

$$3x = 30 \quad \text{Simplify.}$$

$$x = 10 \quad \text{Divide both sides by 3.}$$

The missing length is 10 units. ●

Now Work PROBLEM 43

A.2 Assess Your Understanding

Concepts and Vocabulary

1. A(n) ____ triangle is one that contains an angle of 90 degrees. The longest side is called the __________.
2. For a triangle with base b and altitude h, a formula for the area A is _________.
3. The formula for the circumference C of a circle of radius r is ________.
4. Two triangles are ______ if corresponding angles are equal and the lengths of the corresponding sides are proportional.
5. Which of the following is not a case for determining congruent triangles?
   (a) Angle–Side–Angle (b) Side–Angle–Side
   (c) Angle–Angle–Angle (d) Side-Side-Side
6. Choose the formula for the volume of a sphere of radius r.
   (a) $\frac{4}{3}\pi r^2$ (b) $\frac{4}{3}\pi r^3$ (c) $4\pi r^3$ (d) $4\pi r^2$
7. ***True or False*** In a right triangle, the square of the length of the longest side equals the sum of the squares of the lengths of the other two sides.
8. ***True or False*** The triangle with sides of lengths 6, 8, and 10 is a right triangle.
9. ***True or False*** The surface area of a sphere of radius r is $\frac{4}{3}\pi r^2$.
10. ***True or False*** The triangles shown are congruent.

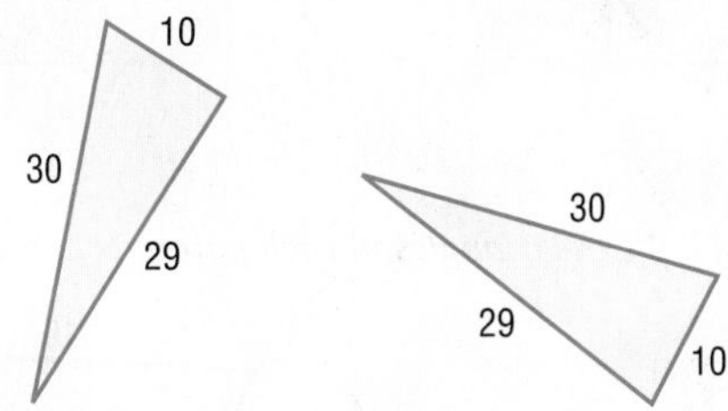

11. ***True or False*** The triangles shown are similar.

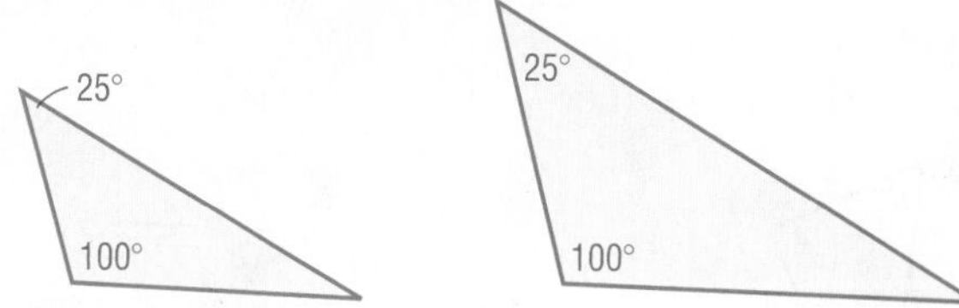

12. ***True or False*** The triangles shown are similar.

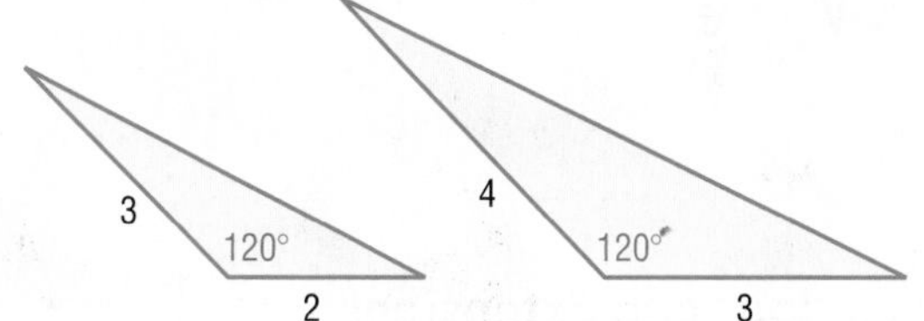

Skill Building

In Problems 13–18, the lengths of the legs of a right triangle are given. Find the hypotenuse.

13. $a = 5, \quad b = 12$
14. $a = 6, \quad b = 8$
15. $a = 10, \quad b = 24$
16. $a = 4, \quad b = 3$
17. $a = 7, \quad b = 24$
18. $a = 14, \quad b = 48$

In Problems 19–26, the lengths of the sides of a triangle are given. Determine which are right triangles. For those that are, identify the hypotenuse.

19. 3, 4, 5
20. 6, 8, 10
21. 4, 5, 6
22. 2, 2, 3
23. 7, 24, 25
24. 10, 24, 26
25. 6, 4, 3
26. 5, 4, 7
27. Find the area A of a rectangle with length 4 inches and width 2 inches.
28. Find the area A of a rectangle with length 9 centimeters and width 4 centimeters.
29. Find the area A of a triangle with height 4 inches and base 2 inches.
30. Find the area A of a triangle with height 9 centimeters and base 4 centimeters.
31. Find the area A and circumference C of a circle of radius 5 meters.
32. Find the area A and circumference C of a circle of radius 2 feet.
33. Find the volume V and surface area S of a closed rectangular box with length 8 feet, width 4 feet, and height 7 feet.
34. Find the volume V and surface area S of a closed rectangular box with length 9 inches, width 4 inches, and height 8 inches.
35. Find the volume V and surface area S of a sphere of radius 4 centimeters.
36. Find the volume V and surface area S of a sphere of radius 3 feet.
37. Find the volume V and surface area S of a closed right circular cylinder with radius 9 inches and height 8 inches.
38. Find the volume V and surface area S of a closed right circular cylinder with radius 8 inches and height 9 inches.

In Problems 39–42, find the area of the shaded region.

39.

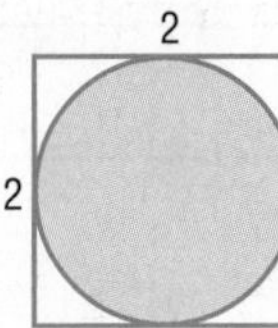

40.

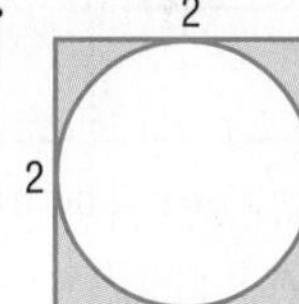

41.

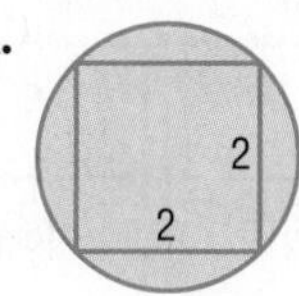

42.

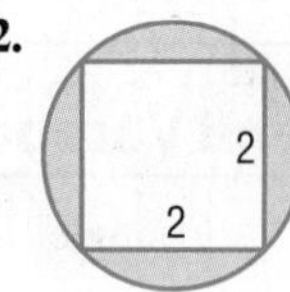

In Problems 43–46, the triangles in each pair are similar. Find the missing length x and the missing angles A, B, and C.

43.

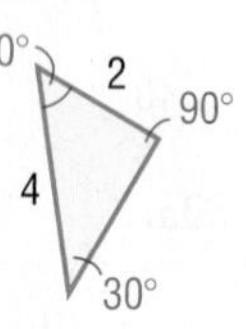

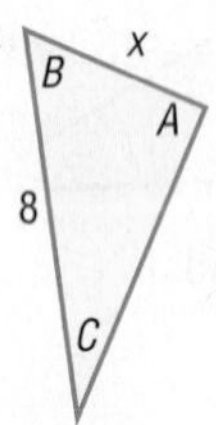

44.

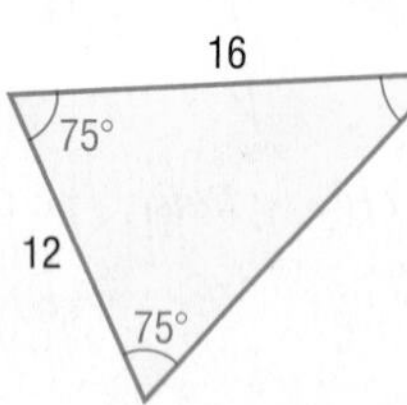

x, B, A, 6, C

45.

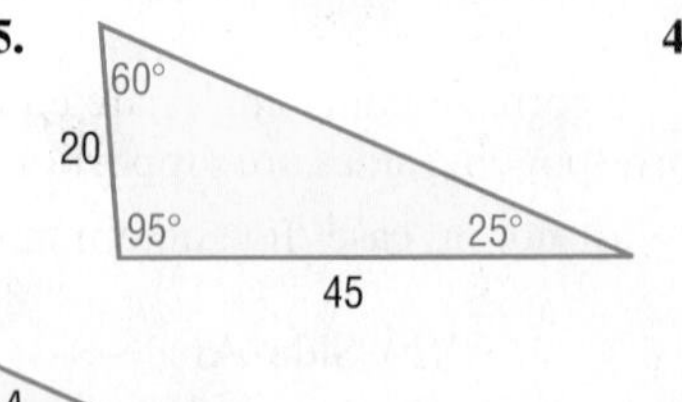

A, 30, B, C, x

46.

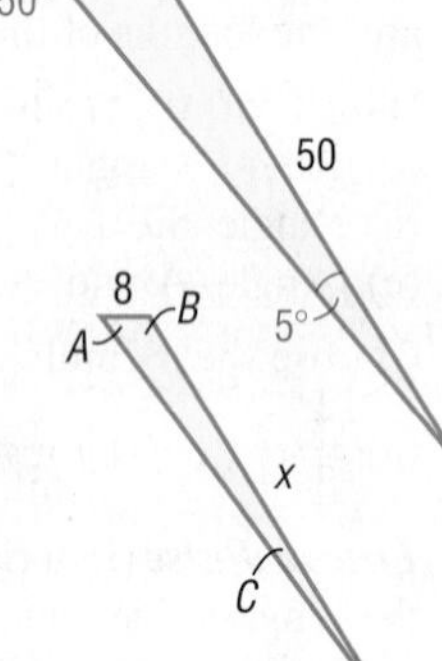

Applications and Extensions

47. How many feet has a wheel with a diameter of 16 inches traveled after four revolutions?

48. How many revolutions will a circular disk with a diameter of 4 feet have completed after it has rolled 20 feet?

49. In the figure shown, $ABCD$ is a square, with each side of length 6 feet. The width of the border (shaded portion) between the outer square $EFGH$ and $ABCD$ is 2 feet. Find the area of the border.

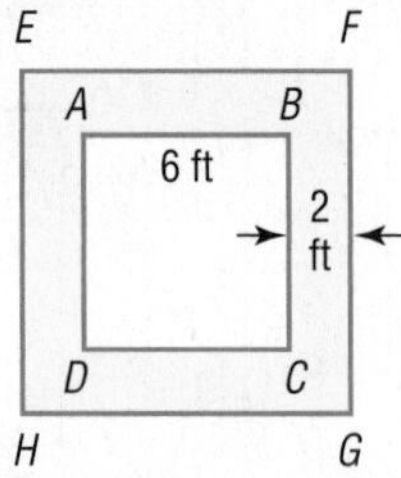

50. Refer to the figure. Square $ABCD$ has an area of 100 square feet; square $BEFG$ has an area of 16 square feet. What is the area of the triangle CGF?

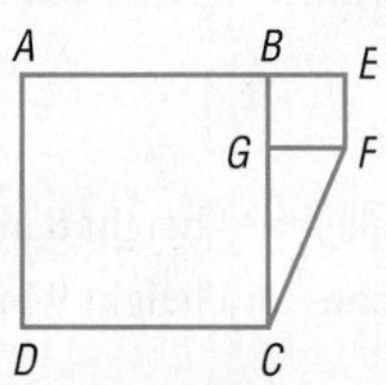

51. Architecture A **Norman window** consists of a rectangle surmounted by a semicircle. Find the area of the Norman window shown in the illustration. How much wood frame is needed to enclose the window?

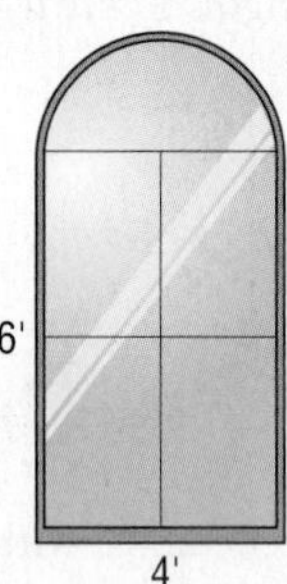

52. Construction A circular swimming pool that is 20 feet in diameter is enclosed by a wooden deck that is 3 feet wide. What is the area of the deck? How much fence is required to enclose the deck?

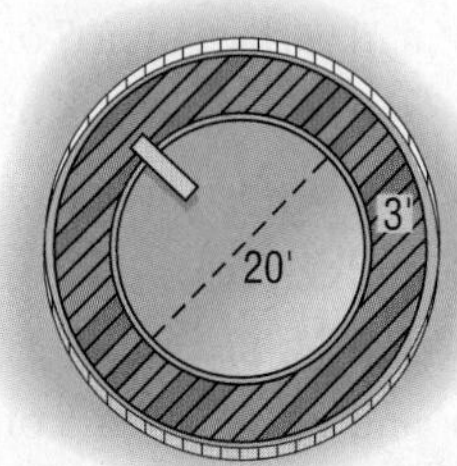

53. How Tall Is the Great Pyramid? The ancient Greek philosopher Thales of Miletus is reported on one occasion to have visited Egypt and calculated the height of the Great Pyramid of Cheops by means of shadow reckoning. Thales knew that each side of the base of the pyramid was 252 paces and that his own height was 2 paces. He measured the length of the pyramid's shadow to be 114 paces and determined the length of his shadow to be 3 paces. See the illustration. Using similar triangles, determine the height of the Great Pyramid in terms of the number of paces.

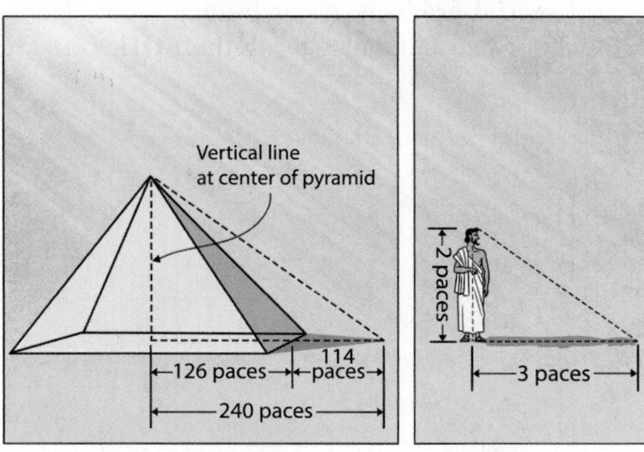

Source: Diggins, Julie E, *String Straightedge and Shadow: The Story of Geometry*, 2003, Whole Spirit Press, http://wholespiritpress.com.

54. The Bermuda Triangle Karen is doing research on the Bermuda Triangle which she defines roughly by Hamilton, Bermuda; San Juan, Puerto Rico; and Fort Lauderdale, Florida. On her atlas Karen measures the straight-line distances from Hamilton to Fort Lauderdale, Fort Lauderdale to San Juan, and San Juan to Hamilton to be approximately 57 millimeters (mm), 58 mm, and 53.5 mm respectively. If the actual distance from Fort Lauderdale to San Juan is 1046 miles, approximate the actual distances from San Juan to Hamilton and from Hamilton to Fort Lauderdale.

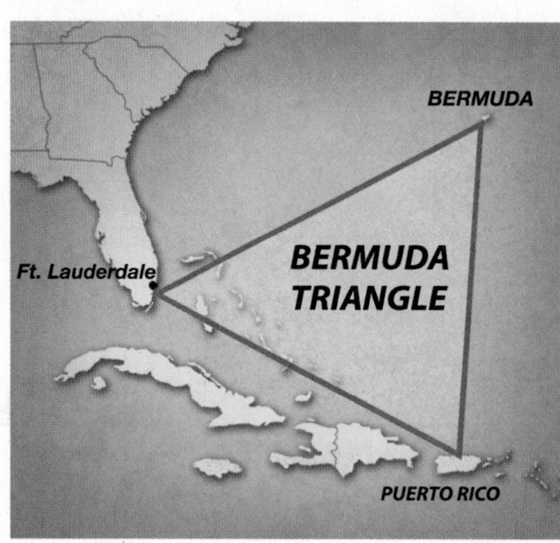

In Problems 55–57, use the facts that the radius of Earth is 3960 miles and 1 mile = 5280 feet.

55. **How Far Can You See?** The conning tower of the U.S.S. *Silversides,* a World War II submarine now permanently stationed in Muskegon, Michigan, is approximately 20 feet above sea level. How far can you see from the conning tower?

56. **How Far Can You See?** A person who is 6 feet tall is standing on the beach in Fort Lauderdale, Florida, and looks out onto the Atlantic Ocean. Suddenly, a ship appears on the horizon. How far is the ship from shore?

57. **How Far Can You See?** The deck of a destroyer is 100 feet above sea level. How far can a person see from the deck? How far can a person see from the bridge, which is 150 feet above sea level?

58. Suppose that m and n are positive integers with $m > n$. If $a = m^2 - n^2$, $b = 2mn$, and $c = m^2 + n^2$, show that a, b, and c are the lengths of the sides of a right triangle. (This formula can be used to find the sides of a right triangle that are integers, such as 3, 4, 5; 5, 12, 13; and so on. Such triplets of integers are called **Pythagorean triples.**)

Explaining Concepts: Discussion and Writing

59. You have 1000 feet of flexible pool siding and intend to construct a swimming pool. Experiment with rectangular-shaped pools with perimeters of 1000 feet. How do their areas vary? What is the shape of the rectangle with the largest area? Now compute the area enclosed by a circular pool with a perimeter (circumference) of 1000 feet. What would be your choice of shape for the pool? If rectangular, what is your preference for dimensions? Justify your choice. If your only consideration is to have a pool that encloses the most area, what shape should you use?

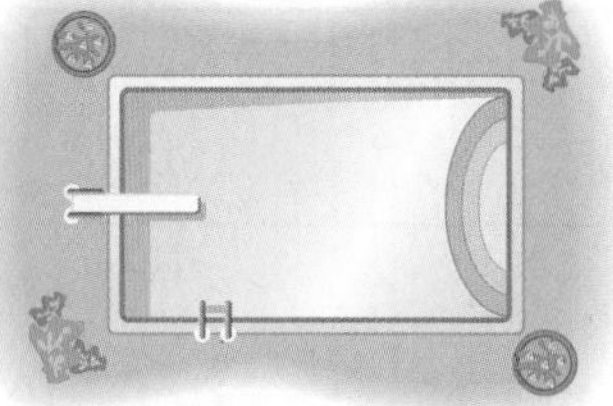

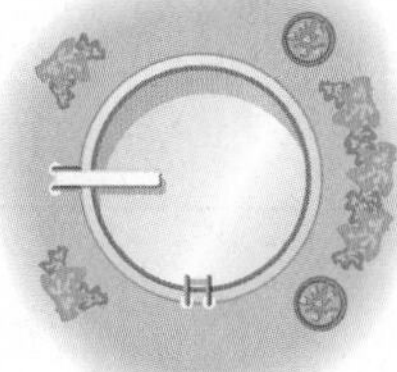

60. **The Gibb's Hill Lighthouse, Southampton, Bermuda,** in operation since 1846, stands 117 feet high on a hill 245 feet high, so its beam of light is 362 feet above sea level. A brochure states that the light itself can be seen on the horizon about 26 miles distant. Verify the accuracy of this information. The brochure further states that ships 40 miles away can see the light and that planes flying at 10,000 feet can see it 120 miles away. Verify the accuracy of these statements. What assumption did the brochure make about the height of the ship?

A.3 Factoring Polynomials; Completing the Square

OBJECTIVES 1 Know Formulas for Special Products (p. A22)
2 Factor Polynomials (p. A23)
3 Complete the Square (p. A24)

1 Know Formulas for Special Products

Certain products, which we call **special products**, occur frequently in algebra. For example, we can find the product of two binomials using the **FOIL** (*F*irst, *O*uter, *I*nner, *L*ast) method.

$$(ax + b)(cx + d) = ax(cx + d) + b(cx + d)$$

$$= \underbrace{ax \cdot cx}_{\text{First}} + \underbrace{ax \cdot d}_{\text{Outer}} + \underbrace{b \cdot cx}_{\text{Inner}} + \underbrace{b \cdot d}_{\text{Last}}$$

$$= acx^2 + adx + bcx + bd$$

$$= acx^2 + (ad + bc)x + bd$$

EXAMPLE 1 **Using FOIL**

(a) $(x - 3)(x + 3) = \underset{F}{x^2} + \underset{O}{3x} - \underset{I}{3x} - \underset{L}{9} = x^2 - 9$

(b) $(x + 2)^2 = (x + 2)(x + 2) = x^2 + 2x + 2x + 4 = x^2 + 4x + 4$

(c) $(x - 3)^2 = (x - 3)(x - 3) = x^2 - 3x - 3x + 9 = x^2 - 6x + 9$

(d) $(x + 3)(x + 1) = x^2 + x + 3x + 3 = x^2 + 4x + 3$

(e) $(2x + 1)(3x + 4) = 6x^2 + 8x + 3x + 4 = 6x^2 + 11x + 4$ ●

Now Work PROBLEM 9

Some products have been given special names because of their form. In the list that follows, x and a are real numbers.

Difference of Two Squares

$$(x - a)(x + a) = x^2 - a^2 \quad \textbf{(1)}$$

Squares of Binomials, or Perfect Squares

$$(x + a)^2 = x^2 + 2ax + a^2 \quad \textbf{(2a)}$$

$$(x - a)^2 = x^2 - 2ax + a^2 \quad \textbf{(2b)}$$

Cubes of Binomials, or Perfect Cubes

$$(x + a)^3 = x^3 + 3ax^2 + 3a^2x + a^3 \quad \textbf{(3a)}$$

$$(x - a)^3 = x^3 - 3ax^2 + 3a^2x - a^3 \quad \textbf{(3b)}$$

Difference of Two Cubes

$$(x - a)(x^2 + ax + a^2) = x^3 - a^3 \qquad (4)$$

Sum of Two Cubes

$$(x + a)(x^2 - ax + a^2) = x^3 + a^3 \qquad (5)$$

Now Work PROBLEMS 13, 17, AND 21

2 Factor Polynomials

Consider the following product:

$$(2x + 3)(x - 4) = 2x^2 - 5x - 12$$

The two polynomials on the left side are called **factors** of the polynomial on the right side. Expressing a given polynomial as a product of other polynomials, that is, finding the factors of a polynomial, is called **factoring**.

We shall restrict our discussion here to factoring polynomials in one variable into products of polynomials in one variable, where all coefficients are integers. We call this **factoring over the integers**.

Any polynomial can be written as the product of 1 times itself or as -1 times its additive inverse. If a polynomial cannot be written as the product of two other polynomials (excluding 1 and -1), then the polynomial is said to be **prime**. When a polynomial has been written as a product consisting only of prime factors, it is said to be **factored completely**. Examples of prime polynomials (over the integers) are

$$2 \quad 3 \quad 5 \quad x \quad x + 1 \quad x - 1 \quad 3x + 4 \quad x^2 + 4$$

COMMENT Over the real numbers, $3x + 4$ factors into $3\left(x + \frac{4}{3}\right)$. It is the noninteger $\frac{4}{3}$ that causes $3x + 4$ to be prime over the integers. ■

The first factor to look for in a factoring problem is a common monomial factor present in each term of the polynomial. If one is present, use the Distributive Property to factor it out.

EXAMPLE 2 Identifying Common Monomial Factors

Polynomial	Common Monomial Factor	Remaining Factor	Factored Form
$2x + 4$	2	$x + 2$	$2x + 4 = 2(x + 2)$
$3x - 6$	3	$x - 2$	$3x - 6 = 3(x - 2)$
$2x^2 - 4x + 8$	2	$x^2 - 2x + 4$	$2x^2 - 4x + 8 = 2(x^2 - 2x + 4)$
$8x - 12$	4	$2x - 3$	$8x - 12 = 4(2x - 3)$
$x^2 + x$	x	$x + 1$	$x^2 + x = x(x + 1)$
$x^3 - 3x^2$	x^2	$x - 3$	$x^3 - 3x^2 = x^2(x - 3)$
$6x^2 + 9x$	$3x$	$2x + 3$	$6x^2 + 9x = 3x(2x + 3)$

●

Notice that, once all common monomial factors have been removed from a polynomial, the remaining factor is either a prime polynomial of degree 1 or a polynomial of degree 2 or higher. (Do you see why?)

The list of special products (1) through (5) given earlier provides a list of factoring formulas when the equations are read from right to left. For example, equation (1) states that if the polynomial is the difference of two squares, $x^2 - a^2$, it can be factored into $(x - a)(x + a)$. The following example illustrates several factoring techniques.

EXAMPLE 3 **Factoring Polynomials**

Factor completely each polynomial.

(a) $x^4 - 16$ (b) $x^3 - 1$ (c) $9x^2 - 6x + 1$
(d) $x^2 + 4x - 12$ (e) $3x^2 + 10x - 8$ (f) $x^3 - 4x^2 + 2x - 8$

Solution

(a) $x^4 - 16 = (x^2 - 4)(x^2 + 4) = (x - 2)(x + 2)(x^2 + 4)$

Difference of squares; Difference of squares

(b) $x^3 - 1 = (x - 1)(x^2 + x + 1)$

Difference of cubes

(c) $9x^2 - 6x + 1 = (3x - 1)^2$

Perfect square

(d) $x^2 + 4x - 12 = (x + 6)(x - 2)$

The product of 6 and −2 is −12, and the sum of 6 and −2 is 4.

(e) $3x^2 + 10x - 8 = (3x - 2)(x + 4)$

$12x - 2x = 10x$; $3x^2$, -8

(f) $x^3 - 4x^2 + 2x - 8 = (x^3 - 4x^2) + (2x - 8)$

Group terms

$= x^2(x - 4) + 2(x - 4) = (x^2 + 2)(x - 4)$

Distributive Property; Distributive Property

COMMENT The technique used in part (f) is called **factoring by grouping.** ■

Now Work PROBLEMS 33, 49, AND 83

3 Complete the Square

The idea behind completing the square in one variable is to "adjust" an expression of the form $x^2 + bx$ to make it a perfect square. Perfect squares are trinomials of the form

$$x^2 + 2ax + a^2 = (x + a)^2 \quad \text{or} \quad x^2 - 2ax + a^2 = (x - a)^2$$

For example, $x^2 + 6x + 9$ is a perfect square because $x^2 + 6x + 9 = (x + 3)^2$. And $p^2 - 12p + 36$ is a perfect square because $p^2 - 12p + 36 = (p - 6)^2$.

So how do we "adjust" $x^2 + bx$ to make it a perfect square? We do it by adding a number. For example, to make $x^2 + 6x$ a perfect square, add 9. But how do we know to add 9? If we divide the coefficient of the first-degree term, 6, by 2, and then square the result, we obtain 9. This approach works in general.

WARNING To use $\left(\frac{1}{2}b\right)^2$ to complete the square, the coefficient of the x^2 term must be 1. ■

Completing the Square of $x^2 + bx$

Identify the coefficient of the first-degree term. Multiply this coefficient by $\frac{1}{2}$ and then square the result. That is, determine the value of b in $x^2 + bx$ and compute $\left(\frac{1}{2}b\right)^2$.

EXAMPLE 4

Completing the Square

Determine the number that must be added to each expression to complete the square. Then factor the expression.

Start	Add	Result	Factored Form
$y^2 + 8y$	$\left(\frac{1}{2}\cdot 8\right)^2 = 16$	$y^2 + 8y + 16$	$(y + 4)^2$
$x^2 + 12x$	$\left(\frac{1}{2}\cdot 12\right)^2 = 36$	$x^2 + 12x + 36$	$(x + 6)^2$
$a^2 - 20a$	$\left(\frac{1}{2}\cdot(-20)\right)^2 = 100$	$a^2 - 20a + 100$	$(a - 10)^2$
$p^2 - 5p$	$\left(\frac{1}{2}\cdot(-5)\right)^2 = \frac{25}{4}$	$p^2 - 5p + \frac{25}{4}$	$\left(p - \frac{5}{2}\right)^2$

Notice that the factored form of a perfect square is either

$$x^2 + bx + \left(\frac{b}{2}\right)^2 = \left(x + \frac{b}{2}\right)^2 \quad \text{or} \quad x^2 - bx + \left(\frac{b}{2}\right)^2 = \left(x - \frac{b}{2}\right)^2$$

Now Work PROBLEM 73

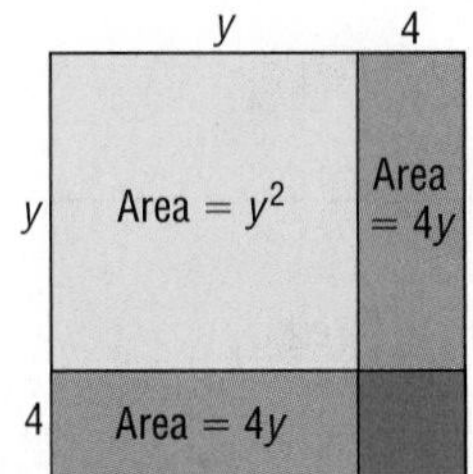

Figure 22

Are you wondering why we refer to making an expression a perfect square as "completing the square"? Look at the square in Figure 22. Its area is $(y + 4)^2$. The yellow area is y^2 and each orange area is $4y$ (for a total area of $8y$). The sum of these areas is $y^2 + 8y$. To complete the square, we need to add the area of the green region: $4 \cdot 4 = 16$. As a result, $y^2 + 8y + 16 = (y + 4)^2$.

A.3 Assess Your Understanding

Concepts and Vocabulary

1. $(x^2 - 4)(x^2 + 4) =$ ______.
2. $(x - 2)(x^2 + 2x + 4) =$ ______.
3. ***True or False*** The polynomial $x^2 + 4$ is prime.
4. ***True or False*** $3x^3 - 2x^2 - 6x + 4 = (3x - 2)(x^2 + 2)$.
5. To complete the square of the expression $x^2 + 5x$, you would ____ the number ____.
6. Choose the best description of $x^2 - 64$.
   (a) Prime (b) Difference of two squares
   (c) Difference of two cubes (d) Perfect Square
7. Choose the complete factorization of $4x^2 - 8x - 60$.
   (a) $2(x + 3)(x - 5)$ (b) $4(x^2 - 2x - 15)$
   (c) $(2x + 6)(2x - 10)$ (d) $4(x + 3)(x - 5)$
8. To complete the square of $x^2 + bx$, use which of the following?
   (a) $(2b)^2$ (b) $2b^2$ (c) $\left(\frac{1}{2}b\right)^2$ (d) $\frac{1}{2}b^2$

Skill Building

In Problems 9–24, multiply out each expression. Express your answer as a single polynomial in standard form.

9. $(x + 2)(x + 4)$
10. $(x + 3)(x + 5)$
11. $(2x + 5)(x + 2)$
12. $(3x + 1)(2x + 1)$
13. $(x - 7)(x + 7)$
14. $(x - 1)(x + 1)$
15. $(2x + 3)(2x - 3)$
16. $(3x + 2)(3x - 2)$
17. $(x + 4)^2$
18. $(x - 5)^2$
19. $(2x - 3)^2$
20. $(3x - 4)^2$
21. $(x - 2)^3$
22. $(x + 1)^3$
23. $(2x + 1)^3$
24. $(3x - 2)^3$

In Problems 25–72, factor completely each polynomial. If the polynomial cannot be factored, say it is prime.

25. $x^2 - 36$
26. $x^2 - 9$
27. $2 - 8x^2$
28. $3 - 27x^2$
29. $x^2 + 11x + 10$
30. $x^2 + 5x + 4$
31. $x^2 - 10x + 21$
32. $x^2 - 6x + 8$
33. $4x^2 - 8x + 32$
34. $3x^2 - 12x + 15$
35. $x^2 + 4x + 16$
36. $x^2 + 12x + 36$
37. $15 + 2x - x^2$
38. $14 + 6x - x^2$
39. $3x^2 - 12x - 36$
40. $x^3 + 8x^2 - 20x$
41. $y^4 + 11y^3 + 30y^2$
42. $3y^3 - 18y^2 - 48y$
43. $4x^2 + 12x + 9$
44. $9x^2 - 12x + 4$
45. $6x^2 + 8x + 2$
46. $8x^2 + 6x - 2$
47. $x^4 - 81$
48. $x^4 - 1$
49. $x^6 - 2x^3 + 1$
50. $x^6 + 2x^3 + 1$
51. $x^7 - x^5$
52. $x^8 - x^5$
53. $16x^2 + 24x + 9$
54. $9x^2 - 24x + 16$
55. $5 + 16x - 16x^2$
56. $5 + 11x - 16x^2$
57. $4y^2 - 16y + 15$
58. $9y^2 + 9y - 4$
59. $1 - 8x^2 - 9x^4$
60. $4 - 14x^2 - 8x^4$
61. $x(x + 3) - 6(x + 3)$
62. $5(3x - 7) + x(3x - 7)$
63. $(x + 2)^2 - 5(x + 2)$
64. $(x - 1)^2 - 2(x - 1)$
65. $(3x - 2)^3 - 27$
66. $(5x + 1)^3 - 1$
67. $3(x^2 + 10x + 25) - 4(x + 5)$
68. $7(x^2 - 6x + 9) + 5(x - 3)$
69. $x^3 + 2x^2 - x - 2$
70. $x^3 - 3x^2 - x + 3$
71. $x^4 - x^3 + x - 1$
72. $x^4 + x^3 + x + 1$

In Problems 73–78, determine the number that should be added to complete the square of each expression. Then factor each expression.

73. $x^2 + 10x$
74. $p^2 + 14p$
75. $y^2 - 6y$
76. $x^2 - 4x$
77. $x^2 - \frac{1}{2}x$
78. $x^2 + \frac{1}{3}x$

Applications and Extensions

In Problems 79–88, expressions that occur in calculus are given. Factor completely each expression.

79. $2(3x + 4)^2 + (2x + 3)\cdot 2(3x + 4)\cdot 3$
80. $5(2x + 1)^2 + (5x - 6)\cdot 2(2x + 1)\cdot 2$
81. $2x(2x + 5) + x^2\cdot 2$
82. $3x^2(8x - 3) + x^3\cdot 8$
83. $2(x + 3)(x - 2)^3 + (x + 3)^2\cdot 3(x - 2)^2$
84. $4(x + 5)^3(x - 1)^2 + (x + 5)^4\cdot 2(x - 1)$
85. $(4x - 3)^2 + x\cdot 2(4x - 3)\cdot 4$
86. $3x^2(3x + 4)^2 + x^3\cdot 2(3x + 4)\cdot 3$
87. $2(3x - 5)\cdot 3(2x + 1)^3 + (3x - 5)^2\cdot 3(2x + 1)^2\cdot 2$
88. $3(4x + 5)^2\cdot 4(5x + 1)^2 + (4x + 5)^3\cdot 2(5x + 1)\cdot 5$
89. Show that $x^2 + 4$ is prime.
90. Show that $x^2 + x + 1$ is prime.

Explaining Concepts: Discussion and Writing

91. Do you prefer to memorize the rule for the square of a binomial $(x + a)^2$ or to use FOIL to obtain the product? Write a brief position paper defending your choice.

92. Make up a polynomial that factors into a perfect square.

93. Explain to a fellow student what you look for first when presented with a factoring problem. What do you do next?

A.4 Solving Equations

PREPARING FOR THIS SECTION *Before getting started, review the following:*

- Factoring Polynomials (Appendix A, Section A.3, pp. A23–A24)
- Zero-Product Property (Appendix A, Section A.1, p. A4)
- Square Roots (Appendix A, Section A.1, pp. A9–A10)
- Absolute Value (Appendix A, Section A.1, pp. A5–A6)
- Complete the Square (Appendix A, Section A.3, pp. A24–A25)

Now Work the **'Are You Prepared?'** problems on page A34.

OBJECTIVES

1 Solve Equations by Factoring (p. A29)
2 Solve Equations Involving Absolute Value (p. A29)
3 Solve a Quadratic Equation by Factoring (p. A30)
4 Solve a Quadratic Equation by Completing the Square (p. A31)
5 Solve a Quadratic Equation Using the Quadratic Formula (p. A32)

An **equation in one variable** is a statement in which two expressions, at least one containing the variable, are equal. The expressions are called the **sides** of the equation. Since an equation is a statement, it may be true or false, depending on the value of the variable. Unless otherwise restricted, the admissible values of the variable are those in the domain of the variable. These admissible values of the variable, if any, that result in a true statement are called **solutions**, or **roots**, of the equation. To **solve an equation** means to find all the solutions of the equation.

For example, the following are all equations in one variable, x:

$$x + 5 = 9 \qquad x^2 + 5x = 2x - 2 \qquad \frac{x^2 - 4}{x + 1} = 0 \qquad \sqrt{x^2 + 9} = 5$$

The first of these statements, $x + 5 = 9$, is true when $x = 4$ and false for any other choice of x. That is, 4 is a solution of the equation $x + 5 = 9$. We also say that 4 **satisfies** the equation $x + 5 = 9$, because, when 4 is substituted for x, a true statement results.

Sometimes an equation will have more than one solution. For example, the equation

$$\frac{x^2 - 4}{x + 1} = 0$$

has $x = -2$ and $x = 2$ as solutions.

Usually, we will write the solution of an equation in set notation. This set is called the **solution set** of the equation. For example, the solution set of the equation $x^2 - 9 = 0$ is $\{-3, 3\}$.

Some equations have no real solution. For example, $x^2 + 9 = 5$ has no real solution, because there is no real number whose square, when added to 9, equals 5.

An equation that is satisfied for every value of the variable for which both sides are defined is called an **identity**. For example, the equation

$$3x + 5 = x + 3 + 2x + 2$$

is an identity, because this statement is true for any real number x.

> One method for solving an equation is to replace the original equation by a succession of **equivalent equations**, equations having the same solution set, until an equation with an obvious solution is obtained.

For example, consider the following succession of equivalent equations:

$$\begin{aligned} 2x + 3 &= 13 \\ 2x &= 10 \\ x &= 5 \end{aligned}$$

We conclude that the solution set of the original equation is $\{5\}$.

How do we obtain equivalent equations? In general, there are five ways.

Procedures That Result in Equivalent Equations

1. Interchange the two sides of the equation:

   Replace $3 = x$ by $x = 3$

2. Simplify the sides of the equation by combining like terms, eliminating parentheses, and so on:

   Replace $(x + 2) + 6 = 2x + (x + 1)$
   by $x + 8 = 3x + 1$

3. Add or subtract the same expression on both sides of the equation:

   Replace $3x - 5 = 4$
   by $(3x - 5) + 5 = 4 + 5$

4. Multiply or divide both sides of the equation by the same nonzero expression:

   Replace $\dfrac{3x}{x-1} = \dfrac{6}{x-1} \qquad x \neq 1$
   by $\dfrac{3x}{x-1} \cdot (x - 1) = \dfrac{6}{x-1} \cdot (x - 1)$

5. If one side of the equation is 0 and the other side can be factored, then we may use the Zero-Product Property* and set each factor equal to 0:

   Replace $x(x - 3) = 0$
   by $x = 0$ or $x - 3 = 0$

WARNING Squaring both sides of an equation does not necessarily lead to an equivalent equation. For example, $x = 3$ has one solution, but $x^2 = 9$ has two solutions, $x = -3$ and $x = 3$. ■

Whenever it is possible to solve an equation in your head, do so. For example,

The solution of $2x = 8$ is $x = 4$.
The solution of $3x - 15 = 0$ is $x = 5$.

Now Work PROBLEM 15

EXAMPLE 1 **Solving an Equation**

Solve the equation: $3x - 5 = 4$

Solution Replace the original equation by a succession of equivalent equations.

$$3x - 5 = 4$$
$$(3x - 5) + 5 = 4 + 5 \quad \text{Add 5 to both sides.}$$
$$3x = 9 \quad \text{Simplify.}$$
$$\frac{3x}{3} = \frac{9}{3} \quad \text{Divide both sides by 3.}$$
$$x = 3 \quad \text{Simplify.}$$

The last equation, $x = 3$, has the single solution 3. All these equations are equivalent, so 3 is the only solution of the original equation, $3x - 5 = 4$.

*The Zero-Product Property says that if $ab = 0$, then $a = 0$ or $b = 0$ or both equal 0.

Check: Check the solution by substituting 3 for x in the original equation.

$$3x - 5 = 3(3) - 5 = 9 - 5 = 4$$

The solution checks. The solution set is $\{3\}$.

Now Work PROBLEMS 29 AND 35

1 Solve Equations by Factoring

EXAMPLE 2 **Solving Equations by Factoring**

Solve the equations: (a) $x^3 = 4x$ (b) $x^3 - x^2 - 4x + 4 = 0$

Solution (a) Begin by collecting all terms on one side. This results in 0 on one side and an expression to be factored on the other.

$$x^3 = 4x$$

$$x^3 - 4x = 0$$

$$x(x^2 - 4) = 0 \quad \text{Factor.}$$

$$x(x - 2)(x + 2) = 0 \quad \text{Factor again.}$$

$$x = 0 \text{ or } x - 2 = 0 \text{ or } x + 2 = 0 \quad \text{Apply the Zero-Product Property.}$$

$$x = 0 \text{ or } x = 2 \text{ or } x = -2 \quad \text{Solve for } x.$$

Check: $x = -2$: $(-2)^3 = -8$ and $4(-2) = -8$ -2 is a solution.

$x = 0$: $0^3 = 0$ and $4 \cdot 0 = 0$ 0 is a solution.

$x = 2$: $2^3 = 8$ and $4 \cdot 2 = 8$ 4 is a solution.

The solution set is $\{-2, 0, 2\}$.

(b) Group the terms of $x^3 - x^2 - 4x + 4 = 0$ as follows:

$$(x^3 - x^2) - (4x - 4) = 0$$

Factor out x^2 from the first grouping and 4 from the second.

$$x^2(x - 1) - 4(x - 1) = 0$$

This reveals the common factor $(x - 1)$, so

$$(x^2 - 4)(x - 1) = 0$$

$$(x - 2)(x + 2)(x - 1) = 0 \quad \text{Factor again.}$$

$$x - 2 = 0 \text{ or } x + 2 = 0 \text{ or } x - 1 = 0 \quad \text{Apply the Zero-Product Property.}$$

$$x = 2 \quad x = -2 \quad x = 1 \quad \text{Solve for } x.$$

Check:

$x = -2$: $(-2)^3 - (-2)^2 - 4(-2) + 4 = -8 - 4 + 8 + 4 = 0$ -2 is a solution.

$x = 1$: $1^3 - 1^2 - 4(1) + 4 = 1 - 1 - 4 + 4 = 0$ 1 is a solution.

$x = 2$: $2^3 - 2^2 - 4(2) + 4 = 8 - 4 - 8 + 4 = 0$ 2 is a solution.

The solution set is $\{-2, 1, 2\}$.

Now Work PROBLEM 39

2 Solve Equations Involving Absolute Value

On the real number line, there are two points whose distance from the origin is 5 units, –5 and 5, so the equation $|x| = 5$ will have the solution set $\{-5, 5\}$.

EXAMPLE 3 **Solving an Equation Involving Absolute Value**

Solve the equation: $|x + 4| = 13$

Solution There are two possibilities.

$$x + 4 = 13 \quad \text{or} \quad x + 4 = -13$$
$$x = 9 \quad \text{or} \quad x = -17$$

The solution set is $\{-17, 9\}$.

Now Work PROBLEM 51

3 Solve a Quadratic Equation by Factoring

DEFINITION

A **quadratic equation** is an equation equivalent to one of the form

$$ax^2 + bx + c = 0 \tag{1}$$

where a, b, and c are real numbers and $a \neq 0$.

A quadratic equation written in the form $ax^2 + bx + c = 0$ is said to be in **standard form**.

Sometimes, a quadratic equation is called a **second-degree equation** because, when it is in standard form, the left side is a polynomial of degree 2.

When a quadratic equation is written in standard form, it may be possible to factor the expression on the left side into the product of two first-degree polynomials. Then, by using the Zero-Product Property and setting each factor equal to 0, the resulting linear equations can be solved to obtain the solutions of the quadratic equation.

EXAMPLE 4 **Solving a Quadratic Equation by Factoring**

Solve the equation: $2x^2 = x + 3$

Solution Put the equation $2x^2 = x + 3$ in standard form by subtracting x and 3 from both sides.

$$2x^2 = x + 3$$
$$2x^2 - x - 3 = 0 \quad \text{Subtract } x \text{ and 3 from both sides.}$$

The left side may now be factored as

$$(2x - 3)(x + 1) = 0 \quad \text{Factor.}$$

so that

$$2x - 3 = 0 \quad \text{or} \quad x + 1 = 0 \quad \text{Use the Zero-Product Property.}$$
$$x = \frac{3}{2} \qquad x = -1 \quad \text{Solve.}$$

The solution set is $\left\{-1, \frac{3}{2}\right\}$.

When the left side factors into two linear equations with the same solution, the quadratic equation is said to have a **repeated solution**. This solution is also called a **root of multiplicity 2**, or a **double root**.

EXAMPLE 5 **Solving a Quadratic Equation by Factoring**

Solve the equation: $9x^2 - 6x + 1 = 0$

Solution This equation is already in standard form, and the left side can be factored.

$$9x^2 - 6x + 1 = 0$$
$$(3x - 1)(3x - 1) = 0 \quad \text{Factor.}$$

so

$$x = \frac{1}{3} \quad \text{or} \quad x = \frac{1}{3} \quad \text{Solve for } x.$$

This equation has only the repeated solution $\frac{1}{3}$. The solution set is $\left\{\frac{1}{3}\right\}$. ●

Now Work PROBLEM 69

The Square Root Method

Suppose that we wish to solve the quadratic equation

$$x^2 = p \qquad (2)$$

where $p \geq 0$ is a nonnegative number. Proceeding as in the earlier examples,

$$x^2 - p = 0 \quad \text{Put in standard form.}$$
$$(x - \sqrt{p})(x + \sqrt{p}) = 0 \quad \text{Factor (over the real numbers).}$$
$$x = \sqrt{p} \quad \text{or} \quad x = -\sqrt{p} \quad \text{Solve.}$$

We have the following result:

> If $x^2 = p$ and $p \geq 0$, then $x = \sqrt{p}$ or $x = -\sqrt{p}$. **(3)**

When statement (3) is used, it is called the **Square Root Method**. In statement (3), note that if $p > 0$ the equation $x^2 = p$ has two solutions, $x = \sqrt{p}$ and $x = -\sqrt{p}$. We usually abbreviate these solutions as $x = \pm\sqrt{p}$, which is read as "x equals plus or minus the square root of p."

For example, the two solutions of the equation

$$x^2 = 4$$

are

$$x = \pm\sqrt{4} \quad \text{Use the Square Root Method.}$$

and, since $\sqrt{4} = 2$, we have

$$x = \pm 2$$

The solution set is $\{-2, 2\}$.

Now Work PROBLEM 83

4 Solve a Quadratic Equation by Completing the Square

EXAMPLE 6 **Solving a Quadratic Equation by Completing the Square**

Solve by completing the square: $2x^2 - 8x - 5 = 0$

Solution First, rewrite the equation so that the constant is on the right side.

$$2x^2 - 8x - 5 = 0$$
$$2x^2 - 8x = 5 \quad \text{Add 5 to both sides.}$$

Next, divide both sides by 2 so that the coefficient of x^2 is 1. (This enables us to complete the square at the next step.)

$$x^2 - 4x = \frac{5}{2}$$

Finally, complete the square by adding 4 to both sides.

$$x^2 - 4x + 4 = \frac{5}{2} + 4 \qquad \text{Add 4 to both sides.}$$

$$(x-2)^2 = \frac{13}{2} \qquad \text{Factor on the left; simplify on the right.}$$

$$x - 2 = \pm\sqrt{\frac{13}{2}} \qquad \text{Use the Square Root Method.}$$

$$x - 2 = \pm\frac{\sqrt{26}}{2} \qquad \sqrt{\frac{13}{2}} = \frac{\sqrt{13}}{\sqrt{2}} = \frac{\sqrt{13}}{\sqrt{2}}\cdot\frac{\sqrt{2}}{\sqrt{2}} = \frac{\sqrt{26}}{2}$$

$$x = 2 \pm \frac{\sqrt{26}}{2}$$

The solution set is $\left\{2 - \frac{\sqrt{26}}{2}, 2 + \frac{\sqrt{26}}{2}\right\}$.

NOTE If we wanted an approximation, say rounded to two decimal places, of these solutions, we would use a calculator to get $\{-0.55, 4.55\}$. ■

Now Work PROBLEM 87

5 Solve a Quadratic Equation Using the Quadratic Formula

We can use the method of completing the square to obtain a general formula for solving any quadratic equation

$$ax^2 + bx + c = 0 \qquad a \neq 0$$

NOTE There is no loss in generality to assume that $a > 0$, since if $a < 0$ we can multiply by -1 to obtain an equivalent equation with a positive leading coefficient. ■

As in Example 6, rearrange the terms as

$$ax^2 + bx = -c \qquad a > 0$$

Since $a > 0$, divide both sides by a to get

$$x^2 + \frac{b}{a}x = -\frac{c}{a}$$

Now the coefficient of x^2 is 1. To complete the square on the left side, add the square of $\frac{1}{2}$ of the coefficient of x; that is, add

$$\left(\frac{1}{2}\cdot\frac{b}{a}\right)^2 = \frac{b^2}{4a^2}$$

to both sides. Then

$$x^2 + \frac{b}{a}x + \frac{b^2}{4a^2} = \frac{b^2}{4a^2} - \frac{c}{a}$$

$$\left(x + \frac{b}{2a}\right)^2 = \frac{b^2 - 4ac}{4a^2} \qquad \frac{b^2}{4a^2} - \frac{c}{a} = \frac{b^2}{4a^2} - \frac{4ac}{4a^2} = \frac{b^2 - 4ac}{4a^2} \qquad (4)$$

Provided that $b^2 - 4ac \geq 0$, we can now use the Square Root Method to get

$$x + \frac{b}{2a} = \pm\sqrt{\frac{b^2 - 4ac}{4a^2}}$$

$$x + \frac{b}{2a} = \frac{\pm\sqrt{b^2 - 4ac}}{2a}$$

The square root of a quotient equals the quotient of the square roots. Also, $\sqrt{4a^2} = 2a$ since $a > 0$.

$$x = -\frac{b}{2a} \pm \frac{\sqrt{b^2 - 4ac}}{2a} \qquad \text{Add } -\frac{b}{2a} \text{ to both sides.}$$

$$= \frac{-b \pm \sqrt{b^2 - 4ac}}{2a} \qquad \text{Combine the quotients on the right.}$$

What if $b^2 - 4ac$ is negative? Then equation (4) states that the left expression (a real number squared) equals the right expression (a negative number). Since this is impossible for real numbers, we conclude that if $b^2 - 4ac < 0$, the quadratic equation has no *real* solution. (We discuss quadratic equations for which the quantity $b^2 - 4ac < 0$ in detail in the next section.)

THEOREM

Quadratic Formula

Consider the quadratic equation

$$ax^2 + bx + c = 0 \qquad a \neq 0$$

If $b^2 - 4ac < 0$, this equation has no real solution.
If $b^2 - 4ac \geq 0$, the real solution(s) of this equation is (are) given by the **quadratic formula**:

$$x = \frac{-b \pm \sqrt{b^2 - 4ac}}{2a} \qquad \textbf{(5)}$$

The quantity $\boldsymbol{b^2 - 4ac}$ is called the **discriminant** of the quadratic equation, because its value tells us whether the equation has real solutions. In fact, it also tells us how many solutions to expect.

Discriminant of a Quadratic Equation

For a quadratic equation $ax^2 + bx + c = 0$, $a \neq 0$:

1. If $b^2 - 4ac > 0$, there are two unequal real solutions.
2. If $b^2 - 4ac = 0$, there is a repeated solution, a double root.
3. If $b^2 - 4ac < 0$, there is no real solution.

When asked to find the real solutions of a quadratic equation, always evaluate the discriminant first to see if there are any real solutions.

EXAMPLE 7

Solving a Quadratic Equation Using the Quadratic Formula

Use the quadratic formula to find the real solutions, if any, of the equation

$$3x^2 - 5x + 1 = 0$$

Solution The equation is in standard form, so compare it to $ax^2 + bx + c = 0$ to find a, b, and c.

$$3x^2 - 5x + 1 = 0$$
$$ax^2 + bx + c = 0 \qquad a = 3, b = -5, c = 1$$

With $a = 3$, $b = -5$, and $c = 1$, evaluate the discriminant $b^2 - 4ac$.

$$b^2 - 4ac = (-5)^2 - 4(3)(1) = 25 - 12 = 13$$

Since $b^2 - 4ac > 0$, there are two real solutions, which can be found using the quadratic formula.

$$x = \frac{-b \pm \sqrt{b^2 - 4ac}}{2a} = \frac{-(-5) \pm \sqrt{13}}{2(3)} = \frac{5 \pm \sqrt{13}}{6}$$

The solution set is $\left\{\frac{5 - \sqrt{13}}{6}, \frac{5 + \sqrt{13}}{6}\right\}$.

EXAMPLE 8

Solving a Quadratic Equation Using the Quadratic Formula

Use the quadratic formula to find the real solutions, if any, of the equation

$$3x^2 + 2 = 4x$$

Solution The equation, as given, is not in standard form.

$$3x^2 + 2 = 4x$$

$$3x^2 - 4x + 2 = 0 \quad \text{Put in standard form.}$$

$$ax^2 + bx + c = 0 \quad \text{Compare to standard form.}$$

With $a = 3$, $b = -4$, and $c = 2$, the discriminant is

$$b^2 - 4ac = (-4)^2 - 4(3)(2) = 16 - 24 = -8$$

Since $b^2 - 4ac < 0$, the equation has no real solution.

Now Work PROBLEMS 93 AND 99

SUMMARY

Procedure for Solving a Quadratic Equation

To solve a quadratic equation, first put it in standard form:

$$ax^2 + bx + c = 0$$

Then:

STEP 1: Identify a, b, and c.

STEP 2: Evaluate the discriminant, $b^2 - 4ac$.

STEP 3: (a) If the discriminant is negative, the equation has no real solution.
(b) If the discriminant is zero, the equation has a repeated real solution.
(c) If the discriminant is positive, the equation has two distinct real solutions.

If you can easily spot factors, use the factoring method to solve the equation. Otherwise, use the quadratic formula or the method of completing the square.

A.4 Assess Your Understanding

'Are You Prepared?' *Answers are given at the end of these exercises. If you get a wrong answer, read the pages listed in red.*

1. Factor $x^2 - 5x - 6$. (pp. A23–A24)

2. To complete the square of $x^2 - 4x$, add _________ to both sides. (pp. A24–A25)

3. The solution set of the equation $(x - 3)(3x + 5) = 0$ is _________. (p. A4)

4. ***True or False*** $\sqrt{x^2} = |x|$. (pp. A9–A10)

Concepts and Vocabulary

5. An equation that is satisfied for every choice of the variable for which both sides are defined is called a(n) ________.

6. ***True or False*** The solution of the equation $3x - 8 = 0$ is $\frac{3}{8}$.

7. ***True or False*** Some equations have no solution.

8. To solve the equation $x^2 + 5x = 0$ by completing the square, you would _____ the number ____ to both sides.

9. The quantity $b^2 - 4ac$ is called the __________ of a quadratic equation. If it is ________, the equation has no real solution.

10. ***True or False*** Quadratic equations always have two real solutions.

11. ***True or False*** If the discriminant of a quadratic equation is positive, then the equation has two solutions that are negatives of one another.

12. An admissible value for the variable that makes the equation a true statement is called a(n) ______ of the equation.
(a) identity (b) solution (c) degree (d) model

13. A quadratic equation is sometimes called a ______ equation.
(a) first-degree (b) second-degree
(c) third-degree (d) fourth-degree

14. Which of the following quadratic equations is in standard form?
(a) $x^2 - 7x = 5$ (b) $9 = x^2$
(c) $(x + 5)(x - 4) = 0$ (d) $0 = 5x^2 - 6x - 1$

Skill Building

In Problems 15–80, solve each equation.

15. $3x = 21$ **16.** $3x = -24$ **17.** $5x + 15 = 0$ **18.** $3x + 18 = 0$

19. $2x - 3 = 5$ **20.** $3x + 4 = -8$ **21.** $\frac{1}{3}x = \frac{5}{12}$ **22.** $\frac{2}{3}x = \frac{9}{2}$

23. $6 - x = 2x + 9$ **24.** $3 - 2x = 2 - x$ **25.** $2(3 + 2x) = 3(x - 4)$ **26.** $3(2 - x) = 2x - 1$

27. $8x - (2x + 1) = 3x - 10$ **28.** $5 - (2x - 1) = 10$ 29. $\frac{1}{2}x - 4 = \frac{3}{4}x$ **30.** $1 - \frac{1}{2}x = 5$

31. $0.9t = 0.4 + 0.1t$ **32.** $0.9t = 1 + t$ **33.** $\frac{2}{y} + \frac{4}{y} = 3$ **34.** $\frac{4}{y} - 5 = \frac{5}{2y}$

35. $(x + 7)(x - 1) = (x + 1)^2$ **36.** $(x + 2)(x - 3) = (x - 3)^2$ **37.** $z(z^2 + 1) = 3 + z^3$

38. $w(4 - w^2) = 8 - w^3$ 39. $x^2 = 9x$ **40.** $x^3 = x^2$

41. $t^3 - 9t^2 = 0$ **42.** $4z^3 - 8z^2 = 0$ **43.** $\frac{3}{2x - 3} = \frac{2}{x + 5}$

44. $\frac{-2}{x + 4} = \frac{-3}{x + 1}$ **45.** $(x + 2)(3x) = (x + 2)(6)$ **46.** $(x - 5)(2x) = (x - 5)(4)$

47. $\frac{2}{x - 2} = \frac{3}{x + 5} + \frac{10}{(x + 5)(x - 2)}$ **48.** $\frac{1}{2x + 3} + \frac{1}{x - 1} = \frac{1}{(2x + 3)(x - 1)}$ **49.** $|2x| = 6$

50. $|3x| = 12$ 51. $|2x + 3| = 5$ **52.** $|3x - 1| = 2$

53. $|1 - 4t| = 5$ **54.** $|1 - 2z| = 3$ **55.** $|-2x| = 8$ **56.** $|-x| = 1$

57. $|-2|x = 4$ **58.** $|3|x = 9$ **59.** $|x - 2| = -\frac{1}{2}$ **60.** $|2 - x| = -1$

61. $|x^2 - 4| = 0$ **62.** $|x^2 - 9| = 0$ **63.** $|x^2 - 2x| = 3$ **64.** $|x^2 + x| = 12$

65. $|x^2 + x - 1| = 1$ **66.** $|x^2 + 3x - 2| = 2$ **67.** $x^2 = 4x$ **68.** $x^2 = -8x$

69. $z^2 + 4z - 12 = 0$ **70.** $v^2 + 7v + 12 = 0$ **71.** $2x^2 - 5x - 3 = 0$ **72.** $3x^2 + 5x + 2 = 0$

73. $x(x - 7) + 12 = 0$ **74.** $x(x + 1) = 12$ **75.** $4x^2 + 9 = 12x$ **76.** $25x^2 + 16 = 40x$

77. $6x - 5 = \frac{6}{x}$ **78.** $x + \frac{12}{x} = 7$ **79.** $\frac{4(x - 2)}{x - 3} + \frac{3}{x} = \frac{-3}{x(x - 3)}$ **80.** $\frac{5}{x + 4} = 4 + \frac{3}{x - 2}$

In Problems 81–86, solve each equation by the Square Root Method.

81. $x^2 = 25$ **82.** $x^2 = 36$ 83. $(x - 1)^2 = 4$

84. $(x + 2)^2 = 1$ **85.** $(2y + 3)^2 = 9$ **86.** $(3x - 2)^2 = 4$

In Problems 87–92, solve each equation by completing the square.

87. $x^2 + 4x = 21$ **88.** $x^2 - 6x = 13$ **89.** $x^2 - \frac{1}{2}x - \frac{3}{16} = 0$

90. $x^2 + \frac{2}{3}x - \frac{1}{3} = 0$ **91.** $3x^2 + x - \frac{1}{2} = 0$ **92.** $2x^2 - 3x - 1 = 0$

In Problems 93–104, find the real solutions, if any, of each equation. Use the quadratic formula.

93. $x^2 - 4x + 2 = 0$ **94.** $x^2 + 4x + 2 = 0$ **95.** $x^2 - 5x - 1 = 0$

96. $x^2 + 5x + 3 = 0$ **97.** $2x^2 - 5x + 3 = 0$ **98.** $2x^2 + 5x + 3 = 0$

99. $4y^2 - y + 2 = 0$ **100.** $4t^2 + t + 1 = 0$ **101.** $4x^2 = 1 - 2x$

102. $2x^2 = 1 - 2x$ **103.** $x^2 + \sqrt{3}x - 3 = 0$ **104.** $x^2 + \sqrt{2}x - 2 = 0$

In Problems 105–110, use the discriminant to determine whether each quadratic equation has two unequal real solutions, a repeated real solution, or no real solution without solving the equation.

105. $x^2 - 5x + 7 = 0$ **106.** $x^2 + 5x + 7 = 0$ **107.** $9x^2 - 30x + 25 = 0$

108. $25x^2 - 20x + 4 = 0$ **109.** $3x^2 + 5x - 8 = 0$ **110.** $2x^2 - 3x - 4 = 0$

Applications and Extensions

In Problems 111–116, solve each equation. The letters a, b, and c are constants.

111. $ax - b = c, \quad a \neq 0$

112. $1 - ax = b, \quad a \neq 0$

113. $\frac{x}{a} + \frac{x}{b} = c, \quad a \neq 0, b \neq 0, a \neq -b$

114. $\frac{a}{x} + \frac{b}{x} = c, \quad c \neq 0$

115. $\frac{1}{x-a} + \frac{1}{x+a} = \frac{2}{x-1}$

116. $\frac{b+c}{x+a} = \frac{b-c}{x-a}, \quad c \neq 0, a \neq 0$

Problems 117–122 list some formulas that occur in applications. Solve each formula for the indicated variable.

117. Electricity $\frac{1}{R} = \frac{1}{R_1} + \frac{1}{R_2}$ for R

118. Finance $A = P(1 + rt)$ for r

119. Mechanics $F = \frac{mv^2}{R}$ for R

120. Chemistry $PV = nRT$ for T

121. Mathematics $S = \frac{a}{1-r}$ for r

122. Mechanics $v = -gt + v_0$ for t

123. Show that the sum of the roots of a quadratic equation is $-\frac{b}{a}$.

124. Show that the product of the roots of a quadratic equation is $\frac{c}{a}$.

125. Find k such that the equation $kx^2 + x + k = 0$ has a repeated real solution.

126. Find k such that the equation $x^2 - kx + 4 = 0$ has a repeated real solution.

127. Show that the real solutions of the equation $ax^2 + bx + c = 0$ are the negatives of the real solutions of the equation $ax^2 - bx + c = 0$. Assume that $b^2 - 4ac \geq 0$.

128. Show that the real solutions of the equation $ax^2 + bx + c = 0$ are the reciprocals of the real solutions of the equation $cx^2 + bx + a = 0$. Assume that $b^2 - 4ac \geq 0$.

Explaining Concepts: Discussion and Writing

129. Which of the following pairs of equations are equivalent? Explain.
(a) $x^2 = 9; \quad x = 3$
(b) $x = \sqrt{9}; \quad x = 3$
(c) $(x-1)(x-2) = (x-1)^2; \quad x - 2 = x - 1$

130. The equation

$$\frac{5}{x+3} + 3 = \frac{8+x}{x+3}$$

has no solution, yet when we go through the process of solving it, we obtain $x = -3$. Write a brief paragraph to explain what causes this to happen.

131. Find an equation that has no solution and give it to a fellow student to solve. Ask the fellow student to write a critique of your equation.

132. Describe three ways you might solve a quadratic equation. State your preferred method; explain why you chose it.

133. Explain the benefits of evaluating the discriminant of a quadratic equation before attempting to solve it.

134. Find three quadratic equations: one having two distinct solutions, one having no real solution, and one having exactly one real solution.

135. The word *quadratic* seems to imply four (*quad*), yet a quadratic equation is an equation that involves a polynomial of degree 2. Investigate the origin of the term *quadratic* as it is used in the expression *quadratic equation*. Write a brief essay on your findings.

'Are You Prepared?' Answers

1. $(x-6)(x+1)$ **2.** 4 **3.** $\left\{-\frac{5}{3}, 3\right\}$ **4.** True

A.5 Complex Numbers; Quadratic Equations in the Complex Number System*

OBJECTIVES 1 Add, Subtract, Multiply, and Divide Complex Numbers (p. A38)
2 Solve Quadratic Equations in the Complex Number System (p. A42)

Complex Numbers

One property of a real number is that its square is nonnegative. For example, there is no real number x for which

$$x^2 = -1$$

To remedy this situation, we introduce a new number called the *imaginary unit.*

DEFINITION

The **imaginary unit**, which we denote by i, is the number whose square is -1. That is,

$$i^2 = -1$$

This should not surprise you. If our universe were to consist only of integers, there would be no number x for which $2x = 1$. This was remedied by introducing numbers such as $\frac{1}{2}$ and $\frac{2}{3}$, the *rational numbers*. If our universe were to consist only of rational numbers, there would be no x whose square equals 2. That is, there would be no number x for which $x^2 = 2$. To remedy this, we introduced numbers such as $\sqrt{2}$ and $\sqrt[3]{5}$, the *irrational numbers*. Recall that the *real numbers* consist of the rational numbers and the irrational numbers. Now, if our universe were to consist only of real numbers, then there would be no number x whose square is -1. To remedy this, we introduce the number i, whose square is -1.

*This section may be omitted without any loss of continuity.

In the progression outlined, each time we encountered a situation that was unsuitable, a new number system was introduced to remedy the situation. The number system that results from introducing the number i is called the **complex number system**.

DEFINITION

Complex numbers are numbers of the form $a + bi$, where a and b are real numbers. The real number a is called the **real part** of the number $a + bi$; the real number b is called the **imaginary part** of $a + bi$; and i is the imaginary unit, so $i^2 = -1$.

For example, the complex number $-5 + 6i$ has the real part -5 and the imaginary part 6.

When a complex number is written in the form $a + bi$, where a and b are real numbers, it is in **standard form**. However, if the imaginary part of a complex number is negative, such as in the complex number $3 + (-2)i$, we agree to write it instead in the form $3 - 2i$.

Also, the complex number $a + 0i$ is usually written merely as a. This serves to remind us that the real numbers are a subset of the complex numbers. Similarly, the complex number $0 + bi$ is usually written as bi. Sometimes the complex number bi is called a **pure imaginary number**.

1 Add, Subtract, Multiply, and Divide Complex Numbers

Equality, addition, subtraction, and multiplication of complex numbers are defined so as to preserve the familiar rules of algebra for real numbers. Two complex numbers are equal if and only if their real parts are equal and their imaginary parts are equal.

Equality of Complex Numbers

$$a + bi = c + di \quad \text{if and only if} \quad a = c \text{ and } b = d \qquad (1)$$

Two complex numbers are added by forming the complex number whose real part is the sum of the real parts and whose imaginary part is the sum of the imaginary parts.

Sum of Complex Numbers

$$(a + bi) + (c + di) = (a + c) + (b + d)i \qquad (2)$$

To subtract two complex numbers, use this rule:

Difference of Complex Numbers

$$(a + bi) - (c + di) = (a - c) + (b - d)i \qquad (3)$$

EXAMPLE 1

Adding and Subtracting Complex Numbers

(a) $(3 + 5i) + (-2 + 3i) = [3 + (-2)] + (5 + 3)i = 1 + 8i$

(b) $(6 + 4i) - (3 + 6i) = (6 - 3) + (4 - 6)i = 3 + (-2)i = 3 - 2i$

Now Work PROBLEM 15

Products of complex numbers are calculated as illustrated in Example 2.

EXAMPLE 2

Multiplying Complex Numbers

$$(5 + 3i) \cdot (2 + 7i) = 5 \cdot (2 + 7i) + 3i(2 + 7i) = 10 + 35i + 6i + 21i^2$$

Distributive Property Distributive Property

$$= 10 + 41i + 21(-1)$$

$i^2 = -1$

$$= -11 + 41i$$

Based on the procedure of Example 2, the **product** of two complex numbers is defined as follows:

Product of Complex Numbers

$$(a + bi) \cdot (c + di) = (ac - bd) + (ad + bc)i \quad (4)$$

Do not bother to memorize formula (4). Instead, whenever it is necessary to multiply two complex numbers, follow the usual rules for multiplying two binomials, as in Example 2, remembering that $i^2 = -1$. For example,

$$(2i)(2i) = 4i^2 = 4(-1) = -4$$

$$(2 + i)(1 - i) = 2 - 2i + i - i^2 = 3 - i$$

Now Work PROBLEM 21

Algebraic properties for addition and multiplication, such as the commutative, associative, and distributive properties, hold for complex numbers. The property that every nonzero complex number has a multiplicative inverse, or reciprocal, requires a closer look.

DEFINITION If $z = a + bi$ is a complex number, then its **conjugate**, denoted by $\overline{z}$, is defined as

$$\overline{z} = \overline{a + bi} = a - bi$$

NOTE The conjugate of a complex number can be found by changing the sign of the imaginary part. ■

For example, $\overline{2 + 3i} = 2 - 3i$ and $\overline{-6 - 2i} = -6 + 2i$.

EXAMPLE 3

Multiplying a Complex Number by Its Conjugate

Find the product of the complex number $z = 3 + 4i$ and its conjugate $\overline{z}$.

Solution Since $\overline{z} = 3 - 4i$, we have

$$z\overline{z} = (3 + 4i)(3 - 4i) = 9 - 12i + 12i - 16i^2 = 9 + 16 = 25$$

The result obtained in Example 3 has an important generalization.

THEOREM The product of a complex number and its conjugate is a nonnegative real number. That is, if $z = a + bi$, then

$$z\overline{z} = a^2 + b^2 \quad (5)$$

Proof If $z = a + bi$, then

$$z\bar{z} = (a + bi)(a - bi) = a^2 - (bi)^2 = a^2 - b^2i^2 = a^2 + b^2$$ ■

To express the reciprocal of a nonzero complex number z in standard form, multiply the numerator and denominator of $\frac{1}{z}$ by $\bar{z}$. That is, if $z = a + bi$ is a nonzero complex number, then

$$\frac{1}{a + bi} = \frac{1}{z} = \frac{1}{z} \cdot \frac{\bar{z}}{\bar{z}} = \frac{\bar{z}}{z\bar{z}} = \frac{a - bi}{a^2 + b^2}$$

↑ Use (5).

$$= \frac{a}{a^2 + b^2} - \frac{b}{a^2 + b^2}i$$

EXAMPLE 4 Writing the Reciprocal of a Complex Number in Standard Form

Write $\frac{1}{3 + 4i}$ in standard form $a + bi$; that is, find the reciprocal of $3 + 4i$.

Solution The idea is to multiply the numerator and denominator by the conjugate of $3 + 4i$, that is, by the complex number $3 - 4i$. The result is

$$\frac{1}{3 + 4i} = \frac{1}{3 + 4i} \cdot \frac{3 - 4i}{3 - 4i} = \frac{3 - 4i}{9 + 16} = \frac{3}{25} - \frac{4}{25}i$$ ●

To express the quotient of two complex numbers in standard form, multiply the numerator and denominator of the quotient by the conjugate of the denominator.

EXAMPLE 5 Writing the Quotient of Two Complex Numbers in Standard Form

Write each of the following in standard form.

(a) $\frac{1 + 4i}{5 - 12i}$ (b) $\frac{2 - 3i}{4 - 3i}$

Solution

(a) $$\frac{1 + 4i}{5 - 12i} = \frac{1 + 4i}{5 - 12i} \cdot \frac{5 + 12i}{5 + 12i} = \frac{5 + 12i + 20i + 48i^2}{25 + 144}$$
$$= \frac{-43 + 32i}{169} = -\frac{43}{169} + \frac{32}{169}i$$

(b) $$\frac{2 - 3i}{4 - 3i} = \frac{2 - 3i}{4 - 3i} \cdot \frac{4 + 3i}{4 + 3i} = \frac{8 + 6i - 12i - 9i^2}{16 + 9} = \frac{17 - 6i}{25} = \frac{17}{25} - \frac{6}{25}i$$ ●

Now Work PROBLEM 29

EXAMPLE 6 Writing Other Expressions in Standard Form

If $z = 2 - 3i$ and $w = 5 + 2i$, write each of the following expressions in standard form.

(a) $\frac{z}{w}$ (b) $\overline{z + w}$ (c) $z + \bar{z}$

Solution (a) $\dfrac{z}{w} = \dfrac{z \cdot \overline{w}}{w \cdot \overline{w}} = \dfrac{(2-3i)(5-2i)}{(5+2i)(5-2i)} = \dfrac{10-4i-15i+6i^2}{25+4}$

$$= \frac{4-19i}{29} = \frac{4}{29} - \frac{19}{29}i$$

(b) $\overline{z+w} = \overline{(2-3i)+(5+2i)} = \overline{7-i} = 7+i$

(c) $z + \overline{z} = (2-3i) + (2+3i) = 4$

The conjugate of a complex number has certain general properties that will be useful later.

For a real number $a = a + 0i$, the conjugate is $\overline{a} = \overline{a+0i} = a - 0i = a$.

THEOREM

The conjugate of a real number is the real number itself.

Other properties that are direct consequences of the definition of the conjugate are given next. In each statement, z and w represent complex numbers.

THEOREM

The conjugate of the conjugate of a complex number is the complex number itself.

$$\overline{(\overline{z})} = z \quad \textbf{(6)}$$

The conjugate of the sum of two complex numbers equals the sum of their conjugates.

$$\overline{z+w} = \overline{z} + \overline{w} \quad \textbf{(7)}$$

The conjugate of the product of two complex numbers equals the product of their conjugates.

$$\overline{z \cdot w} = \overline{z} \cdot \overline{w} \quad \textbf{(8)}$$

The proofs of equations (6), (7), and (8) are left as exercises.

Powers of *i*

The **powers of *i*** follow a pattern that is useful to know.

$$i^1 = i \qquad i^5 = i^4 \cdot i = 1 \cdot i = i$$
$$i^2 = -1 \qquad i^6 = i^4 \cdot i^2 = -1$$
$$i^3 = i^2 \cdot i = -1 \cdot i = -i \qquad i^7 = i^4 \cdot i^3 = -i$$
$$i^4 = i^2 \cdot i^2 = (-1)(-1) = 1 \qquad i^8 = i^4 \cdot i^4 = 1$$

And so on. The powers of i repeat with every fourth power.

EXAMPLE 7 **Evaluating Powers of *i***

(a) $i^{27} = i^{24} \cdot i^3 = (i^4)^6 \cdot i^3 = 1^6 \cdot i^3 = -i$

(b) $i^{101} = i^{100} \cdot i^1 = (i^4)^{25} \cdot i = 1^{25} \cdot i = i$

EXAMPLE 8 Writing the Power of a Complex Number in Standard Form

Write $(2 + i)^3$ in standard form.

Solution Use the special product formula for $(x + a)^3$.

$$(x + a)^3 = x^3 + 3ax^2 + 3a^2x + a^3$$

NOTE Another way to find $(2 + i)^3$ is to multiply out $(2 + i)^2(2 + i)$. ■

Using this special product formula,

$$\begin{aligned}(2 + i)^3 &= 2^3 + 3 \cdot i \cdot 2^2 + 3 \cdot i^2 \cdot 2 + i^3 \\ &= 8 + 12i + 6(-1) + (-i) \\ &= 2 + 11i\end{aligned}$$

●

Now Work PROBLEM 43

2 Solve Quadratic Equations in the Complex Number System

Quadratic equations with a negative discriminant have no real number solution. However, if we extend our number system to allow complex numbers, quadratic equations will always have a solution. Since the solution to a quadratic equation involves the square root of the discriminant, we begin with a discussion of square roots of negative numbers.

DEFINITION If N is a positive real number, we define the **principal square root of $-N$**, denoted by $\sqrt{-N}$, as

$$\sqrt{-N} = \sqrt{N}i$$

where i is the imaginary unit and $i^2 = -1$.

WARNING In writing $\sqrt{-N} = \sqrt{N}i$, be sure to place i outside the $\sqrt{\ }$ symbol. ■

EXAMPLE 9 Evaluating the Square Root of a Negative Number

(a) $\sqrt{-1} = \sqrt{1}i = i$

(b) $\sqrt{-4} = \sqrt{4}i = 2i$

(c) $\sqrt{-8} = \sqrt{8}i = 2\sqrt{2}i$ ●

EXAMPLE 10 Solving Equations

Solve each equation in the complex number system.

(a) $x^2 = 4$ (b) $x^2 = -9$

Solution (a) $x^2 = 4$

$$x = \pm\sqrt{4} = \pm 2$$

The equation has two solutions, -2 and 2. The solution set is $\{-2, 2\}$.

(b) $x^2 = -9$

$$x = \pm\sqrt{-9} = \pm\sqrt{9}i = \pm 3i$$

The equation has two solutions, $-3i$ and $3i$. The solution set is $\{-3i, 3i\}$. ●

Now Work PROBLEMS 51 AND 55

WARNING When working with square roots of negative numbers, do not set the square root of a product equal to the product of the square roots (which can be done with positive real numbers). To see why, look at this calculation: We know that $\sqrt{100} = 10$. However, it is also true that $100 = (-25)(-4)$, so

$$10 = \sqrt{100} = \sqrt{(-25)(-4)} = \sqrt{-25}\sqrt{-4} = (\sqrt{25}i)(\sqrt{4}i) = (5i)(2i) = 10i^2 = -10$$

↑ Here is the error. ■

Because we have defined the square root of a negative number, we can now restate the quadratic formula without restriction.

THEOREM

Quadratic Formula

In the complex number system, the solutions of the quadratic equation $ax^2 + bx + c = 0$, where a, b, and c are real numbers and $a \neq 0$, are given by the formula

$$x = \frac{-b \pm \sqrt{b^2 - 4ac}}{2a} \quad (9)$$

EXAMPLE 11

Solving a Quadratic Equation in the Complex Number System

Solve the equation $x^2 - 4x + 8 = 0$ in the complex number system.

Solution Here $a = 1$, $b = -4$, $c = 8$, and $b^2 - 4ac = (-4)^2 - 4(1)(8) = -16$. Using equation (9), we find that

$$x = \frac{-(-4) \pm \sqrt{-16}}{2(1)} = \frac{4 \pm \sqrt{16}i}{2} = \frac{4 \pm 4i}{2} = \frac{2(2 \pm 2i)}{2} = 2 \pm 2i$$

The equation has two solutions: $2 - 2i$ and $2 + 2i$.

The solution set is $\{2 - 2i, 2 + 2i\}$.

✓**Check:** $2 + 2i$: $(2 + 2i)^2 - 4(2 + 2i) + 8 = 4 + 8i + 4i^2 - 8 - 8i + 8$
$= 4 + 4i^2$
$= 4 - 4 = 0$

$2 - 2i$: $(2 - 2i)^2 - 4(2 - 2i) + 8 = 4 - 8i + 4i^2 - 8 + 8i + 8$
$= 4 - 4 = 0$ ●

Now Work PROBLEM 61

The discriminant $b^2 - 4ac$ of a quadratic equation still serves as a way to determine the character of the solutions.

Character of the Solutions of a Quadratic Equation

In the complex number system, consider a quadratic equation $ax^2 + bx + c = 0$ with real coefficients.

1. If $b^2 - 4ac > 0$, the equation has two unequal real solutions.
2. If $b^2 - 4ac = 0$, the equation has a repeated real solution, a double root.
3. If $b^2 - 4ac < 0$, the equation has two complex solutions that are not real. The solutions are conjugates of each other.

The third conclusion in the display is a consequence of the fact that if $b^2 - 4ac = -N < 0$, then by the quadratic formula, the solutions are

$$x = \frac{-b + \sqrt{b^2 - 4ac}}{2a} = \frac{-b + \sqrt{-N}}{2a} = \frac{-b + \sqrt{N}i}{2a} = \frac{-b}{2a} + \frac{\sqrt{N}}{2a}i$$

and

$$x = \frac{-b - \sqrt{b^2 - 4ac}}{2a} = \frac{-b - \sqrt{-N}}{2a} = \frac{-b - \sqrt{N}i}{2a} = \frac{-b}{2a} - \frac{\sqrt{N}}{2a}i$$

which are conjugates of each other.

EXAMPLE 12 **Determining the Character of the Solutions of a Quadratic Equation**

Without solving, determine the character of the solutions of each equation.

(a) $3x^2 + 4x + 5 = 0$ (b) $2x^2 + 4x + 1 = 0$ (c) $9x^2 - 6x + 1 = 0$

Solution (a) Here $a = 3, b = 4$, and $c = 5$, so $b^2 - 4ac = 16 - 4(3)(5) = -44$. The solutions are two complex numbers that are not real and are conjugates of each other.

(b) Here $a = 2, b = 4$, and $c = 1$, so $b^2 - 4ac = 16 - 8 = 8$. The solutions are two unequal real numbers.

(c) Here $a = 9, b = -6$, and $c = 1$, so $b^2 - 4ac = 36 - 4(9)(1) = 0$. The solution is a repeated real number—that is, a double root. ●

Now Work PROBLEM 75

A.5 Assess Your Understanding

Concepts and Vocabulary

1. ***True or False*** The square of a complex number is sometimes negative.
2. $(2 + i)(2 - i) =$ _____.
3. ***True or False*** In the complex number system, a quadratic equation has four solutions.
4. In the complex number $5 + 2i$, the number 5 is called the _____ part; the number 2 is called the _____ part; the number i is called the _____ _____.
5. ***True or False*** The conjugate of $2 + 5i$ is $-2 - 5i$.
6. ***True or False*** All real numbers are complex numbers.
7. ***True or False*** If $2 - 3i$ is a solution of a quadratic equation with real coefficients, then $-2 + 3i$ is also a solution.
8. Which of the following is the principal square root of -4?
(a) $-2i$ (b) $2i$ (c) -2 (d) 2
9. Which operation involving complex numbers requires the use of a conjugate?
(a) division (b) multiplication
(c) subtraction (d) addition
10. Powers of i repeat every _____ power.
(a) second (b) third (c) fourth (d) fifth

Skill Building

In Problems 11–48, perform the indicated operation, and write each expression in the standard form $a + bi$.

11. $(2 - 3i) + (6 + 8i)$
12. $(4 + 5i) + (-8 + 2i)$
13. $(-3 + 2i) - (4 - 4i)$
14. $(3 - 4i) - (-3 - 4i)$
15. $(2 - 5i) - (8 + 6i)$
16. $(-8 + 4i) - (2 - 2i)$
17. $3(2 - 6i)$
18. $-4(2 + 8i)$
19. $2i(2 - 3i)$
20. $3i(-3 + 4i)$
21. $(3 - 4i)(2 + i)$
22. $(5 + 3i)(2 - i)$
23. $(-6 + i)(-6 - i)$
24. $(-3 + i)(3 + i)$
25. $\frac{10}{3 - 4i}$
26. $\frac{13}{5 - 12i}$
27. $\frac{2 + i}{i}$
28. $\frac{2 - i}{-2i}$
29. $\frac{6 - i}{1 + i}$
30. $\frac{2 + 3i}{1 - i}$
31. $\left(\frac{1}{2} + \frac{\sqrt{3}}{2}i\right)^2$
32. $\left(\frac{\sqrt{3}}{2} - \frac{1}{2}i\right)^2$
33. $(1 + i)^2$
34. $(1 - i)^2$

35. i^{23} **36.** i^{14} **37.** i^{-15} **38.** i^{-23} **39.** $i^6 - 5$

40. $4 + i^3$ **41.** $6i^3 - 4i^5$ **42.** $4i^3 - 2i^2 + 1$ **43.** $(1 + i)^3$ **44.** $(3i)^4 + 1$

45. $i^7(1 + i^2)$ **46.** $2i^4(1 + i^2)$ **47.** $i^6 + i^4 + i^2 + 1$ **48.** $i^7 + i^5 + i^3 + i$

In Problems 49–54, perform the indicated operations, and express your answer in the form $a + bi$.

49. $\sqrt{-4}$ **50.** $\sqrt{-9}$ **51.** $\sqrt{-25}$

52. $\sqrt{-64}$ **53.** $\sqrt{(3 + 4i)(4i - 3)}$ **54.** $\sqrt{(4 + 3i)(3i - 4)}$

In Problems 55–74, solve each equation in the complex number system.

55. $x^2 + 4 = 0$ **56.** $x^2 - 4 = 0$ **57.** $x^2 - 16 = 0$ **58.** $x^2 + 25 = 0$

59. $x^2 - 6x + 13 = 0$ **60.** $x^2 + 4x + 8 = 0$ **61.** $x^2 - 6x + 10 = 0$ **62.** $x^2 - 2x + 5 = 0$

63. $8x^2 - 4x + 1 = 0$ **64.** $10x^2 + 6x + 1 = 0$ **65.** $5x^2 + 1 = 2x$ **66.** $13x^2 + 1 = 6x$

67. $x^2 + x + 1 = 0$ **68.** $x^2 - x + 1 = 0$ **69.** $x^3 - 8 = 0$ **70.** $x^3 + 27 = 0$

71. $x^4 = 16$ **72.** $x^4 = 1$ **73.** $x^4 + 13x^2 + 36 = 0$ **74.** $x^4 + 3x^2 - 4 = 0$

In Problems 75–80, without solving, determine the character of the solutions of each equation in the complex number system.

75. $3x^2 - 3x + 4 = 0$ **76.** $2x^2 - 4x + 1 = 0$ **77.** $2x^2 + 3x = 4$

78. $x^2 + 6 = 2x$ **79.** $9x^2 - 12x + 4 = 0$ **80.** $4x^2 + 12x + 9 = 0$

81. $2 + 3i$ is a solution of a quadratic equation with real coefficients. Find the other solution.

82. $4 - i$ is a solution of a quadratic equation with real coefficients. Find the other solution.

In Problems 83–86, $z = 3 - 4i$ and $w = 8 + 3i$. Write each expression in the standard form $a + bi$.

83. $z + \overline{z}$ **84.** $w - \overline{w}$ **85.** $z\overline{z}$ **86.** $\overline{z - w}$

Applications and Extensions

87. Electrical Circuits The impedance Z, in ohms, of a circuit element is defined as the ratio of the phasor voltage V, in volts, across the element to the phasor current I, in amperes, through the elements. That is, $Z = \frac{V}{I}$. If the voltage across a circuit element is $18 + \ i$ volts and the current through the element is $3 - 4i$ amperes, determine the impedance.

88. Parallel Circuits In an ac circuit with two parallel pathways, the total impedance Z, in ohms, satisfies the formula $\frac{1}{Z} = \frac{1}{Z_1} + \frac{1}{Z_2}$, where Z_1 is the impedance of the first pathway and Z_2 is the impedance of the second pathway. Determine the total impedance if the impedances of the two pathways are $Z_1 = 2 + i$ ohms and $Z_2 = 4 - 3i$ ohms.

89. Use $z = a + bi$ to show that $z + \overline{z} = 2a$ and $z - \overline{z} = 2bi$.

90. Use $z = a + bi$ to show that $\overline{\overline{z}} = z$.

91. Use $z = a + bi$ and $w = c + di$ to show that $\overline{z + w} = \overline{z} + \overline{w}$.

92. Use $z = a + bi$ and $w = c + di$ to show that $\overline{z \cdot w} = \overline{z} \cdot \overline{w}$.

Explaining Concepts: Discussion and Writing

93. Explain to a friend how you would add two complex numbers and how you would multiply two complex numbers. Explain any differences between the two explanations.

94. Write a brief paragraph that compares the method used to rationalize the denominator of a radical expression and the method used to write the quotient of two complex numbers in standard form.

95. Use an Internet search engine to investigate the origins of complex numbers. Write a paragraph describing what you find, and present it to the class.

96. Explain how the method of multiplying two complex numbers is related to multiplying two binomials.

97. What Went Wrong? A student multiplied $\sqrt{-9}$ and $\sqrt{-9}$ as follows:

$$\begin{aligned}\sqrt{-9} \cdot \sqrt{-9} &= \sqrt{(-9)(-9)} \\ &= \sqrt{81} \\ &= 9\end{aligned}$$

The instructor marked the problem incorrect. Why?

A.6 Interval Notation; Solving Inequalities

PREPARING FOR THIS SECTION *Before getting started, review the following:*

- Algebra Essentials (Appendix A, Section A.1, pp. A1–A10)

Now Work the **'Are You Prepared?'** problems on page A52.

OBJECTIVES
1. Use Interval Notation (p. A46)
2. Use Properties of Inequalities (p. A47)
3. Solve Inequalities (p. A49)
4. Solve Combined Inequalities (p. A50)
5. Solve Inequalities Involving Absolute Value (p. A51)

Suppose that a and b are two real numbers and $a < b$. The notation $a < x < b$ means that x is a number *between* a and b. The expression $a < x < b$ is equivalent to the two inequalities $a < x$ and $x < b$. Similarly, the expression $a \le x \le b$ is equivalent to the two inequalities $a \le x$ and $x \le b$. The remaining two possibilities, $a \le x < b$ and $a < x \le b$, are defined similarly.

Although it is acceptable to write $3 \ge x \ge 2$, it is preferable to reverse the inequality symbols and write instead $2 \le x \le 3$ so that the values go from smaller to larger, reading from left to right.

A statement such as $2 \le x \le 1$ is false because there is no number x for which $2 \le x$ and $x \le 1$. Finally, never mix inequality symbols, as in $2 \le x \ge 3$.

1 Use Interval Notation

Let a and b represent two real numbers with $a < b$.

DEFINITION

An **open interval**, denoted by (a, b), consists of all real numbers x for which $a < x < b$.

A **closed interval,** denoted by $[a, b]$, consists of all real numbers x for which $a \le x \le b$.

The **half-open**, or **half-closed**, **intervals** are $(a, b]$, consisting of all real numbers x for which $a < x \le b$, and $[a, b)$, consisting of all real numbers x for which $a \le x < b$.

In Words
The notation $[a, b]$ represents all real numbers between a and b, inclusive. The notation (a, b) represents all real numbers between a and b, not including either a or b.

In each of these definitions, a is called the **left endpoint** and b the **right endpoint** of the interval.

The symbol ∞ (read as "infinity") is not a real number, but notation used to indicate unboundedness in the positive direction. The symbol $-\infty$ (read as "negative infinity") also is not a real number, but notation used to indicate unboundedness in the negative direction. The symbols ∞ and $-\infty$ are used to define five other kinds of intervals:

$[a, \infty)$ Consists of all real numbers x for which $x \ge a$
(a, ∞) Consists of all real numbers x for which $x > a$
$(-\infty, a]$ Consists of all real numbers x for which $x \le a$
$(-\infty, a)$ Consists of all real numbers x for which $x < a$
$(-\infty, \infty)$ Consists of all real numbers

Note that ∞ and $-\infty$ are never included as endpoints, since neither is a real number.

Table 1 summarizes interval notation, corresponding inequality notation, and their graphs.

Table 1

Interval	Inequality	Graph
The open interval (a, b)	$a < x < b$	
The closed interval $[a, b]$	$a \le x \le b$	
The half-open interval $[a, b)$	$a \le x < b$	
The half-open interval $(a, b]$	$a < x \le b$	
The interval $[a, \infty)$	$x \ge a$	
The interval (a, ∞)	$x > a$	
The interval $(-\infty, a]$	$x \le a$	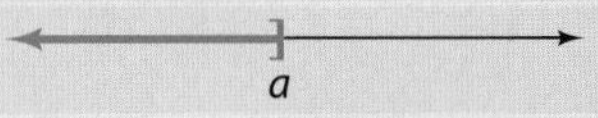
The interval $(-\infty, a)$	$x < a$	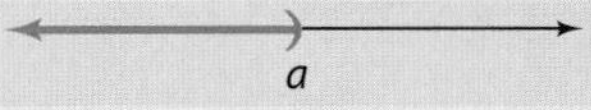
The interval $(-\infty, \infty)$	All real numbers	

EXAMPLE 1 **Writing Inequalities Using Interval Notation**

Write each inequality using interval notation.

(a) $1 \le x \le 3$ (b) $-4 < x < 0$ (c) $x > 5$ (d) $x \le 1$

Solution

(a) $1 \le x \le 3$ describes all real numbers x between 1 and 3, inclusive. In interval notation, we write $[1, 3]$.
(b) In interval notation, $-4 < x < 0$ is written $(-4, 0)$.
(c) In interval notation, $x > 5$ is written $(5, \infty)$.
(d) In interval notation, $x \le 1$ is written $(-\infty, 1]$.

EXAMPLE 2 **Writing Intervals Using Inequality Notation**

Write each interval as an inequality involving x.

(a) $[1, 4)$ (b) $(2, \infty)$ (c) $[2, 3]$ (d) $(-\infty, -3]$

Solution

(a) $[1, 4)$ consists of all real numbers x for which $1 \le x < 4$.
(b) $(2, \infty)$ consists of all real numbers x for which $x > 2$.
(c) $[2, 3]$ consists of all real numbers x for which $2 \le x \le 3$.
(d) $(-\infty, -3]$ consists of all real numbers x for which $x \le -3$.

Now Work PROBLEMS 15, 27, AND 35

2 Use Properties of Inequalities

The product of two positive real numbers is positive, the product of two negative real numbers is positive, and the product of 0 and 0 is 0. For any real number a, the value of a^2 is 0 or positive; that is, a^2 is nonnegative. This is called the **nonnegative property**.

In Words
The square of a real number is never negative.

Nonnegative Property

For any real number a,

$$a^2 \ge 0 \quad (1)$$

When the same number is added to both sides of an inequality, an equivalent inequality is obtained. For example, since $3 < 5$, then $3 + 4 < 5 + 4$ or $7 < 9$. This is called the **addition property** of inequalities.

In Words
The addition property states that the sense, or direction, of an inequality remains unchanged if the same number is added to each side.

Addition Property of Inequalities

For real numbers a, b, and c,

$$\text{If } a < b, \text{ then } a + c < b + c \qquad \textbf{(2a)}$$

$$\text{If } a > b, \text{ then } a + c > b + c \qquad \textbf{(2b)}$$

Now let's see what happens when each side of an inequality is multiplied by a nonzero number. Begin with $3 < 7$ and multiply each side by 2. The numbers 6 and 14 that result obey the inequality $6 < 14$.

Now start with $9 > 2$ and multiply each side by -4. The numbers -36 and -8 that result obey the inequality $-36 < -8$.

Note that the effect of multiplying both sides of $9 > 2$ by the negative number -4 is that the direction of the inequality symbol is reversed.

These results illustrate the following general **multiplication properties** for inequalities:

In Words
Multiplying by a negative number reverses the inequality.

Multiplication Properties for Inequalities

For real numbers a, b, and c,

$$\begin{aligned}&\text{If } a < b \text{ and if } c > 0, \text{ then } ac < bc\\ &\text{If } a < b \text{ and if } c < 0, \text{ then } ac > bc\end{aligned} \qquad \textbf{(3a)}$$

$$\begin{aligned}&\text{If } a > b \text{ and if } c > 0, \text{ then } ac > bc\\ &\text{If } a > b \text{ and if } c < 0, \text{ then } ac < bc\end{aligned} \qquad \textbf{(3b)}$$

In Words
The multiplication properties state that the sense, or direction, of an inequality *remains the same* if each side is multiplied by a *positive* real number, whereas the direction *is reversed* if each side is multiplied by a *negative* real number.

EXAMPLE 3 **Multiplication Property of Inequalities**

(a) If $2x < 6$, then $\frac{1}{2}(2x) < \frac{1}{2}(6)$ or $x < 3$.

(b) If $\frac{x}{-3} > 12$, then $-3\left(\frac{x}{-3}\right) < -3(12)$ or $x < -36$.

(c) If $-4x < -8$, then $\frac{-4x}{-4} > \frac{-8}{-4}$ or $x > 2$.

(d) If $-x > 8$, then $(-1)(-x) < (-1)(8)$ or $x < -8$. ●

Now Work PROBLEM 49

Reciprocal Property for Inequalities

$$\text{If } a > 0, \text{ then } \frac{1}{a} > 0 \qquad \text{If } \frac{1}{a} > 0, \text{ then } a > 0 \qquad \textbf{(4a)}$$

$$\text{If } a < 0, \text{ then } \frac{1}{a} < 0 \qquad \text{If } \frac{1}{a} < 0, \text{ then } a < 0 \qquad \textbf{(4b)}$$

In Words
The reciprocal property states that the reciprocal of a positive real number is positive and that the reciprocal of a negative real number is negative.

3 Solve Inequalities

An **inequality in one variable** is a statement involving two expressions, at least one containing the variable, separated by one of the inequality symbols: $<$, $\leq$, $>$, or $\geq$. To **solve an inequality** means to find all values of the variable for which the statement is true. These values are called **solutions** of the inequality.

For example, the following are all inequalities involving one variable x:

$$x + 5 < 8 \qquad 2x - 3 \geq 4 \qquad x^2 - 1 \leq 3 \qquad \frac{x+1}{x-2} > 0$$

As with equations, one method for solving an inequality is to replace it by a series of equivalent inequalities until an inequality with an obvious solution, such as $x < 3$, is obtained. Equivalent inequalities are obtained by applying some of the same properties that are used to find equivalent equations. The addition property and the multiplication properties for inequalities form the basis for the following procedures.

Procedures That Leave the Inequality Symbol Unchanged

1. Simplify both sides of the inequality by combining like terms and eliminating parentheses:

   Replace $x + 2 + 6 > 2x + 5(x + 1)$
   by $x + 8 > 7x + 5$

2. Add or subtract the same expression on both sides of the inequality:

   Replace $3x - 5 < 4$
   by $(3x - 5) + 5 < 4 + 5$

3. Multiply or divide both sides of the inequality by the same *positive* expression:

   Replace $4x > 16$ by $\dfrac{4x}{4} > \dfrac{16}{4}$

Procedures That Reverse the Sense or Direction of the Inequality Symbol

1. Interchange the two sides of the inequality:

   Replace $3 < x$ by $x > 3$

2. Multiply or divide both sides of the inequality by the same *negative* expression:

   Replace $-2x > 6$ by $\dfrac{-2x}{-2} < \dfrac{6}{-2}$

As the examples that follow illustrate, we solve inequalities using many of the same steps that we would use to solve equations. In writing the solution of an inequality, either set notation or interval notation may be used, whichever is more convenient.

EXAMPLE 4 **Solving an Inequality**

Solve the inequality $4x + 7 \geq 2x - 3$, and graph the solution set.

Solution

$$\begin{aligned} 4x + 7 &\geq 2x - 3 \\ 4x + 7 - 7 &\geq 2x - 3 - 7 && \text{Subtract 7 from both sides.} \\ 4x &\geq 2x - 10 && \text{Simplify.} \\ 4x - 2x &\geq 2x - 10 - 2x && \text{Subtract } 2x \text{ from both sides.} \\ 2x &\geq -10 && \text{Simplify.} \\ \frac{2x}{2} &\geq \frac{-10}{2} && \text{Divide both sides by 2. (The direction of the inequality symbol is unchanged.)} \\ x &\geq -5 && \text{Simplify.} \end{aligned}$$

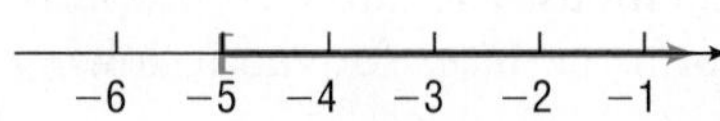

Figure 23 $x \geq -5$

The solution set is $\{x | x \geq -5\}$ or, using interval notation, all numbers in the interval $[-5, \infty)$. See Figure 23 for the graph.

Now Work PROBLEM 61

4 Solve Combined Inequalities

EXAMPLE 5 **Solving a Combined Inequality**

Solve the inequality $-5 < 3x - 2 < 1$, and graph the solution set.

Solution Recall that the inequality

$$-5 < 3x - 2 < 1$$

is equivalent to the two inequalities

$$-5 < 3x - 2 \quad \text{and} \quad 3x - 2 < 1$$

Solve each of these inequalities separately.

$-5 < 3x - 2$		$3x - 2 < 1$
$-5 + 2 < 3x - 2 + 2$	Add 2 to both sides.	$3x - 2 + 2 < 1 + 2$
$-3 < 3x$	Simplify.	$3x < 3$
$\frac{-3}{3} < \frac{3x}{3}$	Divide both sides by 3.	$\frac{3x}{3} < \frac{3}{3}$
$-1 < x$	Simplify.	$x < 1$

The solution set of the original pair of inequalities consists of all x for which

$$-1 < x \quad \text{and} \quad x < 1$$

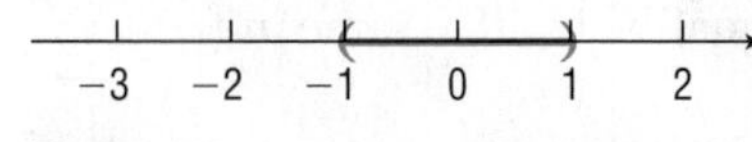

Figure 24 $-1 < x < 1$

This may be written more compactly as $\{x | -1 < x < 1\}$. In interval notation, the solution is $(-1, 1)$. See Figure 24 for the graph.

Observe in the preceding process that solving each of the two inequalities required exactly the same steps. A shortcut to solving the original inequality algebraically is to deal with the two inequalities at the same time, as follows:

$$\begin{aligned} -5 < \quad 3x - 2 \quad &< 1 \\ -5 + 2 < 3x - 2 + 2 &< 1 + 2 && \text{Add 2 to each part.} \\ -3 < \quad 3x \quad &< 3 && \text{Simplify.} \\ \frac{-3}{3} < \quad \frac{3x}{3} \quad &< \frac{3}{3} && \text{Divide each part by 3.} \\ -1 < \quad x \quad &< 1 && \text{Simplify.} \end{aligned}$$

Now Work PROBLEM 77

EXAMPLE 6 **Using the Reciprocal Property to Solve an Inequality**

Solve the inequality $(4x - 1)^{-1} > 0$, and graph the solution set.

Solution Recall that $(4x - 1)^{-1} = \dfrac{1}{4x - 1}$. The Reciprocal Property states that if $\dfrac{1}{a} > 0$, then $a > 0$.

$$(4x - 1)^{-1} > 0$$

$$\frac{1}{4x - 1} > 0$$

$$4x - 1 > 0 \quad \text{Reciprocal Property}$$

$$4x > 1 \quad \text{Add 1 to both sides.}$$

$$x > \frac{1}{4} \quad \text{Divide both sides by 4.}$$

The solution set is $\left\{x \middle| x > \dfrac{1}{4}\right\}$, that is, all x in the interval $\left(\dfrac{1}{4}, \infty\right)$. Figure 25 illustrates the graph. ●

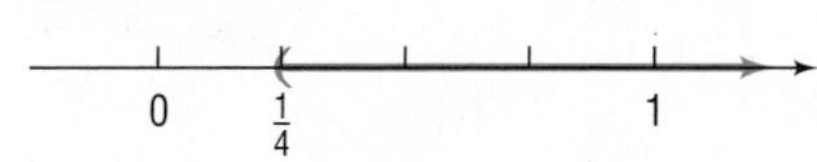

Figure 25 $x > \dfrac{1}{4}$

Now Work PROBLEM 87

5 Solve Inequalities Involving Absolute Value

EXAMPLE 7 **Solving an Inequality Involving Absolute Value**

Solve the inequality $|x| < 4$, and graph the solution set.

Solution We are looking for all points whose coordinate x is a distance less than 4 units from the origin. See Figure 26 for an illustration. Because any x between -4 and 4 satisfies the condition $|x| < 4$, the solution set consists of all numbers x for which $-4 < x < 4$, that is, all x in the interval $(-4, 4)$. ●

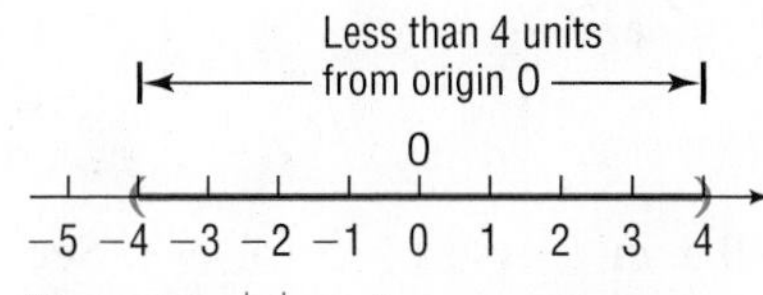

Figure 26 $|x| < 4$

EXAMPLE 8 **Solving an Inequality Involving Absolute Value**

Solve the inequality $|x| > 3$, and graph the solution set.

Solution We are looking for all points whose coordinate x is a distance greater than 3 units from the origin. Figure 27 illustrates the situation. Any number x less than -3 or greater than 3 satisfies the condition $|x| > 3$. The solution set consists of all numbers x for which $x < -3$ or $x > 3$, that is, all x in $(-\infty, -3) \cup (3, \infty)$.* ●

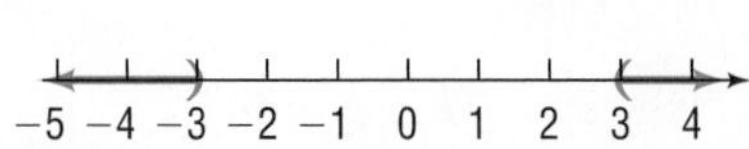

Figure 27 $|x| > 3$

Examples 7 and 8 illustrate the following results:

THEOREM

If a is any positive number, then

$\lvert u\rvert < a$	is equivalent to	$-a < u < a$	**(5)**
$\lvert u\rvert \leq a$	is equivalent to	$-a \leq u \leq a$	**(6)**
$\lvert u\rvert > a$	is equivalent to	$u < -a$ or $u > a$	**(7)**
$\lvert u\rvert \geq a$	is equivalent to	$u \leq -a$ or $u \geq a$	**(8)**

*Recall that the symbol $\cup$ stands for the union of two sets. Refer to page A2 if necessary.

EXAMPLE 9 **Solving an Inequality Involving Absolute Value**

Solve the inequality $|2x + 4| \le 3$, and graph the solution set.

Solution

$$|2x+4| \le 3$$ This follows the form of statement (6); the expression $u = 2x + 4$ is inside the absolute value bars.

$$-3 \le 2x+4 \le 3$$ Apply statement (6).

$$-3-4 \le 2x+4-4 \le 3-4$$ Subtract 4 from each part.

$$-7 \le 2x \le -1$$ Simplify.

$$\frac{-7}{2} \le \frac{2x}{2} \le \frac{-1}{2}$$ Divide each part by 2.

$$-\frac{7}{2} \le x \le -\frac{1}{2}$$ Simplify.

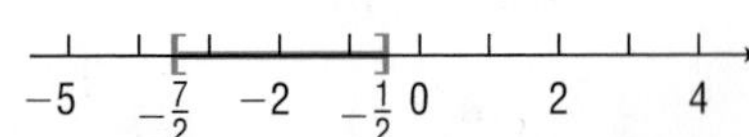

Figure 28 $|2x + 4| \le 3$

The solution set is $\left\{x \middle| -\frac{7}{2} \le x \le -\frac{1}{2}\right\}$, that is, all x in the interval $\left[-\frac{7}{2}, -\frac{1}{2}\right]$. See Figure 28 for a graph of the solution set. ●

Now Work PROBLEM 93

EXAMPLE 10 **Solving an Inequality Involving Absolute Value**

Solve the inequality $|2x - 5| > 3$, and graph the solution set.

Solution $|2x - 5| > 3$ This follows the form of statement (7); the expression $u = 2x - 5$ is inside the absolute value bars.

$$2x-5 < -3 \quad \text{or} \quad 2x-5 > 3$$ Apply statement (7).

$$2x-5+5 < -3+5 \quad \text{or} \quad 2x-5+5 > 3+5$$ Add 5 to each part.

$$2x < 2 \quad \text{or} \quad 2x > 8$$ Simplify.

$$\frac{2x}{2} < \frac{2}{2} \quad \text{or} \quad \frac{2x}{2} > \frac{8}{2}$$ Divide each part by 2.

$$x < 1 \quad \text{or} \quad x > 4$$ Simplify.

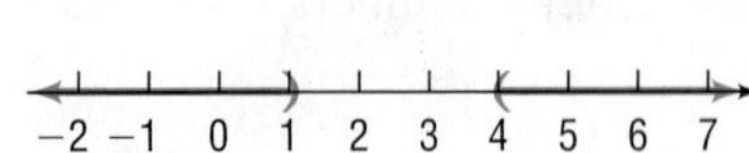

Figure 29 $|2x - 5| > 3$

The solution set is $\{x | x < 1 \text{ or } x > 4\}$, that is, all x in $(-\infty, 1) \cup (4, \infty)$. See Figure 29 for a graph of the solution set. ●

WARNING A common error to be avoided is to attempt to write the solution $x < 1$ or $x > 4$ as the combined inequality $1 > x > 4$, which is incorrect, since there are no numbers x for which $x < 1$ and $x > 4$. Another common error is to "mix" the symbols and write $1 < x > 4$, which makes no sense. ■

Now Work PROBLEM 99

A.6 Assess Your Understanding

'Are You Prepared?' *Answers are given at the end of these exercises. If you get a wrong answer, read the pages listed in red.*

1. Graph the inequality: $x \ge -2$. (pp. A4–A5)
2. ***True or False*** $-5 > -3$ (pp. A4–A5)
3. $|-2| =$ ______. (p. A5)
4. ***True or False*** $|x| \ge 0$ for any real number x. (pp. A5–A6)

Concepts and Vocabulary

5. If each side of an inequality is multiplied by a(n) ________ number, then the sense of the inequality symbol is reversed.
6. A(n) _______ ________, denoted $[a, b]$, consists of all real numbers x for which $a \le x \le b$.
7. The solution set of the equation $|x| = 5$ is {__________}.
8. The solution set of the inequality $|x| < 5$ is {x|__________}.
9. ***True or False*** The equation $|x| = -2$ has no solution.
10. ***True or False*** The inequality $|x| \ge -2$ has the set of real numbers as the solution set.

11. Which of the following will change the direction, or sense, of an inequality?
(a) Dividing each side by a positive number
(b) Interchanging sides
(c) Adding a negative number to each side
(d) Subtracting a positive number from each side

12. Which pair of inequalities is equivalent to $0 < x \le 3$?
(a) $x > 0$ and $x \ge 3$ (b) $x < 0$ and $x \ge 3$
(c) $x > 0$ and $x \le 3$ (d) $x < 0$ and $x \le 3$

13. Which of the following pairs of inequalities is equivalent to $|x| > 4$?
(a) $x > -4$ and $x < 4$ (b) $x < -4$ and $x < 4$
(c) $x > -4$ or $x > 4$ (d) $x < -4$ or $x > 4$

14. Which of the following has no solution?
(a) $|x| < -5$ (b) $|x| \le 0$ (c) $|x| > 0$ (d) $|x| \ge 0$

Skill Building

In Problems 15–20, express the graph shown in blue using interval notation. Also express each as an inequality involving x.

15. 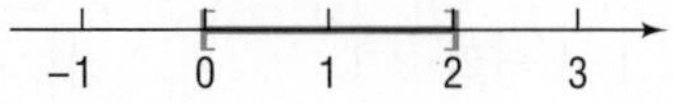
16. 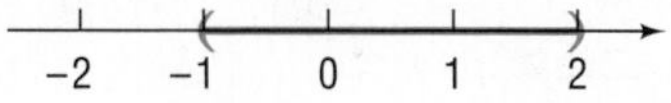
17. 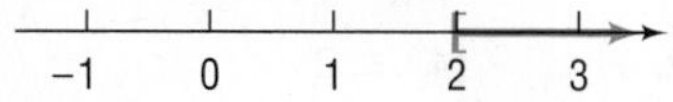
18. 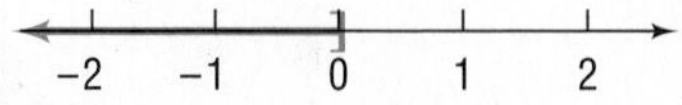
19. 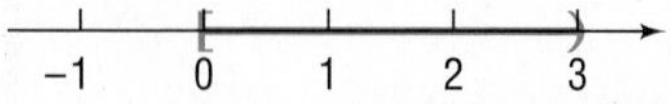
20. 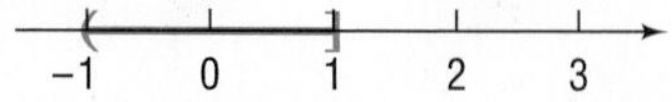

In Problems 21–26, an inequality is given. Write the inequality obtained by:
(a) Adding 3 to each side of the given inequality.
(b) Subtracting 5 from each side of the given inequality.
(c) Multiplying each side of the given inequality by 3.
(d) Multiplying each side of the given inequality by −2.

21. $3 < 5$ 22. $2 > 1$ 23. $4 > -3$ 24. $-3 > -5$ 25. $2x + 1 < 2$ 26. $1 - 2x > 5$

In Problems 27–34, write each inequality using interval notation, and illustrate each inequality using the real number line.

27. $0 \le x \le 4$ 28. $-1 < x < 5$ 29. $4 \le x < 6$ 30. $-2 < x < 0$
31. $x \ge 4$ 32. $x \le 5$ 33. $x < -4$ 34. $x > 1$

In Problems 35–42, write each interval as an inequality involving x, and illustrate each inequality using the real number line.

35. $[2, 5]$ 36. $(1, 2)$ 37. $(-3, -2)$ 38. $[0, 1)$
39. $[4, \infty)$ 40. $(-\infty, 2]$ 41. $(-\infty, -3)$ 42. $(-8, \infty)$

In Problems 43–56, fill in the blank with the correct inequality symbol.

43. If $x < 5$, then $x - 5$ ________ 0.
44. If $x < -4$, then $x + 4$ ________ 0.
45. If $x > -4$, then $x + 4$ ________ 0.
46. If $x > 6$, then $x - 6$ ________ 0.
47. If $x \ge -4$, then $3x$ ________ -12.
48. If $x \le 3$, then $2x$ ________ 6.
49. If $x > 6$, then $-2x$ ________ -12.
50. If $x > -2$, then $-4x$ ________ 8.
51. If $x \ge 5$, then $-4x$ ________ -20.
52. If $x \le -4$, then $-3x$ ________ 12.
53. If $2x > 6$, then x ________ 3.
54. If $3x \le 12$, then x ________ 4.
55. If $-\frac{1}{2}x \le 3$, then x ________ -6.
56. If $-\frac{1}{4}x > 1$, then x ________ -4.

In Problems 57–104, solve each inequality. Express your answer using set notation or interval notation. Graph the solution set.

57. $x + 1 < 5$ 58. $x - 6 < 1$ 59. $1 - 2x \le 3$
60. $2 - 3x \le 5$ 61. $3x - 7 > 2$ 62. $2x + 5 > 1$
63. $3x - 1 \ge 3 + x$ 64. $2x - 2 \ge 3 + x$ 65. $-2(x + 3) < 8$
66. $-3(1 - x) < 12$ 67. $4 - 3(1 - x) \le 3$ 68. $8 - 4(2 - x) \le -2x$
69. $\frac{1}{2}(x - 4) > x + 8$ 70. $3x + 4 > \frac{1}{3}(x - 2)$ 71. $\frac{x}{2} \ge 1 - \frac{x}{4}$
72. $\frac{x}{3} \ge 2 + \frac{x}{6}$ 73. $0 \le 2x - 6 \le 4$ 74. $4 \le 2x + 2 \le 10$
75. $-5 \le 4 - 3x \le 2$ 76. $-3 \le 3 - 2x \le 9$ 77. $-3 < \frac{2x - 1}{4} < 0$
78. $0 < \frac{3x + 2}{2} < 4$ 79. $1 < 1 - \frac{1}{2}x < 4$ 80. $0 < 1 - \frac{1}{3}x < 1$

81. $(x+2)(x-3) > (x-1)(x+1)$

82. $(x-1)(x+1) > (x-3)(x+4)$

83. $x(4x+3) \leq (2x+1)^2$

84. $x(9x-5) \leq (3x-1)^2$

85. $\frac{1}{2} \leq \frac{x+1}{3} < \frac{3}{4}$

86. $\frac{1}{3} < \frac{x+1}{2} \leq \frac{2}{3}$

87. $(4x+2)^{-1} < 0$

88. $(2x-1)^{-1} > 0$

89. $0 < \frac{2}{x} < \frac{3}{5}$

90. $0 < \frac{4}{x} < \frac{2}{3}$

91. $0 < (2x-4)^{-1} < \frac{1}{2}$

92. $0 < (3x+6)^{-1} < \frac{1}{3}$

93. $|2x| < 8$

94. $|3x| < 12$

95. $|3x| > 12$

96. $|2x| > 6$

97. $|2x-1| \leq 1$

98. $|2x+5| \leq 7$

99. $|1-2x| > 3$

100. $|2-3x| > 1$

101. $|-4x| + |-5| \leq 9$

102. $|-x| - |4| \leq 2$

103. $|-2x| \geq |-4|$

104. $|-x-2| \geq 1$

Applications and Extensions

105. Express the fact that x differs from 2 by less than $\frac{1}{2}$ as an inequality involving an absolute value. Solve for x.

106. Express the fact that x differs from -1 by less than 1 as an inequality involving an absolute value. Solve for x.

107. Express the fact that x differs from -3 by more than 2 as an inequality involving an absolute value. Solve for x.

108. Express the fact that x differs from 2 by more than 3 as an inequality involving an absolute value. Solve for x.

109. What is the domain of the variable in the expression $\sqrt{3x+6}$?

110. What is the domain of the variable in the expression $\sqrt{8+2x}$?

111. A young adult may be defined as someone older than 21, but less than 30 years of age. Express this statement using inequalities.

112. Middle-aged may be defined as being 40 or more and less than 60. Express this statement using inequalities.

113. Life Expectancy The Social Security Administration determined that an average 30-year-old male in 2014 could expect to live at least 51.9 more years and that an average 30-year-old female in 2014 could expect to live at least 55.6 more years.

(a) To what age could an average 30-year-old male expect to live? Express your answer as an inequality.

(b) To what age could an average 30-year-old female expect to live? Express your answer as an inequality.

(c) Who can expect to live longer, a male or a female? By how many years?

Source: Social Security Administration, 2014

114. General Chemistry For a certain ideal gas, the volume V (in cubic centimeters) equals 20 times the temperature T (in degrees Celsius). If the temperature varies from 80° to 120° C inclusive, what is the corresponding range of the volume of the gas?

115. Real Estate A real estate agent agrees to sell an apartment complex according to the following commission schedule: \$45,000 plus 25% of the selling price in excess of \$900,000. Assuming that the complex will sell at some price between \$900,000 and \$1,100,000 inclusive, over what range does the agent's commission vary? How does the commission vary as a percent of selling price?

116. Sales Commission A used car salesperson is paid a commission of \$25 plus 40% of the selling price in excess of owner's cost. The owner claims that used cars typically sell for at least owner's cost plus \$200 and at most owner's cost plus \$3000. For each sale made, over what range can the salesperson expect the commission to vary?

117. Federal Tax Withholding The percentage method of withholding for federal income tax (2014) states that a single person whose weekly wages, after subtracting withholding allowances, are over \$753, but not over \$1762, shall have \$97.75 plus 25% of the excess over \$753 withheld. Over what range does the amount withheld vary if the weekly wages vary from \$900 to \$1100, inclusive?

Source: Employer's Tax Guide. Internal Revenue Service, 2014.

118. Exercising Sue wants to lose weight. For healthy weight loss, the American College of Sports Medicine (ACSM) recommends 200 to 300 minutes of exercise per week. For the first six days of the week, Sue exercised 40, 45, 0, 50, 25, and 35 minutes. How long should Sue exercise on the seventh day in order to stay within the ACSM guidelines?

119. Electricity Rates Commonwealth Edison Company's charge for electricity in January 2014 was 8.21¢ per kilowatt-hour. In addition, each monthly bill contains a customer charge of \$15.37. If last year's bills ranged from a low of \$72.84 to a high of \$237.04, over what range did usage vary (in kilowatt-hours)?

Source: Commonwealth Edison Co., 2014.

120. Water Bills The Village of Oak Lawn charges homeowners \$57.07 per quarter-year plus \$5.81 per 1000 gallons for water usage in excess of 10,000 gallons. In 2014 one homeowner's quarterly bill ranged from a high of \$150.03 to a low of \$97.74. Over what range did water usage vary?

Source: Village of Oak Lawn, Illinois, January 2014.

121. Markup of a New Car The markup over dealer's cost of a new car ranges from 12% to 18%. If the sticker price is \$18,000, over what range will the dealer's cost vary?

122. IQ Tests A standard intelligence test has an average score of 100. According to statistical theory, of the people who take the test, the 2.5% with the highest scores will have scores of more than 1.96σ above the average, where σ (sigma, a number called the **standard deviation**) depends on the nature of the test. If $\sigma = 12$ for this test and there is (in principle) no upper limit to the score possible on the test, write the interval of possible test scores of the people in the top 2.5%.

123. Computing Grades In your Economics 101 class, you have scores of 68, 82, 87, and 89 on the first four of five tests. To get a grade of B, the average of the first five test scores must be greater than or equal to 80 and less than 90.

(a) Solve an inequality to find the range of the score that you need on the last test to get a B.

(b) What score do you need if the fifth test counts double?

124. "Light" Foods For food products to be labeled "light," the U.S. Food and Drug Administration requires that the altered product must either contain at least one-third fewer calories than the regular product or it must contain at least one-half less fat than the regular product. If a serving of Miracle Whip® Light contains 20 calories and 1.5 grams of fat, then what must be true about either the number of calories or the grams of fat in a serving of regular Miracle Whip®?

125. Arithmetic Mean If $a < b$, show that $a < \frac{a+b}{2} < b$. The number $\frac{a+b}{2}$ is called the **arithmetic mean** of a and b.

126. Refer to Problem 125. Show that the arithmetic mean of a and b is equidistant from a and b.

127. Geometric Mean If $0 < a < b$, show that $a < \sqrt{ab} < b$. The number $\sqrt{ab}$ is called the **geometric mean** of a and b.

128. Refer to Problems 125 and 127. Show that the geometric mean of a and b is less than the arithmetic mean of a and b.

129. Harmonic Mean For $0 < a < b$, let h be defined by

$$\frac{1}{h} = \frac{1}{2}\left(\frac{1}{a} + \frac{1}{b}\right)$$

Show that $a < h < b$. The number h is called the **harmonic mean** of a and b.

130. Refer to Problems 125, 127, and 129. Show that the harmonic mean of a and b equals the geometric mean squared divided by the arithmetic mean.

131. Another Reciprocal Property Prove that if $0 < a < b$, then $0 < \frac{1}{b} < \frac{1}{a}$.

Explaining Concepts: Discussion and Writing

132. Make up an inequality that has no solution. Make up one that has exactly one solution.

133. The inequality $x^2 + 1 < -5$ has no real solution. Explain why.

134. Do you prefer to use inequality notation or interval notation to express the solution to an inequality? Give your reasons. Are there particular circumstances when you prefer one to the other? Cite examples.

135. How would you explain to a fellow student the underlying reason for the multiplication properties for inequalities (page A48)? That is, the sense or direction of an inequality remains the same if each side is multiplied by a positive real number, whereas the direction is reversed if each side is multiplied by a negative real number?

'Are You Prepared?' Answers

1. −4 −2 0 **2.** False **3.** 2 **4.** True

A.7 *n*th Roots; Rational Exponents

PREPARING FOR THIS SECTION *Before getting started, review the following:*

- Exponents, Square Roots (Appendix A, Section A.1, pp. A8–A10)

Now Work the 'Are You Prepared?' problems on page A61.

OBJECTIVES
1. Work with *n*th Roots (p. A56)
2. Simplify Radicals (p. A57)
3. Rationalize Denominators (p. A58)
4. Solve Radical Equations (p. A59)
5. Simplify Expressions with Rational Exponents (p. A59)

1 Work with *n*th Roots

DEFINITION The **principal *n*th root of a real number *a***, $n \geq 2$ an integer, symbolized by $\sqrt[n]{a}$, is defined as follows:

$$\sqrt[n]{a} = b \quad \text{means} \quad a = b^n$$

where $a \geq 0$ and $b \geq 0$ if n is even and a, b are any real numbers if n is odd.

In Words
The symbol $\sqrt[n]{a}$ means "give me the number that, when raised to the power n, equals a."

Notice that if a is negative and n is even, then $\sqrt[n]{a}$ is not defined. When it is defined, the principal nth root of a number is unique.

The symbol $\sqrt[n]{a}$ for the principal nth root of a is called a **radical**; the integer n is called the **index**, and a is called the **radicand**. If the index of a radical is 2, we call $\sqrt[2]{a}$ the **square root** of a and omit the index 2 by simply writing $\sqrt{a}$. If the index is 3, we call $\sqrt[3]{a}$ the **cube root** of a.

EXAMPLE 1 **Simplifying Principal *n*th Roots**

(a) $\sqrt[3]{8} = \sqrt[3]{2^3} = 2$

(b) $\sqrt[3]{-64} = \sqrt[3]{(-4)^3} = -4$

(c) $\sqrt[4]{\dfrac{1}{16}} = \sqrt[4]{\left(\dfrac{1}{2}\right)^4} = \dfrac{1}{2}$

(d) $\sqrt[6]{(-2)^6} = |-2| = 2$

These are examples of **perfect roots**, since each simplifies to a rational number. Notice the absolute value in Example 1(d). If n is even, then the principal nth root must be nonnegative.

In general, if $n \geq 2$ is an integer and a is a real number, we have

$$\sqrt[n]{a^n} = a \qquad \text{if } n \geq 3 \text{ is odd} \qquad \textbf{(1a)}$$

$$\sqrt[n]{a^n} = |a| \qquad \text{if } n \geq 2 \text{ is even} \qquad \textbf{(1b)}$$

Now Work PROBLEM 11

Radicals provide a way of representing many irrational real numbers. For example, it can be shown that there is no rational number whose square is 2. Using radicals, we can say that $\sqrt{2}$ is the positive number whose square is 2.

EXAMPLE 2

Using a Calculator to Approximate Roots

Use a calculator to approximate $\sqrt[5]{16}$.

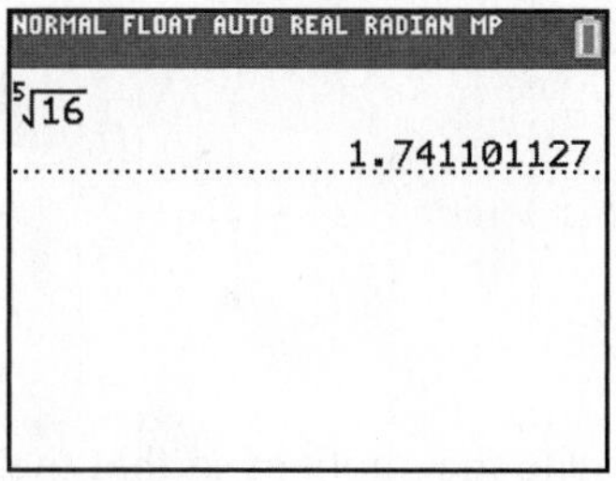

Solution Figure 30 shows the result using a TI-84 Plus C graphing calculator.

NORMAL FLOAT AUTO REAL RADIAN MP

$\sqrt[5]{16}$

1.741101127

Figure 30

Now Work PROBLEM 117

2 Simplify Radicals

Let $n \geq 2$ and $m \geq 2$ denote integers, and let a and b represent real numbers. Assuming that all radicals are defined, we have the following properties:

Properties of Radicals

$$\sqrt[n]{ab} = \sqrt[n]{a}\,\sqrt[n]{b} \tag{2a}$$

$$\sqrt[n]{\frac{a}{b}} = \frac{\sqrt[n]{a}}{\sqrt[n]{b}} \qquad b \neq 0 \tag{2b}$$

$$\sqrt[n]{a^m} = (\sqrt[n]{a})^m \tag{2c}$$

When used in reference to radicals, the direction to "simplify" will mean to remove from the radicals any perfect roots that occur as factors.

EXAMPLE 3

Simplifying Radicals

(a) $\sqrt{32} = \sqrt{16\cdot 2} = \sqrt{16}\cdot\sqrt{2} = 4\sqrt{2}$

Factor out 16, a perfect square. (2a)

(b) $\sqrt[3]{16} = \sqrt[3]{8\cdot 2} = \sqrt[3]{8}\cdot\sqrt[3]{2} = \sqrt[3]{2^3}\cdot\sqrt[3]{2} = 2\sqrt[3]{2}$

Factor out 8, a perfect cube. (2a)

(c) $\sqrt[3]{-16x^4} = \sqrt[3]{-8\cdot 2\cdot x^3\cdot x} = \sqrt[3]{(-8x^3)(2x)}$

Factor perfect cubes inside radical. Group perfect cubes.

$= \sqrt[3]{(-2x)^3\cdot 2x} = \sqrt[3]{(-2x)^3}\cdot\sqrt[3]{2x} = -2x\sqrt[3]{2x}$

(2a)

(d) $\sqrt[4]{\dfrac{16x^5}{81}} = \sqrt[4]{\dfrac{2^4x^4x}{3^4}} = \sqrt[4]{\left(\dfrac{2x}{3}\right)^4\cdot x} = \sqrt[4]{\left(\dfrac{2x}{3}\right)^4}\cdot\sqrt[4]{x} = \left|\dfrac{2x}{3}\right|\sqrt[4]{x}$

Now Work PROBLEMS 15 AND 21

Two or more radicals can be combined, provided that they have the same index and the same radicand. Such radicals are called **like radicals**.

EXAMPLE 4

Combining Like Radicals

(a) $-8\sqrt{12} + \sqrt{3} = -8\sqrt{4\cdot 3} + \sqrt{3}$

$= -8\cdot\sqrt{4}\sqrt{3} + \sqrt{3}$

$= -16\sqrt{3} + \sqrt{3} = -15\sqrt{3}$

$$\text{(b) } \sqrt[3]{8x^4} + \sqrt[3]{-x} + 4\sqrt[3]{27x} = \sqrt[3]{2^3x^3x} + \sqrt[3]{-1\cdot x} + 4\sqrt[3]{3^3x}$$
$$= \sqrt[3]{(2x)^3}\cdot\sqrt[3]{x} + \sqrt[3]{-1}\cdot\sqrt[3]{x} + 4\sqrt[3]{3^3}\cdot\sqrt[3]{x}$$
$$= 2x\sqrt[3]{x} - 1\cdot\sqrt[3]{x} + 12\sqrt[3]{x}$$
$$= (2x + 11)\sqrt[3]{x}$$

Now Work PROBLEM 39

3 Rationalize Denominators

When radicals occur in quotients, it is customary to rewrite the quotient so that the new denominator contains no radicals. This process is referred to as **rationalizing the denominator**.

The idea is to multiply by an appropriate expression so that the new denominator contains no radicals. For example:

If a Denominator Contains the Factor	Multiply by	To Obtain a Denominator Free of Radicals
$\sqrt{3}$	$\sqrt{3}$	$(\sqrt{3})^2 = 3$
$\sqrt{3} + 1$	$\sqrt{3} - 1$	$(\sqrt{3})^2 - 1^2 = 3 - 1 = 2$
$\sqrt{2} - 3$	$\sqrt{2} + 3$	$(\sqrt{2})^2 - 3^2 = 2 - 9 = -7$
$\sqrt{5} - \sqrt{3}$	$\sqrt{5} + \sqrt{3}$	$(\sqrt{5})^2 - (\sqrt{3})^2 = 5 - 3 = 2$
$\sqrt[3]{4}$	$\sqrt[3]{2}$	$\sqrt[3]{4}\cdot\sqrt[3]{2} = \sqrt[3]{8} = 2$

In rationalizing the denominator of a quotient, be sure to multiply both the numerator and the denominator by the expression.

EXAMPLE 5 **Rationalizing Denominators**

Rationalize the denominator of each expression.

(a) $\dfrac{4}{\sqrt{2}}$ (b) $\dfrac{\sqrt{3}}{\sqrt[3]{2}}$ (c) $\dfrac{\sqrt{x} - 2}{\sqrt{x} + 2}$, $x \ge 0$

Solution

(a) $\dfrac{4}{\sqrt{2}} = \dfrac{4}{\sqrt{2}}\cdot\dfrac{\sqrt{2}}{\sqrt{2}} = \dfrac{4\sqrt{2}}{(\sqrt{2})^2} = \dfrac{4\sqrt{2}}{2} = 2\sqrt{2}$

Multiply by $\dfrac{\sqrt{2}}{\sqrt{2}}$.

(b) $\dfrac{\sqrt{3}}{\sqrt[3]{2}} = \dfrac{\sqrt{3}}{\sqrt[3]{2}}\cdot\dfrac{\sqrt[3]{4}}{\sqrt[3]{4}} = \dfrac{\sqrt{3}\,\sqrt[3]{4}}{\sqrt[3]{8}} = \dfrac{\sqrt{3}\,\sqrt[3]{4}}{2}$

Multiply by $\dfrac{\sqrt[3]{4}}{\sqrt[3]{4}}$.

(c) $\dfrac{\sqrt{x} - 2}{\sqrt{x} + 2} = \dfrac{\sqrt{x} - 2}{\sqrt{x} + 2}\cdot\dfrac{\sqrt{x} - 2}{\sqrt{x} - 2} = \dfrac{(\sqrt{x} - 2)^2}{(\sqrt{x})^2 - 2^2}$

$$= \frac{(\sqrt{x})^2 - 4\sqrt{x} + 4}{x - 4} = \frac{x - 4\sqrt{x} + 4}{x - 4}$$

Now Work PROBLEM 53

4 Solve Radical Equations

When the variable in an equation occurs in a square root, cube root, and so on—that is, when it occurs in a radical—the equation is called a **radical equation**. Sometimes a suitable operation will change a radical equation to one that is linear or quadratic. A commonly used procedure is to isolate the most complicated radical on one side of the equation and then eliminate it by raising each side to a power equal to the index of the radical. Care must be taken, however, because apparent solutions that are not, in fact, solutions of the original equation may result. These are called **extraneous solutions**. Therefore, we need to check all answers when working with radical equations, and we check them in the *original* equation.

EXAMPLE 6 **Solving a Radical Equation**

Find the real solutions of the equation: $\sqrt[3]{2x-4} - 2 = 0$

Solution The equation contains a radical whose index is 3. Isolate it on the left side.

$$\sqrt[3]{2x-4} - 2 = 0$$

$$\sqrt[3]{2x-4} = 2 \quad \text{Add 2 to both sides.}$$

Now raise each side to the third power (the index of the radical is 3) and solve.

$$\left(\sqrt[3]{2x-4}\right)^3 = 2^3 \quad \text{Raise each side to the power 3.}$$

$$2x - 4 = 8 \quad \text{Simplify.}$$

$$2x = 12 \quad \text{Add 4 to both sides.}$$

$$x = 6 \quad \text{Divide both sides by 2.}$$

Check: $\sqrt[3]{2(6)-4} - 2 = \sqrt[3]{12-4} - 2 = \sqrt[3]{8} - 2 = 2 - 2 = 0$

The solution set is $\{6\}$. ●

Now Work PROBLEM 63

5 Simplify Expressions with Rational Exponents

Radicals are used to define rational exponents.

DEFINITION

If a is a real number and $n \geq 2$ is an integer, then

$$a^{1/n} = \sqrt[n]{a} \qquad (3)$$

provided that $\sqrt[n]{a}$ exists.

Note that if n is even and $a < 0$, then $\sqrt[n]{a}$ and $a^{1/n}$ do not exist.

EXAMPLE 7 **Writing Expressions Containing Fractional Exponents as Radicals**

(a) $4^{1/2} = \sqrt{4} = 2$ (b) $8^{1/2} = \sqrt{8} = 2\sqrt{2}$

(c) $(-27)^{1/3} = \sqrt[3]{-27} = -3$ (d) $16^{1/3} = \sqrt[3]{16} = 2\sqrt[3]{2}$ ●

DEFINITION

If a is a real number and m and n are integers containing no common factors, with $n \geq 2$, then

$$a^{m/n} = \sqrt[n]{a^m} = \left(\sqrt[n]{a}\right)^m \qquad (4)$$

provided that $\sqrt[n]{a}$ exists.

We have two comments about equation (4):

1. The exponent $\frac{m}{n}$ must be in lowest terms, and $n \geq 2$ must be positive.
2. In simplifying the rational expression $a^{m/n}$, either $\sqrt[n]{a^m}$ or $(\sqrt[n]{a})^m$ may be used, the choice depending on which is easier to simplify. Generally, taking the root first, as in $(\sqrt[n]{a})^m$, is easier.

EXAMPLE 8

Using Equation (4)

(a) $4^{3/2} = (\sqrt{4})^3 = 2^3 = 8$ (b) $(-8)^{4/3} = (\sqrt[3]{-8})^4 = (-2)^4 = 16$

(c) $(32)^{-2/5} = (\sqrt[5]{32})^{-2} = 2^{-2} = \frac{1}{4}$ (d) $25^{6/4} = 25^{3/2} = (\sqrt{25})^3 = 5^3 = 125$

Now Work PROBLEM 67

It can be shown that the Laws of Exponents hold for rational exponents. The next example illustrates using the Laws of Exponents to simplify.

EXAMPLE 9

Simplifying Expressions Containing Rational Exponents

Simplify each expression. Express your answer so that only positive exponents occur. Assume that the variables are positive.

(a) $(x^{2/3}y)(x^{-2}y)^{1/2}$ (b) $\left(\frac{2x^{1/3}}{y^{2/3}}\right)^{-3}$ (c) $\left(\frac{9x^2y^{1/3}}{x^{1/3}y}\right)^{1/2}$

Solution

(a)
$$\begin{aligned}(x^{2/3}y)(x^{-2}y)^{1/2} &= (x^{2/3}y)\left[(x^{-2})^{1/2}y^{1/2}\right]\\ &= x^{2/3}yx^{-1}y^{1/2}\\ &= (x^{2/3}\cdot x^{-1})(y\cdot y^{1/2})\\ &= x^{-1/3}y^{3/2}\\ &= \frac{y^{3/2}}{x^{1/3}}\end{aligned}$$

(b)
$$\left(\frac{2x^{1/3}}{y^{2/3}}\right)^{-3} = \left(\frac{y^{2/3}}{2x^{1/3}}\right)^3 = \frac{(y^{2/3})^3}{(2x^{1/3})^3} = \frac{y^2}{2^3(x^{1/3})^3} = \frac{y^2}{8x}$$

(c)
$$\left(\frac{9x^2y^{1/3}}{x^{1/3}y}\right)^{1/2} = \left(\frac{9x^{2-(1/3)}}{y^{1-(1/3)}}\right)^{1/2} = \left(\frac{9x^{5/3}}{y^{2/3}}\right)^{1/2} = \frac{9^{1/2}(x^{5/3})^{1/2}}{(y^{2/3})^{1/2}} = \frac{3x^{5/6}}{y^{1/3}}$$

Now Work PROBLEM 87

The next two examples illustrate some algebra that you will need to know for certain calculus problems.

EXAMPLE 10

Writing an Expression as a Single Quotient

Write the following expression as a single quotient in which only positive exponents appear.

$$(x^2+1)^{1/2} + x\cdot\frac{1}{2}(x^2+1)^{-1/2}\cdot 2x$$

Solution

$$\begin{aligned}(x^2+1)^{1/2} + x\cdot\frac{1}{2}(x^2+1)^{-1/2}\cdot 2x &= (x^2+1)^{1/2} + \frac{x^2}{(x^2+1)^{1/2}}\\ &= \frac{(x^2+1)^{1/2}(x^2+1)^{1/2}+x^2}{(x^2+1)^{1/2}}\\ &= \frac{(x^2+1)+x^2}{(x^2+1)^{1/2}}\\ &= \frac{2x^2+1}{(x^2+1)^{1/2}}\end{aligned}$$

Now Work PROBLEM 93

EXAMPLE 11 **Factoring an Expression Containing Rational Exponents**

Factor and simplify: $\frac{4}{3}x^{1/3}(2x+1) + 2x^{4/3}$

Solution Begin by writing $2x^{4/3}$ as a fraction with 3 as the denominator.

$$\frac{4}{3}x^{1/3}(2x+1) + 2x^{4/3} = \frac{4x^{1/3}(2x+1)}{3} + \frac{6x^{4/3}}{3} \underset{\text{Add the two fractions}}{=} \frac{4x^{1/3}(2x+1)+6x^{4/3}}{3}$$

$$\underset{\text{2 and } x^{1/3} \text{ are common factors}}{=} \frac{2x^{1/3}[2(2x+1)+3x]}{3} \underset{\text{Simplify}}{=} \frac{2x^{1/3}(7x+2)}{3}$$

Now Work PROBLEM 105

A.7 Assess Your Understanding

'Are You Prepared?' *Answers are given at the end of these exercises. If you get a wrong answer, read the pages in red.*

1. $(-3)^2 =$ __; $-3^2 =$ ___ (pp. A8–A9)
2. $\sqrt{16} =$ __; $\sqrt{(-4)^2} =$ __ (pp. A9–A10)

Concepts and Vocabulary

3. In the symbol $\sqrt[n]{a}$, the integer n is called the ________.
4. We call $\sqrt[3]{a}$ the _______ _______ of a.
5. Let $n \geq 2$ and $m \geq 2$ be integers, and let a and b be real numbers. Which of the following is not a property of radicals? Assume all radicals are defined.
   (a) $\sqrt[n]{\frac{a}{b}} = \frac{\sqrt[n]{a}}{\sqrt[n]{b}}$ (b) $\sqrt[n]{a+b} = \sqrt[n]{a} + \sqrt[n]{b}$
   (c) $\sqrt[n]{ab} = \sqrt[n]{a}\sqrt[n]{b}$ (d) $\sqrt[n]{a^m} = (\sqrt[n]{a})^m$
6. If a is a real number and $n \geq 2$ is an integer, then which of the following expressions is equivalent to $\sqrt[n]{a}$, provided that it exists?
   (a) a^{-n} (b) a^n (c) $\frac{1}{a^n}$ (d) $a^{1/n}$
7. Which of the following phrases best defines like radicals?
   (a) Radical expressions that have the same index
   (b) Radical expressions that have the same radicand
   (c) Radical expressions that have the same index and the same radicand
   (d) Radical expressions that have the same variable
8. To rationalize the denominator of the expression $\frac{\sqrt{2}}{1-\sqrt{3}}$, multiply both the numerator and the denominator by which of the following?
   (a) $\sqrt{3}$ (b) $\sqrt{2}$ (c) $1+\sqrt{3}$ (d) $1-\sqrt{3}$
9. ***True or False*** $\sqrt[5]{-32} = -2$
10. ***True or False*** $\sqrt[4]{(-3)^4} = -3$

Skill Building

In Problems 11–48, simplify each expression. Assume that all variables are positive when they appear.

11. $\sqrt[3]{27}$ 12. $\sqrt[4]{16}$ 13. $\sqrt[3]{-8}$ 14. $\sqrt[3]{-1}$

15. $\sqrt{8}$ 16. $\sqrt[3]{54}$ 17. $\sqrt[3]{-8x^4}$ 18. $\sqrt[4]{48x^5}$

19. $\sqrt[4]{x^{12}y^8}$ 20. $\sqrt[5]{x^{10}y^5}$ 21. $\sqrt[4]{\dfrac{x^9y^7}{xy^3}}$ 22. $\sqrt[3]{\dfrac{3xy^2}{81x^4y^2}}$

23. $\sqrt{36x}$ 24. $\sqrt{9x^5}$ 25. $\sqrt[3]{162x^9y^{12}}$ 26. $\sqrt[3]{-40x^{14}y^{10}}$

27. $\sqrt{3x^2}\sqrt{12x}$ 28. $\sqrt{5x}\sqrt{20x^3}$ 29. $(\sqrt{5}\sqrt[3]{9})^2$ 30. $(\sqrt[3]{3}\sqrt{10})^4$

31. $(3\sqrt{6})(2\sqrt{2})$ 32. $(5\sqrt{8})(-3\sqrt{3})$ 33. $3\sqrt{2} + 4\sqrt{2}$ 34. $6\sqrt{5} - 4\sqrt{5}$

35. $-\sqrt{18} + 2\sqrt{8}$ 36. $2\sqrt{12} - 3\sqrt{27}$ 37. $(\sqrt{3} + 3)(\sqrt{3} - 1)$ 38. $(\sqrt{5} - 2)(\sqrt{5} + 3)$

39. $5\sqrt[3]{2} - 2\sqrt[3]{54}$ 40. $9\sqrt[3]{24} - \sqrt[3]{81}$ 41. $(\sqrt{x} - 1)^2$ 42. $(\sqrt{x} + \sqrt{5})^2$

43. $\sqrt[3]{16x^4} - \sqrt[3]{2x}$ 44. $\sqrt[4]{32x} + \sqrt[4]{2x^5}$ 45. $\sqrt{8x^3} - 3\sqrt{50x}$ 46. $3x\sqrt{9y} + 4\sqrt{25y}$

47. $\sqrt[3]{16x^4y} - 3x\sqrt[3]{2xy} + 5\sqrt[3]{-2xy^4}$ 48. $8xy - \sqrt{25x^2y^2} + \sqrt[3]{8x^3y^3}$

In Problems 49–62, rationalize the denominator of each expression. Assume that all variables are positive when they appear.

49. $\dfrac{1}{\sqrt{2}}$ 50. $\dfrac{2}{\sqrt{3}}$ 51. $\dfrac{-\sqrt{3}}{\sqrt{5}}$ 52. $\dfrac{-\sqrt{3}}{\sqrt{8}}$

53. $\dfrac{\sqrt{3}}{5 - \sqrt{2}}$ 54. $\dfrac{\sqrt{2}}{\sqrt{7} + 2}$ 55. $\dfrac{2 - \sqrt{5}}{2 + 3\sqrt{5}}$ 56. $\dfrac{\sqrt{3} - 1}{2\sqrt{3} + 3}$

57. $\dfrac{5}{\sqrt{2} - 1}$ 58. $\dfrac{-3}{\sqrt{5} + 4}$ 59. $\dfrac{5}{\sqrt[3]{2}}$ 60. $\dfrac{-2}{\sqrt[3]{9}}$

61. $\dfrac{\sqrt{x+h} - \sqrt{x}}{\sqrt{x+h} + \sqrt{x}}$ 62. $\dfrac{\sqrt{x+h} + \sqrt{x-h}}{\sqrt{x+h} - \sqrt{x-h}}$

In Problems 63–66, solve each equation.

63. $\sqrt[3]{2t - 1} = 2$ 64. $\sqrt[3]{3t + 1} = -2$ 65. $\sqrt{15 - 2x} = x$ 66. $\sqrt{12 - x} = x$

In Problems 67–82, simplify each expression.

67. $8^{2/3}$ 68. $4^{3/2}$ 69. $(-27)^{1/3}$ 70. $16^{3/4}$

71. $16^{3/2}$ 72. $25^{3/2}$ 73. $9^{-3/2}$ 74. $16^{-3/2}$

75. $\left(\dfrac{9}{8}\right)^{3/2}$ 76. $\left(\dfrac{27}{8}\right)^{2/3}$ 77. $\left(\dfrac{8}{9}\right)^{-3/2}$ 78. $\left(\dfrac{8}{27}\right)^{-2/3}$

79. $(-1000)^{-1/3}$ 80. $-25^{-1/2}$ 81. $\left(-\dfrac{64}{125}\right)^{-2/3}$ 82. $-81^{-3/4}$

In Problems 83–90, simplify each expression. Express your answer so that only positive exponents occur. Assume that the variables are positive.

83. $x^{3/4}x^{1/3}x^{-1/2}$ 84. $x^{2/3}x^{1/2}x^{-1/4}$ 85. $(x^3y^6)^{1/3}$ 86. $(x^4y^8)^{3/4}$

87. $\dfrac{(x^2y)^{1/3}(xy^2)^{2/3}}{x^{2/3}y^{2/3}}$ 88. $\dfrac{(xy)^{1/4}(x^2y^2)^{1/2}}{(x^2y)^{3/4}}$ 89. $\dfrac{(16x^2y^{-1/3})^{3/4}}{(xy^2)^{1/4}}$ 90. $\dfrac{(4x^{-1}y^{1/3})^{3/2}}{(xy)^{3/2}}$

Applications and Extensions

In Problems 91–104, expressions that occur in calculus are given. Write each expression as a single quotient in which only positive exponents and/or radicals appear.

91. $\dfrac{x}{(1 + x)^{1/2}} + 2(1 + x)^{1/2} \quad x > -1$ 92. $\dfrac{1 + x}{2x^{1/2}} + x^{1/2} \quad x > 0$

93. $2x(x^2+1)^{1/2} + x^2 \cdot \frac{1}{2}(x^2+1)^{-1/2} \cdot 2x$

94. $(x+1)^{1/3} + x \cdot \frac{1}{3}(x+1)^{-2/3} \quad x \neq -1$

95. $\sqrt{4x+3} \cdot \frac{1}{2\sqrt{x-5}} + \sqrt{x-5} \cdot \frac{1}{5\sqrt{4x+3}} \quad x > 5$

96. $\frac{\sqrt[3]{8x+1}}{3\sqrt[3]{(x-2)^2}} + \frac{\sqrt[3]{x-2}}{24\sqrt[3]{(8x+1)^2}} \quad x \neq 2, x \neq -\frac{1}{8}$

97. $\frac{\sqrt{1+x} - x \cdot \frac{1}{2\sqrt{1+x}}}{1+x} \quad x > -1$

98. $\frac{\sqrt{x^2+1} - x \cdot \frac{2x}{2\sqrt{x^2+1}}}{x^2+1}$

99. $\frac{(x+4)^{1/2} - 2x(x+4)^{-1/2}}{x+4} \quad x > -4$

100. $\frac{(9-x^2)^{1/2} + x^2(9-x^2)^{-1/2}}{9-x^2} \quad -3 < x < 3$

101. $\frac{\frac{x^2}{(x^2-1)^{1/2}} - (x^2-1)^{1/2}}{x^2} \quad x < -1 \text{ or } x > 1$

102. $\frac{(x^2+4)^{1/2} - x^2(x^2+4)^{-1/2}}{x^2+4}$

103. $\frac{\frac{1+x^2}{2\sqrt{x}} - 2x\sqrt{x}}{(1+x^2)^2} \quad x > 0$

104. $\frac{2x(1-x^2)^{1/3} + \frac{2}{3}x^3(1-x^2)^{-2/3}}{(1-x^2)^{2/3}} \quad x \neq -1, x \neq 1$

In Problems 105–114, expressions that occur in calculus are given. Factor each expression. Express your answer so that only positive exponents occur.

105. $(x+1)^{3/2} + x \cdot \frac{3}{2}(x+1)^{1/2} \quad x \geq -1$

106. $(x^2+4)^{4/3} + x \cdot \frac{4}{3}(x^2+4)^{1/3} \cdot 2x$

107. $6x^{1/2}(x^2+x) - 8x^{3/2} - 8x^{1/2} \quad x \geq 0$

108. $6x^{1/2}(2x+3) + x^{3/2} \cdot 8 \quad x \geq 0$

109. $3(x^2+4)^{4/3} + x \cdot 4(x^2+4)^{1/3} \cdot 2x$

110. $2x(3x+4)^{4/3} + x^2 \cdot 4(3x+4)^{1/3}$

111. $4(3x+5)^{1/3}(2x+3)^{3/2} + 3(3x+5)^{4/3}(2x+3)^{1/2} \quad x \geq -\frac{3}{2}$

112. $6(6x+1)^{1/3}(4x-3)^{3/2} + 6(6x+1)^{4/3}(4x-3)^{1/2} \quad x \geq \frac{3}{4}$

113. $3x^{-1/2} + \frac{3}{2}x^{1/2} \quad x > 0$

114. $8x^{1/3} - 4x^{-2/3} \quad x \neq 0$

In Problems 115–122, use a calculator to approximate each radical. Round your answer to two decimal places.

115. $\sqrt{2}$

116. $\sqrt{7}$

117. $\sqrt[3]{4}$

118. $\sqrt[3]{-5}$

119. $\frac{2+\sqrt{3}}{3-\sqrt{5}}$

120. $\frac{\sqrt{5}-2}{\sqrt{2}+4}$

121. $\frac{3\sqrt{5}-\sqrt{2}}{\sqrt{3}}$

122. $\frac{2\sqrt{3}-\sqrt[3]{4}}{\sqrt{2}}$

123. **Calculating the Amount of Gasoline in a Tank** A Shell station stores its gasoline in underground tanks that are right circular cylinders lying on their sides. See the illustration. The volume V of gasoline in the tank (in gallons) is given by the formula

$$V = 40h^2\sqrt{\frac{96}{h} - 0.608}$$

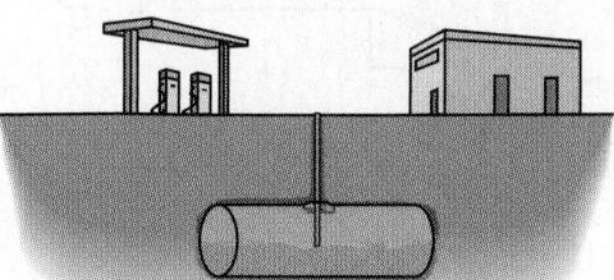

where h is the height of the gasoline (in inches) as measured on a depth stick.
(a) If $h = 12$ inches, how many gallons of gasoline are in the tank?
(b) If $h = 1$ inch, how many gallons of gasoline are in the tank?

124. **Inclined Planes** The final velocity v of an object in feet per second (ft/sec) after it slides down a frictionless inclined plane of height h feet is

$$v = \sqrt{64h + v_0^2}$$

where v_0 is the initial velocity (in ft/sec) of the object.
(a) What is the final velocity v of an object that slides down a frictionless inclined plane of height 4 feet? Assume that the initial velocity is 0.
(b) What is the final velocity v of an object that slides down a frictionless inclined plane of height 16 feet? Assume that the initial velocity is 0.
(c) What is the final velocity v of an object that slides down a frictionless inclined plane of height 2 feet with an initial velocity of 4 ft/sec?

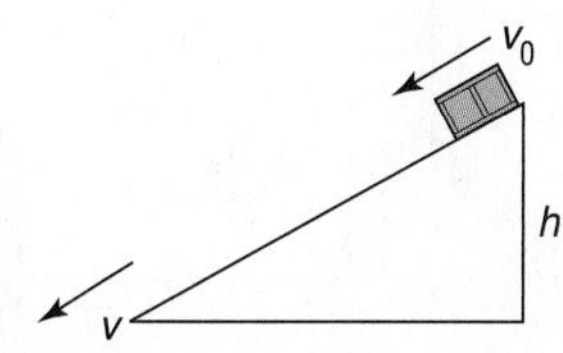

Problems 125 and 126 require the following information.

Period of a Pendulum *The period T, in seconds, of a pendulum of length l, in feet, may be approximated using the formula*

$$T = 2\pi\sqrt{\frac{l}{32}}$$

In Problems 125 and 126, express your answer both as a square root and as a decimal.

125. Find the period T of a pendulum whose length is 64 feet.

126. Find the period T of a pendulum whose length is 16 feet.

Explaining Concepts: Discussion and Writing

127. Give an example to show that $\sqrt{a^2}$ is not equal to a. Use it to explain why $\sqrt{a^2} = |a|$.

'Are You Prepared?' Answers

1. 9; −9 **2.** 4; 4

A.8 Lines

OBJECTIVES
1 Calculate and Interpret the Slope of a Line (A64)
2 Graph Lines Given a Point and the Slope (A67)
3 Find the Equation of a Vertical Line (A67)
4 Use the Point–Slope Form of a Line; Identify Horizontal Lines (A68)
5 Find the Equation of a Line Given Two Points (A69)
6 Write the Equation of a Line in Slope–Intercept Form (A69)
7 Identify the Slope and *y*-Intercept of a Line from Its Equation (A70)
8 Graph Lines Written in General Form Using Intercepts (A71)
9 Find Equations of Parallel Lines (A72)
10 Find Equations of Perpendicular Lines (A73)

In this section we study a certain type of equation that contains two variables, called a *linear equation,* and its graph, a *line.*

1 Calculate and Interpret the Slope of a Line

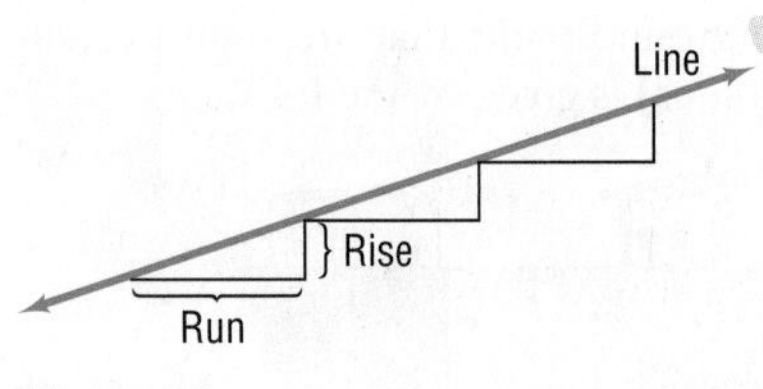

Figure 31

Consider the staircase illustrated in Figure 31. Each step contains exactly the same horizontal **run** and the same vertical **rise**. The ratio of the rise to the run, called the *slope,* is a numerical measure of the steepness of the staircase. For example, if the run is increased and the rise remains the same, the staircase becomes less steep. If the run is kept the same but the rise is increased, the staircase becomes more steep. This important characteristic of a line is best defined using rectangular coordinates.

DEFINITION

Let $P = (x_1, y_1)$ and $Q = (x_2, y_2)$ be two distinct points. If $x_1 \neq x_2$, the **slope *m*** of the nonvertical line L containing P and Q is defined by the formula

$$m = \frac{y_2 - y_1}{x_2 - x_1} \qquad x_1 \neq x_2 \qquad (1)$$

If $x_1 = x_2$, then L is a **vertical line** and the slope m of L is **undefined** (since this results in division by 0).

Figure 32(a) on page A65 provides an illustration of the slope of a nonvertical line; Figure 32(b) illustrates a vertical line.

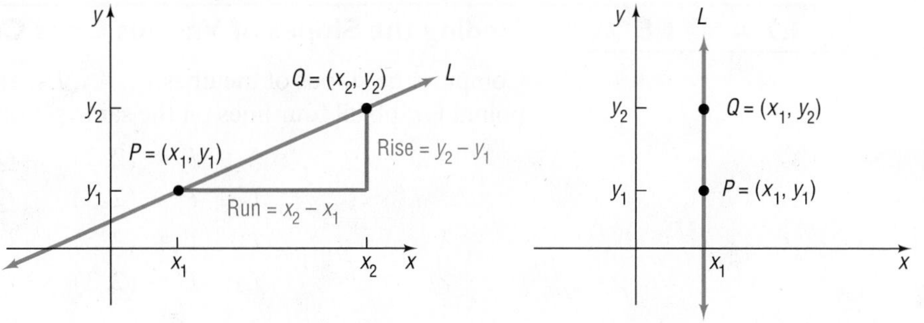

Figure 32

(a) Slope of L is $m = \dfrac{y_2 - y_1}{x_2 - x_1}, x_1 \neq x_2$

(b) Slope is undefined; L is vertical

> **In Words**
> The symbol Δ is the Greek uppercase letter delta. In mathematics, Δ is read "change in," so $\dfrac{\Delta y}{\Delta x}$ is read "change in y divided by change in x."

As Figure 32(a) illustrates, the slope m of a nonvertical line may be viewed as

$$m = \frac{y_2 - y_1}{x_2 - x_1} = \frac{\text{Rise}}{\text{Run}} \quad \text{or as } m = \frac{y_2 - y_1}{x_2 - x_1} = \frac{\text{Change in } y}{\text{Change in } x} = \frac{\Delta y}{\Delta x}$$

That is, the slope m of a nonvertical line measures the amount y changes when x changes from x_1 to x_2. The expression $\dfrac{\Delta y}{\Delta x}$ is called the **average rate of change** of y with respect to x.

Two comments about computing the slope of a nonvertical line may prove helpful:

1. Any two distinct points on the line can be used to compute the slope of the line. (See Figure 33 for justification.) Since any two distinct points can be used to compute the slope of a line, the average rate of change of a line is always the same number.

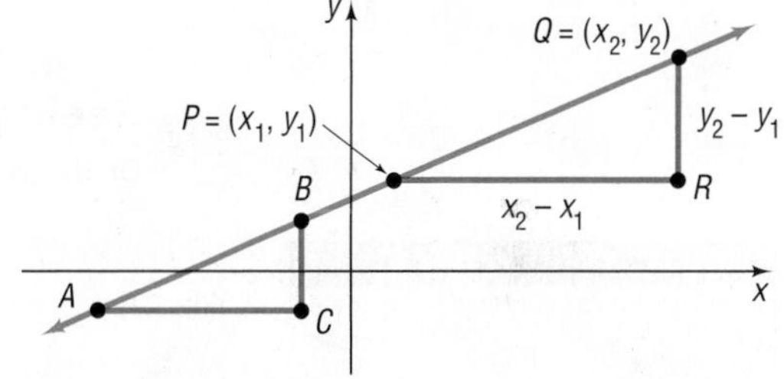

Figure 33
Triangles ABC and PQR are similar (equal angles), so ratios of corresponding sides are proportional. Then
Slope using P and Q $= \dfrac{y_2 - y_1}{x_2 - x_1} = \dfrac{d(B, C)}{d(A, C)} =$ Slope using A and B

2. The slope of a line may be computed from $P = (x_1, y_1)$ to $Q = (x_2, y_2)$ or from Q to P because

$$\frac{y_2 - y_1}{x_2 - x_1} = \frac{y_1 - y_2}{x_1 - x_2}$$

EXAMPLE 1

Finding and Interpreting the Slope of a Line Given Two Points

The slope m of the line containing the points $(1, 2)$ and $(5, -3)$ may be computed as

$$m = \frac{-3 - 2}{5 - 1} = \frac{-5}{4} = -\frac{5}{4} \quad \text{or as} \quad m = \frac{2 - (-3)}{1 - 5} = \frac{5}{-4} = -\frac{5}{4}$$

For every 4-unit change in x, y will change by -5 units. That is, if x increases by 4 units, then y will decrease by 5 units. The average rate of change of y with respect to x is $-\dfrac{5}{4}$. ●

Now Work PROBLEMS 13 AND 19

EXAMPLE 2 Finding the Slopes of Various Lines Containing the Same Point (2, 3)

Compute the slopes of the lines L_1, L_2, L_3, and L_4 containing the following pairs of points. Graph all four lines on the same set of coordinate axes.

$$\begin{aligned} &L_1\colon\ P=(2,3) \quad Q_1=(-1,-2)\\ &L_2\colon\ P=(2,3) \quad Q_2=(3,-1)\\ &L_3\colon\ P=(2,3) \quad Q_3=(5,3)\\ &L_4\colon\ P=(2,3) \quad Q_4=(2,5)\end{aligned}$$

Solution Let m_1, m_2, m_3, and m_4 denote the slopes of the lines L_1, L_2, L_3, and L_4, respectively. Then

$$m_1=\frac{-2-3}{-1-2}=\frac{-5}{-3}=\frac{5}{3} \quad \text{A rise of 5 divided by a run of 3}$$

$$m_2=\frac{-1-3}{3-2}=\frac{-4}{1}=-4$$

$$m_3=\frac{3-3}{5-2}=\frac{0}{3}=0$$

m_4 is undefined because $x_1 = x_2 = 2$

The graphs of these lines are given in Figure 34.

Figure 34

Figure 34 illustrates the following facts:

1. When the slope of a line is positive, the line slants upward from left to right (L_1).
2. When the slope of a line is negative, the line slants downward from left to right (L_2).
3. When the slope is 0, the line is horizontal (L_3).
4. When the slope is undefined, the line is vertical (L_4).

Seeing the Concept

On the same screen, graph the following equations:

$Y_1 = 0$ Slope of line is 0.

$Y_2 = \frac{1}{4}x$ Slope of line is $\frac{1}{4}$.

$Y_3 = \frac{1}{2}x$ Slope of line is $\frac{1}{2}$.

$Y_4 = x$ Slope of line is 1.

$Y_5 = 2x$ Slope of line is 2.

$Y_6 = 6x$ Slope of line is 6.

See Figure 35.

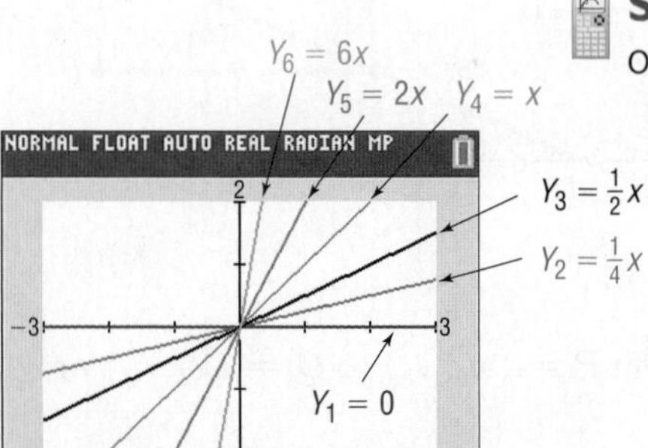

Figure 35

Seeing the Concept

On the same screen, graph the following equations:

$Y_1 = 0$ Slope of line is 0.

$Y_2 = -\frac{1}{4}x$ Slope of line is $-\frac{1}{4}$.

$Y_3 = -\frac{1}{2}x$ Slope of line is $-\frac{1}{2}$.

$Y_4 = -x$ Slope of line is -1.

$Y_5 = -2x$ Slope of line is -2.

$Y_6 = -6x$ Slope of line is -6.

See Figure 36.

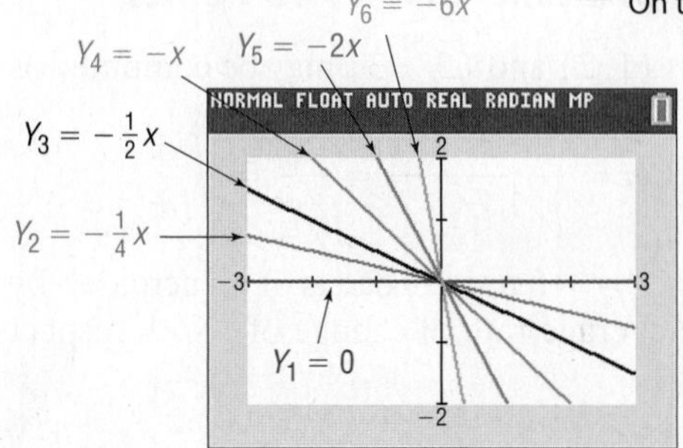

Figure 36

Figures 35 and 36 on page A66 illustrate that the closer the line is to the vertical position, the greater the magnitude of the slope.

2 Graph Lines Given a Point and the Slope

EXAMPLE 3 **Graphing a Line Given a Point and a Slope**

Draw a graph of the line that contains the point $(3, 2)$ and has a slope of:

(a) $\frac{3}{4}$ (b) $-\frac{4}{5}$

Solution

(a) Slope $= \frac{\text{Rise}}{\text{Run}}$. The slope $\frac{3}{4}$ means that for every horizontal movement (run) of 4 units to the right, there will be a vertical movement (rise) of 3 units. Start at the given point $(3, 2)$ and move 4 units to the right and 3 units up, arriving at the point $(7, 5)$. Drawing the line through this point and the point $(3, 2)$ gives the graph. See Figure 37.

Figure 37

(b) The fact that the slope is

$$-\frac{4}{5} = \frac{-4}{5} = \frac{\text{Rise}}{\text{Run}}$$

means that for every horizontal movement of 5 units to the right, there will be a corresponding vertical movement of -4 units (a downward movement). Start at the given point $(3, 2)$ and move 5 units to the right and then 4 units down, arriving at the point $(8, -2)$. Drawing the line through these points gives the graph. See Figure 38.

Alternatively, consider that

$$-\frac{4}{5} = \frac{4}{-5} = \frac{\text{Rise}}{\text{Run}}$$

so for every horizontal movement of -5 units (a movement to the left), there will be a corresponding vertical movement of 4 units (upward). This approach leads to the point $(-2, 6)$, which is also on the graph of the line in Figure 38.

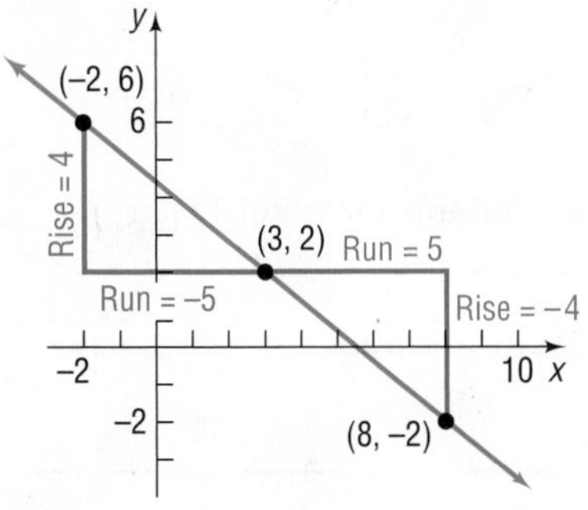

Figure 38

Now Work PROBLEM 25 (Graph the line)

3 Find the Equation of a Vertical Line

EXAMPLE 4 **Graphing a Line**

Graph the equation: $x = 3$

Solution To graph $x = 3$, we find all points (x, y) in the plane for which $x = 3$. No matter what y-coordinate is used, the corresponding x-coordinate always equals 3. Consequently, the graph of the equation $x = 3$ is a vertical line with x-intercept 3 and undefined slope. See Figure 39.

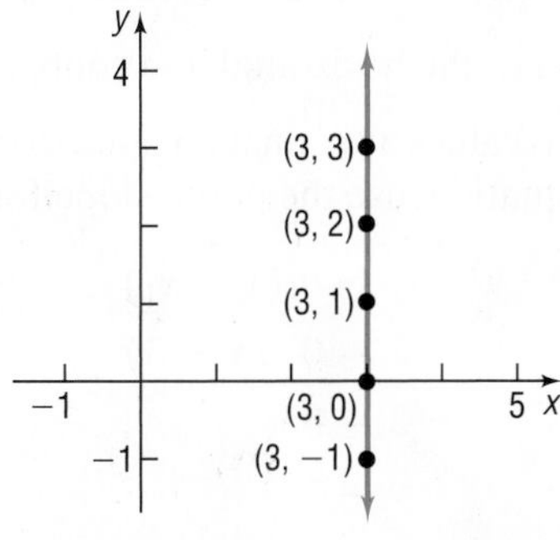

Figure 39 $x = 3$

Example 4 suggests the following result:

THEOREM

Equation of a Vertical Line

A vertical line is given by an equation of the form

$$x = a$$

where a is the x-intercept.

COMMENT To graph an equation using a graphing utility, we need to express the equation in the form $y = \{\textit{expression in } x\}$. But $x = 3$ cannot be put in this form. To overcome this, most graphing utilities have special commands for drawing vertical lines. DRAW, LINE, PLOT, and VERT are among the more common ones. Consult your manual to determine the correct methodology for your graphing utility. ■

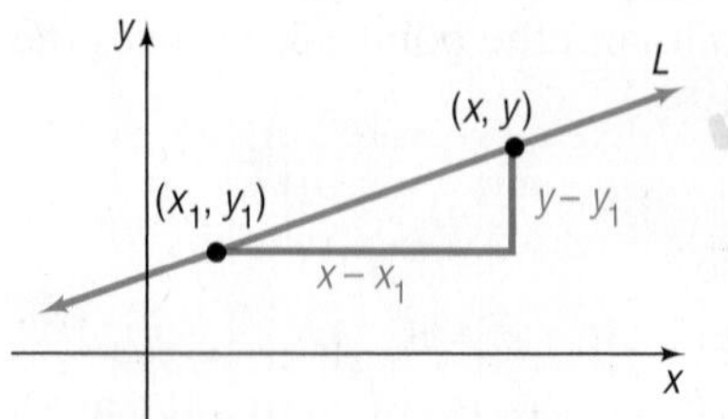

Figure 40

4 Use the Point–Slope Form of a Line; Identify Horizontal Lines

Let L be a nonvertical line with slope m that contains the point (x_1, y_1). See Figure 40. For any other point (x, y) on L, we have

$$m = \frac{y - y_1}{x - x_1} \quad \text{or} \quad y - y_1 = m(x - x_1)$$

THEOREM

Point–Slope Form of an Equation of a Line

An equation of a nonvertical line with slope m that contains the point (x_1, y_1) is

$$y - y_1 = m(x - x_1) \qquad \textbf{(2)}$$

EXAMPLE 5 **Using the Point–Slope Form of a Line**

An equation of the line with slope 4 that contains the point $(1, 2)$ can be found by using the point–slope form with $m = 4$, $x_1 = 1$, and $y_1 = 2$.

$$y - y_1 = m(x - x_1)$$
$$y - 2 = 4(x - 1) \qquad m = 4, x_1 = 1, y_1 = 2$$
$$y = 4x - 2 \qquad \text{Solve for } y.$$

See Figure 41 for the graph. ●

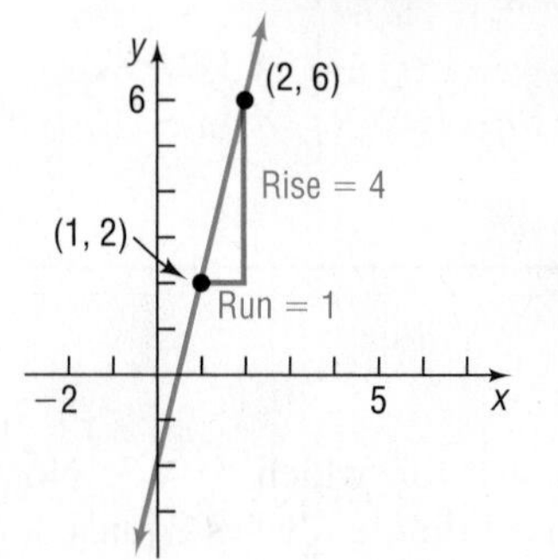

Figure 41 $y = 4x - 2$

Now Work PROBLEM 25 (Find the point-slope form)

EXAMPLE 6 **Finding the Equation of a Horizontal Line**

Find an equation of the horizontal line containing the point $(3, 2)$.

Solution Because all the y-values are equal on a horizontal line, the slope of a horizontal line is 0. To get an equation, use the point–slope form with $m = 0$, $x_1 = 3$, and $y_1 = 2$.

$$y - y_1 = m(x - x_1)$$
$$y - 2 = 0 \cdot (x - 3) \qquad m = 0, x_1 = 3, \text{and } y_1 = 2$$
$$y - 2 = 0$$
$$y = 2$$

See Figure 42 for the graph. ●

Figure 42 $y = 2$

Example 6 suggests the following result:

THEOREM

Equation of a Horizontal Line

A horizontal line is given by an equation of the form

$$y = b$$

where b is the y-intercept.

5 Find the Equation of a Line Given Two Points

EXAMPLE 7 **Finding an Equation of a Line Given Two Points**

Find an equation of the line containing the points $(2, 3)$ and $(-4, 5)$. Graph the line.

Solution First compute the slope of the line.

$$m = \frac{5 - 3}{-4 - 2} = \frac{2}{-6} = -\frac{1}{3} \qquad m = \frac{y_2 - y_1}{x_2 - x_1}$$

Use the point $(2, 3)$ and the slope $m = -\frac{1}{3}$ to get the point–slope form of the equation of the line.

$$y - 3 = -\frac{1}{3}(x - 2) \qquad y - y_1 = m(x - x_1)$$

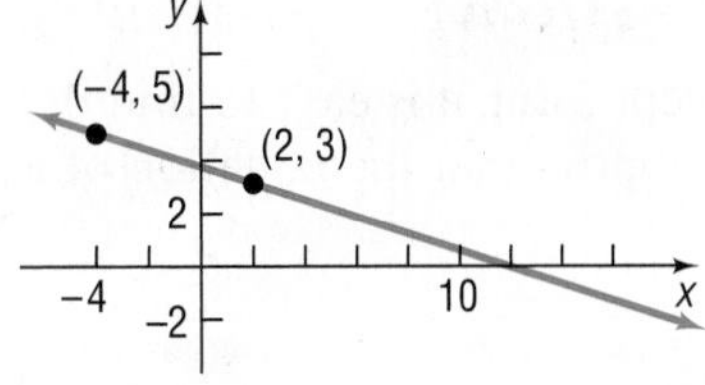

Figure 43 $y - 3 = -\frac{1}{3}(x - 2)$

See Figure 43 for the graph.

In the solution to Example 7, we could have used the other point, $(-4, 5)$, instead of the point $(2, 3)$. The equation that results, although it looks different, is equivalent to the equation that we obtained in the example. (Try it for yourself.)

Now Work PROBLEM 39

6 Write the Equation of a Line in Slope–Intercept Form

Another useful equation of a line is obtained when the slope m and y-intercept b are known. In this event, both the slope m of the line and a point $(0, b)$ on the line are known; then use the point–slope form, equation (2), to obtain the following equation:

$$y - b = m(x - 0) \quad \text{or} \quad y = mx + b$$

THEOREM

Slope–Intercept Form of an Equation of a Line

An equation of a line with slope m and y-intercept b is

$$y = mx + b \tag{3}$$

Now Work PROBLEMS 47 AND 53 (Express answer in slope–intercept form)

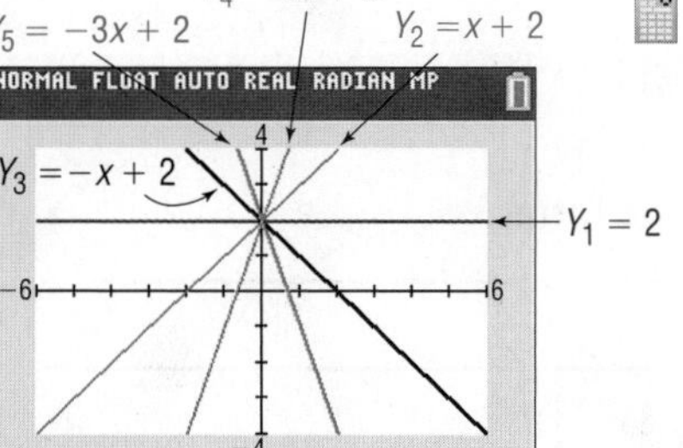

Figure 44 $y = mx + 2$

Seeing the Concept

To see the role that the slope m plays, graph the following lines on the same screen.

$$Y_1 = 2 \qquad m = 0$$
$$Y_2 = x + 2 \qquad m = 1$$
$$Y_3 = -x + 2 \qquad m = -1$$
$$Y_4 = 3x + 2 \qquad m = 3$$
$$Y_5 = -3x + 2 \qquad m = -3$$

See Figure 44. What do you conclude about the lines $y = mx + 2$?

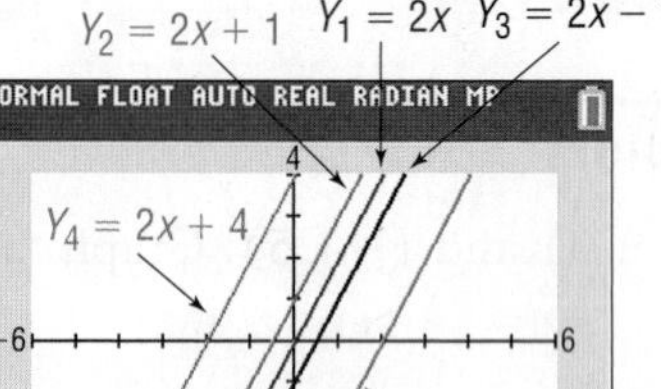

Figure 45 $y = 2x + b$

Seeing the Concept

To see the role of the y-intercept b, graph the following lines on the same screen.

$$Y_1 = 2x \qquad b = 0$$
$$Y_2 = 2x + 1 \qquad b = 1$$
$$Y_3 = 2x - 1 \qquad b = -1$$
$$Y_4 = 2x + 4 \qquad b = 4$$
$$Y_5 = 2x - 4 \qquad b = -4$$

See Figure 45. What do you conclude about the lines $y = 2x + b$?

7 Identify the Slope and y-Intercept of a Line from Its Equation

When the equation of a line is written in slope–intercept form, it is easy to find the slope m and y-intercept b of the line. For example, suppose that the equation of a line is

$$y = -2x + 7$$

Compare this equation to $y = mx + b$.

$$\underset{y\,=\,mx\ \ +\,b}{y = -2x + 7}$$

The slope of this line is -2 and its y-intercept is 7.

Now Work PROBLEM 73

EXAMPLE 8 Finding the Slope and y-Intercept

Find the slope m and y-intercept b of the equation $2x + 4y = 8$. Graph the equation.

Solution To obtain the slope and y-intercept, write the equation in slope–intercept form by solving for y.

$$2x + 4y = 8$$
$$4y = -2x + 8$$
$$y = -\frac{1}{2}x + 2 \qquad y = mx + b$$

The coefficient of x, $-\frac{1}{2}$, is the slope, and the constant, 2, is the y-intercept. Graph the line with y-intercept 2 and with slope $-\frac{1}{2}$. Starting at the point $(0, 2)$, go to the right 2 units and then down 1 unit to the point $(2, 1)$. Draw the line through these points. See Figure 46.

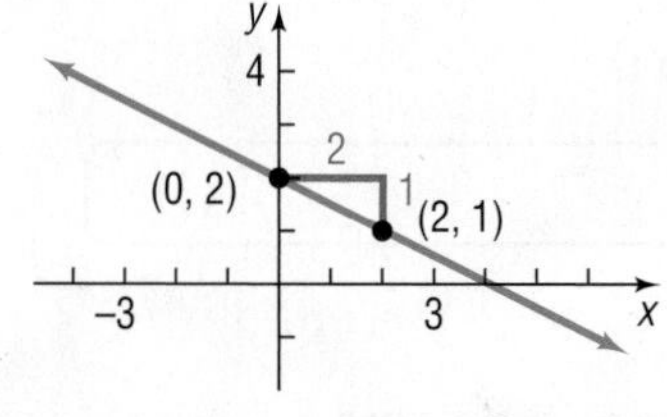

Figure 46 $y = -\frac{1}{2}x + 2$

Now Work PROBLEM 79

8 Graph Lines Written in General Form Using Intercepts

Refer to Example 8. The form of the equation of the line $2x + 4y = 8$ is called the *general form*.

DEFINITION The equation of a line is in **general form*** when it is written as

$$Ax + By = C \tag{4}$$

where A, B, and C are real numbers and A and B are not both 0.

If $B = 0$ in equation (4), then $A \neq 0$ and the graph of the equation is a vertical line: $x = \frac{C}{A}$. If $B \neq 0$ in equation (4), then we can solve the equation for y and write the equation in slope–intercept form as we did in Example 8.

Another approach to graphing equation (4) is to find its intercepts. Remember, the intercepts of the graph of an equation are the points where the graph crosses or touches a coordinate axis.

EXAMPLE 9 Graphing an Equation in General Form Using Its Intercepts

Graph the equation $2x + 4y = 8$ by finding its intercepts.

Solution To obtain the x-intercept, let $y = 0$ in the equation and solve for x.

$$\begin{aligned} 2x + 4y &= 8 \\ 2x + 4(0) &= 8 \quad \text{Let } y = 0. \\ 2x &= 8 \\ x &= 4 \quad \text{Divide both sides by 2.} \end{aligned}$$

The x-intercept is 4, and the point $(4, 0)$ is on the graph of the equation.

To obtain the y-intercept, let $x = 0$ in the equation and solve for y.

$$\begin{aligned} 2x + 4y &= 8 \\ 2(0) + 4y &= 8 \quad \text{Let } x = 0. \\ 4y &= 8 \\ y &= 2 \quad \text{Divide both sides by 4.} \end{aligned}$$

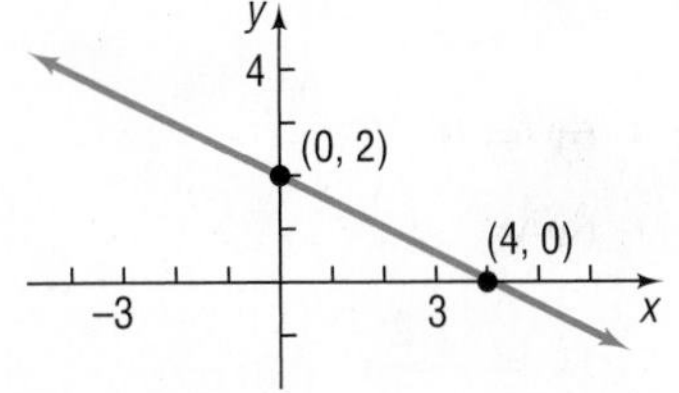

Figure 47 $2x + 4y = 8$

The y-intercept is 2, and the point $(0, 2)$ is on the graph of the equation.

Plot the points $(4, 0)$ and $(0, 2)$ and draw the line through the points. See Figure 47.

Now Work PROBLEM 93

Every line has an equation that is equivalent to an equation written in general form. For example, a vertical line whose equation is

$$x = a$$

can be written in the general form

$$1 \cdot x + 0 \cdot y = a \quad A = 1, B = 0, C = a$$

A horizontal line whose equation is

$$y = b$$

can be written in the general form

$$0 \cdot x + 1 \cdot y = b \quad A = 0, B = 1, C = b$$

*Some texts use the term **standard form**.

Lines that are neither vertical nor horizontal have general equations of the form

$$Ax + By = C \quad A \neq 0 \text{ and } B \neq 0$$

Because the equation of every line can be written in general form, any equation equivalent to equation (4) is called a **linear equation**.

9 Find Equations of Parallel Lines

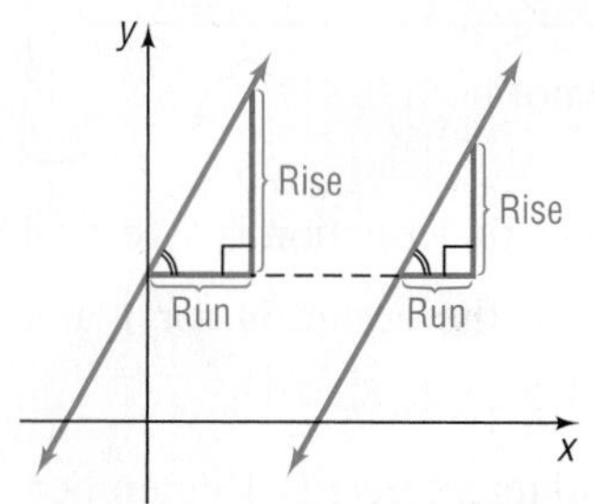

Figure 48 Parallel lines

When two lines (in the plane) do not intersect (that is, they have no points in common), they are **parallel**. Look at Figure 48. There we have drawn two parallel lines and have constructed two right triangles by drawing sides parallel to the coordinate axes. The right triangles are similar. (Do you see why? Two angles are equal.) Because the triangles are similar, the ratios of corresponding sides are equal.

THEOREM Criteria for Parallel Lines

Two nonvertical lines are parallel if and only if their slopes are equal and they have different y-intercepts.

The use of the phrase "if and only if" in the preceding theorem means that actually two statements are being made, one the converse of the other.

If two nonvertical lines are parallel, then their slopes are equal and they have different y-intercepts.

If two nonvertical lines have equal slopes and they have different y-intercepts, then they are parallel.

EXAMPLE 10 **Showing That Two Lines Are Parallel**

Show that the lines given by the following equations are parallel.

$$L_1:\ 2x + 3y = 6 \qquad L_2:\ 4x + 6y = 0$$

Solution To determine whether these lines have equal slopes and different y-intercepts, write each equation in slope–intercept form.

$$L_1:\ 2x + 3y = 6 \qquad\qquad L_2:\ 4x + 6y = 0$$
$$3y = -2x + 6 \qquad\qquad 6y = -4x$$
$$y = -\frac{2}{3}x + 2 \qquad\qquad y = -\frac{2}{3}x$$
$$\text{Slope} = -\frac{2}{3};\ y\text{-intercept} = 2 \qquad\qquad \text{Slope} = -\frac{2}{3};\ y\text{-intercept} = 0$$

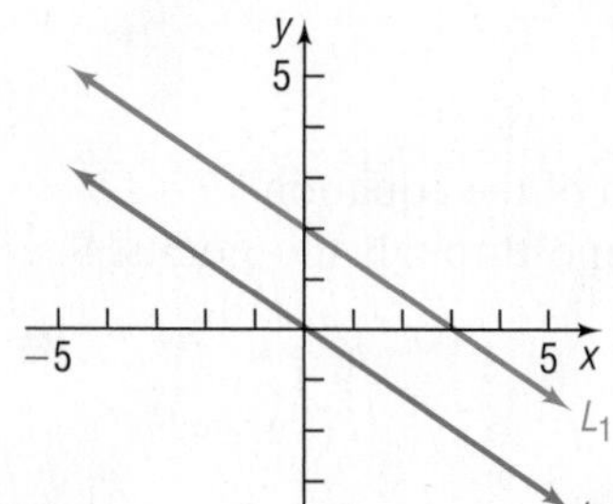

Figure 49

Because these lines have the same slope, $-\frac{2}{3}$, but different y-intercepts, the lines are parallel. See Figure 49.

EXAMPLE 11 **Finding a Line That Is Parallel to a Given Line**

Find an equation for the line that contains the point $(2, -3)$ and is parallel to the line $2x + y = 6$.

Solution Since the two lines are to be parallel, the slope of the line being sought equals the slope of the line $2x + y = 6$. Begin by writing the equation of the line $2x + y = 6$ in slope–intercept form.

$$2x + y = 6$$
$$y = -2x + 6$$

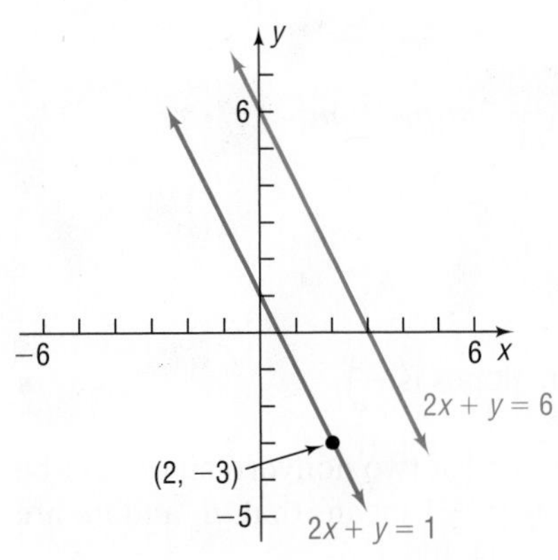

Figure 50

The slope is -2. Since the line being sought also has slope -2 and contains the point $(2, -3)$, use the point–slope form to obtain its equation.

$$y - y_1 = m(x - x_1) \qquad \text{Point–slope form}$$
$$y - (-3) = -2(x - 2) \qquad m = -2, x_1 = 2, y_1 = -3$$
$$y + 3 = -2x + 4 \qquad \text{Simplify.}$$
$$y = -2x + 1 \qquad \text{Slope–intercept form}$$
$$2x + y = 1 \qquad \text{General form}$$

This line is parallel to the line $2x + y = 6$ and contains the point $(2, -3)$. See Figure 50. ●

Now Work PROBLEM 61

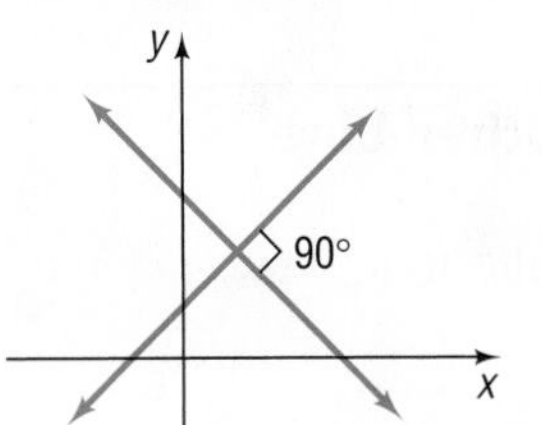

Figure 51 Perpendicular lines

10 Find Equations of Perpendicular Lines

When two lines intersect at a right angle (90°), they are **perpendicular**. See Figure 51.

The following result gives a condition, in terms of their slopes, for two lines to be perpendicular.

THEOREM

Criterion for Perpendicular Lines

Two nonvertical lines are perpendicular if and only if the product of their slopes is -1.

In Problem 130 you are asked to prove the "if" part of the theorem:

If two nonvertical lines have slopes whose product is -1, then the lines are perpendicular.

Below we prove the "only if" part of the statement:

If two nonvertical lines are perpendicular, then the product of their slopes is -1.

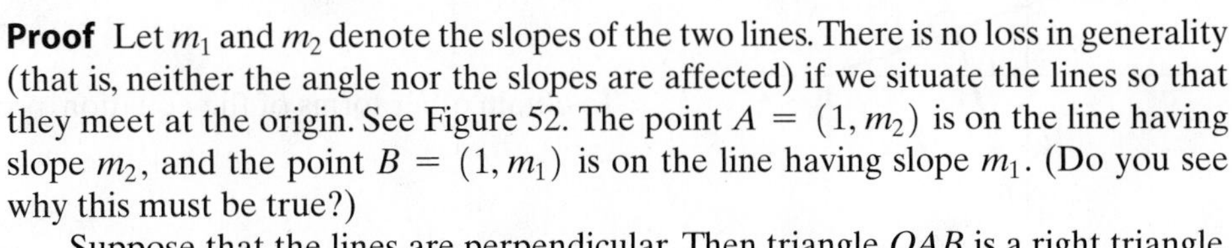

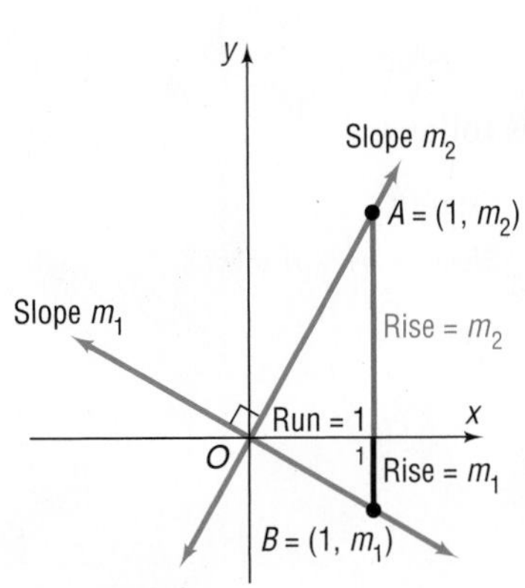

Figure 52

Proof Let m_1 and m_2 denote the slopes of the two lines. There is no loss in generality (that is, neither the angle nor the slopes are affected) if we situate the lines so that they meet at the origin. See Figure 52. The point $A = (1, m_2)$ is on the line having slope m_2, and the point $B = (1, m_1)$ is on the line having slope m_1. (Do you see why this must be true?)

Suppose that the lines are perpendicular. Then triangle OAB is a right triangle. As a result of the Pythagorean Theorem, it follows that

$$[d(O, A)]^2 + [d(O, B)]^2 = [d(A, B)]^2 \qquad (5)$$

Using the distance formula, the squares of these distances are

$$[d(O, A)]^2 = (1 - 0)^2 + (m_2 - 0)^2 = 1 + m_2^2$$
$$[d(O, B)]^2 = (1 - 0)^2 + (m_1 - 0)^2 = 1 + m_1^2$$
$$[d(A, B)]^2 = (1 - 1)^2 + (m_2 - m_1)^2 = m_2^2 - 2m_1m_2 + m_1^2$$

Using these facts in equation (5), we get

$$\left(1 + m_2^2\right) + \left(1 + m_1^2\right) = m_2^2 - 2m_1m_2 + m_1^2$$

which, upon simplification, can be written as

$$m_1m_2 = -1$$

If the lines are perpendicular, the product of their slopes is -1. ■

You may find it easier to remember the condition for two nonvertical lines to be perpendicular by observing that the equality $m_1m_2 = -1$ means that m_1 and m_2 are negative reciprocals of each other; that is, $m_1 = -\dfrac{1}{m_2}$ and $m_2 = -\dfrac{1}{m_1}$.

EXAMPLE 12

Finding the Slope of a Line Perpendicular to Another Line

If a line has slope $\frac{3}{2}$, any line having slope $-\frac{2}{3}$ is perpendicular to it. ●

EXAMPLE 13

Finding the Equation of a Line Perpendicular to a Given Line

Find an equation of the line that contains the point $(1, -2)$ and is perpendicular to the line $x + 3y = 6$. Graph the two lines.

Solution First write the equation of the given line in slope–intercept form to find its slope.

$$x + 3y = 6$$

$$3y = -x + 6 \quad \text{Proceed to solve for } y.$$

$$y = -\frac{1}{3}x + 2 \quad \text{Place in the form } y = mx + b.$$

The given line has slope $-\frac{1}{3}$. Any line perpendicular to this line will have slope 3.

Because the point $(1, -2)$ is on this line with slope 3, use the point–slope form of the equation of a line.

$$y - y_1 = m(x - x_1) \quad \text{Point–slope form}$$

$$y - (-2) = 3(x - 1) \quad m = 3, x_1 = 1, y_1 = -2$$

$$y + 2 = 3(x - 1)$$

To obtain other forms of the equation, proceed as follows:

$$y + 2 = 3x - 3 \quad \text{Simplify.}$$

$$y = 3x - 5 \quad \text{Slope–intercept form}$$

$$3x - y = 5 \quad \text{General form}$$

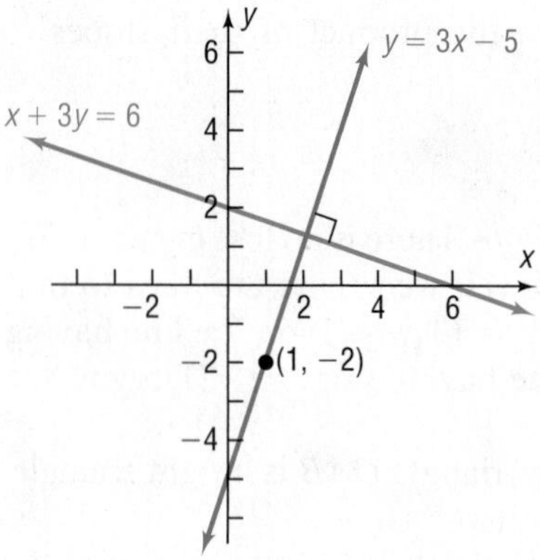

Figure 53

Figure 53 shows the graphs. ●

Now Work PROBLEM 67

WARNING Be sure to use a square screen when you use a graphing calculator to graph perpendicular lines. Otherwise, the angle between the two lines will appear distorted. A discussion of square screens is given in Section B.5 of Appendix B. ■

A.8 Assess Your Understanding

Concepts and Vocabulary

1. The slope of a vertical line is ____________; the slope of a horizontal line is _____.

2. For the line $2x + 3y = 6$, the x-intercept is ____ and the y-intercept is _____.

3. ***True or False*** The equation $3x + 4y = 6$ is written in general form.

4. ***True or False*** The slope of the line $2y = 3x + 5$ is 3.

5. ***True or False*** The point $(1, 2)$ is on the line $2x + y = 4$.

6. Two nonvertical lines have slopes m_1 and m_2, respectively. The lines are parallel if __________ and the ____________ are unequal; the lines are perpendicular if ____________.

7. The lines $y = 2x + 3$ and $y = ax + 5$ are parallel if $a =$ _____.

8. The lines $y = 2x - 1$ and $y = ax + 2$ are perpendicular if $a =$ _____.

9. ***True or False*** Perpendicular lines have slopes that are reciprocals of one another.

10. Choose the formula for finding the slope m of a nonvertical line that contains the two distinct points (x_1, y_1) and (x_2, y_2).

(a) $m = \dfrac{y_2 - x_2}{y_1 - x_1} \quad x_1 \neq y_1$

(b) $m = \dfrac{y_2 - x_1}{x_2 - y_1} \quad y_1 \neq x_2$

(c) $m = \dfrac{x_2 - x_1}{y_2 - y_1} \quad y_1 \neq y_2$

(d) $m = \dfrac{y_2 - y_1}{x_2 - x_1} \quad x_1 \neq x_2$

11. If a line slants downward from left to right, then which of the following describes its slope?
(a) positive (b) zero
(c) negative (d) undefined

12. Choose the correct statement about the graph of the line $y = -3$.
(a) The graph is vertical with x-intercept -3.
(b) The graph is horizontal with y-intercept -3.
(c) The graph is vertical with y-intercept -3.
(d) The graph is horizontal with x-intercept -3.

Skill Building

In Problems 13–16, (a) find the slope of the line and (b) interpret the slope.

13.

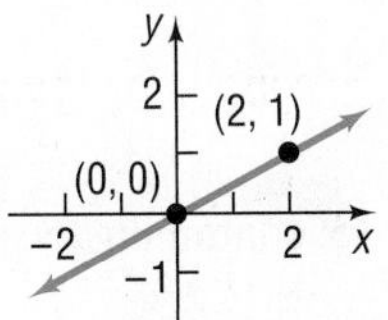

14.

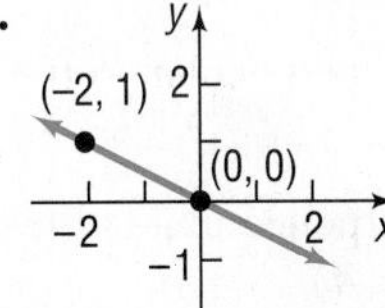

15.

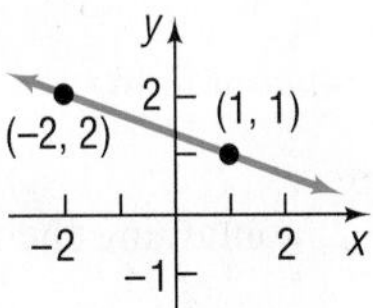

16.

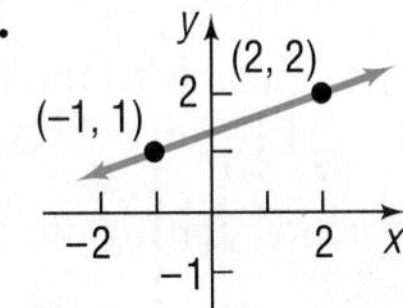

In Problems 17–24, plot each pair of points and determine the slope of the line containing them. Graph the line.

17. $(2, 3); (4, 0)$ **18.** $(4, 2); (3, 4)$ **19.** $(-2, 3); (2, 1)$ **20.** $(-1, 1); (2, 3)$

21. $(-3, -1); (2, -1)$ **22.** $(4, 2); (-5, 2)$ **23.** $(-1, 2); (-1, -2)$ **24.** $(2, 0); (2, 2)$

In Problems 25–32, graph the line that contains the point P and has slope m. In Problems 25–30, find the point-slope form of the equation of the line. In Problems 31 and 32, find an equation of the line.

25. $P = (1, 2); m = 3$ **26.** $P = (2, 1); m = 4$ **27.** $P = (2, 4); m = -\dfrac{3}{4}$ **28.** $P = (1, 3); m = -\dfrac{2}{5}$

29. $P = (-1, 3); m = 0$ **30.** $P = (2, -4); m = 0$ **31.** $P = (0, 3)$; slope undefined **32.** $P = (-2, 0)$; slope undefined

In Problems 33–38, the slope and a point on a line are given. Use this information to locate three additional points on the line. Answers may vary. [**Hint:** *It is not necessary to find the equation of the line. See Example 3.*]

33. Slope 4; point $(1, 2)$ **34.** Slope 2; point $(-2, 3)$ **35.** Slope $-\dfrac{3}{2}$; point $(2, -4)$

36. Slope $\dfrac{4}{3}$; point $(-3, 2)$ **37.** Slope -2; point $(-2, -3)$ **38.** Slope -1; point $(4, 1)$

In Problems 39–46, find an equation of the line L.

39.

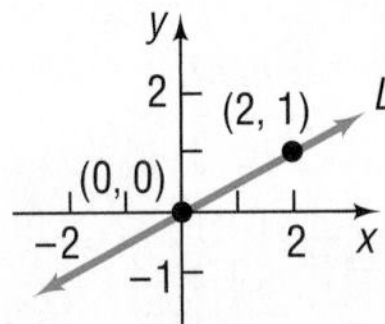

40.

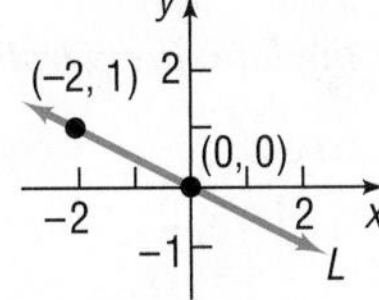

41.

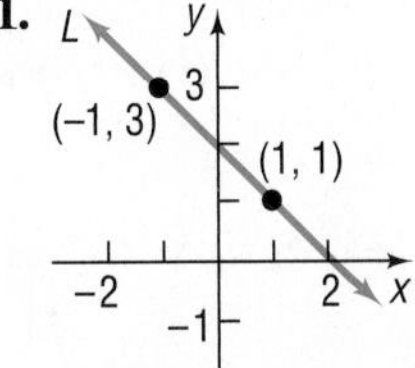

42.

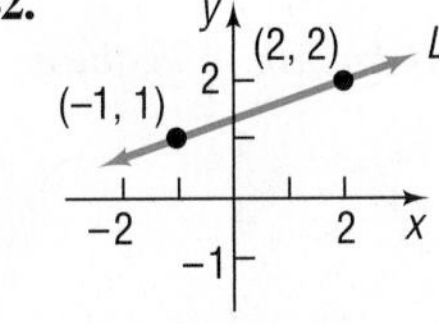

43.
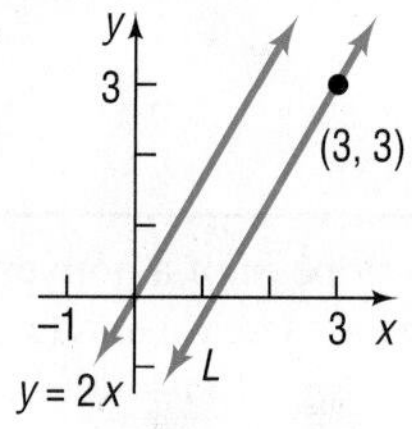

L is parallel to y = 2x

44.
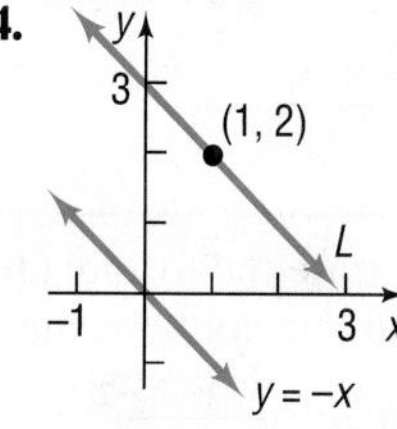

L is parallel to y = −x

45.
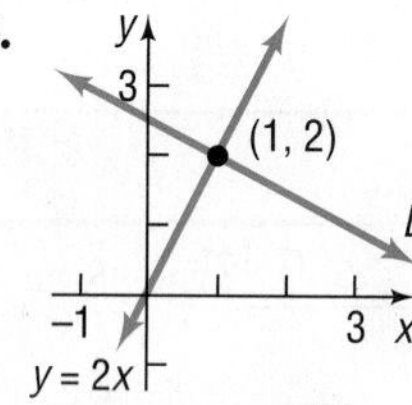

L is perpendicular to y = 2x

46.
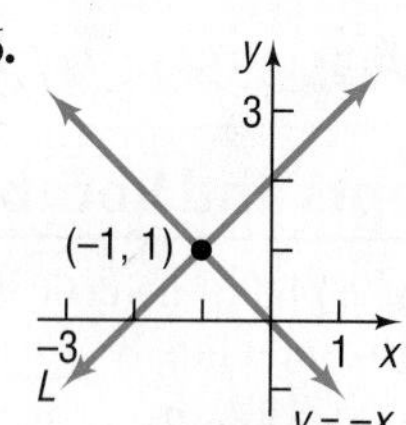

L is perpendicular to y = −x

In Problems 47–72, find an equation for the line with the given properties. Express your answer using either the general form or the slope–intercept form of the equation of a line, whichever you prefer.

47. Slope = 3; containing the point $(-2, 3)$

48. Slope = 2; containing the point $(4, -3)$

49. Slope $= -\frac{2}{3}$; containing the point $(1, -1)$

50. Slope $= \frac{1}{2}$; containing the point $(3, 1)$

51. Containing the points $(1, 3)$ and $(-1, 2)$

52. Containing the points $(-3, 4)$ and $(2, 5)$

53. Slope = −3; y-intercept = 3

54. Slope = −2; y-intercept = −2

55. x-intercept = 2; y-intercept = −1

56. x-intercept = −4; y-intercept = 4

57. Slope undefined; containing the point $(2, 4)$

58. Slope undefined; containing the point $(3, 8)$

59. Horizontal; containing the point $(-3, 2)$

60. Vertical; containing the point $(4, -5)$

61. Parallel to the line $y = 2x$; containing the point $(-1, 2)$

62. Parallel to the line $y = -3x$; containing the point $(-1, 2)$

63. Parallel to the line $2x - y = -2$; containing the point $(0, 0)$

64. Parallel to the line $x - 2y = -5$; containing the point $(0, 0)$

65. Parallel to the line $x = 5$; containing the point $(4, 2)$

66. Parallel to the line $y = 5$; containing the point $(4, 2)$

67. Perpendicular to the line $y = \frac{1}{2}x + 4$; containing the point $(1, -2)$

68. Perpendicular to the line $y = 2x - 3$; containing the point $(1, -2)$

69. Perpendicular to the line $2x + y = 2$; containing the point $(-3, 0)$

70. Perpendicular to the line $x - 2y = -5$; containing the point $(0, 4)$

71. Perpendicular to the line $x = 8$; containing the point $(3, 4)$

72. Perpendicular to the line $y = 8$; containing the point $(3, 4)$

In Problems 73–92, find the slope and y-intercept of each line. Graph the line.

73. $y = 2x + 3$

74. $y = -3x + 4$

75. $\frac{1}{2}y = x - 1$

76. $\frac{1}{3}x + y = 2$

77. $y = \frac{1}{2}x + 2$

78. $y = 2x + \frac{1}{2}$

79. $x + 2y = 4$

80. $-x + 3y = 6$

81. $2x - 3y = 6$

82. $3x + 2y = 6$

83. $x + y = 1$

84. $x - y = 2$

85. $x = -4$

86. $y = -1$

87. $y = 5$

88. $x = 2$

89. $y - x = 0$

90. $x + y = 0$

91. $2y - 3x = 0$

92. $3x + 2y = 0$

In Problems 93–102, (a) find the intercepts of the graph of each equation and (b) graph the equation.

93. $2x + 3y = 6$

94. $3x - 2y = 6$

95. $-4x + 5y = 40$

96. $6x - 4y = 24$

97. $7x + 2y = 21$

98. $5x + 3y = 18$

99. $\frac{1}{2}x + \frac{1}{3}y = 1$

100. $x - \frac{2}{3}y = 4$

101. $0.2x - 0.5y = 1$

102. $-0.3x + 0.4y = 1.2$

103. Find an equation of the x-axis.

104. Find an equation of the y-axis.

In Problems 105–108, the equations of two lines are given. Determine whether the lines are parallel, perpendicular, or neither.

105. $y = 2x - 3$
$y = 2x + 4$

106. $y = \frac{1}{2}x - 3$
$y = -2x + 4$

107. $y = 4x + 5$
$y = -4x + 2$

108. $y = -2x + 3$
$y = -\frac{1}{2}x + 2$

In Problems 109–112, write an equation of each line. Express your answer using either the general form or the slope–intercept form of the equation of a line, whichever you prefer.

109.

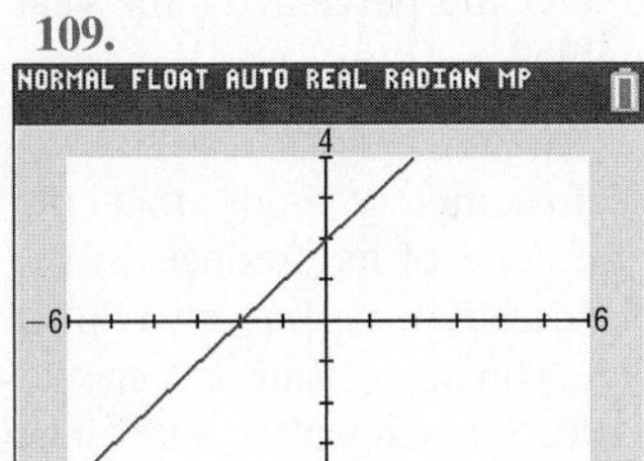

110.

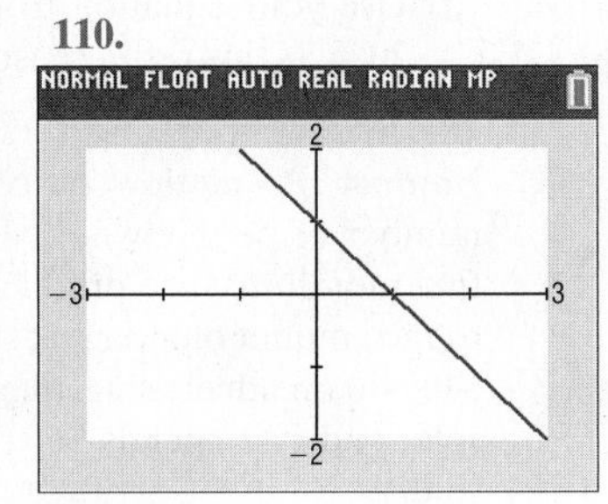

111.

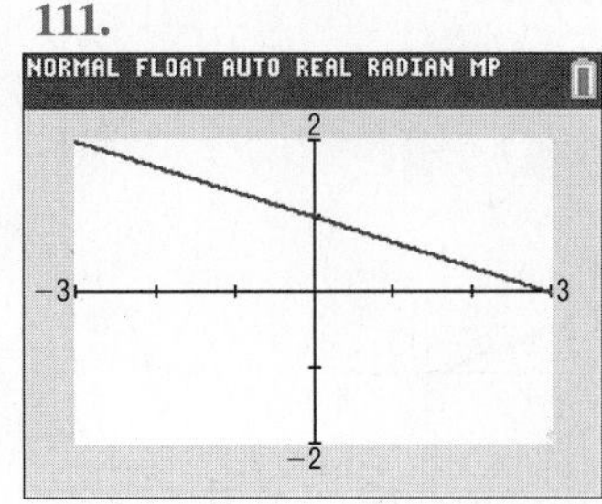

112.

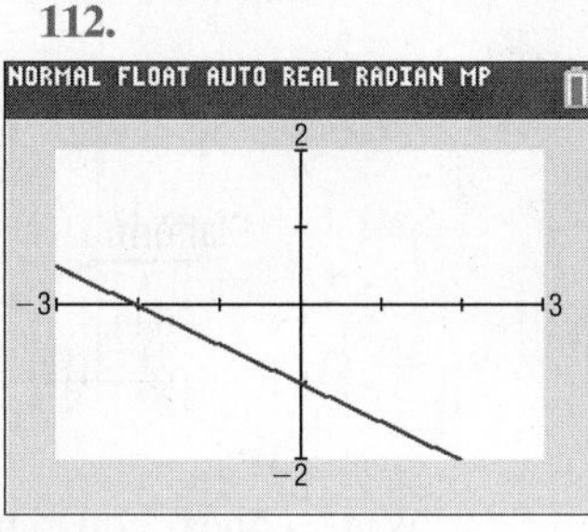

Applications and Extensions

113. Geometry Use slopes to show that the triangle whose vertices are $(-2, 5)$, $(1, 3)$, and $(-1, 0)$ is a right triangle.

114. Geometry Use slopes to show that the quadrilateral whose vertices are $(1, -1)$, $(4, 1)$, $(2, 2)$, and $(5, 4)$ is a parallelogram.

115. Geometry Use slopes to show that the quadrilateral whose vertices are $(-1, 0)$, $(2, 3)$, $(1, -2)$, and $(4, 1)$ is a rectangle.

116. Geometry Use slopes and the distance formula to show that the quadrilateral whose vertices are $(0, 0)$, $(1, 3)$, $(4, 2)$, and $(3, -1)$ is a square.

117. Truck Rentals A truck rental company rents a moving truck for one day by charging \$39 plus \$0.60 per mile. Write a linear equation that relates the cost C, in dollars, of renting the truck to the number x of miles driven. What is the cost of renting the truck if the truck is driven 110 miles? 230 miles?

118. Cost Equation The **fixed costs** of operating a business are the costs incurred regardless of the level of production. Fixed costs include rent, fixed salaries, and costs of leasing machinery. The **variable costs** of operating a business are the costs that change with the level of output. Variable costs include raw materials, hourly wages, and electricity. Suppose that a manufacturer of jeans has fixed daily costs of \$500 and variable costs of \$8 for each pair of jeans manufactured. Write a linear equation that relates the daily cost C, in dollars, of manufacturing the jeans to the number x of jeans manufactured. What is the cost of manufacturing 400 pairs of jeans? 740 pairs?

119. Cost of Driving a Car The annual fixed costs of owning a small sedan are \$4462, assuming the car is completely paid for. The cost to drive the car is approximately \$0.17 per mile. Write a linear equation that relates the cost C and the number x of miles driven annually.

Source: AAA, April 2013

120. Wages of a Car Salesperson Dan receives \$375 per week for selling new and used cars at a car dealership in Oak Lawn, Illinois. In addition, he receives 5% of the profit on any sales that he generates. Write a linear equation that represents Dan's weekly salary S when he has sales that generate a profit of x dollars.

121. Electricity Rates in Illinois Commonwealth Edison Company supplies electricity to residential customers for a monthly customer charge of \$15.37 plus 8.21 cents per kilowatt-hour for up to 800 kilowatt-hours (kW-hr).

(a) Write a linear equation that relates the monthly charge C, in dollars, to the number x of kilowatt-hours used in a month, $0 \le x \le 800$.
(b) Graph this equation.
(c) What is the monthly charge for using 200 kilowatt-hours?
(d) What is the monthly charge for using 500 kilowatt-hours?
(e) Interpret the slope of the line.

Source: Commonwealth Edison Company, January 2014.

122. Electricity Rates in Florida Florida Power & Light Company supplies electricity to residential customers for a monthly customer charge of \$7.24 plus 9.07 cents per kilowatt-hour for up to 1000 kilowatt-hours.

(a) Write a linear equation that relates the monthly charge C, in dollars, to the number x of kilowatt-hours used in a month, $0 \le x \le 1000$.
(b) Graph this equation.
(c) What is the monthly charge for using 200 kilowatt-hours?
(d) What is the monthly charge for using 500 kilowatt-hours?
(e) Interpret the slope of the line.

Source: Florida Power & Light Company, March 2014.

123. Measuring Temperature The relationship between Celsius (°C) and Fahrenheit (°F) degrees of measuring temperature is linear. Find a linear equation relating °C and °F if 0°C corresponds to 32°F and 100°C corresponds to 212°F. Use the equation to find the Celsius measure of 70°F.

124. Measuring Temperature The Kelvin (K) scale for measuring temperature is obtained by adding 273 to the Celsius temperature.

(a) Write a linear equation relating K and °C.
(b) Write a linear equation relating K and °F (see Problem 123).

125. Access Ramp A wooden access ramp is being built to reach a platform that sits 30 inches above the floor. The ramp drops 2 inches for every 25-inch run.

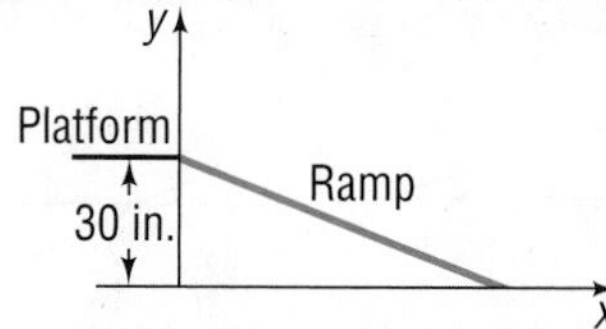

(a) Write a linear equation that relates the height y of the ramp above the floor to the horizontal distance x from the platform.
(b) Find and interpret the x-intercept of the graph of your equation.
(c) Design requirements stipulate that the maximum run be 30 feet and that the maximum slope be a drop of 1 inch for each 12 inches of run. Will this ramp meet the requirements? Explain.
(d) What slopes could be used to obtain the 30-inch rise and still meet design requirements?

Source: www.adaptiveaccess.com/wood_ramps.php

126. Cigarette Use A report in the Child Trends DataBase indicated that in 2000, 20.6% of twelfth grade students reported daily use of cigarettes. In 2012, 9.3% of twelfth grade students reported daily use of cigarettes.
(a) Write a linear equation that relates the percent y of twelfth grade students who smoke cigarettes daily to the number x of years after 2000.
(b) Find the intercepts of the graph of your equation.
(c) Do these intercepts have meaningful interpretation?
(d) Use your equation to predict the percent for the year 2025. Is this result reasonable?

Source: www.childtrendsdatabank.org

127. Product Promotion A cereal company finds that the number of people who will buy one of its products in the first month that the product is introduced is linearly related to the amount of money it spends on advertising. If it spends \$40,000 on advertising, then 100,000 boxes of cereal will be sold, and if it spends \$60,000, then 200,000 boxes will be sold.
(a) Write a linear equation that relates the amount A spent on advertising to the number x of boxes the company aims to sell.
(b) How much expenditure on advertising is needed to sell 300,000 boxes of cereal?
(c) Interpret the slope.

128. Show that the line containing the points (a, b) and (b, a), $a \neq b$, is perpendicular to the line $y = x$. Also show that the midpoint of (a, b) and (b, a) lies on the line $y = x$.

129. The equation $2x - y = C$ defines a **family of lines**, one line for each value of C. On one set of coordinate axes, graph the members of the family when $C = -4$, $C = 0$, and $C = 2$. Can you draw a conclusion from the graph about each member of the family?

130. Prove that if two nonvertical lines have slopes whose product is -1, then the lines are perpendicular. [**Hint:** Refer to Figure 52 and use the converse of the Pythagorean Theorem.]

Explaining Concepts: Discussion and Writing

131. Which of the following equations might have the graph shown? (More than one answer is possible.)
(a) $2x + 3y = 6$
(b) $-2x + 3y = 6$
(c) $3x - 4y = -12$
(d) $x - y = 1$
(e) $x - y = -1$
(f) $y = 3x - 5$
(g) $y = 2x + 3$
(h) $y = -3x + 3$

132. Which of the following equations might have the graph shown? (More than one answer is possible.)
(a) $2x + 3y = 6$
(b) $2x - 3y = 6$
(c) $3x + 4y = 12$
(d) $x - y = 1$
(e) $x - y = -1$
(f) $y = -2x - 1$
(g) $y = -\frac{1}{2}x + 10$
(h) $y = x + 4$

133. The figure shows the graph of two parallel lines. Which of the following pairs of equations might have such a graph?

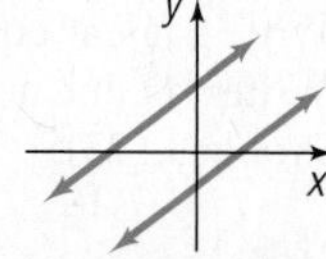

(a) $x - 2y = 3$
$x + 2y = 7$
(b) $x + y = 2$
$x + y = -1$
(c) $x - y = -2$
$x - y = 1$
(d) $x - y = -2$
$2x - 2y = -4$
(e) $x + 2y = 2$
$x + 2y = -1$

134. The figure shows the graph of two perpendicular lines. Which of the following pairs of equations might have such a graph?
(a) $y - 2x = 2$
$y + 2x = -1$
(b) $y - 2x = 0$
$2y + x = 0$
(c) $2y - x = 2$
$2y + x = -2$
(d) $y - 2x = 2$
$x + 2y = -1$
(e) $2x + y = -2$
$2y + x = -2$

135. *m* is for Slope The accepted symbol used to denote the slope of a line is the letter m. Investigate the origin of this practice. Begin by consulting a French dictionary and looking up the French word *monter*. Write a brief essay on your findings.

136. **Grade of a Road** The term *grade* is used to describe the inclination of a road. How is this term related to the notion of slope of a line? Is a 4% grade very steep? Investigate the grades of some mountainous roads and determine their slopes. Write a brief essay on your findings.

137. **Carpentry** Carpenters use the term *pitch* to describe the steepness of staircases and roofs. How is pitch related to slope? Investigate typical pitches used for stairs and for roofs. Write a brief essay on your findings.

138. Can the equation of every line be written in slope–intercept form? Why?

139. Does every line have exactly one x-intercept and one y-intercept? Are there any lines that have no intercepts?

140. What can you say about two lines that have equal slopes and equal y-intercepts?

141. What can you say about two lines with the same x-intercept and the same y-intercept? Assume that the x-intercept is not 0.

142. If two distinct lines have the same slope but different x-intercepts, can they have the same y-intercept?

143. If two distinct lines have the same y-intercept but different slopes, can they have the same x-intercept?

144. Which form of the equation of a line do you prefer to use? Justify your position with an example that shows that your choice is better than another. Have reasons.

145. **What Went Wrong?** A student is asked to find the slope of the line joining $(-3, 2)$ and $(1, -4)$. He states that the slope is $\frac{3}{2}$. Is he correct? If not, what went wrong?

A.9 Building Linear Models from Data

OBJECTIVES
1 Draw and Interpret Scatter Diagrams (A79)
2 Distinguish between Linear and Nonlinear Relations (A80)
3 Use a Graphing Utility to Find the Line of Best Fit (A82)

1 Draw and Interpret Scatter Diagrams

Linear models can be constructed by fitting a linear function to data. The first step is to plot the ordered pairs using rectangular coordinates. The resulting graph is a **scatter diagram**.

EXAMPLE 1 Drawing and Interpreting a Scatter Diagram

In baseball, the on-base percentage for a team represents the percentage of time that the players safely reach base. The data given in Table 2 on page A80 represent the number of runs scored y and the on-base percentage x for teams in the National League during the 2013 baseball season.

(a) Draw a scatter diagram of the data, treating on-base percentage as the independent variable.

(b) Use a graphing utility to draw a scatter diagram.

(c) Describe what happens to runs scored as the on-base percentage increases.

Solution

(a) To draw a scatter diagram, plot the ordered pairs listed in Table 2, with the on-base percentage as the x-coordinate and the runs scored as the y-coordinate. See Figure 54(a) on page A80. Notice that the points in the scatter diagram are not connected.

(b) Figure 54(b) shows a scatter diagram using a TI-84 Plus C graphing calculator.

(c) The scatter diagrams show that as the on-base percentage increases, the number of runs scored also increases.

Table 2

Team	On-Base Percentage, x	Runs Scored, y	(x, y)
Arizona	32.3	685	(32.3, 685)
Atlanta	32.1	688	(32.1, 688)
Chicago Cubs	30.0	602	(30.0, 602)
Cincinnati	32.7	698	(32.7, 698)
Colorado	32.3	706	(32.3, 706)
LA Dodgers	32.6	649	(32.6, 649)
Miami	29.3	513	(29.3, 513)
Milwaukee	31.1	640	(31.1, 640)
NY Mets	30.6	619	(30.6, 619)
Philadelphia	30.6	610	(30.6, 610)
Pittsburgh	31.3	634	(31.3, 634)
San Diego	30.8	618	(30.8, 618)
San Francisco	32.0	629	(32.0, 629)
St. Louis	33.2	783	(33.2, 783)
Washington	31.3	656	(31.3, 656)

Source: espn.go.com

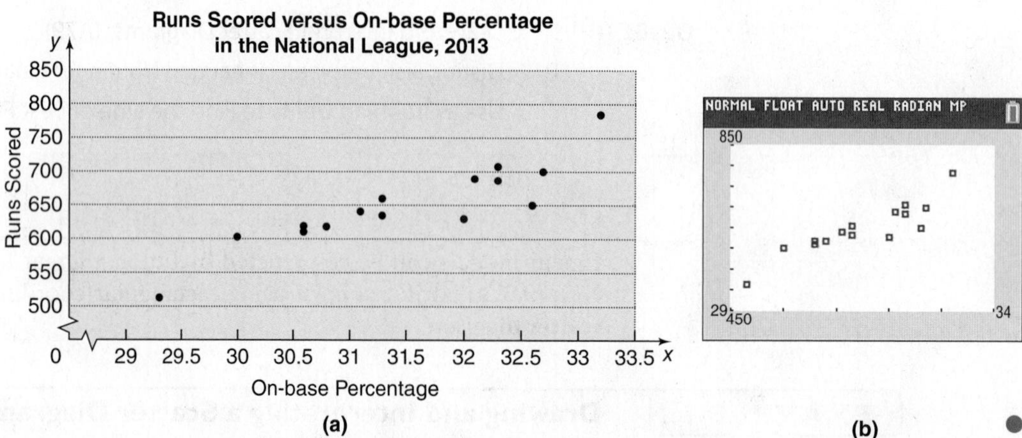

Figure 54 On-base percentage (a) (b)

Now Work PROBLEM 9(a)

2 Distinguish between Linear and Nonlinear Relations

Notice that the points in Figure 54 do not follow a perfect linear relation. However, the data do exhibit a linear pattern. There are numerous possible explanations why the data are not perfectly linear, but one easy explanation is the fact that other variables besides on-base percentage (such as number of home runs hit) play a role in determining runs scored.

Scatter diagrams are used to help us to see the type of relation that exists between two variables. In this text, we will discuss a variety of different relations that may exist between two variables. For now, we concentrate on distinguishing between linear and nonlinear relations. See Figure 55.

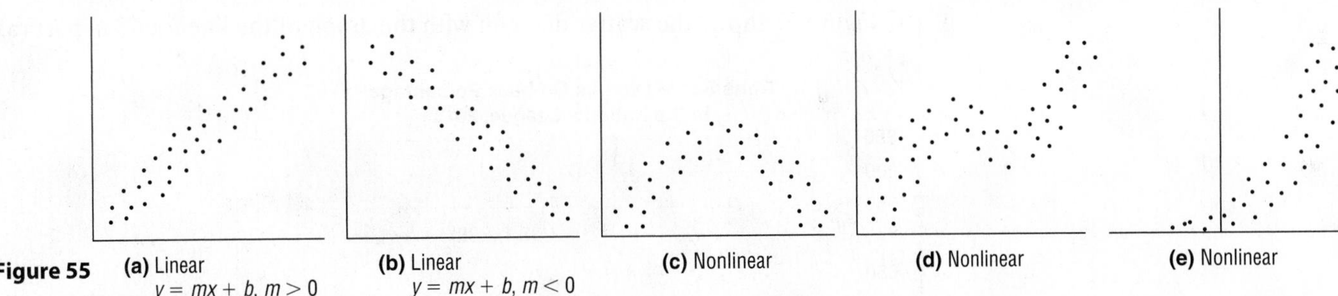

Figure 55 (a) Linear $y = mx + b, m > 0$ (b) Linear $y = mx + b, m < 0$ (c) Nonlinear (d) Nonlinear (e) Nonlinear

EXAMPLE 2 Distinguishing between Linear and Nonlinear Relations

Determine whether the relation between the two variables in each scatter diagram in Figure 56 is linear or nonlinear.

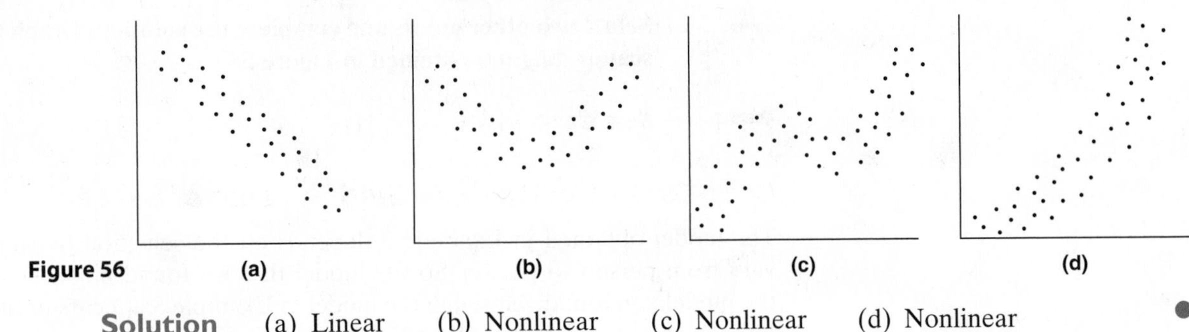

Figure 56 (a) (b) (c) (d)

Solution (a) Linear (b) Nonlinear (c) Nonlinear (d) Nonlinear ●

Now Work PROBLEM 3

This section considers data whose scatter diagrams suggest that a linear relation exists between the two variables.

Suppose that the scatter diagram of a set of data appears to indicate a linear relationship, as in Figure 55(a) or (b). We might want to model the data by finding an equation of a line that relates the two variables. One way to obtain a model for such data is to draw a line through two points on the scatter diagram and determine the equation of the line.

EXAMPLE 3 Finding a Model for Linearly Related Data

Use the data in Table 2 from Example 1.

(a) Select two points and find an equation of the line containing the points.

(b) Graph the line on the scatter diagram obtained in Example 1(a).

Solution (a) Select two points, say $(30.6, 610)$ and $(32.1, 688)$. The slope of the line joining the points $(30.6, 610)$ and $(32.1, 688)$ is

$$m = \frac{688 - 610}{32.1 - 30.6} = \frac{78}{1.5} = 52$$

The equation of the line that has slope 52 and passes through $(30.6, 610)$ is found using the point–slope form with $m = 52$, $x_1 = 30.6$, and $y_1 = 610$.

$$\begin{aligned} y - y_1 &= m(x - x_1) && \text{Point–slope form of a line} \\ y - 610 &= 52(x - 30.6) && x_1 = 30.6,\ y_1 = 610,\ m = 52 \\ y - 610 &= 52x - 1591.2 && \\ y &= 52x - 981.2 && \text{The model} \end{aligned}$$

(b) Figure 57 shows the scatter diagram with the graph of the line found in part (a).

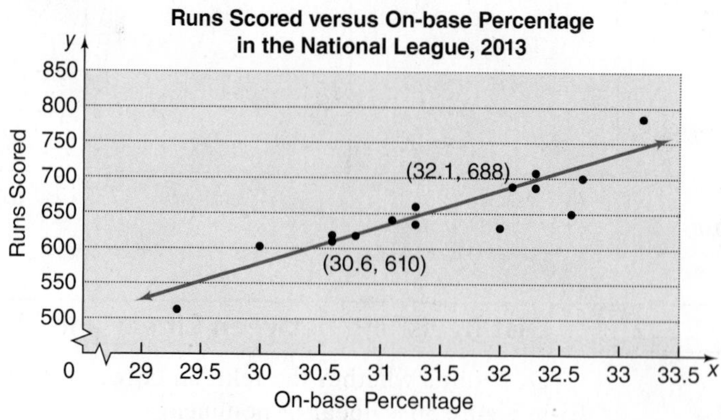

Figure 57

Select two other points and complete the solution. Graph the line on the scatter diagram obtained in Figure 54.

Now Work PROBLEMS 9(b) AND (c)

3 Use a Graphing Utility to Find the Line of Best Fit

The model obtained in Example 3 depends on the selection of points, which will vary from person to person. So the model that we found might be different from the model you found. Although the model in Example 3 appears to fit the data well, there may be a model that "fits them better." Do you think your model fits the data better? Is there a *line of best fit*? As it turns out, there is a method for finding a model that best fits linearly related data (called the **line of best fit**).*

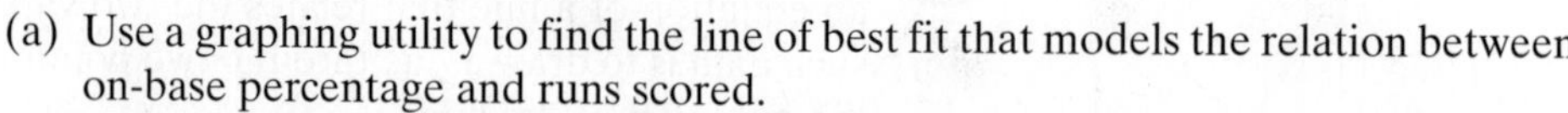

EXAMPLE 4 Finding a Model for Linearly Related Data

Use the data in Table 2 from Example 1.

(a) Use a graphing utility to find the line of best fit that models the relation between on-base percentage and runs scored.
(b) Graph the line of best fit on the scatter diagram obtained in Example 1(b).
(c) Interpret the slope.
(d) Use the line of best fit to predict the number of runs a team will score if their on-base percentage is 33.1.

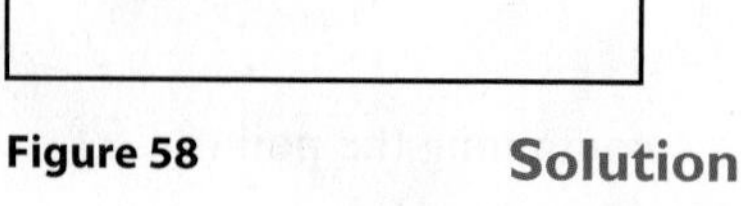

Figure 58

Solution (a) Graphing utilities contain built-in programs that find the line of best fit for a collection of points in a scatter diagram. Executing the LINear REGression program provides the results shown in Figure 58. This output shows the equation $y = ax + b$, where a is the slope of the line and b is the y-intercept. The line of best fit that relates on-base percentage to runs scored may be expressed as the line

$$y = 49.40x - 906.29 \quad \text{The model}$$

(b) Figure 59 shows the graph of the line of best fit, along with the scatter diagram.
(c) The slope of the line of best fit is 49.40, which means that, for every 1 percent increase in the on-base percentage, runs scored increase by 49.40, on average.
(d) Letting $x = 33.1$ in the equation of the line of best fit, we obtain

$$y = 49.40(33.1) - 906.29 \approx 729 \text{ runs}$$

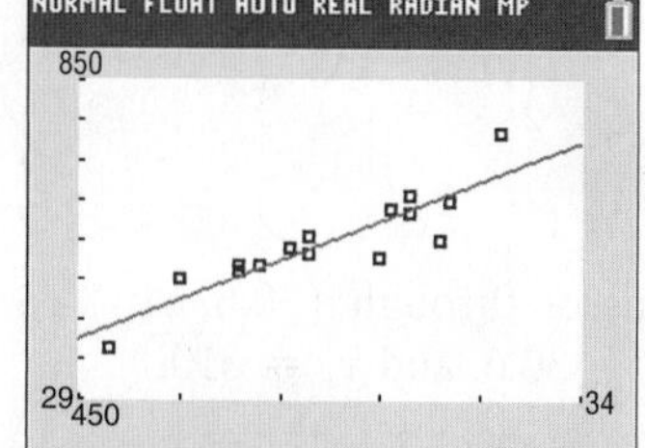

Figure 59

Now Work PROBLEMS 9(d) AND (e)

* We shall not discuss the underlying mathematics of lines of best fit in this text.

Does the line of best fit appear to be a good fit? In other words, does it appear to accurately describe the relation between on-base percentage and runs scored?

And just how "good" is this line of best fit? Look again at Figure 58. The last line of output is $r = 0.896$. This number, called the **correlation coefficient**, r, $-1 \le r \le 1$, is a measure of the strength of the linear relation that exists between two variables. The closer $|r|$ is to 1, the more nearly perfect the linear relationship is. If r is close to 0, there is little or no linear relationship between the variables. A negative value of r, $r < 0$, indicates that as x increases, y decreases; a positive value of r, $r > 0$, indicates that as x increases, y does also. The data given in Table 2, which have a correlation coefficient of 0.896, are indicative of a linear relationship with positive slope.

A.9 Assess Your Understanding

Concepts and Vocabulary

1. A ______________ is used to help us to see what type of relation, if any, may exist between two variables.

2. ***True or False*** The correlation coefficient is a measure of the strength of a linear relation between two variables and must lie between -1 and 1, inclusive.

Skill Building

In Problems 3–8, examine the scatter diagram and determine whether the type of relation is linear or nonlinear.

3.

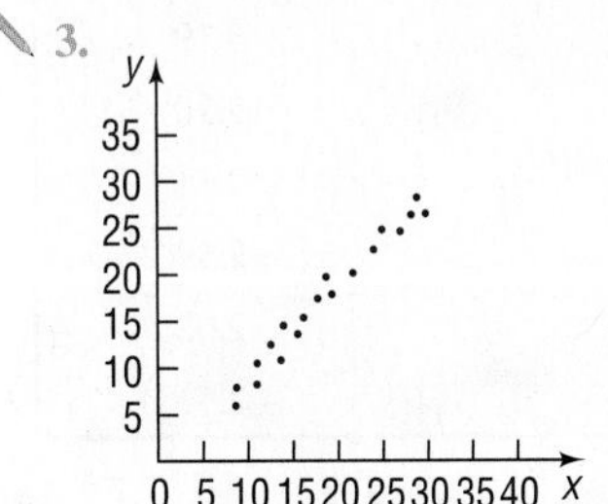

4.

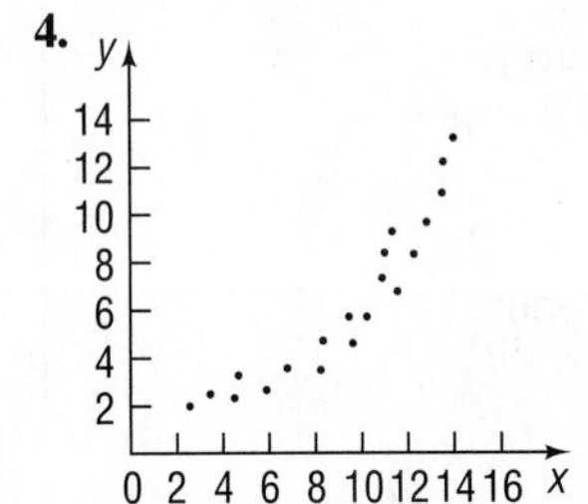

5.

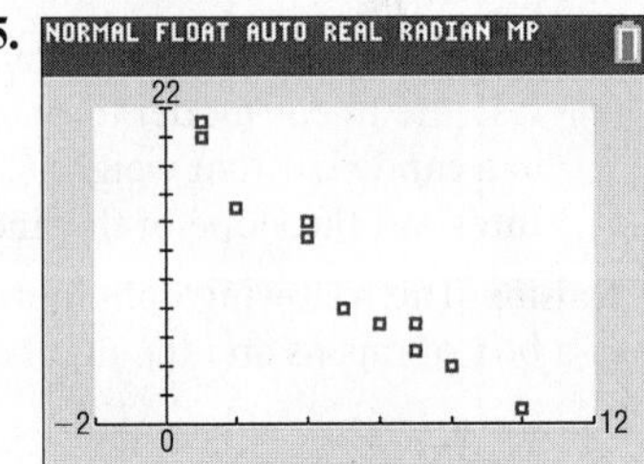

6.

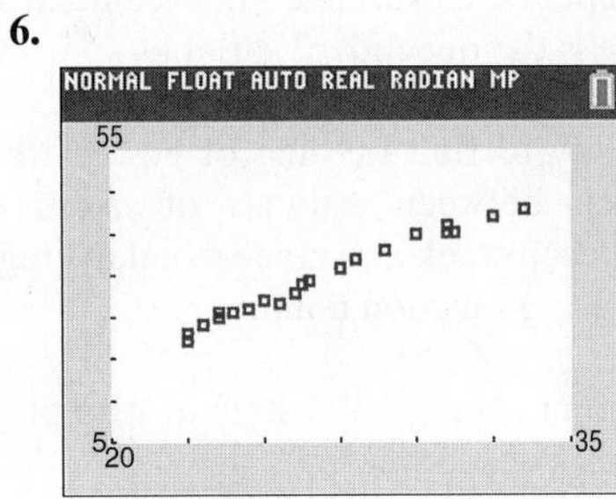

7.

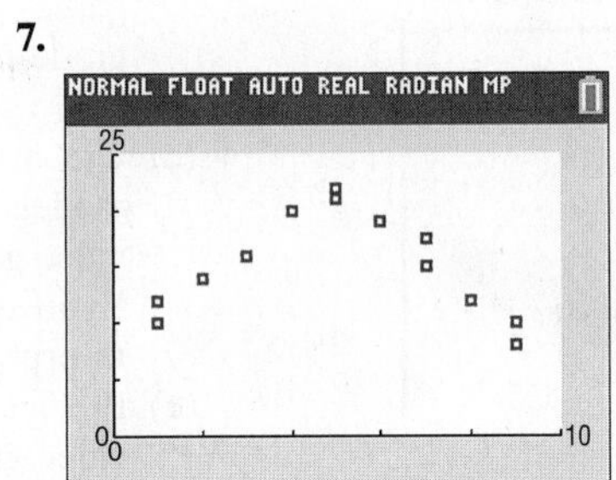

8.

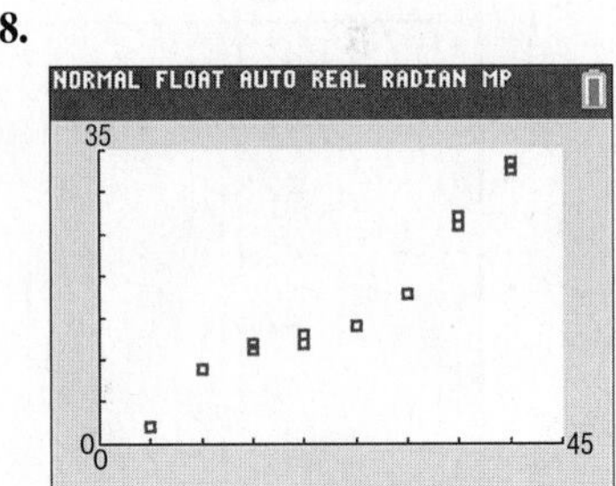

In Problems 9–14:

(a) Draw a scatter diagram.
(b) Select two points from the scatter diagram, and find the equation of the line containing the points selected.
(c) Graph the line found in part (b) on the scatter diagram.
(d) Use a graphing utility to find the line of best fit.
(e) Use a graphing utility to draw the scatter diagram and graph the line of best fit on it.

9.

x	3	4	5	6	7	8	9
y	4	6	7	10	12	14	16

10.

x	3	5	7	9	11	13
y	0	2	3	6	9	11

11.

x	−2	−1	0	1	2
y	−4	0	1	4	5

12.

x	−2	−1	0	1	2
y	7	6	3	2	0

13.

x	−20	−17	−15	−14	−10
y	100	120	118	130	140

14.

x	−30	−27	−25	−20	−14
y	10	12	13	13	18

Applications and Extensions

15. Candy The following data represent the weight (in grams) of various candy bars and the corresponding number of calories.

Candy Bar	Weight, *x*	Calories, *y*
Hershey's Milk Chocolate®	44.28	230
Nestle's Crunch®	44.84	230
Butterfinger®	61.30	270
Baby Ruth®	66.45	280
Almond Joy®	47.33	220
Twix® (with caramel)	58.00	280
Snickers®	61.12	280
Heath®	39.52	210

Source: Megan Pocius, student at Joliet Junior College

(a) Draw a scatter diagram of the data, treating weight as the independent variable.
(b) What type of relation appears to exist between the weight of a candy bar and the number of calories?
(c) Select two points and find a linear model that contains the points.
(d) Graph the line on the scatter diagram drawn in part (a).
(e) Use the linear model to predict the number of calories in a candy bar that weighs 62.3 grams.
(f) Interpret the slope of the line found in part (c).

16. Raisins The following data represent the weight (in grams) of a box of raisins and the number of raisins in the box.

Weight (grams), *w*	Number of Raisins, *N*
42.3	87
42.7	91
42.8	93
42.4	87
42.6	89
42.4	90
42.3	82
42.5	86
42.7	86
42.5	86

Source: Jennifer Maxwell, student at Joliet Junior College

(a) Draw a scatter diagram of the data, treating weight as the independent variable.
(b) What type of relation appears to exist between the weight of a box of raisins and the number of raisins?
(c) Select two points and find a linear model that contains the points.
(d) Graph the line on the scatter diagram drawn in part (b).
(e) Use the linear model to predict the number of raisins in a box that weighs 42.5 grams.
(f) Interpret the slope of the line found in part (c).

17. Video Games and Grade-Point Average Professor Grant Alexander wanted to find a linear model that relates the number of hours a student plays video games each week, h, to the cumulative grade-point average, G, of the student. He obtained a random sample of 10 full-time students at his college and asked each student to disclose the number of hours spent playing video games and the student's cumulative grade-point average.

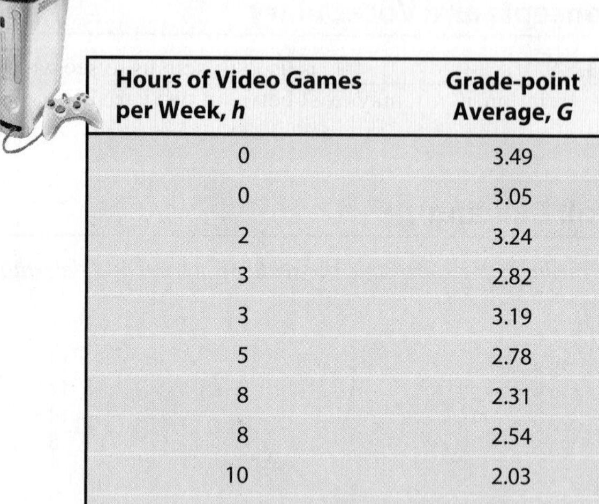

Hours of Video Games per Week, *h*	Grade-point Average, *G*
0	3.49
0	3.05
2	3.24
3	2.82
3	3.19
5	2.78
8	2.31
8	2.54
10	2.03
12	2.51

(a) Explain why the number of hours spent playing video games is the independent variable and cumulative grade-point average is the dependent variable.
(b) Use a graphing utility to draw a scatter diagram.
(c) Use a graphing utility to find the line of best fit that models the relation between number of hours of video game playing each week and grade-point average. Express the model using function notation.
(d) Interpret the slope.
(e) Predict the grade-point average of a student who plays video games for 8 hours each week.
(f) How many hours of video game playing do you think a student plays whose grade-point average is 2.40?

18. Height versus Head Circumference A pediatrician wanted to find a linear model that relates a child's height, H, to head circumference, C. She randomly selects nine children from her practice, measures their height and head circumference, and obtains the data shown. Let H represent the independent variable and C the dependent variable.

(a) Use a graphing utility to draw a scatter diagram.
(b) Use a graphing utility to find the line of best fit that models the relation between height and head circumference. Express the model using function notation.
(c) Interpret the slope.

(d) Predict the head circumference of a child who is 26 inches tall.
(e) What is the height of a child whose head circumference is 17.4 inches?

Height, H (inches)	Head Circumference, C (inches)
25.25	16.4
25.75	16.9
25	16.9
27.75	17.6
26.5	17.3
27	17.5
26.75	17.3
26.75	17.5
27.5	17.5

Source: Denise Slucki, student at Joliet Junior College

19. Flight Time and Ticket Price The following data represent nonstop flight time (in minutes) and one-way ticket price (in dollars) for flying from Chicago to various cities on Southwest Airlines.

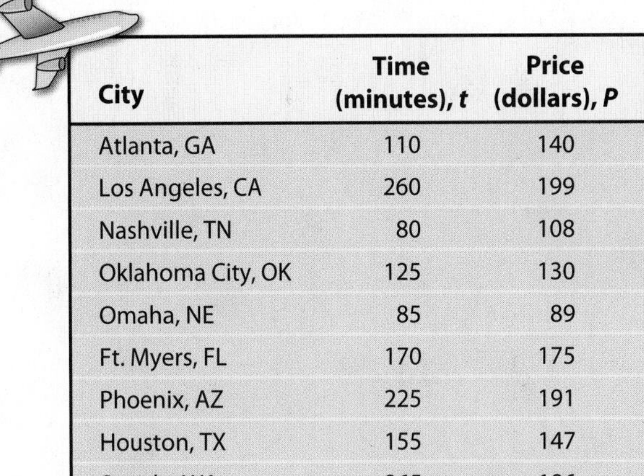

City	Time (minutes), t	Price (dollars), P
Atlanta, GA	110	140
Los Angeles, CA	260	199
Nashville, TN	80	108
Oklahoma City, OK	125	130
Omaha, NE	85	89
Ft. Myers, FL	170	175
Phoenix, AZ	225	191
Houston, TX	155	147
Seattle, WA	265	196
St. Louis, MO	65	93

Source: Southwest.com, for midmorning flights in May 2014

(a) Use a graphing utility to draw a scatter diagram.
(b) Use a graphing utility to find the line of best fit that models the relation between flight time and airfare. Express the model using function notation.
(c) Interpret the slope.
(d) Predict the airfare for a flight from Chicago to Kansas City, Missouri, if the flight time is 85 minutes. Round to the nearest dollar.
(e) Predict the flight time from Chicago to Baltimore, Maryland if the airfare is \$120. Round to the nearest minute.

Explaining Concepts: Discussion and Writing

20. Maternal Age versus Down Syndrome A biologist would like to know how the age of the mother affects the incidence of Down syndrome. The data to the right represent the age of the mother and the incidence of Down syndrome per 1000 pregnancies. Draw a scatter diagram treating age of the mother as the independent variable. Would it make sense to find the line of best fit for these data? Why or why not?

21. Find the line of best fit for the ordered pairs $(1, 5)$ and $(3, 8)$. What is the correlation coefficient for these data? Why is this result reasonable?

22. What does a correlation coefficient of 0 imply?

23. Explain why it does not make sense to interpret the y-intercept in Problem 15.

24. Refer to Problem 17. Solve $G(h) = 0$. Provide an interpretation of this result. Find $G(0)$. Provide an interpretation of this result.

Age of Mother, x	Incidence of Down Syndrome, y
33	2.4
34	3.1
35	4
36	5
37	6.7
38	8.3
39	10
40	13.3
41	16.7
42	22.2
43	28.6
44	33.3
45	50

Source: Hook, E.B., *Journal of the American Medical Association*, 249, 2034-2038, 1983.

Appendix B
Graphing Utilities

Outline

B.1 The Viewing Rectangle
B.2 Using a Graphing Utility to Graph Equations
B.3 Using a Graphing Utility to Locate Intercepts and Check for Symmetry
B.4 Using a Graphing Utility to Solve Equations
B.5 Square Screens
B.6 Using a Graphing Utility to Graph a Polar Equation
B.7 Using a Graphing Utility to Graph Parametric Equations

B.1 The Viewing Rectangle

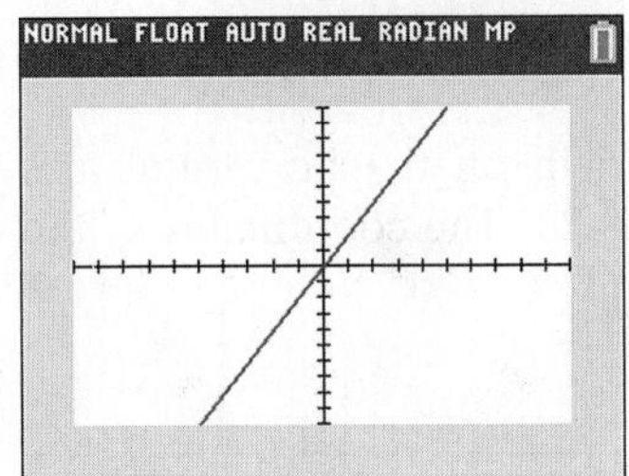

Figure 1 $y = 2x$

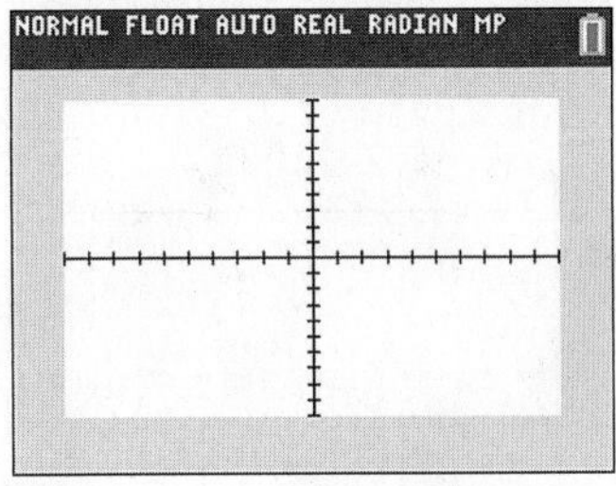

Figure 2 Viewing window

All graphing utilities (that is, all graphing calculators and all computer software graphing packages) graph equations by plotting points on a screen. The screen itself actually consists of small rectangles called **pixels**. The more pixels the screen has, the better the resolution. Most graphing calculators have 50 to 100 pixels per inch; most smartphones have 300 to 450 pixels per inch. When a point to be plotted lies inside a pixel, the pixel is turned on (lights up). The graph of an equation is a collection of pixels. Figure 1 shows how the graph of $y = 2x$ looks on a TI-84 Plus C graphing calculator.

The screen of a graphing utility will display the coordinate axes of a rectangular coordinate system. However, the scale must be set on each axis. The smallest and largest values of x and y to be included in the graph must also be set. This is called **setting the viewing rectangle** or **viewing window**. Figure 2 shows a typical viewing window.

To select the viewing window, values must be given to the following expressions:

Xmin:	the smallest value of x
Xmax:	the largest value of x
Xscl:	the number of units per tick mark on the x-axis
Ymin:	the smallest value of y
Ymax:	the largest value of y
Yscl:	the number of units per tick mark on the y-axis

Figure 3 illustrates these settings and their relation to the Cartesian coordinate system.

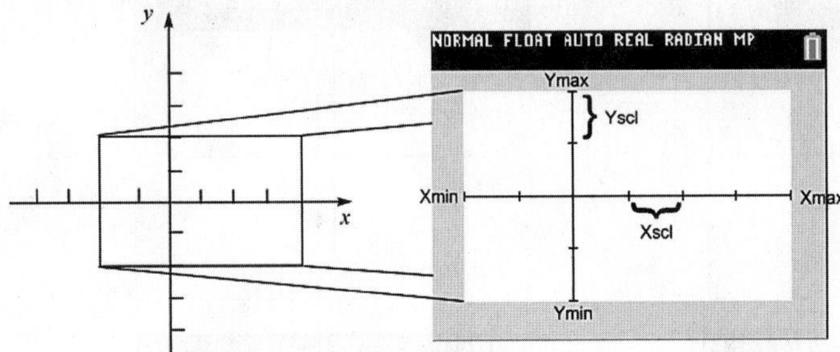

Figure 3

If the scale used on each axis is known, the minimum and maximum values of x and y shown on the screen can be determined by counting the tick marks. Look again at Figure 2. For a scale of 1 on each axis, the minimum and maximum values of x are -10 and 10, respectively; the minimum and maximum values of y are

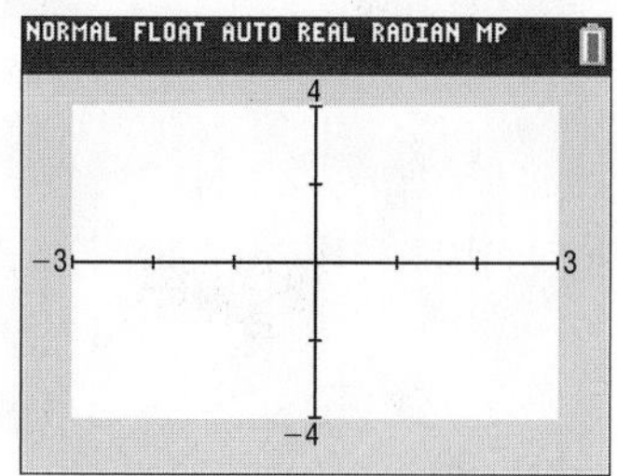

Figure 4

also −10 and 10. If the scale is 2 on each axis, then the minimum and maximum values of x are −20 and 20, respectively; and the minimum and maximum values of y are −20 and 20, respectively.

Conversely, if the minimum and maximum values of x and y are known, the scales can be determined by counting the tick marks displayed. This text follows the practice of showing the minimum and maximum values of x and y in illustrations so that the reader will know how the viewing window was set. See Figure 4. The numbers outside of the viewing window stand for

$$\text{Xmin} = -3 \qquad \text{Ymin} = -4$$
$$\text{Xmax} = 3 \qquad \text{Ymax} = 4$$
$$\text{Xscl} = 1 \qquad \text{Yscl} = 2$$

EXAMPLE 1

Finding the Coordinates of a Point Shown on a Graphing Utility Screen

Find the coordinates of the point shown in Figure 5. Assume that the coordinates are integers.

Solution First note that the viewing window used in Figure 5 is

$$\text{Xmin} = -3 \qquad \text{Ymin} = -4$$
$$\text{Xmax} = 3 \qquad \text{Ymax} = 4$$
$$\text{Xscl} = 1 \qquad \text{Yscl} = 2$$

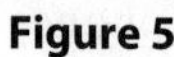

Figure 5

The point shown is 2 tick units to the left of the origin on the horizontal axis $(\text{scale} = 1)$ and 1 tick up on the vertical axis $(\text{scale} = 2)$. The coordinates of the point shown are $(-2, 2)$. ●

B.1 Exercises

In Problems 1–4, determine the coordinates of the points shown. Tell in which quadrant each point lies. Assume that the coordinates are integers.

1.

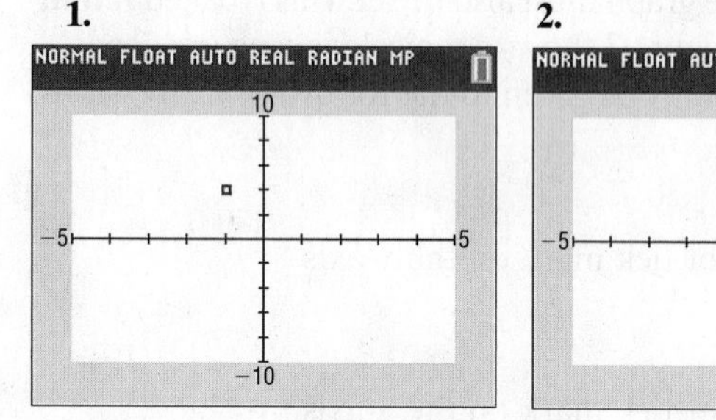

2.

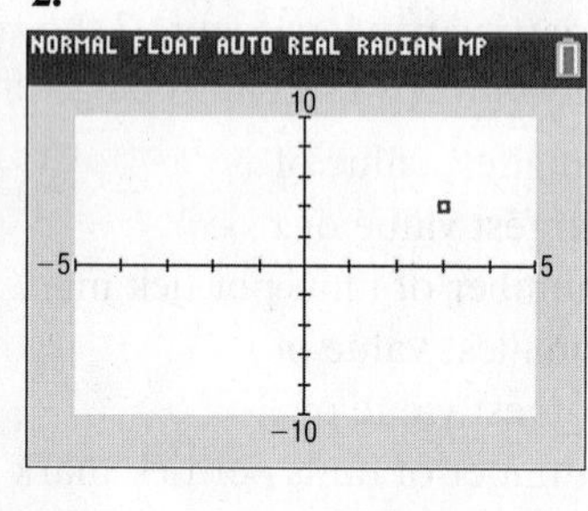

3.

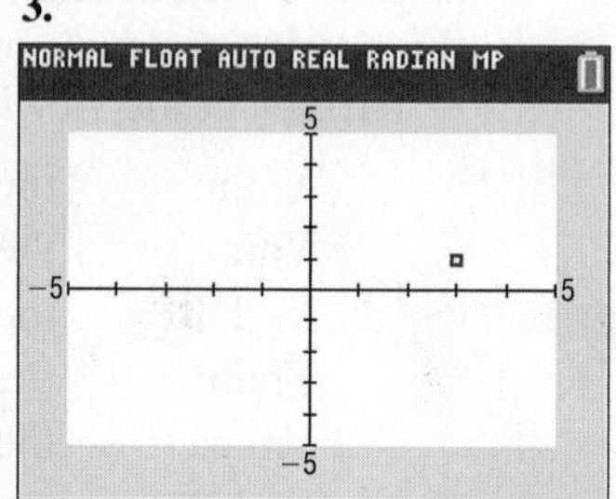

4.

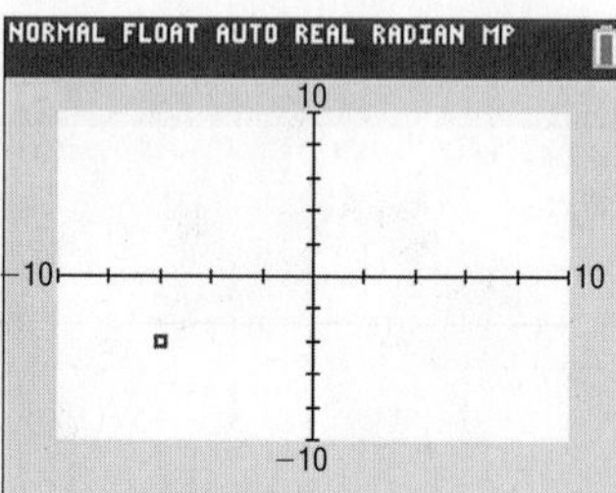

In Problems 5–10, determine the viewing window used.

5.

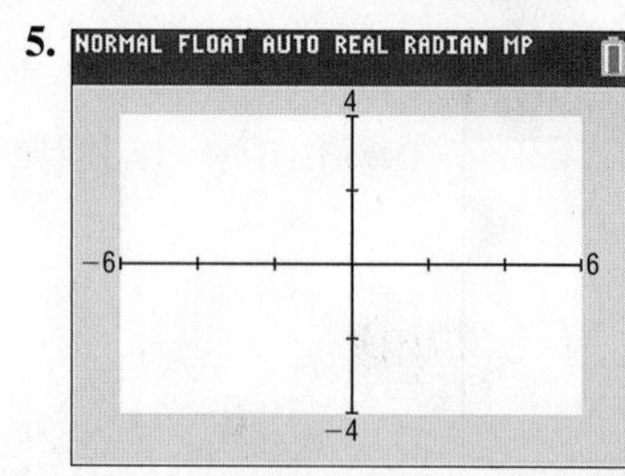

6.

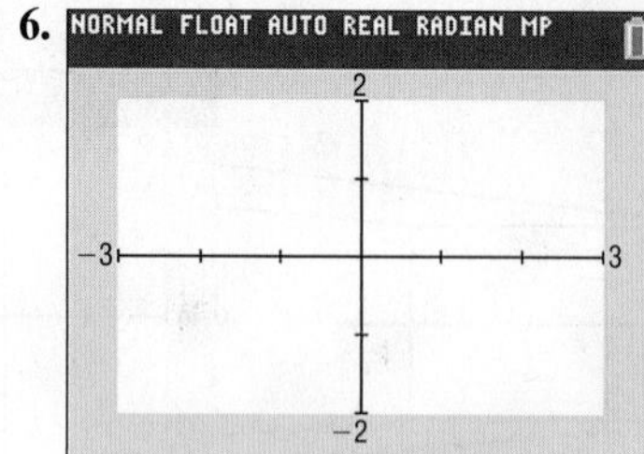

7.

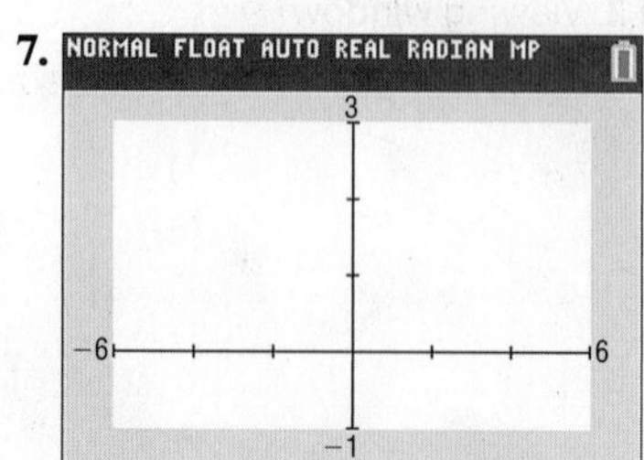

8.

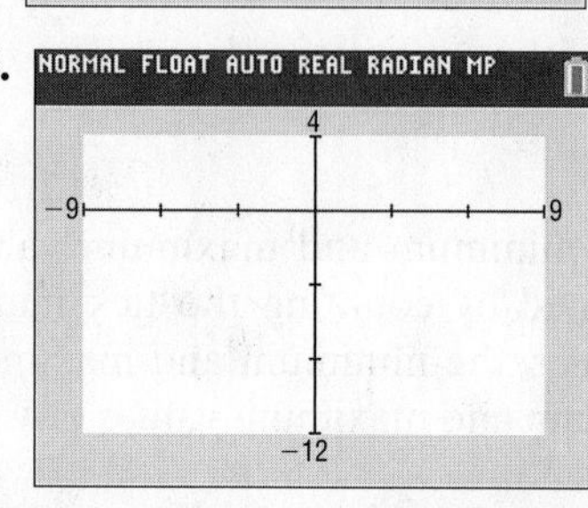

9.

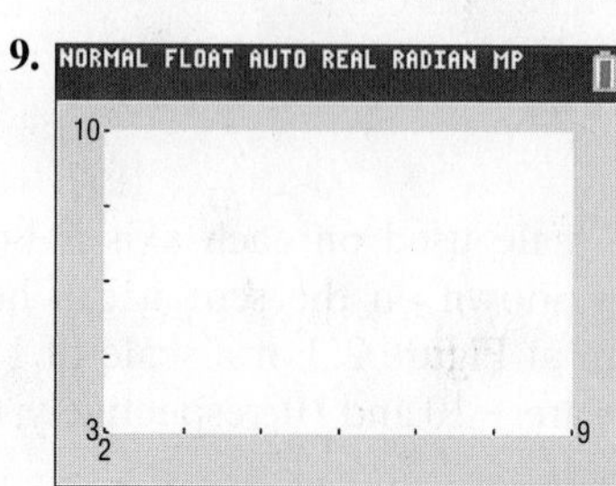

10.

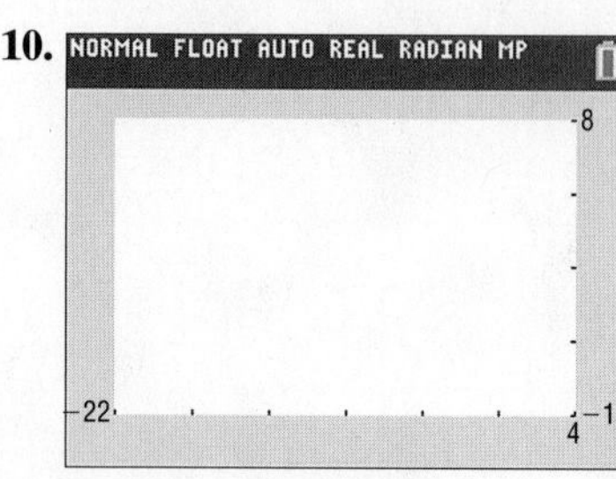

In Problems 11–16, select a setting so that each of the given points will lie within the viewing rectangle.

11. $(-10, 5), (3, -2), (4, -1)$
12. $(5, 0), (6, 8), (-2, -3)$
13. $(40, 20), (-20, -80), (10, 40)$
14. $(-80, 60), (20, -30), (-20, -40)$
15. $(0, 0), (100, 5), (5, 150)$
16. $(0, -1), (100, 50), (-10, 30)$

B.2 Using a Graphing Utility to Graph Equations

From Examples 2 and 3 in Section 1.2, recall that a graph can be obtained by plotting points in a rectangular coordinate system and connecting them. Graphing utilities perform these same steps when graphing an equation. For example, the TI-84 Plus C determines 265 evenly spaced input values,* starting at *X*min and ending at *X*max; uses the equation to determine the output values; plots these points on the screen; and finally (if in the connected mode) draws a line between consecutive points.

To graph an equation in two variables x and y using a graphing utility requires that the equation be written in the form $y = \{expression\ in\ x\}$. If the original equation is not in this form, replace it by equivalent equations until the form $y = \{expression\ in\ x\}$ is obtained.

Steps for Graphing an Equation Using a Graphing Utility

STEP 1: Solve the equation for y in terms of x.

STEP 2: Get into the graphing mode of the graphing utility. The screen will usually display $Y_1 =$, prompting you to enter the expression involving x found in Step 1. (Consult your manual for the correct way to enter the expression; for example, $y = x^2$ might be entered as x^2 or as x*x or as $x\ x^y\ 2$).

STEP 3: Select the viewing window. Without prior knowledge about the behavior of the graph of the equation, it is common to select the **standard viewing window**** initially. The viewing window is then adjusted based on the graph that appears. In this text the standard viewing window is

$$X\text{min} = -10 \qquad Y\text{min} = -10$$
$$X\text{max} = 10 \qquad Y\text{max} = 10$$
$$X\text{scl} = 1 \qquad Y\text{scl} = 1$$

STEP 4: Graph.

STEP 5: Adjust the viewing window until a complete graph is obtained.

EXAMPLE 1

Graphing an Equation on a Graphing Utility

Graph the equation: $6x^2 + 3y = 36$

Solution **STEP 1:** Solve for y in terms of x.

$$6x^2 + 3y = 36$$
$$3y = -6x^2 + 36 \quad \text{Subtract } 6x^2 \text{ from both sides of the equation.}$$
$$y = -2x^2 + 12 \quad \text{Divide both sides of the equation by 3 and simplify.}$$

*These input values depend on the values of *X*min and *X*max. For example, if $X\text{min} = -10$ and $X\text{max} = 10$, then the first input value will be -10 and the next input value will be $-10 + \frac{10 - (-10)}{264} = -9.9242$, and so on.

**Some graphing utilities have a ZOOM-STANDARD feature that automatically sets the viewing window to the standard viewing window and graphs the equation.

STEP 2: From the $Y_1 =$ screen, enter the expression $-2x^2 + 12$ after the prompt.
STEP 3: Set the viewing window to the standard viewing window.
STEP 4: Graph. The screen should look like Figure 6.
STEP 5: The graph of $y = -2x^2 + 12$ is not complete. The value of Ymax must be increased so that the top portion of the graph is visible. After increasing the value of Ymax to 12, we obtain the graph in Figure 7. The graph is now complete.

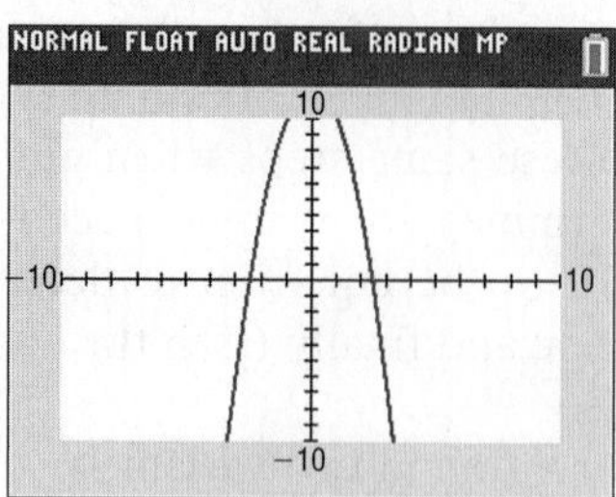

Figure 6 $y = -2x^2 + 12$

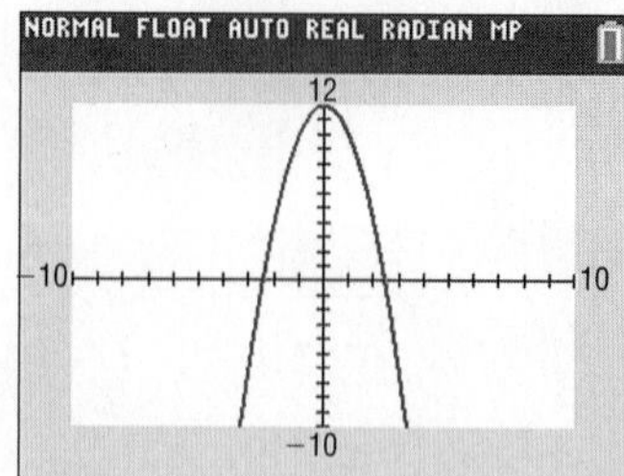

Figure 7 $y = -2x^2 + 12$

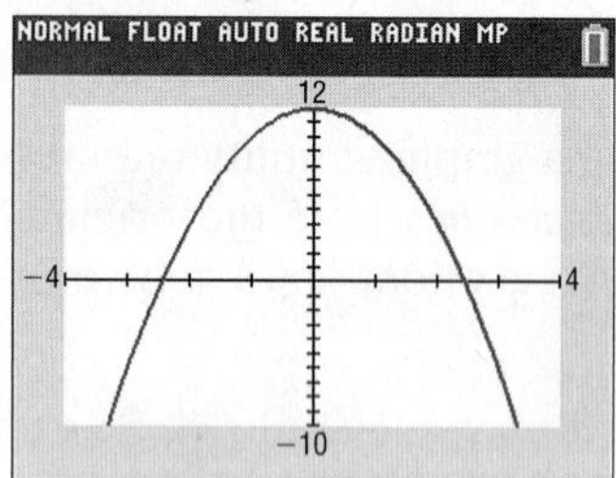

Figure 8 $y = -2x^2 + 12$

Look again at Figure 7. Although a complete graph is shown, the graph might be improved by adjusting the values of Xmin and Xmax. Figure 8 shows the graph of $y = -2x^2 + 12$ using $X\text{min} = -4$ and $X\text{max} = 4$. Do you think this is a better choice for the viewing window?

EXAMPLE 2

Creating a Table and Graphing an Equation

Create a table and graph the equation: $y = x^3$

Solution Most graphing utilities have the capability of creating a table of values for an equation. (Check your manual to see if your graphing utility has this capability.) Table 1 illustrates a table of values for $y = x^3$ on a TI-84 Plus C. See Figure 9 for the graph.

Table 1

NORMAL FLOAT AUTO REAL RADIAN MP
PRESS ENTER TO EDIT

X	Y1
-5	-125
-4	-64
-3	-27
-2	-8
-1	-1
0	0
1	1
2	8
3	27
4	64
5	125

Y1=X³

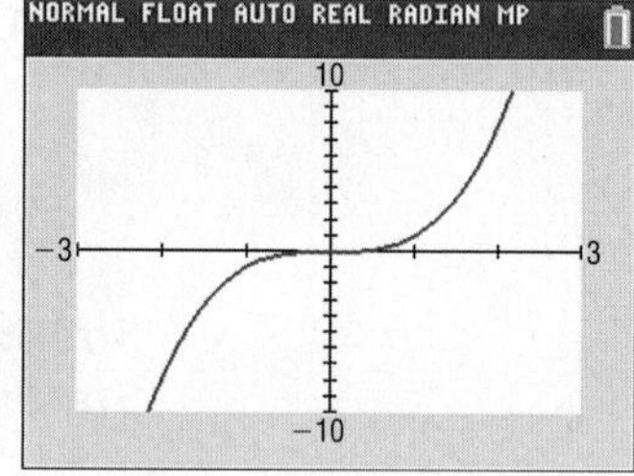

Figure 9 $y = x^3$

B.2 Exercises

In Problems 1–16, graph each equation using the following viewing windows:

(a) $X\text{min} = -5$, $X\text{max} = 5$, $X\text{scl} = 1$, $Y\text{min} = -4$, $Y\text{max} = 4$, $Y\text{scl} = 1$

(b) $X\text{min} = -10$, $X\text{max} = 10$, $X\text{scl} = 2$, $Y\text{min} = -8$, $Y\text{max} = 8$, $Y\text{scl} = 2$

1. $y = x + 2$ **2.** $y = x - 2$ **3.** $y = -x + 2$ **4.** $y = -x - 2$
5. $y = 2x + 2$ **6.** $y = 2x - 2$ **7.** $y = -2x + 2$ **8.** $y = -2x - 2$
9. $y = x^2 + 2$ **10.** $y = x^2 - 2$ **11.** $y = -x^2 + 2$ **12.** $y = -x^2 - 2$
13. $3x + 2y = 6$ **14.** $3x - 2y = 6$ **15.** $-3x + 2y = 6$ **16.** $-3x - 2y = 6$

17–32. *For each of the above equations, create a table, $-5 \le x \le 5$, and list points on the graph.*

B.3 Using a Graphing Utility to Locate Intercepts and Check for Symmetry

Value and Zero (or Root)

Most graphing utilities have an eVALUEate feature that, given a value of x, determines the value of y for an equation. This feature can be used to evaluate an equation at $x = 0$ to determine the y-intercept. Most graphing utilities also have a ZERO (or ROOT) feature that can be used to determine the x-intercept(s) of an equation.

EXAMPLE 1

Finding Intercepts Using a Graphing Utility

Use a graphing utility to find the intercepts of the equation $y = x^3 - 8$.

Solution Figure 10(a) shows the graph of $y = x^3 - 8$.

The eVALUEate feature of a TI-84 Plus C graphing calculator accepts as input a value of x and determines the value of y. Letting $x = 0$, we find that the y-intercept is -8. See Figure 10(b).

The ZERO feature of a TI-84 Plus C is used to find the x-intercept(s). See Figure 10(c). The x-intercept is 2.

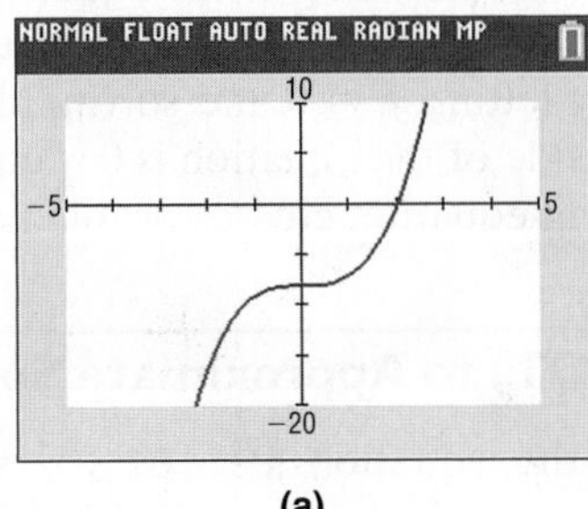

(a)

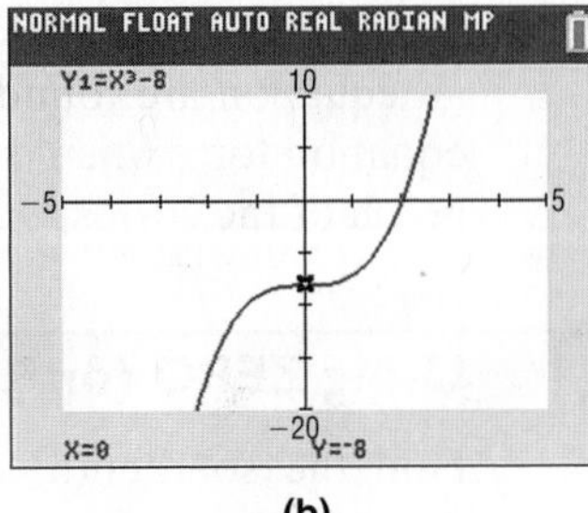

(b)

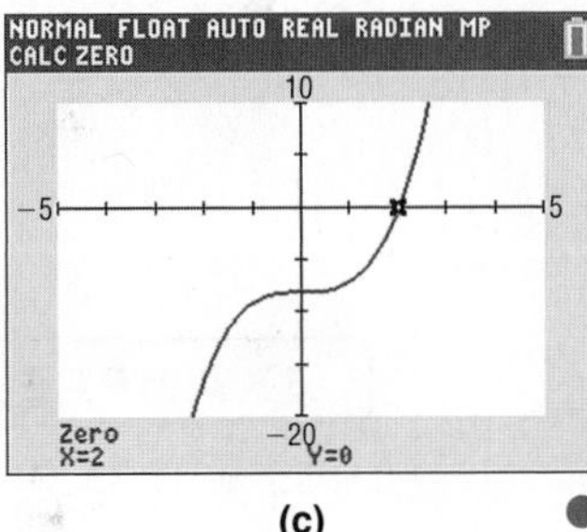

(c)

Figure 10

EXAMPLE 2

Graphing the Equation $y = \frac{1}{x}$

Graph the equation $y = \frac{1}{x}$. Based on the graph, infer information about intercepts and symmetry.

Solution Figure 11 shows the graph. Infer from the graph that there are no intercepts; also infer that symmetry with respect to the origin is a possibility. The TABLE feature on a graphing utility can provide further evidence of symmetry with respect to the origin. Using a TABLE, observe that for any ordered pair (x, y), the ordered pair $(-x, -y)$ is also a point on the graph.

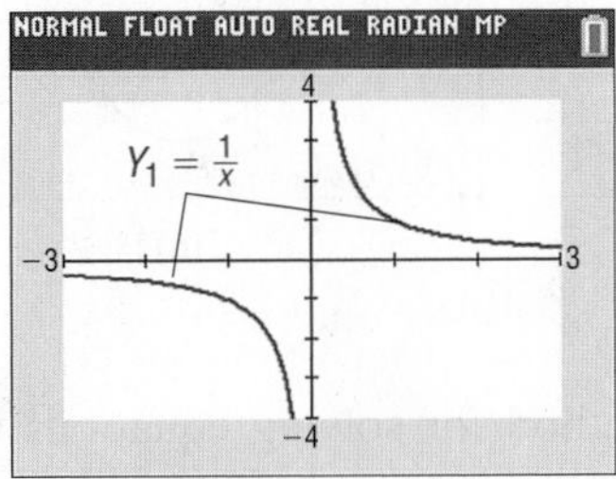

Figure 11

B.3 Exercises

In Problems 1–6, use ZERO (or ROOT) to approximate the smaller of the two x-intercepts of each equation. Express the answer rounded to two decimal places.

1. $y = x^2 + 4x + 2$
2. $y = x^2 + 4x - 3$
3. $y = 2x^2 + 4x + 1$
4. $y = 3x^2 + 5x + 1$
5. $y = 2x^2 - 3x - 1$
6. $y = 2x^2 - 4x - 1$

In Problems 7–12, use ZERO (or ROOT) to approximate the ***positive*** *x-intercepts of each equation. Express each answer rounded to two decimal places.*

7. $y = x^3 + 3.2x^2 - 16.83x - 5.31$
8. $y = x^3 + 3.2x^2 - 7.25x - 6.3$
9. $y = x^4 - 1.4x^3 - 33.71x^2 + 23.94x + 292.41$
10. $y = x^4 + 1.2x^3 - 7.46x^2 - 4.692x + 15.2881$
11. $y = x^3 + 19.5x^2 - 1021x + 1000.5$
12. $y = x^3 + 14.2x^2 - 4.8x - 12.4$

B.4 Using a Graphing Utility to Solve Equations

For many equations, there are no algebraic techniques that lead to a solution. For such equations, a graphing utility can often be used to investigate possible solutions. When a graphing utility is used to solve an equation, *approximate* solutions usually are obtained. Unless otherwise stated, this text follows the practice of giving approximate solutions *rounded to two decimal places.*

The ZERO (or ROOT) feature of a graphing utility can be used to find the solutions of an equation when one side of the equation is 0. In using this feature to solve equations, make use of the fact that the x-intercepts (or zeros) of the graph of an equation are found by letting $y = 0$ and solving the equation for x. Solving an equation for x when one side of the equation is 0 is equivalent to finding where the graph of the corresponding equation crosses or touches the x-axis.

EXAMPLE 1 **Using ZERO (or ROOT) to Approximate Solutions of an Equation**

Find the solution(s) of the equation $x^2 - 6x + 7 = 0$. Round answers to two decimal places.

Solution The solutions of the equation $x^2 - 6x + 7 = 0$ are the same as the x-intercepts of the graph of $Y_1 = x^2 - 6x + 7$. Begin by graphing the equation. See Figure 12(a).

From the graph there appear to be two x-intercepts (solutions to the equation): one between 1 and 2, the other between 4 and 5.

Using the ZERO (or ROOT) feature of the graphing utility, determine that the x-intercepts, and thus the solutions to the equation, are $x = 1.59$ and $x = 4.41$, rounded to two decimal places. See Figures 12(b) and (c).

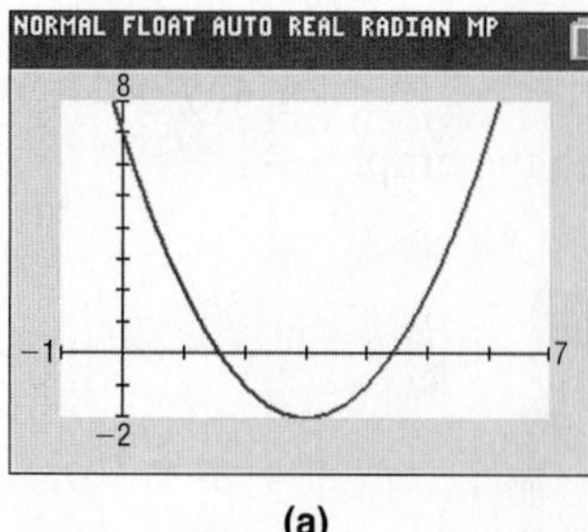

(a)

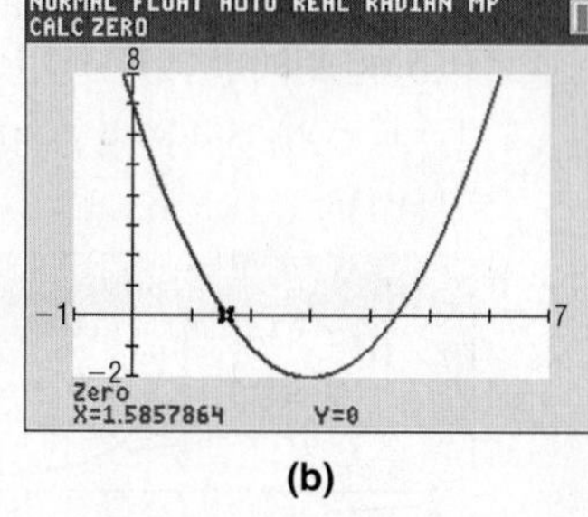

(b)

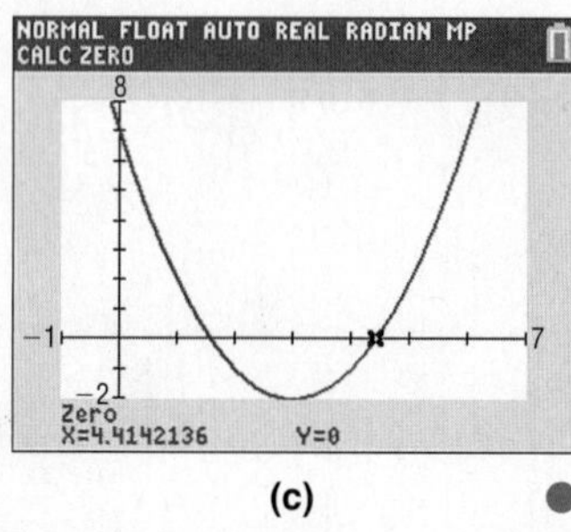

(c)

Figure 12

A second method for solving equations using a graphing utility involves the INTERSECT feature of the graphing utility. This feature is used most effectively when one side of the equation is not 0.

EXAMPLE 2

Using INTERSECT to Approximate Solutions of an Equation

Find the solution(s) of the equation $3(x-2) = 5(x-1)$.

Solution Begin by graphing each side of the equation as follows: graph $Y_1 = 3(x-2)$ and $Y_2 = 5(x-1)$. See Figure 13(a).

At the point of intersection of the graphs, the value of the y-coordinate is the same. Conclude that the x-coordinate of the point of intersection represents the solution of the equation. Do you see why? The INTERSECT feature on a graphing utility determines the point of intersection of the graphs. Using this feature, find that the graphs intersect at $(-0.5, -7.5)$. See Figure 13(b). The solution of the equation is therefore $x = -0.5$.

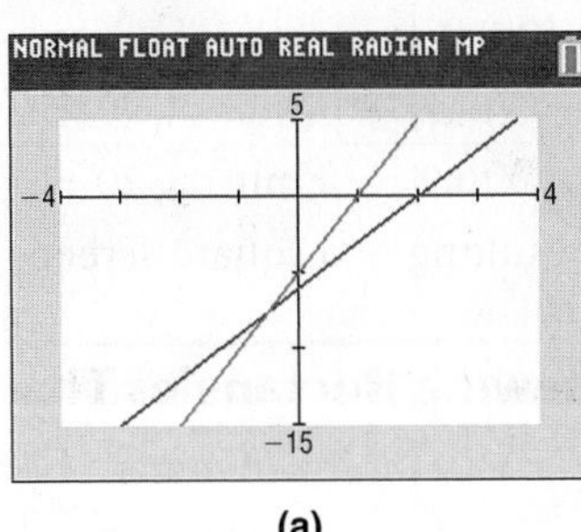

(a)

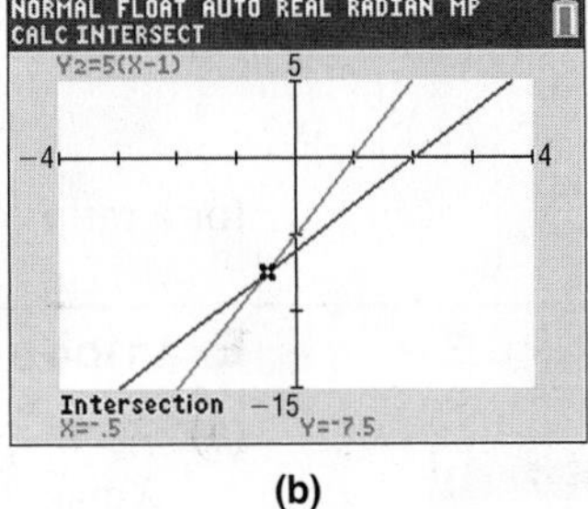

(b)

Figure 13

SUMMARY

The following steps can be used for approximating solutions of equations.

Steps for Approximating Solutions of Equations Using ZERO (or ROOT)

STEP 1: Write the equation in the form $\{expression\ in\ x\} = 0$.

STEP 2: Graph $Y_1 = \{expression\ in\ x\}$.
Be sure that the graph is complete. That is, be sure that all the intercepts are shown on the screen.

STEP 3: Use ZERO (or ROOT) to determine each x-intercept of the graph.

Steps for Approximating Solutions of Equations Using INTERSECT

STEP 1: Graph $Y_1 = \{expression\ in\ x\ on\ the\ left\ side\ of\ the\ equation\}$.
Graph $Y_2 = \{expression\ in\ x\ on\ the\ right\ side\ of\ the\ equation\}$.

STEP 2: Use INTERSECT to determine each x-coordinate of the point(s) of intersection, if any.
Be sure that the graphs are complete. That is, be sure that all the points of intersection are shown on the screen.

EXAMPLE 3

Solving a Radical Equation

Find the real solutions of the equation $\sqrt[3]{2x-4} - 2 = 0$.

Solution Figure 14 shows the graph of the equation $Y_1 = \sqrt[3]{2x-4} - 2$. From the graph, there is one x-intercept near 6. Using ZERO (or ROOT), find that the x-intercept is 6. The only solution is $x = 6$.

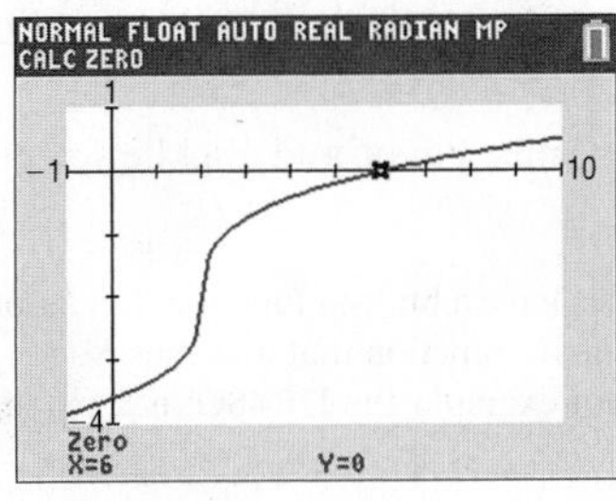

Figure 14

B.5 Square Screens

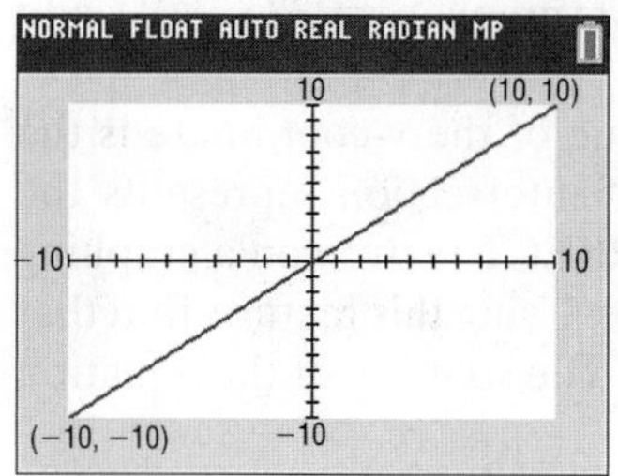

Figure 15 $y = x$

Most graphing utilities have a rectangular screen. Because of this, using the same settings for both x and y will result in a distorted view. For example, Figure 15 shows the graph of the line $y = x$ connecting the points $(-10, -10)$ and $(10, 10)$.

We expect the line to bisect the first and third quadrants, but it doesn't. The selections for Xmin, Xmax, Ymin, and Ymax must be adjusted so that a **square screen** results. On the TI-84 Plus C, this is accomplished by setting the ratio of x to y at 8:5.* For example, if

$$\text{Xmin} = -16 \qquad \text{Ymin} = -10$$
$$\text{Xmax} = 16 \qquad \text{Ymax} = 10$$

then the ratio of x to y is

$$\frac{\text{Xmax} - \text{Xmin}}{\text{Ymax} - \text{Ymin}} = \frac{16 - (-16)}{10 - (-10)} = \frac{32}{20} = \frac{8}{5}$$

for a ratio of 8:5, resulting in a square screen.

EXAMPLE 1 Examples of Viewing Rectangles That Result in Square Screens

(a)	(b)	(c)
Xmin = −8	Xmin = −16	Xmin = −24
Xmax = 8	Xmax = 16	Xmax = 24
Xscl = 1	Xscl = 1	Xscl = 3
Ymin = −5	Ymin = −10	Ymin = −15
Ymax = 5	Ymax = 10	Ymax = 15
Yscl = 1	Yscl = 1	Yscl = 3

●

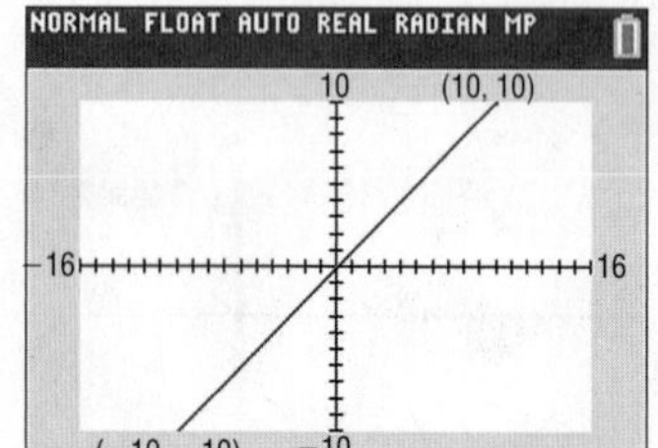

Figure 16 $y = x$

Figure 16 shows the graph of the line $y = x$ on a square screen using the viewing rectangle given in part (b). Notice that the line now bisects the first and third quadrants. Compare this illustration to Figure 15.

B.5 Exercises

In Problems 1–8, determine which of the given viewing rectangles result in a square screen.

1.	2.	3.	4.
Xmin = −8	Xmin = −5	Xmin = 0	Xmin = −16
Xmax = 8	Xmax = 5	Xmax = 16	Xmax = 16
Xscl = 2	Xscl = 1	Xscl = 4	Xscl = 8
Ymin = −5	Ymin = −4	Ymin = −2	Ymin = −10
Ymax = 5	Ymax = 4	Ymax = 8	Ymax = 10
Yscl = 2	Yscl = 1	Yscl = 2	Yscl = 5

5.	6.	7.	8.
Xmin = −6	Xmin = −8	Xmin = −3	Xmin = −10
Xmax = 6	Xmax = 8	Xmax = 5	Xmax = 14
Xscl = 1	Xscl = 4	Xscl = 1	Xscl = 2
Ymin = −2	Ymin = −5	Ymin = −2	Ymin = −7
Ymax = 2	Ymax = 5	Ymax = 3	Ymax = 8
Yscl = 0.5	Yscl = 1	Yscl = 1	Yscl = 3

9. If Xmin = −4, Xmax = 12, and Xscl = 1, how should Ymin, Ymax, and Yscl be selected so that the viewing rectangle contains the point $(4, 8)$ and the screen is square?

10. If Xmin = −6, Xmax = 10, and Xscl = 2, how should Ymin, Ymax, and Yscl be selected so that the viewing rectangle contains the point $(4, 8)$ and the screen is square?

*Some graphing utilities have a built-in function that automatically squares the screen. For example, the TI-84 Plus C has a ZSquare function that does this. Some graphing utilities require a ratio other than 8:5 to square the screen. For example, the HP 48G requires the ratio of x to y to be 2:1 for a square screen. Consult your manual.

B.6 Using a Graphing Utility to Graph a Polar Equation

Most graphing utilities require the following steps in order to obtain the graph of a polar equation. Be sure to be in POLAR mode.

Graphing a Polar Equation Using a Graphing Utility

STEP 1: Set the mode to POLAR. Solve the equation for r in terms of θ.

STEP 2: Select the viewing rectangle in polar mode. Besides setting Xmin, Xmax, Xscl, and so forth, the viewing rectangle in polar mode requires setting the minimum and maximum values for θ and an increment setting for θ (θstep). In addition, a square screen and radian measure should be used.

STEP 3: Enter the expression involving θ that you found in Step 1. (Consult your manual for the correct way to enter the expression.)

STEP 4: Graph.

EXAMPLE 1 **Graphing a Polar Equation Using a Graphing Utility**

Use a graphing utility to graph the polar equation $r \sin \theta = 2$.

Solution **STEP 1:** Solve the equation for r in terms of θ.

$$r \sin \theta = 2$$

$$r = \frac{2}{\sin \theta}$$

STEP 2: From the POLAR mode, select the viewing rectangle.

$$\theta\text{min} = 0 \qquad X\text{min} = -8 \qquad Y\text{min} = -5$$

$$\theta\text{max} = 2\pi \qquad X\text{max} = 8 \qquad Y\text{max} = 5$$

$$\theta\text{step} = \frac{\pi}{24} \qquad X\text{scl} = 1 \qquad Y\text{scl} = 1$$

θstep determines the number of points that the graphing utility will plot. For example, if θstep is $\frac{\pi}{24}$, the graphing utility will evaluate r at $\theta = 0\,(\theta\text{min})$, $\frac{\pi}{24}, \frac{2\pi}{24}, \frac{3\pi}{24}$, and so forth, up to $2\pi\,(\theta\text{max})$. The smaller θstep is, the more points the graphing utility will plot. Experiment with different values for θmin, θmax, and θstep to see how the graph is affected.

STEP 3: Enter the expression $\frac{2}{\sin \theta}$ after the prompt $r_1 =$.

STEP 4: Graph.

The graph is shown in Figure 17. ●

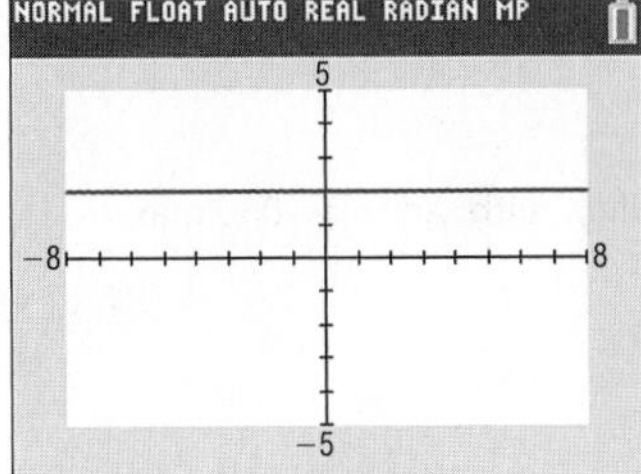

Figure 17 $r \sin \theta = 2$

B.7 Using a Graphing Utility to Graph Parametric Equations

Most graphing utilities have the capability of graphing parametric equations. The following steps are usually required to obtain the graph of parametric equations. Check your owner's manual to see how yours works.

Graphing Parametric Equations Using a Graphing Utility

STEP 1: Set the mode to PARAMETRIC. Enter $x(t)$ and $y(t)$.

STEP 2: Select the viewing window. In addition to setting *X*min, *X*max, *X*scl, and so on, the viewing window in parametric mode requires setting minimum and maximum values for the parameter *t* and an increment setting for *t* (*T*step).

STEP 3: Graph.

EXAMPLE 1 Graphing a Curve Defined by Parametric Equations Using a Graphing Utility

Graph the curve defined by the parametric equations

$$x = 3t^2 \qquad y = 2t \qquad -2 \le t \le 2$$

Solution

STEP 1: Enter the equations $x(t) = 3t^2$, $y(t) = 2t$ with the graphing utility in PARAMETRIC mode.

STEP 2: Select the viewing window. The interval is $-2 \le t \le 2$, so select the following square viewing window:

$$\begin{array}{lll} T\text{min} = -2 & X\text{min} = 0 & Y\text{min} = -5 \\ T\text{max} = 2 & X\text{max} = 16 & Y\text{max} = 5 \\ T\text{step} = 0.1 & X\text{scl} = 1 & Y\text{scl} = 1 \end{array}$$

Choose *T*min $= -2$ and *T*max $= 2$ because $-2 \le t \le 2$. Finally, the choice for *T*step will determine the number of points that the graphing utility will plot. For example, with *T*step at 0.1, the graphing utility will evaluate *x* and *y* at $t = -2, -1.9, -1.8$, and so on. The smaller the *T*step, the more points the graphing utility will plot. Experiment with different values of *T*step to see how the graph is affected.

STEP 3: Graph. Watch the direction in which the graph is drawn. This direction shows the orientation of the curve.

The graph shown in Figure 18 is complete. ●

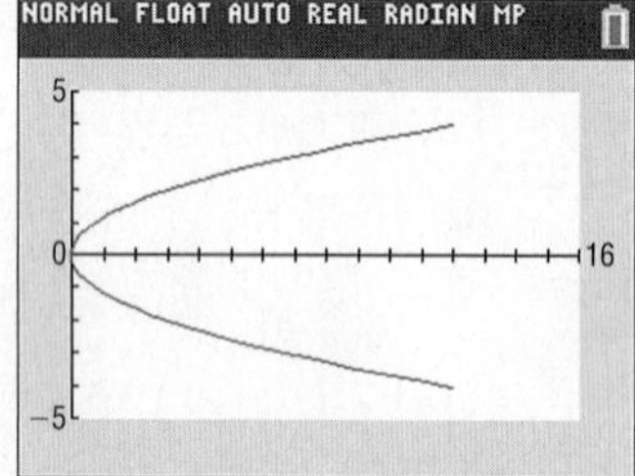

Figure 18
$x = 3t^2, y = 2t, -2 \le t \le 2$

Exploration

Graph the following parametric equations using a graphing utility with *X*min $= 0$, *X*max $= 16$, *Y*min $= -5$, *Y*max $= 5$, and *T*step $= 0.1$.

1. $x = \dfrac{3t^2}{4}, \quad y = t, \quad -4 \le t \le 4$
2. $x = 3t^2 + 12t + 12, \quad y = 2t + 4, \quad -4 \le t \le 0$
3. $x = 3t^{2/3}, \quad y = 2\sqrt[3]{t}, \quad -8 \le t \le 8$

Compare these graphs to Figure 18. Conclude that parametric equations defining a curve are not unique; that is, different parametric equations can represent the same graph.

Exploration

In FUNCTION mode, graph $x = \dfrac{3y^2}{4}\left(Y_1 = \sqrt{\dfrac{4x}{3}} \text{ and } Y_2 = -\sqrt{\dfrac{4x}{3}}\right)$ with *X*min $= 0$, *X*max $= 16$, *Y*min $= -5$, *Y*max $= 5$. Compare this graph with Figure 18. Why do the graphs differ?

Answers

CHAPTER 1 Graphs and Functions

1.1 Assess Your Understanding *(page 5)*

7. x-coordinate or abscissa; y-coordinate or ordinate **8.** quadrants **9.** midpoint **10.** F **11.** F **12.** T **13.** b **14.** a

15. (a) Quadrant II **(b)** x-axis **(c)** Quadrant III **(d)** Quadrant I **(e)** y-axis **(f)** Quadrant IV

17. The points will be on a vertical line that is 2 units to the right of the y-axis.

19. $\sqrt{5}$ **21.** $\sqrt{10}$ **23.** $2\sqrt{17}$ **25.** $\sqrt{85}$ **27.** $\sqrt{53}$ **29.** $\sqrt{a^2 + b^2}$

31. $d(A, B) = \sqrt{13}$
$d(B, C) = \sqrt{13}$
$d(A, C) = \sqrt{26}$
$(\sqrt{13})^2 + (\sqrt{13})^2 = (\sqrt{26})^2$
Area $= \dfrac{13}{2}$ square units

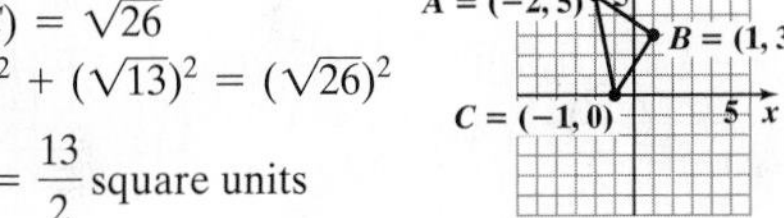

33. $d(A, B) = \sqrt{130}$
$d(B, C) = \sqrt{26}$
$d(A, C) = 2\sqrt{26}$
$(\sqrt{26})^2 + (2\sqrt{26})^2 = (\sqrt{130})^2$
Area $= 26$ square units

35. $d(A, B) = 4$
$d(B, C) = \sqrt{41}$
$d(A, C) = 5$
$4^2 + 5^2 = (\sqrt{41})^2$
Area $= 10$ square units

37. $(4, 0)$ **39.** $\left(\frac{3}{2}, 1\right)$ **41.** $(5, -1)$ **43.** $\left(\frac{a}{2}, \frac{b}{2}\right)$ **45.** $(5, 3)$ **47.** $(3, -13), (3, 11)$

49. $(4 + 3\sqrt{3}, 0); (4 - 3\sqrt{3}, 0)$ **51. (a)** $(-1, 1)$ **(b)** $(0, 13)$ **53.** $(1, 2)$ **55.** $\sqrt{17}; 2\sqrt{5}; \sqrt{29}$ **57.** $\left(\frac{s}{2}, \frac{s}{2}\right)$

59. $d(P_1, P_2) = 6$; $d(P_2, P_3) = 4$; $d(P_1, P_3) = 2\sqrt{13}$; right triangle **61.** $d(P_1, P_2) = 2\sqrt{17}$; $d(P_2, P_3) = \sqrt{34}$; $d(P_1, P_3) = \sqrt{34}$; isosceles right triangle

63. $90\sqrt{2} \approx 127.28$ ft **65. (a)** $(90, 0), (90, 90), (0, 90)$ **(b)** $5\sqrt{2161} \approx 232.43$ ft **(c)** $30\sqrt{149} \approx 366.20$ ft

67. $d = 50t$ mi **69. (a)** $(2.65, 1.6)$ **(b)** Approximately 1.285 units

71. \$21,142; a slight underestimate

1.2 Assess Your Understanding *(page 18)*

3. intercepts **4.** y-axis **5.** $(-3, 4)$ **6.** T **7.** F **8.** radius **9.** T **10.** d **11.** F **12.** a

13. $(0, 0)$ is on the graph. **15.** $(0, 3)$ is on the graph. **17.** $(0, 2)$ and $(\sqrt{2}, \sqrt{2})$ are on the graph

19. $(-2, 0), (0, 2)$

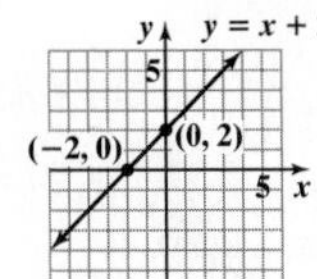

21. $(-4, 0), (0, 8)$

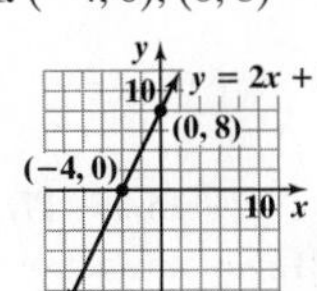

23. $(-1, 0), (1, 0), (0, -1)$

25. $(-2, 0), (2, 0), (0, 4)$

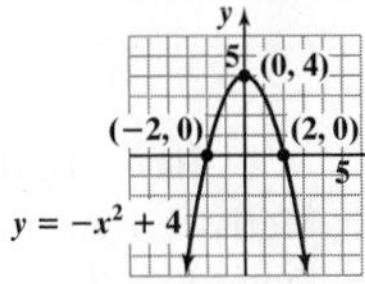

27. $(3, 0), (0, 2)$

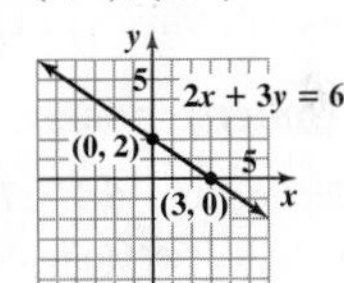

29. $(-2, 0), (2, 0), (0, 9)$

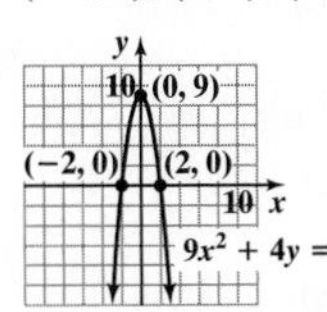

31.

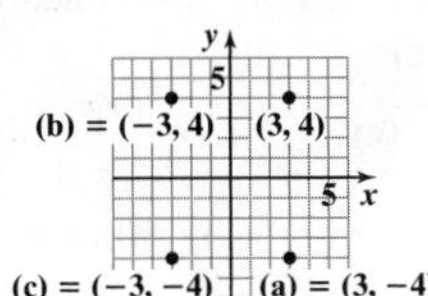

33.

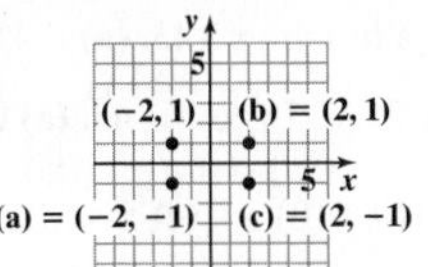

35.

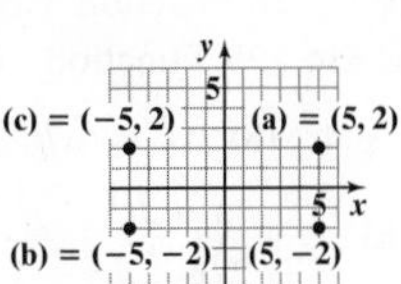

37.

39.

41. (a) $(-1, 0), (1, 0)$ **(b)** Symmetric with respect to the x-axis, the y-axis, and the origin

43. (a) $\left(-\frac{\pi}{2}, 0\right), (0, 1), \left(\frac{\pi}{2}, 0\right)$ **(b)** Symmetric with respect to the y-axis

45. (a) $(0, 0)$ **(b)** Symmetric with respect to the x-axis

47. (a) $(-2, 0), (0, 0), (2, 0)$ **(b)** Symmetric with respect to the origin

49. (a) $(x, 0), -2 \le x \le 1$ **(b)** No symmetry

51. (a) No intercepts **(b)** Symmetric with respect to the origin

53.

55.

57. $(-4, 0), (0, -2), (0, 2)$; symmetric with respect to the x-axis **59.** $(0, 0)$; symmetric with respect to the origin **61.** $(0, 9), (3, 0), (-3, 0)$; symmetric with respect to the y-axis **63.** $(-2, 0), (2, 0), (0, -3), (0, 3)$; symmetric with respect to the x-axis, y-axis, and origin **65.** $(0, -27), (3, 0)$; no symmetry **67.** $(0, -4), (4, 0), (-1, 0)$; no symmetry **69.** $(0, 0)$; symmetric with respect to the origin **71.** $(0, 0)$; symmetric with respect to the origin

73. 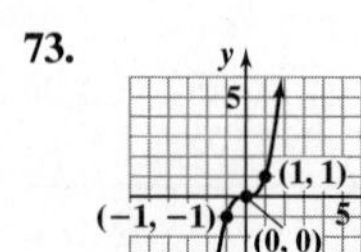

75. 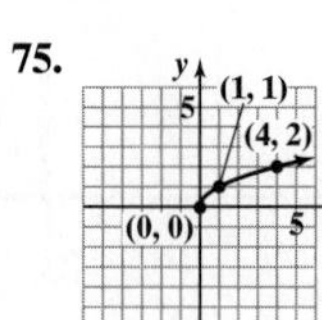

77. $a = -4$ or $a = 1$

79. Center (2, 1); radius = 2; $(x - 2)^2 + (y - 1)^2 = 4$

81. Center $\left(\frac{5}{2}, 2\right)$; radius $= \frac{3}{2}$; $\left(x - \frac{5}{2}\right)^2 + (y - 2)^2 = \frac{9}{4}$

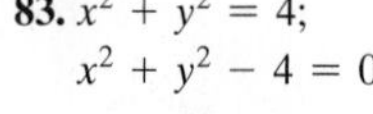

83. $x^2 + y^2 = 4$; $x^2 + y^2 - 4 = 0$

85. $x^2 + (y - 2)^2 = 4$; $x^2 + y^2 - 4y = 0$

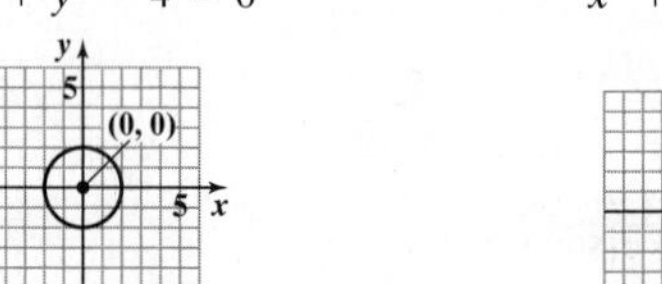

87. $(x - 4)^2 + (y + 3)^2 = 25$; $x^2 + y^2 - 8x + 6y = 0$

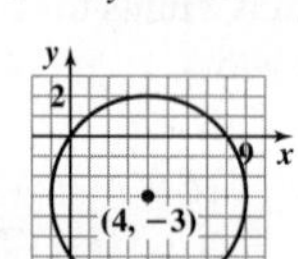

89. $(x + 2)^2 + (y - 1)^2 = 16$; $x^2 + y^2 + 4x - 2y - 11 = 0$

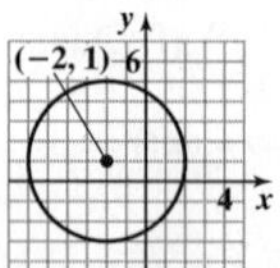

91. $\left(x - \frac{1}{2}\right)^2 + y^2 = \frac{1}{4}$; $x^2 + y^2 - x = 0$

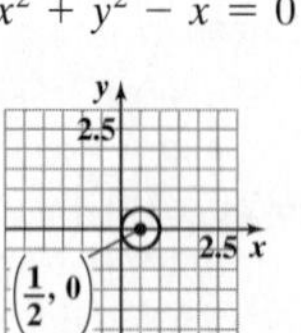

93. **(a)** $(h, k) = (0, 0)$; $r = 2$ **(b)**

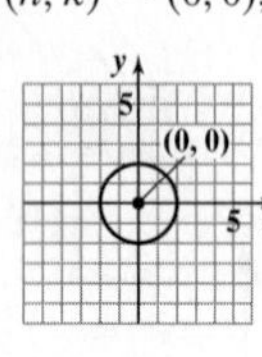

(c) $(\pm 2, 0)$; $(0, \pm 2)$

95. **(a)** $(h, k) = (3, 0)$; $r = 2$ **(b)**

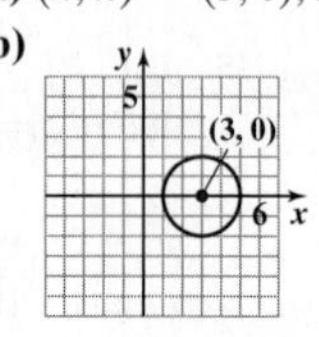

(c) (1, 0); (5, 0)

97. **(a)** $(h, k) = (1, 2)$; $r = 3$ **(b)**

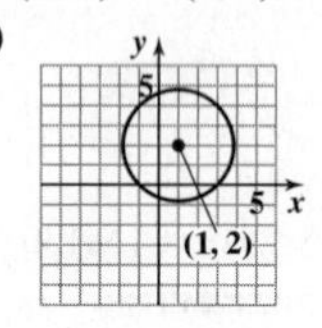

(c) $(1 \pm \sqrt{5}, 0)$; $(0, 2 \pm 2\sqrt{2})$

99. **(a)** $(h, k) = (-2, 2)$; $r = 3$ **(b)**

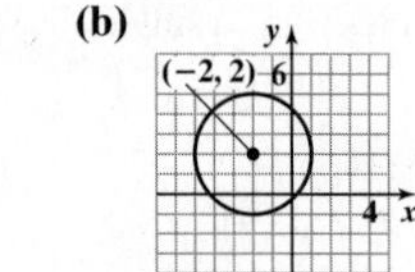

(c) $(-2 \pm \sqrt{5}, 0)$; $(0, 2 \pm \sqrt{5})$

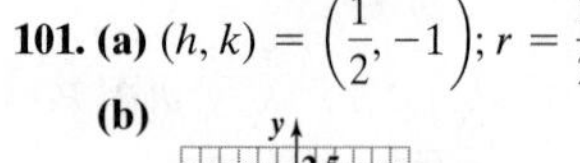

101. **(a)** $(h, k) = \left(\frac{1}{2}, -1\right)$; $r = \frac{1}{2}$ **(b)**

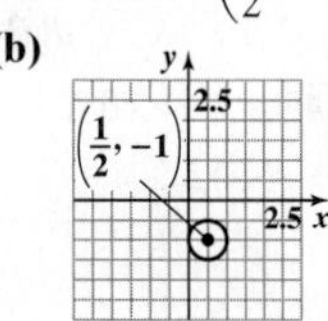

(c) $(0, -1)$

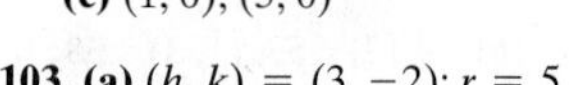

103. **(a)** $(h, k) = (3, -2)$; $r = 5$ **(b)**

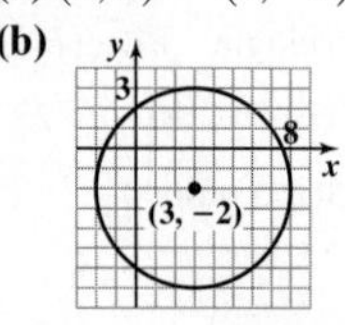

(c) $(3 \pm \sqrt{21}, 0)$; $(0, -6)$, $(0, 2)$

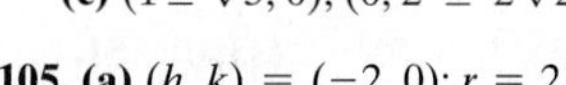

105. **(a)** $(h, k) = (-2, 0)$; $r = 2$ **(b)**

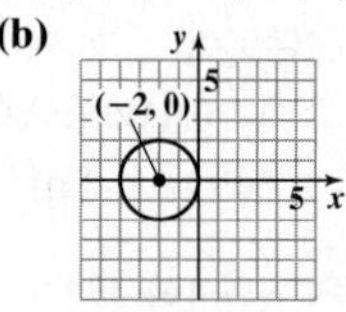

(c) $(0, 0)$, $(-4, 0)$

107. $x^2 + y^2 = 13$ **109.** $(x - 2)^2 + (y - 3)^2 = 9$ **111.** $(x + 1)^2 + (y - 3)^2 = 5$ **113.** $(x + 1)^2 + (y - 3)^2 = 1$
115. $(-1, -2)$ **117.** 4 **119.** **(a)** (0, 0), (2, 0), (0, 1), (0, −1) **(b)** x-axis symmetry **121.** 18 units^2 **123.** $x^2 + (y - 139)^2 = 15{,}625$
125. (b), (c), (e), (g) **127.** **(a)** $y = \sqrt{x^2}$ and $y = |x|$ have the same graph. **(b)** $\sqrt{x^2} = |x|$
(c) $x \ge 0$ for $y = (\sqrt{x})^2$, while x can be any real number for $y = x$. **(d)** $y \ge 0$ for $y = \sqrt{x^2}$

1.3 Assess Your Understanding *(page 35)*

5. independent; dependent **6.** F **7.** F **8.** F **9.** a **10.** c **11.** d **12.** a **13.** vertical **14.** 5; −3 **15.** $a = -2$ **16.** F **17.** T **18.** a
19. Function; Domain: {Elvis, Colleen, Kaleigh, Marissa}; Range: {January 8, March 15, September 17} **21.** Not a function; Domain: {20, 30, 40}; Range: {200, 300, 350, 425} **23.** Not a function; Domain: {−3, 2, 4}; Range: {6, 9, 10} **25.** Function; Domain: {1, 2, 3, 4}; Range: {3} **27.** Not a function; Domain: {−2, 0, 3}; Range: {3, 4, 6, 7} **29.** Function; Domain: {−2, −1, 0, 1}; Range: {0, 1, 4} **31.** Function **33.** Function
35. Not a function **37.** Not a function **39.** Function **41.** Not a function **43.** **(a)** −4 **(b)** 1 **(c)** −3 **(d)** $3x^2 - 2x - 4$ **(e)** $-3x^2 - 2x + 4$
(f) $3x^2 + 8x + 1$ **(g)** $12x^2 + 4x - 4$ **(h)** $3x^2 + 6xh + 3h^2 + 2x + 2h - 4$ **45.** **(a)** 0 **(b)** $\frac{1}{2}$ **(c)** $-\frac{1}{2}$ **(d)** $\frac{-x}{x^2 + 1}$ **(e)** $\frac{-x}{x^2 + 1}$
(f) $\frac{x + 1}{x^2 + 2x + 2}$ **(g)** $\frac{2x}{4x^2 + 1}$ **(h)** $\frac{x + h}{x^2 + 2xh + h^2 + 1}$ **47.** **(a)** 4 **(b)** 5 **(c)** 5 **(d)** $|x| + 4$ **(e)** $-|x| - 4$ **(f)** $|x + 1| + 4$ **(g)** $2|x| + 4$
(h) $|x + h| + 4$ **49.** **(a)** $-\frac{1}{5}$ **(b)** $-\frac{3}{2}$ **(c)** $\frac{1}{8}$ **(d)** $\frac{2x - 1}{3x + 5}$ **(e)** $\frac{-2x - 1}{3x - 5}$ **(f)** $\frac{2x + 3}{3x - 2}$ **(g)** $\frac{4x + 1}{6x - 5}$ **(h)** $\frac{2x + 2h + 1}{3x + 3h - 5}$ **51.** All real numbers
53. All real numbers **55.** $\{x \mid x \ne -4, x \ne 4\}$ **57.** $\{x \mid x \ne 0\}$ **59.** $\{x \mid x \ge 4\}$ **61.** $\{x \mid x > 1\}$ **63.** $\{x \mid x > 4\}$ **65.** $\{t \mid t \ge 4, t \ne 7\}$
67. 4 **69.** $2x + h$ **71.** $2x + h - 1$ **73.** $\frac{-(2x + h)}{x^2(x + h)^2}$ **75.** $\frac{6}{(x + 3)(x + h + 3)}$ **77.** $\frac{1}{\sqrt{x + h - 2} + \sqrt{x - 2}}$
79. **(a)** $f(0) = 3$; $f(-6) = -3$ **(b)** $f(6) = 0$; $f(11) = 1$ **(c)** Positive **(d)** Negative
(e) −3, 6, and 10 **(f)** $-3 < x < 6$; $10 < x \le 11$ **(g)** $\{x \mid -6 \le x \le 11\}$ **(h)** $\{y \mid -3 \le y \le 4\}$ **(i)** −3, 6, 10 **(j)** 3 **(k)** 3 times **(l)** Once
(m) 0, 4 **(n)** −5, 8 **81.** Not a function **83.** Function **(a)** Domain: $\{x \mid -\pi \le x \le \pi\}$; Range: $\{y \mid -1 \le y \le 1\}$ **(b)** $\left(-\frac{\pi}{2}, 0\right)$, $\left(\frac{\pi}{2}, 0\right)$, $(0, 1)$
(c) y-axis **85.** Not a function **87.** Function **(a)** Domain: $\{x \mid 0 < x < 3\}$; Range: $\{y \mid y < 2\}$ **(b)** (1, 0) **(c)** None **89.** Function
(a) Domain: all real numbers; Range: $\{y \mid y \le 2\}$ **(b)** $(-3, 0)$, $(3, 0)$, $(0, 2)$ **(c)** y-axis **91.** Function **(a)** Domain: all real numbers; Range: $\{y \mid y \ge -3\}$ **(b)** (1, 0), (3, 0), (0, 9) **(c)** None **93.** **(a)** Yes **(b)** $f(-2) = 9$; $(-2, 9)$ **(c)** $0, \frac{1}{2}$; $(0, -1)$, $\left(\frac{1}{2}, -1\right)$ **(d)** All real numbers
(e) $-\frac{1}{2}, 1$ **(f)** −1 **95.** **(a)** No **(b)** $f(4) = -3$; $(4, -3)$ **(c)** 14; (14, 2) **(d)** $\{x \mid x \ne 6\}$ **(e)** −2 **(f)** $-\frac{1}{3}$ **97.** **(a)** Yes **(b)** $f(2) = \frac{8}{17}$; $\left(2, \frac{8}{17}\right)$

(c) $-1, 1; (-1, 1), (1, 1)$ **(d)** All real numbers **(e)** 0 **(f)** 0 **99.** $A = -\frac{7}{2}$ **101.** $A = -4$ **103.** $A(x) = \frac{1}{2}x^2$ **105.** $G(x) = 14x$ **107. (a)** P is the dependent variable; a is the independent variable. **(b)** $P(20) = 231.427$ million; In 2012, there were 231.427 million people 20 years of age or older. **(c)** $P(0) = 327.287$ million; In 2012, there were 327.287 million people. **109. (a)** 15.1 m, 14.071 m, 12.944 m, 11.719 m **(b)** 1.01 sec, 1.43 sec, 1.75 sec **(c)** 2.02 sec **111. (a)** \$222 **(b)** \$225 **(c)** \$220 **(d)** \$230

113. (a) Approximately 10.4 ft high
(b) Approximately 9.9 ft high
(c)

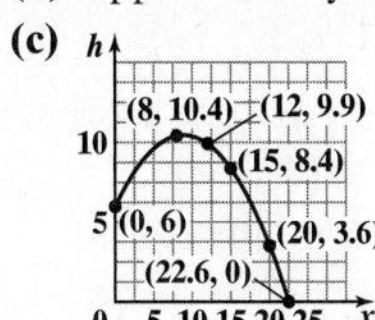

(d) The ball will not go through the hoop; $h(15) \approx 8.4$ ft. If $v = 30$ ft/sec, $h(15) = 10$ ft.

115. (a) About 81.07 ft **(b)** About 129.59 ft **(c)** About 26.63 ft **(d)** About 528.13 ft
(e)

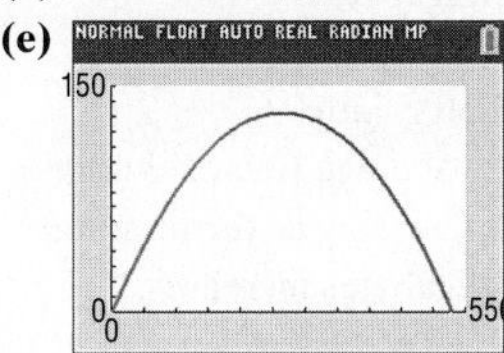

(f) About 115.07 ft and 413.05 ft **(g)** 275 ft; maximum height shown in the table is 131.8 ft **(h)** 264 ft

117. (a) \$30; It costs \$30 if you use 0 gigabytes. **(b)** \$30; It costs \$30 if you use 5 gigabytes. **(c)** \$90; It costs \$90 if you use 15 gigabytes. **(d)** $\{g \mid 0 \le g \le 60\}$. There are at most 60 gigabytes in a month.
121. Yes; $f(x) = 0$, domain $x = a$ **123. (a)** II **(b)** V **(c)** IV **(d)** III **(e)** I
125.

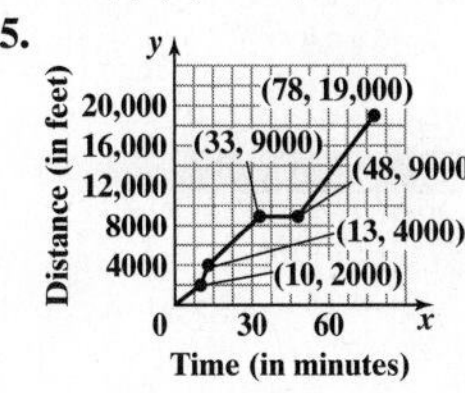

127. (a) (7, 7.4) **(b)** (4.2, 6) **(c)** Increasing from 0 to 30 mi/h **(d)** 0 mi/h **(e)** 50 mi/h **(f)** (2, 4), (4.2, 6), (7, 7.4), (7.6, 8)
129. Yes. $f(x) = 0$ satisfies the condition that $f(x) = -f(x)$.

1.4 Assess Your Understanding *(page 50)*

6. increasing **7.** even; odd **8.** T **9.** T **10.** F **11.** c **12.** d **13.** Yes **15.** No **17.** $(-8, -2)$; $(0, 2)$; $(5, 7)$ **19.** Yes; 10 **21.** $-2, 2; 6, 10$ **23.** $f(-8) = -4$ **25. (a)** $(-2, 0), (0, 3), (2, 0)$ **(b)** Domain: $\{x \mid -4 \le x \le 4\}$ or $[-4, 4]$; Range: $\{y \mid 0 \le y \le 3\}$ or $[0, 3]$ **(c)** Increasing on $(-2, 0)$ and $(2, 4)$; Decreasing on $(-4, -2)$ and $(0, 2)$ **(d)** Even **27. (a)** $(0, 1)$ **(b)** Domain: all real numbers; Range: $\{y \mid y > 0\}$ or $(0, \infty)$ **(c)** Increasing on $(-\infty, \infty)$ **(d)** Neither **29. (a)** $(-\pi, 0), (0, 0), (\pi, 0)$ **(b)** Domain: $\{x \mid -\pi \le x \le \pi\}$ or $[-\pi, \pi]$; Range: $\{y \mid -1 \le y \le 1\}$ or $[-1, 1]$ **(c)** Increasing on $\left(-\frac{\pi}{2}, \frac{\pi}{2}\right)$; Decreasing on $\left(-\pi, -\frac{\pi}{2}\right)$ and $\left(\frac{\pi}{2}, \pi\right)$ **(d)** Odd **31. (a)** $\left(0, \frac{1}{2}\right), \left(\frac{1}{3}, 0\right), \left(\frac{5}{2}, 0\right)$ **(b)** Domain: $\{x \mid -3 \le x \le 3\}$ or $[-3, 3]$; Range: $\{y \mid -1 \le y \le 2\}$ or $[-1, 2]$ **(c)** Increasing on (2, 3); Decreasing on $(-1, 1)$; Constant on $(-3, -1)$ and (1, 2) **(d)** Neither **33. (a)** 0; 3 **(b)** $-2, 2; 0, 0$ **35. (a)** $\frac{\pi}{2}$; 1 **(b)** $-\frac{\pi}{2}$; -1 **37.** Odd **39.** Even **41.** Odd **43.** Neither **45.** Even **47.** Odd **49.** Absolute maximum: $f(1) = 4$; absolute minimum: $f(5) = 1$; local maximum: $f(3) = 3$; local minimum: $f(2) = 2$ **51.** Absolute maximum: $f(3) = 4$; absolute minimum: $f(1) = 1$; local maximum: $f(3) = 4$; local minimum: $f(1) = 1$ **53.** Absolute maximum: none; absolute minimum: $f(0) = 0$: local maximum: $f(2) = 3$; local minimum: $f(0) = 0$ and $f(3) = 2$ **55.** Absolute maximum: none; absolute minimum: none: local maximum: none; local minimum: none

57.

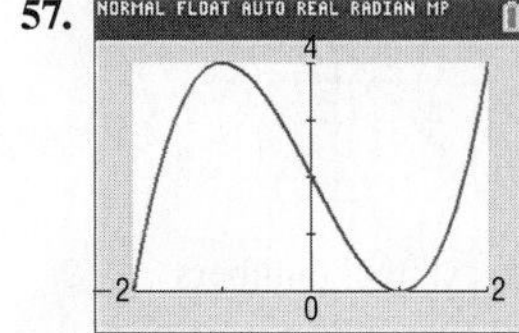

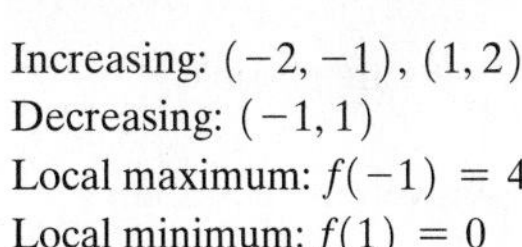

Increasing: $(-2, -1), (1, 2)$
Decreasing: $(-1, 1)$
Local maximum: $f(-1) = 4$
Local minimum: $f(1) = 0$

59.

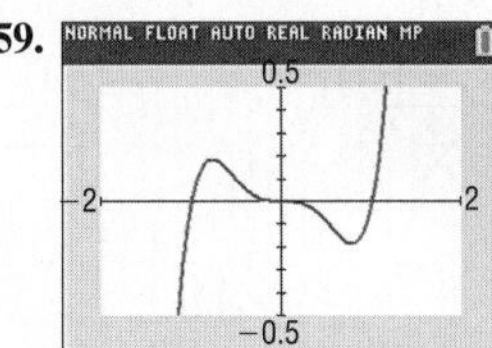

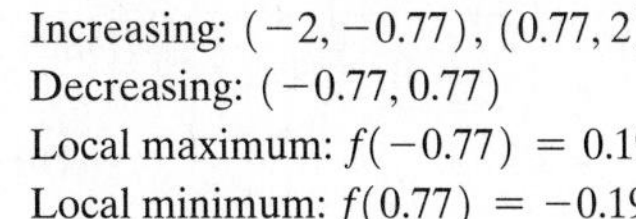

Increasing: $(-2, -0.77), (0.77, 2)$
Decreasing: $(-0.77, 0.77)$
Local maximum: $f(-0.77) = 0.19$
Local minimum: $f(0.77) = -0.19$

61.

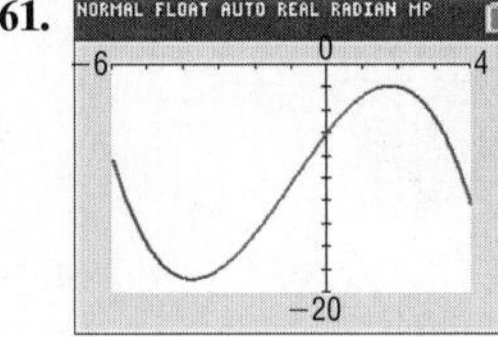

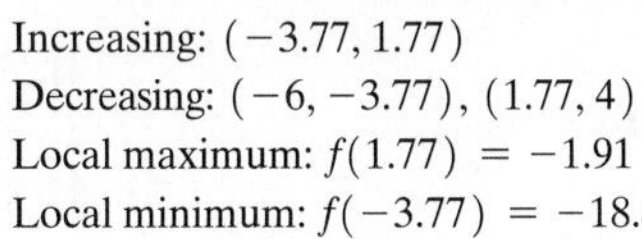

Increasing: $(-3.77, 1.77)$
Decreasing: $(-6, -3.77), (1.77, 4)$
Local maximum: $f(1.77) = -1.91$
Local minimum: $f(-3.77) = -18.89$

63.

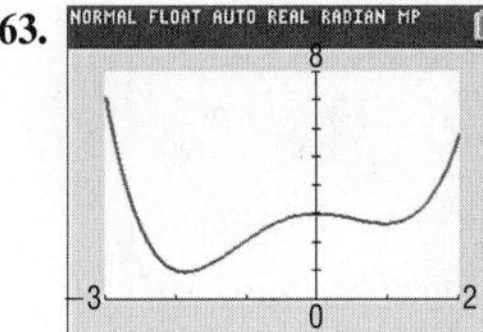

Increasing: $(-1.87, 0), (0.97, 2)$
Decreasing: $(-3, -1.87), (0, 0.97)$
Local maximum: $f(0) = 3$
Local minima: $f(-1.87) = 0.95$, $f(0.97) = 2.65$

65. (a) -4 **(b)** -8 **(c)** -10 **67. (a)** 17 **(b)** -1 **(c)** 11 **69. (a)** 5 **(b)** $y = 5x - 2$ **71. (a)** -1 **(b)** $y = -x$ **73. (a)** 4 **(b)** $y = 4x - 8$ **75. (a)** Odd **(b)** Local maximum value: 54 at $x = -3$ **77. (a)** Even **(b)** Local maximum value: 24 at $x = -2$ **(c)** 47.4 sq. units

79. (a)

(b) 10 riding lawn mowers
(c) \$239/mower

81. (a), (b)

(c) \$5/gigabyte
(d) \$6.25/gigabyte
(e) \$7.50/gigabyte
(f) The average rate of change is increasing as the number of gigabytes increases.

83. (a) On average, the population is increasing at a rate of 0.036 g/h from 0 to 2.5 h. **(b)** On average, from 4.5 to 6 h, the population is increasing at a rate of 0.1 g/h. **(c)** The average rate of change is increasing over time.

85. (a) 1 **(b)** 0.5 **(c)** 0.1 **(d)** 0.01
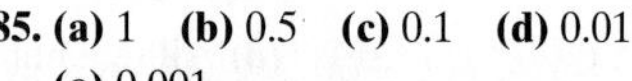
(e) 0.001
(f)
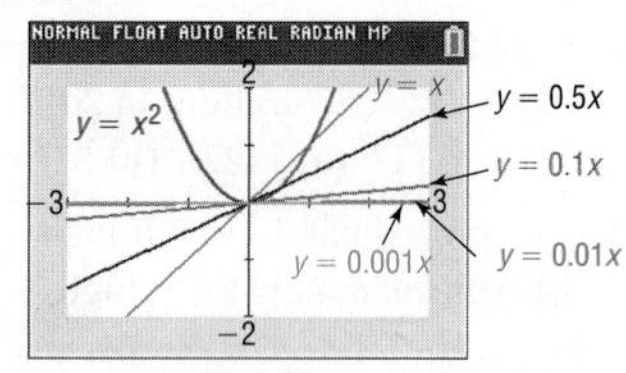

(g) They are getting closer to the tangent line at (0, 0).
(h) They are getting closer to 0.

87. (a) 2
(b) 2; 2; 2; 2
(c) $y = 2x + 5$
(d)
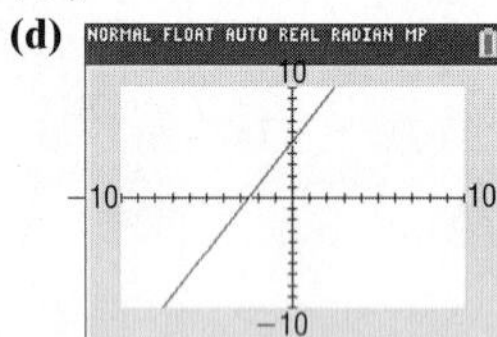

89. (a) $2x + h + 2$
(b) 4.5; 4.1; 4.01; 4
(c) $y = 4.01x - 1.01$
(d)
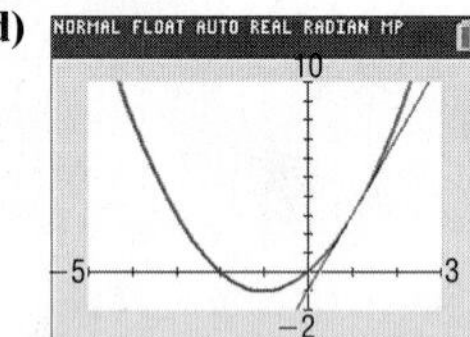

91. (a) $4x + 2h - 3$
(b) 2; 1.2; 1.02; 1
(c) $y = 1.02x - 1.02$
(d)
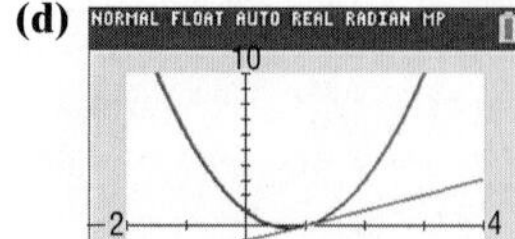

93. (a) $-\frac{1}{(x+h)x}$
(b) $-\frac{2}{3}; -\frac{10}{11}; -\frac{100}{101}; -1$
(c) $y = -\frac{100}{101}x + \frac{201}{101}$
(d)
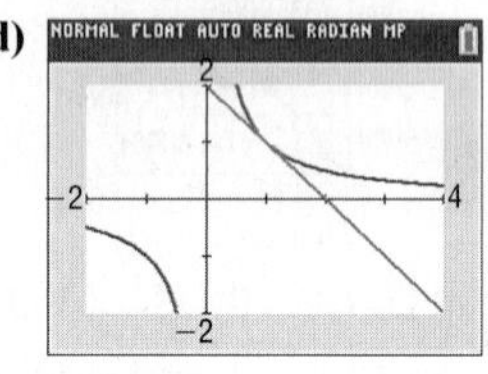

97. At most one **99.** Yes; the function $f(x) = 0$ is both even and odd. **101.** Not necessarily. It just means $f(5) > f(2)$.

1.5 Assess Your Understanding *(page 61)*

4. $(-\infty, 0)$ **5.** piecewise-defined **6.** T **7.** F **8.** F **9.** b **10.** a **11.** C **13.** E **15.** B **17.** F

19.
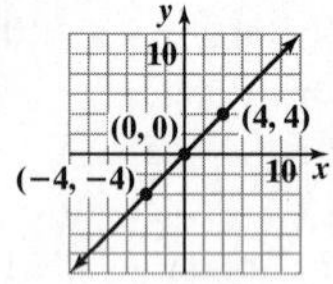

21.
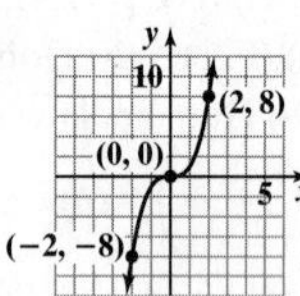

23.
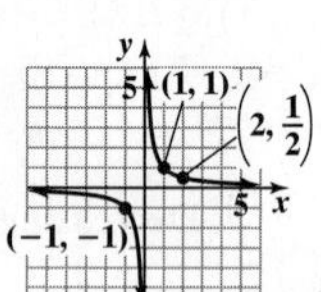

25.
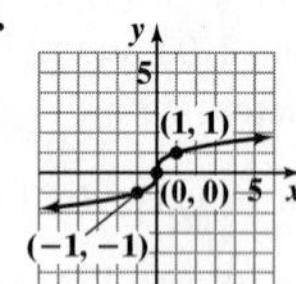

27. (a) 4 **(b)** 2 **(c)** 5 **29. (a)** −4 **(b)** −2 **(c)** 0 **(d)** 25

31. (a) All real numbers
(b) (0, 1)
(c)
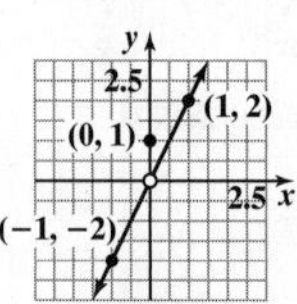

(d) $\{y \mid y \neq 0\}$; $(-\infty, 0) \cup (0, \infty)$
(e) Discontinuous at $x = 0$

33. (a) All real numbers
(b) (0, 3)
(c)
(d) $\{y \mid y \geq 1\}$; $[1, \infty)$
(e) Continuous

35. (a) $\{x \mid x \geq -2\}$; $[-2, \infty)$
(b) (0, 3), (2, 0)
(c)
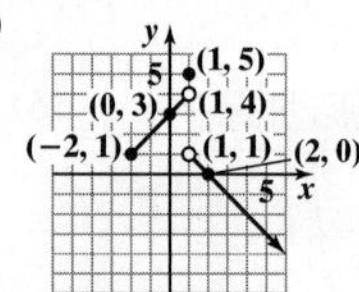

(d) $\{y \mid y < 4, y = 5\}$; $(-\infty, 4) \cup \{5\}$
(e) Discontinuous at $x = 1$

37. (a) All real numbers
(b) $(-1, 0)$, $(0, 0)$
(c)
(d) All real numbers
(e) Discontinuous at $x = 0$

39. (a) $\{x \mid x \geq -2, x \neq 0\}$; $[-2, 0) \cup (0, \infty)$
(b) No intercepts
(c)
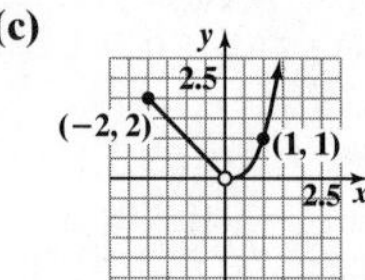

(d) $\{y \mid y > 0\}$; $(0, \infty)$
(e) Discontinuous at $x = 0$

41. $f(x) = \begin{cases} -x & \text{if } -1 \leq x \leq 0 \\ \frac{1}{2}x & \text{if } 0 < x \leq 2 \end{cases}$ (Other answers are possible.)

43. $f(x) = \begin{cases} -x & \text{if } x \leq 0 \\ -x + 2 & \text{if } 0 < x \leq 2 \end{cases}$ (Other answers are possible.)

45. (a) \$34.99 **(b)** \$64.99 **(c)** \$184.99

47. (a) \$44.46 **(b)** \$126.03
(c) $C(x) = \begin{cases} 1.24816x + 19.50 & \text{if } 0 \leq x \leq 30 \\ 0.5757x + 39.6738 & \text{if } x > 30 \end{cases}$
(d)
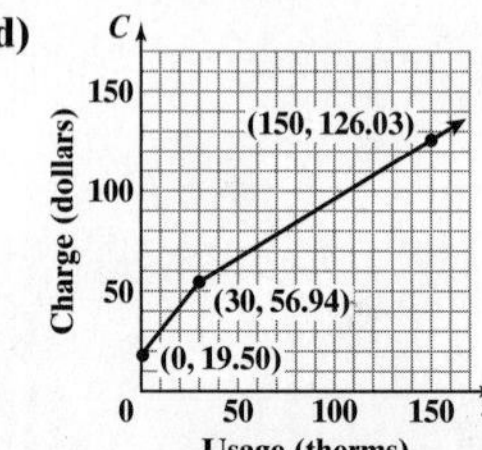

49. $f(x) = \begin{cases} 0.10x & \text{if } 0 < x \le 9075 \\ 907.50 + 0.15(x - 9075) & \text{if } 9075 < x \le 36{,}900 \\ 5081.25 + 0.25(x - 36{,}900) & \text{if } 36{,}900 < x \le 89{,}350 \\ 18{,}193.75 + 0.28(x - 89{,}350) & \text{if } 89{,}350 < x \le 186{,}350 \\ 45{,}353.75 + 0.33(x - 186{,}350) & \text{if } 186{,}350 < x \le 405{,}100 \\ 117{,}541.25 + 0.35(x - 405{,}100) & \text{if } 405{,}100 < x \le 406{,}750 \\ 118{,}188.75 + 0.396(x - 406{,}750) & \text{if } x > 406{,}750 \end{cases}$

51. (a)

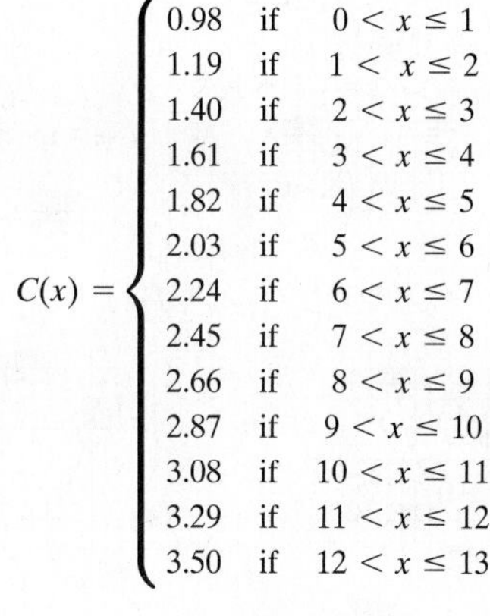

(b) $C(x) = 10 + 0.4x$ **(c)** $C(x) = 70 + 0.25x$

53. (a) $C(s) = \begin{cases} 9000 & \text{if } s \le 659 \\ 7500 & \text{if } 660 \le s \le 679 \\ 5250 & \text{if } 680 \le s \le 699 \\ 3000 & \text{if } 700 \le s \le 719 \\ 1500 & \text{if } 720 \le s \le 739 \\ 750 & \text{if } s \ge 740 \end{cases}$ **(b)** \$1500 **(c)** \$7500

55. (a) 10°C **(b)** 4°C **(c)** −3°C **(d)** −4°C
(e) The wind chill is equal to the air temperature.
(f) At wind speed greater than 20 m/s, the wind chill factor depends only on the air temperature.

57. $C(x) = \begin{cases} 0.98 & \text{if } 0 < x \le 1 \\ 1.19 & \text{if } 1 < x \le 2 \\ 1.40 & \text{if } 2 < x \le 3 \\ 1.61 & \text{if } 3 < x \le 4 \\ 1.82 & \text{if } 4 < x \le 5 \\ 2.03 & \text{if } 5 < x \le 6 \\ 2.24 & \text{if } 6 < x \le 7 \\ 2.45 & \text{if } 7 < x \le 8 \\ 2.66 & \text{if } 8 < x \le 9 \\ 2.87 & \text{if } 9 < x \le 10 \\ 3.08 & \text{if } 10 < x \le 11 \\ 3.29 & \text{if } 11 < x \le 12 \\ 3.50 & \text{if } 12 < x \le 13 \end{cases}$

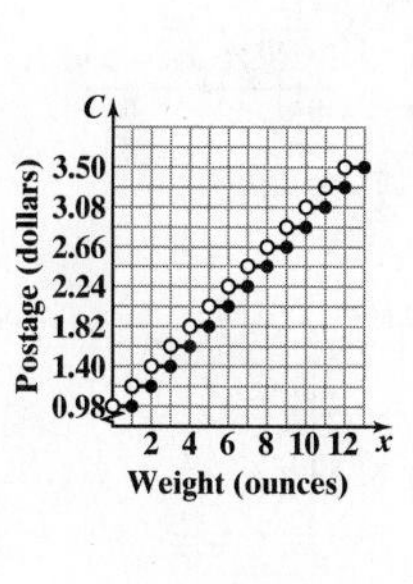

59. Each graph is that of $y = x^2$, but shifted horizontally. If $y = (x - k)^2, k > 0$, the shift is right k units; if $y = (x + k)^2, k > 0$, the shift is left k units. **61.** The graph of $y = -f(x)$ is the reflection about the x-axis of the graph of $y = f(x)$. **63.** Yes. The graph of $y = (x - 1)^3 + 2$ is the graph of $y = x^3$ shifted right 1 unit and up 2 units. **65.** They all have the same general shape. All three go through the points $(-1, -1)$, $(0, 0)$, and $(1, 1)$. As the exponent increases, the steepness of the curve increases (except near $x = 0$).

1.6 Assess Your Understanding *(page 74)*

1. horizontal; right **2.** y **3.** F **4.** T **5.** d **6.** a **7.** F **8.** b **9.** B **11.** H **13.** I **15.** L **17.** F **19.** G **21.** $y = (x - 4)^3$ **23** $y = x^3 + 4$ **25.** $y = -x^3$ **27.** $y = 4x^3$ **29.** $y = -(\sqrt{-x} + 2)$ 31. $y = -\sqrt{x + 3} + 2$ **33.** c **35.** c **37. (a)** −7 and 1 **(b)** −3 and 5 **(c)** −5 and 3 **(d)** −3 and 5 **39. (a)** $(-3, 3)$ **(b)** $(4, 10)$ **(c)** Decreasing on $(-1, 5)$ **(d)** Decreasing on $(-5, 1)$

41.

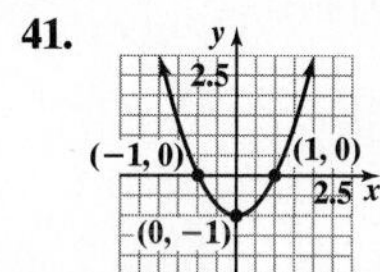

Domain: $(-\infty, \infty)$; Range: $[-1, \infty)$

43.

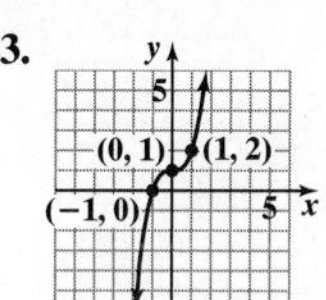

Domain: $(-\infty, \infty)$; Range: $(-\infty, \infty)$

45.

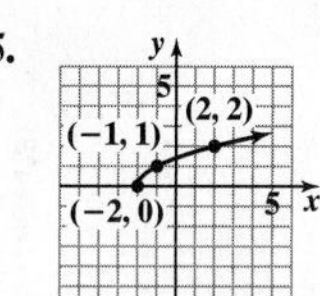

Domain: $[-2, \infty)$; Range: $[0, \infty)$

47.

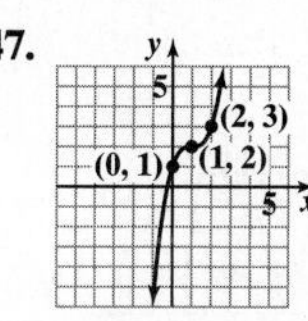

Domain: $(-\infty, \infty)$; Range: $(-\infty, \infty)$

49.

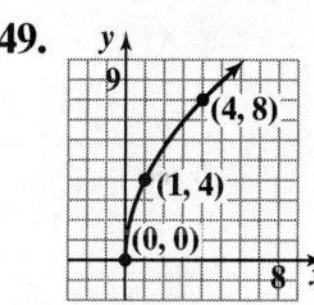

Domain: $[0, \infty)$; Range: $[0, \infty)$

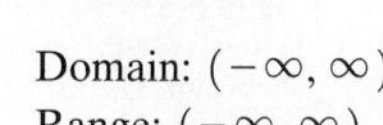

51.

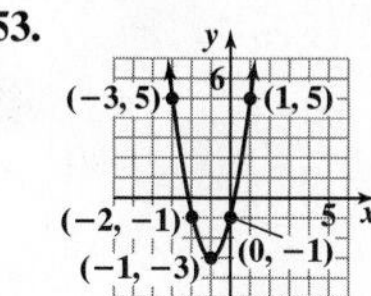

Domain: $(-\infty, \infty)$; Range: $(-\infty, \infty)$

53.

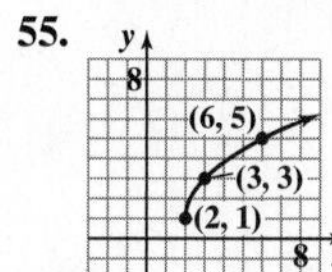

Domain: $(-\infty, \infty)$; Range: $[-3, \infty)$

55.

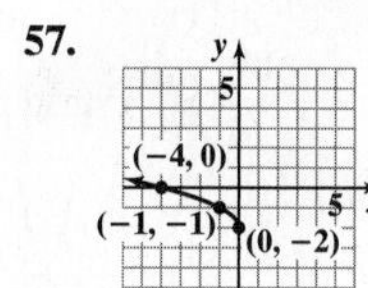

Domain: $[2, \infty)$; Range: $[1, \infty)$

57.

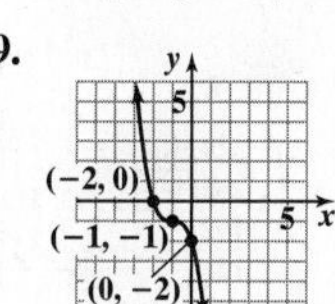

Domain: $(-\infty, 0]$; Range: $[-2, \infty)$

59.

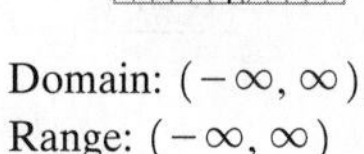

Domain: $(-\infty, \infty)$; Range: $(-\infty, \infty)$

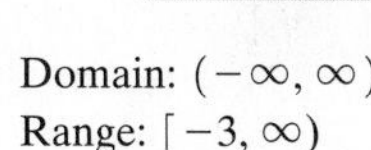
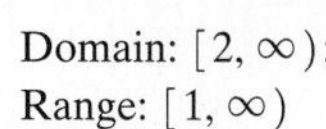
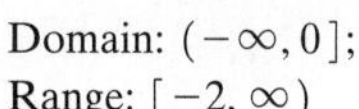

61.

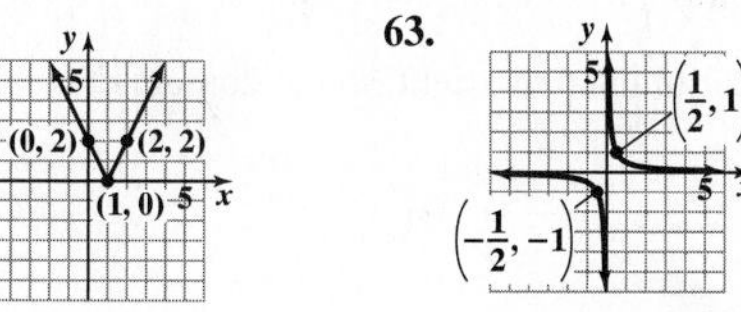

Domain: $(-\infty, \infty)$; Range: $[0, \infty)$

63.

Domain: $(-\infty, 0) \cup (0, \infty)$; Range: $(-\infty, 0) \cup (0, \infty)$

65. (a) $F(x) = f(x) + 3$

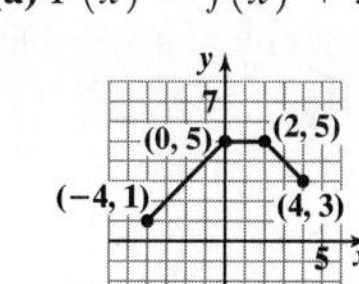

(b) $G(x) = f(x + 2)$

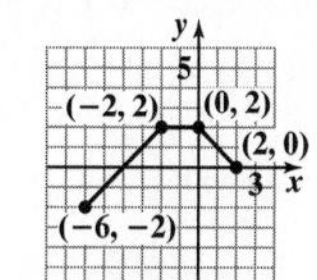

(c) $P(x) = -f(x)$

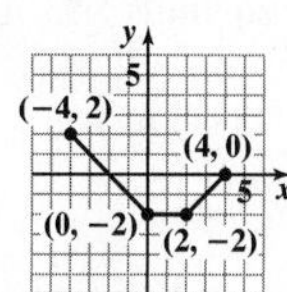

(d) $H(x) = f(x + 1) - 2$

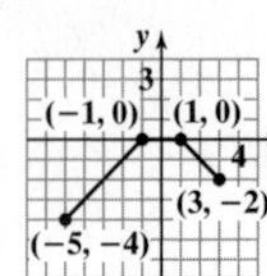

(e) $Q(x) = \frac{1}{2}f(x)$

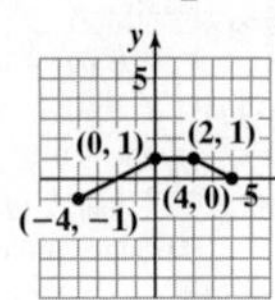

(f) $g(x) = f(-x)$ **(g)** $h(x) = f(2x)$ **67. (a)** $F(x) = f(x) + 3$

(b) $G(x) = f(x + 2)$

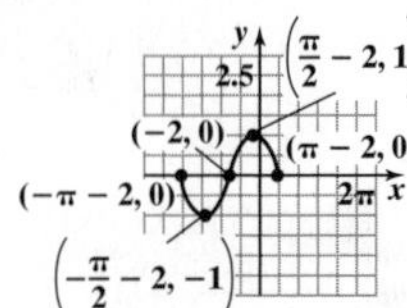

(c) $P(x) = -f(x)$ **(d)** $H(x) = f(x + 1) - 2$

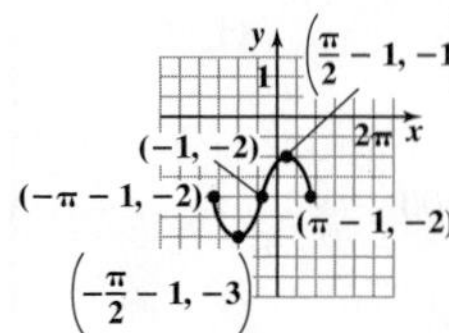

(e) $Q(x) = \frac{1}{2}f(x)$

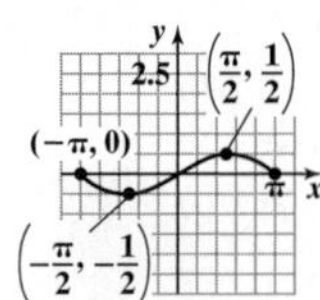

(f) $g(x) = f(-x)$

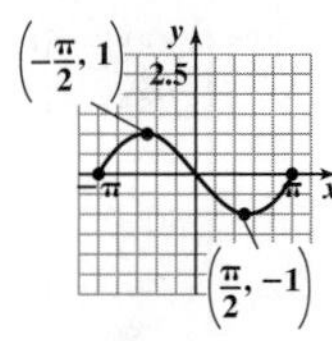

(g) $h(x) = f(2x)$

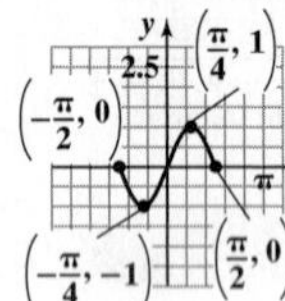

69. $f(x) = (x + 1)^2 - 1$

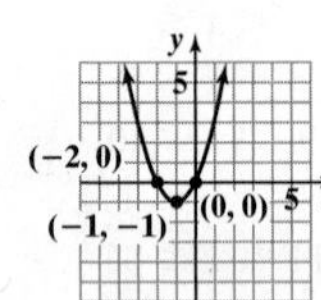

71. $f(x) = (x - 4)^2 - 15$

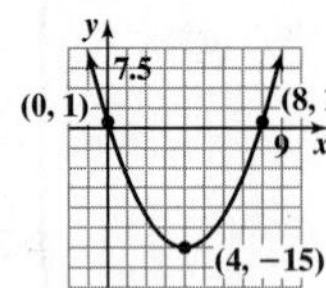

73. $f(x) = 2(x - 3)^2 + 1$

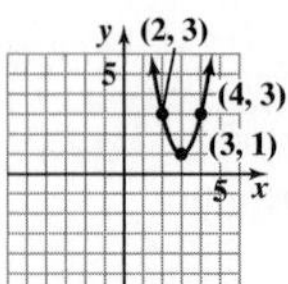

75. $f(x) = -3(x + 2)^2 - 5$

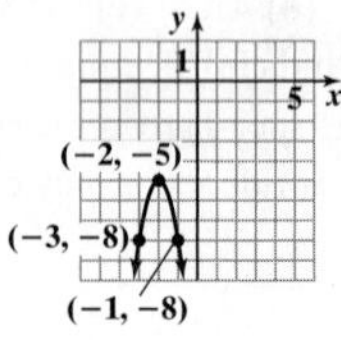

77. (a)

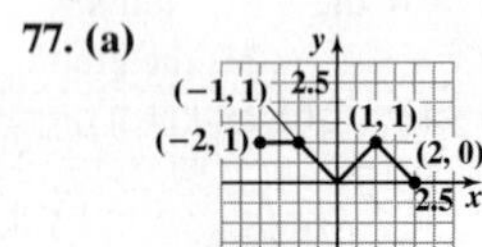

(b)

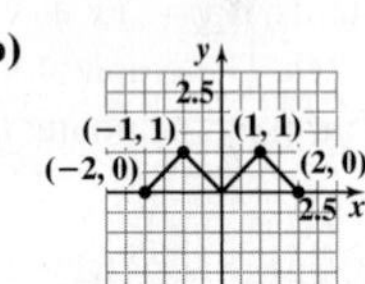

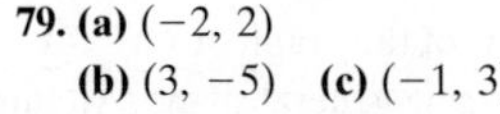

79. (a) $(-2, 2)$ **(b)** $(3, -5)$ **(c)** $(-1, 3)$

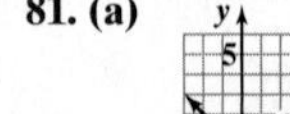

81. (a)

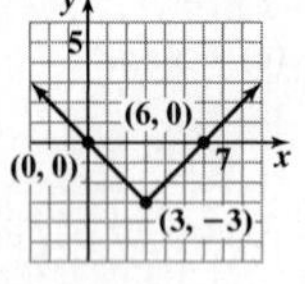

(b) 9 square units

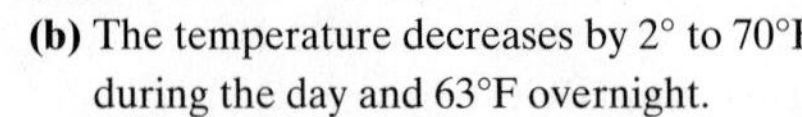

83. (a) 72°F; 65°F **(b)** The temperature decreases by 2° to 70°F during the day and 63°F overnight.

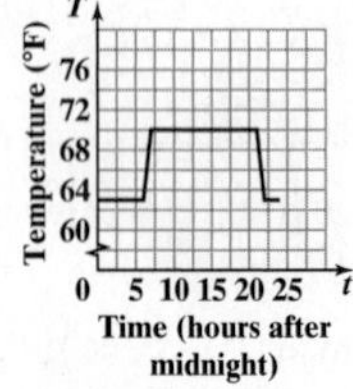

(c) The time at which the temperature adjusts between the daytime and overnight settings is moved to 1 hr sooner. It begins warming up at 5:00 AM instead of 6:00 AM, and it begins cooling down at 8:00 PM instead of 9:00 PM.

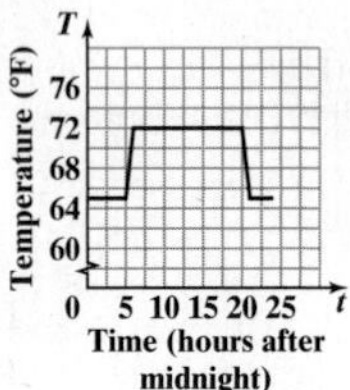

85.

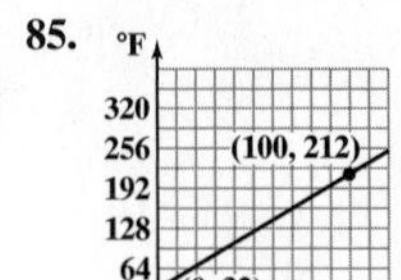

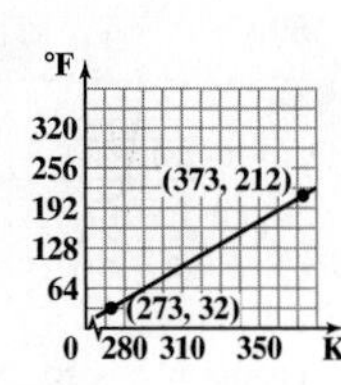

87.

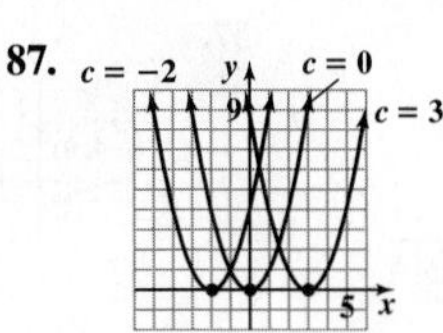

89. The graph of $y = 4f(x)$ is a vertical stretch by a factor of 4. The graph of $y = f(4x)$ is a horizontal compression by a factor of $\frac{1}{4}$.

91. $\frac{16}{3}$ sq. units **93.** The domain of $g(x) = \sqrt{x}$ is $[0, \infty)$. The graph of $g(x - k)$ is the graph of g shifted k units to the right, so the domain of $g(x - k)$ is $[k, \infty)$.

1.7 Assess Your Understanding *(page 87)*

5. $f(x_1) \neq f(x_2)$ **6.** one-to-one **7.** 3 **8.** $y = x$ **9.** $[4, \infty)$ **10.** T **11.** a **12.** d **13.** one-to-one **15.** not one-to-one **17.** not one-to-one **19.** one-to-one **21.** one-to-one **23.** not one-to-one **25.** one-to-one

27.

Annual Rainfall (inches)	Location
49.7	Atlanta, Georgia
43.8	Boston, Massachusetts
4.2	Las Vegas, Nevada
61.9	Miami, Florida
12.8	Los Angeles, California

Domain: {49.7, 43.8, 4.2, 61.9, 12.8}
Range: {Atlanta, Boston, Las Vegas, Miami, Los Angeles}

29.

Monthly Cost of Life Insurance	Age
\$10.59	30
\$12.52	40
\$15.94	45

Domain: {\$10.59, \$12.52, \$15.94}
Range: {30, 40, 45}

31. $\{(5,-3),(9,-2),(2,-1),(11,0),(-5,1)\}$
Domain: $\{5, 9, 2, 11, -5\}$
Range: $\{-3, -2, -1, 0, 1\}$

33. $\{(1,-2),(2,-3),(0,-10),(9,1),(4,2)\}$
Domain: $\{1, 2, 0, 9, 4\}$
Range: $\{-2, -3, -10, 1, 2\}$

35. $f(g(x)) = f\left(\frac{1}{3}(x-4)\right) = 3\left[\frac{1}{3}(x-4)\right] + 4 = (x-4) + 4 = x$

$g(f(x)) = g(3x+4) = \frac{1}{3}[(3x+4) - 4] = \frac{1}{3}(3x) = x$

37. $f(g(x)) = f\left(\frac{x}{4} + 2\right) = 4\left[\frac{x}{4} + 2\right] - 8 = (x+8) - 8 = x$

$g(f(x)) = g(4x-8) = \frac{4x-8}{4} + 2 = (x-2) + 2 = x$

39. $f(g(x)) = f(\sqrt[3]{x+8}) = (\sqrt[3]{x+8})^3 - 8 = (x+8) - 8 = x$

$g(f(x)) = g(x^3 - 8) = \sqrt[3]{(x^3-8)+8} = \sqrt[3]{x^3} = x$

41. $f(g(x)) = f\left(\frac{1}{x}\right) = \frac{1}{\left(\frac{1}{x}\right)} = x; x \neq 0$, $g(f(x)) = g\left(\frac{1}{x}\right) = \frac{1}{\left(\frac{1}{x}\right)} = x, x \neq 0$

43. $f(g(x)) = f\left(\frac{4x-3}{2-x}\right) = \frac{2\left(\frac{4x-3}{2-x}\right) + 3}{\frac{4x-3}{2-x} + 4}$

$= \frac{2(4x-3) + 3(2-x)}{4x - 3 + 4(2-x)} = \frac{5x}{5} = x, x \neq 2$

$g(f(x)) = g\left(\frac{2x+3}{x+4}\right) = \frac{4\left(\frac{2x+3}{x+4}\right) - 3}{2 - \frac{2x+3}{x+4}}$

$= \frac{4(2x+3) - 3(x+4)}{2(x+4) - (2x+3)} = \frac{5x}{5} = x, x \neq -4$

45.

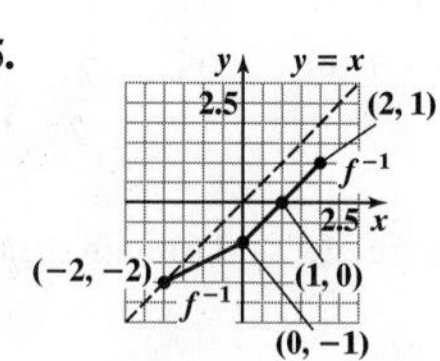

47.

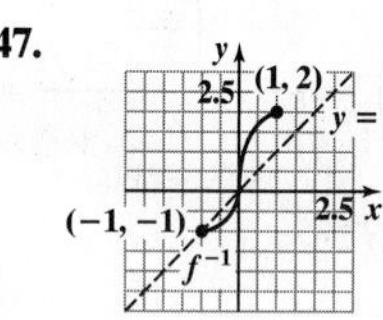

49.

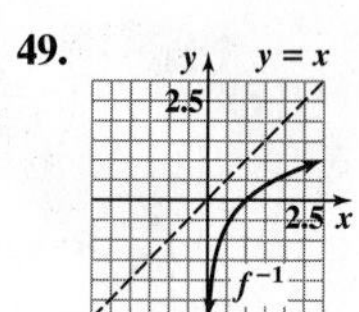

51. (a) $f^{-1}(x) = \frac{1}{3}x$

$f(f^{-1}(x)) = f\left(\frac{1}{3}x\right) = 3\left(\frac{1}{3}x\right) = x$

$f^{-1}(f(x)) = f^{-1}(3x) = \frac{1}{3}(3x) = x$

(b) Domain of f = Range of f^{-1} = All real numbers;
Range of f = Domain of f^{-1} = All real numbers

(c)

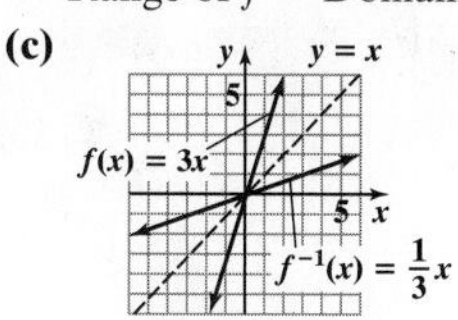

53. (a) $f^{-1}(x) = \frac{x}{4} - \frac{1}{2}$

$f(f^{-1}(x)) = f\left(\frac{x}{4} - \frac{1}{2}\right) = 4\left(\frac{x}{4} - \frac{1}{2}\right) + 2$

$= (x-2) + 2 = x$

$f^{-1}(f(x)) = f^{-1}(4x+2) = \frac{4x+2}{4} - \frac{1}{2}$

$= \left(x + \frac{1}{2}\right) - \frac{1}{2} = x$

(b) Domain of f = Range of f^{-1} = All real numbers;
Range of f = Domain of f^{-1} = All real numbers

(c)

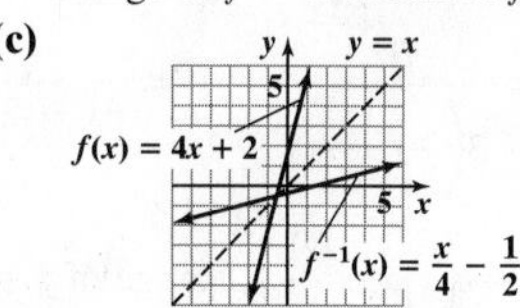

55. (a) $f^{-1}(x) = \sqrt[3]{x+1}$

$f(f^{-1}(x)) = f(\sqrt[3]{x+1})$
$= (\sqrt[3]{x+1})^3 - 1 = x$

$f^{-1}(f(x)) = f^{-1}(x^3 - 1)$
$= \sqrt[3]{(x^3-1)+1} = x$

(b) Domain of f = Range of f^{-1} = All real numbers;
Range of f = Domain of f^{-1} = All real numbers

(c)

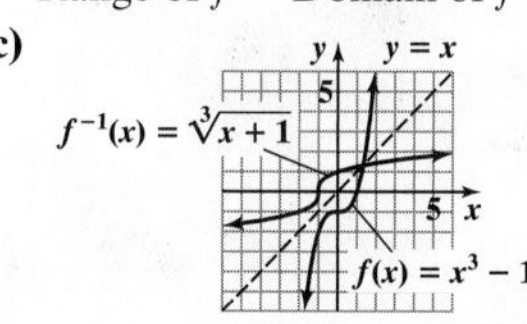

57. (a) $f^{-1}(x) = \sqrt{x-4}, x \geq 4$

$f(f^{-1}(x)) = f(\sqrt{x-4}) = (\sqrt{x-4})^2 + 4 = x$

$f^{-1}(f(x)) = f^{-1}(x^2+4) = \sqrt{(x^2+4) - 4} = \sqrt{x^2} = x, x \geq 0$

(b) Domain of f = Range of f^{-1} = $\{x \mid x \geq 0\}$;
Range of f = Domain of f^{-1} = $\{x \mid x \geq 4\}$

(c)

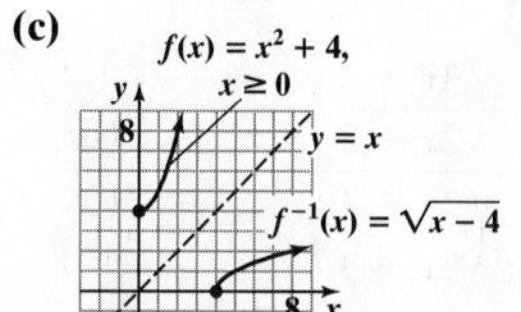

59. (a) $f^{-1}(x) = \dfrac{4}{x}$

$$f(f^{-1}(x)) = f\left(\frac{4}{x}\right) = \frac{4}{\left(\frac{4}{x}\right)} = x$$

$$f^{-1}(f(x)) = f^{-1}\left(\frac{4}{x}\right) = \frac{4}{\left(\frac{4}{x}\right)} = x$$

(b) Domain of f = Range of $f^{-1} = \{x|x \neq 0\}$; Range of f = Domain of $f^{-1} = \{x|x \neq 0\}$

(c)

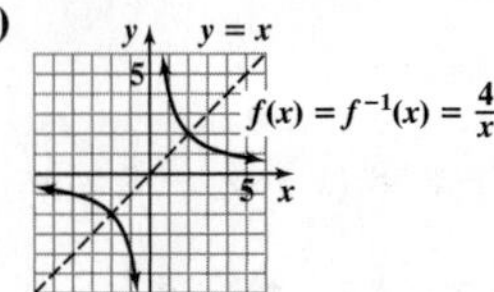

61. (a) $f^{-1}(x) = \dfrac{2x+1}{x}$

$$f(f^{-1}(x)) = f\left(\frac{2x+1}{x}\right) = \frac{1}{\frac{2x+1}{x} - 2} = \frac{x}{(2x+1) - 2x} = x$$

$$f^{-1}(f(x)) = f^{-1}\left(\frac{1}{x-2}\right) = \frac{2\left(\frac{1}{x-2}\right) + 1}{\frac{1}{x-2}} = \frac{2 + (x-2)}{1} = x$$

(b) Domain of f = Range of $f^{-1} = \{x|x \neq 2\}$; Range of f = Domain of $f^{-1} = \{x|x \neq 0\}$

(c)

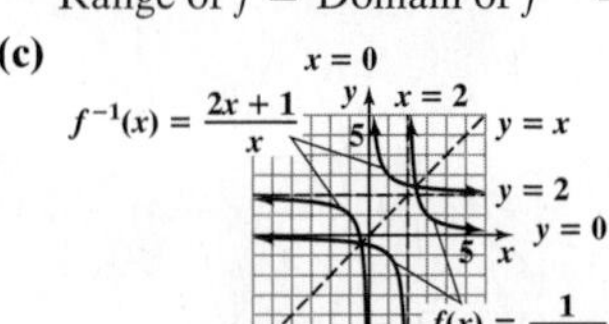

63. (a) $f^{-1}(x) = \dfrac{2-3x}{x}$

$$f(f^{-1}(x)) = f\left(\frac{2-3x}{x}\right) = \frac{2}{3 + \frac{2-3x}{x}} = \frac{2x}{3x + 2 - 3x} = \frac{2x}{2} = x$$

$$f^{-1}(f(x)) = f^{-1}\left(\frac{2}{3+x}\right) = \frac{2 - 3\left(\frac{2}{3+x}\right)}{\frac{2}{3+x}} = \frac{2(3+x) - 3\cdot 2}{2} = \frac{2x}{2} = x$$

(b) Domain of f = Range of $f^{-1} = \{x|x \neq -3\}$; Range of f = Domain of $f^{-1} = \{x|x \neq 0\}$

65. (a) $f^{-1}(x) = \dfrac{-2x}{x-3}$

$$f(f^{-1}(x)) = f\left(\frac{-2x}{x-3}\right) = \frac{3\left(\frac{-2x}{x-3}\right)}{\frac{-2x}{x-3} + 2} = \frac{3(-2x)}{-2x + 2(x-3)} = \frac{-6x}{-6} = x$$

$$f^{-1}(f(x)) = f^{-1}\left(\frac{3x}{x+2}\right) = \frac{-2\left(\frac{3x}{x+2}\right)}{\frac{3x}{x+2} - 3} = \frac{-2(3x)}{3x - 3(x+2)} = \frac{-6x}{-6} = x$$

(b) Domain of f = Range of $f^{-1} = \{x|x \neq -2\}$; Range of f = Domain of $f^{-1} = \{x|x \neq 3\}$

67. (a) $f^{-1}(x) = \dfrac{x}{3x-2}$

$$f(f^{-1}(x)) = f\left(\frac{x}{3x-2}\right) = \frac{2\left(\frac{x}{3x-2}\right)}{3\left(\frac{x}{3x-2}\right) - 1} = \frac{2x}{3x - (3x-2)} = \frac{2x}{2} = x$$

$$f^{-1}(f(x)) = f^{-1}\left(\frac{2x}{3x-1}\right) = \frac{\frac{2x}{3x-1}}{3\left(\frac{2x}{3x-1}\right) - 2} = \frac{2x}{6x - 2(3x-1)} = \frac{2x}{2} = x$$

(b) Domain of f = Range of $f^{-1} = \left\{x \middle| x \neq \frac{1}{3}\right\}$; Range of f = Domain of $f^{-1} = \left\{x \middle| x \neq \frac{2}{3}\right\}$

69. (a) $f^{-1}(x) = \dfrac{3x+4}{2x-3}$

$$f(f^{-1}(x)) = f\left(\frac{3x+4}{2x-3}\right) = \frac{3\left(\frac{3x+4}{2x-3}\right) + 4}{2\left(\frac{3x+4}{2x-3}\right) - 3} = \frac{3(3x+4) + 4(2x-3)}{2(3x+4) - 3(2x-3)} = \frac{17x}{17} = x$$

$$f^{-1}(f(x)) = f^{-1}\left(\frac{3x+4}{2x-3}\right) = \frac{3\left(\frac{3x+4}{2x-3}\right) + 4}{2\left(\frac{3x+4}{2x-3}\right) - 3} = \frac{3(3x+4) + 4(2x-3)}{2(3x+4) - 3(2x-3)} = \frac{17x}{17} = x$$

(b) Domain of f = Range of $f^{-1} = \left\{x \middle| x \neq \frac{3}{2}\right\}$; Range of f = Domain of $f^{-1} = \left\{x \middle| x \neq \frac{3}{2}\right\}$

71. (a) $f^{-1}(x) = \dfrac{-2x+3}{x-2}$

$$f(f^{-1}(x)) = f\left(\frac{-2x+3}{x-2}\right) = \frac{2\left(\frac{-2x+3}{x-2}\right)+3}{\frac{-2x+3}{x-2}+2} = \frac{2(-2x+3)+3(x-2)}{-2x+3+2(x-2)} = \frac{-x}{-1} = x$$

$$f^{-1}(f(x)) = f^{-1}\left(\frac{2x+3}{x+2}\right) = \frac{-2\left(\frac{2x+3}{x+2}\right)+3}{\frac{2x+3}{x+2}-2} = \frac{-2(2x+3)+3(x+2)}{2x+3-2(x+2)} = \frac{-x}{-1} = x$$

(b) Domain of f = Range of $f^{-1} = \{x|x \neq -2\}$; Range of f = Domain of $f^{-1} = \{x|x \neq 2\}$

73. (a) $f^{-1}(x) = \dfrac{2}{\sqrt{1-2x}}$

$$f(f^{-1}(x)) = f\left(\frac{2}{\sqrt{1-2x}}\right) = \frac{\frac{4}{1-2x}-4}{2\cdot\frac{4}{1-2x}} = \frac{4-4(1-2x)}{2\cdot 4} = \frac{8x}{8} = x$$

$$f^{-1}(f(x)) = f^{-1}\left(\frac{x^2-4}{2x^2}\right) = \frac{2}{\sqrt{1-2\left(\frac{x^2-4}{2x^2}\right)}} = \frac{2}{\sqrt{\frac{4}{x^2}}} = \sqrt{x^2} = x, \text{ since } x > 0$$

(b) Domain of f = Range of $f^{-1} = \{x|x > 0\}$; Range of f = Domain of $f^{-1} = \left\{x \middle| x < \frac{1}{2}\right\}$

75. (a) 0 **(b)** 2 **(c)** 0 **(d)** 1 **77.** 7 **79.** Domain of f^{-1}: $[-2, \infty)$; range of f^{-1}: $[5, \infty)$ **81.** Domain of g^{-1}: $[0, \infty)$; range of g^{-1}: $(-\infty, 0]$

83. Increasing on the interval $(f(0), f(5))$ **85.** $f^{-1}(x) = \dfrac{1}{m}(x-b), m \neq 0$ **87.** Quadrant I

89. Possible answer: $f(x) = |x|, x \geq 0$, is one-to-one; $f^{-1}(x) = x, x \geq 0$

91. (a) $r(d) = \dfrac{d+90.39}{6.97}$

(b) $r(d(r)) = \dfrac{6.97r - 90.39 + 90.39}{6.97} = \dfrac{6.97r}{6.97} = r$

$d(r(d)) = 6.97\left(\dfrac{d+90.39}{6.97}\right) - 90.39 = d + 90.39 - 90.39 = d$

(c) 56 miles per hour

93. (a) 77.6 kg

(b) $h(W) = \dfrac{W-50}{2.3} + 60 = \dfrac{W+88}{2.3}$

(c) $h(W(h)) = \dfrac{50 + 2.3(h-60) + 88}{2.3} = \dfrac{2.3h}{2.3} = h$

$W(h(W)) = 50 + 2.3\left(\dfrac{W+88}{2.3} - 60\right)$

$= 50 + W + 88 - 138 = W$

(d) 73 inches

95. (a) $\{g|36{,}900 \leq g \leq 89{,}350\}$

(b) $\{T|5081.25 \leq T \leq 18{,}193.75\}$

(c) $g(T) = \dfrac{T - 5081.25}{0.25} + 36{,}900$

Domain: $\{T|5081.25 \leq T \leq 18{,}193.75\}$

Range: $\{g|36{,}900 \leq g \leq 89{,}350\}$

97. (a) t represents time, so $t \geq 0$.

(b) $t(H) = \sqrt{\dfrac{H-100}{-4.9}} = \sqrt{\dfrac{100-H}{4.9}}$

(c) 2.02 seconds

99. $f^{-1}(x) = \dfrac{-dx+b}{cx-a}$; $f = f^{-1}$ if $a = -d$ **103.** No

Review Exercises *(page 94)*

1. (a) $2\sqrt{5}$ **(b)** $(2, 1)$ **2. (a)** 5 **(b)** $\left(-\frac{1}{2}, 1\right)$ **3. (a)** 12 **(b)** $(4, 2)$

4.

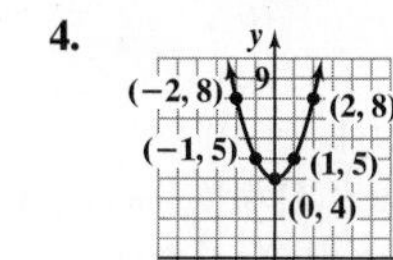

5. $(-4, 0), (0, 2), (0, 0), (0, -2), (2, 0)$ **6.** $(0, 0)$; symmetric with respect to the x-axis **7.** $(\pm 4, 0), (0, \pm 2)$; symmetric with respect to the x-axis, y-axis, and origin **8.** $(0, 1)$; symmetric with respect to the y-axis **9.** $(0, 0), (\pm 1, 0)$; symmetric with respect to the origin **10.** $(0, 0), (-1, 0), (0, -2)$; no symmetry **11.** $(x+2)^2 + (y-3)^2 = 16$ **12.** $(x+1)^2 + (y+2)^2 = 1$

13. Center (0, 1); radius = 2

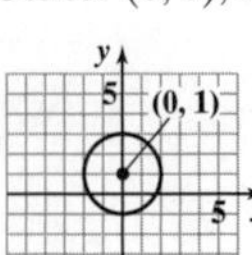

Intercepts: $(-\sqrt{3}, 0), (\sqrt{3}, 0),$ $(0, -1), (0, 3)$

14. Center (1, −2); radius = 3

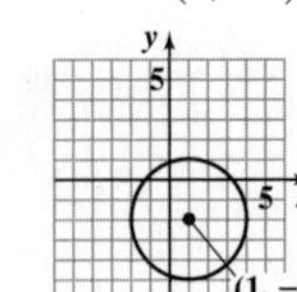

Intercepts: $(1 - \sqrt{5}, 0), (1 + \sqrt{5}, 0),$ $(0, -2 - 2\sqrt{2}), (0, -2 + 2\sqrt{2})$

15. Center (1, −2); radius = $\sqrt{5}$

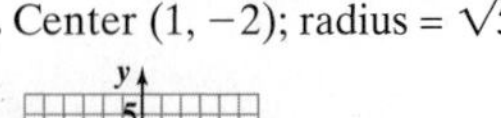

Intercepts: (0, 0), (2, 0), (0, −4)

16. $d(A, B) = \sqrt{(1 - 3)^2 + (1 - 4)^2} = \sqrt{13}$ and
$d(B, C) = \sqrt{(-2 - 1)^2 + (3 - 1)^2} = \sqrt{13}$
17. Center (1, −2); Radius $= 4\sqrt{2}$; $(x - 1)^2 + (y + 2)^2 = 32$

18. Function; domain $\{-1, 2, 4\}$, range $\{0, 3\}$ **19.** Not a function **20.** **(a)** 2 **(b)** −2 **(c)** $-\dfrac{3x}{x^2 - 1}$ **(d)** $-\dfrac{3x}{x^2 - 1}$ **(e)** $\dfrac{3(x - 2)}{x^2 - 4x + 3}$ **(f)** $\dfrac{6x}{4x^2 - 1}$ **21.** **(a)** 0 **(b)** 0 **(c)** $\sqrt{x^2 - 4}$ **(d)** $-\sqrt{x^2 - 4}$ **(e)** $\sqrt{x^2 - 4x}$ **(f)** $2\sqrt{x^2 - 1}$ **22.** **(a)** 0 **(b)** 0 **(c)** $\dfrac{x^2 - 4}{x^2}$ **(d)** $-\dfrac{x^2 - 4}{x^2}$ **(e)** $\dfrac{x(x - 4)}{(x - 2)^2}$ **(f)** $\dfrac{x^2 - 1}{x^2}$

23. $\{x | x \neq -3, x \neq 3\}$ **24.** $\{x | x \leq 2\}$ **25.** $\{x | x \neq 0\}$ **26.** $\{x | x \neq -3, x \neq 1\}$ **27.** $[-1, 2) \cup (2, \infty)$ **28.** $\{x | x > -8\}$, or $(-8, \infty)$

29. $-4x + 1 - 2h$ **30.** **(a)** Domain: $\{x | -4 \leq x \leq 3\}$; Range: $\{y | -3 \leq y \leq 3\}$
(b) (0, 0) **(c)** −1 **(d)** −4 **(e)** $\{x | 0 < x \leq 3\}$

(f)

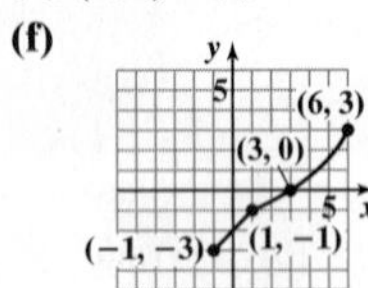

(g)

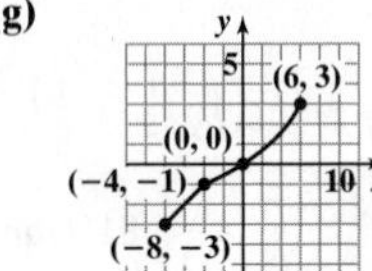

(h)

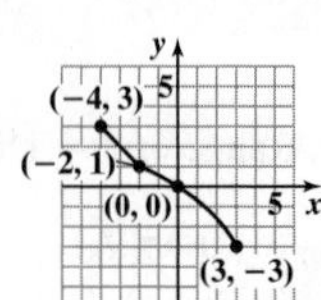

31. **(a)** Domain: $\{x | x \leq 4\}$ or $(-\infty, 4]$
Range: $\{y | y \leq 3\}$ or $(-\infty, 3]$
(b) Increasing on $(-\infty, -2)$ and $(2, 4)$; Decreasing on $(-2, 2)$
(c) Local maximum value is 1 and occurs at $x = -2$.
Local minimum value is −1 and occurs at $x = 2$.
(d) Absolute maximum: $f(4) = 3$
Absolute minimum: none
(e) No symmetry
(f) Neither
(g) x-intercepts: −3, 0, 3 y-intercept: 0

32. Odd **33.** Even **34.** Neither **35.** Odd

36.

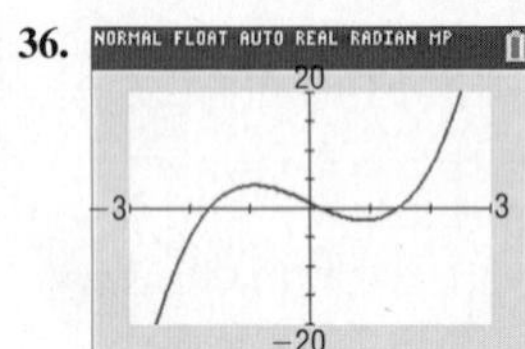

Local maximum value: 4.04 at $x = -0.91$
Local minimum value: −2.04 at $x = 0.91$
Increasing: $(-3, -0.91)$; $(0.91, 3)$
Decreasing: $(-0.91, 0.91)$

37.

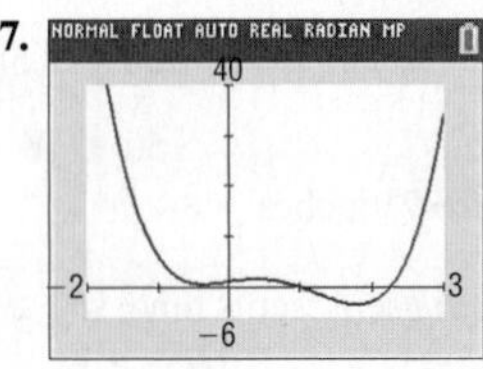

Local maximum value: 1.53 at $x = 0.41$
Local minima values: 0.54 at
$x = -0.34$ and −3.56 at $x = 1.80$
Increasing: $(-0.34, 0.41)$; $(1.80, 3)$
Decreasing: $(-2, -0.34)$; $(0.41, 1.80)$

38. **(a)** 23 **(b)** 7 **(c)** 47 **39.** −5 **40.** −17 **41.** No **42.** Yes

43.

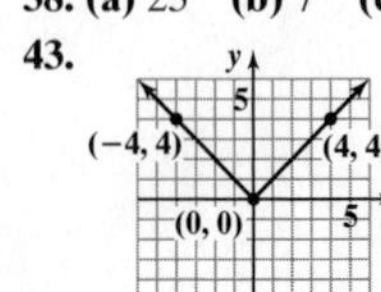

44.

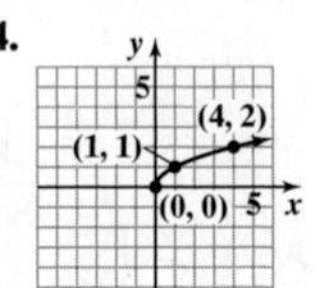

45.

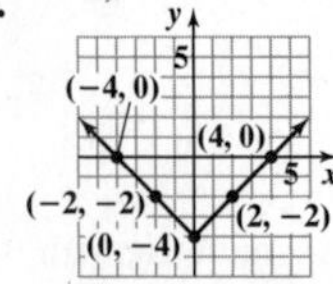

Intercepts: $(-4, 0)$, $(4, 0)$, $(0, -4)$
Domain: all real numbers
Range: $\{y | y \geq -4\}$ or $[-4, \infty)$

46.

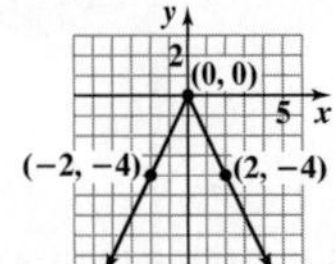

Intercept: (0, 0)
Domain: all real numbers
Range: $\{y | y \leq 0\}$ or $(-\infty, 0]$

47.

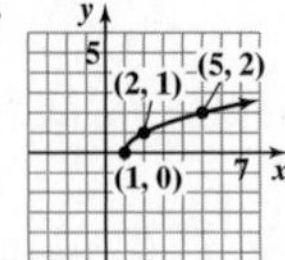

Intercept: (1, 0)
Domain: $\{x | x \geq 1\}$ or $[1, \infty)$
Range: $\{y | y \geq 0\}$ or $[0, \infty)$

48.

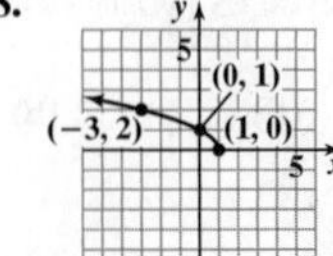

Intercepts: (0, 1), (1, 0)
Domain: $\{x | x \leq 1\}$ or $(-\infty, 1]$
Range: $\{y | y \geq 0\}$ or $[0, \infty)$

49.

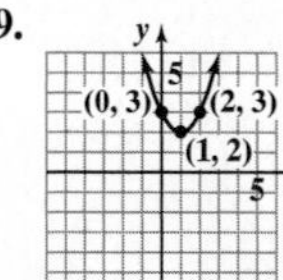

Intercept: (0, 3)
Domain: all real numbers
Range: $\{y|y \geq 2\}$ or $[2, \infty)$

50.

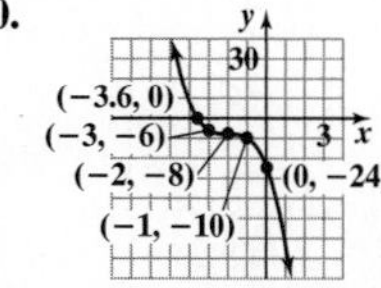

Intercepts: (0, –24), $(-2 - \sqrt[3]{4}, 0)$ or about (–3.6, 0)
Domain: all real numbers
Range: all real numbers

51. (a) $\{x|x > -2\}$ or $(-2, \infty)$
(b) (0, 0)
(c)

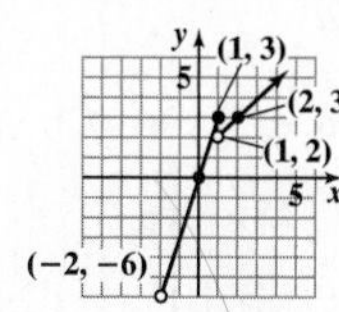

(d) $\{y|y > -6\}$ or $(-6, \infty)$
(e) Discontinuous at $x = 1$

52. (a) $\{x|x \geq -4\}$ or $[-4, \infty)$
(b) (0, 1)
(c)

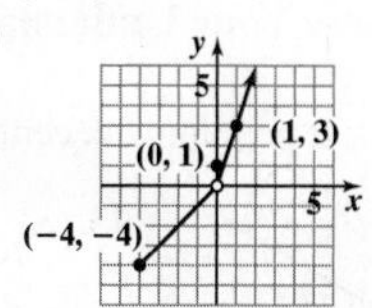

(d) $\{y|-4 \leq y < 0 \text{ or } y > 0\}$ or $[-4, 0) \cup (0, \infty)$
(e) Discontinuous at $x = 0$

53. $A = 11$ **54. (a)** one-to-one **(b)** $\{(2, 1), (5, 3), (8, 5), (10, 6)\}$ **55.**

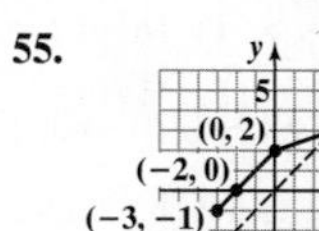

56. $f^{-1}(x) = \dfrac{2x + 3}{5x - 2}$

$$f(f^{-1}(x)) = \frac{2\left(\dfrac{2x+3}{5x-2}\right) + 3}{5\left(\dfrac{2x+3}{5x-2}\right) - 2} = x$$

$$f^{-1}(f(x)) = \frac{2\left(\dfrac{2x+3}{5x-2}\right) + 3}{5\left(\dfrac{2x+3}{5x-2}\right) - 2} = x$$

Domain of f = range of f^{-1} = all real numbers except $\frac{2}{5}$

Range of f = domain of f^{-1} = all real numbers except $\frac{2}{5}$

57. $f^{-1}(x) = \dfrac{x + 1}{x}$

$$f(f^{-1}(x)) = \frac{1}{\dfrac{x+1}{x} - 1} = x$$

$$f^{-1}(f(x)) = \frac{\dfrac{1}{x-1} + 1}{\dfrac{1}{x-1}} = x$$

Domain of f = range of f^{-1} = all real numbers except 1

Range of f = domain of f^{-1} = all real numbers except 0

58. $f^{-1}(x) = x^2 + 2, x \geq 0$
$f(f^{-1}(x)) = \sqrt{x^2 + 2 - 2} = |x| = x, x \geq 0$
$f^{-1}(f(x)) = (\sqrt{x - 2})^2 + 2 = x$
Domain of f = range of f^{-1} = $[2, \infty)$
Range of f = domain of f^{-1} = $[0, \infty)$

59. $f^{-1}(x) = (x - 1)^3$;
$f(f^{-1}(x)) = ((x - 1)^3)^{1/3} + 1 = x$
$f^{-1}(f(x)) = (x^{1/3} + 1 - 1)^3 = x$
Domain of f = range of f^{-1} = $(-\infty, \infty)$
Range of f = domain of f^{-1} = $(-\infty, \infty)$

Chapter Test *(page 96)*

1. $d = 2\sqrt{13}$ **2.** (2, 1)

3.

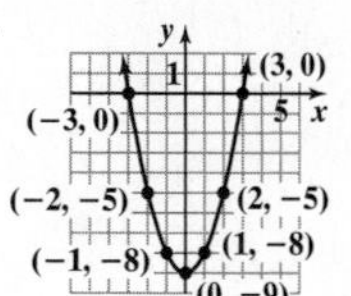

4.

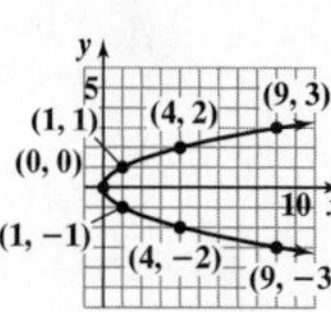

5. Intercepts: (–3, 0), (3, 0), (0, 9); symmetric with respect to the y-axis

6. $x^2 + y^2 - 8x + 6y = 0$

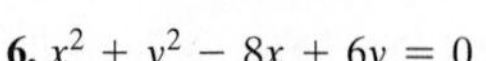

7. Center: (–2, 1); radius: 3

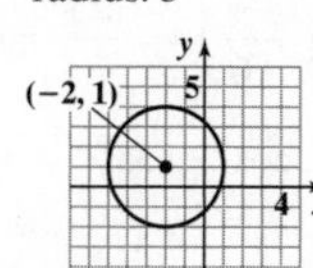

8. (a) Function; domain: {2, 4, 6, 8}; range: {5, 6, 7, 8} **(b)** Not a function **(c)** Not a function **(d)** Function; domain; all real numbers; range: $\{y|y \geq 2\}$ **9.** Domain: $\left\{x \middle| x \leq \frac{4}{5}\right\}$; $f(-1) = 3$ **10.** Domain: $\{x|x \neq -2\}$; $g(-1) = 1$ **11.** Domain: $\{x|x \neq -9, x \neq 4\}$; $h(-1) = \frac{1}{8}$
12. (a) Domain: $\{x|-5 \leq x \leq 5\}$; range: $\{y|-3 \leq y \leq 3\}$ **(b)** (0, 2), (–2, 0), and (2, 0) **(c)** $f(1) = 3$ **(d)** $x = -5$ and $x = 3$
(e) $\{x|-5 \leq x < -2 \text{ or } 2 < x \leq 5\}$ or $[-5, -2) \cup (2, 5]$ **13.** Local maxima values: $f(-0.85) \approx -0.86$; $f(2.35) \approx 15.55$; local minimum value: $f(0) = -2$; the function is increasing on the intervals (–5, –0.85) and (0, 2.35) and decreasing on the intervals (–0.85, 0) and (2.35, 5).

14. (a)

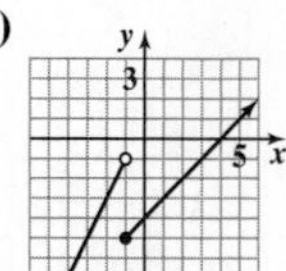

(b) (0, –4), (4, 0)
(c) $g(-5) = -9$
(d) $g(2) = -2$

15. 19 **16. (a)**

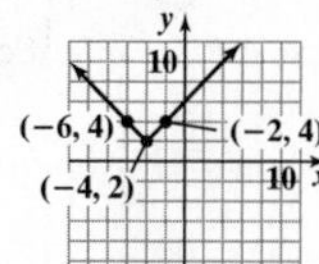

(b)

17. $f^{-1}(x) = \dfrac{2 + 5x}{3x}$, domain of $f = \left\{x \middle| x \neq \frac{5}{3}\right\}$, range of $f = \{y|y \neq 0\}$; domain of $f^{-1} = \{x|x \neq 0\}$; range of $f^{-1} = \left\{y \middle| y \neq \frac{5}{3}\right\}$

18. The point (–5, 3) must be on the graph of f^{-1}.

CHAPTER 2 Trigonometric Functions

2.1 Assess Your Understanding *(page 109)*

3. standard position **4.** central angle **5.** d **6.** $r\theta; \frac{1}{2}r^2\theta$ **7.** b **8.** $\frac{s}{t}; \frac{\theta}{t}$ **9.** T **10.** F

11. **13.**

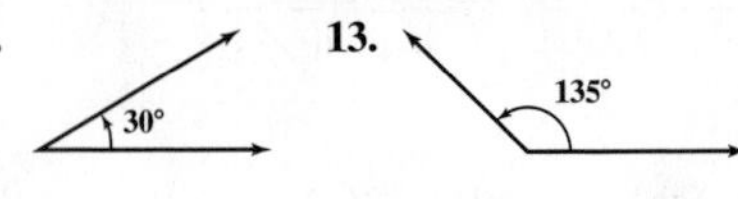

15. **17.**

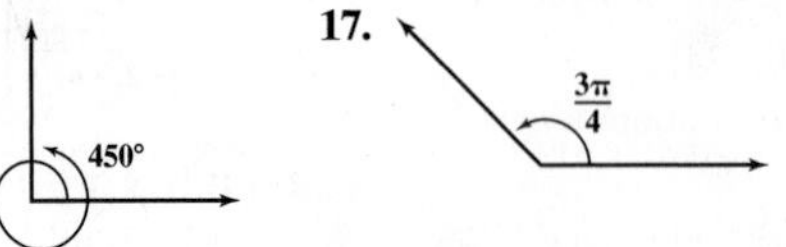

19. **21.**

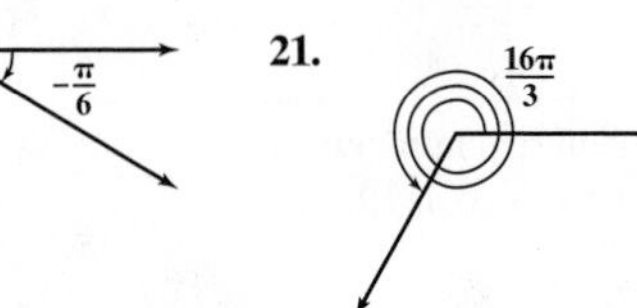

23. 40.17° **25.** 50.24° **27.** 9.15° **29.** 40°19′12″ **31.** 18°15′18″ **33.** 19°59′24″ **35.** $\frac{\pi}{6}$ **37.** $\frac{4\pi}{3}$ **39.** $-\frac{\pi}{3}$ **41.** π **43.** $-\frac{3\pi}{4}$ **45.** $-\frac{\pi}{2}$ **47.** 60° **49.** −225° **51.** 90° **53.** 15° **55.** −90° **57.** −30° **59.** 0.30 **61.** −0.70 **63.** 2.18 **65.** 179.91° **67.** 114.59° **69.** 362.11° **71.** 5 m **73.** 6 ft **75.** 0.6 radian **77.** $\frac{\pi}{3} \approx 1.047$ in. **79.** 25 m^2 **81.** $2\sqrt{3} \approx 3.464$ ft **83.** 0.24 radian **85.** $\frac{\pi}{3} \approx 1.047$ in.^{2} **87.** $s = 2.094$ ft; $A = 2.094$ ft^2 **89.** $s = 14.661$ yd; $A = 87.965$ yd^2 **91.** $3\pi \approx 9.42$ in; $5\pi \approx 15.71$ in. **93.** $2\pi \approx 6.28$ m^2 **95.** $\frac{675\pi}{2} \approx 1060.29$ ft^2 **97.** $\frac{1075\pi}{3} \approx 1125.74$ in.^{2} **99.** $\omega = \frac{1}{60}$ radian/s; $v = \frac{1}{12}$ cm/s **101.** ≈23.2 mph **103.** ≈120.6 km/h **105.** ≈452.5 rpm **107.** ≈359 mi **109.** ≈898 mi/h **111.** ≈2292 mi/h **113.** $\frac{3}{4}$ rpm **115.** ≈2.86 mi/h **117.** ≈31.47 rpm **119.** 63π square feet **121.** ≈1037 mi/h **123.** Radius ≈ 3979 mi; circumference ≈ 25,000 mi **125.** $v_1 = r_1\omega_1$, $v_2 = r_2\omega_2$, and $v_1 = v_2$, so $r_1\omega_1 = r_2\omega_2$ and $\frac{r_1}{r_2} = \frac{\omega_2}{\omega_1}$. **134.** $-\frac{7}{3}$ **135.** $\{x \mid x \neq \pm 3\}$ **136.** $y = -|x + 3| - 4$ **137.** $\left(-\frac{1}{2}, 4\right)$

2.2 Assess Your Understanding *(page 126)*

7. b **8.** (0, 1) **9.** $\left(\frac{\sqrt{2}}{2}, \frac{\sqrt{2}}{2}\right)$ **10.** a **11.** $\frac{y}{r}; \frac{x}{r}$ **12.** F **13.** $\sin t = \frac{1}{2}; \cos t = \frac{\sqrt{3}}{2}; \tan t = \frac{\sqrt{3}}{3}; \csc t = 2; \sec t = \frac{2\sqrt{3}}{3}; \cot t = \sqrt{3}$ **15.** $\sin t = \frac{\sqrt{21}}{5}; \cos t = -\frac{2}{5}; \tan t = -\frac{\sqrt{21}}{2}; \csc t = \frac{5\sqrt{21}}{21}; \sec t = -\frac{5}{2}; \cot t = -\frac{2\sqrt{21}}{21}$ **17.** $\sin t = \frac{\sqrt{2}}{2}; \cos t = -\frac{\sqrt{2}}{2};$ $\tan t = -1; \csc t = \sqrt{2}; \sec t = -\sqrt{2}; \cot t = -1$ **19.** $\sin t = -\frac{1}{3}; \cos t = \frac{2\sqrt{2}}{3}; \tan t = -\frac{\sqrt{2}}{4}; \csc t = -3; \sec t = \frac{3\sqrt{2}}{4}; \cot t = -2\sqrt{2}$

21. −1 **23.** 0 **25.** −1 **27.** 0 **29.** −1 **31.** $\frac{1}{2}(\sqrt{2} + 1)$ **33.** 2 **35.** $\frac{1}{2}$ **37.** $\sqrt{6}$ **39.** 4 **41.** 0 **43.** $2\sqrt{2} + \frac{4\sqrt{3}}{3}$ **45.** 1

47. $\sin\frac{2\pi}{3} = \frac{\sqrt{3}}{2}; \cos\frac{2\pi}{3} = -\frac{1}{2}; \tan\frac{2\pi}{3} = -\sqrt{3}; \csc\frac{2\pi}{3} = \frac{2\sqrt{3}}{3}; \sec\frac{2\pi}{3} = -2; \cot\frac{2\pi}{3} = -\frac{\sqrt{3}}{3}$

49. $\sin 210° = -\frac{1}{2}; \cos 210° = -\frac{\sqrt{3}}{2}; \tan 210° = \frac{\sqrt{3}}{3}; \csc 210° = -2; \sec 210° = -\frac{2\sqrt{3}}{3}; \cot 210° = \sqrt{3}$

51. $\sin\frac{3\pi}{4} = \frac{\sqrt{2}}{2}; \cos\frac{3\pi}{4} = -\frac{\sqrt{2}}{2}; \tan\frac{3\pi}{4} = -1; \csc\frac{3\pi}{4} = \sqrt{2}; \sec\frac{3\pi}{4} = -\sqrt{2}; \cot\frac{3\pi}{4} = -1$

53. $\sin\frac{8\pi}{3} = \frac{\sqrt{3}}{2}; \cos\frac{8\pi}{3} = -\frac{1}{2}; \tan\frac{8\pi}{3} = -\sqrt{3}; \csc\frac{8\pi}{3} = \frac{2\sqrt{3}}{3}; \sec\frac{8\pi}{3} = -2; \cot\frac{8\pi}{3} = -\frac{\sqrt{3}}{3}$

55. $\sin 405° = \frac{\sqrt{2}}{2}; \cos 405° = \frac{\sqrt{2}}{2}; \tan 405° = 1; \csc 405° = \sqrt{2}; \sec 405° = \sqrt{2}; \cot 405° = 1$

57. $\sin\left(-\frac{\pi}{6}\right) = -\frac{1}{2}; \cos\left(-\frac{\pi}{6}\right) = \frac{\sqrt{3}}{2}; \tan\left(-\frac{\pi}{6}\right) = -\frac{\sqrt{3}}{3}; \csc\left(-\frac{\pi}{6}\right) = -2; \sec\left(-\frac{\pi}{6}\right) = \frac{2\sqrt{3}}{3}; \cot\left(-\frac{\pi}{6}\right) = -\sqrt{3}$

59. $\sin(-135°) = -\frac{\sqrt{2}}{2}; \cos(-135°) = -\frac{\sqrt{2}}{2}; \tan(-135°) = 1; \csc(-135°) = -\sqrt{2}; \sec(-135°) = -\sqrt{2}; \cot(-135°) = 1$

61. $\sin\frac{5\pi}{2} = 1; \cos\frac{5\pi}{2} = 0; \tan\frac{5\pi}{2}$ is undefined; $\csc\frac{5\pi}{2} = 1; \sec\frac{5\pi}{2}$ is undefined; $\cot\frac{5\pi}{2} = 0$

63. $\sin\left(-\frac{14\pi}{3}\right) = -\frac{\sqrt{3}}{2}; \cos\left(-\frac{14\pi}{3}\right) = -\frac{1}{2}; \tan\left(-\frac{14\pi}{3}\right) = \sqrt{3}; \csc\left(-\frac{14\pi}{3}\right) = -\frac{2\sqrt{3}}{3}; \sec\left(-\frac{14\pi}{3}\right) = -2; \cot\left(-\frac{14\pi}{3}\right) = \frac{\sqrt{3}}{3}$

65. 0.47 **67.** 1.07 **69.** 0.32 **71.** 3.73 **73.** 0.84 **75.** 0.02 **77.** $\sin\theta = \frac{4}{5}; \cos\theta = -\frac{3}{5}; \tan\theta = -\frac{4}{3}; \csc\theta = \frac{5}{4}; \sec\theta = -\frac{5}{3}; \cot\theta = -\frac{3}{4}$

79. $\sin\theta = -\frac{3\sqrt{13}}{13}; \cos\theta = \frac{2\sqrt{13}}{13}; \tan\theta = -\frac{3}{2}; \csc\theta = -\frac{\sqrt{13}}{3}; \sec\theta = \frac{\sqrt{13}}{2}; \cot\theta = -\frac{2}{3}$

81. $\sin\theta = -\frac{\sqrt{2}}{2}; \cos\theta = -\frac{\sqrt{2}}{2}; \tan\theta = 1; \csc\theta = -\sqrt{2}; \sec\theta = -\sqrt{2}; \cot\theta = 1$

83. $\sin\theta = \frac{3}{5}; \cos\theta = \frac{4}{5}; \tan\theta = \frac{3}{4}; \csc\theta = \frac{5}{3}; \sec\theta = \frac{5}{4}; \cot\theta = \frac{4}{3}$ **85.** 0 **87.** 0 **89.** −0.1 **91.** 3 **93.** 5 **95.** $\frac{\sqrt{3}}{2}$ **97.** $\frac{1}{2}$ **99.** $\frac{3}{4}$ **101.** $\frac{\sqrt{3}}{2}$

103. $\sqrt{3}$ **105.** $-\frac{\sqrt{3}}{2}$ **107.** $\frac{\sqrt{3}}{2}$ **109.** $\frac{\sqrt{2}}{4}$ **111. (a)** $\frac{\sqrt{2}}{2}$; $\left(\frac{\pi}{4}, \frac{\sqrt{2}}{2}\right)$ **(b)** $\left(\frac{\sqrt{2}}{2}, \frac{\pi}{4}\right)$ **(c)** $\left(\frac{\pi}{4}, -2\right)$

113. Answers may vary. One set of possible answers is $-\frac{11\pi}{3}, -\frac{5\pi}{3}, \frac{\pi}{3}, \frac{7\pi}{3}, \frac{13\pi}{3}$.

115.

θ	0.5	0.4	0.2	0.1	0.01	0.001	0.0001	0.00001
$\sin\theta$	0.4794	0.3894	0.1987	0.0998	0.0100	0.0010	0.0001	0.00001
$\frac{\sin\theta}{\theta}$	0.9589	0.9735	0.9933	0.9983	1.0000	1.0000	1.0000	1.0000

$\frac{\sin\theta}{\theta}$ approaches 1 as θ approaches 0.

117. $R \approx 310.56$ ft; $H \approx 77.64$ ft **119.** $R \approx 19{,}541.95$ m; $H \approx 2278.14$ m **121. (a)** 1.20 sec **(b)** 1.12 sec **(c)** 1.20 sec

123. (a) 1.9 hr; 0.57 hr **(b)** 1.69 hr; 0.75 hr **(c)** 1.63 hr; 0.86 hr **(d)** 1.67 hr; tan 90° is undefined **125.** 15.4 ft **127.** 71 ft **129.** $y = -\frac{4\sqrt{3}}{7}$

131. (a) 16.56 ft **(b)**

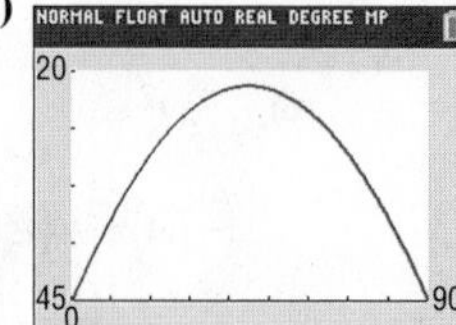

(c) 67.5°

133. (a) values estimated to the nearest tenth: $\sin 1 \approx 0.8$; $\cos 1 \approx 0.5$; $\tan 1 \approx 1.6$; $\csc 1 \approx 1.3$; $\sec 1 \approx 2.0$; $\cot 1 \approx 0.6$; actual values to the nearest tenth: $\sin 1 \approx 0.8$; $\cos 1 \approx 0.5$; $\tan 1 \approx 1.6$; $\csc 1 \approx 1.2$; $\sec 1 \approx 1.9$; $\cot 1 \approx 0.6$
(b) values estimated to the nearest tenth: $\sin 5.1 \approx -0.9$; $\cos 5.1 \approx 0.4$; $\tan 5.1 \approx -2.3$; $\csc 5.1 \approx -1.1$; $\sec 5.1 \approx 2.5$; $\cot 5.1 \approx -0.4$; actual values to the nearest tenth: $\sin 5.1 \approx -0.9$; $\cos 5.1 \approx 0.4$; $\tan 5.1 \approx -2.4$; $\csc 5.1 \approx -1.1$; $\sec 5.1 \approx 2.6$; $\cot 5.1 \approx -0.4$

139. $f^{-1}(x) = \frac{2}{x} + 7$ **140.** $-780°$ **141.** $f(x+3) = \frac{x+2}{x^2+6x+11}$ **142.** $2\sqrt{10}$

2.3 Assess Your Understanding *(page 142)*

5. 2π; π **6.** All real numbers except odd multiples of $\frac{\pi}{2}$ **7.** b **8.** a **9.** 1 **10.** F **11.** $\frac{\sqrt{2}}{2}$ **13.** 1 **15.** 1 **17.** $\sqrt{3}$ **19.** $\frac{\sqrt{2}}{2}$ **21.** 0 **23.** $\sqrt{2}$ **25.** $\frac{\sqrt{3}}{3}$

27. II **29.** IV **31.** IV **33.** II **35.** $\tan\theta = -\frac{3}{4}$; $\cot\theta = -\frac{4}{3}$; $\sec\theta = \frac{5}{4}$; $\csc\theta = -\frac{5}{3}$ **37.** $\tan\theta = 2$; $\cot\theta = \frac{1}{2}$; $\sec\theta = \sqrt{5}$; $\csc\theta = \frac{\sqrt{5}}{2}$

39. $\tan\theta = \frac{\sqrt{3}}{3}$; $\cot\theta = \sqrt{3}$; $\sec\theta = \frac{2\sqrt{3}}{3}$; $\csc\theta = 2$ **41.** $\tan\theta = -\frac{\sqrt{2}}{4}$; $\cot\theta = -2\sqrt{2}$; $\sec\theta = \frac{3\sqrt{2}}{4}$; $\csc\theta = -3$

43. $\cos\theta = -\frac{5}{13}$; $\tan\theta = -\frac{12}{5}$; $\csc\theta = \frac{13}{12}$; $\sec\theta = -\frac{13}{5}$; $\cot\theta = -\frac{5}{12}$ **45.** $\sin\theta = -\frac{3}{5}$; $\tan\theta = \frac{3}{4}$; $\csc\theta = -\frac{5}{3}$; $\sec\theta = -\frac{5}{4}$; $\cot\theta = \frac{4}{3}$

47. $\cos\theta = -\frac{12}{13}$; $\tan\theta = -\frac{5}{12}$; $\csc\theta = \frac{13}{5}$; $\sec\theta = -\frac{13}{12}$; $\cot\theta = -\frac{12}{5}$ **49.** $\sin\theta = \frac{2\sqrt{2}}{3}$; $\tan\theta = -2\sqrt{2}$; $\csc\theta = \frac{3\sqrt{2}}{4}$; $\sec\theta = -3$; $\cot\theta = -\frac{\sqrt{2}}{4}$

51. $\cos\theta = -\frac{\sqrt{5}}{3}$; $\tan\theta = -\frac{2\sqrt{5}}{5}$; $\csc\theta = \frac{3}{2}$; $\sec\theta = -\frac{3\sqrt{5}}{5}$; $\cot\theta = -\frac{\sqrt{5}}{2}$ **53.** $\sin\theta = -\frac{\sqrt{3}}{2}$; $\cos\theta = \frac{1}{2}$; $\tan\theta = -\sqrt{3}$; $\csc\theta = -\frac{2\sqrt{3}}{3}$; $\cot\theta = -\frac{\sqrt{3}}{3}$ **55.** $\sin\theta = -\frac{3}{5}$; $\cos\theta = -\frac{4}{5}$; $\csc\theta = -\frac{5}{3}$; $\sec\theta = -\frac{5}{4}$; $\cot\theta = \frac{4}{3}$ **57.** $\sin\theta = \frac{\sqrt{10}}{10}$; $\cos\theta = -\frac{3\sqrt{10}}{10}$; $\csc\theta = \sqrt{10}$; $\sec\theta = -\frac{\sqrt{10}}{3}$; $\cot\theta = -3$ **59.** $-\frac{\sqrt{3}}{2}$ **61.** $-\frac{\sqrt{3}}{3}$ **63.** 2 **65.** -1 **67.** -1 **69.** $\frac{\sqrt{2}}{2}$ **71.** 0 **73.** $-\sqrt{2}$ **75.** $\frac{2\sqrt{3}}{3}$

77. 1 **79.** 1 **81.** 0 **83.** 1 **85.** -1 **87.** 0 **89.** 0.9 **91.** 9 **93.** 0 **95.** All real numbers **97.** Odd multiples of $\frac{\pi}{2}$ **99.** Odd multiples of $\frac{\pi}{2}$

101. $-1 \le y \le 1$ **103.** All real numbers **105.** $|y| \ge 1$ **107.** Odd; yes; origin **109.** Odd; yes; origin **111.** Even; yes; y-axis

113. (a) $-\frac{1}{3}$ **(b)** 1 **115. (a)** -2 **(b)** 6 **117. (a)** -4 **(b)** -12 **119.** ≈ 15.81 min **121.** Let a be a real number and $P = (x, y)$ be the point on the unit circle that corresponds to t. Consider the equation $\tan t = \frac{y}{x} = a$. Then $y = ax$. But $x^2 + y^2 = 1$, so $x^2 + a^2x^2 = 1$. So $x = \pm\frac{1}{\sqrt{1+a^2}}$ and $y = \pm\frac{a}{\sqrt{1+a^2}}$; that is, for any real number a, there is a point $P = (x, y)$ on the unit circle for which $\tan t = a$. In other words, the range of the tangent function is the set of all real numbers. **123.** Suppose that there is a number p, $0 < p < 2\pi$, for which $\sin(\theta + p) = \sin\theta$ for all θ. If $\theta = 0$, then $\sin(0 + p) = \sin p = \sin 0 = 0$, so $p = \pi$. If $\theta = \frac{\pi}{2}$, then $\sin\left(\frac{\pi}{2} + p\right) = \sin\left(\frac{\pi}{2}\right)$. But $p = \pi$. Thus, $\sin\left(\frac{3\pi}{2}\right) = -1 = \sin\left(\frac{\pi}{2}\right) = 1$. This is impossible. Therefore, the smallest positive number p for which $\sin(\theta + p) = \sin\theta$ for all θ is 2π. **125.** $\sec\theta = \frac{1}{\cos\theta}$; since $\cos\theta$ has period 2π, so does $\sec\theta$. **127.** If $P = (a, b)$ is the point on the unit circle corresponding to θ, then $Q = (-a, -b)$ is the point on the unit circle corresponding to $\theta + \pi$. Thus, $\tan(\theta + \pi) = \frac{-b}{-a} = \frac{b}{a} = \tan\theta$. Suppose that there exists a number p, $0 < p < \pi$, for which $\tan(\theta + p) = \tan\theta$ for all θ. Then, if $\theta = 0$, then $\tan p = \tan 0 = 0$. But this means that p is a multiple of π. Since no multiple of π exists in the interval $(0, \pi)$, this is a contradiction. Therefore, the period of $f(\theta) = \tan\theta$ is π.

129. Let $P = (a, b)$ be the point on the unit circle corresponding to θ. Then $\csc\theta = \frac{1}{b} = \frac{1}{\sin\theta}$; $\sec\theta = \frac{1}{a} = \frac{1}{\cos\theta}$; $\cot\theta = \frac{a}{b} = \frac{1}{b/a} = \frac{1}{\tan\theta}$.

131. $(\sin\theta\cos\phi)^2 + (\sin\theta\sin\phi)^2 + \cos^2\theta = \sin^2\theta\cos^2\phi + \sin^2\theta\sin^2\phi + \cos^2\theta = \sin^2\theta(\cos^2\phi + \sin^2\phi) + \cos^2\theta = \sin^2\theta + \cos^2\theta = 1$

137. Even **138.** **139.** Yes **140.** -27

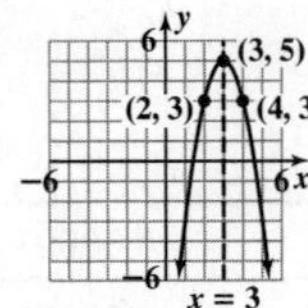

2.4 Assess Your Understanding *(page 155)*

3. $1; \frac{\pi}{2}$ **4.** $3; \pi$ **5.** $3; \frac{\pi}{3}$ **6.** T **7.** F **8.** T **9.** d **10.** d **11. (a)** 0 **(b)** $-\frac{\pi}{2} < x < \frac{\pi}{2}$ **(c)** 1 **(d)** $0, \pi, 2\pi$

(e) $f(x) = 1$ for $x = -\frac{3\pi}{2}, \frac{\pi}{2}$; $f(x) = -1$ for $x = -\frac{\pi}{2}, \frac{3\pi}{2}$ **(f)** $-\frac{5\pi}{6}, -\frac{\pi}{6}, \frac{7\pi}{6}, \frac{11\pi}{6}$ **(g)** $\{x \mid x = k\pi, k \text{ an integer}\}$ **13.** Amplitude $= 2$; period $= 2\pi$

15. Amplitude $= 4$; period $= \pi$ **17.** Amplitude $= 6$; period $= 2$ **19.** Amplitude $= \frac{1}{2}$; period $= \frac{4\pi}{3}$ **21.** Amplitude $= \frac{5}{3}$; period $= 3$

23. F **25.** A **27.** H **29.** C **31.** J

33.

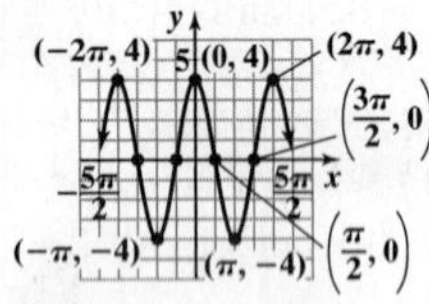

Domain: $(-\infty, \infty)$
Range: $[-4, 4]$

35.

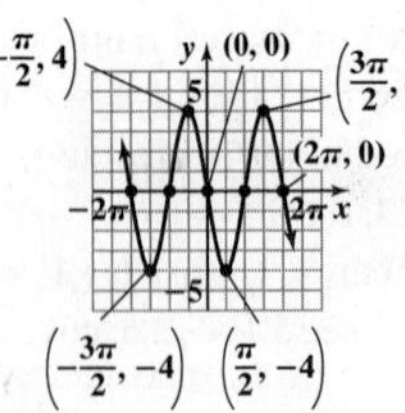

Domain: $(-\infty, \infty)$
Range: $[-4, 4]$

37.

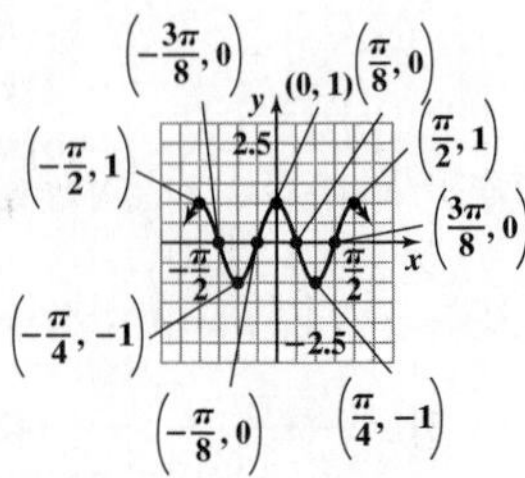

Domain: $(-\infty, \infty)$
Range: $[-1, 1]$

39.

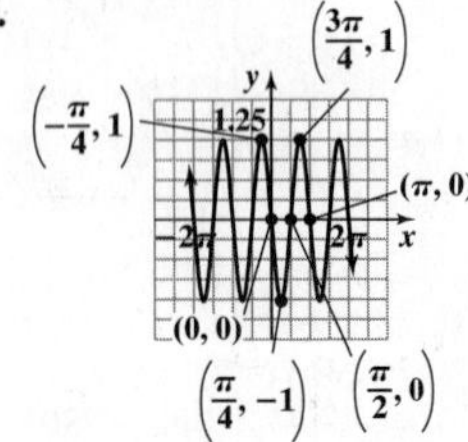

Domain: $(-\infty, \infty)$
Range: $[-1, 1]$

41.

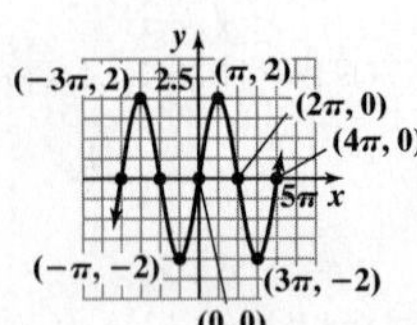

Domain: $(-\infty, \infty)$
Range: $[-2, 2]$

43.

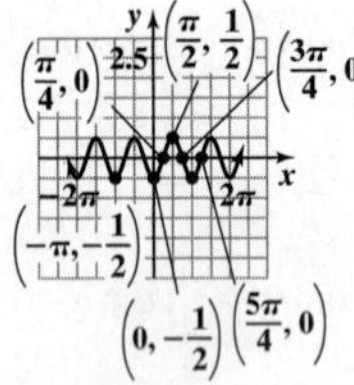

Domain: $(-\infty, \infty)$
Range: $\left[-\frac{1}{2}, \frac{1}{2}\right]$

45.

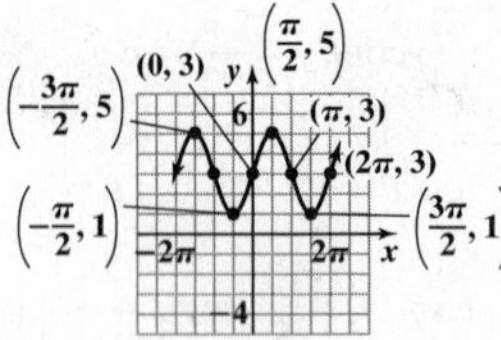

Domain: $(-\infty, \infty)$
Range: $[1, 5]$

47.

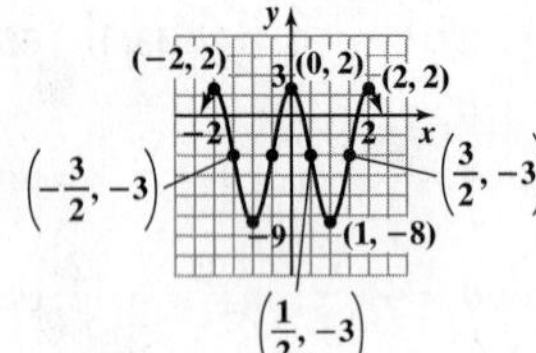

Domain: $(-\infty, \infty)$
Range: $[-8, 2]$

49.

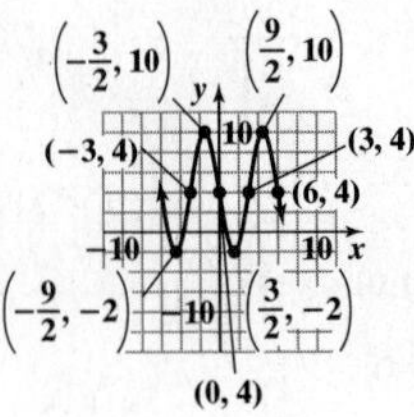

Domain: $(-\infty, \infty)$
Range: $[-2, 10]$

51.

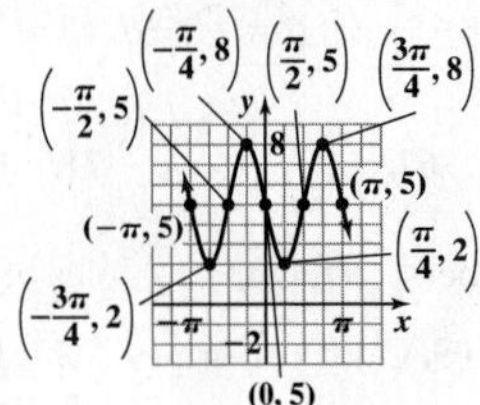

Domain: $(-\infty, \infty)$
Range: $[2, 8]$

53.

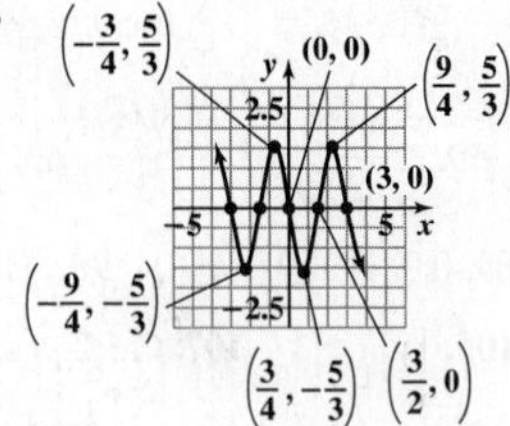

Domain: $(-\infty, \infty)$
Range: $\left[-\frac{5}{3}, \frac{5}{3}\right]$

55.

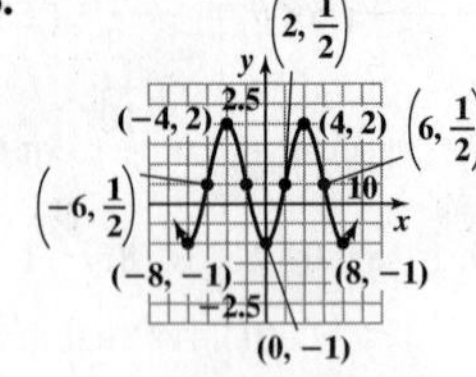

Domain: $(-\infty, \infty)$
Range: $[-1, 2]$

57. $y = \pm 3\sin(2x)$ **59.** $y = \pm 3\sin(\pi x)$ **61.** $y = 5\cos\left(\frac{\pi}{4}x\right)$ **63.** $y = -3\cos\left(\frac{1}{2}x\right)$ **65.** $y = \frac{3}{4}\sin(2\pi x)$ **67.** $y = -\sin\left(\frac{3}{2}x\right)$

69. $y = -\cos\left(\frac{4\pi}{3}x\right) + 1$ **71.** $y = 3\sin\left(\frac{\pi}{2}x\right)$ **73.** $y = -4\cos(3x)$ **75.** $\frac{2}{\pi}$ **77.** $\frac{\sqrt{2}}{\pi}$

79. $f(g(x)) = \sin(4x)$

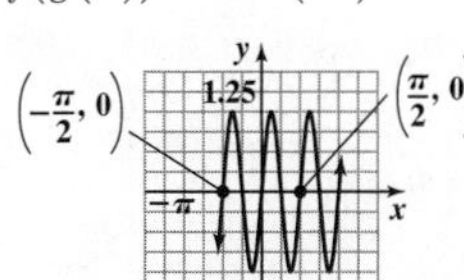

$g(f(x)) = 4\sin x$

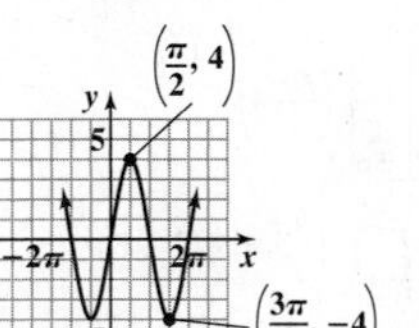

81. $f(g(x)) = -2\cos x$

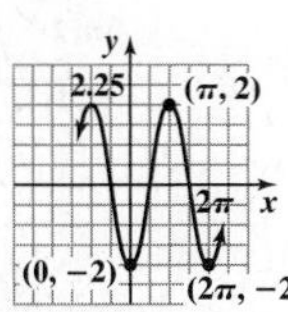

$g(f(x)) = \cos(-2x)$

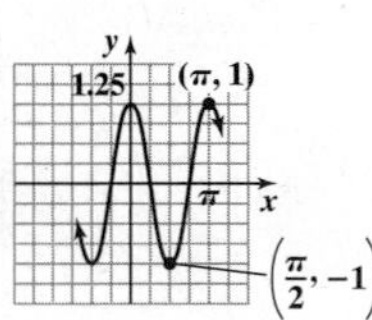

83.

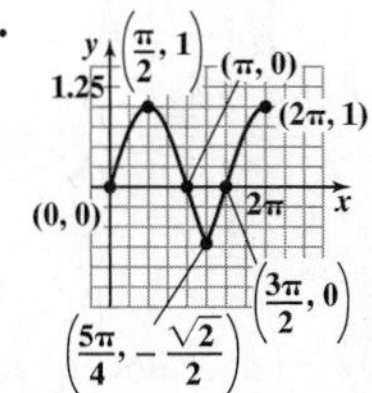

85. Period $= \frac{1}{30}$ s

Amplitude $= 220$ amp

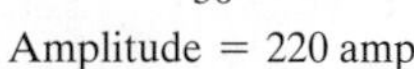

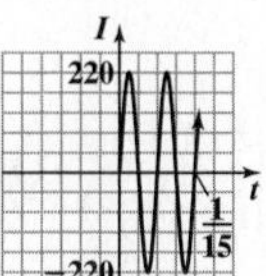

87. (a) Amplitude $= 220$ V

Period $= \frac{1}{60}$ s

(b), (e)

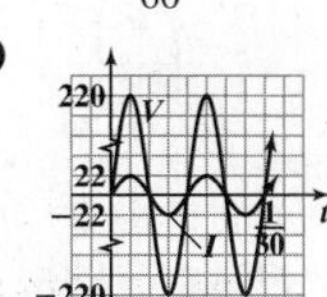

(c) $I(t) = 22\sin(120\pi t)$

(d) Amplitude $= 22$ amp

Period $= \frac{1}{60}$ s

89. (a) $P(t) = \frac{[V_0 \sin(2\pi f t)]^2}{R} = \frac{V_0^2}{R}\sin^2(2\pi f t)$ **(b)** Since the graph of P has amplitude $\frac{V_0^2}{2R}$ and period $\frac{1}{2f}$ and is of the form $y = A\cos(\omega t) + B$, then $A = -\frac{V_0^2}{2R}$ and $B = \frac{V_0^2}{2R}$. Since $\frac{1}{2f} = \frac{2\pi}{\omega}$, then $\omega = 4\pi f$. Therefore, $P(t) = -\frac{V_0^2}{2R}\cos(4\pi f t) + \frac{V_0^2}{2R} = \frac{V_0^2}{2R}[1 - \cos(4\pi f t)]$.

91. (a) Physical potential: $\omega = \frac{2\pi}{23}$; emotional potential: $\omega = \frac{\pi}{14}$; intellectual potential: $\omega = \frac{2\pi}{33}$

(b)

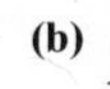

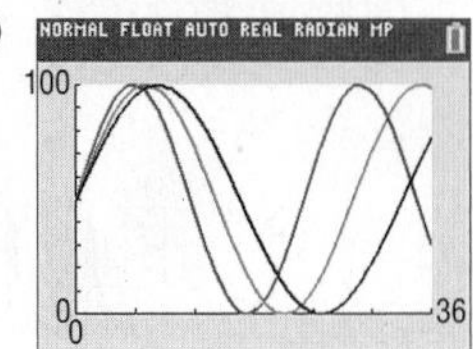

(c) No

(d) Physical potential peaks at 15 days after 20th birthday. Emotional potential is 50 at 17 days, with a maximum at 10 days and a minimum at 24 days. Intellectual potential starts fairly high, drops to a minimum at 13 days, and rises to a maximum at 29 days.

93.

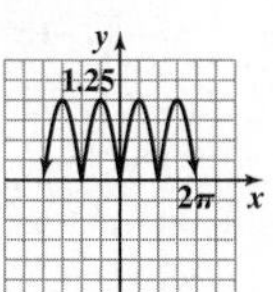

95. Answers may vary. $\left(-\frac{5\pi}{3}, \frac{1}{2}\right), \left(-\frac{\pi}{3}, \frac{1}{2}\right), \left(\frac{\pi}{3}, \frac{1}{2}\right), \left(\frac{5\pi}{3}, \frac{1}{2}\right)$ **97.** Answers may vary. $\left(-\frac{3\pi}{4}, 1\right), \left(\frac{\pi}{4}, 1\right), \left(\frac{5\pi}{4}, 1\right), \left(\frac{9\pi}{4}, 1\right)$

103. $2x + h - 5$ **104.** $-\frac{\sqrt{2}}{2}$ **105.** $(0, 5), \left(-\frac{5}{3}, 0\right), \left(-\frac{7}{3}, 0\right)$ **106.** $\frac{25\pi}{9}$ sq. units

2.5 Assess Your Understanding *(page 165)*

3. origin; odd multiples of $\frac{\pi}{2}$ **4.** y-axis; odd multiples of $\frac{\pi}{2}$ **5.** b **6.** T **7.** 0 **9.** 1

11. $\sec x = 1$ for $x = -2\pi, 0, 2\pi$; $\sec x = -1$ for $x = -\pi, \pi$ **13.** $-\frac{3\pi}{2}, -\frac{\pi}{2}, \frac{\pi}{2}, \frac{3\pi}{2}$ **15.** $-\frac{3\pi}{2}, -\frac{\pi}{2}, \frac{\pi}{2}, \frac{3\pi}{2}$

17.

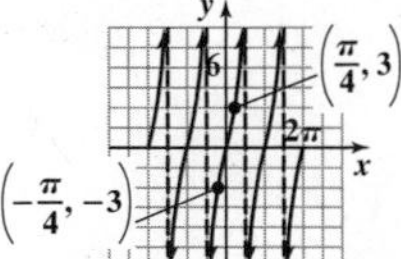

Domain: $\left\{x \middle| x \neq \frac{k\pi}{2}, k \text{ is an odd integer}\right\}$

Range: $(-\infty, \infty)$

19.

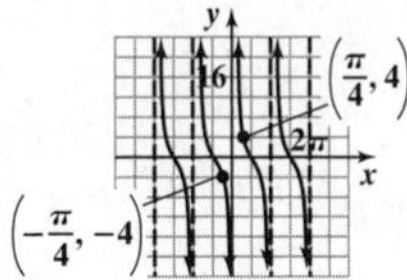

Domain: $\{x | x \neq k\pi, k \text{ is an integer}\}$

Range: $(-\infty, \infty)$

21.

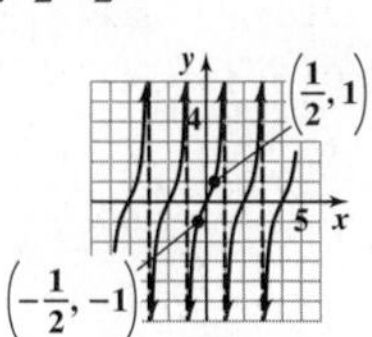

Domain: $\{x | x \text{ does not equal an odd integer}\}$

Range: $(-\infty, \infty)$

23.

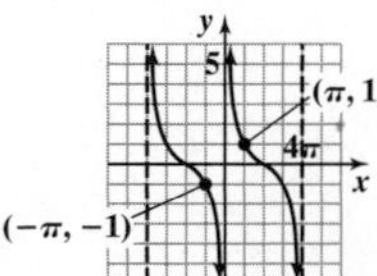

Domain: $\{x | x \neq 4k\pi, k \text{ is an integer}\}$

Range: $(-\infty, \infty)$

25.

Domain: $\left\{x \middle| x \neq \frac{k\pi}{2}, k \text{ is an odd integer}\right\}$

Range: $\{y | y \leq -2 \text{ or } y \geq 2\}$

27.

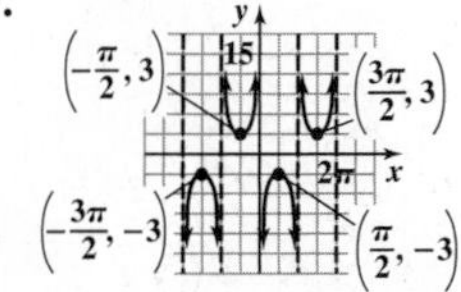

Domain: $\{x | x \neq k\pi, k \text{ is an integer}\}$

Range: $\{y | y \leq -3 \text{ or } y \geq 3\}$

29.

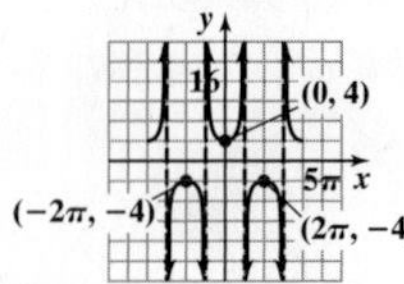

Domain: $\{x \mid x \neq k\pi, k \text{ is an odd integer}\}$
Range: $\{y \mid y \leq -4 \text{ or } y \geq 4\}$

31.

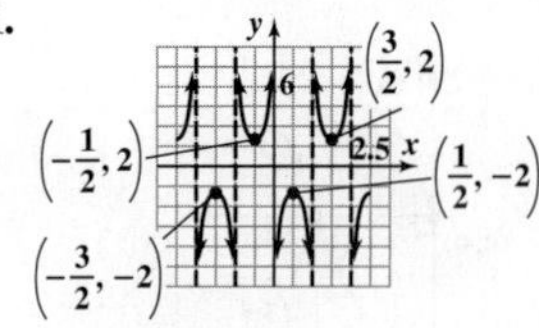

Domain: $\{x \mid x \text{ does not equal an integer}\}$
Range: $\{y \mid y \leq -2 \text{ or } y \geq 2\}$

33.

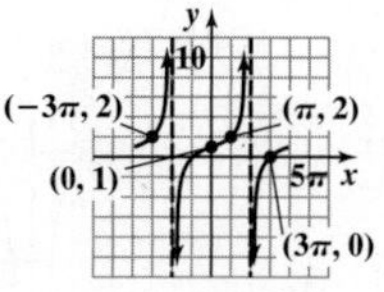

Domain: $\{x \mid x \neq 2\pi k, k \text{ is an odd integer}\}$
Range: $(-\infty, \infty)$

35.

Domain: $\left\{x \middle| x \neq \frac{3}{4}k, k \text{ is an odd integer}\right\}$
Range: $\{y \mid y \leq 1 \text{ or } y \geq 3\}$

37.

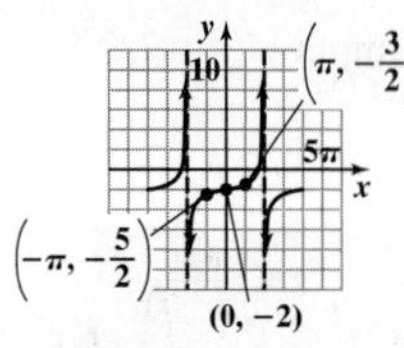

Domain: $\{x \mid x \neq 2\pi k, k \text{ is an odd integer}\}$
Range: $(-\infty, \infty)$

39.

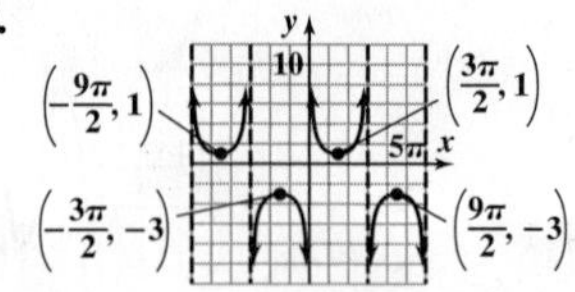

Domain: $\{x \mid x \neq 3\pi k, k \text{ is an integer}\}$
Range: $\{y \mid y \leq -3 \text{ or } y \geq 1\}$

41. $\frac{2\sqrt{3}}{\pi}$ **43.** $\frac{6\sqrt{3}}{\pi}$ **45.** $f(g(x)) = \tan(4x)$ $g(f(x)) = 4\tan x$

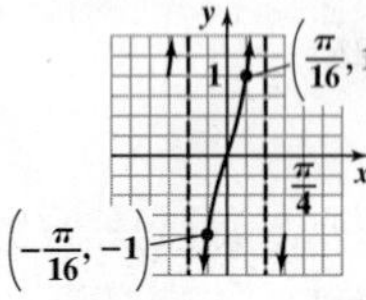

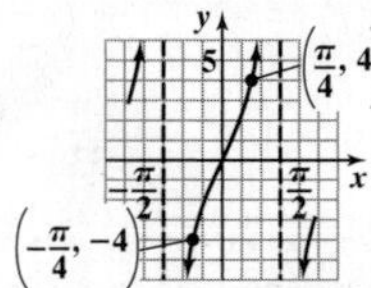

47. $f(g(x)) = -2\cot x$ $g(f(x)) = \cot(-2x)$

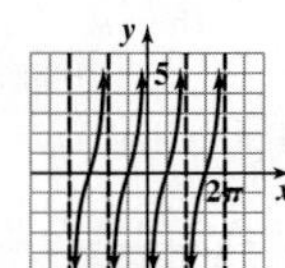

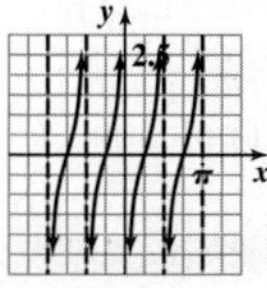

49.

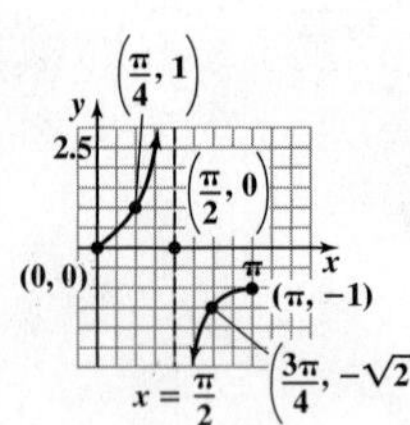

51. (a) $L(\theta) = \dfrac{3}{\cos\theta} + \dfrac{4}{\sin\theta} = 3\sec\theta + 4\csc\theta$

(b)

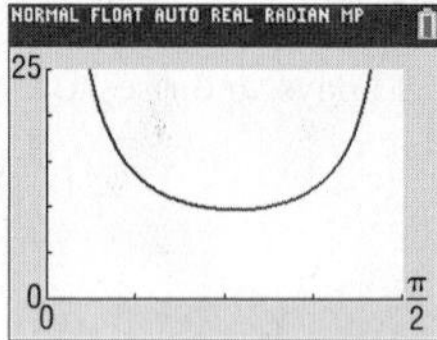

(c) ≈ 0.83 **(d)** ≈ 9.86 ft

53.

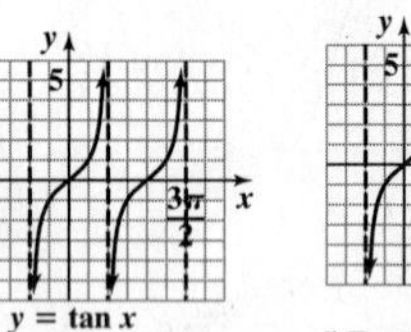

54. x-axis **55.** $(x + 3)^2 + (y - 1)^2 = 9$
56. $h(4) = -\frac{2}{13}$ **57.** Not a function

2.6 Assess Your Understanding *(page 175)*

1. phase shift **2.** False
3. Amplitude $= 4$
Period $= \pi$
Phase shift $= \frac{\pi}{2}$

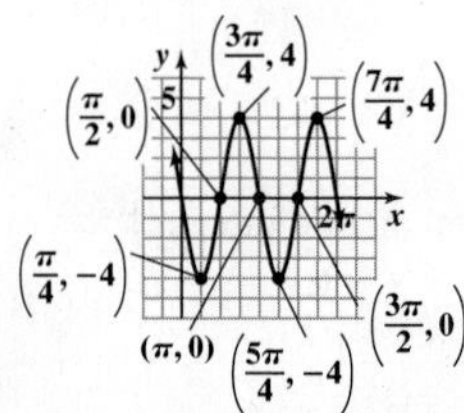

5. Amplitude $= 2$
Period $= \frac{2\pi}{3}$
Phase shift $= -\frac{\pi}{6}$

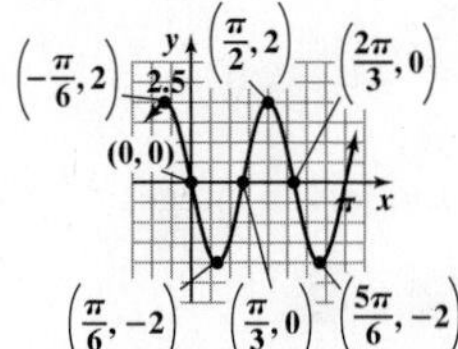

7. Amplitude $= 3$
Period $= \pi$
Phase shift $= -\frac{\pi}{4}$

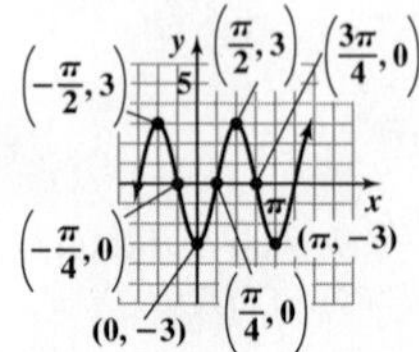

9. Amplitude $= 4$
Period $= 2$
Phase shift $= -\dfrac{2}{\pi}$

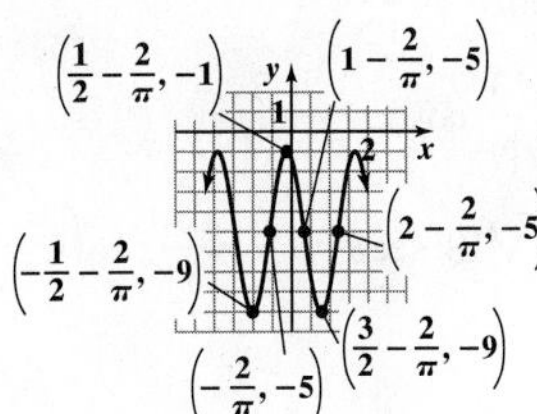

11. Amplitude $= 3$
Period $= 2$
Phase shift $= \dfrac{2}{\pi}$

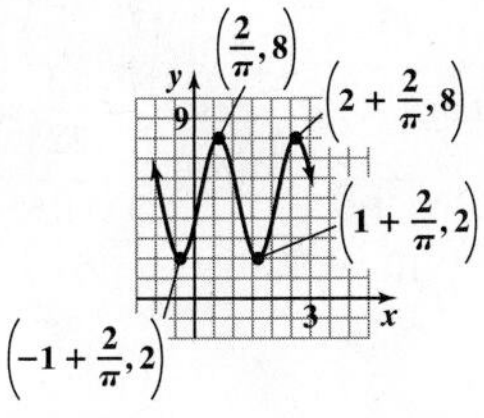

13. Amplitude $= 3$
Period $= \pi$
Phase shift $= \dfrac{\pi}{4}$

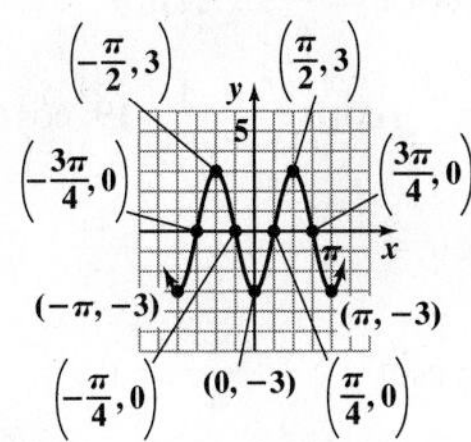

15. $y = 2\sin\left[2\left(x - \frac{1}{2}\right)\right]$ or $y = 2\sin(2x - 1)$

17. $y = 3\sin\left[\frac{2}{3}\left(x + \frac{1}{3}\right)\right]$ or $y = 3\sin\left(\frac{2}{3}x + \frac{2}{9}\right)$

19.

21.

23.

25.

27. Period $= \dfrac{1}{15}$ s
Amplitude $= 120$ amp
Phase shift $= \dfrac{1}{90}$ s

29. (a)

(b) $y = 8.5\sin\left[\frac{2\pi}{5}\left(x - \frac{11}{4}\right)\right] + 24.5$

or

$y = 8.5\sin\left(\frac{2\pi}{5}x - \frac{11\pi}{10}\right) + 24.5$

(c)

(d) $y = 9.46\sin(1.247x + 2.096) + 24.088$

(e)

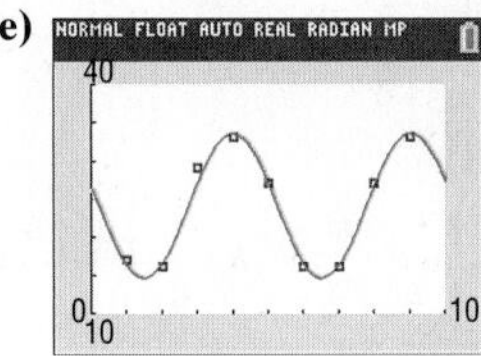

31. (a)

(b) $y = 23.65\sin\left[\frac{\pi}{6}(x - 4)\right] + 51.75$ or $y = 23.65\sin\left(\frac{\pi}{6}x - \frac{2\pi}{3}\right) + 51.75$

(c)

(d) $y = 24.25\sin(0.493x - 1.927) + 51.61$

(e)

33. (a) 6:55 PM **(b)** $y = 3.12\sin\left[\frac{24\pi}{149}(x - 3.3959)\right] + 2.74$ or $y = 3.12\sin\left[\frac{24\pi}{149}x - 1.7184\right] + 2.74$ **(c)** 1.49 ft

35. (a) $y = 1.615\sin\left(\frac{2\pi}{365}x - 1.39\right) + 12.135$
(b) 12.42 h
(c)

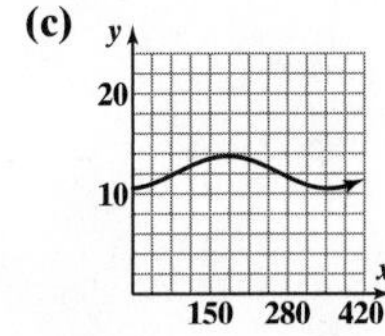

(d) The actual hours of sunlight on April 1, 2014, were 12.43 hours. This is close to the predicted amount of 12.42 hours.

37. (a) $y = 6.96\sin\left(\frac{2\pi}{365}x - 1.39\right) + 12.41$
(b) 13.63 h
(c)

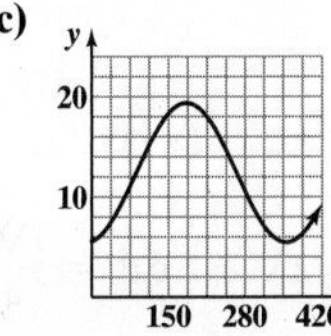

(d) The actual hours of sunlight on April 1, 2014, were 13.37 hours. This is close to the predicted amount of 13.63 hours.

41. Local minima: $f(1.76) \approx 1.53$; Local maxima: $f(-0.76) \approx 17.47$ **42.** -11 **43.** $\theta = 105°$ **44.** $-2, 4$

Review Exercises *(page 182)*

1. $\frac{3\pi}{4}$ **2.** $\frac{\pi}{10}$ **3.** 135° **4.** −450° **5.** $\frac{1}{2}$ **6.** $\frac{3\sqrt{2}}{2}-\frac{4\sqrt{3}}{3}$ **7.** $-3\sqrt{2}-2\sqrt{3}$ **8.** 3 **9.** 0 **10.** 0 **11.** 1 **12.** 1 **13.** 1 **14.** −1 **15.** 1

16. $\cos\theta=\frac{3}{5}$; $\tan\theta=\frac{4}{3}$; $\csc\theta=\frac{5}{4}$; $\sec\theta=\frac{5}{3}$; $\cot\theta=\frac{3}{4}$ **17.** $\sin\theta=-\frac{12}{13}$; $\cos\theta=-\frac{5}{13}$; $\csc\theta=-\frac{13}{12}$; $\sec\theta=-\frac{13}{5}$; $\cot\theta=\frac{5}{12}$

18. $\sin\theta=\frac{3}{5}$; $\cos\theta=-\frac{4}{5}$; $\tan\theta=-\frac{3}{4}$; $\csc\theta=\frac{5}{3}$; $\cot\theta=-\frac{4}{3}$ **19.** $\cos\theta=-\frac{5}{13}$; $\tan\theta=-\frac{12}{5}$; $\csc\theta=\frac{13}{12}$; $\sec\theta=-\frac{13}{5}$; $\cot\theta=-\frac{5}{12}$

20. $\cos\theta=\frac{12}{13}$; $\tan\theta=-\frac{5}{12}$; $\csc\theta=-\frac{13}{5}$; $\sec\theta=\frac{13}{12}$; $\cot\theta=-\frac{12}{5}$ **21.** $\sin\theta=-\frac{\sqrt{10}}{10}$; $\cos\theta=-\frac{3\sqrt{10}}{10}$; $\csc\theta=-\sqrt{10}$; $\sec\theta=-\frac{\sqrt{10}}{3}$; $\cot\theta=3$

22. $\sin\theta=-\frac{2\sqrt{2}}{3}$; $\cos\theta=\frac{1}{3}$; $\tan\theta=-2\sqrt{2}$; $\csc\theta=-\frac{3\sqrt{2}}{4}$; $\cot\theta=-\frac{\sqrt{2}}{4}$

23. $\sin\theta=\frac{\sqrt{5}}{5}$; $\cos\theta=-\frac{2\sqrt{5}}{5}$; $\tan\theta=-\frac{1}{2}$; $\csc\theta=\sqrt{5}$; $\sec\theta=-\frac{\sqrt{5}}{2}$

24.

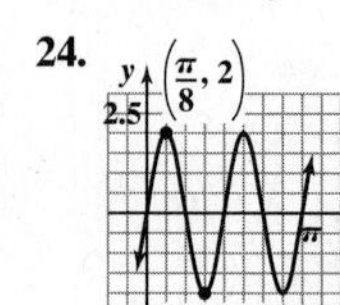

Domain: $(-\infty,\infty)$
Range: $[-2,2]$

25.

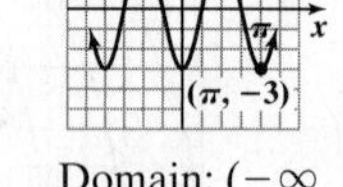

Domain: $(-\infty,\infty)$
Range: $[-3,3]$

26.

Domain: $\left\{x \middle| x\neq\frac{k\pi}{2}, k \text{ is an odd integer}\right\}$
Range: $(-\infty,\infty)$

27.

Domain: $\left\{x \middle| x\neq\frac{\pi}{6}+k\cdot\frac{\pi}{3}, k \text{ is an integer}\right\}$
Range: $(-\infty,\infty)$

28.

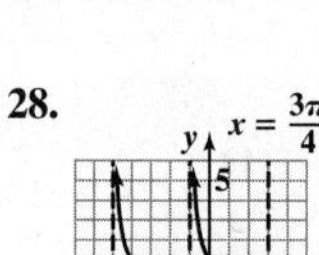

Domain: $\left\{x \middle| x\neq-\frac{\pi}{4}+k\pi, k \text{ is an integer}\right\}$
Range: $(-\infty,\infty)$

29.

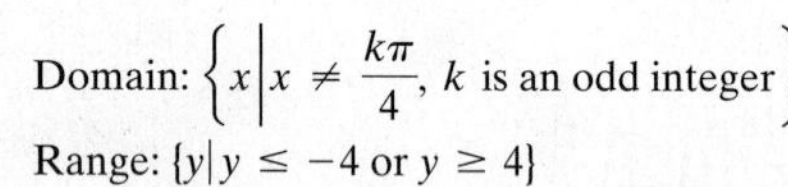

Domain: $\left\{x \middle| x\neq\frac{k\pi}{4}, k \text{ is an odd integer}\right\}$
Range: $\{y \mid y\leq-4 \text{ or } y\geq 4\}$

30.

Domain: $\left\{x \middle| x\neq-\frac{\pi}{4}+k\pi, k \text{ is an integer}\right\}$
Range: $\{y \mid y\leq-1 \text{ or } y\geq 1\}$

31.

Domain: $(-\infty,\infty)$
Range: $[-6,2]$

32.

Domain: $\left\{x \middle| x\neq\frac{3\pi}{4}+k\cdot 3\pi, k \text{ is an integer}\right\}$
Range: $(-\infty,\infty)$

33. Amplitude = 1; Period = π **34.** Amplitude = 2; Period = $\frac{2}{3}$

35. Amplitude = 4
Period = $\frac{2\pi}{3}$
Phase shift = 0

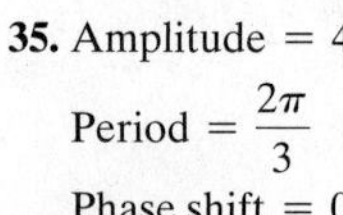

36. Amplitude = 1
Period = 4π
Phase shift = $-\pi$

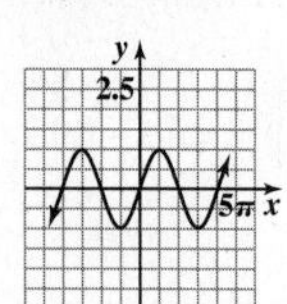

37. Amplitude = $\frac{1}{2}$
Period = $\frac{4\pi}{3}$
Phase shift = $\frac{2\pi}{3}$

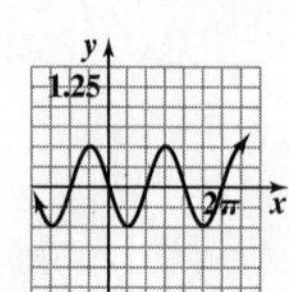

38. Amplitude = $\frac{2}{3}$
Period = 2
Phase shift = $\frac{6}{\pi}$

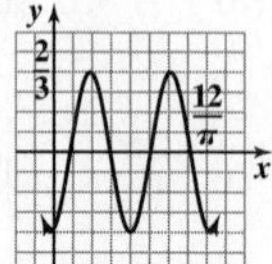

39. $y=5\cos\frac{x}{4}$ **40.** $y=-7\sin\left(\frac{\pi}{4}x\right)$ **41.** 0.38 **42.** 1.02 **43.** Sine, cosine, cosecant, and secant: negative; tangent and cotangent: positive

44. IV **45.** $\sin\theta=\frac{2\sqrt{2}}{3}$; $\cos\theta=-\frac{1}{3}$; $\tan\theta=-2\sqrt{2}$; $\csc\theta=\frac{3\sqrt{2}}{4}$; $\sec\theta=-3$; $\cot\theta=-\frac{\sqrt{2}}{4}$

46. $\sin t=\frac{5\sqrt{29}}{29}$, $\cos t=-\frac{2\sqrt{29}}{29}$, $\tan t=-\frac{5}{2}$ **47.** Domain: $\left\{x \middle| x\neq \text{odd multiple of } \frac{\pi}{2}\right\}$; range: $\{y \mid |y|\geq 1\}$; period = 2π

48. **(a)** 32.34° **(b)** 63°10′48″

49. $\frac{\pi}{3}\approx 1.05$ ft; $\frac{\pi}{3}\approx 1.05$ ft^2 **50.** $8\pi\approx 25.13$ in.; $\frac{16\pi}{3}\approx 16.76$ in. **51.** Approximately 114.59 revolutions/hr

52. 0.1 revolution/sec = $\frac{\pi}{5}$ radian/sec

53. (a) $\frac{1}{15}$ **(b)** 220 **(c)** $-\frac{1}{180}$ **(d)**

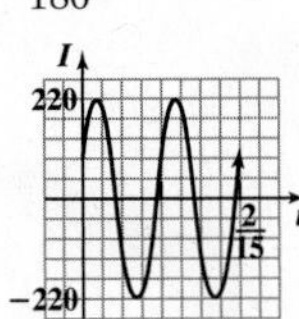

54. (a)

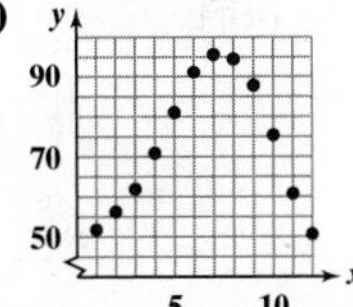

(b) $y = 20\sin\left[\frac{\pi}{6}(x-4)\right] + 75$ or $y = 20\sin\left(\frac{\pi}{6}x - \frac{2\pi}{3}\right) + 75$

(c)

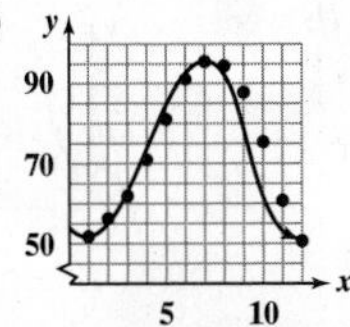

(d) $y = 19.81\sin(0.543x - 2.296) + 75.66$

(e)

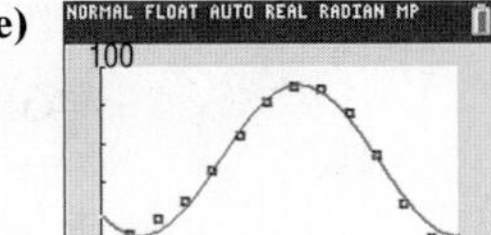

55.

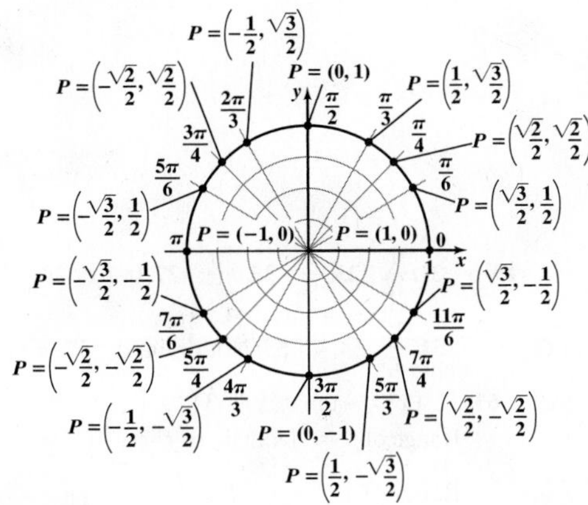

Chapter Test *(page 184)*

1. $\frac{13\pi}{9}$ **2.** $-\frac{20\pi}{9}$ **3.** $\frac{13\pi}{180}$ **4.** $-22.5°$ **5.** 810° **6.** 135° **7.** $\frac{1}{2}$ **8.** 0 **9.** $-\frac{1}{2}$ **10.** $-\frac{\sqrt{3}}{3}$ **11.** 2 **12.** $\frac{3(1-\sqrt{2})}{2}$ **13.** 0.292 **14.** 0.309 **15.** -1.524 **16.** 2.747 **17.**

	$\sin\theta$	$\cos\theta$	$\tan\theta$	$\sec\theta$	$\csc\theta$	$\cot\theta$
θ in QI	+	+	+	+	+	+
θ in QII	+	−	−	−	+	−
θ in QIII	−	−	+	−	−	+
θ in QIV	−	+	−	+	−	−

18. $-\frac{3}{5}$

19. $\cos\theta = -\frac{2\sqrt{6}}{7}$; $\tan\theta = -\frac{5\sqrt{6}}{12}$; $\csc\theta = \frac{7}{5}$; $\sec\theta = -\frac{7\sqrt{6}}{12}$; $\cot\theta = -\frac{2\sqrt{6}}{5}$ **20.** $\sin\theta = -\frac{\sqrt{5}}{3}$; $\tan\theta = -\frac{\sqrt{5}}{2}$; $\csc\theta = -\frac{3\sqrt{5}}{5}$; $\sec\theta = \frac{3}{2}$; $\cot\theta = -\frac{2\sqrt{5}}{5}$ **21.** $\sin\theta = \frac{12}{13}$; $\cos\theta = -\frac{5}{13}$; $\csc\theta = \frac{13}{12}$; $\sec\theta = -\frac{13}{5}$; $\cot\theta = -\frac{5}{12}$ **22.** $\frac{7\sqrt{53}}{53}$ **23.** $-\frac{5\sqrt{146}}{146}$ **24.** $-\frac{1}{2}$

25.

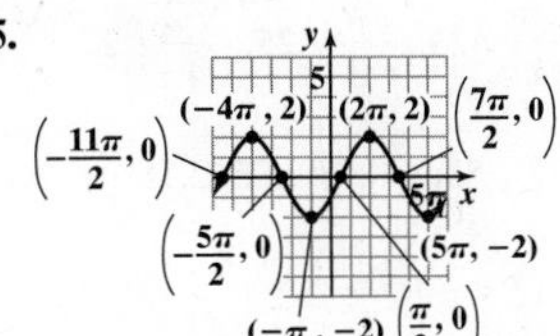

26.

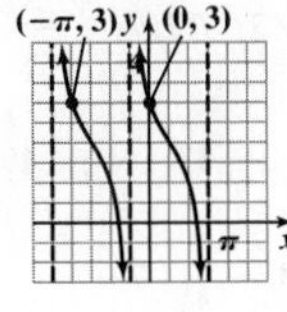

27. $y = -3\sin\left(3x + \frac{3\pi}{4}\right)$ **28.** 78.93 ft^2 **29.** 143.5 rpm

Cumulative Review *(page 185)*

1. $\left\{-1, \frac{1}{2}\right\}$ **2.** $y - 5 = -3(x+2)$ or $y = -3x - 1$ **3.** $x^2 + (y+2)^2 = 16$

4. A line; slope $\frac{2}{3}$; intercepts $(6, 0)$ and $(0, -4)$

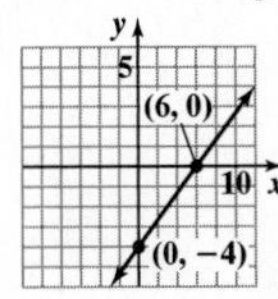

5. A circle; center $(1, -2)$; radius 3

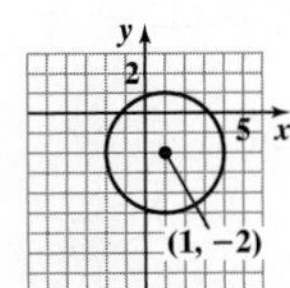

6.

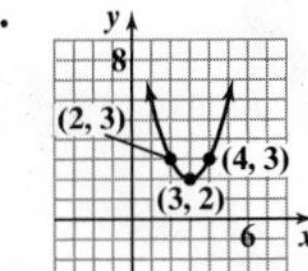

7. (a)
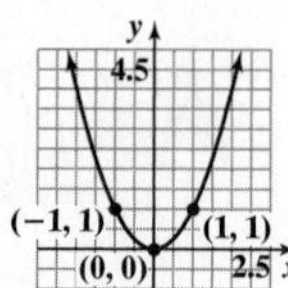

(b)
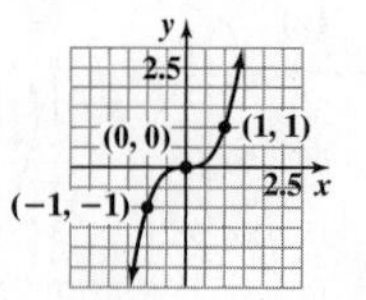

(c)
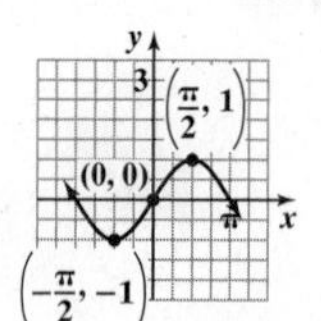

(d)
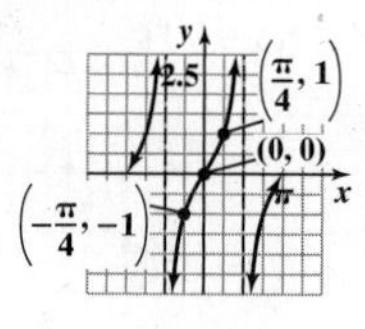

8. $f^{-1}(x) = \frac{1}{3}(x + 2)$ **9.** -2 **10.**
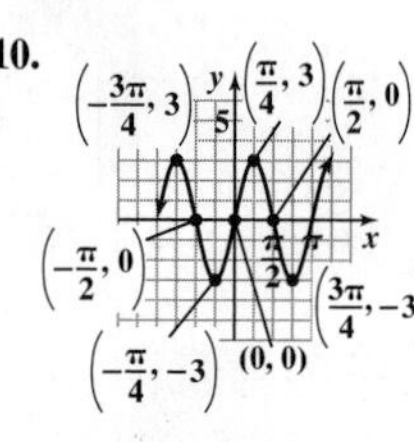

11. $3 - \frac{3\sqrt{3}}{2}$ **12.** $y = 3\cos\left(\frac{\pi}{6}x\right)$

CHAPTER 3 Analytic Trigonometry

3.1 Assess Your Understanding *(page 198)*

7. $x = \sin y$ **8.** $0 \le x \le \pi$ **9.** $-\infty < x < \infty$ **10.** F **11.** T **12.** T **13.** d **14.** a **15.** 0 **17.** $-\frac{\pi}{2}$ **19.** 0 **21.** $\frac{\pi}{4}$ **23.** $\frac{\pi}{3}$ **25.** $\frac{5\pi}{6}$ **27.** 0.10 **29.** 1.37 **31.** 0.51 **33.** -0.38 **35.** -0.12 **37.** 1.08 **39.** $\frac{4\pi}{5}$ **41.** $-\frac{3\pi}{8}$ **43.** $-\frac{\pi}{8}$ **45.** $-\frac{\pi}{5}$ **47.** $\frac{\pi}{4}$ **49.** Not defined **51.** $\frac{1}{4}$ **53.** 4 **55.** Not defined **57.** π

59. $f^{-1}(x) = \sin^{-1}\frac{x-2}{5}$
Range of f = Domain of $f^{-1} = [-3, 7]$
Range of $f^{-1} = \left[-\frac{\pi}{2}, \frac{\pi}{2}\right]$

61. $f^{-1}(x) = \frac{1}{3}\cos^{-1}\left(-\frac{x}{2}\right)$
Range of f = Domain of $f^{-1} = [-2, 2]$
Range of $f^{-1} = \left[0, \frac{\pi}{3}\right]$

63. $f^{-1}(x) = -\tan^{-1}(x + 3) - 1$
Range of f = Domain of $f^{-1} = (-\infty, \infty)$
Range of $f^{-1} = \left(-1 - \frac{\pi}{2}, \frac{\pi}{2} - 1\right)$

65. $f^{-1}(x) = \frac{1}{2}\left[\sin^{-1}\left(\frac{x}{3}\right) - 1\right]$
Range of f = Domain of $f^{-1} = [-3, 3]$
Range of $f^{-1} = \left[-\frac{1}{2} - \frac{\pi}{4}, -\frac{1}{2} + \frac{\pi}{4}\right]$

67. $\left\{\frac{\sqrt{2}}{2}\right\}$ **69.** $\left\{-\frac{1}{4}\right\}$ **71.** $\{\sqrt{3}\}$ **73.** $\{-1\}$
75. (a) 13.92 h or 13 h, 55 min **(b)** 12 h **(c)** 13.85 h or 13 h, 51 min
77. (a) 13.3 h or 13 h, 18 min **(b)** 12 h **(c)** 13.26 h or 13 h, 15 min
79. (a) 12 h **(b)** 12 h **(c)** 12 h **(d)** It is 12 h. **81.** 3.35 min

83. (a) $\frac{\pi}{3}$ square units **(b)** $\frac{5\pi}{12}$ square units **85.** 4250 mi **87.** 6 **88.** The graph passes the horizontal-line test.
89. 7 **90.** $\frac{\sqrt{3}}{4}$

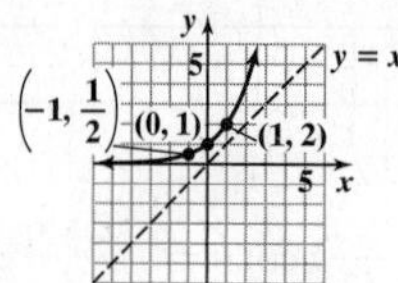

3.2 Assess Your Understanding *(page 205)*

4. $x = \sec y$; ≥ 1; 0; π **5.** cosine **6.** F **7.** T **8.** T **9.** $\frac{\sqrt{2}}{2}$ **11.** $-\frac{\sqrt{3}}{3}$ **13.** 2 **15.** $\sqrt{2}$ **17.** $-\frac{\sqrt{2}}{2}$ **19.** $\frac{2\sqrt{3}}{3}$ **21.** $\frac{3\pi}{4}$ **23.** $-\frac{\pi}{3}$ **25.** $\frac{\sqrt{2}}{4}$ **27.** $\frac{\sqrt{5}}{2}$ **29.** $-\frac{\sqrt{14}}{2}$ **31.** $-\frac{3\sqrt{10}}{10}$ **33.** $\sqrt{5}$ **35.** $-\frac{\pi}{4}$ **37.** $\frac{\pi}{6}$ **39.** $-\frac{\pi}{2}$ **41.** $\frac{\pi}{6}$ **43.** $\frac{2\pi}{3}$ **45.** 1.32 **47.** 0.46 **49.** -0.34 **51.** 2.72 **53.** -0.73 **55.** 2.55 **57.** $\frac{1}{\sqrt{1+u^2}}$ **59.** $\frac{u}{\sqrt{1-u^2}}$ **61.** $\frac{\sqrt{u^2-1}}{|u|}$ **63.** $\frac{\sqrt{u^2-1}}{|u|}$ **65.** $\frac{1}{u}$ **67.** $\frac{5}{13}$ **69.** $\frac{3\pi}{4}$ **71.** $-\frac{3}{4}$ **73.** $\frac{5}{13}$ **75.** $\frac{\pi}{6}$ **77.** $-\sqrt{15}$
79. (a) $\theta = 31.89°$ **(b)** 54.64 ft in diameter **(c)** 37.96 ft high **81. (a)** $\theta = 22.3°$ **(b)** $v_0 = 2940.23$ ft/s

83.
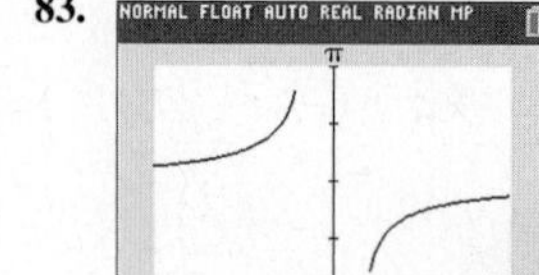

87. $\{x \mid x \ne -5, x \ne 5\}$ **88.** Neither **89.** $\frac{7\pi}{4}$ **90.** $\frac{5\pi}{2} \approx 7.85$ in.

3.3 Assess Your Understanding *(page 213)*

7. F **8.** T **9.** T **10.** F **11.** d **12.** a **13.** $\left\{\frac{7\pi}{6}, \frac{11\pi}{6}\right\}$ **15.** $\left\{\frac{7\pi}{6}, \frac{11\pi}{6}\right\}$ **17.** $\left\{\frac{3\pi}{4}, \frac{7\pi}{4}\right\}$ **19.** $\left\{\frac{2\pi}{3}, \frac{4\pi}{3}\right\}$ **21.** $\left\{\frac{3\pi}{4}, \frac{5\pi}{4}\right\}$ **23.** $\left\{\frac{\pi}{3}, \frac{2\pi}{3}, \frac{4\pi}{3}, \frac{5\pi}{3}\right\}$

25. $\left\{\frac{\pi}{4}, \frac{3\pi}{4}, \frac{5\pi}{4}, \frac{7\pi}{4}\right\}$ **27.** $\left\{\frac{\pi}{2}, \frac{7\pi}{6}, \frac{11\pi}{6}\right\}$ **29.** $\left\{\frac{\pi}{3}, \frac{2\pi}{3}, \frac{4\pi}{3}, \frac{5\pi}{3}\right\}$ **31.** $\left\{\frac{4\pi}{9}, \frac{8\pi}{9}, \frac{16\pi}{9}\right\}$ **33.** $\left\{\frac{3\pi}{4}, \frac{7\pi}{4}\right\}$ **35.** $\left\{\frac{11\pi}{6}\right\}$

37. $\left\{\theta \,\middle|\, \theta = \frac{\pi}{6} + 2k\pi, \theta = \frac{5\pi}{6} + 2k\pi\right\}; \frac{\pi}{6}, \frac{5\pi}{6}, \frac{13\pi}{6}, \frac{17\pi}{6}, \frac{25\pi}{6}, \frac{29\pi}{6}$ **39.** $\left\{\theta \,\middle|\, \theta = \frac{5\pi}{6} + k\pi\right\}; \frac{5\pi}{6}, \frac{11\pi}{6}, \frac{17\pi}{6}, \frac{23\pi}{6}, \frac{29\pi}{6}, \frac{35\pi}{6}$

41. $\left\{\theta \,\middle|\, \theta = \frac{\pi}{2} + 2k\pi, \theta = \frac{3\pi}{2} + 2k\pi\right\}; \frac{\pi}{2}, \frac{3\pi}{2}, \frac{5\pi}{2}, \frac{7\pi}{2}, \frac{9\pi}{2}, \frac{11\pi}{2}$ **43.** $\left\{\theta \,\middle|\, \theta = \frac{\pi}{3} + k\pi, \theta = \frac{2\pi}{3} + k\pi\right\}; \frac{\pi}{3}, \frac{2\pi}{3}, \frac{4\pi}{3}, \frac{5\pi}{3}, \frac{7\pi}{3}, \frac{8\pi}{3}$

45. $\left\{\theta \,\middle|\, \theta = \frac{8\pi}{3} + 4k\pi, \theta = \frac{10\pi}{3} + 4k\pi\right\}; \frac{8\pi}{3}, \frac{10\pi}{3}, \frac{20\pi}{3}, \frac{22\pi}{3}, \frac{32\pi}{3}, \frac{34\pi}{3}$ **47.** $\{0.41, 2.73\}$ **49.** $\{1.37, 4.51\}$ **51.** $\{2.69, 3.59\}$ **53.** $\{1.82, 4.46\}$

55. $\{2.08, 5.22\}$ **57.** $\{0.73, 2.41\}$ **59.** $\left\{\frac{\pi}{2}, \frac{2\pi}{3}, \frac{4\pi}{3}, \frac{3\pi}{2}\right\}$ **61.** $\left\{\frac{\pi}{2}, \frac{7\pi}{6}, \frac{11\pi}{6}\right\}$ **63.** $\left\{0, \frac{\pi}{4}, \frac{5\pi}{4}\right\}$ **65.** $\left\{\frac{\pi}{2}, \frac{2\pi}{3}, \frac{4\pi}{3}, \frac{3\pi}{2}\right\}$ **67.** $\{\pi\}$ **69.** $\left\{\frac{\pi}{4}, \frac{5\pi}{4}\right\}$

71. $\left\{0, \frac{\pi}{3}, \pi, \frac{5\pi}{3}\right\}$ **73.** $\left\{\frac{\pi}{6}, \frac{5\pi}{6}, \frac{3\pi}{2}\right\}$ **75.** $\left\{\frac{\pi}{2}\right\}$ **77.** $\{0\}$ **79.** $\left\{\frac{\pi}{3}, \frac{5\pi}{3}\right\}$ **81.** No real solution **83.** $-1.31, 1.98, 3.84$ **85.** 0.52

87. 1.26 **89.** $-1.02, 1.02$ **91.** 0, 2.15 **93.** 0.76, 1.35 **95.** $\frac{\pi}{3}, \frac{2\pi}{3}, \frac{4\pi}{3}, \frac{5\pi}{3}$

97. (a) $-2\pi, -\pi, 0, \pi, 2\pi, 3\pi, 4\pi$ **(b)**

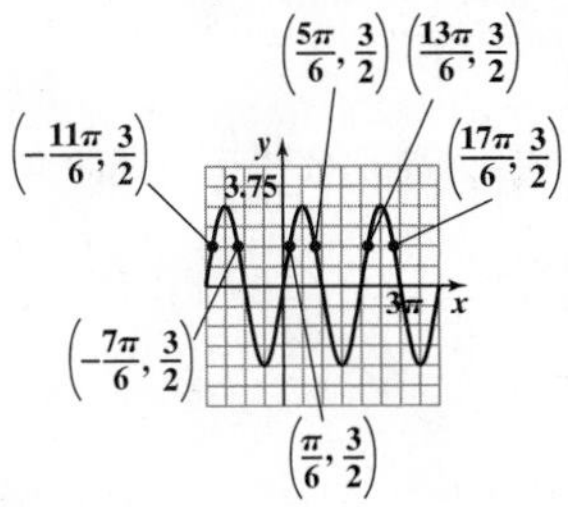

(c) $\left\{-\frac{11\pi}{6}, -\frac{7\pi}{6}, \frac{\pi}{6}, \frac{5\pi}{6}, \frac{13\pi}{6}, \frac{17\pi}{6}\right\}$

(d) $\left\{x \,\middle|\, -\frac{11\pi}{6} < x < -\frac{7\pi}{6} \text{ or } \frac{\pi}{6} < x < \frac{5\pi}{6} \text{ or } \frac{13\pi}{6} < x < \frac{17\pi}{6}\right\}$

99. (a) $\left\{x \,\middle|\, x = -\frac{\pi}{4} + k\pi, k \text{ is any integer}\right\}$ **(b)** $-\frac{\pi}{2} < x < -\frac{\pi}{4}$ or $\left(-\frac{\pi}{2}, -\frac{\pi}{4}\right)$

101. (a), (d)

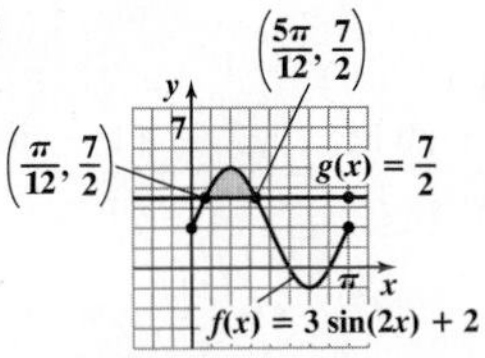

(b) $\left\{\frac{\pi}{12}, \frac{5\pi}{12}\right\}$

(c) $\left\{x \,\middle|\, \frac{\pi}{12} < x < \frac{5\pi}{12}\right\}$ or $\left(\frac{\pi}{12}, \frac{5\pi}{12}\right)$

103. (a), (d)

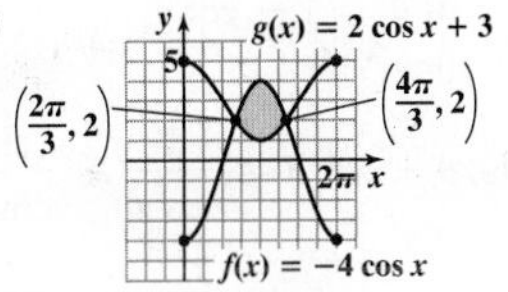

(b) $\left\{\frac{2\pi}{3}, \frac{4\pi}{3}\right\}$

(c) $\left\{x \,\middle|\, \frac{2\pi}{3} < x < \frac{4\pi}{3}\right\}$ or $\left(\frac{2\pi}{3}, \frac{4\pi}{3}\right)$

105. (a) 0 s, 0.43 s, 0.86 s **(b)** 0.21 s **(c)** $[0, 0.03] \cup [0.39, 0.43] \cup [0.86, 0.89]$ **107. (a)** 150 mi **(b)** 6.06, 8.44, 15.72, 18.11 min **(c)** Before 6.06 min, between 8.44 and 15.72 min, and after 18.11 min **(d)** No **109.** 2.03, 4.91

111. (a) 30°, 60° **(b)** 123.6 m **(c)**

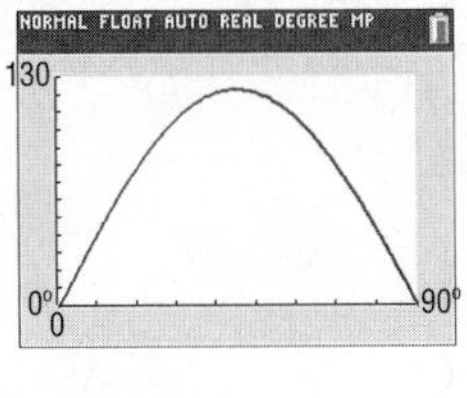

113. 28.90° **115.** Yes; it varies from 1.25 to 1.34. **117.** 1.47

119. If θ is the original angle of incidence and ϕ is the angle of refraction, then $\frac{\sin\theta}{\sin\phi} = n_2$. The angle of incidence of the emerging beam is also ϕ, and the index of refraction is $\frac{1}{n_2}$. Thus, θ is the angle of refraction of the emerging beam.

123. Center $(5, -2)$; radius $= 3$ **124.** $y = \sqrt{-x}$ **125.** $\tan\theta = -\frac{1}{3}$; $\csc\theta = -\sqrt{10}$; $\sec\theta = \frac{\sqrt{10}}{3}$; $\cot\theta = -3$ **126.** Amplitude: 2
Period: π
Phase shift: $\frac{\pi}{2}$

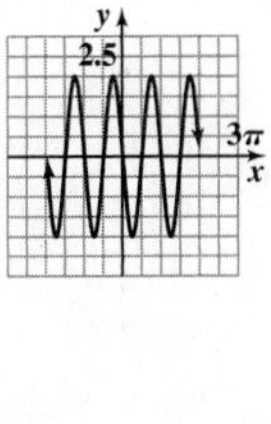

3.4 Assess Your Understanding *(page 223)*

3. identity; conditional **4.** −1 **5.** 0 **6.** T **7.** F **8.** T **9.** c **10.** b **11.** $\frac{1}{\cos\theta}$ **13.** $\frac{1+\sin\theta}{\cos\theta}$ **15.** $\frac{1}{\sin\theta\cos\theta}$ **17.** 2 **19.** $\frac{3\sin\theta+1}{\sin\theta+1}$

21. $\csc\theta\cdot\cos\theta = \frac{1}{\sin\theta}\cdot\cos\theta = \frac{\cos\theta}{\sin\theta} = \cot\theta$ **23.** $1+\tan^2(-\theta) = 1+(-\tan\theta)^2 = 1+\tan^2\theta = \sec^2\theta$

25. $\cos\theta(\tan\theta+\cot\theta) = \cos\theta\left(\frac{\sin\theta}{\cos\theta}+\frac{\cos\theta}{\sin\theta}\right) = \cos\theta\left(\frac{\sin^2\theta+\cos^2\theta}{\cos\theta\sin\theta}\right) = \cos\theta\left(\frac{1}{\cos\theta\sin\theta}\right) = \frac{1}{\sin\theta} = \csc\theta$

27. $\tan u\cot u - \cos^2 u = \tan u\cdot\frac{1}{\tan u} - \cos^2 u = 1-\cos^2 u = \sin^2 u$ **29.** $(\sec\theta-1)(\sec\theta+1) = \sec^2\theta - 1 = \tan^2\theta$

31. $(\sec\theta+\tan\theta)(\sec\theta-\tan\theta) = \sec^2\theta-\tan^2\theta = 1$ **33.** $\cos^2\theta(1+\tan^2\theta) = \cos^2\theta\sec^2\theta = \cos^2\theta\cdot\frac{1}{\cos^2\theta} = 1$

35. $(\sin\theta+\cos\theta)^2+(\sin\theta-\cos\theta)^2 = \sin^2\theta+2\sin\theta\cos\theta+\cos^2\theta+\sin^2\theta-2\sin\theta\cos\theta+\cos^2\theta$
$= \sin^2\theta+\cos^2\theta+\sin^2\theta+\cos^2\theta = 1+1 = 2$

37. $\sec^4\theta-\sec^2\theta = \sec^2\theta(\sec^2\theta-1) = (1+\tan^2\theta)\tan^2\theta = \tan^4\theta+\tan^2\theta$

39. $\sec u-\tan u = \frac{1}{\cos u}-\frac{\sin u}{\cos u} = \frac{1-\sin u}{\cos u}\cdot\frac{1+\sin u}{1+\sin u} = \frac{1-\sin^2 u}{\cos u(1+\sin u)} = \frac{\cos^2 u}{\cos u(1+\sin u)} = \frac{\cos u}{1+\sin u}$

41. $3\sin^2\theta+4\cos^2\theta = 3\sin^2\theta+3\cos^2\theta+\cos^2\theta = 3(\sin^2\theta+\cos^2\theta)+\cos^2\theta = 3+\cos^2\theta$

43. $1-\frac{\cos^2\theta}{1+\sin\theta} = 1-\frac{1-\sin^2\theta}{1+\sin\theta} = 1-\frac{(1+\sin\theta)(1-\sin\theta)}{1+\sin\theta} = 1-(1-\sin\theta) = \sin\theta$

45. $\frac{1+\tan v}{1-\tan v} = \frac{1+\frac{1}{\cot v}}{1-\frac{1}{\cot v}} = \frac{\frac{\cot v+1}{\cot v}}{\frac{\cot v-1}{\cot v}} = \frac{\cot v+1}{\cot v-1}$ **47.** $\frac{\sec\theta}{\csc\theta}+\frac{\sin\theta}{\cos\theta} = \frac{\frac{1}{\cos\theta}}{\frac{1}{\sin\theta}}+\tan\theta = \frac{\sin\theta}{\cos\theta}+\tan\theta = \tan\theta+\tan\theta = 2\tan\theta$

49. $\frac{1+\sin\theta}{1-\sin\theta} = \frac{1+\frac{1}{\csc\theta}}{1-\frac{1}{\csc\theta}} = \frac{\frac{\csc\theta+1}{\csc\theta}}{\frac{\csc\theta-1}{\csc\theta}} = \frac{\csc\theta+1}{\csc\theta-1}$

51. $\frac{1-\sin v}{\cos v}+\frac{\cos v}{1-\sin v} = \frac{(1-\sin v)^2+\cos^2 v}{\cos v(1-\sin v)} = \frac{1-2\sin v+\sin^2 v+\cos^2 v}{\cos v(1-\sin v)} = \frac{2-2\sin v}{\cos v(1-\sin v)} = \frac{2(1-\sin v)}{\cos v(1-\sin v)} = \frac{2}{\cos v} = 2\sec v$

53. $\frac{\sin\theta}{\sin\theta-\cos\theta} = \frac{1}{\frac{\sin\theta-\cos\theta}{\sin\theta}} = \frac{1}{1-\frac{\cos\theta}{\sin\theta}} = \frac{1}{1-\cot\theta}$

55. $(\sec\theta-\tan\theta)^2 = \sec^2\theta-2\sec\theta\tan\theta+\tan^2\theta = \frac{1}{\cos^2\theta}-\frac{2\sin\theta}{\cos^2\theta}+\frac{\sin^2\theta}{\cos^2\theta} = \frac{1-2\sin\theta+\sin^2\theta}{\cos^2\theta} = \frac{(1-\sin\theta)^2}{1-\sin^2\theta} = \frac{(1-\sin\theta)^2}{(1-\sin\theta)(1+\sin\theta)}$
$= \frac{1-\sin\theta}{1+\sin\theta}$

57. $\frac{\cos\theta}{1-\tan\theta}+\frac{\sin\theta}{1-\cot\theta} = \frac{\cos\theta}{1-\frac{\sin\theta}{\cos\theta}}+\frac{\sin\theta}{1-\frac{\cos\theta}{\sin\theta}} = \frac{\cos\theta}{\frac{\cos\theta-\sin\theta}{\cos\theta}}+\frac{\sin\theta}{\frac{\sin\theta-\cos\theta}{\sin\theta}} = \frac{\cos^2\theta}{\cos\theta-\sin\theta}+\frac{\sin^2\theta}{\sin\theta-\cos\theta}$
$= \frac{\cos^2\theta-\sin^2\theta}{\cos\theta-\sin\theta} = \frac{(\cos\theta-\sin\theta)(\cos\theta+\sin\theta)}{\cos\theta-\sin\theta} = \sin\theta+\cos\theta$

59. $\tan\theta+\frac{\cos\theta}{1+\sin\theta} = \frac{\sin\theta}{\cos\theta}+\frac{\cos\theta}{1+\sin\theta} = \frac{\sin\theta(1+\sin\theta)+\cos^2\theta}{\cos\theta(1+\sin\theta)} = \frac{\sin\theta+\sin^2\theta+\cos^2\theta}{\cos\theta(1+\sin\theta)} = \frac{\sin\theta+1}{\cos\theta(1+\sin\theta)} = \frac{1}{\cos\theta} = \sec\theta$

61. $\frac{\tan\theta+\sec\theta-1}{\tan\theta-\sec\theta+1} = \frac{\tan\theta+(\sec\theta-1)}{\tan\theta-(\sec\theta-1)}\cdot\frac{\tan\theta+(\sec\theta-1)}{\tan\theta+(\sec\theta-1)} = \frac{\tan^2\theta+2\tan\theta(\sec\theta-1)+\sec^2\theta-2\sec\theta+1}{\tan^2\theta-(\sec^2\theta-2\sec\theta+1)}$
$= \frac{\sec^2\theta-1+2\tan\theta(\sec\theta-1)+\sec^2\theta-2\sec\theta+1}{\sec^2\theta-1-\sec^2\theta+2\sec\theta-1} = \frac{2\sec^2\theta-2\sec\theta+2\tan\theta(\sec\theta-1)}{-2+2\sec\theta}$
$= \frac{2\sec\theta(\sec\theta-1)+2\tan\theta(\sec\theta-1)}{2(\sec\theta-1)} = \frac{2(\sec\theta-1)(\sec\theta+\tan\theta)}{2(\sec\theta-1)} = \tan\theta+\sec\theta$

63. $\frac{\tan\theta-\cot\theta}{\tan\theta+\cot\theta} = \frac{\frac{\sin\theta}{\cos\theta}-\frac{\cos\theta}{\sin\theta}}{\frac{\sin\theta}{\cos\theta}+\frac{\cos\theta}{\sin\theta}} = \frac{\frac{\sin^2\theta-\cos^2\theta}{\cos\theta\sin\theta}}{\frac{\sin^2\theta+\cos^2\theta}{\cos\theta\sin\theta}} = \frac{\sin^2\theta-\cos^2\theta}{1} = \sin^2\theta-\cos^2\theta$

65. $\frac{\tan u-\cot u}{\tan u+\cot u}+1 = \frac{\frac{\sin u}{\cos u}-\frac{\cos u}{\sin u}}{\frac{\sin u}{\cos u}+\frac{\cos u}{\sin u}}+1 = \frac{\frac{\sin^2 u-\cos^2 u}{\cos u\sin u}}{\frac{\sin^2 u+\cos^2 u}{\cos u\sin u}}+1 = \sin^2 u-\cos^2 u+1 = \sin^2 u+(1-\cos^2 u) = 2\sin^2 u$

67. $\dfrac{\sec\theta+\tan\theta}{\cot\theta+\cos\theta}=\dfrac{\dfrac{1}{\cos\theta}+\dfrac{\sin\theta}{\cos\theta}}{\dfrac{\cos\theta}{\sin\theta}+\cos\theta}=\dfrac{\dfrac{1+\sin\theta}{\cos\theta}}{\dfrac{\cos\theta+\cos\theta\sin\theta}{\sin\theta}}=\dfrac{1+\sin\theta}{\cos\theta}\cdot\dfrac{\sin\theta}{\cos\theta(1+\sin\theta)}=\dfrac{\sin\theta}{\cos\theta}\cdot\dfrac{1}{\cos\theta}=\tan\theta\sec\theta$

69. $\dfrac{1-\tan^2\theta}{1+\tan^2\theta}+1=\dfrac{1-\tan^2\theta+1+\tan^2\theta}{1+\tan^2\theta}=\dfrac{2}{1+\tan^2\theta}=\dfrac{2}{\sec^2\theta}=2\cos^2\theta$

71. $\dfrac{\sec\theta-\csc\theta}{\sec\theta\csc\theta}=\dfrac{\sec\theta}{\sec\theta\csc\theta}-\dfrac{\csc\theta}{\sec\theta\csc\theta}=\dfrac{1}{\csc\theta}-\dfrac{1}{\sec\theta}=\sin\theta-\cos\theta$

73. $\sec\theta-\cos\theta=\dfrac{1}{\cos\theta}-\cos\theta=\dfrac{1-\cos^2\theta}{\cos\theta}=\dfrac{\sin^2\theta}{\cos\theta}=\sin\theta\cdot\dfrac{\sin\theta}{\cos\theta}=\sin\theta\tan\theta$

75. $\dfrac{1}{1-\sin\theta}+\dfrac{1}{1+\sin\theta}=\dfrac{1+\sin\theta+1-\sin\theta}{(1+\sin\theta)(1-\sin\theta)}=\dfrac{2}{1-\sin^2\theta}=\dfrac{2}{\cos^2\theta}=2\sec^2\theta$

77. $\dfrac{\sec\theta}{1-\sin\theta}=\dfrac{\sec\theta}{1-\sin\theta}\cdot\dfrac{1+\sin\theta}{1+\sin\theta}=\dfrac{\sec\theta(1+\sin\theta)}{1-\sin^2\theta}=\dfrac{\sec\theta(1+\sin\theta)}{\cos^2\theta}=\dfrac{1+\sin\theta}{\cos^3\theta}$

79. $\dfrac{(\sec v-\tan v)^2+1}{\csc v(\sec v-\tan v)}=\dfrac{\sec^2 v-2\sec v\tan v+\tan^2 v+1}{\dfrac{1}{\sin v}\left(\dfrac{1}{\cos v}-\dfrac{\sin v}{\cos v}\right)}=\dfrac{2\sec^2 v-2\sec v\tan v}{\dfrac{1}{\sin v}\left(\dfrac{1-\sin v}{\cos v}\right)}=\dfrac{\dfrac{2}{\cos^2 v}-\dfrac{2\sin v}{\cos^2 v}}{\dfrac{1-\sin v}{\sin v\cos v}}=\dfrac{2-2\sin v}{\cos^2 v}\cdot\dfrac{\sin v\cos v}{1-\sin v}$

$=\dfrac{2(1-\sin v)}{\cos v}\cdot\dfrac{\sin v}{1-\sin v}=\dfrac{2\sin v}{\cos v}=2\tan v$

81. $\dfrac{\sin\theta+\cos\theta}{\cos\theta}-\dfrac{\sin\theta-\cos\theta}{\sin\theta}=\dfrac{\sin\theta}{\cos\theta}+1-1+\dfrac{\cos\theta}{\sin\theta}=\dfrac{\sin^2\theta+\cos^2\theta}{\cos\theta\sin\theta}=\dfrac{1}{\cos\theta\sin\theta}=\sec\theta\csc\theta$

83. $\dfrac{\sin^3\theta+\cos^3\theta}{\sin\theta+\cos\theta}=\dfrac{(\sin\theta+\cos\theta)(\sin^2\theta-\sin\theta\cos\theta+\cos^2\theta)}{\sin\theta+\cos\theta}=\sin^2\theta+\cos^2\theta-\sin\theta\cos\theta=1-\sin\theta\cos\theta$

85. $\dfrac{\cos^2\theta-\sin^2\theta}{1-\tan^2\theta}=\dfrac{\cos^2\theta-\sin^2\theta}{1-\dfrac{\sin^2\theta}{\cos^2\theta}}=\dfrac{\cos^2\theta-\sin^2\theta}{\dfrac{\cos^2\theta-\sin^2\theta}{\cos^2\theta}}=\cos^2\theta$

87. $\dfrac{(2\cos^2\theta-1)^2}{\cos^4\theta-\sin^4\theta}=\dfrac{[2\cos^2\theta-(\sin^2\theta+\cos^2\theta)]^2}{(\cos^2\theta-\sin^2\theta)(\cos^2\theta+\sin^2\theta)}=\dfrac{(\cos^2\theta-\sin^2\theta)^2}{\cos^2\theta-\sin^2\theta}=\cos^2\theta-\sin^2\theta=(1-\sin^2\theta)-\sin^2\theta=1-2\sin^2\theta$

89. $\dfrac{1+\sin\theta+\cos\theta}{1+\sin\theta-\cos\theta}=\dfrac{(1+\sin\theta)+\cos\theta}{(1+\sin\theta)-\cos\theta}\cdot\dfrac{(1+\sin\theta)+\cos\theta}{(1+\sin\theta)+\cos\theta}=\dfrac{1+2\sin\theta+\sin^2\theta+2(1+\sin\theta)\cos\theta+\cos^2\theta}{1+2\sin\theta+\sin^2\theta-\cos^2\theta}$

$=\dfrac{1+2\sin\theta+\sin^2\theta+2(1+\sin\theta)(\cos\theta)+(1-\sin^2\theta)}{1+2\sin\theta+\sin^2\theta-(1-\sin^2\theta)}=\dfrac{2+2\sin\theta+2(1+\sin\theta)(\cos\theta)}{2\sin\theta+2\sin^2\theta}$

$=\dfrac{2(1+\sin\theta)+2(1+\sin\theta)(\cos\theta)}{2\sin\theta(1+\sin\theta)}=\dfrac{2(1+\sin\theta)(1+\cos\theta)}{2\sin\theta(1+\sin\theta)}=\dfrac{1+\cos\theta}{\sin\theta}$

91. $(a\sin\theta+b\cos\theta)^2+(a\cos\theta-b\sin\theta)^2=a^2\sin^2\theta+2ab\sin\theta\cos\theta+b^2\cos^2\theta+a^2\cos^2\theta-2ab\sin\theta\cos\theta+b^2\sin^2\theta$

$=a^2(\sin^2\theta+\cos^2\theta)+b^2(\cos^2\theta+\sin^2\theta)=a^2+b^2$

93. $\dfrac{\tan\alpha+\tan\beta}{\cot\alpha+\cot\beta}=\dfrac{\tan\alpha+\tan\beta}{\dfrac{1}{\tan\alpha}+\dfrac{1}{\tan\beta}}=\dfrac{\tan\alpha+\tan\beta}{\dfrac{\tan\beta+\tan\alpha}{\tan\alpha\tan\beta}}=(\tan\alpha+\tan\beta)\cdot\dfrac{\tan\alpha\tan\beta}{\tan\alpha+\tan\beta}=\tan\alpha\tan\beta$

95. $(\sin\alpha+\cos\beta)^2+(\cos\beta+\sin\alpha)(\cos\beta-\sin\alpha)=(\sin^2\alpha+2\sin\alpha\cos\beta+\cos^2\beta)+(\cos^2\beta-\sin^2\alpha)$

$=2\cos^2\beta+2\sin\alpha\cos\beta=2\cos\beta(\cos\beta+\sin\alpha)=2\cos\beta(\sin\alpha+\cos\beta)$

97. $\ln|\sec\theta|=\ln|\cos\theta|^{-1}=-\ln|\cos\theta|$

99. $\ln|1+\cos\theta|+\ln|1-\cos\theta|=\ln(|1+\cos\theta||1-\cos\theta|)=\ln|1-\cos^2\theta|=\ln|\sin^2\theta|=2\ln|\sin\theta|$

101. $g(x)=\sec x-\cos x=\dfrac{1}{\cos x}-\cos x=\dfrac{1}{\cos x}-\dfrac{\cos^2 x}{\cos x}=\dfrac{1-\cos^2 x}{\cos x}=\dfrac{\sin^2 x}{\cos x}=\sin x\cdot\dfrac{\sin x}{\cos x}=\sin x\cdot\tan x=f(x)$

103. $f(\theta)=\dfrac{1-\sin\theta}{\cos\theta}-\dfrac{\cos\theta}{1+\sin\theta}=\dfrac{1-\sin\theta}{\cos\theta}\cdot\dfrac{1+\sin\theta}{1+\sin\theta}-\dfrac{\cos\theta}{1+\sin\theta}\cdot\dfrac{\cos\theta}{\cos\theta}=\dfrac{1-\sin^2\theta}{\cos\theta(1+\sin\theta)}-\dfrac{\cos^2\theta}{\cos\theta(1+\sin\theta)}$

$=\dfrac{\cos^2\theta}{\cos\theta(1+\sin\theta)}-\dfrac{\cos^2\theta}{\cos\theta(1+\sin\theta)}=0=g(\theta)$

105. $1200 \sec\theta\,(2\sec^2\theta - 1) = 1200\dfrac{1}{\cos\theta}\left(\dfrac{2}{\cos^2\theta} - 1\right) = 1200\dfrac{1}{\cos\theta}\left(\dfrac{2}{\cos^2\theta} - \dfrac{\cos^2\theta}{\cos^2\theta}\right) = 1200\dfrac{1}{\cos\theta}\left(\dfrac{2-\cos^2\theta}{\cos^2\theta}\right) = \dfrac{1200\,(1 + 1 - \cos^2\theta)}{\cos^3\theta}$

$= \dfrac{1200\,(1+\sin^2\theta)}{\cos^3\theta}$

111. 13 **112.** $(x+6)^2 + y^2 = 7$ **113.** $\sin\theta = \dfrac{5}{13}$; $\cos\theta = -\dfrac{12}{13}$; $\tan\theta = -\dfrac{5}{12}$; $\csc\theta = \dfrac{13}{5}$; $\sec\theta = -\dfrac{13}{12}$; $\cot\theta = -\dfrac{12}{5}$ **114.** $-\dfrac{2}{\pi}$

3.5 Assess Your Understanding *(page 235)*

5. $-$ **6.** $-$ **7.** F **8.** F **9.** F **10.** T **11.** a **12.** d **13.** $-\frac{1}{4}(\sqrt{2}+\sqrt{6})$ **15.** $2-\sqrt{3}$ **17.** $\frac{1}{4}(\sqrt{6}+\sqrt{2})$ **19.** $\frac{1}{4}(\sqrt{2}-\sqrt{6})$

21. $-\frac{1}{4}(\sqrt{6}+\sqrt{2})$ **23.** $\sqrt{6}-\sqrt{2}$ **25.** $\frac{1}{2}$ **27.** 0 **29.** 1 **31.** -1 **33.** $\frac{1}{2}$ **35. (a)** $\dfrac{2\sqrt{5}}{25}$ **(b)** $\dfrac{11\sqrt{5}}{25}$ **(c)** $\dfrac{2\sqrt{5}}{5}$ **(d)** 2

37. (a) $\dfrac{4-3\sqrt{3}}{10}$ **(b)** $\dfrac{-3-4\sqrt{3}}{10}$ **(c)** $\dfrac{4+3\sqrt{3}}{10}$ **(d)** $\dfrac{25\sqrt{3}+48}{39}$ **39. (a)** $-\dfrac{5+12\sqrt{3}}{26}$ **(b)** $\dfrac{12-5\sqrt{3}}{26}$ **(c)** $\dfrac{-5+12\sqrt{3}}{26}$ **(d)** $\dfrac{-240+169\sqrt{3}}{69}$

41. (a) $-\dfrac{2\sqrt{2}}{3}$ **(b)** $\dfrac{-2\sqrt{2}+\sqrt{3}}{6}$ **(c)** $\dfrac{-2\sqrt{2}+\sqrt{3}}{6}$ **(d)** $\dfrac{9-4\sqrt{2}}{7}$ **43.** $\dfrac{1-2\sqrt{6}}{6}$ **45.** $\dfrac{\sqrt{3}-2\sqrt{2}}{6}$ **47.** $\dfrac{8\sqrt{2}-9\sqrt{3}}{5}$

49. $\sin\left(\dfrac{\pi}{2}+\theta\right) = \sin\dfrac{\pi}{2}\cos\theta + \cos\dfrac{\pi}{2}\sin\theta = 1\cdot\cos\theta + 0\cdot\sin\theta = \cos\theta$

51. $\sin(\pi-\theta) = \sin\pi\cos\theta - \cos\pi\sin\theta = 0\cdot\cos\theta - (-1)\sin\theta = \sin\theta$

53. $\sin(\pi+\theta) = \sin\pi\cos\theta + \cos\pi\sin\theta = 0\cdot\cos\theta + (-1)\sin\theta = -\sin\theta$

55. $\tan(\pi-\theta) = \dfrac{\tan\pi - \tan\theta}{1+\tan\pi\tan\theta} = \dfrac{0-\tan\theta}{1+0\cdot\tan\theta} = -\tan\theta$

57. $\sin\left(\dfrac{3\pi}{2}+\theta\right) = \sin\dfrac{3\pi}{2}\cos\theta + \cos\dfrac{3\pi}{2}\sin\theta = (-1)\cos\theta + 0\cdot\sin\theta = -\cos\theta$

59. $\sin(\alpha+\beta) + \sin(\alpha-\beta) = \sin\alpha\cos\beta + \cos\alpha\sin\beta + \sin\alpha\cos\beta - \cos\alpha\sin\beta = 2\sin\alpha\cos\beta$

61. $\dfrac{\sin(\alpha+\beta)}{\sin\alpha\cos\beta} = \dfrac{\sin\alpha\cos\beta + \cos\alpha\sin\beta}{\sin\alpha\cos\beta} = \dfrac{\sin\alpha\cos\beta}{\sin\alpha\cos\beta} + \dfrac{\cos\alpha\sin\beta}{\sin\alpha\cos\beta} = 1 + \cot\alpha\tan\beta$

63. $\dfrac{\cos(\alpha+\beta)}{\cos\alpha\cos\beta} = \dfrac{\cos\alpha\cos\beta - \sin\alpha\sin\beta}{\cos\alpha\cos\beta} = \dfrac{\cos\alpha\cos\beta}{\cos\alpha\cos\beta} - \dfrac{\sin\alpha\sin\beta}{\cos\alpha\cos\beta} = 1 - \tan\alpha\tan\beta$

65. $\dfrac{\sin(\alpha+\beta)}{\sin(\alpha-\beta)} = \dfrac{\sin\alpha\cos\beta + \cos\alpha\sin\beta}{\sin\alpha\cos\beta - \cos\alpha\sin\beta} = \dfrac{\dfrac{\sin\alpha\cos\beta + \cos\alpha\sin\beta}{\cos\alpha\cos\beta}}{\dfrac{\sin\alpha\cos\beta - \cos\alpha\sin\beta}{\cos\alpha\cos\beta}} = \dfrac{\dfrac{\sin\alpha\cos\beta}{\cos\alpha\cos\beta} + \dfrac{\cos\alpha\sin\beta}{\cos\alpha\cos\beta}}{\dfrac{\sin\alpha\cos\beta}{\cos\alpha\cos\beta} - \dfrac{\cos\alpha\sin\beta}{\cos\alpha\cos\beta}} = \dfrac{\tan\alpha + \tan\beta}{\tan\alpha - \tan\beta}$

67. $\cot(\alpha+\beta) = \dfrac{\cos(\alpha+\beta)}{\sin(\alpha+\beta)} = \dfrac{\cos\alpha\cos\beta - \sin\alpha\sin\beta}{\sin\alpha\cos\beta + \cos\alpha\sin\beta} = \dfrac{\dfrac{\cos\alpha\cos\beta - \sin\alpha\sin\beta}{\sin\alpha\sin\beta}}{\dfrac{\sin\alpha\cos\beta + \cos\alpha\sin\beta}{\sin\alpha\sin\beta}} = \dfrac{\dfrac{\cos\alpha\cos\beta}{\sin\alpha\sin\beta} - \dfrac{\sin\alpha\sin\beta}{\sin\alpha\sin\beta}}{\dfrac{\sin\alpha\cos\beta}{\sin\alpha\sin\beta} + \dfrac{\cos\alpha\sin\beta}{\sin\alpha\sin\beta}} = \dfrac{\cot\alpha\cot\beta - 1}{\cot\beta + \cot\alpha}$

69. $\sec(\alpha+\beta) = \dfrac{1}{\cos(\alpha+\beta)} = \dfrac{1}{\cos\alpha\cos\beta - \sin\alpha\sin\beta} = \dfrac{\dfrac{1}{\sin\alpha\sin\beta}}{\dfrac{\cos\alpha\cos\beta - \sin\alpha\sin\beta}{\sin\alpha\sin\beta}} = \dfrac{\dfrac{1}{\sin\alpha}\cdot\dfrac{1}{\sin\beta}}{\dfrac{\cos\alpha\cos\beta}{\sin\alpha\sin\beta} - \dfrac{\sin\alpha\sin\beta}{\sin\alpha\sin\beta}} = \dfrac{\csc\alpha\csc\beta}{\cot\alpha\cot\beta - 1}$

71. $\sin(\alpha-\beta)\sin(\alpha+\beta) = (\sin\alpha\cos\beta - \cos\alpha\sin\beta)(\sin\alpha\cos\beta + \cos\alpha\sin\beta) = \sin^2\alpha\cos^2\beta - \cos^2\alpha\sin^2\beta$

$= (\sin^2\alpha)(1-\sin^2\beta) - (1-\sin^2\alpha)(\sin^2\beta) = \sin^2\alpha - \sin^2\beta$

73. $\sin(\theta + k\pi) = \sin\theta\cos k\pi + \cos\theta\sin k\pi = (\sin\theta)(-1)^k + (\cos\theta)(0) = (-1)^k\sin\theta$, k any integer

75. $\dfrac{\sqrt{3}}{2}$ **77.** $-\dfrac{24}{25}$ **79.** $-\dfrac{33}{65}$ **81.** $\dfrac{63}{65}$ **83.** $\dfrac{48+25\sqrt{3}}{39}$ **85.** $\dfrac{4}{3}$ **87.** $u\sqrt{1-v^2} - v\sqrt{1-u^2}$: $-1 \le u \le 1$; $-1 \le v \le 1$

89. $\dfrac{u\sqrt{1-v^2} - v}{\sqrt{1+u^2}}$: $-\infty < u < \infty$; $-1 \le v \le 1$ **91.** $\dfrac{uv - \sqrt{1-u^2}\sqrt{1-v^2}}{v\sqrt{1-u^2} + u\sqrt{1-v^2}}$: $-1 \le u \le 1$; $-1 \le v \le 1$ **93.** $\left\{\dfrac{\pi}{2}, \dfrac{7\pi}{6}\right\}$ **95.** $\left\{\dfrac{\pi}{4}\right\}$ **97.** $\left\{\dfrac{11\pi}{6}\right\}$

99. Let $\alpha = \sin^{-1} v$ and $\beta = \cos^{-1} v$. Then $\sin\alpha = \cos\beta = v$, and since $\sin\alpha = \cos\left(\dfrac{\pi}{2}-\alpha\right)$, $\cos\left(\dfrac{\pi}{2}-\alpha\right) = \cos\beta$.

If $v \ge 0$, then $0 \le \alpha \le \dfrac{\pi}{2}$, so $\left(\dfrac{\pi}{2}-\alpha\right)$ and β both lie on $\left[0, \dfrac{\pi}{2}\right]$. If $v < 0$, then $-\dfrac{\pi}{2} \le \alpha < 0$, so $\left(\dfrac{\pi}{2}-\alpha\right)$ and β both lie on $\left(\dfrac{\pi}{2}, \pi\right]$.

Either way, $\cos\left(\dfrac{\pi}{2}-\alpha\right) = \cos\beta$ implies $\dfrac{\pi}{2} - \alpha = \beta$, or $\alpha + \beta = \dfrac{\pi}{2}$.

101. Let $\alpha = \tan^{-1}\frac{1}{v}$ and $\beta = \tan^{-1} v$. Because $v \neq 0, \alpha, \beta \neq 0$. Then $\tan\alpha = \frac{1}{v} = \frac{1}{\tan\beta} = \cot\beta$, and since

$\tan\alpha = \cot\left(\frac{\pi}{2} - \alpha\right)$, $\cot\left(\frac{\pi}{2} - \alpha\right) = \cot\beta$. Because $v > 0$, $0 < \alpha < \frac{\pi}{2}$, and so $\left(\frac{\pi}{2} - \alpha\right)$ and β both lie on $\left(0, \frac{\pi}{2}\right)$.

Then $\cot\left(\frac{\pi}{2} - \alpha\right) = \cot\beta$ implies $\frac{\pi}{2} - \alpha = \beta$, or $\alpha = \frac{\pi}{2} - \beta$.

103. $\sin(\sin^{-1} v + \cos^{-1} v) = \sin(\sin^{-1} v)\cos(\cos^{-1} v) + \cos(\sin^{-1} v)\sin(\cos^{-1} v) = (v)(v) + \sqrt{1-v^2}\sqrt{1-v^2} = v^2 + 1 - v^2 = 1$

105. $\frac{\sin(x+h) - \sin x}{h} = \frac{\sin x\cos h + \cos x\sin h - \sin x}{h} = \frac{\cos x\sin h - \sin x(1-\cos h)}{h} = \cos x \cdot \frac{\sin h}{h} - \sin x \cdot \frac{1-\cos h}{h}$

107. (a) $\tan(\tan^{-1}1 + \tan^{-1}2 + \tan^{-1}3) = \tan((\tan^{-1}1 + \tan^{-1}2) + \tan^{-1}3) = \frac{\tan(\tan^{-1}1 + \tan^{-1}2) + \tan(\tan^{-1}3)}{1 - \tan(\tan^{-1}1 + \tan^{-1}2)\tan(\tan^{-1}3)}$

$$= \frac{\frac{\tan(\tan^{-1}1) + \tan(\tan^{-1}2)}{1 - \tan(\tan^{-1}1)\tan(\tan^{-1}2)} + 3}{1 - \frac{\tan(\tan^{-1}1) + \tan(\tan^{-1}2)}{1 - \tan(\tan^{-1}1)\tan(\tan^{-1}2)}\cdot 3} = \frac{\frac{1+2}{1-1\cdot 2} + 3}{1 - \frac{1+2}{1-1\cdot 2}\cdot 3} = \frac{\frac{3}{-1} + 3}{1 - \frac{3}{-1}\cdot 3} = \frac{-3+3}{1+9} = \frac{0}{10} = 0$$

(b) From the definition of the inverse tangent function, $0 < \tan^{-1}1 < \frac{\pi}{2}$, $0 < \tan^{-1}2 < \frac{\pi}{2}$, and $0 < \tan^{-1}3 < \frac{\pi}{2}$, so $0 < \tan^{-1}1 + \tan^{-1}2 + \tan^{-1}3 < \frac{3\pi}{2}$.

On the interval $\left(0, \frac{3\pi}{2}\right)$, $\tan\theta = 0$ if and only if $\theta = \pi$. Therefore, from part (a), $\tan^{-1}1 + \tan^{-1}2 + \tan^{-1}3 = \pi$.

109. $\tan\theta = \tan(\theta_2 - \theta_1) = \frac{\tan\theta_2 - \tan\theta_1}{1 + \tan\theta_1\tan\theta_2} = \frac{m_2 - m_1}{1 + m_1 m_2}$

111. $2\cot(\alpha - \beta) = \frac{2}{\tan(\alpha-\beta)} = 2\left(\frac{1 + \tan\alpha\tan\beta}{\tan\alpha - \tan\beta}\right) = 2\left(\frac{1 + (x+1)(x-1)}{(x+1) - (x-1)}\right) = 2\left(\frac{1 + x^2 - 1}{x + 1 - x + 1}\right) = \frac{2x^2}{2} = x^2$

113. $\tan\frac{\pi}{2}$ is not defined; $\tan\left(\frac{\pi}{2} - \theta\right) = \frac{\sin\left(\frac{\pi}{2} - \theta\right)}{\cos\left(\frac{\pi}{2} - \theta\right)} = \frac{\cos\theta}{\sin\theta} = \cot\theta$.

114. 17,043 mph **115.** 510° **116.** $\frac{9\pi}{2}\text{ cm}^2 \approx 14.14\text{ cm}^2$ **117.** $\sin\theta = -\frac{2\sqrt{5}}{5}$; $\cos\theta = \frac{\sqrt{5}}{5}$; $\csc\theta = -\frac{\sqrt{5}}{2}$; $\sec\theta = \sqrt{5}$; $\cot\theta = -\frac{1}{2}$

3.6 Assess Your Understanding *(page 245)*

1. $\sin^2\theta$; $2\cos^2\theta$; $2\sin^2\theta$ **2.** $1 - \cos\theta$ **3.** $\sin\theta$ **4.** T **5.** F **6.** F **7.** b **8.** c **9. (a)** $\frac{24}{25}$ **(b)** $\frac{7}{25}$ **(c)** $\frac{\sqrt{10}}{10}$ **(d)** $\frac{3\sqrt{10}}{10}$

11. (a) $\frac{24}{25}$ **(b)** $-\frac{7}{25}$ **(c)** $\frac{2\sqrt{5}}{5}$ **(d)** $-\frac{\sqrt{5}}{5}$ **13. (a)** $-\frac{2\sqrt{2}}{3}$ **(b)** $\frac{1}{3}$ **(c)** $\sqrt{\frac{3+\sqrt{6}}{6}}$ **(d)** $\sqrt{\frac{3-\sqrt{6}}{6}}$

15. (a) $\frac{4\sqrt{2}}{9}$ **(b)** $-\frac{7}{9}$ **(c)** $\frac{\sqrt{3}}{3}$ **(d)** $\frac{\sqrt{6}}{3}$ **17. (a)** $-\frac{4}{5}$ **(b)** $\frac{3}{5}$ **(c)** $\sqrt{\frac{5+2\sqrt{5}}{10}}$ **(d)** $\sqrt{\frac{5-2\sqrt{5}}{10}}$

19. (a) $-\frac{3}{5}$ **(b)** $-\frac{4}{5}$ **(c)** $\frac{1}{2}\sqrt{\frac{10-\sqrt{10}}{5}}$ **(d)** $-\frac{1}{2}\sqrt{\frac{10+\sqrt{10}}{5}}$ **21.** $\frac{\sqrt{2-\sqrt{2}}}{2}$ **23.** $1 - \sqrt{2}$ **25.** $-\frac{\sqrt{2+\sqrt{3}}}{2}$

27. $\frac{2}{\sqrt{2+\sqrt{2}}} = (2 - \sqrt{2})\sqrt{2+\sqrt{2}}$ **29.** $-\frac{\sqrt{2-\sqrt{2}}}{2}$ **31.** $-\frac{4}{5}$ **33.** $\frac{\sqrt{10(5-\sqrt{5})}}{10}$ **35.** $\frac{4}{3}$ **37.** $-\frac{7}{8}$ **39.** $\frac{\sqrt{10}}{4}$ **41.** $-\frac{\sqrt{15}}{3}$

43. $\sin^4\theta = (\sin^2\theta)^2 = \left(\frac{1 - \cos(2\theta)}{2}\right)^2 = \frac{1}{4}[1 - 2\cos(2\theta) + \cos^2(2\theta)] = \frac{1}{4} - \frac{1}{2}\cos(2\theta) + \frac{1}{4}\cos^2(2\theta)$

$= \frac{1}{4} - \frac{1}{2}\cos(2\theta) + \frac{1}{4}\left(\frac{1 + \cos(4\theta)}{2}\right) = \frac{1}{4} - \frac{1}{2}\cos(2\theta) + \frac{1}{8} + \frac{1}{8}\cos(4\theta) = \frac{3}{8} - \frac{1}{2}\cos(2\theta) + \frac{1}{8}\cos(4\theta)$

45. $\cos(3\theta) = 4\cos^3\theta - 3\cos\theta$ **47.** $\sin(5\theta) = 16\sin^5\theta - 20\sin^3\theta + 5\sin\theta$ **49.** $\cos^4\theta - \sin^4\theta = (\cos^2\theta + \sin^2\theta)(\cos^2\theta - \sin^2\theta) = \cos(2\theta)$

51. $\cot(2\theta) = \frac{1}{\tan(2\theta)} = \frac{1 - \tan^2\theta}{2\tan\theta} = \frac{1 - \frac{1}{\cot^2\theta}}{2\left(\frac{1}{\cot\theta}\right)} = \frac{\frac{\cot^2\theta - 1}{\cot^2\theta}}{\frac{2}{\cot\theta}} = \frac{\cot^2\theta - 1}{\cot^2\theta}\cdot\frac{\cot\theta}{2} = \frac{\cot^2\theta - 1}{2\cot\theta}$

53. $\sec(2\theta) = \frac{1}{\cos(2\theta)} = \frac{1}{2\cos^2\theta - 1} = \frac{1}{\frac{2}{\sec^2\theta} - 1} = \frac{1}{\frac{2 - \sec^2\theta}{\sec^2\theta}} = \frac{\sec^2\theta}{2 - \sec^2\theta}$ **55.** $\cos^2(2u) - \sin^2(2u) = \cos[2(2u)] = \cos(4u)$

57. $\dfrac{\cos(2\theta)}{1+\sin(2\theta)} = \dfrac{\cos^2\theta - \sin^2\theta}{1 + 2\sin\theta\cos\theta} = \dfrac{(\cos\theta - \sin\theta)(\cos\theta + \sin\theta)}{\sin^2\theta + \cos^2\theta + 2\sin\theta\cos\theta} = \dfrac{(\cos\theta - \sin\theta)(\cos\theta + \sin\theta)}{(\sin\theta + \cos\theta)(\sin\theta + \cos\theta)} = \dfrac{\cos\theta - \sin\theta}{\cos\theta + \sin\theta}$

$= \dfrac{\frac{\cos\theta - \sin\theta}{\sin\theta}}{\frac{\cos\theta + \sin\theta}{\sin\theta}} = \dfrac{\frac{\cos\theta}{\sin\theta} - \frac{\sin\theta}{\sin\theta}}{\frac{\cos\theta}{\sin\theta} + \frac{\sin\theta}{\sin\theta}} = \dfrac{\cot\theta - 1}{\cot\theta + 1}$

59. $\sec^2\dfrac{\theta}{2} = \dfrac{1}{\cos^2\left(\frac{\theta}{2}\right)} = \dfrac{1}{\frac{1+\cos\theta}{2}} = \dfrac{2}{1+\cos\theta}$

61. $\cot^2\dfrac{v}{2} = \dfrac{1}{\tan^2\left(\frac{v}{2}\right)} = \dfrac{1}{\frac{1-\cos v}{1+\cos v}} = \dfrac{1+\cos v}{1-\cos v} = \dfrac{1 + \frac{1}{\sec v}}{1 - \frac{1}{\sec v}} = \dfrac{\frac{\sec v + 1}{\sec v}}{\frac{\sec v - 1}{\sec v}} = \dfrac{\sec v + 1}{\sec v}\cdot\dfrac{\sec v}{\sec v - 1} = \dfrac{\sec v + 1}{\sec v - 1}$

63. $\dfrac{1 - \tan^2\left(\frac{\theta}{2}\right)}{1 + \tan^2\left(\frac{\theta}{2}\right)} = \dfrac{1 - \frac{1-\cos\theta}{1+\cos\theta}}{1 + \frac{1-\cos\theta}{1+\cos\theta}} = \dfrac{\frac{1+\cos\theta - (1-\cos\theta)}{1+\cos\theta}}{\frac{1+\cos\theta+1-\cos\theta}{1+\cos\theta}} = \dfrac{2\cos\theta}{1+\cos\theta}\cdot\dfrac{1+\cos\theta}{2} = \cos\theta$

65. $\dfrac{\sin(3\theta)}{\sin\theta} - \dfrac{\cos(3\theta)}{\cos\theta} = \dfrac{\sin(3\theta)\cos\theta - \cos(3\theta)\sin\theta}{\sin\theta\cos\theta} = \dfrac{\sin(3\theta - \theta)}{\frac{1}{2}(2\sin\theta\cos\theta)} = \dfrac{2\sin(2\theta)}{\sin(2\theta)} = 2$

67. $\tan(3\theta) = \tan(\theta + 2\theta) = \dfrac{\tan\theta + \tan(2\theta)}{1 - \tan\theta\tan(2\theta)} = \dfrac{\tan\theta + \frac{2\tan\theta}{1-\tan^2\theta}}{1 - \frac{\tan\theta(2\tan\theta)}{1-\tan^2\theta}} = \dfrac{\tan\theta - \tan^3\theta + 2\tan\theta}{1 - \tan^2\theta - 2\tan^2\theta} = \dfrac{3\tan\theta - \tan^3\theta}{1 - 3\tan^2\theta}$

69. $\dfrac{1}{2}(\ln|1 - \cos(2\theta)| - \ln 2) = \ln\left(\dfrac{|1-\cos(2\theta)|}{2}\right)^{1/2} = \ln|\sin^2\theta|^{1/2} = \ln|\sin\theta|$ **71.** $\left\{\dfrac{\pi}{3}, \dfrac{2\pi}{3}, \dfrac{4\pi}{3}, \dfrac{5\pi}{3}\right\}$ **73.** $\left\{0, \dfrac{2\pi}{3}, \dfrac{4\pi}{3}\right\}$

75. $\left\{0, \dfrac{\pi}{3}, \dfrac{\pi}{2}, \dfrac{2\pi}{3}, \pi, \dfrac{4\pi}{3}, \dfrac{3\pi}{2}, \dfrac{5\pi}{3}\right\}$ **77.** No real solution **79.** $\left\{0, \dfrac{\pi}{3}, \pi, \dfrac{5\pi}{3}\right\}$ **81.** $\dfrac{\sqrt{3}}{2}$ **83.** $\dfrac{7}{25}$ **85.** $\dfrac{24}{7}$ **87.** $\dfrac{24}{25}$ **89.** $\dfrac{1}{5}$ **91.** $\dfrac{25}{7}$ **93.** $0, \dfrac{\pi}{3}, \pi, \dfrac{5\pi}{3}$

95. $\dfrac{\pi}{2}, \dfrac{3\pi}{2}$ **97. (a)** $W = 2D(\csc\theta - \cot\theta) = 2D\left(\dfrac{1}{\sin\theta} - \dfrac{\cos\theta}{\sin\theta}\right) = 2D\dfrac{1-\cos\theta}{\sin\theta} = 2D\tan\dfrac{\theta}{2}$ **(b)** $\theta = 24.45°$

99. (a) $R = \dfrac{v_0^2\sqrt{2}}{16}\cos\theta(\sin\theta - \cos\theta)$

$= \dfrac{v_0^2\sqrt{2}}{32}(2\cos\theta\sin\theta - 2\cos^2\theta)$

$= \dfrac{v_0^2\sqrt{2}}{32}[\sin(2\theta) - \cos(2\theta) - 1]$

(b) $\dfrac{3\pi}{8}$ or $67.5°$ **(c)** $32(2 - \sqrt{2}) \approx 18.75$ ft

(d) 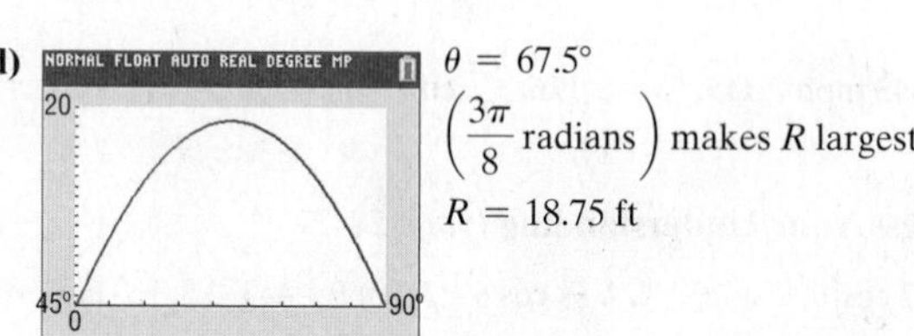

$\theta = 67.5°$ $\left(\dfrac{3\pi}{8}\text{ radians}\right)$ makes R largest. $R = 18.75$ ft

101. $A = \dfrac{1}{2}h(\text{base}) = h\left(\dfrac{1}{2}\text{base}\right) = s\cos\dfrac{\theta}{2}\cdot s\sin\dfrac{\theta}{2} = \dfrac{1}{2}s^2\sin\theta$ **103.** $\sin(2\theta) = \dfrac{4x}{4+x^2}$ **105.** $-\dfrac{1}{4}$

107. $\dfrac{2z}{1+z^2} = \dfrac{2\tan\left(\frac{\alpha}{2}\right)}{1+\tan^2\left(\frac{\alpha}{2}\right)} = \dfrac{2\tan\left(\frac{\alpha}{2}\right)}{\sec^2\left(\frac{\alpha}{2}\right)} = \dfrac{\frac{2\sin\left(\frac{\alpha}{2}\right)}{\cos\left(\frac{\alpha}{2}\right)}}{\frac{1}{\cos^2\left(\frac{\alpha}{2}\right)}} = 2\sin\left(\dfrac{\alpha}{2}\right)\cos\left(\dfrac{\alpha}{2}\right) = \sin\left(2\cdot\dfrac{\alpha}{2}\right) = \sin\alpha$ **109.** y 1.25 2π x

111. $\sin\dfrac{\pi}{24} = \dfrac{\sqrt{2}}{4}\sqrt{4 - \sqrt{6} - \sqrt{2}}$

$\cos\dfrac{\pi}{24} = \dfrac{\sqrt{2}}{4}\sqrt{4 + \sqrt{6} + \sqrt{2}}$

113. $\sin^3\theta + \sin^3(\theta + 120°) + \sin^3(\theta + 240°) = \sin^3\theta + (\sin\theta\cos 120° + \cos\theta\sin 120°)^3 + (\sin\theta\cos 240° + \cos\theta\sin 240°)^3$

$= \sin^3\theta + \left(-\dfrac{1}{2}\sin\theta + \dfrac{\sqrt{3}}{2}\cos\theta\right)^3 + \left(-\dfrac{1}{2}\sin\theta - \dfrac{\sqrt{3}}{2}\cos\theta\right)^3$

$= \sin^3\theta + \dfrac{1}{8}(3\sqrt{3}\cos^3\theta - 9\cos^2\theta\sin\theta + 3\sqrt{3}\cos\theta\sin^2\theta - \sin^3\theta) - \dfrac{1}{8}(\sin^3\theta + 3\sqrt{3}\sin^2\theta\cos\theta + 9\sin\theta\cos^2\theta + 3\sqrt{3}\cos^3\theta)$

$= \dfrac{3}{4}\sin^3\theta - \dfrac{9}{4}\cos^2\theta\sin\theta = \dfrac{3}{4}[\sin^3\theta - 3\sin\theta(1 - \sin^2\theta)] = \dfrac{3}{4}(4\sin^3\theta - 3\sin\theta) = -\dfrac{3}{4}\sin(3\theta)$ (from Example 2)

115. $-\dfrac{1}{2}$ **117.** $\{x \mid x \le 5\}$ **118.** Odd **119.** $\dfrac{\sqrt{3}+1}{2}$ **120.**

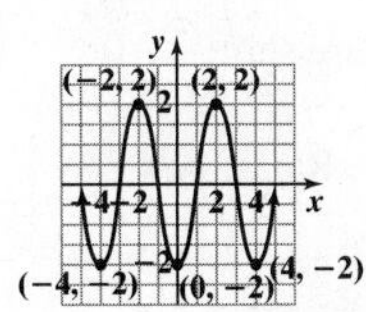

3.7 Assess Your Understanding *(page 250)*

1. $\dfrac{1}{2}\left(\dfrac{\sqrt{3}}{2}-1\right)$ **3.** $-\dfrac{1}{2}\left(\dfrac{\sqrt{3}}{2}+1\right)$ **5.** $\dfrac{\sqrt{2}}{2}$ **7.** $\dfrac{1}{2}[\cos(2\theta)-\cos(6\theta)]$ **9.** $\dfrac{1}{2}[\sin(6\theta)+\sin(2\theta)]$ **11.** $\dfrac{1}{2}[\cos(2\theta)+\cos(8\theta)]$

13. $\dfrac{1}{2}[\cos\theta-\cos(3\theta)]$ **15.** $\dfrac{1}{2}[\sin(2\theta)+\sin\theta]$ **17.** $2\sin\theta\cos(3\theta)$ **19.** $2\cos(3\theta)\cos\theta$ **21.** $2\sin(2\theta)\cos\theta$ **23.** $2\sin\theta\sin\dfrac{\theta}{2}$

25. $\dfrac{\sin\theta+\sin(3\theta)}{2\sin(2\theta)}=\dfrac{2\sin(2\theta)\cos\theta}{2\sin(2\theta)}=\cos\theta$ **27.** $\dfrac{\sin(4\theta)+\sin(2\theta)}{\cos(4\theta)+\cos(2\theta)}=\dfrac{2\sin(3\theta)\cos\theta}{2\cos(3\theta)\cos\theta}=\dfrac{\sin(3\theta)}{\cos(3\theta)}=\tan(3\theta)$

29. $\dfrac{\cos\theta-\cos(3\theta)}{\sin\theta+\sin(3\theta)}=\dfrac{2\sin(2\theta)\sin\theta}{2\sin(2\theta)\cos\theta}=\dfrac{\sin\theta}{\cos\theta}=\tan\theta$

31. $\sin\theta[\sin\theta+\sin(3\theta)]=\sin\theta[2\sin(2\theta)\cos\theta]=\cos\theta[2\sin(2\theta)\sin\theta]=\cos\theta\left[2\cdot\dfrac{1}{2}[\cos\theta-\cos(3\theta)]\right]=\cos\theta[\cos\theta-\cos(3\theta)]$

33. $\dfrac{\sin(4\theta)+\sin(8\theta)}{\cos(4\theta)+\cos(8\theta)}=\dfrac{2\sin(6\theta)\cos(2\theta)}{2\cos(6\theta)\cos(2\theta)}=\dfrac{\sin(6\theta)}{\cos(6\theta)}=\tan(6\theta)$

35. $\dfrac{\sin(4\theta)+\sin(8\theta)}{\sin(4\theta)-\sin(8\theta)}=\dfrac{2\sin(6\theta)\cos(-2\theta)}{2\sin(-2\theta)\cos(6\theta)}=\dfrac{\sin(6\theta)}{\cos(6\theta)}\cdot\dfrac{\cos(2\theta)}{-\sin(2\theta)}=\tan(6\theta)[-\cot(2\theta)]=-\dfrac{\tan(6\theta)}{\tan(2\theta)}$

37. $\dfrac{\sin\alpha+\sin\beta}{\sin\alpha-\sin\beta}=\dfrac{2\sin\dfrac{\alpha+\beta}{2}\cos\dfrac{\alpha-\beta}{2}}{2\sin\dfrac{\alpha-\beta}{2}\cos\dfrac{\alpha+\beta}{2}}=\dfrac{\sin\dfrac{\alpha+\beta}{2}}{\cos\dfrac{\alpha+\beta}{2}}\cdot\dfrac{\cos\dfrac{\alpha-\beta}{2}}{\sin\dfrac{\alpha-\beta}{2}}=\tan\dfrac{\alpha+\beta}{2}\cot\dfrac{\alpha-\beta}{2}$

39. $\dfrac{\sin\alpha+\sin\beta}{\cos\alpha+\cos\beta}=\dfrac{2\sin\dfrac{\alpha+\beta}{2}\cos\dfrac{\alpha-\beta}{2}}{2\cos\dfrac{\alpha+\beta}{2}\cos\dfrac{\alpha-\beta}{2}}=\dfrac{\sin\dfrac{\alpha+\beta}{2}}{\cos\dfrac{\alpha+\beta}{2}}=\tan\dfrac{\alpha+\beta}{2}$

41. $1+\cos(2\theta)+\cos(4\theta)+\cos(6\theta)=[1+\cos(6\theta)]+[\cos(2\theta)+\cos(4\theta)]=2\cos^2(3\theta)+2\cos(3\theta)\cos(-\theta)$
$=2\cos(3\theta)[\cos(3\theta)+\cos\theta]=2\cos(3\theta)[2\cos(2\theta)\cos\theta]=4\cos\theta\cos(2\theta)\cos(3\theta)$

43. $\left\{0,\dfrac{\pi}{3},\dfrac{\pi}{2},\dfrac{2\pi}{3},\pi,\dfrac{4\pi}{3},\dfrac{3\pi}{2},\dfrac{5\pi}{3}\right\}$ **45.** $\left\{0,\dfrac{\pi}{5},\dfrac{2\pi}{5},\dfrac{3\pi}{5},\dfrac{4\pi}{5},\pi,\dfrac{6\pi}{5},\dfrac{7\pi}{5},\dfrac{8\pi}{5},\dfrac{9\pi}{5}\right\}$

47. (a) $y=2\sin(2061\pi t)\cos(357\pi t)$
(b) $y_{\max}=2$
(c)

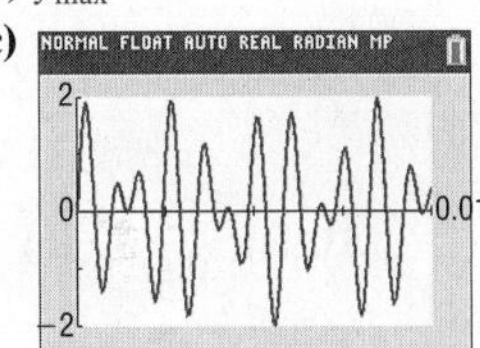

49. $I_u=I_x\cos^2\theta+I_y\sin^2\theta-2I_{xy}\sin\theta\cos\theta=I_x\cos^2\theta+I_y\sin^2\theta-I_{xy}\sin 2\theta$
$=I_x\left(\dfrac{\cos 2\theta+1}{2}\right)+I_y\left(\dfrac{1-\cos 2\theta}{2}\right)-I_{xy}\sin 2\theta$
$=\dfrac{I_x}{2}\cos 2\theta+\dfrac{I_x}{2}+\dfrac{I_y}{2}-\dfrac{I_y}{2}\cos 2\theta-I_{xy}\sin 2\theta$
$=\dfrac{I_x+I_y}{2}+\dfrac{I_x-I_y}{2}\cos 2\theta-I_{xy}\sin 2\theta$

$I_v=I_x\sin^2\theta+I_y\cos^2\theta+2I_{xy}\sin\theta\cos\theta=I_x\left(\dfrac{1-\cos 2\theta}{2}\right)+I_y\left(\dfrac{\cos 2\theta+1}{2}\right)+I_{xy}\sin 2\theta$
$=\dfrac{I_x}{2}-\dfrac{I_x}{2}\cos 2\theta+\dfrac{I_y}{2}\cos 2\theta+\dfrac{I_y}{2}+I_{xy}\sin 2\theta$
$=\dfrac{I_x+I_y}{2}-\dfrac{I_x-I_y}{2}\cos 2\theta+I_{xy}\sin 2\theta$

51. $\sin(2\alpha)+\sin(2\beta)+\sin(2\gamma)=2\sin(\alpha+\beta)\cos(\alpha-\beta)+\sin(2\gamma)=2\sin(\alpha+\beta)\cos(\alpha-\beta)+2\sin\gamma\cos\gamma$
$=2\sin(\pi-\gamma)\cos(\alpha-\beta)+2\sin\gamma\cos\gamma=2\sin\gamma\cos(\alpha-\beta)+2\sin\gamma\cos\gamma=2\sin\gamma[\cos(\alpha-\beta)+\cos\gamma]$
$=2\sin\gamma\left(2\cos\dfrac{\alpha-\beta+\gamma}{2}\cos\dfrac{\alpha-\beta-\gamma}{2}\right)=4\sin\gamma\cos\dfrac{\pi-2\beta}{2}\cos\dfrac{2\alpha-\pi}{2}=4\sin\gamma\cos\left(\dfrac{\pi}{2}-\beta\right)\cos\left(\alpha-\dfrac{\pi}{2}\right)$
$=4\sin\gamma\sin\beta\sin\alpha=4\sin\alpha\sin\beta\sin\gamma$

53.
$$\begin{aligned}\sin(\alpha-\beta)&=\sin\alpha\cos\beta-\cos\alpha\sin\beta\\ \sin(\alpha+\beta)&=\sin\alpha\cos\beta+\cos\alpha\sin\beta\\ \sin(\alpha-\beta)+\sin(\alpha+\beta)&=2\sin\alpha\cos\beta\\ \sin\alpha\cos\beta&=\frac{1}{2}[\sin(\alpha+\beta)+\sin(\alpha-\beta)]\end{aligned}$$

55. $2\cos\dfrac{\alpha+\beta}{2}\cos\dfrac{\alpha-\beta}{2}=2\cdot\dfrac{1}{2}\left[\cos\left(\dfrac{\alpha+\beta}{2}+\dfrac{\alpha-\beta}{2}\right)+\cos\left(\dfrac{\alpha+\beta}{2}-\dfrac{\alpha-\beta}{2}\right)\right]=\cos\dfrac{2\alpha}{2}+\cos\dfrac{2\beta}{2}=\cos\alpha+\cos\beta$

57. $(0, -12), (-4, 0), (4, 0)$ **58.** $\frac{2\sqrt{6}}{7}$ **59.** Amplitude: 5; Period: $\frac{\pi}{2}$; Phase shift: $\frac{\pi}{4}$

60. $f^{-1}(x) = \sin^{-1}\left(\frac{x+5}{3}\right)$; Range of f = Domain of $f^{-1} = [-8, -2]$; Range of $f^{-1} = \left[-\frac{\pi}{2}, \frac{\pi}{2}\right]$

Review Exercises *(page 253)*

1. $\frac{\pi}{2}$ **2.** $\frac{\pi}{2}$ **3.** $\frac{\pi}{4}$ **4.** $-\frac{\pi}{6}$ **5.** $\frac{5\pi}{6}$ **6.** $-\frac{\pi}{3}$ **7.** $\frac{\pi}{4}$ **8.** $\frac{3\pi}{4}$ **9.** $\frac{3\pi}{8}$ **10.** $\frac{3\pi}{4}$ **11.** $-\frac{\pi}{3}$ **12.** $\frac{\pi}{7}$ **13.** $-\frac{\pi}{9}$ **14.** 0.9 **15.** 0.6 **16.** 5 **17.** Not defined

18. $-\frac{\pi}{6}$ **19.** π **20.** $-\sqrt{3}$ **21.** $\frac{2\sqrt{3}}{3}$ **22.** $\frac{4}{5}$ **23.** $-\frac{4}{3}$ **24.** $f^{-1}(x) = \frac{1}{3}\sin^{-1}\left(\frac{x}{2}\right)$; Range of f = Domain of $f^{-1} = [-2, 2]$; Range of $f^{-1} = \left[-\frac{\pi}{6}, \frac{\pi}{6}\right]$

25. $f^{-1}(x) = \cos^{-1}(3 - x)$; Range of f = Domain of $f^{-1} = [2, 4]$; Range of $f^{-1} = [0, \pi]$ **26.** $\sqrt{1 - u^2}$ **27.** $\frac{|u|}{u\sqrt{u^2 - 1}}$

28. $\tan\theta\cot\theta - \sin^2\theta = 1 - \sin^2\theta = \cos^2\theta$ **29.** $\sin^2\theta(1 + \cot^2\theta) = \sin^2\theta\csc^2\theta = 1$

30. $5\cos^2\theta + 3\sin^2\theta = 2\cos^2\theta + 3(\cos^2\theta + \sin^2\theta) = 3 + 2\cos^2\theta$

31. $\frac{1 - \cos\theta}{\sin\theta} + \frac{\sin\theta}{1 - \cos\theta} = \frac{(1 - \cos\theta)^2 + \sin^2\theta}{\sin\theta(1 - \cos\theta)} = \frac{1 - 2\cos\theta + \cos^2\theta + \sin^2\theta}{\sin\theta(1 - \cos\theta)} = \frac{2(1 - \cos\theta)}{\sin\theta(1 - \cos\theta)} = 2\csc\theta$

32. $\frac{\cos\theta}{\cos\theta - \sin\theta} = \frac{\frac{\cos\theta}{\cos\theta}}{\frac{\cos\theta - \sin\theta}{\cos\theta}} = \frac{1}{1 - \frac{\sin\theta}{\cos\theta}} = \frac{1}{1 - \tan\theta}$ **33.** $\frac{\csc\theta}{1 + \csc\theta} = \frac{\frac{1}{\sin\theta}}{1 + \frac{1}{\sin\theta}} = \frac{1}{1 + \sin\theta} = \frac{1}{1 + \sin\theta}\cdot\frac{1 - \sin\theta}{1 - \sin\theta} = \frac{1 - \sin\theta}{1 - \sin^2\theta} = \frac{1 - \sin\theta}{\cos^2\theta}$

34. $\csc\theta - \sin\theta = \frac{1}{\sin\theta} - \sin\theta = \frac{1 - \sin^2\theta}{\sin\theta} = \frac{\cos^2\theta}{\sin\theta} = \cos\theta\cdot\frac{\cos\theta}{\sin\theta} = \cos\theta\cot\theta$

35. $\frac{1 - \sin\theta}{\sec\theta} = \cos\theta(1 - \sin\theta)\cdot\frac{1 + \sin\theta}{1 + \sin\theta} = \frac{\cos\theta(1 - \sin^2\theta)}{1 + \sin\theta} = \frac{\cos^3\theta}{1 + \sin\theta}$ **36.** $\cot\theta - \tan\theta = \frac{\cos\theta}{\sin\theta} - \frac{\sin\theta}{\cos\theta} = \frac{\cos^2\theta - \sin^2\theta}{\sin\theta\cos\theta} = \frac{1 - 2\sin^2\theta}{\sin\theta\cos\theta}$

37. $\frac{\cos(\alpha + \beta)}{\cos\alpha\sin\beta} = \frac{\cos\alpha\cos\beta - \sin\alpha\sin\beta}{\cos\alpha\sin\beta} = \frac{\cos\alpha\cos\beta}{\cos\alpha\sin\beta} - \frac{\sin\alpha\sin\beta}{\cos\alpha\sin\beta} = \cot\beta - \tan\alpha$

38. $\frac{\cos(\alpha - \beta)}{\cos\alpha\cos\beta} = \frac{\cos\alpha\cos\beta + \sin\alpha\sin\beta}{\cos\alpha\cos\beta} = \frac{\cos\alpha\cos\beta}{\cos\alpha\cos\beta} + \frac{\sin\alpha\sin\beta}{\cos\alpha\cos\beta} = 1 + \tan\alpha\tan\beta$

39. $(1 + \cos\theta)\left(\tan\frac{\theta}{2}\right) = (1 + \cos\theta)\cdot\frac{\sin\theta}{1 + \cos\theta} = \sin\theta$

40. $2\cot\theta\cot 2\theta = 2\left(\frac{\cos\theta}{\sin\theta}\right)\left(\frac{\cos 2\theta}{\sin 2\theta}\right) = \frac{2\cos\theta(\cos^2\theta - \sin^2\theta)}{2\sin^2\theta\cos\theta} = \frac{\cos^2\theta - \sin^2\theta}{\sin^2\theta} = \cot^2\theta - 1$

41. $1 - 8\sin^2\theta\cos^2\theta = 1 - 2(2\sin\theta\cos\theta)^2 = 1 - 2\sin^2(2\theta) = \cos(4\theta)$

42. $\frac{\sin(3\theta)\cos\theta - \sin\theta\cos(3\theta)}{\sin(2\theta)} = \frac{\sin(2\theta)}{\sin(2\theta)} = 1$ **43.** $\frac{\sin(2\theta) + \sin(4\theta)}{\cos(2\theta) + \cos(4\theta)} = \frac{2\sin(3\theta)\cos(-\theta)}{2\cos(3\theta)\cos(-\theta)} = \tan(3\theta)$

44. $\frac{\cos(2\theta) - \cos(4\theta)}{\cos(2\theta) + \cos(4\theta)} - \tan\theta\tan(3\theta) = \frac{-2\sin(3\theta)\sin(-\theta)}{2\cos(3\theta)\cos(-\theta)} - \tan\theta\tan(3\theta) = \tan(3\theta)\tan\theta - \tan\theta\tan(3\theta) = 0$

45. $\frac{1}{4}(\sqrt{6} - \sqrt{2})$ **46.** $-2 - \sqrt{3}$ **47.** $\frac{1}{4}(\sqrt{6} - \sqrt{2})$ **48.** $\frac{1}{4}(\sqrt{2} - \sqrt{6})$ **49.** $\frac{1}{2}$ **50.** $\frac{1}{2}$ **51.** $\sqrt{2} - 1$ **52.** $\frac{\sqrt{2 + \sqrt{2}}}{2}$ **53. (a)** $-\frac{33}{65}$ **(b)** $-\frac{56}{65}$

(c) $-\frac{63}{65}$ **(d)** $\frac{33}{56}$ **(e)** $\frac{24}{25}$ **(f)** $\frac{119}{169}$ **(g)** $\frac{5\sqrt{26}}{26}$ **(h)** $\frac{2\sqrt{5}}{5}$ **54. (a)** $-\frac{16}{65}$ **(b)** $-\frac{63}{65}$ **(c)** $-\frac{56}{65}$ **(d)** $\frac{16}{63}$ **(e)** $\frac{24}{25}$ **(f)** $\frac{119}{169}$ **(g)** $\frac{\sqrt{26}}{26}$ **(h)** $-\frac{\sqrt{10}}{10}$

55. (a) $-\frac{63}{65}$ **(b)** $\frac{16}{65}$ **(c)** $\frac{33}{65}$ **(d)** $-\frac{63}{16}$ **(e)** $\frac{24}{25}$ **(f)** $-\frac{119}{169}$ **(g)** $\frac{2\sqrt{13}}{13}$ **(h)** $-\frac{\sqrt{10}}{10}$ **56. (a)** $\frac{-\sqrt{3} - 2\sqrt{2}}{6}$ **(b)** $\frac{1 - 2\sqrt{6}}{6}$ **(c)** $\frac{-\sqrt{3} + 2\sqrt{2}}{6}$

(d) $\frac{8\sqrt{2} + 9\sqrt{3}}{23}$ **(e)** $-\frac{\sqrt{3}}{2}$ **(f)** $-\frac{7}{9}$ **(g)** $\frac{\sqrt{3}}{3}$ **(h)** $\frac{\sqrt{3}}{2}$ **57. (a)** 1 **(b)** 0 **(c)** $-\frac{1}{9}$ **(d)** Not defined **(e)** $\frac{4\sqrt{5}}{9}$ **(f)** $-\frac{1}{9}$ **(g)** $\frac{\sqrt{30}}{6}$

(h) $-\frac{\sqrt{6}\sqrt{3 - \sqrt{5}}}{6}$ **58.** $\frac{4 + 3\sqrt{3}}{10}$ **59.** $\frac{33}{65}$ **60.** $-\frac{48 + 25\sqrt{3}}{39}$ **61.** $\frac{-\sqrt{2}}{10}$ **62.** $-\frac{24}{25}$ **63.** $-\frac{7}{25}$ **64.** $\left\{\frac{\pi}{3}, \frac{5\pi}{3}\right\}$ **65.** $\left\{\frac{2\pi}{3}, \frac{5\pi}{3}\right\}$ **66.** $\left\{\frac{3\pi}{4}, \frac{7\pi}{4}\right\}$

67. $\left\{0, \frac{\pi}{2}, \pi, \frac{3\pi}{2}\right\}$ **68.** $\left\{\frac{\pi}{3}, \frac{2\pi}{3}, \frac{4\pi}{3}, \frac{5\pi}{3}\right\}$ **69.** $\{0.25, 2.89\}$ **70.** $\left\{0, \frac{2\pi}{3}, \pi, \frac{4\pi}{3}\right\}$ **71.** $\left\{0, \frac{\pi}{6}, \frac{5\pi}{6}\right\}$ **72.** $\left\{\frac{\pi}{6}, \frac{\pi}{2}, \frac{5\pi}{6}\right\}$ **73.** $\left\{\frac{\pi}{3}, \frac{5\pi}{3}\right\}$

74. $\left\{\frac{\pi}{4}, \frac{\pi}{2}, \frac{3\pi}{4}, \frac{3\pi}{2}\right\}$ **75.** $\left\{\frac{\pi}{2}, \pi\right\}$ **76.** 0.78 **77.** -1.11 **78.** 1.77 **79.** 1.23 **80.** 2.90 **81.** $\{1.11\}$ **82.** $\{0.87\}$ **83.** $\{2.22\}$ **84.** $\left\{-\frac{\sqrt{3}}{2}\right\}$ **85.** $\{0\}$

86. $\sin 15° = \sqrt{\dfrac{1 - \cos 30°}{2}} = \sqrt{\dfrac{1 - \dfrac{\sqrt{3}}{2}}{2}} = \sqrt{\dfrac{2 - \sqrt{3}}{4}} = \dfrac{\sqrt{2 - \sqrt{3}}}{2};$

$\sin 15° = \sin(45° - 30°) = \sin 45° \cos 30° - \cos 45° \sin 30° = \dfrac{\sqrt{2}}{2}\cdot\dfrac{\sqrt{3}}{2} - \dfrac{\sqrt{2}}{2}\cdot\dfrac{1}{2} = \dfrac{\sqrt{6}}{4} - \dfrac{\sqrt{2}}{4} = \dfrac{\sqrt{6} - \sqrt{2}}{4};$

$\left[\dfrac{\sqrt{2 - \sqrt{3}}}{2}\right]^2 = \dfrac{2 - \sqrt{3}}{4} = \dfrac{4(2 - \sqrt{3})}{4 \cdot 4} = \dfrac{8 - 4\sqrt{3}}{16} = \dfrac{6 - 2\sqrt{12} + 2}{16} = \left(\dfrac{\sqrt{6} - \sqrt{2}}{4}\right)^2$

87. $\cos(2\theta) = 2\cos^2\theta - 1$

Chapter Test *(page 255)*

1. $\frac{\pi}{6}$ **2.** $-\frac{\pi}{4}$ **3.** $\frac{\pi}{5}$ **4.** $\frac{7}{3}$ **5.** 3 **6.** $-\frac{4}{3}$ **7.** 0.39 **8.** 0.78 **9.** 1.25 **10.** 0.20

11. $\dfrac{\csc\theta + \cot\theta}{\sec\theta + \tan\theta} = \dfrac{\csc\theta + \cot\theta}{\sec\theta + \tan\theta}\cdot\dfrac{\csc\theta - \cot\theta}{\csc\theta - \cot\theta} = \dfrac{\csc^2\theta - \cot^2\theta}{(\sec\theta + \tan\theta)(\csc\theta - \cot\theta)} = \dfrac{1}{(\sec\theta + \tan\theta)(\csc\theta - \cot\theta)}$

$= \dfrac{1}{(\sec\theta + \tan\theta)(\csc\theta - \cot\theta)}\cdot\dfrac{\sec\theta - \tan\theta}{\sec\theta - \tan\theta} = \dfrac{\sec\theta - \tan\theta}{(\sec^2\theta - \tan^2\theta)(\csc\theta - \cot\theta)} = \dfrac{\sec\theta - \tan\theta}{\csc\theta - \cot\theta}$

12. $\sin\theta\tan\theta + \cos\theta = \sin\theta\cdot\dfrac{\sin\theta}{\cos\theta} + \cos\theta = \dfrac{\sin^2\theta}{\cos\theta} + \dfrac{\cos^2\theta}{\cos\theta} = \dfrac{\sin^2\theta + \cos^2\theta}{\cos\theta} = \dfrac{1}{\cos\theta} = \sec\theta$

13. $\tan\theta + \cot\theta = \dfrac{\sin\theta}{\cos\theta} + \dfrac{\cos\theta}{\sin\theta} = \dfrac{\sin^2\theta}{\sin\theta\cos\theta} + \dfrac{\cos^2\theta}{\sin\theta\cos\theta} = \dfrac{\sin^2\theta + \cos^2\theta}{\sin\theta\cos\theta} = \dfrac{1}{\sin\theta\cos\theta} = \dfrac{2}{2\sin\theta\cos\theta} = \dfrac{2}{\sin(2\theta)} = 2\csc(2\theta)$

14. $\dfrac{\sin(\alpha + \beta)}{\tan\alpha + \tan\beta} = \dfrac{\sin\alpha\cos\beta + \cos\alpha\sin\beta}{\dfrac{\sin\alpha}{\cos\alpha} + \dfrac{\sin\beta}{\cos\beta}} = \dfrac{\sin\alpha\cos\beta + \cos\alpha\sin\beta}{\dfrac{\sin\alpha\cos\beta}{\cos\alpha\cos\beta} + \dfrac{\cos\alpha\sin\beta}{\cos\alpha\cos\beta}} = \dfrac{\sin\alpha\cos\beta + \cos\alpha\sin\beta}{\dfrac{\sin\alpha\cos\beta + \cos\alpha\sin\beta}{\cos\alpha\cos\beta}}$

$= \dfrac{\sin\alpha\cos\beta + \cos\alpha\sin\beta}{1}\cdot\dfrac{\cos\alpha\cos\beta}{\sin\alpha\cos\beta + \cos\alpha\sin\beta} = \cos\alpha\cos\beta$

15. $\sin(3\theta) = \sin(\theta + 2\theta) = \sin\theta\cos(2\theta) + \cos\theta\sin(2\theta) = \sin\theta\cdot(\cos^2\theta - \sin^2\theta) + \cos\theta\cdot 2\sin\theta\cos\theta = \sin\theta\cos^2\theta - \sin^3\theta + 2\sin\theta\cos^2\theta$
$= 3\sin\theta\cos^2\theta - \sin^3\theta = 3\sin\theta(1 - \sin^2\theta) - \sin^3\theta = 3\sin\theta - 3\sin^3\theta - \sin^3\theta = 3\sin\theta - 4\sin^3\theta$

16. $\dfrac{\tan\theta - \cot\theta}{\tan\theta + \cot\theta} = \dfrac{\dfrac{\sin\theta}{\cos\theta} - \dfrac{\cos\theta}{\sin\theta}}{\dfrac{\sin\theta}{\cos\theta} + \dfrac{\cos\theta}{\sin\theta}} = \dfrac{\dfrac{\sin^2\theta - \cos^2\theta}{\sin\theta\cos\theta}}{\dfrac{\sin^2\theta + \cos^2\theta}{\sin\theta\cos\theta}} = \dfrac{\sin^2\theta - \cos^2\theta}{\sin^2\theta + \cos^2\theta} = \dfrac{(1 - \cos^2\theta) - \cos^2\theta}{1} = 1 - 2\cos^2\theta$ **17.** $\frac{1}{4}(\sqrt{6} + \sqrt{2})$

18. $2 + \sqrt{3}$ **19.** $\frac{\sqrt{5}}{5}$ **20.** $\frac{12\sqrt{85}}{49}$ **21.** $\frac{2\sqrt{13}(\sqrt{5} - 3)}{39}$ **22.** $\frac{2 + \sqrt{3}}{4}$ **23.** $\frac{\sqrt{6}}{2}$ **24.** $\frac{\sqrt{2}}{2}$ **25.** $\left\{\frac{\pi}{3}, \frac{2\pi}{3}, \frac{4\pi}{3}, \frac{5\pi}{3}\right\}$ **26.** $\{0, 1.911, \pi, 4.373\}$

27. $\left\{\frac{3\pi}{8}, \frac{7\pi}{8}, \frac{11\pi}{8}, \frac{15\pi}{8}\right\}$ **28.** $\{0.285, 3.427\}$ **29.** $\{0.253, 2.889\}$

Cumulative Review *(page 256)*

1. $\left\{\frac{-1 - \sqrt{13}}{6}, \frac{-1 + \sqrt{13}}{6}\right\}$ **2.** $y + 1 = -1(x - 4)$, or $x + y = 3$; $6\sqrt{2}$; $(1, 2)$ **3.** x-axis symmetry; $(0, -3)$, $(0, 3)$, $(3, 0)$

4.

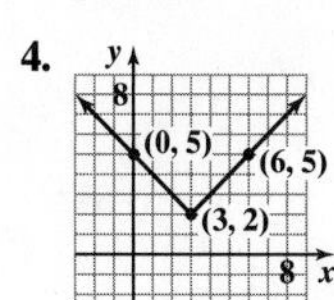

5.

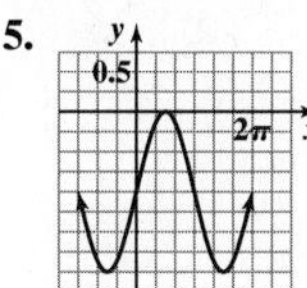

6. (a)

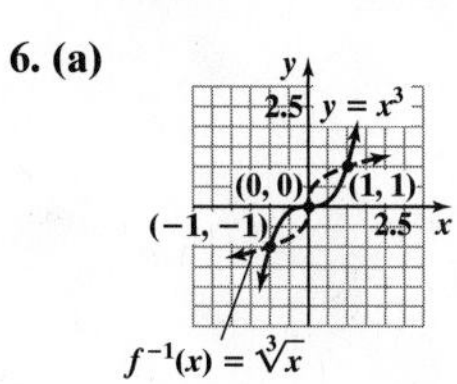

(b)

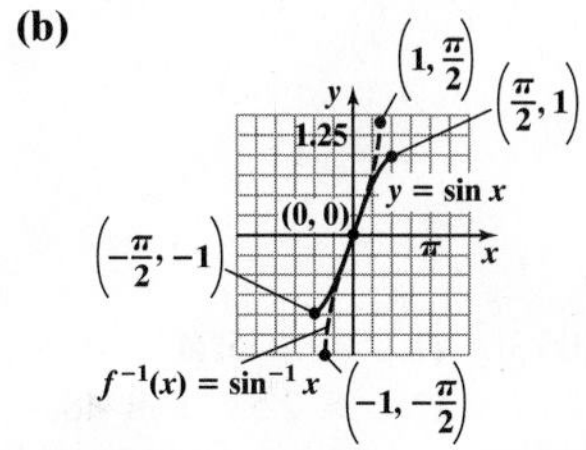

(c)

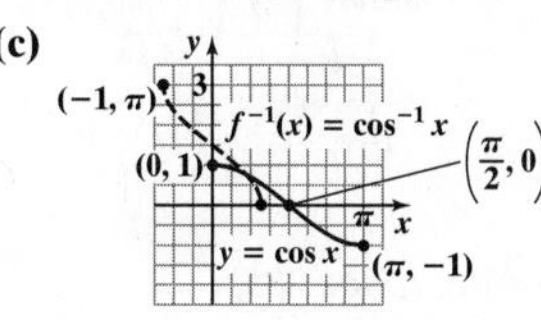

7. (a) $-\frac{2\sqrt{2}}{3}$ **(b)** $\frac{\sqrt{2}}{4}$ **(c)** $\frac{4\sqrt{2}}{9}$ **(d)** $\frac{7}{9}$ **(e)** $\sqrt{\frac{3 + 2\sqrt{2}}{6}}$ **(f)** $-\sqrt{\frac{3 - 2\sqrt{2}}{6}}$ **8.** $\frac{\sqrt{5}}{5}$ **9. (a)** $-\frac{2\sqrt{2}}{3}$ **(b)** $-\frac{2\sqrt{2}}{3}$ **(c)** $\frac{7}{9}$ **(d)** $\frac{4\sqrt{2}}{9}$ **(e)** $\frac{\sqrt{6}}{3}$

CHAPTER 4 Applications of Trigonometric Functions

4.1 Assess Your Understanding *(page 266)*

4. F **5.** b **6.** angle of elevation **7.** T **8.** F **9.** $\sin\theta = \frac{5}{13}$; $\cos\theta = \frac{12}{13}$; $\tan\theta = \frac{5}{12}$; $\cot\theta = \frac{12}{5}$; $\sec\theta = \frac{13}{12}$; $\csc\theta = \frac{13}{5}$

11. $\sin\theta = \frac{2\sqrt{13}}{13}$; $\cos\theta = \frac{3\sqrt{13}}{13}$; $\tan\theta = \frac{2}{3}$; $\cot\theta = \frac{3}{2}$; $\sec\theta = \frac{\sqrt{13}}{3}$; $\csc\theta = \frac{\sqrt{13}}{2}$

13. $\sin\theta = \frac{\sqrt{3}}{2}$; $\cos\theta = \frac{1}{2}$; $\tan\theta = \sqrt{3}$; $\cot\theta = \frac{\sqrt{3}}{3}$; $\sec\theta = 2$; $\csc\theta = \frac{2\sqrt{3}}{3}$

15. $\sin\theta = \frac{\sqrt{6}}{3}$; $\cos\theta = \frac{\sqrt{3}}{3}$; $\tan\theta = \sqrt{2}$; $\cot\theta = \frac{\sqrt{2}}{2}$; $\sec\theta = \sqrt{3}$; $\csc\theta = \frac{\sqrt{6}}{2}$

17. $\sin\theta = \frac{\sqrt{5}}{5}$; $\cos\theta = \frac{2\sqrt{5}}{5}$; $\tan\theta = \frac{1}{2}$; $\cot\theta = 2$; $\sec\theta = \frac{\sqrt{5}}{2}$; $\csc\theta = \sqrt{5}$

19. 0 **21.** 1 **23.** 0 **25.** 0 **27.** 1 **29.** $a \approx 13.74, c \approx 14.62, A = 70°$ **31.** $b \approx 5.03, c \approx 7.83, A = 50°$ **33.** $a \approx 0.71, c \approx 4.06, B = 80°$ **35.** $b \approx 10.72, c \approx 11.83, B = 65°$ **37.** $b \approx 3.08, a \approx 8.46, A = 70°$ **39.** $c \approx 5.83, A \approx 59.0°, B \approx 31.0°$ **41.** $b \approx 4.58, A \approx 23.6°, B \approx 66.4°$ **43.** 23.6° and 66.4° **45.** 4.59 in.; 6.55 in. **47. (a)** 5.52 in. or 11.83 in. **49.** 70.02 ft **51.** 985.91 ft **53.** 137.37 m **55.** 80.5° **57. (a)** 111.96 ft/sec or 76.3 mi/h **(b)** 82.42 ft/sec or 56.2 mi/h **(c)** Under 18.8° **59. (a)** 2.4898×10^{13} miles **(b)** 0.000214° **61.** 554.52 ft **63.** S76.6°E **65.** The embankment is 30.5 m high. **67.** 3.83 mi **69.** 1978.09 ft **71.** 60.27 ft **73.** The buildings are 7984 ft apart. **75.** 69.0° **77.** 38.9° **79.** The white ball should hit the top cushion 4.125 ft from the upper left corner. **84.** Amplitude: 8; period: $\frac{\pi}{3}$

85. $\frac{\sqrt{6}-\sqrt{2}}{4}$ or $\frac{\sqrt{2-\sqrt{3}}}{2}$ **86.** 0.236, 0.243, 0.248 **87.** $\left\{\frac{\pi}{2}, \frac{7\pi}{6}, \frac{11\pi}{6}\right\}$

4.2 Assess Your Understanding *(page 278)*

4. a **5.** $\frac{\sin A}{a} = \frac{\sin B}{b} = \frac{\sin C}{c}$ **6.** F **7.** F **8.** ambiguous case **9.** $a \approx 3.23, b \approx 3.55, A = 40°$ **11.** $a \approx 3.25, c \approx 4.23, B = 45°$ **13.** $C = 95°, c \approx 9.86, a \approx 6.36$ **15.** $A = 40°, a = 2, c \approx 3.06$ **17.** $C = 120°, b \approx 1.06, c \approx 2.69$ **19.** $A = 100°, a \approx 5.24, c \approx 0.92$ **21.** $B = 40°, a \approx 5.64, b \approx 3.86$ **23.** $C = 100°, a \approx 1.31, b \approx 1.31$ **25.** One triangle; $B \approx 30.7°, C \approx 99.3°, c \approx 3.86$ **27.** One triangle; $C \approx 36.2°, A \approx 43.8°, a \approx 3.51$ **29.** No triangle **31.** Two triangles; $C_1 \approx 30.9°, A_1 \approx 129.1°, a_1 \approx 9.07$ or $C_2 \approx 149.1°, A_2 \approx 10.9°, a_2 \approx 2.20$ **33.** No triangle **35.** Two triangles; $A_1 \approx 57.7°, B_1 \approx 97.3°, b_1 \approx 2.35$ or $A_2 \approx 122.3°, B_2 \approx 32.7°, b_2 \approx 1.28$ **37.** 1490.48 ft **39.** 335.16 ft **41.** 153.42 ft; 136.59 ft **43.** The tree is 39.4 ft high. **45.** Adam receives 100.6 more frequent flyer miles. **47. (a)** Station Able is about 143.33 mi from the ship: Station Baker is about 135.58 mi from the ship. **(b)** Approximately 41 min **49.** 84.7°; 183.72 ft **51.** 2.64 mi **53.** 38.5 in. **55.** 449.36 ft **57.** 187,600,000 km or 101,440,000 km **59.** The diameter is 252 ft.

61. $\frac{a-b}{c} = \frac{a}{c} - \frac{b}{c} = \frac{\sin A}{\sin C} - \frac{\sin B}{\sin C} = \frac{\sin A - \sin B}{\sin C} = \frac{2\sin\left(\frac{A-B}{2}\right)\cos\left(\frac{A+B}{2}\right)}{2\sin\frac{C}{2}\cos\frac{C}{2}} = \frac{\sin\left(\frac{A-B}{2}\right)\cos\left(\frac{\pi}{2}-\frac{C}{2}\right)}{\sin\frac{C}{2}\cos\frac{C}{2}} = \frac{\sin\left(\frac{A-B}{2}\right)}{\cos\frac{C}{2}}$

63. $\frac{a-b}{a+b} = \frac{\frac{a-b}{c}}{\frac{a+b}{c}} = \frac{\frac{\sin\left[\frac{1}{2}(A-B)\right]}{\cos\frac{C}{2}}}{\frac{\cos\left[\frac{1}{2}(A-B)\right]}{\sin\frac{C}{2}}} = \frac{\tan\left[\frac{1}{2}(A-B)\right]}{\cot\frac{C}{2}} = \frac{\tan\left[\frac{1}{2}(A-B)\right]}{\tan\left(\frac{\pi}{2}-\frac{C}{2}\right)} = \frac{\tan\left[\frac{1}{2}(A-B)\right]}{\tan\left[\frac{1}{2}(A+B)\right]}$

69. $(h, k) = (-1, 3); r = 4$ **70.** (graph: $(-3\pi, 4)$, $(\pi, 4)$, $(-4\pi, 0)$, $(2\pi, 0)$, $(4\pi, 0)$, 5π, $(-2\pi, 0)$, $(-\pi, -4)$, $(3\pi, -4)$, $(0, 0)$) **71.** $-\frac{\sqrt{15}}{7}$ **72.** $3\sqrt{5} \approx 6.71$

4.3 Assess Your Understanding *(page 285)*

3. Cosines **4.** a **5.** b **6.** F **7.** F **8.** T **9.** $b \approx 2.95, A \approx 28.7°, C \approx 106.3°$ **11.** $c \approx 3.75, A \approx 32.1°, B \approx 52.9°$ **13.** $A \approx 48.5°, B \approx 38.6°, C \approx 92.9°$ **15.** $A \approx 127.2°, B \approx 32.1°, C \approx 20.7°$ **17.** $c \approx 2.57, A \approx 48.6°, B \approx 91.4°$ **19.** $a \approx 2.99, B \approx 19.2°, C \approx 80.8°$ **21.** $b \approx 4.14, A \approx 43.0°, C \approx 27.0°$ **23.** $c \approx 1.69, A = 65.0°, B = 65.0°$ **25.** $A \approx 67.4°, B = 90°, C \approx 22.6°$ **27.** $A = 60°, B = 60°, C = 60°$ **29.** $A \approx 33.6°, B \approx 62.2°, C \approx 84.3°$ **31.** $A \approx 97.9°, B \approx 52.4°, C \approx 29.7°$ **33.** $A = 85°, a = 14.56, c = 14.12$ **35.** $A = 40.8°, B = 60.6°, C = 78.6°$ **37.** $A = 80°, b = 8.74, c = 13.80$

39. Two triangles: $B_1 = 35.4°, C_1 = 134.6°, c_1 = 12.29; B_2 = 144.6°, C_2 = 25.4°, c_2 = 7.40$ **41.** $B = 24.5°, C = 95.5°, a = 10.44$
43. 165 yd **45. (a)** 26.4° **(b)** 30.8 h **47. (a)** 63.7 ft **(b)** 66.8 ft **(c)** 92.8° **49. (a)** 492.6 ft **(b)** 269.3 ft
51. 342.33 ft **53.** The footings should be 7.65 ft apart.
55. Suppose $0 < \theta < \pi$. Then, by the Law of Cosines, $d^2 = r^2 + r^2 - 2r^2\cos\theta = 4r^2\left(\frac{1-\cos\theta}{2}\right) \Rightarrow d = 2r\sqrt{\frac{1-\cos\theta}{2}} = 2r\sin\frac{\theta}{2}$.

Since, for any angle in $(0, \pi)$, d is strictly less than the length of the arc subtended by θ, that is, $d < r\theta$, then $2r\sin\frac{\theta}{2} < r\theta$, or $2\sin\frac{\theta}{2} < \theta$. Since $\cos\frac{\theta}{2} < 1$, then, for $0 < \theta < \pi$, $\sin\theta = 2\sin\frac{\theta}{2}\cos\frac{\theta}{2} < 2\sin\frac{\theta}{2} < \theta$. If $\theta \ge \pi$, then, since $\sin\theta \le 1$, $\sin\theta < \theta$. Thus $\sin\theta < \theta$ for all $\theta > 0$.

57. $$\sin\frac{C}{2} = \sqrt{\frac{1-\cos C}{2}} = \sqrt{\frac{1-\frac{a^2+b^2-c^2}{2ab}}{2}} = \sqrt{\frac{2ab-a^2-b^2+c^2}{4ab}} = \sqrt{\frac{c^2-(a-b)^2}{4ab}} = \sqrt{\frac{(c+a-b)(c+b-a)}{4ab}}$$
$$= \sqrt{\frac{(2s-2b)(2s-2a)}{4ab}} = \sqrt{\frac{(s-a)(s-b)}{ab}}$$

64. $(2, 5)$ **65.** $-\frac{3\sqrt{3}}{\pi}$ **66.** $\sin\theta = \frac{2\sqrt{6}}{7}$; $\csc\theta = \frac{7\sqrt{6}}{12}$; $\sec\theta = -\frac{7}{5}$; $\cot\theta = -\frac{5\sqrt{6}}{12}$ **67.** $y = -3\sin(4x)$

4.4 Assess Your Understanding *(page 291)*

2. $\frac{1}{2}ab\sin C$ **3.** $\sqrt{s(s-a)(s-b)(s-c)}$; $\frac{1}{2}(a+b+c)$ **4.** T **5.** c **6.** c **7.** 2.83 **9.** 2.99 **11.** 14.98 **13.** 9.56 **15.** 3.86 **17.** 1.48 **19.** 2.82
21. 30 **23.** 1.73 **25.** 19.90 **27.** $K = \frac{1}{2}ab\sin C = \frac{1}{2}a\sin C\left(\frac{a\sin B}{\sin A}\right) = \frac{a^2\sin B\sin C}{2\sin A}$ **29.** 0.92 **31.** 2.27 **33.** 5.44 **35.** 9.03 sq ft **37.** \$5446.38
39. The area of home plate is about 216.5 in.^{2} **41.** $K = \frac{1}{2}r^2(\theta + \sin\theta)$ **43.** The ground area is 7517.4 ft^2.

45. (a) Area $\Delta OAC = \frac{1}{2}|OC||AC| = \frac{1}{2}\cdot\frac{|OC|}{1}\cdot\frac{|AC|}{1} = \frac{1}{2}\sin\alpha\cos\alpha$

(b) Area $\Delta OCB = \frac{1}{2}|BC||OC| = \frac{1}{2}|OB|^2\frac{|BC|}{|OB|}\cdot\frac{|OC|}{|OB|} = \frac{1}{2}|OB|^2\sin\beta\cos\beta$

(c) Area $\Delta OAB = \frac{1}{2}|BD||OA| = \frac{1}{2}|OB|\frac{|BD|}{|OB|} = \frac{1}{2}|OB|\sin(\alpha+\beta)$

(d) $$\frac{\cos\alpha}{\cos\beta} = \frac{\frac{|OC|}{1}}{\frac{|OC|}{|OB|}} = |OB|$$

(e) Area ΔOAB = Area ΔOAC + Area ΔOCB
$$\frac{1}{2}|OB|\sin(\alpha+\beta) = \frac{1}{2}\sin\alpha\cos\alpha + \frac{1}{2}|OB|^2\sin\beta\cos\beta$$
$$\sin(\alpha+\beta) = \frac{1}{|OB|}\sin\alpha\cos\alpha + |OB|\sin\beta\cos\beta$$
$$\sin(\alpha+\beta) = \frac{\cos\beta}{\cos\alpha}\sin\alpha\cos\alpha + \frac{\cos\alpha}{\cos\beta}\sin\beta\cos\beta$$
$$\sin(\alpha+\beta) = \sin\alpha\cos\beta + \cos\alpha\sin\beta$$

47. 31,145 ft^2 **49. (a)** The perimeter and area are both 36. **(b)** The perimeter and area are both 60.
51. $K = \frac{1}{2}ah = \frac{1}{2}ab\sin C \Rightarrow h = b\sin C = \frac{a\sin B\sin C}{\sin A}$

53. $\angle POQ = 180° - \left(\frac{A}{2} + \frac{B}{2}\right) = 180° - \frac{1}{2}(180° - C) = 90° + \frac{C}{2}$, and $\sin\left(90° + \frac{C}{2}\right) = \cos\left(-\frac{C}{2}\right) = \cos\frac{C}{2}$, since cosine is an even function.

Therefore, $r = \frac{c\sin\frac{A}{2}\sin\frac{B}{2}}{\sin\left(90° + \frac{C}{2}\right)} = \frac{c\sin\frac{A}{2}\sin\frac{B}{2}}{\cos\frac{C}{2}}$.

55. $\cot\frac{A}{2} + \cot\frac{B}{2} + \cot\frac{C}{2} = \frac{s-a}{r} + \frac{s-b}{r} + \frac{s-c}{r} = \frac{3s-(a+b+c)}{r} = \frac{3s-2s}{r} = \frac{s}{r}$
60. $\frac{\pi}{10}$ **61.** $\{x \mid x \ne -3, 3\}$

62. $\sin t = \frac{\sqrt{2}}{3}$, $\cos t = -\frac{\sqrt{7}}{3}$, $\tan t = -\frac{\sqrt{14}}{7}$, $\csc t = \frac{3\sqrt{2}}{2}$, $\sec t = -\frac{3\sqrt{7}}{7}$, $\cot t = -\frac{\sqrt{14}}{2}$

63. $\csc\theta - \sin\theta = \frac{1}{\sin\theta} - \sin\theta = \frac{1-\sin^2\theta}{\sin\theta} = \frac{\cos^2\theta}{\sin\theta} = \cos\theta \cdot \frac{\cos\theta}{\sin\theta} = \cos\theta\cot\theta$

4.5 Assess Your Understanding *(page 301)*

2. Simple harmonic; amplitude **3.** Simple harmonic; damped **4.** T **5.** $d = -5\cos(\pi t)$ **7.** $d = -6\cos(2t)$ **9.** $d = -5\sin(\pi t)$ **11.** $d = -6\sin(2t)$ **13. (a)** Simple harmonic **(b)** 5 m **(c)** $\frac{2\pi}{3}$ sec **(d)** $\frac{3}{2\pi}$ oscillation/sec **15. (a)** Simple harmonic **(b)** 6 m **(c)** 2 sec **(d)** $\frac{1}{2}$ oscillation/sec **17. (a)** Simple harmonic **(b)** 3 m **(c)** 4π sec **(d)** $\frac{1}{4\pi}$ oscillation/sec **19. (a)** Simple harmonic **(b)** 2 m **(c)** 1 sec **(d)** 1 oscillation/sec

21.

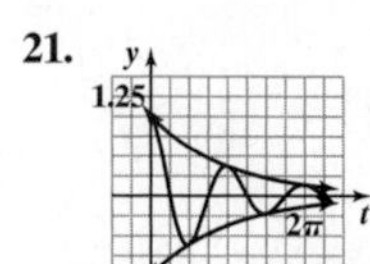

23.

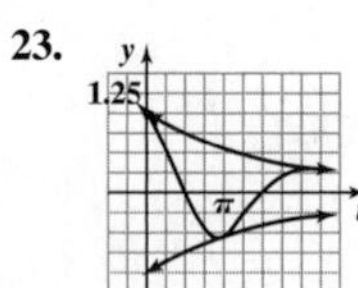

25.

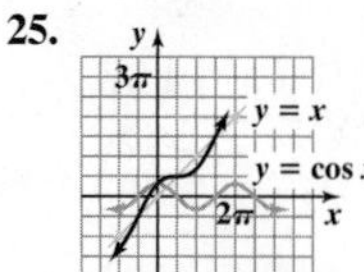

27.

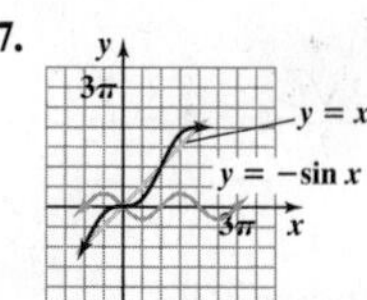

29.

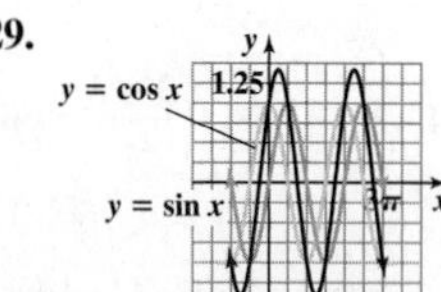

31.

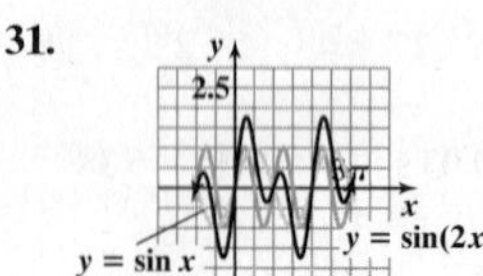

33. (a) 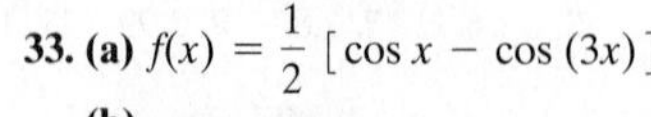 $f(x) = \frac{1}{2}[\cos x - \cos(3x)]$

(b)

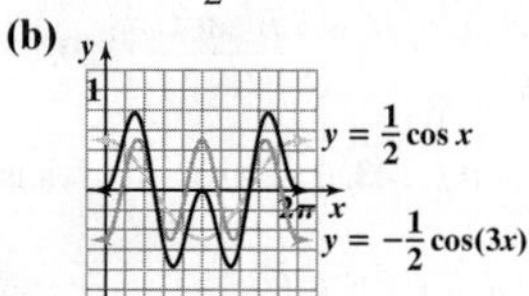

35. (a) 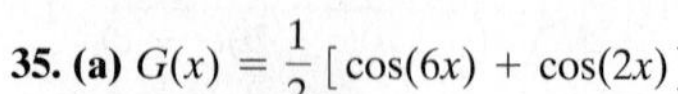 $G(x) = \frac{1}{2}[\cos(6x) + \cos(2x)]$

(b)

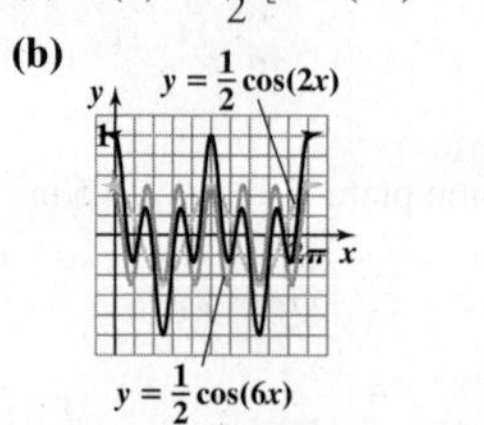

37. (a) 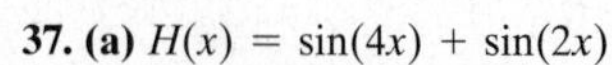 $H(x) = \sin(4x) + \sin(2x)$

(b)

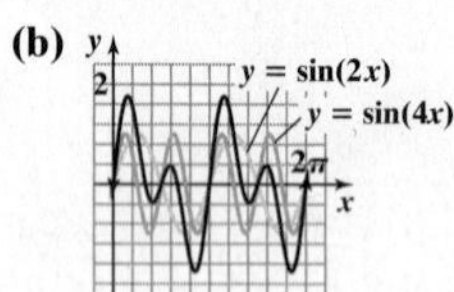

39. (a) 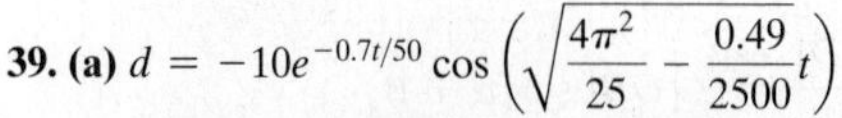 $d = -10e^{-0.7t/50}\cos\left(\sqrt{\frac{4\pi^2}{25} - \frac{0.49}{2500}}\,t\right)$

(b)

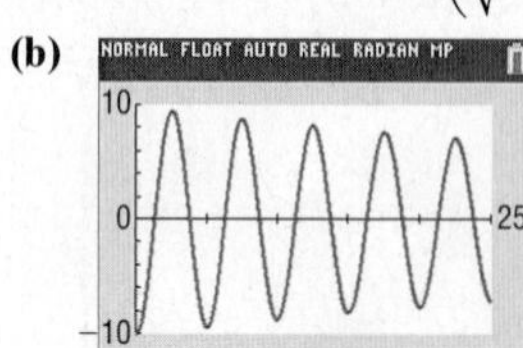

41. (a) 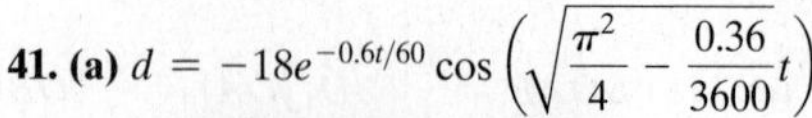 $d = -18e^{-0.6t/60}\cos\left(\sqrt{\frac{\pi^2}{4} - \frac{0.36}{3600}}\,t\right)$

(b)

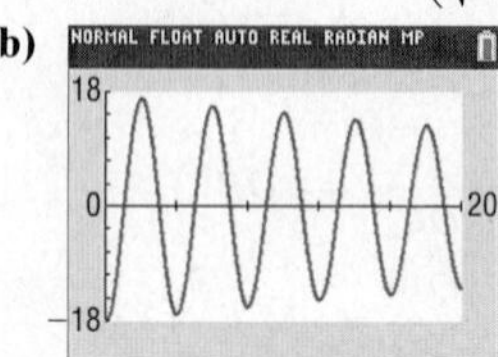

43. (a) $d = -5e^{-0.8t/20}\cos\left(\sqrt{\frac{4\pi^2}{9} - \frac{0.64}{400}}\,t\right)$

(b)

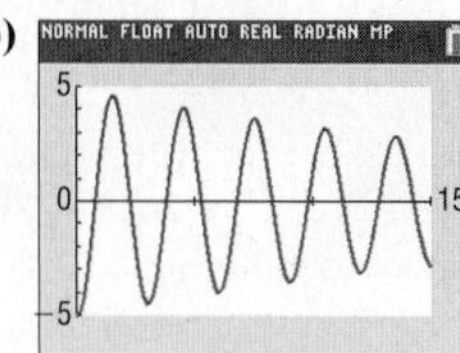

45. (a) The motion is damped. The bob has mass $m = 20$ kg with a damping factor of 0.7 kg/sec.
(b) 20 m leftward
(c)

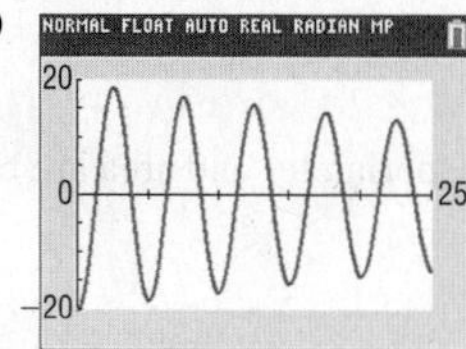

(d) 18.33 m leftward **(e)** $d \to 0$

47. (a) The motion is damped. The bob has mass $m = 40$ kg with a damping factor of 0.6 kg/sec.
(b) 30 m leftward
(c)

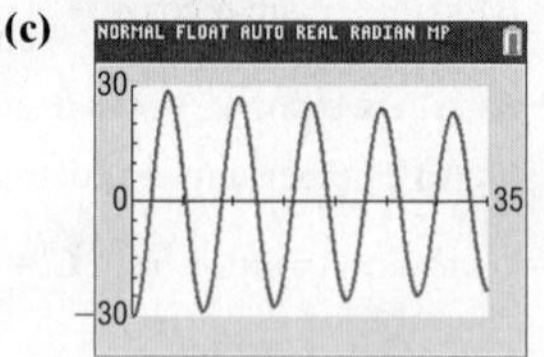

(d) 28.47 m leftward **(e)** $d \to 0$

49. (a) The motion is damped. The bob has mass $m = 15$ kg with a damping factor of 0.9 kg/sec.
(b) 15 m leftward
(c)

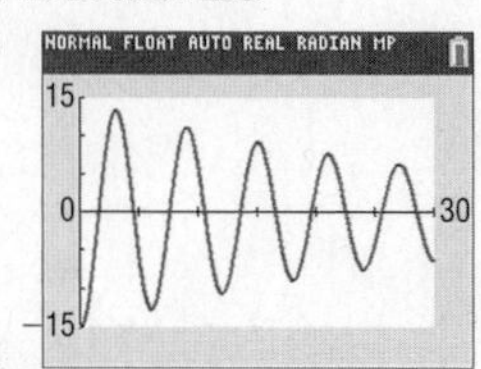

(d) 12.53 m leftward **(e)** $d \to 0$

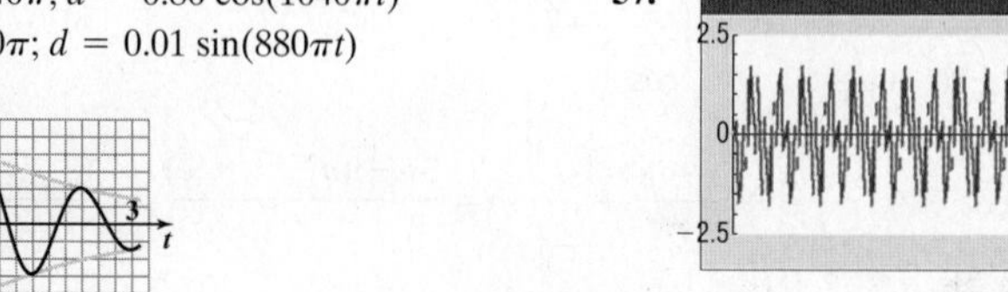

51. $\omega = 1040\pi$; $d = 0.80\cos(1040\pi t)$
53. $\omega = 880\pi$; $d = 0.01\sin(880\pi t)$
55. (a)

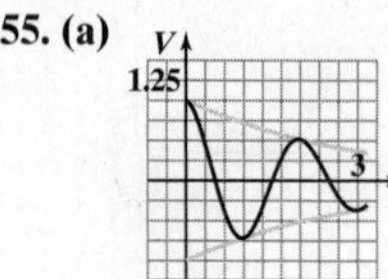

(b) At $t = 0, t = 2$; at $t = 1, t = 3$
(c) During the approximate intervals $0.35 < t < 0.67$, $1.29 < t < 1.75$, and $2.19 < t \le 3$

57.

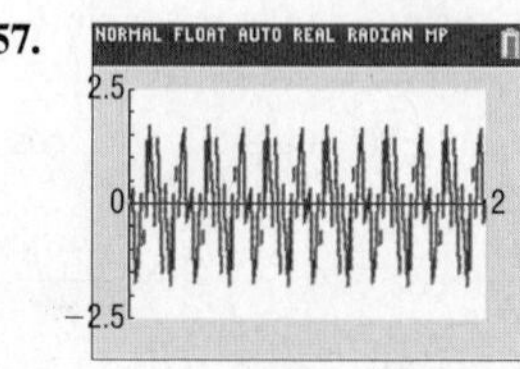

61. $f(x) = \dfrac{\sin x}{x}$

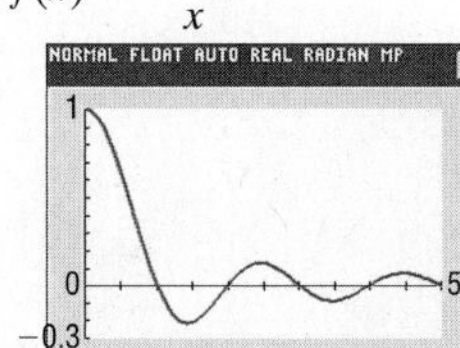

63. $y = \dfrac{1}{x}\sin x$

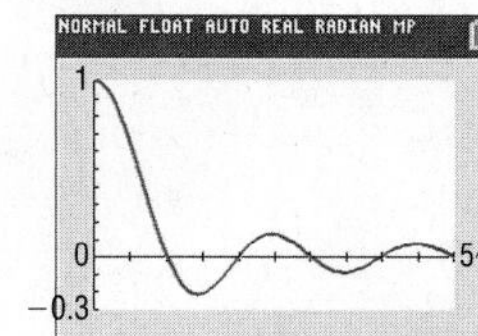

$y = \dfrac{1}{x^2}\sin x$

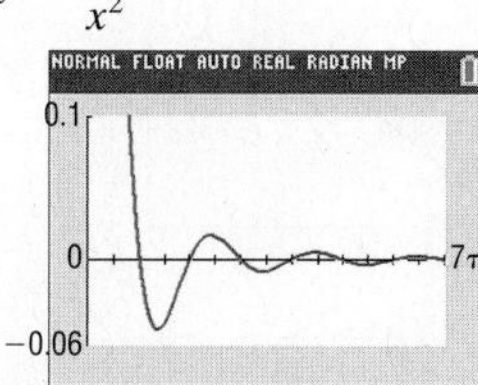

$y = \dfrac{1}{x^3}\sin x$

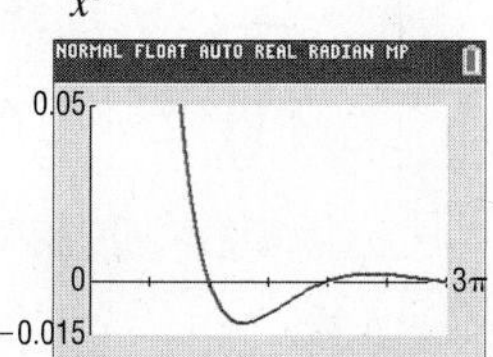

65. $f^{-1}(x) = \dfrac{4x - 3}{x - 1}$ **66.** $\dfrac{5\sqrt{21}}{21}$ **67.** $A \approx 20°, C \approx 50°, c \approx 44.85$ **68. (a)** $\dfrac{3\sqrt{10}}{10}$ **(b)** $\dfrac{\sqrt{10}}{10}$ **(c)** $\dfrac{1}{3}$

Review Exercises *(page 305)*

1. $\sin\theta = \dfrac{4}{5}$; $\cos\theta = \dfrac{3}{5}$; $\tan\theta = \dfrac{4}{3}$; $\cot\theta = \dfrac{3}{4}$; $\sec\theta = \dfrac{5}{3}$; $\csc\theta = \dfrac{5}{4}$ **2.** $\sin\theta = \dfrac{\sqrt{3}}{2}$; $\cos\theta = \dfrac{1}{2}$; $\tan\theta = \sqrt{3}$; $\cot\theta = \dfrac{\sqrt{3}}{3}$; $\sec\theta = 2$; $\csc\theta = \dfrac{2\sqrt{3}}{3}$
3. 0 **4.** 1 **5.** 1 **6.** $A = 70°, b \approx 3.42, a \approx 9.40$ **7.** $a \approx 4.58, A \approx 66.4°, B \approx 23.6°$ **8.** $C = 100°, b \approx 0.65, c \approx 1.29$
9. $B \approx 56.8°, C \approx 23.2°, b \approx 4.25$ **10.** No triangle **11.** $b \approx 3.32, A \approx 62.8°, C \approx 17.2°$ **12.** $A \approx 36.2°, C \approx 63.8°, c \approx 4.55$
13. No triangle **14.** $A \approx 83.3°$, B $\approx 44.0°, C \approx 52.6°$ **15.** $c \approx 2.32, A \approx 16.1°, B \approx 123.9°$ **16.** $B \approx 36.2°, C \approx 63.8°, c \approx 4.55$
17. $A \approx 39.6°, B \approx 18.6°, C \approx 121.9°$ **18.** Two triangles: $B_1 \approx 13.4°, C_1 \approx 156.6°, c_1 \approx 6.86$ or $B_2 \approx 166.6°, C_2 \approx 3.4°, c_2 \approx 1.02$
19. $b \approx 11.52, c \approx 10.13, C \approx 60°$ **20.** $a \approx 5.23, B \approx 46.0°, C \approx 64.0°$ **21.** 1.93 **22.** 18.79 **23.** 6 **24.** 3.80 **25.** 0.32 **26.** 1.92 in.^{2}
27. 48.2° and 41.8° **28.** 23.32 ft **29.** 2.15 mi **30.** 132.55 ft/min **31.** 12.7° **32.** 29.97 ft **33.** 6.22 mi **34. (a)** 131.8 mi **(b)** 23.1° **(c)** 0.21 hr
35. 8798.67 sq ft **36.** S4.0°E **37.** 76.94 in. **38.** 79.69 in. **39.** $d = -3\cos\left(\dfrac{\pi}{2}t\right)$
40. (a) Simple harmonic **(b)** 6 ft **(c)** π s **(d)** $\dfrac{1}{\pi}$ oscillation/s
41. (a) Simple harmonic **(b)** 2 ft **(c)** 2 s **(d)** $\dfrac{1}{2}$ oscillation/s
42. (a) $d = -15e^{-0.75t/80}\cos\left(\sqrt{\dfrac{4\pi^2}{25} - \dfrac{0.5625}{6400}}\,t\right)$
(b)

43. (a) The motion is damped. The bob has mass $m = 20$ kg with a damping factor of 0.6 kg/s.

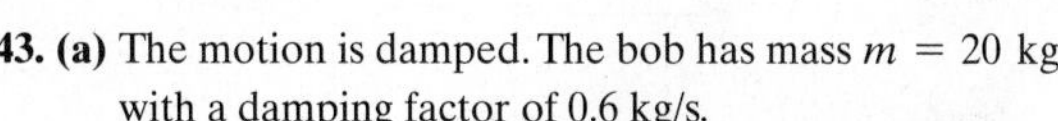

(b) 15 m leftward

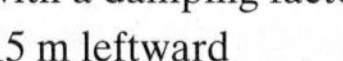

(c)

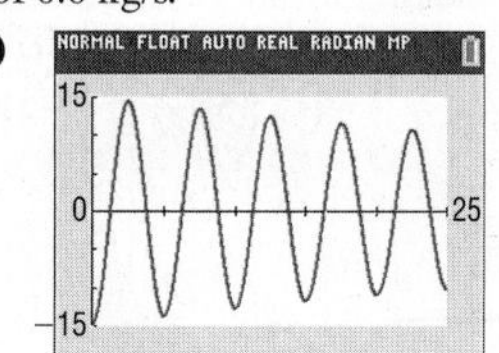

(d) 13.92 m leftward
(e) $d \to 0$

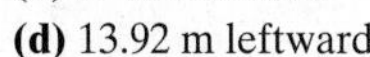

44.

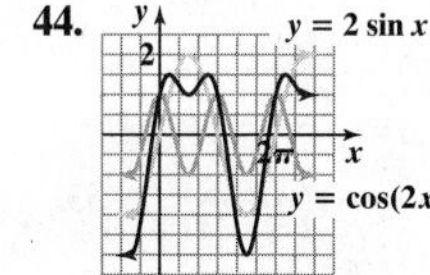

Chapter Test *(page 307)*

1. $\sin\theta = \dfrac{\sqrt{5}}{5}$; $\cos\theta = \dfrac{2\sqrt{5}}{5}$; $\tan\theta = \dfrac{1}{2}$; $\csc\theta = \sqrt{5}$; $\sec\theta = \dfrac{\sqrt{5}}{2}$; $\cot\theta = 2$ **2.** 0 **3.** $a = 15.88, B \approx 57.5°, C \approx 70.5°$
4. $b \approx 6.85, C = 117°, c \approx 16.30$ **5.** $A \approx 52.4°, B \approx 29.7°, C \approx 97.9°$ **6.** $b \approx 4.72, c \approx 1.67, B = 105°$ **7.** No triangle
8. $c \approx 7.62, A \approx 80.5°, B \approx 29.5°$ **9.** 15.04 square units **10.** 19.81 square units **11.** 61.0° **12.** 1.3° **13.** The area of the shaded region is 9.26 cm^2.
14. 54.15 square units **15.** Madison will have to swim about 2.23 miles. **16.** 12.63 square units **17.** The lengths of the sides are 15, 18, and 21.
18. $d = 5(\sin 42°)\sin\left(\dfrac{\pi t}{3}\right)$ or $d \approx 3.346\sin\left(\dfrac{\pi t}{3}\right)$

Cumulative Review *(page 308)*

1. $\left\{\dfrac{1}{3}, 1\right\}$ **2.** $(x + 5)^2 + (y - 1)^2 = 9$

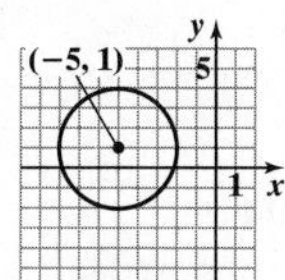

3. $\{x \mid x \le -1 \text{ or } x \ge 4\}$ **4.**

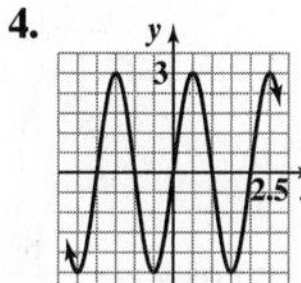

5.

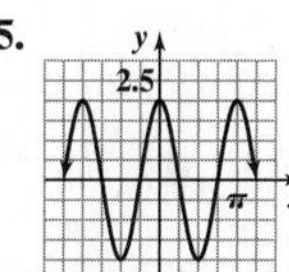

6. (a) $-\dfrac{2\sqrt{5}}{5}$ **(b)** $\dfrac{\sqrt{5}}{5}$ **(c)** $-\dfrac{4}{5}$ **(d)** $-\dfrac{3}{5}$ **(e)** $\sqrt{\dfrac{5 - \sqrt{5}}{10}}$ **(f)** $-\sqrt{\dfrac{5 + \sqrt{5}}{10}}$

7. (a) 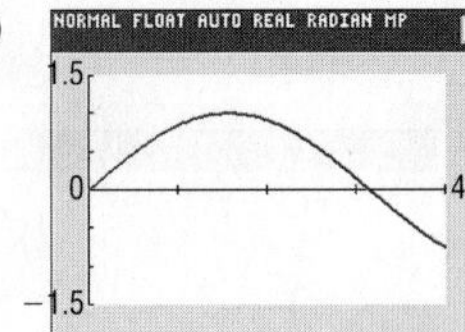

(b) 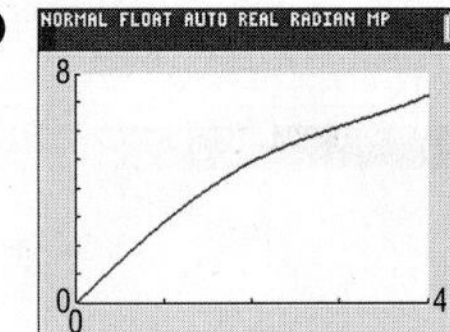

8. (a) 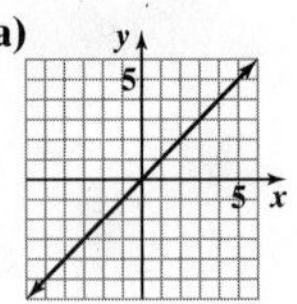

(b) 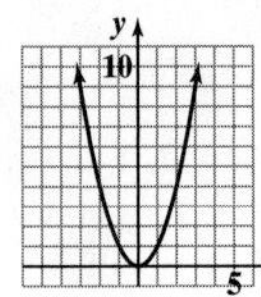

(c) 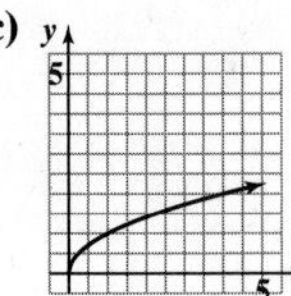

(d) 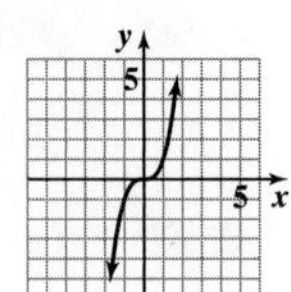

(e) 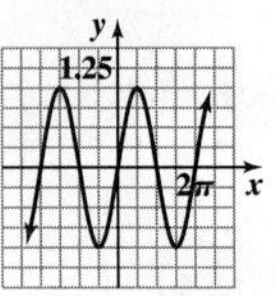

(f) 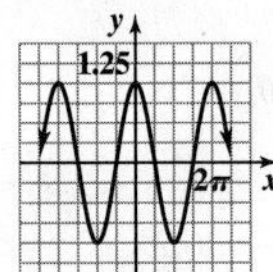

(g) 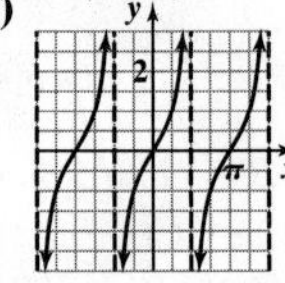

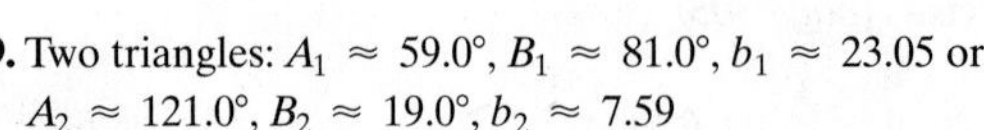

9. Two triangles: $A_1 \approx 59.0°, B_1 \approx 81.0°, b_1 \approx 23.05$ or $A_2 \approx 121.0°, B_2 \approx 19.0°, b_2 \approx 7.59$

10. $\left\{0, \frac{2\pi}{3}, \frac{4\pi}{3}\right\}$ **11.** π ft ≈ 3.14 ft

CHAPTER 5 Polar Coordinates; Vectors

5.1 Assess Your Understanding *(page 319)*

5. pole; polar axis **6.** $r\cos\theta$; $r\sin\theta$ **7.** b **8.** d **9.** T **10.** F **11.** *A* **13.** *C* **15.** *B* **17.** *A*

19.

21. 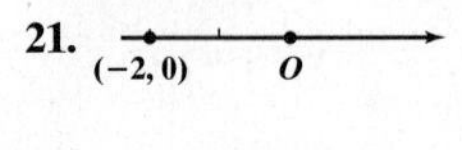

23. 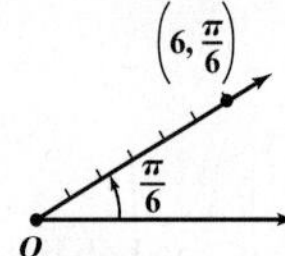

25. 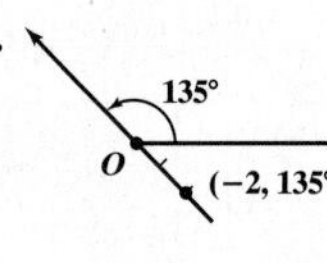

27. 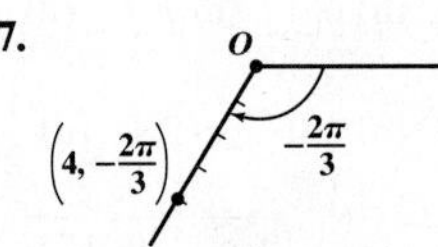

29. 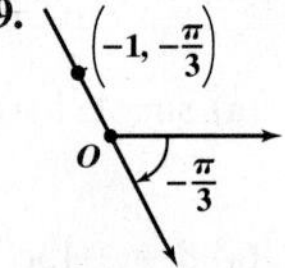

31.

33. 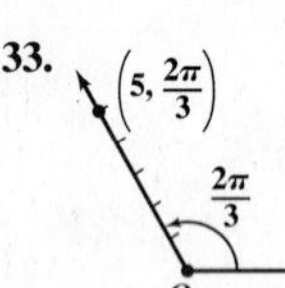

(a) $\left(5, -\frac{4\pi}{3}\right)$ **(b)** $\left(-5, \frac{5\pi}{3}\right)$ **(c)** $\left(5, \frac{8\pi}{3}\right)$

35. **(a)** $(2, -2\pi)$ **(b)** $(-2, \pi)$ **(c)** $(2, 2\pi)$

37. **(a)** $\left(1, -\frac{3\pi}{2}\right)$ **(b)** $\left(-1, \frac{3\pi}{2}\right)$ **(c)** $\left(1, \frac{5\pi}{2}\right)$

39. 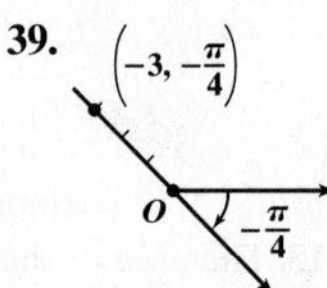

(a) $\left(3, -\frac{5\pi}{4}\right)$ **(b)** $\left(-3, \frac{7\pi}{4}\right)$ **(c)** $\left(3, \frac{11\pi}{4}\right)$

41. $(0, 3)$ **43.** $(-2, 0)$ **45.** $(-3\sqrt{3}, 3)$ **47.** $(\sqrt{2}, -\sqrt{2})$ **49.** $\left(-\frac{1}{2}, \frac{\sqrt{3}}{2}\right)$ **51.** $(2, 0)$ **53.** $(-2.57, 7.05)$ **55.** $(-4.98, -3.85)$ **57.** $(3, 0)$ **59.** $(1, \pi)$ **61.** $\left(\sqrt{2}, -\frac{\pi}{4}\right)$ **63.** $\left(2, \frac{\pi}{6}\right)$ **65.** $(2.47, -1.02)$ **67.** $(9.30, 0.47)$ **69.** $r^2 = \frac{3}{2}$ or $r = \frac{\sqrt{6}}{2}$ **71.** $r^2\cos^2\theta - 4r\sin\theta = 0$ **73.** $r^2\sin 2\theta = 1$ **75.** $r\cos\theta = 4$ **77.** $x^2 + y^2 - x = 0$ or $\left(x - \frac{1}{2}\right)^2 + y^2 = \frac{1}{4}$ **79.** $(x^2 + y^2)^{3/2} - x = 0$ **81.** $x^2 + y^2 = 4$ **83.** $y^2 = 8(x + 2)$

85. (a) $(-10, 36)$ **(b)** $\left(2\sqrt{349}, 180° + \tan^{-1}\left(-\frac{18}{5}\right)\right) \approx (37.36, 105.5°)$ **(c)** $(-3, -35)$ **(d)** $\left(\sqrt{1234}, 180° + \tan^{-1}\left(\frac{35}{3}\right)\right) \approx (35.13, 265.1°)$

90. $\left(-\frac{5}{4}, \frac{9}{2}\right)$ **91.** $(0, -11)$ **92.** $\sin\theta = \frac{15}{17}$; $\cos\theta = -\frac{8}{17}$; $\tan\theta = -\frac{15}{8}$; $\csc\theta = \frac{17}{15}$; $\sec\theta = -\frac{17}{8}$; $\cot\theta = -\frac{8}{15}$

93. Amplitude $= 4$; Period $= 3\pi$

5.2 Assess Your Understanding *(page 333)*

7. polar equation **8.** F **9.** $-\theta$ **10.** $\pi - \theta$ **11.** T **12.** $2n; n$ **13.** c **14.** b

15. $x^2 + y^2 = 16$; circle, radius 4, center at pole

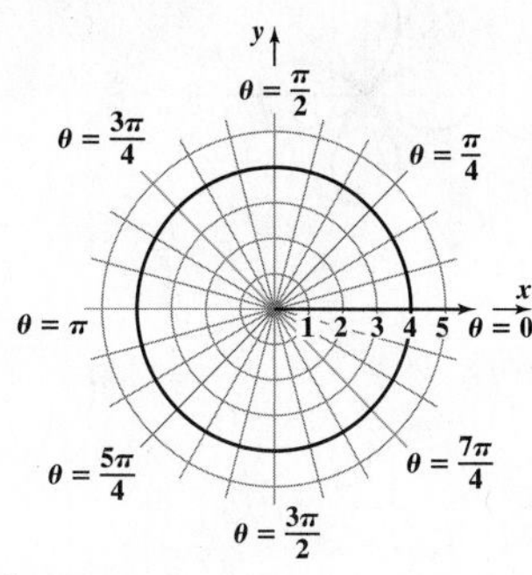

17. $y = \sqrt{3}\,x$; line through pole, making an angle of $\frac{\pi}{3}$ with polar axis

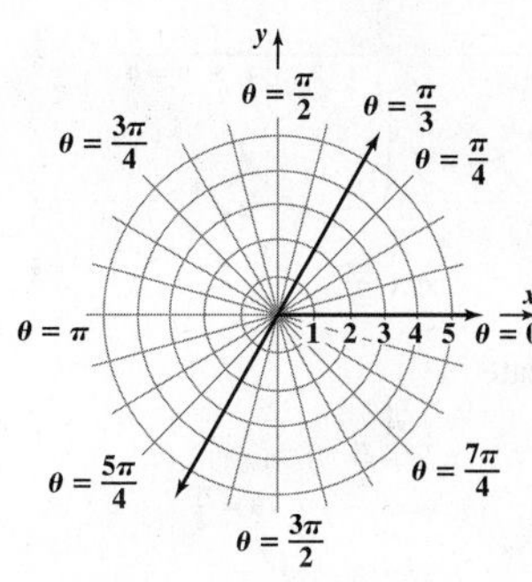

19. $y = 4$; horizontal line 4 units above the pole

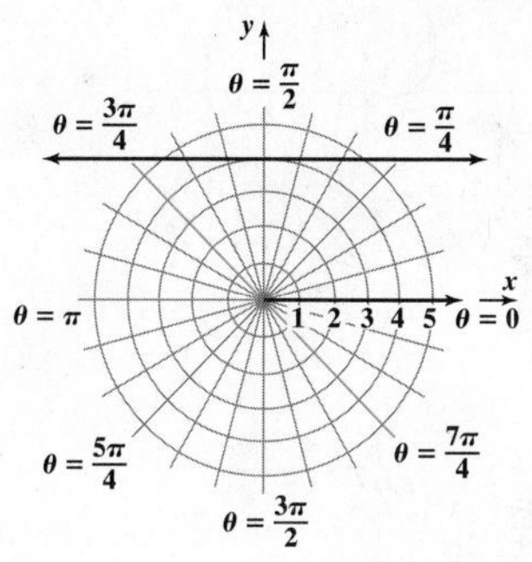

21. $x = -2$; vertical line 2 units to the left of the pole

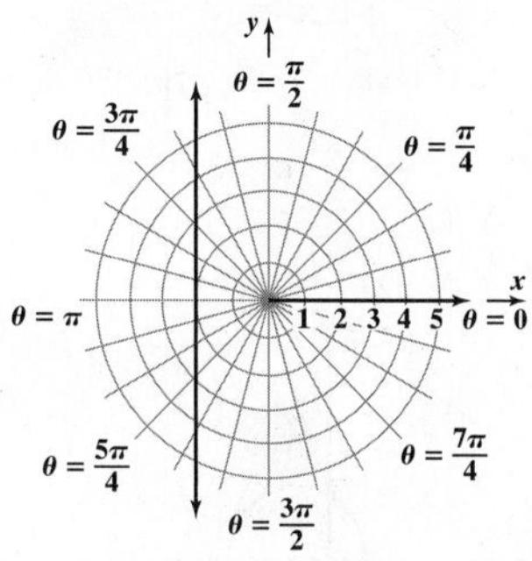

23. $(x - 1)^2 + y^2 = 1$; circle, radius 1, center $(1, 0)$ in rectangular coordinates

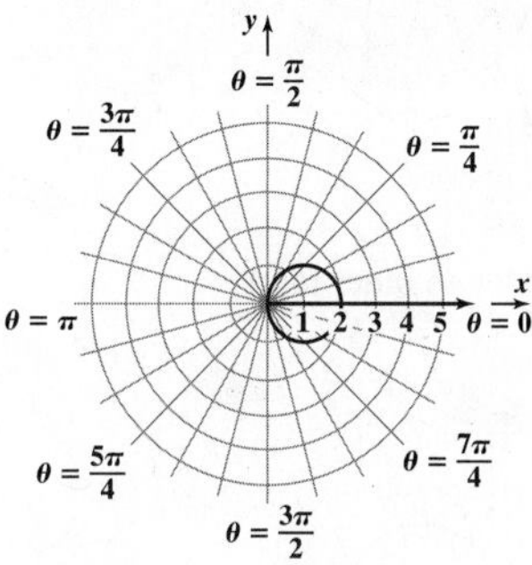

25. $x^2 + (y + 2)^2 = 4$; circle, radius 2, center at $(0, -2)$ in rectangular coordinates

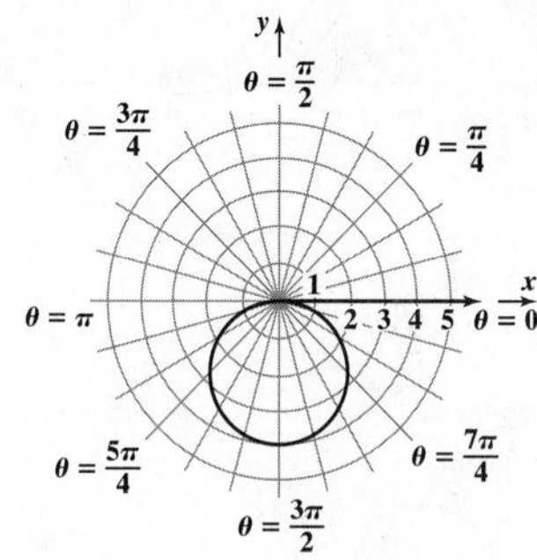

27. $(x - 2)^2 + y^2 = 4, x \neq 0$; circle, radius 2, center at $(2, 0)$ in rectangular coordinates, hole at $(0, 0)$

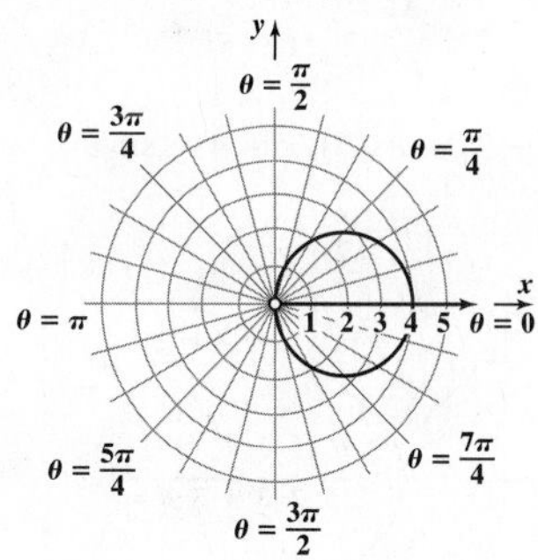

29. $x^2 + (y + 1)^2 = 1, x \neq 0$; circle, radius 1, center at $(0, -1)$ in rectangular coordinates, hole at $(0, 0)$

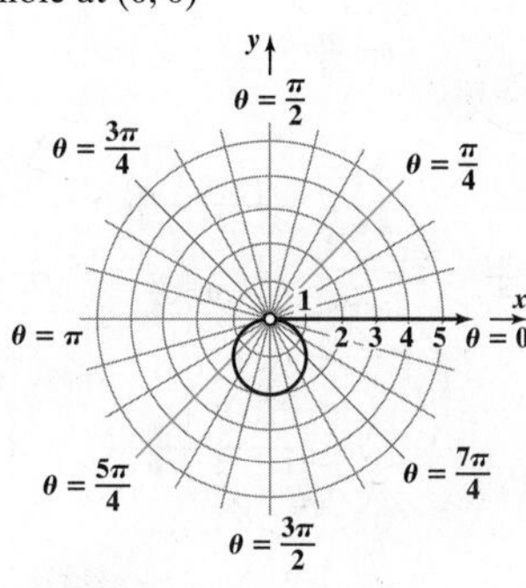

31. *E* **33.** *F* **35.** *H* **37.** *D*

39. Cardioid

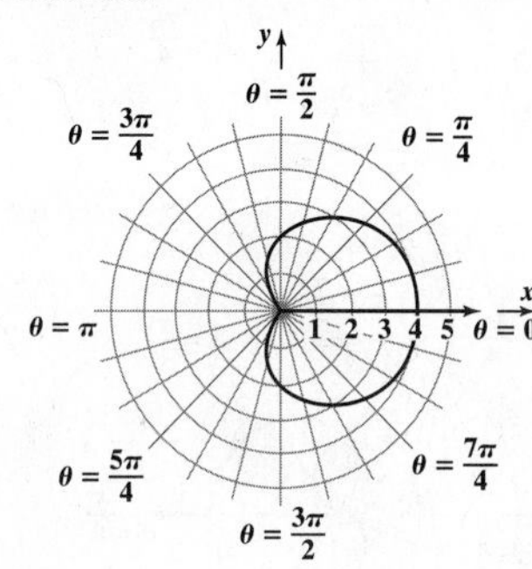

41. Cardioid

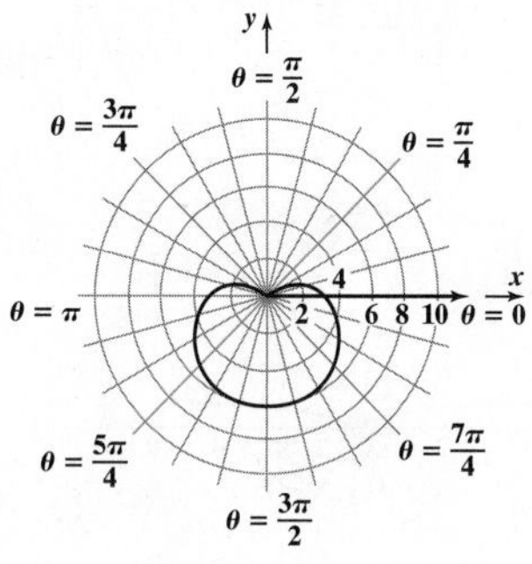

43. Limaçon without inner loop

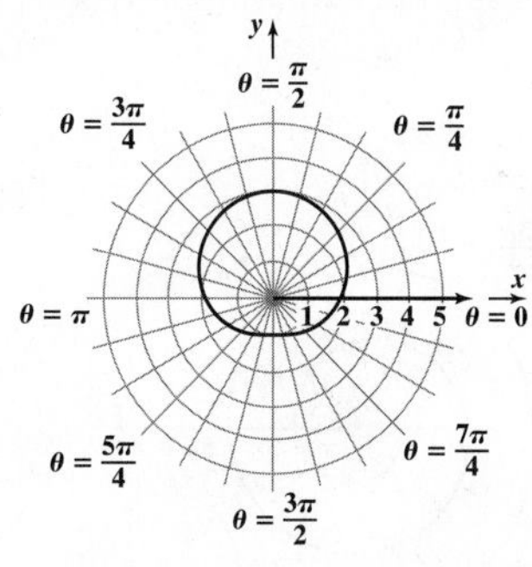

45. Limaçon without inner loop

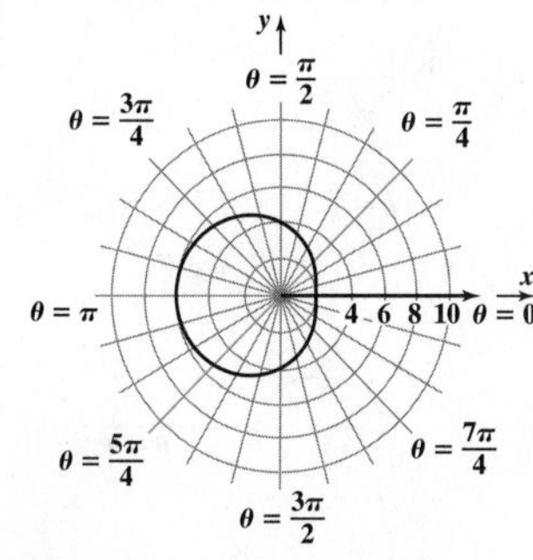

47. Limaçon with inner loop

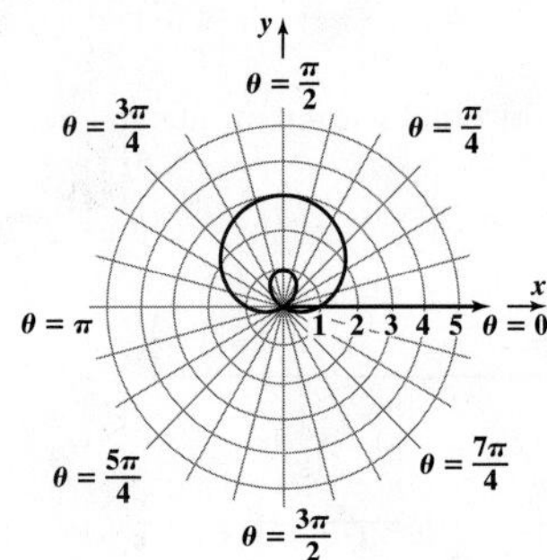

49. Limaçon with inner loop

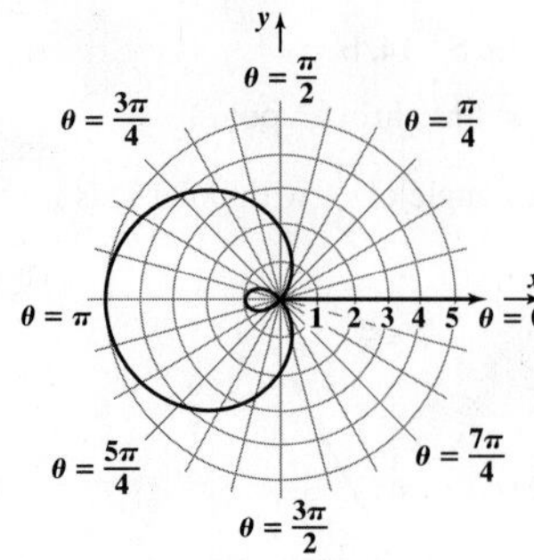

51. Rose

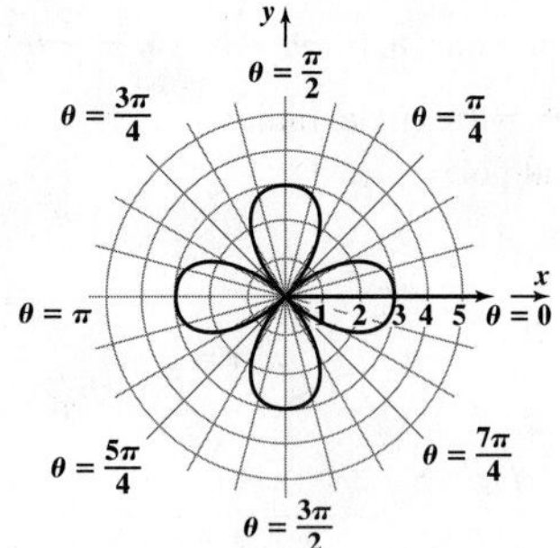

53. Rose

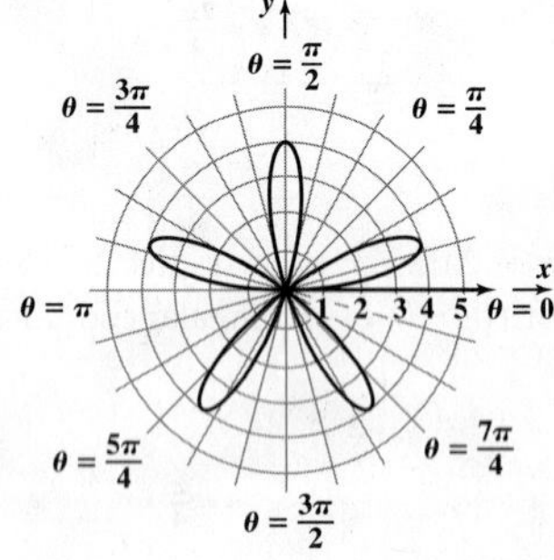

55. Lemniscate

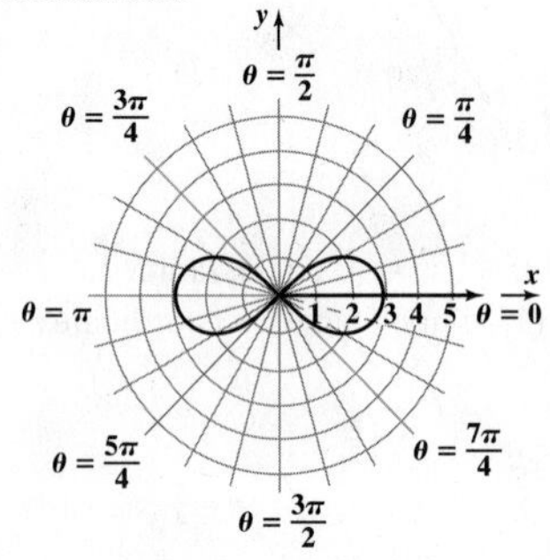

57. Spiral

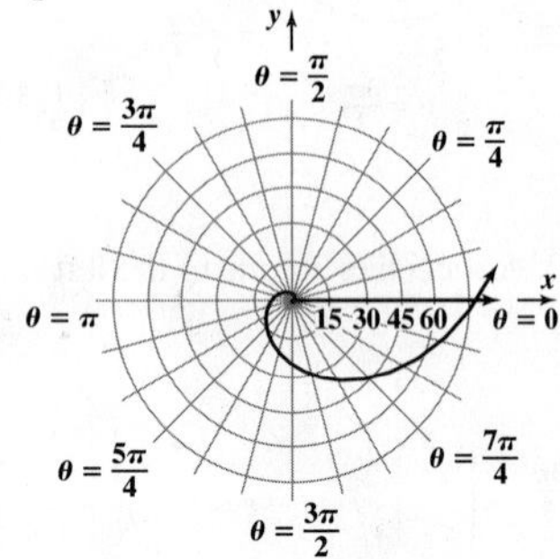

59. Cardioid

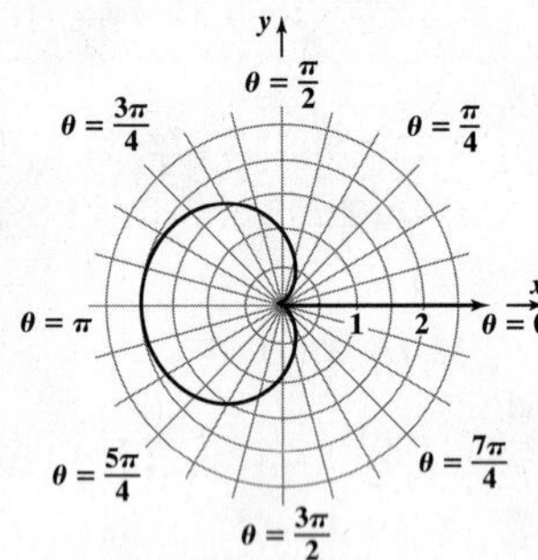

61. Limaçon with inner loop

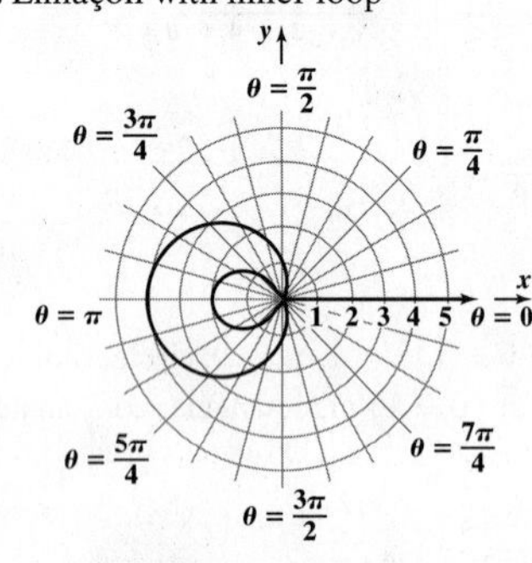

63.

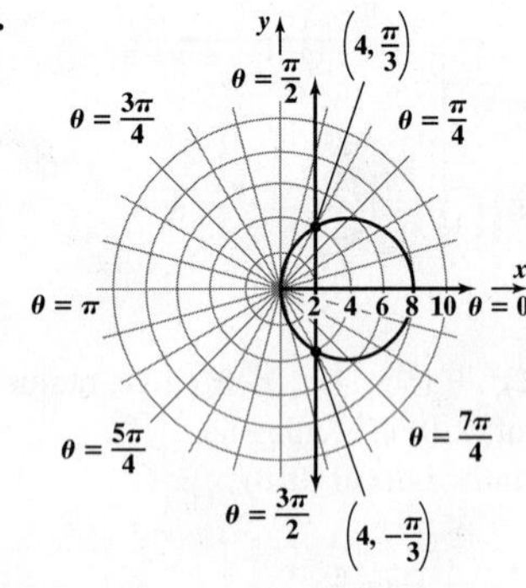

65.

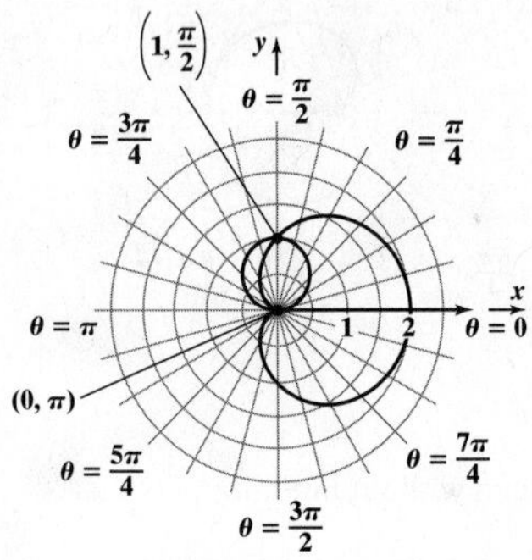

67.

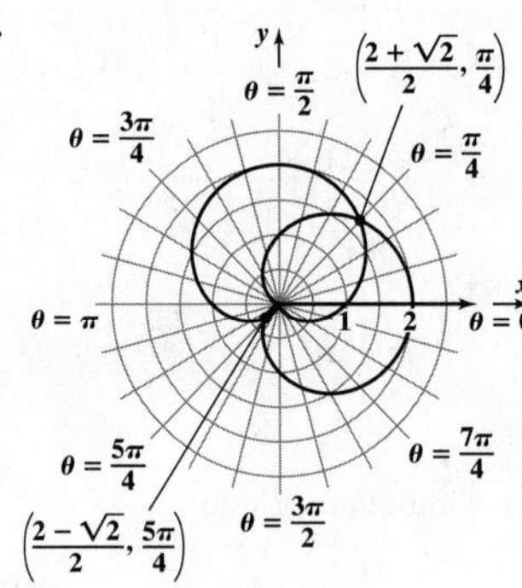

69. $r = 3 + 3 \cos \theta$ **71.** $r = 4 + \sin \theta$

73.

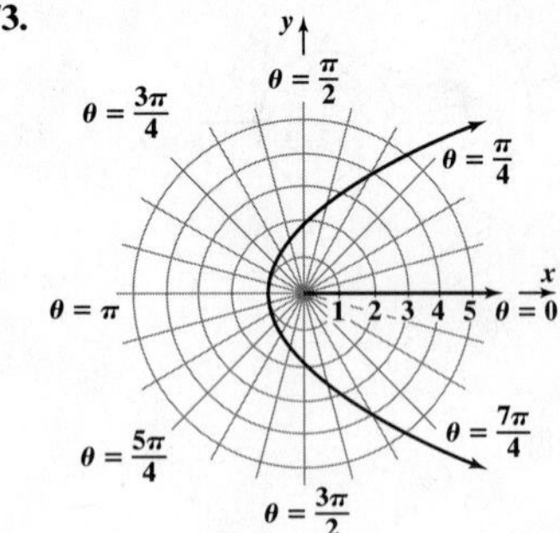

75.

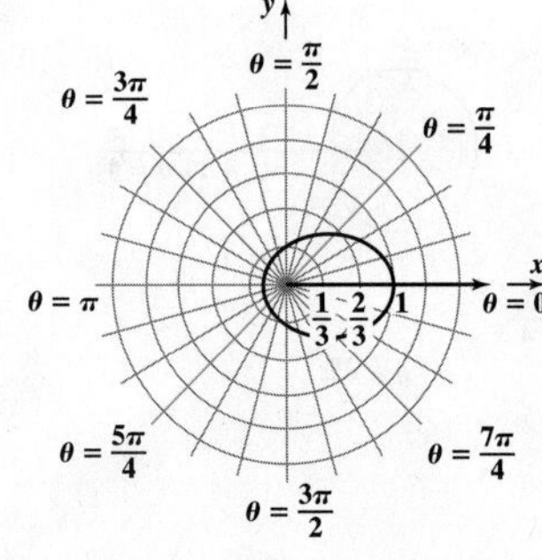

77.

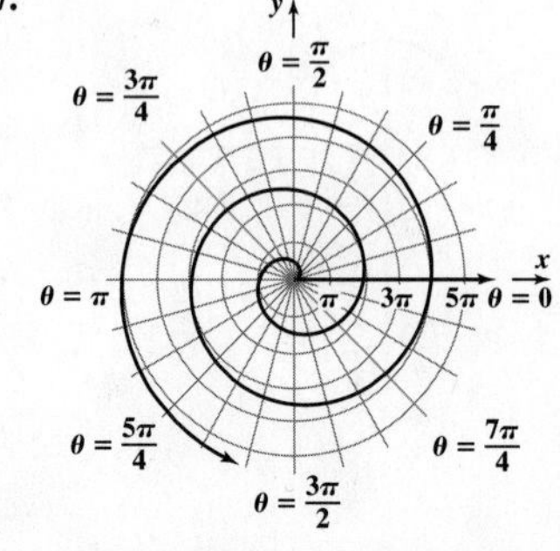

79.

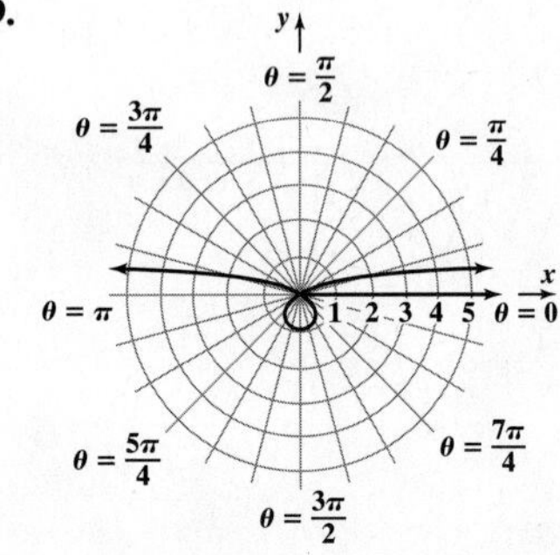

81.

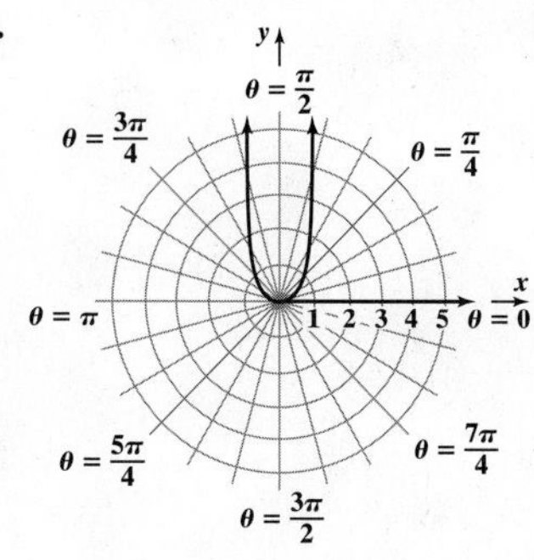

83. $r \sin\theta = a$
$y = a$

85.
$$\begin{aligned} r &= 2a\sin\theta \\ r^2 &= 2ar\sin\theta \\ x^2 + y^2 &= 2ay \\ x^2 + y^2 - 2ay &= 0 \\ x^2 + (y-a)^2 &= a^2 \end{aligned}$$
Circle, radius a, center at $(0, a)$ in rectangular coordinates

87.
$$\begin{aligned} r &= 2a\cos\theta \\ r^2 &= 2ar\cos\theta \\ x^2 + y^2 &= 2ax \\ x^2 - 2ax + y^2 &= 0 \\ (x-a)^2 + y^2 &= a^2 \end{aligned}$$
Circle, radius a, center at $(a, 0)$ in rectangular coordinates

89. (a) $r^2 = \cos\theta$; $r^2 = \cos(\pi - \theta)$
$r^2 = -\cos\theta$
Not equivalent; test fails.
$(-r)^2 = \cos(-\theta)$
$r^2 = \cos\theta$
New test works.

(b) $r^2 = \sin\theta$: $r^2 = \sin(\pi - \theta)$
$r^2 = \sin\theta$
Test works.
$(-r)^2 = \sin(-\theta)$
$r^2 = -\sin\theta$
Not equivalent; new test fails.

93. $f(x + 2) = x^2 + x - 2$ **94.** $420°$ **95.** Amplitude $= 2$; period $= \frac{2\pi}{5}$ **96.** $b \approx 6.81, C = 85°, c \approx 8.09$

Historical Problems *(page 341)*

1. (a) $1 + 4i, 1 + i$ **(b)** $-1, 2 + i$

5.3 Assess Your Understanding *(page 342)*

5. real; imaginary **6.** magnitude; modulus; argument **7.** $r_1 r_2$; $\theta_1 + \theta_2$; $\theta_1 + \theta_2$ **8.** r^n; $n\theta$; $n\theta$ **9.** three **10.** T **11.** c **12.** a

13.

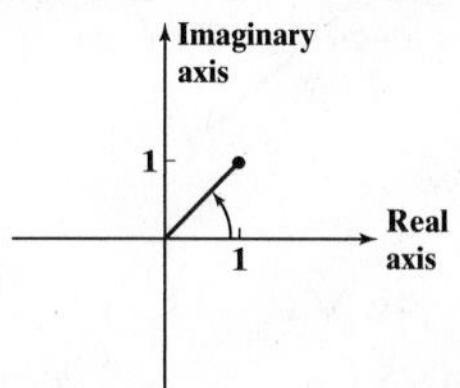

$\sqrt{2}\,(\cos 45° + i \sin 45°)$

15.

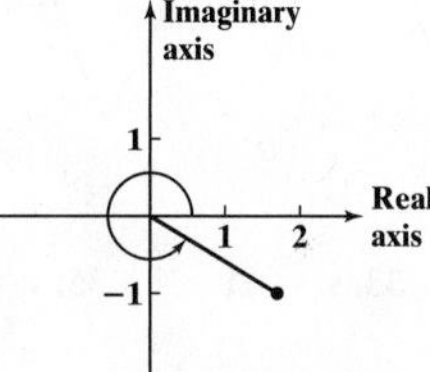

$2(\cos 330° + i \sin 330°)$

17.

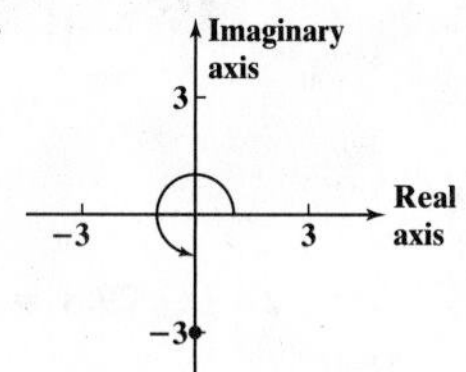

$3(\cos 270° + i \sin 270°)$

19.

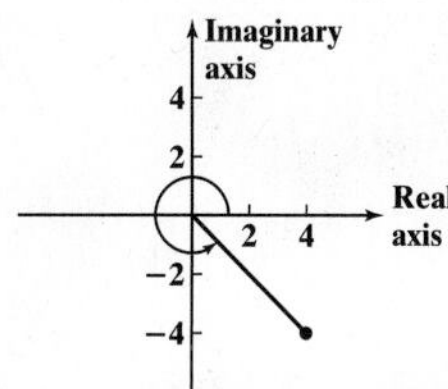

$4\sqrt{2}\,(\cos 315° + i \sin 315°)$

21.

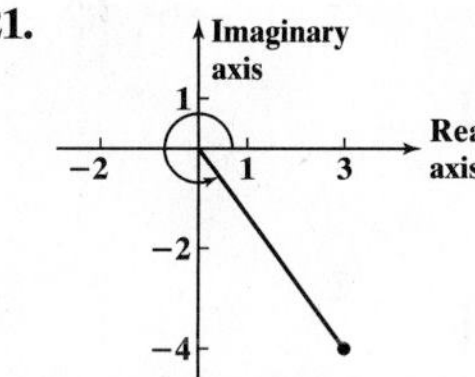

$5(\cos 306.9° + i \sin 306.9°)$

23.

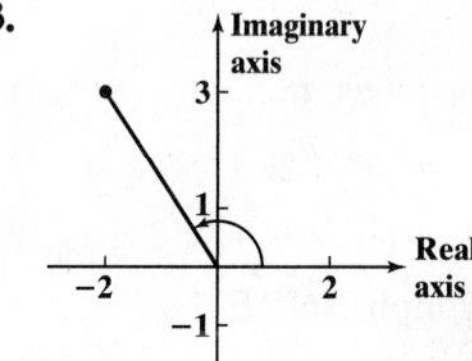

$\sqrt{13}\,(\cos 123.7° + i \sin 123.7°)$

25. $-1 + \sqrt{3}i$ **27.** $2\sqrt{2} - 2\sqrt{2}i$ **29.** $-3i$ **31.** $-0.035 + 0.197i$ **33.** $1.970 + 0.347i$

35. $zw = 8(\cos 60° + i \sin 60°)$; $\frac{z}{w} = \frac{1}{2}(\cos 20° + i \sin 20°)$ **37.** $zw = 12(\cos 40° + i \sin 40°)$; $\frac{z}{w} = \frac{3}{4}(\cos 220° + i \sin 220°)$

39. $zw = 4\left(\cos\frac{9\pi}{40} + i\sin\frac{9\pi}{40}\right)$; $\frac{z}{w} = \cos\frac{\pi}{40} + i\sin\frac{\pi}{40}$ **41.** $zw = 4\sqrt{2}\,(\cos 15° + i \sin 15°)$; $\frac{z}{w} = \sqrt{2}\,(\cos 75° + i \sin 75°)$

43. $-32 + 32\sqrt{3}i$ **45.** $32i$ **47.** $\frac{27}{2} + \frac{27\sqrt{3}}{2}i$ **49.** $-\frac{25\sqrt{2}}{2} + \frac{25\sqrt{2}}{2}i$ **51.** $-4 + 4i$ **53.** $-23 + 14.142i$

55. $\sqrt[6]{2}(\cos 15° + i \sin 15°)$, $\sqrt[6]{2}(\cos 135° + i \sin 135°)$, $\sqrt[6]{2}(\cos 255° + i \sin 255°)$

57. $\sqrt[4]{8}(\cos 75° + i \sin 75°)$, $\sqrt[4]{8}(\cos 165° + i \sin 165°)$, $\sqrt[4]{8}(\cos 255° + i \sin 255°)$, $\sqrt[4]{8}(\cos 345° + i \sin 345°)$

59. $2(\cos 67.5° + i \sin 67.5°)$, $2(\cos 157.5° + i \sin 157.5°)$, $2(\cos 247.5° + i \sin 247.5°)$, $2(\cos 337.5° + i \sin 337.5°)$

61. $\cos 18° + i \sin 18°$, $\cos 90° + i \sin 90°$, $\cos 162° + i \sin 162°$, $\cos 234° + i \sin 234°$, $\cos 306° + i \sin 306°$

63. $1, i, -1, -i$

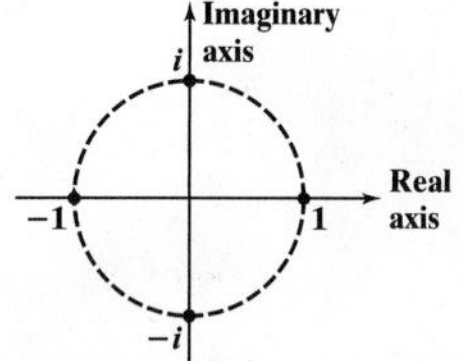

65. Look at formula (8). $|z_k| = \sqrt[n]{r}$ for all k.

67. Look at formula (8). The z_k are spaced apart by an angle of $\frac{2\pi}{n}$.

69. Assume the theorem is true for $n \geq 1$.

For $n = 0$:

$z^0 = r^0[\cos(0 \cdot \theta) + i\sin(0 \cdot \theta)]$

$1 = 1 \cdot [\cos(0) + i\sin(0)]$

$1 = 1 \cdot [1 + 0]$

$1 = 1$ True

For negative integers:

$$\begin{aligned} z^{-n} &= (z^n)^{-1} = (r^n[\cos(n\theta) + i\sin(n\theta)])^{-1} \quad \text{with } n \geq 1 \\ &= \frac{1}{r^n[\cos(n\theta) + i\sin(n\theta)]} \\ &= \frac{1}{r^n[\cos(n\theta) + i\sin(n\theta)]} \cdot \frac{\cos(n\theta) - i\sin(n\theta)}{\cos(n\theta) - i\sin(n\theta)} \\ &= \frac{\cos(n\theta) - i\sin(n\theta)}{r^n(\cos^2(n\theta) + \sin^2(n\theta))} \\ &= \frac{\cos(n\theta) - i\sin(n\theta)}{r^n} \\ &= r^{-n}[\cos(n\theta) - i\sin(n\theta)] \\ &= r^{-n}[\cos(-n\theta) + i\sin(-n\theta)] \end{aligned}$$

Thus, De Moivre's Theorem is true for all integers.

71. ≈ 40.50 **72.** $\frac{4}{3}\pi$ **73.** $5\sqrt{2}$ **74.** $\sin\theta = -\frac{\sqrt{15}}{4}$; $\tan\theta = -\sqrt{15}$; $\csc\theta = -\frac{4\sqrt{15}}{15}$; $\sec\theta = 4$; $\cot\theta = -\frac{\sqrt{15}}{15}$

5.4 Assess Your Understanding *(page 354)*

1. vector **2.** 0 **3.** unit **4.** position **5.** horizontal; vertical **6.** resultant **7.** T **8.** F **9.** a **10.** b

11.

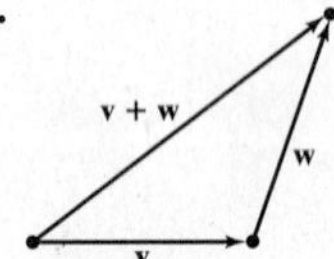

13. 3v

15.

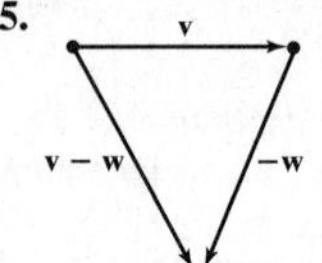

17. 3v, u, 3v + u − 2w, −2w

19. T **21.** F **23.** F **25.** T **27.** 12 **29.** $\mathbf{v} = 3\mathbf{i} + 4\mathbf{j}$ **31.** $\mathbf{v} = 2\mathbf{i} + 4\mathbf{j}$ **33.** $\mathbf{v} = 8\mathbf{i} - \mathbf{j}$ **35.** $\mathbf{v} = -\mathbf{i} + \mathbf{j}$ **37.** 5 **39.** $\sqrt{2}$ **41.** $\sqrt{13}$ **43.** $-\mathbf{j}$

45. $\sqrt{89}$ **47.** $\sqrt{34} - \sqrt{13}$ **49.** $\mathbf{i}$ **51.** $\frac{3}{5}\mathbf{i} - \frac{4}{5}\mathbf{j}$ **53.** $\frac{\sqrt{2}}{2}\mathbf{i} - \frac{\sqrt{2}}{2}\mathbf{j}$ **55.** $\mathbf{v} = \frac{8\sqrt{5}}{5}\mathbf{i} + \frac{4\sqrt{5}}{5}\mathbf{j}$, or $\mathbf{v} = -\frac{8\sqrt{5}}{5}\mathbf{i} - \frac{4\sqrt{5}}{5}\mathbf{j}$

57. $\{-2 + \sqrt{21}, -2 - \sqrt{21}\}$ **59.** $\mathbf{v} = \frac{5}{2}\mathbf{i} + \frac{5\sqrt{3}}{2}\mathbf{j}$ **61.** $\mathbf{v} = -7\mathbf{i} + 7\sqrt{3}\mathbf{j}$ **63.** $\mathbf{v} = \frac{25\sqrt{3}}{2}\mathbf{i} - \frac{25}{2}\mathbf{j}$ **65.** 45° **67.** 150° **69.** 333.4° **71.** 258.7°

73. $\mathbf{F} = 20\sqrt{3}\mathbf{i} + 20\mathbf{j}$ **75.** $\mathbf{F} = (20\sqrt{3} + 30\sqrt{2})\mathbf{i} + (20 - 30\sqrt{2})\mathbf{j}$

77. (a) $\mathbf{v}_a = 550\mathbf{j}$; $\mathbf{v}_w = 50\sqrt{2}\mathbf{i} + 50\sqrt{2}\mathbf{j}$

(b) $\mathbf{v}_g = 50\sqrt{2}\mathbf{i} + (550 + 50\sqrt{2})\mathbf{j}$

(c) $\|\mathbf{v}_g\| = 624.7$ mph; N6.5°E

79. $\mathbf{v} = (250\sqrt{2} - 30)\mathbf{i} + (250\sqrt{2} + 30\sqrt{3})\mathbf{j}$; 518.8 km/h; N38.6°E

81. Approximately 4031 lb **83.** 8.6° left of direct heading across the river; 1.52 min

85. (a) N7.05°E **(b)** 12 min **87.** Tension in right cable: 1000 lb; tension in left cable: 845.2 lb

89. Tension in right part: 1088.4 lb; tension in left part: 1089.1 lb **91.** $\mu = 0.36$

93. 13.68 lb **95.** The truck must pull with a force of 4635.2 lb.

97. (a) $(-1, 4)$

(b)

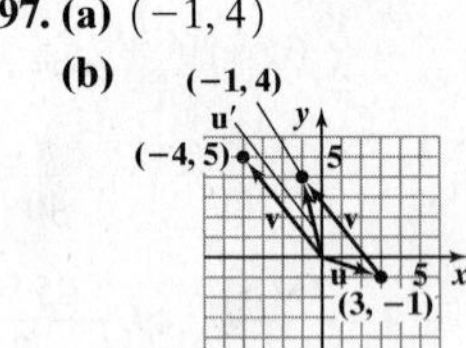

99.

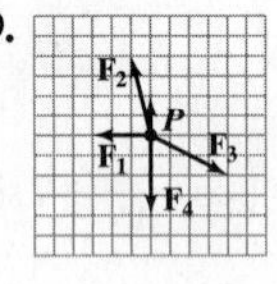

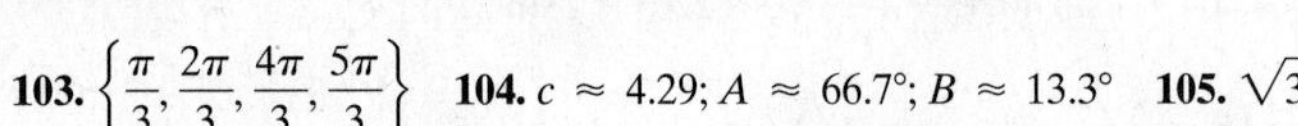

103. $\left\{\frac{\pi}{3}, \frac{2\pi}{3}, \frac{4\pi}{3}, \frac{5\pi}{3}\right\}$ **104.** $c \approx 4.29$; $A \approx 66.7°$; $B \approx 13.3°$ **105.** $\sqrt{3}$

106. Amplitude $= \frac{3}{2}$; period $= \frac{\pi}{3}$

Phase shift $= -\frac{\pi}{2}$

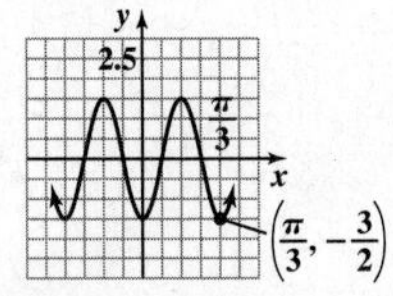

Historical Problem *(page 363)*

$(a\mathbf{i} + b\mathbf{j}) \cdot (c\mathbf{i} + d\mathbf{j}) = ac + bd$

Real part $[(\overline{a + bi})(c + di)] = $ real part$[(a - bi)(c + di)] = $ real part$[ac + adi - bci - bdi^2] = ac + bd$

5.5 Assess Your Understanding *(page 363)*

2. dot product **3.** orthogonal **4.** parallel **5.** T **6.** F **7.** d **8.** b **9. (a)** 0 **(b)** 90° **(c)** orthogonal **11. (a)** 0 **(b)** 90° **(c)** orthogonal

13. (a) $\sqrt{3} - 1$ **(b)** 75° **(c)** neither **15. (a)** -50 **(b)** 180° **(c)** parallel **17. (a)** 0 **(b)** 90° **(c)** orthogonal

19. $\frac{2}{3}$ **21.** $\mathbf{v}_1 = \frac{5}{2}\mathbf{i} - \frac{5}{2}\mathbf{j}, \mathbf{v}_2 = -\frac{1}{2}\mathbf{i} - \frac{1}{2}\mathbf{j}$ **23.** $\mathbf{v}_1 = -\frac{1}{5}\mathbf{i} - \frac{2}{5}\mathbf{j}, \mathbf{v}_2 = \frac{6}{5}\mathbf{i} - \frac{3}{5}\mathbf{j}$ **25.** $\mathbf{v}_1 = \frac{14}{5}\mathbf{i} + \frac{7}{5}\mathbf{j}, \mathbf{v}_2 = \frac{1}{5}\mathbf{i} - \frac{2}{5}\mathbf{j}$ **27.** Approximately 1.353
29. 9 ft-lb **31. (a)** $\|\mathbf{I}\| \approx 0.022$; the intensity of the sun's rays is approximately 0.022 W/cm^2. $\|\mathbf{A}\| = 500$; the area of the solar panel is 500 cm^2.
(b) $W = 10$; ten watts of energy is collected. **(c)** Vectors **I** and **A** should be parallel with the solar panels facing the sun.
33. Force required to keep the Sienna from rolling down the hill: 737.6 lb; force perpendicular to the hill: 5248.4 lb
35. Timmy must exert 85.5 lb. **37.** 60° **39.** Let $\mathbf{v} = a\mathbf{i} + b\mathbf{j}$. Then $\mathbf{0} \cdot \mathbf{v} = 0a + 0b = 0$.
41. $\mathbf{v} = \cos\alpha\mathbf{i} + \sin\alpha\mathbf{j}, 0 \le \alpha \le \pi$; $\mathbf{w} = \cos\beta\mathbf{i} + \sin\beta\mathbf{j}, 0 \le \beta \le \pi$. If θ is the angle between **v** and **w**, then $\mathbf{v} \cdot \mathbf{w} = \cos\theta$, since $\|\mathbf{v}\| = 1$ and $\|\mathbf{w}\| = 1$. Now $\theta = \alpha - \beta$ or $\theta = \beta - \alpha$. Since the cosine function is even, $\mathbf{v} \cdot \mathbf{w} = \cos(\alpha - \beta)$. Also, $\mathbf{v} \cdot \mathbf{w} = \cos\alpha\cos\beta + \sin\alpha\sin\beta$. So $\cos(\alpha - \beta) = \cos\alpha\cos\beta + \sin\alpha\sin\beta$.
43. (a) If $\mathbf{u} = a_1\mathbf{i} + b_1\mathbf{j}$ and $\mathbf{v} = a_2\mathbf{i} + b_2\mathbf{j}$, then, since $\|\mathbf{u}\| = \|\mathbf{v}\|, a_1^2 + b_1^2 = \|\mathbf{u}\|^2 = \|\mathbf{v}\|^2 = a_2^2 + b_2^2$,
$(\mathbf{u} + \mathbf{v}) \cdot (\mathbf{u} - \mathbf{v}) = (a_1 + a_2)(a_1 - a_2) + (b_1 + b_2)(b_1 - b_2) = (a_1^2 + b_1^2) - (a_2^2 + b_2^2) = 0.$
(b) The legs of the angle can be made to correspond to vectors $\mathbf{u} + \mathbf{v}$ and $\mathbf{u} - \mathbf{v}$.
45. $(\|\mathbf{w}\|\mathbf{v} + \|\mathbf{v}\|\mathbf{w}) \cdot (\|\mathbf{w}\|\mathbf{v} - \|\mathbf{v}\|\mathbf{w}) = \|\mathbf{w}\|^2\mathbf{v}\cdot\mathbf{v} - \|\mathbf{w}\|\|\mathbf{v}\|\mathbf{v}\cdot\mathbf{w} + \|\mathbf{v}\|\|\mathbf{w}\|\mathbf{w}\cdot\mathbf{v} - \|\mathbf{v}\|^2\mathbf{w}\cdot\mathbf{w} = \|\mathbf{w}\|^2\mathbf{v}\cdot\mathbf{v} - \|\mathbf{v}\|^2\mathbf{w}\cdot\mathbf{w} = \|\mathbf{w}\|^2\|\mathbf{v}\|^2 - \|\mathbf{v}\|^2\|\mathbf{w}\|^2 = 0$
47. $\|\mathbf{u} + \mathbf{v}\|^2 - \|\mathbf{u} - \mathbf{v}\|^2 = (\mathbf{u} + \mathbf{v}) \cdot (\mathbf{u} + \mathbf{v}) - (\mathbf{u} - \mathbf{v}) \cdot (\mathbf{u} - \mathbf{v}) = (\mathbf{u}\cdot\mathbf{u} + \mathbf{u}\cdot\mathbf{v} + \mathbf{v}\cdot\mathbf{u} + \mathbf{v}\cdot\mathbf{v}) - (\mathbf{u}\cdot\mathbf{u} - \mathbf{u}\cdot\mathbf{v} - \mathbf{v}\cdot\mathbf{u} + \mathbf{v}\cdot\mathbf{v})$
$= 2(\mathbf{u}\cdot\mathbf{v}) + 2(\mathbf{v}\cdot\mathbf{u}) = 4(\mathbf{u}\cdot\mathbf{v})$

49. 12 **50.** $\frac{9}{2}$ **51.** $(1 - \sin^2\theta)(1 + \tan^2\theta) = (\cos^2\theta)(\sec^2\theta) = \cos^2\theta \cdot \frac{1}{\cos^2\theta} = 1$ **52.** $A \approx 27.2°, B \approx 39.0°, C \approx 113.8°$

5.6 Assess Your Understanding *(page 372)*

2. *xy*-plane **3.** components **4.** 1 **5.** F **6.** T **7.** All points of the form $(x, 0, z)$ **9.** All points of the form $(x, y, 2)$ **11.** All points of the form $(-4, y, z)$
13. All points of the form $(1, 2, z)$ **15.** $\sqrt{21}$ **17.** $\sqrt{33}$ **19.** $\sqrt{26}$ **21.** (2, 0, 0); (2, 1, 0); (0, 1, 0); (2, 0, 3); (0, 1, 3); (0, 0, 3)
23. (1, 4, 3); (3, 2, 3); (3, 4, 3); (3, 2, 5); (1, 4, 5); (1, 2, 5) **25.** (−1, 2, 2); (4, 0, 2); (4, 2, 2); (−1, 2, 5); (4, 0, 5); (−1, 0, 5) **27.** $\mathbf{v} = 3\mathbf{i} + 4\mathbf{j} - \mathbf{k}$
29. $\mathbf{v} = 2\mathbf{i} + 4\mathbf{j} + \mathbf{k}$ **31.** $\mathbf{v} = 8\mathbf{i} - \mathbf{j}$ **33.** 7 **35.** $\sqrt{3}$ **37.** $\sqrt{22}$ **39.** $-\mathbf{j} - 2\mathbf{k}$ **41.** $\sqrt{105}$ **43.** $\sqrt{38} - \sqrt{17}$ **45.** **i** **47.** $\frac{3}{7}\mathbf{i} - \frac{6}{7}\mathbf{j} - \frac{2}{7}\mathbf{k}$
49. $\frac{\sqrt{3}}{3}\mathbf{i} + \frac{\sqrt{3}}{3}\mathbf{j} + \frac{\sqrt{3}}{3}\mathbf{k}$ **51.** $\mathbf{v}\cdot\mathbf{w} = 0; \theta = 90°$ **53.** $\mathbf{v}\cdot\mathbf{w} = -2, \theta \approx 100.3°$ **55.** $\mathbf{v}\cdot\mathbf{w} = 0; \theta = 90°$ **57.** $\mathbf{v}\cdot\mathbf{w} = 52; \theta = 0°$
59. $\alpha \approx 64.6°; \beta \approx 149.0°; \gamma \approx 106.6°; \mathbf{v} = 7(\cos 64.6°\mathbf{i} + \cos 149.0°\mathbf{j} + \cos 106.6°\mathbf{k})$
61. $\alpha = \beta = \gamma \approx 54.7°; \mathbf{v} = \sqrt{3}(\cos 54.7°\mathbf{i} + \cos 54.7°\mathbf{j} + \cos 54.7°\mathbf{k})$ **63.** $\alpha = \beta = 45°; \gamma = 90°; \mathbf{v} = \sqrt{2}(\cos 45°\mathbf{i} + \cos 45°\mathbf{j} + \cos 90°\mathbf{k})$
65. $\alpha \approx 60.9°; \beta \approx 144.2°; \gamma \approx 71.1°; \mathbf{v} = \sqrt{38}(\cos 60.9°\mathbf{i} + \cos 144.2°\mathbf{j} + \cos 71.1°\mathbf{k})$ **67. (a)** $\mathbf{d} = \mathbf{a} + \mathbf{b} + \mathbf{c} = \langle 7, 1, 5 \rangle$ **(b)** 8.66 ft
69. $(x - 3)^2 + (y - 1)^2 + (z - 1)^2 = 1$ **71.** Radius = 2, center (−1, 1, 0) **73.** Radius = 3, center (2, −2, −1) **75.** Radius = $\frac{3\sqrt{2}}{2}$, Center (2, 0, −1)
77. 2 newton-meters = 2 joules **79.** 9 newton-meters = 9 joules **80.** $f^{-1} = \cos^{-1}\left(\frac{x - 5}{3}\right)$
Range of f = Domain of $f^{-1} = [2, 8]$
Range of $f^{-1} = [0, \pi]$
81. $3\sqrt{5}$ **82.** $\frac{1}{2}$
83. $c = 3\sqrt{5} \approx 6.71$; $A \approx 26.6°; B \approx 63.4°$

5.7 Assess Your Understanding *(page 379)*

1. T **2.** T **3.** T **4.** F **5.** F **6.** T **7.** 2 **9.** 4 **11.** $-11A + 2B + 5C$ **13.** $-6A + 23B - 15C$ **15. (a)** $5\mathbf{i} + 5\mathbf{j} + 5\mathbf{k}$ **(b)** $-5\mathbf{i} - 5\mathbf{j} - 5\mathbf{k}$
(c) **0** **(d)** **0** **17. (a)** $\mathbf{i} - \mathbf{j} - \mathbf{k}$ **(b)** $-\mathbf{i} + \mathbf{j} + \mathbf{k}$ **(c)** **0** **(d)** **0** **19. (a)** $-\mathbf{i} + 2\mathbf{j} + 2\mathbf{k}$ **(b)** $\mathbf{i} - 2\mathbf{j} - 2\mathbf{k}$ **(c)** **0** **(d)** **0** **21. (a)** $3\mathbf{i} - \mathbf{j} + 4\mathbf{k}$
(b) $-3\mathbf{i} + \mathbf{j} - 4\mathbf{k}$ **(c)** **0** **(d)** **0** **23.** $-9\mathbf{i} - 7\mathbf{j} - 3\mathbf{k}$ **25.** $9\mathbf{i} + 7\mathbf{j} + 3\mathbf{k}$ **27.** **0** **29.** $-27\mathbf{i} - 21\mathbf{j} - 9\mathbf{k}$ **31.** $-18\mathbf{i} - 14\mathbf{j} - 6\mathbf{k}$ **33.** **0** **35.** −25
37. 25 **39.** **0** **41.** Any vector of the form $c(-9\mathbf{i} - 7\mathbf{j} - 3\mathbf{k})$, where c is a nonzero scalar **43.** Any vector of the form $c(-\mathbf{i} + \mathbf{j} + 5\mathbf{k})$, where c is a nonzero scalar **45.** $\sqrt{166}$ **47.** $\sqrt{555}$ **49.** $\sqrt{34}$ **51.** $\sqrt{998}$ **53.** $\frac{11\sqrt{19}}{57}\mathbf{i} + \frac{\sqrt{19}}{57}\mathbf{j} + \frac{7\sqrt{19}}{57}\mathbf{k}$ or $-\frac{11\sqrt{19}}{57}\mathbf{i} - \frac{\sqrt{19}}{57}\mathbf{j} - \frac{7\sqrt{19}}{57}\mathbf{k}$

55. 98 cubic units **57.** $\mathbf{u} \times \mathbf{v} = \begin{vmatrix} \mathbf{i} & \mathbf{j} & \mathbf{k} \\ a_1 & b_1 & c_1 \\ a_2 & b_2 & c_2 \end{vmatrix} = (b_1c_2 - b_2c_1)\mathbf{i} - (a_1c_2 - a_2c_1)\mathbf{j} + (a_1b_2 - a_2b_1)\mathbf{k}$

$\|\mathbf{u} \times \mathbf{v}\|^2 = \sqrt{(b_1c_2 - b_2c_1)^2 + (a_1c_2 - a_2c_1)^2 + (a_1b_2 - a_2b_1)^2)}^2$
$= b_1^2c_2^2 - 2b_1b_2c_1c_2 + b_2^2c_1^2 + a_1^2c_2^2 - 2a_1a_2c_1c_2 + a_2^2c_1^2 + a_1^2b_2^2 - 2a_1a_2b_1b_2 + a_2^2b_1^2$
$\|\mathbf{u}\|^2 = a_1^2 + b_1^2 + c_1^2, \|\mathbf{v}\|^2 = a_2^2 + b_2^2 + c_2^2$
$\|\mathbf{u}\|^2\|\mathbf{v}\|^2 = (a_1^2 + b_1^2 + c_1^2)(a_2^2 + b_2^2 + c_2^2) = a_1^2a_2^2 + a_1^2b_2^2 + a_1^2c_2^2 + b_1^2a_2^2 + b_1^2b_2^2 + b_1^2c_2^2 + a_2^2c_1^2 + b_2^2c_1^2 + c_1^2c_2^2$
$(\mathbf{u}\cdot\mathbf{v})^2 = (a_1a_2 + b_1b_2 + c_1c_2)^2 = (a_1a_2 + b_1b_2 + c_1c_2)(a_1a_2 + b_1b_2 + c_1c_2)$
$= a_1^2a_2^2 + a_1a_2b_1b_2 + a_1a_2c_1c_2 + b_1b_2c_1c_2 + b_1b_2a_1a_2 + b_1^2b_2^2 + b_1b_2c_1c_2 + a_1a_2c_1c_2 + c_1^2c_2^2$
$= a_1^2a_2^2 + b_1^2b_2^2 + c_1^2c_2^2 + 2a_1a_2b_1b_2 + 2b_1b_2c_1c_2 + 2a_1a_2c_1c_2$
$\|\mathbf{u}\|^2\|\mathbf{v}\|^2 - (\mathbf{u}\cdot\mathbf{v})^2 = a_1^2b_2^2 + a_1^2c_2^2 + b_1^2a_2^2 + a_2^2c_1^2 + b_2^2c_1^2 + b_1^2c_2^2 - 2a_1a_2b_1b_2 - 2b_1b_2c_1c_2 - 2a_1a_2c_1c_2$, which equals $\|\mathbf{u} \times \mathbf{v}\|^2$.

59. By Problem 58, since $\mathbf{u}$ and $\mathbf{v}$ are orthogonal, $\|\mathbf{u} \times \mathbf{v}\| = \|\mathbf{u}\|\|\mathbf{v}\|$. If, in addition, $\mathbf{u}$ and $\mathbf{v}$ are unit vectors, $\|\mathbf{u} \times \mathbf{v}\| = 1 \cdot 1 = 1$.

61. Assume that $\mathbf{u} = a\mathbf{i} + b\mathbf{j} + c\mathbf{k}$, $\mathbf{v} = d\mathbf{i} + e\mathbf{j} + f\mathbf{k}$, and $\mathbf{w} = l\mathbf{i} + m\mathbf{j} + n\mathbf{k}$. Then $\mathbf{u} \times \mathbf{v} = (bf - ec)\mathbf{i} - (af - dc)\mathbf{j} + (ae - db)\mathbf{k}$, $\mathbf{u} \times \mathbf{w} = (bn - mc)\mathbf{i} - (an - lc)\mathbf{j} + (am - lb)\mathbf{k}$, and $\mathbf{v} + \mathbf{w} = (d + l)\mathbf{i} + (e + m)\mathbf{j} + (f + n)\mathbf{k}$.
Therefore, $(\mathbf{u} \times \mathbf{v}) + (\mathbf{u} \times \mathbf{w}) = (bf - ec + bn - mc)\mathbf{i} - (af - dc + an - lc)\mathbf{j} + (ae - db + am - lb)\mathbf{k}$ and $\mathbf{u} \times (\mathbf{v} + \mathbf{w})$
$= [b(f + n) - (e + m)c]\mathbf{i} - [a(f + n) - (d + l)c]\mathbf{j} + [a(e + m) - (d + l)b]\mathbf{k}$
$= (bf - ec + bn - mc)\mathbf{i} - (af - dc + an - lc)\mathbf{j} + (ae - db + am - lb)\mathbf{k}$, which equals $(\mathbf{u} \times \mathbf{v}) + (\mathbf{u} \times \mathbf{w})$.

64. $\frac{\pi}{4}$ **65.** $(17, 4.22), (-17, 1.08)$ **66.** $\left\{\frac{\pi}{12}, \frac{5\pi}{12}, \frac{3\pi}{4}, \frac{13\pi}{12}, \frac{17\pi}{12}, \frac{7\pi}{4}\right\}$ **67.** $\cos\theta = -\frac{21}{29}$; $\csc\theta = \frac{29}{20}$; $\sec\theta = -\frac{29}{21}$; $\cot\theta = -\frac{21}{20}$

Review Exercises *(page 382)*

1. 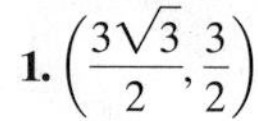 $\left(\frac{3\sqrt{3}}{2}, \frac{3}{2}\right)$

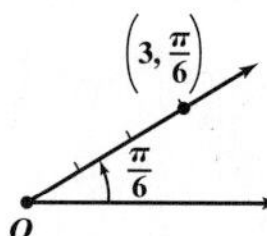

2. $(1, \sqrt{3})$

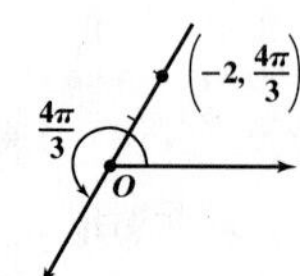

3. $(0, 3)$

4. $\left(3\sqrt{2}, \frac{3\pi}{4}\right), \left(-3\sqrt{2}, -\frac{\pi}{4}\right)$

5. $\left(2, -\frac{\pi}{2}\right), \left(-2, \frac{\pi}{2}\right)$

6. $(5, 0.93), (-5, 4.07)$

7. (a) $x^2 + (y - 1)^2 = 1$ **(b)** circle, radius 1, center $(0, 1)$ in rectangular coordinates

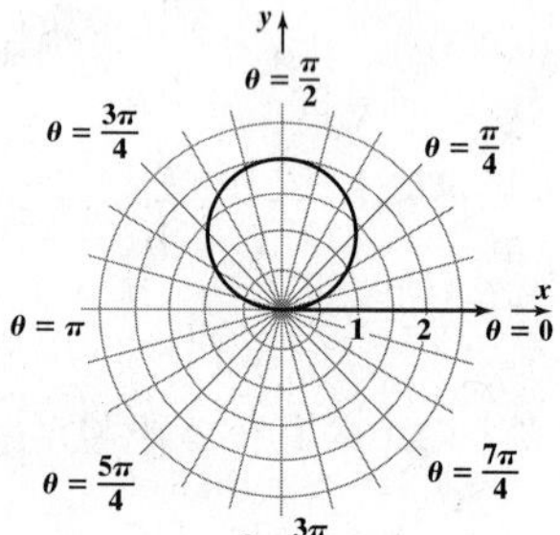

8. (a) $x^2 + y^2 = 25$ **(b)** circle, radius 5, center at pole

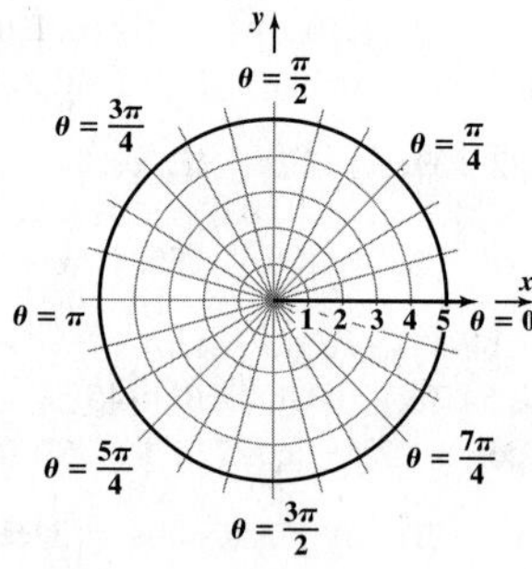

9. (a) $x - y = 0$ **(b)** line through pole, making an angle of $\frac{\pi}{4}$ with polar axis

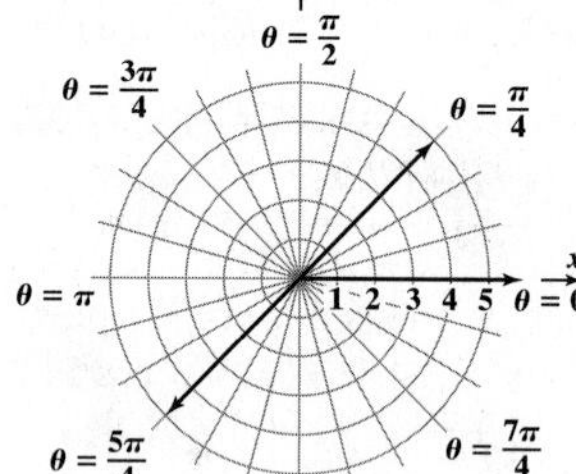

10. (a) $(x - 4)^2 + (y + 2)^2 = 25$ **(b)** circle, radius 5, center $(4, -2)$ in rectangular coordinates

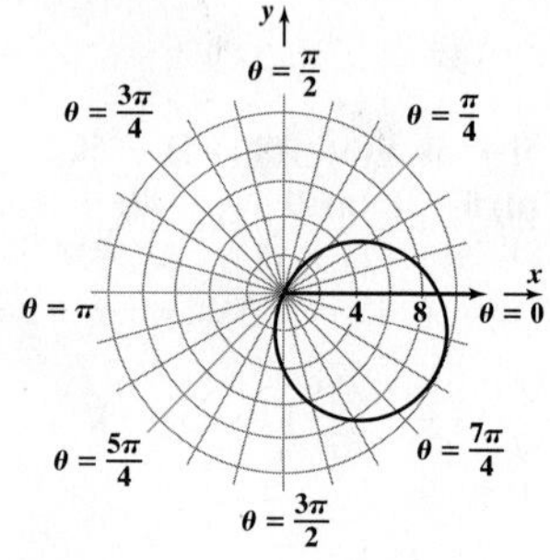

11. Circle; radius 2, center at $(2, 0)$ in rectangular coordinates; symmetric with respect to the polar axis

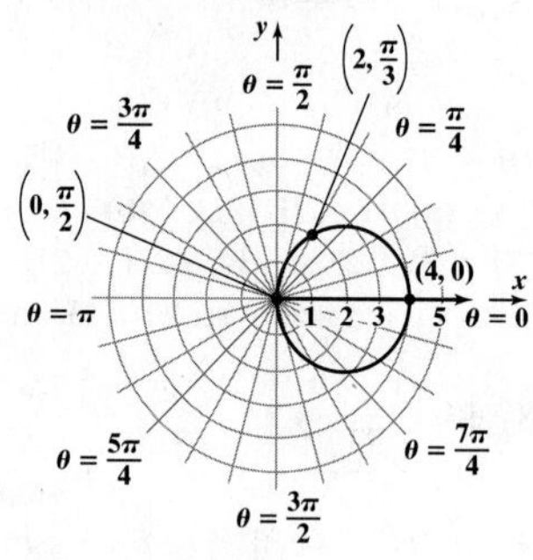

12. Cardioid; symmetric with respect to the line $\theta = \frac{\pi}{2}$

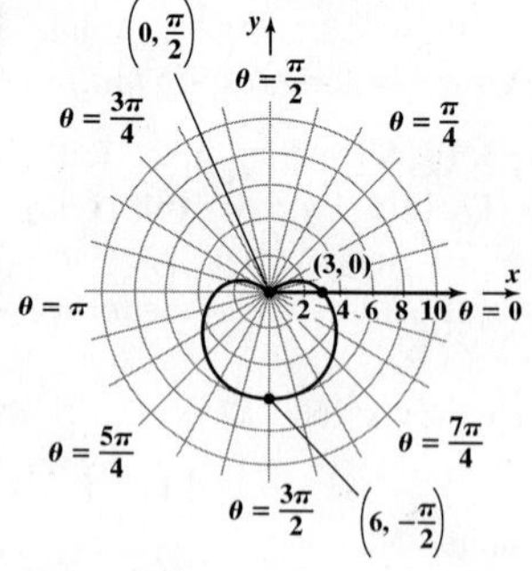

13. Limaçon without inner loop; symmetric with respect to the polar axis

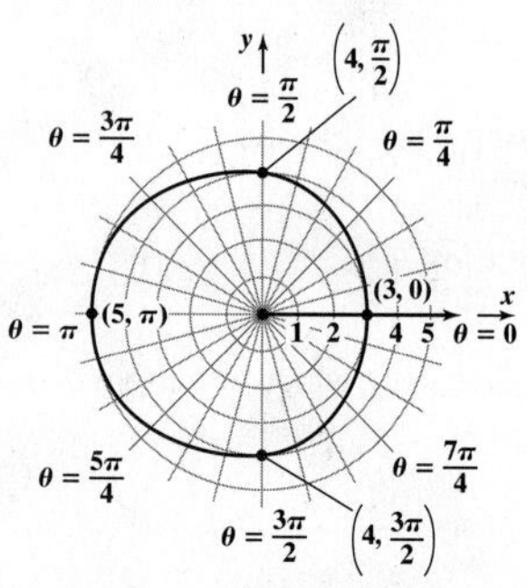

14. $\sqrt{2}\,(\cos 225° + i \sin 225°)$ **15.** $5(\cos 323.1° + i \sin 323.1°)$

16. $-\sqrt{3} + i$

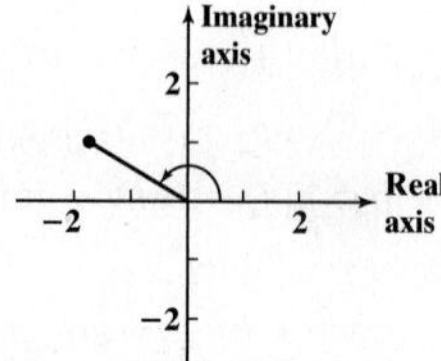

17. $-\frac{3}{2} + \frac{3\sqrt{3}}{2}i$

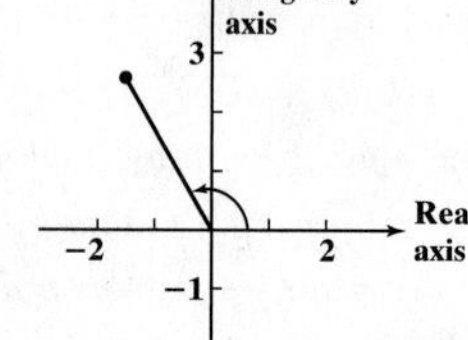

18. $0.10 - 0.02i$

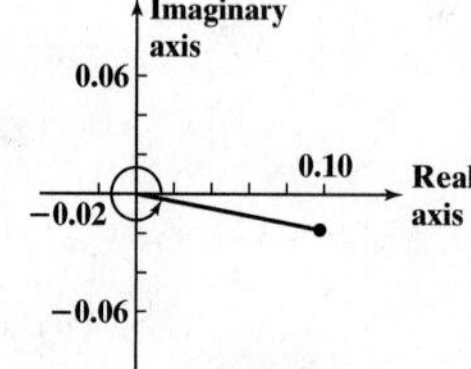

19. $zw = \cos 130° + i \sin 130°; \dfrac{z}{w} = \cos 30° + i \sin 30°$ **20.** $zw = 6(\cos 0 + i \sin 0) = 6; \dfrac{z}{w} = \dfrac{3}{2}\left(\cos \dfrac{8\pi}{5} + i \sin \dfrac{8\pi}{5}\right)$

21. $zw = 5(\cos 5° + i \sin 5°); \dfrac{z}{w} = 5\,(\cos 15° + i \sin 15°)$ **22.** $\dfrac{27}{2} + \dfrac{27\sqrt{3}}{2}i$ **23.** $4i$ **24.** 64 **25.** $-527 - 336i$

26. $3, 3(\cos 120° + i \sin 120°), 3(\cos 240° + i \sin 240°)$ or $3, -\dfrac{3}{2} + \dfrac{3\sqrt{3}}{2}i, -\dfrac{3}{2} - \dfrac{3\sqrt{3}}{2}i$

27.

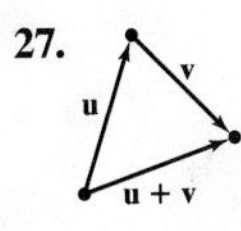

28.

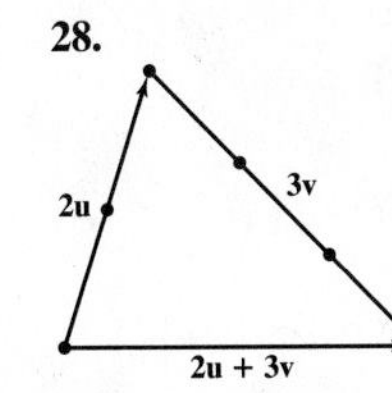

29. $\mathbf{v} = 2\mathbf{i} - 4\mathbf{j}; \|\mathbf{v}\| = 2\sqrt{5}$ **30.** $\mathbf{v} = -\mathbf{i} + 3\mathbf{j}; \|\mathbf{v}\| = \sqrt{10}$ **31.** $2\mathbf{i} - 2\mathbf{j}$ **32.** $-20\mathbf{i} + 13\mathbf{j}$ **33.** $\sqrt{5}$ **34.** $\sqrt{5} + 5 \approx 7.24$ **35.** $-\dfrac{2\sqrt{5}}{5}\mathbf{i} + \dfrac{\sqrt{5}}{5}\mathbf{j}$ **36.** $\mathbf{v} = \dfrac{3}{2}\mathbf{i} + \dfrac{3\sqrt{3}}{2}\mathbf{j}$ **37.** 120° **38.** $\sqrt{43} \approx 6.56$ **39.** $\mathbf{v} = 3\mathbf{i} - 5\mathbf{j} + 3\mathbf{k}$ **40.** $21\mathbf{i} - 2\mathbf{j} - 5\mathbf{k}$ **41.** $\sqrt{38}$ **42.** 0 **43.** $3\mathbf{i} + 9\mathbf{j} + 9\mathbf{k}$ **44.** 0 **45.** $\dfrac{\sqrt{19}}{19}\mathbf{i} + \dfrac{3\sqrt{19}}{19}\mathbf{j} + \dfrac{3\sqrt{19}}{19}\mathbf{k}$ or $-\dfrac{\sqrt{19}}{19}\mathbf{i} - \dfrac{3\sqrt{19}}{19}\mathbf{j} - \dfrac{3\sqrt{19}}{19}\mathbf{k}$ **46.** $\mathbf{v}\cdot\mathbf{w} = -11; \theta \approx 169.7°$ **47.** $\mathbf{v}\cdot\mathbf{w} = -4; \theta \approx 153.4°$ **48.** $\mathbf{v}\cdot\mathbf{w} = 1; \theta \approx 70.5°$ **49.** $\mathbf{v}\cdot\mathbf{w} = 0; \theta = 90°$ **50.** Parallel **51.** Neither **52.** Orthogonal **53.** $\mathbf{v}_1 = \dfrac{4}{5}\mathbf{i} - \dfrac{3}{5}\mathbf{j}; \mathbf{v}_2 = \dfrac{6}{5}\mathbf{i} + \dfrac{8}{5}\mathbf{j}$ **54.** $\mathbf{v}_1 = \dfrac{9}{10}(3\mathbf{i} + \mathbf{j}); \mathbf{v}_2 = -\dfrac{7}{10}\mathbf{i} + \dfrac{21}{10}\mathbf{j}$ **55.** $a \approx 56.1°; \beta \approx 138°; \gamma \approx 68.2°$ **56.** $2\sqrt{83}$ **57.** $-2\mathbf{i} + 3\mathbf{j} - \mathbf{k}$ **58.** 0 **59.** $\sqrt{29} \approx 5.39$ mi/h 0.4 mi **60.** Left cable: 1843.21 lb; right cable: 1630.41 lb **61.** 50 ft-lb **62.** A force of 697.2 lb is needed to keep the van from rolling down the hill. The magnitude of the force perpendicular to the hill is 7969.6 lb.

Chapter Test *(page 384)*

1–3.

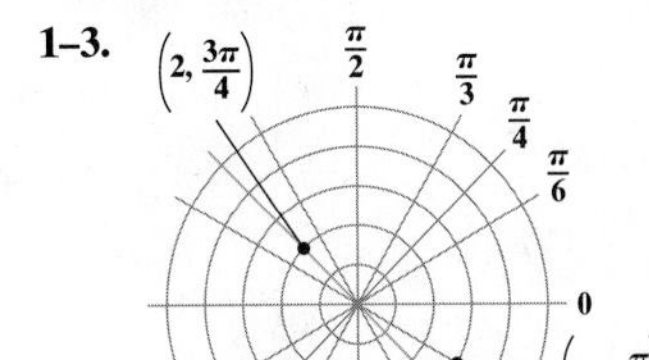

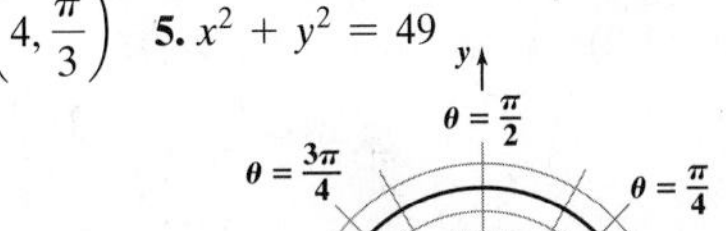

4. $\left(4, \dfrac{\pi}{3}\right)$ **5.** $x^2 + y^2 = 49$

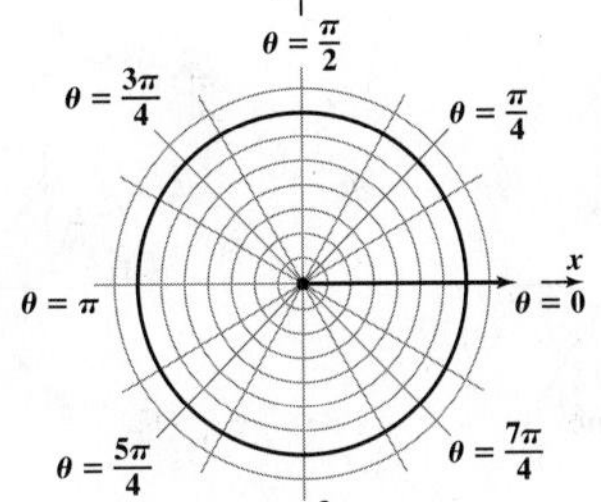

6. $\dfrac{y}{x} = 3$ or $y = 3x$

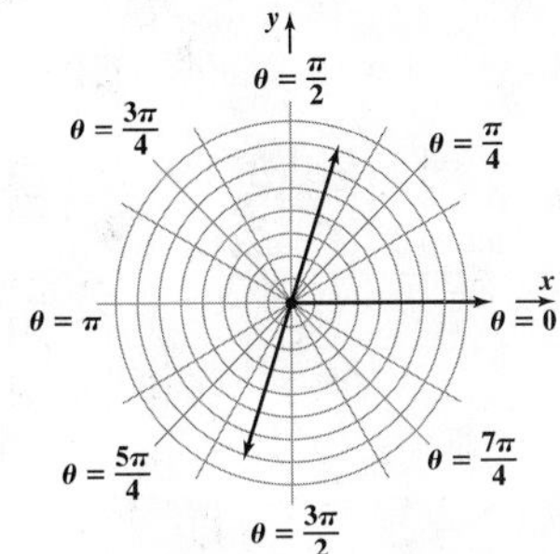

7. $8y = x^2$

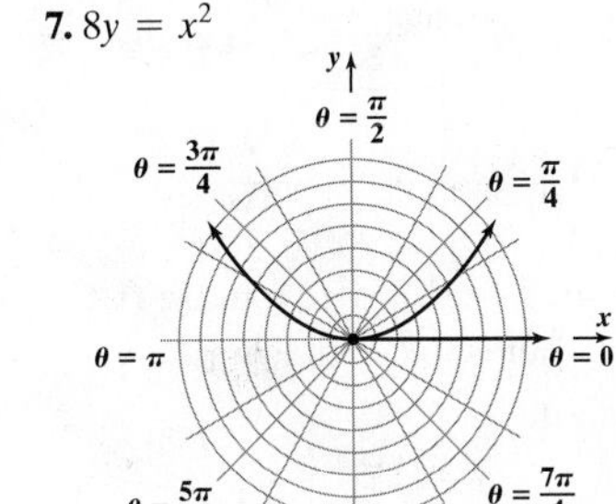

8. $r^2 \cos \theta = 5$ is symmetric about the pole, the polar axis, and the line $\theta = \dfrac{\pi}{2}$.
9. $r = 5 \sin \theta \cos^2 \theta$ is symmetric about the line $\theta = \dfrac{\pi}{2}$. The tests for symmetry about the pole and the polar axis fail, so the graph of $r = 5 \sin \theta \cos^2 \theta$ may or may not be symmetric about the pole or the polar axis.

10. $z \cdot w = 6(\cos 107° + i \sin 107°)$ **11.** $\dfrac{w}{z} = \dfrac{3}{2}(\cos 297° + i \sin 297°)$
12. $w^5 = 243(\cos 110° + i \sin 110°)$
13. $z_0 = 2\sqrt[3]{2}(\cos 40° + i \sin 40°), z_1 = 2\sqrt[3]{2}(\cos 160° + i \sin 160°), z_2 = 2\sqrt[3]{2}(\cos 280° + i \sin 280°)$

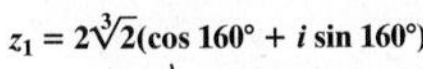

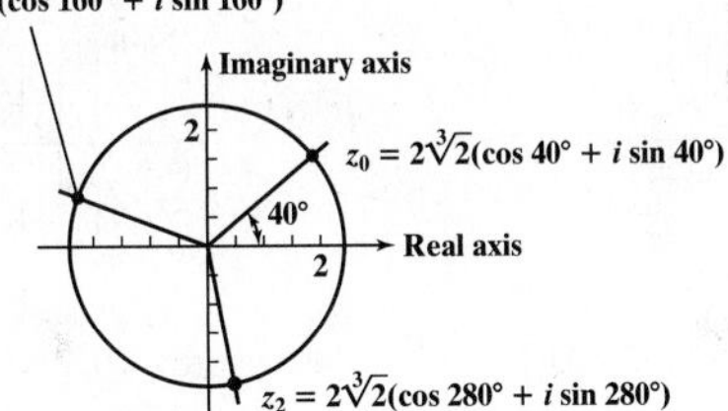

14. $\mathbf{v} = \langle 5\sqrt{2}, -5\sqrt{2} \rangle$ **15.** $\|\mathbf{v}\| = 10$ **16.** $\mathbf{u} = \dfrac{\mathbf{v}}{\|\mathbf{v}\|} = \left\langle \dfrac{\sqrt{2}}{2}, -\dfrac{\sqrt{2}}{2} \right\rangle$ **17.** 315° off the positive x-axis **18.** $\mathbf{v} = 5\sqrt{2}\mathbf{i} - 5\sqrt{2}\mathbf{j}$ **19.** $\mathbf{v}_1 + 2\mathbf{v}_2 - \mathbf{v}_3 = \langle 6, -10 \rangle$ **20.** Vectors $\mathbf{v}_1$ and $\mathbf{v}_4$ are parallel. **21.** Vectors $\mathbf{v}_2$ and $\mathbf{v}_3$ are orthogonal. **22.** 172.87° **23.** $-9\mathbf{i} - 5\mathbf{j} + 3\mathbf{k}$ **24.** $\alpha \approx 57.7°, \beta \approx 143.3°, \gamma \approx 74.5°$ **25.** $\sqrt{115}$ **26.** The cable must be able to endure a tension of approximately 670.82 lb.

Cumulative Review *(page 385)*

1. $\{-3, 3\}$ **2.** $y = \dfrac{\sqrt{3}}{3}x$ **3.** $x^2 + (y - 1)^2 = 9$

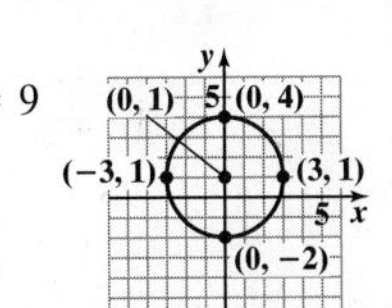

4. $\left\{x \middle| x \neq \dfrac{1}{2}\right\}$ **5.** Symmetry with respect to the y-axis

6.

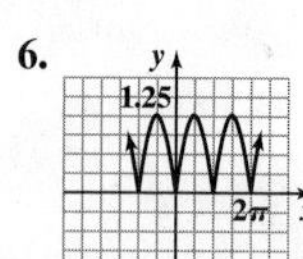

7.

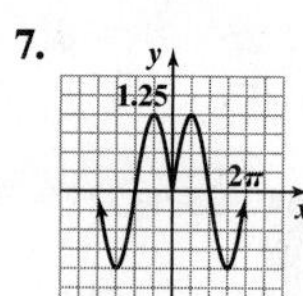

8. $-\frac{\pi}{6}$

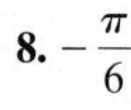

9.

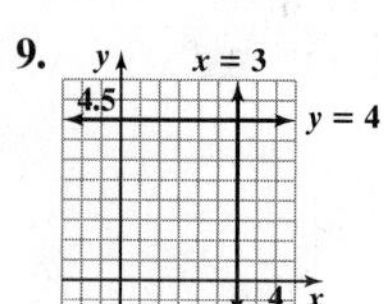

10.

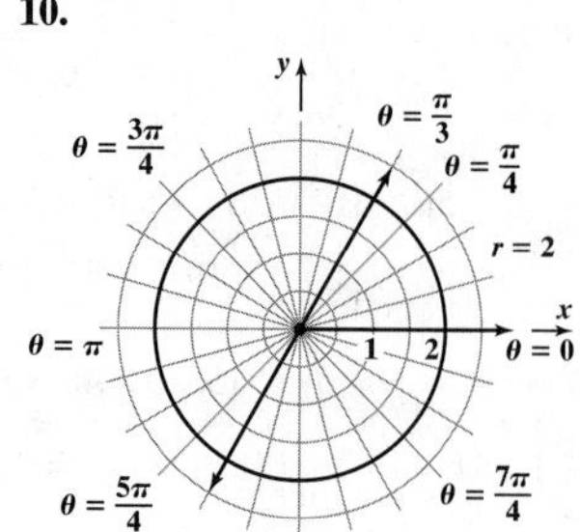

11. Amplitude: 4; period: 2

CHAPTER 6 Analytic Geometry

6.2 Assess Your Understanding *(page 394)*

6. parabola **7.** axis of symmetry **8.** latus rectum **9.** c **10.** (3, 2) **11.** d **12.** c **13.** B **15.** E **17.** H **19.** C

21. $y^2 = 16x$

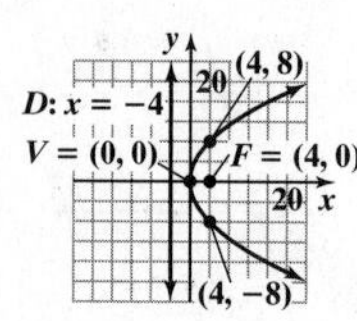

23. $x^2 = -12y$

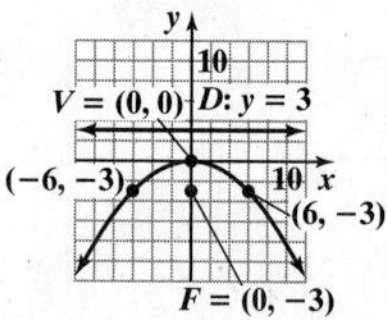

25. $y^2 = -8x$

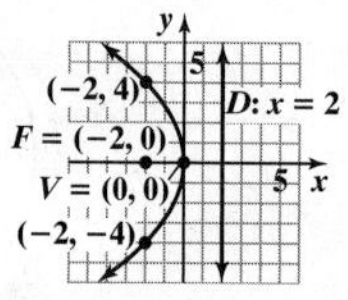

27. $x^2 = 2y$

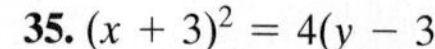

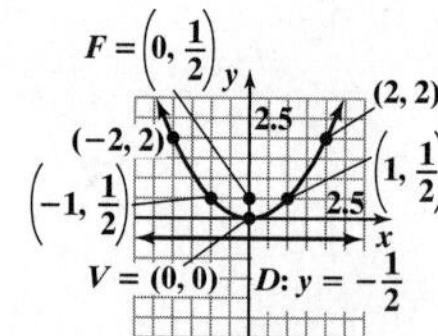

29. $x^2 = \frac{4}{3}y$

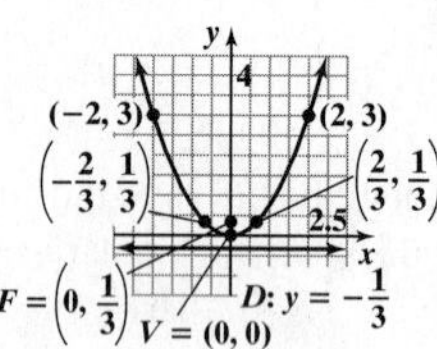

31. $(x - 2)^2 = -8(y + 3)$

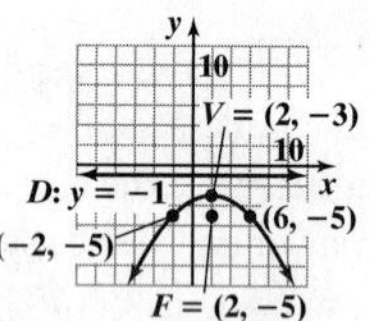

33. $(y + 2)^2 = 4(x + 1)$

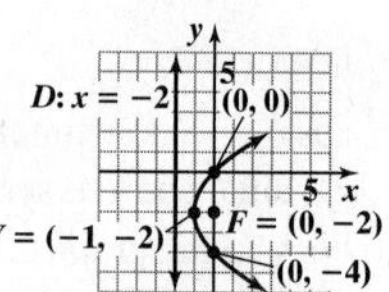

35. $(x + 3)^2 = 4(y - 3)$

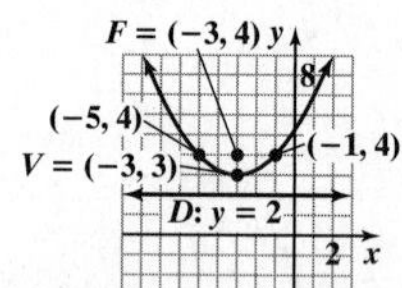

37. $(y + 2)^2 = -8(x + 1)$

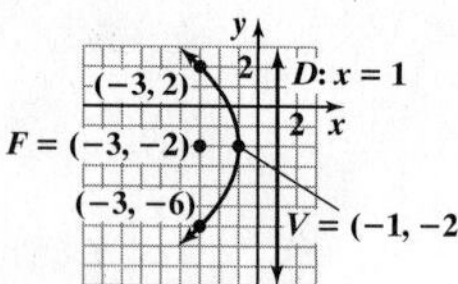

39. Vertex: (0, 0); focus: (0, 1); directrix: $y = -1$

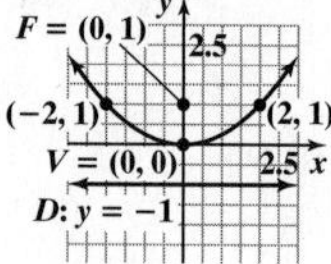

41. Vertex: (0, 0); focus: $(-4, 0)$; directrix: $x = 4$

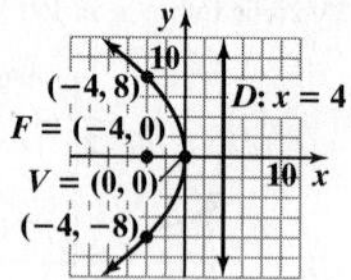

43. Vertex: $(-1, 2)$; focus: (1, 2); directrix: $x = -3$

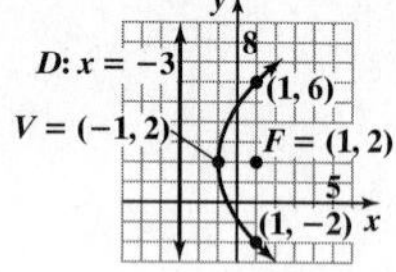

45. Vertex: $(3, -1)$; focus: $\left(3, -\frac{5}{4}\right)$; directrix: $y = -\frac{3}{4}$

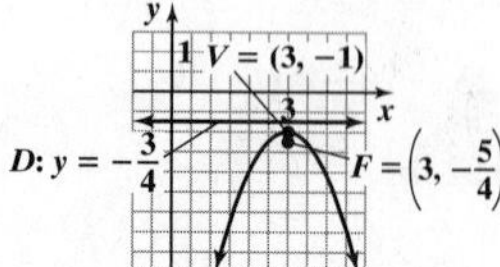

47. Vertex: $(2, -3)$; focus: $(4, -3)$; directrix: $x = 0$

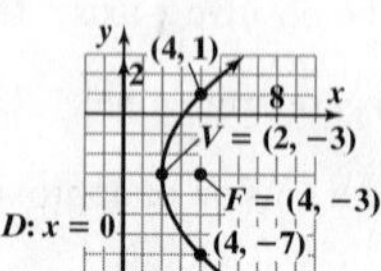

49. Vertex: (0, 2); focus: $(-1, 2)$; directrix: $x = 1$

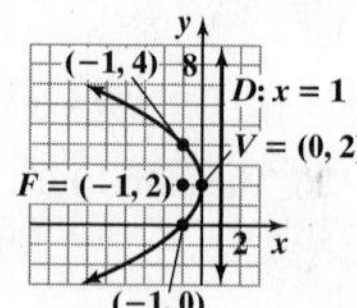

51. Vertex: $(-4, -2)$; focus: $(-4, -1)$; directrix: $y = -3$

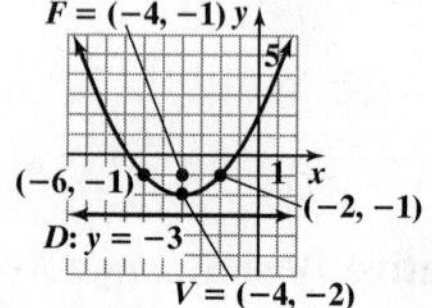

53. Vertex: $(-1, -1)$; focus: $\left(-\frac{3}{4}, -1\right)$; directrix: $x = -\frac{5}{4}$

55. Vertex: $(2, -8)$; focus: $\left(2, -\frac{31}{4}\right)$; directrix: $y = -\frac{33}{4}$

57. $(y - 1)^2 = x$ **59.** $(y - 1)^2 = -(x - 2)$

61. $x^2 = 4(y - 1)$ **63.** $y^2 = \frac{1}{2}(x + 2)$

65. 1.5625 ft from the base of the dish, along the axis of symmetry

67. 1 in. from the vertex, along the axis of symmetry

69. 20 ft **71.** 0.78125 ft

73. 4.17 ft from the base, along the axis of symmetry

75. 24.31 ft, 18.75 ft, 7.64 ft

77. (a) $y = -\frac{2}{315}x^2 + 630$

(b) 567 ft: 119.7 ft; 478 ft: 267.3 ft; 308 ft: 479.4 ft **(c)** No

79. $Cy^2 + Dx = 0, C \neq 0, D \neq 0$

$$Cy^2 = -Dx$$
$$y^2 = -\frac{D}{C}x$$

This is the equation of a parabola with vertex at $(0, 0)$ and axis of symmetry the x-axis. The focus is $\left(-\frac{D}{4C}, 0\right)$; the directrix is the line $x = \frac{D}{4C}$. The parabola opens to the right if $-\frac{D}{C} > 0$ and to the left if $-\frac{D}{C} < 0$.

81. $Cy^2 + Dx + Ey + F = 0, C \neq 0$

$$Cy^2 + Ey = -Dx - F$$
$$y^2 + \frac{E}{C}y = -\frac{D}{C}x - \frac{F}{C}$$
$$\left(y + \frac{E}{2C}\right)^2 = -\frac{D}{C}x - \frac{F}{C} + \frac{E^2}{4C^2}$$
$$\left(y + \frac{E}{2C}\right)^2 = -\frac{D}{C}x + \frac{E^2 - 4CF}{4C^2}$$

(a) If $D \neq 0$, then the equation may be written as

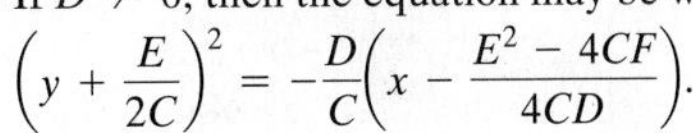

$$\left(y + \frac{E}{2C}\right)^2 = -\frac{D}{C}\left(x - \frac{E^2 - 4CF}{4CD}\right).$$

This is the equation of a parabola with vertex at $\left(\frac{E^2 - 4CF}{4CD}, -\frac{E}{2C}\right)$ and axis of symmetry parallel to the x-axis.

(b)–(d) If $D = 0$, the graph of the equation contains no points if $E^2 - 4CF < 0$, is a single horizontal line if $E^2 - 4CF = 0$, and is two horizontal lines if $E^2 - 4CF > 0$.

82. $(0, 2), (0, -2), (-36, 0)$; symmetric with respect to the x-axis. **83.** $-70°$

84. $\sin\theta = \frac{5\sqrt{89}}{89}$; $\cos\theta = -\frac{8\sqrt{89}}{89}$; $\csc\theta = \frac{\sqrt{89}}{5}$; $\sec\theta = -\frac{\sqrt{89}}{8}$; $\cot\theta = -\frac{8}{5}$ **85.** $-\frac{2\sqrt{10}}{3}$

6.3 Assess Your Understanding *(page 404)*

7. ellipse **8.** b **9.** $(0, -5)$; $(0, 5)$ **10.** 5; 3; x **11.** $(-2, -3)$; $(6, -3)$ **12.** a **13.** C **15.** B

17. Vertices: $(-5, 0), (5, 0)$
Foci: $(-\sqrt{21}, 0), (\sqrt{21}, 0)$

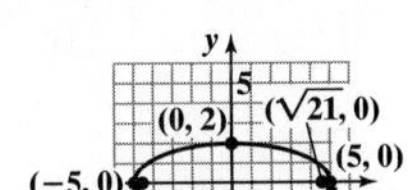

19. Vertices: $(0, -5), (0, 5)$
Foci: $(0, -4), (0, 4)$

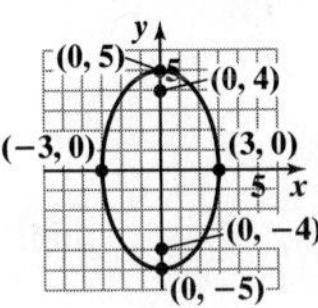

21. $\frac{x^2}{4} + \frac{y^2}{16} = 1$
Vertices: $(0, -4), (0, 4)$
Foci: $(0, -2\sqrt{3}), (0, 2\sqrt{3})$

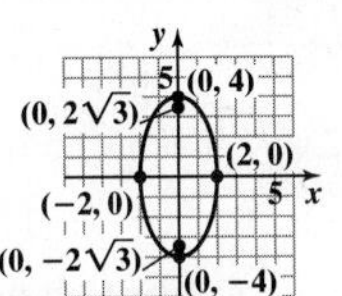

23. $\frac{x^2}{8} + \frac{y^2}{2} = 1$
Vertices: $(-2\sqrt{2}, 0), (2\sqrt{2}, 0)$
Foci: $(-\sqrt{6}, 0), (\sqrt{6}, 0)$

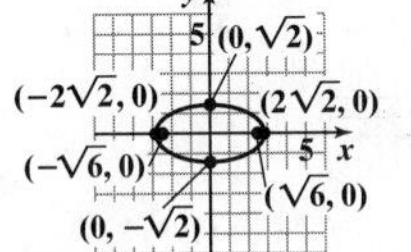

25. $\frac{x^2}{16} + \frac{y^2}{16} = 1$

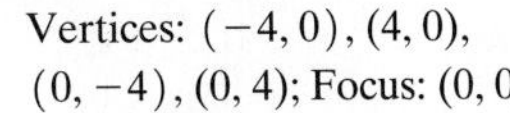

Vertices: $(-4, 0), (4, 0), (0, -4), (0, 4)$; Focus: $(0, 0)$

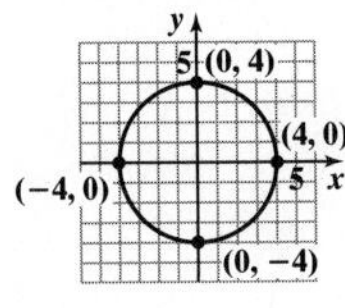

27. $\frac{x^2}{25} + \frac{y^2}{16} = 1$

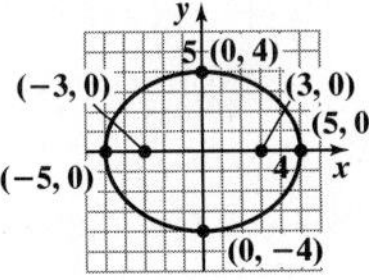

29. $\frac{x^2}{9} + \frac{y^2}{25} = 1$

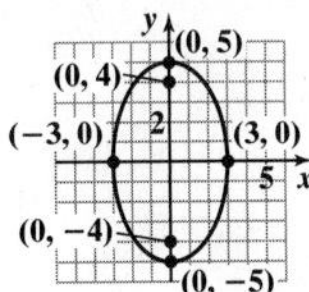

31. $\frac{x^2}{9} + \frac{y^2}{5} = 1$

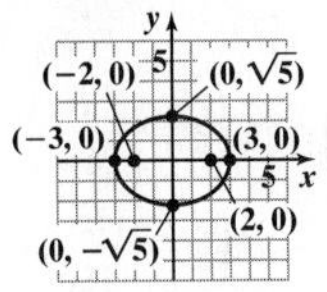

33. $\frac{x^2}{25} + \frac{y^2}{9} = 1$

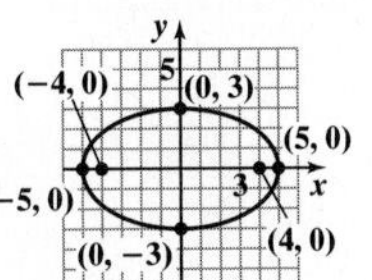

35. $\frac{x^2}{4} + \frac{y^2}{13} = 1$

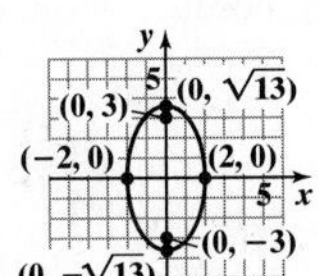

37. $x^2 + \frac{y^2}{16} = 1$

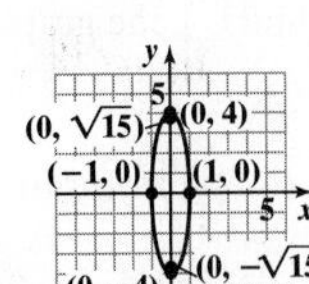

39. $\frac{(x + 1)^2}{4} + (y - 1)^2 = 1$

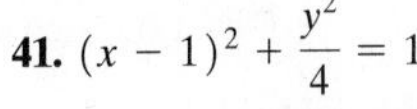

41. $(x - 1)^2 + \frac{y^2}{4} = 1$

43. Center: $(3, -1)$; vertices: $(3, -4), (3, 2)$; foci: $\left(3, -1 - \sqrt{5}\right), \left(3, -1 + \sqrt{5}\right)$

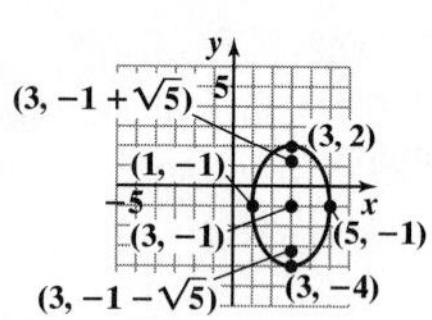

45. $\dfrac{(x + 5)^2}{16} + \dfrac{(y - 4)^2}{4} = 1$

Center: $(-5, 4)$; vertices: $(-9, 4), (-1, 4)$; foci: $\left(-5 - 2\sqrt{3}, 4\right), \left(-5 + 2\sqrt{3}, 4\right)$

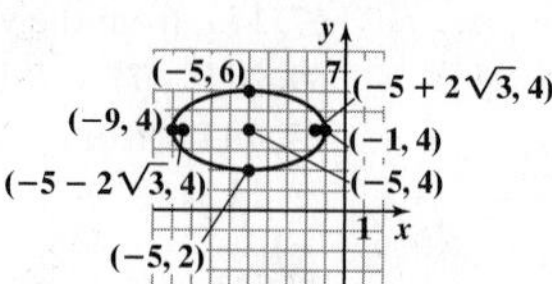

47. $\dfrac{(x + 2)^2}{4} + (y - 1)^2 = 1$

Center: $(-2, 1)$; vertices: $(-4, 1), (0, 1)$; foci: $\left(-2 - \sqrt{3}, 1\right), \left(-2 + \sqrt{3}, 1\right)$

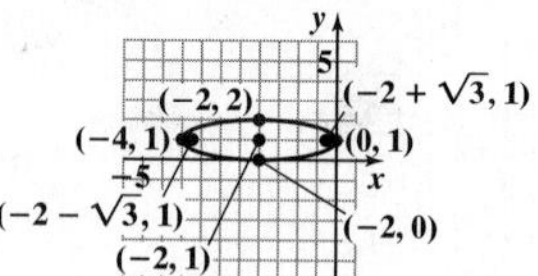

49. $\dfrac{(x - 2)^2}{3} + \dfrac{(y + 1)^2}{2} = 1$

Center: $(2, -1)$; vertices: $\left(2 - \sqrt{3}, -1\right)$; $\left(2 + \sqrt{3}, -1\right)$; foci: $(1, -1), (3, -1)$

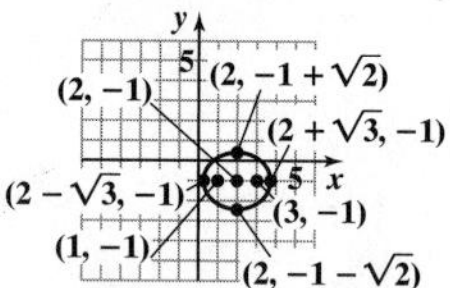

51. $\dfrac{(x - 1)^2}{4} + \dfrac{(y + 2)^2}{9} = 1$

Center: $(1, -2)$; vertices: $(1, -5), (1, 1)$; foci: $\left(1, -2 - \sqrt{5}\right), \left(1, -2 + \sqrt{5}\right)$

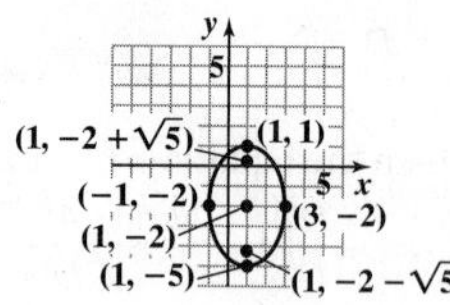

53. $x^2 + \dfrac{(y + 2)^2}{4} = 1$

Center: $(0, -2)$; vertices: $(0, -4), (0, 0)$; foci: $\left(0, -2 - \sqrt{3}\right), \left(0, -2 + \sqrt{3}\right)$

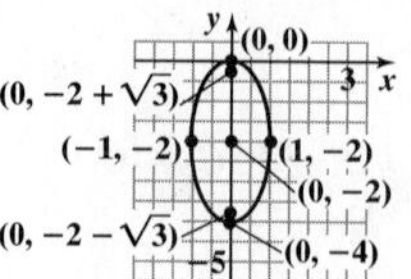

55. $\dfrac{(x - 2)^2}{25} + \dfrac{(y + 2)^2}{21} = 1$

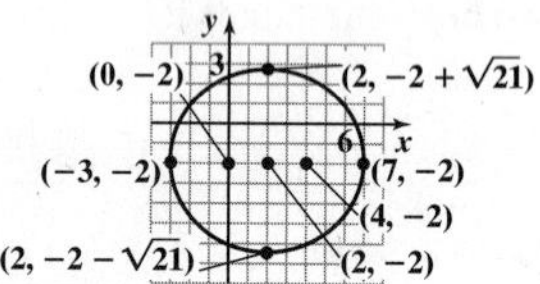

57. $\dfrac{(x - 4)^2}{5} + \dfrac{(y - 6)^2}{9} = 1$

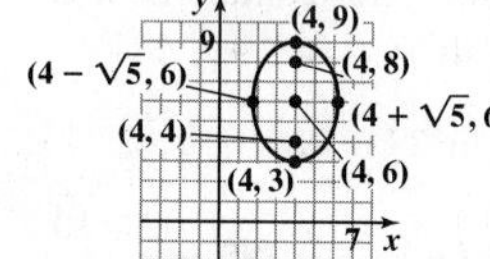

59. $\dfrac{(x - 2)^2}{16} + \dfrac{(y - 1)^2}{7} = 1$

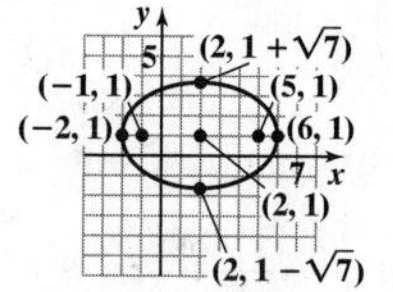

61. $\dfrac{(x - 1)^2}{10} + (y - 2)^2 = 1$

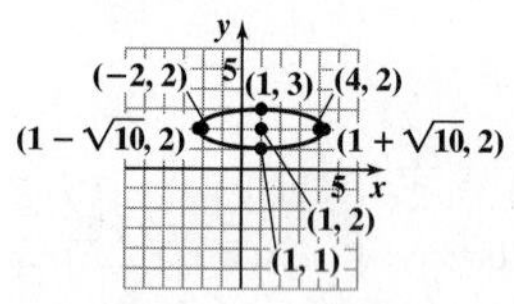

63. $\dfrac{(x - 1)^2}{9} + \dfrac{(y - 2)^2}{9} = 1$

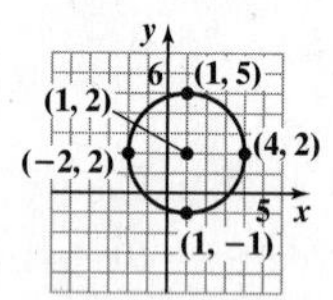

65.

67.

69. $\dfrac{x^2}{100} + \dfrac{y^2}{36} = 1$ **71.** 43.3 ft **73.** 24.65 ft, 21.65 ft, 13.82 ft **75.** 30 ft **77.** The elliptical hole will have a major axis of length $2\sqrt{41}$ in. and a minor axis of length 8 in. **79.** 91.5 million mi; $\dfrac{x^2}{(93)^2} + \dfrac{y^2}{8646.75} = 1$

81. Perihelion: 460.6 million mi; mean distance: 483.8 million mi; $\dfrac{x^2}{(483.8)^2} + \dfrac{y^2}{233{,}524.2} = 1$

83. 35 million mi **85.** $5\sqrt{5} - 4$

87. $Ax^2 + Cy^2 + Dx + Ey + F = 0 \quad A \neq 0, C \neq 0$

$$Ax^2 + Dx + Cy^2 + Ey = -F$$

$$A\left(x^2 + \frac{D}{A}x\right) + C\left(y^2 + \frac{E}{Cy}\right) = -F$$

$$A\left(x + \frac{D}{2A}\right)^2 + C\left(y + \frac{E}{2C}\right)^2 = -F + \frac{D^2}{4A} + \frac{E^2}{4C}$$

(a) If $\dfrac{D^2}{4A} + \dfrac{E^2}{4C} - F$ is of the same sign as A (and C), this is the equation of an ellipse with center at $\left(-\dfrac{D}{2A}, -\dfrac{E}{2C}\right)$.

(b) If $\dfrac{D^2}{4A} + \dfrac{E^2}{4C} - F$, the graph is the single point $\left(-\dfrac{D}{2A}, -\dfrac{E}{2C}\right)$.

(c) If $\dfrac{D^2}{4A} + \dfrac{E^2}{4C} - F$ is of the sign opposite that of A (and C), the graph contains no points, because in this case, the left side has the sign opposite that of the right side.

89. $5 - 2\sqrt{3}, 5 + 2\sqrt{3}$ **90.** Domain: $\{x \mid x \neq 5\}$; Horizontal asymptote: $y = 2$; Vertical asymptote: $x = 5$

91. -0.28 **92.** $b \approx 10.94, c \approx 17.77, B = 38°$

6.4 Assess Your Understanding *(page 417)*

7. hyperbola **8.** transverse axis **9.** b **10.** (2, 4); (2, −2) **11.** (2, 6); (2, −4) **12.** c **13.** 2; 3; x **14.** $y = -\frac{4}{9}x$; $y = \frac{4}{9}x$ **15.** B **17.** A

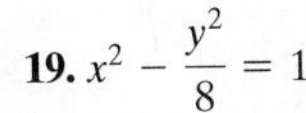

19. $x^2 - \frac{y^2}{8} = 1$

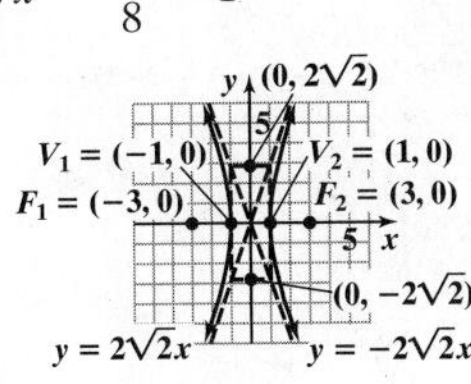

21. $\frac{y^2}{16} - \frac{x^2}{20} = 1$

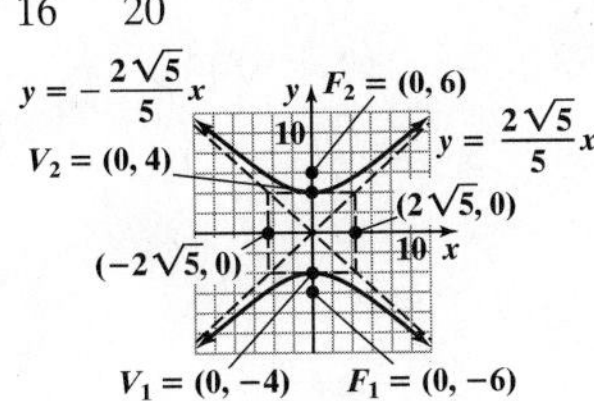

23. $\frac{x^2}{9} - \frac{y^2}{16} = 1$

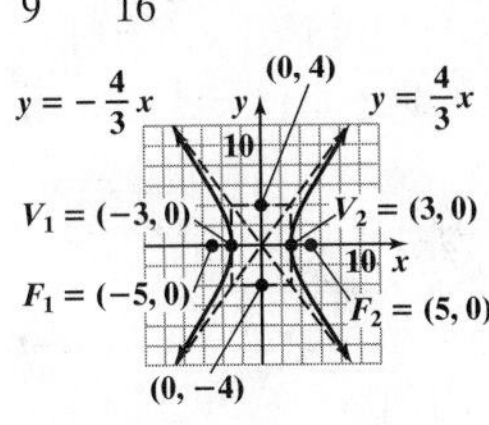

25. $\frac{y^2}{36} - \frac{x^2}{9} = 1$

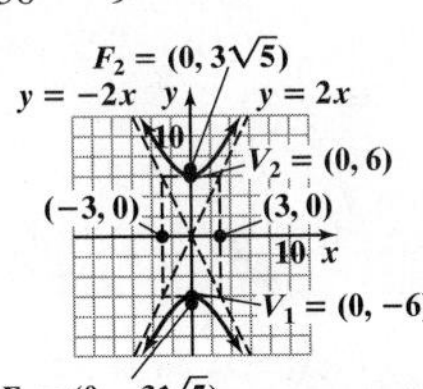

27. $\frac{x^2}{8} - \frac{y^2}{8} = 1$

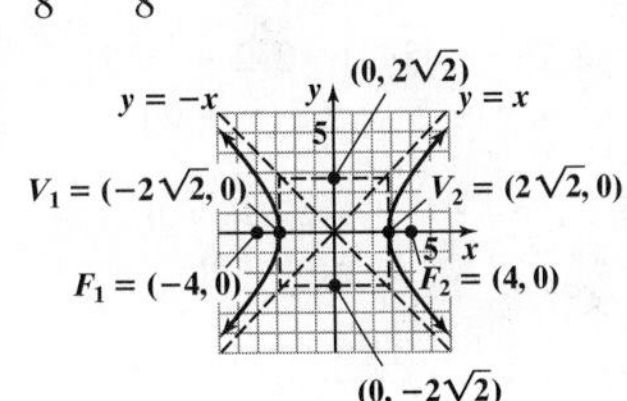

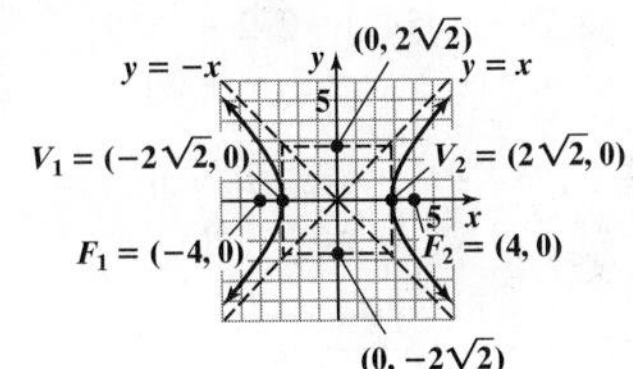

29. $\frac{x^2}{25} - \frac{y^2}{9} = 1$

Center: (0, 0)
Transverse axis: x-axis
Vertices: $(-5, 0), (5, 0)$
Foci: $(-\sqrt{34}, 0), (\sqrt{34}, 0)$
Asymptotes: $y = \pm\frac{3}{5}x$

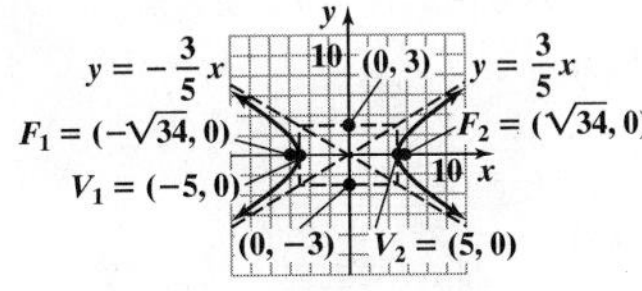

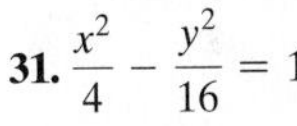

31. $\frac{x^2}{4} - \frac{y^2}{16} = 1$

Center: (0, 0)
Transverse axis: x-axis
Vertices: $(-2, 0), (2, 0)$
Foci: $(-2\sqrt{5}, 0), (2\sqrt{5}, 0)$
Asymptotes: $y = \pm 2x$

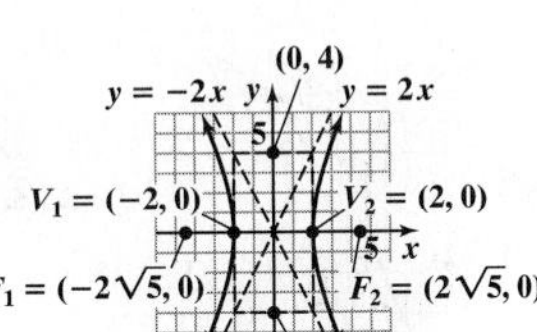

33. $\frac{y^2}{9} - x^2 = 1$

Center: (0, 0)
Transverse axis: y-axis
Vertices: $(0, -3), (0, 3)$
Foci: $(0, -\sqrt{10}), (0, \sqrt{10})$
Asymptotes: $y = \pm 3x$

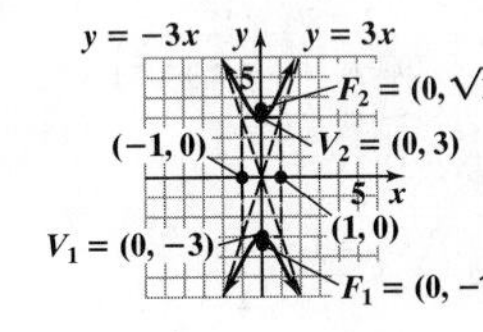

35. $\frac{y^2}{25} - \frac{x^2}{25} = 1$

Center: (0, 0)
Transverse axis: y-axis
Vertices: $(0, -5), (0, 5)$
Foci: $(0, -5\sqrt{2}), (0, 5\sqrt{2})$
Asymptotes: $y = \pm x$

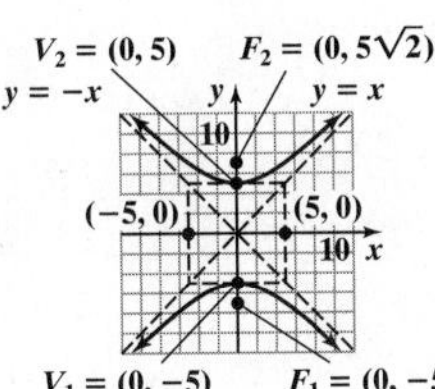

37. $x^2 - y^2 = 1$

39. $\frac{y^2}{36} - \frac{x^2}{9} = 1$

41. $\frac{(x-4)^2}{4} - \frac{(y+1)^2}{5} = 1$

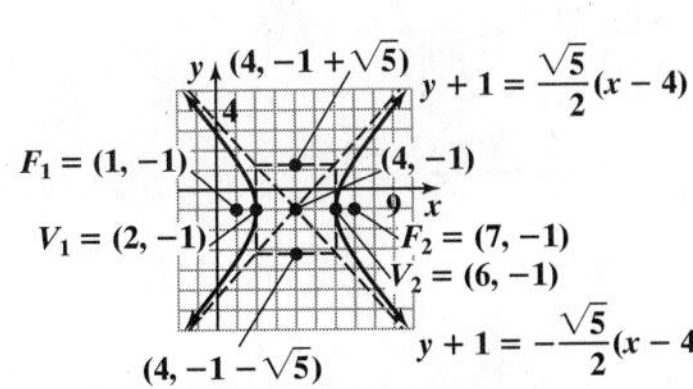

43. $\frac{(y+4)^2}{4} - \frac{(x+3)^2}{12} = 1$

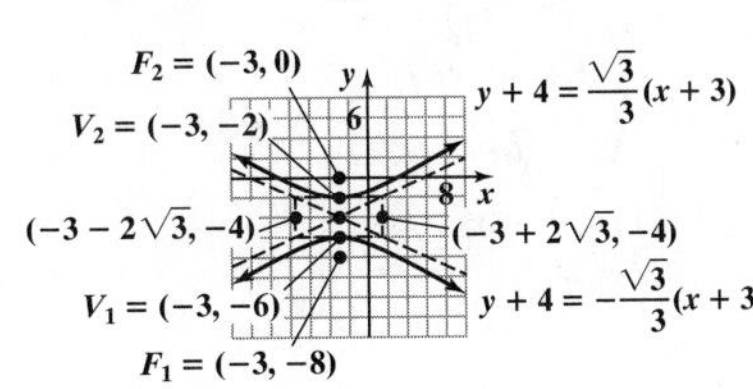

45. $(x-5)^2 - \frac{(y-7)^2}{3} = 1$

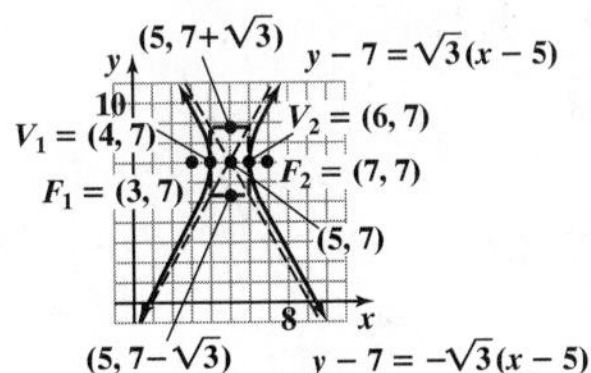

47. $\dfrac{(x-1)^2}{4} - \dfrac{(y+1)^2}{9} = 1$

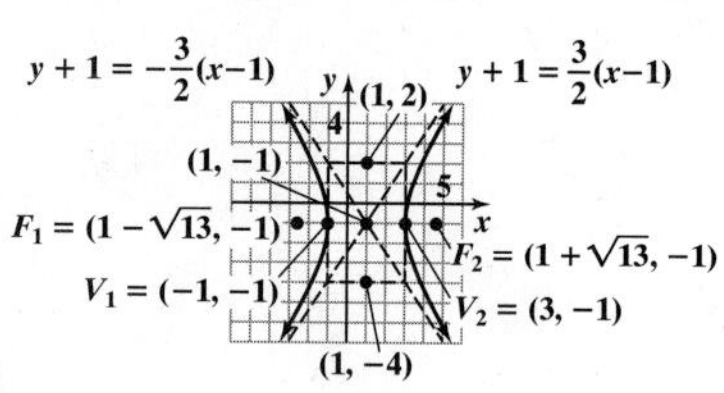

49. $\dfrac{(x-2)^2}{4} - \dfrac{(y+3)^2}{9} = 1$
Center: $(2, -3)$
Transverse axis: parallel to x-axis
Vertices: $(0, -3)$, $(4, -3)$
Foci: $\left(2 - \sqrt{13}, -3\right)$, $\left(2 + \sqrt{13}, -3\right)$
Asymptotes: $y + 3 = \pm\dfrac{3}{2}(x - 2)$

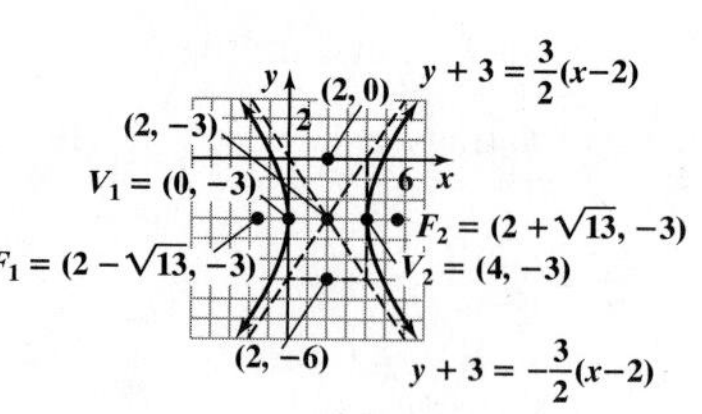

51. $\dfrac{(y-2)^2}{4} - (x+2)^2 = 1$
Center: $(-2, 2)$
Transverse axis: parallel to y-axis
Vertices: $(-2, 0)$, $(-2, 4)$
Foci: $\left(-2, 2 - \sqrt{5}\right)$, $\left(-2, 2 + \sqrt{5}\right)$
Asymptotes: $y - 2 = \pm 2(x + 2)$

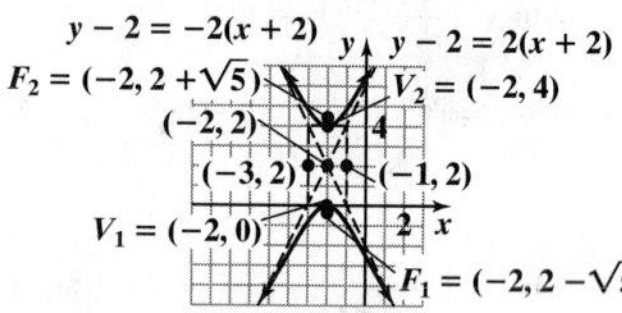

53. $\dfrac{(x+1)^2}{4} - \dfrac{(y+2)^2}{4} = 1$
Center: $(-1, -2)$
Transverse axis: parallel to x-axis
Vertices: $(-3, -2)$, $(1, -2)$
Foci: $\left(-1 - 2\sqrt{2}, -2\right)$, $\left(-1 + 2\sqrt{2}, -2\right)$
Asymptotes: $y + 2 = \pm(x + 1)$

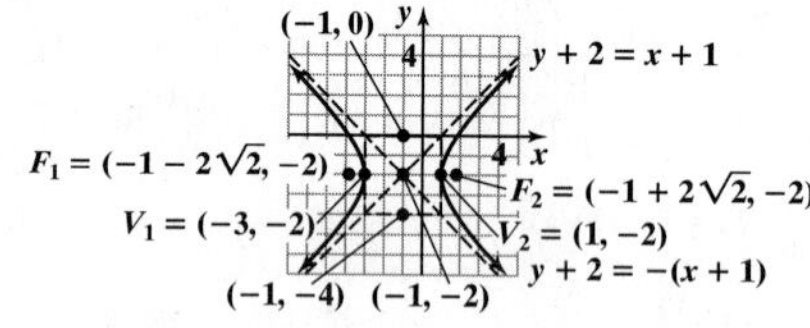

55. $(x-1)^2 - (y+1)^2 = 1$
Center: $(1, -1)$
Transverse axis: parallel to x-axis
Vertices: $(0, -1)$, $(2, -1)$
Foci: $\left(1 - \sqrt{2}, -1\right)$, $\left(1 + \sqrt{2}, -1\right)$
Asymptotes: $y + 1 = \pm(x - 1)$

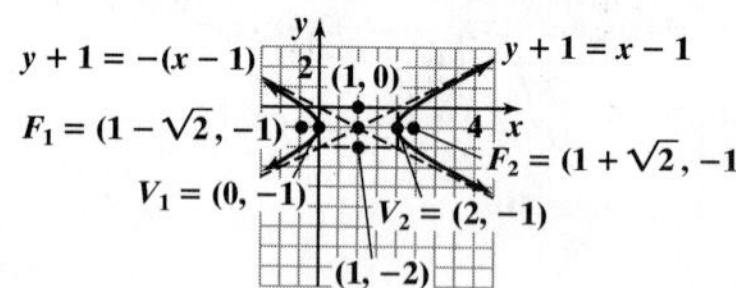

57. $\dfrac{(y-2)^2}{4} - (x+1)^2 = 1$
Center: $(-1, 2)$
Transverse axis: parallel to y-axis
Vertices: $(-1, 0)$, $(-1, 4)$
Foci: $\left(-1, 2 - \sqrt{5}\right)$, $\left(-1, 2 + \sqrt{5}\right)$
Asymptotes: $y - 2 = \pm 2(x + 1)$

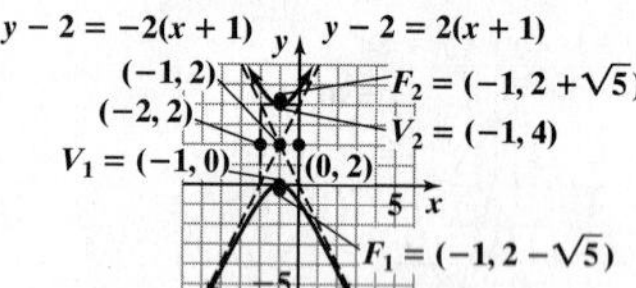

59. $\dfrac{(x-3)^2}{4} - \dfrac{(y+2)^2}{16} = 1$
Center: $(3, -2)$
Transverse axis: parallel to x-axis
Vertices: $(1, -2)$, $(5, -2)$
Foci: $\left(3 - 2\sqrt{5}, -2\right)$, $\left(3 + 2\sqrt{5}, -2\right)$
Asymptotes: $y + 2 = \pm 2(x - 3)$

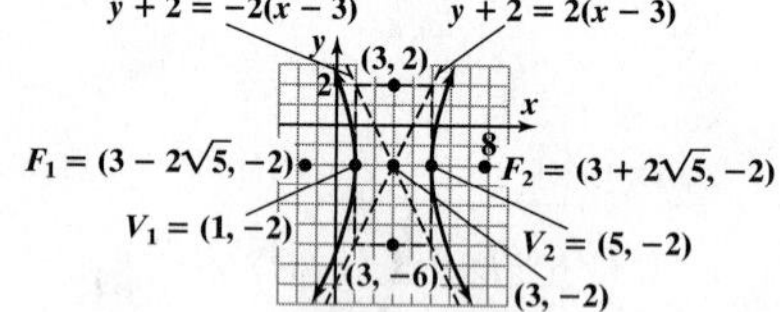

61. $\dfrac{(y-1)^2}{4} - (x+2)^2 = 1$
Center: $(-2, 1)$
Transverse axis: parallel to y-axis
Vertices: $(-2, -1)$, $(-2, 3)$
Foci: $\left(-2, 1 - \sqrt{5}\right)$, $\left(-2, 1 + \sqrt{5}\right)$
Asymptotes: $y - 1 = \pm 2(x + 2)$

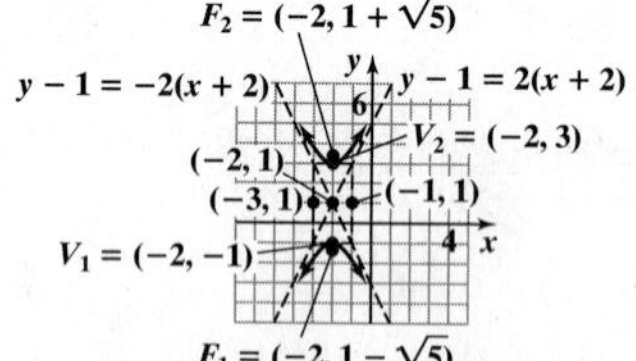

63.

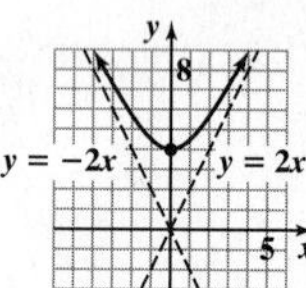

65.

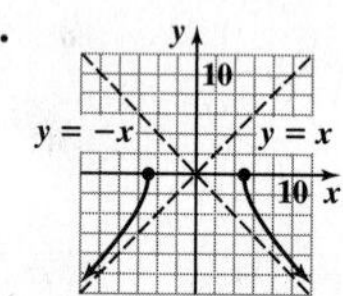

67. Center: (3, 0)
Transverse axis: parallel to x-axis
Vertices: (1, 0), (5, 0)
Foci: $(3 - \sqrt{29}, 0)$, $(3 + \sqrt{29}, 0)$
Asymptotes: $y = \pm\dfrac{5}{2}(x - 3)$

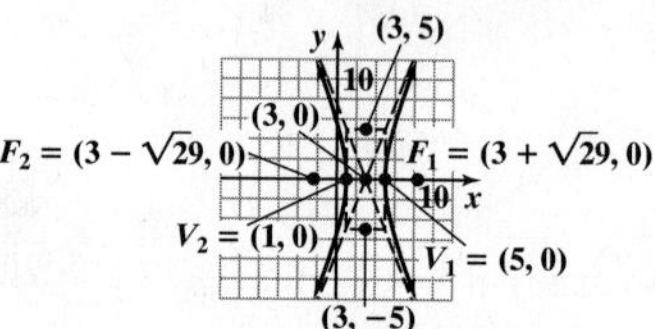

69. Vertex: (0, 3); focus: (0, 7); directrix: $y = -1$

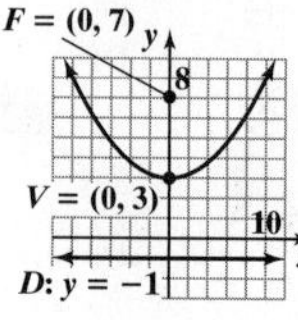

71. $\dfrac{(x-5)^2}{9} + \dfrac{y^2}{25} = 1$
Center: (5, 0); vertices: (5, 5), (5, −5); foci: (5, −4), (5, 4)

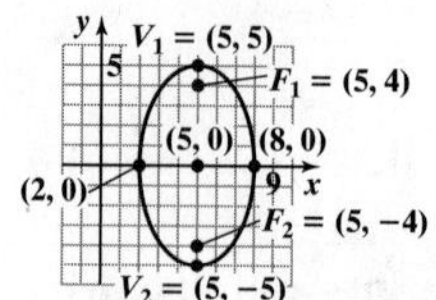

73. $(x-3)^2 = 8(y+5)$
Vertex: (3, −5); focus: (3, −3); directrix: $y = -7$

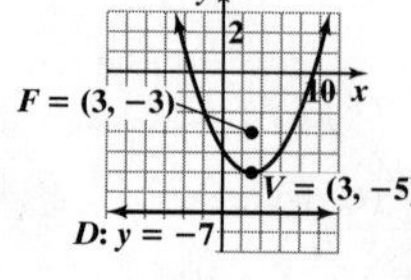

75. The fireworks display is 50,138 ft north of the person at point A. **77.** The tower is 592.4 ft tall. **79. (a)** $y = \pm x$ **(b)** $\frac{x^2}{100} - \frac{y^2}{100} = 1, x \geq 0$

81. If the eccentricity is close to 1, the "opening" of the hyperbola is very small. As e increases, the opening gets bigger.

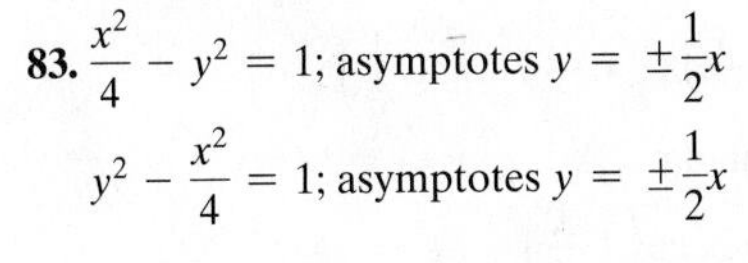

83. $\frac{x^2}{4} - y^2 = 1$; asymptotes $y = \pm\frac{1}{2}x$

$y^2 - \frac{x^2}{4} = 1$; asymptotes $y = \pm\frac{1}{2}x$

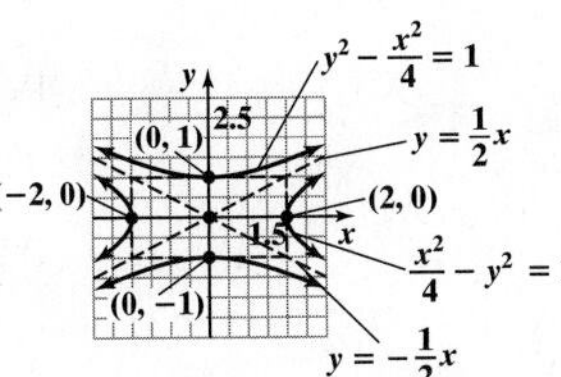

85. $Ax^2 + Cy^2 + F = 0$

$Ax^2 + Cy^2 = -F$

If A and C are of opposite sign and $F \neq 0$, this equation may be written as $\frac{x^2}{\left(-\frac{F}{A}\right)} + \frac{y^2}{\left(-\frac{F}{C}\right)} = 1$, where $-\frac{F}{A}$ and $-\frac{F}{C}$ are opposite in sign. This is the equation of a hyperbola with center (0, 0). The transverse axis is the x-axis if $-\frac{F}{A} > 0$; the transverse axis is the y-axis if $-\frac{F}{A} < 0$.

87. Amplitude $= \frac{3}{2}$; Period $= \frac{\pi}{3}$; Phase shift $= -\frac{\pi}{2}$

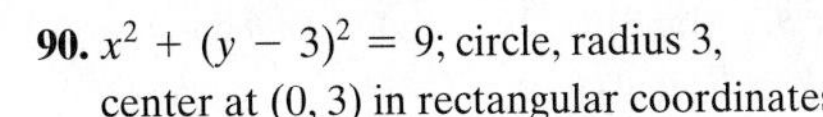

90. $x^2 + (y - 3)^2 = 9$; circle, radius 3, center at (0, 3) in rectangular coordinates

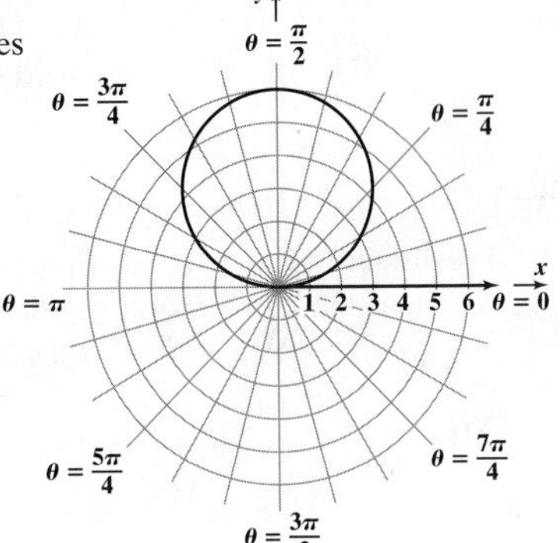

88. $c \approx 13.16, A \approx 31.6°, B = 48.4°$ **89.** $(6, -6\sqrt{3})$

6.5 Assess Your Understanding *(page 426)*

5. $\cot(2\theta) = \frac{A - C}{B}$ **6.** parabola **7.** T **8.** $B^2 - 4AC < 0$ **9.** T **10.** F **11.** Parabola **13.** Ellipse **15.** Hyperbola

17. Hyperbola **19.** Circle **21.** $x = \frac{\sqrt{2}}{2}(x' - y'), y = \frac{\sqrt{2}}{2}(x' + y')$ **23.** $x = \frac{\sqrt{2}}{2}(x' - y'), y = \frac{\sqrt{2}}{2}(x' + y')$

25. $x = \frac{1}{2}(x' - \sqrt{3}y'), y = \frac{1}{2}(\sqrt{3}x' + y')$ **27.** $x = \frac{\sqrt{5}}{5}(x' - 2y'), y = \frac{\sqrt{5}}{5}(2x' + y')$ **29.** $x = \frac{\sqrt{13}}{13}(3x' - 2y'), y = \frac{\sqrt{13}}{13}(2x' + 3y')$

31. $\theta = 45°$ (see Problem 21)

$x'^2 - \frac{y'^2}{3} = 1$

Hyperbola
Center at origin
Transverse axis is the x'-axis.
Vertices at $(\pm 1, 0)$

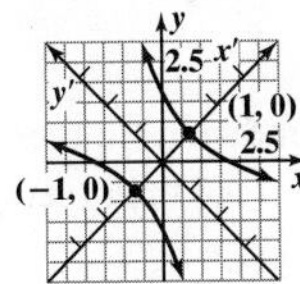

33. $\theta = 45°$ (see Problem 23)

$x'^2 + \frac{y'^2}{4} = 1$

Ellipse
Center at (0, 0)
Major axis is the y'-axis.
Vertices at $(0, \pm 2)$

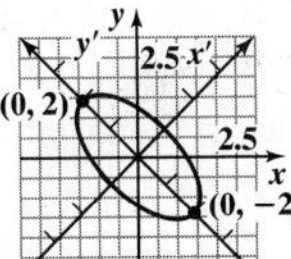

35. $\theta = 60°$ (see Problem 25)

$\frac{x'^2}{4} + y'^2 = 1$

Ellipse
Center at (0, 0)
Major axis is the x'-axis.
Vertices at $(\pm 2, 0)$

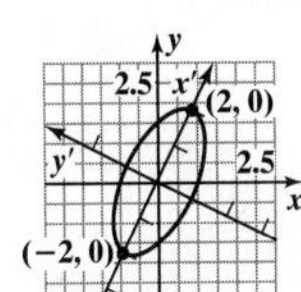

37. $\theta \approx 63°$ (see Problem 27)
$y'^2 = 8x'$
Parabola
Vertex at $(0, 0)$
Focus at $(2, 0)$

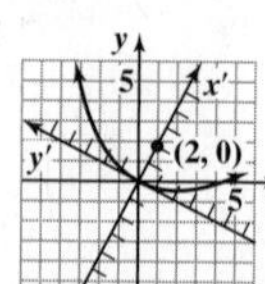

39. $\theta \approx 34°$ (see Problem 29)
$\dfrac{(x' - 2)^2}{4} + y'^2 = 1$
Ellipse
Center at $(2, 0)$
Major axis is the x'-axis.
Vertices at $(4, 0)$ and $(0, 0)$

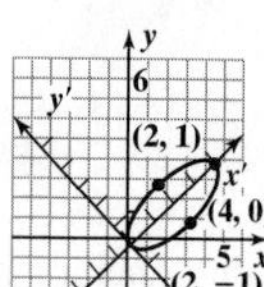

41. $\cot(2\theta) = \dfrac{7}{24}$;
$\theta = \sin^{-1}\left(\dfrac{3}{5}\right) \approx 37°$
$(x' - 1)^2 = -6\left(y' - \dfrac{1}{6}\right)$
Parabola
Vertex at $\left(1, \dfrac{1}{6}\right)$
Focus at $\left(1, -\dfrac{4}{3}\right)$

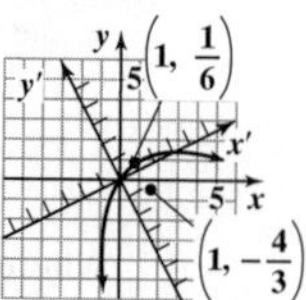

43. Hyperbola **45.** Hyperbola **47.** Parabola **49.** Ellipse **51.** Ellipse

53. Refer to equation (6):
$$\begin{aligned} A' &= A\cos^2\theta + B\sin\theta\cos\theta + C\sin^2\theta \\ B' &= B(\cos^2\theta - \sin^2\theta) + 2(C - A)(\sin\theta\cos\theta) \\ C' &= A\sin^2\theta - B\sin\theta\cos\theta + C\cos^2\theta \\ D' &= D\cos\theta + E\sin\theta \\ E' &= -D\sin\theta + E\cos\theta \\ F' &= F \end{aligned}$$

55. Use Problem 53 to find $B'^2 - 4A'C'$. After much cancellation, $B'^2 - 4A'C' = B^2 - 4AC$.

57. The distance between P_1 and P_2 in the $x'y'$-plane equals $\sqrt{(x_2' - x_1')^2 + (y_2' - y_1')^2}$.
Assuming that $x' = x\cos\theta - y\sin\theta$ and $y' = x\sin\theta + y\cos\theta$, then
$$\begin{aligned}(x_2' - x_1')^2 &= (x_2\cos\theta - y_2\sin\theta - x_1\cos\theta + y_1\sin\theta)^2 \\ &= \cos^2\theta(x_2 - x_1)^2 - 2\sin\theta\cos\theta(x_2 - x_1)(y_2 - y_1) + \sin^2\theta(y_2 - y_1)^2, \text{ and}\end{aligned}$$
$$(y_2' - y_1')^2 = (x_2\sin\theta + y_2\cos\theta - x_1\sin\theta - y_1\cos\theta)^2 = \sin^2\theta(x_2 - x_1)^2 + 2\sin\theta\cos\theta(x_2 - x_1)(y_2 - y_1) + \cos^2\theta(y_2 - y_1)^2.$$
Therefore,
$$\begin{aligned}(x_2' - x_1')^2 + (y_2' - y_1')^2 &= \cos^2\theta(x_2 - x_1)^2 + \sin^2\theta(x_2 - x_1)^2 + \sin^2\theta(y_2 - y_1)^2 + \cos^2\theta(y_2 - y_1)^2 \\ &= (x_2 - x_1)^2(\cos^2\theta + \sin^2\theta) + (y_2 - y_1)^2(\sin^2\theta + \cos^2\theta) = (x_2 - x_1)^2 + (y_2 - y_1)^2.\end{aligned}$$

61. $A \approx 39.4°, B \approx 54.7°, C \approx 85.9°$ **62.** 38.5 **63.** $r^2\cos\theta\sin\theta = 1$ **64.** $\sqrt{29}(\cos 291.8° + i\sin 291.8°)$

6.6 Assess Your Understanding *(page 432)*

3. conic; focus; directrix **4.** 1; <1; >1 **5.** T **6.** T **7.** Parabola; directrix is perpendicular to the polar axis, 1 unit to the right of the pole.

9. Hyperbola; directrix is parallel to the polar axis, $\dfrac{4}{3}$ units below the pole.

11. Ellipse; directrix is perpendicular to the polar axis, $\dfrac{3}{2}$ units to the left of the pole.

13. Parabola; directrix is perpendicular to the polar axis, 1 unit to the right of the pole; vertex is at $\left(\dfrac{1}{2}, 0\right)$.

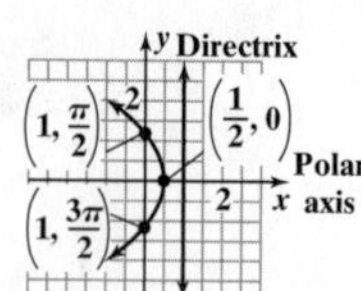

15. Ellipse; directrix is parallel to the polar axis, $\dfrac{8}{3}$ units above the pole; vertices are at $\left(\dfrac{8}{7}, \dfrac{\pi}{2}\right)$ and $\left(8, \dfrac{3\pi}{2}\right)$.

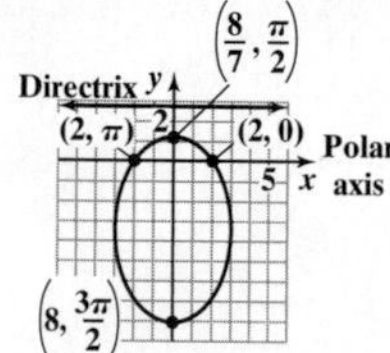

17. Hyperbola; directrix is perpendicular to the polar axis, $\dfrac{3}{2}$ units to the left of the pole; vertices are at $(-3, 0)$ and $(1, \pi)$.

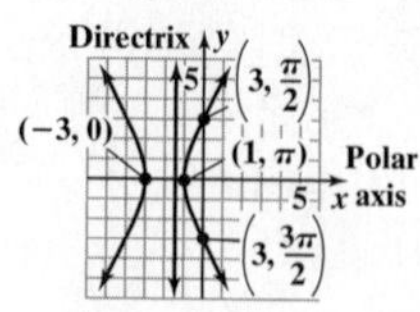

19. Ellipse; directrix is parallel to the polar axis, 8 units below the pole; vertices are at $\left(8, \dfrac{\pi}{2}\right)$ and $\left(\dfrac{8}{3}, \dfrac{3\pi}{2}\right)$.

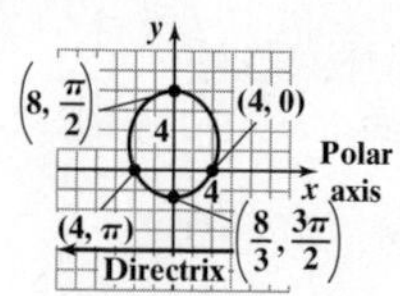

21. Ellipse; directrix is parallel to the polar axis, 3 units below the pole; vertices are at $\left(6, \dfrac{\pi}{2}\right)$ and $\left(\dfrac{6}{5}, \dfrac{3\pi}{2}\right)$.

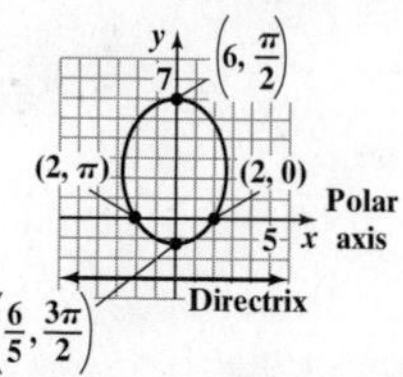

23. Ellipse; directrix is perpendicular to the polar axis, 6 units to the left of the pole; vertices are at $(6, 0)$ and $(2, \pi)$.

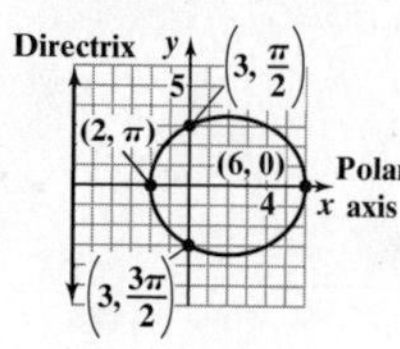

25. $y^2 + 2x - 1 = 0$ **27.** $16x^2 + 7y^2 + 48y - 64 = 0$ **29.** $3x^2 - y^2 + 12x + 9 = 0$ **31.** $4x^2 + 3y^2 - 16y - 64 = 0$

33. $9x^2 + 5y^2 - 24y - 36 = 0$ **35.** $3x^2 + 4y^2 - 12x - 36 = 0$ **37.** $r = \dfrac{1}{1 + \sin\theta}$ **39.** $r = \dfrac{12}{5 - 4\cos\theta}$ **41.** $r = \dfrac{12}{1 - 6\sin\theta}$

43. Use $d(D, P) = p - r\cos\theta$ in the derivation of equation (a) in Table 5.

45. Use $d(D, P) = p + r\sin\theta$ in the derivation of equation (a) in Table 5.

47. 27.81 **48.** Amplitude = 4; Period = 10π **49.** $\left\{\dfrac{\pi}{3}, \pi, \dfrac{5\pi}{3}\right\}$ **50.** 26

6.7 Assess Your Understanding *(page 443)*

2. plane curve; parameter **3.** ellipse **4.** cycloid **5.** F **6.** T

7. 
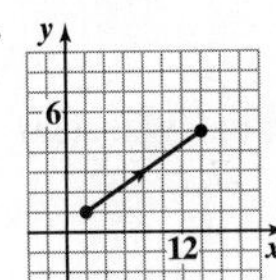

$x - 3y + 1 = 0$

9. 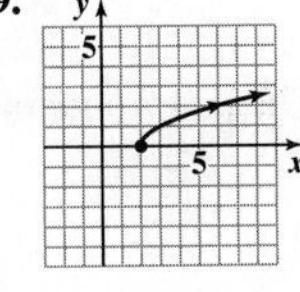

$y = \sqrt{x - 2}$

11. 
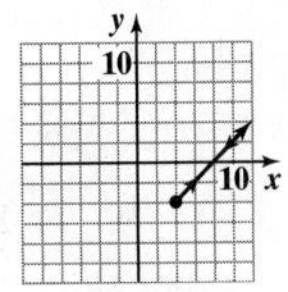

$y = x - 8$

13. 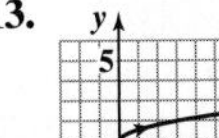

$x = 3(y - 1)^2$

15. 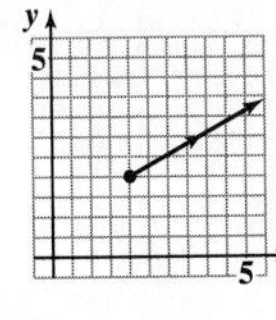

$2y = 2 + x$

17. 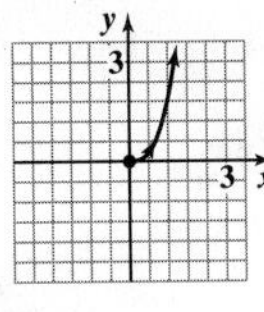

$y = x^3$

19. 
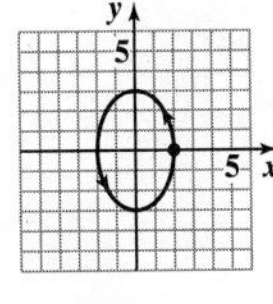

$\dfrac{x^2}{4} + \dfrac{y^2}{9} = 1$

21. 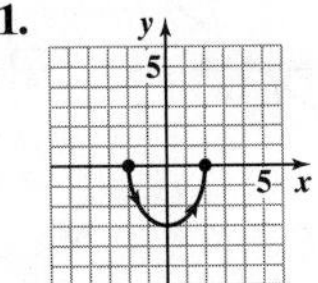

$\dfrac{x^2}{4} + \dfrac{y^2}{9} = 1$

23. 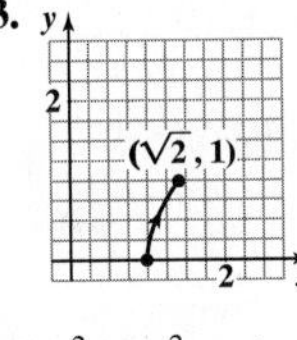

$x^2 - y^2 = 1$

25. back and forth twice

$x + y = 1$

27. $x = t$, $y = 4t - 1$ or $x = \dfrac{t + 1}{4}$, $y = t$

29. $x = t$, $y = t^2 + 1$ or $x = t^3$, $y = t^6 + 1$

31. $x = t$, $y = t^3$ or $x = \sqrt[3]{t}$, $y = t$

33. $x = t$, $y = t^{2/3}, t \ge 0$ or $x = t^3$, $y = t^2, t \ge 0$

35. $x = t + 2, y = t, 0 \le t \le 5$ **37.** $x = 3\cos t, y = 2\sin t, 0 \le t \le 2\pi$

39. $x = 2\cos(\pi t), y = -3\sin(\pi t), 0 \le t \le 2$

41. $x = 2\sin(2\pi t), y = 3\cos(2\pi t), 0 \le t \le 1$

43. 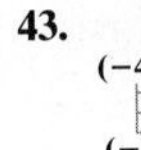
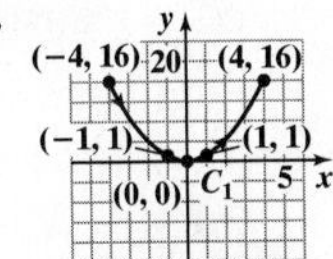

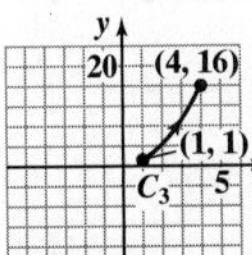

45. 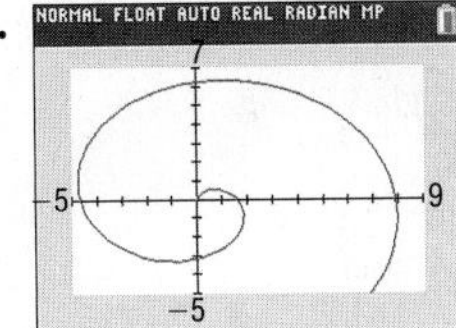

47.

49. (a) $x = 3$
$y = -16t^2 + 50t + 6$
(b) 3.24 s
(c) 1.56 s; 45.06 ft
(d) 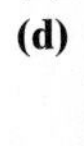
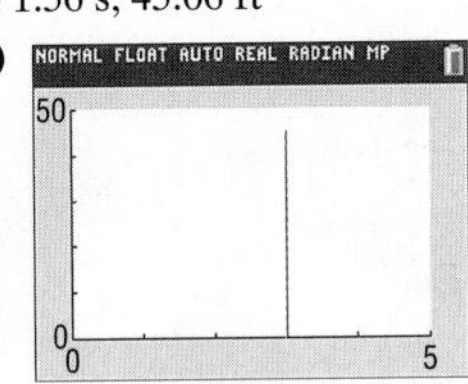

51. (a) Train: $x_1 = t^2, y_1 = 1$;
Bill: $x_2 = 5(t - 5), y_2 = 3$
(b) Bill won't catch the train.
(c) 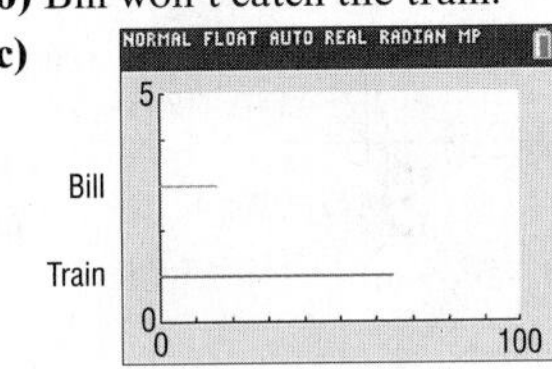

53. (a) $x = (145\cos 20°)t$
$y = -16t^2 + (145\sin 20°)t + 5$
(b) 3.20 s
(c) 435.65 ft
(d) 1.55 s; 43.43 ft
(e) 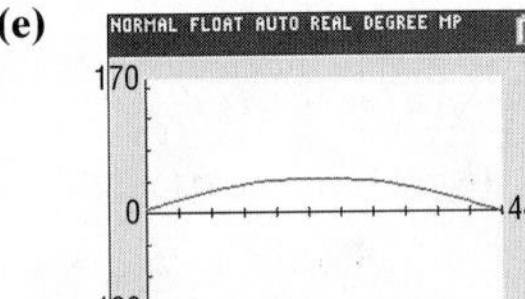

55. (a) $x = (40 \cos 45°)t$
$y = -4.9t^2 + (40 \sin 45°)t + 300$
(b) 11.23 s
(c) 317.52 m
(d) 2.89 s; 340.82 m
(e)

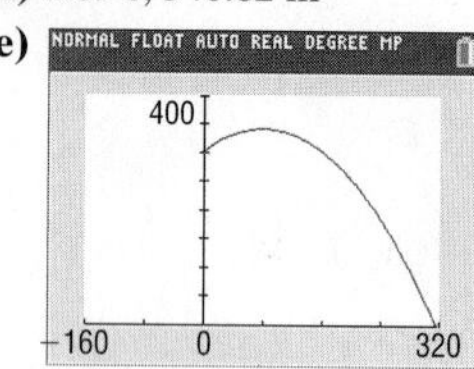

57. (a) Camry: $x = 40t - 5, y = 0$; Chevy Impala: $x = 0, y = 30t - 4$
(b) $d = \sqrt{(40t - 5)^2 + (30t - 4)^2}$
(c)

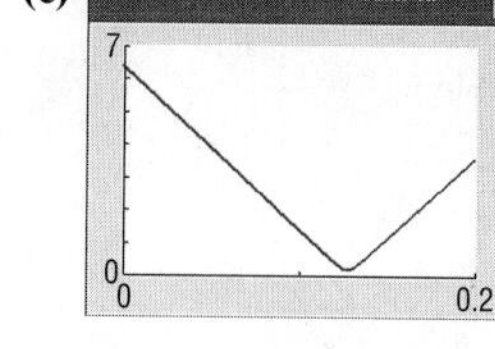

(d) 0.2 mi; 7.68 min
(e)

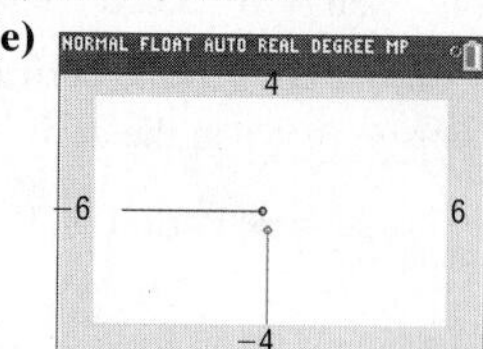

59. (a) $x = \frac{\sqrt{2}}{2}v_0 t,\ y = -16t^2 + \frac{\sqrt{2}}{2}v_0 t + 3$ **(b)** Maximum height is 139.1 ft. **(c)** The ball is 272.25 ft from home plate. **(d)** Yes, the ball will clear the wall by about 99.5 ft. **61.** The orientation is from (x_1, y_1) to (x_2, y_2).

65. **66.**

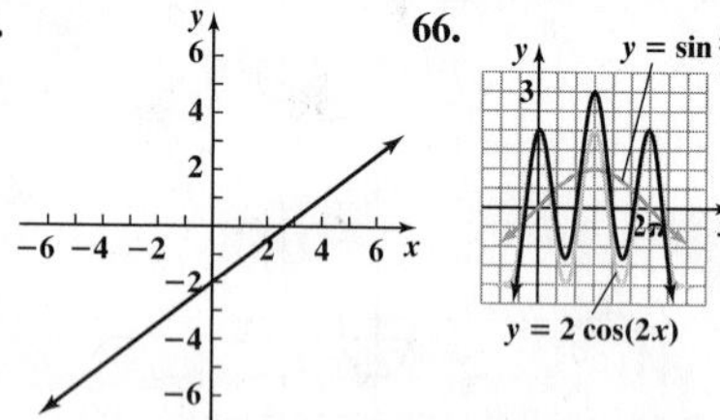

67. Approximately 2733 miles **68. (a)** Simple harmonic **(b)** 2 m **(c)** $\frac{\pi}{2}$ s **(d)** $\frac{2}{\pi}$ oscillations/s

Review Exercises *(page 447)*

1. Parabola; vertex (0, 0), focus $(-4, 0)$, directrix $x = 4$ **2.** Hyperbola; center (0, 0), vertices (5, 0) and $(-5, 0)$, foci $(\sqrt{26}, 0)$ and $(-\sqrt{26}, 0)$, asymptotes $y = \frac{1}{5}x$ and $y = -\frac{1}{5}x$ **3.** Ellipse; center (0, 0), vertices (0, 5) and $(0, -5)$, foci (0, 3) and $(0, -3)$

4. $x^2 = -4(y - 1)$: Parabola; vertex (0, 1), focus (0, 0), directrix $y = 2$ **5.** $\frac{x^2}{2} - \frac{y^2}{8} = 1$: Hyperbola; center (0, 0), vertices $(\sqrt{2}, 0)$ and $(-\sqrt{2}, 0)$, foci $(\sqrt{10}, 0)$ and $(-\sqrt{10}, 0)$, asymptotes $y = 2x$ and $y = -2x$ **6.** $(x - 2)^2 = 2(y + 2)$: Parabola; vertex $(2, -2)$, focus $\left(2, -\frac{3}{2}\right)$, directrix $y = -\frac{5}{2}$

7. $\frac{(y - 2)^2}{4} - (x - 1)^2 = 1$: Hyperbola; center (1, 2), vertices (1, 4) and (1, 0), foci $(1, 2 + \sqrt{5})$ and $(1, 2 - \sqrt{5})$, asymptotes $y - 2 = \pm 2(x - 1)$

8. $\frac{(x - 2)^2}{9} + \frac{(y - 1)^2}{4} = 1$: Ellipse; center (2, 1), vertices (5, 1) and $(-1, 1)$, foci $(2 + \sqrt{5}, 1)$ and $(2 - \sqrt{5}, 1)$

9. $(x - 2)^2 = -4(y + 1)$: Parabola; vertex $(2, -1)$, focus $(2, -2)$, directrix $y = 0$

10. $\frac{(x - 1)^2}{4} + \frac{(y + 1)^2}{9} = 1$: Ellipse; center $(1, -1)$, vertices (1, 2) and $(1, -4)$, foci $(1, -1 + \sqrt{5})$ and $(1, -1 - \sqrt{5})$

11. $y^2 = -8x$

(−2, 4)
F = (−2, 0)
V = (0, 0)
(−2, −4)
D: x = 2

12. $\frac{y^2}{4} - \frac{x^2}{12} = 1$

$V_2 = (0, 2)$
$F_2 = (0, 4)$
$y = -\frac{\sqrt{3}}{3}x$
$y = \frac{\sqrt{3}}{3}x$
$(-2\sqrt{3}, 0)$
$(2\sqrt{3}, 0)$
$V_1 = (0, -2)$
$F_1 = (0, -4)$

13. $\frac{x^2}{16} + \frac{y^2}{7} = 1$

$(0, \sqrt{7})$
$V_1 = (-4, 0)$
$V_2 = (4, 0)$
$F_1 = (-3, 0)$
$F_2 = (3, 0)$
$(0, -\sqrt{7})$

14. $(x - 2)^2 = -4(y + 3)$

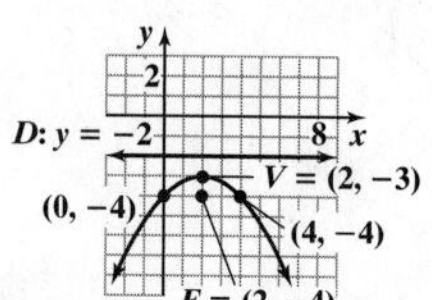

15. $(x + 2)^2 - \frac{(y + 3)^2}{3} = 1$

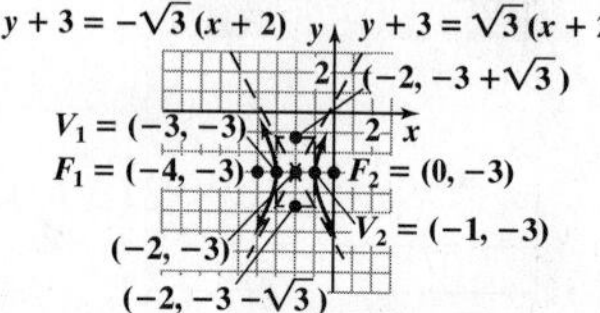

16. $\frac{(x + 4)^2}{16} + \frac{(y - 5)^2}{25} = 1$

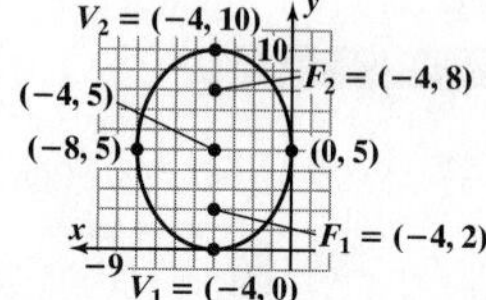

17. $\dfrac{(x+1)^2}{9} - \dfrac{(y-2)^2}{7} = 1$

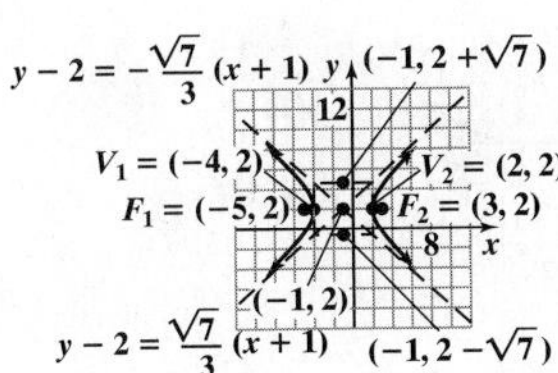

18. $\dfrac{(x-3)^2}{9} - \dfrac{(y-1)^2}{4} = 1$

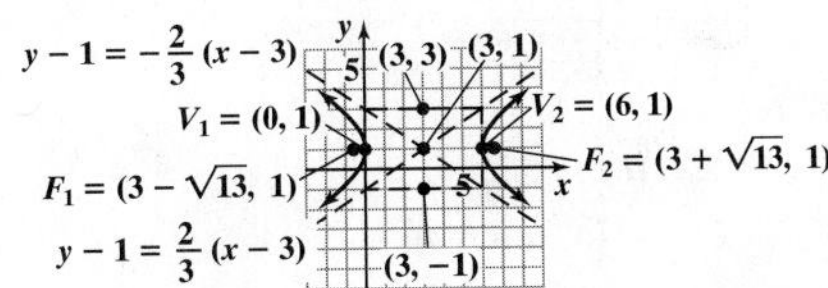

19. Parabola **20.** Ellipse
21. Parabola **22.** Hyperbola
23. Ellipse

24. $x'^2 - \dfrac{y'^2}{9} = 1$

Hyperbola
Center at the origin
Transverse axis the x'-axis
Vertices at $(\pm 1, 0)$

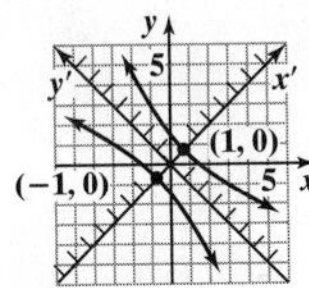

25. $\dfrac{x'^2}{2} + \dfrac{y'^2}{4} = 1$

Ellipse
Center at origin
Major axis the y'-axis
Vertices at $(0, \pm 2)$

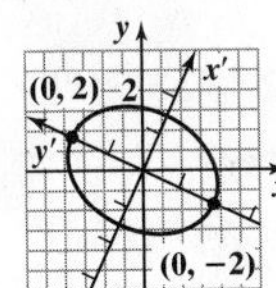

26. $y'^2 = -\dfrac{4\sqrt{13}}{13}x'$

Parabola
Vertex at the origin
Focus on the x'-axis at $\left(-\dfrac{\sqrt{13}}{13}, 0\right)$

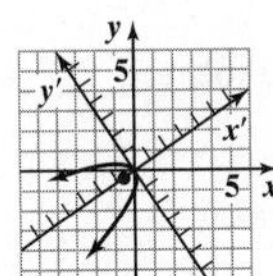

27. Parabola; directrix is perpendicular to the polar axis 4 units to the left of the pole; vertex is $(2, \pi)$.

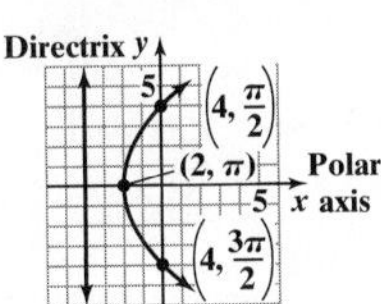

28. Ellipse; directrix is parallel to the polar axis 6 units below the pole; vertices are $\left(6, \dfrac{\pi}{2}\right)$ and $\left(2, \dfrac{3\pi}{2}\right)$.

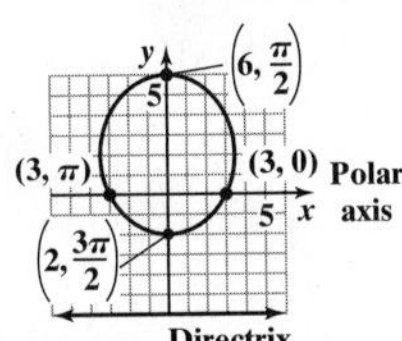

29. Hyperbola; directrix is perpendicular to the polar axis 1 unit to the right of the pole; vertices are $\left(\dfrac{2}{3}, 0\right)$ and $(-2, \pi)$.

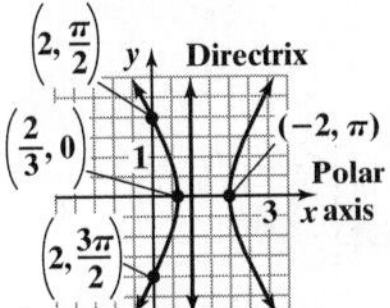

30. $y^2 - 8x - 16 = 0$ **31.** $3x^2 - y^2 - 8x + 4 = 0$

32.

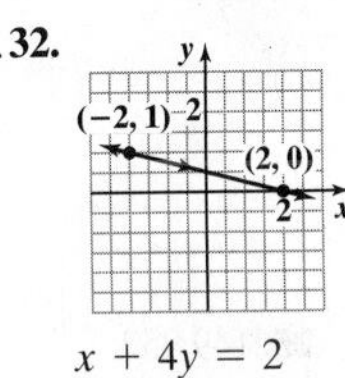

$x + 4y = 2$

33.

(0, 6) 7 (0, 2) (−3, 2) (3, 2) 5 (0, −2)

$\dfrac{x^2}{9} + \dfrac{(y-2)^2}{16} = 1$

34.

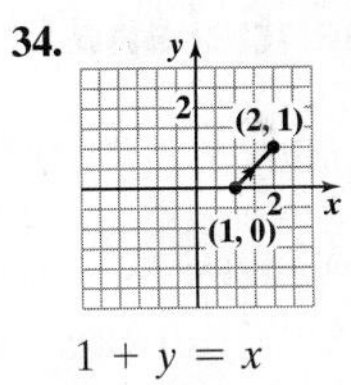

$1 + y = x$

35. $x = t, y = -2t + 4, -\infty < t < \infty$ **36.** $x = 4\cos\left(\dfrac{\pi}{2}t\right), y = 3\sin\left(\dfrac{\pi}{2}t\right), 0 \le t \le 4$ **37.** $\dfrac{x^2}{5} - \dfrac{y^2}{4} = 1$ **38.** The ellipse $\dfrac{x^2}{16} + \dfrac{y^2}{7} = 1$

$x = \dfrac{t-4}{-2}, y = t, -\infty < t < \infty$

39. $\dfrac{1}{4}$ ft or 3 in. **40.** 19.72 ft, 18.86 ft, 14.91 ft **41.** 450 ft

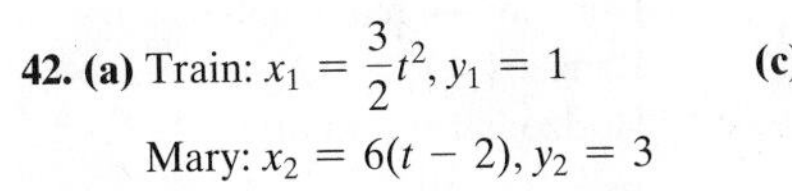

42. (a) Train: $x_1 = \dfrac{3}{2}t^2, y_1 = 1$

Mary: $x_2 = 6(t - 2), y_2 = 3$

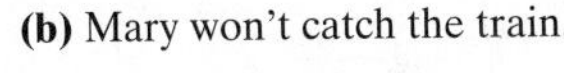

(b) Mary won't catch the train.

(c)

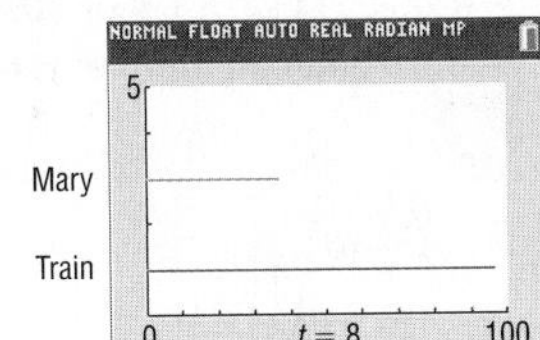

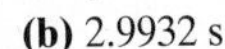

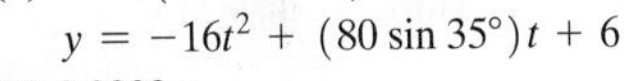

43. (a) $x = (80\cos 35°)t$

$y = -16t^2 + (80\sin 35°)t + 6$

(b) 2.9932 s

(c) 1.4339 s; 38.9 ft

(d) 196.15 ft

(e)

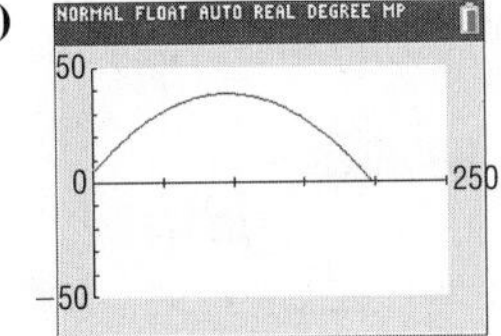

Chapter Test *(page 449)*

1. Hyperbola; center: $(-1, 0)$; vertices: $(-3, 0)$ and $(1, 0)$; foci: $(-1 - \sqrt{13}, 0)$ and $(-1 + \sqrt{13}, 0)$; asymptotes: $y = -\dfrac{3}{2}(x+1)$ and $y = \dfrac{3}{2}(x+1)$

2. Parabola; vertex: $\left(1, -\dfrac{1}{2}\right)$; focus: $\left(1, \dfrac{3}{2}\right)$; directrix: $y = -\dfrac{5}{2}$

3. Ellipse; center: $(-1, 1)$; foci: $(-1 - \sqrt{3}, 1)$ and $(-1 + \sqrt{3}, 1)$; vertices: $(-4, 1)$ and $(2, 1)$

4. $(x+1)^2 = 6(y-3)$

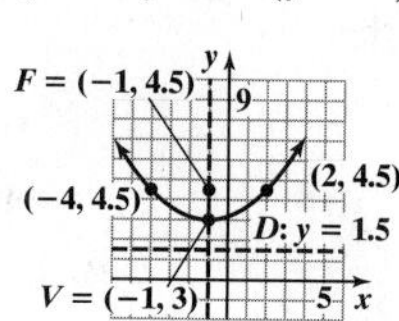

5. $\dfrac{x^2}{7} + \dfrac{y^2}{16} = 1$

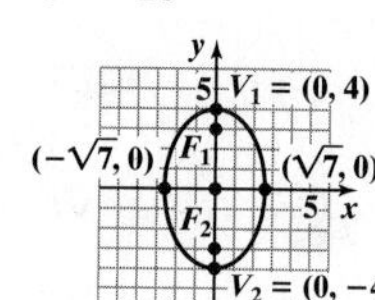

6. $\dfrac{(y-2)^2}{4} - \dfrac{(x-2)^2}{8} = 1$

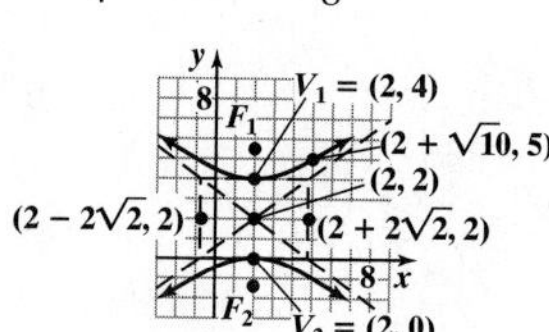

7. Hyperbola **8.** Ellipse **9.** Parabola

10. $x'^2 + 2y'^2 = 1$. This is the equation of an ellipse with center at $(0, 0)$ in the $x'y'$-plane. The vertices are at $(-1, 0)$ and $(1, 0)$ in the $x'y'$-plane.

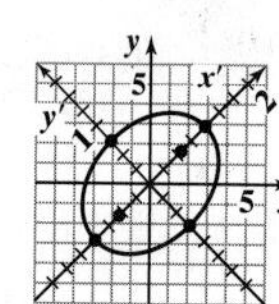

11. Hyperbola; $(x+2)^2 - \dfrac{y^2}{3} = 1$ **12.** $y = 1 - \sqrt{\dfrac{x+2}{3}}$

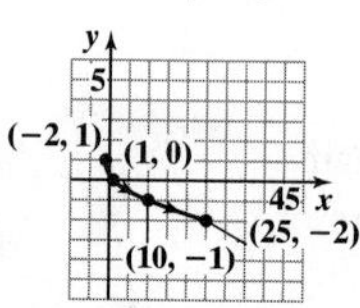

13. The microphone should be located $\frac{2}{3}$ ft from the base of the reflector, along its axis of symmetry.

Cumulative Review *(page 449)*

1. $-6x + 5 - 3h$

2. (a) $y = 2x - 2$ **(b)** $(x-2)^2 + y^2 = 4$ **(c)** $\dfrac{x^2}{9} + \dfrac{y^2}{4} = 1$ **(d)** $y = 2(x-1)^2$ **(e)** $y^2 - \dfrac{x^2}{3} = 1$

3. $\theta = \dfrac{\pi}{12} \pm \pi k$, k is any integer; $\theta = \dfrac{5\pi}{12} \pm \pi k$, k is any integer **4.** $\theta = \dfrac{\pi}{6}$

5. $r = 8 \sin \theta$ **6.** $\left\{x \,\middle|\, x \neq \dfrac{3\pi}{4} \pm \pi k, k \text{ is an integer}\right\}$ **7.** $\{22.5°\}$ **8.** $y = \dfrac{x^2}{5} + 5$

CHAPTER 7 Exponential and Logarithmic Functions

7.1 Assess Your Understanding *(page 464)*

6. Exponential function; growth factor; initial value **7.** a **8.** T **9.** T **10.** $\left(-1, \dfrac{1}{a}\right)$; $(0, 1)$; $(1, a)$ **11.** 4 **12.** F **13.** b **14.** c **15. (a)** 8.815 **(b)** 8.821 **(c)** 8.824 **(d)** 8.825 **17. (a)** 21.217 **(b)** 22.217 **(c)** 22.440 **(d)** 22.459 **19.** 1.265 **21.** 0.347 **23.** 3.320 **25.** 149.952 **27.** Neither **29.** Exponential; $H(x) = 4^x$ **31.** Exponential; $f(x) = 3(2^x)$ **33.** Linear; $H(x) = 2x + 4$ **35.** B **37.** D **39.** A **41.** E

43.

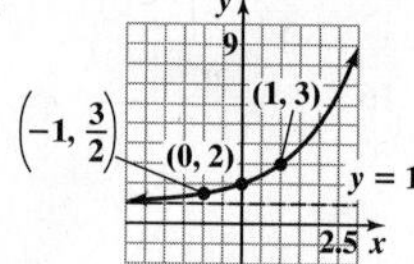

Domain: All real numbers
Range: $\{y \mid y > 1\}$ or $(1, \infty)$
Horizontal asymptote: $y = 1$

45.

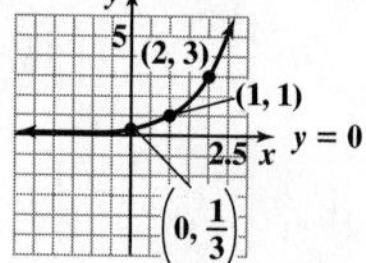

Domain: All real numbers
Range: $\{y \mid y > 0\}$ or $(0, \infty)$
Horizontal asymptote: $y = 0$

47.

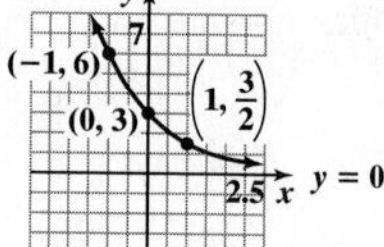

Domain: All real numbers
Range: $\{y \mid y > 0\}$ or $(0, \infty)$
Horizontal asymptote: $y = 0$

49.

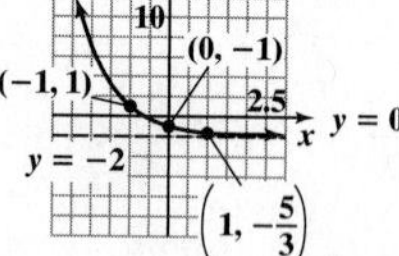

Domain: All real numbers
Range: $\{y \mid y > -2\}$ or $(-2, \infty)$
Horizontal asymptote: $y = -2$

51.

Domain: All real numbers
Range: $\{y \mid y > 2\}$ or $(2, \infty)$
Horizontal asymptote: $y = 2$

53.

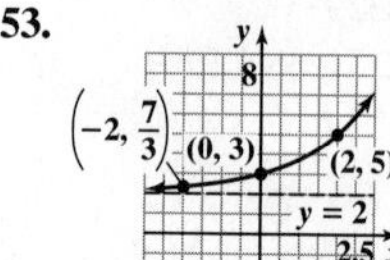

Domain: All real numbers
Range: $\{y \mid y > 2\}$ or $(2, \infty)$
Horizontal asymptote: $y = 2$

55.

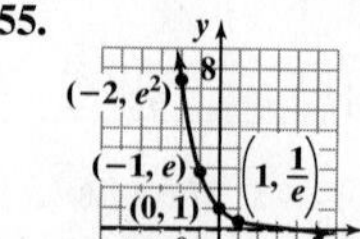

Domain: All real numbers
Range: $\{y \mid y > 0\}$ or $(0, \infty)$
Horizontal asymptote: $y = 0$

57.

Domain: All real numbers
Range: $\{y \mid y > 0\}$ or $(0, \infty)$
Horizontal asymptote: $y = 0$

59.

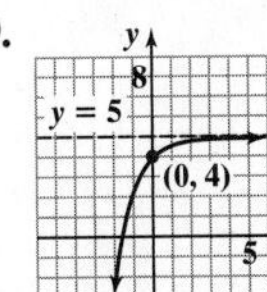

Domain: All real numbers
Range: $\{y \mid y < 5\}$ or $(-\infty, 5)$
Horizontal asymptote: $y = 5$

61.

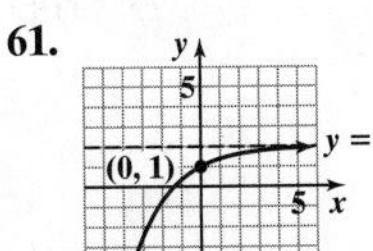

Domain: All real numbers
Range: $\{y \mid y < 2\}$ or $(-\infty, 2)$
Horizontal asymptote: $y = 2$

63. $\{3\}$ **65.** $\{-4\}$ **67.** $\{2\}$ **69.** $\left\{\frac{3}{2}\right\}$ **71.** $\{-\sqrt{2}, 0, \sqrt{2}\}$ **73.** $\{6\}$ **75.** $\{-1, 7\}$ **77.** $\{-4, 2\}$ **79.** $\{-4\}$ **81.** $\{1, 2\}$ **83.** $\frac{1}{49}$ **85.** $\frac{1}{4}$ **87.** 5 **89.** $f(x) = 3^x$ **91.** $f(x) = -6^x$ **93.** $f(x) = 3^x + 2$ **95. (a)** 16; (4, 16) **(b)** -4; $\left(-4, \frac{1}{16}\right)$ **97. (a)** $\frac{9}{4}$; $\left(-1, \frac{9}{4}\right)$ **(b)** 3; (3, 66) **99. (a)** 60; $(-6, 60)$ **(b)** -4; $(-4, 12)$ **(c)** -2

101.

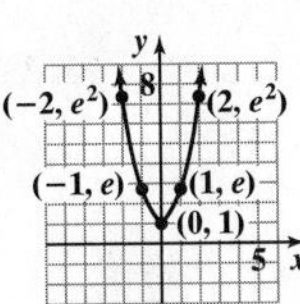

Domain: $(-\infty, \infty)$
Range: $[1, \infty)$
Intercept: (0, 1)

103.

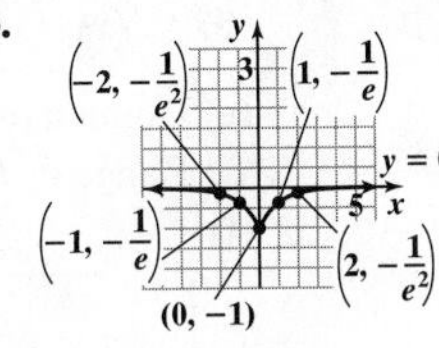

Domain: $(-\infty, \infty)$
Range: $[-1, 0)$
Intercept: $(0, -1)$

105. (a) 74% **(b)** 47% **(c)** Each pane allows only 97% of light to pass through. **107. (a)** \$16,231 **(b)** \$8626 **(c)** As each year passes, the sedan is worth 90% of its value the previous year. **109. (a)** 30% **(b)** 9% **(c)** Each year only 30% of the previous survivors survive again. **111.** 3.35 mg; 0.45 mg **113. (a)** 0.632 **(b)** 0.982 **(c)** 1 **(d)**

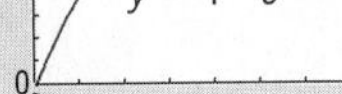

(e) About 7 min

115. (a) 0.0516 **(b)** 0.0888 **117. (a)** 70.95% **(b)** 72.62% **(c)** 100%
119. (a) 5.41 amp, 7.59 amp, 10.38 amp **(b)** 12 amp
(d) 3.34 amp, 5.31 amp, 9.44 amp
(e) 24 amp
(c), (f)

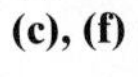
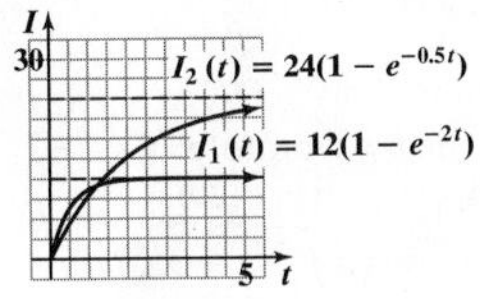

121. 36
123.

Final Denominator	Value of Expression	Compare Value to $e \approx$ 2.718281828
1 + 1	2.5	$2.5 < e$
2 + 2	2.8	$2.8 > e$
3 + 3	2.7	$2.7 < e$
4 + 4	2.721649485	$2.721649485 > e$
5 + 5	2.717770035	$2.717770035 < e$
6 + 6	2.718348855	$2.718348855 > e$

125. $f(A + B) = a^{A+B} = a^A \cdot a^B = f(A) \cdot f(B)$ **127.** $f(\alpha x) = a^{\alpha x} = (a^x)^\alpha = [f(x)]^\alpha$

129. (a) $f(-x) = \frac{1}{2}(e^{-x} + e^{-(-x)})$
$= \frac{1}{2}(e^{-x} + e^x)$
$= \frac{1}{2}(e^x + e^{-x})$
$= f(x)$

(b)

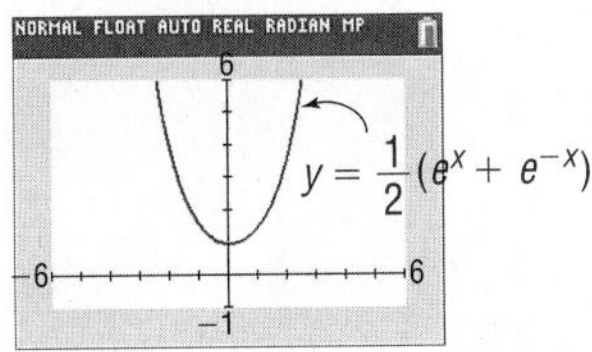

(c) $(\cosh x)^2 - (\sinh x)^2$
$= \left[\frac{1}{2}(e^x + e^{-x})\right]^2 - \left[\frac{1}{2}(e^x - e^{-x})\right]^2$
$= \frac{1}{4}[e^{2x} + 2 + e^{-2x} - e^{2x} + 2 - e^{-2x}]$
$= \frac{1}{4}(4) = 1$

131. 59 minutes **135.** $a^{-x} = (a^{-1})^x = \left(\frac{1}{a}\right)^x$ **136.** $\frac{5\pi}{9}$ **137.** Amplitude = 4; period = $\frac{\pi}{3}$ **138.** $\frac{3\pi}{10}$ **139.** $(-4\sqrt{3}, 4)$

7.2 Assess Your Understanding *(page 478)*

4. $\{x \mid x > 0\}$ or $(0, \infty)$ **5.** $\left(\frac{1}{a}, -1\right)$, $(1, 0)$, $(a, 1)$ **6.** 1 **7.** F **8.** T **9.** a **10.** c **11.** $2 = \log_3 9$ **13.** $2 = \log_a 1.6$ **15.** $x = \log_2 7.2$ **17.** $x = \ln 8$ **19.** $2^3 = 8$ **21.** $a^6 = 3$ **23.** $3^x = 2$ **25.** $e^x = 4$ **27.** 0 **29.** 2 **31.** -4 **33.** $\frac{1}{2}$ **35.** 4 **37.** $\frac{1}{2}$ **39.** $\{x \mid x > 3\}$; $(3, \infty)$
41. All real numbers except 0; $\{x \mid x \neq 0\}$; $(-\infty, 0) \cup (0, \infty)$ **43.** $\{x \mid x > 10\}$; $(10, \infty)$ **45.** $\{x \mid x > -1\}$; $(-1, \infty)$
47. $\{x \mid x < -1 \text{ or } x > 0\}$; $(-\infty, -1) \cup (0, \infty)$ **49.** $\{x \mid x \geq 1\}$; $[1, \infty)$ **51.** 0.511 **53.** 30.099 **55.** 2.303 **57.** -53.991 **59.** $\sqrt{2}$

61.

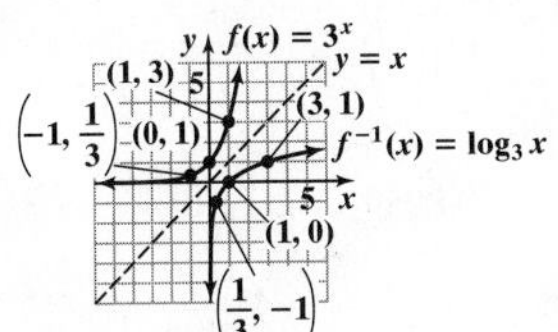

63.

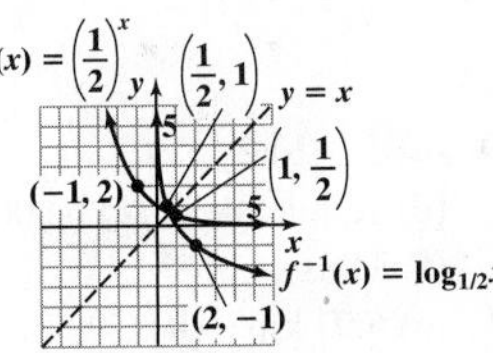

65. B **67.** D **69.** A **71.** E

73. **(a)** Domain: $(-4, \infty)$
(b)
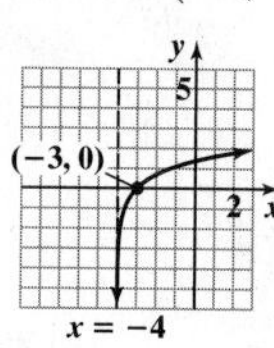

(c) Range: $(-\infty, \infty)$
Vertical asymptote: $x = -4$
(d) $f^{-1}(x) = e^x - 4$
(e) Domain of f^{-1}: $(-\infty, \infty)$
Range of f^{-1}: $(-4, \infty)$
(f)
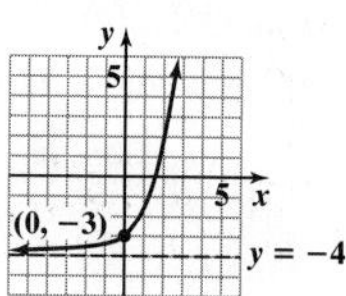

75. **(a)** Domain: $(0, \infty)$
(b)
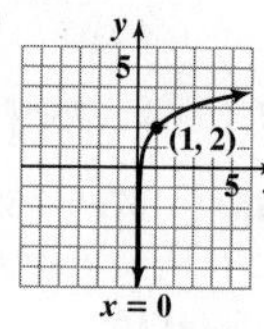

(c) Range: $(-\infty, \infty)$
Vertical asymptote: $x = 0$
(d) $f^{-1}(x) = e^{x-2}$
(e) Domain of f^{-1}: $(-\infty, \infty)$
Range of f^{-1}: $(0, \infty)$
(f)
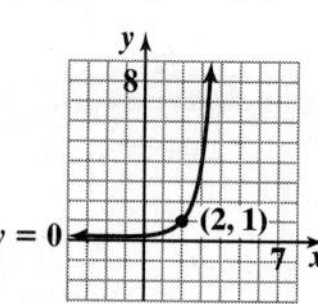

77. **(a)** Domain: $(0, \infty)$
(b)
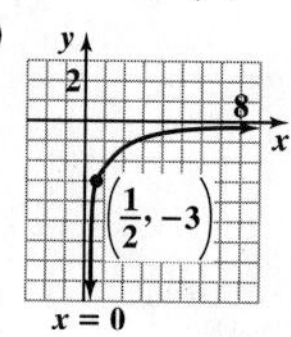

(c) Range: $(-\infty, \infty)$
Vertical asymptote: $x = 0$
(d) $f^{-1}(x) = \dfrac{1}{2}e^{x+3}$
(e) Domain of f^{-1}: $(-\infty, \infty)$
Range of f^{-1}: $(0, \infty)$
(f)
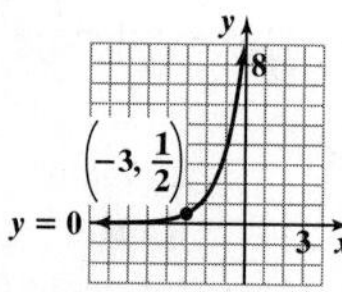

79. **(a)** Domain: $(4, \infty)$
(b)
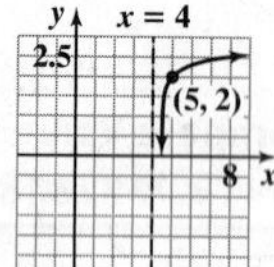

(c) Range: $(-\infty, \infty)$
Vertical asymptote: $x = 4$
(d) $f^{-1}(x) = 10^{x-2} + 4$
(e) Domain of f^{-1}: $(-\infty, \infty)$
Range of f^{-1}: $(4, \infty)$
(f)
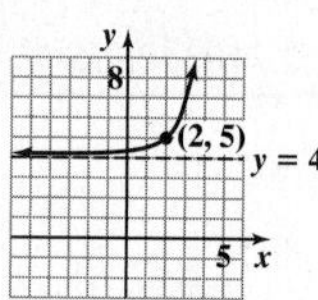

81. **(a)** Domain: $(0, \infty)$
(b)
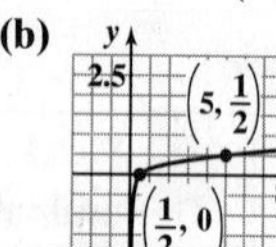

(c) Range: $(-\infty, \infty)$
Vertical asymptote: $x = 0$
(d) $f^{-1}(x) = \dfrac{1}{2} \cdot 10^{2x}$
(e) Domain of f^{-1}: $(-\infty, \infty)$
Range of f^{-1}: $(0, \infty)$
(f)
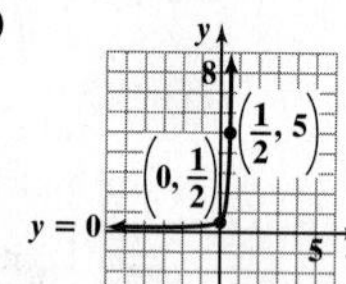

83. **(a)** Domain: $(-2, \infty)$
(b)
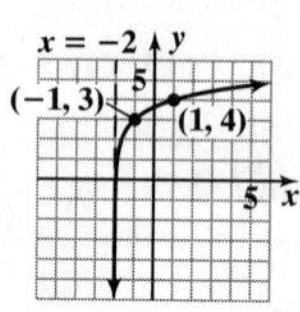

(c) Range: $(-\infty, \infty)$
Vertical asymptote: $x = -2$
(d) $f^{-1}(x) = 3^{x-3} - 2$
(e) Domain of f^{-1}: $(-\infty, \infty)$
Range of f^{-1}: $(-2, \infty)$
(f)
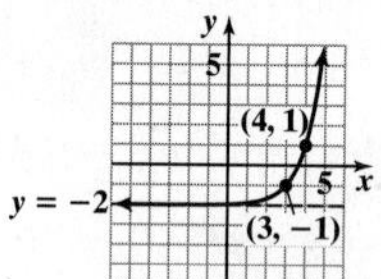

85. **(a)** Domain: $(-\infty, \infty)$
(b)
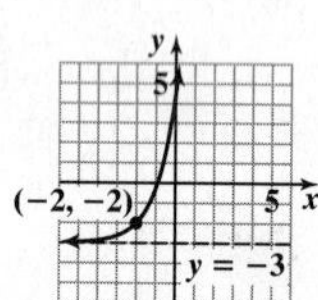

(c) Range: $(-3, \infty)$
Horizontal asymptote: $y = -3$
(d) $f^{-1}(x) = \ln(x + 3) - 2$
(e) Domain of f^{-1}: $(-3, \infty)$
Range of f^{-1}: $(-\infty, \infty)$
(f)
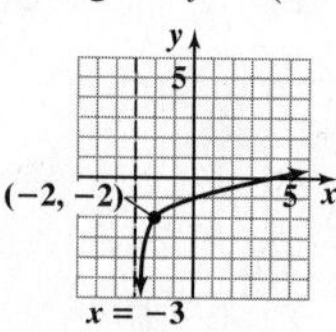

87. **(a)** Domain: $(-\infty, \infty)$
(b)
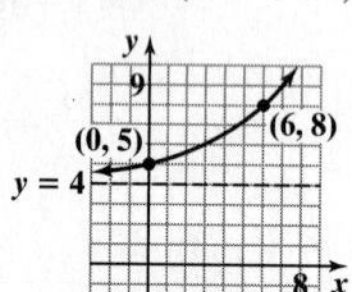

(c) Range: $(4, \infty)$
Horizontal asymptote: $y = 4$
(d) $f^{-1}(x) = 3\log_2(x - 4)$
(e) Domain of f^{-1}: $(4, \infty)$
Range of f^{-1}: $(-\infty, \infty)$
(f)
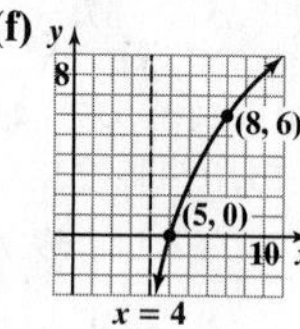

89. $\{9\}$ **91.** $\left\{\dfrac{7}{2}\right\}$ **93.** $\{2\}$ **95.** $\{5\}$ **97.** $\{3\}$
99. $\{2\}$ **101.** $\left\{\dfrac{\ln 10}{3}\right\}$ **103.** $\left\{\dfrac{\ln 8 - 5}{2}\right\}$
105. $\{-2\sqrt{2}, 2\sqrt{2}\}$ **107.** $\{-1\}$
109. $\left\{5 \ln \dfrac{7}{5}\right\}$ **111.** $\left\{2 - \log \dfrac{5}{2}\right\}$
113. **(a)** $\left\{x \,\middle|\, x > -\dfrac{1}{2}\right\}; \left(-\dfrac{1}{2}, \infty\right)$
(b) 2; (40, 2) **(c)** 121; (121, 3) **(d)** 4

115.
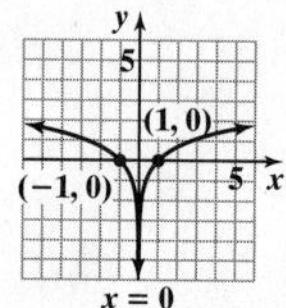

Domain: $\{x \mid x \neq 0\}$
Range: $(-\infty, \infty)$
Intercepts: $(-1, 0)$, $(1, 0)$

117.
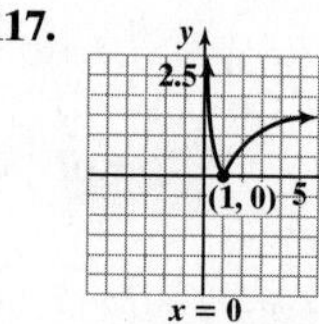

Domain: $\{x \mid x > 0\}$
Range: $\{y \mid y \geq 0\}$
Intercept: $(1, 0)$

119. **(a)** 1 **(b)** 2 **(c)** 3
(d) It increases. **(e)** 0.000316
(f) 3.981×10^{-8}
121. **(a)** 5.97 km **(b)** 0.90 km
123. **(a)** 6.93 min **(b)** 16.09 min
125. $h \approx 2.29$, so the time between injections is about 2 h, 17 min.

127. 0.2695 s
0.8959 s
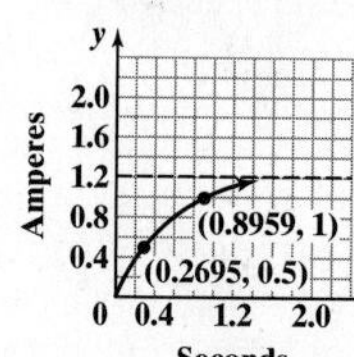

129. 50 decibels (dB) **131.** 90 dB **133.** 8.1 **135.** **(a)** $k \approx 11.216$ **(b)** 6.73 **(c)** 0.41% **(d)** 0.14%

137. Because $y = \log_1 x$ means $1^y = 1 = x$, which cannot be true for $x \neq 1$

139. 12 **140.** -1 **141.** $\left\{\frac{\pi}{3}, \pi, \frac{5\pi}{3}\right\}$ **142.** $\mathbf{v} \cdot \mathbf{w} = 5; \theta = 45°$

7.3 Assess Your Understanding *(page 489)*

1. 0 **2.** M **3.** r **4.** $\log_a M; \log_a N$ **5.** $\log_a M; \log_a N$ **6.** $r \log_a M$ **7.** 7 **8.** F **9.** F **10.** F **11.** b **12.** b **13.** 71 **15.** -4 **17.** 7 **19.** 1 **21.** 1 **23.** 3 **25.** $\frac{5}{4}$ **27.** 4 **29.** $a + b$ **31.** $b - a$ **33.** $3a$ **35.** $\frac{1}{5}(a + b)$ **37.** $2 + \log_5 x$ **39.** $3 \log_2 z$ **41.** $1 + \ln x$ **43.** $\ln x - x$ **45.** $2 \log_a u + 3 \log_a v$ **47.** $2 \ln x + \frac{1}{2}\ln(1 - x)$ **49.** $3 \log_2 x - \log_2(x - 3)$ **51.** $\log x + \log(x + 2) - 2 \log(x + 3)$ **53.** $\frac{1}{3}\ln(x - 2) + \frac{1}{3}\ln(x + 1) - \frac{2}{3}\ln(x + 4)$ **55.** $\ln 5 + \ln x + \frac{1}{2}\ln(1 + 3x) - 3 \ln(x - 4)$ **57.** $\log_5 u^3v^4$ **59.** $\log_3\left(\frac{1}{x^{5/2}}\right)$ **61.** $\log_4\left[\frac{x - 1}{(x + 1)^4}\right]$ **63.** $-2 \ln(x - 1)$ **65.** $\log_2[x(3x - 2)^4]$ **67.** $\log_a\left(\frac{25x^6}{\sqrt{2x + 3}}\right)$ **69.** $\log_2\left[\frac{(x + 1)^2}{(x + 3)(x - 1)}\right]$ **71.** 2.771 **73.** -3.880 **75.** 5.615 **77.** 0.874

79. $y = \frac{\log x}{\log 4}$ **81.** $y = \frac{\log(x + 2)}{\log 2}$ **83.** $y = \frac{\log(x + 1)}{\log(x - 1)}$

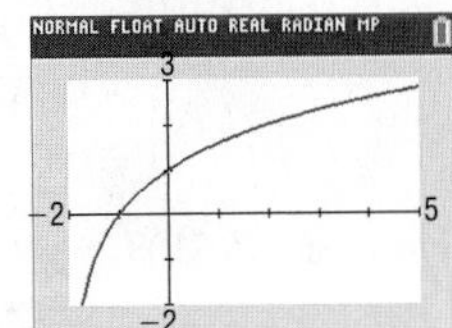

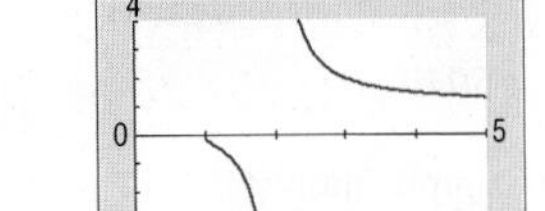

85. **(a)** $(f \circ g)(x) = x$; $\{x \mid x \text{ is any real number}\}$ or $(-\infty, \infty)$
(b) $(g \circ f)(x) = x$; $\{x \mid x > 0\}$ or $(0, \infty)$ **(c)** 5
(d) $(f \circ h)(x) = \ln x^2$; $\{x \mid x \neq 0\}$ or $(-\infty, 0) \cup (0, \infty)$ **(e)** 2

87. $y = Cx$ **89.** $y = Cx(x + 1)$ **91.** $y = Ce^{3x}$ **93.** $y = Ce^{-4x} + 3$

95. $y = \frac{\sqrt[3]{C}(2x + 1)^{1/6}}{(x + 4)^{1/9}}$ **97.** 3 **99.** 1

101. $\log_a(x + \sqrt{x^2 - 1}) + \log_a(x - \sqrt{x^2 - 1}) = \log_a[(x + \sqrt{x^2 - 1})(x - \sqrt{x^2 - 1})] = \log_a[x^2 - (x^2 - 1)] = \log_a 1 = 0$

103. $\ln(1 + e^{2x}) = \ln[e^{2x}(e^{-2x} + 1)] = \ln e^{2x} + \ln(e^{-2x} + 1) = 2x + \ln(1 + e^{-2x})$

105. $y = f(x) = \log_a x$; $a^y = x$ implies $a^y = \left(\frac{1}{a}\right)^{-y} = x$, so $-y = \log_{1/a} x = -f(x)$.

107. $f(x) = \log_a x$; $f\left(\frac{1}{x}\right) = \log_a \frac{1}{x} = \log_a 1 - \log_a x = -f(x)$

109. $\log_a \frac{M}{N} = \log_a(M \cdot N^{-1}) = \log_a M + \log_a N^{-1} = \log_a M - \log_a N$, since $a^{\log_a N^{-1}} = N^{-1}$ implies $a^{-\log_a N^{-1}} = N$; that is, $\log_a N = -\log_a N^{-1}$.

115.

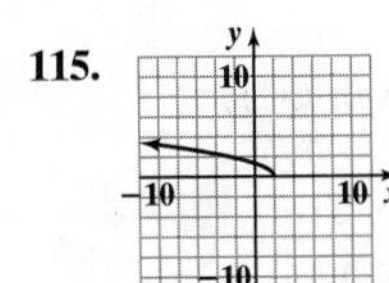

Domain: $\{x \mid x \leq 2\}$ or $(-\infty, 2]$
Range: $\{y \mid y \geq 0\}$ or $[0, \infty)$

116. $\sin\theta = -\frac{5}{13}$; $\cos\theta = \frac{12}{13}$; $\tan\theta = -\frac{5}{12}$; $\csc\theta = -\frac{13}{5}$; $\sec\theta = \frac{13}{12}$; $\cot\theta = -\frac{12}{5}$ **117.** $\frac{5\pi}{6}$ **118.** $r = 2\sin\theta$

7.4 Assess Your Understanding *(page 495)*

5. $\{16\}$ **7.** $\left\{\frac{16}{5}\right\}$ **9.** $\{6\}$ **11.** $\{16\}$ **13.** $\left\{\frac{1}{3}\right\}$ **15.** $\{3\}$ **17.** $\{5\}$ **19.** $\left\{\frac{21}{8}\right\}$ **21.** $\{-6\}$ **23.** $\{-2\}$ **25.** $\{-1 + \sqrt{1 + e^4}\} \approx \{6.456\}$ **27.** $\left\{\frac{-5 + 3\sqrt{5}}{2}\right\} \approx \{0.854\}$ **29.** $\{2\}$ **31.** $\left\{\frac{9}{2}\right\}$ **33.** $\{7\}$ **35.** $\{-2 + 4\sqrt{2}\}$ **37.** $\{-\sqrt{3}, \sqrt{3}\}$ **39.** $\left\{\frac{1}{3}, 729\right\}$ **41.** $\{8\}$

43. $\{\log_2 10\} = \left\{\frac{\ln 10}{\ln 2}\right\} \approx \{3.322\}$ **45.** $\{-\log_8 1.2\} = \left\{-\frac{\ln 1.2}{\ln 8}\right\} \approx \{-0.088\}$ **47.** $\left\{\frac{1}{3}\log_2 \frac{8}{5}\right\} = \left\{\frac{\ln\frac{8}{5}}{3 \ln 2}\right\} \approx \{0.226\}$

49. $\left\{\frac{\ln 3}{2 \ln 3 + \ln 4}\right\} \approx \{0.307\}$ **51.** $\left\{\frac{\ln 7}{\ln 0.6 + \ln 7}\right\} \approx \{1.356\}$ **53.** $\{0\}$ **55.** $\left\{\frac{\ln \pi}{1 + \ln \pi}\right\} \approx \{0.534\}$ **57.** $\left\{\frac{\ln 3}{\ln 2}\right\} \approx \{1.585\}$

59. $\{0\}$ **61.** $\left\{\log_4(-2 + \sqrt{7})\right\} \approx \{-0.315\}$ **63.** $\{\log_5 4\} \approx \{0.861\}$ **65.** No real solution **67.** $\{\log_4 5\} \approx \{1.161\}$ **69.** $\{2.79\}$

71. $\{-0.57\}$ **73.** $\{-0.70\}$ **75.** $\{0.57\}$ **77.** $\{0.39, 1.00\}$ **79.** $\{1.32\}$ **81.** $\{1.31\}$ **83.** $\{1\}$ **85.** $\{16\}$ **87.** $\left\{-1, \frac{2}{3}\right\}$ **89.** $\{0\}$

91. $\{\ln(2 + \sqrt{5})\} \approx \{1.444\}$ **93.** $\left\{e^{\frac{\ln 5 \cdot \ln 3}{\ln 15}}\right\} \approx \{1.921\}$ **95. (a)** $\{5\}$; $(5, 3)$ **(b)** $\{5\}$; $(5, 4)$ **(c)** $\{1\}$; yes, at $(1, 2)$ **(d)** $\{5\}$ **(e)** $\left\{-\frac{1}{11}\right\}$

97. (a), (b)

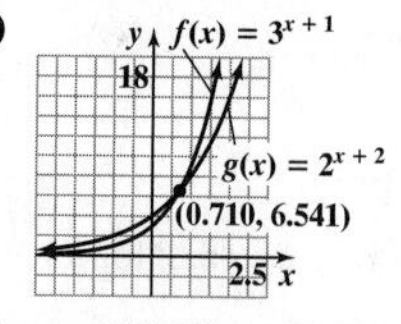

(c) $\{x \mid x > 0.710\}$ or $(0.710, \infty)$

99. (a), (b), (c)

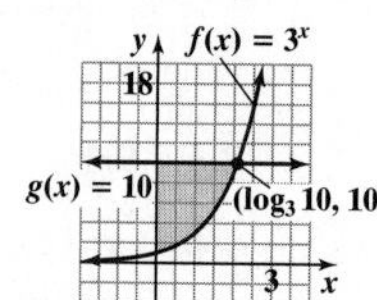

101. (a), (b), (c)

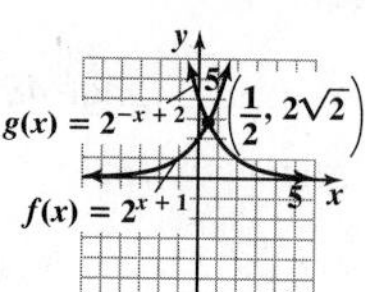

103. (a)

(b) 2 **(c)** $\{x \mid x < 2\}$ or $(-\infty, 2)$

105. (a) 2047 **(b)** 2059
107. (a) After 4.2 yr **(b)** After 6.5 yr **(c)** After 12.8 yr

110. one-to-one
111. 642.6 rpm
112. $4\sqrt{2}(\cos 135° + i \sin 135°)$
113. $2x - 5y = 10$

7.5 Assess Your Understanding *(page 504)*

1. \$15 **2.** $13\frac{1}{3}\%$ **3.** principal **4.** *I*; *Prt*; simple interest **5.** 4 **6.** effective rate of interest **7.** \$108.29 **9.** \$609.50 **11.** \$697.09 **13.** \$1246.08 **15.** \$88.72 **17.** \$860.72 **19.** \$554.09 **21.** \$59.71 **23.** 5.095% **25.** 5.127% **27.** $6\frac{1}{4}\%$ compounded annually **29.** 9% compounded monthly **31.** 25.992% **33.** 24.573% **35. (a)** About 8.69 yr **(b)** About 8.66 yr **37.** 6.823% **39.** 10.15 yr; 10.14 yr **41.** 15.27 yr or 15 yr, 3 mo **43.** \$104,335 **45.** \$12,910.62 **47.** About \$30.17 per share or \$3017 **49.** Not quite. Jim will have \$1057.60. The second bank gives a better deal, since Jim will have \$1060.62 after 1 yr. **51.** Will has \$11,632.73; Henry has \$10,947.89. **53. (a)** \$63,449 **(b)** \$44,267 **55.** About \$1019 billion; about \$232 billion **57.** \$940.90 **59.** 2.53% **61.** 34.31 yr **63. (a)** \$3686.45 **(b)** \$3678.79 **65.** \$6439.28

67. (a) 11.90 yr **(b)** 22.11 yr **(c)**
$$mP = P\left(1 + \frac{r}{n}\right)^{nt}$$
$$m = \left(1 + \frac{r}{n}\right)^{nt}$$
$$\ln m = \ln\left(1 + \frac{r}{n}\right)^{nt} = nt \ln\left(1 + \frac{r}{n}\right)$$
$$t = \frac{\ln m}{n \ln\left(1 + \frac{r}{n}\right)}$$

69. (a) 1.59% **(b)** In 2029 or after 21 yr **71.** 22.7 yr
76. $f^{-1}(x) = \frac{2x}{x - 1}$
77. $\tan\theta = 3\sqrt{7}$; $\csc\theta = -\frac{8\sqrt{7}}{21}$; $\sec\theta = -8$; $\cot\theta = \frac{\sqrt{7}}{21}$
78. $C = 80°$, $a \approx 7.83$, $b \approx 10.55$ **79.** $\{6\}$

7.6 Assess Your Understanding *(page 516)*

1. (a) 500 insects **(b)** $0.02 = 2\%$ per day **(c)** About 611 insects **(d)** After about 23.5 days **(e)** After about 34.7 days **3. (a)** $-0.0244 = -2.44\%$ per year **(b)** About 391.7 g **(c)** After about 9.1 yr **(d)** 28.4 yr **5. (a)** $N(t) = N_0e^{kt}$ **(b)** 5832 **(c)** 3.9 days **7. (a)** $N(t) = N_0e^{kt}$ **(b)** 25,198 **9.** 9.797 g **11.** 9953 yr ago **13. (a)** 5:18 PM **(b)** About 14.3 min **(c)** The temperature of the pan approaches 70°F. **15.** 18.63°C; 25.1°C **17.** 1.7 ppm; 7.17 days, or 172 h **19.** 0.26 M; 6.58 h, or 395 min **21.** 26.6 days

23. (a) In 1984, 91.8% of households did not own a personal computer.
(b)

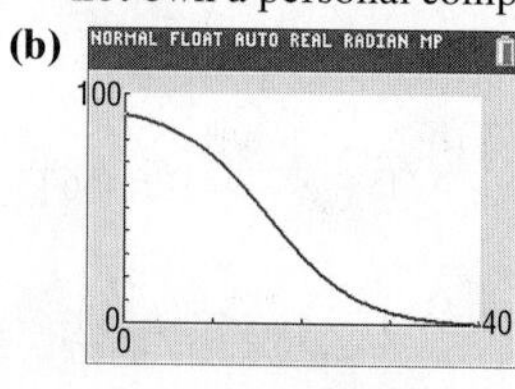

(c) 70.6% **(d)** During 2011

25. (a)

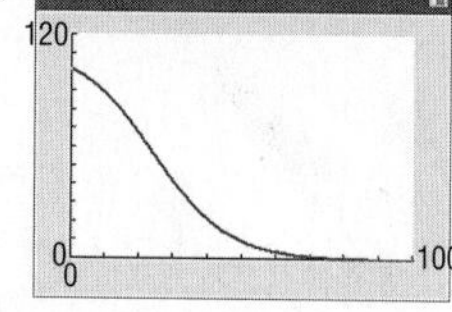

(b) 0.78, or 78%
(c) 50 people
(d) As n increases, the probability decreases.

27. (a) 9.23×10^{-3}, or about 0
(b) 0.81, or about 1
(c) 5.01, or about 5
(d) 57.91°, 43.99°, 30.07°

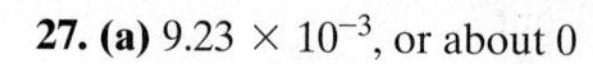

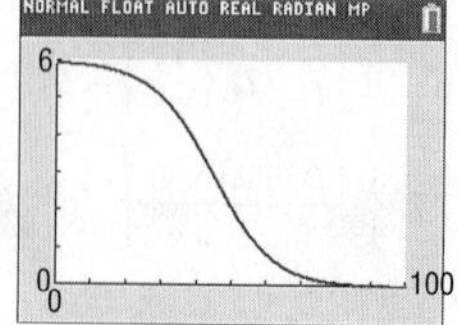

29. (a) $P(t) = 50(3)^{t/20}$ **(b)** 661 **(c)** In 48 days **(d)** $P(t) = 50e^{0.055t}$

30. $\cos\theta = -\frac{\sqrt{7}}{4}$; $\tan\theta = -\frac{3\sqrt{7}}{7}$; $\csc\theta = \frac{4}{3}$; $\sec\theta = -\frac{4\sqrt{7}}{7}$; $\cot\theta = -\frac{\sqrt{7}}{3}$ **31.** $c \approx 7.39$, $A \approx 39.4°$, $B \approx 30.6°$ **32.** 43.82 square units

33. $2\ln x + \frac{1}{2}\ln y - \ln z$

7.7 Assess Your Understanding *(page 523)*

1. (a) 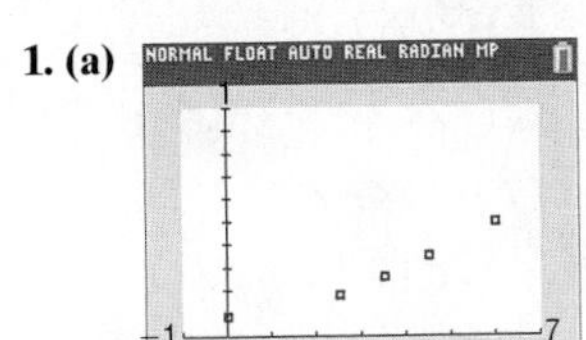

(b) $y = 0.0903(1.3384)^x$
(c) $N(t) = 0.0903e^{0.2915t}$

(d) 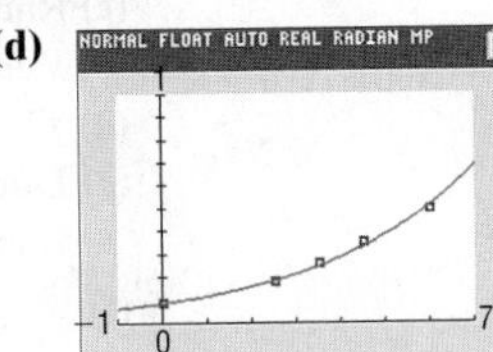

(e) 0.69
(f) After about 7.26 h

3. (a) 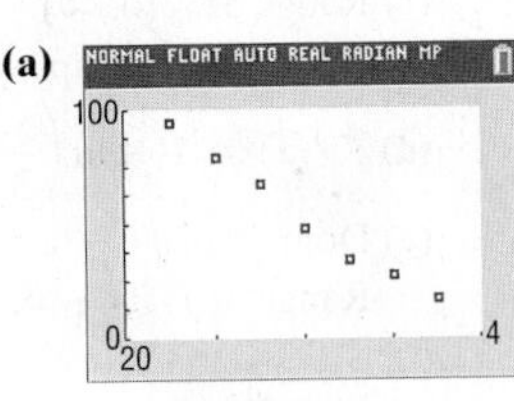

(b) $y = 118.7226(0.7013)^x$
(c) $A(t) = 118.7226e^{-0.3548t}$

(d) 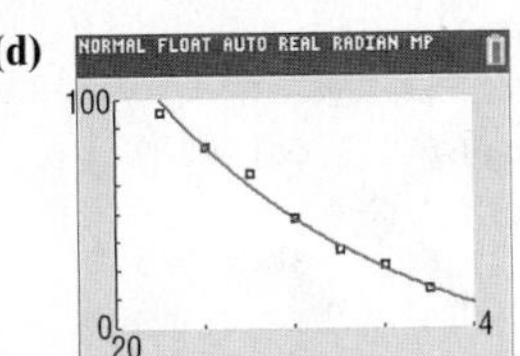

(e) 28.7%
(f) $k = -0.3548 = -35.48\%$ is the exponential growth rate. It represents the rate at which the percentage of patients surviving advanced-stage breast cancer is decreasing.

5. (a) 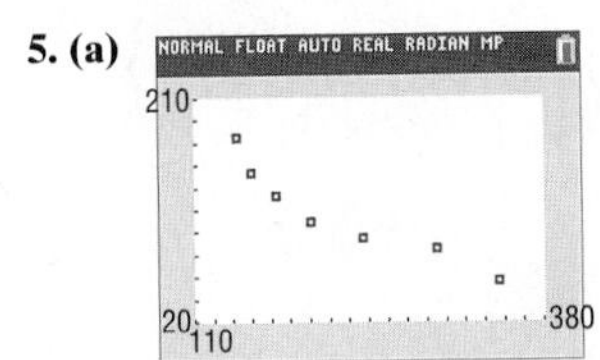

(b) $y = 330.0549 - 34.5008\ln x$

(c) [graph]
(d) 185 billion pounds
(e) Under by 5 billion pounds

7. (a) 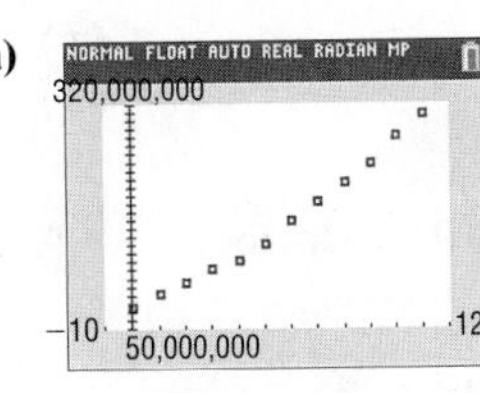

(b) $y = \dfrac{762{,}176{,}844.4}{1 + 8.7428e^{-0.0162x}}$

(c) 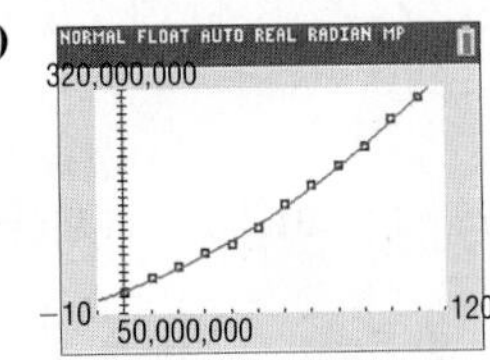

(d) 762,176,844
(e) Approximately 315,203,288 **(f)** 2023

9. (a) 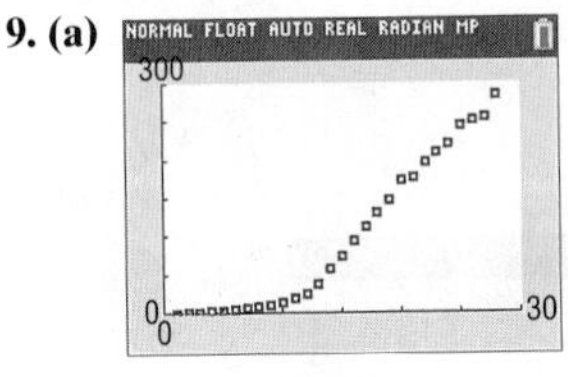

(b) $y = \dfrac{286.2055}{1 + 226.8644e^{-0.2873x}}$

(c) 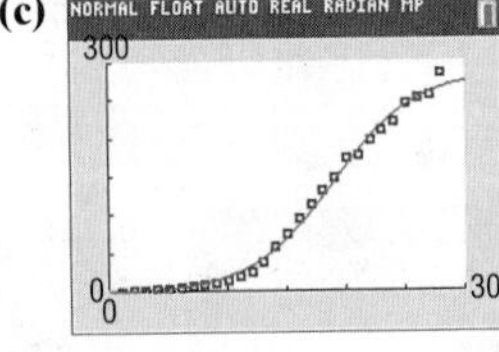

(d) 286.2 thousand cell sites
(e) 281.3 thousand cell sites

11. (a) 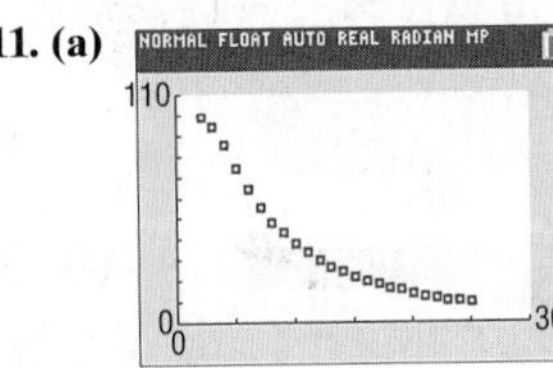

(b) Exponential
(c) $y = 115.5779(0.9012)^x$

(d) 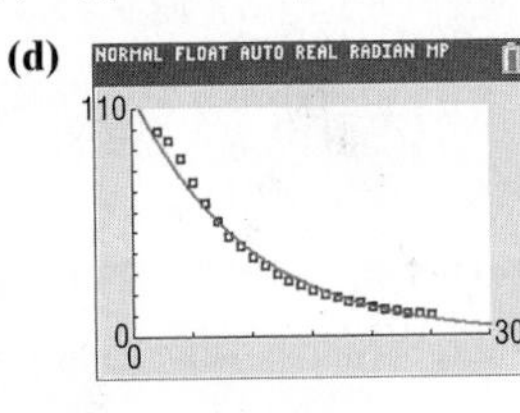

(e) 5.1%

12. 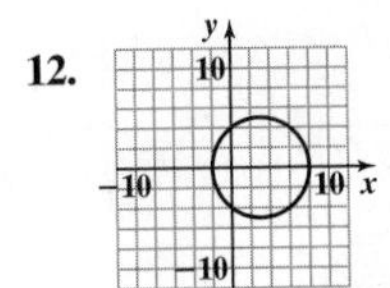

13. 1 **14.** $\left\{\frac{\pi}{12}, \frac{5\pi}{12}, \frac{13\pi}{12}, \frac{17\pi}{12}\right\}$ **15.** {133}

Review Exercises *(page 528)*

1. (a) 81 **(b)** 2 **(c)** $\frac{1}{9}$ **(d)** -3 **2.** $\log_5 z = 2$ **3.** $5^{13} = u$ **4.** $\left\{x \middle| x > \frac{2}{3}\right\}$; $\left(\frac{2}{3}, \infty\right)$ **5.** $\{x \mid x < 1 \text{ or } x > 2\}$; $(-\infty, 1) \cup (2, \infty)$

6. -3 **7.** $\sqrt{2}$ **8.** 0.4 **9.** $\log_3 u + 2\log_3 v - \log_3 w$ **10.** $8\log_2 a + 2\log_2 b$

11. $2\log x + \frac{1}{2}\log(x^3 + 1)$ **12.** $2\ln(2x + 3) - 2\ln(x - 1) - 2\ln(x - 2)$ **13.** $\frac{25}{4}\log_4 x$ **14.** $-2\ln(x + 1)$ **15.** $\ln\left[\frac{16\sqrt{x^2 + 1}}{\sqrt{x(x - 4)}}\right]$ **16.** 2.124

17. 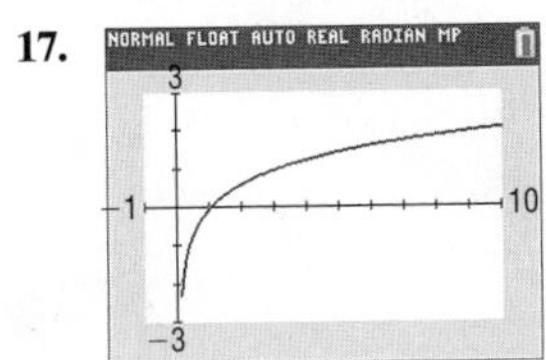

18. (a) Domain of f: $(-\infty, \infty)$
(b) 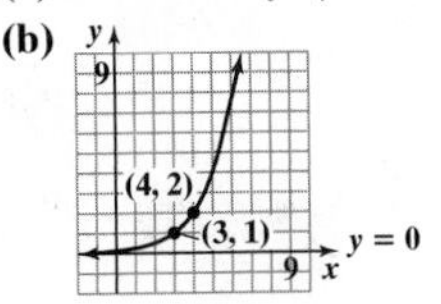

(c) Range of f: $(0, \infty)$
Horizontal asymptote: $y = 0$
(d) $f^{-1}(x) = 3 + \log_2 x$
(e) Domain of f^{-1}: $(0, \infty)$
Range of f^{-1}: $(-\infty, \infty)$
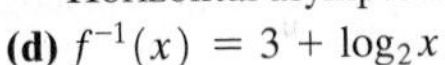

(f) 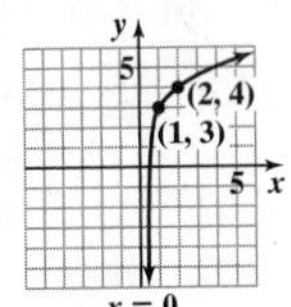

19. (a) Domain of f: $(-\infty, \infty)$

(b)

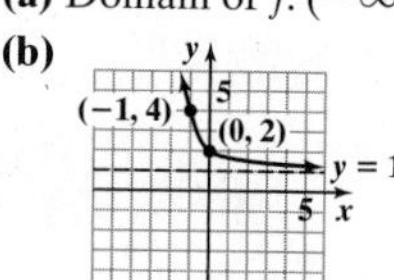

(c) Range of f: $(1, \infty)$
Horizontal asymptote: $y = 1$

(d) $f^{-1}(x) = -\log_3(x - 1)$

(e) Domain of f^{-1}: $(1, \infty)$
Range of f^{-1}: $(-\infty, \infty)$

(f)

y 5 (2, 0) 5 x (4, −1) x = 1

20. (a) Domain of f: $(-\infty, \infty)$

(b)

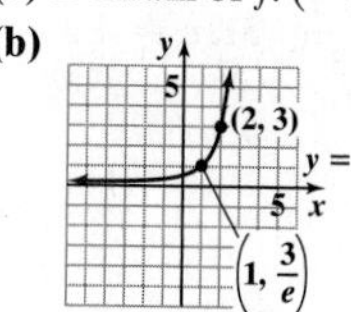

(c) Range of f: $(0, \infty)$
Horizontal asymptote: $y = 0$

(d) $f^{-1}(x) = 2 + \ln\left(\dfrac{x}{3}\right)$

(e) Domain of f^{-1}: $(0, \infty)$
Range of f^{-1}: $(-\infty, \infty)$

(f)

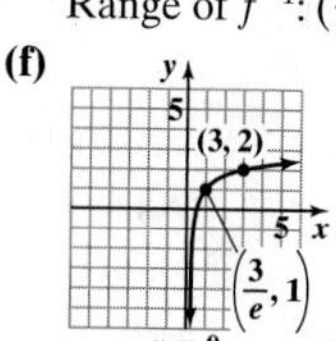

21. (a) Domain of f: $(-3, \infty)$

(b)

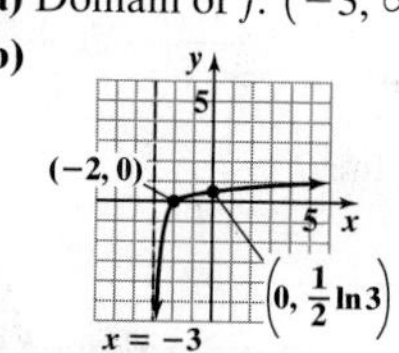

(c) Range of f: $(-\infty, \infty)$
Vertical asymptote: $x = -3$

(d) $f^{-1}(x) = e^{2x} - 3$

(e) Domain of f^{-1}: $(-\infty, \infty)$
Range of f^{-1}: $(-3, \infty)$

(f)

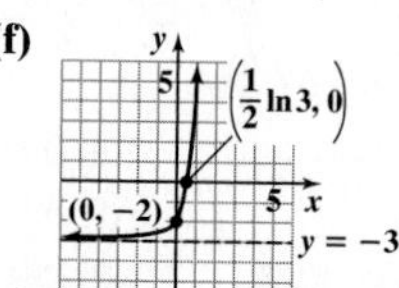

22. $\left\{-\dfrac{16}{9}\right\}$ **23.** $\left\{\dfrac{-1 - \sqrt{3}}{2}, \dfrac{-1 + \sqrt{3}}{2}\right\} \approx \{-1.366, 0.366\}$ **24.** $\left\{\dfrac{1}{4}\right\}$ **25.** $\left\{\dfrac{2 \ln 3}{\ln 5 - \ln 3}\right\} \approx \{4.301\}$ **26.** $\{-2, 6\}$ **27.** $\{83\}$ **28.** $\left\{\dfrac{1}{2}, -3\right\}$

29. $\{1\}$ **30.** $\{-1\}$ **31.** $\{1 - \ln 5\} \approx \{-0.609\}$ **32.** $\left\{\log_3(-2 + \sqrt{7})\right\} = \left\{\dfrac{\ln(-2 + \sqrt{7})}{\ln 3}\right\} \approx \{-0.398\}$

33. (a), (e)

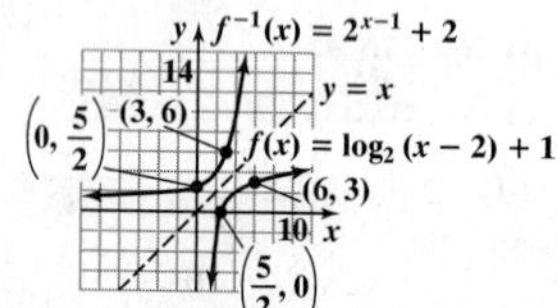

(b) 3; (6, 3)
(c) 10; (10, 4)
(d) $\left\{x \,\middle|\, x > \dfrac{5}{2}\right\}$ or $\left(\dfrac{5}{2}, \infty\right)$
(e) $f^{-1}(x) = 2^{x-1} + 2$

34. (a) 37.3 W **(b)** 6.9 dB **35. (a)** 11.77 **(b)** 9.56 in.
36. (a) 9.85 yr **(b)** 4.27 yr **37.** \$20,398.87; 4.04%; 17.5 yr
38. \$41,668.97 **39.** 24,765 yr ago **40.** 55.22 min, or 55 min, 13 sec
41. 7,967,521,519 **42.** 7.204 g; 0.519 g

43. (a) 0.3 **(b)** 0.8
(c)

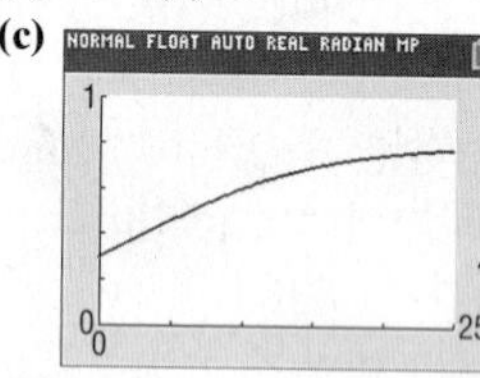

(d) In 2026

44. (a)

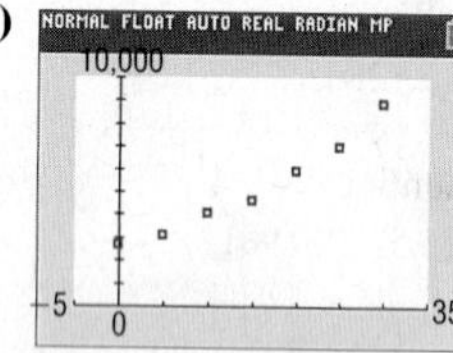

(b) $y = 2638.26(1.0407)^x$
(c) $A(t) = 2638.26e^{0.0399x}$
(d)

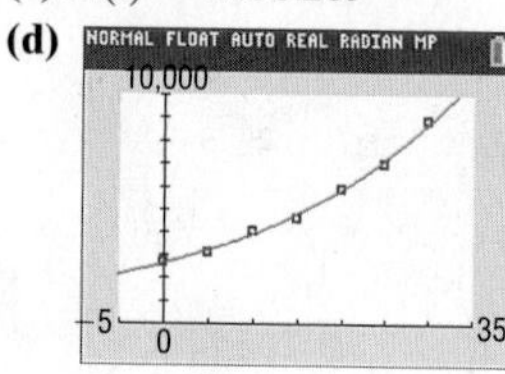

(e) 2021–22

45. (a)

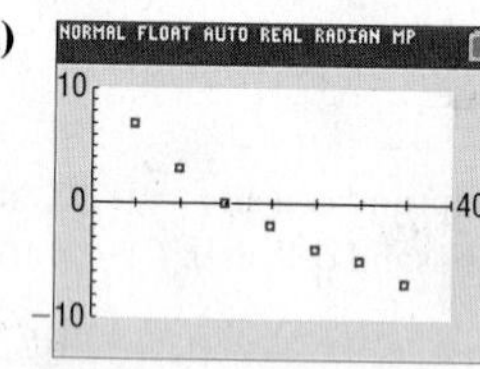

(b) $y = 18.921 - 7.096 \ln x$
(c)

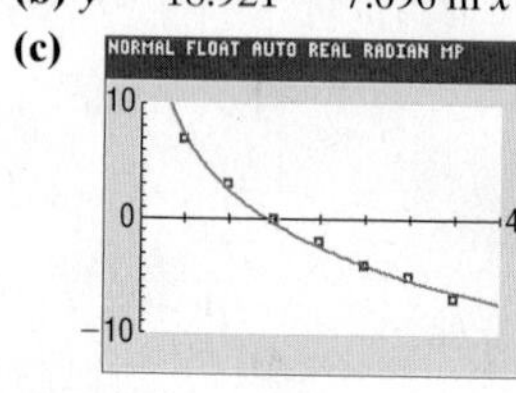

(d) Approximately −3°F

46. (a)

NORMAL FLOAT AUTO REAL RADIAN MP 50 −1 0 9

(b) $C = \dfrac{46.9292}{1 + 21.2733e^{-0.7306t}}$
(c)

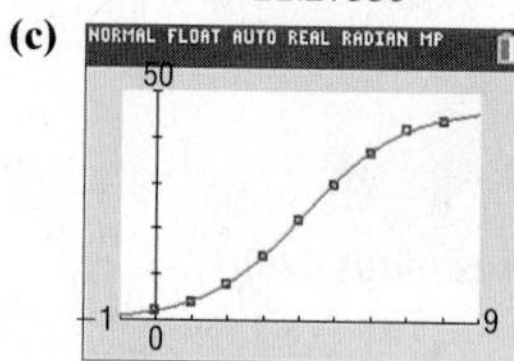

(d) About 47 people; 50 people
(e) 2.4 days; during the tenth hour of day 3
(f) 9.5 days

Chapter Test *(page 530)*

1. $x = 5$ **2.** $b = 4$ **3.** $x = 625$ **4.** $e^3 + 2 \approx 22.086$ **5.** $\log 20 \approx 1.301$

6. $\log_3 21 = \dfrac{\ln 21}{\ln 3} \approx 2.771$ **7.** $\ln 133 \approx 4.890$

8. (a) Domain of f: $\{x|-\infty < x < \infty\}$ or $(-\infty, \infty)$

(b)

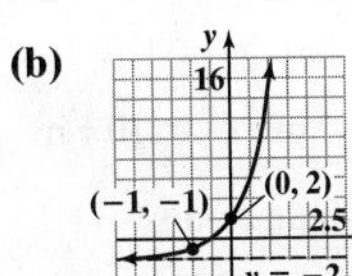

(c) Range of f: $\{y|y > -2\}$ or $(-2, \infty)$; Horizontal asymptote: $y = -2$
(d) $f^{-1}(x) = \log_4(x + 2) - 1$
(e) Domain of f^{-1}: $\{x|x > -2\}$ or $(-2, \infty)$
Range of f^{-1}: $\{y|-\infty < y < \infty\}$ or $(-\infty, \infty)$
(f)

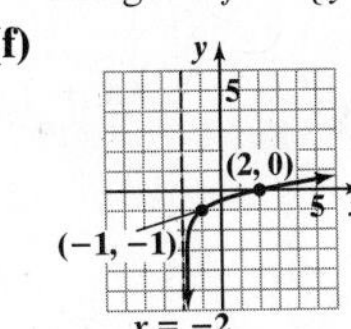

9. (a) Domain of f: $\{x|x > 2\}$ or $(2, \infty)$
(b)

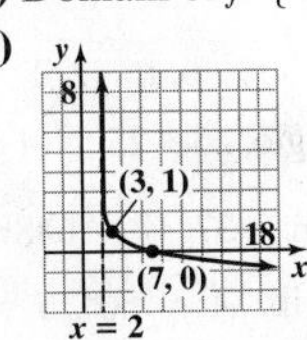

(c) Range of f: $\{y|-\infty < y < \infty\}$ or $(-\infty, \infty)$; vertical asymptote: $x = 2$
(d) $f^{-1}(x) = 5^{1-x}+2$
(e) Domain of f^{-1}: $\{x|-\infty < x < \infty\}$ or $(-\infty, \infty)$
Range of f^{-1}: $\{y|y > 2\}$ or $(2, \infty)$
(f)

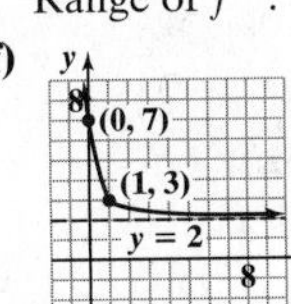

10. $\{1\}$ **11.** $\{91\}$ **12.** $\{-\ln 2\} \approx \{-0.693\}$ **13.** $\left\{\frac{1-\sqrt{13}}{2}, \frac{1+\sqrt{13}}{2}\right\} \approx \{-1.303, 2.303\}$ **14.** $\left\{\frac{3\ln 7}{1-\ln 7}\right\} \approx \{-6.172\}$
15. $\{2\sqrt{6}\} \approx \{4.899\}$ **16.** $2 + 3\log_2 x - \log_2(x-6) - \log_2(x+3)$ **17.** About 250.39 days **18. (a)** \$1033.82 **(b)** \$963.42 **(c)** 11.9 yr
19. (a) About 83 dB **(b)** The pain threshold will be exceeded if 31,623 people shout at the same time.

Cumulative Review *(page 531)*

1. Yes; no **2. (a)** 10 **(b)** $2x^2 + 3x + 1$ **(c)** $2x^2 + 4xh + 2h^2 - 3x - 3h + 1$ **3.** $\left(\frac{1}{2}, \frac{\sqrt{3}}{2}\right)$ is on the graph. **4.** $\{-26\}$

5.

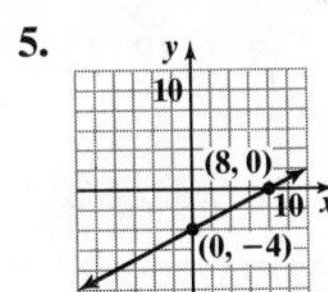

6.

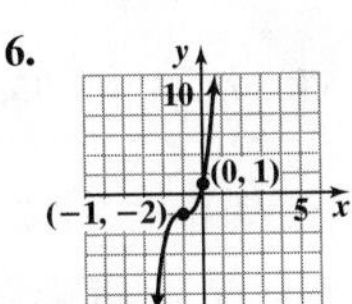

7. (a), (c)

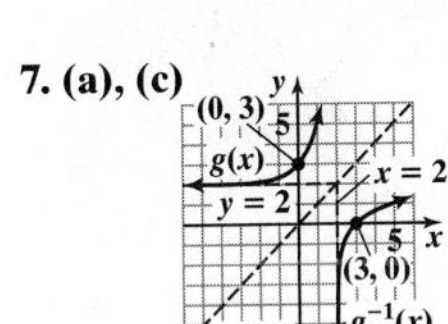

Domain g = range $g^{-1} = (-\infty, \infty)$
Range g = domain $g^{-1} = (2, \infty)$
(b) $g^{-1}(x) = \log_3(x - 2)$

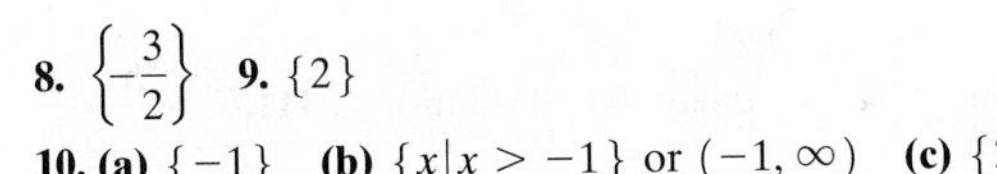

8. $\left\{-\frac{3}{2}\right\}$ **9.** $\{2\}$
10. (a) $\{-1\}$ **(b)** $\{x|x > -1\}$ or $(-1, \infty)$ **(c)** $\{25\}$
11. (a)

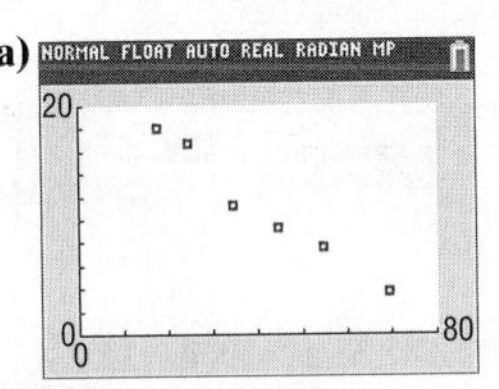

(b) Logarithmic; $y = 49.293 - 10.563 \ln x$
(c) Highest value of $|r|$

APPENDIX A Review

A.1 Assess Your Understanding *(page A10)*

1. variable **2.** origin **3.** strict **4.** base; exponent or power **5.** d **6.** b **7.** T **8.** T **9.** F **10.** F **11.** {1, 2, 3, 4, 5, 6, 7, 8, 9}
13. {4} **15.** {1, 3, 4, 6} **17.** {0, 2, 6, 7, 8} **19.** {0, 1, 2, 3, 5, 6, 7, 8, 9} **21.** {0, 1, 2, 3, 5, 6, 7, 8, 9}
23.

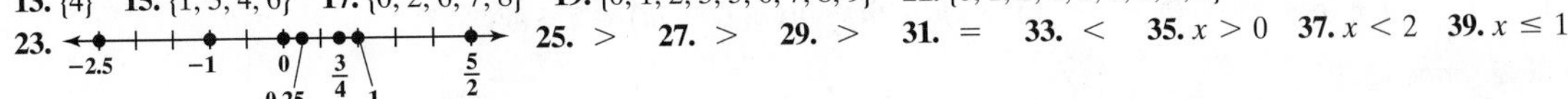

25. > **27.** > **29.** > **31.** = **33.** < **35.** $x > 0$ **37.** $x < 2$ **39.** $x \le 1$

41. (number line, −2) **43.** (number line, −1)

45. 1 **47.** 2 **49.** 6 **51.** 4 **53.** −28 **55.** $\frac{4}{5}$ **57.** 0 **59.** 1 **61.** 5 **63.** 1
65. 22 **67.** 2 **69.** $x = 0$ **71.** $x = 3$ **73.** None **75.** $x = 0, x = 1, x = -1$ **77.** $\{x|x \ne 5\}$ **79.** $\{x|x \ne -4\}$ **81.** 0°C **83.** 25°C **85.** 16
87. $\frac{1}{16}$ **89.** $\frac{1}{9}$ **91.** 9 **93.** 5 **95.** 4 **97.** $64x^6$ **99.** $\frac{x^4}{y^2}$ **101.** $\frac{x}{y}$ **103.** $-\frac{8x^3z}{9y}$ **105.** $\frac{16x^2}{9y^2}$ **107.** −4 **109.** 5 **111.** 4 **113.** 2 **115.** $\sqrt{5}$ **117.** $\frac{1}{2}$ **119.** 10; 0
121. 81 **123.** 304,006.671 **125.** 0.004 **127.** 481.890 **129.** 0.000 **131.** $A = lw$ **133.** $C = \pi d$ **135.** $A = \frac{\sqrt{3}}{4}x^2$ **137.** $V = \frac{4}{3}\pi r^3$ **139.** $V = x^3$
141. (a) \$6000 **(b)** \$8000 **143.** $|x - 4| \ge 6$ **145. (a)** $2 \le 5$ **(b)** $6 > 5$ **147. (a)** Yes **(b)** No **149.** No; $\frac{1}{3}$ is larger; 0.000333... **151.** No

A.2 Assess Your Understanding *(page A19)*

1. right; hypotenuse **2.** $A = \frac{1}{2}bh$ **3.** $C = 2\pi r$ **4.** similar **5.** c **6.** b **7.** T **8.** T **9.** F **10.** T **11.** T **12.** F **13.** 13 **15.** 26 **17.** 25 **19.** Right triangle; 5 **21.** Not a right triangle **23.** Right triangle; 25 **25.** Not a right triangle **27.** 8 in.^{2} **29.** 4 in.^{2} **31.** $A = 25\pi$ m^2; $C = 10\pi$ m **33.** $V = 224$ ft^3; $S = 232$ ft^2 **35.** $V = \frac{256}{3}\pi$ cm^3; $S = 64\pi$ cm^2 **37.** $V = 648\pi$ in.^{3}; $S = 306\pi$ in.^{2} **39.** π square units **41.** 2π square units **43.** $x = 4$ units; $A = 90°$; $B = 60°$; $C = 30°$ **45.** $x = 67.5$ units; $A = 60°$; $B = 95°$; $C = 25°$ **47.** About 16.8 ft **49.** 64 ft^2 **51.** $24 + 2\pi \approx 30.28$ ft^2; $16 + 2\pi \approx 22.28$ ft **53.** 160 paces **55.** About 5.477 mi **57.** From 100 ft: 12.2 mi; From 150 ft: 15.0 mi

A.3 Assess Your Understanding *(page A25)*

1. $x^4 - 16$ **2.** $x^3 - 8$ **3.** T **4.** F **5.** add; $\frac{25}{4}$ **6.** b **7.** d **8.** c **9.** $x^2 + 6x + 8$ **11.** $2x^2 + 9x + 10$ **13.** $x^2 - 49$ **15.** $4x^2 - 9$ **17.** $x^2 + 8x + 16$ **19.** $4x^2 - 12x + 9$ **21.** $x^3 - 6x^2 + 12x - 8$ **23.** $8x^3 + 12x^2 + 6x + 1$ **25.** $(x + 6)(x - 6)$ **27.** $2(1 + 2x)(1 - 2x)$ **29.** $(x + 1)(x + 10)$ **31.** $(x - 7)(x - 3)$ **33.** $4(x^2 - 2x + 8)$ **35.** Prime **37.** $-(x - 5)(x + 3)$ **39.** $3(x + 2)(x - 6)$ **41.** $y^2(y + 5)(y + 6)$ **43.** $(2x + 3)^2$ **45.** $2(3x + 1)(x + 1)$ **47.** $(x - 3)(x + 3)(x^2 + 9)$ **49.** $(x - 1)^2(x^2 + x + 1)^2$ **51.** $x^5(x - 1)(x + 1)$ **53.** $(4x + 3)^2$ **55.** $-(4x - 5)(4x + 1)$ **57.** $(2y - 5)(2y - 3)$ **59.** $-(3x - 1)(3x + 1)(x^2 + 1)$ **61.** $(x + 3)(x - 6)$ **63.** $(x + 2)(x - 3)$ **65.** $(3x - 5)(9x^2 - 3x + 7)$ **67.** $(x + 5)(3x + 11)$ **69.** $(x - 1)(x + 1)(x + 2)$ **71.** $(x - 1)(x + 1)(x^2 - x + 1)$ **73.** 25; $(x + 5)^2$ **75.** 9; $(y - 3)^2$ **77.** $\frac{1}{16}$; $\left(x - \frac{1}{4}\right)^2$ **79.** $2(3x + 4)(9x + 13)$ **81.** $2x(3x + 5)$ **83.** $5(x + 3)(x - 2)^2(x + 1)$ **85.** $3(4x - 3)(4x - 1)$ **87.** $6(3x - 5)(2x + 1)^2(5x - 4)$ **89.** The possibilities are $(x \pm 1)(x \pm 4) = x^2 \pm 5x + 4$ or $(x \pm 2)(x \pm 2) = x^2 \pm 4x + 4$, none of which equals $x^2 + 4$.

A.4 Assess Your Understanding *(page A34)*

5. identity **6.** F **7.** T **8.** add; $\frac{25}{4}$ **9.** discriminant; negative **10.** F **11.** F **12.** b **13.** b **14.** d **15.** $\{7\}$ **17.** $\{-3\}$ **19.** $\{4\}$ **21.** $\left\{\frac{5}{4}\right\}$ **23.** $\{-1\}$ **25.** $\{-18\}$ **27.** $\{-3\}$ **29.** $\{-16\}$ **31.** $\{0.5\}$ **33.** $\{2\}$ **35.** $\{2\}$ **37.** $\{3\}$ **39.** $\{0, 9\}$ **41.** $\{0, 9\}$ **43.** $\{21\}$ **45.** $\{-2, 2\}$ **47.** $\{6\}$ **49.** $\{-3, 3\}$ **51.** $\{-4, 1\}$ **53.** $\left\{-1, \frac{3}{2}\right\}$ **55.** $\{-4, 4\}$ **57.** $\{2\}$ **59.** No real solution **61.** $\{-2, 2\}$ **63.** $\{-1, 3\}$ **65.** $\{-2, -1, 0, 1\}$ **67.** $\{0, 4\}$ **69.** $\{-6, 2\}$ **71.** $\left\{-\frac{1}{2}, 3\right\}$ **73.** $\{3, 4\}$ **75.** $\left\{\frac{3}{2}\right\}$ **77.** $\left\{-\frac{2}{3}, \frac{3}{2}\right\}$ **79.** $\left\{-\frac{3}{4}, 2\right\}$ **81.** $\{-5, 5\}$ **83.** $\{-1, 3\}$ **85.** $\{-3, 0\}$ **87.** $\{-7, 3\}$ **89.** $\left\{-\frac{1}{4}, \frac{3}{4}\right\}$ **91.** $\left\{\frac{-1 - \sqrt{7}}{6}, \frac{-1 + \sqrt{7}}{6}\right\}$ **93.** $\{2 - \sqrt{2}, 2 + \sqrt{2}\}$ **95.** $\left\{\frac{5 - \sqrt{29}}{2}, \frac{5 + \sqrt{29}}{2}\right\}$ **97.** $\left\{1, \frac{3}{2}\right\}$ **99.** No real solution **101.** $\left\{\frac{-1 - \sqrt{5}}{4}, \frac{-1 + \sqrt{5}}{4}\right\}$ **103.** $\left\{\frac{-\sqrt{3} - \sqrt{15}}{2}, \frac{-\sqrt{3} + \sqrt{15}}{2}\right\}$ **105.** No real solution **107.** Repeated real solution **109.** Two unequal real solutions **111.** $x = \frac{b + c}{a}$ **113.** $x = \frac{abc}{a + b}$ **115.** $x = a^2$ **117.** $R = \frac{R_1R_2}{R_1 + R_2}$ **119.** $R = \frac{mv^2}{F}$ **121.** $r = \frac{S - a}{S}$ **123.** $\frac{-b + \sqrt{b^2 - 4ac}}{2a} + \frac{-b - \sqrt{b^2 - 4ac}}{2a} = \frac{-2b}{2a} = \frac{-b}{a}$ **125.** $k = -\frac{1}{2}$ or $\frac{1}{2}$ **127.** The solutions of $ax^2 - bx + c = 0$ are $\frac{b + \sqrt{b^2 - 4ac}}{2a}$ and $\frac{b - \sqrt{b^2 - 4ac}}{2a}$. **129.** b

A.5 Assess Your Understanding *(page A44)*

1. T **2.** 5 **3.** F **4.** real; imaginary; imaginary unit **5.** F **6.** T **7.** F **8.** b **9.** a **10.** c **11.** $8 + 5i$ **13.** $-7 + 6i$ **15.** $-6 - 11i$ **17.** $6 - 18i$ **19.** $6 + 4i$ **21.** $10 - 5i$ **23.** 37 **25.** $\frac{6}{5} + \frac{8}{5}i$ **27.** $1 - 2i$ **29.** $\frac{5}{2} - \frac{7}{2}i$ **31.** $-\frac{1}{2} + \frac{\sqrt{3}}{2}i$ **33.** $2i$ **35.** $-i$ **37.** i **39.** -6 **41.** $-10i$ **43.** $-2 + 2i$ **45.** 0 **47.** 0 **49.** $2i$ **51.** $5i$ **53.** $5i$ **55.** $\{-2i, 2i\}$ **57.** $\{-4, 4\}$ **59.** $\{3 - 2i, 3 + 2i\}$ **61.** $\{3 - i, 3 + i\}$ **63.** $\left\{\frac{1}{4} - \frac{1}{4}i, \frac{1}{4} + \frac{1}{4}i\right\}$ **65.** $\left\{\frac{1}{5} - \frac{2}{5}i, \frac{1}{5} + \frac{2}{5}i\right\}$ **67.** $\left\{-\frac{1}{2} - \frac{\sqrt{3}}{2}i, -\frac{1}{2} + \frac{\sqrt{3}}{2}i\right\}$ **69.** $\{2, -1 - \sqrt{3}i, -1 + \sqrt{3}i\}$ **71.** $\{-2, 2, -2i, 2i\}$ **73.** $\{-3i, -2i, 2i, 3i\}$ **75.** Two complex solutions that are conjugates of each other **77.** Two unequal real solutions **79.** A repeated real solution **81.** $2 - 3i$ **83.** 6 **85.** 25 **87.** $2 + 3i$ ohms **89.** $z + \bar{z} = (a + bi) + (a - bi) = 2a$; $z - \bar{z} = (a + bi) - (a - bi) = 2bi$ **91.** $\overline{z + w} = \overline{(a + bi) + (c + di)} = \overline{(a + c) + (b + d)i} = (a + c) - (b + d)i = (a - bi) + (c - di) = \bar{z} + \bar{w}$

A.6 Assess Your Understanding *(page A52)*

5. negative **6.** closed interval **7.** $-5, 5$ **8.** $-5 < x < 5$ **9.** T **10.** T **11.** b **12.** c **13.** d **14.** a **15.** $[0, 2]$; $0 \le x \le 2$ **17.** $[2, \infty)$; $x \ge 2$ **19.** $[0, 3)$; $0 \le x < 3$ **21. (a)** $6 < 8$ **(b)** $-2 < 0$ **(c)** $9 < 15$ **(d)** $-6 > -10$ **23. (a)** $7 > 0$ **(b)** $-1 > -8$ **(c)** $12 > -9$ **(d)** $-8 < 6$ **25. (a)** $2x + 4 < 5$ **(b)** $2x - 4 < -3$ **(c)** $6x + 3 < 6$ **(d)** $-4x - 2 > -4$

27. $[0, 4]$

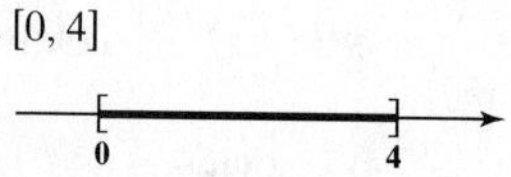

0 4

29. $[4, 6)$

4 6

31. $[4, \infty)$

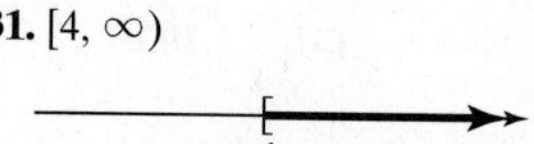

4

33. $(-\infty, -4)$

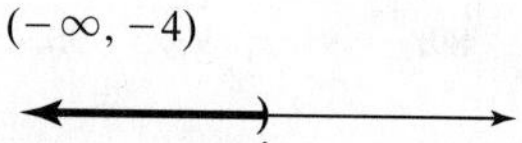

−4

35. $2 \le x \le 5$ **37.** $-3 < x < -2$ **39.** $x \ge 4$ **41.** $x < -3$

43. $<$ **45.** $>$ **47.** $\ge$ **49.** $<$ **51.** $\le$ **53.** $>$ **55.** $\ge$

57. $\{x|x < 4\}$ or $(-\infty, 4)$ **59.** $\{x|x \ge -1\}$ or $[-1, \infty)$ **61.** $\{x|x > 3\}$ or $(3, \infty)$ **63.** $\{x|x \ge 2\}$ or $[2, \infty)$

65. $\{x|x > -7\}$ or $(-7, \infty)$ **67.** $\left\{x\middle|x \le \frac{2}{3}\right\}$ or $\left(-\infty, \frac{2}{3}\right]$ **69.** $\{x|x < -20\}$ or $(-\infty, -20)$ **71.** $\left\{x\middle|x \ge \frac{4}{3}\right\}$ or $\left[\frac{4}{3}, \infty\right)$

73. $\{x|3 \le x \le 5\}$ or $[3, 5]$ **75.** $\left\{x\middle|\frac{2}{3} \le x \le 3\right\}$ or $\left[\frac{2}{3}, 3\right]$ **77.** $\left\{x\middle|-\frac{11}{2} < x < \frac{1}{2}\right\}$ or $\left(-\frac{11}{2}, \frac{1}{2}\right)$ **79.** $\{x|-6 < x < 0\}$ or $(-6, 0)$

81. $\{x|x < -5\}$ or $(-\infty, -5)$ **83.** $\{x|x \ge -1\}$ or $[-1, \infty)$ **85.** $\left\{x\middle|\frac{1}{2} \le x < \frac{5}{4}\right\}$ or $\left[\frac{1}{2}, \frac{5}{4}\right)$ **87.** $\left\{x\middle|x < -\frac{1}{2}\right\}$ or $\left(-\infty, -\frac{1}{2}\right)$

89. $\left\{x\middle|x > \frac{10}{3}\right\}$ or $\left(\frac{10}{3}, \infty\right)$ **91.** $\{x|x > 3\}$ or $(3, \infty)$ **93.** $\{x|-4 < x < 4\}; (-4, 4)$ **95.** $\{x|x < -4 \text{ or } x > 4\}; (-\infty, -4) \cup (4, \infty)$

97. $\{x|0 \le x \le 1\}; [0, 1]$ **99.** $\{x|x < -1 \text{ or } x > 2\}; (-\infty\ -1) \cup (2, \infty)$ **101.** $\{x|-1 \le x \le 1\}; [-1, 1]$ **103.** $\{x|x \le -2 \text{ or } x \ge 2\}; (-\infty, -2] \cup [2, \infty)$

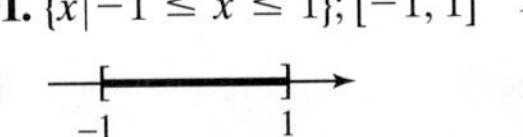

105. $|x - 2| < \frac{1}{2}; \left\{x\middle|\frac{3}{2} < x < \frac{5}{2}\right\}$ **107.** $|x + 3| > 2; \{x|x < -5 \text{ or } x > -1\}$ **109.** $\{x|x \ge -2\}$ **111.** $21 < \text{Age} < 30$ **113. (a)** Male $\ge$ 81.9 years **(b)** Female $\ge$ 85.6 years **(c)** A female can expect to live 3.7 years longer. **115.** The agent's commission ranges from \$45,000 to \$95,000, inclusive. As a percent of selling price, the commission ranges from 5% to 8.6%, inclusive. **117.** The amount withheld varies from \$134.50 to \$184.50, inclusive. **119.** The usage varies from 700 kW · hr to 2700 kW · hr, inclusive. **121.** The dealer's cost varies from \$15,254.24 to \$16,071.43, inclusive.

123. (a) You need at least a 74 on the fifth test. **(b)** You need at least a 77 on the fifth test. **125.** $\frac{a+b}{2} - a = \frac{a+b-2a}{2} = \frac{b-a}{2} > 0$; therefore, $a < \frac{a+b}{2}$. $b - \frac{a+b}{2} = \frac{2b-a-b}{2} = \frac{b-a}{2} > 0$; therefore, $b > \frac{a+b}{2}$. **127.** $(\sqrt{ab})^2 - a^2 = ab - a^2 = a(b-a) > 0$; thus $(\sqrt{ab})^2 > a^2$ and $\sqrt{ab} > a$. $b^2 - (\sqrt{ab})^2 = b^2 - ab = b(b-a) > 0$; thus $b^2 > (\sqrt{ab})^2$ and $b > \sqrt{ab}$.

129. $h - a = \frac{2ab}{a+b} - a = \frac{ab - a^2}{a+b} = \frac{a(b-a)}{a+b} > 0$; thus $h > a$. $b - h = b - \frac{2ab}{a+b} = \frac{b^2 - ab}{a+b} = \frac{b(b-a)}{a+b} > 0$; thus $h < b$.

131. Since $0 < a < b$, then $a - b < 0$ and $\frac{a-b}{ab} < 0$. So $\frac{a}{ab} - \frac{b}{ab} < 0$, or $\frac{1}{b} - \frac{1}{a} < 0$. Therefore, $\frac{1}{b} < \frac{1}{a}$. And $0 < \frac{1}{b}$ because $b > 0$.

A.7 Assess Your Understanding *(page A61)*

3. index **4.** cube root **5.** b **6.** d **7.** c **8.** c **9.** T **10.** F **11.** 3 **13.** -2 **15.** $2\sqrt{2}$ **17.** $-2x\sqrt[3]{x}$ **19.** x^3y^2 **21.** x^2y **23.** $6\sqrt{x}$ **25.** $3x^2y^3\sqrt[4]{2x}$ **27.** $6x\sqrt{x}$ **29.** $15\sqrt[3]{3}$ **31.** $12\sqrt{3}$ **33.** $7\sqrt{2}$ **35.** $\sqrt{2}$ **37.** $2\sqrt{3}$ **39.** $-\sqrt[3]{2}$ **41.** $x - 2\sqrt{x} + 1$ **43.** $(2x-1)\sqrt[3]{2x}$ **45.** $(2x-15)\sqrt{2x}$ **47.** $-(x+5y)\sqrt[3]{2xy}$ **49.** $\frac{\sqrt{2}}{2}$ **51.** $-\frac{\sqrt{15}}{5}$ **53.** $\frac{(5+\sqrt{2})\sqrt{3}}{23}$ **55.** $\frac{8\sqrt{5}-19}{41}$ **57.** $5\sqrt{2}+5$ **59.** $\frac{5\sqrt[3]{4}}{2}$ **61.** $\frac{2x+h-2\sqrt{x^2+xh}}{h}$ **63.** $\left\{\frac{9}{2}\right\}$ **65.** $\{3\}$ **67.** 4 **69.** -3 **71.** 64 **73.** $\frac{1}{27}$ **75.** $\frac{27\sqrt{2}}{32}$ **77.** $\frac{27\sqrt{2}}{32}$ **79** $-\frac{1}{10}$ **81.** $\frac{25}{16}$ **83.** $x^{7/12}$ **85.** xy^2 **87.** $x^{2/3}y$ **89.** $\frac{8x^{5/4}}{y^{3/4}}$ **91.** $\frac{3x+2}{(1+x)^{1/2}}$ **93.** $\frac{x(3x^2+2)}{(x^2+1)^{1/2}}$ **95.** $\frac{22x+5}{10\sqrt{x-5}\sqrt{4x+3}}$ **97.** $\frac{2+x}{2(1+x)^{3/2}}$ **99.** $\frac{4-x}{(x+4)^{3/2}}$ **101.** $\frac{1}{x^2(x^2-1)^{1/2}}$ **103.** $\frac{1-3x^2}{2\sqrt{x}(1+x^2)^2}$ **105.** $\frac{1}{2}(5x+2)(x+1)^{1/2}$ **107.** $2x^{1/2}(3x-4)(x+1)$ **109.** $(x^2+4)^{1/3}(11x^2+12)$ **111.** $(3x+5)^{1/3}(2x+3)^{1/2}(17x+27)$ **113.** $\frac{3(x+2)}{2x^{1/2}}$ **115.** 1.41 **117.** 1.59 **119.** 4.89 **121.** 2.15 **123. (a)** 15,660.4 gal **(b)** 390.7 gal **125.** $2\sqrt{2\pi} \approx 8.89$ sec

A.8 Assess Your Understanding *(page A75)*

1. undefined; 0 **2.** 3; 2 **3.** T **4.** F **5.** T **6.** $m_1 = m_2$; y-intercepts; $m_1 m_2 = -1$ **7.** 2 **8.** $-\frac{1}{2}$ **9.** F **10.** d **11.** c **12.** b

13. (a) Slope $= \frac{1}{2}$ **(b)** If x increases by 2 units, y will increase by 1 unit.

15. (a) Slope $= -\frac{1}{3}$ **(b)** If x increases by 3 units, y will decrease by 1 unit.

17. Slope $= -\frac{3}{2}$

19. Slope $= -\frac{1}{2}$

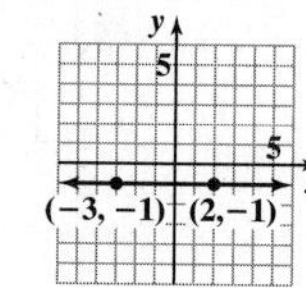

21. Slope $= 0$

23. Slope undefined

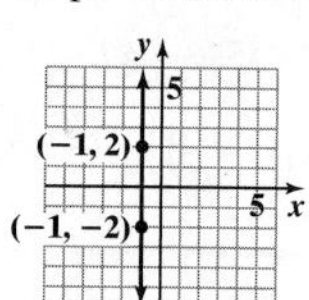

25.

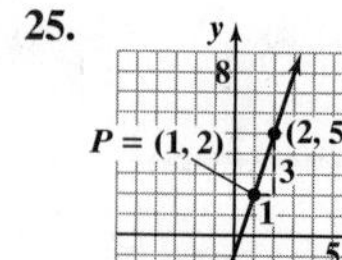

27. $y - 4 = -\frac{3}{4}(x - 2)$

29. $y - 3 = 0$

31. $x = 0$

33. (2, 6); (3, 10); (4, 14) **35.** (4, −7); (6, −10); (8, −13) **37.** (−1, −5); (0, −7); (1, −9) **39.** $x - 2y = 0$ or $y = \frac{1}{2}x$

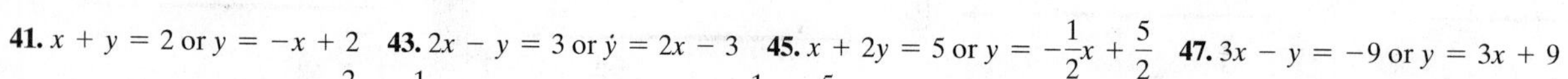

41. $x + y = 2$ or $y = -x + 2$ **43.** $2x - y = 3$ or $y = 2x - 3$ **45.** $x + 2y = 5$ or $y = -\frac{1}{2}x + \frac{5}{2}$ **47.** $3x - y = -9$ or $y = 3x + 9$ **49.** $2x + 3y = -1$ or $y = -\frac{2}{3}x - \frac{1}{3}$ **51.** $x - 2y = -5$ or $y = \frac{1}{2}x + \frac{5}{2}$ **53.** $3x + y = 3$ or $y = -3x + 3$ **55.** $x - 2y = 2$ or $y = \frac{1}{2}x - 1$ **57.** $x = 2$; no slope–intercept form **59.** $y = 2$ **61.** $2x - y = -4$ or $y = 2x + 4$ **63.** $2x - y = 0$ or $y = 2x$ **65.** $x = 4$; no slope–intercept form **67.** $2x + y = 0$ or $y = -2x$ **69.** $x - 2y = -3$ or $y = \frac{1}{2}x + \frac{3}{2}$ **71.** $y = 4$

73. Slope $= 2$; y-intercept $= 3$

75. Slope $= 2$; y-intercept $= -2$

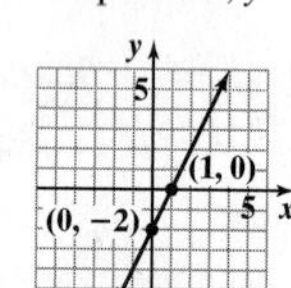

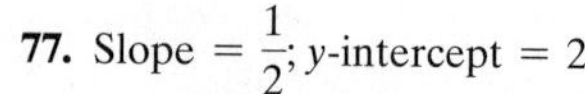

77. Slope $= \frac{1}{2}$; y-intercept $= 2$

79. Slope $= -\frac{1}{2}$; y-intercept $= 2$

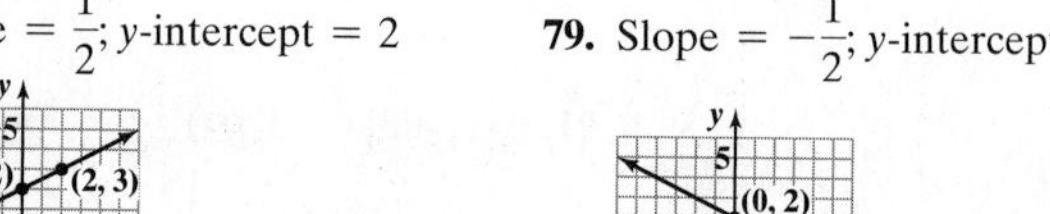

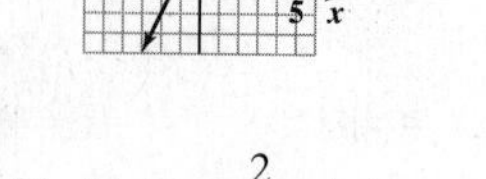

81. Slope $= \frac{2}{3}$; y-intercept $= -2$

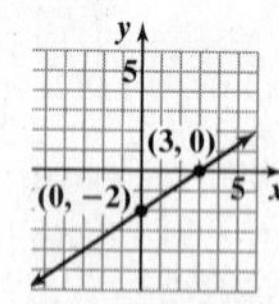

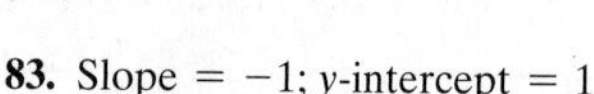

83. Slope $= -1$; y-intercept $= 1$

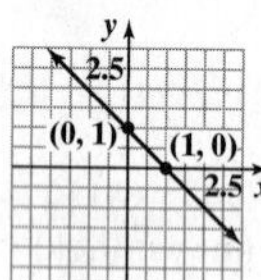

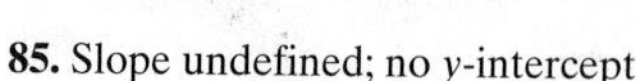

85. Slope undefined; no y-intercept

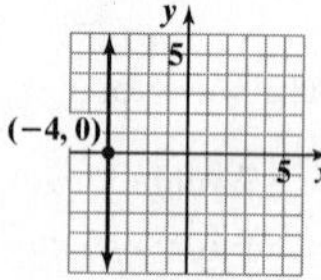

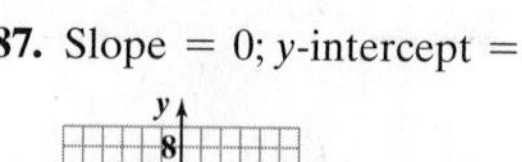

87. Slope $= 0$; y-intercept $= 5$

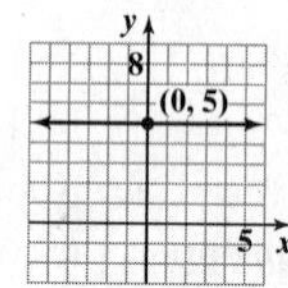

89. Slope $= 1$; y-intercept $= 0$

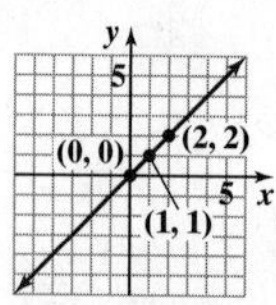

91. Slope $= \frac{3}{2}$; y-intercept $= 0$

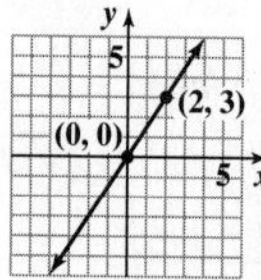

93. (a) x-intercept: 3; y-intercept: 2 **(b)**

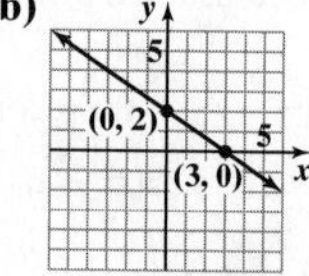

95. (a) x-intercept: −10; y-intercept: 8 **(b)**

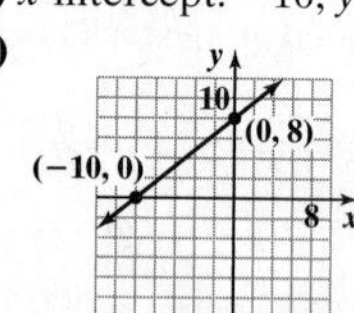

97. (a) x-intercept: 3; y-intercept: $\frac{21}{2}$ **(b)**

99. (a) x-intercept: 2; y-intercept: 3 **(b)**

101. (a) x-intercept: 5; y-intercept: −2 **(b)**

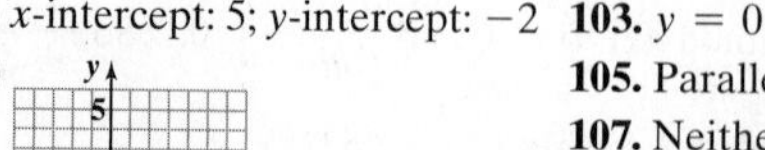

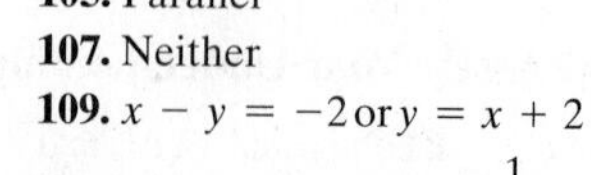

103. $y = 0$

105. Parallel

107. Neither

109. $x - y = -2$ or $y = x + 2$

111. $x + 3y = 3$ or $y = -\frac{1}{3}x + 1$

113. $P_1 = (-2, 5)$, $P_2 = (1, 3)$, $m_1 = -\frac{2}{3}$; $P_2 = (1, 3)$, $P_3 = (-1, 0)$, $m_2 = \frac{3}{2}$; because $m_1 m_2 = -1$, the lines are perpendicular and the points (−2, 5), (1, 3), and (−1, 0) are the vertices of a right triangle; thus, the points P_1, P_2, and P_3 are the vertices of a right triangle.

115. $P_1 = (-1, 0)$, $P_2 = (2, 3)$, $m = 1$; $P_3 = (1, -2)$, $P_4 = (4, 1)$, $m = 1$; $P_1 = (-1, 0)$, $P_3 = (1, -2)$, $m = -1$; $P_2 = (2, 3)$, $P_4 = (4, 1)$, $m = -1$; opposite sides are parallel, and adjacent sides are perpendicular; the points are the vertices of a rectangle.

117. $C = 0.60x + 39$; \$105.00; \$177.00 **119.** $C = 0.17x + 4462$

121. (a) $C = 0.0821x + 15.37, 0 \le x \le 800$
(b)

(c) \$31.79 **(d)** \$56.42
(e) Each additional kW-h used adds \$0.0821 to the bill.

123. $°\text{C} = \frac{5}{9}(°\text{F} - 32)$; approximately 21.1°C **125. (a)** $y = -\frac{2}{25}x + 30$ **(b)** x-intercept: 375; The ramp meets the floor 375 in. (31.25 ft) from the base of the platform. **(c)** The ramp does not meet design requirements. It has a run of 31.25 ft. **(d)** The only slope possible for the ramp to comply with the requirement is for it to drop 1 in. for every 12-in. run.

127. (a) $A = \frac{1}{5}x + 20{,}000$ **(b)** \$80,000 **(c)** Each additional box sold requires an additional \$0.20 in advertising. **129.** All have the same slope, 2; the lines are parallel.

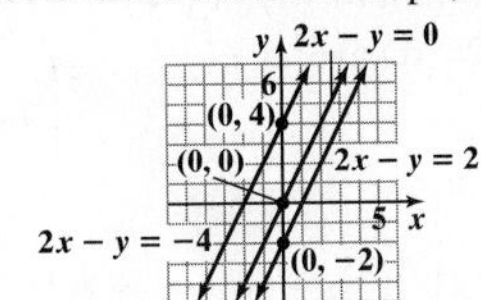

131. (b), (c), (e), (g) **133.** (c) **139.** No; no **141.** They are the same line.
143. Yes, if the y-intercept is 0.

A.9 Assess Your Understanding *(page A83)*

1. scatter diagram **2.** T **3.** Linear relation, $m > 0$ **5.** Linear relation, $m < 0$ **7.** Nonlinear relation

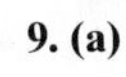

9. (a)

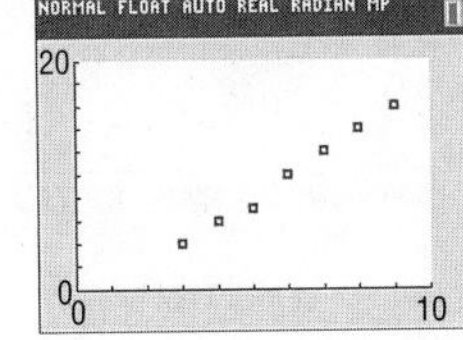

(b) Answers will vary. Using (4, 6) and (8, 14), $y = 2x - 2$.

(c)

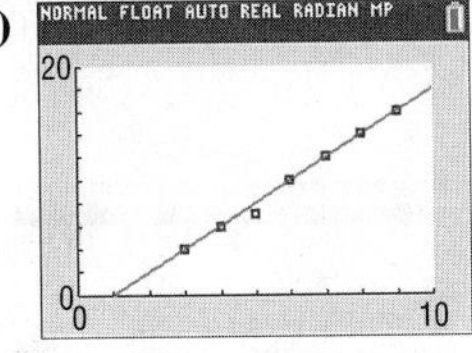

(d) $y = 2.0357x - 2.3571$

(e)

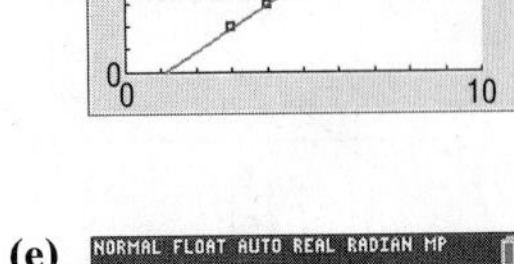

11. (a)

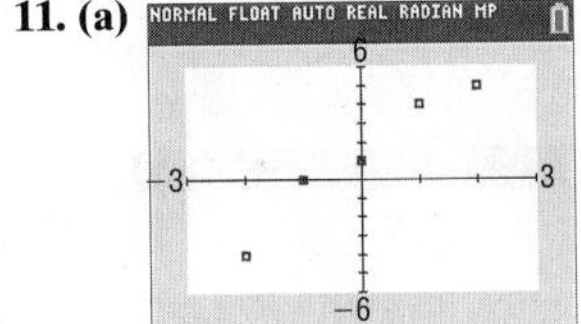

(b) Answers will vary. Using $(-2, -4)$ and $(2, 5)$, $y = \frac{9}{4}x + \frac{1}{2}$.

(c)

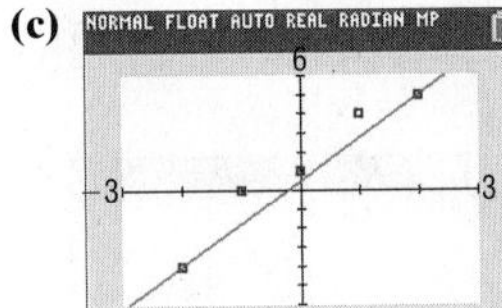

(d) $y = 2.2x + 1.2$

(e)

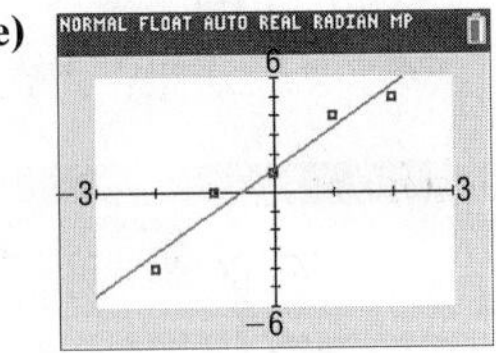

13. (a)

(b) Answers will vary. Using $(-20, 100)$ and $(-10, 140)$, $y = 4x + 180$.

(c)

(d) $y = 3.8613x + 180.2920$

(e)

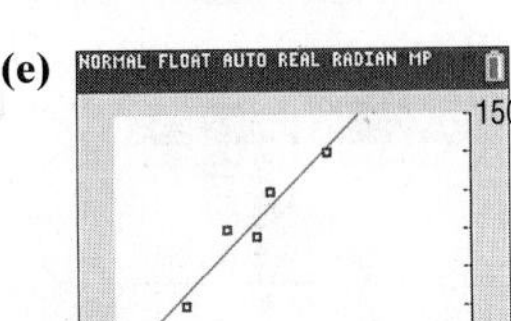

15. (a)

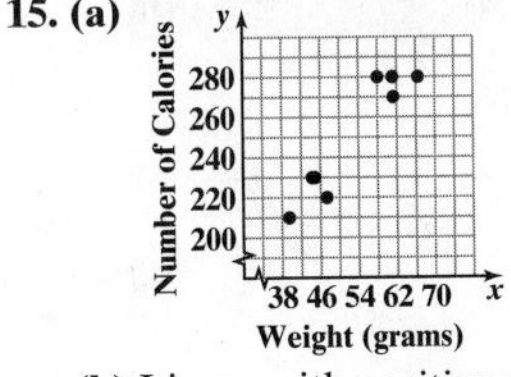

(b) Linear with positive slope

(c) Answers will vary. Using the points (39.52, 210) and (66.45, 280), $y = 2.599x + 107.288$.

(d)

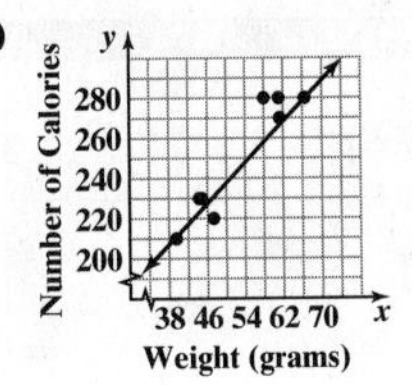

(e) 269 calories
(f) If the weight of a candy bar is increased by 1 gram, the number of calories will increase by 2.599, on average.

17. (a) The independent variable is the number of hours spent playing video games, and cumulative grade-point average is the dependent variable, because we are using number of hours playing video games to predict (or explain) cumulative grade-point average.

(b)

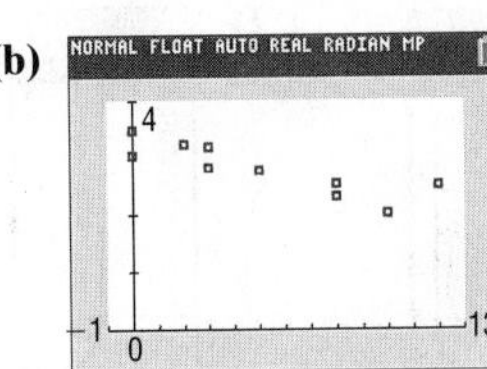

(c) $G(h) = -0.0942h + 3.2763$
(d) If the number of hours playing video games in a week increases by 1 hour, the cumulative grade-point average decreases 0.09, on average.
(e) 2.52
(f) Approximately 9.3 hours

19. (a) 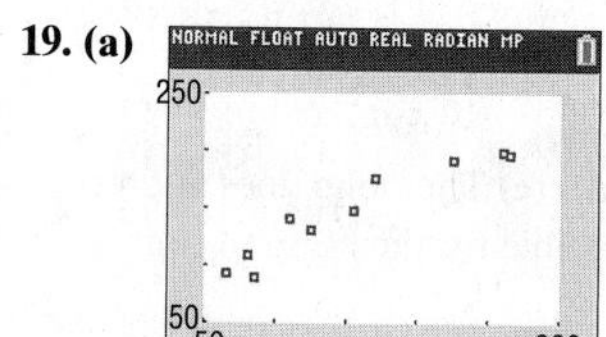

(b) $y = 0.5426x + 63.2336$
(c) If the flight time increases by 1 minute, then the ticket price increases by about \$0.5426, on average.
(d) \$109 **(e)** 105 minutes

21. $y = 1.5x + 3.5$; $r = 1$; The linear relationship between two points is perfect.
23. A value of $x = 0$ does not make sense.

APPENDIX B Graphing Utilities

B.1 Exercises *(page B2)*

1. $(-1, 4)$; II **3.** $(3, 1)$; I **5.** $Xmin = -6$, $Xmax = 6$, $Xscl = 2$, $Ymin = -4$, $Ymax = 4$, $Yscl = 2$
7. $Xmin = -6$, $Xmax = 6$, $Xscl = 2$, $Ymin = -1$, $Ymax = 3$, $Yscl = 1$ **9.** $Xmin = 3$, $Xmax = 9$, $Xscl = 1$, $Ymin = 2$, $Ymax = 10$, $Yscl = 2$
11. $Xmin = -11$, $Xmax = 5$, $Xscl = 1$, $Ymin = -3$, $Ymax = 6$, $Yscl = 1$ **13.** $Xmin = -30$, $Xmax = 50$, $Xscl = 10$, $Ymin = -90$, $Ymax = 50$, $Yscl = 10$ **15.** $Xmin = -10$, $Xmax = 110$, $Xscl = 10$, $Ymin = -10$, $Ymax = 160$, $Yscl = 10$

B.2 Exercises *(page B4)*

1. (a) 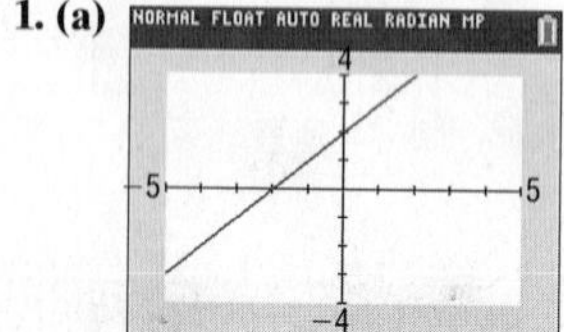
(b) 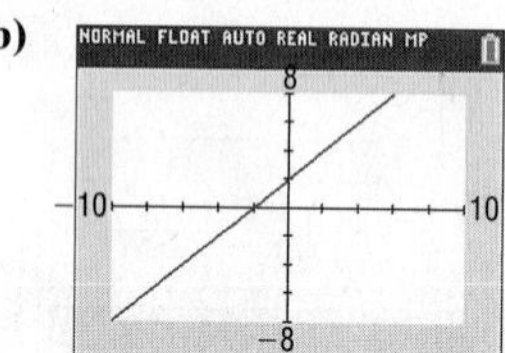
3. (a) 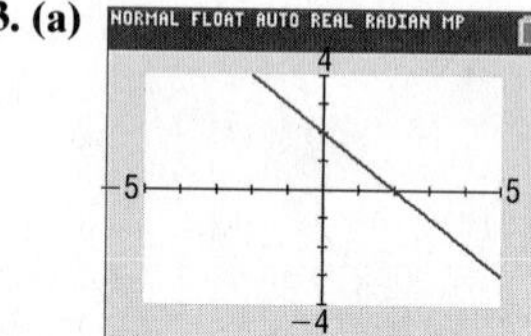
(b) 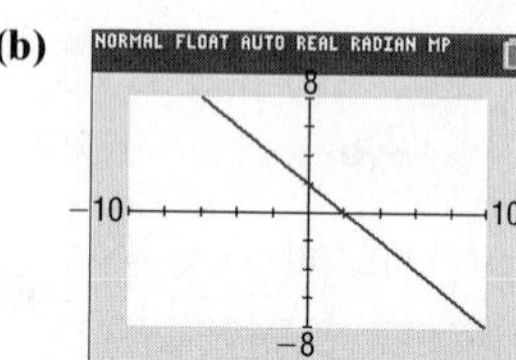

5. (a) 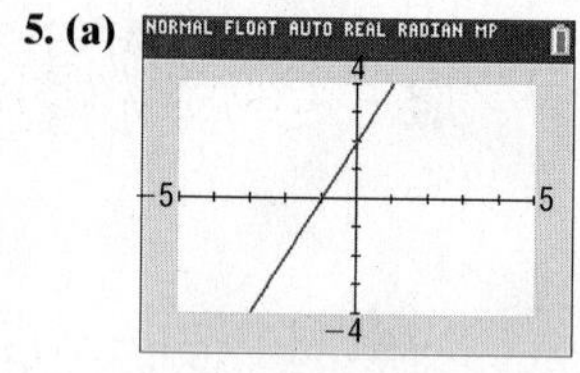
(b) 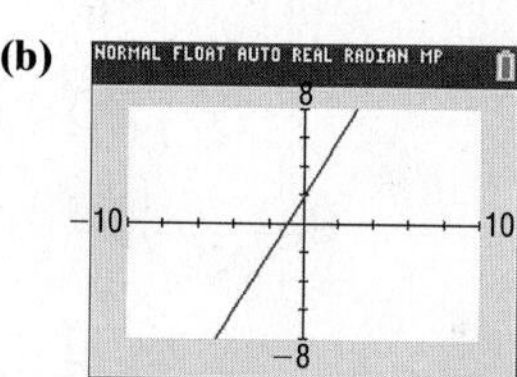
7. (a) 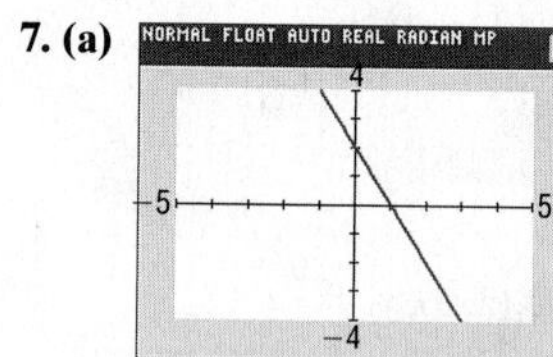
(b) 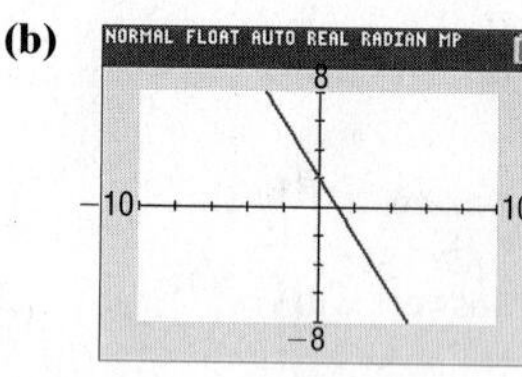

9. (a) 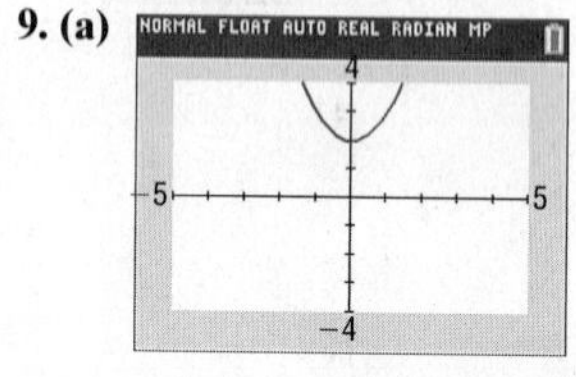
(b) 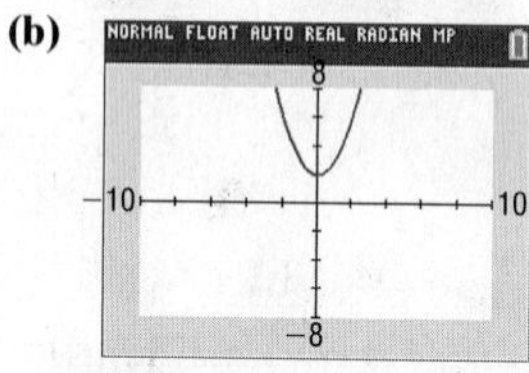
11. (a) 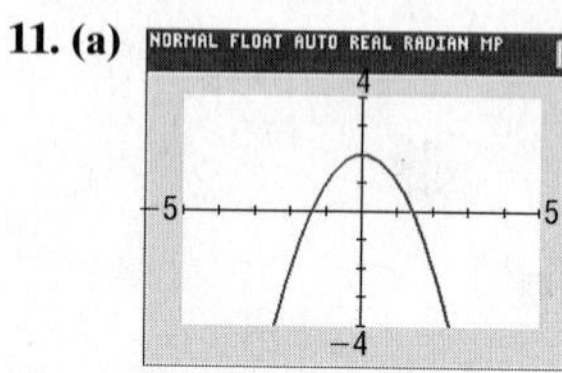
(b) 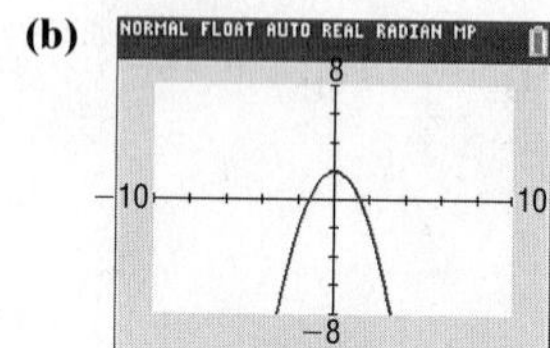

13. (a) 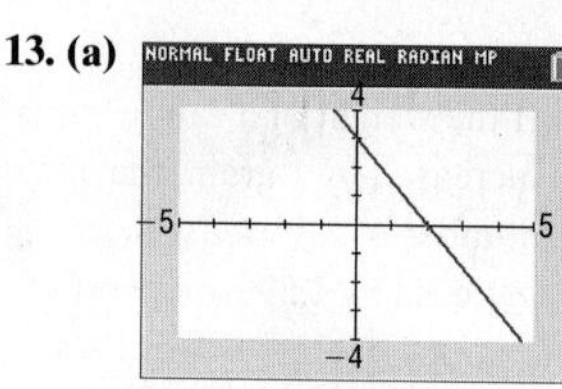
(b) 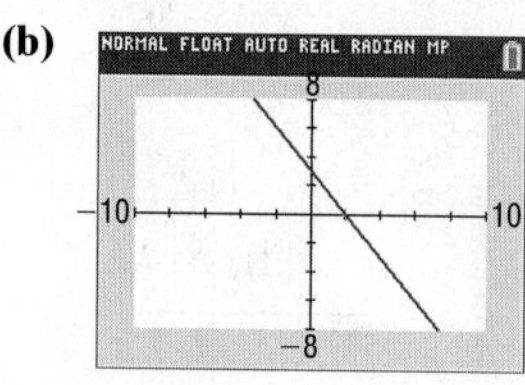
15. (a)
(b)

17. NORMAL FLOAT AUTO REAL RADIAN MP
PRESS ENTER TO EDIT

X	Y1
-5	-3
-4	-2
-3	-1
-2	0
-1	1
0	2
1	3
2	4
3	5
4	6
5	7

Y1=X+2

19. NORMAL FLOAT AUTO REAL RADIAN MP
PRESS ENTER TO EDIT

X	Y1
-5	7
-4	6
-3	5
-2	4
-1	3
0	2
1	1
2	0
3	-1
4	-2
5	-3

Y1=-X+2

21. NORMAL FLOAT AUTO REAL RADIAN MP
PRESS ENTER TO EDIT

X	Y1
-5	-8
-4	-6
-3	-4
-2	-2
-1	0
0	2
1	4
2	6
3	8
4	10
5	12

Y1=2X+2

23. NORMAL FLOAT AUTO REAL RADIAN MP
PRESS ENTER TO EDIT

X	Y1
-5	12
-4	10
-3	8
-2	6
-1	4
0	2
1	0
2	-2
3	-4
4	-6
5	-8

Y1=-2X+2

25. NORMAL FLOAT AUTO REAL RADIAN MP
PRESS ENTER TO EDIT

X	Y1
-5	27
-4	18
-3	11
-2	6
-1	3
0	2
1	3
2	6
3	11
4	18
5	27

Y1=X²+2

27. NORMAL FLOAT AUTO REAL RADIAN MP
PRESS ENTER TO EDIT

X	Y1
-5	-23
-4	-14
-3	-7
-2	-2
-1	1
0	2
1	1
2	-2
3	-7
4	-14
5	-23

Y1=-X²+2

29. NORMAL FLOAT AUTO REAL RADIAN MP
PRESS ENTER TO EDIT

X	Y1
-5	10.5
-4	9
-3	7.5
-2	6
-1	4.5
0	3
1	1.5
2	0
3	-1.5
4	-3
5	-4.5

Y1=-(3/2)X+3

31. NORMAL FLOAT AUTO REAL RADIAN MP
PRESS ENTER TO EDIT

X	Y1
-5	-4.5
-4	-3
-3	-1.5
-2	0
-1	1.5
0	3
1	4.5
2	6
3	7.5
4	9
5	10.5

Y1=(3/2)X+3

B.3 Exercises *(page B6)*

1. −3.41 **3.** −1.71 **5.** −0.28 **7.** 3.00 **9.** 4.50 **11.** 1.00, 23.00

B.5 Exercises *(page B8)*

1. Yes **3.** Yes **5.** No **7.** Yes **9.** Answers may vary. A possible answer is $Y\text{min} = 0$, $Y\text{max} = 10$, and $Y\text{scl} = 1$.

Credits

Photos

Front Matter Achieve Your Potential (Retain Your Knowledge), Mike Flippo/Shutterstock; Resources for Success, Sturti/E+/Getty Images

Chapter 1 Pages 1 and 97, Leigh Prather/Shutterstock/Asset Library; Page 12, Snapper/Fotolia; Page 21(L), Stockyimages/Fotolia; Page 21(R), Department of Energy; Page 22, Tom Donoghue/Polaris/Newscom; Page 39, NASA; Page 40, Exactostock/SuperStock; Page 78, Kenzee/Dreamstime LLC.

Chapter 2 Pages 99 and 186, NLSA; Page 111, Ryan McVay/Digital Vision/Getty Images; Page 159, Srdjan Draskovic/Dreamstime.

Chapter 3 Pages 188 and 257, Sebastian Kaulitzki/Fotolia.

Chapter 4 Pages 258 and 309, Jennifer Thermes/Photodisc/Getty Images; Page 270, Sergey Karpov/Shutterstock; Page 293, Alexandre Fagundes De Fagundes/Dreamstime; Page 295, Anton Ignatenco/iStock/360/Getty Images.

Chapter 5 Pages 311 and 385, Aviator70/Fotolia; Page 333, Pearson Education, Inc.; Page 341, Science & Society Picture Library/Contributor/Getty Images; Page 353, Hulton Archive/Stringer/Getty Images.

Chapter 6 Pages 386 and 450, Marcel Clemens/Shutterstock; Page 404, Thomas Barrat/Shutterstock.

Chapter 7 Pages 452 and 532, Rawpixel/Fotolia; Page 488, Hulton Archive/Handout/Getty Images; Page 497(L), Stockbyte/Getty Images; Page 497(R), Pascal Saez/ Alamy; Page 502, Lacroix Serge/iStock/360/Getty Images; Page 517, Jupiterimages/Photos.com/360/Getty Images.

Appendix A Page A15, Feraru Nicolae/Shutterstock; Page A77, Tetra Images/Alamy.

Text

TI 84 Plus C screenshots courtesy of Texas Instruments. Pages 98, 186, and 532: Screenshots from Microsoft® Excel®. Used by permission of Microsoft Corporation. Chapter 1, Page 25: Diamond price. Used with permission of Diamonds.com. Copyright © Martin Rapaport. All rights reserved. Chapter 3, Pages 247 and 251: Product of Inertia and Moment of Inertia from *Engineering Mechanics: Dynamics* by Russell C. Hibbeler. Published by Pearson Education, © 2013. Chapter 7, Page 525: Cell Phone Towers by CTIA-The Wireless Association. Copyright © 2013 by CTIA-The Wireless Association. Used by permission of CTIA-The Wireless Association®.

Subject Index

Abscissa, 2
Absolute maximum and minimum of functions, 46–47
Absolute value, 336, A5–A6
  inequalities involving, A51–A52
  solving equations involving, A29–A30
Absolute value function, 57, 59
Acute angles, 259–261
  complementary, 261
  trigonometric functions of, 259–261
Addition. *See also* Sum
  of complex numbers, A38
  of vectors, 348–349
    geometrically, 345
    in space, 368
Addition property of inequalities, A48
Aerodynamic forces, 385
Airplane wings, 311
Algebra essentials, A1–A14
  distance on the real number line, A5–A6
  domain of variable, A7
  evaluating algebraic expressions, A6–A7
  evaluating exponents, A10
  graphing inequalities, A5
  Laws of Exponents, A8–A9
  real number line, A4
  sets, A1–A4
  to simplify trigonometric expressions, 219–220
  to solve geometry problems, 4
  square roots, A9–A10
Algebraic vector, 346–347
Alpha particles, 419
Altitude of triangle, A15
Ambiguous case, 274
Amplitude
  of simple harmonic motion, 295
  of sinusoidal functions, 149–151, 168–171
Analytic trigonometry, 188–257
  algebra to simplify trigonometric expressions, 219–220
  Double-angle Formulas, 238–242
    to establish identities, 239–242
    to find exact values, 239
  Half-angle Formulas, 242–244
    to find exact values, 242–244
    for tangent, 244
  inverse functions. *See* Inverse functions
  Product-to-Sum Formulas, 248–249
  Sum and Difference Formulas, 226–238
    for cosines, 226–227
    defined, 226
    to establish identities, 228
    to find exact values, 227, 229–230
    involving inverse trigonometric function, 232
    for sines, 228–229
    for tangents, 231
  Sum-to-Product Formulas, 249–250
  trigonometric equations, 208–218
    calculator for solving, 211
    graphing utility to solve, 213
    identities to solve, 212–213
    involving single trigonometric function, 208–211
    linear, 209–210
    linear in sine and cosine, 233–234
    quadratic in from, 211–212
    solutions of, defined, 208
  trigonometric identities, 218–225
    basic, 219
    establishing, 220–223, 230–231, 239–242
    Even-Odd, 219
    Pythagorean, 219
    Quotient, 219
    Reciprocal, 219
Angle(s), 100–113. *See also* Trigonometric functions
  acute, 259–261
    complementary, 261
    trigonometric functions of. *See* Right triangle trigonometry
  central, 103
  complementary, 261
  defined, 100
  of depression, 263–264
  direction, 350
    of vector, 370–372
  drawing, 101–102
  of elevation, 263–264
  elongation, 281
  Greek letters to denote, 100
  of incidence, 217
  inclination, 130
  initial side of, 100
  measurement of, 101–107
    arc length, 103–104
    degrees, 101–103, 104–107
    to find the area of a sector of a circle, 107
    to find the linear speed of an object traveling in circular motion, 108–109
    radians, 103, 104–107
  negative, 100
  optical (scanning), 247
  positive, 100
  quadrantal, 101
  of refraction, 217
  of repose, 207
  right, 101, A14
  in standard position, 100–101
  straight, 101
  terminal side of, 100
  between vectors, 359
    in space, 369–370
  viewing, 200
Angle–angle case of similar triangle, A18
Angle–side–angle case of congruent triangle, A17
Angular speed, 108
Aphelion, 406, 433, 450
Apollonius of Perga, 386
Approximate decimals, A3
Araybhata the Elder, 125
Arc length, 103–104
Area
  formulas for, A15–A16
  of parallelogram, 378–379
  of sector of circle, 107
  of triangle, 289–295, A15
    SAS triangles, 289–290
    SSS triangles, 290–291
Argument
  of complex number, 337
  of function, 27
Arithmetic calculator, A10
Arithmetic mean, A55
ASA triangles, 272, 273–274
Associative property of vector addition, 345
Asymptote(s), 73–74, 412–414
  horizontal, 73, 74
  vertical, 74
Average rate of change, A65
  of function, 48–50
    exponential functions, 456–457
    finding, 48–49
    linear functions, 456–457
    slope of secant line, 49–50
Axis/axes
  of complex plane, 336
  of cone, 387
  coordinate, 2
  of ellipse, 397, 428
  of hyperbola
    conjugate, 407
    transverse, 407, 428
  polar, 312
  rotation of, 421–423
    analyzing equation using, 423–425
    formulas for, 422
    identifying conic without, 425–426
  of symmetry, of parabola, 388
Azimuth, 266

Barry, Rick, 39
Base of exponent, A8
Basic trigonometric identities, 219
Bearing (direction), 266
Bernoulli, Jakob, 333
Bernoulli, Johann, 442
Bessel, Friedrich, 269
Best fit, line of, A82–A83
Binomial(s)
  cubing, A22
  squares of (perfect squares), A22
Blood alcohol concentration (BAC), 477
Bonds, zero-coupon, 506
Bounding curves, 298
Box, volume and surface area of, A16
Brachistochrone, 442
Brancazio, Peter, 39
Branches of hyperbola, 407
Brewster's Law, 218
Briggs, Henry, 488
Bürgi, Joost, 488

Calculator(s), A10. *See also* Graphing utility(ies)
  approximating roots on, A57
  converting between decimals and degrees, minutes, seconds on, 102
  converting from polar coordinates to rectangular coordinates, 315
  to evaluate powers of 2, 453–454
  functions on, 28
  inverse sine on, 191–192
  kinds of, A10
  logarithms on, 487
  trigonometric equations solved using, 211
Calculus
  area under curve, 201
  area under graph, 53
  difference quotient in, 29–30, 54, 237, 469, 490
  double-angle formulas in, 240
  e in, 461
  exponential equations in, 494
  Extreme Value Theorem in, 47
  functions and
    exponential, 453
    increasing, decreasing, or constant, 44, 515
    local maxima and local minima in, 45–46
    secant line, 49–50
  independent variable in, 145
  integral, 242
  logarithms and, 484, 494
  polar equations and, 332
  projectile motion, 437–439
  radians in, 128
  simplifying expressions with rational exponents in, A60

Calculus (*continued*)
Snell's Law and, 217
tangent line and, 293
trigonometric functions and equations in, 213, 216, 240, 247, 248
trigonometric identities useful in, 242
of variations, 442
Carbon dating, 510
Cardano, Girolamo, 341
Cardioid, 326–327, 332
Carlson, Tor, 522
Carrying capacity, 513
Cartesian (rectangular) coordinates
converted to polar coordinates, 316–317
polar coordinates converted to, 314–315
polar coordinates vs., 312
polar equations graphed by converting to, 322–325
in space, 365–366
Cartesian (rectangular) form of complex number, 337–338
Catenary, 396
Ceilometer, 264
Cell division, 508, 513
Cellular telephones, 1
Center
of circle, 15
of hyperbolas, 407
of sphere, 374
Central angle, 103
Change-of-Base Formula, 487–488
Circle(s), 15–18, 387
arc length of, 103–104
area of, A16
area of sector of, 107
center of, 15
central angle of, 103
circumference of, A16
defined, 15
general form of equation of, 17–18
graphing, 16–17, 331
inscribed, 294
intercepts of, 17
polar equation of, 322, 324
radius of, 15
standard form of equation of, 15–16
unit, 16, 114–117
Circular functions, 116
Circular motion, 108–109
simple harmonic motion and, 296
Circumference, A16
Clark, William, 258, 309
Clinton, Bill, 54
Clock, Huygens's, 442
Closed interval, A46–A47
Coefficient
correlation, A83
damping, 298
Cofunctions, 261
names of, 126
Common logarithms (log), 474, 487, 488
Commutative property
of dot products, 358, 369
of vector addition, 345
Complementary angles, 261
Complementary Angle Theorem, 261
Complement of set, A2
Complete graph, 10
Completing the square, A24–A25, A31–A32
identifying conics without, 420–421
Complex number(s), 353, 363
argument of, 337
conjugates of, 337
De Moivre's Theorem and, 339–340
geometric interpretation of, 336
magnitude (modulus) of, 336
in polar form
converting from rectangular form to, 337–338
converting to rectangular form, 337–338
products and quotients of, 338
product of, 338
quotient of, 338
Complex numbers, A37–A42
addition, subtraction, and multiplication of, A38–A40
conjugates of, A39–A40
definition of, A38
equality of, A38
imaginary part of, A37
real part of, A38
in standard form, A38, A40–A41
power of, A42
reciprocal of, A40
Complex number system, A38
quadratic equations in, A42–A44
Complex plane, 336–338
defined, 336
imaginary axis of, 336
plotting points in, 336–338
real axis of, 336
Complex roots, 340–341
Components of vectors, 346, 348
in space, 367
Compound interest, 498–504
computing, 498–500
continuous, 501
defined, 498
doubling or tripling time for money, 503–504
effective rates of return, 501–502
formula, 499–500
future value of lump sum of money, 498–501
present value of lump sum of money, 502–503
Compressions, 68–70, 71
Conditional equation, 218
Cone
axis of, 387
generators of, 387
right circular, 387
vertex of, 387
Congruent triangles, A16–A18
Conics
defined, 428
degenerate, 387
directrix of, 428
eccentricity of, 428
ellipse, 387, 397–407
with center at (h, k), 401–402
with center at the origin, 397–401
with center not at origin, 402–403
center of, 397
defined, 397, 428
eccentricity of, 406, 429, 430
foci of, 397
graphing of, 399–402
length of major axis, 397
major axis of, 397, 428
minor axis of, 397
solving applied problems involving, 403–404
vertices of, 397
focus of, 428
general form of, 420–421
hyperbolas, 386, 387, 407–419
asymptotes of, 412–414
branches of, 407
with center at (h, k), 414–415
with center at the origin, 407–412
with center not at the origin, 414–415
center of, 407
conjugate, 419
conjugate axis of, 407
defined, 407, 428
eccentricity of, 419, 430
equilateral, 419
foci of, 407
graphing equation of, 409–410
solving applied problems involving, 415–416
transverse axis of, 407, 428
vertices of, 407
identifying, 420–421
without a rotation of axes, 425–426
names of, 387
parabola, 387, 388–396
axis of symmetry of, 388
defined, 388, 428
directrix of, 388
eccentricity of, 430
focus of, 388
graphing equation of, 389
solving applied problems involving, 393–394
with vertex at (h, k), 391–392
with vertex at the origin, 388–391
vertex of, 388
paraboloids of revolution, 386, 393
parametric equations, 434–446
applications to mechanics, 442
for curves defined by rectangular equations, 440–442
cycloid, 441–442
defined, 434
describing, 436–437
graphing using graphing utility, 434–435
rectangular equation for curve defined parametrically, 435–437
time as parameter in, 437–439
polar equations of, 428–433
analyzing and graphing, 428–431
converting to rectangular equation, 432
focus at pole; eccentricity e, 429–431
rotation of axes to transform equations of, 421–423
analyzing equation using, 423–425
formulas for, 422
Conjugate
of complex number, A39–A40
of conjugate of complex number, A41
of product of two complex numbers, A41
of real number, A41
of sum of two complex numbers, A41
Conjugate axis, 407
Conjugate hyperbola, 419
Conjugate of complex numbers, 337
Constant(s), A6
Constant functions, 44–45, 46, 58
Consumer Price Index (CPI), 507
Continuous compounding, 501
Cooling, Newton's Law of, 511–512
Coordinates, 2. *See also* Rectangular (Cartesian) coordinates
of ordered triple, 365
of point on number line, A4
Copernicus, 109
Correlation coefficient, A83
Correspondence between two sets, 23
Cosecant
defined, 259
periodic properties of, 134
Cosecant function, 115
domain of, 132, 133
graph of, 164–165
inverse, 203
approximate value of, 204
calculator to evaluate, 204–205
definition of, 203
exact value of, 204
range of, 132, 133
Cosine(s)
defined, 259
direction, 371–372
exact value of, 227
Law of, 282–288
in applied problems, 284–285

defined, 282
historical feature on, 285
proof of, 283
Pythagorean Theorem as special case of, 283
SAS triangles solved using, 283–284
SSS triangles solved using, 284
periodic properties of, 134
Sum and Difference Formula for, 226–227
trigonometric equations linear in, 233–234
Cosine function, 115
domain of, 132, 133, 148
graphs of, 145–160
amplitude and period, 149–151
equation for, 154–155
key points for, 151–154
hyperbolic, 469
inverse, 193–195
defined, 193
exact value of, 194–195
exact value of expressions involving, 202–203
implicit form of, 193
properties of, 148
range of, 132, 133, 148
Cost(s)
fixed, A77
variable, A77
Cotangent
defined, 259
periodic properties of, 134
Cotangent function, 115
domain of, 132, 133
graph of, 163–164
inverse, 203
approximating the value of, 204–205
calculator to evaluate, 204–205
definition of, 203
range of, 133
Counting numbers (natural numbers), A3
Cross (vector) product, 353, 375–381
defined, 375
determinants to find, 375–376
to find the area of a parallelogram, 378–379
to find vector orthogonal to two given vectors, 378
properties of, 376–378
algebraic, 376–377
geometric, 377–378
of two vectors in space, 375–376
Cube(s)
of binomials (perfect cubes), A22
difference of two, A23, A24
sum of two, A23
Cube function, 27, 58
Cube root, 56, 59, A56
complex, 340–341
Curve(s)
bounding, 298
defined by rectangular equations, 440–442
defined parametrically, 435–437
graphing utility to graph parametrically-defined, B10
of quickest descent, 442
sawtooth, 303
Curve fitting
sinusoidal, 171–175
hours of daylight, 174–175
sine function of best fit, 175
temperature data, 171–174
Curvilinear motion, 437
Cycle of sinusoidal graph, 146, 151
Cycloid, 441–442

Damped motion, 298–299
Damping factor (damping coefficient), 298
Data
fitting exponential functions to, 520
linear models from, A79–A85
sinusoidal model from, 171–174
Day length, 99
Decay, Law of, 510–511. *See also* Exponential growth and decay
Decimals, A3
approximate, A3
converting between degrees, minutes, seconds and, 102–103
repeating, A3
Declination of the Sun, 200
Decomposition, 360–362
Decreasing functions, 44–45, 46, 48
Deflection, force of, 311
Degenerate conics, 387
Degree of polynomial, A23
Degrees, 101–103
converting between decimals and, 102–103
converting between radians and, 104–107
historical note on, 101
De Moivre, Abraham, 339
De Moivre's Theorem, 339–340
Denominator, rationalizing the, A58
Dependent variable, 27
Depreciation, 452
Depression, angle of, 263–264
Descartes, René, 1
Descartes's Law. *See* Snell's Law of Refraction
Determinants, 375–376
Difference(s). *See also* Subtraction
of complex numbers, A38
first, 167
of logarithms, 485
of two cubes, A23, A24
of two squares, A22, A23, A24
of vectors, 345
Difference quotient, 29–30, 54, 237, 469
Directed line segment, 344
Direction angle, 350
Direction angles of vector, 370–372
Direction (bearing), 266
Direction cosines, 371–372
Direction of vectors, 344, 349–351
Directrix, 428
of parabola, 388
Dirichlet, Lejeune, 1
Discriminant, A33, A43
Disjoint sets, A3
Distance, mean, 406, 450
Distance formula, 3–4
proof of, 3
in space, 366
using, 4
Distributive Property
of dot products, 358, 369
of real numbers, A4
Division. *See also* Quotient(s)
of complex numbers, A40–A41
Domain, 23, 24, 30–32
of absolute value function, 59
of constant function, 58
of cosecant function, 132, 133
of cosine function, 132, 133, 148
of cotangent function, 132, 133
of cube function, 58
of cube root function, 59
defined by an equation, 31
of identity function, 58
of inverse function, 82
of logarithmic function, 471–472
of logistic models, 513
of one-to-one function, 79
of reciprocal function, 59
of secant function, 132, 133
of sine function, 132, 133, 146
of square function, 58
of square root function, 58
of tangent function, 132, 133, 162
of the trigonometric functions, 131–132
unspecified, 35
of variable, A7
Domain-restricted function, 86
Dot product, 353, 358–365
angle between vectors using, 359
to compute work, 362
defined, 358
finding, 359
historical feature on, 363
orthogonal vectors and, 360–362
parallel vectors and, 360
properties of, 358–359, 369
of two vectors, 358–359
in space, 368, 369
Double-angle Formulas, 238–242
to establish identities, 239–242
to find exact values, 239
Double root (root of multiplicity 2), A30
Drag, 385

e, 460–462, 469
defined, 461
Earthquakes, magnitude of, 481
Eccentricity, 428
of ellipse, 406, 429–431
of hyperbola, 419, 430
of parabola, 430
Eddin, Nasir, 109, 285
Effective rates of return, 501–502
Elements (Euclid), 285
Elements of sets, A1
Elevation, angle of, 263–264
Ellipse, 387, 397–407
with center at (h, k), 401–402
with center at the origin, 397–401
major axis along x-axis, 398–399
major axis along y-axis, 400
with center not at origin, 402–403
center of, 397
defined, 397, 428
eccentricity of, 406, 429–431
foci of, 397
graphing of, 399–402
major axis of, 397, 428
length of, 397
minor axis of, 397
solving applied problems involving, 403–404
vertices of, 397
Ellipsis, A3
Elliptical orbits, 386
Elongation angle, 281
Empty (null) sets, A1
Epicycloid, 446
Equality
of complex numbers, A38
of sets, A2
of vectors, 344, 348
in space, 367
Equation(s)
conditional, 218
domain of a function defined by, 31
equivalent, A27–A28, B3
even and odd functions identified from, 43–44
exponential, 462–464, 476, 493–495
quadratic in form, 494
as function, 26
graphing utility to graph, B3–B4
intercepts from, 11
inverse function defined by, 84–86
linear. *See* Linear equation(s)
polar. *See* Polar equations
quadratic. *See* Quadratic equation(s)
satisfying the, 9, A27
of secant line, finding, 50
second-degree, A30
sides of, 9, A27
solution set of, A27

Equation(s) (*continued*)
  solving, A27–A37
    by factoring, A29, A30–A31
    with graphing calculator, B6–B7
    involving absolute value, A29–A30
  in two variables, graphs of, 9–15
    intercepts from, 10–11
    by plotting points, 9–10
    symmetry test using, 12–13
    $x = y^2$, 14
    $y = 1 \div x$, 15
    $y = x^3$, 14
Equilateral hyperbola, 419
Equilateral triangle, 7
Equilibrium, static, 352–353
Equilibrium (rest) position, 295
Equivalent equations, A27–A28, B3
Eratosthenes of Cyrene, 113
Error triangle, 8
Euclid, 285
Euler, Leonhard, 1, 109
Even functions, 150
  determining from graph, 42–43
  identifying from equation, 43–44
Evenness ratio, 480
Even-Odd identity, 219
Even-Odd Properties, 141
Explicit form of function, 29
Exponent(s), A8
  Laws of, 454, 463, A8–A9
  logarithms related to, 470
Exponential equations, 462–464
  defined, 462
  solving, 462–464, 476, 493–494
    equations quadratic in form, 494
    using graphing utility, 494–495
Exponential expressions, changing between logarithmic expressions and, 470–471
Exponential functions, 453–469
  defined, 454
  e, 460–462, 469
  evaluating, 453–457
  fitting to data, 520
  graph of, 457–460
    using transformations, 460, 461–462
  identifying, 455–457
  power function vs., 455
  properties of, 458, 460, 464
  ratio of consecutive outputs of, 455–457
Exponential growth and decay, 454, 508–518
  law of decay, 510–511
  logistic models, 513–515
    defined, 513
    domain and range of, 513
    graph of, 513
    properties of, 513
  uninhibited growth, 508–510
Exponential law, 508
Extraneous solutions, A59
Extreme values of functions, 46–47
Extreme Value Theorem, 47

Factored completely, A23
Factoring
  defined, A23
  equations solved by, A29, A30–A31
  of expression containing rational exponents, A61
  over the integers, A23
  polynomials, A23–A24
    by grouping, A24
Factors, A23
Family of lines, A78
Family of parabolas, 78
Fermat, Pierre de, 469
Ferris, George W., 22, 216
Financial models, 498–507
  compound interest, 498–504
  doubling time for investment, 503–504
  effective rates of return, 501–502
  future value of a lump sum of money, 498–501
  present value of a lump sum of money, 500, 502–503
  tripling time for investment, 504
Finck, Thomas, 109, 125
First-degree equation. *See* Linear equation(s)
First differences, 167
Fixed costs, A77
Focus/foci, 428
  of ellipse, 397
  of hyperbola, 407
  of parabola, 388
FOIL method, A22
Foot-pounds, 362
Force(s), 295
  aerodynamic, 385
  of deflection, 311
  resultant, 351
Force vector, 350
Formulas, geometry, A15–A16
Frequency, 158
  in simple harmonic motion, 296
Function(s), 1–98. *See also* Exponential functions; Inverse functions; Linear functions; Trigonometric functions
  absolute value, 57, 59
  argument of, 27
  average rate of change of, 48–50
    finding, 48–49
    slope of secant line, 49–50
  on calculators, 28
  circular, 116
  constant, 44–45, 46, 58
  continuous, 47n
  cube, 27, 58
  cube root, 56, 59
  decreasing, 44–45, 46, 48
  defined, 24
  difference quotient of, 29–30
  domain of, 24, 30–32
    unspecified, 35
  domain-restricted, 86
  equation as, 26
  even and odd, 150
    determining from graph, 42–43
    identifying from equation, 43–44
  explicit form of, 29
  graph of, 32–35, 64–78
    combining procedures, 67–68, 72–73
    determining odd and even functions from, 42–43
    determining properties from, 44–45
    identifying, 32–33
    information from or about, 33–35
    using compressions and stretches, 68–70, 71
    using reflections about the *x*-axis or *y*-axis, 70–71
    using vertical and horizontal shifts, 65–68, 71
  identically equal, 218
  identity, 58
  implicit form of, 29
  important facts about, 29
  increasing, 44–45, 46, 48
  library of, 55–59
  local maxima and local minima of, 45–48
  one-to-one, 78–81
  periodic, 134
  piecewise-defined, 59–60
  range of, 24
  reciprocal, 59, 164. *See also* Cosecant function; Secant function
  relation as, 23–26
  square, 58
  square root, 55, 58
  sum of two, graph of, 299–301
  value (image) of, 24, 26–28
Function keys, A10
Function notation, 35
Fundamental identities of trigonometric functions, 136–138
  quotient, 136
  reciprocal, 136
Fundamental period, 134
Future value, 498–501

Gauss, Karl Friedrich, 341
General form
  of conics, 420–421
  of equation of circle, 17–18
  linear equation in, A71–A72
Generators of cone, 387
Geometric mean, A55
Geometric vectors, 344
Geometry essentials, A14–A21
  congruent and similar triangles, A16–A18
  formulas, A15–A16
  Pythagorean Theorem and its converse, A14–A15
Geometry problems, algebra to solve, 4
Gibbs, Josiah, 353
Grade, A78
Graph(s)/graphing
  bounding curves, 298
  of circles, 16–17, 331
  complete, 10
  of cosecant function, 164–165
    using transformations, 165
  of cosine function, 147–149
  of cotangent function, 163–164
  of ellipse, 399–402
  of equations in two variables, 9–15
    intercepts from, 10–11
    by plotting points, 9–10
    symmetry test using, 12–13
    $x = y^2$, 14
    $y = 1 \div x$, 15
    $y = x^3$, 14
  of exponential functions, 457–460
    using transformations, 460, 461–462
  of function, 32–35, 64–78
    combining procedures, 67–68, 72–73
    determining odd and even functions from, 42–43
    determining properties from, 44–45
    identifying, 32–33
    information from or about, 33–35
    in library of functions, 55–59
    using compressions and stretches, 68–70, 71
    using reflections about the *x*-axis or *y*-axis, 70–71
    using vertical and horizontal shifts, 65–68, 71
  of inequalities, A5
  of inverse functions, 83–84
  of lines
    given a point and the slope, A67
    using intercepts, A71–A72
  to locate absolute maximum and absolute minimum of function, 46–47
  to locate local maxima and local minima of function, 45–46
  of logarithmic functions, 472–475
    base not 10 or *e*, 488
    inverse, 473–475
  of logistic models, 513–515
  of parabola, 389
  of parametric equations, 434–435, B9–B10
  of piecewise-defined functions, 59–60
  of polar equations, 321–335
    cardioid, 326–327, 332
    circles, 331
    of conics, 429–431
    by converting to rectangular coordinates, 322–325

defined, 322
lemniscate, 330, 332
limaçon with inner loop, 328–329, 332
limaçon without inner loop, 327–328, 332
by plotting points, 326–331
polar grids for, 321
rose, 329–330, 332
sketching, 332–333
spiral, 330–331
using graphing utility, 323, B9
of secant function, 164–165
using transformations, 165
of sine and cosine functions, 145–160, 171, 300–301
amplitude and period, 149–151
equation for, 154–155
key points for, 151–154
of vectors, 346
Graphing calculator(s), A10
Graphing utility(ies), B1–B10
coordinates of point shown on, B2
eVALUEate feature, B5
to fit exponential function to data, 519
to fit logarithmic function to data, 521
to fit logistic function to data, 522
functions on, 48
to graph a circle, 18
to graph equations, B3–B4
to graph parametric equations, B9–B10
to graph polar equations, B9
identity established with, 221
INTERSECT feature, B6–B7
line of best fit from, A82–A83
to locate intercepts and check for symmetry, B5–B6
logarithmic and exponential equations solved using, 494–495
MAXIMUM and MINIMUM features, 48
PARametric mode, 438
polar equations using, 323
REGression options, 519
sine function of best fit on, 175
sinusoidal function on, 153
to solve equations, B6–B7
square screens, B8
tables on, B4
trigonometric equations solved using, 213
viewing rectangle, B1–B3
setting, B1
ZERO (or ROOT) feature, B5, B6
Grassmann, Hermann, 353, 363
Greek letters, to denote angles, 100
Greeks, ancient, 109, 113
Grouping, factoring by, A24
Growth, uninhibited, 508–510
Growth factor, 454

Hale-Bopp comet, orbit of, 386, 450
Half-angle Formulas, 242–244
to find exact values, 242–244
for tangent, 244
Half-life, 510
Half-line (ray), 100
Half-open/half-closed intervals, A46–A47
Hamilton, William Rowan, 353
Harmonic mean, A55
Heron of Alexandria, 290, 291
Heron's Formula, 290–291
historical feature on, 291
proof of, 290–291
Horizontal asymptote, 73, 74
Horizontal component of vector, 348
Horizontal compression or stretches, 69–70
Horizontal lines, 323, 331, A68–A69
Horizontal-line test, 80
Horizontal shifts, 66–68, 71
Huygens, Christiaan, 442
Huygens's clock, 442

Hyperbolas, 386, 387, 407–419
asymptotes of, 412–414
branches of, 407
with center at (h, k), 414–415
with center at the origin, 407–412
transverse axis along x-axis, 409–410, 414
transverse axis along y-axis, 410–411, 414
with center not at the origin, 414–415
center of, 407
conjugate, 419
conjugate axis of, 407
defined, 407, 428
eccentricity of, 419, 430
equilateral, 419
foci of, 407
graphing equation of, 409–410
solving applied problems involving, 415–416
transverse axis of, 407, 428
vertices of, 407
Hyperbolic cosine function, 469
Hyperbolic sine function, 469
Hyperboloid, 418
Hypocycloid, 445
Hypotenuse, 259, A14

i, A41–A42
Identically equal functions, 218
Identity(ies), A27
definition of, 218
polarization, 364
Pythagorean, 137, 219
trigonometric, 218–225
basic, 219
establishing, 220–223, 228, 239–242
Even-Odd, 219
Pythagorean, 219
Quotient, 219
Reciprocal, 136, 219
trigonometric equations solved using, 212–213
Identity function, 58
Image (value) of function, 24, 26–28
Imaginary axis of complex plane, 336
Imaginary unit, A37
Implicit form of function, 29
Incidence, angle of, 217
Inclination, 130
Increasing functions, 44–45, 46, 48
Independent variable, 27
in calculus, 145
Index/indices
of radical, A56
of refraction, 217
Inequality(ies)
combined, A50–A51
interval notation to write, A47
involving absolute value, A51–A52
nonstrict, A5
in one variable, A49
properties of, A47–A48, A51
sides of, A5
solving, A49–A52
strict, A5
Inequality symbols, A5
Inertia
moment of, 251
product of, 247
Infinity, 73–74, A46–A47
Inflation, 506
Inflection point, 513
Initial point of directed line segment, 344
Initial side of angle, 100
Initial value of exponential function, 454
Input to relation, 23
Inscribed circle, 294
Integers, A3
factoring over the, A23
Integrals, 242

Intercept(s)
of circle, 17
from an equation, 11
from a graph, 10–11
graphing an equation in general form using, A71–A72
graphing utility to find, B5–B6
from graph of linear equation, 14
graph of lines using, A71–A72
Interest
compound, 498–504
computing, 498–500
continuous, 501
defined, 498
doubling or tripling time for money, 503–504
effective rates of return, 501–502
formula, 499–500
future value of lump sum of money, 498–501
present value of lump sum of money, 502–503
rate of, 498
effective, 501–502
simple, 498
Internal Revenue Service Restructuring and Reform Act (RRA), 54
Intersection of sets, A2
Intervals, A46
closed, A46–A47
endpoints of, A46
half-open, or half-closed, A46–A47
open, A46–A47
writing, using inequality notation, A47
Invariance, 427
Inverse functions, 81–86, 189–207. *See also* Logarithmic functions
cosine, 193–195
defined, 193
exact value of, 194–195
exact value of expressions involving, 202–203
implicit form of, 193
defined by a map or an ordered pair, 81–83
domain of, 82
of domain-restricted function, 84, 86
finding, 81–82, 197–198
defined by an equation, 84–86
graph of, 83–84
range of, 82
secant, cosecant, and cotangent, 203–204
approximating the value of, 204–205
calculator to evaluate, 204–205
definition of, 203
sine, 189–193
approximate value of, 191–192
defined, 190
exact value of, 190–191
exact value of expressions involving, 202–203, 232
implicit form of, 190
properties of, 192–193
solving equations involving, 198
Sum and Difference Formulas involving, 232
tangent, 195–197
defined, 196
exact value of, 196–197
exact value of expressions involving, 202–203
implicit form of, 196
verifying, 82–83
written algebraically, 205
Inverse trigonometric equations, 198
Irrational numbers, A3, A37
decimal representation of, A3
Isosceles triangle, 7

Jība, 125
Jīva, 125
Joules (newton-meters), 362

Latus rectum, 389, 390
Law of Cosines, 282–288
  in applied problems, 284–285
  defined, 282
  historical feature on, 285
  proof of, 283
  Pythagorean Theorem as special case of, 283
  SAS triangles solved using, 283–284
  SSS triangles solved using, 284
Law of Decay, 510–511. *See also* Exponential growth and decay
Law of Sines
  in applied problems, 276–278
  defined, 272
  historical feature on, 285
  proof of, 277–278
  SAA or ASA triangles solved using, 273–274
  SSA triangles solved using, 274–276
Law of Tangents, 281, 285
Laws of Exponents, 454, 463, A8–A9
Left endpoint of interval, A46
Legs of triangle, 259, A14
Leibniz, Gottfried Wilhelm, 1
Lemniscate, 330, 332
Length of arc of a circle, 103–104
Lewis, Meriwether, 258, 309
Lift, 311, 385
Light detector, 264
Light projector, 264
Like radicals, A57
Limaçon
  with inner loop, 328–329, 332
  without inner loop, 327–328, 332
Line(s), A64–A79. *See also* Linear equation(s)
  of best fit, A82–A83
  family of, A78
  graphing
    given a point and the slope, A67
    using intercepts, A71–A72
  horizontal, 323, 331, A68–A69
  point-slope form of, A68–A69
  polar equation of, 322–323, 331
  slope of, A64–A67, A70
    containing two points, A65
    from linear equation, A70
  vertical, 323, 331, A64
  y-intercept of, A70
Linear equation(s). *See also* Line(s)
  defined, A72
  in general form, A71–A72
  given two points, A69
  for horizontal line, A68–A69
  in one variable, A27
  for parallel line, A72–A73
  for perpendicular line, A73–A74
  slope from, A70
  in slope-intercept form, A69–A70
  for vertical line, A67–A68
Linear functions
  building from data, A79–A85
  graphing utility to find the line of best fit, A82–A83
  identifying, 455–457
  nonlinear relations vs., A80–A82
  scatter diagrams, A79–A80
Linear models from data, A79–A85
Linear speed, 108–109
Linear trigonometric equation, 209–210
Line segment, 344
  midpoint of, 5
Local maxima and local minima of functions, 45–48
Logarithmic equations, 491–497
  defined, 476
  solving, 476–477, 491–493
Logarithmic functions, 470–482
  changing between logarithmic expressions and exponential expressions, 470–471
  defined, 470
  domain of, 471–472
  evaluating, 471
  fitting to data, 521
  graph of, 472–475
    base not 10 or e, 488
  properties of, 472, 478
  range of, 471
Logarithmic spiral, 331
Logarithms, 482–490
  on calculators, 487
  common (log), 474, 487, 488
  evaluating, with bases other than 10 or e, 487–488
  historical feature on, 488
  logarithmic expression as single, 485–486
  logarithmic expression as sum or difference of, 485
  natural (ln), 473, 487, 488
  properties of, 482–490
    establishing, 483
    proofs of, 483–484
    summary of, 488
    using, with even exponents, 493
  relating to exponents, 470
Logistic functions, fitting to data, 522
Logistic models, 513–515
  defined, 513
  domain and range of, 513
  graph of, 513
  properties of, 513
Loudness, 481

Magnitude
  of earthquake, 481
  vector in terms of direction cosines and, 372
  of vectors, 344, 346, 348, 349, 350–351
    in space, 368
Magnitude (modulus), 336, 337, 338
Major axis, 428
Mandelbrot sets, 343
Map, inverse function defined by, 81
Mapping, 23
Maxima of functions
  absolute, 46–47
  local, 45–48
Mean
  arithmetic, A55
  geometric, A55
  harmonic, A55
Mean distance, 406, 450
Mechanics, parametric equations applied to, 442
Medians of triangle, 7
Menelaus of Alexandria, 109
Metrica (Heron), 291
Midpoint formula, 5
Mind, mapping of, 188, 257
Mindomo (software), 257
Minima of functions
  absolute, 46–47
  local, 45–48
Minutes, 102–103
Model(s)
  linear, from data, A79–A85
  sinusoidal, 171–175
    best-fit, 175
    daylight hours, 174–175
    temperature data, 171–174
Modulus (magnitude), 336, 337, 338
Mollweide, Karl, 281
Mollweide's Formula, 281
Moment of inertia, 251
Monomial, common factors, A23
Monter, A78
Motion
  circular, 108–109, 296
  curvilinear, 437
  damped, 298–299
  Newton's second law of, 345
  projectile, 437–439
  simple harmonic, 295–297
Multiplication. *See also* Product(s)
  of complex numbers, A39
  of vectors, by numbers geometrically, 345–346
Multiplication properties for inequalities, A48

Napier, John, 488
Nappes, 387
Natural logarithms (ln), 473, 487, 488
Natural numbers (counting numbers), A3
Nautical miles, 112
Negative angle, 100
Negative numbers
  real, A4
  square root of, A9, A42–A43
Newton-meters (joules), 362
Newton's Law of Cooling, 511–512, 516
Newton's Law of Heating, 516
Newton's Second Law of Motion, 295, 345, 385
Niccolo of Brescia (Tartaglia), 341
Nonlinear relations, A80–A82
Nonnegative property of inequalities, A47
Nonstrict inequalities, A5
nth roots, A56–A57
  rationalizing the denominator, A59
  simplifying, A56
  simplifying radicals, A57–A58
Null (empty) sets, A1
Numbers
  irrational, A3
  natural (counting), A3
  rational, A3

Oblique triangle, 272
Odd functions, 150
  determining from graph, 42–43
  identifying from equation, 43–44
One-to-one functions, 78–81
  defined, 79
  horizontal-line test for, 80
Open interval, A46–A47
Optical (scanning) angle, 247
Orbits
  elliptical, 386
  planetary, 406
Ordered pair(s), 2
  inverse function defined by, 82
  as relations, 23–24
Ordinary (statute) miles, 112
Ordinate (y-coordinate), 2
Orientation, 434
Origin, 2, 365
  of real number line, A4
  symmetry with respect to, 12–13
Orthogonal vectors, 360–362
Output of relation, 23

Parabola(s), 387, 388–396
  axis of symmetry of, 388
  defined, 388, 428
  directrix of, 388
  eccentricity of, 430
  family of, 78
  focus of, 388
  graphing equation of, 389
  solving applied problems involving, 393–394
  with vertex at (h, k), 391–392
  with vertex at the origin, 388–391
    finding equation of, 390–391
    focus at $(a, 0)$, $a > 0$, 389–390
  vertex of, 388
Paraboloids of revolution, 386, 393
Parallax, 268
Parallelepiped, 380
Parallel lines, A72–A73
Parallelogram, area of, 378–379

Parallel vectors, 360
Parameter, 434
  time as, 437–439
Parametric equations, 434–446
  for curves defined by rectangular equations, 440–442
    applications to mechanics, 442
    cycloid, 441–442
  defined, 434
  describing, 436–437
  graphing, 434–435
    using graphing utility, B9–B10
  rectangular equation for curve defined parametrically, 435–437
  time as parameter in, 437–439
Pascal, Blaise, 442
Payment period, 498
Pendulum, period of, A64
Perfect cubes, A22
Perfect roots, A56
Perfect squares, A22, A24
Perfect triangle, 294
Perihelion, 406, 433, 450
Perimeter, formulas for, A15
Period
  fundamental, 134
  of simple harmonic motion, 295
  of sinusoidal functions, 150, 151, 168–171
  of trigonometric functions, 133–135
Periodic functions, 134
Period of pendulum, A64
Phase shift, 168–171
  to graph $y = A\sin(\omega x - \varphi) + B$, 168–171
Phones, cellular, 1
Physics, vectors in, 344
Piecewise-defined functions, 59–60
Pitch, A79
Pixels, B1
Plane(s)
  complex, 336–338
    defined, 336
    imaginary axis of, 336
    plotting points in, 336–338
    real axis of, 336
Plane curve, 434
Planets, orbit of, 406
Plotting points, 2, 312–314
  graph equations by, 9–10
Point(s)
  coordinates of
    on graphing utility, B2
    on number line, A4
  distance between two, 3
  inflection, 513
  initial, 344
  plotting, 2, 312–314
    graph equations by, 9–10
  polar coordinates of, 313–314
  terminal, 344
Point-slope form of equation of line, A68–A69
Polar axis, 312
Polar coordinates, 312–321
  conversion from rectangular coordinates, 316–317
  conversion to rectangular coordinates, 314–315
  defined, 312
  plotting points using, 312–314
  of a point, 313–314
  polar axis of, 312
  pole of, 312
  rectangular coordinates vs., 312
Polar equations
  calculus and, 332
  classification of, 331–332
  of conics, 428–433
    analyzing and graphing, 429–431
    converting to rectangular equation, 432
    focus at pole; eccentricity e, 429–431
  defined, 322
  graph of, 321–335
    cardioid, 326–327, 332
    circles, 331
    by converting to rectangular coordinates, 322–325
    defined, 322
    lemniscate, 330, 332
    limaçon with inner loop, 328–329, 332
    limaçon without inner loop, 327–328, 332
    by plotting points, 326–331
    polar grids for, 321
    rose, 329–330, 332
    sketching, 332–333
    spiral, 330–331
    using graphing utility, 323, B9
  historical feature on, 333
  identifying, 322–325
  testing for symmetry, 325
  transforming rectangular form to, 318–319
  transforming to rectangular form, 318
Polar form of complex number, 337–338
Polar grids, 321
Polarization identity, 364
Pole, 312
Polynomial(s)
  degree of, A23
  factoring, A23–A24
    by grouping, A24
  prime, A23
  special products formulas, A22–A23
Position vector, 346–348
  in space, 367
Positive angle, 100
Positive real numbers, A4
Power(s), 158. *See also* Exponent(s)
  of i, A41–A42
  log of, 484
Power functions, exponential function vs., 455
Present value, 500, 502–503
Prime polynomials, A23
Principal, 498
Principal nth root of real number, A56
Principal square root, A9, A42
Product(s). *See also* Dot product; Multiplication
  of complex numbers, A39
    in polar form, 338
  of inertia, 247
  log of, 484
  special, A22–A23
  vector (cross), 353
Product-to-Sum Formulas, 248–249
Projectile motion, 437–439
Projection, vector, 360–361
Projection of P on the x-axis, 296
Projection of P on the y-axis, 296
Prolate spheroid, 406
Ptolemy, 217, 285
Pure imaginary number, A38
Pythagorean Identities, 137, 219
Pythagorean Theorem, 259, A14–A15
  applying, A15
  converse of, A14–A15
  as special case of Law of Cosines, 283

Quadrant, angle lying in, 101
Quadrantal angles, 101
  trigonometric functions of, 117–119
Quadrants, 2
Quadratic equation(s)
  character of the solutions of, A43–A44
  in the complex number system, A42–A44
  definition of, A30
  factoring, A30–A31
  solving
    completing the square, A31–A32
    procedure for, A34
    quadratic formula, A32–A34, A43
    Square Root Method, A31
  in standard form, A30
Quadratic formula, A32–A34, A43
Quaternions, 353
Quotient(s). *See also* Division
  of complex numbers in polar form, 338
  difference, 29–30, 54
  difference, 237, 469
  log of, 484
Quotient identity, 136, 219

Radians, 103
  converting between degrees and, 104–107
Radical equations, A59
  defined, A59
  graphing utility to solve, B7
  solving, A59
Radicals, A56
  fractional exponents as, A59–A60
  index of, A56
  like, A57
  properties of, A57
  rational exponents defined using, A59
  simplifying, A56–A57
Radical sign, A9
Radicand, A56
Radioactive decay, 510–511
Radius, 15
  of sphere, 374
Range, 23, 24
  of absolute value function, 59
  of constant function, 58
  of cosecant function, 132, 133
  of cosine function, 132, 133, 148
  of cotangent function, 133
  of cube function, 58
  of cube root function, 59
  of identity function, 58
  of inverse function, 82
  of logarithmic function, 471
  of logistic models, 513
  of one-to-one function, 79
  of projectile, 242
  of reciprocal function, 59
  of secant function, 133
  of sine function, 132, 133, 146
  of square function, 58
  of square root function, 58
  of tangent function, 133, 162
  of the trigonometric functions, 132–133
Rate of change, average, 48–50, A65
  of linear and exponential functions, 456–457
Rate of interest, 498
Rates of return, effective, 501–502
Rational exponents, A59–A61
Rationalizing the denominator, A58
Rational numbers, A3, A37
Rays (half-lines), 100
  of central angle, 103
  vertex of, 100
Real axis of complex plane, 336
Real number(s), A3–A6, A37
  approximate decimals, A3
  conjugate of, A41
  defined, A3
  principal nth root of, A56
  principal nth root of, A56
Real number line, A4
Real part of complex numbers, A38
Reciprocal function, 59, 164. *See also* Cosecant function; Secant function
Reciprocal identities, 136, 219
Reciprocal property for inequalities, A48, A51
Rectangle, area and perimeter of, A15

Rectangular (Cartesian) coordinates, 2
converted to polar coordinates, 316–317
polar coordinates converted to, 314–315
polar coordinates vs., 312
polar equations graphed by converting to, 322–325
in space, 365–366
Rectangular (Cartesian) form of complex number, 337–338
Rectangular equations
for curve defined parametrically, 435–437
polar equations converted to, 318–319, 432
transforming to polar equation, 318–319
Rectangular grid, 321
Reflections about x-axis or y-axis, 70–71
Refraction, 217
Regiomontanus, 109, 285
Relation(s), 23. *See also* Function(s)
defined, 23
as function, 23–26
input to, 23
nonlinear, A80–A82
ordered pairs as, 23–24
Relative maxima and minima of functions, 45–48
Repeated solution, A30
Repeating decimals, A3
Repose, angle of, 207
Rest (equilibrium) position, 295
Resultant force, 351
Review, A1–A86
of algebra, A1–A14
distance on the real number line, A5–A6
domain of variable, A7
evaluating algebraic expressions, A6–A7
evaluating exponents, A10
graphing inequalities, A5
Laws of Exponents, A8–A9
sets, A1–A4
square roots, A9–A10
complex numbers, A37–A42
of geometry, A14–A21
congruent and similar triangles, A16–A18
formulas, A15–A16
Pythagorean theorem and its converse, A14–A15
inequalities
combined, A50–A51
properties of, A47–A48
solving, A49–A52
interval notation, A46–A47
of nth roots, A56–A57
rationalizing the denominator, A58
simplifying, A56
simplifying radicals, A57–A58
of polynomials
factoring, A23–A24
special products formulas, A22–A23
of rational exponents, A59–A61
Revolutions per unit of time, 108
Rhaeticus, 109
Richter scale, 481
Right angle, 101, A14
Right circular cone, 387
Right circular cylinder, volume and surface area of, A16
Right endpoint of interval, A46
Right-hand rule, 365
Right triangles, 259, 261–266, A14
applications of, 262–266
solving, 261–266
Right triangle trigonometry, 259–271
Complementary Angle Theorem, 261
fundamental identities, 136–138
values of trigonometric functions of acute angles, 259–261
Rise, A64
Root(s), A27. *See also* Solution(s)
complex, 340–341
of multiplicity 2 (double root), A30
perfect, A56
Rose, 329–330, 332
Roster method, A1
Rotation of axes, 421–423
analyzing equation using, 423–425
formulas for, 422
identifying conic without, 425–426
Rounding, A10
Round-off errors, 262
Run, A64
Rutherford, Ernest, 419

SAA triangles, 272, 273–274
SAS triangles, 272, 283–284, 289–290
Satisfying equations, 9, A27
Sawtooth curve, 303
Scalar, 345
Scalar product. *See* Dot product
Scale of number line, A4
Scanning (optical) angle, 247
Scatter diagrams, 171–172, A79–A80
Scientific calculators, A10
Secant
defined, 259
periodic properties of, 134
Secant function, 115
domain of, 132, 133
graph of, 164–165
inverse, 203
approximating the value of, 204
calculator to evaluate, 204–205
definition of, 203
range of, 133
Secant line, 49–50
finding equation of, 50
slope of, 49–50
Second-degree equation. *See* Quadratic equation(s)
Seconds, 102–103
Seed, 343
Set(s), A1–A4
complement of, A2
correspondence between two, 23
disjoint, A3
elements of, A1
empty (null), A1
equal, A2
intersection of, A2
Mandelbrot, 343
of numbers, A1–A4
universal, A2
Set-builder notation, A1–A2
Shannon's diversity index, 480
Shifts, graphing functions using vertical and horizontal, 65–68, 71
Side–angle–side case of congruent triangle, A17
Side–angle–side case of similar triangle, A18
Sides
of equation, 9, A27
of inequality, A5
Side–side–side case of congruent triangle, A17
Side–side–side case of similar triangle, A18
Similar triangles, A16–A18
Simple harmonic motion, 295–297
amplitude of, 295
analyzing, 297
circular motion and, 296
defined, 295
equilibrium (rest) position, 295
frequency of object in, 296
model for, 295–297
period of, 295
Simple interest, 498
Simplifying
expressions with rational exponents, A59–A61
nth roots, A56
radicals, A57–A58
Sine
defined, 259
historical feature on, 125
Law of
in applied problems, 276–278
defined, 272
historical feature on, 285
proof of, 277–278
SAA or ASA triangles solved using, 273–274
SSA triangles solved using, 274–276
periodic properties of, 134
Sum and Difference Formula for, 228–229
trigonometric equations linear in, 233–234
Sine function, 115
of best fit, 175
domain of, 132, 133, 146
graphs of, 145–160
amplitude and period, 149–151
equation for, 154–155
key points for, 151–154
hyperbolic, 469
inverse, 189–193
approximate value of, 191–192
defined, 190
exact value of, 190–191
exact value of expressions involving, 202–203
implicit form of, 190
properties of, 192–193
properties of, 146
range of, 132, 133, 146
Sinusoidal graphs, 145–160, 300–301
amplitude and period, 149–151
equation for, 154–155
key points for, 151–154
steps for, 171
Sinusoidal models, 171–175
best-fit, 175
daylight hours, 174–175
temperature data, 171–174
Six trigonometric functions
exact values of, 125
of quadrantal angles, 117–118
of t, 115
Slope, A64–A67, A70
containing two points, A65
graphing lines given, A67
from linear equation, A70
of secant line, 49–50
Slope-intercept form of equation of line, A69–A70
Snell, Willebrord, 217
Snell's Law of Refraction, 217
Solution(s), A27
extraneous, A59
of inequality, A49–A52
repeated, A30
of trigonometric equations, 208
Solution set of equation, A27
Special products, A22–A23
Speed
angular, 108
linear, 108–109
Sphere, 374
volume and surface area of, A16
Spherical trigonometry, 309
Spheroid, prolate, 406
Spiral, 330–331
Square(s)
of binomials (perfect squares), A22, A24
difference of two, A22, A23, A24
perfect, A22, A24
Square function, 58
Square root(s), A9–A10, A56
complex, 340
of negative number, A9, A42–A43
principal, A9, A42
Square root function, 55, 58
Square Root Method, A31

SSA triangles, 272, 274–276
SSS triangles, 272, 284, 290–291
Standard deviation, A55
Standard form
  complex number in, A38, A40–A41
    power of, A42
    quotient of two, A40–A41
    reciprocal of, A40
  of equation of circle, 15–16
  quadratic equation in, A30
Standard position, angle in, 100–101
Static equilibrium, 352–353
Statute (ordinary) miles, 112
Straight angle, 101
Stretches, graphing functions using, 68–70, 71
Strict inequalities, A5
Subsets, A2
Subtraction. *See also* Difference(s)
  of complex numbers, A38
  of vectors, 348–349
    in space, 368
Sum. *See also* Addition
  of logarithms, 485
  of two cubes, A23
  of two functions, graph of, 299–301
Sum and Difference Formulas, 226–238
  for cosines, 226–227
  defined, 226
  to establish identities, 228, 230–231
  to find exact values, 227, 229–230
  involving inverse trigonometric function, 232
  for sines, 228–229
  for tangents, 231
Sum-to-Product Formulas, 249–250
Sun, declination of, 200
Surface area, formulas for, A16
Symmetry, 12–15
  axis of, of parabola, 388
  graphing utility to check for, B5–B6
  of polar equations, 325
  with respect to origin, 12–13
  with respect to the line $\theta = \frac{\pi}{2}$ (y-axis), 325
  with respect to the polar axis (x-axis), 325
  with respect to the pole (origin), 325
  with respect to the x-axis, 12, 13
  with respect to the y-axis, 13

Tables, on graphing utility, B4
Tangent(s)
  defined, 259
  graph of, 160–163
  Half-angle Formulas for, 244
  historical feature on, 125–126
  Law of, 281, 285
  periodic properties of, 134
  Sum and Difference Formulas for, 231
Tangent function, 115
  domain of, 132, 133, 162
  inverse, 195–197
    defined, 196
    exact value of, 196–197
    exact value of expressions involving, 202–203
    implicit form of, 196
  properties of, 162
  range of, 133, 162
Tartaglia (Niccolo of Brescia), 341
Tautochrone, 442
Terminal point of directed line segment, 344
Terminal side of angle, 100
Terminating decimals, A3
3 by 3 determinants, 375
Thrust, 385
TI-84 Plus C, B3
Time, as parameter, 437–439
Transformations, 64–78, 391, 402, 414
  combining, 67–68, 72–73
  compressions and stretches, 68–70, 71
  cosecant and secant graphs using, 165
  of cosine function, 148–149
  defined, 64
  graphs using, of exponential functions, 460, 461–462
  reflections about the x-axis or y-axis, 70–71
  of sine function, 146–147
  vertical and horizontal shifts, 65–68, 71
Transverse axis, 407, 428
Triangle(s). *See also* Law of Sines
  area of, 289–295, A15
  ASA, 272, 273–274
  congruent, A16–A18
  equilateral, 7
  error, 8
  isosceles, 7
  legs of, 259, A14
  medians of, 7
  oblique, 272
  perfect, 294
  right, 259, 261–266, A14
    applied problems involving, 262–266
    solving, 261–266
  SAA, 272, 273–274
  SAS, 272, 283–284, 289–290
  similar, A16–A18
  SSA, 272, 274–276
  SSS, 272, 284, 290–291
Trigonometric equations, 208–218
  calculator for solving, 211
  graphing utility to solve, 213
  identities to solve, 212–213
  involving single trigonometric function, 208–211
  linear, 209–210
  linear in sine and cosine, 233–234
  quadratic in from, 211–212
  solutions of, defined, 208
Trigonometric expressions, written algebraically, 205, 232
Trigonometric functions
  of acute angles, 259–261
  applications of, 258–310
    damped motion, 298–299
    graphing sum of two functions, 299–301
    involving right triangles, 262–266
    Law of Cosines, 284–285
    Law of Sines, 283
    Law of Tangents, 281, 285
    simple harmonic motion, 295–297
  calculator to approximate values of, 124
  circle of radius r to evaluate, 125
  cosecant and secant graphs, 164–165
  domain and the range of, 131–133
  exact value of
    of $\frac{\pi}{4} = 45°$, 119–120, 122–123
    of $\frac{\pi}{6} = 30°$ and $\frac{\pi}{3} = 60°$, 120–124
    given one function and the quadrant of the angle, 138–140
    using even-odd properties, 141
  fundamental identities of, 136–138
    quotient, 136
    reciprocal, 136
  of general angles, signs of, in a given quadrant, 135–136
  historical feature on, 109, 125–126
  period of, 133–135
  phase shift, 168–171
    to graph $y = A\sin(\omega x - \varphi) + B$, 168–171
  properties of, 131–145
    Even-Odd, 141
  of quadrantal angles, 117–119
  right triangle trigonometry, 259–271
    Complementary Angle Theorem, 261
    fundamental identities, 136–138
  sine and cosine graphs, 145–160
    amplitude and period, 149–151, 168–171
    equation for, 154–155
    key points for, 151–154
  sinusoidal curve fitting, 171–175
  of t, 115
  tangent and cotangent graphs, 160–164
  unit circle approach to, 114–131
Trigonometric identities, 218–225
  basic, 219
  establishing, 220–223
    Double-angle Formulas for, 239–242
    Sum and Difference Formulas for, 228, 230–231
  Even-Odd, 219
  Pythagorean, 219
  Quotient, 219
  Reciprocal, 219
Truncation, A10
2 by 2 determinants, 375

Umbra versa, 126
Unbounded in the negative direction, 73
Unbounded in the positive direction, 74
Uninhibited growth, 508–510
Unit circle, 16, 114–117
Unit vector, 346, 349–350
  in space, 368–369
Universal sets, A2

Value (image) of function, 24, 26–28
Variable(s), A6
  dependent, 27
  domain of, A7
  independent, 27
    in calculus, 145
Variable costs, A77
Vector(s), 344–357
  adding, 345, 348–349
  algebraic, 346–347
  angle between, 359
  components of, 346, 348
  decomposing, 360–362
  defined, 344
  difference of, 345
  direction of, 344, 349–351
  dot product of two, 358–359
  equality of, 344, 348
  finding, 350–351
  force, 350
  geometric, 344
  graphing, 346
  historical feature on, 353
  magnitudes of, 344, 346, 349, 350–351
  modeling with, 351–353
  multiplying by numbers geometrically, 345–346
  objects in static equilibrium, 352–353
  orthogonal, 360–362
  parallel, 360
  in physics, 344
  position, 346–348
  scalar multiples of, 345, 349, 358
  in space, 365–374
    angle between two vectors, 369–370
    cross product of two, 375–376
    direction angles, 370–372
    distance formula, 366
    dot product, 369
    operations on, 367–369
    orthogonal to two given vectors, 378
    position vectors, 367
    rectangular coordinates, 365–366
  subtracting, 348–349
  unit, 346, 349–350
  velocity, 350–351
  zero, 344
Vector product. *See* Cross (vector) product
Vector projection, 360–361

Velocity vector, 350–351
Venn diagrams, A2
Vertex/vertices
  of cone, 387
  of ellipse, 397
  of hyperbola, 407
  of parabola, 388
  of ray, 100
Vertical asymptote, 74
Vertical component of vector, 348
Vertical line, 323, 331, A64
Vertical-line test, 32–33, 35
Vertically compressed or stretched graphs, 68–69
Vertical shifts, 65–68, 71
Viète, François, 285
Viewing angle, 200
Viewing rectangle, 3, B1–B3
  setting, B1
Volume, formulas for, A16

Wallis, John, 341
Waves, traveling speeds of, 217
Weight, 385
Whispering galleries, 403–404
Wings, airplane, 311, 385
Work, 374
  dot product to compute, 362

x-axis, 2
  projection of P on the, 296
  reflections about, 70–71
  symmetry with respect to, 12, 13
x-coordinate, 2
x-intercept, 10
xy-plane, 2, 366
xz-plane, 366

y-axis, 2
  projection of P on the, 296
  reflections about, 71
  symmetry with respect to, 13
y-coordinate (ordinate), 2
y-intercept, 10, A70
  from linear equation, A70
yz-plane, 366

Zero-coupon bonds, 506
Zero-level earthquake, 481
Zero-Product Property, A4
Zero vector, 344

LIBRARY OF FUNCTIONS

Identity Function

$f(x) = x$

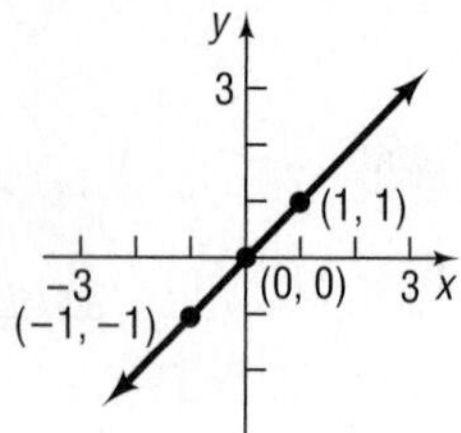

Square Function

$f(x) = x^2$

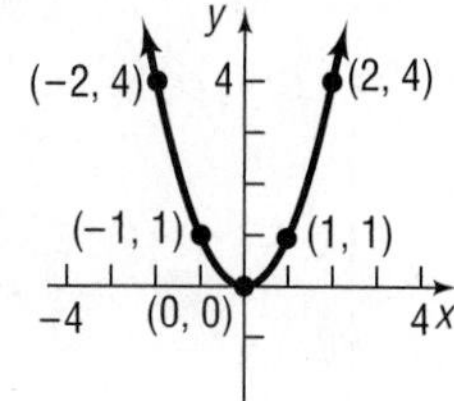

Cube Function

$f(x) = x^3$

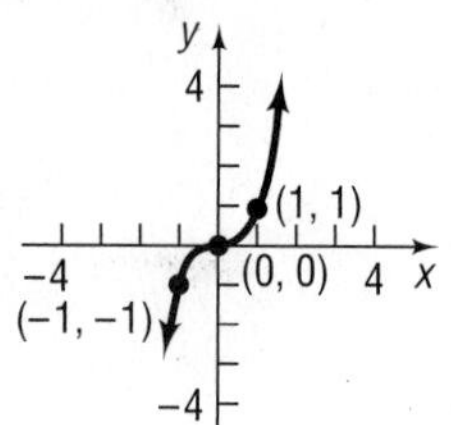

Square Root Function

$f(x) = \sqrt{x}$

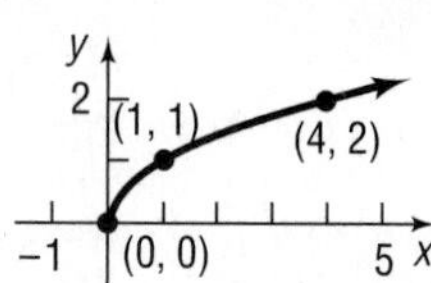

Reciprocal Function

$f(x) = \frac{1}{x}$

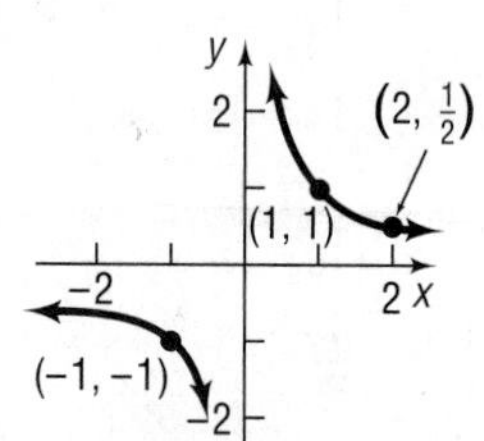

Cube Root Function

$f(x) = \sqrt[3]{x}$

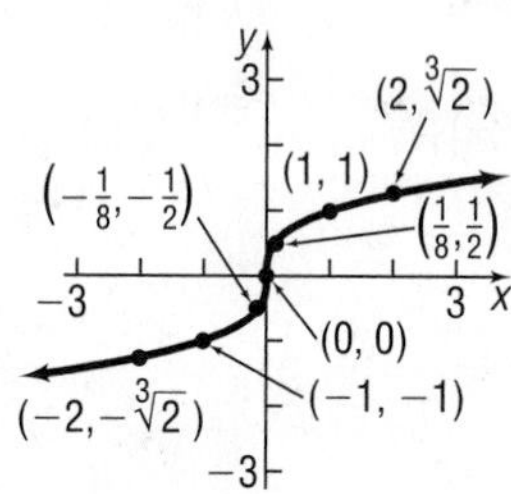

Absolute Value Function

$f(x) = |x|$

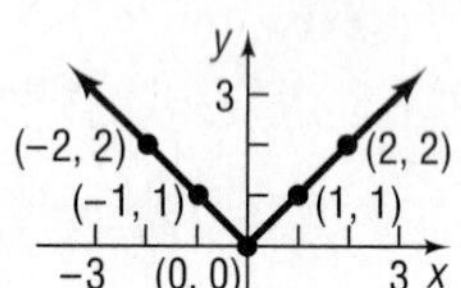

Exponential Function

$f(x) = e^x$

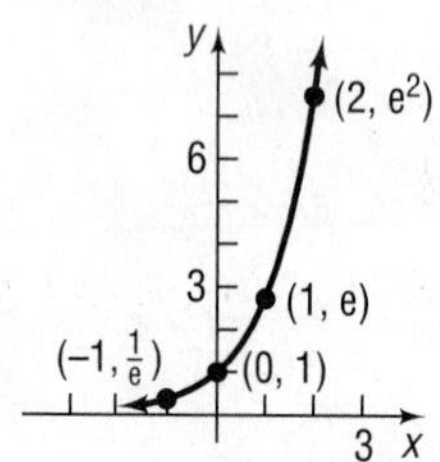

Natural Logarithm Function

$f(x) = \ln x$

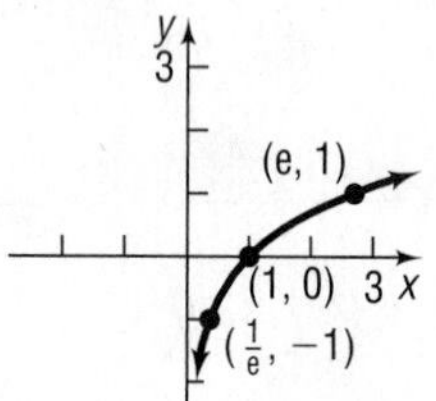

Sine Function

$f(x) = \sin x$

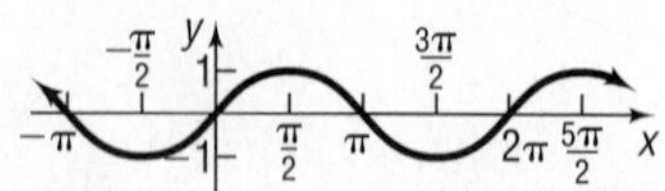

Cosine Function

$f(x) = \cos x$

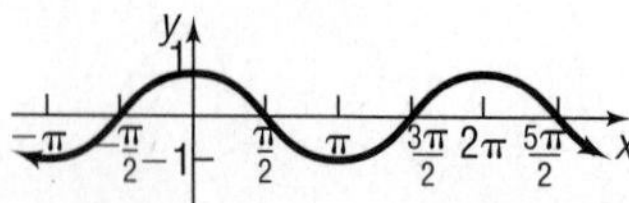

Tangent Function

$f(x) = \tan x$

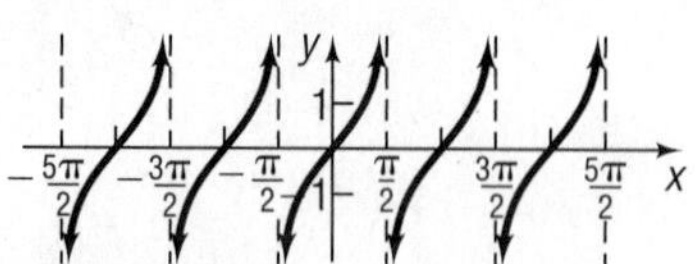

Cosecant Function

$f(x) = \csc x$

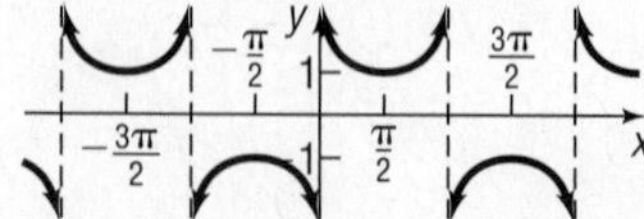

Secant Function

$f(x) = \sec x$

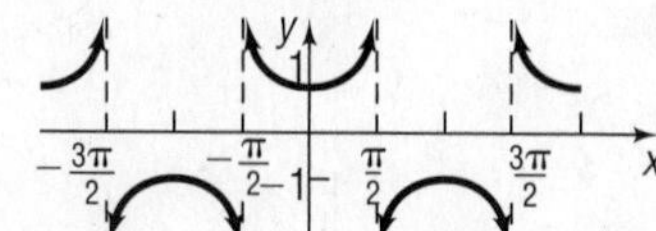

Cotangent Function

$f(x) = \cot x$

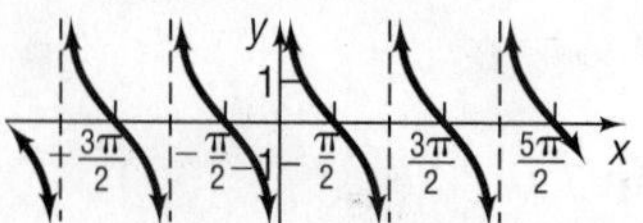

FORMULAS/EQUATIONS

Distance Formula

If $P_1 = (x_1, y_1)$ and $P_2 = (x_2, y_2)$, the distance from P_1 to P_2 is

$$d(P_1, P_2) = \sqrt{(x_2 - x_1)^2 + (y_2 - y_1)^2}$$

Standard Equation of a Circle

The standard equation of a circle of radius r with center at (h, k) is

$$(x - h)^2 + (y - k)^2 = r^2$$

Slope Formula

The slope m of the line containing the points $P_1 = (x_1, y_1)$ and $P_2 = (x_2, y_2)$ is

$$m = \frac{y_2 - y_1}{x_2 - x_1} \qquad \text{if } x_1 \neq x_2$$

$$m \text{ is undefined} \qquad \text{if } x_1 = x_2$$

Point–Slope Equation of a Line

The equation of a line with slope m containing the point (x_1, y_1) is

$$y - y_1 = m(x - x_1)$$

Slope–Intercept Equation of a Line

The equation of a line with slope m and y-intercept b is

$$y = mx + b$$

Quadratic Formula

The solutions of the equation $ax^2 + bx + c = 0, a \neq 0$, are

$$x = \frac{-b \pm \sqrt{b^2 - 4ac}}{2a}$$

If $b^2 - 4ac > 0$, there are two unequal real solutions.
If $b^2 - 4ac = 0$, there is a repeated real solution.
If $b^2 - 4ac < 0$, there are two complex solutions that are not real.

GEOMETRY FORMULAS

Circle

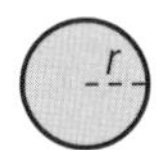

r = Radius, A = Area, C = Circumference

$$A = \pi r^2 \qquad C = 2\pi r$$

Triangle

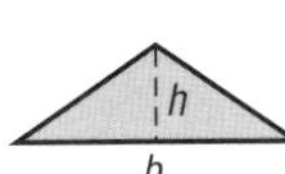

b = Base, h = Altitude (Height), A = area

$$A = \tfrac{1}{2} bh$$

Rectangle

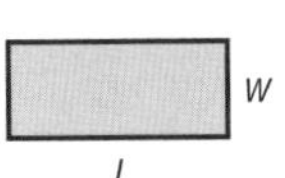

l = Length, w = Width, A = area, P = perimeter

$$A = lw \qquad P = 2l + 2w$$

Rectangular Box (closed)

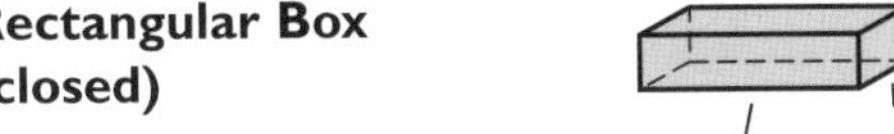

l = Length, w = Width, h = Height, V = Volume, S = Surface area

$$V = lwh \qquad S = 2lw + 2lh + 2wh$$

Sphere

r = Radius, V = Volume, S = Surface area

$$V = \tfrac{4}{3}\pi r^3 \qquad S = 4\pi r^2$$

Right Circular Cylinder (closed)

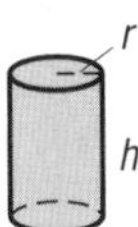

r = Radius, h = Height, V = Volume, S = Surface area

$$V = \pi r^2 h \qquad S = 2\pi r^2 + 2\pi rh$$

CONICS

Parabola

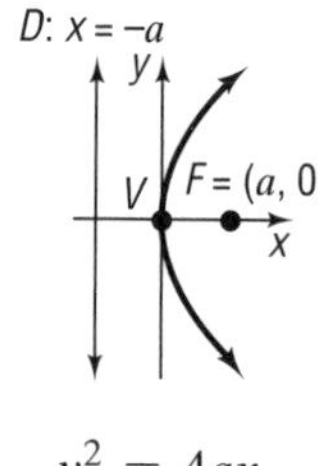

$y^2 = 4ax$

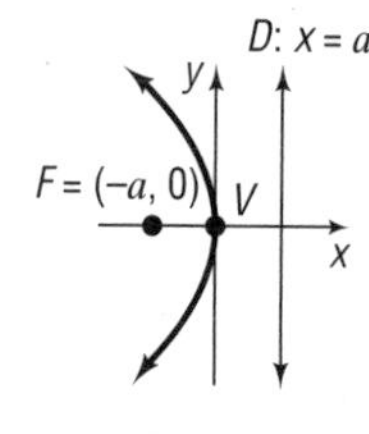

$y^2 = -4ax$

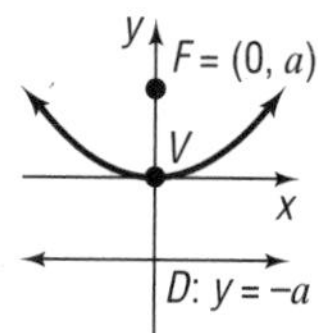

$x^2 = 4ay$

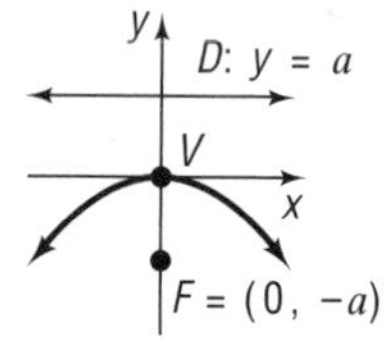

$x^2 = -4ay$

Ellipse

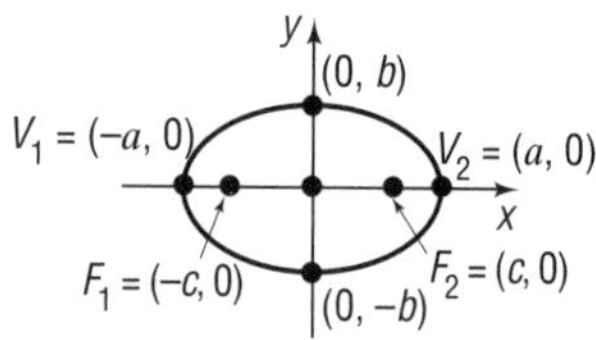

$$\frac{x^2}{a^2} + \frac{y^2}{b^2} = 1, \quad a > b, \quad c^2 = a^2 - b^2$$

y
$V_2 = (0, a)$
$F_2 = (0, c)$
$(b, 0)$
$(-b, 0)$
x
$F_1 = (0, -c)$
$V_1 = (0, -a)$

$$\frac{x^2}{b^2} + \frac{y^2}{a^2} = 1, \quad a > b, \quad c^2 = a^2 - b^2$$

Hyperbola

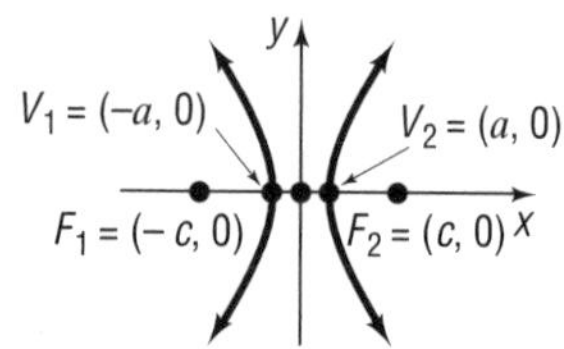

$$\frac{x^2}{a^2} - \frac{y^2}{b^2} = 1, \quad c^2 = a^2 + b^2$$

Asymptotes: $y = \frac{b}{a}x, \quad y = -\frac{b}{a}x$

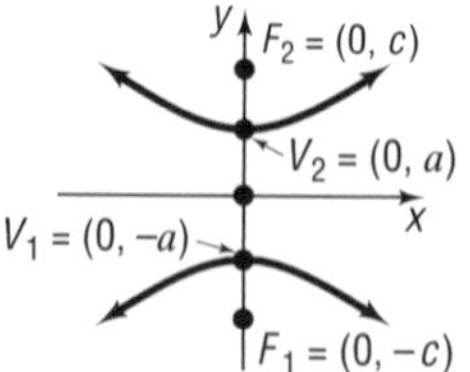

$$\frac{y^2}{a^2} - \frac{x^2}{b^2} = 1, \quad c^2 = a^2 + b^2$$

Asymptotes: $y = \frac{a}{b}x, \quad y = -\frac{a}{b}x$

PROPERTIES OF LOGARITHMS

$$\log_a(MN) = \log_a M + \log_a N$$

$$\log_a\left(\frac{M}{N}\right) = \log_a M - \log_a N$$

$$\log_a M^r = r\log_a M$$

$$\log_a M = \frac{\log M}{\log a} = \frac{\ln M}{\ln a}$$

$$a^r = e^{r \ln a}$$